APPLIED THREE-DIMENSIONAL SUBSURFACE GEOLOGICAL MAPPING

WITH STRUCTURAL METHODS

Third Edition

Daniel J. Tearpock
Richard E. Bischke

With James Brenneke and David C. Metzner

Pearson

Boston • Columbus • New York • San Francisco • Amsterdam • Cape Town
Dubai • London • Madrid • Milan • Munich • Paris • Montreal • Toronto • Delhi • Mexico City
São Paulo • Sydney • Hong Kong • Seoul • Singapore • Taipei • Tokyo

For information about buying this title in bulk quantities, or for special sales opportunities (which may include electronic versions; custom cover designs; and content particular to your business, training goals, marketing focus, or branding interests), please contact our corporate sales department at corpsales@pearsoned.com or (800) 382-3419.

For government sales inquiries, please contact governmentsales@pearsoned.com.

For questions about sales outside the U.S., please contact intlcs@pearson.com.

Visit us on the Web: informit.com

Library of Congress Control Number: 2020932225

Cover images: David C. Metzner

ISBN-13: 978-0-13-485961-3
ISBN-10: 0-13-485961-8

ScoutAutomatedPrintCode

This book is dedicated to the many geoscientists and engineers around the world who convert chaos into logic, the abstract into reality, and the uninterpretable into prospects, thus providing our world with a continuing supply of natural resources to meet our increasing global demands.

CONTENTS

CHAPTER 6 CROSS SECTIONS 197

CHAPTER 9 INTERPRETATION OF THREE-DIMENSIONAL SEISMIC DATA 479

CHAPTER 13 GROWTH STRUCTURES 775

FOREWORDS

FOREWORD TO THE FIRST EDITION

The petroleum geologist is vital to the economic security of oil-producing nations. The geologist's ability to effectively explore and find oil and gas in the future will have a profound impact on oil-producing nations. Therefore, there is going to be a vital need in the years to come for well-trained, well-educated petroleum geologists.

Applied Subsurface Geological Mapping is a complete and thorough exposition on the subject of subsurface petroleum geologic methods. It explains and illustrates graphically the principles necessary to successfully search for and develop deposits of oil and gas. It covers a wide range of topics in subsurface mapping and is written and illustrated in an easy-to-understand manner. The material not only is designed for the student of petroleum geology but also is appropriate for the experienced geologist who needs to sharpen his tools.

The authors of the book are uniquely qualified to write a comprehensive work on applied subsurface geological mapping. Dan Tearpock has extensive practical petroleum geological and engineering experience, having worked as a development geologist for several companies and as a consultant. In addition, he has organized training courses and taught subsurface mapping to many geologists, geophysicists, and engineers. The co-author, Dr. Richard E. Bischke, has a PhD in Geology from Columbia University. He served as Chief Geophysicist for International Exploration for over twelve years and has most recently served on the Research Staff of Princeton University. Through the expertise and experience of both, they have combined to write a text that will quickly become the "Bible" of subsurface mapping.

Frank W. Harrison, Jr.
Consulting Geologist
Past President AAPG (1981–82)

FOREWORD TO THE SECOND EDITION

The daily life of geoscientists includes only a few truly fundamental activities, but among them is making maps. Indeed, we mark the birth of modern geology by the publication in 1815 of William Smith's *Geologic Map of England.* That map was the scientific progeny originating from an economic incentive to identify and locate in the subsurface the best seams of coal. This coupling of science and economics has continued over the intervening two centuries with the publication of this, the second edition of *Applied Subsurface Geological Mapping With Structural Methods.*

Today the scope of subsurface mapping goes well beyond the original goal to map with spatial accuracy, and it now includes many technologies that, when used together, yield insight to both predictive relationships and geological history. Toward that end of integration, the authors have expanded their treatment of geological settings and included new structural geology chapters on compressional, strike-slip, and extensional settings, as well as growth structures. They have also added a chapter on 3D seismic interpretation and expanded the emphasis on computing tools to include those used to contour, correlate, build cross sections, and analyze fault seal.

Through the years, this constantly increasing complexity has led to a greater appreciation of the value of the integrated solution. Accordingly, Tearpock and Bischke have produced a book with an unusual breadth of practical applications gathered from their extensive teaching, consulting, and resultant research. This mix has produced an analytically rigorous work and a selection of illustrative applications that help the novice to the experienced become both a better specialist and a better integrator.

Our profession is fortunate to have the skills of Dan Tearpock and Richard Bischke because both of these authors are passionate about their subject and dedicated to sharing their ideas with others.

Dr. Richard S. Bishop
Past President AAPG (1998–99)

FOREWORD TO THE THIRD EDITION

Geoscientists must visualize the unknown from limited data and extrapolate renderings of surfaces and rock volumes to predict the location of valuable recoverable solids, fluids, and gases. We communicate these 3D solutions using maps, cross sections, and conceptual models so that integrated teams can work together to provide prosperity to an energy-hungry world. That world has changed dramatically in the last several decades. Geoscientists must dedicate themselves to the task of lifelong learning as rapid advances in technology continually provide us with new tools.

Applied Three-Dimensional Subsurface Geological Mapping is a valuable treatise on Mother Nature's rule book. The third edition includes an expanded tool kit of computing, horizontal drilling, unconventional resources, 3D seismic, new structural solutions, and enhanced seismic imaging. Discussions include how traditional techniques such as restored tops and the Wharton method of net pay mapping can be adapted to the workstation, how measuring depth in the subsurface varies from wireline to LWD logs, and a discussion of horizontal wells.

Dan Tearpock and Richard Bischke have built a strong foundation. Contributions from Jim Brenneke, Dick Bischke, and David Metzner enhance our subsurface understanding to help the reader cultivate the venerable habits of the highly successful oil finder. On a personal note, it was a great honor for me to work with Dan Tearpock to fulfill the promise of the 2006 AAPG Convention. Our theme, "Perfecting the Search—Delivering on Promises," celebrated good science serving human needs in a business context.

Dr. Charles A. Sternbach
Consulting Geologist
Past President AAPG (2017–18)

PREFACE

Many textbooks cover numerous geologic subjects; however, since Margaret S. Bishop's classic textbook (1960), and the first edition of this textbook, no complete and detailed book on the subject of *subsurface mapping and structural methods* has been published.

Subsurface geological maps are the most important and widely used vehicle to explore for and develop hydrocarbon reserves. Geologists, geophysicists, and engineers are expected to understand the many aspects of subsurface mapping and to be capable of preparing accurate subsurface maps. Yet, the subject of subsurface mapping is probably the least taught of all petroleum-related subjects. Many colleges and universities do not teach applied subsurface mapping courses, and over the past decade, many company-sponsored training programs have been curtailed or eliminated.

In this early part of the new millennium, we must become more aware of our limitations. This involves questioning our methods and thinking more about our interpretation techniques. We need to consider the tools we have at our disposal to support our interpretations and to generate a better-quality product. Inaccurate procedures, unjustified shortcuts, and limited mapping and structural skills will result in a poor product. During the past decade, the petroleum industry has experienced sweeping changes; new technologies have emerged requiring new skills.

In today's petroleum exploration and development activities, a geoscientist spends a great deal of his or her time in front of a workstation or computer correlating, interpreting, and mapping with the ultimate goal of generating viable interpretations resulting in economic prospects. The computers provide increased speed and efficiency in nearly every aspect of exploration and development. However, problems have developed with the computer-based activities. What we often see today is the computers, not the interpreters, driving interpretations and maps, with blind acceptance of the results. Too often, this procedure results in contour maps that violate geological principles, interpretations that are unlikely in three dimensions, and prospects that depict subsurface geology and geometry that are outright impossible.

We cannot accept computers driving interpretations, nor can we blindly accept the resultant maps (Tearpock and Brenneke 2001a, 2001b). In the hands of geoscientists properly educated in the basics and fundamentals of geology, including field experience, workstations and personal computers are powerful tools. This text contains many of the techniques and methods required to generate sound geologic interpretations and prospects. It can help you become a more productive and successful geoscientist or engineer.

It has been estimated that 30% or more of future reserve additions will be found in areas that are maturely developed. These reserves will come from redeveloping older oil and gas fields. This future potential is to be found in proved producing reservoirs, reservoirs with proved reserves behind pipe, various types of accumulations (attic, infill, and untested fault blocks), and wildcatting in and around the fields and in deeper stratigraphic sections than the current limit within the area.

The techniques presented in this book, from correctly mapping well data to structural balancing, are ideally suited for both exploration and development activities but especially valuable when conducting detailed geoscience projects involving the development or redevelopment of a mature area.

We have expanded this third edition to include new methods or techniques developed during the last 10 years, the understanding and applications of computer-based log correlation, cross section construction, and 3D seismic interpretation within several productive shale basins. Also stressed are new approaches for horizonal drilling and mapping techniques. We have organized a wealth of information that exists in the literature, in addition to material that has never been published. This third edition builds on the new advances and our experience in petroleum exploration and exploitation, as well as our extensive experience in teaching subsurface exploration and development mapping and structural geology courses to geologists, geophysicists, and engineers in the energy industry (worldwide) and at the college and university level.

We present a variety of subsurface mapping and structural techniques applicable in the four major petroleum-related tectonic settings: extensional, compressional, strike-slip, and diapiric. The detailed techniques presented throughout the book are intended to expand your knowledge and improve your skills in preparing geologic interpretations. The knowledge of the principles and techniques presented, plus a burning desire to explore the unknown, will ensure your success.

This textbook is specifically designed for geologists, geophysicists, and engineers who prepare subsurface geological interpretations with accompanying maps. This book should also be beneficial to supervisors, managers, technical assistants, investors, and other persons who have a requirement for the use, preparation, or evaluation of subsurface interpretations and maps.

See the information below for information about registering your copy on the InformIT site for electronic access to color versions of select figures.

Good luck and good prospecting.

Daniel J. Tearpock
Richard E. Bischke

A NOTE FROM THE PUBLISHER

Color figures are available online. To access these, please go to one of the websites listed below and sign up for access:

- For instructors: https://www.pearson.com/us/higher-education/subject-catalog/download-instructor-resources.html
- For readers: http://www.informit.com/register

Register your copy of *Applied Three-Dimensional Subsurface Geological Mapping, Third Edition,* on the InformIT site for convenient access to updates and/or corrections as they become available. To start the registration process, go to informit.com/register and log in or create an account. Enter the product ISBN (9780134859613) and click Submit. Look on the Registered Products tab for an Access Bonus Content link next to this product, and follow that link to access any available bonus materials. If you would like to be notified of exclusive offers on new editions and updates, please check the box to receive email from us.

ACKNOWLEDGMENTS

REVIEWERS—FIRST, SECOND, AND THIRD EDITIONS

We wish to extend our appreciation to the many persons who provided assistance and encouragement during these three projects. We are grateful to the following persons for their review of the text, either in part or in its entirety. Their corrections, suggestions, and insight proved invaluable during the preparation of the textbook, and their suggestions were liberally incorporated into the text. The reviewers are Bob Bass, petrophysicist; Margaret S. Bishop, professor of geology (University of Houston, retired); Robert Bodziak, geophysicist (Pioneer); William G. Brown, consultant and professor of geology (Baylor University); John C. Crowell, professor emeritus (University of California–Santa Barbara); Wendalin Frantz, professor of geology (Bloomsburg University, retired); James Harris, geological engineering consultant, retired; Frank Harrison, consultant and past president of AAPG (1981–82); Ron Hartman, geophysicist and past president of International Exploration; Martin Link (Core Laboratories, Inc.); Brian Lock, professor of geology (University of Louisiana at Lafayette); Donald Dorn Lopez; Miguel Morales, geophysicist (Chevron); John Mosar, geologist (Geological Survey of Norway); Wayne Orlowski, geoscience applications manager (Silicon Graphics, Inc.); Gary Rapp, manager, offshore exploitation and development (Amerada Hess Corporation); Earl Roberts, geophysicist (Pioneer); Mike Sewell, reservoir engineer (Newfield Exploration Company); John H. Shaw (Harvard University); Emery Steffenhagen, petroleum geologist (retired); Ken Thies; Scott Wilkerson; Hongbin Xiao, structural geologist (Saudi Aramco); and David Watso (Unocal Corp.).

CONTRIBUTORS—FIRST, SECOND, AND THIRD EDITIONS

We are grateful to the numerous companies, organizations, professional societies, and especially to our colleagues and friends for their contributions, which have improved the quality of all three editions of this textbook: American Association of Petroleum Geologists; American Geophysical Union;

American Journal of Science; Anadarko Petroleum Corporation; Margaret S. Bishop, professor of geology (University of Houston, retired); Wayne Boeckelman Sr., drafting supervisor (Atwater Consultants, Ltd.); Joseph L. Brewton, Subsurface Consultants & Associates, LLP.; William G. Brown, consultant and professor of geology (Baylor University); Bureau of Economic Geology, University of Texas; Canadian Society of Petroleum Geologists; CGG Land; Chinese Petroleum Institute; Colorado School of Mines; Denver Geophysical Society; Earth Resources Foundation (University of Sydney); Eastman Christensen; Elsevier Science Journal of Structural Geology; Gardes Directional Drilling; Geological Society of America; Gulf Coast Association of Geological Societies; Gulf Publishing Company; Gyrodata, Inc.; Clay Harmon, geophysicist and principal (Juniper Energy, L.P.); James Harris, geological engineering consultant (retired); Steve Hook, advanced geologist (Texaco USA); Houston Geological Society; IHRDC; Jebco Seismic, Inc.; Journal of Petroleum Technology; Brian Lock, professor of geology (University of Louisiana at Lafayette); Lafayette Geological Society; Don Medwedeff, senior research geologist (Arco Oil and Gas); Merlin Profilers, Inc.; John Mosar, geologist (Geological Survey of Norway); Van Mount, senior research geologist (Arco Oil and Gas); Muzium Brunei; National Research Council of Canada; New Mexico Bureau of Mines and Mineral Resources; New Orleans Geological Society; Petrôleos de Venezuela, S.A.; Harvey Pousson, associate professor of mathematics (University of Louisiana at Lafayette); Friends at Pioneer Natural Resources; Pearson Inc.; Princeton University; Rocky Mountain Association of Geologists; Dietrich Roeder, consultant; Royal Society of London; Sandefer Oil and Gas, Inc.; Seitel Data Ltd.; John Shaw, chaired professor (Harvard University); Ted Snedden, geologist, Basin Systems Research (Texaco E&P Technology Division); Sociedad Venezolana de Geôlogos; Society for Sedimentary Geology; Subsurface Consultants & Associates LLC; John Suppe, Blair professor of geology (Princeton University, University of Houston); Swiss Geological Society; Tectonophysics; Tenneco Oil Company; Texaco USA, Eastern E&P Region; TGS Offshore Geophysical Company; United States Geological Survey; Mike Welborne, reservoir engineer and operations manager (Gas Transportation Corporation); W. H. Freeman and Company; W. W. Norton and Company, Inc.; and Hongbin Xiao, structural geologist (Saudi Aramco).

DRAFTING—FIRST EDITION

A textbook is enhanced by the quality of the figures contained in it. We are indebted to several outstanding drafts people for their understanding, cooperation, and especially their conscientiousness in the preparation of the hundreds of line drawings contained in this text. Sharon Light (Chevron USA) has superior drafting skills and did an outstanding job in the preparation of numerous figures; she was responsible for the special illustrations. Steve Nelson (independent draftsman) was extremely patient with our many changes; the figures he prepared illustrate his exceptional drafting talent. Thank you both.

DRAFTING—SECOND EDITION

From the first edition to the second, we went from ink and graph paper to powerful computer programs for drafting many of the new figures. We wish to extend our appreciation to Oanh Nguyen for his talent and dedication in the preparation of the figures and the layout of the book in the second edition.

SUPPORT PERSONNEL—FIRST EDITION

We are grateful to the many individuals who provided support in the form of typing, word processing, reproduction, data collection, organization, and secretarial assistance: Elsie Bischke, Karen Davis, Susie Melacen, Nicole Tearpock-McMorris, Danielle Tearpock-Hitt, Paula Hebert, and Laura Washispack.

SUPPORT PERSONNEL—SECOND EDITION

For several reasons, this second edition was more difficult to prepare than the first edition. Dr. Larry Walker served as our internal editor for this second edition. But Larry was much more than an editor. He helped us improve our ideas, made many sound recommendations to both the text and figures, provided guidance in frustrating circumstances, and brought a calm, steady organized plan to the development of the second edition. Much thanks goes out to Larry Walker.

We owe a special thanks to Karen Hundl for her diligent work in the conversion of files, word processing, proofreading, reproduction, and many other important services she has provided to us in the preparation of the second edition. Karen has been an inspiration and a great person with whom to work. Thanks, Karen.

CONTRIBUTING AUTHORS—SECOND EDITION

We greatly expanded the depth and scope of this second edition, calling on the expertise of several of our friends in the industry, who are experts in their own right, to contribute to the writing of certain sections.

We wish to thank David Metzner, geophysicist, for his hard work, knowledge, dedication, and contributions to Chapter 9, Interpretation of Three-Dimensional Seismic Data. David is an exceptionally talented geophysicist. He not only understands the philosophical doctrine presented in Chapter 1 and throughout the book, but is dedicated to its rigorous application. His contribution provides a significant enhancement to the textbook. Thank you, Dave.

Geologists have lagged in their use of computers for several reasons, including the lack of digital log databases, limited computer proficiency, and primitive software design. This is all changing. William C. Ross is one of the true pioneers in sound log correlation software. With the advances that Bill and others are making, geologists stand on the brink of a significant increase in productivity with the use of the computer.

Bill Ross and Nancy Ash-Shofner, of A2D Technologies, are contributing authors to several sections of this second edition. Their contribution centers around computer-based log correlation in Chapter 4 and computer-generated cross sections in Chapter 6. Bill has developed software that closely adheres to the hand-correlation procedures presented in this text. His contribution should provide the impetus for all geologists to take advantage of the efficiency of computers. Computer log correlation captured in a digital database allows both the geologist and geophysicist to share a single database, in effect moving toward a true Shared Earth Model. Thank you, Bill and Nancy, for your contributions.

Richard Banks was, and still is, a trailblazer when it comes to computer-based mapping. Dick has written mapping programs and worked with many of the programs in use today. The goal of computer-based mapping is to use the correct methods that must fulfill the needs of the users. In general, a computer-generated map should not only look like a hand-contoured map, but must be both geologically and geometrically valid in three dimensions. Richard Banks and Dr. Joseph K. Sukkar, of Scientific Computer Applications, have provided a wealth of knowledge as contributing authors to the computer-based contouring section in Chapter 2. We thank you for your insight and contribution.

Joe Brewton has been a colleague and friend for many years. He is a co-author of our 1994 book entitled *Quick Look Techniques for Prospect Evaluation*. Joe was kind enough to tackle two key subjects for this second edition: fault seal analysis and horizontal wells. Joe is an expert in both areas of study. His presentation of these subjects will be of great benefit to geologists, geophysicists and engineers. Thanks, Joe.

Randy Etherington is well known for his technical skills in structural geology and structural balancing software. Like many other aspects of geologic work, the computer is now playing a major role in allowing us to conduct structural geologic analysis and to evaluate alternative solutions rapidly. We thank Randy for his contribution of the section on computer-based fault seal analysis in Chapter 7.

SPECIAL RECOGNITION AND THANKS FROM DANIEL J. TEARPOCK

First Edition

I would especially like to thank Emery Steffenhagen for his continued support and encouragement during the preparation of this manuscript. He has served as an exceptional reviewer and mentor. I thank my friend and colleague Dick Bischke for agreeing to co-author this book, and for his many contributions throughout the text, especially his mathematical derivation of the equations relating vertical separation to throw. I wish to thank Margaret S. Bishop for her excellent reviews and her contributions to the text from her previously published textbook on subsurface mapping; Clay Harmon for his conscientious contributions to Chapter 5 (Integration of Geophysical Data in Subsurface Mapping); Jim Harris for his critical reviews, suggestions, extensive help, and preparation of a number of the figures; and Jeff Sandefer and the entire staff of Sandefer Oil and Gas, Inc., for their extensive support during the preparation of the manuscript.

Second Edition

I want to thank the great staff at Subsurface Consultants & Associates LLC (SCA), and especially Hines Austin, who continuously provided encouragement and support throughout the preparation of this second edition. A special thanks goes out to my daughters Nicole and Danielle and my grandsons Justin, Jonathan, Tyler, and Jesse for their understanding and patience with their busy dad and "Pappy." Finally, to my parents John (deceased) and Laura Tearpock who sacrificed much in their lives to make sure that I received a college education, which without this geoscience education, this book would not have become a reality. Thank you, both.

SPECIAL RECOGNITION AND THANKS FROM RICHARD E. BISCHKE

First Edition

I would like to thank Ron Hartman, who, through his compassion and wisdom, has acted as an excellent reviewer and mentor. Particular thanks go to John Suppe and Hongbin Xiao for all their help and for allowing us to publish critical portions of their unpublished data. Thanks to Don Medwedeff, Steve Hook, Van Mount, and Ted Snedden for allowing us to publish portions of their unpublished work. Lastly, I would like to thank Dan Tearpock and my wife Elsie for their friendship and hard work over the years.

Second Edition

I would like to thank John Suppe, Peter Verrall, Jeffrey Milnes, Joe Brewton and Dan Tearpock for helping me to understand how the earth works. I also thank Larry Walker for his meticulous review of this manuscript.

Third Edition

I would like to thank Dan Tearpock who introduced me to the concept that "geometric impossibilities on subsurface maps creates mapping errors and probable mistakes in the generation of prospects." Also, I thank Joe Brewton, whose practical knowledge of subsurface mapping techniques was an inspiration to many geologists and geophysicists who worked with him. Lastly, my thanks go to David Metzner and Jim Brenneke, whose practical knowledge of 3D workstation methods and horizontal drilling techniques and methods made the updated version of the text possible.

SPECIAL RECOGNITION AND THANKS FROM DAVID C. METZNER

Second and Third Editions

I wish to thank Dan Tearpock and Dick Bischke for the opportunity to be part of the second edition by contributing Chapter 9, and Dick Bischke especially for his friendship and confidence in asking me to update Chapter 9 for the third edition. I would also like to thank my wife, Marcia, for her support and understanding throughout the extended period of long days at the office. I also wish to thank the management team at Pioneer Natural Resources who gave permission for me to use my workstation after hours to compose text and figures and for their review of the completed chapter. I couldn't have done it without a support team at Pioneer. Thank you, Cindy, Shi Chen, and David, for all you did to help me. Finally, I wish to thank my professors, mentors, and coworkers who helped feed my desire to know more about Earth and how to apply geoscience to finding oil and gas.

SPECIAL RECOGNITION AND THANKS FROM JAMES C. BRENNEKE

Third Edition

I would like to thank Bob Bass and Gary Rogers, who first taught me the Applied Subsurface Geological Mapping class, along with Sia Agah and Bob Shoup, who worked with me through many sessions of the Applied Subsurface Geological Mapping class until I was finally ready to teach the class on my own. I would also like to thank Sia and Bob for their many discussions of how the third edition might be updated.

ABOUT THE AUTHORS

IN MEMORIAM
DANIEL J. TEARPOCK

Daniel J. Tearpock was a noted scientist, author, and entrepreneur. Without Dan, this book would not be possible. He was the past chairman and CEO of Subsurface Consultants & Associates LLC (SCA), headquartered in Houston, Texas. SCA is an international upstream petroleum consultancy and training company. In addition to managing the overall company, he conducted and supervised a variety of consulting projects in the areas of exploration and development, acquisitions and divestitures, as well as serving as project manager or advisor to various oil companies. He also conducted certain industry training courses and seminars around the world.

Mr. Tearpock authored or co-authored three books and more than 30 technical articles in the areas of project management, petroleum subsurface mapping, shared earth models, structural geology, and geothermal energy. He generated numerous exploration and development projects, either as the sole generator or as part of an organized team. The largest development discovery was 4.5 million BO in the offshore Gulf of Mexico in 1980. The largest exploration discovery was more than 250 BCF of gas in the offshore Gulf of Mexico in 1988.

Some of Dan Tearpock's recognitions and honors include Honored Professional in Marquis' *Who's Who in Science and Engineering, 1998–1999 Edition*; 1998 Recipient of the Distinguished Service Award of the Bloomsburg University Alumni Association; Entrepreneur of the Year Finalist, 1996 and 1998 (sponsored by USA Today, NASDAQ, Ernst & Young LLP, and the Kauffman Foundation); and Honored Professional of National Directory of *Who's Who in Executives and Professionals, 1995-1996 Edition*. In 2014, the Gulf Coast Association Geological Societies dedicated the *GCAGS Transactions* volume to Mr. Tearpock. In recent years, he also received multiple honors and awards commemorating his professional achievements, including the AAPG Division of Professional Affairs' Heritage Award, the AAPG Honorary Member Award, and nomination to the Advisory Board of his alma mater, Bloomsburg University College of Science and Technology.

Mr. Tearpock served as an adjunct associate professor at Tulane University (1984–85), teaching in the Master of Petroleum Engineering program. He was responsible for teaching several geoscience courses. He was also a geology instructor at Montgomery College (1974–75).

Mr. Tearpock received his bachelor's degree in earth science from Bloomsburg University (1970) and his master's degree in geology from Temple University (1977). He was a certified petroleum geologist (certification no. 4114) and a member of various organizations, including the American Association of Petroleum Geologists, American Geological Institute, American Petroleum Institute, European Association of Geoscientists and Engineers, Geological Society of America, Houston Geological Society, Indonesian Petroleum Association, Lafayette Geological Society, New Orleans Geological Society, Society of Exploration Geophysicists, and Society of Petroleum Engineers.

RICHARD E. BISCHKE
(PGCS, retired Subsurface Consultants & Associates LLC)

Dr. Richard (Dick) Bischke was Chief Structural Geophysicist for Subsurface Consultants & Associates LLC (SCA). He works primarily on worldwide projects for major oil companies and large independents. He is a 3D-workstation geophysicist specializing in the interpretation of and prospect generation within complexly deformed structures, including salt-related structures.

Dr. Bischke was an Associate Professor of Geology and Geophysics at Temple University (1973–82) and an adjunct Professor of Engineering at Drexel University. During this time, he became associated with the consulting firm International Exploration (INTEX). He consulted mainly with major oil companies and large independents, primarily in frontier basins, from 1975 to 1988 as Chief Exploration Geophysicist. After participating in an extensive structural study in the Philippines on a basin evaluation project sponsored by the World Bank for the Philippine National Oil Company, he worked with Dr. John Suppe on the prospect generation within several Philippine fold belts. He later joined Dr. Suppe at Princeton University to conduct applied research on extensional and compressional structures (1988–91).

During the 1991–2015 period, he conducted extensive studies throughout the world. He assisted in the exploration and development of the super-giant El Furrial trend and in the exploration of the offshore Orinoco Delta Complex in Venezuela, and he has worked on other projects throughout Venezuela and other regions of South America. Dr. Bischke has expertise in all the tectonic provinces, including compressional, extensional, strike-slip, and diapiric. In the United States, he has conducted studies in the eastern and western overthrust belts, including the Appalachians, the Northern and Southern Rockies, and the Ouchita Fold Belt, as well as in the North Slope of Alaska, offshore and onshore California, shallow and deep-water Gulf of Mexico, and the Gulf Coast and Midland Basins.

He was involved with several significant oil and gas discoveries in Louisiana, Venezuela, and the offshore Palawan, Philippines. He has published on a variety of topics that include earthquake research, seismic structural interpretation, structural balancing and subsurface mapping and interpretation techniques. With Dan Tearpock, Dr. Bischke coauthored the first and second editions of this title, which has become the standard guide for structural mapping interpretations, relied on worldwide by both industry and governments. He taught industry schools to coworkers and management from this text.

Dr. Bischke is also heavily involved in teaching courses in structural geology and balancing in compressional, extensional, and strike-slip regimes. He has authored many papers and several books on tectonics and earthquakes; on extensional, compressional, and strike-slip structures; and on subsurface mapping and prospecting.

He is a member of several societies, including the American Association of Petroleum Geologists, Houston Geological Society, and Lafayette Geophysical Society. Dr. Bischke and Dan Tearpock, as co-authors, won Best Paper Awards from the New Orleans Geological Society and the Gulf Coast Association of Geological Societies.

Dr. Bischke is now the team leader of the third edition of this text.

JAMES C. BRENNEKE
(Retired, BP and Subsurface Consultants & Associates LLC)

Jim Brenneke is a geologist with more than 40 years of experience in the petroleum industry. He joined Shell Oil in 1977 with a master of science plus additional PhD work from the University of Illinois and worked for various Shell subsidiaries until 1993. During that time, he worked on reservoir diagenesis studies, U.S. and international exploration geology, U.S. production geology, and U.S. operations geology. Jim joined SCA in 1994 where he worked with Dan Tearpock, Dick Bischke, and David Metzner. With SCA he worked on various U.S. and international consulting projects as well as held management responsibilities, including Vice President of Geoscience and Engineering. While at SCA, Jim specialized in prospect generation, prospect appraisal, and production geology.

Jim left SCA in 2006 to join BP to work on the development of new fields in the Gulf of Mexico Deepwater. He retired from BP in 2016 as Geology Discipline Lead for BP's Gulf of Mexico Deepwater production department. He currently teaches Applied Subsurface Geological Mapping for SCA.

Jim is a member of AAPG and Houston Geological Society.

DAVID C. METZNER

David Metzner began his career in the petroleum industry in 1982 working as a geophysicist primarily involved in the exploration and development of oil and gas prospects through the integration and mapping of seismic and well data. David is a highly skilled interpreter with applied workstation experience dating back to 1985 and the onset of regional 3D coverage in the deep-water Gulf of Mexico.

He authored Chapter 9, Interpretation of Three-Dimensional Seismic Data, in the second edition of *Applied Subsurface Geological Mapping* (2003), which lays out a practical workstation-based interpretation methodology. In this latest edition, he continues building that methodology based on current practices and workstation technology.

David has been involved in many significant plays throughout his career, including the early exploration of the deep-water Gulf of Mexico and the subsequent development of several of the large oil and gas discoveries. He was involved in the exploration efforts that led to the western extension of the super-giant El Furrial trend in Venezuela. He has worked unconventional reservoirs since 2006, including the early evaluation and development of the Haynesville shale in northern Louisiana and eastern Texas and also explored the Pearsall shale in the Maverick Basin.

From 2014 to 2019, David was employed as a Senior Staff Geophysicist at Pioneer Natural Resources in Irving, Texas, focused on the 3D seismic characterization of the Wolfcamp and Spraberry shales as well as other producing intervals in the Midland Basin in West Texas. Past experience has been with Placid Oil, Arco Oil and Gas, Subsurface Consultants and Associates, EOG Resources, and Encana.

David retired from full-time industry employment in mid-2019 to start Crossroads Geoscience Solutions LLC, a consulting company based in Emory, Texas. He has been a member of the American Association of Petroleum Geologists and the Society of Exploration Geophysicists since 1982.

CHAPTER 1

INTRODUCTION TO SUBSURFACE MAPPING

TEXTBOOK OVERVIEW

This textbook begins where many others end. It focuses on subsurface structural and mapping methods and techniques and their application to the petroleum industry. These techniques are also important and applicable to other fields of study, and geologists, geophysicists, engineers, and students in related fields, such as mining, groundwater, environmental, or waste disposal, should benefit from this text as well.

The objectives of subsurface petroleum geology are to find and develop oil and gas reserves. These objectives are best achieved by the use and integration of all available data and the correct application of these data. This textbook covers various aspects of geoscience interpretation and the construction of subsurface maps and cross sections based on data obtained from well logs, seismic sections, and outcrops. It is concerned with correct structural interpretation and mapping techniques and how to use them to generate the most reasonable subsurface interpretation that is consistent with all the data.

Subsurface geological maps are perhaps the most important vehicle used to explore for undiscovered hydrocarbons and to develop proven hydrocarbon reserves. However, the subject of subsurface mapping is probably the least discussed, yet most important, aspect of petroleum exploration and development. As a field is developed from its initial discovery, a large volume of well, seismic, and production data are obtained. With these data, the accuracy of the subsurface interpretation is improved through time. The most accurate interpretation for any specific oil or gas field can be prepared only after the field has been extensively drilled and most of the hydrocarbons have been depleted. However, accurate and reliable subsurface interpretations and maps are required throughout all exploration and development activities.

From regional exploration to a field discovery and through the life of a producing field, many management decisions are based on the interpretations geoscientists present on subsurface maps.

These decisions involve investment capital to purchase leases, permit and drill wells, build production facilities, and work over or recomplete wells, just to name a few. An exploration or development prospect generator must employ the best and most accurate methods available to find and develop hydrocarbon reserves at the lowest cost per net equivalent barrel. Therefore, when preparing subsurface interpretations, it is essential to use all the available data, evaluate all possible alternate solutions, use valid structural interpretation methods, and use the most accurate mapping techniques to arrive at a finished product that is consistent with correct geological models.

Subsurface geoscientists have the formidable task of mapping unseen structures that may exist thousands of meters beneath the earth's surface. In order to accurately interpret and map these structures, the geoscientist must have a good understanding of the basic principles of structural geology, stratigraphy, sedimentation, and other related geological disciplines. The geoscientist must also be thoroughly familiar with the structural style of the region being worked. Since all subsurface interpretations and the accompanying maps are based on **limited data**, the geologist, geophysicist, or engineer must use (1) his or her educational background, (2) field and work experience, (3) imagination, (4) an understanding of local structures, (5) an ability to visualize in three dimensions in order to evaluate the various possible alternate interpretations and decide on the most reasonable, and (6) correct subsurface structural and mapping methods and techniques.

There are many textbooks on structural geology, tectonics, stratigraphy, sedimentology, structural styles, petroleum geology, and other related geological subjects. Since Bishop's classic work (1960), our original text published in 1991 and revised and expanded in 2003 was the only single-source textbook on applied subsurface mapping techniques. Many new developments occurred during the 1990s. In particular, the field of structural geology advanced to the extent that structural methods and techniques have enhanced subsurface interpretation and mapping. As we entered the twenty-first century, the objective of the second edition of *Applied Subsurface Geological Mapping* was to present a variety of subsurface structural, mapping, and cross-section methods and techniques applicable in various geological settings, including extensional, compressional, strike-slip, and diapiric tectonic areas. The detailed structural and mapping techniques illustrated throughout the book are intended to expand your knowledge and improve your skills in preparing geological interpretations using a variety of maps and cross sections.

This third edition of *Applied Subsurface Geological Mapping* is to update subsurface structural mapping and cross-section methods and techniques with new understanding based on recent technical work, with an emphasis on shale play basins. This edition is also an effort to recognize the significant changes in workflows associated with the widespread use of computer workstations and 3D seismic and the growing importance of *unconventional resources* in the petroleum industry.

All energy companies expect positive economic results through their exploration and development efforts. Some companies are more successful than others. Many factors lead to success, including advanced technology, aggressive management, experience, and serendipity. A significant underlying cause of success that is often overlooked or taken for granted, however, is the **quality** of subsurface structural and mapping methods. The application of the numerous techniques presented in this book should improve the quality of any subsurface interpretation. This improved quality should positively affect any company's economic picture. Quality is improved by:

1. Developing the most reasonable subsurface interpretation for the area being studied, even in areas where the data are sparse or absent.
2. Generating more accurate and reliable exploration and exploitation prospects (thereby reducing associated risk).
3. Correctly integrating geological, geophysical, and engineering data to establish the best development plan for a new field discovery.

4. Optimizing hydrocarbon recoveries through accurate volumetric reserve estimates.
5. Planning a more successful exploration or development drilling program, or preparing a recompletion and workover depletion plan for a mature field.
6. Accurately evaluating and developing any required secondary recovery programs.

THE PHILOSOPHICAL DOCTRINE OF ACCURATE SUBSURFACE INTERPRETATION AND MAPPING

The philosophical doctrine of accurate subsurface interpretation and mapping presented here is designed to provide geologists, geophysicists, and engineers with the tools necessary to prepare the most reasonable subsurface interpretations. In our quest for hydrocarbons, we are always searching for excellence. The material contained in this book can serve as a teaching medium as well as a source of reference for conducting subsurface investigations beginning with the initial stages of exploration, continuing through field development, and ending with enhanced recovery and field depletion.

Have you ever wondered what makes the difference between a great, successful oil and gas prospector and one who is mediocre or below average? Have you ever wondered why one geoscience team or company has a much greater success rate than others working within the same area? When you place your investment dollars into exploration or development prospects or in the purchase of a producing field, do you ask or even consider what methodology or philosophy was used by the generating geoscientist or team (Tearpock 1998)? Remember, people, not computers or workstations, find oil and gas!

Is the methodology or philosophy used by geoscientists an important factor? Well, decades of research, observation, and analyses indicate the primary reason why some individuals, teams, or companies are more successful than others is the direct result of the philosophy and methods used. It is not serendipity, luck, or guesswork that finds hydrocarbons. It is solid scientific work.

The Philosophical Doctrine of accurate subsurface interpretation and mapping presented here provides the best-proven process for finding and developing hydrocarbons. It requires common sense, a certain technical background, experience, logic, and the application of proven scientific methods. The key points of this philosophy are:

1. All subsurface interpretations must be geologically and geometrically valid in three dimensions.
2. An interpreter must have a fundamental, classic education in geology and a strong background in structural geology for the tectonic setting being worked.
3. Sufficient planning, time, and detail are required to generate reliable prospects. Haste makes waste.
4. All subsurface data must be used to develop a reasonable and accurate subsurface interpretation.
5. Accurate correlations (well log and seismic) are required for reliable geological interpretations.
6. The use of correct mapping techniques and methods is essential to generate reasonable and correct subsurface interpretations.
7. All important and relevant geological surfaces must be mapped and the maps integrated to arrive at a reasonable and accurate subsurface picture.
8. The mapping of multiple horizons is essential to develop reasonably correct 3D interpretations of complexly faulted areas.

9. Balanced cross sections are required to prepare a reasonably correct interpretation of complexly deformed structures.
10. All work should be documented.

The following text provides more detail, defining the ten points of the Philosophical Doctrine.

1. All subsurface interpretations must be geologically and geometrically valid in three dimensions. Most subsurface data are either 1D (well log) or 2D (well log cross sections and conventional seismic sections); however, these data are used to generate a 3D depiction of the subsurface. Even though it is intuitive that all interpretations must be valid in three dimensions, too often subsurface structure maps, cross sections, and seismic interpretations are made without much consideration given to establishing a 3D framework or verifying that the interpretation is even possible in three dimensions (Tearpock et al. 1994). There are a number of methods and techniques that can be used to validate an interpretation before investment dollars are committed to a prospect, lease, concession, or field purchase. Management should require that such work be done prior to investment decisions.

2. An interpreter must have a fundamental, classic education in geology, including field experience, and a strong background in structural geology for the tectonic setting being worked. When an interpretation is made in a particular tectonic setting, the interpreter must know as much as possible about the geology of the area so that the interpretation represents geology that is known to fit the style of the area. This places the requirement on an interpreting geoscientist initially to have a fundamental, classic education in geology. Without knowledge and understanding of geology, the applications of geophysics, petrophysics, workstation activities, and even computer mapping will be questionable and potentially inconsistent with geological principles.

Of special interest is the area of structural geology. With few exceptions, interpreters are working in tectonic areas where a knowledge of structural geology is important. A limited understanding of structural geology is one of the shortcomings in numerous geological interpretations that result in unrealistic or even impossible interpretations in three dimensions.

3. Sufficient planning, time, and detail are required to generate reliable prospects. Haste makes waste. Do not be too anxious to drill that next dry hole. There are not many shortcuts to good prospecting. Initially, it is essential to develop a program or project plan designed to meet the objectives and provide the needed deliverables. With limited time available to complete a project, alternate solutions may not be analyzed, all the data may not be used, unjustified shortcuts might be taken, or incorrect techniques may be applied. When you consider the cost of a dry hole, an unsuccessful exploration program, or the loss of investor confidence, the time taken to ***do the project right the first time*** is time well spent. Remember Murphy's law: "If something can go wrong, it will."

Management must also realize that each project is in some ways unique. Therefore, the detail required for any given project to achieve the objectives should be specific to that project. Consider two projects: one on a simple four-way closure and the other on a complex compressional duplex. It is obvious that these studies will require different plans, timetables, and technical details. Use common sense in evaluating the time, plan, and detail required.

This habit does not mean that external time constraints will never dictate the scope of work. A prospect map completed after a lease sale or a reserve estimate completed after an acquisition may be worth much less than a slightly less accurate map available before a lease sale or a less accurate reserve estimate before an acquisition price is agreed. In these and other cases, some shortcuts may be necessary. It is important, however, that these shortcuts be taken deliberately and with a recognition of the additional risk that is involved. It is especially important that management be aware of the additional risk involved. If the additional risk is unacceptable, perhaps management can make additional resources available to do the required work.

4. All subsurface data must be used to develop a reasonable and accurate subsurface interpretation. The data available for interpretation are limited and are typically 1D and 2D (including a seismic section displayed on a 3D conventional workstation). A body of data itself can be confusing with respect to true subsurface relationships. For example, cross sections and seismic sections can misrepresent true 3D subsurface relationships by the simple nature of their orientations. All data (well log, seismic, production, paleontologic, etc.) must be integrated into an interpretation if it is to be considered sound and viable.

Important thoughts to keep in mind while compiling and using subsurface data are the physical limits and accuracy of these data. A degree of confidence must be built into any final interpretation, but the end result can be only as good as the data that are used. The reasons for failing to recognize inaccuracies of data can range from blindly accepting autocorrelations to not investigating questionable directional surveys, mud logs, electric logs, core reports, velocity functions, production data, and so on. Inclusion of incorrect data may be unavoidable due to inaccuracies inherent in tools and procedures that generate data or to historical methods of recordkeeping. Many times, however, it is possible to determine whether the data are out of date, incomplete, or subject to error.

Questionable data may represent the only data that are currently available. In other cases, more reliable data are available and should be sought. The physical limits and possible inaccuracies of the data should always be noted when presenting completed work. Always acknowledge questionable findings and possible discrepancies in a final interpretation. Also, all data used must be used correctly in order to ensure accuracy in any interpretation.

5. Accurate correlations (well log and seismic) are required for reliable geological interpretations. An interpretation that properly integrates all data, such as well log, seismic, and production data, is always more accurate than an interpretation that ignores one of these sources (Tearpock and Bischke 1991). Likewise, the correlations must be accurate because geological interpretations have their foundation in correct correlations. Consider that all aspects of subsurface interpretation and prospecting are based on correlations. These correlations are used to prepare cross sections and fault, unconformity, salt, structure, and isochore/isopach maps. Workstations have given geoscientists the ability to work immensely large databases in a relatively short time. They have also provided, however, a method of autopicking correlations. These correlations are at times blindly accepted and therefore not verified by the interpreter before plunging into map generation. This can be a blessing as well as a curse to the less knowledgeable.

Eventually, a geoscientist's correlations, right or wrong, are incorporated into the final interpretation. Incorrect correlations can be costly; they can result in a dry hole, an unsuccessful workover or recompletion, the purchase of an uneconomic property, or the sale of a producing property that has significant, unrecognized potential.

6. The use of correct mapping techniques and methods is essential to generate reasonable and correct subsurface interpretations. Maps and cross sections are in most cases the primary vehicles used to organize, interpret, and present available subsurface information. The reliability of any subsurface interpretation presented on maps and cross sections is directly related to the use of accurate and correct mapping methods and techniques.

Geoscientists who have a good understanding of the mapping methods applicable in the area of study prepare the most accurate geological interpretations. There is no substitute for correct mapping techniques. A poor understanding of mapping techniques can result in incorrect procedures, unjustified shortcuts, and costly inaccurate interpretations. We illustrate interpretive contouring as an example of the use of a correct mapping technique. Interpretive contouring is the most acceptable method of contouring subsurface features (Bishop 1960). Unlike other contouring methods, interpretive contouring allows the geoscientist to use knowledge of the structural and depositional style in the tectonic setting being worked, the ability to think in three dimensions, field and work experience, imagination, and geological license to generate an interpretation that is geologically sound.

In today's world of computer-contoured maps, can you, as a geoscientist, define which gridding or triangulation procedure should be used to generate a map that honors the data and reflects a mapped interpretation in a way similar to a hand-contoured map made by an experienced geoscientist? This is one example of a technique with which a geoscientist must be familiar in order to generate a good subsurface interpretation.

7. All important and relevant geological surfaces must be mapped and the maps integrated to arrive at a reasonable and accurate subsurface picture. These include surfaces such as tops of formations, stratigraphic markers, faults, unconformities, and salt. For example, in faulted areas it is typically the faults that form the structures, such as rollovers, fault bend folds, and fault propagation folds. To develop a good understanding of any faulted structure, one must analyze, interpret, and map the faults. We cannot overemphasize the importance of mapping faults. In addition, faults play an important role in both the migration and trapping of hydrocarbons. Therefore, a reasonable structural interpretation in most faulted areas is dependent on an accurate 3D understanding of the faults in the area.

The next step in the interpretation process is to integrate the faults with the structure to arrive at an accurate interpretation. If you want to drill more than your share of dry holes, do not map faults.

8. The mapping of multiple horizons is essential to develop reasonably correct 3D interpretations of complexly faulted areas. It allows the interpreter to establish a 3D structural framework prior to generating prospects. The mapping of at least three horizons (shallow, intermediate, and deep) allows the geoscientist to determine whether the interpretation is plausible and fits at all levels from shallow to deep. Remember, almost any set of fault and structural data can be forced to fit on one horizon. The true test of an interpretation is to have the data fit at all structural levels. Therefore, *no prospect should be accepted for review without verification that multiple levels have been interpreted and mapped.* The multiple horizon mapping must show that the interpreted structural framework is geologically sound and that it conforms to 3D spatial and temporal relationships.

9. Balanced cross sections are required to prepare a reasonably correct interpretation of complexly deformed structures. In complexly deformed areas, the application of sound structural methods and the generating of structurally balanced cross sections are necessary to prepare admissible interpretations. Structural interpretations are not cast in stone. Any given structure was not always as it is today. If an interpreted structure does not volume-balance or a cross section does not area-balance, then the interpretation cannot be correct from a simple geometric point of view. Thus, balancing can direct the interpreter toward the correct interpretation. Structural balancing provides a better understanding of the present configuration of structures; how structures form; how, when, and where fluids may have entered the structures; and where hydrocarbons may presently exist. Structural balancing should be an integral part of geological interpretations in complex areas.

10. All work should be documented. Significant volumes of data are collected, evaluated, used, and manipulated during a project. Good, accurate recording of data and a description of procedures taken by an interpreter make everyone's work go more smoothly and accurately. The documentation of work is an integral part of that work. All subsurface projects will provide better results if the data collected and generated are recorded in some format that can easily be referenced, used, or revised. These data include information on formation tops and bases, faults, unconformities, net sand counts, correction factors, and more. All completed work needs to be supported by all the raw data, whether obtained from commercial services or internally generated.

Many persons, including the interpreter, supervisors, managers, other members of an organized study team, and persons inheriting the area of study, may at some time need to review the subsurface data. Numerous types of data sheets are available for documenting geological activities. Everyone should become familiar with the forms used in their company and use them on a regular basis.

The Philosophical Doctrine presented here has been employed by many successful geoscientists. Hydrocarbons are not found by luck, serendipity, or guesswork but instead by solid scientific work. Historical data show that those geoscientists who practice this philosophy on a regular basis are more successful finders and developers of hydrocarbons than those who do not. Workstations and computers have enabled geoscientists to evaluate more data in a short amount of time and to apply the philosophy and methods faster than by hand. However, we must remember that people, not workstations, find oil and gas, and the philosophy and methods used have a great impact on success.

MENTORS AND MENTORING

One additional characteristic of most highly effective subsurface mappers is the existence of at least one strong mentor early in their career. Much of the learning about subsurface mapping is derived from solving your own problems working with your own data. Historically, more experienced geoscientists have worked closely with less experienced geoscientists, sharing knowledge, methods, and techniques. One can even state that advanced mappers and oil finders are typically **only as good as their mentors**.

Some companies have recognized the importance of mentors in developing the careers of younger geoscientists. These companies explicitly encourage experienced geoscientists to establish mentor–mentee relationships with less experienced geoscientists.

Mentoring does not necessarily involve older geoscientists transferring experience and expertise to younger geoscientists. A young geologist who is experienced in geosteering, for example, may well serve as a mentor to an older geologist who is not experienced in geosteering. Similarly, a young sedimentologist or structural geologist may serve as a mentor to an older geologist who is less experienced in these specialties.

TYPES OF SUBSURFACE MAPS AND CROSS SECTIONS

When conducting any detailed subsurface geological study, a variety of maps and cross sections may be required. The numerous techniques available to use in the preparation of these maps and sections are discussed in subsequent chapters.

As mentioned earlier, the primary focus of this book centers on structural and mapping methods and the maps and cross sections used to find and develop hydrocarbons. However, the techniques are applicable to many other related geological fields. The following is a list of the types of maps and sections discussed in this book.

SUBSURFACE MAPS. Structure, structural shape, porosity top and base, fault surface, unconformity, salt, net sand, net hydrocarbon, net oil, net gas, isochore, interval isopach, facies, and palinspastic maps.

CROSS SECTIONS. Structure, stratigraphic, problem-solving, final illustration, balanced, and correlation sections. Conventional and isometric fence diagrams and 3D models are also presented.

CHAPTER 2

CONTOURING AND CONTOURING TECHNIQUES*

INTRODUCTION

A wide variety of subsurface maps are discussed in the following chapters. Each map presents a specific type of subsurface data obtained from one or more sources. The purposes of these maps are to present data in a form that can be understood and used to explore for, develop, or evaluate energy resources such as oil and gas.

It might seem elementary to have a chapter on contouring, since most geologists are taught the basics of contouring in several introductory geology courses. However, there are two good reasons for this chapter. First, part of our audience includes members of the geophysical and petroleum engineering disciplines. They may have had little, if any, training in basic contouring principles and methods. Second, because the understanding and correct application of contouring and contouring techniques is of paramount importance in establishing a solid foundation in subsurface mapping, a review of contouring is appropriate.

The majority of subsurface maps use the *contour line* as the vehicle to convey the various types of subsurface data. By definition, a contour line is a line that connects points of equal value. Usually this value is compared to some chosen reference, such as sea level in the case of structure contour maps. In preparing subsurface maps, we are dealing with data beneath the earth's surface, which cannot be seen or touched directly. Therefore, the preparation of a geologically reasonable subsurface map requires an in-depth knowledge of geology, interpretation skills, imagination, an understanding of 3D geometry, and the use of correct mapping techniques.

* For all figures in this chapter (in the printed book only), see the preface for information about registering your copy on the InformIT site for access to the electronic versions in color.

Any map that uses the contour line as its vehicle for illustration is called a *contour map.* A contour map illustrates a 3D surface or solid in 2D plan (map) view. Any set of data that can be expressed numerically can be contoured.

The following list shows examples of contourable data and the associated contour map.

Data	***Type of Map***
Elevation	Structure, Fault, Salt
Thickness of sediments	Interval Isopach
Percentage of sand	Percent Sand
Feet or meters of pay	Net Pay Isochore
Pressure	Isobar
Temperature	Isotherm
Time	Isochron
Lithology	Isolith

If the same set of data points to be contoured is given to several interpreters, the individually contoured maps generated would likely be different. Differences in an interpretation are the result of educational background, the amount of geological training, field and work experience, imagination, and interpretive abilities (such as visualizing in three dimensions). Yet the use of all the available data and an understanding and application of the basic principles and techniques of contouring should be the same. These principles and actual techniques are fundamental to the construction of a mechanically correct map.

In the first part of this chapter, the importance of visualizing in three dimensions and the basic rules of contouring are discussed. In addition, various techniques for contouring by hand are illustrated and certain important guidelines identified. Later, computer-based contouring is discussed.

THREE-DIMENSIONAL PERSPECTIVE

In this section, we show how 3D surfaces are represented by contours in map view. A good understanding of the geometry within our subsurface geological world, tectonics, and the principles of 3D spatial relationships are essential to any attempt at constructing a picture of the subsurface. Some geoscientists have a stronger educational background in the geological sciences than others. In addition, some geoscientists have an innate ability to visualize in three dimensions, whereas others do not. One of the best ways to develop this ability is to practice perceiving objects in three dimensions.

In its simplest form, we can view a plane dipping in the subsurface with respect to the horizontal. (The reference datum for all the examples shown is sea level.) Figure 2-1 shows an isometric view of a plane dipping at an angle of 45 deg with respect to the horizontal and a projection of that dipping plane upward onto a horizontal surface to form a contour map. This dipping plane intersects an infinite number of horizontal planes; but, for any contour map, only a finite set of evenly spaced horizontal plane intersections can be used to construct the map. (For a subsurface structure contour map, the intersections used may be 50 ft or m, 100 ft or m, or even 500 ft or m apart.) By choosing evenly spaced finite values, we have established the contour interval for the map.

Next, it is important to choose values that are easy to use for the contour lines. For example, if a 100-ft contour interval is chosen, then the contour line values selected to construct the map should be in even increments of 100 ft, such as 7000 ft, 7100 ft, and 7200 ft. Any increment of 100 ft could be chosen, such as 7040 ft, 7140 ft, and 7240 ft. This approach, however, makes the map more difficult to construct and harder to read and understand. In Figure 2-1, a 100-ft contour interval was chosen for the map (the minus sign in front of the depth value indicates the value is below sea level). The intersection of each horizontal plane with the dipping plane results in a line of intersection projected into map view on the contour map above the isometric view. This contour map is a 2D representation of the 3D dipping plane.

Now we complicate the picture by introducing a dipping surface that is not a plane but is curved (Fig. 2-2). The curved surface intersects an infinite number of horizontal planes, as did

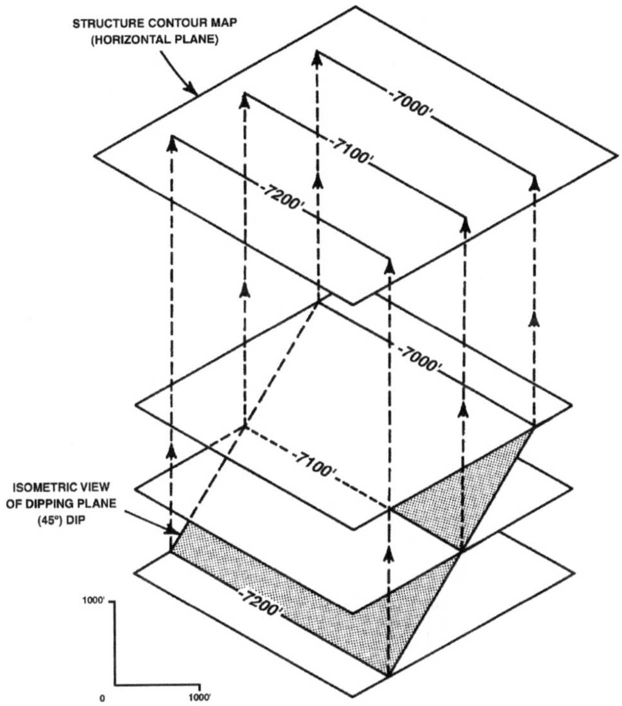

Figure 2-1 Isometric view of dipping plane intersecting three horizontal planes. (Modified from Appelbaum. Geological & Engineering Mapping of Subsurface: A workshop course by Robert Appelbaum. Published by permission of Pearson Education, Inc.)

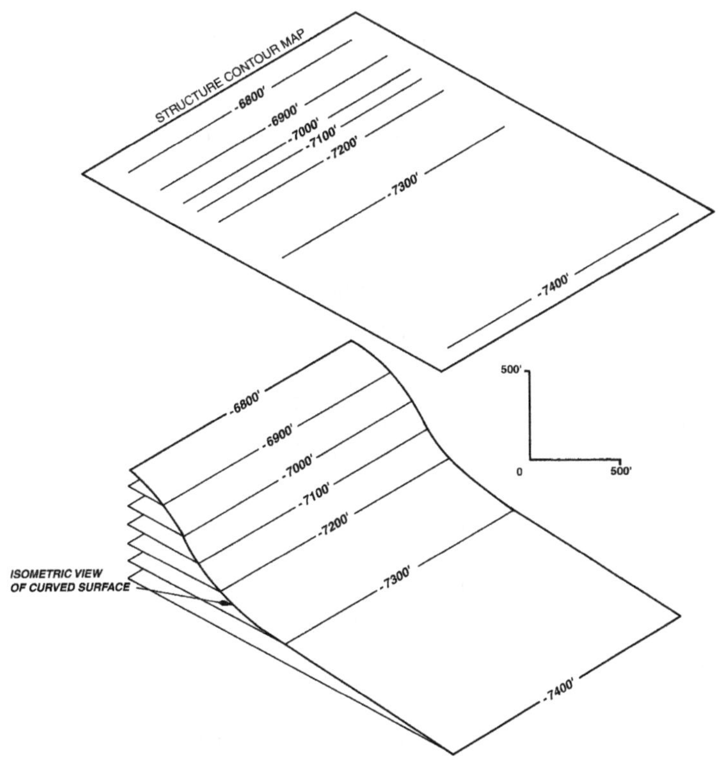

Figure 2-2 Isometric view of a curved surface intersecting a finite number of evenly spaced horizontal planes. (Modified from Appelbaum. Geological & Engineering Mapping of Subsurface: A workshop course by Robert Appelbaum. Published by permission of Pearson Education, Inc.)

the plane in the first example. Each intersection of two surfaces results in a line of intersection, which everywhere has the same value. By projecting these lines into plan view onto a contour map, a 3D surface is represented in two dimensions. If we consider the curved surface as a surface with a changing slope, then the spacing of the contours on the map is representative of the change in slope of the curved surface. In other words, steep slopes are represented by closely spaced contours, and gentle slopes are represented by widely spaced contours (Figs. 2-2 and 2-3). This relationship of contour spacing to change in slope angle assumes that the contour interval for the map is constant.

Finally, we must emphasize that any person generating a subsurface map must have the geological background to understand whether or not the map produced truly represents what is possible in the subsurface. Too often, contour maps violate geological principles or depict structures that are unlikely or even impossible in the subsurface. All maps generated, whether done by hand or computer, must be driven by sound geological principles. Computers are fast, but the accuracy of the resultant interpretation or map is dependent on many factors, such as the algorithm used for gridding, distribution of data, and other factors (refer to section Computer-Based Contouring Concepts and Applications). The computer is a tool, as are an engineer's scale and ten-point spacing dividers. Yes, it is a powerful tool, but nonetheless a tool. We cannot accept computers driving interpretations, nor can we blindly accept the resultant maps. Our educational background, experience, and geological principles should control any interpretation or generated map. The workstations and personal computers in wide usage today are powerful, but they are

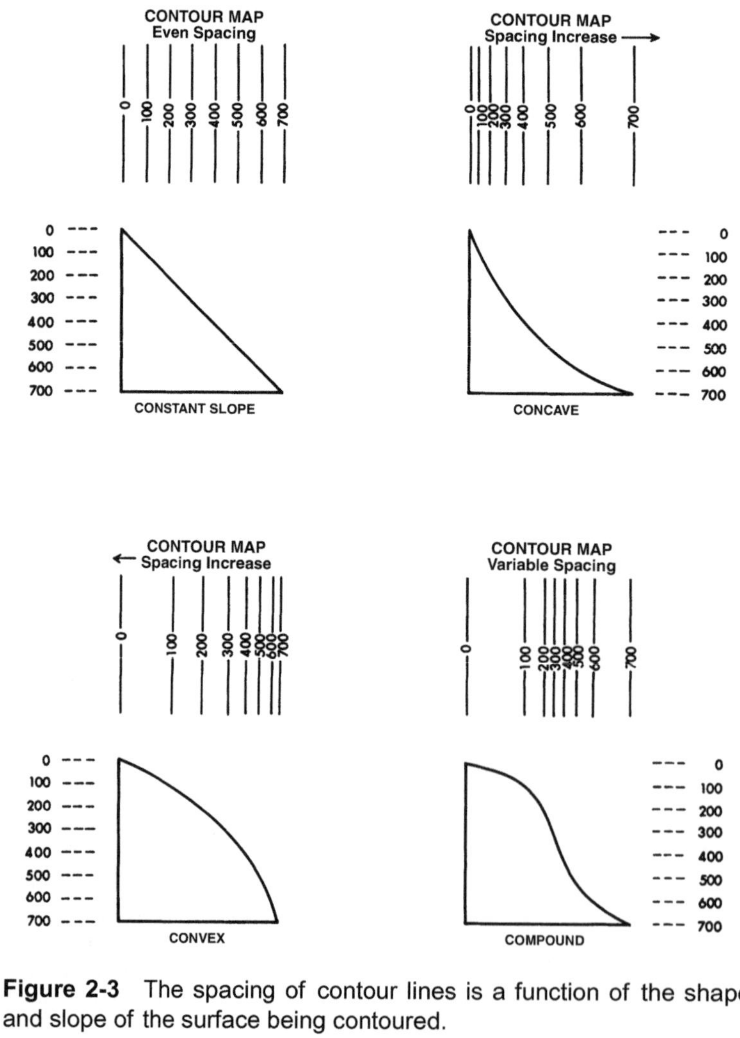

Figure 2-3 The spacing of contour lines is a function of the shape and slope of the surface being contoured.

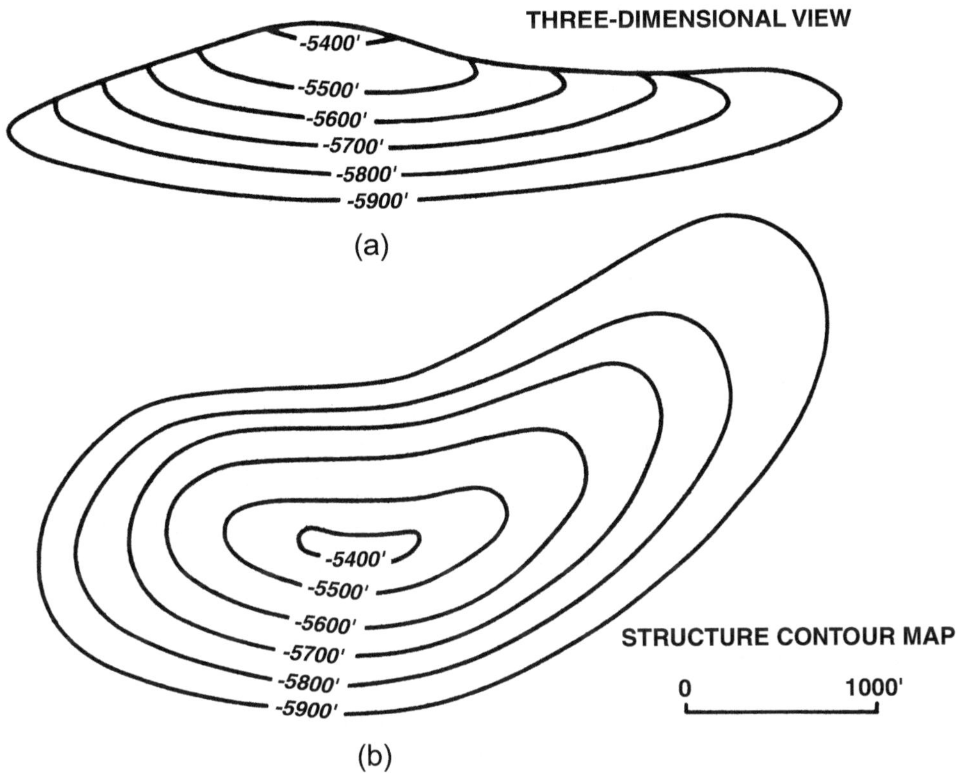

Figure 2-4 (a) A 3D view of an anticlinal structure and (b) the contour map representing the structure.

not artificial intelligence capable of generating geologically and geometrically reasonable interpretations or maps.

Look at a 3D subsurface formation that is similar in shape to a topographic elongated anticline and the contour map representing that surface (Fig. 2-4). The contour map graphically illustrates the subsurface formation in the same manner that a topographic map depicts the surface of the earth. By using your ability to think in three dimensions, it is possible to look at the contour map and visualize the formation in its true subsurface 3D form.

CALCULATION OF BED DIP

Contour spacing not only provides a qualitative measure of bed dip, with closely spaced contours representing steeply dipping surfaces and widely spaced contours representing gently dipping surfaces, it also allows for the quantitative calculation of bed dip. The contour interval measures the elevation change on a surface. The contour spacing, measured from the scale of the map, reflects the horizontal distance over which the depth of the surface changes by the contour interval. The elevation change determined from the contour interval is typically called the *rise*, and the contour spacing is called the *run*. From basic trigonometry,

$$\tan\phi = C.I./C.S. \tag{2-1}$$

where

C.I. = contour interval = rise
C.S. = contour spacing = run
ϕ = the dip of the surface being contoured

Rearranging,

$$\phi = \arctan(C.I./C.S.) = \arctan\frac{\text{rise}}{\text{run}} \tag{2-2}$$

where

Rise = elevation change
Run = horizontal distance from map scale

To get true bed dip, the contour spacing must be measured perpendicular to the strike of the contours. Contour spacing measured at any other angle will yield an apparent dip that is less than the true dip. A graph of apparent dip and the angle between the line of section and the true dip direction is presented in Suppe (1985).

The preceding discussion applies to depth structure maps. Bed dip can be determined qualitatively from contour spacing on time structure maps. Bed dip cannot be calculated quantitatively from time structure (isochron) maps without converting the time contours into depth. The depth interval between time contours is frequently not constant over the area of a structure map, as interval velocity is generally a function of depth and may also vary laterally.

RULES OF CONTOURING

Several rules must be followed in order to draw mechanically correct contour maps. This section lists these rules and discusses a few exceptions.

1. *A contour line cannot cross itself or any other contour except under special circumstances* (see rule 2). Since a contour line connects points of equal value, it cannot cross a line of the same value or lines of different values.
2. *A contour line cannot merge with contours of the same value or different values.* Contour lines may appear to merge or even cross where there is an overhang, overturned fold, or vertical surface (Fig. 2-5). With these exceptions, the key word is *appear*. Consider a vertical cliff that is being mapped. In map view the contours appear to merge, but in 3D space these lines are above each other. For the sake of clarity, contours should be dashed on the underside of an overhang or overturned fold.
3. *A contour line must pass between points whose values are lower and higher than its own value* (Fig. 2-6). In other words, there must be a contour line between points whose values are lower and higher than the value of the contour line.
4. *A contour line of a given value is repeated to indicate reversal of slope direction.* Figure 2-6 illustrates the application of this rule across a structural high (anticline) and a structural low (syncline).
5. *A contour line on a continuous surface must close within the mapped area or end at the edge of the map.* Geoscientists often finesse this rule by preparing what is commonly referred to as a *postage stamp* map. This is a map that covers a very small area when compared to the areal extent of the structure. It is very easy to close a contour line at the edge of a postage stamp map.

These five contouring rules are simple. If they are followed during mapping, the result will be a map that is mechanically correct. Some computer mapping software may be programmed to obey these five rules, but not all mapping software is so programmed. In addition to these rules are other guidelines to contouring that make a map easier to construct, read, and understand.

1. All contour maps should have a chosen reference to which the contour values are compared. A structure contour map, as an example, typically uses mean sea level as the chosen reference.

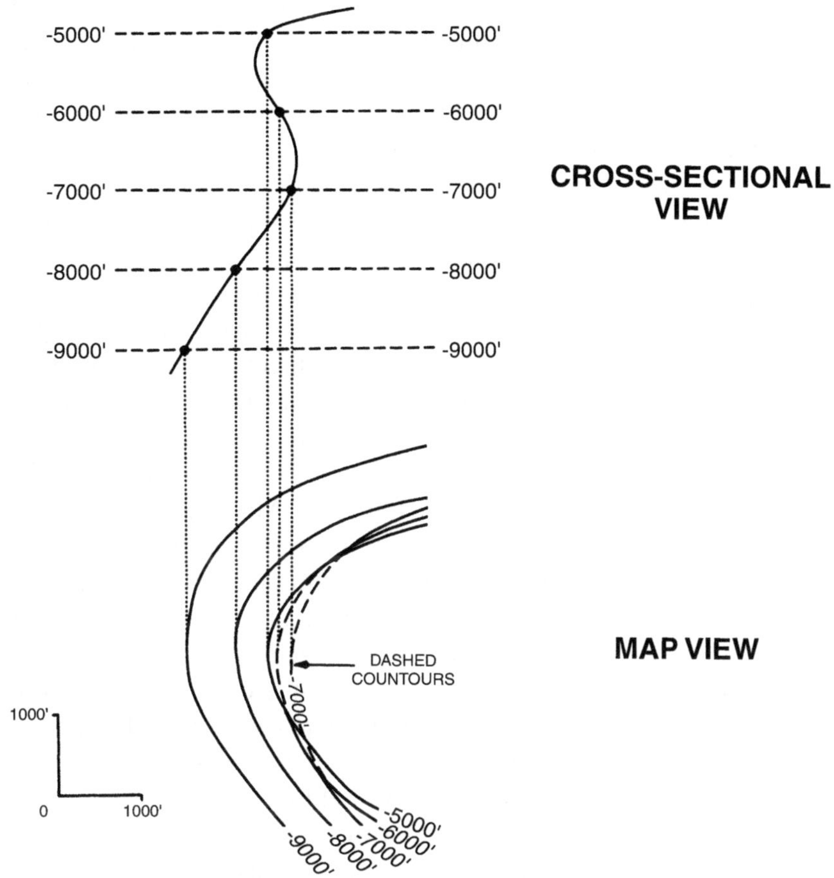

Figure 2-5 To clearly illustrate a 3D overhang or overturned fold, dash the contours on the underside of the structure. (From Tearpock and Harris 1987. Published by permission of Tenneco Oil Company.)

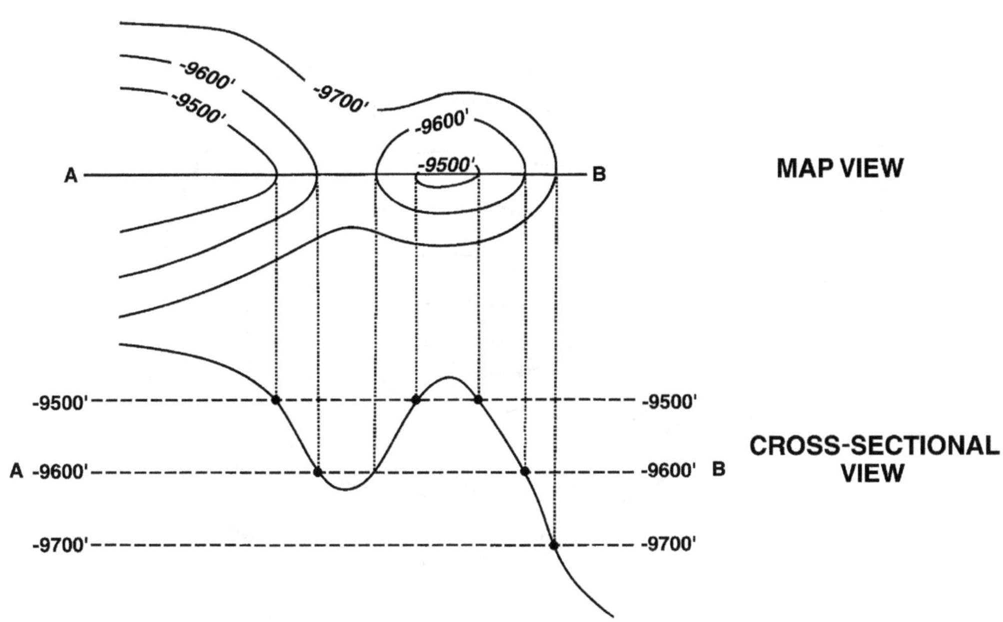

Figure 2-6 A contour line must be repeated to show reversal of slope direction. (From Tearpock and Harris 1987. Published by permission of Tenneco Oil Company.)

Therefore, the elevations on the map can be referenced as being above or below mean sea level. A negative sign in front of a depth value means the elevation is below sea level (e.g., −7000 ft).

2. The contour interval on a map should be constant. The use of a constant contour interval makes a map easier to read and visualize in three dimensions because the distance between successive contour lines has a direct relationship to the steepness of slope. Remember, steep slopes are represented by closely spaced contours and gentle slopes by widely spaced contours (see Fig. 2-3). If for some reason the contour interval is changed on a map, it should be clearly indicated. This can occur where a mapped surface contains both very steep and gentle slopes, such as those seen in areas of salt diapirs. The choice of a contour interval is an important decision. Several factors must be considered in making such a choice. These factors include the density of data, the practical limits of data accuracy (e.g., directional surveys), the steepness of slope, the scale of the map, and its purpose. If the contour interval chosen is too large, small closures with less relief than the contour interval may be overlooked. If the contour interval is too small, however, the map can become too cluttered and reflect inaccuracies of the basic data.
3. All maps should include a graphic scale (Fig. 2-7). Many people may eventually work with or review a map. A graphic scale provides an exact reference and gives the reviewer an idea of the areal extent of the map and the magnitude of the features shown. Also, it is not uncommon for a map to be reproduced, projected, or viewed on a computer screen. During this process, the map may be reduced or enlarged. Without a graphic scale, the values shown on the map may become useless. Because it is easy to stretch or squeeze a map on a computer, we recommend that *scale bars have both an X and a Y graphic scale* (Fig. 2-7).
4. Every fifth contour should be thicker than the other contours, and it should be labeled with the value of the contour. This fifth contour is referred to as an *index contour*. For example, with a structure contour map using a 100-ft contour interval, it is customary to thicken and label the contours every 500 ft. At times it may be necessary to label other contours for clarity (Fig. 2-7).
5. Hachured lines should be used to indicate closed depressions (Fig. 2-7).
6. Start contouring in areas with the maximum number of control points (Fig. 2-8).
7. Construct the contours in groups of several lines rather than one single contour at a time (Fig. 2-8). This should save time and provide better visualization of the surface being contoured.

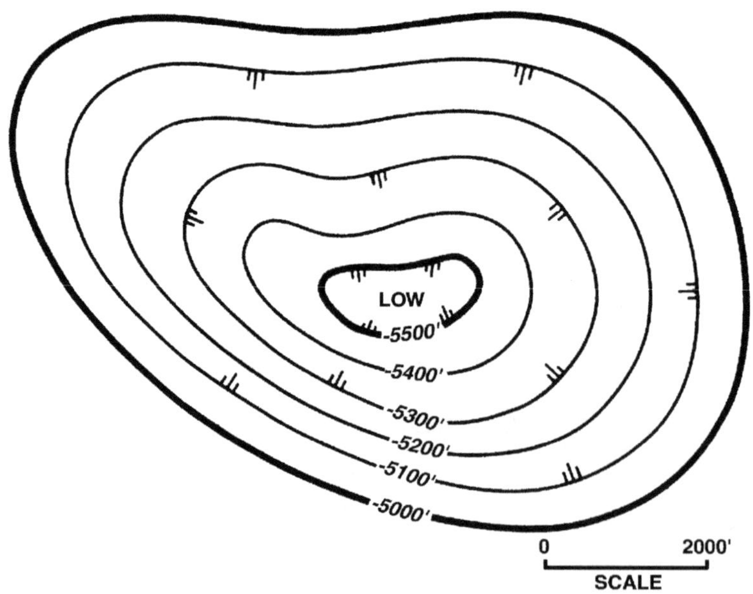

Figure 2-7 Closed depressions are indicated by hachured lines.

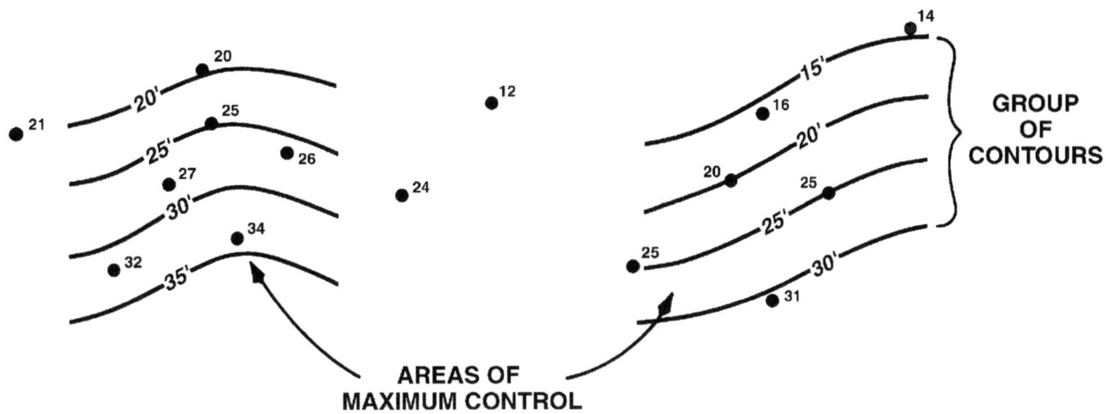

Figure 2-8 Begin contouring in areas of maximum control using groups of contour lines.

8. Initially, choose the simplest contour solution that honors the control points and provides a realistic subsurface interpretation.
9. Use a smooth rather than undulating style of contouring unless the data indicate otherwise (Fig. 2-9).
10. Initially, a hand-drawn map should be contoured in pencil with the lines lightly drawn so they can be erased as the map requires revision.
11. If possible, prepare hand-contoured maps on some type of transparent material such as mylar or vellum. Often, several individual maps have to be overlaid one on top of the other (see Chapter 8). The use of transparent material makes this type of work easier and faster.

METHODS OF CONTOURING BY HAND

As mentioned previously, different contoured interpretations can be constructed from the same set of values. The differences in the finished maps may be the result of the geoscientists' educational background, experience levels, interpretive abilities, or other individual factors. This section establishes that the differences can also be the result of the method of contouring used by each geoscientist (see section Computer-Based Contouring).

Unlike topographic data, which are usually obtainable in whatever quantity needed to construct very accurate contour maps, data from the subsurface is scarce. Therefore, any subsurface map is subject to individual interpretation. The amount of data, the areal extent of that data, and the purpose for which a map is being prepared may dictate the use of a specific method of contouring. There are four distinct methods of hand contouring: (1) mechanical, (2) equal-spaced, (3) parallel, and (4) interpretive (Rettger 1929; Bishop 1960; and Dennison 1968).

1. **Mechanical Contouring**. By using this method of contouring, one may assume that the slope or angle of dip of the surface being contoured is uniform between points of control and that any change occurs at the control points. Figure 2-10 is an example of a mechanically contoured map. With this approach, the spacing of the contours is mathematically (mechanically) proportioned between adjacent control points. We use line A-A' in Figures 2-10 and 2-11 to illustrate this method of contouring. Wells No. 2, 4, and 5 lie on line A-A' with depths of 600 ft, 400 ft, and 200 ft, respectively. The contour interval used for this map is 100 ft. First, we can see that the 600-ft, 400-ft, and 200-ft contour lines pass through Wells No. 2, 4, and 5. Next, we need to determine the position of the 500-ft and 300-ft contour lines. Remembering that this method assumes a uniform slope or dip between control

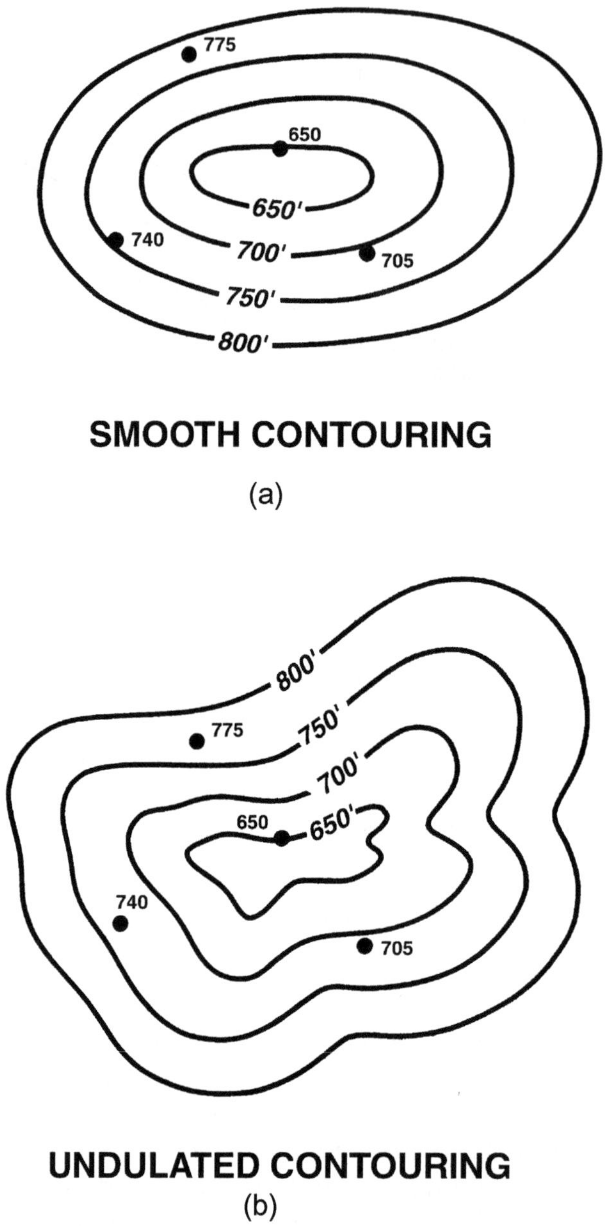

Figure 2-9 A smooth style of contouring is preferred over an undulating style.

points, we can use ten-point spacing dividers or an engineer's scale to interpolate the location of these two contour lines. The 500-ft contour line lies midway between the 600-ft and 400-ft contour lines. Likewise, the 300-ft contour line is placed midway between the 400-ft and 200-ft contour lines. When this procedure is repeated for all adjacent control points, the result is a mechanically contoured map that is geometrically accurate.

Mechanical contouring allows for little, if any, geological interpretation. Even though the map is mechanically correct, the result may be a map that is geologically unreasonable, especially in areas of sparse control.

Although mechanical contouring is not recommended for most contour mapping, it does have application in a few areas and may be a good first step when beginning work in a new geographic area. When there is a sufficient amount of seismic or well control, such as in a densely drilled mature oil or gas field, this method may provide reasonable results, since there is little

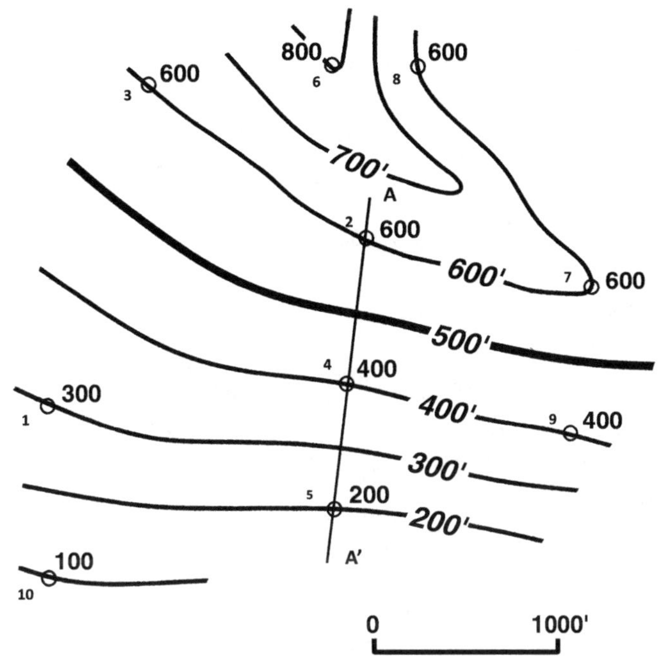

Figure 2-10 Mechanical contouring method. (Modified from Bishop 1960. Published by permission of author.)

room for interpretation. This method is at times employed in litigation, equity determinations, and unitization because it supposedly minimizes individual bias in the contouring. However, although individual bias may be minimized, the method does not allow for true geological interpretation. The method is therefore not recommended for the activities listed here.

2. **Parallel Contouring**. With this method of contouring, the contour lines are drawn parallel or nearly parallel to each other. This method does not assume uniformity of slope or angle of dip as in the mechanical contouring method. Therefore, the spacing between contours may vary (Fig. 2-12).

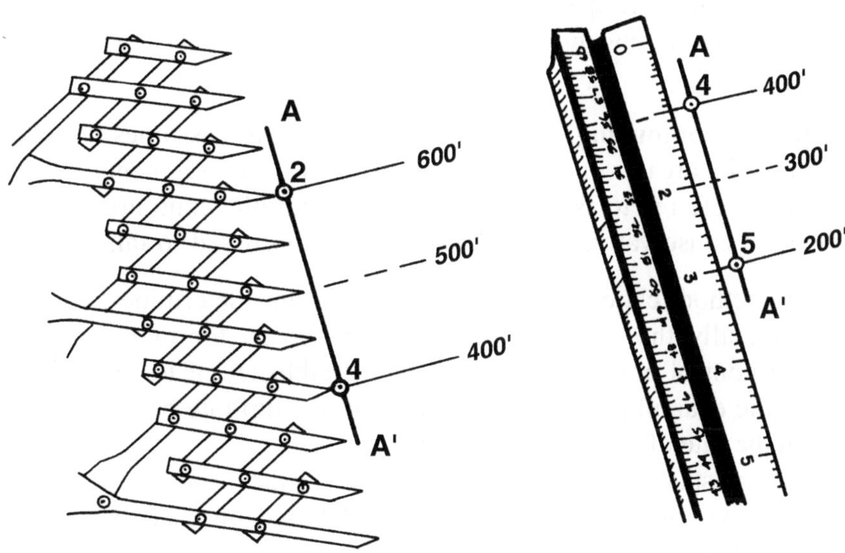

Figure 2-11 Two different methods of establishing contour spacing.

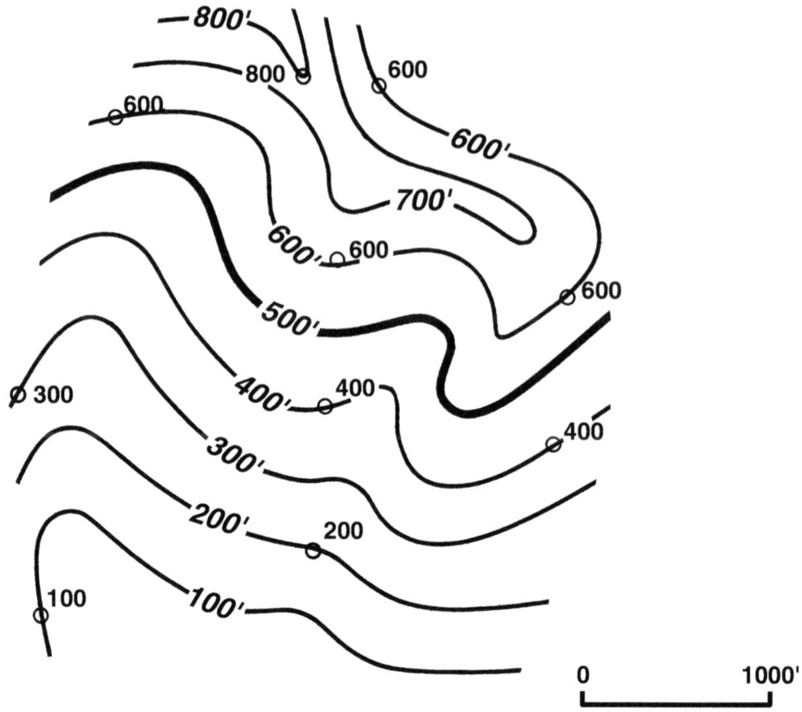

Figure 2-12 Parallel contouring method. (Modified from Bishop 1960. Published by permission of author.)

As with the previous method, if honored exactly, parallel contouring may yield an unrealistic geological picture. Figure 2-13 shows a map that has been contoured using this method. Notice that the highs appear as bubble-shaped structures with the adjoining synclines represented as sharp cusps. This map depicts an unreasonable geological picture.

Although this method may yield an unrealistic map, it does have advantages over mechanical contouring. First, the method allows some *geological license* to draw a more realistic map than one constructed using the mechanical method, because there is no assumption of uniform dip between control points. Also, this method is not as conservative as true mechanical contouring. Therefore, it may reveal features that would not be represented on a mechanically contoured map.

Parallel contouring is frequently used for fault surface maps (e.g., Figs. 7-12 and 7-47) where the dip of the fault may change with depth but the strike of the fault is relatively constant with depth.

3. **Equal-Spaced Contouring**. This method of contouring assumes uniform slope or angle of dip over an entire area, over an individual flank, or over a segment of a structure. Sometimes this method is referred to as a special version of parallel contouring. Equal-spaced contouring is the least conservative of the three methods discussed so far.

To use this method, choose closely spaced data and determine the slope or angle of dip between them. Usually the slope or angle of dip chosen for mapping is the steepest found between adjacent control points. Once the dip is established, it is held constant over the entire mapped area. In the example shown in Figure 2-14, the dip rate between Wells No. 2 and 4 was used to establish the rate of dip for the entire map.

Since the equal-spaced method of contouring is the least conservative, it may result in numerous highs, lows, or undulations that are not based on established points of control but are the result of maintaining a constant dip rate or slope. The advantage to this method, in the early stages of mapping, is that it may indicate a *maximum* number of structural highs and lows

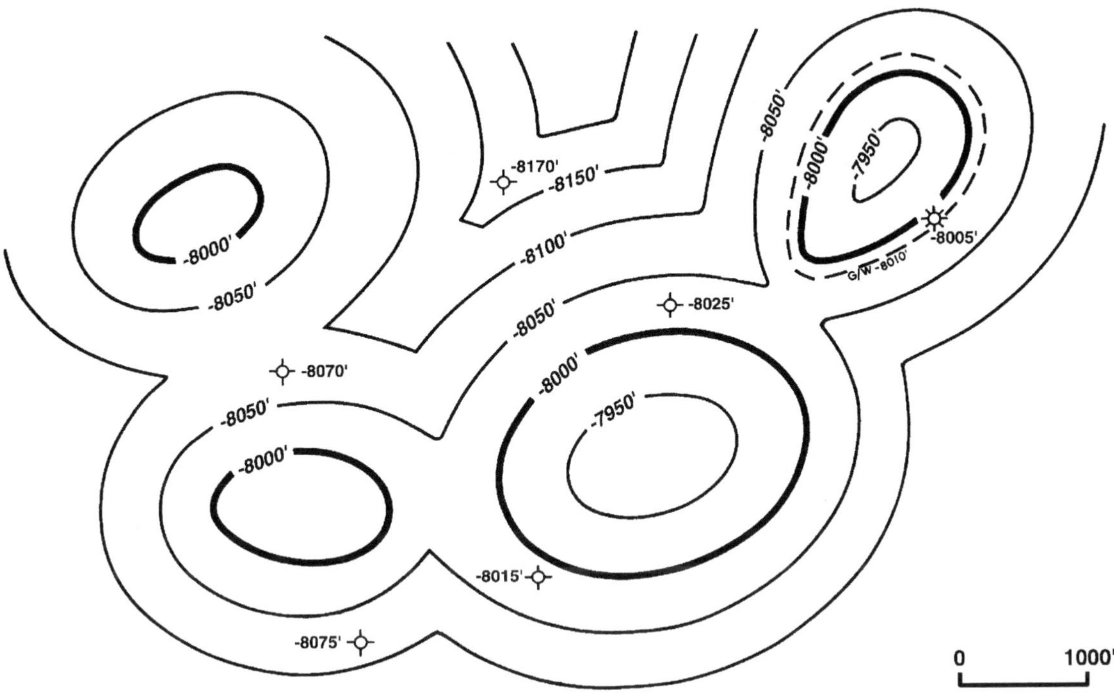

Figure 2-13 An example of an unrealistic structure map constructed using the parallel contouring method.

expected in the study area. One assumption that must be made in using this method is that the data used to establish the slope or rate of dip are not on opposite sides of a nose or on opposite flanks of a fold. In Figure 2-14, Wells No. 6 and 8 are on the northern flank of this structure; therefore, neither well can be used with a well on the southern flank to establish the rate of dip. These two wells can be and were used to establish the rate of dip for contouring the northern flank of this southeast trending structural nose.

This method does have application on the flanks of structures that have a uniform dip, such as kink band folds. For example, the back limb of a fault bend fold, which typically is parallel to the dip of the thrust fault that formed the fold, may have a constant dip between axial surfaces (Fig. 10-36). In such a case, the equal-spaced method of contouring may have some applicability in contouring this back limb.

4. **Interpretive Contouring**. With this method of contouring, a geoscientist has extreme geological license to prepare a map to reflect the best interpretation of the area of study while honoring the available control points (Fig. 2-15). No assumptions, such as constant bed dip or parallelism of contours, are made when using this method. Therefore, the geoscientist can use experience, imagination, ability to think in three dimensions, and an understanding of the structural and depositional style in the geological region being worked to develop a realistic interpretation. Interpretive contouring is the most acceptable and the most commonly used method of hand contouring.

As mentioned earlier, the specific method chosen for contouring may be dictated by such factors as the number of control points, the areal extent of these points, and the purpose of the map. It is essential to remember that no matter which method is used in making a subsurface map, *the map is not correct.* No one can really develop a correct interpretation of the subsurface with the same accuracy as that of a topographic map. What is important is to develop the

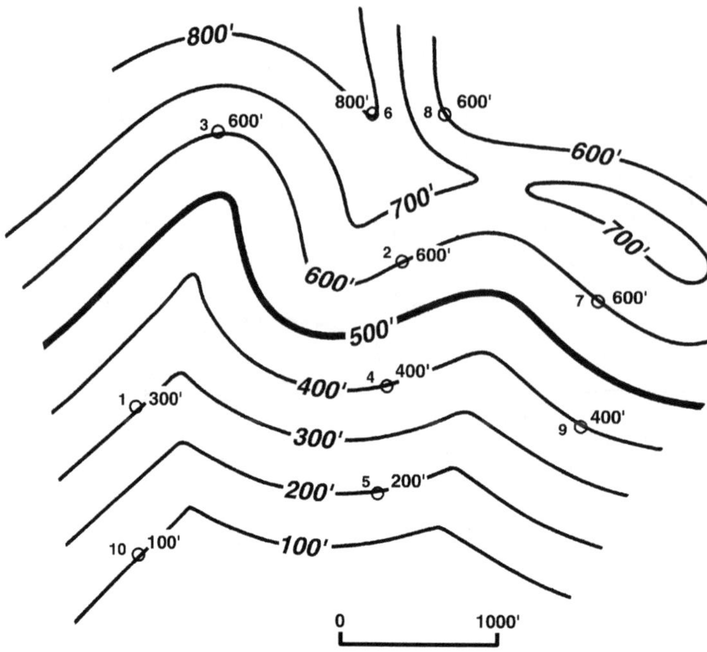

Figure 2-14 Equal-spaced contouring method. (Modified from Bishop 1960. Published by permission of author.)

most *reasonable and realistic interpretation* of the subsurface with the available data, whether the maps are constructed by hand or with the use of a computer.

As an exercise in contouring methods, use vellum to contour the data points in Figure 2-16, using all four methods, and compare the results. Which method results in the most optimistic picture? The most pessimistic?

Special guidelines are used in contouring fault, structure, and isochore maps. Additional guidelines for these maps are discussed in the appropriate chapters. When using computers for mapping, there are other guidelines that should be used or at least considered.

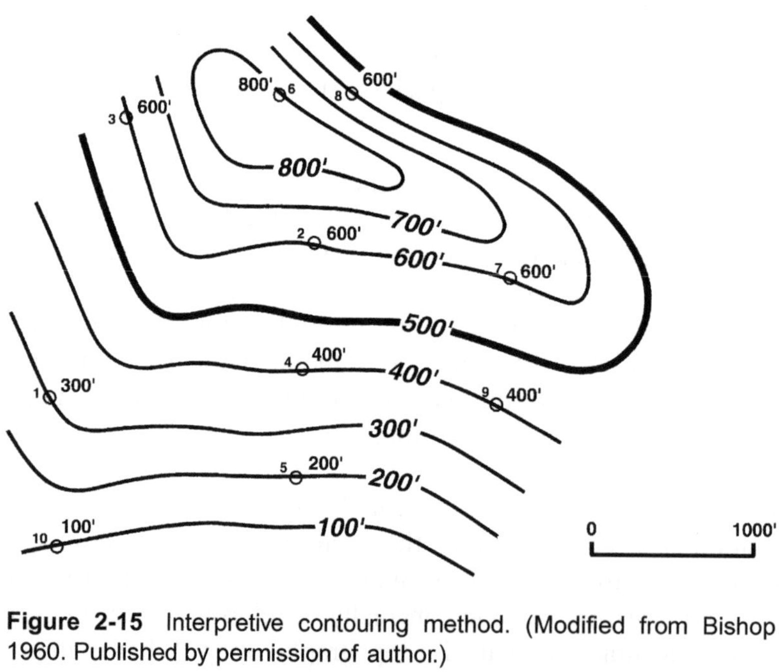

Figure 2-15 Interpretive contouring method. (Modified from Bishop 1960. Published by permission of author.)

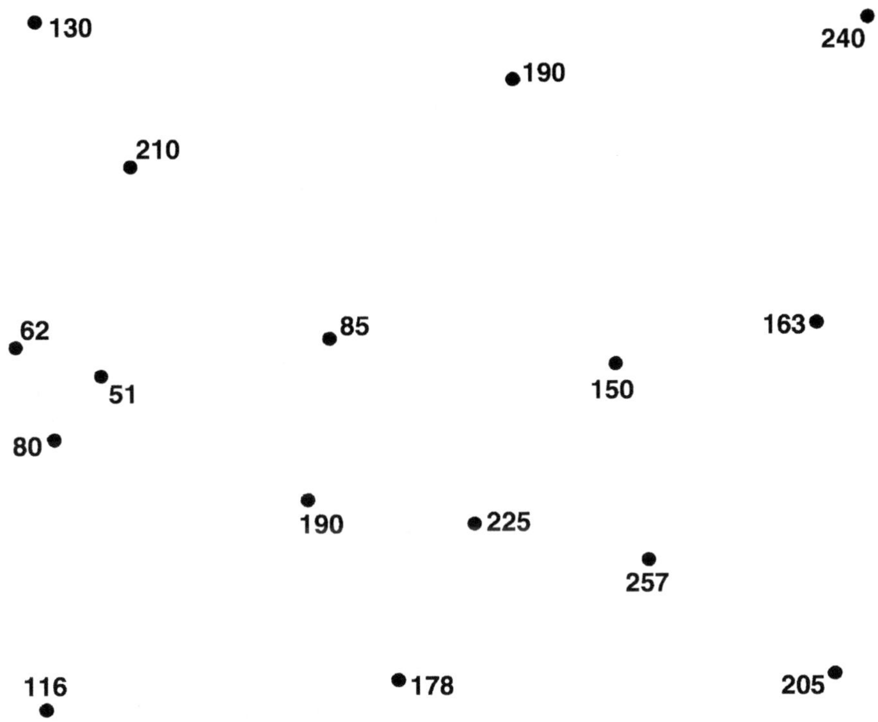

Figure 2-16 Data points to be contoured using all four methods of contouring. (Reproduced from Analysis of Geologic Structures by John M. Dennison, by permission of W. W. Norton & Company, Inc. Copyright 1968 by W. W. Norton & Company, Inc.)

COMPUTER-BASED CONTOURING CONCEPTS AND APPLICATIONS

Computers have altered the way we make geological and geophysical maps. They allow us to quickly create a map without having to think about the surface that is being contoured. They give us the ability to generate a map to the point where the actual geology may be overlooked. Computers have made it easy to skip tried-and-tested techniques that ensure accurate maps because those techniques take too long or are not available in the computer program. *This is the downside of computer mapping*, the side that lacks the interpretive chemistry that occurs when a geoscientist draws contours by hand and thinks about a surface being contoured. That interpretive thought process in many instances has been replaced with concerns about transferring data from one program to another and learning which parameters will actually create a surface map. There is an upside to computer mapping, as well. With the speed and power of computers, the geoscientist can quickly test many interpretations, easily check two surfaces to see whether they cross, use colors to see whether faults reverse direction along strike, and view in three dimensions the created surface to understand the reasonableness and validity of its form. With computers, just as with hand contouring, if the correct methods and proper quality control are not used, then the generated map will likely be wrong.

In this section we discuss how the concepts used in contouring by hand may be implemented on a computer. What is simple for the human brain to accomplish may be extremely difficult for the computer. The contouring we discuss is limited to data sets where we do not have an unlimited number of data points, as would be available for topographic data. Instead, we cover computer contouring of data that represents a surface that has been "sampled" at a limited number of locations (e.g., wells, seismic bins, gravity or magnetic stations). We do not have a precise mathematical equation for the surface, nor do we have aerial photographs. This process of contouring is sometimes called *surface modeling*.

Surface Modeling

We start from a table of X, Y, and Z, where X and Y define the locations of our *samples* and Z is the *height* of the surface at that location. The problem is to position the contour lines so as to depict a reasonable geological surface. Mathematically, this is an interpolation problem.

It can be said that interpolation is the mathematical art of estimating. An interpolation process calculates the value (height) of a surface at locations where it is unknown, based on the values at locations where it is known. Interpolation is an art in the sense that there is no limit to the number of mathematical formulae that may be conceived to make the estimates, and the choice of formulae includes subjective and aesthetic criteria: Does the map look geologically reasonable? Does it come close to how you would do it by hand? Is it pleasing?

The great mathematicians of the past, Newton, Sterling, Chebyshev, Gauss, among others, who dealt with interpolation methods, did not address (or at least did not publish) satisfactory methods to deal with 3D, randomly distributed data. This fact, plus the subjective and aesthetic requirements of contour maps, explains the proliferation of computer contouring methods that have evolved in the past few decades. Two approaches to deal with random data distribution have emerged: indirect (gridded) and direct (nongridded).

Indirect Technique (Gridding). Computer contouring methods that use the gridding technique typically start by using the primary (original) data points to generate a set of secondary (calculated) points. These secondary points, which are traditionally along an orderly geometric pattern (grid), then replace the primary data in steps that generate the contour maps. All subsequent calculations to locate the positions of the contour lines use the secondary data set (the grid). The purpose of using this technique is to simplify the subsequent steps by making the geometry more manageable.

It is easy to ensure that any contour line drawn through a data grid honors its given grid of data points. However, contour lines that honor the data grid *cannot be guaranteed* to honor the original (primary) data points.

Direct Technique (Triangulation). The triangulation technique is the most common of the direct contouring techniques that interpolate values along a pattern which need not be regular but which is derived from the pattern of the original data. The pattern includes the locations of the original data, which are kept throughout the subsequent processing, thus providing the opportunity that all contour lines will honor all the original data.

For both techniques, gridding *and* triangulation, many ways exist to solve the basic problem. Our goal is to choose a technique that most nearly fulfills the needs of the user: geologist, geophysicist, or engineer. In general, the computer-contoured map is more acceptable to the user if it is geologically reasonable and looks as if it has been contoured by hand.

Steps Involved in Gridding

Gridding has come to mean using the original data points to estimate values at the calculated points, or *grid nodes*. Gridding always involves three steps:

1. Select a grid size and origin.
2. Select neighboring data points to be used in calculating a value at each grid node.
3. Estimate the value at that grid node using values from the neighboring points.

The last two steps are where numerous schemes have been developed to make maps that are aesthetically pleasing and that honor the data points as much as possible.

Selecting Neighbors and Estimating Values at Grid Nodes. We now describe some of the methods used for selecting neighbors and estimating values at grid nodes. When selecting neighbors, two criteria are important: (1) neighbors should be evenly distributed around the grid node, and (2) only data points near the grid node should be considered neighbors. As we discuss some of the methods for selecting neighbors, certain data distributions will make it difficult to meet these criteria. We now describe several (out of many) schemes to select neighbors.

1. **Nearest "n" Neighbors**. This is the simplest method of selecting neighbors (Fig. 2-17a). Although this method gives acceptable results for evenly distributed data, it has problems when applied to an uneven data set. For example, where data have been gathered along a straight line (e.g., 2D seismic data, as in Fig. 2-17b), all data points may lie on one side of the grid node. Linear data are unsuitable for surface fitting. To ensure that data are chosen on all sides of a grid node, the domain around the node may be broken into sectors, or segments, such as quadrant, octant, or other pie segments. One method of ensuring a better distribution of data around each grid node is by sector search. Eight sectors are used in Figure 2-18. In the sector search, a certain number, "n," of nearest neighbors are selected from each segment. Obviously, the neighborhood has to expand in order to find some

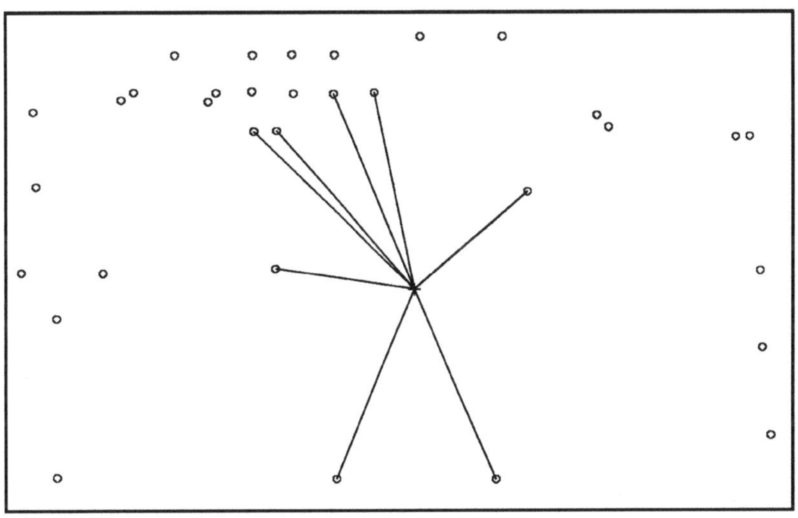

(a)

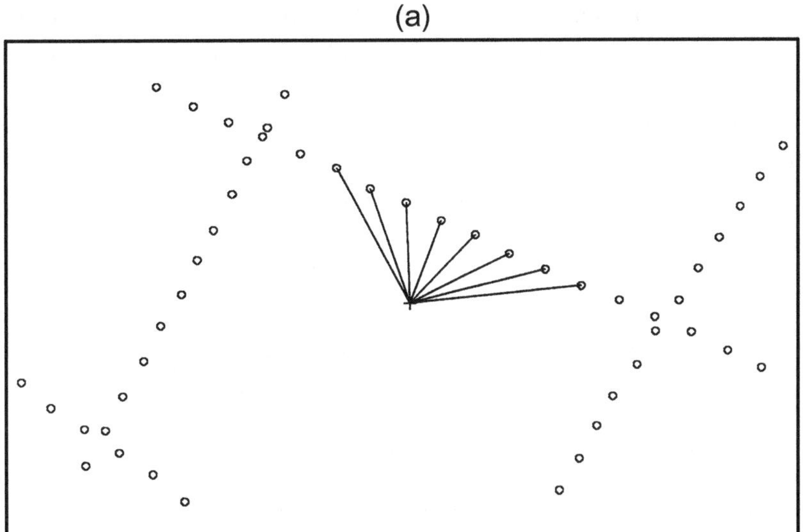

(b)

Figure 2-17 (a) "n" nearest neighbors. This is the simplest method of selecting neighbors. (b) "n" nearest neighbors may all be on one side of grid node for 2D seismic data. (AAPG©1991, reprinted by permission of the AAPG whose permission is required for further use.)

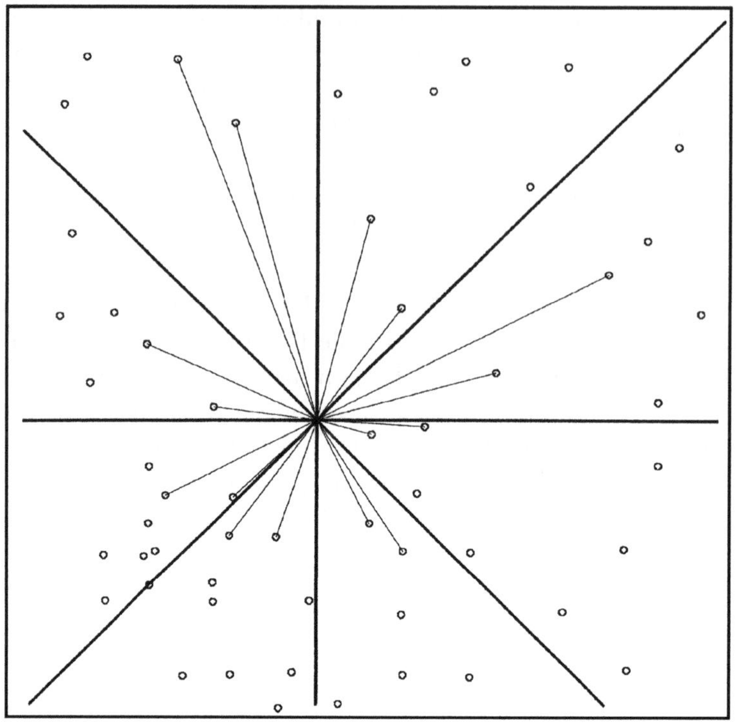

Figure 2-18 Two nearest neighbors in each octant. (AAPG©1991, reprinted by permission of the AAPG whose permission is required for further use.)

neighbors in each segment. This may mean ignoring some nearby control points in order to satisfy the limitation that only "n" points be taken from each segment.

One obvious drawback to any of the sector search procedures is their lack of uniqueness; that is, if the data are rotated, neighbors and neighborhoods change. It would be preferable to find a method that is invariant under rotation.

2. **Natural Neighbors**. On a flat surface that contains more than two data points, any three data points that lie on the circumference of a circle that contains no other data points form what is called a *Delaunay triangle*. These three points are also defined as natural neighbors.

Estimating Values at Grid Nodes. Once we have selected neighbors of a grid node, we proceed to use the Z values of these neighbors to estimate a value (and perhaps a slope) at the grid node. The following are some of the methods used:

1.	Weighted average	6.	Minimum curvature
2.	Least squares	7.	Polynomial fit
3.	Tangential	8.	Double Fourier
4.	Spline	9.	Triangle plane
5.	Hyperbolic		

Each of these methods (and other schemes) has its advocates and adversaries. Any of them works well if the data points are well distributed and well behaved. Each method has problems under certain circumstances.

Most gridded contouring programs give their users an opportunity to select the method of choosing neighbors and the method for estimating values at grid nodes. All gridded contouring programs require users to select grid size.

The pertinent points about indirect techniques (gridding) are:

1. Gridding *can never guarantee* maps that honor all of the data points. On the other hand, the nonhonoring of data may be acceptable if the data are noisy or if the calculated value and the observed value at a data point location differ by an amount that is within the accuracy of the data. In fact, if the data are particularly noisy, the maps may be more pleasing if all of the data are not honored.
2. Sparse data sets that contain clusters of closely spaced data can be troublesome for computer contouring systems, gridded or nongridded. An example of such clustered data distribution is in oil and gas exploration areas, which include wildcat areas (sparse data) and some oil and gas fields (clustered data).
3. Changing the grid size *often produces a different map* because the neighbors of grid nodes change with changes in grid size.
4. The user must choose a method of selecting neighbors and a method of estimating values at grid nodes (interpolation).

Steps Involved in Triangulation. Triangulation in contouring is almost instinctive. Most of us, consciously or subconsciously, were triangulating when we were learning to contour in college. We connected data points with straight lines and subdivided the lines according to contour intervals. At the end of the process, the straight lines connecting the data points were a fairly good approximation of Delaunay triangles. It is therefore not surprising that triangulation, which involves subdividing the map area into triangles (leaving no gaps and creating no overlap), was one of the earlier proposed first steps toward computer contouring.

As in gridding, triangulation requires the selection of neighbors for each data point. However, in triangulation the task of determining neighbors is finished when the data set has been triangulated. The resulting triangulation has the following characteristics:

1. The X-Y data set containing "n" data points has been broken down into a set of (2n-m-2) triangles, where "m" is the number of data points on the convex hull. The convex hull is defined as the smallest convex polygon that encloses all of the data.
2. If the triangles are Delaunay triangles, they are as nearly equilateral as possible for any data distribution, and they are invariant under rotation.
3. In large data sets, each data point will have an average of six natural neighbors.

Figure 2-19 is a sample data set with Delaunay triangles connecting natural neighbors. If no interpolation were performed, and if the surface being contoured were treated as flat in each triangle, the map would have the characteristics of mechanical contouring and would look like Figure 2-20.

Mechanical contouring, however, tends to be very angular and unrealistic. It is usually necessary to interpolate values within the Delaunay triangles (i.e., create more and smaller triangles) based on curved mathematical models in order to overcome the angular appearance. Figure 2-21 shows a set of the subtriangles formed when each leg of the original Delaunay triangles is divided into 16 segments.

The pertinent points about triangulation are:

1. Triangulation always honors every data point because the original data points always remain in the data set being contoured.
2. The interpretation is essentially the same, regardless of the number of triangles or smoothing.
3. The user does not have to worry about data distribution.

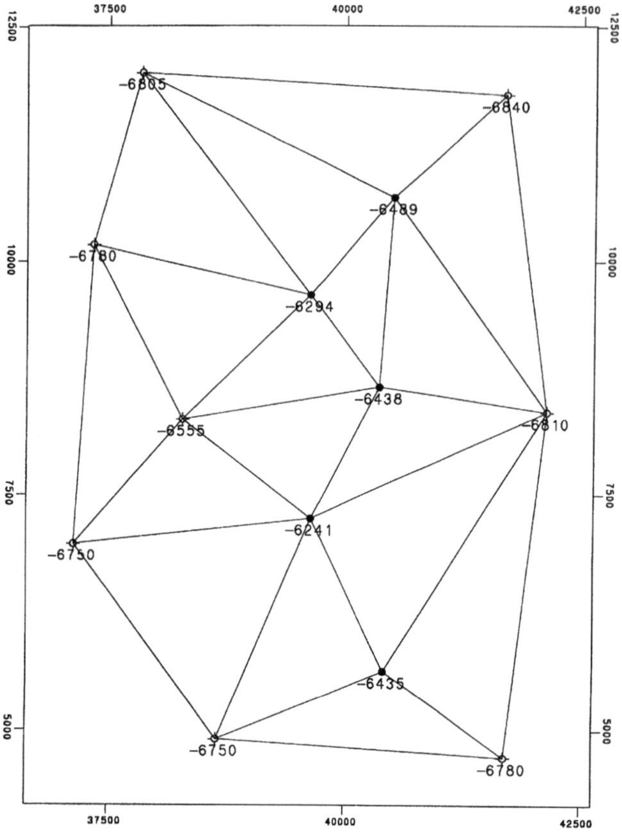

Figure 2-19 Delaunay triangles and natural neighbors. (AAPG©1991, reprinted by permission of the AAPG whose permission is required for further use.)

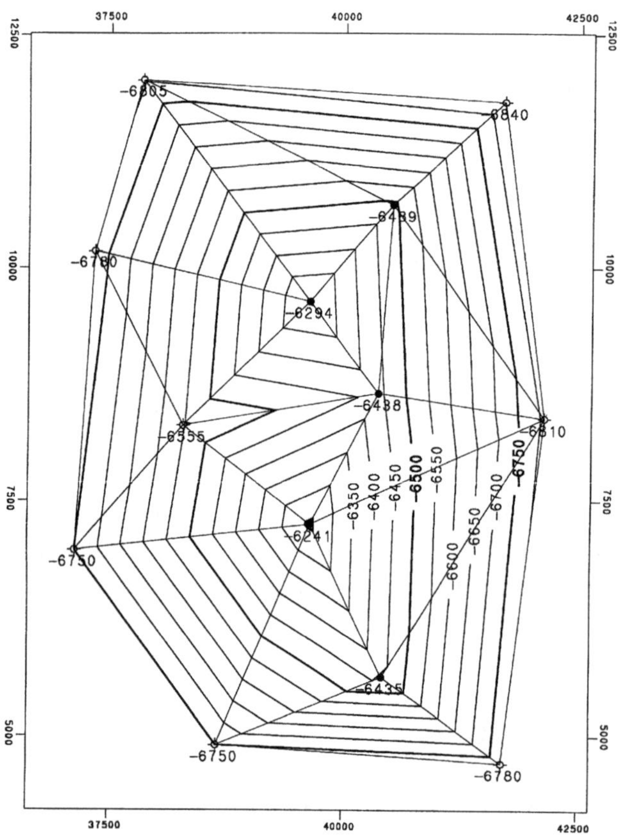

Figure 2-20 Mechanical contouring in basic Delaunay triangles. (Published by permission of Scientific Computer Applications, Inc.)

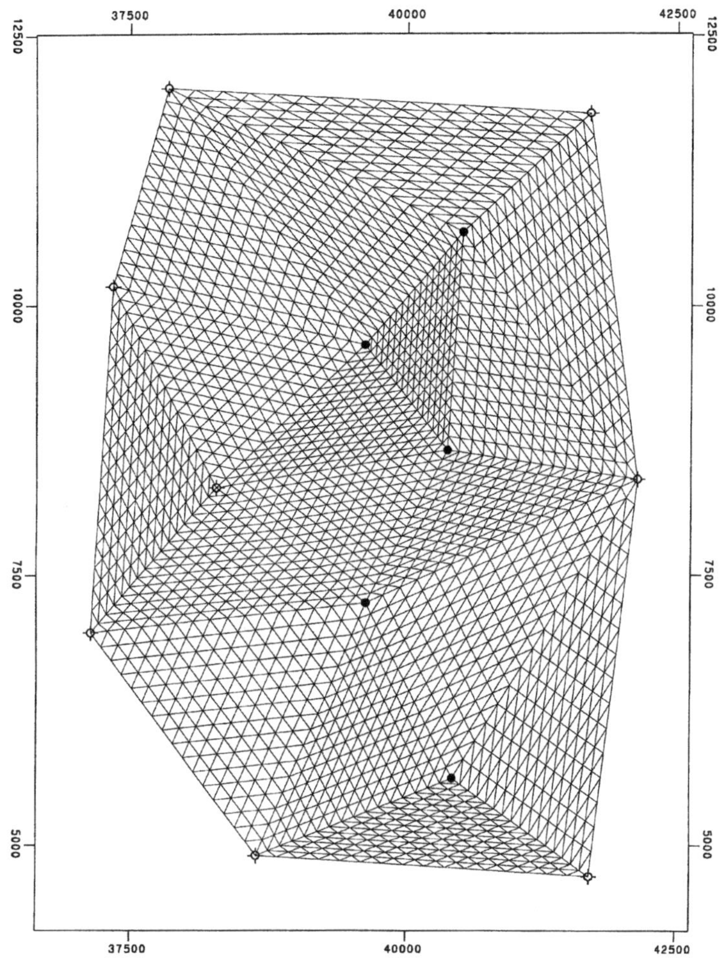

Figure 2-21 Subtriangles forming each side of Delaunay triangles are divided into 16 segments. (AAPG©1991, reprinted by permission of the AAPG whose permission is required for further use.)

Sample Data Set. Figure 2-22 shows a small (13-point) data set representing a structural surface. We encourage you to contour these data yourself before reviewing the computer-contoured maps. These were the data used to generate Figures 2-19, 2-20, and 2-21. Figures 2-23a through f are montages of contour maps of these data generated using various gridded contouring programs, which in turn used various methods of (1) selecting neighbors and (2) calculating values at grid nodes. Figure 2-24 is a contour map of these same data generated using Delaunay triangles, where a value was interpolated at each vertex of the subtriangles shown in Figure 2-21.

Reviewing all these gridded and triangulated maps of the same data set shows that vast differences exist in map interpretation provided by various methods.

Conformable Geology and Multisurface Stacking

So far, we have been discussing single-surface contouring; that is, contouring in which there is only one Z value at each X-Y point. But in oil and gas exploration and production there typically are many surfaces (tops of formations and other horizons) that have been penetrated by the bit or the seismic wavelet. Most of these surfaces are related to each other, for they represent stratigraphic units deposited progressively on a relatively flat sea floor. So, we need to address multisurface contouring.

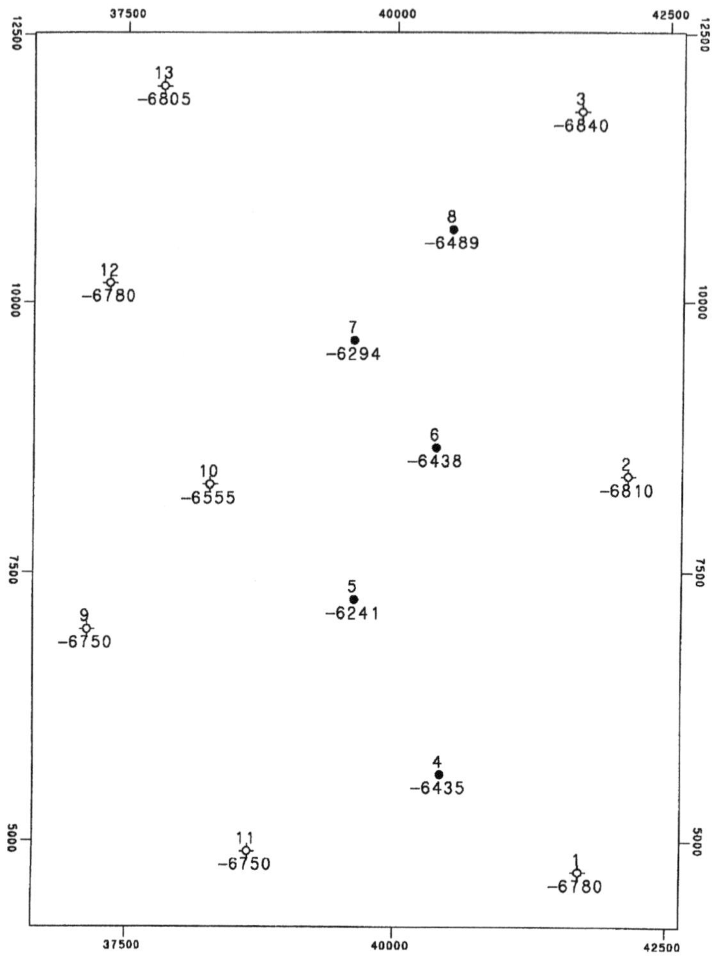

Figure 2-22 Sample data set. (AAPG©1991, reprinted by permission of the AAPG whose permission is required for further use.)

In nature, many associated surfaces resemble one another. When material is deposited on an old surface, the resulting new surface tends to exhibit the same or similar features as the old surface, although the new features tend to be attenuated. Forces may reshape the structure after deposition. Deformation applied to suites of surfaces results in similar features for neighboring surfaces that reflect their common history. It is intuitively expected and empirically verifiable that under these circumstances, the *true vertical thickness* (isochore) between adjacent surfaces tends to be less complicated than the surfaces themselves. The process by which we take advantage of these facts to get improved interpretation is called *stacking,* or conformable mapping. When done by hand, the procedure is usually as follows:

1. Contour the shallowest structure, which is usually the one that has the most data. Designate it as Surface 1.
2. Prepare an isochore contour map of the interval between Surface 1 and Surface 2, using all the data points that penetrate Surfaces 1 and 2.
3. Create and contour estimated structural points on Surface 2. First, on a light table, overlay the Surface 1 structure map and the interval isochore map. Then overlay the Surface 2 base map on those maps. Using points where Surface 1 structure contours cross the interval isochore contours, add the thickness to the depth of Surface 1 and plot the calculated depth for each of those points on Surface 2. Then contour the structure on Surface 2.

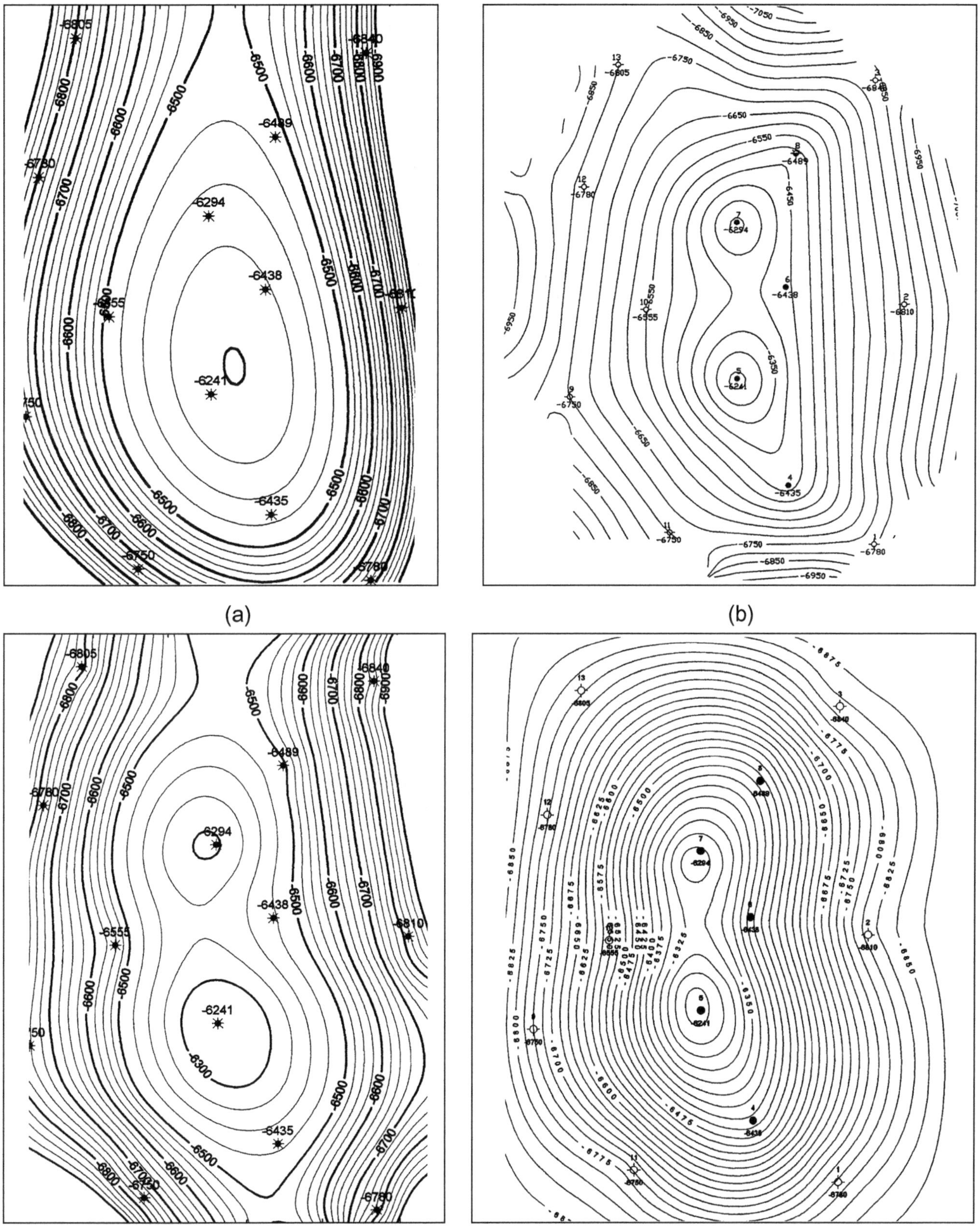

Figure 2-23 (a)–(f) Six structural contour maps using various gridding methods for the same data set. (Published by permission of Scientific Computer Applications, Inc.)

Figure 2-23 *(continued)*

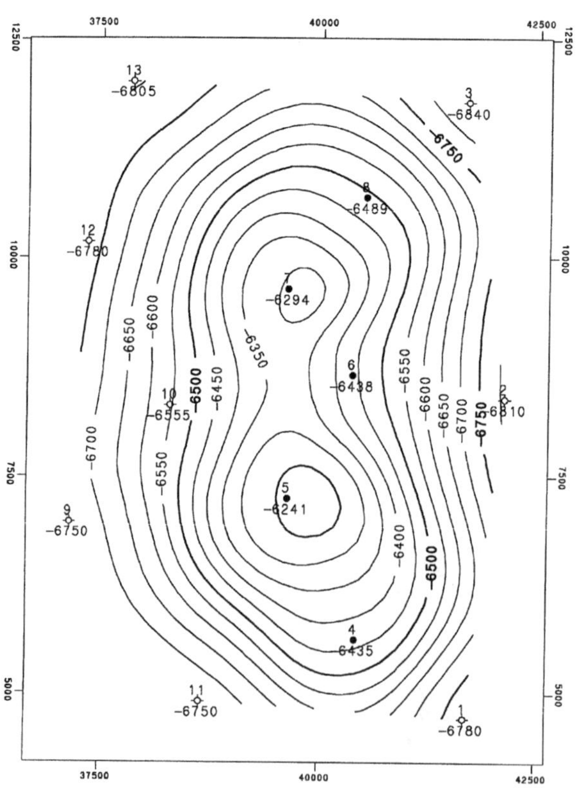

Figure 2-24 Contour map of sample data generated using subtriangles of Delaunay triangles as illustrated in Figure 2-21. (AAPG©1991, reprinted by permission of the AAPG whose permission is required for further use.)

4. Contour a second isochore map between Surfaces 2 and 3. Repeat steps 1 to 3, working down through the stack of surfaces to copy structural shapes from shallow surfaces onto deeper surfaces.

To illustrate the value of stacking, consider the case shown in Figure 2-25, which shows in cross section the stacking process for two surfaces. Of the 11 wells, only five penetrate both surfaces. These are designated by the lowercase letter b. Six wells penetrate only Surface A and are designated by the letter a. If data from the wells that penetrate Surface B were used alone, the surface would tend to be interpreted as shown by the dashed cross-section line.

Instead, stacking recognizes that there is a *true vertical thickness* (isochore) value at each of these b points, and the computer program uses them to calculate (interpolate or extrapolate) estimated true vertical thickness values at all of the a points. These calculated thicknesses are subtracted from the elevation of the known Surface A values to get a calculated elevation of Surface B at all of the a points.

A real case will show the benefit of stacking. Figures 2-26 through 2-28 represent a Top-of-Unit structure map, a Unit isochore map, and a Base-of-Unit structure map for the same 13-point sample data set that was used earlier. These maps were made using multisurface stacking. The isochore map was contoured using the five wells that penetrate the Base-of-Unit, interpolating or extrapolating as necessary to cover the entire map area. Then the elevations for the Base-of-Unit at the eight other wells were derived by subtracting the isochore values from the Top-of-Unit elevations. Notice that the Base-of-Unit mimics features of the Top-of-Unit. Highs are shifted in the direction of thinning isochores.

Figure 2-29 shows the same Top-of-Unit and Base-of-Unit and Unit isochore contours all plotted on the same map. It illustrates precisely how a geoscientist would generate a Base-of-Unit map by hand, using the Top-of-Unit map and the Unit isochore map. Note the three-contour crossing points (e.g., wherever the two surfaces, Top-of-Unit and Base-of-Unit, are 100 ft apart, that is a point on the 100-ft isochore).

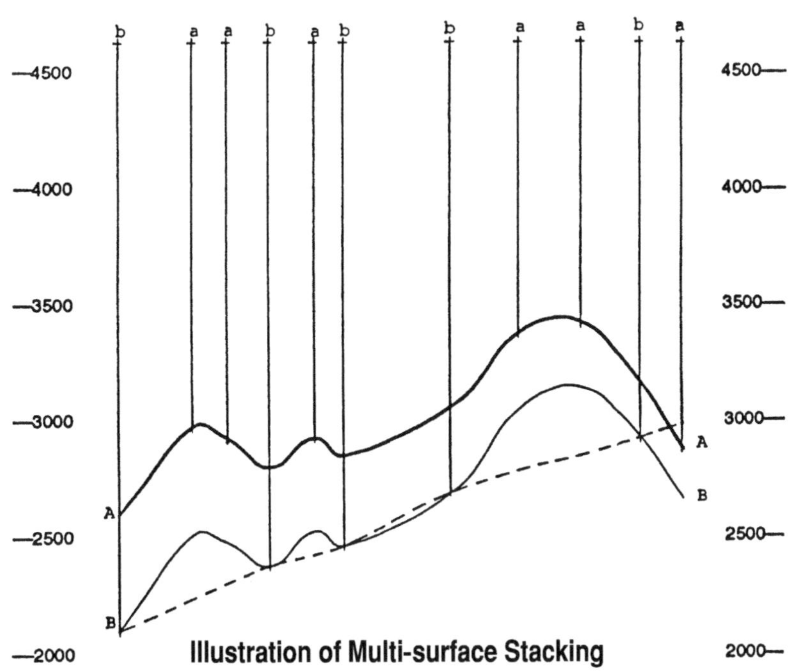

Figure 2-25 Cross section showing stacking process for two surfaces. (AAPG©1991, reprinted by permission of the AAPG whose permission is required for further use.)

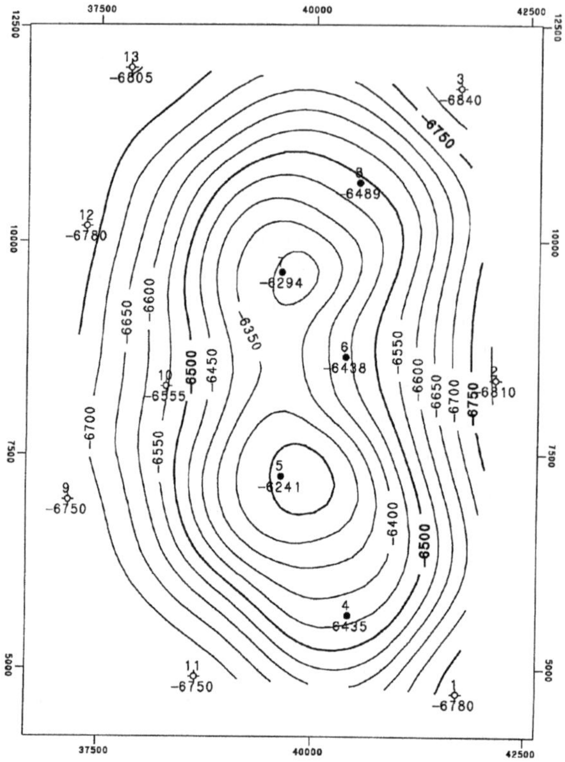

Figure 2-26 Structure contour map on Top-of-Unit. (Published by permission of Scientific Computer Applications, Inc.)

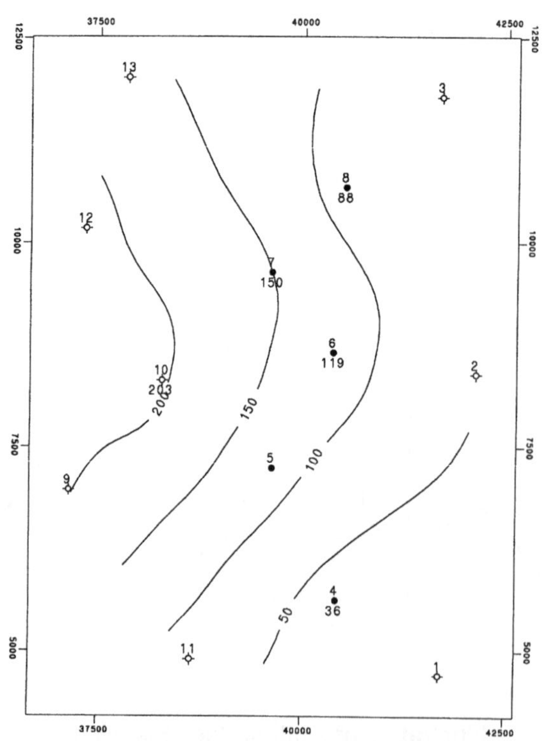

Figure 2-27 Unit isochore map. (Published by permission of Scientific Computer Applications, Inc.)

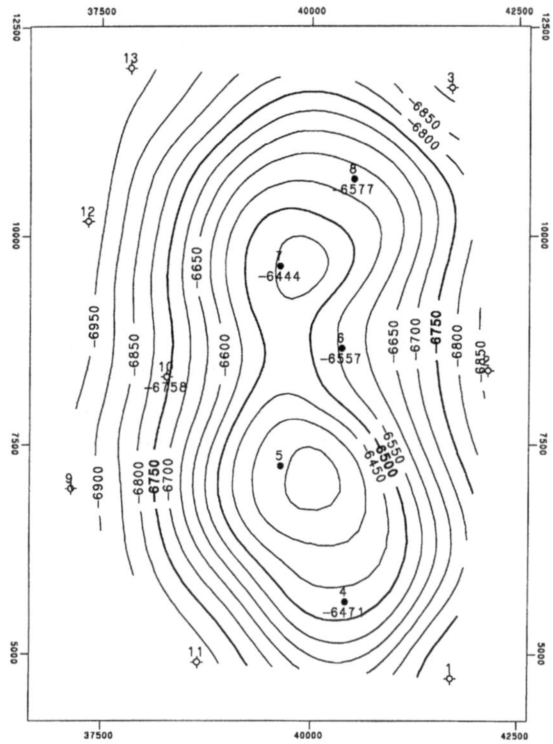

Figure 2-28 Structure contour map on Base-of-Unit with stacking. (Published by permission of Scientific Computer Applications, Inc.)

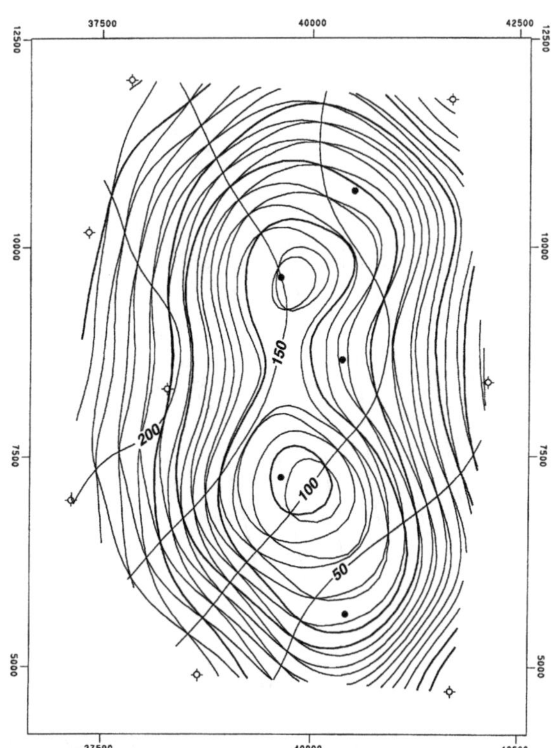

Figure 2-29 Top-of-Unit and Base-of-Unit maps and Unit isochore map overlaid. (Published by permission of Scientific Computer Applications, Inc.)

Base-of-Unit has been penetrated by five wells only. If it were processed without the information derived from the 13 wells that penetrate the Top-of-Unit and the knowledge that the surfaces are similar, that is, if it were processed *without stacking,* we would get a structure contour map as shown in Figure 2-30. If we proceed to make a Unit Isochore by subtracting Figure 2-30 from Figure 2-26, it would be as shown in Figure 2-31. Note that all Unit isochore values are honored, but the interpretation is much different than that shown in Figure 2-28 and is unreasonable if Top-of-Unit and Base-of-Unit are conformable surfaces. The value of stacking is clearly seen when maps with and without stacking are compared.

Some software programs utilize the concept of multisurface stacking without explicitly creating isochore maps using a concept called *conformable gridding.* In conformable gridding, data points on the surface being mapped are honored, but the shape of a better-constrained surface is honored in areas away from control points. Conformable gridding mimics multisurface stacking, and isochore maps constructed from conformably gridded surfaces should match traditional isochore maps.

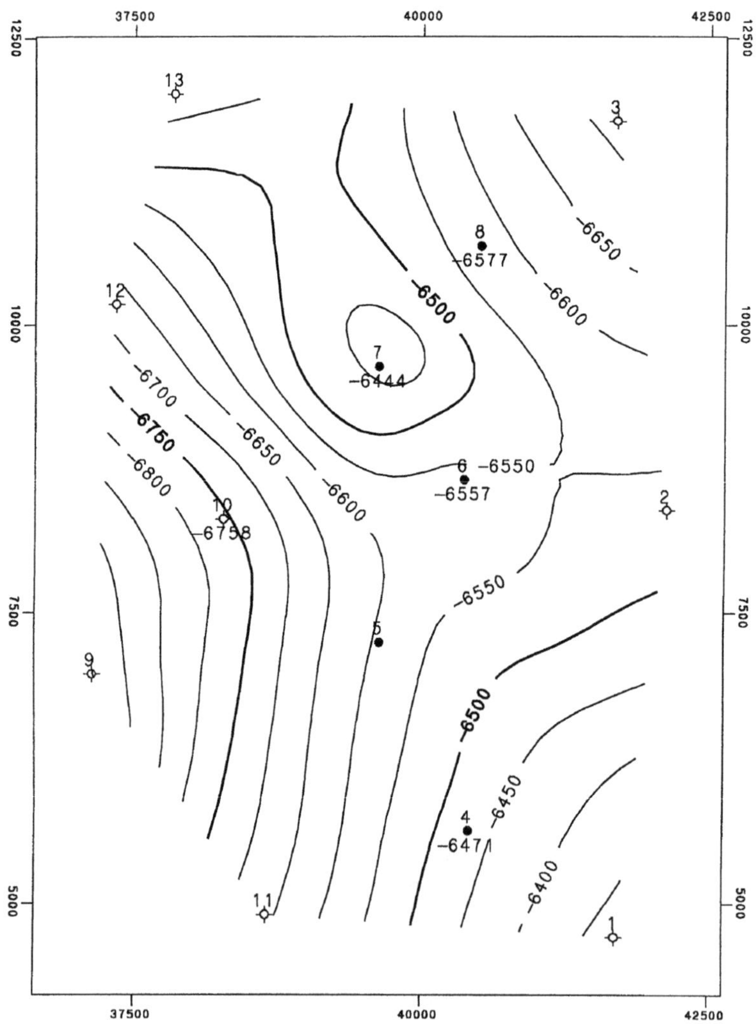

Figure 2-30 Structure contour map on Base-of-Unit without stacking. (Published by permission of Scientific Computer Applications, Inc.)

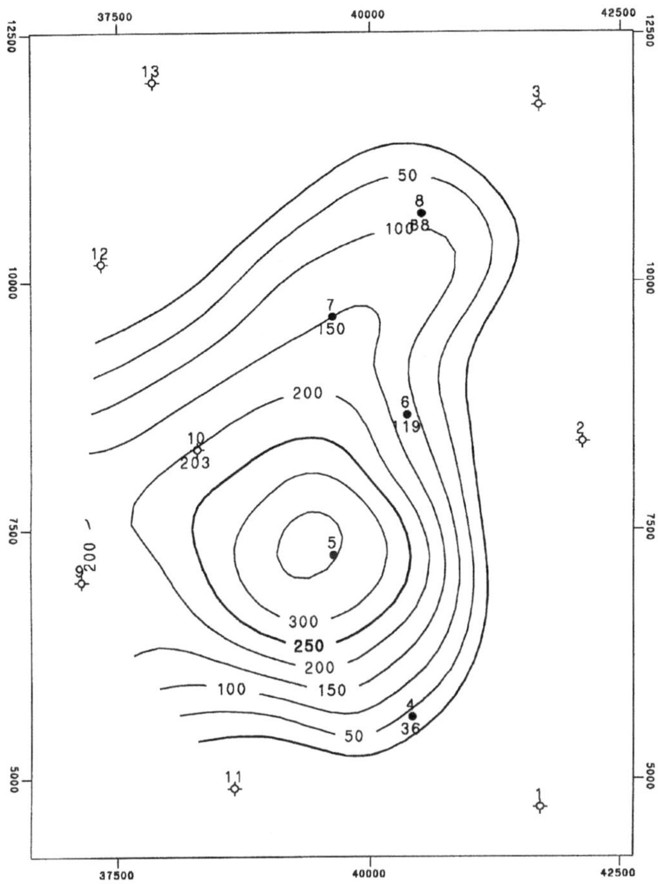

Figure 2-31 Unit isochore map without stacking. (Published by permission of Scientific Computer Applications, Inc.)

The recent shift to drilling increasing numbers of horizontal or highly deviated wells has allowed us to expand on the concept of multisurface mapping or conformable mapping. With vertical wells, mappers tend to project data points from shallower wells down to deeper horizons. With horizontal wells, we can augment this process by projecting data points from deeper horizons onto shallower horizons. Thompson et al. (2017) recently reported on a workflow, called 2D Conformal Mapping (2DCM), that they use to refine structure maps and help with real-time adjustments to directional wells.

In the workflow outlined by Thompson, tops from multiple surfaces in horizontal wells are compared to structural models constructed using conformable mapping (Fig. 2-32). One horizon is designated a reference horizon, the Horizon A in the case of Figure 2-32. Other horizons are constructed to be conformable with the reference horizon (Thompson et al. 2017). The reference horizon should be carefully selected not only on the availability of well control but also on the presence of a good, mappable seismic event if seismic is available (Vogt, personal communication, 2018). The seismic data provides structural control away from well control, such as on the flanks of a field or in portions of the field not yet developed by drilling. Construction of the reference horizon should be done very carefully, as it influences the mapping of all other horizons. Particular care should be paid to any velocity anomalies that might influence depth conversion of the seismic horizon away from well control, as is discussed in Chapter 5.

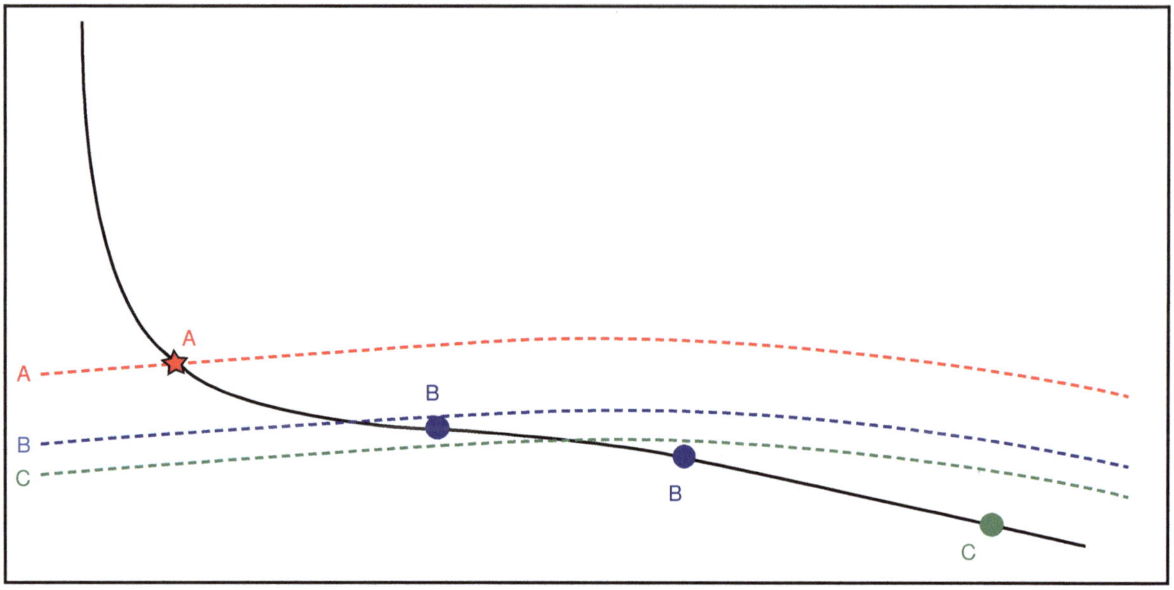

Figure 2-32 New horizontal well drilled on existing structural model. The horizon pick at the reference Horizon A matches the existing model, but the two picks of Horizon B and one pick of Horizon C do not match the existing model. (From Thompson et al. 2017. Published by permission of Society of Petroleum Engineers.)

Figure 2-32 shows that the A Horizon pick of the reference horizon ties the existing structural model very well but that picks of Horizons B and C do not tie the existing model. Control points for the reference Horizon A are constructed using the picks on Horizons B and C (Fig. 2-33). These control points are located directly above the horizon picks for Horizons B and C and are shifted from the existing structural model by the same true vertical depth distance that horizon picks B and C are offset from the structural model (Thompson et al. 2017). In Figure 2-33, all of the picks for Horizons B and C are below the existing structural model, but in other cases, the horizon picks could be below or above the existing structural model.

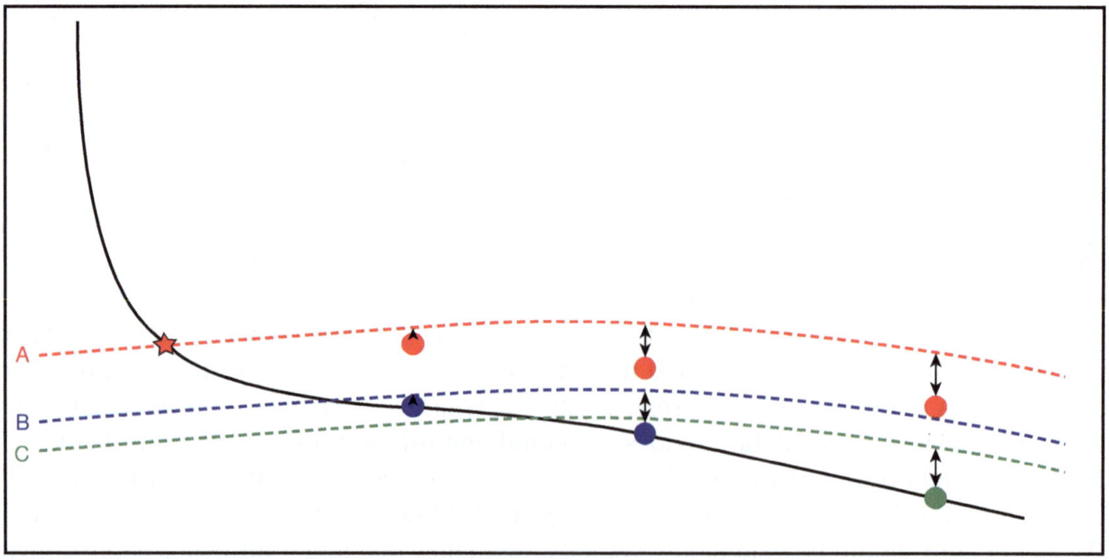

Figure 2-33 Control points for the reference horizon are constructed on the basis of the vertical distance between the deeper horizon picks and the existing structural model. (From Thompson et al. 2017. Published by permission of Society of Petroleum Engineers.)

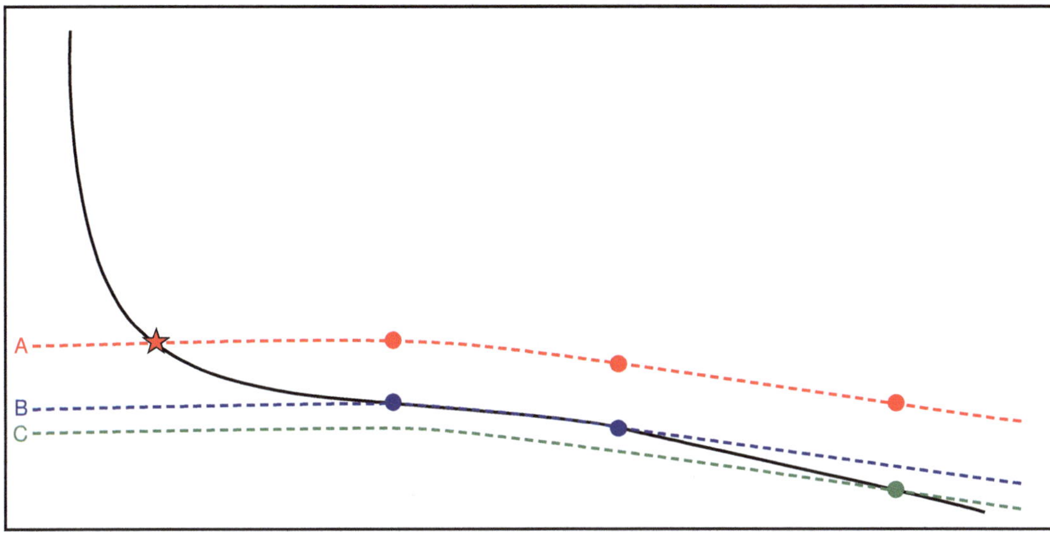

Figure 2-34 The reference Horizon A is remapped using existing data plus the new control points. Deeper Horizons B and C are remapped using the updated reference Horizon A and conformable mapping. (From Thompson et al. 2017. Published by permission of Society of Petroleum Engineers.)

The reference Horizon A is remapped using all existing data, including the new control points. Horizons B and C are then remapped using conformable mapping with the reference horizon. The actual picks for Horizons B and C from the new horizontal well are excluded from the database when the deeper horizons are remapped (Thompson et al. 2017). The new structural model ties all of the horizon picks in the new horizontal well (Fig. 2-34). With the correct computer software, this remapping can be done in real time as a well is being drilled, allowing real-time adjustment to directional plans for horizontal wells (Thompson et al. 2017).

In this example, one horizontal well results in four control points for the reference Horizon A, one actual penetration point, and three control points generated from penetration points of other horizons. Over time, the reference horizon may contain many more control points than actual well penetration points, resulting in a significantly more accurate structure map than using actual well penetrations alone.

Contouring Faulted Surfaces on the Computer

Faults historically have made contouring by hand or computer more complicated because faults destroy the continuity of surfaces being contoured. When contouring faulted surfaces by hand, geologists compensate mentally for fault **vertical separation** (missing or repeated section) and copy shapes across faults that have structural compatibility. Vertical separation is described in Chapters 7 and 8.

Various schemes have been used in an attempt to teach the computer to be able to contour and display faulted surfaces. One of the simplest is to define the fault by a group of connected vectors that separate data on one side of the fault from data on the other side. No fault surfaces are used in the construction. The mapping of this type of fault information on a structural horizon is sometimes called a **trace fault**. Data on each side of the fault are contoured separately and extrapolated to the fault. The vertical separation resulting from the fault is implied from data on each side of the trace fault. The fault is displayed as a vertical fault without any fault gap, and shapes are not necessarily copied across the fault. Figure 2-35 shows an example of a trace fault. Note that fault vertical separation is implicit from the data and changes radically. Also note that shape is not copied across the trace fault.

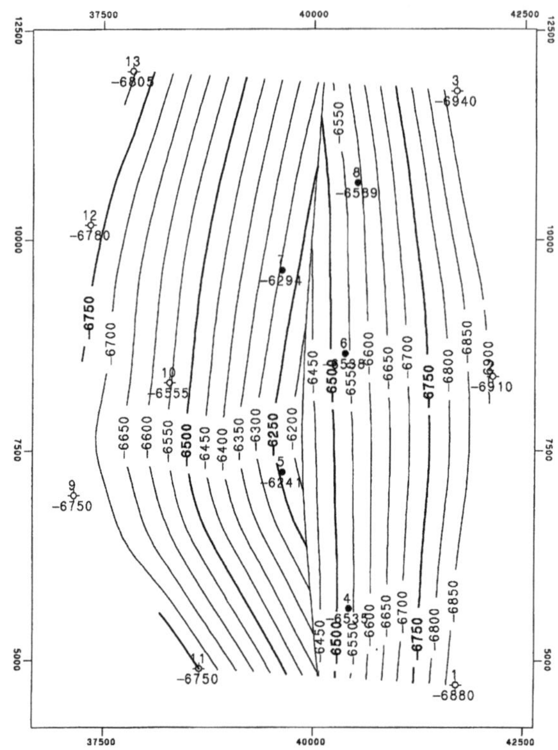

Figure 2-35 Example of trace fault. (Published by permission of Scientific Computer Applications, Inc.)

Another procedure that is used to allow computers to handle faulted surfaces is based on **fault polygons**. In this procedure the faulted surface is divided into a series of polygons that describe individual fault blocks. Data in each fault block are contoured separately, one surface at a time. Fault vertical separation is implicit and is not treated as an explicit variable. Figure 2-36 is an example of a map contoured using fault polygons.

Another procedure for contouring faulted surfaces on the computer is known as the **restored surface method** (fault/structure map integration), which is further discussed in Chapters 7 and 8. It is based on contouring both the fault surfaces and their vertical separations. Hence vertical separation (missing or repeated section) is explicit rather than implicit. Figure 2-37 shows a faulted structure map made using the restored surface method and the same data shown in Figures 2-35 and 2-36. Figure 2-38 is contoured on the fault, which has a constant 100 ft of vertical separation.

In this restored surface method, faulted systems are treated as what they are: sets of 3D fault blocks containing mappable strata, which once were continuous surfaces. The boundaries of these fault blocks are the fault surfaces, and they are contourable. The restored surface method is essentially the procedure that is used when contouring faulted surfaces by hand.

In the restored surface method, faulted systems are processed in three steps designed to honor continuity of shape across faults:

1. Move (i.e., restore palinspastically) the fault blocks, together with the contained geological horizons, to their prefaulted positions.
2. Having restored the "continuous surface" attribute to the geological horizons, perform all the stacking (discussed earlier) and interpolations needed to obtain a smooth map or cross section.
3. Rebreak (i.e., reverse the first step) and return the fault blocks and their contents to their faulted positions and display contour maps or cross sections.

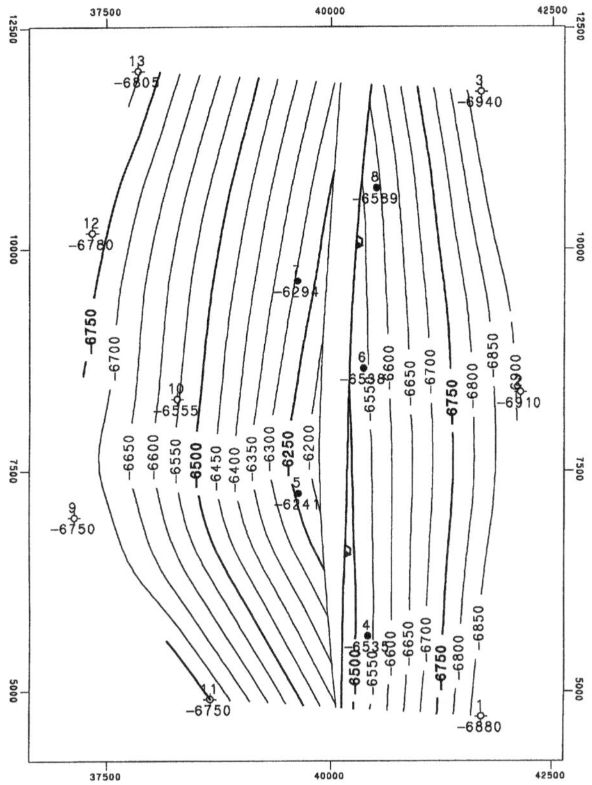

Figure 2-36 Example of contouring using fault polygons. (Published by permission of Scientific Computer Applications, Inc.)

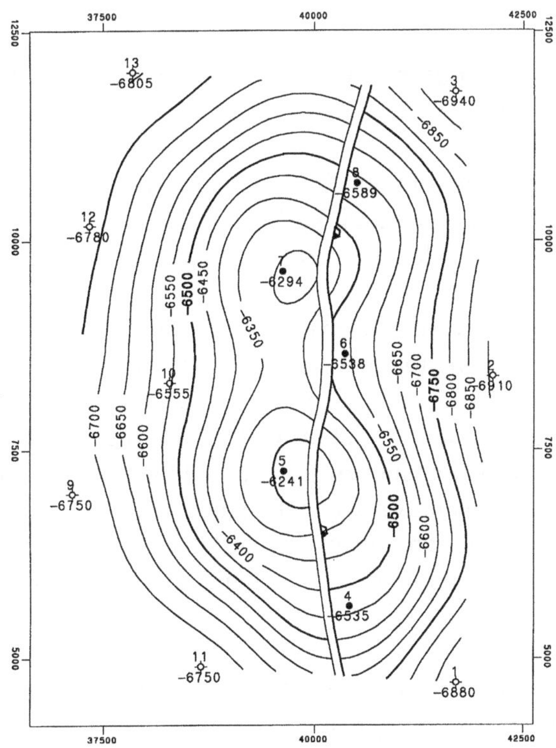

Figure 2-37 Faulted structure using restored surface method. (Published by permission of Scientific Computer Applications, Inc.)

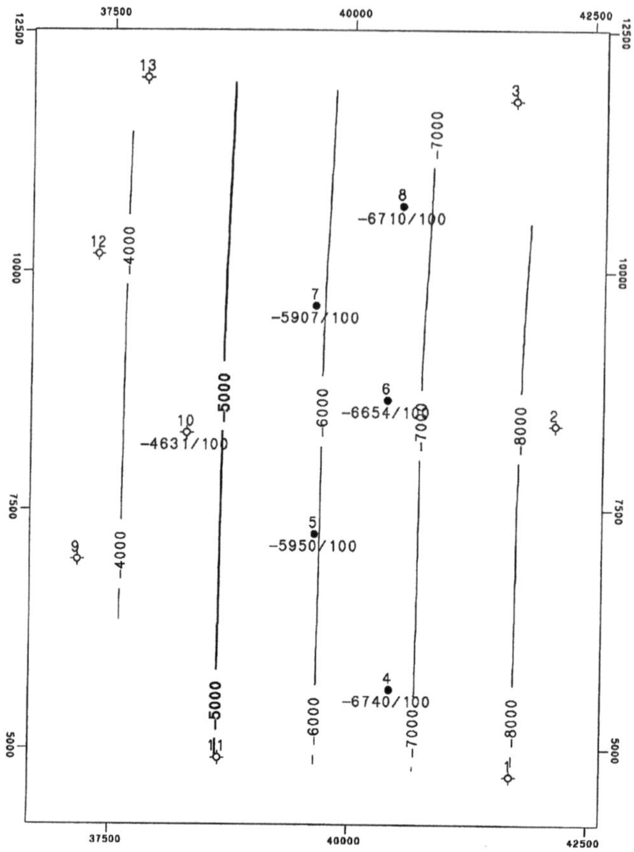

Figure 2-38 Fault surface for 100-ft fault. (Published by permission of Scientific Computer Applications, Inc.)

Procedure for Contouring Faulted Surfaces. To accomplish the steps of the restored surface method, certain data are needed. These data consist of:

1. XYZ data for all horizons (in their faulted position).
2. Three or more XYZ points for each fault.
3. Three or more XYZ points for the vertical separation for each fault.

The missing section, or vertical separation, to the computer, is just another mathematical surface, which can vary over the mapped area and can be contoured. Vertical separation is designated as positive for normal faults and negative for reverse faults.

It is valuable to make a mental picture of the faulting process. In order to restore the fault blocks to their prefaulted position successfully, it is helpful to have a reasonable hypothesis about the order in time of the various faulting events, since the most successful restoration to unfaulted positions needs to be done in steps in reverse order to the original faulting. The faulted system normally is analyzed first by making contour maps and/or cross sections of all fault surfaces in order to:

1. Test the faults for reasonableness. Do observed fault cuts assigned to the same fault result in a fault surface that makes geological sense?

2. Infer and/or ascertain the hierarchy of the faults with respect to age. Where two faults meet in space and do not cross one another, which one extends beyond that junction? Which one is therefore older? The analysis lends itself to "what if" games regarding the faulting sequence.

To perform its task, the computer can use a set of **restore** commands, each of which describes a fault block and instructs the program to move it to its prefaulted or restored position. Restore commands take us "backward in time." The first restore reverses the most recent faulting event; the last restore reverses the oldest faulting event. *Geological knowledge must be used to make decisions regarding fault analysis.*

We use as an example a sand unit offset by a bifurcating fault system, and we create structure maps of the top and base of the unit. The system consists of two merging faults, Fault A and Fault B, contoured in Figures 2-39 and 2-40. The maps also show the line of bifurcation where Fault B merges with Fault A. Data for these faults were obtained by correlating logs, picking horizon tops, locating faults, and measuring missing sections (vertical separations). Fault A has observed cuts on both sides of the line of bifurcation and Fault B does not, so we conclude that Fault A is the older fault and Fault B is the younger fault. Hence Fault B will be restored first, and then Fault A will be restored. Also, we note that vertical separation of Fault A east of the line of bifurcation is the sum of the vertical separations of Faults A and B west of the line of bifurcation because Fault B no longer exists. Vertical separation balance *must be maintained* around the line of bifurcation.

Next, we restore the fault blocks to their original unfaulted positions and then contour the paleosurface. Figures 2-41 through 2-43 display conceptually the steps taken by the computer program in restoring the faults. Notice the changes in well depths from figure to figure. Figure 2-41a shows the depths of the Top-of-Unit in the three fault blocks, as separated by the approximate locations of Fault A and Fault B. Figure 2-41b is a north-south cross section before any restoration has occurred. Figure 2-42a is a base map on the Top-of-Unit after Fault B has been restored, and Figure 2-42b is a cross section after Fault B has been restored.

Next, we restore the block moved during fault event A (the block downthrown to Fault A), which incidentally contains restored Fault B. That restoration produces an unfaulted surface, which is the paleosurface prior to all faulting. We then contour that surface. Figure 2-43a is a structure map of the Top-of-Unit that was contoured after Fault A was restored. Figure 2-43b shows the paleosurfaces in cross section.

Now that the faulted system has been restored to its prefaulting configuration, the structural surfaces are continuous and can be *stacked*, as described earlier in this chapter.

The final step in processing faulted surfaces is to rebreak and move geological surfaces to their true (postfaulted) positions. This step is the mathematical inverse of restoration. In the restored surface method, the computer program generates hanging wall and footwall fault traces as the intersection between structural horizons and faults. Rigorous vertical separation balance should be maintained at all fault intersections. In our example, the structure contour maps for Top-of-Unit and Base-of-Unit are completed and shown in Figures 2-44a and b.

Limitations. Limitations of the restored surfaces method as implemented by the computer have their origin in mathematical simplifications, the most important of which is the use of purely vertical movement in the fault restoration process. This implies that the method becomes less applicable when fault movement is essentially horizontal, for example, strike-slip or very low-angle thrust faults.

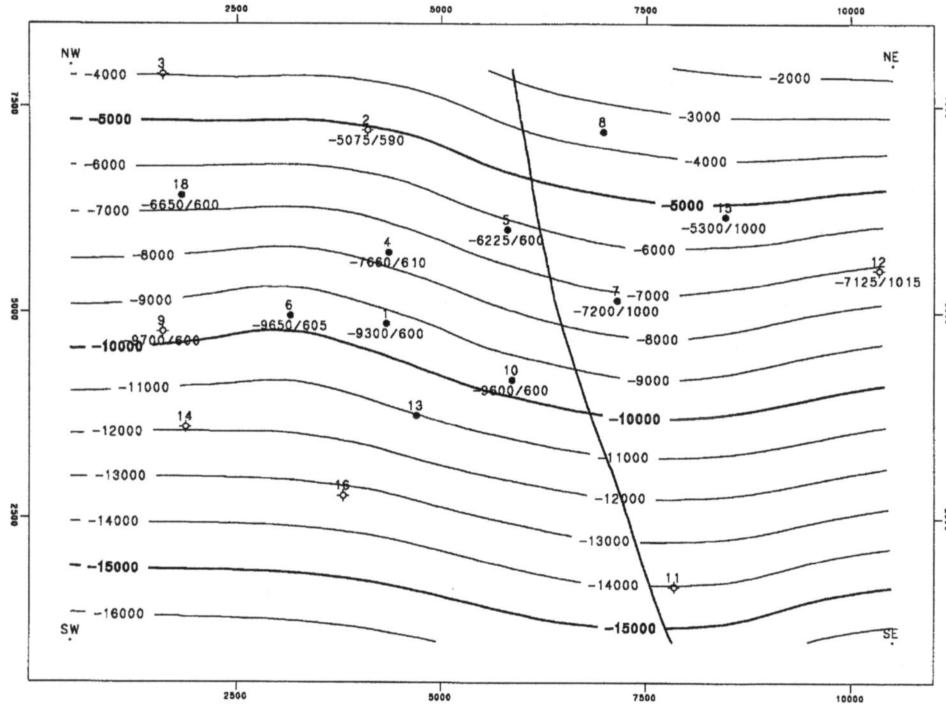

Figure 2-39 Fault surface map of Fault A with vertical separation (missing section) posted. Line of bifurcation with Fault B is shown. (Published by permission of Subsurface Consultants & Associates, LLC.)

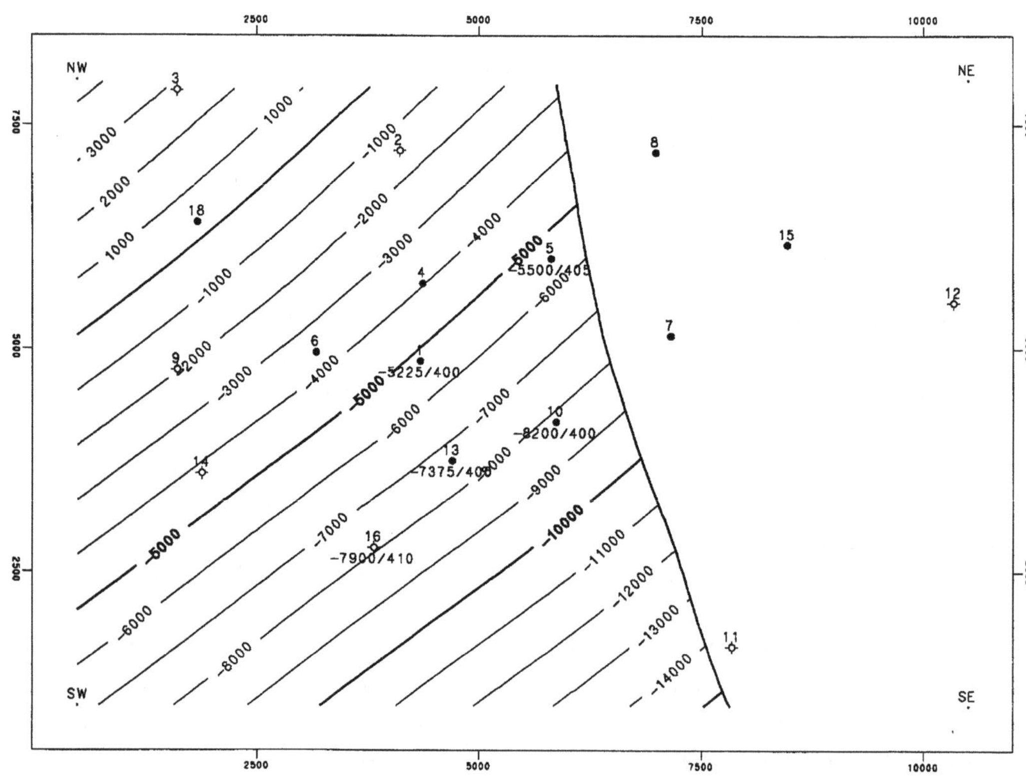

Figure 2-40 Fault surface map of Fault B with vertical separation (missing section) posted. Line of bifurcation with Fault A is shown. (Published by permission of Subsurface Consultants & Associates, LLC.)

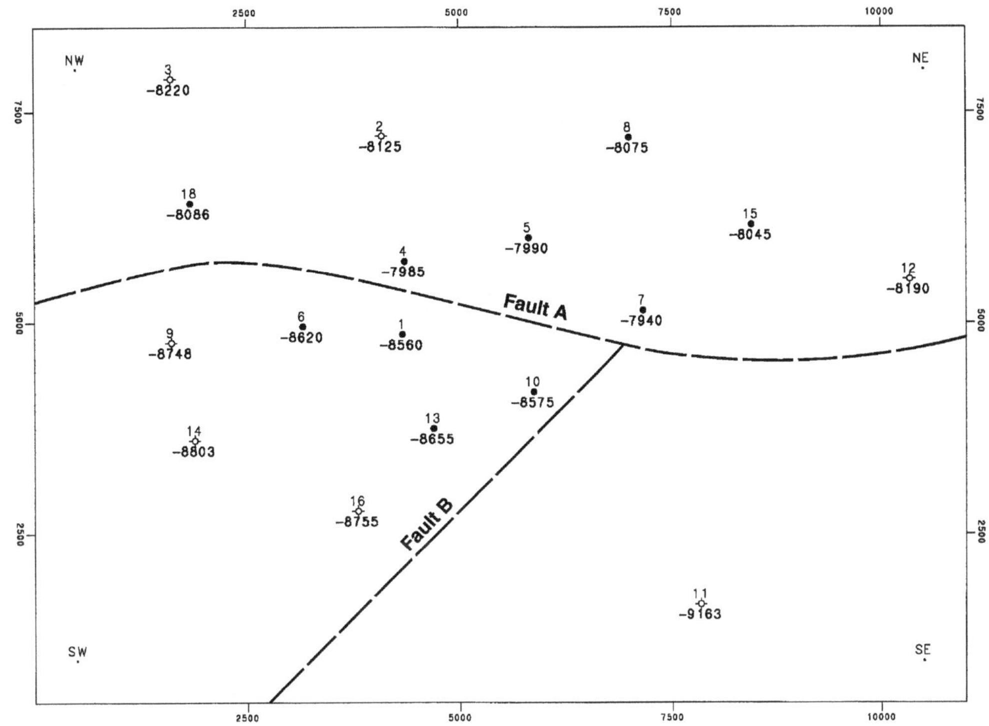

(a)

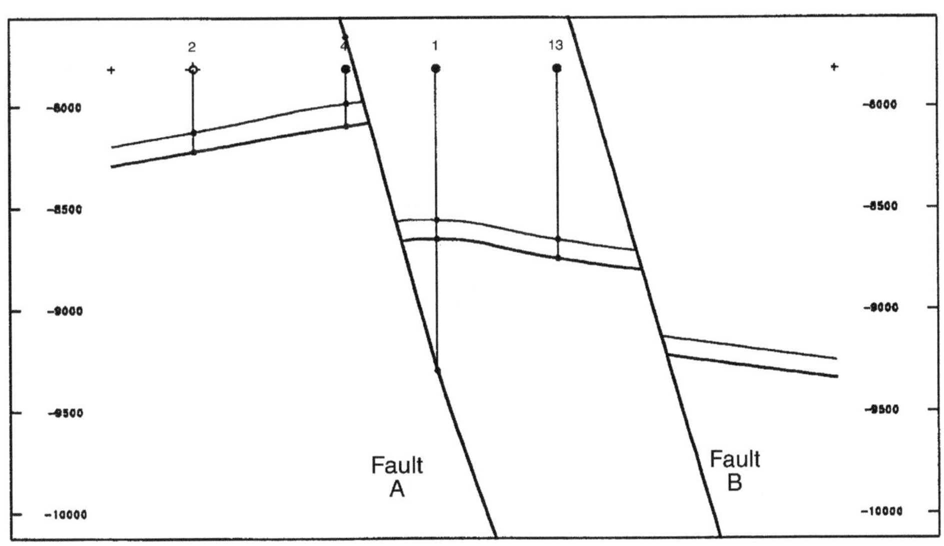

(b)

Figure 2-41 (a) Base map with Top-of-Unit elevations and approximate traces of Faults A and B. (b) North-south cross section before restoration of faults. (Published by permission of Subsurface Consultants & Associates, LLC.)

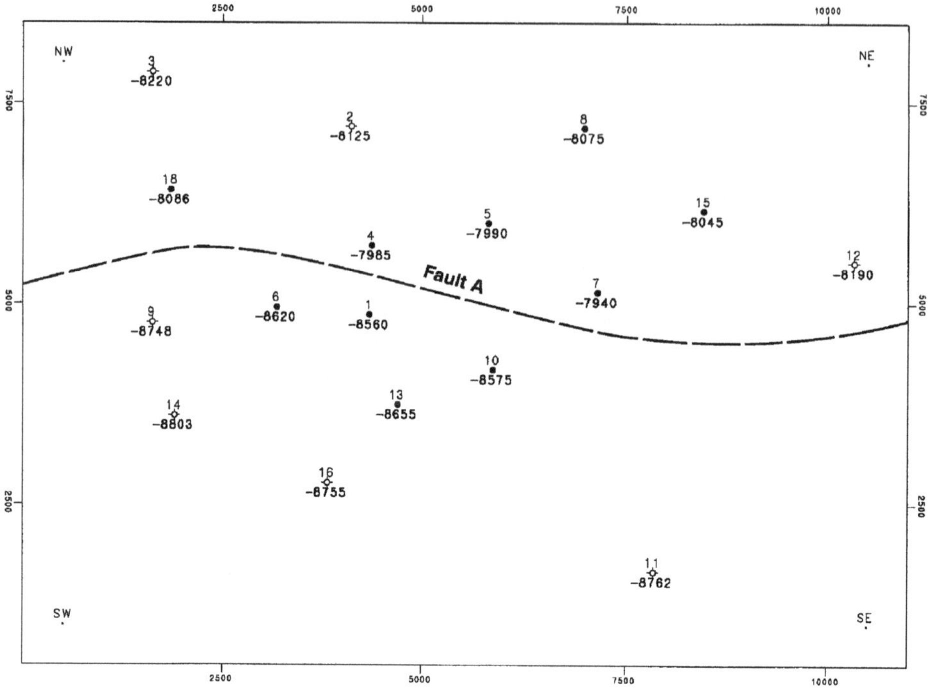

(a)

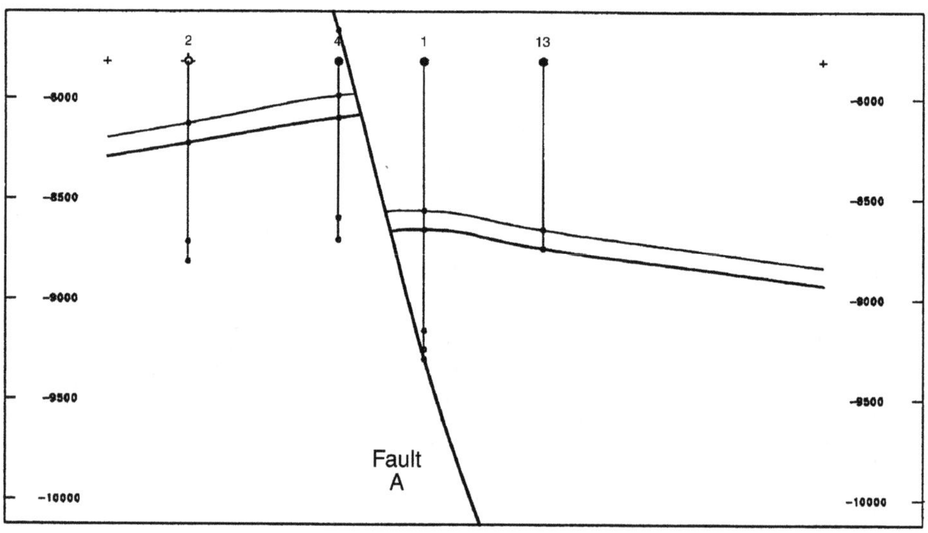

(b)

Figure 2-42 (a) Base map of Top-of-Unit after Fault B has been restored. (b) North-south cross section after Fault B has been restored. (Published by permission of Subsurface Consultants & Associates, LLC.)

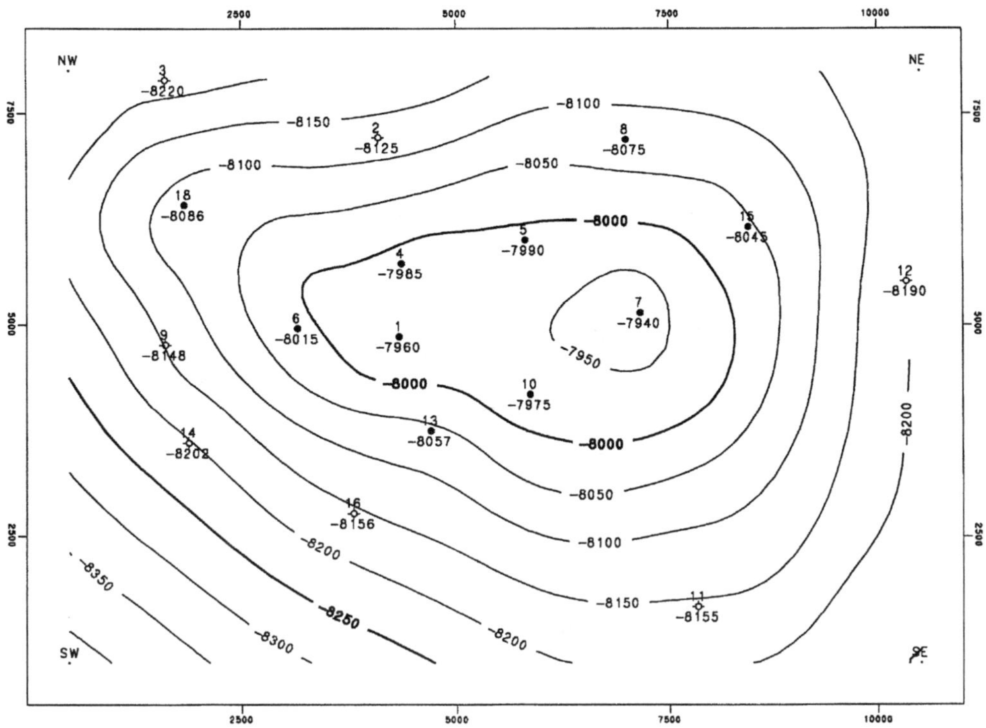

(a)

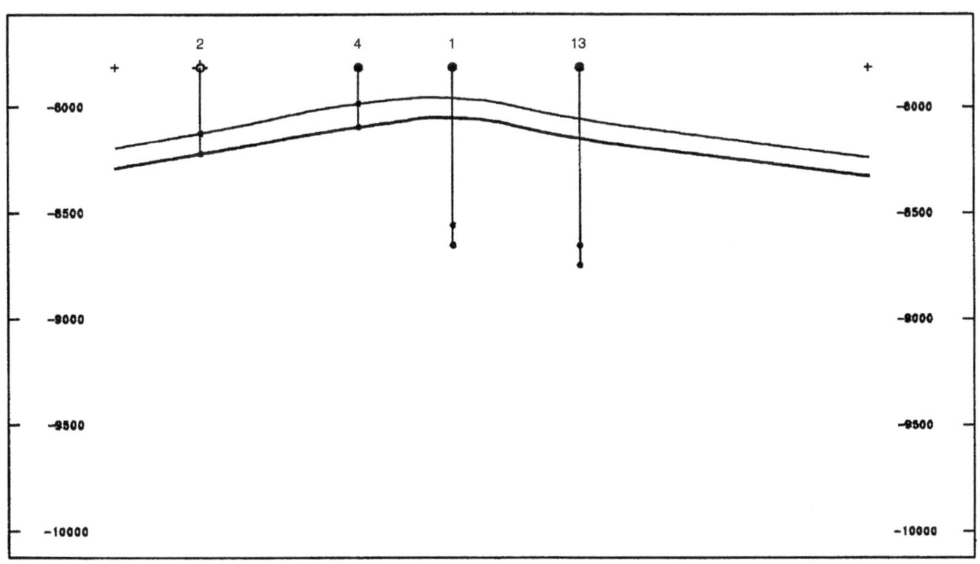

(b)

Figure 2-43 (a) Structure map of Top-of-Unit contoured after Fault A has been restored. This is the paleosurface prior to faulting. (b) North-south cross section after Fault A has been restored. (Published by permission of Subsurface Consultants & Associates, LLC.)

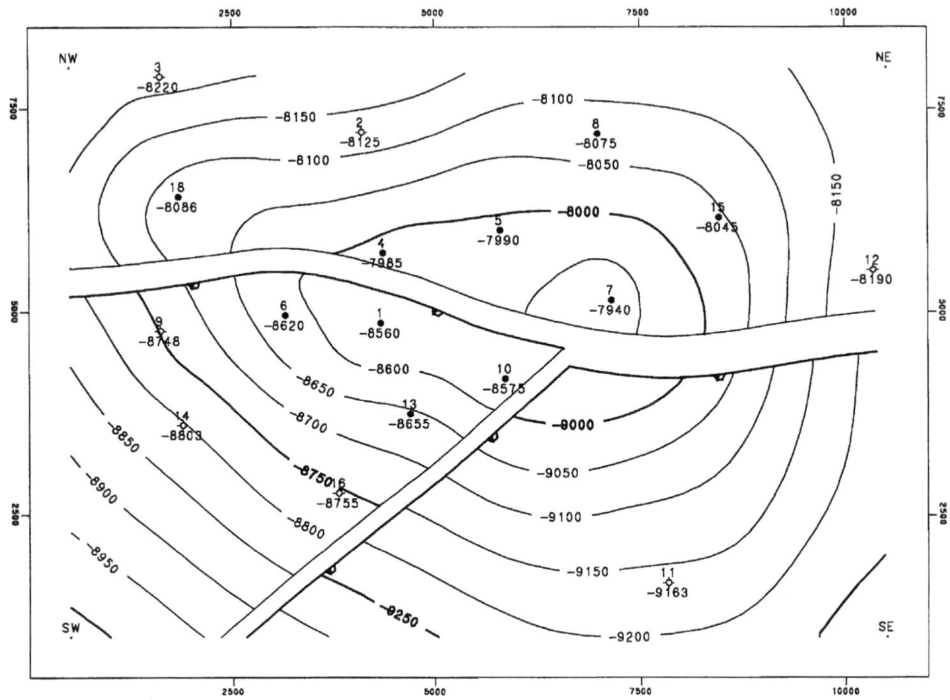

(a)

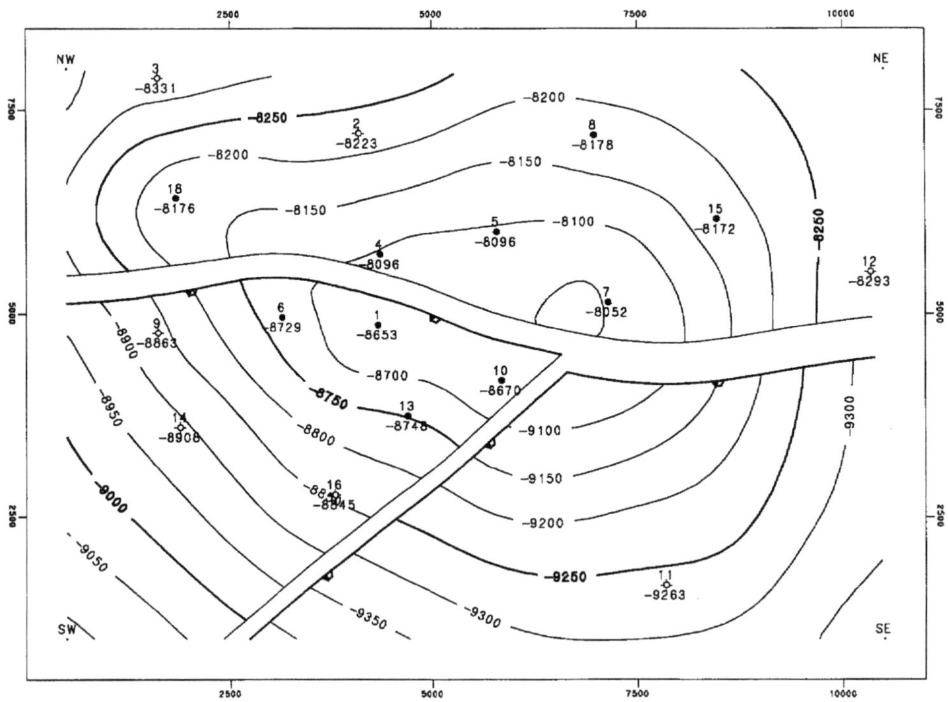

(b)

Figure 2-44 Completed structure maps of (a) Top-of-Unit, and (b) Base-of-Unit, after surfaces have been moved to their true postfaulted positions. (Published by permission of Subsurface Consultants & Associates, LLC.)

CHAPTER 3

MEASURING DEPTH, DIRECTIONALLY DRILLED WELLS, AND DIRECTIONAL SURVEYS*

INTRODUCTION

Accurate subsurface maps require that the data used to construct those maps be placed in the correct X (latitude) and Y (longitude) position. For many maps, especially depth structure maps but also net pay, isotherm, and isobar maps, it is important that data also be properly located in Z (depth) as well. There are significant uncertainties in all of these measurements (X, Y, and Z). It is important that geoscientists understand the sources and magnitudes of these uncertainties so that maps can be made as accurately as possible and important business decisions can be made on the basis of those maps with the explicit recognition of the uncertainty contained within the maps.

X and Y uncertainty derives from any uncertainty associated with the surface location along with the added uncertainty related to the directional path of the well. Depth uncertainty similarly is associated with uncertainties in the measured depth of the well along with the added uncertainty of converting measured depth to true vertical depth (TVD). The goal of this chapter is not to instruct the reader in how to recognize and minimize these uncertainties. The primary goal is to make the reader aware that these uncertainties exist and that they must be considered in any subsurface map.

There are two distinct types of depth uncertainty to keep in mind. The first is relative depth uncertainty. This is the depth uncertainty of different log markers within a single well. The second depth uncertainty is absolute depth uncertainty. This is the uncertainty relative to a specific

---

* For all figures in this chapter (in the printed book only), see the preface for information about registering your copy on the InformIT site for access to the electronic versions in color.

reference, usually mean sea level. Absolute uncertainty is most important when comparing depths from different wells.

VERTICAL WELLS

Early in the development of the oil industry, all wells were drilled vertically because vertical wells are easier, cheaper, and safer to drill. That situation began to change in the late 1920s as operators struggled to develop oil fields that straddled the California coastline (LeRoy et al. 1977, Chilingarian and Vorabutr 1981). As recently as 1990, about 80% of all wells being drilled in the United States were vertical, and the remaining 20% were evenly split between deviated wells and horizontal wells (Baker Hughes Weekly Rig Count as of July 2018). By the middle of 2018, the picture had changed completely with almost 90% of the wells being drilled in the United States being horizontal and the remaining 10% being evenly split between vertical and directional wells (Baker Hughes Weekly Rig Count as of July 2018).

Determining the X, Y, and Z values for a vertical well is relatively straightforward. The X and Y values are taken from the surveyed location of the rig drilling the well. The Z is determined by direct measurement in the well. The situation for deviated and horizontal wells is more complicated and is discussed in more detail later in this chapter.

Surface Location Uncertainty

Although determining the X and Y location of a well is relatively straightforward, the available data is not always accurate. A recent attempt to locate several hundred wells in southwestern Wyoming revealed that about a quarter of them were mislocated in the state's database by up to one mile (Joyce 2016). A number of studies encompassing 12,000 wells found that 40% of the surface locations were off by over 100 ft (Stigant 2012). There are numerous reasons why the surface location of wells can be incorrect in a database. In the early days of petroleum exploration, well locations were not always recorded or were recorded only very generally (Frischkneckt et al. 1985). Older survey methods (pre-1990) were not as accurate as modern methods using GPS satellites, especially in frontier areas and in difficult terrain such as the swamps of southern Louisiana. During the 1990s, the U.S. government deliberately degraded civilian accuracy of GPS data, but this practice was ended in May 2000. Current commercial GPS surveys are usually accurate within 1 ft or less.

Positional uncertainty of the surface location of carefully surveyed older wells is frequently ± 66 to 164 ft (± 20 to 50 m) (Williamson 1999). Even when the surface location of a well is accurately surveyed, the location may be incorrectly labeled on the drilling permit, scout ticket, or well header. Accurate surface locations may be entered incorrectly into computer databases. Finally, the surface location may be reported in the wrong geodetic datum and ellipsoid or the wrong map projection, causing the well to appear in the wrong location on a map (Williamson 1999, Jamieson 2012). A single point on the earth's surface will have different latitude and longitude depending on the geodetic datum and ellipsoid referenced. Taking latitude and longitude from one datum and using it incorrectly with a different datum can produce hundreds of feet of error in the surface location (Stigant 2012, Jamieson 2012).

Jamieson (2012) lists several key facts pertinent to surface locations, including that latitude and longitude, elevations/heights, units, and headings are not unique unless qualified with a datum or reference; that every time data is handled or transferred, there is a chance that the references are not properly interpreted; and that most databases have incomplete metadata describing the references. Most geoscientists do not understand geodesy, the science of positioning on the earth's surface. Unless you understand geodesy, you should always consult someone who does when setting up a mapping project (Jamieson 2012).

In many areas with a long history of drilling, it may be necessary to confirm the surface location of old wells. In some cases, it may be possible to do this from geographically registered air photos or even satellite images (Barnes 2005). In other cases, it may be necessary to use airborne magnetometers carried by either low-flying planes or helicopters to locate surface locations from the magnetic signatures of casing cut off below ground level as part of the abandonment process (Frischkneckt et al. 1985, Hammack et al. 2016). In difficult cases, it may be necessary to check surface locations on foot. In a few cases, it may not be possible to locate or confirm the location of old wells, especially if the well's casing was cut off below the surface and pulled, leaving no equipment or surface magnetic anomaly to find.

Even when the surface location of a well is accurate, the data from that well can be assigned to the wrong wellbore. For example, data from a sidetrack may be incorrectly stored in a database as being from the original hole or from a different sidetrack. Log data can also be incorrectly stored under a different well with a similar name (Storey 2014). When data does not fit the map, it is important to go back and check that the data is assigned to the correct well at the right surface location.

Well Depth

Depth in a well is usually measured either with drill pipe, called *driller's depth*, or with wireline, called *logger's depth*. The reference elevation of the well has traditionally been the kelly bushing, a large piece of the drilling rig. Other common reference elevations are the rig floor (RF) or derrick floor (DF), the rotary table (RT), the bradenhead flange (BHF), and ground level (GL). In some offshore wells, a well may be started with one rig and finished with another rig. The same reference point on the different rigs may be at significantly different elevations relative to mean sea level (MSL). Thus, the zero depth for a shallow log run may be at a different elevation than the zero depth for a deeper log run.

We begin by discussing measured depth (MD), also known as *along hole depth* (e.g., Brooks et al. 2005) or *true along hole depth* (TAH) (Forsyth et al. 2013). For vertical wells MD is equal to TVD (which we discuss later in the chapter). In deviated wells, MD is not equal to TVD and must be corrected for the directional path of the well. Before we deal with the directional path of the well, however, it is important to understand the uncertainty in MD, as TVD can never be known more accurately than we know MD.

Depth is the most fundamental measurement made in a well, and yet it is a measurement for which there are few, if any, industry standards or accepted auditable practices (Forsyth et al. 2013, Bolt 2015). Bolt (2015) recently listed some of the costs of depth uncertainties (Table 3-1). These costs range from minor, thousands or tens of thousands of dollars, to major, tens of millions or hundreds of millions of dollars. Depth uncertainty is one of the largest uncertainties in assessing the value of oil and gas fields (Fig. 3-1), but it is rarely considered. It is important to get depth values as accurately as possible.

The largest investment decision that many geoscientists will ever work on is the development plan for a new discovery. These decisions can involve billions of dollars of investment capital in

Table 3-1 Potential cost of depth uncertainties (after Bolt 2015)

Event	*Cost ($)*
Lost time repeat surveys	Thousands
Interval relogging	Tens of thousands
Field depth resurveys	Hundreds of thousands
Field studies and sidetracks	Millions
Equity value discrepancies	Tens of millions
Incorrect reserves and geological models	Hundreds of millions

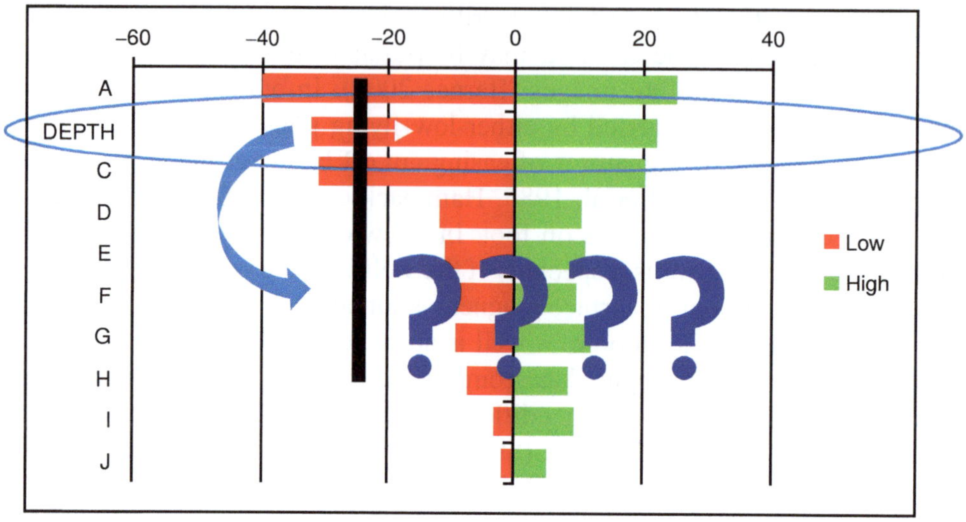

Figure 3-1 Depth is one of the key uncertainties in determining the value of oil and gas fields. (After Bolt 2015. Published with permission from Harald Bolt.)

frontier discoveries, for example. Suboptimal development plans can result in the unnecessary spending of tens or hundreds of millions of dollars by installing larger facilities or drilling more wells than needed to develop a field. Alternatively, fields may be undercapitalized, requiring retrofitting or additional facilities at a much higher cost than if the initial field development plan was more appropriate. Worst of all would be the failure to develop a profitable discovery because the cost of the development was inaccurately assessed. Failure to accurately recognize field complexity and compartmentalization due to a lack of understanding of depth uncertainty associated with fluid contacts, for example, can contribute to very expensive business mistakes (Forsyth et al. 2013; Bolt 2015).

All structure maps are created relative to a horizontal reference plane, usually MSL. The elevation of MSL is constant over short distances but it is not constant over large distances. MSL in the United States and Canada used to be the National Geodetic Vertical Datum of 1929 (NGVD 29). NGVD 29 assumed that 26 tidal gages in the United States and Canada were the same elevation. More accurate surveying has revealed that these 26 gages actually differ in elevation by up to 5 ft (1.5 m) (FEMA 2014). NGVD 29 was supplanted with a new datum in the early 1990s, the North American Vertical Datum of 1988 (NAVD 88). Elevations relative to MSL for pre-1991 wells in the United States and Canada are likely to be relative to the NGVD 29 datum, whereas elevations relative to MSL for wells drilled after 1991 are likely to be referenced to NAVD 88. Although reference elevation differences for nearby wells referenced to the different datums are likely to be very small, it is possible to check elevations relative to the two different standards using a program (VERTCON) on the U.S. government's National Oceanic and Atmospheric Administration (NOAA) website. Different areas of the world have different datums for MSL, but these differences are unlikely to affect any local mapping project.

Driller's depth is measured with joints of drill pipe, which are generally measured on a pipe rack. Traditionally, joints of drill pipe have been measured from pipe shoulder to pipe shoulder using steel measuring tape under ambient surface conditions, although in recent years they are frequently measured using lasers (Jamieson, 2012). A joint of pipe will have a slightly different length when measured in North Dakota in the winter than if the same joint of pipe were measured in the Permian Basin in August. Driller's depth is determined by summing the length of every joint of pipe in the drill string along with any bottomhole assembly (BHA) that is in the hole below the reference elevation of the well. The reference for the length of every joint of pipe run in the drill string is known as the driller's tally. Traditionally, mudlogs, casing points, and total depth (TD) of a well have been based on driller's depth.

Since Schlumberger introduced the resistivity log in France in 1927, most well depth measurements that are used in subsurface mapping are based on various types of wireline logs. These logs are generally measured with cable or wireline depths. Recently, more and more logs have been acquired using logging while drilling (LWD). LWD logs are measured, somewhat indirectly, using driller's depth. Actual log readings are referenced to time. The depth of the bit is also referenced to time. Cross-referencing the time of a log reading and the depth of the bit at that time, along with knowledge of the position of tool sensors relative to the bit, allows the depth of log measurements to be determined.

In some situations, pipe-conveyed logs (PCL), also referred to as *tough logging conditions* (TLC or TCL) logs, are acquired using pipe-conveyed wireline tools. Some logs are also obtained on slickline, but these logs are almost invariably obtained within cased holes. Their depth is usually tied to an open hole log and are not considered further in this book. It is important that geoscientists and engineers making and working with subsurface maps understand that depth measurements of different types of logs, especially wireline and LWD logs, are made with different measuring tools and frequently give different depths for the same subsurface marker (Fig. 3-2). To produce accurate subsurface maps, it is important that mappers understand how depth is measured for different types of logs, how and why these depths may differ, and how these differences can be minimized.

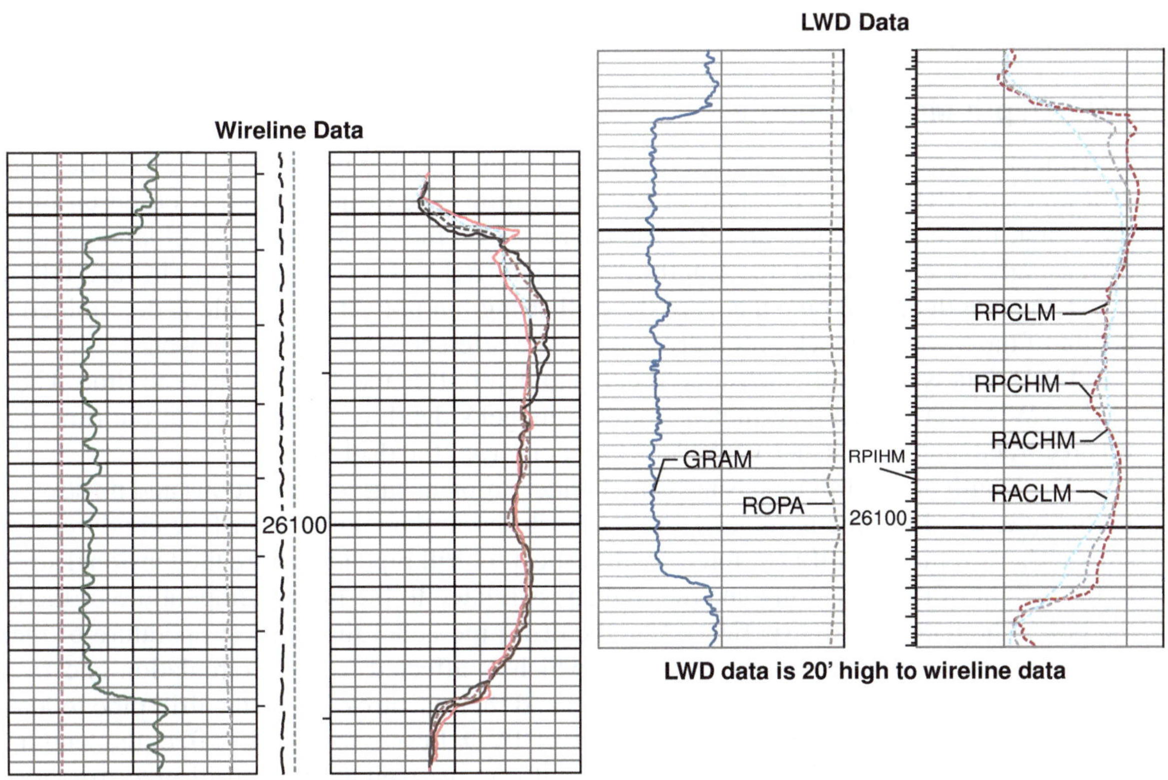

Figure 3-2 Wireline and LWD logs from a single Gulf of Mexico well showing a hydrocarbon-bearing sand that is at different depths on the two different logs. The 20-ft depth difference occurs because wireline depth and LWD depth are measured using different tools. (Published with permission of J. Brenneke.)

Wireline Depth Measurements

Since the 1930s, most formation depths in wells have been measured using wireline tools. Wireline depths are measured by determining the position of various logging tools on the logging sonde and measuring the length of wireline cable lowered into the hole. In most cases the length of cable is measured using some type of wheel device. A number of factors can affect the measurement of cable length, but in modern wells, these factors are relatively minor and are not discussed in detail in this book. Interested readers are referred to extensive discussions of these factors in publications such as Bolt (2015).

The main factors affecting wireline length changes in the hole are cable stretch and thermal expansion/contraction. Most service companies have relatively accurate algorithms that allow depth correction for cable stretch (Bolt 2015, 2016), although these algorithms continue to be improved (e.g., Fitzgerald and Pedersen 2007). Wellbore deviation affects the stretch of wireline cables as friction on the borehole wall in deviated wells decreases cable tension (Fitzgerald and Pedersen 2007). Older algorithms are probably less accurate than more recent ones. The thermal expansion/contraction of modern logging cables is complicated, as they are composed of several different materials that react differently to temperature. The reaction to temperature may vary from one cable type to another. Some workers report that increased temperatures lengthen cables, whereas other workers report that increased temperatures shorten cables (Bolt 2015).

Wireline data may be measured either while running into the hole or when pulling out of the hole. Different companies prefer one method or the other, but the depths from the two methods are different (Pedersen and Constable 2006; Fitzgerald and Pedersen 2007). Most final logs are recorded while pulling out of the hole, although the depth on these logs may be adjusted to the depth recorded running into the hole.

LWD Depth Measurements

LWD depth measurements are based on driller's depths, which in turn are based on the length of drill pipe and BHA in the borehole at any given time. The driller's depth is based on three factors. First, the length of pipe and BHA is measured on the surface using steel tape measures or, increasingly, lasers. Second, the length of pipe and BHA is recorded in the driller's tally book. Third, the length of drill pipe that is in the drill string but not yet in the hole is determined by the height of the traveling block or top drive above the reference location of the well.

LWD data is not really measured relative to depth. It is measured relative to time. The depth of the LWD data is then determined from the logger's depth of the drill bit at the time the data was recorded and the distance of the LWD sensors from the drill bit.

The length of the drill pipe is measured under ambient surface conditions and no stress. Once in the hole, the drill pipe is subjected to forces different than ambient surface conditions and no stress. The two most significant forces with respect to drill pipe length are the tensional stress due to the weight of the drill string and thermal stress related to the in situ temperature of the pipe in the hole. The temperature experienced by the drill pipe is not the static temperature of the earth but the actual temperature in the borehole, which is affected by mud circulation and frictional heating from the drill bit. Both tensional stress and thermal stress cause the drill pipe to lengthen, and these forces tend to increase as a function of TVD. Lesser forces that affect the length of the drill pipe are the mud weight, which partially offsets the weight of the drill pipe; friction along the borehole wall, weight on bit (WOB), and borehole geometry, which also affect the tension on the drill string and thus its stretch; pump pressure, which can cause the drill pipe to balloon slightly, shortening the pipe; and torque, which can also shorten the pipe (Chia et al. 2006).

The length of each section of drill pipe is recorded in the driller's tally book. Occasionally, a joint or stand is left out of the tally book due to human error. Also occasionally, one joint may

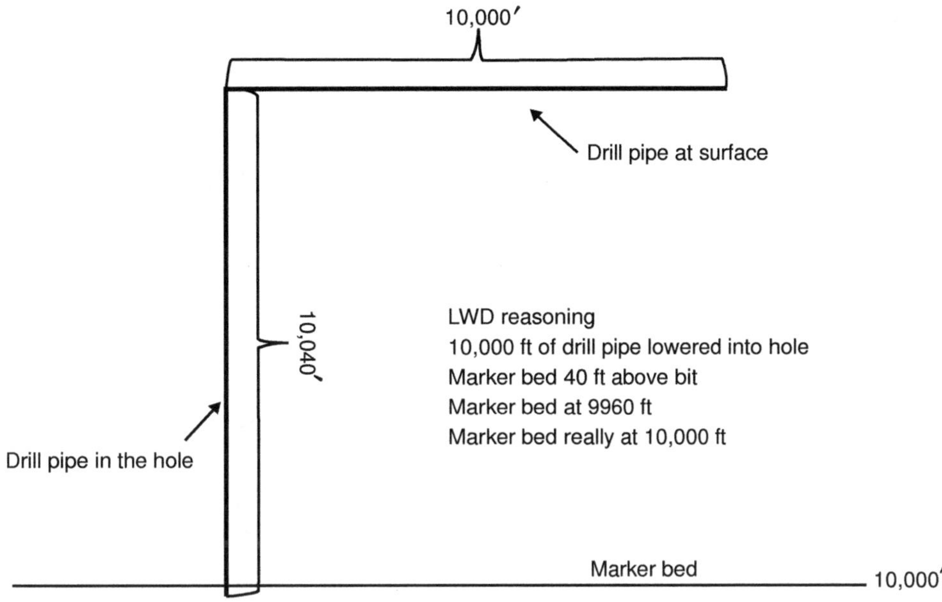

Figure 3-3 Diagram explaining why LWD logs are almost invariably shallower than wireline logs in deeper wells. The length of the drill pipe is measured under ambient conditions at the surface. Tensional stress and thermal stress result in pipe stretch, which is generally not corrected for on LWD logs. (Published with permission of J. Brenneke.)

be switched out for another of a slightly different length, perhaps because the first joint's threads were scored or some other imperfection in the joint was recognized. If the length of the first joint is already recorded in the tally book, it might not be replaced with the length of the second joint. The final accuracy of driller's depth cannot exceed the accuracy of the driller's tally book.

Once a well is drilled as deeply as can be done with a given amount of drill pipe, the drill pipe is set in the slips at the rig floor and a new joint or joints of drill pipe is attached to the existing drill string. The top drive or traveling block is then attached to the top of the new joint or joints of drill pipe and drilling is resumed. The distance the top drive or traveling block moves down from its position in the top of the derrick after drilling resumes is used to determine the depth of the bit in the well.

As a well is drilled, the main forces acting on the drill pipe usually result in elongating the drill string. This elongation results in events logged in the well to be recorded at shallower depths than they really occur (Fig. 3-3).

Pipe-Conveyed Log Depth Measurement

Pipe-conveyed logging (PCL) is usually performed when wireline logs are required but the conditions of the wellbore do not allow the wireline sonde to go down in the wellbore on its own. The wireline sonde is attached to the end of the drill pipe, and the wireline cable is run outside of the drill pipe through a side entry sub. The wireline operator plays out cable as the drill pipe is lowered into the hole. Prior to logging the interval in question, the PCL log is depth tied to an existing log, which may be either a wireline log or an LWD log. During the logging operation, the wireline operator attempts to synchronize the movement of the cable with the movement of the drill pipe. This synchronization is frequently not perfect, resulting in depth errors that are difficult to quantify (Wilson et al. 2004). LWD logs are also usually acquired with some WOB, whereas PCL logs are acquired with no WOB, resulting in some difference in driller's depth between the two logs (Wilson et al. 2004). PCL logs are relatively uncommon and are not considered further in this book.

Relative Depth Uncertainty

Relative depth uncertainty is depth uncertainty within a single well. Both wireline and LWD logs generally have good relative depth uncertainty, so the distance between markers within a single well is usually relatively accurate. On some occasions, however, LWD relative depth uncertainty becomes less accurate (Fig. 3-4) (Storey 2013). The example in Figure 3-4 is somewhat extreme, but smaller variations are not uncommon (e.g., Forsyth et al. 2013, Fig. 5). The cause of

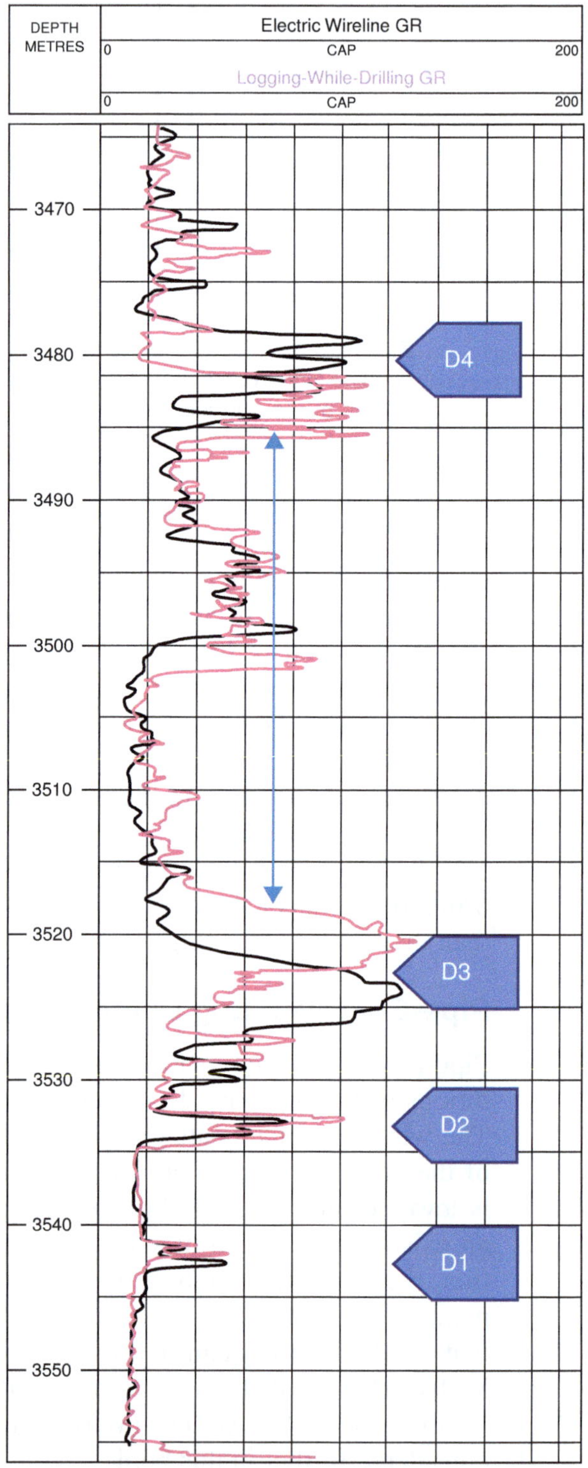

Figure 3-4 Relative depth uncertainty between an LWD log and a wireline log. The two gamma ray (GR) logs are on depth near both the top and the bottom of this section, but they are off depth by several meters near the center of the interval. The interval thickness between the D3 and D4 markers is 32.6 m (107 ft) based on the wireline log and 40.3 m (132 ft) based on the LWD log. This thickness difference, approximately 25%, would have a significant impact on volumetrics. (From Storey 2013. Published with permission of Oilfield Technology.)

these relative depth uncertainties in the LWD logs is not addressed by Storey (2013) or Forsyth et al. (2013) but could be related to poor calibration of the hanging block in the middle of a stand (R. Wylie, personal communication, 2018) or by variations in WOB. The relative depth uncertainty in the LWD data can be seen when both LWD and wireline logs are available but may go unrecognized if only LWD depth is available.

It is important to check the WOB to minimize the possibility of relative depth uncertainty in an LWD log. If the WOB decreases while the drill pipe is rotating and the hanging block is stationary, the drill bit is probably drilling a new hole, and the LWD sensors are recording values from the new section. These readings may not be displayed because LWD logs generally display only the first readings from a given depth, and if the hanging block is not moving, the depth is assumed to be constant. Similarly, if the WOB is increasing and the hanging block is moving, the bit may not be moving as much as the hanging block. In extreme cases, if the drill string goes from extension into compression, the drill pipe could buckle or corkscrew, resulting in significant movement of the hanging block with no movement of the drill bit (Fig. 3-5).

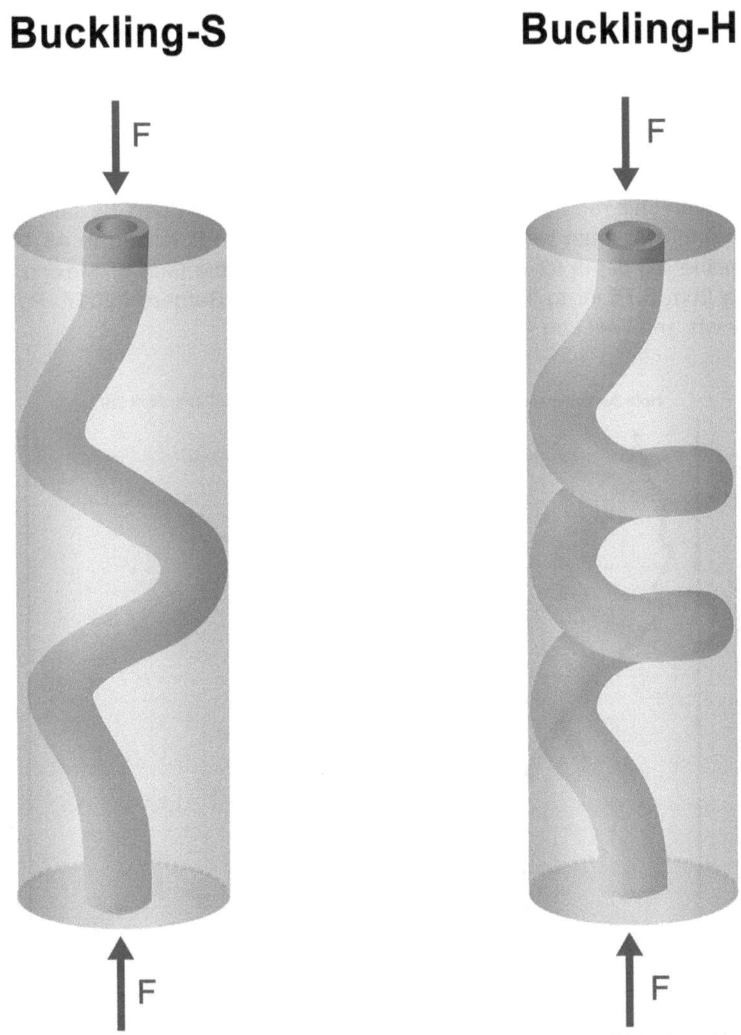

Figure 3-5 Drill pipe in compression can produce sinusoidal buckling (Buckling-S) or helical buckling (Buckling-H), which can cause driller's depth to be greater than TAH. In a deviated well, some parts of the drill string may be in compression while other parts are in tension. (From Jamieson 2012. Image courtesy of Pegasus Vertex, Inc.)

Absolute Depth Uncertainty

Although the relative depth uncertainty within a single well is generally small, the uncertainty in the absolute depth of any well is greater (Fig. 3-2). This uncertainty produces a depth uncertainty

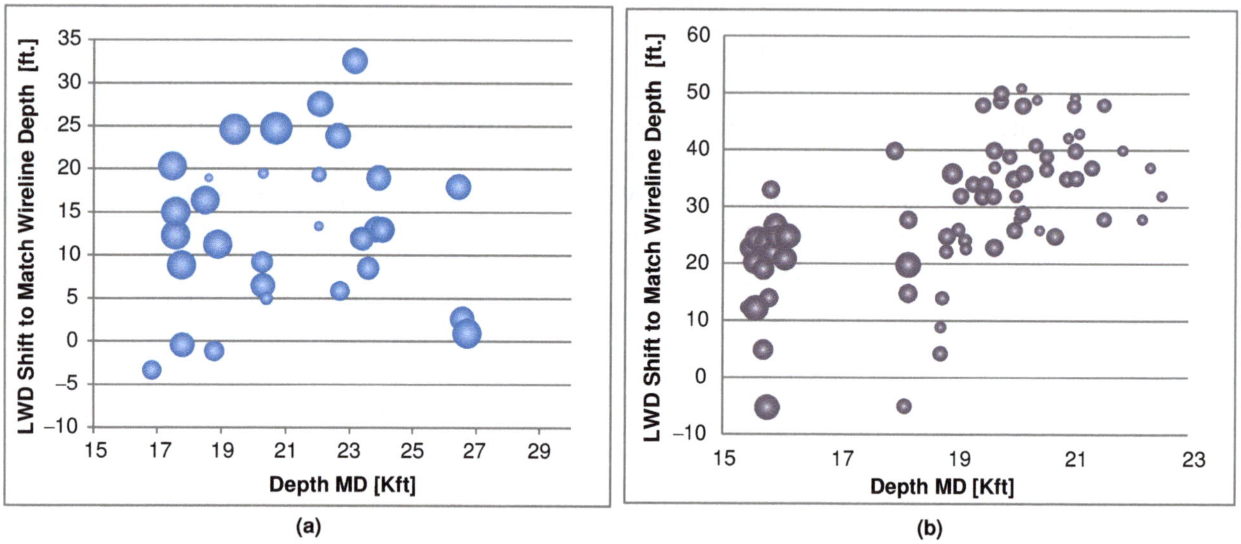

Figure 3-6 (a) LWD depth shift required to match wireline log depth for deep, deviated wells in the Gulf of Mexico. Dot size is proportional to maximum well deviation that ranged from 5.38 deg to 48.47 deg. (b) LWD depth shift required to match wireline log depth for deep, near-vertical wells in the Gulf of Mexico. Dot size is proportional to hole diameter ranging from 5.75 in. to 14.75 in. (From Pineda and Bergeron 2017. Published with permission of Society of Petrophysicists and Well Log Analysts.)

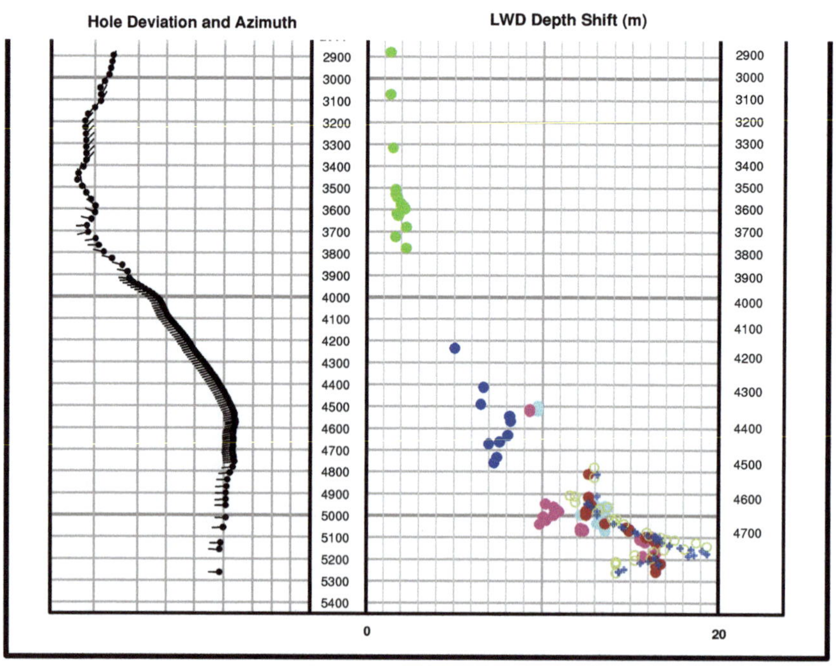

Figure 3-7 Depth shift required to tie an LWD log to a wireline log in a deviated well from Kristin Field, Norwegian Sea. The required log shift reaches a maximum of 19.5 m (64 ft) between 5100 m and 5200 m MD (16,730 ft to 16,060 ft). This shift decreases to just over 14 m (46 ft) at TD. (From Pedersen and Constable 2006. Published with permission of Society of Petrophysicists and Well Log Analysts.)

between wells. Because drill pipe stretch is rarely corrected for in LWD logs, LWD logs must almost invariably be shifted deeper to match wireline logs (Fig. 3-6).

When both wireline and LWD logs are run in a well, the LWD logs can be depth shifted to tie the wireline log. This depth shift is not constant, however, but varies with depth, and not always in a constant direction (Fig. 3-7). The real problem arises when only LWD logs are run in a well and it is compared to wells with wireline logs. There are no correction tables or rules of thumb that allow LWD logs to be depth shifted to mimic wireline depths, but in wells deeper than 15,000 ft, depth shifts of 20 ft to 50 ft are not uncommon.

LWD logs can be depth shifted to correct for drill pipe stretch if sufficient drilling data is available (Chia et al. 2006). These corrections allow the length of each joint of drill pipe to be corrected for mechanical stretch and thermal expansion and can dramatically reduce the depth differences between LWD and wireline logs (Fig. 3-8). The required drilling data are almost always available to the service company running the log but may not be available to other companies who wish to use the log in mapping.

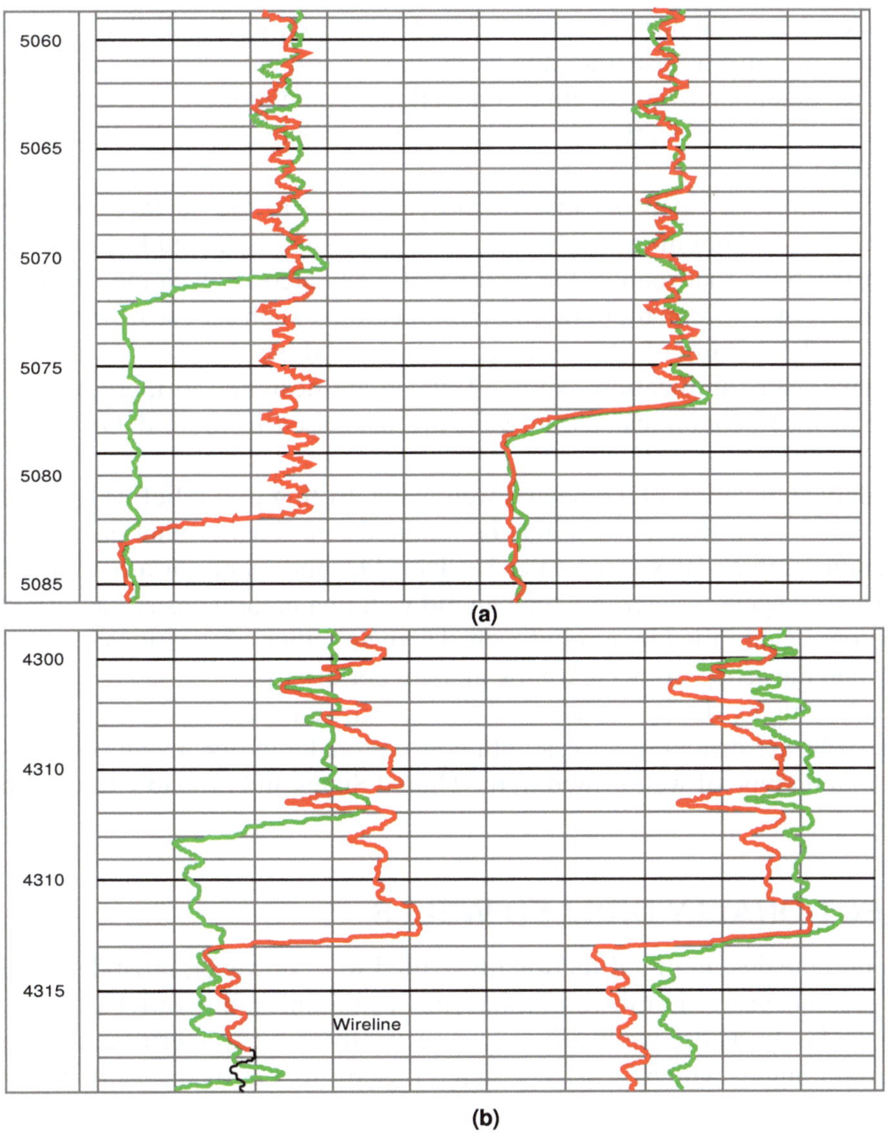

Figure 3-8 (a) LWD and wireline GR logs from a deep, highly deviated well. The left side shows the curves differ in depth by 11 m (36 ft) as supplied in final form by the service company. The right side of the figure shows the depth agreement between the two logs after both the LWD and wireline depths have been corrected. (b) LWD and wireline GR logs from a well in a different field with a simpler geometric profile. Again, the logs on the left side of the figure are the final results supplied by the service company. Here the depth difference is 5.3 m (17 ft). The logs on the right side of the figure have been depth corrected. Here the depth correction to the wireline log is very small, but the depth shift to the LWD log is substantial. (From Chia et al. 2006. Published with permission of Society of Petroleum Engineers.)

DIRECTIONALLY DRILLED WELLS

A directionally drilled, or deviated, well is defined as a well drilled at an angle less than 90 deg to the horizontal (Fig. 3-9). Wells are normally deviated intentionally in response to a predetermined plan; however, straight holes often deviate from the vertical due to bit rotation and natural deviation tendencies related to rock types and structure.

The technique of controlled directional drilling began in the late 1920s on the U.S. Pacific coast (LeRoy et al. 1977, Chilingarian and Vorabutr 1981). Through the use of controlled directional drilling, a wellbore is deviated along a preplanned course to intersect a subsurface target horizon(s) at a specific location (Fig. 3-10). Our primary interest in directionally drilled wells in this textbook centers around their application to subsurface mapping.

APPLICATION OF DIRECTIONALLY DRILLED WELLS

There are a number of reasons to drill a directional well. Some of the more common applications are shown in Figure 3-11. For many years, the most common application was the drilling of offshore wells from a single platform location (Fig. 3-11a). The use of a single platform or subsea manifold from which multiple wells are drilled improves economics and simplifies production facilities. Onshore, many unconventional plays are now being drilled from well pads that include multiple wells. This approach minimizes lost time due to rig moves and can also minimize the amount of surface disturbance and the cost of pipeline connections. Some modern rigs can actually "walk" short distances from surface location to surface location without being broken down for transport (e.g., Wethe 2015).

Onshore, wells are commonly deviated because of inaccessibility to the surface location directly over the subsurface target. Buildings, towns, cities, rivers, shorelines, and mountains are the kinds of surface obstructions that require the drilling of a deviated well. Some offshore wells are also located away from seafloor topographic complications such as escarpments or unstable conditions, requiring the drilling of deviated wells. Horizontal wells are a special type of directional well. They have many applications, most of which are designed to increase productivity rate and improve project economics. These wells are discussed in more detail in the next section.

One very important safety application of a deviated well is the drilling of a relief well to kill a well that has blown out (Fig. 3-11f). Relief wells may be required to kill a well that has blown out during drilling/completion, such as BP's Deepwater Horizon Macondo well, or a completed well such as California's Aliso Canyon disaster in which an old well in a natural gas storage field flowed gas for 118 days before being killed by a relief well. There are other applications of deviated wells, but they are beyond the scope of this textbook and are not discussed.

Over the last two decades, highly deviated wells and horizontal wells have become increasingly common. Horizontal wells are drilled to access both conventional and unconventional reservoirs and are the most common type of well drilled in the United States and probably the world today.

COMMON TYPES OF DIRECTIONALLY DRILLED WELLS

Many complex factors go into the design of a directionally drilled well; however, most deviated wells fall into one of three types. For years, the most common type was a simple ramp well (Fig. 3-9a), sometimes called an L-shaped hole. These wells are drilled vertically to a predetermined depth and then deviated to a certain angle, which is usually held constant to TD of the well. Many

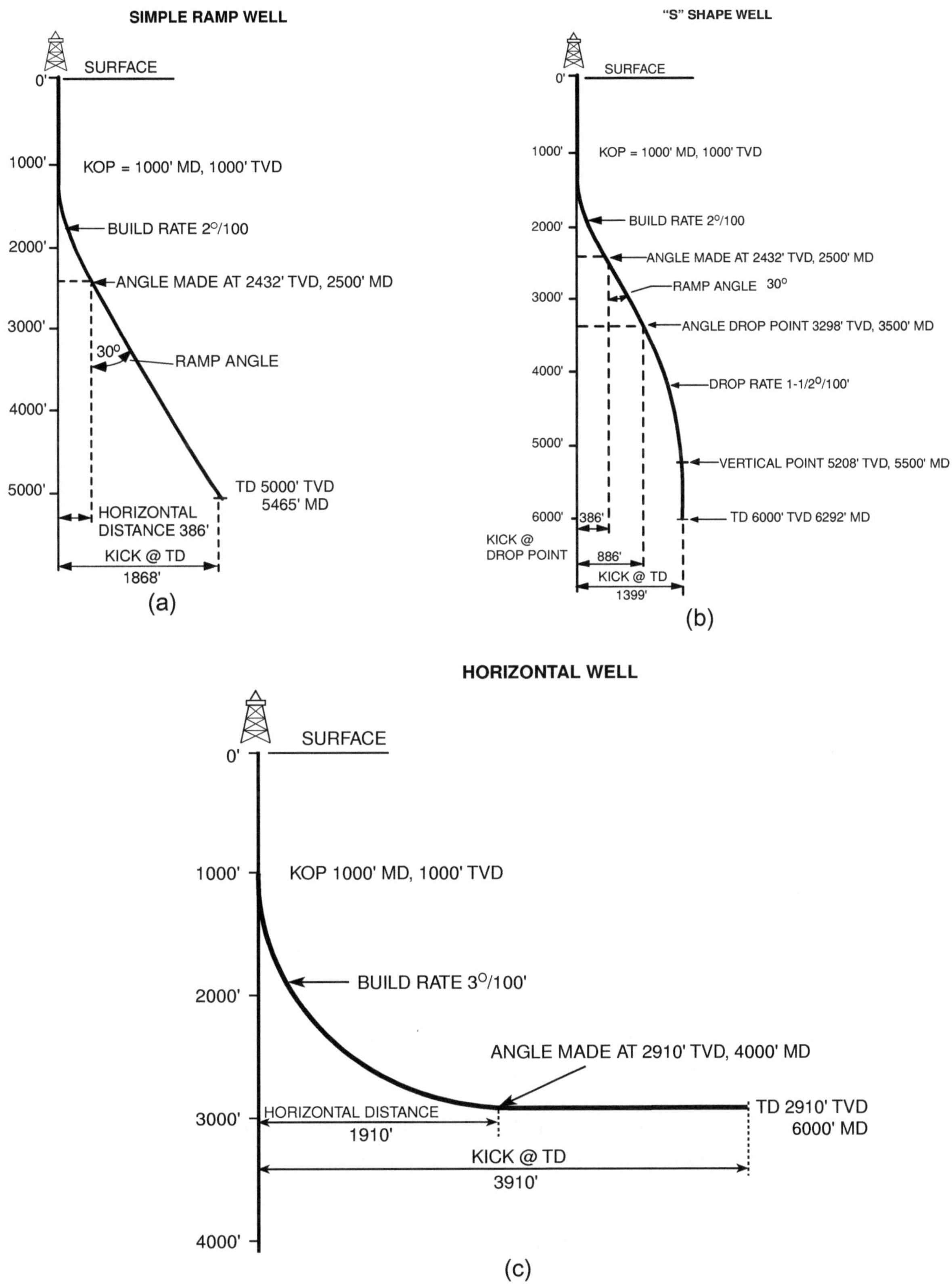

Figure 3-9 Diagrammatic cross sections of (a) a simple ramp or L-shaped well, (b) a more complicated S-shaped well, and (c) a horizontal well. (Figures [a] and [b] published by permission of Tenneco Oil Company; [c] published by permission of J. Brewton.)

wells are drilled with an S-shaped design. For an S-shaped hole, the well begins as a vertical hole and then builds to a predetermined angle, maintains this angle to a designated depth, and then the angle is lowered again, often going back to vertical (Fig. 3-9b). Finally, horizontal wells are configured by continuously building the deviation angle until the desired near-horizontal orientation is reached (Fig. 3-9c). Under very rare circumstances, such as in the tar sands of Canada, wells may be spudded at an angle. This allows for larger step-out distances at very shallow depths.

General Terminology

The terms used to describe various aspects of a directionally drilled well are defined in Table 3-2 and illustrated in Figure 3-9.

Table 3-2 Aspects of directionally drilled wells

KOP	Kick-off point. Depth of initial deviation from vertical measured as measured depth (MD), true vertical depth (TVD), or subsea true vertical depth (SSTVD).
Build rate	Build angle. Rate at which the angle changes during deviation. It is usually expressed in degrees per 100 ft drilled. Example: 2 deg per 100 ft.
Ramp angle	Hole angle, drift angle, angle of deviation. Angle from the vertical that a well maintains from the end of the build through the ramp segment of the well.
BHL	Bottomhole location. Horizontal and vertical coordinates to the total depth point usually measured from the surface location.
Drop rate	Rate at which the ramp angle changes in degrees per 100 ft. Measured in S-shaped holes.
Vertical point	The depth where the well is back to vertical, measured as MD, TVD, or SSTVD.

Horizontal Wells

Horizontal wells are typically considered to be wells with the borehole drilled within about 3 deg of bed dip or wells drilled nearly horizontally. These include extended-reach wells with long horizontal displacements, as well as long- and medium-radius horizontal wells. Short-radius horizontal wells are often called *drain hole wells*.

Extended-reach wells can be similar to the S-shaped well but with very high ramp angles in the 80 deg range. Although they are nearly horizontal, they might not be considered true horizontal wells. These wells are generally drilled when the surface location is necessarily a great distance from the target. Long-radius horizontal wells have build rates in the 3 deg per 100-ft range, and generally the horizontal part of the borehole is several thousand feet in length. Medium-radius horizontal wells have build rates in the 30 deg per 100-ft range and are usually drilled for shallow objectives. Short-radius horizontal wells are borehole segments drilled from a vertical borehole that penetrated the objective interval, with the deviation from vertical to horizontal made within a vertical interval of about 20 ft. These wells usually have horizontal segments of only a few hundred feet, but several horizontal segments may be drilled from the same vertical wellbore. These drain hole wells are used in low permeability reservoirs and enhanced recovery projects.

The purpose of drilling most horizontal wells is to improve the economics of a project by increasing production rates. For example, horizontal wells can improve production rates from (1) reservoirs containing heavy oil, such as the Orinoco heavy oil belt (Venezuela); (2) reservoirs with mostly fracture porosity, such as the Austin Chalk (onshore Gulf of Mexico); and

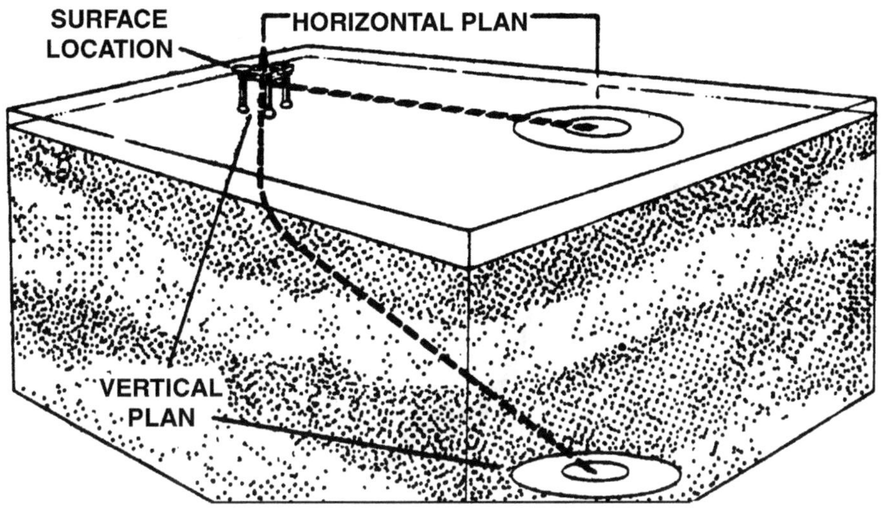

Figure 3-10 Block diagram showing the vertical and horizontal plan views of a well directionally drilled to a predetermined subsurface target. (Published by permission of Eastman Christensen.)

(3) reservoirs of low permeability. This latter feature is responsible for the rapid development of unconventional reservoirs in the United States and the rapid shift of well type from predominantly vertical in 1990 to dominantly horizontal in 2018 (Fig. 3-12).

Horizontal wells can also be used to penetrate and produce from multiple reservoirs that are laterally discontinuous. For example, separate fluvial channel sands that are in the same stratigraphic interval but are laterally discontinuous can be penetrated by a single horizontal well to drain multiple sand bodies. Also, for multilobed reservoirs with attic reserves above the highest wells on the structure, a horizontal well can be designed to encounter all the lobes, whereas a vertical well might miss some of the lobes that truncate up-dip. Another application of horizontal wells is to drill for thin oil zones over water. These zones are subject to coning of water into vertical wells. A horizontal well can significantly reduce the problem of water coning. There are other applications of horizontal wells, but most well plans are based on increasing production rates and shortening well life to improve project economics.

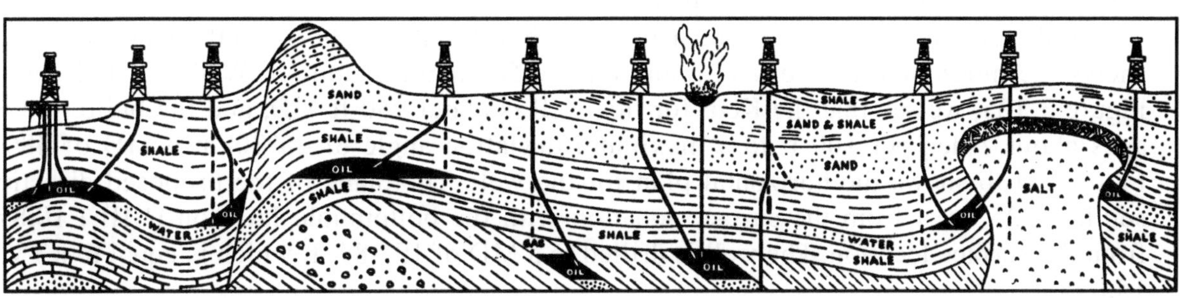

Figure 3-11 Applications of directional drilling. (a) Multiple wells offshore or from artificial islands. (b) Shoreline drilling. (c) Fault control. (d) Inaccessible surface location. (e) Stratigraphic trap. (f) Relief well control. (g) Straightening hole and side tracking. (h, i, j) Saltdome drilling. (From LeRoy et al. 1977. Published by permission of the Colorado School of Mines.)

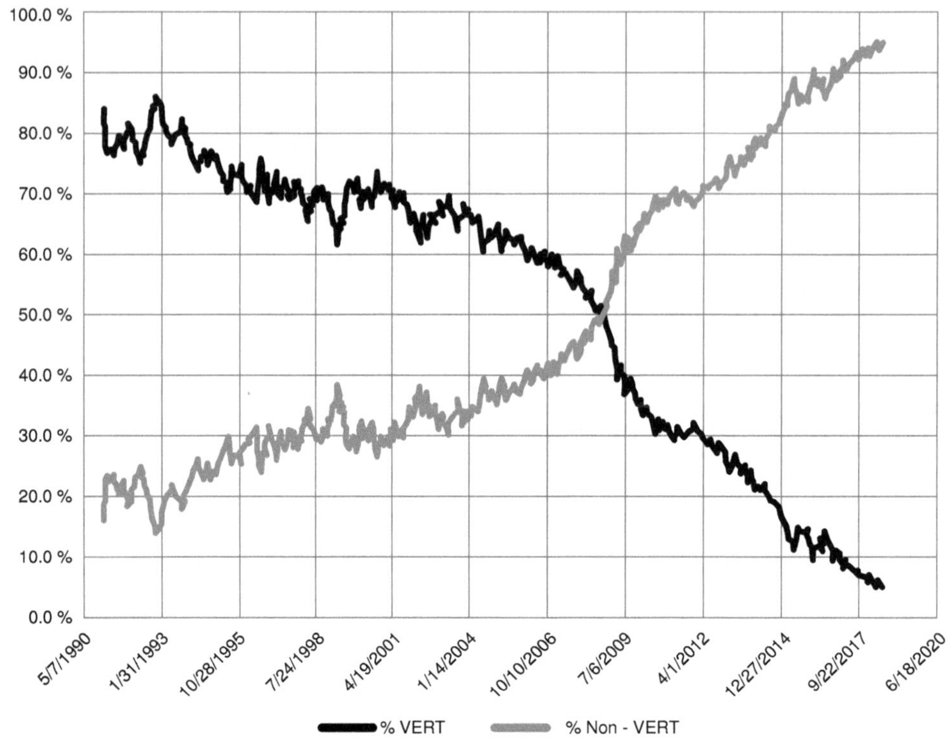

Figure 3-12 Percentage of wells of various well types drilled in the United States. Most nonvertical wells drilled after 2007 are horizontal wells. By 2018, 88% of wells drilled in the United States were horizontal. Data from Baker Hughes website. (Published with permission of J. Brenneke.)

DIRECTIONAL WELL PLAN

A variety of data go into the design of a directionally drilled well, including the depth and distance from the surface location to each subsurface target, diameter of the target, KOP, build rate, platform location, lease lines, hole size, and total depth of the well. Once preliminary studies indicate the need for a deviated well, most companies rely on a directional drilling service company to prepare the final directional plan.

A directional well design consists of both vertical and horizontal plans. Figure 3-13 shows the horizontal and vertical plans for the Diamond Shamrock Well No. 1. Reviewing Figure 3-13a, we see that the KOP for this well is about 950 ft MD, and the build rate averages about 2 deg per 100 ft to a maximum deviation angle of 45°30' at a MD of 4800 ft. Figure 3-13b shows the horizontal plan for the well. The plane of the proposed direction is south 46°25' west from the surface location with the bottomhole location (BHL) 10,873 ft from the surface location. The well is drilled to a MD of 19,484 ft, which is equal to a TVD of 15,695 ft.

A grid reference for north is shown in Figure 3-13b. Several coordinate systems exist for industry's use and, unfortunately, it is common for the data sources (e.g., directional well surveys and seismic surveys) and maps used in a project to be in different coordinate systems. It is critical that all of them be converted to the same system to reduce errors in location.

Figure 3-14 illustrates the vertical plan for a well that has a maximum deviation angle of 94 deg and a maximum build angle of 14 deg per 100 ft. The wellbore is vertical near the surface, and at a depth of 1659 ft (TVD), it is horizontal.

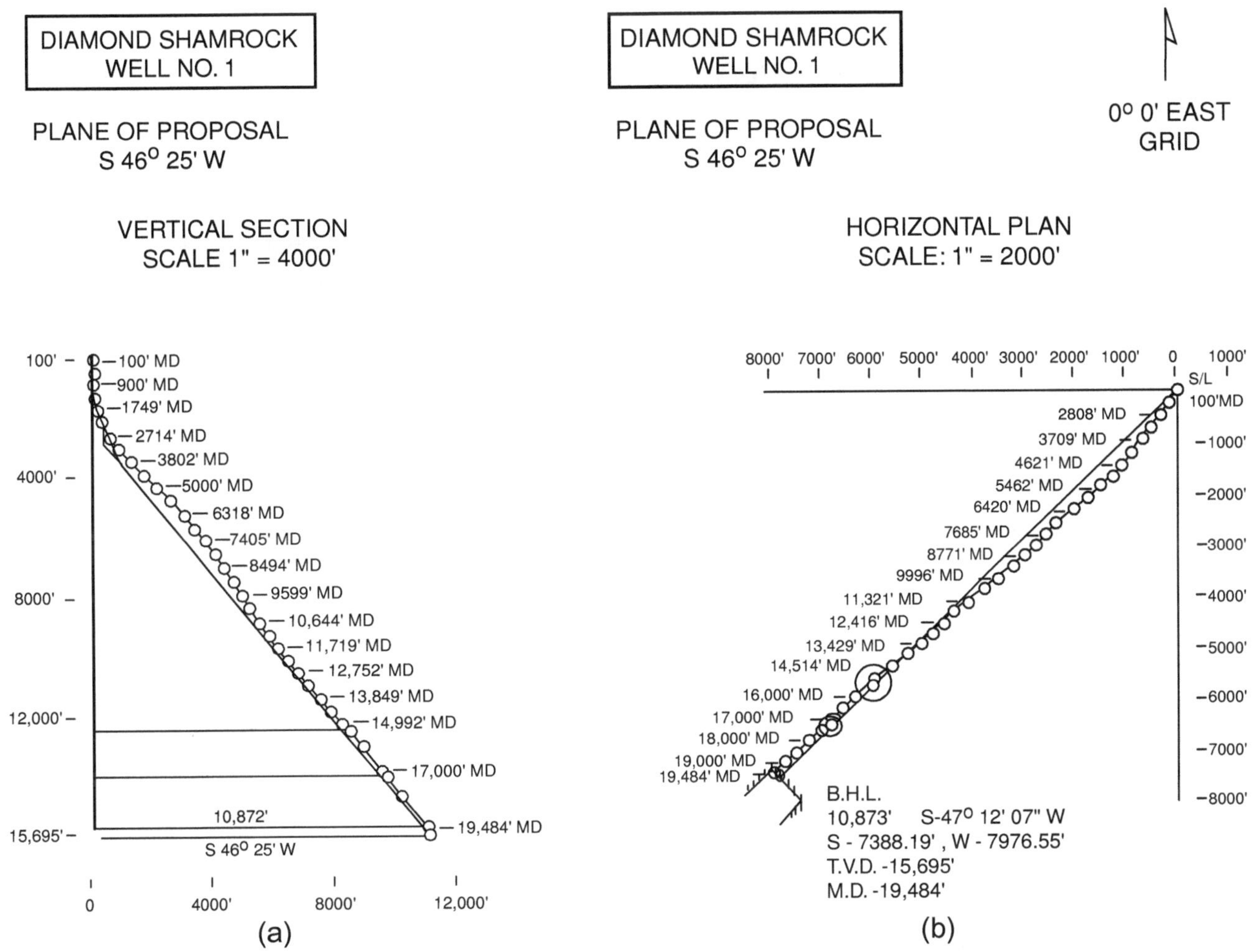

Figure 3-13 (a) Vertical section plan for a directional well. (b) Horizontal plan for the same directional well shown in Figure 3-13a. (Published by permission of Gardes Directional Drilling.)

Commonly, deviated wells are drilled with a build rate of 2 deg per 100 ft of hole drilled. Figure 3-15 shows the scaled chart for a 2 deg per 100 ft build rate. Such charts are used to make a quick estimate of well design after structure maps have been made on target horizons. For example, from the chart, a target horizon located at a TVD of 10,000 ft and a horizontal distance of 4000 ft from the platform location requires the drilling of a well with a deviation angle of approximately 23 deg to a MD of 10,800 ft. Such charts are available from directional service companies for build rates ranging from 1 deg to 5 deg per 100 ft. In most cases today, however, these charts have been replaced with computer software that can calculate preliminary directional plans quickly.

DIRECTIONAL TOOLS USED FOR MEASUREMENTS

Three features of a directional wellbore are measured at given points within the well: (1) measured depth, which is the distance from the surface to a given point, measured along the wellbore; (2) angle of inclination from the vertical (drift angle or deviation angle); and (3) drift direction, or the directional path of the wellbore. These parameters are the basis for calculations of the position of each point in the subsurface, and all this information is included in a directional survey. Drift angle and drift direction are measured by a survey tool conveyed in drill pipe or by wireline, and MD is determined by length of drill pipe or wireline. The various tools that are used fall into two categories: magnetic and nonmagnetic.

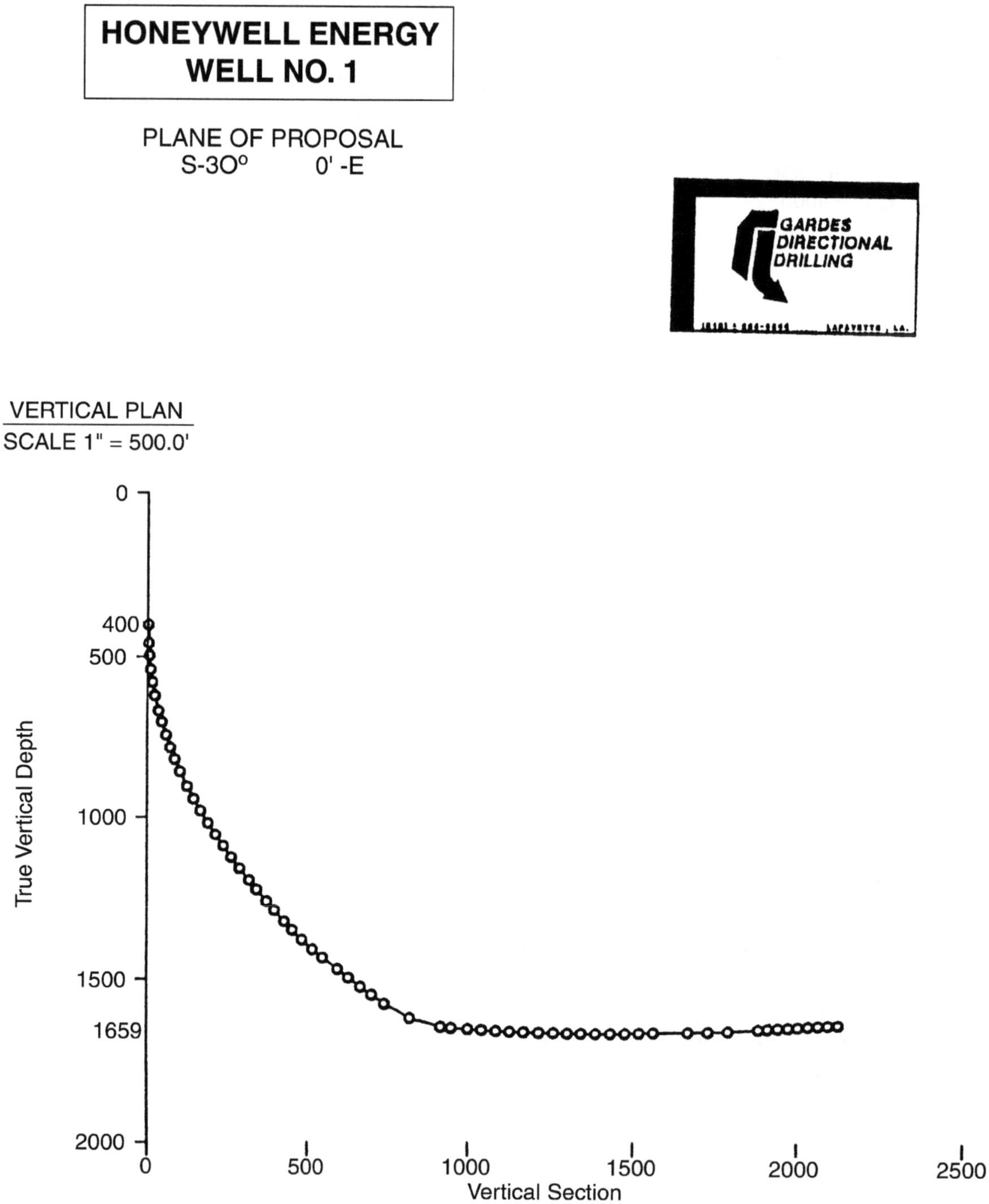

Figure 3-14 Vertical plan for a nearly horizontal well with a maximum deviation angle of 94 deg. (Published by permission of Gardes Directional Drilling.)

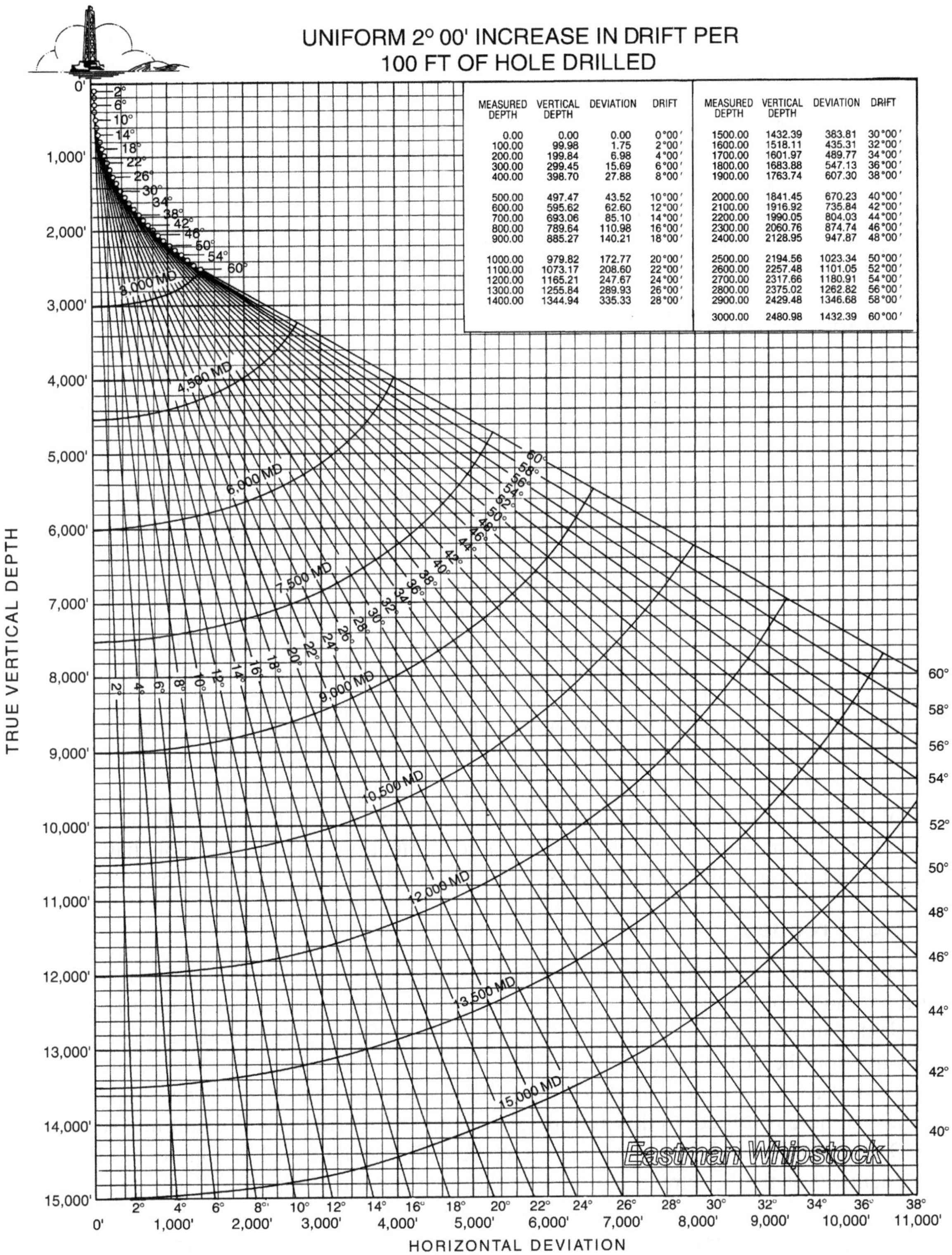

MEASURED DEPTH	VERTICAL DEPTH	DEVIATION	DRIFT	MEASURED DEPTH	VERTICAL DEPTH	DEVIATION	DRIFT
0.00	0.00	0.00	0°00′	1500.00	1432.39	383.81	30°00′
100.00	99.98	1.75	2°00′	1600.00	1518.11	435.31	32°00′
200.00	199.84	6.98	4°00′	1700.00	1601.97	489.77	34°00′
300.00	299.45	15.69	6°00′	1800.00	1683.88	547.13	36°00′
400.00	398.70	27.88	8°00′	1900.00	1763.74	607.30	38°00′
500.00	497.47	43.52	10°00′	2000.00	1841.45	670.23	40°00′
600.00	595.62	62.60	12°00′	2100.00	1916.92	735.84	42°00′
700.00	693.06	85.10	14°00′	2200.00	1990.05	804.03	44°00′
800.00	789.64	110.98	16°00′	2300.00	2060.76	874.74	46°00′
900.00	885.27	140.21	18°00′	2400.00	2128.95	947.87	48°00′
1000.00	979.82	172.77	20°00′	2500.00	2194.56	1023.34	50°00′
1100.00	1073.17	208.60	22°00′	2600.00	2257.48	1101.05	52°00′
1200.00	1165.21	247.67	24°00′	2700.00	2317.66	1180.91	54°00′
1300.00	1255.84	289.93	26°00′	2800.00	2375.02	1262.82	56°00′
1400.00	1344.94	335.33	28°00′	2900.00	2429.48	1346.68	58°00′
				3000.00	2480.98	1432.39	60°00′

Figure 3-15 Scaled chart for a build rate of 2 deg per 100 ft of hole drilled. (Published by permission of Eastman Christensen.)

Magnetic Surveys

Magnetic is a generic term for describing several survey tools that use a magnetic compass for direction and therefore must be run inside a special nonmagnetic drill collar to negate the effects of the drill pipe. An example of such a survey is the *single-shot magnetic survey*. This device records, on a heat-resistant film disc, the magnetic direction and inclination angle of the wellbore at specific depth intervals. A *multishot survey* uses a filmstrip to record several readings of hole angle and direction at different depth intervals. Newer wireline tools provide real-time survey data at the surface. Measurement accuracy of magnetic survey tools has been improved within the last 30 years. Modern magnetic tools are used for measurement-while-drilling (MWD) surveying. These surveys provide real-time data for more efficient directional drilling.

Nonmagnetic Surveys

Nonmagnetic survey tools are of two types: those with no direction-finding device and those with a gyroscopic mechanism for determining direction. A drift indicator tool (e.g., a Totco tool) measures only drift angle and is usually run in vertical wells or shallow vertical sections of deviated wells where directional information is not required. This tool generally consists of a housing or barrel, a motion indicator, a timer, a punch, and a printed paper disc. The unit is either run on a wireline or dropped on a drill bit. When the motion sensor determines that the tool is no longer moving, the timer is activated, and after a predetermined interval of time the punch is released. The punch, which is allowed to swing freely and act as a plumb bob, drops vertically and punches a hole in the paper disc, which is marked in degrees. Figure 3-16 shows a Totco disc scaled to a maximum of 8 deg. The hole punched in the disc indicates an inclination angle of 4.5 deg, but the drift direction is unknown.

Gyroscopic survey tools are widely used and are capable of providing more accurate data than magnetic survey tools. Because the magnetic compass is replaced by a gyrocompass, the system can be run in both cased and uncased holes and run where cased holes are nearby, as in a platform well cluster. The gyro system can be set up as a single-shot or multishot instrument. Conventional tools, referred to as *free* gyros, are less accurate than the newer *rate* gyros, which are of different construction. A gyro survey is sometimes run after an MWD magnetic survey to provide additional information for well path determination. Gyroscopic tools have recently been developed for use while drilling (gyroscope while drilling [GWD]) (e.g., Weston et al. 2014).

Survey Frequency

Survey frequency can affect the accuracy of a survey, especially if the well is drilled with the traditional mud motor and bent housing. Deviated wells drilled using a mud motor and bent housing are drilled using a combination of sliding and drilling. During sliding, the bent sub is oriented in the direction the driller wants to turn the well, and the mud motor rotates the bit while the drill pipe is not rotated. After a short section of hole is drilled this way, the drill pipe is rotated and additional hole is drilled. In this system, sliding and then drilling will result in a slightly different TVD than drilling and then sliding, even if the MD, borehole inclination, and borehole azimuth are identical at the beginning and end of the hole section (Jamieson 2012). Wells drilled with rotary steerable assemblages may be less susceptible to errors due to survey frequency, but it is still best practice to survey deviated wells at frequent intervals, generally every 100 ft or less.

DIRECTIONAL SURVEY CALCULATIONS

More than 20 methods have been developed to calculate the directional survey, which provides the 3D location of a directional wellbore anywhere along its entire length. The more common methods include (1) tangential, which is the least accurate and is no longer used; (2) trapezoidal, also

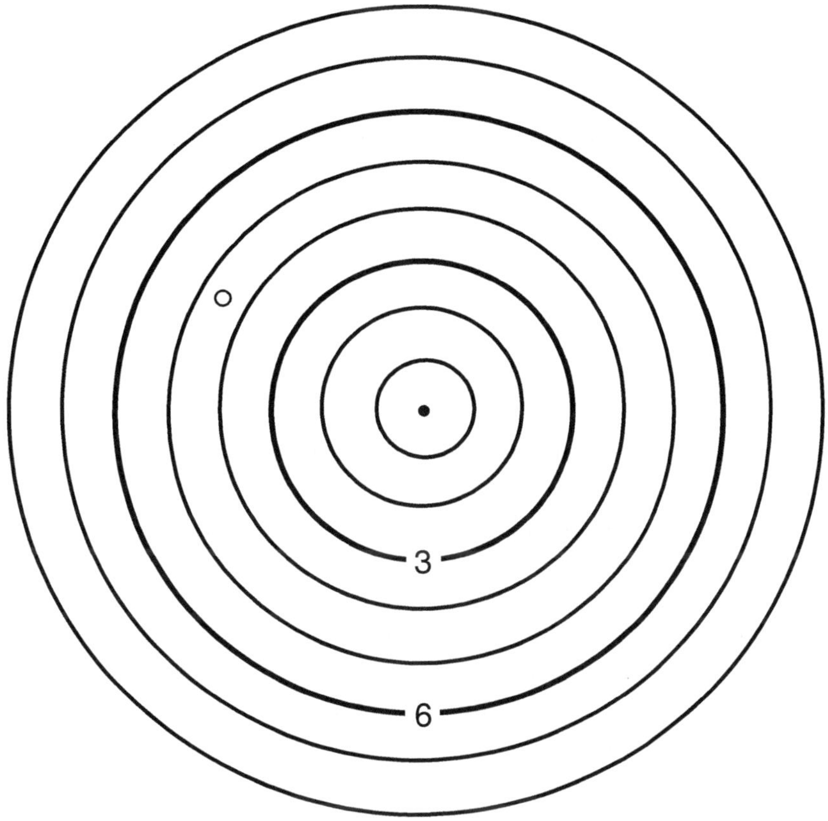

Figure 3-16 Example of a Totco survey. Well deviation angle is 4.5 deg.

called *balanced tangential*; (3) average angle; (4) radius of curvature; and (5) minimum curvature. Many companies have their own methods, and new algorithms are occasionally published. The accuracy of any method varies with the configuration of the wellbore. The radius of curvature and minimum curvature methods are the most widely used today, and the latter is considered slightly more accurate for most wells. The method of calculation is typically noted on the survey. If you work in an area with deviated wells and surveys of different vintages, you can reduce the wellbore position uncertainty due to calculation method by using the same algorithm, such as minimum curvature, to recalculate all surveys. Computer software for this is readily available. The TVD calculated for a survey can vary by tens of feet between an older method, such as the tangential method, and a more modern method, such as the minimum curvature method.

If you are working an area with surveys of different vintages and using a computer workstation, all of the directional survey data for the wells should be entered as the basic survey measurements, that is, MD, deviation angle, and direction of deviation. Allow the computer software to calculate the directional survey results using the radius of curvature or minimum curvature method. Some computer software allows you to enter calculated data from directional surveys, such as TVD and rectangular coordinates, rather than MD. If you enter these calculated data from old surveys rather than the MD, you are introducing unnecessary positioning errors associated with older calculation methods into your database.

Three measurements go into the directional survey: (1) measured depth, (2) deviation angle, and (3) direction of deviation. These measurements, taken at specific depth intervals, are used to calculate the directional survey of a well. Figure 3-17 shows a portion of the directional survey from a deviated well. The tabular printout for this directional survey has nine columns of data for each survey point in the well (Table 3-3).

Table 3-3 Explanation of data shown in Figure 3-17

Column	*Data*
1	Subsea depth of wellbore in feet
2	Measured depth of wellbore in feet
3	True vertical depth of wellbore in feet
4	Angle of wellbore deviation (inclination angle)
5	Planned direction of wellbore (true bearing)
6	Distance in feet from the surface location along the proposed directional path
7	True bearing and distance of each survey point from the surface location in rectangular coordinates
8	True bearing and distance from surface location directly to each survey point
9	Maximum change in hole angle in degrees per 100 ft

If the cased or surface section of the hole was surveyed with a nonmagnetic survey tool such as a drift indicator, the angle of the cased portion of the hole will be displayed on the survey along with an estimate of the maximum possible deviation of this portion of the hole, as shown in Figure 3-18. The 146.75-ft maximum deviation, prominently shown on the survey, indicates that if the wellbore drift in the cased portion of the hole were all in the same direction, the well at a depth of 3513 ft could be as much as 146.75 ft from the surface location. Notice at the end of the survey that it indicates that the bottom of the hole lies within a circle of radius 146.75 ft with its center located 203.94 ft south 63°27' west of the surface location. Such information may be important in fault, structure, and isochore mapping.

Directional surveys are used to plot wellbores on base maps. The applications of these plots are discussed later in this chapter.

DIRECTIONAL SURVEY

(1) SUBSEA DEPTH FEET	(2) MEAS. DEPTH FEET	(3) TRUE VERTICAL DEPTH FEET	(4) DRIFT ANGLE D M	(5) DRIFT DIREC D M	(6) VERTICAL SECTION FEET	(7) TOTAL RECTANGULAR COORDINATES FEET	(7) FEET	(8) CLOSURES DISTANCE FEET	(8) DIRECT D M	(9) DOG LEG SEVERITY DEG/100FT
5299.88	5700	5330.88	31 30	N 37 0 E	1490.00	1390.73 N	538.16 E	1491.22	N 21 9 E	1.04
5385.14	5800	5416.14	31 30	N 32 0 E	1541.27	1433.77 N	567.75 E	1542.09	N 21 36 E	2.61
5470.52	5900	5501.52	31 15	N 25 0 E	1593.10	1479.50 N	592.57 E	1593.76	N 21 50 E	3.65
5556.01	6000	5587.01	31 15	N 20 0 E	1644.95	1527.41 N	612.42 E	1645.61	N 21 51 E	2.59
5641.95	6100	5672.95	30 15	N 14 0 E	1695.73	1576.28 N	627.36 E	1696.54	N 21 42 E	3.23
5728.23	6200	5759.23	30 30	N 10 0 E	1745.28	1625.73 N	637.87 E	1746.39	N 21 25 E	2.04
5814.50	6300	5845.50	30 15	N 8 0 E	1794.24	1675.67 N	645.78 E	1795.81	N 21 5 E	1.04
5900.88	6400	5931.88	30 15	N 8 0 E	1842.79	1725.56 N	652.79 E	1844.91	N 20 43 E	0.00
5987.59	6500	6018.59	29 30	N 10 0 E	1891.02	1774.76 N	660.59 E	1893.71	N 20 25 E	1.25
6075.26	6600	6106.26	28 0	N 12 0 E	1937.98	1821.97 N	669.76 E	1941.17	N 20 11 E	1.78
6164.16	6700	6195.16	26 30	N 12 0 E	1982.85	1866.75 N	679.28 E	1986.50	N 19 60 E	1.50
6254.23	6800	6285.23	25 0	N 13 0 E	2025.50	1909.17 N	688.68 E	2029.58	N 19 50 E	1.56
6344.86	6900	6375.86	25 0	N 16 0 E	2067.24	1950.08 N	699.26 E	2071.66	N 19 44 E	1.27
6435.40	7000	6466.40	25 15	N 18 0 E	2109.43	1990.68 N	711.68 E	2114.07	N 19 40 E	0.89
6525.75	7100	6556.75	25 30	N 20 0 E	2152.15	2031.20 N	725.63 E	2156.92	N 19 40 E	0.89
6616.10	7200	6647.10	25 15	N 20 0 E	2194.92	2071.47 N	740.29 E	2199.77	N 19 40 E	0.25
6706.73	7300	6737.73	24 45	N 20 0 E	2237.11	2111.18 N	754.74 E	2242.03	N 19 40 E	0.50
6797.82	7400	6828.82	24 0	N 20 0 E	2278.30	2149.96 N	768.86 E	2283.30	N 19 41 E	0.75
6889.17	7500	6920.17	24 0	N 20 0 E	2318.90	2188.18 N	782.77 E	2323.98	N 19 41 E	0.00
6980.53	7600	7011.53	24 0	N 20 0 E	2359.50	2226.40 N	796.68 E	2364.65	N 19 41 E	0.00

Radius of Curvature Calculation

Figure 3-17 Part of the directional survey for a deviated well. (Published by permission of Gardes Directional Drilling.)

DIRECTIONAL SURVEY UNCERTAINTIES

The error introduced by the method of calculation becomes almost academic when the other directional survey uncertainties are considered. Uncertainty in measurement can be directly related to the type of survey tool and its operating condition and procedures. The diagrams in Figures 3-19 and 3-20 are derived from a predictive mathematical model that was based on extensive research on North Sea wells by Wolff and de Wardt (1981), who showed that position uncertainties of boreholes are controlled by systematic errors. Vertical uncertainty (Fig. 3-19) refers to inconsistencies in determination of depths. Lateral uncertainty (Fig. 3-20) applies to inconsistencies in determination of lateral position. Magnetic surveys are compared to gyroscopic surveys. The magnetic and free gyroscope surveys display a significant range of potential inaccuracy, which can be due to the quality of survey equipment, operating procedures, and wellbore position. Note that more highly deviated wellbores are subject to more uncertainty in position. The actual predicted amounts of inaccuracy are not directly applicable to areas other than the North Sea, circa 1980, for a variety of reasons. For example, a higher latitudinal position produces more uncertainty in magnetic surveys. Also, the mathematical model does not apply to horizontal wells, including extended-reach wells. Many companies have detailed proprietary information about degree of error in particular types of survey tools, as used under certain operating conditions in given areas.

DIRECTIONAL SURVEY DATA
BAYOU FER BLANC FIELD, LAFOURCHE REALTY CO. WELL NO. B-3

MEASURED	DRIFT	TVD	COURSE DEVIATION		COURSE COORDINATES				TOTAL COORDINATES			
DEPTH	ANGLE	DEPTH	FEET	DIRECTION	NORTH	SOUTH	EAST	WEST	NORTH	SOUTH	EAST	WEST
1000.00	1-45	999.53	30.54	CASING								
2000.00	2-30	1998.58	43.62	CASING								
3000.00	2-45	2997.43	47.98	CASING								
3513.00	2-45	3509.84	24.61	CASING								
		MAXIMUM POSSIBLE DEVIATION OF CASED HOLE AT THIS POINT IS 146.75 FEET.										
4000.00	2- 0	3996.54	17.00	S 39 W	0.	13.21	0.	10.70	0.	13.21	0.	10.70
4500.00	1- 0	4496.47	8.73	S 36 E	0.	7.06	5.13	0.	0.	20.27	0.	5.57
5000.00	0-30	4996.45	4.36	N 82 E	0.61	0.	4.32	0.	0.	19.66	0.	1.25
5500.00	0-30	5496.43	4.36	S 64 W	0.	1.91	0.	3.92	0.	21.57	0.	5.17
6000.00	0-45	5996.39	6.54	S 26 W	0.	5.88	0.	2.87	0.	27.46	0.	8.04
6500.00	1- 0	6496.31	8.73	S 20 W	0.	8.20	0.	2.98	0.	35.66	0.	11.02
7000.00	1- 0	6996.23	8.73	S 85 W	0.	0.76	0.	8.69	0.	36.42	0.	19.71
7500.00	1-15	7496.11	10.91	S 77 W	0.	2.45	0.	10.63	0.	38.87	0.	30.34
8000.00	2- 0	7995.81	17.45	S 24 W	0.	15.94	0.	7.10	0.	54.81	0.	37.44
8500.00	2- 0	8495.73	8.33	S 30 W	0.	0.	0.	3.00	0.	30.45	0.	45.00
9000.00	1-45	8995.50	15.27	S 25 W	0.	13.84	0.	6.45	0.	64.29	0.	51.45
9500.00	2-15	9495.11	19.63	S 33 W	0.	16.46	0.	10.69	0.	80.75	0.	62.14
10000.00	2-30	9994.64	21.81	S 58 W	0.	11.56	0.	18.50	0.	92.31	0.	80.64
10500.00	1-45	10494.41	15.27	S 75 W	0.	3.95	0.	14.75	0.	96.26	0.	95.39
11000.00	2-30	10993.93	21.81	N 85 W	1.90	0.	0.	21.73	0.	94.36	0.	117.11
11500.00	1-45	11493.70	15.27	S 75 W	0.	3.95	0.	14.75	0.	98.31	0.	131.86
12000.00	2- 0	11993.39	17.45	N 38 W	13.75	0.	0.	10.74	0.	84.56	0.	142.60
12500.00	2- 0	12493.09	17.45	S 40 W	0.	13.37	0.	11.22	0.	97.93	0.	153.82
13000.00	2- 0	12992.78	17.45	S 82 W	0.	2.43	0.	17.28	0.	100.36	0.	171.10
13500.00	1-30	13492.61	13.09	N 64 W	5.74	0.	0.	11.76	0.	94.62	0.	182.86
13700.00	1- 0	13692.58	3.49	N 7 E	3.46	0.	0.43	0.	0.	91.15	0.	182.44

THE BOTTOM OF THE HOLE LIES WITHIN A CIRCLE OF RADIUS 146.75 FEET WITH ITS CENTER LOCATED 203.94 FEET SOUTH 63 DEGREES 27 MINUTES WEST OF THE SURFACE LOCATION

Figure 3-18 A directional survey from a well in which the surface casing was surveyed with a Totco tool providing deviation angle but not direction. Notice, at 3513 ft, that the maximum possible deviation of the cased hole is 146.75 ft. This assumes that the wellbore deviation in this portion of the hole was in one direction.

Tenneco Oil Company conducted a detailed study of directional survey uncertainties in 1980. An important conclusion from the study indicates that there is a 90% certainty that any directional well will have an error of 35 ft or less TVD and 140 ft or less in departure. This conclusion is drawn irrespective of MD, hole angle, or survey type. This means that wells with hydrocarbon contacts that vary up to 35 ft TVD may well be in the same reservoir, with the variations due merely to survey error rather than such geological factors as permeability barriers or faults.

Today's directional surveys are potentially more accurate than those described above, provided that the tools and operating procedures are of high quality. However, that does nothing to mitigate the errors inherent in surveys of older wells. Always remember that data based on directional wells, including new wells, should not be taken at face value. Apparent discrepancies in data used in mapping may be due to errors in depth and lateral position of points in wells. For example, the depth of a mapped surface or a water contact in a deviated well may not match satisfactorily with depths at nearby points. The reason could be an error in the directional survey or perhaps in the spotting of the wellbore. The same caution applies even to data discrepancy in a vertical well, which may have a significant but unrecognized deviation from the vertical.

Jamieson (2012) gives some typical current uncertainties in modern directional surveys. He states that a typical uncertainty in MD is 2 m/1000 m, a typical uncertainty in inclination is

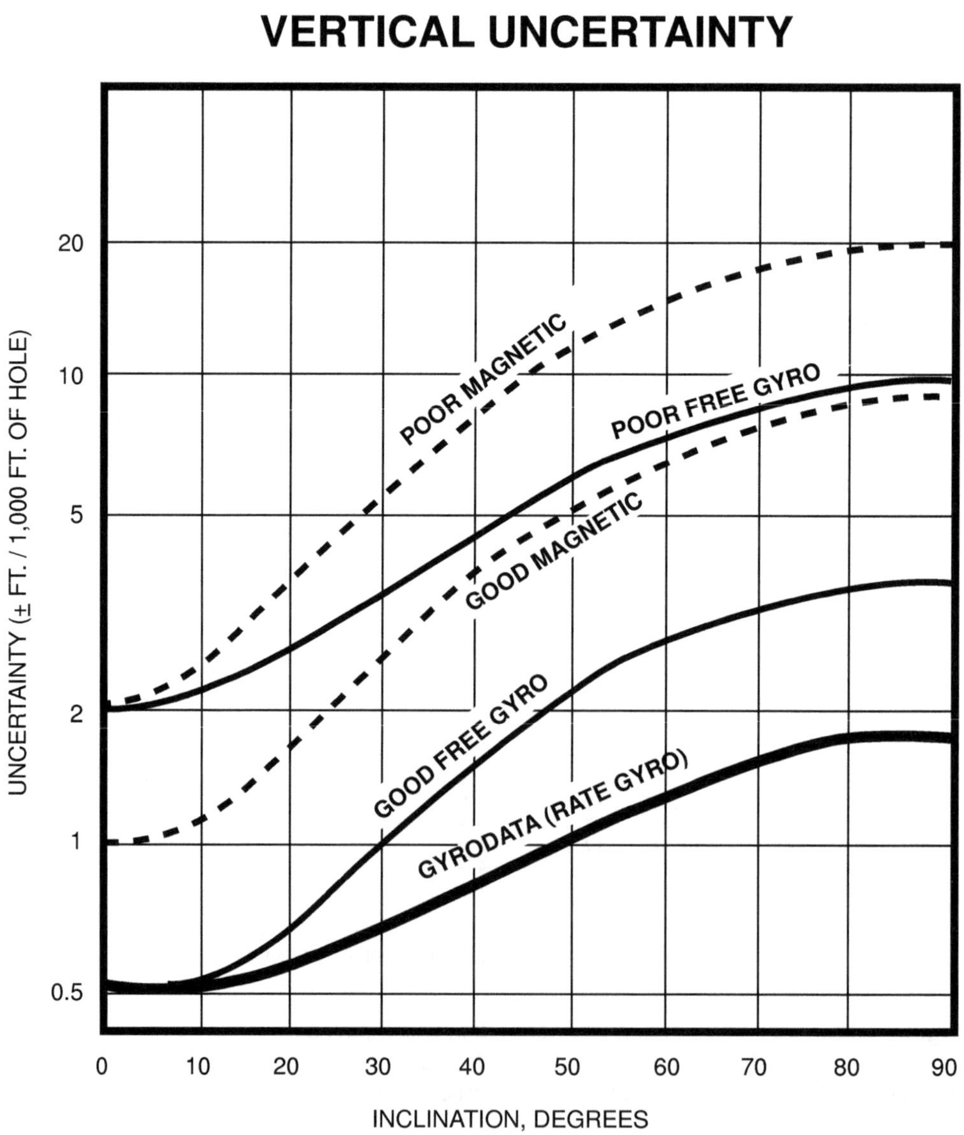

Figure 3-19 Expected vertical uncertainty in a deviated well considering various types of surveys. (Modified from Wolff and de Wardt 1981. Published by permission of the Journal of Petroleum Technology and Gyrodata, Inc.)

0.3 deg, and a typical uncertainty in azimuth is 1 deg. These uncertainties result in a cone of location uncertainty along the wellbore and ellipses of uncertainty at TD (Fig. 3-21).

Survey errors can be corrected in some cases in fields that have hydrocarbon/water contacts. Wells are often "corrected" to fit the contact. This is usually done by selecting a contact that fits most of the wells and then adjusting the depth of the well(s) that does not fit. An example of how to adjust a well is shown in Figure 3-22. The water contact in Wells No. 1 and 2 is at a depth of −9738 ft, whereas the contact in Well No. 3 is at a depth of −9748 ft. As data from two wells are in agreement with a water contact at −9738 ft, Well No. 3 is adjusted upward 10 ft to correct the water level from −9748 ft to −9738 ft. Not only is the water level corrected, but the structural depths of the sand (top and base) are also corrected upward 10 ft. Therefore, the top of the sand at a depth of −9720 ft becomes −9710 ft. In some cases, with stacked pays, depth shifting wells to produce a flat-water contact at one horizon will also significantly reduce water contact depth variation at other horizons. An understanding of directional survey errors can at times eliminate the need for a "production fault" or "permeability barrier" to explain discrepancies in water levels. Avoiding such production faults or permeability barriers can have a significant impact on field development plans.

Sag Correction in MWD Directional Surveys

The BHA of the tool string has to be flexible enough to pass through curves in the wellbore. When wells are drilled at high inclinations, the flexibility of the BHA causes it to sag, which can cause an error in the inclination reading of the sensor (Jamieson 2012). Wells drilled at inclinations above 45 deg should have sag corrections applied to the MWD directional survey. If a well with an MWD survey does not seem to fit the map or if production is disappointing, it could be due to a failure to apply a sag correction (Jamieson, 2012).

DIRECTIONAL WELL PLOTS

Directional survey data are used to determine the depth and lateral position of the borehole along its entire length. These data are normally plotted on a base map in one of two ways: (1) as straight lines from the surface to BHL or (2) as detailed directional plots.

The simplified straight-line method of plotting a directional well is shown in Figure 3-23a. The only data required and plotted on the base map are the surface location and BHL, which may be the only data available. The MD to TD may be recorded next to the BHL. A straight dashed line is usually drawn between the surface location and BHL. This directional well plot provides absolutely no information about the position or depth of the wellbore in the subsurface between the surface and bottomhole locations. Such a plot is not helpful in the interpretation, construction, and evaluation of fault, structure, or isochore maps. Some computer databases use this type of map to show what wells have been drilled in a given area. Data in the database is then tied to the wells on the screen and is available with the click of a mouse.

When directional survey data are actually plotted to provide detail about the lateral position and subsea depth of the wellbore, as shown in Figure 3-23b, the plot has real value. Such a plot provides a visual guide (in map view) to the location and subsea depth of the wellbore anywhere along its path. It saves time in preparing subsurface maps and is extremely helpful in the interpretation, construction, and evaluation of fault, structure, and isochore maps. Later chapters examine several important benefits to fault surface and structure mapping derived from plotting on a base map the actual location and subsea depth, at fixed increments (usually 500 ft or 1000 ft), of all directional wells. This type of basemap is seldom used these days, as computer workstations plot the location of formation tops and fault cuts on directional well paths.

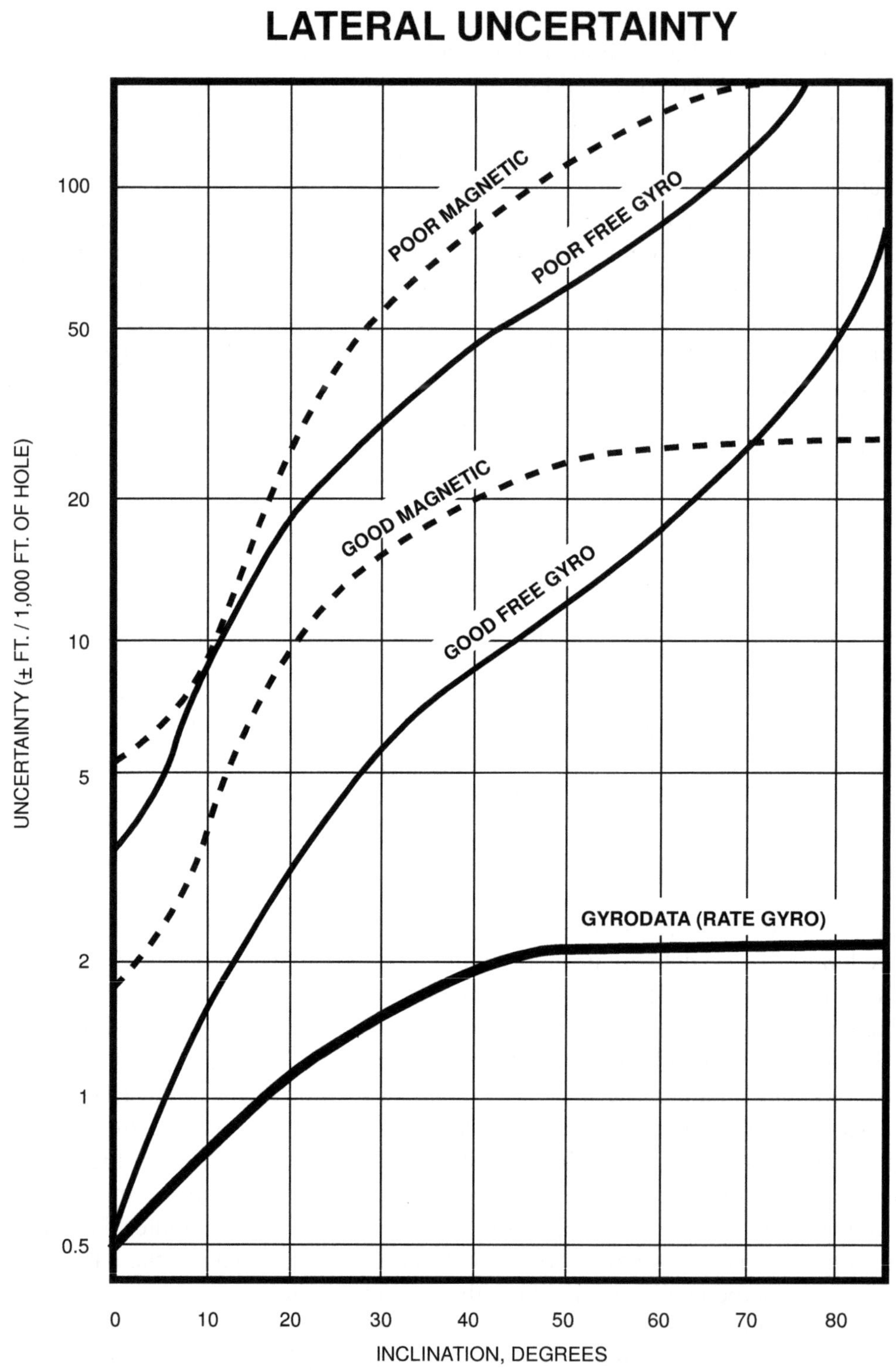

Figure 3-20 Expected lateral uncertainty in a deviated well considering various types of survey tools. (Modified from Wolff and de Wardt 1981. Published by permission of the Journal of Petroleum Technology and Gyrodata, Inc.)

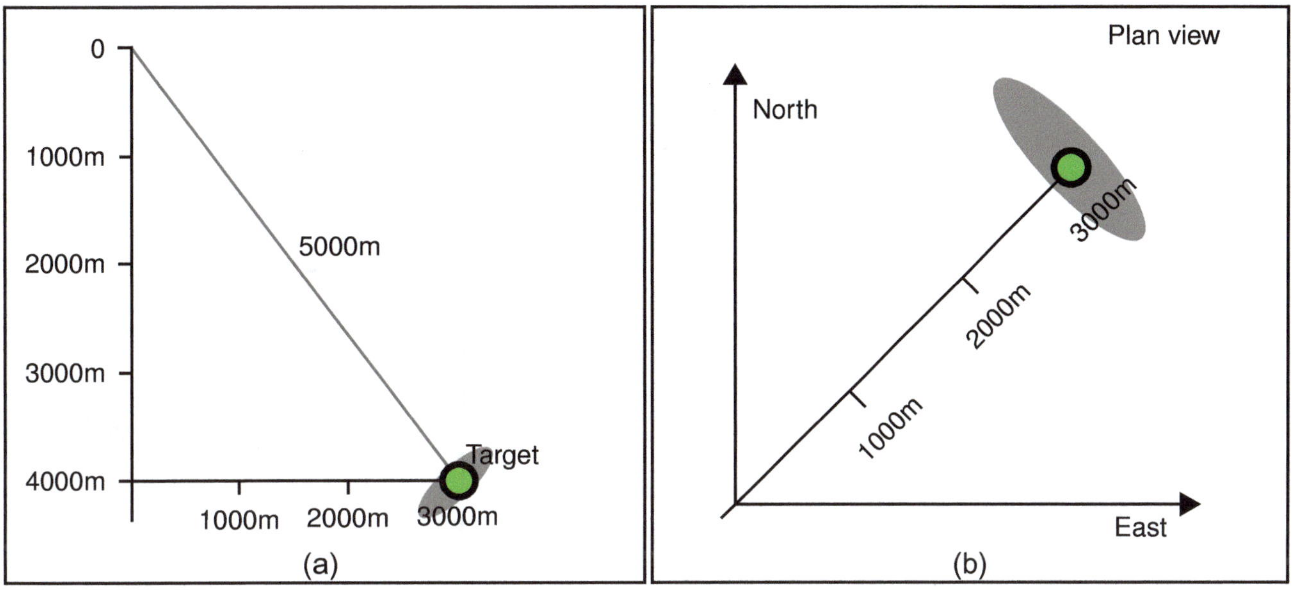

Figure 3-21 Ellipses of uncertainty in a hypothetical well drilled 4000 m vertically and then 3000 m horizontally and surveyed with modern directional survey tools. (a) Vertical section showing vertical and lateral uncertainty in a vertical plane. (b) Plan view showing lateral uncertainty. (Jamieson 2012. Published with permission of the author.)

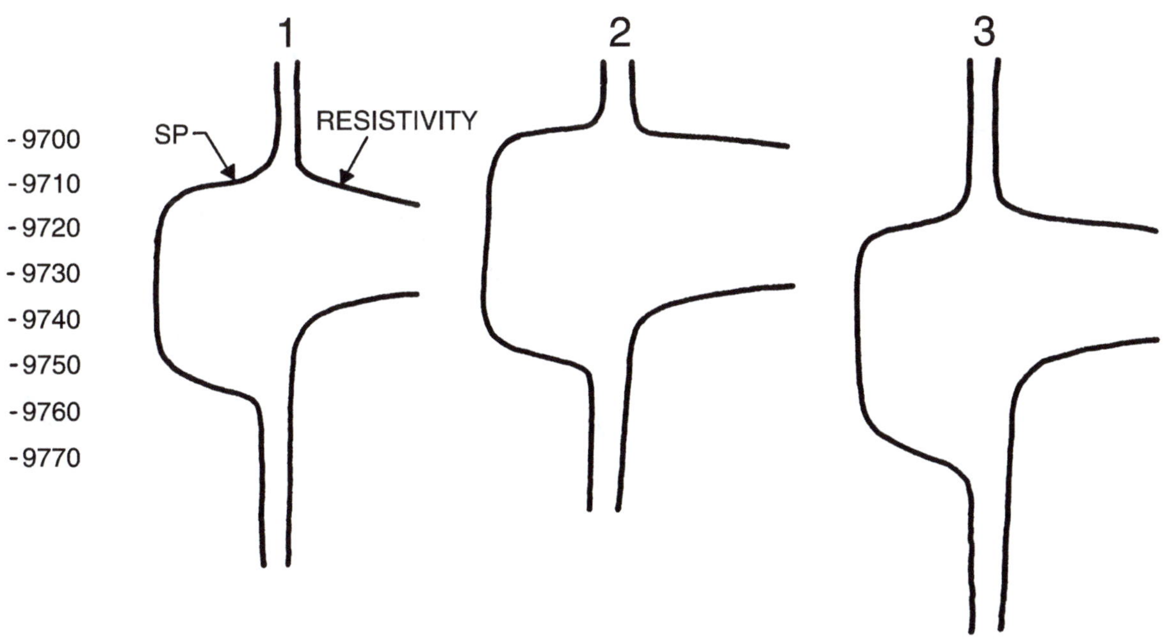

Figure 3-22 Different hydrocarbon/water contacts caused by directional survey errors.

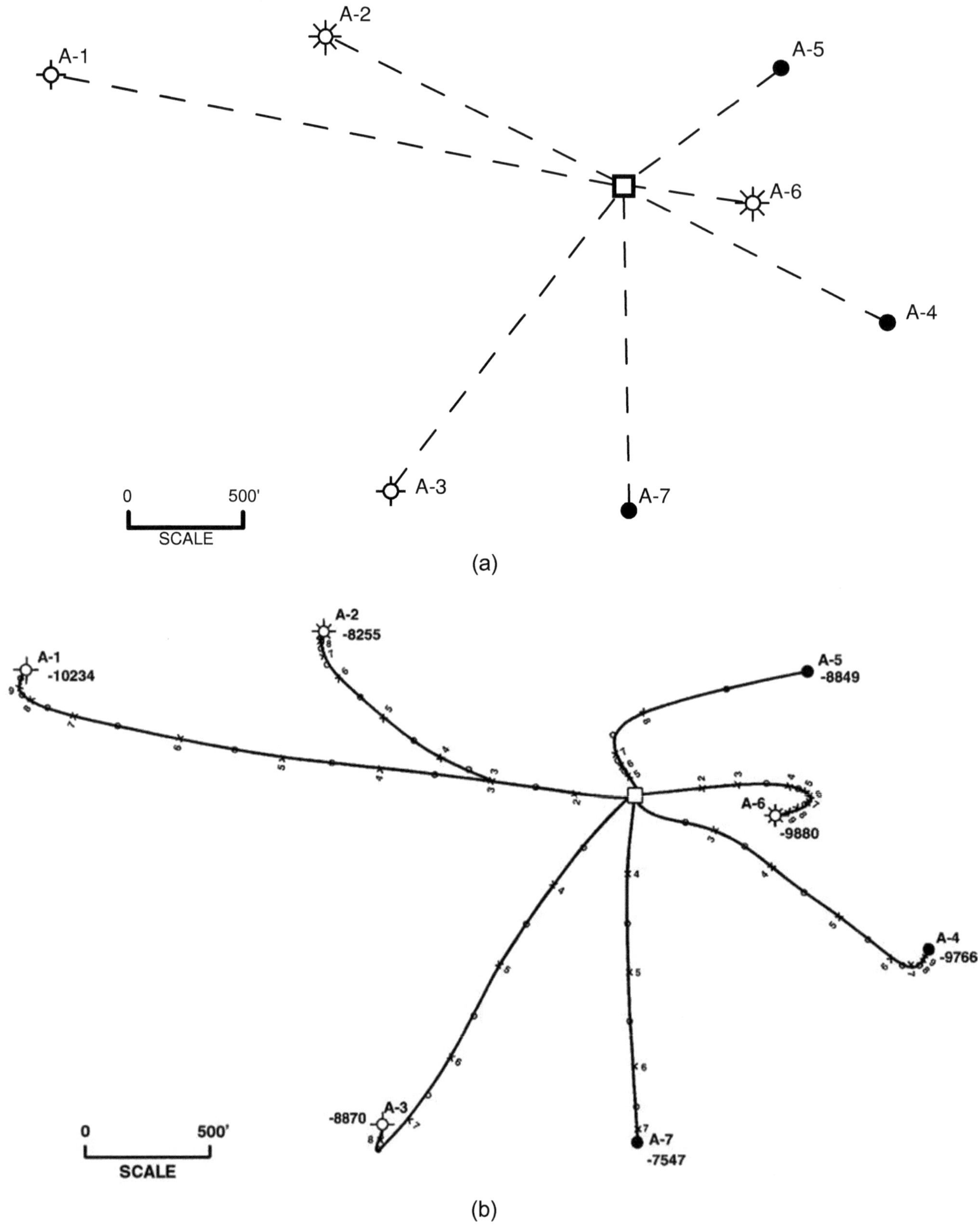

Figure 3-23 (a) Straight-line method of plotting directional wells in map view. (b) Detailed plot of directional survey data indicating the location and subsea depth of the wellbores along their entire length. Compare this plot to that in Figure 3-13a.

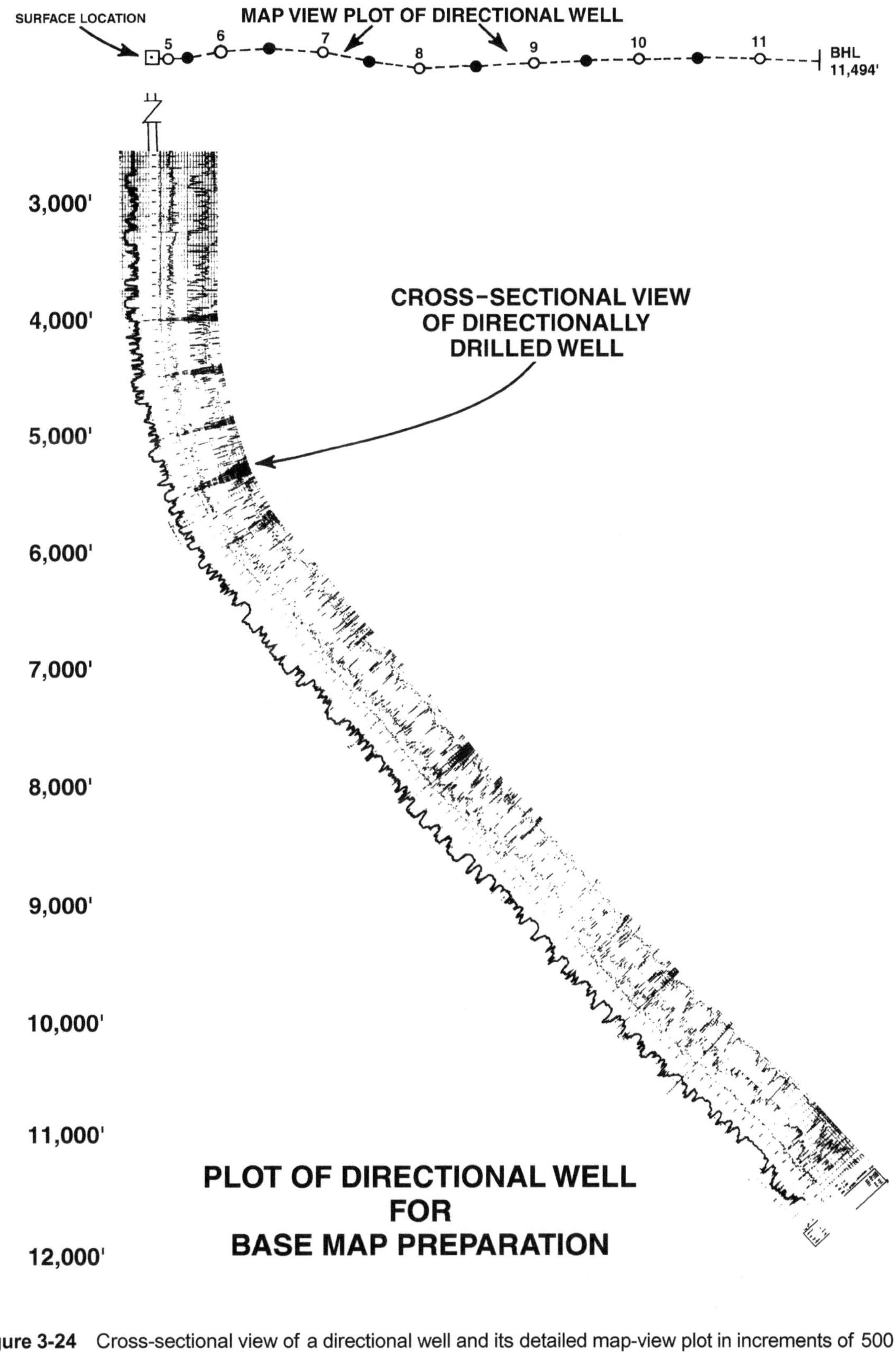

Figure 3-24 Cross-sectional view of a directional well and its detailed map-view plot in increments of 500 ft.

Figure 3-24 illustrates the cross-sectional view of a directionally drilled well and the detailed map-view plot of the directional data in 500-ft increments of subsea TVD along the actual well path. If subsea data are to be used in mapping, the directional well data are corrected to subsea before being plotted on the base map.

Figure 3-25 shows the map-view well plot for the Cognac "A" Platform in the Mississippi Canyon Block 194 in the U.S. Gulf of Mexico. The platform is the largest multiple-well platform in the world. The platform is located in 1000 ft of water and has a total height of 1260 ft. Wells were deviated with high angles, up to 75 deg. Horizontal displacements up to 11,500 ft result in a well pattern that covers an area with a diameter of more than 4 mi. Total cost for the project was over *$1 billion.*

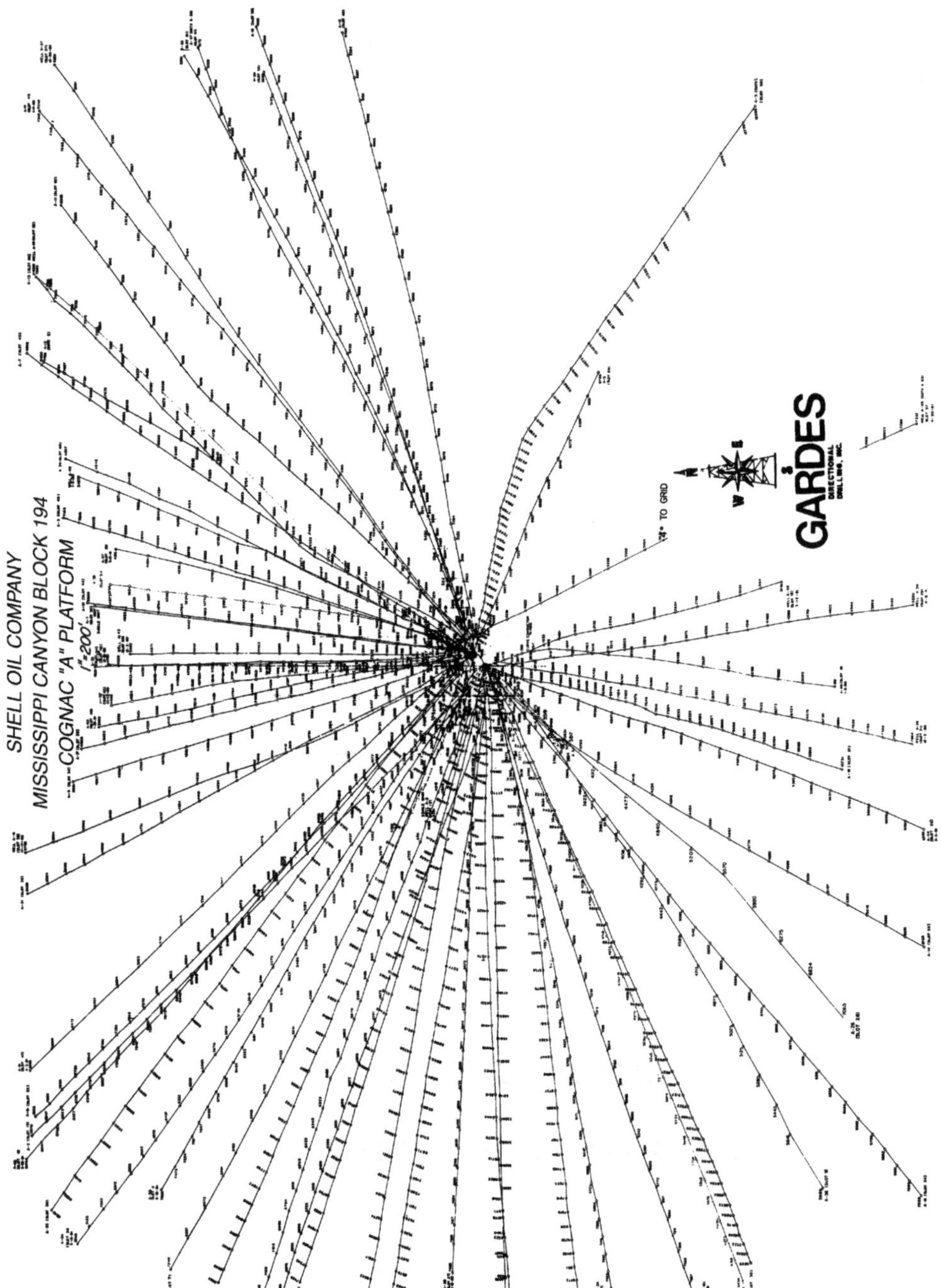

Figure 3-25 Spider directional well plot for the Cognac "A" Platform, Offshore Gulf of Mexico. (Published by permission of Gardes Directional Drilling.)

WELLS WITHOUT DIRECTIONAL SURVEYS

Many wells that are planned to be vertical are drilled without running a directional survey. Wells that are reported to be vertical but do not have a directional survey are frequently not vertical. Anderson (1929) surveyed many "vertical" wells in California and showed that the wells frequently deviated significantly from the vertical. This situation is still true (Fig. 3-26) (Stigant 2012). If tops do not seem to match surrounding wells, it is frequently desirable to run a new directional survey if the wellbore is still available. New directional surveys should always be run in wells that are being sidetracked, and these directional surveys should be run as deep as possible.

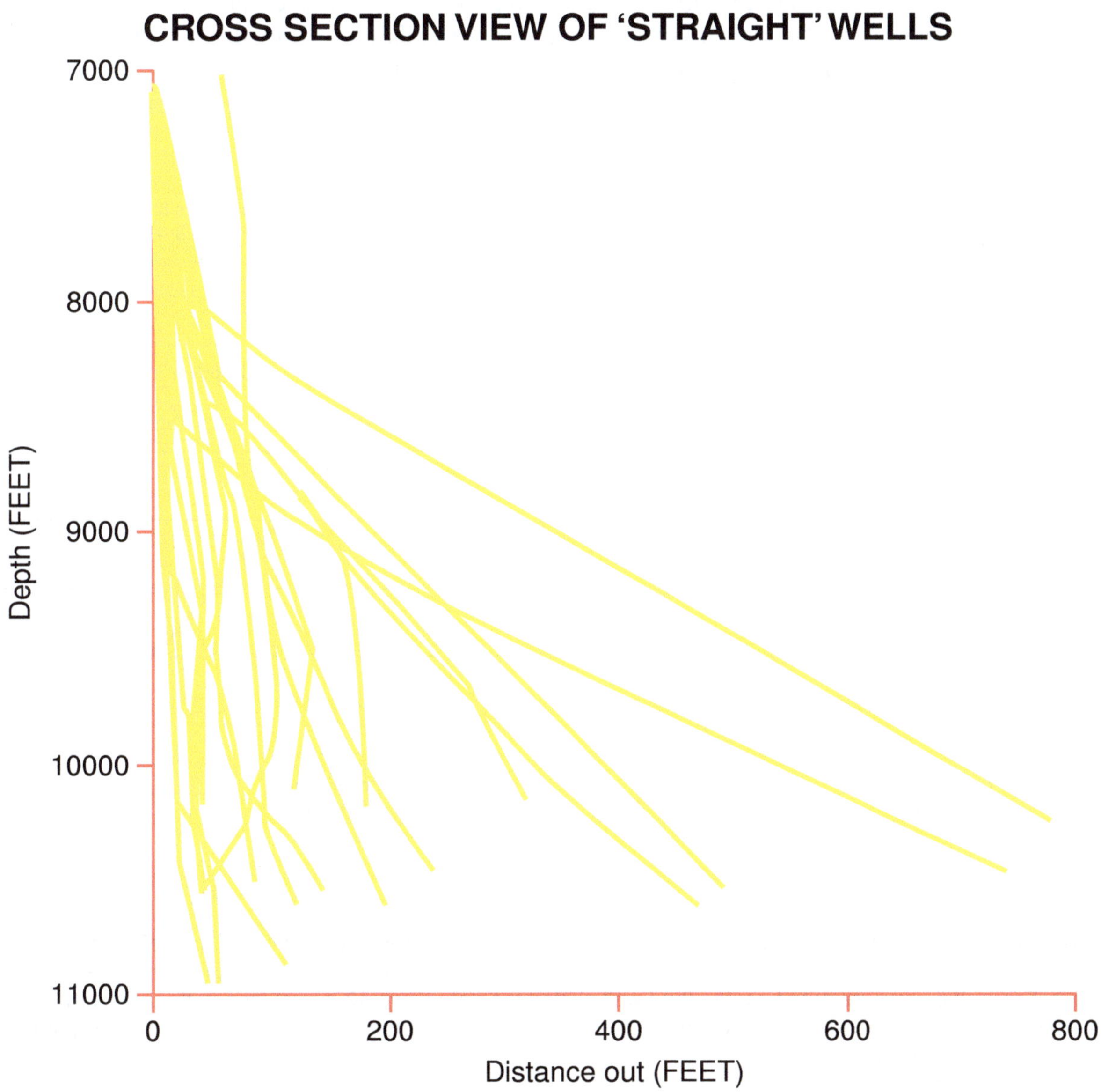

Figure 3-26 Cross sectional profile of a number of wells in West Texas that were thought to be vertical before directional surveys were run in them. (Modified after Stigant 2012. Used with permission of American Association of Petroleum Geologists [AAPG].)

Reservoir Pressures

In onshore or shallow water fields, where well costs are relatively low, there are frequently multiple wells that penetrate the hydrocarbon/water contact in a reservoir compartment. Figure 3-22 shows how directional survey uncertainties can result in different contact depths. In deep water and frontier fields, where well costs are much higher, there may be few if any penetrations of the hydrocarbon/water contact. In these areas, wells may rely on reservoir pressure data from repeat formation testers to identify hydrocarbon contacts and recognize separate reservoir compartments (e.g., Brown 2003; Chen 2014). Differences in depth measurement between LWD and wireline logs can result in pressure data that suggests reservoir compartmentalization where none may exist. Carefully corrected depth data can identify where the risk of compartmentalization may have been overestimated (Fig. 3-27).

Just as water contacts that differ due to directional survey uncertainties can be corrected, resulting in corrected formation tops, pressure data trends that differ due to directional survey issues or along hole depth measurement issues can be shifted to align, shifting formation tops as well. Before shifting the depth of tops, it is important to have convincing evidence that wells are in the same pressure compartment.

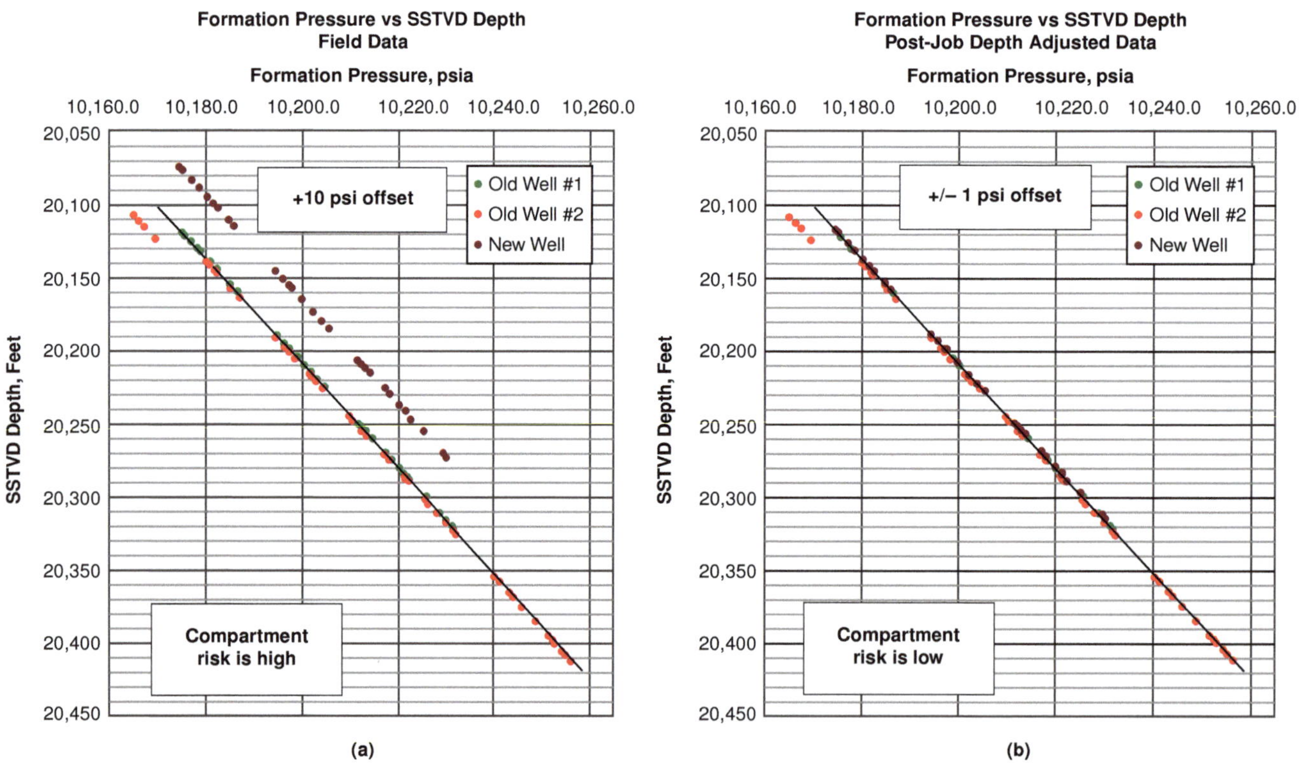

Figure 3-27 Pressure data from three wells in a discovery. (a) One well's pressure data lies 10 psi off the pressure trend from the other two wells, suggesting that the third well lies in a different fault block than the first two wells. (b) Post-job depth correction of the pressure data from the third well shifted the data down 42 ft, resulting in the pressure data being within 1 psi of the pressure trend from the other two wells, suggesting that the risk of the wells being in separate faults blocks is less. (From Cribbs 2020.)

CHAPTER 4

LOG CORRELATION TECHNIQUES*

INTRODUCTION

Correlation can be defined as the determination of structural or stratigraphic units that are equivalent in time, age, or stratigraphic position. For the purpose of preparing subsurface interpretations, including maps and cross sections, the two general sources of correlation data are electric wireline or LWD (logging-while-drilling) logs and seismic sections. In this chapter, we discuss basic procedures for correlating well logs, introduce plans for performing various phases of well log correlation, and present fundamental concepts and techniques for correlating well logs from vertical as well as directionally drilled wells.

Fundamentally, electric well log curves are used to delineate the boundaries of subsurface units for the preparation of a variety of subsurface maps and cross sections (Doveton 1986). These maps and cross sections are used to develop an interpretation of the subsurface for the purpose of exploring for and exploiting natural resources, including hydrocarbons.

After the preparation of an accurate well and seismic basemap, electric log and seismic correlation work is the next step in the process of conducting a detailed geological/geophysical study. No geological interpretation can be prepared without detailed correlations. ***Accurate correlations are paramount for reliable geological interpretations.***

* For all figures in this chapter (in the printed book only), see the preface for information about registering your copy on the InformIT site for access to the electronic versions in color.

General Log Measurement Terminology

An understanding of several log depth measurements is important for converting log depths to depths used in mapping. The following is a list of measurements, their abbreviations, and definitions of depth terminology. These terms are illustrated in Figure 4-1.

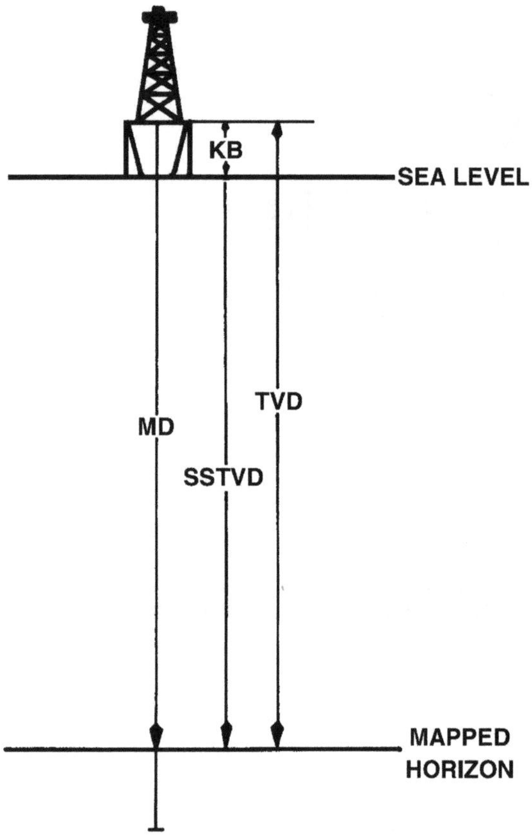

Figure 4-1 Diagram showing general log measurement terminology.

- **KB** Vertical distance from kelly bushing to sea level. The kelly bushing is the most common reference for measuring logs. The rig floor (RF) or derrick floor (DF) or rotary table (RT) are other common reference points.
- **MD** Measured depth. Measured distance along the path of a wellbore from the kelly bushing to TD (total depth of the well) or any correlation point in between. May also be called along hole depth (AHD) (Brooks et al. 2005) or true along hole depth (TAH) (Forsyth et al. 2013).
- **TVD** True vertical depth. Vertical distance from the kelly bushing to any point in the subsurface.
- **SSTVD** Subsea true vertical depth. Vertical distance from sea level to any point in the subsurface. SSTVD may either be positive (above sea level) or negative (below sea level).
- **Vertical Wellbore** A well drilled 90 deg to a horizontal reference, usually sea level (also called a straight hole).

The SSTVD measurement is the only depth measurement from a common reference datum, sea level. Therefore, SSTVD is the depth most often used for mapping. Logging depths measured from a vertical or directionally drilled well for mapping are usually corrected to SSTVD. For vertical wells, the SSTVD = KB − TVD. The measurements for directionally drilled wells are discussed in Chapter 3.

ELECTRIC LOG CORRELATION PROCEDURES AND GUIDELINES

What is well log correlation? *Electric log correlation is pattern recognition.* It is often debated whether this pattern recognition is more of an art or a science, but we believe both play a part in correlation work. Professionals involved with log correlation must be well versed in sound geological principles, including depositional processes and environments, and have an understanding of the tectonic setting under study. They should also be familiar with the principles of logging tools and measurements, general reservoir engineering fundamentals, and basic qualitative and quantitative log analyses.

The best way to develop log correlation ability is by actually performing correlation work. A geologist should become more proficient with increased experience in correlation. Proficiency in correlating well logs in one tectonic setting or depositional environment does not always ensure similar competence in other settings. In other words, someone who is an expert at correlating well logs in the South American Andes Mountains Overthrust may not be equally competent when working in, for example, Offshore West Africa. Just as it took time to become proficient at correlating logs in the South American Andes, so too will it take time and familiarity in the new area to become proficient.

When geologists correlate one log to another, they are attempting to match the pattern of curves on one log to the pattern of curves found on the second log. A variety of curves may be represented on a log. For correlation work, it is best to correlate well logs that have the same type of curves; however, this is not always possible. A geologist may be required to correlate logs that have different curves. And at times, even if the logs have the same curves, the character or magnitude of the fluctuations of the curves may be different from one log to the next. Therefore, the correlation work must be independent of the magnitude of the fluctuations and the variety of curves on the individual well logs. Figure 4-2 shows sections from two electric logs. The patterns of curves on Well No. A-1 are very similar to the patterns on Well No. A-2. We can say that these two logs have a high degree of correlation.

The data presented on a well log are representative of the subsurface formations found in the wellbore. A correlated log provides information about the subsurface, such as stratigraphic markers, tops and bases of stratigraphic units, depth and amount of missing or repeated section resulting from faults, lithology, depth to and thickness of hydrocarbon-bearing zones, porosity and permeability of productive zones, and depth to unconformities. The information obtained from correlated logs is the raw data used to prepare subsurface interpretations and maps. The maps may include fault, structure, stratigraphic, salt, unconformity, and a variety of isochore maps. These data can also be used to prepare a variety of cross sections. Accurate correlation is the foundation of reliable subsurface interpretations. Subsurface geological maps based on log correlation are only as reliable as the correlations used in their construction. Eventually, a geologist's correlations, right or wrong, are incorporated into the construction of subsurface geological maps. An incorrect correlation can be costly in terms of a dry hole or an unsuccessful workover or recompletion; therefore, it is essential that extreme care be taken in correlating well logs.

In this section, we introduce you to a general correlation procedure and discuss some guidelines for electric log correlation. The process of correlating logs varies from one individual to the next. As geologists gain experience, they modify and eventually establish a correlation procedure that works best for them. If you have no experience in log correlation or want to improve your skills, you can begin by using the procedures and guidelines discussed in this section.

Electric logs can be correlated by hand or with the assistance of a computer. When correlated by hand, the electric logs are arranged on a worktable in one of two ways (Fig. 4-3). The arrangement shown in Figure 4-3a is preferred over that shown in Figure 4-3b by most geologists because more log section can be viewed at one time and the logs are easier to slide during correlation. The manipulation of logs on a computer screen is described in the section Computer-Based Log Correlation.

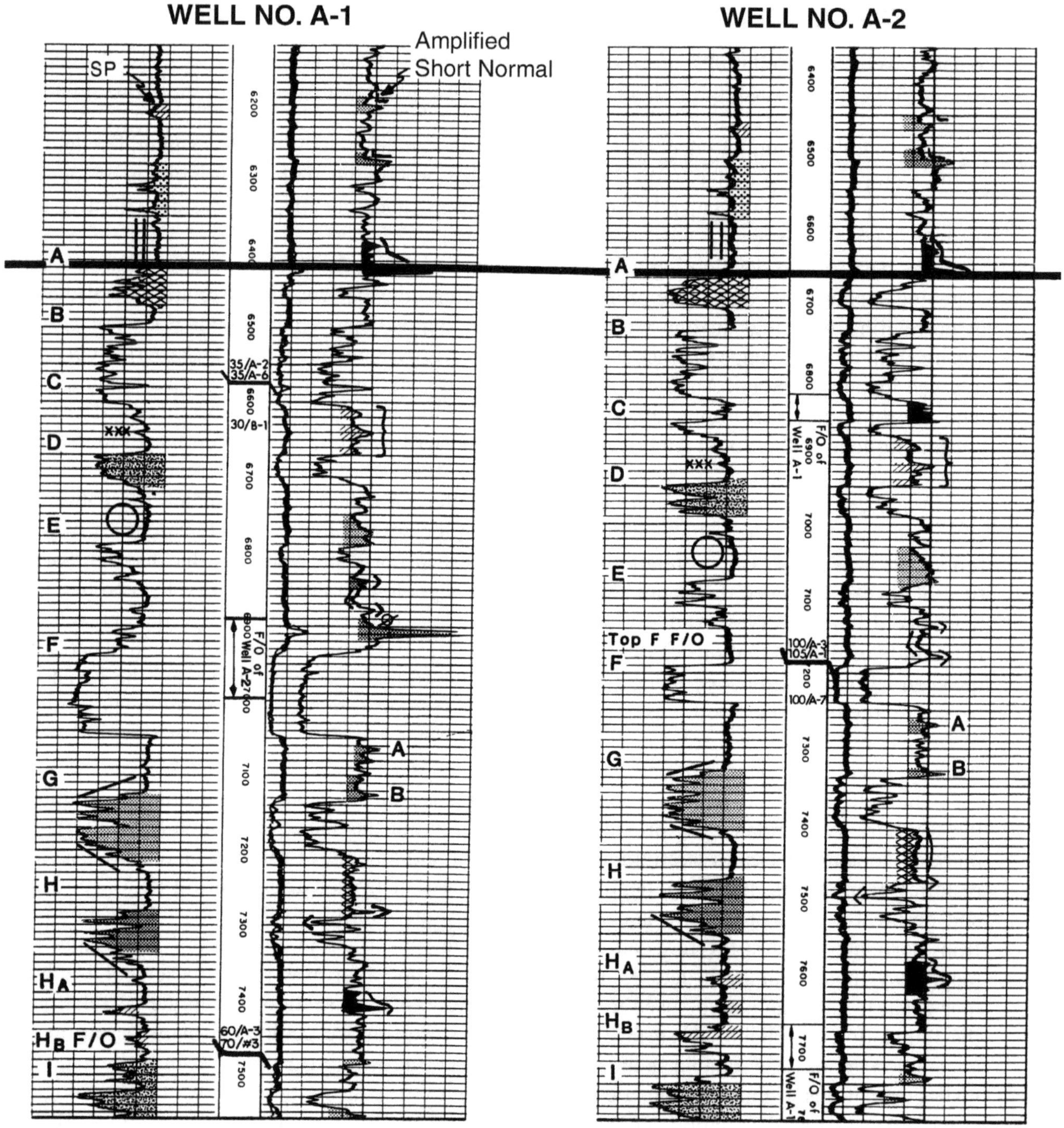

Figure 4-2 Portion of two electric logs illustrating methods of annotating recognizable correlation patterns on well logs.

As a starting point, align the depth scale of the logs and look for correlation, as shown in Figure 4-2. If no correlation is evident, begin to slide one of the logs until a good correlation point is found, and mark it. Continue this process over the entire length of each log until all recognized correlations have been identified. This process may seem relatively easy, but it can be complicated by such factors as stratigraphic thinning, bed dip, faulting, unconformities, lateral facies changes, poor log quality, and directionally drilled wells. If seismic data is available near the wells being correlated, it should be consulted prior to beginning correlating. Many of the complicating factors listed above may be apparent on the seismic, which will make correlating easier and more efficient.

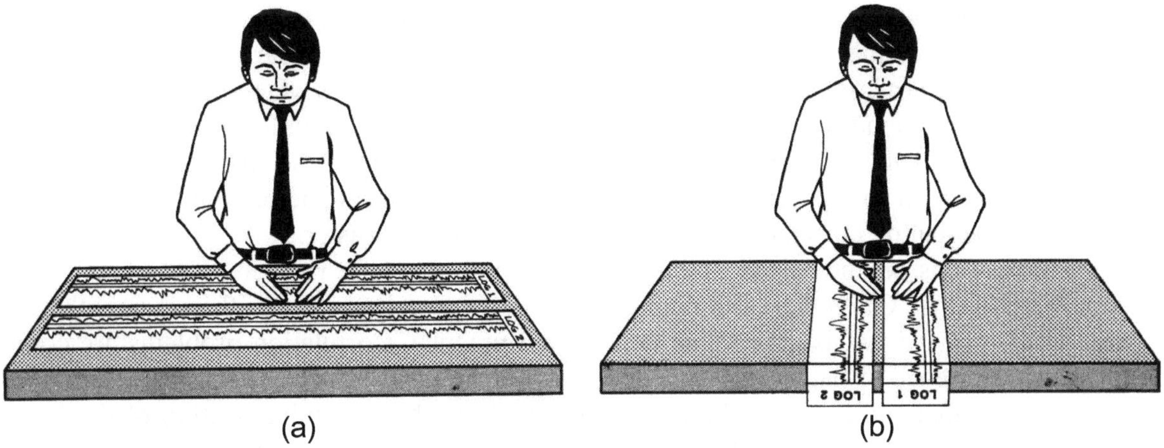

Figure 4-3 (a) Preferred method of arranging logs for correlation. (b) Alternate method of arranging logs for correlation.

There are some basic, universally valid guidelines, which are useful in the log correlation process. If followed, these guidelines should improve your correlation efficiency and minimize correlation problems.

1. For initial quick-look correlation, review major sand or carbonate bodies using the spontaneous potential (SP) curve or gamma ray (GR) curve.
2. For detailed correlation work, first correlate shale sections.
3. Initially, use the amplified shallow resistivity curve (focused, short normal, or shallowest reading array induction), which usually provides the most reliable shale correlations.
4. Use colored pencils to identify specific correlation points.
5. Always begin correlation at the top of the log, not the middle.
6. Correlate the entire log.
7. Do not force a correlation.
8. In highly faulted areas, first correlate down the log and then correlate up the log.

After an initial quick look using the SP curve or GR curve to identify the major sand or carbonate bodies, concentrate your correlation work on shale sections. There are three good reasons for this. First, the clay and silt particles that make up shales are deposited in low-energy regimes. These low-energy environments responsible for shale deposition commonly cover large geographic areas. Therefore, the log curves (sometimes referred to as *log signatures*) in shales are often highly correlatable from well to well and can be recognized over long distances. Second, prominent sand bodies are often not good correlation markers because they commonly exhibit significant variation in thickness and character from well to well and are often laterally discontinuous. Finally, the resistivity curves for the same sand on two well logs being correlated may be different. Variations in fluid content in a sand bed may cause pronounced resistivity differences (e.g., water versus gas).

Individual shale beds exhibit distinctive resistivity characteristics over large areas. Therefore, when all log curves are considered, the amplified shallow-investigation resistivity curve provides the most reliable shale correlations. Although all log curves should be used for correlation work, the amplified resistivity curve is five times more sensitive than the unamplified curve and exhibits patterns that are easier to recognize and correlate from well to well. The amplified curve should be the initial curve used for correlation (Fig. 4-2).

The liberal use of colored pencils is an excellent way to identify and mark correlation patterns on well logs. The correlation patterns might be peaks, valleys, or groups of wiggles that are recognizable in many or all of the well logs being correlated (Fig. 4-2). The colored pencils should be erasable in the event that correlations are changed. *Do not mark on original logs.* A blueline or blackline copy of the original logs should be used for marking during correlation.

In general, structures become less complicated toward the surface because of several factors. Many faults tend to die upward toward the surface and are either small or nonexistent in the upper part of the logs. This makes for easier correlations. Also, in many geological provinces, especially in soft rock basins, the structural dip, both local and regional, decreases upward. Therefore, beginning correlation at the top of a log is usually easier.

Correlations are not always straightforward and everyone runs into correlation problems from time to time. Often there is a tendency to force a correlation rather than bypass the problem area until further work is done. This is not good practice. Correlation problems are commonly due to the presence of faults, high bed dips, unconformities, and facies changes. It is best to pass the problem area and continue the correlation work on the remaining section of the log. Later, when the remainder of the problem log and other logs have been correlated, the questionable correlations can be reviewed again with this new information.

In highly faulted areas it is advantageous to approach a recognized fault from two directions. First, correlate down the log to the fault and then correlate up the log to the fault. By taking this approach, determination of the amount of missing or repeated section and depth of the fault in the correlated well will be more accurate (Figs. 4-2 and 4-10). This method is discussed in detail later in this chapter.

CORRELATION TYPE LOG

*A **correlation type log** is defined as a log that exhibits a complete stratigraphic section in a field or regional area of study.* The type log should reflect the deepest and thickest stratigraphic section penetrated. Because of faults, unconformities, and variations in stratigraphy affecting the sedimentary section, a correlation type log is often composed of sections from several individual logs and is referred to as a *composite type log*.

Do not confuse a correlation type log with other kinds of type logs, such as stratigraphic type logs, composite sand type logs, or show logs. A stratigraphic type log is usually prepared to depict the depositional environments that existed in a particular field or area of study (Fig. 4-4). Although it may include portions of several logs to depict the entire stratigraphic section, it is usually not prepared in the strict sense of a correlation type log. Therefore, it may contain faults or unconformities and include sections of wells near the crest of the structure, which do not represent the thickest sedimentary section.

Composite pay logs or show logs are prepared to illustrate the potential productivity within a field or area of study that have shows, contain hydrocarbons, or have the potential to be hydrocarbon bearing (Fig. 4-5). These logs are not prepared for use as a correlation aid and therefore are not prepared in the rigid manner of a correlation type log.

When beginning geological work in a new area of study in which a type log has already been prepared, it is important to carefully review the log to see that it meets with the requirements of a correlation type log. If the type log has an incomplete stratigraphic section, its use will result in correlation errors. The type log must have the complete stratigraphic section if it is to be a useful tool for correlation.

Figure 4-6 shows a cross section through a complex diapiric salt structure. We use this figure to illustrate the procedure for preparing a correlation type log. This structure exhibits a number of complexities, including a salt overhang, several faults, an unconformity, diapiric shale, stratigraphic thinning, and stratigraphic pinchouts in the up-structure position near the salt. We consider each of the four wells that have penetrated the structure and evaluate the applicability of each as a type log.

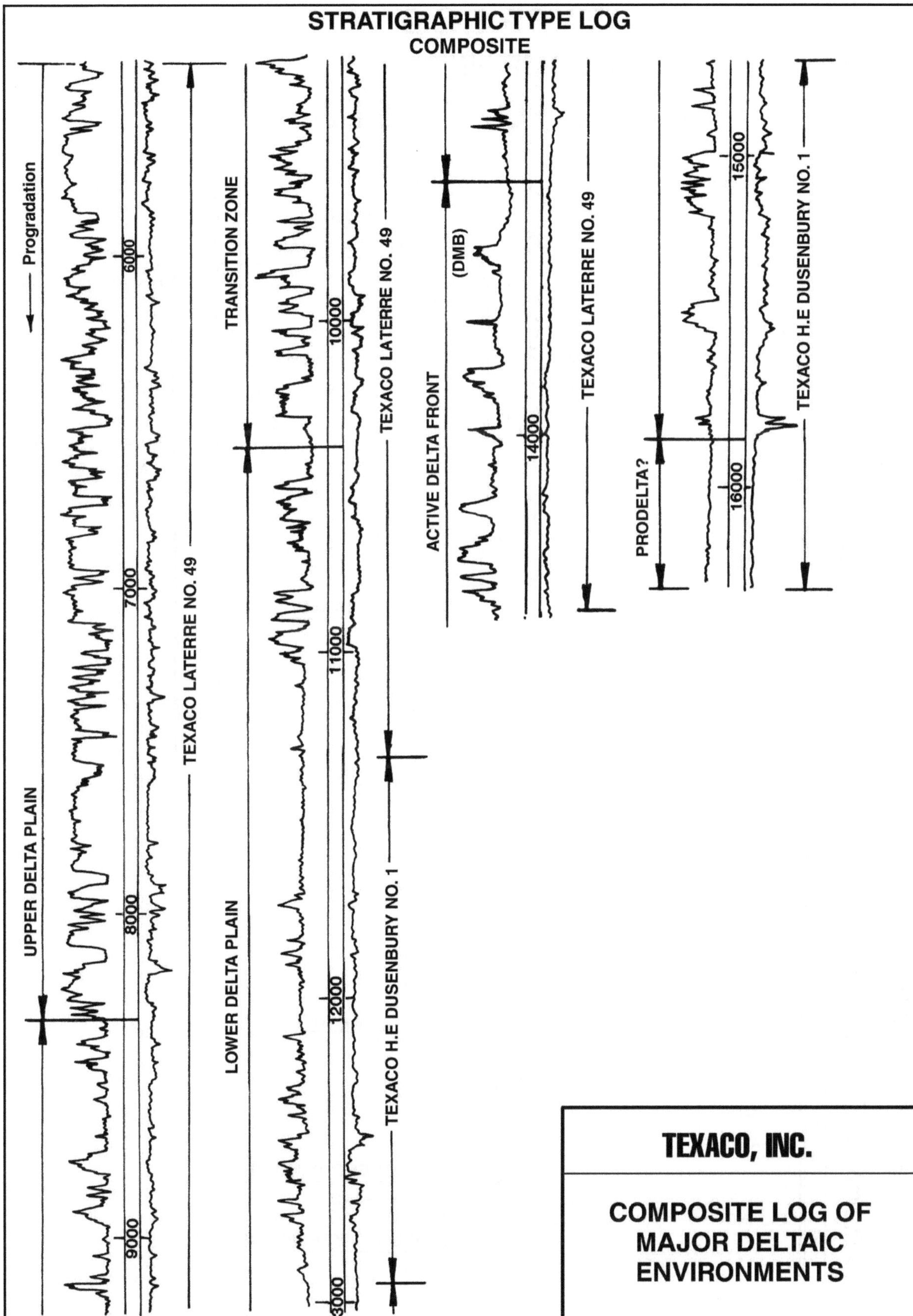

Figure 4-4 Composite stratigraphic type log. (Published by permission of Texaco USA.)

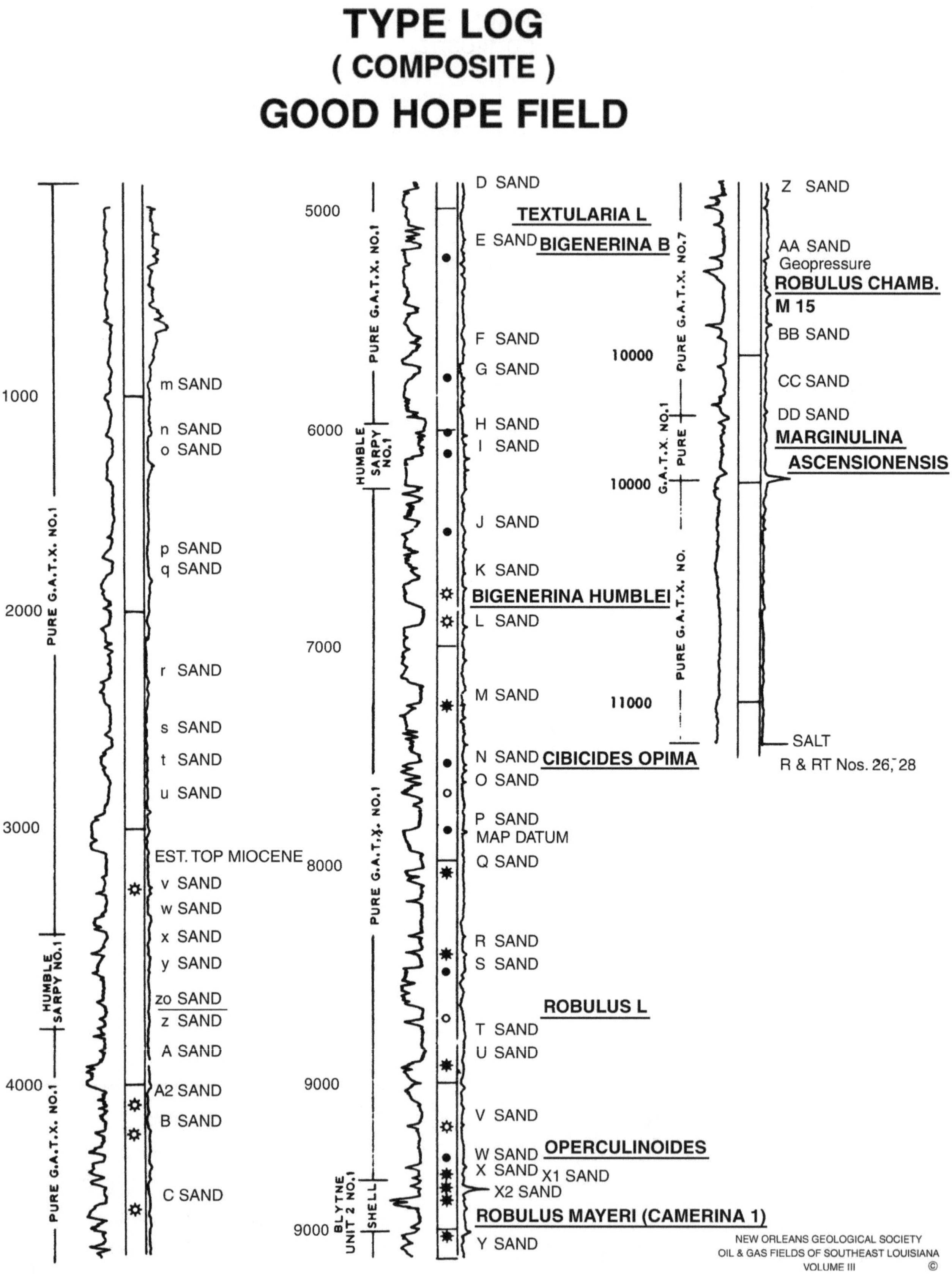

Figure 4-5 Composite show log from Good Hope Field, St. Charles Parish, Louisiana. (Published by permission of the New Orleans Geological Society.)

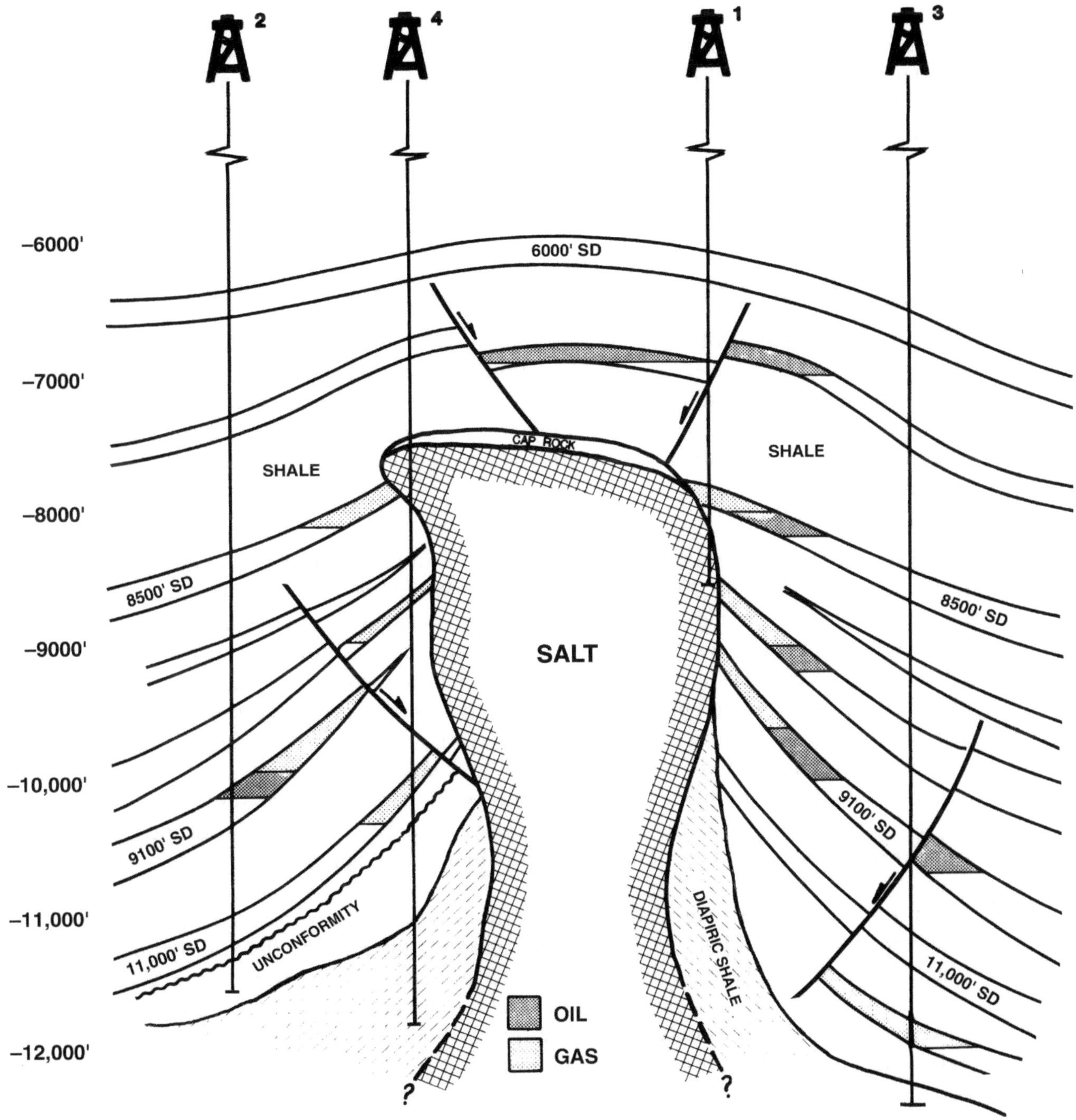

Figure 4-6 A cross section through a complex diapiric salt structure, penetrated by four vertical wells.

Well No. 1 is not a good candidate as a type log for several reasons: it reaches a depth of only −8700 ft, crosses a crestal fault, encounters salt at a shallow depth, and does not penetrate a complete section. Well No. 2 is drilled off the flank of the structure and penetrates a thick, nearly complete stratigraphic section. However, it crosses an unconformity at about −11,300 ft. Well No. 3 is also drilled in a down-dip position and penetrates the entire stratigraphic section before encountering diapiric shale near the total depth (TD) of the well. However, it crosses a fault at about −10,500 ft in the 9100-ft Sand. Well No. 4 drilled in a crestal position is not suitable as a type log because it penetrates the salt overhang, encounters a thinner stratigraphic section than that penetrated by Wells No. 2 and 3, crosses a fault and an unconformity, and does not penetrate a complete stratigraphic section.

For this particular example, the correlation type log must be a composite log including sections from Wells No. 2 and 3. The best type log consists of Well No. 2 from the surface down to a correlation marker just below the 11,000-ft Sand and Well No. 3 from the same marker to TD in diapiric shale. This composite type log meets all the requirements in the definition shown earlier and is an excellent standard for all other well log correlation work on this structure.

Normally, faults are not included on a correlation type log. However, if a major decollement, such as a thrust or listric growth fault, serves as the deepest limit of prospective section, it is advisable to place the fault on the type log.

ELECTRIC LOG CORRELATION—VERTICAL WELLS

We begin the discussion of actual correlation work by reviewing electric log correlation in vertical wells. In general, electric log correlation is easier and more straightforward in vertical wells than in wells that are directionally drilled. Later in this chapter, after we discuss the fundamental concepts and techniques of correlation in vertical wells, we review the same concepts and techniques as they apply to directionally drilled (deviated) wells.

Log Correlation Plan

When given the task of correlating logs in a specific field or area of interest, you might ask yourself one of several questions: "Where do I start?" or "Which log do I correlate first, second, third, etc.?" Before starting the log correlation in an area, a general *log correlation plan* needs to be developed. In this section, we illustrate a log correlation plan that provides an answer to the questions asked and establishes a preferred order in which to correlate electric logs from vertically drilled wells. This correlation plan can be adapted to most geological settings. For the purpose of illustration, we use a structure map of the 8000-ft Horizon on a normally faulted anticlinal structure in an extensional geological setting (Fig. 4-7). Considering the faults on the anticline in the up-dip position, the structure becomes more complex in the up-dip direction.

The following log correlation plan is intended to make correlation work more systematic and easier to conduct and to result in fewer difficulties in correlation.

Step 1. First, prepare a correlation type log. Remember that a correlation type log must show a complete unfaulted interval of sediments representative of the thickest and deepest sedimentary section in the field. For the structure in Figure 4-7, the wells furthest off structure, such as Well No. 5 or 7 or a composite of the two, are good candidates for a type log if these wells are drilled deeply enough. These wells, positioned off the crest of the structure, should show the thickest and most complete sedimentary section.

Step 2. A good correlation plan involves the correlation of each well with a minimum of two other wells. To ensure good correlation efficiency over the entire area, the plan should be established to correlate by means of closed loops. The correlation plan in Figure 4-7 illustrates a sequence of closed loops. The recommended order of correlation is represented by "billiard ball" type numbers for correlation sequence. Using this procedure, the log correlation work within a loop begins and ends with the same log, eliminating correlation mis-ties and reducing the chance of other correlation errors.

Step 3. First correlate wells expected to exhibit the most complete and thickest stratigraphic section. On a structure such as the one shown in Figure 4-7, the structurally lowest wells usually have the thickest section. These include wells represented by the billiard ball type correlation sequence number 1.

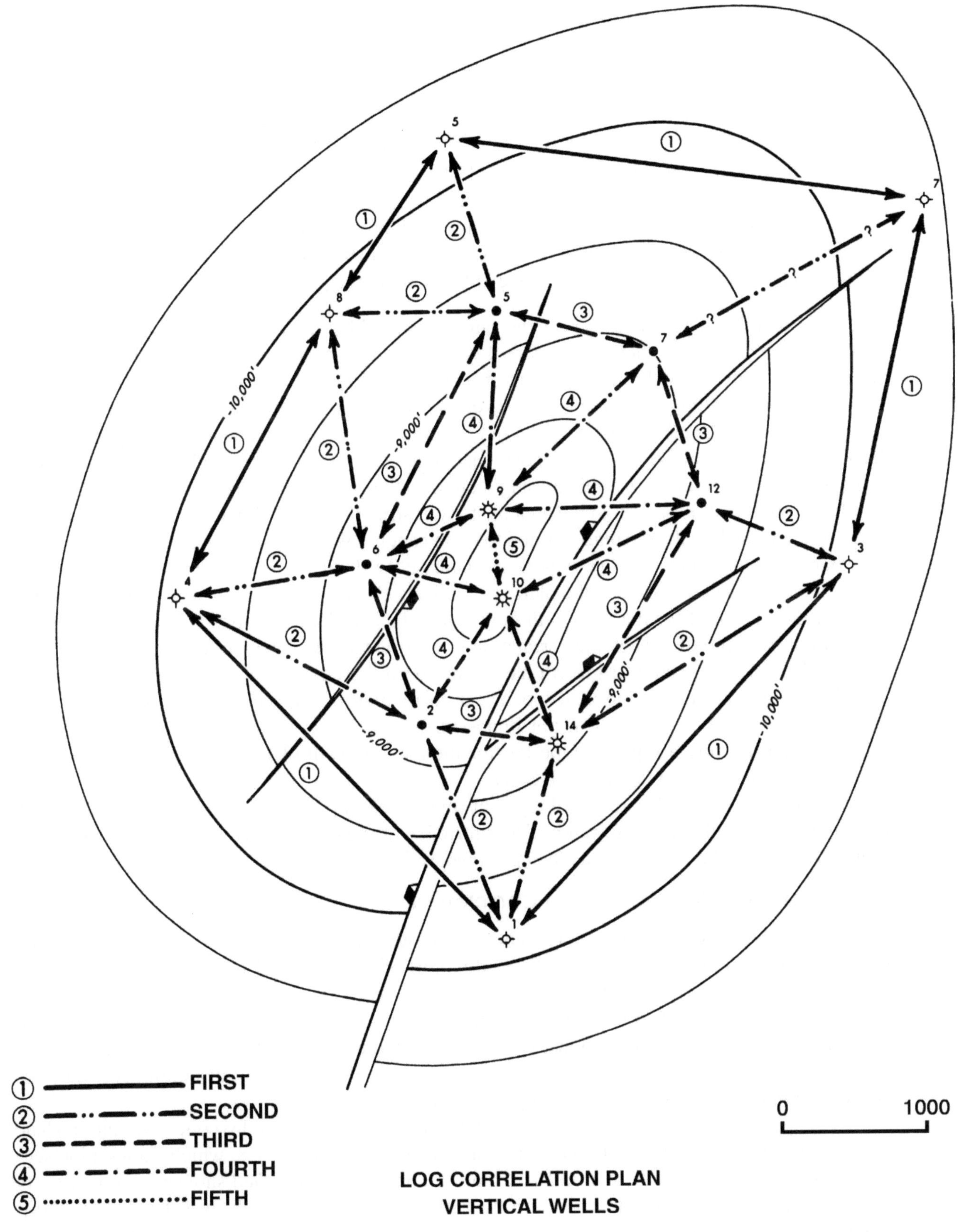

Figure 4-7 An example of a log correlation plan for vertical wells. Notice that there is a hierarchy in the correlation sequence and that all the wells are correlated in closed loops.

Step 4. Continue log correlation, progressing from wells in a down-structure position to wells in an up-structure position (see the billiard ball type correlation sequence numbers 2, 3, 4, and 5). These numbers indicate the recommended correlation sequence for the example.

Step 5. Generally, correlate wells located nearest each other. In most cases, closely spaced wells should have a similar stratigraphic section and so correlation is usually easier.

Step 6. In many geological areas, rapid changes in stratigraphy (particularly changes in thickness) occur over short distances. Where possible, correlate wells anticipated to have a similar stratigraphic interval thickness. In extensional or diapiric tectonic areas, wells at or near the same structural position often exhibit similar stratigraphy. For areas involving plunging folds, similar stratigraphy may be exhibited along the plunge of the fold.

Also, if wells that exist at or near the same structural bend are connected together, as in the first order correlation as shown in Figure 4-7, the dip rate of the structure should be similar in these various wells, thus eliminating correlation problems related to changing structural dip.

In geological settings containing *syndepositional faults* (commonly referred to as growth faults), some special considerations must be given in preparing a log correlation plan. For our purposes, we define a growth normal fault as a syndepositional fault resulting in an expanded stratigraphic section in the hanging wall fault block, with missing section on the growth fault changing with depth.

If a growth fault is present in the area of study, restrict the correlation to wells within one fault block of the growth fault. Keeping in mind that the hanging wall block of a major growth fault has an expanded stratigraphic section, which can increase the difficulty in correlation, start the correlation in the footwall block using the plan just outlined. Once the correlations are completed in the footwall fault block of the major growth fault, carry the correlations, if possible, into the hanging wall fault block. Initially, check the correlation in wells located in down-structure positions.

If a significant amount of growth has occurred on the fault, the thickness of the sediments can be so great in the hanging wall block that correlation of the section from hanging wall to footwall blocks may be difficult, if not impossible. In such a case, the best correlations can be achieved by preparing a separate type log for the hanging wall block and correlating this fault block independently from the footwall block.

We note here that other terminology is at times used in place of hanging wall and footwall. For a normal fault, the hanging wall is also referred to as the *downthrown fault block*. As well, the footwall block can be referred to as the *upthrown fault block*.

Basic Concepts in Electric Log Correlation

Now that we have established a plan of correlation, we examine some basic concepts of electric log correlation. Figure 4-8 shows the SP and amplified short normal resistivity curve from the electric logs of two vertical wells. Initial *quick-look* correlations can be made by reviewing major sands. Sands are the dominant and most obvious features seen on the SP curve or GR curve and serve as good quick-look correlations. Because sand bodies commonly exhibit significant variation in thickness and character from well to well and are often laterally discontinuous, they are not recommended for detailed electric log correlations.

We suggest that all detailed electric log correlation be undertaken by concentrating on shale sections. We apply this approach to electric log correlation in Figure 4-8, which shows a log segment from two vertical wells (No. A-1 and No. 3). The SP and amplified short normal resistivity curves are shown for each log. There are two major sands seen in each well, labeled 10,000-ft Sand and 10,300-ft Sand.

First we review these two logs using the tops of the major sands as the primary vehicle for correlation (Fig. 4-8a). Imagine that we prefer to correlate major sand bodies and we are correlating Wells No. A-1 and 3 in Figure 4-8a. By correlating the sands, we see that the interval from the

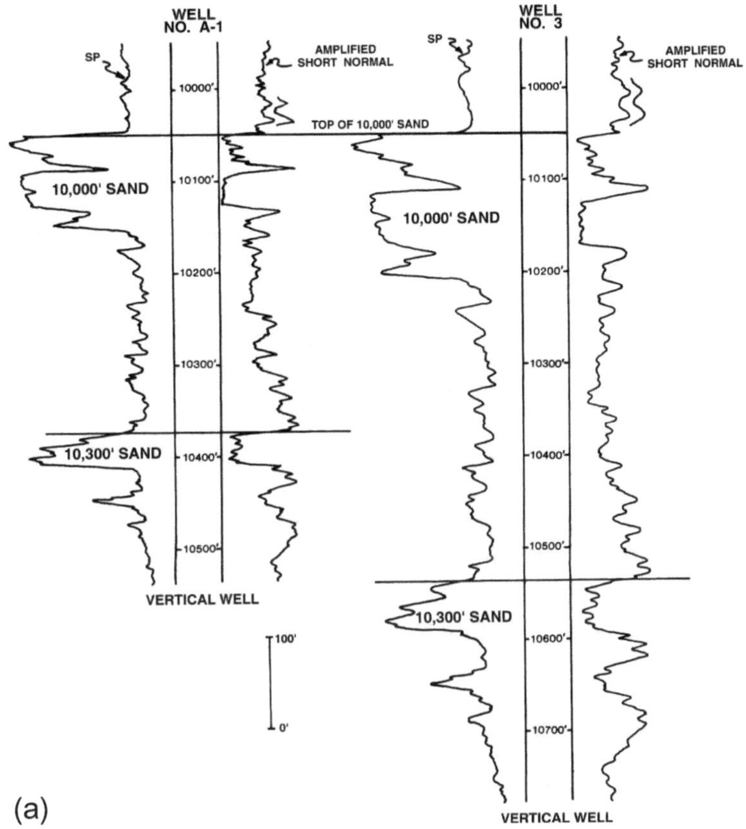

(a)

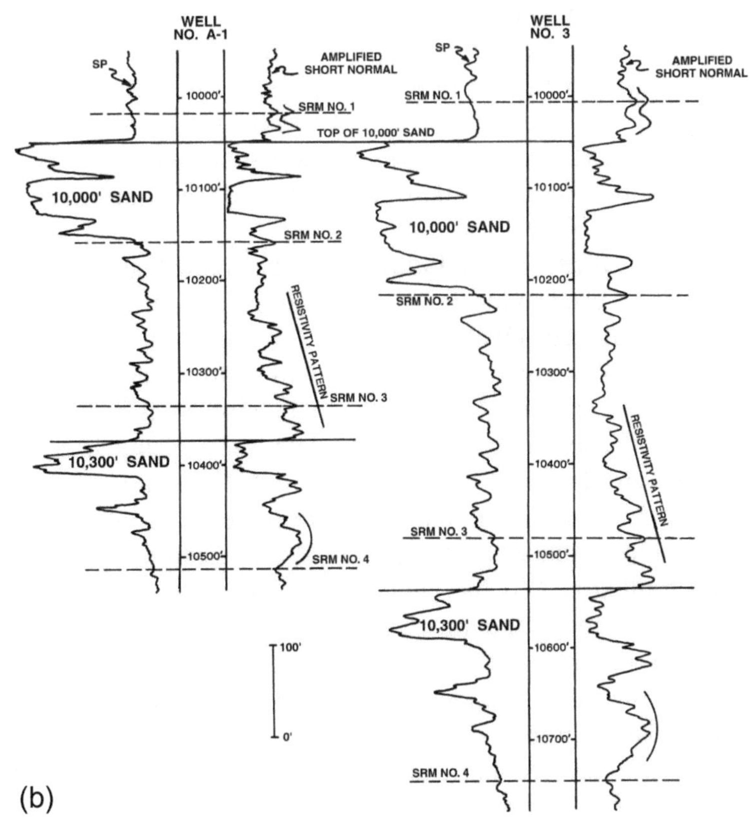

(b)

Figure 4-8 (a) Correlation of two vertical wells using the major sands as the primary vehicle for correlation. (b) Detailed correlation of the two vertical wells shown in Figure 4-8a using all the reliable shale and sand correlation markers. (SRM = shale resistivity marker.)

top of the 10,000-ft Sand to the top of the 10,300-ft Sand is about 325 ft thick in Well No. A-1 and 480 ft in Well No. 3. The interval in Well No. A-1 is short between the two sand tops by 155 ft. This short interval, based on the sand correlations, suggests the possibility of a 155-ft normal fault in Well No. A-1.

Now we correlate these same two logs, shown in Figure 4-8b, using the guidelines outlined earlier. The guidelines recommend that detailed correlations be conducted in the shale sections using all the electric log curves with an initial emphasis on the amplified resistivity curve. This curve provides the most reliable shale correlations.

Through detailed correlations of the shale sections and the sands, a number of correlation markers are identified on the two logs. These include a series of shale resistivity markers (SRMs) labeled SRM No. 1 through SRM No. 4, with certain diagnostic resistivity correlation patterns highlighted on the resistivity side of the logs, in addition to the two major sands. All these correlation markers indicate that both log segments have a high degree of correlation and that no fault is present in Well No. A-1.

It appears that the stratigraphic section in Well No. A-1 is uniformly thin relative to Well No. 3. The thickness ratio for the intervals between each of the four shale markers shows a consistency in the stratigraphic thinning in Well No. A-1 when compared to Well No. 3. This uniform thinning supports the idea that although Well No. A-1 is short to No. 3 as a result of stratigraphic thinning, the two logs exhibit correlation.

Faults Versus Variations in Stratigraphy

We begin the log correlation section to determine faults in an extensional setting where faults typically create a missing section on well logs. The differentiation between recognizing faults versus variations in stratigraphic thickness in well log correlation is very important. We stated earlier that reliable interpretations presented on maps and cross sections are bedrocked in accurate correlations. If a stratigraphically thin section is correlated incorrectly as a fault, this erroneous fault data will be incorporated into the construction of a fault surface map and later integrated into the structural interpretation. The purpose of this section is to outline procedures that are effective during correlation to help differentiate between faults and variations in stratigraphic thickness.

Fault Determinations: Depth and Missing Section. Now that we have a basic understanding of how shale markers are used to aid in log correlation, look at the log segment from two other electric logs run in vertical wells (Wells No. 1 and 3 in Fig. 4-9). Some geologists prefer to correlate major sands. If we do that and use the 8600-ft and 9000-ft Sands as the principal correlations, we see that the section in Well No. 3 between the two major sands is 80 ft short and a fault appears possible in the well. With the limited correlation data, the missing section and depth of the fault is uncertain. Also, the correlation of the top of the 9000-ft Sand in Well No. 3 is questionable. Is there a fault in Well No. 3, and is the fault (1) within the shale interval between the base of the 8600-ft Sand and the top of the 9000-ft Sand or (2) at the top of the 9000-ft Sand, or (3) is part of the 9000-ft Sand faulted out? If the fault is above or at the top of the 9000-ft Sand, then the interval from the 8600-ft Sand to the top of the 9000-ft Sand is 80 ft short. If part of the top of the 9000-ft Sand is faulted out, then the interval is short by some amount greater than 80 ft. With the major sand correlation methodology, the nature of the short section in Well No. 3 is not apparent, and so we have a correlation problem.

Now we will follow the recommended correlation procedures illustrated in Figure 4-10. These procedures provide a number of correlation markers, including SRMs 1 through 7 and specific resistivity characteristics highlighted on the resistivity curve of each log. These detailed correlation markers show that the interval between each correlation marker is comparable in both wells except for the short section identified between SRM 5 and SRM 7. Notice that SRM 6 is

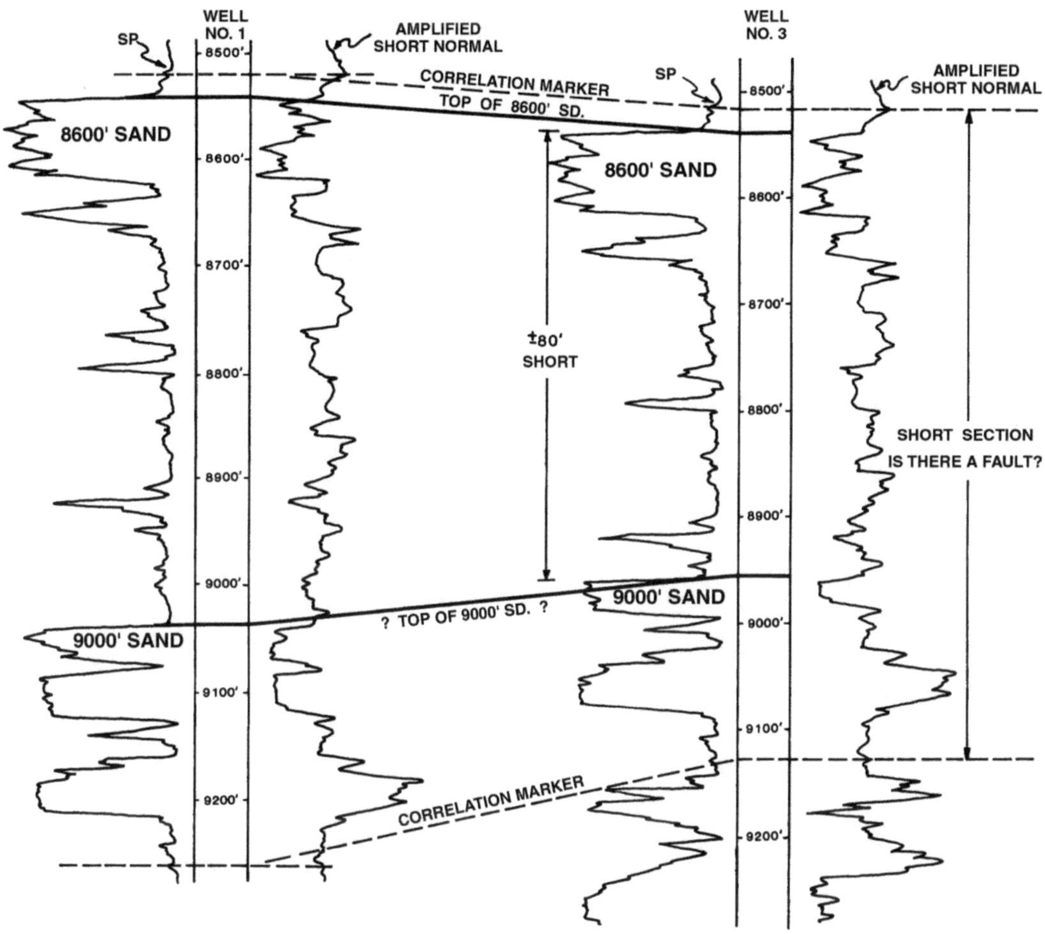

Figure 4-9 Correlation of the major sands in two vertical wells may provide insufficient correlation data to accurately determine the depth and size of a suspected fault.

missing in Well No. 3, as is the lower segment of the resistivity character highlighted through SRM 6 in Well No. 1. Finally, these detailed shale correlations and the correlation data contained within the 9000-ft Sand indicate that the upper portion of the 9000-ft Sand is also missing in Well No. 3. We have isolated the short section in Well No. 3 to a specific interval 135 ft thick. **The isolation of this short section to one particular location indicates that the short section is the result of a fault rather than a variation in stratigraphy**. The location of the short section provides the depth of the fault in this well. By measuring the amount of section in Well No. 1 that is missing in Well No. 3, we determine the missing section due to the fault (135 ft) by correlation with Well No. 1. The missing section is highlighted in Figure 4-10. In order to ensure confidence in the fault, Well No. 3 should be correlated with at least one more nearby well.

For each fault recognized in a well, there are three important pieces of data that must be obtained for documentation and later use in mapping: (1) the amount of missing section, (2) the depth of the fault in the log, and (3) the well or wells correlated to identify the fault. The fault data (135 ft/8957 ft/Well No. 1) and information regarding the faulted out (F/O) top of the 9000-ft Sand are annotated next to the fault symbol on the log. Refer to Figure 4-10 again for an example of how these data are annotated on a log. For convenience, in most of our examples we use measured well depths for faults and other points. For your mapping purposes, we recommend that you annotate subsea depths.

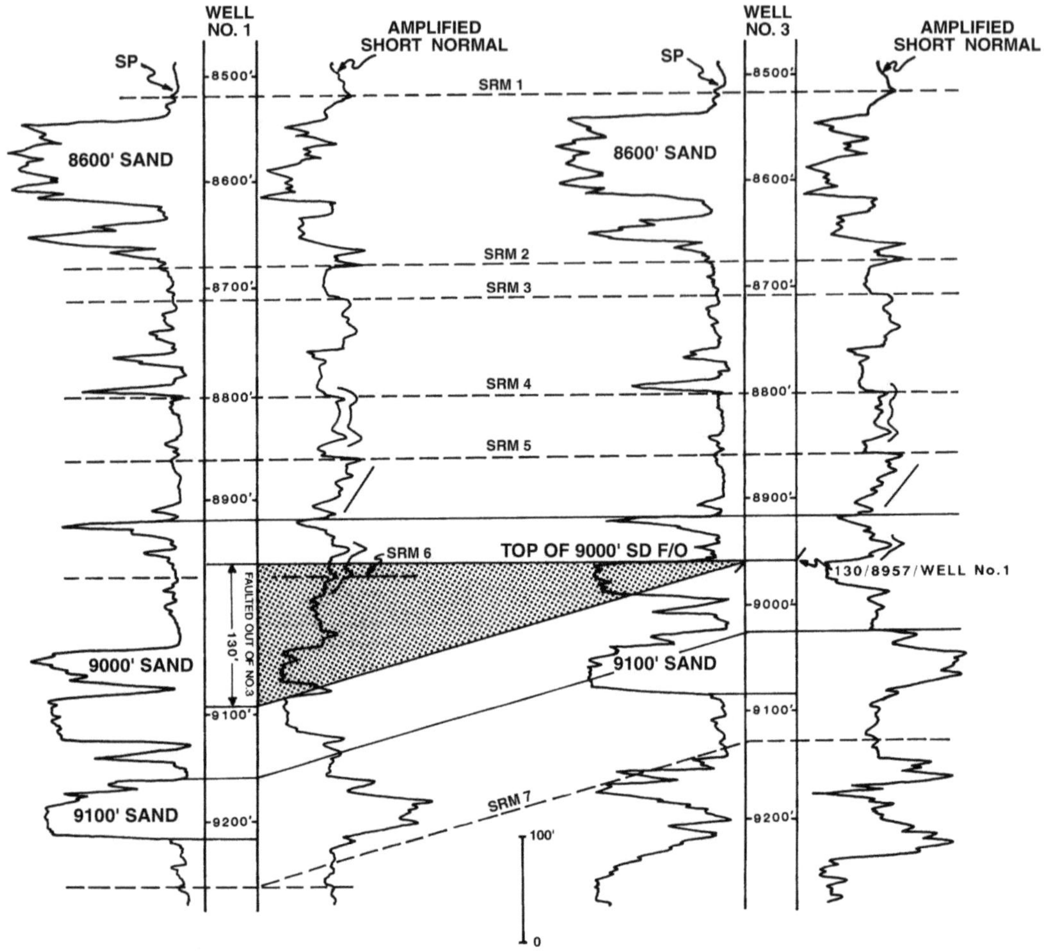

Figure 4-10 Detailed correlation of the two vertical wells shown in Figure 4-9 using all recognizable correlation markers to determine the depth and missing section for a fault in Well No. 3. Notice that the top of the 9000-ft Sand and SRM 6 are faulted out of Well No. 3.

Most computer-based log interpretation systems automatically record the depth of the fault and provide a specific place in the computer database to record the amount of missing section associated with the fault. Few computer-based log interpretations systems provide a specific place in the computer database to record the well or wells correlated to identify the fault. The geoscientist correlating the logs will usually have to enter this information as an unstructured annotation in the database. We strongly recommend that interpreters take the additional time to add this information, as it makes checking the fault interpretation much easier in the future.

The accuracy of identifying the depth of a fault in a well and determining the amount of missing section is directly related to (1) the detail to which the logs are correlated, (2) the number of logs used for correlation, and (3) variations in stratigraphic thickness seen in the wells. Obviously, the smaller the interval between established correlation markers, the more precise the correlation in pinpointing the depth of the fault in the well and the amount of missing section. Missing section can be incorrectly estimated if the reference well is of different thickness than the faulted well. A thickness ratio can be used to appropriately adjust the amount of missing section that is measured in the reference well log (see following section, Stratigraphic Variations).

The correlation detail and accuracy required are often dictated by the type of geological study being conducted. For example, if you are involved in a regional geological study, pinpointing the depth of a fault within several hundred feet on a well log may be sufficient. Also, you may be interested only in the larger faults (i.e., faults greater than 100 ft). If the study is to be detailed for field development or enhanced recovery, however, it may be necessary to *locate the depth of all recognizable faults to within ± 20 ft*. The same variation in accuracy applies to missing section as well.

Stratigraphic Variations. Figure 4-11 shows a log segment from Wells No. A-1 and 3. In this section, we use the correlation procedures to establish specific correlation markers to recognize stratigraphic variations so that such thickness changes are not mistaken for faults.

In Figure 4-11, two correlation markers are identified in each well. Based on these markers, Well No. A-1 is 155 ft short to Well No. 3. Is the short section in Well No. A-1 the result of a fault or variations in stratigraphy? With the limited correlation data shown in the figure, it is impossible to determine why the section in Well No. A-1 is short. We could use the major sands in each well to aid the correlation work, but this added information provides little help in determining the nature of the short section.

So far, we have shown that it is important to identify as many correlation markers as possible, especially in questionable log intervals. *Closely spaced correlations generally improve the accuracy of the correlation, help differentiate faults from stratigraphic variations, and improve the estimate for the amount of missing section and depth of identified faults.* Therefore, in order to accurately correlate Wells No. A-1 and 3, additional correlation markers are required.

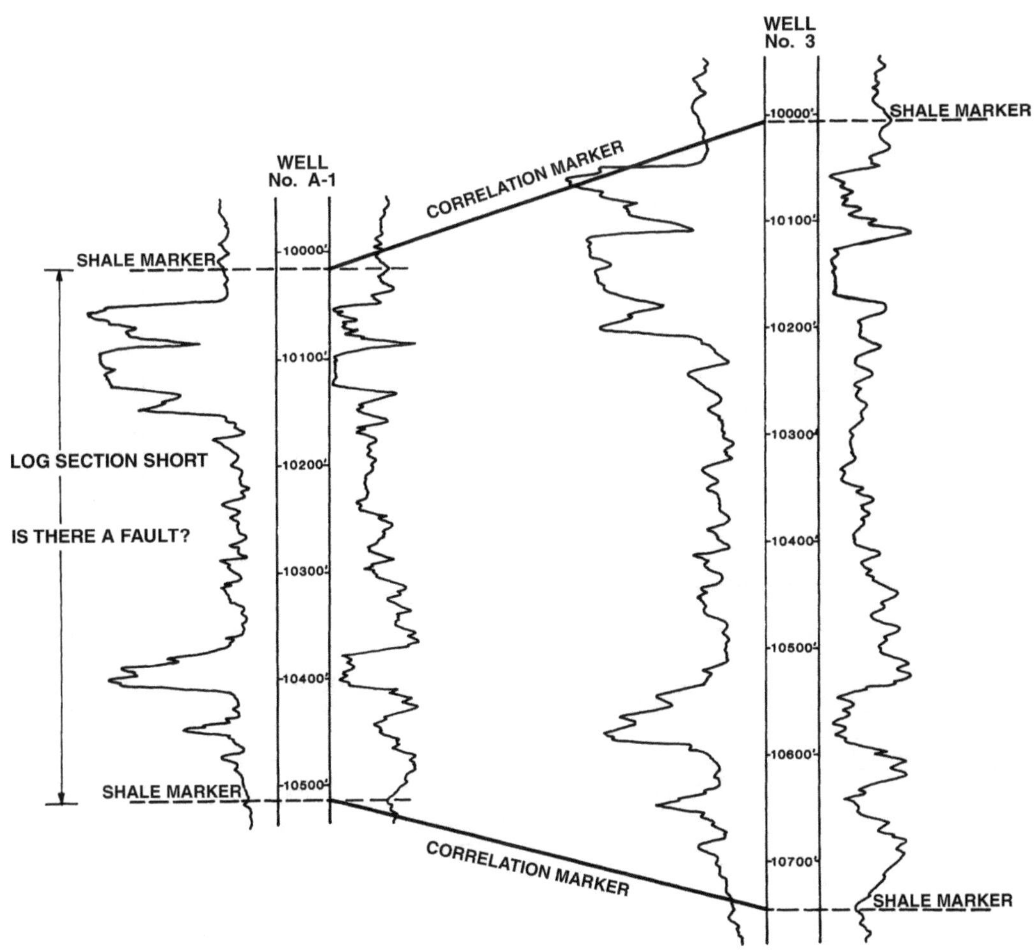

Figure 4-11 Correlation of Wells No. A-1 and 3 using limited correlation markers. Is there a fault in Well No. A-1?

Figure 4-12 shows the same two logs with additional correlation markers identified. The correlation process is improved with these additional markers. Notice that the shortening in Well No. A-1 is not isolated to one specific interval but is present in all the intervals in the well between SRM 1 and SRM 4. This evidence strongly suggests that the thickness variations in Well No. A-1 are stratigraphic and not the result of faulting. If necessary, interval thickness ratios can be calculated between correlation markers to provide further evidence to support the conclusion.

$$T_r = \frac{T_s}{T_l} \tag{4.1}$$

where

T_r = thickness ratio

T_s = interval thickness in short log

T_l = interval thickness in long log

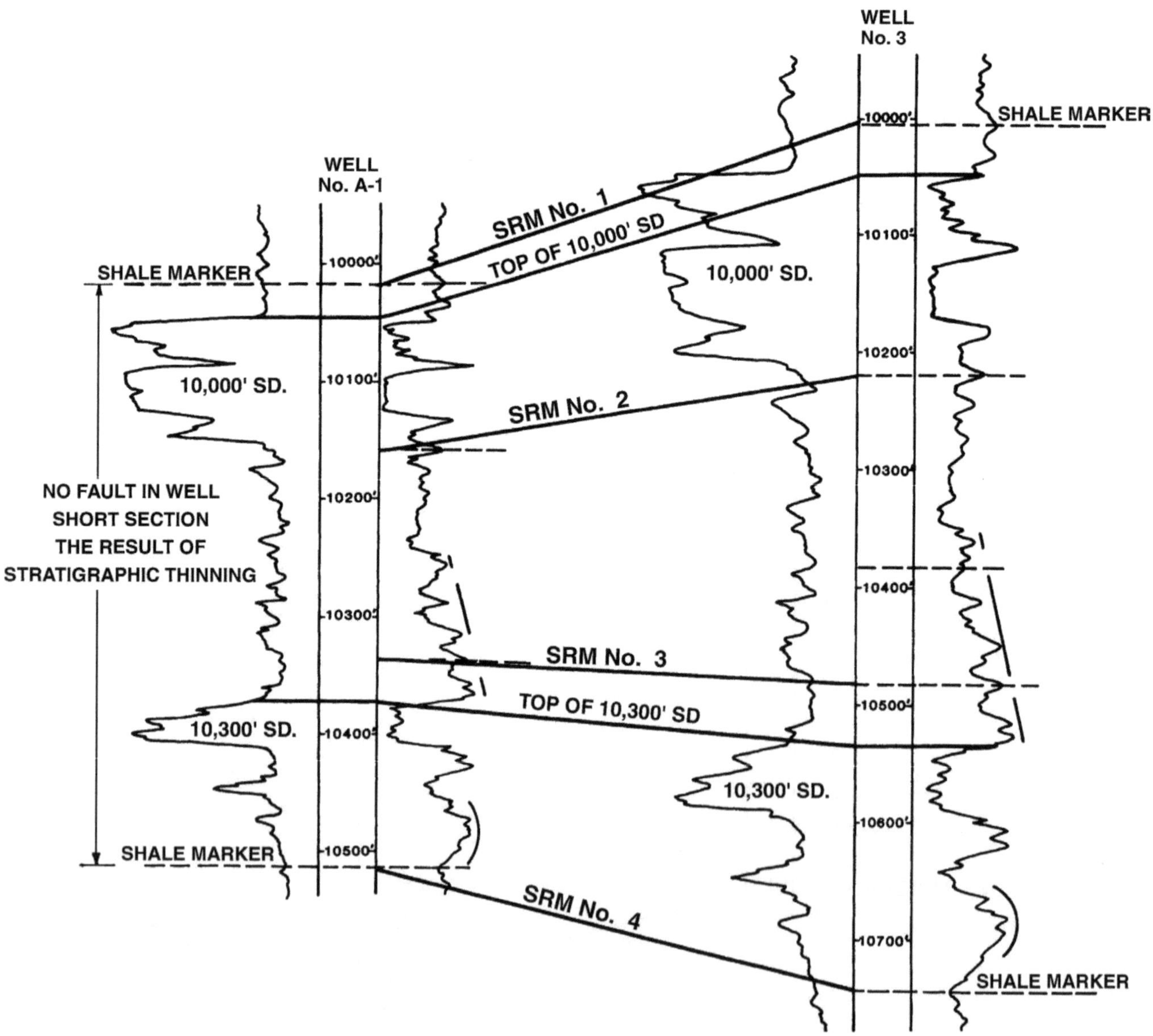

Figure 4-12 Correlation of Wells No. A-1 and 3 using all recognizable correlation markers indicates that there is no fault in Well No. A-1. The short log section is the result of variations in stratigraphy.

Pitfalls in Vertical Well Log Correlation

As a final topic on well log correlation in vertical wells, we look at some pitfalls caused by changes in formation dip. Figure 4-13a shows an electric log section from four vertical wells. Using the detailed correlation procedures, we determine that the stratigraphic section shown in each well is the same section. From right to left in the figure, the well logs show an increasing thickness in the section from 100 ft in Well No. 4 to 200 ft in Well No. 1. What is the cause of thickness change in this section: variations in stratigraphic thickness, faulting, or something else?

General structure mapping and dipmeter data show that the bed dip is different in the vicinity of each well: 0 deg in Well No. 4, 20 deg in Well No. 3, 40 deg in Well No. 2, and 60 deg in Well No. 1. All four wells lie in a line perpendicular to bed strike (parallel to bed dip). Analysis of the dip data and logs suggests an increase in thickness of the stratigraphic section in the up-structure direction. Normally, however, we expect to see a constant or reduced thickness of a stratigraphic section in the up-structure direction. So, are these thickness changes seen on each log due to faulting, stratigraphic variations, or a geometric problem resulting from changing bed dip?

In Figure 4-13b, the logs are hung in their true structural position with the dip of the formation at each well location shown. The formation dips and the relationships between true bed or stratigraphic thickness and log or vertical thickness are shown on the figure. Notice that even though the log thickness in Well No. 1 is twice that seen in Well No. 4, the stratigraphic thickness is identical at both locations. This example illustrates that caution must be taken when correlating logs on a structure with significant changes in bed dip. Changing bed dip can result in changing log thickness in vertical wells, even though the section is not faulted and the stratigraphic thickness is constant.

To better understand the stratigraphy and growth history of a structure, and to resolve some of the geometric problems caused by changes in bed dip, the true stratigraphic variables required to calculate true stratigraphic thickness (TST) are the true vertical thickness (TVT) of the section as seen in a vertical well and the bed dip.

$$\text{TST} = \text{TVT}(\cos\phi) \tag{4.2}$$

where

TST = true stratigraphic thickness
TVT = true vertical thickness
ϕ = true bed dip

Logs cannot be correlated in a vacuum. The correlation plan shown earlier illustrated the need to know the structural relationship of logs being correlated. This can be accomplished by having a well-log basemap available during correlation that shows the general structure and the location and structural position of each well being correlated.

Finally, let's consider a situation as shown in Figure 4-14. In this case, the stratigraphic section identified in the two wells has a decreasing TST in the up-structure direction. We can say that the structure was actively growing during the time of deposition of the section, resulting in stratigraphic thinning toward the crest of the structure. However, by log correlation, *the vertical log thickness of the stratigraphic section in each well is exactly the same.* If you recognized the same interval thickness in each well irrespective of structural position, you might make the incorrect assumption that since the thicknesses are equal, the structure was not active during deposition, or that the TST of the stratigraphic section as seen in both wells is the same.

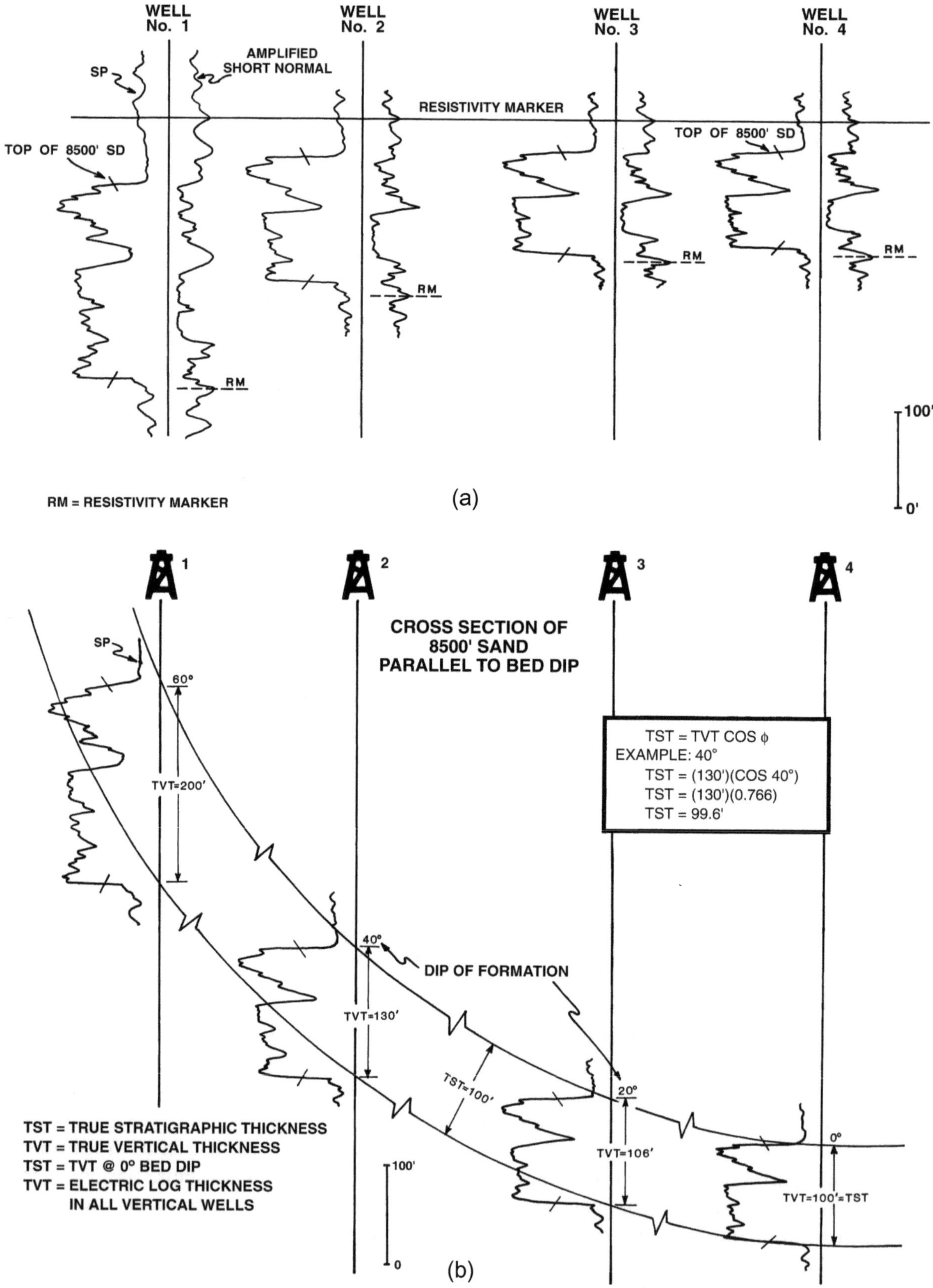

Figure 4-13 (a) Electric log stratigraphic section laid out perpendicular to the strike of the 8500-ft Sand (parallel to dip). (b) Electric log section showing the relationship of TST to TVT with changing bed dip. The TVT is that thickness seen in a vertical or straight hole.

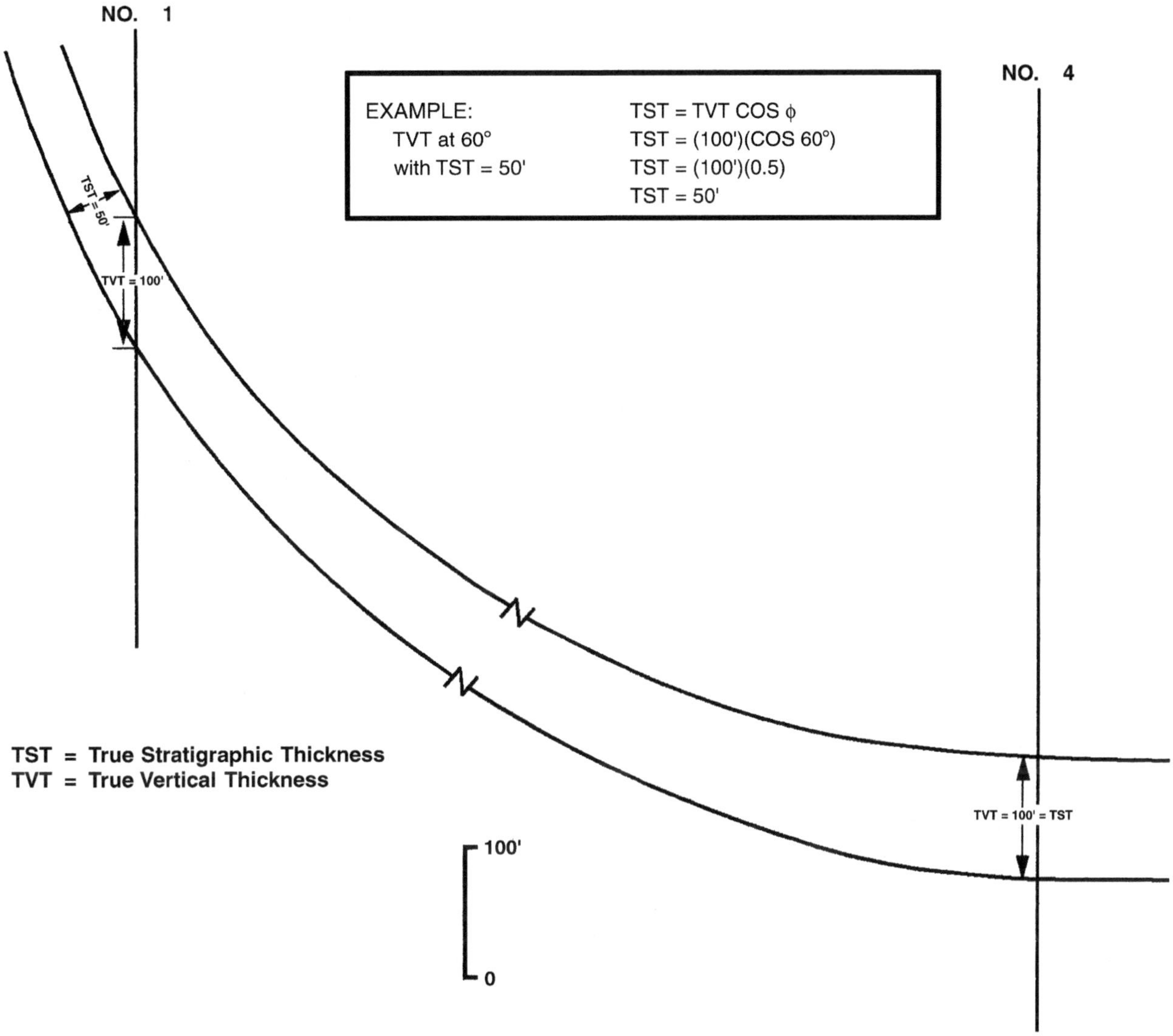

Figure 4-14 The cross section shows the effect of changing bed dip on the true vertical thickness of a stratigraphic section.

A review of the wells in cross section in Figure 4-14 shows that the stratigraphic thickness of the section actually decreases in the up-dip direction such that the stratigraphic thickness in Well No. 1 is only one-half the thickness found in Well No. 4. *It is the changing bed dip that causes the vertical log thickness to be the same in each well.* Equation (4-2) can be used to calculate the TST in each well to develop a better understanding of the stratigraphic thicknesses as seen in each well.

These examples show that log thickness in vertical wells varies with changes in structural dip and is equal to the TST only when the dip of the unit is zero (Fig. 4-14, Well No. 4). They also emphasize that log correlation is not an isolated process. Structural and stratigraphic relationships and geological knowledge of the area of study must always be kept in mind during correlation. Errors in correlation and incorrect assumptions on such aspects as the local growth history of a structure may be incorporated into the geological work. Such pitfalls can be prevented if the structural and stratigraphic framework of the area is considered during correlation.

ELECTRIC LOG CORRELATION—DIRECTIONALLY DRILLED WELLS

In this portion of the chapter we discuss fundamental concepts and techniques for correlating directionally drilled wells. Additional complexities in correlation arise when working with logs from wells deviated from the vertical. We also look at the correlation of vertical wells with directionally drilled wells (often referred to as deviated wells).

What is a directionally drilled well? We discussed earlier that a vertical well is one drilled 90 deg to the horizontal reference, usually sea level. A directionally drilled well can be defined as a well drilled at an angle less than 90 deg to the horizontal reference, as shown in Figure 4-15. Some general directional well terminology was discussed in Chapter 3. These terms are again illustrated in Figure 4-15 for ease of reference. Other terminology discussed earlier in this chapter for vertical wells is also applicable to deviated wells.

Most wells drilled in an offshore environment and many wells onshore are drilled directionally. The most common well is a simple ramp well (Fig. 4-15a), sometimes called an *L-shaped hole*. These wells are deviated to a certain angle, which is usually held constant to TD of the well. Many wells are drilled with an S-shaped design. With an S-shaped hole, the well builds to one angle, maintains this angle to a designated depth, and then the angle is lowered again, often going back to vertical (Fig. 4-15b). Today we see a large number of horizontal wells, which are shaped by continually building the angle until the desired near-horizontal orientation is reached (Fig. 4-15c). By 2018, almost 90% of the wells being drilled in the United States were classified as horizontal wells, and another 5% were classified as deviated wells according to the weekly Baker Hughes rig count.

Log Correlation Plan

Just as with vertical wells, there must be some system to log correlation of directionally drilled wells. Due to the nature of deviated wells, a good correlation plan is critical to accurate correlations. For this log correlation plan, we once again use the structure map on the 8000-ft Sand on a normally faulted anticlinal structure (Fig. 4-16). The correlation plan outlined here is intended to make correlation systematic, provide a logical method for correlating directionally drilled wells with other directionally drilled wells or with vertical wells, and reduce correlation problems.

Step 1. Construct a correlation type log. Refer to the section on correlation type logs for the complete definition of a type log. Do not use a deviated well in the construction of a type log because a log from a directionally drilled well does not represent the true vertical stratigraphic section. Wells farthest off structure serve as good type log candidates. If no vertical wells are available for the correlation type log, a TVD log from a deviated well drilled in an area of low bed dip may be an acceptable substitute.

Step 2. Correlate all the vertical wells before correlating the deviated wells, since the vertical wells are usually easier to correlate. For the vertical wells, use the same plan outlined in Figure 4-7.

Step 3. Once the vertical wells have been correlated, begin correlating the deviated wells. To begin directional well log correlation, first organize the wells according to their direction of deviation with respect to structural strike. *Deviated wells are classified into one of three groups: (1) wells drilled down-dip, (2) wells drilled along strike, and (3) wells drilled up-dip.*

Step 4. Begin correlation of these three groups with the wells drilled generally down-dip. First correlate the wells with the least amount of deviation, and where possible, correlate in closed loops with each well log correlated with a minimum of two other wells. The wells with the least amount of deviation will have a log section thickness closer to that seen in a vertical well than other wells drilled down-dip. Looking at the wells drilled from Platform B in Figure 4-16, the first directional wells correlated are those represented by a billiard ball type

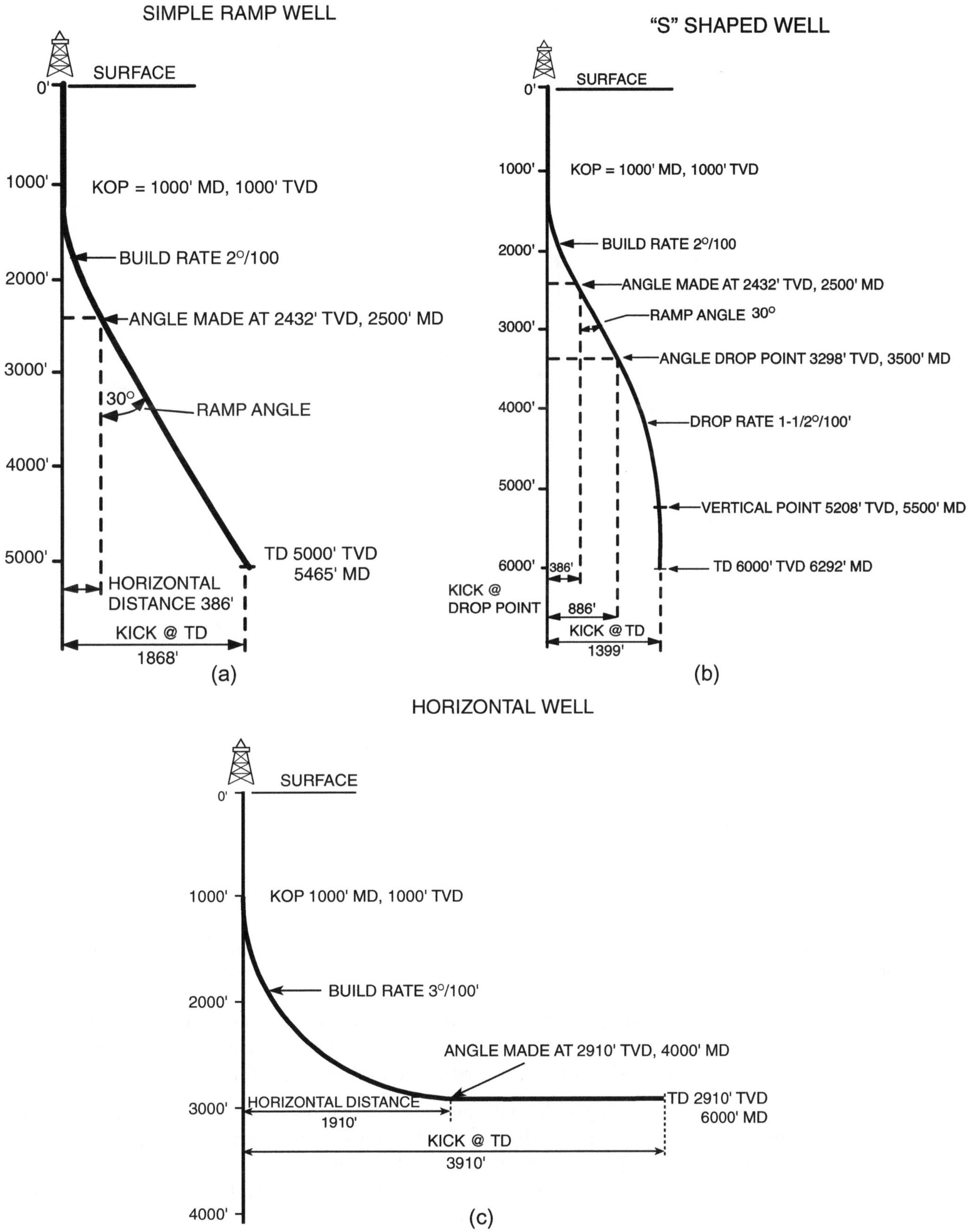

Figure 4-15 Diagrammatic cross sections illustrating (a) a simple ramp or L-shaped well; (b) a more complicated S-shaped well; (c) a horizontal well. [(a) and (b) published by permission of Tenneco Oil Company; (c) published by permission of J. Brewton.]

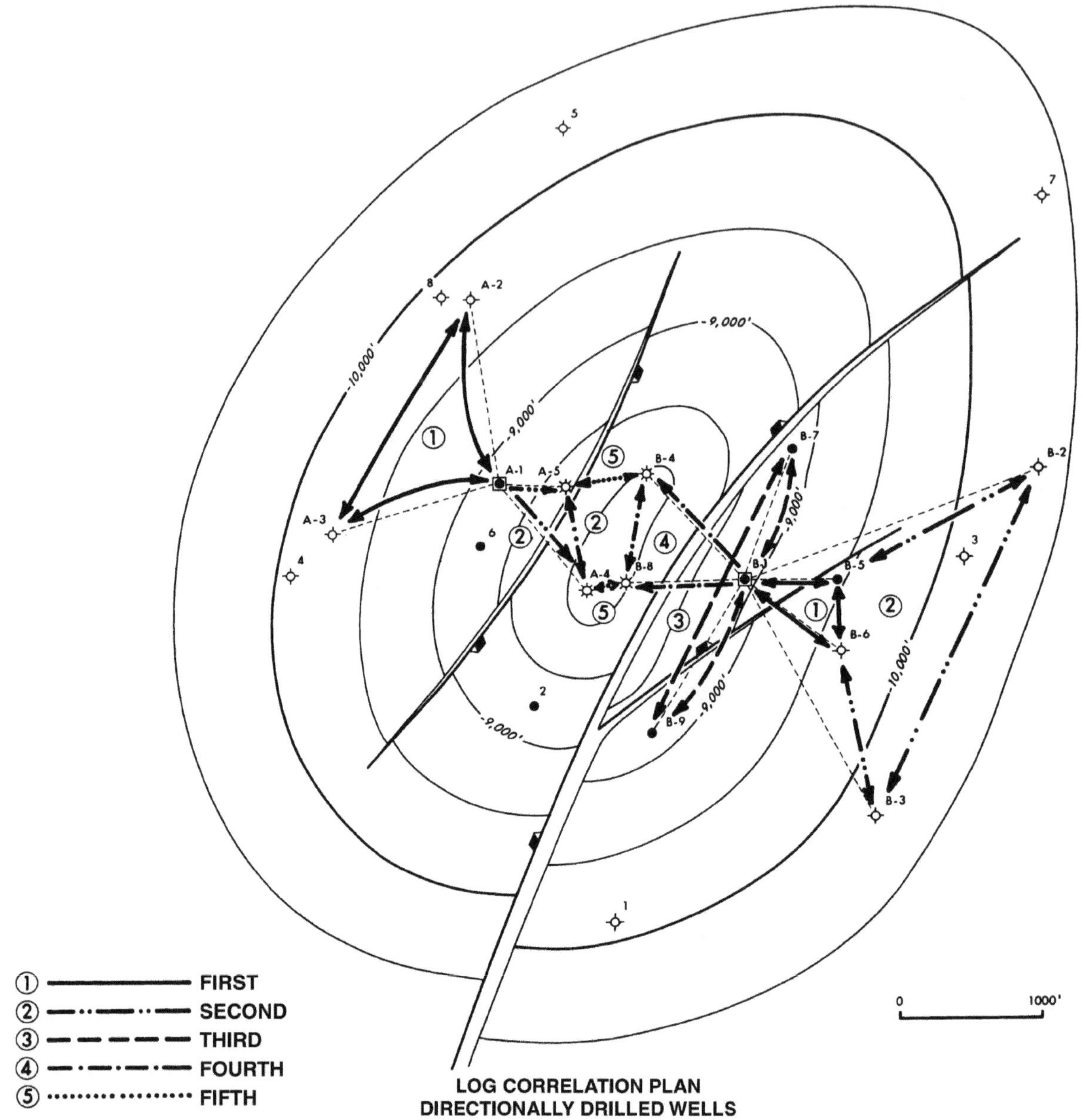

Figure 4-16 An example of a log correlation plan for directionally drilled wells. The plan shows the hierarchy of the log correlation sequence and illustrates how to correlate deviated wells in closed loops.

correlation sequence number 1. There are two wells drilled with a minimum down-dip deviation (Wells No. B-5 and B-6). These wells can be correlated to each other and then with the straight hole, Well No. B-1, drilled as a vertical well from the platform.

Step 5. Continue correlating wells with increased deviation in the down-dip direction. For this example, these are Wells No. B-2 and B-3, indicated by correlation sequence number 2. These two highly deviated wells can be correlated with each other and then with Wells No. B-5 and B-6. Also, the vertical Well No. 3 may be used to correlate B-2 and B-3, since it is an off-structure well exhibiting a thick stratigraphic section.

Step 6. When all wells classified as being deviated down-dip are correlated, the next group to correlate are those wells deviated along structural strike. From Platform B, Wells No. B-7 and B-9 fall into this category. These wells can be correlated to each other and then with straight hole B-1 to close the loop. When correlating wells drilled along strike, the effect of wellbore deviation can be removed from the representative thickness of the directionally drilled wells by using a TVD log. In other words, the TVD log of a well drilled along strike will be similar to a vertical well drilled in the same dip location. The TVD log will still be affected by any true stratigraphic variations as well as apparent stratigraphic variations due to bed dip. Using a TVD log of a well drilled along strike can often simplify correlation.

Step 7. Correlate the wells deviated up-dip. Those wells drilled closest to the crest of the structure usually are complicated by stratigraphic thinning, faulting, and unconformities. The correlation of these wells can be most difficult; therefore, they are normally correlated last when all other correlation information is available and you can recognize the best correlation markers. Wells No. B-4 and B-8 drilled from the B Platform fall into this category. They are labeled as correlation sequence number 4. Wells drilled in an up-dip direction can have variable log section thickness due to the geometric relationship between a wellbore and structural dip. A log section from a well drilled in an up-dip direction can be thicker, thinner, or equal to the thickness of a log section from a nearby vertical well drilled through the same stratigraphic section. This potential complexity can add to the difficulty of correlating wells drilled in an up-dip direction. Because of these complexities, we recommend that these wells be correlated last, after significant knowledge is gained from other correlation work.

Step 8. Generally, it is best to correlate wells located nearest each other, especially in areas where significant changes in stratigraphic thickness are probable. Wells nearest each other and approximately in the same structural position usually are expected to have the most comparable interval thicknesses.

Step 9. After correlating the wells from one platform, begin correlation of wells on any additional platforms in the area. In Figure 4-16, the A Platform wells in the northwest portion of the field should be correlated next. It is not necessary, however, to isolate correlation to a single platform. Often, wells from one platform are drilled in a direction toward another platform. If wells from separate platforms are in close proximity to one another, they should be correlated to each other. Notice that correlation sequence number 5 illustrates the correlation of B-4 with A-5, and B-8 with A-4. Wells No. A-1 and B-1 are straight holes drilled from separate platforms, but since they are located in a similar structural position, they can also be correlated to each other.

The primary focus of this correlation plan is to provide a logical method for correlating all vertical and deviated wells in an area of study. The plan outlined is by no means the only one that can be used. *The important point is to have a plan*. Without one, log correlation becomes a random process, often resulting in some type of correlation problem or in miscorrelations.

CORRELATION OF VERTICAL AND DIRECTIONALLY DRILLED WELLS

In this section, we discuss general procedures for correlating vertical wells with directionally drilled wells. Directional wells have a measured log thickness (MLT) that can be less than, greater than, or equal to the log thickness in a vertical well drilled through the same stratigraphic section. These different MLTs result in additional complexities that must be considered when undertaking correlation work using well logs from both vertical and deviated wells.

Now we look at the correlation of a vertical well with a deviated well. Figure 4-17 shows a portion of an electric log from vertical Well No. A-1 and the electric log from directionally drilled Well No. A-2. The wells are in close proximity to each other. The detailed electric log correlation

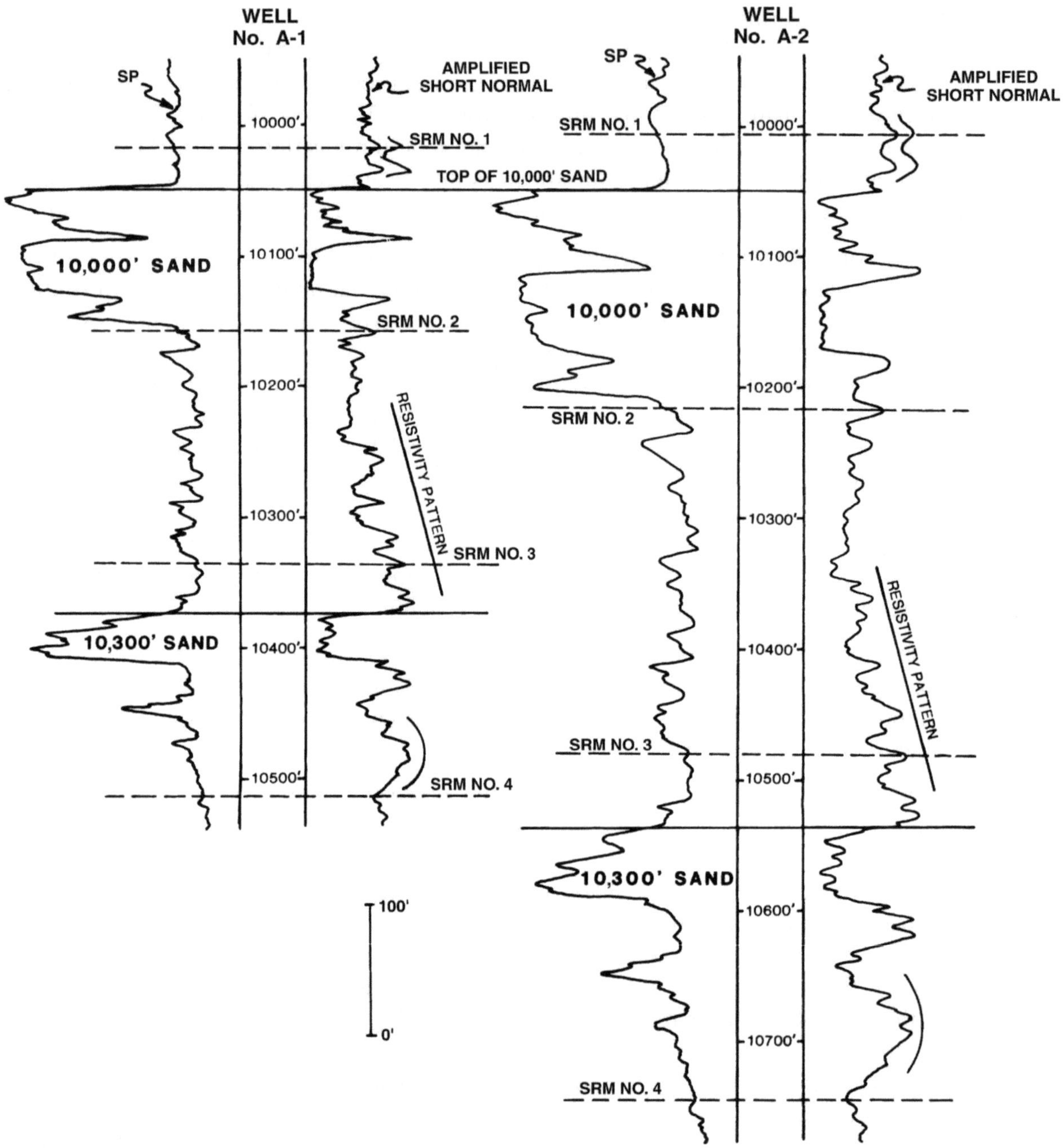

Figure 4-17 Portion of an electric log from a vertical well (A-1) and a directionally drilled well (A-2). The electric log sections show detailed correlations.

(sand and shale sections) for both wells indicates that they have penetrated the same stratigraphic section. Although both wells have a high degree of correlation (see SRM 1 through SRM 4), the stratigraphic section in Well No. A-2 is much thicker than the same section seen in Well No. A-1. The log section in Well No. A-1 from SRM 1 to SRM 4 is 490 ft thick. The same section in Well No. A-2 is 735 ft. Earlier in the chapter, in the discussion on vertical wells, we mentioned that a short section in one well with respect to another might be the result of stratigraphic changes or a fault. If the short section is isolated to one particular location, the short section is most likely the result of a fault rather than variations in stratigraphy. Conversely, if the short section is uniformly distributed over a series of intervals, the short section is probably due to stratigraphic variations rather than a fault.

Based on correlation criteria, the thinner section in Well No. A-1 appears to be the result of stratigraphic thinning rather than a fault. In this example, however, we introduce another possible explanation for the shortening. Since Well No. A-2 is directionally drilled, the thickness seen in the well with respect to Well No. A-1 may be completely the result of the wellbore deviation. Figure 4-18 shows vertical Well No. A-1 and deviated Well No. A-2 in its true orientation with respect to the vertical. Well No. A-2 is drilled due west at a deviation angle of 48 deg (from the vertical). The correlation markers in each well show that the strata are horizontal, and the thick section seen in Well No. A-2 is solely the result of wellbore deviation. We have now introduced another complexity in correlation that must be considered when both vertical and deviated wells are present in the area of study.

Here are several procedures that can be used to help correlate a vertical well with a directionally drilled well.

1. Mark the angle of deviation for the directional well on the log at least every 1000 ft. This provides a reminder that the well is deviated and indicates the angle of deviation at 1000-ft intervals on the actual log.
2. To compare interval thicknesses, slide the vertical well log as you correlate from marker to marker. This allows you to compensate during correlation for the expanded or reduced section in the directional well as a result of its deviation.
3. Calculate a thickness ratio for certain correlation intervals of interest to help evaluate whether any short section is the result of faulting, stratigraphic thinning, or just wellbore deviation (Fig. 4-18).
4. If a copy machine with a reduction mode is available, calculate the correction factor required to convert the deviated (stretched) log section to a vertical log section, and then reduce the log by the appropriate reduction factor. Use the reduced log for correlation.
5. *In areas of horizontal beds or low relief*, the MD log from a deviated well can be corrected for wellbore deviation and converted into a TVD log to use for correlation.
6. In areas with bed dips greater than 5 deg to 10 deg, if dip data are available from a dipmeter log or previously constructed structure maps, these data can be used to convert the deviated log to a TVT log. A TVT log is one in which the measured thickness has been corrected for wellbore deviation and bed dip to the thickness represented in a vertical well. In areas of significant dip, a TVD log provides little aid, if any, in correlation and can actually cause correlation problems (see section on MLT, TVDT, TVT, and TST) unless the well is drilled along the strike of the beds.

Estimating the Missing Section for Normal Faults

Earlier in this chapter, we discussed the procedure for estimating the depth and missing section for a fault in a vertical well by correlation with another vertical well. Now we present the method for estimating the depth and missing section for a fault when deviated wells are considered. First, we look at the situation involving an area with horizontal beds.

Horizontal Beds. We begin with a fault in the deviated Well No. A-2 (Fig. 4-19a). By correlation with vertical Well No. A-1, this well cuts a fault near the 10,000-ft Sand level. To determine the depth and missing section, we correlate the logs in the same manner as previously outlined in this chapter. First, correlate down the logs starting with SRM 1. We can say that correlation is lost at points A in both wells. Mark this location on the two logs. Next, find a correlation point below this section on the logs, such as SRM 4, and correlate up the logs. We now lose correlation in the wells at points B. By detailed correlation of the shale markers and sands, we have determined

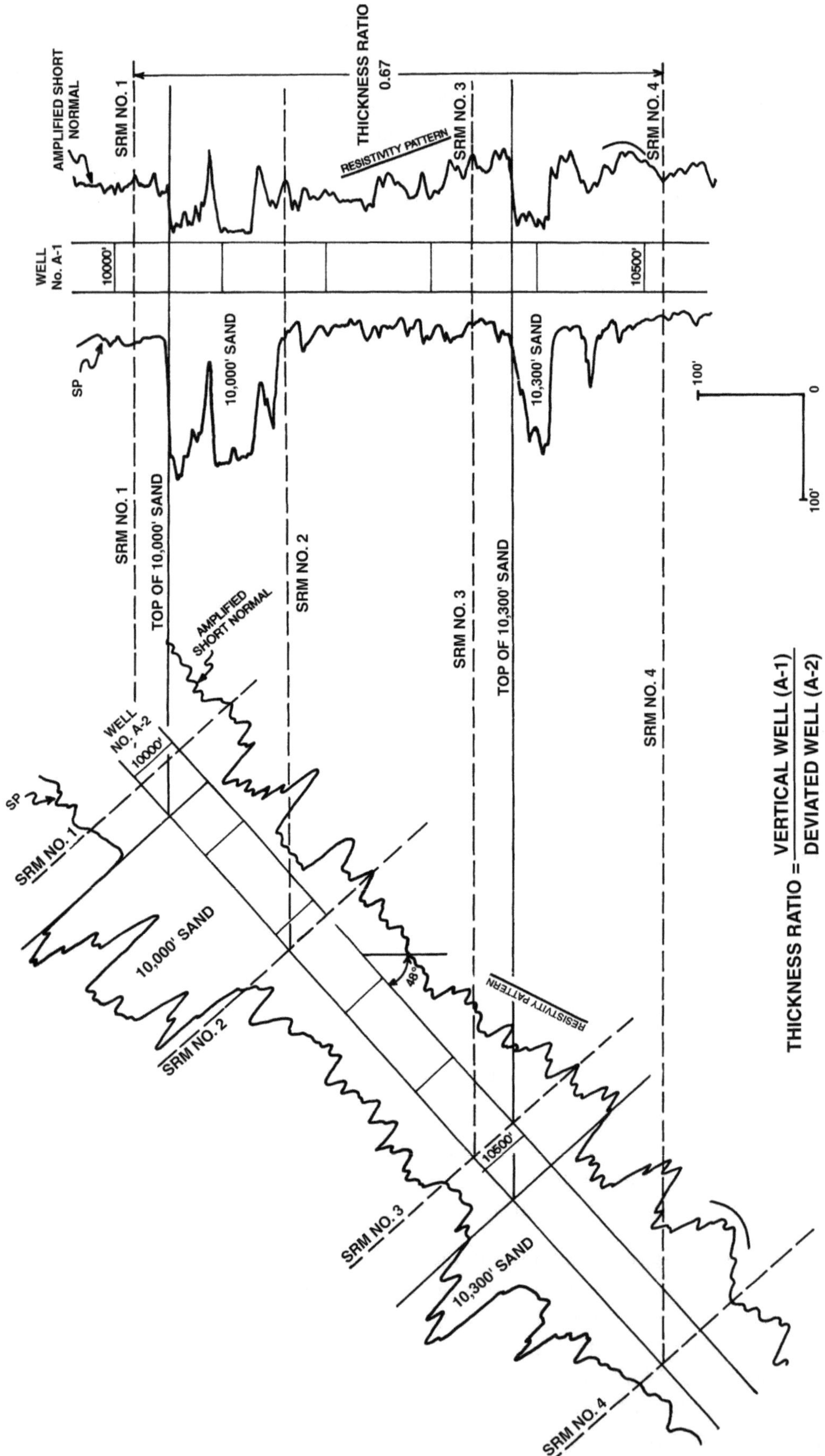

Figure 4-18 Vertical Well No. A-1 and deviated Well No. A-2 (shown in its true orientation with respect to vertical). The correlation markers show that the thicker section in Well No. A-2 is a direct result of its deviation from the vertical.

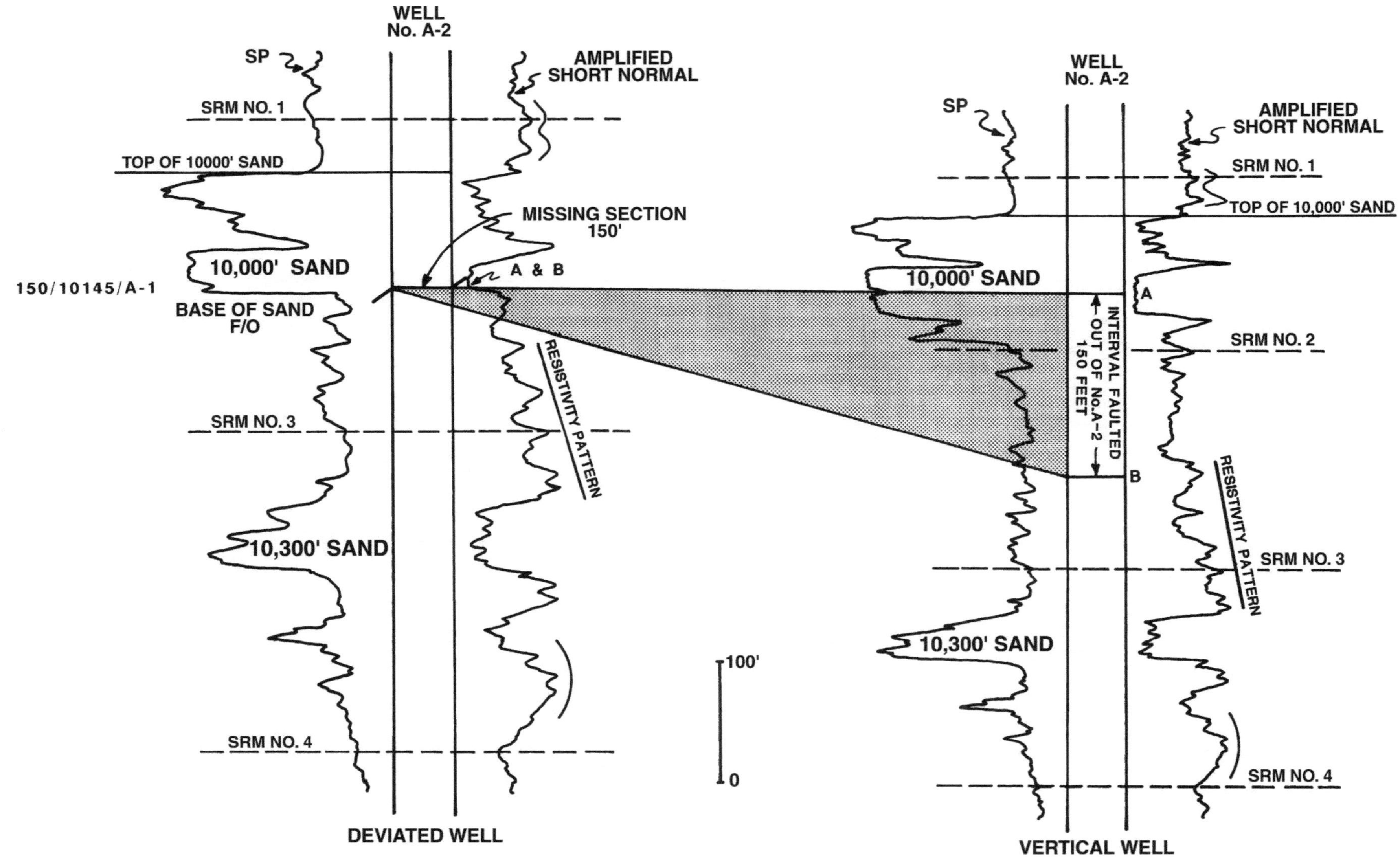

(a)

Figure 4-19(a) Detailed correlation of a deviated well with a vertical well to locate the depth and the missing section for a fault in the deviated well. The base of the 10,000-ft Sand is faulted out.

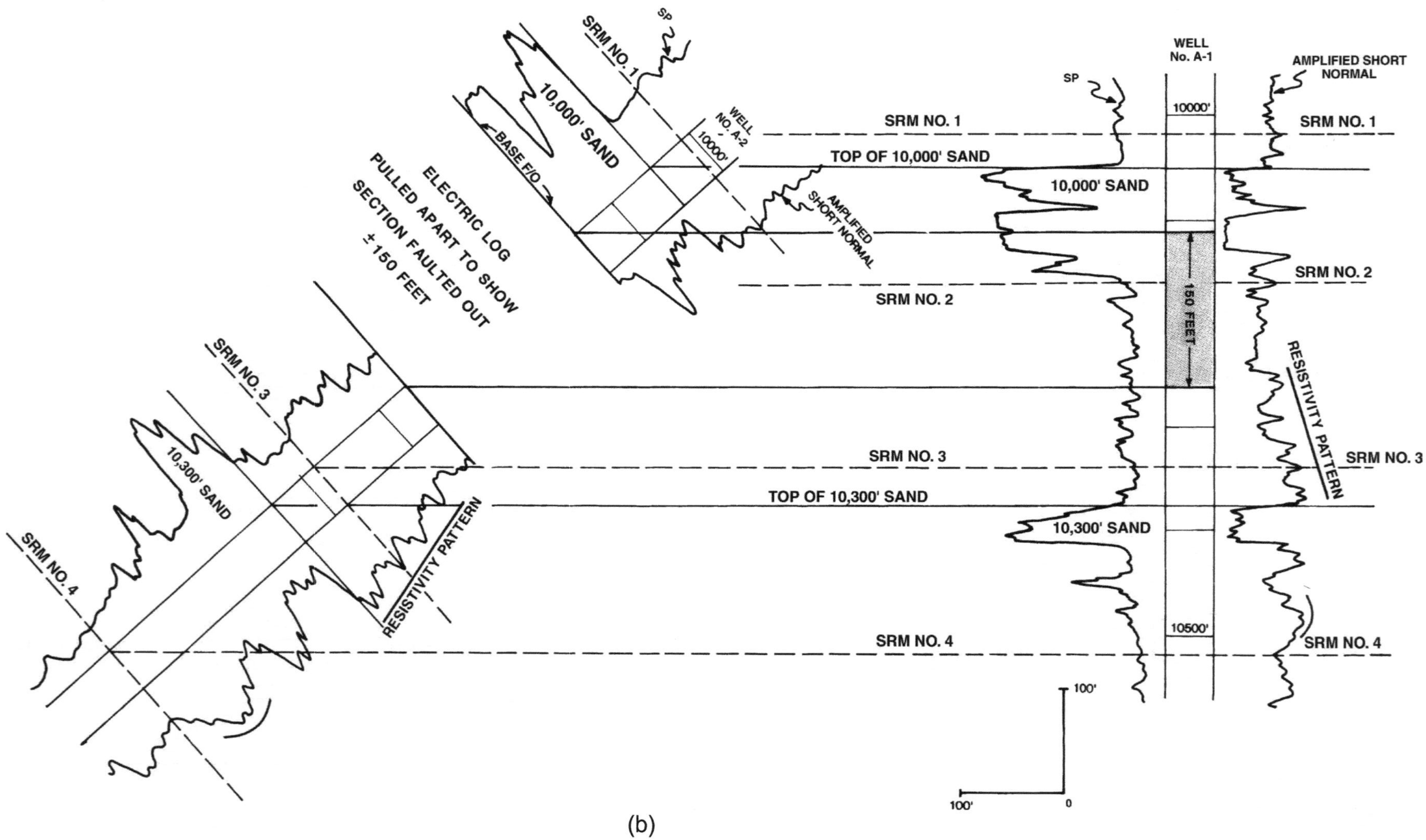

(b)

Figure 4-19(b) The simplified stratigraphic section through Wells No. A-1 and A-2 illustrates that the missing section in Well No. A-2 is equivalent to the vertical section highlighted in Well No. A-1. No thickness correction factor is required in this example.

that Well No. A-2 is faulted, and the section in Well No. A-1 that is stratigraphically equivalent to the missing section in Well No. A-2 is highlighted in Figure 4-19a–b.

The faulted out or missing section in Well No. A-2 is equal to 150 ft by correlation with Well No. A-1. Notice that the base of the 10,000-ft Sand is faulted out of Well No. A-2. This information is annotated on the log along with the amount of missing section, the depth of the fault, and the well(s) used to correlate the fault.

The missing section in directional Well No. A-2 is determined by correlation with Well No. A-1, which is a vertical well. In a vertical well, the measured log thickness and vertical thickness are the same. Since *missing section* is expressed as *the vertical thickness of the stratigraphic interval faulted out of a well,* the vertical thickness of the missing section in Well No. A-2 is 150 ft. The 150 ft represents the missing section for the fault. This information will be used in future fault and structure mapping.

Figure 4-19b is a simplified stratigraphic section showing Wells No. A-1 and A-2 positioned in their true orientation with respect to the vertical. Well No. A-2, which is deviated at 48 deg from the vertical, is pulled apart at the fault to show the restoration of the faulted-out section. This cross section clearly illustrates that the missing section in Well No. A-2 is equal to the 150 ft of vertical section highlighted in Well No. A-1.

Now consider a fault in vertical Well No. A-1 correlated with deviated Well No. A-2 (Fig. 4-20a). Well No. A-1 has a fault near the base of the 10,000-ft Sand. Detailed correlation, as shown in the figure, identifies a 225-ft section in deviated Well No. A-2 that is faulted out of Well No. A-1. The faulted-out section is highlighted in the figure. Since the missing section for the fault is determined as the TVT of the stratigraphic interval faulted out of the well, the estimate of 225 ft of missing section based on the deviated log thickness must be corrected to express the missing section in terms of TVT.

Figure 4-20b is a stratigraphic section showing Wells No. A-1 and A-2 positioned in their true orientation relative to vertical. The log section of Well No. A-1 is pulled apart at the fault to show the restoration of the faulted out section. Since we are working in an area with horizontal beds, the correction of the measured log thickness in Well No. A-2 to TVT is determined by the simple trigonometric solution of a right triangle. The insert in the center of the figure shows that the TVT of the missing section is equivalent to the vertical side of a right triangle whose hypotenuse is equal to the log thickness of the missing section in deviated Well No. A-2.

Where

$$\mathrm{TVT} = (\mathrm{MLT})(\cos\psi) \tag{4-3}$$

$$\mathrm{TVT} = \text{true vertical thickness}$$

$$\mathrm{MLT} = \text{measured log thickness in deviated well}$$

$$\psi = \text{angle of wellbore deviation from vertical}$$

Therefore,

$$\begin{aligned}\mathrm{TVT} &= (225\,\mathrm{ft})(\cos 48^\circ)\\ &= (225\,\mathrm{ft})(0.669)\\ \mathrm{TVT} &= 151\,\mathrm{ft}\end{aligned}$$

The actual (corrected) missing section for the fault in Well A-1 determined by correlation with deviated Well No. A-2 is 150 ft.

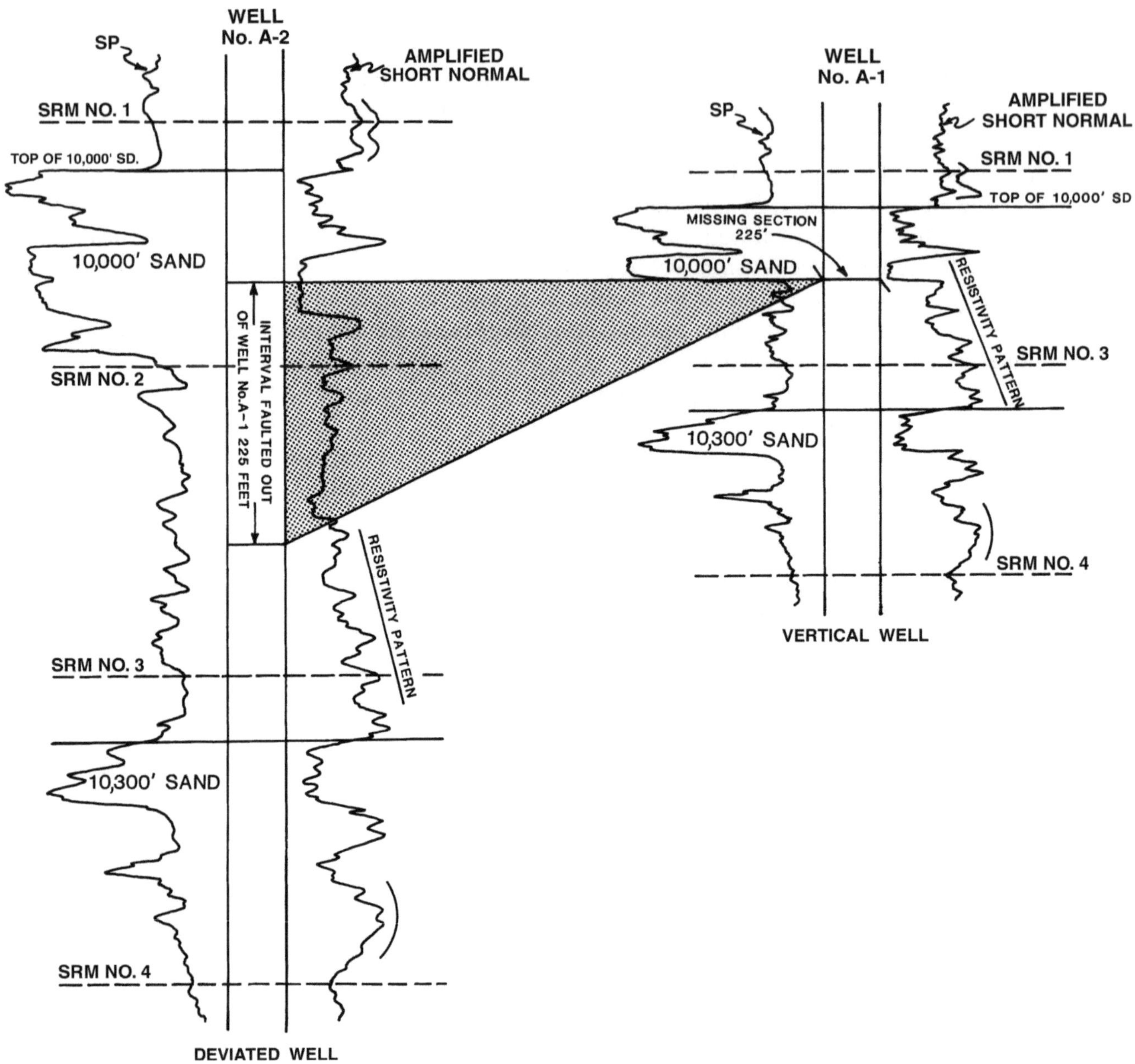

(a)

Figure 4-20(a) Detailed correlation of a vertical well with a deviated well to locate the depth and determine the missing section for a fault in the vertical well.

Dipping Beds. The procedure for correlating deviated wells in an area of significant dip is, in effect, the same as that presented thus far in this chapter. *The primary difference occurs in estimating the actual missing section resulting from a normal fault.* Since the missing section due to a fault is defined as the TVT of the stratigraphic interval faulted out of a well, any missing section value determined by correlation with a deviated well exhibiting a measured log thickness must be converted to TVT. In the previous section, we defined a simple trigonometric relationship for calculating the correction factor applicable in areas with horizontal beds. Where dipping beds are present, the mathematical correction factor becomes somewhat more complex.

There are several equations available for calculating a correction factor to convert MLT in a deviated well to TVT. We present two methods for computing the correction factor.

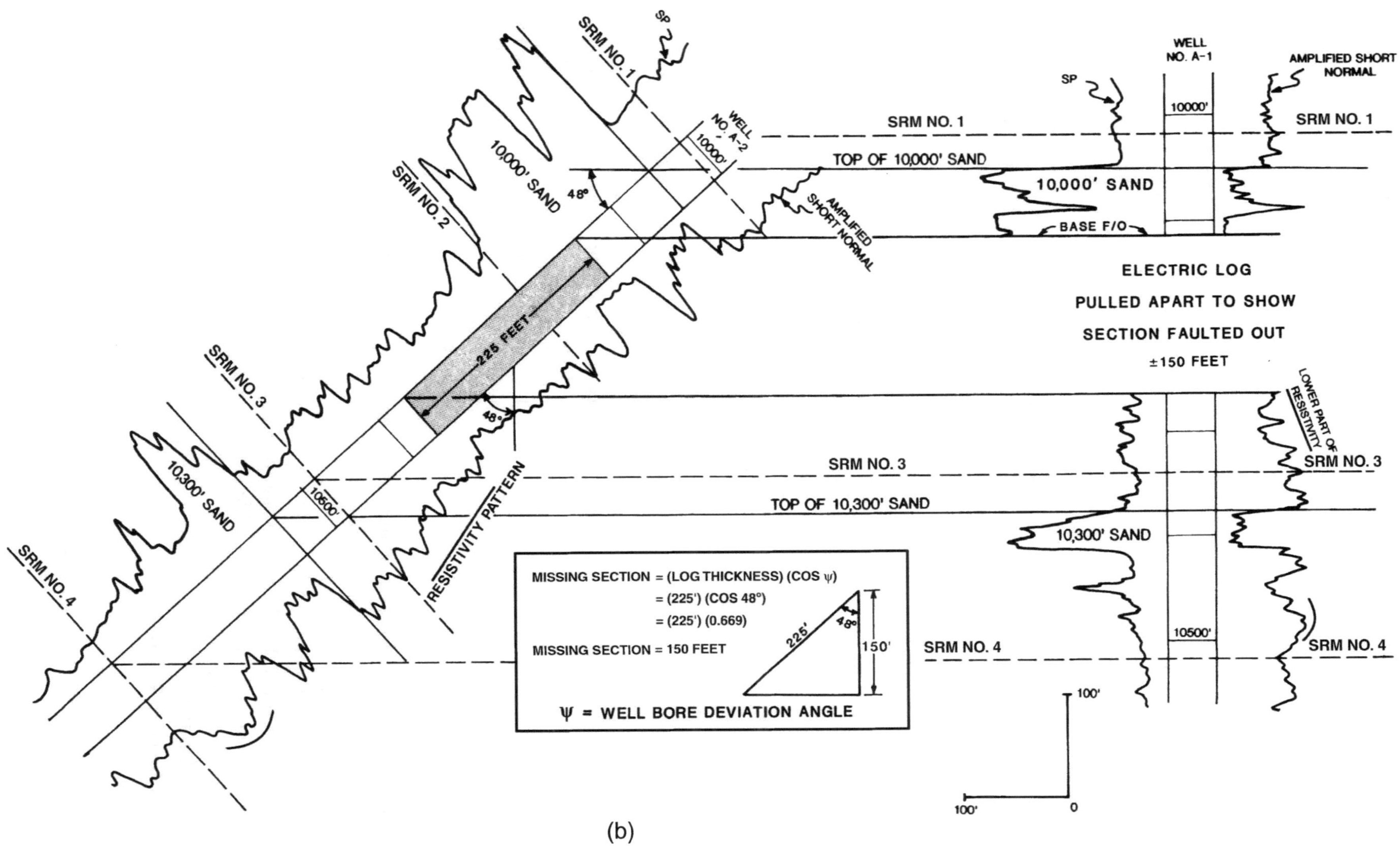

Figure 4-20(b) A simplified stratigraphic section illustrating the relationship of the missing section in vertical Well No. A-1 to the exaggerated sections seen in deviated Well No. A-2. The exaggerated section in Well No. A-2 must be corrected for wellbore deviation to determine the true amount of missing section.

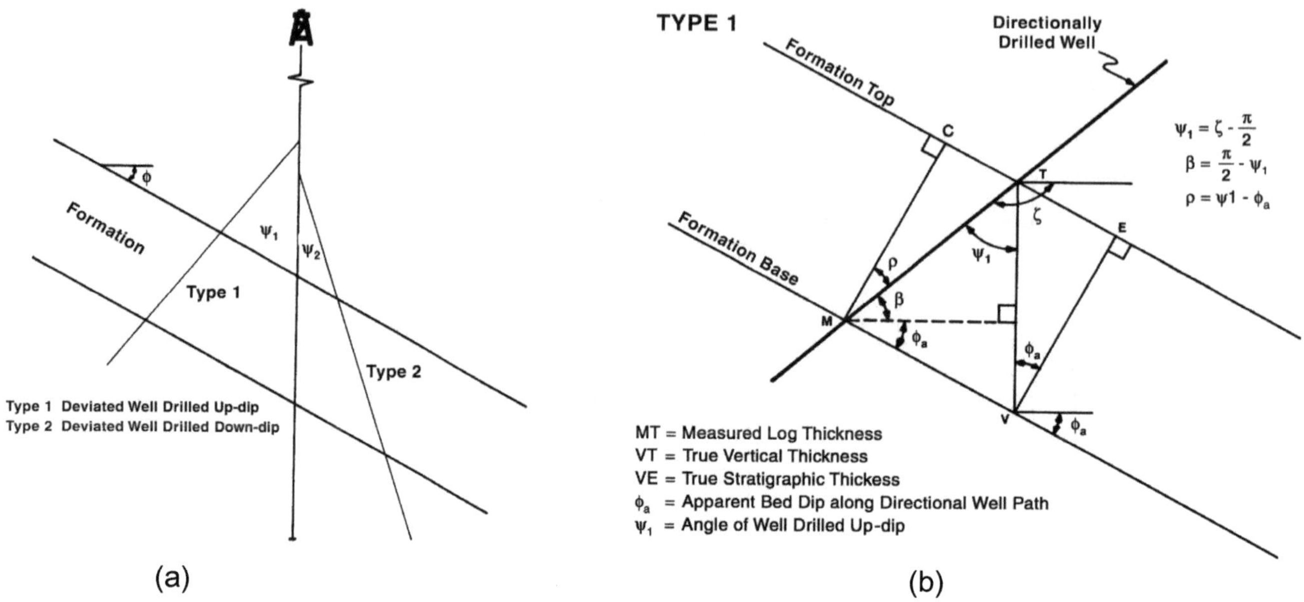

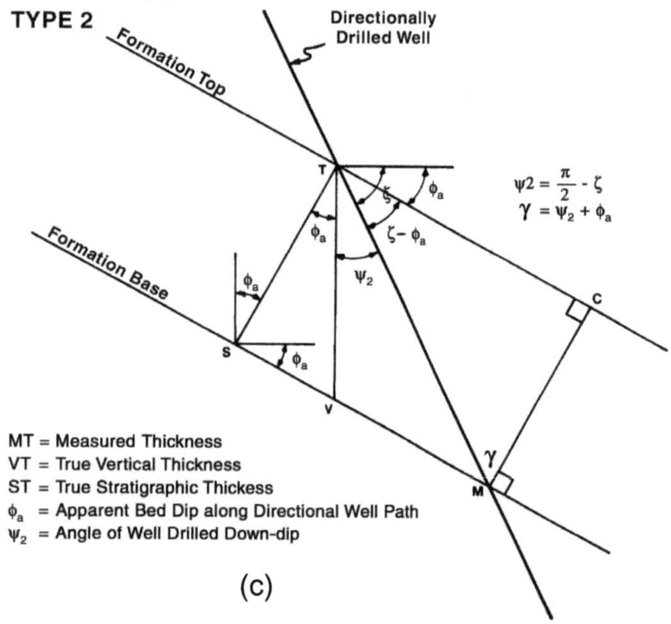

Figure 4-21 (a) Cross section illustrating the formation/wellbore deviation relationship for the Type 1 and 2 correction factors. (b) The cross section shows the detailed wellbore/formation geometry used to derive the Type 1 correction factor. (c) The cross section shows the detailed wellbore/formation geometry used to derive the Type 2 correction factor. (Published by permission of D. Tearpock and R. Bischke.)

Method 1—Two-Dimensional Correction Factor. For the 2D correction factor, there are two correction factor equations, assuming that the well is drilled at a constant inclination and azimuth over the interval in question. The first, called *Type 1*, is used where a deviated well is drilled in a direction that is general opposite to that of the bed dip. In other words, the well is deviated in the up-dip direction (Fig. 4-21a). The bed dip would be an *apparent* dip in most cases. The second, called *Type 2*, is used where a deviated well is drilled in the same general direction as that of the bed dip; the well is deviated in the down-dip direction (Fig. 4-21a). Again, the bed dip is typically apparent dip.

Type 1 Equation. The derivation of the Type 1 equation (well deviated up-dip) is shown here and illustrated in Figure 4-21b.

$$\cos\rho = MC/MT \quad (1a) \qquad \cos\phi_a = VE/VT \quad (2a)$$

$$MC = MT\cos\rho \quad (1b) \qquad VE = VT\cos\phi_a \quad (2b)$$

$$MC = VE$$

Therefore, equating 1b and 2b,

$$\mathrm{MT}\cos\rho = \mathrm{VT}\cos\phi_a$$

Rearranging,

$$\mathrm{VT} = \frac{\mathrm{MT}\cos\rho}{\cos\phi_a}$$

Substituting $\rho = \psi - \phi_a$ and relabeling VT as TVT, and MT as MLT,

$$\mathrm{TVT} = \mathrm{MLT}\frac{\cos(\psi_1 + \phi_a)}{\cos\phi_a} \qquad (4\text{-}4)$$

Type 2 Equation. The derivation of Type 2 (well deviated down-dip) is shown here and illustrated in Figure 4-21c.

$$\cos\phi_a = \mathrm{ST/VT} \quad (1a) \quad \cos\gamma = \mathrm{MC}/\mathrm{MT} \quad (2a)$$

$$\mathrm{ST} = \mathrm{VT}\cos\phi_a \quad (1b) \quad \mathrm{MC} = \mathrm{MT}\cos\gamma \quad (2b)$$

$$\mathrm{MC} = \mathrm{ST}$$

Therefore, using 1b and 2b and relabeling VT as TVT, and MT as MLT,

$$\mathrm{TVT}\cos\phi_a = \mathrm{MLT}\cos\gamma$$

Rearranging,

$$\mathrm{TVT} = \mathrm{MLT}\frac{\cos\gamma}{\cos\phi_a}$$

Substituting $\gamma = \psi_2 + \phi_a$

$$\mathrm{TVT} = \mathrm{MLT}\frac{\cos(\psi_2 + \phi_a)}{\cos\phi_a} \qquad (4.5)$$

With Eqs. (4-4) or (4-5), the data required to calculate the correction factor are (1) ϕ_a = wellbore deviation from the vertical, (2) ϕ_a = apparent bed dip (bed dip in the direction of wellbore deviation), and (3) MLT = measured log thickness in the deviated well. The apparent bed dip is the most difficult data to obtain for these equations. The only source of apparent bed dip is from an existing structure map. If an existing structure map is available the apparent bed dip along the well path can easily be calculated using the rise/run equation discussed in Chapter 2 (Eq. 2-2).

Method 2—Three-Dimensional Correction Factor. In this section, we present an exact 3D correction factor equation. A version of the equation was first presented by J. Setchell (1958) and has been used successfully for more than 60 years. We consider this 3D correction factor equation preferable because this equation can be used to calculate the correction factor regardless of the direction of wellbore deviation, and the true dip of the beds is used instead of the apparent dip, which is used in the 2D equations.

To derive the general 3D equation, we introduce a 3D spherical coordinate system (Fig. 4-22) with one horizontal axis in the direction of dip (called the x-axis), one axis perpendicular to the first and also horizontal (called the y-axis), and one axis perpendicular to the other two (called the z-axis). In this equation we assume that the well is straight within the interval of interest (i.e., the well has both a constant deviation and azimuth over this interval). If the well has a significant change in deviation and/or azimuth over the bed of interest, you can calculate the straight-line distance, inclination and azimuth from the point of entry into a unit to the point of exit from the unit and use these values in Eq. (4.6) rather than measured log thickness and information from the well directional survey. We also assume that the top and base of the bed are parallel to one another. The case where the top and the base of the bed is not parallel is discussed in Chapter 14. The origin is the point (T) at which the wellbore first penetrates the bed.

Where

MLT = measured log thickness

TVT = true vertical thickness

ϕ = dip of bed

α = Δ azimuth (minimum angle, well azimuth to bed dip)

ψ = wellbore drift angle

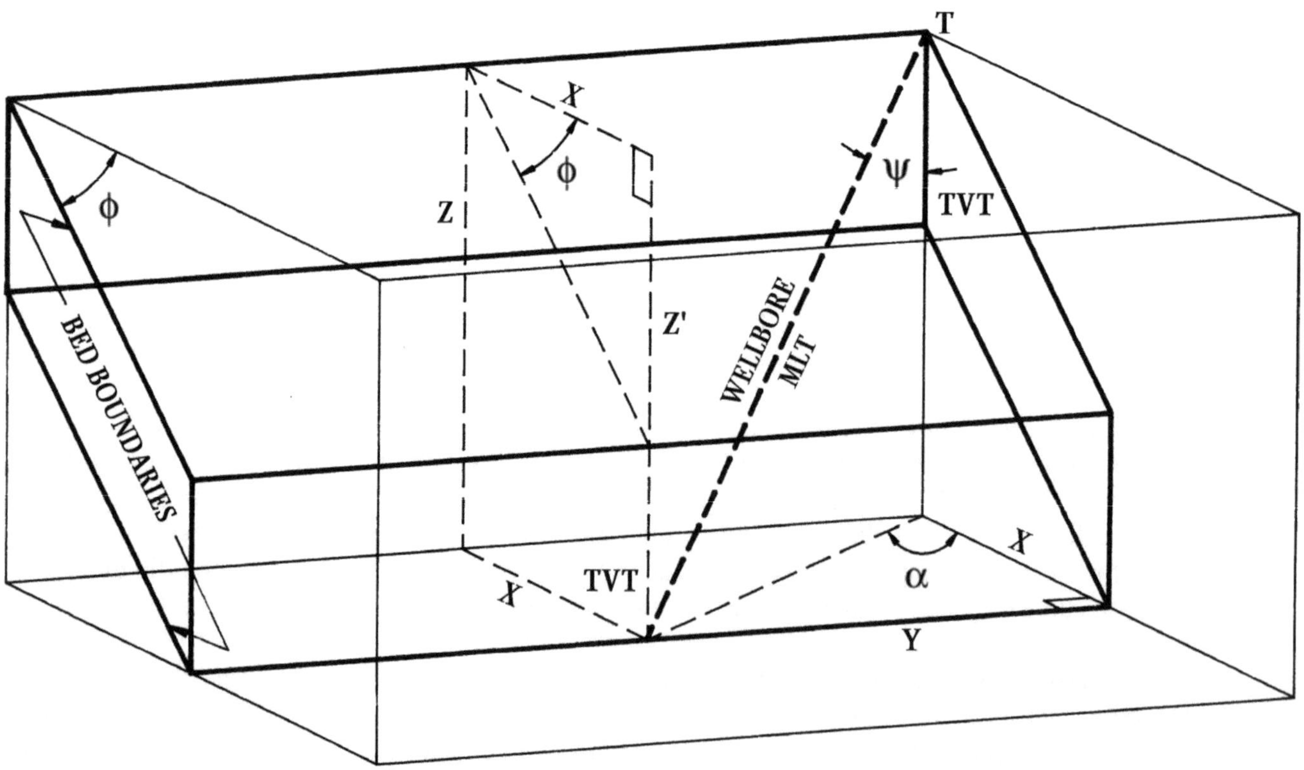

Figure 4-22 A 3D spherical coordinate system with one horizontal axis in the direction of bed dip (x-axis), one axis perpendicular to the first and also horizontal (y-axis), and one axis perpendicular to the other two (z-axis). (Published by permission of D. Tearpock.)

Spherical coordinate system equations:	Trigonometric definition of Z:
$X = MLT \sin\psi \cos\alpha$	$Z' = Z' + TVT\, Z = X \tan\phi$
$Z = MLT \cos\psi$	$Z = X \tan\phi + TVT$

Substituting the spherical coordinate definition for Z, X, and the alternate trigonometric definition of Z easily yields Eq. (4-6).

$$\begin{aligned} Z &= MLT \cos\psi \\ X \tan\phi + TVT &= MLT \cos\psi \\ TVT &= MLT \cos\psi - X \tan\phi \\ TVT &= MLT \cos\psi - MLT \sin\psi \cos\alpha \tan\phi \\ &\text{or} \\ TVT &= MLT[\cos\psi - (\sin\psi \cos\alpha \tan\phi)] \end{aligned} \tag{4.6}$$

Application of Methods 1 and 2. Now apply these methods using the data shown in Figure 4-23. Wells No. A-1, A-2, and A-3 are drilled from a single location and penetrate a section that includes Bed A, which is of constant thickness and dips at 35 deg due east. Well No. A-1, a vertical well, cuts a fault that completely faults out Bed A. Well No. A-2 is drilled in a down-dip direction with an average deviation angle of 36 deg through Bed A. Well No. A-3 is drilled directly up-dip with a deviation angle of 35 deg through Bed A. For simplicity, the wells are all assumed to be drilled in a vertical plane parallel to the dip of Bed A.

Through detailed correlation of the three wells, a fault is identified in Well No. A-1. Based on the correlations, Bed A is completely faulted out of the well. Assuming that Wells No. A-2 and A-3 are the only wells available for correlation, the amount of missing section for the fault in Well No. A-1 must be estimated from these deviated wells.

Considering again that missing section is defined as the true vertical thickness of the stratigraphic interval faulted out of a well, and the fault in Well No. A-1 completely faults out Bed A, the missing section resulting from the fault in Well No. A-1 is equal to the true vertical thickness of Bed A. A review of Figure 4-23 shows that Bed A has a true vertical thickness of 200 ft. When we correlate Well No. A-1 with the two deviated wells, however, we obtain a missing section in terms of deviated log thickness rather than true vertical thickness. Therefore, these deviated log thicknesses must be converted to true vertical thickness to estimate the amount of missing section. The missing section (Bed A) in Well No. A-1 has a logged thickness in Well No. A-2 of 504 ft Zand a logged thickness in Well No. A-3 of 164 ft. The logged thickness of 504 ft in Well No. A-2 is over two and one-half times greater than the TVT. The logged thickness of 164 ft in Well No. A-3 is less than the TVT (0.82).

By log correlation, the amount of missing section for the fault in Well No. A-1 ranges from 164 ft, based on correlation with Well No. A-3, to 504 ft, by correlation with Well No. A-2. What is the actual missing section for the fault? In order to determine the actual missing section, the log thickness measured in Wells No. A-2 and A-3 must be corrected to true vertical thickness.

TVT Calculation Using Method 1

Type 1: Well Drilled Up-Dip. In order to use Eq. (4-4) to calculate the correction factor, three pieces of data are required: (1) the wellbore deviation angle (ψ_1), which can be obtained from the directional survey of Well No. A-3, (2) the formation dip (ϕ_a), which is true dip in this case, obtained from a dip meter log or a completed structure map, and (3) the MLT in Well No. A-3 that is equivalent to the section missing in Well A-1.

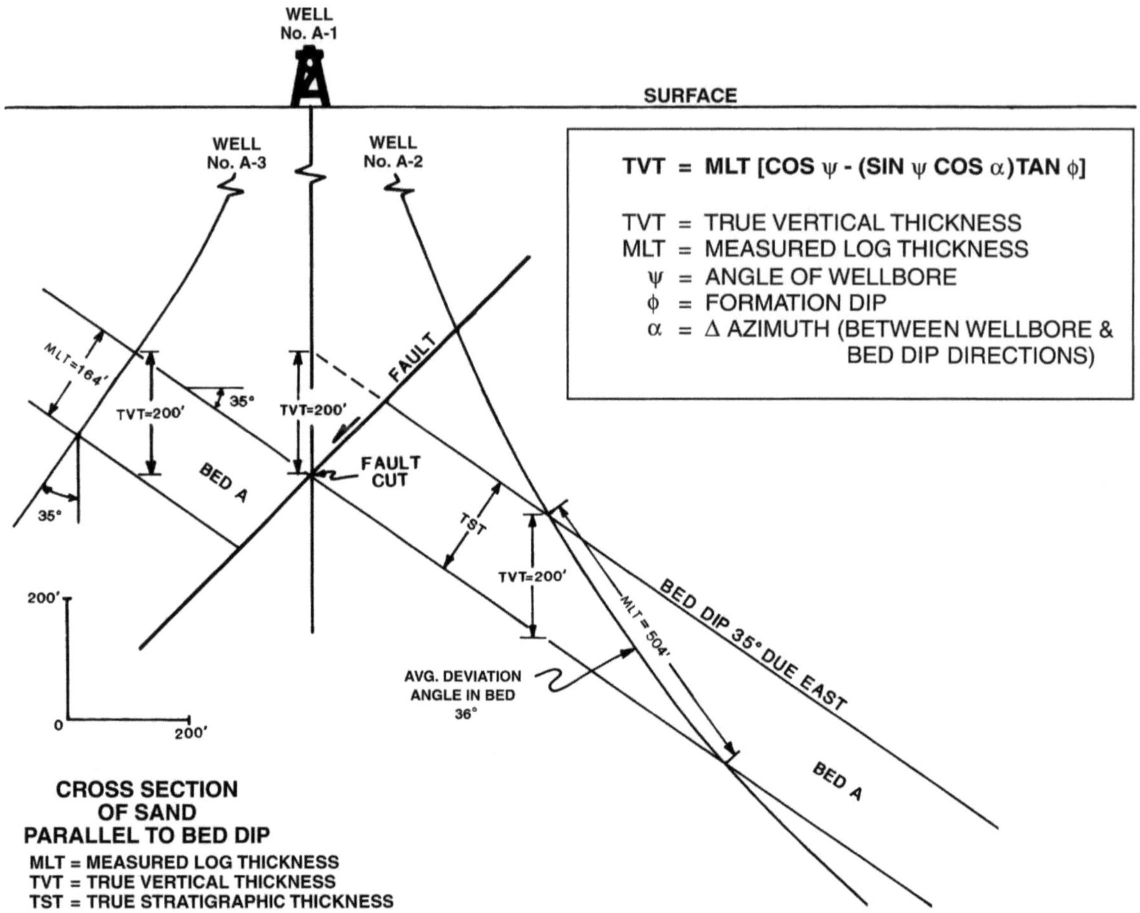

Figure 4-23 One vertical and two deviated wells penetrate Bed A. In Well No. A-1, Bed A is faulted out. The cross section shows the relationship of the missing section in Well No. A-1 to the MLT of the equivalent interval in deviated Wells No. A-2 and A-3. The equation shown is used to correct MLT to TVT.

Data:

$$\psi_1 = 35\deg$$
$$\phi_a = 35\deg$$
$$MLT = 164\,ft$$
$$TVT = MLT\frac{\cos(\psi_1 + \phi_a)}{\cos\phi_a}$$
$$= 164'\frac{\cos 0^\circ}{\cos 35^\circ}$$
$$= 164'\frac{1}{0.8192}$$
$$= 164'(1.2207)$$
$$\mathbf{TVT = 200\,ft}$$

Type 2: Well Drilled Down-Dip. In order to use Eq. (4-5), the same parameters used in Eq. (4-4) are required.

Data:

$$\psi_1 = 36\,\text{deg}$$
$$\phi_a = 35\,\text{deg}$$
$$\text{MLT} = 504\,\text{ft}$$

$$\begin{aligned}
\text{TVT} &= \text{MLT}\frac{\cos(\psi_1 + \phi_a)}{\cos\phi_a} \\
&= 504'\frac{\cos(36° + 35°)}{\cos 35°} \\
&= 504'\frac{\cos 71°}{\cos 35°} \\
&= 504'\frac{0.3256}{0.8192} \\
&= 504'(0.3975)
\end{aligned}$$

$$\mathbf{TVT = 200\,ft}$$

Since Wells No. A-2 and A-3 are drilled directly down-dip and up-dip respectively, the value for bed dip in Eqs. (4-4) and (4-5) is equal to the true bed dip (ϕ). This can be obtained from a structure map or a dipmeter if available. If the direction of wellbore deviation is not parallel to bed dip, however, then an apparent bed dip in the direction of wellbore deviation must be determined. This apparent bed dip cannot come from a dipmeter log because this log calculates true bed dip. This is one of the main drawbacks to the 2D equations. Apparent bed dip can be derived from a structure map by using Eq. 2-2 where rise/run is measured along the path of the deviated well and not perpendicular to bed strike.

TVT Calculation Using Method 2 The required data to solve the general 3D equation are (1) wellbore deviation angle (ψ) obtained from a directional survey, (2) wellbore deviation azimuth (α_w) obtained from a directional survey, (3) true bed dip (ϕ) measured from a completed structure map or dipmeter if run in the wellbore, (4) bed dip azimuth (α_a) measured from a completed structure map or obtained from a dipmeter, and (5) MLT that is equivalent to the missing section.

TVT for Well No. A-2:

Data:

$$\psi = 36\,\text{deg}$$
$$\alpha_w = 90\,\text{deg}$$
$$\phi = 35\,\text{deg}$$

$$\text{MLT} = 504\,\text{ft}$$

$$\alpha = 0\,\text{deg}$$

$$\begin{aligned}
\text{TVT} &= \text{MLT}\,[\cos\psi - (\sin\psi\cos\alpha\tan\phi)] \\
&= 504'\,[\cos 36° - (\sin 36° \cos 0° \tan 35°)] \\
&= 504'\,[0.809 - (0.5878)(1)(0.70)] \\
&= 504'\,[0.809 - 412]
\end{aligned}$$

$$\text{TVT} = 504'\,[0.397]$$
$$\mathbf{TVT = 200\,ft}$$

TVT for Well No. A-3: The data for this calculation are exactly the same as for Well No. A-2, with two exceptions. The azimuth (α_w) for Well No. A-3 is due west, or 270 deg, and therefore the Δ azimuth (α) is 180 deg and ψ is 35 deg.

$$= 164'\left[\cos 35° - \left(\sin 35° \cos 180° \tan 35°\right)\right]$$

$$= 164'[0.8192 - (0.5736)(-1)(0.7002)]$$

$$= 164'[0.8192 - (-0.4016)]$$

$$= 164'[0.8192 + 0.4016]$$

$$\text{TVT} = 164'[1.2208]$$

$$\mathbf{TVT = 200\,ft}$$

Through the use of the 2D and 3D equations, we have successfully calculated the amount of missing section for the fault in Well No. A-1. Notice the close agreement between the different equations. The missing section estimated at 200 ft can now be used for all future fault surface mapping and structure map integration as well as other geological work, such as cross-fault drainage analysis. The actual procedure for integrating a fault and structure map is detailed in Chapter 8. We have now defined two methods for obtaining the correct values for missing section when correlating with deviated wells. *Anytime you are correlating in an area with deviated wells and dipping beds, missing and repeated section are defined in terms of TVT.* Significant errors in structure maps, net pay maps, and even proposed well locations can occur if these corrections are not made or if missing section is not understood in terms of vertical separation (see Chapter 7).

MLT, TVDT, TVT, AND TST

The thickness of any given interval on a log is referred to as the *measured log thickness* (MLT). In a vertical well, the MLT for any given interval is equal to the TVT of the interval. We know from the previous discussion in this chapter, however, that the MLT in a directionally drilled well is normally not equal to TVT because of the wellbore deviation and to bed dip in areas of dipping beds. However, remember that MLT will approximate TVT in the near-vertical part(s) of a directional well.

True vertical depth thickness (TVDT) is defined as the MLT between two specific points in a deviated well, corrected only for wellbore deviation. The *true vertical thickness* (TVT) is defined as the thickness of an interval measured in the vertical direction. It is the thickness seen in a vertical well. For a directionally drilled well, the TVT can be calculated using the equations introduced in the previous section. The *true stratigraphic thickness* (TST) is defined as the thickness of a given interval measured at a right angle to the bedding surface in a vertical cross section. It can be calculated by multiplying the TVT by the cosine of bed dip.

These various thicknesses are graphically illustrated in Figure 4-24. In Figure 4-24a, a well deviated up-dip at an angle of 50 deg penetrates strata dipping 35 deg due west. The MLT of the strata is 127 ft. To correct the MLT to TVDT, the MLT (127 ft) is multiplied by the cosine of the wellbore deviation angle (50 deg). The resultant TVDT is 82 ft.

Figure 4-24b shows the same strata penetrated by a well drilled down-dip at an angle of 40 deg. The MLT of the strata in the well is 476 ft. Corrected to TVDT, it is 357 ft. The correction factor equations developed in the previous section are used to calculate a TVT of 150 ft for the penetrated strata in Figure 4-24b. The TST of the strata calculated by multiplying the TVT (150 ft) by the cosine of the bed dip (35 deg) is 123 ft.

Notice that for the well drilled down-dip, the MLT is 3.17 times greater than the TVT, and the TVDT is 2.38 times greater than the TVT. For the well drilled up-dip, the MLT is less than the TVT, and the TVDT is about one-half the TVT.

The understanding of these various measurements is very important in log correlation and the determination of the value for the missing section. Very often, when a well is directionally drilled, a TVD log is automatically prepared as part of the logging program. The TVD log is then used to aid in correlation, determine missing or repeated section resulting from a fault, and

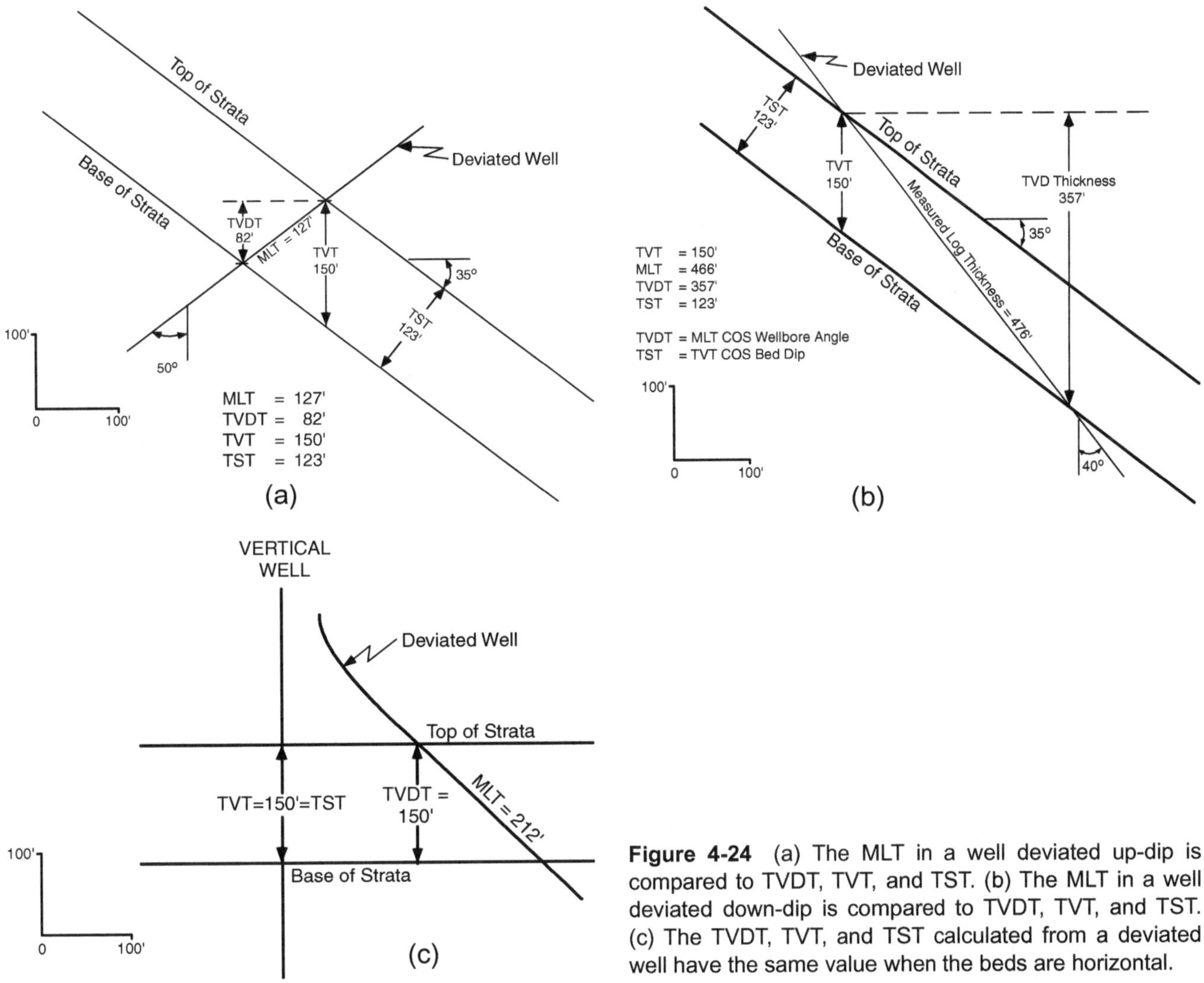

Figure 4-24 (a) The MLT in a well deviated up-dip is compared to TVDT, TVT, and TST. (b) The MLT in a well deviated down-dip is compared to TVDT, TVT, and TST. (c) The TVDT, TVT, and TST calculated from a deviated well have the same value when the beds are horizontal.

count net sand and net pay. Figure 4-24 illustrates that in areas of significant dip, a TVD log of a well not drilled along strike may provide little, if any, advantage over the deviated well log for correlation and can actually complicate the estimation of missing section. Observe in Figure 4-24b that the TVDT, which is the thickness seen in a TVD log, is still 2.38 times greater than the TVT.

Consider a fault in this section (Fig. 4-24b) that exactly faults the entire unit out of the well. The amount of missing section in that well would be 150 ft. By correlation with the well deviated down-dip, the missing section would be estimated to be 476 ft; with a TVD log, the estimated missing section for the fault would be 357 ft. We can conclude that the TVD log could provide a considerable error in the estimation of the amount of missing section. If the data are available, we recommend that a TVT log be prepared. This log can be used to aid in correlation and to estimate the amount of missing section due to a fault, since it represents the TVT of the interval logged.

The preparation of a TVD log for the well deviated up-dip in Figure 4-24a could actually result in additional correlation problems. Notice that the measured thickness of the strata in the deviated well is 127 ft. Converting this MLT to a TVD log thickness actually reduces the thickness of the interval to 82 ft. This reduced thickness could be mistakenly interpreted as stratigraphic thinning. If the TVD log were used to estimate the missing section for the fault, it would result in an underestimate. For example, if we again use a fault in this section with a missing section of 150 ft, then by correlation with a TVD log for the well drilled up-dip, the missing section for the fault would be estimated at 82 ft, or nearly one-half the actual size.

In areas of horizontal or nearly horizontal beds, TVDT is equal to or nearly equal to TVT (Fig. 4-24c) and can be of significant help in correlation and estimating the value for missing section. *In areas of dipping beds, however, a TVD log may provide little help and can actually cause additional correlation problems that may result in mapping errors.*

ELECTRIC LOG CORRELATION—HORIZONTAL WELLS

Horizontal wells require some additional techniques in log correlation. Ordinary methods for correlating deviated wells can be used before the wellbore reaches an angle close to the bed dip angle. However, where the borehole angle is very close to the bed dip, the log curves are extremely stretched compared to a vertical well log. Also, the response of the logging tool is affected by its eccentric position in the borehole and lack of symmetry of the beds around the borehole. In vertical wells a logging tool is more or less in the center of the hole and the bed boundaries are near horizontal or at some reasonable angle to the borehole, so the tool is reading the same lithology around the borehole at a given depth. In horizontal boreholes the bed boundaries are nearly parallel to the borehole, so it is possible to have sand on one side of the hole and shale on the other. The tool would then be reading some averaged value for the two lithologies. Next, we briefly discuss some of the basic methods of recognizing bed boundaries and correlating horizontal wells.

Direct Detection of Bed Boundaries

When a logging tool crosses a bed boundary, there is some change in the value recorded on the log. In vertical wells this change is fairly abrupt, so specific log curve shapes can be matched from well to well. In horizontal wells these curve shapes are stretched out due to the low angle between borehole and bed boundary. For example, a gamma ray curve might change from indicating shale to indicating sand within a foot or two in a vertical well; but in a horizontal well this change may be stretched out over tens of feet or more as the borehole gradually crosses from shale into sand. Although the gamma ray log is directly indicating the bed boundary, the stretched curve may be difficult to correlate without rescaling or shrinking its length. We discuss methods for doing this in the sections on TVD and TSD methods.

Electromagnetic propagation resistivity logging tools, such as those used in logging-while-drilling (LWD) systems, record spikes on the log curve at boundaries where significant changes in resistivity occur. Sand-shale bed boundaries, limestone-shale bed boundaries, and hydrocarbon-water contacts are examples of where these spikes would occur. These types of boundaries would be directly detectable with this type of tool, and the spikes could be directly correlated to the equivalent boundary in a nearby vertical well. Fagin et al. (1991) give an example of recognizing an oil/water contact using this method, as well as modeling log response to the proximity of an oil/water contact before it is penetrated.

Modeling Log Response of Bed Boundaries and Fluid Contacts

Logging service companies can compute the response of their tools to model stratigraphic sections and proximity to fluid contacts for horizontal boreholes. Using a nearby vertical well for control, resistivities and thicknesses are assigned to the model section. The response of the tool to the vertical model is computed and compared to the actual log from the vertical well. If it is a good match, then a set of log curves are computed for multiple angles of penetration of the stratigraphic section (i.e., various apparent bed dip angles to the borehole orientation). These model curves are then compared to the log from a horizontal borehole. If a match is found, this correlation to the vertical well determines the position of the horizontal borehole in the stratigraphic section or its proximity to a fluid contact. Leake and Shray (1991) give an LWD example using the technique to maintain a horizontal well within a 40-ft sand. This approach is well illustrated by Amer et al. (2013).

True Vertical Depth Cross Section

The previous methods are widely used to recognize specific bed boundaries and fluid contacts. But how can a geologist correlate a long section of a horizontal well log to a log from a vertical well? The geologist can build a cross section to aid in making the correlations using a vertical section plot from the directional survey and a TVD log from the horizontal well. If the TVD of the horizontal borehole stays constant for long sections of the hole, a measured depth log may be used with the TVD and vertical section (through the wellbore from the surface location to total depth) for correlation points posted on the log. The cross section can help you determine whether your correlation picks are at least geometrically consistent. TVD logs have highly distorted thicknesses in horizontal wellbores, but if the target zone is thick and you have been able to correlate the log as it approaches horizontal, you may be able to recognize the target zone bed boundary. Typically, the gamma ray and resistivity logs are used for TVD log correlation. However, if multiple bed boundaries are crossed or the borehole goes in and out of the target zone, a satisfactory correlation of a TVD log is unlikely. A TVT log would more accurately represent the thickness seen in the vertical well with which you are correlating, but the correction factor or reduction in length can be extreme. Most computer-based log correlation software can construct and display a TVT log if the bed dip is known.

The TVD cross section can be constructed with standard cross-section methods and a few additional considerations.

1. **1.** Plot the vertical section of the well on cross-section paper with a suitable vertical exaggeration (e.g., 10:1).
2. **2.** Plot the bed boundaries for the part of the well as it approaches the horizontal segment of the well and estimate the vertical thickness to the top of the objective zone using a nearby vertical well. To determine the top of the objective zone, a pilot hole may be drilled vertically to the objective stratigraphic section before drilling the horizontal part of the hole.
3. **3.** Estimate the apparent bed dip in the plane of the cross section from maps, seismic data, or correlations among nearby vertical wells. Draw the bed boundaries on the cross section and try to pick the top of the objective zone where the projected top intersects the plotted borehole.
4. **4.** Adjust the dip of the bed boundary if your correlation indicates a different penetration point.
5. **5.** Continue projecting the beds and thicknesses along the cross section and look for correlations where the wellbore would be intersecting the plotted bed boundary. Note that if the wellbore goes up through a stratigraphic section, the log curve will be inverted (Fig. 4-25).
6. **6.** If the wellbore cuts a fault, use the TVT in the vertical well to estimate the missing section for the fault. That is, if your correlation shows the wellbore is 10 ft below the top of the target zone before crossing the fault and 20 ft above the top after crossing the fault, then the fault is a 30-ft fault. This is basically the estimated restored tops method described later in this chapter.

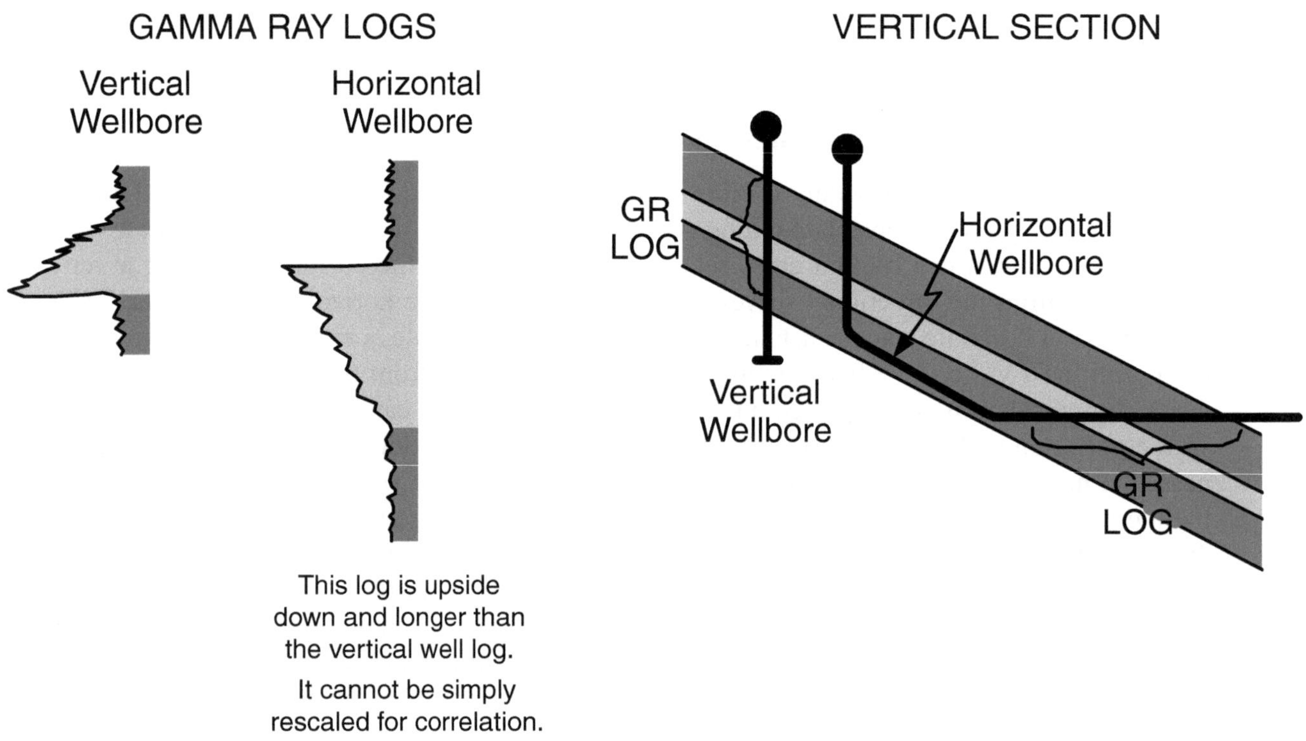

Figure 4-25 A gamma ray log in a wellbore that goes up stratigraphic section is upside down compared to a log from a vertical well. (From Tearpock 1999. Published by permission of Anadarko Petroleum Corporation.)

In general, the TVD cross section requires that bed boundaries are recognizable on a measured depth log recorded in the horizontal borehole or at least on the computed TVD log. Modeling of log response is often used to recognize the bed boundaries, and a cross section is constructed to illustrate the correlations. Singer (1992) gives an example of using modeling and illustrating the correlations with a cross section. Amer et al. (2013) also illustrate matching the LWD measurements in a horizontal well with modeled results from a nearby vertical well in order to correlate the horizontal well. This correlation work is most easily done on a computer.

True Stratigraphic Depth Method

Kyte et al. (1994) describe a method for correlating logs and building cross sections from horizontal wells. They call it the true stratigraphic depth, or TSD, method. Figure 4-26 illustrates the relationship of TVD, TVT, TST, and TSD. TVD is the actual vertical depth of a point in the wellbore or on the log. TSD is the stratigraphic position of a point in the wellbore or on the log. So any place on the same bed boundary would be considered to have the same TSD, and therefore, these points would correlate back to the same stratigraphic point in the vertical well. The idea is to use a computer program to rescale the log (usually the GR log) from the horizontal

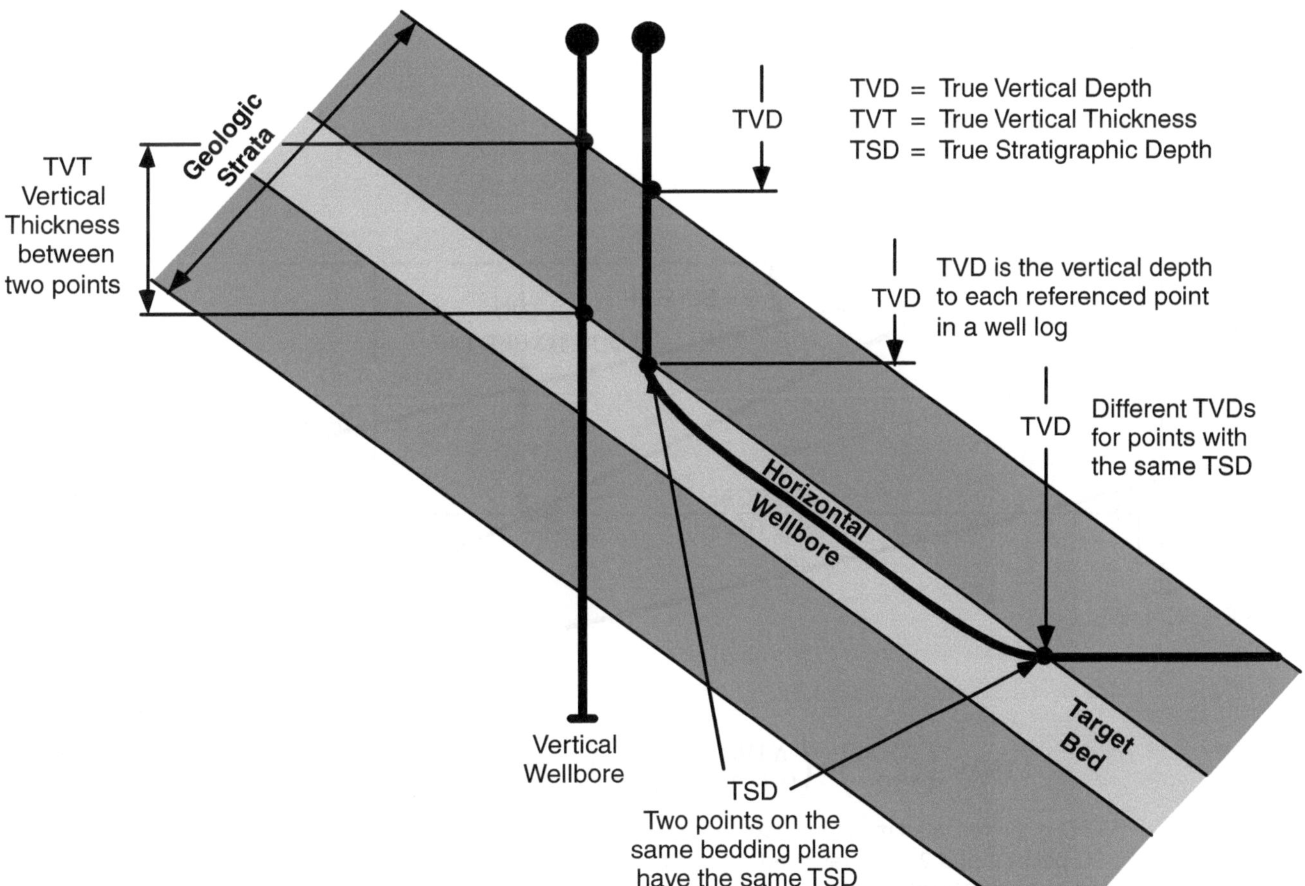

Figure 4-26 A schematic vertical section along a horizontal wellbore with vertically exaggerated bed dip illustrating TVD, TVT, and TSD. Points in a wellbore that are in the same stratigraphic position have the same TSD. (From Tearpock 1999. Published by permission of Anadarko Petroleum Corporation.)

well for correlation with a nearby vertical well. One such program, developed by Union Pacific Resources Corporation, is called GRNAV. The technique uses the TVDs for a log from the horizontal well and an estimated change in TVD due to bed dip for the bed or correlation point. This allows an estimate of the stratigraphic position of a point on the log compared to a nearby vertical well. To make the correlation, a part of the log curve is rescaled based on this estimate and compared to the vertical well. The next segment along the wellbore is then correlated the same way until the whole log is correlated. In practice, the method is computer-based to correlate as the well is being drilled to provide real-time geological information for steering and maintaining the wellbore in the target zone.

Figure 4-27 illustrates the method for determining the rescaled log length or thickness for correlating a stratigraphic interval between two points in a borehole going down section. TSD point 1 has already been correlated with the vertical well. TSD point 2 is a point in the wellbore selected by the interpreter. From TSD point 1, the bed dip is plotted and a ΔTVD for TSD point 1 is calculated at a point vertically above TSD point 2. Subtracting this ΔTVD value from the total TVD difference between the two TSD points leaves ΔTSD, which will be the rescaled log thickness of the horizontal well to use for correlation (Fig. 4-27). In the computer program a slider bar is used interactively to change the estimated dip and change the displayed log thickness. The log

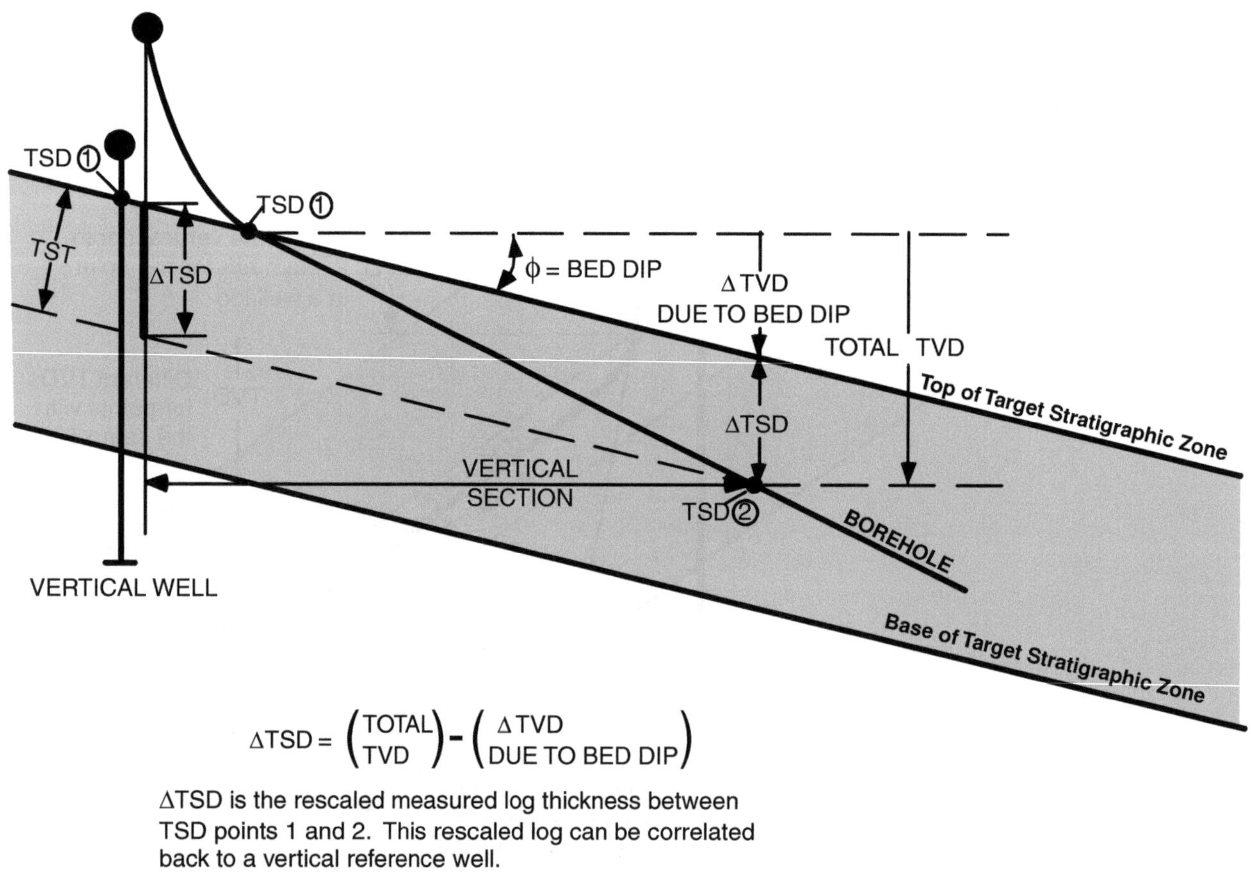

Figure 4-27 A schematic example of determining ΔTSD from TVD values to rescale a horizontal well log for correlation with a log from a vertical well. This example shows a wellbore or borehole going down section. (From Tearpock 1999. Published by permission of Anadarko Petroleum Corporation.)

is then stretched or squeezed until the curve is matched to the vertical well to make the correlation. The program also generates a TVD cross section as correlations are made.

Figure 4-28 illustrates the method for determining rescaled log thickness for a wellbore directed up section, which might occur if the wellbore extends too shallow relative to the target zone. A deeper horizon, such as the base of the target zone, is correlatable with the vertical well, and TSD point 2 is established and the apparent dip is drawn (Fig. 4-28). For the part of the wellbore going up section to the selected TSD point 3, the total TVD difference between TSD points 2 and 3 and the ΔTVD for TSD point 2 *add* to give ΔTSD. This vertical thickness will be the rescaled log thickness to use for correlations of the log section between TSD points 2 and 3. Of course, the log curve will also be inverted compared to the vertical well if the wellbore is deviated up stratigraphic section (Fig. 4-25). The GRNAV program can correct an inverted log curve.

Figure 4-29 is a generic illustration of the correlation of a horizontal well log and a vertical well log from Kyte et al. (1994). Figure 4-29a is a gamma ray log from a nearby vertical well or offset log with A, B, and C identifying three correlation markers. The wellbore geometry is shown in Figure 4-29b in a TVD cross section along the vertical section of the wellbore. A, B, and C are the bed boundaries identified by the correlation markers in the vertical well.

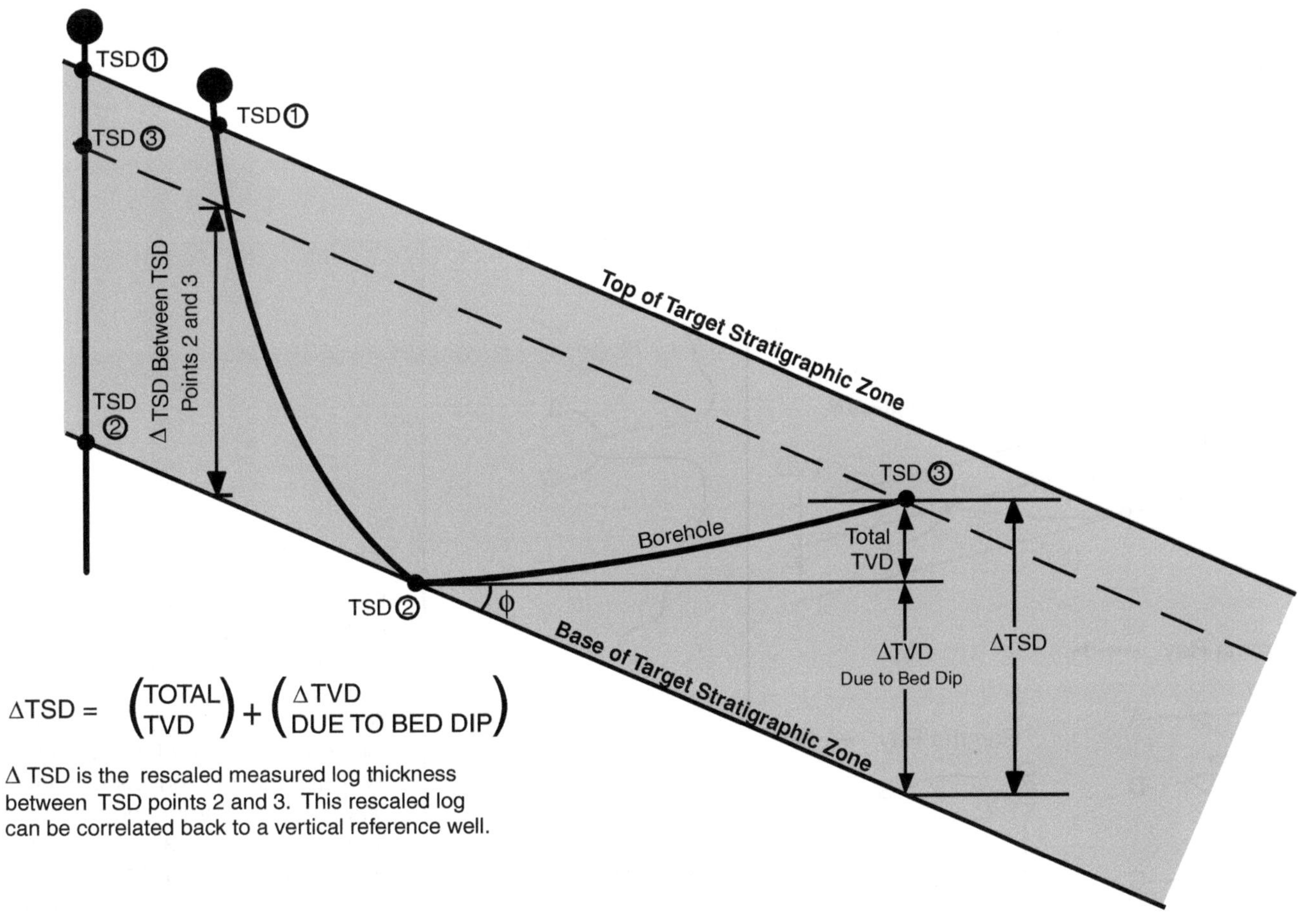

Figure 4-28 A schematic example of determining ΔTSD from TVD values to rescale a horizontal well log for correlation with a log from a vertical well. This example shows a wellbore or borehole going up section. (From Tearpock 1999. Published by permission of Anadarko Petroleum Corporation.)

The wellbore goes down stratigraphic section and up section with numbers indicating selected points illustrated in Figure 4-29e, the horizontal well log. Figures 4-29c and d are horizontal well log segments rescaled with the TSD method. The horizontal well log has been rescaled in Figure 4-29c from point 1 to point 3 in the horizontal well log, Figure 4-29e, to make the correlation to the vertical well. From point 3 on the horizontal well log to point 4, the log has been rescaled and *inverted* in Figure 4-29d, because the wellbore is going up section from points 3 to 4 (Fig. 4-29b).

The TSD method is an effective way to rescale a log from a horizontal well with a computer program and correlate it to a nearby well. Good log character is required, and even with a gamma ray log, the response may be more gradual than in a vertical well due to the low angle at which the wellbore cuts across the bed boundary.

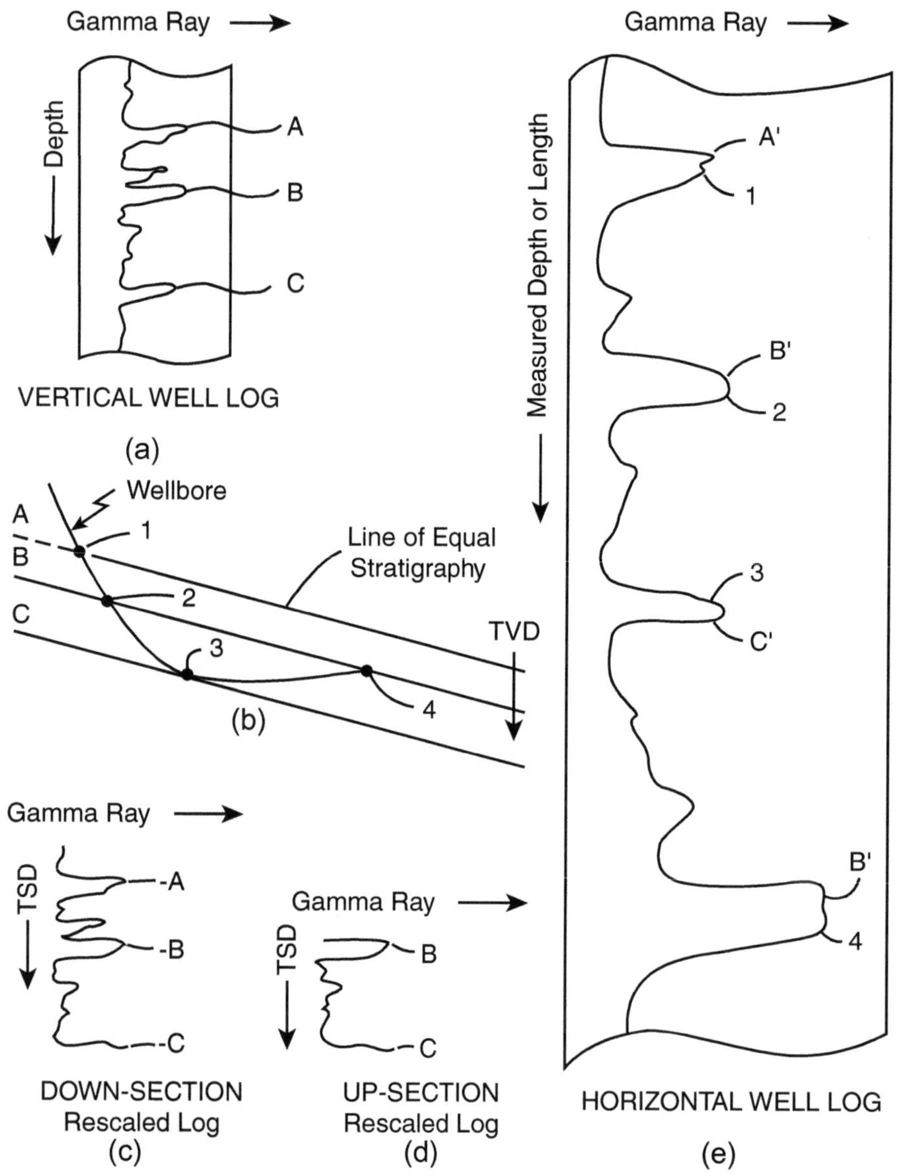

Figure 4-29 A generic illustration of the correlation of a horizontal well log with a vertical well log. (a) Vertical well log. (b) A TVD cross section drawn at true scale along the vertical section of the wellbore. Two gamma ray log curves have been rescaled from the horizontal well log for (c), a down-section part, and (d), an up-section part. (e) Horizontal well log. (From Kyte et al. 1994. Published by permission of Anadarko Petroleum Corporation.)

COMPUTER-BASED LOG CORRELATION

As we have discussed so far in this chapter, paper-based log correlation techniques have been the basic tools utilized by geologists for more than 60 years. The transfer of these techniques onto the computer screen has resulted in significant advantages both in efficiency (i.e., saving time) and in the application of high-resolution stratigraphic analysis. This was shown in the previous section on horizontal well log correlation. Despite these promises, geologists were initially slow to migrate onto the computer. In this section of the chapter, we address the traditional problems geologists have encountered in their attempt to use the computer for log-correlation work, describe some of the solutions to these problems, and provide examples of the significant advantages gained from computer-based log correlation in subsurface interpretation.

Geologists often cite two significant reasons they cannot or will not work with well logs on the computer: (1) the lack of an inexpensive digital well log database, and (2) software tools that do not reflect geologists' actual workflow with well logs. The digital-data availability issue is being actively addressed. Since the late 1990s, there has been a dramatic increase in the rate of well log digitization, including the advent of digital raster well log formats (Montgomery 1997a and b). The newly developed raster well log formats provide companies with an affordable option for accessing thousands of well logs for basin-wide studies. True digital log data is also becoming more widely available.

The second issue, that software tools did not reflect geologists' workflow, has also been addressed with several new software offerings devoted to well log interpretation/correlation. The next section utilizes one of these new applications and addresses the specific tools and techniques necessary to transfer paper-based log-correlation techniques onto the computer screen.

Well Log Correlation: The Transition from Paper-Based to Screen-Based

Log correlation is a basic skill of the geologist, as discussed earlier in this chapter. A geologist uses this skill to perform the often-repeated tasks of picking correlation markers, formation tops, faults, and unconformities, along with net sand, porosity, and pay analysis. Advanced applications include electric log facies analysis and sequence stratigraphy. In the hands of an experienced geologist, these techniques can be used to correlate subtle stratigraphic features, faults, and unconformities with great accuracy over large distances.

In paper-based log correlation, we first overlay two logs to place the two well log curves immediately adjacent to one another, as previously shown in Figure 4-3. We then begin slipping (sliding) the paper logs past one another to look for similarities in log pattern or log character that we interpret as correlations. We start marking the logs where we see similarities (or correlations) using color and all manner of annotations, such as boxes, squares, curlicues, asterisks, and so on, as shown in Figure 4-2. The identification of consistent log character from well to well convinces the geologist of an accurate log correlation.

As we mentioned before, a geologist will attempt to begin such an exercise using a well log considered to have a reasonably complete geological section—in other words, a type log that lacks significant missing section due to unconformities or faults. The detailed annotation of log character on such a well defines the normal succession of basin fill in the study area. Detailed correlation of all well logs to the normal succession at the type well allows geologists to identify missing section that may be related to normal faults or unconformities and to identify repeated section resulting from reverse faults.

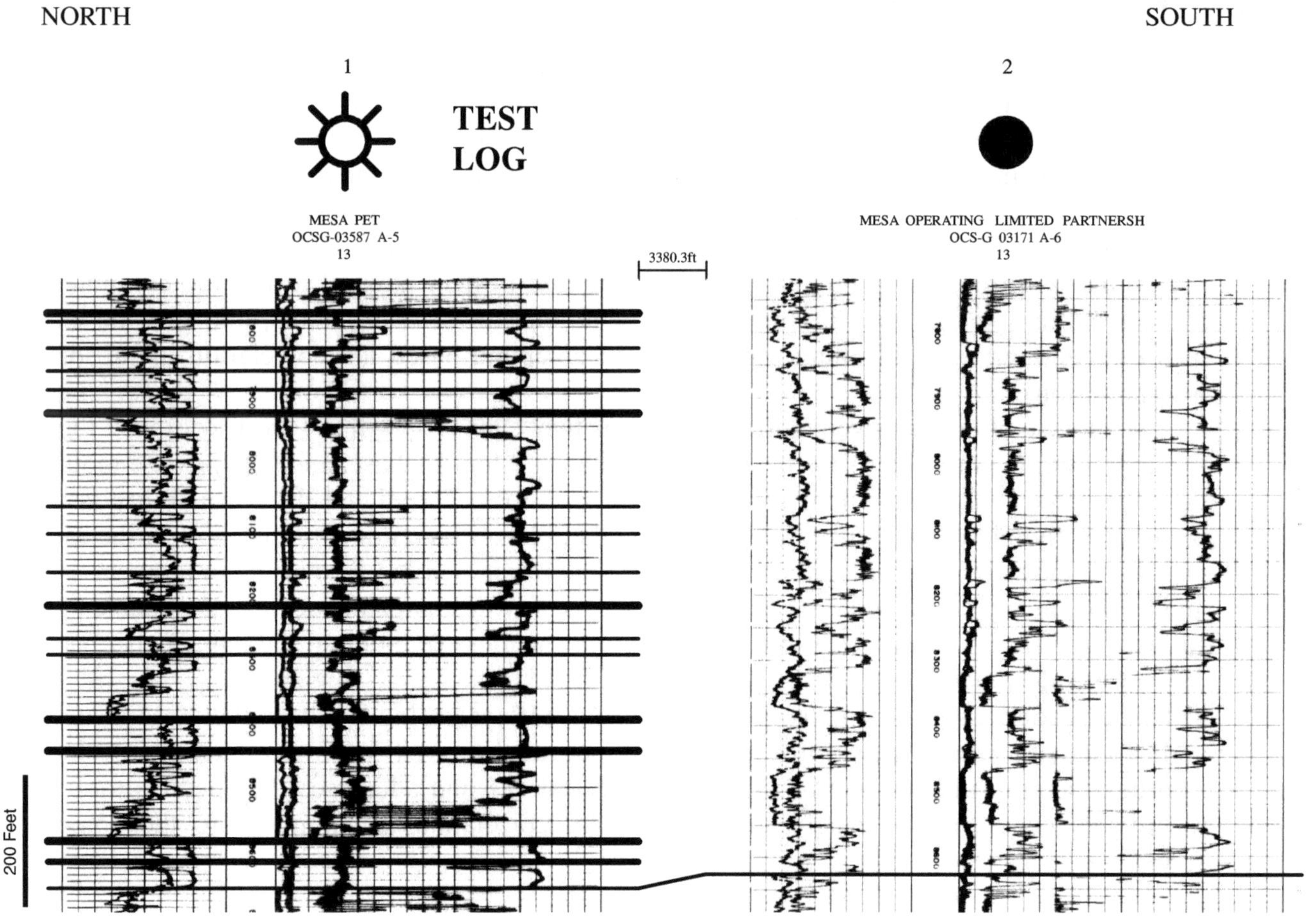

Figure 4-30 Two full-width electric logs, displayed side by side, with a series of arbitrarily named markers placed on the "type log." Both logs are digital raster well log images with depth intelligence added to the depth track. (Published by permission of W. C. Ross and A2D Technologies.)

On-Screen Log Correlation

The logs we use for illustration purposes are from the South Pelto area in the offshore Gulf of Mexico. The stratigraphy in this area consists of a series of sandstones and shales of Pliocene age. Figure 4-30 shows two well logs side by side. Each log is displayed at full width. For these types of full-width log displays, log-character correlation is impeded on the computer screen by the wide separation of the respective log curves in each well log window. To simulate paper-based log correlation procedures, we can create a display with multiple well logs so that the geologist is able to overlay adjacent logs, slip (or scroll) the logs vertically to find a character match, and then annotate points or zones of correlation from log to log.

Configuring the width of each well log window to approximately two inches (for a 1 in. = 100 ft log scale) simulates overlaying two paper well logs for correlation by placing the two curves of greatest interest for correlation immediately adjacent to one another (Fig. 4-31a). The two-inch width of the log windows was chosen in order to display just one log curve from the full-width log. Choosing which well log curve is displayed in the two-inch wide log window (e.g., SP, GR, resistivity, conductivity) depends on which log curve is most useful for correlation. In our example, we place the SP and GR curves in this window. By creating narrow windows with single log curves (or single log tracks), a computer-screen display can simulate paper well-log overlay techniques and effectively increase the number of well logs that can be displayed on the screen at one time.

Computer log correlation can now proceed by simply pointing and clicking on each well log with the correlation tool at the various points of log correlation. Log slipping (or vertical scrolling) can be performed to help line up each of the logs prior to correlation itself.

To simulate the log annotation techniques used in paper-based correlation, the following procedure can be used. First, nominate a type well for the starting point in your correlation exercise. Next, place on the type log a series of markers (Fig. 4-31a). They are designated by an arbitrary naming convention, such as GM1 (generic marker 1), GM2, GM3, and their purpose is analogous to manual annotations on paper well logs. As with paper-based annotation, the markers should be placed with as much frequency as is needed to capture all the log character deemed necessary for detailed correlation.

Once the markers are placed on the type well (the starting point), the geologist can begin transferring each marker from well log to well log across a panel of wells (Fig. 4-31b and c). The correlation work proceeds by correlating all the wells in the study area using this level of detail. When creating a correlation section such as shown in Figure 4-31, the type log may be placed at the beginning of the section and repeated at the end of the section, rather than in the middle of the section, as in Figure 4-31. This procedure automatically loop-ties the correlations.

Whereas the level of detail in computer correlation may not be any higher than the paper-based log annotation techniques, computer-based correlation of markers carries the advantage that each marker top is automatically placed into a digital database of tops for each well. Consequently, as you build arbitrary cross sections through your 3D database of well control, all the marker correlation work will carry forward onto these cross sections. The power of this technology for geometric cross-sectional analysis, fault picking, and unconformity recognition is considerable, as we illustrate in the next section.

Example of Unconformity Identification

To illustrate the power of computer-based correlation for unconformity identification, we use an example of a correlation exercise from the Lewis Shale Formation within the eastern Green River basin in southern Wyoming (Fig. 4-32). The Lewis Shale overlies the non-marine and shallow marine Almond Formation (Winn et al. 1985).

A subtle unconformity was discovered by careful well log correlation within the transgressive Lewis Shale, just above the Almond Formation. To illustrate how this unconformity was detected, we begin with a 10-well, west-to-east cross section constructed parallel to depositional dip across the Almond and Lewis Formations (Fig. 4-32a). The shallow-marine Almond Sandstone and the marine Lewis Shale represent the transgression of the Lewis seaway from east to west (right to left) on the section.

The on-screen correlation involved the use of arbitrarily named markers to perform high-resolution, time-stratigraphic correlations, following the sequence stratigraphic methodology of Van Wagoner et al. (1990). Following this strategy, we established a type well at the Davis Oil Armor State #1 (fifth well in cross section) and began correlating the initial markers across the panel of well logs. In the early stages of log correlation, it became apparent that there were distinct packages of rock within the Lewis Shale that could easily be distinguished and shown to be discordant with respect to one another along an unconformity surface (Fig. 4-32a). The discordant surface coincides with the top of a regional bentonite marker denoted as B1. The geometric discordance is enhanced through extreme vertical exaggeration (V.E. = 166:1). The vertical exaggeration is facilitated on a 21-inch computer screen by using narrow log windows, close well spacing, and by changes in correlation draw style. Figure 4-32a and b illustrate the visual advantages gained by changing the log correlation draw style from a conventional style where correlations are drawn straight across the log (i.e., edge-to-edge) to a style where the correlations are connected from the center point of each log (i.e., center-to-center). The use of center-to-center correlation drawing style smooths the connections

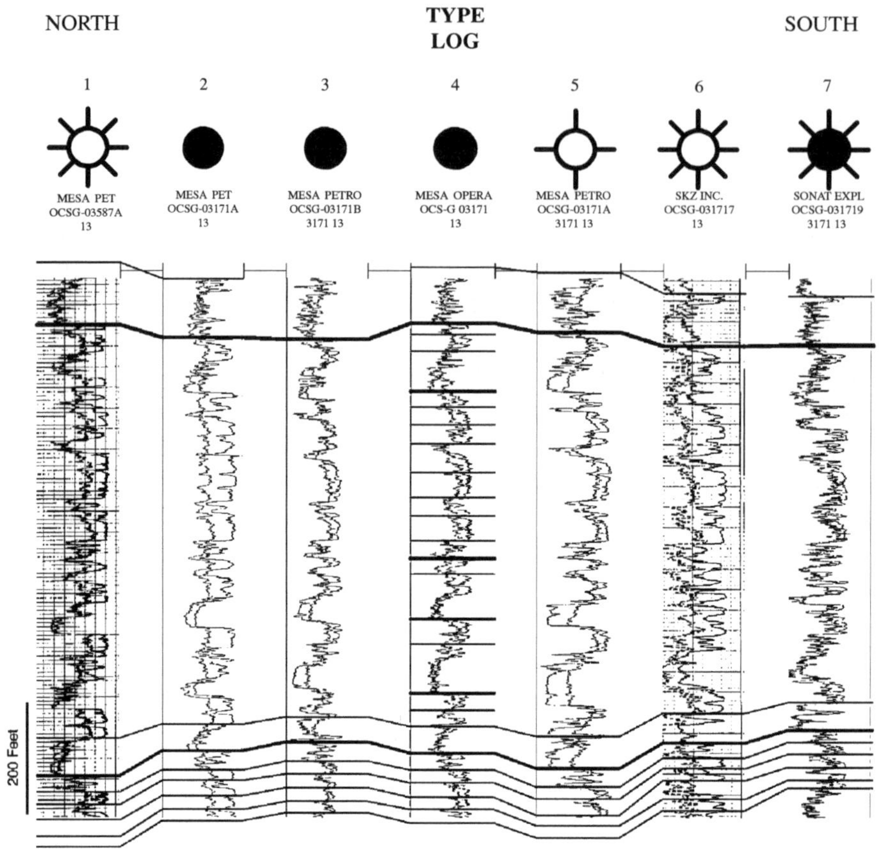
NORTH
TYPE LOG
SOUTH
1
2
3
4
5
6
7
MESA PET OCSG-03587A 13
MESA PET OCSG-03171A 13
MESA PETRO OCSG-03171B 3171 13
MESA OPERA OCS-G 03171 13
MESA PETRO OCSG-03171A 3171 13
SKZ INC. OCSG-031717 13
SONAT EXPL OCSG-031719 3171 13
200 Feet

(a)

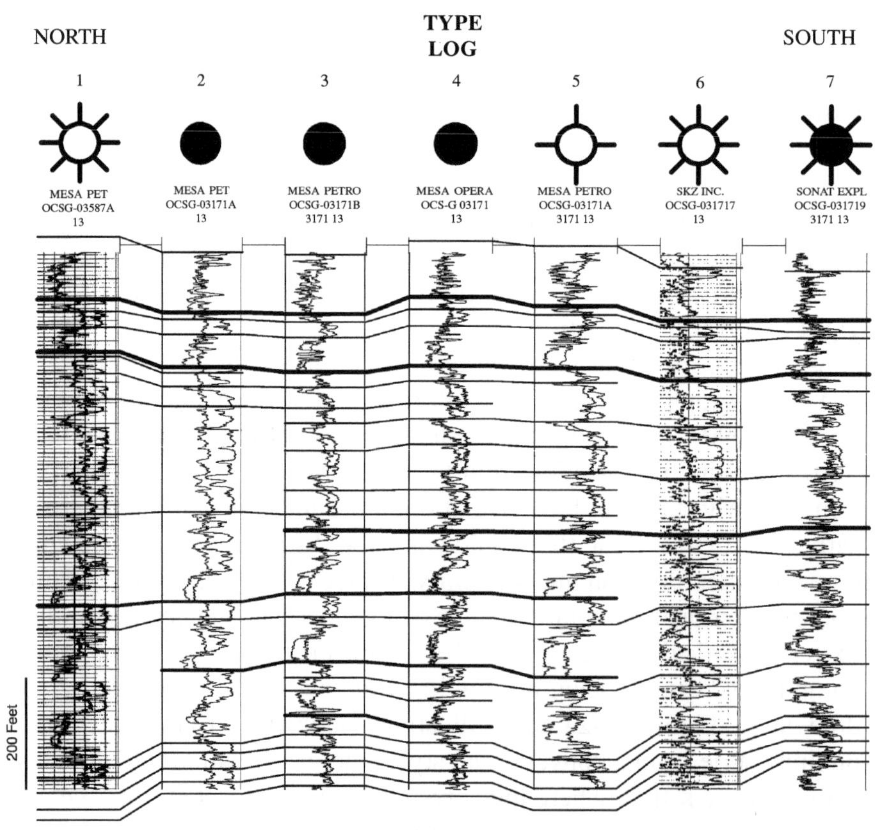
NORTH
TYPE LOG
SOUTH
1
2
3
4
5
6
7
MESA PET OCSG-03587A 13
MESA PET OCSG-03171A 13
MESA PETRO OCSG-03171B 3171 13
MESA OPERA OCS-G 03171 13
MESA PETRO OCSG-03171A 3171 13
SKZ INC. OCSG-031717 13
SONAT EXPL OCSG-031719 3171 13
200 Feet

(b)

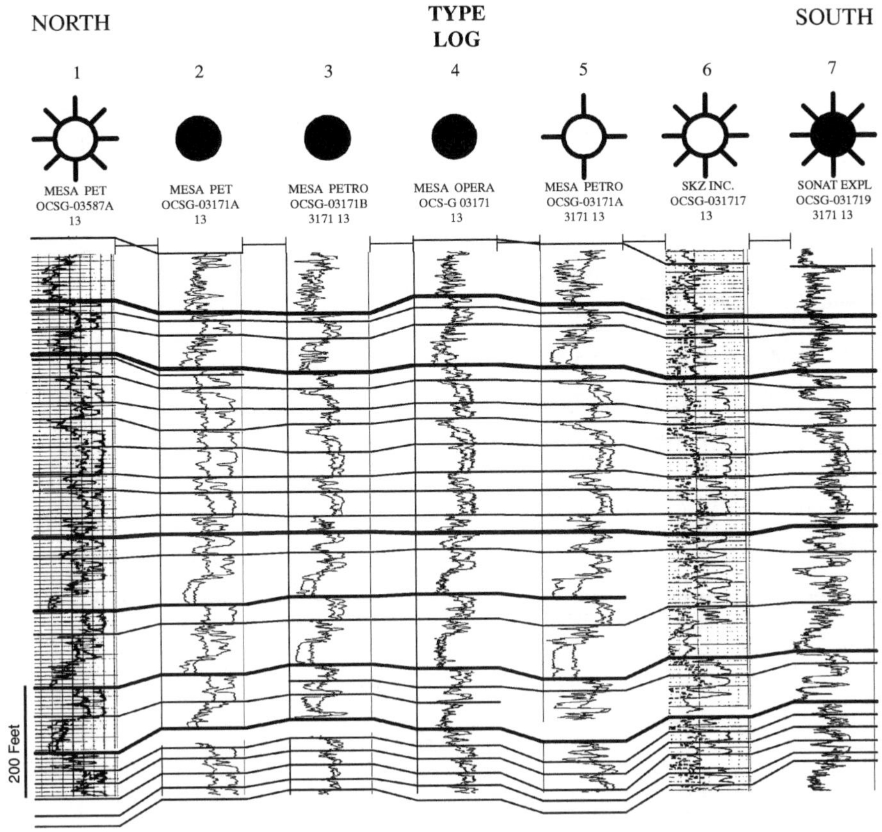

(c)

Figure 4-31 (a) Cross-section window configured to simulate paper-based log correlation. Closely spaced log windows, which here contain spontaneous potential and gamma ray curves centered within them, simulate the log overlay techniques used in paper-based methods. The type log contains a series of arbitrarily named markers to be used as starting points in a log correlation exercise. The wells in the cross section have an average spacing of less than one-quarter mile and were constructed with a combination of raster and digital (Log ASCII Standard) well log data. (b) Partially correlated panel of well logs illustrating the progression of correlation work from the central type log. (c) Completed correlation panel of well logs. (Published by permission of W. C. Ross and A2D Technologies.)

between correlation points and provides a clearer view of both discordant relationships and subtle basin-fill geometries. Upon initial inspection, the discordant relationship (correlation markers terminating in an easterly or basinward direction against a common surface) was interpreted as depositional downlap (Fig. 4-32b). Computer-based cross-sectional analysis allows the interpreter to quickly choose an alternative cross-section datum to evaluate alternative interpretations of unconformity type. For example, in Figure 4-32c a higher marker (B2) is utilized as the stratigraphic datum, providing a view that would suggest the unconformity is an onlap unconformity.

Example of Fault Identification

For our second example, we illustrate the power of computer software to identify faults and interactively interpret missing section. We start with another cross section from the South Pelto area in the offshore Gulf of Mexico (Fig. 4-33a). The 6-well cross section contains numerous marker horizons that can be identified and correlated throughout the area. As discussed earlier in the chapter, using numerous markers in log correlation enables the geologist to quickly identify missing section due to normal faulting or erosional truncation. In Figure 4-33b we have added a well (Mesa Well No. A-6) and, due to the established high-resolution correlation framework, we can immediately determine

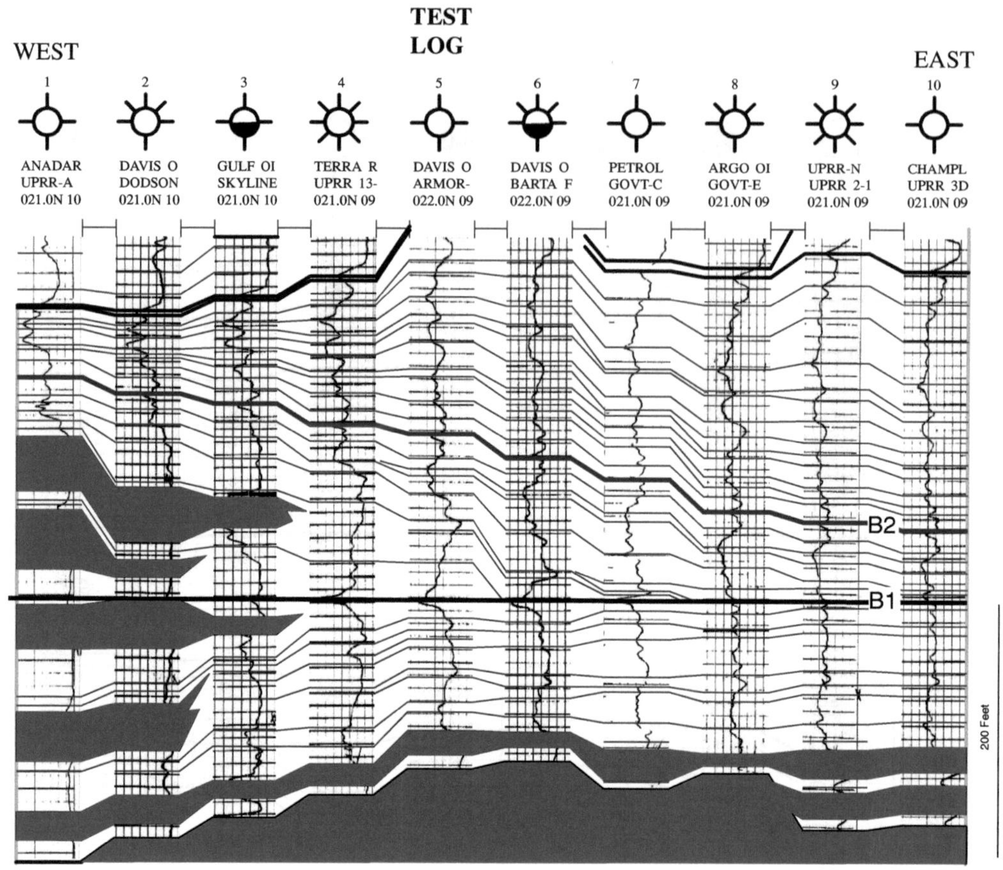
TEST LOG
WEST
EAST
1
2
3
4
5
6
7
8
9
10
ANADAR UPRR-A 021.0N 10
DAVIS O DODSON 021.0N 10
GULF OI SKYLINE 021.0N 10
TERRA R UPRR 13- 021.0N 09
DAVIS O ARMOR- 022.0N 09
DAVIS O BARTA F 022.0N 09
PETROL GOVT-C 021.0N 09
ARGO OI GOVT-E 021.0N 09
UPRR-N UPRR 2-1 021.0N 09
CHAMPL UPRR 3D 021.0N 09
B2
B1
200 Feet

(a)

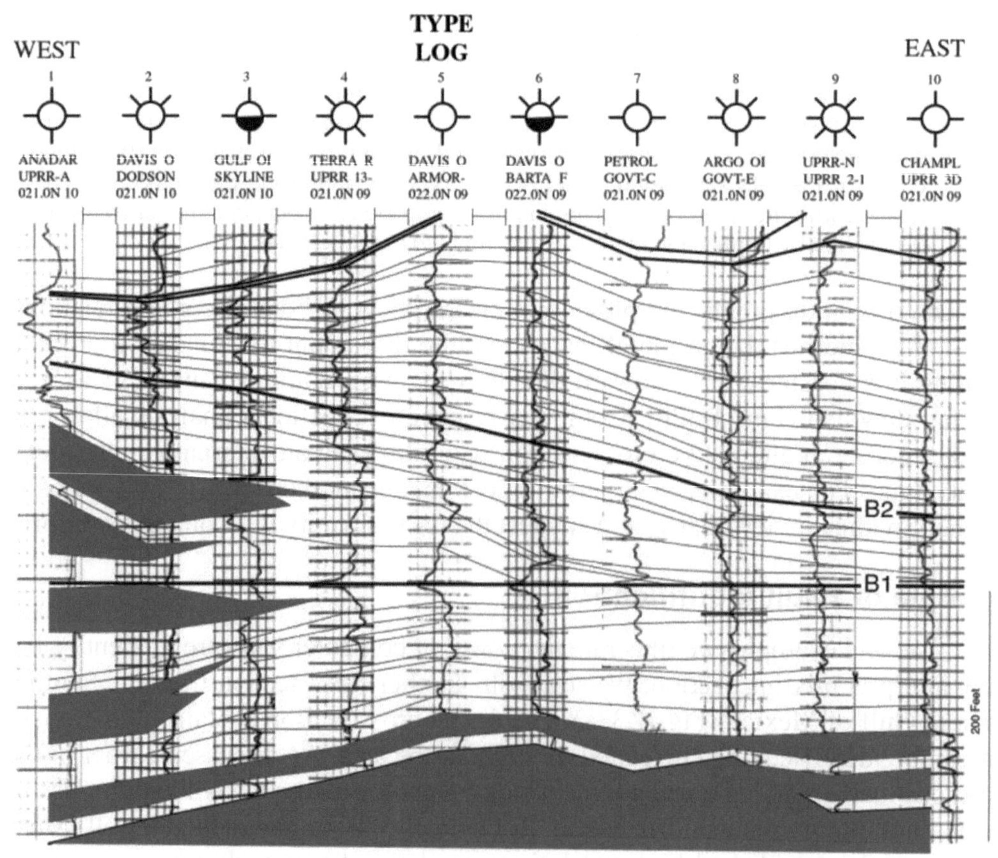
TYPE LOG
WEST
EAST
1
2
3
4
5
6
7
8
9
10
ANADAR UPRR-A 021.0N 10
DAVIS O DODSON 021.0N 10
GULF OI SKYLINE 021.0N 10
TERRA R UPRR 13- 021.0N 09
DAVIS O ARMOR- 022.0N 09
DAVIS O BARTA F 022.0N 09
PETROL GOVT-C 021.0N 09
ARGO OI GOVT-E 021.0N 09
UPRR-N UPRR 2-1 021.0N 09
CHAMPL UPRR 3D 021.0N 09
B2
B1
200 Feet

(b)

(c)
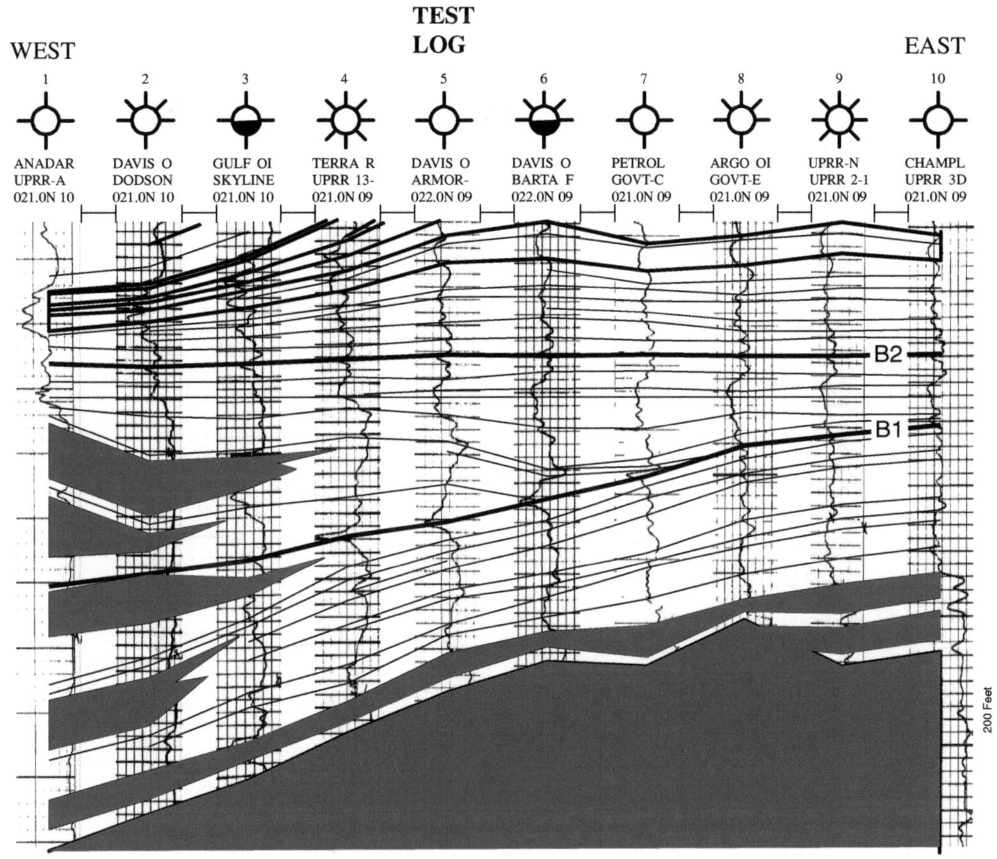

Figure 4-32 (a) West-east stratigraphic cross section, hung on the B1 datum, constructed with conductivity curves centered within narrow well log windows. The light gray shading depicts sands within the Almond Formation. The darker gray pattern represents sands and shales of the non-marine Almond Formation, and the white region depicts shale facies within the Lewis Shale Formation. This cross section was constructed with well log images. Style of drawing correlations is from log edge to log edge. The cross section is eighteen miles in length with an average well spacing of less than two miles. The vertical exaggeration is 166:1. (b) Same cross section with correlation draw style changed to center-to-center of logs, which accentuates basin-fill geometries. Downlap on surface B1 is suggested. (c) Same cross section hung on the B2 stratigraphic datum. The higher datum changes the geometry of the unconformable relationship and suggests onlap versus downlap. (Published by permission of W. C. Ross and A2D Technologies.)

that correlation markers A, B, and C are obviously missing in the A-6 well due to faulting. To determine the actual amount of section missing (in feet) due to faulting, we can pick a fault location and use a program function to "split" or "gap" the well log. The gap in the well log represents the amount of section missing due to normal faulting. The size of the gap is interpreted by interactively "dragging" down, or opening up, the boundary of the gap (Fig. 4-33c). Using one or more of the cross-section wells for reference, the geologist can reconstruct the amount of missing section by pulling down, or opening, the gap until the correlations above and below the gap are approximately parallel (Fig. 4-33d). In this example, we have interpreted the missing section to be approximately 215 ft thick. This value is also the amount of vertical separation, which is discussed in Chapter 7.

The log-splitting tool allows you to create correlation cross sections hung on stratigraphic markers, with missing section (due to faulting or erosional truncation) displayed as missing section gaps. Once the gaps are interpreted, you can pick tops within the missing section to allow the cross section to be hung on missing top markers (Fig. 4-33d). The hand-generated equivalent to this computer method is shown in Figure 4-20b, for which a paper log was cut and pulled apart by the amount of missing section.

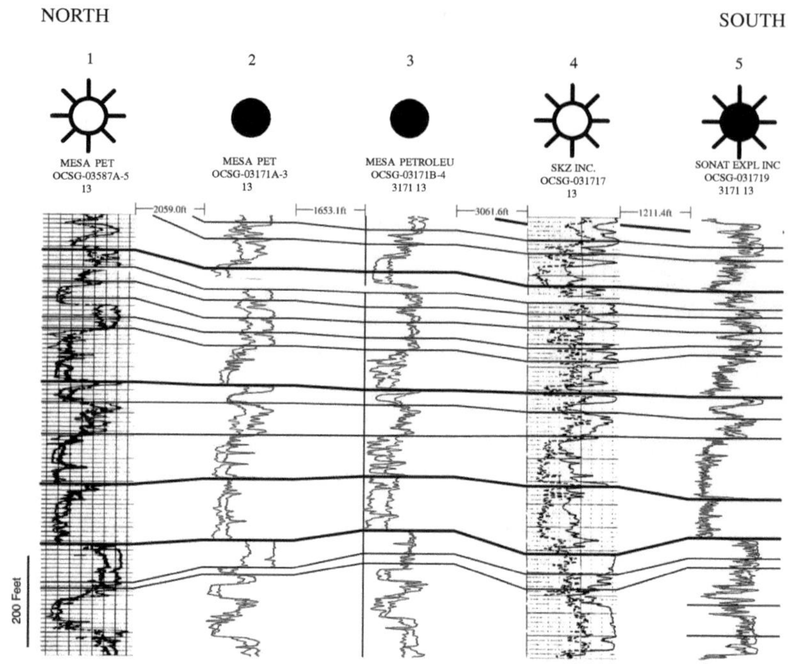

NORTH
SOUTH
1
2
3
4
5
MESA PET
OCSG-03587A-5
13
MESA PET
OCSG-03171A-3
13
MESA PETROLEU
OCSG-03171B-4
3171 13
SKZ INC.
OCSG-031717
13
SONAT EXPL INC
OCSG-031719
3171 13
2059.0ft
1653.1ft
3061.6ft
1211.4ft
200 Feet

(a)

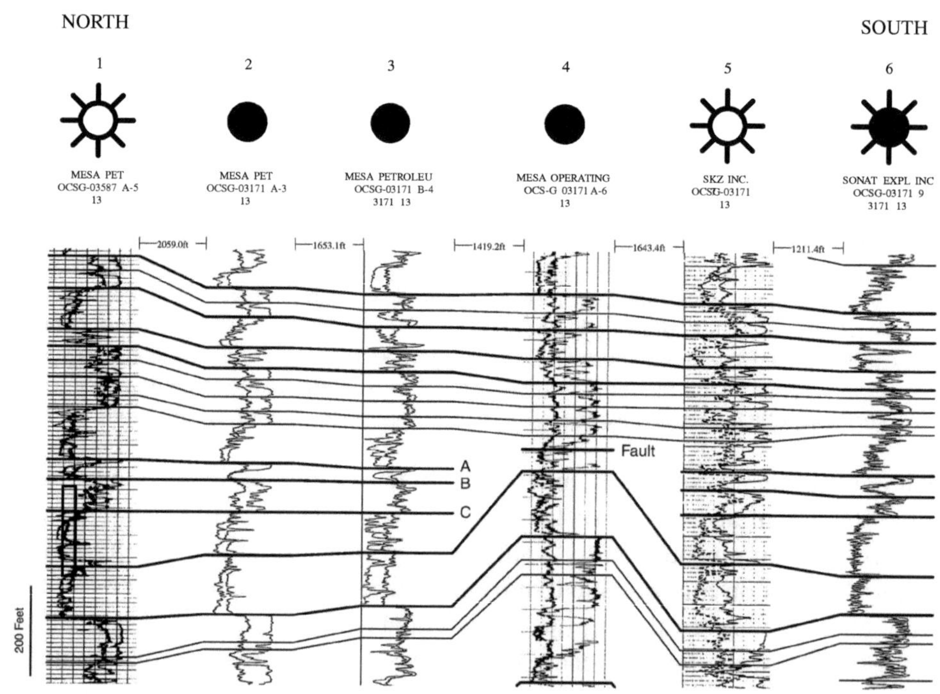

NORTH
SOUTH
1
2
3
4
5
6
MESA PET
OCSG-03587 A-5
13
MESA PET
OCSG-03171 A-3
13
MESA PETROLEU
OCSG-03171 B-4
3171 13
MESA OPERATING
OCS-G 03171 A-6
13
SKZ INC.
OCSG-03171
13
SONAT EXPL INC
OCSG-03171 9
3171 13
2059.0ft
1653.1ft
1419.2ft
1643.4ft
1211.4ft
Fault
A
B
C
200 Feet

(b)

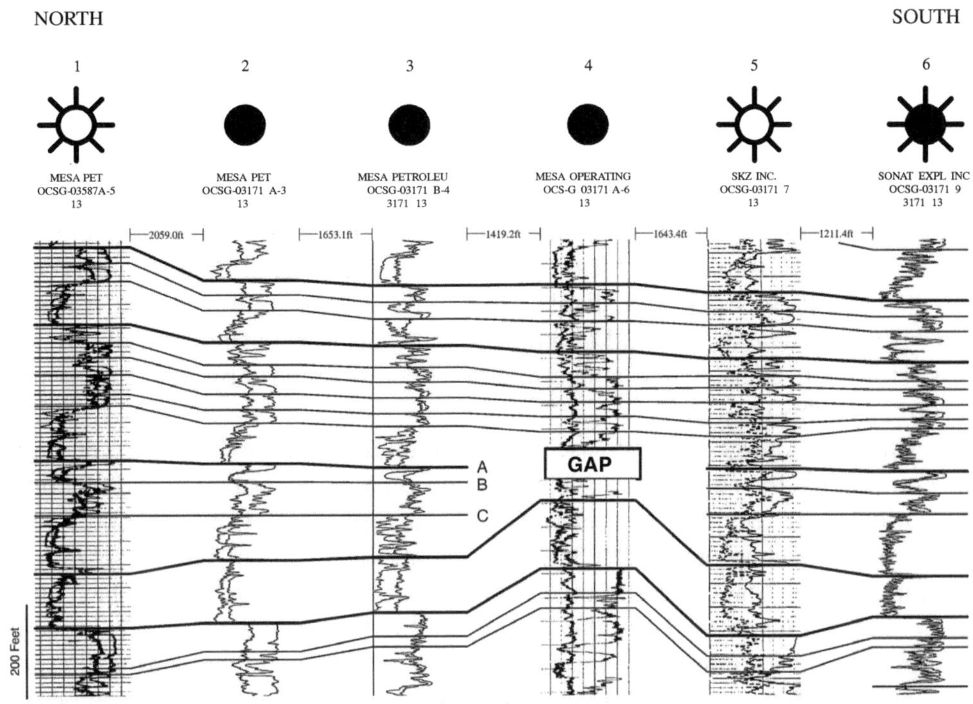

(c)

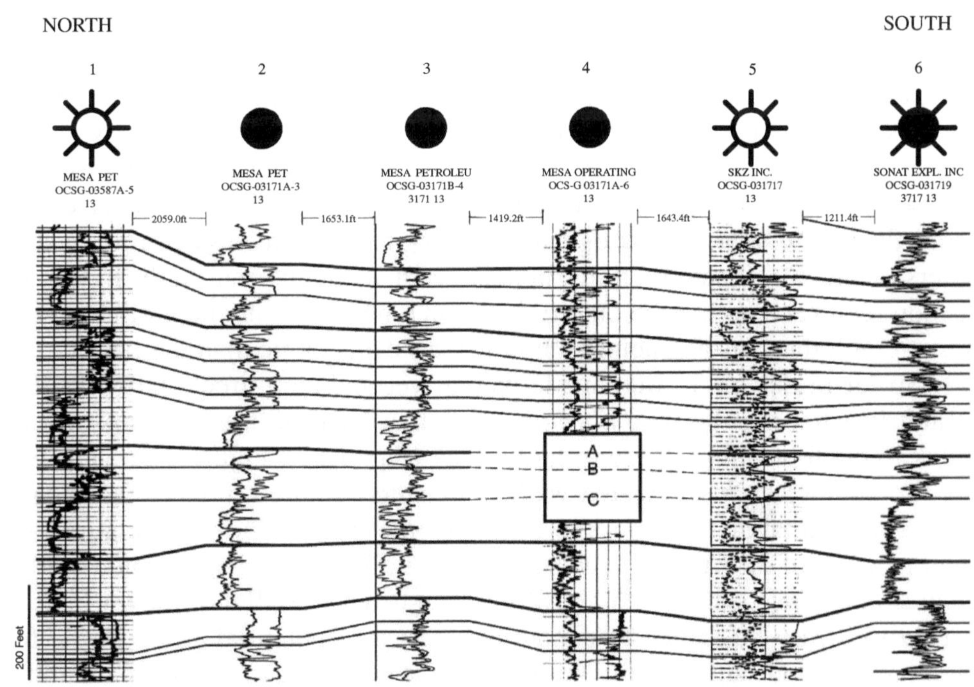

(d)

Figure 4-33 Cross section illustrating the technique to determine missing section. (a) Cross section with numerous markers demonstrating a continuous, unfaulted stratigraphic section. Average well spacing is less than one-quarter mile. (b) Same cross section with an additional well (Well No. A-6) that has missing section. The detailed correlation work demonstrates that the A, B, and C correlations are cut out by a fault (i.e., where correlations converge). (c) Same cross section with fault picked and log partially pulled apart (gapped) for Well No. A-6. (d) Same cross section with missing section interval fully interpreted and missing tops (A, B, and C correlations) placed within it. (Published by permission of W. C. Ross and A2D Technologies.)

The use of high-resolution marker correlations, narrow window correlation displays, and interactive fault picking/gapping tools allows geologists to correlate hundreds of well logs within a short period of time. Correlation exercises are significantly aided by this fault-gapping technology. By picking and gapping faults during a correlation exercise, significant amounts of distortion related to missing section can be eliminated (compare Fig. 4-33b and d).

There are many advantages of computer-based log correlation over paper-based techniques. Perhaps the most significant are the capabilities for instant access to log data from a computerized database, automatic rescaling of images and digital data, rapid redatuming capabilities, and the ability to track and display hundreds of correlations/markers within a computerized database. For many traditional geologists, the transition to the computer has been a painful process. However, many of the concerns have been overcome by adopting some of the techniques described in this section.

Many geologists had expressed concern about being unable to see enough of the well log on the computer screen at one time. This is a legitimate concern. However, technology has improved dramatically in recent years, providing both high-resolution computer monitors and image-enhancement technologies, which allow several thousands of feet of section to be viewed on the computer screen for reconnaissance log review. Ultimately, geologists have agreed that the advantages of migrating to the computer outweigh the disadvantages of learning a new set of skills. As with geophysicists, who made the transition to geophysical workstations in the mid-1980s, geologists have rapidly made the transition to geological workstations. The enormous increase in productivity for geologists is on par with their geophysicist colleagues.

REPEATED SECTION

A repeated section in a well is defined as a part of the stratigraphic section that appears twice on a log as the result of a fault, consequently lengthening the log section. A repeated section is commonly thought of as a compressional tectonic phenomenon, occurring as the result of a reverse fault pushing the stratigraphic section in the hanging wall up and over the same section in the footwall. Figure 4-34 illustrates the geometry required to result in a repeated section due to a reverse fault. A repeated section can also occur with a normal fault. In this situation, a repeated section requires a specific geometry between a normal fault and a directionally drilled well, or a normal fault that cuts a structure dipping at a steeper angle than the fault.

So far in this book, we have looked at vertical and directionally drilled wells penetrating a fault in what is called the *normal sense*; that is, from the hanging wall (downthrown) fault block to the footwall (upthrown) fault block (Fig. 4-35). Figure 4-36 is a cross section illustrating the geological and deviated well parameters required to cause a repeated section in a log for a situation involving a normal fault. Geometries, as shown in Figure 4-36, are not uncommon in areas of directionally drilled wells. In offshore areas, deviated wells are drilled from a central platform location and are often designed to parallel a known fault (Fig. 4-37). In this way, a well may be drilled to penetrate a series of potential strata in the footwall (upthrown) block of an important trapping fault. If care is not taken to precisely map the fault and drill the well, it is possible for the well to cross the fault and enter the hanging wall (downthrown) block of the fault. In Figure 4-37, the well deviated toward the salt crosses a normal fault *backwards*, resulting in a repeated section. Also notice that by crossing the fault, the well does not penetrate productive Sand B, which is upthrown to the fault, but instead penetrates the nonproductive Sand B downthrown.

Figure 4-38 is an example of a repeated section resulting from an extensional (normal) fault. The amount of wellbore deviation is shown on the right of the log section. Notice that the deviation angle ranges from 40°45′ to 42°15′. The well was drilled to parallel a fault and to stay in

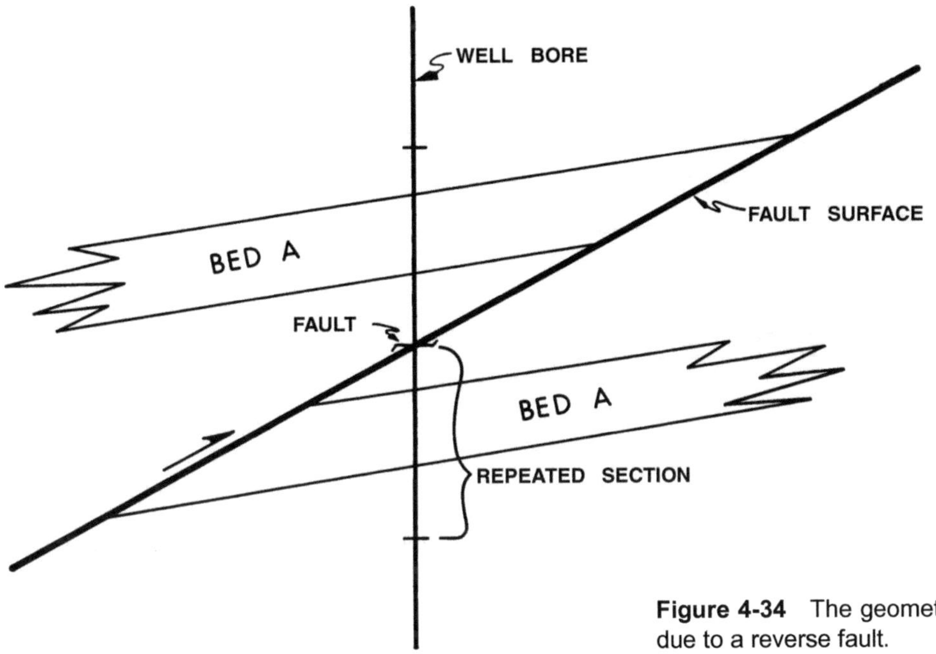

Figure 4-34 The geometry required for a repeated section due to a reverse fault.

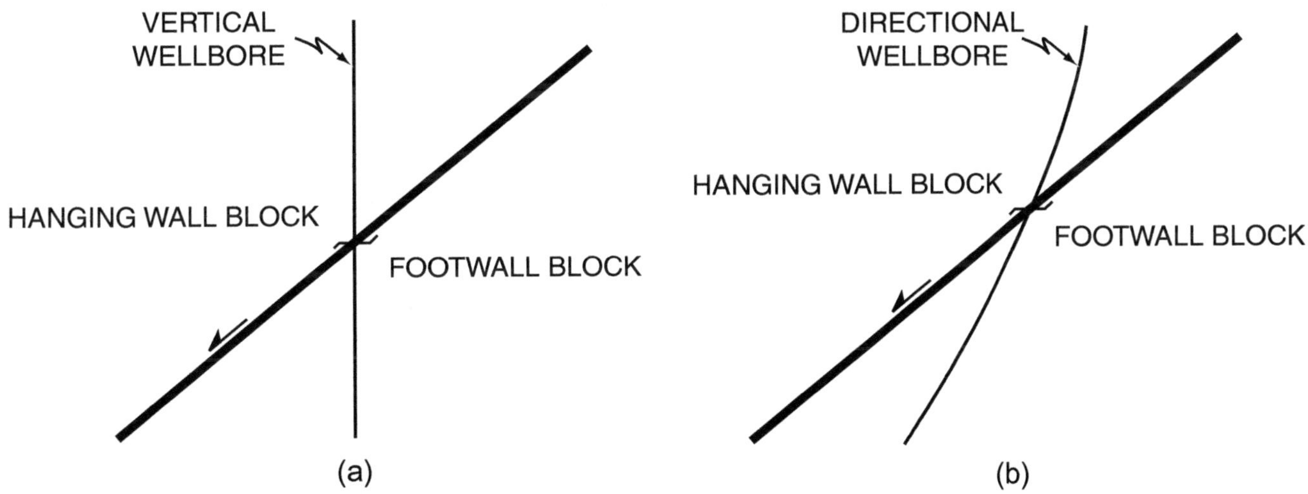

Figure 4-35 (a) A vertical well intersecting a fault in the normal sense (from hanging wall to footwall fault blocks). (b) A deviated well intersecting a fault in the normal sense (from hanging wall to footwall fault blocks).

the footwall fault block. Since wellbore deviation is measured from the vertical and fault dip from the horizontal, it is necessary to subtract the hole angle from 90 deg to compare its "dip," or plunge, angle to that of the fault. By doing this, the well angle measured from the horizontal, for the section shown, ranges from 47°45′ min to 49°15′ min. Based on the fault surface map, the fault dip is approximately 52 deg. Therefore, the well inclines downward at an angle less than the fault dip, establishing a well/fault geometry similar to that shown in Figure 4-36. The result is the repeated section shown in Figure 4-38.

Figure 4-39 is another excellent example of a repeated section resulting from a normal fault cut backwards by a deviated well. Observe that the footwall fault block is productive of hydrocarbons, whereas the hanging wall block is not.

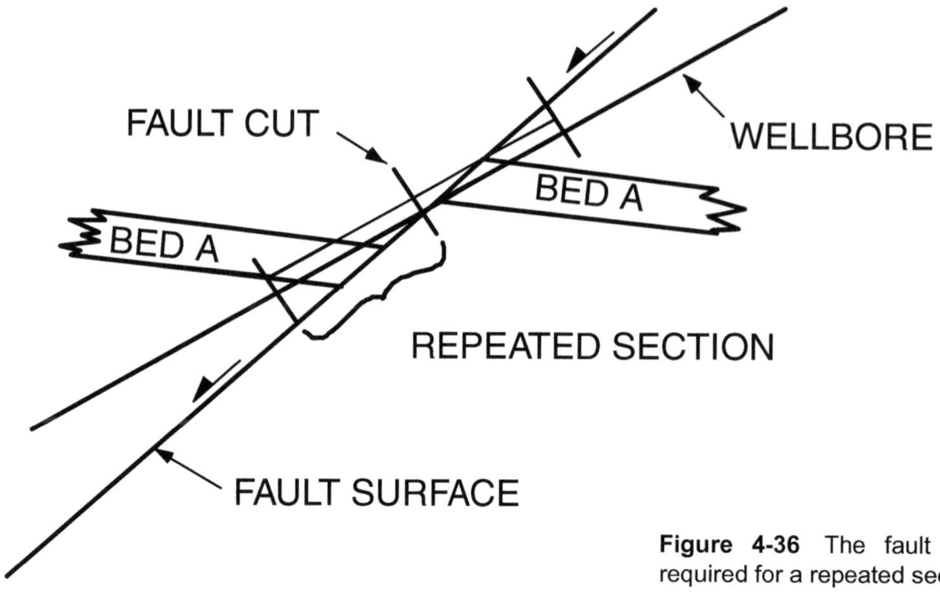

Figure 4-36 The fault and deviated well parameters required for a repeated section resulting from a normal fault.

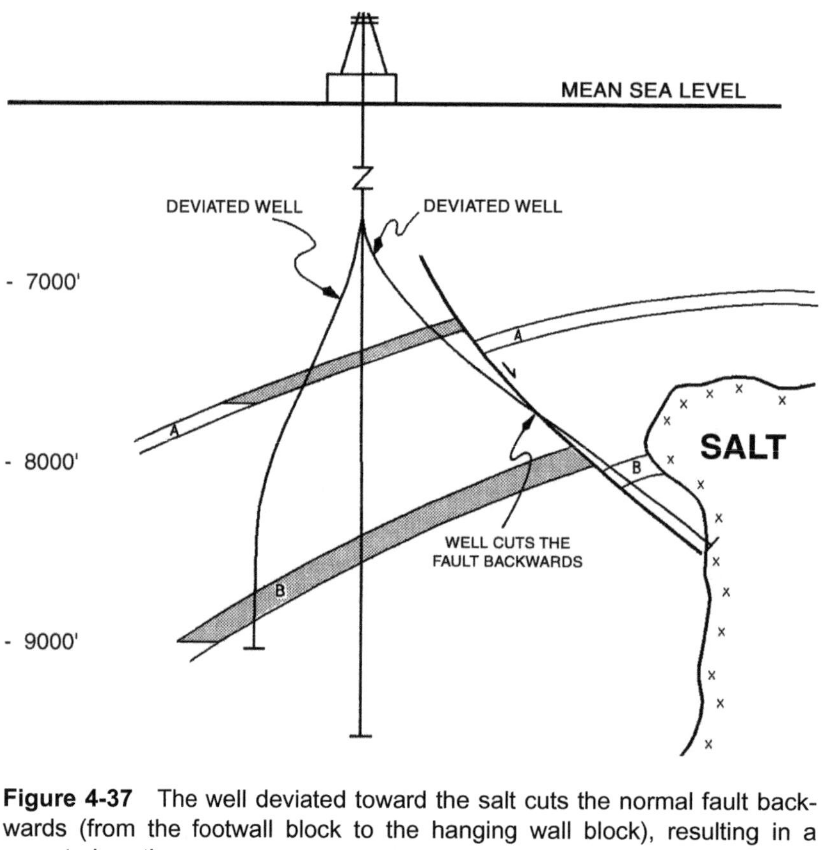

Figure 4-37 The well deviated toward the salt cuts the normal fault backwards (from the footwall block to the hanging wall block), resulting in a repeated section.

The identification of a repeated section that has previously gone unrecognized can result in excellent prospect potential. Look at the generic situation shown in Figure 4-40. In this case, a well was designed to stay in the footwall of an important trapping fault. Good control on the fault was available at the time of drilling, but a fault surface map was not prepared. The well meandered across the fault, resulting in an unrecognized repeated section on the well log. The stratigraphic intervals penetrated in the footwall of the fault are productive of hydrocarbons, whereas the hanging wall interval from A to B are wet. The repeated section went unrecognized and,

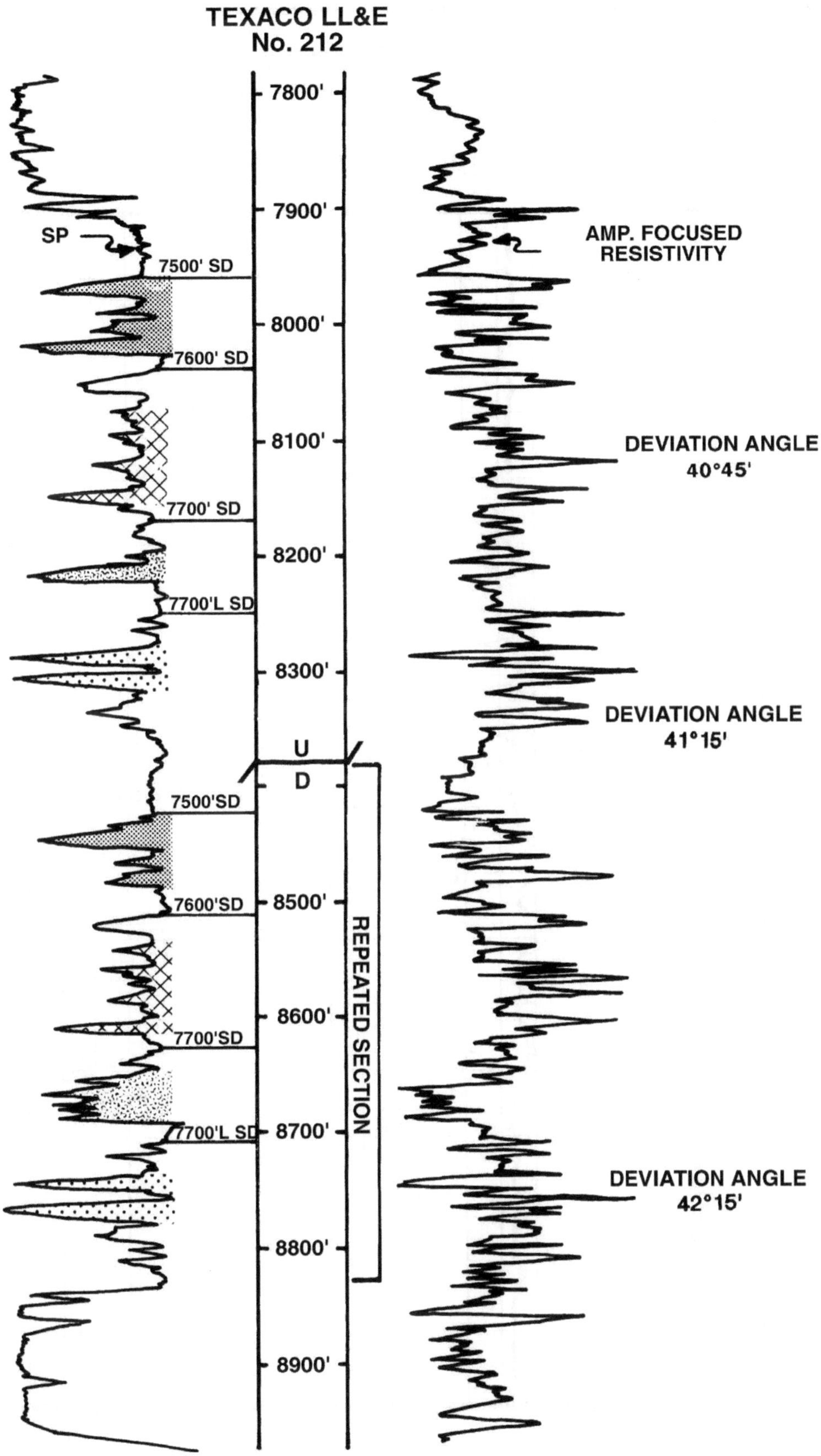

Figure 4-38 Repeated section in the Texaco LL&E Well No. 212 due to the well cutting a normal fault backwards. (Published by permission of Texaco, USA.)

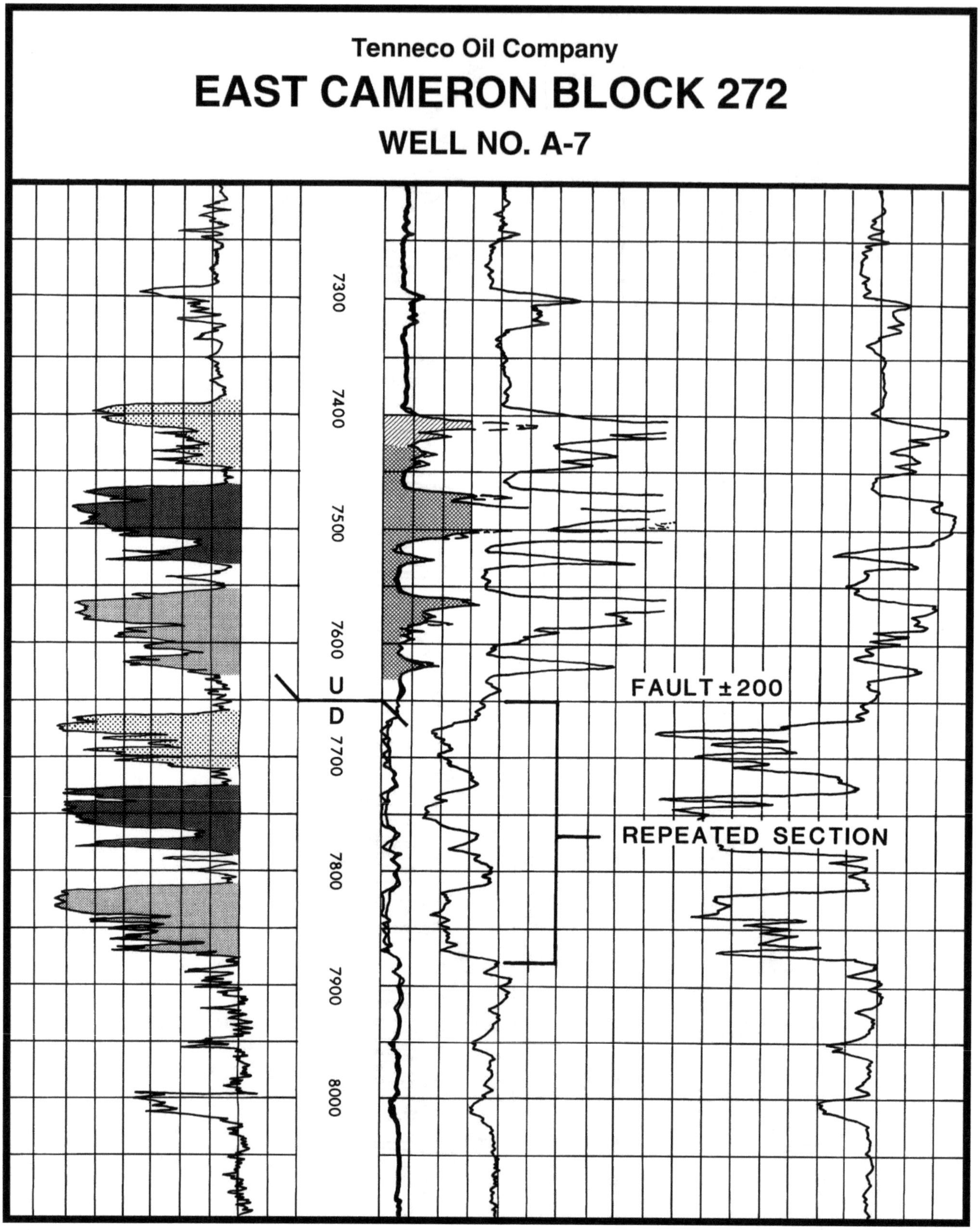

Figure 4-39 Repeated section in a deviated well due to the well cutting a normal fault backwards. Notice that the footwall (upthrown) fault block contains hydrocarbons, while the hanging wall (downthrown) block is wet. (From Tearpock and Harris 1987. Published by permission of Tenneco Oil Company.)

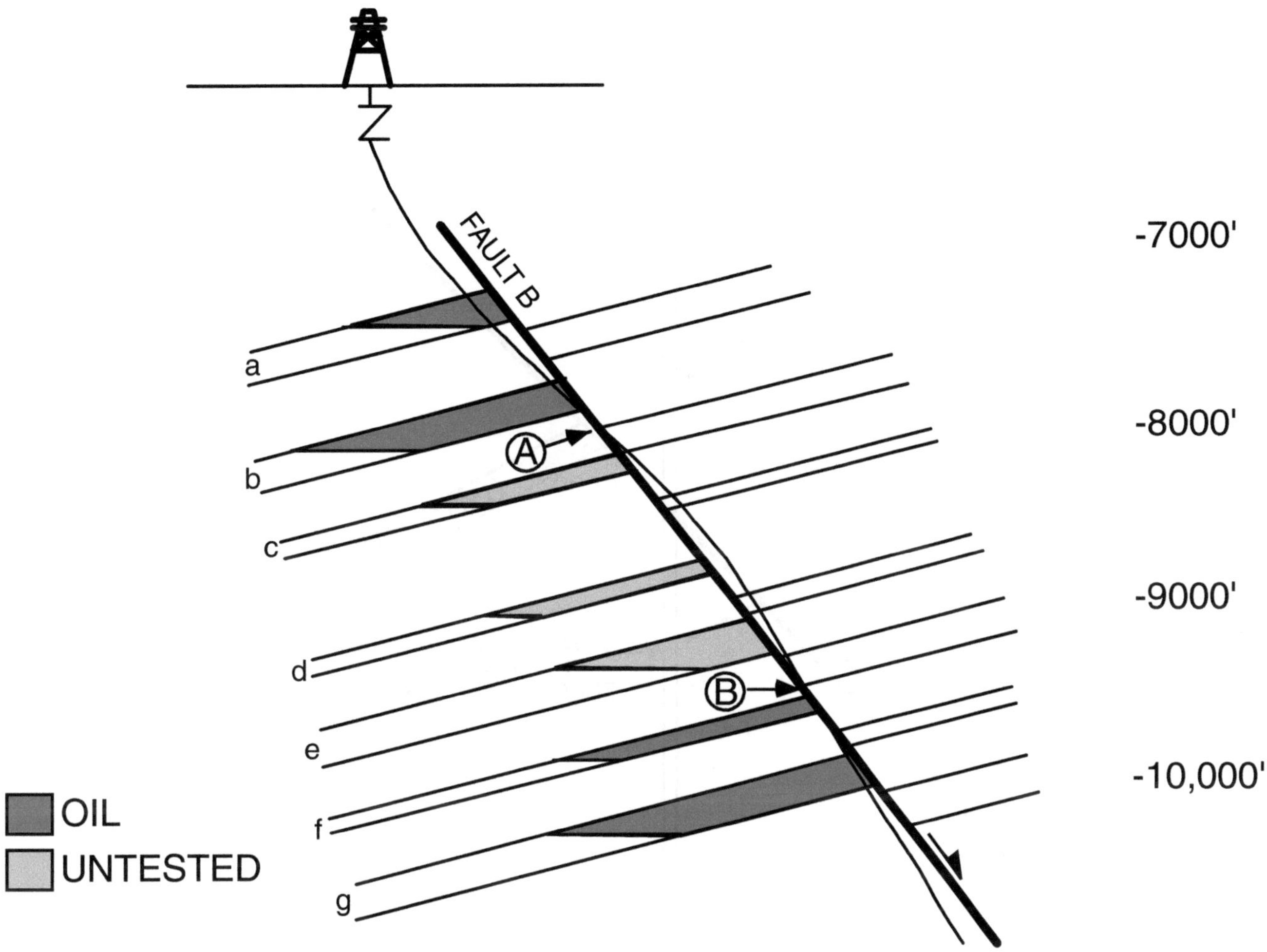

Figure 4-40 The identification of a previously unrecognized repeated section can result in a potential prospect. In this example, Sands c through e are untested upthrown to Fault B.

consequently, the wet intervals penetrated in the hanging wall were thought to be in the footwall and were condemned as nonproductive.

If an observant geoscientist recognizes the repeated section and understands that the non-productive log section represents the well penetration in the hanging wall block, a new prospect can be generated in an area previously condemned. In Figure 4-40 the new prospect includes the stratigraphic sections c through e, present in the footwall block of the fault and not seen in the deviated well.

ESTIMATING RESTORED TOPS

A particular marker or stratum being mapped may be faulted out of one or more wells in the area of study. If this situation exists, it is often possible to estimate an upthrown (footwall) and downthrown (hanging wall) restored top for the missing marker or stratum in the faulted well(s). *A restored top is defined as an estimated depth at a well for a specific marker or stratum that is faulted out of that well.* Figure 4-41 illustrates the concept of restoring tops. An upthrown (footwall) restored top is the estimated depth of a marker or bed if the well were in the upthrown block and the marker not faulted out (i.e., if the fault were farther to the left in Fig. 4-41). Similarly, the downthrown (hanging wall) restored top is the estimated depth of the marker or bed if the well

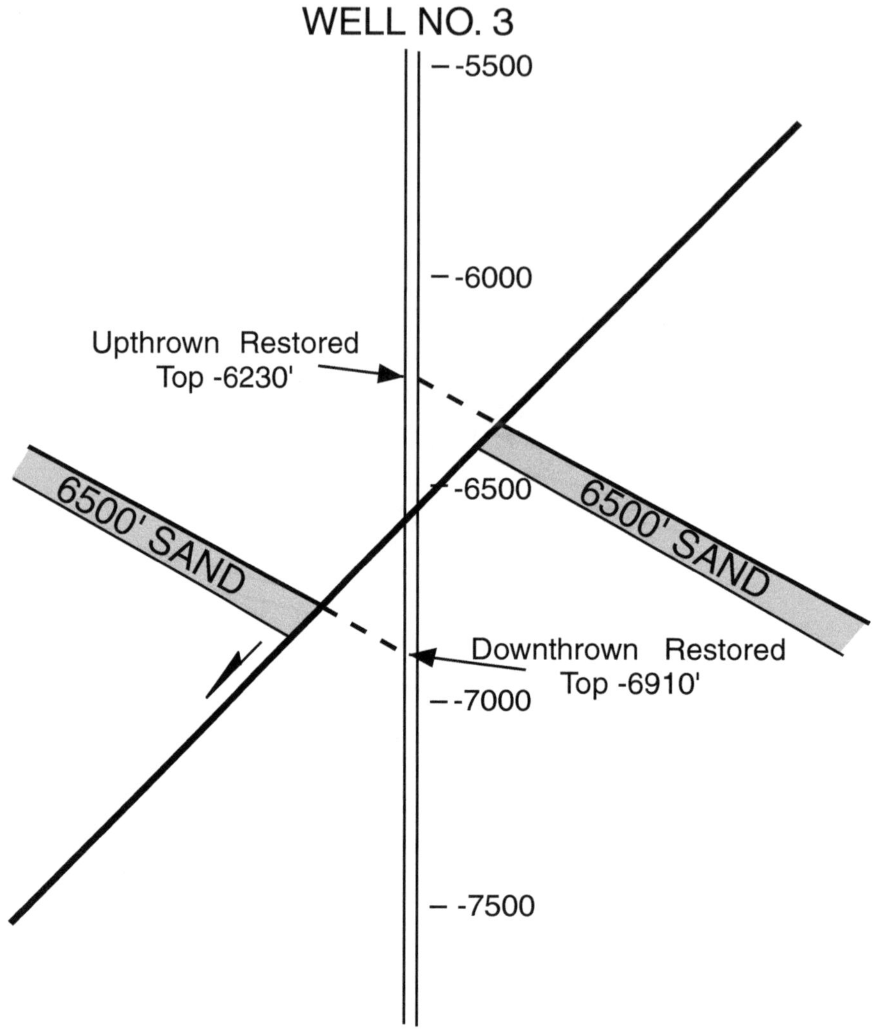

Figure 4-41 Basic concept of restoring tops. The upthrown restored top is the estimated depth in the well for the mapped horizon if the well were in the upthrown fault block. The downthrown restored top is the estimated depth in the well for the horizon if the well were in the downthrown block.

were in the downthrown block and the marker not faulted out (if the fault were farther to the right in Fig. 4-41). We strongly recommend that these data points be honored in structure mapping just as any other top obtained from well log data. This is important for maintaining the integrity of a structural interpretation and it provides additional information that aids in the actual structure contouring, as described in Chapter 8.

Vertical Wells

The procedure for determining the upthrown and downthrown restored tops (sometimes referred to as *equivalent points*) in vertical wells is discussed here and illustrated in Figure 4-42.

1. Figure 4-42 shows three vertical well logs positioned side by side with no horizontal scale. By correlation, we determine that there is a 150-ft fault in Well No. 2 and that the objective Unit C is faulted out. Well No. 1 is in the upthrown fault block and Well No. 3 is in the downthrown block. To determine the upthrown and downthrown restored tops, these two nearby wells are used.

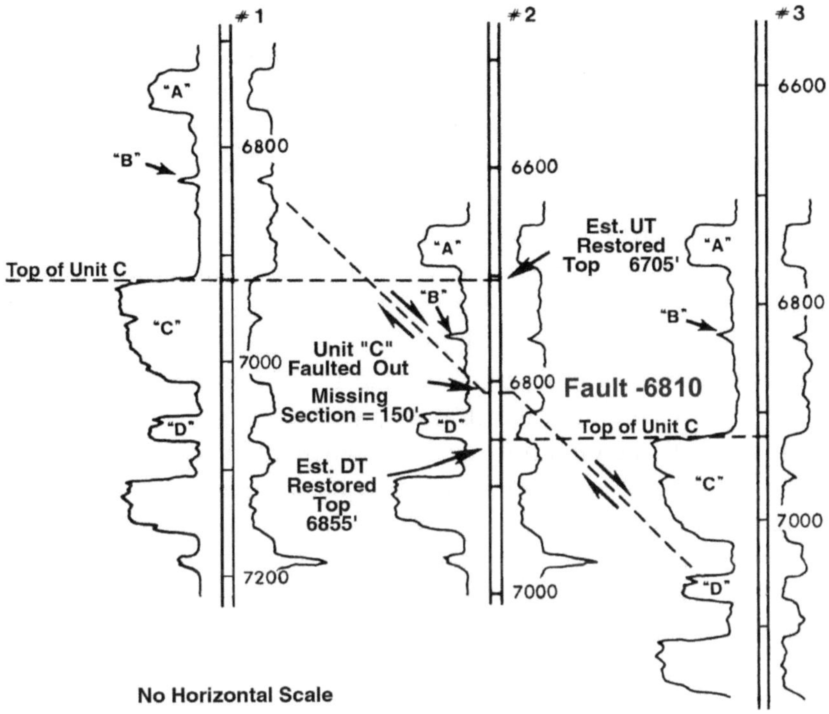

Figure 4-42 The method for estimating restored tops in vertical wells is shown by special correlation of the three vertical wells.

2. *Downthrown Restored Top for Unit C.* To estimate the downthrown restored top, Well No. 3 in the downthrown fault block is used to correlate with Well No. 2. Identify a good marker, such as a field-wide sand or resistivity marker, ***in both logs above*** the faulted-out section. In Figure 4-42, the "B" marker is used as the marker in the wells. Line up the marker on both logs and correlate ***down*** the logs until the top of "C" is reached in Well No. 3. The equivalent measured depth reading for the top of "C" in Well No. 2 is 6855 ft. This represents the estimated depth for the top of "C" if the well were in the downthrown (hanging wall block).
3. *Upthrown Restored Top for Unit C.* To estimate the upthrown restored top, Well No. 1, which is in the upthrown block, is used to correlate with Well No. 2. Identify a good marker ***in both wells below*** the faulted-out section. In this case it is the "D" marker. Line up the "D" marker on both logs and correlate ***up*** the logs until the top of "C" is reached in Well No. 1. This equivalent depth reading for the top of "C" in Well No. 2 is 6705 ft. This represents the estimated depth for the top of "C" if the well were in the upthrown block.

The accuracy of the restored top estimates can be checked by subtracting the upthrown restored top from the downthrown restored top and then comparing the difference to the estimated value of the missing section for the fault. In this case, 6855 ft − 6705 ft = 150 ft. The difference in the restored tops and the amount of missing section in Well No. 2 are exactly the same. Therefore, we conclude that the estimated depths for the restored tops are reasonable.

A similar method to restore a top is to calculate the vertical distance that the top should be above or below the fault in the well. In the example shown in Figure 4-42, a point in upthrown Well No. 1 can be picked that is correlative to the fault pick in Well No. 2; it would be just above "D." The true vertical thickness between that point and the top of Unit C can be determined. That thickness of 105 ft is equal to the vertical distance between the fault and the upthrown restored top of "C" in Well No. 2. In fact, that TVT can be used to calculate the restored top 105 ft above the fault, at 6705 ft, in Well No. 2. Similarly, the depth of the downthrown restored top in Well No. 2 can be determined by using the 45-ft TVT of the interval between the fault-correlative point and the top of Unit C in downthrown Well No. 3. In the next section, we use the TVTs of such reference intervals to restore the top of a unit faulted out of a deviated well.

In the example illustrated in Figure 4-42, the section in the three wells is approximately equal in thickness. What if a reference well is of different thickness than the well with a fault? That could result in an incorrect restoration of a top, so an adjustment to the thickness measured in the reference well is necessary. Simply apply a thickness ratio. Determine an average thickness ratio between the wells for stratigraphic intervals immediately above and below the fault in the faulted well. Then use that ratio to modify the TVT of each reference interval as measured in the reference well. For example, if the stratigraphic section in the faulted well is 0.80 as thick as that in the reference well, then multiply each reference interval measured in the reference well by 0.80 to provide a more accurate estimate for restoring each top.

The two restored tops should be honored in structure contouring of any mapped surface, whether it is a marker, sand top, particular stratum, and so on. If the map is referenced to sea level, measured log depths must be converted to subsea depths for structure mapping. The specific technique to apply the restored tops in structure mapping is discussed in Chapter 8.

Deviated Wells

All the wells are vertical in the example shown in Figure 4-42. Where deviated wells are present, the estimation of restored tops requires additional steps because TVTs must be used instead of the MLT in the deviated wells.

There are *three separate cases* for log correlation involving deviated wells: (1) a fault in a deviated well correlated with a vertical well; (2) a fault in a vertical well correlated with a deviated well; and (3) a fault in a deviated well correlated with another deviated well. Previously, we showed that missing section is defined in terms of TVT, and we also used TVT in restoring tops in vertical wells. When estimating restored tops in deviated wells, TVT must be used. Probably the most accurate way to estimate restored tops in deviated wells is to convert the deviated logs to TVT logs. In order to make a TVT log, bed dip and azimuth data are required in addition to the well deviation angle and well azimuth. If time, costs, or other factors prohibit the preparation of TVT logs, the restored tops can still be estimated with reasonable accuracy using the deviated well logs and applying a correction factor to derive TVTs. The correction of MLT to TVT was described in the section Estimating the Missing Section for Normal Faults.

For case No. 1, in which a fault in a deviated well is correlated with a vertical well, the same basic procedure for estimating a restored top using two vertical wells can be applied, with one significant variation. The depth of a restored top for the faulted-out unit cannot be transferred (projected) directly to the deviated well log from the top of the unit in the vertical well log, as it was in Figure 4-42, because the deviated well log depths are not vertical depths and the thickness of a given interval is not a TVT. A calculation for depth of each restored top must be made, and it is based on the TVT of a reference interval in the vertical well.

The procedure for restoring a top faulted out of a deviated well is shown in Figure 4-43. A fault is identified in the deviated Well No. 3 at a depth of −6810 ft, with a missing section of 100 ft (Fig. 4-43a). The B Sand is faulted out of the well. We wish to restore the upthrown and downthrown tops for the B Sand. Well No. 1 in the downthrown block and Well No. 2 in the upthrown block are available to estimate the restored tops.

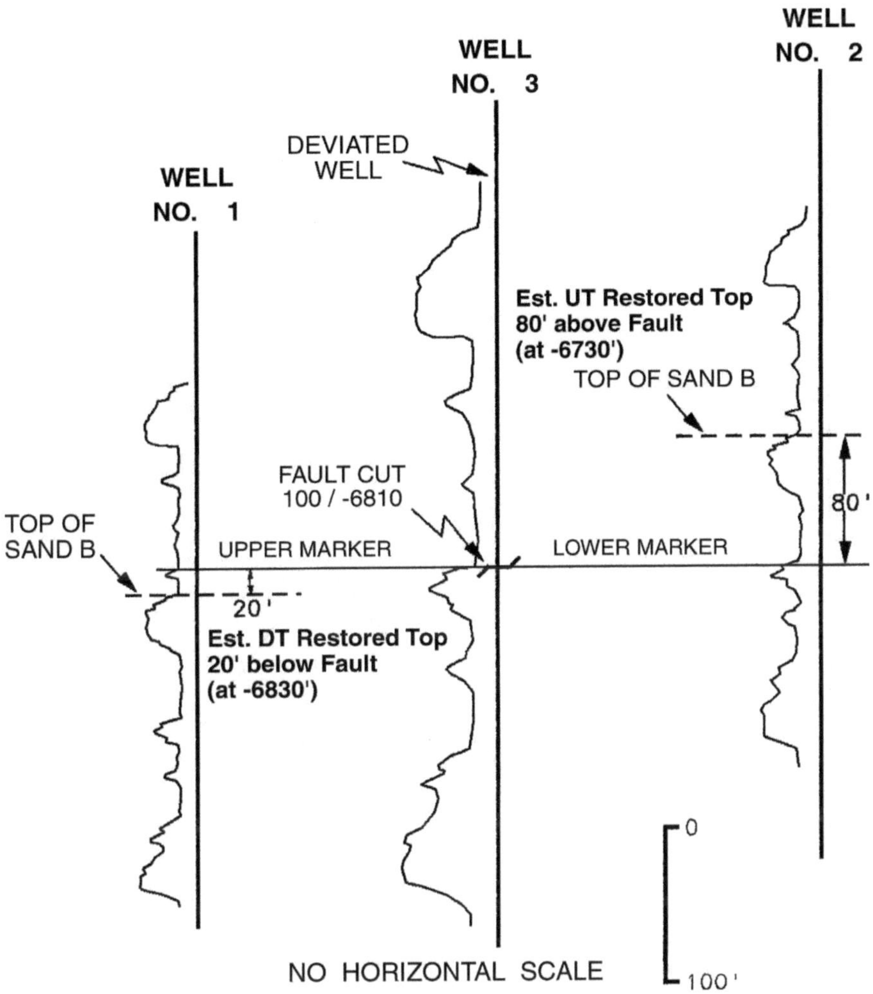

(a)

Figure 4-43 (a) The two vertical wells (Nos. 1 and 2) and the deviated well (No. 3) illustrate case No. 1, estimating restored tops for a unit faulted out of a deviated well. (b) A structural cross section containing the three wells shown in (a); it illustrates the precise positions of the restored tops with respect to Well No. 3: directly above (upthrown restored top) and directly below (downthrown restored top) the actual fault pick in the well. The top of Sand B in each fault block is projected to the appropriate restored top to establish the structural dip shown in the cross section.

1. *Upthrown Restored Top for the B Sand.* To estimate the upthrown restored top, identify a marker ***below*** the B Sand in Well No. 2 that is also in the faulted well. For this example, the lower marker used and shown in Well No. 2 is exactly at the fault in Well No. 3. By correlating up the logs, we determine that the top of the B Sand is 80 ft above the lower marker in Well No. 3. Since the marker is exactly at the fault in Well No. 3, the upthrown restored top for the B Sand in the well is 80 ft above the subsea depth of the fault. The subsea depth of the fault is −6810 ft; therefore, the upthrown restored top is −6730 ft. It is positioned vertically above the fault in the deviated well rather than within the inclined wellbore (Fig. 4-43b).
2. *Downthrown Restored Top for the B Sand.* The same procedure is followed to estimate the downthrown restored top. Using Well No. 1 in the downthrown fault block, identify a marker ***above*** but as close to the fault cut as possible. In this case, the upper marker used in

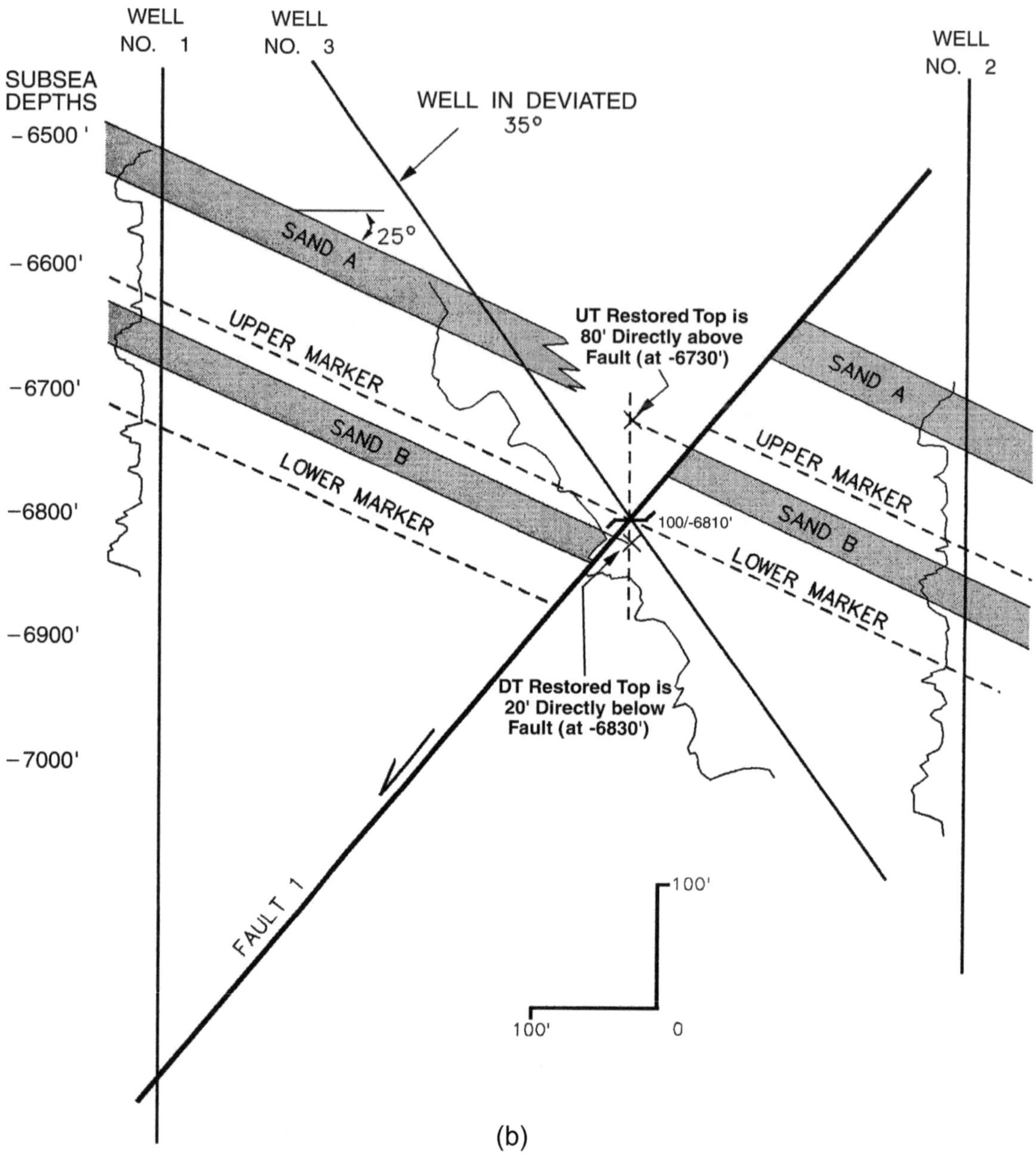

(b)

Figure 4-43 *(continued)*

Well No. 1 is exactly at the fault in Well No. 3. Correlate down the two logs from the upper marker until the top of the B Sand is reached in Well No. 1. Using the TVT of that interval, the restored top is calculated to be 20 ft vertically beneath the fault in Well No. 3, at −6830 ft (Fig. 4-43b).

The depths for the restored tops were not projected directly from faulted Well No. 3, as was done in the vertical well example (Fig. 4-42). Instead, for the upthrown restored top, the ***vertical distance*** from the fault-correlative point to the top of the sand in unfaulted Well No. 2 was *subtracted* from the subsea depth of the fault to estimate the upthrown restored top. Similarly, the vertical distance from the fault-correlative point to the top of the sand in unfaulted Well No. 1 was *added* to the subsea depth of the fault to estimate the downthrown restored top.

Observe that the correlation marker used to determine each restored top was chosen as the closest marker to the fault. It is recommended that, whenever possible, you choose a correlation marker as close to the fault as possible in order to minimize any errors due to correlation or due to stratigraphic variations between the wells being used.

In Figure 4-43b, which is a structural cross section through the three wells, an X is placed 80 ft vertically above and 20 ft vertically below the fault in Well No. 3. These X's show the precise position of the upthrown and downthrown restored tops for the B Sand. It is important to understand that the restored tops are located ***vertically above and below*** the location of the fault in the well and ***not within*** the deviated well either in cross-section view or in map view. This understanding is critical when using these restored tops in structure mapping.

The fact that restored tops in a deviated well are plotted directly above the location of the fault in the well can cause complications when using a workstation. We are not aware of any interpretation software that will plot a restored top directly above the fault cut. If the restored top is stored with other tops from the well in the software database, the restored tops will plot along the wellbore, not directly above the fault cut. This will cause the restored tops to be plotted in the wrong spot on a map. It is possible that an upthrown restored top may be plotted in the downthrown fault block and vice versa. One possible solution is to store the restored tops in a pseudo-well as digital control points at the same X-Y position as the fault cut. The challenge of using restored tops in structure mapping using a workstation is discussed in more detail in Chapter 8.

The 100-ft difference between the downthrown and upthrown restored tops agrees with the 100 ft estimated for the missing section in the well. Therefore, we conclude that the depths for the estimated restored tops are reasonable. The depth difference between the downthrown and upthrown restored tops should always agree with the amount of missing section estimated for the fault.

For cases 2 and 3, in which a fault in a vertical well is correlated with a deviated well or a fault in a deviated well is correlated with another deviated well, the procedure is basically the same as for case 1 with one exception. When an unfaulted deviated well is used as a reference, the ***TVT*** from a marker to the mapped (faulted) horizon ***must be calculated*** to accurately estimate the restored tops. The measured (deviated) log thickness cannot be used to properly estimate restored tops. For example, if Well No. 2 in Figure 4-43a were a deviated well, the interval thickness of 80 ft measured from the upper marker to the top of Sand B would have to be corrected to TVT before calculating the subsea depth of the upthrown restored top.

UNCONFORMITIES

Unconformities are present in all geological settings, especially on steeply dipping growth structures. Excellent hydrocarbon traps can occur at unconformities where overlying impermeable rocks form seals for reservoirs. Therefore, it is important to recognize, analyze, and often map unconformities in the subsurface.

There are a host of interrelated versions of the broad term *unconformity*. Some are primarily erosional, some are nondepositional, and others are combinations of both. The subject of unconformities is extensive and beyond the scope of this book. We present important information on the recognition of unconformities during electric log correlation.

An unconformity appears on an electric log as missing section. Since missing section can be due to an unconformity or to a normal fault, care must be taken during correlation so that an unconformity is not mistaken as a fault. In this section we discuss several general guidelines to follow during correlation to recognize an unconformity.

1. Structural dip commonly is different above and below an unconformity. Dipmeter data can be used to indicate this change in dip. The structural dip below an unconformity is usually steeper (Fig. 4-44).
2. If missing section is recognized in two or more wells at the same or nearly the same correlative horizon, such as "C" in Figure 4-44a, an unconformity, rather than a fault, should be suspected.
3. The amount of missing section resulting from an unconformity increases in the up-structure direction. This is illustrated in Figure 4-44. The missing section in Wells No. 1, 2, and 3 increases in the up-structure direction; Well No. 3 has the least amount of section missing and Well No. 1 has the greatest amount missing.
4. The stratigraphic section below an unconformity is truncated in a younger sequence in down-structure wells than in up-structure wells. The truncated sequence becomes older in the up-structure direction. In Figure 4-44a, the sedimentary sequence just below the unconformity in Well No. 3 (the G interval) is younger than the K interval in Well No. 1. To recognize the sequence trend, you must correlate **up** the well logs.
5. If the depositional environment results in an onlap sedimentary sequence (Fig. 4-44b) rather than the sequence shown in Figure 4-44a, the missing section in logs correlated in the up-structure position increases above and below the unconformity. The sedimentary sequence just above the unconformity is younger in the up-structure direction.
6. Unconformities must be mapped. Since an excellent hydrocarbon trap can occur at an unconformity, a map of the unconformity is vital. In order to identify the intersection of the unconformity and the underlying sedimentary sequence, an unconformity map must be integrated with structure maps of relevant surfaces within that sequence. The mapping techniques required are discussed in Chapter 8.

Figure 4-45 illustrates the use of a dipmeter to aid in the identification of an angular unconformity. The figure also shows a dipmeter response to beds affected by the proximity of salt with an overhang. The dipmeter reaches a maximum dip of 62 deg at about −6500 ft; thereafter, the dip slightly flattens to 40 deg at TD. This change in dip is in response to the proximity to the salt face. Such dipmeter responses can be used in evaluating a salt feature for possible overhang, with potential stratigraphic section beneath it.

In the mid-1990s, Dr. Richard Bischke developed a powerful new technique for evaluating structural and stratigraphic information from well log as well as seismic data. This technique, today known as the Multiple Bischke Plot Analysis (MBPA), has numerous applications (Bischke 1994b; Bischke et al. 1999). One application of the technique is to assist in the differentiation of unconformities from faults in well logs where missing section is present. It is a rapid, robust, and powerful method for application with unconformities, as well as validation of general log correlations (see Chapter 13).

ANNOTATION AND DOCUMENTATION

The importance of good, accurate documentation cannot be overemphasized. The generation of good quality maps depends on a volume of accurate data. These data include faults, horizons, tops and bases of stratigraphic units, net sand, net pay, and so on. In this section, we illustrate a recommended method of annotating 1-in. and 5-in. electric logs and documenting the log data.

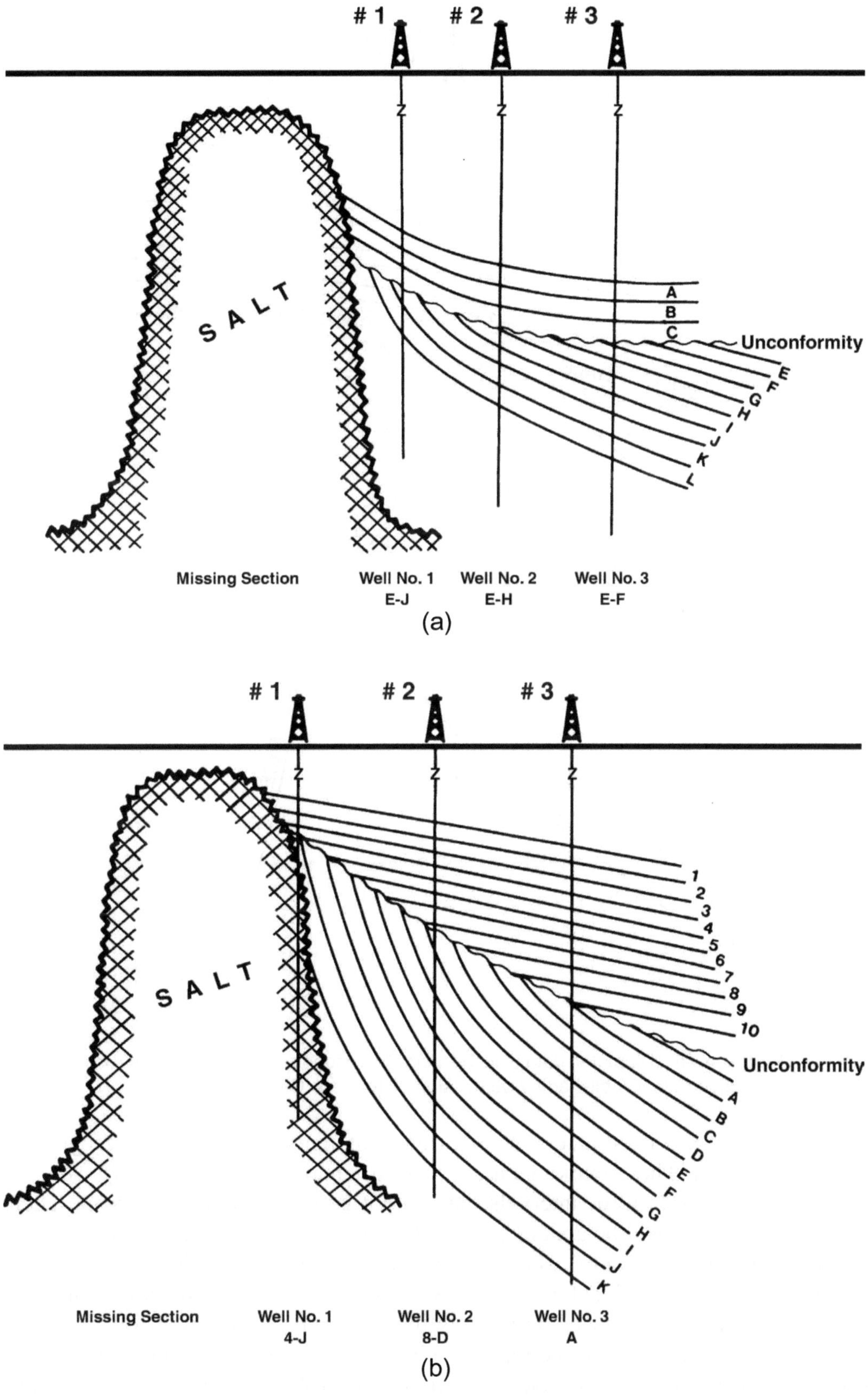

Figure 4-44 (a) Well log correlation can be used to recognize an angular unconformity. (b) If an onlap sedimentary sequence is deposited above an angular unconformity, certain log correlation guidelines can be used to recognize the unconformity.

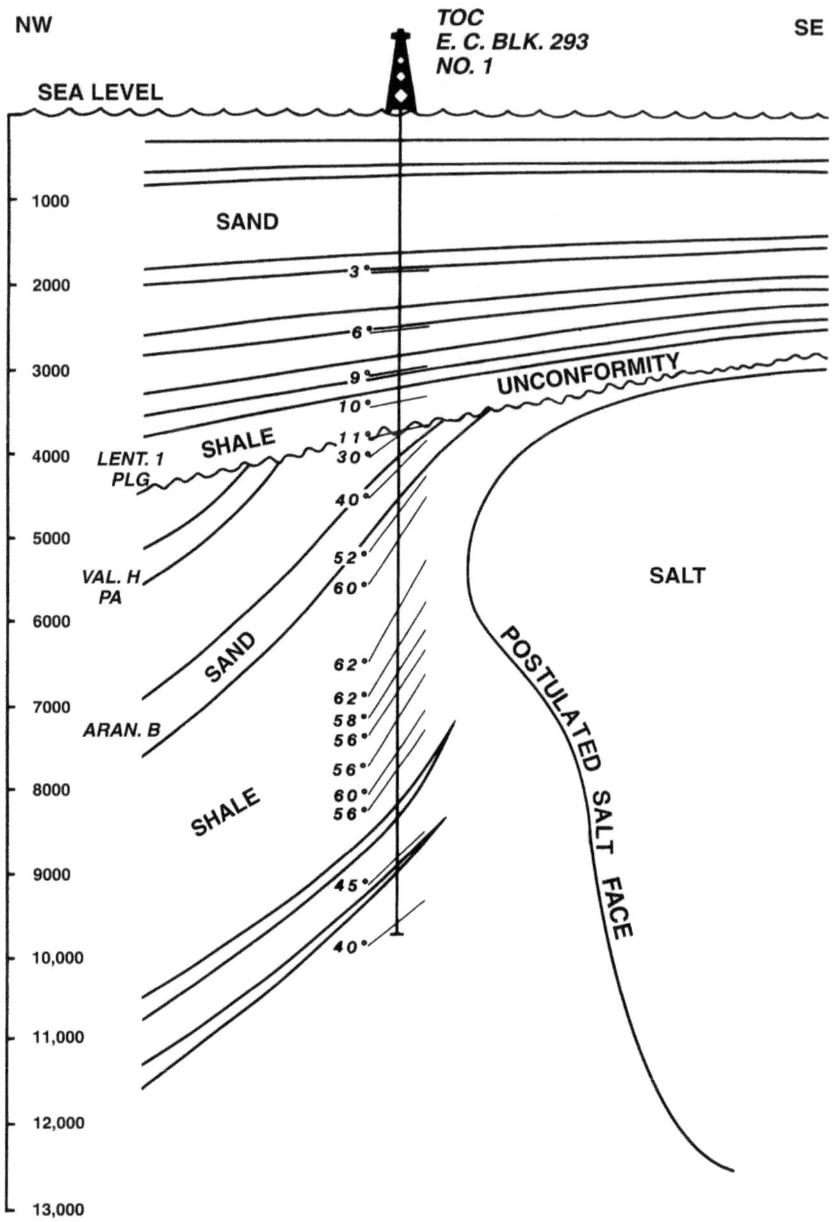

Figure 4-45 Dipmeter data can be used to recognize an unconformity in the subsurface. (Published by permission of Tenneco Oil Company.)

In Figure 4-2 we illustrated the importance of marking logs with recognizable symbols and the use of color. These markings are a form of annotation. *They identify your correlations.* The logs in Figure 4-2 are of intervals that do not contain any hydrocarbons. There are some additional data that should be annotated on a 1-in. or 5-in. log with recognized pay. Figure 4-46 illustrates the additional data that should be annotated in the pay section on a 1-in. log. The annotation includes the name of each pay zone, perforation intervals, well status, cumulative production from each interval produced and a note on why each interval went off production, and the measured and subsea depth for the top of each important productive interval. Any intervals that appear productive on the log but have not been produced should be noted, such as Sands 2 and 6A in Figure 4-46. If sidewall cores are taken, the results of analyzing these cores should be posted. Finally, any recognized faults must be indicated on the log, such as Fault F, which faults out Sand 3 at a log depth of 8350 ft. It is important to note that the fault cut in Figure 4-46 includes the amount of missing section and the well or wells from which the missing section was determined.

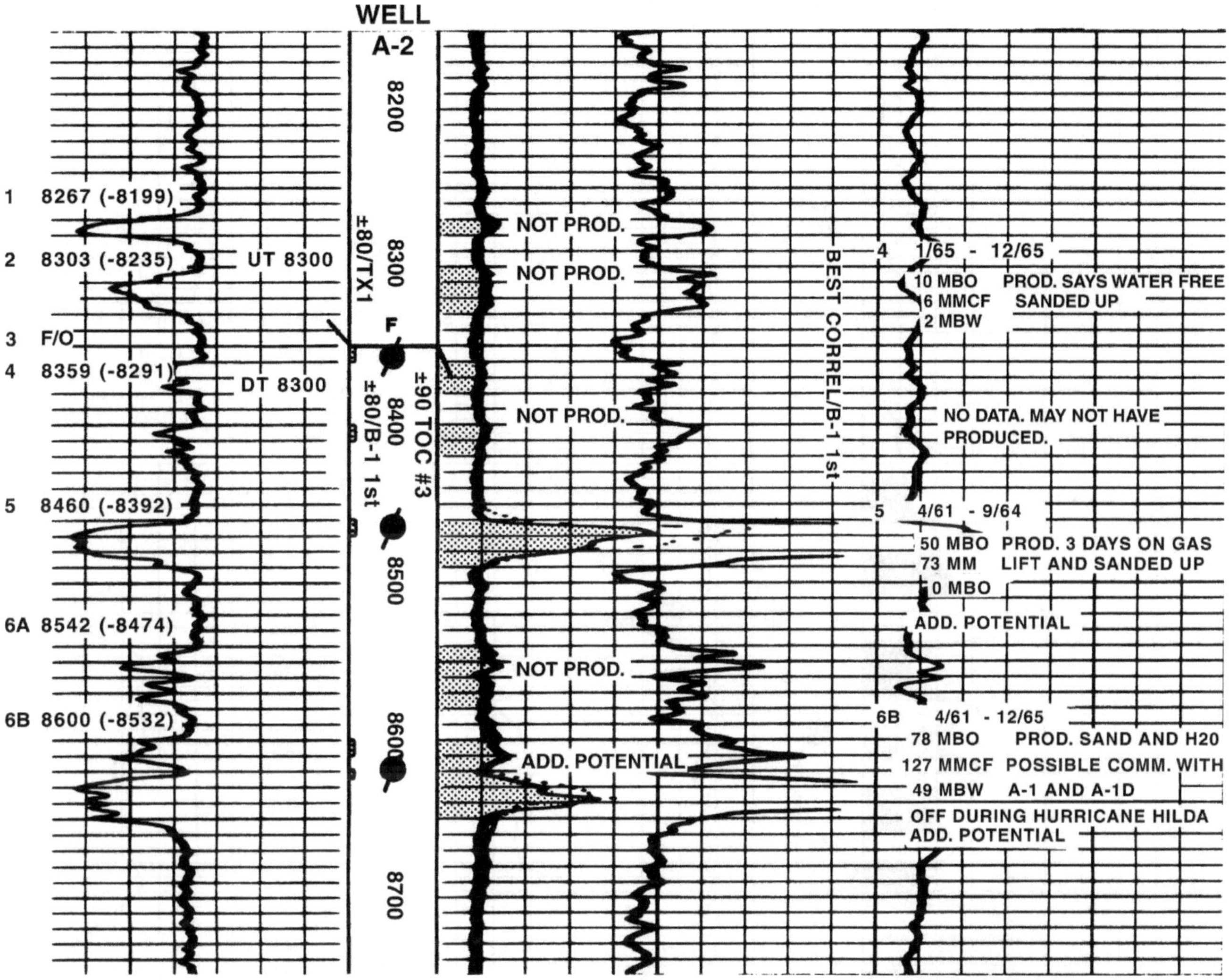

Figure 4-46 Electric log annotation in a hydrocarbon-bearing (pay) section. (From Tearpock and Harris 1987. Published by permission of Tenneco Oil Company.)

Annotating data when correlating logs on a workstation is, in many cases, easier as the software automatically enters tops into a database. Some data, such as missing section associated with a fault, is usually not included in the database automatically. Most software interpretation packages include places where additional notes, such as missing section, can be added to the database. It is important that interpreters use these features and record all important data. If picking up an interpretation project from a colleague, it is also important to check the note section of a database. Unlike a note handwritten on a paper log, you will never see a note in a computer database unless you look for it.

The annotation of a detailed 5-in. log is crucial, and an example is shown in Figure 4-47. The information annotated includes the name of productive zone, measured and subsea depth of the correlated tops, perforation intervals and corresponding production, well status, net pay counts, limit of pay (full to base of interval or water contact), and basic core data (at least the porosity and permeability data). Notice that the net pay on the 5-in. log in Figure 4-47 is assigned per 10-ft intervals on the left side of the log. This annotation is used to support the net pay count and later in preparation of net pay isochore maps.

Finally, the documentation of well data means recording it in some format that can be easily used. There are various types of manual and computer data sheets available for documenting mapping parameters. These data sheets should be used at all times to document the log correlation data.

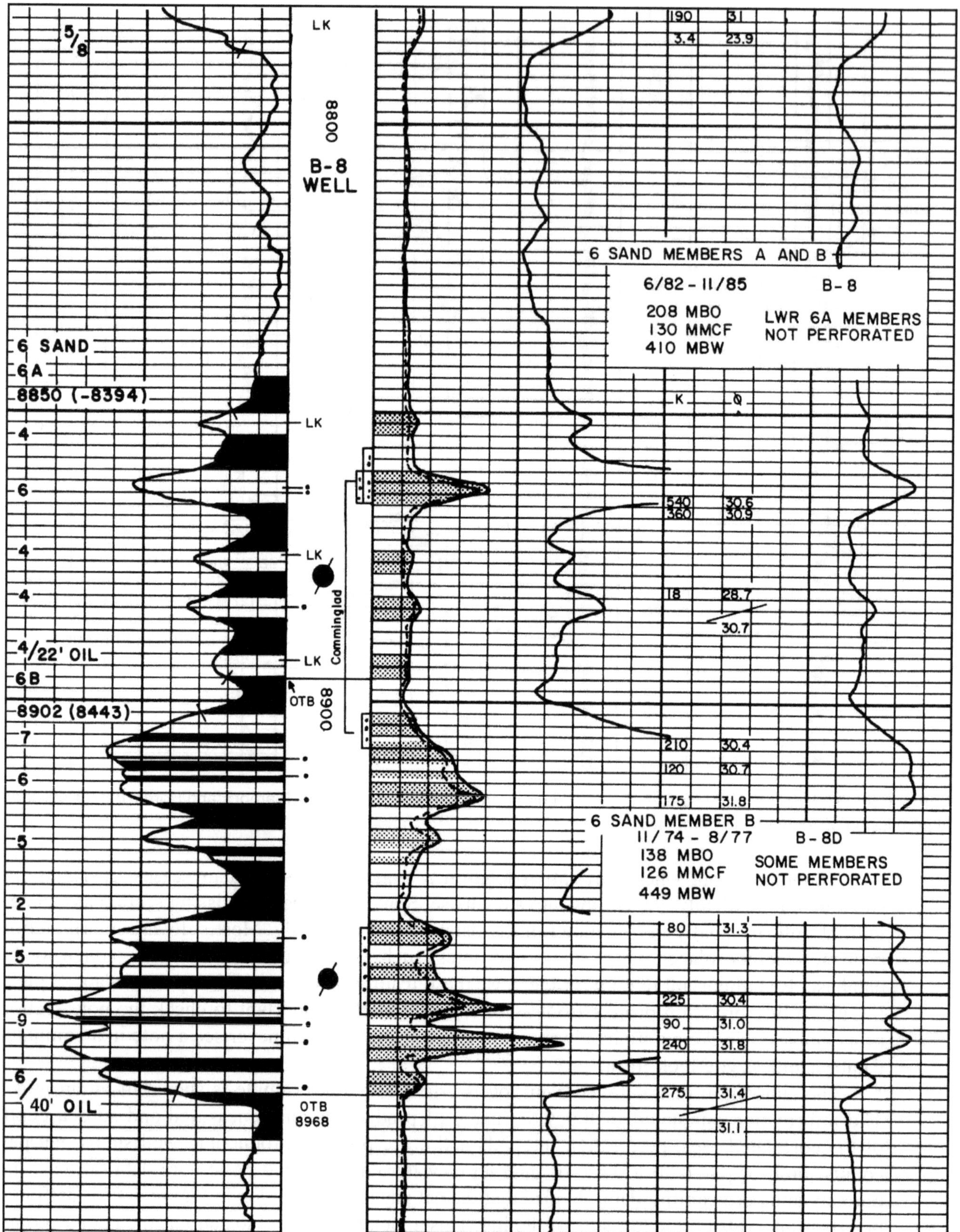

Figure 4-47 Annotation of a detailed 5-in. log. (Modified from Tearpock and Harris 1987. Published by permission of Tenneco Oil Company.)

CHAPTER 5

INTEGRATION OF GEOPHYSICAL DATA IN SUBSURFACE MAPPING*

INTRODUCTION AND PHILOSOPHY

Since the publication of the first edition of this book, seismic interpretation and mapping techniques have been revolutionized through the introduction of sequence stratigraphy (Wilgus et al. 1988; Van Wagoner et al. 1990) and 3D-migrated data sets (Brown 1999). Since the publication of the second edition of this book, 3D seismic data sets, interpreted exclusively on computer workstations, have become increasingly common. Chapter 9 covers the subject of 3D seismic interpretation methods and techniques. In this chapter, we review the *basic principles* of seismic interpretation and the integration of seismic data with well log data, along with the philosophy of geophysical mapping and the introduction of correct mapping procedures, techniques, and methods. These subjects are just as relevant today as they were prior to the development of the seismic workstation. These basic principles can determine the ultimate success of an interpretation and mapping project. More information on basic seismic interpretation from a geophysical perspective is provided by Herron (2011).

Some geoscientists may encounter interpretation projects that do not involve the seismic workstation, and others may wish to review the basic principles of seismic interpretation and integration. This chapter is designed to accomplish these tasks and to help the geologist and the new geophysicist take their first steps into the exciting but demanding world of applied geophysics.

This chapter also discusses the general principles and the details involved in the use of 2D geophysical data as applied to subsurface maps. More specifically, the discussion centers on the

* For all figures in this chapter (in the printed book only), see the preface for information about registering your copy on the InformIT site for access to the electronic versions in color.

use of reflection seismic data both to aid in the visualization of the subsurface geology and to extract data useful in the creation of accurate maps. The first section contains a general discussion of the integration of well-log and seismic data as applied to both the 2D workstation and paper sections, as well as the benefits and limitations of the seismic method. The second section is a more detailed discussion of some of the techniques and procedures for integrating seismic data into subsurface geological maps.

The discussions in this chapter are intended to benefit the individual who may not be familiar with seismic data and, indeed, may not understand how seismic data are acquired and processed. We focus on practical approaches to using seismic data in the search for hydrocarbon traps. The technical details of seismic acquisition and processing are beyond the scope of this book, as is the topic of theoretical geophysics. These are very important subjects that a working geoscientist must understand. Many interpreters who have access to seismic data are not geophysicists. However, it is our intent to illustrate techniques that will make the non-geophysicist comfortable with using these data in the construction of subsurface maps.

This chapter should make it obvious that valuable information is present in seismic data and that an interpretation that properly integrates the subsurface geological data with the seismic data is always more accurate than an interpretation that ignores one of these data sets. It will soon become apparent that the discussion has a strong regional bias in that most of the examples are from the offshore Gulf of Mexico. There are several reasons for this. Perhaps the most obvious reason is that this region has a greater abundance of high-quality seismic data than anywhere else in the world. This fact means that (1) it is easier to get good examples from this region than from most others, and (2) this region is highly prospective, which increases the likelihood that North American geoscientists will work in this region at some point during their careers.

This regional orientation does not mean that the techniques outlined are limited to the Gulf of Mexico. In fact, the techniques presented here can be used to establish the three-dimensional geometric validity for subsurface maps in any tectonic environment, anywhere in the world.

Seismic Data Applied to Subsurface Interpretations

On a fundamental level, seismic data can provide two major benefits. First, it can be acquired in frontier areas or over areas that have sparse well control. An interpretation and generated maps can thus be extended with some confidence into areas that have little or no well control. This is an important benefit, especially when one considers that few wildcat prospects actually have wells in the immediate area of the prospect. The second benefit is that seismic data can provide explicitly 2D and 3D data as opposed to the one-dimensional nature of a wellbore. The 2D and 3D character of seismic data as opposed to well data is illustrated in Figure 5-1. It should be added that in most cases, the 2D appearance of a seismic section is an artifact of the data being reduced to a flat sheet of paper or computer screen. The data on the line may actually represent a very complex 3D subsurface geological world!

Where a geological structure is complex in three dimensions, the most insidious and potentially dangerous pitfall is assuming that all the data on a section represent a planar slice through the earth directly underneath the line. In complex areas, the data on a line may not represent the geological structure directly beneath the line, as illustrated in Figure 5-2. Methods for handling some of these effects, called *sideswipe* (Sheriff 1973), are discussed later in the chapter, although sometimes sideswipe cannot be fully corrected on 2D seismic data and can just be recognized and ignored (Fig. 5-3). However, the two-dimensional nature of 2D seismic data often does mean, when compared to the use of well data, that one less dimension must be inferred in order to construct an accurate subsurface structural interpretation.

Assumptions and Limitations

The techniques outlined in this chapter assume that the data used are properly acquired and processed up to and including migration of the data. The techniques also assume that the geology of the subsurface beneath the line permits the acquisition of good quality data. Of course, there are many areas where the subsurface does not cooperate and does not yield good seismic data. Some possible major problems include severe horizontal velocity gradients, high noise areas, high bed dips, and extremely complex geology, all of which may invalidate many assumptions necessary for the acquisition of good data. If these problems are present, they may pose a challenge for even the best geophysicist. In such instances, the expertise of a geophysicist and structural geologist may be necessary to solve the complexities. In areas of complex faulting and structure, a 3D data set may help resolve the structural complexities. However, even in these complex areas, 2D seismic data may contain valuable information that can be used in creating a reasonable subsurface interpretation. Examples of the usefulness of seismic data in these complex areas are presented in the section that covers structural balancing (Chapter 10).

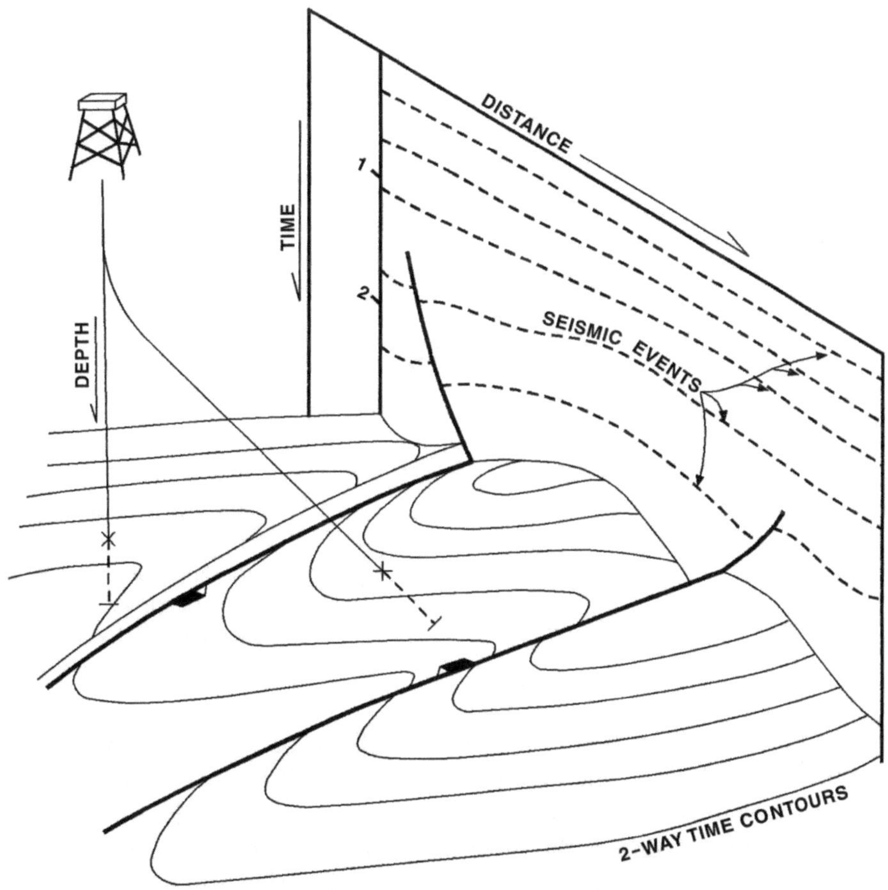

Figure 5-1 Sampling differences between wells and surface seismic events. Well data are single points, and correlations of points between wells must be inferred. Seismic events, however, explicitly demonstrate horizon continuity.

The techniques and parameters employed in field acquisition of seismic data can influence the quality of data, so they should be based on knowledge of the geology of the area and the acquisition procedures that are most suitable. Careful planning can result in achieving the best possible data set given the geological constraints.

THE PROCESS

The procedure for making subsurface maps from seismic data is similar to the sequence of steps used in constructing interpretations from subsurface well log information. The first step is one of data validation; that is, analysis of what the seismic data represent. Do the seismic data actually have some relationship to the geology in the subsurface? This procedure is similar to the checking one does when a log is first used. In the case of log data, decisions about the validity and meaning of the log response must be made before the data can be used to form an interpretation.

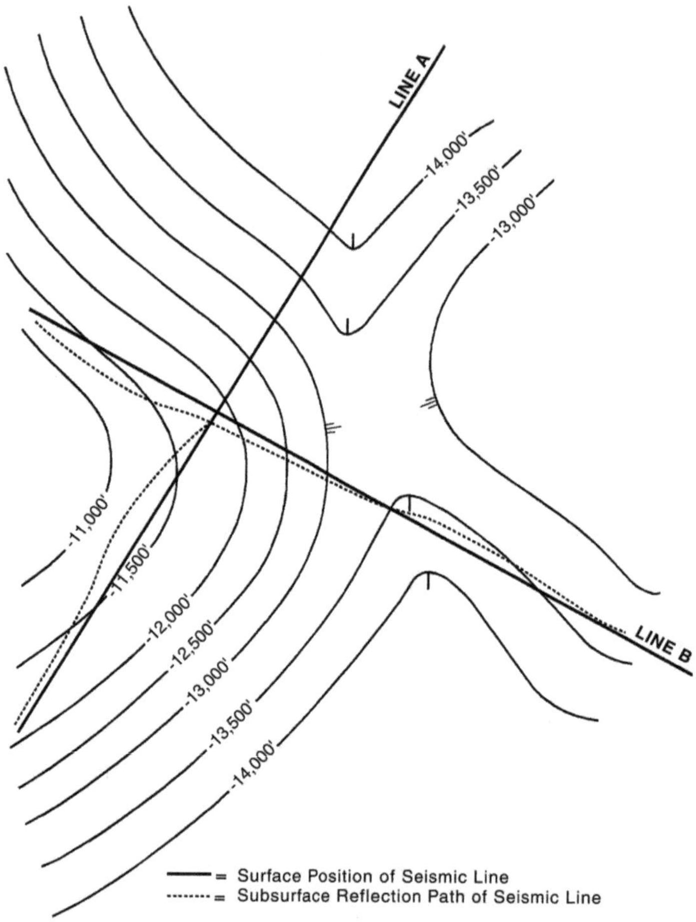

Figure 5-2 Effect of subsurface structure on actual subsurface reflection path of line. Subsurface reflection points do not occur vertically beneath the surface location of the lines where reflected from dipping subsurface horizons.

The second step is the actual interpretation of the seismic section. This step is analogous to the correlation of well logs when using subsurface well log information. Because the validity of the remaining work rests on having an accurate and geologically correct interpretation of the seismic data, validating the data is the most important part of the process.

Some aspects of seismic interpretation as they relate to the construction of subsurface maps are covered in this chapter. However, we do not attempt to cover the subject exhaustively. Several excellent books on seismic interpretation are listed in the references and are recommended to those who may be unfamiliar with seismic interpretation techniques (Badley 1985; Herron 2011). Just as a basic knowledge of well logs is needed to use log data properly, a basic understanding of the reflection seismic method is needed before seismic data can be interpreted correctly.

The third step involves extracting the information from the seismic data and transferring it onto the map so that it can be used effectively. The 2D seismic workstation process collects data in relation to a basemap. Usually, transferring the data to a map is referred to as *posting*. This

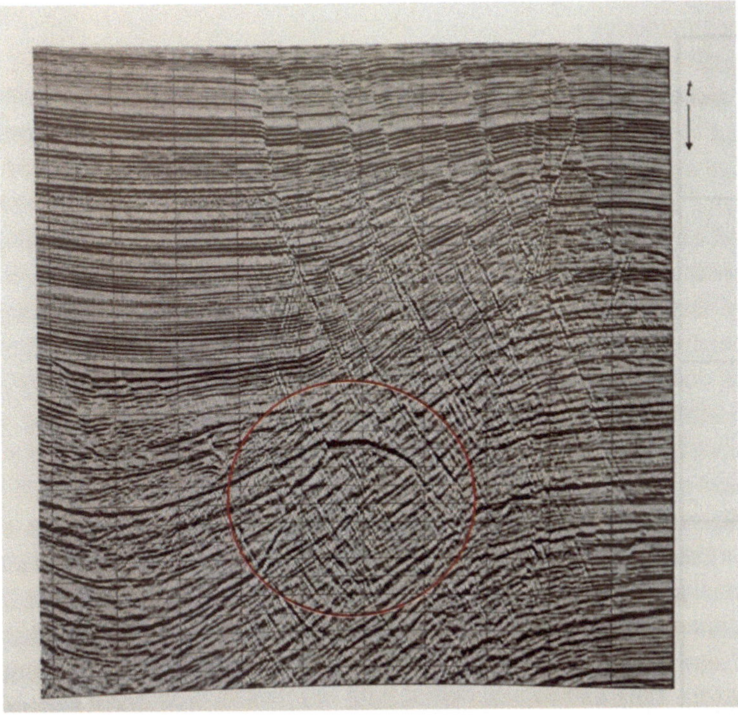

Figure 5-3 The antiformal feature within the red circle on this 2D time-migrated seismic line is an example of an out-of-plane reflection called side-swipe. (Herron 2011. Published by permission of Society of Economic Geophysicists.)

procedure is practically identical to that used when recording subsurface well log data. As seismic sections have a 2D aspect that well log data do not possess, there are some unique aspects to this step when using seismic data. This step also represents the merger of the subsurface well log and the seismic information. Both types of data should be posted and used to construct the final interpretation. If you do not understand the seismic data and require assistance, experienced interpreters are usually available. There is typically some usable information on even the worst seismic data that can add to the confidence and validity of your final subsurface interpretation and accompanying maps. *A valid interpretation should agree with and satisfy all types of information.* If 3D data are available, all the data should be used.

We appeal to geophysicists as well as geologists to work in a synergistic manner. Very often, there are two sets of interpretations and maps: the *geological* one and the *geophysical* one. A subsurface interpretation and maps should accurately represent the subsurface geology, incorporating both well log and seismic data into a seamless interpretation. There is only one configuration of the subsurface, and it is the job of the interpreters (geologists and geophysicists) to create an *integrated and reconciled interpretation* ***using all the data available.***

The last step is the construction of the subsurface geological maps. This step represents the culmination of all previous work, and in many instances, it will be the result by which your work is measured. Any subsurface map is only as good as the information it contains, so do not rush to begin this step before the previous steps are completely finished.

In practice, however, constructing a map is never the last step. Several iterations of validation, interpretation, and mapping are typically necessary before a satisfactory subsurface map is completed. Figure 5-4 is a conceptual flow diagram of this process. At some point, it will become apparent that most of the major questions have been resolved and a satisfactory interpretation and maps have been made. While pride and satisfaction in the result are deserved, keep in mind that additional data from either drilling or additional seismic acquisition will almost always change some of your ideas. Furthermore, as seismic profiles are not geological profiles, subsurface maps can never perfectly represent the structural configuration within the earth. We lack perfect and complete velocity functions over our data sets, and we lack the high-frequency

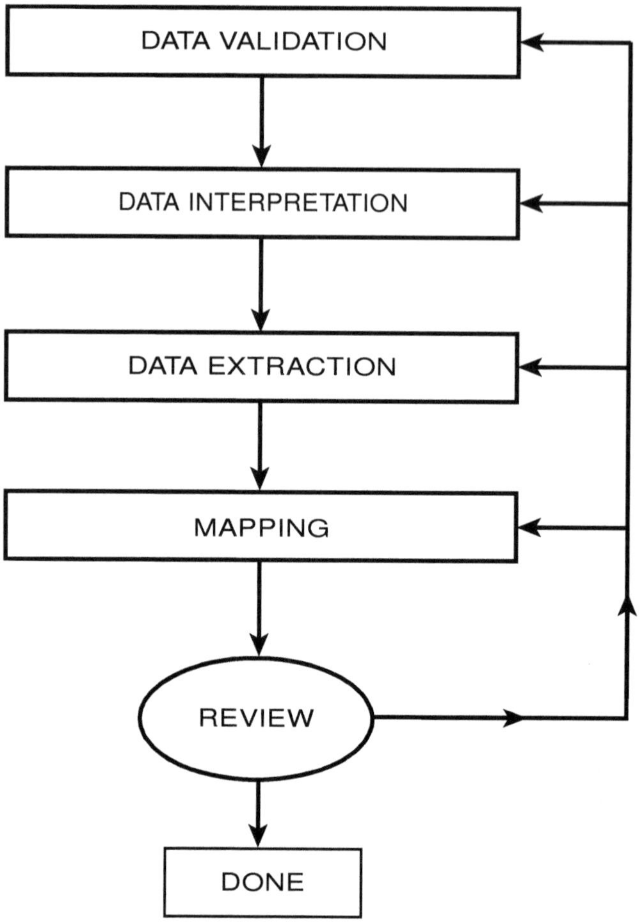

Figure 5-4 Flow diagram of subsurface mapping procedures.

wavelengths required to resolve geological subtleties. The more your ideas are actually tested, the more obvious it becomes that interpretation and mapping are both an art and a science. Ideally, you will asymptotically approach the truth as more data and better interpretation techniques become available. The measure of an interpreter is his or her ability to approach the truth quickly with the limited data available.

DATA VALIDATION AND INTERPRETATION

Examining the Seismic Sections

The first step toward obtaining the information you need from seismic data is to examine the 2D or 3D lines on the workstation. Start the process by deciding what the data represents. The majority of seismic data that are used for subsurface interpretations and mapping are seismic time sections, although seismic depth sections are becoming increasingly common. Figure 5-5 is a seismic time section over a simply deformed area. It is very tempting to think, "This is easy; all those dark lines are the rock layers, and at shallow depths, there is little difficulty picking the fault that dips toward the left part of the section. The fault trace on the line is concave upward, so this must be one of those common listric faults that everyone writes about." Without realizing it, you have made assumptions about the data and the geology that may or may not be justified. In many cases, these assumptions are close enough to the truth that it really doesn't matter. In other situations, these assumptions, while not completely without merit, may bias your interpretation in a way that leads you completely down the wrong path.

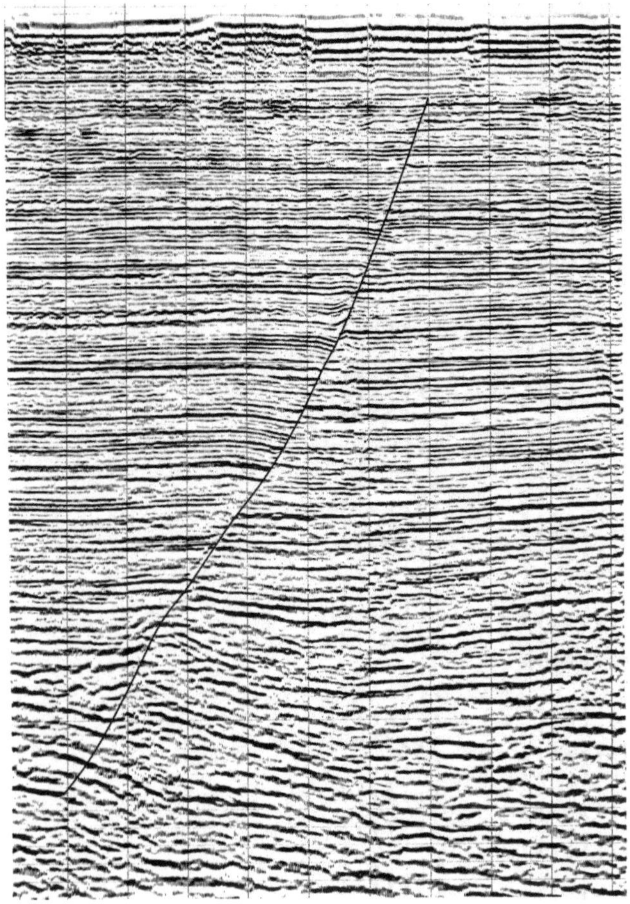

Figure 5-5 Seismic line over faulted area, Gulf of Mexico. (Published by permission of TGS/GECO.)

The first incorrect assumption is that the reflections represent discrete layers of rock. A reflection seen on a section may or may not represent a discrete sedimentary boundary. The vertical complexity of the sedimentary sequence and the frequency content of the recorded and processed seismic signal determine the appearance of the seismic wiggles. Figure 5-6 is a *synthetic seismogram* illustrating the relative "size" of seismic wiggles in an average velocity Tertiary section, such as that in the Gulf of Mexico, in comparison to a well log curve. It is obvious that the *vertical resolution* of a well log is vastly superior to that of a seismic trace. The seismic wiggle trace is a composite of waveforms from reflections from many boundaries in the subsurface. Figure 5-7 illustrates how a series of interfaces can combine or convolve their reflections to produce a simple seismic reflection.

At this point, you may be overwhelmed by the potential complexity of the seismic waveform. Let us say that in most cases, it is safe to assume that the *individual reflections represent mappable, isochronous, sedimentary unit boundaries or sequence boundaries* (Vail and Mitchum 1977; Wilgus et al. 1988). This assumption usually will not sacrifice the integrity of the final map in the least. In areas where there is no radical thinning or thickening of the sedimentary section, it is reasonable to assume that the reflections, at the very least, parallel the sedimentary units.

The exception is noted in the situation where you may be forced to map a horizon that causes no seismic event but is located in an interval between two diverging reflectors. In some cases, the most likely position of the horizon is not parallel to either reflector. Mapping a nonevent is often referred to as *phantoming* (Sheriff 1973).

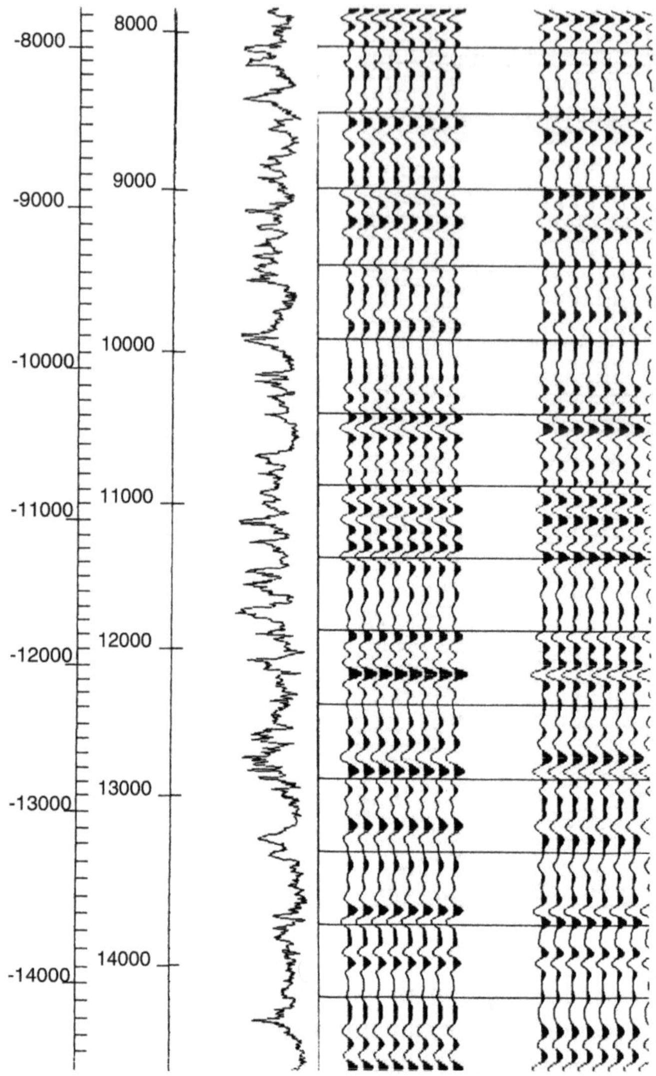

Figure 5-6 Synthetic seismogram illustrating lack of vertical resolution in seismic data.

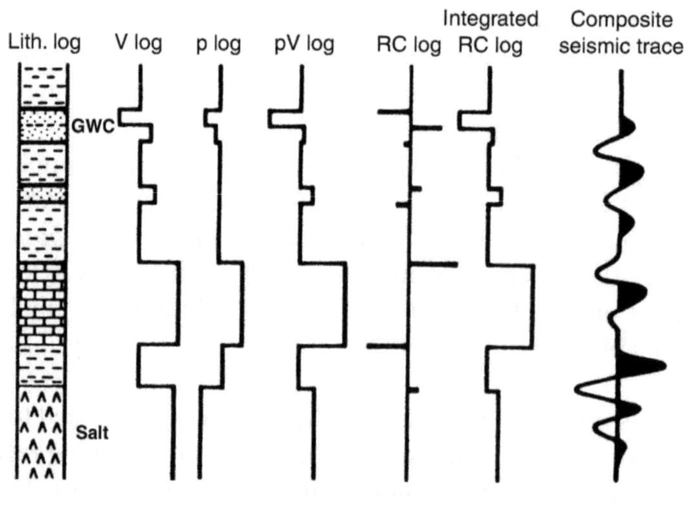

Figure 5-7 Composition of seismic trace from velocity and density contrasts. (After Anstey 1982. Published by permission of the International Human Resources Development Corporation Press.)

Keep in mind that the vertical or horizontal resolution of seismic data will never be as good as that of well log data, but experience has shown that the seismic reflections typically represent isochronous geological surfaces (Payton 1977). This fact makes it possible to map 2D seismic data between wells.

A second incorrect assumption, which is shown in Figure 5-5, is that a curved fault trace seen on a seismic section represents a listric fault. Indeed, the fault shown in the figure is slightly listric, but the reason for stating this is not because of the curved expression of the fault trace on the seismic section; rather, it is the presence of the rollover seen on the seismic *time* section. A perfectly linear feature in the subsurface may look curved when plotted on a seismic *time* section—a situation often encountered when plotting directional well paths on seismic sections. You cannot rely on the linearity or nonlinearity of a feature on a seismic time section to be a reliable indicator of its actual geometry in the subsurface without first converting the feature from time to depth and displaying the section with equivalent vertical and horizontal scales. Similarly, you cannot determine the amount of fault or bed dip from a seismic time section without converting your interpretation into depth.

The most insidious assumption you can make regarding a seismic section is that the section is really just a *geological cross section of the earth* directly under the line. You must always keep in mind that a seismic section is displayed in two very different dimensions: *space and time, and not in the geological realm of space and depth.*

A **time section** is simply a series of traces displayed next to one another on a piece of paper or on a computer screen (ignoring variable density displays). The distance along the line is a physical distance and represents a distance along the surface of the earth. Therefore, looking horizontally along a section requires only that you understand the scale. If you are not working on a workstation, a typical full-scale paper section might be 5 in. to the mile along the top of the section. Figure 5-8 shows a typical time section with annotation illustrating the accepted working terminology for the various parts of its display. On the workstation, the horizontal scale is typically posted along the top or base of the section, or it can be obtained in one of the windows.

It is the *vertical dimension* on seismic time sections that can lead you astray. It sometimes seems reasonable to assume that the vertical dimension also translates directly into a scalable physical distance. *The vertical section is displayed in two-way time.* It represents the amount of time it takes for a seismic signal or wave front to travel from the surface, down through the earth to the reflector, and back to the surface. It would be simple if the seismic velocity field remained constant throughout the earth, but this is *not* the case.

Even in "geologically simple" areas, velocity changes with depth. In general, the deeper the rock, the *higher* its velocity. Figure 5-9 is a time–depth table from *checkshot* data taken in a well in Pliocene sediments. (A checkshot measures the actual time for a surface seismic source to travel to a receiver lowered down a wellbore. This one-way time is converted to two-way time by doubling the times for any given depth.) Underlined in the figure are the subsea depths at 1 sec, 2 sec, and 3 sec. The depth represented by 1 sec of two-way time is 3227 ft. The depth at 2 and 3 sec is 6996 and 11,642 ft, respectively. In this example, then, depending on the depth, an incremental 1 sec of two-way time may represent 3227, 3769, or 4646 ft. Therefore, before you make conclusions concerning the listric shape of faults from a time section, you must convert the two-way times to depth and display the depth section at true scale.

One expensive way to convert the two-way times is to have all the seismic section's depth converted. This may not be necessary, but this depth conversion process is routinely conducted during 3D-depth migration (Chapter 9). An easier and less costly method is to convert all time points to depth, using a valid checkshot, before constructing a geometric interpretation.

To demonstrate how different the perspective can become after converting everything to the same dimension, look at Figure 5-10, which shows a depth-converted fault trace plotted at the

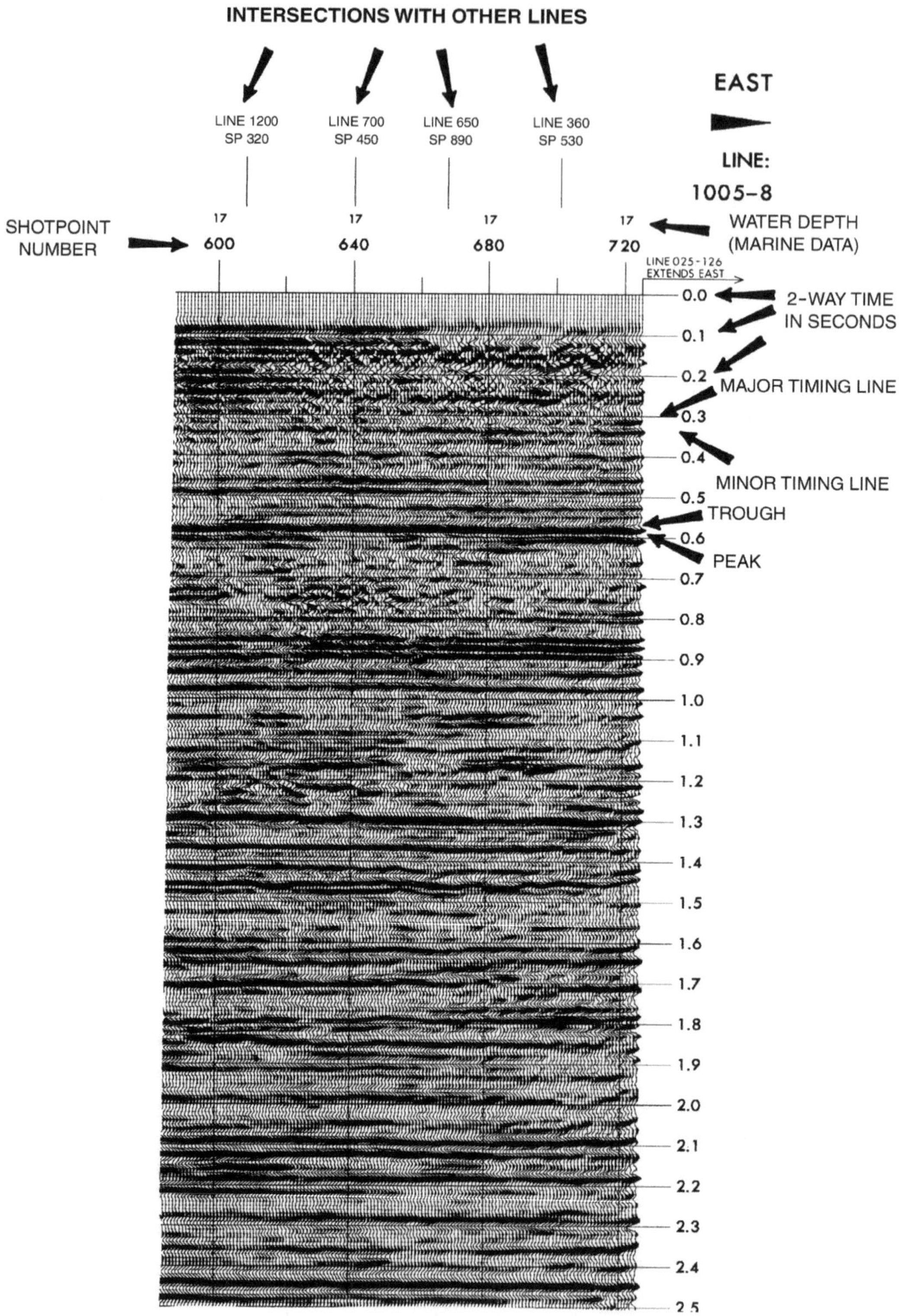

Figure 5-8 Descriptive nomenclature of a typical seismic section. (Published by permission of TGS/GECO.)

same horizontal scale as the seismic profile in Figure 5-5. Does the fault look as curved as the trace on the time section? This example illustrates the effect time sections can have in *distorting the true geometry of a geological feature.*

Furthermore, when considering the listric nature of the fault, you cannot assume that this seismic section orientation is **perpendicular** to the strike of the fault surface, since fault strike cannot be determined on the basis of one line. Essentially, the only statement you can make is that the trace of the fault on this section *appears* to be concave upward in time. Figure 5-11 shows a hypothetical fault surface and seismic line showing its fault trace on the section. Observe that the fault surface is curved, but the dip of the surface itself maintains a fairly constant angle. It is enlightening to note that because of the orientation of the line with respect to the fault surface, the trace of the fault, even on the depth section, *appears* to represent a listric fault, when in fact this is not the case. The easiest mistake to make in seismic interpretation is to *infer 3D geology from observations based on a single seismic section, from either a 2D or a 3D data set.*

Concepts in Tying Seismic Data

Rationale for Tying Loops. How do we take into account the 3D nature of the earth when using 2D seismic data? Keep in mind our earlier contention that a 2D seismic section rarely represents a true planar slice through the earth. Furthermore, the interpretation of 3D workstation data is made through the collation of individual 2D profiles. An advantage of 3D data sets, as opposed to 2D data sets, is that the interpreter of 3D workstation data can select the location and direction of any 2D profile. Thus, the process used to aid in the construction of a valid interpretation is called *tying* the profiles together. Anyone who has worked with geophysicists has heard the phrases "tie the data," or "tie the loop." What tying the data does for the interpreter is build a 3D picture of the

	0	10	20	30	40	50	60	70	80	90
0	0	31	63	94	125	157	188	220	251	282
100	314	345	377	408	439	471	502	534	565	597
200	628	659	691	722	754	785	817	849	880	912
300	943	975	1007	1038	1070	1102	1133	1165	1197	1229
400	1260	1292	1324	1356	1388	1420	1452	1483	1515	1547
500	1579	1612	1644	1676	1708	1740	1772	1804	1837	1869
600	1901	1934	1966	1998	2031	2063	2096	2129	2161	2194
700	2226	2259	2292	2325	2358	2390	2423	2456	2489	2522
800	2555	2588	2622	2655	2688	2721	2755	2788	2822	2855
900	2889	2922	2956	2989	3023	3057	3091	3125	3159	3193
1000	3227	3261	3295	3329	3363	3398	3432	3467	3501	3536
1100	3570	3605	3640	3674	3709	3744	3779	3814	3849	3885
1200	3920	3955	3990	4026	4061	4097	4133	4168	4204	4240
1300	4276	4312	4348	4384	4420	4456	4493	4529	4566	4602
1400	4639	4676	4712	4749	4786	4823	4860	4898	4935	4972
1500	5010	5047	5085	5122	5160	5198	5236	5274	5312	5350
1600	5388	5427	5465	5504	5542	5581	5620	5659	5698	5737
1700	5776	5815	5855	5894	5934	5973	6013	6053	6093	6133
1800	6173	6213	6253	6294	6334	6375	6415	6456	6497	6538
1900	6579	6620	6662	6703	6745	6786	6828	6870	6912	6954
2000	6996	7038	7081	7123	7166	7209	7252	7295	7338	7382
2100	7426	7470	7515	7559	7604	7649	7694	7739	7784	7829
2200	7874	7919	7964	8009	8054	8099	8144	8189	8233	8278
2300	8323	8368	8413	8457	8502	8547	8592	8637	8682	8727
2400	8772	8817	8862	8908	8953	8999	9045	9090	9136	9183
2500	9229	9276	9322	9369	9416	9463	9510	9557	9605	9652
2600	9699	9746	9793	9840	9887	9934	9982	10029	10076	10123
2700	10170	10217	10264	10312	10359	10407	10455	10503	10551	10599
2800	10648	10697	10746	10796	10845	10895	10945	10995	11045	11095
2900	11145	11195	11245	11295	11344	11394	11444	11493	11543	11592
3000	11642	11692	11741	11791	11841	11890	11940	11990	12041	12091
3100	12142	12193	12244	12295	12347	12398	12449	12500	12550	12600
3200	12650	12699	12748	12796	12843	12890	12937	12983	13029	13074
3300	13118	13162	13206	13249	13292	13334	13377	13419	13461	13504
3400	13546	13588	13631	13674	13717	13759	13802	13845	13888	13931
3500	13974	14017	14060	14103	14145	14188	14231	14274	14316	14359
3600	14402	14445	14488	14531	14574	14617	14659	14702	14745	14788
3700	14831	14874	14917	14960	15003	15046	15088	15131	15174	15217
3800	15260	15303	15346	15389	15432	15475	15517	15560	15603	15646
3900	15689	15732	15775	15818	15861	15904	15946	15989	16032	16075
4000	16118	16161	16204	16247	16290	16333	16375	16418	16461	16504
4100	16547	16590	16633	16676	16719	16762	16804	16847	16890	16933
4200	16976	17019	17062	17105	17148	17191	17233	17276	17319	17362
4300	17405	17448	17491	17534	17577	17620	17662	17705	17748	17791
4400	17834	17877	17920	17963	18006	18049	18091	18134	18177	18220
4500	18263	18306	18349	18392	18435	18478	18520	18563	18606	18649
4600	18692	18735	18778	18821	18864	18907	18949	18992	19035	19078
4700	19121	19164	19207	19250	19293	19336	19378	19421	19464	19507
4800	19550	19593	19636	19679	19722	19765	19807	19850	19893	19936
4900	19979	20022	20065	20108	20151	20194	20236	20279	20322	20365
5000	20408	20451	20494	20537	20580	20623	20665	20708	20751	20794
5100	20837	20880	20923	20966	21009	21052	21094	21137	21180	21223
5200	21266	21309	21352	21395	21438	21481	21523	21566	21609	21652
5300	21695	21738	21781	21824	21867	21910	21952	21995	22038	22081
5400	22124	22167	22210	22253	22296	22339	22381	22424	22467	22510

Figure 5-9 Time–depth table from offshore Gulf of Mexico well.

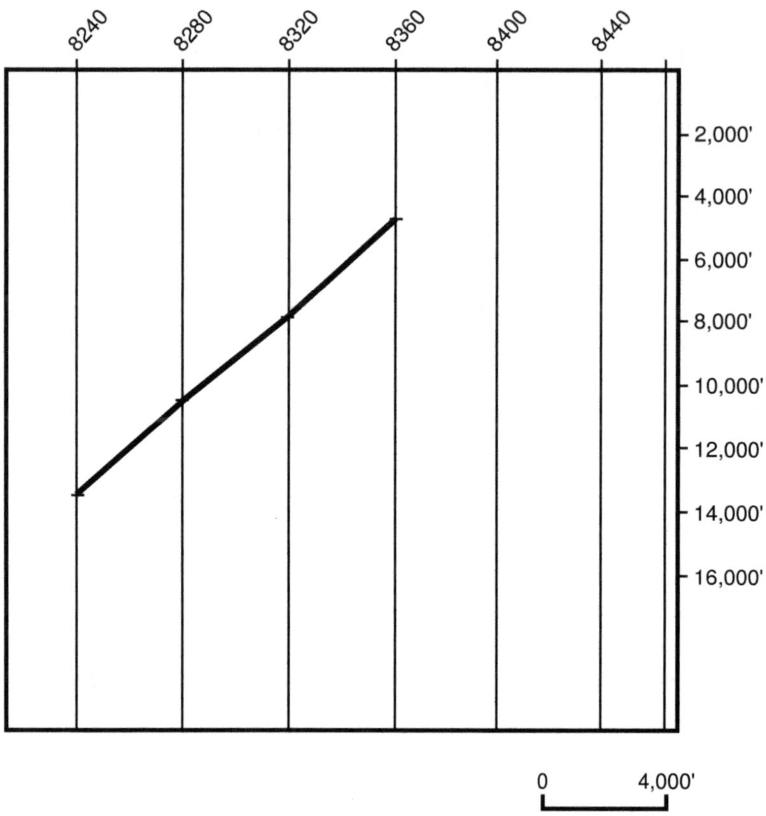

Figure 5-10 Depth-converted fault trace from seismic line in Figure 5-5. Notice that the fault does not appear as curved on the depth-converted section as it does on the time section in Figure 5-5.

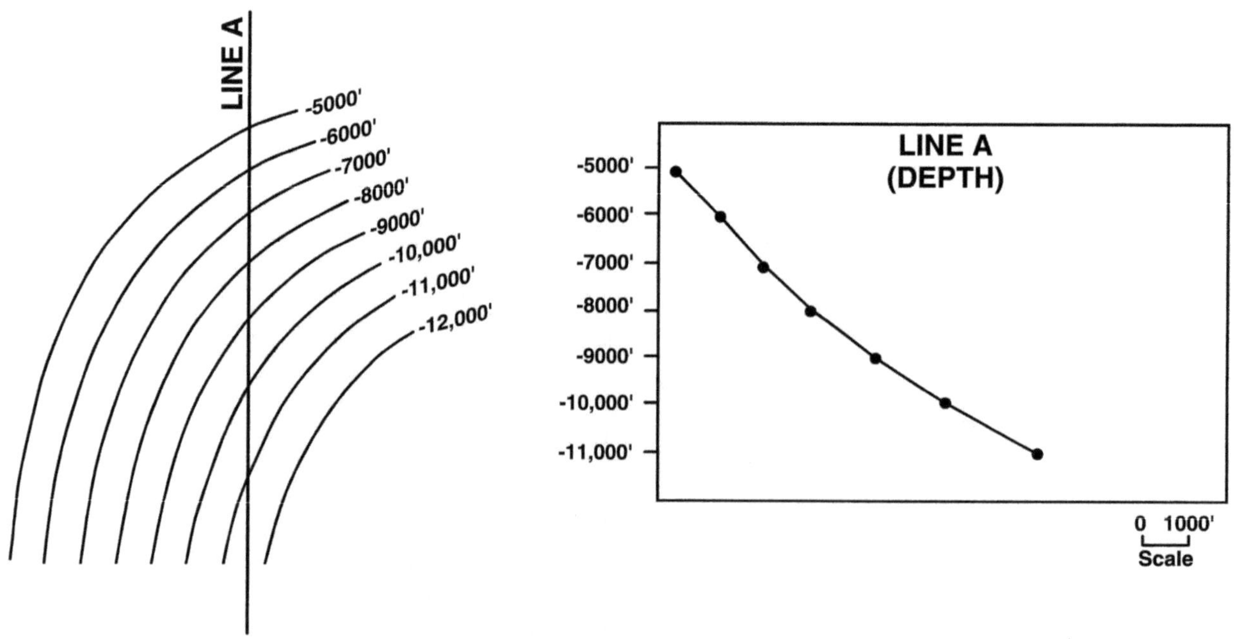

Figure 5-11 Line orientation causing an apparent listric fault on seismic line. The dip on the fault plane is uniform; however, the line orientation makes the fault appear listric, which it is not.

subsurface. Both structure contour maps and fault surface contour maps are 2D approximations of 3D geological surfaces. It follows that two vertical sections intersecting a common surface (i.e., geological log cross sections or seismic sections) will show the intersection of that surface at the same elevation on both profiles. This is illustrated in Figure 5-12.

Even though this seems self-evident, the most common error in seismic interpretation is failure to ensure that *all geological surfaces* that affect an interpretation have been tied around a loop along the lines. This includes tying the faults from line to line. For example, our experience with 2D and 3D data sets demonstrates that the *failure to loop-tie fault surfaces* can result in mapping two faults as one. The failure to loop-tie fault surfaces on 3D data can result in so-called trapping faults that do not exist. This problem is most important where, in the strike direction, one fault replaces another. This area is called the *fault ramp* or *bridge* (Chapter 11). The only cases where tying faults is difficult are in areas where the fault surfaces are near vertical or in areas of complex deformation where the strike lines are poorly imaged. This may seem laborious (and often is), but the ability to *tie surfaces* by following a laterally continuous seismic event is one of the major advantages that seismic data has over well data. Well data forces the interpreter to *infer* a continuous surface from point information, whereas seismic data shows explicit continuity for the horizons and faults being mapped. By tying surfaces, you can eliminate some of the ambiguity that may arise when just using point information from well data. In effect, the act of tying both horizons and faults on a network of lines continually extends the surface and eliminates a number of possible surface configurations that may arise from the point data in wells.

Figure 5-13 is a set of diagrams illustrating the utility of tying surfaces in order to eliminate this 3D ambiguity. Figure 5-13a is a *fault surface map that was constructed solely from fault depth data* derived from well log correlation. Figure 5-13b shows a different fault surface configuration that satisfies the same set of well log fault data. It is apparent that in some cases the point data from the wells are *insufficient* to uniquely define a geological surface. An interpretation that satisfies the data may not be a unique solution. Figure 5-13c shows the same area with a set of two seismic lines, and Figure 5-13d is a representation of these two lines, tied at their intersection and interpreted. To satisfy the requirement that *all nonvertical surfaces should tie*, it is easy to see that all but one of the possible fault surfaces cannot be justified with the grid of seismic data. Figure 5-13e shows a completed fault surface map, which is different than the other two but more accurate because both well data and seismic data are integrated into the fault surface interpretation.

Tying seismic data serves two important purposes. First, it establishes a relationship between the traces of surfaces seen on seismic profiles. In other words, by tying the data, we can assure ourselves that a given interpretation of a geological surface on one line is indeed the same surface as interpreted on an intersecting line. This principle applies to both 2D and 3D data sets. The second benefit of tying seismic data is the ability to project the horizon being mapped into areas where *well control may not exist.* This forms the basis for many wildcat prospects. As previously mentioned, few wildcat prospects have wells near them. Seismic data allow you to extend a mapped horizon into areas with little subsurface control.

Loop-Tying as a Proof of Correctness. Do seismic interpretations have to tie between the grid of lines? The answer is a qualified yes. Ignoring 3D imaging problems for a moment, we can state that a *valid interpretation must tie to be correct.* All faults and horizons must be related and understood in the framework of a spatial grid of lines. The traces of geological surfaces, as seen on seismic sections (the seismic events), must intersect at the tie points between lines.

Be aware of the tendency, however, to believe that a given interpretation must be correct because it ties between lines. Any nonvertical surface being interpreted on a section can be drawn

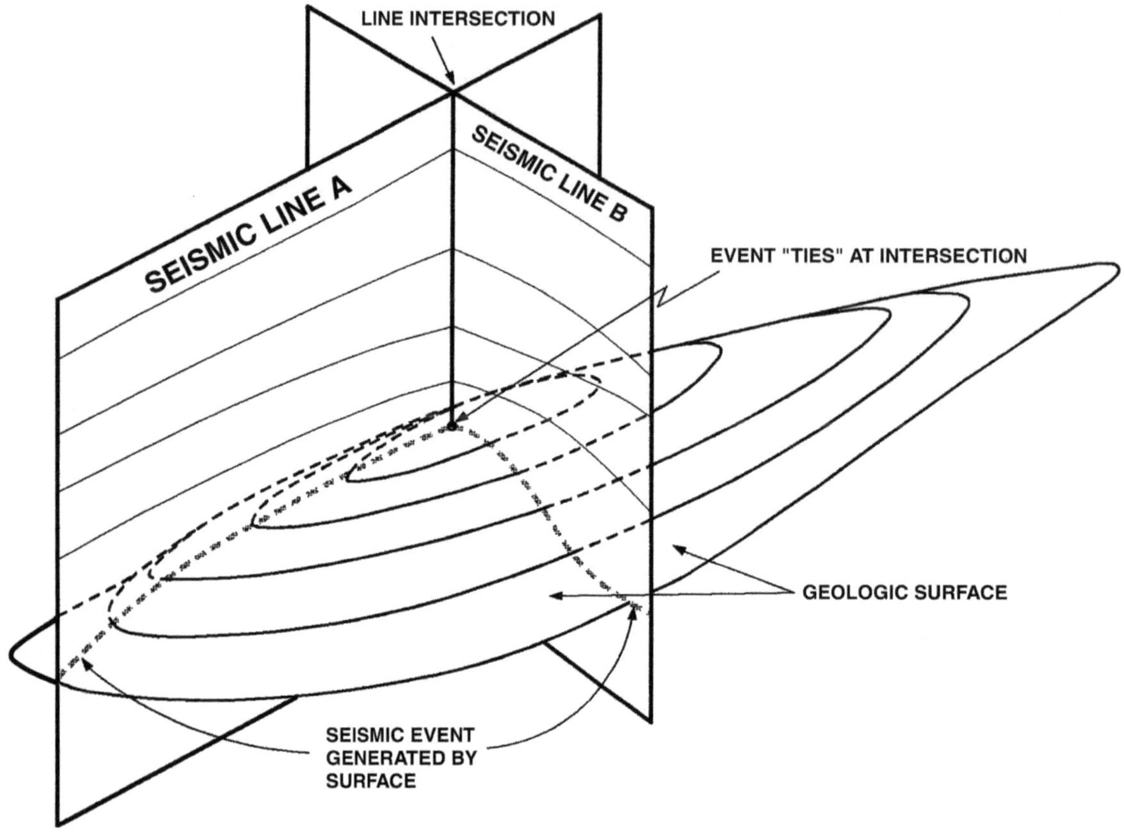

Figure 5-12 Intersection of seismic sections through a subsurface structure. Events must tie at line intersections because they represent the same subsurface point.

to tie between any set of lines. Of course, when it is observed that seismic profiles show interpreted horizons crossing the actual seismic events, there is good cause not to believe in the interpretation. An exception occurs in reverse-faulted and thrust-faulted terranes where horizons may overlap because of faulting. More subtle problems can creep into the interpretation during the process of tying faults. Fault traces on seismic sections are sometimes difficult to see, and tying an array of closely spaced faults can present a near-impossible task, even on 3D data sets. An insufficiently spaced grid of seismic data in an area of dense faulting can present problems of *aliasing*, particularly if there is little data oriented along the strike of the faults. Figure 5-14 illustrates how aliasing can be a problem in highly faulted areas. Line Y is a strike line on which the faults are poorly imaged.

The geometry of the faults in Figure 5-14 may help with the tying problem. Notice that the maximum upper limits of two of the faults on line Z are the same as the maximum upper limits of the two faults on line X. It would be a *reasonable first guess* that the faults that die or are buried at the same elevation are the same pair of faults. This supposition, however, should be consistent with a believable tie and the construction of a reasonable set of fault surface maps.

Perhaps the most useful advice about tying data is to always think in terms of geological surfaces and to regard what is seen on seismic sections as merely the trace of this surface intersecting the surface of the seismic line. Remember, however, that the time sections are not depth sections. Making geological interpretations based on one seismic profile defies the 3D nature of the earth. Tying the data is a visualization tool that helps create a 3D representation of the subsurface.

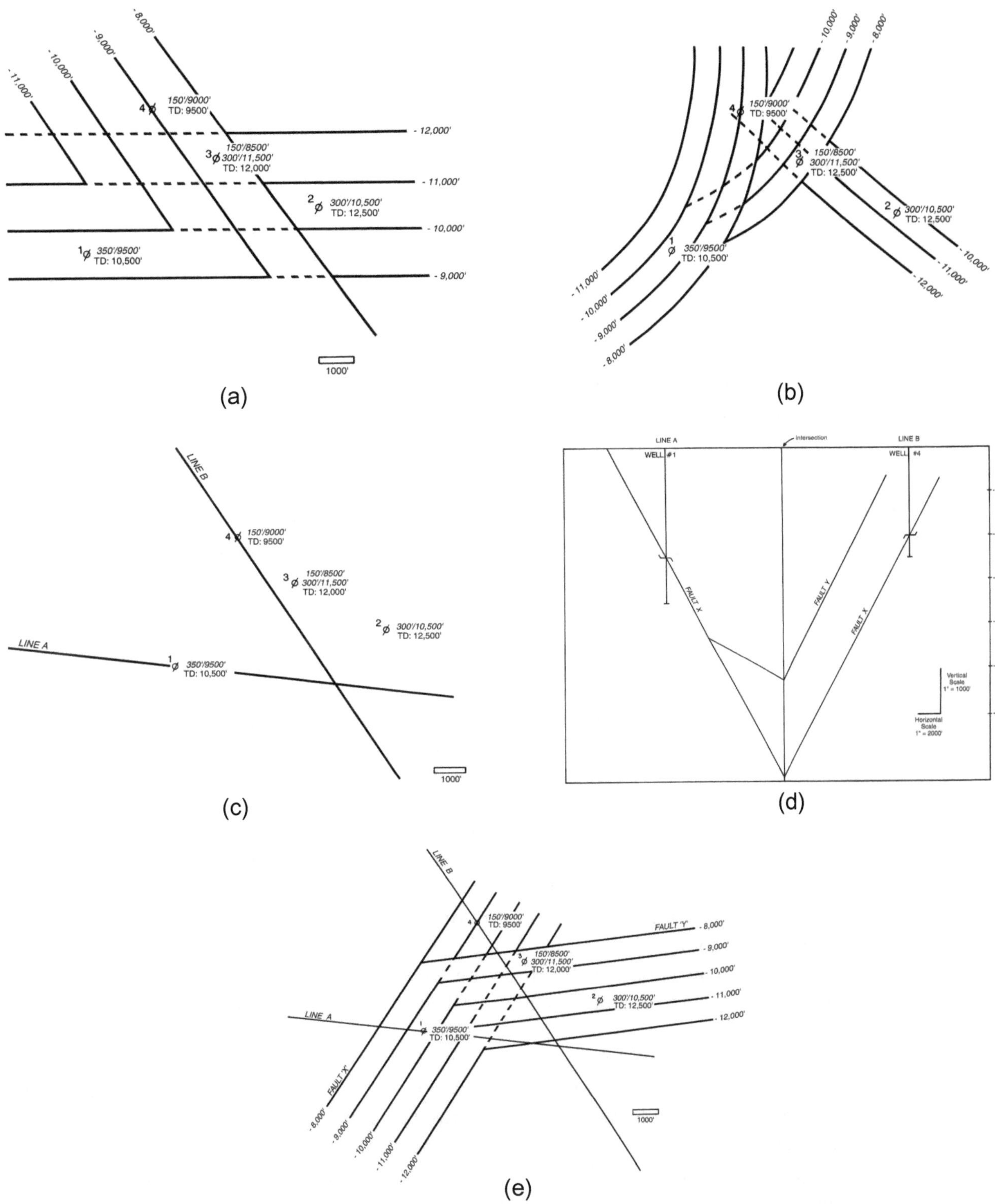

Figure 5-13 (a) Possible interpretation of hypothetical fault data observed in a set of wells. (b) Another possible interpretation of the same fault data. Multiple fault surface interpretations are possible from this set of data. (c) Hypothetical seismic grid through faulted area. (d) Appearance of hypothetical seismic sections when interpreted and tied. Note that fault traces on profiles meet (tie) at the intersection of the two lines. (e) Final fault surface interpretation, integrating both seismic and well data.

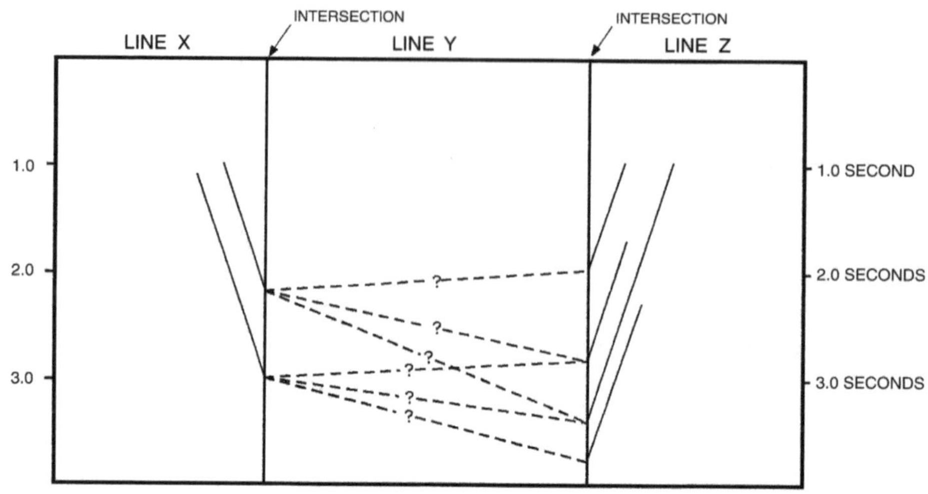

Figure 5-14 Ambiguity when tying closely spaced faults. Line Y is a strike line. Multiple interpretations may be possible.

Procedures in Tying Seismic Data

Contemplating the Data. Certain organizational procedures can save you unnecessary work during an interpretation. When you receive a set of paper seismic sections, it is tempting to immediately begin interpreting and mapping. A lot of unnecessary rework can be avoided by first taking the time to arrange the paper sections into stacks of data with similar orientations. For example, put all your north-south data into one pile, arranged consecutively from west to east or east to west. Similarly, put all your east-west oriented data into a separate pile, ordered consecutively going north to south or south to north. If your lines are arranged randomly, separate the lines into groups of roughly similar orientations. When you use a workstation, similar procedures can be conducted by scrolling through the 3D data set or by selecting parallel lines on the workstation.

Begin by selecting a line that is oriented in the same direction as the dip of the predominant geological features you are interpreting. In areas of folds and thrusts, pick lines that are oriented perpendicular to the strike of the fold axes and thrust faults. In listric fault regions, choose lines oriented perpendicular to the strike of the listric faults, and so on. With each line, ask yourself questions about the structure. Which faults are the dominant faults? Where do the horizons show changes in apparent dip direction? Where are the crests of the highs and the bottoms of the lows? (Tearpock and Bischke 1991; see the section Long Wavelength Domain Mapping in Chapter 9.)

If using paper sections, unfold the next line and put it beside the first line. If at a workstation, bring up a new window and compare a new line to the existing line. Look at both lines and ask yourself more questions. Are the same faults present on this line as on the previous line? If a fault is almost certainly the same as one on the previous line, create a fault name, pick the fault, and, when using a computer, assign the picked faults to the same file. Do additional faults appear on this second line? If the dips change dramatically across the fault surface and the horizons lack compatibility, then the fault is likely to be large and will extend over large horizontal distances. Do the crests of the highs and bottoms of the lows change between the lines? Does the structural style change, or is it similar? Do the seismic dip rates at similar seismic times differ between the lines? What could be causing these differences? Pull up a third line and ask the same questions.

Continue this process of looking at two lines at a time until you have looked at all the similarly oriented lines. Make notes about major changes that you observe but cannot adequately explain at the moment. Now follow the same procedure with the strike lines. This may seem somewhat tedious and simple-minded, but we have found that a modest amount of time taken to do this sort of work is well worth the effort. It can save enormous amounts of time spent correcting mistakes made because you made too cursory an interpretation or became biased. We have found that when we are failing and having problems with an interpretation, doing this little exercise methodically and with a critical and questioning eye can help create an interpretation that is more likely to be correct.

We always peruse or scroll through a 3D data set before embarking on an interpretation. This procedure allows the interpreter to get acquainted with the structural style of an area and to get a better feel for and understanding of the data set. Do not do this perusal too quickly, and do be sure to make notes on interesting features that deserve further investigation during your interpretation project.

Picking a Reflection to Interpret and Map. Seismic profiles (e.g., Fig. 5-5) contain numerous reflections, and it is obvious that it would not be possible or practical to interpret and map every event. The interpreter should look critically at the sections and decide which seismic horizons are best to interpret and to map. Typically, the chosen events correspond to selected stratigraphic horizons, although reflection strength or continuity can also influence the decision.

Today, interpreters typically map sequence boundaries because not only are these boundaries commonly the most laterally continuous events on a section, but they are also directly related to rock type and the geological history of an area. *Sequence boundaries* are geological unconformities, or surfaces of erosion or nondeposition, and represent approximate isochronous surfaces (Payton 1977). They can generally be located on seismic sections by observing where reflections converge or are truncated against a (usually) strong event.

Sequence stratigraphic analysis is an exciting use of geology in geophysical interpretation. Covering this subject in any detail is a book in itself, and therefore we direct you to the references in the bibliography (in particular, Payton 1977; Berg and Woolverton 1985; Wilgus et al. 1988). You should be aware of the implications of this growing body of knowledge, as it has had and will continue to have a major impact on petroleum exploration.

Annotating the Well Information. Now that you are confident that your geological interpretation is underway, mark the position of any wells that intersect the seismic lines. Straight holes require only that you have reasonably correct checkshot data close to the line. Directional wells require that a directional survey be available. Directional surveys and the projection of deviated wells into a seismic line are covered in Chapters 3 and 6. Remember that you must convert the depth points to their equivalent two-way travel times in order to annotate them correctly on a time seismic section.

It is important to remember that any projection is a *compromise* and will commonly cause some confusion. Figure 5-15a is an illustration of a projection of a directional well onto a seismic section. Notice on Figure 5-15a and 5-15b that an orthogonal projection of the well into the seismic line suggests that the well penetrated the footwall of the interpreted fault. However, the fault surface map (Fig. 5-15b) clearly shows that the well never crossed the fault from hanging wall to footwall; it is entirely within the downthrown block. This illusion occurs because of the compromises inherent in projecting a 3D entity onto a 2D profile. The routine orthogonal projection of a well into a seismic line can also cause significant mis-ties of horizons and faults. The tendency to cause mis-ties increases with the distance a well is projected, especially in areas of moderate to steeply dipping beds. If 3D seismic data is available, it is always advisable to tie an arbitrary seismic line directly along the directional path of a well.

Tying Well Data to Seismic with Checkshot Information. Once the well position is annotated, the information from the well data, in the form of geological tops and fault cuts, must be located and marked on the time sections or loaded into the computer and annotated on the profile. How do you know where to find the event that corresponds to the geological horizons? There are basically two methods used to tie the geological control into the seismic data: (1) using a time–depth function calculated from checkshot data or (2) tying into the seismic data with a synthetic seismogram.

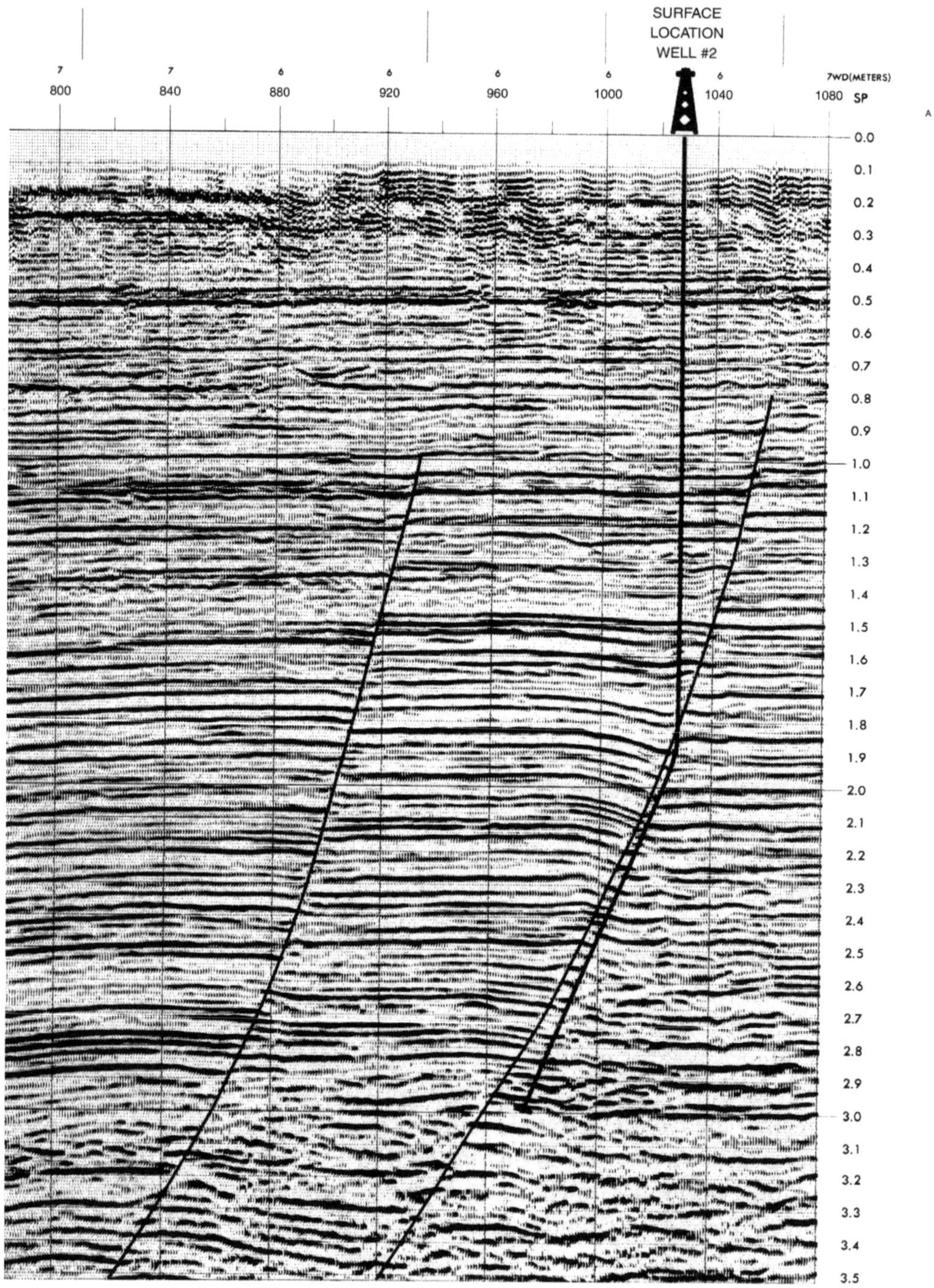

Figure 5-15a Seismic section with directional well projected onto section (straight-line perpendicular projection used). (Seismic line published by permission of TGS/GECO.)

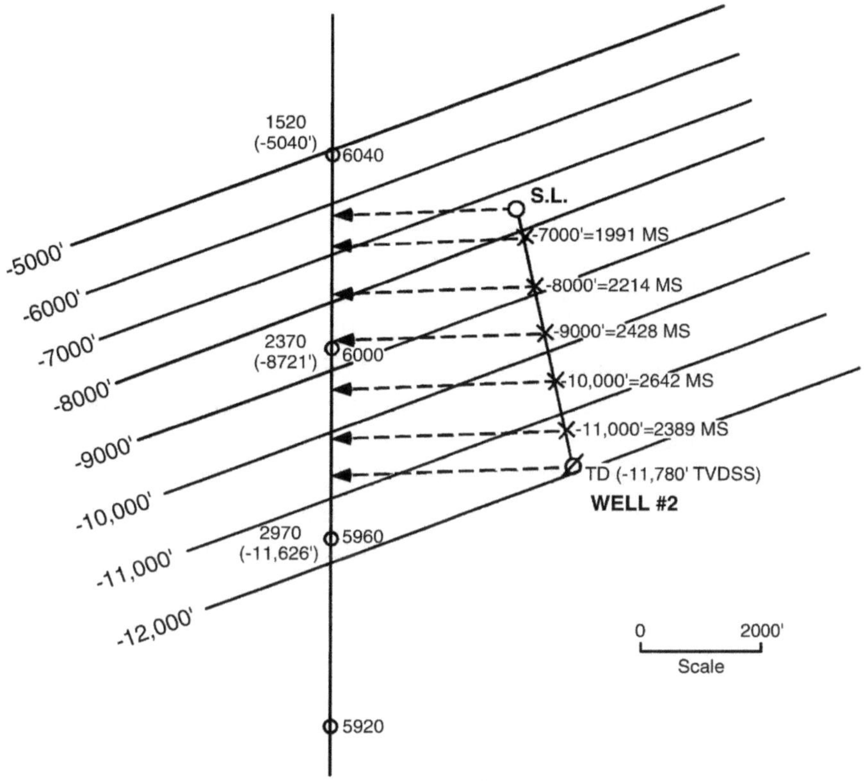

Figure 5-15b Fault surface map of fault observed in Figure 5-15a. Projection onto line makes well appear to have penetrated the upthrown block. Fault surface and the well's directional survey clearly indicate that the well remained in the downthrown block. Care must be exercised in projecting wells onto seismic lines.

The simplest but least accurate method of tying well data to seismic is to use the checkshot data to convert the tops from the log data from depth to time, and post the equivalent horizons on the seismic section at the proper times. The problem with this method is that you never know what kind of assumptions may have been made in the processing of the seismic line to correct to the proper datum. This is why data from different contractors may have static shifts between lines when tied together. "Ground truth" is hard to ascertain in these circumstances.

There is also a temptation to place unwarranted faith in the checkshot data and to believe it over all other information. We have seen cases where an interpreter tied a sand occurring in the middle of a 1000-ft shale interval to a level within a no-data zone and above a distinct event, simply because a nearby checkshot indicated such a tie. Particularly when sonic and density information indicates that a sand should generate a strong seismic response, it is likely that the sand ties to the strong event in the middle of the mostly reflection-free shale zone. This is not to say that shale intervals cannot generate strong seismic events. However, if a sand is present and fairly close to tying an event on the line, then correlate the sand to the reflection and do not be misled by the checkshot listing.

Tying Well Data to Seismic with Synthetic Seismograms. Tying the well data into the seismic data with a synthetic seismogram is the preferred method, as it will usually provide reasonable results. Its usefulness, however, depends almost exclusively on the availability of good quality sonic and bulk density log data from the wells. In some areas, these logs are run as a matter of course, whereas in other regions they are the exception rather than the rule. Particularly in older basins, there may be a shortage of high quality sonic and density log data. It is imperative that the geoscientist determines the quality of the data used to make the synthetic seismogram. We have seen synthetics made from sonic logs obtained from washed-out wellbores that have recorded mostly mud arrivals and cycle skipping. Needless to say, the synthetic seismograms from such logs are useless.

Figure 5-16 shows a high-quality synthetic seismogram and its tie to a seismic profile. This seismogram is shown adjacent to a seismic section through the actual well location. As you can see, the match is good, though not perfect. The procedure for tying the proper event is to locate the chosen horizon on the log plotted next to the synthetic, then draw a horizontal line over to the synthetic trace. Lay the synthetic seismogram over the seismic profile at the appropriate location, then shift the synthetic up or down to determine whether it matches the seismic data. The horizon line drawn earlier will show where the actual seismic event, corresponding to the log horizon, is located on the seismic line.

Always be wary of *forcing* yourself to see correlations between the seismic data and the synthetic seismogram. If there are problems with either the seismic data or the synthetic data, it may be impossible to make a valid correlation. As a rule of thumb, a shift of the synthetic by more than about *one hundred milliseconds* should be highly suspect. Also, if you can turn the synthetic upside down and get equally good correlations, you should be suspicious of the validity of this method for tying well horizons into seismic data.

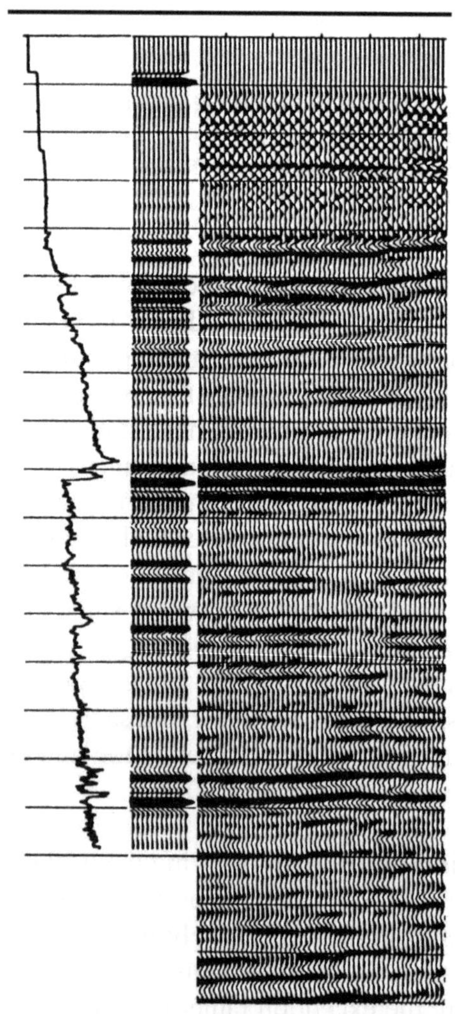

Figure 5-16 Tie of seismic data to synthetic seismogram. Events correlate well from synthetic seismogram to seismic profile. (From Badley 1985, provided by Merlin Profilers, Ltd.)

If applicable, this method of tying in the well data is preferable over tying into a particular seismic event with just checkshot data. If a correlation exists between the synthetic and the seismic profiles, then the synthetic seismogram method will ensure, with a reasonable degree of certainty, that the event being mapped is the intended geological horizon. This is particularly valuable in areas of abrupt stratigraphic thinning and thickening. Figure 5-17 illustrates how an incorrect pick for the mapping horizon can have a profound effect on the depth of the horizon away from the well control. As shown, a small error in the thinner stratigraphic section will cause a much larger error to occur where the stratigraphic section is thicker.

A *vertical seismic profile* (usually abbreviated VSP) derived from a synthetic trace is also an excellent tool for tying into the seismic data. The methods for using the VSP to tie into the seismic line are the same as those used for synthetic seismograms. Establish a correlation between the traces on the VSP and the seismic, and use the plotted log as a guide for tying into the seismic section. An added benefit to both the synthetic seismogram and the vertical seismic profile is the ability to use these data to analyze the relationship between the lithology in the well and the seismic character.

The subject of tying well information into seismic data is deceptively simple. There is a dangerous tendency to believe that your first tie from a well is the correct one. Make sure that it is, because being tied into the wrong horizon can cause you to miss the relationship between a given log response and its correlative seismic response. The only other advice we can give is to avoid a "railroad track" mentality (looking strictly at a narrow 80-millisecond strip of the seismic data), or you may totally miss what is present above and below the horizon you are mapping. It is frequently advisable to map multiple horizons at the same time (Herron 2011) to help avoid focusing on a single horizon to the exclusion of other horizons.

Tying the Faults. We strongly recommend that fault surfaces be loop-tied prior to loop-tying horizons. We have found that the time spent mapping faults is recovered when trying to identify the location of mis-ties. Plus, you have a higher-quality product. Geophysicists loop-tie surfaces for the same reasons that geologists loop-tie stratigraphic markers. That is to make sure that you are on the same surface where you began and not on a different surface. Faults are geological surfaces just as horizons are and should be loop-tied just as horizons are (Herron 2011). Mapping faults first has distinct advantages. First, in areas of complex structure the style of faulting helps determine the type of structures that exist in the area (Chapter 10). How can you determine how the structure formed if you do not understand how the faults behave in three dimensions? Second, problems may arise when mapping horizons on portions of a profile that contain a lower seismic frequency or where the data are not perfectly coherent. If the position of recognized *faults has not been identified in three dimensions, then geoscientists may map horizons right through fault surfaces*. This will result in a horizon mis-tie, as described in the next section. Our examination of the horizon mis-tie problem on many data sets indicates that the *geoscientist may not even suspect that an existing fault produced the mis-tie*. After all, the assumption was made that the faulting is understood or that fault surface mapping is not required. The result is that the geoscientist may create a nonexistent fault to solve the mis-tie problem. The nonexistent fault constructed through semi-coherent data or data that abruptly changes reflection character creates additional mis-ties and more nonexistent faults and mis-picked horizons. We have seen cases where prospects have been generated as a result of the nonexistent faults. Needless to say, the wells were dry.

There is another common problem with the picking of nonexistent faults in semi-coherent data or data that rapidly changes reflection character. These nonexistent faults may actually be *axial surfaces* or changes in bed dip that are common to compressional and salt-related folds (Chapter 10). Where the geoscientist encounters a nonexistent fault, the horizons must be offset,

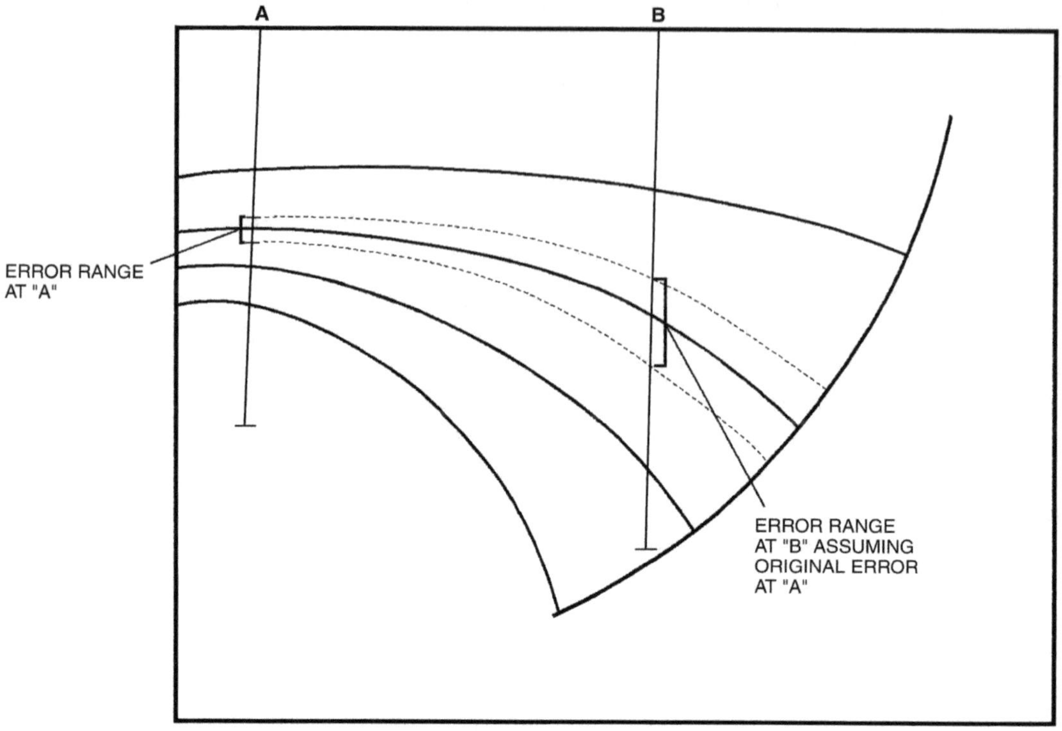

Figure 5-17 Increase in magnitude of error in areas with rapid changes in interval thicknesses. Any uncertainty in tying formation top to seismic event can introduce a greater uncertainty where stratigraphic section thickens away from well control.

which results in a horizon mis-pick. A mis-picked horizon results in mis-ties and in other nonexistent faults that are created to solve the mis-tie problem.

However, if fault surface maps are constructed for each of these nonexistent faults, the fault may not map as a smooth curved surface but will contain offsets and kinks. The presence of an offset or kink typically implies more than one fault. Also, the overall geometry of a mapped fault surface may be simply unreasonable. These types of fault surfaces are not viable geological surfaces and should be rejected. Faults that map as smooth surfaces are considered more plausible. Furthermore, with 2D data sets, questions typically arise as to which faults link to form a continuous fault surface. Fault surface maps can help resolve the fault correlation problem.

We find that mapping fault surfaces not only results in better interpretations and in higher-quality prospects, but the process saves time. The time taken to construct quality fault surface maps is justified when you consider the time required to attempt to solve existent and nonexistent mis-tie problems, the reworking of an interpretation that proves incorrect, and the costs of an unnecessary dry hole.

Tying the Lines and Horizons. If you have followed the procedures so far, you now have a tentative interpretation and well data annotated on each profile. To this point, we have not described loop-tying the data. That is the next step. On a good day, portions of your preliminary interpretation will probably be wrong when you finish loop-tying the data. Again, methodically tying the loops will improve the chances of finishing the task correctly and in a timely manner. The use of a workstation has advantages in that any errors encountered during the interpretation can be readily corrected and alternative ideas can be easily tested.

We find it much easier to interpret the most obvious geological features first and to tie them together on all the lines. After the large features tie, begin another iteration of tying through the data volume, concentrating on the smaller "second-order" features. As the size of the features being tied together decreases, the number of lines required to tie them also decreases, so the work goes more quickly toward the end of the process.

It is important to pick a loop-tying scheme that will allow you to make the smallest number of assumptions while carrying your surface around the map area. Tying a path that crosses the fewest faults and that crosses faults at their location of smallest displacement will more likely be correct.

The initial task for tying the loops is to post all the intersections of the seismic data on all the lines. With paper sections, depending on the number of lines that are being tied, this process can take anywhere from a morning to several weeks. Figure 5-18 shows a seismic basemap with two seismic lines. The corresponding profiles are shown in Figures 5-19 and 5-20. Line A intersects line B at a location just north of shotpoint 480 on line B. If interpreting on a workstation, the computer should mark the line intersections for you.

Line intersections are seldom cooperative enough to fall on a *downline* (a downline, shown in Fig. 5-8, merely refers to the dark vertical line printed on the time section at the shotpoint locations annotated on the maps). Depending on the precision required, you can either make a rough estimate of the intersection or use a scale to determine that the intersection is exactly 150 ft, or 1.829 traces, to the right of shotpoint 480 on line B. All the intersections that you intend to tie together must be marked.

The next step, when using paper sections, is to fold one of the lines at the marked intersection. *This fold must be vertical.* Use a straightedge to ensure that the section is folded vertically. Align the folded section with the unfolded section at the appropriate intersection. Figure 5-21 demonstrates a tie between the two paper sections. On the workstation, the horizon intersection picks are posted automatically.

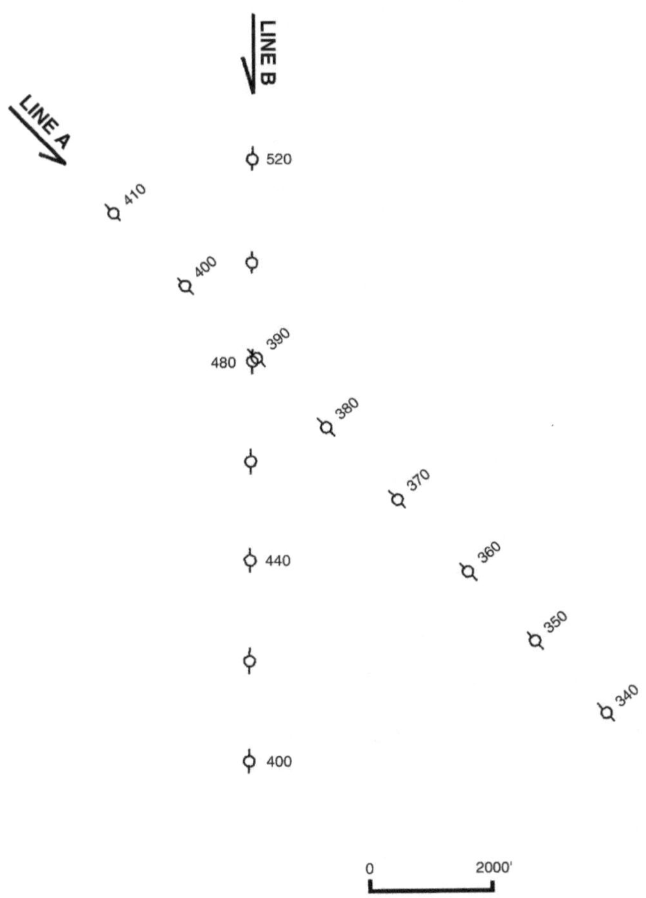

Figure 5-18 Seismic basemap showing positions of seismic lines A and B.

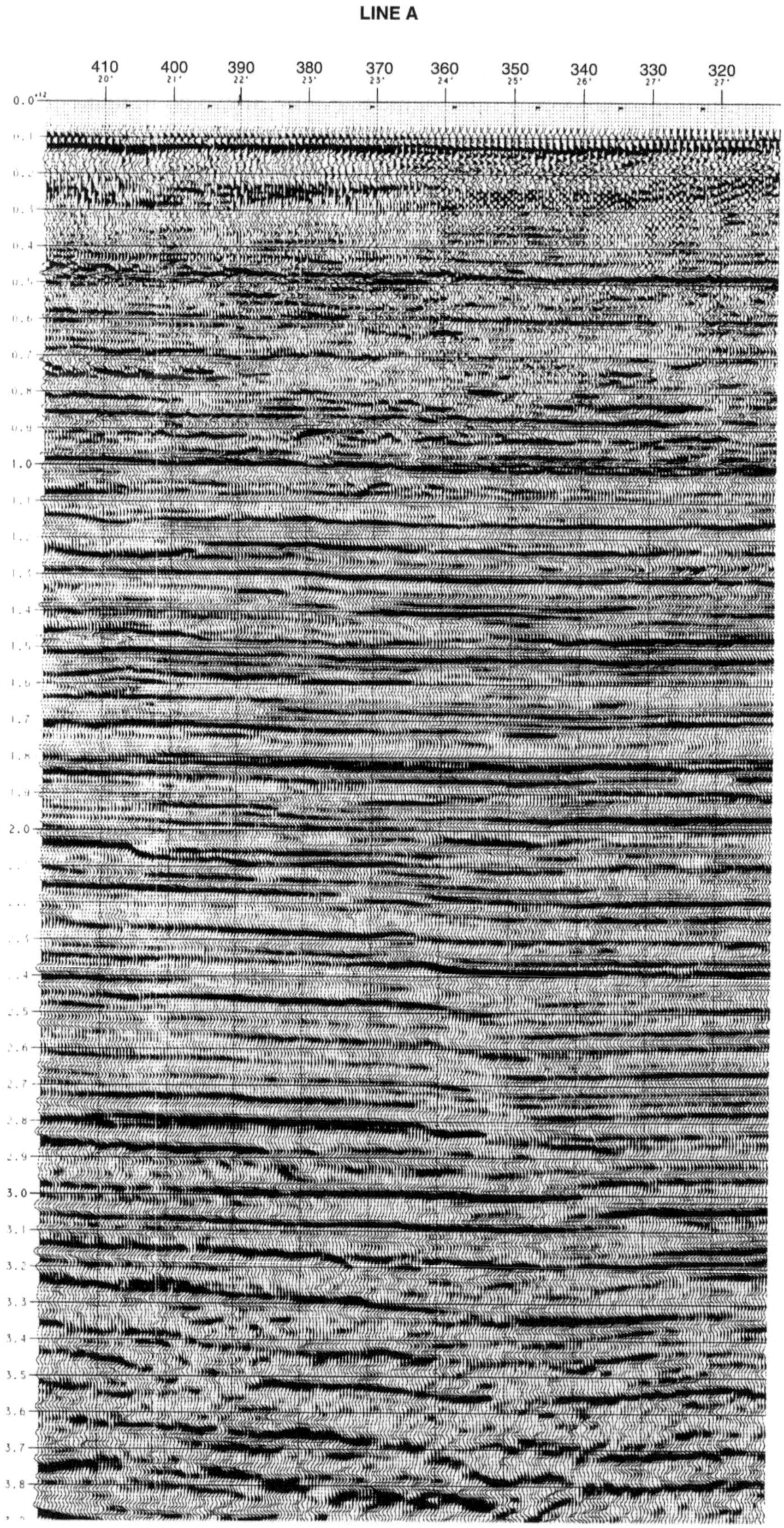

Figure 5-19 Seismic line A. Profile shows gently dipping reflectors cut by a small growth fault. (Seismic data published by permission of TGS/GECO.)

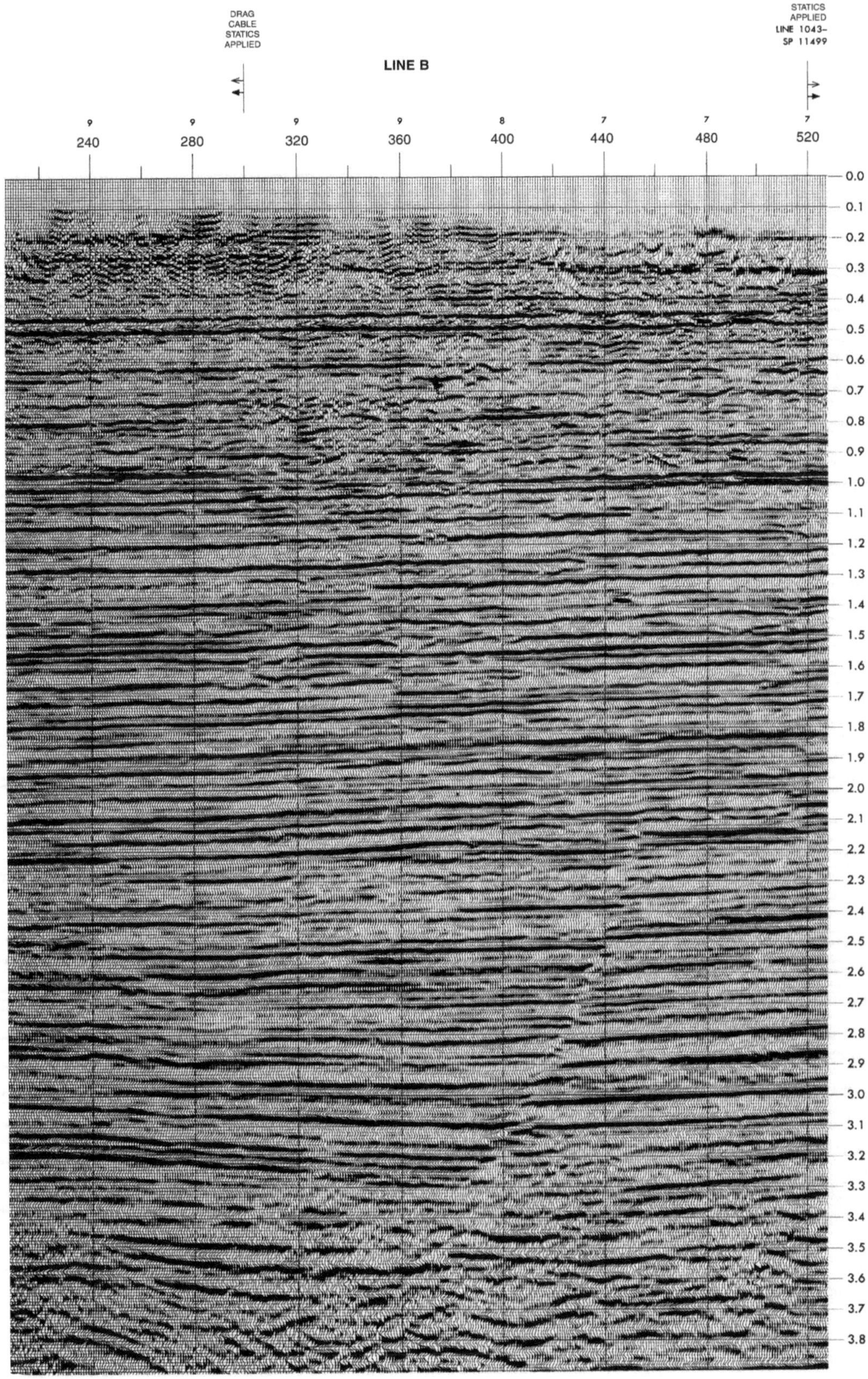

Figure 5-20 Seismic line B. Profile shows gently dipping reflectors cut by a small growth fault. (Seismic data published by permission of TGS/GECO.)

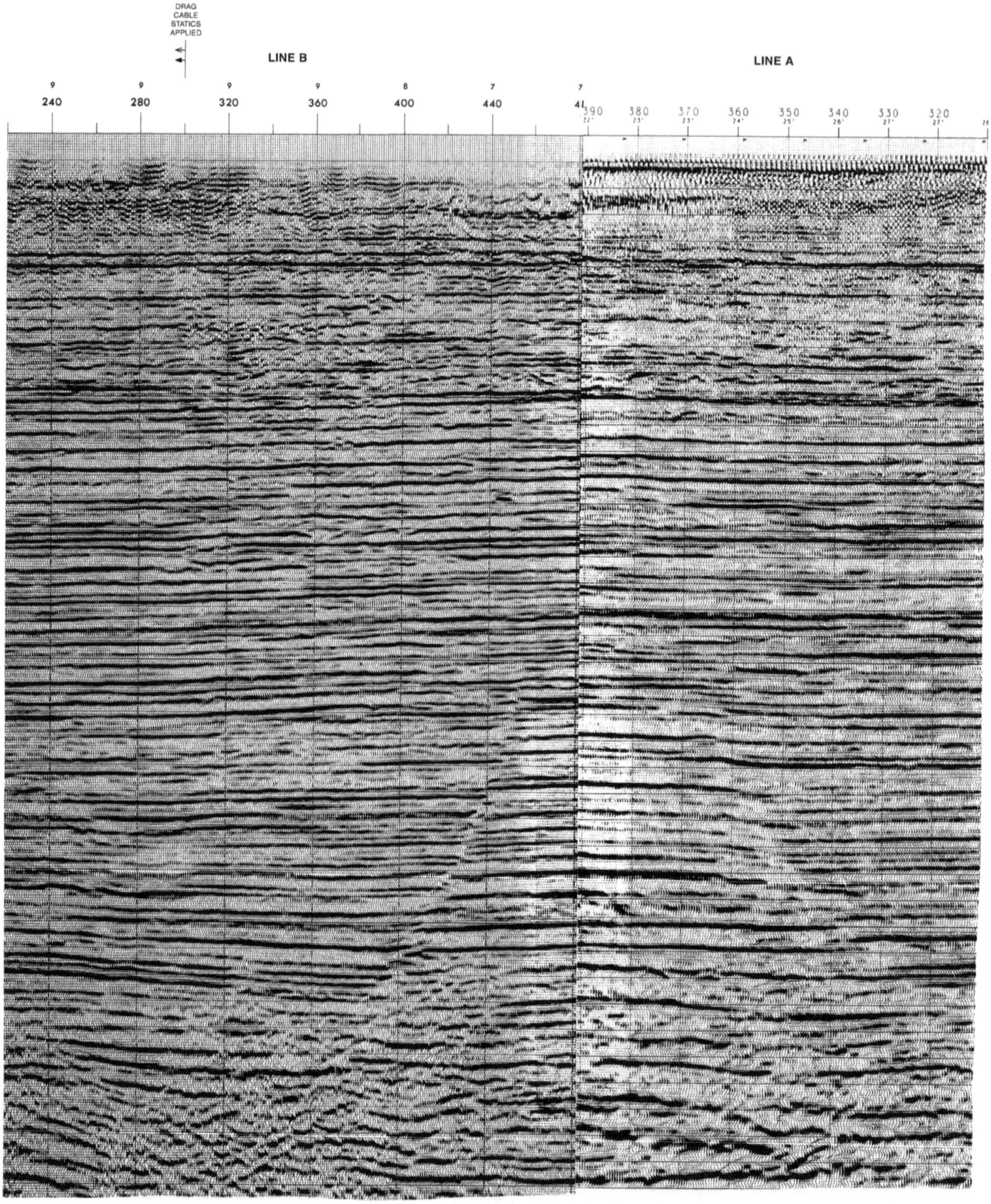

Figure 5-21 Seismic lines A and B intersected with each other. Notice line A appears to be too “shallow” in relation to line B when the timing lines are aligned. (Seismic data published by permission of TGS/GECO.)

The first thing you will probably notice is that the lines may not tie perfectly when the intersecting lines are aligned at a common two-way time (Fig. 5-21). Sometimes they will match perfectly, but more likely, they will not match at a common time. At least one section will have to be slightly shifted vertically to establish a good correlation between the lines. Figure 5-22 shows how the two lines have to be shifted relative to one another to "tie" the events. Which line has been shifted? Has line A been shifted down 10 milliseconds (+10 ms), or has line B been shifted up 10 milliseconds (−10 ms)? You now decide which line shows the "real" two-way time to the event.

In an interpretation and mapping project of any complexity, it is necessary to pick a reference seismic line so that all the other data may be posted with the appropriate time shifts relative to the reference line. In effect, the procedure for choosing a proper reference line is similar to that for picking a correlation type log (Chapter 4). When using wells, you are picking the log that best demonstrates the geological section of the mapping area. When choosing a reference line, pick a line that has as many of the following characteristics as possible.

1. The reference line should cross as many of the other lines as possible. This is required to reduce the number of indirect calculations of static shifts.
2. The reference line should be a dip line or as close to it as possible.
3. The reference line should be of high quality relative to the rest of your data. In short, it must be one of the most believable lines in your collection.

Mis-Ties

There are two kinds of mis-ties that must be corrected before posting data on a basemap: *static* mis-ties and *migration* mis-ties. The static mis-ties are reflection-time invariant corrections made to the event times, and they are the easiest to recognize and correct. The migration mis-ties are corrections that vary with the two-way time of the events being mapped, and they are more difficult to correct. Real problems occur when both static and migration mis-ties are present in a data set. The static component must be recognized and corrected first, and the migration correction is made after the static solution is determined.

Static Mis-Ties. Static mis-ties can be recognized because they cause "bulk" shifts of the intersecting line, either up or down to achieve a good correlation (Sheriff 1973). They commonly occur between data sets of varying vintages and contractors because of different datum corrections and assumptions. Figure 5-21 shows a static mis-tie between two intersecting lines. The easiest way to determine a static mis-tie problem is to search for any shift between flat-lying events that are normally present in the shallow part of most basins. In Figure 5-21, line A ties line B perfectly with a static shift. We do not know, however, if the times on line A are too large or if the times on line B are too small. There is no absolute answer. This is the reason for picking a reference line; it establishes a reference or datum. The rest of the lines can then have the time picks for a given event adjusted to the datum established by the reference line. If a line does not directly intersect the reference line, then its relative mis-tie with a line that does intersect the reference line is added together with the adjustment value from the line intersecting the reference line. This is harder to describe than to illustrate, so Figure 5-22 shows an example for keeping track of static mis-ties on lines that do not directly intersect the reference line.

Once a reference line is chosen, annotate the rest of the sections with their respective static mis-tie values relative to the chosen datum. An important point to note about this process is that it should be carried out only where you are tying events that are relatively low-dip (probably less than 8 to 10 deg at most). High dip rates cause an effect called migration mis-tie, which is discussed next.

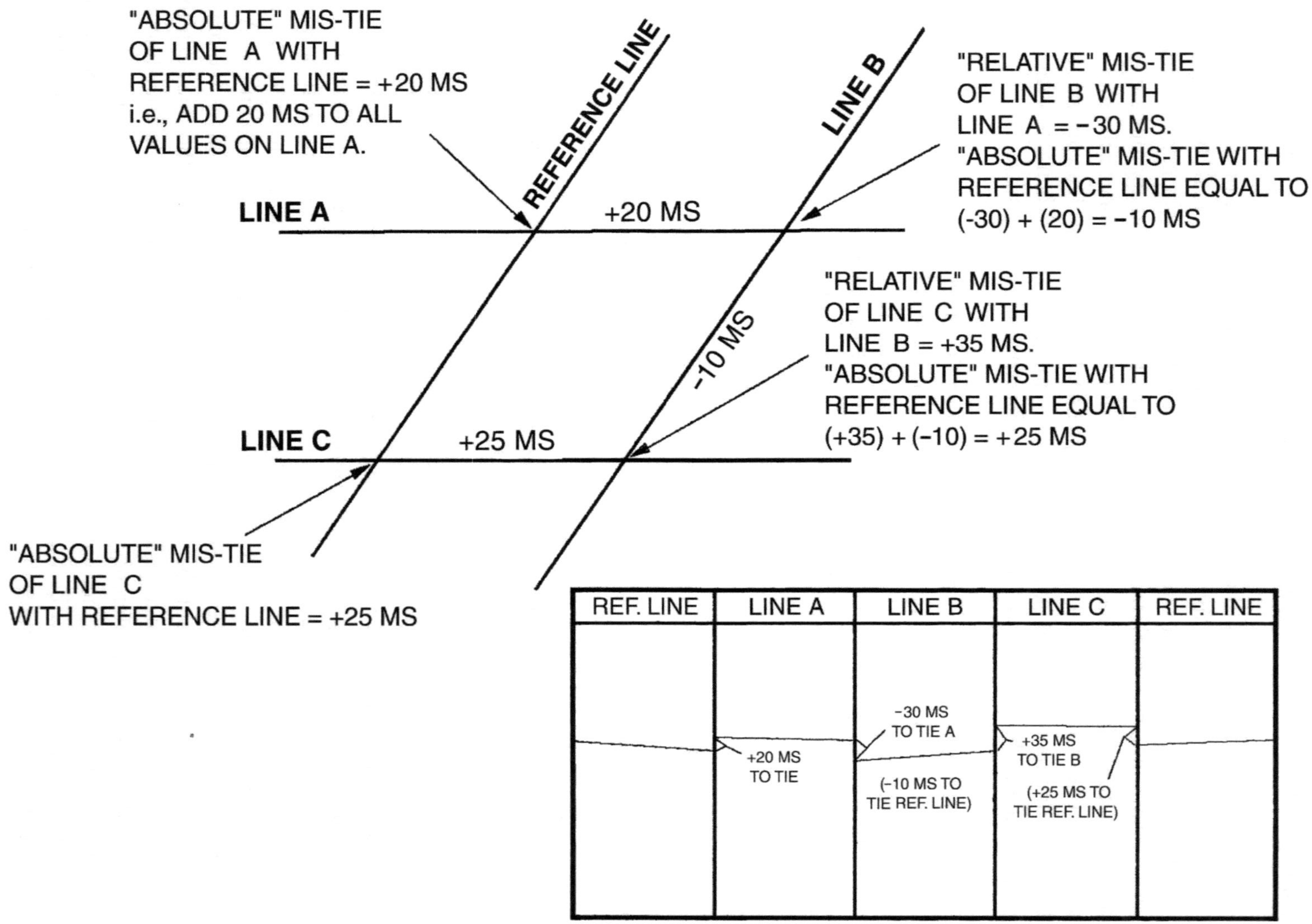

Figure 5-22 Keeping track of relative mis-tie through a grid of seismic data. All absolute mis-ties are in reference to reference line.

Migration Mis-Ties. A migration mis-tie is one of the more difficult aspects of interpretation of 2D seismic data, and it becomes problematic in areas of high bed dips. As noted previously, you may encounter varying amounts of mis-tie with different time ranges of events. In general, the shallow, low-dip events will tie reasonably well with a static shift. But as the dip increases with increasing depth (greater time on the section), you may notice that events on the dip lines appear to be too deep (too large a time) relative to the intersecting strike line. Figure 5-23 is a sketch of this phenomenon. This mis-tie problem is present because of the limits of 2D seismic data in imaging a 3D surface. A concise and readable description of migration issues is provided by Herron (2011).

To understand this problem, it is first important that you understand what the migration process does to seismic data. Figure 5-24 is a simplified illustration of a 2D seismic line shot in the dip direction. Two simple, normal incidence raypaths are drawn on the section from surface positions A and B. By definition, a normal incidence ray will intersect the reflector Z at a right angle. Raypaths drawn to satisfy this condition are A-A2 and B-B2. Assume the two-way travel time for a reflection from point A2 is 2.0 sec and the two-way travel time for a reflection from point B2 is 2.1 sec. Both these reflection points are being recorded at the surface at positions A and B, respectively. So, on an unmigrated seismic section, the events appear to have the positions shown by dashed line Z2. The seismic lines are recording data at surface locations A and B from subsurface reflection points A2 and B2, which are located up-dip of surface locations A and B.

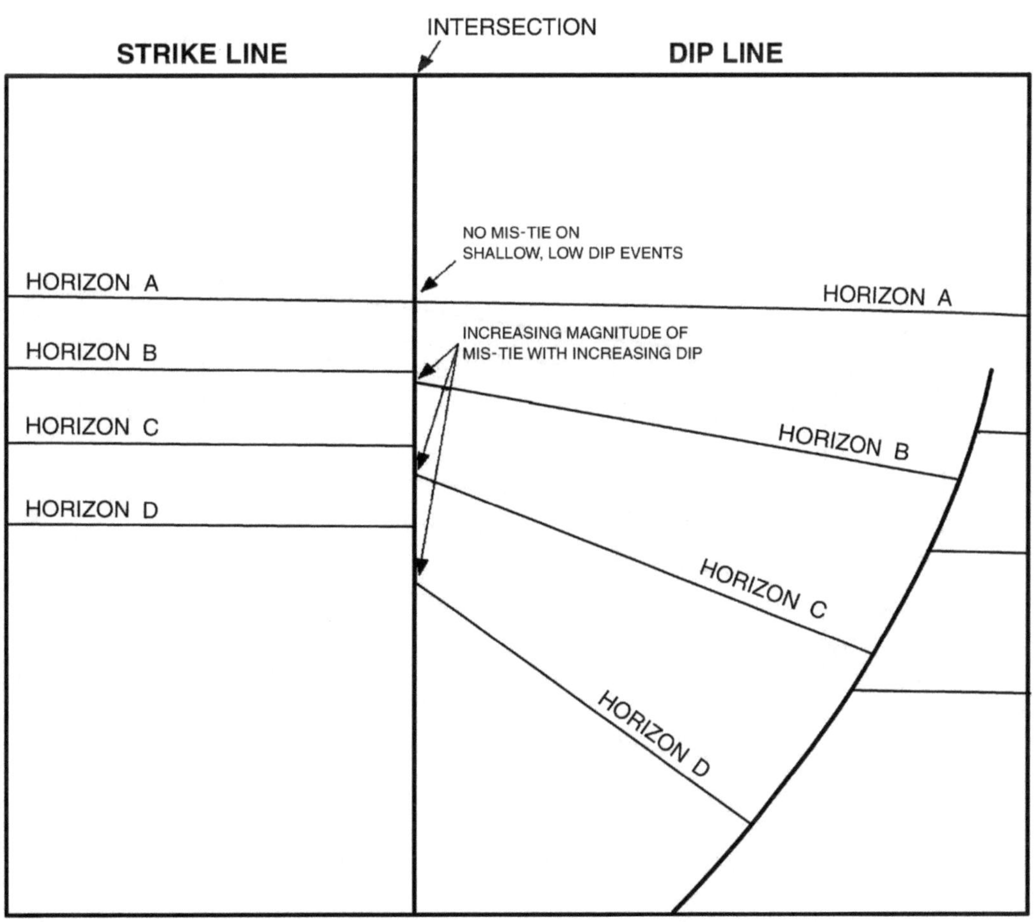

Figure 5-23 Appearance of migrated strike and dip lines at their intersection. Notice events on strike line are too "high" (too small a time).

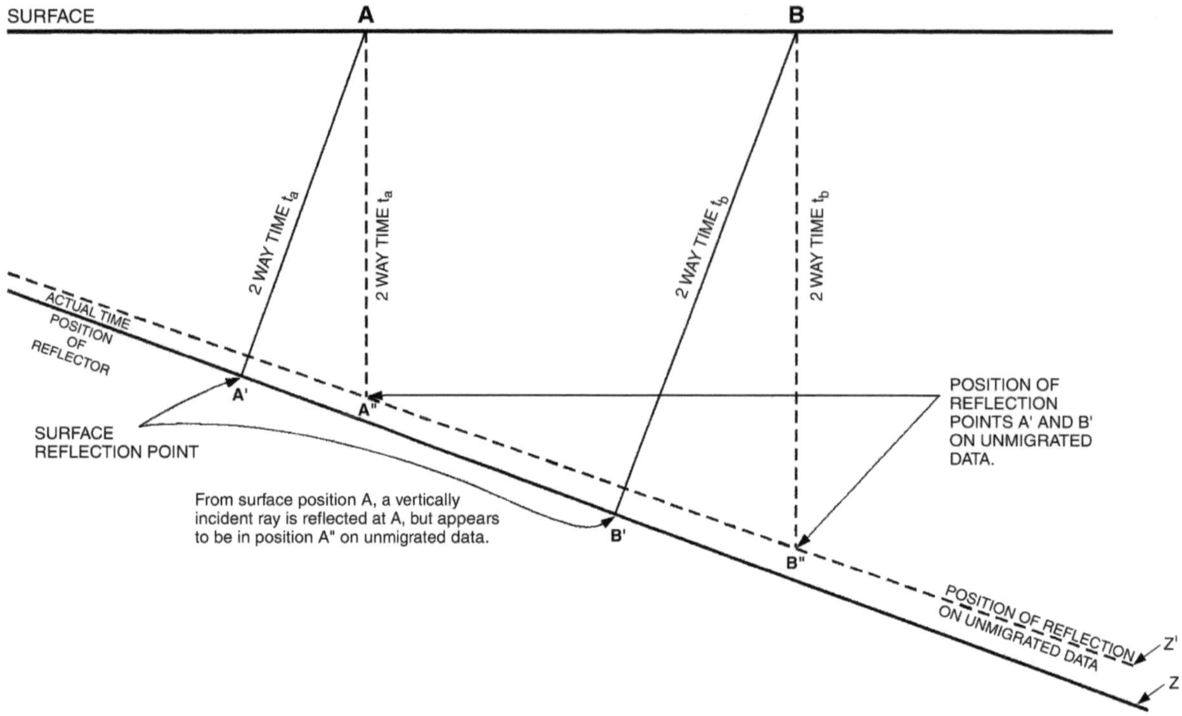

Figure 5-24 Diagram illustrating why data need to be migrated. Subsurface reflection points A2 and B2 are positioned at A3 and B3 on unmigrated data. Migration moves points A3 and B3 back to their actual time positions A2 and B2.

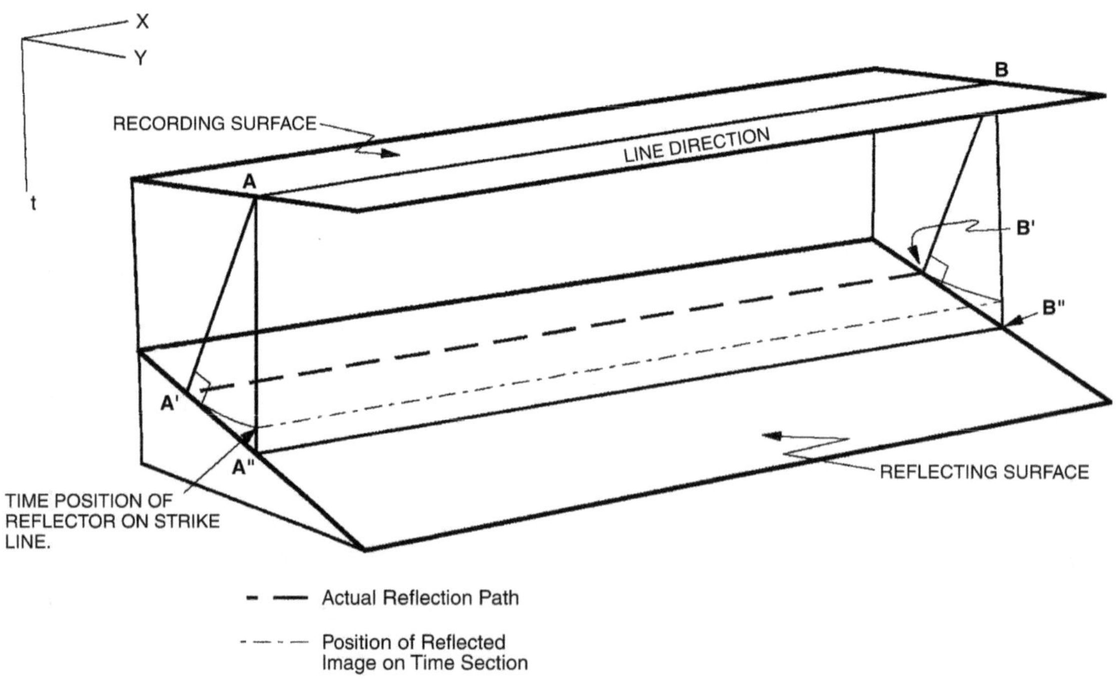

Figure 5-25 Strike line imaging. Reflection path for strike line is not underneath the line. Data is being recorded from up-dip.

Migration is a fairly complex process that corrects the data by moving it back to its proper time position relative to the surface location. In other words, after migration, a reflection point should be positioned correctly with respect to the surface recording points. Migrating the example would move the event Z2 to a position coincident with the actual event Z. Migration will always *steepen events or reflectors and cause a given event to appear "deeper"* (occur at a larger time) for any given shotpoint when compared to the unmigrated data. Migration of the seismic time data is critically important in obtaining a reasonably accurate interpretation of the subsurface.

The problem with 2D migration is that it can fully correct only a **true dip line**. Because the 2D migration algorithm cannot move data from out of the plane of the line, a line that has an apparent dip and not true dip will not be fully corrected. Figure 5-25 shows the extreme case: a true strike line that is oriented perpendicular to dip direction and thus contains no component of dip whatsoever. The data are still being recorded from up-dip, but the migration algorithm cannot move data out of the 2D plane of the line. The *apparent dip is zero*, so the migration algorithm really has no effect on the line. Event Z is positioned on the line at the two-way time of A-A2 and B-B2, much shallower (smaller two-way time) than is really the case beneath the surface line A-B. Figure 5-23 shows what occurs at the intersection of a true dip and true strike line. All the events on the strike line appear to come in too shallow (too small a time). This is important to remember when tying data. Only in very peculiar circumstances will a strike line event intersect a dip line deeper (greater time) than the corresponding event on the dip line. (This assumes that any static mis-tie has already been taken into account.)

Figure 5-26 illustrates what effect a migration mis-tie can have on a line intersection in areas where there is increasing dip with depth: the mis-tie becomes larger and larger in the deeper (greater time) part of the section. In an interesting twist to this problem, notice that the strong event at about 2.6 sec on the strike line is actually "deeper" than its correlative event on the dip line. How can this be possible if the data are coming from up-dip? Notice that all the events are too shallow on the strike line until about 2.4 sec, when they begin to appear deeper. By tracing horizontally (isotime) from the strike line to the dip line, we can see that the events are deeper because they are being recorded from up-dip, which happens to be downthrown to a buried growth fault, whereas the section ties the dip line in the upthrown block. Notice also that the fault intersection point on the strike line is actually 300 ms deeper than a "mechanical" tie would indicate. In growth fault areas, this problem is not uncommon. Many growth faults have larger dips in the nongrowth sediments located beneath the fault at depth.

The problem can be handled in either of two ways. The first and easiest method is to use the strike lines only to tie the events among the dip lines, but ignore them as valid sources of data points for drawing maps. In areas where there is already abundant well control and adequate density of dip-line coverage, this may be a viable option. The strike lines can still be used to tie events among the lines, but their actual time values are ignored. The disadvantage of this method is that the strike lines contain information on cross structures that is often critical to development of a potential hydrocarbon trap or reservoir size (see Chapter 11).

The best option (and the most time consuming) is to explicitly correct the strike line data by moving the data to their proper position relative to the surface locations. There is an easy graphical way to accomplish this task that the 2D seismic workstation does not automatically provide. Figure 5-27a shows hypothetical dip and strike lines posted on a structure map. Figure 5-27b shows an intersection of migrated lines A and B, as they would appear on migrated time sections. If you assume that the dip line A is properly migrated, then the data on strike line B at the intersection is actually coming from a position that is up-dip of the intersection with the dip line.

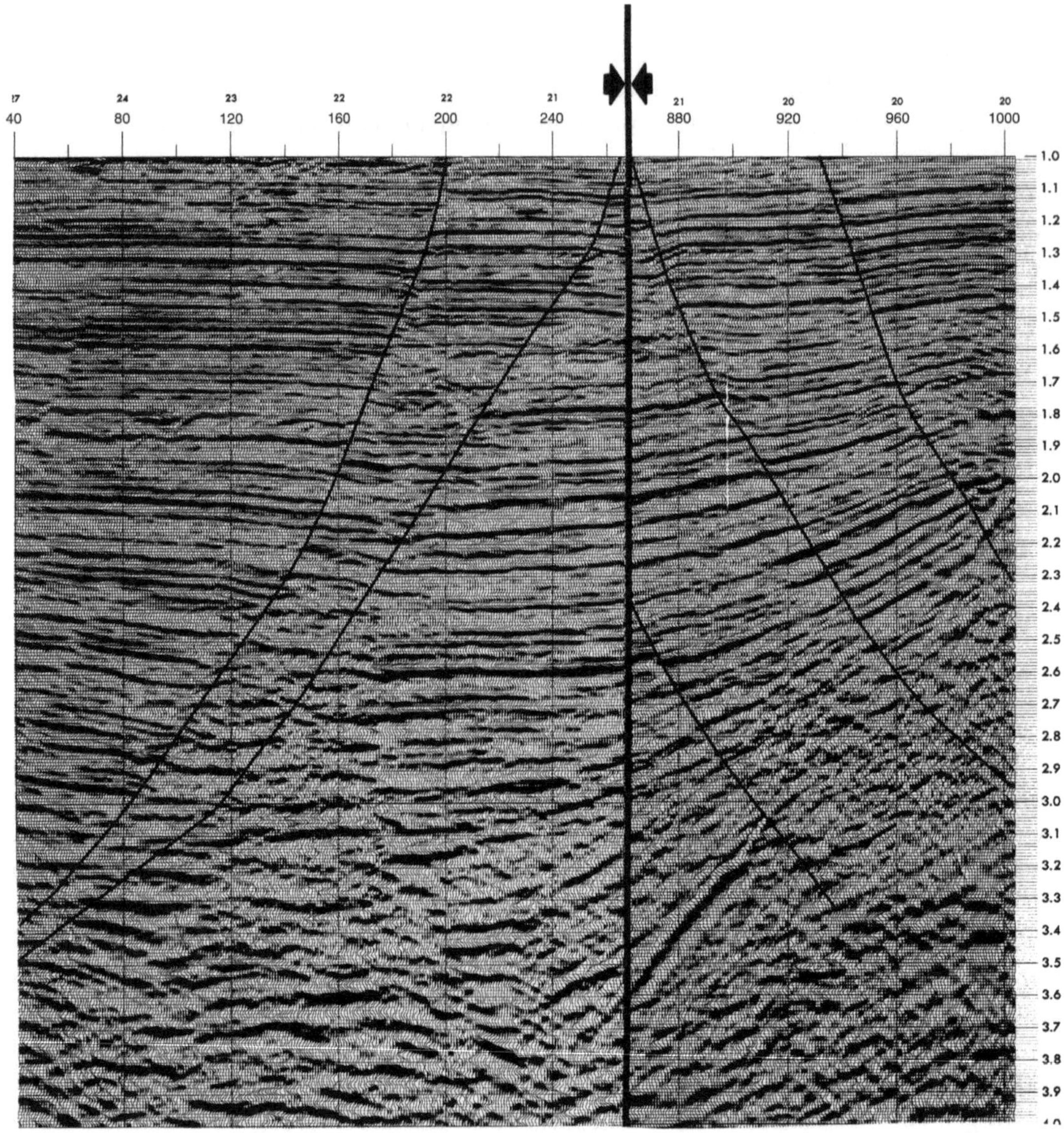

Figure 5-26 An unusual case of migration mis-tie. Strike section events occur earlier ("shallower") than same events on dip section until approximately 2.4 sec. At the bold event at 2.6 sec, strike line continues to record up-dip; however, this is also downthrown to an expanding growth fault. (Seismic lines published by permission of TGS/GECO.)

This concept is easier to illustrate than to explain, so Figure 5-27b shows an imaginary horizontal line drawn from the event intersection on the strike line to its real reflection point on the dip line. The process for correcting a series of lines is to find the actual reflection points for all the intersections of the strike lines with dip lines, and to mark these points on the map. These points locate positions near where the reflected event actually originated (Point A on Fig. 5-27b). If this procedure is carried out over a series of intersections, then a "corrected" basemap can be made that shows the approximate *reflection path of the strike line.* The dashed shotpoint display in Fig. 5-27a illustrates how to "relocate" the strike line up-dip before posting the data points to use in making a map. Once this corrected basemap is made, interpolate the individual shotpoints and post the times from the strike lines at their approximate subsurface locations. One important point about this technique: any correction is only valid for the particular event being interpreted and loop-tied. For example, another event that is deeper may require another corrected basemap

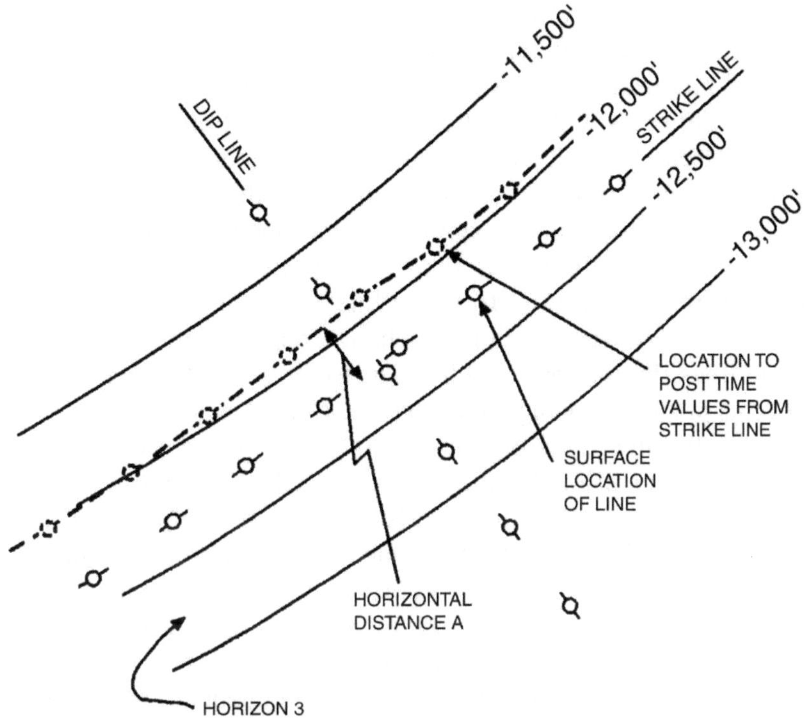

(a)

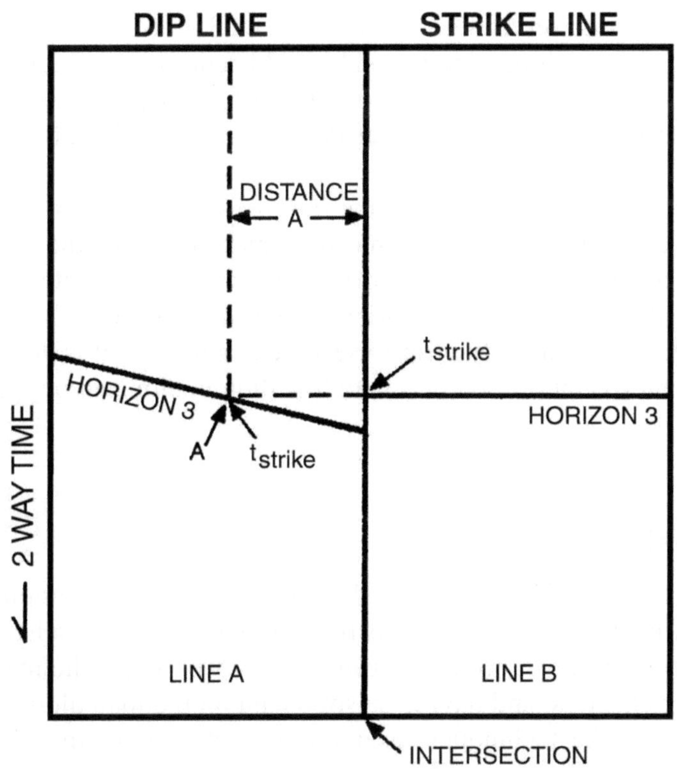

(b)

Figure 5-27 (a) Approximate method for dealing with migration mis-tie involves relocating strike line data points up-dip as determined from a true dip line. (b) Method for calculating the distance that strike data must be moved to account for migration mis-tie. Find intersection of time observed on strike line (at line intersection) with the same event on dip line.

to be constructed for its structure map. In other words, the correction is not a fixed value, but rather, it varies depending on the depth of the event being mapped.

This discussion has been directed toward tying actual seismic events among lines. Faults must also be tied, and in areas with high bed dips, the migration mis-tie problem can make mapping fault surfaces extremely difficult. It is vital to remember that if the fault surface reflections are coming from somewhere other than beneath the line, then the fault surface reflection is subject to migration effects. What makes mapping faults especially difficult is the spatial aspect of the migration mis-tie problem. It is entirely possible to have data points for a given fault surface coming from further and further away from the line, simply because the profile crosses the fault at an oblique angle. This causes the fault to change depth relative to the profile. Figure 5-28a shows two seismic lines that intersect each other. Line A is a dip line and shows increasing dip of the seismic events with increasing time (depth). Line B is a strike line with an obvious fault cutting the events on the line. It is obvious from the tie that the events on line B are actually coming from up-dip of the surface location of the line. The discontinuity in the events is caused by fault X, and thus the image of fault X is also coming from up-dip of the surface location of the line. Point X on Line A in Figure 5-28a is the actual tie point for fault X. Figure 5-28b is an illustration of the method used to construct a corrected fault surface map by moving the times for the fault trace further and further up-dip with increasing depth of the fault. The complex nature of the migration process illustrates the advantages of 3D migration. In areas with complex velocity fields, prestack time migration (PSTM) may not be adequate and prestack depth migration (PSDM) may be required. Fortunately, the increased availability of 3D data and increases in computing power mean that 3D PSTM and 3D PSDM data are becoming much more common.

It is important for interpreters to realize that even 3D PSDM data may not accurately depict the subsurface (Fig. 5-29). Three-dimensional PSDM relies on a velocity model and an algorithm to work. An inaccurate velocity model or selection of an inappropriate algorithm can produce a seismic dataset that appears reasonable but does not accurately reflect the subsurface geology (Herron 2011).

One additional complication that seismic interpreters must consider is illumination. In areas of steep dips and complex geology, seismic data may return to the surface far away from the surface shotpoint. Seismic surveys need to be designed so that data required for processing is acquired, although this requires some a priori knowledge of the geological structure. If it is not possible to do an illumination study prior to seismic acquisition, it is good practice to do one after an initial round of interpretation, so areas with low illumination can be identified (Herron 2011).

DATA EXTRACTION

Picking and Posting

After the seismic lines have been interpreted, transfer all the information to a basemap and begin the process of making a subsurface map. As pointed out earlier, the seismic data should be posted along with all the subsurface information from electric well logs. The mapping process is covered extensively in other chapters, and having seismic data on the map along with the subsurface well data should not affect the techniques used for the actual mapping. The following discussion pertains to posting data from interpreted sections. Most of these data are automatically recorded during interpretation on a workstation.

Types of Data from Seismic. When interpreting by hand, the most obvious type of data to post from seismic sections are the actual two-way travel times for the events that correspond to the geological horizons being mapped. This is analogous to posting formation tops on the map

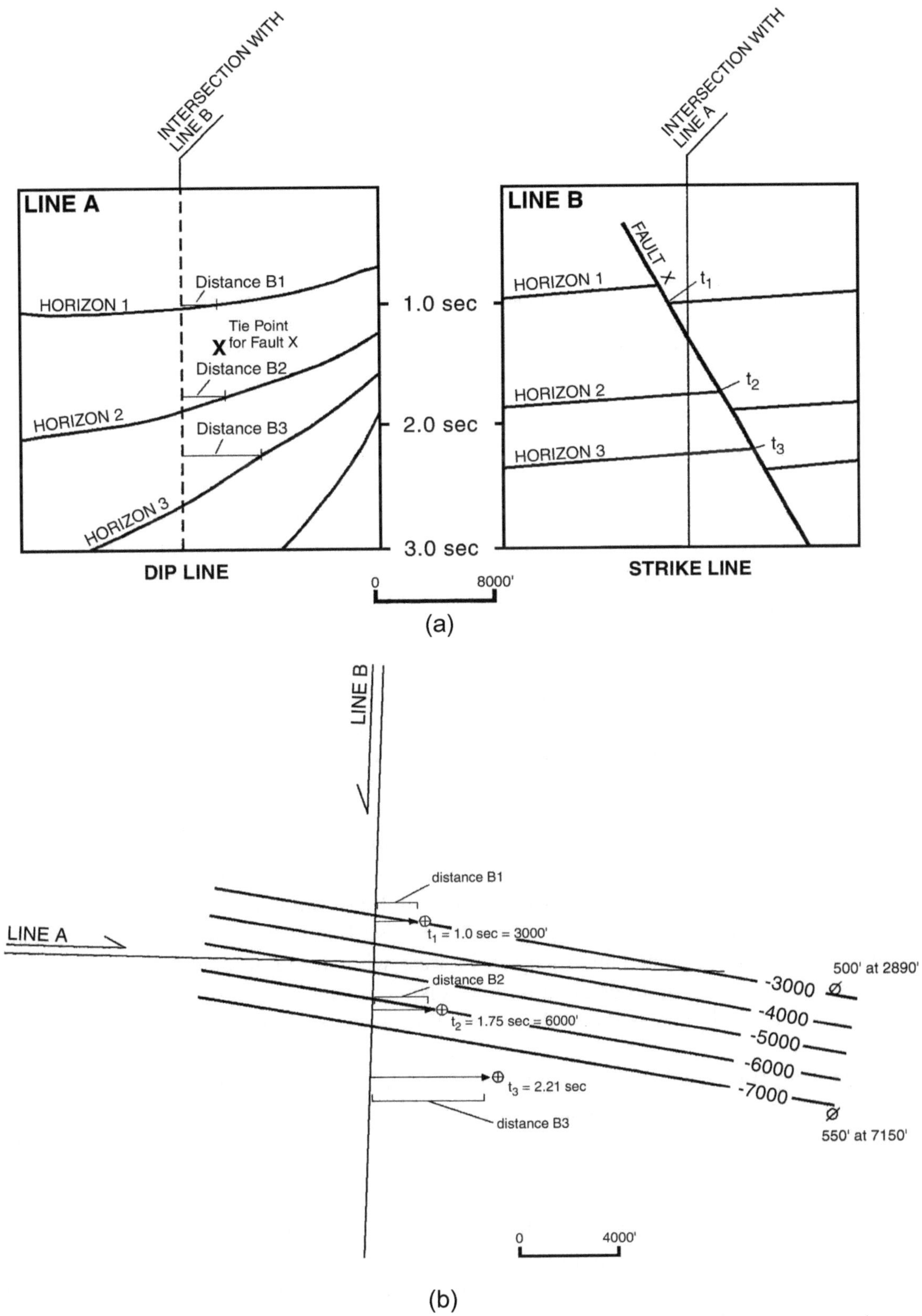

Figure 5-28 (a) Hypothetical dip and strike lines with fault observed on strike line. Notice the effect that an increasing amount of dip, with increasing depth on the dip line, has on the correlative events on the strike line. (b) Relocation of fault surface points to proper spatial location, using method outlined in text. Notice that the change in the distance the points must be shifted actually changes as the record time (depth) increases.

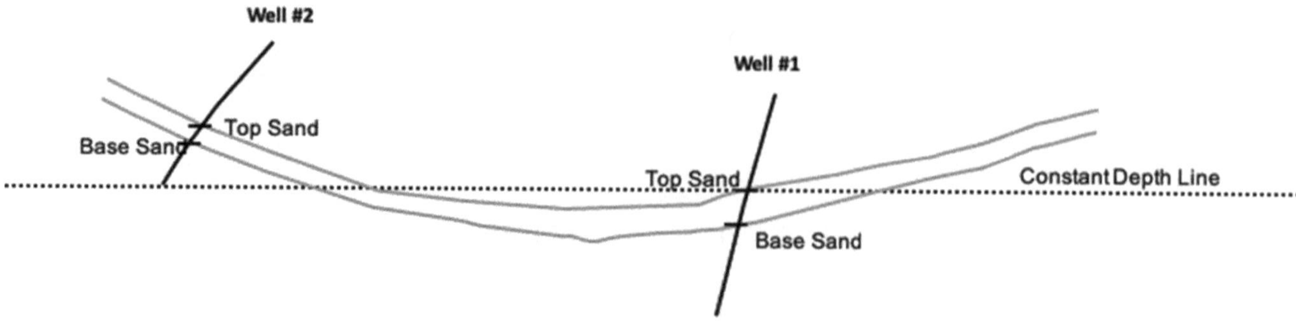

Figure 5-29 Two wells tied to a pay sand interpreted on well-imaged 3D PSDM seismic data. Although the seismic data clearly shows that Well No. 2 is approximately 100 ft high to Well No. 1, in reality the two wells are flat to each other. The elevation gain on the left side of the figure is a velocity anomaly that was unrecognized prior to drilling. (Published by permission of J. Brenneke.)

when using well data. The same two-way travel times can be posted for any fault surfaces being mapped. There are several other types of mapping information that can be extracted from the seismic data. One type of information that is extremely useful is the upthrown and downthrown intersections (cutout points) of the horizon being mapped with the surface of a fault. These intersection points have both a vertical datum and a location associated with them. The significance of these points, assuming the data is reasonably high quality, is that they can be used to help position the upthrown and downthrown traces of the fault and the approximate width of the fault gap. In practice, seismic methods overstate gap width by about a factor of 2, and the correct width of the gap is finally derived from structural horizon and fault surface map integration (Chapter 8).

Many interpreters post a solid bar on the map in an easily identifiable color to indicate the fault trace identification. In complex areas, you can assign a unique color code to each fault being tied on the seismic sections and post the trace on a basemap in the same color.

Depending on the area, there may be some other useful information that can be posted on the map. Inflection points of anticlines and synclines can be posted on the map to assist in contouring. If the seismic event being mapped has an amplitude anomaly associated with a hydrocarbon-bearing sand, then the areal extent of the amplitude anomaly can be posted. In areas adjacent to saltdomes, you may be fortunate to have data sufficient to identify a salt/sediment interface. This contact can be posted and mapped. In areas with stratigraphic discontinuity in the objective sands, you may be able to detect a unique seismic response indicating where the sand is present and where it is not. The extent of the potential reservoir body can thus be mapped.

Extracting the Data. When tying the data around a loop by hand, use a colored pencil to color in the troughs of the seismic data. At each downline or shotpoint marked on the line, mark with a pencil a consistent part of the waveform, using either the maximum trough (which is easy to see), the maximum peak, or the crossover.

Posting the Information. Now that the data have been interpreted, they have to be transferred from the seismic lines to the basemap. With an engineer's scale, measure the two-way time in milliseconds to the event being mapped. Next, add or subtract the constant value in milliseconds that represents the amount of static mis-tie between this seismic line and the reference seismic line. Post the two-way time on the map at the actual reflection point for the horizon being mapped. If you constructed "pseudo" basemaps with all the strike lines repositioned up-dip, post two-way times at the "corrected" shotpoint locations. Remember that the adjusted strike line location may vary, depending on the depth of the seismic event that is being mapped. Mark all the intersections of the seismic events associated with fault surfaces, and post the upthrown and downthrown intersection (cutout) points. Finally, record any other valuable information and post it on the map.

Converting Time to Depth

You are now at the point where the actual mapping is about to take place. One problem remains: converting the two-way travel time values on your map to depth data. This is done automatically on a workstation. If you are using paper sections, there are several different ways to accomplish this task. As in all the previous tasks discussed, there are both simple and detailed methods. Deciding which method to use depends on the complexity of the time–depth relationship in the area in which you are working. Some areas are so complex that attempting to make a valid map without the assistance of a geophysicist on a workstation is a mistake. Some areas are well-behaved in the time–depth domain, and the process for converting time to depth is trivial.

Brute Conversion with a Time–Depth Table. The first method for converting time to depth is extremely easy. The procedure for converting the values involves determining the depth values corresponding to the posted time values in a time–depth table that has been generated from checkshot data. Checkshot data is acquired during the evaluation phase of drilling a well. A checkshot measures the amount of time it takes for the first arrival of a seismic wave to travel from a surface source near the well to a receiver lowered down the wellbore. After the data are acquired, they are usually interpreted to generate a set of one-way travel times for specific depths in the well.

Conversion of these one-way times to two-way times involves multiplying the time values by two. (The receiver only measures the time it takes for a wave to travel to a given depth, whereas a seismic line measures the time required for a wave to travel to a reflecting horizon and back to the surface.) The two-way time–depth pairs are then interpolated from the usually sparse set of data points to generate a table with the time–depth pairs calculated at even increments of time. Such a table was shown in Figure 5-9. A plot of a particular checkshot can be shown graphically by constructing a time versus depth graph. The actual times are shown as data points, and a line is fitted to the data points using a cubic spline curve-fitting computer program.

Once the correct depth for a given two-way travel time has been found, you can post the depth value beside the time value on the map. Using a contrasting color pencil to post depth information is a good method for keeping the time values distinct from the depth values you will use to contour the map.

Where should this method be used? Use this simple technique in areas where there does not appear to be any large lateral variations in the relationship between seismic time and depth. The method for determining whether this is appropriate is as follows. First, use a time–depth table generated from the closest well to the mapping area (or a well in the middle of the area) to convert all the times to depth. Second, examine your map, looking for obvious discrepancies between the well tops posted on your map and the converted time values. If, for example, the depth from a converted time value is 11,200 ft subsea, and a well top at the same location is 10,700 ft subsea, then there is an obvious problem to be addressed.

Recognizing Velocity Problems. If the velocity information does not tie accurately to the wells, there are two possible sources for the differences in the values posted. One possibility, often overlooked, is that the *wrong horizon* was interpreted on the seismic line, and thus inappropriate time values are being converted to depth. This problem could arise from an incorrect pick across a fault or from an incorrect tie to a well. The error can be caused by using too large a shift to tie to a synthetic seismogram. Perhaps the synthetic seismogram contains problems related to log quality or some other factor. There exists a whole multitude of possible causes similar to these.

Another possibility is that there is a *strong horizontal velocity gradient* in the area that causes the time–depth relationship to change laterally. Lateral variations are often easy to determine if you have logs and synthetic seismic events that are easy to correlate in two wells (Fig. 5-29). In this case, you absolutely know that both the correlations and time picks on seismic data are correct and the correlations in the wells are correct. However, the correlations between the wells do not agree with one another. In other cases, the problem may not be so easy to identify. If either the well correlations or the seismic event ties are ambiguous, the presence of a gradient may not be obvious because of the difficulty in deciding what interpretation to rely on initially. It may be difficult to recognize a horizontal velocity gradient lacking well control in the proper locations.

Lateral velocity changes can be extremely difficult to manage. It is beyond the scope of this book to attempt a complete discussion of the methods of handling velocity gradients. The key question is when to recognize the need for expert help. A simple rule of thumb is this: if the gradient in the area being mapped is severe enough to cause the depth uncertainty for a given two-way time value to exceed the average amount of closure on the features you are mapping, you should be careful. In particular, if this depth uncertainty can be observed to occur between checkshots in wells that are as physically close as the average dimensions of a prospective closure, the likelihood of correctly mapping a structural closure is small. If a problem exists, seek the assistance of a senior geophysicist who understands all aspects of handling velocity gradients.

Accounting for Small Velocity Problems. If the magnitude of velocity changes over your mapped area is not severe, you can often "eyeball correct" a map to account for a gradient. To correct for gentle gradients, interpreters may make ad hoc adjustments to the map to account for the horizontal velocity gradients. In practice, interpreters will attempt to find the geological reason for the velocity differences and use a different time–depth table on either side of the geological "boundary" causing the anomaly. For example, in the Gulf Coast tertiary section, a large growth fault will commonly have a downthrown section that has a different velocity field than the upthrown section. In some areas, the seismic velocities may be faster on the downthrown side because of the increased amount of sand in the downthrown block. In other cases, the seismic velocities may be slower in the downthrown block because the thickened section was deposited rapidly and is undercompacted and slightly overpressured, and therefore the velocities are slower. Whatever the situation, if you can determine the reason for the gradient, you can often adjust the contours to honor the well control.

We have found that the easiest way to handle velocity gradients that do not appear to have definite boundaries is to use the following technique, illustrated in Figures 5-30a to 5-30d. A basemap with posted information obtained from both well logs and seismic sections is shown in Figure 5-30a.

First, prepare a pure-time map. Map the isotime contours of the time values, as shown in Figure 5-30b. Next, determine the average velocity in the depth range being mapped. Simply determine the number of milliseconds of two-way time that the contour interval, in depth, represents at the depth range you are mapping. A typical Tertiary value might be about 22 ms per 100 ft. At each point of well control, use the time map as a guide and begin contouring in depth, using the distance between each time contour as a rough indicator of the magnitude of bed dip. Carry the contours about halfway to the next well and then start at that well and contour away from that well until you meet the previous contours (Fig. 5-30c). The discrepancy in the depth values

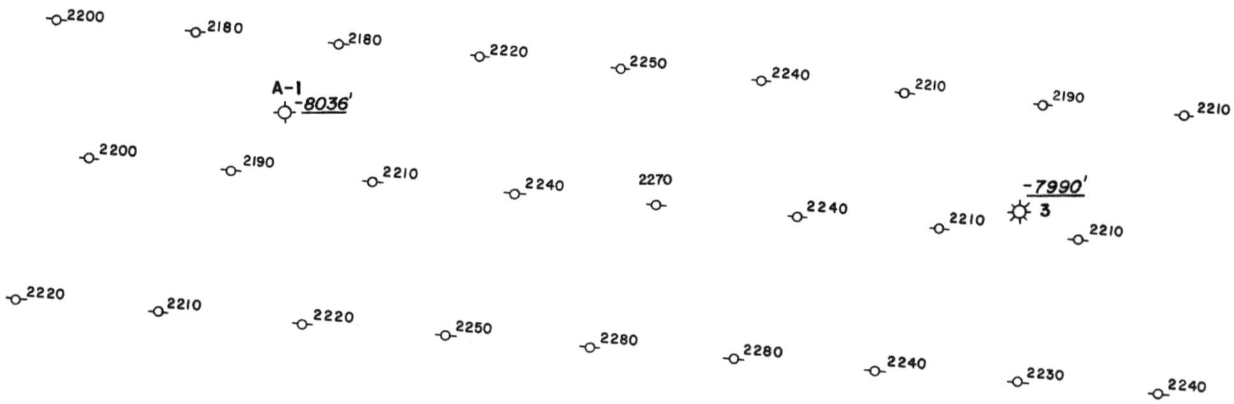

(a)

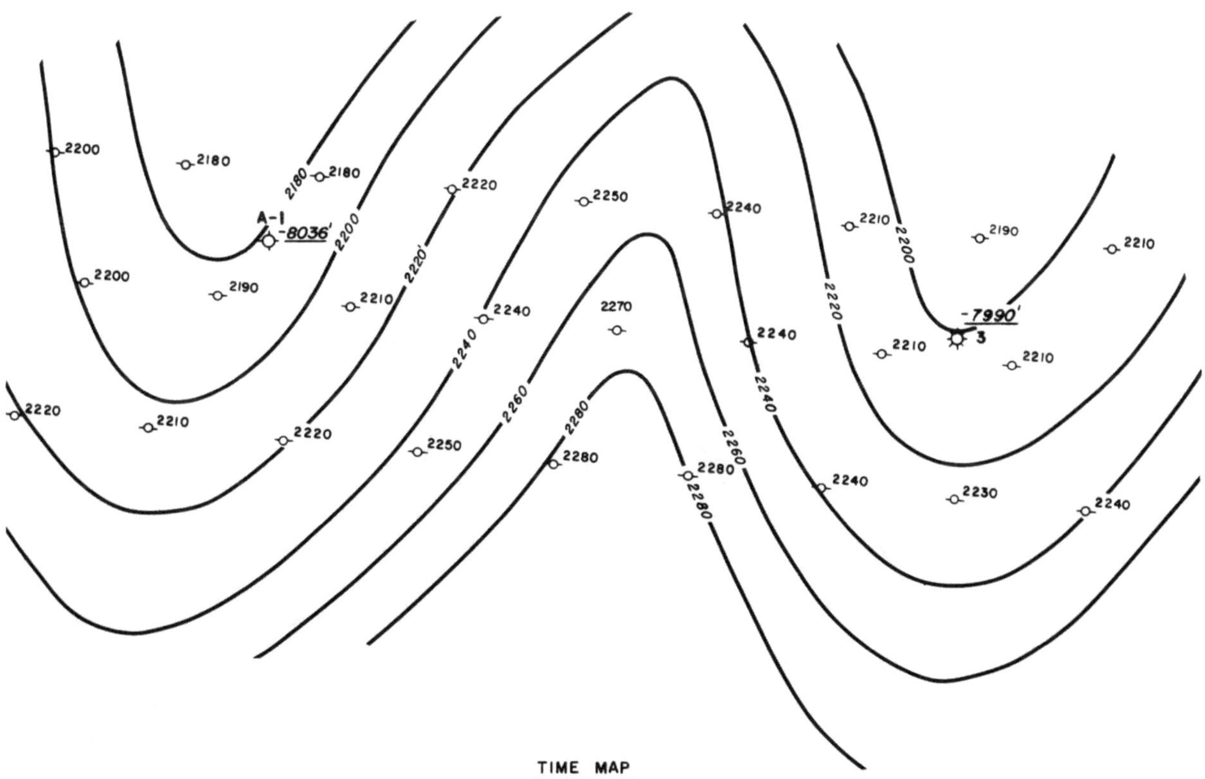

(b)

Figure 5-30 (a) Basemap showing posted seismic two-way times and formation tops from well data for a mapping horizon. (b) Isotime map of seismic two-way data. (c) Preliminary depth map constructed from each point of control. Contouring begins at each point of well control and moves outward. (d) Final depth map with horizontal velocity gradient "lost" in the syncline between the two plunging noses. The difference in depth values has been adjusted in the contouring of the syncline between the well control points.

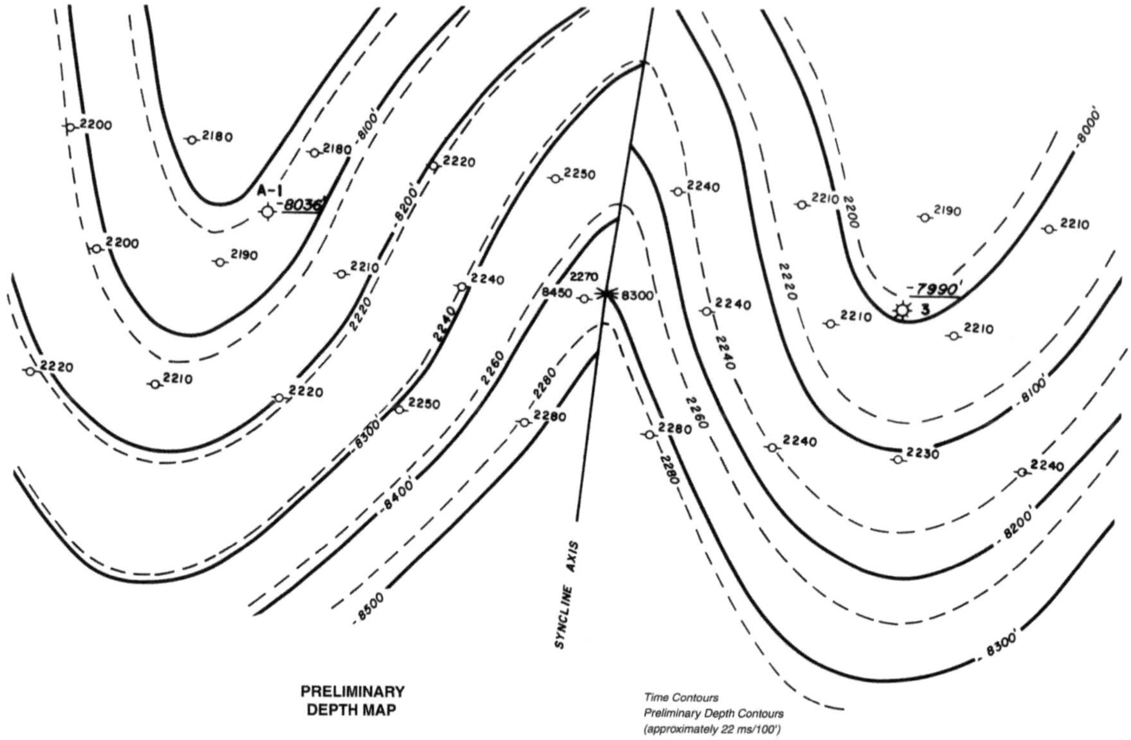

(c)

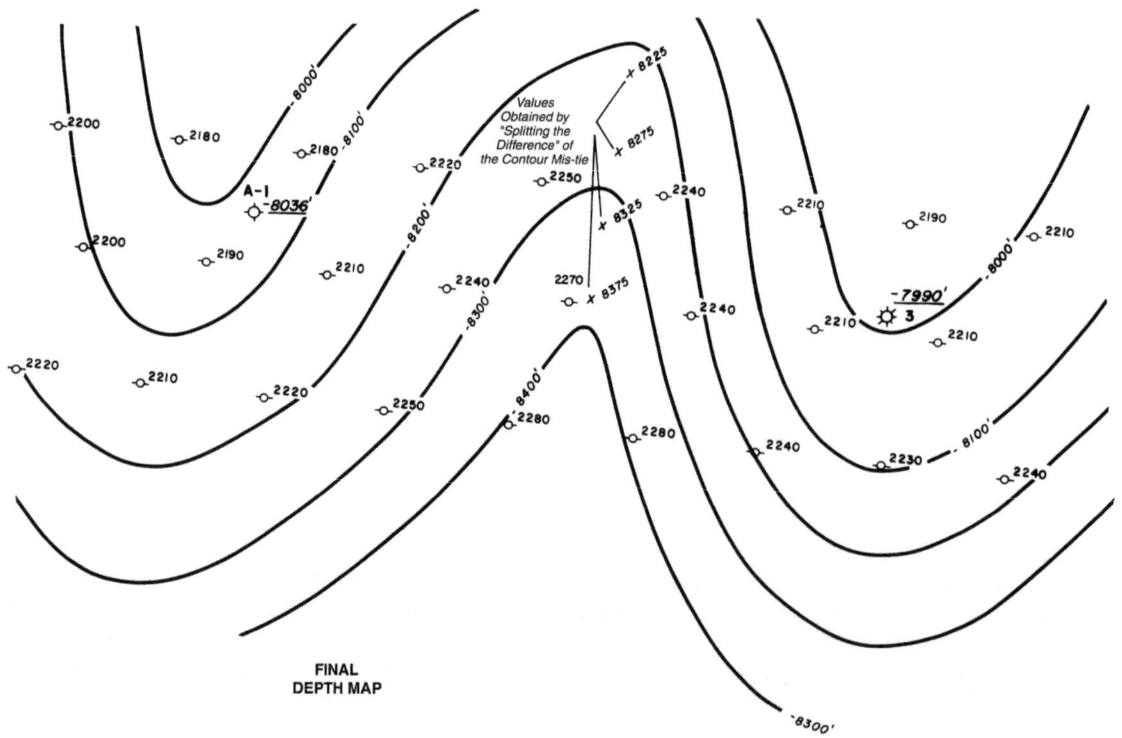

(d)

Figure 5-30 *(continued)*

can then be adjusted by splitting the difference between the two sets of contours and gradually adjusting the mis-tie in the spacing of the depth contours (Fig. 5-30d).

The points to remember about this technique are, (1) it is quick but imprecise, (2) it is appropriate when applied to minor velocity problems over large areas, and (3) it is useful when the map must be finished quickly. A caveat: Never use this technique when the gradient is severe and is present over a single structure. It is an appropriate and useful technique when you are mapping on a large scale and need a method of "absorbing" the mis-ties in the synclines between the major closures. Remember that seismic data do not have the vertical resolution of well data. A seismic line sampled at 4 ms and picked to an accuracy of 10 ms will give you about 40 to 60 ft of error in an average Gulf of Mexico Tertiary section. Using this technique to account for a 200-ft mis-tie problem seems acceptable.

The most important point to remember about velocity problems is that it is very easy to exceed your expertise with simplistic solutions. Look critically at your data and get assistance if it is needed.

SOME FINAL THOUGHTS ON SEISMIC MAPPING

In this section, we discuss several additional problems that yield to common sense and intuition when making maps using seismic data. The main "rule" is to stand back from the work at various stages during a project and ask yourself some hard questions about the quality of the work. These thoughts are presented in the form of questions that the geoscientist must ask himself or herself during the interpretation project.

Have you picked the right event to map? This is meant as more than just a verification to determine whether you are tied correctly into a specific horizon. In a more general sense, have you picked the correct geological event to interpret and tie through your data volume? If there is a choice as to what event to map, always pick the strongest, most laterally continuous event in your data set. In many areas, this will be a sequence boundary. If there is no inherent reason to pick one event in an interval over another, always opt for the one that has the most lateral continuity. Your chances of incorrectly interpreting the correlation across a fault or mis-tying a seismic line are much reduced when mapping these obvious sequence boundaries.

Does your interpretation make geological sense? Such geologically unreasonable features as a radically thinner downthrown section on a growth fault, or horizons that cut across reflectors, are warning signs that the interpretation is suspect. Check the angles of the fault surfaces on your data. Are the angles reasonable for the tectonic regime being mapped? Look for impossibilities on your interpreted section before posting the data on the map. We once asked a colleague, who found over 2 billion barrels of oil, "What was the most important technique used to find this oil?" The reply was, "I try not to do anything stupid." This can save a lot of potential rework. Several minutes of self-criticism each hour can be a very valuable quality-control technique for your work.

Has the interpretation taken the path of working from the known to the unknown or from good data to bad data? It is very easy to become absorbed in solving small interpretive problems before understanding the large-scale features. The best way to work is to solve the obvious, incontestable problems before tackling the subtleties.

Seismic interpretation and mapping is a process that can be learned only by practice. The more you interpret, the easier it is to recognize the pitfalls and sources of error in your visualization of the subsurface world from seismic data.

CHAPTER 6

CROSS SECTIONS*

INTRODUCTION

A contour map depicts the horizontal plan view configuration of a single attribute of a stratigraphic unit, such as structure, thickness, and percent porosity. In contrast, a cross section depicts the configuration of many units as typically viewed in a *vertical plane*. Since a map or cross section alone cannot represent the complete subsurface geological picture, both must be used to conduct a complete and detailed study.

Geological cross sections constitute a very important geological exploration and exploitation tool. They are useful in all phases of subsurface geology as well as in reservoir engineering. Cross sections are used for solving structural and stratigraphic problems in addition to being employed as finished illustrations for display or presentation. Used in conjunction with maps, they provide another viewing dimension that is helpful in visualizing a geological picture in three dimensions.

If a cross section is oriented perpendicular to the strike of the structure, it is termed a *dip section*. If the section is oriented parallel to the strike of the structure, it is called a *strike section*. Finally, if the orientation is oblique to the structural axis, it is termed an *oblique section*.

Various data can be used to construct a vertical cross section. It can be based on surface data (dips), electric well log data (markers, unit tops and bases, dips, and faults), seismic data, or entirely from completed subsurface maps. As mentioned in Chapter 1, all available data should be used in the preparation of a subsurface interpretation, whether it is in the form of a structure map or cross section.

* For all figures in this chapter (in the printed book only), see the preface for information about registering your copy on the InformIT site for access to the electronic versions in color.

PLANNING A CROSS SECTION

Prior to making a cross section, ask yourself a number of questions regarding the planned section construction. The answers to these questions facilitate the preparation of the section and improve its value as an aid to solving problems or illustrating the final geological picture.

What is the purpose of the cross section? Is the section going to be used as a structural or stratigraphic aid to solving problems? Is it going to be used as a communication device to illustrate the final geological picture? Will the section show the gross geological framework or be designed to show significant detail? What sources of data are to be used in the construction? Should the section be prepared true to scale (with the same vertical and horizontal scales) or at an exaggerated scale? What datum is to be used? The answers to these questions provide insight into the planning and preparation of any proposed cross section.

Initially, you must *determine the specific objective* for preparing a cross section. If it is to be prepared to aid in the interpretation of the structural framework and solve problems related to faulting and structural dip, then the section required is a structural cross section. If the intent of the section is to solve stratigraphic problems relating to detailed correlations, permeability barriers, unconformities, facies changes, or changes in depositional environments, then a stratigraphic cross section is needed. A structural or stratigraphic section can be used as a visual aid to communicate or illustrate, as well as to solve, problems. The intent will affect the preparation of the section.

The next step in preparing a cross section is to *choose the orientation of the line of section*. The choice is dependent first on the type of section you intend to prepare (structure or stratigraphic); second, on the type of geological structure (i.e., diapiric, extensional, compressional, or strike-slip); and third, on the data to be used in the section (well logs, seismic data, structure maps, or surface data).

Finally, *the scales of the proposed section must be selected*. Two separate scales must be considered: the vertical and the horizontal. The scales used are dependent on the type of section being prepared, the actual length of the section, data used, and desired detail. Whenever practical, *use the same horizontal and vertical scales*. However, there may be a special consideration that requires the section to have different vertical and horizontal scales. Often, the vertical scale is larger than the horizontal, and where this situation occurs, the section is said to have *vertical exaggeration*. Each of the specific conditions that go into the planning and preparation of a line of section are discussed in detail in this chapter.

STRUCTURAL CROSS SECTIONS

Structural cross sections illustrate structural features such as dips, faults, and folds (Silver 1982). They are usually prepared to study structural problems related to subsurface units, fault geometry, and general correlations. Such problem-solving is accomplished by enabling you to visualize the subsurface structure in a vertical plane. Electric well logs (Fig. 6-1) or well log sticks (Fig. 6-2) can be used in the construction of structural cross sections. There are times when electric logs are not available and other data must be used. These data could include drill-time logs, core data, and lithologic logs prepared from cuttings descriptions.

Structural cross sections are drawn in the direction of interest. The section can be oriented perpendicular, parallel, or oblique to structural strike. For solving structural problems, it is common for the line of section to be laid out in the dip direction or over the crest of a structure. A line of section parallel to the dip of a fault is best for solving fault problems. In a complex area such as that of a faulted diapiric salt structure (Fig. 6-3), the evaluation of the structure, salt, and fault geometry may require a number of cross sections to be laid out in the direction of structural dip and perpendicular to fault strike. Each structure and the problems to be solved must be evaluated individually as to the planned direction and the number of sections required to adequately study the geological feature.

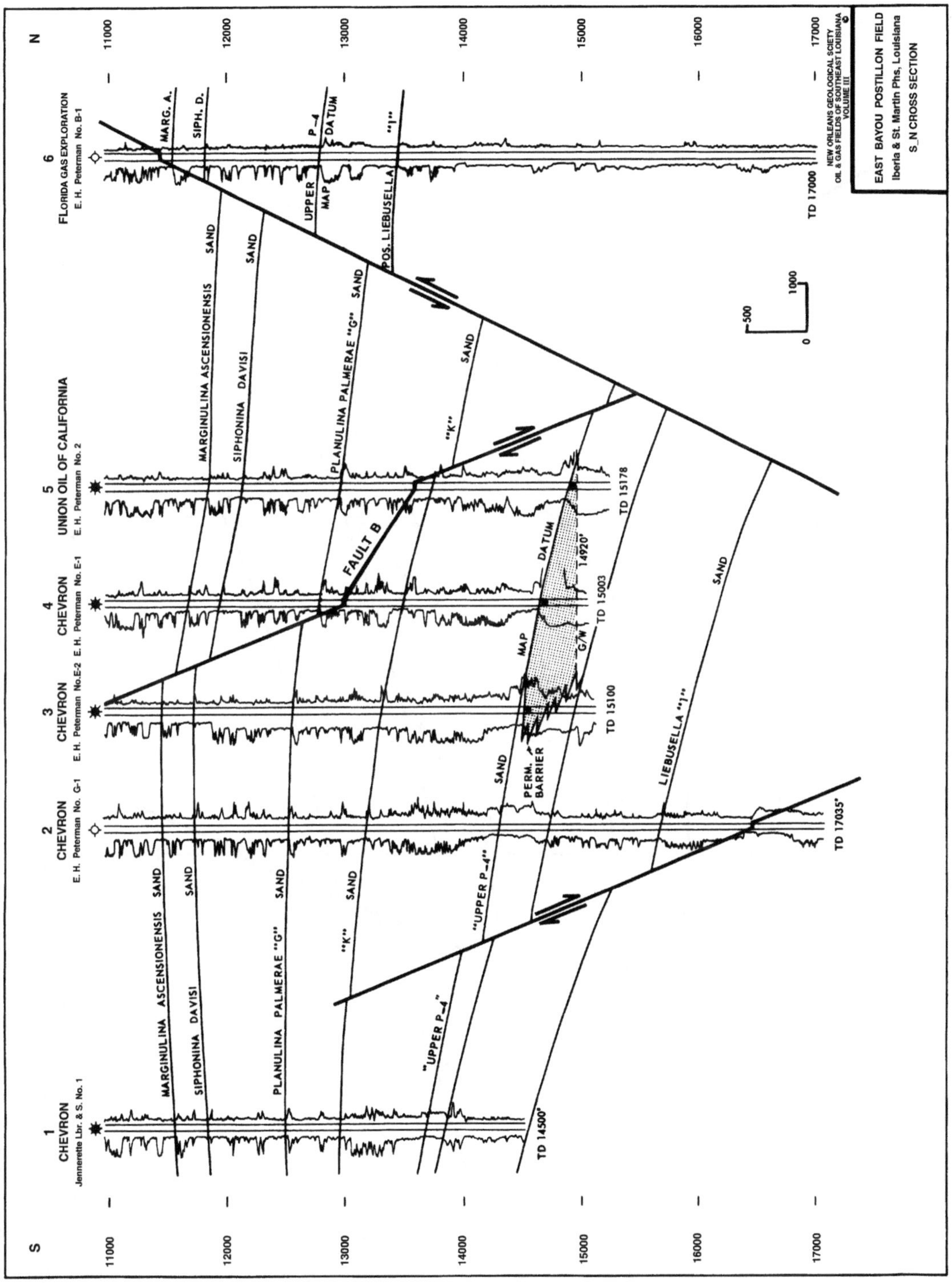

Figure 6-1 Structural cross section prepared from electric log data. (Modified from Oil and Gas Fields of Southeast Louisiana, v. 3, 1983. Published by permission of the New Orleans Geological Society.)

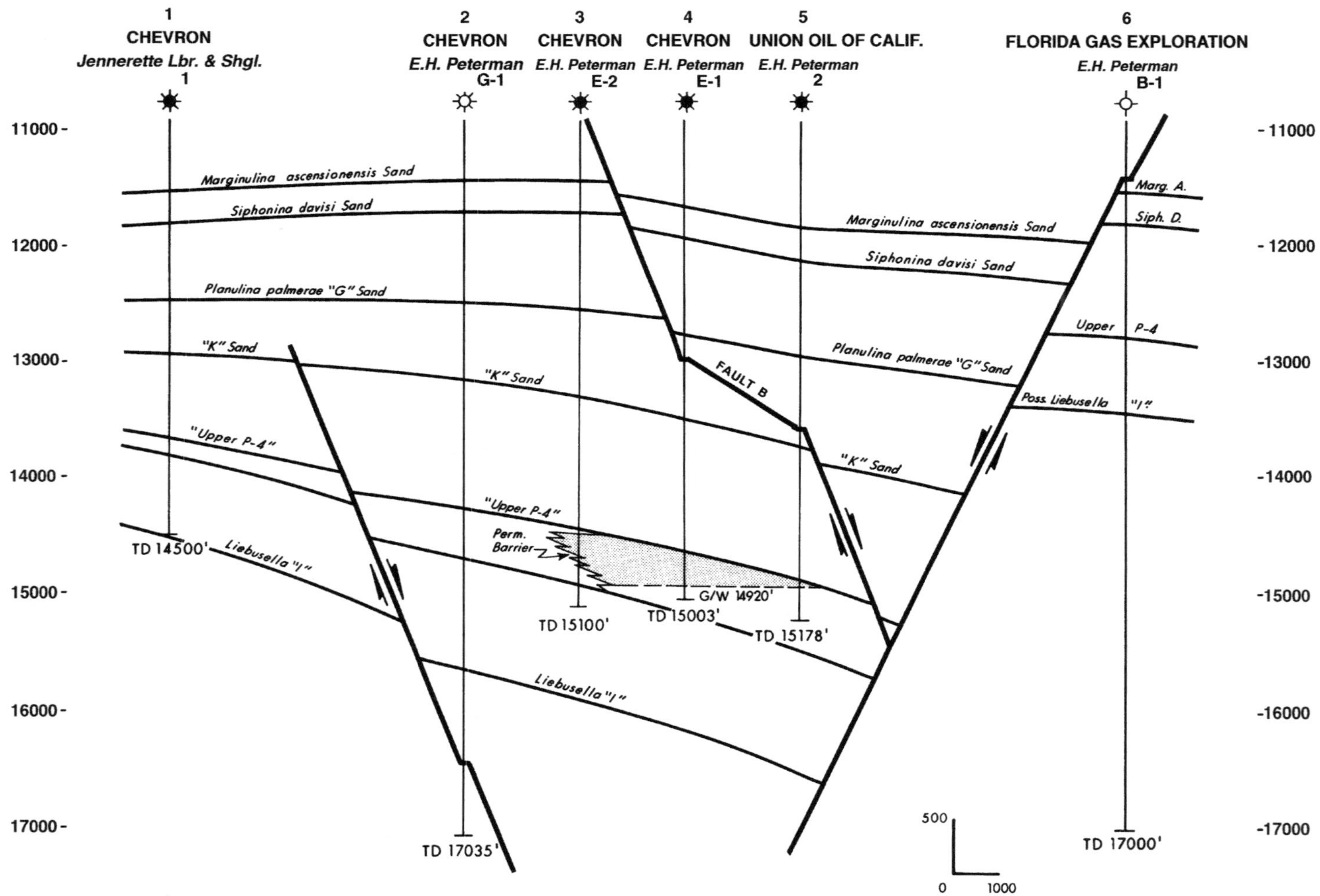

Figure 6-2 Structural cross section prepared using well log sticks. (Modified from Oil and Gas Fields of Southeast Louisiana, v. 3, 1983. Published by permission of the New Orleans Geological Society.)

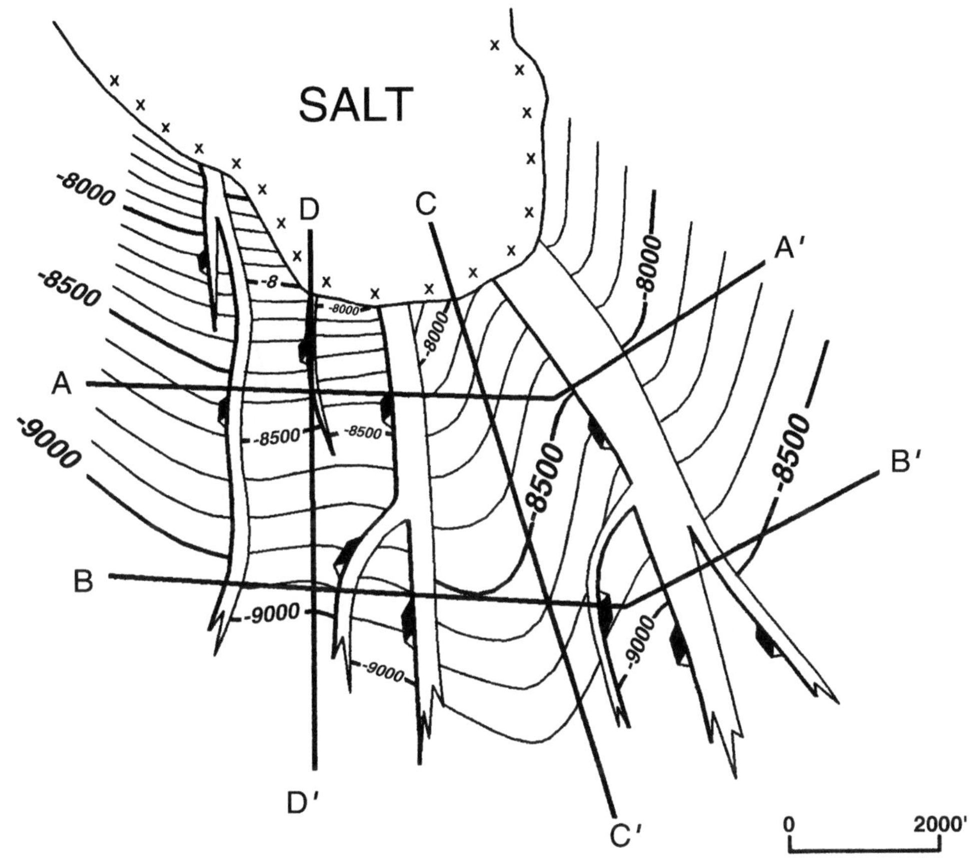

Figure 6-3 Typical cross section layout for a complex piercement salt structure.

Oil and gas wells do not normally lie in a straight line, as shown in Figure 6-4a, and so the direction of a planned cross section may not be in a straight line. Instead, line segments between adjacent wells may vary in length and direction, giving the line of section a zigzag appearance (Fig. 6-4b). With a zigzag section, wherever the direction of the line of section changes, the apparent dip of horizons and faults on the section also changes. An example of such a change can be seen in Figure 6-1 for Fault B. Notice that the dip of the fault changes from 46 deg between Wells No. 3 and 4 to about 18 deg between Wells No. 4 and 5. This change indicates that the line of section between Wells No. 3 and 4 is perpendicular or nearly perpendicular to the strike of Fault B; but between Wells No. 4 and 5, the line of section is more parallel to the strike of the fault. These apparent changes in fault dip are illusions of subsurface geometry caused by the orientation of the line of section. Such illusions must be considered when laying out any line of section in order to prevent confusion to an observer and also to make sure that the section shows what it was intended to show.

We recommend that structural cross sections be *drawn with the same horizontal and vertical scales* whenever possible. With the same scales, the cross section is prepared true to scale with no vertical or horizontal exaggeration. At times, however, exaggeration is required to permit legible vertical detail. The effects of vertical exaggeration are discussed in detail later in this chapter.

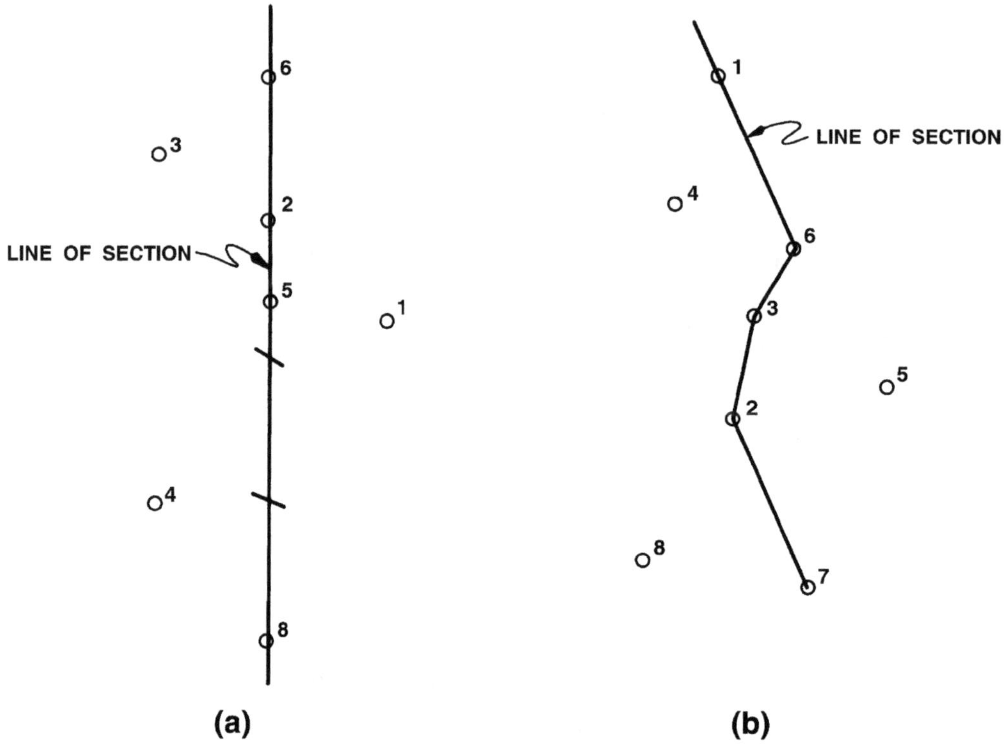

Figure 6-4 (a) A cross section oriented through wells that form a straight line. (b) A zigzag line of section formed by line segments between adjacent wells.

Electric Log Sections

When preparing cross sections with electric logs, certain procedures are helpful in maximizing the usefulness of the section. These procedures will vary depending on whether you are making a cross section using paper logs or making a cross section on a workstation. We discuss using paper logs first.

1. The logs that are chosen for a section must fit the scale of the cross section. The logs may need to be enlarged or reduced to make the scale equal to that of the cross section.
2. If possible, use the same vertical and horizontal scale. This is usually convenient for field studies; however, regional or semiregional sections may require exaggerated vertical scales to get them down to a manageable size.
3. Data must be legible. Log heading data, correlations, depths, and so on, may have to be posted after the section is laid out in order for them to be legible.

How are electric logs mounted on a manually generated cross section? The procedure usually begins with the preparation of a film positive for each of the original logs to be used in the section. If required, the logs are reduced to the appropriate vertical scale to accommodate the section. The film positives are normally taped on coordinate grid cross-section paper with the correct proportional horizontal spacing between wells. It is good practice to plot your line of section on a well basemap or structure map (Fig. 6-13) so the spacing between well logs for the section can be measured directly from the basemap. For a structural section, the logs are normally hung with sea level as the reference datum. Therefore, the measured log depths must be converted to

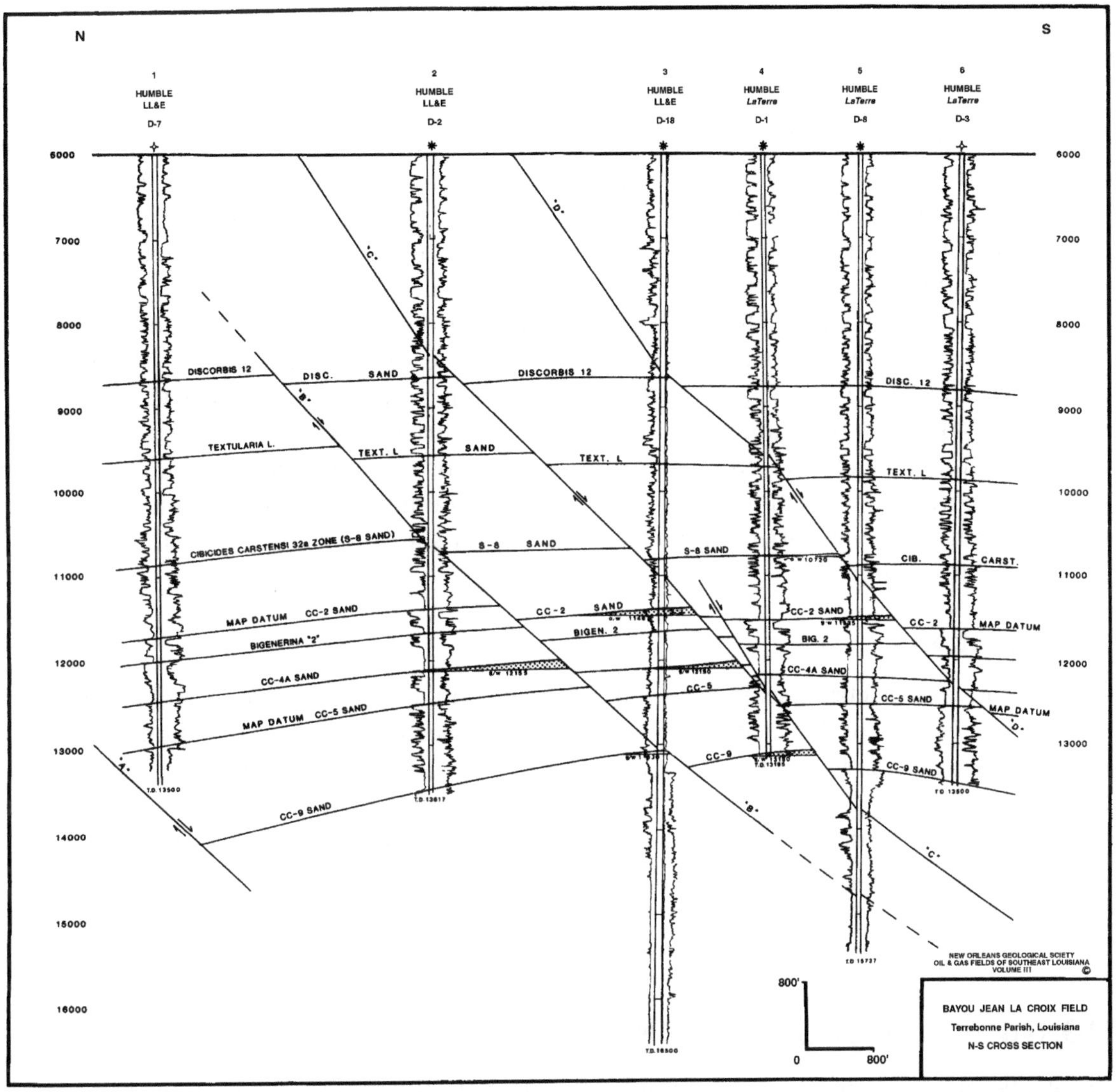

Figure 6-5 Completed structural cross section from Bayou Jean La Croix Field, Terrebonne Parish, Louisiana. (Reprinted from Oil and Gas Fields of Southeast Louisiana, v. 3, 1983. Published by permission of the New Orleans Geological Society.)

subsea depths for vertical position on the section. Finally, an ozalid or xerographic print of the cross-section base is made, and you are now ready to begin the cross-section interpretation.

Cross sections are more easily prepared if you use a computer. In a few simple steps, a line of section is laid out with proportionally spaced wells that can be hung on a given datum. The datum is readily changed and the logs are easily manipulated in making interpretations. The procedures are described in the section Cross-Section Construction Using a Computer.

As the first step in an interpretation, we recommend that all recognized correlation markers, as well as locations of faults in each well, be indicated on the logs used in the section. The faults can be connected from well to well to reflect the proposed fault interpretation. Lastly, in developing your interpretation, lines can be drawn from well to well connecting the correlation markers, with the lines being offset across faults (Fig. 6-5). Straight-line sections are typically used to initially evaluate a structure (Fig. 6-8a). They portray the dip in straight-line segments, so such features as variations in dip between wells will not be apparent. Although such a section does not represent the true attitude of the structure, during the initial phases of a study it provides significant information on the general structure, fault geometry, and correlations.

Cross sections usually go through several stages of revisions, with each such revision improving the accuracy and reasonableness of the interpretation. Remember that the final cross-sectional interpretation must agree with the completed geological maps, be geologically reasonable, have 3D geometric validity, and conform to the structural style of the area. If possible, cross sections should be retrodeformable or structurally balanced (see Chapter 9).

Figure 6-5 is a completed structural cross section from the Bayou Jean La Croix Field in Terrebonne Parish, Louisiana. It is a good example of what is called a *finished illustration* (Langstaff and Morrill 1981) structural cross section prepared from electric well logs (1 in. = 100 ft) and finished structure maps on the various horizons shown. The section is geologically reasonable for the tectonic setting, illustrating the subsurface structural geology, including fault geometry, correlations, and areas of hydrocarbon accumulation. The section was prepared true to scale with the same horizontal and vertical scales shown graphically in the lower right-hand corner of the section.

Stick Sections

An alternative to electric logs is the use of log sticks in the preparation of a cross section. A *stick* is defined as a vertical or deviated line that represents an electric log. A stick section has several advantages over the electric log section, including simplicity, clarity, and ease of construction (Lock 1989).

Since sticks do not show any correlation data (stratigraphic correlations or faults), it is necessary to record the depth of all pertinent correlations and faults obtained from the actual electric logs. Stick sections are often used to solve structural problems because of their simplicity and lack of clutter. A typical stick structure cross section is shown in Figure 6-2. This is the same cross section shown in Figure 6-1, which incorporates the actual electric well logs.

STRATIGRAPHIC CROSS SECTIONS

Problems related to changing stratigraphy require a stratigraphic cross section. They are drawn to illustrate stratigraphic correlations, unconformities, permeability barriers, stratigraphic thickness changes, facies changes, and other stratigraphic characteristics. Many of the comments made about structural sections also apply to stratigraphic sections. Vertical and horizontal scales must be assigned, the line of section laid out based on the intent of the section, a datum chosen, and the logs prepared to place on the cross section. The datum for a stratigraphic cross section is normally chosen as some stratigraphic marker with the section set up so that the chosen datum is horizontal. By using a horizontal datum, the distorting effects of structure (folds and faults) are eliminated. This is equivalent to unfolding and unfaulting the strata. Figure 6-6 is an example of two stratigraphic cross sections from southwest Kansas/northwest Oklahoma, each hung on a specific stratigraphic datum. These sections were prepared as part of a detailed study to evaluate the complexities of the structural and stratigraphic factors controlling the trapping of hydrocarbons in the Morrowan Sandstones (Mannhard and Busch 1974).

In preparing structural cross sections, recall that you had the choice of using the actual electric well logs or stick representations of the logs. For stratigraphic cross sections, the actual well logs are usually used, since the work involves solving problems related to the lithology. Changes in lithology from well to well can be evaluated only with *real* log data, although in special cases stick sections may be appropriate.

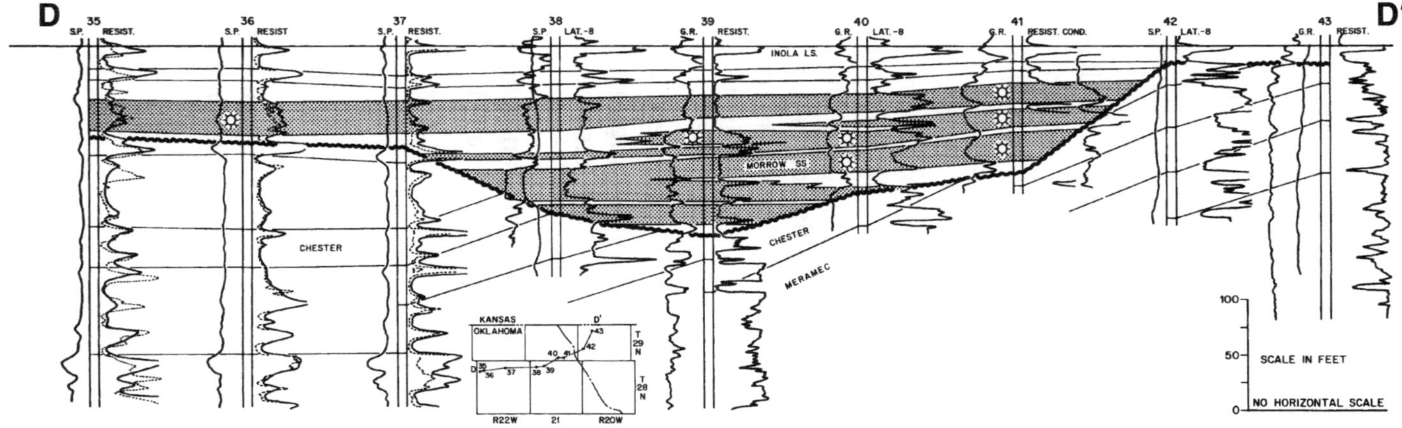

Cross Section D-D′.

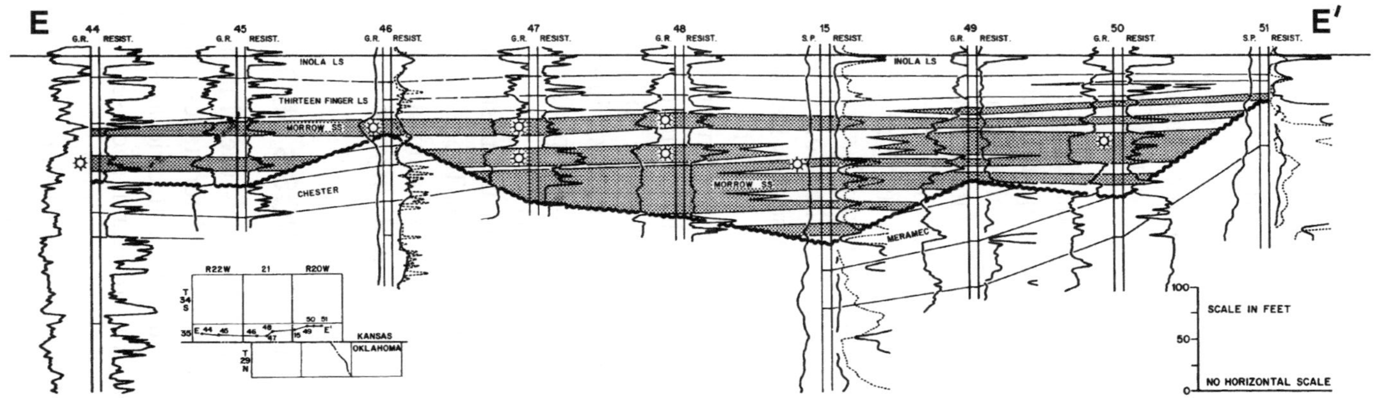

Cross Section E-E′.

Figure 6-6 Two stratigraphic correlation sections prepared to evaluate stratigraphic complexities in the Morrowan Sandstones. (From Mannhard and Busch 1974; AAPG©1974, reprinted by permission of the American Association of Petroleum Geologists (AAPG) whose permission is required for further use.)

In the preparation of a stratigraphic section, the choice of an appropriate marker bed to use as the datum is extremely important. The choice of a poor datum can pose problems, such as incorrectly illustrating the original configuration of the stratigraphic units under study. For example, a poorly chosen datum may incorrectly depict an actual channel sand as a bar.

Remember, one of the primary objectives in laying out a stratigraphic section is to reconstruct the sand geometry at the time of deposition or shortly thereafter. When working in areas of predominant sand/shale deposition, keep in mind that sands and shales compact to differing degrees. The effects of differential compaction as the result of sediment burial are recorded on the electric logs. If you have a good idea of the environment of deposition for the sands being evaluated, the choice of the datum may be relatively easy. If the environments are in question, however, you should ask yourself which marker is most likely to have been close to horizontal at the time of deposition.

Figure 6-7 illustrates three cross sections of the Pennsylvanian Anvil Rock Sandstone, with each section having a different reference datum. In the upper section, the No. 7 coal seam above the sand is chosen as the datum. Figure 6-7 is one of those special cases where log sticks are appropriate for a stratigraphic section. With this reconstruction, the cross section does not show the effects of draping over the sand. In the middle section, the top of the Anvil Sandstone is used as the datum. By using this datum, the sand may give the appearance of a channel-fill sand regardless of the sand's actual original configuration. The bottom section using the No. 5 coal seam below the sand is the best choice in this particular situation because this last section depicts the Anvil Sandstone as a channel-fill sand that shows the effects of differential compaction. The channel sand interpretation is also supported by the lateral truncation of the 5A and 6 coal seams against the sand. If the intent was to reflect the geometry at the time the channel was active and the Anvil Rock Sandstone was deposited, the middle section hung on the sandstone itself would come closest to showing this geometry. Figure 6-7 illustrates the fact that stratigraphic sections may be highly exaggerated to emphasize a specific point, as this figure has a vertical exaggeration of greater than 50:1.

We emphasize that care must always be taken when choosing the stratigraphic datum for a stratigraphic cross section. There are no actual rules of thumb to apply when choosing the datum. However, the answers to the initial questions that should be posed prior to constructing a section often can serve as a guide in your choice of the best datum.

PROBLEM-SOLVING CROSS SECTIONS

Cross sections can be very useful in helping to solve structural and stratigraphic problems from the earliest through the later stages of a project. We call such sections *problem-solving cross sections*. As stated earlier, electric logs are usually required if the problem is stratigraphic. If the basic problem is structural, stick sections may be useful in solving fault and structural geometry problems.

Two different stick sections using the same data are shown in Figure 6-8. Figure 6-8a is an example of a problem-solving cross section utilizing the straight-line method of illustrating the geological interpretation. With this method, straight lines are drawn from well to well representing the horizon and fault correlations. Obviously, the straight lines between wells may not illustrate the true geological picture, such as changes in bed dip between control points. However, in the initial stage of a geological study, this type of section is very helpful in evaluating alternate correlations and fault and structural interpretations. Remember, the interpretation must be geologically reasonable and have 3D geometric validity.

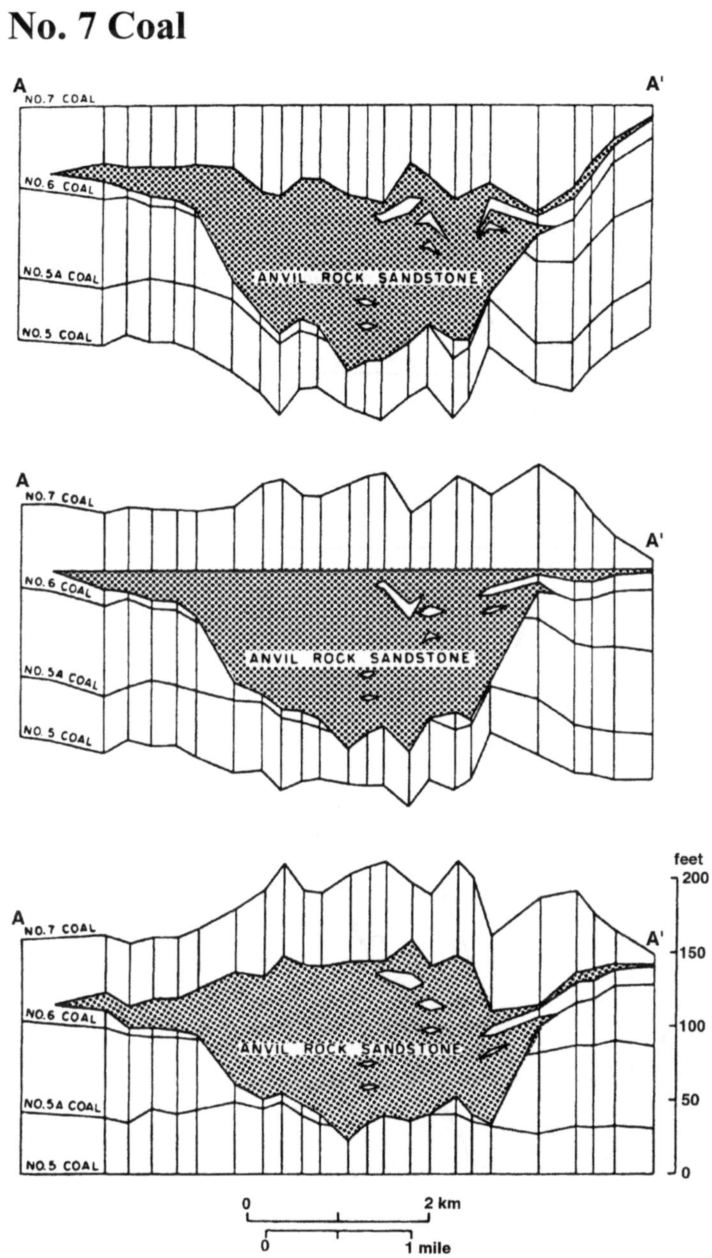

Figure 6-7 The choice of a reference datum is very important in the construction of a stratigraphic cross section. Three stratigraphic cross-sectional interpretations of the Anvil Rock Sandstone are shown based on three different reference data (Potter 1963).

FINISHED ILLUSTRATION (SHOW) CROSS SECTIONS

A *finished illustration (show) cross section* illustrates the final interpretation. It is constructed after all the fault and structure maps have been prepared, and it is used to complement the fault and structure maps. Finished illustration cross sections also serve as visual aids to communicate and present the final geological interpretation.

Figure 6-8b shows a completed finished illustration cross section for the initial interpretation shown in Figure 6-8a. Observe that the correlations between wells reflect the true geometry of the horizons and faults as opposed to the initial straight line interpretation.

Figure 6-9a and b present the details of constructing a finished illustration cross section from electric well log control and completed subsurface maps (in this case, a structure and fault map). The section is a dip section crossing a normally faulted anticlinal structure. The figures depict the data used in constructing the 6850-ft Sand and Faults A, B, C, and D on the finished

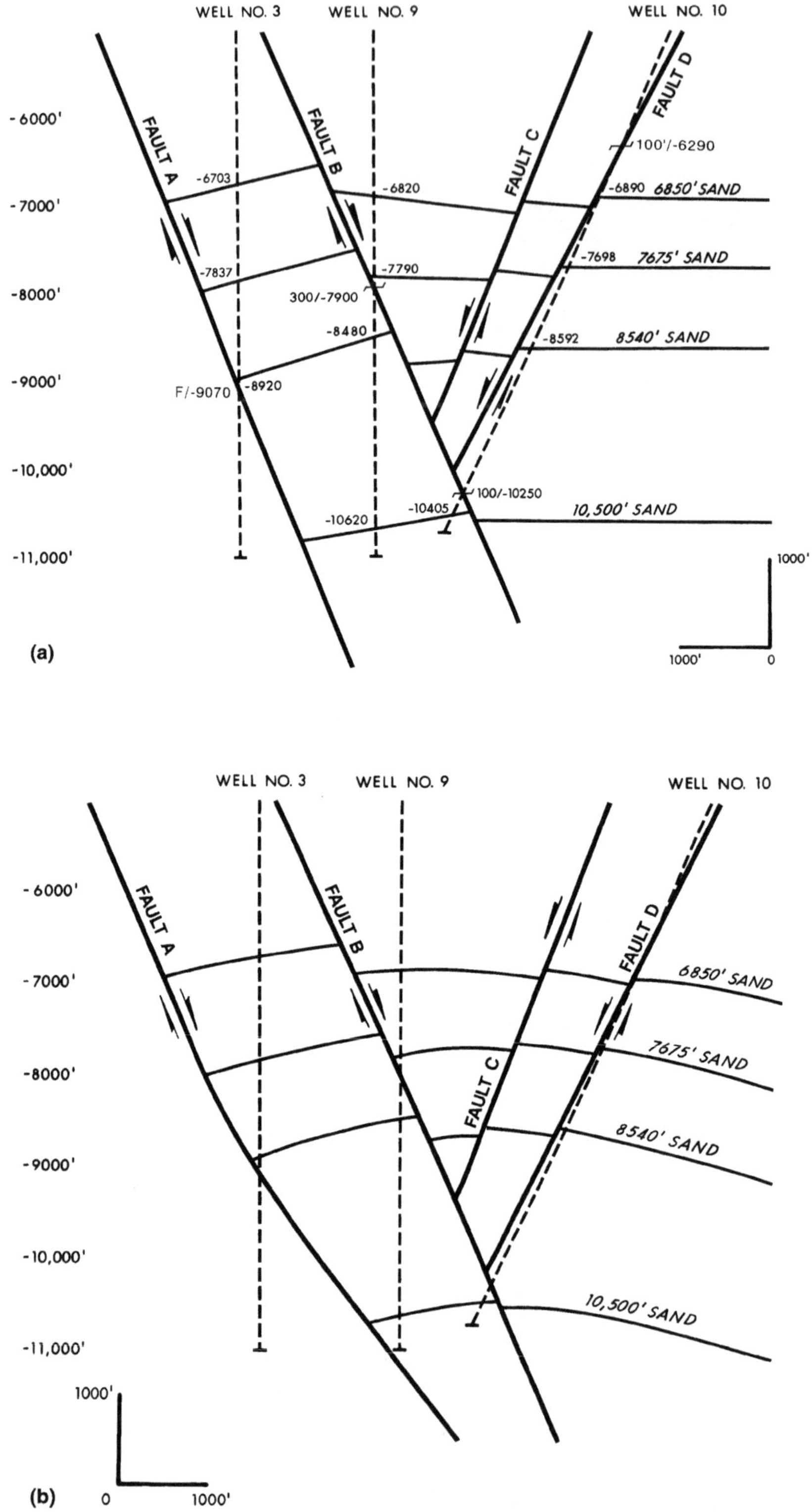

Figure 6-8 (a) Straight-line problem-solving cross section. (b) Finished illustration cross section constructed from completed fault and structure maps (compare this with Fig. 6-8a).

illustration cross section. The final illustration cross section constructed from all the fault and structure maps is shown in Figure 6-8b.

We can review Figure 6-9a and b to illustrate the actual procedure for constructing this type of cross section by hand. The upper part of Figure 6-9a is a structure contour map on the 6850-ft Sand. The line of section, drawn on the structure map, starts at the downthrown trace of Fault A, continues through Wells No. 3, 9, and 10, and terminates at the −7100-ft contour line upthrown to Fault D.

It is always a good idea to review the planning and preparation procedures presented earlier before beginning any cross section until you feel that you are thoroughly familiar with them. The first task for a manually drawn cross section is to prepare the cross-section coordinate grid paper, with the logs proportionately spaced, and then begin plotting all appropriate available data points. Since all the data are acquired from completed maps and accepted log correlations, it is not necessary to place the actual electric logs on the section; however, this is a matter of personal preference. The first points to plot are those for the top of the 6850-ft Sand, taken from the structure map in Figure 6-9a (upper). Starting with Well No. 3, the first data point to place on the section is the top of the sand in this well, which is at a depth of −6703 ft. This point, shown at Well No. 3 on the structure map, is plotted at the appropriate depth on the log stick for Well No. 3 on the cross section (position A). The second data point is the intersection of the section line and the −6700-ft structure contour line on the structure map, 100 ft southeast of Well No. 3. This −6700-ft subsea depth point is plotted on the cross section 100 ft from Well No. 3 (position B). The third data point is the intersection of the section line with the −6600-ft contour line on the structure map. This point is 725 ft southeast of Well No. 3 and is plotted on the cross section at position C.

What is the fourth data point? It is the intersection of the section line and the upthrown trace of Fault B, which is at an estimated subsea depth of −6585 ft and measures 900 ft from Well No. 3. This point is plotted on the cross section at position D.

This procedure is continued for each measurable data point along the line of section on the structure map. Although there are no wells north of Well No. 3, the section extends to Fault A using all contour lines as data points. It should be noted that measurable data points include such items as gas/oil contacts, oil/water contacts, and upthrown and downthrown fault traces. You will also notice that there is an obvious structural crest about 150 ft south of Well No. 9. The crestal position and its subsea depth have been estimated and incorporated into the cross section. Remember that this section is being constructed after the structure maps are in their final stages of completion, and all data shown on the maps should be used in the construction of the cross section.

Each point for this cross section is identified on the structure map along the section line and plotted on the cross section. When all the data points are plotted and connected with a correlation line, a detailed interpretation of the 6850-ft Sand is illustrated in cross-sectional view (lower part of Fig. 6-9a). This procedure should be completed for all the sands planned for the cross section.

Turning to Figure 6-9b upper, we see a contour map for Fault B with the line of section drawn on the map. Following the same procedure outlined for the 6850-ft Sand, the location of each fault data point is identified on the fault surface map and plotted on the cross section in the lower part of the figure. The same procedure is followed for Faults A, C, and D.

We now have completed the plotting of all the available data points for the 6850-ft Sand and all the faults for the finished illustration cross section in Figure 6-8b. The interpretations of the other sands were constructed using the same procedure just outlined. The result is a detailed

(a)

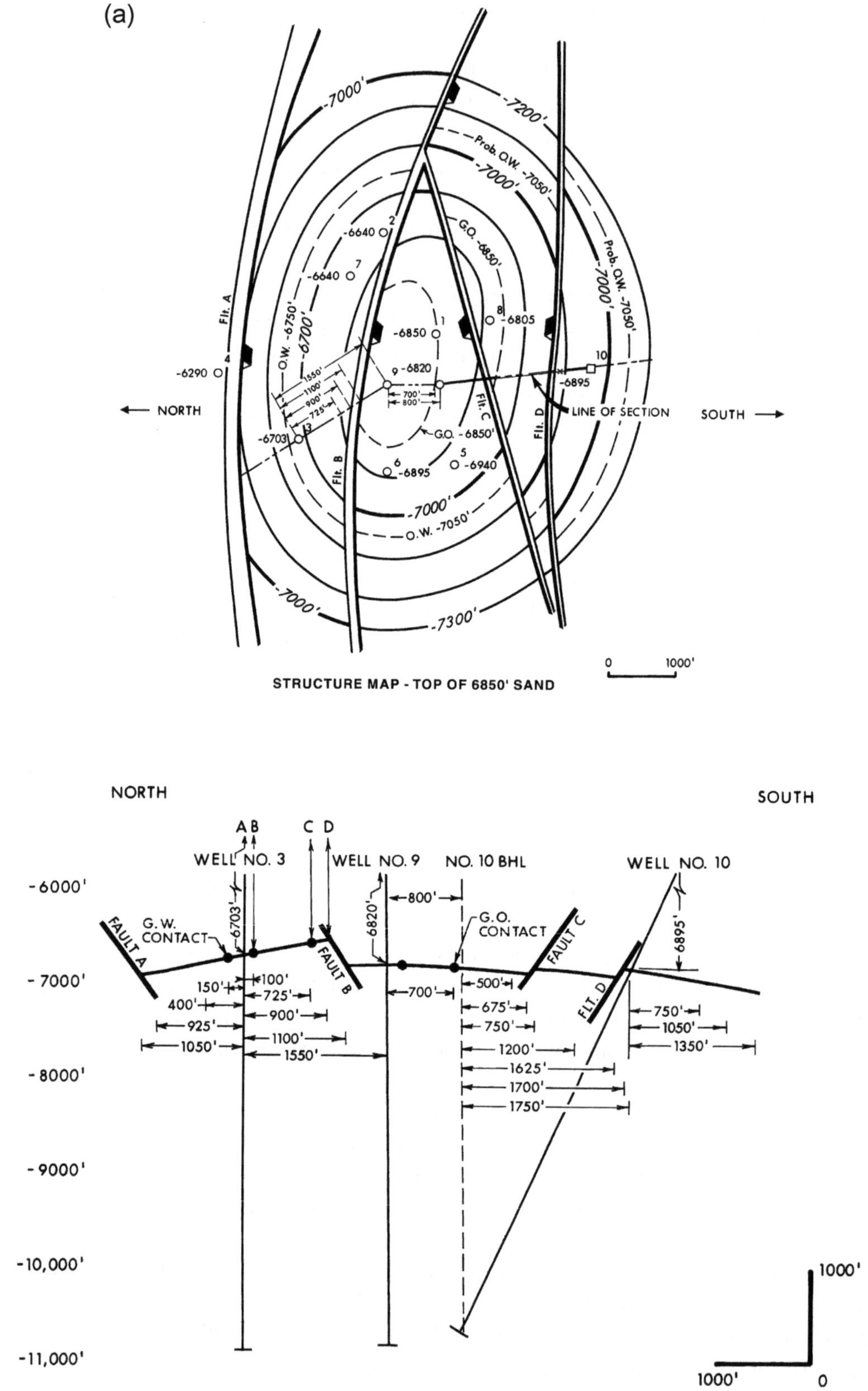

Figure 6-9 (a, upper) Structure map on the top of the 6850-ft Sand. Line of cross section shown on map.(a, lower) Details for construction of a finished illustration section for the 6850-ft Sand. (b, upper) Fault surface map for Fault B. Line of section shown on map. (b, lower) Detailed construction for Fault B for incorporation into finished illustration cross section.

(b)

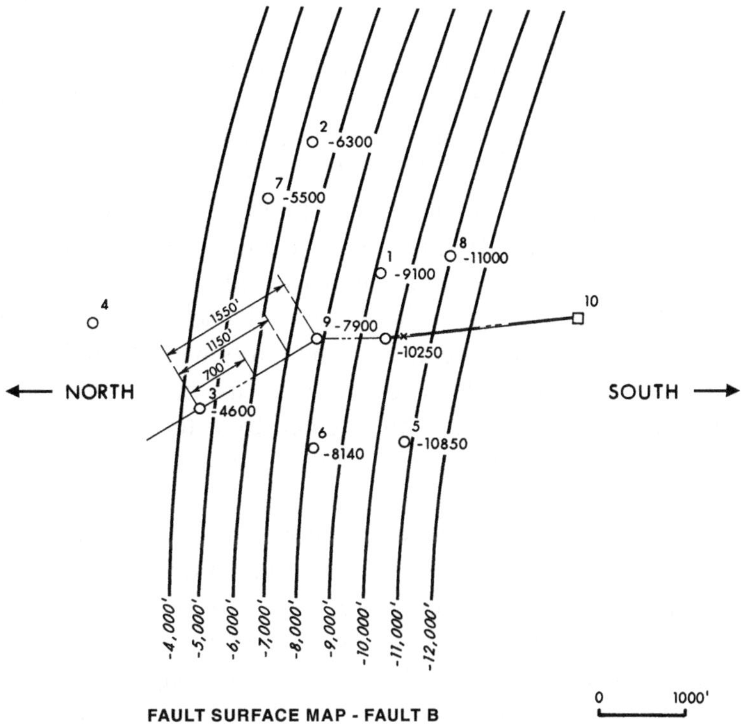

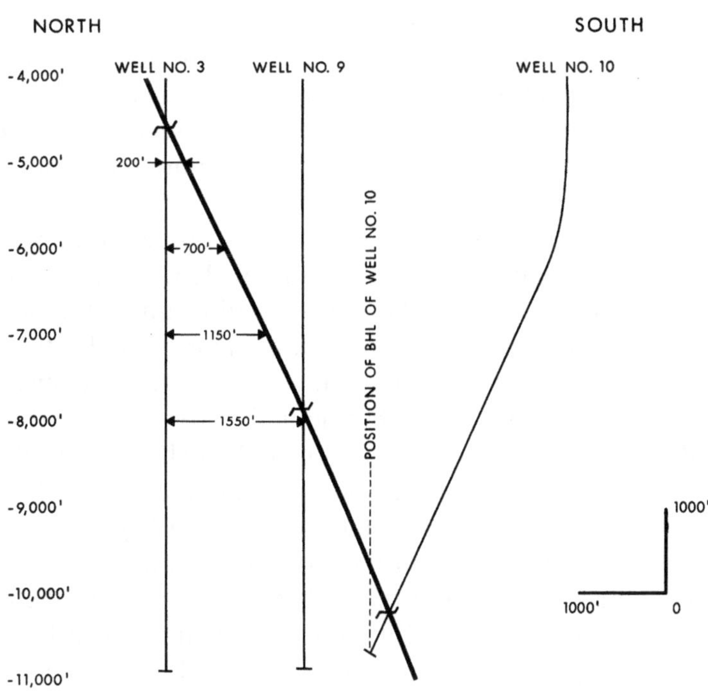

Figure 6-9 *(continued)*

cross section representing an accurate picture of the structural interpretation of the horizons and faults in a vertical plane.

In summary, a finished illustration cross section is important for several reasons. It serves as a visual aid for presenting a completed geological interpretation and can also be helpful in identifying mapping problems that might otherwise go undetected, particularly in fields with closely spaced productive sands. An accurately detailed cross section might indicate any number of mapping "busts," such as areas where thickness compatibility between horizons has not been honored. The cross section in Figure 6-10, prepared to evaluate cross-fault drainage, is an example of a detailed finished illustration cross section constructed almost exclusively from final structure maps for each of the horizons shown on the section. The section shows that the interpretation for each horizon, the faults, and the intersection of the faults with the individual horizons is geologically reasonable for the most part. However, a few areas on the cross section appear to indicate some minor geological busts, which are highlighted by asterisks. Although none of these problems seem to be serious, the fact that they are visible shows the sensitivity of this detailed cross section to interpretation error.

When mapping several very closely spaced horizons, it is not too difficult to cross contours from one horizon to the next, if extreme care is not taken during mapping. It is good practice to underlay the structure map currently being prepared with one already constructed on a horizon immediately above or below. Such practice ensures compatibility in the structural interpretation for a series of horizons being mapped and prevents major mapping busts. This topic is discussed in detail in Chapter 8, but it is introduced here to show how detailed cross sections are used to identify errors in structure contouring.

At times it is even possible to cross structure contours from two mapped horizons that are actually hundreds of feet apart, resulting in a mapping bust. Figure 6-11 illustrates just such a situation. Figure 6-11a shows structure maps on two different horizons. The prospective stratigraphic section for an exploratory play lies between these mapped horizons (P-5 and P-7), which are separated vertically by nearly 800 ft of stratigraphic section in the off-flank position labeled "T" on the structure maps. A cursory review of the structure maps indicates that they appear geologically reasonable.

Based on a study of the growth activity of a salt dome on the west, it was evident that the salt mass was a positive feature during the time of deposition of the prospective section. So the section between the P-5 and P-7 horizons should thin in the up-structure position to the west, and it does. The stratigraphic thinning was incorporated into the structural picture, but because care was not taken during the mapping process, the two structure maps depict the two mapped horizons not only converging but actually crossing up-structure. The detailed cross section of the P-5 and P-7 horizons in Figure 6-11b shows the effect of this mapping bust very clearly. Notice that the two mapped horizons intersect at the location marked C and then cross, creating a negative structural volume, which is an *impossible* geological situation. This mapping bust is due to carelessness in not using the technique of underlaying the completed P-5 structure map when constructing the P-7 map. Such a mistake can put a question in the minds of management as to the reliability of the work and jeopardize what might otherwise be a great exploration prospect. The preparation of a quick cross section would have caught this mapping bust.

A precise finished illustration cross section can identify many mapping problems and provide the time to correct the work before the final interpretation and maps are prepared. Also, these cross sections are excellent aids for reviewing the possibility of juxtaposed reservoirs across a fault (Fig. 6-10). Studies of such sand occurrences are important in evaluating the possibility for

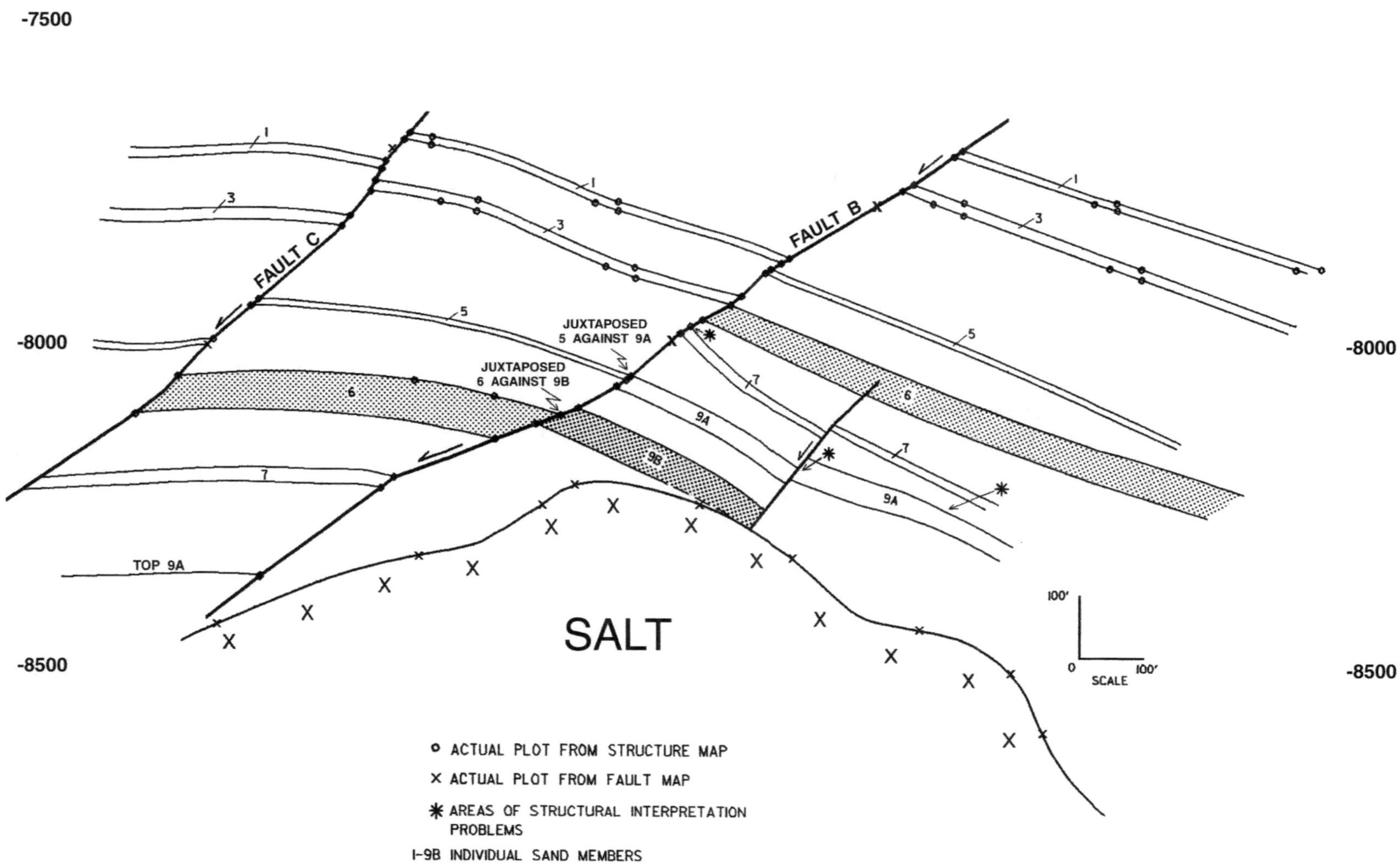

Figure 6-10 A detailed finished illustration cross section constructed almost exclusively from completed maps, including fault maps, structure maps on the top and base of each of the seven stratigraphic units shown, and a salt map. Notice that the detailed section indicates that the 6 Sand downthrown to Fault B is in juxtaposition with the 9B Sand upthrown to Fault B. Cross-fault drainage was evident from production data and confirmed by this detailed section. Asterisks indicate locations of possible mapping errors. The original scale of the cross section was 1 in. = 100 ft; therefore, detail on the order of tens of feet could be incorporated into the section.

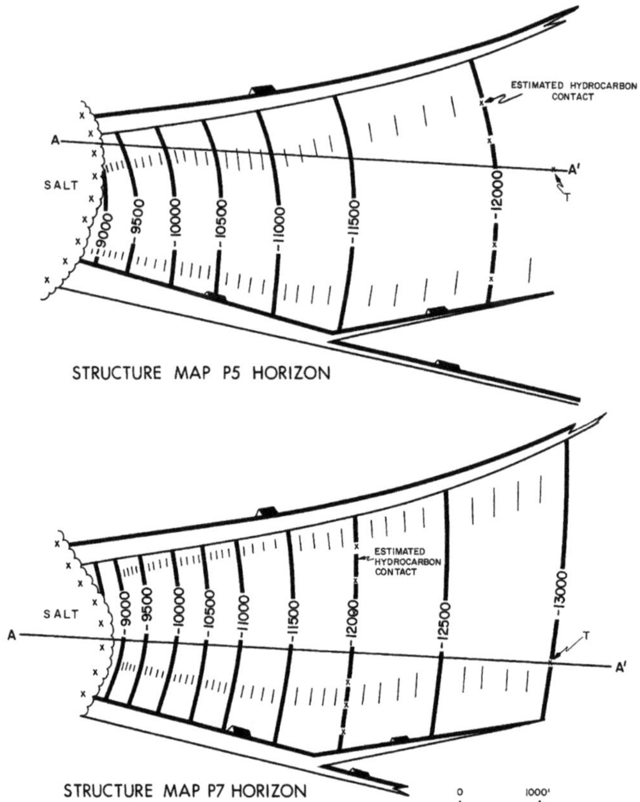

(a)

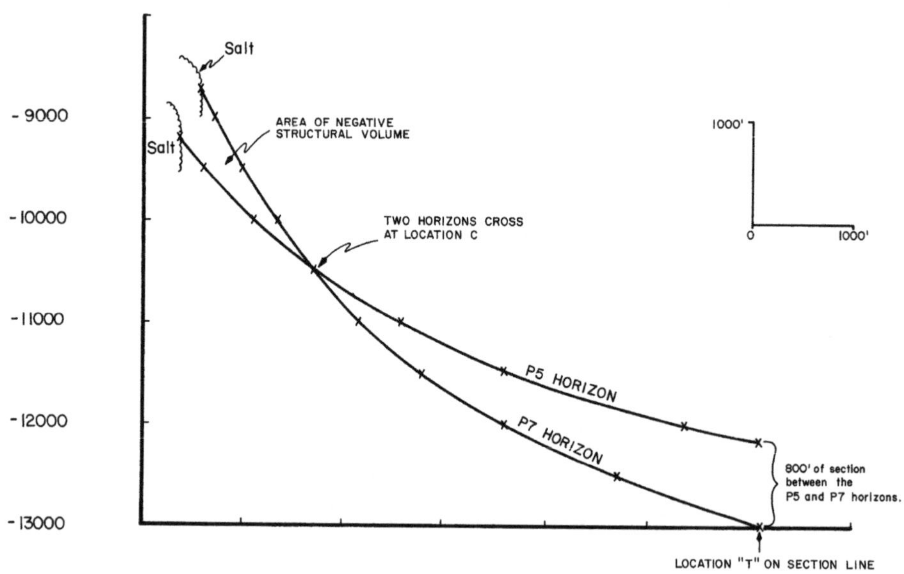

(b)

Figure 6-11 (a) Completed structure maps on the P5 and P7 horizons. These two horizons bracket a prospective section trapped by salt to the west and faults to the north and south. (b) Structure cross section prepared from the completed structure maps on the P5 and P7 horizons. The cross section clearly indicates that the completed structure maps were not constructed correctly.

cross-fault drainage from one fault block to another (Smith 1980). Another type of cross section used for evaluation of juxtaposition across a fault is described in the section Fault-Seal Analysis.

CORRELATION SECTIONS

In this segment of the chapter, we discuss a special type of cross section called a *correlation section*. This section is a special type of stratigraphic cross section that is primarily used as a detailed correlation aid. There are several important guidelines helpful in preparing this type of section.

1. Choose a stratigraphic datum that best serves the intended purpose of the correlation cross section (see Fig. 6-7). A reliable shale marker is usually a good choice for a stratigraphic datum (Sneider et al. 1977).
2. Limit the section to a short vertical log interval in order to show significant correlation detail.
3. Position the logs as closely as possible with no horizontal scale to include as many logs as needed in the section.

A correlation section can serve as an excellent correlation aid in defining the lateral and vertical continuity of permeable units within a specific area and stratigraphic interval. These sections can also be used as good prospecting tools to evaluate and illustrate the potential for hydrocarbons.

Figure 6-12 shows the layout for a typical correlation section. The actual electric log showing all the curves can be used; however, to reduce clutter and provide sufficient space for the inclusion of many logs, the spontaneous potential (SP) or gamma ray (GR) and the amplified resistivity curves are often sufficient. As mentioned in Chapter 4, the amplified resistivity curve is usually the most helpful for correlation work. This discussion of layout and correlation procedure applies to both manually prepared and computer-based correlation sections (see the section Cross-Section Construction Using a Computer).

In preparing a correlation section, follow the same correlation procedures outlined earlier in Chapter 4. Since a correlation section is a type of stratigraphic section, hang it on a well-defined datum such as the one shown in Figure 6-12a. Next, correlate the shale sections using all correlatable shale markers indicated by the SP and resistivity curves. Figure 6-12a shows the shale correlations for this section. Observe the parallel to semiparallel trend of the shale markers, which indicates a fairly uniform stratigraphic thickness. A time-stratigraphic framework now exists within which you can correlate permeable units. We use sands in this example.

We begin the sand correlations as shown in Figure 6-12b. As can be seen, the sands are not as vertically and laterally continuous and uniform as the shale sections. Lateral changes in depositional environment can cause sudden variations in sand thicknesses even within an area of limited lateral extent. Because the clays and silt that make up the shales are usually deposited in quiet waters over large areas, abrupt changes in stratigraphic thicknesses in shales are not common. Such extensive and consistent deposition of shale usually provides for good lateral correlation continuity from well to well.

The 8300-ft, 8500-ft, and 9200-ft Sands are separated vertically from the other sands, but they appear laterally continuous from well to well. The stippled pattern for the 9000-ft Sand is representative of the rest of the section. Within this overall gross sand interval are several distinct sand members in three of the wells. Based on this correlation section, it is speculative whether there is vertical or lateral continuity of individual sand members from well to well. As for the

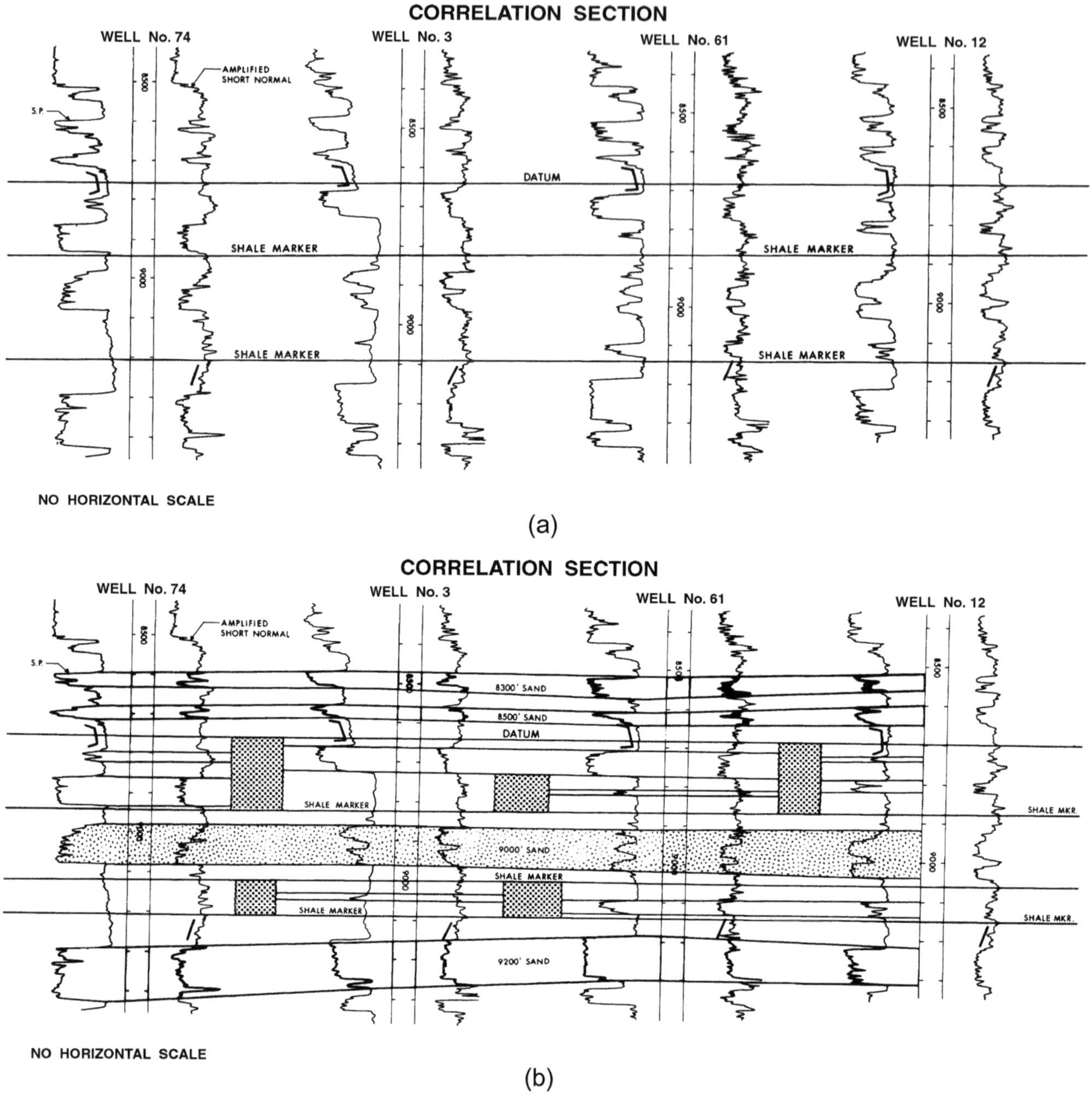

Figure 6-12 (a) Correlation section is first laid out by hanging the logs on a reference datum and correlating all recognizable shale markers. (b) Completed correlation section shows the lateral and vertical continuity (or lack of continuity) of the individual sands seen in each well.

other individual sands shown in each log, they appear to be laterally discontinuous from well to well, indicating rapid changes in depositional environments and laterally limited sand bodies. The rectangular areas between wells represent discontinuity of sands somewhere between the wells.

This type of information regarding the continuity of sands is most important to the development geologist and reservoir engineer. The layout of one such section or a number of sections can often aid in making critical decisions, such as the following:

1. Which well or wells must be perforated to maximize the drainage efficiency within a reservoir?
2. Which sand interval or intervals should be perforated within a single wellbore to optimize hydrocarbon recovery?

3. Is a specific reservoir competitive with an adjoining lease operator? In other words, can another operator's well on an adjacent lease drain your reserves because of possible lateral continuity of sands across the lease? If so, what action must be taken to protect your reserves?
4. Are any additional development wells required to optimize field production?
5. Can remaining reserves, identified in an abandoned well, be recovered with other existing wellbores? By this we mean, is there continuity of the hydrocarbon sand from the location of the abandoned well to a well capable of being completed?

We mentioned earlier that correlation sections can be used as an excellent exploitation tool. Figure 6-13 shows part of a geological and engineering prospect package designed to justify the drilling of a development oil well into what was considered a depleted/nonproducible oil reservoir.

Figure 6-13a is a structure map on the top of a prospective sand called the "9300-Foot Sand." An oil reservoir is present upthrown to Fault A. Six wells penetrated the oil reservoir with Wells No. 4 and 10 having produced minor amounts of oil and gas. Due to the minimum amount of oil production from Well No. 10 and the minimum gas production and pressure decline in Well No. 4, the reservoir was considered depleted. However, further study, with the use of detailed maps, log correlations, perforation data, production data, and the correlation section A-A′ (Fig. 6-13c), revealed that the reservoir consisted of three distinct sands: (1) a small, highly calcareous fringe complex, such as that seen in Wells No. 7 and 10; (2) a major cut-and-fill channel sand seen in Wells No. 3 and 4; and (3) an upper transgressive sand member separated from the fringe and channel sands by a shale break. This transgressive sand member is seen in all wells.

The correlation section (Fig. 6-13c) shows that the oil production in Well No. 10 was from the small fringe complex, and the gas production in Well No. 4 from the upper transgressive sand member. Reserves were calculated volumetrically, from net sand and net hydrocarbon isochore maps, and they were quite large compared to the actual oil and gas production. This suggested that the fringe complex and transgressive sand are separate members not in communication with the main channel sand (see delineation of major channel sand on the structure map [Fig. 6-13a], shown as a permeability barrier, and on the net sand map [Fig. 6-13b], highlighted as the limit of major channel sand). Therefore, significant reserves remain to be produced in this channel sand, which can be recovered by a recompletion in Well No. 3, if possible, or through the drilling of a new well. This prospect used the correlation section as an integral part of the prospecting process, as well as a final illustration to present the idea to management.

CROSS-SECTION DESIGN

In this section, we discuss the specific procedures for laying out cross sections. We review the layout of sections for four different tectonic settings: (1) extensional (normally faulted), (2) diapiric salt, (3) compressional (reverse or thrust faulted), and (4) strike-slip fault settings.

Before reviewing the section design for the four different tectonic settings, we summarize some design guidelines for cross sections.

1. Cross sections can be run from well to well in a zigzag pattern or as a straight line with data from each well not on the line projected into the line of section. Both types of sections have inherent problems. Zigzag sections tend to distort the subsurface geology due to changes in the strike direction of the section. If you treat a zigzag section as a series of two well straight sections, interpretation problems can be minimized. Straight-line sections may require the

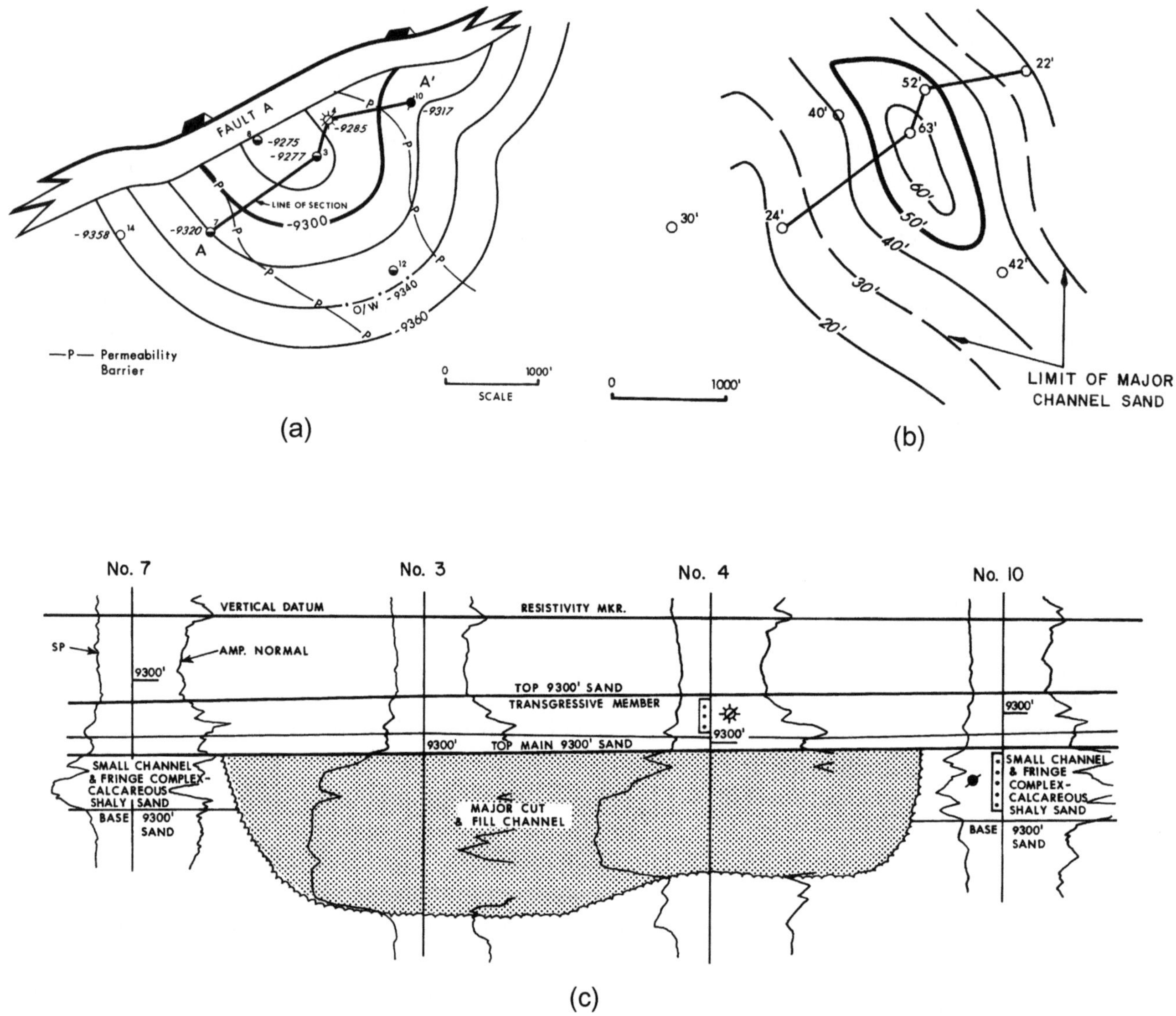

Figure 6-13 (a) Structure map on the top of the 9300-ft (hydrocarbon-bearing) Sand trapped upthrown to Fault A. (b) Net sand isochore map of the 9300-ft Sand delineating the limit of good quality major sand development. (c) A detailed correlation section through the 9300-ft Sand Reservoir trapped upthrown to Fault A. The section clearly illustrates that the upper transgressive member, the fringe complex, and the channel sand are separate and distinct members of the 9300-ft Sand package. Production data indicate minor oil production from Well No.10 and that the transgressive member perforated in Well No. 4 pressure depleted (it is a closed system separated from the main 9300-ft Sand by a continuous shale break).

projection of well data over long distances, resulting in well projection problems. If you understand the various methods and limitations of projecting well data, however, straight-line sections can be used very effectively.

2. Deviated wells can be included in a cross section if the line of section runs along the plan view path of the deviated well (Fig. 6-15).

3. If a line of section being laid out intersects two closely spaced wells, include the well that penetrates the deepest section if the total depths are significantly important. Spacing may not be the critical factor; instead, similar structural geometry may be critical. In such cases, if two closely spaced wells reflect different geometry, choose the well that illustrates the geometry expected in the cross section.
4. When preparing both strike and dip sections in the same field, it is good practice to tie the sections together with a specific well placed on both sections (Fig. 6-14).

Use common sense in laying out all sections. Remember, a cross section is intended to help you visualize the structure in three dimensions, give you another perspective view of the structure, and serve as an aid in solving a variety of problems related to general correlations, fault geometry, or the structural interpretation.

Extensional Structures

Figure 6-14 shows the layout for two dip sections and one strike section for a typical extensional structure. The dip sections, which are perpendicular to the strike of the faults, provide the best information for studying the faults. For single fault systems, these sections can be balanced to help develop the best structural interpretation. Sections involving bifurcating or compensating fault systems are difficult to balance due to out-of-the-plane motion, which is often difficult to account for in balancing (see Chapter 11). Initially, the cross sections can be used as problem-solving sections to help delineate the fault and structural geometry for the area under study. At the initial stages of the geological study, the straight-line method of construction for faults, horizons, salt bodies, unconformities, and other features is recommended. During the later stages of mapping, the sections can be upgraded or revised to help develop and illustrate the final interpretation. Also, cross sections such as section A-A′ in Figure 6-14 can be useful in resolving possible correlation problems in deviated wells.

Diapiric Salt Structures

Diapiric salt structures are in general structurally complex and therefore often require the layout of a number of sections to develop a reasonable geological interpretation. For an example of the cross-section design, we use the salt structure shown in Figure 6-15. A typical cross-section layout must be designed to incorporate both straight and deviated wells, since both types are commonly drilled on these structures. There are two basic directions for the cross sections: (1) strike sections (sometimes referred to as *peripheral* sections), which parallel or semiparallel structural strike, and (2) dip sections (sometimes referred to as *radial* sections), which parallel structural dip. For salt structures, we recommend that dip sections be constructed to continue past the last up-dip well control to include the salt in the cross section (see sections C-C′ and D-D′ in Fig. 6-15).

For the structure in Figure 6-15, we show the layout of two strike and three dip cross sections. Due to the nature of the structure and the position of the wells, there is only one straight-line section, E-E′. All other sections have a zigzag pattern. The actual procedure for laying out the sections is the same as that outlined in the previous section. Observe how each line of section that includes a deviated well follows the path of that deviated well from the surface to total depth. The portion of the line of section that follows a deviated well path is dashed on the figure for clarity.

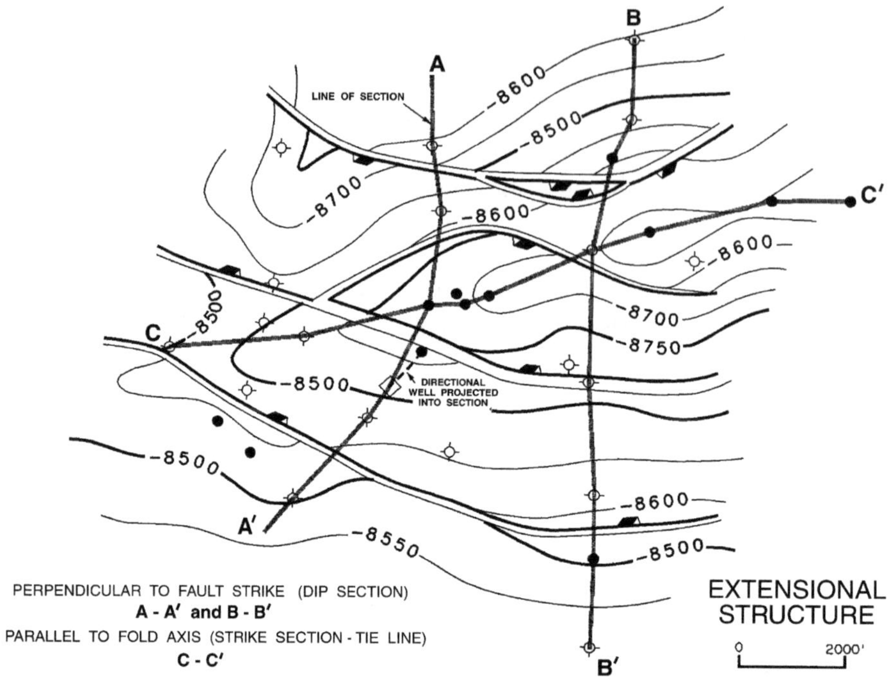

Figure 6-14 Typical layout of cross sections in an extensional tectonic setting. (Map was computer-drafted.)

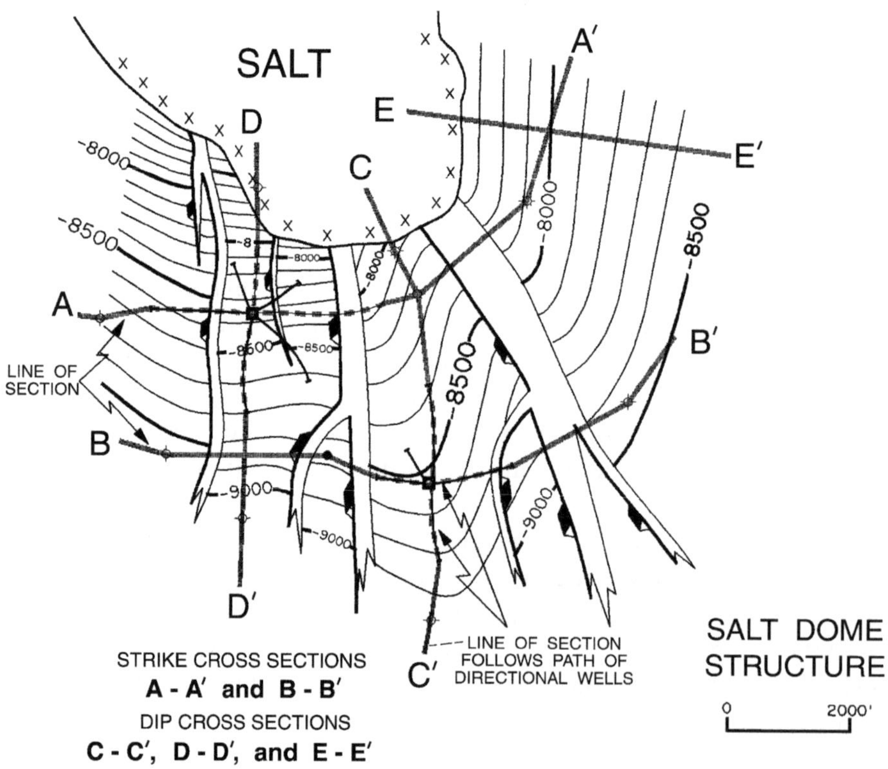

Figure 6-15 Typical layout of cross sections for a complex piercement salt structure including both straight and vertical wells. (Map was computer-drafted.)

Once the structural interpretation is complete, the sections can be upgraded to serve as displays to illustrate the final interpretation. The structure maps represent the horizontal view and the cross sections show the vertical view of the 3D geometry of the structural interpretation.

Look at sections A-A′ and B-B′ in Figure 6-15. With these sections, the general fault geometry in the southern portion of this field can be studied because the sections are laid out roughly perpendicular to the strike direction of the faults. These sections are actually laid out perpendicular to the fault traces, not the strike of the faults themselves. As discussed in Chapter 8 (e.g., Fig. 8-32), fault traces can differ significantly from the strike of a fault if beds are steeply dipping, as they frequently are near salt domes. These sections can also be used to study the juxtapositioning of sands to evaluate the possibility of cross-fault drainage. The cross sections C-C′, D-D′, and E-E′, which are dip sections, provide information on the correlations, structural dip, sediment/salt interface, growth characteristics of the structure, and data on the down-dip extent of any hydrocarbon accumulations.

Finally, notice that section E-E′ is laid out with no well control. This section may be important in evaluating the structural interpretation developed for the eastern flank of the field. Such a section is constructed using the detailed illustration cross-section techniques discussed earlier. It is constructed solely from completed structure, fault, and salt maps.

Compressional Structures

The most common hydrocarbon trap in compressional areas is found in hanging wall anticlines, which form as a direct result of thrust faulting and include such structures as fault propagation folds, fault bend folds, and duplexes (Chapter 10). The anticlines commonly exhibit an asymmetry with steep frontal limbs and elongated longitudinal axes (B-axes) perpendicular to transport direction. In order to study the internal geometry of these plunging compressional folds, cross sections are normally laid out perpendicular to the B-axis (i.e., parallel to transport direction).

Figure 6-16 is a structure map on top of the Upper Triassic Nugget Sandstone in the Painter Reservoir and East Painter Reservoir Fields. Cross sections A-A′ and B-B′ shown on the map are laid out parallel to transport direction and perpendicular to the strike direction of the thrust faults. In compressional settings it is best to construct straight-line sections and plunge-project well data into the line of section to preclude the possibility of physically moving structural geometry within the fold. Such a section will also provide the best information for studying the thrust faults and balancing the structural interpretation.

Strike-Slip Faulted Structures

The cross-section layout for structures associated with strike-slip faults is basically the same as discussed in the previous section. Strike-slip fault systems are commonly associated with faulted or unfaulted elongated anticlines. If we wish to study the geometry of an associated anticline, dip cross sections must be laid out perpendicular to the elongated fold axis. It is also recommended to tie the dip sections with at least one strike section laid out parallel or subparallel to the axis of the fold.

Figure 6-17 shows a strike-slip fault system with associated faulted anticlines on either side of the strike-slip fault. Cross sections A-A′ and B-B′ are laid out perpendicular to the fold axes. These sections can be constructed as zigzag sections in which the section line passes through each well, or as straight-line sections with the well data projected into the line of section. Cross section C-C′ is the tie line to sections A-A′ and B-B′. If internal faults such as those shown in Figure 6-17 exist on the structure, cross sections perpendicular to the fault can help in the evaluation.

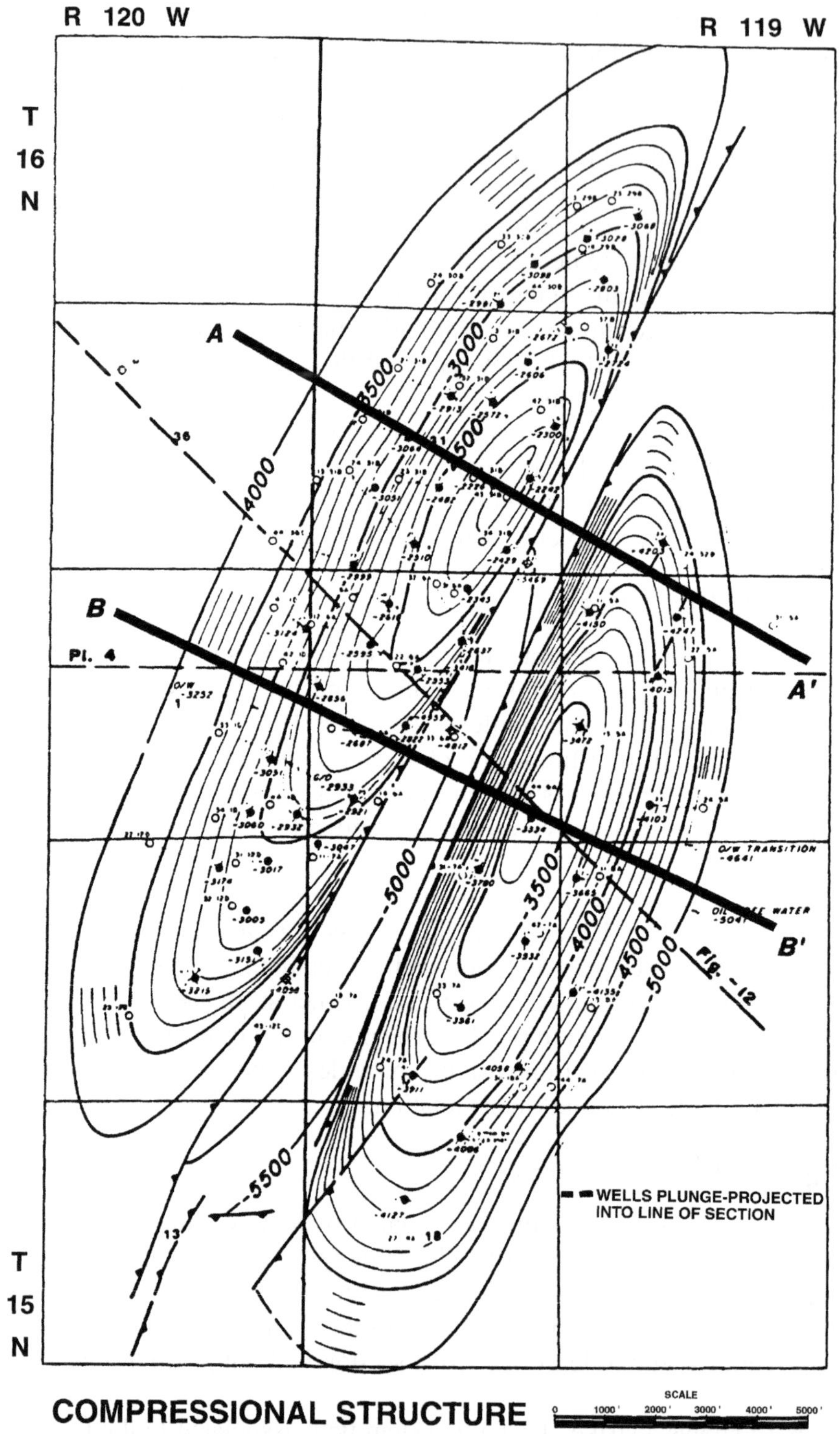

SECTIONS A-A′ AND B-B′ ARE PARALLEL TO TRANSPORT DIRECTION AND PERPENDICULAR TO STRIKE OF THRUST FAULTS.

Figure 6-16 Cross-section layout for a typical compressional structure. Structure map is on the Upper Nugget Sandstone in the Painter Reservoir and East Painter Reservoir Fields, Wyoming. (Modified from Lamerson 1982. Published by permission of the Rocky Mountain Association of Geologists.)

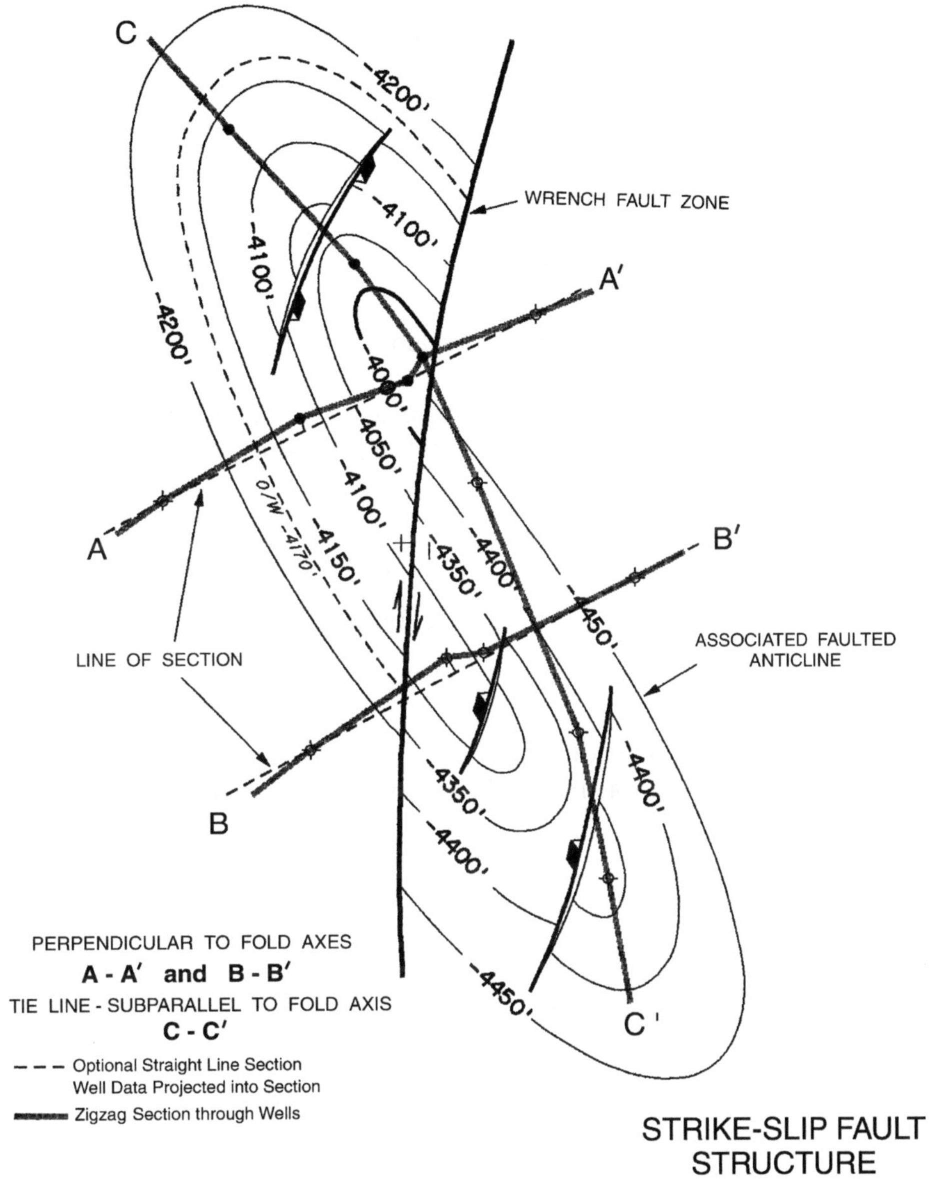

Figure 6-17 Cross section layout for a strike-slip fault system with associated faulted anticlines.

VERTICAL EXAGGERATION

Earlier in this chapter, we mentioned that, whenever possible, cross sections should be constructed using the same horizontal and vertical scales (true-scale sections). Special considerations may require that a section be prepared with different (exaggerated) scales, particularly when constructing large regional or semiregional sections. Typically, it is the vertical scale that is exaggerated. Vertical exaggeration can be incorporated into both structural and stratigraphic cross sections. Keep in mind that with the use of a vertically exaggerated scale comes various types of *distortion*, such as that of stratigraphic unit thickness and dip angles of horizons and faults.

The degree of vertical exaggeration is defined as

$$v_E = \frac{H_L}{V_L} \qquad (6\text{-}1)$$

where

V_E = Vertical exaggeration

V_L = Length of unit distance on the vertical scale

H_L = Length of unit distance on the horizontal scale

Say, for example, that you wish to prepare a cross section with a horizontal scale of 1 in. = 10,000 ft and a vertical scale of 1 in. = 2000 ft. By using Eq. (6-1), the vertical exaggeration for this cross section is

$$V_E = \frac{10{,}000\,\text{ft}}{2000\,\text{ft}}$$

$$V_E = 5$$

If a map has a horizontal scale of 1 in. = 4000 ft and you wish to prepare a cross section with a vertical exaggeration equal to **4**, Eq. (6-1) can be rearranged to determine the vertical scale required for this exaggeration.

$$V_L = \frac{H_L}{V_E}$$

Therefore,

$$V_L = \frac{4000\,\text{ft}}{4}$$

$$V_L = 1000\,\text{ft}$$

We mentioned at the beginning of this section that there are certain situations in which a cross section with a vertical exaggeration is required; in fact, it may have some advantages over a cross section with equal scales. Several situations that may require a cross section with a vertical exaggeration are (1) the preparation of a cross section in an area of low structural relief, (2) the construction of a section in an area of gently dipping beds, (3) a section that would be unreasonably long with equal scales, or (4) the need for extensive vertical detail. Consideration must also be given to the cost and size limitations of available reproduction equipment. Figure 6-18 shows two cross sections across the Uinta Basin, Colorado, which are the same except for scales. The upper section has a vertical exaggeration of 12; the lower section is true scale with the same horizontal and vertical scales. Notice how much detail can be shown on the upper, vertically exaggerated cross section as compared to the true-scale cross section.

When a cross section is prepared with vertical exaggeration, the dip angle of beds, faults, or any other line is exaggerated. The exaggerated dip angle is not simply the product of true dip

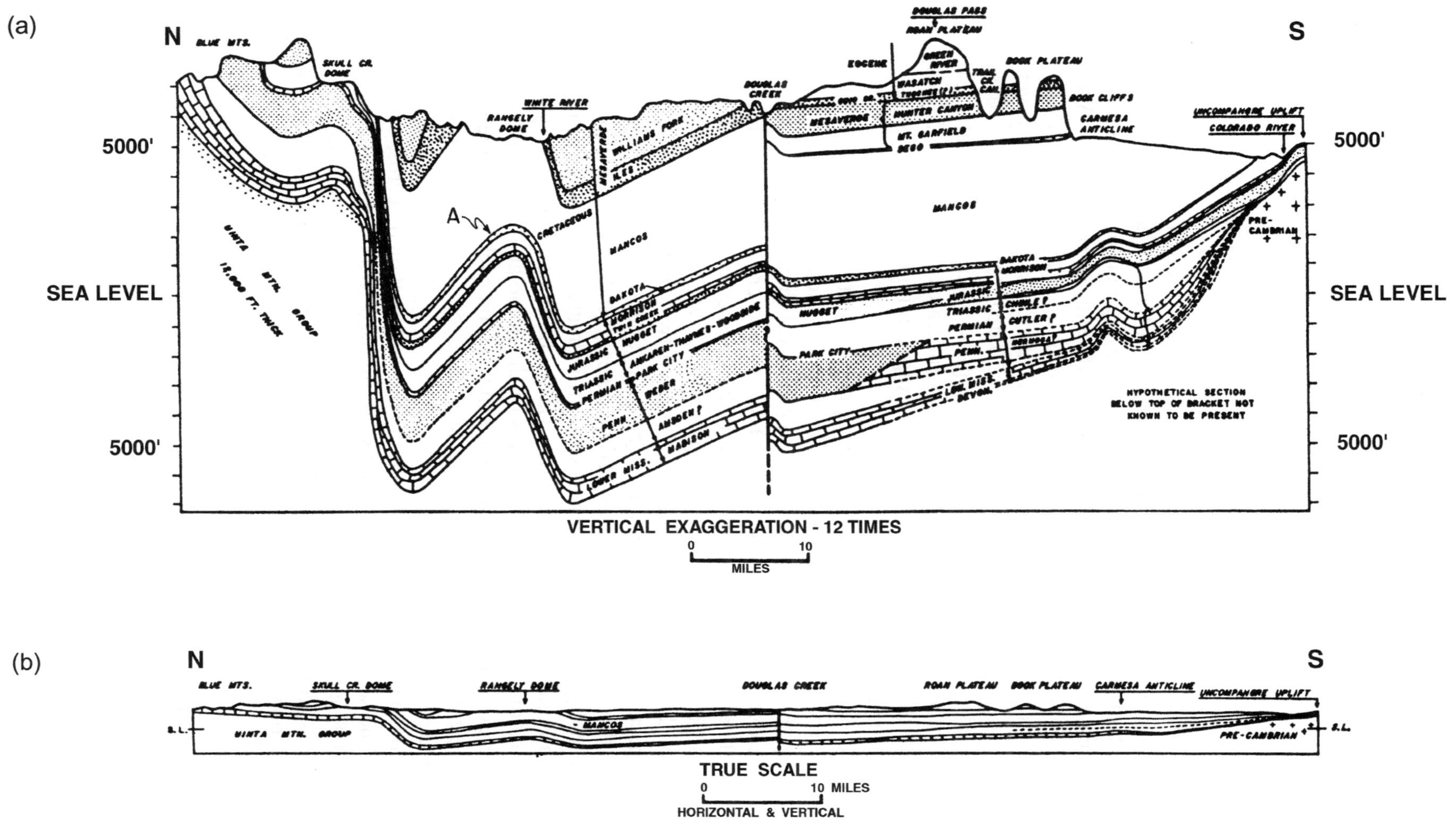

Figure 6-18 (a) A generalized cross section across the Uinta Basin, Colorado, showing the effect of vertical exaggeration (12 times). (b) Same general cross section constructed to true scale. (Modified from Suter 1947; AAPG©1947, reprinted by permission of the AAPG whose permission is required for further use.)

multiplied by exaggeration. Equation (6-2) defines the relationship between the true dip and the exaggerated dip (see Dennison 1968; Langstaff and Morrill 1981).

$$V_E = \frac{\tan[\text{exaggerateddip}(\delta_E)]}{\tan[\text{truedip}(\delta)]}$$

Therefore,

$$\tan[\text{exaggerateddip}(\delta_E)] = (V_E)\tan[\text{truedip}(\delta)] \qquad (6\text{-}2)$$

Using Eq. (6-2), the exaggerated dip for any cross section can be calculated if the true dip and vertical exaggeration are known. Likewise, if the exaggerated dip and vertical exaggeration for a horizon or fault are known, the true dip can be calculated. Figure 6-19 is a graphic representation of Eq. (6-2) from Langstaff and Morrill (1981). We can look at an example to illustrate the use of the graph. Referring once again to Figure 6-18, the Dakota Formation has a dip of 50 deg at location A, and the cross section has a vertical exaggeration of 12. By entering the graph on the X-axis at 50 deg, representing the exaggerated dip of 50 deg, and entering the Y-axis at 12, representing the vertical exaggeration, the intersection of the two lines generated from these data points indicates a true dip of 5.7 deg. You can see by this example that there can be significant exaggeration of dip on a cross section with a large vertical exaggeration.

As pointed out earlier, thickness of stratigraphic units is also affected by vertical exaggeration. If we consider that the exaggeration in the vertical direction on a cross section is a function of Eq. (6-1), it becomes apparent that interval thickness in cross section varies as a function of the exaggerated dip. Figure 6-20 shows two cross sections. Figure 6-20a is plotted true to scale and shows a unit of constant thicknesses "T." Figure 6-20b shows the effect on the unit's thickness with a vertical exaggeration of two. Notice that in the area of low dip, the thickness of the interval increased by a factor of two. In the area of steep dip, the interval thickness has been attenuated as a result of the exaggerated scale. It is interesting to note, however, that even though the apparent true bed thickness (stratigraphic thickness) is attenuated in the area of steep dip, the vertical thickness is still increased by a factor of two. This effect can also be seen in Figure 6-18 across the Uinta Basin. In the area of Rangely Dome, we see the effect of attenuation of the units on the flanks of the dome and the apparent thickening of the units over the crest of the structure and in the synclines on each side of the dome. In the preparation and review of cross sections with vertical exaggeration, be sure to keep such thickness changes in mind.

The decision to use vertical exaggeration in the construction of a cross section must be based on the scale of the section and its intended use. Employed wisely, vertical exaggeration can be a very useful tool. Finally, we emphasize that the use of both a vertical and horizontal graphic scale is necessary in all constructed cross sections, whether they are true scale or incorporate some type of exaggeration.

PROJECTION OF WELLS

It is best to construct straight cross sections directly through wells; however, for various reasons this may not be possible, so it sometimes becomes necessary to project a well into a line of section.

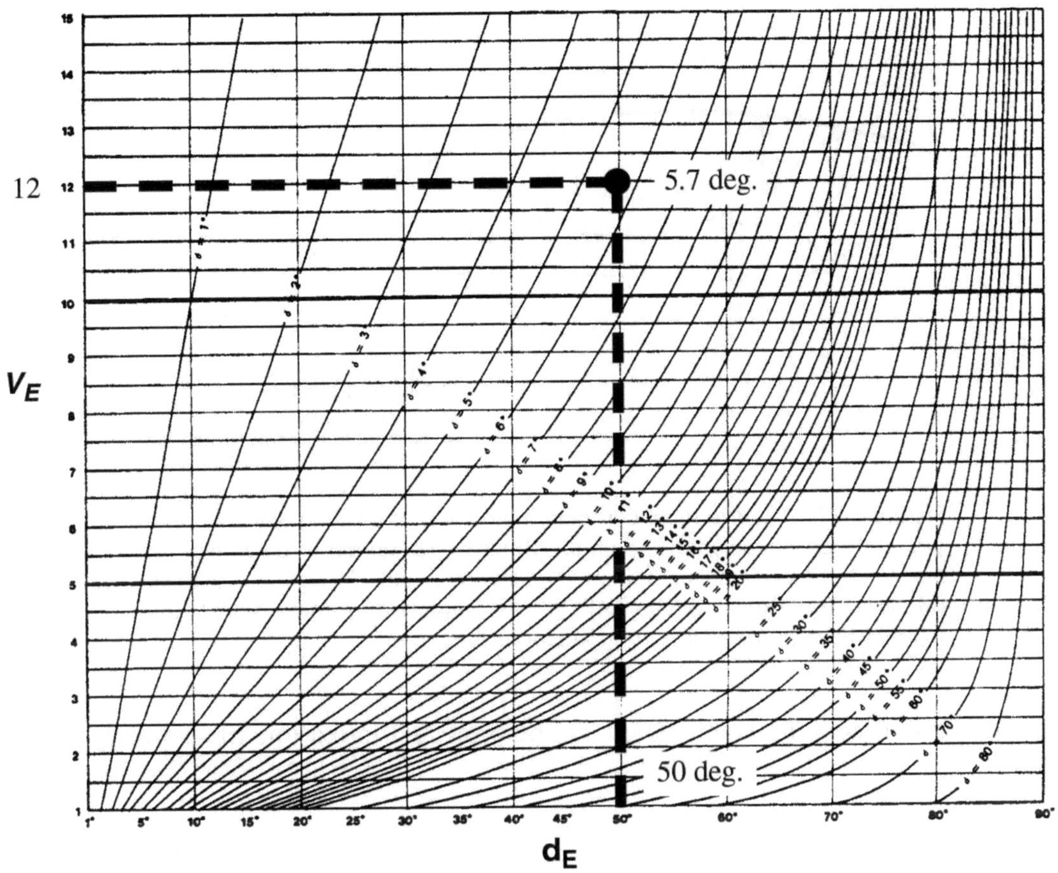

Figure 6-19 Solutions to Eq. (6-2) for different values of V_E (vertical exaggeration), and (true dip). True dip occurs where V_E = 1. (From Langstaff and Morrill 1981. Published by permission of the International Human Resources Development Corporation Press.)

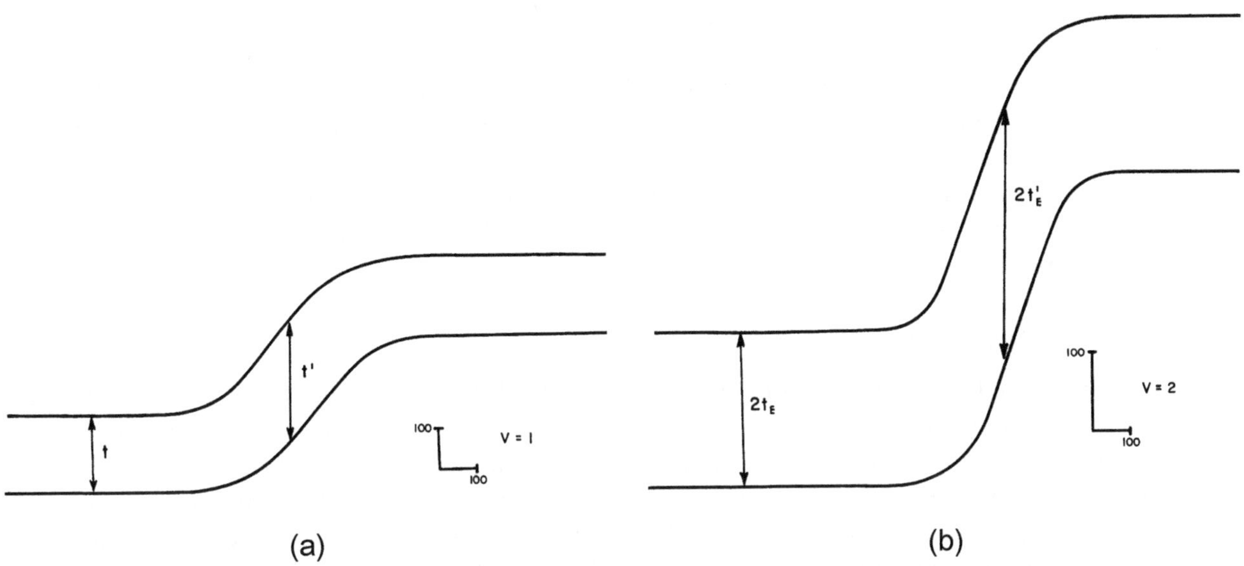

Figure 6-20 (a) is constructed to true scale. (b) shows apparent attenuation due to vertical exaggeration. Attenuation is greatest where true dip is steepest. Notice in the area of steep dip that, although the stratigraphic thickness is attenuated, the vertical thickness is twice the vertical thickness of that shown in (a), which is drawn with a true scale. (Modified from Langstaff and Morrill 1981. Published by permission of the International Human Resources Development Corporation Press.)

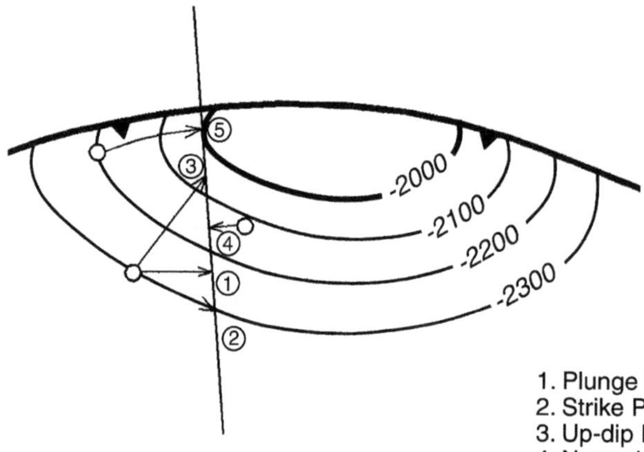

Figure 6-21 The five most common methods of projecting well data into a line of section include (1) plunge, (2) strike, (3) up-dip or down-dip, (4) normal-to-the-section-line (minimum distance), and (5) parallel to fault. (Modified from Brown 1984a. AAPG©1984, reprinted by permission of the AAPG whose permission is required for further use.)

There are several ways in which a well can be projected into a line of section (Fig. 6-21). These include

1. Plunge projection
2. Strike projection
3. Up-dip or down-dip projection
4. Normal-to-the-section-line projection (minimum distance method)
5. Parallel-to-fault-strike projection

All these methods can be used to project well data into a cross section. The best method to use, however, depends on various factors, including (1) the structural style of the area being worked, (2) the orientation of the section to the axis of the structure, (3) the horizontal distance of a well from the line of section, (4) whether there are faults on the section line, and (5) the general dip of the structure. The most commonly used methods for projecting wells into a line of section constructed by hand are the *plunge* and *strike* projections. As discussed later in this chapter, the most commonly used method of projecting wells into a line of section constructed on a computer is the normal-to-the-section-line.

Plunge Projection

In structural settings such as compressional fold-and-thrust belts containing plunging structures, the preferred method for projection of well data into a line of section is parallel to the B-axis of the fold in the hanging wall, either up-plunge or down-plunge. Figure 6-22 is a map view of a cylindrical fold. The longitudinal or B-axis of the fold is shown as B-B′, which is coincident with the axial line (Brown 1984a). The dashed lines represent plunge lines (fold elements) which are, by definition, parallel to the B-axis in a cylindrical fold. Notice that the dip rates are the same along each individual fold element, including the axial line. This is true only where the plunge rate is constant for the portion of the fold depicted.

One of the main objectives in plunge-projecting well data into a line of section is to preclude the possibility of *physically moving structural geometry* within the fold. By definition, a

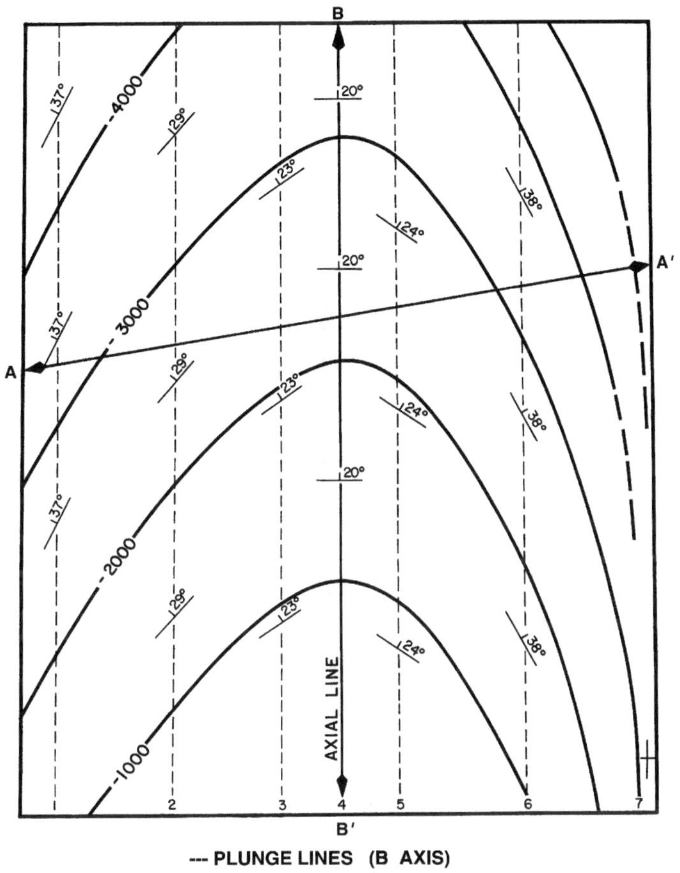

Figure 6-22 Map view of a plunging cylindrical fold. The longitudinal fold axis (B-axis) is shown as B-B' The dashed lines represent plunge lines that parallel the B-axis. (Modified from Brown 1984a. AAPG©1984, reprinted by permission of the AAPG whose permission is required for further use.)

cylindrical fold is one in which all fold elements are parallel to one another and to the B-axis, which is parallel to the direction of plunge. A good way to visualize this is to equate a cylindrical fold to a cylinder of pipe. The edges of the pipe represent fold elements and are parallel to one another for the entire length of the pipe. Thus, the shape of the pipe's cross section is the same at each end and everywhere in between. Similarly, a cylindrical fold contains fold elements that are everywhere parallel, and the cross-section geometry is the same throughout the length of the cylindrical portion of the fold. Therefore, projection of fold geometry into cross section A-A'(Fig. 6-22) along a line that is not parallel to the plunge lines (e.g., sides of the pipe) would result in the attempted *physical movement of geometry to an improper position within the fold* (e.g., projection of a 38-deg dip from plunge line No. 6 to plunge line No. 5, which has only 24 deg of dip).

We now consider the application of plunge projection. Figure 6-23 is a structure map of a plunging cylindrical fold cut by a reverse fault. We want to construct cross section A-A′ as shown on the structure map. As only Wells No. 1 and 7 actually lie on the line of section and we wish to include more data than that available from the two wells, we have no choice but to project the additional wells into the line of section.

Well No. 1 is drilled off the flank of the structure and Well No. 7 is in the footwall in the opposite fault block. Wells No. 2 through 6 can be projected into the line of section in several different ways: most commonly, either parallel to structural strike (dashed arrow lines) or parallel to plunge (solid arrow lines). In the former case, Wells No. 2 through 5 are projected into

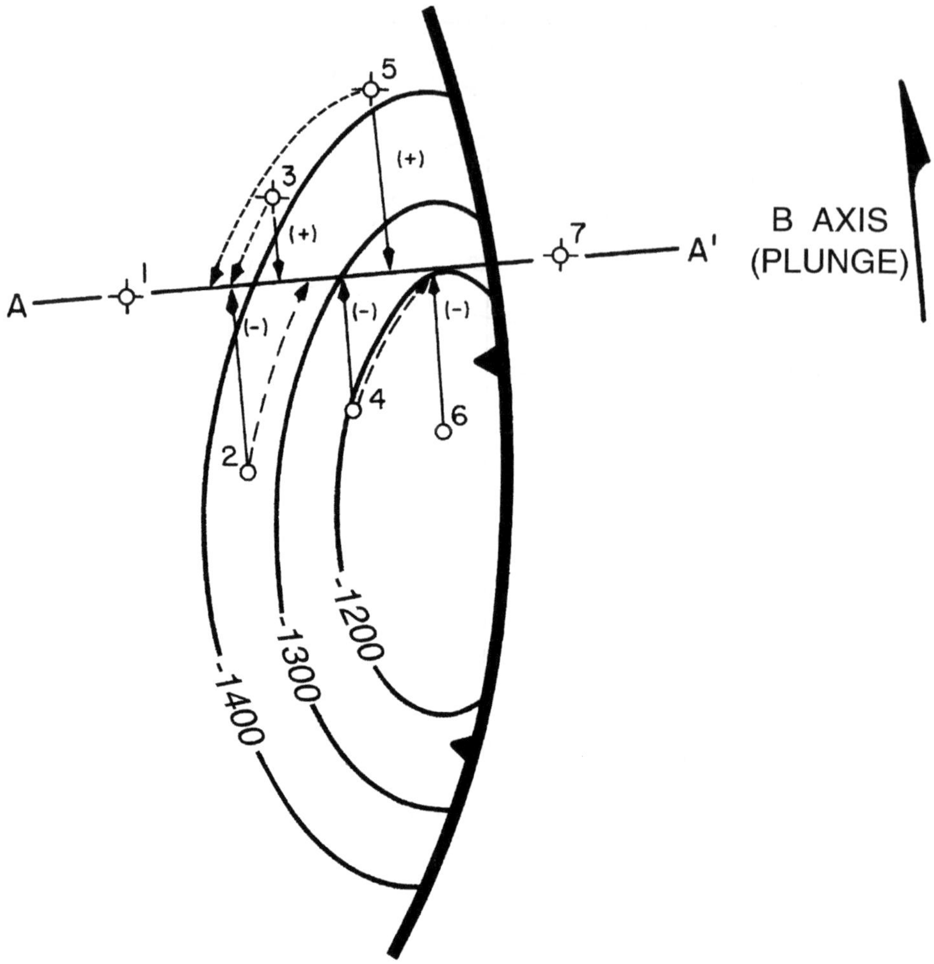

Figure 6-23 Structure map of a plunging cylindrical fold cut by a west-dipping reverse fault. (From Brown 1984a. AAPG©1984, reprinted by permission of the AAPG whose permission is required for further use.)

A-A′ along structural strike (parallel to the contours), and Well No. 6 cannot be projected at all. Figure 6-24, the cross section obtained by projection along structural strike, is obviously very confusing and results in an *unreasonable depiction* of the fault shape, as well as the relationship of hanging wall and footwall rocks.

The direction of plunge of the B-axis is indicated on the structure map in Figure 6-23 to show the direction in which the wells must be plunge-projected. The B-axial direction can be obtained by analysis of any form of dip data. The solid arrows show the planned projection path of the wells into the section. Next to each projection is a + or − sign. These signs indicate whether the structural elevation for a specific horizon or structural marker (fault, dip rate, etc.) in the wells must be adjusted positively or negatively for the direction and rate of plunge over the distance projected into the section line. For example, Well No. 2 has a subsea depth for the horizon of −1340 ft. When we project this well along plunge, the top of the horizon intersects the cross section at a subsea depth of −1410 ft. Therefore, the top must be adjusted downward by 70 ft to correct for the change in elevation. This correction in elevation can actually be calculated with trigonometry by using the angle of plunge and the length of the projection from the well site to

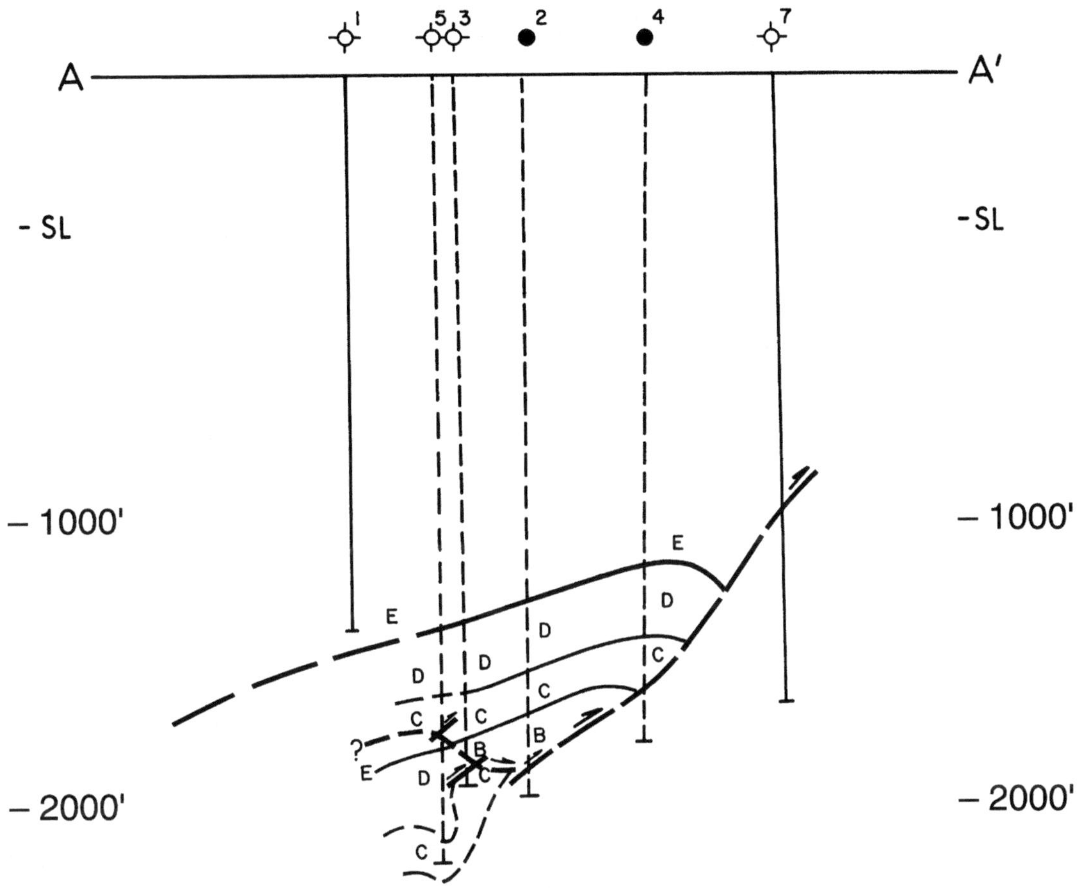

Figure 6-24 Structural interpretation along cross section A-A′ (Fig. 6-23) based on strike-projecting well data into the line of section. The strike-projected data results in an unacceptable interpretation. (Modified from Brown 1984a. AAPG©1984, reprinted by permission of the AAPG whose permission is required for further use.)

the line of section. Each well must be corrected in this manner, with (+) values indicating elevations raised because the well is projected up-plunge, and (−) values indicating elevations lowered because the well is projected down-plunge.

Figure 6-25 is the completed cross section using all the data from the wells projected into the line of section parallel to plunge, or the B-axis (in the hanging wall). In order to differentiate between the wells that actually lie on the section line from those that have been projected, the wellbore stick for each projected well is dashed. Notice also that the numerical sequence of wells (1 through 7) is now in proper order.

In conclusion, for plunging folds, the plunge-projection method offers the most accurate means of projection of structural geometry for the hanging wall block into a cross section. This projection may be up-plunge or down-plunge, or both, and structural elevations likewise are adjusted up or down. The projections are made in order to ascertain the true cross-sectional shape of the structure.

Strike Projection

If you are mapping in a geological setting involving diapiric structures or in a tectonic area with nonplunging folds, projection along strike is often more beneficial than other types of

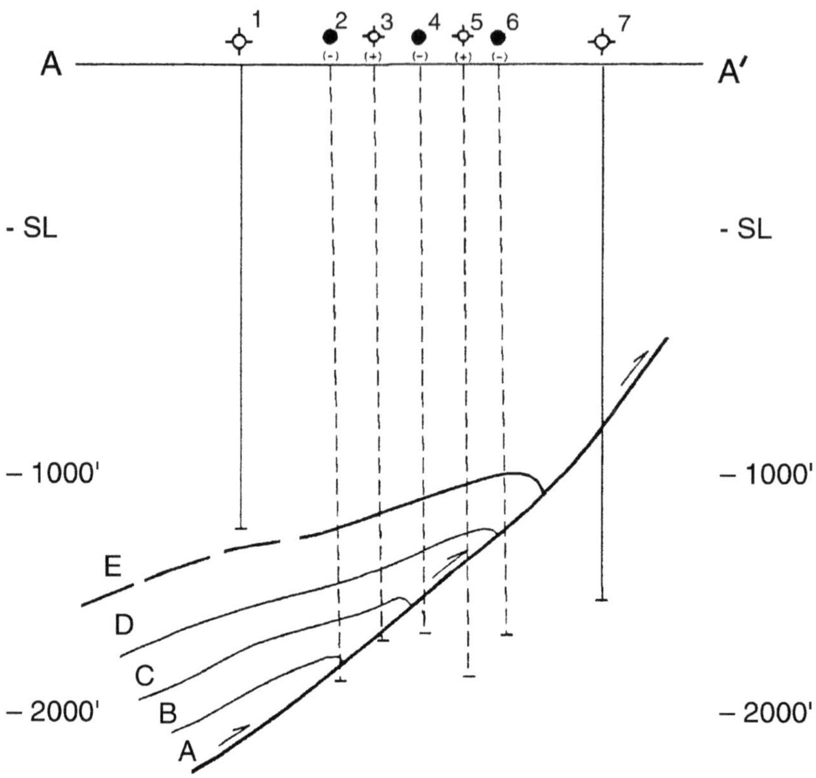

Figure 6-25 Structural interpretation along cross section A-A′ (Fig. 6-23) based on the well data projected into the line of section parallel to plunge of the B-axis. The projection of data along plunge provides the most accurate method of projecting the fold geometry into the cross section. (Modified from Brown 1984a. AAPG©1984, reprinted by permission of the AAPG whose permission is required for further use.)

projections. In the areas of extensional or diapiric tectonics, the dip rate on structures is typically constant along structural strike (Fig. 6-26). Therefore, in order to preserve structural geometry as well as stratigraphic relationships (syndepositional structures), wells should be projected parallel to strike.

As you look at each example cross section, imagine each to be a seismic section into which wells are projected. Consider whether the well data would tie closely with the seismic interpretation.

We first look at an example of projecting along strike in an area without faults (Fig. 6-27). In Figure 6-27a, Well No. 2 is projected parallel to strike into the line of Section A-A′. In cross section, the M Horizon point projects into the correct structural position on the section and the rate of dip at that point would be proper for the cross section. Using this method, the projection does not require any correction factors for elevation or dip rate. In Figure 6-27b, Well No. 2 is projected along strike and again projects correctly into the line of section B-B′. We can conclude that in this type of structural situation, projection along strike is preferred.

Now we look at some examples of projecting a well along strike of a horizon cut by faults (Fig. 6-28). Figure 6-28a is a portion of a structure map showing a west-dipping structure cut by an east-dipping normal fault. East-west cross section A-A′ is plotted. Wells No. 1, 2, and 4 lie on the section line. Since data from Well No. 3 is important to the interpretation, the well must be projected into the line of section. Considering the specific geological conditions, Well No. 3 is projected parallel to strike into the A-A′ cross section, as shown below the structure

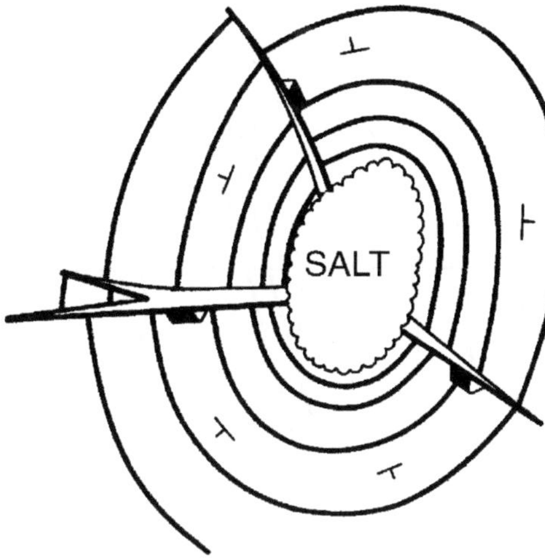

Figure 6-26 Nonplunging diapiric salt structure exhibiting constant dip along structural strike.

map in the figure. The top of the horizon projects into the section in the correct structural position and maintains the correct structural dip, but the depth of the fault in the projected well is incorrect and must be adjusted. This adjustment is required because as Well No. 3 is projected from its actual position into the line of section, the distance from the well to the fault changes.

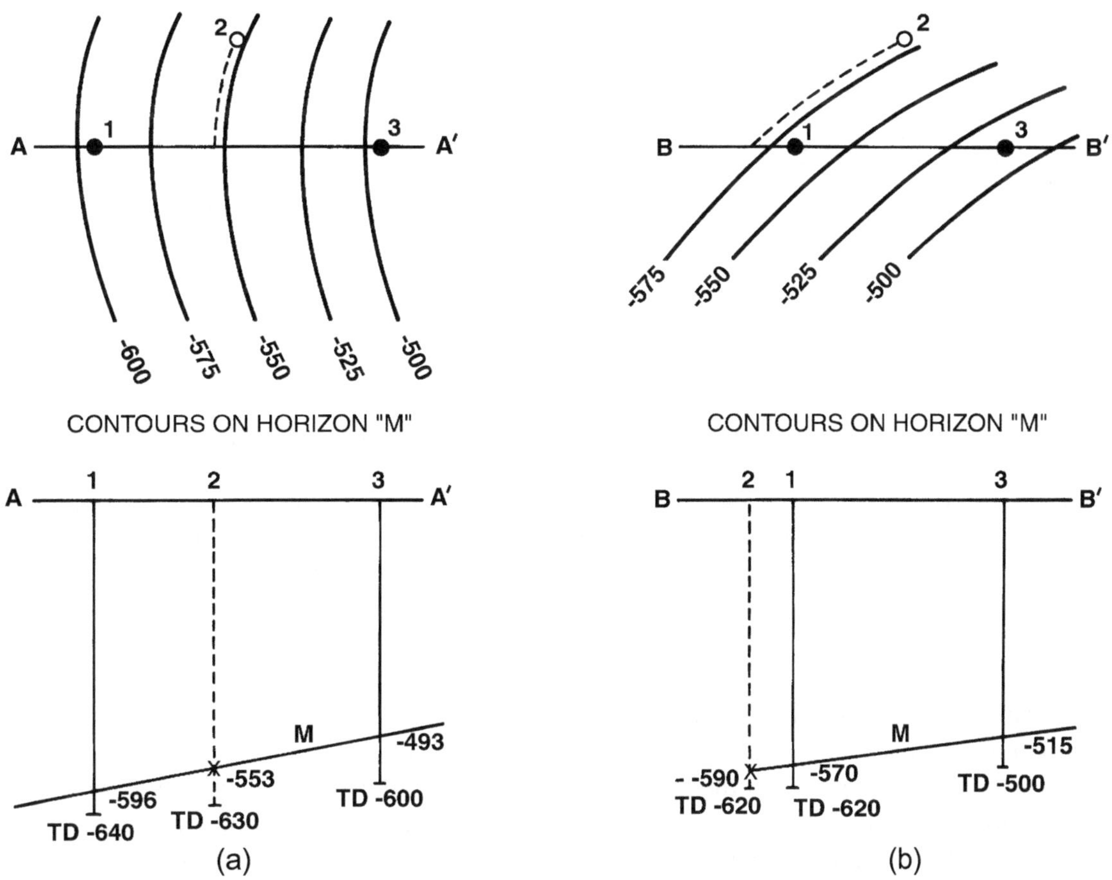

Figure 6-27 (a) Projection of well data along strike. Upper figure is a portion of a structure contour map. The lower figure is cross section A-A′ illustrating the projected data for Well No. 2. (b) Projection of well data along strike. Upper figure is a portion of a structure contour map. The lower figure is cross section B-B′ illustrating the projected data for Well No. 2. (Published by permission of Tenneco Oil Company.)

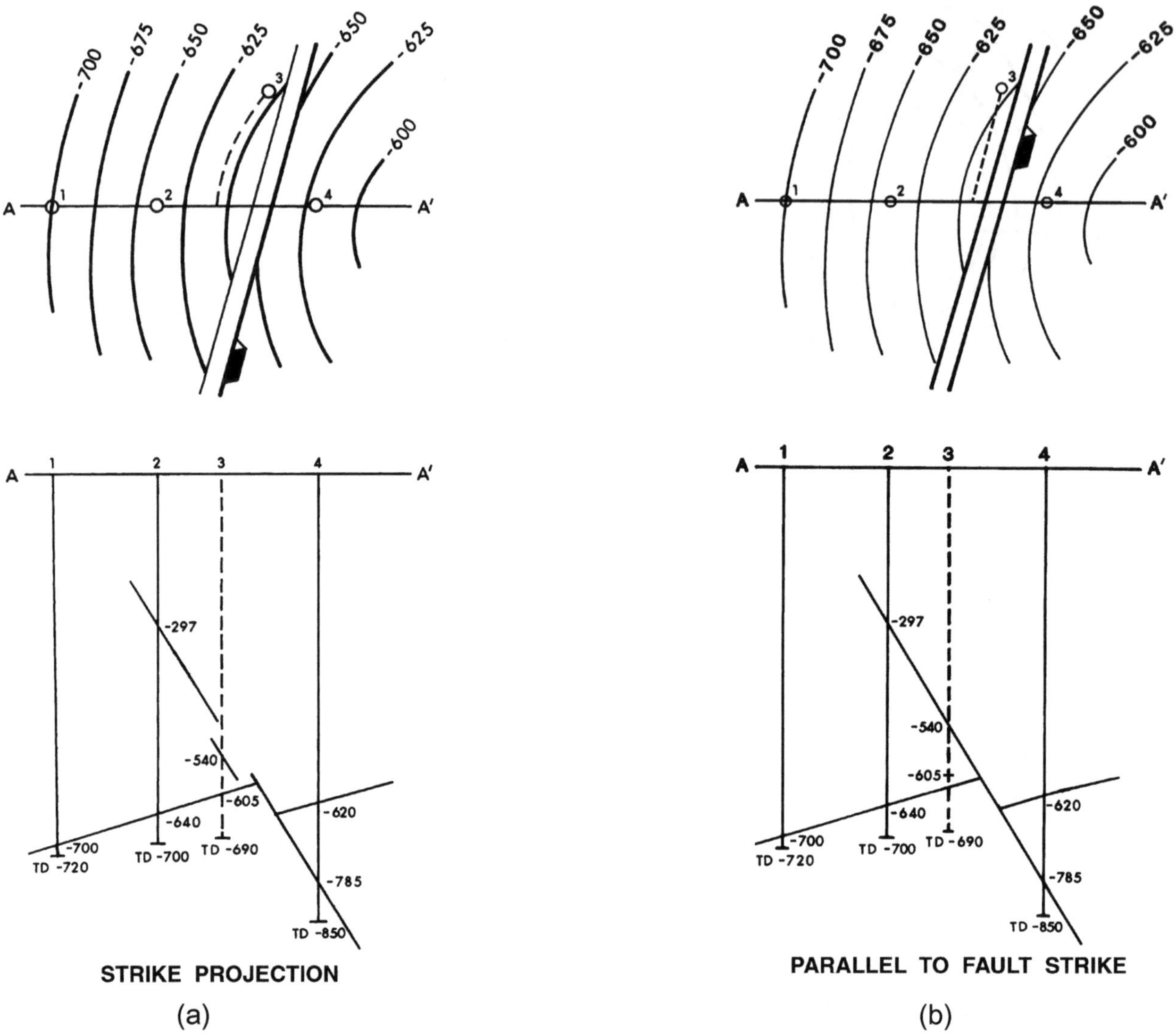

Figure 6-28 (a) Strike projection of well data on a horizon cut by a normal fault. The strike direction of the fault does not parallel the strike of the mapped horizon. Upper figure is the structure contour map. In the lower figure, cross section A-A′ illustrates that the well depth point for the mapped horizon fits the cross section; however, the fault location in Well No. 3 is too deep and must be adjusted. (b) Parallel-to-fault-strike projection. The well data are projected parallel to the strike of the fault surface. The fault projects correctly into the line of section, but the horizon depth must be adjusted. (Published by permission of Tenneco Oil Company.)

You might already have asked yourself the question, if the well is projected parallel to the fault, how will the fault and horizon project into the line? Figure 6-28b depicts this situation. Well No. 3 is projected into the section line parallel to the *strike of the fault surface*. With this projection, the fault location in the well fits the section correctly, but the horizon point is projected too shallow, requiring an elevation adjustment from −590 ft to −605 ft.

When projecting wells along strike of a horizon cut by faults, either the horizons or the faults usually will require an elevation adjustment. Therefore, we caution that projected wells often have limited use in a cross section and can, at times, cause more confusion than clarity. If a well must be projected into a line of section, be sure to identify the projected electric log or wellbore stick with a dashed line and clearly note any required corrections or adjustments.

Other Types of Projections

Figure 6-29 illustrates two other types of projection: one normal to the line of section and one parallel to the bed dip direction. Well No. 2 in Figure 6-29 is projected normal or at a right angle to the section line. This is sometimes referred to as the minimum distance method (Brown 1984a). Also, it is the typical method for projecting a well into a seismic line. Using this method for well projection, the horizon does not plot correctly on the section. In addition, it is possible that the dip of the horizon may be different at the actual well location than at its projected location on the line of section, since the well is projected into a higher structural position. This can add confusion to the section and may also require adjustments in the stratigraphic interval thicknesses in the projected well. This is the default method for projecting wells into cross sections when working on a computer.

The second projection shown in Figure 6-29 is a down-dip projection of Well No. 4. With this projection, the horizon is projected incorrectly. Also, a problem may exist with the projected horizon dip rate, since the well is being projected into a down-dip position. Neither method is recommended for projecting well data into a cross section.

Projection of Deviated Wells

If a directionally drilled well is to be included in a cross section, it is best to have the line of section follow the plan view path of the deviated well (Fig. 6-15). If the line of section is not coincident with the well path, the well will have to be projected into the cross section. The procedure for such data projection is shown in Figure 6-30, which shows the map view of directional Well No. 6 and its projection into the line of section A-A′. The deviated well path is almost coincident with the section line, but not exactly. Therefore, directional survey data points must be projected into the line of section. This is accomplished by first determining the approximate direction of structural strike, either from a structure map or seismic data, and then projecting each data point parallel to the horizon strike direction as shown in the figure.

If the electric log is represented by a stick, each directional survey data point is plotted individually in cross section and then the points are connected with a smoothly curved line representing the electric log in stick form. Notice in Figure 6-30 that the projected data points for the 4000-ft, 5000-ft, 6000-ft, and 7000-ft true vertical depths are not the same distance from the surface location as the actual depths plotted along the directional path. This correction must be

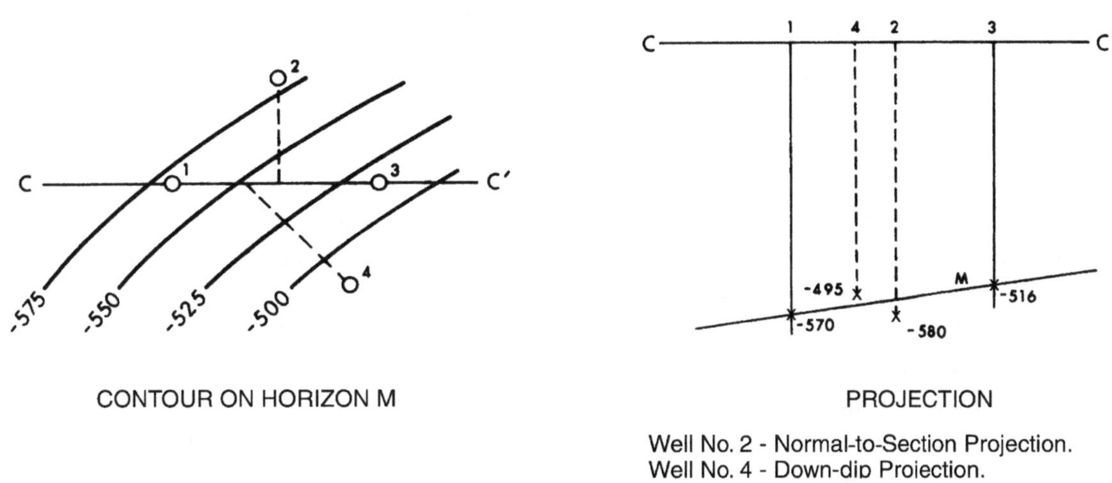

Figure 6-29 Projection of well data into a line of section, using the normal-to-section and down-dip projection methods. Neither method is recommended for projecting well data. (Published by permission of Tenneco Oil Company.)

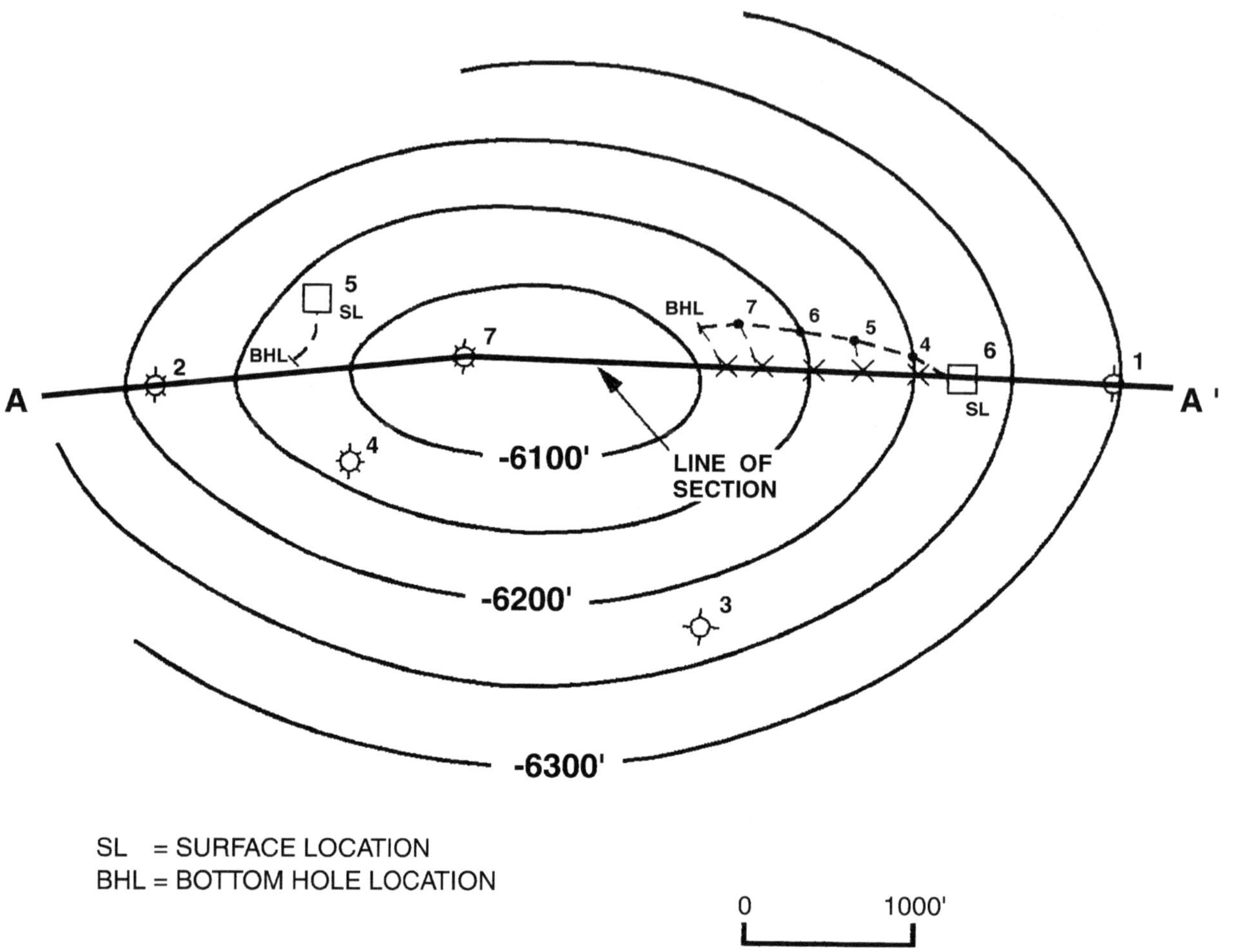

Figure 6-30 Structure contour map with line of section A-A′. Deviated Well No. 6 is projected into the line of section parallel to horizon strike.

made when the individual directional survey points are plotted in cross section. If each survey data point is projected correctly and plotted correctly in cross section, the projected directional wellbore stick will depict the true representation of the directionally drilled well in cross section.

When the actual electric log for a deviated well is used to project into a cross section being constructed by hand, the log must be cut at various locations to retain true vertical or subsea true vertical depth, whichever is used. In Figure 6-30, we see that the projected depth points plotted on the map are not the same distance from the surface location as the actual plotted depths. Therefore, adjustments must be made to the log in order to include it in the cross section. These adjustments are made by cutting out small sections of the log at various depths, as needed, to maintain true vertical or subsea depth on the section after projection. Computer software can display a deviated well projected onto a plane but most software does not project the deviated well along strike. In part this is because the software does not know the strike of the bed.

Figure 6-31 shows a deviated well log that is cut at various depths to retain its true depth in cross section after projection. It is best to delete sections of the log that do not contain any important correlations. If possible, shale sections should be used to make these adjustments.

Figure 6-32 is a north-south structural cross section through West Cameron Block 192 Field, northern Gulf of Mexico. The section is composed from six vertical and six directionally drilled wells. The structure map on the U Sand is placed as an insert on the cross section. Notice that the line of section has a zigzag pattern from north to south. Apparently, this zigzag pattern was used

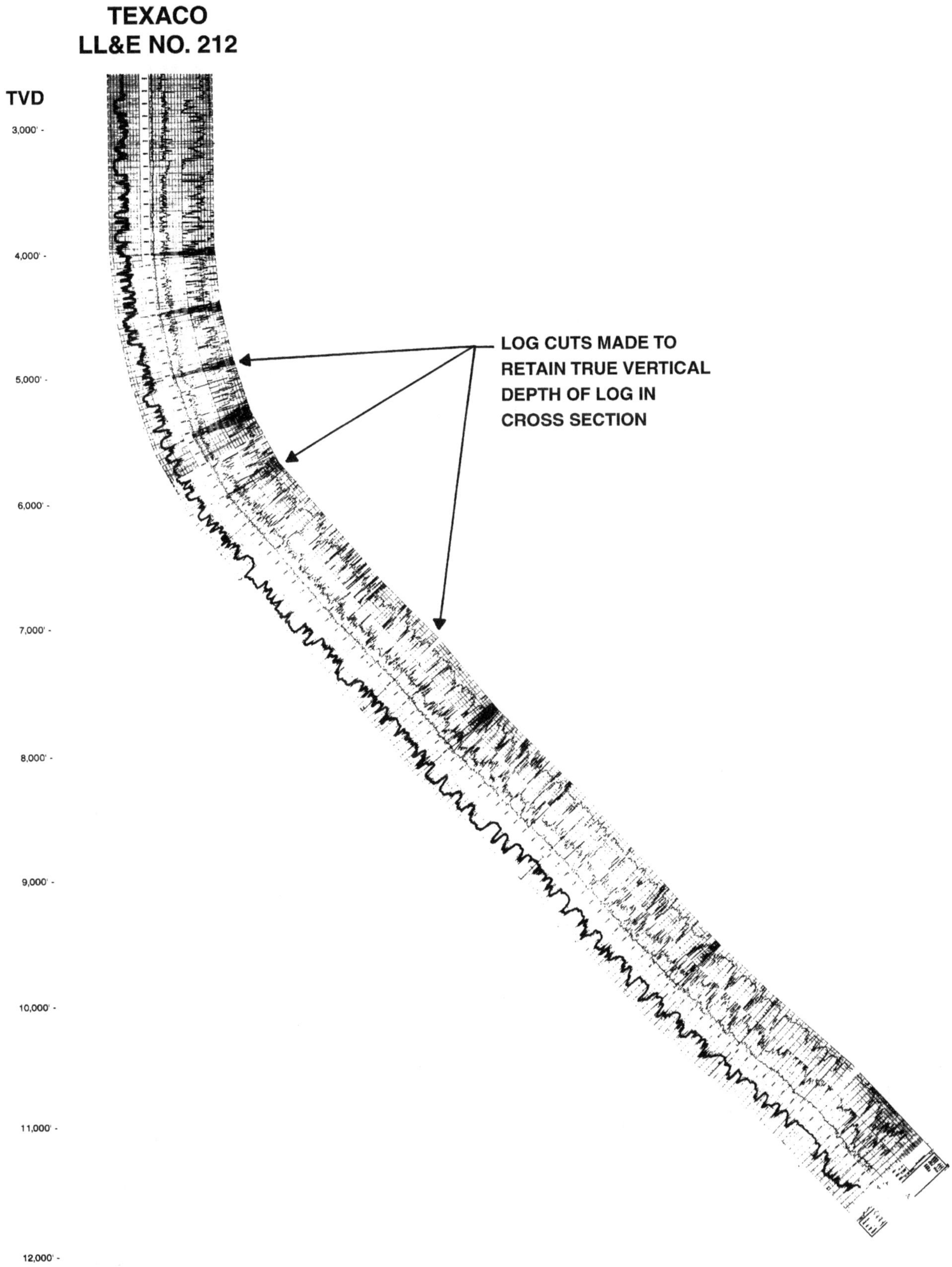

Figure 6-31 The film of the log for Well No. 212 is cut at various depths to retain true vertical depth of the entire log in cross section.

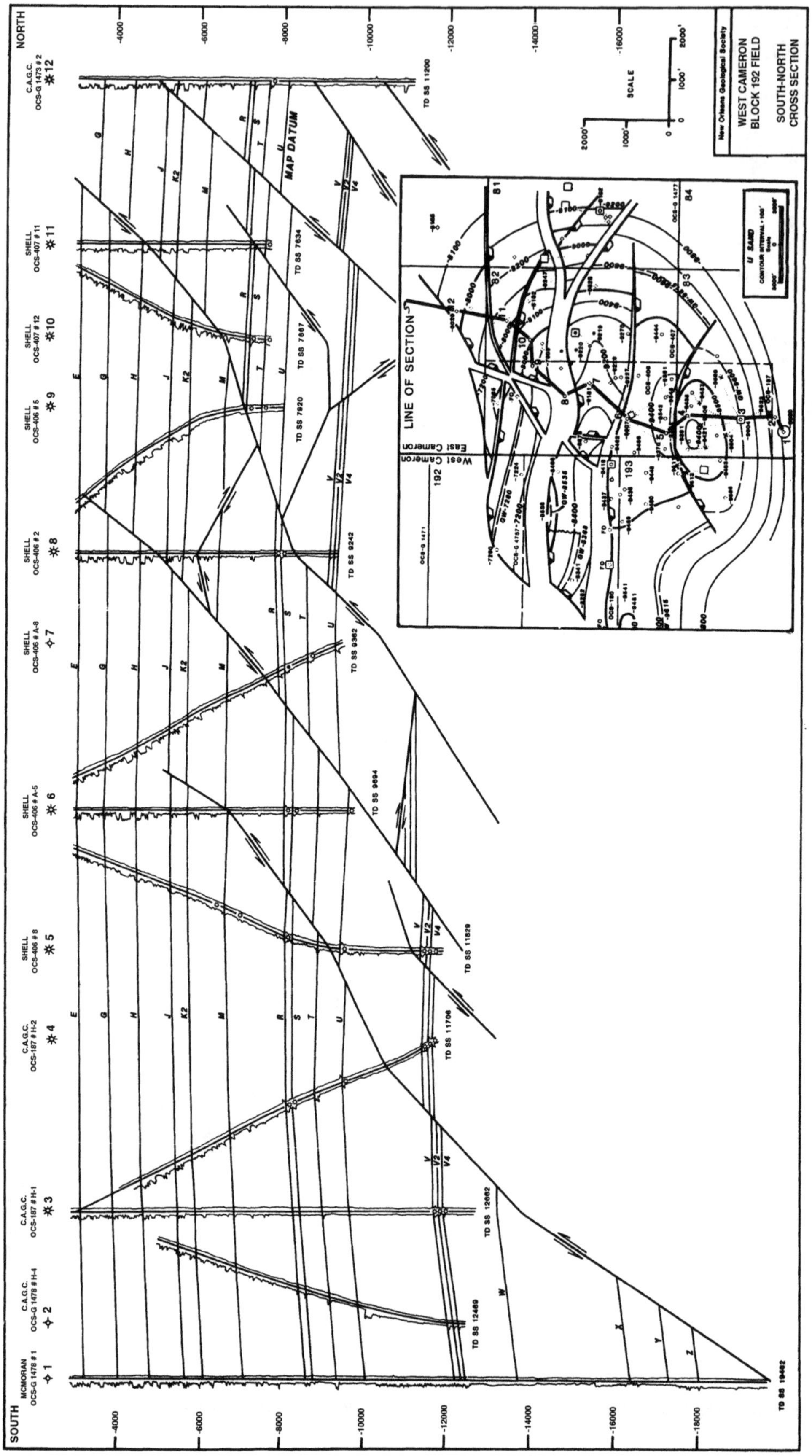

Figure 6-32 North-south structural cross section in West Cameron Block 192, northern Gulf of Mexico. The section line (see structure map insert) has a zigzag pattern to accommodate the directionally drilled wells, thus minimizing the projection of well data into the line. (From Offshore Louisiana Oil and Gas Fields, v. II, 1988. Published by permission of the New Orleans Geological Society.)

to accommodate directionally drilled wells. In other words, the path of the section line changes direction to parallel as closely as possible the path of the directional wells. By doing so, the projection of data from each deviated well into the line of section is minimized, thereby providing the most accurate representation of the directionally drilled well data in the cross section.

Projecting a Well into a Seismic Line

A well path projected into a seismic line (Fig. 5-15a and b) can often cause more confusion than clarification to an observer unfamiliar with the projection or details of the geological structure. Off-depth projections of horizon and fault positions in wells result in a mis-tie of the seismic with the well data. Therefore, care must be taken to properly project a well into a seismic section and label clearly the projected well data. If you are working with 3D seismic data, then the creation of an arbitrary line through a well can avoid the problems of projection into a line. However, if the velocity function used to convert the well depths to time for plotting the well is inaccurate, or if the well and/or seismic lines are incorrectly spotted, then a mis-tie can still occur.

The techniques outlined for projecting a well into a cross section are also valid for projecting well data into a seismic section. The following are a few additional suggestions that may help when projecting a well into a seismic section.

1. Be certain that the velocity function is the most accurate one for the area. Incorrect velocity information can cause some serious errors when projecting a well, particularly if it involves a directionally drilled well. Even when working with seismic displayed in depth it is important to remember that the velocity function used to convert time seismic into depth seismic may not be completely accurate (Fig. 5-29). Also, remember that a seismic section *is not* a geological cross section and that the apparent angle of a well on a seismic section may not resemble the angle of the well on the equivalent geological cross section. It is also important to know both the horizontal and vertical scales of the seismic section being used.
2. Consider dashing all projected wells posted on a seismic section. If the well path actually crosses the plane of the line, mark the point clearly with a different color solid line or a horizontal dash across the well path. Many computer systems use this or a similar convention when projecting wells onto seismic sections.
3. Remember that when a well is projected along structural strike, many of the faults may appear to intersect the projected well path at different depths than the actual intersections in the well. Be ready to field questions when a line like this is used for presentation purposes.
4. Too often, the *minimum distance* (normal-to-the-line) projection method is used for projecting well data into a seismic section. Whenever possible, do not use this method, because the data frequently will be projected incorrectly, resulting in a conflict between the well and seismic data (Fig. 5-15a and b). The minimum distance projection method is the default method for projecting wells onto seismic lines for all interpretation software systems of which we are aware. Many software systems may not allow an alternative method of projection.
5. It is not good practice to project well data over long distances into a seismic section. Remember the geometry of any structure changes laterally even over short horizontal distances. Projections of well data over long distances can and often do cause great confusion in the interpreted seismic section.

CROSS-SECTION CONSTRUCTION ACROSS FAULTS

In Chapter 7 we discuss the importance of the fault component *vertical separation* and define the missing or repeated section in a wellbore, resulting from a fault, as being equal to this fault component. Therefore, when constructing a cross section with faults, the vertical separation must be used to correctly displace a horizon or stratigraphic unit from one fault block to another. In this section, the method for using the vertical separation in the preparation of a cross section containing a fault is discussed in detail. The concept is basic and the correct technique of construction very simple, yet the technique is often misused. Before reading this section, it may be necessary to become familiar with the fault component *vertical separation* and the difference between this component and fault *throw*. These subjects are covered in the beginning of Chapter 7, particularly in the sections titled Definition of Fault Displacement and Fault Data Determined from Well Logs.

If the missing section in an electric log is mistakenly used as throw in the construction of a cross section involving a normal fault, chances are the displacement of horizons across the fault will be incorrect. The values for vertical separation and throw are the same only for a vertical fault and for a fault that cuts through horizontal beds. With increasing bed dip, however, the values are different and can be significantly so with steeply dipping beds. Therefore, the bed displacement error in a cross section, using the missing section as throw, will usually be greater with increasing structural dip.

Figure 6-33 graphically depicts the correct method for using vertical separation (missing section) to obtain the proper bed displacement across a normal fault. The incorrect method using throw as the missing section is also shown for comparison. In this example, assume there is sufficient control from well data or a structure map in the downthrown fault block to establish the dip of the horizon (Bed A) and its intersection with Fault 1 in the downthrown block at −7100 ft, as shown in Figure 6-33a. The missing section for Fault 1 determined by log correlation is 300 ft. With this information, the next step is to determine where Bed A intersects the fault in the upthrown fault block. This is accomplished by correctly plotting the 300 ft of missing section as *vertical separation* across the fault into the opposite fault block. *Vertical separation* is defined as the measured vertical distance between a bed projected from one fault block across the fault to a point where the projection is vertically over or under the same bed in the opposite block. *The missing or repeated section observed in a wellbore as the result of a fault is equal in value to vertical separation and not throw.* Applying this definition to the normal fault in the cross section in Figure 6-33b, the dip of Bed A is projected from the downthrown block through the fault, as if the fault were not there, to the upthrown block until it is 300 ft vertically beneath the fault. The depth at this point on the fault (−6900 ft) is the upthrown intersection of Bed A with the fault, and Bed A is then constructed from that point.

Now look at the missing section plotted incorrectly as throw (Fig. 6-33c). *Throw* is defined as the difference in vertical depth between the fault intersection with the bed in one fault block and the fault intersection with the same bed in the opposing fault block, determined in a direction perpendicular to the strike of the fault. Applying this definition to the cross section in Figure 6-33c, which is perpendicular to the strike of Fault 1, the depth of intersection of Bed A with the fault in the upthrown fault block is −6800 ft.

There is a 100-ft vertical difference between the two methods in the estimated depth of the upthrown intersection of Bed A with the fault. Incorrectly using the missing section as throw places the upthrown block 100 ft too high. This mistake could be costly, particularly if a well were planned for the upthrown block based on the incorrect interpretation, which plotted the amount of vertical separation as throw.

Using the missing section incorrectly as throw has an additional problem if the line of section is not perpendicular to the strike of the fault. Throw is measured perpendicular to the strike of the fault. If the line of section is not perpendicular to the strike of the fault, it illustrates apparent throw, which is always less than true throw (Suppe 1985). The best way to determine the strike of the fault is to map the surface of the fault as described in Chapter 7. If the missing section is used correctly as vertical separation, the orientation of the line of section relative to the strike of the fault is irrelevant as vertical separation is independent of the strike of the fault as the bed is projected across the fault using the dip of the bed along the line of the cross section.

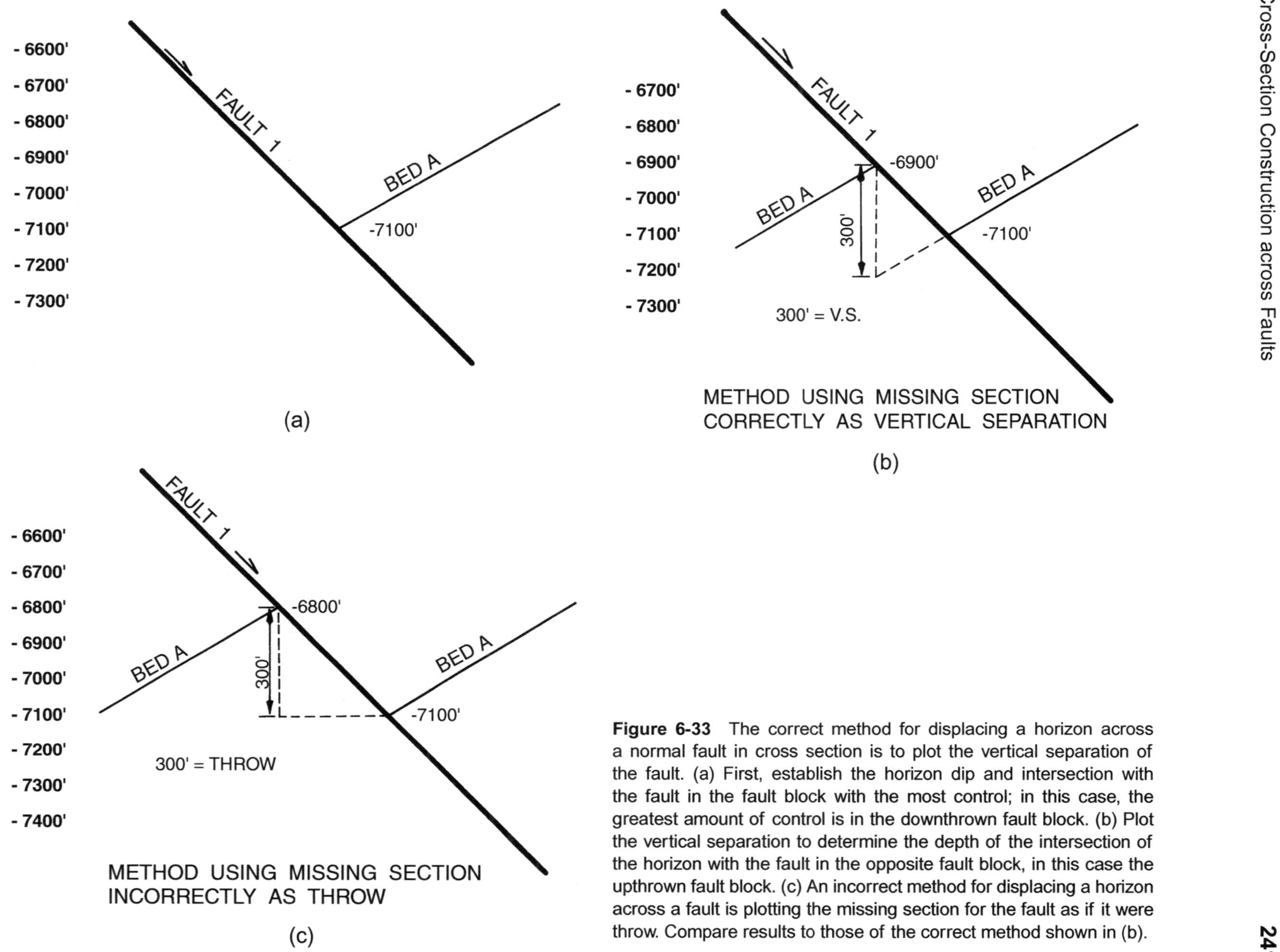

Figure 6-33 The correct method for displacing a horizon across a normal fault in cross section is to plot the vertical separation of the fault. (a) First, establish the horizon dip and intersection with the fault in the fault block with the most control; in this case, the greatest amount of control is in the downthrown fault block. (b) Plot the vertical separation to determine the depth of the intersection of the horizon with the fault in the opposite fault block, in this case the upthrown fault block. (c) An incorrect method for displacing a horizon across a fault is plotting the missing section for the fault as if it were throw. Compare results to those of the correct method shown in (b).

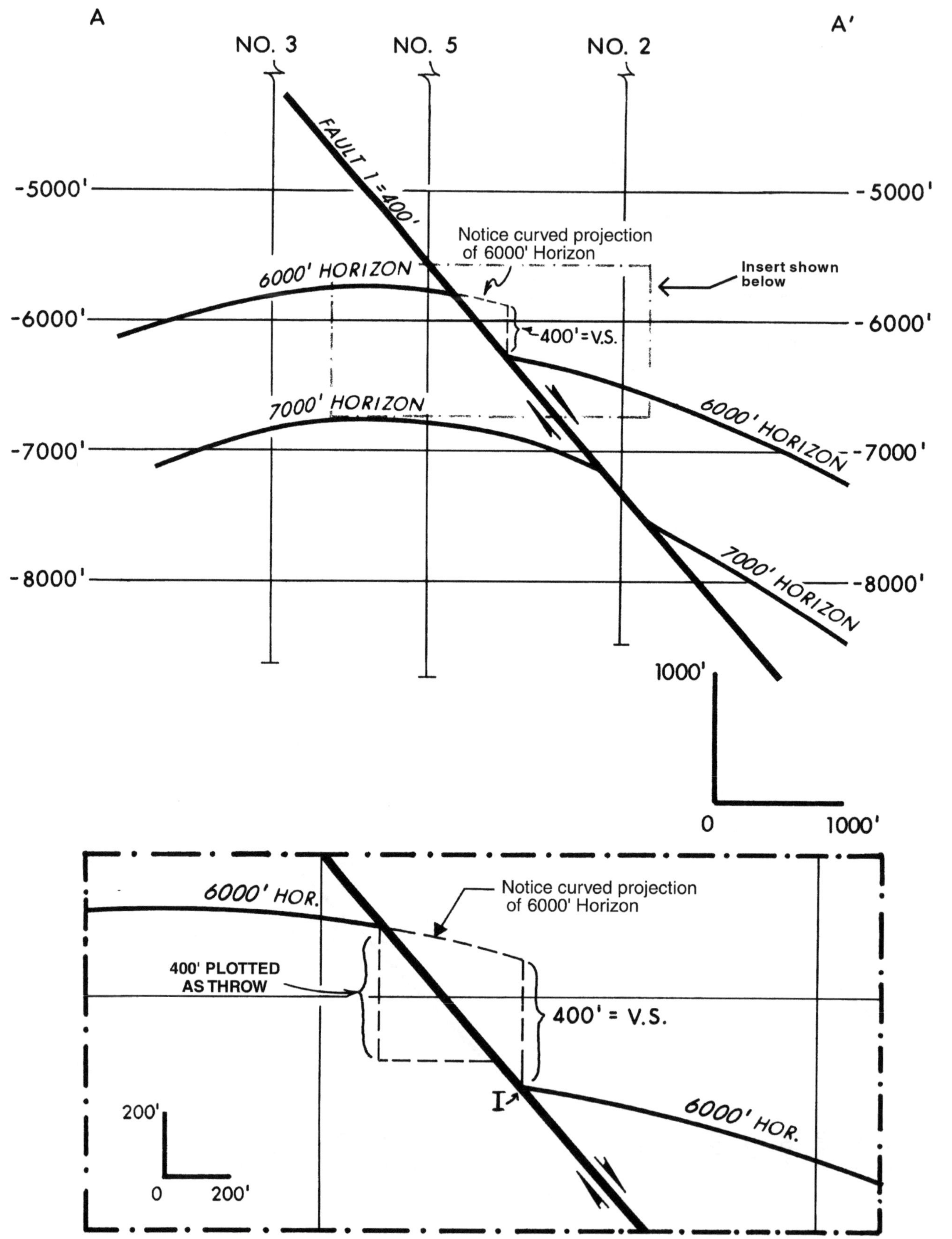

Figure 6-34 An illustration using the correct method for displacing a horizon across a normal fault.

Because this technique is so important, we will review it with two other geological examples, one involving a normal fault and the other a reverse fault. Figure 6-34 shows cross section A-A′ and an insert enlargement of a portion of the section. The missing section in Wells No. 2, 3, and 5 for the normal Fault 1 is 400 ft. The missing section is equal to vertical separation, so we use this fault component value to prepare the cross section construction across Fault 1.

The structural attitude of the 6000-ft horizon in the upthrown block of Fault 1 is established by using the available well control in that block and a completed structure map. Well No. 5 is sufficiently close to the intersection of the 6000-ft horizon with the fault to establish the depth at which this intersection occurs. Once this depth has been established, we must determine the position (depth) at which the horizon intersects the fault in the downthrown block. The depth chosen again depends on your understanding of missing section. You must use the definition of vertical separation to project the dip of the 6000-ft horizon from the upthrown block, through the fault as if the fault were not there, into the downthrown block until the projection is 400 ft vertically above the fault. The point labeled I in the insert is the correct intersection of the horizon with the fault in the downthrown block. Notice that the projection of the 6000-ft horizon from the upthrown to downthrown blocks is curved to follow the actual changing dip of the unit.

Remember, the missing section is not equal to throw in areas involving dipping beds. Therefore, if the mistake is made of using 400 ft of missing section as throw, then the estimated depth of the intersection of the horizon with the fault in the downthrown block would be incorrect. In this case, shown in the insert in Figure 6-34, the intersection point would be too shallow by approximately 75 ft.

We now look at one example involving a reverse fault. The vertical separation for a reverse fault is equal to the repeated section in a well cut by the fault. Figure 6-35 shows cross section A-A′ and an enlarged insert of a portion of the cross section. The repeated section resulting from Fault 1, as correlated with all surrounding well control, is 400 ft, which is the vertical separation of the fault.

Assume, for this example, that due to more well control, the attitude of the 5000-ft horizon and its intersection depth with Fault 1 is first established in the hanging wall block. The task now is to determine the displaced position of the horizon in the footwall block and the depth of its intersection with the fault. With reverse fault geometry, the technique is actually easier than with a normal fault and can be used to establish two control points for the 5000-ft horizon in the opposing fault block (in this case the footwall). Working with the enlarged portion of the figure, follow along the horizon in the hanging wall until the bed is at a point that is 400 ft vertically above the fault surface. This point, labeled I, is the footwall intersection of the horizon with the fault. A second point of control for the 5000-ft horizon can be established by measuring 400 ft vertically down from the intersection of the 5000-ft horizon with the fault in the hanging wall. This point, labeled I′, corresponds to the depth to the top of the horizon in the footwall at the point vertically beneath the intersection of the sand with the fault in the hanging wall.

As an academic exercise, estimate the intersection of the 5000-ft horizon with the fault in the footwall using the incorrect assumption that the 400 ft of repeated section corresponds to the throw of the fault. As a second exercise, estimate the size of the actual fault throw considering the correct geometry shown in Figure 6-35.

It is important to note again that the preciseness of the projection of horizon dip, through a fault, determines the accuracy of the displacement across the fault. If the bed dip changes in the area where the projection is made, the projection must follow this changing bed dip (observe again the curved projection of the 6000-ft horizon in the enlargement insert in Fig. 6-34). In other words, the projection of a horizon across a fault need not be a straight line. If the bed dip is not constant across the fault, then the projection must follow the changing bed dip for you to accurately construct the vertical separation.

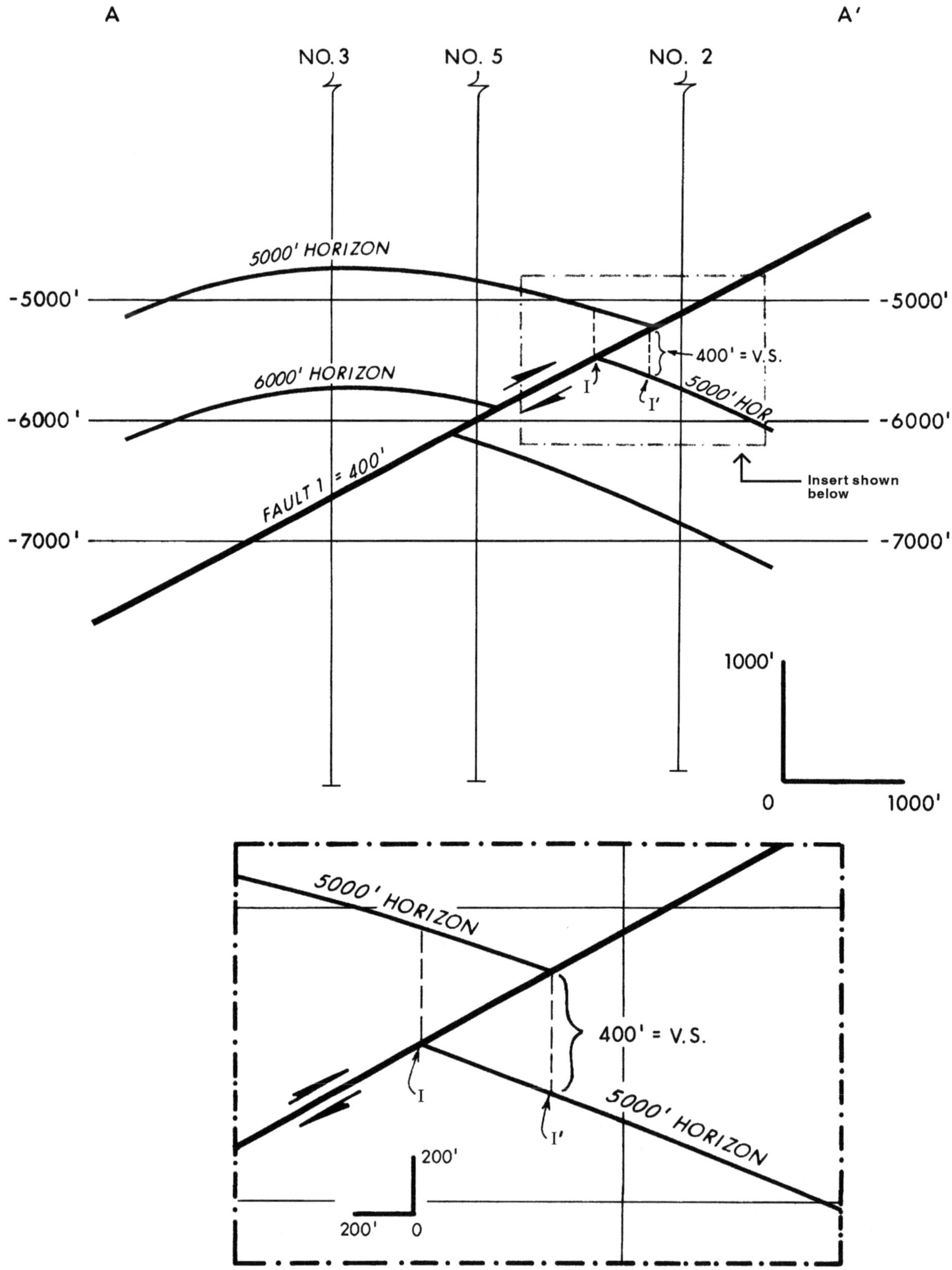

Figure 6-35 An illustration using the correct method for displacing a horizon across a reverse fault.

The precision of determining the intersection of a horizon with a fault using vertical separation is also dependent on knowing the depth of the fault along the line of section. The best way for determining the depth of the fault along the line of section is to create a structure map of the surface of the fault as discussed in Chapter 7.

THREE-DIMENSIONAL VIEWS

Many types of computer software provide 3D views of mapped surfaces and rock volumes. These are more easily generated and preferable to hand-drawn 2D representations of 3D geology. However, the latter are sometimes appropriate or necessary, so we present here some methods that help in 3D visualization of the geology.

In petroleum exploration and exploitation, a map or cross section alone cannot always represent the complete subsurface geological picture, because each is limited to two dimensions. Since neither a map nor cross section presents data in three dimensions, a detailed study may require the use of additional techniques that aid in visualizing the geology in three dimensions. Such additional aids include *log maps, stick or fence diagrams, isometric projections, 3D structural models*, and *block diagrams.*

Log Maps

Log maps are the simplest means of combining plan view information with some vertical dimensional data. Figures 6-36 and 6-37 illustrate two different types of log maps. In Figure 6-36, electric log sections and data charts for the Miocene Middle Cruse Sand Member in Trinidad are superimposed on a structure map of the same sand member. This type of map allows you to quickly review the vertical sand conditions and producing characteristics of the sand in each well with respect to the well's structural position within the reservoir. Notice that only the SP and the amplified short normal resistivity curves were used in the log sections. By looking at this map, you can very quickly see the complexity of the one or more sand members compromising this reservoir and the relationship of sand quality to structural position and well location.

Figure 6-37 is another type of log map in which the SP curve for each electric log is placed next to the appropriate well location in an attempt to portray the 3D paleogeographic conditions during the time of sand deposition. In this field, the sands appear to be of two types: (1) a major cut-and-fill channel sand, and (2) a very poor-quality, highly calcareous fringe complex. The calcareous nature of the sand was determined from cores taken in a number of wells.

If the actual structure map for the 10,500-ft Sand were superimposed on this log map in a manner similar to that shown in Figure 6-36, this additional step would improve the evaluation of the sand conditions with respect to the localization of the major channel sand and calcareous fringe complex sand in comparison to the structural position of the individual wells, production characteristics of the local reservoirs, and hydrocarbon-trapping conditions. Such information can lead to the identification of overlooked workovers and recompletions, in addition to possible development drilling locations.

Fence Diagrams

Fence diagrams, also called *panel diagrams*, consist of a 3D network of cross sections drawn in two dimensions. They are designed to illustrate the areal relationship among several wells that are located in close proximity to each other. Fence diagrams can be either structural or stratigraphic. Figure 6-38 is a typical example of a fence diagram in which actual electric log segments have been traced for use at each well location. Fence diagrams are often prepared using only log sticks, although log sticks do not provide the correlation detail usually needed for detailed work. Fence diagrams are also usually diagrammatic because they seldom have a common horizontal and vertical scale.

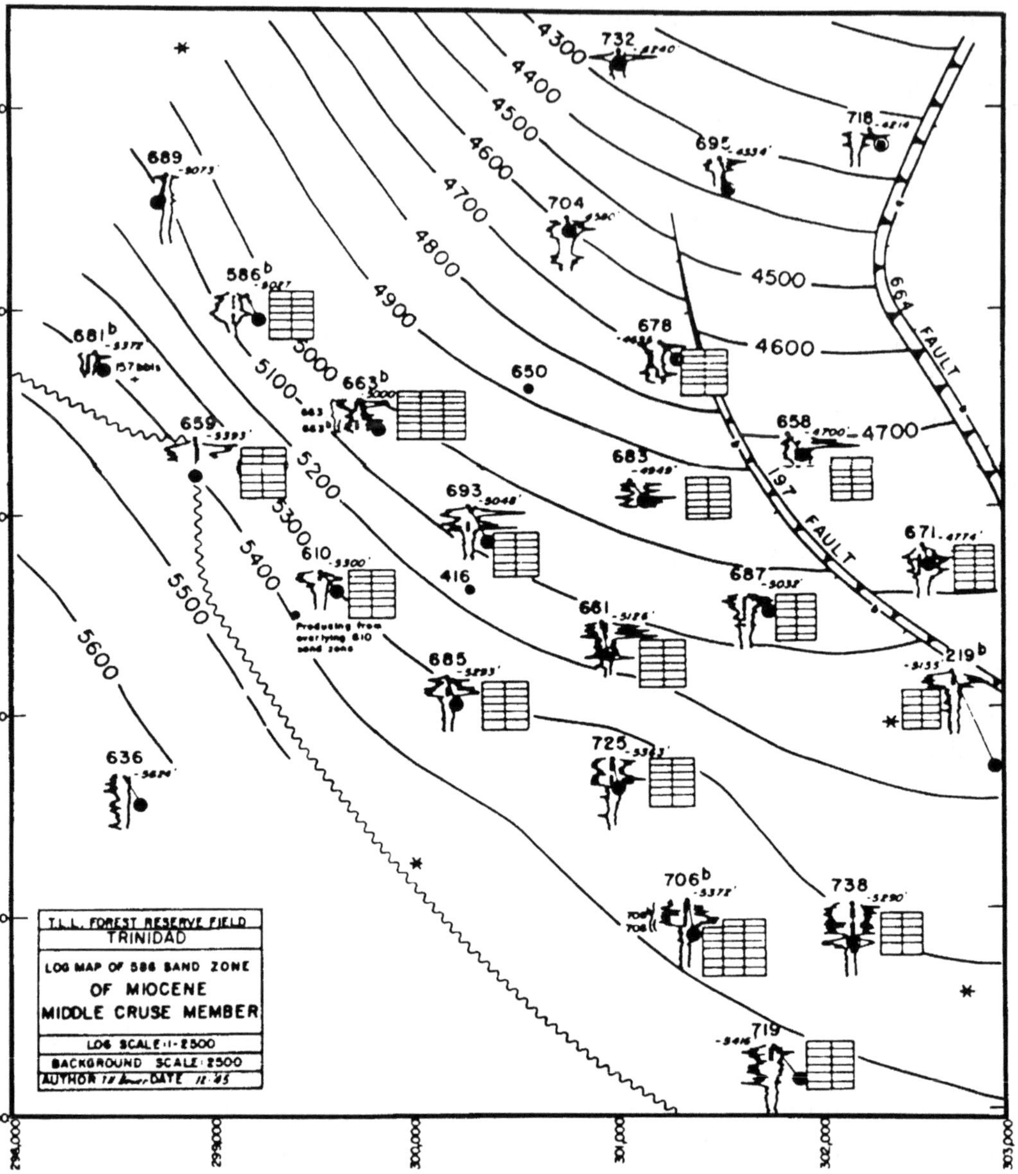

Figure 6-36 Log map of the Miocene Middle Cruse Sand Member (Trinidad). The SP and amplified short normal log curves are superimposed on the structure map of the sand. The relationship of sand quality to structural position within the reservoir can quickly be evaluated. (Modified after Bower 1947; AAPG©1947, reprinted by permission of the AAPG whose permission is required for further use.)

Fence Diagram Construction

The construction of all fence or panel diagrams begins with a well basemap. The plane of the basemap represents a chosen datum plane and each well location is taken to be the point where the well intersects the datum plane. A vertical line is drawn at each well control point and the well data are either hand-plotted along this line or a film of the actual log curves, reduced to the proper scale, is attached to the line. For example, in Figure 6-38, the SP and resistivity curves were hand-traced from reductions of the actual log. Figure 6-39 is a fence diagram constructed using the actual reduced log curves.

In the preparation of a structural fence diagram, the reference plane of the map is a chosen datum, which can be sea level or any elevation above or below. For example, the chosen datum

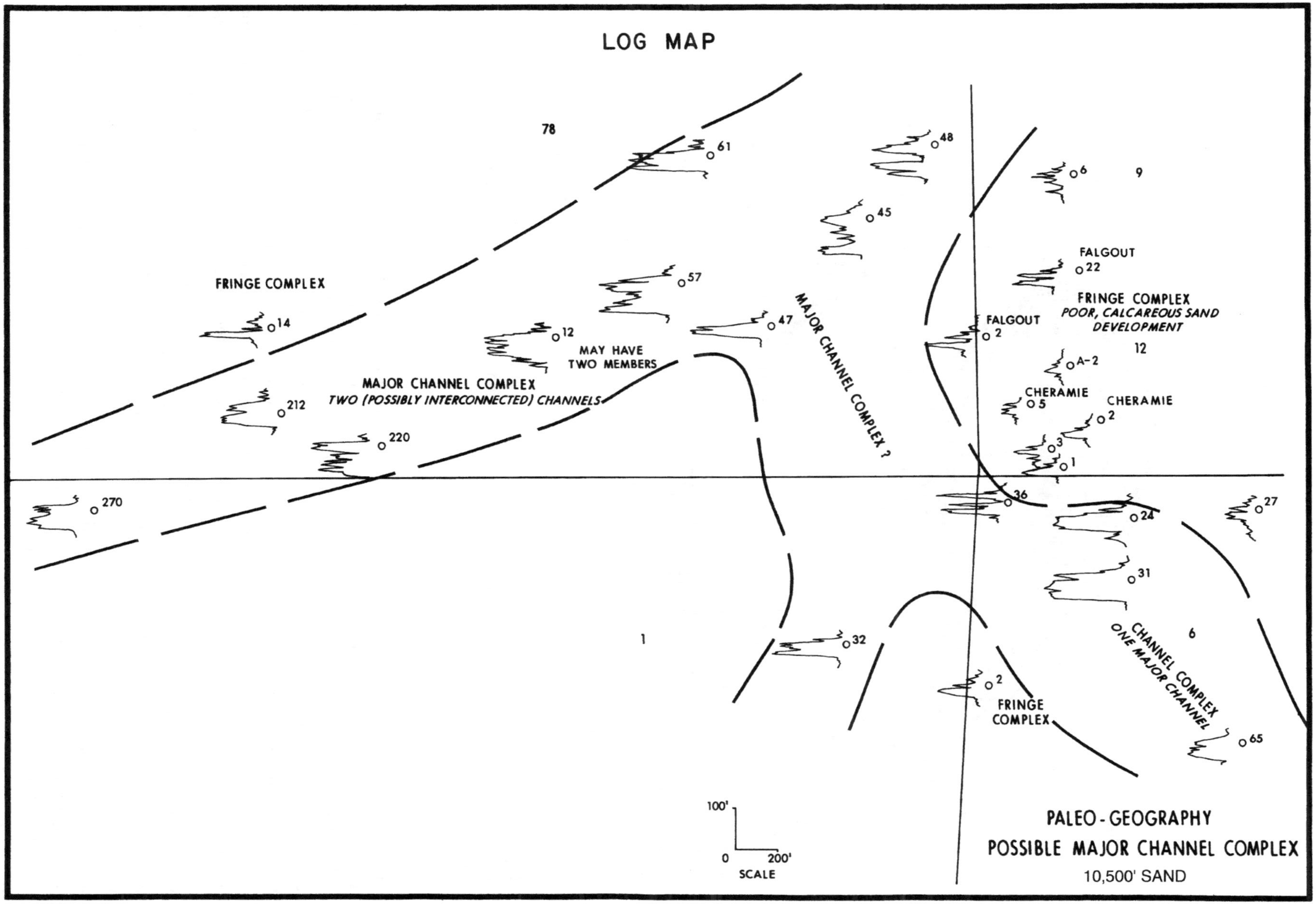

Figure 6-37 Another type of log map. The SP curve for each well is placed at the surface location for the well in an attempt to visualize the 3D characteristics of the 10,500-ft Sand. (Published by permission of Texaco USA.)

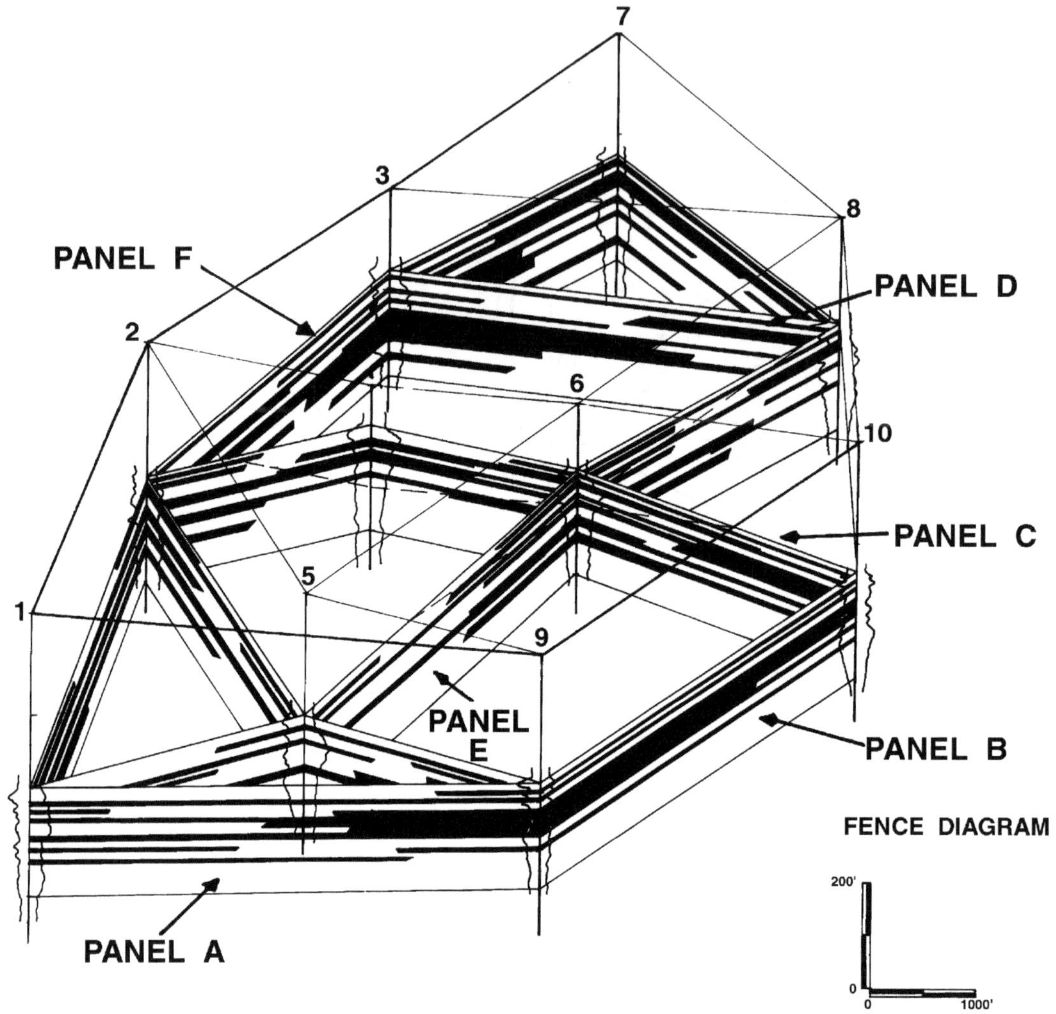

Figure 6-38 Layout of a typical stratigraphic fence diagram in which the actual electric log segment for each well was traced at the well location. (Modified from Boeckelman, unpublished. Published by permission of author.)

for the fence diagram in Figure 6-40 is −1500 ft, which is a subsea datum equivalent to 1500 ft below sea level. The log sections or hand-traced curves are hung on the vertical line at each well location and are referenced to the chosen datum. The structural profile of the bed(s) on the section is drawn by initially connecting the major correlation markers from each well. The number of correlation markers connected depends on the detail desired for the diagram.

It is important to try to minimize the interference of one panel with another to reduce the effect of masking data. It is a good idea to begin to complete the panels starting with the east-west sections first, and in particular, the panels nearest the lower or front end of the map (see panels A, B, and C in Fig. 6-38). The northwest-southeast or northeast-southwest panels are completed next, up to a point where they intersect the east-west panels from the front or where they disappear behind these panels from the rear, such as panels E and F in Figure 6-38. Once all the correlation markers have been connected on all the panels and the desired labeling added, the diagram is complete. If dips are most prominent in the north-south direction, an isometric projection may be more helpful, or the diagram may be drawn with north toward the right instead of toward the top of the map, as is more common.

For the construction of a stratigraphic fence diagram, the plane of the map is considered to be either the bedding plane or an unconformity that is the chosen datum horizon. Just as with the structure fence diagrams, a vertical line is drawn at each well location. Since most stratigraphic fence diagrams are prepared to illustrate correlations, facies changes, permeability barriers, sediment wedging, or some other stratigraphic feature, the actual log sections are normally required. Connect all the correlation markers to complete the fence diagram (Figs. 6-38 or 6-39).

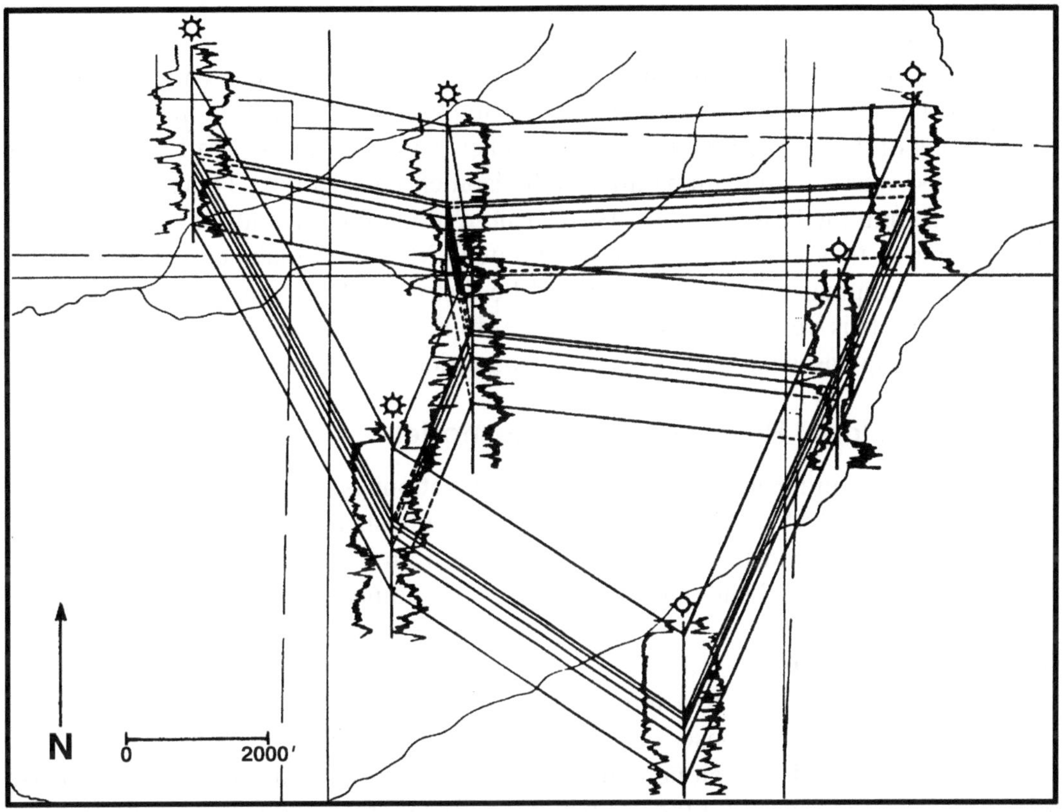

Figure 6-39 Layout of a stratigraphic fence diagram using the actual reduced log curves. (From Langstaff and Morrill 1981. Published by permission of the International Human Resources Development Corporation Press.)

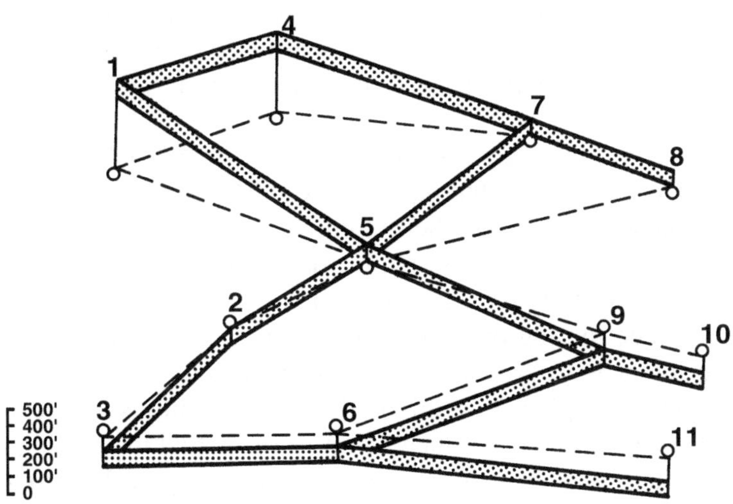

Figure 6-40 Structural fence diagram: reference plane (dashed line) is 1500 ft below sea level. (From Bishop 1960. Published by permission of author.)

Isometric Projections

Isometric projections are sometimes used to emphasize certain stratigraphic variations that might not be apparent on the conventionally oriented fence diagram. In an isometric projection, the map base is shown as if it were turned at an angle and tilted toward the front, which transforms the projection from orthogonal to nonorthogonal axes (Langstaff and Morrill 1981). Figure 6-41 shows the difference between a standard fence diagram and its transfer into an isometric projection. Notice that the rectangular basemap becomes a parallelogram in the isometric projection.

The basemap used in Figure 6-41 has a coordinate grid system with both north-south and east-west scales provided. Any isometric projection will have distortion in any direction except in the directions of the original grid system. In other words, all lines parallel to the original north-south and east-west axes remain parallel in the isometric projection. Therefore, any data points to be measured from the basemap must be measured along one or both of these coordinated grids, and by doing so they can be measured accurately. Measured distances in any direction except parallel to the original grid coordinates will be distorted. They may be greater or less than the actual distance. It is very important, therefore, to always include some type of coordinate grid system on all map bases that are planned as isometric projections.

Three-Dimensional Reservoir Analysis Model

One of the most accurate and detailed types of projections is what we call a *3D reservoir analysis model*, abbreviated RAM (Wayne Boeckelman, unpublished). Using the basic rules of descriptive geometry, we can construct a 3D of a structure from which true length and true slope can be measured. This is accomplished by constructing the folding plane at 90 deg to dip (Fig. 6-42).

Construction Procedure

The construction of a 3D RAM must begin with a completed structure map such as the one shown in Figure 6-43a. Choose a grid that is 90 deg to the dip of the structure (Fig. 6-43b) so that true length and true slope can be plotted and measured from the model. Be sure that grid lines, which later become the 3D panels, go through all the wells on the structure map, as shown in Figure 6-43b, to ensure the use of all the geological control. In this example, there are six wells to include in the grid. Notice, however, that there will eventually be more control points than just those from the six well locations. The model shows that there are actually 18 grid line intersections in addition to those through the six wells. Each of these intersections serves as a data control point. This means that there are a total of 24 geological control points for the construction of this example model. You can immediately see that one of the benefits of this type model is to provide a more accurate interpretation because of the large number of control points.

Once the grid system is laid out, extend a line vertically straight down from each well and grid intersection, as shown in Figure 6-43c. As with all the other types of 3D diagrams, a datum must be chosen. For this example, we chose −9100 ft as the datum, so all data are referenced from this datum. Once the datum is chosen, we are ready to plot data for the specific stratigraphic interval to be studied. In this model we study the 9200-ft Sand.

We begin with a review of how to place the log or hand-drawn curves on the vertical line with reference to the datum. Since the intent is to evaluate the entire 9200-foot Sand package, the representation of this sand package must be placed on the model, and in doing so there are several choices. The most accurate method is to make a reduced film of the actual log section to accommodate the scale of the model and hang the log section on the structural top of the 9200-ft Sand at each well location. As mentioned before, it is best to use a limited number of log curves

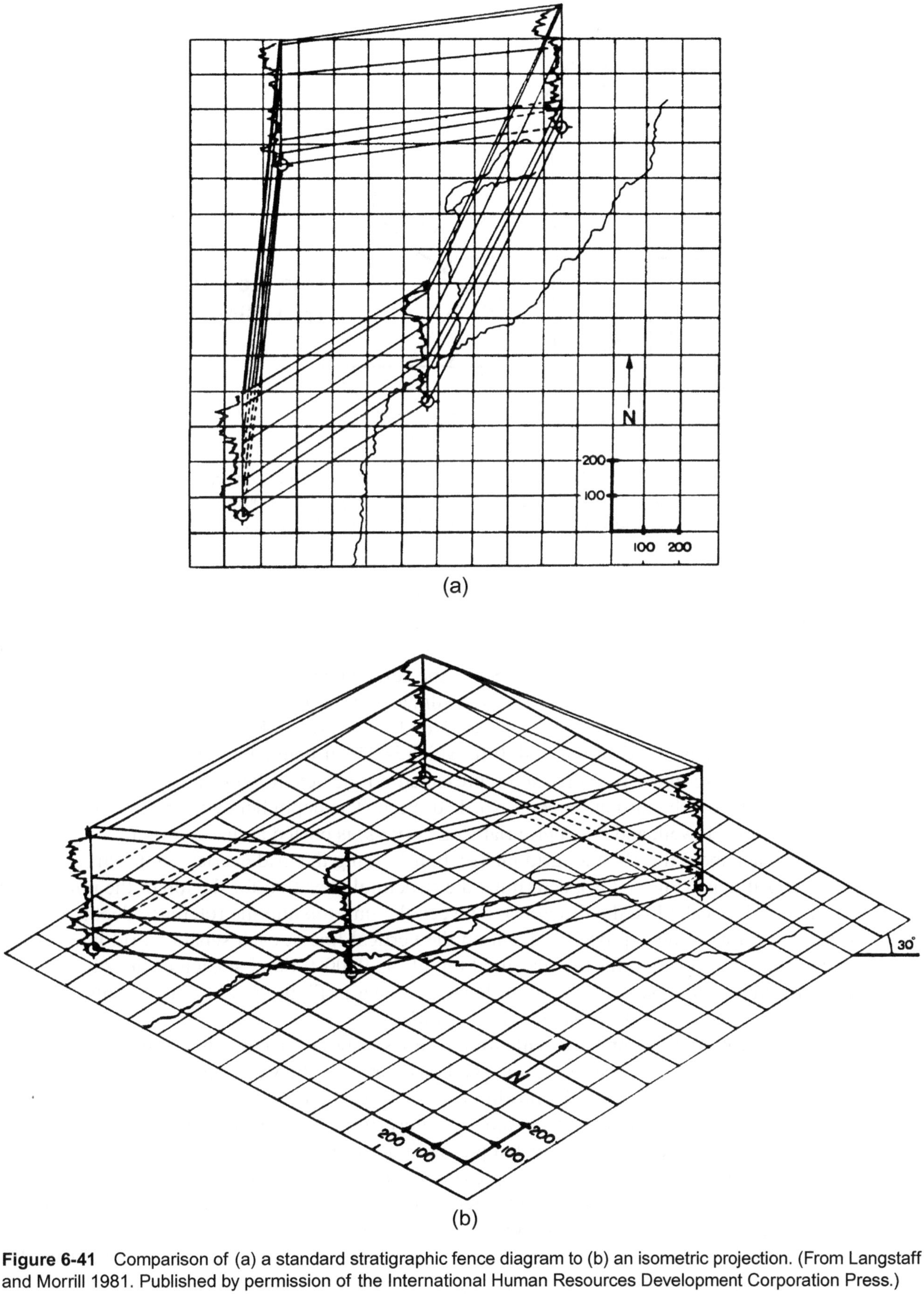

Figure 6-41 Comparison of (a) a standard stratigraphic fence diagram to (b) an isometric projection. (From Langstaff and Morrill 1981. Published by permission of the International Human Resources Development Corporation Press.)

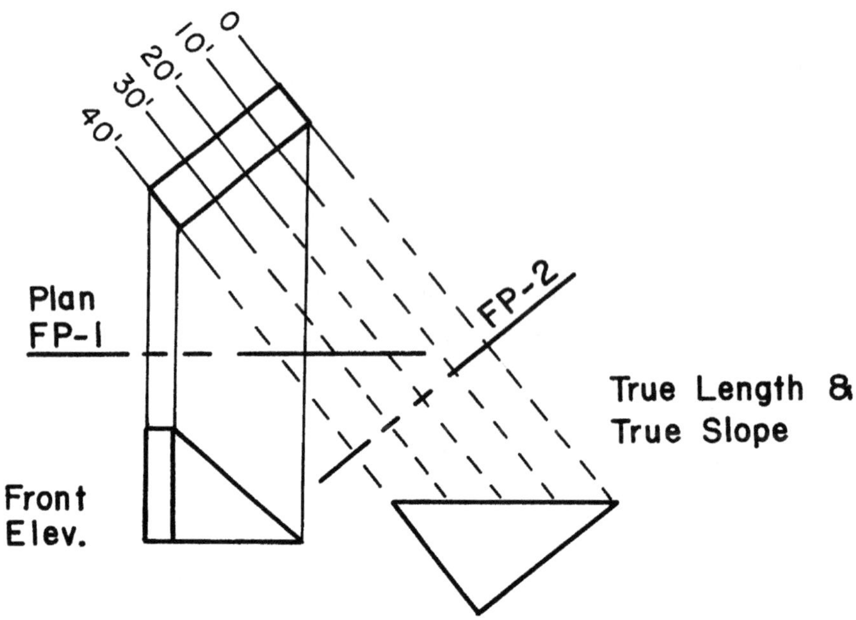

Figure 6-42 A 3D RAM (reservoir analysis model) can be constructed to measure true length and slope by constructing the folding plane at 90 deg to dip. (From Boeckelman, unpublished. Published by permission of author.)

to reduce clutter on the model. The second choice is to hand-trace the reduced log curves on the vertical line at each well location. The choice is more a matter of preference.

Our interest, in this case, is to hang the section on the top of the 9200-ft Sand in order to study and review the entire 9200-ft Sand package. At Well No. 3, for example, the top of the 9200-ft Sand is at a depth of −9150 ft. Since the referenced datum is −9100 ft, the first depth point is 50 ft below the datum. The log or hand-drawn curve(s) is positioned so that the top of the sand is 50 ft below the reference grid at Well No. 3, as shown in Figure 6-43d. For Well No. 1, the top of the sand is at a depth of −9450 ft, which is 350 ft below the referenced grid datum of −9100 ft. Therefore, the top of the sand represented on the actual log or hand-drawn curve(s) is placed 350 ft below the grid at the Well No. 1 location. The depth below the referenced datum is calculated for each well, and the appropriate log curve(s) is posted at that location.

Now what about the 18 other grid intersections? Using the grid overlay on the structure map of the 9200-ft Sand (Fig. 6-43b), determine the structural elevation for the top of the unit at each grid intersection, calculate the difference between the unit top and the reference depth, and post the structural top at the appropriate depth on the vertical line at each separate grid intersection. After posting the elevation for the structural top of the 9200-ft Sand at each of the 24 grid intersections, connect the points with straight lines as shown in Figure 6-43d. These connected lines represent the top of the 9200-ft Sand on the 3D model.

The control points connected with straight lines may not accurately represent the attitude of the sand top, since the true top of the sand has curvature between many control points. The process can be improved further by posting the depth values for all structure contour/grid intersections and connecting these depths. Approximately 30 depth control points for contouring the top of the sand are added using this technique. With these additional control points, the top of the sand can be contoured to more closely represent the true attitude of the sand. Also, grid lines can be added at strategic locations to refine the RAM as desired. The amount of detail depends on the individual interpreter and the intent of the reservoir model.

From the log sections for the six well locations, the base of the sand can be recognized. But what about the base of sand at the other 18 grid intersections? We must now repeat the process

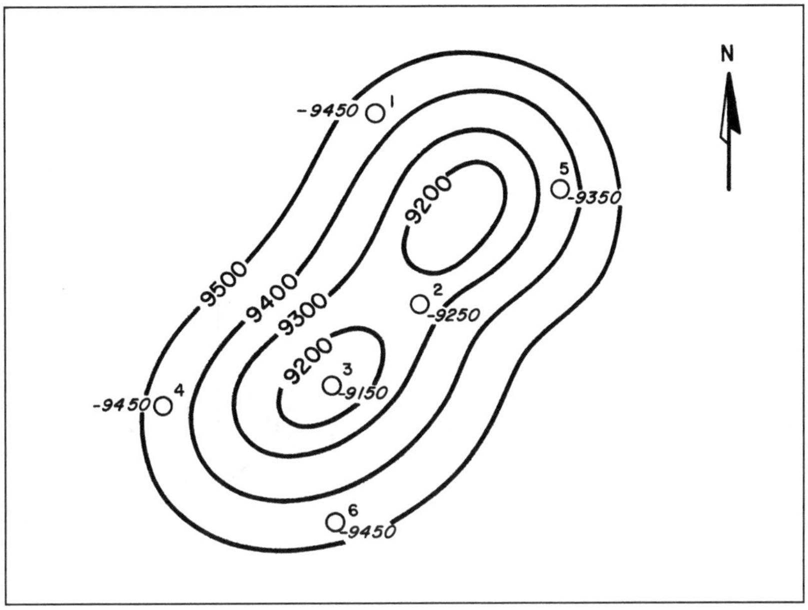

(a)

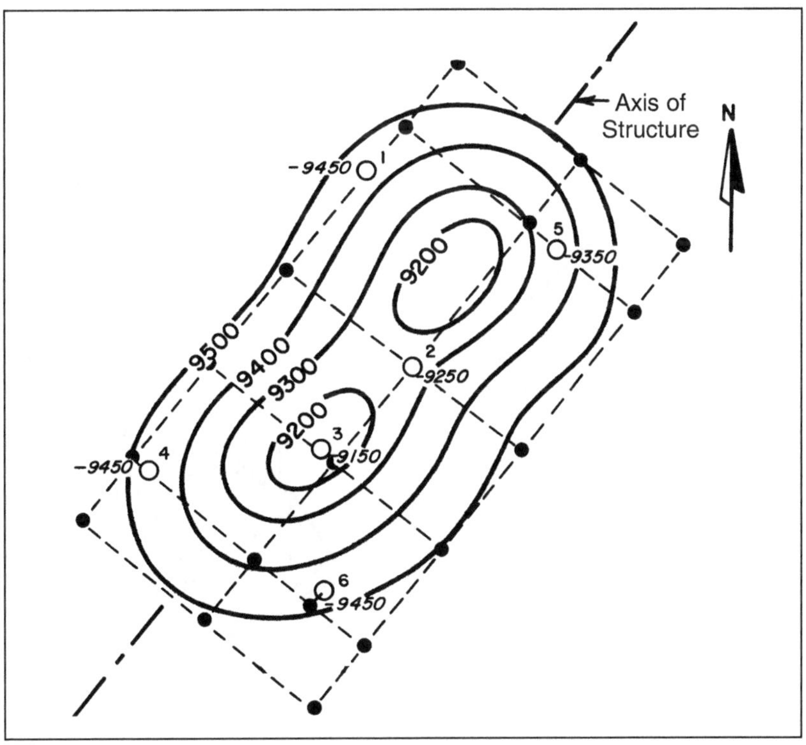

(b)

Figure 6-43 (a) Completed structure map on the top of the 9200-ft Sand. (b) RAM grid positioned at 90 deg to dip. Grid lines must intersect each well to ensure the use of all geological data. (c) Choose a datum and construct lines from the datum vertically straight down from each well and grid location. (d) Post the elevation for the top of the 9200-ft Sand at each grid intersection and connect the points with straight lines representing the top of the 9200-ft Sand. (e) Overlay the grid system on the base of the sand map and post the base of the sand at each grid intersection and each well location. (f) Begin detailed construction of the sand and shale data for each panel starting at each well location, working outward in all panel directions. (From Boeckelman, unpublished. Published by permission of author.)

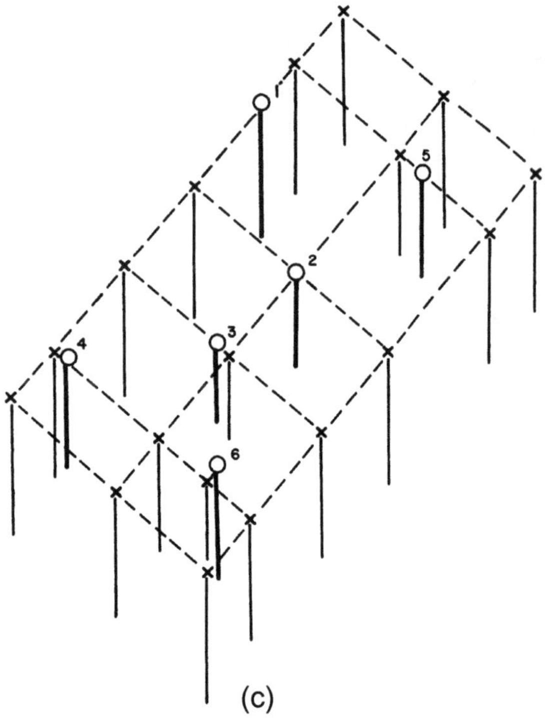

(c)

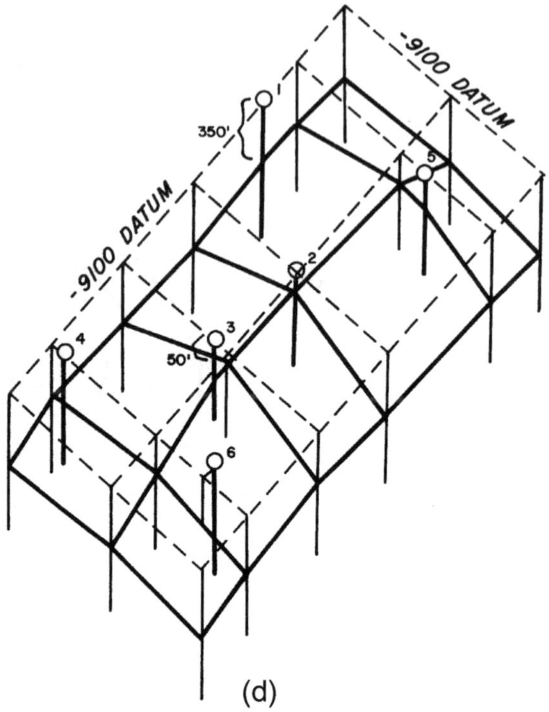

(d)

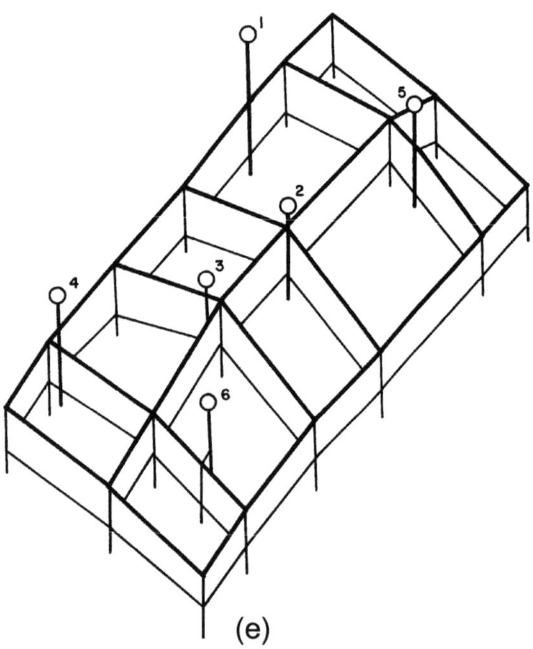

(e)

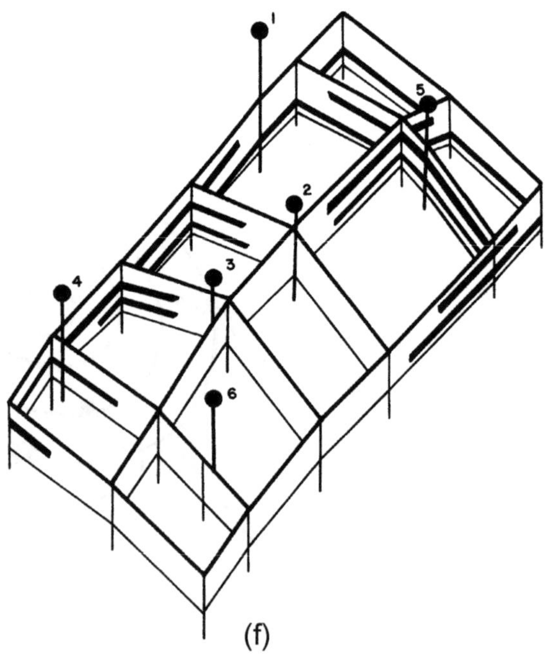

(f)

Figure 6-43 *(continued)*

outlined above for the structural base of the sand. Overlay the grid over a structure map on the base of the sand, determine the structural elevation of the base of the sand at each grid intersection, and mark this base on each vertical intersection line. Connect all the depth points for the base of the sand, and the 3D panels for the 9200-ft Sand are complete, as shown in Figure 6-43e. Notice that all the panels are at right angles to one another, so very little data on any particular panel are hidden by other panels.

Finally, using the net sand and shale data from each of the electric logs, begin construction of the model panel by panel. Start at each well location and work outward in all panel directions to prepare a complete interpretation of the sand/shale distribution for the 9200-ft Sand. This sand/shale work can be done in a general way or detailed based on each foot of analyzed electric log (Fig. 6-43f).

This 3D RAM can be used to accurately evaluate the distribution of sand and shale and to aid in developing a depositional model for the overall sand package as well as the individual members. If hydrocarbons are present, water levels can be plotted and an entire color scheme developed to represent sand, shale, hydrocarbons, and water. There are advantages to this model not associated with other types of cross sections, fence diagrams (Fig. 6-44), or isometric projections.

1. True length and slope measurements can be made.
2. More panels are available for study, providing more detailed and accurate analysis.
3. Because of the right-angle grid system, less data are obscured in this model as compared to other types of panel diagrams.
4. By measuring total net sand at each grid intersection and well location, it is possible to construct a more accurate net sand and net hydrocarbon isochore map.
5. The overall appearance of the model is more pleasing to the eye and is easier to understand and follow.

Figure 6-45 is another example of a completed 3D reservoir model. Compare this model with the fence diagram in Figure 6-38, which is constructed from the same well data. Notice how clearly the structure is represented in the 3D model (Fig. 6-45) compared to the fence diagram (Fig. 6-38). Also observe how much more data are presented with minimal obstruction in the 3D model. Finally, true length and slope can be measured on the model, which cannot be done with the fence diagram.

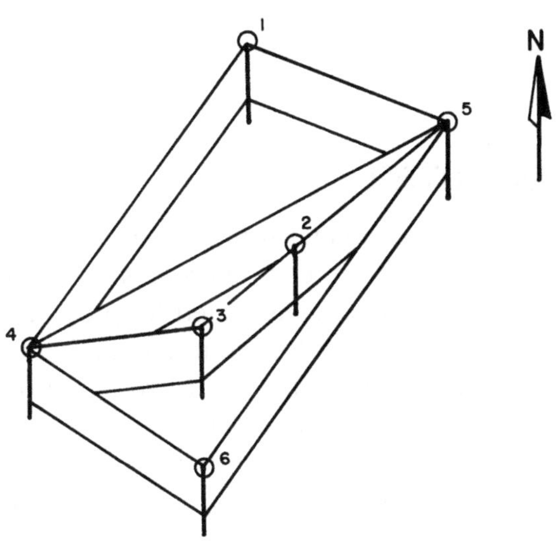

Figure 6-44 Typical fence diagram layout for the 9200-ft Sand. (From Boeckelman, unpublished. Published by permission of author.)

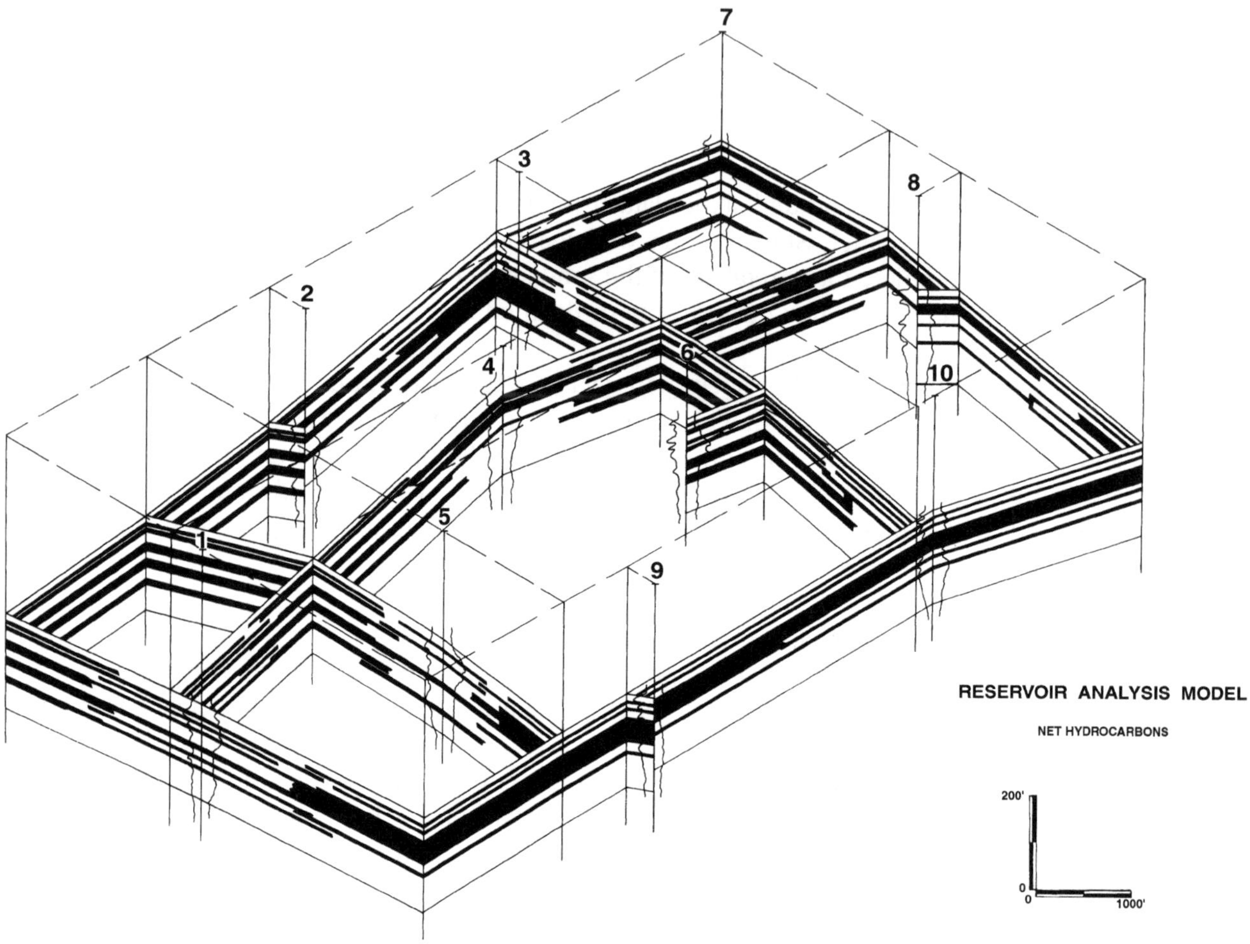

Figure 6-45 Compare this 3D RAM with the fence diagram in Figure 6-38, which was prepared from the same well data.

The use of 3D reservoir analysis models has become less common in recent years as the increased availability of computing power has made the construction of 3D geocellular models practical. These geocellular models not only capture the distribution of lithologies and reservoir and fluid properties in three dimensions, they also allow for the simulation of fluid flow within petroleum reservoirs.

CROSS-SECTION CONSTRUCTION USING A COMPUTER

The most significant advantage of building cross sections using a computer is that the geologist is actively viewing and interacting with the cross section during the interpretation process. In other words, rather than viewing the cross-section display as a final product, constructed for presentation purposes, cross-sectional views can be created at all phases of a geological interpretation to help guide the final result. For example, on the computer the geologist can set up cross-section displays that are designed primarily for correlation purposes and quickly re-datum and respace these cross sections to view the geometric implications of the evolving correlation interpretation. Cross sections can be constructed for many purposes, and with the computer, they can be created on-the-fly in a matter of seconds.

Correlation Cross Sections

During the log correlation phase of subsurface interpretation, the geologist can build cross-section displays, optimized for log correlation, that contain several dozen wells each. Figure 6-46 shows an eight-well correlation cross section through Pliocene sediments. Correlation sections can be set up to show any track of a series of electric logs. By limiting the view to just one track, several dozen closely spaced logs can be viewed on a computer screen at one time, and log picks can be rapidly correlated from the type well across these specialized cross-section panels. The track chosen for Figure 6-46 is the SP/GR track. Note the missing sections, due to faults, displayed in Wells No. 2, 3, 5, and 7 in the stratigraphically datumed cross section. Picking and "gapping" faults during the correlation phase of a study greatly facilitates sorting out the stratigraphy in complexly faulted regions (see section Computer-Based Log Correlation in Chapter 4). Deviated wells need to be corrected to a true vertical depth display utilizing the directional survey.

Stratigraphic Cross Sections

Stratigraphic cross sections are designed to reconstruct original depositional geometries prior to structural deformation. Computer-based cross-section software facilitates the construction of stratigraphic cross sections in a number of ways. The ability to quickly select any correlation and rehang the section on the selected datum allows the geologist to reconstruct, with the click of a button, depositional geometries for successive correlations. To properly reconstruct accurate stratigraphic geometries, wells can be spaced proportional to map distance (or projected to a line of section). For wells with deviated wellbores, the program used should be able to adjust the

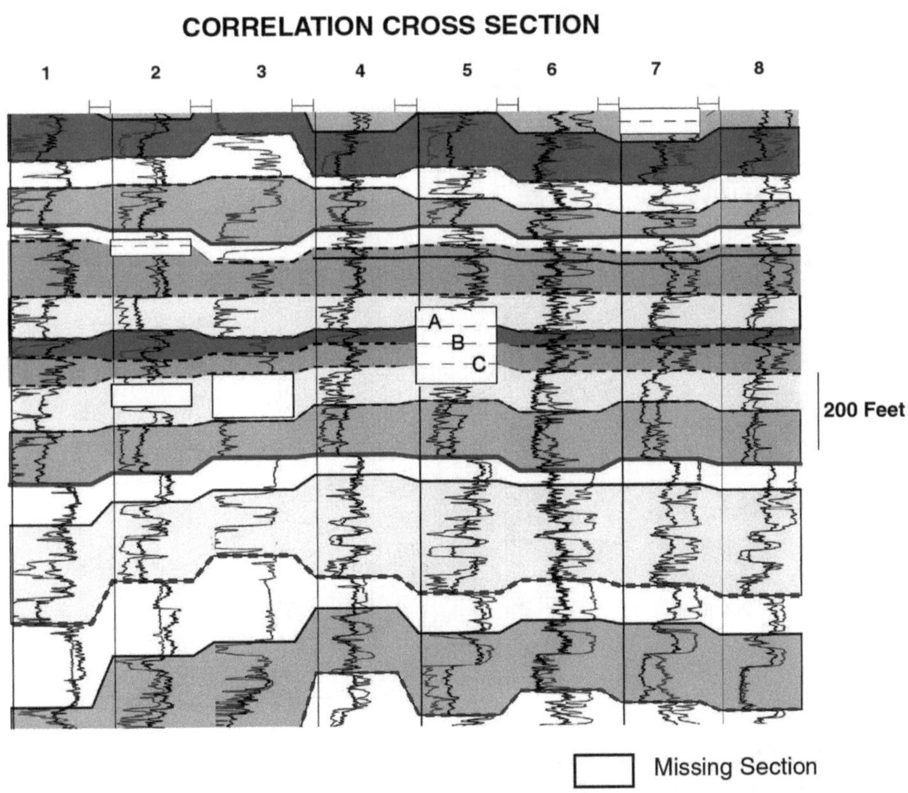

Figure 6-46 Eight-well, closely spaced correlation cross section set up to show the SP/GR track of a series of electric logs. The orientation of the section is constantly changing as all wells in the field study were included for correlation purposes. The missing section due to faulting is restored using a fault-gapping tool. Wells numbered 1 through 8 at the top of the section are all Mesa Petroleum-operated wells. (Published by permission of W. C. Ross and A2D Technologies.)

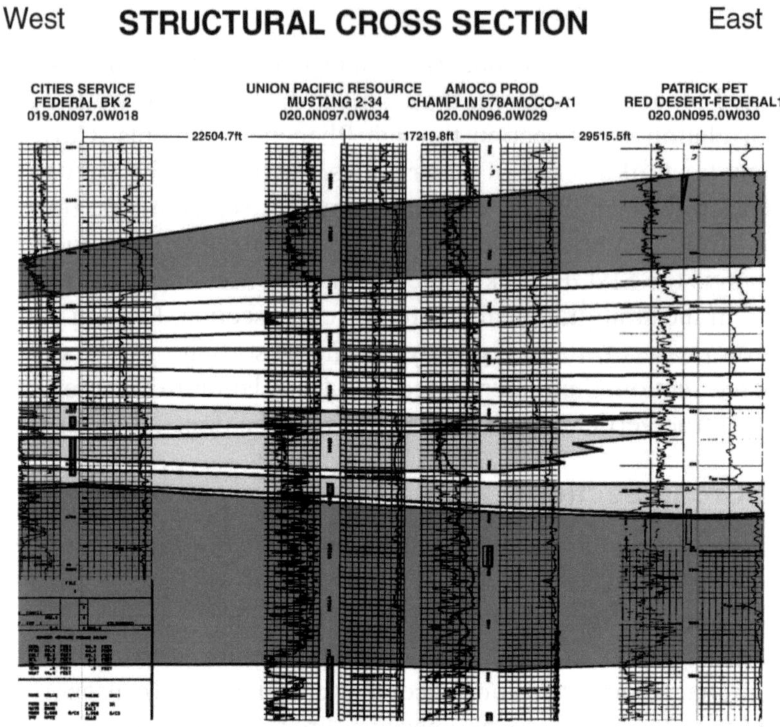

(a)

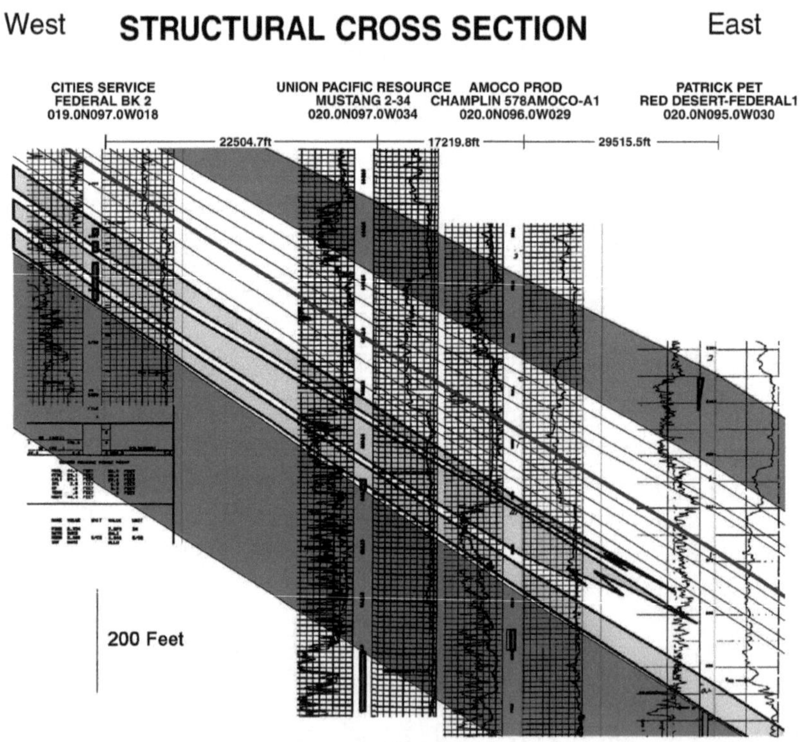

(b)

Figure 6-47 Stratigraphic (a) and structural (b) cross sections of Almond Formation and Lewis Formation stratigraphic relationships in the Green River Basin of Wyoming. Wells are spaced proportional to map distance. Facies changes and sandstone pinchouts, interpreted in the inter-well space in stratigraphic mode (a), are automatically repositioned properly in structural mode (b), and tops picked in the time-stratigraphic gap are used to determine and draw the position of onlap and erosional truncation terminations above and below the sequence boundary, respectively. (Published by permission of W. C. Ross and A2D Technologies.)

spacing of the wells to be proportional to map distance using the latitude and longitude positions along the chosen datum. Cross-section software should support the picking and termination of correlations, facies boundaries, and unconformities in the space between wells (Figs. 6-47 and 6-48). As wells are reordered or the section re-datumed from stratigraphic to structural views, the inter-well interpretation shifts appropriately (Fig. 6-47a and b).

Missing Section in Stratigraphic Cross Sections

Building stratigraphic cross sections can be enhanced by restoring those portions of the cross section that are missing due to erosion, nondeposition, or faulting. Unconformity/fault tools allow geologists to split or gap logs proportionally to reflect the amount of section lost due to erosion or faulting. An example of restoring a missing section due to channel erosion is illustrated in Figure 6-48a, where a siliciclastic package of marine shales and shelf sandstones is truncated by an incised valley system. Figure 6-48b shows the same system where the logs have been split, or gapped, at an unconformity (sequence) boundary. The gap space represents missing time due to erosion (in the valley) and nondeposition (in the interfluves). Correlations from the marine shale interfluve are extended into the missing section gap and represent an estimate of stratigraphic geometry prior to erosion. The ability to pick tops within these gaps allows geologists to hang cross sections on tops that have been cut out by erosion. Tops picked within the channel feature are extended into the interfluve space. When the gaps are closed, the software automatically determines the appropriate onlap and truncation geometries in the inter-well space, above and below the unconformity surface, respectively (Fig. 6-48a).

Structural Cross Sections

Structural cross sections are built by geologists as a complement to structure maps. Creating structural cross sections using a computer is simply a matter of selecting a structural datum and a well spacing proportional to map distance or projected to a line of section. These sections can be produced on-the-fly to check the validity of an evolving interpretation.

To include fault interpretations in a structural interpretation, the geologist first picks faults using an arbitrary, generic fault-naming scheme (Fig. 6-49). In this phase of interpretation, fault picks and missing section estimates can be interactively interpreted in "fault gap mode" by splitting the log and dragging down the fault gaps with a mouse-controlled fault tool (Fig. 4-33c and d in Computer-Based Log Correlation in Chapter 4). Once the generic fault locations in the wells are in place, the geologist can correlate or assign them with the same fault name (Fig. 6-50). These correlated faults each possess a missing section or vertical separation value interpreted by the geologist when the fault was first picked and after the gap was "pulled down" in gap mode. When the faults are correlated, these vertical separation values are interpolated from well to well to create an overall vertical separation fault function along the length of the fault. When the geologist instructs the software to create a structural cross section, the fault gaps are automatically closed and correlations intersecting the fault are automatically offset utilizing the vertical separation value at the fault-correlation intersection point (Fig. 6-51). The appropriate offset is determined using logic described in Chapter 4.

In addition to being a significant time-saver in the construction of structural cross sections, the automated offsetting of correlations provides a powerful constraint on the attitude of beds in the inter-well area, including beds cut out by the fault (see offsets of correlations A, B, and C, which are missing at cross section Well No. 3 in Fig. 6-51).

The ability to interactively interpret fault location (depth) in a well and the amount of missing section (vertical separation), and then to pick markers within these missing-section gaps, provides the geologist with a powerful set of tools for subsurface interpretations in regions dominated by normal faults. Figure 6-52a and b show stratigraphic (with missing-section gaps) and structural versions of a cross section from the Wilcox of South Texas (Edwards 2001). The large faults have

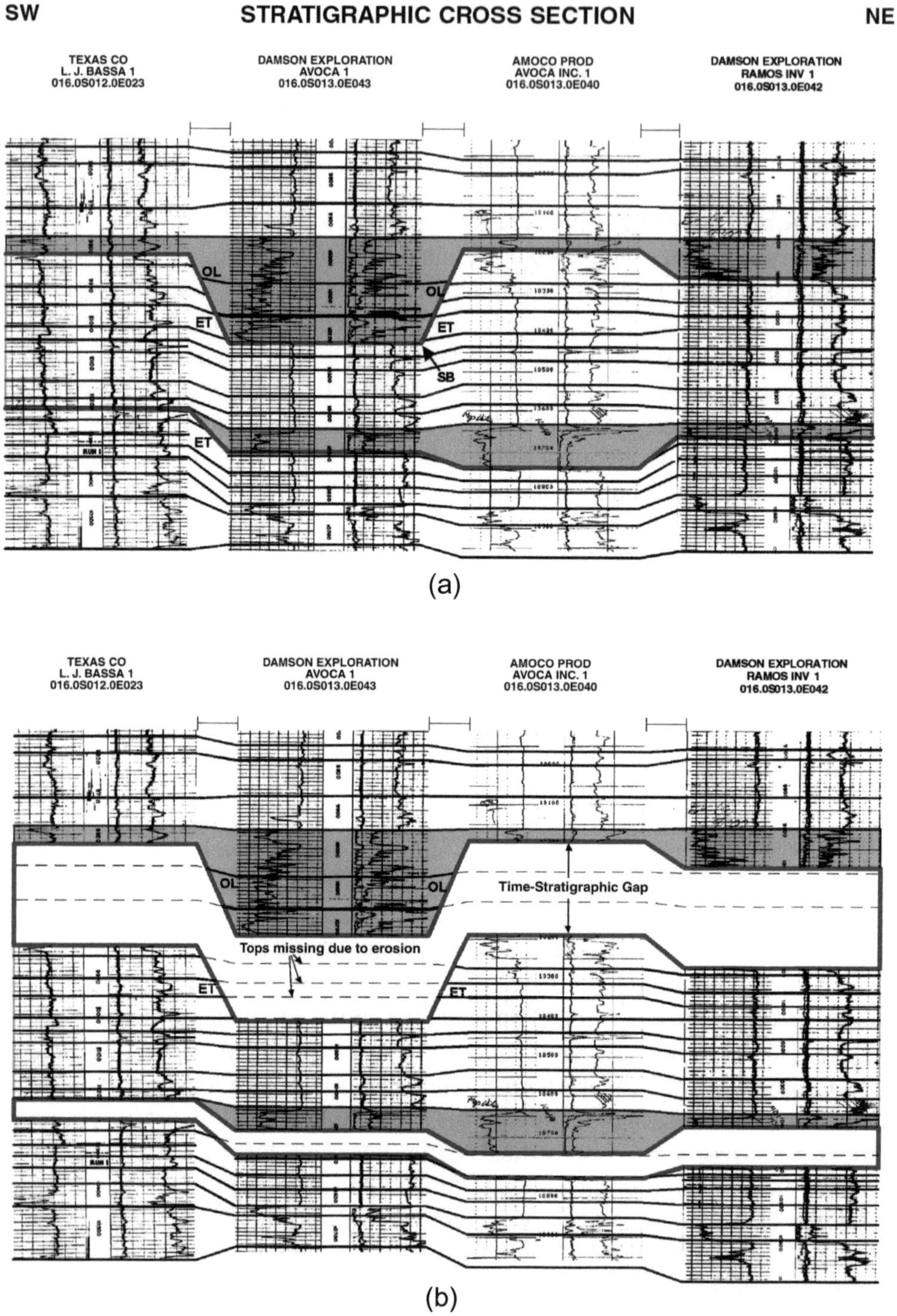

Figure 6-48 Stratigraphic cross sections of shallow marine and incised valley-fill sediments. (a) Shallow marine shelf facies are shown truncated by an incised-valley system. The base of the incised-valley system is a sequence boundary (SB). Erosional truncation (ET) of shelf strata below the sequence boundary and onlap (OL) of channel-fill strata above the unconformity are automatically positioned using the timeline geometries shown in (b). (b) The logs have been pulled apart, or "gapped," at the sequence boundary to represent a missing section due to erosion and/or nondeposition. The intersection of the sequence boundary and tops picked in the time-stratigraphic gap are used to determine and draw the position of onlap and erosional truncation terminations above and below the sequence boundary, respectively. (Published by permission of W. C. Ross and A2D Technologies; and AAPG©1990, reprinted by permission of the AAPG whose permission is required for further use.)

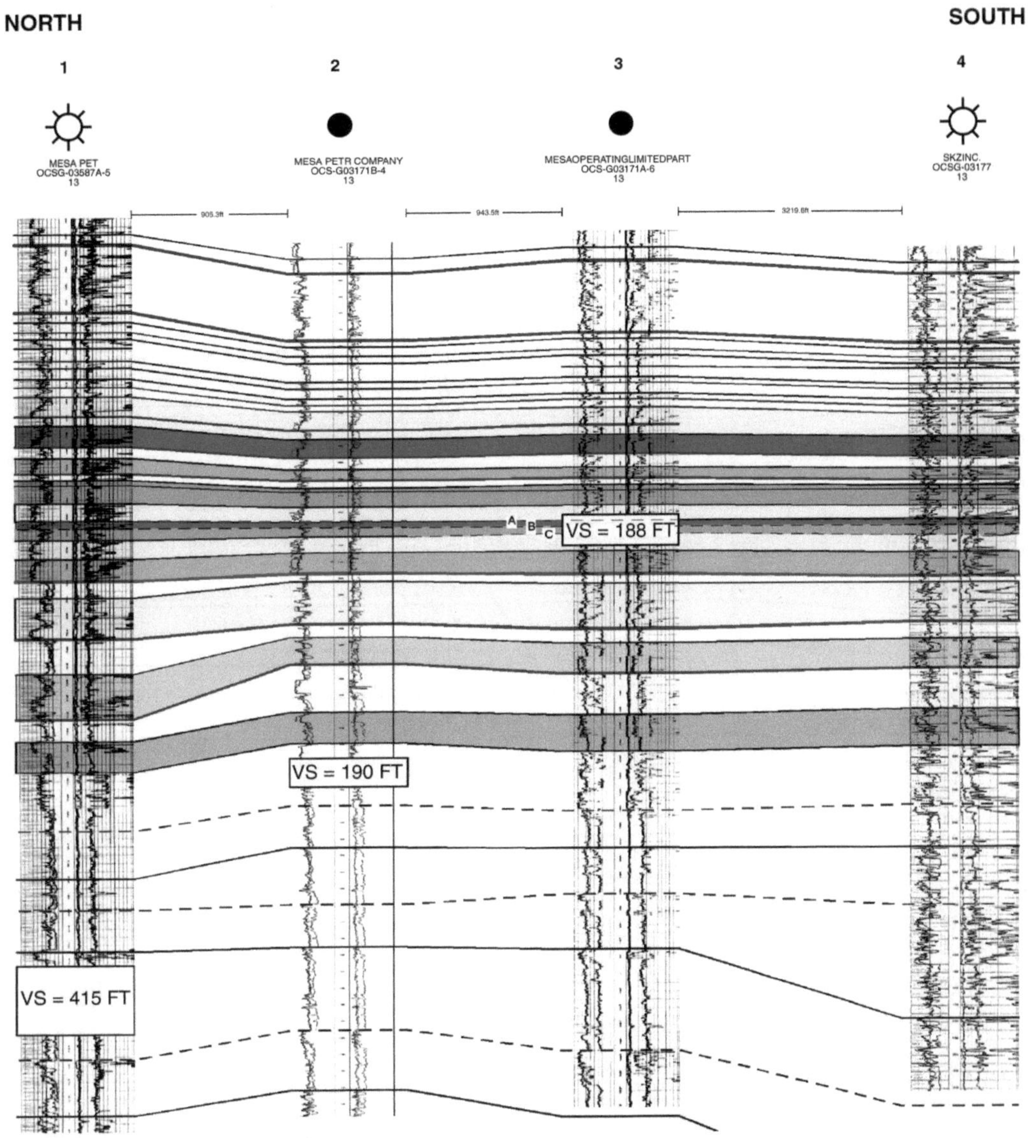

Figure 6-49 Stratigraphic cross section of Pliocene sediments. The missing section fault gaps show the results of fault identification through detailed correlation and vertical separation (VS) analysis. Estimates of a missing section (VS), due to faulting, have been performed at three locations and are shown within each missing section fault gap. Fault names are generic. As a result, no fault correlation is required at this stage of fault interpretation. Correlations A, B, and C are picked within the missing fault gap on the Mesa Petroleum OCS-G 03587 A-6 well. (Published by permission of W. C. Ross and A2D Technologies.)

cut out hundreds of feet of section. Tops cut out by the faults have been picked within the missing-section gaps across the entire cross section. Missing-section stratigraphic cross sections are useful for removing structure and providing an undistorted view of the correlation framework (Fig. 6-52b).

One challenge with constructing cross sections on a computer is controlling the projection of wells into the section. Most if not all software systems project wells into a section using the normal-to-section projection method. As discussed earlier in this chapter, this projection method is not the preferred method in many cases. Some software systems allow the interpreter to manually project wells using alternative projection systems such as plunge projection, structural strike projection, fault-strike projection, and so on. Using these alternative projections, which are

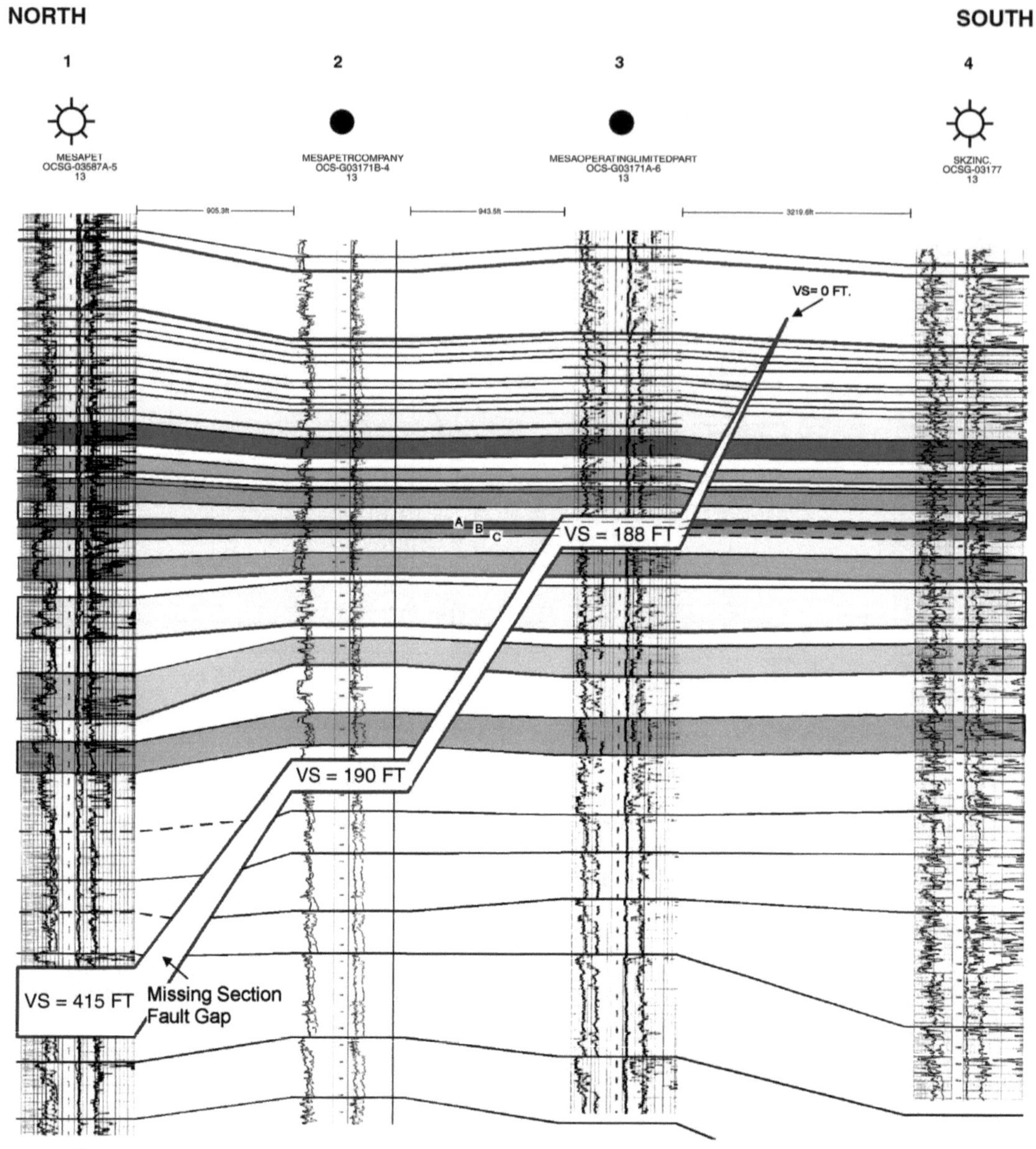

Figure 6-50 Stratigraphic cross section from Figure 6-49 showing fault geometry after geologist assigns a specific fault name to each generically named fault. (Published by permission of W. C. Ross and A2D Technologies.)

preferred in many instances, requires a geologist to manually control how wells are projected, assuming that the software package being used allows this manual control, as the software does not know the plunge or strike of a structure. Having to project wells by hand obviates some, but not all, of the advantages of constructing cross sections on the computer.

FAULT-SEAL ANALYSIS

The evaluation of potential hydrocarbon seal along a fault surface can be critical to exploration and development prospects and to reservoir evaluation. Some aspects of this analysis that are typically considered are (1) juxtaposition of rock units, which we can classify as a structural or geometric aspect; (2) lithologic characteristics of the fault zone such as shale smear, fault

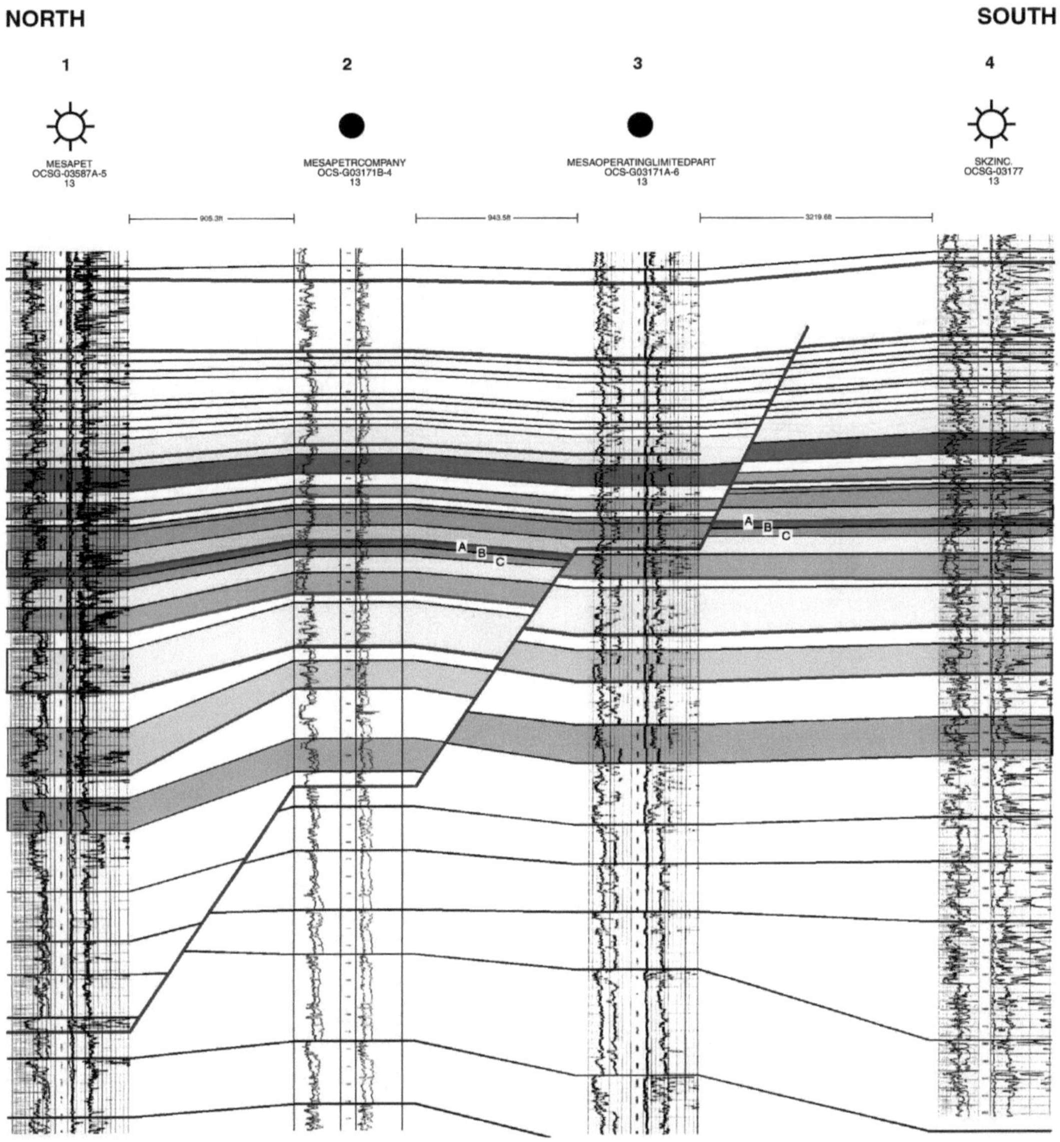

Figure 6-51 Structural cross section with fault and bed-offset geometry resulting from application of vertical-separation fault function to each fault-correlation intersection. Note upthrown and downthrown bed offsets of missing correlations A, B, and C to the left and right of cross-section Well No. 3. (Published by permission of W. C. Ross and A2D Technologies.)

gouge, and the rock types that are juxtaposed along the fault; (3) pressure differentials across the fault zone; and (4) possible vertical conductivity of the fault. Given the context of this book, we address only techniques by which you can determine juxtaposition of rock units across a fault. This determination provides the structural basis for a complete fault-seal analysis. We describe in detail a manual structural method and summarize a computer-based evaluation of juxtaposition.

Studies of faulted reservoirs in the Gulf Coast Tertiary (Smith 1980) indicate that sands that are juxtaposed against themselves across a fault are very unlikely to be sealed at that fault. Sands that are juxtaposed across a fault against a sealing lithology are very likely to be sealed at the fault. Sands that are juxtaposed across a fault against a sand of a different age are sometimes sealed at the fault. Analysis of the sealing potential of a fault generally begins with a study of juxtaposition of beds across the fault.

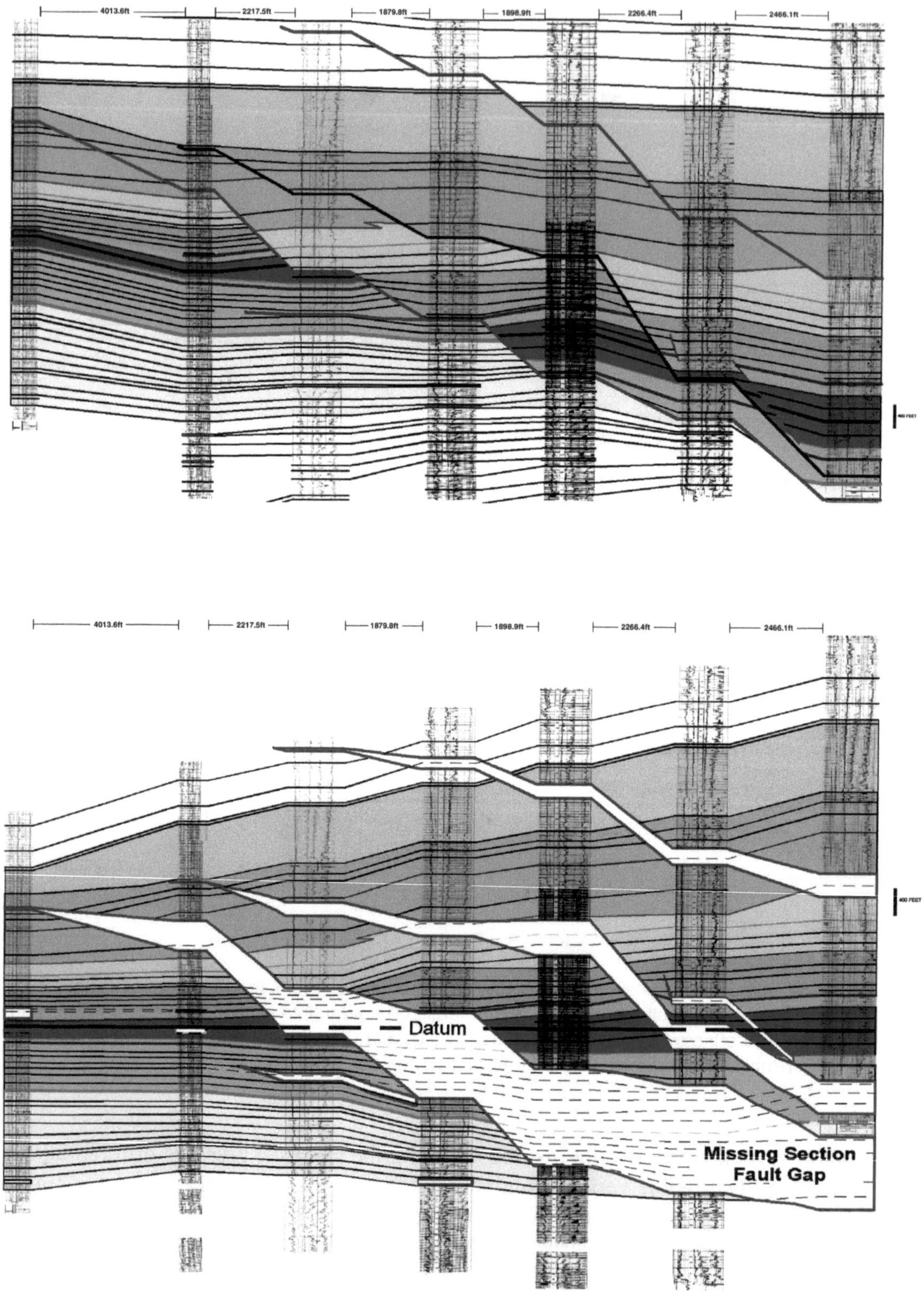

Figure 6-52 (a) Structural cross section of faulted Wilcox strata in Zapata County, South Texas. Correlation offsets were calculated using interpolated vertical separation information interpreted at fault positions in wells. (b) Stratigraphic cross section with a missing section, due to faulting, shown as gaps. Note increasing size of missing-section fault gap in a down-dip direction for lowermost fault. This pattern depicts increasing amounts of vertical separation along the fault with depth. (Published by permission of W. C. Ross and A2D Technologies.)

Analysis of juxtaposition typically involves a depiction of the fault surface showing the traces of stratigraphic units on the footwall and hanging wall. The interpreted fault surface itself may be used, or a horizontal projection of the traces can be made from the interpreted fault surface onto a vertical plane parallel to the strike of the fault. Either method portrays the juxtaposition of units across the fault. Allan (1980, 1989) is generally credited with introducing the technique into the oil industry, although the first widely available description of Allan's technique was by Downey (1984). Rippon (1985) described the construction of fault plane sections for use in the coal mining industry in the United Kingdom. Bouvier et al. (1989) apply a technique known as *fault slicing* (seismic sections parallel to the trace of a fault) to examine the seal characteristics of faults in offshore Nigeria. Fault surface sections are used extensively to evaluate sealing characteristics of faults (e.g., Broussard and Lock 1995; Yielding et al. 1997).

Fault Surface Sections Constructed by Hand

Fault surface sections, variously referred to as fault plane sections, fault profile sections, juxtaposition diagrams, or Allan sections, are simply cross sections with the fault surface as the plane of section. Allan (1980, 1989) proposed this idea to show the juxtaposition of permeable units along the length of a fault. A standard cross section shows the juxtaposition of units only at a single place along the fault. The objective is to build a cross section that shows where reservoir rocks might communicate across the fault and use this information to evaluate the trapping potential of the fault. Plotting permeable intervals (based on well logs and seismic data) for adjacent fault blocks on the same cross section shows the separation and juxtaposition of the permeable units across the fault. For convenience, the fault surface section typically represents a *vertical plane* onto which the rock units are projected from the fault surface.

Structure maps on the top and base of permeable units are the basic data needed (for simplicity, we refer herein to such units as "sands"). The method requires structure maps with the fault traces accurately mapped, as described in Chapter 8, because the fault traces show the location and depth of the intersection of the sand and the fault surface. To build a fault surface section, you could just overlay cross-section paper on a top-of-sand structure map and trace the fault traces, both upthrown and downthrown, from the map onto the cross section. This would be the intersection of the top of sand with the fault surface along the length of the fault. Then overlay the same cross-section paper on the base-of-sand map in exactly the same position and copy the fault traces from the map. Any overlap of a top of sand with a base of sand would show juxtaposition of sand across the fault. The difficulty with this method is that the vertical scale of the cross section is the same as the horizontal scale. At a map scale of 1 in. = 1000 ft, a 100-ft sand would appear only about one-tenth of an inch thick on the cross section. This would make interpretation difficult. Many cross sections use some vertical exaggeration, and this is generally necessary for fault surface sections. As discussed earlier in this chapter, this distorts the true geometry of the units, but the purpose of a fault surface section is only to show juxtaposition of units, and such distortions do not interfere with this purpose. Also, the curvature of the fault surface introduces other distortions in the cross section, but they do not adversely affect fault surface sections either. Finally, the method itself introduces some thickness distortion as structure maps are maps of surfaces projected into a horizontal plane; thus the top and base of sand on the cross section are not in the plane of the fault.

In practice, the vertical scale of a fault surface section is in depth, so the top and base of sand are projected into a vertical plane that is more or less parallel to the strike of the fault surface. Broussard and Lock (1995) referred to this vertical plane as the Allan plane. Figure 6-53 shows a profile of this vertical plane on a standard cross section, with the projection of the sands onto it from the intersections of the top and base of sand with the fault surface. No juxtaposition exists in this case. Figure 6-54 demonstrates, with the same structure, a *pitfall of using sand thickness* to determine the base-of-sand intersection with the fault surface. Note that the apparent sand

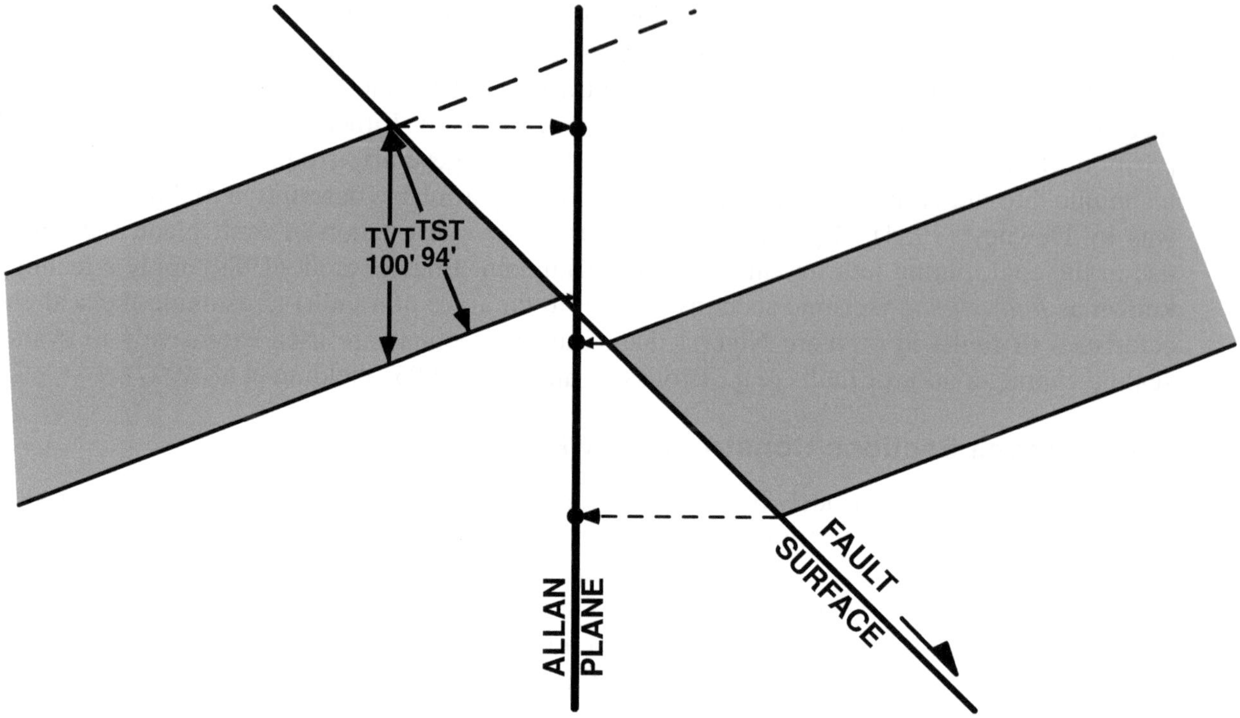

Figure 6-53 A cross section of a faulted sand bed showing the correct projection of the sand from the fault surface to the vertical Allan plane. (After Broussard and Lock 1995. Published by permission of the Gulf Coast Association of Geological Societies.)

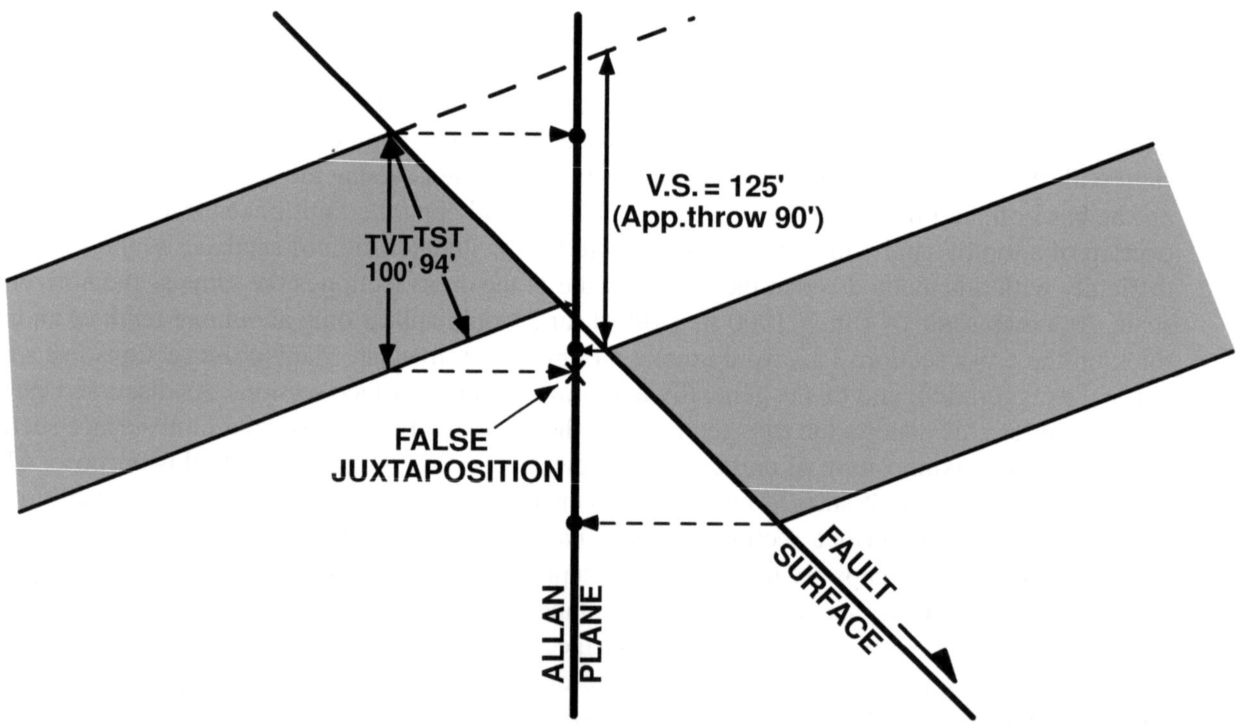

Figure 6-54 A cross section of a faulted sand bed showing an incorrect projection of the sand from the fault surface to the Allan plane. False juxtaposition can be indicated by using bed thickness for the base of sand or by comparing apparent throw to bed thickness. (After Broussard and Lock 1995. Published by permission of the Gulf Coast Association of Geological Societies.)

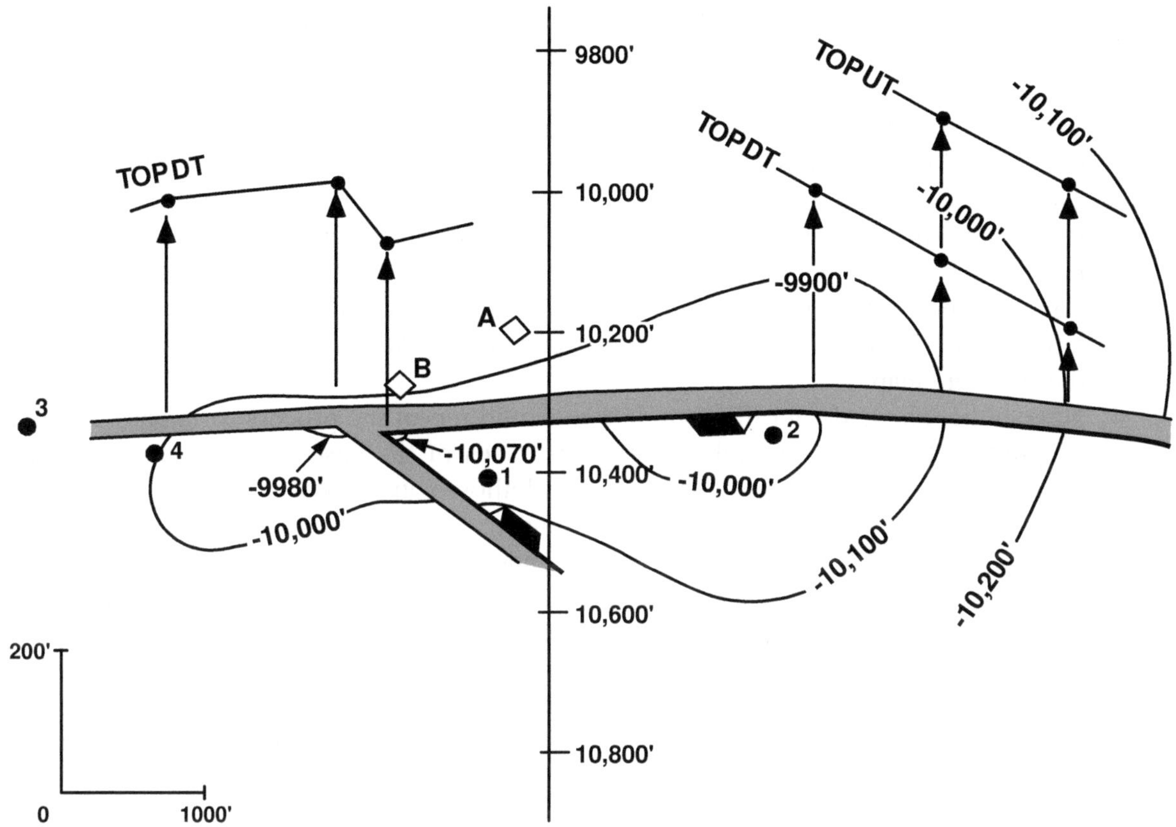

Figure 6-55 A fault surface section is constructed by overlaying depth-scaled cross-section paper on a structure map of the top of sand, with the depth scale perpendicular to the fault trace. The depths of the upthrown and downthrown traces for the top of sand are plotted on the cross section. Although not illustrated here, the traces of the base of sand are then plotted from a base-of-sand map. Wells 1, 2, 3, and 4 and Platforms A and B are used to register the cross section in the same place on both maps. (After Broussard and Lock 1995. Published by permission of the Gulf Coast Association of Geological Societies.)

thickness as projected onto the Allan plane is smaller than the TVT and the stratigraphic thickness of the sand. If, as a shortcut, the TVT thickness were used to determine base of sand within dipping beds at the fault, then it could indicate juxtaposition where there is none (Fig. 6-54). Also, if apparent throw is compared to TVT of the sand and found to be less than the TVT, a false juxtaposition could be inferred. To avoid these pitfalls, always use both the top and base of sand to portray the sand face correctly on the fault surface section.

Figure 6-55 illustrates the process of constructing a fault surface section or Allan section from structure maps. The steps are similar to building a finished illustration (show) cross section, as described elsewhere in this chapter, except the line of section is the fault surface projected into a vertical plane. To illustrate the method, we use one sand offset by a fault and evaluate the juxtaposition of the sand.

1. For the fault surface section, select a convenient depth scale that covers the range of depths along the fault for the top-of-sand and base-of-sand maps. Overlay the cross-section paper on the top-of-sand structure map with the vertical depth scale more or less perpendicular to the fault trace on the map. Mark some registration points, such as well locations, so you can locate the cross section paper in exactly the same place from map to map. In the example, the vertical exaggeration is 5:1.

2. Determine the depth of the fault trace for the top of sand in the upthrown block and plot it on the cross-section paper. For example, from the structure map, find where the −10,000-ft contour intersects with the fault trace upthrown and project parallel to the depth scale to the −10,000-ft depth on the cross section and mark a point. Continue along the upthrown fault trace, plotting the depths onto the cross section (interpolating the depth where necessary). Then connect the points with a line to plot the top of the sand at the fault.
3. Plot the depth of the fault trace for the top of sand in the downthrown block in the same manner. At intersections with other faults, estimate the depth in the corners of the intersection, plot these points on the cross section, and connect them with a straight line. This is the top of sand as it would be on the subject fault surface at the intersecting fault, according to the structure map.
4. Next, although we do not illustrate it, you would repeat the process for a base-of-sand map by plotting the upthrown and downthrown traces of the base of sand. The same registration points used for the top-of-sand map will allow you to locate the cross-section paper in exactly the same place on the base of sand map.
5. Finally, as shown in Figure 6-56, color or shade any overlap or juxtaposition of the sand, or sands if you are plotting more than one. Minor variations in contouring between the top-and base-of-sand maps usually should not affect the juxtaposition interpretation, but they should be evaluated for their significance to the accuracy of the maps.

Commonly, the fault displacement is sufficient to juxtapose different sands. For this situation, choose a depth interval suitable to the sand(s) of interest in one fault block and then

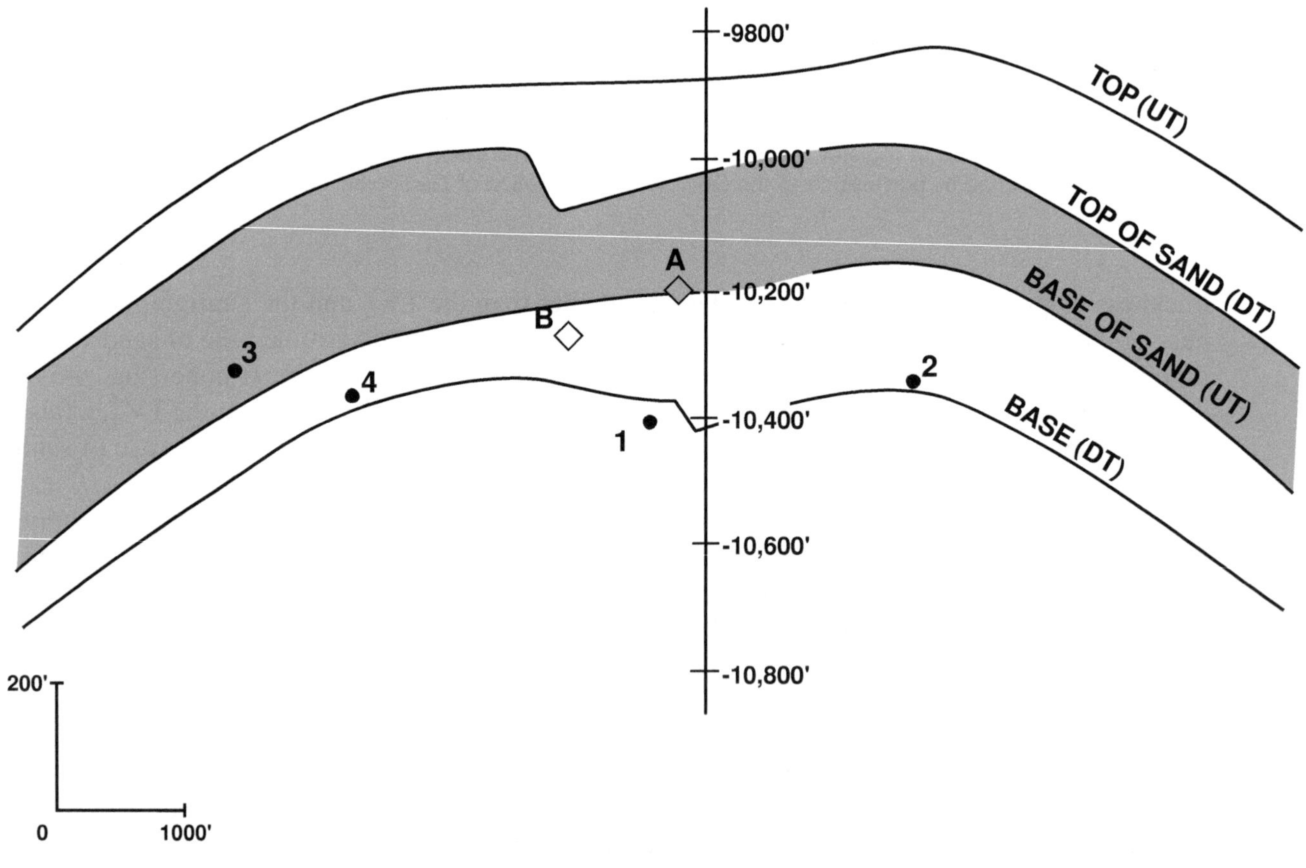

Figure 6-56 A fault surface section of a thick sand with the area shaded gray where the sand is juxtaposed to itself. (After Broussard and Lock 1995. Published by permission of the Gulf Coast Association of Geological Societies.)

determine which sand(s) in the other block fall within that depth range. Follow the steps above, using the structure maps of the top and base of each sand within its respective fault block.

Plotting the depth of the fault traces for the tops and bases of permeable intervals on cross-section paper gives a graphical display of the faces of the intervals against the fault surface that can easily be interpreted for juxtaposition of the intervals. Accurate fault surface maps and top- and base-of-sand maps are necessary for a valid fault surface section.

Computer-Based Fault-Seal Analysis

Modern visualization software provides a means of viewing a fault surface that was interpreted within a data volume on the basis of seismic and well data. An interpreter can examine the fault surface in isometric view, usually looking toward the footwall and in a direction perpendicular to the fault surface. The presentation considerably enhances the interpreter's ability to evaluate whether the fault surface is reasonable, based on aspects such as fault geometry and displacement patterns, which can be plotted on the surface (Needham et al. 1996). Furthermore, juxtaposition of stratigraphic units can be readily examined if horizon data, such as tops and bases of permeable units, at the footwall and hanging wall are plotted as traces on the fault surface. The displayed data are derived directly from the interpreted data volume.

Application of 3D Modeling and Visualization Techniques. The computer-based method in which the fault surface is modeled in three dimensions eliminates the projection errors that are common in the traditional method for constructing fault plane sections. More importantly, the 3D representation portrays the fault surface in a true structural context within the prospect itself and in relation to other faults.

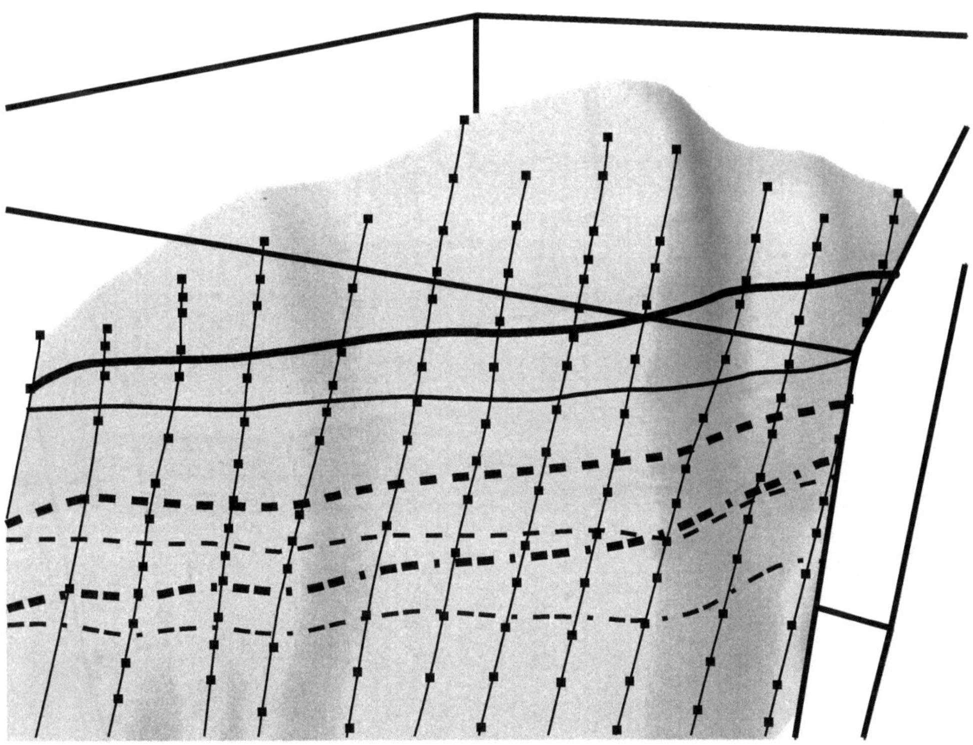

Figure 6-57 Footwall (thick lines and dashes) and hanging wall (thin lines and dashes) projected onto a normal fault surface. Vertical lines with nodes represent the fault surface intersection with the vertical seismic lines. (Published by permission of Badley Earth Sciences.)

A 3D fault surface is constructed by modeling the fault segments picked on vertical (in-line or cross-line) or horizontal (time slice) sections, either on a grid or by using triangulation algorithms (Fig. 6-57). The surface can then be viewed in perspective to provide an assessment of the plausibility of the interpreted fault geometry. Horizon terminations at their respective fault segments are used to plot footwall and hanging wall horizon traces on the perspective view of the fault surface, allowing the interpreter to observe in detail how reservoir layers are juxtaposed by the fault. The horizon separation information can also be used to calculate displacement-related values (throw, heave, vertical separation), which may be gridded or triangulated over the entire fault surface in order to produce displacement maps on the 3D representation of the fault surface. Fault displacement can also be used to calculate other attributes, such as fault-seal potential, which can be posted on the perspective view of the fault surface.

Fault-Seal Potential. Fault seal can be due to juxtaposition of reservoir rock against nonreservoir rock or to fault-zone material of high entry-pressure (e.g., clay smear in the fault zone) (Yielding et al. 1997). Both seal types are dependent on fault displacement and the nature and distribution of lithologies in the wall rocks (Figs. 6-58 and 6-59). The 3D representation of the entire fault surface allows fault-seal attributes to be mapped onto the fault surface. A detailed description of the various methods to predict fault-seal potential, with examples using fault surface sections, can be found in Yielding et al. 1997 and Yielding et al. 2010.

Identification of juxtaposition of reservoir units across the fault surface (Fig. 6-59) uses horizons mapped from seismic and well data and a detailed reservoir stratigraphy defined by isochores. Probable seals due to juxtaposition (sands on one side of the fault in contact with impermeable lithologies, commonly shales) can be clearly identified along the fault surface (Fig. 6-59). Similarly, overlaps of juxtaposed reservoir units (e.g., sand against sand) can be explicitly displayed (Fig. 6-60).

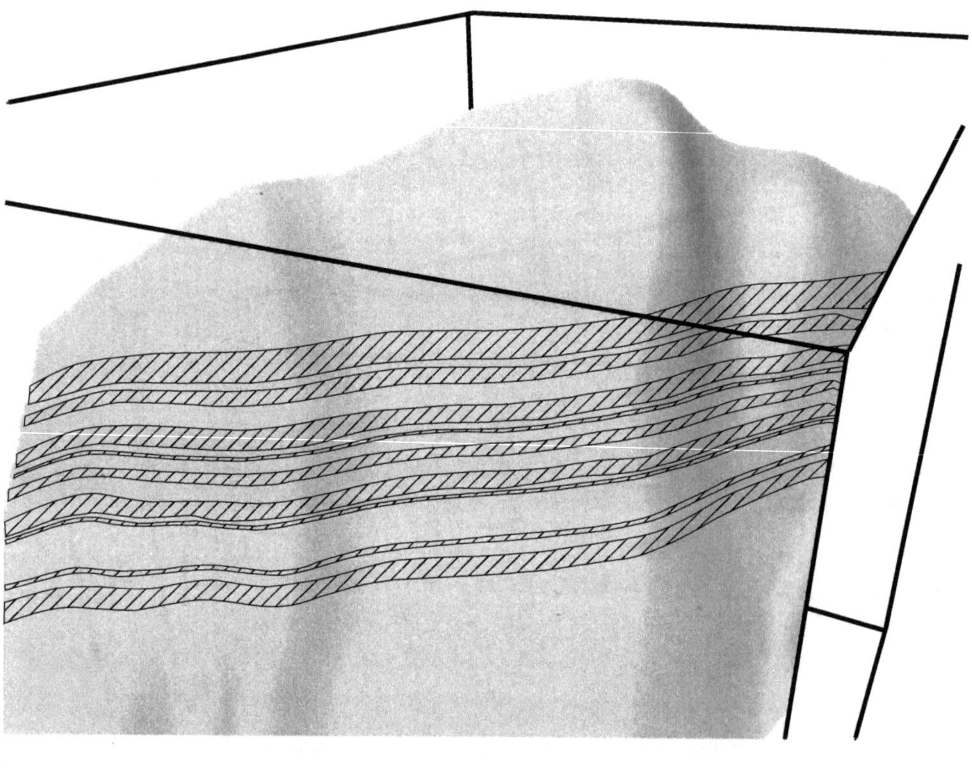

Figure 6-58 Fault surface showing sands in the footwall of a normal fault. (Published by permission of Badley Earth Sciences.)

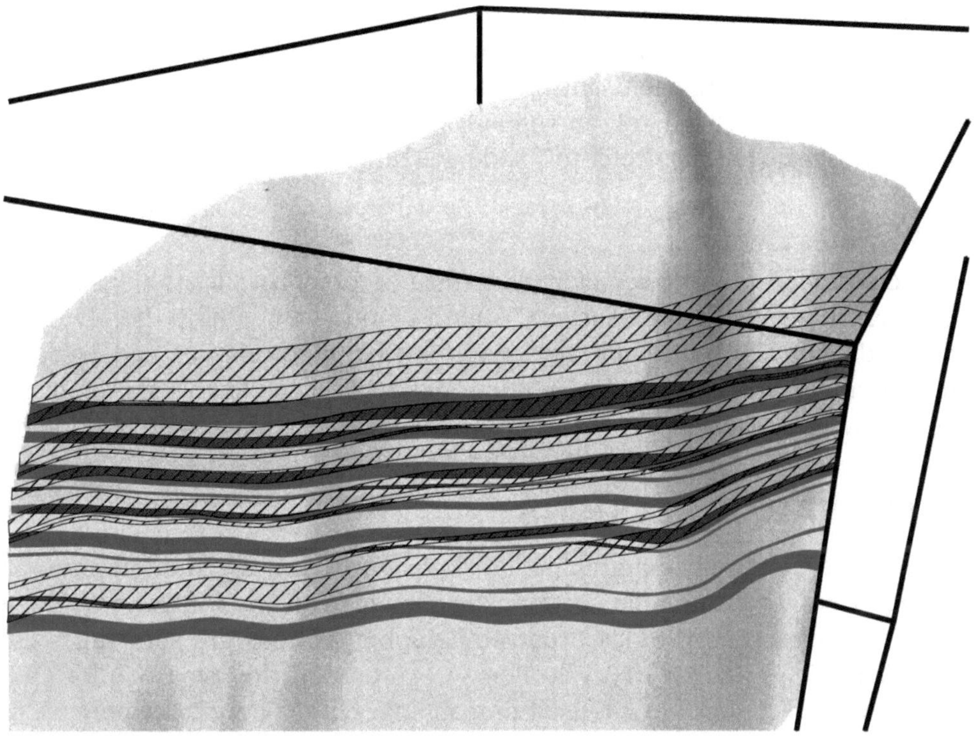

Figure 6-59 Fault surface showing sands in the footwall (hachure pattern) and the hanging wall (solid pattern), and sand juxtaposition where the two patterns overlap. (Published by permission of Badley Earth Sciences.)

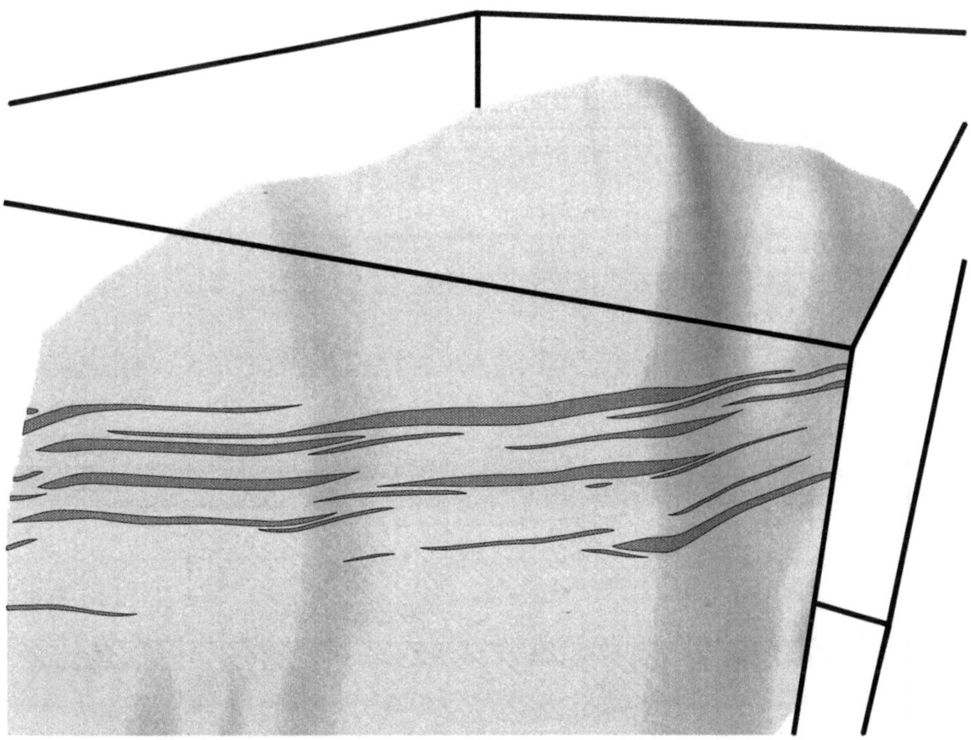

Figure 6-60 Fault surface showing the overlaps of juxtaposed reservoir units. (Published by permission of Badley Earth Sciences.)

Remember that *the accuracy of horizon interpretation is paramount* to the precision of this juxtaposition display and analysis and therefore to the ultimate depiction of the reservoirs.

Once the overlap of permeable units is determined, as in Figure 6-60, an analysis can be made of whether a given contact is likely to support a pressure difference due to lithologic attributes (e.g., gouge ratio, shale smear). Refer to Yielding et al. (1997, 2010) for a detailed discussion.

Conclusions

Fault surface sections, whether depicted as vertical sections parallel to the strike of the fault or as 3D representations of the fault surface, offer a unique view within the 3D data volume. The sections provide data about the fault, the quality of the interpretation, and especially the fault's potential for sealing hydrocarbons. The routine construction and analysis of fault surface sections should be adopted as a crucial part of the workflow in exploration and production projects.

CHRONOSTRATIGRAPHIC SECTIONS

We describe here a chronostratigraphic cross section used by stratigraphers and sequence stratigraphers. Chronostratigraphic sections are plotted with the y-axis in geological time and the x-axis in distance (Fig. 6-61). Chronostratigraphic sections are sometimes called Wheeler diagrams (Sloss 1984) after Harry Wheeler, who proposed this type of plot (Wheeler 1958). Chronostratigraphic sections are not used in mapping, but they are extremely useful in exploration because they *illustrate the history of a basin*. Also important is that the sections can show the distribution of facies across the basin at various times. Source, seal, and reservoir units can be highlighted. Chronostratigraphic sections are usually accompanied with a map that illustrates the line of section. Several chronostratigraphic sections are shown in Figures 6-62 and 6-63.

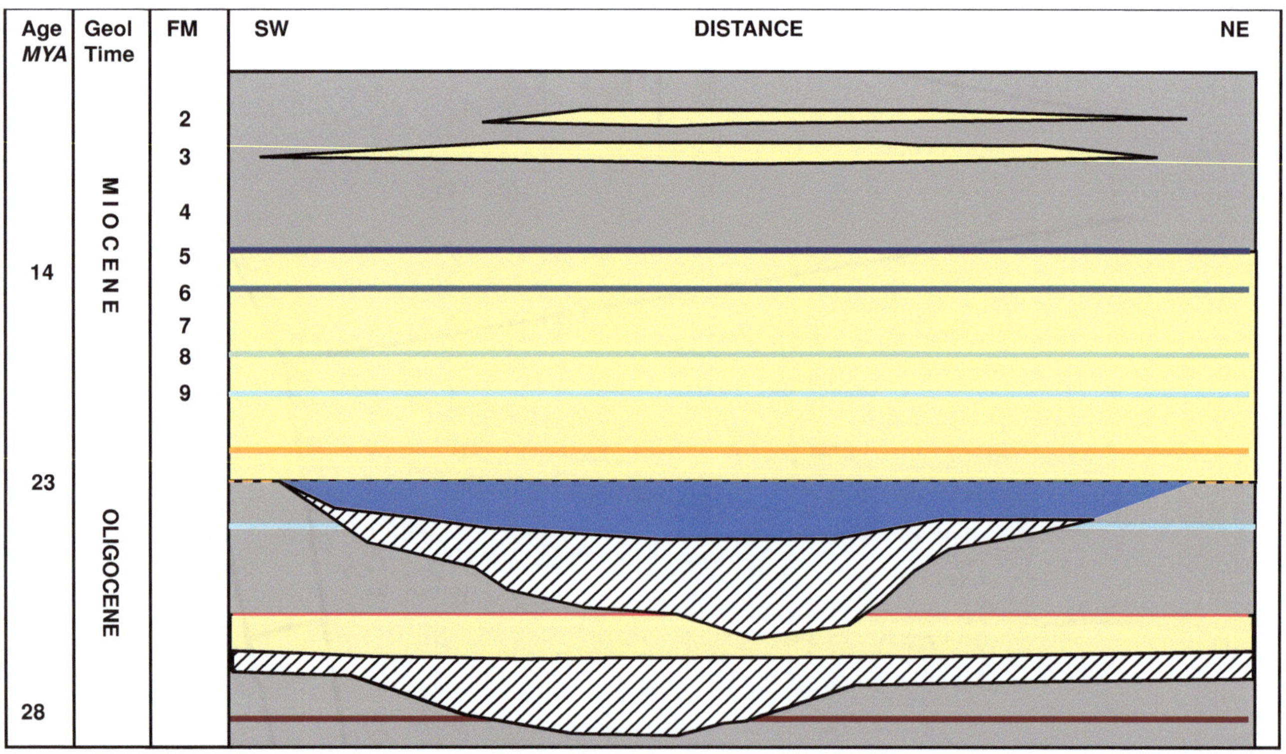

Figure 6-61 Generic chronostratigraphic section illustrating the change in lithology with time and along the line of the section. Hachured intervals equal times of erosion and nondeposition. (Modified from Shoup 2017. Used with permission of R. Shoup.)

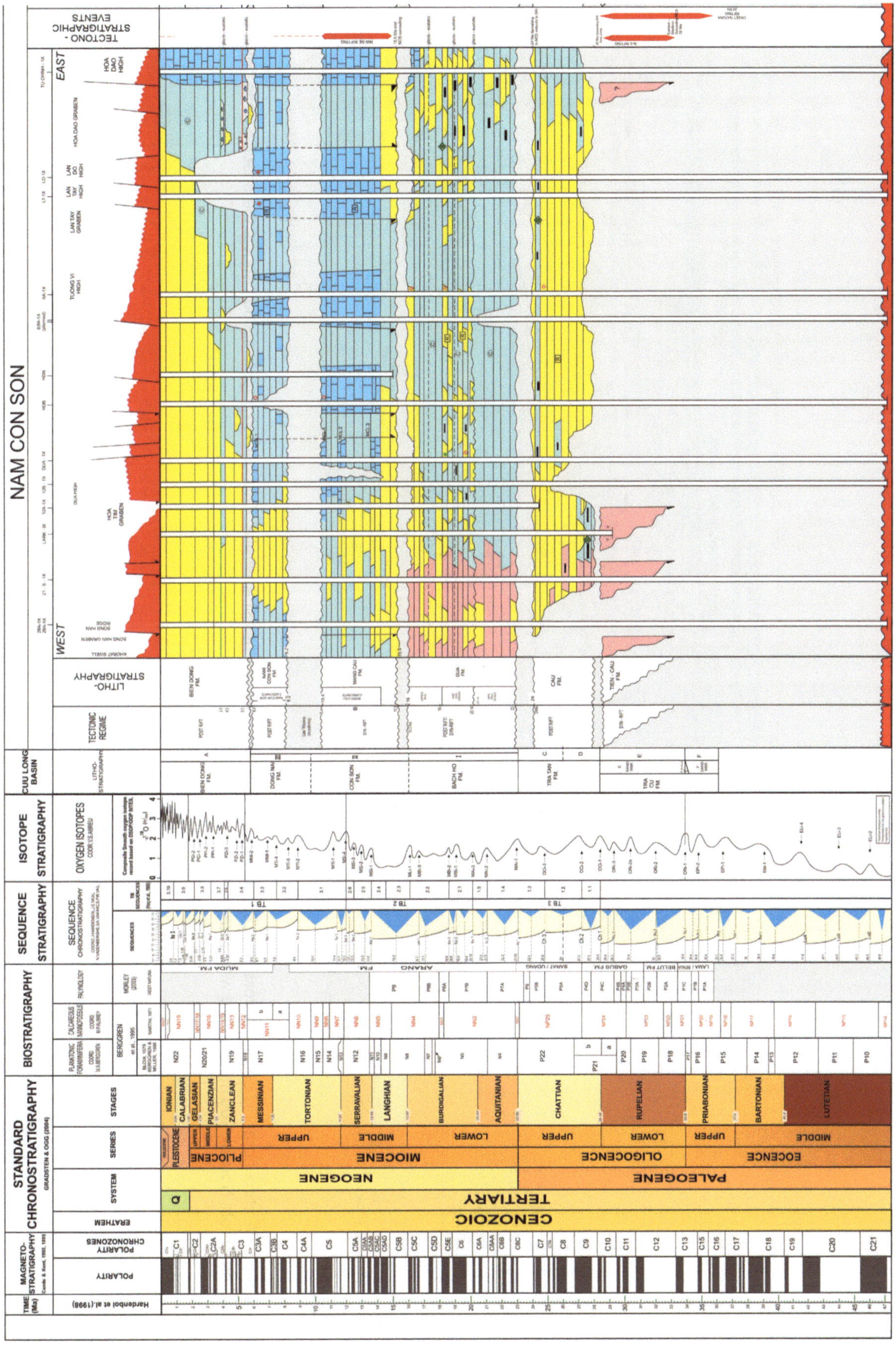

Figure 6-62 A chronostratigraphic section from the Nam Con Son Basin in Southeast Asia. This chronostratigraphic section illustrates the geological history of a basin. Various additional pieces of information, such as seafloor magnetic reversals, sea level curve, biostratigraphy, formation names, and orogenic events, are plotted on the vertical axis along with geological time. Lithology is based on stratigraphy seen in individual wells across the basin. (Modified from Swiecicki and Maynard, 2009; used with permission of Dr. Tony Swiecicki.) Register on the InformIT site (see preface) to view the full-sized color figure.

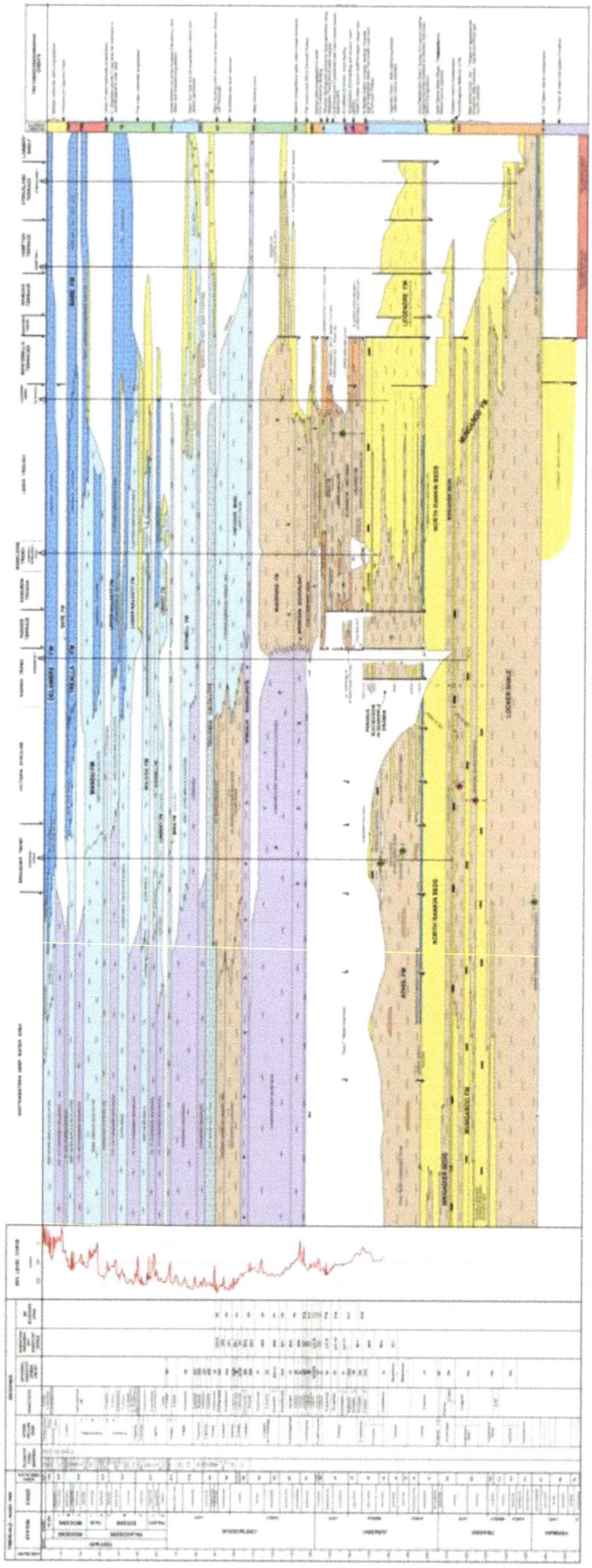

Figure 6-63 A chronographic section across the Dampier Sub-Basin, offshore northwest Australia. This chronographic section shows the specific wells going into the section. (From Norvick 2001. Published with permission from M. S. Norvick.)

CHAPTER 7

FAULT MAPS

INTRODUCTION

Faulted structures play a very significant role in the trapping of hydrocarbons. Therefore, it is imperative that anyone involved in the exploration for, or exploitation of, hydrocarbons should have a significant understanding of faults within the area of study, including their origin and relationship to the formation of structures. Detailed interpretation and mapping of major faults is critical in the process of hydrocarbon exploration and development. This chapter presents the correct subsurface interpretation and mapping techniques required to prepare fault surface maps. Chapter 8 presents the techniques for integrating faults into structural interpretations and maps. Faults themselves are vital to structural development and to hydrocarbon migration and entrapment. A reasonable structural interpretation, in faulted areas, begins with an accurate fault interpretation resulting from the construction of fault surface maps and the proper integration of these fault maps into the structural interpretation. We refer to constructed fault maps as **fault surface maps** rather than the more commonly used term "fault plane maps," since most fault surfaces are not true planes.

The data required to construct a fault surface map are obtained from the correlation of well logs, interpretation of seismic sections, and at times from outcrops. In Chapter 4, we present the methods and procedures for recognizing a fault in a well log and determining its missing or repeated section. In this chapter, we discuss the importance of mapping faults and present the methods for constructing fault surface maps with fault data acquired from electric well logs and seismic sections.

The preparation of accurate fault surface interpretations and maps requires a strong geological background, 3D thinking, and a good understanding of the structural style of the area being worked. When we make reference to the understanding of structural style, we are referring to that specific assemblage of geological structures common to a particular petroleum province (Fig. 7-1). In order to prepare geologically reasonable maps, one must be familiar with the tectonic setting being worked, the fault and structural patterns expected, their origins, and, at times, the process of development. Many of the basic concepts, methods, and techniques for interpreting and mapping faults are universally valid. However, their recognition, interpretation,

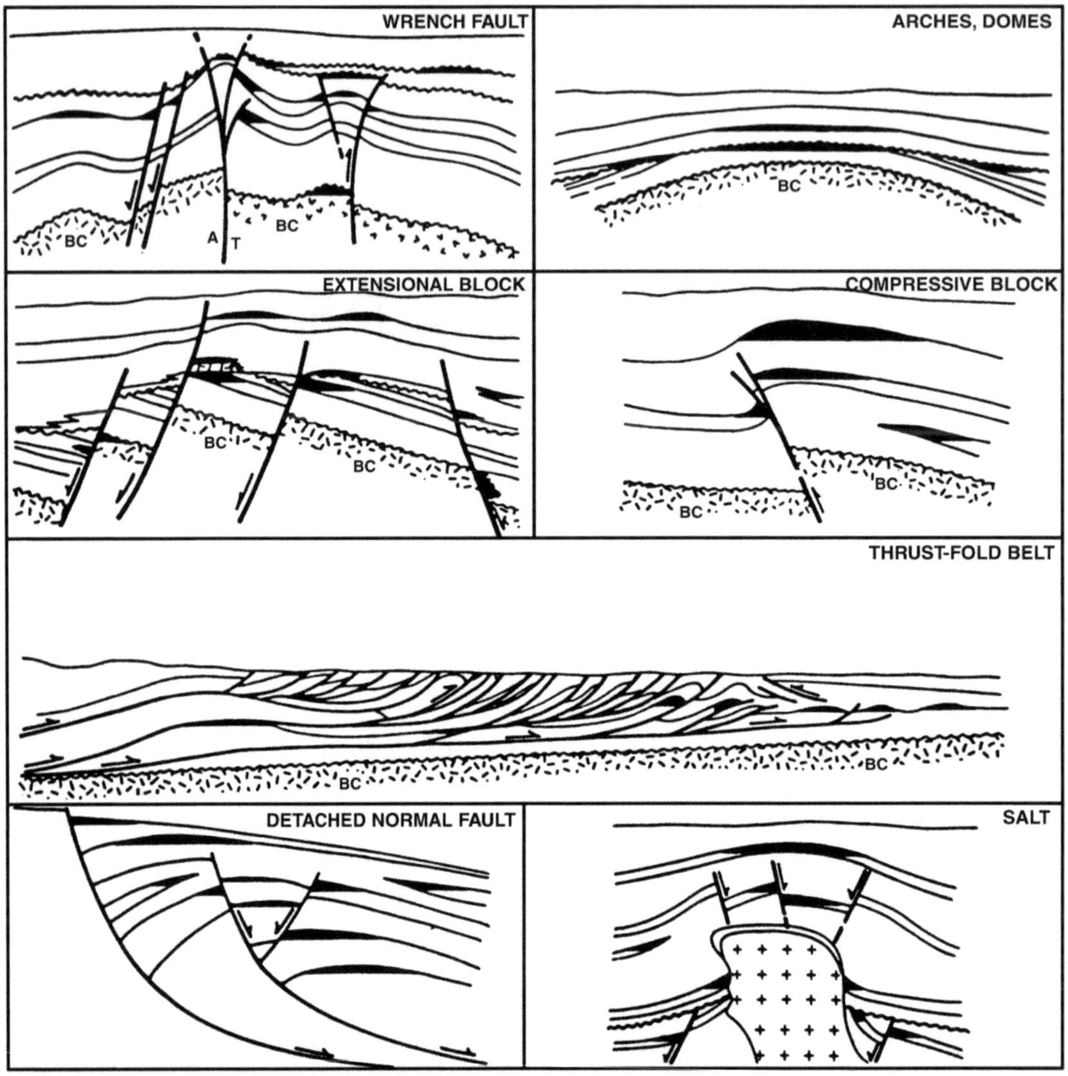

Figure 7-1 Schematic diagrams of hydrocarbon traps associated with various structural styles. BC, basement complex; T, displacement toward viewer; A, away from viewer. (From Harding and Lowell 1979; AAPG©1979, reprinted by permission of the AAPG whose permission is required for further use.) (Salt-related closures modified from Salt Domes by Michel T. Halbouty. Copyright 1979 by Gulf Publishing Company, Houston TX. Used with permission. All rights reserved.)

map construction, and application very much depend on the geoscientist's background and understanding of the kinds of geological structures being worked (see Chapters 10, 11, and 12).

Detailed discussion of structural geology or structural styles is beyond the scope of this text, although many aspects of this subject are presented in several chapters, including Chapters 9, 10, 11, and 12. We consider this book to be an advanced level text with the focus on structural and mapping techniques, and we make the assumption that you have a general understanding of the fundamental principles of classic geological study, including structural geology as outlined in such texts as Billings (1972), Harding and Lowell (1979), and Suppe (1985).

Faulted structures can be simple or complex. To provide the best structural interpretation with the available data, the integrity of the structure must be shown to be sound and geologically reasonable. *To provide the soundest and most accurate geological interpretation in faulted areas, the interpretation, mapping, and validation of the faults is the first step.* The construction of fault surface maps as a fundamental part of any geological study is *absolutely* necessary. The integration of fault surface maps with structure maps is also essential to support the structural interpretation, to prepare accurate maps, to identify prospects, to design wells to be drilled, and to determine the volume of potential hydrocarbons. The integration of fault and structure maps is discussed in detail in Chapter 8.

Too often, geological interpretations and the accompanying maps and cross sections are prepared without giving much consideration to the 3D geometric validity of the interpretation. Testing the validity of geological interpretations is discussed in some detail in Chapter 10 under structural balancing, but it needs to be stated here that the proper construction of fault surface maps and their correct structural integration can go a long way toward providing 3D validity or consistency to any interpretation.

It is not sufficient to rely solely *on what the well logs are indicating or what is seen on the seismic sections*. Cross sections and seismic sections can in themselves misrepresent true subsurface relationships by the simple nature of their orientation as well as by other factors. A good understanding of 3D geometry is essential to any attempt at reconstructing a subsurface picture.

No hydrocarbons have ever been trapped by a fault trace. The trap is along the fault surface itself. Therefore, the mapping of the surfaces of all-important faults is an integral part of any subsurface interpretation, particularly in areas involving multiple faults, where extremely complicated structural relationships can exist. Attempting to reconstruct a complicated structure by using isolated fault data from electric well logs or seismic sections without the benefit of reasonable fault surface maps and their integration with various structural horizons can provide erroneous geological interpretations. Shortcuts are often taken in our preparation of subsurface maps. Such shortcuts include failure to construct fault surface maps, the preparation of a structure map of only one horizon, or the use of a limited number of seismic sections to generate an interpretation. In this chapter and in Chapter 8, we show how such shortcuts can often lead to structural interpretations that are misleading, unreasonable, and therefore costly to any exploration or development program.

The basic concepts and techniques discussed in this chapter apply to the use of data obtained from both vertical and deviated wells in addition to data from seismic sections. The discussions, illustrations, and practice problems deal principally with extensional and compressional faulting that reflect mainly dip-slip movement, but the methods are broadly applicable to all styles of faulting.

FAULT TERMINOLOGY

"Probably no portion of geological literature has a more confused terminology than that dealing with faults." This is a profound statement that is as applicable today as it was when made by H. W. Straley in 1932. A literature search on the subject of fault nomenclature, or terminology, shows that as far back as the turn of the century, there was great inconsistency in the use of fault component terminology.

In 1908 the Council of the Geological Society of America appointed the Committee on the Nomenclature of Faults. This committee was charged with establishing proper fault nomenclature. Since that time, there have been numerous papers on the subject of fault components and their related terminology, usage, and nomenclature. However, despite these numerous publications, many geoscientists and engineers are still confused when it comes to the definitions and usage of various fault component terms.

The correct understanding and usage of fault terminology with respect to certain fault components is essential to the preparation of correct subsurface fault and structural interpretations and maps. Therefore, in this section we discuss the fault components that are important to subsurface mapping in the petroleum industry. Many others are not discussed, not because they are less important but because they do not apply to the interpretation and mapping techniques that are presented in this text. Figure 7-2 graphically defines several of the fault components of interest to us.

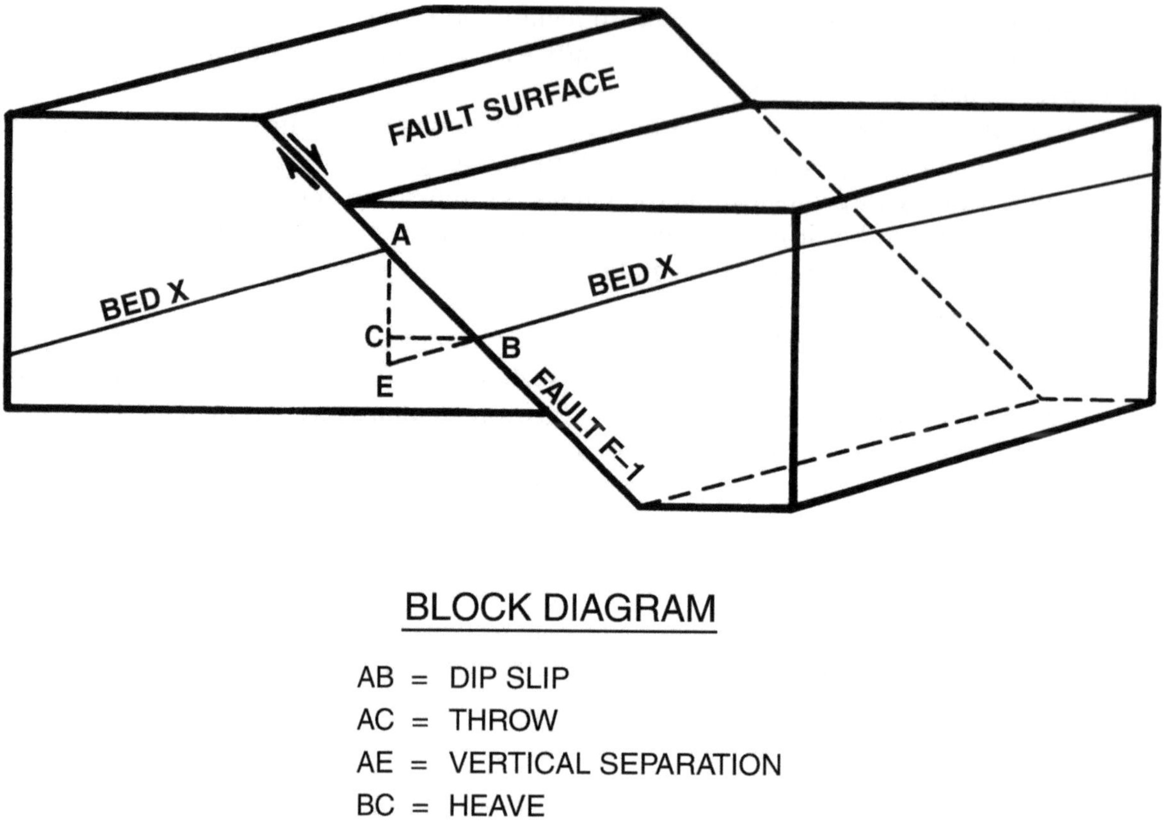

Figure 7-2 Block diagram of Bed X displaced by a normal fault, illustrating four different fault components. The front panel is perpendicular to the strike of Fault F-1. (Modified from Tearpock and Harris 1987. Published by permission of Tenneco Oil Company.)

In the literature, the definitions and uses of the terms *vertical separation, throw*, and *heave* are inconsistent and confusing. The correct usage of fault component terms by all geoscientists and engineers may never be achieved. However, in order to correctly conduct interpretations and construct subsurface maps, we must use the fault component terms vertical separation, throw, and heave *in a consistent manner*. We choose to use the terms as follows.

Vertical separation (AE) *is the vertical component of bed displacement*. It is measured as the vertical distance between a horizon (such as the top of a stratigraphic unit) projected from one fault block across a fault to a point where the projection is vertically over or under the same horizon in the opposite fault block. It is that separation seen in vertical wellbores, vertical shafts, and vertical cross sections (Dennis 1972).

Throw (AC) *is the difference in vertical depth between the fault intersection with a horizon in one fault block and the fault intersection with the same horizon in the opposing fault block, determined in a direction perpendicular to the strike of the fault surface.*

Heave (BC) *is the horizontal distance between the fault intersection with a horizon in one fault block and the fault intersection with the same horizon in the opposing fault block, determined in a direction perpendicular to the strike of the fault surface.*

Missing section *is the vertical thickness of the stratigraphic section faulted out of a wellbore as a direct result of a normal fault cutting through the section*. Missing section is sometimes referred to as fault cut. **Repeated section** *is the vertical thickness of the stratigraphic section repeated in a wellbore as the direct result of a reverse fault cutting through the section.*

The missing or repeated section is determined by correlation of an electric log from one well with other electric logs from nearby wells, as presented in Chapter 4. ***Technically, missing or repeated section is equal in value to vertical separation***. Vertical separation for a fault can also be determined from seismic data, as shown later in this chapter.

We cannot overemphasize the importance for all geoscientists and engineers to understand that ***the missing or repeated section recognized in well logs is NOT throw, nor is it equal to throw***, but rather that it is equal to the fault component vertical separation. A misunderstanding of the technical point can create, and has caused, significant interpretation and mapping errors resulting in millions of lost dollars from dry holes, failed recompletions, workovers, and more. Of special interest is the integration of well log with seismic data. Major interpretation errors occur regarding fault displacement when seismic data are misinterpreted by using throw, or apparent throw, determined from seismic sections as if it were equivalent to the missing section from nearby wells.

Notice in Figure 7-2 that the value for the throw of the fault is not equal to the value for the vertical separation. The fact that they are different components of a fault is of significance when fault data obtained from well logs are used for integrating faults into a structural interpretation (Tearpock and Bischke 1990). The terms *throw* and *vertical separation* are commonly misunderstood and, more importantly, often misused in the preparation of subsurface interpretations and the accompanying fault and structure maps.

Because the understanding of these terms is so important to correct interpretation and construction of subsurface fault and structure maps, we attempt to clarify the issue without causing more confusion than currently exists. We discuss the terms in a general way and review them with respect to subsurface mapping techniques. For a complete review of the subject of fault nomenclature, refer to the references at the end of the book.

DEFINITION OF FAULT DISPLACEMENT

We apply a definition to the word displacement similar to that given by Reid et al. (1913). It is here applied to the relative movement of the two sides of a fault, measured in any specified direction, or to the change in position of a marker or horizon caused by fault movement. There are two ways to estimate displacement resulting from a fault. The first is the *actual relative displacement* of the two sides of a fault, and the other is the *apparent relative displacement*.

Slip is the *actual relative displacement* of a fault (Hill 1959). It is defined as the measurement of the distance of the actual relative motion between two formerly adjacent points on opposite sides of a fault, measured on the fault surface. **Separation** is the *apparent relative displacement* of a fault (Hill 1959). It is defined as the distance, measured in any specified direction, between two parts of a displaced surface on opposite sides of a fault. Separation is apparent movement on a fault with respect to a reference horizon cut by the fault (Dennison 1968). Separations are measurable, whereas slip is usually calculated. Numerous authors (including Reid et al. 1913; Hill 1947; Crowell 1959; Dennis 1972; Tearpock and Harris 1987; Tearpock and Bischke 1990; and others) have emphasized the importance of distinguishing between fault components related to slip and those components related to separation.

We emphasize that components of the actual *slip* cannot routinely be measured in the subsurface due to a lack of conventional sources of data from which the measurements can be made. Therefore, slip components are not routinely mappable. Some *separation components*, on the other hand, are measurable fault components with conventional subsurface data and therefore are mappable. They can be measured in a vertical shaft, an electric log from a wellbore, or on a seismic section, *regardless of orientation with respect to fault strike*. Of these various separation components, vertical separation is the most important parameter for constructing subsurface fault and structure maps (Tearpock and Harris 1987).

Throw and heave cannot be measured by correlation of well logs, as we will demonstrate. Using a seismic interpretation, measurements of throw and heave can be made only if the interpreted seismic profile is oriented perpendicular to the strike of the fault surface. The terms throw and heave have limited practical application in subsurface petroleum interpretation and mapping. In addition, they cause confusion and significant interpretation and mapping errors. We discuss their application and relationship to other fault components later in this chapter and again in Chapters 8.

We cannot leave this subject with the idea that fault slip is not important. Fault slip and related components are important, particularly in mining operations. Slip generally can be determined in mines where the actual fault surface is visible. In fact, the usage of the terms throw and heave originally came from the coal fields of Great Britain where the strata are nearly horizontal or the faults are strike faults (Reid et al. 1913). As mentioned earlier, these terms have found their way into the petroleum industry and are probably here to stay. However, they have limited practical application in subsurface petroleum interpretation and mapping.

MATHEMATICAL RELATIONSHIP OF THROW TO VERTICAL SEPARATION

Throw can be measured in cross sections drawn perpendicular to fault surface strike (or parallel to maximum dip), such as Figure 7-3.

$$\text{Throw} = AC = AB \sin\theta$$

where

$$AB = \text{dip separation}$$
$$\theta = \text{fault dip}$$

Thus, as the fault changes dip, the value of throw must also change. We emphasize at this point that throw (which is related to fault dip and displacement) cannot be directly measured from electric well logs. We do, however, present methods that enable you to calculate the amount of throw, if desired, knowing the vertical separation and other properties, such as fault and bed dips. However, *throw does not normally enter into proper subsurface mapping techniques* (Tearpock and Bischke 1990).

The vertical separation *(AE)* in Figures 7-2 and 7-3 is defined as the distance that a bed has been vertically displaced during faulting (Hill 1947). This distance is of primary importance to us because the vertical separation is recognized and determined from correlated electric well logs, as described in Chapter 4. To illustrate this point, consider the following example. Assume that a structure exists that contains beds that dip uniformly to the west (Fig 7-4). The SP from two wells drilled into these beds is shown on the figure. The dashed line in Figure 7-4 represents a future normal fault. The beds will be displaced in such a manner that the hanging wall portion of Well No. 1 is placed in juxtaposition with the footwall portion of Well No. 2, as shown in Figure 7-5.

The geometric configuration produces the following observations in Figure 7-5. As the hanging wall block is displaced, the top of Bed B in the hanging wall is brought into contact with the top of Bed C in the footwall. Therefore, the missing section in Well No. 1 of the hanging wall includes the stratigraphic section from the top of Bed B to the top of

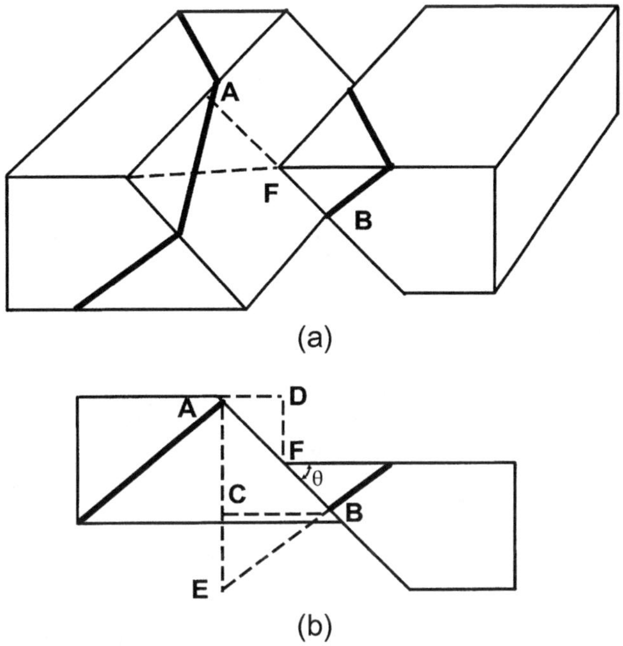

Figure 7-3 (a) Block diagram showing surface displaced by normal fault. (b) Vertical cross section perpendicular to strike of fault; in same plane as front of hanging wall block of block diagram. DF is vertical slip; AD is horizontal dip slip; AC is the throw; BC is the heave; AE is the vertical separation. DF, AC, and AE are all used by some geologists as throw. (Modified from Billings 1972. Published by permission of Pearson Education, Inc.)

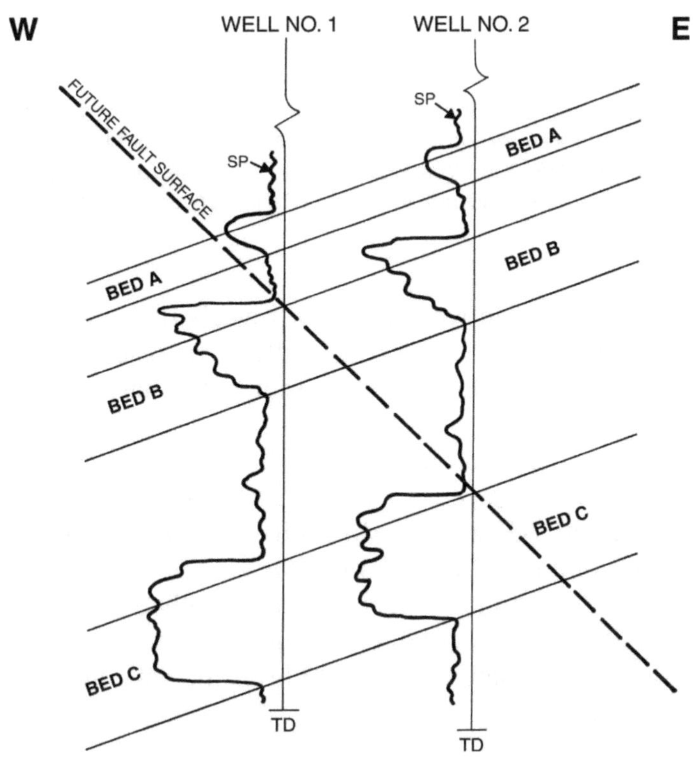

Figure 7-4 Hypothetical example. Unfaulted structure with beds dipping uniformly to the west. Dashed line shows location of future normal fault perpendicular to plane of cross section. (Published by permission of D. Tearpock and R. Bischke.)

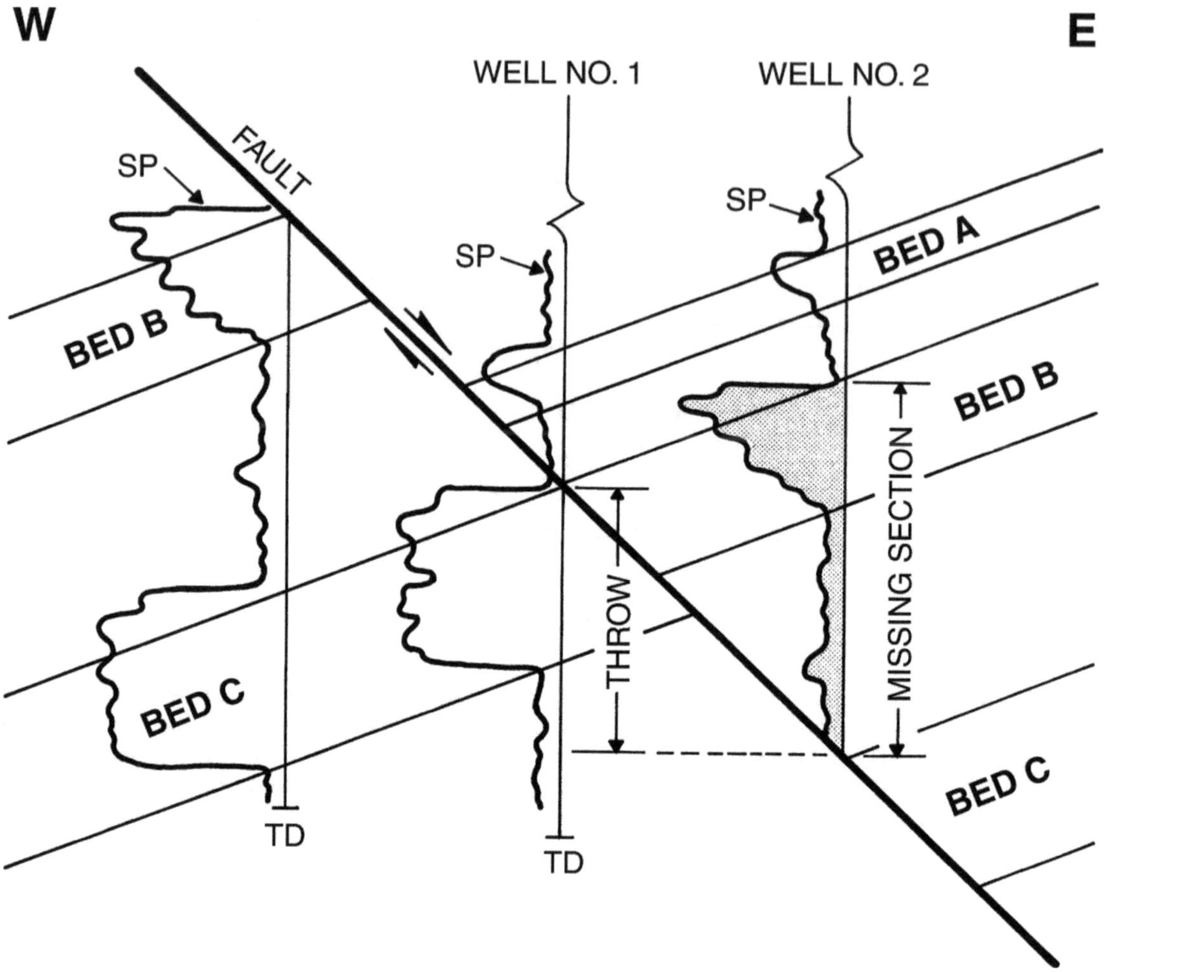

Figure 7-5 Beds are displaced such that the upper portion of Well No. 1 in the hanging wall fault block is juxtaposed with the lower portion of Well No. 2 in the footwall block. Missing section in Well No. 1 by correlation with Well No. 2 is highlighted on the SP curve for Well No. 2. The missing section is equal to vertical separation and not to throw. (Published by permission of D. Tearpock and R. Bischke.)

Bed C. Inspection of the electric well logs now reveals that the missing section as the result of the fault in Well No. 1 (in the hanging wall) is represented by the coarsening upward sand sequence and the lower shale section present in the hanging wall portion of Well No. 2 (shaded section in Fig. 7-5).

This example clearly demonstrates that the throw of the fault is not equal to the missing section in the faulted well. However, the missing section is equal to the vertical separation as defined in Figures 7-2 and 7-3. *We therefore have shown that electric well logs record vertical separation and not throw and that throw does not directly enter into subsurface mapping techniques* (Tearpock and Harris 1987; Tearpock and Bischke 1990). Vertical separation (as well as throw) varies laterally and with depth on a fault. Expect these variations and be prepared to recognize them in your interpretation of well and seismic data, and honor the variations in your maps of faulted structures.

Quantitative Relationship

Vertical separation can be related to throw (Tearpock and Bischke 1990) through two equations, which change with bed dip (Eqs. 7-1a and b), and these equations are derived from Figures 7-6 and 7-2. Performing some trigonometry and using the law of sines, the following relationship is developed:

$$AE\sin\left(\frac{\pi}{2}-\phi_a\right)=AB\sin(\phi-\theta_a)$$

Substituting

$$AB=\frac{AC}{\sin\theta}$$

yields

$$\frac{AE}{AC}=\frac{\sin(\phi_a-\theta)}{\left[\sin\theta\cdot\sin\left(\frac{\pi}{2}\right)-\phi_a\right]}$$

Utilizing trigonometric identities yields

$$\frac{AE}{AC}=\left|\frac{\tan\phi_a}{\tan\theta}-1\right| \qquad \text{(7-1a)}$$

where

AE/AC is taken relative to the absolute value

ϕ_a = apparent bed dip measured in dip direction of fault

θ = fault dip

ϕ_a and θ are taken clockwise from 0 deg to 180 deg

If derived from Figure 7-2, and there is missing section between the hanging wall and the footwall, then

$$AE\ \sin(\pi/2-\phi_a)=AB\ \sin(\phi+\theta_a)=AE/AC=|\ \tan(\phi)/\tan(\theta)+1.0\ | \qquad \text{(7-1b)}$$

Equations (7-1a) and (7-1b) have application in regard to the evaluation of subsurface structure maps (Tearpock and Bischke 1990). This 2D equation gives us the ability to check completed structure maps for accuracy of construction when the missing or repeated section is used to prepare an integrated subsurface structure map (see the section Contouring Faulted Surfaces in Chapter 8 for definition of *integrated structure map*). In Chapters 7 and 8, we discuss the application of Eq. (7-1a) to test the validity of structure maps constructed using well log or seismic data, and we present a method for analyzing the magnitude of error if mapped incorrectly.

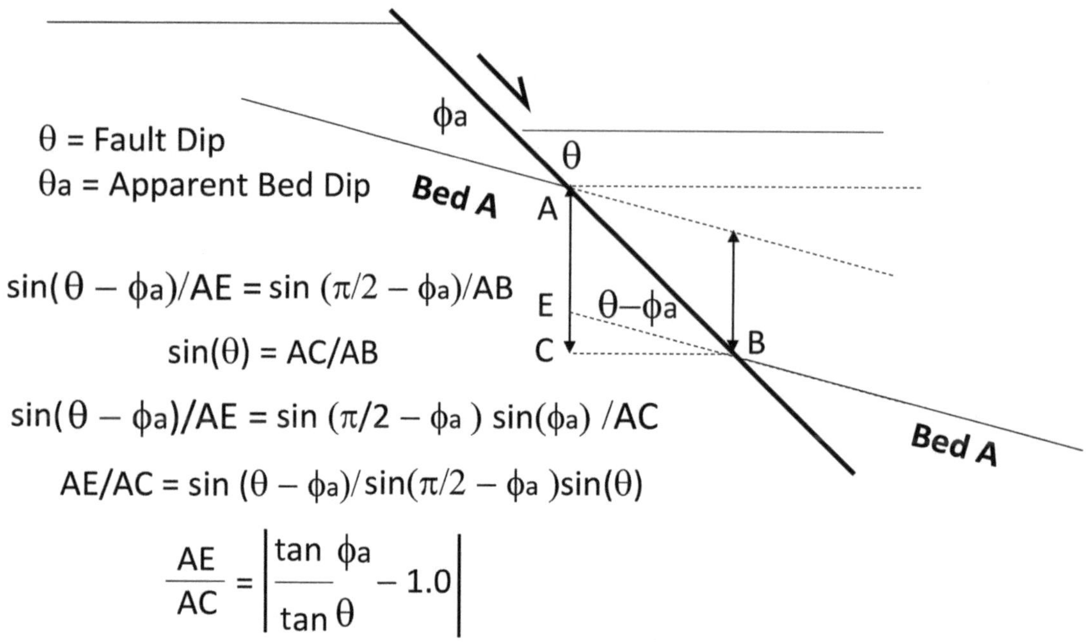

Figure 7-6 Using trigonometry and the law of sines, the relationship of vertical separation to throw is shown graphically in this figure and mathematically in Eq. (7-1a). (Published by permission of D. Tearpock and R. Bischke.)

Using data from the first edition of *Applied Subsurface Geological Mapping* (Tearpock and Bischke 1991), J. R. Sonnad generated a 3D equation for determining throw. The geometric relationships here are similar to those presented in Chapter 4 regarding the Setchell equation. Therefore, the data are used to establish the 3D equation for calculation of throw when the required fault and structural data are available.

$$AC = \frac{AE}{1 - \dfrac{\tan\phi\cos\alpha}{\tan\theta}} \qquad (7\text{-}2)$$

where

ϕ = true bed dip

α = azimuth between bed dip direction and fault dip direction

θ = true fault dip

FAULT DATA DETERMINED FROM WELL LOGS

As a standard practice in reviewing subsurface structural interpretations and accompanying maps with faults, two questions should be asked: (1) What fault data were used to estimate amounts of missing or repeated section for the fault or faults in the preparation of the structure maps? (2) What technique was used to contour across the fault(s)? If the interpreter responds that throw was mapped across the faults and the source of the fault throw data was subsurface well logs or seismic sections, a review of the maps can easily be made to determine whether the use of the word *throw* is simply a verbal substitution for vertical separation or whether the interpreter actually used the vertical separation data incorrectly as if the data were throw. If fault data from electric logs were used as throw for a fault in the construction of an integrated subsurface interpretation, the structure map prepared will probably be incorrect and require revision. The use of well log fault data as *throw* in mapping across faults on structure maps is an incorrect technique. *Herein lies one of the most basic problems with the construction of many subsurface structure maps*—a misunderstanding of what fault data are actually obtained from electric well logs for use in subsurface structure mapping.

Determination of fault data begins with well log and seismic correlations. Although there have been numerous publications on fault component terminology covering the subject of throw and vertical separation, we were unable to find one figure that diagrammatically illustrates the geometric relationship of missing or repeated section, as seen on an electric log from a vertical wellbore, with respect to the fault components throw and vertical separation. Figure 7-3, from Billings (1972), illustrates a vertical cross section perpendicular to the strike of a fault surface showing such fault components as throw, heave, and vertical separation. Notice that the figure shows at least three separate fault components defined by various geoscientists as throw, demonstrating the confusion that surrounds the use of these fault component terms. Although this figure correctly illustrates the difference between throw and vertical separation, it does not relate this geometry to what is seen in a well log.

Since fault data are in part derived from well log correlation, it is very important to understand that the true vertical thickness of missing or repeated section, determined in a well by correlation of electric logs, is actually a measurement of the fault component vertical separation. The cross section in Figure 7-7 diagrammatically shows the geometric relationship between the missing section in a wellbore and the vertical separation of the fault. The east-west structural cross section, which is perpendicular to the strike of the fault surface, shows two beds (A and B) that have been displaced by the normal Fault F-1. This normal fault cuts Well No. 1 at −5230 ft and is dipping at an angle of 45 deg to the east. The beds are dipping at 30 deg to the west. By correlation with Well No. 2, the missing section in Well No. 1 is determined to be 100 ft and the fault is shown to have entirely faulted out Bed A in Well No. 1. As shown in the cross section, the throw of Fault F-1 is represented by the vertical line AC, which is equal to 63 ft. The vertical separation, represented by the vertical line AE, is equal to 100 ft. Thus, by the correlation of Well No. 1 with Well No. 2 in Figure 7-7, we have diagrammatically shown that the missing section obtained for the fault in Well No. 1 is not throw nor equal to throw but rather is a measurement of the fault component vertical separation. With this particular set of conditions, the throw of the fault is only 63 percent of the *vertical separation.* As shown in Figure 7-7, there can be a significant difference between the throw and vertical separation of a fault. If mapped incorrectly, this difference can result in significant error in an integrated structural interpretation and therefore the generated structure maps.

For the most part, an understanding of fault terms and their application to subsurface mapping comes from academic studies and company-sponsored training programs. Textbook discussions on faults commonly use very simplistic examples showing faults cutting horizontal beds. These examples using horizontal beds lead to the misconception that throw and vertical separation are the same fault component or have the same value. Throw and vertical separation, however, have the same value in only three specific circumstances: (1) where the strata being faulted are horizontal, (2) where the fault is vertical, or (3) in a cross section perpendicular to the strike of the fault surface, where the fault strike is at a right angle to the strike of the strata.

In the latter situation, the strata will have an apparent dip of zero deg. Figure 7-8a shows the situation involving horizontal beds where the values for throw and vertical separation are the same. The use of models where faults cut dipping beds (Fig. 7-8b and c) should eliminate the misconception that missing section is always throw, an idea that leads to the preparation of incorrect maps.

The first set of conditions from our list of situations, where vertical separation equals throw, is discussed in greater detail here because it is the most common situation presented in textbooks and the one that has resulted in more misunderstanding of the fault component terminology than any other. Because of exposure only to simplistic examples using horizontal beds, many geoscientists and engineers have failed to recognize that the values for these two different fault components vary from each other depending on the structural attitude of the formation. This major misunderstanding can result in numerous mapping errors in a structural interpretation.

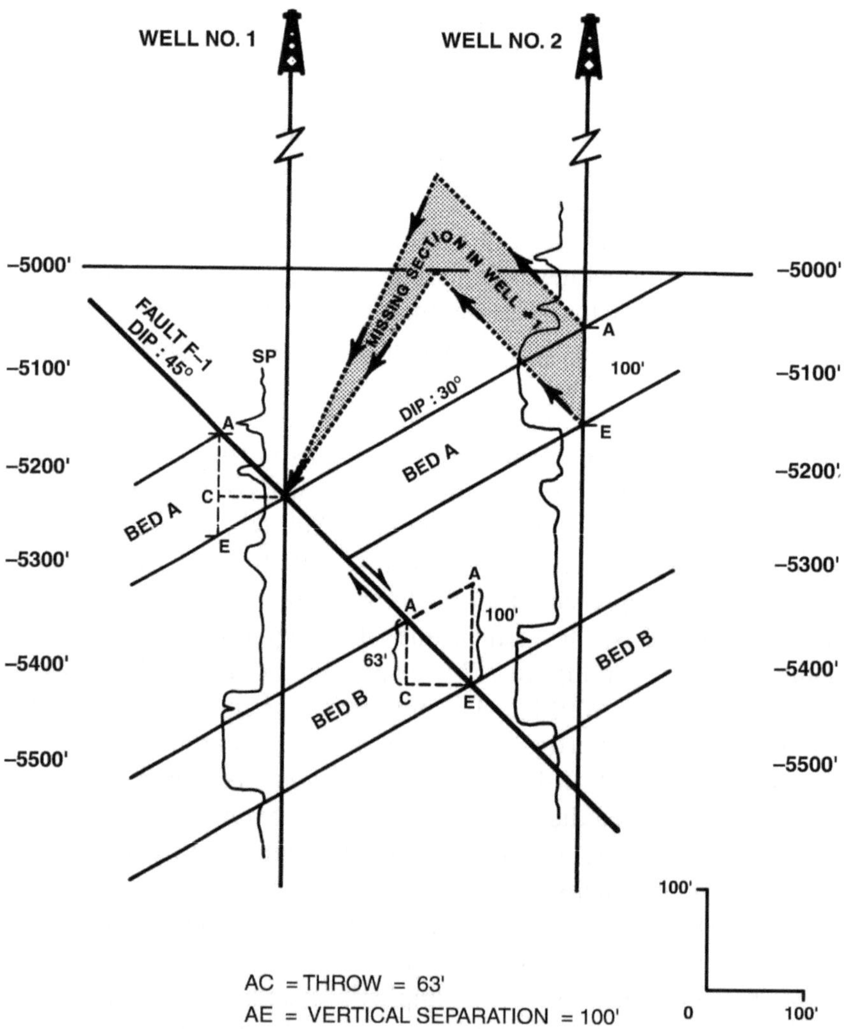

Figure 7-7 Diagrammatic cross section illustrates the geometric relationship between the missing section in a wellbore and the vertical separation of a normal fault. The 63-ft value of throw was calculated mathematically using Eq. (7-1). (Published by permission of D. Tearpock.)

Figures 7-8b and c illustrate the discrepancy in values of vertical separation and throw where bed dip is considered. They further show the relationship where the horizons are dipping in the same general direction as the fault and where the horizons are dipping in the opposite direction to the fault. For example, as shown in Figure 7-8b, the missing section in Well No. 1 is 1000 ft, although the throw of the fault is 1500 ft. In Figure 7-8c, the missing section in Well No. 1 is 1400 ft, whereas the throw is only 1000 ft. We point out again that with the well logs, only vertical separation (missing section) data can be determined. Throw cannot be determined from well log correlation, but it is not important to know in most cases.

Vertical separation is directly measurable from correlation of well logs and is equivalent to the missing section or repeated section caused by a fault, which is valid regardless of the apparent attitude of any horizon considered. Throw and heave are dependent fault variables that change

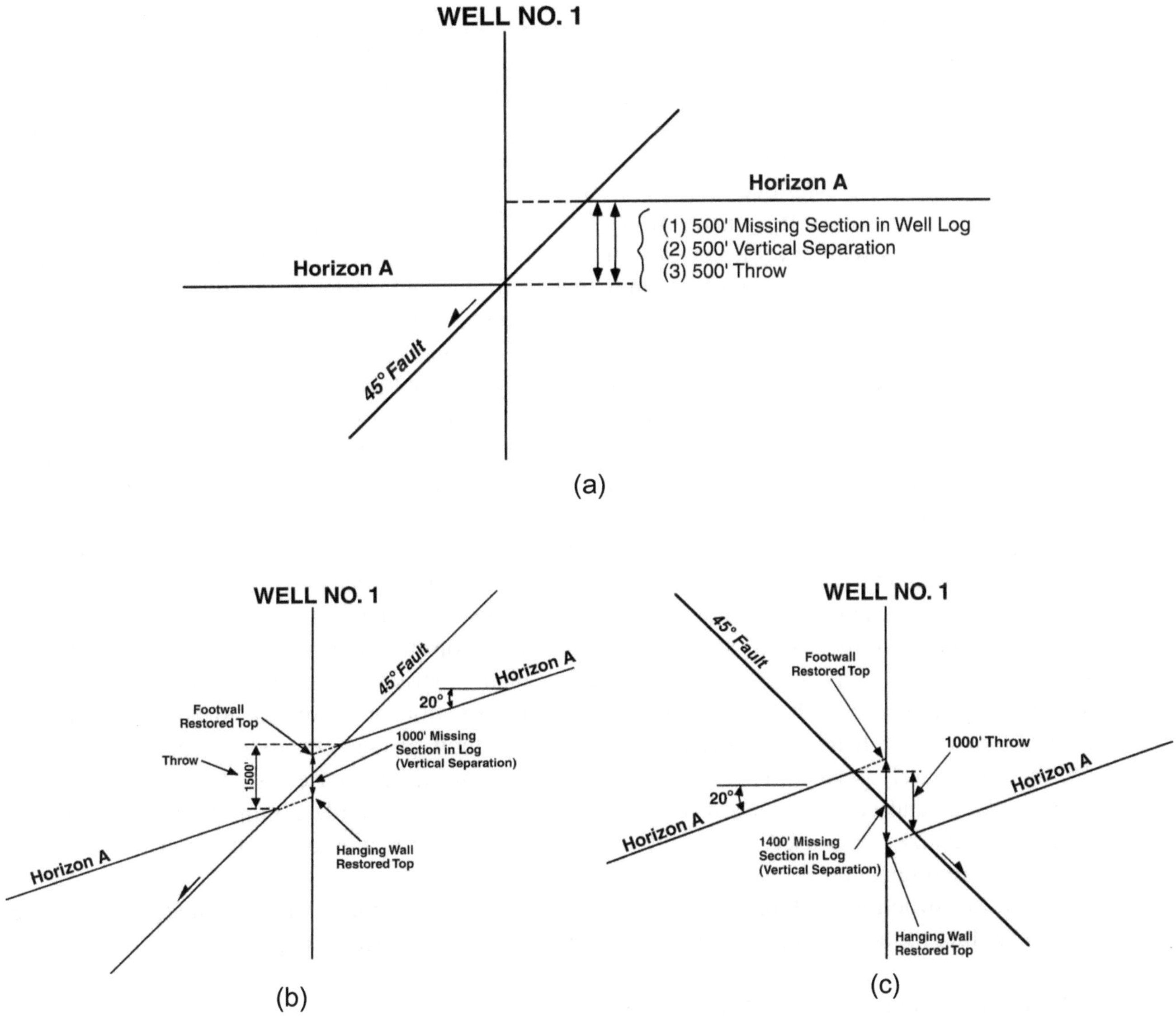

Figure 7-8 (a) The values for throw and vertical separation are the same where the displaced beds are horizontal. (b) Where both the strata and the fault dip in the same general direction, throw is greater than vertical separation. (c) Where the strata and the fault dip in generally opposite directions, throw is less than vertical separation. (Published by permission of D. Tearpock.)

with variations in the apparent attitude of the fault and horizon. For most petroleum-related interpretation and mapping, the estimates for throw and heave have mainly academic value. They can be measured only in a cross section or seismic section that is oriented perpendicular to the strike of the fault surface, or on a structure map after the map has been completed using fault data correctly as vertical separation to construct a technically and structurally reasonable map. By using Eq. (7-2), however, the measured throw across a fault on a completed structure map can be used to check the accuracy of the map. This is discussed in detail in Chapter 8.

FAULT SURFACE MAP CONSTRUCTION

A fault surface map is a type of contour map. It differs from a structure contour map in that the contours are on the surface of a fault rather than on some stratigraphic marker or horizon. The contouring of a faulted horizon presents numerous complex problems, in the contouring of both key horizons and fault surfaces. Although the contouring of key horizons may be the main objective in a mapping project, the contouring of the fault or faults provides essential information about the geology being studied. We present more details on this subject in Chapter 9, which includes 3D diagrams and presentations.

In an area where a fault serves as the boundary limit of a hydrocarbon reservoir, the trap is along the fault surface itself. Construction of the geological picture involves the integration of all fault surface maps with several key structural horizons. Therefore, construction of an accurate fault surface map for each important fault, using all available data, is usually *the first step* in generating a structural interpretation where faults are present.

Earlier in this chapter, we mentioned that the preparation of accurate fault surface maps requires 3D thinking and a good understanding of the regional tectonics being studied. This is so because each tectonic setting has its own characteristic patterns of faulting. For example, most of the faults in areas like offshore Nigeria or the northern Gulf of Mexico Basin are normal faults typically downthrown to the basin and, although they may strike in any direction, the preferred strike direction is roughly parallel to the present or historical coastline. Regional knowledge is very important in developing a geological interpretation, comparing alternative geological solutions, and generating final subsurface maps that are geologically reasonable.

Note that we use the term fault *surface* map, or just fault map, instead of the more common usage of fault *plane* map. Fault surfaces tend to differ from true mathematical planes. They may increase or decrease in dip with depth, as well as change strike direction reflecting a sinuous or angular appearance, which may trend in a specific direction or represent an arcuate shape. In profile they may be listric, antilistric, or kinked. Some fault surfaces are deformed. Fault surfaces are therefore rarely perfect planes. However, on a very localized or field scale, some faults may appear planar and can be mapped as such. Some of the basic fault examples in this textbook represent idealized data used to present and teach a specific technique. In these cases, the fault examples are often simplified as true planes.

The construction of fault surface maps has numerous benefits in the interpretation and understanding of the development of faulted structures. Fault surface maps

1. aid in solving 3D structural problems;
2. define the location of a fault in space in both the horizontal and vertical dimensions;
3. help delineate complex fault patterns;
4. can be integrated with structure maps to delineate accurately the
   a. upthrown and downthrown fault traces
   b. fault gap or overlap
   c. hydrocarbon reservoir limits;

5. are required to evaluate potential cross-fault drainage;
6. are required to construct fault surface sections (Allan diagrams);
7. can be used at times as an indication of the changing stratigraphy (sand/shale) in the footwall by means of a change in dip of the fault (see the section Inverting Fault Dips to Determine Sand/Shale Ratios or Percent Sand in Chapter 11);
8. eliminate the distortion of a fault as seen on a cross section with zigzag well spacing;
9. can be used to estimate the dip and strike of a fault at any location along the fault;
10. aid in the designing of well plans, particularly for directionally drilled wells; and
11. help identify prospects that otherwise might be overlooked.

In the subsurface, faults can be recognized in one of three ways: (1) through the correlation and interpretation of electric logs, (2) by the interpretation of seismic sections, and (3) by inference. In this section, we discuss the use of electric logs to obtain fault data required to construct a fault surface map. The fault information begins with fault data points in electric logs. These data points, which represent the intersection of a drilled well with a fault surface, establish the actual presence of faulting (refer to Chapter 4). For normal faults, the fault is usually represented by a loss of stratigraphic section, whereas a repeat of section is associated with reverse faults, as shown in Figure 7-9. There are exceptions to this generalization, however, such as the special case of steeply dipping beds cut by a normal fault, resulting in a repeated section, as shown in Figure 7-9c.

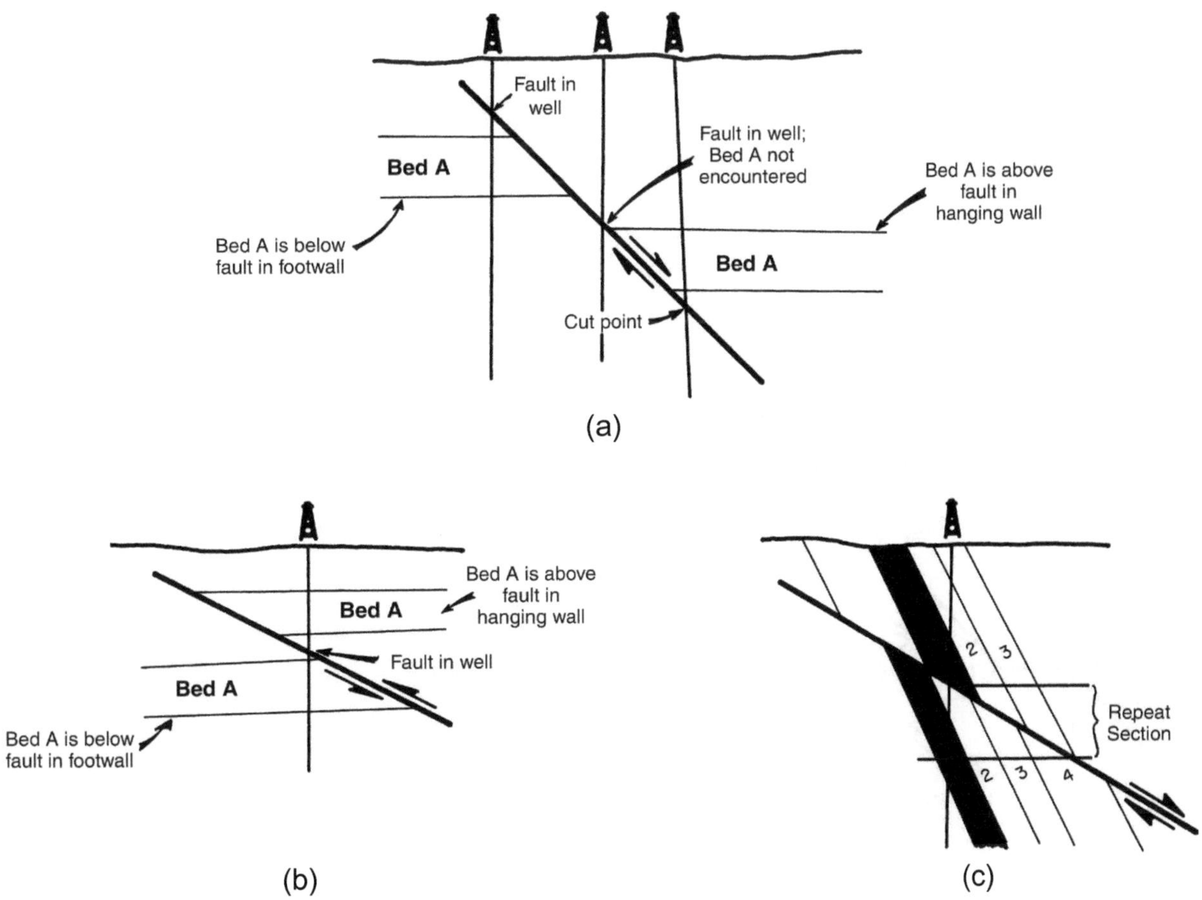

Figure 7-9 (a) Normal fault resulting in a missing section. (b) Reverse fault resulting in a repeated section. (c) Normal fault resulting in a repeated section. In example (c), the beds are dipping at a steeper angle than the fault. (Modified from Bishop 1960. Published by permission of author.)

For a normal fault, any key horizon encountered in a well above the fault is typically in the hanging wall (downthrown) fault block, and any marker encountered below the fault is in the footwall (upthrown) fault block (Fig. 7-9a and c). The only exception is in a deviated well crossing a fault backwards (Fig. 4-36), or on the flanks of salt domes. For a reverse fault, any horizon encountered in a well above the fault is in the hanging wall (upthrown) fault block; a marker encountered below the fault is in the footwall (downthrown) block (Fig. 7-9b). This relationship between the fault pick and any particular horizon thus indicates whether a well is in the footwall or hanging wall fault block for any particular horizon being mapped.

For each fault point in a well, two values are required for use in the interpretation and construction of a fault map: (1) an estimate of the amount of missing or repeated section for the fault, which we define as the vertical separation, and (2) an estimate of the subsea depth of the fault in the well. In Figure 7-10, the recognized fault data in the log of Well No. 2 are clearly marked to indicate the amount of missing section as a result of the fault (150 ft), the depth of the fault (a measured depth of 7280 ft), and the well(s) used in the correlation (Well No. 1). If you are still not sure how to identify faults in electric well logs, refer again to Chapter 4 on electric log correlation.

Fault data from at least three wells, not in a straight line, are required to accurately begin to contour a fault surface in the vicinity of the well control. However, if you are familiar enough with the area and if data from one or more seismic sections are available, then an accurate fault map may be constructed with data from just one well. Obviously, the more fault data available, the better the interpretation of the fault surface. Fault maps also can be prepared from seismic data alone if the coverage is sufficient. This topic is presented later in this chapter.

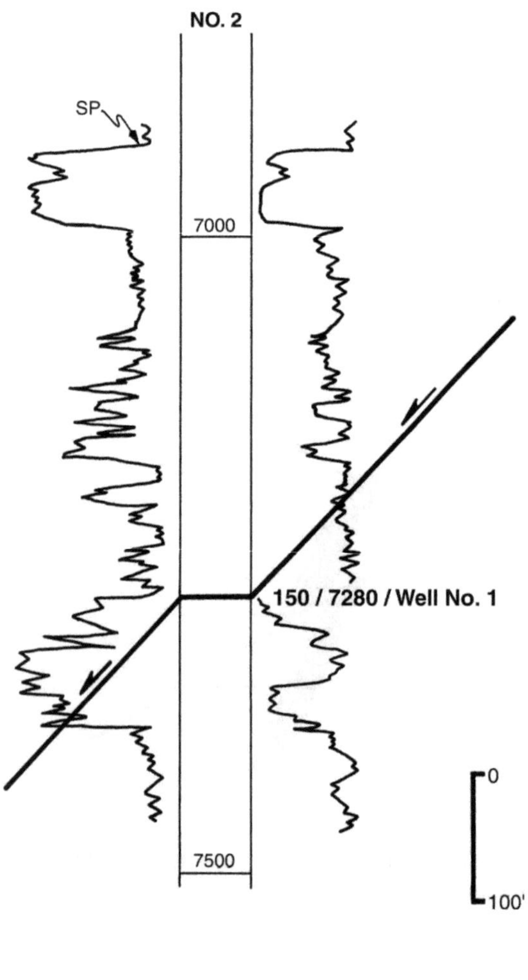

Figure 7-10 Fault information is documented on the electric log. It indicates the missing section, depth of the fault, and the well used for correlation.

Contouring Guidelines

In preparing fault surface maps, certain general guidelines should be followed. If sufficient fault data are available, the fault surface can be contoured in the same way as the elevations of a key horizon (Reiter 1947). Inasmuch as faults result from breaks rather than bends in the strata, they pose some special problems in contouring. The rules differ from the general rules for contouring, in that angular relationships may exist between two intersecting fault surfaces or between a fault surface and a horizon. The general guidelines for contouring a fault surface are as follows:

1. Contours of a fault surface may be open-ended. They do not have to close upon themselves. This is true because faults terminate laterally in the subsurface.
2. Changes in either fault strike or dip are assumed to be gradual unless evidence indicates otherwise (cross-faulting). An exception to this guideline might occur in the case of mapping a reverse-faulted ramp and flat surface (see Chapter 10); in these cases, the changes in dip can be abrupt.
3. Changes in fault strike for normal faults are usually represented by smooth curves rather than by sharp angles. Exceptions to this guideline are deformed fault surfaces and the effect of cross structures.
4. Changes in dip are generally mapped as smooth curves rather than plane segments deflecting at sharp angles. Again, there are exceptions to this guideline, including those listed in guideline 3 and when mapping some thrust faults.
5. Use the interpretive method of contouring outlined in Chapter 2 for preparing fault surface maps.
6. Several fault surfaces may be contoured on a single basemap. Contours of individual faults may merge inasmuch as faults intersect one another in nature. Note: When constructing compensating, bifurcating, or intersecting faults, denote the lines of termination, bifurcation, or intersection on the fault maps.
7. Fault surface maps must be geologically reasonable for the area being mapped.
8. Fault surface maps are normally contoured with a 500-ft or 1000-ft vertical contour interval, since fault surfaces are usually relatively steep. Thrust faults may dip at a low angle, so a smaller contour interval may be appropriate.

The fault surface map will be integrated later with a structure horizon map to generate a completed structure map. This procedure is described in Chapter 8. Three key concepts regarding fault contours need to be remembered for the integration process.

1. The contours of a fault surface join those of a given mapping horizon at points of intersection of fault contours and horizon contours of the same value.
2. Faults commonly dip at different angles than a key horizon being mapped and, consequently, only some of the fault surface contours intersect with those of a given datum (see Chapter 8).
3. The fault surface intersects horizons above and below a given horizon unless the fault is extremely limited in vertical extent.

Fault surface map construction actually involves some subjective interpretation; thus, the more data available for mapping, the less uncertainty in the interpretation. As each of us has a different idea of the geological picture of an area being worked, it is possible in areas with limited well or seismic control to generate several fault surface interpretations using a single or similar sets of fault data. Fault maps, like many other subsurface geological maps, tend to change with time as new well and seismic data become available. Therefore, a fault surface interpretation is never complete until the last well is drilled and all the seismic data to be shot have been shot and interpreted.

Fault Surface Map Construction Techniques

We shall begin with a relatively simple fault contour map involving a single fault, illustrated in Figure 7-11. There are 18 vertical wells in this example from which fault data have been obtained (Fig. 7-11a). First, the amount of missing section and depth for each fault pick in a well are posted next to the appropriate well in which the fault was observed, as shown in the figure. The most common way of posting these data are to indicate the missing section first and then the subsea depth of the fault (e.g., 325 ft/−9240 ft). The minus sign in front of the depth number indicates that the fault in the well is below sea level.

The well control in this example is located in three separate areas on the basemap. Seven wells are located in the western portion of the map, three in the central portion, and eight to the east. As discussed under the general contouring rules, begin contouring in the area or areas of *maximum control*. In this case, we first contour the eastern area, where there are eight wells, followed by the western area, with seven points of control, and finally, the central area with three fault cuts. Use a freehand style of contouring or ten-point proportional dividers to initially establish the contour spacing for the map.

Figure 7-11b shows the contours established for the fault in the three areas of well control. Based on the fault data, we assume that the three fault segments are parts of the same fault, so the final step is to extend the contours into the area of no control and connect the fault segments. There are two possible ways to do this. One method is to extend the contours from each segment toward one another as straight lines until they intersect. This method, illustrated in Figure 7-11c, appears to be a more unreasonable or unlikely interpretation. The second and preferred method is to use an interpretive form of contouring (Fig. 7-11d). With this method, some geological license is used in the interpretation to reflect the expected geometry of the fault surface in this tectonic setting. We gradually change the strike direction of the fault connecting the adjacent segments with a smooth curve.

Now that the fault map is complete, we can estimate the dip of the fault at any location. At a depth of around 5000 ft subsea, the dip of the fault is 65 deg, decreasing to 55 deg at 8500 ft subsea, and finally flattening to about 40 deg between −10,000 ft and −11,000 ft. This type of fault shape is common for growth faults (see the Growth Faults section in this chapter). The fault in Figure 7-11d is contoured as a listric (curvilinear concave-upward surface) growth fault; that is, a fault whose dip decreases with depth, whereas the vertical separation or missing section increases with depth.

Figure 7-11e shows a cross section (A-A′) laid out in a northwest-southeast direction perpendicular to the strike of the fault. Three wells lie on the section with fault data for each well posted. An interpretive method of contouring was used to contour the fault surface with depth. This is the preferred method, which provides the most reasonable interpretation. Other methods could have been used, including the mechanical contouring method, in which the dip rate is constant between each pair of well control points but changes at each well. The equal-spaced contouring method also could have been used. Both methods provide a less reasonable interpretation of the fault surface.

Three examples of the same fault data that were contoured by a computer-based mapping program are shown in Figure 7-11f through h. The projected slope gridding algorithm was used for Figure 7-11f. The result is a map similar to the hand-drawn map in Figure 7-11d in that the fault surface is listric but has a smoother curving lateral bend. A least squares gridding algorithm generated a comparable map except that the extrapolated 5000-ft contour did not conform well to the trend of the other contours and the surface was not so smoothly listric. A closest point algorithm was used to generate the fault surface map in Figure 7-11g, resulting in a fault surface that is not geologically reasonable considering the data, which indicate that the fault is a large growth fault. The mapped surface is not listric and the contours are not credible at the limits of

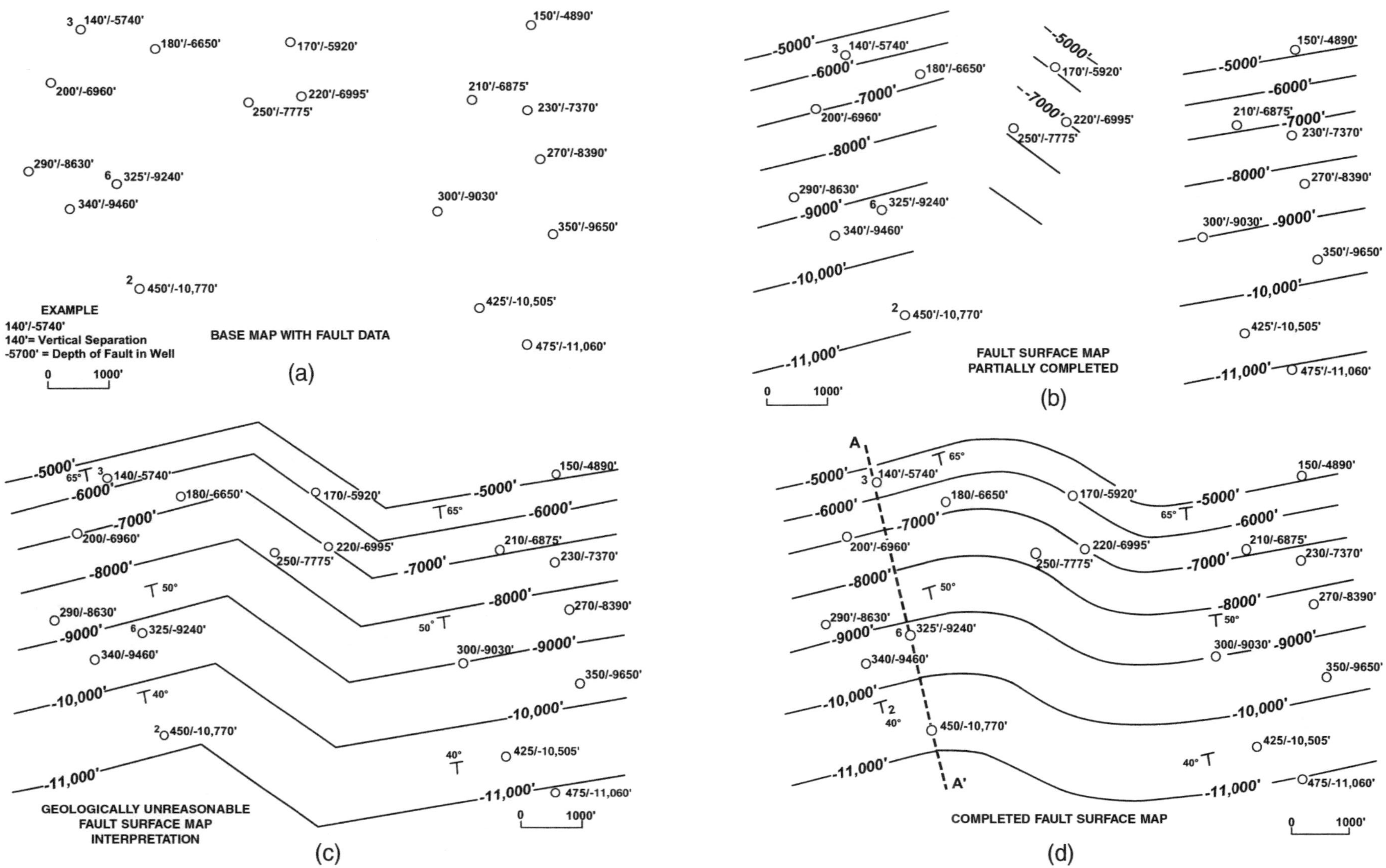

Figure 7-11 (a) Fault surface basemap showing the missing section and depth of the fault in each well. (b) Fault contours established in three areas of well control. Each fault segment is part of the same fault. (c) Unrealistic fault surface interpretation results from connecting each fault segment with straight lines. This is a mechanical approach to contouring. (d) Completed fault surface map using the interpretive form of contouring to reflect the expected geometry of the fault surface. (e) Cross section A-A′ passes directly through Wells No. 3, 6, and 2 and is laid out perpendicular to fault strike. Use of an interpretative approach to contouring results in a gradual, rather than abrupt, change in fault dip with depth. (f) through (h) are computer-contoured maps based on the same fault data as (d). They differ from each other because different gridding algorithms were used: (f) projected slope; (g) closest point; (h) point density.

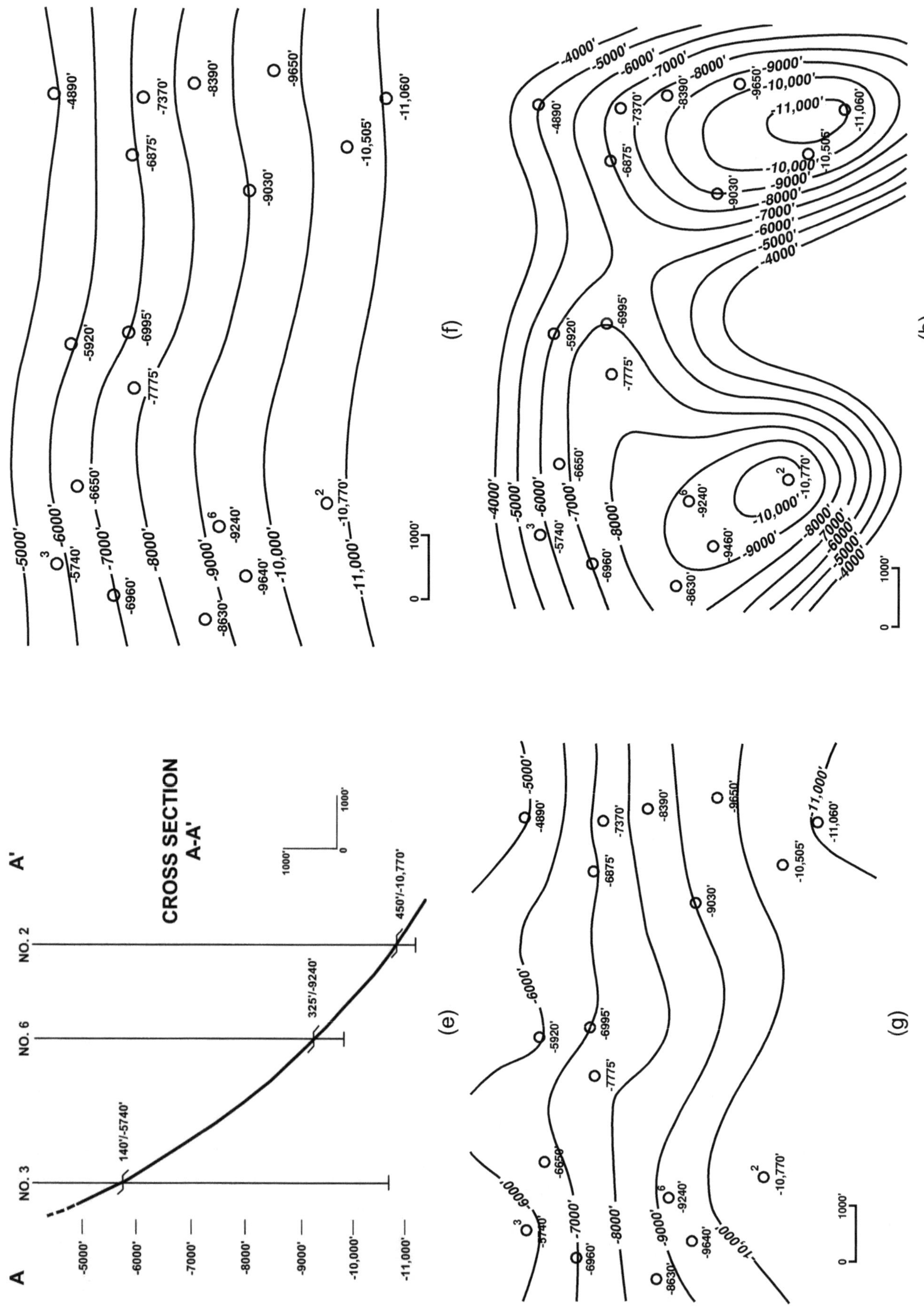

Figure 7-11 *(continued)*

the data, both shallow and deep. The closest point algorithm is best suited to a data set with more numerous and more evenly distributed data points than in our example. Lastly, Figure 7-11g is an extreme example of a bogus fault surface map generated by an algorithm (point density) that is unsuitable to the data set. These three examples demonstrate how critical it is to select an appropriate gridding algorithm (including suitable gridding parameters) for generation of a fault surface interpretation or for integration with a structural horizon map.

Comparison to the hand-contoured map in Figure 7-11d indicates that the projected slope algorithm was the best choice among four algorithms to generate the most reasonable fault surface interpretation. But that does not mean the projected slope algorithm is always the most suitable for mapping a listric fault surface with a lateral bend. The best algorithm is dependent on the number and distribution of data points, among other things. Beware of habitually choosing the same gridding algorithm in computer-based mapping. The interpreter must be sufficiently familiar with all the various algorithms and gridding parameters in order to choose the one most suitable to the data set and the geological surfaces in the area of study.

In developing a final fault surface map interpretation, keep in mind that a fault need not remain constant in strike direction, dip, or vertical separation over its entire extent. Along the strike, the vertical separation may increase, decrease, or remain constant, and the strike direction may change. The vertical separation might increase with depth, decrease to zero up-section, or even decrease with depth. A fault may die laterally, have its displacement transferred to other faults or to folds, combine with other faults, or intersect with or terminate against another fault. In areas of salt diapirs, a fault may terminate against salt, extend through it, or even be deformed due to strata draping around the diapir or by salt movement. We again emphasize that a good interpretation of a fault surface must have 3D validity and comply with the tectonic characteristics of the region being mapped, and you must use correct mapping techniques in its construction.

Figures 7-12 and 7-13 are examples of completed fault surface maps. The fault map in Figure 7-12 is that of the "S" Fault in the Indigo Bayou area, Iberville Parish, Louisiana. Although there is some variation in the amount of missing section for this fault, for the most part it appears to be a post-depositional (nongrowth) fault with a vertical separation ranging from 220 ft to 400 ft. The fault exhibits little, if any, growth with depth. This fault surface map is contoured in accordance with the guidelines and rules outlined in this section.

Contoured fault surfaces of selected normal faults in the Ivanhoe Field, U.K. North Sea, are shown in Figure 7-13 (Hooper et al. 1992). The map is an essential component of a computer-generated set of fault, structure, and isochore maps that were successfully combined in developing more accurate structural and volumetric models of reservoir units than existed at the time. The fault surface maps were used to improve accuracy in determining the intersections of mapped horizons with the faults, and that in turn was the basis for more precise net pay maps. The use of fault surface maps is essential in generating the most accurate maps possible. We present in Chapters 8 and 9 our methodology for integrating fault surface maps with horizon structure maps to generate accurate structure maps, and in Chapter 14 we describe their use in developing the most precise net pay maps.

TYPES OF FAULT PATTERNS

Extensional Faulting

Normal faulting may be defined as motion along a dipping fault surface on which the hanging wall block moves down relative to the footwall block. Normal faults commonly occur as a set with more or less parallel strikes but opposing dips, referred to as a *conjugate fault system*. Typically, each fault has a different amount of slip, with the fault having the major displacement called the **master fault**, and the fault with the relatively minor displacement called an **antithetic**,

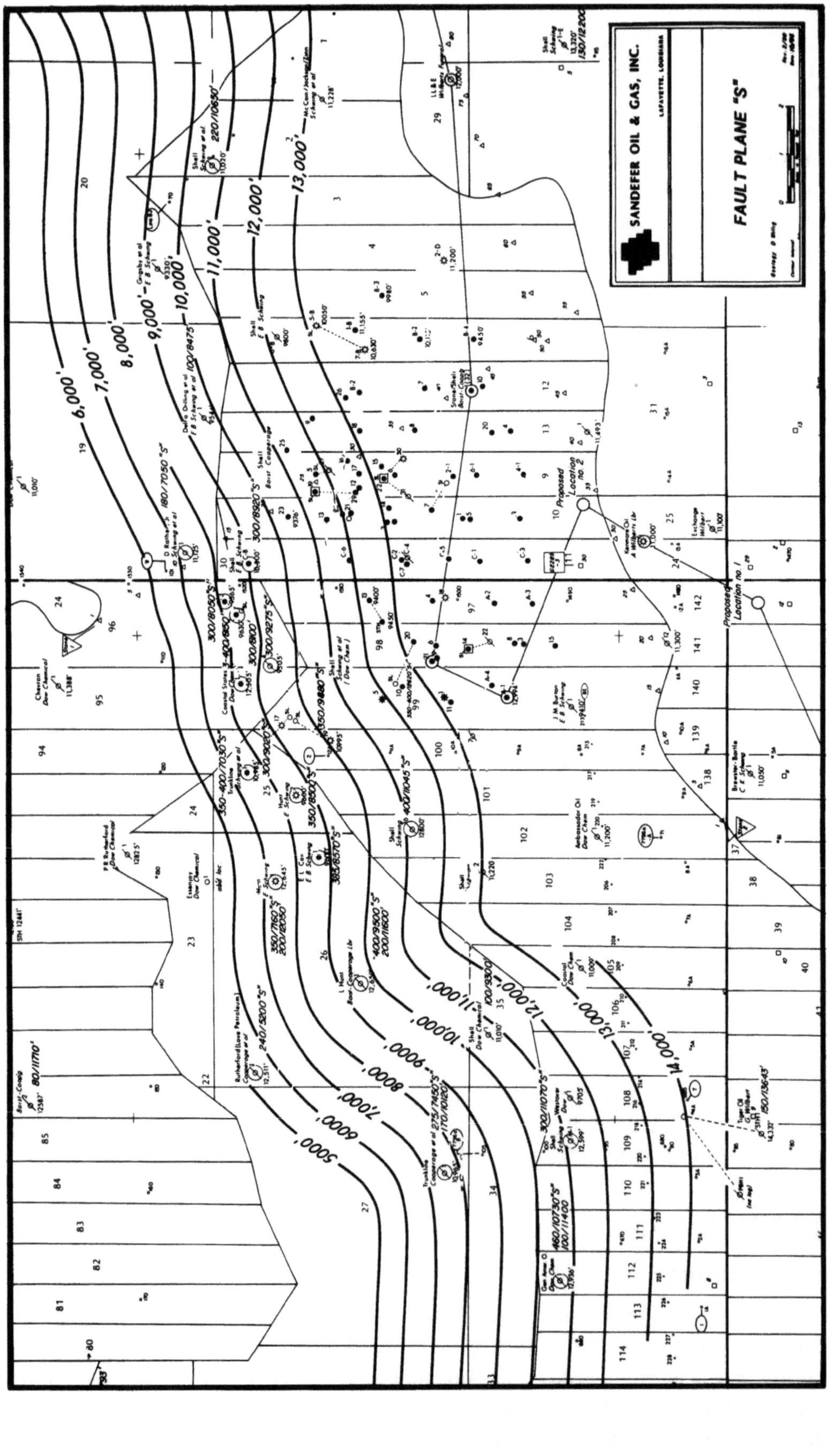

Figure 7-12 Fault surface map on Fault "S" at Indigo Bayou, Iberville Parish, Louisiana. Vertical separation varies slightly from well to well, ranging in size from 220 ft to 400 ft. The fault shows very little change in dip with depth. (Published by permission of Sandefer Oil and Gas, Inc.)

or **compensating, fault**. Normal faults are typically steeply dipping; however, the dips of normal faults may range from almost horizontal to vertical. A normal fault typically results in a missing stratigraphic section in electric well logs and a gap on a structure map between the intersections of the fault and the mapped horizon in the upthrown (footwall) and downthrown (hanging wall) fault blocks. This was illustrated in Figures 7-2 and 7-7. Normal faults can be growth (synsedimentary) or nongrowth in nature; can be isolated or display complex patterns; can be virtually planar, listric, or antilistric in profile; can die downward; or can even exhibit deformation.

Extensional basins associated with salt tectonics can have very complex fault patterns. The maximum principal stress axis in extensional basins is vertical, resulting in normal faulting with initial dips of about 60 deg or greater due to extension. Salt masses are commonly associated with crestal grabens, as well as radial and peripheral faulting. Antithetic faults, also referred to as *compensating faults*, are common in extensional areas.

In addition to single normal faults (structural and growth), there are three principal patterns of normal fault intersections and terminations common in areas of extensional tectonics. These patterns or systems, illustrated in Figures 7-14, 7-17, and 7-20, are (1) bifurcating, (2) compensating, and (3) intersecting. For each of the fault patterns discussed, a fault surface map, one or more cross sections, and a block model are provided to explain and illustrate the pattern.

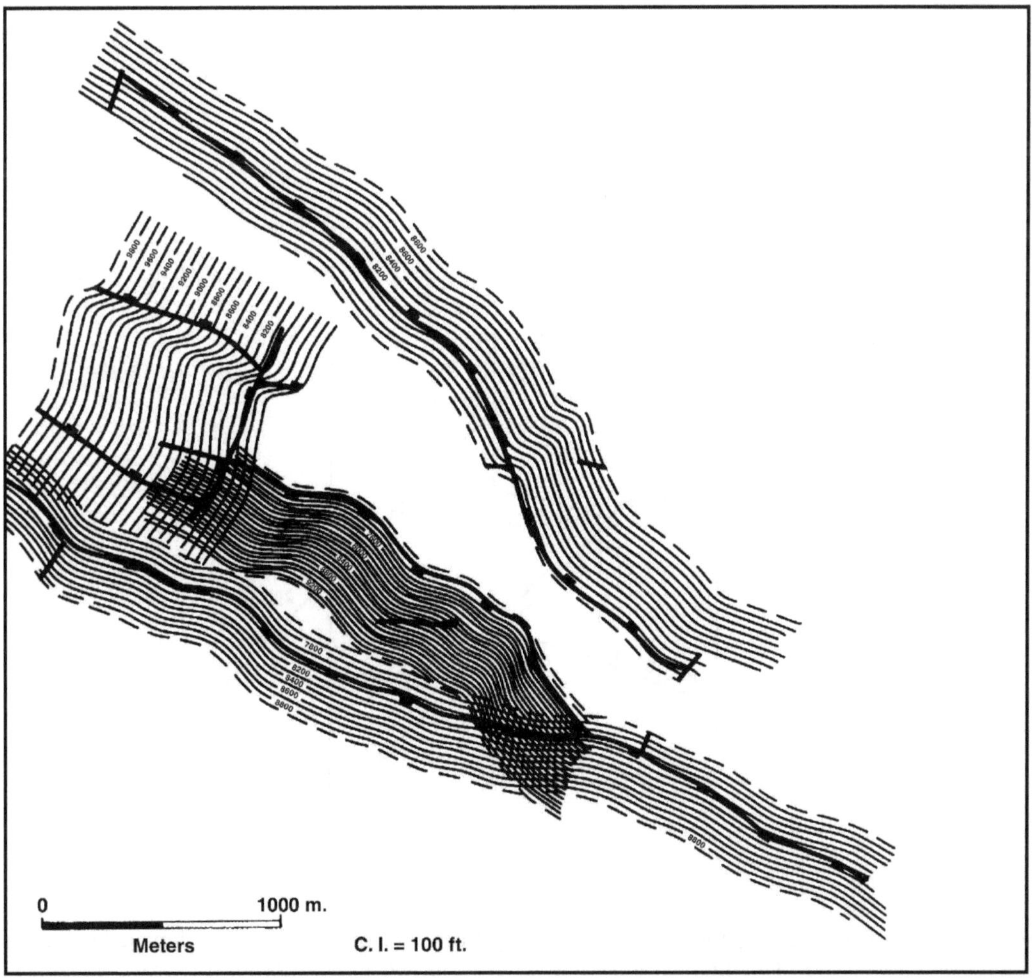

Figure 7-13 Computer-generated fault surface maps of selected normal faults in Ivanhoe Field, U.K. North Sea. Dashed lines indicate that a fault surface extends beyond the contours shown on this map. Wide lines are upthrown fault traces at one horizon. (Modified from Hooper et al. 1992; AAPG©1992, reprinted by permission of the AAPG whose permission is required for further use.)

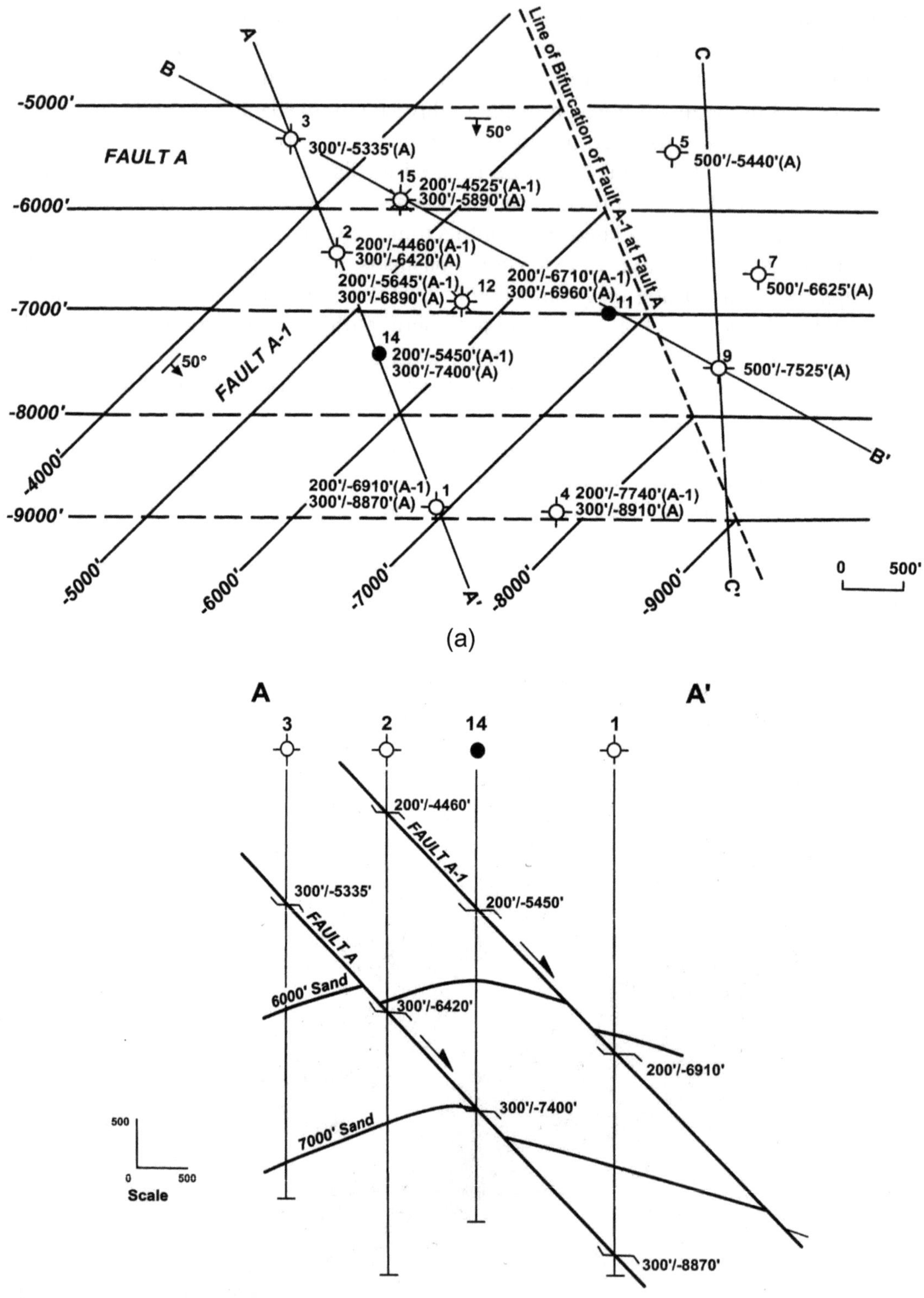

Figure 7-14 (a) Bifurcating fault pattern resulting from two merging faults, dipping in the same general direction. Line of bifurcation indicates where the two faults merge. (b) Cross section A-A′ bisects Faults A and A-1 in such a way that the two faults do not appear on the cross section as merging faults, but instead appear as two parallel faults. (c) Cross section B-B′ is laid out almost perpendicular to Fault A-1 and at an oblique angle to Fault A. In cross section, the faults appear to merge with depth rather than to merge laterally.

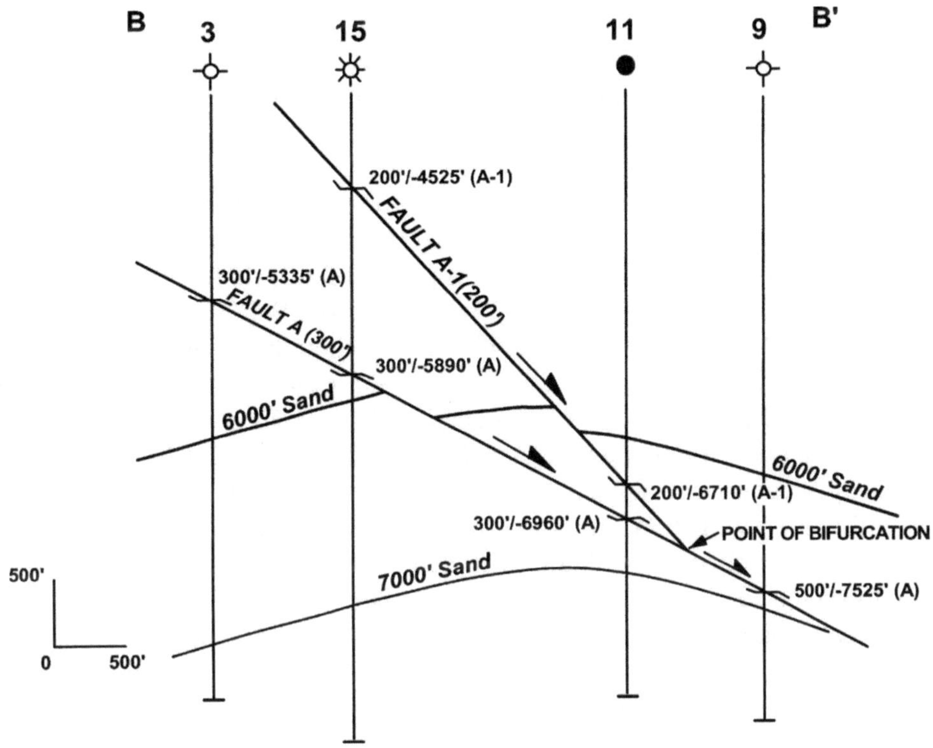

(c)

Figure 7-14 *(continued)*

Bifurcating Fault Pattern. A bifurcating fault pattern or system results from two normal faults that dip in the same general direction, as shown in Figure 7-14. The strike direction of each fault is such that the two faults merge laterally in the subsurface and continue on as one fault. The line along which the two faults *merge* is called the **line of bifurcation** or **intersection**. The total vertical separation of the fault across the line of bifurcation must be conserved. This means that the vertical separation of the single fault, where only one fault exists, is equal to or nearly equal to the sum of the vertical separations of the two faults, where two faults are present. The contoured fault surface map in Figure 7-14a shows two intersecting fault surfaces dipping in the same general direction. This interpretation was made using fault data from 11 wells, the contouring guidelines, and an understanding of the geological setting.

Using Figure 7-14, we review the fault system in detail and illustrate the specific characteristics that classify this as a bifurcating fault pattern. Fault A is striking east-west and dipping 50 deg to the south. Fault A-1 is striking northeast-southwest with a dip of 50 deg to the southeast. The two faults are dipping in the same general direction.

Fault A-1 merges with Fault A where the two faults intersect, as indicated by the dashed line of bifurcation. It is the result of the intersection of contours of the same value on the two faults. There are two faults present west of this line. Fault A has a vertical separation of 300 ft, and the vertical separation for Fault A-1 is 200 ft. East of the line of intersection only one fault (Fault A) exists, with a vertical separation of 500 ft. These vertical separation values across the line of bifurcation satisfy the conservation of vertical separation, also referred to as the *additive property* of faults (see Chapter 8). Notice that the contours for Fault A are dashed west of the line of intersection, indicating that the contour values are deeper than those for Fault A-1. This is a good contouring practice that helps reduce confusion on maps where more than one fault surface is constructed on the same base.

Figure 7-14b and c are two cross sections with a different orientation to the two fault surfaces for each line of section. Remember, cross sections used in conjunction with maps provide another viewing dimension that can be helpful in visualizing the geological picture and solving structural problems. The orientation of the section line, however, is very important. Chosen incorrectly, the line of section can be more confusing then informative.

In the two cross sections through the bifurcating fault pattern, the fault geometry appears different in each section. In cross section A-A′ (Fig. 7-14b), the fault pattern does not appear to be bifurcating. Instead, the two faults appear as parallel faults. Is this real or an optical illusion as a result of the line of section? In Figure 7-14c showing the B-B′ cross section, Fault A-1 appears to merge with Fault A with depth. Real or illusion? Although the two cross sections are geologically and technically correct, they can pose problems for those unfamiliar with fundamental geological principles. When laying out a cross section, be sure to consider the purpose of the cross section and your audience (Chapter 6). As illustrated by Figure 7-14b, a line of section parallel to the line of bifurcation will appear to show two parallel faults. A line of section that crosses the line of bifurcation will show either two faults that merge with depth, as in Figure 7-14c, or a single fault that separates into two faults with depth, as would be illustrated by section C-C′. It is worthwhile to sketch section C-C′ as an exercise to help reinforce this point.

The vertical separation values for the faults have been incorporated into each cross section to represent correctly the offset of the 6000-ft and 7000-ft Sands by Faults A and A-1. The term **bed offset** means that the horizon has been displaced by the fault, and the displacement is defined in terms of vertical separation. Earlier in this chapter, we showed that the missing section in a wellbore as the direct result of a normal fault is the measurement of the displacement in terms of vertical separation. Since this understanding is very important when preparing cross sections, we detail the procedure for using vertical separation in the preparation of cross sections in Chapter 6. Therefore, in the two cross sections in Figure 7-14b and c, the offset for the beds is constructed using the vertical separation from the wellbore fault data. We cannot overemphasize that wellbore fault data are not throw; therefore, we cannot construct a cross section using the fault data as throw.

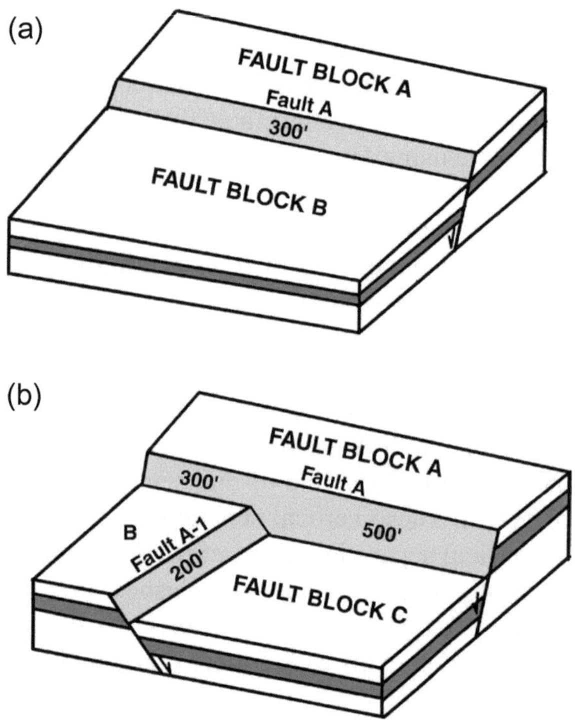

Figure 7-15 Block models show the development of a bifurcating fault pattern. (a) Fault 1 develops with a vertical separation of 300 ft. (b) Fault A-1 develops (vertical separation of 200 ft) and terminates against Fault A. West of the intersection of the two faults, the movement of Fault Block C is accommodated by an additional vertical separation of 200 ft on Fault A.

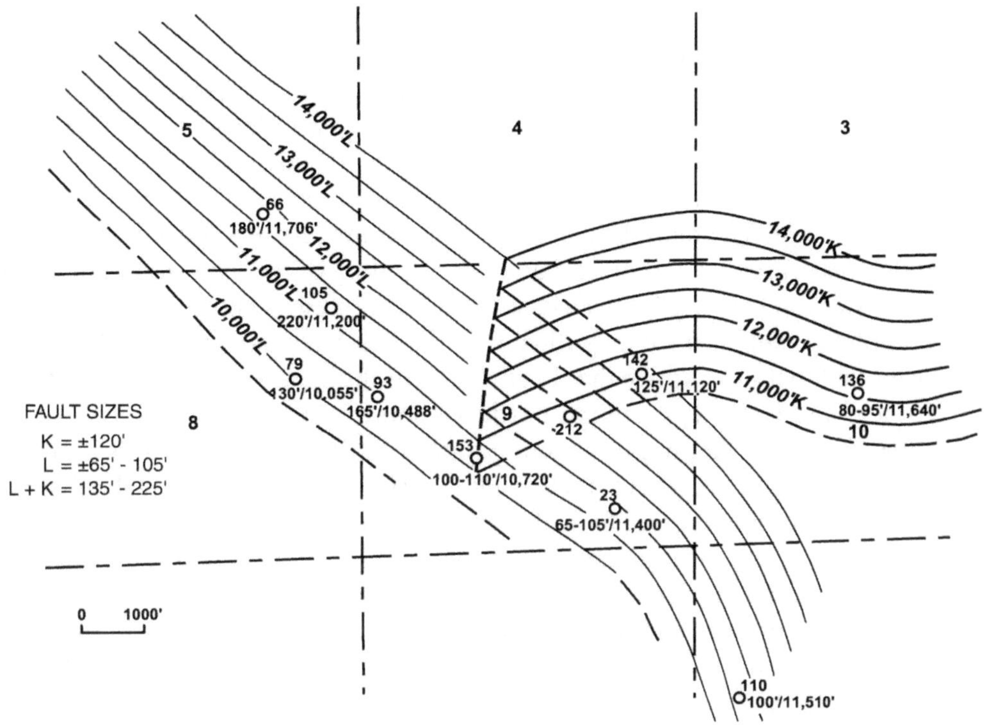

Figure 7-16 Example of a bifurcating fault system. Note the conservation of vertical separation on either side of the line of bifurcation. (Published by permission of Texaco, USA.)

Figure 7-15 is a block diagram of a bifurcating fault pattern. A review of Figure 7-15a and b illustrates the geological development of the fault pattern and individual fault blocks. Fault 1 develops first as the rocks fracture and Fault Block B moves downward, creating a fault with 300 ft of vertical separation. Then Fault Block C moves and creates Fault A-1 with a vertical separation of 200 ft. Because Fault A-1 terminates at Fault A, the surface of Fault A to the left of the intersection must accommodate the displacement of Fault Block C. Therefore, the vertical separation increases to 500 ft on Fault A to the left of the intersection. The term *bifurcating fault pattern* is somewhat unfortunate, as it implies that a single fault splits into two faults. As illustrated by Figure 7-15, this implication is not correct. A bifurcating fault pattern forms when two faults dipping in the same general direction merge to form a single fault, not when one fault splits into two faults.

Figure 7-16 is an example of a bifurcating fault pattern. The fault system shown is that of two faults dipping in the same general direction and merging laterally as indicated by the line of bifurcation. The sum of the vertical separations in the area where two faults are present, east of the line of bifurcation, is equal to or nearly equal to the vertical separation of the one fault west of the intersection.

Fault L dipping to the north-northeast is contoured between −9500 ft and −14,000 ft. The available well control east of the intersection of the two faults indicates that the missing section for Fault L ranges from 65 ft to 105 ft. Fault K, which dips to the north, is contoured between −10,500 ft and −14,000 ft. Based on the well control, the missing section for this fault appears to be about 120 ft. The sum of the vertical separations for Faults K and L east of the line of bifurcation is ±185 ft to 225 ft; west of the bifurcation line, where both faults have laterally merged into one, the vertical separation of Fault L is 165 ft to 225 ft. The nearly equal values for the vertical separation on both sides of the fault intersection show that the vertical separation across the line of bifurcation has generally been conserved.

Compensating Fault Pattern. A compensating fault pattern or system consists of two normal faults dipping in generally opposite directions toward one another (Fig. 7-17) with an acute angle between the strike directions of the two faults. At the line of intersection, one of the two faults terminates against the other. Conservation of vertical separation is maintained on either side of the line of termination, as demonstrated in the following discussion.

We can look in detail at the example fault surface map for the compensating fault pattern shown in Figure 7-17a. The fault data were obtained from 11 wells. Based on these fault data, the general guidelines presented earlier, and an understanding of the expected fault surface geometry in this setting, two intersecting fault surfaces were contoured as shown.

The completed fault surface map illustrates the specific characteristics that classify this as a compensating fault pattern. Notice that Fault A is striking east-west and dipping to the south with a dip of 50 deg, and Fault B is striking northeast-southwest and dipping at 50 deg to the northwest. The two faults are dipping in generally opposite directions and toward one another. Fault B terminates against Fault A where the two fault surfaces intersect at equal subsea elevations, as indicated by a dashed line referred to as the **line of termination**.

Southeast of this line of termination are two faults (Faults A and B). Fault A has a vertical separation of 100 ft, and the vertical separation of Fault B is 200 ft. Northwest of the termination line, only Fault A is present and has a vertical separation of 300 ft. These displacement values satisfy the conservation of vertical separation. Therefore, we say that Fault B is compensating with respect to Fault A.

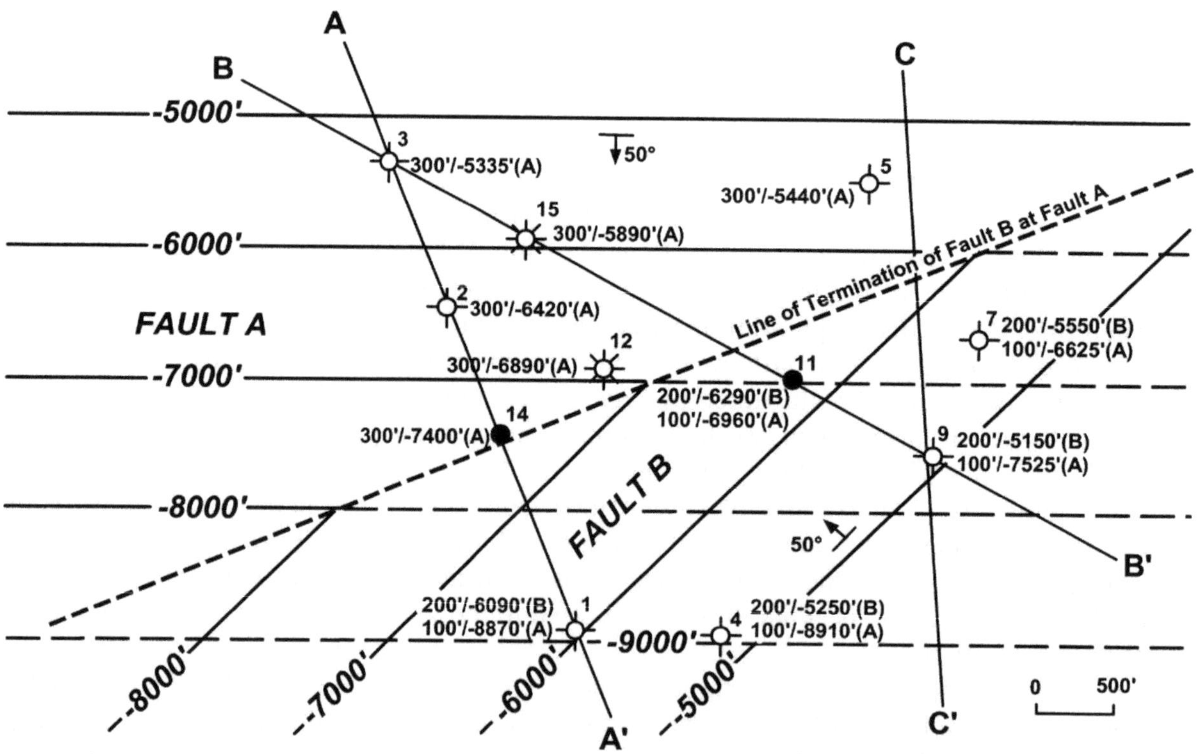

Figure 7-17 (a) Compensating fault pattern resulting from two intersecting faults dipping in generally opposite directions. Line of termination of Fault B at Fault A indicates the intersection of the two faults. (b) Cross section A-A′ illustrates the termination of Fault B at Fault A. Northwest of the intersection, Fault A has a vertical separation of 300 ft, whereas southeast of the intersection, Fault A is 100 ft. Fault B has 200 ft of vertical separation. The vertical separation is conserved across the line of termination.

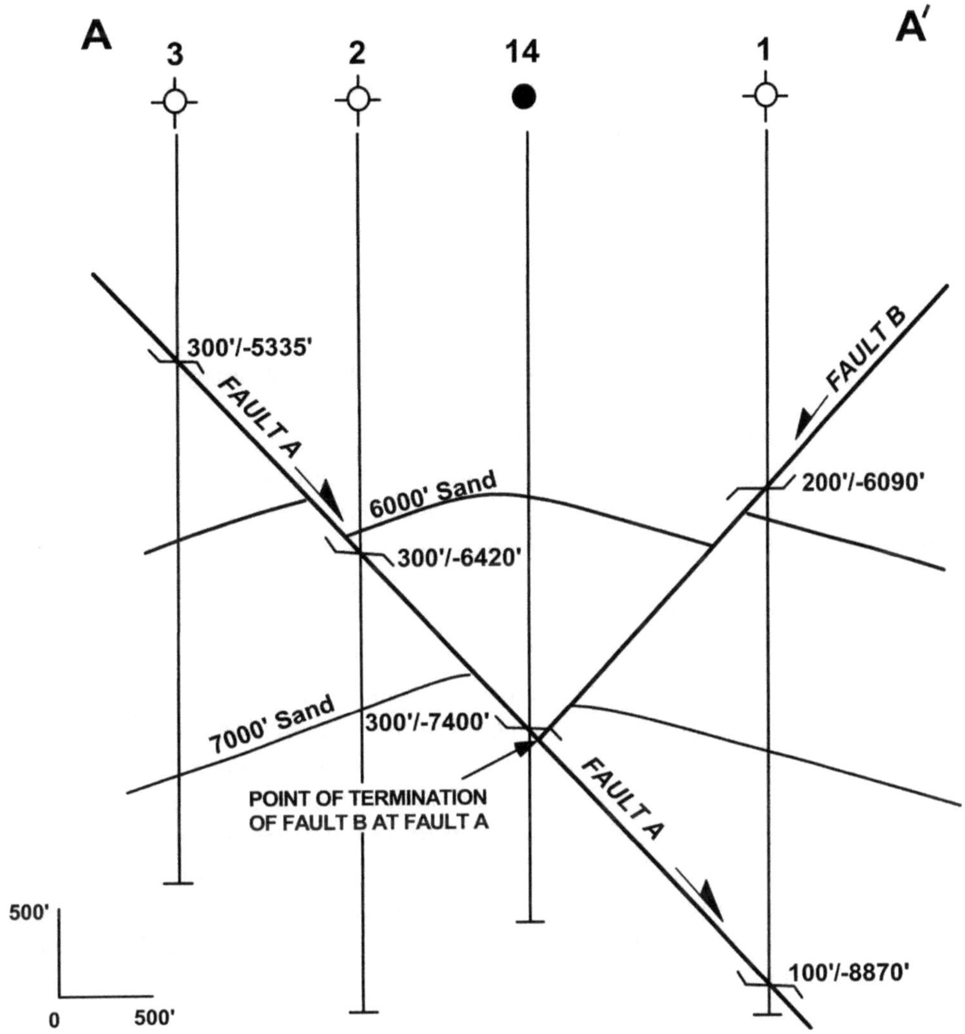

(b)

Figure 7-17 *(continued)*

Figure 7-17b is northwest-southeast stick cross section A-A′ shown in plan view on the fault contour map in Figure. 7-17a. The fault data from Wells No. 3, 2, 14, and 1, which lie directly on the cross section, are posted on the section. Fault B terminates against Fault A at a depth of −7460 ft, which corresponds to the point on the fault map (Fig. 7-17a) where the termination line for Fault B intersects the cross section. In the area where Fault A and Fault B are present, Fault A has a vertical separation of 100 ft, and the vertical separation of Fault B is 200 ft. This is shown in Well No. 1 by the fault cut point for Fault A of 100 ft at −8870 ft and 200 ft at−6090 ft for Fault B. Northwest of the termination of Fault B, only Fault A is present with a vertical separation of 300 ft shown in the fault cuts in Wells No. 2, 3, and 14. In Well No. 14, for example, the 300-ft fault cut is at a depth of −7400 ft. The vertical separation values for the faults have been incorporated into the cross section to correctly represent the offset of the 6000-ft and 7000-ft Sands by Faults A and B.

Figure 7-18 is a block diagram of a compensating fault pattern. At times, you may hear the following as an explanation for the fault displacements within a compensating system: “Think of the system in this way. Northwest of the fault intersection, Fault 1 has a missing section of 300 ft. Since Fault 2 is 200 ft, it *takes away* 200 ft of displacement from Fault 1 southeast of the intersection of the two faults, leaving 100 ft of displacement for Fault 1.” This explanation may

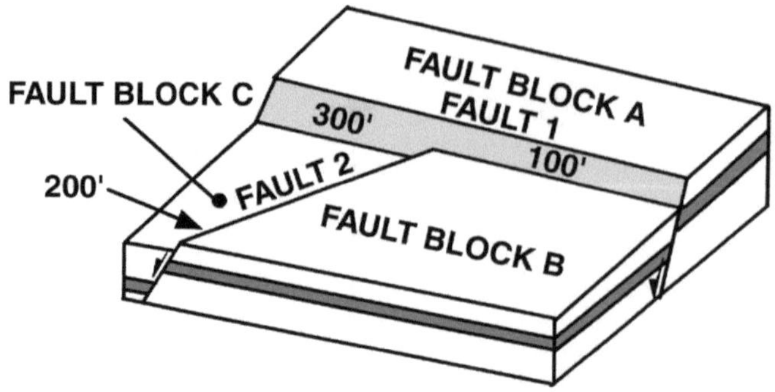

Figure 7-18 Block model of a compensating fault system.

provide you with a visual understanding of the missing section for each fault on both sides of the termination line, but technically it is incorrect and can lead to confusion. *One fault cannot take displacement away from another fault unless there is active inversion.*

Looking at Figure 7-19, think of the geological development of the fault pattern and individual fault blocks in the way they formed. Fault 1 develops as the rocks fracture and Fault Block B moves downward, creating a fault with a vertical separation of 100 ft. Next, Fault Block C moves and creates Fault 2 with a vertical separation of 200 ft. Fault 2 terminates at Fault 1, so Fault 1 to the left of the intersection must accommodate the displacement of Fault Block C. Therefore, the vertical separation increases to 300 ft on Fault 1 to the left of the intersection. Some geologists refer to such movement as a "reactivation of the older fault surface by the younger fault." If we check for conservation of vertical separation, we see that the 200 ft for Fault 2 plus the initial 100 ft for Fault 1 to the left of the intersection equal the final 300 ft of vertical separation for Fault 1 to the left of the intersection. Comparing Figures 7-15 and 7-19, can you see that the orientation of the younger fault is the only fundamental difference between the bifurcating and compensating fault patterns? For simplicity, we use examples of fault systems in which one fault is implied to be younger. It is also possible that the two faults are contemporaneous.

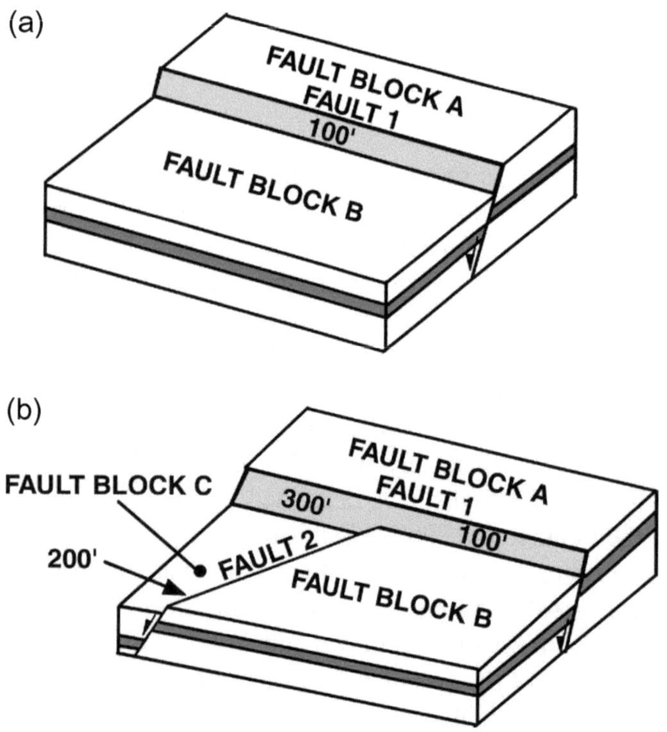

Figure 7-19 Block models show the development of a compensating fault pattern. (a) Fault 1 develops with a vertical separation of 100 ft. (b) Fault 2 develops (vertical separation of 200 ft) and terminates against Fault 1. The portion of Fault 1 west of the intersection of the two faults accommodates the additional 200 ft of movement by Fault Block C, and its vertical separation increases to 300 ft.

Intersecting Fault Pattern. So far, we have discussed two types of fault patterns in which one fault merges or terminates against another fault at their intersection. Now we look at the intersecting fault pattern, which results from two faults (*normal* or *reverse*) dipping in such a manner as to intersect in the subsurface; unlike the bifurcating system, in which the two faults merge, or the compensating system, in which one fault terminates against the other, both faults continue downward. The geometric relationship of intersecting faults is very difficult to visualize. Block models can help illustrate this pattern in three dimensions, and 3D seismic data are at times a fantastic data source from which to view these patterns.

Because of the complexity of this fault pattern, a correct interpretation is rarely achieved, even in areas of adequate well and 2D seismic control (Dickinson 1954). When considering the three fault patterns discussed in this section, the intersecting fault pattern presents the most complexities and the solutions are not at all straightforward. With limited available data, a decision must be made whether the intersecting faults formed contemporaneously (Horsfield 1980) or are of two different ages. Without good seismic control, such as a 3D survey, it is often difficult, if not impossible, to determine if the faults formed contemporaneously or at different times. A possible determination is the strike of the two faults. If the material is homogeneous and contains no significant pre-existing discontinuities, then contemporaneous faults will form in response to a single regional stress field, and the strike of these faults may be parallel or subparallel. The strike of normal faults will be perpendicular to the minimum horizontal stress; the strike of reverse faults will be parallel to the minimum horizontal stress. A second hint to whether the faults are contemporaneous is the dip of the faults. Contemporaneous intersecting faults should have the roughly same dip unless the faults have been rotated by subsequent tectonic action.

If the conclusion is that the faults are of two different ages, the next step is to determine which fault formed first and which was second. Such conclusions affect the construction of the fault surface maps, as well as completed structure maps. Today it is frequently possible in areas with good 3D seismic to determine which fault formed first and which was second.

Because of the complexities and uncertainties surrounding this fault pattern, *we recommend that the intersecting faults be mapped as if there is no offset of one fault by the other unless the data is adequate to determine which fault is younger and which fault is older*. This assumption does result in some error around the intersection of the faults, as even contemporaneous faults do not pass through each other without offset (Ferrill et al. 2000; Ferrill et al. 2009), but it does save time, and the actual interpretation may be impossible to determine from data available. This error extends to any resulting integrated structure map. This compromise should usually introduce less error than an incorrect guess as to the age of faulting. This subject is further discussed in Chapters 8 and 9.

For the intersecting fault example shown in Figure 7-20a, we make one of two assumptions: (1) the faults are contemporaneous, or (2) the relative age of the faults is unknown. With these assumptions, we construct a fault map with both faults meeting at their intersection and continuing downward, unaffected at and past their intersection. The fault maps were prepared using fault data from the 11 wells shown on the map. Fault A is striking east-west and dipping to the south at 50 deg, and Fault B is striking northeast-southwest with a dip of 50 deg to the northwest. The intersection of the two faults is shown as a dashed line. Beneath their intersection, the two faults continue downward with no change in vertical separation. These values are not affected by the intersection of the faults, as they are in the bifurcating and compensating fault systems.

Around the area of fault intersection, a *chaotic zone* may exist in which both the strata and fault surfaces are disrupted (Ferrill et al. 2000; Ferrill et al. 2009). However, well control and seismic data rarely can identify such disruption and, therefore, the mapping around these intersections is inaccurate. This must be kept in mind when mapping horizons affected by intersecting faults.

The cross section A-A′ shown in Figure 7-20b illustrates the simplified (or compromised) method of preparing the fault surface maps as if the two faults were contemporaneous. Although neither fault surface is offset on the fault map, when the fault surface map is integrated with a structure map, all four resultant fault traces may be offset. This is covered in detail in Chapter 8.

Figure 7-21 is a block diagram of an intersecting fault pattern resulting from two different ages of faulting. Fault 1 developed first, followed by Fault 2. Notice in the figure that Fault 2 cuts through Fault 1, displacing the older fault surface and causing a gap in this displaced surface. Since both faults carry to depth, no change in vertical separation occurs for either fault below their intersection.

An interpreted seismic line from a 3D survey over an offshore Gulf of Mexico field is shown in Figure 7-22. It shows very clearly an intersecting fault pattern in which both faults appear to pass through one another as if the faulting were contemporaneous, similar to the example used in Figure 7-20. Intersecting faults dipping in generally opposite directions, as shown in the figure, are given the special name **intersecting horst-graben** faults. Looking at the figure, we can see how this pattern gets its name. Above the fault intersection, the faults form a central graben block; below the intersection is a horst block, thus the name horst-graben fault system.

In areas where there is significant seismic data (3D data), it is sometimes possible to determine the relative ages of the faults. If this is possible, fault surface maps can be constructed for both faults, showing displacement of the older fault by the younger one. Integration of these fault maps with structure maps results in a more accurate representation of the fault intersection.

One final note on the different fault patterns: These patterns can be very complex, involving numerous faults in a single area (e.g., Figure 9 in Ferrill et al. 2009). Also, a fault need not remain as one pattern over its lateral or vertical extent. In other words, a fault that is part of a compensating fault pattern in one area can be part of an intersecting or bifurcating pattern in another (see Fig. 7-22).

Combined Vertical Separation. The term *combined vertical separation* applies to the relationship that results where two faults of different ages intersect. If a fault surface map similar to Figure 7-20a were constructed for intersecting faults of different ages, the fault surface of the older fault would be offset by the younger fault. The zone of combined vertical separation applies to that segment of the intersecting (younger) fault which lies between the offset surfaces of the displaced (older) fault. This zone is called the "zone of combined throw" by Dickinson (1954); however, his use of the word *throw* is a substitution for vertical separation.

Figure 7-23a and b illustrate a sequence of faulting involving two normal faults that results in a combined vertical separation. If a well penetrates the area of combined vertical separation, only one of the two intersecting fault surfaces will be crossed *(only one fault pick in the well)*, but the interval shortening or missing section will be equal to the sum of the vertical separations for both faults. The example in Figure 7-23 shows two intersecting faults of different ages dipping in generally opposite directions. The initial Fault 1 has 100 ft of vertical separation. The younger Fault 2, which has 200 ft of vertical separation, has displaced Fault 1 in a manner similar to a fault displacing a horizon. A review of the stratigraphic section in the area affected by both faults, penetrated by Well No. 2 in Figure 7-23b, shows a vertical shortening (or missing section in the well) of 300 ft, which is equal to the combined vertical separation for Faults 1 and 2.

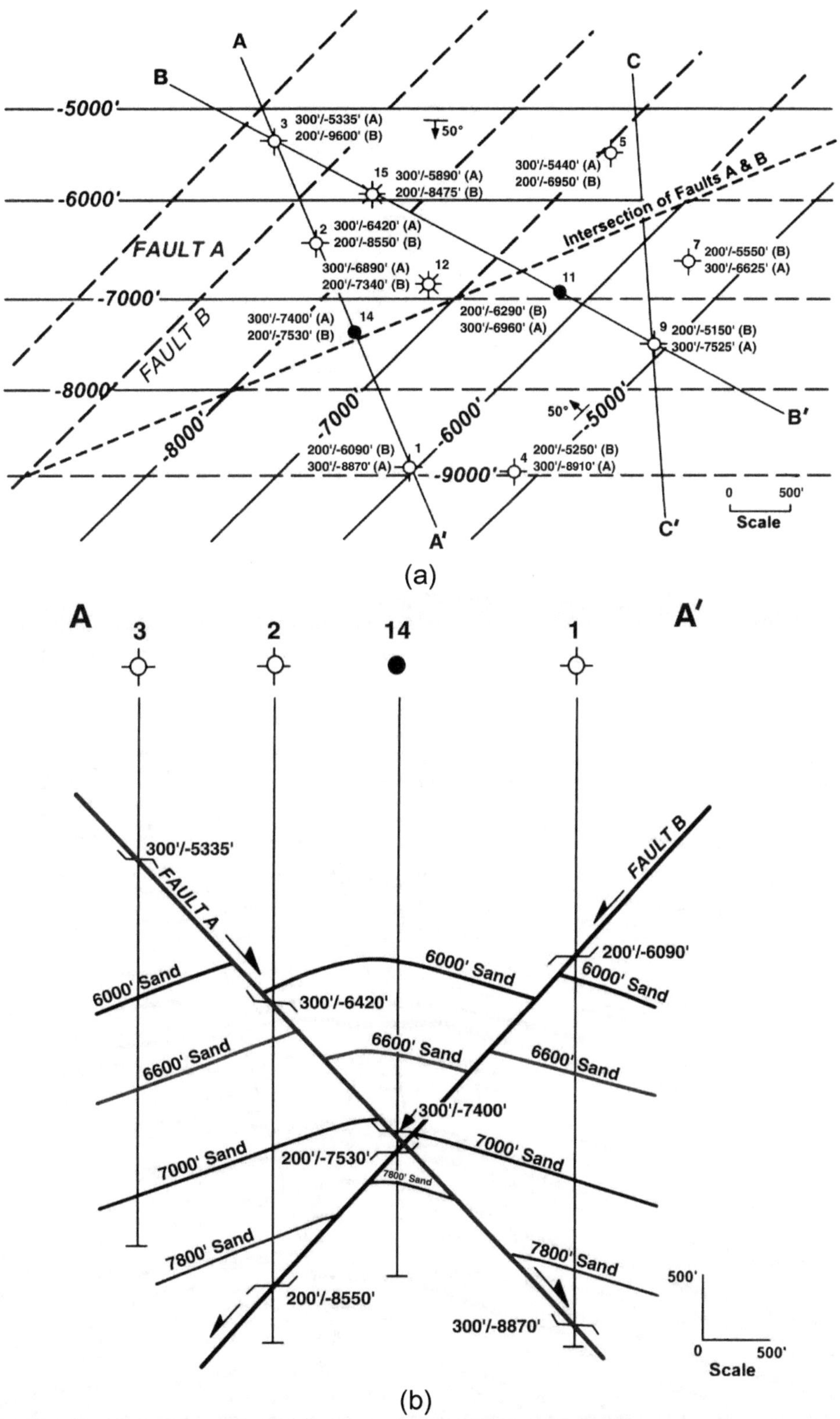

Figure 7-20 (a) Intersecting fault pattern resulting from two faults dipping in opposite directions. Unlike the compensating fault pattern, both faults continue beneath the intersection. This fault map was prepared using the simplified method of assuming neither fault surface is offset. (b) Cross section A-A′ illustrates this intersecting fault pattern. Observe that the faults form a central "graben" block above the intersection and a central "horst" block below the intersection.

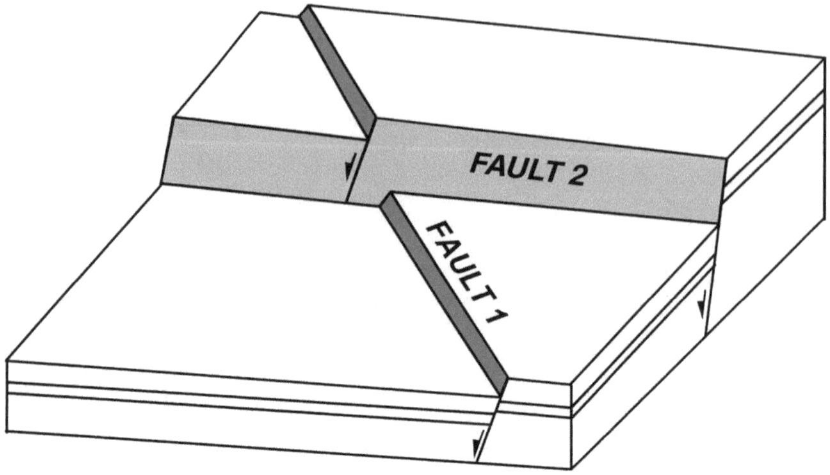

Figure 7-21 Block model of an intersecting fault pattern.

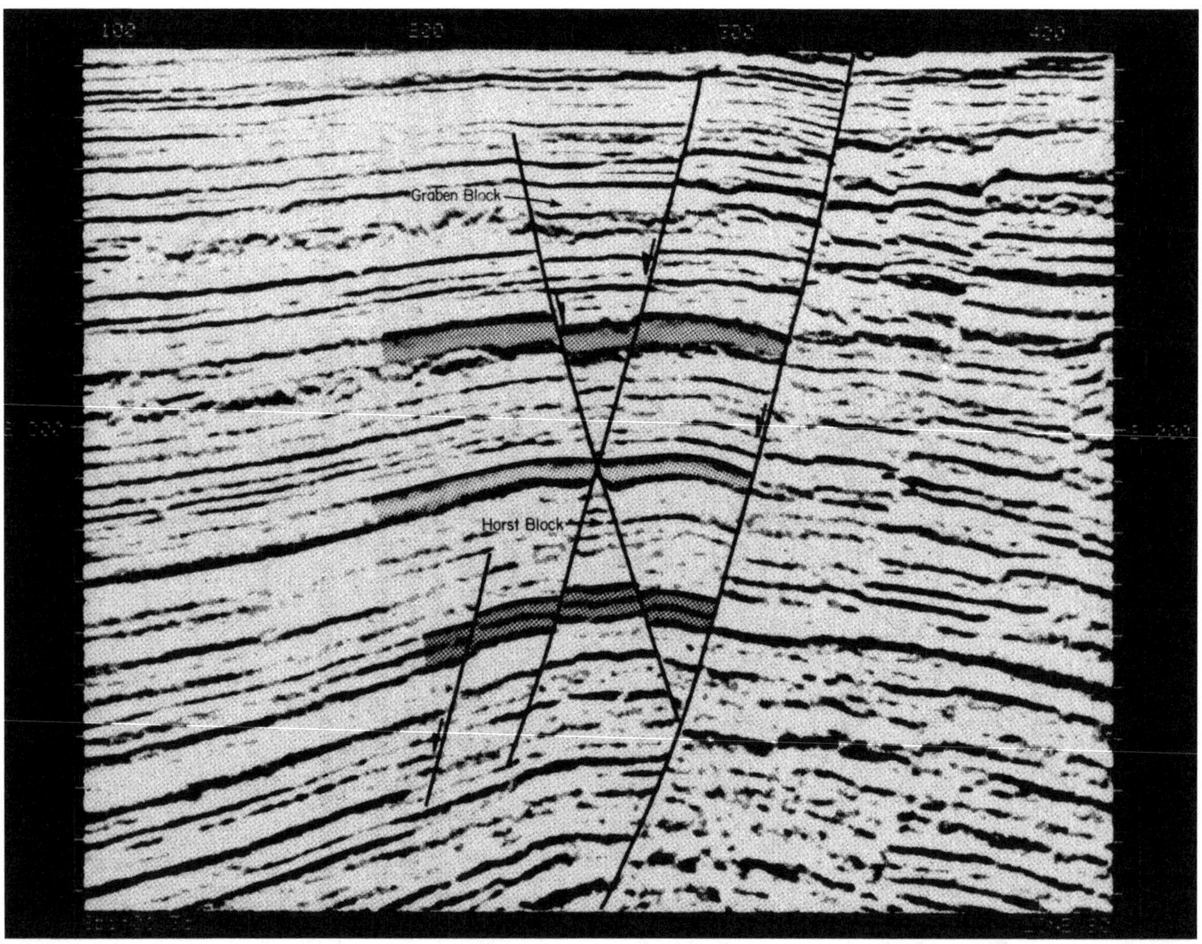

Figure 7-22 Seismic line from a 3D survey shows an intersecting fault pattern. Both faults appear to have formed contemporaneously, since neither fault is offset by the other. Notice that the east-dipping fault intersects and terminates against a second west-dipping fault, forming a compensating fault pattern. (Modified from Tearpock and Harris 1987. Published with permission of Tenneco Oil Company.)

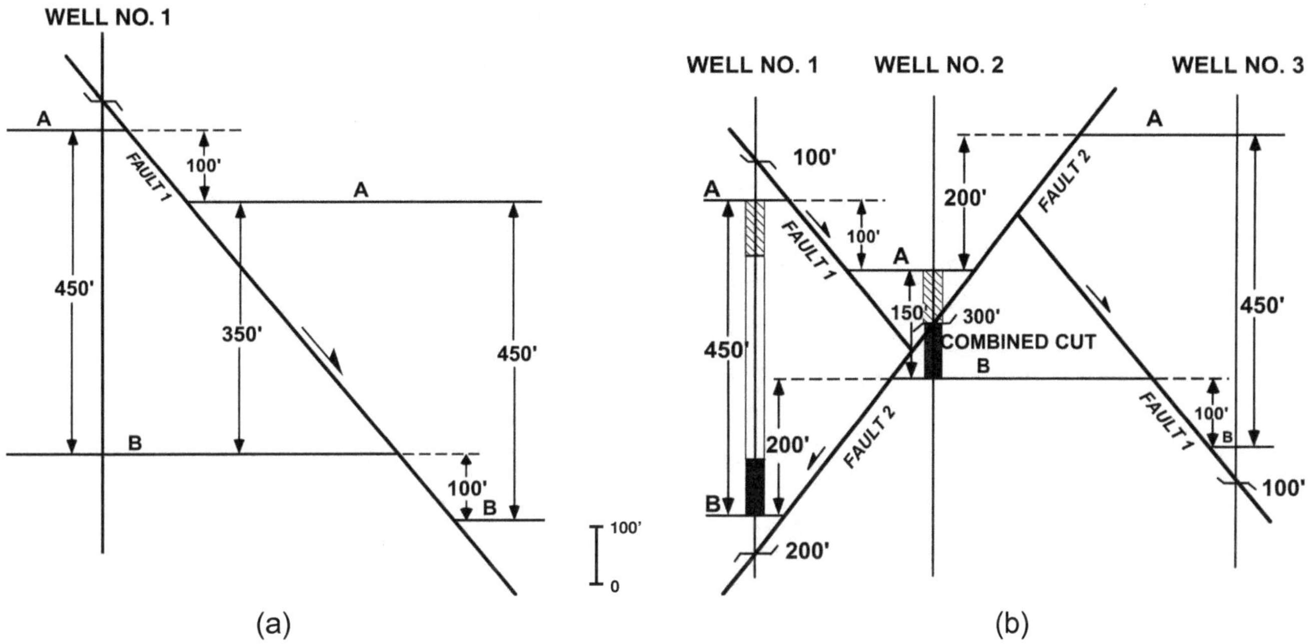

Figure 7-23 Schematic cross sections illustrating the zone of combined vertical separation (zone of combined fault cut), which develops from two intersecting normal faults. (a) Fault 1 with vertical separation of 100 ft. (b) Younger Fault 2, with vertical separation of 200 ft, offsets Fault 1. Well No. 2 penetrates one fault and has 300 ft of missing section. (Published by permission of D. Tearpock and J. Brewton.)

Figure 7-24 is a fault surface map for Fault J. This fault is the north-dipping component of an intersecting fault system composed of two faults of different ages and dipping in opposite directions. The north-dipping fault has a vertical separation of about 80 ft, and the vertical separation of the south-dipping Fault D is about 150 ft. Notice along the line of fault intersection that the fault cuts in Wells No. 2 and 109 are unusually large (235 to 250 ft) compared to the vertical separation of Faults D or J. These two larger fault cuts result from a combined vertical separation.

When working in an extensional area of complex or intersecting faults where an unusually large fault is present in one or more wells, keep the idea of combined vertical separation in mind. An unusually large cut could be the result of a new, previously unrecognized fault, a bifurcating or compensating fault pattern, a buried fault, or a combined vertical separation resulting from intersecting faults of different ages.

The effects of intersecting normal and reverse faults are illustrated in Figure 7-25a through d. The upper part of each figure shows a structure contour map depicting the interruption of a dipping horizon "O" by various combinations of intersecting normal and reverse faults. The cross section in the lower part of each figure shows the structural effects of the intersecting faults on interval "O-P," which has a constant vertical thickness defined as "c." Although the vertical separation for the fault in the zone of combined vertical separation is a function of many variables, including horizon dip, fault dip, vertical separation for each fault, and the relative movements of the individual faults, it is usually equal to the algebraic sum of the vertical separation of both faults.

Where the intersecting (younger) fault is normal (Fig. 7-25a and b), the vertical separation of the intersecting fault in the zone of combined vertical separation is equal to the algebraic sum of individual vertical separations; that is, it is equal to (−b) plus (+a). The figures illustrate that where both faults are normal, the vertical separation across the intersected fault in the zone of combined vertical separation will also be normal. Where the intersected fault is reverse, however, the vertical separation will be normal only if the intersecting fault has the greater vertical

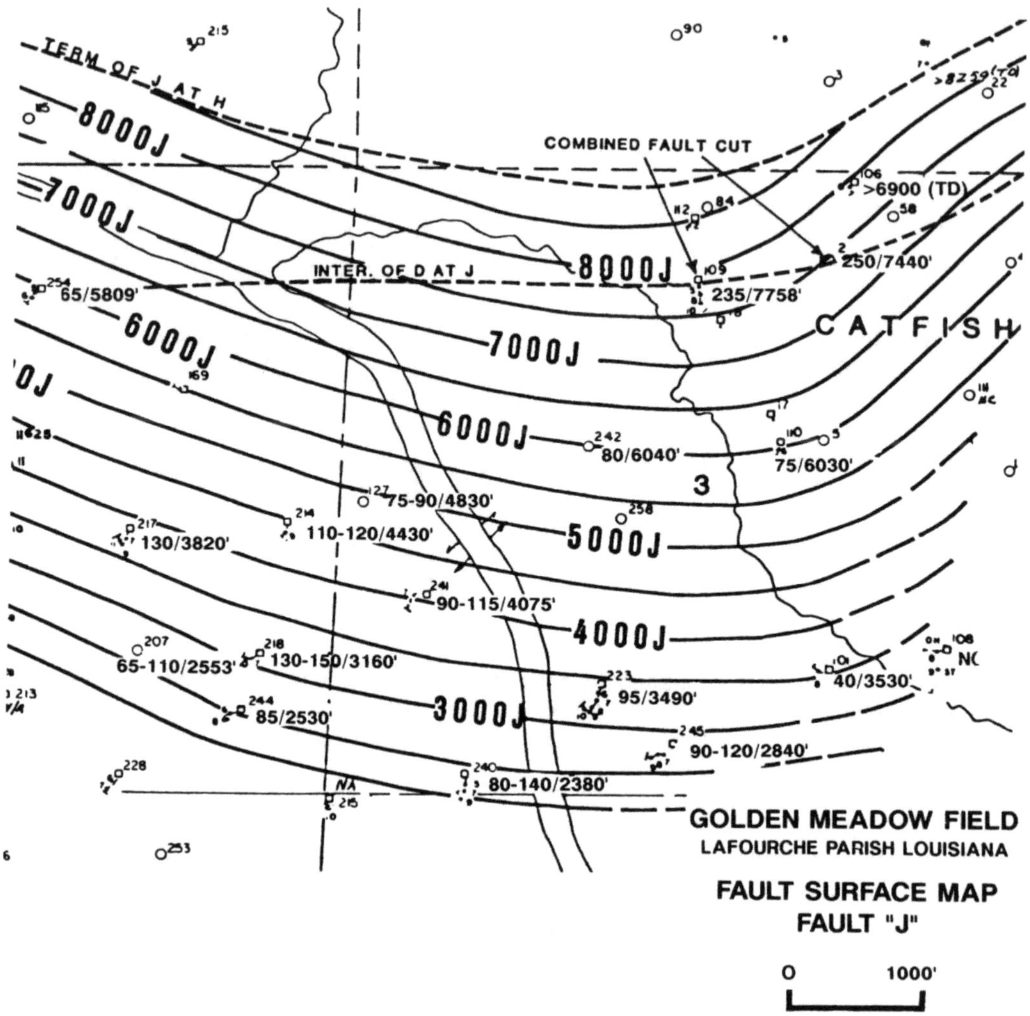

Figure 7-24 Wells No. 2 and 109 each have combined fault cuts as the result of the intersection of Faults D and J, Golden Meadow Field, Lafourche Parish, Louisiana. (Published by permission of Texaco, USA.)

separation and will be reverse only if the intersected reverse fault has the greater vertical separation (Dickinson 1954).

Figure 7-25c and d show the resulting fault geometry where the intersecting fault is reverse. In these two cases, the vertical separation across the intersecting fault in the zone of combined vertical separation is equal to the algebraic difference between the vertical separations of the intersecting and intersected faults; that is, it equals (+b) minus (+a). The vertical separation across the zone of combined vertical separation will be normal where the intersected fault is reverse and has a vertical separation greater than that of the intersecting fault. Where the intersected fault is reverse and has a vertical separation smaller than that of the intersecting fault, or where it is normal, the vertical separation across the zone of combined vertical separation is always reverse for a reverse intersecting fault. Unlike the geometry involving normal faults where only one fault is seen in the zone of combined vertical separation, the geometry in Figure 7-25c and d results in *three faults* (i.e., a well in that zone will penetrate three faults).

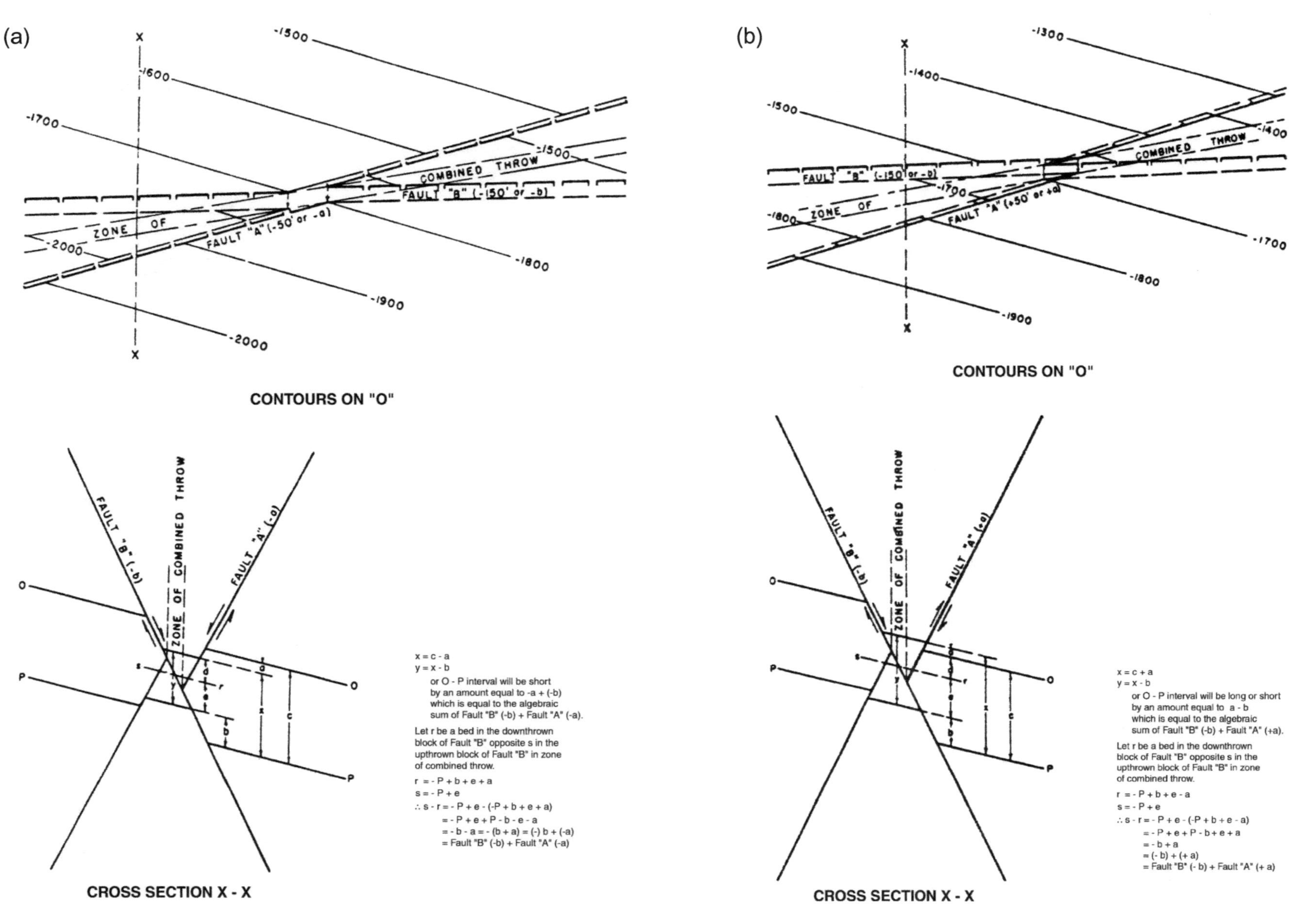

Figure 7-25 (a) Displacement across zone of combined vertical separation (referred to as combined throw by Dickinson [1954]) for intersecting normal faults. (b) Displacement across zone of combined vertical separation for a reverse fault intersected by a normal fault. (c) Displacement across zone of combined vertical separation for intersecting reverse faults. (d) Displacement across zone of combined vertical separation for a normal fault intersected by a reverse fault. (Dickinson 1954; AAPG©1954, reprinted by permission of the AAPG whose permission is required for further use.)

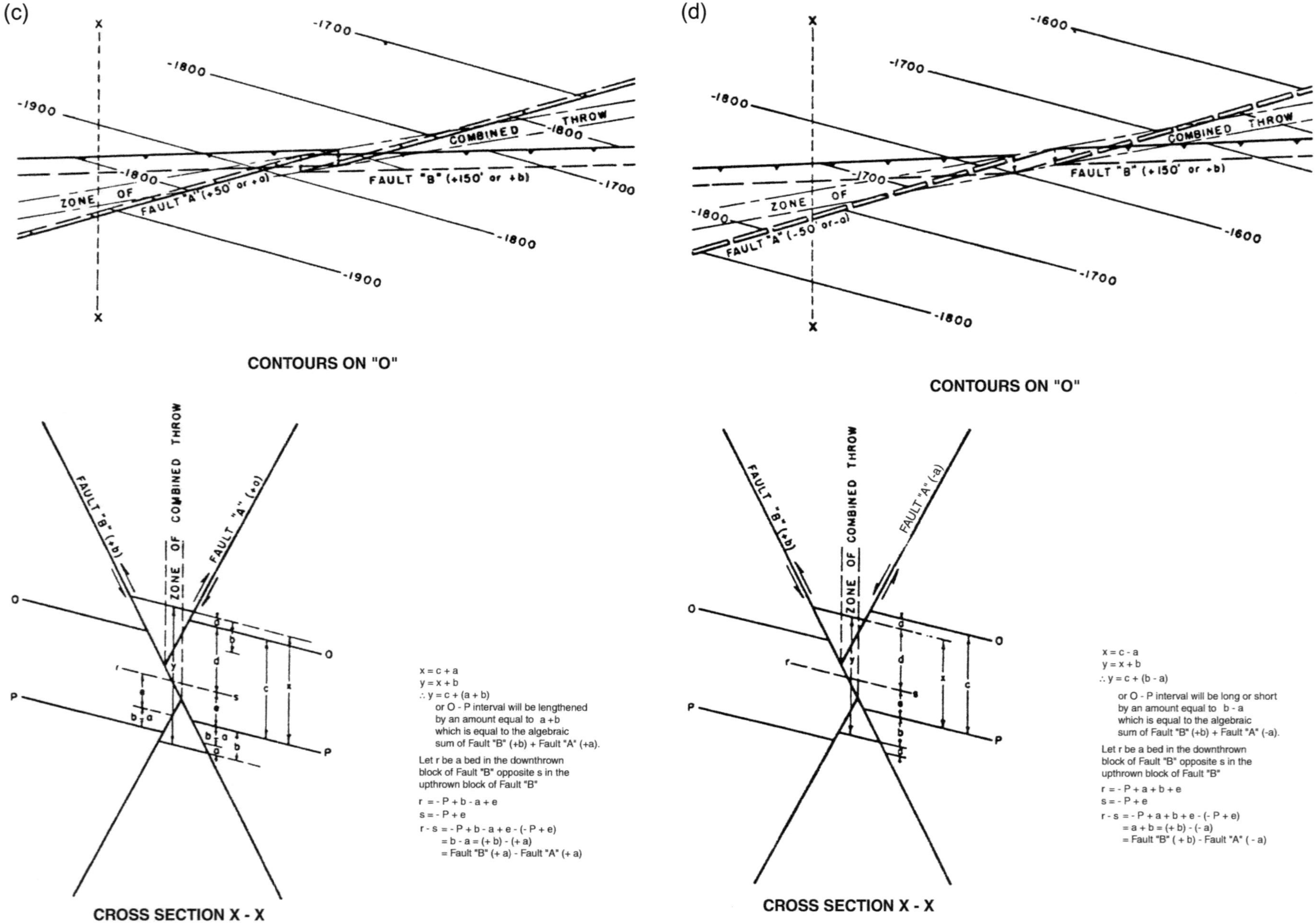

Figure 7-25 *(continued)*

Compressional Faulting

Reverse faulting is defined as motion along a dipping fault surface on which the hanging wall block rises relative to the footwall block. End members of the reverse fault spectrum consist of vertical and horizontal fault surfaces, as is the case with normal faults. An idealized reverse fault is shown in Figure 7-26. As in Figure 7-2, for normal faults, this figure graphically defines the common fault components used in mapping. Depending on relative amounts of fault and formation dips and wellbore deviation angles, normal faults usually omit section, whereas reverse faults usually repeat section in electric logs. Exceptions occur where the bed dips are near vertical, where wells deviate across the main normal fault may contain repeated section, and a reverse fault can have missing section.

However, in most cases dealing with vertical wells, the terms *normal* and *reverse* are in wide use to indicate the presence of missing or repeated section. We should recognize the origin of section omission and repetition during well log correlations and, in doing so, use the genetic terms *extensional* and *compressional* faulting. The previous section of this chapter dealt with the practical aspects of extensional faulting. In this section we discuss principles and practical aspects of compressional faulting.

In analogy with Figure 7-1 for normal faults, we derive from the frontal portions of Figure 7-27 a formula for *repeated section versus throw* (also see Chapter 10 for more complex examples of calculating repeated section with different hanging wall and footwall dips). The formulas are in the dip direction relative to the reverse fault. This formulation applies only to *straight reverse fault segments and not curved reverse faults*, and here the beds across the fault surface dip at the same angle. Using the trigonometry and equating to the dip slip (S), it follows from Figure 7-27 that

$$\sin(\theta) = AC/S, \text{ or}$$
$$S = AC/\sin(\theta)$$

where

S = dip slip = AB
AC = throw
ϕ_a = apparent bed dip measured in dip direction of fault
θ = fault dip
AE = repeated section or vertical separation (Fig. 7-27; also see Fig. 7-6).

From the law of sines, AC and AE are derived relative to S:

$$AC/\sin(\theta) = AE(\sin(\phi + \pi/2))/(\sin(\phi + \theta)) \quad (7\text{-}3)$$

And using trigonometric identities yields and rearranging yields

$$AE/AC = \sin(\phi + \theta)/(\sin(\theta))(\cos(\phi))$$
$$AE/AC = \tan(\phi)/\tan(\theta) + 1.0 \quad (7\text{-}4)$$

The AE/AC ratio can be calculated or determined from the repeated section equation Eq. (7-3) or (7-4) or by measured AE/AC with a scale. For Figure 7-27, the calculated AE/AC = 1.1466.

For Figure 7-28, the fault dips to the left and is in a different quadrant. Thus,

$$AE/AC = 1.0 - \tan(\phi)/\tan(\theta) \quad (7\text{-}5)$$

For Figure 7-28, using Eq. (7-5), $AE/AC = 0.48$ and as $AC = 200$ ft, $AE = 96$ ft (1 m = 3.28 ft).

In general, our study found that irrespective of the fault dip, either dipping to the left or to the right, if the bed dips in the opposite direction of the fault, then use Eq. (7-4). If the bed dips in the same direction as the fault, then use Eq. (7-5). Measurements of *AE*/*AC* can confirm the calculations.

In compressional areas, the determination of fault data begins with well log correlation, just as it does in extensional areas. For a reverse fault, the repeated section, recognized on an electric log by correlation with surrounding well logs, is equal in value to the vertical separation (Tearpock and Harris 1987). Figure 7-28 diagrammatically represents the determination of fault data by the correlation of well logs. This east-west structural cross section is perpendicular to the strike of the fault surface and shows two beds that have been cut by a reverse fault (F-1). At depth, the reverse fault cuts Well No. 1 near −5250 ft and is dipping at an angle of 35 deg to the west, whereas the strata are dipping at 20 deg to the west. By correlation with Well No. 2, the repeated section in Well No. 1 is calculated to be equal to 96 ft and is shown to have completely repeated Bed A in both wells. As shown in the cross section, the throw of Fault F-1 is represented by the vertical line AC, which is equal to 200 ft. Vertical separation, represented by the vertical line AE, is equal to the repeated section, which is 96 ft. This correlation example demonstrates that the repeated section obtained for the fault due to a reverse fault is equal to vertical separation. Notice, however, that the throw is not determined at Well No. 2, but rather from an accurately drawn cross section constructed from Wells No.1 and No. 2 along with the bed and fault dips.

In summary, Figure 10-28 shows a cross section with two wells, each of which has intersected reverse Fault F-1. The fault and beds are dipping in the same direction. By log correlation, we see that the repeated section in each well is equal to 100 ft, which is the vertical separation (AE). Unlike the first example in Figure 7-27, in which the throw was about one-half the size of the vertical separation, here a measurement of throw reveals it to be over twice as large as the vertical separation. Once again, we diagrammatically show that the throw of Fault F-1 cannot be measured accurately by log correlation alone but by cross section, and also that the throw is not equal to the repeated section.

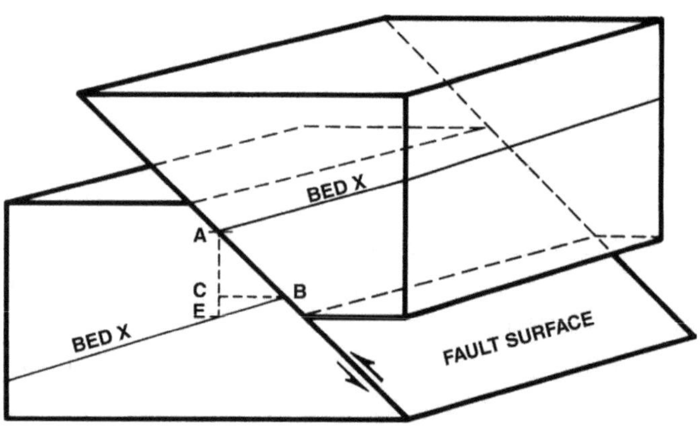

Figure 7-26 Block diagram of Bed X displaced by a reverse fault, showing four different fault components. The front panel is perpendicular to the strike of the fault. (Modified from Tearpock and Harris 1987. Published by permission of Tenneco Oil Company.)

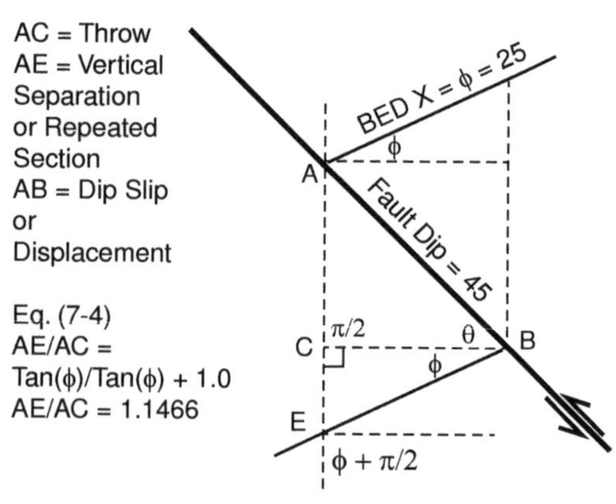

Figure 7-27 Dip cross section used to determine the relationship between repeated section shown by the vertical separation of the bed, which is the vertical line AE, versus the throw (vertical line AC). The AE/AE ratio is calculated from Eq. (7-4). (Published by permission of R. Bischke.)

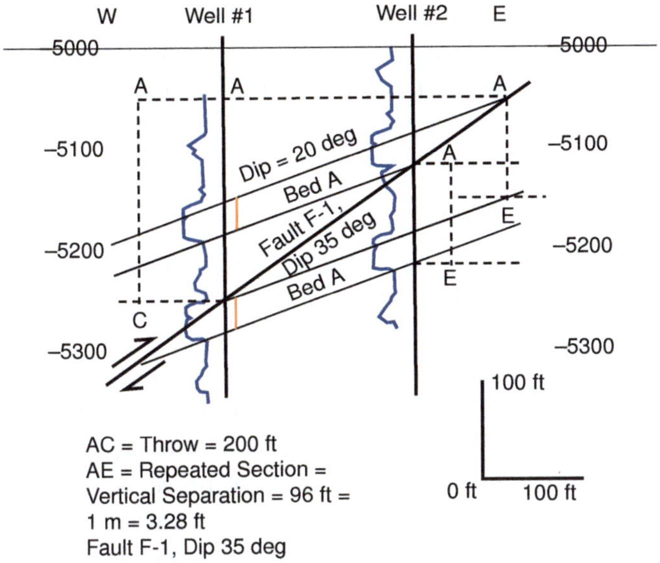

Figure 7-28 Diagrammatic cross section illustrating the relationship between a repeated section in a wellbore and the vertical separation where the beds are dipping in the same direction as the reverse fault. Eq. (7-4) is the formula used for the AE/AC calculation. Compare the relationship in this figure to those shown in Figure 7-27. (Published by permission of R. Bischke.)

Thrust faults are dip-slip faults in which the hanging wall has moved up relative to the footwall. Isolated thrust faults generally dip less than 30 deg during active slip and commonly dip between 10 and 20 deg at their time of formation (Suppe 1985). Many thrust faults lie along bedding planes over part of their length.

Thrust faults develop in two primary settings: (1) compressional plate boundaries and (2) secondary faulting in response to folding. The foreland of fold-and-thrust belts have been the most extensively studied areas of thrust faulting because of significant potential for both petroleum and coal reserves. There are two basic types of fold-and-thrust belts: (1) those dominated by thrust faulting such as the southern Appalachians and the Cordilleran belt in western Canada and the United States, and (2) fold-dominated belts such as the central Appalachians and Jura mountains of Switzerland (Suppe 1985).

Individual thrust faults commonly cut up the stratigraphic section in a sequence of discrete crosscutting segments called **ramps** and **flats**. The thrust fault separates the deformed hanging wall structure from the substrate along a bedding plane zone called a *decollement* (from the French word meaning "unglue"). Thrust and reverse faults, as with normal faults, can bifurcate in the ramp areas with total displacement distributed among the many fault splays and also absorbed in folds. In many cases the hanging wall blocks of thrust faults are folded; this results from bending of the fault blocks as they slip over nonplanar fault surfaces. These folds are called *fault bend folds* (Suppe 1985). The best-known fault bend folds are those associated with the stepping-up of thrust faults from lower to higher decollement surfaces.

Three names are commonly applied to compressional faults: (1) high-angle thrust faults, usually referred to as reverse faults (steeply dipping), (2) low-angle thrust faults (moderately dipping, sometimes defined as less than 45 deg of dip), and (3) overthrust faults (very shallow dip, usually with long horizontal transport).

How do we map the surface of a reverse fault? Most of the fault surface map and cross-section techniques already described for extensional faulting apply to compressional fault surface mapping. For the most part, the differences lie in the complexities of the compressional fault patterns and the limit of good-quality well and seismic data. Probably more than in any other tectonic setting, we must have a thorough understanding of the principles of structural geology and the structural styles at play, and we must be able to think and visualize in three dimensions.

Single Compressional Faults. Figure 7-29 is an example of a single compressional reverse fault surface map. The interpreted fault surface is based on data from seven wells. The techniques used to construct the fault surface map are the same as those used for normal faults. Notice in this example that the depth for each fault cut is expressed as a positive number (300 ft/5245 ft). This means that these fault picks are above sea level. The vertical separation of 300 ft refers to the vertical thickness of the repeated section.

Intersecting Compressional Faults. A compressional fault pattern of two reverse faults dipping in the same general direction is shown in Figure 7-30a. The data from 12 wells, the general guidelines presented earlier, and an understanding of the compressional tectonic setting were used to construct the fault surface map interpretation for the two reverse faults A and B. Fault A is striking east-west and dipping to the south at 35 deg, and Fault B is striking northeast-southwest and dipping 35 deg to the southeast. The two faults merge laterally as indicated by the dashed line of intersection. We can see by reviewing the fault surface map that the vertical separation is conserved across the line of intersection; this means that the combined vertical separations for Fault A (125 ft) and Fault B (150 ft) west of this line equal the vertical separation of Fault A (275 ft) east of the intersection.

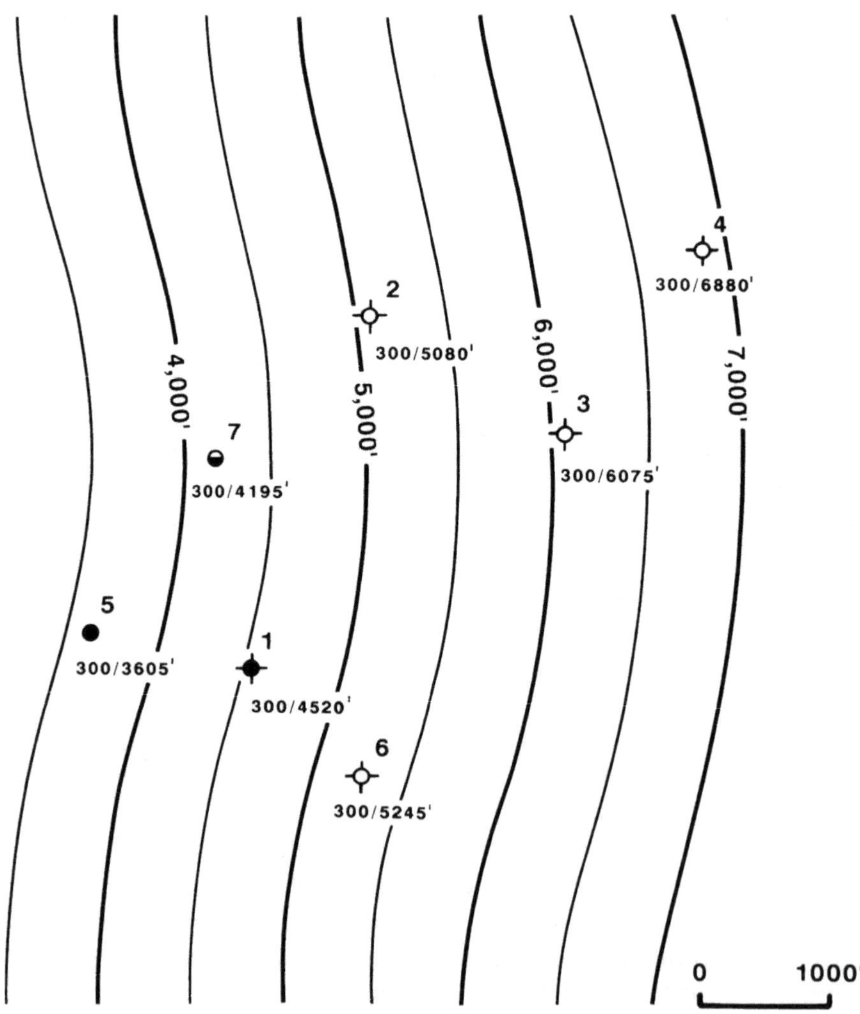

Figure 7-29 Fault surface map for a reverse fault with a vertical separation (repeated section) of 300 ft.

The change in displacement for these reverse faults across the intersection line is the result of their intersection. Do not mistake the change in displacement in this example with changes in fault displacement resulting from its transfer from one structure to another. In other words, this example is not related to what Dahlstrom (1969) refers to as the compensatory mechanism for reverse faults (thrusts) wherein fault displacement that is diminishing in one fault is replaced by increasing displacement on an echelon fault. Those displacement changes occur in a **transfer zone** and result from all the associated faults being rooted in a common sole fault. This type of faulting and change in displacement is discussed in Chapter 10.

Figure 7-30b shows cross section A-A′. The displacement for each fault has been incorporated into the offset of each stratum shown on the cross section. Remember, just as in the extensional faulting examples, the displacement of the strata *is described in terms of vertical separation*. Therefore, the footwall and hanging wall intersections for each stratum with Faults A and B are constructed with the vertical separation technique used to displace a stratum correctly from one fault block to another in cross section. This is the same technique used to construct the cross sections for the extensional and compressional fault patterns, which was introduced in Chapter 6.

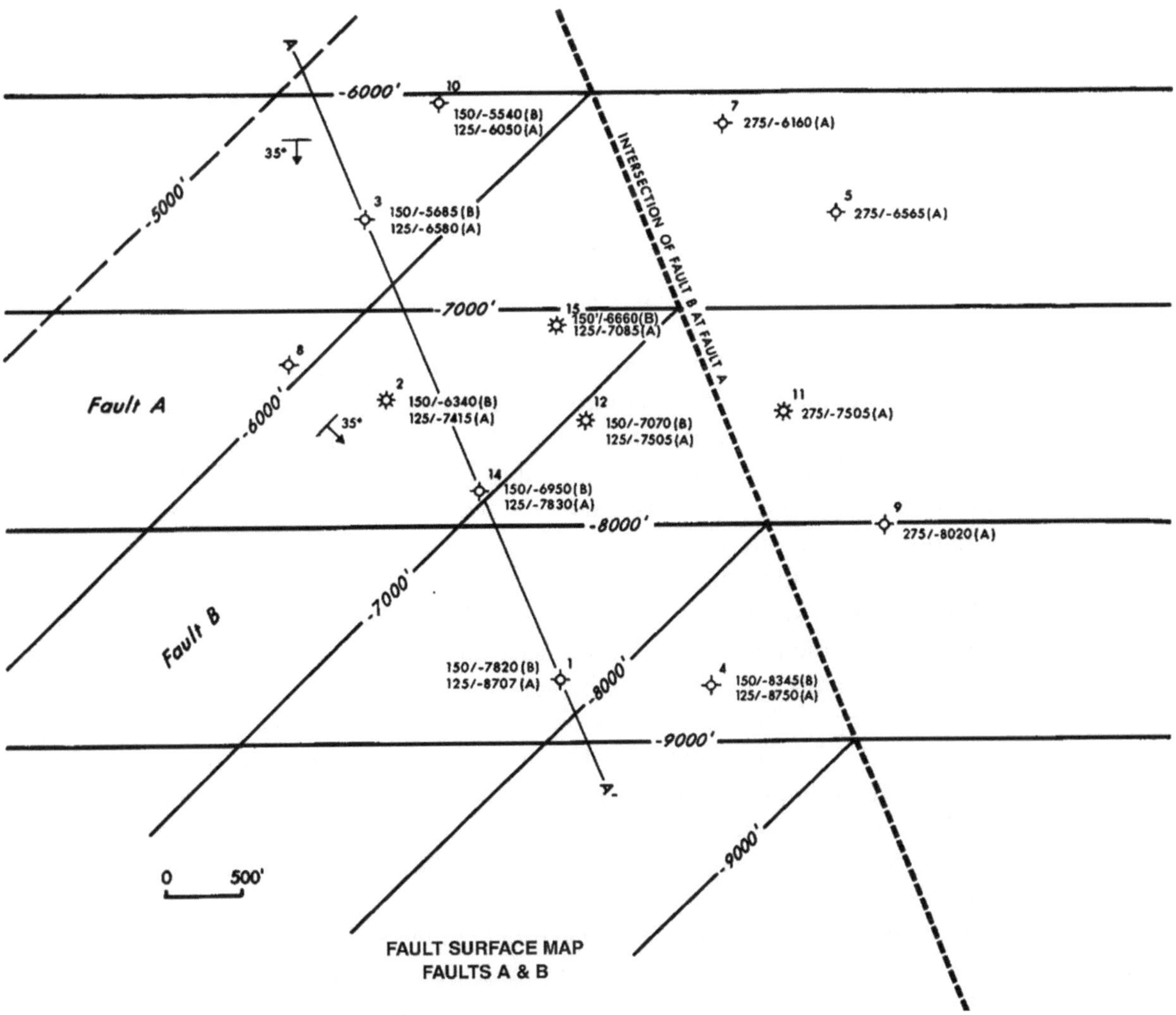

(a)

Figure 7-30 (a) Fault surface map of two intersecting reverse faults. (b) Cross section A-A′ crosses the two faults at an orientation that makes the two intersecting reverse faults appear to be parallel.

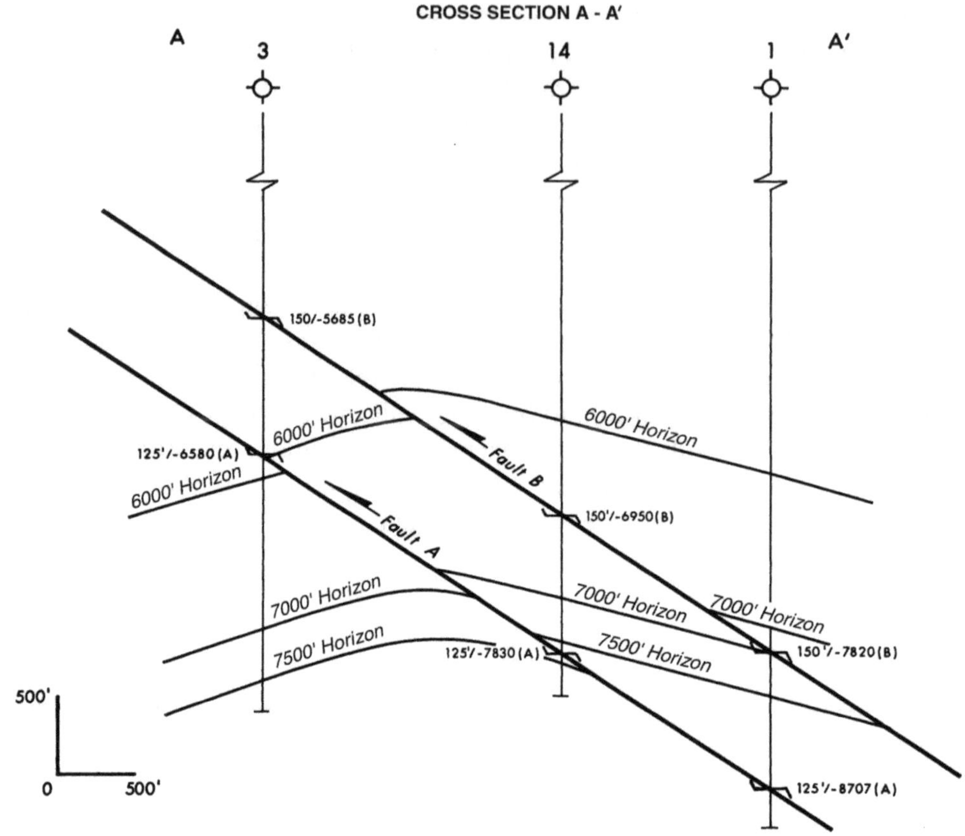

(b)

Figure 7-30 *(continued)*

Ramp and Flat Thrust Faults. Ramp and flat fault surfaces develop such that the fault surface moves subparallel to bedding in incompetent strata and ramps to higher structural levels in competent rock. If the ramp intersects another incompetent rock layer, then the ramp may transform again into another flat. This geometry is illustrated in Figure 7-31. Figure 7-31a is a fault surface map of an idealized ramp and flat thrust fault. As with all fault surface maps, the interpretive method of contouring is recommended. This gives you the geological license to construct the fault surface map to reflect the expected geometry in the area of study.

Figure 7-31b is cross section A-A′, which shows the fault surface and the disrupted beds. Notice how the strata in the hanging wall are thrust up and over the same strata in the footwall. The geometry of these faults is detailed in Chapter 10.

A fault surface map of the Absaroka thrust fault in the Fossil Basin of Utah and Wyoming is shown in Figure 7-32. Twenty oil and gas fields have been found in the hanging wall block of the Absaroka thrust fault in the southern Fossil Basin. The fault surface map in general reflects a steep ramp to the west from about −20,000 ft to −12,000 ft, a flattening or gentle rise in the central portion of the fault map between −12,000 ft and −5000 ft, and then another steepening ramp to the east (Lamerson 1982).

A portion of the Hogsback ramp and flat thrust fault is shown in cross-sectional view in Figure 7-33. This cross section clearly illustrates the features of the ramp and flat portions of this type of thrust fault. Observe that the flat parallels the Cambrian rocks, which serve as the zone of weakness for the basal detachment zone. As the Hogsback thrust fault cuts up-section across Paleozoic and Mesozoic rocks in the footwall, it creates a ramp with west dip of 20 to 30 deg in the hanging wall rocks of the thrust.

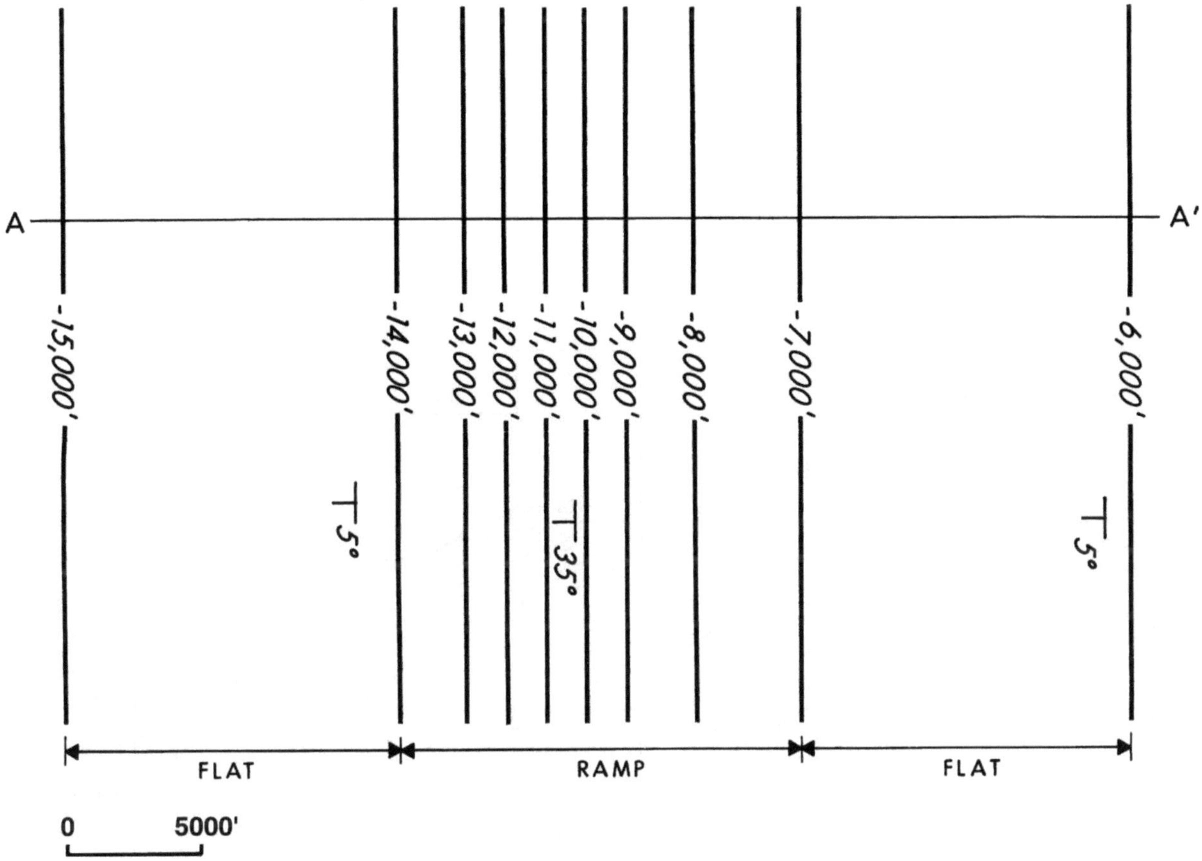

(a)

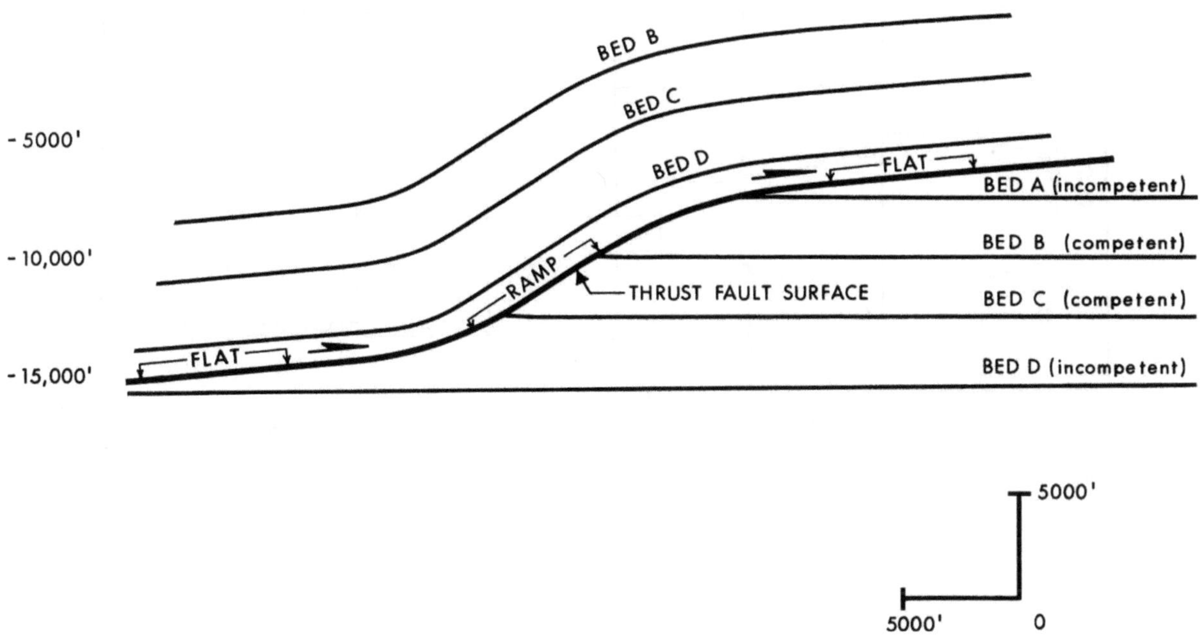

(b)

Figure 7-31 (a) Fault surface map for an idealized ramp and flat thrust fault. (b) Diagrammatic cross section, parallel to the transport direction of the thrust fault shown in Figure 7-31a.

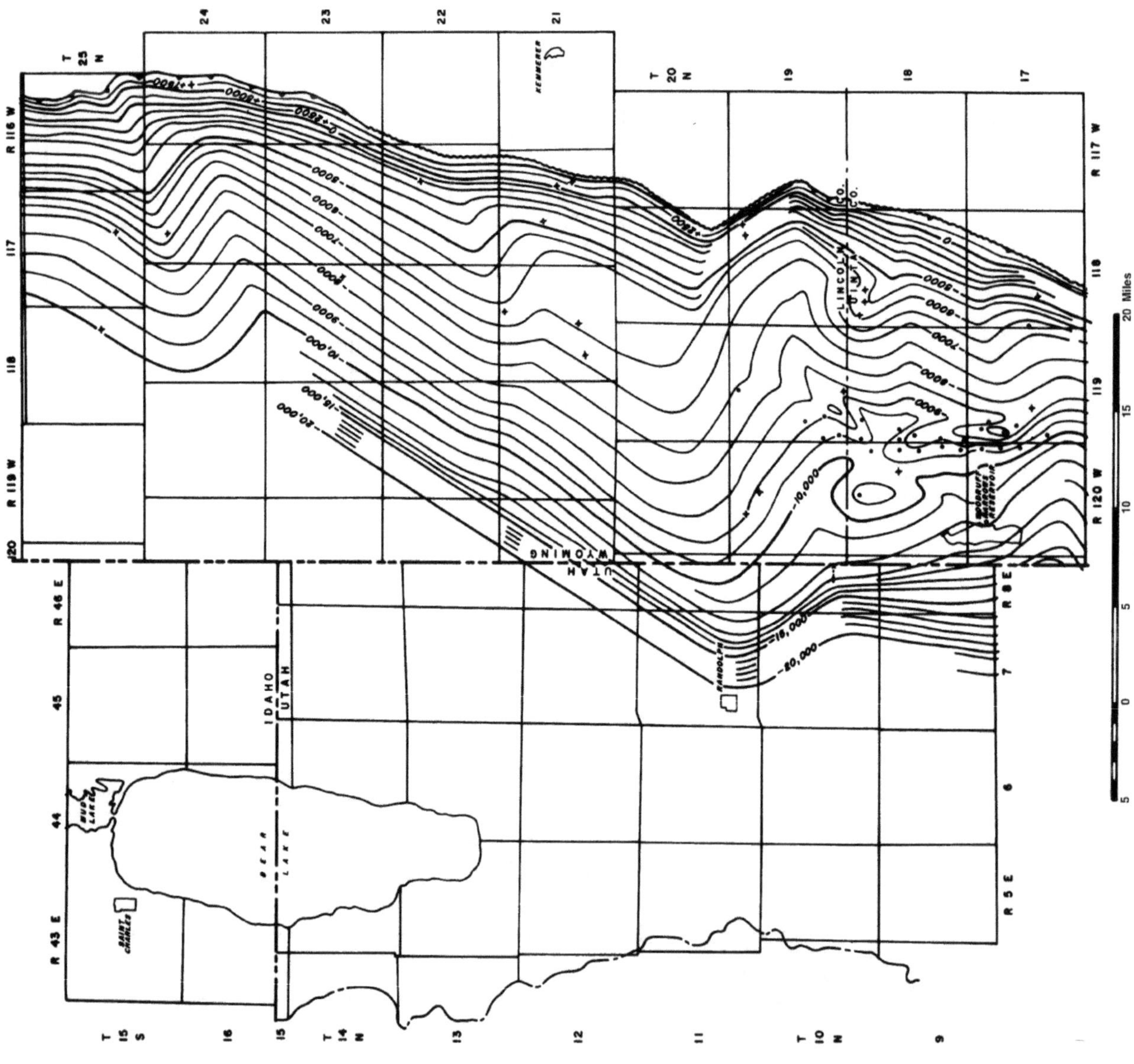
WYOMING
UTAH
IDAHO
UTAH
KEMMERER
LINCOLN CO.
UINTA CO.
RANDOLPH
BEAR LAKE
R 116 W
R 117 W
R 119 W
R 120 W
R 43 E
R 46 E
R 5 E
R 8 E
T 25 N
T 20 N
T 15 S
T 14 N
T 10 N
20 Miles

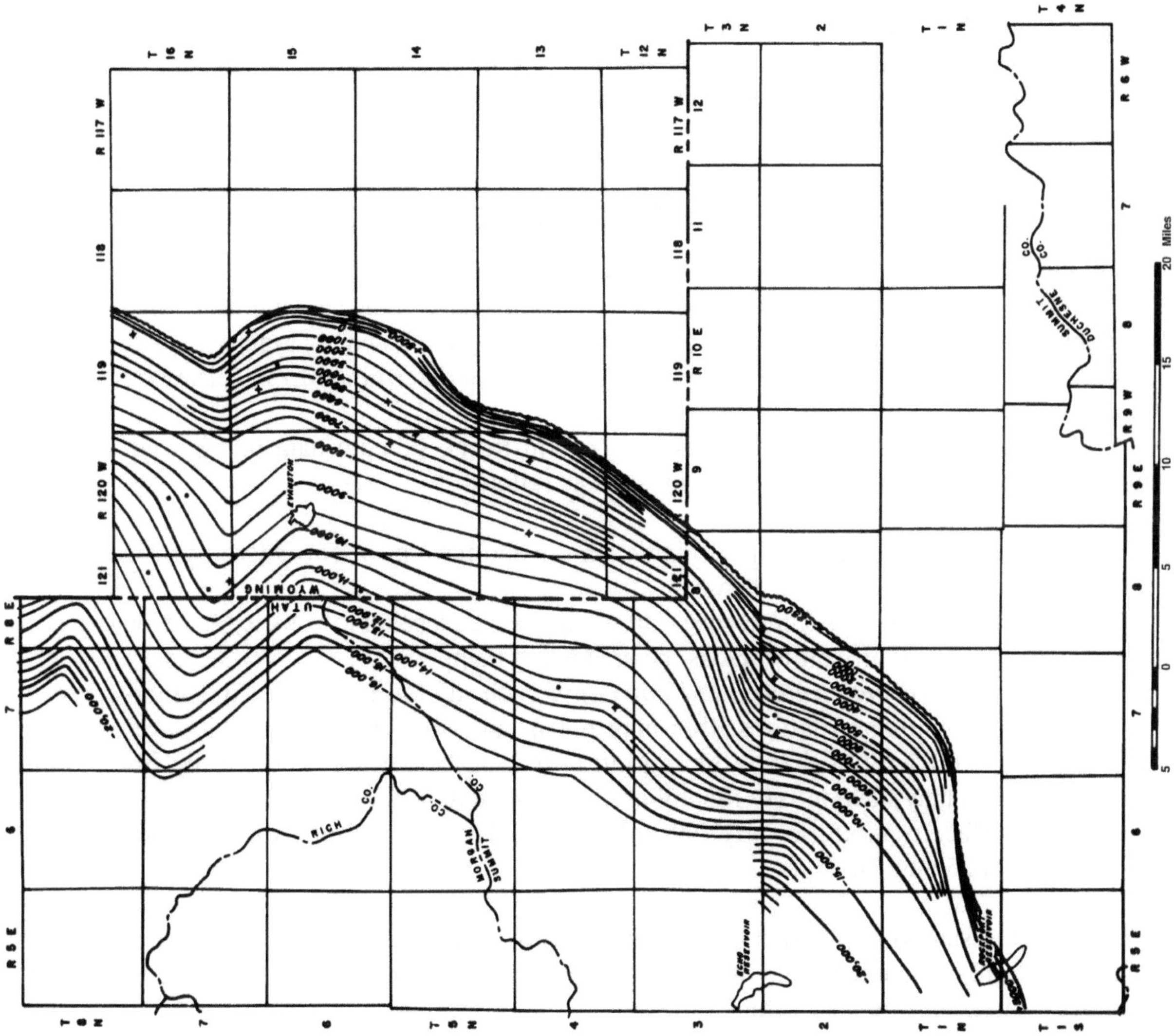

Figure 7-32 Fault surface map of the Absaroka thrust fault in the Fossil Basin of Utah and Wyoming. Note the changes in fault dip from east to west. (From Lamerson 1982. Published by permission of the Rocky Mountain Association of Geologists.)

Fault surface mapping is not commonly done in compressional areas, although there is *no reason* why it should not be used and, indeed, it is often required to aid the understanding of complex thrust fault geometry. Hydrocarbon traps are not common along thrust faults, but instead the anticlines that form in the hanging wall blocks act as the primary trapping mechanism. These structural anticlines are often sufficient for trap delineation with little, if any, fault interaction. We strongly believe, however, that the mapping of faults is an integral part of any geological study regardless of the tectonic setting and can significantly add to the understanding of the geology.

Figure 7-34a is the fault surface map for the Aguaclara ramp and flat thrust fault from the southeastern thrust front of the Eastern Condillera of Colombia, South America (Rowan and Linares 2000). This fault surface interpretation and map were generated from both well log and seismic data. The fault surface in this case is mapped in time (msec) rather than depth. Observe the classical ramp, flat, ramp geometry of the fault surface. One of the seismic sections used in this interpretation is shown in Figure 7-34b. Observe how the hanging wall beds have been folded due to the effects of thrust faulting.

FAULT DATA DETERMINED FROM SEISMIC INFORMATION

Up to this point, most of the information reviewed on fault surface mapping has related primarily to fault data determined from well logs and the use of these data. Seismic data can also provide valuable fault information to aid the mapping process, as shown in Figure 7-35. This section covers some basic techniques for integration of fault data determined from seismic sections with data obtained from well control.

First, we make a few assumptions with regard to the seismic data being used to assist in mapping fault surfaces. These assumptions are as follows:

1. The data being used are of *reasonable quality* and have been correctly processed up to and including migration of the data.
2. *The two-way time-to-depth conversion is known* (i.e., a reliable velocity function exists) and does not vary significantly across the area being mapped.
3. *The lines have been interpreted correctly*, with all faults tied at line intersections. In areas of complex structures, this may not be possible. Where 2D data are used, steep dips pose difficulties due to the migration mis-tie between strike and dip lines. In these cases, you may have to rely mostly on the dip-oriented lines. This sort of problem makes fault surface mapping a very helpful tool in identifying faults on successive dip lines (Chapters 5 and 9).

If these assumptions are met, seismic data can provide *three* very useful types of fault information that aid in the integration of fault data from wells to generate a fault surface interpretation and map. First, seismic data offers a method of establishing a fault identification (ID). Second, it provides the location of a fault surface in space. Third, the amount of vertical separation for a fault can be measured.

The tying process provides us with a method of establishing a *fault ID*. In short, by tying a fault on a series of lines, we can be sure that the fault seen on various lines is formed by the same fault surface. A fault interpretation that does not tie between lines cannot be correct. Merely tying the faults, however, *does not* ensure that the fault interpretation is valid, since any nonvertical surface can be tied between a series of lines. To establish validity, a fault surface map must be constructed to confirm that both the fault dip and the changes in strike of fault contours are geologically reasonable.

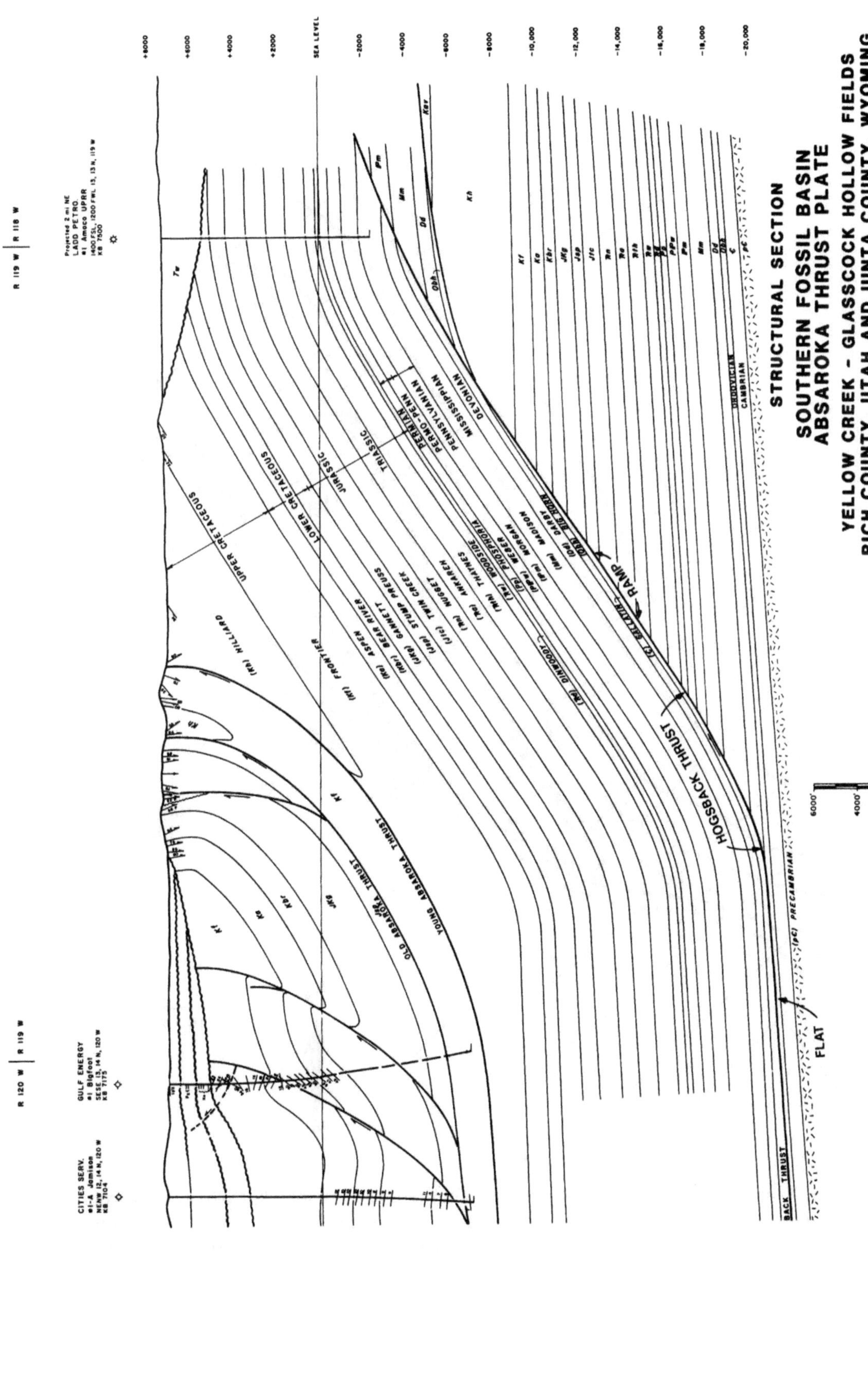

Figure 7-33 Cross section illustrating the geometry of the Hogsback ramp and flat thrust fault. (Lamerson 1982. Published by permission of the Rocky Mountain Association of Geologists.)

Seismic data offer a method of visualizing a fault interpretation through a series of lines. By tying the fault traces from these seismic sections, we can be assured that a given fault seen on one section is the same fault seen on other profiles. A schematic illustration of this is shown in Figure 7-35.

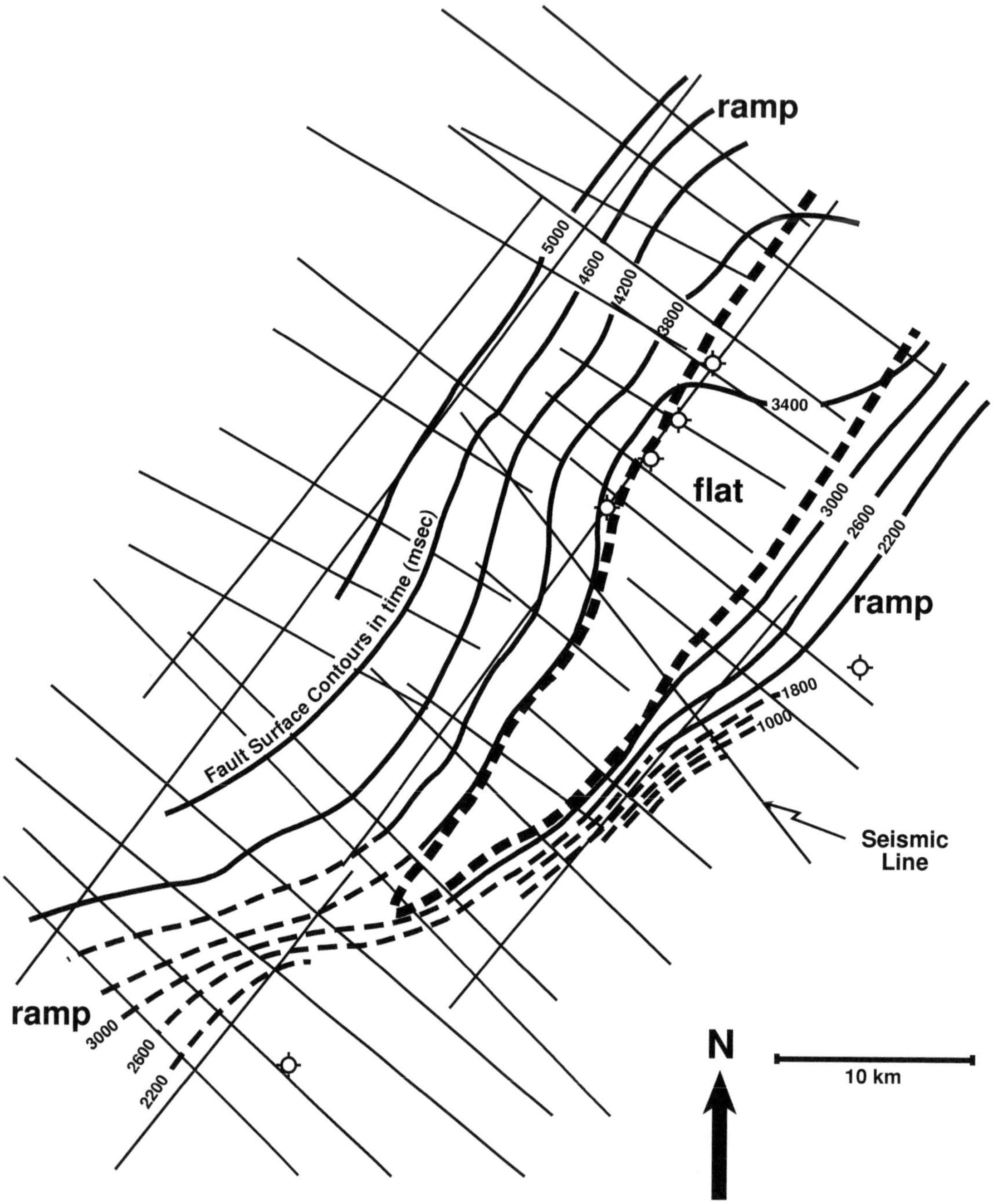

(a)

Figure 7-34 (a) Fault surface map for the Aguaclara ramp and flat thrust fault, Colombia. Contour interval is 400 msec, relative to an arbitrary datum near the surface. (b) Interpreted seismic profile; location shown in (a). Solid lines are axial surfaces. (Modified after Rowan and Linares 2000. AAPG©2000, reprinted by permission of the AAPG whose permission is required for further use.)

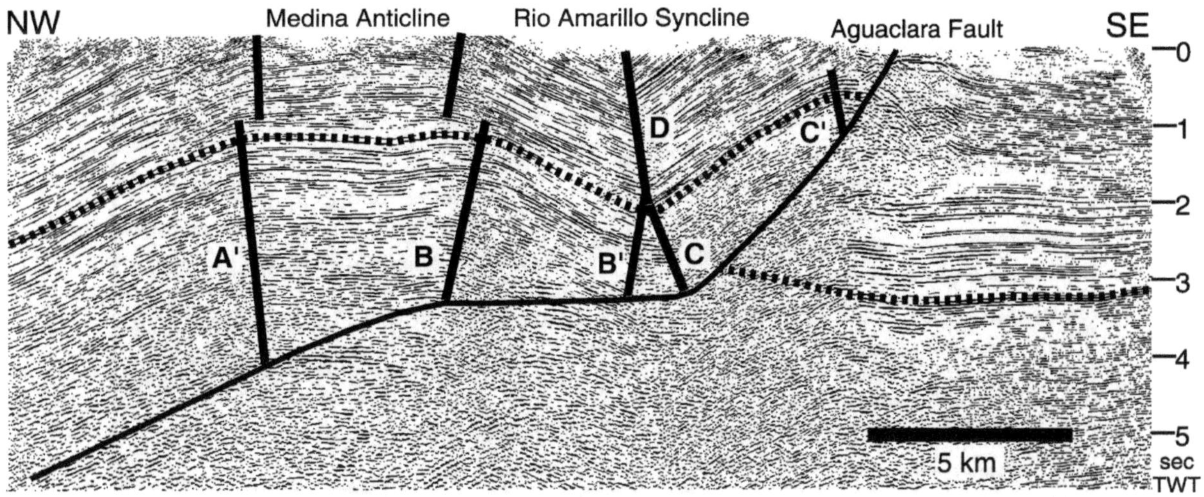

(b)

Figure 7-34 *(continued)*

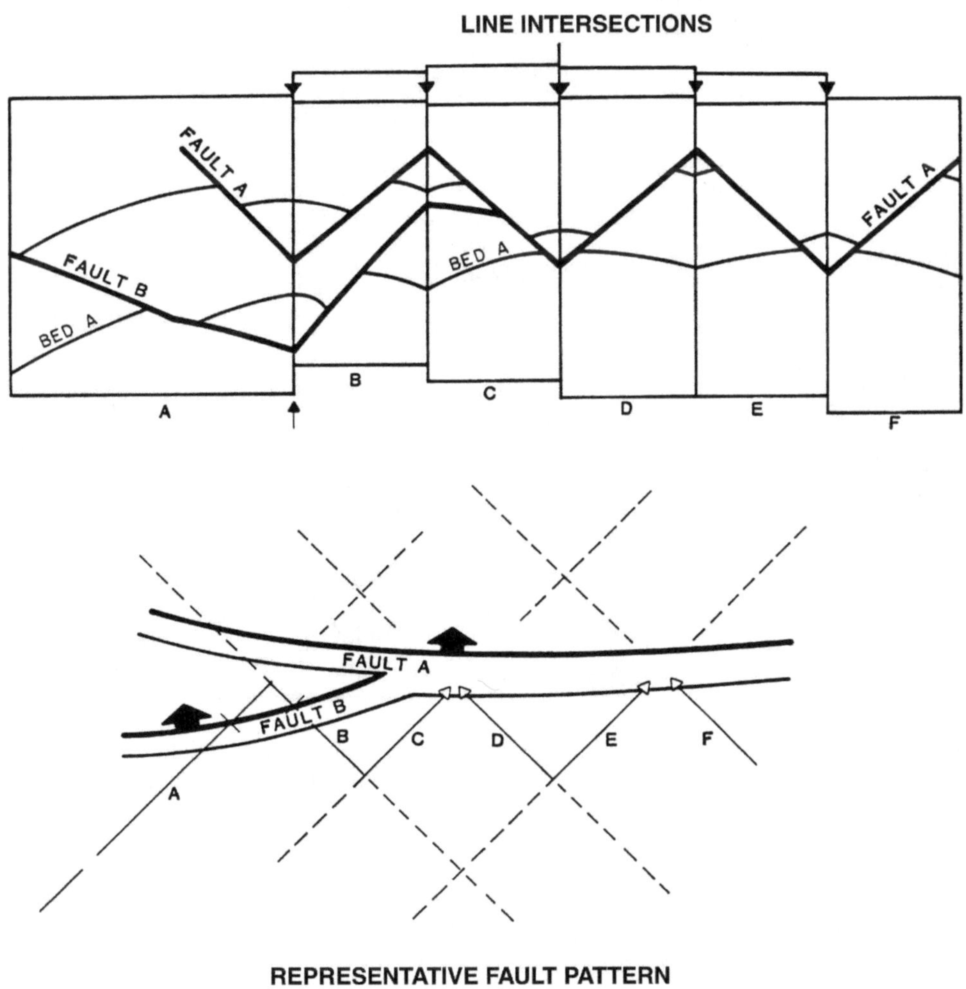

Figure 7-35 Illustration of the usefulness of fault-tying in propagating a fault surface over a series of seismic lines. The lower diagram is a map of the fault pattern at the level of Bed A. (Prepared by C. Harmon. From Tearpock and Harris 1987. Published by permission of Tenneco Oil Company.)

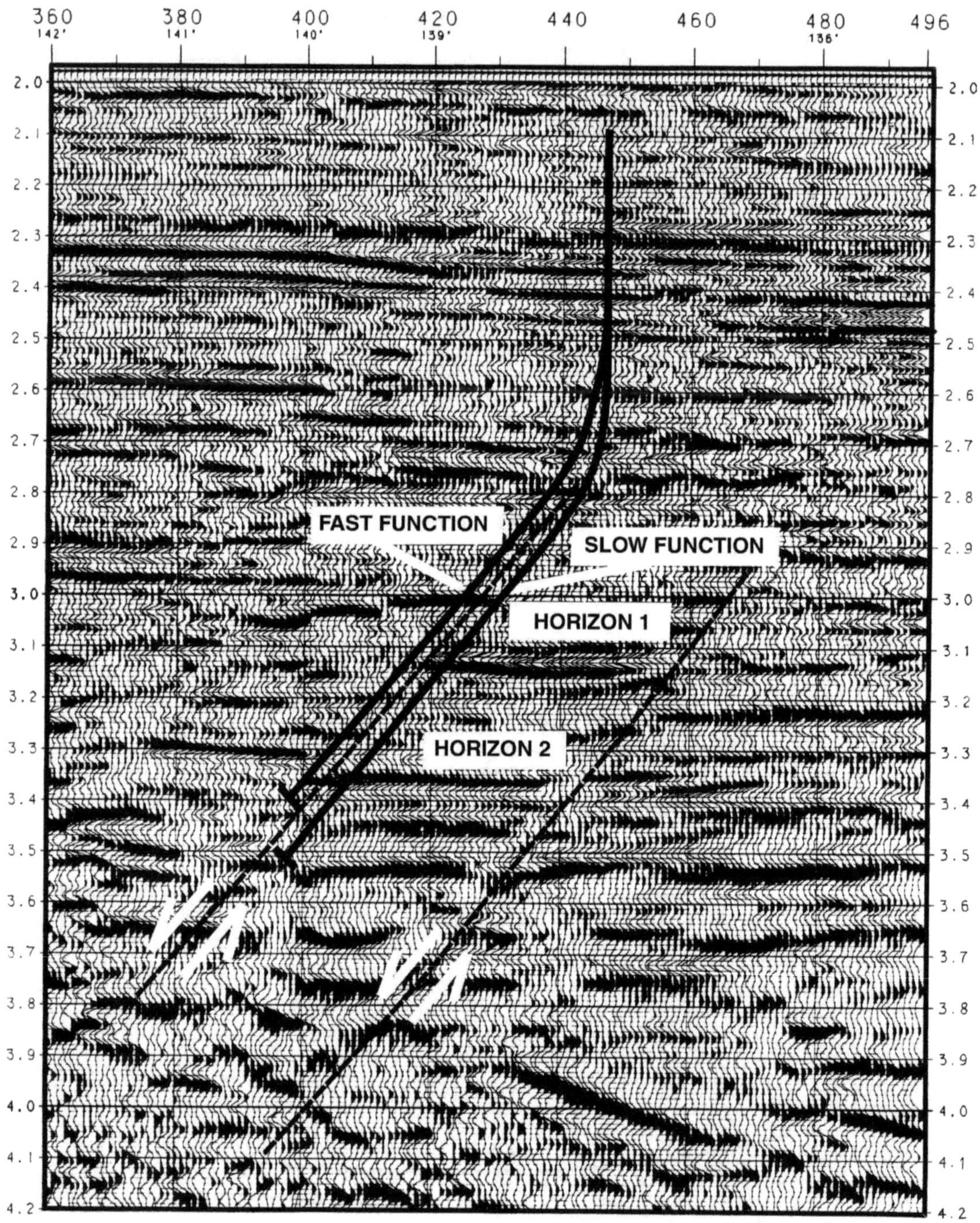

Figure 7-36 Seismic section shows the effect of velocity function uncertainty on the location of a directional well. (Prepared by C. Harmon. Modified from Tearpock and Harris 1987. Published by permission of Tenneco Oil Company.)

Seismic data give us a *location of the fault surface in space*, and unlike well data, data from seismic sections are over a series of points along the profile of the line. This enables us to map fault surfaces over a greater area and in areas where no well control exists.

It is important to remember that accuracy of points plotted from seismic data is dependent on the accuracy (appropriateness) of the velocity function used to convert seismic time to a depth value. Figure 7-36 illustrates what influence a velocity function can have on dipping surfaces. The figure shows two positions for a proposed directional well plotted on a seismic line based on two different velocity functions. Both functions are possibly valid for this prospect. As the illustration shows, the use of the slow velocity function places the directional well in the upthrown fault block; however, the fast function places the well in the downthrown fault block. The upthrown block is the target block to test horizons 1 and 2. The use of the wrong velocity would cause the exploratory well to be drilled in the wrong fault block. In this particular situation, since the correct velocity was not known, a straight hole design was chosen rather than risk missing the prospective fault block with a directional well. The same problem can apply to fault interpretations. In some cases, there may be some uncertainty as to the two-way seismic time corresponding to a particular depth. If the position of a fault is especially critical, error bars can be placed around a posted point that will graphically illustrate spatial uncertainty for a particular datum.

The third type of information seismic data can provide is a *measurement of the vertical separation* of a fault. There is often disagreement between geologists and geophysicists with regard to the amount of missing or repeated section for a fault estimated from well log correlation as compared to that estimated from a seismic section. In many cases, the difference between what the geophysicist is calling displacement and what the geologist is calling displacement is the result of a misunderstanding of fault component terminology. If we assume the vertical resolution limitation of seismic data is known, then the answer to this discrepancy is that two different values for displacement are being measured. As discussed earlier in this chapter, the amount of missing section determined from correlation of an electric log of a well is *not equal to throw*, but rather, equal to vertical separation (unless, of course, the strata in the area have zero dip or the fault is vertical). Then the values for throw and vertical separation are the same. What is commonly measured on a seismic line for fault displacement is not vertical separation but throw and, more commonly, apparent throw, which is a measurement of fault displacement dependent on the orientation of the seismic line in relation to both the strata and fault dip. The only case where we can actually measure the true throw of a fault, at a particular depth, is where the seismic line is oriented *perpendicular* to the strike of the fault surface (not perpendicular to the fault trace as seen on a structural map). This means that if several seismic lines, oriented at different angles to the strike of a fault surface, are used to estimate the throw of that fault, each line will yield a different value for the apparent throw for the same fault. Even if it were possible to shoot all seismic lines perpendicular to the strike of each fault surface being mapped, the throw of the fault is *not* the fault component used in the construction of fault and integrated structure maps, and it is not the displacement value determined from well logs. The added work and expense does not seem justified.

Fault displacement problems from seismic data can be easily reconciled by estimating the vertical separation of a fault from seismic lines, instead of throw or apparent throw. If a seismic line has good correlations across a fault, it is very simple to measure the vertical separation for that fault. Remember, the amount of missing section for a fault in a well log is the measurement of the fault's vertical separation. In order not to confuse components in our fault displacement measurements and constructed maps, the vertical separation for a fault is a required measurement on a seismic section. These depth-converted vertical separation values can then be used to tie the fault data from wells and aid in the construction of the fault and structure maps.

Figure 7-37 illustrates the proper method for measuring the vertical separation from a seismic line. The line is an east-west line shot in the central Gulf of Mexico Basin by TGS Offshore Geophysical Company. It is an excellent line to use to define the procedure for measuring vertical separation because the dip of the structure changes from east to west. At the eastern end of the line, the structure is essentially flat, with zero dip; to the west, the dip increases. Therefore, we can show the procedure for measuring the vertical separation in flat-lying beds as well as those with significant dip. We also show the comparison of the values for vertical separation and apparent throw to reinforce the understanding that these fault components are not the same and that the use of throw in place of vertical separation can cause significant mapping errors.

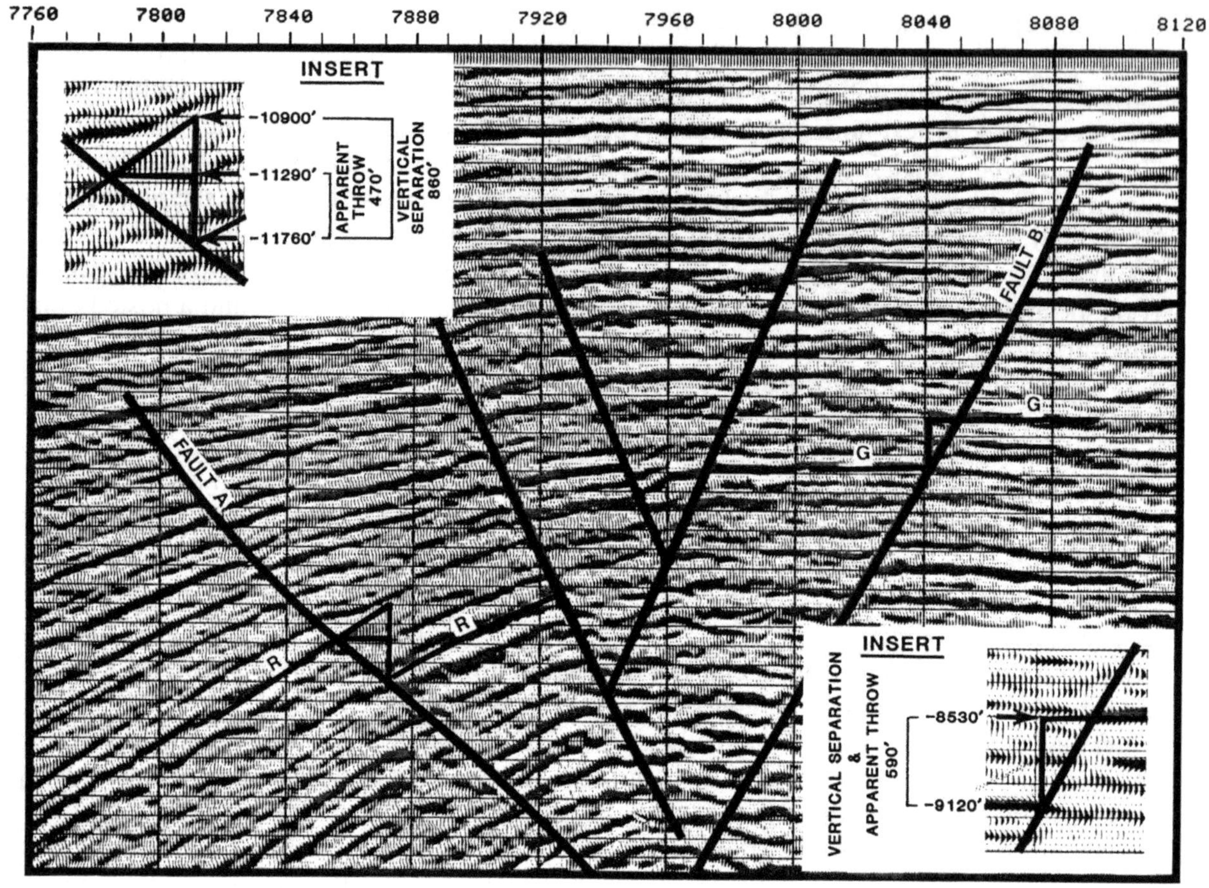

Figure 7-37 Seismic line offshore Gulf of Mexico. The figure inserts illustrate the correct method for estimating the vertical separation of a fault from a seismic line where the events are dipping (Horizon R) and where the events are horizontal (Horizon G). In order to contour the correct displacement, based on seismic data, across a fault, the vertical separation must be estimated. Throw is not used in subsurface mapping. (Modified from Tearpock and Harris 1987. Published by permission of TGS Offshore Geophysical Company and Tenneco Oil Company.)

We begin with the displacement for Fault A at Horizon R. The procedure is more easily followed by using the insert blowup in the upper left corner of the figure. Horizon R is dipping at about 16 deg (depth-corrected) to the west. To measure the displacement of the fault in terms of vertical separation, project the dip of the horizon from the upthrown fault block through the fault, as if the fault were not there, into the downthrown block until the projection is vertically over the intersection of Horizon R with the fault in the downthrown block. Then convert time to depth. The vertical difference in depth from the projection, at −10,900 ft, to the downthrown intersection of the horizon and fault, at −11,760 ft, is 860 ft. This is the measurement of the vertical separation of the fault at this horizon. If a well were drilled through the fault at shotpoint 7865, so that Horizon R is faulted out of the well, the missing section in the well log would be approximately 860 ft.

The measurement of the apparent throw of the fault at this horizon is the vertical difference in depth from the intersection of Horizon R with the fault in the upthrown fault block (−11,290 ft) to the intersection of the horizon with the fault in the downthrown block (−11,760 ft), and this is 470 feet. *The difference between the apparent throw and the vertical separation of the fault at Horizon R is a significant 390 ft.*

Turning to the right side of the figure, we conduct the same procedure and measure the displacement of Horizon G by Fault B, in an area in which the structural bed dip is essentially zero deg. Refer to the insert blowup in the lower right corner of the figure. First, measure the vertical separation by projecting the dip of the horizon in the upthrown block through the fault into the downthrown block until the projection is vertically directly over the intersection of Horizon G with the fault in the downthrown block. The vertical distance between the depth of Horizon G at the fault intersection in the upthrown block at −8530 ft, and the depth of the horizon/fault intersection in the downthrown block, of −9120 ft, is 590 ft.

The apparent throw, measured as the vertical distance from the intersection of the horizon with the fault in the upthrown block to the intersection of the same horizon and fault in the downthrown block, is 590 ft. This is the same value obtained for the vertical separation. Since the strata dip at this location is zero deg, we should expect these values to be the same, based on the definitions and discussion presented so far in this chapter.

If necessary, go back and review these procedures until you thoroughly understand how to measure the vertical separation and clearly see how it differs from apparent throw. It is essential to remember that whether the fault data are derived from well logs or seismic lines, the same fault component for displacement must be used in mapping faults and structures, and that component is *vertical separation*. Of course, vertical separation changes laterally along a fault and varies with depth, so you need to determine it on an appropriate number of seismic sections at depths suitable for the mappable horizons.

Figure 7-38 illustrates the accuracy of the technique outlined above in reconciling well log data with seismic data. The seismic line is a Tenneco Oil Company line in South Marsh Island (SMI), offshore Gulf of Mexico. The seismic line intersects the Signal Well No. 1 in SMI Block 67. A fault in the Signal Well produces a missing section equal to 720 ft at a log depth of 10,940 ft, and it faults out Horizon A. Based on this seismic line, the apparent throw of the fault at Horizon A, marked on the line, is 505 ft, and the vertical separation is 695 ft. Notice that the measurement for the vertical separation agrees very closely with the missing section of 720 ft obtained in the well. The estimate of the apparent throw, however, differs by nearly 200 ft from the estimate of fault displacement by well log correlation. By this example, we again illustrate the importance of measuring the same fault component to obtain the value for fault displacement, which will later be used for fault and structure mapping, whether that value is derived from well log or seismic data.

Seismic and Well Log Data Integration—Fault Surface Map Construction

Since the measurement of fault displacement data can be made from seismic and well log data, both can be used in the construction of fault surface maps. Figure 7-39 illustrates the integration of fault data from two well logs and two seismic lines to construct a fault surface map. The fault data obtained from the well logs are posted next to the appropriate well, and the fault data from the two seismic lines are posted next to the shotpoints at which the data were obtained. Unlike fault data from a well log, which provides fault information from a single depth at a specific location, a seismic line presents continuous fault data over a series of shotpoints along the profile of the line. This permits the fault surface to be mapped over a greater area than would be possible with well log data alone. The seismic also provides a tie to each well to confirm whether the fault cut observed in the well is the same fault surface seen on the seismic line.

Fault-Displacement Mapping. A review of each fault surface map presented in this text reveals that the vertical separation and subsea depth for each well log or seismic fault pick is posted on the basemaps (e.g., 310 ft/−6621 ft). So far, we have used only the subsea depth data to prepare fault surface contour maps. What about the posted vertical separation data? Can these data be used for mapping and, if so, what is their importance? The vertical separation of the fault can and should be contoured where sufficient data are available. Such a map, called a *vertical separation*

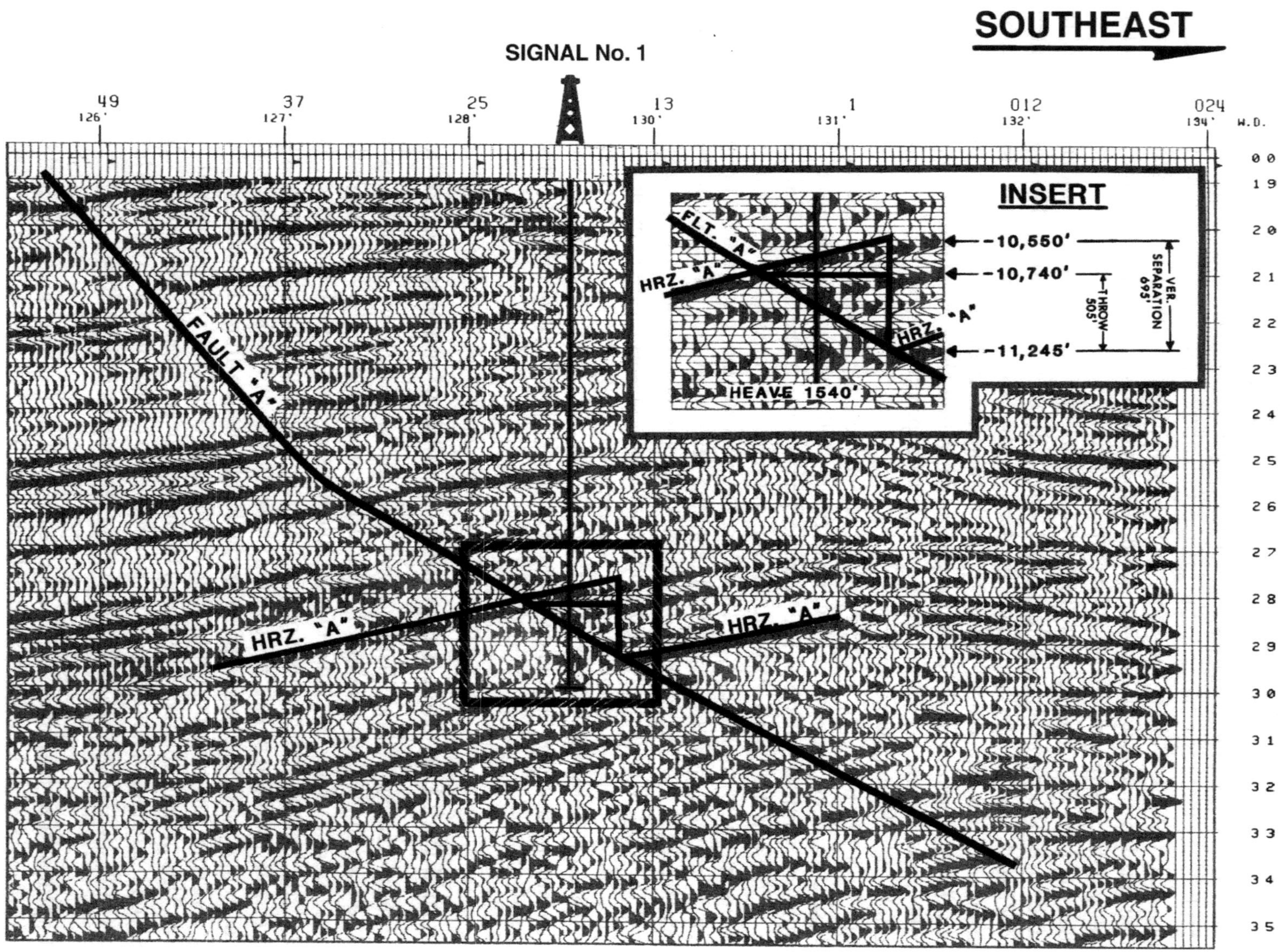

Figure 7-38 Seismic section illustrates the accuracy with which the vertical separation can be estimated, as compared to fault data from a well log directly on the section line. The missing section is 720 ft in the Signal No. 1 Well by correlation with surrounding wells. This compares favorably with the vertical separation of 695 ft estimated from the seismic section. The apparent throw of the fault is only 505 ft. (From Tearpock and Harris 1987. Published by permission of Tenneco Oil Company.)

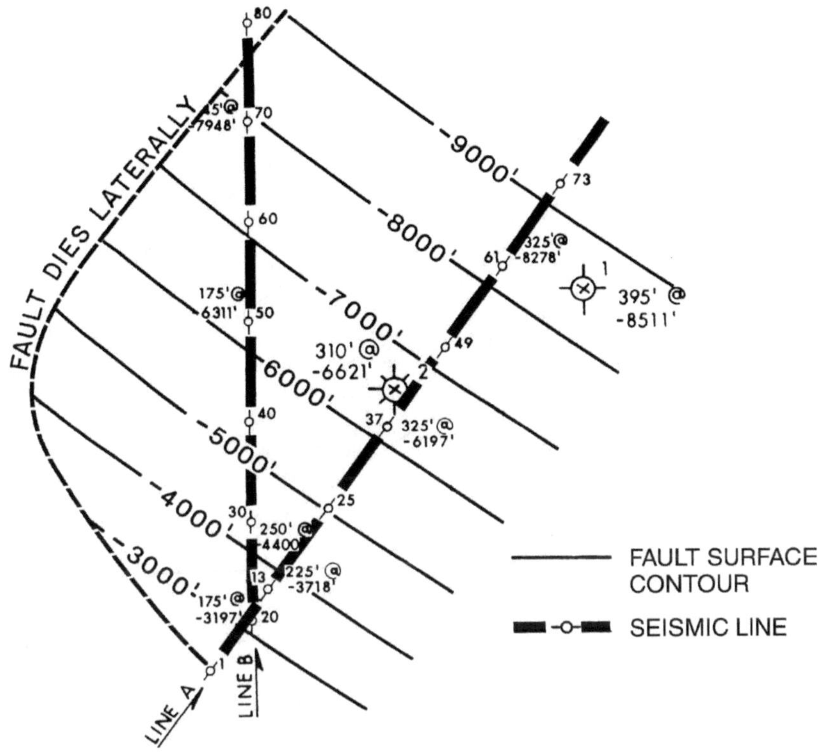

Figure 7-39 Fault surface map prepared using data from well logs and seismic sections. (Modified from Tearpock and Harris 1987. Published by permission of Tenneco Oil Company.)

map, or *throw map* by some, provides additional information on the growth history of the fault and additional data for the construction of integrated structure maps (see Chapter 8).

In Figure 7-40a, vertical separation contours are superimposed on the fault surface contours. This vertical separation is contoured in a dashed pattern with a contour interval of 200 ft. Notice that the vertical separation contours are *perpendicular* to the fault surface contours at depths greater than −5000 ft. This type pattern indicates that the fault is post-depositional and is therefore termed a **structural (nongrowth) fault**. In other words, the fault is not a syndepositional growth fault; it does not show an increase in displacement with depth. However, the deviation in trend of the vertical separation contours, at a depth from about −3000 ft to −5000 ft, shows an increase in vertical separation with depth and indicates the time during which the fault was active.

This information has application in evaluating a fault trap because it graphically illustrates the depth of the geological interval affected by post-depositional faulting. A late trap formed by a post-depositional fault can be evaluated as a prospect as long as migration of hydrocarbons occurred during or after the initiation of fault movement. Displacement mapping for growth faults is also very important and is covered later in this chapter.

Seismic Pitfalls

As we have already mentioned, this text is not intended to instruct in all areas of seismic interpretation, nor is there any intent to discuss in detail the pitfalls of seismic interpretation. However, several seismic pitfalls can be avoided if fault surface and vertical separation mapping are undertaken. We address these subjects to emphasize the importance of constructing fault surface maps as part of any subsurface geological study.

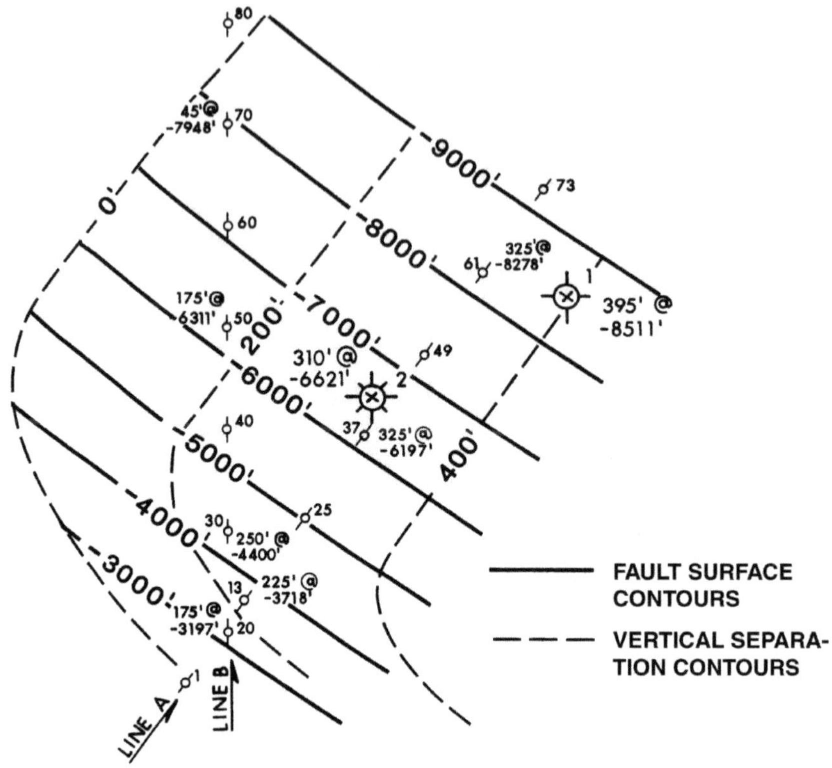

(a)

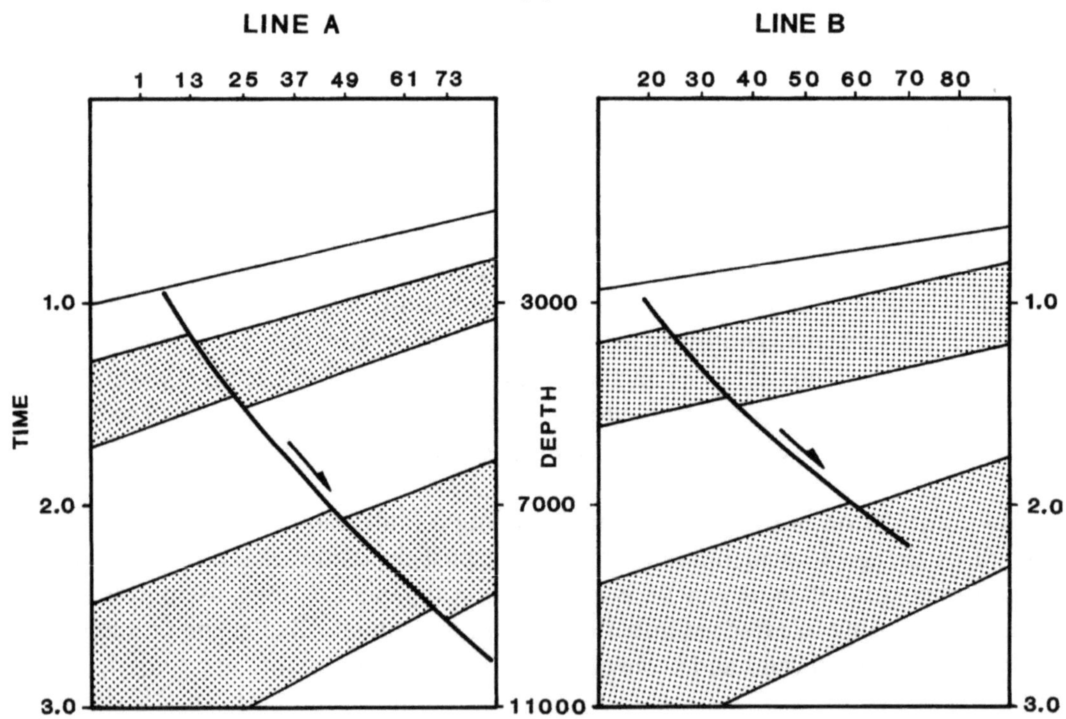

(b)

Figure 7-40 (a) Vertical separation contours added to fault surface contour map provide additional information that is helpful during fault and structure map integration (see Chapter 8). (Modified from Tearpock and Harris 1987. Published by permission of Tenneco Oil Company.) (b) Seismic lines A and B shown in map view in Figure 7-40a. (From Tearpock and Harris 1987. Published by permission of Tenneco Oil Company.)

In extensional basins, do faults die with depth? The answer to this question is obviously yes, but when a fault is interpreted as dying with depth, there has to be a geological explanation for this occurrence (Chapter 10). We have seen faults interpreted on a single seismic section or a limited number of sections as dying with depth, where they actually do not. A fault may *appear* to die with depth on a seismic profile, but this does not mean that it actually dies with depth in the subsurface. What it does indicate is that an interpretation based on one line or a few selected lines can be suspect, requiring additional subsurface work that may include evaluating additional seismic or well control, if it is available.

Seismic line B in Figure 7-40b illustrates something commonly seen on seismic—a fault "dying with depth." Is it real or just an illusion? Line A shows the same fault; however, on this line the fault continues with depth rather than appearing to die at about 2.1 sec as it does on line B. The fault shown on these two seismic lines is the fault contoured in Figure 7-40a. Referring to this figure, we see that the fault does not die with depth. The fault does, however, *die laterally* to the northwest. Notice that Line B is oriented north-south in the general direction in which the fault is laterally dying. At shotpoint 50 on line B, the vertical separation is 175 ft, it is 45 ft at shotpoint 70, and 0 ft (fault dies) at about shotpoint 77. Therefore, on line B the fault *appears* to die with depth. The key word is *appears*. This effect is simply a function of the orientation of the line in relation to the fault surface of this laterally dying fault.

In this particular case, the dilemma of incorrectly interpreting a fault dying with depth can be avoided if a detailed evaluation of the fault surface is conducted. First, Lines A and B can be tied at shotpoint 20. This fault tie confirms that the fault continues with depth. Second, the construction of a fault surface map, incorporating the well and seismic data, shows that the fault continues with depth. Third, as mentioned earlier in this section, whenever possible, contour the vertical separation on the fault map as part of the geological evaluation of that fault. In this case, the vertical separation contoured in Figure 7-40a, using the seismic and well log data, clearly shows that the fault continues with depth but dies laterally to the northwest. These three steps illustrate how correct and detailed fault surface mapping can present a better and more complete geological interpretation of a fault and prevent incorrect subsurface interpretations that could later prove to be costly.

Figure 7-41 is a seismic line that illustrates an authentic example of a pitfall that could have been avoided with fault surface and vertical separation mapping. The profile shows two faults, both coincidentally lining up to look as if one fault surface could be drawn; however, there actually are two faults. This line was a regional line used to illustrate a prospect. As shown in Figure 7-41b, a fault correlation was forced where no fault exists (dashed line). This is significant, because by forcing a one-fault interpretation, a footwall trap is inferred between 2.5 and 2.9 sec that really does not exist. Figure 7-42a is a structure map interpreted from the forced correlation on that line, as well as the interpretation on three other lines of what the geophysicist considered to be the same fault. The result is a curved fault that forms a trap. Notice that the structure map identifies 500 ft of closure in the footwall fault block. Figure 7-42b is the fault surface map implied from the structure map. However, the prospect was generated without this fault surface map being made and evaluated. The prospect was later presented with the intent of placing a future bid on the offshore sale block in which it is located.

The concept of an implied fault surface map is described briefly in Chapter 8 and illustrated in Figure 8-54b; it is discussed in additional detail in Tearpock et al. 1994. It is tempting to create an implied fault surface from Figure 7-42a by connecting points of equal elevation along the fault trace with horizontal lines. Such a fault surface map, however, would imply that the fault is dipping toward the bottom of Figure 7-42a, whereas the seismic line in Figure 7-41 and the fault trace symbols on Figure 7-42a indicate the fault is dipping toward the top of Figure 7-42a.

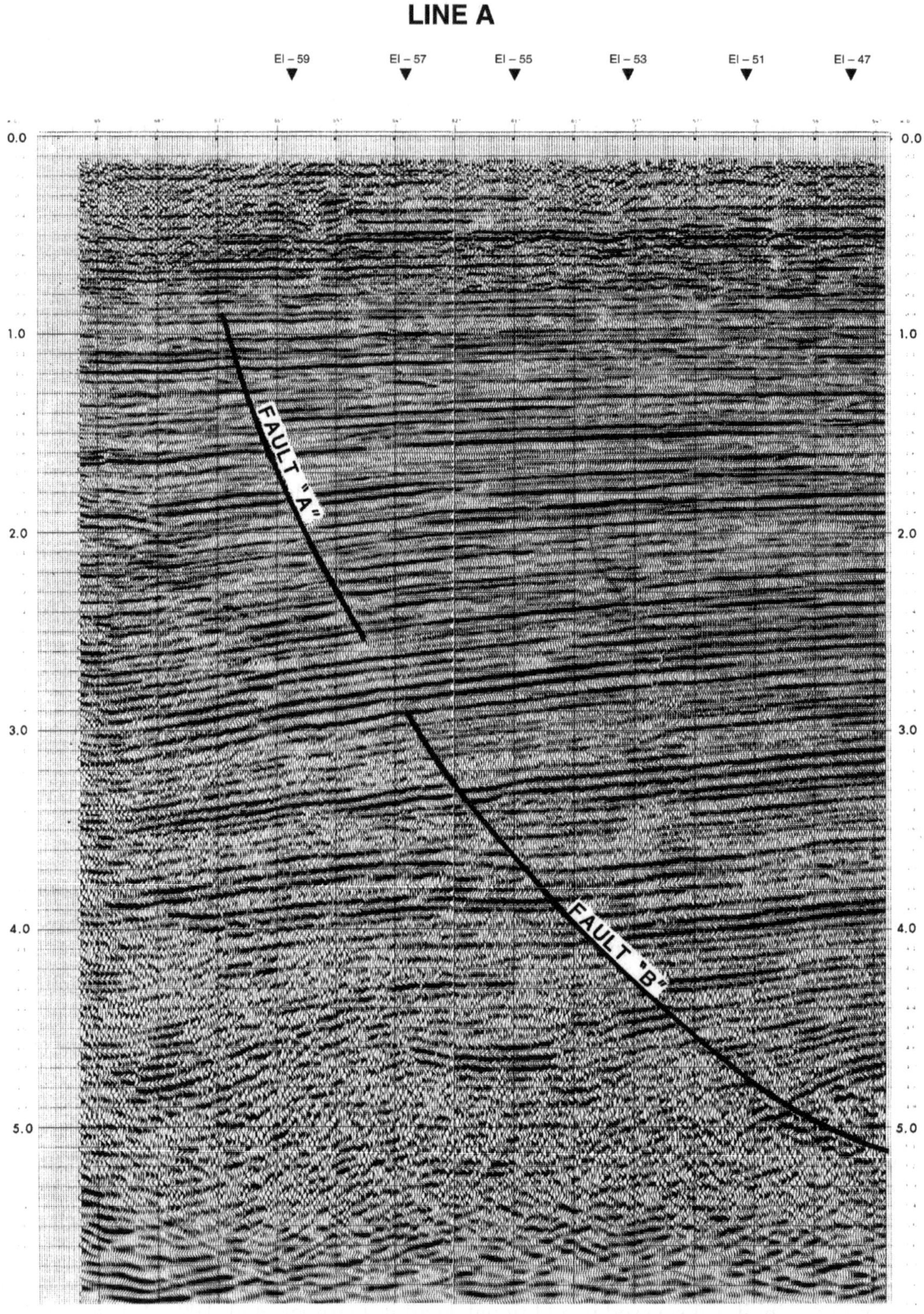

(a)

Figure 7-41 (a) The orientation of this seismic line is such that Faults A and B coincidentally line up and appear as if a one-fault trace could be drawn. (b) A one-fault interpretation forced through continuous reflectors (see dashed line). (From Tearpock and Harris 1987. Published by permission of Tenneco Oil Company.)

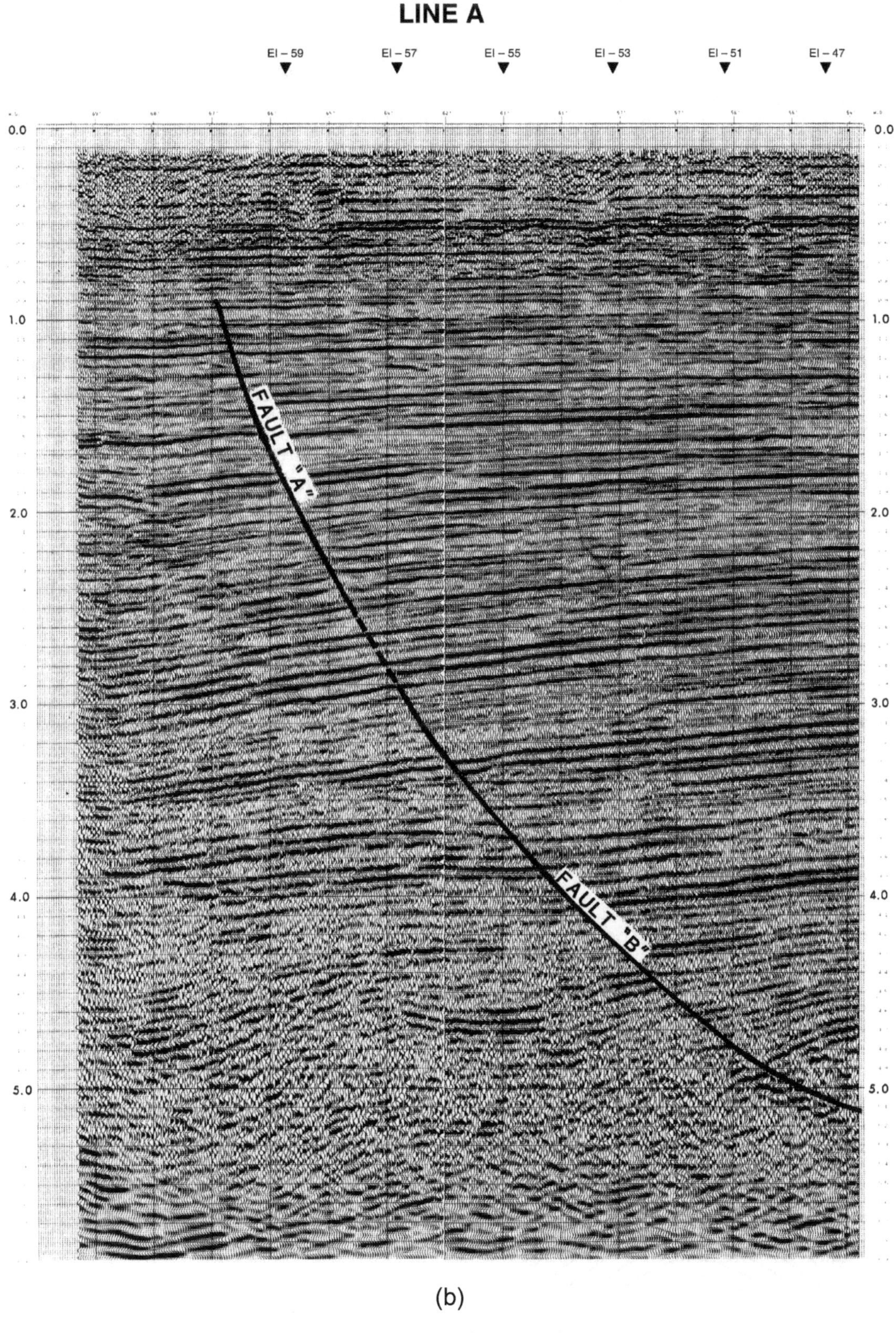

(b)

Figure 7-41 *(continued)*

The prospect was reviewed using the seismic line in Figure 7-41 and the additional lines in the surrounding area. First, based on the structure map in Figure 7-42a, the implied fault surface map in Figure 7-42b was prepared. Second, using all the available lines, the fault and structural interpretations in Figure 7-43a and b were made, which show two faults. The shallower fault is interpreted as a fault that is dying laterally away from an anticline to the west. It appears to die with depth on the seismic line (Fig. 7-41a), but actually the fault surface is simply migrating out of the plane of the line. The deeper fault is a buried growth fault that is also dying laterally, but in a generally opposite direction to that of the shallower fault. Seismic line 120A just happens to be in the correct orientation to make the two fault traces appear to line up on the profile. Figure 7-43 illustrates the correct fault surface and structural interpretation based on all the available data. Notice that based on this new and correct interpretation, there is *no fault trap* and therefore *no prospect*.

The generation of this nonexistent prospect would not have occurred had the procedures outlined in this chapter been followed. First, a fault surface map should always be prepared. Had a fault surface map been prepared based on the one-fault interpretation and map, the sharp convex bend (in the downthrown direction) of the fault surface (Fig. 7-42b) should have been suspect and further review of the fault undertaken. Second, the construction of a vertical separation map would have shown a fault losing displacement with depth, then abruptly gaining displacement again, a very questionable and unlikely situation.

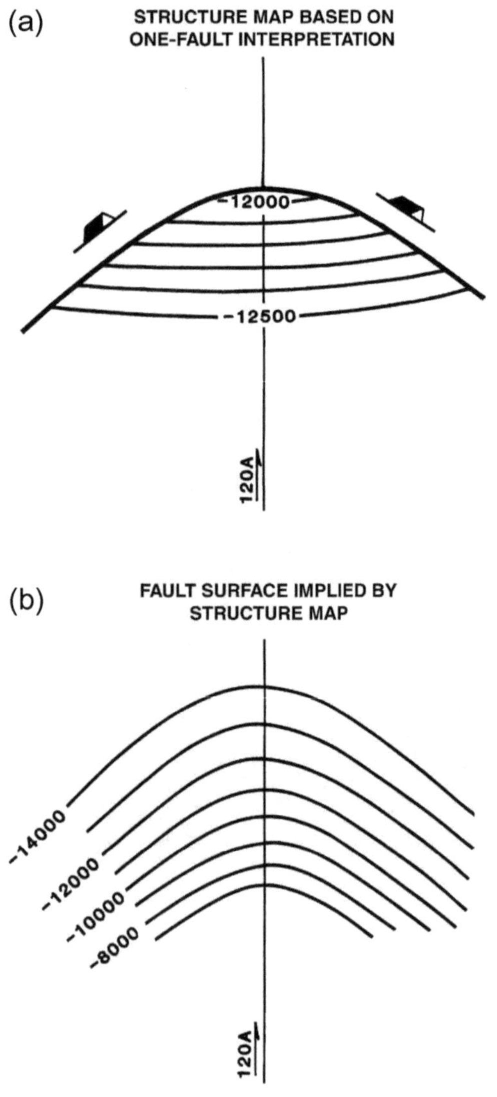

Figure 7-42 (a) Structure map of fault trap created by the forced interpretation. (b) Fault surface map generated from the one-fault interpretation (Fig. 7-41b). (From Tearpock and Harris 1987. Published by permission of Tenneco Oil Company.)

No prospect should be based on a single seismic line or even a few lines where more data are available. Too many *illusions* of geometry can occur on any given line. A gross underuse of costly 3D seismic data, frequently done to "save time," is to use every fifth or tenth line to generate an interpretation and prospect. This means that only 10 to 20 percent of the seismic data are used. Companies spend millions of dollars to acquire and process 3D seismic surveys. The purpose is to provide a dense grid of data to improve the interpretation process. The use of every fifth or tenth line to generate an interpretation is at times a costly, unjustified shortcut that should not be accepted.

In Chapter 1, we stated that the only certainty in a subsurface interpretation, prepared from seismic and well log data, is that it is most likely incorrect. ***The best geoscientist is the one who comes up with the most accurate and geologically reasonable subsurface interpretation***.

Remember the two significant causes of errors in geological interpretations and maps: the failure to use all the data available and the incorrect application of the data. If we fail to make fault surface and vertical separation maps, where the data are available, how can we expect our interpretation and resultant maps to be accurate, geologically reasonable, and representative of the best interpretation of the subsurface?

(a) CORRECT FAULT SURFACE MAP

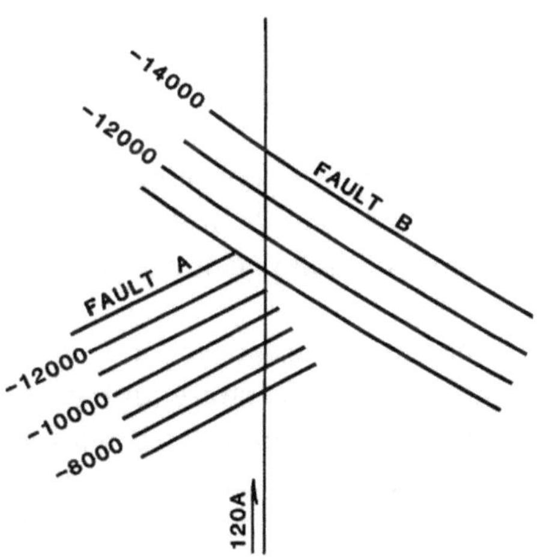

(b) CORRECT MAP SHOWING NON-TRAPPING CONDITIONS

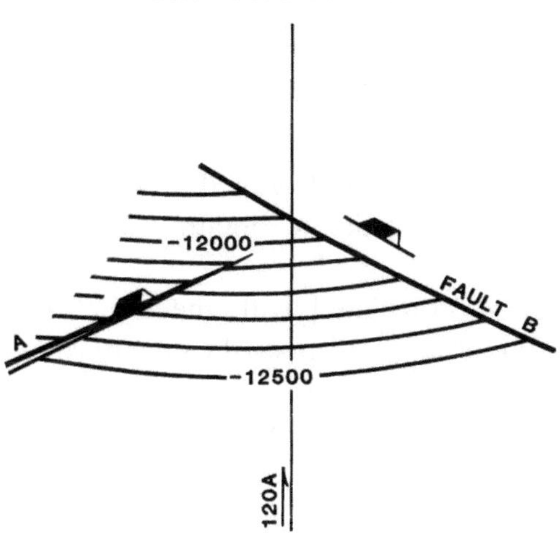

Figure 7-43 (a) Fault surface map for Faults A and B. (b) Completed structure map constructed by integration of fault and structure maps shows no hydrocarbon trap. (From Tearpock and Harris 1987. Published by permission of Tenneco Oil Company.)

GROWTH FAULTS

One of the principal petroleum-related settings for normal faults is that of major large-scale gravity slide structures. The normal faults in this category decrease in dip with depth to low-angle faults, commonly becoming bedding-plane faults. These faults slide along a decollement and are called *detachment faults* (Chapter 11). Such fault systems are well known in areas such as the U.S. Gulf of Mexico, Brunei, and the Niger Delta. These faults are called **listric growth normal faults**, based on the Greek word *listron*, or shovel, because of their curved appearance in profile. The major listric fault on a structure is commonly referred to as the *master* fault if it has the largest relative displacement. Numerous synthetic and antithetic (secondary) faults are often associated with the master fault.

These listric growth faults are syndepositional, commonly exhibiting a significant increase in stratigraphic thickening in the downthrown (hanging wall) block and an increase in displacement with depth. This relationship indicates that there was movement along the fault surface while the sediments were being deposited (Chapter 13). Growth faults also are called syndepositional and contemporaneous faults because fault movement was occurring while the adjacent sediments were being deposited.

Growth faults are commonly associated with hanging wall anticlines, called *rollovers*, which develop as a result of the hanging wall block bending as it conforms to the curved fault surface (Suppe 1985) (see Chapter 10). These rollover anticlines are one of the most important hydrocarbon traps associated with listric growth faults.

Estimating the Vertical Separation for a Growth Fault

The method for estimating the missing section in a well due to a post-depositional fault, and thereby estimating vertical separation for the fault, was outlined in Chapter 4. In this section, we present two methods for estimating the vertical separation for a growth fault penetrated by a wellbore.

The Restored Top Method. The thickness of the stratigraphic section in the footwall (upthrown) block of a growth fault is thinner than the equivalent stratigraphic section in the hanging wall (downthrown) block. Therefore, if a faulted well is correlated with a well in the upthrown block, the missing section due to the growth fault will be estimated to be smaller than if the correlation is made with a well in the downthrown block. So the procedure for estimating the vertical separation for a growth fault is not as straightforward as that for nongrowth faults. In most cases, at least two correlation well logs are needed: one located in the upthrown fault block and the other in the downthrown block. With the available electric well logs, the vertical separation at any horizon faulted out of a well can be estimated by the restored tops method, which is described in detail in Chapter 4.

1. Choose a horizon, faulted out of the subject well, for which you wish to estimate the vertical separation.
2. The log of a well in the upthrown block is then correlated with the deeper log section (the interval below the fault) in the faulted well in order to obtain an estimate of the upthrown restored top for the chosen horizon.
3. The log of a well in the downthrown block is correlated with the upper log section (the interval above the fault) in the faulted well in order to estimate the downthrown restored top for the horizon.
4. The estimate of vertical separation for the growth fault at the horizon is the difference in true vertical depths of the upthrown and downthrown restored tops for the horizon in the

faulted well. If the wells are vertical, the difference in measured depths of the upthrown and downthrown restored tops can be used.

Using the cross section in Figure 7-44, we illustrate the use of the restored tops method. Fault 1 is a growth fault downthrown to the east. In Well No. 1, Beds A, B, and most of C are faulted out. Bed B is a horizon of interest to be mapped; therefore, we want to estimate the vertical separation for the growth fault at the level of Bed B. Three wells are on the cross section: Well No. 2 in the upthrown fault block, Well No. 3 in the downthrown fault block, and Well No. 1, which has the horizon of interest faulted out. If necessary, go back to Chapter 4 and review the section on estimating restored tops before continuing this section.

First, estimate the missing section due to the fault in Well No. 1 using the other two wells (Fig. 7-44). By correlation with Well No. 2 in the upthrown fault block, the missing section at the fault in Well No 1 is 190 ft, but by correlation with Well No. 3 in the downthrown block, the missing section at the fault is 315 ft. Since these two estimates for the missing section are different, which one should be used as vertical separation in mapping Bed B? The answer is neither estimate. The 190-ft value is too small, since it is based on correlation with a well in the stratigraphically thin upthrown block. The 315-ft value is too large because it was obtained by correlation with a well in the stratigraphically thick downthrown block.

How do we obtain an estimate of the vertical separation for the growth fault to be used in the structure mapping of Bed B? The restored tops method outlined earlier must be used. The upthrown restored top for Bed B in Well No. 1, by correlation with Well No. 2 in the upthrown fault block, is at 7705 ft measured log depth (Fig. 7-44). The downthrown restored top in Well No. 1 for Bed B, by correlation with Well No. 3 in the downthrown Block, is at 7980 ft measured log depth. Using the restored tops method, the vertical separation for the growth fault at Bed B is estimated as the difference in restored top depths for Bed B (7980 ft − 7705 ft = 275 ft). Therefore, in the preparation of a structure map on Bed B, the vertical separation value of 275 ft is used to contour across Growth Fault 1. Also, in using this method, two additional control points for mapping are estimated, these being the upthrown and downthrown restored tops in Well No. 1. Remember, these tops must be corrected to true vertical subsea depths before they are used in the structure mapping of the horizon.

The estimated vertical separation of 275 ft for the fault cut for Bed B lies between the low value of 190 ft and the high value of 315 ft for missing section derived by correlation with wells on either side of the fault. The restored tops method is the most accepted procedure for estimating the vertical separation (missing section) on a growth fault at any particular horizon.

The Single-Well Method. Given certain limitations, it is possible to use only one reference well to closely approximate the vertical separation for a growth fault. This method is referred to as the *single-well method.* It is applicable only if the given horizon is close to the top or to the base of the stratigraphic section missing in the faulted well. This method is illustrated in cross-sectional view in Figure 7-45.

Three wells are in the cross section in Figure 7-45: Well No. 3 is in the upthrown fault block, Well No. 2 is in the downthrown fault block, and Well No. 5 is the well of interest. The growth fault has faulted out a significant stratigraphic section from Bed B through Bed F in Well No. 5. For mapping purposes, we are interested in the vertical separation for the growth fault at Beds B, D, and F.

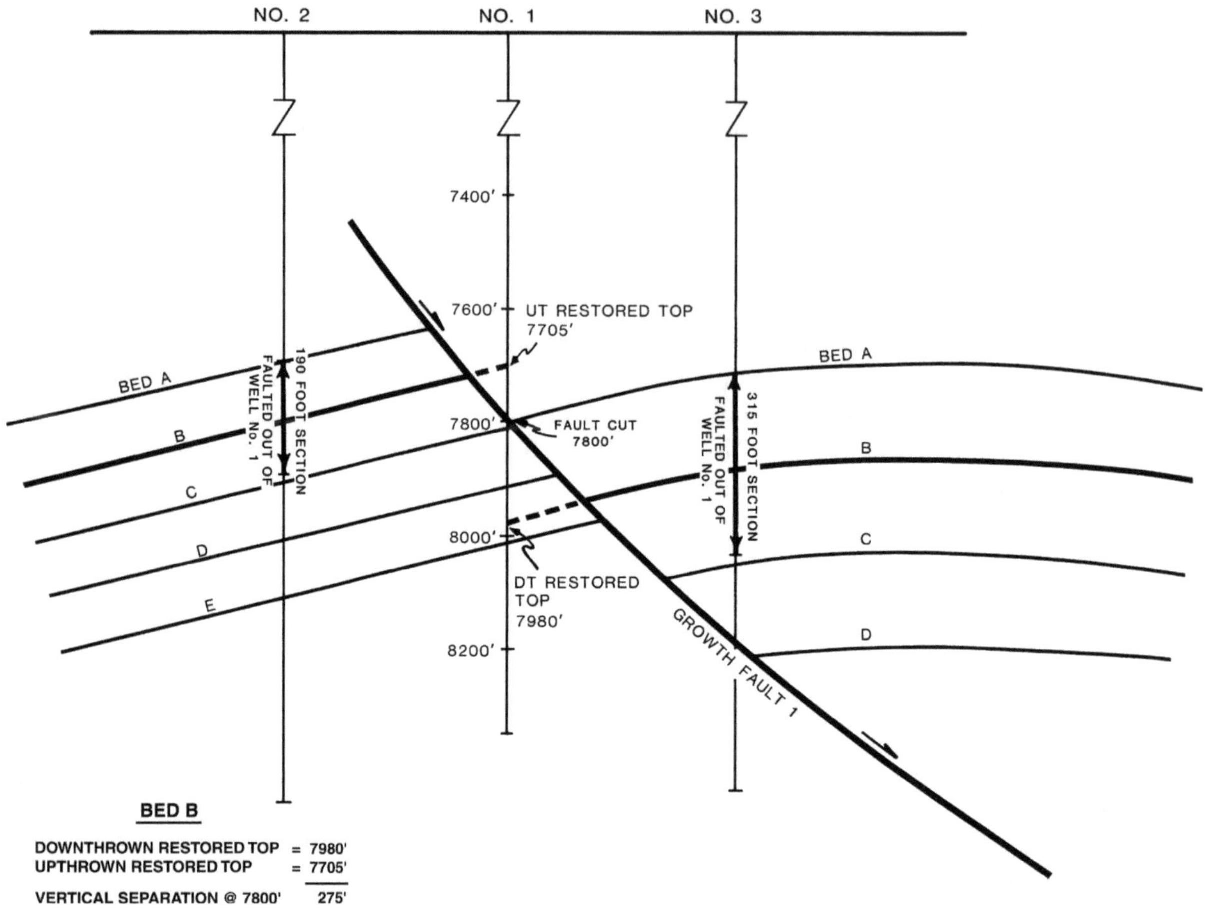

Figure 7-44 Cross section illustrates the restored tops method for estimating the vertical separation for a growth fault. The restored tops are estimates using the procedures discussed in Chapter 4.

First, we estimate the missing section in Well No. 5 with the other two wells (Fig. 7-45). By correlation with Well No. 3 in the upthrown block, the estimated missing section due to the growth fault is 200 ft, and with Well No. 2 in the expanded downthrown fault block, the estimated missing section is 400 ft. There is a two-to-one difference in the estimated missing section at the fault by correlation with these two wells. Which one is correct, or are they both correct? Which one, if any, should be used for vertical separation in the structure mapping of the various horizons? We know that this is a growth fault, which means by definition that the vertical separation increases with depth. Therefore, it is reasonable to assume that vertical separation increases from Bed B to Bed F. But by how much, and what is the vertical separation at each horizon?

The vertical separation at any horizon can be estimated by using the restored tops method, whereas the vertical separation at only two horizons can be estimated by using the single-well method. We begin by using restored tops to estimate the vertical separation at Bed B (Fig. 7-45). The upthrown restored top for Bed B is at 8050 ft, and the downthrown restored top is at 8250 ft. The difference between these two restored top values is 200 ft, which is an estimate of the vertical separation for the fault at Bed B. This is recorded within the inset in Figure 7-45, which contains a summary of the data pertinent to Beds B, D, and F.

The single-well method can be used to estimate the vertical separation for a growth fault, but it can be applied only under certain restraints. Notice in the cross section (Fig. 7-45) that Bed B is at the top of the section missing in Well No. 5; therefore, we can say that most of the interval

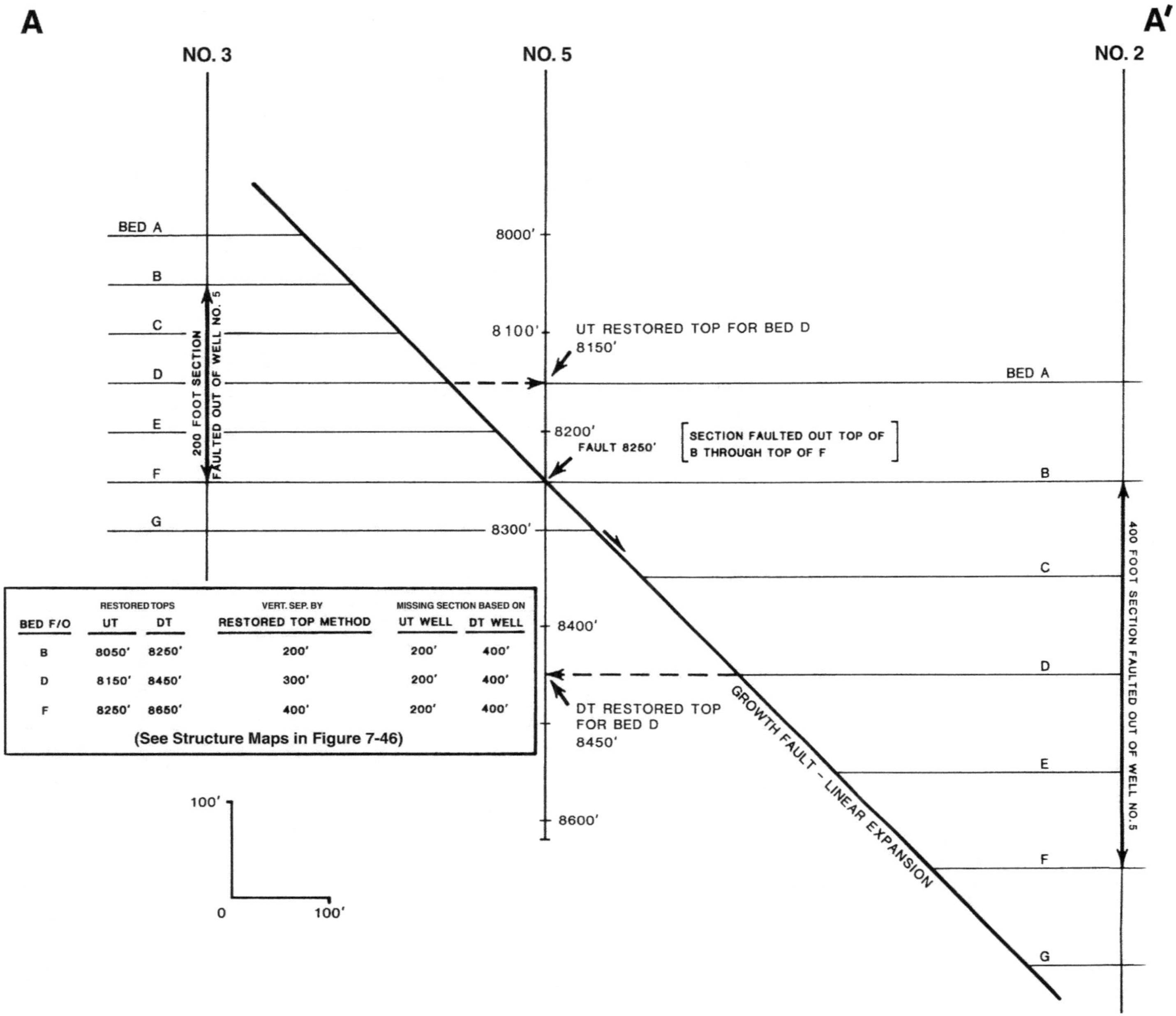

BED F/O	RESTORED TOPS UT	RESTORED TOPS DT	VERT. SEP. BY RESTORED TOP METHOD	MISSING SECTION BASED ON UT WELL	MISSING SECTION BASED ON DT WELL
B	8050′	8250′	200′	200′	400′
D	8150′	8450′	300′	200′	400′
F	8250′	8650′	400′	200′	400′

(See Structure Maps in Figure 7-46)

Figure 7-45 The vertical separation is estimated for this growth fault using both the restored tops and single-well methods. The displacement is determined at the levels of Beds B, D, and F.

faulted out of Well No. 5 can be represented by the thin upthrown section seen in Well No. 3. Therefore, the vertical separation at Bed B can be closely estimated simply by correlating Well No. 5 only with the thin upthrown Well No. 3. By making this correlation, we obtain an estimate of 200 ft for the missing section due to the fault. Observe the 200 ft of section faulted out of Well No. 5, as illustrated on the well log stick for Well No. 3. We can check the accuracy of this simplified technique by comparing the result with that obtained from the restored tops method (Fig. 7-45). They are the same. Using the two methods independently, the vertical separation of the growth fault at Bed B is estimated to be 200 ft.

Now estimate the vertical separation at Horizon F. First, using the restored tops method, the estimated vertical separation at this horizon is 400 ft. Again, the results are summarized in the insert in Figure 7-45. Reviewing the log sticks in cross section, notice that Bed F is at the base of the missing section in Well No. 5. Therefore, most of the section faulted out of Well No. 5 is represented by the thick downthrown section seen in Well No. 2. So, we can use the simplified single-well method and closely estimate the vertical separation at Bed F by correlating Well No. 5 with just one well, the thick downthrown Well No. 2. With this correlation, the vertical separation at Bed F is estimated to be 400 ft. Again, the estimate for the vertical separation is the same using the more detailed restored tops method or the simplified single-well method.

The last horizon requiring an estimated vertical separation is Bed D. Since Bed D is not near the top or bottom of the missing section in Well No. 5, the simplified single-well method cannot be used. Therefore, the restored tops method for estimating the vertical separation at Bed D must be used, and it is illustrated in Figure 7-45. The upthrown restored top is at 8150 ft, and the downthrown restored top is at 8450 ft, resulting in an estimated vertical separation of 300 ft at Bed D.

Figure 7-46 shows a portion of the structure map for each of the three horizons, Beds B, D, and F. For mapping purposes, well depths are converted to subsea depths by using a KB = 50 ft for each well. The beds are dipping, and their apparent dip is horizontal in the cross section in Figure 7-45. The maps were accurately constructed by the integration of the growth-fault surface map with each horizon map, as described in Chapter 8. Note how the restored tops for Well No. 5 are honored by the structure contours. The maps collectively demonstrate that the fault gap widens with the increase in vertical separation with depth, as is typical for any growth fault delineated on a set of horizon maps. Also, from map to map, the position of Well No. 5, within the fault gap, changes relative to the upthrown and downthrown traces of the fault. That also is typical for any well in which mapped horizons are faulted out by the same fault. Compare the three maps with the cross section in Figure 7-45 to visualize why the position of Well No. 5 changes relative to the traces of the fault. If the horizon being mapped is *at or near the top* of the missing section in the faulted well, the fault size vertical separation can be closely approximated by using the single-well method and correlating the faulted well with a nearby well in the thin upthrown fault block. See Bed B in our example in Figure 7-44. In map view, this is represented in Figure 7-45a, which is a portion of the structure map on Bed B.

If the horizon being mapped is *at or near the bottom* of the missing section in the faulted well, the missing section fault size can be closely approximated by using the single-well method and correlating the faulted well with a nearby well in the thick downthrown fault block. This was done for Bed F, as shown in Figure 7-46. In map view, this is represented in Figure 7-47c, which is a portion of the structure map on Bed F. The faulted Well No. 5 is near the upthrown trace of the growth fault. If the horizon being mapped falls clearly within the missing section in the well log (D Horizon, Fig. 7-44), then the restored tops method must

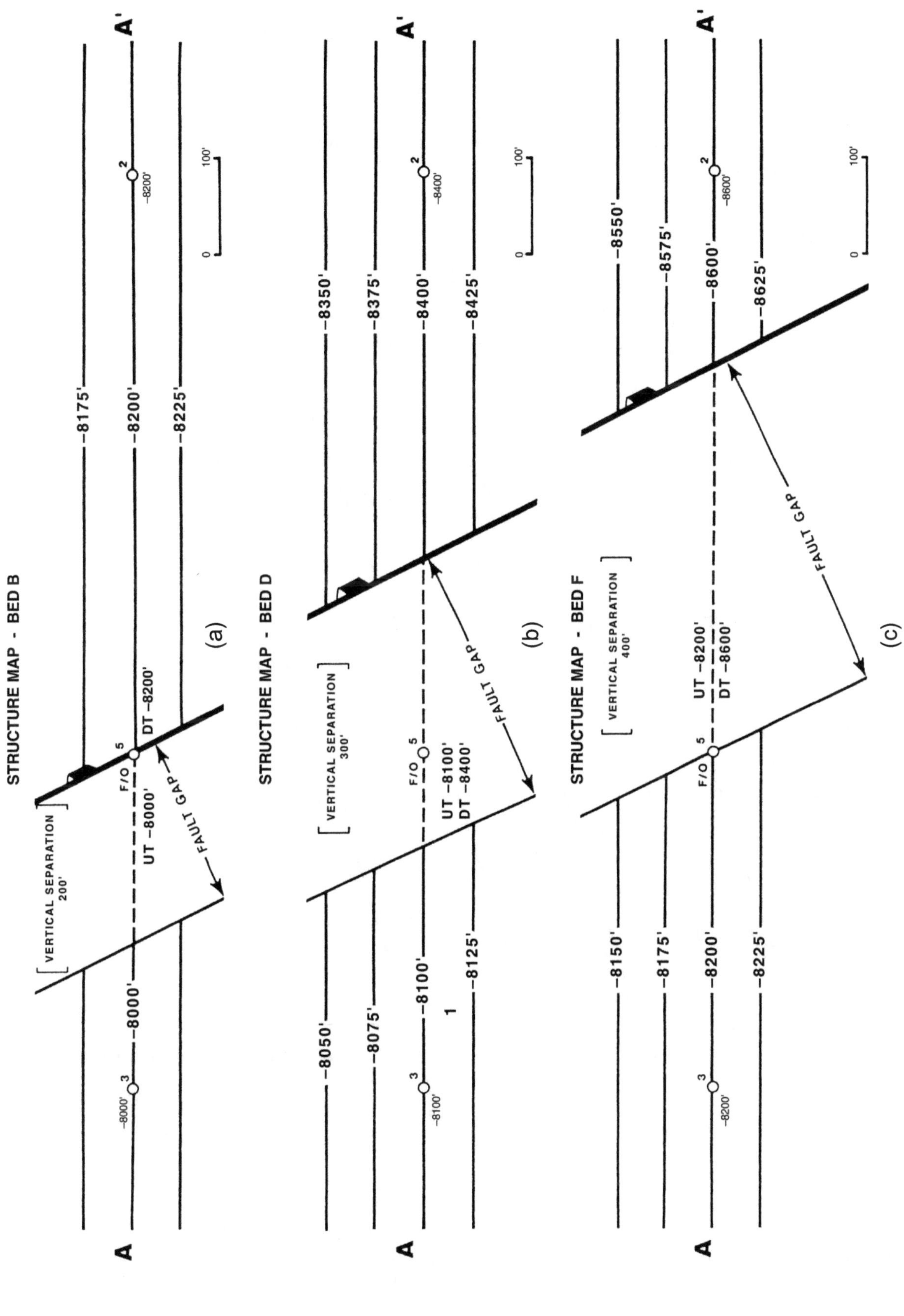

Figure 7-46 Segments of the integrated structure maps for Beds B, D, and F. Notice how the fault gap width changes with depth in response to the increase in vertical separation. Well depths are converted to subsea depths (KB = 50 ft for each well).

be used. The restored tops method is always preferable over the single-well method if time is not a factor. In addition to providing an estimate of vertical separation for the fault at any horizon, it also contributes two control points for contouring (the upthrown and downthrown restored tops).

The faults used to illustrate the methods for estimating the vertical separation for a growth fault do possess linear expansion, and the beds are gently dipping in Figure 7-46. However, these methods are just as applicable for growth faults with nonlinear expansion and in areas with steeply dipping beds.

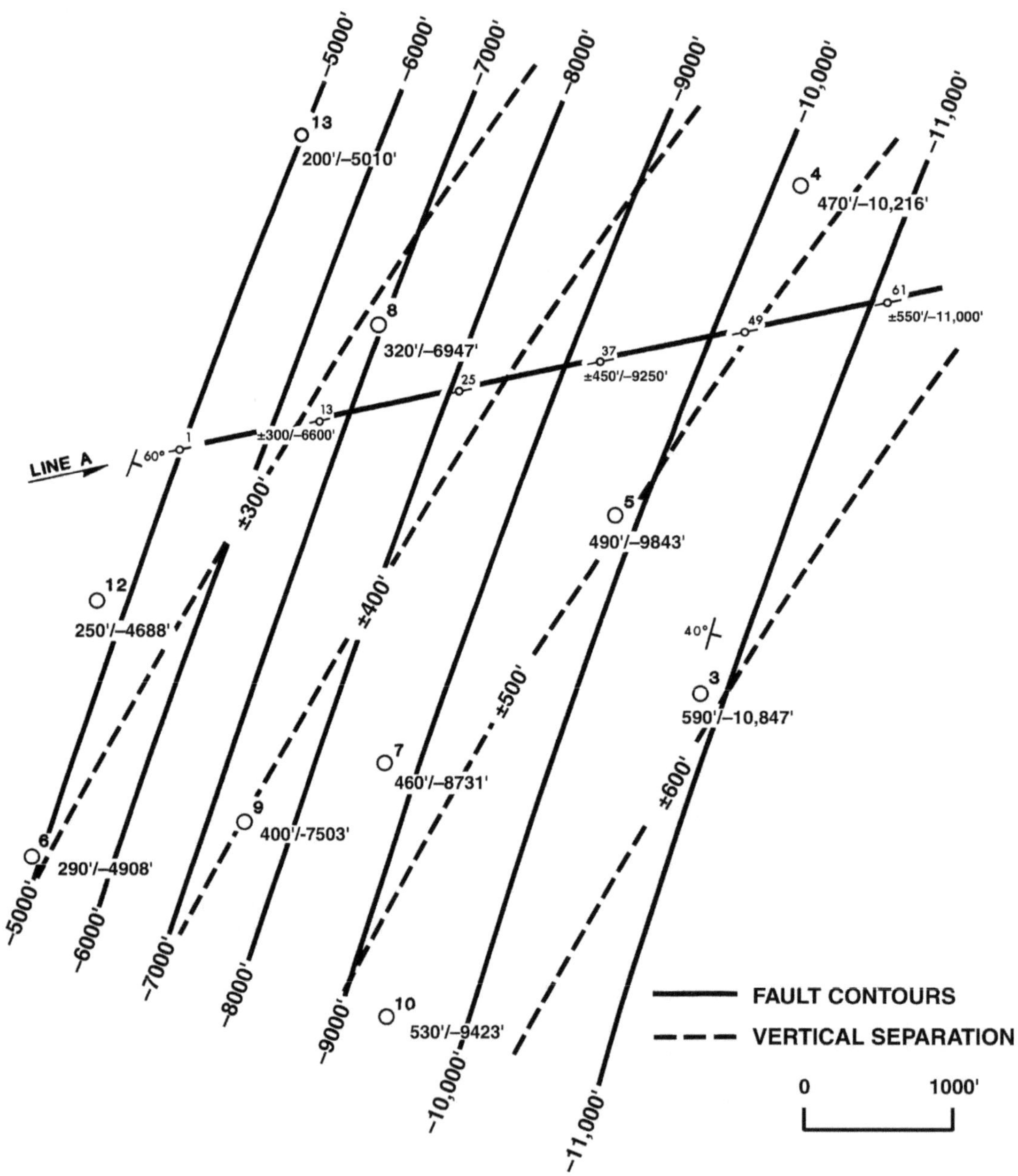

Figure 7-47 (a) Fault surface map of a growth fault. Vertical separation contours are placed on the fault map, adding valuable mapping information pertaining to the growth fault. (b) Vertical separation values are not contoured, but strategically placed on the fault surface map, which is a shortcut method. (From Tearpock and Harris 1987. Published by permission of Tenneco Oil Company.)

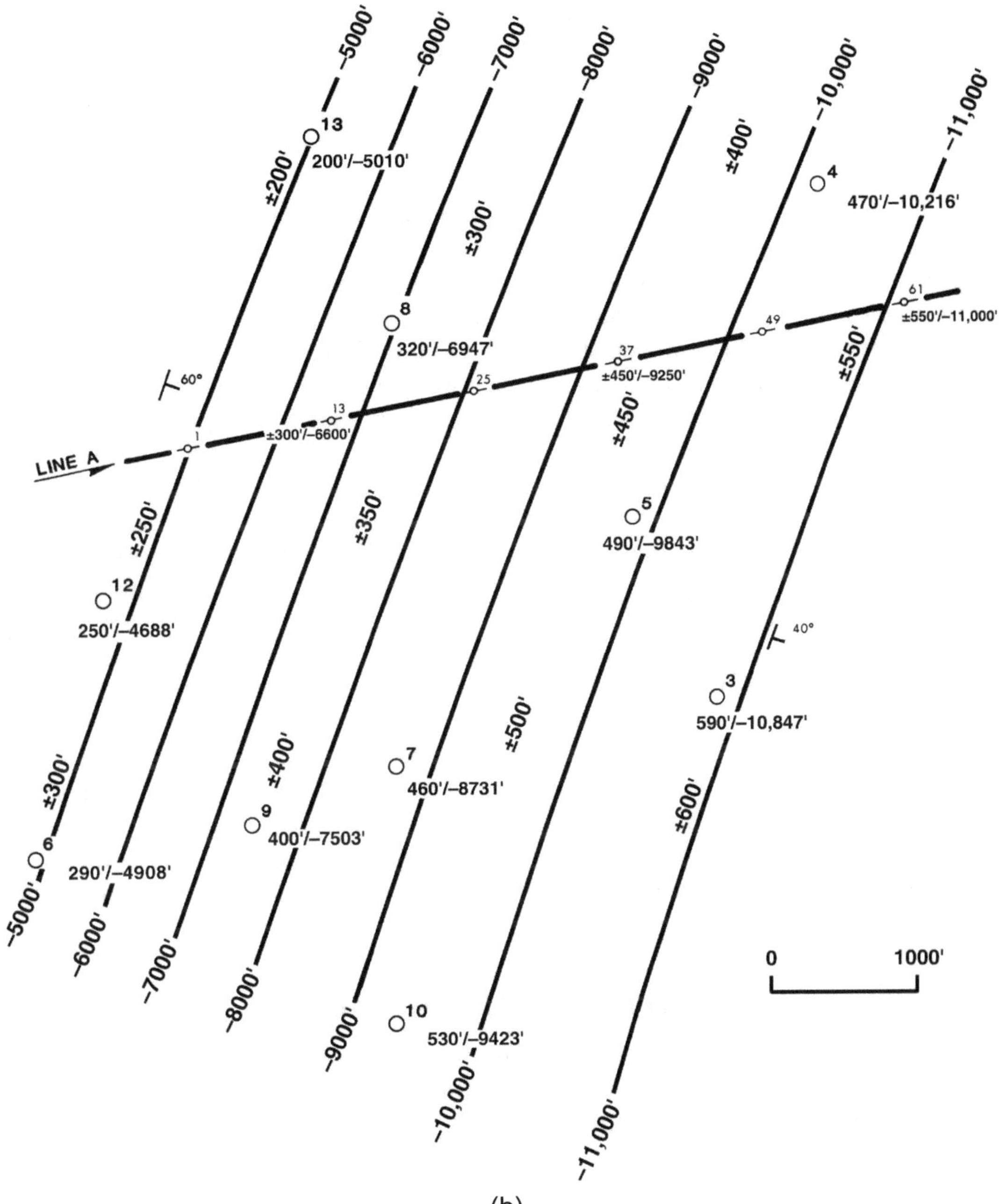

(b)

Figure 7-47 *(continued)*

Growth-Fault Surface Map Construction

Earlier in the chapter, we discussed the importance of preparing a fault surface map for all faults. The procedures and guidelines for constructing fault surface maps as outlined earlier are applicable to growth faults, with one possible addition. Recall we stated that a good geological mapping practice is to contour the vertical separation for the fault, as well as the fault surface (Fig. 7-40a). With a growth fault, however, it is absolutely necessary to contour both the fault surface and vertical separation, whenever possible. The vertical separation for a growth fault varies with depth and changes laterally, and in some cases rather abruptly. Therefore, in order to prepare an accurate interpretation of the fault for future mapping and prospecting, both the fault surface and vertical separation should be mapped.

The vertical separation data for a fault can come from electric well logs or seismic sections. Once the vertical separation data are gathered, there are at least two ways to map the data. The vertical separation values can be contoured on the fault surface map as shown in Figure 7-47a, or they can be recorded at strategic locations on the fault surface map as shown in Figure 7-47b. The actual contouring of the vertical separation is by far the more accurate method and the one we recommend.

Looking at the fault surface and vertical separation contour map in Figure 7-47a (which is only a segment of the entire fault), we can quickly observe a lot of valuable information about this growth fault. The sources of the fault surface and vertical separation data are both well logs and seismic sections. The fault surface contour spacing indicates that the fault has a listric shape with a maximum dip rate of 60 deg at around −5000 ft, decreasing to 40 deg at about −11,000 ft. Both the vertical separation values next to each well and seismic shotpoint and the vertical separation contours indicate that the vertical separation increases with depth from 250 ft at −4688 ft in Well No. 12 to 590 ft at −10,847 ft in Well No. 3. Based on the spacing of the vertical separation contours, the fault growth appears to be generally uniform over this area of the fault. Also, the contours show that the vertical separation increases laterally from the northeast to the southwest. Taking the −9000 fault contour as an example, in the southernmost portion of the map, the vertical separation at this depth is 500 ft. Following along this −9000-ft contour to the northeast, this displacement decreases steadily to only 400 ft at the northern end of the map. Lastly, the contours point to the direction of maximum growth fault activity (major sedimentary depocenter for this fault) as being to the southwest. All this information is important for future integration of this fault interpretation with structure and stratigraphic data to prepare detailed structure and isochore maps for use in the exploration and exploitation of localized hydrocarbon accumulations.

The pattern of fault surface and vertical separation contours is different for nongrowth (structural) faults than it is for growth faults. Refer once again to Figure 7-40a and the discussion. Compare this map with that for the growth fault in Figure 7-47a. Identify the differences observed and establish criteria that can be used to distinguish a growth fault from a post-depositional fault using the fault maps.

DIRECTIONAL SURVEYS AND FAULT SURFACE MAPS

In Chapter 3 we discuss the importance of using the directional survey from a deviated well to calculate the position and depth of the borehole throughout its entire length. Recall that we illustrate the two basic ways in which directional well paths are presented on a map base, shown here again as Figure 7-48a and b.

Using the simplified method of plotting a directional well shown in Figure 7-48a, the only data required and plotted on the basemap are the surface and bottomhole locations. At times the measured depth to total depth may be recorded next to the bottomhole location. Between the surface and bottomhole locations, a straight dashed line is usually drawn. This directional well plot provides absolutely no information about the position or depth of the wellbore in the subsurface between the surface and bottomhole locations. Such a plot is not helpful in the interpretation, construction, and evaluation of fault, structure, or isochore maps.

Where directional survey data are actually plotted to provide detail regarding the position and subsea depth of the wellbore throughout its entire length, as shown in Figure 7-48b, the plot has real value. Such a plot provides a visual guide (in map view) to the location and subsea depth of the wellbore anywhere along its path. Such a plot saves time in preparing subsurface maps and is extremely helpful in the interpretation, construction, and evaluation of fault, structure, and isochore maps. In this section, we look at several important benefits of fault surface mapping derived from plotting the actual position and subsea depth, at evenly spaced increments (usually 500 ft or

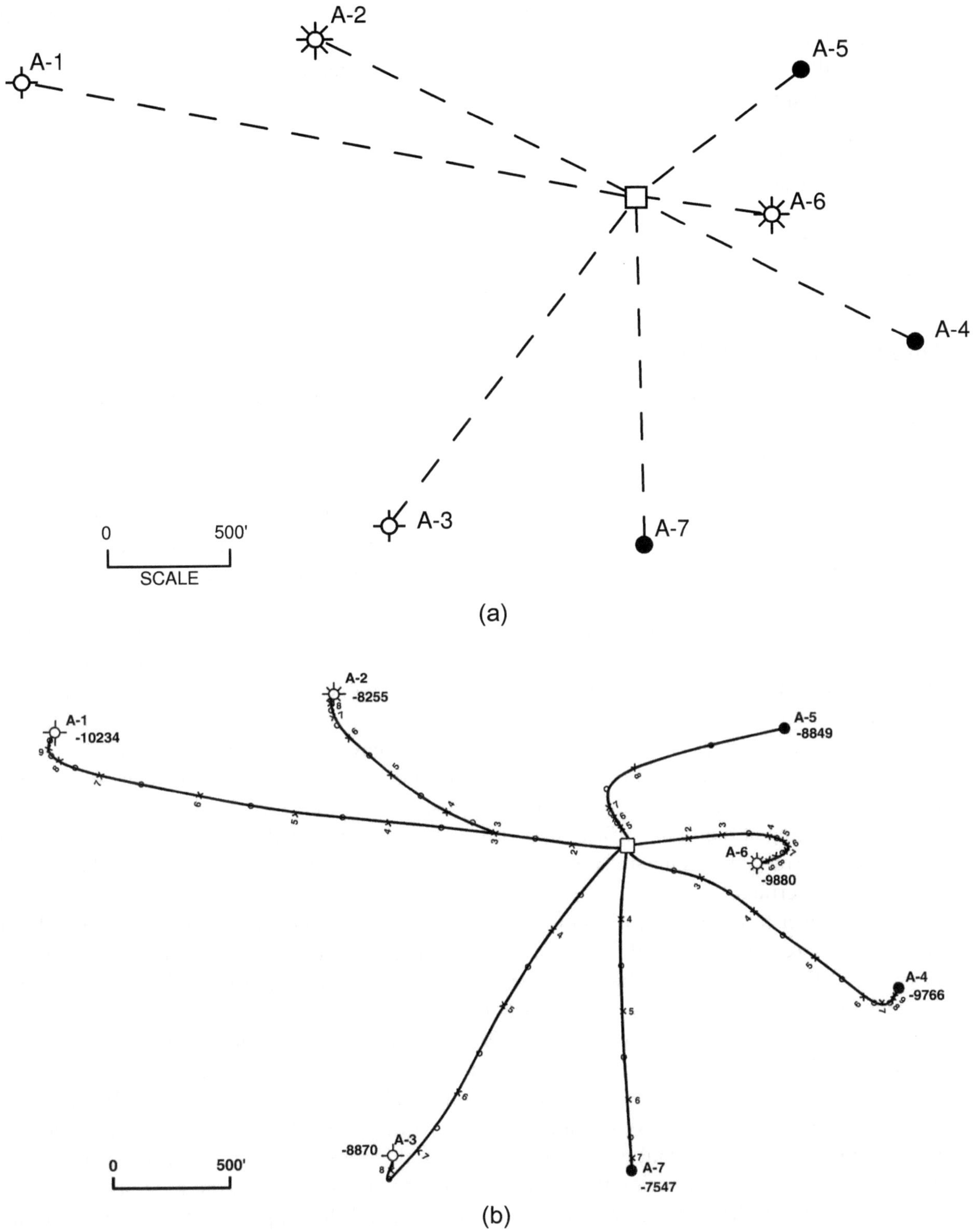

Figure 7-48 (a) Straight-line method of plotting directional wells in map view. (b) Detailed plot of directional survey data indicating the location and subsea depth of each wellbore along its entire length.

1000 ft), of all directional wells on a basemap. The benefits to structure and isochore mapping are covered in the appropriate chapters.

The advent of computer mapping systems has rendered the issue of basemaps largely moot, as computers have no problem in correctly plotting fault cuts and horizon picks in their proper X-Y location. Some database programs, such as Lexco's OWL for the Gulf of Mexico, still use the simplified method illustrated in Figure 7-48a because of the speed of mapping. All software mapping programs with which we are familiar use the more accurate method illustrated in Figure 7-48b. Tearpock and Bischke (1991, 2003) discussed the relative merits of the two methods in some detail, and interested readers are referred to these references.

Directional Well Pitfalls

When is a straight hole not a straight hole? When it is deviated. Many so-called vertical wells are not vertical at all but have been drilled at a deviated angle, as discussed in Chapter 3. This deviation usually is not planned but is the result of natural wellbore drift. As we discuss in Chapter 3, the deviation angle of a drilling well may be monitored in some way, such as with a Totco survey. Remember that a Totco records the deviation angle of the well but not its direction. Most countries have regulations allowing a maximum deviation before an actual directional survey is required by law. Not all wells that are designed to be vertical, however, are monitored for deviation.

The fault surface maps in Figure 7-49a and b illustrate the kinds of problems that can arise when wells are assumed to be vertical or when no directional well data are available. The fault surface map in Figure 7-49a was prepared using the available well data shown on the map. Each depth value represents the subsea depth at which this mapped fault was encountered in the adjacent well. The map does not look too bad, but it does have some unusual variations in contour spacing (dip of fault) and indicates a problem in the northeast part of the map. According to the available data on the basemap and electric logs, all the wells are identified as being vertical. However, looking at the fault surface map, it is apparent that something is wrong. Either the fault picks are incorrect, the wells are spotted at the wrong surface location, or some of these vertical wells are not vertical.

The answer to the problem is shown in Figure 7-49b. Upon additional review of well files, five wells were found to be deviated, as shown in the figure. By replotting the wellbores as deviated wells and respotting the fault picks to correspond to the true wellbore deviation, the fault surface map shown in this figure was constructed.

Well No. 6 is of special interest. The only available directional data on this well was a Totco survey, which measures angle but not direction. In the case of Well No. 6, the assumption was made that the deviation was all in the same direction. Therefore, a circle whose radius is equal to the distance the well could have deviated from the vertical to a depth of −7200 ft was plotted. If the wellbore deviation were all in the same direction, the fault pick should fall somewhere on this circle. The direction of deviation that best fit the fault was used to spot the pick, as shown in the figure. Notice that the contour spacing has evened out and the major contouring problem in the northeast was resolved.

Keep this example in mind, especially when mapping in areas with very old wells. A problem that might at first appear to be that of wrong correlations or an incorrect interpretation might turn out to be nothing more than a problem of a vertical well not being vertical. If the issue is important and the wellbore is still available, it is worthwhile to re-enter the well and run a new, modern directional survey.

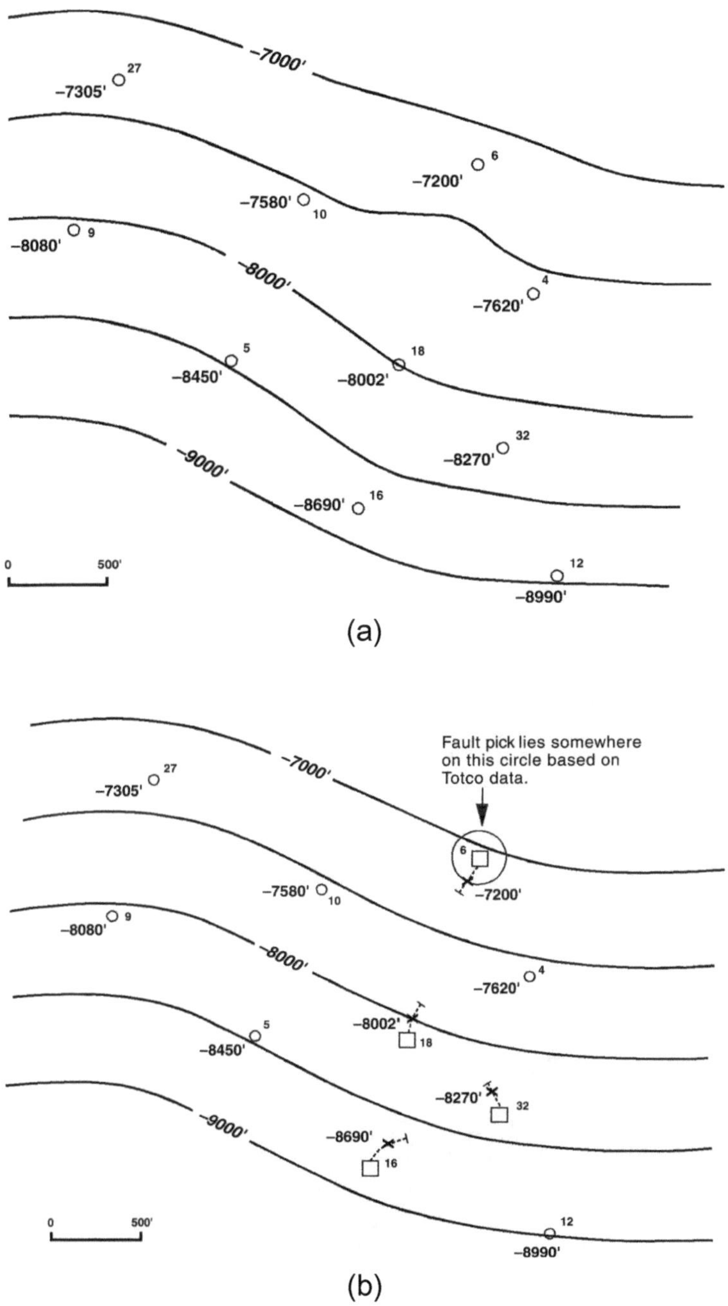

Figure 7-49 (a) Fault surface map prepared with the assumption that all the wells are vertical. Notice the unrealistic variations in dip in the northeast portion of the map. (b) Revised fault surface map based on new directional well data obtained for a number of wells. The new interpretation of the fault surface appears reasonable.

Fault Maps, Directional Wells, and Repeated Sections

In Chapter 4 we discuss the specific geometry between a normal fault and a directional well required for a repeated section (Fig. 4-36). We also look at several examples of repeated sections, such as the one in the Texaco LL&E Well No. 212 (Fig. 4-38).

In the preparation of a fault surface map, a fault resulting in a repeated section is treated in the same manner as a fault resulting in a missing section. Figure 7-50 shows the Texaco LL&E Well No. 212 plotted in cross section along the wellbore deviation path. The well/fault geometry is quite unique in that the well crosses the same Fault H at least four times: twice in the normal sense, resulting in missing section, and twice in the reverse sense, causing two repeated sections. Therefore, the well actually penetrates the downthrown fault block three times and the upthrown block twice.

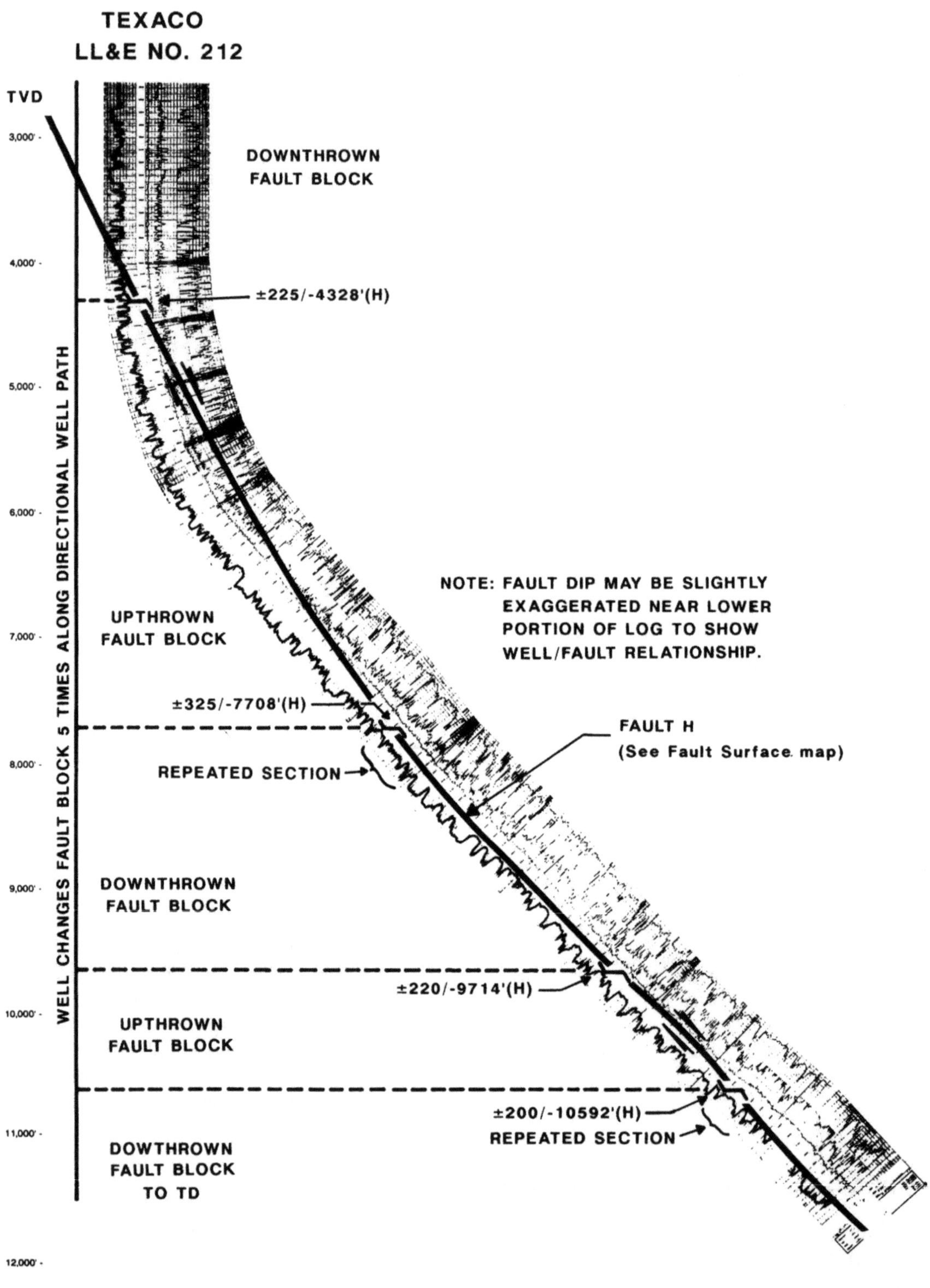

Figure 7-50 Directionally drilled Well No. 212 intersects Fault H four times, resulting in two missing sections and two repeated sections. The well changes fault blocks four times. (Published by permission of Texaco, USA.)

Figure 7-51 shows a portion of the fault surface map for Fault H, in the vicinity of Well No. 212. The four fault picks in the well and fault control data from numerous straight holes and other deviated wells were used to prepare the fault surface interpretation shown. The fault surface has a generally listric shape but displays little if any growth. Near the total depth of the well, Fault H intersects and terminates against Fault C, which is a down-to-the-north fault. Faults E, E-1, and B-1 are interpreted as terminating against Fault H, as shown in Figure 7-51. These faults are not shown in the cross section in Figure 7-50. This example shows the unusual geometries that can occur where directional wells encounter fault surfaces.

VERTICAL SEPARATION—CORRECTION FACTOR AND DOCUMENTATION

The vertical separation for a *nongrowth fault*, as discussed in Chapter 4, is equal to the vertical thickness of the missing or repeated section in a well. In vertical wells the thickness represented on a log is the *true vertical thickness*. Therefore, when correlating a fault in a well log with another log from a vertical well, the vertical separation for the fault is the measured amount of missing or repeated stratigraphic section determined from the vertical well log.

For directionally drilled wells, the log thickness can be greater or less than true vertical thickness. Therefore, a correction must be applied to the log thickness to obtain its true vertical equivalent. The equations used for this conversion are presented in Chapter 4.

The vertical separation for a *growth fault* penetrated by a well should be determined at specific horizons by using the difference in restored tops for each horizon faulted out of the well. Where a deviated well is the reference well for determining the interval thicknesses to be used to restore tops, each thickness must be converted to true vertical thickness.

As discussed in Chapter 1, good-quality work must be backed up by good documentation. This means recording all the data in some format that can be easily used or revised by the person conducting the work, supervisors or managers, other team members, or persons inheriting an area for future study. Fault data must be documented, particularly when deviated wells are used for correlation and require correction factors for log thickness in order to determine the correct value for missing or repeated section. This need for documentation is irrespective of whether correlating and mapping is being done by hand or on a computer workstation. If being done on a workstation, fault data must be documented in the computer database if the software allows it, as most interpretation packages do. If the interpretation software being used does not allow recording the fault data, this data must be recorded separately where it is easily accessible to other workers.

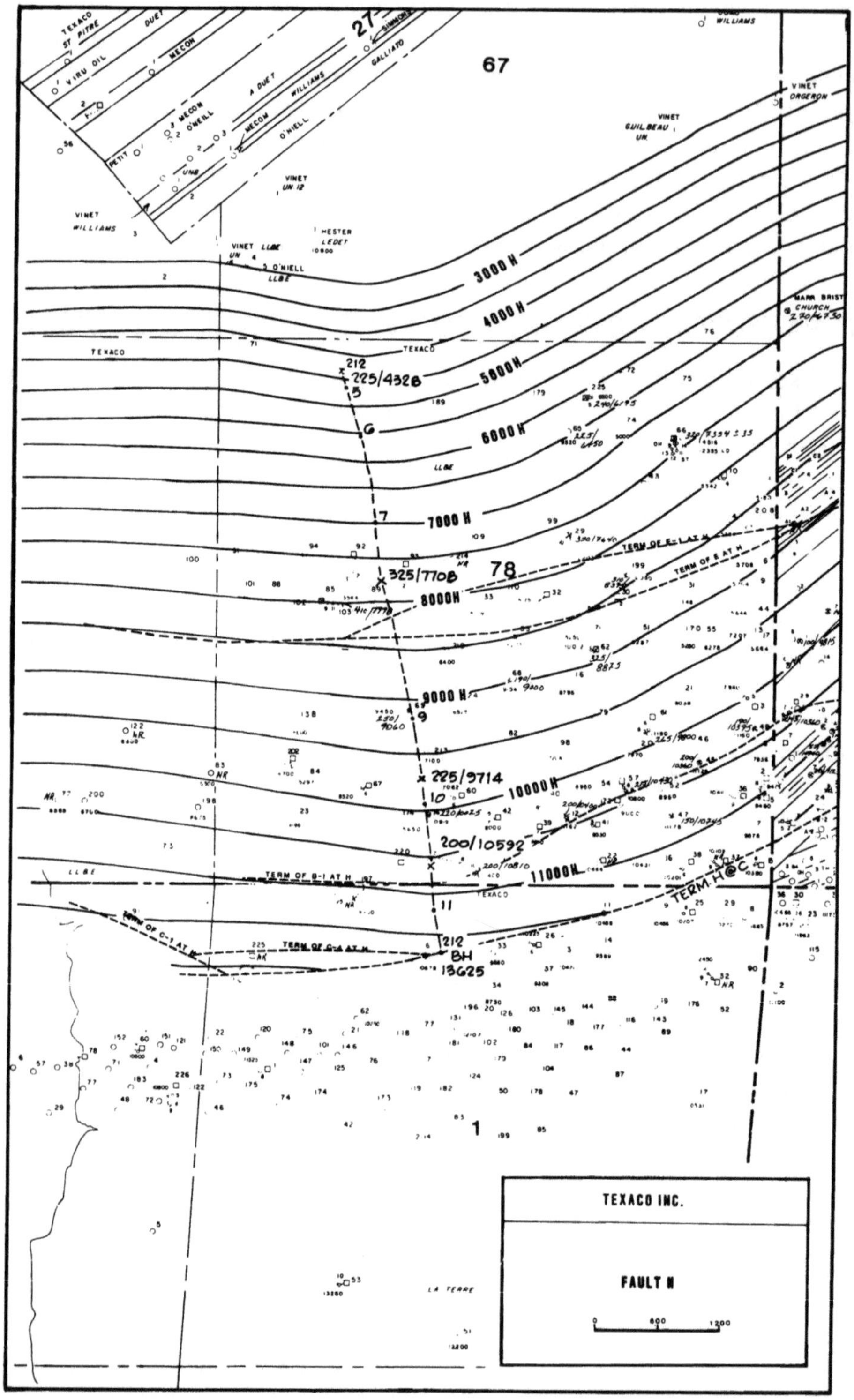

Figure 7-51 Portion of the fault map for Fault H. Note the four fault data points from Well No. 212. (Published by permission of Texaco, USA.)

CHAPTER 8

STRUCTURE MAPS*

INTRODUCTION

The subsurface structure map is one of the primary vehicles used by geoscientists to find and produce hydrocarbons from the initial stage of exploration through the complete development of a field. Each subsurface structure map is a geological or geophysical interpretation based on limited data, technical proficiency, creative imagination, 3D visualization, and experience. We consider the construction of a structure map to be an interpretive and creative process. No two geoscientists will construct a map exactly the same, even with the same data, because each uses the factors just mentioned in addition to educational background and field experience to develop his or her interpretation. However, the interpretation must incorporate *sound geological principles, correct and accurate mapping techniques, and be valid in three dimensions*.

The importance and reliability of subsurface structure mapping increase with advancing stages of field development and depletion. Many management decisions are based on the interpretations presented on subsurface structure maps. These decisions involve investment capital to purchase leases, permit and drill wells, and to work over or recomplete wells, to name a few examples. A geoscientist must employ the best and most accurate methods to find and develop hydrocarbons at the lowest cost per net equivalent barrel.

Since faulted structures play such a significant role in the trapping of hydrocarbons, we devote a considerable portion of this chapter to the correct and accurate subsurface mapping techniques required to integrate fault surface map interpretations into the structural interpretation to construct completed structure maps. A reasonable structural interpretation, in most faulted areas, begins with an accurate fault picture developed from the interpretation of faults and the construction of fault surface maps using fault data from well logs and seismic sections (Chapter 7), followed by the integration of these fault surface maps into the structural interpretation.

---

* For all figures in this chapter (in the printed book only), see the preface for information about registering your copy on the InformIT site for access to the electronic versions in color.

Many petroleum provinces involve multiple faults that result in extremely complicated structural relationships. The attempted reconstruction of a complex structure with isolated fault data from well logs or seismic sections can result in erroneous geological interpretations. Too often, subsurface interpretations and the accompanying structure maps are prepared without giving much consideration to the 3D geometric validity of the interpretation. The most accurate and sound structural interpretation in a faulted area requires (1) the interpretation and construction of fault surface maps for all important structure-forming and trapping faults, (2) the integration of the fault surface maps with the structural horizon maps, and (3) mapping of multiple horizons at various depths (shallow, intermediate, and deep) to justify and support the integrity of any structural interpretation (Tearpock and Harris 1987).

The exploration for and exploitation of hydrocarbons is interpretive and creative work. Most of the time, a geoscientist is dealing with geological structures that are not visible on the surface. The formidable challenge of interpreting these unseen structures can be accomplished only with a clear understanding of basic geological principles, familiarity with the geometry of structural and fault surface relationships, analysis of all available data, use of all technical capabilities, application of technical knowledge and skills, and imagination.

In this chapter, we concentrate on the technical knowledge and skills necessary to develop a geologically reasonable structural interpretation. Technical knowledge and skills fall into two categories: (1) a good understanding of the tectonic setting being worked, and (2) understanding and application of correct interpretation and mapping techniques. The primary focus of this chapter is on the broad range of important structural mapping techniques; however, since the application of many techniques depends on the tectonic style (type of structure and trap), we discuss and illustrate techniques as they apply to different tectonic settings and review a number of real-world examples.

Subsurface structure maps usually are constructed for specific stratigraphic horizons to show, in plan view, the geometric shapes of these horizons. These maps are constructed using correlation data from well logs, interpretations of seismic sections, and in some cases, outcrop data. Remember that *accurate correlations are paramount for reliable subsurface interpretation and mapping*. Subsurface structure maps are no more reliable than the correlations used in their construction. Incorrect correlations will find their way, at some point, into the final interpretation. They may be incorporated into the fault, structure, isochore, or isopach maps and result in serious mapping problems. Therefore, it is essential that utmost care be taken in correlating logs and interpreting seismic sections.

Not every horizon within a stratigraphic sequence is suitable for structure mapping. A horizon that is not correlatable over a large area or one that is limited in areal extent may not be suitable. Maps on stratigraphic horizons of limited extent, if important, can be prepared after the overall structural interpretation has been developed from fieldwide or regional correlations and structure maps.

A structure map is a form of contour map. As discussed in Chapter 4, marine shales exhibit distinctive characteristics over large areas. Therefore, they serve as good correlatable horizons for fieldwide or regional structure mapping. A structure contour map presents a 2D interpretation of the 3D shape of a specific stratigraphic horizon. Each contour connects points of equal elevation above or below sea level for a given stratigraphic horizon. A good structural interpretation requires 3D thinking, as illustrated in the simplified block diagram in Figure 8-1.

Broadly interrelated assemblages of geological structures constitute the fundamental structural styles of petroleum provinces. These assemblages generally are repeated in regions of similar deformation, and the associated types of hydrocarbon traps can be anticipated (Harding and Lowell 1979). There are a number of petroleum-related tectonic habitats around the world; each results, to varying degrees, in different kinds of hydrocarbon traps that may require modified or different mapping techniques. In the first part of this chapter, we discuss numerous subsurface structure mapping techniques. These techniques are then reviewed as they apply to the following tectonic habitats and their associated hydrocarbon traps.

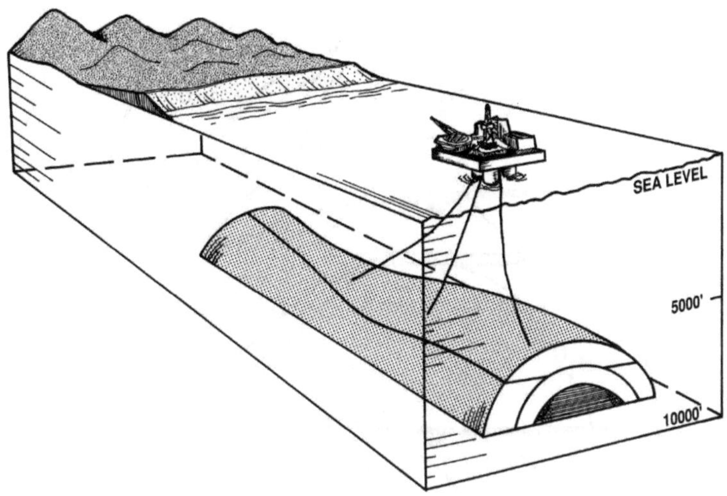

Figure 8-1 A 3D view of an anticlinal structure 7000 ft below sea level.

1. Extensional terranes, including normal faulting and detached listric growth fault systems.
2. Compressional terranes, including reverse faulting and fold and thrust belts.
3. Diapiric salt terranes.
4. Strike-slip fault terranes.

GUIDELINES TO CONTOURING

Review the five basic rules of contouring presented in Chapter 2. In addition to these basic rules, the following guidelines to contouring make a map easier to construct, read, and understand; they also help to ensure the technical accuracy and correctness of the completed map. Some guidelines covered in Chapter 2 are repeated here; many have been expanded, and additional guidelines are presented.

1. All contour maps must have a *chosen reference* to which the contour values are compared. A structure contour map usually uses sea level as the chosen reference. Therefore, the elevations on the map can be referenced above or below mean sea level. A negative sign in front of a depth value indicates that the elevation is below sea level.
2. The *contour interval* on a map should be constant. The use of a constant contour interval makes a map easier to read and visualize in 3D because the distance between successive contour lines has a direct relationship to the steepness of slope. Steep dips are represented by closely spaced contours, gentler dips by contours with a wider spacing. Figure 8-2 illustrates the confusion and difficulty involved in trying to visualize a contoured surface in 3D where the contour interval is not constant over the mapped area. From fault block to fault block, the contour interval changes from 100-ft to 50-ft contours with no consistency. Notice upthrown to Fault A that the contour interval is 50 ft, and downthrown it is 100 ft, yet the contour spacing is about the same. This indicates that the area downthrown to Fault A has a much steeper dip than the area upthrown. However, when we look at this area of the map, the contour spacing gives the illusion that the dip rate in both fault blocks is about the same. The contour interval changes even within the fault block northeast of Fault A.

   The choice of a contour interval is an important decision. Factors to be considered include the density of data, the practical limits of data accuracy, the steepness of dip, the scale, and the purpose of the map.

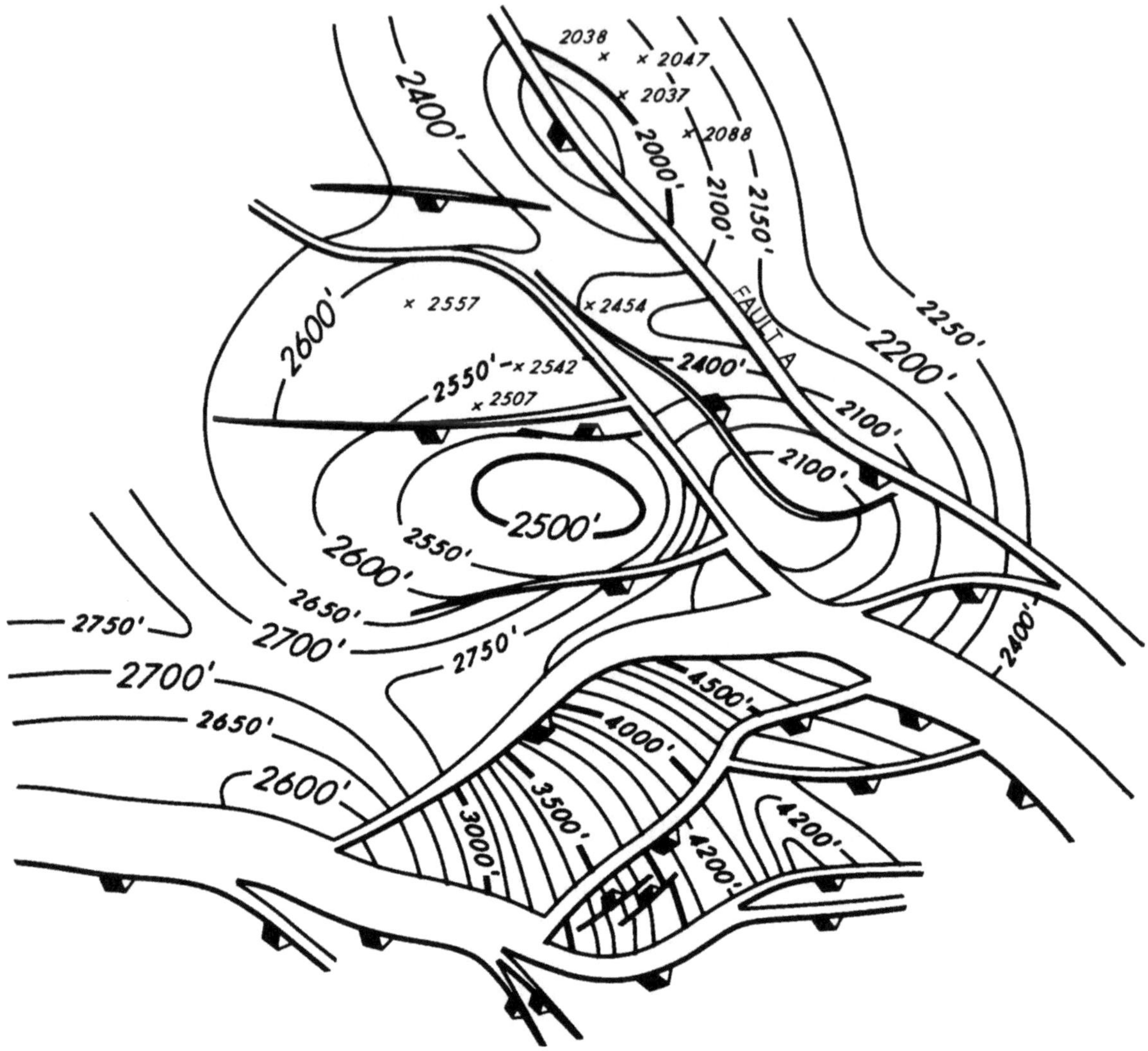

Figure 8-2 This structure map has an inconsistent contour interval randomly changing from a 50-ft to 100-ft contour interval from fault block to fault block. Such inconsistency in the contour interval makes a map difficult to visualize in three dimensions. Observe the change in contour interval upthrown and downthrown to Fault A, and even in the fault block upthrown to Fault A. (From Tearpock and Harris 1987. Published by permission of Tenneco Oil Company.)

3. *Contour spacing* depends on the dip of the structure being mapped. For any given structure, the spacing of contours will vary at different locations unless the equal spacing method of contouring is used. Several graphs are designed for convenient use to compute contour spacing when the dip is known; likewise, the dip on a completed map can be determined by measuring the contour spacing [see Eq. (2-2)]. Figure 8-3 is a graph that relates the dip of beds to the horizontal distance between 100-ft contours. It can be used for determining dip or contour spacing for fault maps as well as structure maps.
4. All maps should include an *X-Y graphic scale* (Fig. 8-4). A graphic scale gives the viewer an idea of the areal extent of the map and the magnitude of the features shown. Maps are commonly reduced or enlarged for various reasons; without a graphic scale, the values shown on the map become meaningless. This is especially true for maps that might be displayed as slides during meetings. The ability of computers to cut, paste, stretch, and squeeze maps requires the use of X-Y scales.

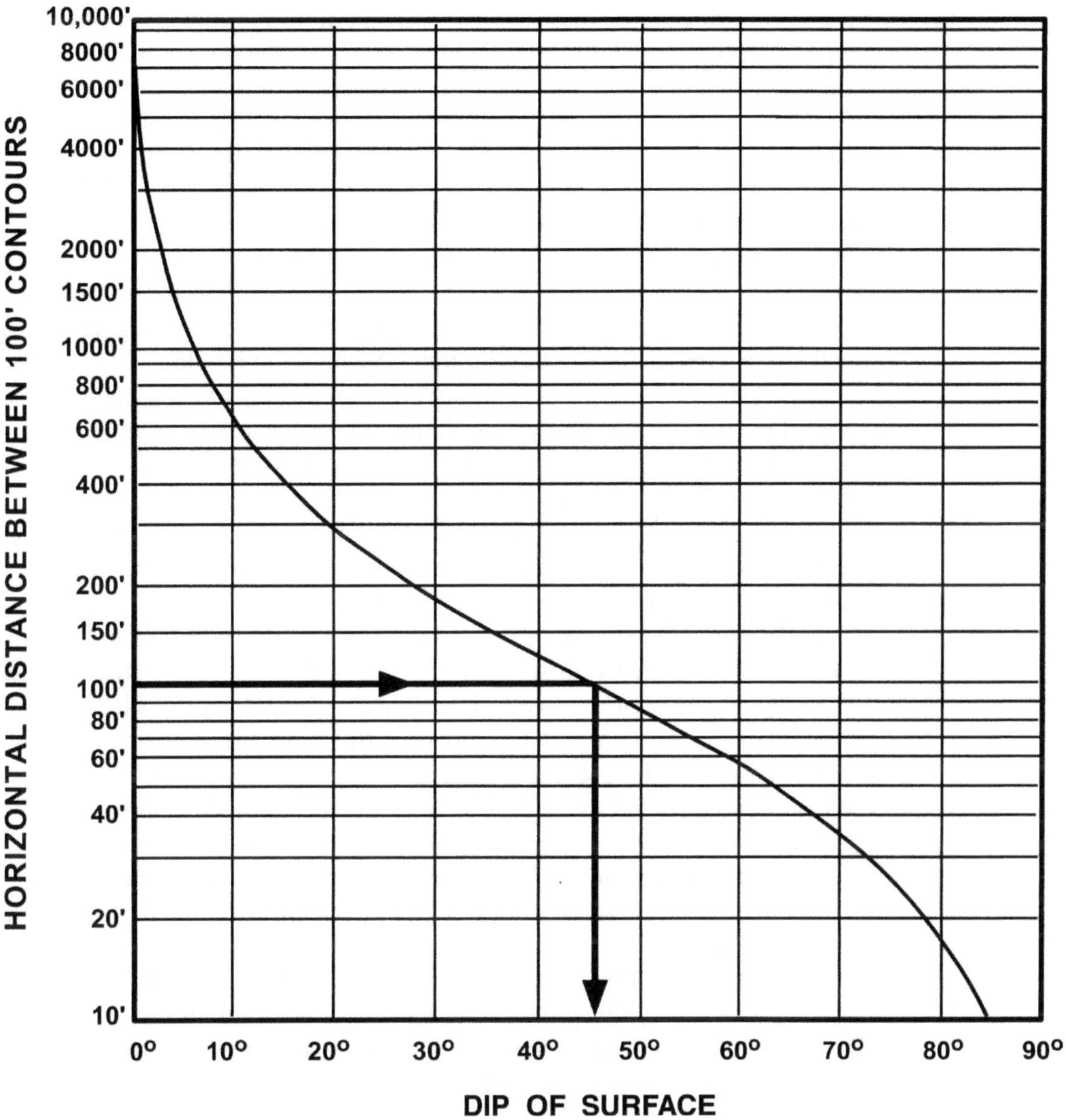

Figure 8-3 Graph of dip versus horizontal distance (in feet) between 100-ft contours. This graph is derived from Equation 2-2, bed dip = arctan (contour interval/contour spacing), also described as bed dip = arctan (rise/run).

5. Every fifth contour is an *index contour*. It should be bolder or thicker than the other contours and labeled with its value (Fig. 8-4).
6. *Hachured lines* should be used to indicate a closed depression (Fig. 8-5).
7. Contouring should be started in areas with the maximum number of control points (Chapter 2, Fig. 2-8). The area or areas of maximum control commonly occur around structural highs or lows.
8. *Construct contours in groups of several lines* rather than one single contour at a time (Fig. 2-8). This method provides better visualization of the surface being contoured and results in more consistent contouring.
9. Initially, choose the simplest contour solution that honors the control and provides a realistic interpretation. The simplest solution may be the best (Occam's razor), and it is usually easy to test. If problems arise with this solution, a more complex interpretation can be prepared.

10. Use a *smooth rather than undulating style* of contouring unless the data indicate otherwise. Some geoscientists argue that a smoothly contoured structure is not likely to occur in nature (Chapter 2, Fig. 2-9). This may be true; however, it is better to keep the structure simple with smooth contours until the data indicate otherwise. It is possible to present a significant misinterpretation by placing unjustified wiggles in contours (Silver 1982).

11. Initially, a hand-contoured map should be *contoured in pencil* with lines lightly drawn so that they can be erased as the map requires revision.

12. *Establish regional dip* whenever possible. Regional dip is the general direction of dip for any given area. Regional dip may not be constant over a large area, but changes should be gradual. In areas of regional dip, contour lines have a certain degree of parallelism along regional strike. Any change in the dip rate may be an indication of local structures. In areas of minor or localized structures, contours extend away from regional dip. Such indications are important because in many petroleum provinces minor anomalous highs that break regional dip commonly are productive of hydrocarbons (Fig. 8-4).

If regional dip is interrupted by a localized structural high, reentrants occur on each side of the minor uplift (Fig. 8-4 or 8-5). If the axis of the localized uplift parallels regional strike, the magnitude of the reentrants may be small compared to reentrants adjacent to a high that is perpendicular to regional strike (Fig. 8-5).

Any flattening or reversal of normal dip is a possible clue to local structures. Therefore, changes of this kind are extremely important. Local uplifts may have their axes perpendicular or parallel to the regional strike. When the axis of a local fold is perpendicular to strike, contours flare outward in a down-dip direction and the distance between contour lines increases as the rate of dip decreases. A nosing or U-shaped projection results in the bottom of the U pointing basinward (Fig. 8-5). Nosings are flanked by reentrants, the axes of which may be perpendicular or oblique to the regional strike. The reentrants begin where the contours start to widen out and are less pronounced down-dip until eventually they disappear.

If the axis of a structurally high area is parallel to regional strike, reentrants are also parallel to strike. As the direction of regional dip reverses at the axis of the reentrants, a high area results down-dip, as shown in Figure 8-5 (Bishop 1960).

13. *Contouring can be optimistic or pessimistic* depending on your experience, corporate guidelines, and exploration philosophy. All contouring, however, must be governed by sound geological principles and the general tectonic style, and optimism must be kept within geologically reasonable limits. Pessimistic contouring can condemn potentially prospective

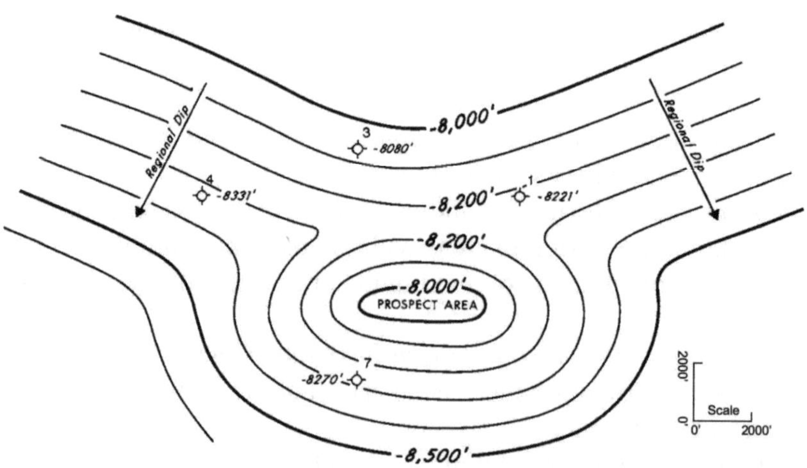

Figure 8-4 An example of a localized structural high indicated by a change in regional dip.

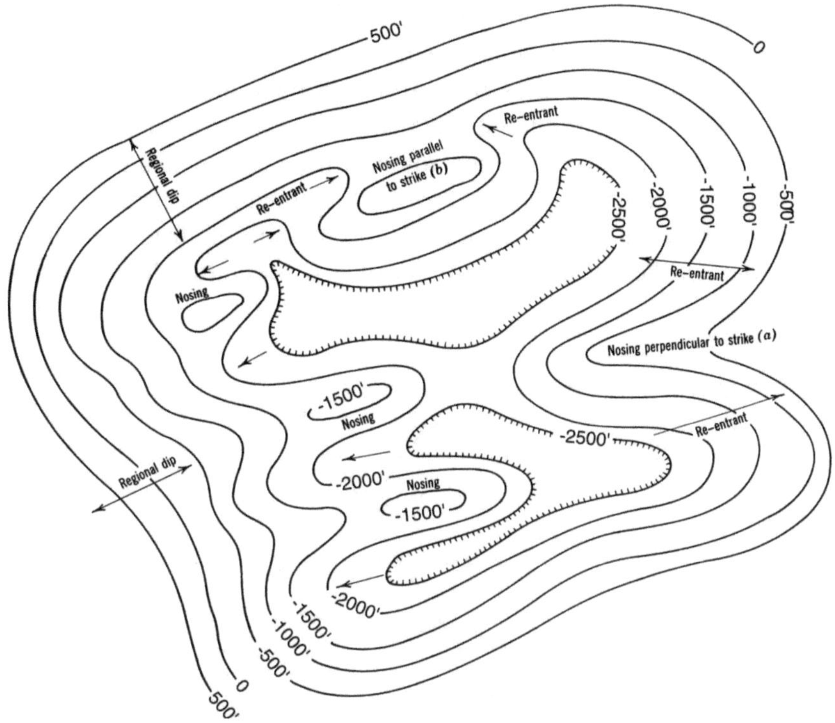

Figure 8-5 A diagrammatic structure contour map of a basin illustrating several important contouring guidelines. (From Bishop 1960. Published by permission of author.)

areas to the point that no exploratory drilling is undertaken. A good mapping philosophy to follow is to map neither optimistically nor pessimistically, but instead use all of your technical expertise to map *realistically*. Structure maps used to determine reserves (Chapter 14) should generally be contoured conservatively.

14. In areas of either limited subsurface control or vertical faults, it is important to contour the limited data to reflect as simple a geological interpretation as possible, rather than just to connect points of equal elevation. Therefore, any *radical change that occurs in the strike of the contours* may suggest faulting even though no fault has been recognized by well control. Figure 8-6a depicts such a situation. In these cases, all available data need to be reviewed, including production and pressure data to help resolve the geological problem. In the example shown in Figure 8-6, notice a significant change in contour strike in the area marked as a possible fault, although no fault is recognized in the wells. An interpretation that fits all the geological and hydrocarbon data includes a vertical fault not intersected by the wells (Fig. 8-6b).

An abrupt increase in the rate of dip is a good indication of faulting. An increase in the rate of dip accompanied by an abrupt change in strike is very strong evidence of faulting (Bishop 1960). Increased dip might, alternatively, result from folding, but in most cases the increase is more abrupt where faulting is responsible.

15. *A change or reversal in the direction of dip* suggests the crossing of a fold axis (Fig. 8-5 or 8-7). Reversal of dip may occur over the crest of an anticline or the trough of a syncline. The amount of dip reversal is often a guide to whether the reversal is due to a regional change or local structure. An excessively steep dip may indicate the presence of a fault or steep fold, while a dip that flattens may be indicative of the crest of a fold or the bottom of a syncline.

16. *Structures may or may not have structural attitude compatibility (contour compatibility) across a fault*. The compatibility of structural attitude on opposite sides of a fault depends primarily on the size and type of fault. For example, within many, if not most, structures, structural compatibility exists across normal and reverse nongrowth faults (Fig. 8-8a). In contrast, many large listric normal faults (such as growth faults) and thrust faults, with significant displacements, result in structures that are not compatible across the large fault (Fig. 8-8b). Large listric normal faults and large thrust faults create structures such as rollover anticlines, fault bend folds, and fault propagation folds. If the fault creates the structure, you should not expect contour compatibility across the fault, although there may be compatibility across smaller faults within the created structure. The method of structure contouring across a fault depends on whether there is a compatibility of structural attitude on both sides of the fault.

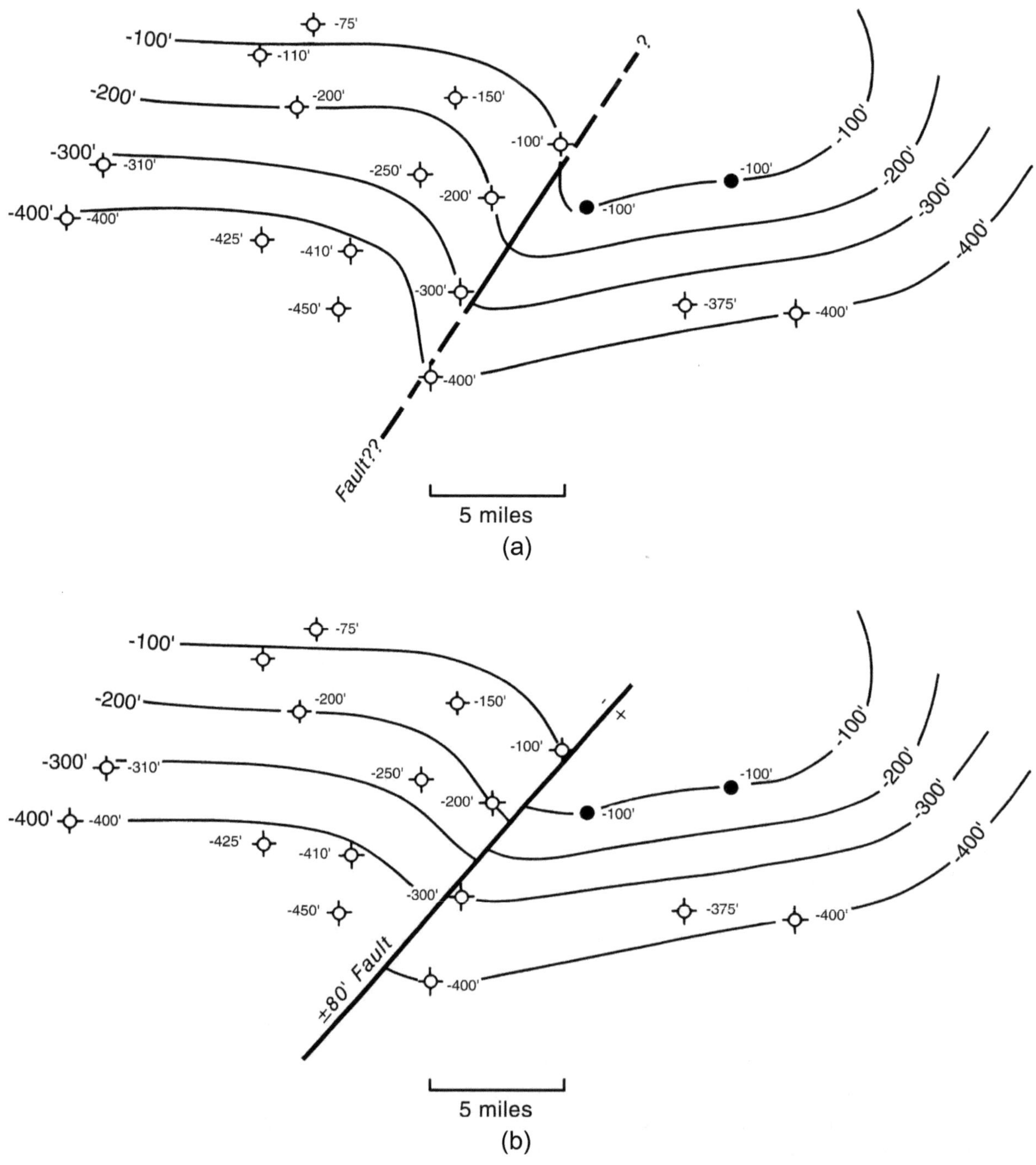

Figure 8-6 (a) An abrupt change in the strike direction of contours suggests the possibility of faulting. (b) Another interpretation of the data in Figure 8-6a that fits the geological and hydrocarbon data includes a vertical fault not intersected by the wells. (Modified after Bishop 1960. Published by permission of author.)

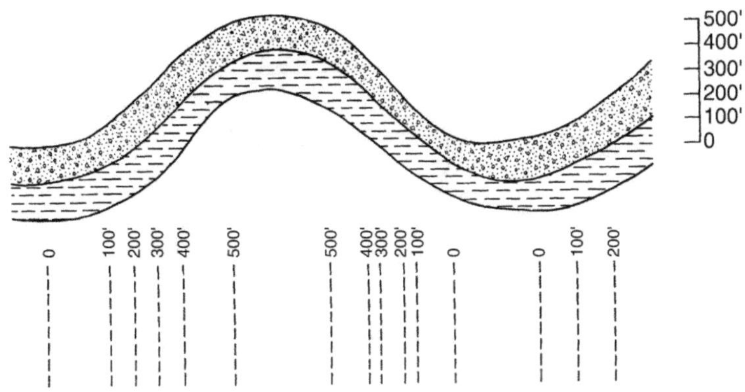

Figure 8-7 Structural highs and lows can sometimes be recognized by their effect on contour spacing. (From Bishop 1960. Published by permission of author.)

17. Structural highs, regardless of origin, usually tend to *flatten across the axis* with gentle dips across the top of the structure (Fig. 8-7). Exceptions do occur, such as the structure in the inner core of a fault propagation fold (Chapter 10). Contour line spacing widens across the crest of the structure compared to spacing on the flanks. Synclines, like anticlines, also tend to flatten across their axes. The widening of contours is often even more pronounced in a syncline. If the data indicate a continuously steep slope up to the crestal high with little if any flattening of dip, this may indicate that the surface is one that has been affected by erosion (the presence of an unconformity).

18. *Closed structural lows* are not common. If possible, avoid closing lows with contours unless the data require it. The presence of closed lows often suggests an eroded surface or the presence of faulting. If the closed low is elongated, faulting is likely, and the greater the size of the closed low, the greater the probability of faulting.

19. *Contour license* refers to the geologist's right to contour a structure in a way that best fits the geological, geophysical, and engineering data, and that best represents the types of structures present in the tectonic setting. The interpretive method of contouring as defined by Bishop (1960) best corresponds to what we call contour license. The use of geological license requires a solid educational background and extensive experience. Contour license does not allow a geologist or geophysicist to ignore or misrepresent valid data.

20. Specific structural highs may be in the form of domes, anticlines, and noses. *Domal structures* are usually the result of local positive features, such as diapirs, that provide relative uplift. On the flanks of domes, the direction of dip is away from the central high and the dip rates are commonly constant along strike, at least within each major fault block around the structure. Therefore, the contour spacing is commonly uniform along strike. In the dip direction, the structurally highest areas on the flank typically have the steeper dips, with a gradual decrease in dip away from the uplift. Contour spacing is close high on the flank of the structure, but widens with distance down-dip.

Anticlinal structures generally appear as elongated domes. Their origin can be the result of compressional forces (e.g., fault-propagation folds associated with reverse faults), extensional forces (e.g., rollover anticlines associated with growth normal faults), or strike-slip forces. In general, the direction of dip is away from the crestal area in two opposing directions. Since anticlines are commonly asymmetric, with inclined axial surfaces, the dip rate and resulting contour spacing may vary around the anticline.

Structural noses that trend off local structures show dip away from the crest in three directions. Contour lines widen, indicating flatter dips in the area of a nose or an associated reentrant. Contours become closer together immediately down-dip of the local high until regional dip is attained again.

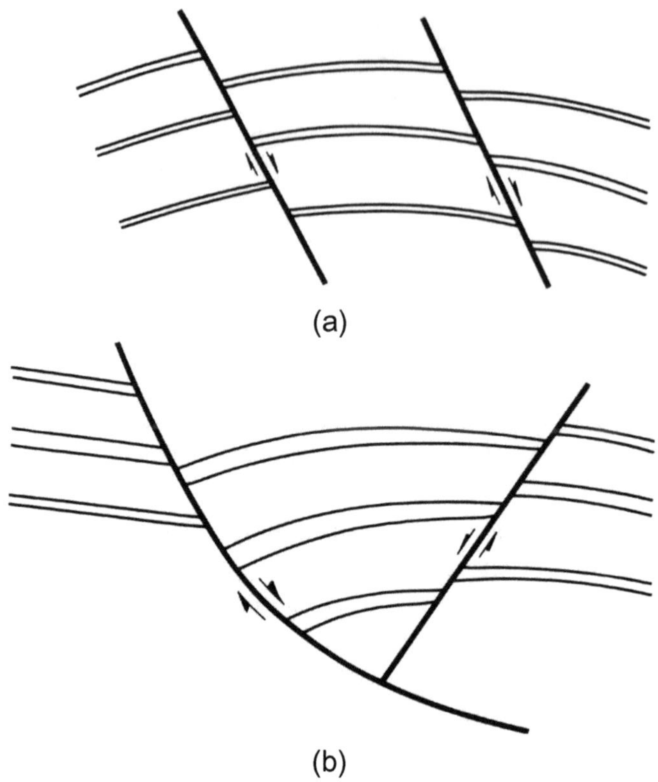

Figure 8-8 (a) Cross section shows structural compatibility across faults. (b) Cross section shows no structural compatibility across major growth fault, but structural compatibility across the smaller antithetic fault.

The best test of the 3D geometric validity of any structure contour map is its predictability. How well does the interpretation hold together with additional data from the drilling of new wells or the shooting of new seismic lines? If the interpretation and maps require major revision each time new data are obtained, there should be serious concern regarding the validity of the interpretation and accompanying maps. On the other hand, if only minor adjustments are required and hydrocarbon traps predicted by the mapping are successfully found, the interpretation may be considered reasonable. Remember, we always work with a limited amount of subsurface data to be interpreted. Each geoscientist must have imagination, an understanding of local structures, the ability to visualize in three dimensions, a sound geological education, field experience, and technical knowledge and skills to evaluate any number of possible (alternative) interpretations. Finally, a geoscientist must decide which interpretation, in his or her judgment, is the most reasonable. The 20 guidelines presented in this section should help you construct more accurate and reasonable structure contour maps.

SUMMARY OF THE METHODS OF CONTOURING BY HAND

As discussed in Chapter 2, four distinct methods of contouring by hand or combinations of methods are commonly used in the preparation of structure contour maps. These are (1) *mechanical*, (2) *equal spaced*, (3) *parallel*, and (4) *interpretive* (see Rettger 1929; Bishop 1960; Dennison 1968).

1. *Mechanical Contouring*. In using this method of contouring, the assumption is made that the slope or angle of dip of the surface being contoured is uniform between points of control and that any change occurs at the control points. With this approach, the spacing of the contours is mathematically (mechanically) proportioned between adjacent control points.

Mechanical contouring allows for little, if any, geological interpretation. Even though the map is mechanically correct, the result may be a map that is geologically unreasonable, especially in areas of sparse control (Fig. 8-9a).

2. *Parallel Contouring*. With this method, the contour lines are drawn parallel or nearly parallel to each other. This method does not assume uniformity of slope or angle of dip; therefore, the contour spacing can vary. Like the previous method, if honored exactly, parallel contouring yields an unrealistic geological picture. It allows for some geological license to draw a map a little closer to the real world as there is no assumption of uniform dip (Fig. 8-9b).
3. *Equal-Spaced Contouring*. This method assumes uniform slope or angle of dip over an entire area or at least over an individual flank of a structure. Sometimes this method is referred to as a special version of parallel contouring. The advantage to this method, in the early stages of mapping, is that it can indicate the maximum number of structural highs and lows expected in an area of study (Fig. 8-9c).
4. *Interpretive Contouring*. With this method, the geoscientist has extreme geological license to prepare a map to reflect the best interpretation of the area of study, while honoring the available control. No assumptions, such as constant bed dip or parallelism of contours, are made. Therefore, the geoscientist can use his or her experience, imagination, ability to think in three dimensions, and an understanding and knowledge of the structural and stratigraphic styles of the geological region to develop an accurate and realistic interpretation. This method is the most acceptable and the most commonly used method of contouring (Fig. 8-9d).

The specific method or combination of methods chosen for hand-contouring may be dictated by such factors as the number of control points, the areal extent of these points, and the purpose of the map. No individual can develop an exact interpretation of the subsurface with the same accuracy as that displayed on a topographic map. What is important is to develop the most reasonable and realistic interpretation of the subsurface with the available data.

CONTOURING FAULTED SURFACES

The contouring of faulted surfaces adds complications in the contouring of both structural horizons and faults. A completed structure contour map on one or more horizons is usually the main objective in any mapping project. In order to construct a completed structure map, however, the faults themselves must be contoured and the fault surface maps integrated with the structure maps. This integration is required to support a reasonable geological interpretation and to prepare accurate maps. In terms of map accuracy, this integration does the following (see Fig. 8-10):

1. Delineates the position of the footwall (upthrown) and hanging wall (downthrown) cutoffs or traces of the fault in map view
2. Provides for the proper contouring of the mapped horizon across the fault
3. Depicts the vertical separation of the fault for any particular mapped horizon
4. Defines the limits of fault-bounded reservoirs

Fault surface mapping is covered in detail in Chapter 7. In this section, we present the proper techniques for integrating a fault surface map with a structure contour map. Included in this section are (1) techniques for positioning the upthrown and downthrown traces of a fault on a structure map; (2) construction of the fault gap or overlap; (3) the mapping of vertical separation versus throw; (4) the use of restored tops in structure mapping; (5) the application of contour compatibility across faults; and (6) the exceptions to contour compatibility.

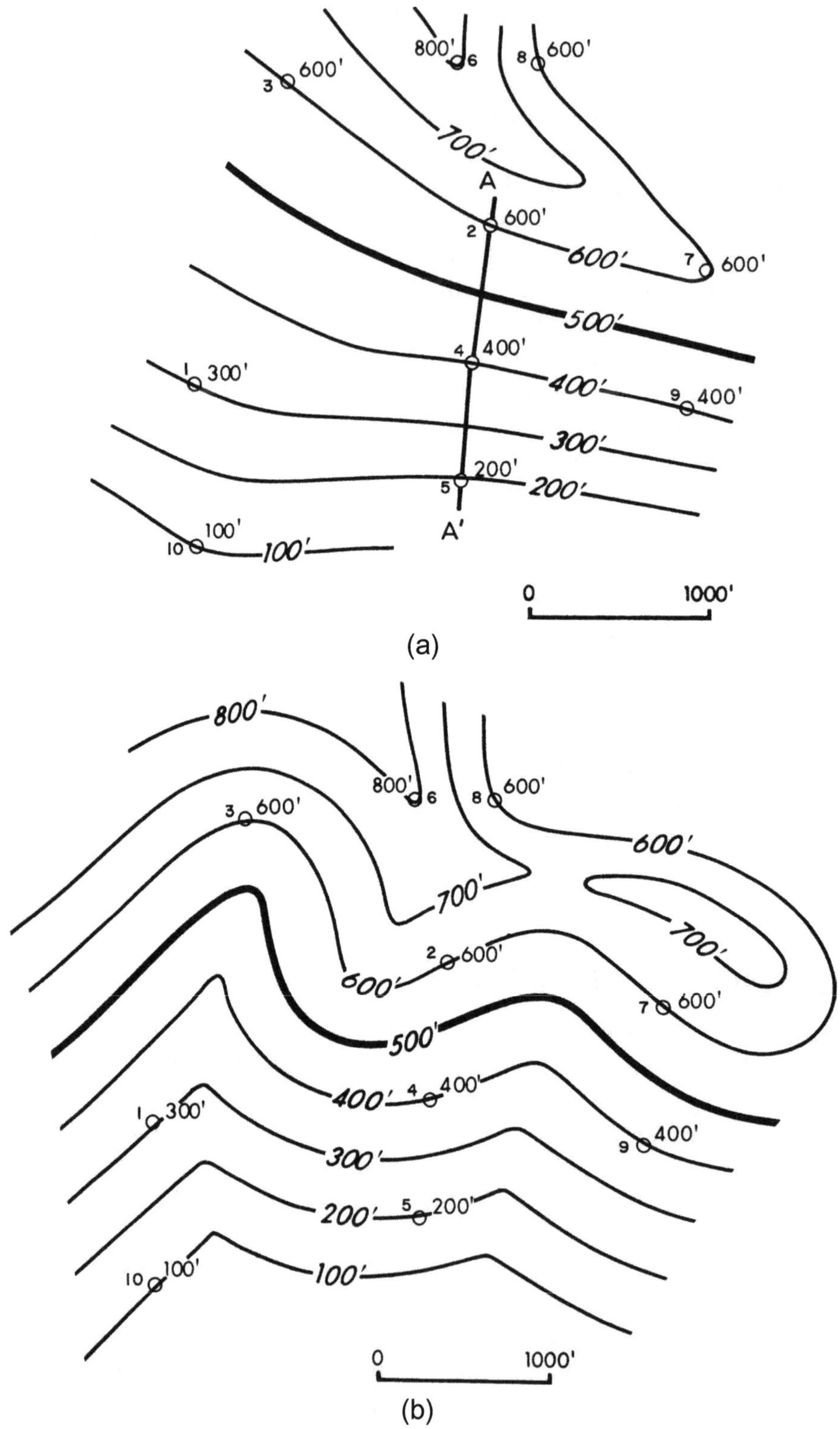

Figure 8-9 (a) Mechanical contouring method. (b) Equal-spaced contouring method. (c) Parallel contouring method. (d) Interpretive contouring method. (Modified after Bishop 1960. Published by permission of author.)

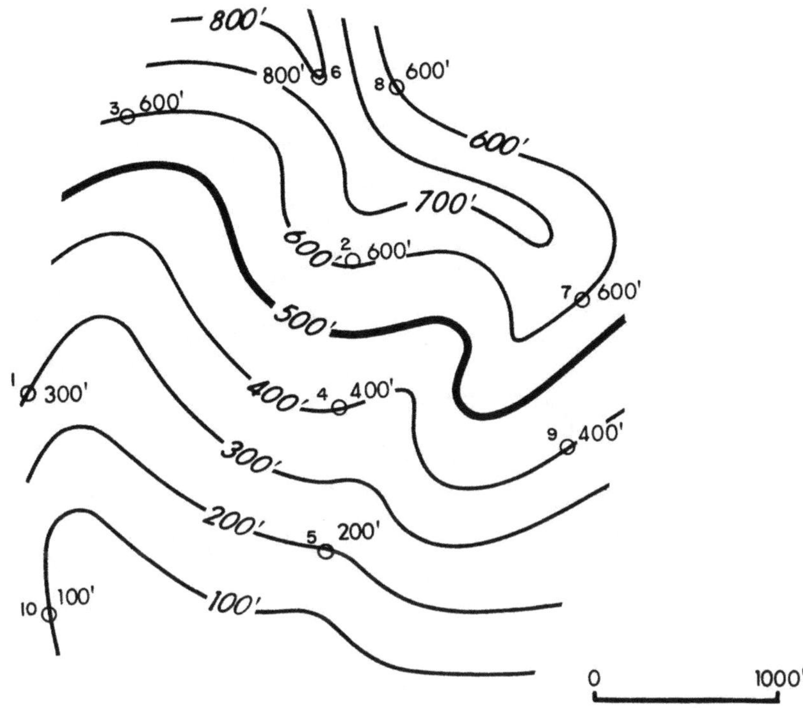

(c)

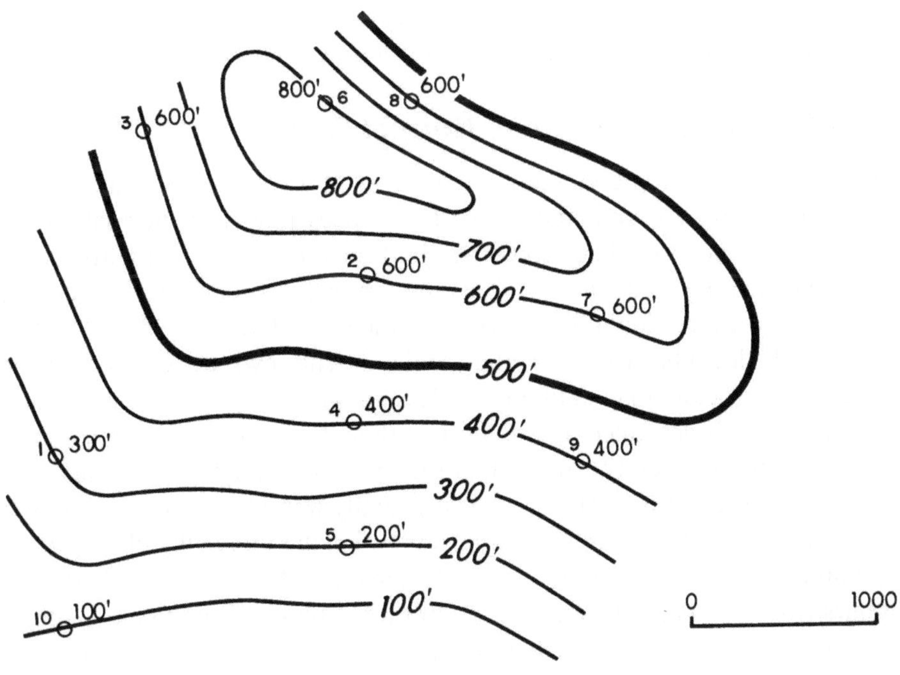

(d)

Figure 8-9 *(continued)*

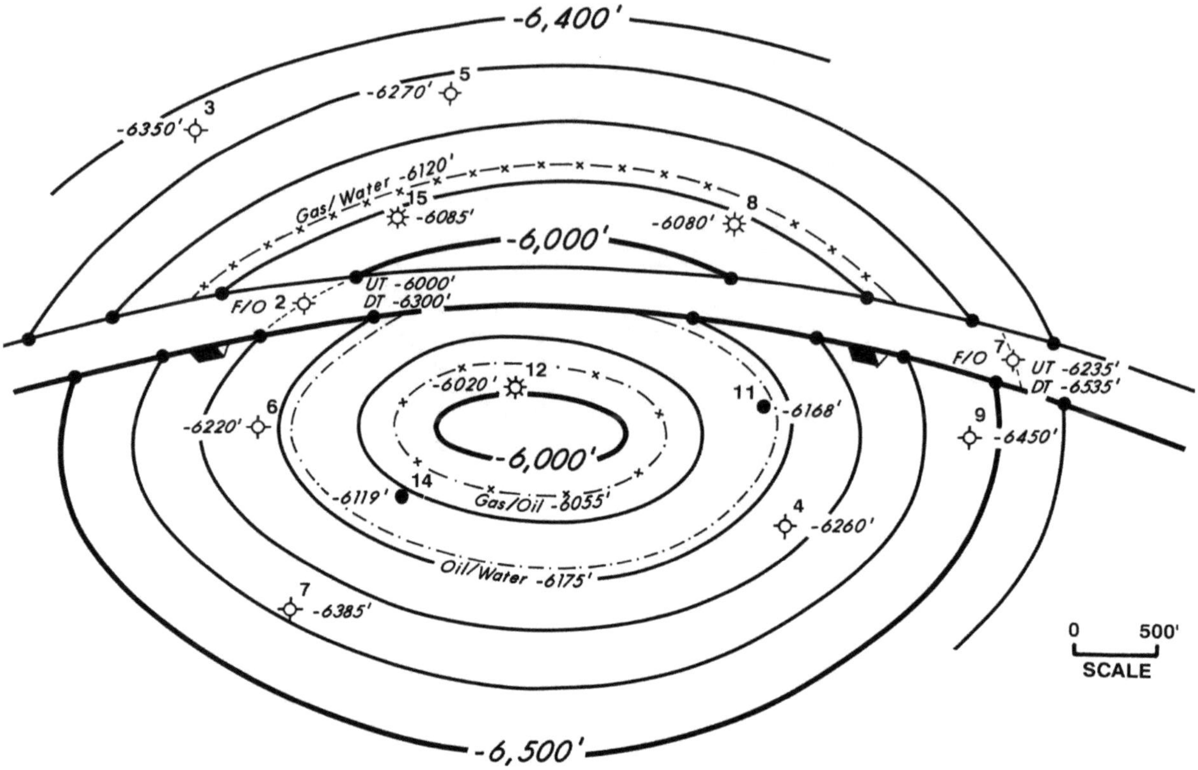

Figure 8-10 Integrated fault and structure map for the 6000-ft Horizon. The darkened circles delineate the intersection of each structure contour with the fault contour of the same elevation.

Techniques for Contouring across Normal Faults

A **fault trace** is a line that represents the intersection of a fault surface and a structural horizon; it is sometimes referred to as a fault cutoff. Two fault traces (lines) are normally required to delineate a fault on a structure map. One line represents the footwall cutoff, or upthrown trace, and the other line represents the hanging wall cutoff or downthrown trace of the fault. Two conventions have been designed to indicate the direction of fault dip: (1) some type of symbol, like a "tent" or triangle, on the hanging wall cutoff (downthrown trace), and (2) the downthrown trace is heavier or thicker than the upthrown trace. The structure map in Figure 8-11a shows a fault displacing a contoured surface, using the conventional symbols described. It is important that all mappers at a given company use the same convention and use it consistently. This will aid managers in quickly reviewing maps from different individuals and subsurface teams. This issue is especially important when mapping on a computer, as computer software does not generally distinguish between the footwall and hanging wall traces.

The techniques presented in this section demonstrate the correct method for projecting established contours from one fault block across a fault into another fault block. Using the available data (Fig. 8-11a), contours are first established for the block with the best control, which in this case is the upthrown block with four wells. These contours are extended to the upthrown trace of Fault 1. To contour across the fault, project the contours from the upthrown block through the fault into the downthrown block. This is shown in Figure 8-11a by a set of dashed contour lines continued across the fault gap indicating what the structural attitude of the horizon would be if the fault were not there. *In other words, where would the contours be drawn if the fault were not there?* Once the contours are projected through the fault gap to the downthrown fault trace, they are adjusted relative to the upthrown contour values by *using the amount of vertical separation*, which in this case is 400 ft. The downthrown block is then contoured. For example, the −8400-ft contour in the upthrown block, when projected into the downthrown block, becomes the −8800-ft contour.

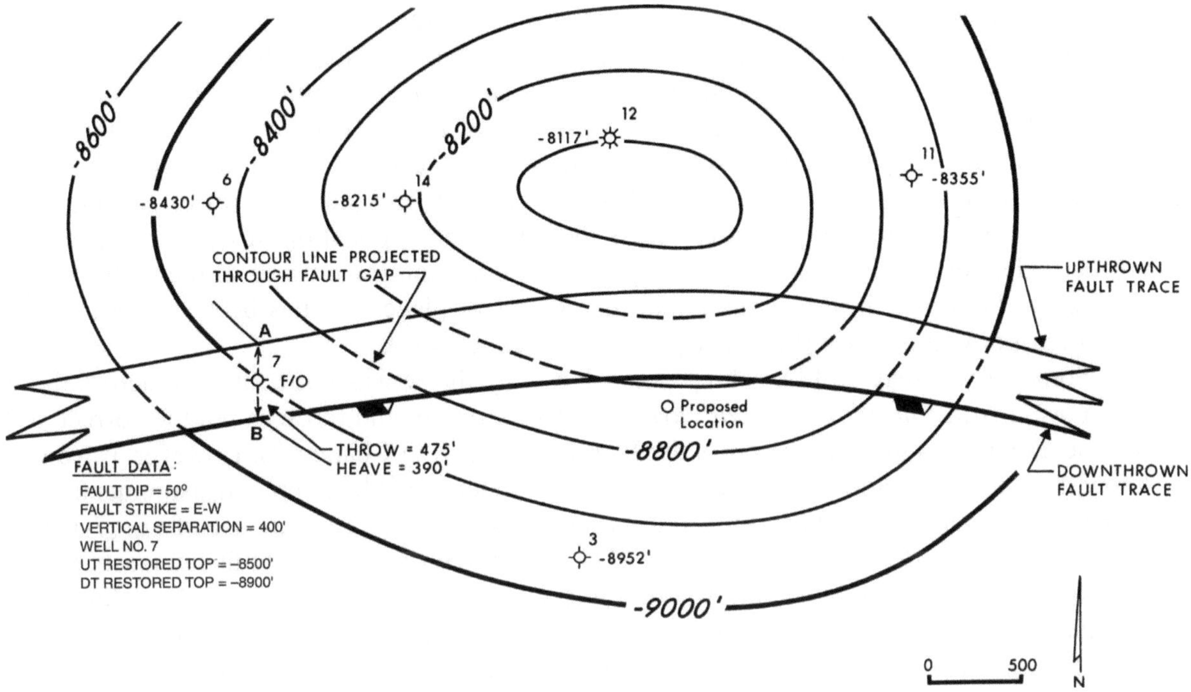

(a)

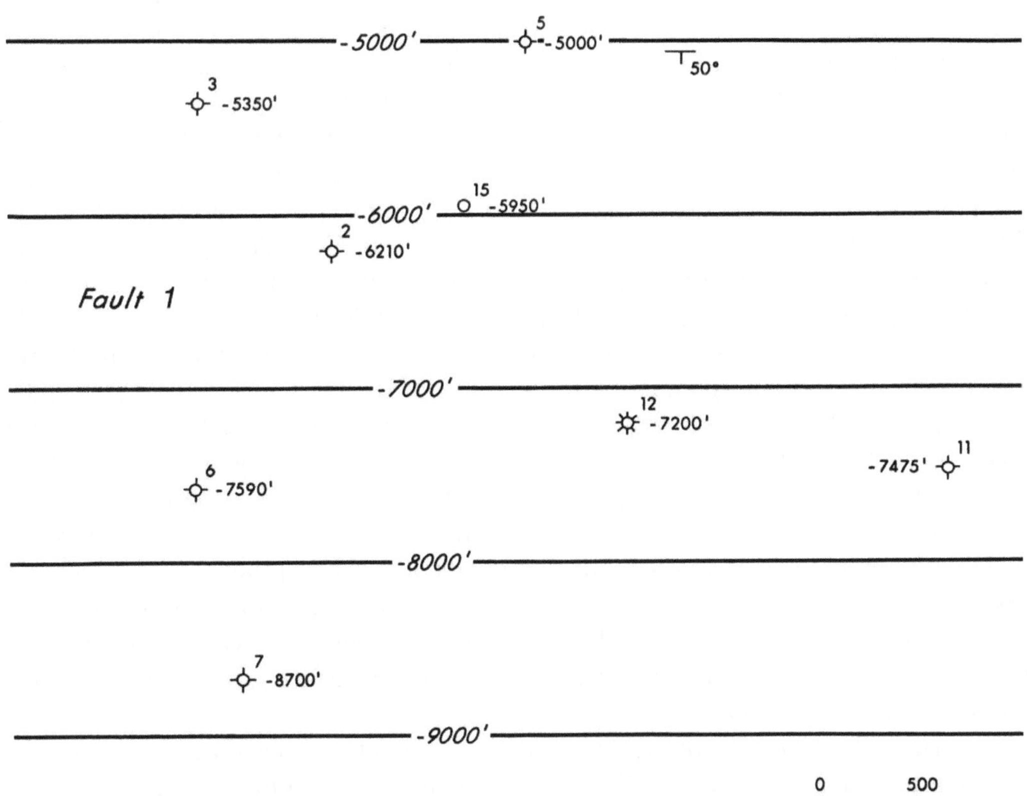

(b)

Figure 8-11 (a) Faulted structure map on the 8000-ft Horizon. The structure is cut by a 400-ft fault. The correct method for contouring vertical separation (missing section) is illustrated by the dashed contour lines. (b) Fault surface map for Fault 1, which strikes east-west and dips at 50 deg to the south. (Published by permission of D. Tearpock.)

Contours may be projected for some distance within the fault gap. Notice how the −8300-ft contour is projected from the upthrown trace of the fault for some distance through the fault gap before it intersects the downthrown trace and enters the downthrown block as an −8700-ft contour. The mechanics of projecting contours, such as the −8300-ft contour, through the fault, as shown in this figure, is the correct technique for contouring across a normal fault, using the vertical separation as obtained from well log correlation (missing section) or seismic sections where there is contour compatibility across the fault. The application of this technique leads to correctly delineating the position of the upthrown and downthrown traces of the fault, thus establishing the fault gap. It also assures that the correct displacement has been mapped across the fault (Tearpock and Harris 1987; Tearpock and Bischke 1990).

Some of you may still be asking yourselves why throw was not contoured across the fault. One reason is that *no throw data are available for mapping*, and throw is not the correct vertical displacement we want to map across the fault (refer again to Chapter 7). However, if we want to know the fault throw and heave, their values can be determined by simple measurements once the structure contour map has been prepared, as shown in Figure 8-11a or by use of Eq. (7-1a).

Throw is the *difference in the vertical depth* between where the fault intersects a specific horizon in the upthrown block and where it intersects that same datum in the downthrown block, *measured perpendicular to the strike of the fault surface* (not perpendicular to the strike of the fault trace). Therefore, a fault surface contour map must be available in order to calculate the throw from a completed structure contour map, as shown in Figure 8-11a. The fault shown in Figure 8-11b strikes east-west; therefore, the throw is determined in a north-south direction (see arrows in fault gap through Well No. 7). Using the points A and B on the map (Fig. 8-11a), the upthrown depth at point A is −8465 ft and the downthrown depth at point B is −8940 ft. The throw of the fault at this location is the difference between these two depths, or 475 ft.

We can calculate the throw mathematically by applying Eq. (7-1a), and knowing that the missing section in Well No 7 is 400 ft, the apparent bed dip, that is, the bed dip perpendicular to the strike of the fault, is about 10 deg at point A, and the fault dip is 50 deg. Equation (7-1a) yields a ratio of vertical separation (*AE*) to throw (*AC*) of 0.85, the absolute value of [(tan10 deg/tan50 deg) − 1]. The throw is thus (400 ft/0.85), or 470 ft. Considering the accuracy of graphical measurements on a contoured map, these two estimates for throw are in excellent agreement. We can see for this particular example that the throw is about 70 ft greater than the vertical separation.

Heave, which is the *horizontal distance* across the fault gap from the upthrown to downthrown traces, measured perpendicular to the strike of the fault surface (not perpendicular to the fault trace), is 390 ft. Therefore, a fault surface map must be available to determine heave as well.

For subsurface petroleum-related structure mapping, the measurements of throw or heave are usually measured for academic purposes and have little application in the actual contouring of a structure map. However, the throw and heave can be used to check a completed structure map using the graphical and mathematical methods described in this chapter and in Chapter 7. If the estimates for throw or heave determined by both methods (graphical and mathematical) are reasonably close, you can conclude that the map construction is reasonable. We point out here that the mapping of throw and heave are important in subsurface mining, the mapping of subsurface mineral deposits and outcrop mapping.

To further illustrate the proper construction of contours across a fault, we review Figure 8-12, using the same data given in Figure 8-11, with one exception. In this case, we map a horizon about 2000 ft shallower, and at this level the fault trace falls on the northern flank of the structure. Here the fault is dipping in the opposite direction as the horizon. The horizon, dipping generally to the north, is displaced by the south-dipping fault. The fault has a vertical separation of 400 ft. From the data points available, the structure contours were first established in the upthrown block and extended to the upthrown trace of Fault 1. As shown in the figure, to contour across the fault it is necessary to project the contours from the upthrown block through the fault gap to the downthrown trace of the fault, and then into the downthrown block. Ask yourself *where the contour would be drawn had the structure not been faulted.* The construction from one block to the other is shown by a series of dashed contour lines placed from the upthrown trace, through the gap, to the downthrown trace. Once the contours have been projected into the downthrown fault block, they are adjusted in depth from the upthrown contour values by the amount of vertical separation, which in this case is 400 ft. As an example, the −6000-ft contour in the upthrown block projected into the downthrown block becomes a −6400-ft contour. As in Figure 8-11, the same technique is followed for construction of all structure contour lines.

Now that the structure contours have been established in both blocks, the fault throw and heave can be calculated and measured. We graphically determine the throw of the fault by estimating the upthrown and downthrown structural depths using points A and B on the map. Notice that the values for throw and heave are different than those estimated in Figure 8-11a. The value for throw is now 325 ft, as compared with 475 ft in Figure 8-11a. The value for heave is 280 ft, as compared with 390 ft in Figure 8-11a. It is important to note that although the vertical separation does not change, there is a difference in the values for throw and heave. The missing section due to the fault has not changed, as it is equal to vertical separation and not throw. Remember, throw and heave are dependent fault slip variables that change with variation in the apparent attitude of the fault or horizon. In this case (Fig. 8-12), the fault intersects the horizon where the horizon dip is generally in the opposite direction to that of the fault. At the map level shown in Figure 8-11a, the fault and horizon are dipping in the same general direction. It is this change in relative dip direction (fault strike is unchanged) that causes the different values for throw and heave shown in the two figures, even though the vertical separation has not changed. Using Eq. (7-1a) and the data on the map near Well No. 2 (average bed dip is 14.5 deg), estimate the throw across the fault at points A and B and compare this estimate with that obtained graphically from the map.

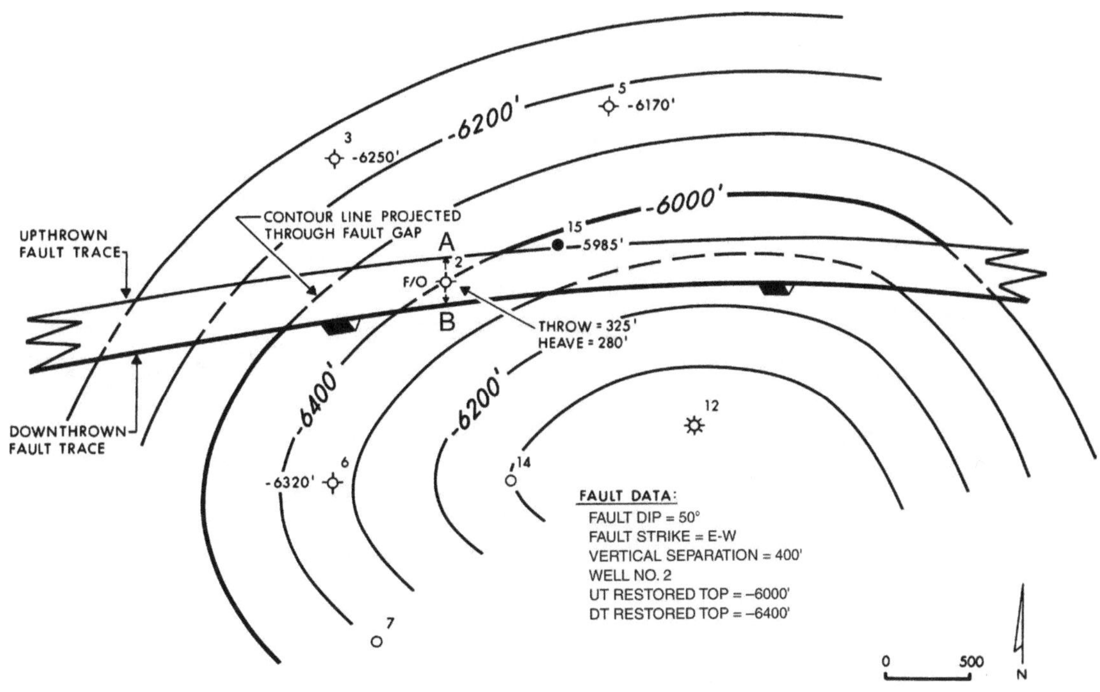

Figure 8-12 Portion of the structure map for the 6000-ft Horizon, showing the method for contouring vertical separation across a fault. (Published by permission of D. Tearpock.)

Mapping Throw across a Fault. Previously, we mentioned that missing section and repeated section due to a fault are equal to vertical separation. Let us assume for a moment, however, that a geoscientist *incorrectly considers the fault data as throw* and contours across the fault as if the missing section were throw. In Figure 8-13, the fault and structural data are exactly the same as that in Figure 8-11; therefore, we can compare the results of this (throw-contoured) map with the map in Figure 8-11a.

Using the available data, the contours are first established in the upthrown fault block and extended to the upthrown trace of Fault F-1. When contouring throw across a fault, the strike direction of a contour changes at this point and becomes perpendicular to the strike of the fault surface (Fig. 8-13). The contour is then projected through the fault gap to the downthrown trace of the fault. The strike direction of the contour is again changed to conform to its trend in the upthrown block.

Follow the −8500-ft contour through its construction in order to gain a good understanding of the technique. The strike direction of the −8500-ft contour in the upthrown block is established by the surrounding well control. At the intersection with the upthrown trace of the fault, the strike direction of the contour line changes abruptly to a strike direction that is perpendicular to the strike of the fault surface. In this example, as in the one in Figure 8-11a, the strike direction of the fault surface is east-west. Therefore, treating the vertical separation as if it were throw, the contour is projected through the fault gap in a north-south direction, perpendicular to the strike of the fault surface. At the intersection of the contour and the downthrown trace of the fault, the contour strike direction changes again to conform to the strike direction in the upthrown fault block. Once the contour is projected into the downthrown fault block, its depth is adjusted from the upthrown contour value by the amount of missing section (or in this case, assumed throw) of the fault (400 ft), so it becomes a 8900-ft contour.

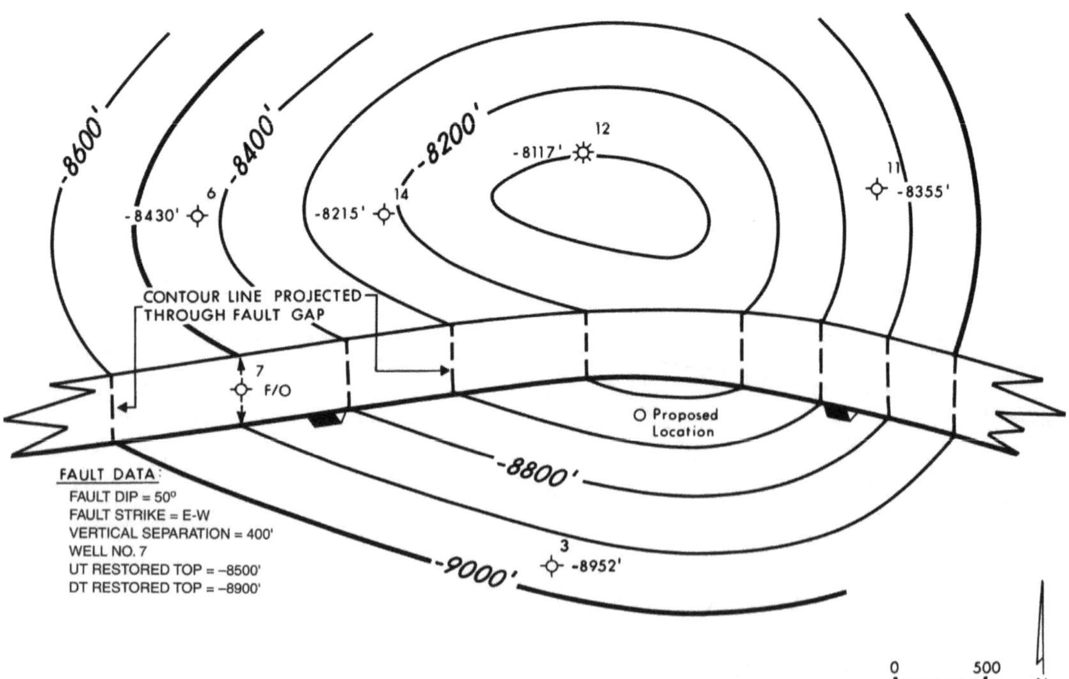

Figure 8-13 Fault and structure data are the same as shown in Figure 8-11a. For this interpretation, the missing section is contoured across the fault incorrectly as if it were throw. Compare the downthrown fault block with that shown in Figure 8-11a. (Published by permission of D. Tearpock.)

A proposed well location is in the downthrown fault block in Figs. 8-11a and 8-13. Considering the correctly contoured map (Fig. 8-11a), the depth at which the proposed well is estimated to penetrate this horizon is −8720 ft, whereas at the same location on the incorrectly contoured map (Fig. 8-13), the well is estimated to penetrate the horizon at −8640 ft. The depth to the horizon is mapped 80 ft shallower on the incorrect map. Contoured depth differences of 80 ft can make the difference between a successful well and a dry hole, or result in a well that is not drilled in the optimum position on the structure.

Based on the nature of this erroneous contouring technique (mapping vertical separation incorrectly as throw), the magnitude of error becomes greater near or on the crest of a structure, as well as becoming larger with increasing structural dip. This is very critical since hydrocarbons are often trapped near the crest of structures and we commonly map these areas.

Over the years, we have reviewed hundreds of maps that have been contoured incorrectly as shown here. This is a common error based on a misconception of the meaning of throw. This error has caused petroleum companies hundreds of millions of dollars in dry holes or poorly positioned wells. The information presented here should provide you the basic knowledge needed to avoid similar or costly errors (Tearpock et al. 1994). Furthermore, good field training in which maps are made on faulted structures using outcrop and well log data can provide the 3D understanding of these various fault components, their important differences, and the impact of mapping vertical separation incorrectly as if it were throw.

Error Analysis—Procedure for Checking Structure Maps. The best way to check whether a map has incorrectly used vertical separation as throw is to determine the vertical separation on the map by projecting the contours through the fault gap, as illustrated in Figure 8-11a and comparing it with the missing section in nearby wells. The missing section in Well No. 7 in Figure 8-11a is 400 ft. This matches the vertical separation shown by the map in Figure 8-11a. The missing section in Well No. 7 in Figure 8-13 is still 400 ft, but the vertical separation determined from the map in Figure 8-13 by projecting the upthrown −8500-ft contour through the fault gap to line up with the downthrown −8800-ft contour is 300 ft (Fig. 8-13). The lack of agreement between the missing section in the well and the vertical separation determined from the map demonstrates that the map in Figure 8-13 is incorrect. Without knowledge of the missing section in a well or the vertical separation measured from a seismic line, there is no way to know whether the map in Figure 8-11a was correct or the map in Figure 8-13 was correct.

The values for the vertical separation, fault dip, and bed dip can be determined from normal mapping parameters. With these data, Eq. (7-2) can be used to calculate the throw at any point along a fault. If the calculated value for throw agrees fairly well with the value determined graphically from the map, we can conclude that the interpretation is reasonable. If the missing or repeated section were mapped as throw instead of vertical separation, the value for throw determined mathematically will not compare favorably with that determined graphically, indicating that the map is in error.

The graph presented in Figure 8-14, which is derived from Eq. (7-1a), can be used during daily operations to check the consistency of structure contour maps and to ensure that a structure map has been contoured correctly across any existing faults. Equation (7-1a) is accurate for the situation in which the bed dip direction is more or less perpendicular to the strike of the fault surface (regardless of whether the beds dip toward the fault or in the same direction as the fault). So the graph in Figure 8-14, and also the graph in Figure 8-16, should be applied to that structural condition. That relationship is common in prospective closures against faults, and prospects are where you most want to check the accuracy of structure maps. For structural configurations where true bed dip is not perpendicular to the strike of the fault, use apparent bed dip measured perpendicular to the strike of the fault.

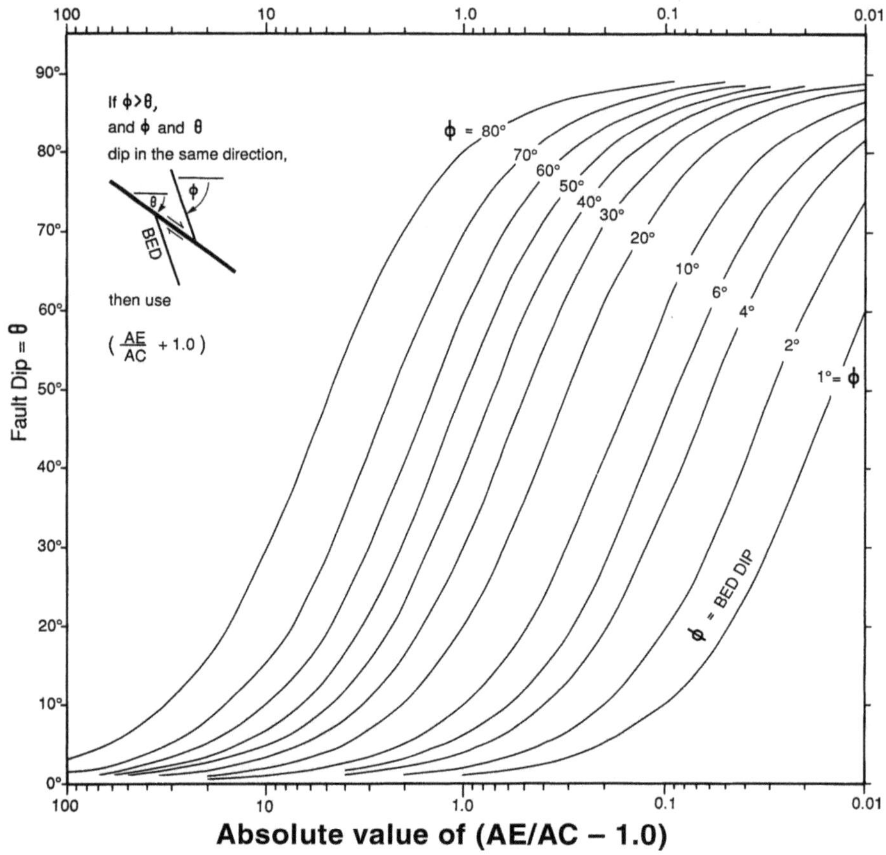

Figure 8-14 Graph used to check contouring across a fault. See text for explanation. (Published by permission of D. Tearpock and R. Bischke.)

In Figure 8-14, the fault and bed dips are taken to be clockwise. If the beds and fault dip in the same direction, the ratio AE/AC (vertical separation/throw) must be added to 1.0 in the figure. This is the case of a repeated section due to a normal fault.

As a practice exercise, use the data in Figure 8-11a, near the proposed location, and the graph (Fig. 8-14) to verify the dip of Fault 1 shown in Figure 8-11b.

An example of how to use Figure 8-14, for steeply dipping beds encountered around a salt dome, is as follows. On Figure 8-14 the fault dip θ is on the y-axis, the bed dip φ are the dipping lines in the central portion of the graph, and the absolute value of $(AE/AC - 1.0)$ is plotted on the x-axis. Assume a fault has a dip of 50 deg. If the vertical separation AE is 400 ft and the throw AC is 200 ft, then the ratio of $AE/AC = 2.0$. The absolute value of $(AE/AC - 1.0) = 1.0$. Next, construct a vertical line from the abscissa of Figure 8-14 at the value of 1.0 and a horizontal line, across the graph from the ordinate, for a fault that dips at $\theta = 50$ deg. The two lines cross the steeply dipping bed-dip curves at about $\phi = 50$ deg. Thus, a correctly contoured map is consistent with a bed that dips at 50 deg into a fault that also dips at 50 deg (in opposite directions), for a vertical separation of 400 ft and a throw of 200 ft.

The relationship shown in Figure 8-15 is used to conduct error analyses on incorrectly contoured maps (i.e., how much error is introduced into a map that assumes the missing or repeated section in a well is throw). By this time it should be readily apparent that if vertical separation is mapped as if it were throw, the errors involved could result in a well that totally misses its target. The graph in Figure 8-16, derived from the relationship in Figure 8-15, can be used to analyze the magnitude of error (Tearpock and Bischke 1990). Figure 8-15 illustrates a faulted horizon for which the footwall position is determined incorrectly by mapping vertical separation as throw. From Figure 8-15:

$$AE = A'C$$

Therefore, from the law of sines,

$$BA \sin(\theta - \phi_a) = A'C \sin(\pi/2 - \theta)$$

As

$$BA = TA / \sin\phi_a$$

$$\frac{TA}{A'C} = \frac{\sin\phi_a \cos\theta}{\sin(\theta - \phi_a)} \tag{8-1}$$

The distance *TA* is the error in depth between the correct (*B*) and incorrect (*A'*) points for the horizon projected onto the footwall. The ratio *TA*/*A'C* provides a percent error relative to the vertical separation value *AE*, which was incorrectly used as throw. Thus, Figure 8-16 can be used to measure the error that is introduced by improper contouring techniques.

On Figure 8-16 the percent error *TA*/*A'C* is plotted on the y-axis against the fault dip q on the x-axis. The bed dips f plot across the body of the graph as the gently to steeply dipping curved lines. The plot is symmetric across a horizontal line at which percent error = 0, to distinguish between beds that dip in the same direction of the fault (plotted on lower half of Fig. 8-16) from beds that dip in the opposite direction of the fault (upper half of plot). On the plot the errors range from 0% to 500%.

An examination of Figure 8-16 shows that unacceptably large errors of greater than 50% are introduced for faults dipping at angles of less than 30 deg (or greater than 150 deg) and for

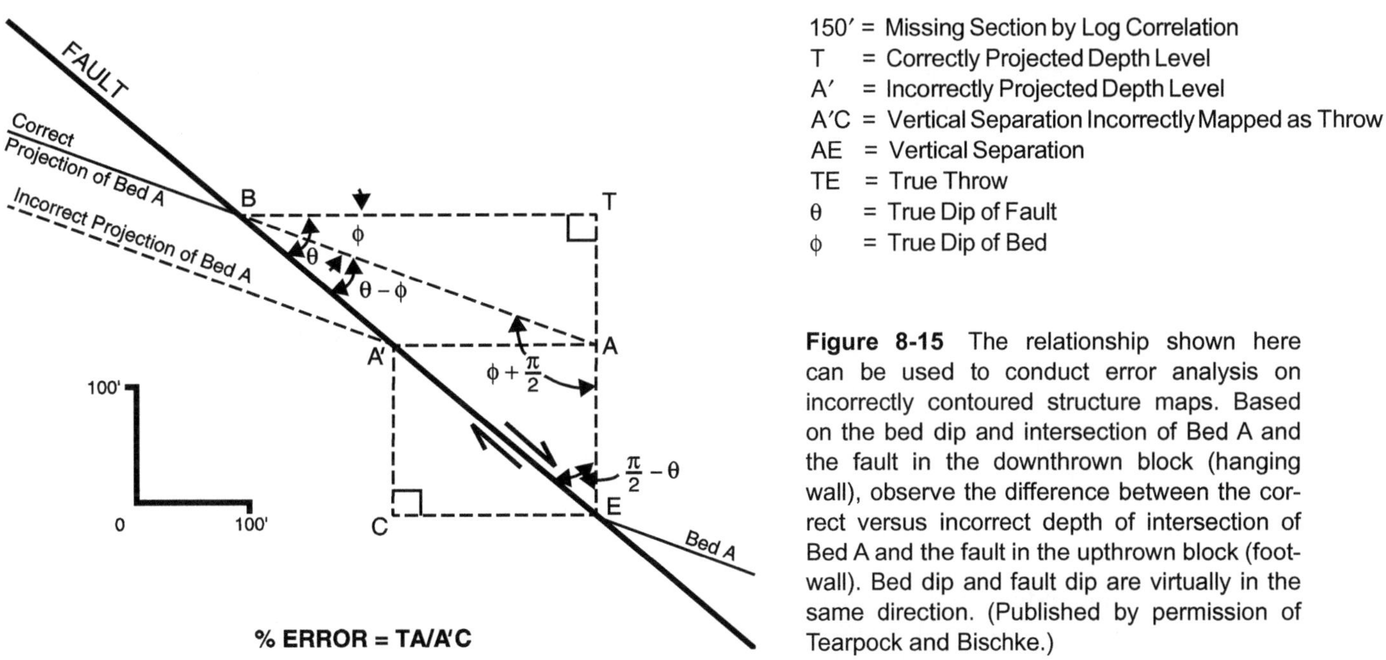

Figure 8-15 The relationship shown here can be used to conduct error analysis on incorrectly contoured structure maps. Based on the bed dip and intersection of Bed A and the fault in the downthrown block (hanging wall), observe the difference between the correct versus incorrect depth of intersection of Bed A and the fault in the upthrown block (footwall). Bed dip and fault dip are virtually in the same direction. (Published by permission of Tearpock and Bischke.)

beds dipping at less than 5 deg (or greater than 175 deg, recognizing that 180 deg is a lower dip than 175 deg). For faults dipping at 45 deg, bed dips of 10 deg or more can introduce unacceptable errors. Notice that if the beds are dipping in the same direction as the fault, then the errors encountered are correspondingly larger (see lower half of the graph) than a situation in which the beds are dipping in the direction opposite to the fault. If the bed dip is about equal to the fault dip, as is often the case along the flanks of salt domes, then the errors involved in mapping vertical separation as throw can readily exceed several hundred percent.

In Figure 8-16, the error is relative to the depth of an incorrectly contoured horizon or a horizon mapped as throw. To calculate the error encountered when mapping vertical separation as throw, consider a bed dipping at 135 deg (or 45 deg to the west) into a fault that dips at 45 deg to the east. On the plot, a 45-deg fault dipping to the east is located along the 0% error line at 45 deg. From this value, project a vertical line from the x-axis into the upper portion of the plot. This portion of the plot defines beds that dip in the direction opposite to the fault. A bed dipping at 135 deg is located between the gently sloping 130-deg to 140-deg dipping lines (see X on plot). The percent error in incorrectly mapping a vertical separation as throw can now be determined by projecting the bed and fault dip intersection at X over to the y-axis.

Thus, a bed that dips at 135 deg (or 45 deg to the west) into a fault dipping at 45 deg (to the east) will have an error of 50%, which means that the correct depth to the bed under consideration is 50%, or one-half the vertical distance away from the incorrect level. Thus, relative to the

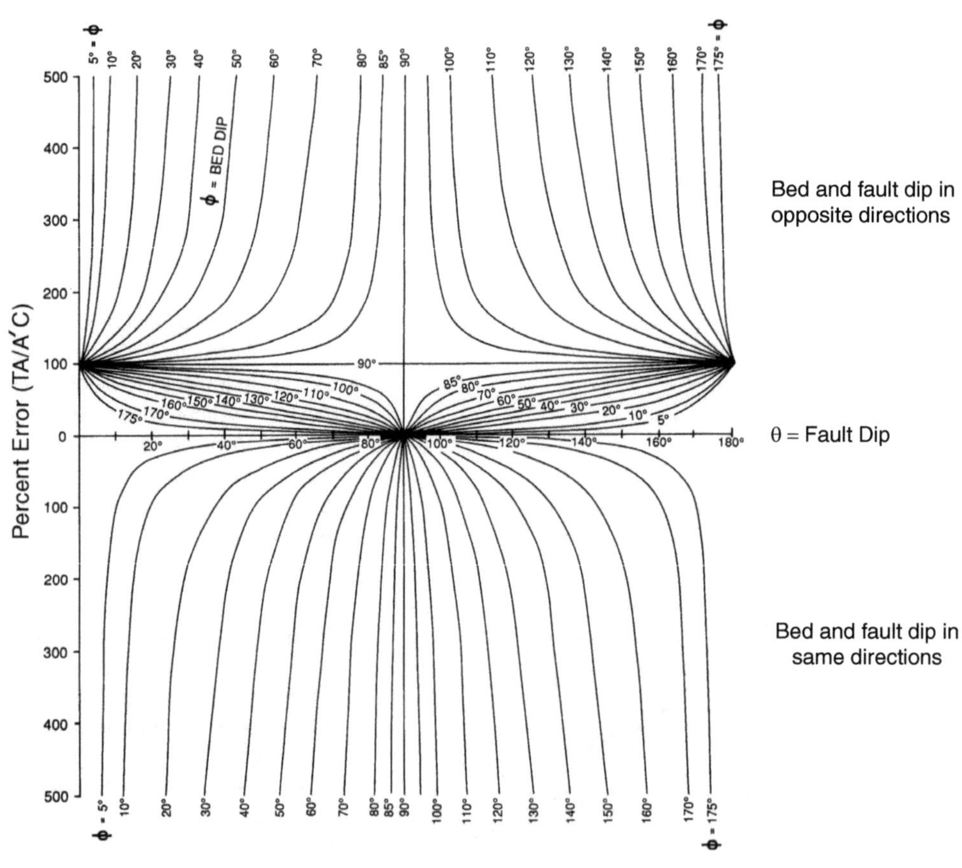

Figure 8-16 Graph used to measure the percent error on an incorrectly contoured structure map. Fault and bed dips are taken to be clockwise from 0 deg to 180 deg. (Published by permission of Tearpock and Bischke.)

correctly contoured bed, the error relative to the depth that should have been correctly contoured as vertical separation, is

$$\text{Error} = \frac{(\text{incorrect depth} - \text{correct depth})}{\text{correct depth}}$$

or

$$\frac{(1-0.5)}{0.5} = \text{a}\,100\%\ \text{error}$$

From this discussion, we can conclude that the errors involved in substituting throw for vertical separation are larger than most interpreters may realize or be willing to accept. Equations (7-1) and (8-1) and the graphs in Figures 8-14 and 8-16 are powerful tools that can be used on a daily basis to help generate correctly contoured maps and to evaluate the maps of others. Such analysis can be routinely conducted when evaluating prospect maps. Use Figure 8-16 to analyze the magnitude of error in the incorrectly contoured map in Figure 8-13, in the area of the proposed location.

The Legitimate Contouring of Throw. The contouring of throw is a legitimate technique. Where all fault data are in terms of throw, such as in mining or outcrop geology, the construction of contours across the fault by mapping throw is an accepted technique. *With regard to petroleum subsurface mapping, however, the technique is not valid, for most cases.* First, well log fault data are not throw. Second, most seismic fault data given as throw are actually apparent throw. Third, if throw is measured from a seismic section, it cannot be tied to fault data from well logs for mapping, since measurements of fault displacement, using logs, are of vertical separation. Finally, throw often varies significantly along the strike of a fault; therefore, if throw is to be mapped, an almost infinite number of fault throw values are required to integrate a fault with a structure contour map.

When throw is the desired fault component to map, as in mining or outcrop mapping, the technique is sometimes incorrectly used. We shall look at one of the main errors made in mapping throw. Remember from previous discussions that *true throw is measured perpendicular to the strike of the fault surface and not necessarily perpendicular to a fault trace shown on a structure map*. Therefore, to project the structure contours through a fault gap, the fault surface map must be available to determine the strike of the fault. Since the strike of a fault can change across the mapped area, it is good practice when mapping throw to place the fault surface map under the structure map to obtain the strike direction of the fault so that all the contours can be projected through the fault gap perpendicular to the strike of the fault surface at any location along the fault.

At times, contours are projected through a fault *perpendicular to the fault trace rather than the strike of the actual fault surface. This can result in serious mapping errors.* If the strike directions of the fault itself and the fault trace are extremely close, the trace may be used to project contours with minimal error. But this is not always the case. *The line of intersection between two inclined planes with different strikes (such as a horizon and fault) is not parallel to the strike of the horizon or the fault.* Therefore, fault traces are normally not parallel to the strike of either the horizon or fault. Figure 8-17 shows such an example. The map in Figure 8-17 is contoured incorrectly by projecting the structure contours from the upthrown fault block through the fault gap *perpendicular to the trace* of the fault. In Figure 8-18, the contours are projected correctly through the fault gap *perpendicular to the strike of the fault surface* (see line indicating fault strike). Notice that the contour value of the point labeled **X** in the downthrown fault block is mapped 95 ft deeper using the incorrect technique. Depending on the amount of vertical separation and the difference in attitudes between the fault surface and horizon, the magnitude of contour errors on completed maps can vary from being insignificant to being very significant. When mapping throw, take the time to use the technique correctly.

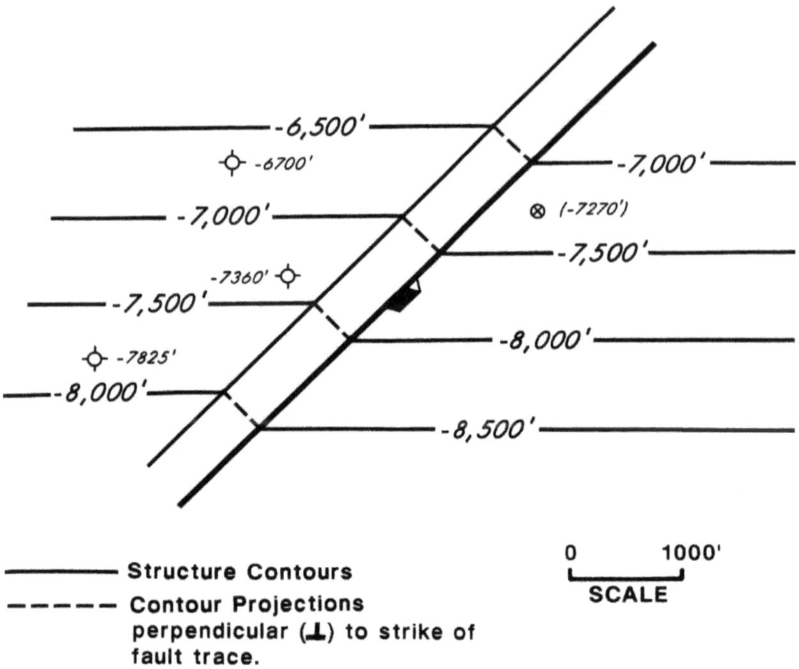

Figure 8-17 Structure contours illustrate the incorrect method for mapping throw perpendicular to the strike of the fault trace (see dashed contours in fault gap).

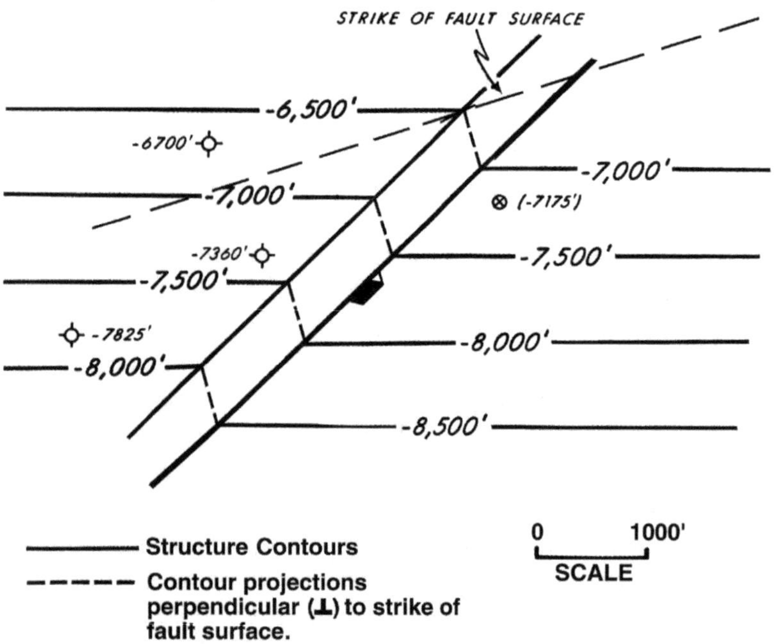

Figure 8-18 Structure contours illustrate the correct method for mapping throw perpendicular to the strike of the fault trace (see dashed contours in fault gap).

For petroleum geology mapping, you might consider this discussion on the correct technique for mapping throw as more of an academic exercise than one having some practical importance. But remember that in mining geology and outcrop mapping, where the actual fault surface can be touched and throw values physically measured, this mapping technique is valid. You should, however, consider one important point. *If the fault data consist of both actual throw measurements from a mine or an outcrop, and well log fault data, they cannot be used together in structure mapping because the well log fault data represent vertical separation.* Equation (7-1), however, can be used to convert the throw data to vertical separation, or vice versa, which can then be used for mapping.

Only if the dip of the horizon being mapped is zero or nearly so, or if the fault is vertical, are the values for vertical separation and throw essentially the same. In these cases, they can be used together in fault and integrated structure mapping.

Later in this chapter, we examine a petroleum-related generic case study, which further illustrates the importance of mapping the missing section correctly as vertical separation rather than as throw in petroleum subsurface mapping. It can make the difference between drilling a successful well and a dry hole.

Technique for Contouring across Reverse Faults

The technique presented for contouring across a normal fault is also applicable for contouring across a reverse fault. Reverse faults and overthrusts, however, produce a fault overlap rather than a fault gap. Figure 8-19 shows the technique for contouring across a reverse fault. The 4500-ft Horizon, which dips generally to the west-northwest, is displaced by a southwest-dipping reverse fault. The vertical separation or repeated section determined from well log correlation is 450 ft.

The technique for contouring across a reverse fault is usually easier than that for a normal fault because there is no projection of contours through a fault gap. With a reverse fault, the hanging wall is thrust up and over the footwall, resulting in an overlap of structural horizons. Therefore, the hanging wall is contoured right up to the hanging wall cutoff at the upthrown fault trace. Likewise, the footwall is contoured up to the footwall cutoff at the downthrown trace of the fault. Since the fault blocks overlap, the strike direction of the contours established in the block that is contoured first (the block with the most control) serves as a guide to the contouring of the other fault block. As with the normal fault example, to guide the strike direction of the contours, consider how the structure would be contoured if the fault were not there. Wells positioned in the fault overlap serve as a guide to the contouring of both fault blocks.

Referring back to Figure 7-29 in Chapter 7, the thickness of the repeated section or vertical separation resulting from a reverse fault can be calculated by measuring the vertical distance from the top of the mapped horizon in the hanging wall to the top of the same horizon in the footwall. Notice that Well No. 2 in Figure 8-19, in the fault overlap, has penetrated the top of this horizon twice: the first time at 4700 ft and the second time at 5150 ft. The vertical difference between these two tops is equivalent to the repeated section or vertical separation, which is equal to 450 ft. *The vertical separation can therefore be seen directly in the fault overlap.*

There is one problem with the construction of a reverse fault overlap; it results in significant clutter, which can be confusing. Some confusion is eliminated by dashing the contours on the footwall beneath the fault, within the zone of fault overlap, as illustrated in Figure 8-19. Another method of eliminating clutter is to pull the fault blocks apart and present each fault block separately. This is a good way to construct a structure map with a reverse fault, especially if hydrocarbons are present and isochore maps are required. The method is shown later in this chapter.

There is an additional complication in using a workstation to map structures cut by reverse faults. Consider Well No. 2 in Figure 8-19. This well contains two tops for the same horizon as it penetrated both the hanging wall block and the footwall block. Most computer software packages do not handle repeated section very well, neither repeated tops in a well nor repeated horizons on a seismic line. With most computer software packages, it may be necessary to map the hanging wall block separately from the footwall block. Digital control points can be added to the map if needed to ensure contour compatibility across the fault.

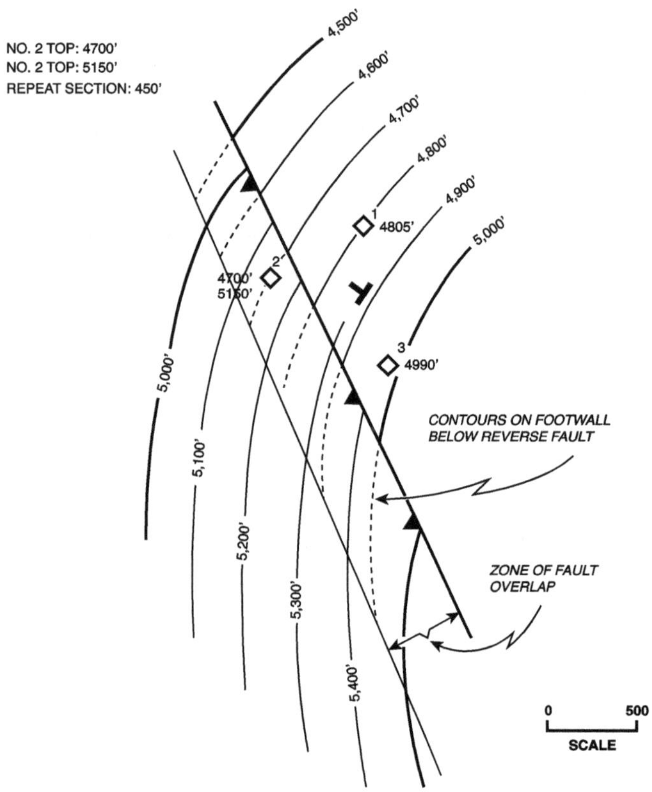

Figure 8-19 The correct method for contouring repeated section (vertical separation) across a reverse fault is illustrated by the solid and dashed contours in the fault overlap. This reverse fault has 450 ft of vertical separation (repeated section) as evidenced by the repeated tops in Well 2 that are 450 ft apart.

MANUAL INTEGRATION OF FAULT AND STRUCTURE MAPS

In this section, we present the technique for manually integrating a fault surface map with a structural horizon map. The correct application of this technique is essential for accurate structure mapping in faulted areas. The technique provides the following important contributions to the structure contour map interpretation.

1. An accurate delineation of the fault location for any mapped horizon;
2. The precise construction of the upthrown and downthrown traces of the fault;
3. The correct width of the fault gap or overlap; and
4. The proper projection of structure contours across the fault based on fault data (vertical separation) from well logs or seismic sections.

Normal Faults

The step-by-step method of integrating a fault surface map of a normal fault with a structural horizon map is illustrated in Figures 8-20 and 8-21. Figure 8-20 is a fault surface map constructed from fault data from the wells shown on the figure. The fault strikes generally north-south, is slightly convex to the east with a dip of 45 deg, and has a vertical separation of 400 ft. The stratum being contoured is called the 7000-ft Horizon. The subsea tops for this horizon in each well are shown on the partly completed structure map in Figure 8-21a. In Well No. 3, for example, the top of the 7000-ft Horizon is at a depth of −7045 ft.

The 7000-ft Horizon is cut by Fault A. A review of the depth of the fault picks and the structure tops for each well indicates that five of the seven wells are in the upthrown block and that only Wells No. 4 and 5 are in the downthrown fault block. For a structure cut by a normal fault, wells in which the fault cut is above the structure top are in the upthrown (footwall) block and wells in which the structure top is above the fault cut are in the downthrown (hanging wall) block, unless the well drills

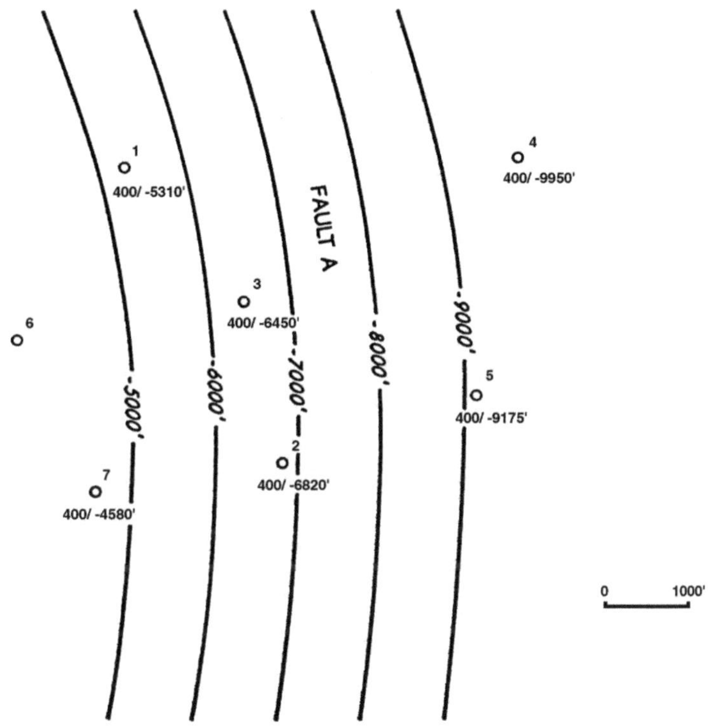

Figure 8-20 Fault surface map for Fault A constructed from well log fault data from seven wells. The fault map has a 1000-ft contour interval.

the fault "backwards" and penetrates the fault from the upthrown (footwall) into the downthrown (hanging wall). Employing the general contouring guideline of beginning the structure contouring in the area or fault block with the most control, the contouring of the 7000-ft Horizon originates in the upthrown fault block. The initial contouring indicates an anticlinal structure with a slightly elongated east-west axis (Fig. 8-21a). The contours are not continued across the entire map area because at some point the contours in this upthrown fault block intersect Fault A and terminate at the upthrown fault trace. By underlaying the fault surface map beneath the structure map, the approximate location of this intersection can be determined and the contouring stopped in the vicinity of this intersection.

When doing this method by hand, the next step is to overlay the structure map (Fig. 8-21a) onto the fault map (Fig. 8-20), as shown in Figure 8-21b, to determine the upthrown trace of the fault. The upthrown trace occurs where structure contours in the upthrown block intersect fault contours of the same elevation. These intersections are highlighted on the map by a small mark placed at the end of each contour line, as shown in the figure. Because the fault surface map has a contour interval of 1000 ft and the structure contour map has a 200-ft contour interval, the precise position of the intersections for each contour line should be located by interpolation, using 10-point proportional dividers or a scale. Dividers or a scale allow subdivision of the 1000-ft fault contours into 200-ft intervals anywhere on the map without the clutter of actually drawing the extra contours (see Fig. 2-11). For a small portion of the fault map, 200-ft contours are shown between the −7000-ft and −8000-ft depths. These contours serve two purposes: (1) to illustrate the accuracy of the intersection of the fault and structure contours for the −7000-ft, −7200 ft, −7400-ft, −600-ft, −7800-ft, and −8000-ft elevations; and (2) to show that the construction of 200-ft fault contours over the entire map would result in unnecessary clutter. Two-hundred-foot (200-ft) contours can, however, be placed as marks in localized areas, as shown in the figure, to determine the intersections precisely. Once used for this purpose, the additional contours are then erased. After all the intersections have been identified, the upthrown trace of the fault is accurately constructed simply by connecting all the marks with a smooth line, as shown in Figure 8-21c.

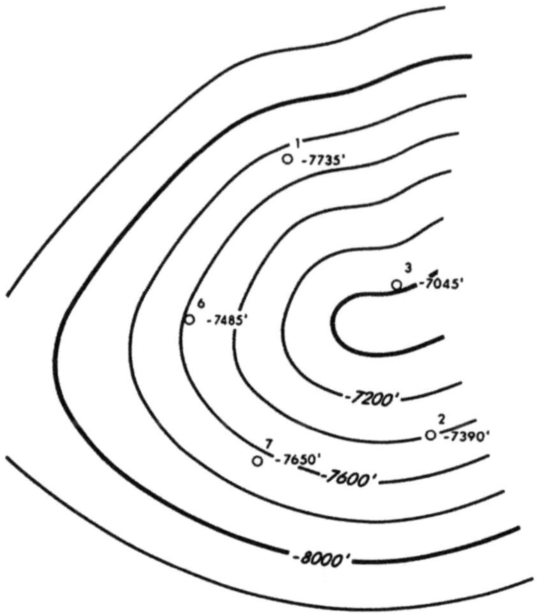

(a)

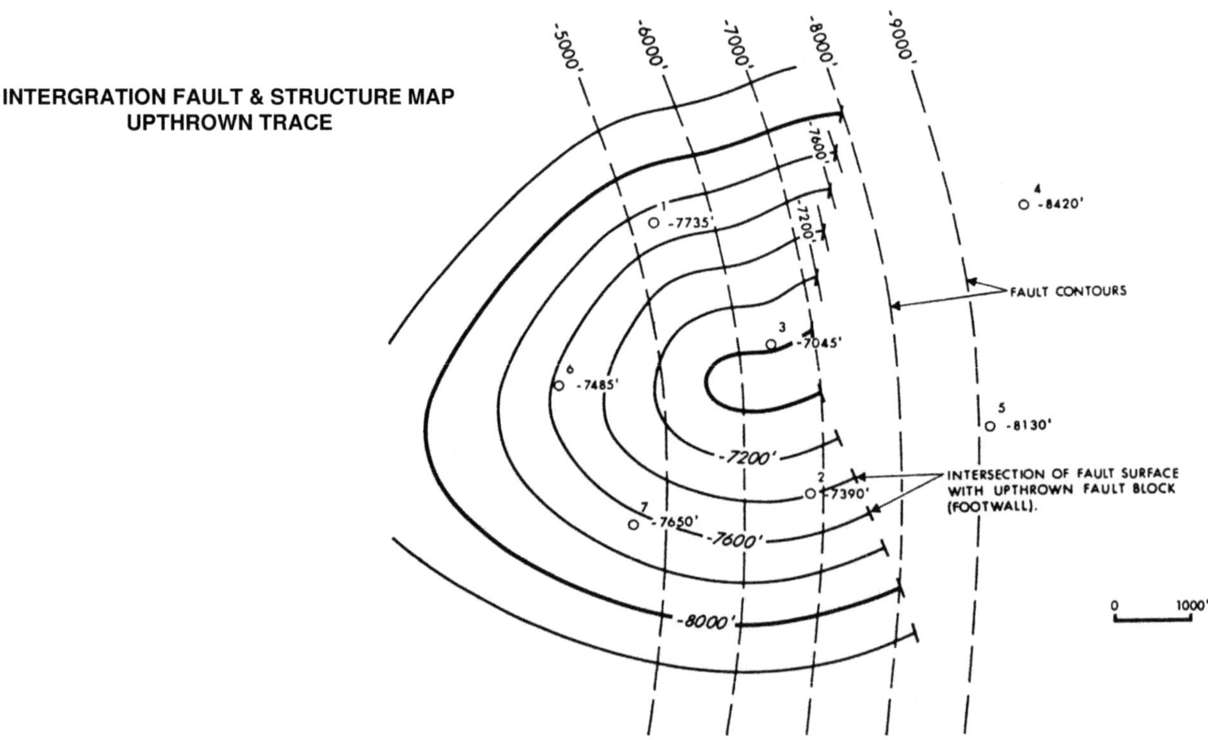

(b)

Figure 8-21 (a) Partially completed structure map on the 7000-ft Horizon. (b) Integration of the fault and structure maps to identify the intersection of the fault surface with the upthrown fault block of the 7000-ft Horizon. (c) Structure map shows the delineation of the upthrown trace of Fault A and the projection of form contours into the downthrown fault block. (d) Integration of the fault and structure maps to identify the intersection of the fault surface with the downthrown fault block of the 7000-ft Horizon. (e) The final, integrated structure map for the 7000-ft Horizon.

PREFAULTED MODEL

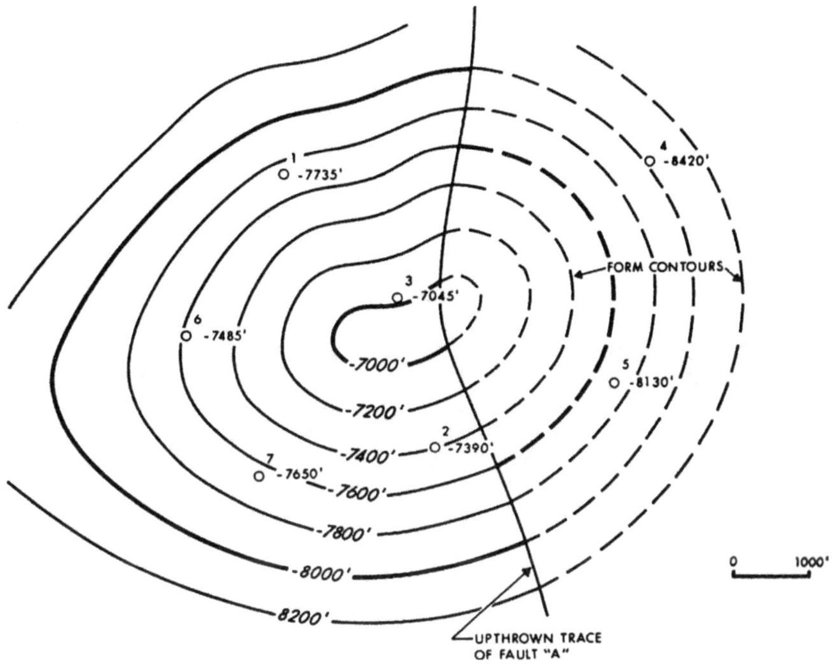

(c)

INTERGRATION FAULT & STRUCTURE MAP
DOWNTHROWN TRACE

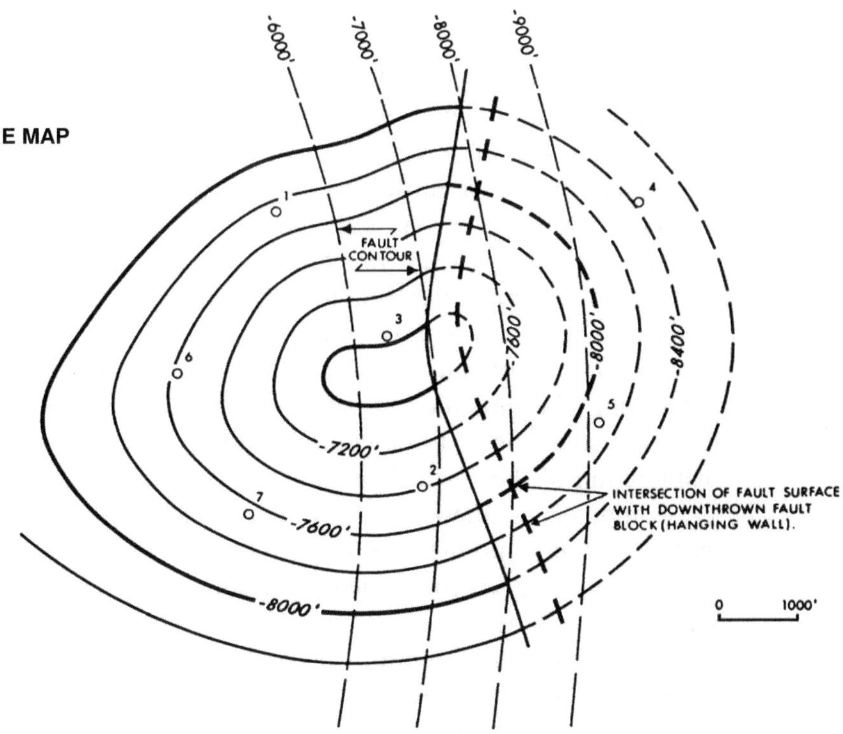

(d)

Figure 8-21 (*continued*)

COMPLETED STRUCTURE MAP
7000' SAND

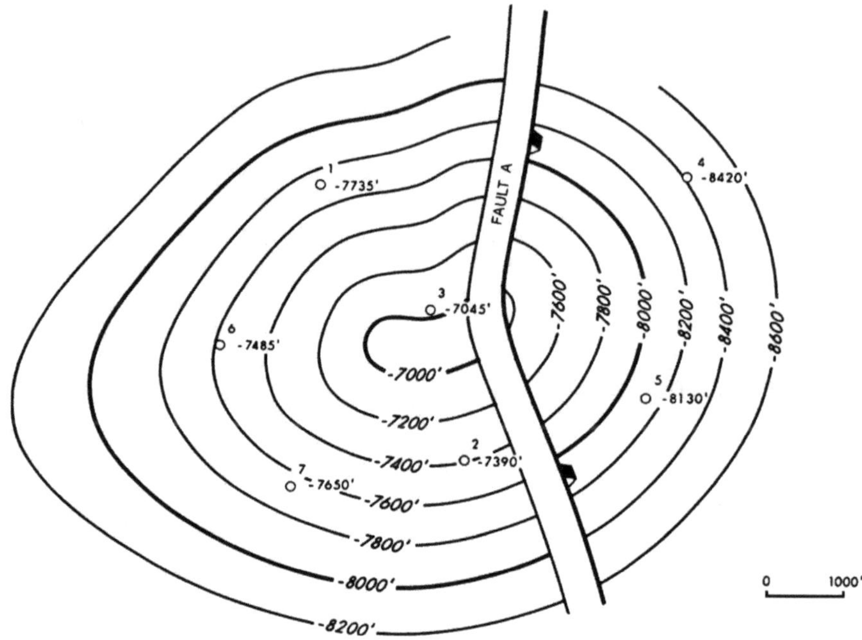

(e)

Figure 8-21 *(continued)*

The next step is to project the structure contours through the upthrown trace of the fault into the downthrown fault block. There are two points of control in the downthrown fault block that must be honored: the horizon elevation in Well No. 4 is −8420 ft, and in Well No. 5 it is −8130 ft. Earlier in this chapter, we presented the technique for projecting contours from one fault block across a fault into the adjacent fault block. This technique is used to complete the construction of the anticline in the downthrown fault block and delineate the downthrown trace of the fault. An easy way to remember the technique *to guide your contouring through the fault is to consider how the structure would be contoured if the fault were not there.* One method is to restore the faulted block to an unfaulted condition and relabel the depth values in the wells to account for the restoration. The unfaulted structure can then be mapped. Once completed, the map can be refaulted or restored again to its faulted condition. The structure contours must be adjusted again in consideration of the vertical separation. This is a time-consuming task when done by hand, whereas computers are more capable of using such methodology. For simplicity, often it is best to project the contours into the opposing fault block as form contours (contours without values) in order to complete the structural picture, as shown in Figure 8-21c.

Once contours are projected through the fault into the downthrown block, their values are adjusted from those in the upthrown block by the amount of vertical separation, which in this case is 400 ft. Thus, the projection of the −7400-ft contour from the upthrown block becomes a −7800-ft contour in the downthrown block. Continue this procedure for all contours projected through the fault. In Figure 8-21d, the downthrown contours have been assigned structural elevation values and now the downthrown trace of the fault can be determined. Again, the fault surface map is placed under the structure map and the intersections of the structure contours in the downthrown block with the fault contours of the same elevation are identified and indicated by small marks.

Finally, connect all the marks with a heavy smooth line to accurately delineate the downthrown trace of the fault. Place a symbol on the downthrown trace to show the direction of fault dip, erase the contours in the fault gap, and the integrated structure contour map for the faulted anticlinal structure is complete (Fig. 8-21e). By correctly integrating the fault and structure maps, we have (1) accurately delineated the position of the fault trace on the structure for a particular horizon; (2) precisely constructed the upthrown and downthrown traces of the fault; (3) established the actual width of the fault gap; and (4) projected the structure contours correctly across the fault. *An understanding of this technique is paramount to the correct and precise integration of a fault map and a structure map;* therefore, take the time to review and master this process.

For the example shown in Figure 8-21, the process was relatively easy because the structural pattern is a simple anticline cut by only one fault. But the procedure is basically the same for a more complex structure and pattern of faults. Figure 8-22 shows the complexly faulted anticlinal structure of an oil and gas field. A fault surface map was prepared for each of the 13 individual normal faults, and the integration technique was used to prepare the completed structure map. With a complex structure such as the one shown here, the logistics are more involved and it takes more time to prepare the fault maps and integrated structural interpretation, but the methodology is basically the same.

Some software packages allow a geoscientist to create a surface from a horizon in a fault block and a surface from a fault interpretation. These software packages allow the horizon surface to be projected to or clipped at the fault surface. The intersection of these two surfaces forms a fault trace.

If the software package you are using does not allow you to create an intersection of the horizon surface and the fault surface, you can use the same procedure used for mapping a faulted structure by hand. First, create a polygon on your horizon map that roughly corresponds to the

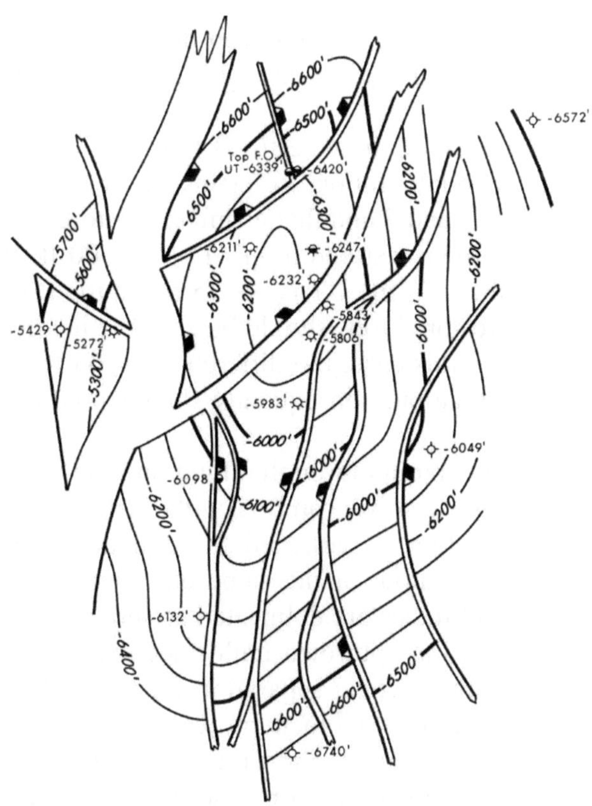

Figure 8-22 An integrated structure map of a very complexly faulted anticlinal structure. Each fault was integrated with the structural interpretation as shown in Figure 8-21.

fault gap. Second, contour both the horizon and the fault. Third, display both the horizon contours and the fault surface contours on the screen at the same time. Fourth, adjust the polygon to match the intersection of the horizon contours in the fault block where you have the best data with the fault surface contours. As you are likely to use different contour intervals for the horizon and the fault surface, you will have to interpolate between fault contours, just as you would if contouring by hand. Fifth, adjust the horizon contours in the other fault block to reflect contour compatibility and the vertical separation of the fault, if necessary. Finally, display both the horizon and fault surface contours again and adjust the polygon so that it ties the intersection of the horizon and fault surface contours.

The details of how to accomplish these methods of integration with the various workstation programs are beyond the scope of this book. Chapter 9 on 3D seismic interpretation provides some additional guidance for this method in a general sense. Each geoscientist must learn the individual mapping program being used and determine the best method for combining the use of applications within each package in order to achieve this type of subsurface integration accuracy.

Restored Tops—An Aid to Structural Integration. In Chapter 4 we mention that in dealing with normal faults, a given horizon or marker being mapped may be faulted out of one or more wells in the area of study. It is often possible, however, to estimate an upthrown and downthrown restored top for any horizon missing from the faulted well(s). A restored top is an estimated elevation for a specific marker or horizon that is faulted out of a well. In other words, *a restored top is an estimate of the depth of a marker or horizon in a given fault block were it not faulted out of the well.* The procedures for estimating restored tops for both straight and deviated wells were discussed in detail in Chapter 4. In Chapter 7, we also showed the importance of restored tops in estimating the amount of vertical separation for a growth fault.

In this section, we discuss how these restored tops are used to provide additional control points for fault and horizon map integration. Referring again to Figure 8-10, notice that the horizon is faulted out in Wells No. 2 and 7 and that estimated restored tops are indicated next to each well. For Well No. 2, the upthrown restored top (UT) is −6000 ft, and the downthrown restored top (DT) is −6300 ft; for Well No. 7, the UT is −6235 ft and the DT is −6535 ft. The vertical difference between DT and UT restored tops should be equal or nearly equal to the section missing in the wells due to the fault. In this case, the difference is 300 ft, which is equal to the missing section determined by log correlation; consequently, the restored tops appear consistent with the available data.

Procedures for projecting contours through a fault and the integration of a fault map with a structure map have been discussed, so we can review the structure map in Figure 8-10 and determine the amount of missing section. A darkened circle marks the intersection of each structure contour with the fault surface contour at the same elevation. By connecting these marks, the upthrown and downthrown fault traces are delineated as shown on the map. Any contour may be taken from its intersection with the upthrown fault trace and projected through the fault to the downthrown trace to estimate the vertical separation for the fault at this horizon. For example, the −6100-ft contour in the upthrown block projected through the fault gap intersects the −6400-ft contour in the downthrown block, indicating the fault displacement (vertical separation) of 300 ft, which was based on the restored tops and used in the mapping. The restored tops in Wells No. 2 and 7 provide important control points for contouring the structure map.

Restored tops are located in a vertical well itself (Fig. 4-42), or vertically above and below the fault pick in a deviated well (Fig. 4-43). They are not placed at the upthrown and downthrown traces of a fault. Using Figure 8-10 as an example, the restored tops for Wells No. 2 and

7 are placed right at the well locations. The dashed lines through each of the two wells illustrate the honoring of the restored tops in the wellbores. In Well No. 7, an interpolated contour of −6235 ft projects from the upthrown fault block into the well, and an interpolated contour of −6535 ft projects from the downthrown fault block into the well. Similarly for Well No. 2, the −6000-ft contour in the upthrown block and the −6300-ft contour in the downthrown block project into the well.

How are these restored tops used as an aid in structure mapping? Depths for restored tops are honored in the same way as any other well control point during structure mapping. For example, the UT restored top for Well No. 2 is −6000 ft; therefore, the −6000-ft structure contour in the upthrown block honors this control point at Well No. 2. Likewise, the −6300-ft contour in the downthrown fault block is projected through the fault gap to intersect with Well No. 2, whose DT restored top is −6300 ft. Each UT and DT restored top provides two additional points of control to aid in structure contouring in and around a fault. The contouring of the structure map in Figure 8-10 confirms that the four restored tops were used as control points in the construction of the final structure map.

In areas of limited well control, restored tops in faulted wells provide significant structural information which is often necessary for the preparation of a realistic and accurate structure map. Considering the well control in Figure 8-23a, how would you contour these data points? The data can be contoured in a number of ways. Figure 8-23b shows one interpretation which appears to be reasonable. The map does honor all the established well control and was used to propose the two drilling locations shown on the map. The first location is upthrown to Fault C in Reservoir C-3. An oil show in Well No. 14 establishes the down-dip limit of oil at −9245 ft. The proposed well is designed to penetrate the reservoir in the optimum position for maximizing the well's drainage efficiency. Based on this interpretation, the volume of anticipated recoverable oil up-dip of the oil/water contact is 6480 acre-feet. Considering a reasonable recovery factor for this area of 450 barrels per acre-foot, this prospect represents 2,916,000 barrels of potentially recoverable oil. Downthrown to Fault C, a development location is proposed to maximize the drainage efficiency for the east-west elongated Reservoir C-5. Based on this structural interpretation, the volume of estimated reserves is 1,107,000 barrels of recoverable oil.

The interpretation appears reasonable except for the two wells within the fault gap in which the 9000-ft Horizon has been faulted out. For whatever reason, no restored tops were estimated for these wells to incorporate into the structural interpretation. Figure 8-23c is a structure contour map for the same two reservoirs using the UT and DT restored tops for the 9000-ft Horizon in Wells No. 3 and 11 in the interpretation. It is obvious that the four restored tops are very important data needed to develop a more accurate interpretation of this structure. The addition of the restored top data has a significant impact on the interpretation of the overall geometry of the structure and the proposed well locations. For the upthrown Reservoir C-3, the volume of recoverable hydrocarbons has been reduced to 3480 acre-feet or 1,566,000 barrels of recoverable oil, a reduction in volume of 46%. The map for Reservoir C-5 in the downthrown block, using the restored tops, indicates that the volume of Reservoir C-5 is smaller by 41% compared to the map in Figure 8-23b. Potentially recoverable hydrocarbons decrease from 1,107,000 barrels to 648,000 barrels. More important, the proposed development location for Reservoir C-5 is actually down-dip of the oil/water contact, and if drilled, the well would result in a dry hole.

Information provided by the UT and DT restored tops was critical in this example. The tops are valuable mapping data that should not be ignored, particularly in areas of limited well control. These extra data points guide the structure contours into and through the fault gap. Figure 8-23 makes this point very clear. Upthrown and downthrown restored tops improve the accuracy of a structure contour map, and in this case reduced the size of the two prospective reservoirs.

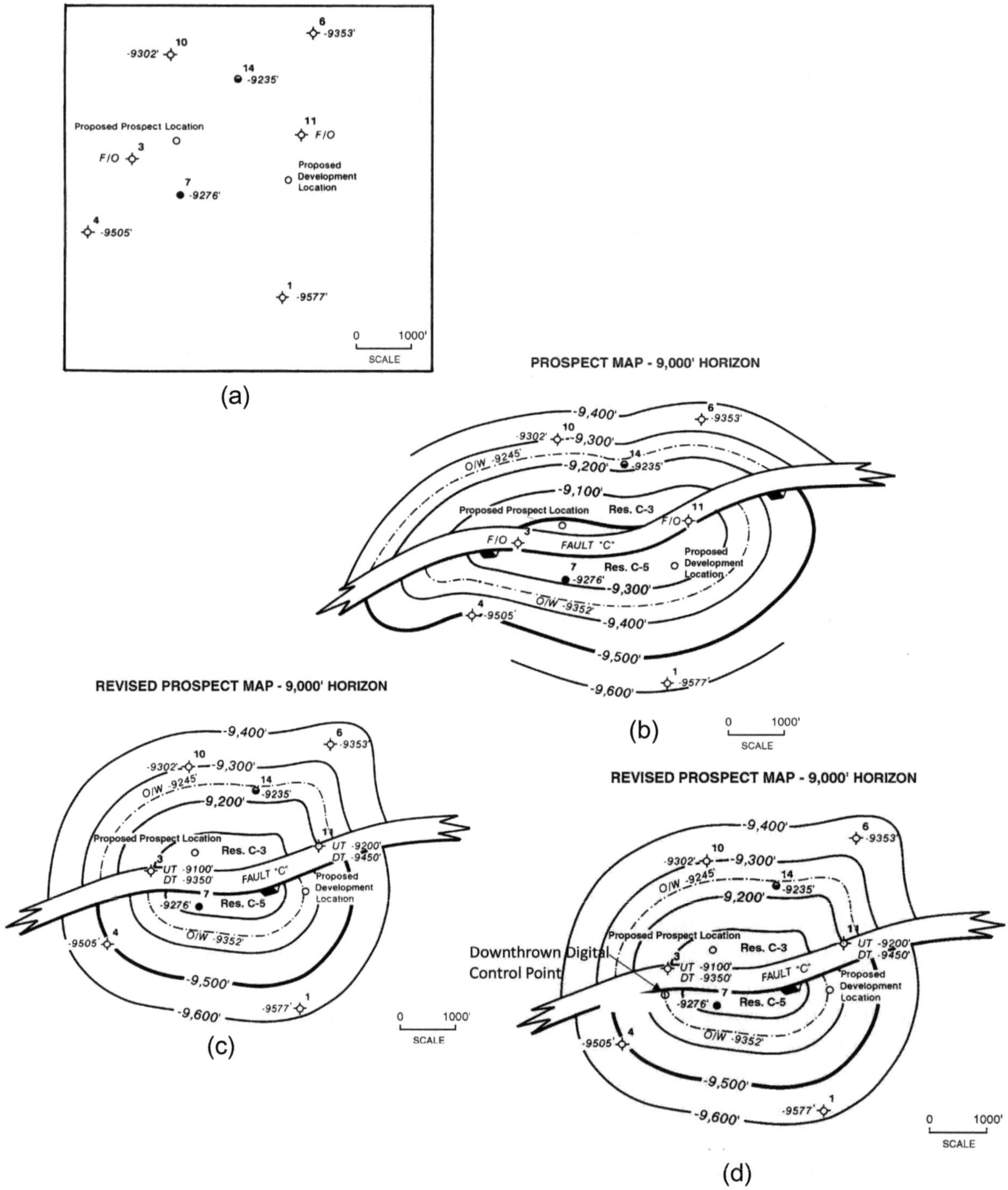

Figure 8-23 (a) Basemap with posted data for the 9000-ft Horizon. How would you contour the data? (b) Structural interpretation of the 9000-ft Horizon using all the well data except for the restored tops in Wells No. 3 and 11. Two development wells are proposed based on this interpretation. (c) Revised structural interpretation of the 9000-ft Horizon using all the available well data including the restored tops for Wells No. 3 and 11. Compare this interpretation with that shown in Figure 8-23b. (d) Revised structural interpretation with the DT restored top from Well No. 3 plotted just outside the fault gap as a digital control point for contouring on a computer. This figure illustrates only a single digital control point. When contouring on the computer, digital control points for both upthrown and downthrown restored tops would be used for both Well No. 3 and Well No. 11. Digital control points could also be used to project contours across the fault gap and constrain contouring of the downthrown fault block. (Figure 23d published by permission of J. Brenneke.)

Restored tops are also helpful in generating prospects where none previously existed. Figure 8-24a shows a portion of a structure contour map showing a potential oil reservoir. Notice Well No. 22 in the western portion of the map, upthrown to Fault A, has an oil show with an oil/water contact at 6458 ft TVD (true vertical depth). Well No. 25 to the east is wet and the mapping horizon is faulted out in Well No. 65 by Fault A. Based on the structural interpretation shown in this figure, the volume of potentially recoverable oil is insufficient to justify the drilling of an up-dip development well into Reservoir A-1. The structural closure is too small, and the target area between the oil/water contact and the upthrown trace of the fault too small to risk the drilling of a well. During the preparation of this structure map, the trace of Fault A was constructed based on isolated fault data without the use of a fault surface map and without estimated restored tops for the faulted out Well No. 65.

The structure map in Figure 8-24b was constructed using a fault surface map for Fault A to integrate with the structure and to contour across the fault. The UT and DT restored tops for this horizon in Well No. 65 were calculated and used, and maps on shallower and deeper horizons, in which the tops in Well No. 65 were present, were also made. These additional mapping data change both the configuration of Reservoir A-1 and the volume of potentially recoverable hydrocarbons. This accurately *integrated structure contour map* delineates a reservoir of sufficient size (569,000 barrels of oil) and closure to be reevaluated as a potential development location. The UT and DT restored tops are used in Figure 8-24b to guide the strike direction of the contours into the fault and to honor the vertical separation, thus changing the structural configuration of the reservoir.

We cannot overemphasize the importance of using all available data when preparing subsurface maps of any type. Geoscientists always work with limited data, so to ignore or not use valuable data, for any reason, is unthinkable. Our job is to develop the most reasonable and accurate interpretation possible from available data. The use of upthrown and downthrown restored tops is a necessary part of preparing a structure contour map. The time required to estimate these tops and incorporate them into the integrated structural picture is minimal, but it can have a significant impact on the final interpretation, as shown in the examples presented.

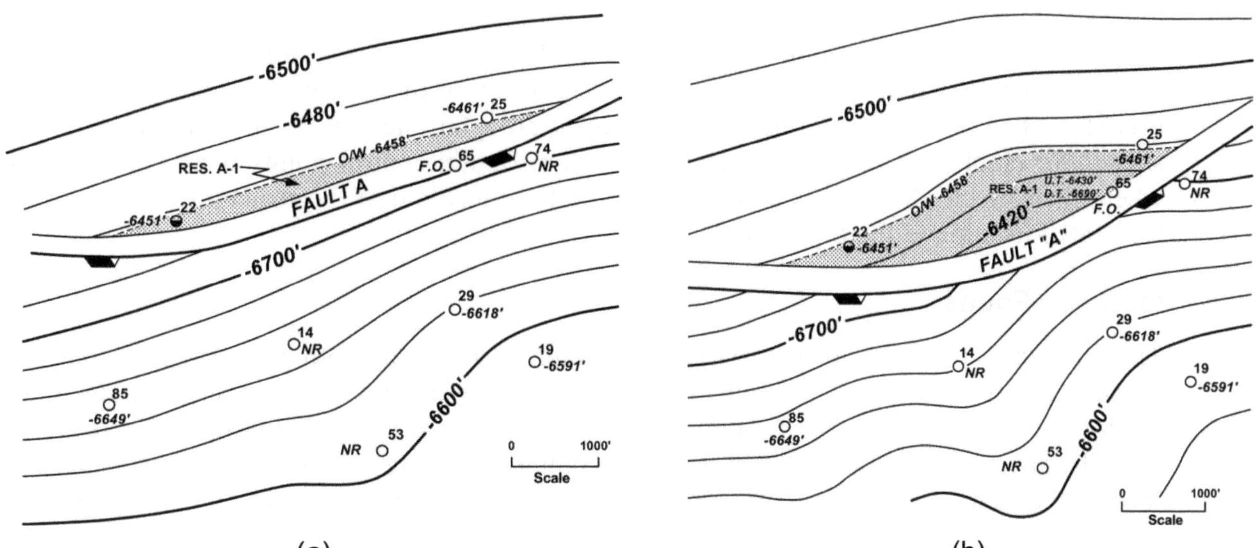

Figure 8-24 (a) Structure map on the 6450-ft Horizon shows Reservoir A-1 of insufficient size to justify a development well. This map was constructed without the integration of a fault map, and the restored tops for Well No. 65 were not used. (b) Revised structure map on the 6450-ft Horizon, Reservoir A-1. The revised interpretation was prepared using the restored top data from Well No. 65 and the integration of the fault surface map for Fault A with the structural interpretation. Also, shallower and deeper structure maps were used to guide the structure contouring. Reservoir A-1 is now of sufficient size to propose a development well location.

It is relatively simple to use restored tops when contouring by hand, but it is more difficult to use them when contouring on a workstation. This difficulty is for three reasons. First, restored tops need to be named differently than tops that are actually seen in a well. If the top of the J Sand is being mapped, tops actually seen in a well may be labeled J_t, whereas an upthrown restored top might be labeled J_t_ut. We recommend that tops be named with the shortest reasonably descriptive name because top names can obscure features on a computer screen. The shorter the name, the less information is obscured. This issue can be solved fairly simply in most computer systems by telling the system that J_t_ut is an alias of J_t, so the software will treat the two tops as equivalent. Shorter names also slightly minimize the issue of multiple names for the same top. To a computer, the J Sand, J sand, J-Sand, and J_Sand are four unique tops.

This solution runs into the second difficulty in using restored tops on a workstation. Most software programs struggle if there are two tops of the same stratigraphic marker in the same well. A third problem arises because restored tops occur in the fault gap, which is usually represented on a workstation by a polygon within which data is excluded. Finally, some computer systems expect tops to be in a stratigraphic order. Restored tops can violate this order. These last three issues can be finessed by placing digital control points representing the restored tops just outside the exclusion polygon (Fig. 8-23d). These digital control points are placed by estimating where they would project outside the exclusion polygon to match contours projected into the exclusion polygon to tie the restored tops. It may be necessary to move these digital control points to make sure the contours project through the faulted out well and reflect contour compatibility across the fault.

Restored tops in deviated wells are a special challenge when mapping with a computer. As discussed in Chapter 4, restored tops are placed directly above and below the fault cut in a deviated well (Fig. 4-43). They are not placed at their subsea depth along the deviated well path. We are not familiar with any computer mapping system that plots restored tops correctly on deviated well paths, although there may well be such systems available. Go back and review Figure 4-43. If the upthrown restored top for Sand B is plotted on the directional path of Well No. 3, it will plot out of stratigraphic sequence as it will be above the upper marker. It will also plot outside of the fault gap in the downthrown fault block. This issue may be resolved by initially entering restored tops as digital control points at the X-Y position of the fault cut and later moving the digital control points outside the fault exclusion polygon once the form of the contours takes shape. We emphasize that placing restored tops as digital control points is an iterative process where their precise location is dependent on the final location of the structural contours and the integrated location of the fault traces.

Reverse Faults

The method of manually integrating a reverse fault surface map with a structure contour map is illustrated in Figures 8-25 and 8-26. We do not describe the procedure in as much detail as we did for the integration of normal faults, since the procedure is basically the same. There are seven wells from which fault and horizon data were obtained. Figure 8-25 shows the reverse fault surface map constructed using the data from the seven wells. The fault strikes generally north-south with a dip of 35 deg to the west and has a displacement (vertical separation) of 300 ft, as shown by the values of repeated section in the well data.

The technique for integrating a reverse fault surface map with a horizon map is often easier than that for a normal fault because there is no projection of contours through a fault gap. In the case of a reverse fault, the hanging wall is thrust up and over the footwall, resulting in an overlap of structural horizons. Therefore, the hanging wall is contoured up to the hanging wall cutoff at the upthrown trace of the fault, and the footwall is contoured up to the footwall cutoff at the

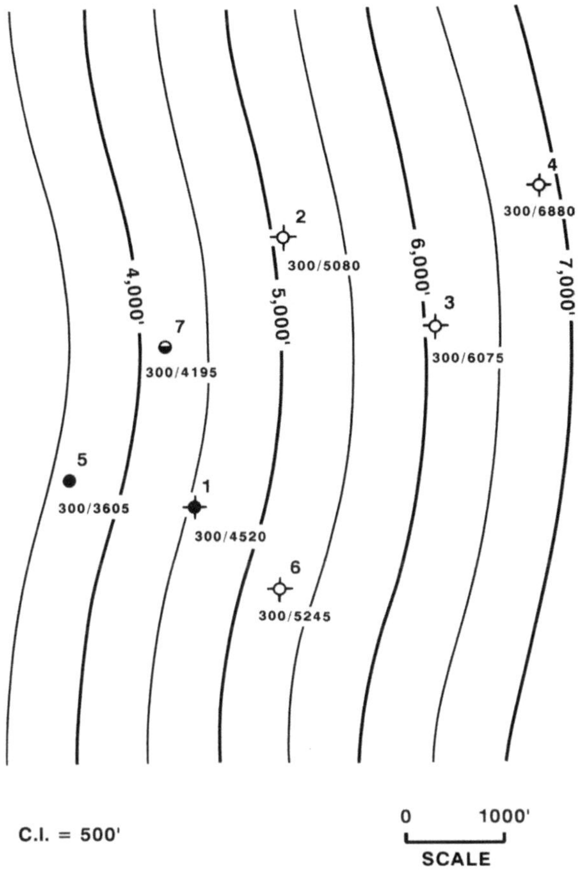

Figure 8-25 Fault surface map for reverse Fault 1 based on well control from seven wells. Depth values are positive, indicating they are above sea level.

downthrown trace of the fault. Since the fault blocks overlap, the strike direction of the contours established in one fault block can often be used to guide the direction of the contours in the opposing block.

The faulted structure for this example (Fig. 8-26) is the nose of a plunging anticline that is penetrated by seven wells. Wells No. 2, 3, and 6 penetrated the footwall, Wells No. 5 and 7 penetrated the hanging wall, and Well No. 1 is in the fault overlap, thereby penetrating the horizon twice: first at 4760 ft in the hanging wall, and second at 4460 ft in the footwall. Notice that the contour values are positive, indicating this structure is above sea level.

We begin contouring this structure in the hanging wall fault block. Place the fault surface map under the structure map as shown in Figure 8-26a. The fault trace on the hanging wall occurs where the structure contours in the hanging wall intersect fault surface contours of the same elevation. These intersections are highlighted on the map as small marks placed at the end of each contour. Likewise, as we contour the footwall horizon, the fault trace is located where the structure contours in the footwall intersect fault contours of the same value. Again, these intersections are highlighted by small marks placed at the end of each contour (Fig. 8-26a). For accuracy, 10-point dividers or a scale may be used to locate the exact points of intersection between structure and fault contours.

The difference in elevation between contours in the footwall and those in the hanging wall equals 300 ft, which is the value of the vertical separation or repeated section determined from well log correlation. For example, the 4000-ft contour in the footwall within the zone of overlap is positioned directly under the 4300-ft contour in the hanging wall. This same 300-ft difference in contour elevation is maintained for each of the structure contours in the mapped area.

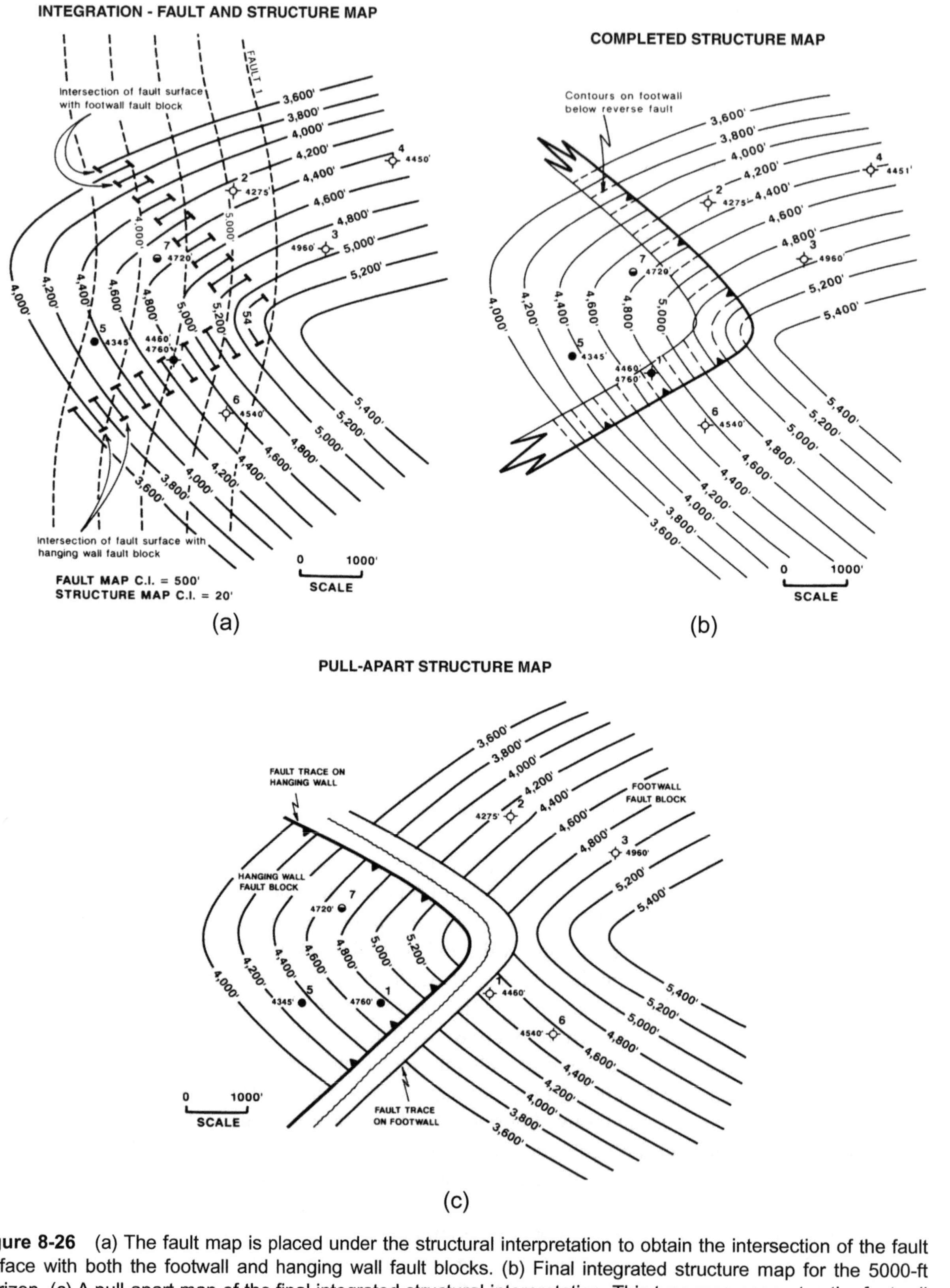

Figure 8-26 (a) The fault map is placed under the structural interpretation to obtain the intersection of the fault surface with both the footwall and hanging wall fault blocks. (b) Final integrated structure map for the 5000-ft Horizon. (c) A pull-apart map of the final integrated structural interpretation. This type map separates the footwall and hanging wall fault blocks to show them individually, without the clutter of overlapping contours.

Finally, connect all the marks representing the intersections of the fault surface with the footwall and hanging wall structure contours with smooth lines to accurately delineate the fault traces (Fig. 8-26b). Place a symbol on the hanging wall fault trace to show the direction of fault dip and dash the structure contours in the footwall, under the fault overlap for clarity. Having completed these various steps, the integrated structure contour map for the faulted plunging anticlinal nose is complete.

By correctly integrating the fault and structure maps, the following is accomplished: (1) the position of the fault on the mapped horizon is accurately delineated; (2) the traces for the fault are precisely constructed; (3) the actual width of the fault overlap is established; and (4) the contours are carried correctly in the fault overlap. Although the actual fault surface map is striking north-south, the trace of the fault on this horizon strikes northwest-southeast on the northern flank of the structure and southwest-northeast on the southern flank of the structure. *The curved shape and position of the trace for Fault 1 on this structure map could not be intuitively constructed without the integration of the fault and structure maps.*

We mentioned earlier that a structure map with a reverse fault overlap is cluttered and can be confusing because of the double set of contours in the overlap. One way to minimize the clutter and confusion when contouring by hand is to dash the contours in the footwall in the zone of overlap (Fig. 8-26b). Another solution to the clutter and confusion is to prepare a *pull-apart map*, as shown in Figure 8-26c. With this map, the two separate fault blocks are pulled apart or mapped separately to show each block separately without any overlap. Such a map should be constructed only after the fault and structural integration is complete in order to eliminate any possibility of contouring errors. In this example, the hanging wall is productive of hydrocarbons. Eventually, net sand and net oil isochore maps will be constructed for the hanging wall reservoir. These maps can be prepared more easily if the fault blocks are separated as shown in Figure 8-26c.

There is an additional complication in using a workstation to map structures cut by reverse faults. Consider Well No. 1 in Figure 8-26. This well contains two tops for the same horizon, as it penetrated both the hanging wall block and the footwall block. Most computer software packages do not handle repeated section well, either with repeated tops in a well or with repeated horizons on a seismic line. With most computer software packages, it may be necessary to map the hanging wall block separately from the footwall block. Digital control points can be added to the map to insure contour compatibility across the fault.

FAULT TRACES AND GAPS—SHORTCUTS AND PITFALLS

The following discussion refers primarily to fault traces and gaps formed by normal faults; however, the geometric relations are also applicable to fault overlaps formed by reverse faults. Fault traces are the two lines on a structure contour map representing the intersection of a fault surface with the structure-contoured horizon in the upthrown and downthrown fault blocks. Together, the fault traces form the overall trace for a fault on any particular mapped horizon. Both **fault gap** and **fault overlap** are each defined as the horizontal distance between the upthrown and downthrown fault traces *measured perpendicular to the fault traces*, as depicted on a completed structure contour map. When the integration technique is used, the fault gap or overlap is *automatically* constructed, since it is the horizontal distance between upthrown and downthrown traces.

With regard to the trace of a fault and the resultant gap or overlap, there are some common misunderstandings that we intend to clarify. Also, we present some shortcuts for delineating fault traces and gaps, and the pitfalls associated with these shortcuts.

There is widespread belief that the horizontal width of a fault gap equals the amount of vertical separation, as obtained from well logs or seismic data. It is thought that the fault gap may therefore be scaled off, using the horizontal reference scale on the map. Thus, for a fault with 500 ft of vertical separation, the width of the fault gap could be mechanically constructed by scaling off a 500-ft Horizontal gap between the upthrown and downthrown traces of the fault, using an engineer's scale or dividers. *This technique is often incorrect and should be employed only under special conditions.* The technique assumes that the positions of the upthrown and downthrown traces of the fault are known. However, fault trace positions are very difficult to define without the use of the fault/structure map integration technique. The width of a fault gap is also a function of the strike and dip of the fault, as well as the strike and dip of the horizon. Therefore, on a single structure contour map, variations in the attitude of either a fault or the horizon may result in changes in the width of the fault gap from place to place, although the vertical separation for the fault itself has not changed.

There is a general rule that should be kept in mind regarding fault gap. A fault dipping in the opposite direction of bed dip will have a narrower fault gap than a fault (with the same displacement) dipping in the same direction as the bed dip. Look again at Figures 8-11a and 8-12 to see the application of this general rule. In Figure 8-11a, the fault is dipping in the same general direction as the mapped horizon, so the fault gap is wider than that shown in Figure 8-12, in which the fault is dipping in the opposite direction to the dip of the mapped horizon. In Figure 8-27, we see the same effect. The fault cuts this horizon north of the crest and is dipping in the opposite direction to the mapped horizon; therefore, the fault gap is thinner in this area than on the flank of the structure. On the flanks, where the fault cuts this horizon south of the structural axis, the fault is dipping at almost a right angle to the mapped horizon, resulting in a wider fault gap. If a deeper horizon were mapped, one in which the fault were entirely to the south of the crest of the structure, the width of the fault gap near the crest would be greater. As a simple exercise, project the well control to a horizon 1000 ft deeper (e.g., change the depth in Well No. 3 from −5300 ft to −6300 ft). Using the contours in Figure 8-27 as a guide, contour that deeper horizon on tracing paper and then integrate the fault map with that horizon map. What observations can you make about the change in width of the fault gap and direction of the fault trace?

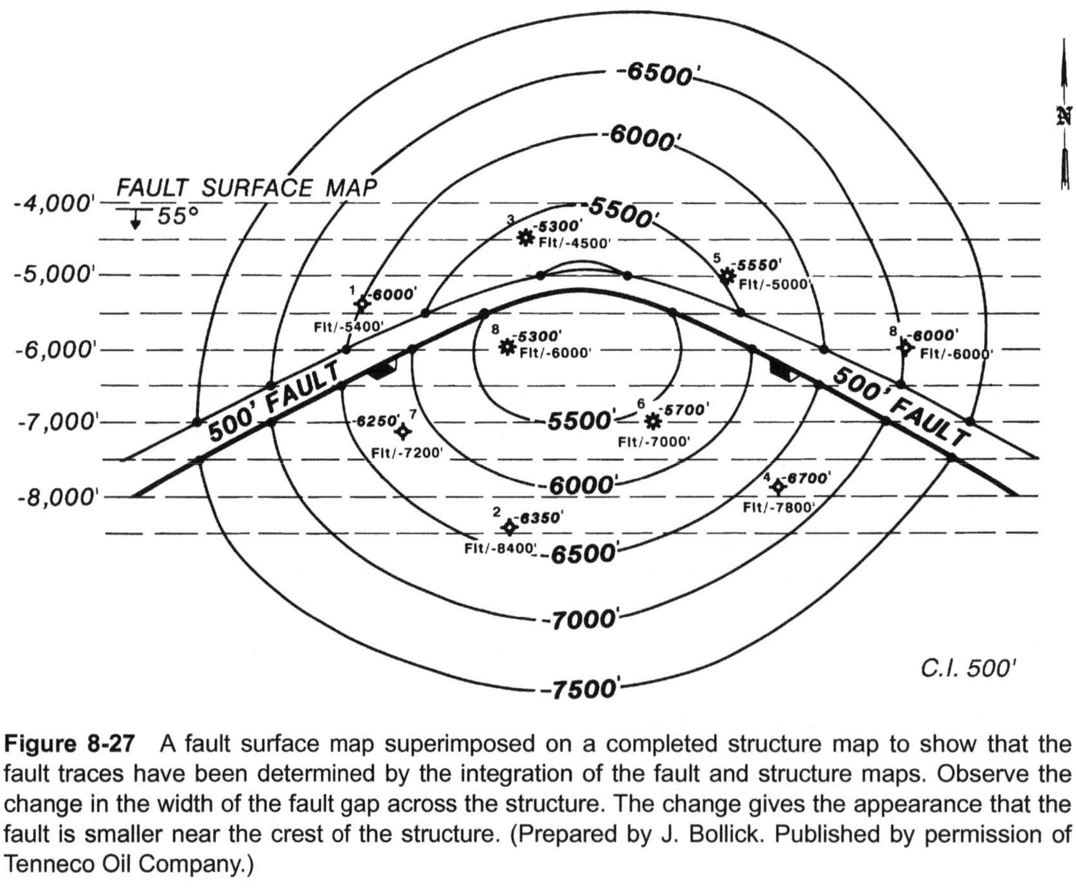

Figure 8-27 A fault surface map superimposed on a completed structure map to show that the fault traces have been determined by the integration of the fault and structure maps. Observe the change in the width of the fault gap across the structure. The change gives the appearance that the fault is smaller near the crest of the structure. (Prepared by J. Bollick. Published by permission of Tenneco Oil Company.)

In Figure 8-10, the curvature of the fault trace and width of the fault gap change across the structure. The gap changes from 350 ft on the flanks to 250 ft near the crest, whereas the fault has a constant vertical separation of 300 ft. Figures 8-11a and 8-12 show a significant difference in the width of the fault gap, which ranges from an average 420 ft in Figure 8-11a to 290 ft in Figure 8-12, despite that the fault on each map is the same fault, with a vertical separation of 400 ft. Finally, examine Figure 8-27, which shows a map of a fault surface, striking east-west and dipping to the south at 55 deg, superimposed on the structure map of the 5500-ft Horizon. The upthrown and downthrown fault traces, and therefore the fault gap, were constructed using the integration technique. Notice how the fault gap thins near the crest of the structure, causing the vertical separation of the fault to *appear* smaller on the crest than on the flanks. The vertical separation of the fault only *appears* to change across the structure because the width of the fault gap changes.

The width of a fault gap depends on the amount of vertical separation and the angles and directions of the fault dip and the horizon dip. Therefore, the *width of a fault gap for any given fault on a structure map cannot be intuitively determined.* It is extremely difficult at best, regardless of vertical separation and our ability to see in three dimensions, to predict the position of fault traces and the width of a fault gap by looking at fault data, or even at a fault surface map and a structure map independently. Examine Figure 8-27 again and ask yourself whether you could have predicted the curvature of the overall fault trace and the variation in width of the fault gap on this final structure map simply by reviewing the fault surface map and structure map data separately without integration. Most likely, your answer is no. *We emphasize that if the construction of the upthrown and downthrown fault traces is done properly by integrating the fault and structure maps, the width of the fault gap is automatically constructed without any need for guesses or estimations* (Tearpock and Harris 1987).

Rule of 45

Figure 8-28 shows a special condition in which the fault traces can be delineated with reasonable accuracy, without integrating a fault and structure map, and where the fault gap is equal in width to the amount of missing section in the well. The technique used for this situation is known as the rule of 45. *If a fault is inclined at 45 deg and the beds are approximately horizontal (zero degree dip)*, the position of the traces can be identified with reasonable accuracy without integration. The width of the fault gap is equal to the amount of vertical separation for the fault, measured off using the map scale as reference (see map insert in Fig. 8-28).

Consider Figure 8-29a. It is a prospect map illustrating a hydrocarbon trap upthrown to a fault and limited down-dip by a water contact in Well No. 3. The interpretation is based on both well log and seismic data. The completed structure map shows the horizon tops, fault data from the wells, and the subsea depths of the fault at various shotpoints on the seismic lines. The prospect map with a proposed location looks very attractive at first glance.

Without detailed knowledge of how the interpretation was generated, management might approve the drilling of such a location. However, upon detailed investigation, it is learned that the completed trace for Fault B was prepared using the position of the fault on the horizon from the two seismic lines and the rule of 45 to position the trace relative to the well locations. Does the application of this method provide a reasonable interpretation in this example? The answer is no. One major structural feature *not considered* in using the rule of 45 was the structural dip: the horizon is far from flat. As we have shown, bed dip is one of the structural features that affects the position of a fault on a completed structure map.

The structure map shown in Figure 8-29b was prepared using the same structural and fault data. In this case, however, a fault surface map was first constructed (Fig. 8-29c) and then integrated (cross-contoured) with the structure map to arrive at the correct positioning of the upthrown and downthrown fault traces.

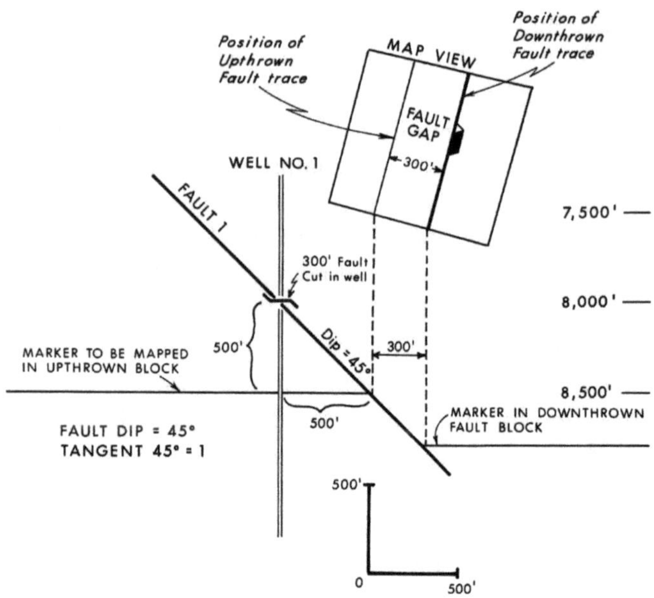

Figure 8-28 The rule of 45. On any given mapped horizon, the horizontal distance from a well to the nearest fault trace is equal to the vertical difference in depth between the fault and the horizon in the well, if the fault is dipping at 45 deg and the mapped horizon is horizontal. In these specific circumstances, the width of the fault gap will be equal to the missing section in the well, which is equal to the vertical separation of the fault.

The correct integration illustrates that the original prospect map has an enormous error regarding the position of the fault and the prospect potential. This error would result in the proposed well being drilled downthrown to Fault B, missing the oil reservoir entirely. How often can a petroleum company afford such mistakes? A number of prospect maps reviewed worldwide have shown similar mistakes resulting in costly dry holes. And in one particular case, one company drilled two such dry holes in one year. The company's CEO stated that no additional prospects would be drilled without constructing fault surface maps and integrating the fault with the structural interpretation.

Tangent or Circle Method

Another technique, referred to as the *tangent* or *circle method* (Lyle 1951; Bishop 1960), can be used to quickly delineate the approximate position of the fault traces in areas where the beds are nearly horizontal. Lyle introduced this technique in 1951 for mapping faults in southwest Texas and Bishop (1960) diagrammatically illustrated its use.

Figure 8-30a is a base map showing the depth of a normal fault and the depth to the top of a key horizon in each well. More than one fault is represented. The tangent or circle method is used here to quickly prepare a structure map including the fault traces. This technique is good for rapidly checking a completed structure map for accuracy of the positioning of the fault traces. The use of this method requires a low to flat bed dip and knowledge of the dip angle of each fault.

1. The horizontal distance between the well and the fault trace on the horizon equals the *cotangent* of the dip angle of the fault multiplied by the vertical distance between the fault pick and the horizon in the well (Fig. 8-30b).
2. The well serves as the center point of a circle, the radius of which equals the distance between the fault and the well as determined in step 1 (Fig. 8-30c).
3. A tangent common to two or more circles determines a fault trace. The exterior tangent is the fault trace for wells on the same side of the fault; the interior tangent is the fault trace for wells on the opposite sides of the fault (Fig. 8-30c). Refer again to Chapter 7, Figure 7-9, and the accompanying text for determining in which fault block a well is located for any mapped horizon.
4. Finally, contour the horizon tops assuming continuity of structural attitude across the faults (Fig. 8-30d).

If the beds are horizontal, the fault gap width can be determined as shown in Figure 8-30b. Then the fault gap can be drawn on the structure map rather than representing the fault as one line, as shown in Figure 8-30d. This method of estimating the fault trace is useful if the dip of the fault surface can be determined or at least assumed with some degree of certainty and if the horizon being mapped is approximately horizontal. Compare the mapping results in Figure 8-30d with those shown in Figure 8-30e and f. Figure 8-30e is a fault map for the two faults, and Figure 8-30f shows the fault traces constructed by the integration of the fault and structure maps. The position of the fault in Figure 8-30d compares very favorably with the position of the fault in Figure 8-30f, showing that in an area of relatively flat beds, the tangent method can provide a good approximation for positioning the fault on a structure map. However, we caution that *this technique does not work in areas of dipping beds.*

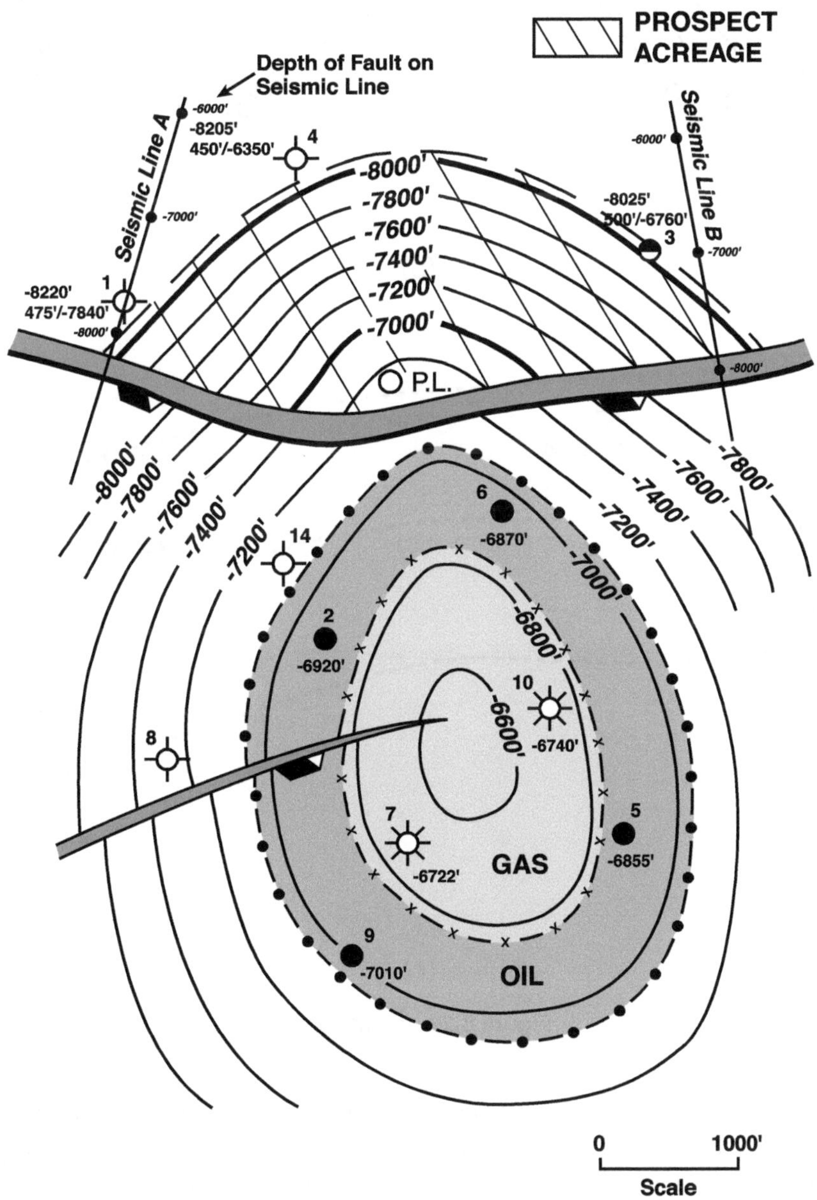

(a)

Figure 8-29 (a) Prospect map based on well log and seismic data. Position of the fault trace based on the rule of 45 relative to the well locations and on the fault position as interpreted on the seismic profiles. (b) Reinterpretation of prospect map in (a), based on integration of the fault surface map with the horizon map. Fault trace moves about 900 ft north. (c) Fault surface map. (Published by permission of D. Tearpock.)

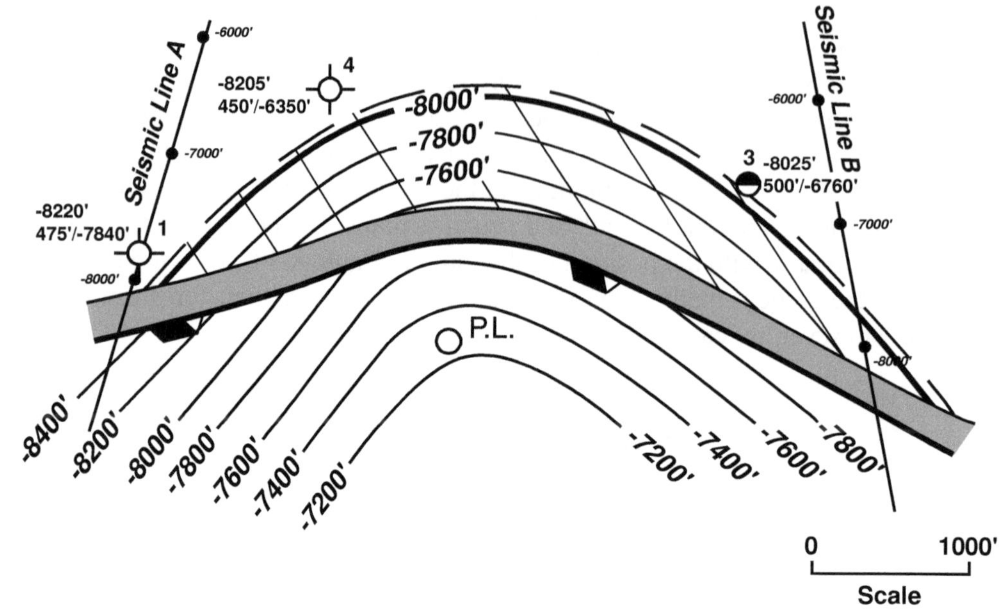

(b)

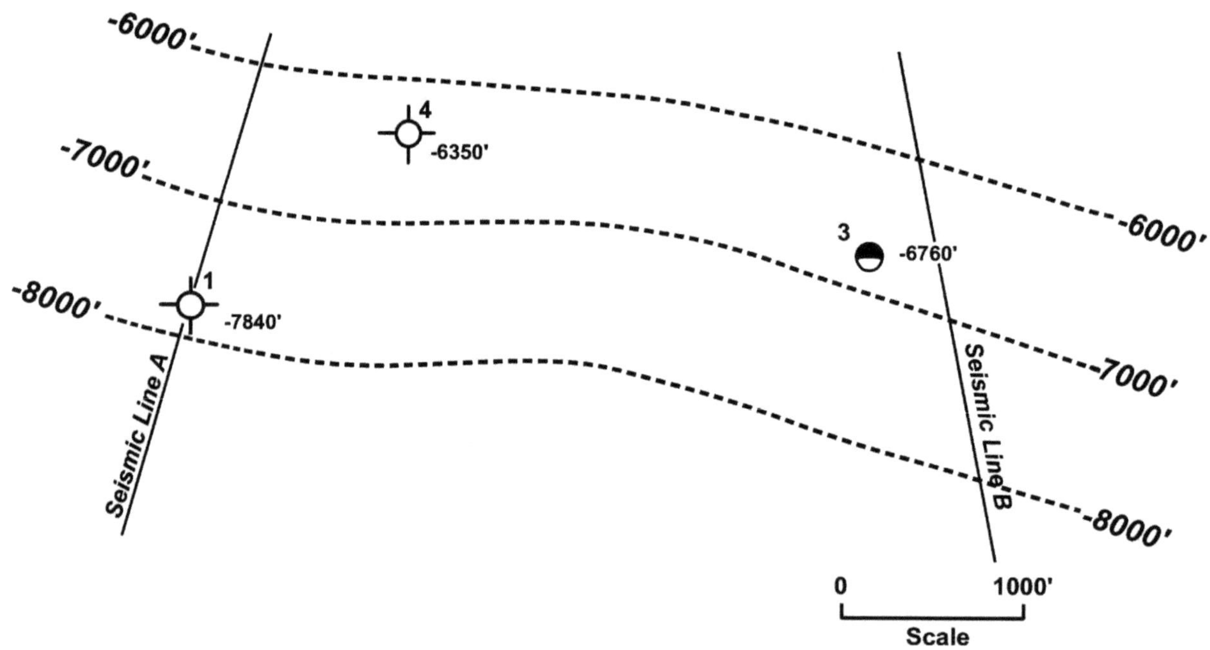

(c)

Figure 8-29 *(continued)*

In Figure 8-30c and d, the fault traces are constructed as straight-line segments with sharp angular changes in strike direction. In order to prepare a more accurate map, the change in strike direction can be smoothed to best fit the overall pattern of the faults and the data in the area being mapped. The fault surface map for Figure 8-30e and the integrated structure map in Figure 8-30f were also constructed as straight-line segments to make the results comparable with those in Figure 8-30d.

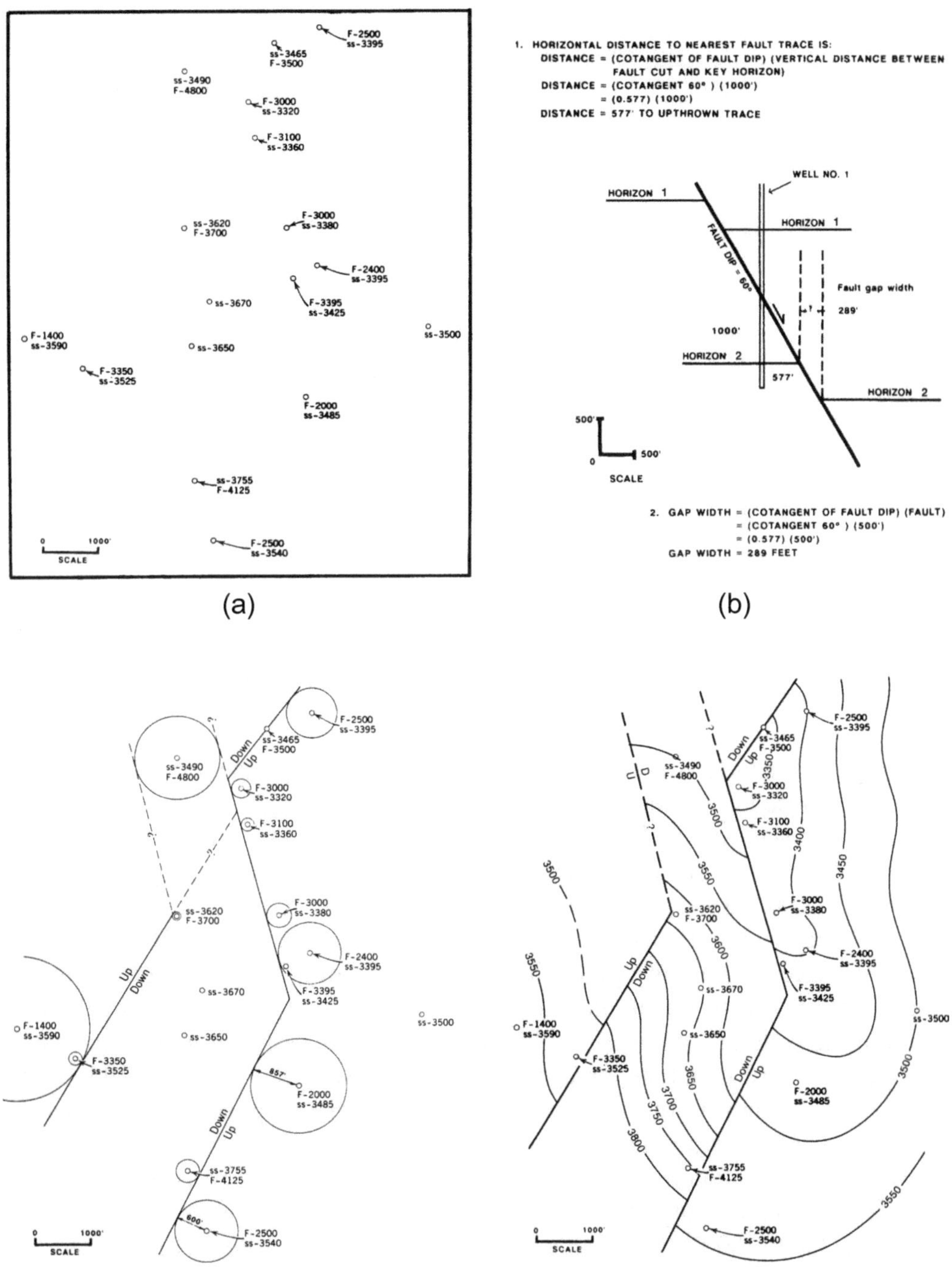

Figure 8-30 (a) Basemap with the depth to fault picks (F−3100 ft) and sand tops (ss -3320 ft) posted. (Modified after Bishop 1960. Published by permission of author.) (b) Determination of the horizontal distance from a well to the nearest fault trace using the cotangent rule and the width of the fault gap. (c) Assumed angle of fault = 60 deg. The position of the fault trace is determined by the circle method. (Modified after Bishop 1960. Published by permission of author.) (d) Structure contour map. Position of the fault determined by the circle method. (e) Integration of mapped horizon and fault surface maps to determine the position and width of the fault gaps. (f) Final integrated structure map. Compare position of fault traces with those shown in Figure 8-30d. (From Bishop 1960. Published by permission of author.)

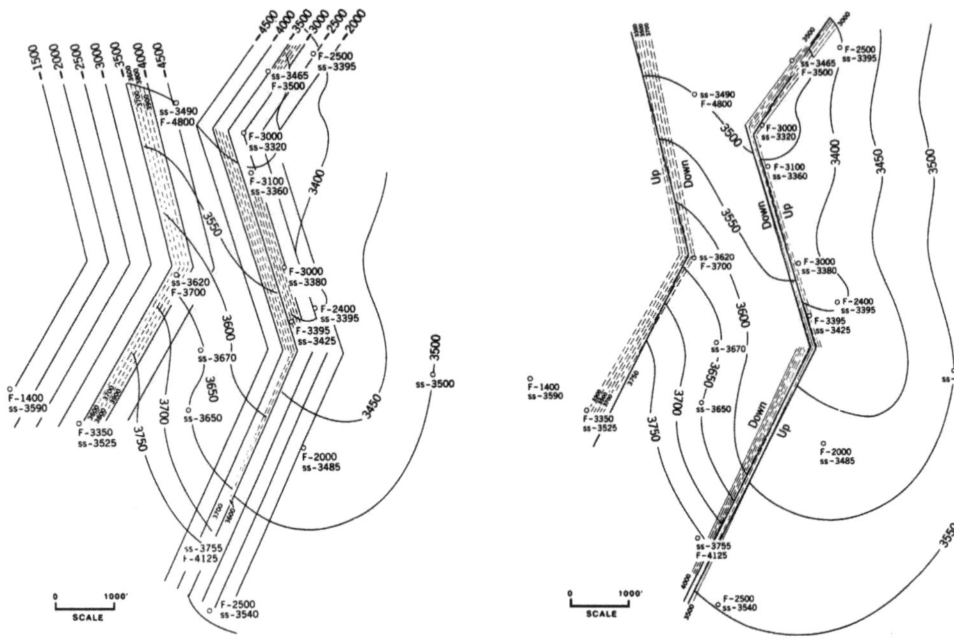

(e) (f)

Figure 8-30 (*continued*)

The tangent method is not accurate if the horizon being mapped has significant dip (more than 5 deg) because the wellbore is then not perpendicular to the bedding. The correct construction of the fault traces and fault gap for faults on a dipping structure is especially critical because the fault position has a significant impact on any proposed location for an exploration or development well. Figure 8-31a shows a portion of a structure map on the flank of a piercement salt dome. The depth to the mapped horizon and the missing section and depth of the faults are shown next to each well. Since the fault is a perfect plane with a dip of 45°, the structure map in Figure 8-31a was prepared (incorrectly) using the rule of 45 to determine the position of the fault traces and the width of the fault gap. As an example, look at Well No. 6 shown in the figure insert. The fault cuts the well at −9510 ft and the top of the horizon intersects the well at −9395 ft. From the figure insert, we see that Fault A crosses the well 115 ft below the mapped horizon. Therefore, based on the 45-degree trigonometric relationship of 1 to 1, the downthrown trace of the fault was located 115 ft west of the location of Well No. 6, as the horizon in the well is in the downthrown block. Also, since the fault has a dip of 45 deg, the width of the fault gap was scaled to equal the missing section of the fault (400 ft). Using this method, the fault cut in Well No. 3 does not fit as part of the same fault identified in Wells No. 2, 5, 6, and 10; therefore, a second fault, Fault B, was postulated as shown on the figure, resulting in an untested fault trap up-dip of Well No. 5.

The fault in this example has a constant dip of 45 deg and, in this respect, the above methods were appropriate. However, the beds are *not* horizontal; they dip quite steeply (about 40 deg in the up-structure position). Therefore, neither the rule of 45 nor the tangent method is appropriate for the preparation of the structure contour map in Figure 8-31a. Figure 8-31b shows the fault surface map and the correctly contoured structure map for this horizon. The fault is a peripheral fault paralleling the face of the salt; the beds are dipping at about 40 deg at −7000 ft and flatten to about 32 deg between −11,000 ft and −12,000 ft. Since the beds are steeply dipping, the fault map was required for integration of the fault and structure maps to delineate accurately the fault traces on the structure map. The upthrown and downthrown traces of the fault, and therefore the resulting fault gap, were constructed using the integration technique discussed earlier in the chapter.

Observe that although the fault has a constant vertical separation of 400 ft (as determined from well log correlation), and the fault is a perfect plane dipping at 45 deg, the trace of the fault is curved and nowhere along its trace is the gap 400 ft wide. Measuring across the fault gap at the location where the −8000-ft structure contour in the upthrown block intersects the upthrown trace of the fault, the width of the fault gap is approximately 1700 ft. At the intersection of the −10,000-ft structure contour and the upthrown trace of the fault, the width of the fault gap has decreased to 1200 ft. If a well were drilled, on the basis of the map in Figure 8-31a, in the "untested fault block," then the mapped horizon would probably be faulted out of that well (as shown by the correct map in Figure 8-31b). Study this figure very carefully, for it shows how important it is to integrate the fault and structure maps on steeply dipping structures to accurately position the fault traces and delineate the fault gap.

As a general rule of thumb, we consider any structure with a dip greater than 5 deg as steeply dipping for purposes of fault and structure map integration. However, to prepare a structure map as accurately as possible regardless of the structural dip, the integration technique should always be used to delineate the position of the fault traces and the width of the fault gap.

Figure 8-32 shows three different sets of integrated fault and structure contour maps that result in significantly different fault traces and gaps. For example, the set on the left side of

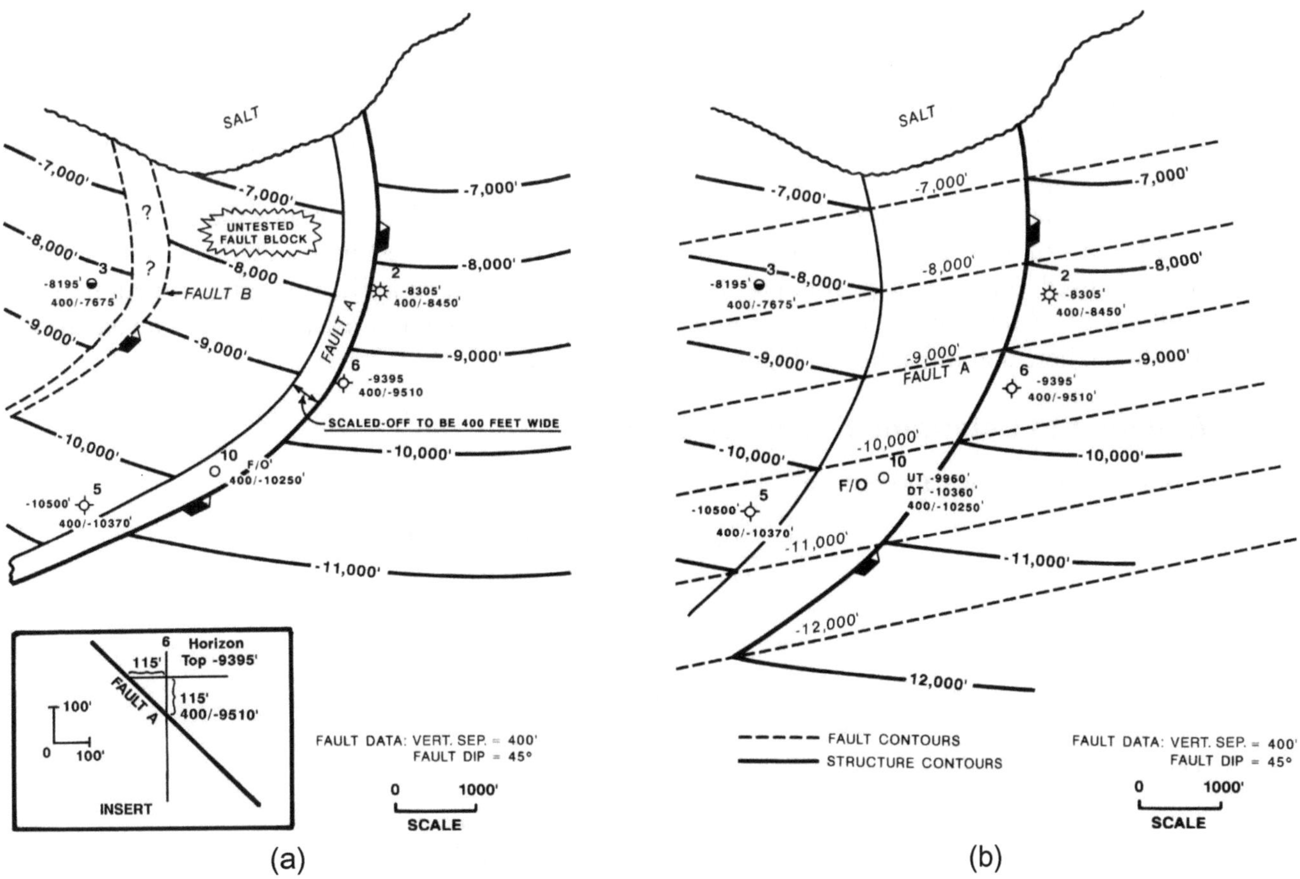

Figure 8-31 (a) Portion of a structure map on a steeply dipping horizon on the flank of a piercement salt dome. The map was incorrectly prepared using the rule of 45 to position the fault traces. Although the fault is dipping at 45 deg, the rule of 45 is not applicable because the beds are not horizontal. (b) Correctly contoured structure map based on the integration of the structural interpretation with the fault surface map. Compare this map to that shown in Figure 8-30a.

the figure consists of completed structure maps A(1), B(1), and C(1), all of which use the same oblique down-dip fault with 100 ft of vertical separation to integrate with the structure. A(1) shows a horizon with a dip that is constant, B(1) shows a steeply dipping horizon which increases in dip in the upstructure direction, and C(1) is similar to B(1) except that the horizon dip is less. Despite the constant fault surface in all three cases, each integrated structure map results in a strikingly different fault trace pattern and fault gap width. Take a few minutes to review the different fault traces and gaps for sets 2 and 3 in Figure 8-32.

The nine separate patterns shown in Figure 8-32 and those in Figures 8-10, 8-26, 8-27, and 8-31 reinforce our statements that the final fault traces and fault gap for any particular set of fault and structural conditions are not intuitively predictable and, in fact, are often impossible to predict without actually integrating the two surfaces. The precise positioning of the fault traces and delineation of the width of the gap or overlap (in the case of a reverse fault) can mean the difference between generating and not generating a prospect, the difference between drilling a successful well versus a dry hole, or the difference between a reservoir whose calculated volume matches its performance versus a reservoir that appears to either have overproduced or underproduced the volumetric estimate of reserves. Taking shortcuts to identify the positions of fault traces and to determine the widths of fault gaps is a risky business unless the beds are relatively flat-lying and the fault dips are very close to 45 deg.

New Circle Method

We have emphasized the importance of integrating a fault and structure map in areas of dipping beds to accurately delineate a fault trace. We have also developed a mathematical relationship that can be used in areas of dipping beds to estimate the *approximate position* of the fault traces, if there is sufficient well control. These equations can be used as a quick check when reviewing structure maps.

Two equations are required for this method. The first equation estimates the distance from a well (radius of a circle) to the nearest fault trace, and the second estimates the heave of the fault, which is used to locate the position of the other trace of the same fault. The strike direction of the fault *must be known* in order to determine the direction in which to measure heave. The equations are shown here and illustrated in Figures 8-33 and 8-34.

Equation to Determine Radius of Circle

From the law of sines and Figure 8-33,

$$\frac{\sin(\pi/2-\theta)}{DX} = \frac{\sin(\theta+\pi-\phi_a)}{AX}$$

and

$$\cos(\pi-\phi_a) = BD/DX$$

Therefore,

$$\frac{\sin(\pi/2-\theta)}{BD/\cos(\pi-\phi_a)} = \frac{\sin(\theta+\pi-\phi_a)}{AX}$$

or

$$BD = \frac{AX\cos(\pi-\phi_a)\sin(\pi/2-\phi_a)}{\sin(\theta+\pi-\phi_a)}$$

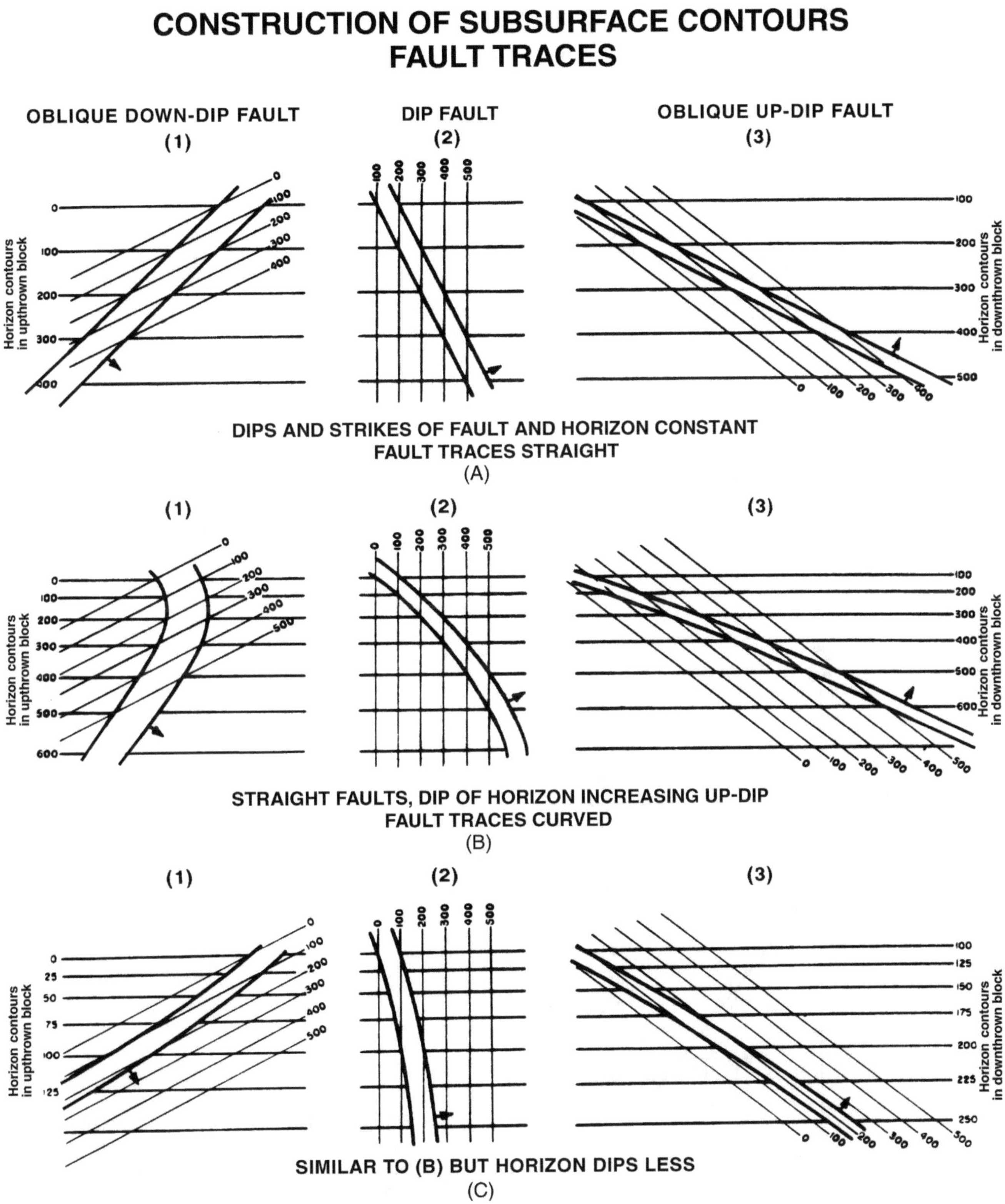

Figure 8-32 Three different sets of integrated fault and structure maps illustrating significantly different fault traces. (Published by permission of Tenneco Oil Company.)

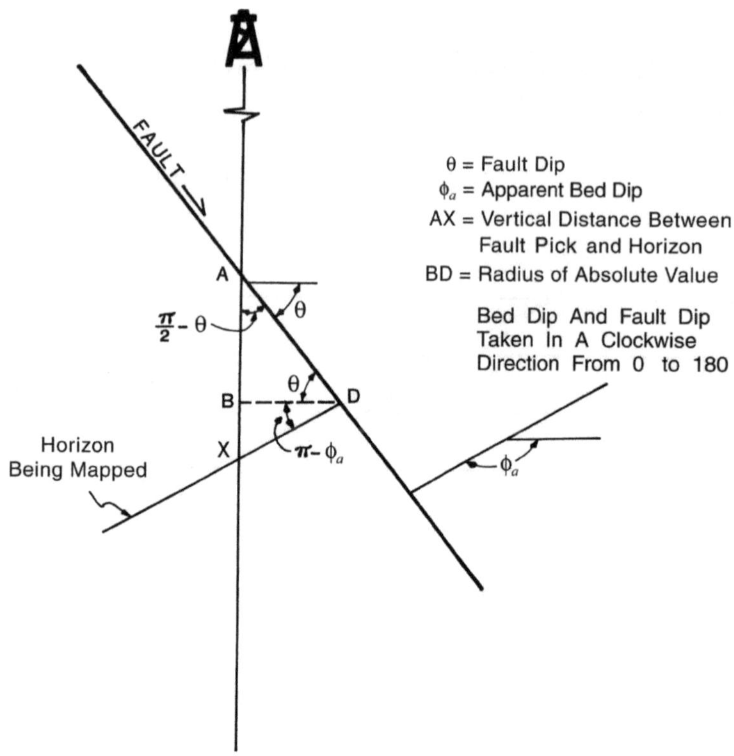

Figure 8-33 Diagrammatic illustration of the fault and formation parameters used to derive Eq. (8-2). View is perpendicular to strike of fault surface.

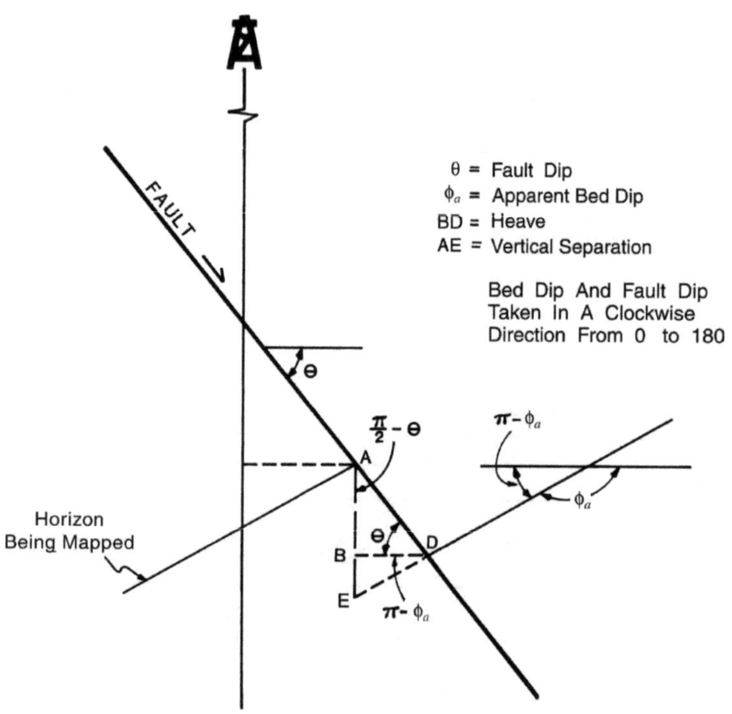

Figure 8-34 Diagrammatic illustration of the fault and formation parameters used to derive Eq. (8-3). View is perpendicular to strike of fault surface.

Therefore,

$$\text{Radius} = \left| \frac{-AX \cos\phi_a \cos\theta}{\sin(\theta - \phi_a)} \right| \tag{8-2}$$

where

ϕ_a = apparent bed dip
θ = fault dip
θ and ϕ_a are taken clockwise from 0 deg to 180 deg

If $\phi_a = \pi$, then Eq. (8-2) becomes

$$\text{Radius} = \frac{-AX \cos(\pi)\cos\theta}{\sin\theta}$$
$$= \frac{AX \cos\theta}{\sin\theta}$$
$$\text{Radius}* = \text{AX} \operatorname{cotan} q$$

Equation to Determine Heave

Utilizing the law of sines and Figure 8-34,

$$\frac{\sin(\pi/2 - \phi_a)}{DE} = \frac{\sin(\theta + \pi - \phi_a)}{AE}$$

and

$$\cos(\pi - \phi_a) = BD/DE$$
$$DE = \frac{BD}{\cos(\pi - \phi_a)}$$

Therefore,

$$BD\ (\text{Heave}) = \frac{\text{AE} \sin(\pi/2 - \theta)\cos(\pi - \phi_a)}{\sin(\theta - \phi_a)}$$
$$BD = \left| \frac{AE \cos\theta(-\cos\phi_a)}{\sin(\theta - \phi_a)} \right| \tag{8-3}$$

which is the same form as the radius Eq. (8-2). For example,

$$\theta = 60 \text{ deg}$$
$$\phi_a = 45 \text{ deg}$$
$$AE = 1000 \text{ ft}$$
$$BD\ (\text{Heave}) = \left| \frac{1000 \cos 60^\circ \cos 45^\circ}{\sin(60^\circ - 45^\circ)} \right|$$
$$BD\ (\text{Heave}) = \left| \frac{1000(0.5)(0.707)}{0.2588} \right|$$
$$BD = 353.5/0.2588$$
$$\boldsymbol{BD = 1366 \text{ ft}}$$

In order to use these equations, *the strike direction of the fault surface must be known*. These equations should be used only as a quick look or check; they do not take the place of the fault/structure integration technique.

Fault Gap Versus Fault Heave

Before leaving the subject of fault gap, we briefly discuss the fault component heave, which is often misused or mistaken for fault gap. **Fault gap** is the horizontal distance between the upthrown trace and downthrown trace of a fault *measured perpendicular to the strike of the fault traces*. **Fault heave** is the horizontal distance between the upthrown and downthrown traces of a fault *measured perpendicular to the strike of the fault surface itself*. The width of a fault gap is equal to fault heave only where the beds are horizontal, or nearly so, and the strike of the fault is parallel to the strike of the formation. Considering most other situations, the width of the fault gap and the fault heave are not the same. Therefore, the term *fault gap* (or *overlap* in the case of a reverse fault) is *not synonymous* with *fault heave*.

The completed structure map in Figure 8-31b clearly illustrates the difference between fault gap and fault heave. Using the completed structure map, measure the heave for Fault A at the location where the −9000-ft contour in the upthrown block intersects the upthrown trace of the fault. First, draw a reference line perpendicular to the strike of the fault surface at the intersection of the −9000-ft structure contour with the −9000-ft fault contour. This line intersects the downthrown trace of the fault at some point. Using the horizontal reference scale on the map, measure the length of this reference line, which is the measurement of fault heave at this location along the fault. The fault heave is equal to about 1825 ft. Now the distance across the gap at the same −9000-ft point, measured perpendicular to the upthrown trace, is equal to 1425 ft. Therefore, we can see that the heave is not equal to gap and they are measured in two completely different directions.

Now, take a minute to determine the throw of the fault, using the same reference line. If done correctly, the line intersects the downthrown trace of the fault at a depth of about −10,825 ft so the fault throw is 1825 ft (10,825 ft − 9000 ft = 1825 ft). Since the fault has a dip of 45 deg, the throw of the fault must equal the heave, or 1825 ft. We can easily check the measured result for fault throw by using Eq. (7-1). To use the equation, the bed dip must be averaged for the interval between the −9000-ft contour in the upthrown block to the −10,825-ft contour in the downthrown block because the gap is too wide and bed dip changes rapidly in this area. For the equation, use 38 deg as an average bed dip. Also, use Eq. (8-3) to estimate the fault heave mathematically and compare the results with the measurement shown in Figure 8-34.

These measurements are shown in Figure 8-35. Notice for this example that there is a significant difference between the strike direction of the fault gap and that for the fault heave. Fault heave and fault gap are different fault components that cannot be referred to interchangeably. As another exercise, measure the fault overlap and heave at some point on the reverse fault shown in Figure 8-26b and compare the results.

Subsurface geoscientists should, as a general rule, avoid using the terms throw *and* heave *and should always refer to measurements derived from log correlations and map construction as* vertical separation *and* fault gap width (Tearpock and Harris 1987).

STRUCTURE MAP—GENERIC CASE STUDY

In the section Techniques for Contouring across Normal Faults, we present the technique for projecting established contours from one fault block across a fault into the opposing fault block to contour correctly the displacement resulting from a fault. The displacement to which we refer is vertical separation, as determined from well log correlations or seismic data. In the same section, the incorrect technique of contouring displacement as throw is also presented. Now we look at a generic case study that illustrates the impact of preparing a map with the incorrect, as opposed to correct, contouring technique.

This field example shows an anticline cut by a large 600-ft down-to-the-south normal Fault A and a smaller 100-ft antithetic (compensating) normal Fault B (Fig. 8-36). The structure map in Figure 8-37a was prepared with the incorrect assumption that the missing section in the wellbores is equal to throw (note the dashed contours in the fault gap drawn perpendicular to the strike of the fault). Based on this interpretation, two development wells are proposed. Location X, upthrown to Fault A in Reservoir A, is estimated to penetrate the reservoir up-dip and to the east of Well No. 12 at a depth of −4960 ft. The purpose of the well is to improve the drainage efficiency of this reservoir. Location Y, upthrown to Fault B, is estimated to penetrate Reservoir B at a depth of −5360 ft and is designed to be drilled in the optimum structural position to drain the attic reserves in this reservoir.

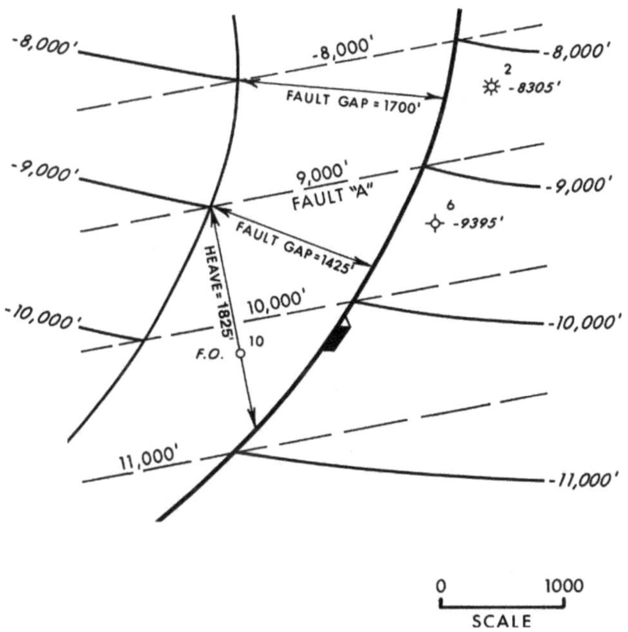

Figure 8-35 A part of Figure 8-31b illustrating the measurements of fault gap and fault heave.

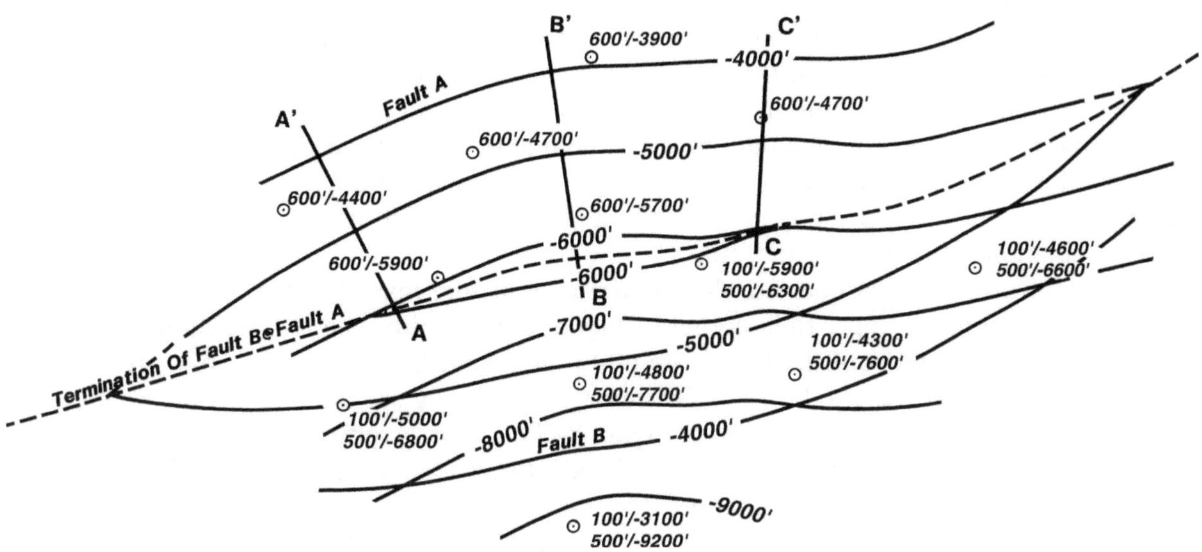

Figure 8-36 Fault surface map (Faults A and B) for the generic case study. (Published by permission of Tenneco Oil Company.)

The structure map in Figure 8-37b was constructed correctly, using the missing section in the wellbores as vertical separation. Observe the dashed contours in the fault gap showing the trend projection of the contours from one fault block to the next. This correct interpretation shows a different structural picture. First, Well X is estimated to penetrate the formation at a subsea depth of −5030 ft, which is the depth of the oil/water contact in Reservoir A. Therefore, Well X, if drilled, would probably be a dry hole. The penetration point for Well Y, in Reservoir B, is estimated at −5415 ft, or 55 ft deeper than that shown in Figure 8-37a. Based on the correctly contoured map, the proposed well is about 800 ft west of the optimum position to efficiently drain the remaining reserves in Reservoir B.

The correctly contoured map has the following impact on these two reservoirs and proposed wells.

Reservoir A (upthrown to Fault A)

1. It eliminates the drilling of a dry hole.
2. It improves the volumetric reserve estimate, in this case, a 36% reduction in reserves.

Reservoir B (upthrown to Fault B)

1. It improves the volumetric reserves estimate for the reservoir (a reduction of 11%).
2. It improves the configuration of the reservoir, affecting the location of the proposed development well.

The magnitude of error created by incorrectly mapping vertical separation as if it were throw is greatest near or on the crest of a structure, as illustrated in this generic case. On both the east and west flanks of the structure, where the strike direction of the contours is nearly perpendicular to the strike of the fault, the projection of contours through the fault is almost identical for both structure map interpretations. Where the strike direction of the structure contours is perpendicular to the strike of the fault, lines parallel to the structure contour strike, or a cross section in this orientation, give the appearance that the formation has an apparent dip of zero degrees (horizontal). Therefore, whether the contours are projected through the fault as if the fault were not there (mapping vertical separation), or the strike direction of the contours is projected through the fault perpendicular to the strike of the fault (mapping throw), the direction of contour projection through the fault is similar. However, as the crest of the structure is approached, the projection of the contours is radically different for each structure map. Review the −5000-ft contour starting in the upthrown block of Fault A and follow the projection across the fault into the downthrown fault block on both maps; notice the significant difference in contour projections. Here again we emphasize that it is very important to contour correctly over the entire map, but it is especially critical in and around the crest of a structure where hydrocarbon accumulations are expected to occur.

Three cross-section lines are shown on each structure map. The B-B′ cross section for each map is shown in the upper right corner of Figure 8-37a and b. The B-B′ section in Figure 8-36a shows that if the throw of the fault were 600 ft as mapped, then the vertical separation in cross section B-B′ would have to be 725 ft; therefore, the missing section in the wells would also have to be 725 ft. The B-B′ cross section in Figure 8-37b shows a vertical separation of 600 ft, which agrees with the missing section in each well and a throw across the fault at this location of 500 ft. This cross-section geometry is supported by the well log data and the completed structure map. As a cross-section exercise, complete sections A-A′ and C-C′ for both maps and evaluate the results. If you are interested, use Eq. (7-1) again to test the validity of the structure contour map in Figure 8-37b.

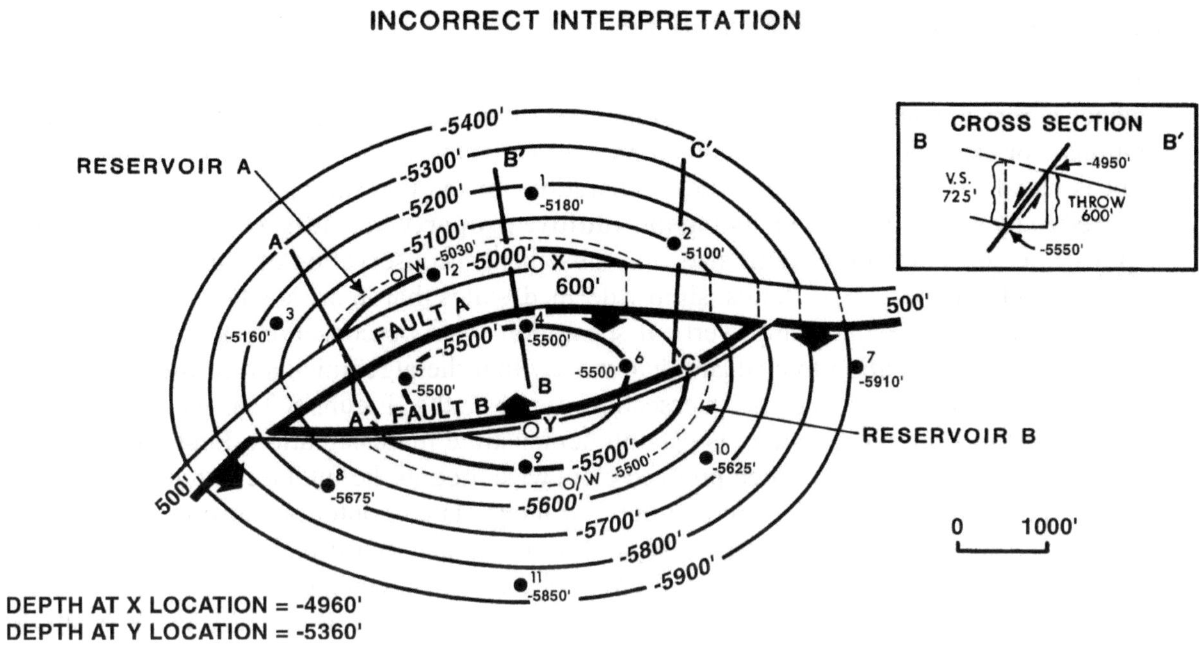

(a)

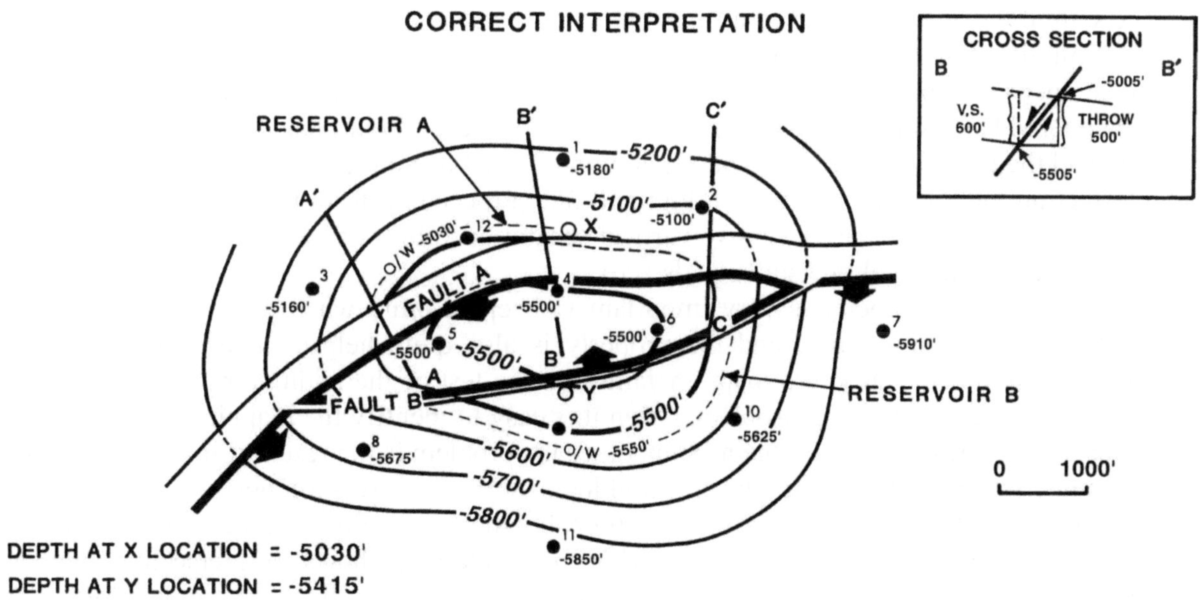

(b)

Figure 8-37 (a) Generic case study: structure map prepared using missing section (vertical separation) incorrectly as if it were throw. (b) Generic case study: structure map prepared using vertical separation to correctly contour across Faults A and B. (Published by permission of Tenneco Oil Company.)

THE ADDITIVE PROPERTY OF FAULTS

In the preparation of accurate structure maps, the vertical separations of intersecting faults will balance across the intersection. This is sometimes called the additive property of the vertical separations. For example, where two faults merge into one, the vertical separation of the "surviving" fault is equal to the sum of the vertical separations of the two merging faults. This property is referred to by several names, including the **additive property of faults, conservation of vertical separation, conservation of fault size**, and (mistakenly) conservation of fault throw.

The additive property of faults, simply defined, states that in areas where two or more faults have an intersecting relationship, vertical separation must be conserved at the intersection. To create a simple calculation to evaluate that, we say that the algebraic sum of the vertical separations must equal zero around the intersection of the faults. Figure 8-38a, which is a structure map of a bifurcating fault system, is used to demonstrate the calculation. First, imagine a circle around the fault intersection. Then proceed clockwise around the intersection and note the vertical separations as determined from contours, or interpolated contours, close to the intersection. Assign a positive value to the vertical separation if the clockwise path across the fault is from the upthrown to the downthrown block. Conversely, assign a negative value if the path across the fault is from the downthrown to the upthrown block. Beginning at the 12 o'clock position, the first fault encountered has a vertical separation value of +400 ft (−7800 ft versus −7400 ft). The next fault has a vertical separation value of −150 ft, using an interpolated contour of −7550 ft versus the downthrown contour of −7700 ft. The third fault has a vertical separation of −250 ft. The sum of the three values is zero.

Simplified maps of fault patterns in Figure 8-38b and c further illustrate the method. Using the compensating pattern in Figure 8-38b and beginning at the 12 o'clock position, the first fault encountered has a vertical separation of +1200 ft. The second fault has a −400-ft vertical separation, and the third fault has a vertical separation of −800 ft. So the algebraic sum equals zero. An intersecting fault pattern is shown in Figure 8-38c. Proceeding from the 12 o'clock position, we cross four faults that have successive vertical separation values of −200 ft, +300 ft, +200 ft, and −300 ft. Again, the algebraic sum equals zero.

This additive property is very important to keep in mind when preparing structure maps involving intersecting faults, and the property is also quite helpful in quickly evaluating or reviewing completed structure maps. A failure in the test of the additive property of faults can indicate a structure mapping problem. Often it is easy to identify the mapping bust, but difficult to determine why there is a problem. Sometimes the problem is the result of incorrect contouring, a misunderstanding of the property, or a well log or seismic correlation bust.

The example shown in Figure 8-39a is probably the result of incorrect contouring and failure to integrate the fault and structure maps. This completed map shows an intersecting compensating fault pattern with Fault A dipping to the north and having a vertical separation of 400 ft at the intersection (−12,400 ft versus −12,000 ft) and Fault B dipping to the southwest and intersecting Fault A. Fault B becomes smaller to the south so that near the intersection, the vertical separation is only 100 ft. Therefore, the vertical separation of the resultant Fault A east of the intersection must be 300 ft (400 ft − 100 ft = 300 ft); however, it is contoured as 450 ft on the map.

There appear to be three separate problems with this structure map. (1) Although this is a compensating fault pattern, the map is prepared incorrectly, depicting a bifurcating pattern where the individual vertical separations for Faults A and B are added. The 400 ft for Fault A added to the 100 ft for Fault B equal 500 ft. Notice that the vertical separations do not even add correctly, which is another problem related to the completed map. (2) It is obvious that this map was prepared without being integrated with fault maps. When fault and structure maps are integrated, the fault traces and gaps are properly constructed. Thus, an error of constructing fault traces that

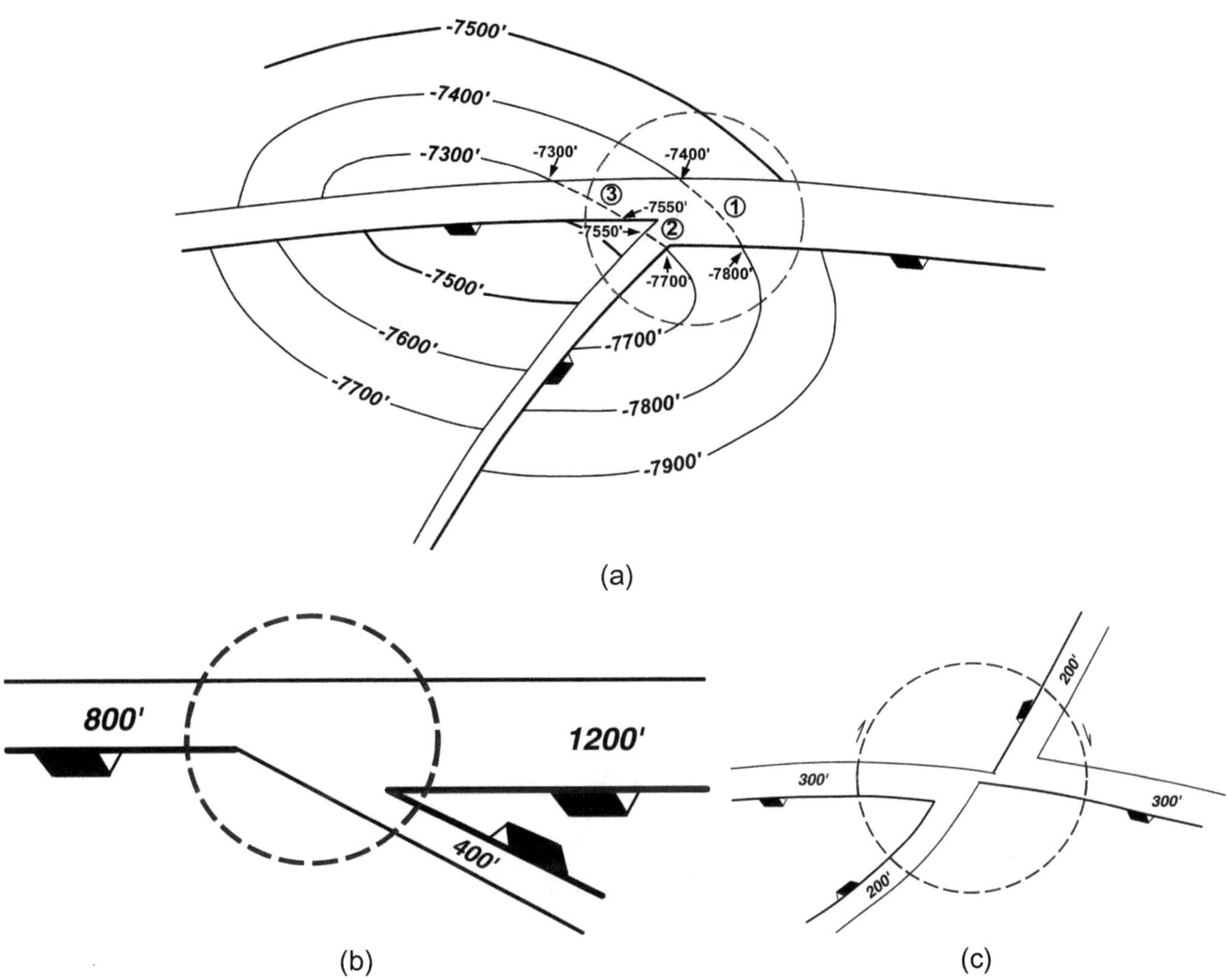

Figure 8-38 Conservation of vertical separation at fault intersections. (Published by permission ofTearpock et al. 1994.)

depict the addition of fault vertical separation rather than their subtraction, such as in the case of this compensating fault pattern, would not occur. (3) The contouring across Fault A is incorrect at the intersection. The contours on opposite sides of Fault A east of the intersection should have an elevation difference of 350 ft rather than the 450 ft contoured on the map, since this is a compensating fault pattern.

Using tracing paper, recontour this map (Fig. 8-39a) to reflect an acceptable structural interpretation around the fault intersection. Finally, examine the structure map further for mapping busts in contouring across Fault A-2 and the intersection of Fault A-2 with A-1. See if you can identify and correct these problems.

Figure 8-39b shows another example of a completed structure map, generated from seismic data, that fails to honor the conservation of vertical separation. Unlike the cited error in Figure 8-39a, which is minor as errors go, the error in Figure 8-39b is quite significant. A test of the additive property of faults around the intersection of Faults A, A-1, and A-2 indicates an error on the order of 1800 ft. Such a significant error is not due to incorrect contouring, but is most likely the result of a seismic correlation mis-tie across one or more of the intersecting faults. Look at the structural dip direction for the westernmost fault block versus the fault block upthrown to Faults A and A-1. What are your thoughts?

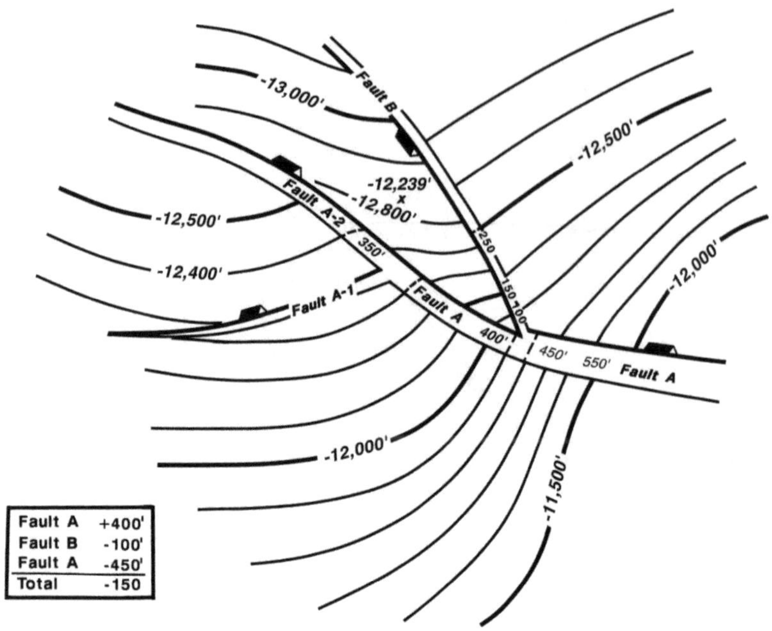

(a)

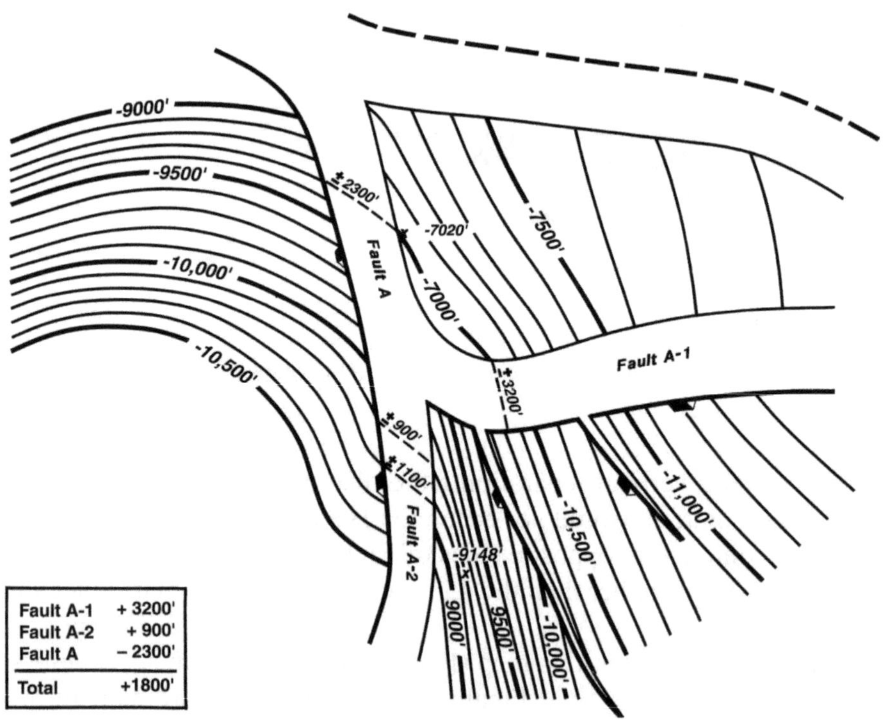

(b)

Figure 8-39 (a) Vertical separation is not conserved around the intersection of Faults A and B. This mapping bust appears to be the result of incorrect contouring and the failure to integrate the fault and structure maps. (b) The failure to conserve the vertical separation around the intersection of Faults A-1 and A-2 in this example is significant. This mapping bust is probably due to a seismic correlation mis-tie across one or both of the intersecting faults. (Modified from Tearpock and Harris 1987. Published by permission of Tenneco Oil Company.)

INTEGRATION OF SEISMIC AND WELL DATA FOR STRUCTURE MAPPING

In Chapters 5 and 7 we cover the basic techniques for integration of fault data determined from seismic sections. In this section, we discuss the additional data provided by seismic sections to assist in the construction of subsurface structure maps. The lateral resolution of seismic data contributes information away from well control and helps define a larger, more continuous surface than is mappable from well data alone.

This discussion assumes that several conditions have been met before the seismic data are used to make a structure map. *The first condition* is that the seismic data have been correctly processed. *The second condition* is that the seismic lines have been approximately tied into the event that is correlative with the actual horizon being mapped. In some areas this may be simply a matter of using the correct velocity function to locate the proper time event that corresponds to the horizon. In more complex areas, the tie to seismic information may require the use of a synthetic seismogram or vertical seismic profile to locate the proper event to map.

The third condition is that there should be no significant velocity changes over the area being mapped. If a lateral velocity gradient appears to be present, a gradient map may be required to deal with this problem. If the seismically derived horizons do not agree with the horizons in the wells across the area being mapped, there are several potential reasons for this disagreement. First, a velocity gradient may be present. These are common in some areas and rare in others. Second, the well log or seismic correlations may be wrong. Look hard at both sets of data before modifying or discarding one of them. Third, the wrong event may be interpreted as corresponding to the horizon being mapped. This can cause problems, especially in areas of extreme thinning and thickening of the stratigraphic intervals.

If the velocity field is uniform over the area, or if appropriate adjustments have been made for a nonuniform velocity field, then seismic data can be extremely valuable in subsurface structure mapping. Seismic data provides (1) estimates of structural elevations for the horizon being mapped; (2) continuous dip information along the seismic line; and (3) accurate structural data points at the intersection of faults with the horizon along the seismic line. The use of a 3D data set allows the geoscientist to actually interpret and map a 3D data volume.

Figure 8-40 shows an example of seismically derived data that can be posted to aid in the construction of a subsurface structure map. The first and most obvious data points are the horizon depths posted at the shotpoint locations. These values are seismic time picks converted to depth with the appropriate velocity function. Second, fault data are posted along the intersection of the faults with the horizon. For example, line D illustrates how to post the upthrown and downthrown elevations for the mapping horizon at the intersection of the horizon with the fault along the line of section. For Fault B, the upthrown elevation of the intersection of the horizon with the fault is −9460 ft, and the downthrown elevation at the fault/horizon intersection is −9815 ft. Finally, the vertical separation data estimated across the fault are posted and used to aid in the mapping (e.g., 450 ft for Fault B on Line D).

When using seismic data, be aware of the difference between the apparent throw values and the vertical separation, and label the latter on the map. The estimate for apparent throw is the difference in elevation between the upthrown and downthrown intersections of a horizon with the fault along the line of section. Notice the changing values for apparent throw along fault A; these changing values are a function of the orientation of the line of section crossing the fault (lines B, D, and E). However, the values for the vertical separation along this portion of the fault do not change. For example, based on the upthrown and downthrown values measured along lines E, B, and D, the apparent throw of Fault A is 520 ft at line E, 425 ft at line B, and 375 ft across line D. Considering these apparent throw values, the fault could be incorrectly interpreted as decreasing in size or dying to the east. But the estimate of the vertical separation at each line shows that vertical separation for Fault A is a constant 500 ft in the area east of the intersection with Fault B. Contrast this with Fault B, which shows a decreasing vertical separation as the fault dies laterally to the east; however, the

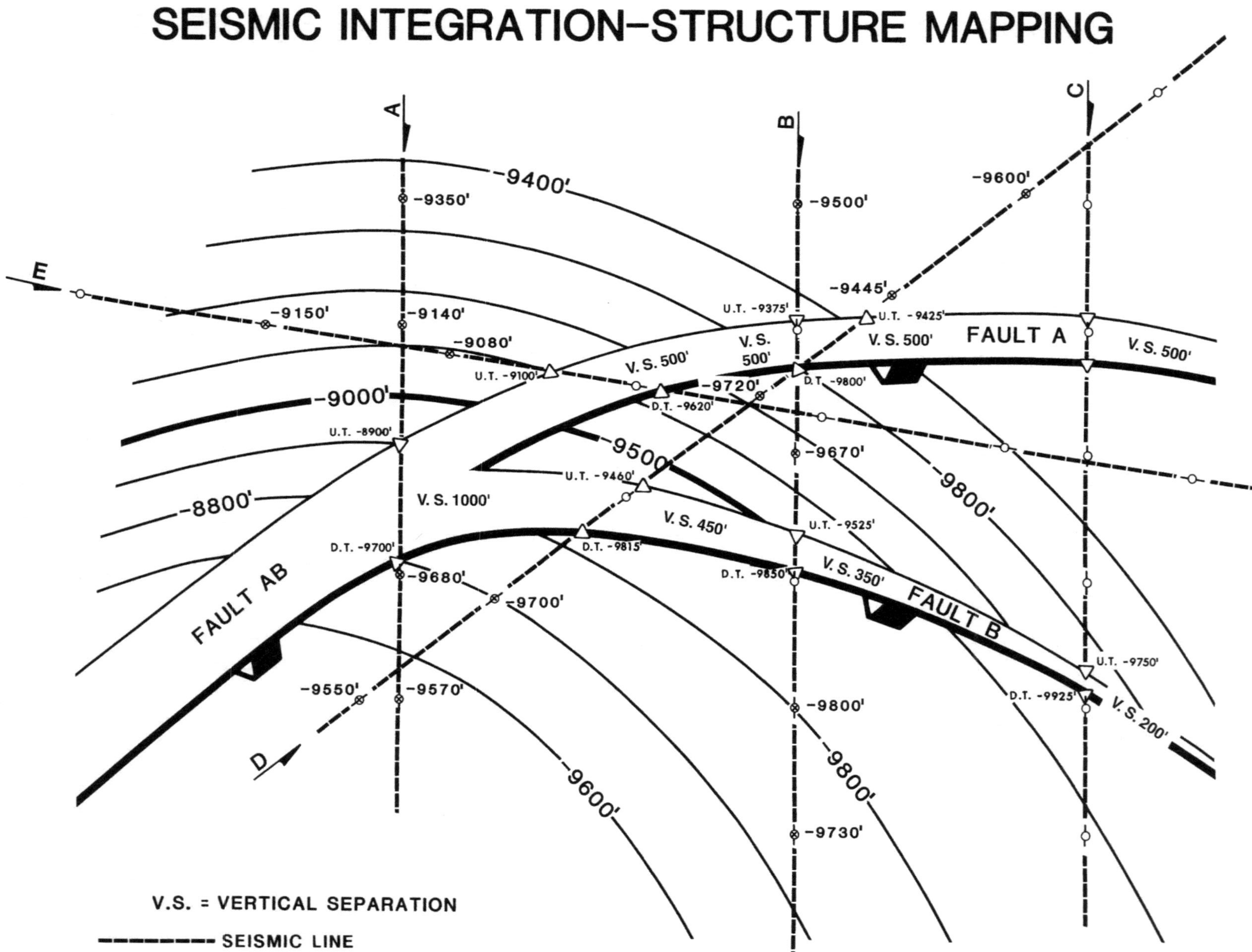

Figure 8-40 Completed structure map illustrating seismically derived subsurface data used to prepare the map. Depths are in feet. (Prepared by C. Harmon. Modified from Tearpock and Harris 1987. Published by permission of Tenneco Oil Company.)

apparent throw, as measured from line D, is 355 ft, and it is 325 ft from line B, suggesting a fault of almost constant apparent throw. It is most common to use the vertical separations for mapping displacement across faults, whether the data are from well logs or seismic sections.

The contouring techniques presented at the beginning of this chapter apply to the use of seismically derived data as well as log data. Seismic data offer some additional advantages that are lacking in well data, such as the ability to post upthrown and downthrown mapping control points at fault/horizon intersections. These additional data give two more control points to help in contouring the final structure map.

In summary, seismic data allow us to extend the interpretation beyond the limits of well control. It helps in the construction of a continuous subsurface picture of the horizon being mapped and provides additional, more complete information about the characteristics of the faults that intersect these horizons.

OTHER MAPPING TECHNIQUES

Mapping Unconformities

An **unconformity** is a surface of erosion and/or nondeposition that separates strata of different ages. The development of an unconformity involves several stages of activity. The first is the deposition of the initial sediments; second, the area is subject to uplift and subaerial erosion and/or nondeposition, developing an unconformable surface; and third, the deposition of younger sediments above the unconformity. Unconformities are present in all geological settings, but are especially common on steeply dipping structures.

Unconformities alone, or in combination with faults or stratigraphic anomalies, such as sand pinchouts, serve as locations of excellent hydrocarbon traps (Fig. 8-41). It is therefore important to recognize and map unconformities in the subsurface. In Chapter 4, we present the guidelines to recognizing unconformities by log correlation and dipmeters, and to distinguishing faults from unconformities. In this section we discuss general techniques for mapping an unconformity in the subsurface and integrating that unconformity map with a structure map.

An unconformity map is a type of surface contour map. It differs from a typical structure map in that the contours are on an unconformable surface rather than some stratigraphic marker or horizon. Unconformity maps are not normally made to stand alone; instead, they are integrated with a structural horizon map to delineate the intersection of the unconformity with the horizon being mapped. This intersection marks the termination of the horizon at the unconformity. If a hydrocarbon trap is present beneath the unconformity, the termination of the reservoir against the unconformity defines a boundary of the reservoir (Fig. 8-41).

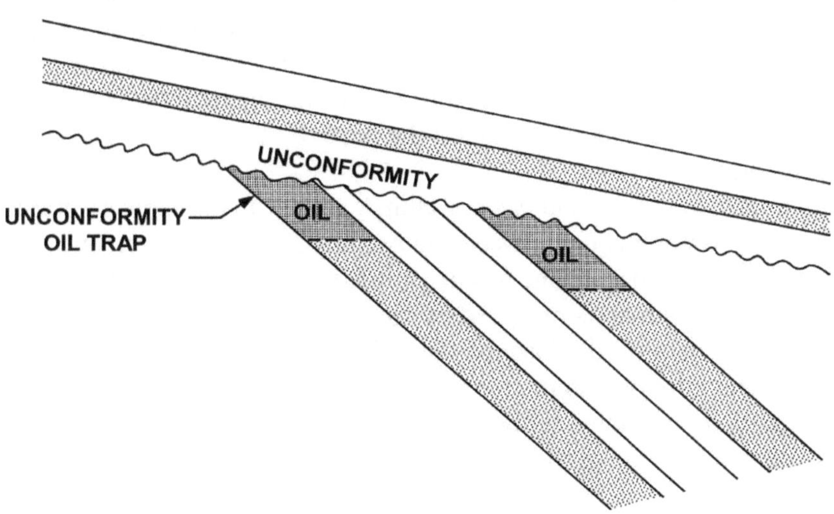

Figure 8-41 Typical hydrocarbon trap beneath an angular unconformity.

Although there are various kinds of unconformities, for petroleum exploration and exploitation we are primarily interested in two types: (1) angular unconformity and (2) disconformity. An *angular unconformity* is one in which the rocks above and below the unconformity are not parallel with one another, with the rocks below the unconformable surface typically dipping at a steeper angle than those above. Figure 8-42a is a north-south electric log cross section showing a good example of an angular unconformity determined by log correlation (Lock and Voorhies 1988). Good, easily traceable resistivity markers in the section below the unconformity are seen to be progressively cut out at the unconformity surface. Notice, in the interval above the unconformity, that by correlating down the logs, the section is seen to be interrupted in each well at the same stratigraphic marker. From the Tribal Oil No. 1 to the Sinclair No. 5, as much as 150 ft of section is missing as a result of the unconformity. Angular unconformities and disconformities can often be mapped accurately on seismic data (Fig. 8-42b)

A *disconformity* is an unconformity in which the rocks above and below the unconformity are approximately parallel. In the subsurface, disconformities are particularly difficult to recognize. In some cases, paleontological data may be helpful, in the absence of evidence based on well log correlation or seismic data.

Unconformities occur on a variety of scales ranging from very local in extent, such as those associated with meandering stream channels across a flood plain, to those covering many square miles. A major unconformity in the area around the western flank of the Sabine Uplift (located along the Louisiana-Texas border in the United States) serves as the trap for the giant (5-billion-barrel) East Texas Oil Field (Fig. 8-43). Actually, this giant field results from the intersection of two unconformities creating the up-dip oil trap in the Woodbine Sandstones. Unconformities are very common around steeply dipping structures, especially salt domes. An excellent example of a major angular unconformity associated with a salt dome is shown in Figure 4-45.

Mapping Techniques. Figure 8-44a is a map of a deformed unconformity surface based on log correlations and seismic data. Figure 8-44b shows the integration of the top of Stratigraphic Unit A, which is oil-bearing, with the unconformity, thus delineating the western limit of the horizon against the unconformity. The termination of the mapped horizon against the unconformity is determined by overlaying the structure map on the top of Unit A onto the unconformity map. The intersection of each structure contour for the top of the unit with each unconformity contour at the same elevation delineates the position of the unconformity on the top of Unit A (Fig. 8-44b). This integration is basically the same as that for fault and structure maps.

Depending on the scale of the map, thickness of the stratigraphic unit, dip of the beds, and other considerations, a map on the top of a particular stratigraphic unit may not be sufficient to accurately show the effect of the unconformity on that unit. At times, a structure map on the base of the stratigraphic unit is required in addition to a map showing the unconformity wedge between the top and base of the unit. Figure 8-44c is a structure map on the base of Unit A, and it delineates the position of the intersection of the unconformity with the mapped horizon. The procedure for determining the position of the unconformity on the base of the Unit A is the same as that just shown for the top of the unit.

Notice that the unconformity traces on the top and base of Unit A are not in the same position, indicating that there is an unconformity wedge zone between the two traces. This wedge zone can be mapped in combination with the structure map on the top of Unit A in order to depict the top of the reservoir, as shown in Figure 8-44d. This map is prepared by first overlaying the structure map on the top of the unit onto the structure map on the base of the unit and transferring the trace of the unconformity on the base to the structure map on the top of the unit. There are now two unconformity traces on the top of structure map and they delineate the wedge zone. Between these two traces, the reservoir varies from full thickness to zero thickness. Since the structure in the wedge

(a)

(b)

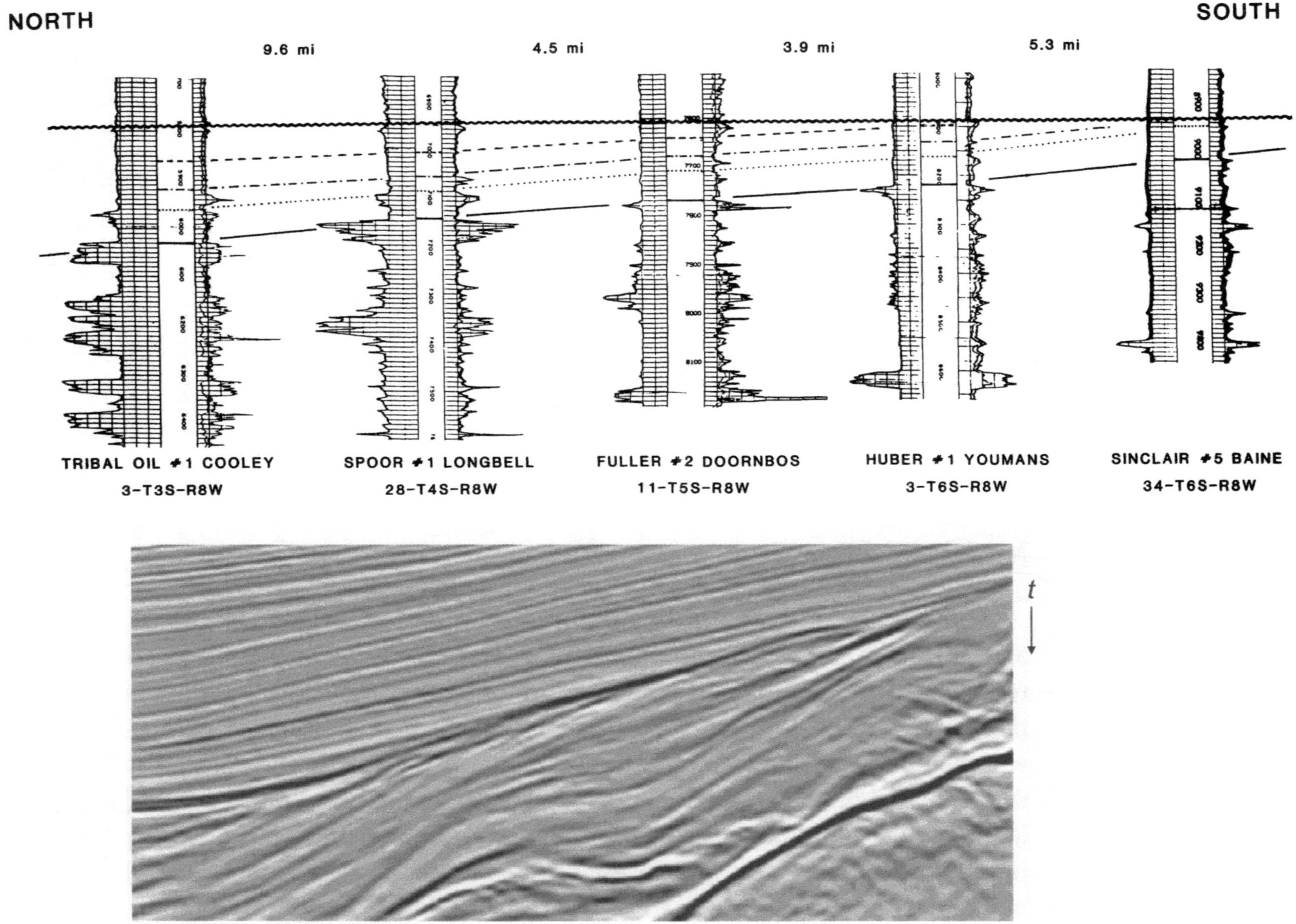

Figure 8-42 (a) Example of an angular unconformity recognized by electric log correlation. (From Lock and Voorhies 1988. Published by permission of the Gulf Coast Association of Geological Societies.) (b) Example of an angular unconformity recognized on seismic data (From Herron 2011. Published with permission from Society of Economic Geophysicists, seismic courtesy of PGS.)

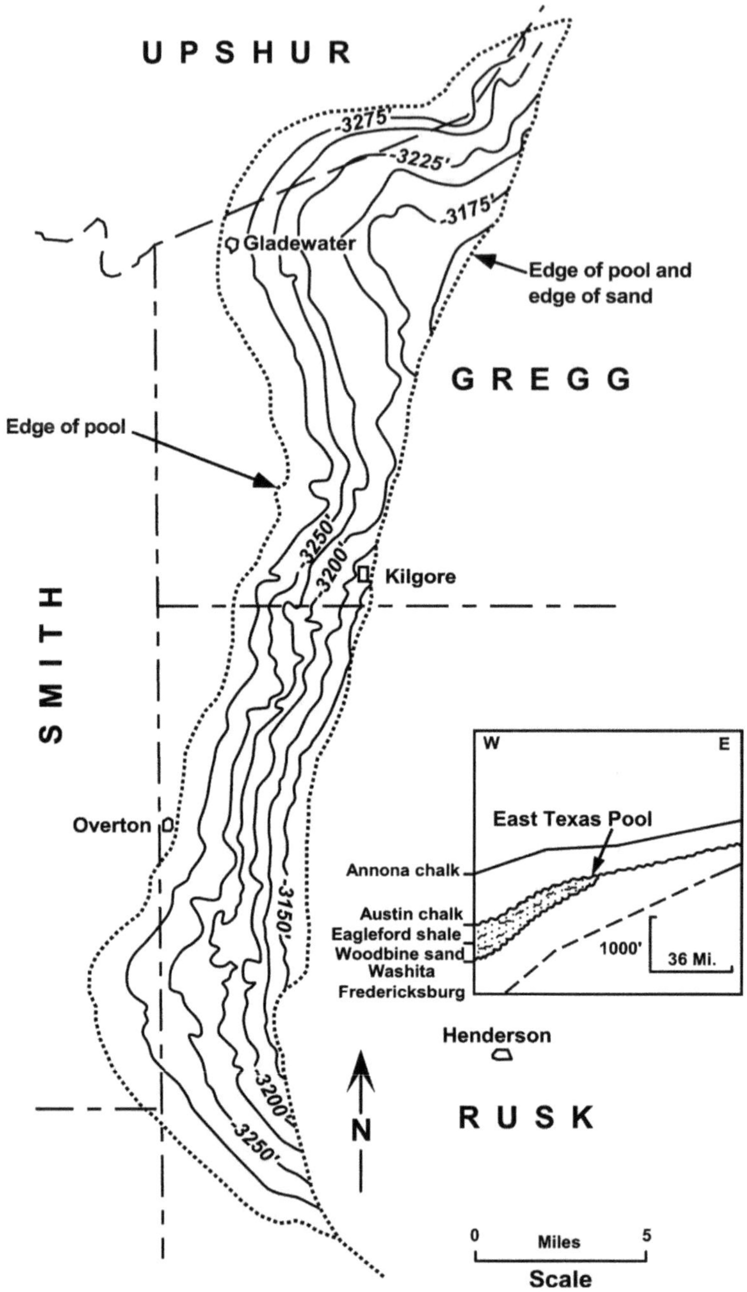

Figure 8-43 Structure map on top of the Woodbine Sand in the East Texas pool. As shown in the cross-section insert, the intersection of two unconformity surfaces marks the eastern boundary of this unconformity trap. (From Geology of Petroleum, first ed. By A. I. Levorsen, Copyright 1954 by W. H. Freeman and Company. Reprinted by permission.)

zone is actually that of the unconformity itself, next overlay the top of the structure map onto the unconformity map and trace the structure contours in the wedge zone from the unconformity map. For example, the −6200-ft contour on the top of Unit A, at the trace of the unconformity, continues as the −6200-ft contour on the unconformity in the wedge zone until it intersects the western limit of the wedge zone. Notice that the contour changes strike direction in the wedge. This procedure is followed for all the contours on the map. The result is a structure map on the top of the reservoir. Observe that the unconformity wedge zone for this interval comprises a significant area of the oil reservoir; therefore, in such situations the wedge zone must be contoured.

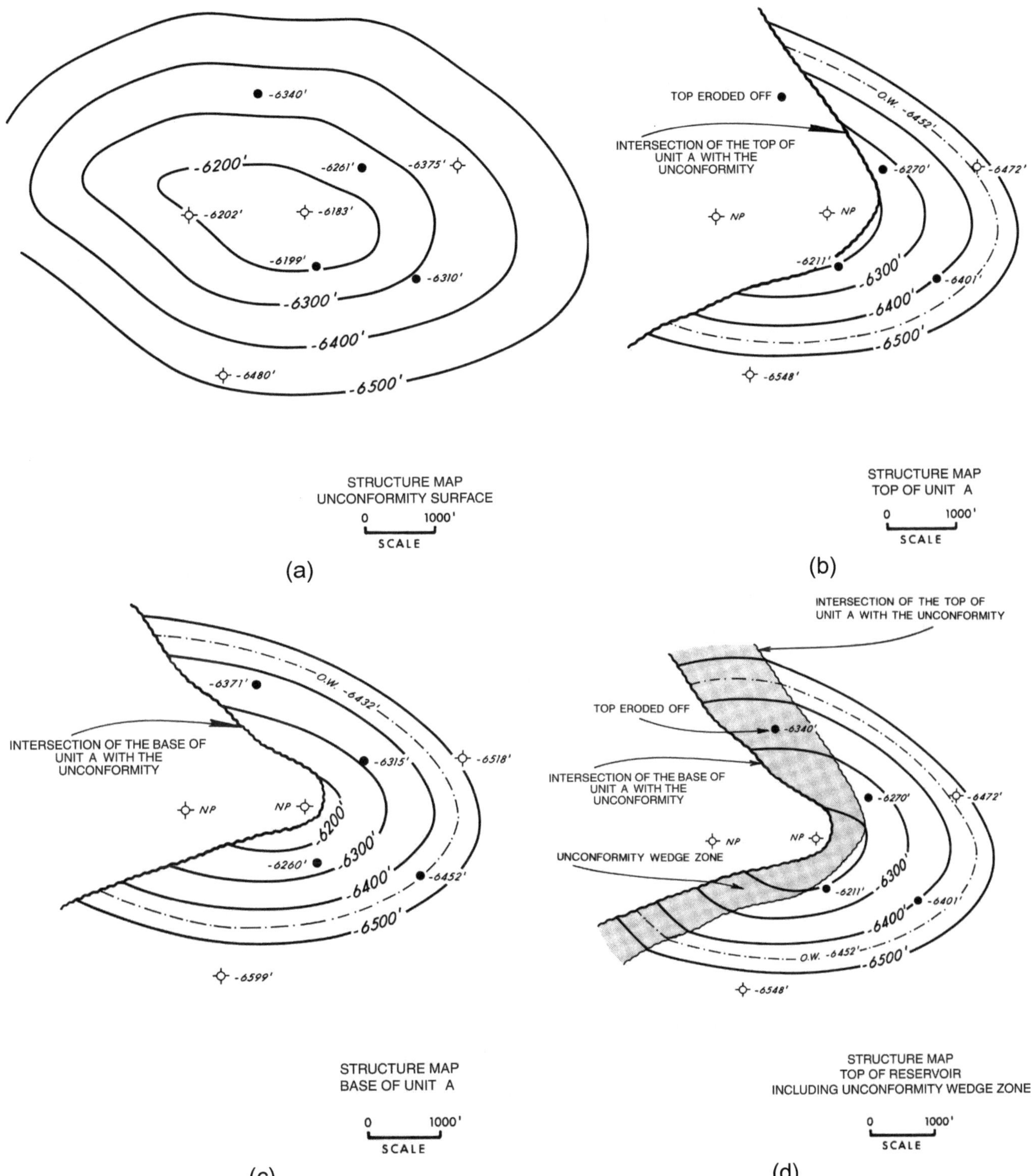

Figure 8-44 Construction of a map of the top of a reservoir trapped beneath an unconformity. (a) Contour map on an unconformity surface based on both electric log and seismic correlation data. (b) The termination of the top of Unit A is defined by the intersection of the unconformity with the top of Unit A. (c) The intersection of the base of Unit A with the unconformity surface defines the western limit of this oil reservoir Unit A. (d) Map of the top of the reservoir. An unconformity wedge zone is defined as the area between the unconformity traces on the top and base of the sand unit. The map of the wedge zone is a copy of the map of the unconformity in that area. The maps of the wedge zone and the top of Unit A are combined to form the top of reservoir map.

Mapping across Vertical Faults

High angle and vertical faults, particularly those with small displacements relative to the overall structure, tend to offset the structure without much if any change in the structural attitude across the fault. *For a vertical fault, the vertical separation equals throw.* Figure 8-45 shows that the throw is 400 ft and the vertical separation is also 400 ft. They are the same because there is no projection of bed dip across a fault gap or overlap. Therefore, we need only concern ourselves with one value for fault displacement when contouring across a vertical fault. For many cases, there is continuity of structural attitude, so the strike direction of the contours is maintained when mapping across the fault. In very complexly faulted areas, however, the strike direction of the contours may be affected from fault block to fault block, particularly if any rotation of the fault block or horizontal displacement occurs.

Figure 8-46 depicts a structure cut by two vertical faults. The contours from one block are projected across the fault into the next fault block. The contours are projected across the fault as if the fault were not there, representing a continuity of structure across the fault, although each contour is adjusted for the vertical separation. This technique assumes no horizontal displacement. Since the fault is vertical, there is no fault gap or overlap, so the fault is depicted as a single line on the finished structure map.

One of the biggest problems with a vertical fault, besides recognizing the presence of the fault, is estimating the vertical separation of the fault for contouring. The fault cannot be intersected by a vertical well. An estimate of the vertical separation must come from intersections of the fault with deviated wells, from seismic data, or from sufficient well control within each fault block to contour both the upthrown and downthrown blocks separately and use the offset contours to determine the vertical separation for the fault.

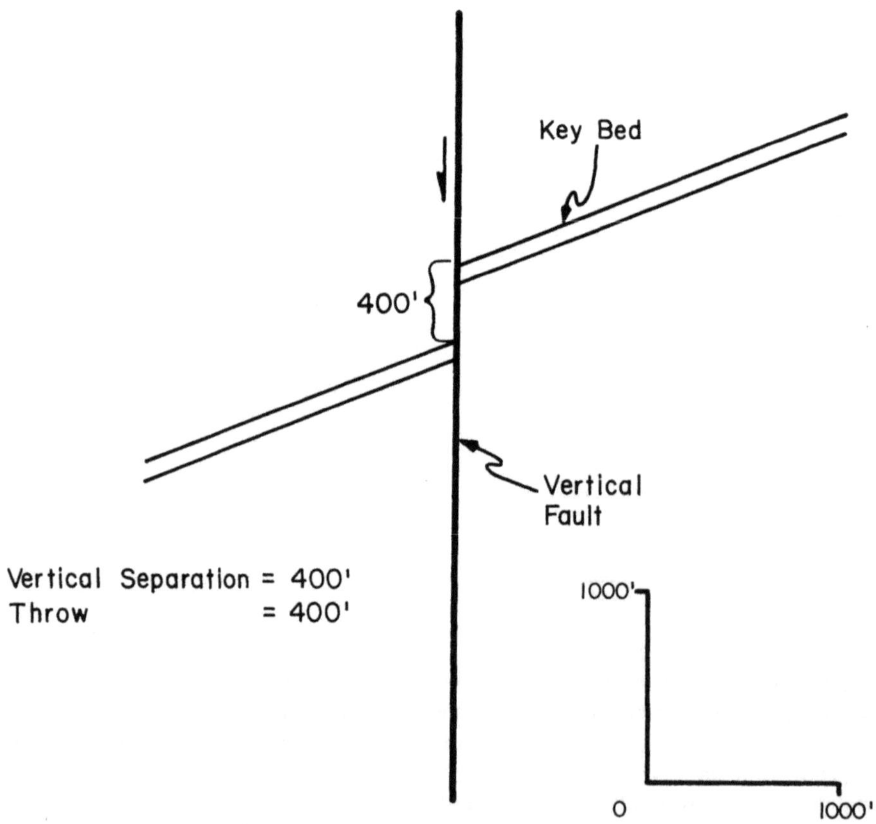

Figure 8-45 Cross section shows that the values for both vertical separation and throw are the same across a vertical fault.

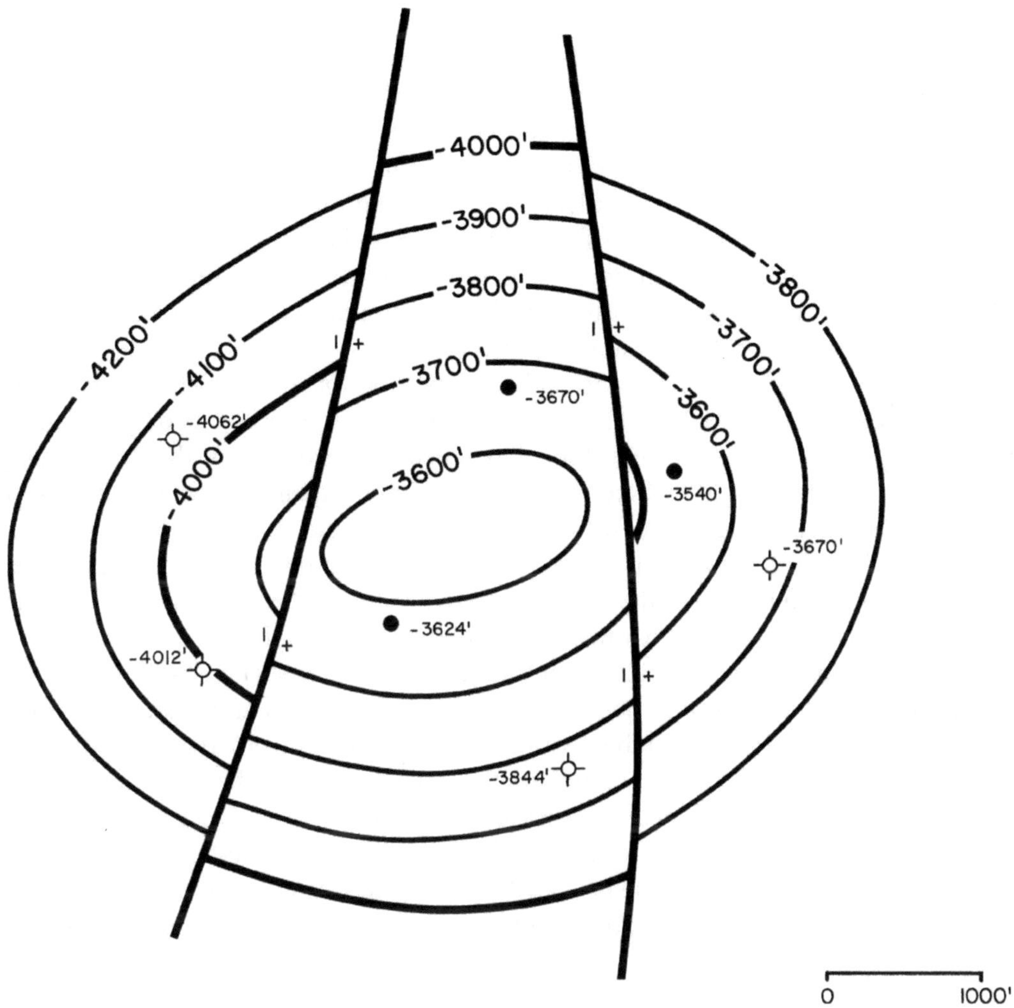

Figure 8-46 Completed structure contour map on the 3600-ft Horizon cut by two vertical faults. The structure contours are drawn from one fault block to the next assuming structural continuity across the faults.

Top of Structure Versus Top of Porosity

Subsurface structure maps are made on specific stratigraphic units to depict the 3D geometric shape of the geological structures being mapped. Once the geometry of the structure has been determined, the primary effort is focused on the mapping of all hydrocarbon-bearing stratigraphic units.

At times, for various reasons, a structure map is prepared on a good seismic event or resistivity marker that is correlatable in all or most of the wells in a region or field, instead of mapping an actual hydrocarbon-bearing unit. In some cases this may be done because the hydrocarbon-bearing unit is discontinuous or has great vertical variation not reflecting the true shape of the structure. Therefore, it is necessary to prepare a structure contour map first on a stratigraphically equivalent marker in order to construct a map that conforms to the true structure of the field or region. This marker may be a few feet or several hundred feet above the actual hydrocarbon-bearing unit(s). Once the structural framework is prepared by contouring the data from the stratigraphically equivalent marker, a second map, called a **top-of-porosity map**, is required on the top of any productive unit for the purpose of delineating the actual configuration and limits of the reservoir (Fig. 8-47).

It is common for the upper portion of a particular stratigraphic unit to be composed of nonreservoir-quality rock. This nonreservoir-quality rock is often referred to as a *tight zone* or *tight streak*. Although the top of the unit may represent an actual stratigraphically equivalent horizon, it is underlain everywhere by impermeable rock. Therefore, the structure maps prepared to interpret the true structure commonly cannot be used to evaluate the reservoir itself.

Once a structure map is completed, the next step is to prepare a top-of-porosity map for accurate delineation of the reservoir, and for later use in the construction of net hydrocarbon isochore maps. Two parameters are considered in evaluating the importance of separately mapping the top of porosity: (1) the thickness of the tight zone, and (2) the relief of the structure. A thick tight zone has a greater effect than one that is thin. Low-relief structures introduce greater error in delineating the limits of a reservoir than steeply dipping structures, particularly if the low-relief structure contains a bottom water reservoir.

Figure 8-48a shows a structure map and cross section for the 6000-ft reservoir. This unit consists of nonreservoir-quality rock in the upper 75 ft. The same reservoir is mapped on the top of the porous rock or porosity top in Figure 8-48b. Notice in cross section A-A′ that by mapping on the top of the unit, in which the upper 75 ft consists of nonreservoir-quality rock, the limit of the reservoir (gas/water contact) is extended beyond the true gas/water contact as mapped on the top of porosity (also see Figure 8-48c). Even though no net pay is assigned to the tight zone, the productive area of the reservoir mapped on the top of the nonproductive portion of the unit is larger. In turn, the volume of the reservoir is also larger than that mapped on the porosity top. In this case, the volume, based on net gas isochore maps, is larger by 32%. This added reservoir area (Fig. 8-48c) created by mapping on the top of the stratigraphic unit does not contain hydrocarbons and therefore is not productive; consequently, the volume of recoverable hydrocarbons based on this map is overestimated.

The decision to prepare a separate map on the top of porosity, where the upper portion of a unit is not productive, needs to be made on a reservoir-by-reservoir basis. Depending on the geometry of the reservoir and thickness of the tight zone, the difference in volume between a map on the top of the unit and a map on the top of porosity may be too insignificant to warrant additional mapping. In Chapter 14, we discuss in greater detail the impact of nonproductive zones in isochore mapping.

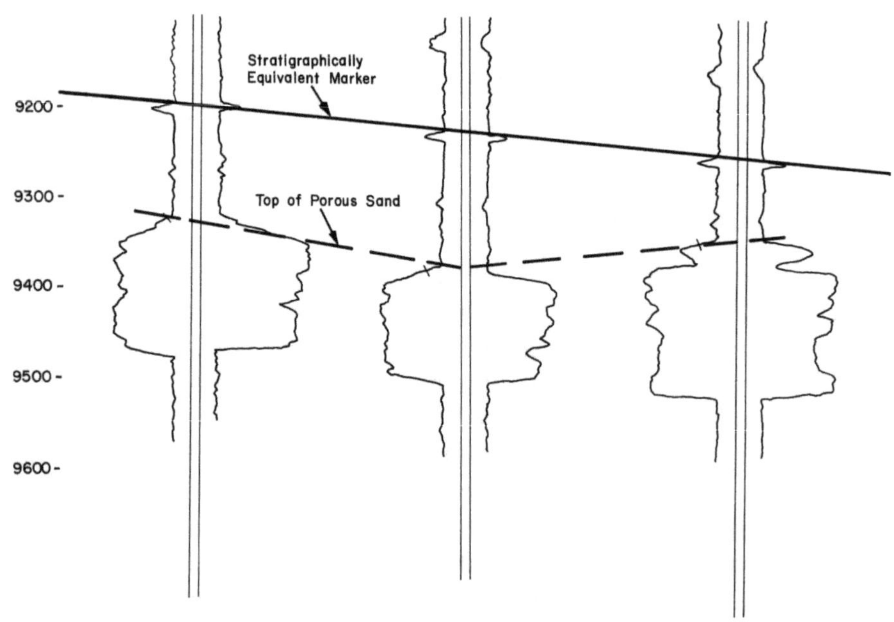

Figure 8-47 Electric logs from three wells. The upper stratigraphic marker conforms to true structure and is used to construct a map representing the true structural framework of the area. The top of the thick productive sand member does not conform to structure, but it represents a porosity top. It must be mapped separately to delineate the actual reservoir configuration.

Contour Compatibility—Closely Spaced Horizons

Structure maps on vertically close horizons should exhibit, in most cases, compatible contour configurations. In order to maintain compatibility during manual structure contour mapping of two or more vertically consecutive horizons, a completed map should be used as a guide for the next horizon to be mapped. The simplest way to do this is to lay the completed map under the base map for the next horizon to be mapped, and use it as a guide for contouring the new horizon map. This technique usually ensures that consecutively mapped horizons possess structural compatibility or contour compatibility.

Figure 8-49 illustrates the use of this technique and also shows the problems that can arise if vertically consecutive horizon maps are not overlaid during structure contouring. In Figure 8-49a, the completed structure map for Horizon A was used to guide the contouring of Horizon B. Notice the compatibility in the structural configuration of the two horizons. Figure 8-49b illustrates the kind of contouring errors that can arise when a completed map is not used to guide the structure contouring of another, vertically close horizon. Notice that in the area of good well control, the contours are similar to those in Figure 8-49a. The major problem occurs in the areas of limited control on Horizon B. Observe in Figure 8-49b that several contours on Horizon A actually cross contours of the same value on Horizon B (e.g., the −5950-ft contour), which incorrectly indicates a stratigraphic thickness

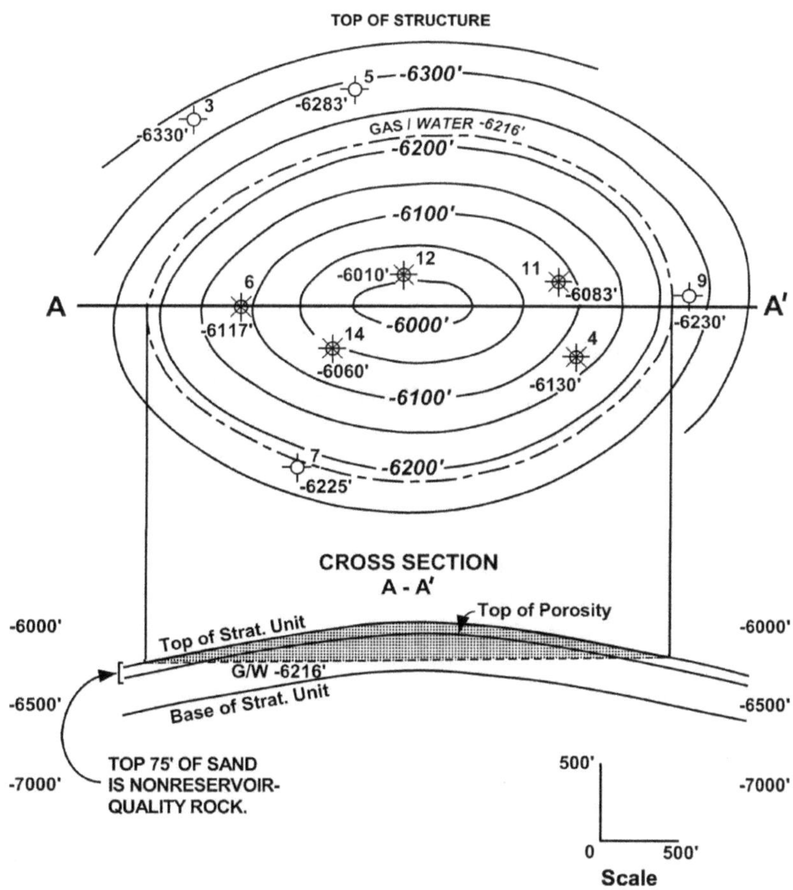

(a)

Figure 8-48 (a) Structure map on top of the 6000-ft unit, with a gas/water contact at a depth of −6216 ft, and cross section A-A′ illustrating (1) the top of the unit, (2) top of porosity, and (3) base of unit. (b) Structure map on the top of porosity for the 6000-ft unit, with the gas/water contact at a depth of −6216 ft, and cross section A-A′. (c) Mapping on top of structure versus top of porosity results in a 32% increase in volume.

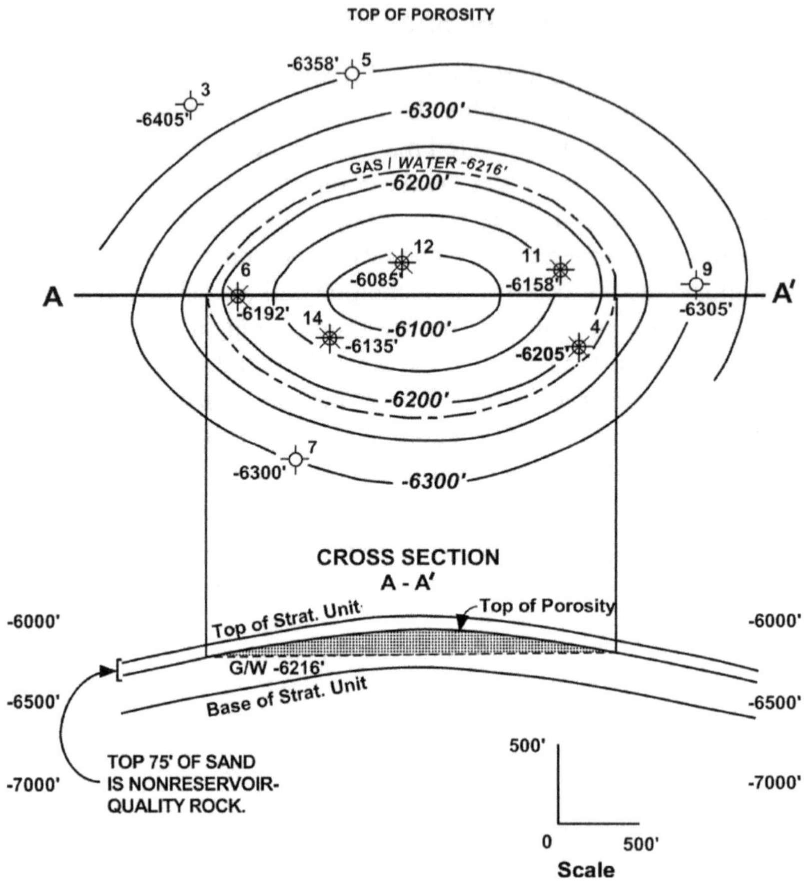

(b)

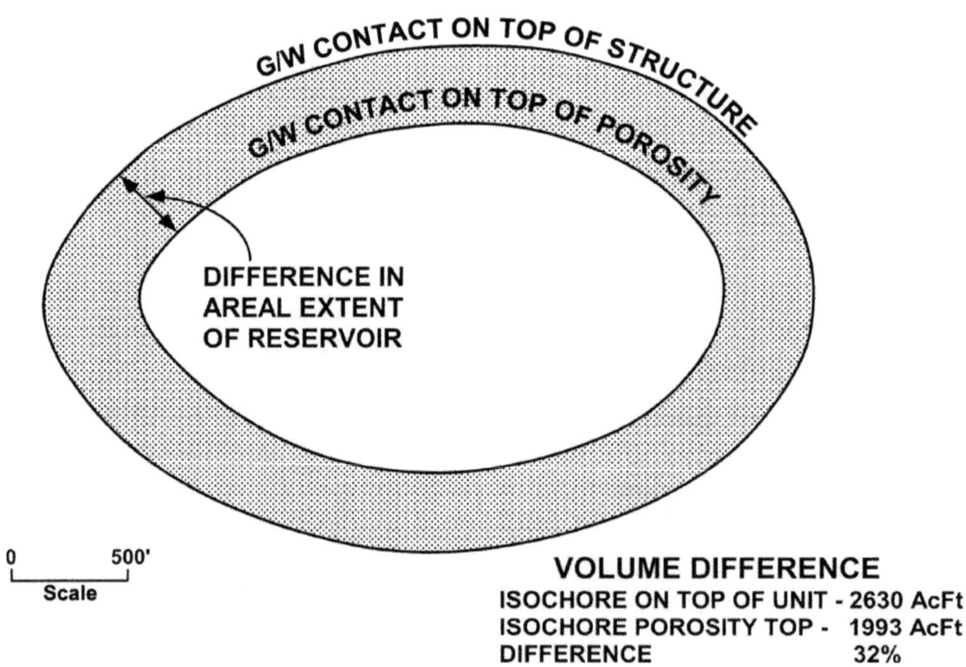

(c)

Figure 8-48 (*continued*)

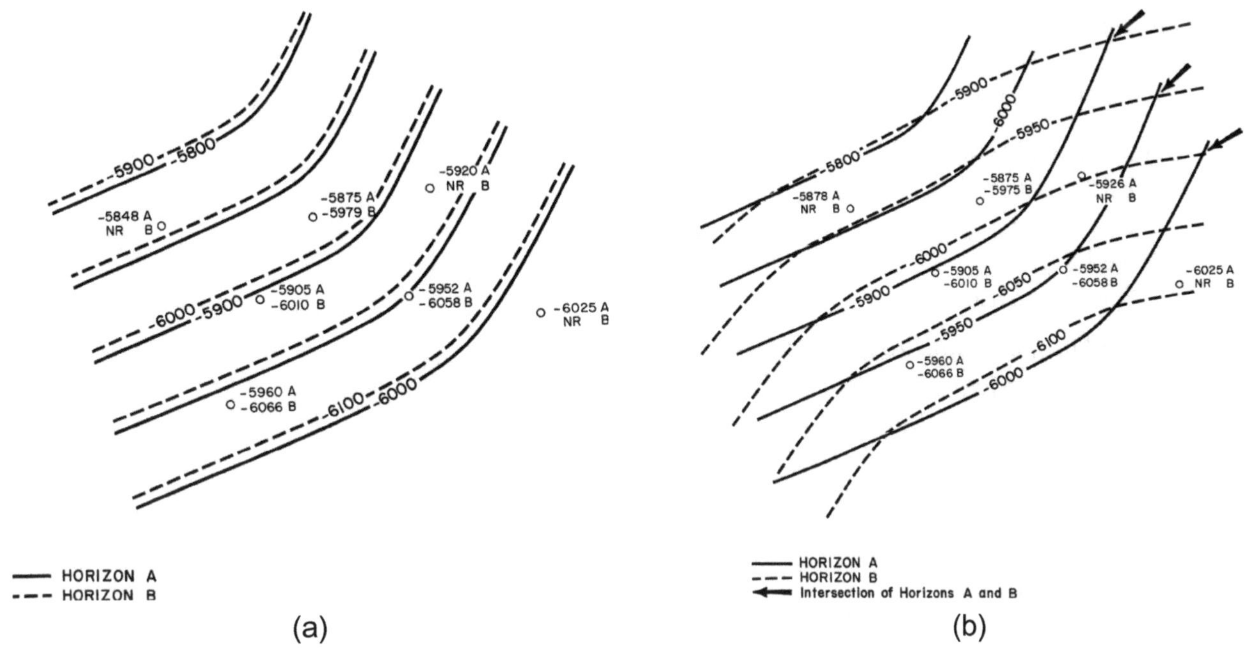

Figure 8-49 (a) Structure contour maps on Horizons A and B showing structural compatibility. The configuration of Horizon A was used to guide the contouring of Horizon B, especially in areas of limited or no well control. (b) An alternate interpretation of the structure contour map on Horizon B drawn independently and not taking advantage of the configuration of Horizon A, which has additional subsurface well control. A geologically impossible situation is shown; a contour of a given value (e.g., −5950) on Horizon A crosses a contour of the same value on Horizon B.

between Horizons A and B of zero feet, as well as a crossing of the horizons. We know, however, from the available well data, that there are about 100 ft gross between these two horizons.

One contouring rule states that contours of the same value cannot cross on separately mapped horizons. If contours of the same value, mapped on different horizons, intersect or cross, then the interval between the two horizons would have zero or negative thickness, which is a physical impossibility. This is a serious mapping error that occurs because completed structure maps are not used to guide the contouring of other, vertically close horizons.

A major mapping bust is shown in Figure 6-11a and b in Chapter 6. In the situation shown, the two horizons are not vertically close (separated by 800 ft of section in the off-structure position) and yet the structure contour maps for the P5 and P7 Horizons cross in the up-structure position. To avoid such errors, we recommend that completed structure maps always be used to guide the contouring of other horizons, whether they are above or below the completed map.

Many software packages are designed to mimic this approach through conformal gridding. After a structure map has been made on one horizon, the software uses the shape of that horizon to guide the contouring of a nearby horizon while still honoring all of the control points of the second horizon (Thompson et al. 2017). This concept is also discussed in Chapter 2 under Conformable Geology and Multisurface Stacking.

APPLICATION OF CONTOUR COMPATIBILITY ACROSS FAULTS

The method of contouring across faults using the vertical separation requires the structure (individual horizons being mapped) to have the same or similar attitude on opposite sides of the fault, as shown in Figures 6-5, 6-8, 8-10, 8-22, and 8-23. This means that the use of the technique requires *structural compatibility*, or *contour compatibility*, across the fault. For many tectonic settings, the compatibility of structural attitude across faults is more the rule than the exception; for others, it is not. Use of vertical separation for contouring across faults, shown earlier in this chapter, does have application on a worldwide basis in nearly every tectonic habitat, including some of the areas listed here.

Thin-Skinned Extensional Areas
Gulf of Mexico—onshore and offshore
West Coast of Africa— Gabon and Niger Delta areas

Crystalline Basement-Involved Extensional Areas
East African Rift area
North Sea
Red Sea
South America

Thin-Skinned Compressional Areas
Southern Mexican Ridges
South American Andes
Western United States and Canada

Diapiric Salt Tectonic Areas
Gulf of Mexico
North Sea
Northern German Salt Basin

Strike-Slip Fault Tectonic Areas
Western United States

The use of seismic data is very helpful in determining whether there is structural (contour) compatibility across faults in any area being worked (Figs. 7-37 and 7-38). To be a good interpreter and mapper, you must have a good understanding of the tectonic setting being worked and detailed knowledge of the individual structures being mapped. It is very important to know where to apply a particular technique, such as mapping across faults, and where the structure being mapped is an exception to the rule, thus requiring a different technique.

Exceptions to Contour Compatibility across Faults

Four key geological situations usually result in differing structural attitudes and display structural incompatibility across a fault:

1. Normal *listric* growth faults exhibiting large vertical separation;
2. Thrust or large reverse faults and large strike-slip faults;
3. Intermediate to late-stage strike-slip faults; and
4. Ramps relating to rapidly dying thrust faults.

A listric normal growth fault, depending on its vertical separation and the local amount of contemporaneous deposition, may or may not possess structural (contour) compatibility across the fault. Usually, the larger the fault, the less chance of compatibility. For example, Figure 7-37 shows a growth fault (Fault A) with compatible beds across the fault. However, Figure 8-50 illustrates a large growth fault (Fault A) that does not have compatibility from the upthrown to downthrown fault blocks of the main fault. Figure 8-51 is a cross section from the Gulf of Mexico that shows a fault with both characteristics. The structural attitude is compatible across the fault in the shallow section, gradually becoming incompatible with depth. There are several reasons for this change in structural attitude, including the activity of the large Miocene growth fault shown to the far left in the cross section. The seismic section in Figure 11-12 illustrates a faulted rollover anticline. Structural compatibility is lacking across the large fault on the left, which is the fault that formed the anticlinal structure on the right. On the other hand, all faults within the anticlinal structure do exhibit structural compatibility.

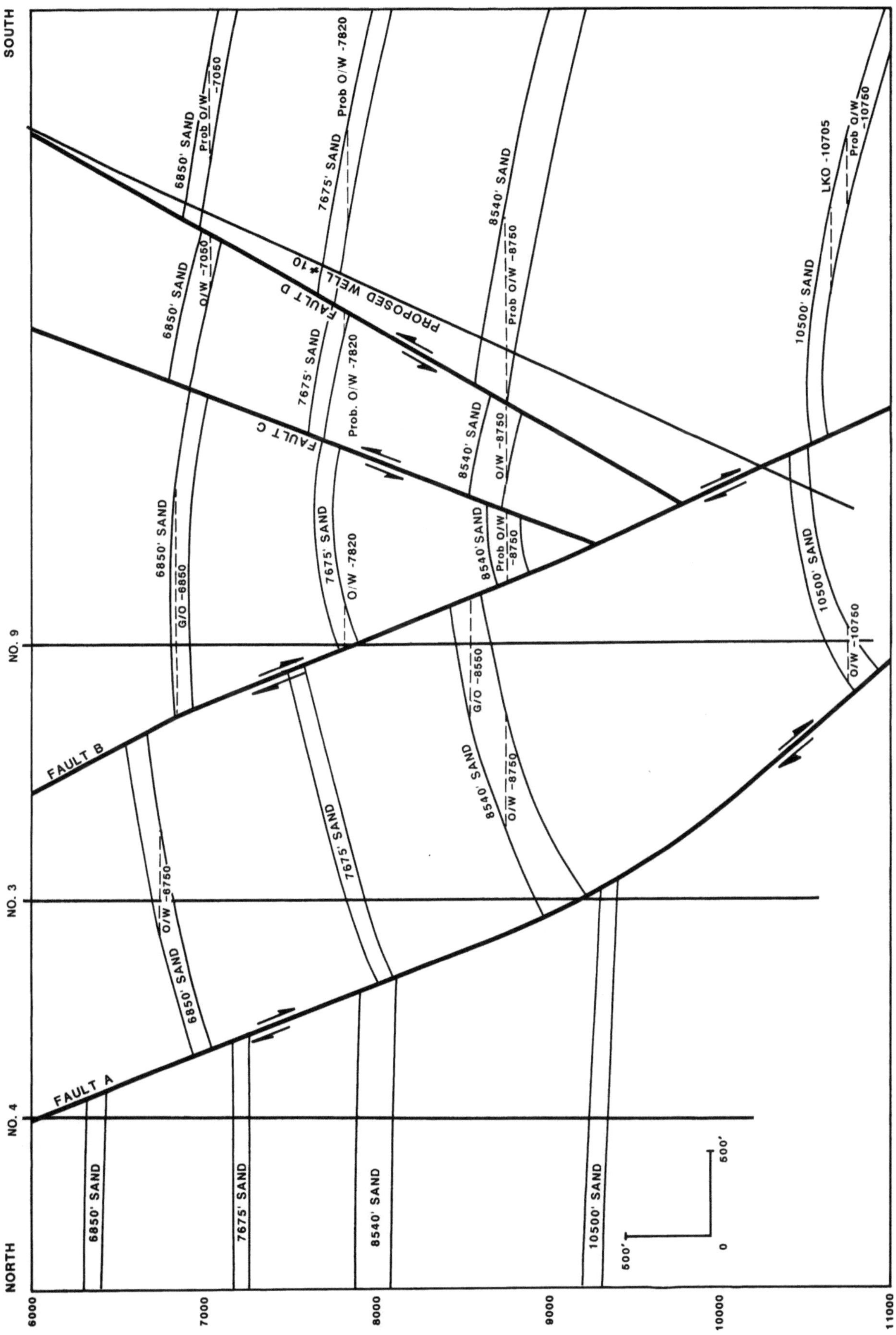

Figure 8-50 Observe the differing bed attitude upthrown and downthrown to the major growth Fault A. The rollover anticline, containing all the hydrocarbon accumulations, displays a compatibility of bed attitude across Faults B, C, and D. (Modified from Tearpock and Harris 1987. Published by permission of Tenneco Oil Company.)

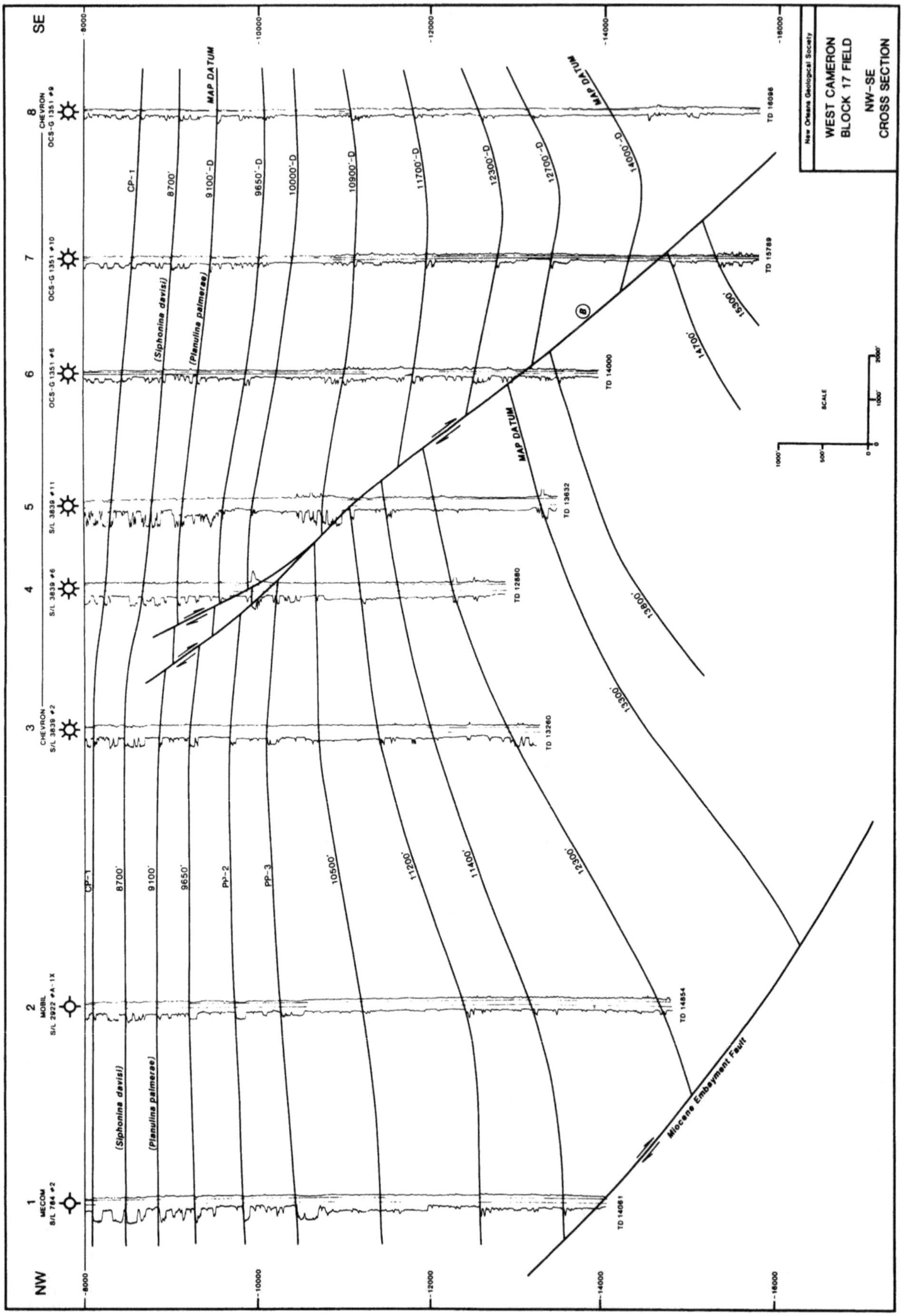

Figure 8-51 The cross section from an offshore Gulf of Mexico field exhibits structural compatibility in the shallow section and structural incompatibility in the deeper section across the same normal fault. (Published by permission of the New Orleans Geological Society.)

Figure 8-26, discussed earlier, is an example of an area cut by a reverse fault that exhibits structural (contour) compatibility across the fault, so the vertical separation (the repeated section in this case) is used to map across the fault. In Figure 6-25, however, there are significantly different structural attitudes on opposite sides of the blind thrust fault; in this case, the technique of contouring the vertical separation across the fault cannot be used.

Vertical separation cannot easily be used to contour across a fault where there is no compatibility of structural attitude across the fault. For such structures, it is necessary to *map each fault block separately and independently using the available well or seismic control. Each contoured fault block must be individually integrated with the fault surface map for the displacing fault to determine the upthrown and downthrown fault traces* (Tearpock and Harris 1987).

Observe in Figure 8-50 that although the vertical separation mapping technique cannot be used across the major growth Fault A, the faulted hanging wall anticline that contains all the reservoirs does exhibit structural compatibility across all faults. Vertical separation must be used to map across all the faults that cut the rollover anticline in the hanging wall block of the major growth fault. This type of situation is quite common in both extensional and compressional areas. Major growth normal, thrust, or strike-slip faults may downdrop, thrust, or slide one fault block with respect to the other to such a degree that there is no compatibility of structural attitude across these major faults. However, the associated structures formed by these faults commonly possess structural compatibility across the faults within the structure. This is shown clearly in the cross section in Figure 8-50. Figure 8-22 is a structure map of a similar anticline downthrown to a very large growth fault (which is not shown on the map). Note the structural compatibility across the large fault (on the west), even though the vertical separation is about 1100 ft.

A good example of an area exhibiting both compatible and incompatible structural attitudes across faults, within the same field, is seen in the Cactus-Nispero Field in the Southern Zone, Mexico. Figure 8-52a is a structure map on the middle Cretaceous showing this salt-cored structure cut by both normal and reverse faults. Figure 8-52b and c are cross sections through the field, both of which are shown in plan view (F-F2 and G-G2) on the structure map. A careful review of the structure map and cross sections shows that the structure exhibits contour compatibility across most of the faults, and therefore the vertical separation should be used to map across these faults. Two areas provide major exceptions. In cross section F-F2, structural compatibility is lost across the reverse fault cutting Well No. C-87. On the structure map, the large normal fault just north of the salt body does not exhibit contour compatibility from the upthrown to downthrown blocks. To contour these two areas accurately, each block must be contoured separately and then integrated with the fault map for each associated fault.

Although many ramps relating to rapidly dying faults do not exhibit contour compatibility, some do. The complication here is that the vertical separation across the fault changes rapidly along the ramp, so the contour spacing on opposite sides of the ramp will be different.

All these examples stress the importance of understanding the 3D geometry of any structure being interpreted and mapped. The accuracy of the mapping depends on the use of the correct mapping techniques, which often depends on the geometry of the structure itself. Different techniques may be needed within the same field.

MAPPING TECHNIQUES FOR VARIOUS TECTONIC HABITATS

In this section, we review the mapping techniques applicable to four major tectonic habitats and their associated hydrocarbon traps. These habitats include (1) extensional, (2) compressional, (3) strike-slip fault, and (4) diapiric settings. In addition to discussing the various types of hydrocarbon traps in each tectonic setting, we illustrate the fault surface map and the integrated contour construction of fault traces for each of the fault patterns observed in these settings, and present a field example where possible.

Extensional Tectonics

For extensional tectonics, we include nongrowth normal fault systems and detached listric growth fault systems, since both commonly occur together within this setting. In many areas of the world, hydrocarbons are trapped upthrown to normal faults. Important exceptions to this are found in the Gulf of Mexico and the Niger Delta, where hydrocarbons are commonly trapped downthrown as well as upthrown to normal faults. Also, reservoir-bearing rollover anticlines commonly occur in blocks downthrown to listric growth normal faults.

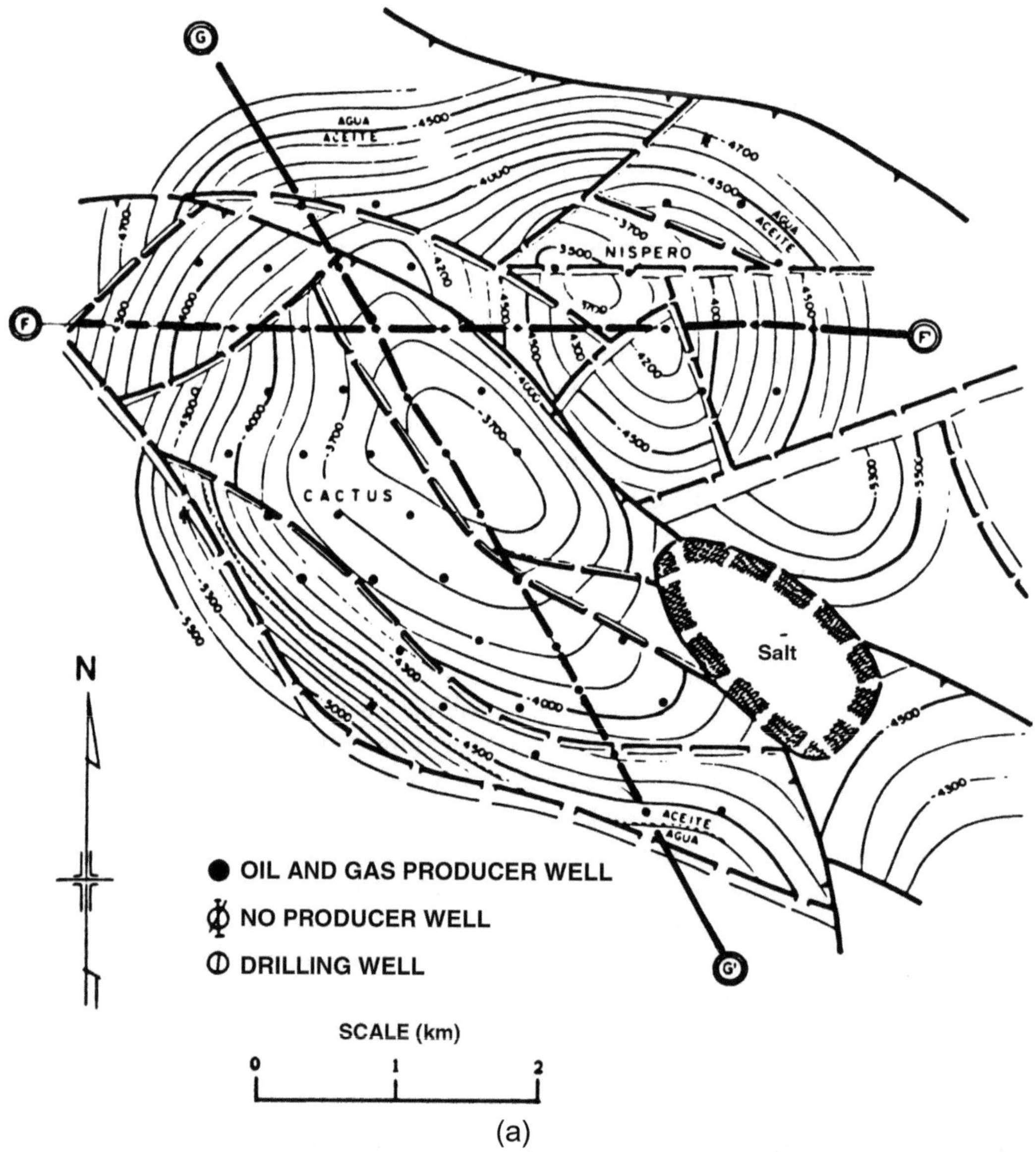

Figure 8-52 (a) Structure map of the Middle Cretaceous, Cactus-Nispero Field, Southern Zone, Mexico. The structure is a salt-cored structure cut by normal and reverse faults. (b) Geological cross section F-F′ through the Cactus-Nispero Field. (c) Geological cross section G-G′ through the Cactus Field. (From Santiago-Acevedo 1980. AAPG©1980, reprinted by permission of the AAPG whose permission is required for further use.)

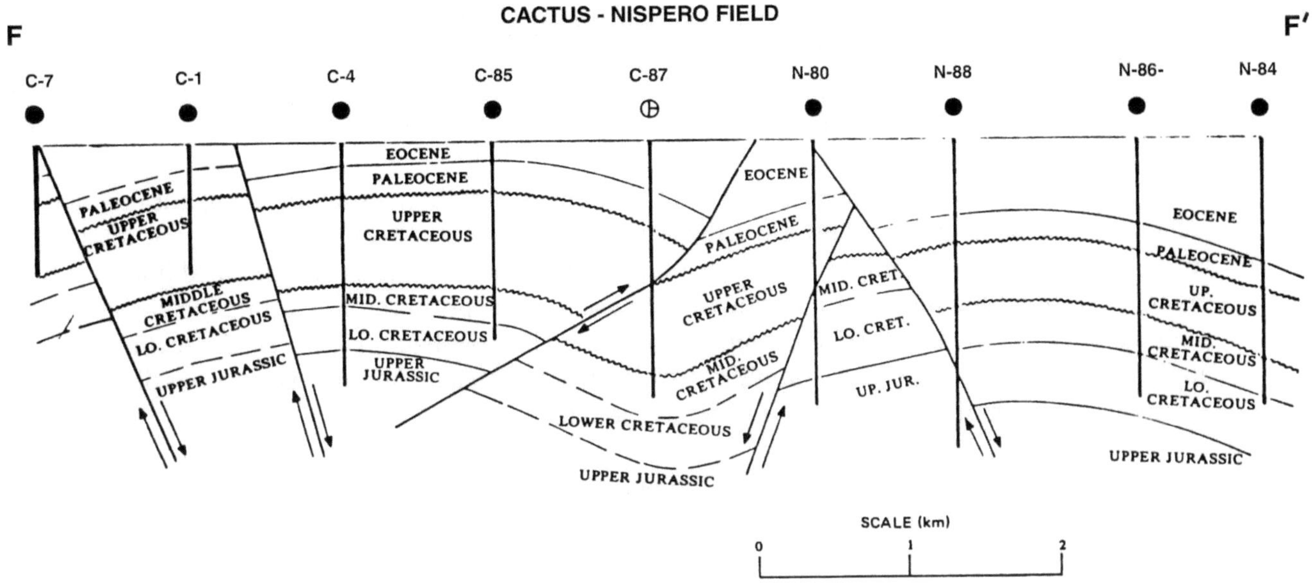

(b)

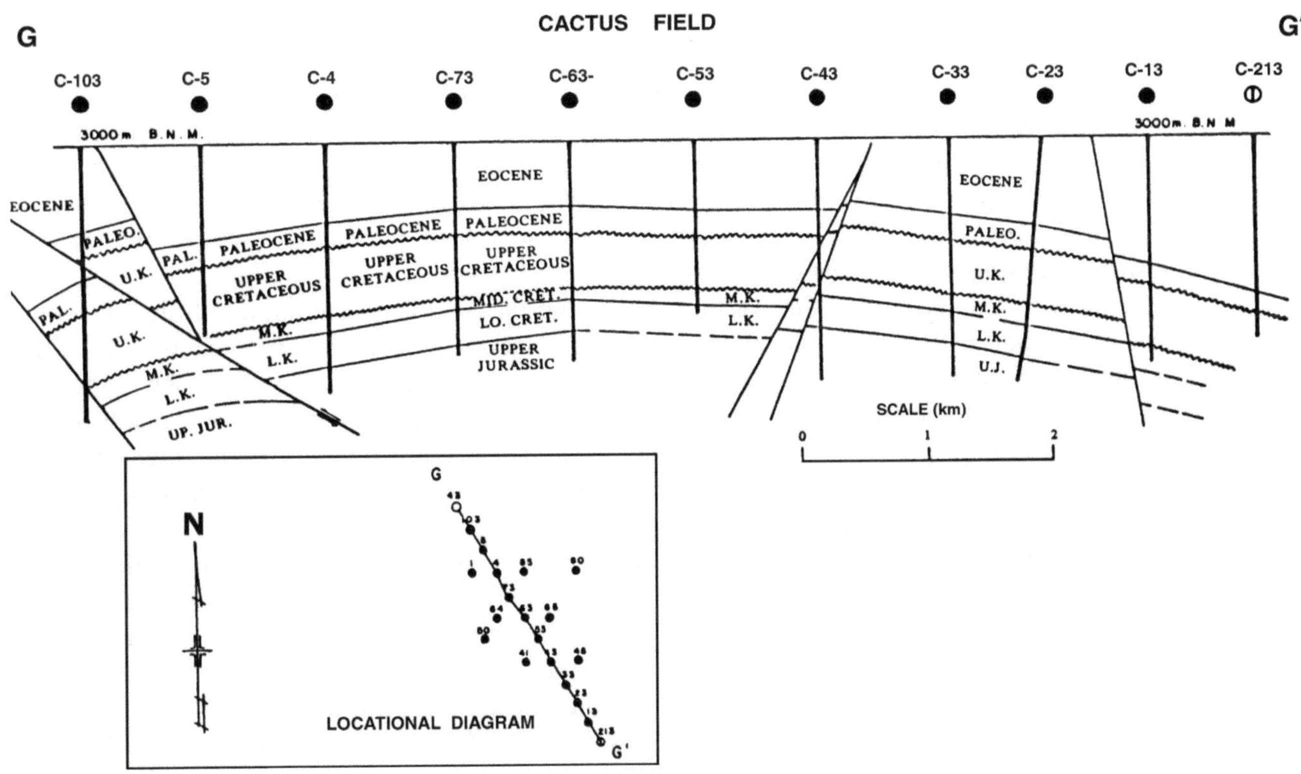

(c)

Figure 8-52 *(continued)*

Faults are the primary trapping mechanism for hydrocarbons in extensional habitats, although four-way closures, unconformities, and stratigraphic traps are also important. Our primary discussion centers around five different types of fault patterns: (1) single nongrowth (structural) faults, (2) compensating, (3) bifurcating, (4) intersecting, and (5) growth fault systems.

The construction techniques for all fault patterns that exhibit structural compatibility across a fault are the same as those used for the working model (Fig. 8-21) to arrive at an integrated structure map. For the examples of each fault pattern, we show at least two mapped horizons that bound a stratigraphic interval of constant thickness. For the purpose of detailing the correct mapping techniques, the data for the examples are somewhat idealized. However, the techniques are applicable in both simple and complex petroleum geological mapping. Where possible, actual examples for each of the fault patterns are presented.

Compensating Fault Pattern. Figure 8-53a shows the fault surface contour map for Faults A and B, which together represent a typical compensating fault pattern. If necessary, refer again to Chapter 7 and review the major aspects of this fault pattern. Several key points must be reemphasized here. First, the line of termination must be shown on the fault surface map; its importance will become clear as we review the structure maps on the 6000-ft and 7000-ft Horizons. Second, vertical separation is conserved across fault intersections. Observe that throughout the area where two faults are present (i.e., the area south of the line of termination of Fault B at Fault A in our example), Fault A has 100 ft of vertical separation and Fault B has a vertical separation of 200 ft. Only Fault A is present north of the termination line, and there its vertical separation is 300 ft.

Figure 8-53b is a completed structure contour map on the 6000-ft Horizon. The structure is a faulted anticline. Figure 8-53c shows the fault surface map for Faults A and B superimposed onto the completed structure map for the 6000-ft Horizon. Using this figure, *we review the construction of the upthrown and downthrown traces for each fault, the construction of the intersection of the two faults, and the proper horizon contour construction across each fault using the vertical separation (missing section from each wellbore).*

Begin with the northern upthrown fault block and the −6200-ft contour, and follow its complete construction over the entire structure. The intersection of this structure contour with the fault contour of the same value defines the upthrown trace of Fault A at point A. The contour is projected through the fault, as if the fault were not there, and adjusted in the downthrown block by the amount of the vertical separation (300 ft) to become the −6500-ft contour. The intersection of this −6500-ft structure contour with the fault contour of the same value defines the downthrown trace of Fault A at point B. The −6500-ft contour downthrown to Fault A crosses the axis of the structure and intersects with the −6500-ft contour for Fault B at point C, defining the downthrown trace of Fault B at this location. The contour is now projected through Fault B, as if the fault were not there, and then adjusted by 200 ft (representing the vertical separation for Fault B) in the upthrown block of Fault B to become the −6300-ft contour. This −6300-ft structure contour intersects the −6300-ft fault surface contour at point D, defining the upthrown trace of Fault B at this location. This contour once again crosses the axis of the structure, this time on the eastern side of the anticline, and intersects with the −6300-ft contour of Fault A at point E, defining the downthrown trace of Fault A. Project the contour through the fault gap of Fault A and adjust its value by 100 ft for the vertical separation at point F, defining the upthrown trace of Fault A. Finally, this contour continues in the upthrown block of Fault A until it intersects with point A, which was our starting point. This technique is applied for each structure contour on the map to arrive at the integrated structure map shown.

The heavy dashed line (Fig. 8-53c) striking southwest-northeast is the line of termination of Fault B at Fault A. *The intersection of the fault traces on the completed structure map for both faults* falls on this line of termination. For this map, there are two trace intersections: (1) the intersection of the downthrown traces of Faults A and B shown at point G, and (2) the intersection of the upthrown trace of Fault B with the downthrown trace of Fault A shown at point H. Since all trace intersections must fall on the line of termination, it is very important to place this line on the fault surface map. By following the line, we can predict the location of the fault intersections for other horizons either shallower or deeper in the subsurface, because all intersections must fall on this line. Observe that north of the line of termination, only Fault A is present, with a vertical separation of 300 ft, whereas south of the line, both Faults A and B are present, with vertical separations of 100 ft and 200 ft respectively. Take a minute and test the additive property of faults around the fault intersection on the completed structure map.

This construction technique of integrating the fault and structure maps removes all the guesswork with regard to: (1) fault trace construction; (2) the location of all faults on the structure map; (3) the location of all fault intersections; and (4) the width of the fault gaps. The integration of the two maps constructs all these features automatically without the need for guesses or estimates. Take one or two additional structure contour lines and go through the same procedure just described to make sure you fully understand the technique for integrating a fault map with a structure map.

Figure 8-53d is a structure map on the deeper 7000-ft Horizon, and it shows that the faults have migrated with depth. The technique of construction is exactly the same as presented for the 6000-ft Horizon. In Figure 8-53e, we again superimpose the fault map onto the completed structure map. Choose several structure contours and follow each across the mapped area to review the construction techniques as discussed for the preparation of the completed structure map for the 6000-ft Horizon.

Finally, the completed fault traces for both the 6000-ft and 7000-ft Horizons, and the general trend of the structure contours is superimposed onto the fault map for Faults A and B in Figure 8-53f. Looking at each set of fault traces (those for the 6000-ft and 7000-ft Horizons), can you detect any difference between them? Use 10-point proportional dividers or an engineer's scale to measure the width of the fault gap for Fault A on both horizons. Notice that the gap is wider for Fault A on both sides of the line of termination on the structure map for the 7000-ft Horizon, when compared with the fault gap on the structure map for the 6000-ft Horizon. Remember the general rule of fault gap width presented earlier. *A fault dipping up-dip (in the opposite direction of bed dip) will have a narrower fault gap than a fault of the same or similar vertical separation dipping down-dip (in the same direction as the bed dip).* Notice that Fault A on the 6000-ft Horizon is on the northern side of the structural axis and is dipping generally in the opposite direction to the formation. On the deeper structure map for the 7000-ft Horizon, Fault A has crossed the axis of the structure and is now dipping more or less in the same general direction as the formation. Therefore, the fault gap is wider as mapped on the structure map for the 7000-ft Horizon.

With the technique of integrating the fault and structure maps, *this change in fault gap width is automatically built into the integration*. It is unlikely that such a change in gap width could have been intuitively incorporated into the completed structure contour maps had the fault and structure map integration technique not been used. For this example, the change in gap width is significant, but depending on the configuration of the structure and the geometry of the fault, the change in width can be minor. Let the integration technique be the tool used to establish fault gap widths.

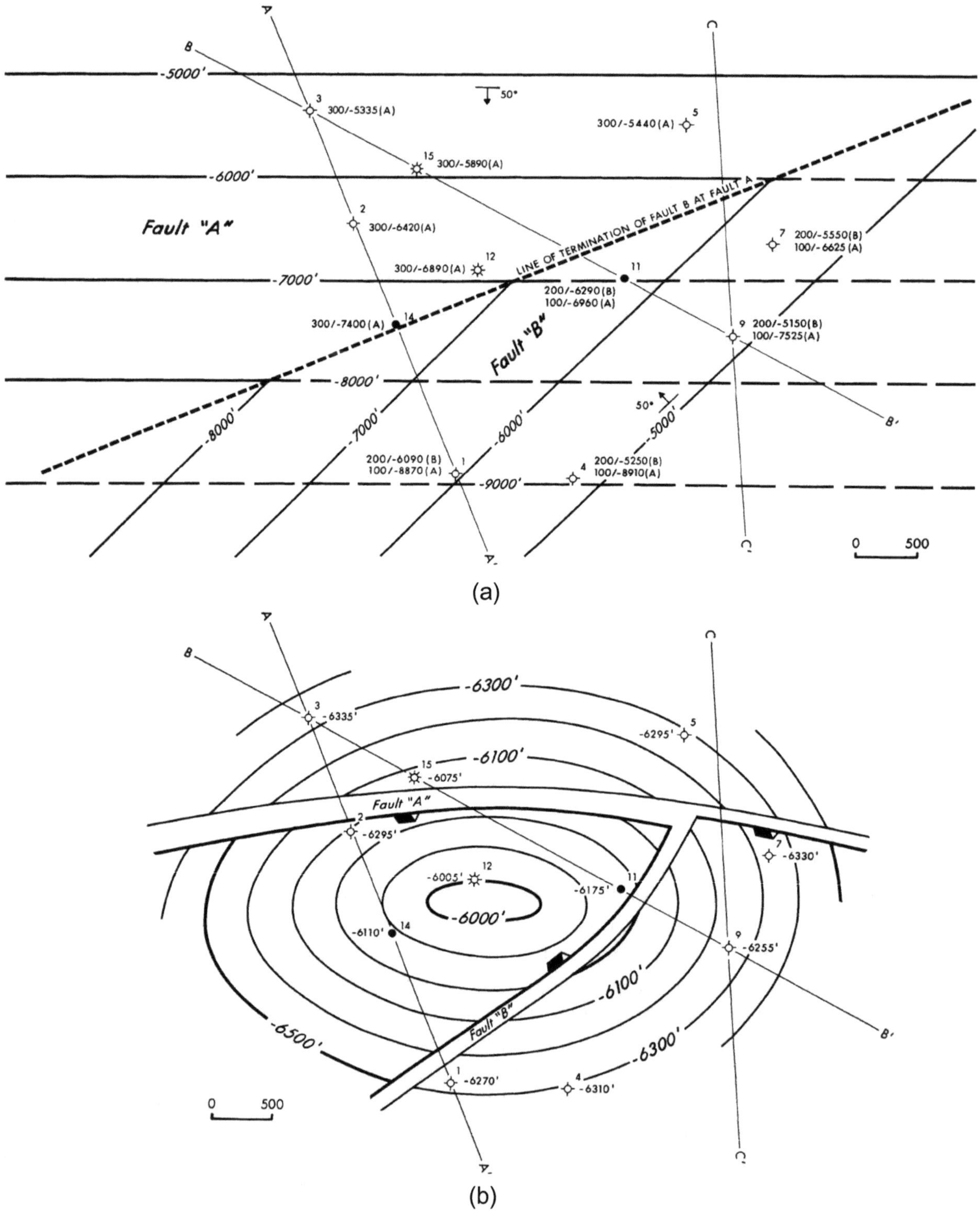

Figure 8-53 (a) Fault surface map for a compensating fault system including Faults A and B. (b) Integrated structure map on the 6000-ft Horizon. (c) Fault and structure maps (6000-ft Horizon) superimposed to illustrate the details of the integration technique. (d) Integrated structure map on the 7000-ft Horizon. Upthrown and downthrown restored tops in Wells No. 9 and 14 were used to aid in the structural interpretation. (e) Fault and structure maps (7000-ft Horizon) superimposed to show the accurate construction of the faulted 7000-ft Horizon map using the integration technique. (f) Overlay of fault traces for both the 6000-ft and 7000-ft Horizons. The figure shows that the intersections of all fault traces fall on the line of termination. The location of the fault intersections on any structural horizon can be predicted from the strike of the line of termination.

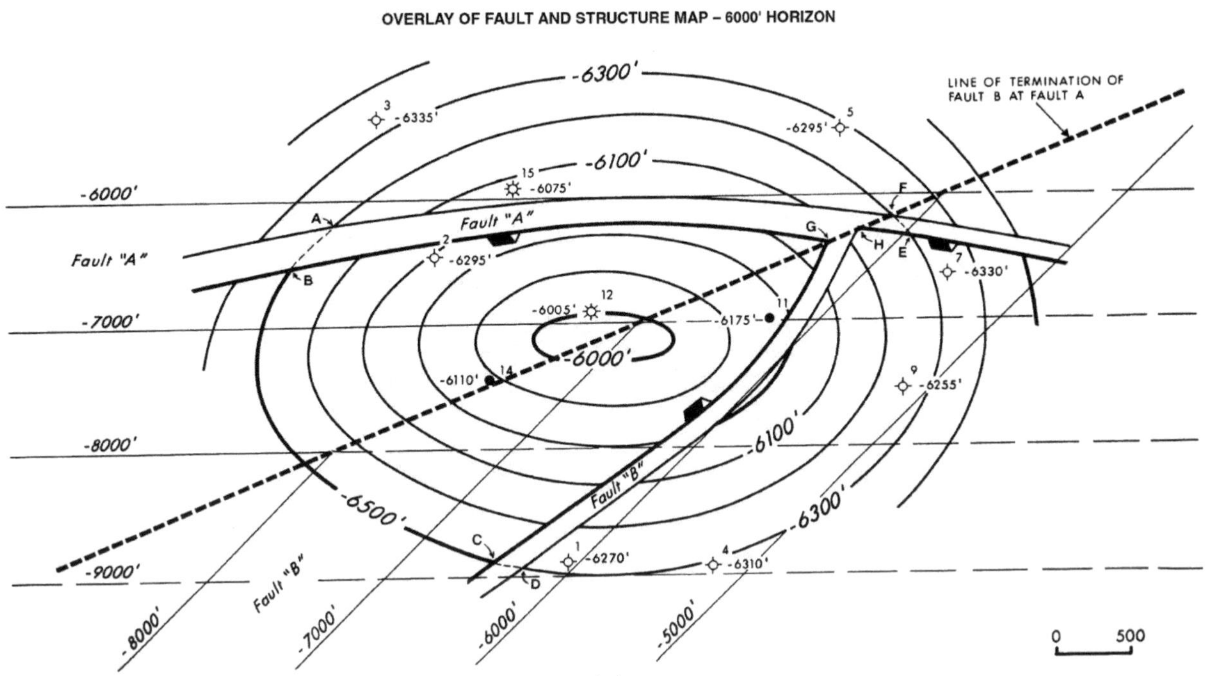

(c)

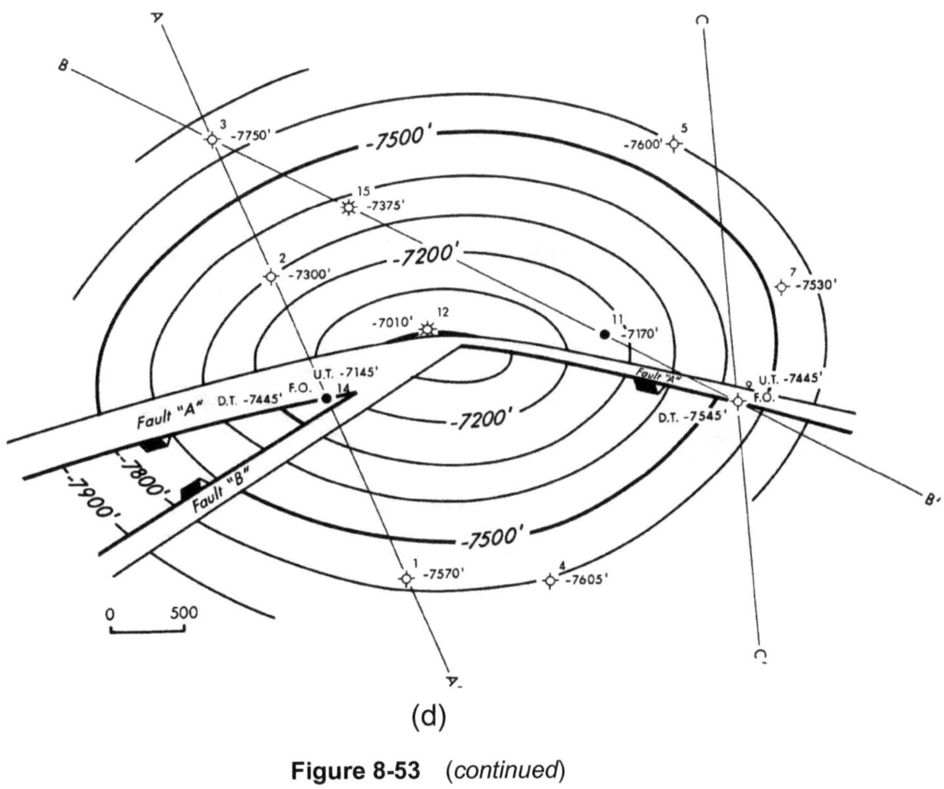

(d)

Figure 8-53 *(continued)*

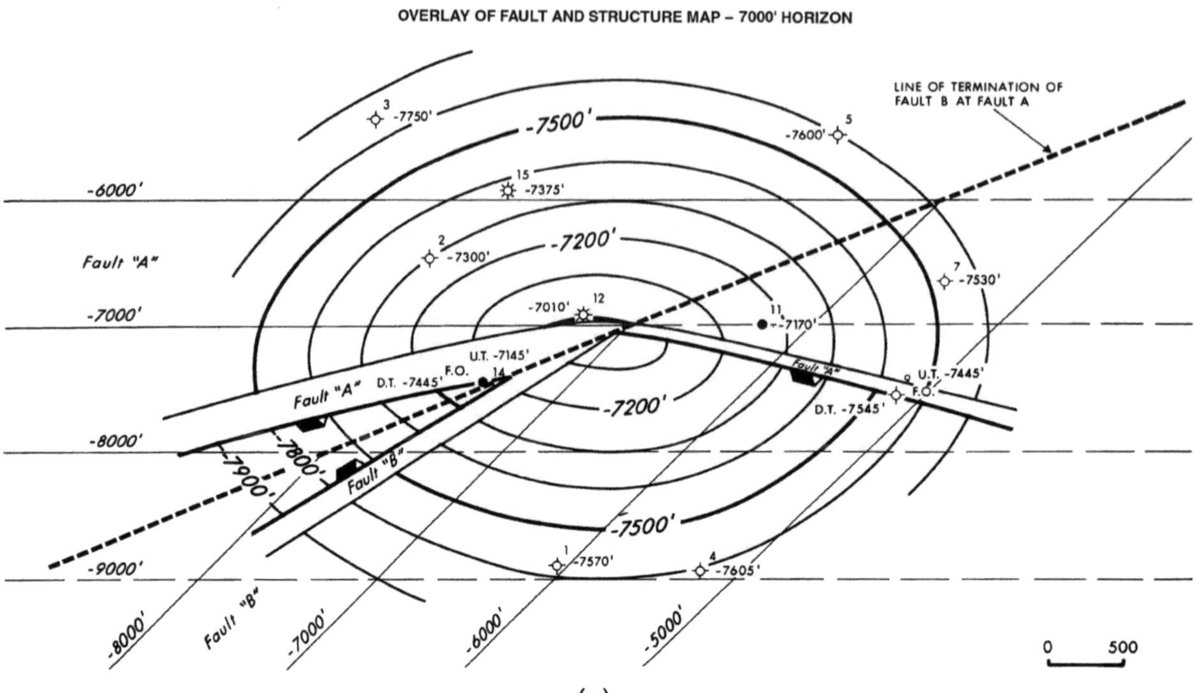

(e)

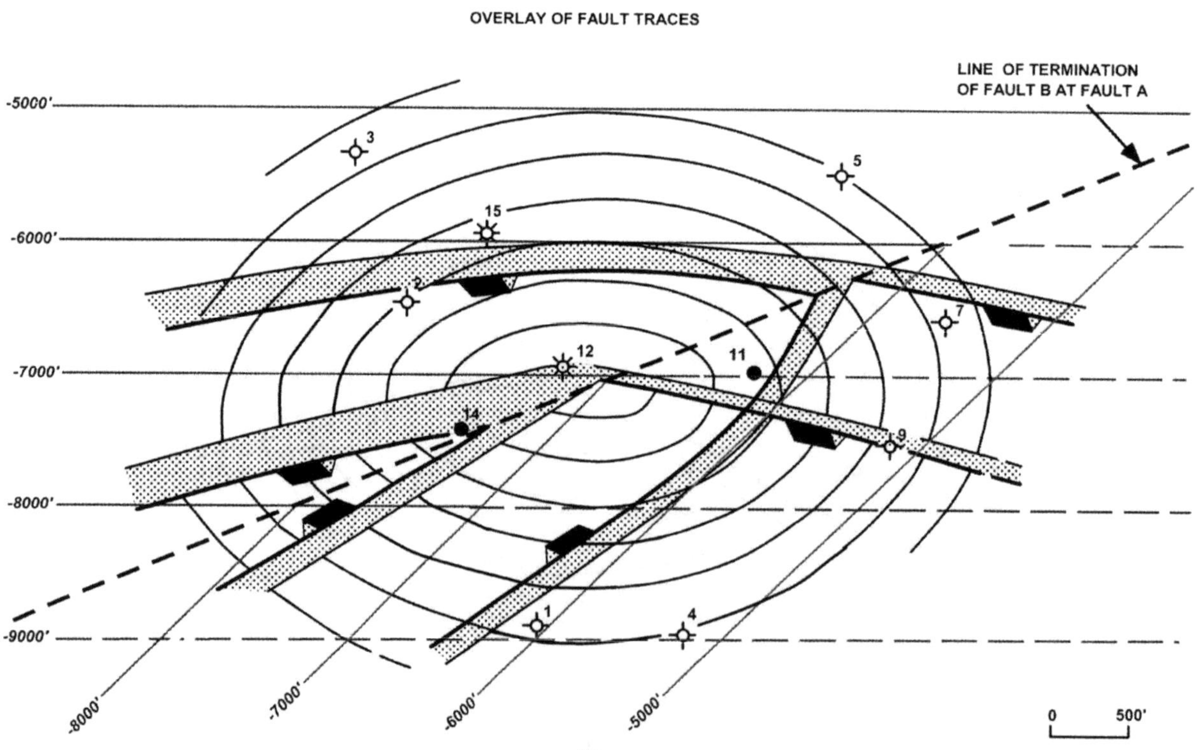

(f)

Figure 8-53 *(continued)*

The completed structure map in Figure 8-22 illustrates two compensating faults to the large down-to-the-east normal fault. Fault surface maps were prepared for each fault, and they were integrated with the structure map to arrive at the completed map.

Look at the completed structure map in Figure 8-54a. What type of fault pattern is represented by the intersection of Faults A and D? Is it a compensating fault pattern, or is it a mapping error? Can you find any other errors on this map?

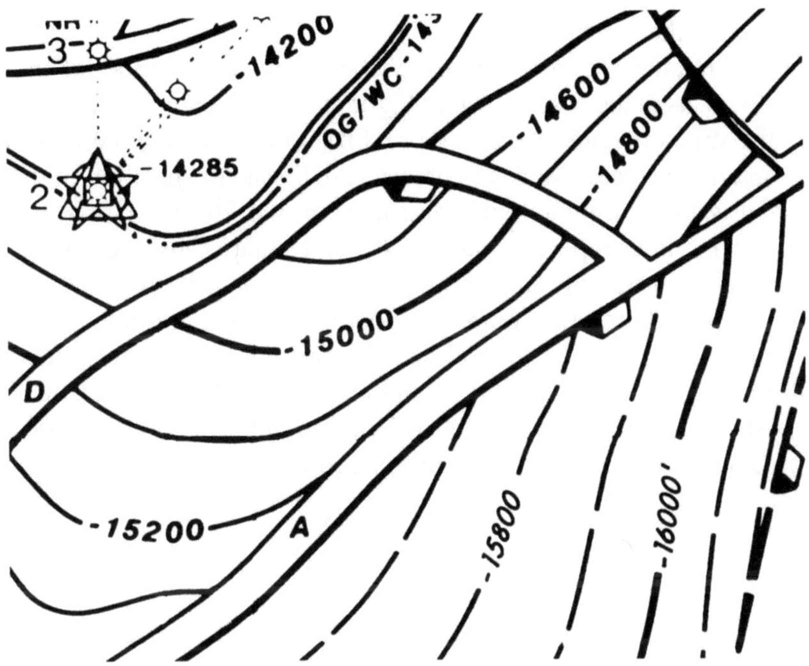

(a)

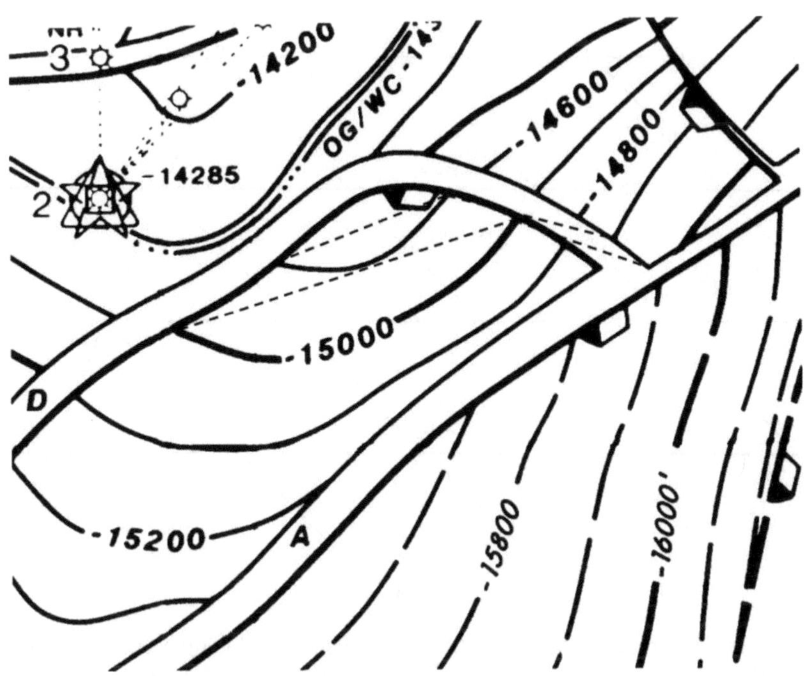

(b)

Figure 8-54 (a) Compensating fault system or mapping bust? (b) Implied strike of Fault D (dashed) based on where the −14,900-ft′ contour intersects the fault trace.

Bifurcating Fault Pattern. The fault map in Figure 8-55a shows a fault contour map for Faults A and A-1, which together form a bifurcating fault pattern. Refer again to Chapter 7, if necessary, to refresh your memory on the characteristics of a bifurcating fault pattern. Figure 8-55b is the completed structure map on the 6000-ft Horizon. Figure 8-55c shows the fault surface map for Faults A and A-1 superimposed onto the structure map for the 6000-ft Horizon. Choose several structure contours upthrown to Fault A and go through the integration technique to check the position of the fault traces and the fault intersections, the width of the fault gaps, and the contour construction for the 6000-ft Horizon across the two faults.

Fault traces are relatively easy to delineate when there are a number of structure contours intersecting fault contours, as in the case shown in Figure 8-55; however, in areas where fault contours are semiparallel to the structure contours, it is more difficult to delineate the fault traces. In these situations, you must interpolate depth values to precisely delineate the position of the fault traces.

Figure 8-55d is the completed structure map on the 7000-ft Horizon. Figure 8-55e shows the fault surface map superimposed onto the structure map. Observe that the trace intersections for Faults A and A-1 fall directly on the line of bifurcation. Figure 8-55f shows the fault traces, as seen on the completed structure maps for both the 6000-ft and the 7000-ft Horizons with the fault surface map superimposed. Notice how the fault intersection migrates from north to south along the path of the line of bifurcation. Also observe that the width of the fault gap for Fault A is wider on the structure map for the 7000-ft Horizon than it is on the 6000-ft Horizon structure map. This occurs here for the same reason discussed for the compensating fault example.

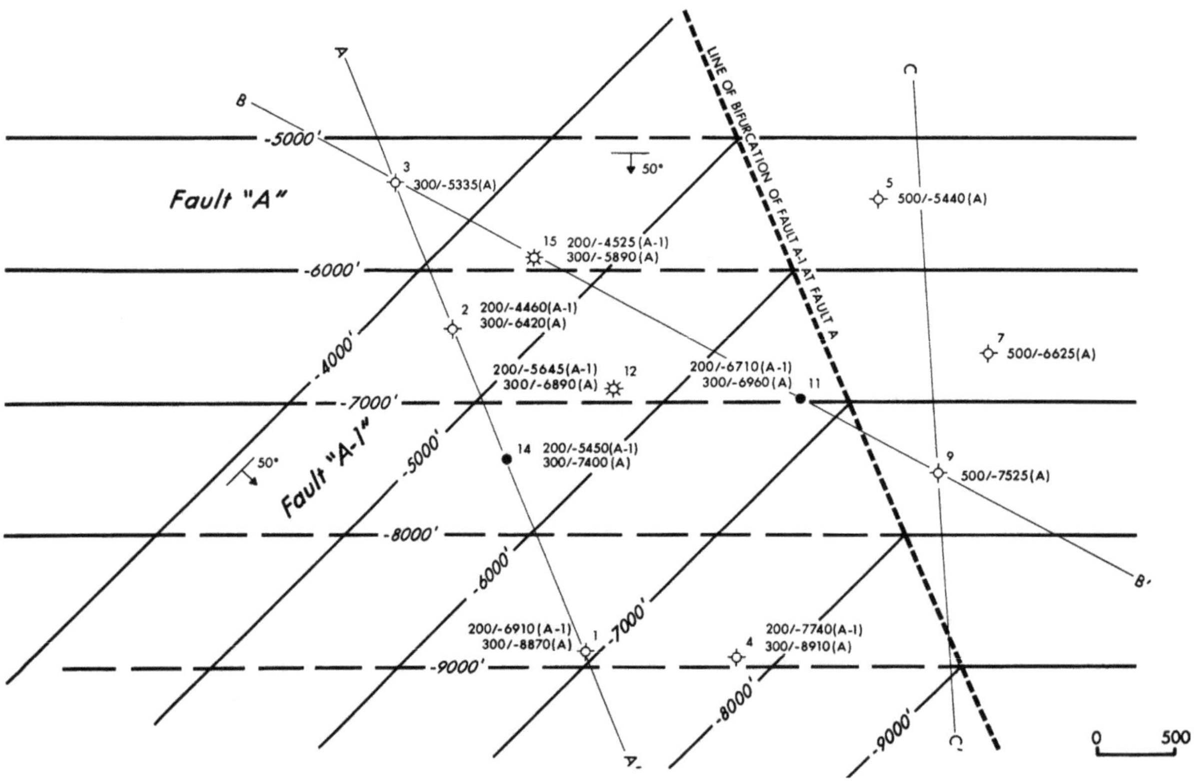

Figure 8-55 (a) Fault surface map for a bifurcating fault system including Faults A and A-1. (b) Integrated structure map on the 6000-ft Horizon. (c) Fault and structure maps (6000-ft Horizon) superimposed to show the accuracy of the integration technique: to position the fault traces and intersections and to determine the width of the fault gap. (d) Integrated structure map for the 7000-ft Horizon. (e) The fault map superimposed onto the 7000-ft structure map. (f) Overlay of fault traces on both the 6000-ft and 7000-ft Horizons. As with the compensating fault pattern, all intersections of fault traces fall on the line of bifurcation.

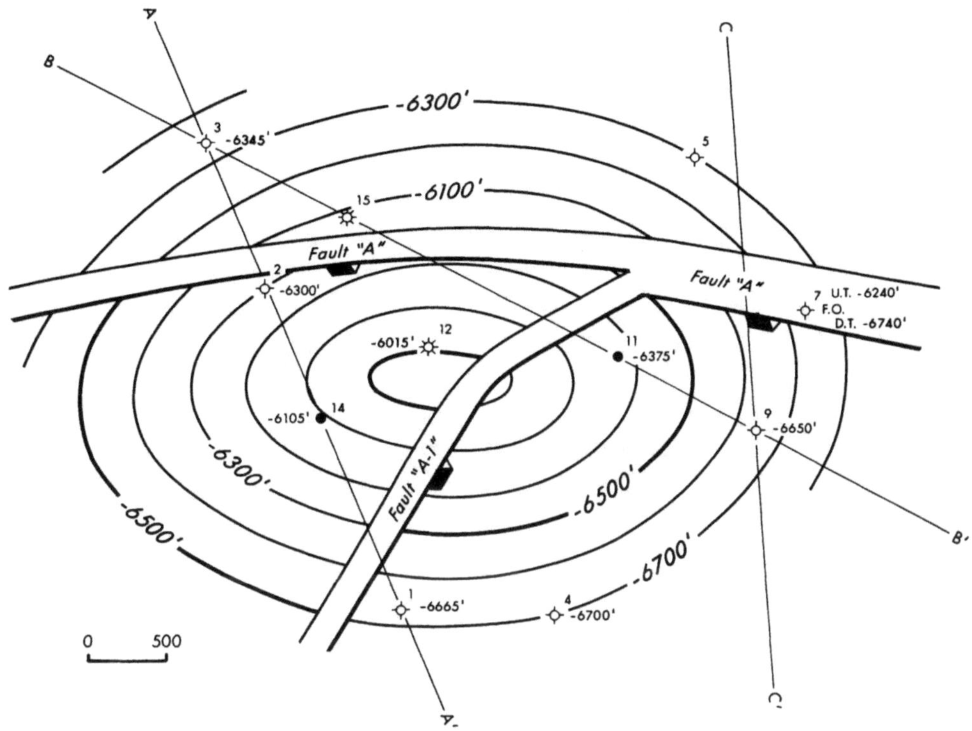

(b)

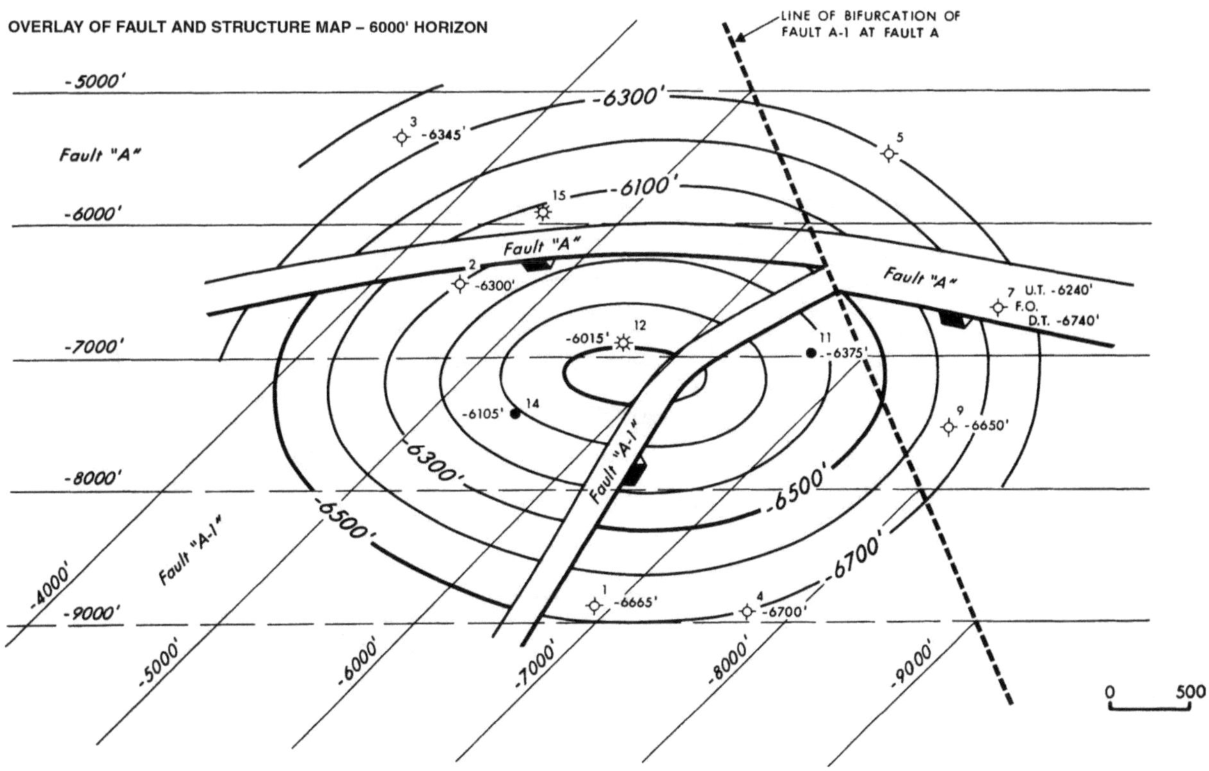

(c)

Figure 8-55 *(continued)*

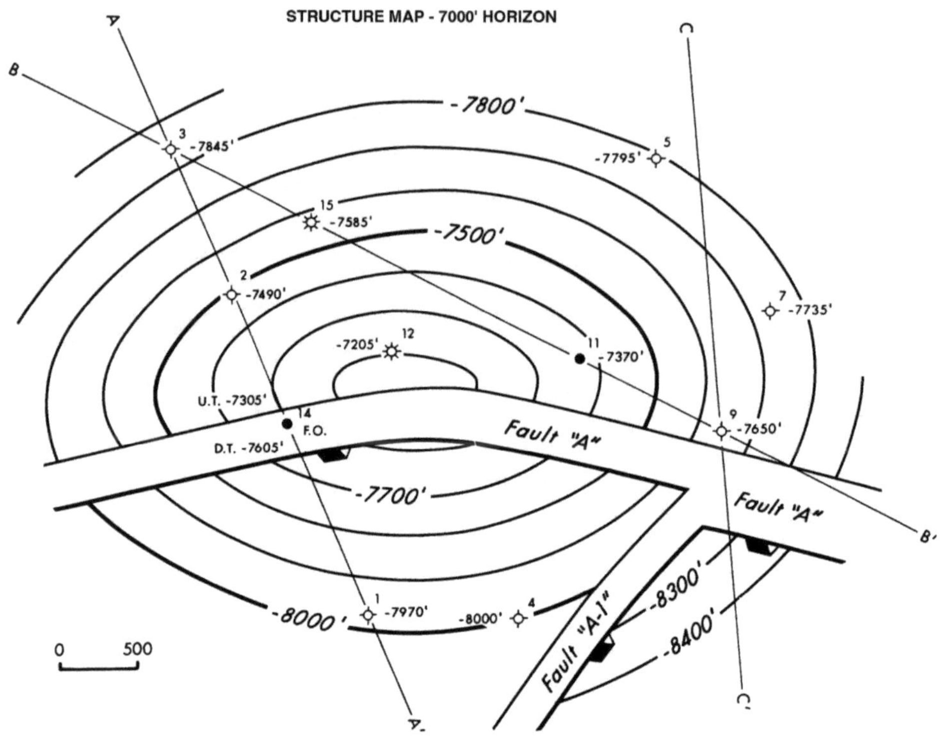

(d)

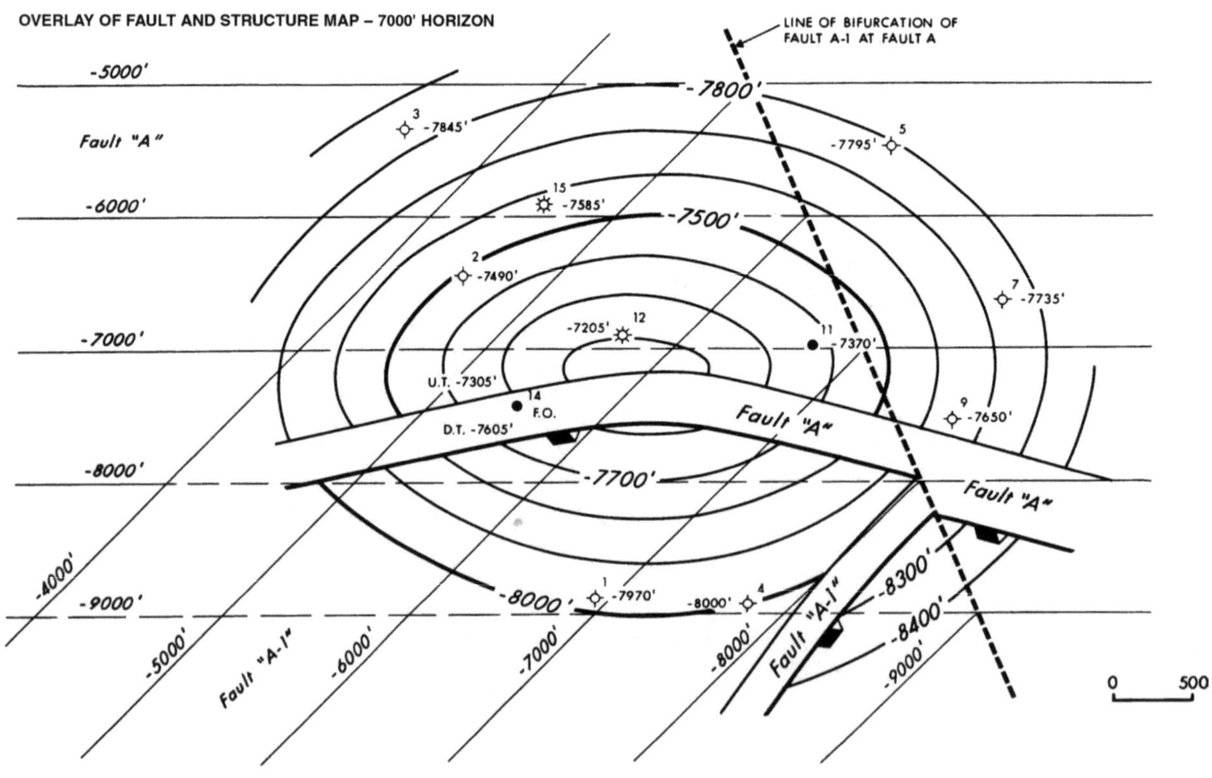

(e)

Figure 8-55 (*continued*)

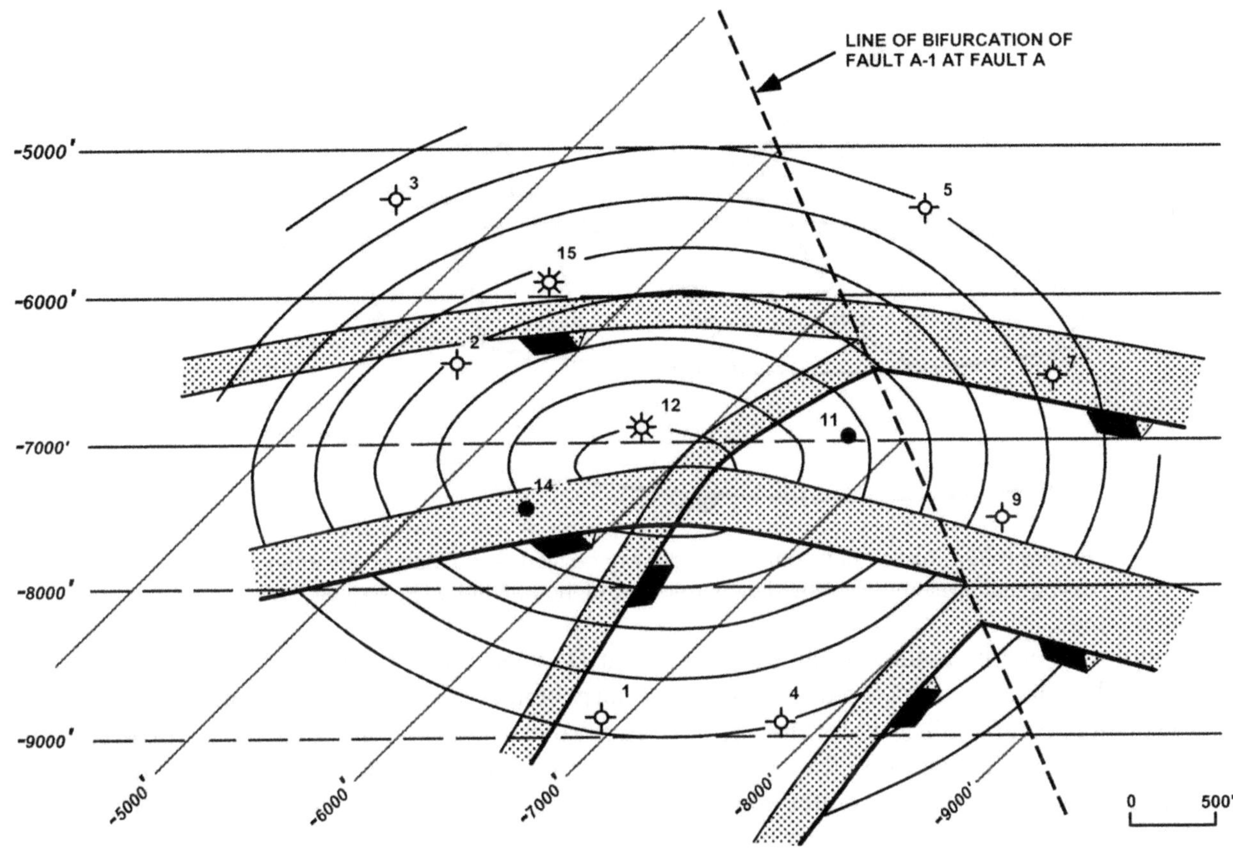

(f)

Figure 8-55 *(continued)*

Figure 8-56 is a portion of the structure map on the Nodosaria Marker, and Figure 8-57 is a portion of the U Sand structure map from the offshore Gulf of Mexico. Both figures show good examples of the mapping of a bifurcating fault system. The fault system in Figure 8-56 actually has two bifurcations. Take a minute to check the construction of these completed maps in the area around the bifurcations by applying the additive property of faults.

Now we revisit Figure 8-54. Both of the faults that merge dip in the same direction, so this is a bifurcating system. The gap for the fault that extends to the east of the point of merger would be expected to be larger than the gap on either of the two faults, thus reflecting the additive property of faults. However, the gap is much narrower than that of either fault. Actually, the gap is of the size expected if this were a compensating fault system, so the interpretation is incorrect. The only way that the gap could be correctly constructed to be so narrow is if the fault were to abruptly become almost vertical right at the point of merger of the two faults, and that is certainly unreasonable. In further review of the map, we can use the implied strike technique (Tearpock et al. 1994) on the northern fault, whose trace bends sharply. The implied strike technique takes advantage of the fact that at each point where a structure contour intersects a fault trace, we know the depth of both the horizon and the fault surface. Connecting the depth points from the fault surface along one or more fault traces allows us to approximate the fault surface map that is implied by the fault trace on the structure map. Using the two points where the −14,900-ft contour intersects the downthrown fault trace of Fault D and the one point where the −14,900-ft

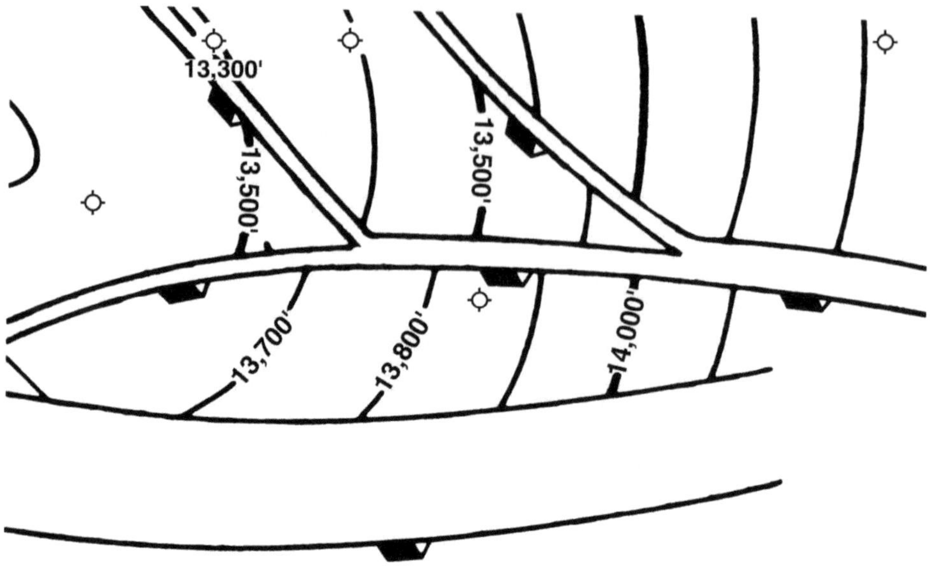

Figure 8-56 Portion of the structure map on the Nodosaria Marker illustrating a bifurcating fault system. (Published by permission of the Lafayette Geological Society.)

contour meets the upthrown fault trace of Fault D, we see that the implied fault strike of Fault D abruptly changes (Fig. 8-54b). Is such a sharp bend reasonable for a normal fault? Finally, structural compatibility is lacking across the northern fault. Notice that the vertical separation decreases rapidly from 600 ft to 300 ft in an easterly direction. The poor contour compatibility is apparently due to the interpreter honoring that displacement. Again, do you think that this abrupt change in vertical separation is reasonable along this single fault, or could something else be happening? Notice that it occurs near the sharp bend in the implied fault surface. Could Fault D actually be two faults?

Intersecting Fault Pattern. Figure 8-58a illustrates a typical intersecting fault pattern composed of Faults A and B. Cross section A-A′ is shown in Figure 8-58b. In the section in Chapter 7 dealing with the intersecting fault pattern, the fault surface maps were contoured with the assumption that neither fault surface was offset (Fig. 8-58b). In other words, the two faults were considered contemporaneous rather than of two different ages. The fault traces for the structure maps shown in Figure 8-58 were constructed with this assumption. Notice, however, in Figure 8-58c through g, that *all the fault traces are offset.* Figure 8-58d shows, for example, that although the fault surfaces themselves are not offset, the fault traces on the completed structure maps show *significant offsets for all the fault traces* at the fault intersections. The positions of the offset traces are accurate for the constructed fault surface maps. If the faults are of two different ages, however, then the completed maps are subject to error around the intersection. If the faulting is contemporaneous, then the traces are accurate as shown on the completed structure maps.

Compare the completed structure maps in Figure 8-58d and f. There is a pronounced change in the configuration of the fault traces shown on these two maps. The change is the result of the intersecting geometries of the two horizons with Faults A and B at different depths. Without the integration of the fault and structure maps, the accurate delineation of these fault traces, particularly for the 7000-ft Horizon, would be difficult, if not impossible.

Finally, observe that all the fault trace intersections fall on the line of intersection. In this case of two intersecting faults, there are four fault trace intersections (Fig. 8-58e and g). This construction again emphasizes the necessity of placing the line of intersection on the completed fault maps.

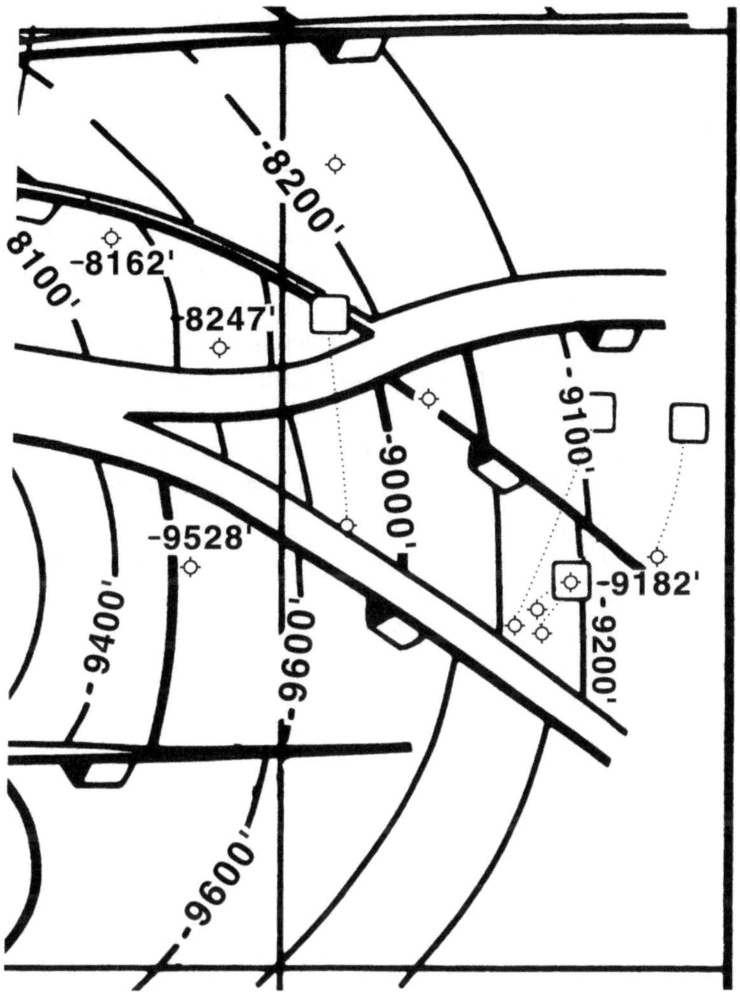

Figure 8-57 Example of a bifurcating fault system from the offshore Gulf of Mexico. (Published by permission of the New Orleans Geological Society.)

Where faults of two different ages intersect, the construction of accurate fault surface maps showing the offset of an earlier fault by a later one may require seismic data to determine the relative ages of the faults. Also, a graphical solution to the construction of the fault surfaces and their effects on stratigraphic horizons can be undertaken. A graphical technique is presented by Dickinson (1954). The method is not detailed here; however, if it is absolutely necessary to prepare precise fault and structure maps reflecting the accurate geometry of intersecting faults of different ages, then the Dickinson method should be considered. In most cases, the relative ages of different intersecting faults can be determined if reasonable quality seismic data is available.

Figure 8-59, from Dickinson's paper, shows a completed structure map for a horizon above a deep-seated salt dome in Texas. Such a complex intersecting fault pattern is not uncommon in areas of salt tectonics. In this example, Faults 3 and 5 are intersecting. A test was planned for the upthrown block to the east. In order to test the block at an optimum structural position, it was necessary to accurately locate the position of the intersection of Faults 3 and 5, so the graphical solution outlined by Dickinson was used to prepare the fault and structure maps.

Intersecting faults of different ages are quite common in certain geological settings, especially where the stress field has shifted over time. Ferrill et al. (2009) illustrate this very well from the Jean d'Arc Basin offshore Newfoundland, which has a set of early faults that formed when the least principal stress was east-west. These faults strike predominantly north-south.

During a later phase of rifting, the least principal stress was north-south and the faults strike predominantly east-west. Contemporaneous intersecting normal faults tend to have similar strikes as they form in the same regional stress field. Intersecting normal faults of different ages may have similar or different strikes, depending on the regional stress field when the faults formed.

In addition to the intersecting normal faults, some areas around the world, such as western Texas, California, southern Oklahoma, the North Sea, and the southern zone of Mexico, contain intersecting normal and reverse faults. The Dickinson method is also very helpful in developing a structural picture of these intersecting faults. For many situations, however, particularly those involving small faults, the shortcut method of assuming contemporaneous faulting, shown earlier in this section, is often adequate for the construction of the fault surface maps and completed structure maps.

We cannot leave this subject of intersecting faults without covering an important misunderstanding with regard to intersections of fault traces on completed structure maps. There seems to be a popular belief that where two faults of *different* ages intersect, the fault trace of the intersected or older fault is not offset on a completed structure map, as shown in Figure 8-60. We strongly emphasize that *this idea is incorrect* for most cases. *The traces of each of the intersecting faults on any mapped horizon always offset, except in two special cases.* These exceptions, which are covered in detail by Dickinson (1954), occur where (1) the beds are horizontal, the faults intersect at right angles, and the movement on both faults is entirely dip-slip (Fig. 8-60);

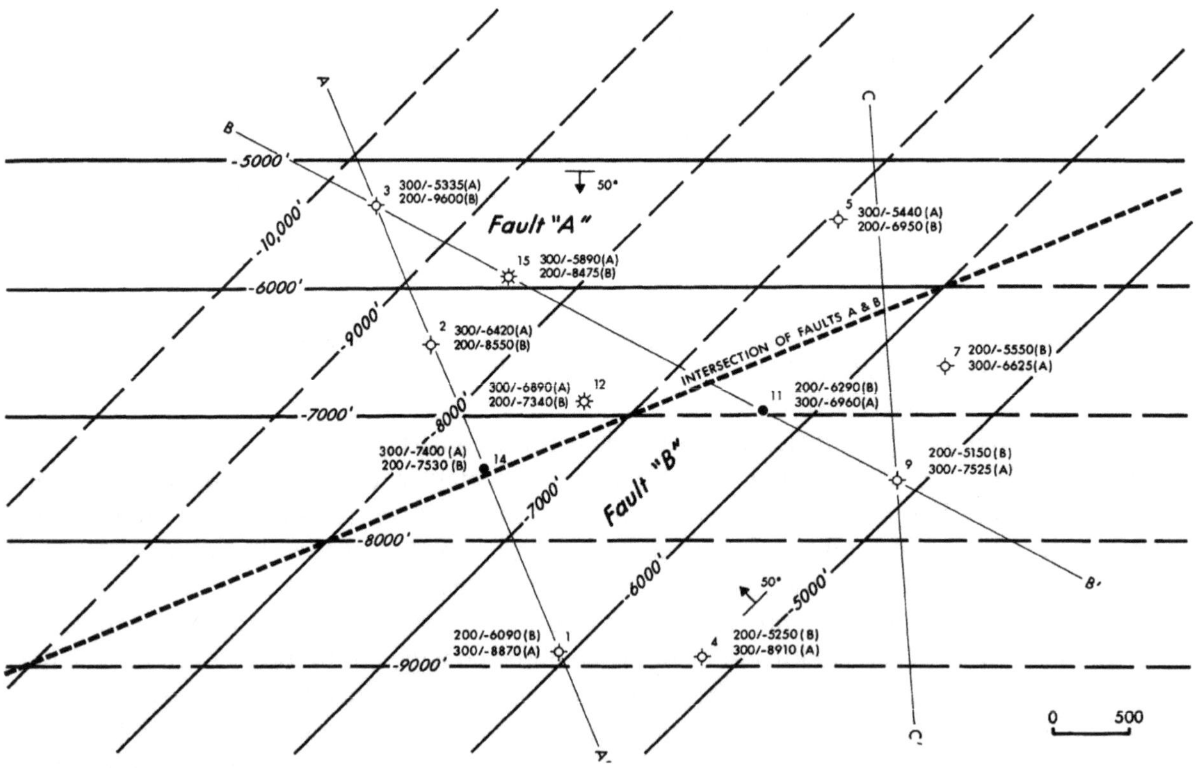

(a)

Figure 8-58 (a) Fault surface map for an intersecting fault system including Faults A and B. Fault B does not terminate at the line of intersection. (b) Cross section A-A′ illustrates the shortcut method of constructing the fault map as if the fault surfaces are not offset. (c) Integrated structure map on the 6000-ft Horizon. (d) Completed structure map on the 6600-ft Horizon. Observe that all fault traces are offset at the intersection. (e) Fault surface map and the structure map on the 6600-ft Horizon superimposed to show the accuracy of fault trace construction. (f) Completed structure map on the 7000-ft Horizon. (g) Overlay of the fault surface map and the 7000-ft structure map. Notice that all fault trace intersections fall directly on the line of intersection.

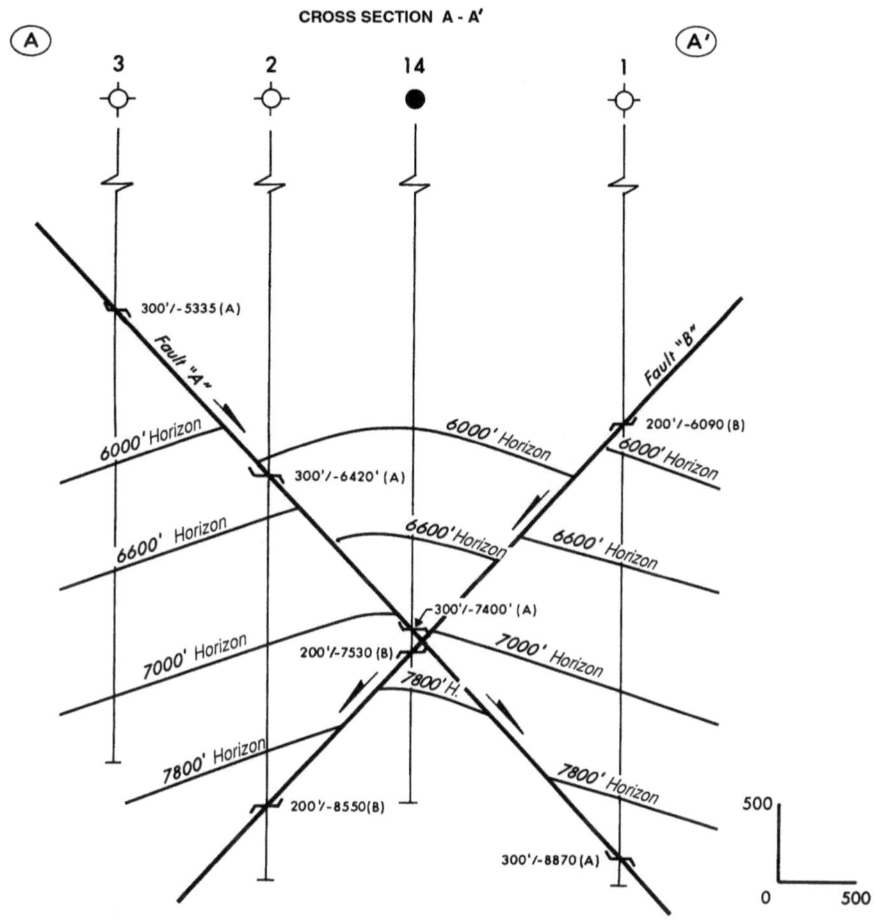

(b)

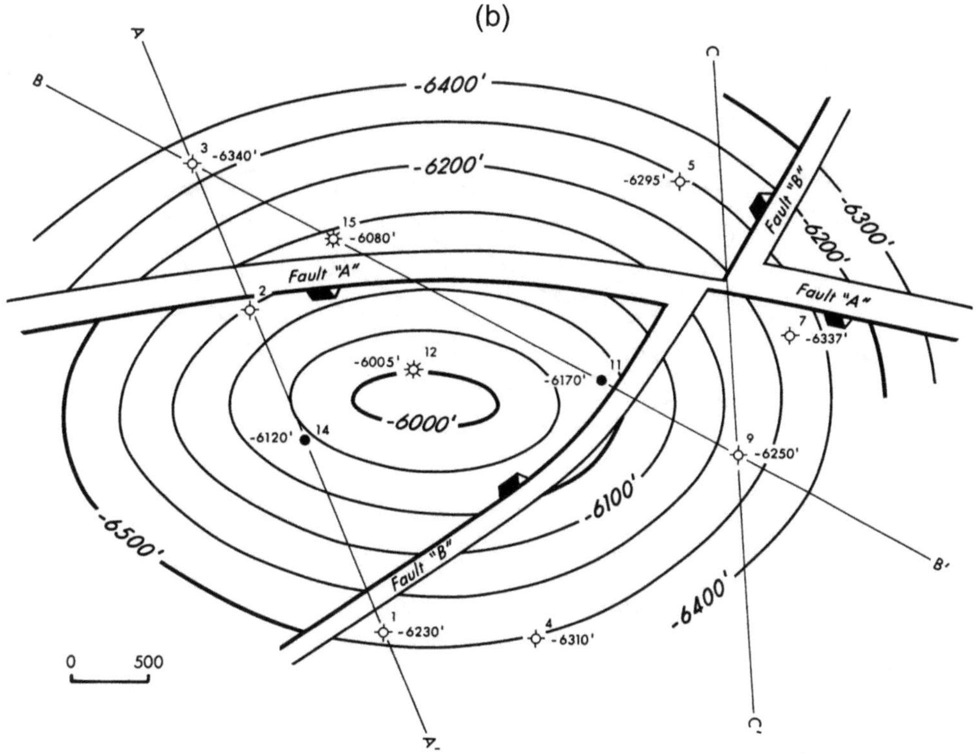

(c)

Figure 8-58 (*continued*)

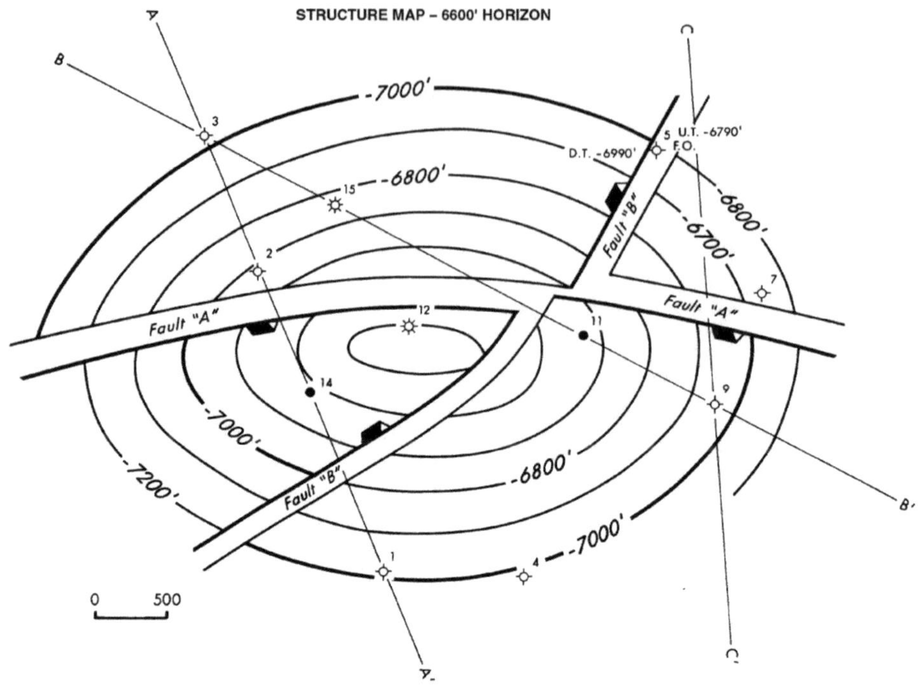

(d)

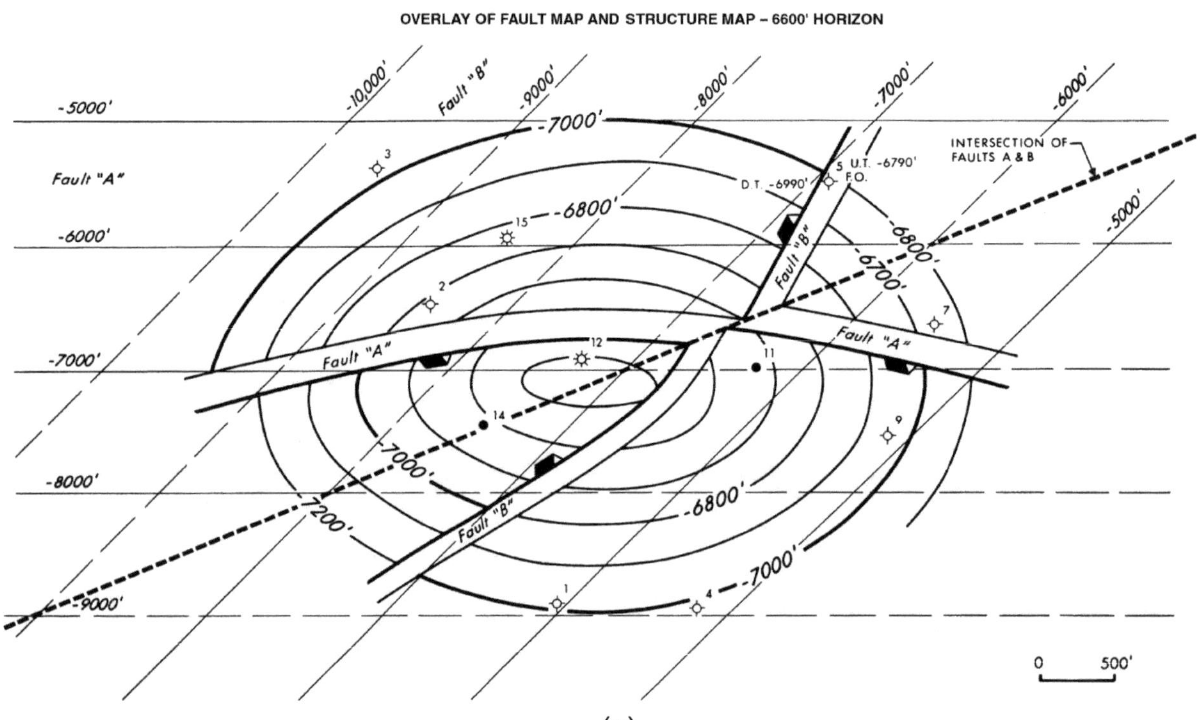

(e)

Figure 8-58 *(continued)*

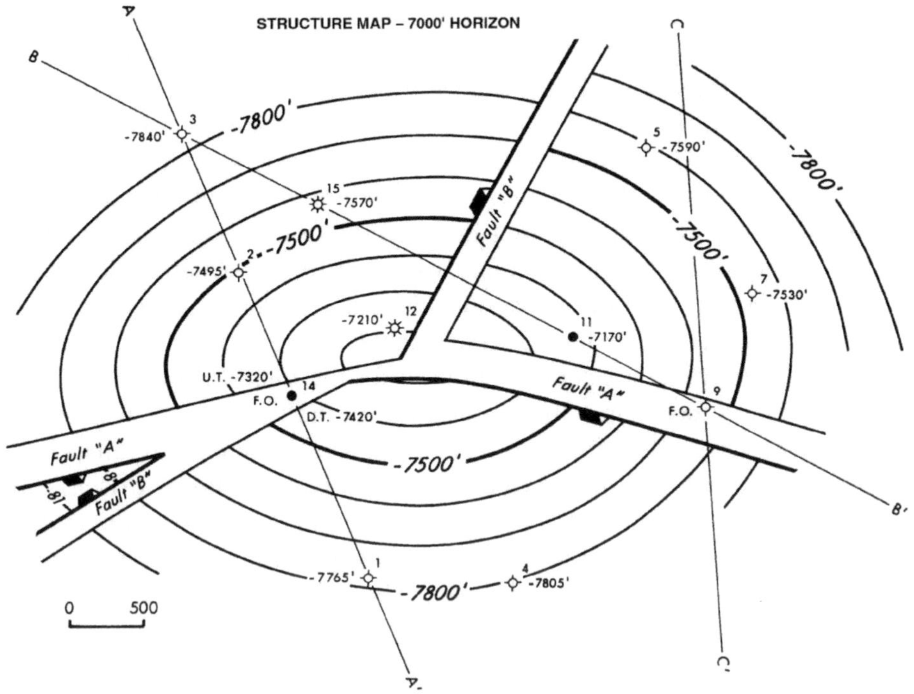

(f)

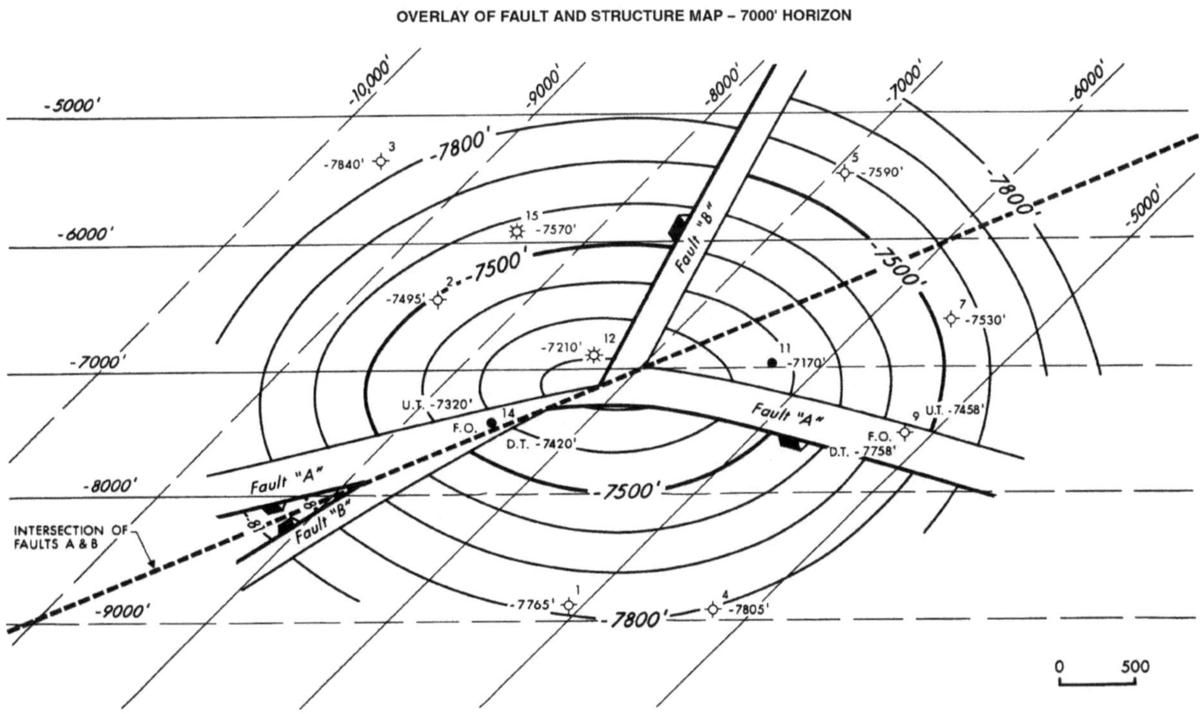

(g)

Figure 8-58 (*continued*)

or (2) the intersection of the (older) intersected fault and the mapped horizon is parallel to the direction of slip on the younger fault. Obviously, these two exceptional situations are rare. In both cases, the trace of the (older) intersected fault will not be offset, but the trace of the (younger) intersecting fault will be. So, *if neither of the faults traces is offset, then the structure map is incorrect (no exceptions)*. These guidelines apply to intersecting normal faults and intersecting reverse faults. Dickinson (1954) also discusses intersections of normal faults with reverse faults.

A structure map with intersecting faults showing the intersected (older) fault traces with no offset must meet one of the two special cases, or it is constructed incorrectly. Usually, this type of error occurs because (1) the interpreter does not understand the relationships of intersecting fault construction, or (2) no fault surface maps were prepared and used to integrate with the structure map to arrive at an accurate solution for the intersecting faults. Figure 8-61 is a portion of a completed structure map with an intersecting fault pattern. A review of the map shows that only one fault trace is offset for each pair of crossing faults, so the map is immediately suspect. Consider the two exceptions to the rule that both traces should be offset. The map shows that the beds are dipping, so exception (1), based on horizontal beds, does not apply. However, exception (2) might be valid if the movement on Fault B was in exactly the same direction as the trace of Fault A. However, we cannot determine that. The only way to determine the accuracy of the fault traces is to have a fault surface map and integrate it with the horizon map. However, for the map in Figure 8-61, the fault traces were prepared showing no offset for the assumed older Fault A.

If a pair of intersecting fault traces is shown on a completed structure map with neither trace offset and the traces resembling a *cross* (Fig. 8-62), the construction is incorrect. Such errors as the ones shown in Figures 8-61 and 8-62 can place suspicion on the rest of the structural interpretation.

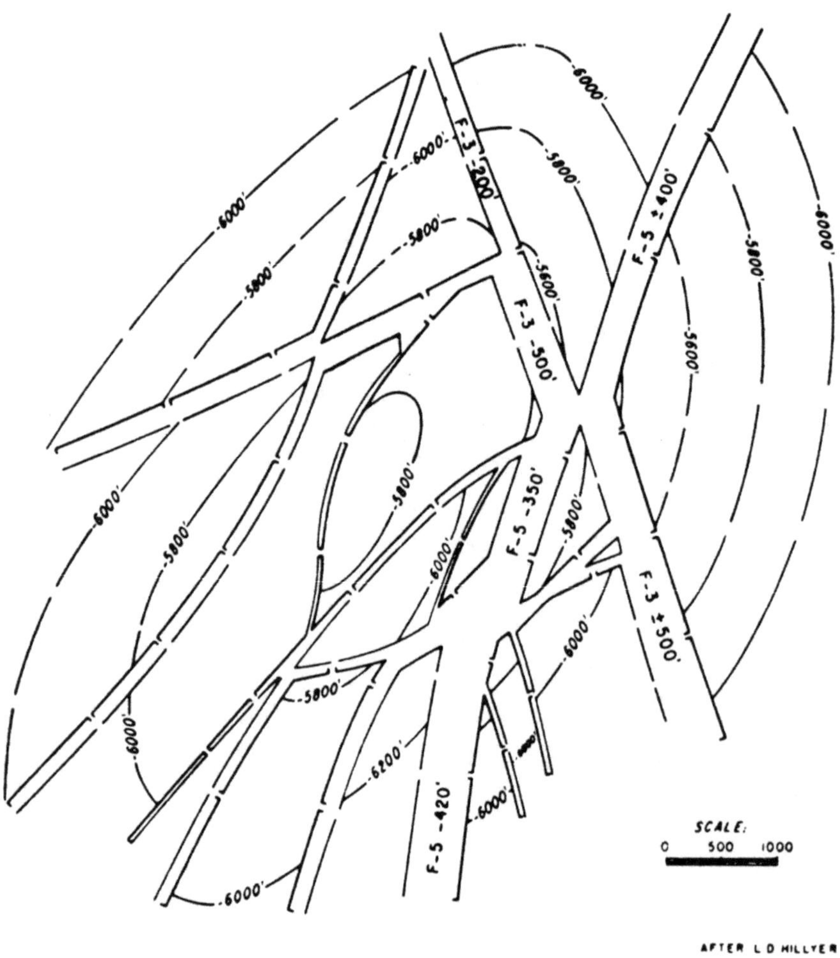

Figure 8-59 Structure contour map over a deep-seated salt dome in Texas illustrating several intersecting faults. (From Dickinson 1954. AAPG©1954, reprinted by permission of the APPG whose permission is required for further use.)

If the area you are working is complexly faulted with intersecting faults and extreme accuracy is required, it is advisable to become familiar with the Dickinson method for resolving the geometry of fault intersections.

Combined Vertical Separation. In Chapter 7, we discussed the combined vertical separation that results where two faults of different ages intersect. In Figure 8-60, the zone of combined vertical separation at a given horizon is shown in map view for this example of intersecting normal faults. It is the area between points J, E′, N, and M. A well drilled within this zone will cross only one of the two intersecting faults, but the interval shortening, or missing section, is equal to the sum of the vertical separations for both faults. For this example, the earlier fault has a vertical separation of 150 ft and the later fault is 250 ft; therefore, the vertical separation for a fault in the area of combined vertical separation is 400 ft.

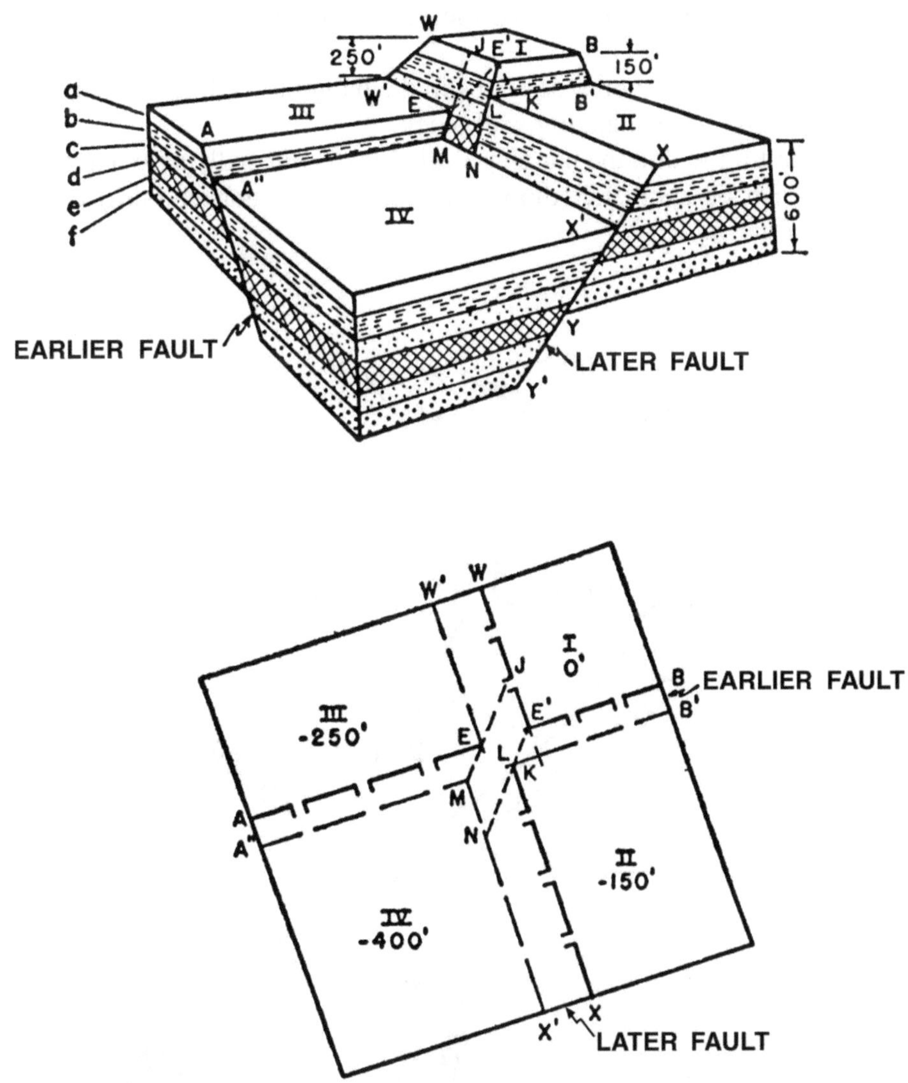

SUBSURFACE MAP ON TOP OF UPPER LAYER OF BLOCK

Figure 8-60 Block model and map view of intersecting normal faults in horizontal strata. Fault dip is 60 deg and the movement is all dip-slip. This result is only applicable in this special case of right-angle faults and flat beds. (Modified from Dickinson 1954. AAPG©1954, reprinted by permission of the APPG whose permission is required for further use.)

Growth Faults. A growth fault is a special type of normal fault. A growth fault is a normal fault that is contemporaneous with deposition, and it is often referred to as a *syndepositional fault*. There are several primary characteristics of growth faults.

1. They are commonly arcuate in lateral extent and concave toward the basin.
2. Fault dip typically decreases with depth, commonly becoming a bedding plane fault at depth (Chapter 11). We refer to this shape as *listric*, based on the Greek word *listron*, or shovel, due to its curved shape in cross section. However, fault dip can locally increase with depth (see the section Compaction Effects along Growth Normal Faults, in Chapter 11).
3. The vertical separation for the fault normally increases with depth. Displacements of several thousands of feet or kilometers are not uncommon. Growth faults can also include displacements that decrease with depth as well (see Chapter 11).
4. The time-stratigraphic intervals in the hanging wall blocks are thicker than the equivalent intervals in the footwall blocks (see the section Estimating the Vertical Separation for a Growth Fault, in Chapter 7).
5. A rollover anticline commonly develops in the hanging wall fault block of a growth fault as a result of collapse of the hanging wall block.
6. Secondary faulting within the hanging wall block is normally associated with growth faults. These faults can be synthetic (dipping parallel to the master fault) or antithetic (dipping toward the master fault).

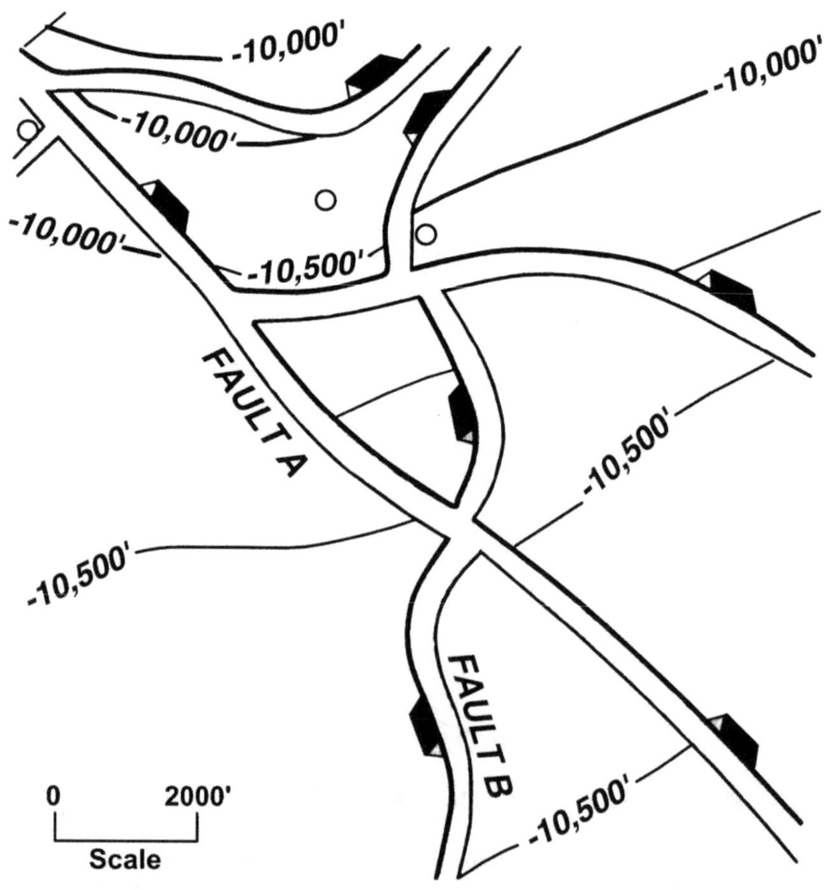

Figure 8-61 Portion of a structure map showing incorrectly interpreted and constructed intersecting faults.

A majority of growth faults dip toward the current basin or the paleo-basin and strike parallel or semiparallel to the coastline. However, growth faults can dip in any direction, including toward the margin of the basin. Along strike, most growth faults are generally concave in the direction of dip.

Faults are not perfect planar surfaces of slip; they generally have undulations of some type. As two fault blocks slip past one another, there must be deformation in at least one fault block, because rocks are not strong enough to support large voids. This is the main reason why many major rollover folds exist within the hanging wall fault blocks, formed by bending of the fault blocks as they slip over nonplanar (listric) fault surfaces. This mechanism of folding is called *collapse folding* (Hamblin 1965) (see Chapter 11 for details of the collapse mechanism). The rollover anticline, associated with listric growth faults, is one of the most widely recognized petroleum-producing folds, as seen in such areas as the Gulf of Mexico, Brunei, and the Niger Delta.

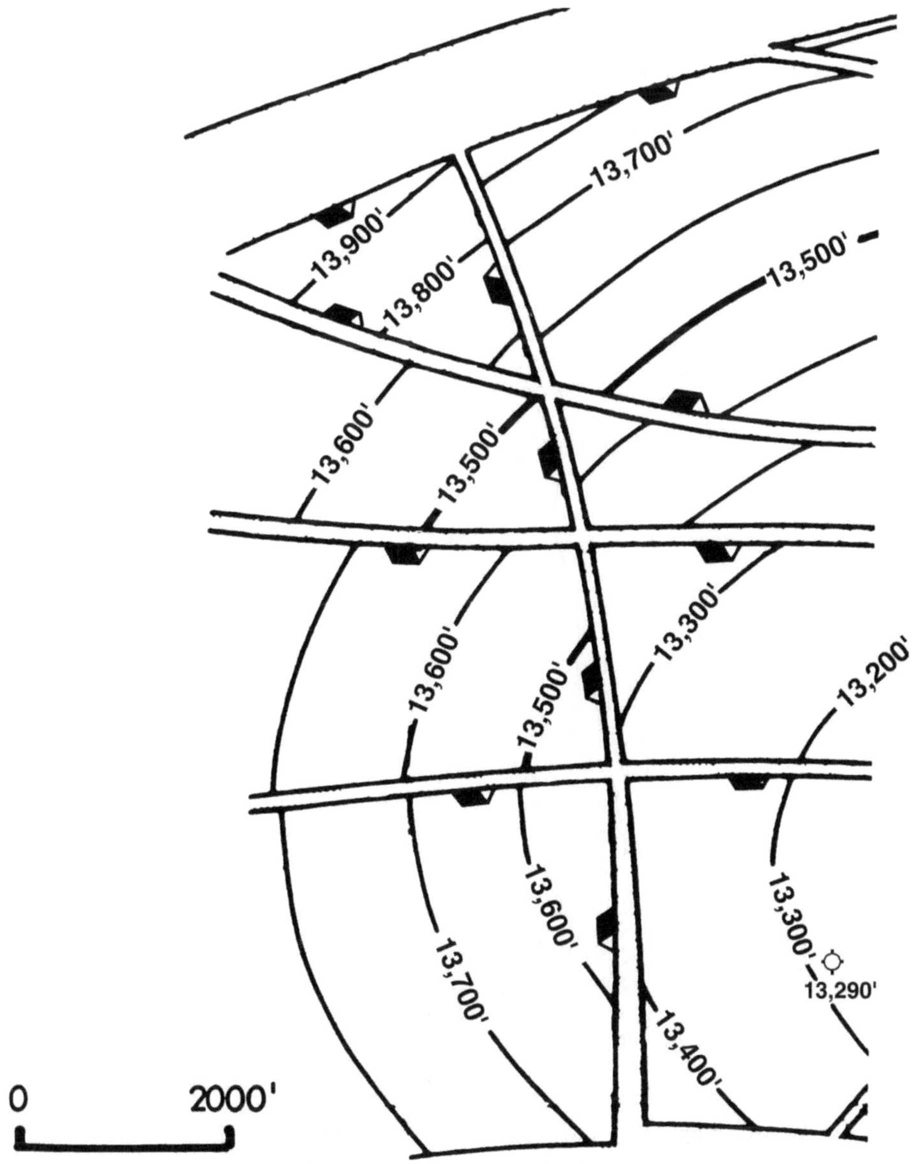

Figure 8-62 Completed structure map showing several intersecting faults with each of the trace intersections forming a "cross." (The final construction is incorrect. No fault traces are offset.)

For contouring purposes, the hanging wall and footwall may or may not have structural compatibility across a growth fault. The primary factors are the vertical separation of the fault and the amount of deposition in the downthrown block. Usually, the larger the fault, the less the compatibility. If structural compatibility does exist, the vertical separation can be contoured across the fault. If compatibility is absent, then each fault block must be contoured separately. At times, only the upthrown or downthrown block of a growth fault is mapped, usually because (1) good control or correlation is lacking across the fault, or (2) only one block is productive and that is the only block mapped.

Figure 8-63 is a structure map on the G Sand in East Cameron Block 270 Field, Gulf of Mexico. This field is productive from 19 separate horizons trapped both upthrown and downthrown to the west-dipping growth Fault A. Rollover into Fault A has formed a north-south trending anticline, which is the primary hydrocarbon trap. This field has produced more than 38 million barrels of oil and condensate and nearly 1 trillion cu ft of gas.

Superimposed on the southern half of the structure map is a portion of the fault surface map for Fault A from −8000 ft to −9000 ft. The structure contours drawn upthrown and downthrown to the fault, in this area, indicate reasonable compatibility in structure across this growth fault, which has a vertical separation of about 800 ft at this level. The existence of this structural

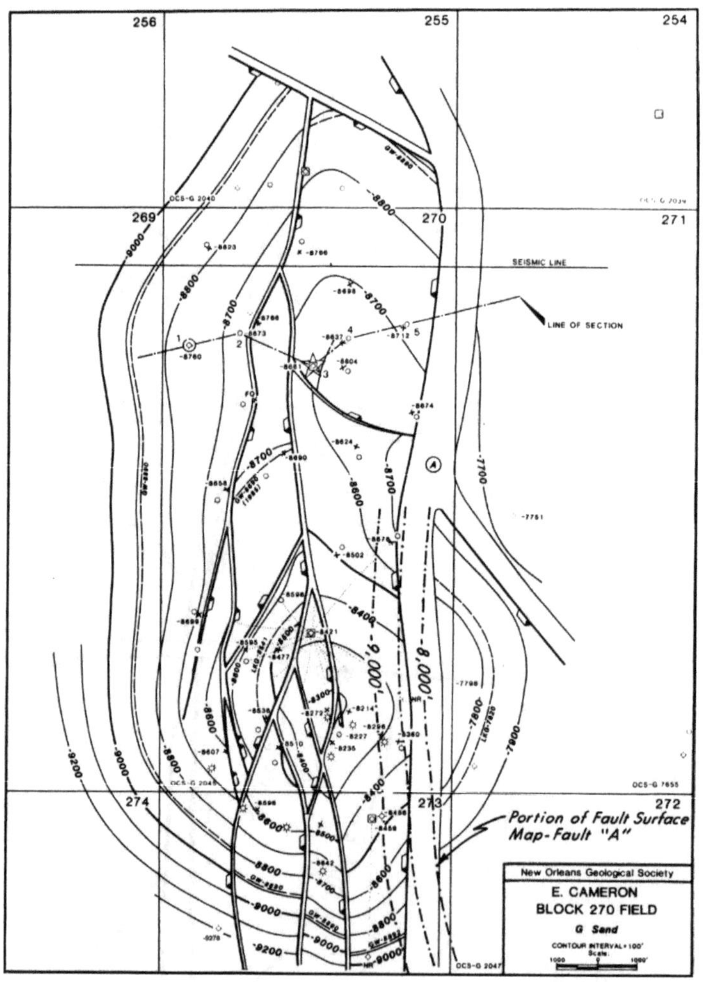

Figure 8-63 Structure map on the G Sand Unit, East Cameron Block 270 Field, Gulf of Mexico. A portion of the fault surface map for Fault A is superimposed on the structure map. The primary hydrocarbon trap is a faulted rollover anticline formed in response to Growth Fault A. (Published by permission of the New Orleans Geological Society.)

compatibility does not, however, ensure such compatibility along fault strike nor with depth (recall the cross section in Figure 8-51). Notice that the rollover anticline is complexly faulted with secondary synthetic and antithetic faults. Although the fault maps for these faults are not published, it is easy to recognize that there is good structural compatibility across all the secondary faults; therefore, the vertical separation must have been used to contour across these faults.

One final note on growth faults. It is commonly observed that the most productive interval in a growth fault complex is the stratigraphic section deposited during the most active fault movement. Therefore, when prospecting around growth faults, it is important to understand the history of the fault movement. The technique of plotting a growth fault's history, the Multiple Bischke Plot Analysis (MBPA) discussed in Chapter 13, is most helpful in evaluating the fault history and its potential for hydrocarbon accumulations.

Diapiric Salt Tectonics

A diapir is a mass of rock that has flowed ductilely and appears to have discordantly pierced or intruded overlying rocks and sediments (after Jackson and Talbot 1991). The word comes from the Greek verb diaperein, meaning "to pierce," and was first used in the Carpathian Mountains of Romania. The most common use of the word is in association with salt and shale masses. Although this may not be the best term to describe a salt body or to correctly describe the motion of salt movement, it is in common use. Most salt diapirs probably formed by a mainly lateral movement of the salt into the diapir, with the movement being caused by differential loading upon the salt layer (Jackson and Talbot 1991; Diegel et al. 1995; Hudec and Jackson 2007), often called *down building*. In other words, the sediments move down into the salt, forcing the salt upwards, often forming a tear-shaped salt body (Fig. 8-72). The diapir increases in relief by growing downward relative to the sedimentary surface as the base of the sediments subsides into the mobile salt. Actual upward movement may be minimal for many domes. The extent of upward movement depends in part on the buoyancy of the salt mass and the strength of the overburden.

Diapiric salt masses are associated with large accumulations of hydrocarbons in many areas around the world, including the U.S. Gulf Coast, the southern zone of Mexico, Gabon, Senegal, the Canadian Arctic, the North Sea and adjacent Europe, Romania, the Zagros Mountains of Iran, and the Ukraine. These *salt domes* are commonly structurally and stratigraphically very complex. They come in all shapes and sizes from small, needle-like spines less than 2000 ft across, such as the Rabbit Island Salt Spine (Fig. 8-64), to very large salt massifs such as the Marchand-Timbalier-Caillou Island Salt Massif, which is more than 27 mi long and 13 mi wide at a depth of 20,000 ft (Fig. 8-65). Regardless of the actual size of the salt mass, most are associated with significant quantities of hydrocarbons. For example, the Rabbit Island Field has estimated reserves of 1.5 trillion cu ft of gas and 55 million barrels of condensate and oil. The fields associated with the Marchand-Timbalier-Caillou Island Salt Massif have combined ultimate reserves of over 500 million barrels of oil. Figure 8-66 shows the extent of salt structures in the northwestern Gulf of Mexico and adjacent interior basins. Salt masses in the Southern Permian Basin of the North Sea area are shown in Figure 8-67. Notice the general clustering of salt pillows versus piercement salt bodies.

An important development in petroleum exploration has been the recognition during the past 35 years that some irregular salt masses are sheets of salt that are detached from the source layer (Diegel et al. 1995). The salt often rides up ramps and forms sheets on different structural layers. Their size varies from small to hundreds of square miles in area. Typical sequences of strata and trapping structures exist beneath these salt sheets, and significant discoveries have been made.

Several types of normal faulting are commonly associated with most diapirs, as well as with many deep-seated salt structures. If the regional stress field is nearly isotropic, the normal faults develop in an outward radiating pattern from the salt structure, and these are called *radial faults*.

If the regional stress field is more anisotropic, or the structure exhibits a variable growth history in different areas around the dome, the fault pattern may have a more preferred orientation, resulting in *peripheral faults* that are parallel to the face of the salt or ride down the flank of the salt. Rowan et al. (1999) discuss faults associated with various salt structures in much more detail.

Growth of a salt structure can develop in different stages. Graben blocks often form in the sediments over the salt mass due to extension. The dip of associated faults can be toward the dome as well as away from it. Although not directly associated with the salt dome development, regional growth faults are often found on or near salt structures. In certain areas of the world, salt diapirs are associated with reverse faults.

Hydrocarbon Traps. Salt structures have an extremely varied and complex geometry resulting in numerous types of hydrocarbon traps. Any single salt structure may have more than one type of trap, depending on the history of the salt structure and surrounding sediments. Figure 8-68 (Halbouty 1979) shows an idealized section through a salt dome illustrating the more common types of hydrocarbon traps, including (1) a simple domal anticline, (2) graben fault traps over the dome, (3) porous caprock (limestone or dolomite), (4) up-dip pinchouts of permeable units, (5) traps beneath an overhang, (6) traps against the salt itself, (7) unconformities, (8) traps along faults downthrown away from the dome, and (9) traps along faults downthrown toward the dome. Fault traps like 8 and 9 can also be considered first-fault out-of-the-basin type traps. In addition to the traps shown in Figure 8-68, radial and peripheral faults also serve as excellent hydrocarbon traps.

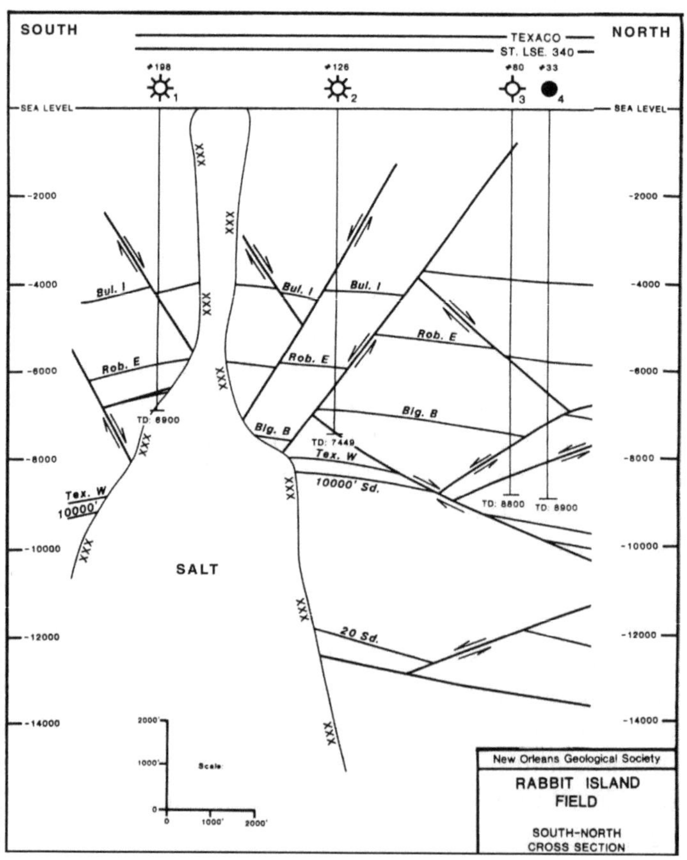

Figure 8-64 Cross section showing the needle-like Rabbit Island Salt Spine, which is less than 2000 ft across to a depth of about 7000 ft. (Published by permission of the New Orleans Geological Society.)

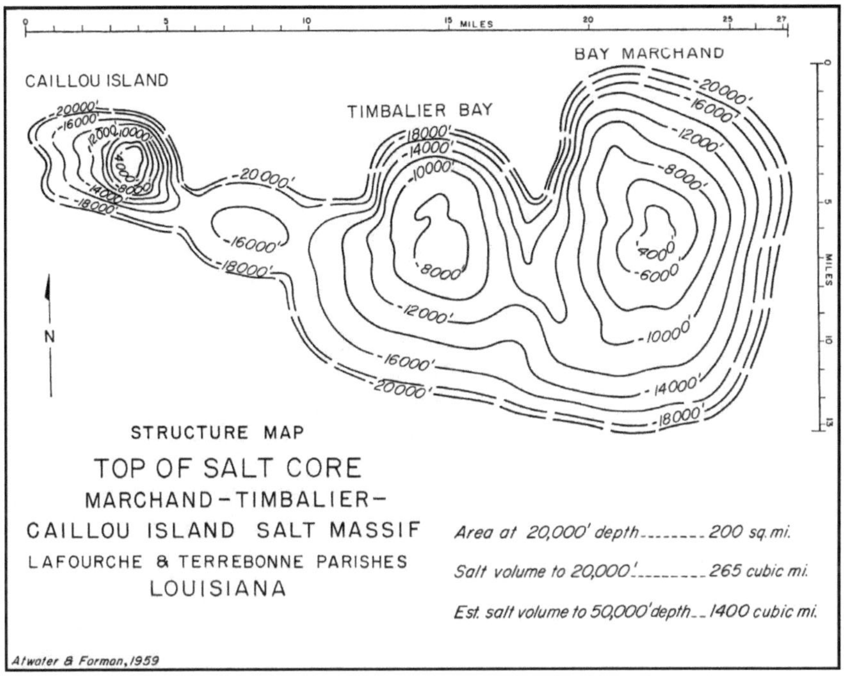

Figure 8-65 Structure map on the top of the Marchard-Timbalier-Caillou Island Salt Massif. (From Atwater and Forman 1959. AAPG©1959, reprinted by permission of the APPG whose permission is required for further use.)

Considering their complexity, salt structures require very precise interpretation and detailed mapping in order to exploit all the hydrocarbon potential. When working with salt-related structures, maps are needed for the salt itself and for salt/sediment, salt/fault, and fault/sediment interfaces, in addition to associated unconformities and top of shale, if present.

Contouring the Salt Surface. No special techniques are required to contour the salt surface. The key to making a good salt map is having sufficient data. Typically, only a limited number of wells penetrate salt, and commonly the quality of seismic data near steeply dipping salt is poor. Figure 8-69 is the salt surface contour map for Main Pass Block 299 Field, which is of special interest because the salt dome has an overhang on the south and southwest flanks. Notice how the contours under the overhang are dashed, making the map easier to use when preparing structure contour maps, as well as making the salt map easier to read. Some of the deviated wells were drilled to traps beneath the overhang.

Although the steep flanks of salt bodies are frequently poorly imaged on seismic data, there are other geophysical techniques that can provide additional data to assist in mapping a salt body. A salt proximity survey can consist of either a source above the salt and receivers in a well near the salt (e.g., Barker and Guzman 1992; Li et al. 2003; O'Brien 2004; Lou et al. 2012). The technique takes advantage of difference in velocity between salt and sediment to identify the location of the salt face. Another version of a salt proximity uses a source exterior to the salt body and receivers in a well within the salt. This arrangement is technically more challenging and frequently identifies only a small segment of the salt body (O'Brien 2004). An additional geophysical technique is a walkaway or walkaround vertical seismic profile (VSP) (Barker and Guzman 1992). These surveys can record reflections from the salt face that are not recorded on some surface seismic. In addition, properly designed conventional 3D seismic surveys with very long offsets can do an excellent job of delineating a very steeply dipping salt flank (e.g., Figure 1 in Hartman et al. 2017). Data from all of these geophysical sources, where available, should be combined with well log data to make the most accurate map of the salt surface possible (Li et al. 2003).

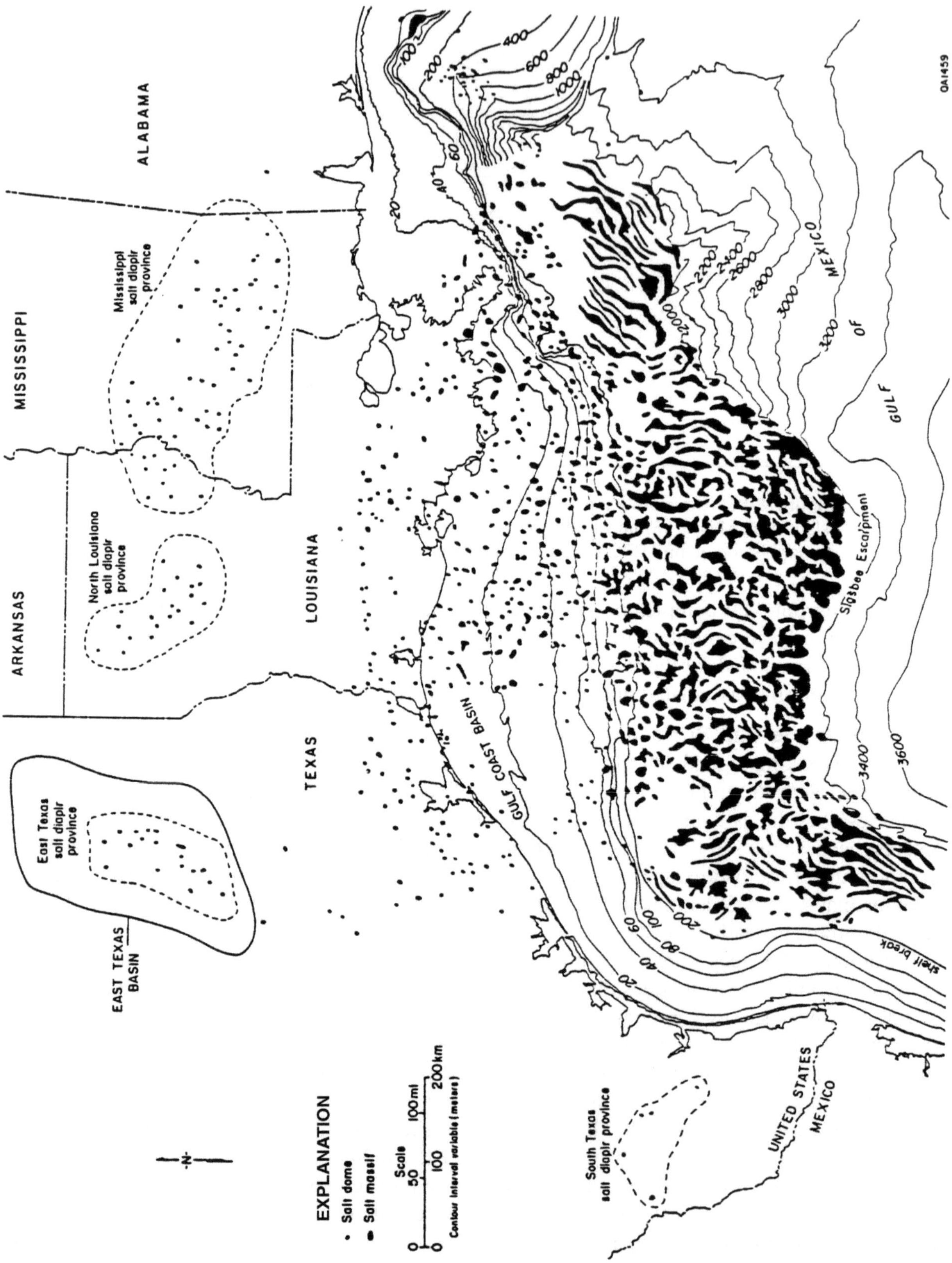

Figure 8-66 Salt structures in the Gulf of Mexico and adjacent interior basins. (From Seni and Jackson 1983. AAPG©19831983, reprinted by permission of the AAPG whose permission is required for further use.)

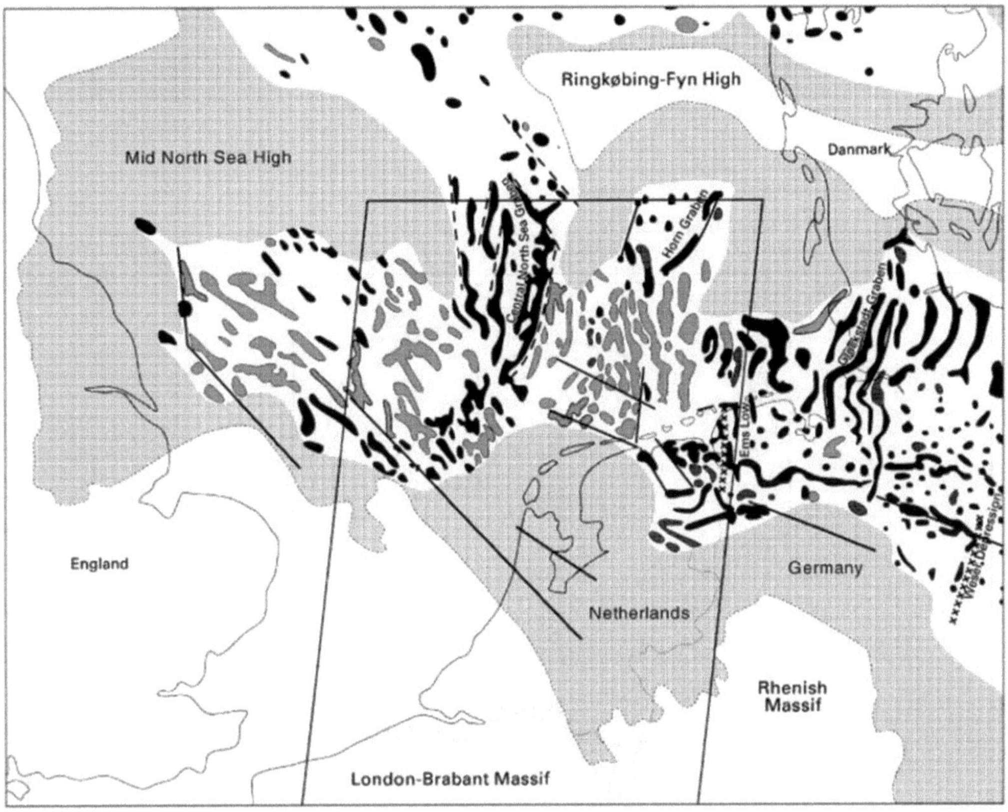

Figure 8-67 Map of salt pillows and piercement salt bodies in the Southern Permian Basin, North Sea Area. (From Remmelts 1995. AAPG©1995, published by permission of the AAPG whose permission is required for further use.)

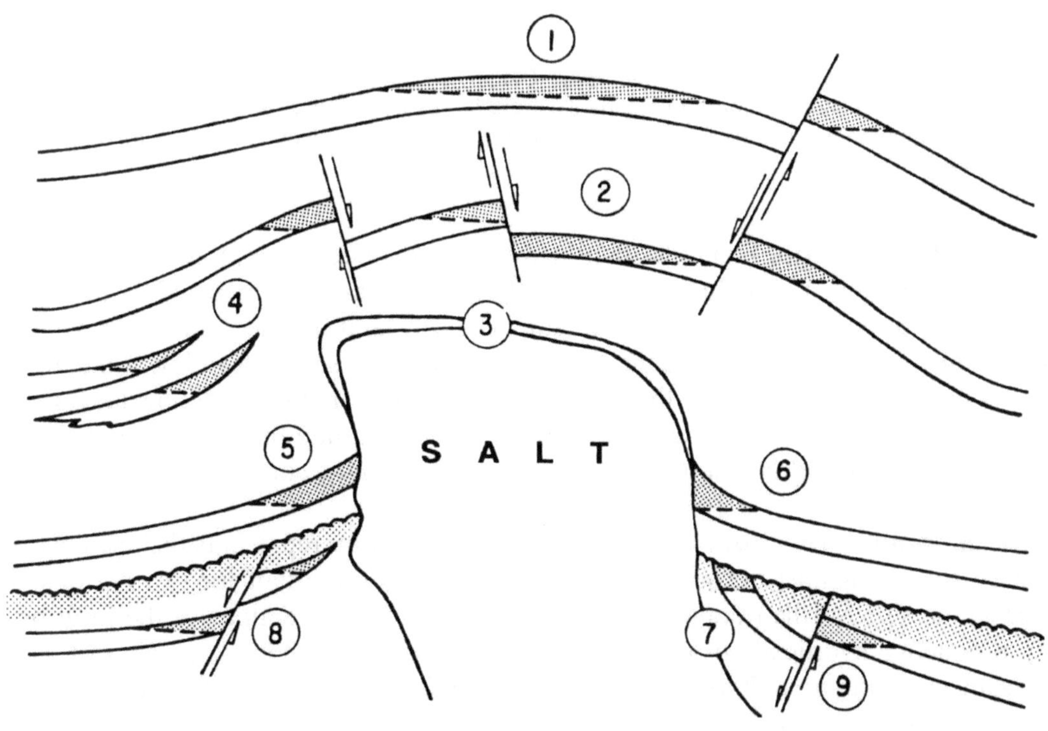

Figure 8-68 An idealized diapiric salt structure showing common types of hydrocarbon traps. (From Salt Domes, by Michel T. Halbouty. Copyright 1979 by Gulf Publishing Company, Houston, Texas. Used with permission. All rights reserved.)

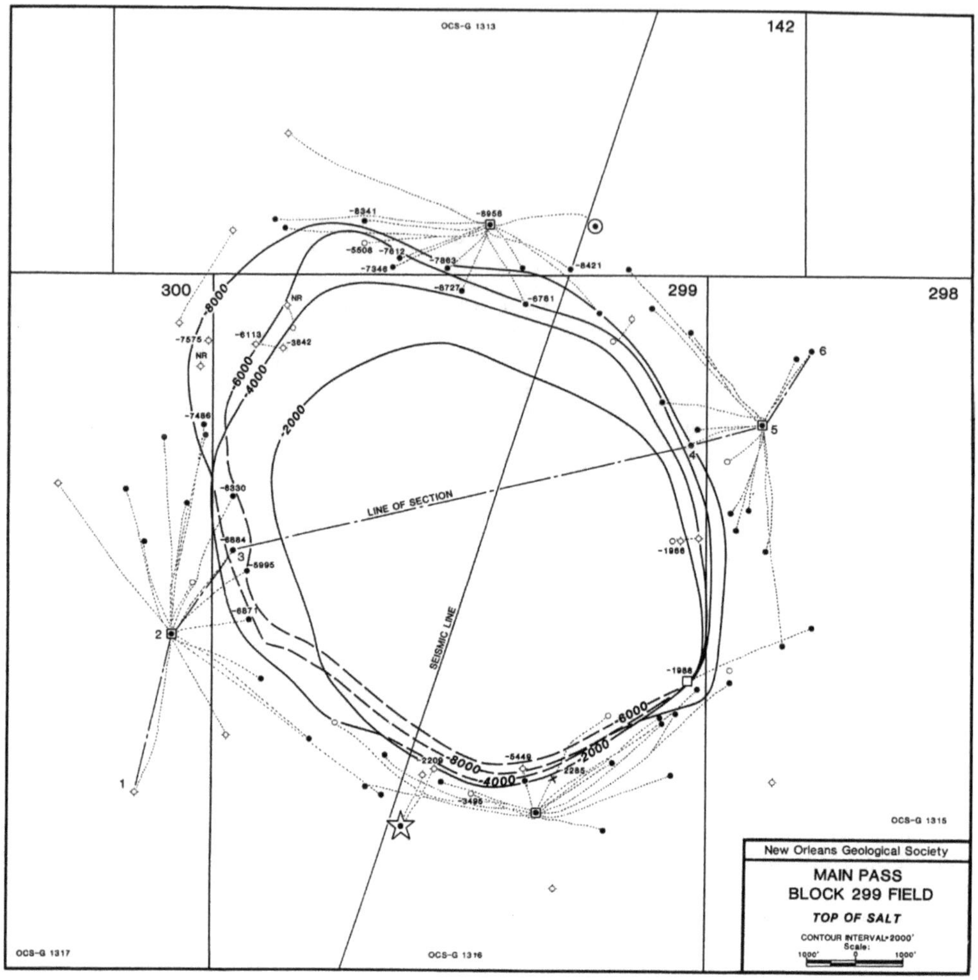

Figure 8-69 Surface of salt, Main Pass Block 299 Field, Offshore, Gulf of Mexico. Observe the dashed contours to clearly illustrate the salt overhang. Some deviated wells extend beneath the overhang. (Published by permission of the New Orleans Geological Society.)

Salt/Fault Intersection. Figure 8-70a illustrates the method for contouring the intersection of a salt mass and a fault. The technique is basically the same as that presented earlier for contouring the upthrown or downthrown traces of a fault. The salt/fault intersection occurs where the structure contours on the salt intersect the fault contours of the same elevation. Like all intersecting surfaces presented in map view, the salt/fault intersection should be delineated as shown in the figure. This line of intersection can serve as a directional guide for planning a deviated well that tests potential reservoir units in the structural trap between the surfaces.

Salt/Sediment Intersection. The salt/sediment intersection represents the termination of a sedimentary surface against the salt mass. This interface is important for two reasons: (1) it often serves as the seal for the trapping of hydrocarbons; and (2) it delineates the limit of the potentially productive hydrocarbon-bearing horizon. Figure 8-70b shows a portion of a salt contour map (same as in Figure 8-70a) and the line of termination of the 8000-ft Horizon against the salt. The termination of the horizon occurs where the structure contours intersect salt contours of the same elevation. The termination line bends sharply above 800 ft. This was located accurately by integrating interpolated contours on the two surfaces.

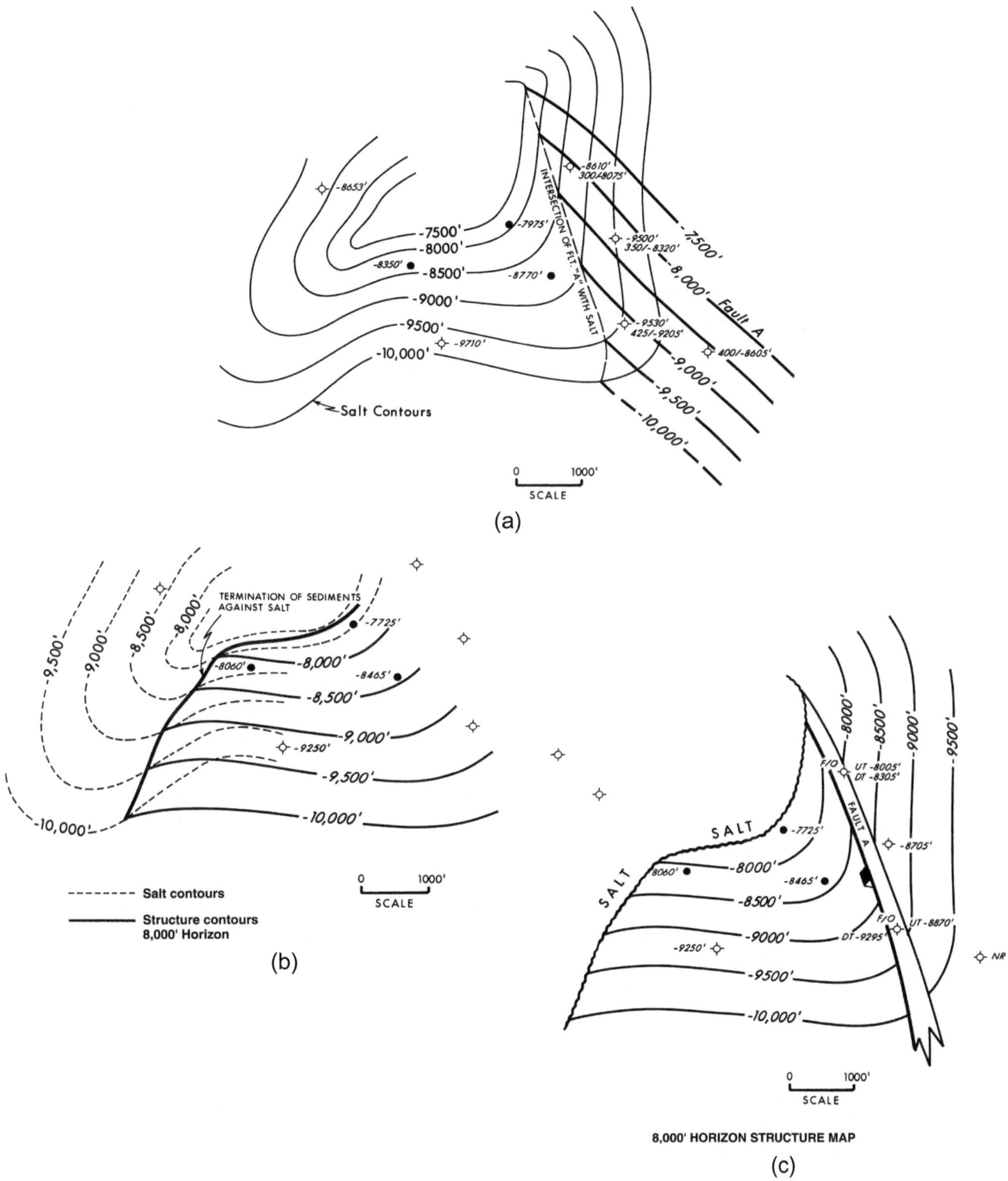

Figure 8-70 (a) Integration of the contour maps on the salt surface and on Fault A delineates the intersection of the fault with salt. (b) Integration of the contour maps on the salt surface and on the structure contours for the 8000-ft Horizon. The salt/sediment boundary is located where the salt contours intersect the fault contours of the same elevation. (c) Completely integrated structure map on the 8000-ft Horizon, Reservoir A, in the southeast portion of this piercement salt structure. This oil reservoir is bounded to the north and west by salt, to the east by Fault A, and to the south by an oil/water contact at a depth of −8605 ft.

In some cases, sediments continue all the way to the salt face as with the shallow horizons, as illustrated in Figure 8-72. In these cases, the structure map on the surface of the salt can be integrated with the structure map of the reservoir to identify the termination of the reservoir unit. In other cases, the salt face is encased in a “sheath” of shale and impermeable rock, as at White Castle dome (Johnson and Bredeson 1971) and the reservoir sands terminate against this sheath, prior to reaching the salt. In still other cases, the salt structure was a positive feature during deposition and porous sediments pinch out prior to reaching the salt face. This situation is common in many turbidite fields in the deepwater Gulf of Mexico (e.g., Thunder Horse North Field, Reitz and Hartman 2017). In these latter two cases, it can be difficult to map a surface that can be integrated with the reservoir horizon to delineate the edge of the reservoir because the edge of a sheath of sediment is frequently not as visible on geophysical data as is the edge of salt.

Completed Structural Picture. The mapping of all intersections and the integration of the fault, salt, and structure maps results in a completed structural interpretation, such as that shown in Figure 8-70c for the 8000-ft Horizon. This example is relatively simple, but it does illustrate the use of the mapping techniques. Salt features are normally associated with highly faulted, highly complex structures such as the one shown in Figure 8-71, which is a structure map on the Grand Isle Ash at Grand Isle Block 16 Field, northern Gulf of Mexico. This structure map is unique in that it contains contours both on the salt and on the structural horizon. It is the intersection of these salt and structure contours that delineates the salt/sediment boundary.

As mentioned in Chapter 6, steeply dipping structures, such as those related to salt tectonics, require the layout of a number of cross sections to help develop an accurate geological interpretation and aid in the structure map construction. Typical cross sections for steeply dipping structures such as a salt dome are designed to incorporate both straight and deviated wells. Initially, problem-solving type cross sections are laid out to help resolve correlation problems and aid in developing the structural interpretation. Later, during the advanced stages of mapping, these sections can be converted to finished illustration sections for display and presentation. For further information on the use of cross sections as an aid to structure interpretation and mapping of salt structures, refer to Chapter 6.

A salt diapir commonly forms the core of a large structural high. In response to the draping of strata around the diapir as the structure develops, the strike of the strata is typically parallel or subparallel to the face of the salt. So, the structural contours on a given horizon tend to be parallel or subparallel to contours on the salt surface. This is illustrated in Figure 8-71 for Grand Isle Block 16 Field.

Around certain salt structures, subsurface data show structure contours intersecting salt at a sharp angle. Where this happens, it may be an indication of several possible situations, including (1) the possibility of a salt overhang, such as the one shown in Figure 8-72, or (2) an unrecognized peripheral fault sliding down or located near the face of the salt, causing the strike direction of the contours to *appear* to turn into the salt. Actually, the contours are striking into the downthrown side of the peripheral fault. Figure 8-73 is an example of such a situation. Notice on the southwestern portion of the salt dome that the contours appear to strike directly into the salt. Due to the complex nature of the faulting pattern, Faults C, D, and E, which appear as radial faults shallow in the section, become peripheral faults with depth, paralleling the southwest flank of this salt structure. As a result, the structure contours tend to strike at various angles into the fault. In contrast, observe the general conformity of the structure contours with the salt on the western and eastern flanks of the structure not affected by peripheral faulting.

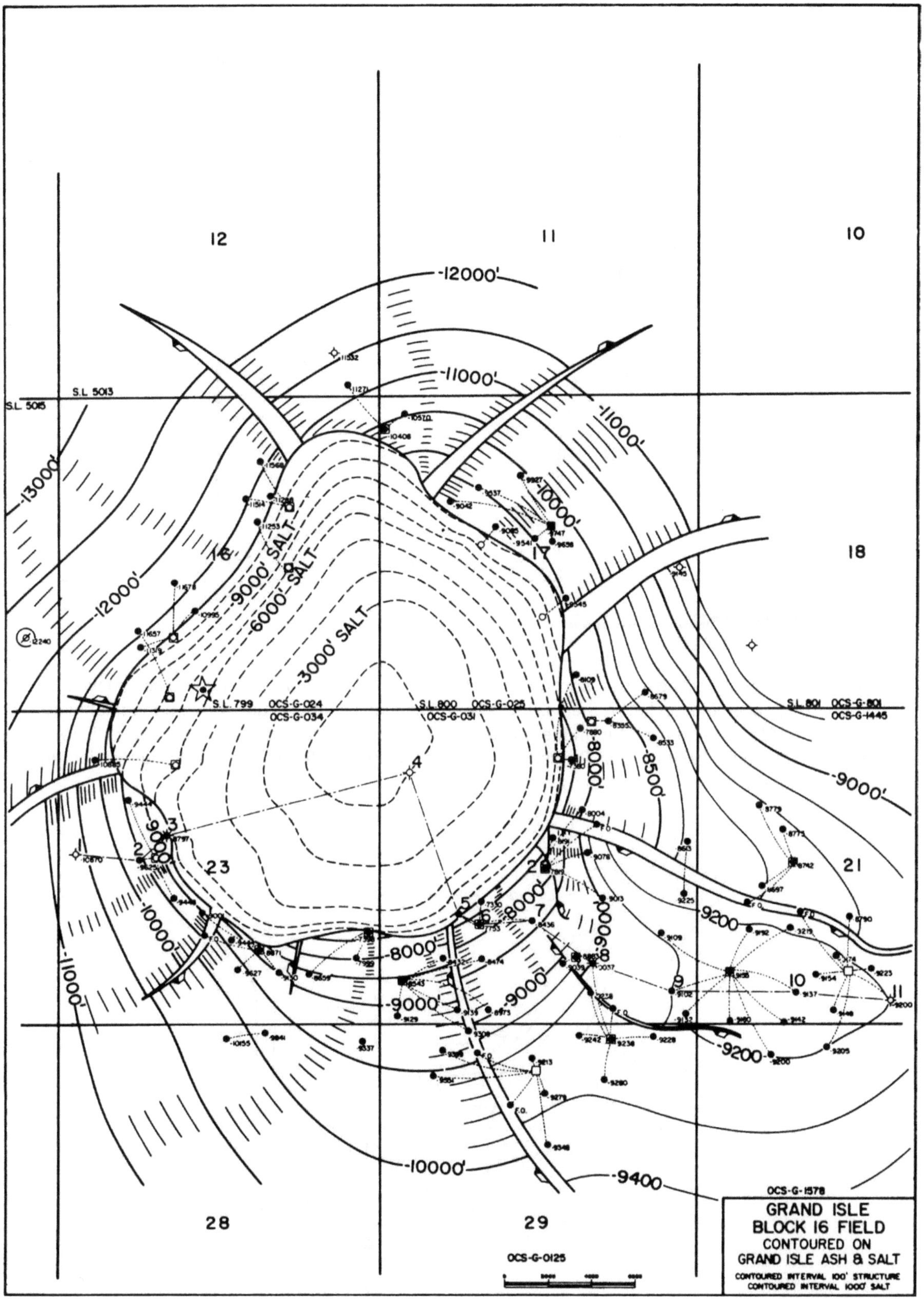

Figure 8-71 Structure contour map on the Grand Isle Ash and on the salt at Grand Isle Block 16 Field, northern Gulf of Mexico. The outline of the salt at this map level is defined by the intersection of the salt and structure contours at the same elevation. (Published by permission of the Lafayette Geological Society.)

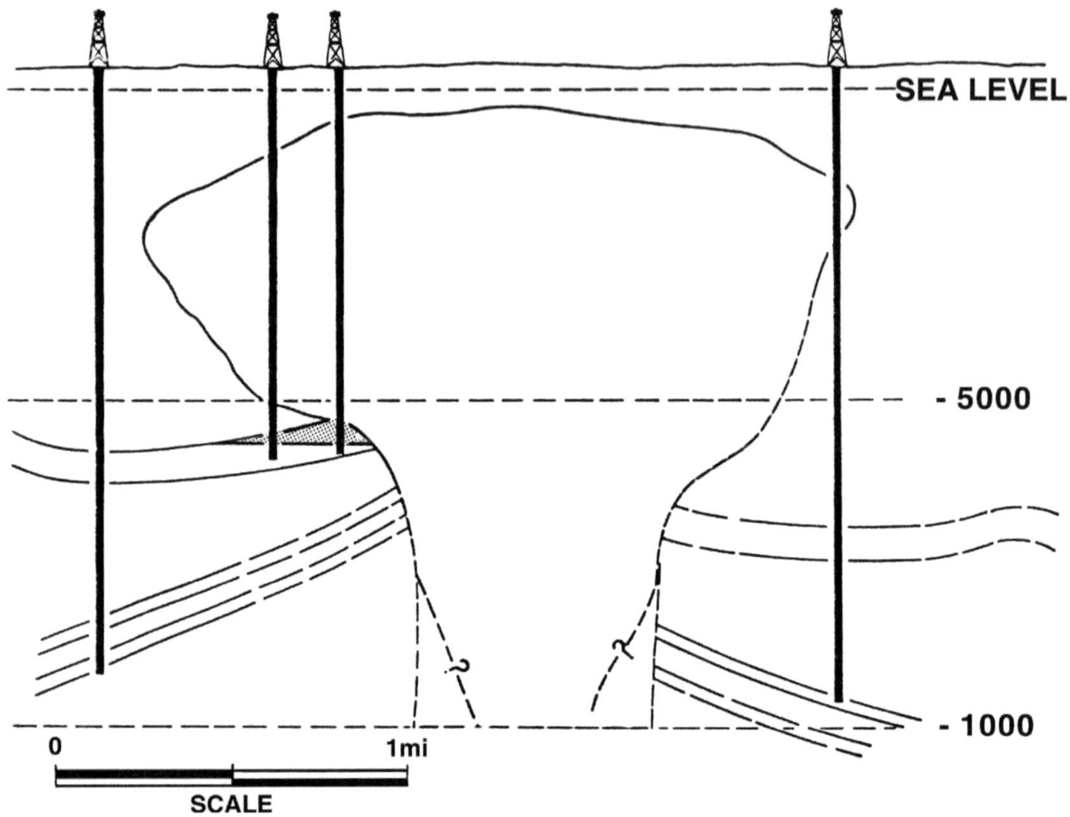

Figure 8-72 Generalized cross section of Bethel Dome, Anderson County, Texas, showing a hydrocarbon accumulation below the salt overhang. The significant amount of overhang might indicate that the dome is detached from the salt source bed at depth. (From Salt Domes, by Michel T. Halbouty. Copyright 1979 by Gulf Publishing Company, Houston, Texas. Used with permission. All rights reserved.)

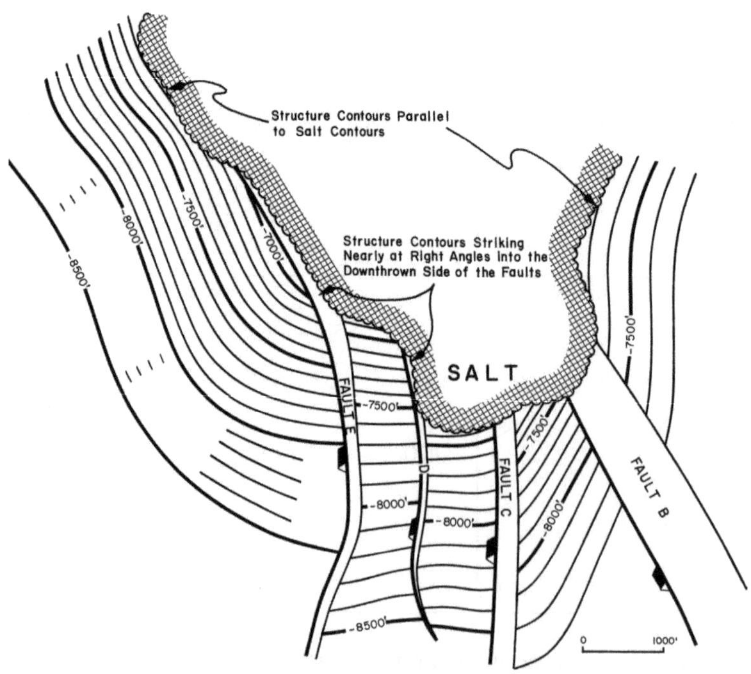

Figure 8-73 Structure contours striking into the downthrown side of Faults E and D give the appearance that the contours are striking into the salt.

Strike-Slip Fault Tectonics

Strike-slip faults are high-angle to vertical faults that form under horizontal compression. A strike-slip fault system may have great linear extent, such as the San Andreas Fault complex, where major lithospheric plates are involved, or it may occur within a local or subregional area as a limited system of finite length. These local strike-slip systems are sometimes referred to as *compartmental* because of the independent deformation on either side of the fault (Fig. 8-74a). In this type of deformation, a structure that seems to be cut by a strike-slip fault need not have a *severed* portion offset on the other side of the fault, since the faulting and deformation are contemporaneous and independent on either side of the fault (Fig. 8-74b). In addition to horizontal motion, vertical movement of varying degrees is commonly associated with strike-slip fault systems, as well as fault block rotation. Methods of interpretation and mapping of strike-slip areas are described in Chapter 12.

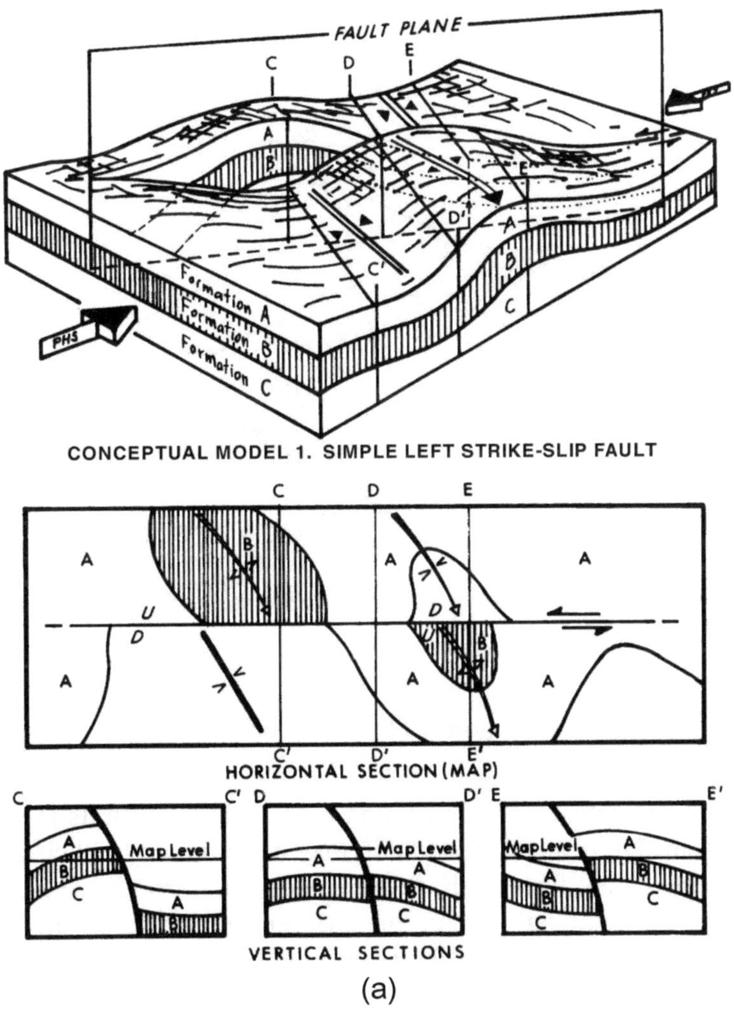

Figure 8-74 (a) Conceptual model of simple left strike-slip fault: Top—block diagram; middle—plan view; bottom—cross-sectional views. (From Stone 1969. Reprinted by permission of the Rocky Mountain Association of Geologists.) (b) Block diagram illustrating different deformational patterns on opposite sides of a finite strike-slip fault. (From Bell 1956. AAPG©1956, reprinted by permission of the AAPG whose permission is required for further use.) (c) The change in direction of asymmetry or fold frequency across the fault is due to a different response to the compressional forces. (From Brown 1982. AAPG©1982, reprinted by permission of the AAPG whose permission is required for further use.)

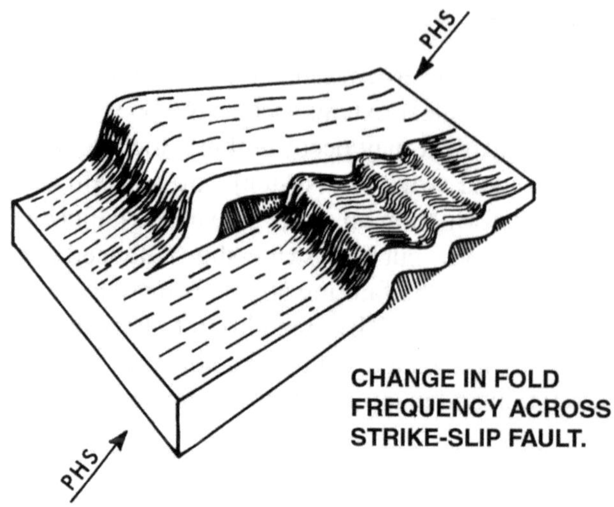

(b)

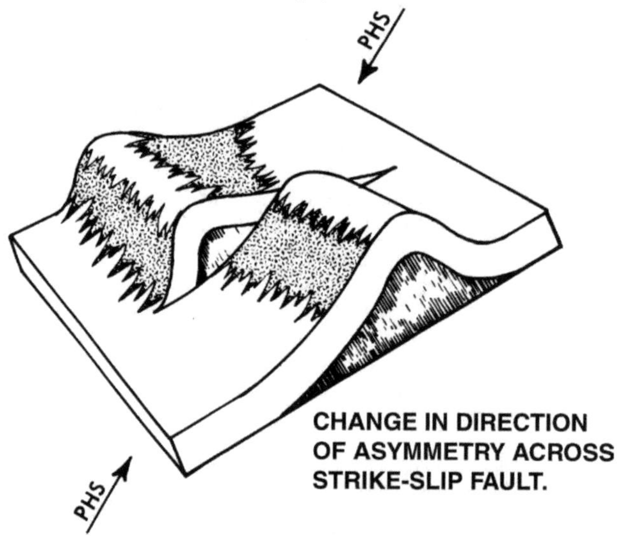

(c)

Figure 8-74 (*continued*)

The primary hydrocarbon traps associated with strike-slip faults are anticlines that straddle the strike-slip system. These anticlines, which may be faulted by either normal or reverse faults, are good traps because they form early and commonly develop large closures sufficient to trap economic quantities of hydrocarbons. In this chapter, our interest is in the mapping techniques that are applicable to these strike-slip fault systems. Figure 8-75 shows an example of a strike-slip fault system with offset faulted anticlines on each side of the fault. In many cases, the faults cutting the anticlines are small and simply offset the structures with little, if any, change in structural attitude across the fault.

Two different types of mapping techniques are required to map these strike-slip faulted structures. Normally, the structures on either side of the main strike-slip fault must be mapped independently (Fig. 8-75) because structures tend to terminate against the strike-slip fault (Bischke et al. 1990). As for the individual anticlines that form on either side of the main strike-slip fault, the vertical separation technique is commonly applicable for mapping across the faults within these anticlines. In other words, the faulted anticlines commonly exhibit contour (structural) compatibility across the small normal or reverse faults that cut the anticlines, unless some type of rotation occurs along the fault.

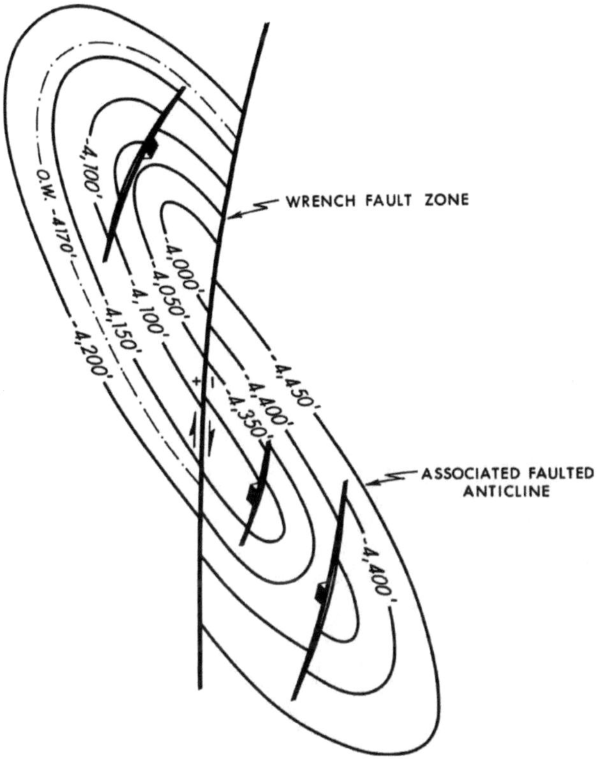

Figure 8-75 Typical strike-slip fault system with associated faulted anticlines.

Figure 8-76 is a structure map from the Rosecrans Oil Field, California, USA. Evidence from seismic well log correlations and focal mechanism solutions shows that the Newport/Inglewood trend is a transpressional system (Wright 1991) (also consult Figs. 12-12 to 12-14). The Rosecrans Oil Field is one structure along this tread that exists along the western portions of the Los Angeles Basin. Rosecrans Field can be used to illustrate the mapping techniques applicable for a strike-slip fault system. Notice that the anticlines adjacent to the Inglewood Fault are cut by small reverse faults. The cross-section insert and the structure map show that there appears to be structural compatibility across most of these small faults. With the use of well control and seismic data, the vertical separation for each fault can be determined and used to contour across the faults.

Notice the two faults labeled C and D on the structure map. Based on this structure map and available cross sections, some fault block rotation, in addition to dip-slip motion, occurred on these faults and resulted in a loss of structural compatibility across the faults. Therefore, it may be difficult, if not impossible, to contour across these faults using the vertical separation. The structures on either side of Faults C (Compton Thrust Fault) and D may require independent contouring. However, north of Fault D the structure again has good contour compatibility across the northern two faults shown on the map. Another possible interpretation for Faults C and D is that these faults do not exist, and instead the apparent faults are actually axial surfaces that bisect kinks in the strata (see Chapter 10).

Figure 8-77 is a structure map on the top of the Ranger Zone in the Wilmington Field in the Los Angeles Basin, California. The anticlinal structure that forms the Wilmington Field is 11 mi long and 3 mi wide. This giant field has produced over 1.2 billion barrels of oil. During the Middle Miocene, compressive stresses formed a north-south couple, folding the strata and establishing the present northwest-southeast Wilmington structure. During its complex structural history, the anticline developed a series of normal faults as a result of tensional forces acting along the structural axis. These normal faults are small, ranging from 100 ft to 400 ft of vertical separation and dipping at angles between 45 deg and 65 deg. Many of the faults are sealing and therefore divide the field into at least seven major production blocks (Mayuga 1970).

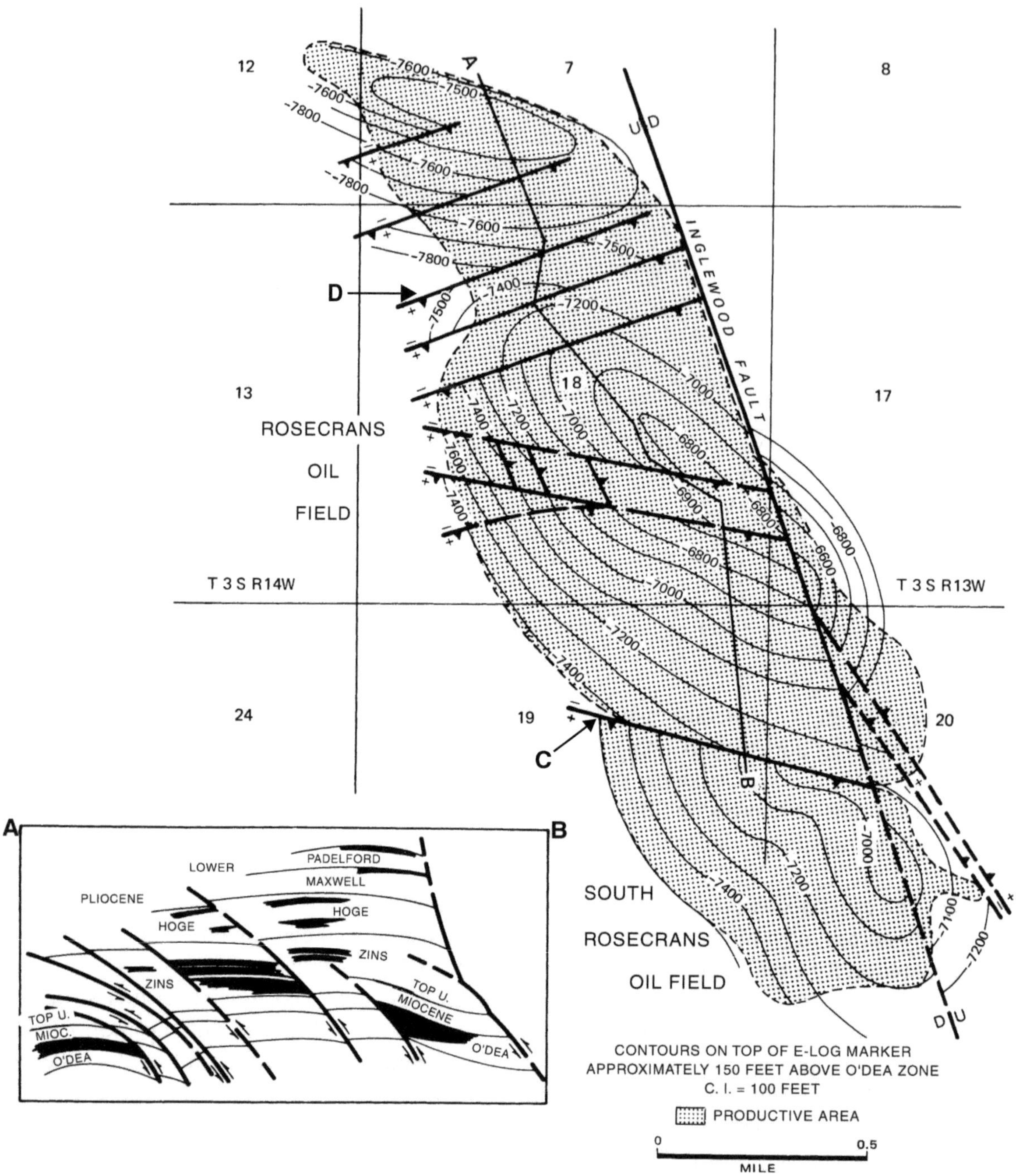

Figure 8-76 Rosecrans Field structure (after California Div. Oil & Gas, 1961) shows a distinct pattern of reverse-faulted anticlinal folds oriented obliquely to the Inglewood Fault. (Modified from Harding 1973. AAPG©1973, reprinted by permission of the APPG whose permission is required for further use.)

A review of the structure map in Figure 8-77 suggests very good structural (contour) compatibility across all the faults in the field. Therefore, all structure and fault map integration in this field would use the vertical separation technique for contouring across the faults. The Wilmington Field is another example of a strike-slip fault-associated structure that exhibits good internal structural compatibility across the faulted anticline.

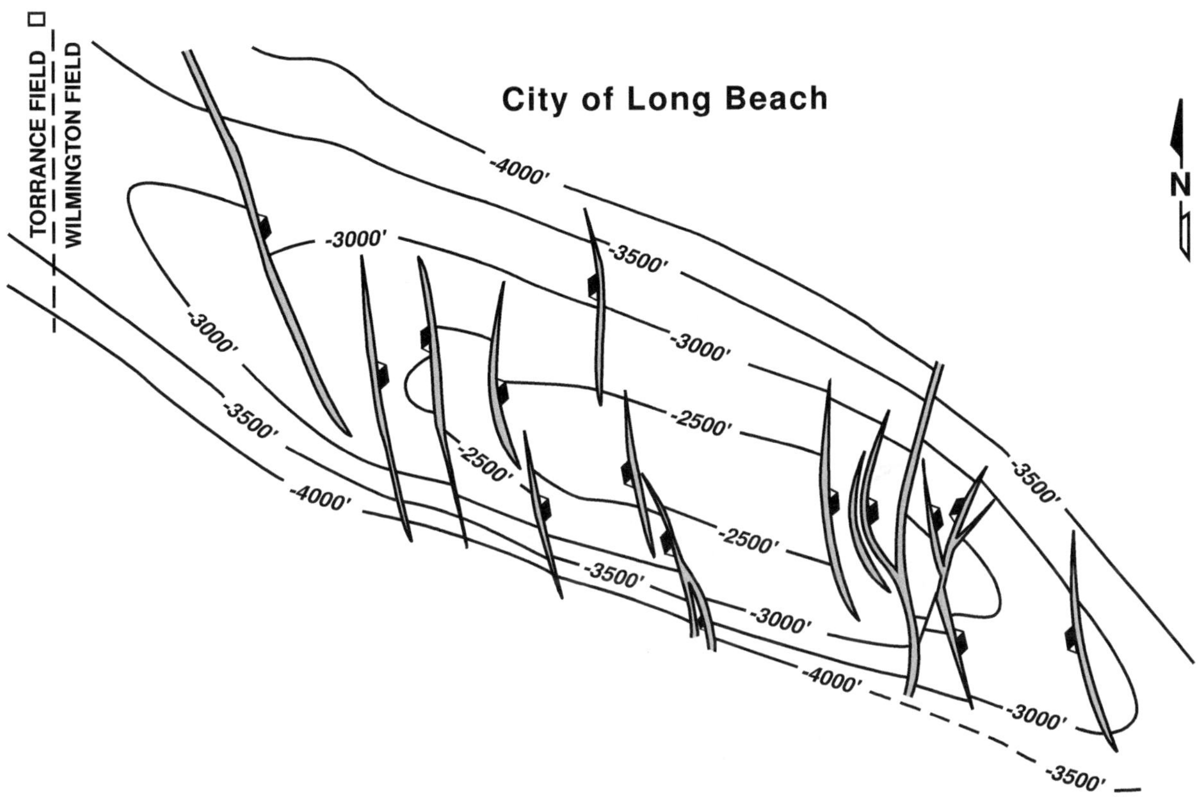

Figure 8-77 Structure contour map on the top of the Ranger Zone, Wilmington Oil Field, southern California. The normal faults in the field exhibit several patterns: (1) single, (2) compensating, (3) bifurcating, and (4) intersecting. Length of the mapped area is about 10 miles. (After Mayuga 1970. AAPG©1970, reprinted by permission of the AAPG whose permission is required for further use.)

Compressional Tectonics

Our discussion of compressional tectonic settings includes both high-angle reverse and thrust faults because they commonly occur together in this setting. Also, this section centers on fold-and-thrust belts, which include forearc, backarc, and collisional belts, as they make up the most prolific compressional habitat. The most common hydrocarbon trap is the hanging wall anticline, which includes such structures as fault propagation folds (snakeheads), fault bend folds, and duplex structures (see Chapter 10). For example, in the Wyoming-Utah backarc fold-and-thrust belt fields in the Rocky Mountains, USA (Fig. 8-78), nearly all the hydrocarbons are trapped in the hanging wall of the Absaroka Thrust. They include such fields as Painter Reservoir, Whitney Canyon, Ryckman Creek, and Anschutz Ranch Fields. Nearly all these fields are found in asymmetric anticlinal folds with the steep limb to the east. Collisional zones, such as the Zagros collisional belt of Iran, are some of the most prolific of all fold-and-thrust belt types. At one time, the Zagros belt accounted for 75% of the world's fold-and-thrust belt production.

An extensive literature search yields very few examples of contoured fault surface maps and integrated structure maps for compressional structures. There are volumes of published fault and integrated structure maps for extensional petroleum areas, but they appear to be scarce for compressional areas. The inverse is true for balanced cross sections. A significant number of balanced cross sections have been published for compressional structures, but very few are published in extensional areas.

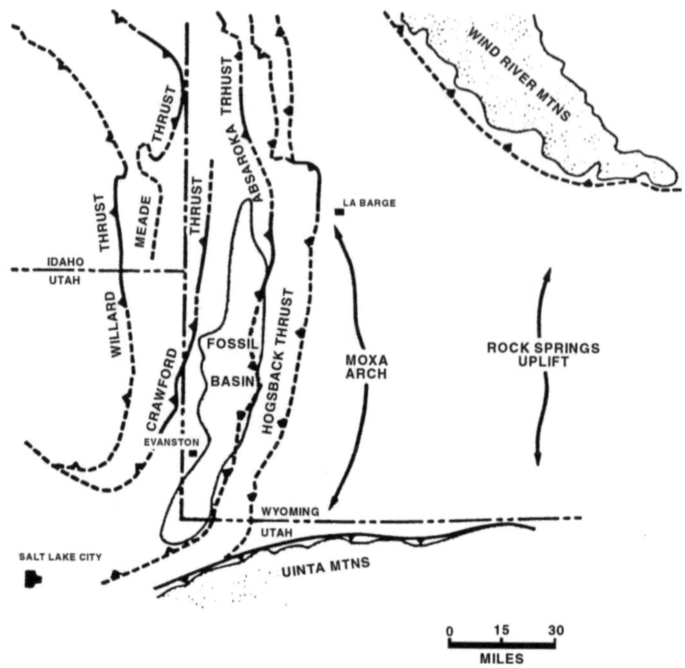

Figure 8-78 Maps showing the position of the Fossil Basin in the hanging wall of the Absaroka Thrust Plate. (From Lamerson 1982. Reprinted by permission of the Rocky Mountain Association of Geologists.)

Fault surface mapping is not commonly done in compressional areas, although there is *no* reason why it should not be; indeed, the construction of fault surface maps would aid in the understanding of the geology. One possible reason for not making fault surface maps and integrating them with structure maps is that, in many cases, hydrocarbons are trapped in the hanging wall anticline, only slightly disrupted or controlled by reverse or thrust faulting. Therefore, fault surface maps are not prepared. Another reason is that balanced sections are often constructed; therefore, the construction of fault surface maps is assumed unnecessary. We believe, however, that the mapping of all related faults is an *essential* part of any structural interpretation. In previous sections of this text, we showed that cross sections and seismic sections can misrepresent the actual structure because of the orientation of the line of section. We also mentioned in Chapter 7, with regard to the tying of seismic lines, that any nonvertical event can tie. The test of a valid fault interpretation is the preparation of a fault surface map that is geologically reasonable based on the available data.

We have shown that fault trace construction for any mapped horizon, particularly in areas of steeply dipping beds, *cannot be done intuitively*. The accurate delineation of a fault trace for any mapped horizon requires the integration of a fault map with the structure map for that horizon. Remember, a fault trace on a specific horizon may be misleading as to the trend and shape of the fault surface itself. The construction of a fault trace on a structure map based on a few well picks and seismic lines is guesswork and is often incorrect. We therefore encourage the construction of fault surface maps in compressional areas.

Reverse Faults. Earlier in this chapter, we showed the techniques for construction of a reverse-fault surface map and the integration of this map with a structure map (Figs. 8-25 and 8-26). In this section, we look at intersecting reverse faults. Unlike the case involving normal faults, where the fault appears as a horizon discontinuity on the map (a gap), no plan view horizon discontinuity is present on a structure map with a reverse fault. The contours on the upthrown block (hanging wall) overlie the contours on the downthrown block (footwall).

The example in Figure 8-79 is drawn in a form similar to the normal fault examples except that fault transport is reversed to show what happens due to compression. The pattern of intersecting faults illustrates the mapping techniques, but the geometry may not be an accurate predictive model. In other words, reverse faults on a plunging anticlinal structure may result from a volume problem created as planar beds are forced into the compound curvature of the nose. Such faults may not necessarily intersect, because the compression may be accommodated on different stratigraphic levels by separate faults. Intersecting faults do occur in compressional areas, however, and therefore understanding the mapping techniques for these faults is important.

Figure 8-79a shows the fault contour map for reverse Faults A and B. Since we are dealing with intersecting faults, the line of intersection must be shown on the fault surface map. Also,

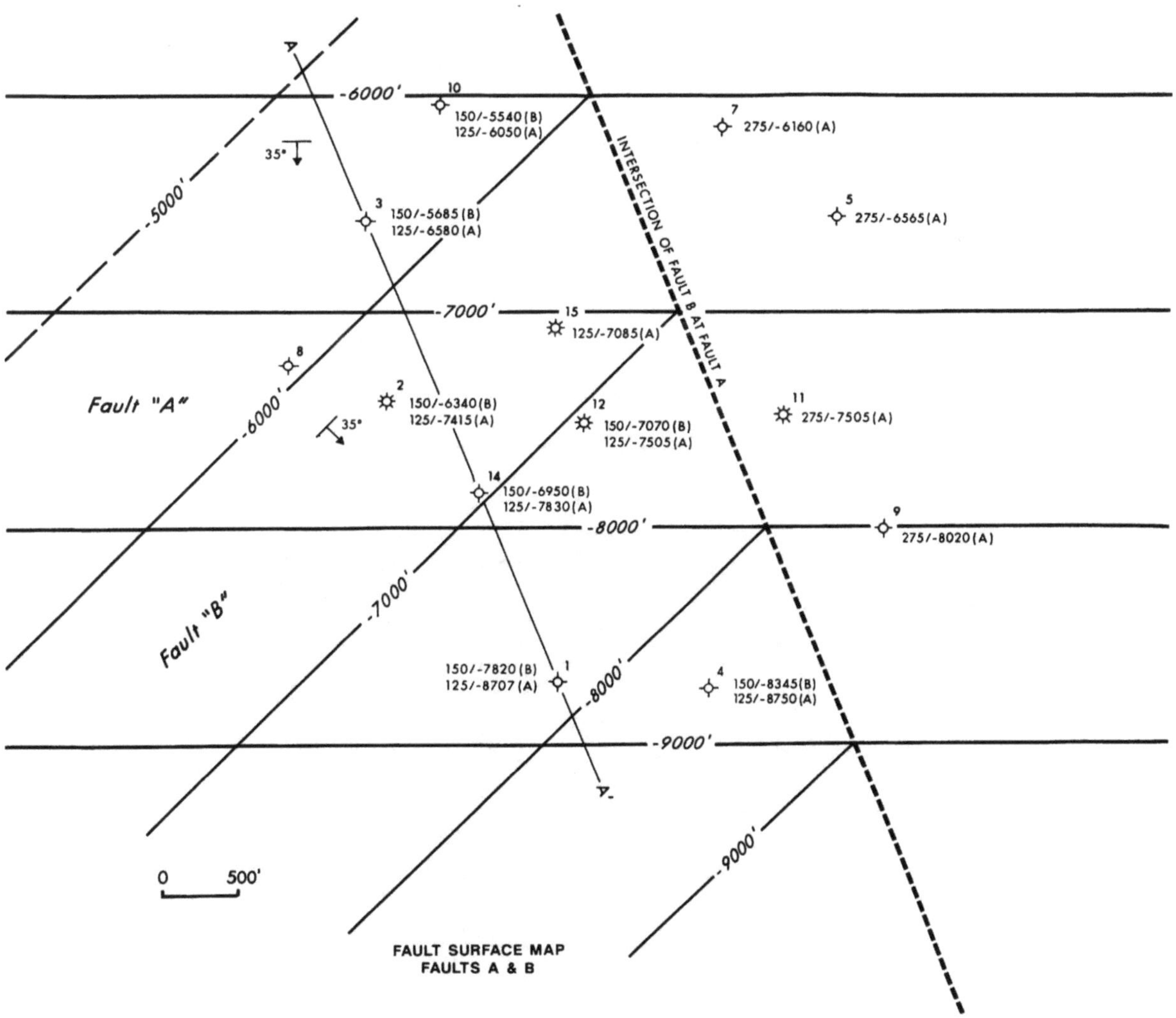

(a)

Figure 8-79 (a) Fault surface map for reverse Faults A and B. (b) Integrated structure map on the 6000-ft Horizon. Contours are dashed on the footwall block in the area of fault overlap for clarity. (c) Fault map superimposed onto the structure map to show the accuracy that is achieved by the integration of the two maps regarding: (1) fault trace construction, (2) position of faults, (3) fault intersections, and (4) proper bed contour construction across each fault. (d) Completed structure map on the 7500-ft Horizon. Contours dashed on footwall in area of fault overlap.

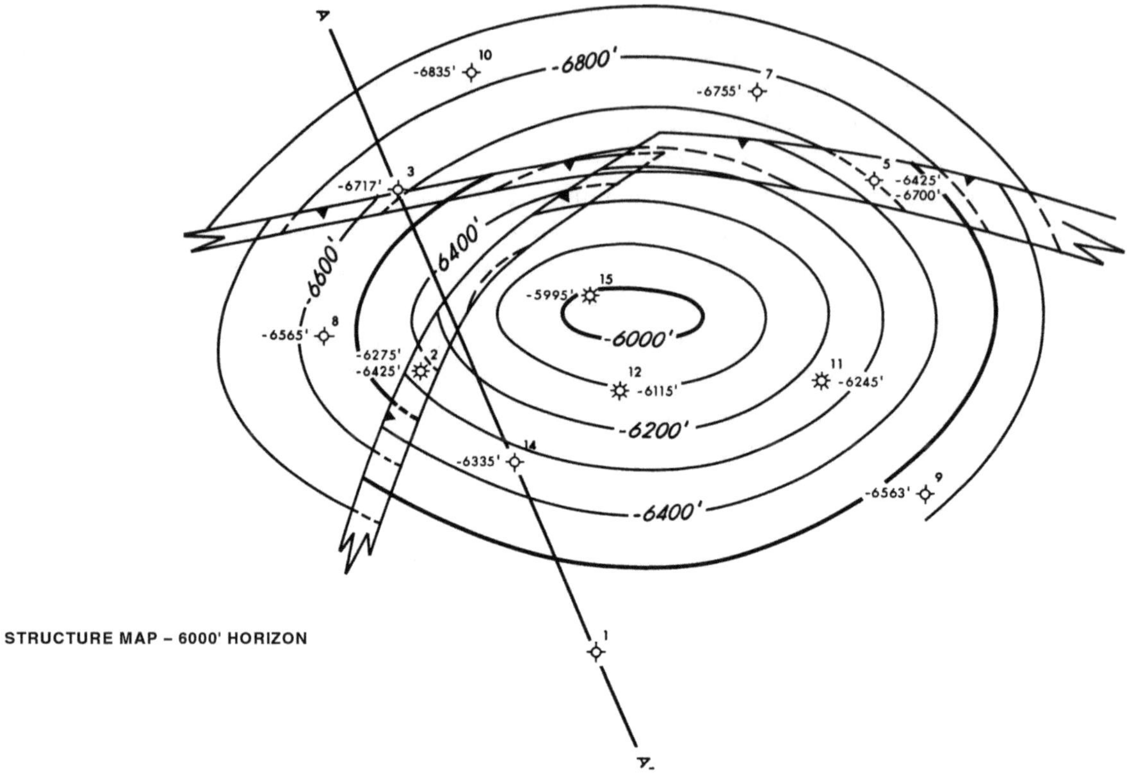

(b)

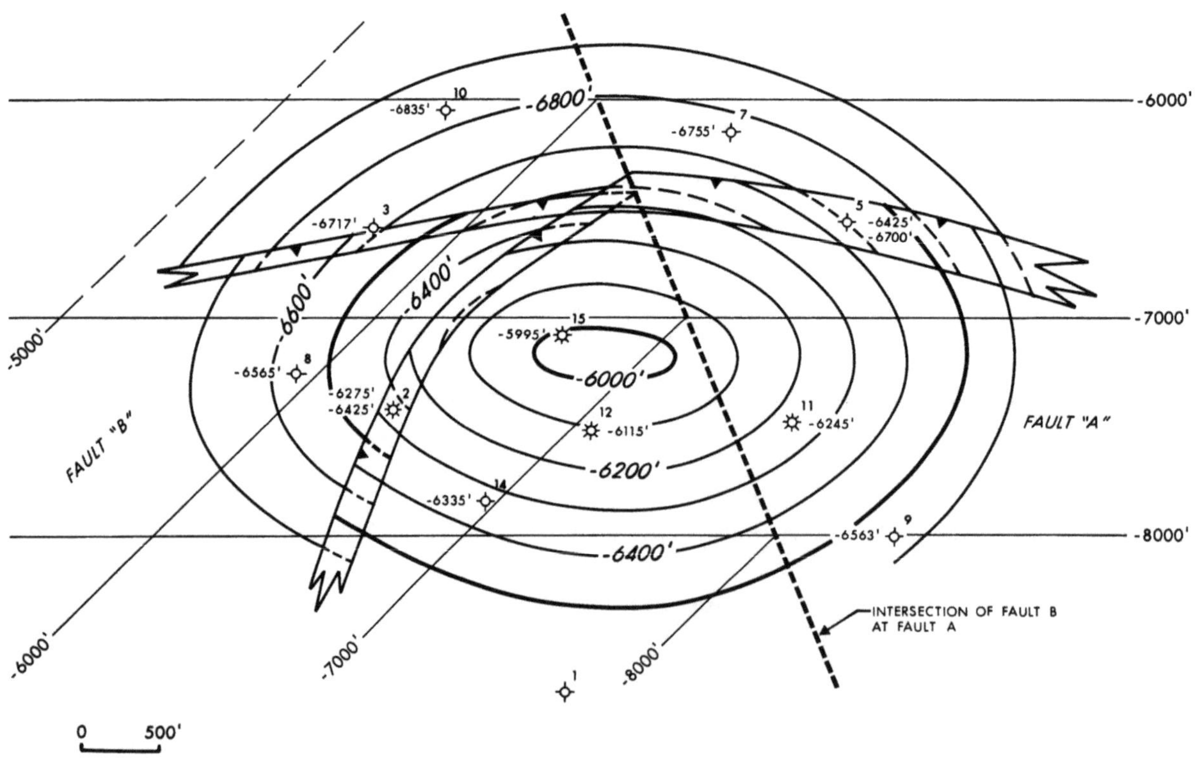

(c)

Figure 8-79 (*continued*)

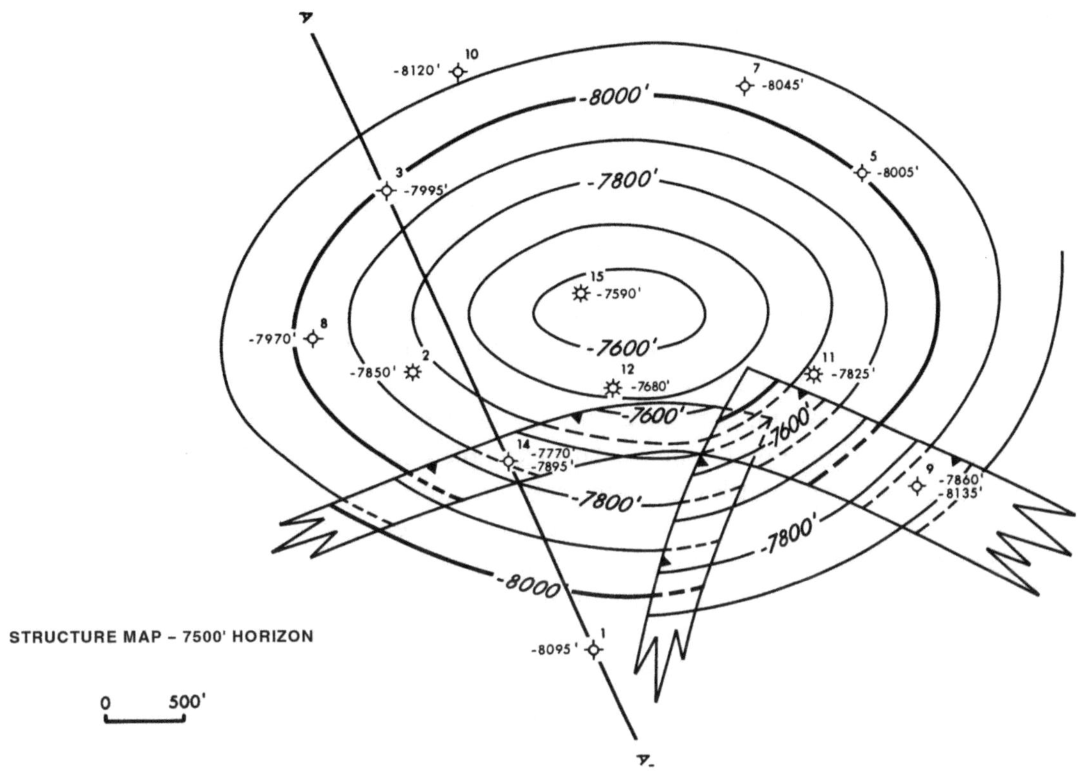

(d)

Figure 8-79 (*continued*)

the additive property of faults must again be considered; throughout the area where two faults are present (west of the intersection line in our example), Fault A has 125 ft of vertical separation (repeated section in wells) and Fault B has 150 ft of vertical separation. Only Fault A with a vertical separation of 275 ft (150 ft + 125 ft = 275 ft) is present east of the line of intersection. Therefore, the vertical separation is conserved across the line of intersection.

Figure 8-79b is a completed structure map on the 6000-ft Horizon faulted by Faults A and B. Figure 8-79c shows the fault map superimposed on the completed structure map. This figure demonstrates the construction of the upthrown trace (trace on the hanging wall) and downthrown trace (trace on the footwall) for each fault, the construction of the intersection of the traces of the two faults, and the proper bed contour construction across each fault using the vertical separation. We have discussed in detail this type of construction with a number of other examples, so we do not detail the construction here. Instead, we recommend that you pick at least two contours and review the construction techniques used to complete this structure map. Notice that the fault intersections fall on the line of intersection. Figure 8-79d is the completed structure map on the deeper 7500-ft Horizon; take a moment and review its construction. The fault overlaps are wider on this horizon because the faults and the horizon are dipping in the same general direction.

As with normal faults, the construction technique of integrating a fault surface map with a structure map removes all the guesswork with regard to (1) the location of the fault on the mapped horizon, (2) fault trace construction, (3) the location of all fault intersections, and (4) the width of the fault overlaps. The integration of the two maps constructs these features automatically without the need for guesses or estimates.

Thrust Faults. Most thrust faults have displacements large enough to offset the two fault blocks so that there is no continuity of structure (contour continuity) across the fault. In such cases, each fault block must be contoured independently and then integrated with the fault map or maps.

Figure 8-80a is a diagrammatic cross section of the Mount Tobin Thrust Fault, Nevada. Notice that there has been significant movement of the upper plate over the thrust fault. Therefore, the upper and lower plates of the Ph Bed require independent contouring and fault integration. Figure 8-80b shows the Mount Tobin Thrust Fault map superimposed on the structure map of the top of the Ph Bed in the upper plate. Observe in the upper portion of the figure that the shape of the fault trace on the top of the Ph Bed is shown as a closed oval (in plan view). This type of configuration is not arrived at intuitively; it requires the integration of the fault and structure maps for accurate delineation.

Figure 8-80c shows the integration of the fault map with the lower plate (the trace on the upper plate is also shown on the map). This example demonstrates the detail required to contour the Ph Bed in the upper and lower plates and integrate these plates with the fault map to accurately determine the position of all the fault traces. Take a few minutes and study the map to be sure you understand how the two maps were integrated to arrive at the finished structure map.

Figure 8-81 is an integrated structure contour map, in two-way time, on the top of the Mirador Formation within the Medina Anticline in the Eastern Cordillera, Colombia (Rowen and Linares 2000). The Medina Anticline was interpreted and mapped based on surface outcrop, well log, and seismic data.

The Medina Anticline is an excellent example of the integration of all data sources to (1) generate a fault surface interpretation and map (Fig. 7-34), and (2) integrate the fault surface with the structural interpretation. The Medina Anticline is a fault bend fold in the hanging wall of the Aguaclara Fault in the southeastern thrust belt of the Eastern Cordillera in Colombia. Active exploration in this area of Colombia has resulted in the discovery of several large hydrocarbon accumulations such as Cusiana and Cupiagna. Although the Medina Anticline has well-defined axial surfaces and nearly planar limbs, the fold appears more rounded than kink in shape. In order to complete the interpretation and mapping for prospect generation, the fault surface and structure map should be converted from time to depth.

Finally, remember we discussed the technique of preparing a *pull-apart map* for compressional areas, or the use of a *flap map*, when mapping by hand. Such a map allows the contouring of a given horizon all the way to its footwall cutoff, which may be a considerable distance from

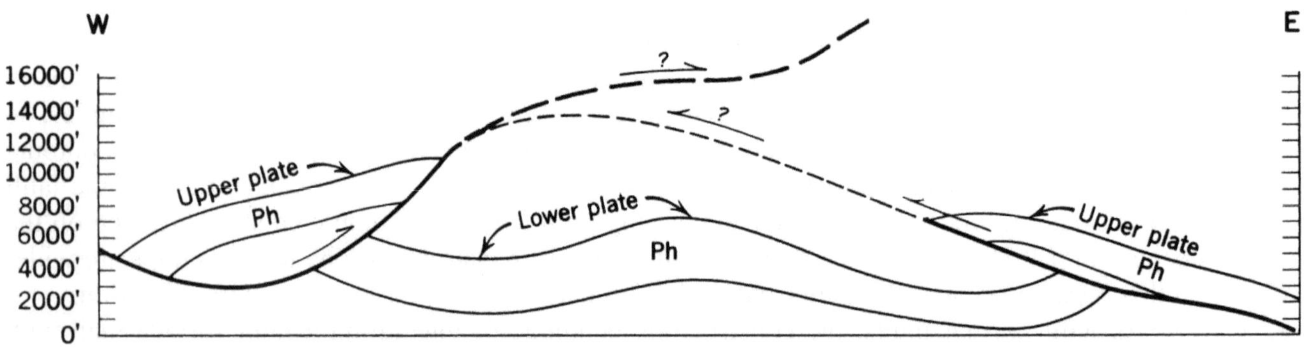

(a)

Figure 8-80 (a) Diagrammatic cross section of thrust fault (see Mount Tobin, Nevada, USGS map series GQ7). (Modified from Bishop 1960. Published by permission of author.) (b) Part of upper plate and fault surface map. Delineates the intersection of the fault with the upper plate. (c) Overlay of fault map and lower plate to show the intersection of the fault with the lower plate. (From Bishop 1960. Published by permission of author.)

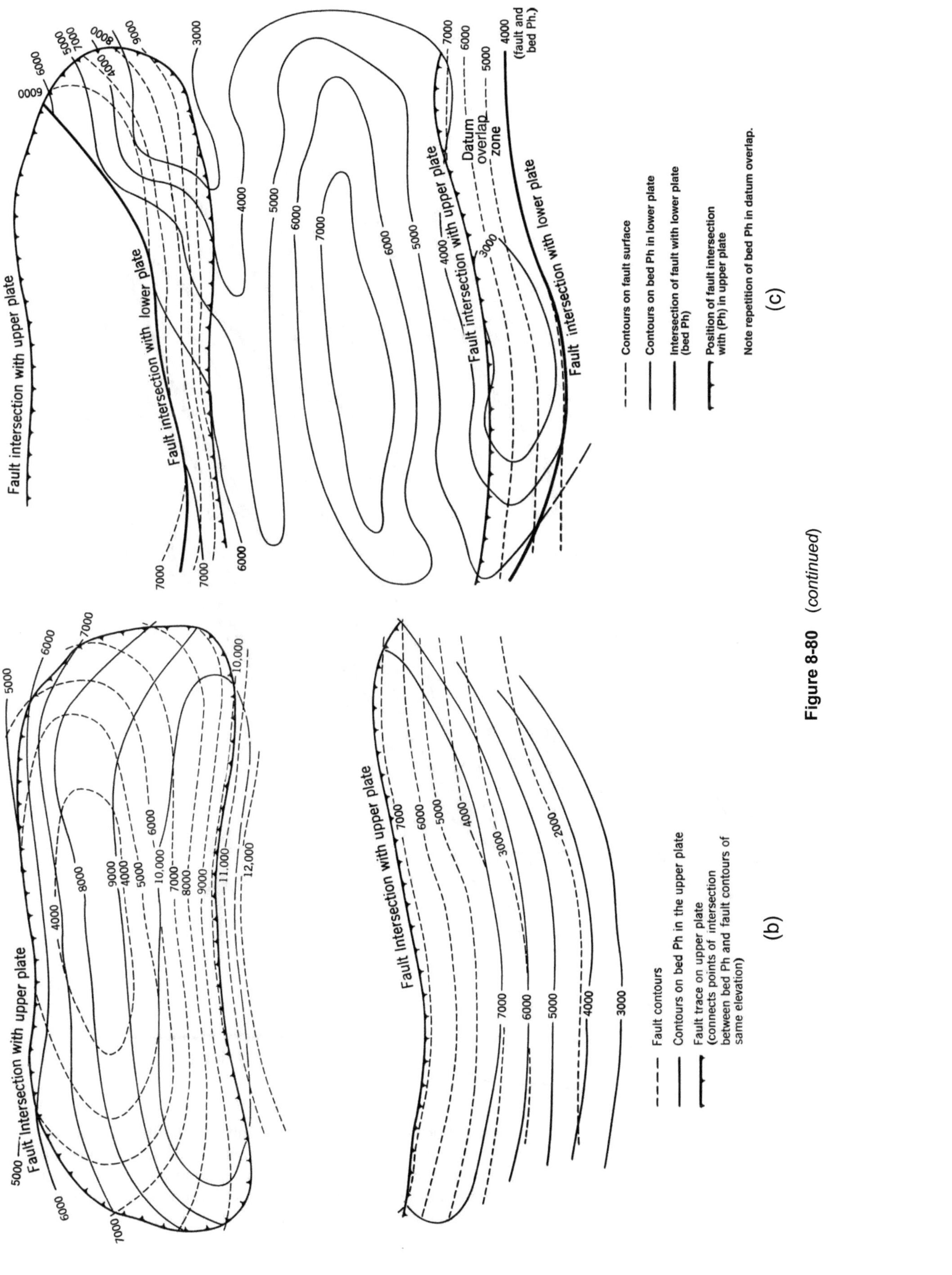

Figure 8-80 *(continued)*

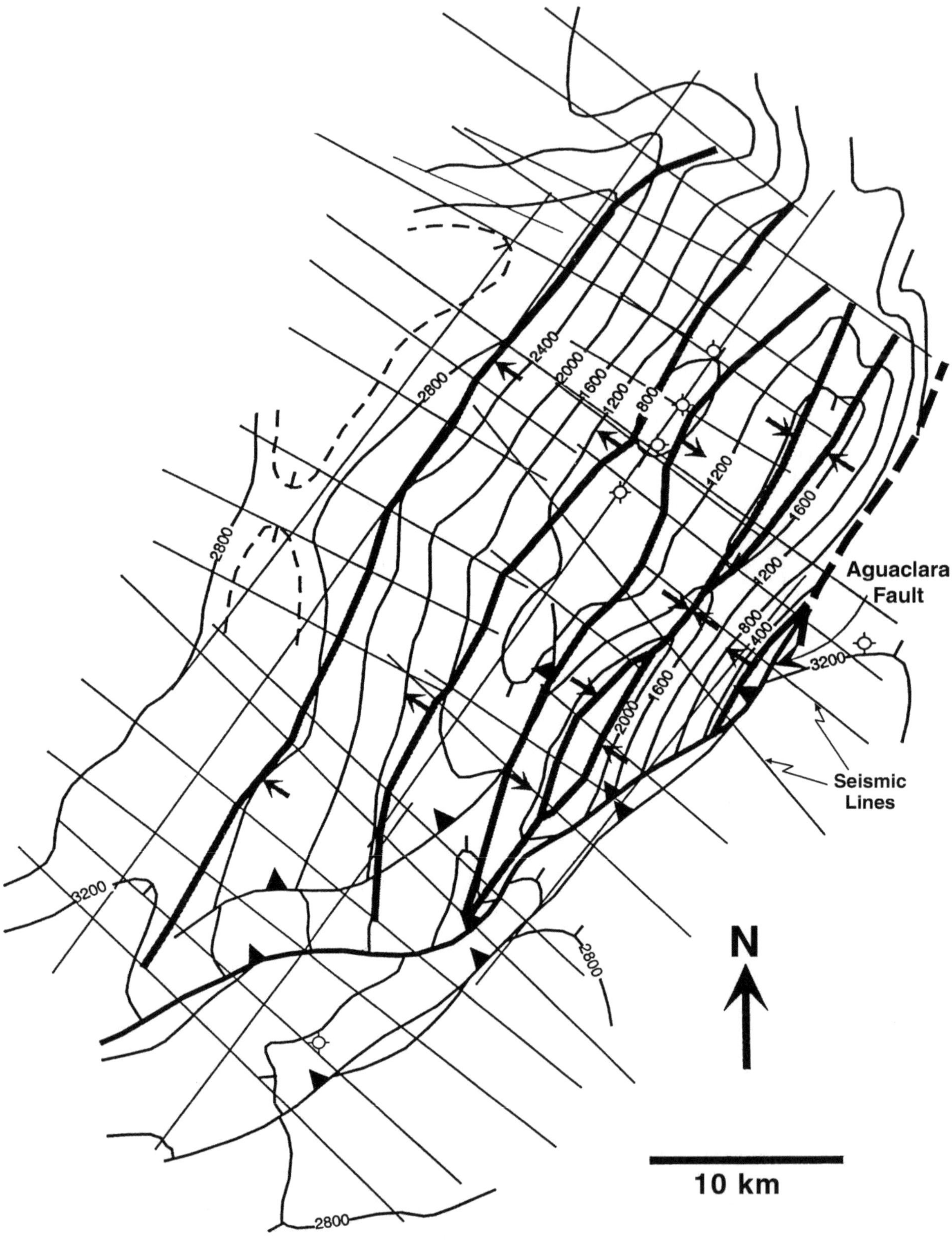

Figure 8-81 Structure contour map of the top Mirador Formation in the Medina Anticline, Eastern Cordillera, Colombia. Thick black lines are axial surface traces interpreted on seismic dip profiles, with arrows pointing in the dip direction. Thick, dashed black line represents erosional truncation at the ground surface. Contour interval is 400 msec, relative to an arbitrary datum near the ground surface. (From Rowan and Linares 2000. AAPG©2000, reprinted by permission of the AAPG whose permission is required for further use.)

the leading edge of the hanging wall block. With this technique, the mylar or paper used for mapping is cut along the leading edge of the hanging wall block, and new material is spliced in underneath to allow mapping of the subthrust surface to its fault cutoff. In complex areas with substantial thrust transport, this eliminates the confusion of dashed versus non-dashed contour lines. The maps can be separated to show the whole surface of the mapped horizon in both the upper block and subthrust. They are also very helpful during isochore mapping.

REQUIREMENTS FOR A REASONABLE STRUCTURAL INTERPRETATION AND COMPLETED MAPS

At the end of any interpretation project, the accuracy of the reconstructed subsurface geology depends on the completeness of the work undertaken. We cite 10 requirements that should be met to ensure the best and most accurate subsurface interpretation and final generated maps.

1. Three-dimensional validity of the interpretation.
2. A clear understanding of basic geological principles, including the intersecting geometry of horizons and faults.
3. A good understanding of the tectonic setting being worked.
4. The use of *all* the available data.
5. An understanding of the accuracy of the data.
6. Correct correlations (well log, seismic, and outcrop where available).
7. The use of correct and accurate mapping techniques.
8. The construction and integration of all required maps.
9. Multiple horizon mapping.
10. Construction of cross sections (balanced, if possible).

If all these requirements are met, there should be a high degree of confidence that the interpretation and completed maps are reasonable and accurate with respect to the available data.

Multiple Horizon Mapping

Almost any set of fault and structural data can be forced to fit on one particular horizon. The test of the fault and structural interpretation, and whether the interpretation fits on a series of horizons at various depths, lies in structural framework mapping. Therefore, one of the most important of the 10 requirements listed above is the interpretation and preparation of integrated multiple horizon structure maps. Multiple horizon interpretation and mapping means the preparation of structure maps on several horizons to ensure that the interpreted structural framework maintains continuity at all levels. Normally, at least three horizons are required to establish confidence in the interpretation. However, depending on the size of the area being mapped and the number and vertical distribution of pay horizons, additional integrated structure maps at various depths may be required.

Once the structural framework (fault and structure pattern) is established by constructing several fieldwide or regional structure maps at various depths, the mapping of all pay horizons between these initially mapped horizons becomes a much easier task.

Discontinuity of Structure with Depth. If the structural geology of an area maintains continuity with depth, the mapping of at least three horizons at various depths establishes and supports the structural interpretation. However, in many instances the structural continuity is interrupted with depth by one or more of three primary geological events: (1) unconformities, (2) thrust faults, or (3) listric growth faults. In these cases, the structural geology above the interrupting

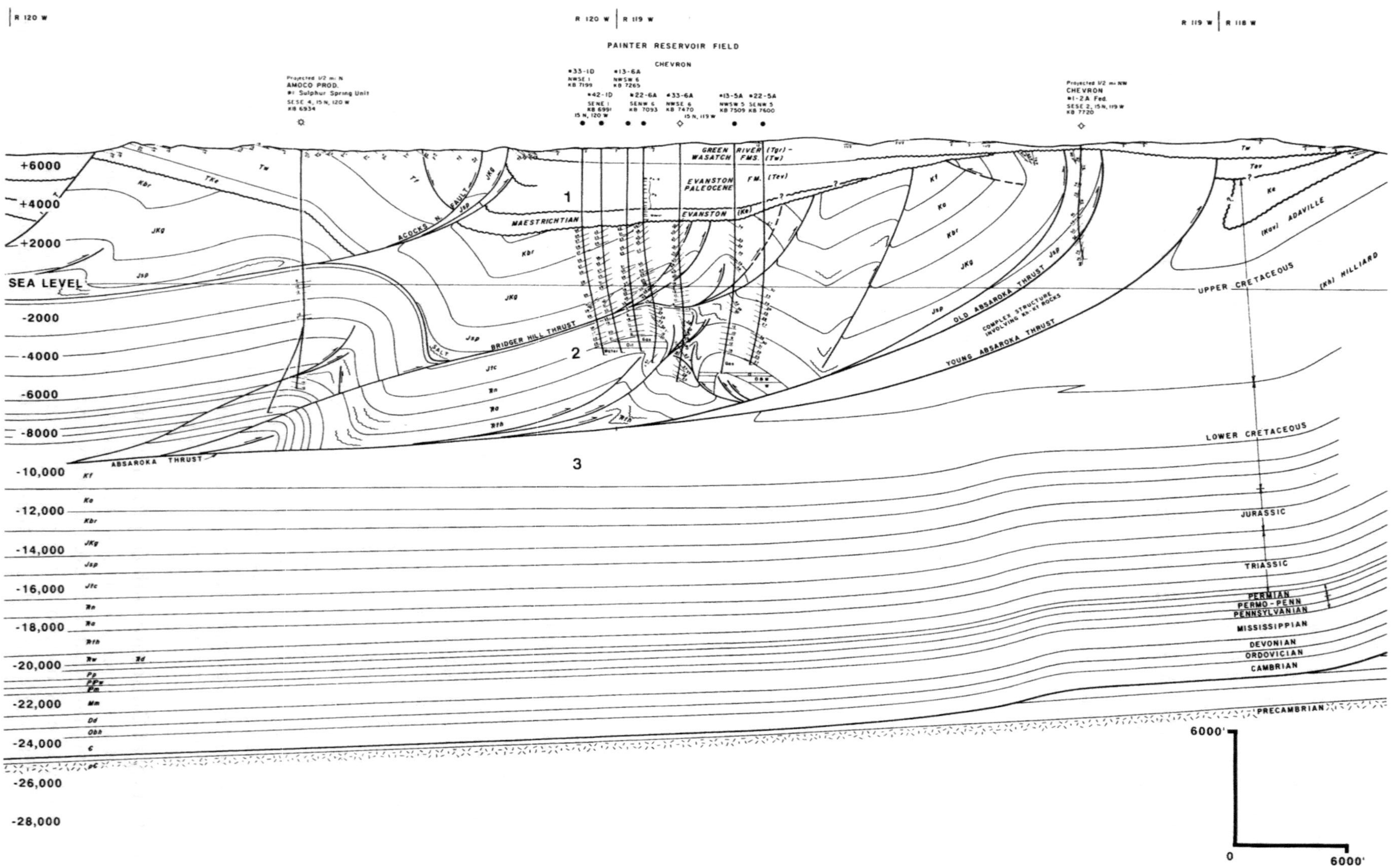

Figure 8-82 Cross section through Painter Reservoir Field. The Sub-Evanston Unconformity and the Absaroka Thrust Fault are interrupting events that result in different structural geometry above and below these events. (From Lamerson 1982. Reprinted by permission of the Rocky Mountain Association of Geologists.)

event most likely will not conform to the structure below. Therefore, separate interpretations are required above and below the interrupting event, in addition to multiple framework horizon mapping.

Figure 8-82 is a cross section through the Painter Reservoir Field in Wyoming. Observe that the structural continuity is interrupted by two different events, first by the Sub-Evanston Unconformity at about 3500 ft, and second at the Absaroka Thrust Fault. Notice how the structure of the area changes above and below each interrupting event. Such major events must be recognized to develop a good structural interpretation and are important in the search for hydrocarbons.

PERCHED WATER

We have emphasized mapping the top of reservoir beds in this chapter as the structure of the top of the reservoirs controls the trapping of oil and natural gas. It is also important to map the base of the reservoir. A good map on the base of the reservoir is required for accurate volumetric determination of edge water reservoirs, as is discussed in detail in Chapter 14. A base of reservoir map may be necessary for constructing an Allan section, as described in Chapter 4. A base of reservoir map is also helpful in understanding the occurrence and distribution of perched water. Perched water was defined by Gaafar et al. (2016) as "movable water above the transition zone that is not directly connected to the actual free-water table of a given hydrocarbon column." The perched hydrocarbon/water contact may be a few feet or tens of feet about the field hydrocarbon/water contact, as at Grane Field in the North Sea (Rolfsvag and Danielsen 2016) or hundreds of feet above the field contact as in the 8500-ft Sand in Garden Banks 191 Field (Fugitt et al. 2000) and in the J3 Sand at Holstein Field in the Gulf of Mexico.

Perched water is trapped in closed lows on the base of the reservoir unit. These closed lows could be the irregularities on the base of the reservoir unit in an anticline (Fig. 8-83; see also Figure 5 in Rolfsvag and Danielsen 2016), erosional lows in a stratigraphic trap (Fig. 8-84), or the faulted limb of an anticline (Fig. 8-85). Perched water is in pressure communication with the hydrocarbon accumulation across the hydrocarbon/water contact, but it is not in pressure communication with the regional aquifer (Fig. 8-86).

In most cases, the amount of perched water in the reservoir is rather small and does not affect the reservoir volumetrics significantly, although some have suggested that significant impact on volumetrics is possible (Gaafar et al. 2016). As the volume of perched water is relatively small, it does not cone into producing wells. Although the perched water is movable, the compressibility of water is small, and reducing the pressure in the reservoir through production does not cause the perched water to expand appreciably. Because perched water is trapped in structural lows by hydrocarbons, it is isolated from the regional aquifer. If the composition of the regional aquifer changes after the hydrocarbons trap the perched water, the composition of the perched water can differ significantly from that of the aquifer. Perched water may, for example, be highly saline, whereas the regional aquifer salinity may be closer to normal marine salinity. A good structure map on the base of the reservoir should allow the distribution of perched water to be determined, potentially identifying economic volumes of hydrocarbons downdip of the perched water.

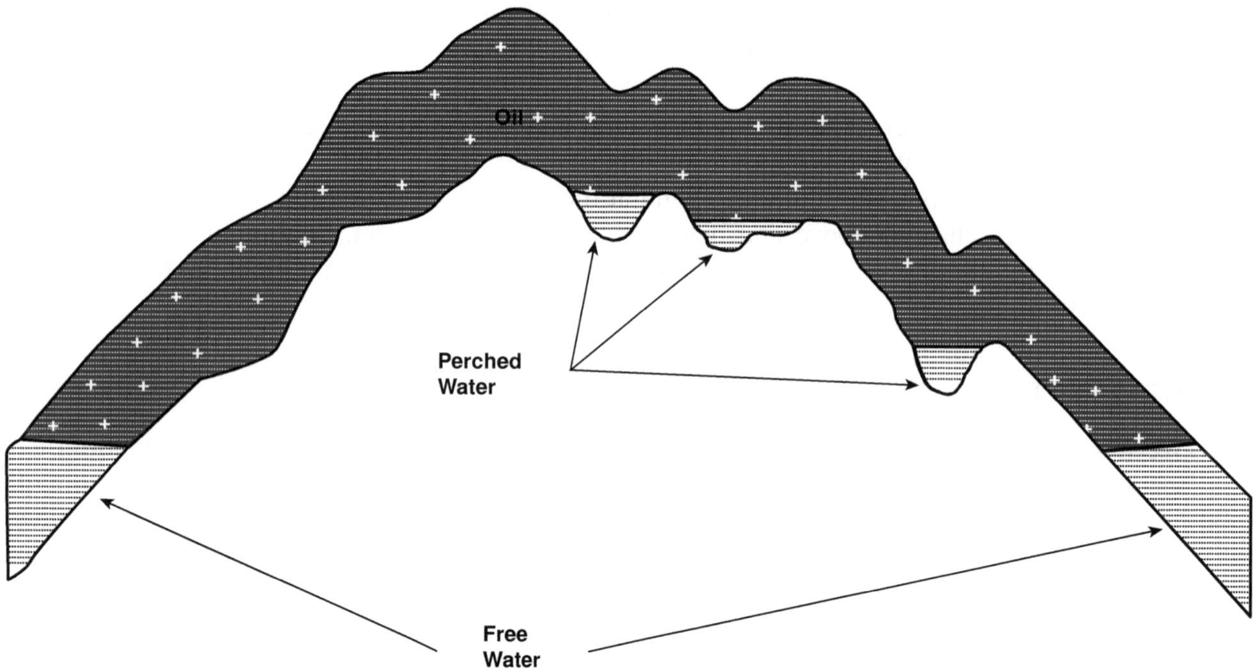

Figure 8-83 Anticline with closed structural lows on the base of the reservoir unit resulting in perched water. (Modified after Weber 1995.)

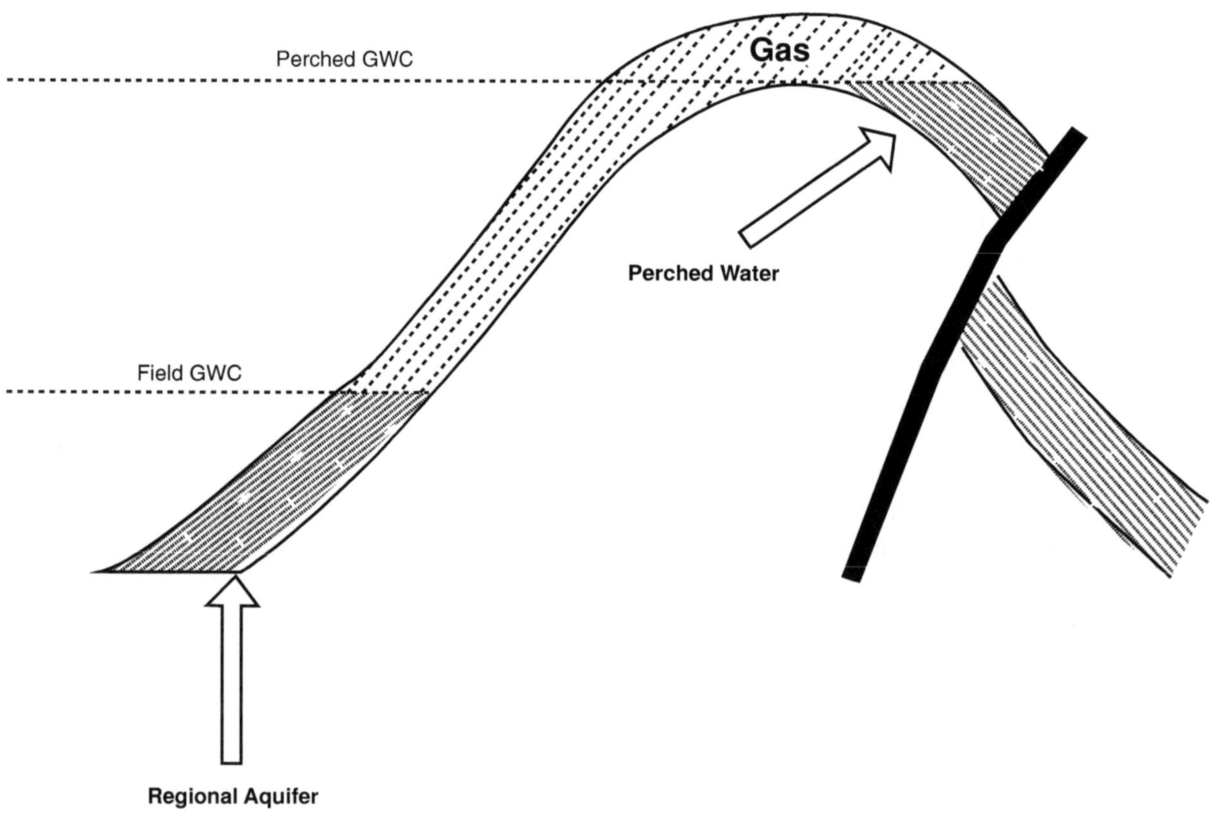

Figure 8-84 Faulted anticline illustrating perched water trapped against a fault. This figure illustrates water trapped against a reverse fault but the figure could just as easily have been drawn with the perched water trapped against a normal fault. (Modified after Weber 1995.)

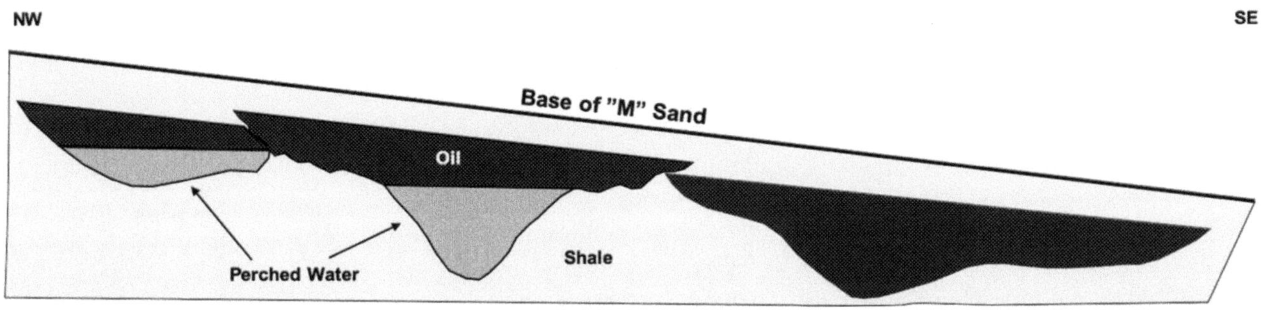

Figure 8-85 Perched water trapped in erosional lows in a system of partially stacked sand bodies inside of a larger erosional scour in the "N" Sand, Ram-Powell Field, Gulf of Mexico. Oil within the individual sands is in pressure communication but the perched water accumulations are isolated from each other. (Modified after Kendrick 2000.)

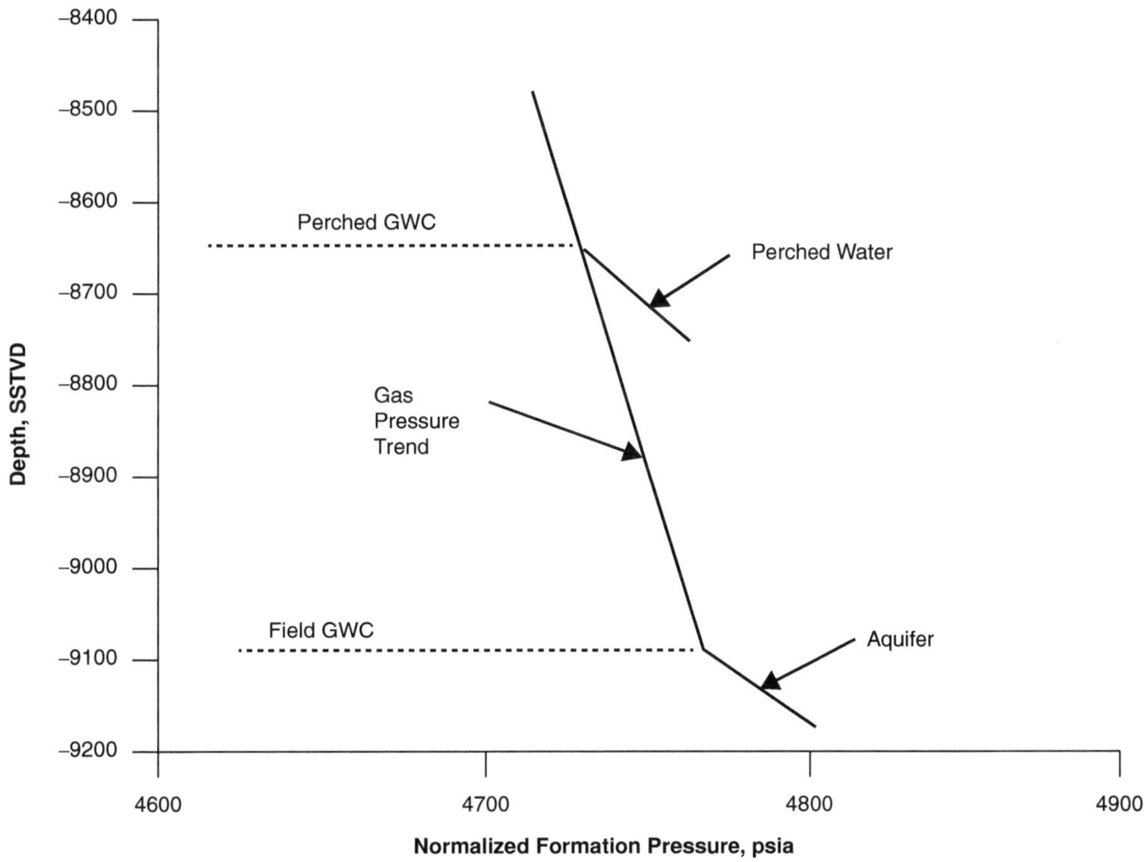

Figure 8-86 Reservoir pressure data from members 3 to 5 of the 8500-ft Sand in four wells in Garden Banks 191 Field in the Gulf of Mexico. One well encountered perched water in this sand. (Modified after Fugitt et al 2000.)

CHAPTER 9

INTERPRETATION OF THREE-DIMENSIONAL SEISMIC DATA*

By David C. Metzner

INTRODUCTION AND PHILOSOPHY

This chapter focuses on the techniques employed in the interpretation and structure mapping of 3D seismic data on a workstation. The focus is on unconventional reservoirs with examples from the Eagle Ford, Wolfcamp, Marcellus, and Haynesville plays and shows how a principled approach to 3D interpretation applies in *all geological settings*. Space does not allow this chapter to be a comprehensive work on all aspects of seismic interpretation, but it does provide fundamental methods that are compatible with the other writings in this textbook (such as Chapters 2, 5, 7, and 8) and with textbooks and memoirs published by other authors. For a more comprehensive discussion and examples of many case histories using 3D seismic data in subsurface mapping, we refer the reader to Brown (2011), Hart (2011), and Krantz et al. (2016) as a starting point. Data examples used in this chapter were provided by the companies cited with each figure except where the providers desired to remain anonymous. Finally, the choice of unconventional plays described in the text and figures was determined by data availability, interesting geology, and relevance to current unconventional production and reserves in North America. During this review of structural features, 10,300 Km^2 of seismic data were viewed or mapped. For more

* For all figures in this chapter (in the printed book only), see the preface for information about registering your copy on the InformIT site for access to the electronic versions in color.

detail on each play, we refer the reader to the following: Eagle Ford, Breyer (2016); Haynesville, Hammes and Gale (2013); and Midland Basin, Sinclair et al. (2017), Blomquist (2016). For an overview of all major unconventional plays refer to Stoneburner (2017).

The Philosophical Doctrine Relative to the Workstation

Chapter 1 of this book reviews the Philosophical Doctrine of Accurate Subsurface Interpretation and Mapping utilized to construct viable and accurate maps. Great economic value is achieved by providing high-quality subsurface interpretations with accompanying maps. The proper construction and use of these maps has a direct positive impact on a company's financial bottom line. *This holds true whether you are working conventional or unconventional reservoirs.* It does not matter which basin, structural style, or stratigraphic complexity is being imaged: 3D seismic has proven its value for decades. Technological innovation in seismic acquisition, data processing, and of course, workstation functionality is moving very quickly. For workstations, this includes faster algorithms, faster processors, statistical analysis, reservoir characterization, and machine learning. We cite several areas of performance improvement discussed in this textbook that apply today to workstation-based interpretations.

1. **Speed.** High-speed hardware and software enable interpreters to complete the basic framework interpretation in just minutes for relatively uncomplicated areas. More complex areas may require several iterations and usually some manual editing and contouring to be considered final.
2. **Accuracy.** Modern 3D seismic data have imaged increasingly finer details of the subsurface geology, allowing more accurate and reliable interpretations.
3. **Integration.** Workstations in use today allow unprecedented integration of all types of data at a high rate of delivery. Proper integration of geological, geophysical, and engineering data leads to optimal development plans for conventional and unconventional plays.
4. **Optimization.** Optimizing hydrocarbon recoveries through accurate reservoir models is attainable. A common practice in use today is geocellular modeling made possible with modern workstations populated with 3D seismic volumes and geological and/or geomechanical properties.

The Philosophical Doctrine is valid when work is done with or without a workstation. Take time to review the basic principles before reading this chapter because they provide the context and background to the techniques and workflows described in the following sections. Even though most of the 10 points in the Philosophical Doctrine are self-explanatory, point 4 requires a bit of clarification relative to the workstation. Point 4 states that "all subsurface data must be used to develop a reasonable and accurate subsurface interpretation." In the 3D seismic world, this does not mean that you interpret by hand every line, crossline, time slice, and arbitrary line in the 3D data set. Methods are discussed in this chapter on how to use all available data to make sound 3D interpretations in a time-efficient manner, allowing geoscientists and engineers the time to do the quality geoscience or engineering that we *need to do*.

Data Optimization

From the preliminary design to the delivery of the 3D seismic survey and loading of the final processed seismic data to a workstation, it is beneficial for the eventual interpreters to be informed about the acquisition and processing of the data volume. Seismic interpreters who are involved from the start of a project potentially have a *much better understanding of the limitations and resolution of the final 3D volume*. Interpreters can provide insight, in advance, as to where the targets are and how to best image them. If interpreters are involved early in the processing flow, they can assist in quality control of the results as work progresses. They can recommend preservation of certain intermediate data sets along the way in case another workflow is applied to the latter stages of the processing. Finally, interpreters need to specify which products are to be delivered. Many options are available as final products; however, most commonly, a prestack time migration is delivered. Poststack processing such as filtering, noise suppression, scaling, attributes, and inversion can be performed either by the vendor or in-house by the interpreters themselves or coworkers.

Once the seismic volume has been finalized and delivered from processing, the next step is to preview the data on the workstation. The interpreter should scroll through the volume and determine how the data look and optimize views and display parameters. The process gives the interpreter an idea of the general structural style of the area. Most interpreters have guidelines or standards as to what, if any, filtering, scaling, or noise suppression is applied. Today's workstations give the interpreter complete control over how the data look and allow the interpreter to select parameters based on how the data are used. Whatever the case may be, it is the interpreter's responsibility to understand as much as possible about the data he or she is interpreting.

Work Environment

Subsurface interpretation and mapping, data integration, and rock property characterization have, with a few exceptions, all migrated to the digital world. Geoscientists and engineers now collaborate as team members from their offices or small workrooms that have wide-screen visualization capabilities. Worldwide real-time sharing of data and interpretations is made possible with high-speed Internet connections and cloud computing. Artificial intelligence, machine learning, and big data analytics enable the ability to analyze multiple types of engineering, geological, and geophysical data to find optimal solutions to today's challenges (Chopra et al. 2018). With all these capabilities in the hands of highly trained professionals working as a team, more is accomplished in less time than in previous generations, and we can understand the reservoir more clearly.

Optimizing Interpretation in Three-Dimensional Space

Many interpreters work in 3D visualization software environments (Fig. 9-1), which allow for any scale, orientation, and color scheme imaginable. Within this type of display environment, multiple seismic attributes can be individually selectable by slice or vertical seismic displays (VSD). Figure 9-2 contains a solid cube display of a coherence attribute volume. This is typically used for fault interpretation or identifying subtle stratigraphic features. Multiple seismic volumes can also be co-rendered as a single volume. Figure 9-3 is a co-rendering of reflectivity and coherence. Using this combination of volumes or other volumes of choice in the same 3D scene is an effective reconnaissance and well planning tool. Applying an artificial light source creates highlights and shadows to enhance subtle features in mapped surfaces (Fig. 9-4). Of course, color is still a great way to bring out the details in the data you need to see. Figures 9-5 through 9-8 show several examples of color schemes, although the options are not limited to those shown.

In addition to color, horizontal and vertical scale settings can enhance or suppress structural or stratigraphic features. Vertical exaggeration (VE) is the ratio of vertical scale to horizontal scale. To enhance faults on a vertical section, expand the VE. Figure 9-9 shows an Eagle Ford example of a dip profile at a typical regional scale. Figure 9-10 is a VE version of Figure 9-9 showing how the faults are much easier to interpret at this scale. Stratigraphic features such as channels, prograding sequences, and unconformities can also be interpreted in this manner. An example of the San Andres formation in the Midland Basin is presented in Figure 9-11. The prograding sequences are interpreted in yellow. The use of alternative color schemes and VE allow stratigraphic details to become more visible to the interpreter. How much of a seismic profile is displayed is a balance between how much of the line you need to see to get the big picture and how much detail is

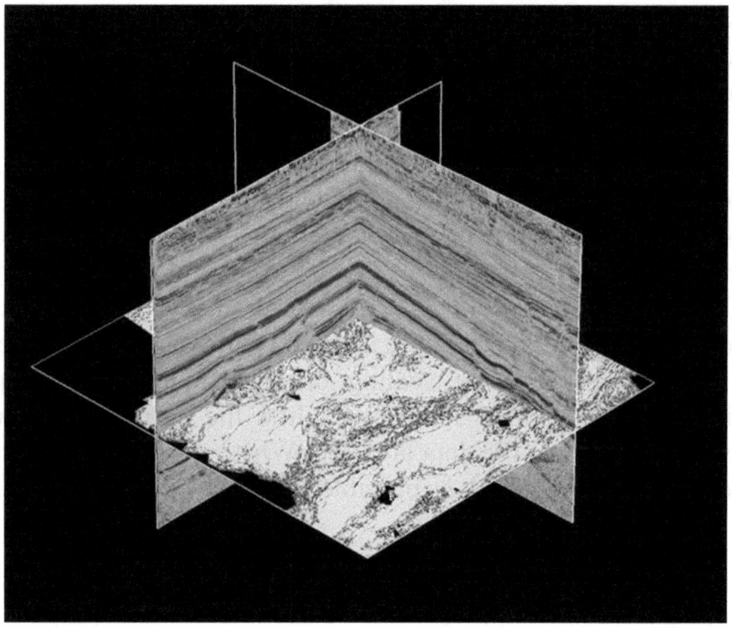

Figure 9-1 This example from the Eagle Ford play in South Texas shows a 3D color visualization view with inline and crossline vertical panels displaying reflectivity and a horizontal slice view displaying the coherence attribute. Interpreters can select any panel and move it to any location within the volume, rotate the volume, zoom in, make an interpretation on the panel, and repeat as necessary to build a 3D interpretation. (Data published with permission from Seitel Data Ltd.)

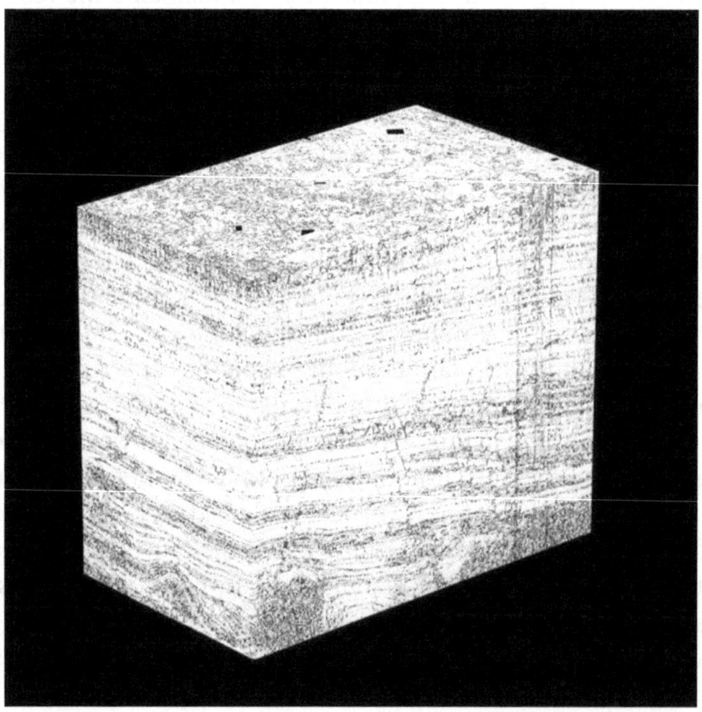

Figure 9-2 Eagle Ford: Example of a 3D coherence attribute volume used to enhance subtle structural and stratigraphic features. Interpreters can slice this volume in any orientation and rotate to view any side. (Data published with permission from Seitel Data Ltd.)

Figure 9-3 Eagle Ford: Multiple volumes are co-rendered and independently colored to enhance features in the volume. In this example, reflectivity in color is co-rendered with coherence in black and white to enhance faulting and stratigraphic layering. (Data published with permission from Seitel Data Ltd.)

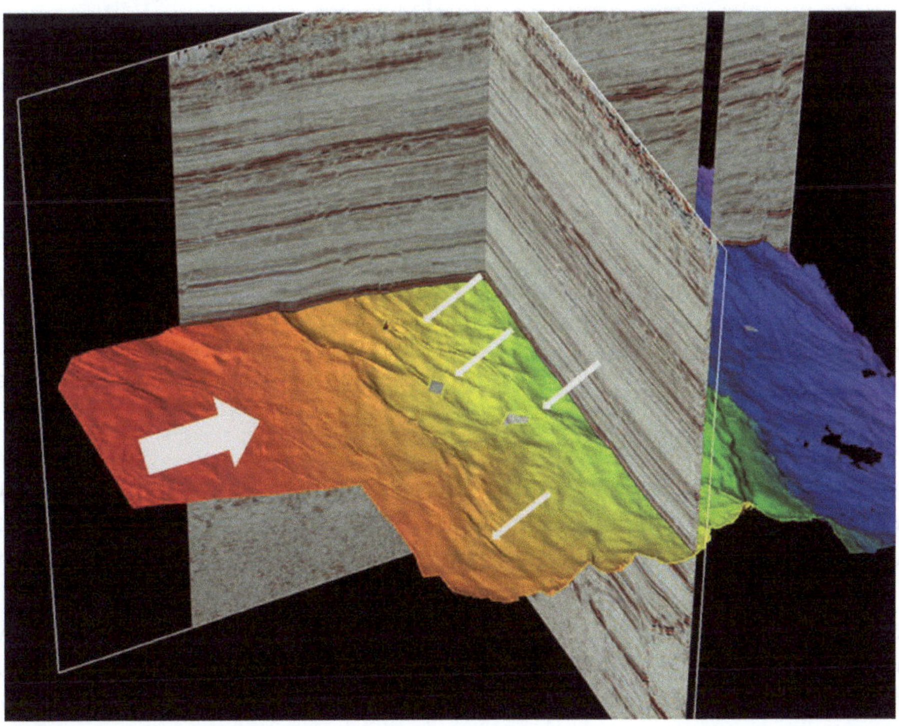

Figure 9-4 Eagle Ford: An artificial light source from the left (large arrow) creates a shadow effect on the interpreted horizon. Note how the fault offsets are enhanced including the subtle features (a few are noted with small arrows). (Data published with permission from Seitel Data Ltd.)

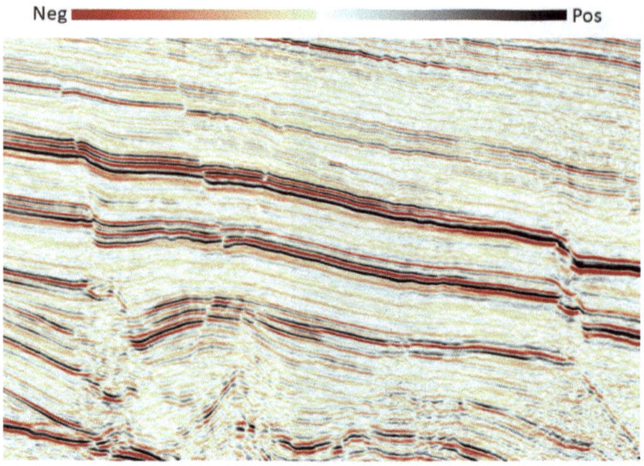

Figure 9-5 Color scheme with high-intensity colors at the extremes. Subtle colors in the low range. Polarity is preserved. (Data published with permission from Seitel Data Ltd.)

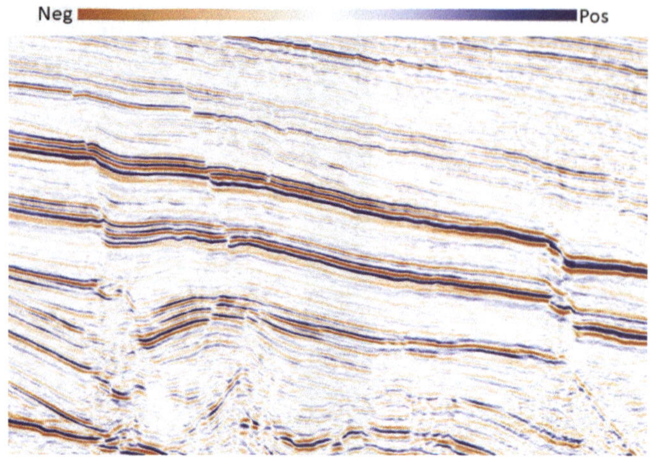

Figure 9-6 Color scheme with high-intensity colors at the extremes. Subtle colors in the low range. Polarity is preserved. (Data published with permission from Seitel Data Ltd.)

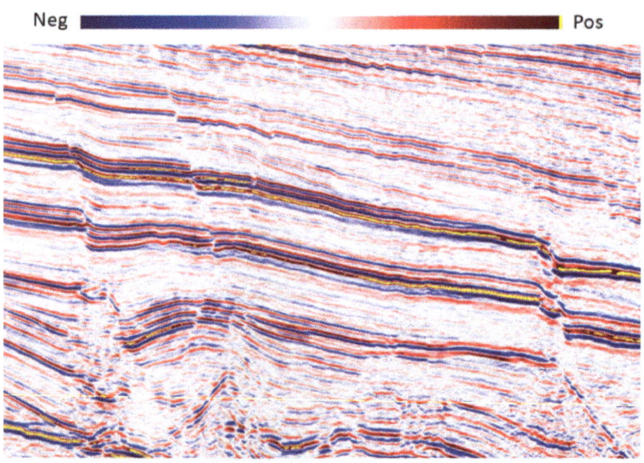

Figure 9-7 Color scheme with dark colors at the extremes. Bright colors in the low range. Polarity is preserved. (Data published with permission from Seitel Data Ltd.)

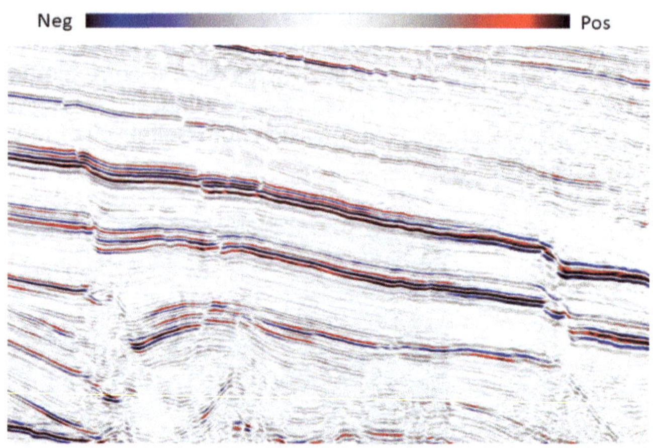

Figure 9-8 Color scheme with dark colors at the extremes. Bright colors in the high range. Subtle colors in the low range. Polarity is preserved in the high range only. (Data published with permission from Seitel Data Ltd.)

necessary to make an accurate interpretation. For example, when working in unconventional plays, where wells are drilled horizontally, one geoscientist may be mapping trends on a regional scale, while another is focused on the wellbore scale and tracking a drilling target horizon. Figure 9-12a shows a typical horizontal well plan with seismic from the Midland Basin Wolfcamp play. The actual well path, as in Figure 9-12b, shows the drilled well path and is displayed with the gamma ray (GR) log used to geosteer the horizontal part of the well. The seismic scale has a VE of 10x, which allows the simultaneous viewing of well path details against a seismic backdrop. Drillers of horizontal wells that encounter structural changes along the well path are challenged to keep the drill bit in the target zone. Using 3D seismic and nearby wells with proper directional drilling and geosteering techniques, these wells can be successfully drilled within the target (Fig. 9-13).

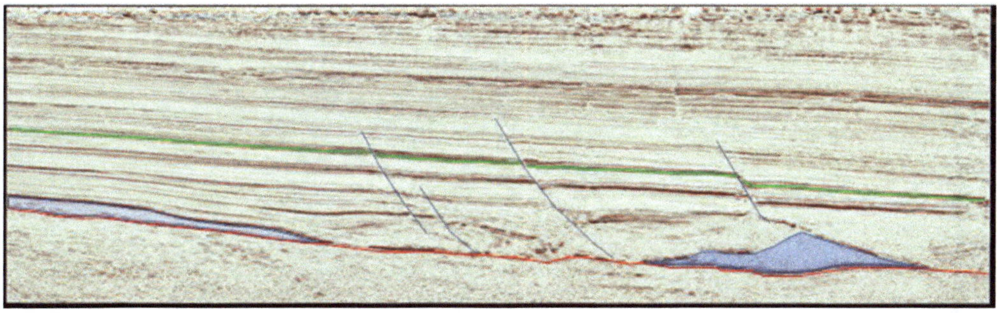

Figure 9-9 Eagle Ford: Regional seismic line for reconnaissance (with slight VE). Regional scale faults are dark blue, Base Eagle Ford horizon is green, basement is red, and salt is shown in light blue. Note smaller faults are difficult to see at this scale. (Data published with permission from Seitel Data Ltd.)

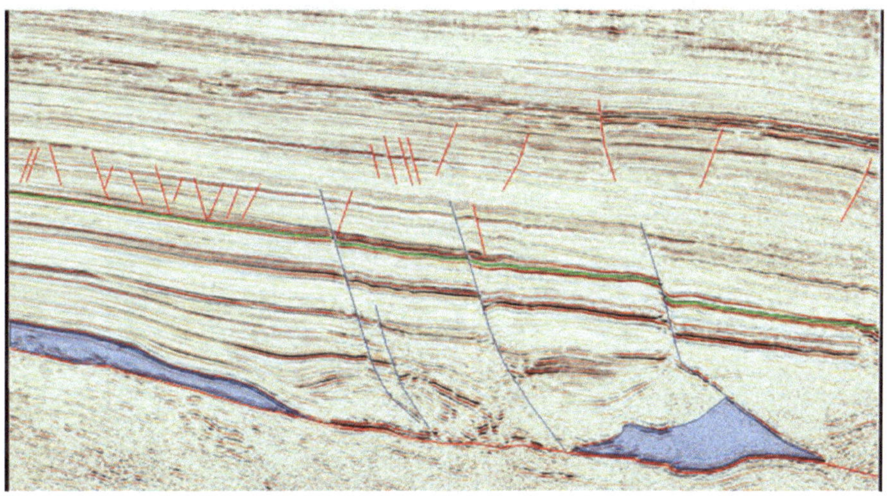

Figure 9-10 Eagle Ford: Expanded vertical exaggeration of Figure 9-9 shows how increased VE improves fault identification. Regional faults are shown in blue, and smaller faults are in red. The Base Eagle Ford horizon is green. (Data published with permission from Seitel Data Ltd.)

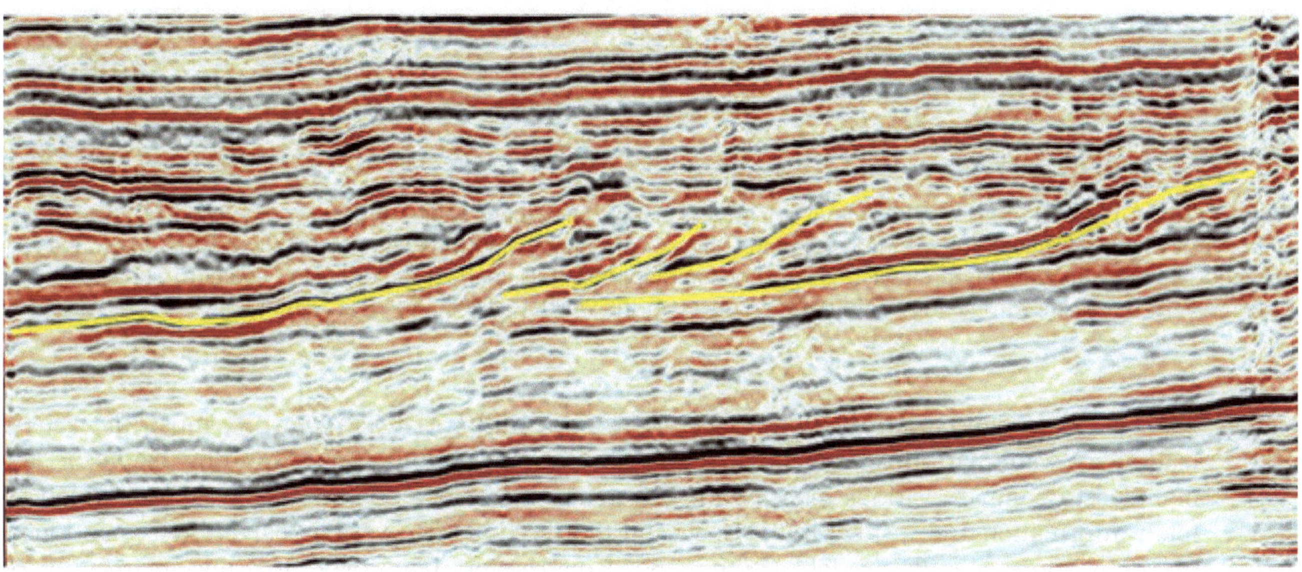

Figure 9-11 Midland Basin: Prograding sequences in the San Andres formation. Stratigraphic detail is shown using a 10x vertical exaggeration. Sequences bounded in yellow are deposited from right to left.

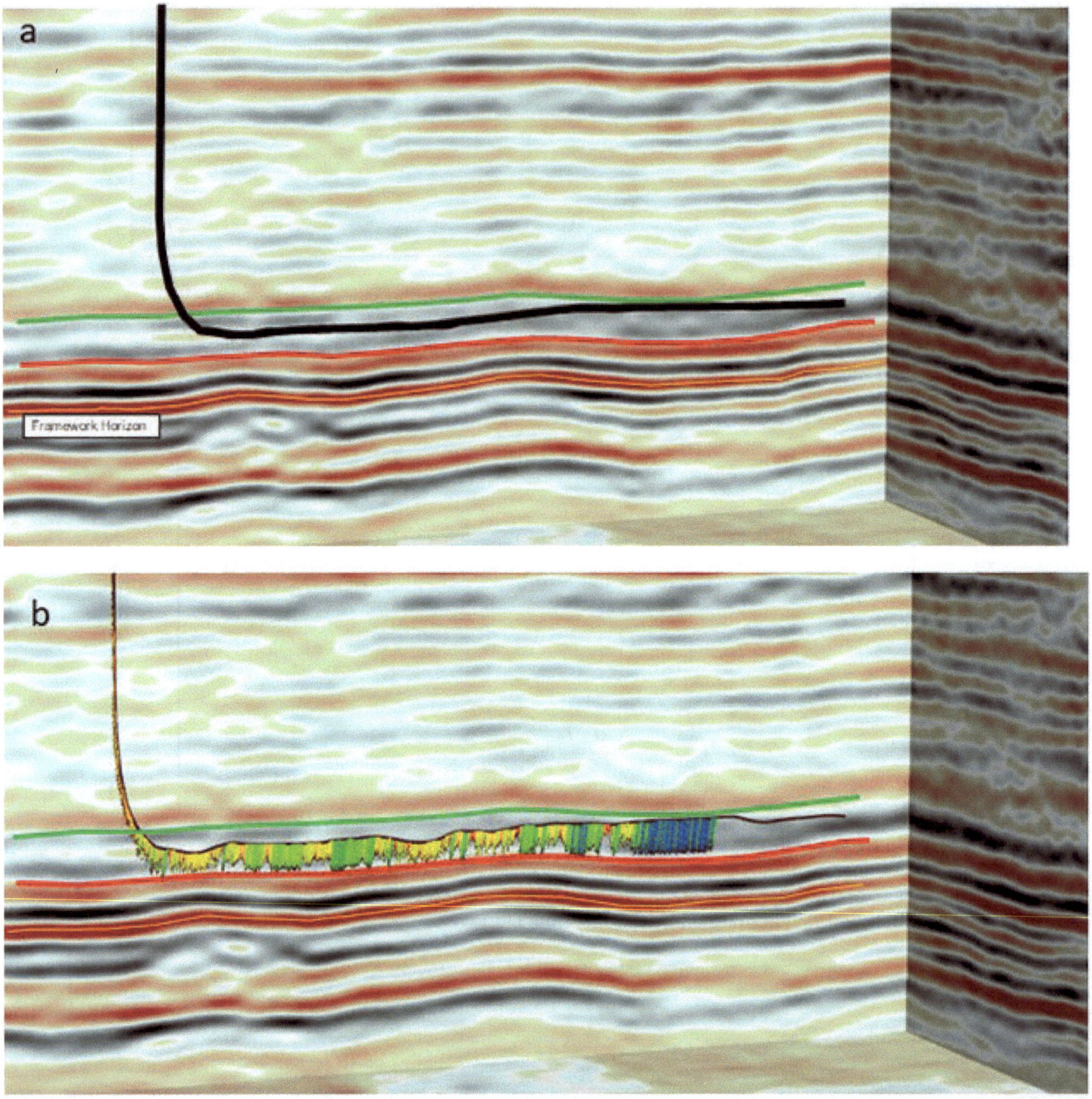

Figure 9-12 (a) Midland Basin: Wolfcamp example of a planned horizontal wellbore displayed on a depth seismic line at 10x vertical exaggeration. The planned well path is black. The target window is 50 ft thick between the green and red horizons. Note that the only interpreted framework horizon mapped in this image is the orange horizon. (b) The same horizontal well, post drill, shown with the GR log used to geosteer the wellbore within the target window.

For *basic interpretation and mapping of a fault and horizon framework*, most 3D visualization software platforms are generally very similar in capability; therefore, this chapter is not intended to compare various interpretation systems. All have a central database from which to fetch information and display in 3D space on the desktop. The various capabilities mentioned previously are generally available. Experienced interpreters likely have had the opportunity to

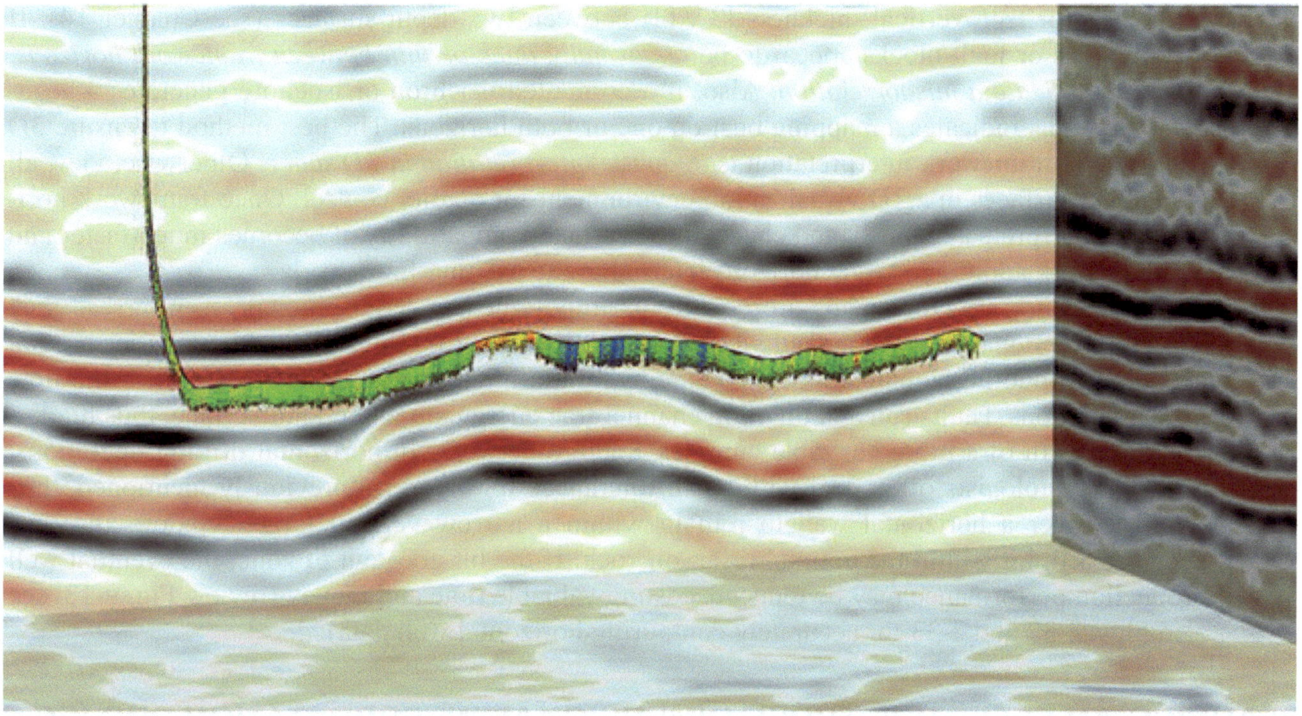

Figure 9-13 Midland Basin: Wolfcamp example of a horizontal well in a deeper target showing structure along the well path. Even with structural variability, using 3D seismic and nearby well control, this well was successfully drilled within the target interval.

use many different software platforms over time. The best platforms give geoscientists the freedom to create interpretations in an accurate and efficient way while sharing that information with the team for better decision making and better well results.

Framework Interpretation and Mapping

Philosophical Doctrine point 8 states, "The mapping of multiple horizons is essential to develop reasonably correct, 3D interpretations of complexly faulted areas." Framework horizon mapping is the practical application of this philosophy. A framework horizon is a seismic or geological correlation marker that is laterally extensive but may not necessarily be a target horizon. Target horizons should be "framed" above and below by framework horizons to ensure the interpretation is valid in 3D. At first, it may seem excessive to interpret and map horizons that are not prospective; however, *not all geologic markers, as correlated on well log data, have a corresponding seismic horizon*. All fault and horizon surfaces should be interpreted so that they intersect correctly in 3D space. In particular, fault interpretations benefit from interpreting some vertical sections in the strike direction to control the shape of the fault surface. Making the effort to create geological framework surfaces that are valid in three dimensions may seem to be a waste of time, but the time invested is typically saved by not having to clean up these surfaces when the structure maps are made.

Framework horizons are effective quality-control tools. They can be used to create *isochron* or *isochore maps* with target horizons or other framework horizons. Isochron maps represent the difference (or thickness in time) between two time surfaces. Isochore maps represent the difference (or true vertical thickness [TVT]) between two depth surfaces (see Chapter 14). If certain areas or fault blocks appear to be *abnormally thick or thin between framework horizons, this may indicate a miscorrelation*. Also, if a computed isochron or isochore generates a negative value, this indicates a mapping bust on one or both horizons. The best method to ensure 3D validity of a framework interpretation is to *do the interpretation in 3D space*. Otherwise, view all completed surfaces in a 3D visualization environment for quality control. Once framework horizons are interpreted, mapped, and checked for vertical compatibility, it becomes much faster and easier to work additional target horizons internal to the existing framework.

Framework structure maps are equally as important in unconventional reservoirs as they are in conventional reservoirs. Faults large enough to cut across multiple bed boundaries are not desirable around horizontal wellbores where hydraulic fracturing is needed to establish the flow of hydrocarbons. Faults can and do often channel frac energy along the fault surface, thus hindering the growth of a more desirable complex fracture network. In horizontal drilling operations, operations geologists often deal with small tolerances for error. As previously mentioned, they often geosteer a horizontal well to remain within a narrow vertical target window and thus keep the drill bit within the optimal reservoir rock. If the bit goes out of zone, the eventual well production performance may be reduced, costing valuable time and money (review Fig. 9-13; see also Figs. 2-32 and 2-33 for accurately contouring horizontal well data).

PLANNING AN EFFECTIVE THREE-DIMENSIONAL INTERPRETATION PROJECT

The Role of the Geoscientist

The development of conventional or unconventional reservoirs, sometimes referred to as *resource plays* or *shale plays*, takes a lot of effort to position resources such as people, materials, and equipment efficiently. The success of bringing a new producing well online is the culmination of *8 to 14 months* of planning and coordination among many groups, both internal and external to the corporation.

Geoscientists play an integral role in that process. From the early stages of mapping a prospective area to the design of well plans to the geosteering of the wellbore during drilling operations, the geoscientist must develop and use the skills necessary to accomplish the task. To begin, geoscientists interpret and map existing well tops and seismic to provide the structure maps for the framework and target horizons. They then team with engineers (drilling, completion, reservoir, and facilities), land negotiators, regulatory personnel, and data analysts to determine areas for development. The geoscientists and the rest of the team lay out a plan of horizontal wells and surface drilling sites (called *pads*) and determine which reservoirs to target for each area.

The next phase in the well cycle is well planning. Structure maps and cross sections based on nearby wells and 3D seismic are the primary tools used to plan a new well. The geoscientist is tasked with determining the vertical depth to the target horizon and the structural dip to follow to keep the drill bit in the target zone throughout the lateral portion of the well. Often, more than one well is planned from a pad, in which case, multiple wells are designed to optimize drilling and completion operations (i.e. where most of the capital is spent).

Once drilling begins, the operations geologists are tasked with interpreting the geology *as it is encountered* by the drill bit. Well logs (usually GR logs) are acquired while drilling and, throughout the day (note 24-hour operations), the logs are electronically transmitted to the surface and relayed to the geologists for interpretation. The operations geologists correlate the incoming GR log with other GR logs from nearby wells to determine the stratigraphic position of the bit and whether it is in the target zone. They then relay to the directional drillers guidance on adjustments to the borehole trajectory. This process continues until a well has reached total depth and drilling stops. Once a well reaches total depth, the rest of the well cycle primarily involves engineering operations.

The process of communication between the rig and home office operations and back to the rig is becoming increasingly automated. Data flow happens in real time, enabling key personnel to process information and make decisions faster. Recent developments in automated geosteering will prove valuable as operations geologists are able to increase their efficiency, allowing companies to increase the number of wells drilled without proportionally increasing personnel.

As operations scale up to grow the number of wells drilling in the same basin, it is important to have a comprehensive planning tool in place. Tools such as a Gantt chart, spreadsheet, or database are necessary to coordinate the timing of well design, land surveys, regulatory permits, rig movements, fracture stimulation crews, and facilities construction. Managing this plan ensures that critical phases of the well cycle are executed on time. Idle time for rigs and crews will negatively impact a company's bottom line. As has been stated many times, "Teams don't plan to fail, they fail to plan" (Tearpock 1997).

Developing an Interpretation Workflow

Workflows are developed to accomplish the 10 objectives listed in our Philosophical Doctrine. How does a workflow differ from a plan? A *workflow* is the detailed step-by-step actions taken to accomplish a task or series of tasks in a *plan*; for example, a workflow might comprise the steps required to interpret faults throughout a data set, to construct maps, or to create a velocity model for depth conversion. By describing, step-by-step, *how* a task is to be done, the workflow provides a level of consistency when similar tasks are done on the same project by different people. The same workflows can also be used in different basins, either unchanged or, if necessary, modified. Seismic interpretation workflows are usually specific to one type of interpretation software and can be difficult to translate to other systems. For this reason, this chapter does not discuss software but rather lays out the philosophical basis for developing or improving your workflow on whatever software is in use.

In practice, following a plan usually requires some tasks to be dependent on completion of other tasks. Each task has a workflow specific to the software in use. For example, the task to design a well plan involves the use of correlated well tops and structure maps that are seismic in depth. Immediately, it becomes obvious that to complete the task of making an accurate well plan, a lot has to happen before reaching this step. Included in this process are a fault interpretation workflow if faults are present, a horizon interpretation workflow, a depth conversion workflow, and a mapping workflow.

Workflows should be developed with the Philosophical Doctrine as a backdrop within the context of the project data and time constraints. Allow enough time to do the work correctly. Computer mapping is a great tool, but it should not have to hide an incorrect interpretation.

Organizing a Workstation Project

Workflows can be made even more effective if the *output* of interpretation and mapping is well organized. An efficient naming convention allows interpreters to find faults, horizons, maps, or other displays quickly. Workstation databases can store large amounts of data and interpretation. Usually, lists can be sorted by name, interpreter, or last modified date, so these attributes can be used to group similar items. For example, faults can be grouped by their location and color (e.g., NE red, S blue), appending a number if the selection of colors runs out (e.g., NE red00, NE red01, …). Some interpreters prefer to use A through Z, then AA through ZZ, and so on. Consistency is the most important aspect of a naming convention to ensure that everyone involved can quickly find what they need.

Following are two additional actions to take that are known to most experienced interpreters but usually are learned the hard way:

1. Export or back up a copy of the data, interpretations, and databases on a routine basis. Protect the interpreted work from a system crash, power outage, file corruption, and of course, user error.
2. Clean up old versions of interpretation. If every version is saved, eventually no one will know the good from the bad. It may not be clear whether a specific map or interpreted horizon is the final result or some older version.

Documenting Work

One way to keep work organized is to document the work on paper or digitally (Philosophical Doctrine point 10). Doing so will benefit all who follow. Even if an area has a history, such as a previous interpretation, most people, understandably, want to do their own interpretation. Nonetheless, knowing what previous interpreters have done can save time and effort in the current work. Avoiding issues noted by predecessors saves valuable time, and building on what they have accomplished increases efficiency during the next phase of interpretation.

Handwritten notes are not as critical today as they once were, but it is often helpful to take a few seconds to jot down key observations or the steps taken to generate a particular display. Most software enables users to save a *template* that can be used to re-create a display in a certain way. Templates promote consistency and reduce the amount of work entailed in creating new workflows. Making a screen capture of interesting features is another way to catalog important aspects of the interpretation. The work performed by employees of a company has value. We recommend keeping a record of progress and sharing knowledge when needed.

Key Point: A systematic approach and well-organized interpretations are effective time-saving strategies.

FAULT INTERPRETATION

Introduction

The principles discussed in the following sections apply to any workstation software and are relevant to extensional or compressional tectonic environments as well as conventional and unconventional reservoirs. The importance and technique of making fault surface maps is discussed in detail in Chapter 7. This section presents traditional methods of fault interpretation and mapping for the workstation. A three-dimensionally correct fault interpretation provides the solid foundation upon which all subsequent horizon interpretation and mapping are built. It is not enough simply to pick fault segments on a series of parallel vertical and/or horizontal seismic profiles. The resulting fault surface should make sense geologically and geometrically and be valid in three dimensions. Therefore, *faults should not only be interpreted as individual picks on the data but also viewed as fault surfaces in order to be deemed geologically valid in three dimensions.*

As of this writing, several of the 3D interpretation software packages are capable of some form of autotracking of faults in a 3D seismic volume. Some autotrackers are designed to work on one fault surface at a time, while others autotrack multiple faults throughout the entire volume. Parameters can be optimized to obtain a meaningful result. However, because these autotracking capabilities and results vary widely based on the software, input data quality, and geologic complexity, a detailed discussion is not provided here. These are exciting developments in workstation capabilities and will undoubtedly gain more acceptance with time. Interpreters of 3D seismic data are encouraged to test this technology on their data and see if the promise of quality and efficiency holds true. As with any machine-driven interpretation, quality control and geologic validity are paramount. The following discussion provides methods of interpreting faults as well as methods of quality control and visualization of results.

Generally, *fault surfaces tend to be smooth, nearly planar, or arcuate surfaces*. They typically do not change radically in strike or dip unless the fault surface has been deformed. For most faults, throw varies in a systematic manner along the strike of a fault surface and can increase or decrease with depth. Fault throw and horizontal extent are related. The larger the throw, the greater the lateral extent will be. A rapid change of throw along a fault likely indicates a fault intersection or bifurcation.

One question that arises during framework interpretation is how much detail is necessary to define a good framework. The answer is usually based on the complexity of the geology or, in some cases, the quality of the seismic. If the fault interpretation is started on widely spaced parallel lines and not tied via intersecting lines or horizontal slices, we may interpret a fault segment on one profile that appears to belong to the same fault as segments picked on other profiles. However, as we *tie* additional lines, the fault picks may be found to be either a series of

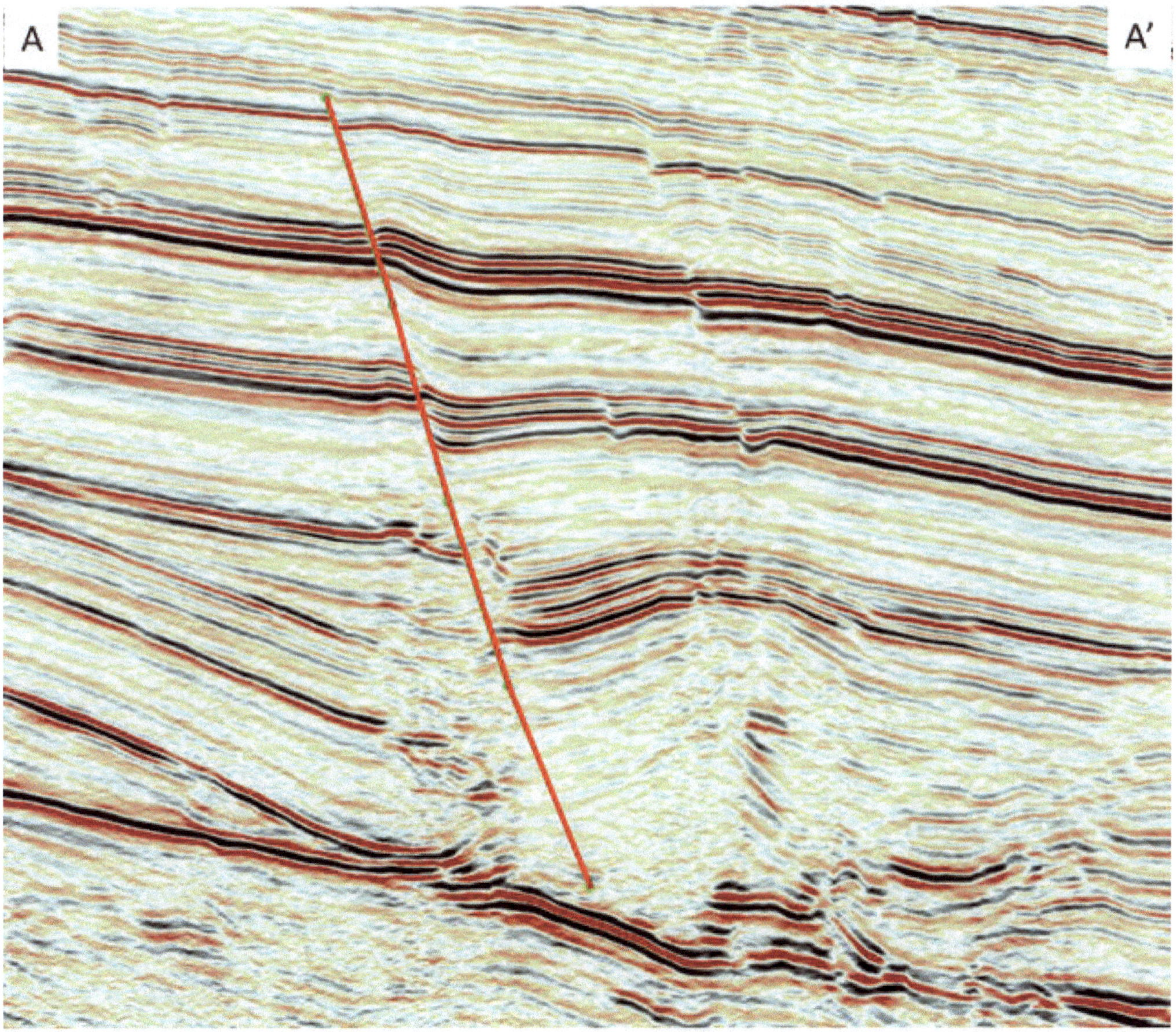

Figure 9-14 Eagle Ford: Dip line A-A′. Parallel to Figure 9-15. The red fault is the most northerly major fault encountered in the down dip area. (Data published with permission from Seitel Data Ltd.)

unconnected faults or several faults that coalesced to form a fault system. What may be initially interpreted to be a single fault surface may in fact be separate faults with different timing and orientation. Figures 9-14 through 9-17 demonstrate how this would look on 3D seismic data. Figures 9-14 and 9-15 are parallel dip lines from a 3D data set. Individual fault picks are color-coded to identify them in subsequent figures. The red and blue fault picks are the first fault encountered in the down dip direction on each dip line, but *the true relationship between the faults emerges when viewed on a* **strike line** (Fig. 9-16). Note how each fault has a unique shape and that the two faults are not connected. Figure 9-17 shows the same relationship in a horizontal slice view using the coherence attribute.

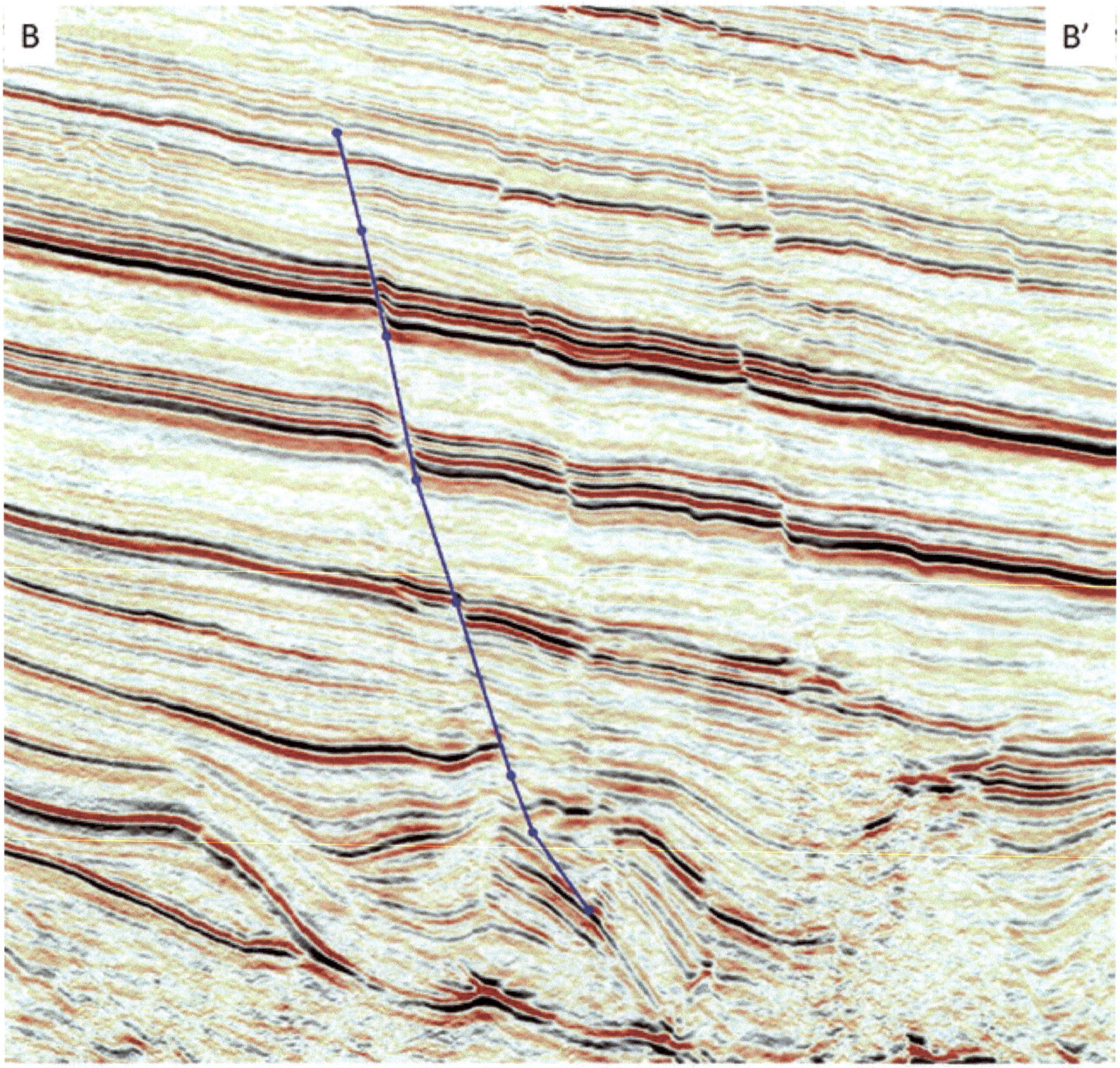

Figure 9-15 Eagle Ford: Dip line B-B′. Profile is located 2 miles southwest of Figure 9-14. The blue fault is also the most northerly fault encountered in the down dip area. (Data published with permission from Seitel Data Ltd.)

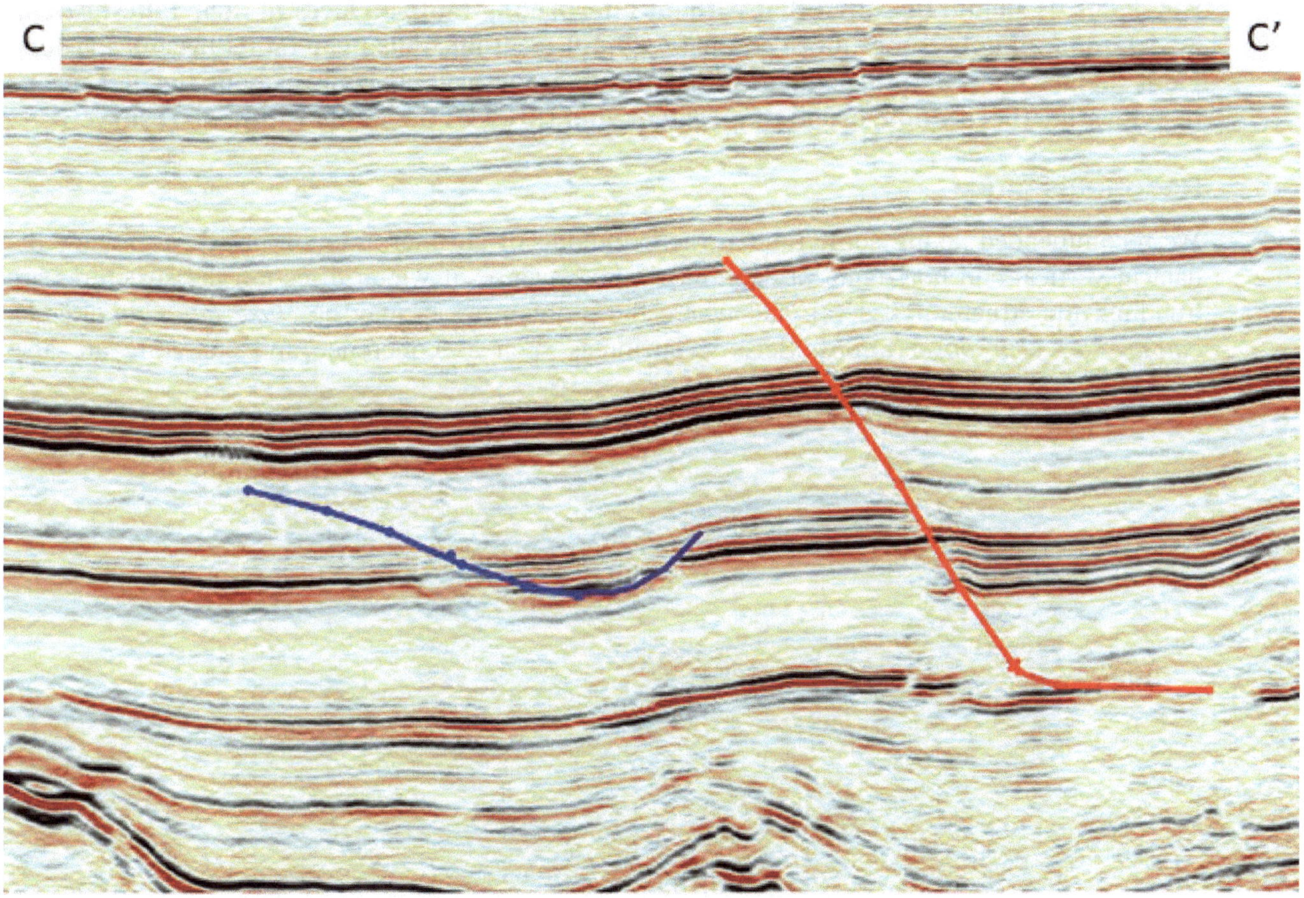

Figure 9-16 Eagle Ford: Strike line C-C′ demonstrates that the red and blue fault picks do not belong to the same fault. (Data published with permission from Seitel Data Ltd.)

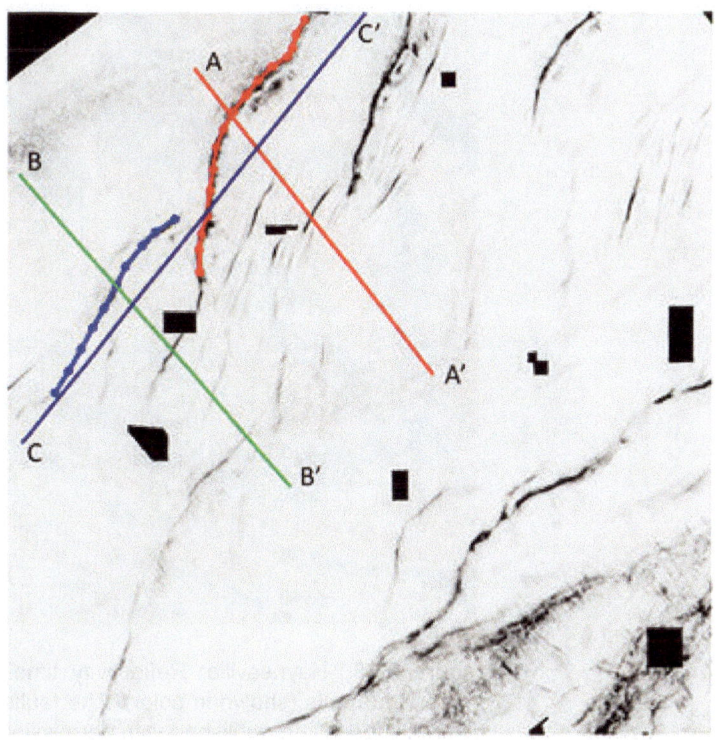

Figure 9-17 Eagle Ford: A horizontal coherence attribute slice view confirms that the red and blue faults are similar but are not the same fault. Profiles from the previous three figures are shown. (Data published with permission from Seitel Data Ltd.)

Key Point: It should be stated that without interpreting fault surfaces to their fullest extent, horizon mapping becomes more challenging and locating wells near faults becomes riskier. Fault surface maps and their integration with the structural horizons provide the best means to determine the location, orientation, extent, and throw of the faults within a 3D survey.

Reconnaissance

We recommended taking some time at the beginning of an interpretation project to review the data in vertical and horizontal orientations to get a sense of which direction the structures are trending, the structural style(s), and where the problem areas may be. *Take notes and screen captures of your observations*. Seismic attributes such as coherence and curvature are also

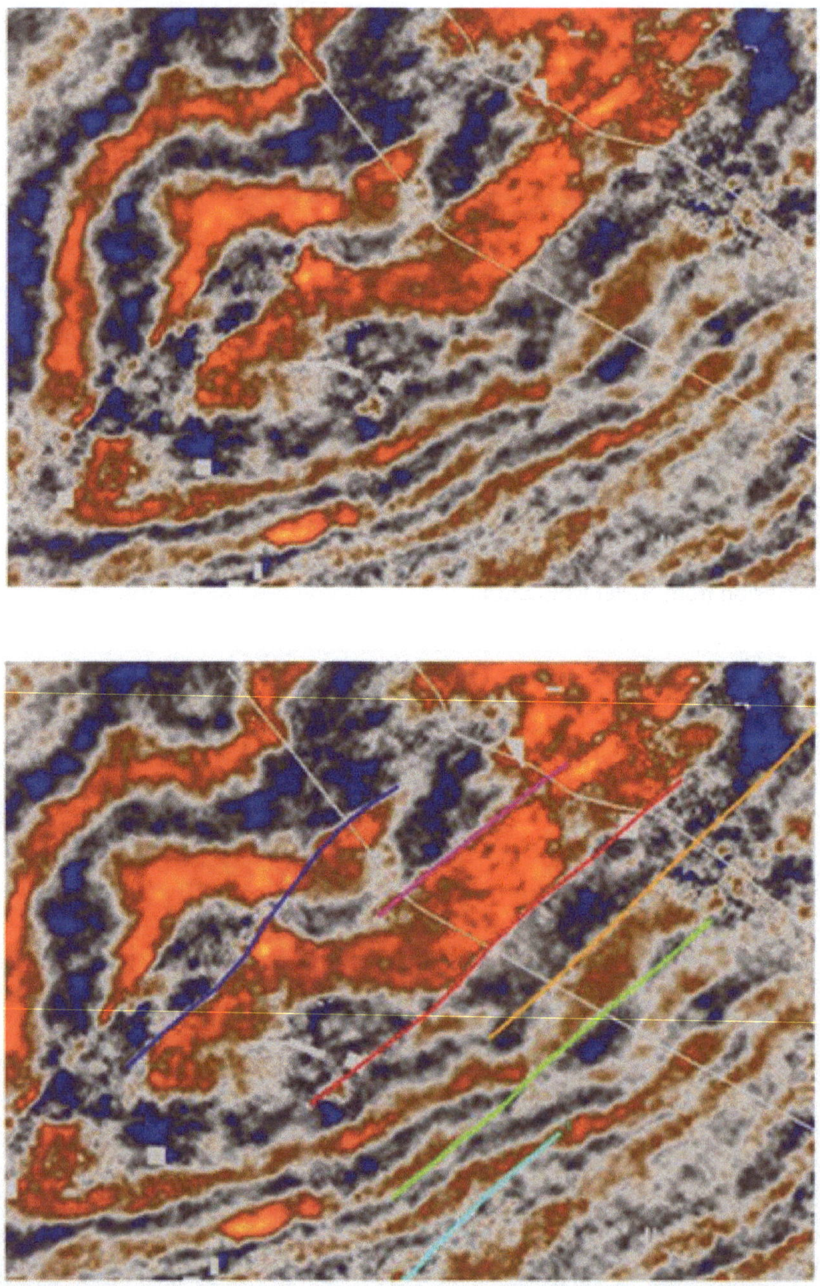

Figure 9-18 Haynesville: Reflectivity time slice shown with and without interpreted faults (shown in color). The faults are difficult to see without the colored lines. (Data published with permission from CGG Land.)

valuable reconnaissance tools. Figure 9-18 demonstrates how difficult it can be to identify faults on reflectivity time slices through a 3D data volume. Figure 9-19 shows the same time slices, but they are now displayed using the coherence attribute to enhance discontinuities. Fault trends and relationships are more easily determined using this type of attribute display.

Creative color bars (palettes) used on variable density displays can enhance discontinuities in the data and allow better fault identification. Some ideas for color schemes were shown previously, but interpreters are not limited to those few. The goal is to identify faults as quickly as possible, so use creativity when it comes to use of color and various display scales. From this reconnaissance, develop a plan to approach the fault interpretation. Consider well locations, any missing or repeated section-correlated in the well logs, and how much detail is necessary to correctly map the faults in your 3D data set.

One of the most effective ways to recon a seismic volume is to autotrack one or more of the most coherent horizons. In Figure 9-20, hundreds of square miles of data have been autotracked in

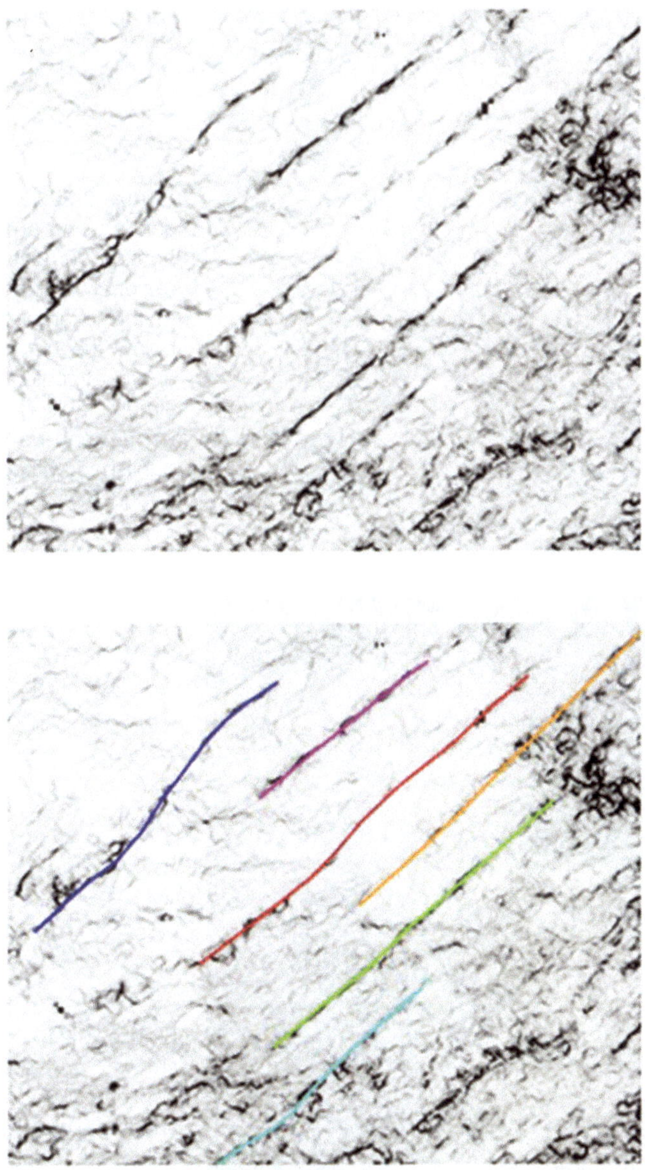

Figure 9-19 The same time slice as in Figure 9-18 now extracted from a coherence attribute volume. The faults are highly visible on this display. (Data published with permission from CGG Land.)

the Haynesville play and are displayed with a dip attribute. Note how quickly the complex fault pattern (A) in the Haynesville play can be identified. In addition, more simple faults (B) and also subtle lineations likely to be faults (C) are highlighted. A similar effect can be seen in Figure 9-21 where the same horizon is displayed as a light-colored time structure. Fault trends stand out in this view.

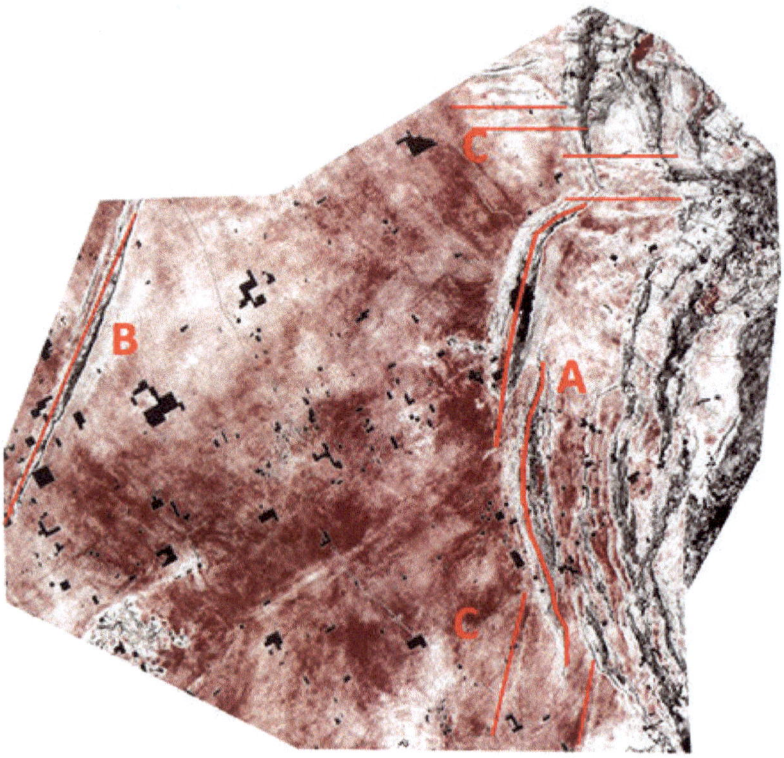

Figure 9-20 Haynesville: Reconnaissance using a preliminary autotracked horizon demonstrates how quickly the regional fault trend can be identified. In this example, the Smackover (just below the Haynesville) is shown with a dip attribute. Complex faulting (A), simple faulting (B), and subtle faulting (C) can be easily identified. (Data published with permission from CGG Land.)

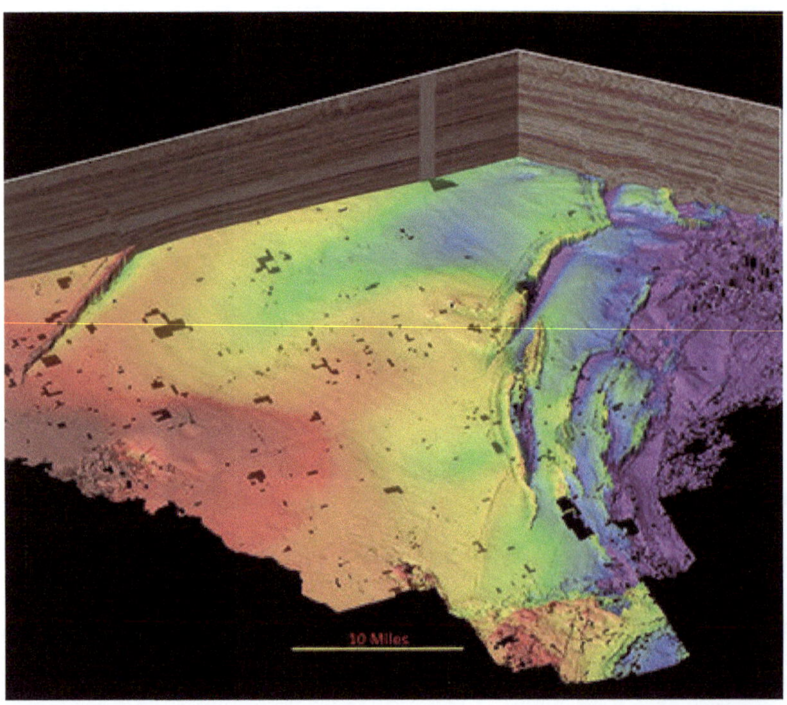

Figure 9-21 Haynesville: This view shows an autotracked Smackover horizon colored by time structure with shading. This is another effective reconnaissance tool. Note the horizon has not been rigorously quality checked and is used only for recon of the data volume. (Data published with permission from CGG Land.)

Similarly, the Marcellus play can be quickly evaluated using the same technique. Figure 9-22 captures the essence of the complex reverse fault trends and crosscutting structures of the seismic reflector nearest the Marcellus shale. Figure 9-23 is a view from above with light shading from the side. In this view, both the primary faults and crosscutting structures are clearly identified.

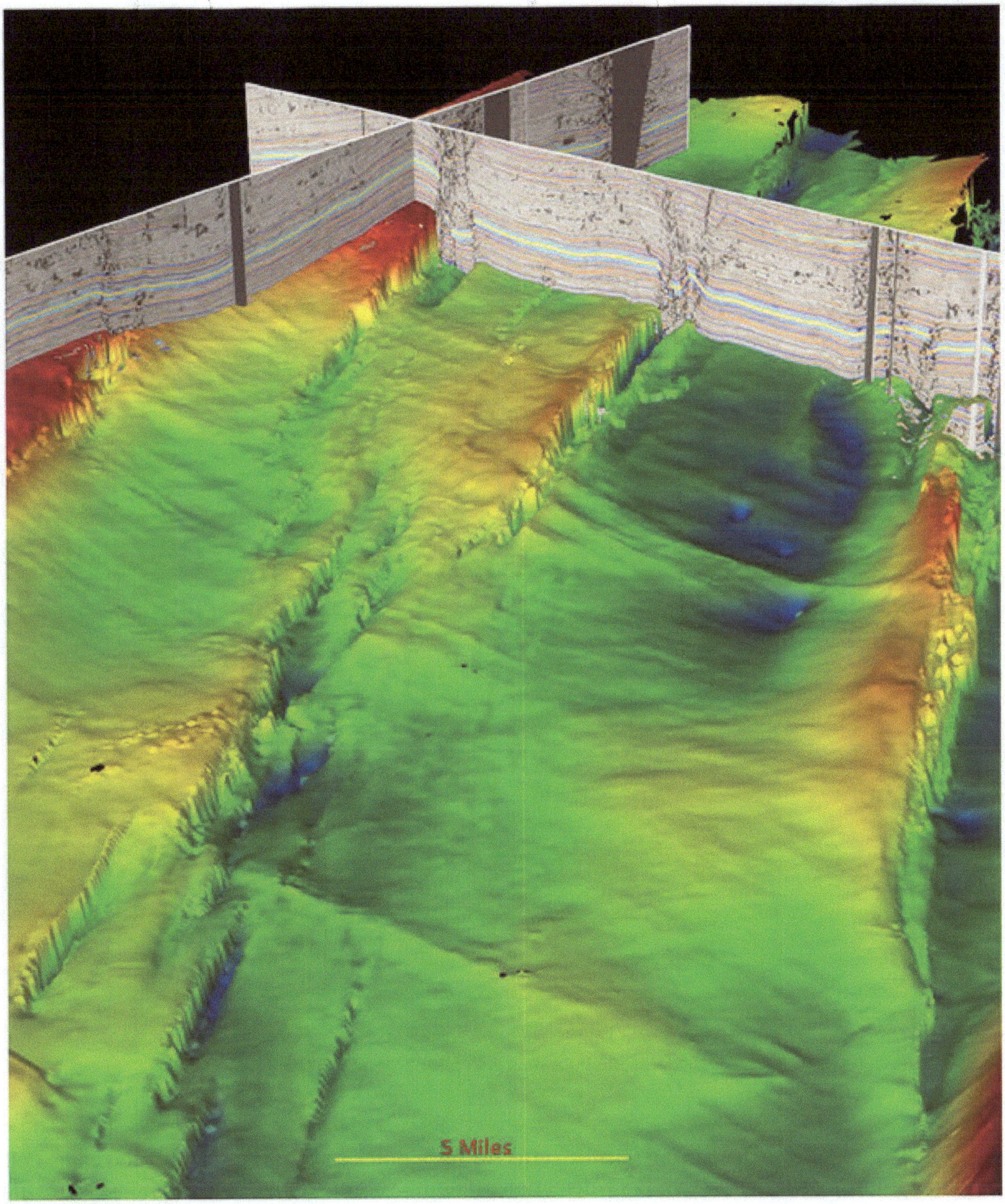

Figure 9-22 Marcellus: Reconnaissance using a 3D perspective view captures the complex reverse fault trends and crosscutting structures of the seismic reflector nearest the Marcellus shale. (Data published with permission from CGG Land.)

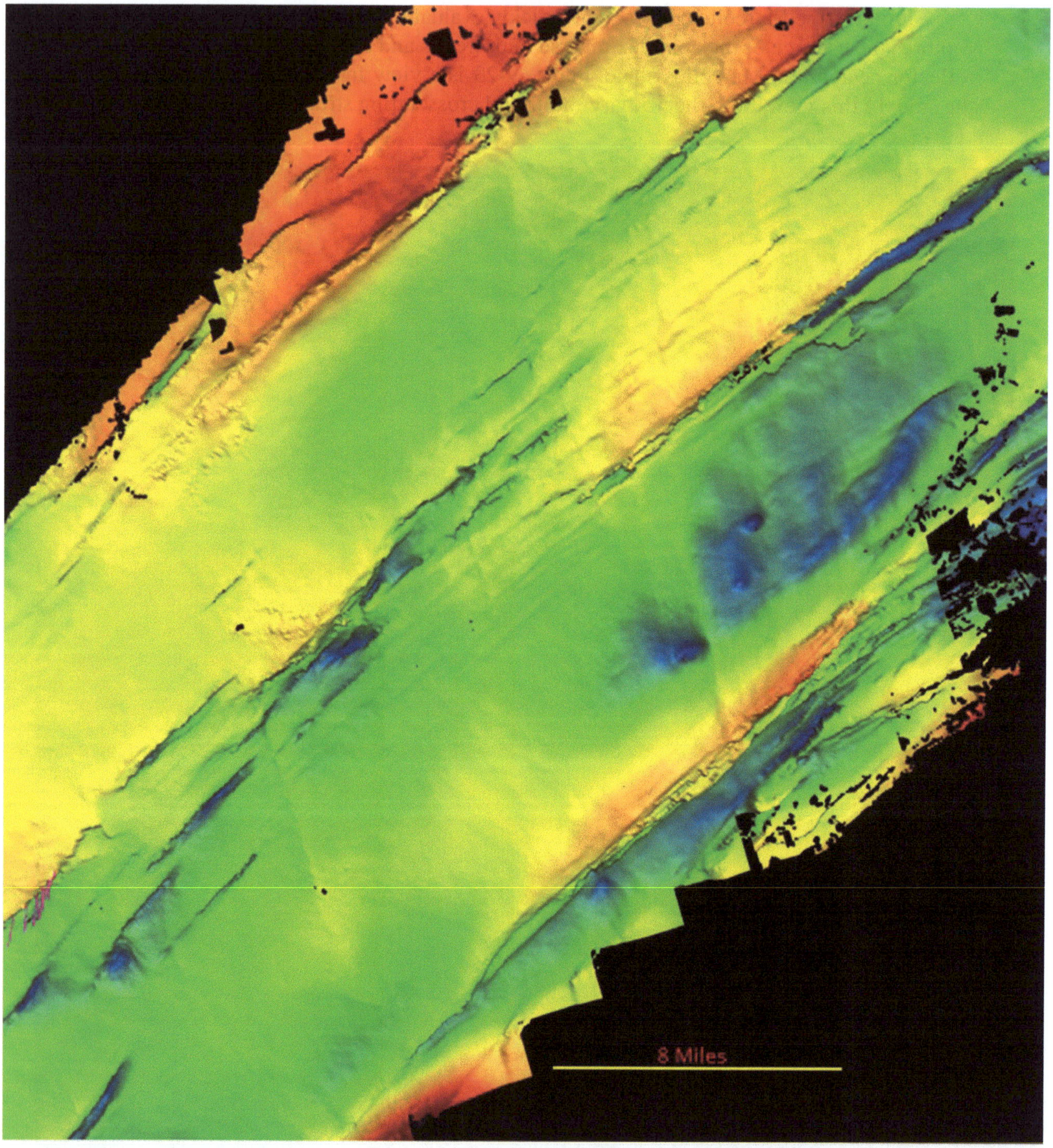

Figure 9-23 Marcellus: View from above with light shading from the side. In this view, both the primary faults and the crosscutting structures are clearly identified. (Data published with permission from CGG Land.)

Key Point: Autotrack a strong seismic reflector for a quick look at the structural trends. Follow up with using slice views of a fault-detection attribute volume to perform reconnaissance along with vertical displays of a reflectivity volume. Use color and scale to enhance the view.

Tying Wells Using Synthetic Seismograms

One of the most important steps to take early in the interpretation of an area is to create synthetic seismograms. The process of tying well data in depth to seismic data in time using a velocity trend is fundamental to the integration of all data into an interpretation. Prior to tying the seismic, a well log (usually a GR log) is correlated with logs from surrounding wells to establish a formation tops framework. This includes wells that have missing or repeated section, which means there is either a normal fault, unconformity, or reverse fault cutting the well. In cases where well control is sparse, the *seismic* can be used to help locate faults in wells. The method to do this is described later in the text.

Wells that have sonic and/or density logs, the basic ingredients to a synthetic seismogram, are good candidates to begin with. The quality of the log curve data should be reviewed and edited if necessary, and the logs should extend over a large depth range for best results. Synthetic seismograms are generated by first computing impedance by multiplying sonic velocity times bulk density. The impedance log is then used to compute a series of reflection coefficients, which are then convolved with a wavelet to generate the synthetic seismogram. The velocity trend used to convert log depth to seismic time can be obtained from integration of the sonic log velocities, vertical seismic profile, checkshot survey, or from time and velocity picks used in data processing, to name a few of the key sources. The quality of the tie between a synthetic seismogram and seismic data can range from very good to very poor. Techniques to adjust the synthetic traces to tie the seismic involve bulk shifts, phase rotation, stretching, or squeezing. Care must be taken not to *overcorrect* the synthetic and cause an unrealistic velocity. The synthetic seismogram in Figure 9-24 shows an example of a good tie to a 3D seismic volume in the Spraberry-Wolfcamp play. The goal is to *match reflection patterns (peaks and troughs) in the synthetic traces (red) to similar patterns in the 3D data at the well (blue)*. From this we determine which horizons are the best to be mapped as the framework. Differences in reflectivity observed between the synthetic and seismic (such as between 1150 and 1350) can be a result of either (1) a higher sample rate for log data than for seismic data, allowing finer detection of geologic beds in the synthetic, or (2) events caused by multiples that exist in the seismic being nongeologic and therefore not seen on the synthetic. The topic of framework horizon selection is discussed later in this chapter.

Integrating Well Control

When working in an area where wells are sparse and there is a suspected fault present in a well, create a synthetic seismogram at the well and then:

1. Display two or three seismic sections that pass through the well location but without the well information displayed on the seismic data. Choose orientations that are based on your reconnaissance of the dip direction of the fault. These sections should intersect in map view.
2. Look for seismic event terminations that indicate the presence of a fault and interpret a fault segment that would represent a *most likely* location for the fault. If no fault is obvious, interpret several possible segments or choose another line orientation.
3. Now redisplay the seismic section showing the well data posted along with the interpreted fault segment. How closely do the seismic fault segment and fault cut from log correlation agree? If they do not agree, consider the following four possible reasons for the mis-tie:
   a. The fault pick in the well needs adjustment.
   b. The seismic interpretation needs adjustment.

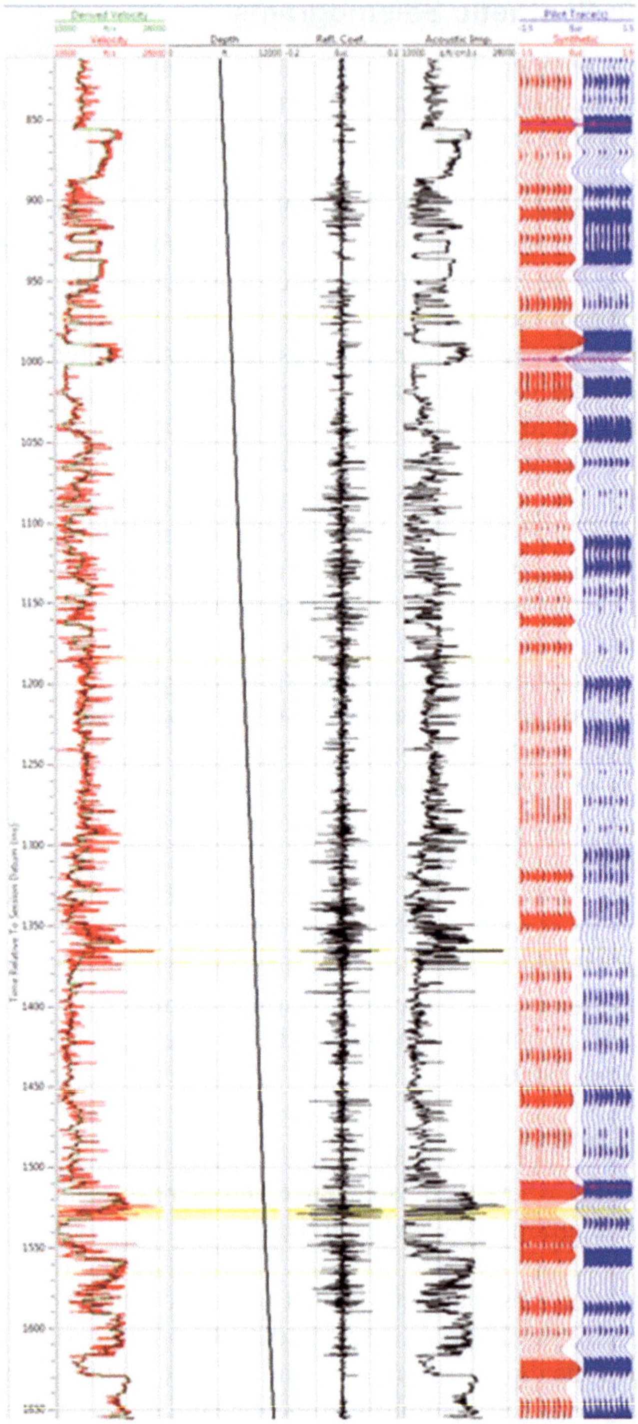

Figure 9-24 Typical synthetic seismogram in the Midland Basin. The left column shows the velocity log (in red) derived from the sonic log. The center column is the reflection coefficient series. The red traces on the right are the synthetic traces, and the blue traces on the far right are from the seismic volume at the well location.

c. The time versus depth relationship between the well and seismic (i.e., the synthetic seismogram) needs adjustment.

d. The well location is inaccurate (incorrect surface location or erroneous directional survey).

Any one of these explanations is possible. In the case of sparse well data, begin by reviewing the well log correlation. Attempt to adjust the fault cut in the well to match the seismic interpretation. If the time-to-depth relationship appears reasonable from the synthetic tie (or other means), the seismic fault and subsurface pick in the well should be very close when displayed on the time seismic volume. If not, other issues could be present. Check for errors in the deviation survey, or in the case of a vertical hole, it may be that it is not really vertical or that the surface location of the well may be misspotted. *Well location problems are more common than one would suspect.* Because fault surfaces typically dip at much higher angles than bedding surfaces, errors in well location generate much larger discrepancies in fault ties than they do in horizon ties. Also, make sure the vertical separation interpreted on the 3D seismic sections agrees with the vertical separation as determined from log correlation (refer to Figure 7-38).

The key point is that *any discrepancies between the seismic data and the subsurface control should be understood and resolved if possible*. After tying the well control, continue working this fault with one of the following strategies.

Fault Interpretation Strategies

There are many ways to approach fault interpretation and mapping in the workstation environment. Some are more efficient than others. As software evolves, techniques also evolve to take advantage of new functionality. Even with more powerful hardware and software, however, the underlying objective in fault interpretation does not change. That objective is to create an accurate 3D representation of all fault surfaces, which leads to better structure maps and better well placement. As mentioned previously, the topic of automated fault interpretation is not covered in this chapter, but the techniques discussed here can be used to validate and correct the machine driven fault interpretation.

A sound fault interpretation strategy is based on the following principles:

1. Define the preliminary fault surfaces quickly with a minimum of fault picks.
2. Integrate the well control.
3. Add fault picks and validate the fault surfaces in 3D.
4. Complete the interpretation of the primary framework faults before completing the final horizon interpretation.

Following are three basic strategies:

1. Traditional method using *tied vertical sections* verified in map view and horizontal slice view.
2. Alternate method to interpret a fault pick on a coherence or symmetry horizontal slice then add picks on vertical displays.
3. Three-dimensional visualization method verified in 3D space.

Which approach to take depends on available software and interpreter preference. Traditional methods are effective in most areas and available on all workstations. Three-dimensional visualization is effective in all areas including structurally complex areas. When possible, use 3D visualization software as a validation tool for the fault surfaces and to refine a preliminary fault interpretation. As visualization capabilities expand, so does the routine use by interpreters.

Strategy 1 is a more traditional approach to fault interpretation. Figures 9-25 through 9-34 demonstrate the workflow for the single fault method. The first step is to interpret a fault segment on the first vertical line (Fig. 9-25). Next, choose two additional lines that tie the first line (Fig. 9-26 basemap) and interpret the fault based on the tie with the previous line. The fault segments interpreted on the two tied seismic lines are shown in Figures 9-27 and 9-28. Figure 9-29

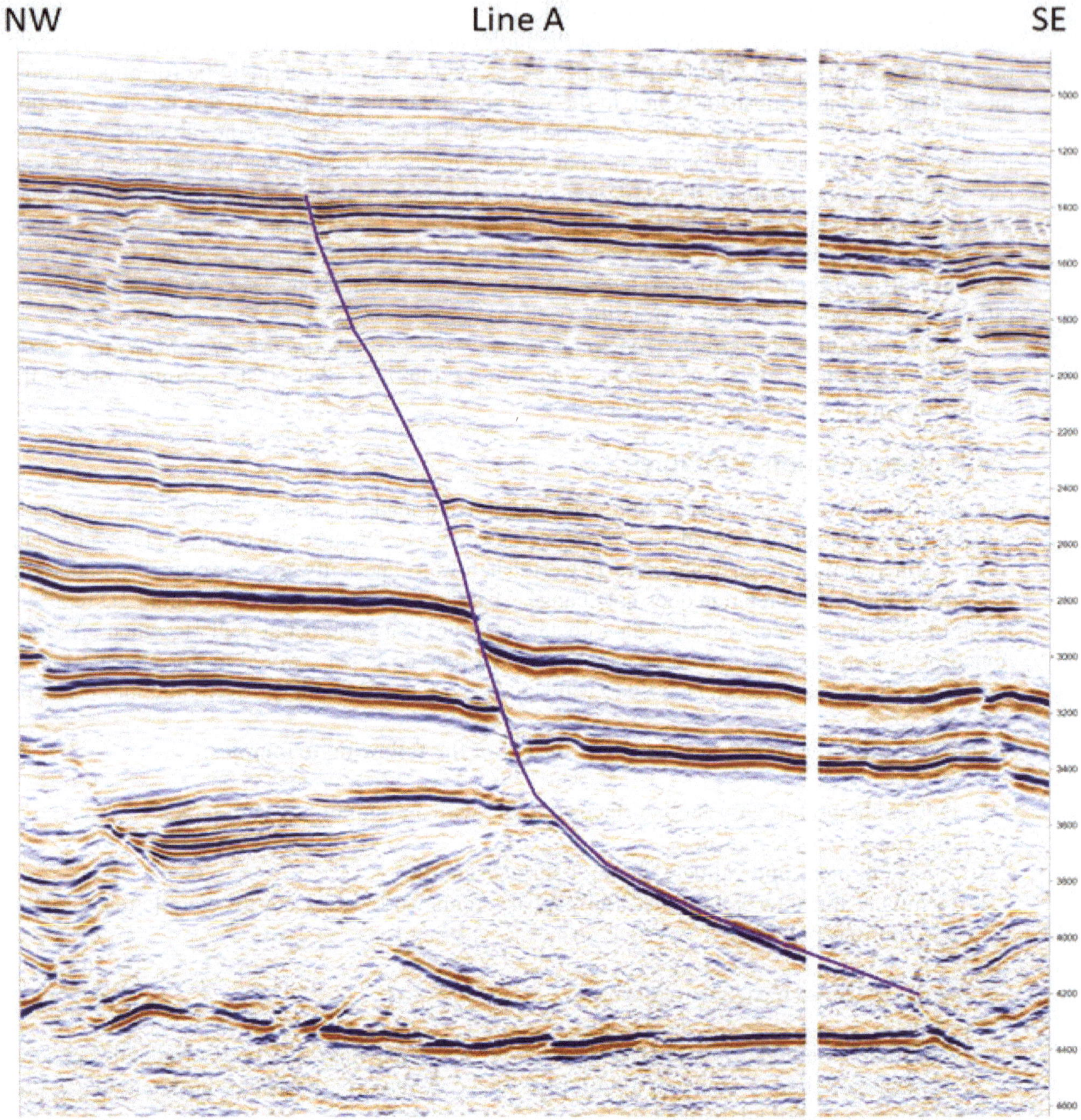

Figure 9-25 Eagle Ford: Fault strategy 1. Line A. First interpreted line for the purple fault. (Data published with permission from Seitel Data Ltd.)

shows the interpretation based on just three intersecting interpreted lines. Coloring the interpretation gives a sense of the fault strike. This is all it takes to begin a preliminary fault interpretation using strategy 1. Continue the interpretation by moving the three intersecting vertical profiles until the fault surface is completely defined (Fig. 9-30). The goal is to accurately define the fault surface in as few segments as possible, so it is acceptable to clean up an interpretation by "pruning" out redundant segments as long as the surface remains valid in three dimensions. The fault surface generated after interpreting 14 intersecting lines is reasonable (Fig. 9-31). The fault surface should also be validated in strike view (Fig. 9-32) and in 3D (Fig. 9-33). These views help

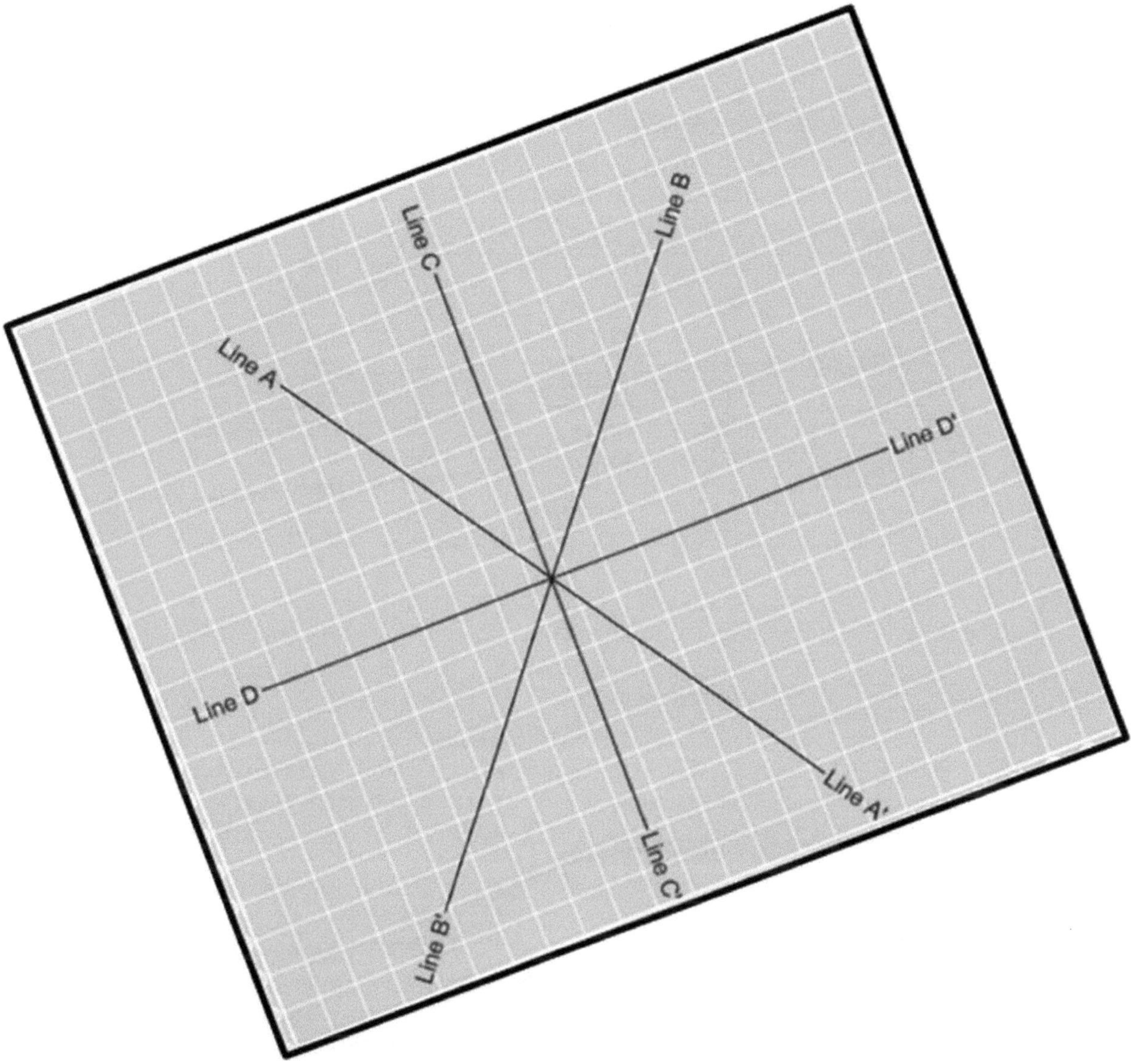

Figure 9-26 Eagle Ford: Fault strategy 1. Basemap showing the initial interpretation plan for the purple fault. (Data published with permission from Seitel Data Ltd.)

visualize the fault shape and compatibility with the seismic. Finally, Figure 9-34 demonstrates how deformed rocks in the footwall can affect the shape of a fault surface. In this case, salt movement has created enough instability to warp the fault surface.

Using a series of intersecting seismic displays and interpreting one fault at a time, such as in strategy 1, is generally an efficient and effective methodology. It is important to note that these are near final fault surfaces that may need to be refined after completion of the framework horizon interpretation. The resulting surface should be a good representation of the geometry of the fault, how far it extends, and in what direction. It should be noted that as faults are interpreted on seismic data, the map view of the resulting contoured fault surface should be updated. This allows quality control on the fly to catch any big errors, such as abnormal changes in fault dip, strike, or miscorrelation, before getting too deep into the interpretation. We recommend not to pick more fault segments than are needed to define the fault surface. The key points are to determine where the major faults occur and how they intersect in the subsurface, if indeed they intersect.

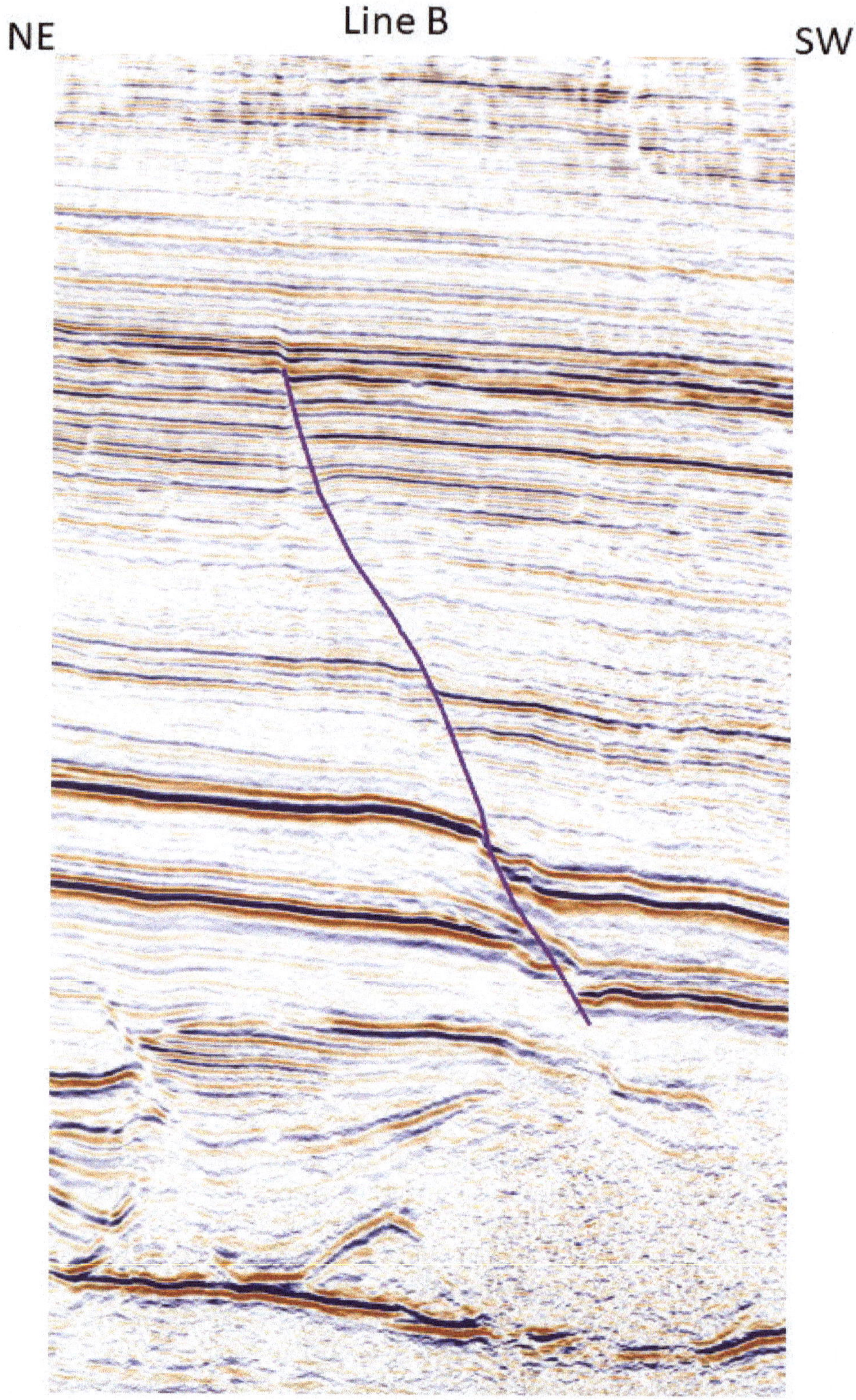

Figure 9-27 Eagle Ford: Fault strategy 1. Second interpreted line for the purple fault tied to the first line. Line B. (Data published with permission from Seitel Data Ltd.)

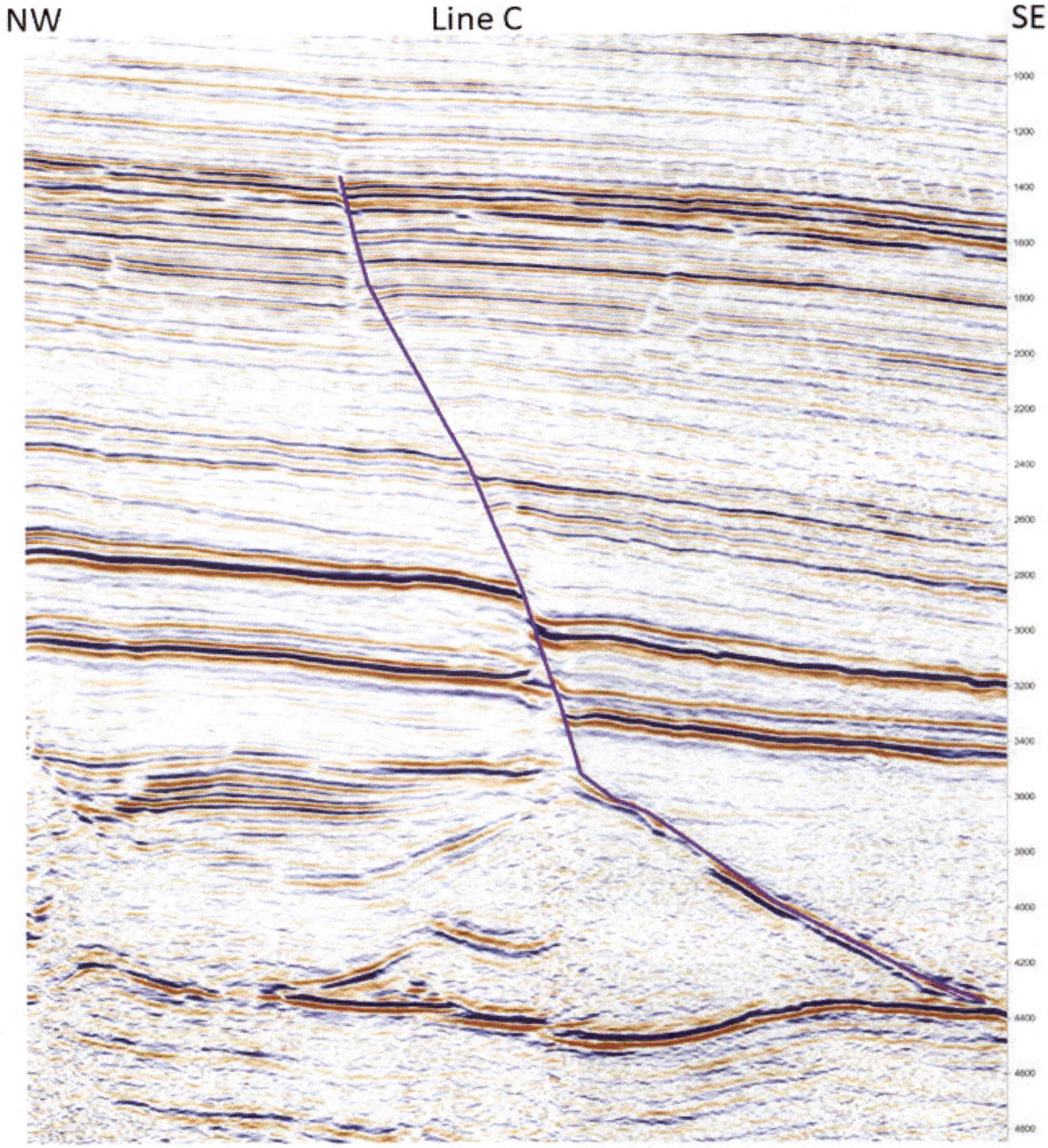

Figure 9-28 Eagle Ford: Fault strategy 1. Third interpreted line for the purple fault tied to the previous lines. Line C. (Data published with permission from Seitel Data Ltd.)

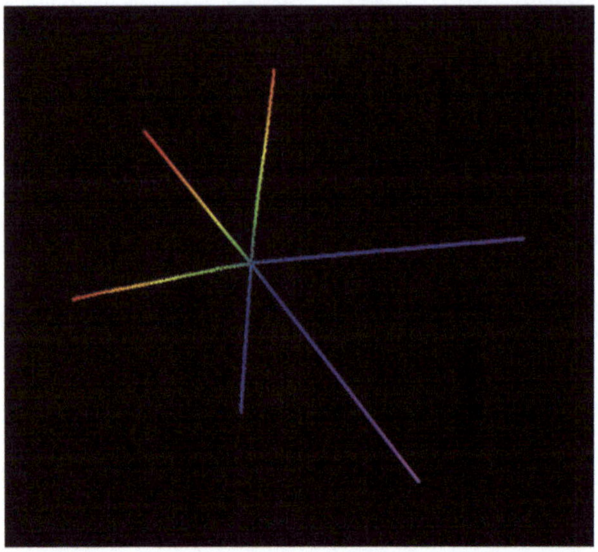

Figure 9-29 Eagle Ford: Fault strategy 1. Perform a quick quality check of three tied interpreted lines for the purple fault colored by z value. (Data published with permission from Seitel Data Ltd.)

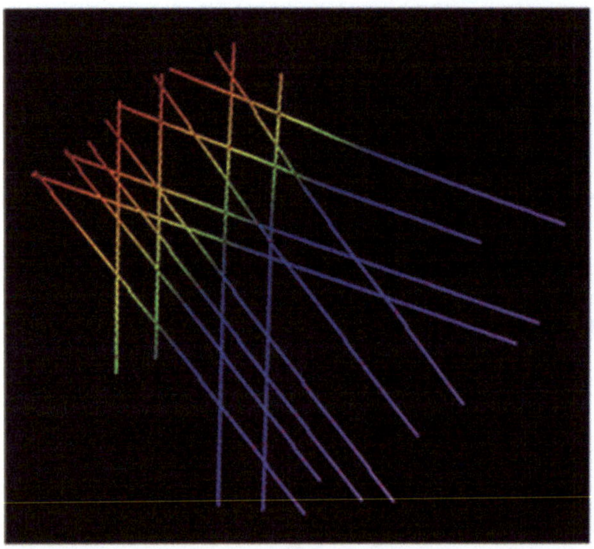

Figure 9-30 Eagle Ford: Fault strategy 1. Fourteen interpreted lines for the purple fault colored by z value. Each line is tied to at least one other line. (Data published with permission from Seitel Data Ltd.)

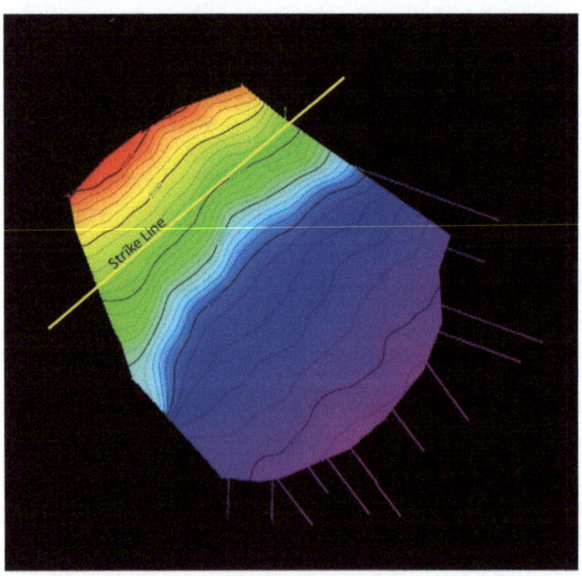

Figure 9-31 Eagle Ford: Fault strategy 1. Final fault surface map for the purple fault. The strike line is shown in Figure 9-32. (Data published with permission from Seitel Data Ltd.)

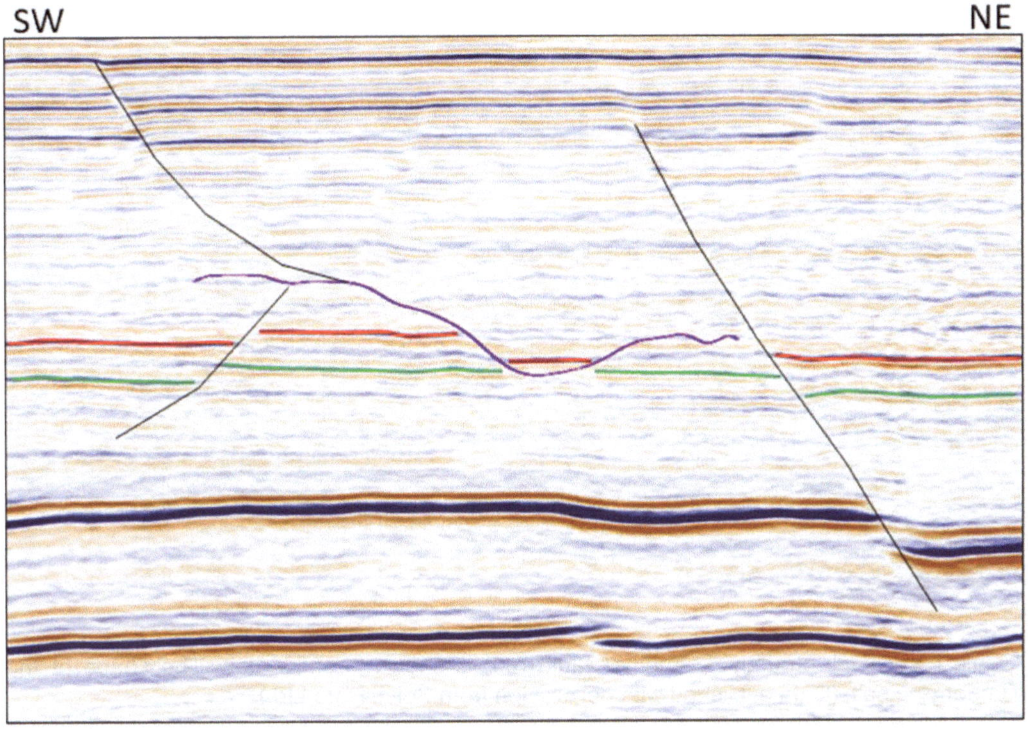

Figure 9-32 Eagle Ford: Fault strategy 1. Strike view for the purple fault. Also shown are other nearby faults (black) and interpreted horizons showing the fault offset (red and green). (Data published with permission from Seitel Data Ltd.)

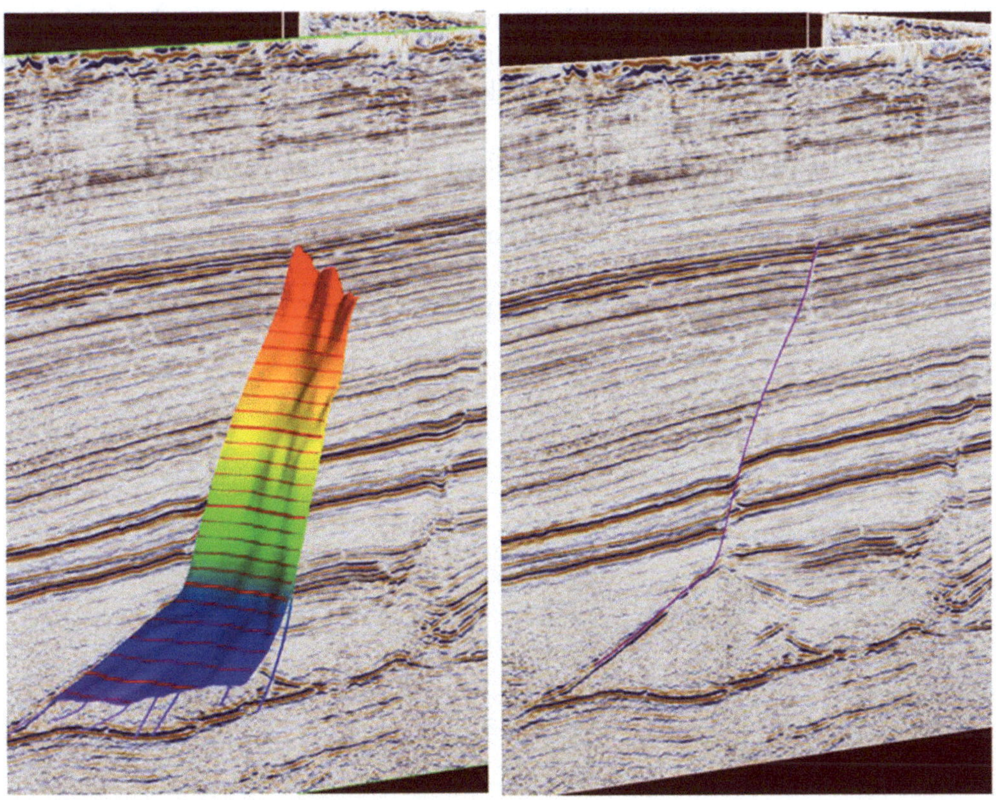

Figure 9-33 Eagle Ford: Fault strategy 1. Three-dimensional view for the purple fault. Left: Full surface cut by a dip seismic line. Right: Intersection of the fault with the seismic line. (Data published with permission from Seitel Data Ltd.)

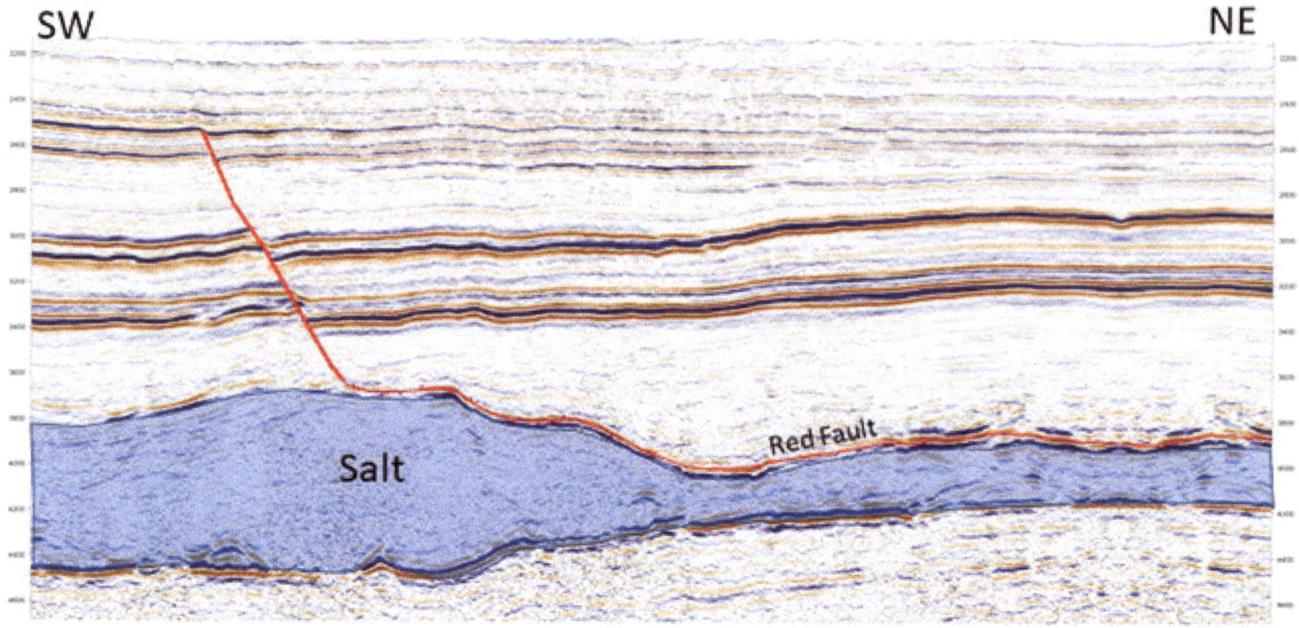

Figure 9-34 Eagle Ford: Fault strategy 1. Another fault (red) is shown in strike view. The shape of the red fault is affected by the underlying salt. (Data published with permission from Seitel Data Ltd.)

Strategy 2 is one of the most time-efficient fault interpretation methods available. It begins with a horizontal slice from any fault detection attribute volume, such as coherence.

1. Use the slice view to identify and interpret a single segment for each fault identified.
2. Display a dip-oriented vertical seismic display from a standard reflectivity volume.
3. Using the cross-posting of the faults picked on the slice, interpret all visible fault segments that were identified previously.
4. Move the vertical panel and repeat the fault picking process. Interpret in this way until the edge of the area of interest is reached.
5. View each fault surface in 3D view and adjust or add picks as needed.

Figures 9-35 through 9-38 show how the process is done.

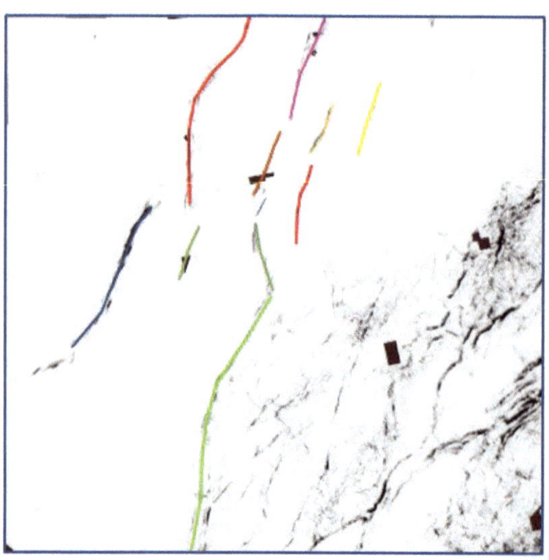

Figure 9-35 Eagle Ford: Fault strategy 2. Eleven faults are identified and interpreted in color on a coherence slice. (Data published with permission from Seitel Data Ltd.)

Strategy 3 is similar to strategies 1 and 2 except that the entire process is done in a 3D view. In areas where it is important to sort out the details of a fault interpretation and the data quality is sufficient, it is best to use a 3D visualization display. This allows the data to be viewed in rapid succession in any orientation. Putting the data in motion, so to speak, allows you to see a single fault surface or the entire fault framework. This validation process takes very little time, and it is well worth the effort when planning and drilling wells near faults.

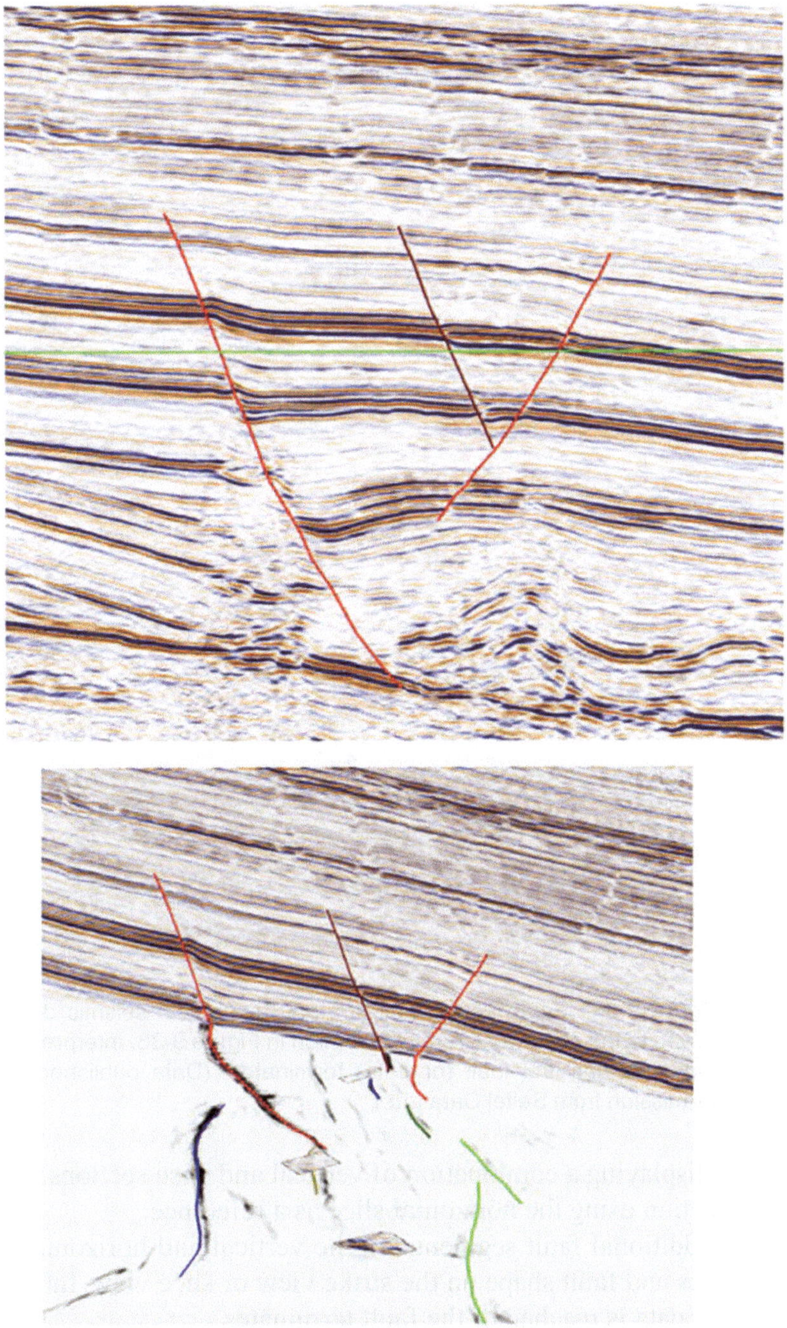

Figure 9-36 Eagle Ford: Fault strategy 2. Initial fault picks from the slice are tied to a vertical line. Each fault intersected is interpreted on the vertical display. (Data published with permission from Seitel Data Ltd.)

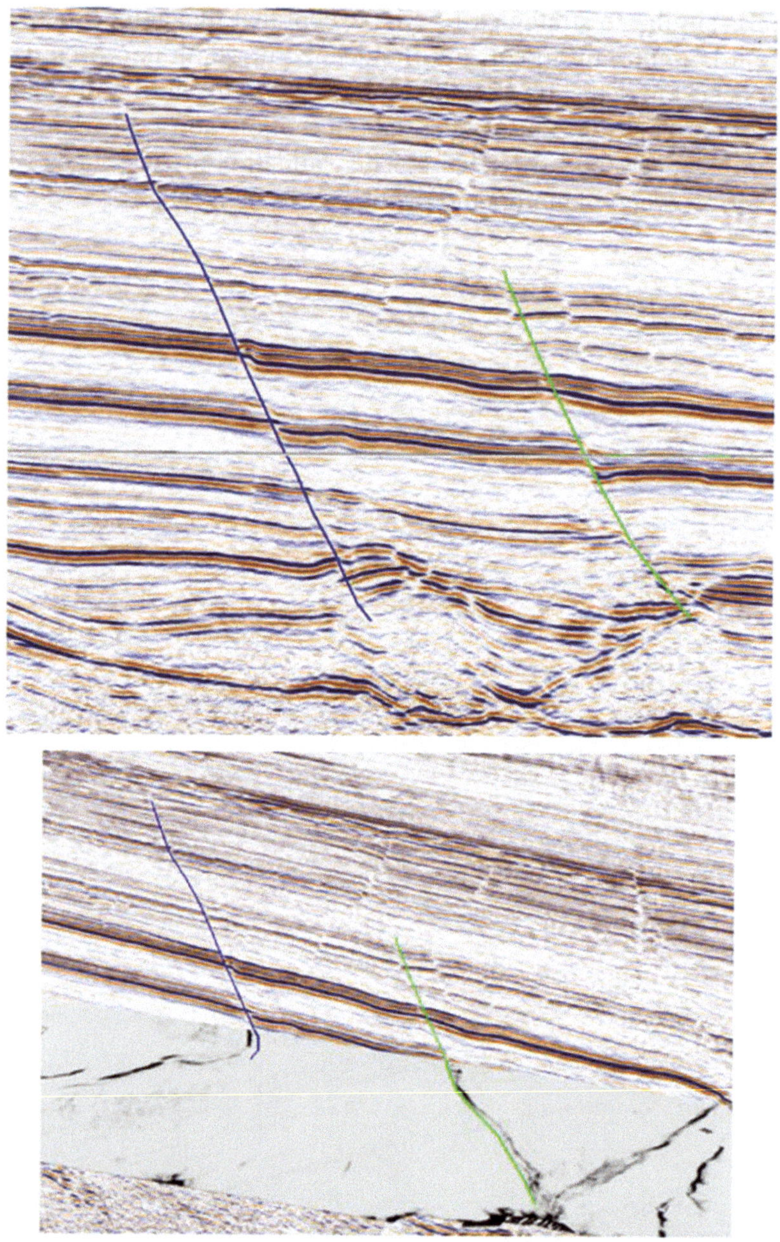

Figure 9-37 Eagle Ford: Fault strategy 2. Vertical seismic display moved to the southwest from the location in Figure 9-36. Interpretation continues until the fault (or data) terminates. (Data published with permission from Seitel Data Ltd.)

1. Begin by displaying a combination of vertical and slice sections. Interpret the first fault on a vertical section using the horizontal slice as a reference.
2. Interpret additional fault segments on the vertical and horizontal slices. Periodically check the fault ties and fault shape on the strike view or slice view. Interpret the fault surface until the edge of data is reached or the fault terminates.

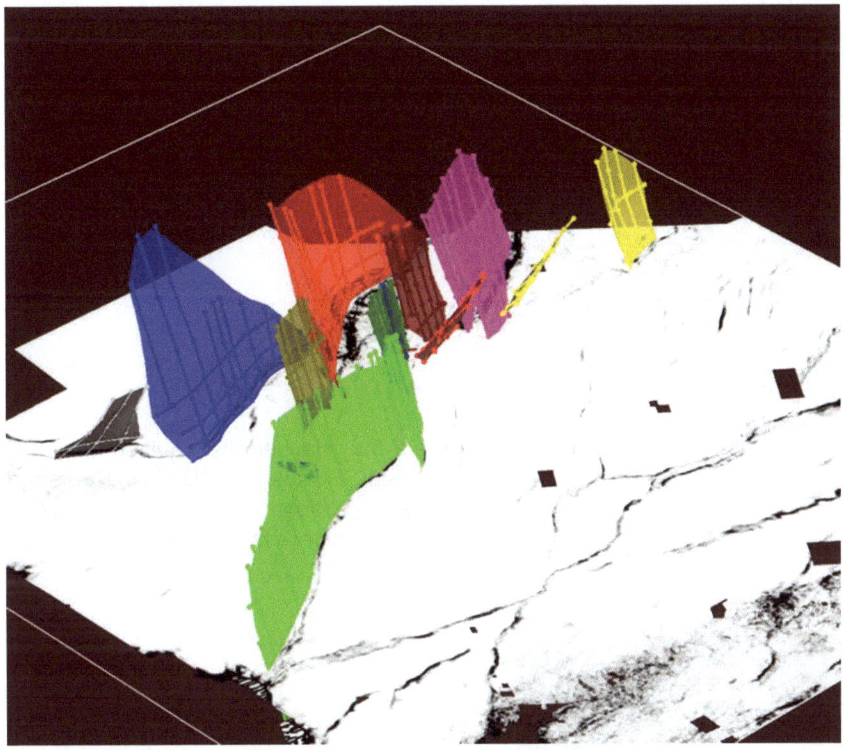

Figure 9-38 Eagle Ford: Fault strategy 2. View each fault surface in 3D and quality check each surface. (Data published with permission from Seitel Data Ltd.)

3. Interpret additional faults in the same way until the framework of major faults is complete. Numerous small or subtle faults can be left for a later time, possibly after the horizon interpretation framework is completed.

Figures 9-39 through 9-43 demonstrate the method. Two faults (Red SE and Blue SE) are interpreted in 3D view. Note the fault surfaces intersect but are clearly not the same fault.

Strategy 3 can be applied to any situation, whether it be conventional or unconventional reservoirs in extensional or compressional terrains. Figures 9-44 through 9-48 demonstrate the same methodology for the reverse faults in the Marcellus play.

HORIZON INTERPRETATION

Selecting the Framework Horizons

The choice of which framework horizons to map and how many are needed is a function of the geologic complexity, reservoir targets, and seismic volume used. Some of the framework horizons may be needed to provide a reference to create target horizons later, and some may be needed only as control for the depth conversion to be performed once the framework interpretation is complete. We recommend a *minimum* of three framework horizons: one shallow, one intermediate, and one deep. The intermediate horizon should be near the primary objective(s)

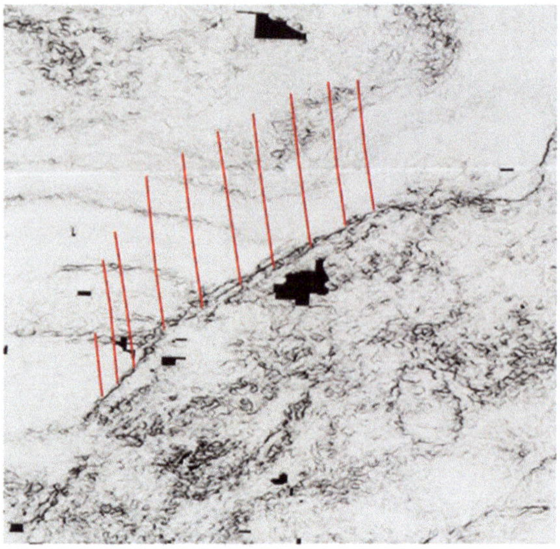

Figure 9-39 Haynesville: Fault strategy 3. Begin the interpretation of the Red SE fault in a 3D view. Red segments were interpreted on vertical sections using the coherence slice as a guide. (Data published with permission from CGG Land.)

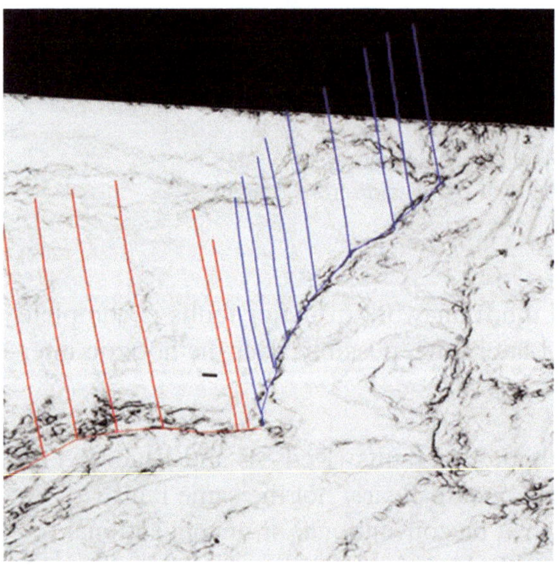

Figure 9-40 Haynesville: Fault strategy 3. Interpret the second fault (blue). Both the Red SE and Blue SE faults are tied along the intersection with the horizontal slice. The coherence attribute clearly shows the red and blue faults are not the same fault. (Data published with permission from CGG Land.)

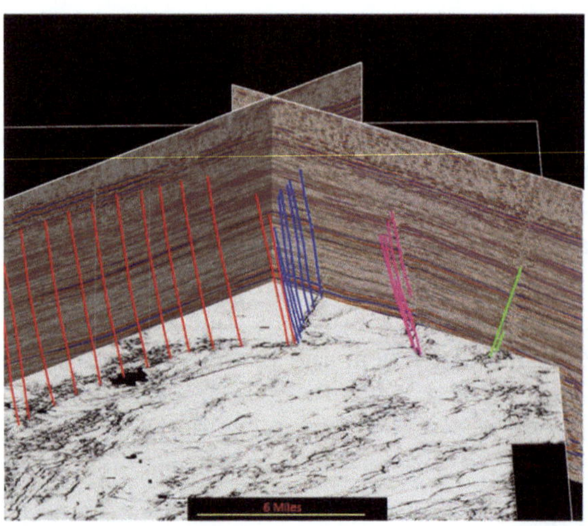

Figure 9-41 Haynesville: Fault strategy 3. Example of an initial fault interpretation for four faults (red, blue, magenta, and green lines) in a 3D view. (Data published with permission from CGG Land.)

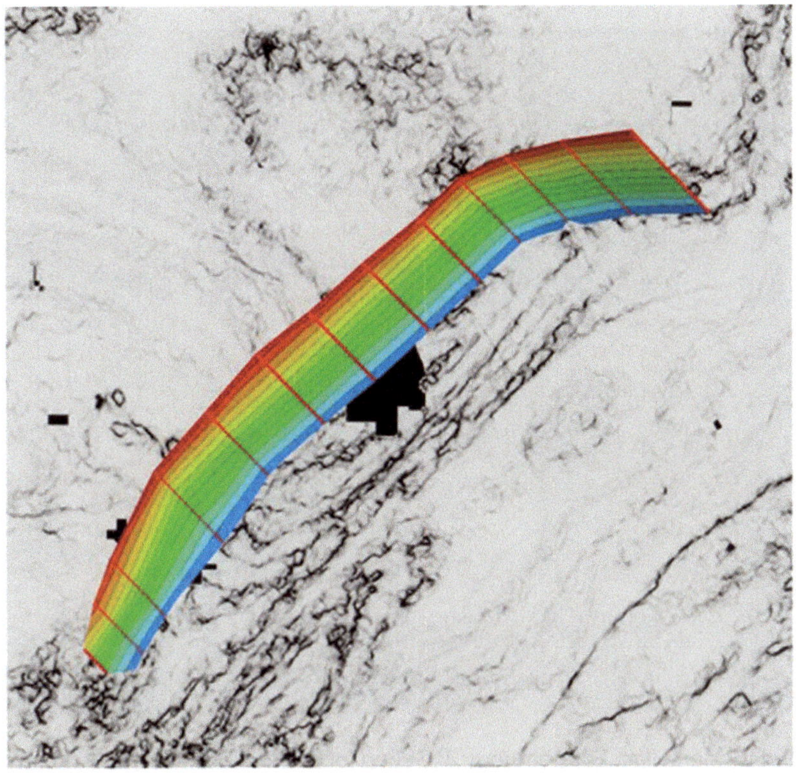

Figure 9-42 Haynesville: Fault strategy 3. Quality check the final fault surface for the Red SE fault. (Data published with permission from CGG Land.)

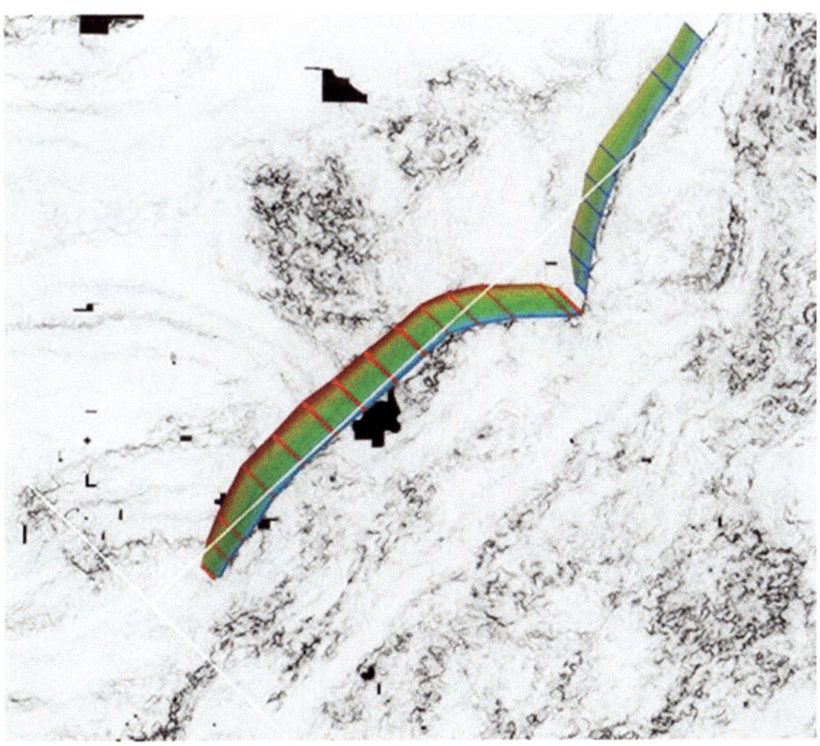

Figure 9-43 Haynesville: Fault strategy 3. Final fault surface for both faults. (Data published with permission from CGG Land.)

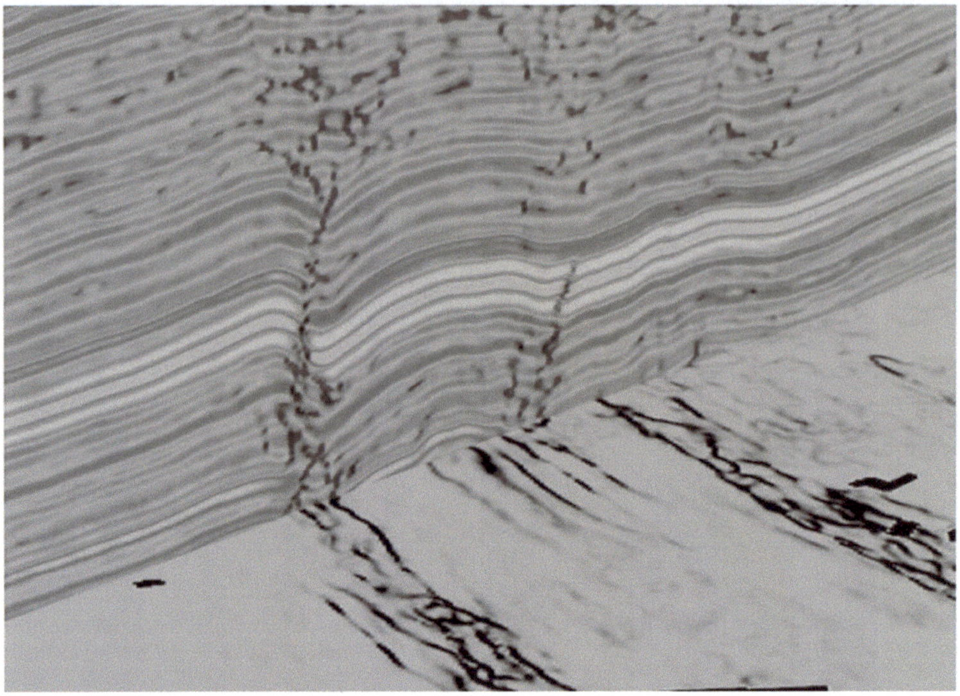

Figure 9-44 Marcellus: Fault strategy 3. Set up for the red fault interpretation. Horizontal slice is a coherence attribute, and the vertical slice is reflectivity data co-rendered with coherence. (Data published with permission from CGG Land.)

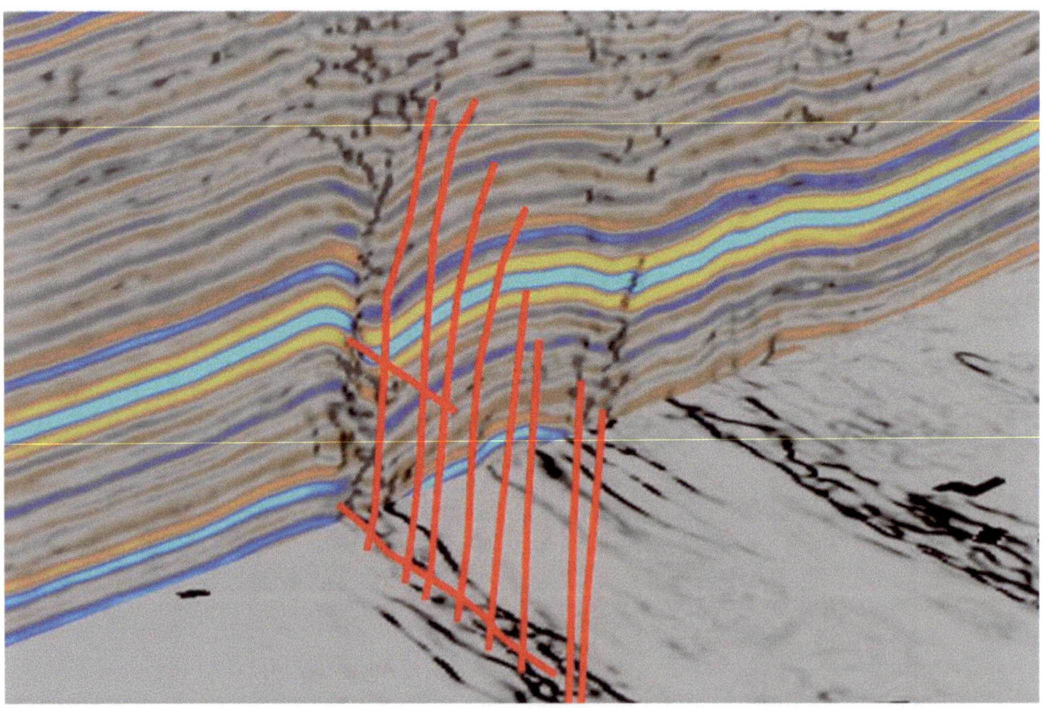

Figure 9-45 Marcellus: Fault strategy 3. Interpreted fault picks on a series of vertical and horizontal slices for the red fault. (Data published with permission from CGG Land.)

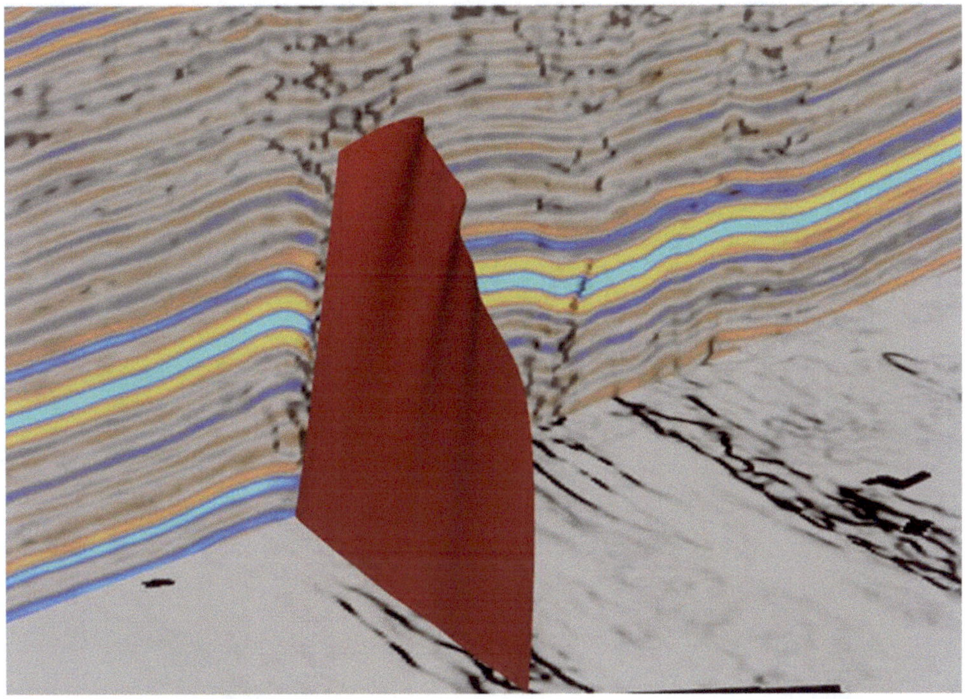

Figure 9-46 Marcellus: Fault strategy 3. Complete the red fault interpretation. (Data published with permission from CGG Land.)

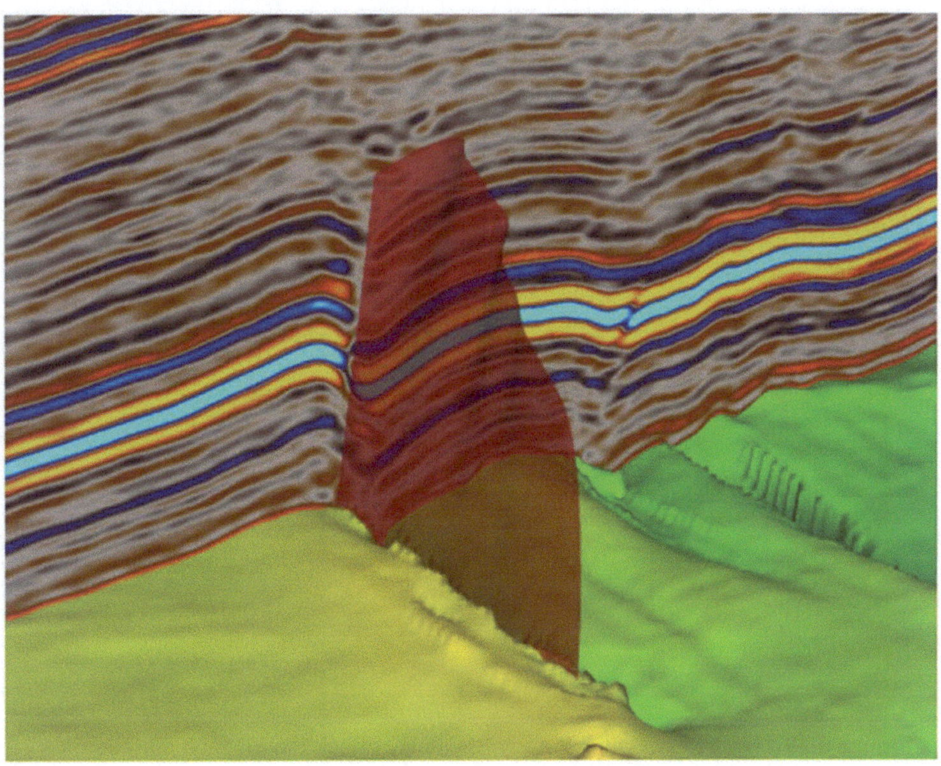

Figure 9-47 Marcellus: Fault strategy 3. Quality check the red fault surface against the autotracked Marcellus horizon. (Data published with permission from CGG Land.)

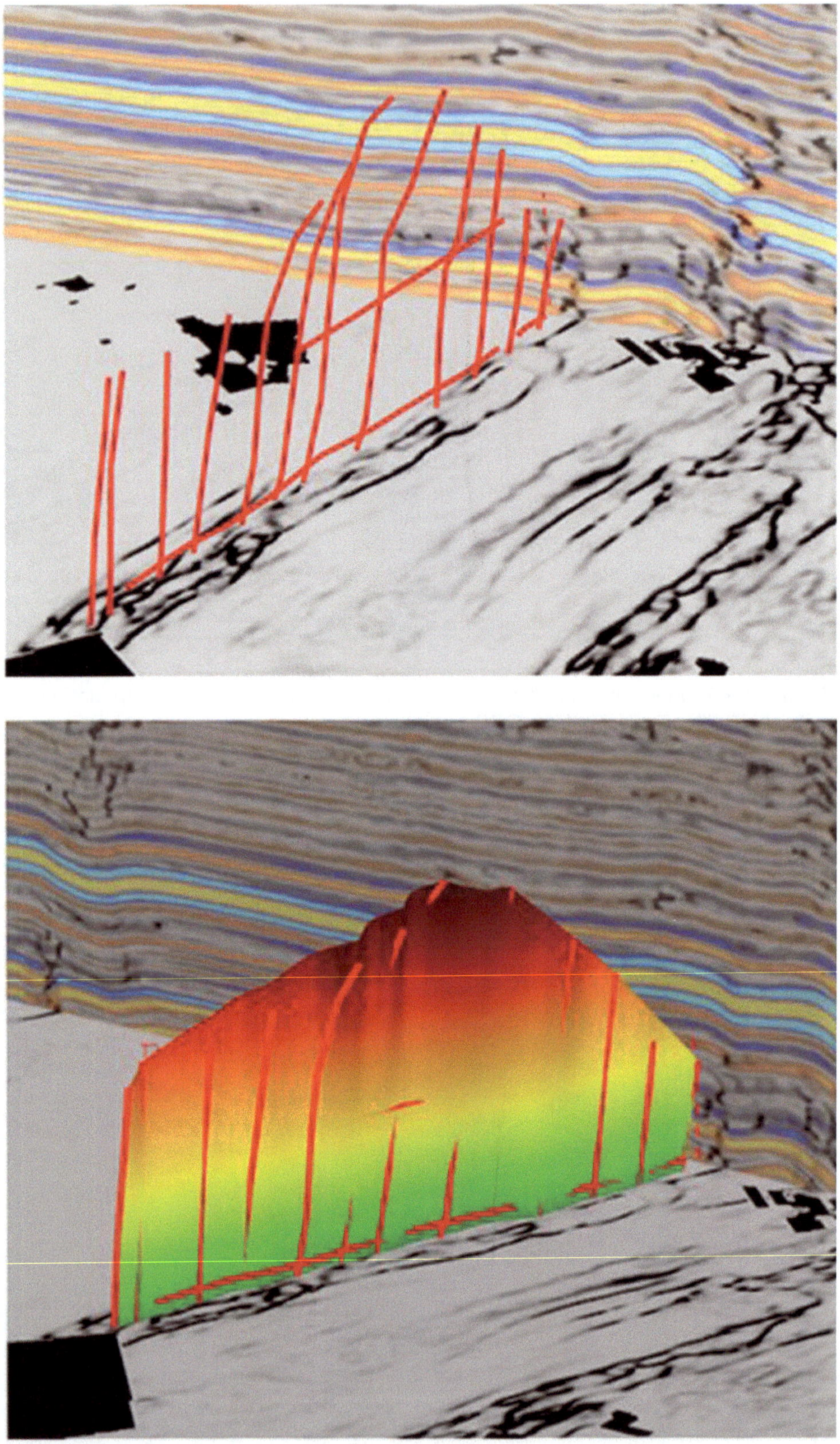

Figure 9-48 Marcellus: Fault strategy 3. Quality check the full fault surface colored by z value. (Data published with permission from CGG Land.)

whenever possible. In this section, we use examples from the Eagle Ford to demonstrate how to map horizons with fault gap polygons.

As mentioned previously, synthetic seismograms are used to tie the well data to the seismic. Figure 9-49 is a synthetic for a typical well tie with framework and target horizons noted. In most cases, the best framework horizons are those that show an acoustic impedance contrast between the shales or tight siltstones and thick limestones. Some target horizons within shale plays are generally subtle or not imaged at all on seismic and therefore need to be mapped with reference to a nearby framework horizon. This is sometimes known as a phantom horizon. For more discussion on synthetic seismograms and phantom horizons, refer back to Chapter 5.

Unconformities must often be mapped as one of the framework horizons because the structure and rock properties are likely to be different above and below the unconformity. Angular unconformities, onlaps, and downlaps sometimes generate complex reflections that are easily identified but difficult to interpret as a horizon. Such an interface may need to be interpreted entirely by hand, which could be a time-consuming process. The techniques described in the following section are intended to make the process of horizon interpretation more effective and efficient.

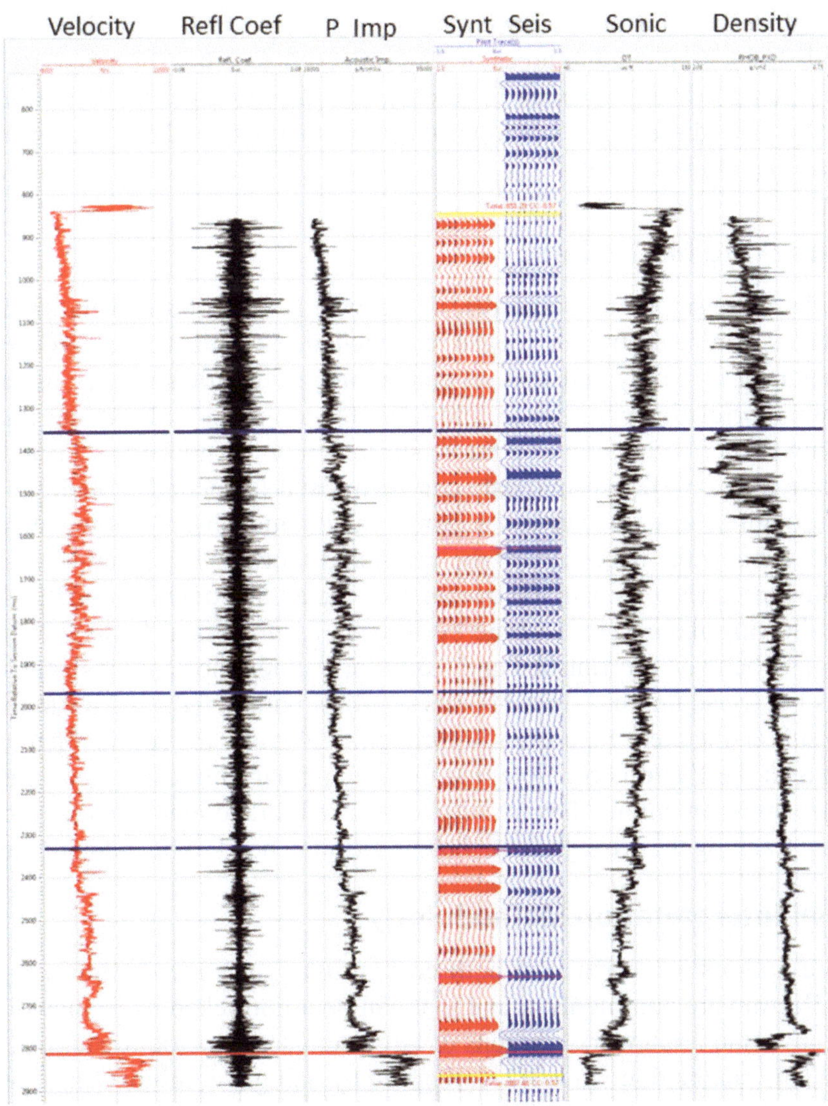

Figure 9-49 Eagle Ford: Synthetic seismogram showing the tie between model traces (Synt) and the seismic trace nearest the wellbore (Seis). Framework horizons and tops are shown in blue, and the Base Eagle Ford in red. (Seismic data published with permission from Seitel Data Ltd.)

Interpretation Strategy

Horizon interpretation is a rather straightforward process, but there are things that can be done to make a more accurate interpretation in a reasonable amount of time. One important time saver has already been mentioned: *complete the preliminary fault interpretation prior to starting the framework horizon interpretation*. This is very important because it significantly reduces the possibility of getting off track if an autotracking method is used in a faulted area. Also, we must remember that most hanging wall structures are formed by large faults. As discussed in Chapters 10 and 11, a geometric relationship commonly exists between fault shape and fold shape. Therefore, the greater the geoscientist's understanding of the faults, the greater the accuracy and geological validity of the horizon interpretations.

One major change implied earlier in this chapter, and since the second edition of this book was published, is the emergence and maturation of unconventional plays such as the Barnett, Eagle Ford, Wolfcamp, Marcellus, Haynesville, and many others. Correspondingly, workstation and software technology have advanced significantly. With this combination of geology and technology came the increased use of autotracked horizons, which is now commonly used by most geoscientists working in unconventional play areas. That said, there is still a place for manual methods to be employed in areas of low reflectivity or poorly imaged areas. These areas, in particular, are notoriously difficult to autotrack, and sometimes the best solution is to interpret manually. We first discuss autotracking and then manual methods that work to fill in areas where autotracking is ineffective.

Autotracking Methodology

A synthetic seismogram helps determine whether the framework horizon is a peak (positive impedance contrast), trough (negative impedance contrast), or zero crossing. Set the tracking parameters accordingly and also set constraints on the tracker, such as the maximum dip change, minimum correlation confidence, and when to stop picking. Next, starting near the tied well, pick a starting point (a seed point) on the seismic event to be tracked and run the autotracker. Hundreds of square miles of 3D seismic data can be tracked in seconds. Quality check the results by stepping through a series of vertical seismic panels and observe how well the autopicks match the seismic. Review the map view of the result and look for large jumps or skips in the tracking. Ask the following questions: Did the tracker pick the areal extent desired? Did the tracker stay on track (i.e., following the correct seismic event even as it crossed faults)? If the answer is yes to both questions, then you are finished and ready for the next horizon. If the answer to one or both questions is no, try adding more seed points or adjust parameters, and rerun the tracker until a satisfactory result is obtained. Figures 9-50 through 9-54 show the results of four autotracked framework horizons in the Eagle Ford. Note that the rectangular untracked areas are *gaps in 3D coverage*. If there are areas remaining that are still not tracked due to low coherency, complex faulting, or consistently tracking the wrong event, those areas may need to be manually interpreted.

Manual Interpretation Methodology

It is the authors' experience in both conventional and unconventional plays that coherent seismic reflectors are commonly found that can be autotracked as framework horizons. However, on occasion, some framework horizons or targeted reservoirs require a great deal of manual interpretation. This can be done in a 3D workspace by using a combination of manual picking in a mesh pattern and interpolation. Another way is to use manual picking in a mesh pattern and gridding. Gridding can create a better-looking surface but may have limitations on editing (depending on the software used). These infill strategies are discussed later in this section.

Our basic philosophy for manual picking can be summed up this way. Horizons should be interpreted in such a way as to be easily revised. Workstations allow selective areal deletion on interpreted surfaces, consequently making repicking a loose mesh pattern more efficient than a line-to line correction. Having a 3D survey on a workstation *does not change the basic need to tie lines to crosslines and resolve mis-ties in a manual interpretation*, such as would be done with a pencil and eraser on 2D paper sections (refer to Chapter 5).

Figure 9-50 Eagle Ford: Autotracked shallow framework time horizon. Note the rectangular untracked areas are gaps in seismic coverage. (Published with permission from Seitel Data Ltd.)

Figure 9-51 Eagle Ford: Autotracked intermediate framework time horizon. (Published with permission from Seitel Data Ltd.)

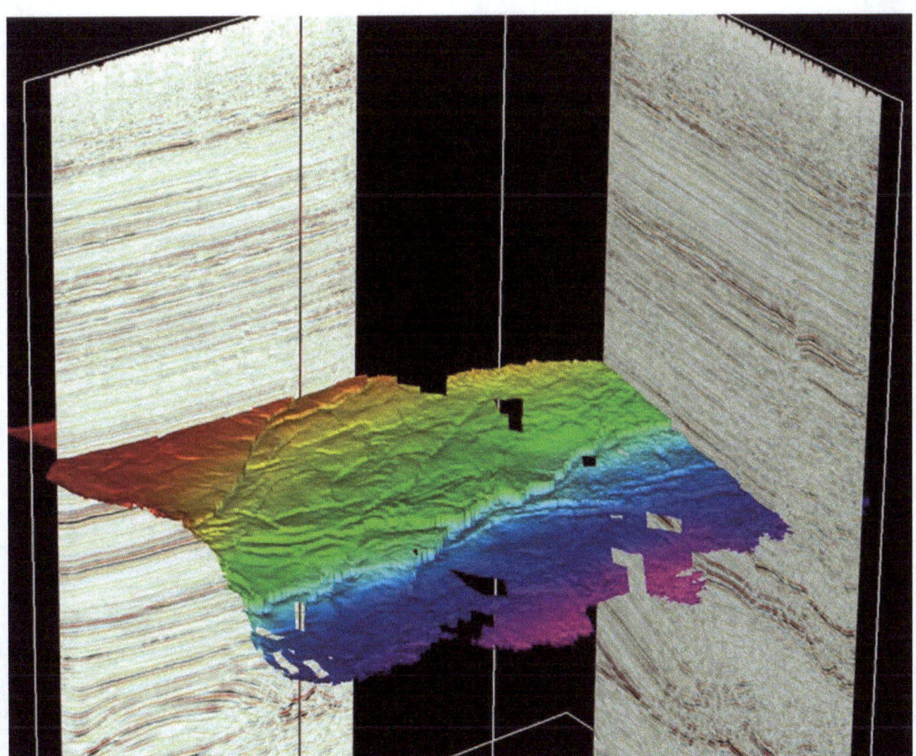

Figure 9-52 Eagle Ford: Autotracked deep framework time horizon. (Published with permission from Seitel Data Ltd.)

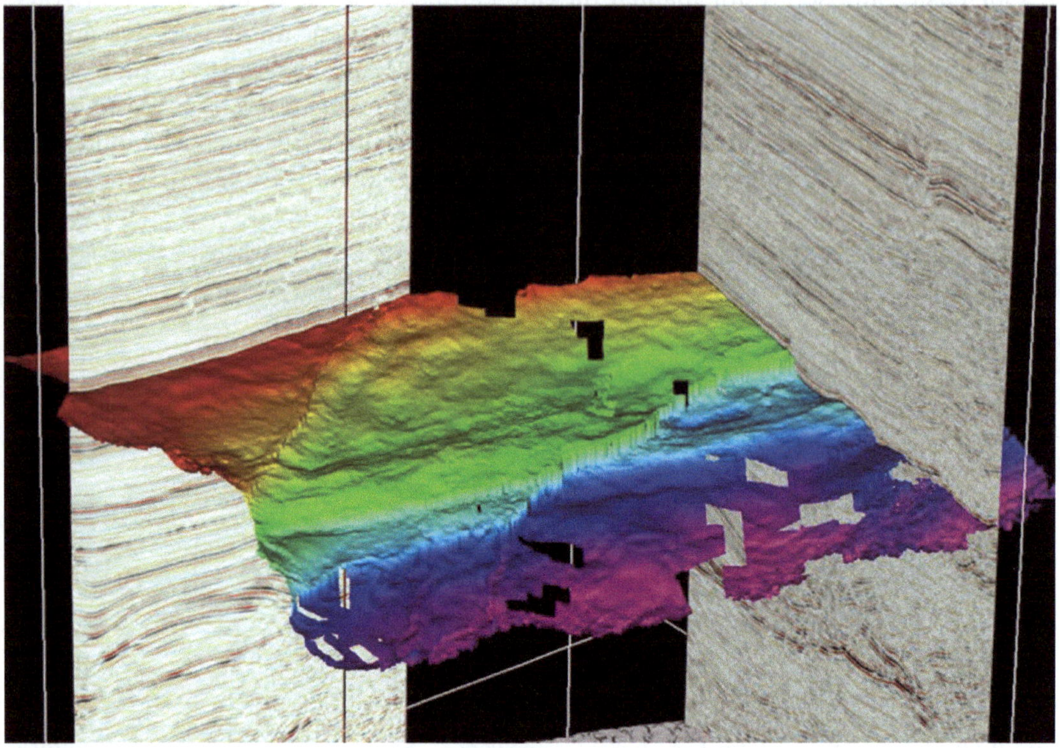

Figure 9-53 Eagle Ford: Autotracked deepest framework time horizon. (Published with permission from Seitel Data Ltd.)

Figure 9-54 Eagle Ford: All four autotracked framework time horizons. (Published with permission from Seitel Data Ltd.)

If starting from a blank surface, it is best to begin near a well (or wells) tied by a synthetic seismogram and work outward from there. When working areas between wells, either of two approaches will work. The first is to use a combination of lines and crosslines to tie data between the wells. The second is to use an arbitrary line directly in line with the wells in question. In either case, picks from intersecting lines can be used as a guide to tie other lines and crosslines. When using arbitrary lines to initially seed an area with interpretation, it is advisable to keep that horizon separate from the line/crossline horizon that will be used for mapping. The reason for this is that, for some of the software in use, editing horizons on arbitrary lines is inefficient once the focus has moved to other arbitrary lines. Keeping them separate allows for fast editing and cleanup later. They can always be merged later, if necessary, for mapping. Figure 9-55 shows an example of an arbitrary seismic line that intersects several wells in a 3D data set. Note that the red correlation marker in the wells corresponds to the magenta horizon that has been manually picked on the section. Figure 9-56 is a map view of the horizon from the arbitrary line and the tied wells with the red marker depths noted. Figure 9-57 shows an example inline with the same magenta horizon interpreted and tied to the red marker. Figure 9-58 is a well-tie horizon interpretation done on lines and crosslines only. There are advantages to both methods, and both can be done quickly. These ties should then be used to work outward from the well control to complete the interpretation. In the case of highly directional wells, be sure to locate the seismic tie at the horizon penetration point in the wellbore.

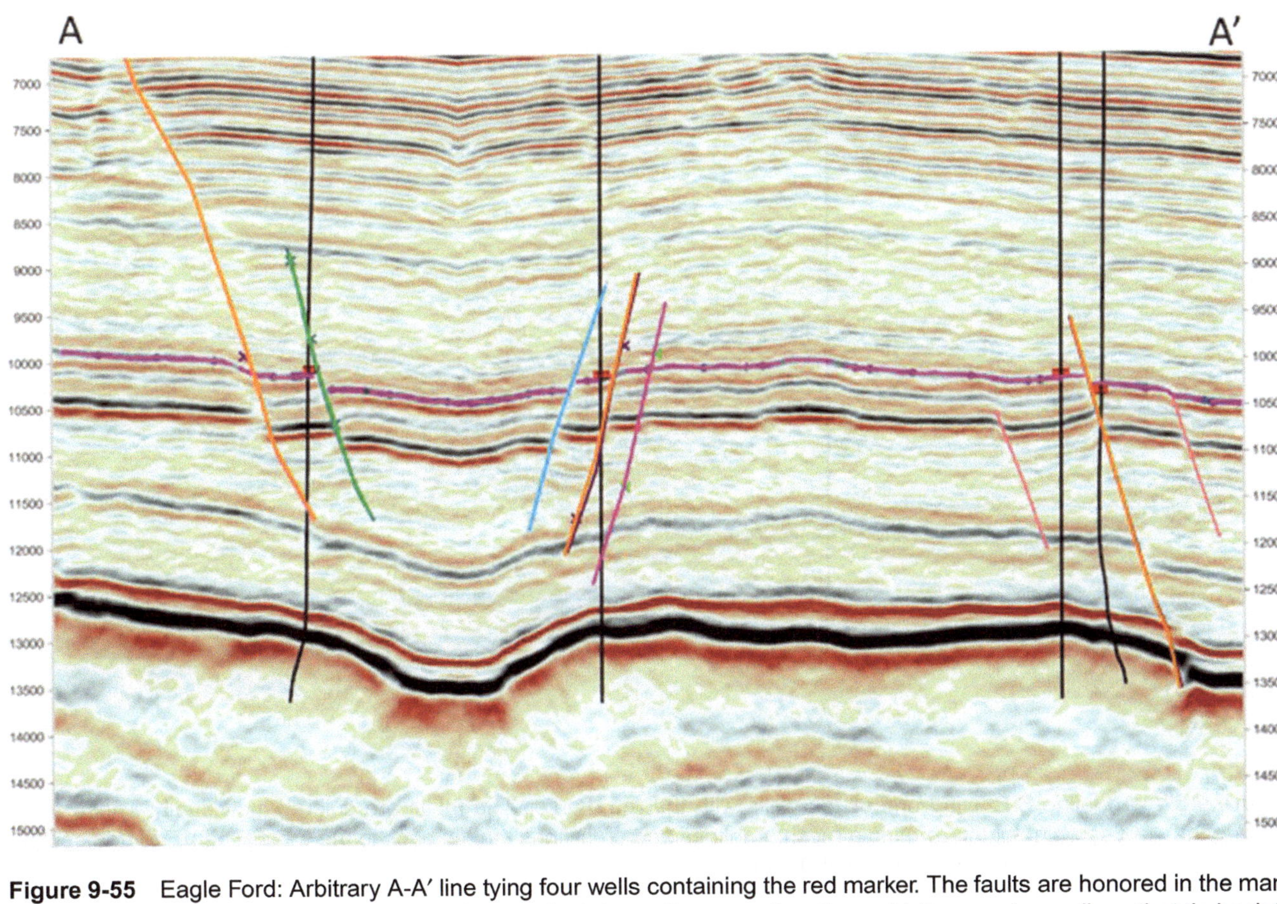

Figure 9-55 Eagle Ford: Arbitrary A-A′ line tying four wells containing the red marker. The faults are honored in the manual interpretation. These picks provide a reference for interpreting a mesh pattern of inlines and crosslines that tie back to the wells. The VE for the section is 3x. (Published with permission from Seitel Data Ltd.)

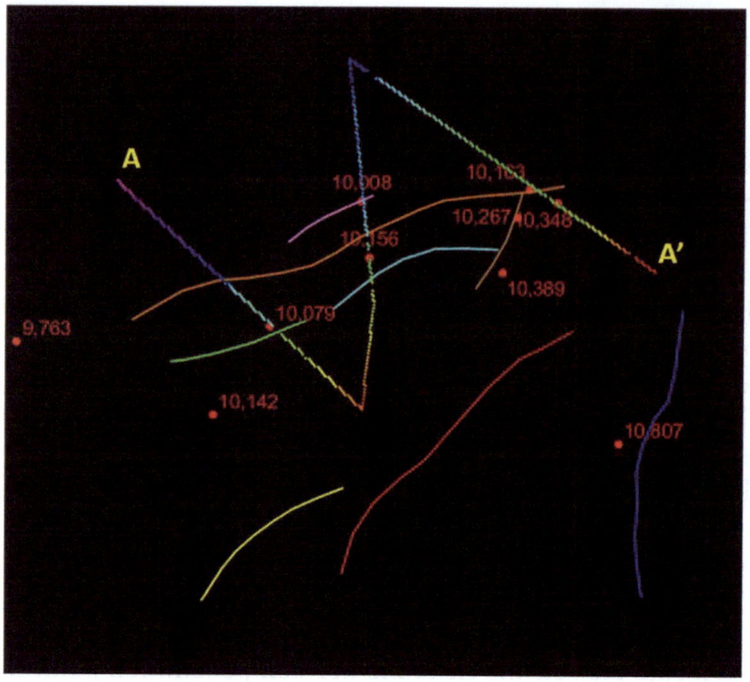

Figure 9-56 Eagle Ford: Map view showing interpretation from arbitrary line A-A′. The faults are colored individually in map view as in the vertical panel. These picks provide a reference for interpreting a mesh of inlines and crosslines that tie back to the wells. (Published with permission from Seitel Data Ltd.)

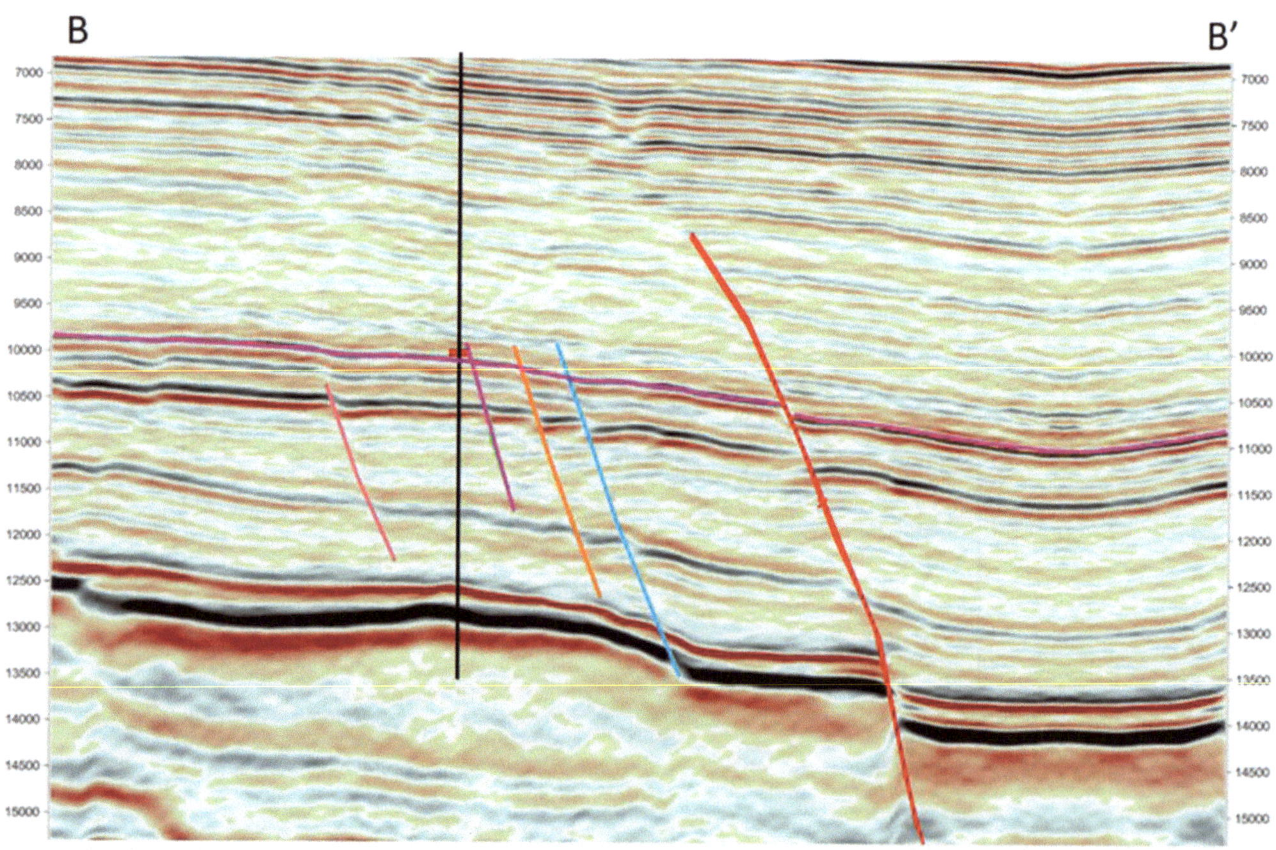

Figure 9-57 An inline B-B′ tying one of ten wells with the red marker. The faults are colored individually. These picks provide the starting point for interpreting a mesh of inlines and crosslines that tie back to the wells. The VE for the section is 3x. (Published with permission from Seitel Data Ltd.)

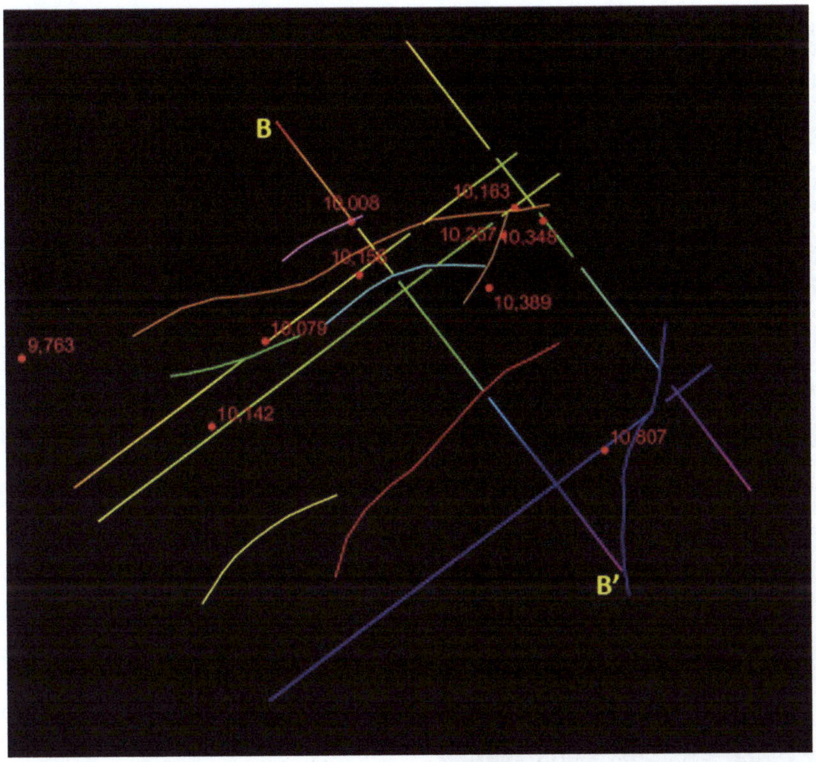

Figure 9-58 Map view showing interpretation from two inlines and three crosslines tying seven wells. These picks provide the starting point for interpreting a mesh of inlines and crosslines that tie back to the wells. (Published with permission from Seitel Data Ltd.)

How dense should the interpretation be when manually interpreting seismic horizons for structure mapping? The answer is similar to the fault interpretation philosophy: *only as dense as needed to accurately define the surface*. In our Philosophical Doctrine, point 4 states, "All subsurface data must be used to develop a reasonable and accurate subsurface interpretation." It is not necessary to interpret every line, crossline, time slice, and arbitrary line. *All orientations should be used as needed to ensure the interpretation is valid in 3D*. Well data should be tied to the seismic data. The manual interpretation performed on selected sections should be preserved and not altered by automated processes such as autotracking or interpolation. The latter should be saved separately so that results can be compared to the manually picked interpretation. If editing becomes necessary, it is much easier to work on a relatively loose mesh of interpreted data than to fix every inline or crossline after the horizon is filled in.

So, what is the optimum density of manual interpretation? That has already been answered to some extent, but for example, if a 20-inline by 20-crossline interpretation yields an accurately contoured structure map on a simple unfaulted structure, then it is not necessary to do a 10 × 10 or 5 × 5. The level of detail at which to begin is dependent on your data and the geology, so start with a wide spacing and move toward increasing levels of detail as complex areas are encountered. Figures 9-59 and 9-60 compare a 20 × 20 line interpretation versus a 10 × 10 line interpretation of the same area. The 10 × 10 shows a higher level of fault detail is attained.

Given what we have said about the density of interpretation, if a framework horizon is a high-impedance contrast, laterally continuous seismic event, it may require only a few picks in an area or a fault block to allow autotracking to infill the rest. *If this is the case, then autotracking is always the better choice*. If for some reason the picks drift off the event, it is better to spend your time adding manual interpretation to the input and rerunning the autotracker rather than trying to clean up afterward. However, autotracked horizon surfaces are notorious for being noisy and generally do not contour well without smoothing.

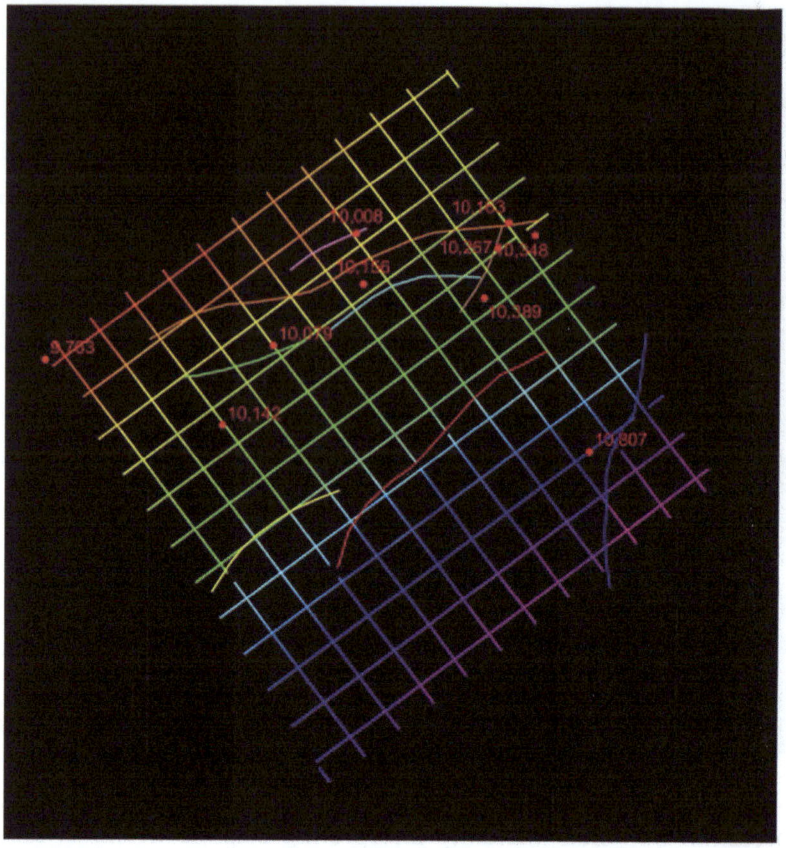

Figure 9-59 Manual interpretation on a 20 × 20 inline-crossline mesh pattern. In faulted areas, this increment lacks detail around the fault intersections, fault trend, and terminations. (Published with permission from Seitel Data Ltd.)

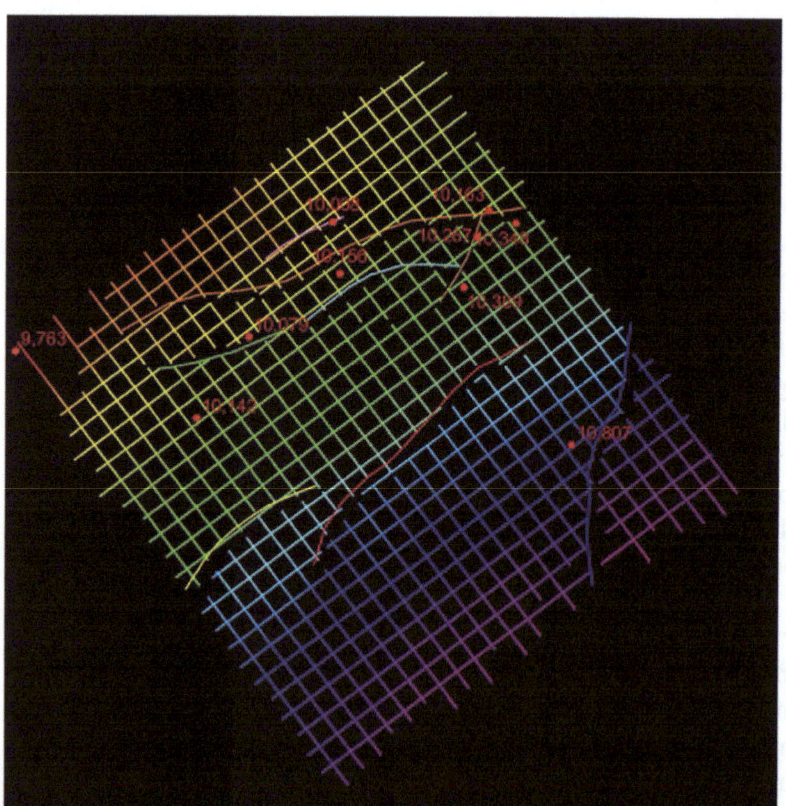

Figure 9-60 Manual interpretation on a 10 × 10 inline-crossline mesh pattern. The detail (lacking in the previous figure) is enhanced. (Published with permission from Seitel Data Ltd.)

Infill Strategies in the Absence of Autotracking

There are basically four methods to infill a surface. The acceptability of the result depends on *data quality, structural complexity, desired objectives*, and *which software parameters are being used*. The four methods are autotracking (discussed previously), manual picking, interpolation, and gridding. Before looking at the methods, these questions should be asked: Why infill a horizon surface at all? Can an accurate structure contour map be constructed without infilling a horizon? If that is the only objective, then yes, it can be done without additional interpretation. There are many reasons, however, why an infilled surface may be necessary. It can be used as a reference for seismic attribute extraction, such as amplitude or coherence. It can be used for detailed structure mapping where there are subtle complexities. Infilled surfaces are essential when performing layer computations such as isochores, isochrons, and depth conversion. Finally, these surfaces can be used as a reference for analysis, targeting horizontal wells, flattening cross sections (or seismic), or a myriad of other uses. The objective for using infilled horizon surfaces will determine which method is best suited to the task.

Manually picking each line is an option that should be used sparingly when needed. For example, when working (1) complexly deformed or difficult areas such as very small fault blocks, (2) near fault intersections, (3) around faults of limited extent that could affect pressure communication between wells, (4) within intersecting fault patterns, (5) in areas of stratigraphic complexity such as angular unconformities or facies changes, or (6) poorly imaged areas. These types of situations usually do not autotrack well, not to mention the difficulties they create in gridding and contouring mapped surfaces later. Detailed manual picking to fill in small areas is one of those tedious tasks that should be used as a *last resort* but is necessary from time to time. Figure 9-60 discussed previously, is the starting point for comparing interpolation, gridding, and autotracking.

Interpolation is faster than manually picking every line. Simply stated, with linear interpolation, the computer uses two interpreted lines and calculates horizon values between them. This is usually tied to the 3D survey geometry, inlines, and/or crosslines and therefore, is *unique for that 3D survey*. The prerequisite for this to work is that interpretation must be done closely enough to reasonably represent the interpreted seismic event. Interpolation can be used to bridge gaps in the data or across sparse interpretation where no changes are expected to occur. It can be used as a quick and dirty infill strategy. Figure 9-61 is an example of an interpolated horizon. The surface is reasonable, but some issues are noted around faults.

Gridding is the most common infill technique besides autotracking. Gridding takes data from sparse and/or irregularly spaced interpretation and computes a surface in XYZ space. The X-Y dimension is known as the *grid cell size* and is not tied to a particular 3D inline/crossline geometry. Gridding algorithms and parameters can be selected to show every nuance of a structural surface or can act as a smoothing filter to remove noise in the interpretation. Parameters for gridding techniques may be specific to an area or play; therefore, we do not discuss each gridding algorithm in detail. Suffice to say that in specific cases, one gridding method is more applicable than another. Experimentation to find the right combination of gridding algorithm and parameters that work best for the data set is acceptable. Computing a gridded surface will usually yield the best-looking map over the other infill methods. Figure 9-62 is a gridded version of Figure 9-60 and, when compared to the autotracked version (Fig. 9-63), the cleaner appearance can easily be seen. Results from all three infill techniques will improve significantly with smoothing. But as with any interpretation, the result should *always be validated with the seismic* to ensure compatibility between the two. Most workstation software allows infilled horizons and gridded surfaces to be used interchangeably; however, a few techniques require one or the other exclusively.

Accurate well log correlation and accurate interpretation of framework and target horizons are the *most important elements* when making reliable structure maps. Using efficient

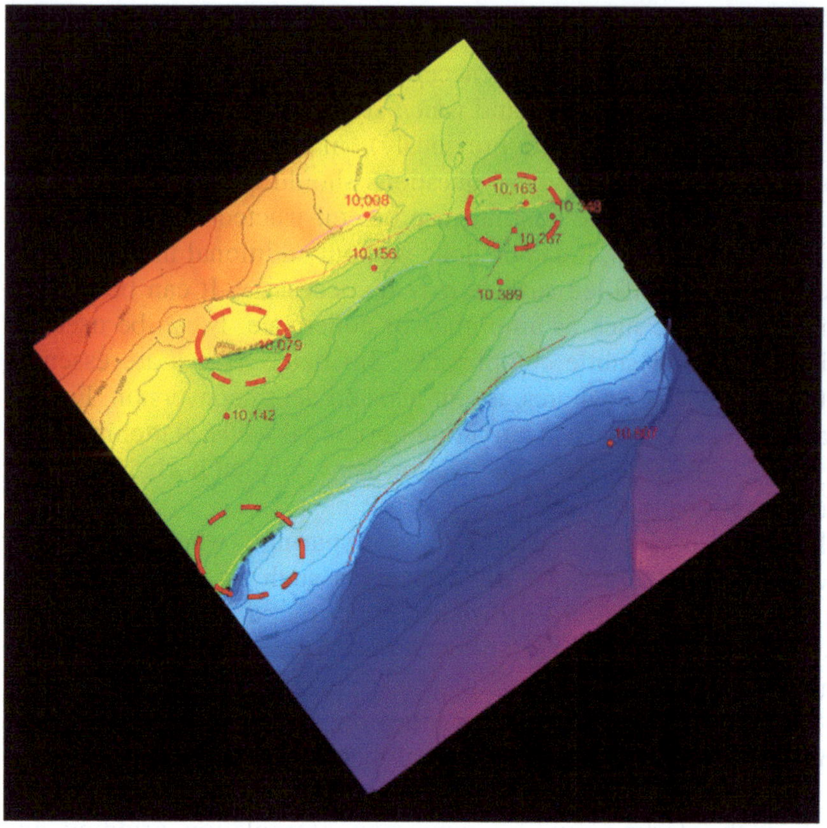

Figure 9-61 Interpolated horizon from manual interpretation on a 10 × 10 inline-crossline mesh pattern. Contouring around faults is poor due to edge effects of the interpolation. (Published with permission from Seitel Data Ltd.)

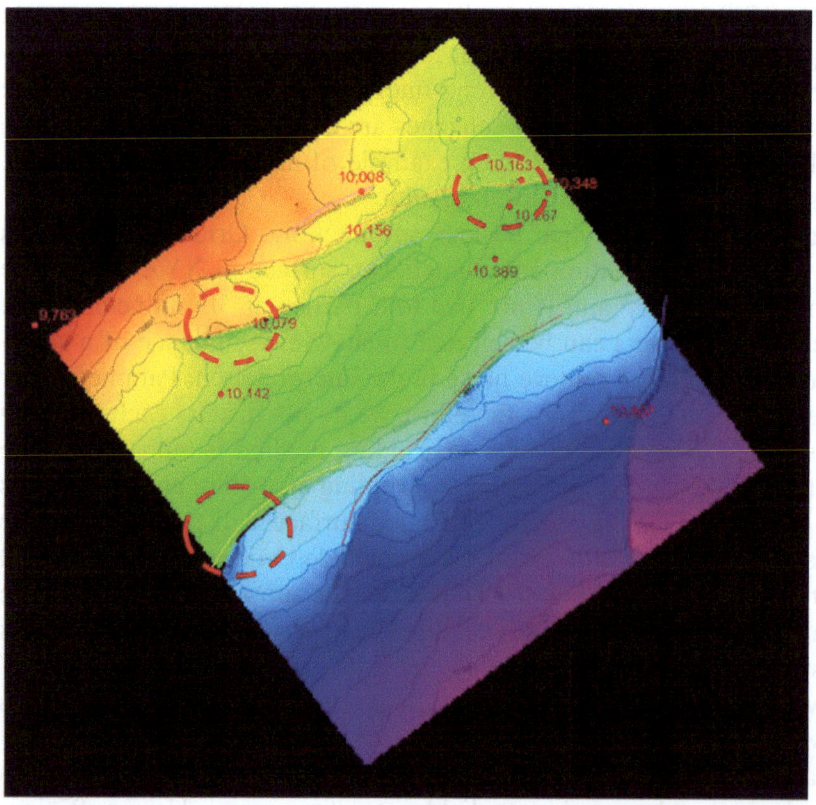

Figure 9-62 Gridded surface from manual interpretation on a 10 × 10 inline-crossline mesh pattern. Contouring around faults is significantly improved. (Published with permission from Seitel Data Ltd.)

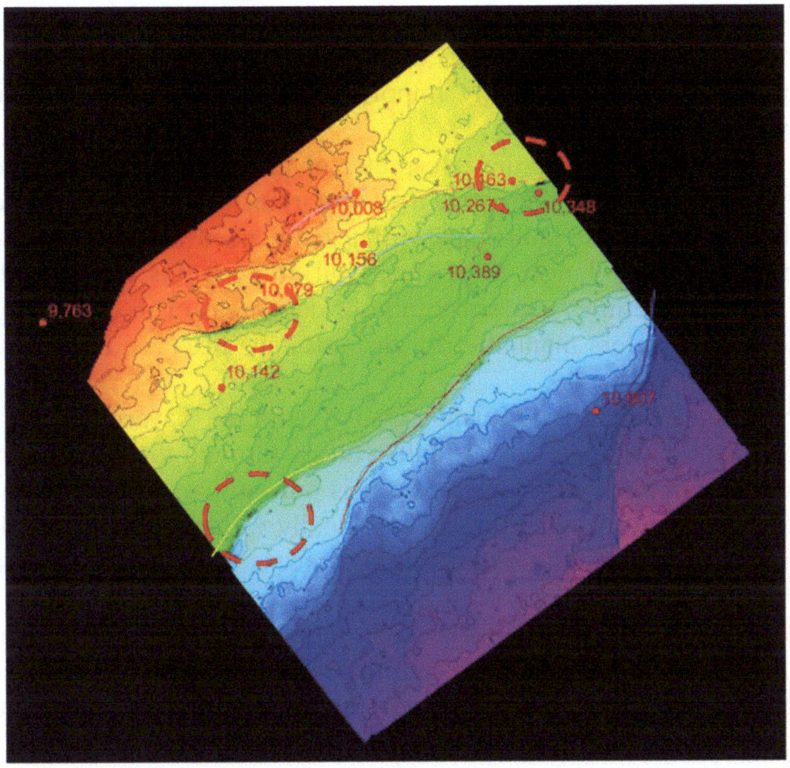

Figure 9-63 Autotracked surface tracked in just seconds from one seed point. However, this result shows the noisy nature of auto-tracked horizons and some issues with contouring near faults. The maps made by each of the three infill techniques will improve significantly when smoothing is applied. (Published with permission from Seitel Data Ltd.)

interpretation techniques allows large data sets with many framework and target horizons to be evaluated in a reasonable amount of time. By using autotracking technology and other infilling strategies, interpreters can have maps ready at any stage for quick evaluation while they continue to refine areas of higher interest. In the next section, we discuss a common technique to convert time-scaled horizons, grids, faults, and seismic volumes into depth-scaled objects in preparation for creating the final structure maps.

STRUCTURE MAPPING

The final phase after the interpretation of the framework horizons and faults is to make a time structure map on each horizon and any non-framework target horizon maps. Structure maps are made by gridding and contouring each surface and accounting for any faults that intersect that surface. We previously mentioned that it is helpful to make preliminary structure maps early in the project to quality check each surface for potential pitfalls such as interpretation mis-ties, fault location, and agreement with existing well control. They are also useful to test the 3D validity with other framework horizons. However, once the depth conversion process is complete, it is best to work completely in the depth domain. The following sections describe the process of making final time-structure maps, although these methods also apply to depth surfaces.

Drawing Accurate Fault Gaps and Overlaps

Drawing accurate fault gaps and overlaps on maps is one area in which good software goes a long way. The many ways to accomplish this task depend on which software is used. Some tools allow the interpreter to manually draw the fault gaps, whereas others compute gaps based on the integration of the fault and horizon surfaces. Some tools work in 2D, while others work in 3D. Construction of fault gaps and overlaps is described in this book in Chapter 8, but generally stated, they are defined as the map representations of the footwall and hanging wall horizon cutoffs where the horizon is intersected by a fault surface. Normal faults create a gap, and reverse faults create an overlap in the horizon. The locations of horizon cutoffs and the width of fault gaps and overlaps in complex areas are extremely important with regard to well design and drilling and completion practices.

Fault gaps are either manually drawn by the interpreter or autogenerated. For speed, it is usually best to autogenerate the fault gaps; therefore, it is important to have well-defined fault surfaces and horizons. The autogenerated fault gaps will usually need some manual editing. Manually drawn fault gaps take slightly longer to produce, but doing it this way provides total control for the interpreter and is sometimes necessary in poor data areas or in complex geology where faults intersect. The author prefers manually drawn fault gaps because, once completed, they rarely need updating.

Figure 9-64 is a typical Eagle Ford vertical seismic display with the Navarro Formation and several faults interpreted. In map view, the series of faults cutting through the horizon provide a guide for manually drawn fault gap polygons (Fig. 9-65). It is our opinion that manually drawn fault gap polygons are one of the best ways to ensure a geologically valid map. In the workstation environment, they also serve to control the horizon infilling process (discussed previously) and computer contouring.

The position and heave of the fault gap can be checked in the workstation environment. One way is to display contours from each surface and locate the points of intersection of contours of equal value. Figure 9-66 is a good example of 3D integration of fault and horizon surfaces. Another way to do this is to simply *subtract* the fault surface grid from the horizon surface grid.

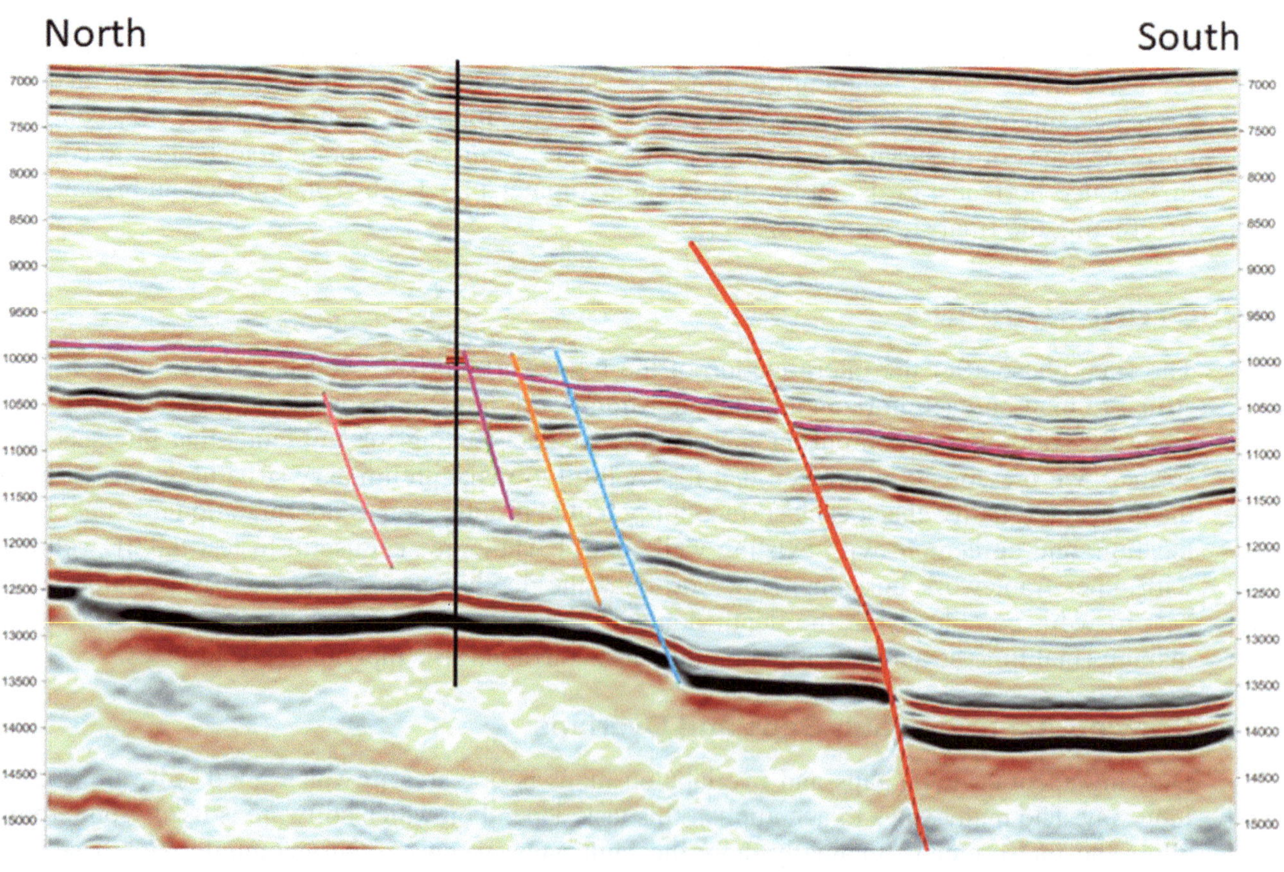

Figure 9-64 Eagle Ford: Typical seismic line showing a series of small faults cutting up through the Navarro Formation (magenta horizon is near the top of Navarro). (Published with permission from Seitel Data Ltd.)

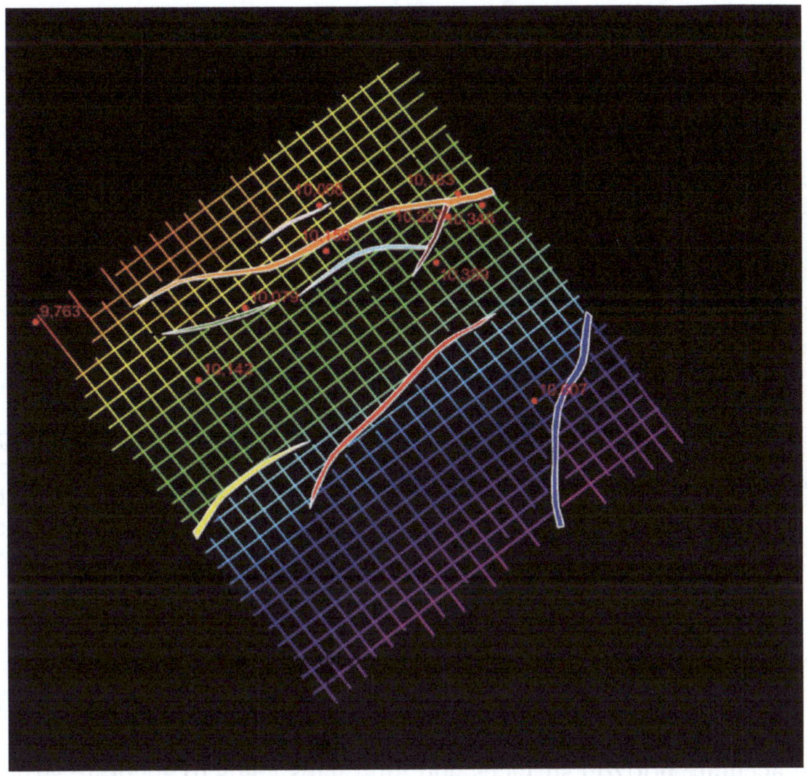

Figure 9-65 Eagle Ford: In map view, the series of faults cutting up through the Navarro Formation provide a guide for hand-drawing the fault gaps as shown. (Published with permission from Seitel Data Ltd.)

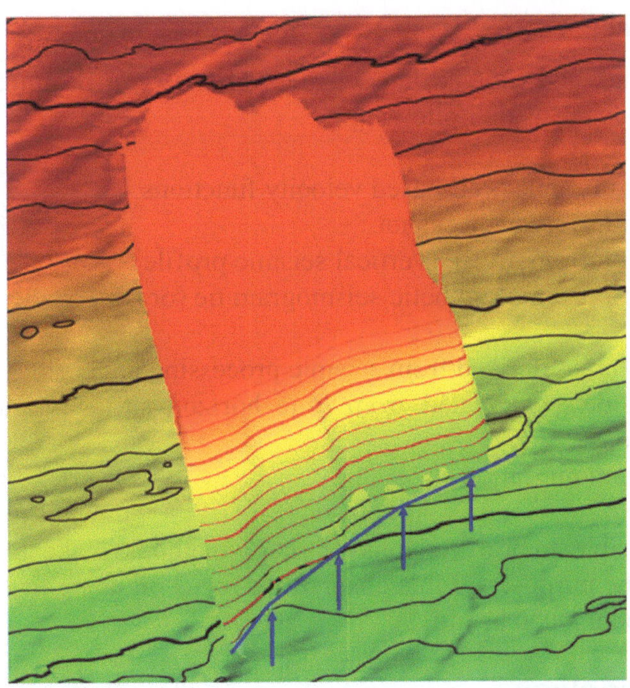

Figure 9-66 Eagle Ford: Each fault gap can be validated by the intersection of the fault surface and horizon surface. The hanging wall fault cutoff is drawn where the contours of equal value intersect (blue line). The same process is repeated for the footwall. (Published with permission from Seitel Data Ltd.)

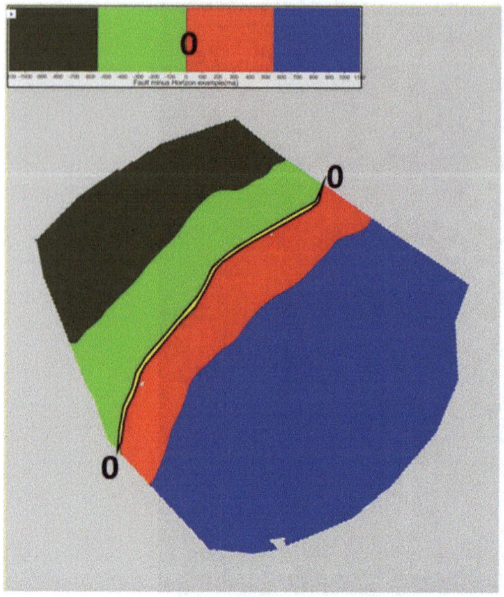

Figure 9-67 Eagle Ford: Subtracting the two intersecting surfaces results in a good indication of where the fault gap should be. In this figure, the green-red color boundary depicts the location of the zero line where the surfaces intersect. The fault gap is shown in yellow. (Published with permission from Seitel Data Ltd.)

Where the two grids intersect, the values should be at or very near zero. Set the map color scale to highlight the very low values near the intersection of the two surfaces. The fault gap should correspond closely to the computed intersection (Fig. 9-67).

Creating accurate horizon surfaces and fault gaps leads to accurate structure maps. With the framework horizons complete, we are now able to build a velocity model to rescale our time-scaled data into depth.

DEPTH CONVERSION

Time horizons, faults, and time-scaled seismic volumes can be transformed into depth-scaled data in several ways, such as the following:

A. Single-well- or multiple-well-based velocity functions
  1. Time to depth from a checkshot
  2. Time to depth from a VSP (vertical seismic profile)
  3. Time to depth from a synthetic seismogram tie (or ties)
B. Volume based
  1. Velocity volumes derived from seismic processing analysis
  2. Velocity models calculated from time horizon surfaces and corresponding well log formation tops
  3. Prestack depth migration

The many possible hybrid versions of these methods make a huge number of possible combinations. For the purposes of this chapter, we focus on using time surfaces and well tops to create a velocity model (such as in B2).

The Basic Methodology

The two essential ingredients to using this method are:

1. Infilled time surfaces from seismic measured in two-way time in seconds (*twt*). Note that two-way time is divided by 2 to calculate one-way time for velocity calculations.

2. Formation tops picked in the wells measured in true vertical subsea depth (*d*) (positive below sea level).

Figure 9-68 is a typical synthetic seismogram from the Midland Basin. Four horizons are chosen as framework horizons, and the corresponding seismic peak or trough is shown on the right. These horizons are tracked on the corresponding seismic reflectors throughout the area of interest. In Figure 9-69, the completed time horizons are shown with the time-scaled seismic volume. The depth control in Figure 9-70 are the formation tops corresponding to the time surfaces. These data together comprise the input data for the velocity model and the subsequent depth scaling of the seismic and time surfaces. Even if the available data were limited to one time surface and one formation top, a single average velocity point (*Vavg*) can be calculated at the well using Eq. (9-1):

$$d / (twt/2) = Vavg \quad (9\text{-}1)$$

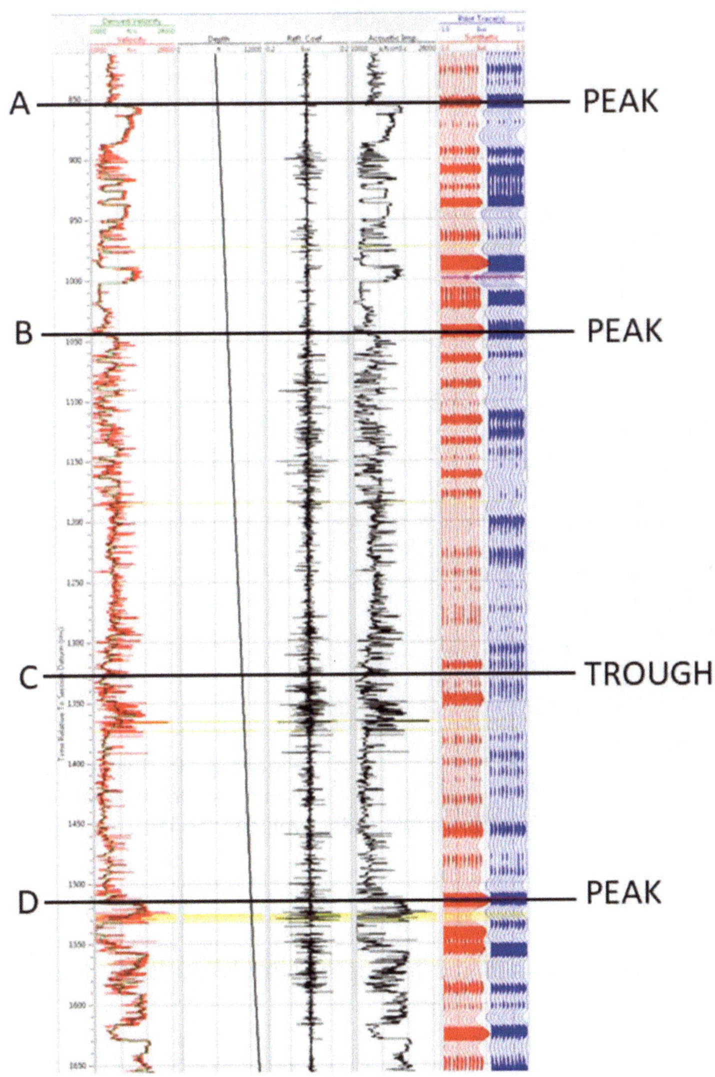

Figure 9-68 Midland Basin: A synthetic seismogram is used to choose the framework horizons for depth conversion. Formation tops are correlated to seismic traces using the character tie between the synthetic from the impedance log and the actual seismic traces at the well.

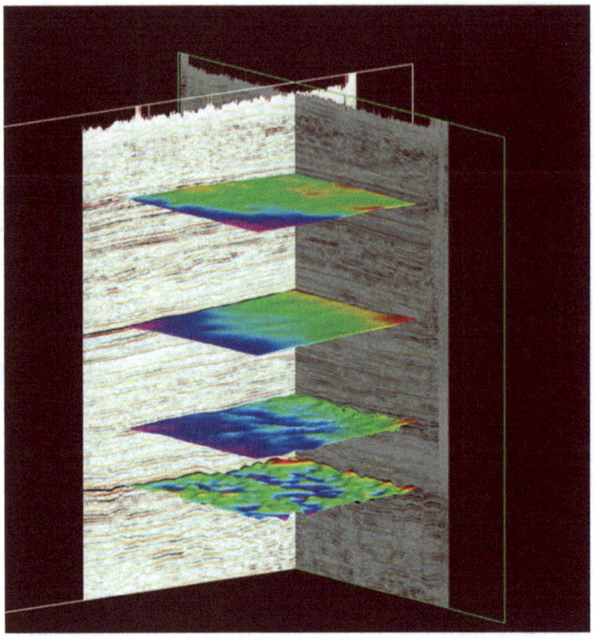

Figure 9-69 Midland Basin: Three-dimensional visualization of the framework horizons and seismic scaled in time.

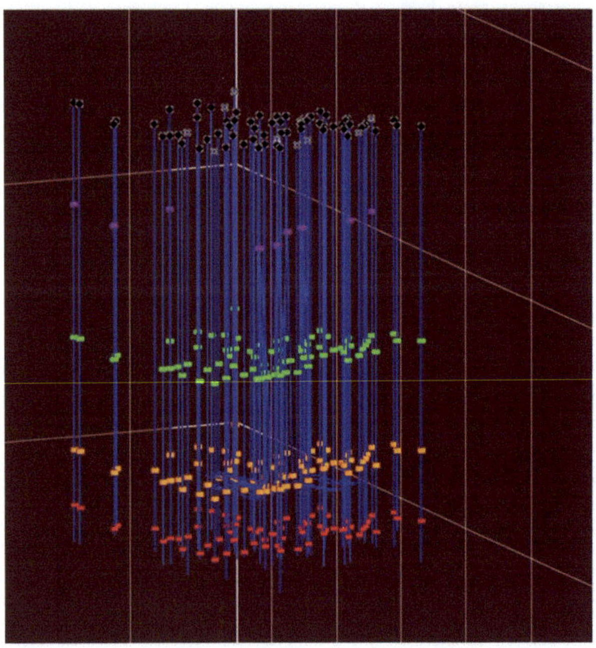

Figure 9-70 Midland Basin: All 80 wells and formation tops shown (four per well) are used as input to the velocity model.

To convert the entire time surface to depth, simply multiply the time surface by the computed velocity using Eq. (9-2):

$$Vavg\ x\ (twt/2) = d \tag{9-2}$$

In practice, this would be the simplest case for this method. More likely, a 3D model would be computed using multiple tops and horizons, as is shown in Figures 9-69 and 9-70. How much well control that is available for velocity calibration varies by location and target horizon. For instance, in the current example in the Midland Basin Wolfcamp play, there are 80 wells within a 40-square-mile area. Each well containing the formation top being mapped contributes a point of velocity control. These *average velocity* points are combined into a velocity cube model and then used to scale a time surface into a depth surface. By definition, this method creates a depth surface that ties each framework formation top exactly. Since the depth calibration depends on accurate formation tops *and* time surfaces, the result should be checked for anomalous velocity values. A rigorous quality control workflow should catch the largest anomalies.

An alternative method of computing a velocity model is one that uses interval velocities. Either method works, but there are pros and cons for each. For average velocity, each surface is computed independently, so there is no reference to isochore or isochron computations, which are a good quality-control tool. Of course, isochore and isochron maps can be computed for quality control prior to the average velocity computation, but they are not needed for the velocity model. When interval velocities are used, isochrons computed from time surfaces and isochores computed from tops are the key ingredients and provide a tool for quality control of the resulting velocity and depth maps. If a well does not have each of the tops used in the model, it cannot contribute to the portion of the model where the top is missing. Also, if multiple time horizons are cut by large listric growth faults or low-angle thrust faults, calculating an accurate isochron is challenging. A positive aspect to using interval velocities is that if lateral gradients exist within an interval, they can be discovered.

Final Steps

Once all of the framework well tops and time horizons have been quality checked, the next step is to build a velocity model and compute the depth grids and depth seismic volume. The actual workflow to accomplish this varies depending on which software is used. One method is to simply calculate the depth map from the time and velocity maps. However, the method of using a velocity model described in this chapter is one of many available, but one thing it does is *tie all the well control* (for better or worse), making a depth map as accurate as the formation tops and shaped by the time horizon. Most workstations are able to use the same information (tops and time horizons) to create 3D velocity models.

It is highly recommended that the velocity volume be reviewed in detail to ensure there are no obvious anomalous velocities that would cause errors in the depth scaling. Figure 9-71 is a 3D visualization of the seismic data co-rendered with the velocity model and framework horizons. Scrolling through the data reveals any areas needing closer scrutiny. More discussion on quality control is covered in the next section. Figure 9-72 is the visualization of all data scaled to depth. The value of having all data scaled in depth has been discussed previously, but to summarize, it is the most effective way to integrate seismic with well logs, petrophysical, whole core, and engineering data such as drilling, completion, and production data.

Quality Control of the Results

We recommend checking the quality of the results of the velocity modeling and depth scaling. Figure 9-73 is an example of average velocity in color draped on top of a depth horizon with depth-scaled seismic in the background. Note the area containing an anomalously low velocity (magenta) at one of the wells (red arrow). Investigation of the well top and vertical distance from kelly bushing to sea level (KB) showed normal values; however, the time horizon has some localized mis-picks from autotracking. Figure 9-74 is the same scene but now with average velocity draped on the time horizon. While the same autotrack mis-picks are present on the horizon (green arrows), the horizon at large is much flatter, meaning that the velocity is miscalculated around the well (red arrow). By correcting the time horizon, as in Figure 9-75, and rescaling the horizons and seismic to depth, the anomalous velocity at the well is mitigated (large arrow). Figure 9-76 shows the corrected depth data after fixing the time horizon and rescaling to depth.

Figure 9-71 Midland Basin: The computed velocity model is co-rendered with the seismic reflectivity data. The vertical coloring shows the velocity change with increasing two-way time.

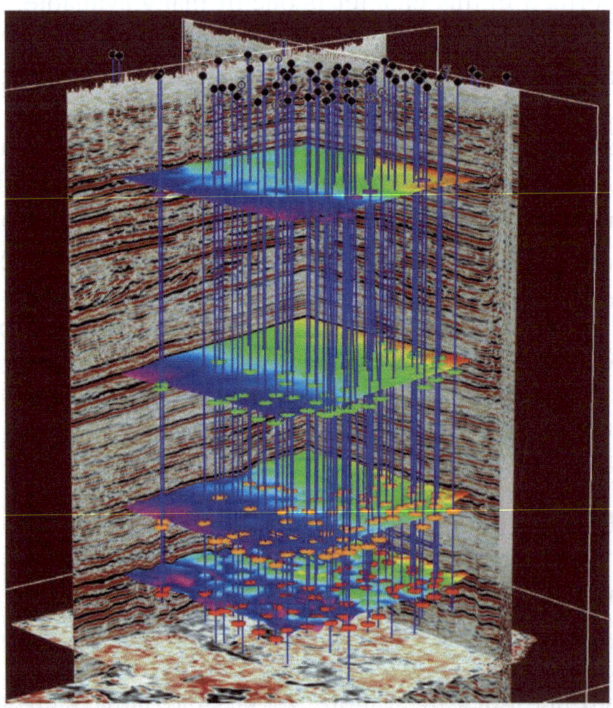

Figure 9-72 Midland Basin: The final computed depth framework showing seismic, formation tops, and horizons. With all data scaled in depth, the analysis and integration of these data with well log and petrophysical data and engineering data is far more effective.

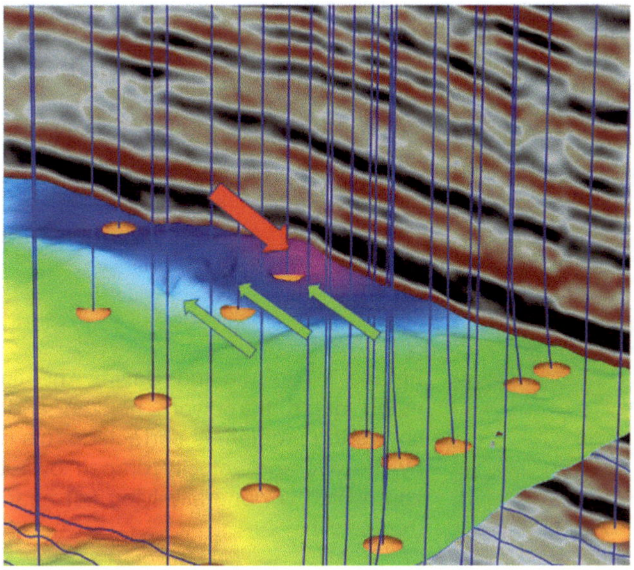

Figure 9-73 Midland Basin: Velocity data (in color) draped over a depth horizon shows an anomalously low velocity (magenta) at a well (large arrow). The well data checks out, but the time horizon has some mis-picks in the autotracking (thin arrows).

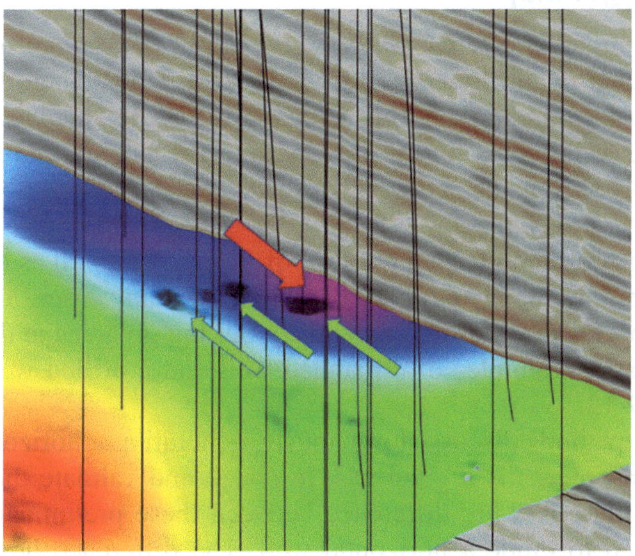

Figure 9-74 Midland Basin: Same scene as in Figure 9-73. However, the velocity data (in color) is draped on the time horizon. The same mis-picks exist in the autotracked horizon (thin arrows), but the horizon is generally flatter, meaning that the velocity is miscalculated around the well (large arrow).

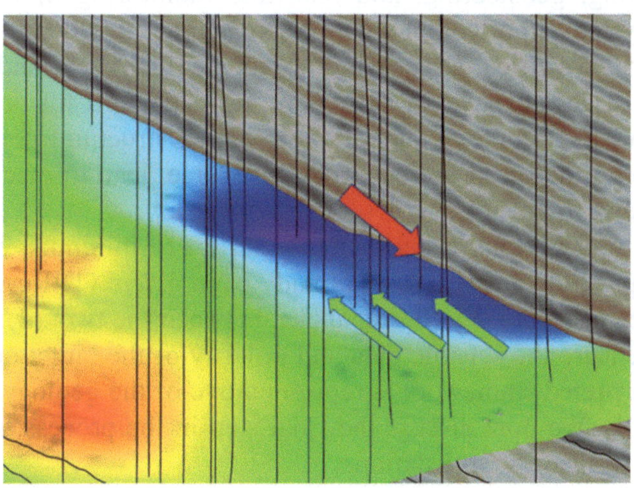

Figure 9-75 Midland Basin: By correcting the time horizon and recomputing the velocity model, the impact should be mitigated in the depth domain.

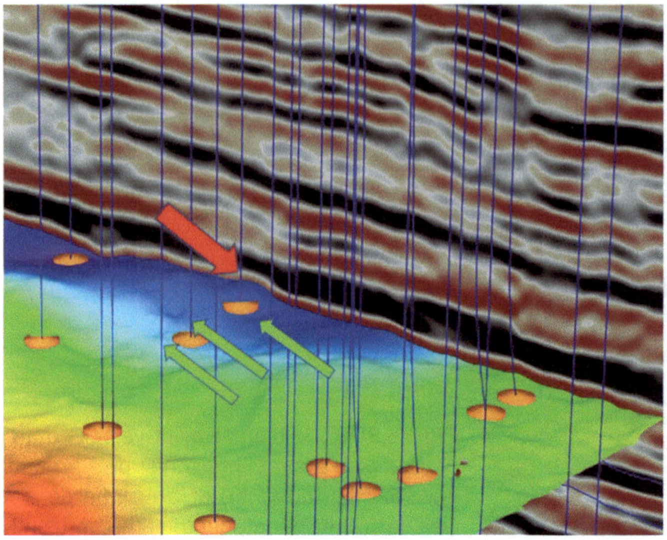

Figure 9-76 Midland Basin: The final computed depth framework horizon and seismic shows the problem is corrected. This is verified by the average velocity draped over the horizon, which is generally lying flat near the well (red arrow).

Creating and Contouring the Target Map

A preliminary depth map is shown in Figure 9-77. *This map is **not** scaled and contoured with enough detail to be used in the horizontal well planning process.* A geoscientist should assess the structural validity of the depth map surrounding the well being planned. This involves taking a close look at all nearby wells and the well depths for the specific interval targeted in the well. Though computed contours are generally acceptable and widely used, a *hand-drawn contour map is highly recommended as the target structure map.*

The landing depth in the horizontal well requires a high level of accuracy as well as the projected dip of the formation and whether any dip changes are predicted. In many cases, a nearby drilled horizontal well can be an analog for the well being planned, but the greater the distance away, the less value the analog will have. Examples of a planned well and a drilled well were shown previously in Figures 9-12 and 9-13.

Similar to the fault interpretation methods discussed previously, generating a horizon time-structure grid during the interpretation process is a good way to check and validate the accuracy of your interpretation. However, there is a big difference between these preliminary time grids and a final target depth grid used to guide a horizontal well. The difference lies in the level of detail and amount of seismic integration with the well data, for example, as in the depth conversion process. Advances in drilling, geosteering, and real-time monitoring of borehole conditions while drilling have routinely led to horizontal wellbores reaching 10,000 ft of lateral length or more while staying within a 50-ft vertical target window (review Figs. 9-12 and 9-13). The final target grids used in well path design have an extremely high standard for accuracy. They are usually contoured at a 10- or 20-ft contour interval. Unplanned structural deviations of ±10 ft or more can jeopardize keeping the bit in the best rock, or worse, lead to mechanical failure and sidetrack—a very costly situation indeed.

Let's step back for a minute and review what a geoscientist can do to minimize error in the mapmaking process. For the most part, the answers have already been given: (1) Review the time- and depth-scaled seismic volumes and note any discrepancies between the two. (2) Tie the wells using synthetic seismograms to ensure the seismic data and polarity of the target horizons are zero phase. (3) Ensure that the interpreted framework is representative of the geology, which includes compatibility with nearby wells. We now discuss the final step in the process: making structure maps for targeting horizontal wells on a workstation.

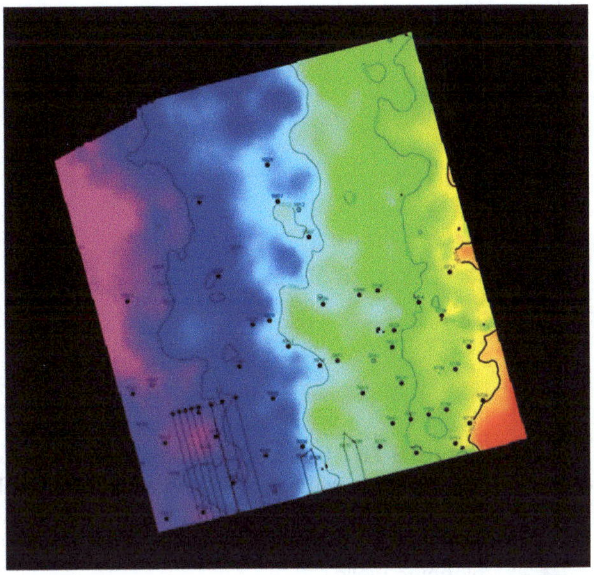

Figure 9-77 Preliminary depth structure map with a computer contour interval of 100 ft. Well tops used in the velocity model are posted in black. Note this map is not detailed enough to be used in the well planning process.

In a previous section regarding depth conversion, it was noted that there is not always an optimal time horizon or well control when it comes to velocity mapping and subsequent depth scaling of the target horizons. Some average velocity maps have a high degree of relative variability. *A velocity anomaly of more than 30 ft per second can result in a depth variation of 10 to 20 ft on the depth map.* If the velocity anomaly lies near the planned well path, it will affect the depth map and consequently the predrill directional plan. This can be mitigated by a diligent geoscientist verifying the structure of the target horizon along the planned well path.

The first step is to verify the formation tops in each of the local wells. Wells found with depth issues are either fixed or removed from the grid. The second step is to verify compatibility with the seismic framework depth grid. A target grid in the Midland Basin Wolfcamp play is rarely the same as a mapped framework horizon, but it can be within a few hundred feet vertically. One solution is to create an isochore map between the framework and the target formation tops and then add (for targets below) or subtract (for targets above) the isochore map to/from the framework depth map. The result is a map that ties the target formation tops and generally has the shape of the framework depth map. Further verification of the target horizon map is needed. For example, many nearby drilled horizontal wells with a similar target will have a point of penetration with the top of the target, which can be mapped as well as additional control points interpreted from the geosteering solution along the well path. However, use of these horizontal wells as additional control should be done with care. For any given horizontal well, there will be increasing uncertainty in the well position along the drilled path. As a final step, a geoscientist involved in planning a horizontal well may opt to use computer contouring (Fig. 9-78) or to manually contour the final target map to remove any anomalous structure not previously removed from the quality-control process. Figure 9-79 shows two example horizontal well paths. Well A, which is planned primarily along strike, should not anticipate more than 20 ft of structural change along the path. The second, Well B, crosses a structural nose with about 50 ft of elevation change along the path. The dip change along the well path is anticipated to be 0.6 deg up then 0.6 deg down across the nose. These dip changes are well within the range experienced in the Midland Basin.

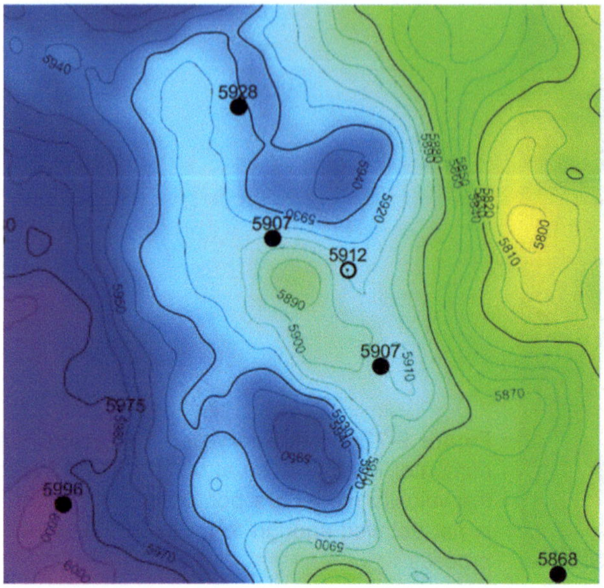

Figure 9-78 Computer gridded and contoured target horizon map with a contour interval of 10 ft. Regional structural dip is generally to the west; however, there is a lot of structural variation to account for when planning wells.

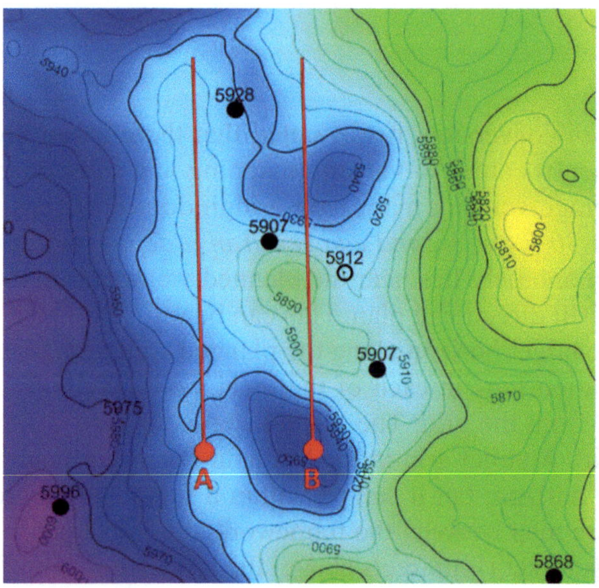

Figure 9-79 Target horizon map with planned wells shown (A and B). Wells planned within several hundred feet of each other can still require different well paths. Well A is generally flat with about 20 ft of vertical structure, while Well B is expected to encounter about 50 ft of structure in the well path.

CONCLUSION

In this chapter, we presented 3D seismic interpretation and mapping techniques consistent with the overlying principles discussed throughout this book. Using a consistent, principled workflow is critical to achieve repeated success in the current and future development of conventional and unconventional oil and gas resources. We discussed ways to increase efficiency through appropriate planning and organization of the geoscience work so that important milestones are met and the overall development plan remains on schedule. Constructing accurate maps to guide the planning and drilling of horizontal wells is a must for oil and gas companies in order to achieve the highest possible return on capital. Workstation technology provides unprecedented integration of geophysical, geological, and engineering data, leading to more robust solutions to some of the industry's most challenging technical hurdles. The philosophy and techniques described in this chapter are meant to inspire geoscientists to apply their creativity using proper tools and skills to make accurate and timely interpretations leading to a positive career path and the success of their company in a competitive world market.

CHAPTER 10

COMPRESSIONAL STRUCTURES: BALANCING AND INTERPRETATION*

INTRODUCTION

The first edition of ASGM contained one chapter on structural geological methods. Since knowledge of structural geology plays a key role in interpretation and mapping, as discussed in number 2 of the Philosophical Doctrine, we believed the chapter was important to the overall content of the textbook. Because of advances in structural geology and balancing during the following decades, in this third edition we have expanded the one chapter into four separate chapters covering compressional, extensional, strike-slip, and growth structures. Knowledge of structural methods in these various tectonic settings will improve your ability to generate viable, valid 3D interpretations, maps, and prospects as well as improve your ability to develop field discoveries. In addition, a number of significant structural advancements occurred since the publication of the second edition, and we have added new sections to the compressional and extensional chapters.

In the age of computer-generated work products, geoscience companies have developed a variety of software programs that can balance most of the common geological structures. It is not our intent to review or discuss these software programs and their various capabilities. Rather, our intent is to implant on the interpreter of geological and seismic data an *understanding of the geometric processes that form the basic structural styles*. This approach has value, as the interpreter benefits from a knowledge of the geological processes that are inherent to the algorithms that form the basis of the software, thus removing the black-box aspect of these programs. The approach should result in better interpretations of the data.

We begin the structural geology section of the third edition with a review of compressional techniques and methods. Much of modern structural geological analysis began with the study of compressional tectonics, and therefore it is appropriate to start here. These four structural

* For all figures in this chapter (in the printed book only), see the preface for information about registering your copy on the InformIT site for access to the electronic versions in color.

chapters center around specific structural methods and techniques. A basic understanding of rock mechanics and structural geology, presented in such textbooks as Billings (1972); Suppe (1985); Woodward, Boyer, and Suppe (1985); and Marshak and Mitra (1988), is a prerequisite to understanding and applying the techniques presented in this chapter.

STRUCTURAL GEOLOGY AND BALANCING

Compressional structures contain extensive proven petroleum reserves in many areas of the world. But even more accumulations remain undiscovered, and existing fields are insufficiently exploited, because the typical complexity of compressional structures and inadequate seismic images inhibit reasonable and accurate interpretation and mapping. Critical to the best possible analysis of the data is the interpreter's knowledge of compressional structural geology and the application of techniques that lead to geologically reasonable interpretations and accurate maps.

One of the most important of the compressional structural geological techniques is structural balancing. The ultimate goals of balancing are to restore complexly deformed rock to its initial state or to its correct palinspastic restoration and to determine the geological sequence of events. Such information can be very useful to the geologist or geophysicist. Not only is the geometry of the structure better understood, resulting in better and more accurate prospect and reservoir maps, but geological trends such as sand patterns can be more accurately located. An understanding of the timing of the structural events should aid in oil migration studies and define how and where fluids may have entered the structure. If the geometry of the structure is understood, then this knowledge can be used to more accurately process seismic data, which in turn results in an even better understanding of the geometries. Balancing can also be effectively used to *check* assumptions and interpretations (Tearpock et al. 1994). Lastly, balancing tends to keep the interpreter more focused. If the section does not balance, then perhaps it is time to reconsider the interpretation. Why drill a well to determine that the interpretation does not balance when restoration can determine a misinterpretation prior to the drilling? Our experience with balancing, as well as that of our colleagues, indicates that *balanced*, geologically possible interpretations can discover significant additional reserves. In short, balancing works.

Structural balancing is based on the intuitively satisfying concept that the interpreter must neither create nor destroy volume during the interpretation process (Goguel 1962). Interpreters may inadvertently introduce a volume imbalance anytime a fault is mis-picked or a horizon is miscorrelated. Fortunately, balancing can detect volume problems prior to the drill bit. Thus, it follows that an interpreted map or cross section, whether it be a geological or seismic section, should volumetrically restore without overlaps or voids in the stratigraphic section. Faulted and folded beds should be restorable to their initial subhorizontal state (Tearpock et al. 1994). Thereby, a structural interpretation may be tested for admissibility. An analogy might be a child who removes a new block puzzle from a box and places it on the floor. Once all the pieces of the puzzle have been removed from its container, the puzzle can be restored to its initial state by placing each block back into its proper position. The first attempt by the child at restoring the puzzle may result in most of the pieces being placed into the box, with one or two pieces remaining on the floor. A second attempt could result in all of the pieces being placed in the box, but with some of the pieces being tilted at various angles or forced to fit.

The geoscientist experiences similar problems when attempting to retrodeform (restore) geological and/or geophysical data. Of course, the correct solution to a puzzle is one that has been perfectly restored to its initial position. There are two types of interpretations: interpretations that are admissible, or geologically possible, and interpretations that are inadmissible, or not geologically possible (Elliot 1983). A balanced interpretation is an admissible interpretation in which the horizons can be restored to their initial subhorizontal position by unfolding the horizons and rotating the beds back to a subhorizontal position along the interpreted faults.

The benefits of balancing are fundamental to correct geological interpretations. The earth's subsurface contains no voids or mass overlaps; thus, a section that does not balance cannot be geologically reasonable on simple geometric grounds. Unfortunately, a balanced section, although physically reasonable, need not necessarily result in the correct geological interpretation. Balancing is *not unique*, and two geoscientists can produce two balanced sections that are not alike. Obviously, the more complete the data set and the better the interpretive techniques, the more likely that the balanced section will reflect reality.

Balancing is still a developing science, and new techniques and interpretations are progressively being introduced. Nevertheless, an interpretation tempered by a concept of mass conservation is the key to admissible geological interpretations and constructions. If the structural interpretation is correct, then balancing techniques can be used to quantify the interpretation.

Balancing can be subdivided into two disciplines: classical balancing, which was primarily developed by Goguel (1962), Bally et al. (1966), and Dahlstrom (1969) and his coworkers; and nonclassical balancing, which was primarily developed by Suppe (1983, 1985) and his students and coworkers. Most of the concepts presented in this introduction can be attributed to Goguel and Dahlstrom.

MECHANICAL STRATIGRAPHY

For many years, structural geologists have argued about the mechanical properties of the upper crust. Does it exhibit elastic and/or frictional behavior as indicated by earthquakes, or is it viscoelastic or viscoplastic, as indicated by the bent strata in the hinge zone of folds? Could time be a factor? Do the sedimentary strata buckle out (Biot 1961), or do the strata follow faults within the sedimentary section (Rich 1934)? Although all of these mechanisms are possible, the evidence now strongly suggests that the deformation that occurs in petroleum basins is primarily controlled by brittle (low temperature) deformation processes, and that the viscous deformation expressed by fold trains (Fig. 10-1) is confined to metamorphic belts (Tearpock and Bischke 1980). The fold style depicted in Figure 10-1, with its near constant wavelength, is not commonly observed in petroleum basins, and thus another deformation mechanism is required to explain the folds that trap hydrocarbons. This mechanism appears to be frictional deformation. Davis et al. (1983), Dahlen et al. (1984), and Dahlen and Suppe (1988) formulate a frictional, or brittle, theory of crystal deformation that applies to both compressional and extensional regimes. The theory resolves the overthrust paradox (Smoluchowski 1909; Hubbert and Rubey 1959) and is consistent with the geological and seismic information collected from petroleum basins. Our intention here is to apply this theory and its observations to our areas of interest. Readers who maintain an interest in mechanics can consult the references listed at the end of this textbook.

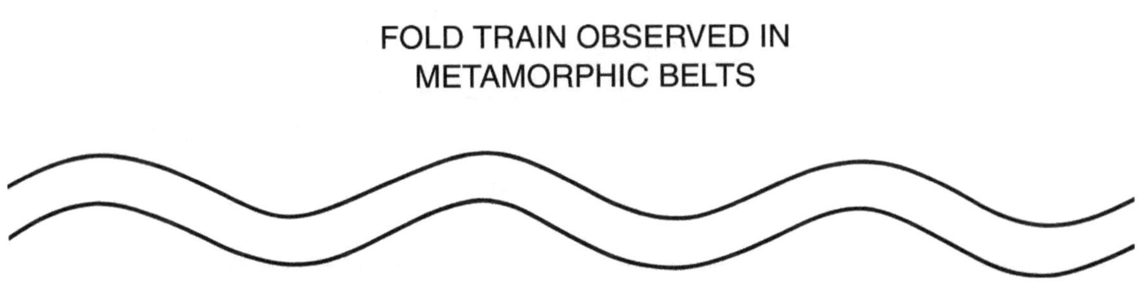

Figure 10-1 Example of a fold train commonly observed in metamorphic belts.

The frictional theory of crystal deformation states that when folds form, the maximum principal stress (σ_1) is inclined slightly to the bedding surfaces (Fig. 10-2). The rock will then fracture along angles that are dependent on the pore pressure and the intrinsic strength of the rock. The weaker the rock, the lower the angle between σ_1 and the fracture.

For example, consider an alternating sequence of limestone and shale layers (Fig. 10-2). Intuitively, shale layers seem to be weaker than better consolidated limestone layers, and it is well known that shales can contain abnormally high fluid pressures that drastically weaken these rocks. The theory states that because shales are weaker than limestone, the angle (α_1) between σ_1 and the fractures in shales must be smaller than the angle (α_2) between σ_1 and the fractures in limestones (Fig. 10-2). As σ_1 is slightly inclined to the bedding, the fractures in the shales are more subhorizontal than the fractures in the limestones. This leads to the primary conclusion of this section: in more competent or stronger rocks, the fractures will form at a high angle to bedding, and in the incompetent rocks, such as overpressured shales, the fractures tend to form parallel or subparallel to bedding.

If motion along these fractures causes them to coalesce, then a decollement, or zone of detachment, will form along the *flat-lying* bedding and may follow incompetent (shale or evaporite) horizons for tens or even hundreds of kilometers (Davis and Engelder 1985). In areas where the weaker layers gain strength or are pinched or faulted out, the decollement may ramp to a higher structural level (Fig. 10-2). As these ramps must pass through rocks that are stronger and have lower pore pressures than shales, the angle (α_2) between σ_1 and the fractures will be larger. Thus, ramps have higher angles with respect to the bedding than do the flatter portions of thrust faults (Fig. 10-2).

Where the ramp connects to a weaker layer on a higher structural level, the ramp transforms into a flat once again. Once a network of ramps and flats forms and a large force is applied to the back of the wedge-shaped region in Figure 10-2, the strata above the flats and ramps will begin to move along the fault. Material will begin to slide along the flats and up the ramps, forming a fold in the hanging wall block. Eventually, large folds will form in a manner that was initially described by Rich (1934), but this process is the subject of a later section.

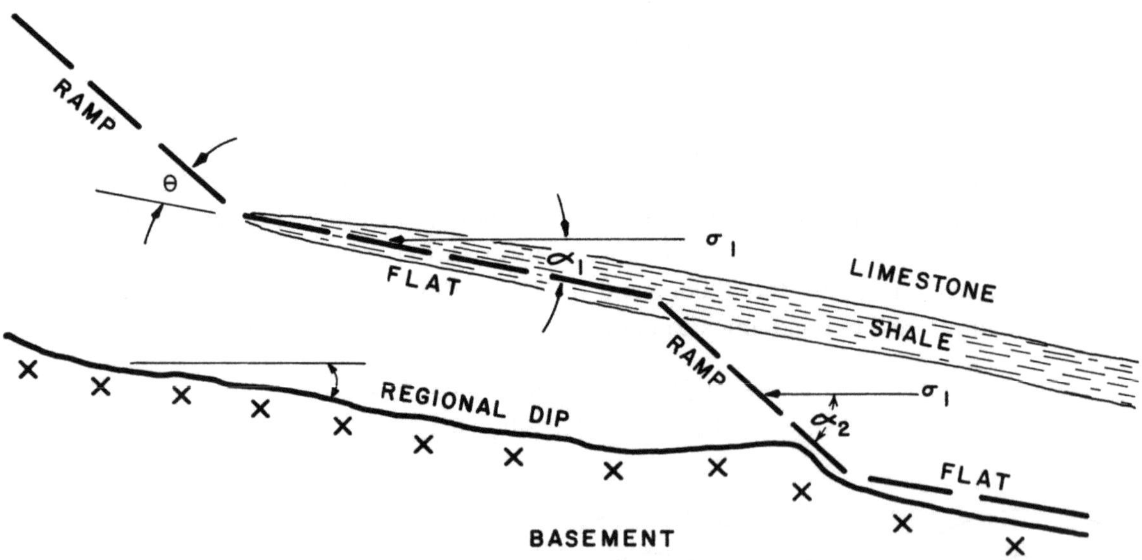

Figure 10-2 Cross section of ramp geometry. For explanation, see text. (Modified after Rich 1934; AAPG©19341934, reprinted by permission of the American Association of Petroleum Geologists (AAPG) whose permission is required for further use.)

The angle at which the ramp steps up from the bedding is called the cutoff, or step-up, angle (θ in Fig. 10-2). This angle is often characteristic of, or fundamental to, a particular fold-thrust belt and depends on both the pore pressure in the rock and the rock type. Similar relationships may exist in extensional terrains. The characteristic cutoff angle in certain fold-thrust belts is generally less than 20 deg and tends to vary within several degrees of its mean value. For example, in Taiwan the characteristic step-up angle is 13.3 deg ± 2.4 deg (Suppe and Namson 1979; Dahlen et al. 1984). An attempt must be made to determine this angle prior to a balancing study. This step-up angle will be used to balance your structures. Note that *the step-up angle is measured relative to regional dip rather than to the horizontal.*

There appear to be at least three methods that give insight into estimating the characteristic step-up angle. Field studies or a literature search can be conducted in the area of interest. As the step-up angle is the angle between the flat and the ramp, field measurements or a description of this relationship will provide the required answer. A second, less direct measurement technique is to observe a well-imaged ramp and flat on a seismic section that is perpendicular to the strike of the fault surface. The section must first be depth-converted to make this measurement. The strata above the ramp will parallel the ramp, and thereby the step-up angle can be determined relative to regional dip (Fig. 10-34). Therefore, a study of the dips across an area may give insight into the characteristic step-up angle. For this method, it is first necessary to know the regional or undeformed dip of the area. For example, suppose that an area has no regional dip. It therefore follows that the nontilted beds will have zero dip. Strata that have moved up ramps, and are deformed, may dip at 12 deg. The characteristic step-up angle is therefore 12 deg. We might, however, be faced with a situation in which 20% of the dips are near zero, 30% are 3 deg, and 50% are about 9 deg or greater. The problem here is attempting to decide whether the regional dip is zero or 3 deg and whether the step-up angle is 9 deg or 12 deg or greater. This matter is often resolved by finding that one of these choices simply works better than the other during the restoration, or balancing, process.

CLASSICAL BALANCING TECHNIQUES

In previous sections we introduced the concepts of *volume conservation* and *brittle deformation*, which we apply to petroleum basins and not to metamorphic belts, which often lie adjacent to our areas of interest. Here we develop these concepts in a manner that can lead to the interpretation and mapping of structures that better define prospects.

The volume conservation concepts that are developed in this section, although rigorous in their general application, do not precisely specify how this volume is to be conserved. Thus, a significant degree of artistic license is left to the interpreter. For this reason, the classical techniques developed by Goguel (1962) and Dahlstrom (1969) are ultimately qualitative in their approach. No formula or graph constrains the interpretation.

Volume Accountability Rule

The basic principle behind all balancing techniques is that nature, and not the interpreter, can create or destroy rock units, and that the interpreter should account for all of the present or pre-existing volume. Engineers are familiar with this concept as one of mass or volume balance, or of volume accountability. Most geoscientists are quick to point out that geological compaction, particularly in growth structures, changes volume with time. In addition, fluid flow through limestone can remove volume by pressure solution, and this volume reduction can be significant (Groshong 1975; Engelder and Engelder 1977). Arguments of this type, although correct, should not be substituted for lazy thinking. We have discovered that even thinking about growth structures in terms of strict volume conservation has forced the development of new balancing and interpretation techniques. If the structure does not balance volumetrically, then what process is causing the imbalance? The conservation of volume principle at least brackets the error or helps define the amount of volume reduction due to compaction or pressure solution. In the case of widespread volume removal, regional balancing and structural analysis may indicate that another process is occurring and to what extent. We normally find, however, that these volume reduction processes are not a major concern and that the interpreter normally can think in terms of volume conservation while being prepared for alternatives.

The economic issue that needs to be addressed here is much more practical and much more likely to confront the interpreter on a daily basis than is pressure solution. Interpreters often unknowingly have a tendency to introduce mass overlaps and gaps into their interpretations (Tearpock et al. 1994). Often, these gaps or overlaps are confined to a particular region of their cross sections or to a particular structure. For example, a given seismic-based cross section and prospect, upon retrodeformation, has twice as much volume between sp (shotpoint) 320 and sp 420 (at about 1.5 sec to 2.2 sec) and no volume between sp 285 and sp 400 (at about 2.8 sec to 3.1 sec). An obvious question thus arises: Does this volume incompatibility affect the viability of the prospect, and would a better interpretation enhance or detract from the prospectivity of the area? Therefore, balancing literally attempts to take the "holes" out of our interpretations, as is shown in the Retrodeformation section in this chapter.

Area Accountability

In the Mechanical Stratigraphy section, we describe the petroleum basin as a low-temperature regime subject to brittle (i.e., frictional) deformation. In such an environment, flow, elongation, and flattening are not of primary importance, and thus the 3D volume problem can be reduced to 2D. In other words, we shall assume that material is not entering or leaving the plane of the geological cross section, and therefore the problem can be reduced to 2D (Goguel 1962). Notable exceptions to this rule would be shale and salt diapirs, which are typically 3D phenomena. These salt structures, which are associated with withdrawal and rim synclines surrounding the diapir (Trusheim 1960), contain a wealth of information that defines salt flowage and can be used to balance salt diapirs in 3D. Another exception is the bifurcating normal fault structure, which moves material out of the plane of cross section. Techniques for studying this type of deformation are briefly addressed in Chapter 11. In the meantime, however, and as long as the deformation is brittle and the transport direction is subperpendicular to fault strike, the 3D problem can be reduced to a 2D cross section that is subperpendicular to the strike of the fault.

Bed Length Consistency

If we accept the premise that petroleum-bearing rocks are brittle and deform at temperatures within the hydrocarbons window, then the 2D problem can be linearized (Goguel 1962). In other words, if there is no large-scale material flow within or across the plane of the 2D cross section, then the seismic reflection or bed length before deformation will remain the same after deformation (Fig. 10-3). This logic will also hold true for the thickness of each bed involved in the deformation, which means that the folding will be of the parallel type. Thus, bed length can be utilized to balance cross sections. If a sedimentary sequence is 2 km long before deformation, it must remain 2 km long after the deformation. The bed may be bent and it may be broken, but it should still be 2 km long.

Although the logic inherent in the above statement may seem self-evident, it appears to be one of the primary causes of the so-called balanced cross section, which is prevalent throughout the literature. The above logic implies that if one measures the bed lengths across a cross section, and the bed lengths are equal on all levels, then the cross section will balance. In practice, however, small changes in the lengths of lines can result in significant volume changes that result from inaccuracies in, or a lack of, subsurface dip data. This follows from the trigonometric relationship that at low angles the length of the adjacent line is about equal to the hypotenuse (Fig. 10-4). Consequently, we can see that the line segment AB is about equal to AC, even though the thickness AX is not equal to the thickness CZ. Therefore, we can often check existing cross sections by simply observing whether beds or formations are subject to unexplained or nonuniform thickness variations. If these thickness variations are not due to logical variations in stratigraphic thickness, then the interpretation should be subjected to further analysis.

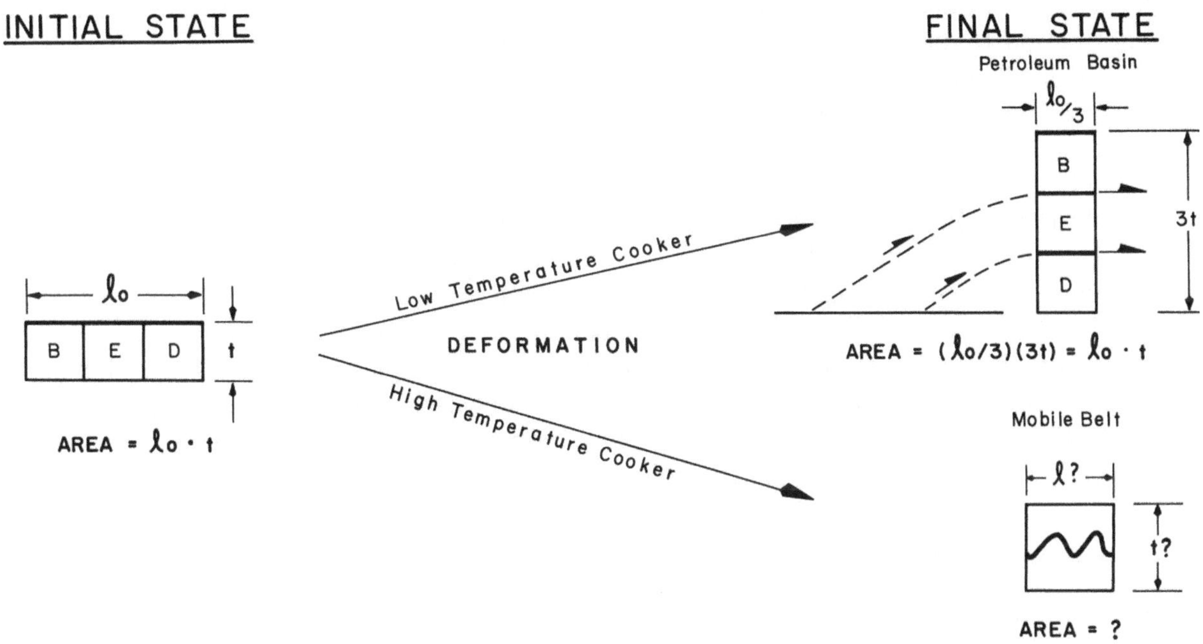

Figure 10-3 Deformation map of petroleum basins versus metamorphic belts. Low temperatures tend to preserve cross-sectional volume, whereas in metamorphic belts, material will flow in and out of the plane of cross section.

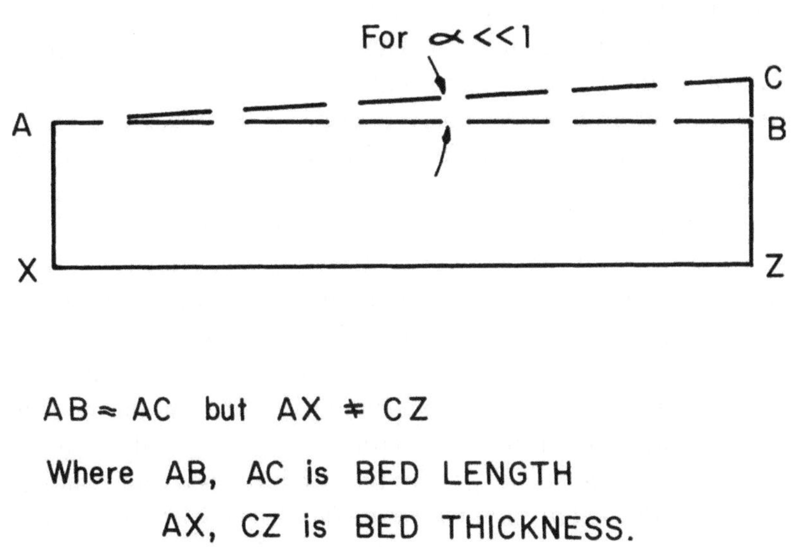

Figure 10-4 Noticeable changes in bed thickness result in small changes in bed length.

Pin Lines

A significant development was made by Dahlstrom (1969), who realized that you can check the validity of any cross section by measuring bed lengths, while keeping an eye out for variations in the thickness of units. This is accomplished through the use of pin lines (Dahlstrom 1969). In this procedure, one attempts to locate regions that are not subject to deformation (such as shear or bedding plane slip) and then affix these regions to the basement by driving an *imaginary* pin vertically through the cross section. Pins are used as a basis for measurement, and bed length consistency is then measured relative to these pin lines (Fig. 10-5). Dahlstrom realized that bed length consistency must be preserved on all structural levels in both 2D and 3D, and that if the bed length consistency does not hold from one section to another, then the interpretation is likely to be in error. Figure 10-5 is modified from Dahlstrom (1969) with Figure 10-5a signifying the undeformed pin state. If the unit is concentrically folded and displaced a distance S, then the bed length (l_o) within the concentric fold after deformation should be the same length (l_o) as it was before deformation (Fig. 10-5a and b).

In Figure 10-5b, the bed length (l_0) within the folded unit is not the same as the pin length (l). This follows, as the folded unit has been shortened a distance S (compare Fig. 10-5b and c). In Figure 10-5c, dipping beds overlie flat beds, which is the classic indication of a geometric discontinuity or decollement (thrust fault). We call this method for picking thrust faults Dahlstrom's rule, and the thrust fault exists between the steeper dipping and the flatter dipping beds (Fig. 10-5c). Thus, when picking thrust faults on seismic data, simply look for steeply dipping beds over more gently dipping beds. These steeply dipping beds must be structurally deformed and typically are inclined at more than 5 deg to regional dip.

Line-Length Exercise

Line-length balancing can be a powerful quick-look tool (Tearpock et al. 1994). We present an example of how line-length balancing may find additional oil in producing fields. Figure 10-6a represents two dip profiles that are similar to those in a large producing trend in South America. The two profiles are from the same field, traverse the same anticlinal structure, and are a short distance from each other. Good to fair quality seismic data from the field image the top of the structure but do not clearly image Faults A, B, and C. Well No. M-5 on profile A and other wells in the field cut Fault A, but Fault B is inferred from the relatively dense well control (Fig. 10-6a). Notice that Wells No. M-1 and M-3, which penetrate the front of the structure on profiles A and B, encounter the reservoir section at a greater depth than do the structurally higher Wells No. M-2 and M-4. Seismic data from an adjoining field on the same structure are of good to excellent quality and clearly image Fault C, which is a bedding plane thrust. Fault C was mapped into the area of profiles A and B from the adjoining field.

The interpretation shown on profile A contains three imbricate blocks formed by Faults A, B, and C. Faults A and B link to large Fault C. The footwall reservoir section has the same bed length along profiles A and B, so pin the structure at the hanging wall cutoff position located in the structurally lowest imbricate blocks (left-hand pin). The pin on the right penetrates the syncline in an off-structural position. In the hanging wall portions of the fold, use a balancing program or a ruler to *measure the bed lengths of the reservoir bed along its top.* The beds are cut by the faults, *so the top of the bed in each imbricate block terminates at the faults.* Therefore, do not include as bed length the distance along a fault. On profile A, the hanging wall bed lengths in the three imbricates are about 11.8 km total.

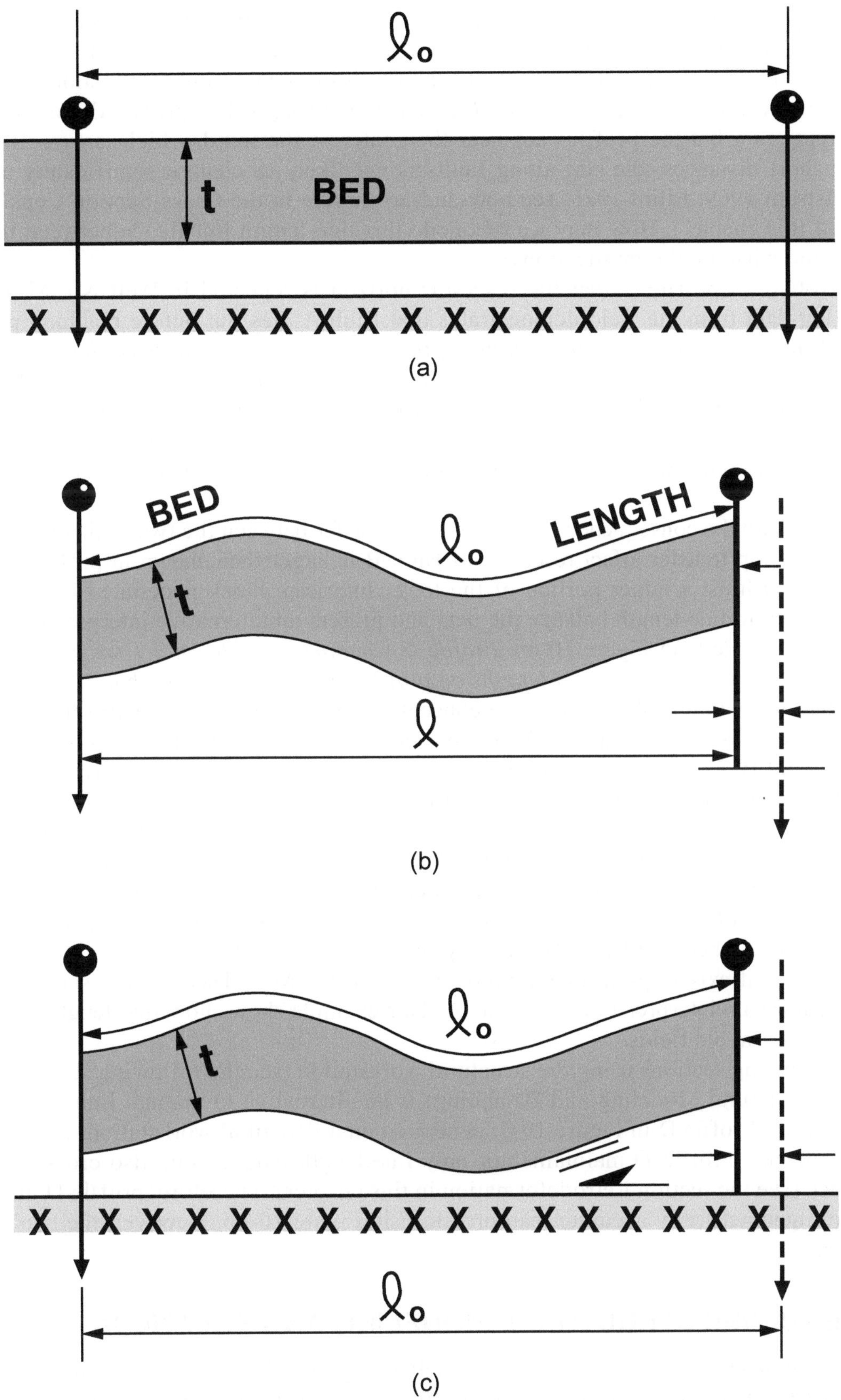

Figure 10-5 Pin lines and bed length consistency. (a) Undeformed bed state. (b) Bed length differs from pin length. (c) Deformed bed. (Modified after Dahlstrom 1969. Published by permission of the National Research Council of Canada.)

Repeating the bed length measurements on profile B, located a short distance from profile A, results in a hanging wall bed length, at the top of the reservoir horizon, of about 10.8 km. Thus, between the two profiles there is a line-length imbalance along the top of the reservoir bed, and profile A contains *1 km more bed length* than profile B. Perhaps the faults are dying out, but the profiles are near the center of the trend, which is over 100 km long. Over short distances, the slip along faults is not likely to change significantly along strike (Dahlstrom 1969; Elliot 1976; see bow-and-arrow rule in the Cross-Section Consistency section of this chapter). How may we reconcile this line-length imbalance between the two profiles, and what are the implications?

Notice on profile A that the reservoir horizon is repeated in Well No. M-5. Abundant well log data from the field demonstrates that Fault A dies out before reaching profile B. In fold-thrust belts and *over short distances, the slip along thrust faults is about constant along strike* (see Cross-Section Consistency). Thus, it is unlikely that Faults A and B would both grow smaller over such a short distance. Alternatively, *slip transfer between faults is common in fold and thrust belts* (Dahlstrom 1969) (Fig. 10-14). The slip on Fault A may transfer to Fault B. In other words, as Fault A dies out, the slip on Fault B increases at the expense of fault A.

What are the consequences of a 1 km slip transfer between the two fault surfaces, and how could this slip transfer affect reserves? If Fault B is larger than shown in profile B, then Fault B may overthrust a larger portion of the lower imbricate block penetrated by Well No. M-3. We proceed to line-length balance the data and present an alternative interpretation of the data shown in profile C in Figure 10-6b. *Profile C contains an additional 1 km of bed length relative to profile B, so that the bed lengths on profiles A and C are both about 11.8 km.* The interpretation shown on profile C uses the concept of a ramp-flat fold geometry that is common to fold-thrust belts (Bally et al. 1966), rather than the upward-listric reverse fault shown on profiles A and B. Upward-listric fault surfaces are common to *extensional* terranes (Chapter 11). As line-length balancing concepts suggest that the bed length should be about 11.8 km on the two profiles, and as we must honor the existing well control, we consider the solution shown in profile C. Profile C contains an additional 1 km of slip on imbricate Fault B. This increase in slip creates more repeated section in the lowermost imbricate block beneath Well No. M-4. This interpretation of the data is exciting, as the new interpretation extends the reservoir horizon in the lower block between Faults B and C by about 1 km to the right, introducing upside potential. This potential exists up-dip of the producing Well No. M-3. The solution shown in profile C may require a reinterpretation of profile A. This example shows how line-length balancing may find new oil in old fields.

Balancing sections using the structural workstation (see the following section, Computer-Aided Structural Modeling and Balancing) is an alternative to manual line-length balancing procedures. Profile D in Figure 10-7, generated on a structural workstation, uses area-balancing concepts. Profile D maintains not only line-length balance but also cross-sectional area balance (see the section Retrodeformation in this chapter). Therefore, profile D in Figure 10-7 is more geometrically accurate than profile C in Figure 10-6b. However, the two profiles are similar.

COMPUTER-AIDED STRUCTURAL MODELING AND BALANCING

Structural analysis, interpretation, and modeling rely heavily on the graphical representation of structural horizon and fault surface geometry. Using structural workstation software, the end product of this graphical representation results in the construction of cross sections. Graphical methods of structural analysis can be applied to geological data to determine the viability of

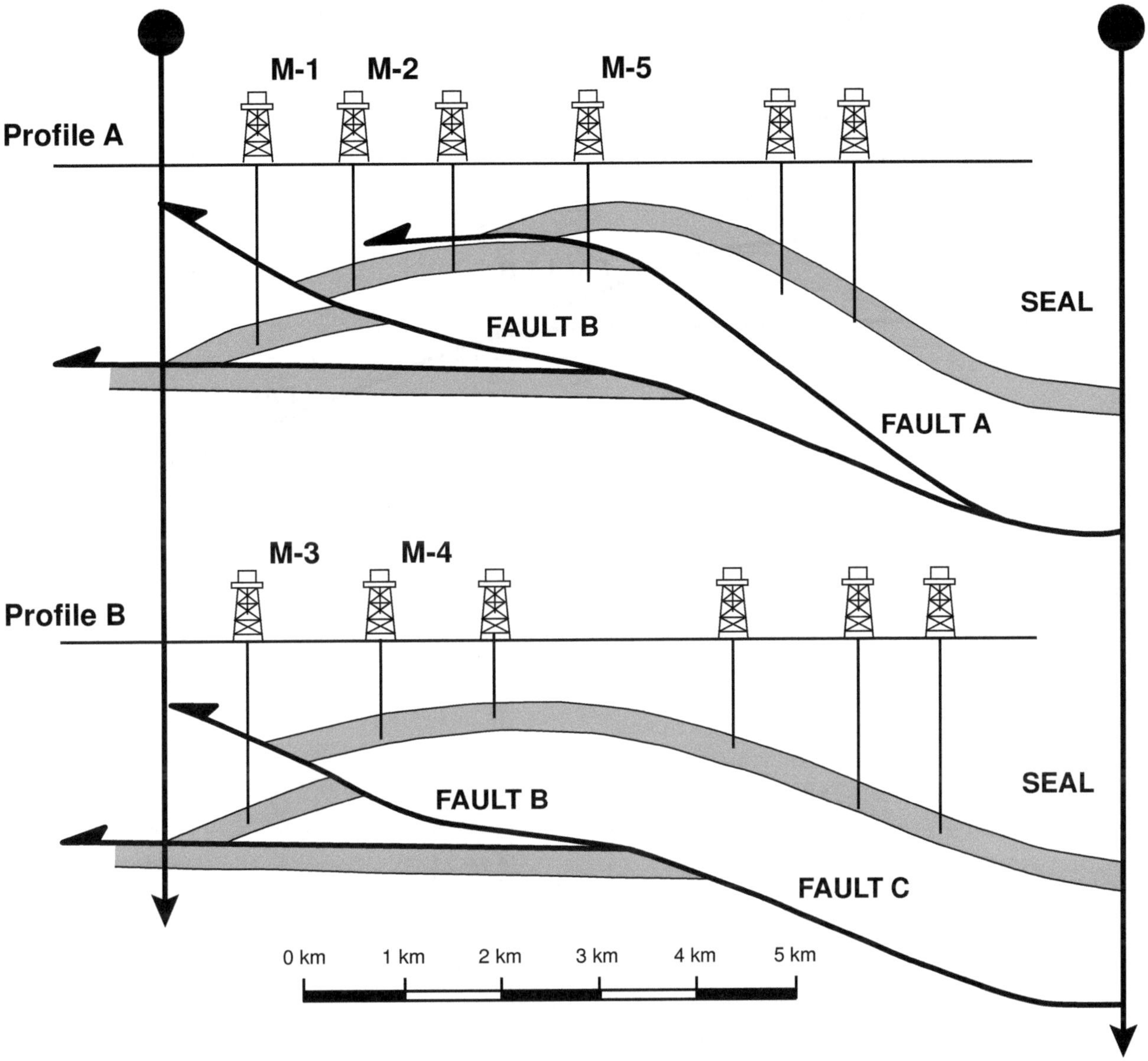

(a)

Figure 10-6 (a) Profiles A and B constructed across an anticline that forms a producing field. The slip imbalance between the two profiles creates a line-length imbalance, as described in text. (b) Profile C represents a reinterpretation of profile B using line-length balancing concepts. Profile C, which uses a ramp-flat thrust fault geometry common to fold-thrust belts, introduces additional potential in the reservoir in the lowermost imbricate block. (Published by permission of R. Bischke.)

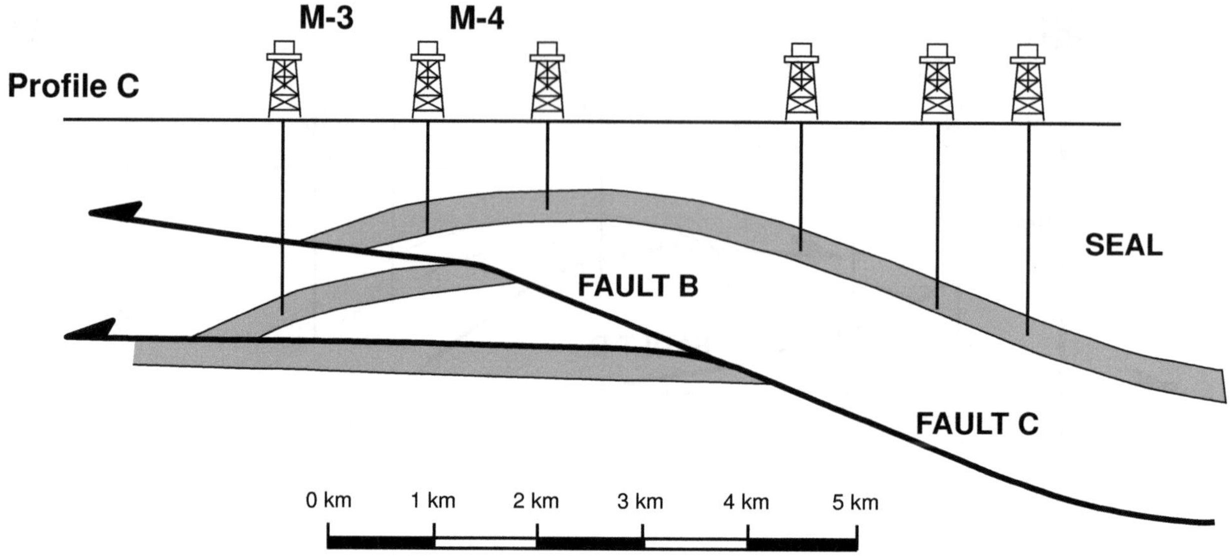

(b)

Figure 10-6 *(continued)*

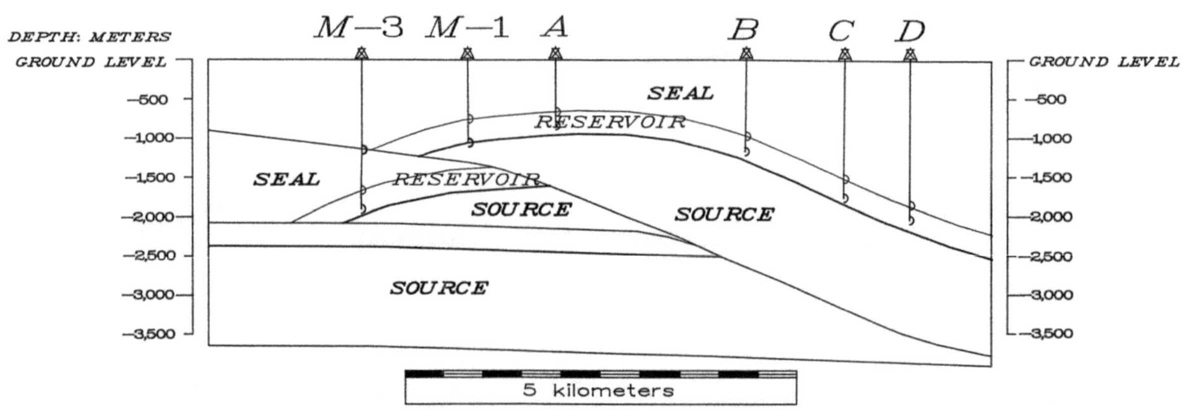

Figure 10-7 Profile D is a reinterpretation of profile C in Figure 10-6b, using structural workstation methods based on balancing concepts. Profile D is similar to the line-length balanced profile C. (Published by permission of R. Bischke.)

cross sections. Historically, structural modeling relied heavily on manual drafting to create cross sections. The emergence and enhancement of computer workstations during the 1990s provided a powerful tool for 2D and 3D structural evaluation. The workstation facilitates the visualization and modeling of structural data and allows interpreters to attack more complicated structural problems (Fig. 10-8). Utilizing workstation software, it is possible to move quickly from the time domain of seismic data into the depth domain of structural visualization. Depth visualization by geoscientists enables the creation of a more complete and accurate depiction of the subsurface structural geology. The technical and economic benefits of computer-aided structural analysis are important, if not key, to the success of petroleum exploration and production in structurally complex areas.

After reviewing the different structural styles presented in Chapters 10 through 12 and their associated algorithms, one may ask, "What is the best and most effective method of applying structural information?" One important approach is the proper use of structural workstation software.

Seismic data are the primary subsurface information; therefore, it is critical to translate seismic time models into seismic depth models. Once depth intervals are selected and assigned respective velocities, the structural workstation software should provide a means to readily move between the time sections and the related depth domains. Data quality and knowledge of related acoustic interval velocities determines the accuracy of the time-depth transition. Again, the workstation is an excellent tool for testing different time-depth pairs. *Iteration of structural models utilizing an array of alternative concepts* helps to refine and perfect the interpretation, which is another strong justification for the implementation of computer-aided structural analysis.

Animated models of fault bend folds, fault propagation folds, and so on, are possible on the workstation. These animated models are helpful when visualizing and constructing forward models of simple structures and illustrating the origin of structures. *The identification and accurate depiction of fault surfaces from seismic data sets is one of the most important steps of seismic structural interpretation.* There is a direct relationship between the geometry of the fault surfaces and the geometry of structure horizons related to the fault surfaces. The *relationship between fold shape and fault shape* is often overlooked by many geoscientists during the seismic interpretation phase of a project. We believe this is often due to a limited structural background by geoscientists, which restricts their understanding of fault–fold relationships. Interpretation errors related to the geometry of faults and horizons become obvious when viewed in the form of a balanced cross section. Risk can be reduced significantly by using comprehensive, balanced 3D structural models.

A validated 3D structural model not only is kinematically correct but also helps to eliminate any errors of interpreted displacement along selected fault surfaces. The elimination of displacements that are kinematically incorrect creates higher quality interpretations. **Whereas a balanced 3D structural interpretation may not be unique, it does add substantially to the validity of any interpretation.** From structural workstation analysis, it can be readily seen that the term *balancing* encompasses *validation*, *retrodeformation*, and/or *restoration* (see Retrodeformation in this chapter). The complexity of retrodeforming a structural cross section manually may be difficult if not impossible in many cases, yet it can be readily and accurately completed with a computer.

Two-dimensional and three-dimensional structural workstation software can significantly expand the interpretive capacity and accuracy of the geoscientist. Software links provide direct communication between structural applications and other geophysical and geological software programs. Accuracy, efficiency, and completeness are improved by the sharing of data in a workstation environment.

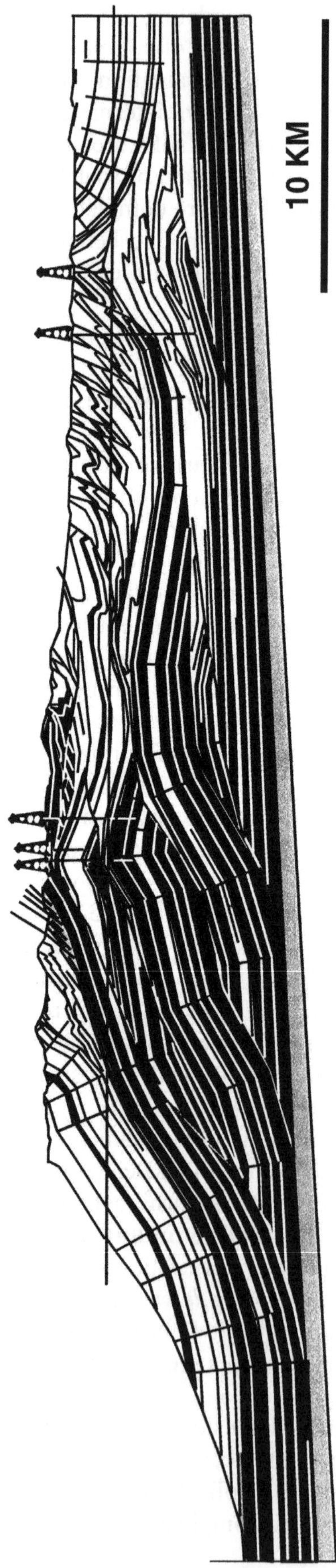

Figure 10-8 Structural cross section across Savanna Creek Duplex and Canadian Rockies. (Published by permission of D. Roeder 1983.)

A comprehensive structural model incorporates all the available geological and geophysical data for a given area. In most cases, structural analysis forces the geoscientist to "fill in the blanks" beyond the limited available information. The good data areas can be readily projected into the poor data areas. Workstations can access and store volumes of data beyond the reasonable capacity or efficiency of manual manipulation.

Accurate dip analysis, sonic logs, lithology logs, deviation surveys, and all other well data are incorporated into an accurate structural interpretation. Detailed surface geology maps, including topography, provide a wealth of information for land-based study areas. All stratigraphic data are an integral part of a comprehensive structural interpretation. Computer-aided structural analysis enables you to *analyze all your data accurately and completely.* The accuracy and reliability of subsurface maps are enhanced and perfected with a detailed, computer-generated structural model.

Structural modeling is the keystone to subsurface modeling and visualization. Therefore, from an industrial point of view, technical and economic success ties directly to the accuracy and effectiveness of the subsurface structural interpretation. The structural analysis not only provides the framework for detailed production activities but also drives frontier exploration. Preseismic structural models are cost-effective prospecting tools during the initial phases of a study. Structural models can help in planning and guiding a seismic acquisition program and can aid in improving the quality of acquired seismic data. Digital cross sections and assigned interval velocities lend themselves to ray-tracing programs and resultant models to help facilitate the planning, acquisition, and interpretation of seismic data. The economic success of a new discovery or the cost of a dry hole dwarfs the cost of a proper structural evaluation. The process of structural modeling and restoration forces the geoscientist to critically think about the interpretation, to question the data, and to understand the hydrocarbon potential of the region. *The computer-based structural interpretation allows geoscientists to quickly and accurately converge on viable geological solutions to complex structure problems.*

Retrodeformation

In a previous section on classical balancing techniques, we introduced a number of powerful rules and constraints to check interpretations. These rules concerning preservation of line length and bed thickness can be quickly applied to cross sections to insure cross-section viability. We now demonstrate that bed line-length and thickness preservation is an important first step in a two-step operation of retrodeformation.

In the introduction to this chapter, we emphasize that, with time, structures move and that structural interpretations should be restorable. The process is called **retrodeformation**, or palinspastic reconstruction. *Any interpretation of subsurface data should be restorable to an initial undeformed state* because the stratigraphic units were deposited parallel to regional dip. Faults induced by compressional forces may cut the strata, causing the hanging wall beds to move over footwall beds. The structure is *thrust forward* and into its present position. Let us assume that this structure is presently imaged on seismic profiles. The retrodeformation process is the *reverse* of the forward-thrusting process. Any interpretation of the faults contained in this seismic data set should be compatible with the hanging wall beds moving back along the fault surface into their undeformed state. The pieces of the seismic puzzle should be *restorable without mass overlaps or voids.* These principles apply to every tectonic regime, but they are most easily applied to compressional and extensional regimes. However, the retrodeformation principle is an *excellent consistency check* on interpretations of compressional, extensional, strike-slip, and salt structures. We apply line-length and bed-thickness preservation concepts to a seismic line to show how these concepts can improve prospect integrity.

Examine Figure 10-9, which is taken from Bally's (1983) classic monograms on seismic interpretation, "Seismic Expressions of Structural Styles." In the forward to his monograms Dr. Bally states, "As to the interpretations presented, the reader will have frequent occasion to disagree or to be unconvinced of the interpretation offered. This properly reflects the fact that seismic reflection profiles are not easily interpreted in a unique way. Because the marked seismic lines are frequently supporting published papers, less critical readers often feel that *such illustrations constitute geological proof*, while in reality they are much more like *drawings on a seismic background* that illustrate an author's concept" (our emphasis). Dr. Bally's statement has many important consequences to industry, so let's examine his statement in more detail.

Dr. Bally makes several important points that management, accountants, economists, and working teams should remember every time geoscientists propose a multimillion dollar well. Economics dictates that wells are expensive and that geoscientists are cheap, and not the other way around. Money should always be available to test the viability of all prospects prior to drilling (Tearpock et al. 1994).

The other concept inherent in Dr. Bally's forwarding statement is that there are two sets of interpretations: those that constitute "geologic proof" and those that constitute "drawings." We call the first type of interpretation an *admissible interpretation* (Elliot 1983). An admissible interpretation maintains 3D structural validity and is a geologically possible interpretation. The second type of interpretation is the *inadmissible interpretation* that does not maintain 3D structural validity and is therefore impossible on simple geometric grounds. Chapters 10, 11, and 12 concentrate on admissible interpretations as applied to prospects and prospect evaluation. With this in mind, we next test Figure 10-9 for its admissibility.

Often, during a prospect review and evaluation of compressional structures, we first check for apparent horizon thickness changes. For example, in nongrowth environments horizons should not change thickness across fault surfaces. The eye is very sensitive to vertical thickness changes and, with a little practice, can readily detect problems in the time domain. Notice on the time profile in Figure 10-9, within the front limb of the structure between sp 125 and sp 175, that the section between the top of Pierre and the top Permo-Pennsylvanian strata apparently thickens. Could this thickness variation result from higher velocity rocks thrust over lower velocity rocks or, alternatively, from imbricate thrusting? *Time profiles are not geological profiles and are subject to geometric distortions.* In order to remove the geometric distortions, the time section needs to be digitized and depth-corrected on a workstation.

Notice on depth-corrected Figure 10-10a that the thickness variations within and beneath the fault zone are exaggerated in the depth domain. These thickness changes are more pronounced within the "fault zone" (refer to Fig. 10-9) that was interpreted in order to retrodeform the structure. The bed dips in the fault zone exceed 40 deg. An interpretation of the depth-corrected section strongly suggests that the fault zone in Figure 10-9 results from high bed dips that are common to compressional terranes. ***In compressional regimes, high bed dips can result in time sections that dramatically distort structures, and we strongly recommend that all interpretations be analyzed in the depth domain.*** The time section in Figure 10-9 bears little resemblance to the depth section in Figure 10-10a.

Figure 10-10b, which represents a restoration of the depth interpretation in Figure 10-10a, shows regions of area imbalance and contains voids in the undeformed state. The horizons change thickness across the restored faults, particularly in the Pierre (Kp) and Niobrara (Kn) units, which indicates a violation of the bed-thickness conservation rule. On a properly restored thrust fault, the beds will maintain approximately constant thickness across the restored structure. This follows because the sedimentary units were deposited parallel to a gentle regional dip.

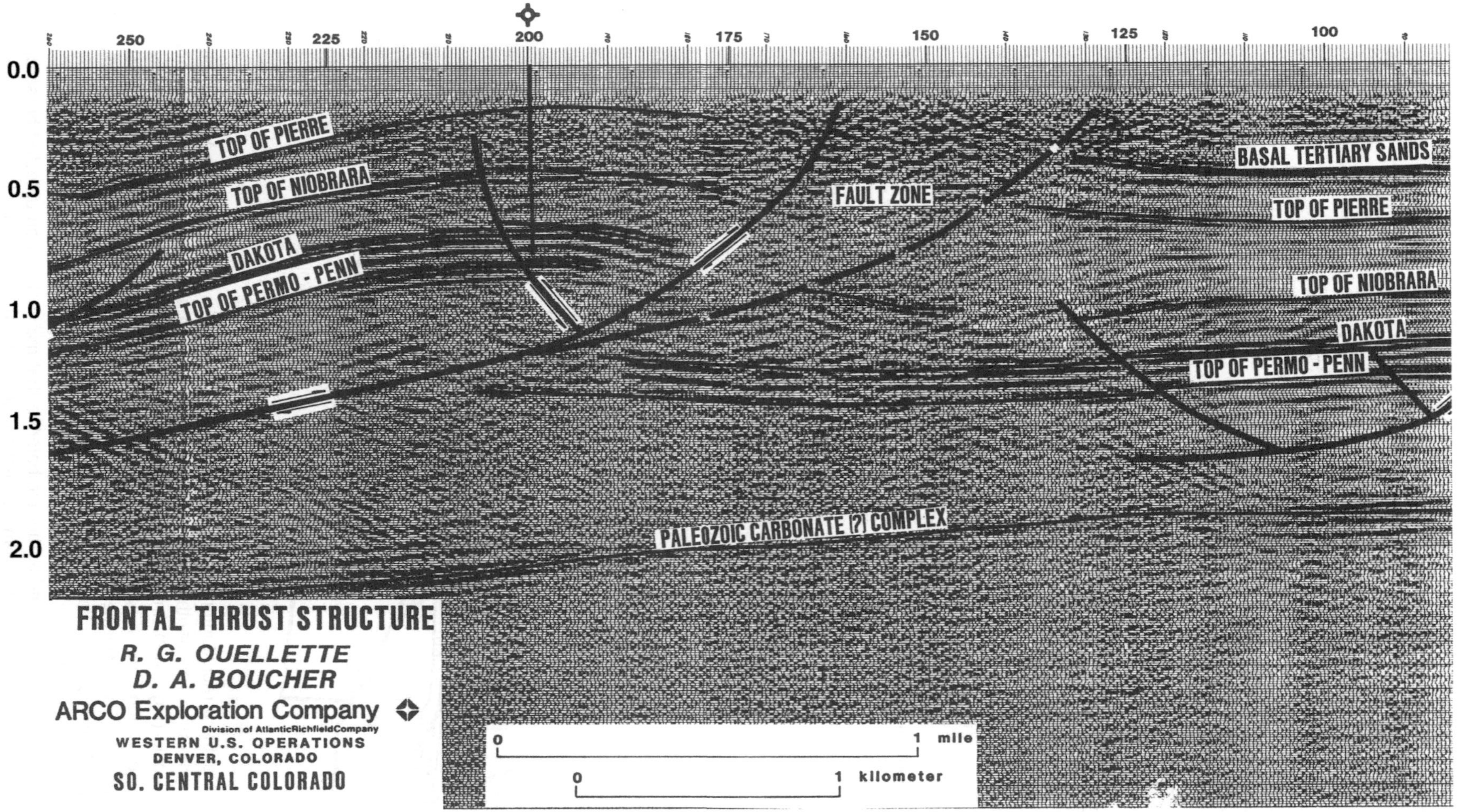

Figure 10-9 Time profile of a fold from the Colorado Rocky Mountains. Beneath the "fault zone," dipping Niobrara reflections over flatter Dakota sandstone reflections may indicate a detachment near the level of the Dakota. (From Bally 1983; AAPG©1983, reprinted by permission of the AAPG whose permission is required for further use.)

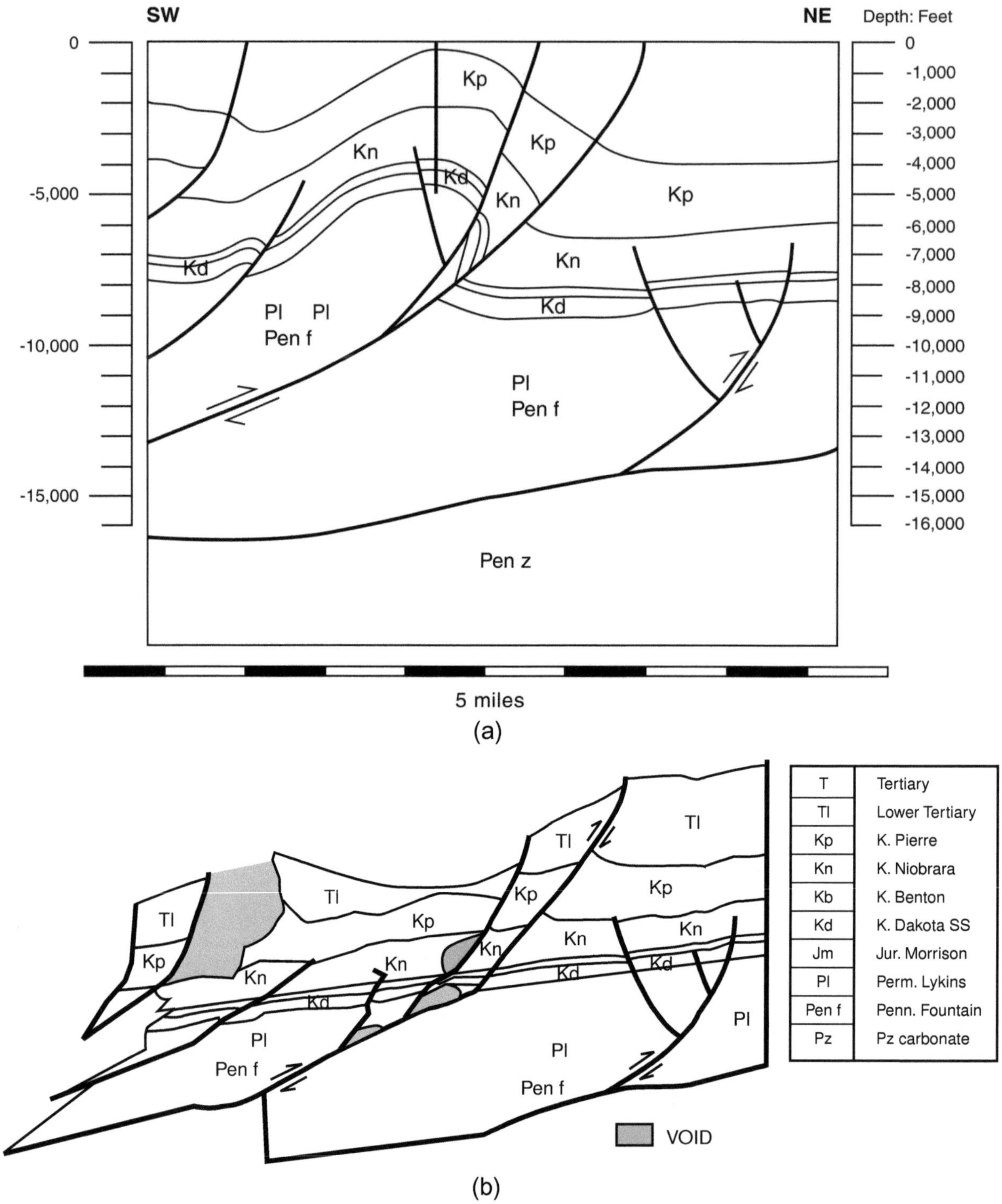

Figure 10-10 (a) Depth-corrected interpretation of time profile shown in Figure 10-9, generated using structural interpretation software. The depth-corrected figure suggests a much tighter fold than the horizontally stretched seismic profile (Fig. 10-9). In the depth domain, the frontal limb fold geometry contains unusual thickness changes above the Dakota sandstone. (b) Fault zone on Figure 10-9 correlates to region of high bed dips in this figure. (b) Retrodeformed Figure 10-10a contains voids and formation thicknesses that do not match or are not uniform across the interpreted faults. This mismatch indicates area and thickness imbalances. (c) Reinterpretation of Figure 10-9 using workstation software and structural principles. Unnatural thickness changes shown in Figure 10-10a indicated an area imbalance that may contain an untested horse block. This figure area-balances and is restorable. (d) Balanced section Figure 10-10c converted to the time domain. This figure can be compared to Figure 10-9 to check for consistency. (Published by permission of R. Bischke.)

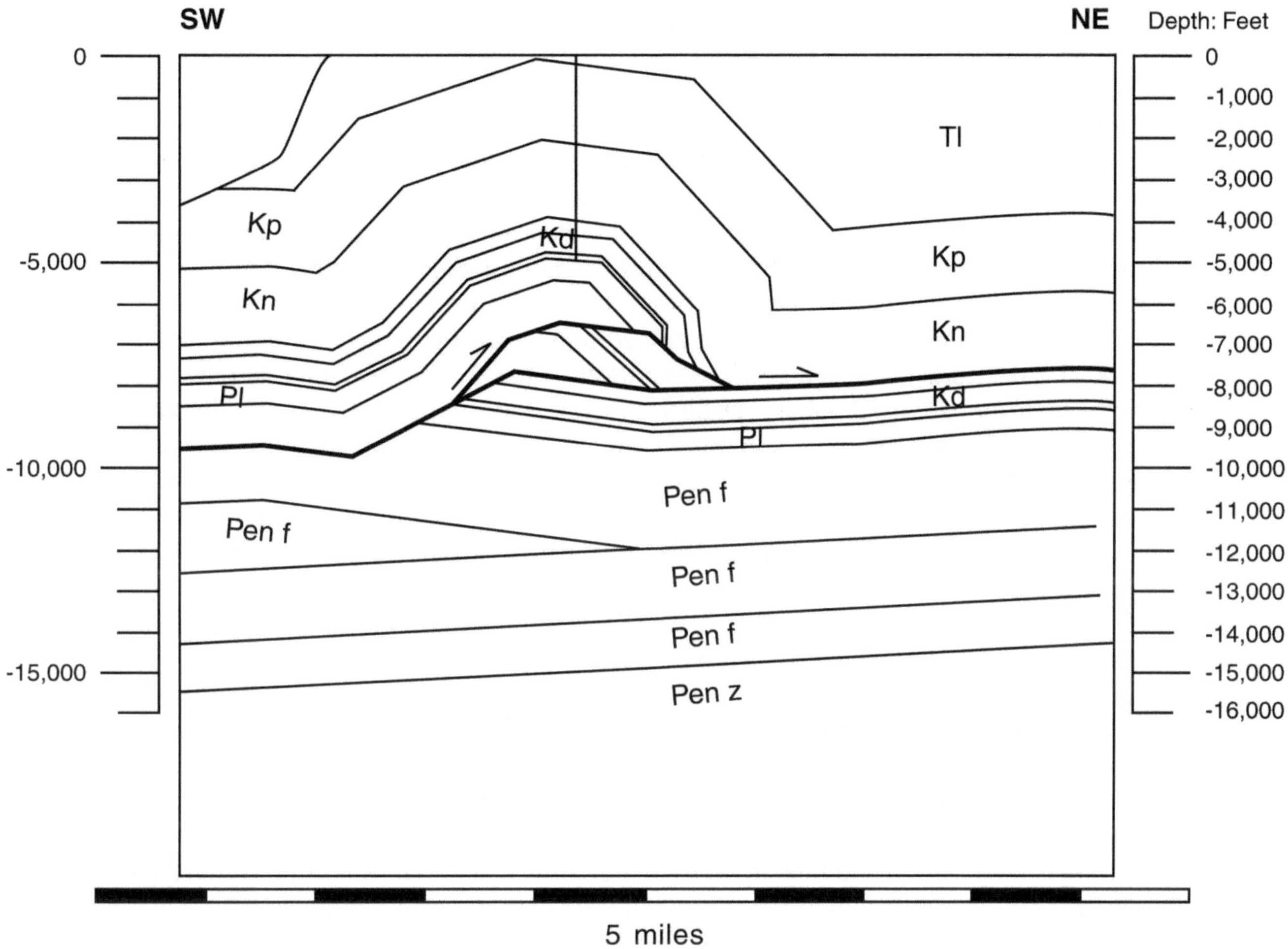

(c)

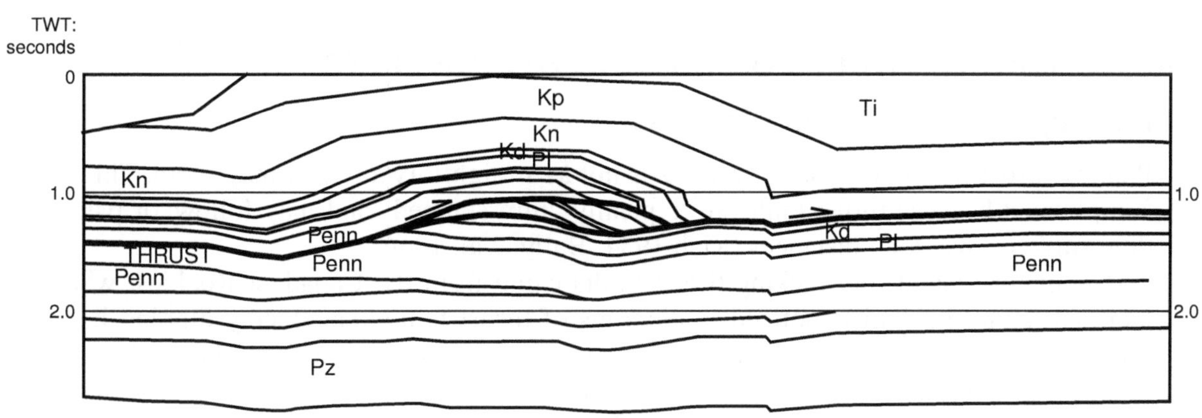

(d)

Figure 10-10 *(continued)*

One of the reasons the structure does not area-balance is that no detachment exists to produce the dipping beds above the "flat" Dakota and top Permo-Pennsylvanian strata (between sp 125 to 175 on Fig. 10-9).

How can we improve the interpretation? Refer to Figure 10-11, a profile from the Canadian Rocky Mountains (Bally et al. 1966). In the Moose Mountain sheet and in the lower central portions of the profile beneath Bow Valley is a structure that resembles the one in Figures 10-9 and 10-10. In Figure 10-11 the thrust fault is observed to ramp beneath the western limb of the fold and flatten beneath the structure's eastern limb. This ramp-flat fault geometry is consistent with high-quality seismic data and is observed in outcrops (Boyer 1986). We use this geometry to reinterpret and balance the structure in Figure 10-9. As mentioned previously, the structure does not balance due to the lack of a detachment located between the level of the dipping Niobrara and the flatter Permo-Pennsylvanian formations. Applying Dahlstrom's rule for picking thrust faults (dipping beds over flatter beds) to the time or depth section, we proceed to balance the structure. The results, shown in Figure 10-10c, require an imbricate fault block, or horse, which is common to fold-thrust belts (Boyer and Elliot 1982). This solution is interesting in that the structure could possess additional hydrocarbons on the level of the repeated Dakota sandstone within the horse. A ramp-flat fault geometry, when applied to Figure 10-9, results in an admissible interpretation, as shown in Figure 10-10c.

Lastly, we convert Figure 10-10c back to the time domain in Figure 10-10d. You can now compare Figure 10-10d to the original time section (Fig. 10-9).

All interpretations of prospects have consequences, which may influence the success of a project and the interpretation of the petroleum system. In Figure 10-9 a possible fault trap exists beneath the fault zone in the upturned beds of the Dakota sandstone. Figure 10-10c indicates that the trapping fault may not exist and that the Dakota strata maintain stratigraphic thickness and may not turn up beneath the proposed fault. This affects prospect risk. The balancing software also predicts the position and thickness of the horizons that are missing from Figure 10-9.

Figure 10-10c predicts that the thrust fault beneath the fold continues toward the northeast to possibly link to other prospects in the petroleum system (Boyer and Elliot 1982). Figure 10-9 suggests that no such link exists in the system, which also affects migration risk.

LOCATING THRUST FAULTS ON DIP AND STRIKE LINES

Picking Thrust Faults on Dip Lines

Picking thrust faults on seismic sections is not as straightforward as it may seem. This subject is complicated because thrust faults are typically "thin skinned" and may follow, or parallel, bedding surfaces over long distances (Rich 1934; Bally et al. 1966).

A major insight into picking thrust faults came from the Canadian Rockies, where petroleum structural geologists noticed in outcrops of thrust faults that steeply dipping beds overlie flatter dipping beds (Bally et al. 1966). In the discussion of pin lines, we show this bed dip discordance or discontinuity in Figure 10-5c and called this method for picking faults Dahlstrom's rule (Dahlstrom 1969). The method works for both dip lines and strike lines.

In Figure 10-12, we can observe a thrust ramping to the left of the fold hinge. The dashed line represents an axial surface, which bisects the limbs of the syncline. The outcrop is perpendicular to the strike of the fault and therefore in the dip direction. To the left of the synclinal axial surface, steeply dipping beds overlie flatter dipping beds, showing a discontinuity and a thrust fault. The thrust in the left part of Figure 10-12 represents a dip ramp, or the area near axial surface BY in Figure 10-37. Alternatively, dipping beds over flat beds can also be observed at the front of the fold, or the region between axial surfaces AX and A′X′ along the upper flat in Figure 10-37c. A similar relationship exists on seismic lines in the strike direction of the fault.

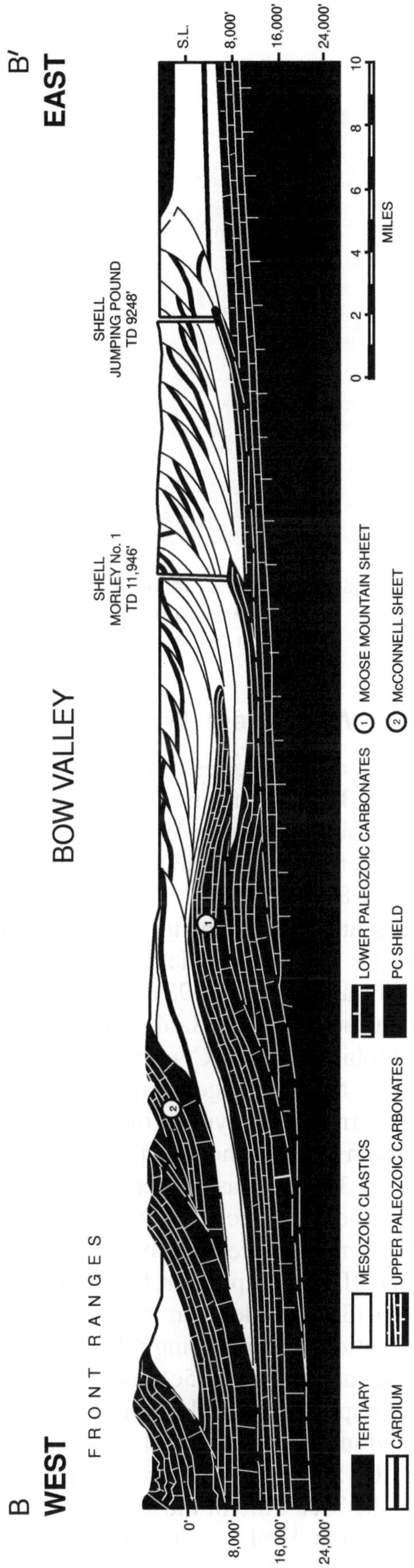

Figure 10-11 Balanced cross section of Canadian Rockies showing ramp-flat fault geometry (From Bally et al. 1966. Published by permission of the Canadian Society of Petroleum Geologists.)

Figure 10-12 Ramp in a thrust fault from the Canadian Rocky Mountains. Dipping beds over flatter beds and the synclinal axial surface define the structural ramp. (From Boyer 1986. Published by permission of the Journal of Structural Geology.)

The 3D Interpretation of the Waha Field, West Texas

The Waha and Coyanosa Fields lie along a series of north-south striking thrust faults that uplift the western edges of the Central Basin Platform, Texas, and along the generally deeper eastern boundaries of the Permian Basin (Shumaker 1992). The Coyanosa Field is 10 mi south of the Waha field and contains distinct overturned (inverted) beds constrained by well control and described by Mitra and Mount (1998).

Figure 10-13 is a 3D north-south striking line through the Lower Ordovician, Devonian, and Mississippian beds of the Waha gas field, perpendicular to the dip of the thrust faults (Shumaker 1992; Hardage et al. 1998; Mitra and Mount 1998). This field is well known not only for its Ellenberger Group production but, according to Hardage et al. (1998), for an east-west cross section originally constructed by Mobil (Railroad Commission of Texas, Case No. 8-83) and later reworked by Kosters et al. (1989) and (Hardage et al. 1998). This cross section shows highly faulted and folded, vertical, overturned beds, typical of fault propagation folds (Fig. 10-14a and b). Twenty wells penetrate the main body of the structure, thus providing adequate data for well log correlations (Fig. 10-16). This 3D line was chosen because it clearly images a lateral ramp structure on the northern flank of the Waha Field. Another lateral ramp, from the El Furrial Field, North Monagas, Venezuela, is shown in Figure 10-51b.

On the northern flank of the 3D line, dipping reflections from 1750 ms to 2750 ms overlie flat dips below 2750 ms (Fig. 10-13). Provided that the dips exceed about 5 deg (see discussion at the end of this section), this relationship of dipping reflections over flat reflections is typical of deformed horizons above reverse faults (Fig. 10-5c) (Dahlstrom 1970). Structural geophysicists use these discontinuous relationships to pick and accurately map thrust faults. In this case, the base of the faulting is picked and interpreted at about 2750 ms. This interpretation is consistent with Kosters et al. (1989) (Fig. 10-15) cross section, which terminates near Well No. 75 (see Fig. 10-16). The 3D strike line is on the back limb of the structure and ties to Well No. 74 on Kosters et al. (1989) cross section (Fig. 10-16). Well No. 78, located 4000 ft (1200 m) south of the Well No. 72, contains more than 2000 ft of overturned Ordovician Simpson section (Fig. 10-15).

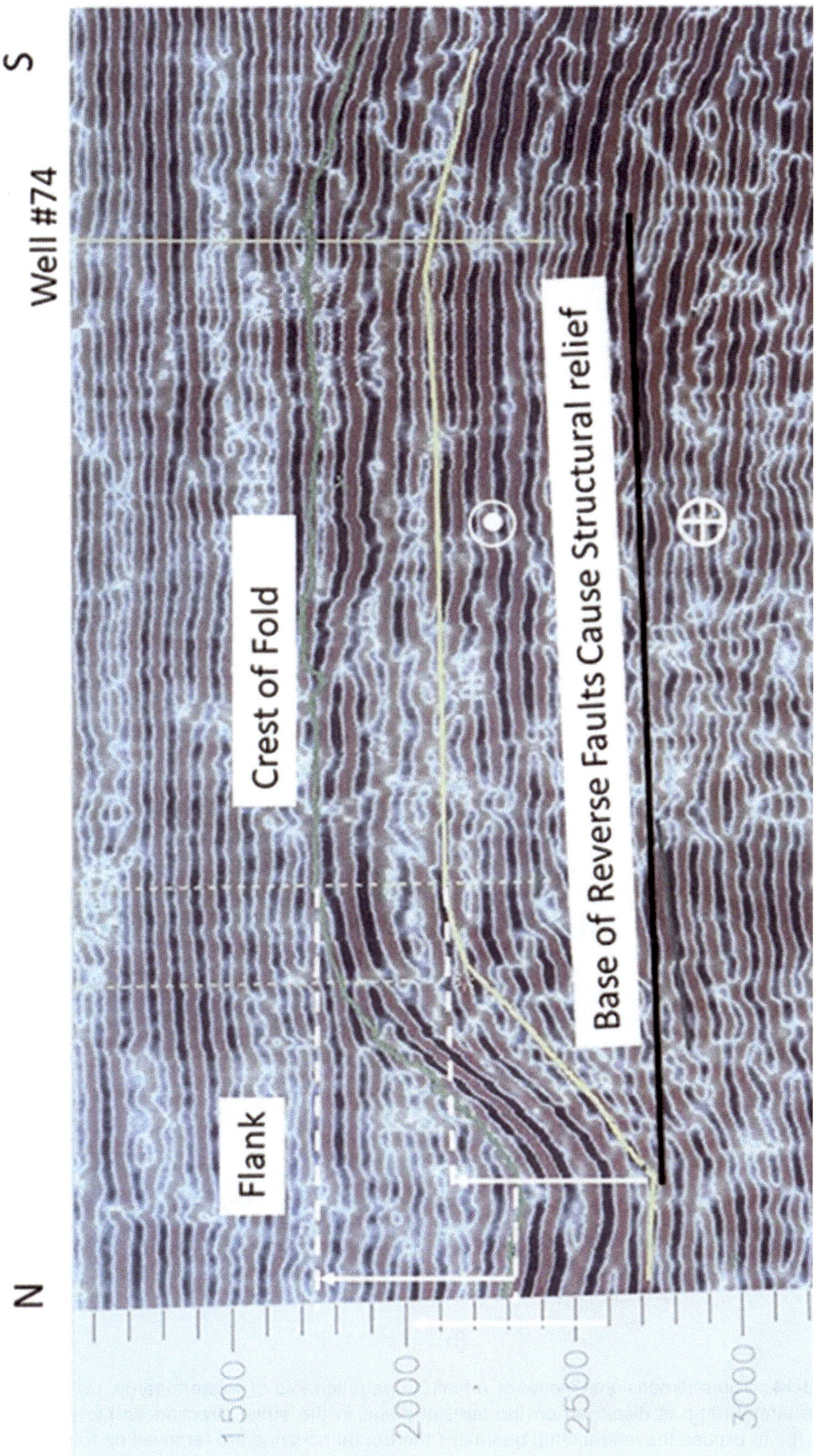

Figure 10-13 Three-dimensional strike line imaging a lateral ramp across the Waha Field, Permian Basin, Texas. This line is tied to the cross section shown in Figure 10-15. (Published by permission of the Bureau of Economic Geology, University of Texas at Austin; edited by Hardage et al. 1998.)

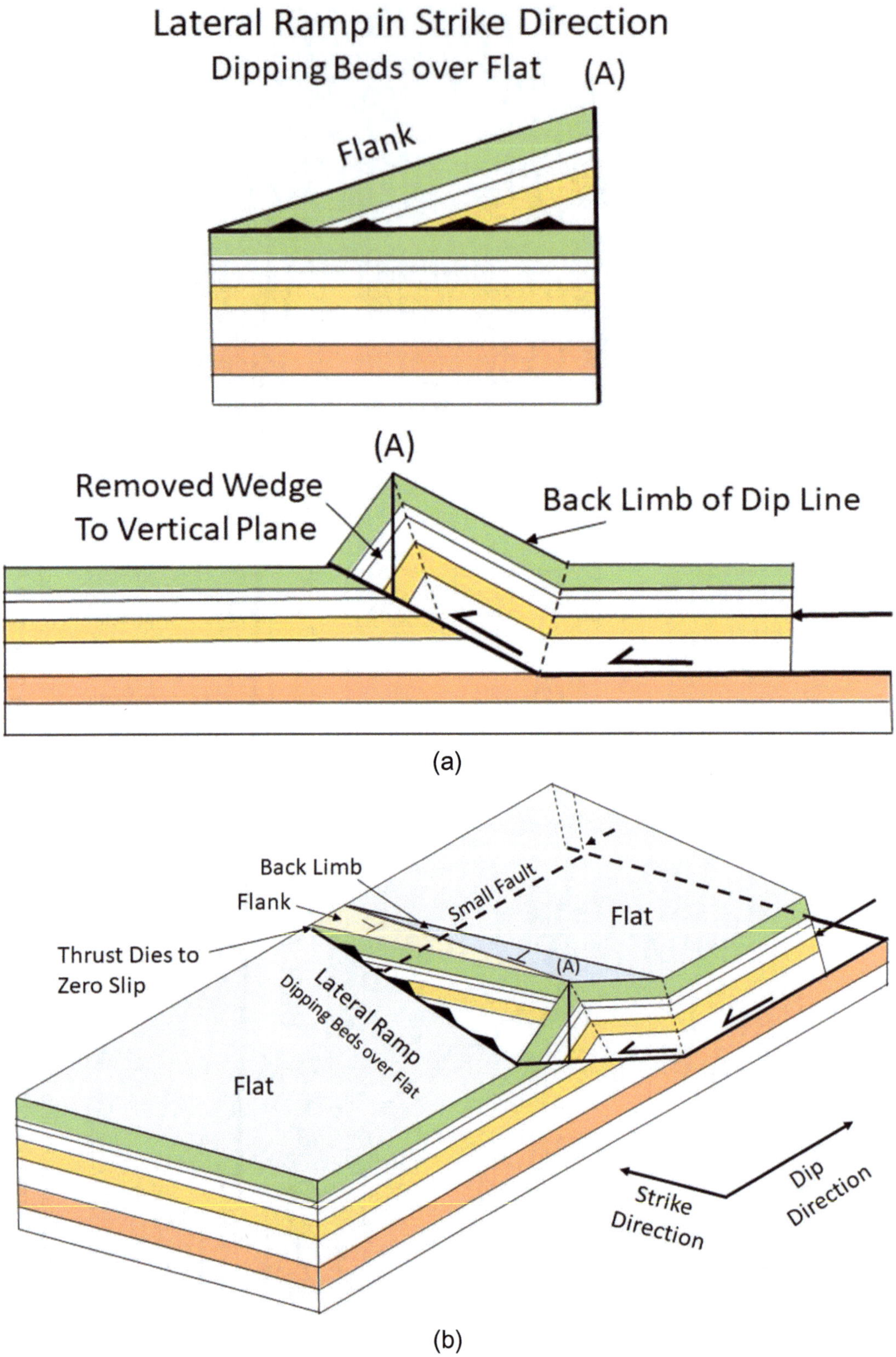

Figure 10-14 Three-dimensional model of a fault propagation fold of a lateral ramp. (a) On this figure, the lateral ramp is displayed on the vertical plane in the strike direction on the flank of structure. (b) To expose the lateral ramp geometry, the frontal horizons are removed up to the vertical plane, exposing the thrust fault and the horizons dipping onto the thrust fault. (Published by permission of R. Bischke.)

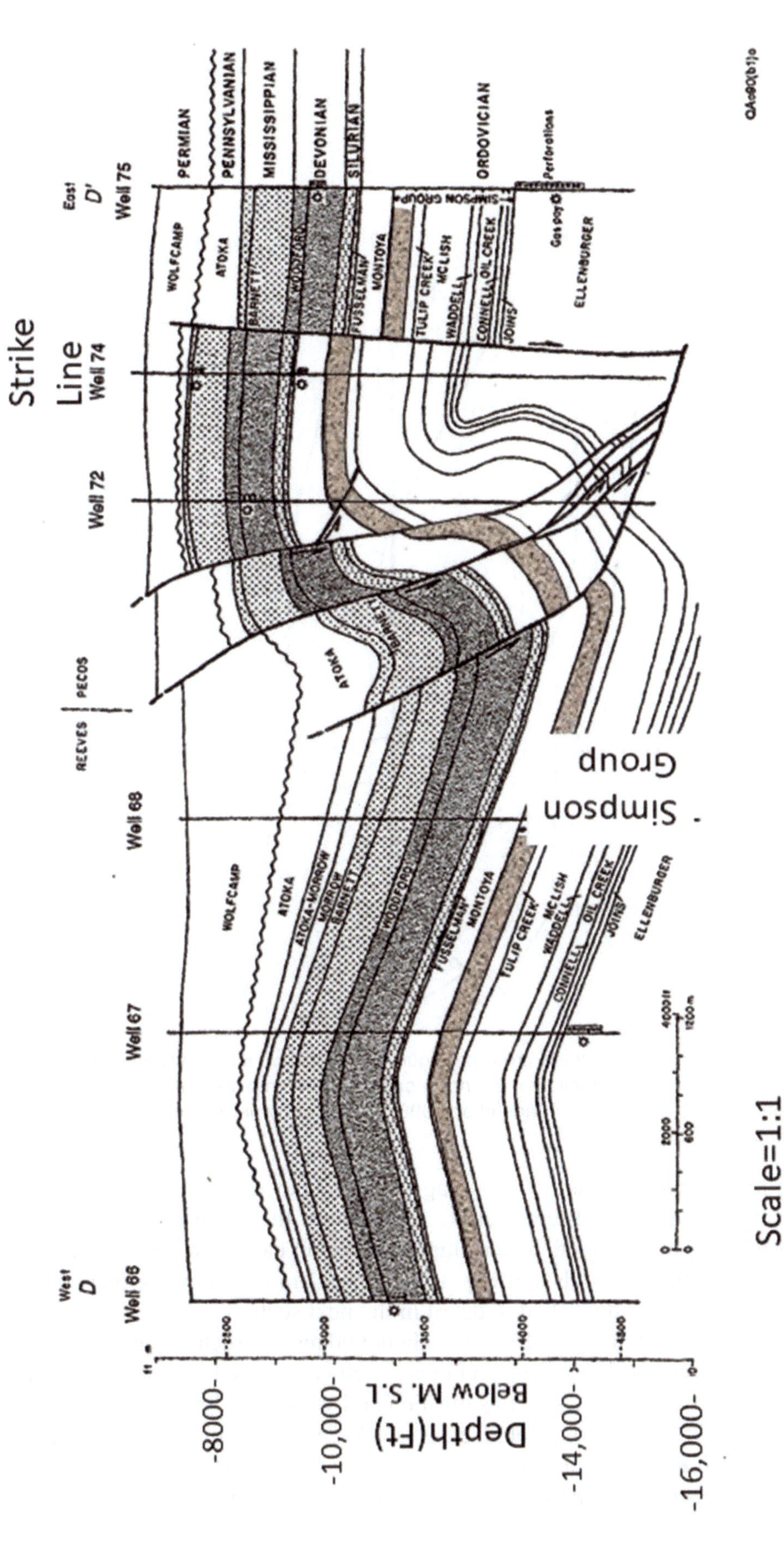

Figure 10-15 East-west cross section of the Waha Field, Permian Basin Texas, in the vicinity of Wells No. 72 and No. 74. This complex cross section is based primarily on well control; the Waha Field contains several large reverse faults and a 3000 ft (914 m) long section of near vertical beds in Well No. 72. (Modified from Kosters et al. 1989. Published by permission of the Bureau of Economic Geology, University of Texas at Austin; edited by Hardage et al. 1998.)

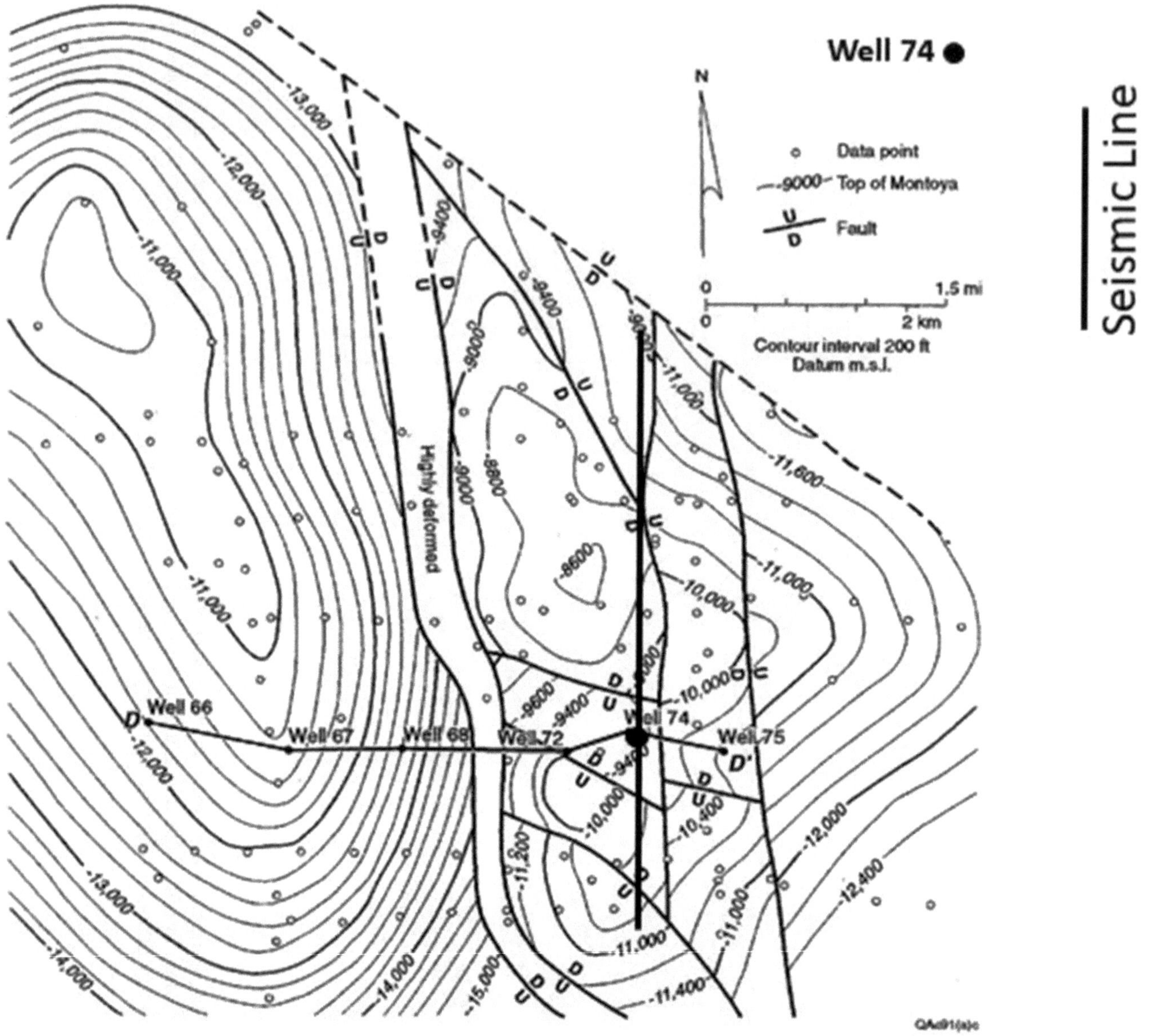

Figure 10-16 Structural-depth map showing wells crossing Figure 10-15, which is the east-west cross section across the Waha Field, Permian Basin, Texas. The direction of the north-south lateral ramp shown on seismic line (Fig. 10-13) is drawn on the figure through Well No. 74. (From Hardage et al. 1998. Published by permission of the Bureau of Economic Geology, University of Texas at Austin.)

The strike line shown in Figure 10-13 images what is defined by industry as a *lateral ramp*. In other words, the north-south directed strike line images the north dipping reflections on the flank of the structure. These reflections terminate against an east-west dipping thrust ramp (consult section Fault Propagation Folds and Fig. 10-43). A 3D model showing these relationships is shown in Figure 10-14 and is described in detail in the next section.

The lateral ramp imaged in Figure 10-13 has about 600 ms of structural relief. Seventeen wells in or near this structure have checkshot data, and thus a well-constrained velocity/depth function exists for the area (Hardage et al. 1998). Using this function and Well No. 72 between 1750 ms to 2350 ms yields about 4500 ft (1370 m) of structural relief. This value of 4500 ft of structural relief on the lateral ramp is consistent with Kosters et al.'s (1989) cross section (Fig. 10-15), as the well log data suggest that the thrust fault deepens as it dips to the east beneath the crest of structure, which is flat.

Furthermore, on the crest of structure is Well No. 72 that encounters five reverse faults (Fig. 10-15). Kosters et al.'s (1989) interpretation of this well log data is that the logs contain 4300 ft (1310 m) of log offset (repeated section and vertical and overturned beds) (Hardage et al. 1998). The reverse faults located at 1800 to 2100 ms are small and repeat 140 to 280 ft of section. The horizontal black-colored fault interpreted on Figure 10-13 is the largest fault with about 850 ft (260 m) of repeated section and is located near the base the of the structure in Well No. 72. Also, the McLish formation in the well has 3000 ft (915 m) of near vertical to overturned beds (Fig. 10-15).

Three-Dimensional Models of Lateral Ramps

Three-Dimensional Fault Propagation Fold Model. Figure 10-13 images a 2D seismic line. Three-dimensional geometric images of lateral ramps maybe difficult to visualize but are important for correctly defining thrust fault surfaces and constructing accurate maps. The reader may want to study the 3D models of lateral ramps shown in the color Figures 10-14 and 10-17. These figures do not include the bed on the top or crest of structure. We discuss a 3D fault propagation model before studying a 3D fault bend fold model.

Figure 10-14a is a modeled fault propagation fold displayed in two dimensions but missing the crestal bed, which is flat at the top of structure. Figure 14a (top) is a *modeled lateral ramp* and is similar to the left-hand side of Figure 10-13 in that Figure 10-14a (top) trends in the strike direction. Notice that the laterally dipping beds terminate on a horizontal trust fault, as do the less coherent beds, just below the yellow line in Figure 10-13 (also consult Fig. 11-51b for a coherent example). Two-dimensional Figure 10-14a (bottom) trends in the dip direction and is similar to Figure 10-15.

Figure 10-14b is a fault-propagation fold model showing the *flank, thrust fault, and horizon dip relationships in three dimensions*. The model is bound on the right-hand side by a 2D profile of a fault-propagation fold (see Fig. 10-14a [bottom]). The crest of structure occurs on the horizon above the green horizon (consult Fig. 10-43). The model shown in Figure 10-14b does not contain the beds above the crest, as the beds above the crest would not be faulted, and the thrust would be buried by the shallower horizons. This buried thrust fault would be called a *blind* fault. The small fault located at depth and near the termination of the flank (on the far-left side of the structure) has less slip on its surface and is blind, or buried by the green and clear horizons. The crest of structure would be to the right of the 2D fault propagation fold shown in Figure 10-14b.

Next, we discuss the 3D geometric properties of the lateral ramp. On Figure 10-14b, the *vertical plane* trends across the flank of the structure, displaying the green to orange horizons and the lateral ramp thrust geometry (compare Fig. 10-14a and 10-14b [top]). The green to orange horizons to the front of the vertical plane have been removed, thus displaying a horizontal thrust fault on the vertical plane and the dipping flank horizons. In two-dimensions, this is the upper figure in Figure 10-14a. This vertical plane, which contains the green to orange horizons, is on the flank of the structure and is in the strike direction.

Notice that the horizons intersect and terminate onto the thrust fault. This vertical plane is perpendicular to the dip direction. The termination is most obvious on the thin clear and orange horizons. As these lateral or flank horizons terminate at the thrust, the dipping beds create a discontinuity in the horizons. *This discontinuity occurs where dipping beds overlie a flat horizon. By definition, this displacement discontinuity defines the lateral fault geometry, and industry has called these geometric relationships* **Dahlstrom's rule** (Fig. 10-5c, Dahlstrom 1970).

Three-Dimensional Fault Bend Fold Model. The 3D model for lateral ramps on fault bend folds is not as complicated as the 3D model for fault propagation folds (Fig. 10-17a and b). As with the fault propagation fold model (Fig. 10-14b), the right-hand side of the model contains a 2D model of the structure, but in this case, a 2D fault bend fold (see Fig. 10-17b [bottom]).

The horizons to the left of the frontal limb of the 2D fault bend fold are removed (i.e., the horizons to the left of the frontal, dipping 2D green horizon in Fig. 10-17a). The frontal and dipping green horizon above the upper flat terminates onto the upper flat, creating a discontinuity

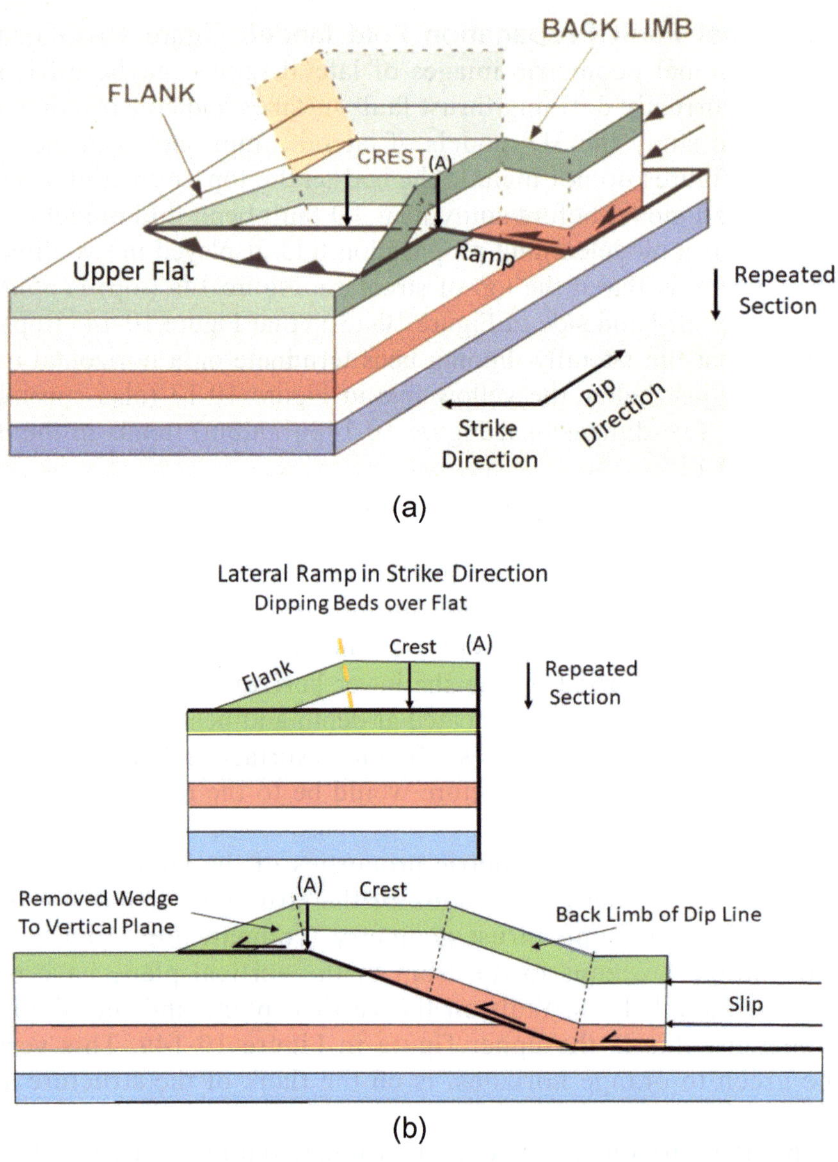

Figure 10-17 (a) Three-dimensional model of a fault bend fold with the frontal limb removed. The barbed line on the footwall, upper flat, depicts the contact of the thrust fault with the removed frontal limb. The repeated section from the crest of structure in the hanging wall to the footwall is the vertical arrow, which points to the thrust fault at the base of the arrow. (b) Two-dimensional model of a fault bend fold moving up a ramp in the dip and strike direction. Strike direction is the top figure, and dip is the bottom figure. (Published by permission of R. Bischke.)

that defines the thrust (Fig. 10-17a). At the top of upper flat, the horizontal fault follows the upper flat to the top of the thrust ramp. The horizons above the upper flat were removed to expose the uppermost portion of the ramp, thus exposing the thrust fault at the top of the ramp. The exposed portion of the thrust fault is the horizontal line in Figure 10-17a, which terminates on the upper flat. Also shown on the upper flat is the frontal position of the thrust, which dies out at the left-hand side or flank of the structure. The slip on the thrust increases to the right, with the *slip reaching a maximum* at the dipping green horizon (the curved hashed line).

The lateral ramp is defined by the dipping green horizon, which forms the flank of the structure in Figure 10-17a and in 2D Figure 10-17b (top). Its significance is that the dipping lateral horizon forms a displacement discontinuity with the upper flat and thus by definition forms a fault. To reiterate, this *discontinuity terminates the dipping horizons onto flat horizons, defining Dahlstrom's rule, that thrusts exist where dipping beds terminate against flatter dipping beds* (Fig. 10-5c, Dahlstrom 1970).

Distinguishing Structural from Stratigraphic Dips

A question may arise as to how to distinguish stratigraphic dips from structural dips. We first refer to Rich (1951), who found that clinoforms in the steeply dipping portions of deltas rarely exceed 5 deg. Second, clinoforms reflect downlap or toplap. Thus, if the seismic reflections are folded and dip at angles exceeding about 5 deg, then the dips are most likely structural and not stratigraphic. Oceanic bed dips tend to be small, and the continental slope tends to dip at less than two degrees. Also, remember that interpreters construct horizontally compressed (vertically extended) seismic sections in order to image subtle changes in dip. Thus, it is always wise to construct several seismic lines at a scale of about one to one from regionally generated velocity functions to view the structure on a less distorted scale.

We also recommend that you use the workstation to scan the strike direction looking for lateral ramps along the flanks of folds. If such a strike ramp is located in the data, pick the fault on several strike lines and construct an initial fault surface map. To complete the fault interpretation and fault surface map, tie the fault surface to the dip lines.

CROSS-SECTION CONSISTENCY

So far, we have generalized the concept of brittle or frictional deformation to a 2D cross section. As deformation is three-dimensional, the brittle deformation interpreted on one cross section imposes constraints on interpretation of the adjoining cross sections, such that the interpreted folds or faults must not terminate abruptly. However, the deformation can be dissipated gradually. In other words, fault slip must be consistent, although not necessarily conserved, from cross section to cross section. Also, the slip can decrease to zero as the result of deformation in the cores of folds.

For example, if a cross section of a complex structure exhibits three thrust faults with a total of 3 mi of slip, then it is very likely that a nearby cross section will also contain three thrust faults of similar shape and form that also contain about 3 mi of slip. If these three thrust faults radically change position and/or shape, then some intervening transverse structure must exist to accommodate the deformation. Such intervening structures are called **transfer structures**, and these structures exist in compressional (Dahlstrom 1969) as well as tensile extensional environments (Gibbs 1984). Transfer structures often occur as tear faults, or cross faults, which form at high angles to the major structural trend. Furthermore, these transverse structures are often responsible for changes in the trends and shapes of structures from cross section to cross section. Figure 10-18 illustrates a transfer by lateral shear from one fault bend fold to another. In Figure 10-18a, the displacement on Fault 1 is compensated by displacement on Fault 2 (see left side of diagram, Fig. 10-18a). The sum of the displacements on Fault 1 and Fault 2 remain constant; thus, as the slip on Fault 1 decreases, the slip on Fault 2 increases. The amplitude of the folds above the faults also change in a like manner. On profile F, the slip on Fault 1 is equal to the slip on Fault 2, and the folds that form above the two faults have the same amplitude. The resulting structures caused by the lateral shear are shown in map view in Fig. 10-18b. The result is that the fold on Fault 1 plunges to the south and is replaced by the fold on Fault 2, which plunges to the north. This slip transfer between folds is very common in fold-thrust belts (Fig. 10-19).

Therefore, we see that small changes are permissible from cross section to cross section, but how much change is possible? Elliott (1976) was the first geoscientist to attempt an answer to this question with the **bow-and-arrow rule** (Fig. 10-20). Additional studies suggest a linear scaling between displacement and fault length (Davis et al. 2005; Bergen and Shaw 2010). Elliot's rule states that the amount of displacement can vary along a fault zone, but at an amount equal to 7% to 12% of its strike length. For example, suppose you mapped deformation along a large thrust fault zone that has a total length of 10 mi. From the bow-and-arrow rule, one would predict that the *maximum dip-slip* motion on the fault would be on the order of 0.7 mi to 1.2 mi. Next, assume that the amount of displacement along another fault is known to *increase to a maximum* along a 10-mi portion of the fault zone. We can now predict not only that the fault is at least 20 mi long but also that there are at least 1.4 mi to 2.4 mi of dip-slip motion on this fault. The bow-and-arrow rule is based on scaling laws, and it follows that laterally restricted faults have small displacements, whereas only laterally extensive or long faults have large displacements.

Bergen and Shaw (2010) used high-quality seismic data to map fault surfaces in order to predict a fault length (L) versus a maximum displacements (D_{max}). Their work suggests that a near linear relationship exists in the form of a power law relationship, where

$$D_{max} = kL^n$$

The constant $k = 0.13$

$$n = 0.87$$

from three different data sets with an $R^2 = 0.89$. For $n = 1.0$, the equation simplifies to $D_{max}/L = k$, and thus the Bergen-Shaw equation is near linear.

Bergen and Shaw carefully studied their displacement-length profiles and from this hard data came to a conclusion that many geoscientists who *map faults* would support. They concluded that if two growing faults are roughly on strike with each other, the two faults may merge and that "folding and fault linkage [are] the probable causes of much of the variation in displacement-length scaling ratios observed in our data sets" (Bergen and Shaw 2010). Our fault surface mapping results based on numerous field and exploration studies supports Bergen and Shaw's conclusions that "fault interactions play a major role in the ultimate displacement profile geometries for thrust faults." Furthermore, we believe that any scaling equation not based on tied fault surface mapping and the fault linkage concept is questionable (Chapter 9, section Fault Interpretation Strategies).

Elliott (1976) developed the bow-and-arrow rule for thrust faults, and a similar relationship may exist for normal faults, particularly for faults in excess of 10 mi in length (Morley 1999). However, no consensus exists for normal faults, as different studies based on different approaches have generated different conclusions.

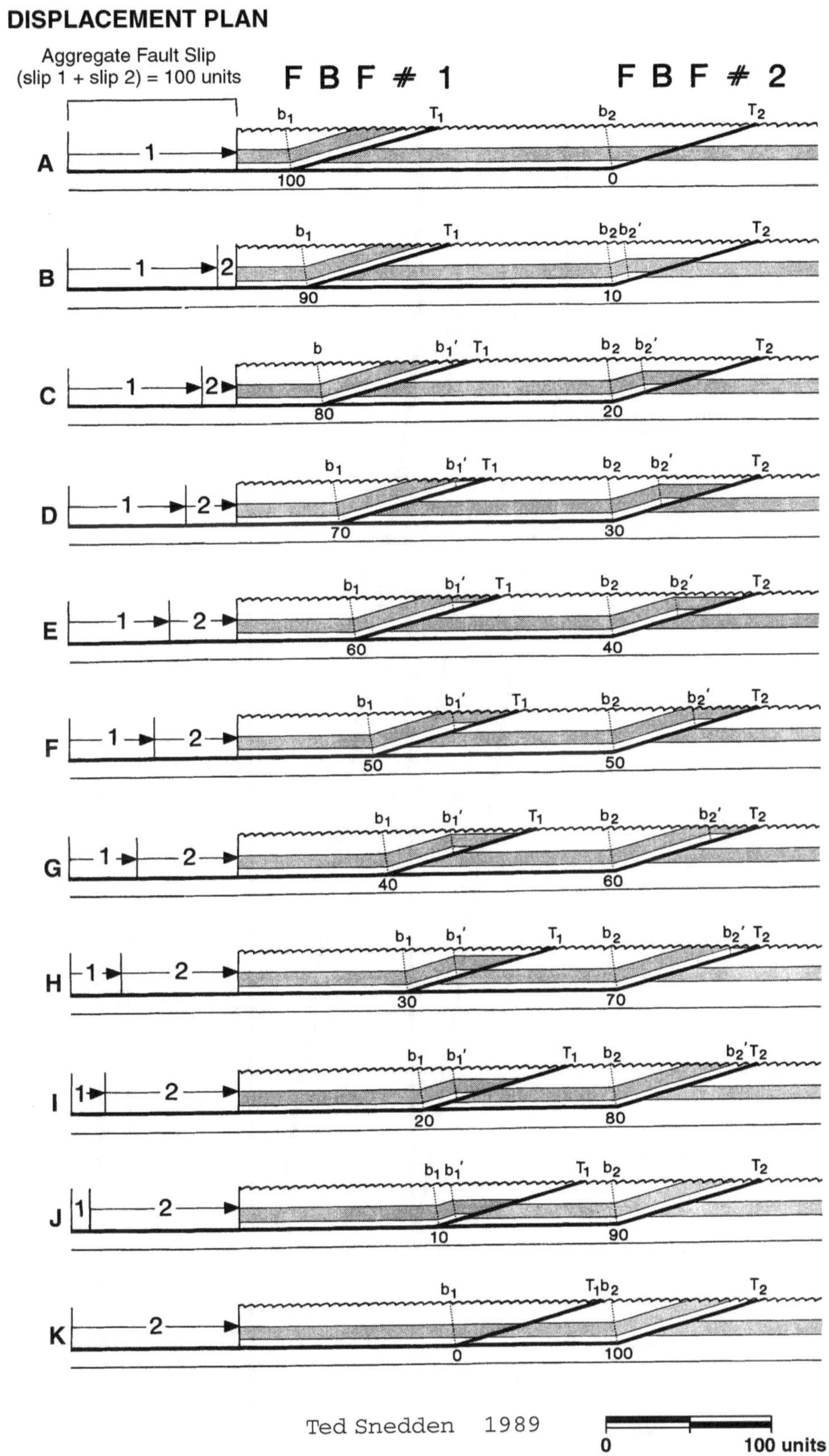

(a)

Figure 10-18 (a) A cross section of a transfer zone from one fault bend fold to another. (b) A top to down view of (a), similar to the aerial radar aperture image of the Appalachian fold-thrust belt shown in Figure 10-19. (Published by permission of Ted Snedden.)

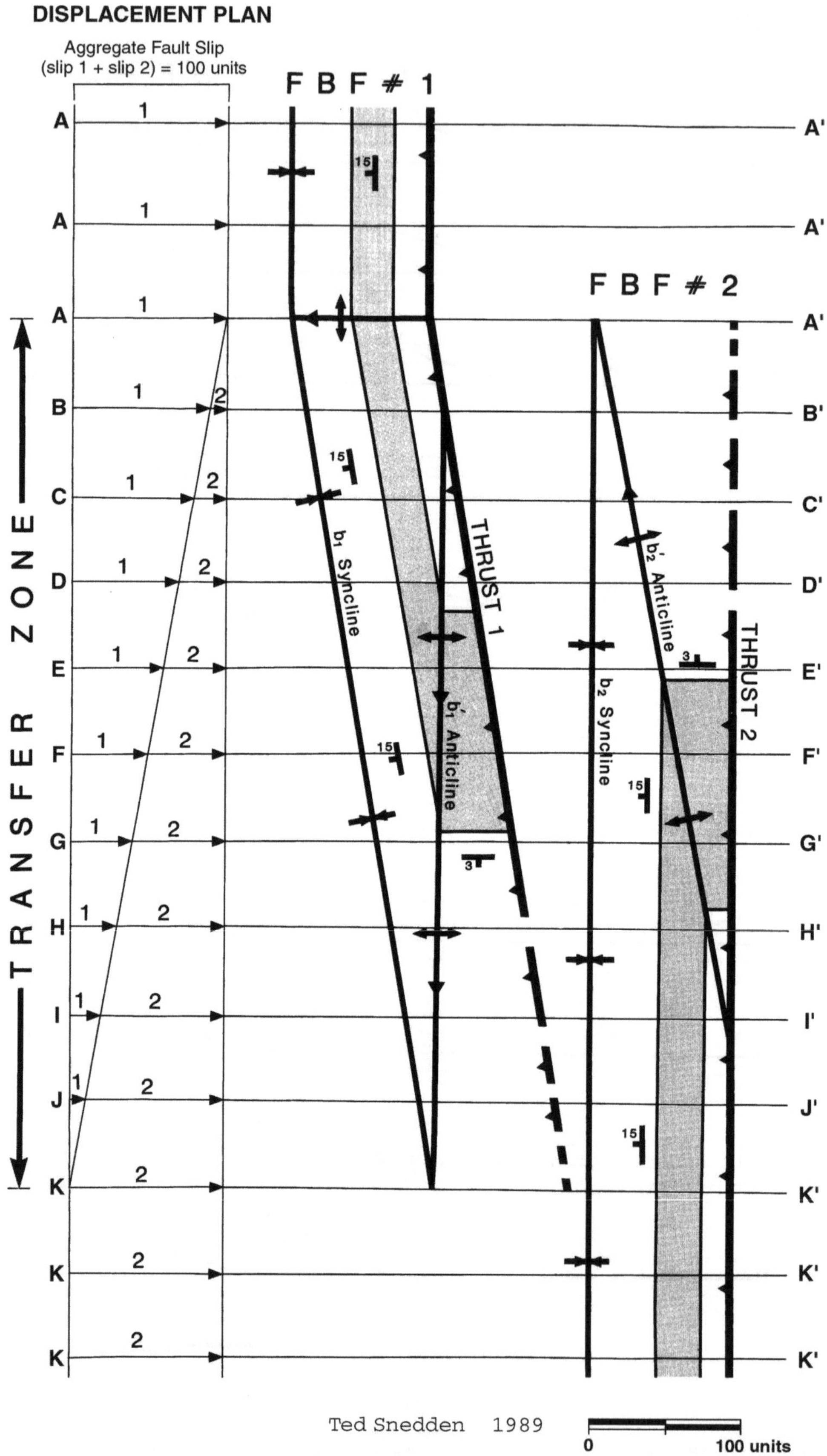

(b)

Figure 10-18 *(continued)*

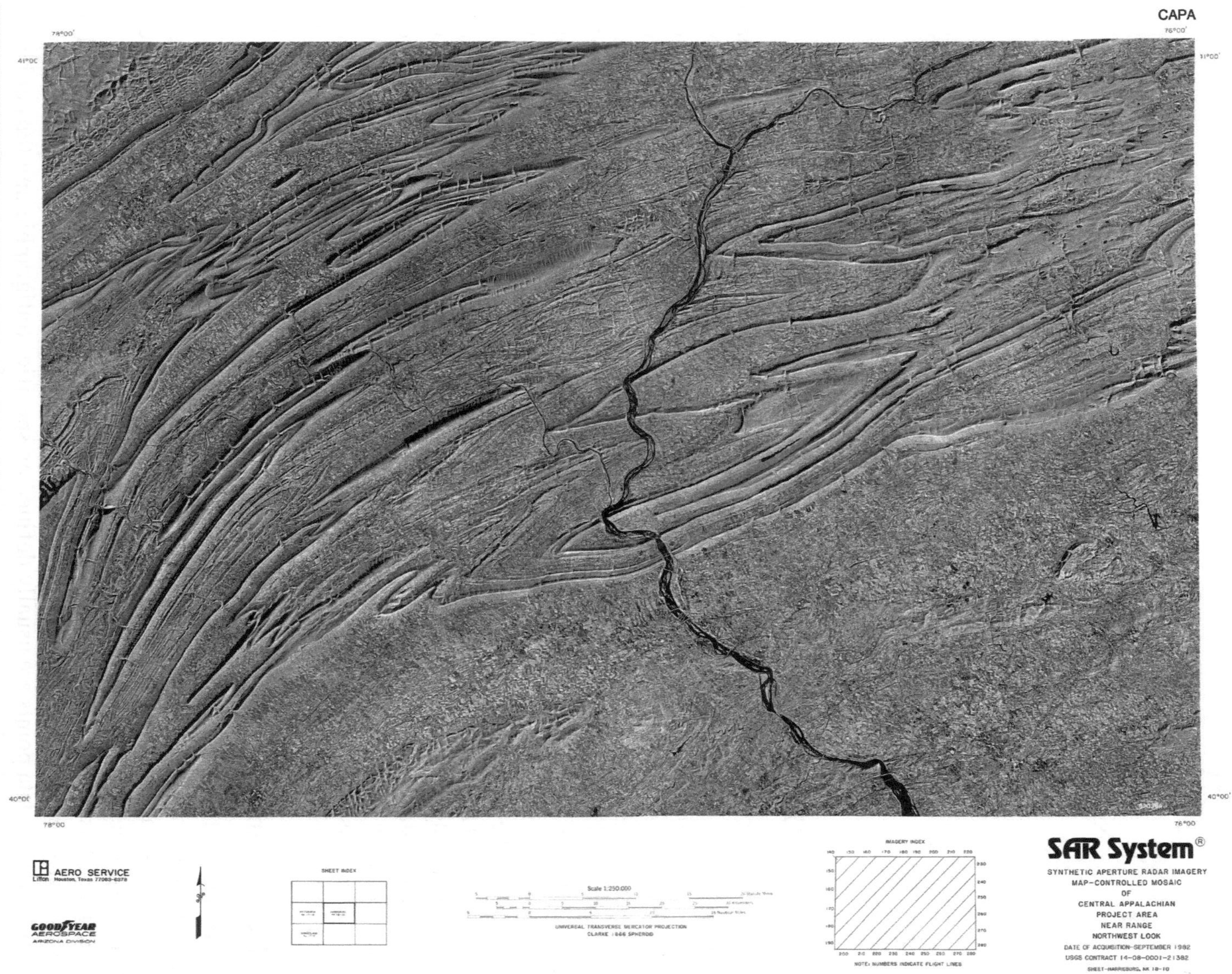

Figure 10-19 Radar aperture image of Appalachian fold-thrust belt near Harrisburg, Pennsylvania, showing an echelon arrangement of plunging anticlines. This displacement transfer is common to fold-thrust belts. Although repeated section exists in the well logs from this area, notice the near absence of surface faulting. The absence of surface faulting is common to portions of many fold-thrust belts, where the deep thrusts occur as blind or as bedding plane thrust faults. (Published by permission of the United States Geological Survey.)

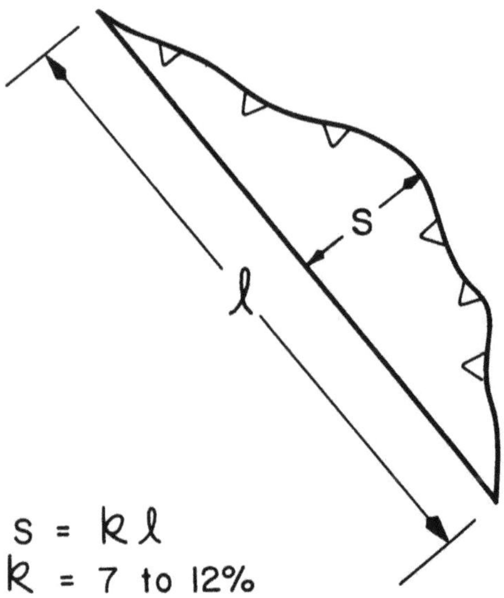

Figure 10-20 Bow-and-arrow rule. Slip perpendicular to fault strike is approximately 10% of the fault length. (Modified after Elliott 1976. Published by permission of the Royal Society of London.)

CROSS-SECTION CONSTRUCTION

Extrapolation of dip data to depth is a critical aspect of interpreting structures that may contain hydrocarbons and of accurately predicting wellbore results. The data can be in the form of outcrop dips, stratigraphic unit tops and bases taken from outcrop or well logs, dipmeter data, and depth-corrected seismic data. There are presently two methods available for extrapolating dip data to depth: the Busk method of segmented circular arcs (Busk 1929) and the kink method, which stresses the long planar limbs exhibited by most folds (Faill 1969, 1973; Laubscher 1977; Suppe 1985; Boyer 1986). Both methods assume that the folding is parallel; that is, stratigraphic unit thickness remains constant (in the absence of more detailed information). The Busk or the kink method can be used to extrapolate any type of dip data. *It is important, however, to be consistent in the use of the data.* For example, the top of a stratigraphic unit is projected to the top of an adjacent unit only if the units being mapped do not change thickness, which is commonly the case over short distances. A dipmeter recording within a stratigraphic unit is not projected to a dipmeter reading in an adjacent well unless these recordings are on the same stratigraphic level. In other words, it is important to understand that you are projecting *time-stratigraphic surfaces* across the structure.

Busk Method Approximation

The **Busk method** (Busk 1929) assumes that the folds are parallel (constant-thickness of stratigraphic units) and that they are concentric; that is, the folds consist of segments of circular arcs. These arc segments are used to project data to depth. Normally, dip data measured from surface outcrops, well logs, or seismic sections will not lie along the plane of cross section. Thus, the data must be projected to the plane using the methods discussed in Chapter 6. Let us assume for simplicity that the data, measured from outcrop, are shown in Figure 10-21a. The data points are usually defined on specific stratigraphic unit tops or bases. Normals (lines perpendicular to dip) are drawn downward from the position of the dip measurement data. These normals intersect at a point that represents a radius of curvature for an arc (point O in Fig. 10-21b), which is used to project the stratigraphic data in the area between the two data points A and B. A compass centered at point O is extended so that it has a radius OA, and then an arc is constructed from point A to line D (Fig. 10-21c). This procedure is then repeated for point B, using radius OB. The results of this exercise are two concentric arc segments, AE and FB, which define a curved layer AE-FB, of constant thickness AF, or EB. If another data point G is introduced (Fig. 10-21d), the normal to this adjoining data point will intersect line segment OB at a different location, point O, and now several different radii (O′B, O′G, OI) are used to complete the stratigraphic extrapolation. In Figure 10-21e, a well with dipmeter data is added, and more normals and arcs are drawn to depict a more complete fold.

The method can be visualized as consisting of several adjoining regions, or domains, in which *the curvature of the beds is constant*, and at the intersection of these domains, the curvature of the beds changes. The Busk method is therefore a *curved dip domain method.* It suffers from an inability to retrodeform easily and to correctly project the front limb of a fold into the adjoining syncline.

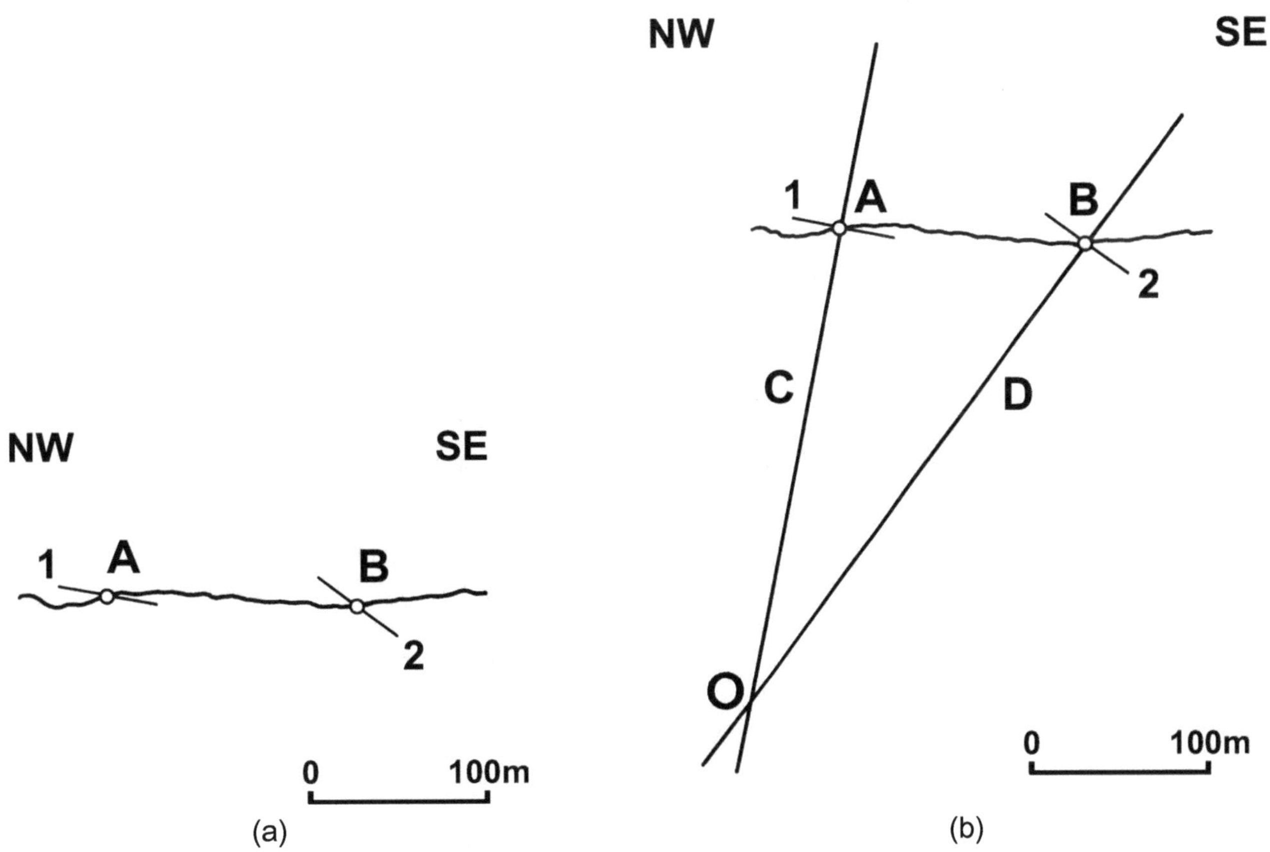

Figure 10-21 (a)–(e) Busk method approximation. The strata are projected to depth along segments of circular arcs. (Modified from Marshak and Mitra 1988.)

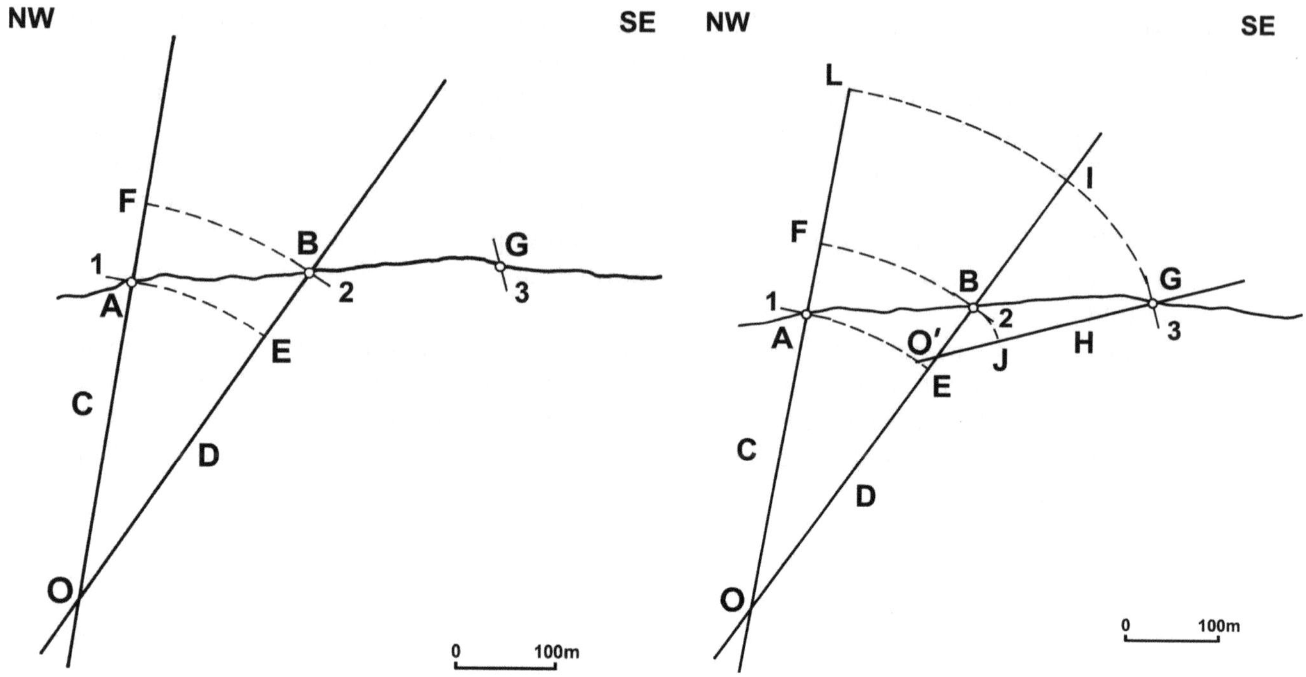

(c)

(d)

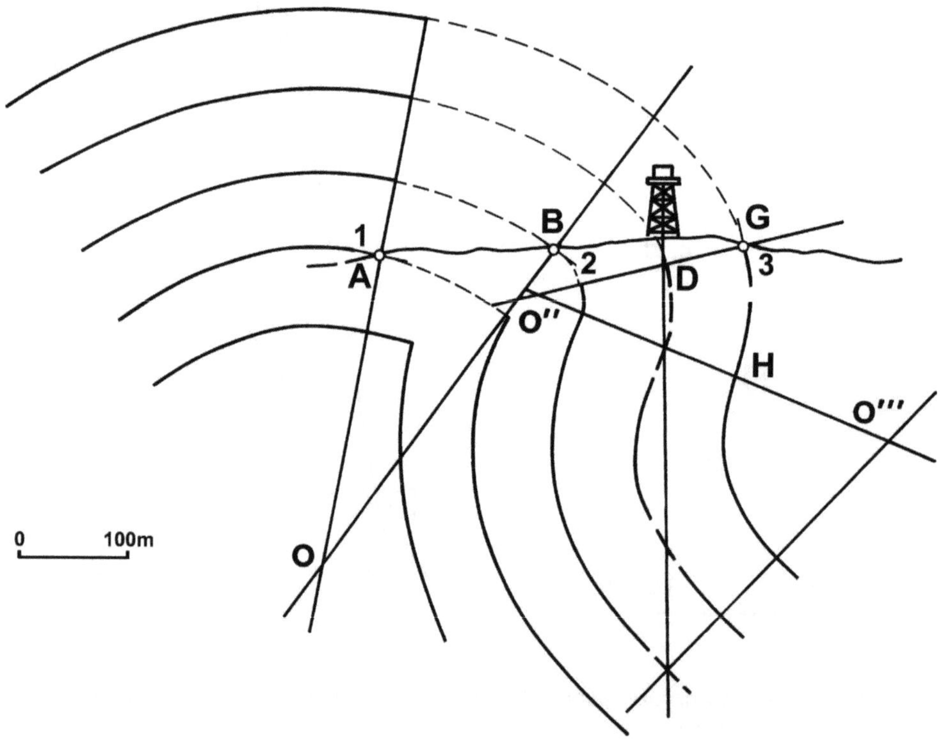

(e)

Figure 10-21 *(continued)*

Kink Method Approximation

The next method that has proven extremely useful for extrapolating data to depth or along a cross section is the **kink method**, or *constant dip domain method* (Faill 1969, 1973; Laubscher 1977; Suppe and Chang 1983). In the Busk method, bed dips that are mutually related are assumed to represent a common curvature domain. However, we could have just as readily bisected the angle between the dips from two adjacent dip data points and created two regions of constant dip related to the two data points. In the limit, or where the data are closely spaced, both methods would be identical.

As shown in Figure 10-22a, the first task in the kink method is to project the bed dip data in cross section. For example, the dip at point B is projected in the direction of bed dip data point A. Next, place two triangles adjacent to each other so that the upper triangle (X) is parallel to bed dip A and can be moved over the lower triangle (Y). (If preferred, a parallel glider can be used in place of two triangles.) Now move the upper triangle upward past the bed dip data point B and construct a line CD so that point D is approximately halfway between bed dip points A and B (Fig. 10-22b). When working with real data, point D need not be halfway between points A and B, and its position will depend on where the beds change dip. This position can often be determined from outcrop or depth-corrected seismic data. *Bisect the angle* between lines CD and DB with a protractor or compass, and then project the dip data at A to the dip domain boundary line with the triangle (line AE, Fig. 10-22c). Move the triangles to a new position so that one of them is parallel to dip data point B, and move this triangle down to continue line AE into the domain of dip data point B (line EF, Fig. 10-22c). The projection process results in two dip domains with each domain containing a constant dip and a theoretical interval of constant thickness (DE). Repeat the process as additional data are introduced (Fig. 10-22d). Notice that in Fig. 10-22d, dip domain B converges and terminates at point O, which is called a *branch point*. A branch point occurs at the intersection of two axial surfaces. A dip domain is eliminated at a branch point, in this case dip domain B. Only two dip domains exist beneath the branch point, whereas three domains exist above the branch point. Notice that the axial surfaces bisect the *bed dip domains* both above and below the branch points. It is important to remember to *bisect the angle between the fold limbs* and not the angle between the axial surfaces.

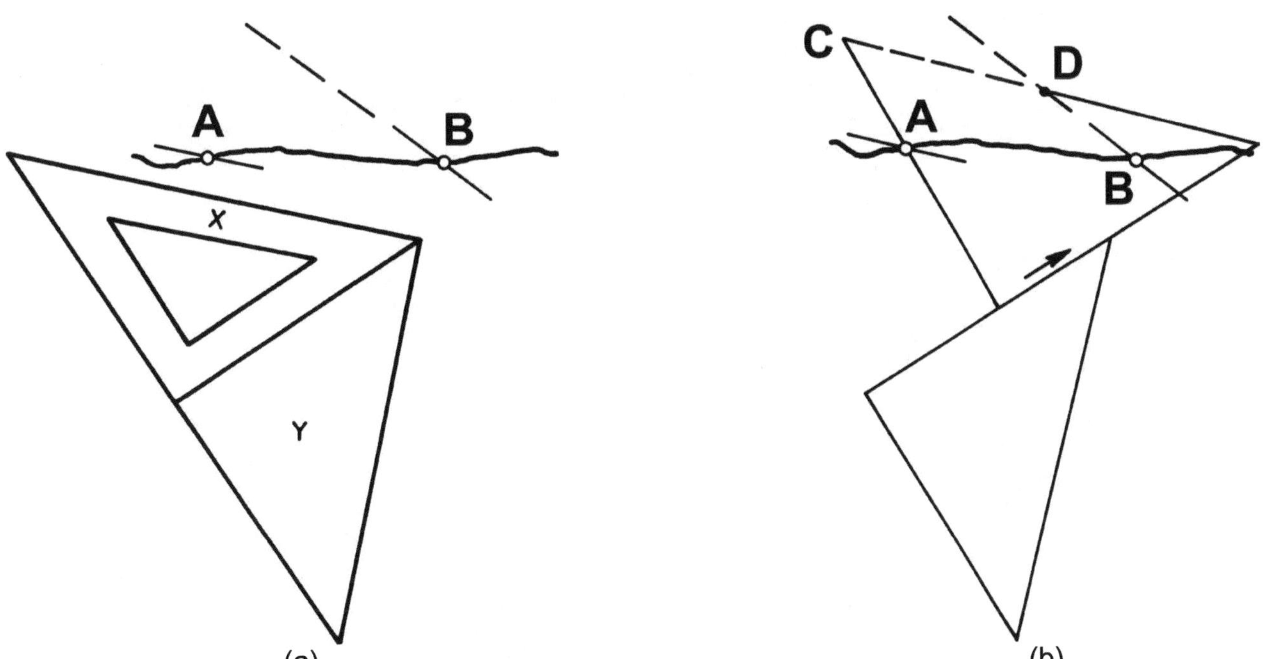

Figure 10-22 (a)–(d) Kink method approximation. The sedimentary beds are projected to depth along planar surfaces. The method applies to the majority of folds, which possess subplanar limbs.

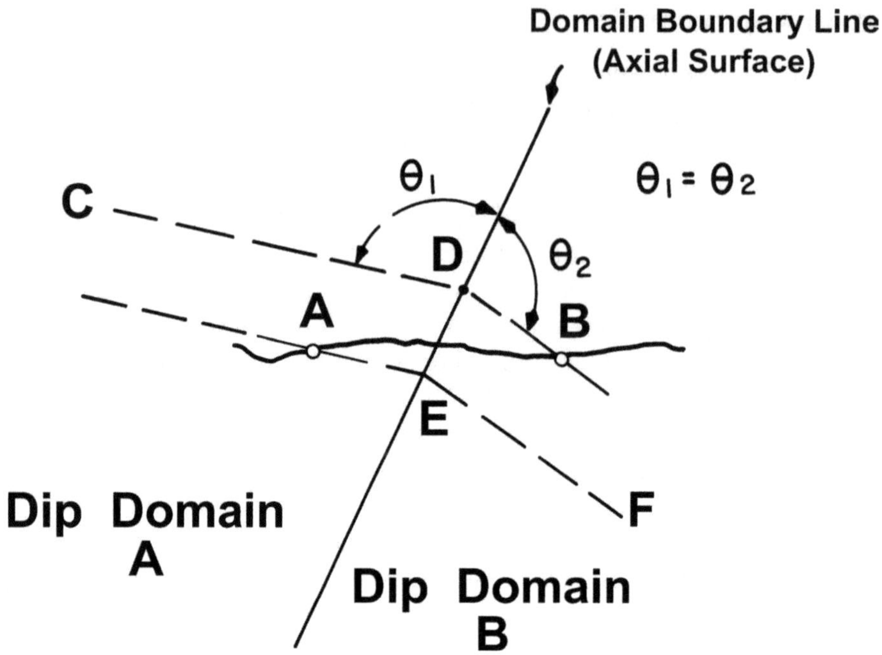

(c)

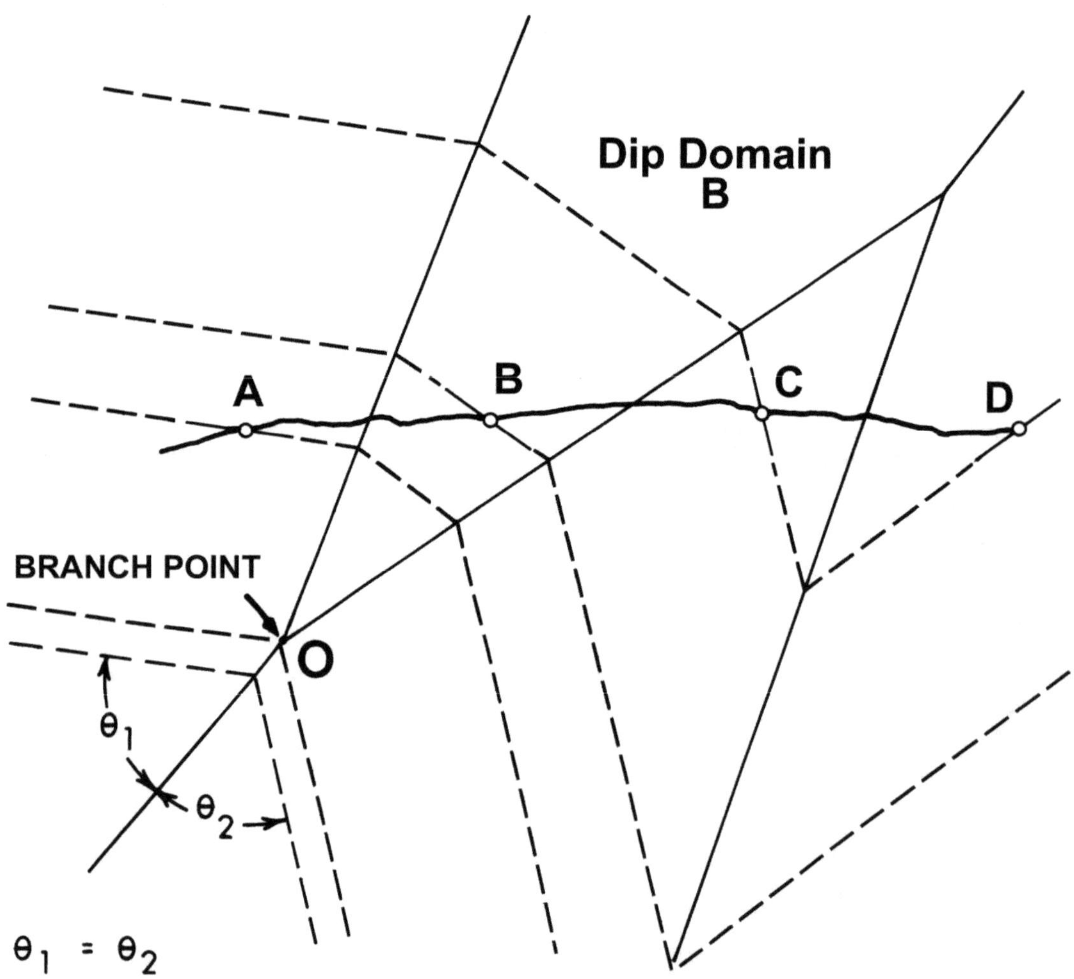

(d)

Figure 10-22 *(continued)*

In many folded areas, extensive regions of relatively constant dip adjoin smaller regions of rapidly changing dip. This is commonly seen on seismic sections. These relationships suggest that many folds possess limbs that have a uniform or near-constant dip but have hinge zones that are curved. As a result of this uniformity in dip, the kink method is readily adapted to work in low-temperature fold belts.

When applying the constant dip domain method, always remember to bisect the angle between the bed dips, thereby creating two adjoining and individual dip domains. Usually, the data are generalized or averaged to eliminate aberrant data points. This can be accomplished by taking two triangles and aligning them so that the top triangle can be passed across the data. In this manner, the triangle can be used as a filter to generalize or average the data. Areas of different generalized dip are defined as individual or separate dip domains, and the dip is then assumed to be approximately constant within each domain. The method also works very well with depth-corrected seismic or well data. The bisection procedure is in fact the continuity principle as applied to balancing (Suppe 1988) (Fig. 10-23):

$$t_1/t_2 = \sin(\alpha_1)/\sin(\alpha_2)$$

Notice that if the fold does not change thickness across the axial surface, then $t_1 = t_2$ and $\alpha_1 = \alpha_2$. If this procedure is judiciously applied, the cross section is more likely to line-length balance and area-balance.

When mapping using the kink method, you will find that as the stratigraphic intervals change thickness, the theoretical structural level of the interval as predicted by the method will deviate from the observed level. Thus, periodic adjustments in bed thickness must be made, usually at the position of the axial surface, which is the dip domain boundary line (Fig. 10-22c). Our preference is to follow the observed stratigraphic unit or sequence boundary in regions of onlap, and so on, even though this results in a divergence of once-parallel lines. If units above the unconformity do not change thickness dramatically, little harm is done by accurately representing the strata.

In areas of good data, the bisected dip domain data will ensure proper line-length and area balancing. In regions where the data are poor or nonexistent, the kink method can be used to project the units being mapped. Even under these conditions, the uniform thickness assumption can be a very powerful tool. Assume, for example, that you are mapping units A and B in Figure 10-24 from the north but that you encounter a region where no data exist. Mapping toward the no-data area from the south results in a good match on unit A but a poor match on unit B. What would you conclude in this case? The mismatch could result from either a dramatic change in thickness or an unrecognized fault in the south-central area that ceased growth prior to the deposition of unit A.

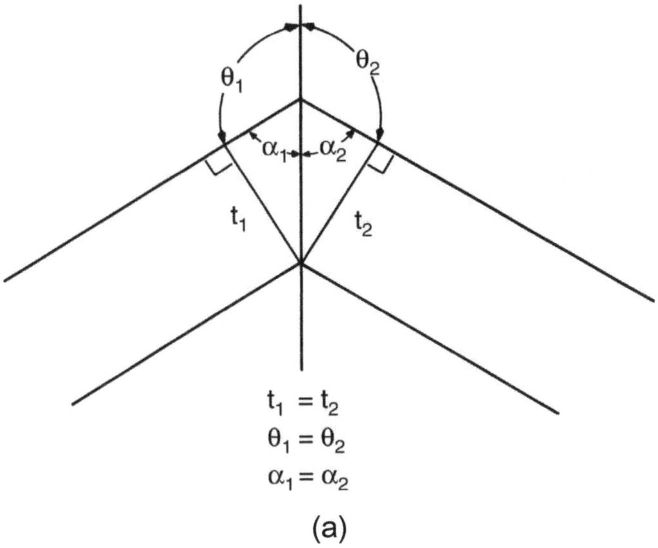

Figure 10-23 (a) Kink method geometry. (After Suppe 1980, 1985.) (b) An outcrop showing a kink band geometry. (b) A kink axial surface in outcrop.

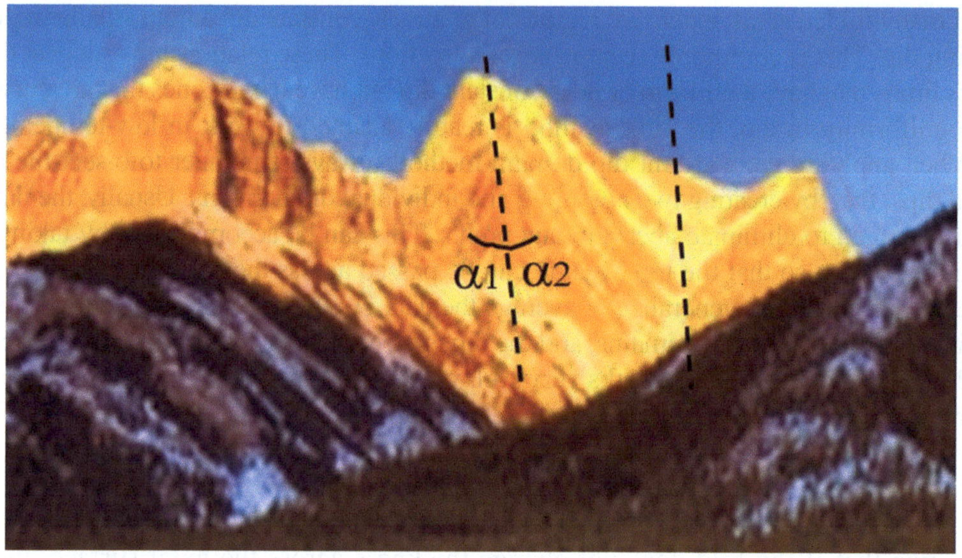

(b)

Figure 10-23 *(continued)*

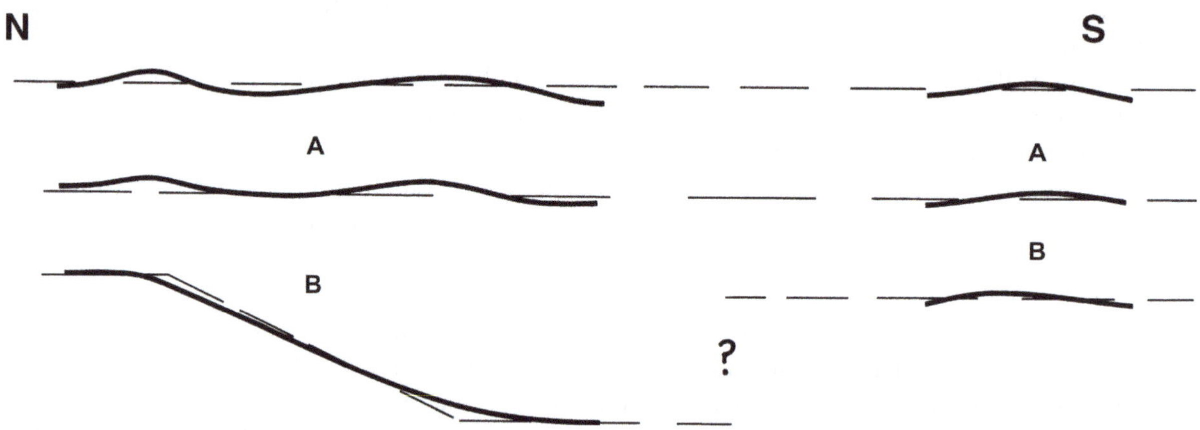

Figure 10-24 Example utilizing the uniform thickness approximation. Major change in the thickness of unit B, but not in unit A, implies that a structure or a stratigraphic change is present in the region that lacks data.

Kink Method Applications

An immediate application of the kink method arises when drilling the crests of the symmetric monoclinal or asymmetric folds that are common to fold-thrust belts worldwide (see the Fault Bend Folds and Fault Propagation Folds sections in this chapter). The improper positioning of wells on the crests of anticlines can result in *drilling wells off-structure* or wells into synclines (Bischke 1994a). *This is particularly true when drilling into an asymmetric fold (fault propagation fold).* In order to avoid costly mistakes, the compressional regime requires a good understanding of structural styles and geometry.

Figure 10-25 shows two different interpretations of an asymmetric fold based on the same bed dip data and seismic data. The steeply dipping limb of the fold was not imaged. Notice that the *crests* of the folds near the surface are positioned the same (use the dip data points as a reference). However, proposed wells are spudded at different locations based on the anticipated structural high at the reservoir level. Which well is more likely to be successful?

Figure 10-25a illustrates a well, positioned near the crest of the fold, which is interpreted to have a steeply dipping front limb. Developing folds verge or move in the direction of steeper bed dips (Fox 1959; Suppe 1985), so the steeper fold limb is defined as the frontal limb in a fold train. The front limb is interpreted in Figure 10-25a to be thinner relative to the more gently dipping back limb. This style of folding is common to high-temperature mobile belts, which do not contain petroleum reserves.

A different type of fold is the **parallel**, or **constant-thickness**, **fold** (Ramsey 1967) interpreted in Figure 10-25b. The beds do not significantly change true stratigraphic thickness from the back limb to the front limb. This type of fold is common to the low-temperature petroleum regime. The interpreters positioned the well on the gently dipping back limb of the structure, in a position farther left than the well in Figure 10-25a.

On seismic time profiles, stratigraphic intervals of constant thickness maintain about the same vertical time thickness. In our example, the depth profile shown in Figure 10-25a is similar to a time profile on which the interpreters attempted to maintain the same vertical time thickness of the intervals. The result is a *thin-limb fold.* On the other hand, the geoscientists who constructed Figure 10-25b made their interpretation on a time profile and then properly depth-corrected it to generate the depth profile shown in the figure. True stratigraphic thickness was maintained, and the result is a *parallel fold.*

Geoscientists who work fold-thrust belts know a majority of the folds within hydrocarbon-producing regions *approximate parallel folds rather than thin-limb folds* (Suppe and Medwedeff 1990; Tearpock et al. 1994). Unless data exists in support of a thin frontal limb, *the parallel fold interpretation is likely to be the better interpretation.*

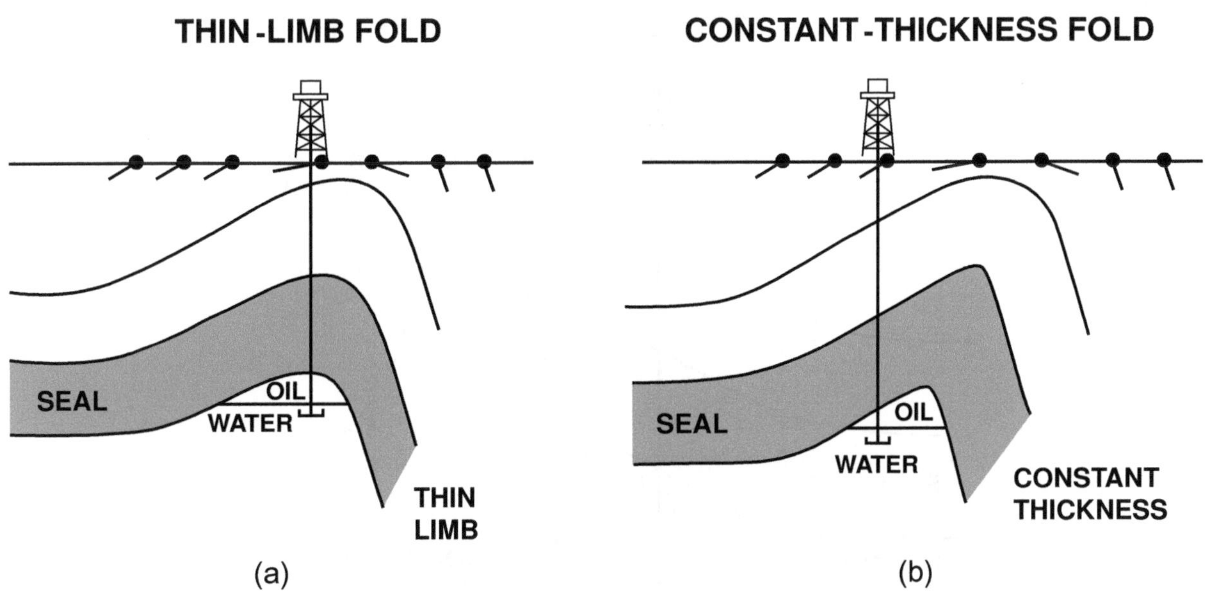

Figure 10-25 Different fold interpretations can result in different proposed well locations. (a) Interpretation of a fold based on surface dip and seismic data, but not using the kink method. An attempt was made to maintain the **vertical thickness** of the beds within the steeply dipping front limb of the fold (see Fig. 10-26). (b) Interpretation of a folded structure based on surface dip and seismic data and using the kink method. An attempt was made to maintain the stratigraphic thickness of the beds within the steeply dipping front limb of the fold. (Published by permission of R. Bischke.)

If the fold is a constant-thickness fold, then the likely result after drilling the two wells is shown in Figure 10-26. In the figure, the two well positions shown in Figures 10-25a and b are redrawn on the constant-thickness fold shown in Figure 10-25b. The well on the right is positioned using the thin frontal limb interpretation. This well is likely to encounter steeply dipping beds in the seal horizon and never test the reservoir. Perhaps the geoscientists who generated the profile shown in Figure 10-25a believed that a seismic time profile is a geological profile. *Time profiles distort geometry, and the distortion increases with increasing bed dip* (Chapter 5).

Notice on the profile shown in Figure 10-25b that if the well were drilled deeper, it might have crossed the axial surface and entered the front limb of the structure. When drilling asymmetric folds, there is always the *risk of crossing the axial surface* that separates the gently dipping back limb from the steeply dipping front limb. If the front limb of the fold is slightly overturned, then beneath the axial surface, the stratigraphic units penetrated by a well will become *younger* with increasing depth (Fig. 10-44). Drilling the syncline in front of the fold is also possible when attempting to exploit asymmetric folds. The fault propagation fold is the second-most common type of fold in fold-thrust belts, so interpreters should be aware of the pitfalls associated with asymmetric folding (Tearpock et al. 1994).

The profile shown in Figure 10-25b illustrates the kink law. The kink law states that if the beds do not change thickness, then *the axial surface bisects the limbs of the fold.* In other words, on constant-thickness folds, the angles between the two fold limbs and the axial surface are about equal, or $\gamma_1 = \gamma_2$, as in Figure 10-26. Most petroleum-related folds come close to obeying a constant-thickness relationship and the kink law (Tearpock et al. 1994). At the correct well position, shown on Figure 10-26, the well was positioned so that it did not cross the axial surface at the reservoir level. This well intersects the reservoir horizon, whereas the dry hole (Fig. 10-25a) crosses the axial surface (Fig. 10-26). A well that crosses an axial surface can even penetrate vertically dipping or overturned beds.

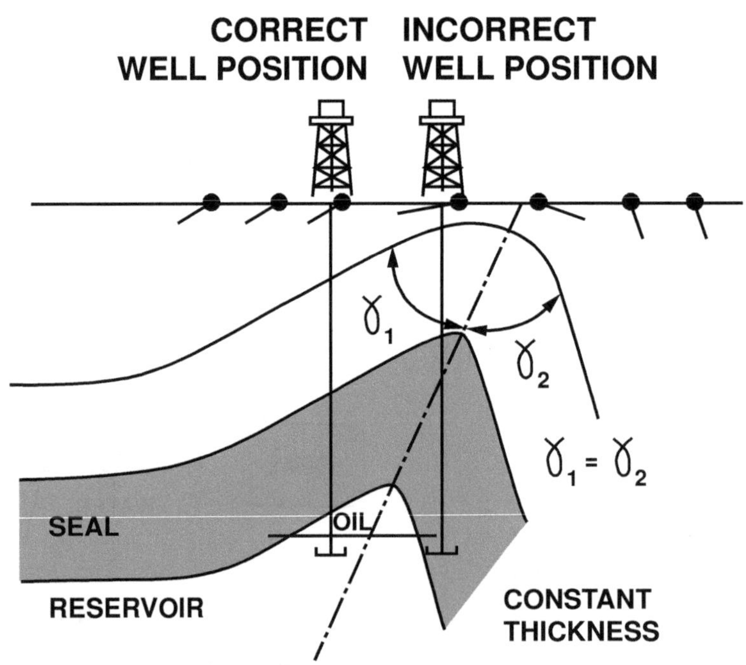

Figure 10-26 As most **post-depositional folds** are constant-thickness folds and obey the kink method, wells spudded on the crests of asymmetric folds will typically intersect steeply dipping beds within the front limbs of these folds. On asymmetric folds, wells spudded on the back limbs are more likely to discover hydrocarbons. This cross section was generated using the kink method, and thus the axial surface bisects the fold limbs. (Modified from Tearpock et al. 1994.)

Figure 10-27 is redrawn from Figure 10-25a to demonstrate that the angles between the two fold limbs and the axial surface are *not* equal, or γ_1 is *not* equal to γ_2. This is an indication that the fold was constructed as a thin-limb fold, and therefore the front limb may be incorrectly located. On the other hand, if the fold is actually a constant-thickness fold, then the well will cross the axial surface and penetrate the steeply dipping beds in the front limb of the fold, as shown in Figure 10-26.

The kink law is a powerful tool when constructing cross sections. Remember to *construct the cross section on a scale of one to one.* To apply the method, simply *bisect the angle between the fold limbs.* These procedures eliminate geometric distortions and provide a clearer picture of the complex relationships concerning folded structures.

As a final exercise, examine the three profiles of folded structures from three different fold-thrust belts in Asia, shown in Figure 10-28. Seismic and surface dip data constrain the profiles. Using the kink law, can you recognize which one of the three wells was drilled on structure, and why? Remember to bisect the angle between the fold limbs. The kink method is easy to use and rapid to apply, and it often generates cross sections that accurately predict wellbore results. The method deteriorates if the projection crosses a large fault. The method assumes that the bed dips remain about constant within each dip domain. However, bed dip may change significantly where crossing large faults and thus violate the constant bed dip assumption. The target depths of the three wells are between 1500 ft and 3500 ft, and the cross sections are drawn at a scale of one to one.

Figure 10-28a shows a well spudded into the front limb of a symmetric monoclinal-type fold. A high-quality seismic line crosses the fold that images a steeply dipping west limb, a flat crestal area, and a more gently dipping east limb. Surface bed dips exist to aid the interpretation. Were high or low bed dips encountered in the well? Using the outcrop data, we know that an axial surface would bisect the angle between the 30-deg to 35-deg front limb dips and the flat crestal dips. We position the axial surface along the change in bed dips that image on the seismic profile, which results in an axial surface that dips steeply to the east (Fig. 10-29a). If seismic data

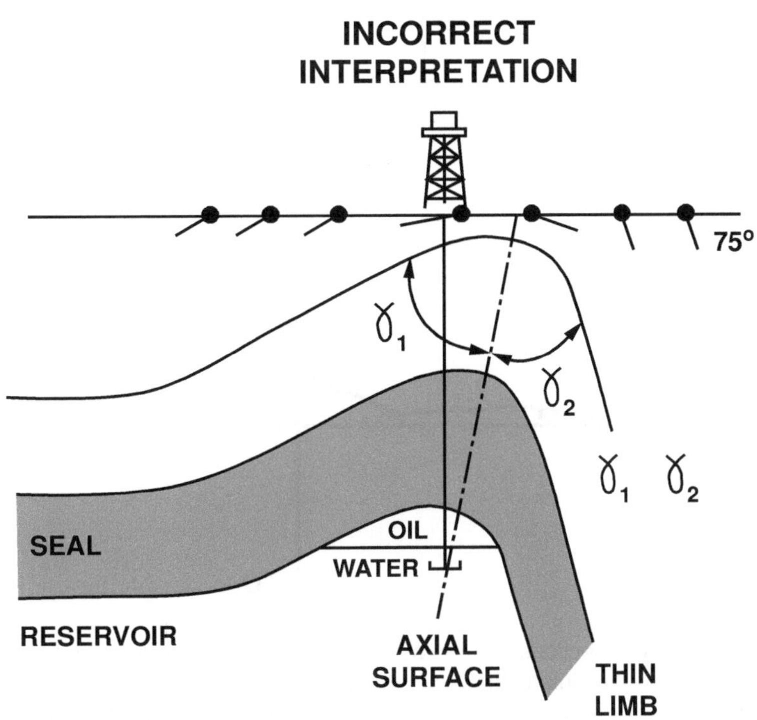

Figure 10-27 On a thin-limb fold, the axial surface does not bisect the limbs of the structure. This geometry contrasts with Figure 10-26 in which the axial surface bisects the fold limbs. (Modified from Tearpock et al. 1994.)

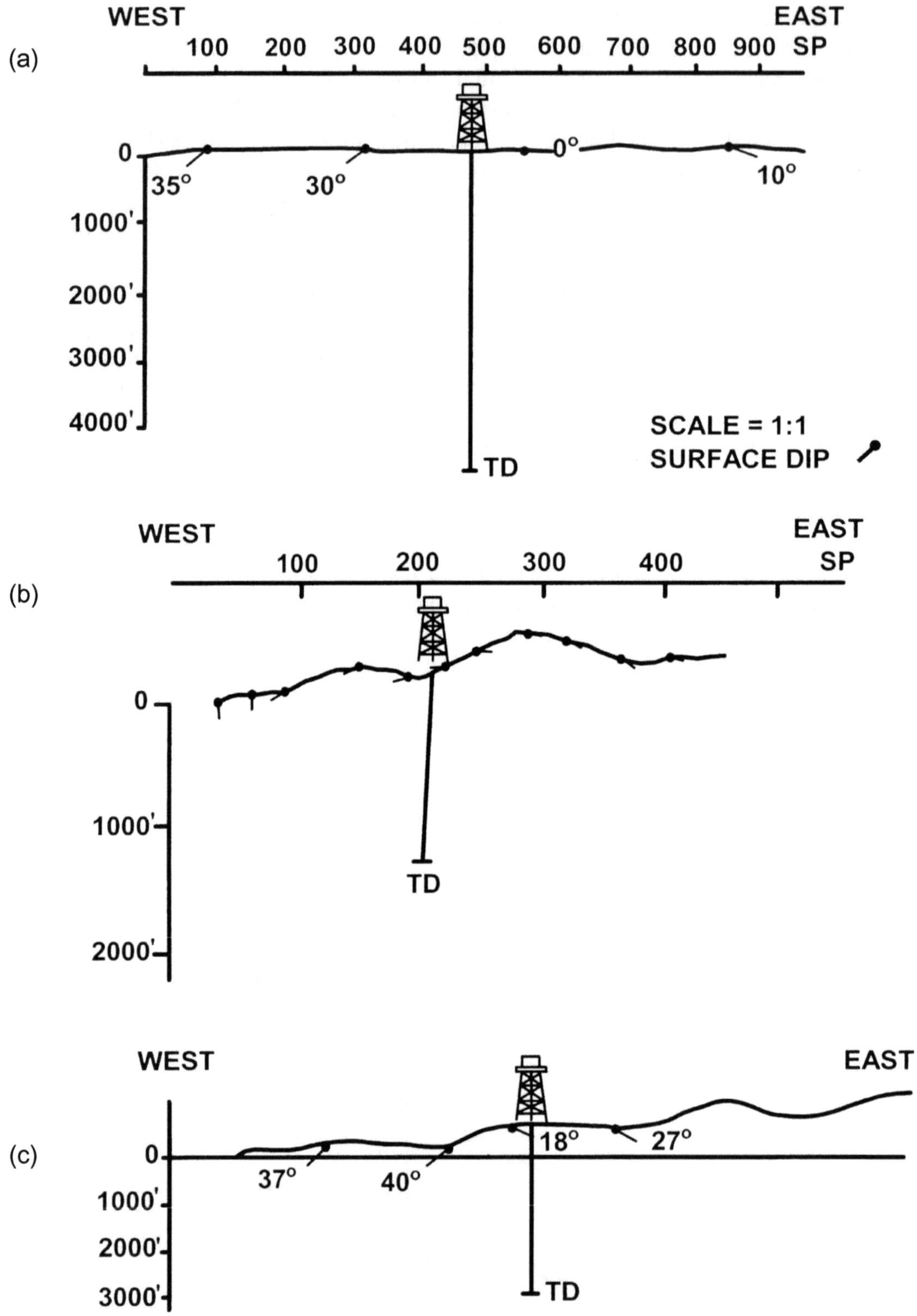

Figure 10-28 (a)–(c) Cross sections of three structures constrained by surface dip and seismic data of varying qualities. Using the kink method, can you predict which one of the three wells discovered hydrocarbons and which wells encountered steeply dipping beds? (From Bischke 1994a. Published by permission of the Houston Geological Society.)

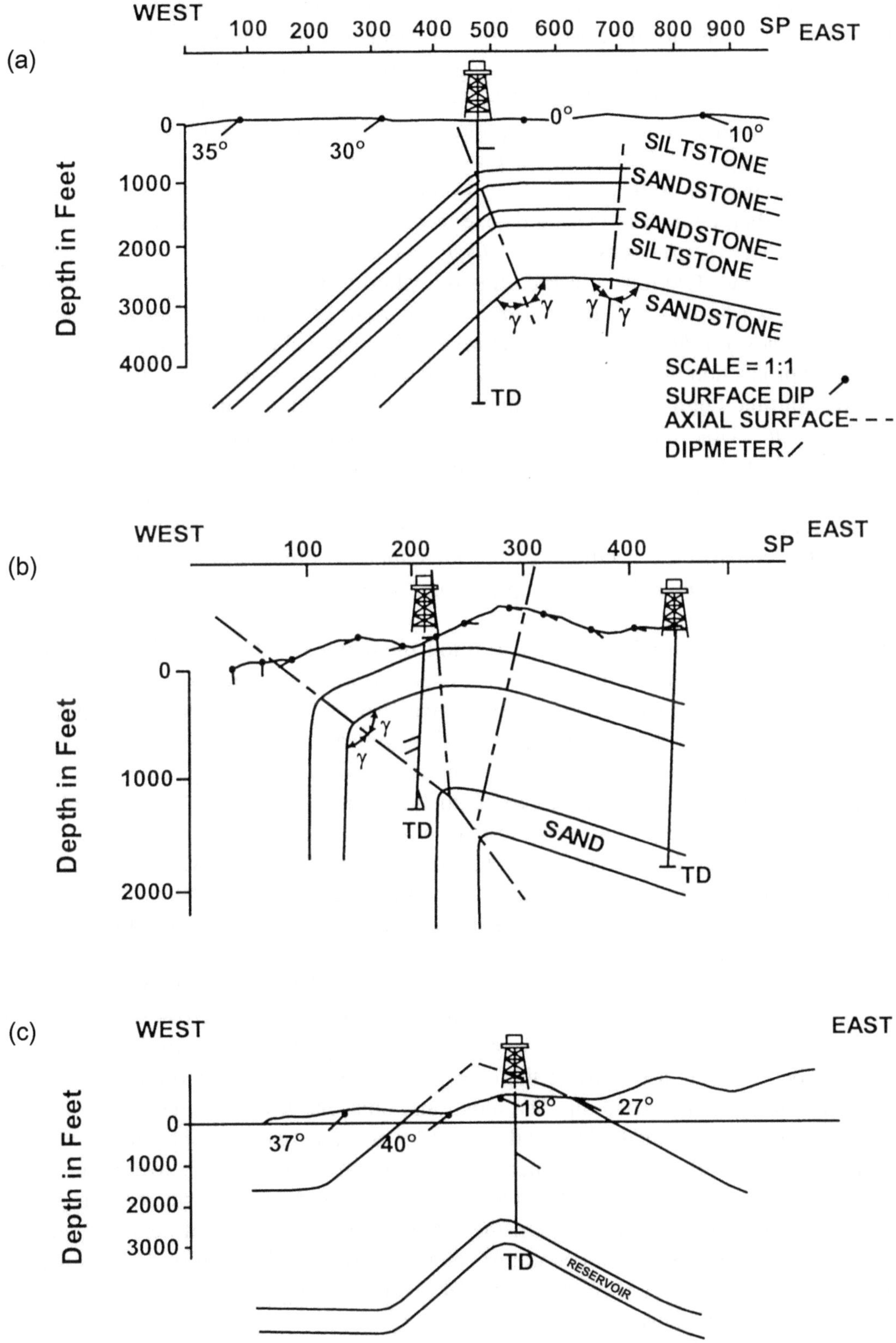

Figure 10-29 (a) The well crosses an axial surface at approximately the 1000-ft level to intersect 25-deg to 40-deg dipping beds. (b) The well is drilled on the crest of an asymmetric fold and encounters near-vertical beds. (c) The well is drilled into the back limb of a tightly folded structure and encounters hydrocarbons. (From Bischke 1994a. Published by permission of the Houston Geological Society.)

do not exist, position the axial surface halfway between the surface data points. Dips exceeding 30 deg exist below and to the west of the axial surface. The kink method predicts that the well should encounter bed dips in excess of 30 deg at depths exceeding 1000 ft (Fig. 10-29a).

Data from the well, shown in Figure 10-29a, confirm the accuracy of the kink method. At a depth of 1200 ft, tadpole dips on the dipmeter log range between 25 deg and 40 deg. The well encountered a sand horizon below 3000 ft that requires more than a 500-ft hydrocarbon column for a discovery in this well. Notice that if a well were positioned on the back limb of the structure, near sp 700, then a smaller hydrocarbon column in that sandstone nevertheless would result in a discovery. *Thus, wells spudded at the back of structures are more likely to encounter hydrocarbons than wells spudded on the front of structures.* Old-timers learned this rule after drilling numerous wells. Furthermore, *wells drilled into the back-limb axial surface not only are more likely to encounter hydrocarbons but are commonly the most productive wells.* We return to this empirical observation and suggest a cause for the increased production in this chapter's section on the kinematics of fault bend folds.

Next, examine Figure 10-28b, which contains a well spudded into an asymmetric *fold at the crest of structure.* This fold, constrained by surface bed dips, exhibits a *near-vertical* front limb. Seismic data do not image these steep bed dips. Applying the kink method to the surface bed dips results in the interpretation shown in Figure 10-29b. The kink method predicts that the well should encounter near-vertical bed dips below 1000 ft. Tadpole dips obtained from dipmeter data below 1000 ft confirm the accuracy of the kink method prediction. The kink method solution suggests that a well spudded between sp 250 and sp 300 would encounter the sand horizon at a depth of 1200 ft.

Last, examine Figure 10-28c, constrained by surface bed dips and poor-quality seismic data. The interpreters who drilled this well used the kink method to constrain the interpretation. The well was spudded off the crest of structure on the more gently dipping back limb. Dipmeter data confirms the interpreted dip. This well resulted in a hydrocarbon discovery and appears to intersect the subsurface crest of the reservoir horizon (Fig. 10-29c).

We have used the recognition of axial surfaces on seismic data and applied them to fold interpretation. When interpreting seismic data, you must always be aware of the probable existence of axial surfaces. *Seismic interpreters commonly mistake axial surfaces for faults because of the abrupt changes in dip.* You can avoid that mistake if you understand the geometry of the possible structures in the area and follow some of the common-sense methodology discussed.

In conclusion, the kink method is relatively easy to use and makes accurate predictions when applied to depth-corrected seismic data and outcrop bed dip data. If the kink projection method does not cross a large fault, then the method typically generates accurate cross sections of subsurface geometry. Bed dips can change across large faults, causing the solution to deteriorate. Remember that surface bed dip data are some of the cheapest data available to interpreters. When employing 2D seismic data, collect the bed dip data along and adjacent to the seismic survey lines. On well-constrained structures, the method typically generates accurate results (Suppe and Medwedeff 1990).

Structures typically contain steeper dipping frontal limbs relative to gentler dipping back limbs. Wells positioned on the back limb of symmetric and asymmetric structures increase the odds of encountering hydrocarbons. On the other hand, wells spudded near the steeply dipping frontal limbs of structures often encounter steep bed dips. If the well penetrates the overturned limb of an asymmetric fold, then the beds will become younger as the well deepens, potentially missing the prospective horizons.

Tearpock et al. (1994) discuss additional pitfalls concerning fault propagation folds and other complex structure styles. A strong structural geological background is key to exploring in these areas. The understanding of compressional structural styles, including the types of faults and folds and their inseparable relationship, is paramount when exploring in fold-thrust belts (Bischke 1994a).

DEPTH-TO-DETACHMENT CALCULATIONS

A method to determine the depth at which folding terminates can be attributed to Chamberlin (1910) and to Bucher (1933), who applied the method to determine the depth to detachment in the Jura Mountains. If the sequence that you are studying consists of a number of folds, then each fold must be isolated and studied separately. In this method, you measure the length (l_o) of a marker or reference bed, the present pin length (l), and the *average* amount that the marker bed has been uplifted ($\overline{U}$) above the undeformed level of the bed, as shown in Figure 10-30. The average uplift ($\overline{U}$) is calculated using the same methods engineers use to calculate reserves (Fig. 10-30). The amount of shortening (S) that the unit has experienced is defined as

$$S = lo - l$$

The average amount of uplift times the present length ($\overline{U}$) equals the average area of uplift, which is then equated to the amount of material that enters the structure from the sides ($S \times d$), where d is the depth to detachment (Fig. 10-31).

It therefore follows (Bucher 1933) that

$$d = l \times \overline{U} / S$$

Alternatively, if the depth to detachment is known, then the method can be used to check fold shape.

A closely related method employed by Laubscher (1961) and described by Goguel (1962) has been used in the petroleum industry. This method also assumes that no material is entering the structure from below, as in a duplex (see the section on duplex structures), and that all the material in the core of the structure is derived from the sides of the structure. Mitra and Namson (1989) point out

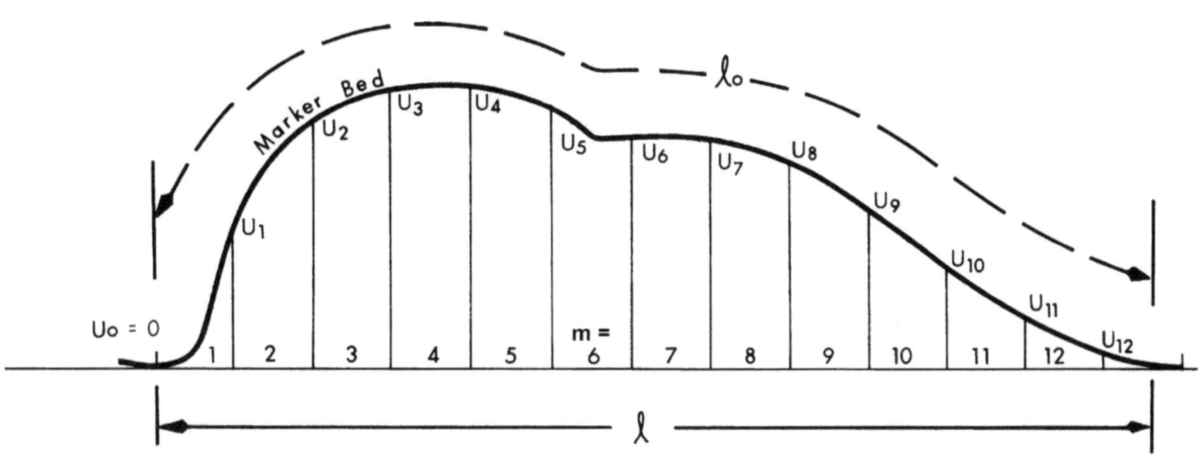

$$\text{AVERAGE UPLIFT} \quad \overline{U}_n = \frac{\sum_{m=o}^{n} U_m}{n+1}$$

Figure 10-30 The average amount that a marker bed has been uplifted can be determined by measuring equally spaced line segments that are drawn between a base level and the marker bed and then averaging the line lengths.

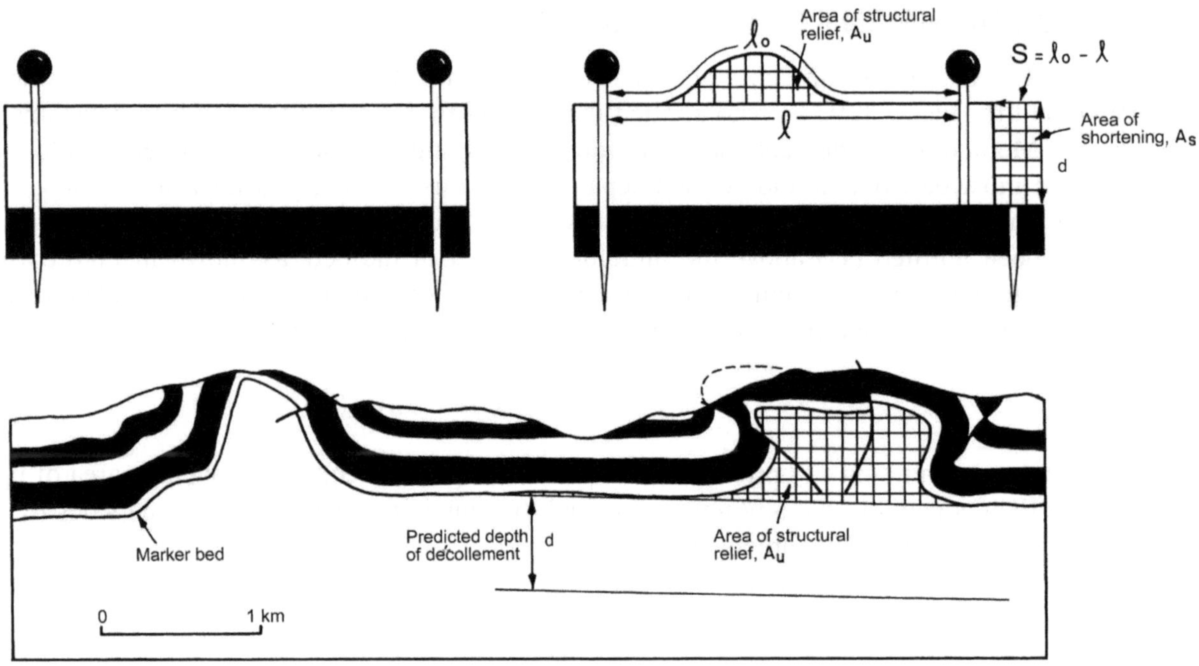

Figure 10-31 Depth-to-detachment calculation. The amount of material entering the cross section from the sides is equal to the material that has been uplifted above base level. (Modified after Laubscher 1961; Suppe 1985. Published by permission of the Swiss Geological Society.)

that these assumptions are invalid if there is interbed shear (i.e., distortion of the vertical pin line) or if material is transferred out of the area of the cross section, as occurs in fault bend folds.

If the material enters the structure from the sides, then the area within the core of a structure (A_u) at a given reference level is measured, as are the final pin length (l) and the initial length (l_o) of a reference or marker bed (Fig. 10-31). As before, the shortening at the reference level is

$$S = lo - l$$

The area (A_u) within the core of the structure beneath the reference bed is assumed to be equal to an equivalent volume that comes in from the side (A_s) (Fig. 10-31). The area can be obtained by planimetry.

Therefore,

$$A_s = (S)(d)$$

where d = depth to detachment, and as

$$A_u = A_s$$

it follows that

$$A_u = (l_o - l)\, d$$

and

$$d = A_u/(l_o - l)$$

NONCLASSICAL METHODS

Introduction

Newer methods of structural interpretation are more precise and more robust than the classical balancing techniques, which have distinct limitations. For example, Dahlstrom (1969) emphasized that within a given area, only a limited number of geological structures are likely to exist. He also realized that these *structures must area-balance and line-length balance, but exactly how does the interpreter accomplish these tasks?* One obvious method is to measure formation bed lengths to check for balance, but this can only be accomplished after the interpretation is *finished*. In addition, two geoscientists given the same data set are very likely to place lines of equal length at different positions within a cross section, although both products may be line-length balanced sections! How are we to evaluate which of the two sections is "correct," and how could the interpretations be *improved*? One problem is that line-length balancing has no rules associated with the method, other than that the bed lengths must be consistent and that the structural styles are limited.

This problem becomes particularly acute when the data in an area are under-constrained, as is often the case and leads to what John Suppe has referred to as the "blank paper" problem (Woodward et al. 1985). For example, you are studying an area in which the only data available are at shallow depths, and these data strongly suggest that the structures continue with depth. Classical balancing lacks constraints, so any attempt to continue the interpretation to depth is likely to result in as many interpretations as there are interpreters. Furthermore, line-length balance can be conducted only after an interpretation is finished. Do methods exist to more directly balance a section, either by hand or with the use of a workstation? More direct methods would certainly be friendlier to the working environment.

Those of you who have worked with various tectonic settings know that several relationships are recurrent from area to area. In extensional terranes, the faults are commonly listric and rollover structures are present, which typically contain antithetic and synthetic minor faults and perhaps a keystone structure. This suggests that some fundamental process controls the development and formation of normal faults and associated structures. In the compressional regime, **folds** are either symmetric, as described by Gwinn (1964) in his work on the Appalachians, or they are asymmetric, such as many of the folds in the Rocky Mountains (Link 1949). Geologists have noticed that where folds are present, faults also seem to exist in association with the folding (Bally et al. 1966; Jones 1971; Woodward et al. 1985). Many different regions around the world possess thrust belts that contain within them symmetric and asymmetric folds, so fundamental processes seem to control the formation of orogenic belts. If we could develop realistic models of these fold-and-thrust belt folds, then the petroleum industry would have powerful tools in which to aid interpretation.

In the section Kink Method Applications, we presented three examples of how surface dip data could reflect symmetric or asymmetric styles (Fig. 10-29). In particular, the surface dips could reflect off-structure problems and the resulting pitfalls. For this reason, we have added Determining Structural Styles to Reduce Risk Using Qualitative and Quantitative Techniques, a section based on recent research presented in this chapter. This section uses measurements of the displacements on thrust faults, throw, structural relief, and repeated section to quantify structural styles. Several mathematical formulas are developed to quantify the process and to check interpretations for structural validity.

Hence, we enter the world of nonclassical methods that utilize mathematical formulas, graphs, and models. Perhaps a word of caution is required at this time for the new geology student. Although models can be very powerful tools (e.g., the plate tectonic model), the improper application of a correct model to the wrong situation will only result in error. To make matters worse, model balancing is nonunique. Different geoscientists, applying the same model to a given structure, are likely to generate similar results, as we shall see. However, the skeptic may point out that this is merely an artifact of being schooled in the same interpretation techniques.

Before we enter the exciting world of *kinematic processes*, we restate that this book is designed primarily to present subsurface mapping techniques and is not a complete reference on interpretation per se. In the balancing sections of this book, the mapping techniques are commonly difficult to separate from the interpretation because you must choose which technique to apply to a given structure, and this choice involves interpretation. Let us caution you that other interpretation techniques exist that do not involve any particular mapping technique, such as growth sedimentary patterns and structures (Medwedeff and Suppe 1986). These growth patterns are often extremely helpful in determining which model or technique to apply to the structure, so we recommend that the serious geoscientist gain familiarity with all the approaches to structural interpretation.

Suppe's Assumptions and Dahlstrom's Rules

When presented with the problem of a poor or nonexistent data set, several approaches are open to the geoscientist. Solutions to this problem seem to involve the following.

1. Collect more and/or better data. This subject is left to the data contractors.
2. Make more assumptions in order to solve the structural problem. If data are lacking or are unobtainable, it is still possible to solve the structural problem, providing you can extrapolate known data, using known geological principles, into the area of interest. For example, if we assume that the kink method is appropriate, we can extrapolate units within the limb of a fold to depths beyond the control data. In this sense, assumptions can substitute for data.

We recommend that you employ the following assumptions and rules credited to Suppe (1988) and Dahlstrom (1970).

Suppe's Assumptions

1. Thrust faults step up abruptly from a decollement and (unless deformed) do not have continuously curved listric shapes.
2. All thrust faults (that produce a given structural style) in a given area step up at approximately the same angle.
3. Layer-parallel slip in a thrust sheet is limited to that caused by changes in dip. This is another way of stating that the kink method applies at all times.

Dahlstrom's Rules

1. Dipping beds over flatter beds define decollements or thrust faults (Fig. 10-5c).
2. Thrust faults cut up, and not down, stratigraphic section.
3. Invent more powerful interpretation methods and techniques so that you can extrapolate existing data into the no-data areas. This is the subject of the remaining sections in this chapter.

Fault Bend Folds

Our examination of seismic sections from various areas of the world (Australia and through the Pacific rim to Alaska, western and eastern United States, western Europe, Argentina, Venezuela, etc.) indicate that there are two commonly recurring fold styles within the low-temperature portions of thrust belts: the **symmetric**, or **fault bend fold type** (Figs. 10-32a and b and 10-34) (Rich 1934; Suppe 1983) and the **asymmetric**, or **fault propagation fold type** (Fig. 10-42) (Link 1949; Suppe 1985). The simplest, end-member structural styles are shown in Figures 10-34 and 10-37 for fault bend folds and in Figure 10-43 for fault propagation folds. See Figure 10-33 for the front planal of a fault bend fold. We stress here that complications in these structures, such as multiple and back thrusts, often exist and that other thrust-related geometries are present (Fig. 10-34). For a more complete discussion of variations in structural style, consult Suppe (1983), Suppe and Medwedeff (1990), and Shaw et al. (2005). We wish to emphasize, however, that these two structural styles are the simplest types of compressional folds that are commonly present in petroleum basins. Fault bend folds appear to be the most common of the two structural styles, and a case history of the El Furrial Fault-Bend Fold Trend is presented in the next paragraph. Also, in the section on Quantitative Methods for Determining Structural Styles is a discussion on qualitative and quantitative methods to determine the two major structural styles.

The following paragraphs present a short case history of the El Furrial Fault-Bend Fold Trend (Fig. 10-51a and b). Fault bend folds and their related structures (see shear fold example in Fig. 10-53), are some of the world's largest producers. Some of the examples that we are familiar with include fault bend style folds in the deep-water Nigeria, such as the Agbami Field, which has reserves reported in excess of 2.0 billion barrels of oil. Another example, this one in the United States, is the line-length balanced Whitney Canyon cross sections associated with the Absaroka fault presented by Lamerson (1982).

A good color 3D seismic example is the El Furrial Fault-Bend Fold, north Monagas, Venezuela, which is a balanced structure (Morales et al. 2005) (Fig. 10-51a) similar to Figure 10-43. Three fault bend folds lie along this trend for 45 km (28 mi), although some of the structures are imbricated. From east to west, these fields consist of the El Furrial, Carito, and Tejero/Santa Barbara structures, that contain about 11 billion barrels of oil in place (Morales et al. 2005). The interpreted and depth-corrected sections shown in Figures 10-51a and b are area-balanced sections generated from software. Figure 10-51b is a strike line that crosses the dip line (Fig. 10-51a) along the vertical dashed line "B" on the front limb of the structure. The orange horizon at the high-impedance contrast is the productive Naricual sands that are about 500 m (16,000 ft) thick.

The El Furrial fault is at a depth of about 5.7 km and repeats about 3.5 km of Naricual section from the crest of structure. Following Dahlstrom's rule 1 from the previous section, that the fault exists where dipping reflections terminate against flatter dipping reflections, we interpret that the fault on Figure 10-51a lies above the black line at the horizontal, high-impedance reflection (the black line is positioned lower so as not to cover the high-impedance contrast). This dipping-over-flat relationship for picking faults is more obvious on the dip line in Figure 10-51a, to the right of vertical line "B," which defines where the strike line crosses the dip line. At this location, the dipping hanging wall reflections terminate against a very bold flat reflection. This bold, flat refection, at about 5.7 km, is caused by a strong impedance contrast caused by over-thrusting denser and higher-velocity rocks over less dense and lower-velocity rocks. Again, the fault is located at the top of the bold reflection, but was placed below the high-impedance contrast in order not to cover the seismic data.

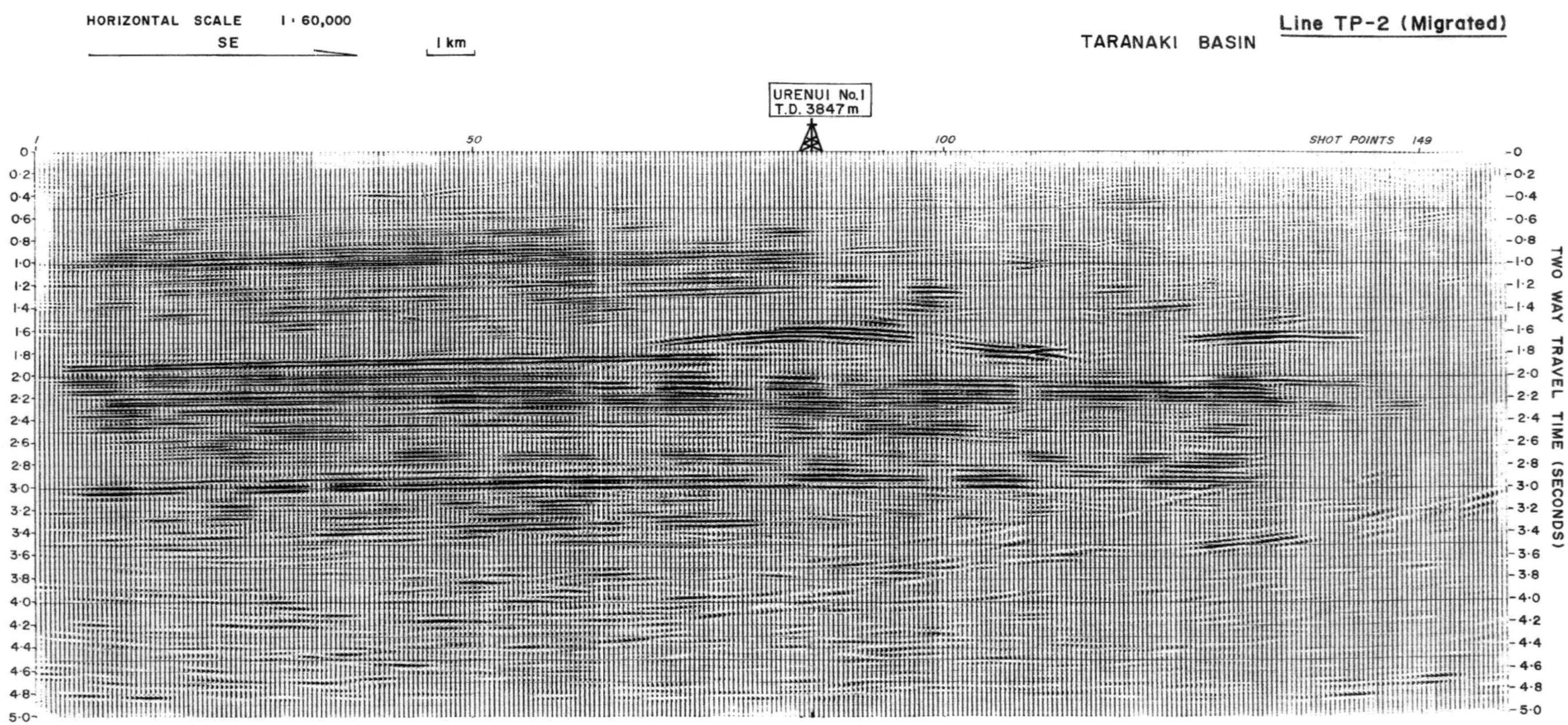

Figure 10-32 (a) Migrated 2D seismic line of fault bend fold from the Taranaki Basin, New Zealand. A symmetric fold is imaged in the vicinity of the well and sp 100, between the two-way travel times at 1.5 sec to 1.9 sec. In the vicinity of sp 75 and sp 100, dipping beds overlie flat beds, indicating a decollement. (From Seismic Atlas of Australian and New Zealand Basins, Skilbeck and Lennox 1984. Published by permission of Earth Resources Foundation, University of Sydney.) (b) For a color 3D fault bend fold example from the El Furrial Field, Monagas Venezuela, see Figure 10-51. (Reprinted by permission of the AAPG whose permission is required for further use.)

Figure 10-33 Frontal limb of a fault bend fold in Hudson Valley, New York, USA, located on Route 23 about 300 meters west of the New York State Thruway. (Compliments of Jon Mosar.)

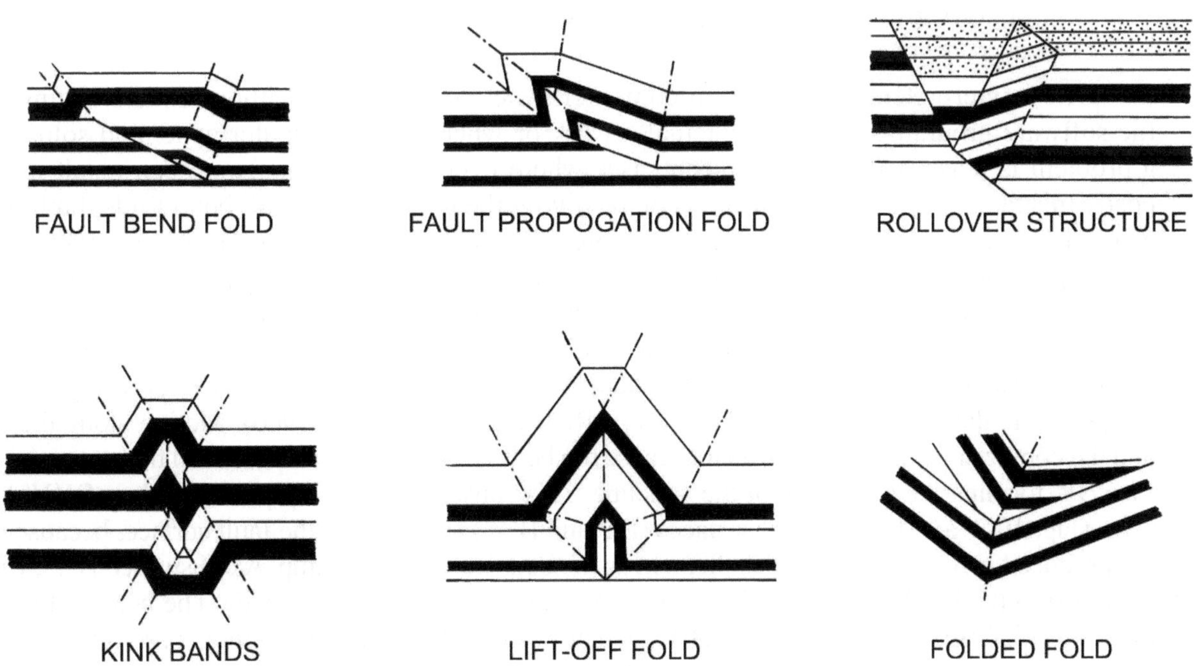

Figure 10-34 Examples of fault-related fold types. (Published by permission of John Suppe.)

Fault bend folds were first described by Rich (1934) in the Pine Mountain thrust region of the Appalachians, where he recognized that this fold style consisted of symmetric anticlines (Fig. 10-35). Rich also recognized that these folds were associated with thrust faults, and he postulated that the folds were the result of *thin-skinned* deformation. Notice that if motion were to occur along the decollement in Figure 10-2, hanging wall material would ride up the ramp and onto the flat. Rich recognized that if this occurs, anticlines and synclines would form (Fig. 10-36). This example was eventually modeled utilizing a mathematical volume conservation concept (Suppe and Namson 1979; Suppe 1980, 1983).

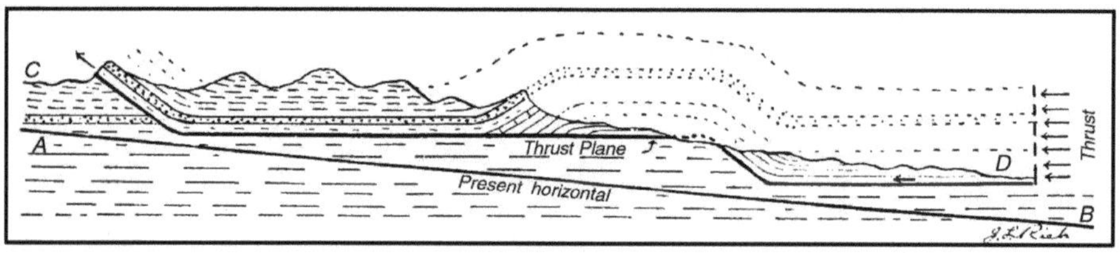

Figure 10-35 Fault bend fold forming over a step-up on a thrust fault. (From Rich 1934. AAPG©1934, reprinted by permission of the AAPG whose permission is required for further use.)

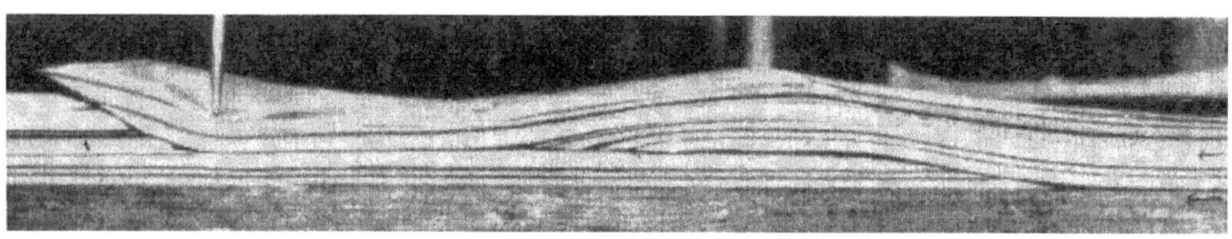

Figure 10-36 Model of fault bend fold constructed from paper sheets. (From Rich 1934. AAPG©19341934, reprinted by permission of the AAPG whose permission is required for further use.)

Often, interpreters rely solely on computer-generated solutions without understanding how the software program's algorithms actually work or whether the computer-generated solution to a problem is reasonable. We have seen cases where overreliance on software can result in very costly mistakes. The following sections on resolving the geometry of fault bend folds, fault propagation folds, and imbricate structures are presented (1) to give the interpreter a better understanding of, and intuition about, the results from balancing software; and (2) to illustrate what a geometrically reasonable computer-generated area-balanced solution looks like.

The kinematics of the simplified fault bend fold modeling process are as follows. Folds form along nonplanar thrust faults where a decollement on a lower structural level (Y level, Fig. 10-37a) ramps to a higher stratigraphic level (X level) (Rich 1934; Bally et al. 1966). Motion along the fault and the conservation of volume principle cause the beds to ride up the ramp and roll through axial surface BY, forming the back limb of the anticline. This causes the back dip panel (or flap) BYY′ B′ to form (Fig. 10-37a). The two axial surfaces (BY and B′Y′) terminate at the fault surface, because they are produced by the bend in the decollement as the beds move up the ramp. Axial surface B′Y′, which is pinned to the bend in the fault, is actively deforming the hanging wall beds. The bend in the fault causes the deformation. Axial surface B′Y′, which was initially at the BY position, passively moves up the ramp as material moves through the bend in the fault. Similarly, the beds moving up the ramp and onto the flat must roll through axial surface AX, forming the frontal dip panel AXX′A′ (Fig. 10-37a). Axial surface AX, which is also pinned to a bend in the fault, is a locus of active deformation and rotation of the hanging wall beds. The beds roll down at axial surface AX and form the front limb of the anticline. As the beds roll through axial surface AX, they experience *bedding plane slip*. This slip produces shear in the frontal limb of the fold (Fig. 10-37b) and causes the frontal limb to dip at a higher angle than the back dip panel. This point is emphasized here because it will be applied to the

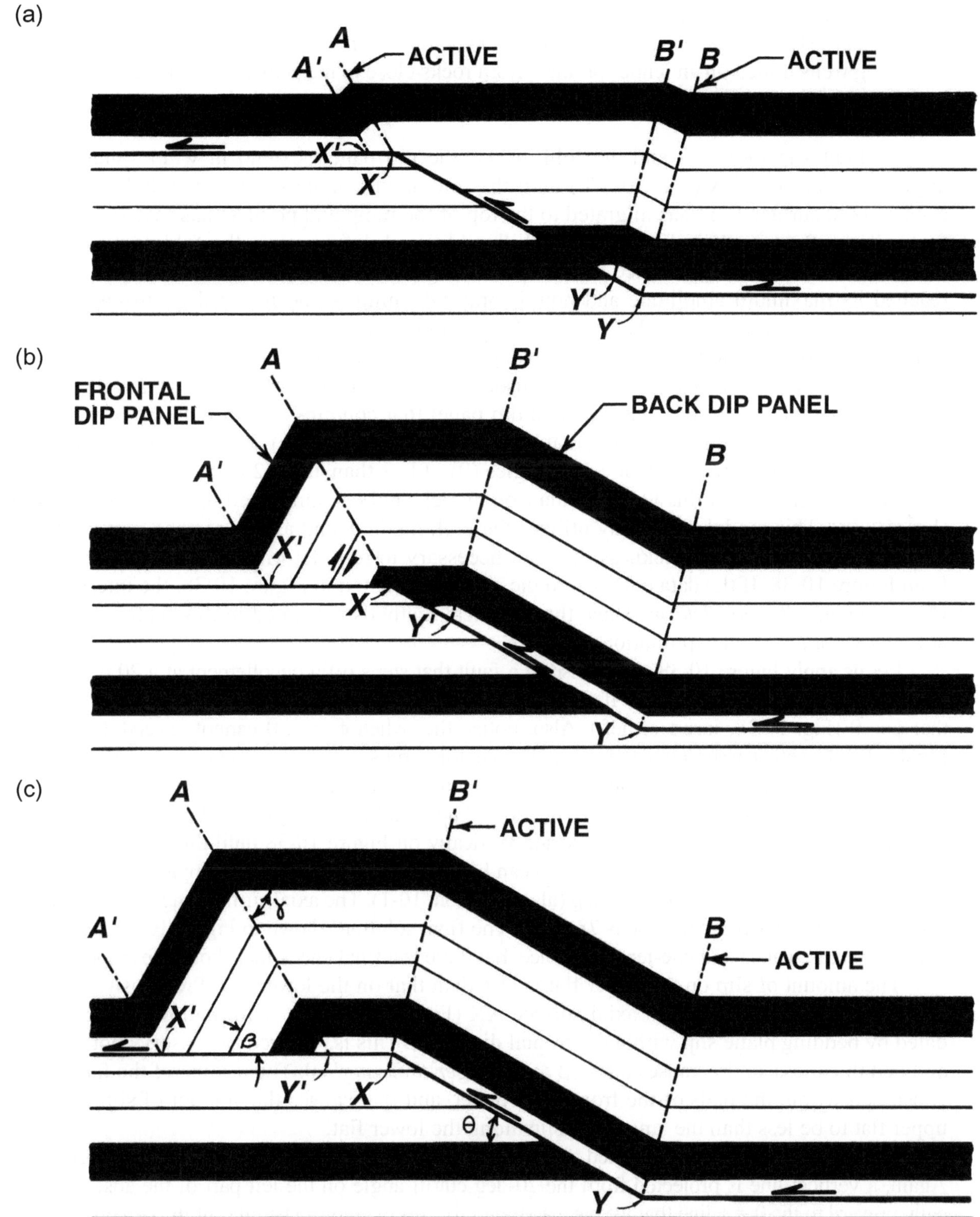

Figure 10-37 Fault bend fold kinematics illustrating the progressive development of beds riding up a thrust ramp. The beds are deformed by the active axial surfaces. (Modified after Suppe 1983, 1985. Published by permission of the American Journal of Science.)

solution of more complicated problems in the section on duplexes. As the beds *roll through the active axial surfaces* BY and AX, a *fracture porosity* is likely to form in the deformed beds. Some of the best producing wells drilled on anticlines produce from rocks close to the active axial surface B′X′, shown in Figure 10-37c. Bending of the strata along this active surface apparently imparts an excellent permeability in some folds.

As fault slip increases and the fold grows, the dip panels extend in width, and point Y′ migrates toward point X (Fig. 10-37b) until the fold attains its maximum amplitude. When this occurs, axial surface B′Y′ has migrated to the top of the ramp and point Y′ has reached the upper footwall cutoff (point X in Fig. 10-37b). With additional deformation, the fold now extends by the lateral motion of axial surface AY′ away from axial surface B′X (Fig. 10-37c). The fold has reached its maximum amplitude and now is only widening, so no material is currently rolling through the AY′ axial surface. This surface has become inactive. However, material continues to roll through the B′X and BY surfaces, probably further fracturing the rock.

The resulting *idealized* fold shape, caused by simple step-up of the hanging wall material along a ramp and onto a flat, has a frontal dip panel that contains a slightly higher dip than the back dip panel (β is usually slightly greater than θ, as in Fig. 10-37c). Thus, the fold geometry is roughly symmetric, particularly at cutoff angles (θ) of less than about 20 deg.

The mathematics of this volume-balanced model can be summarized in the form of a graph (Fig. 10-38). This model vigorously utilizes the kink method, and as this method conserves volume, line length, and bed thickness, it is not necessary to retrodeform a solution that is derived from Figure 10-38. If the data conform to the angles presented in Figure 10-38, the interpretation will *automatically retrodeform*. Thus, the graphical methods presented in this section are useful in both exploration and exploitation activities.

Let us apply Figure 10-38 to the case of a fault that steps off a decollement at a 20 deg (initial) cutoff angle and ramps to an upper flat that parallels the lower decollement (Fig. 10-39a). This means that $\phi = \theta$ (Fig. 10-38, insert on left). Also, notice that when $\phi = \theta$, θ cannot exceed 30 deg (see Fig. 10-38). The other assumption that we shall make for purposes of demonstration is that the amount of slip on the lower decollement is equal to the ramp length. This means that the axial surface (B′Y′ in Fig. 10-39b) has moved up to the top of the ramp. The initial cutoff angle of 20 deg can now be read off the left part of the abscissa and projected vertically on Figure 10-38 until this line intersects the $\theta = \phi$ line. Next, the dip of the front flap (β) can be read off the more steeply dipping lines on Figure 10-37, which in this case is about 23 deg (also see Table 10-1). The axial surface angle (γ) can be read off the ordinate, which in this case is 78.5 deg. The final solution, shown in Figure 10-39b, will automatically area-balance and line-length balance, but there is a final check that should be made.

The amount of slip on the upper flat is less than that on the lower flat. Previously, we stated that as the beds rolled through axial surface AX (Fig. 10-37a), the deformation was accommodated by bedding plane slip within the frontal dip panel. This is required to conserve both volume and bed thickness, and it causes *angle* β *to be larger than angle* θ. Thus, some of the fault slip is consumed within the beds of the frontal dip panel, and that causes the amount of slip along the upper flat to be less than the amount of slip along the lower flat.

The amount of slip to be expected along the upper flat can be determined by using Figure 10-40. Again, a vertical line is projected from the 20-deg cutoff angle on the left part of the abscissa vertically upward to the $\theta = \phi$ line that we have assumed for this example. The ratio of the slip on the upper flat relative to the slip on the lower flat can now be read from the R lines on the diagram, which in this case is about 0.87. Therefore, the slip along the upper flat must be 0.87 of the slip along the lower flat. Field geologists have often observed that the slip on faults dies or decreases within the cores of folds, and this is one reason why fold belts die toward the foreland. This exercise for checking slip is most useful when experimenting with structures that exhibit unusual geometries or complicated shapes. Another check on the solution would be to measure bed lengths on more than one structural level.

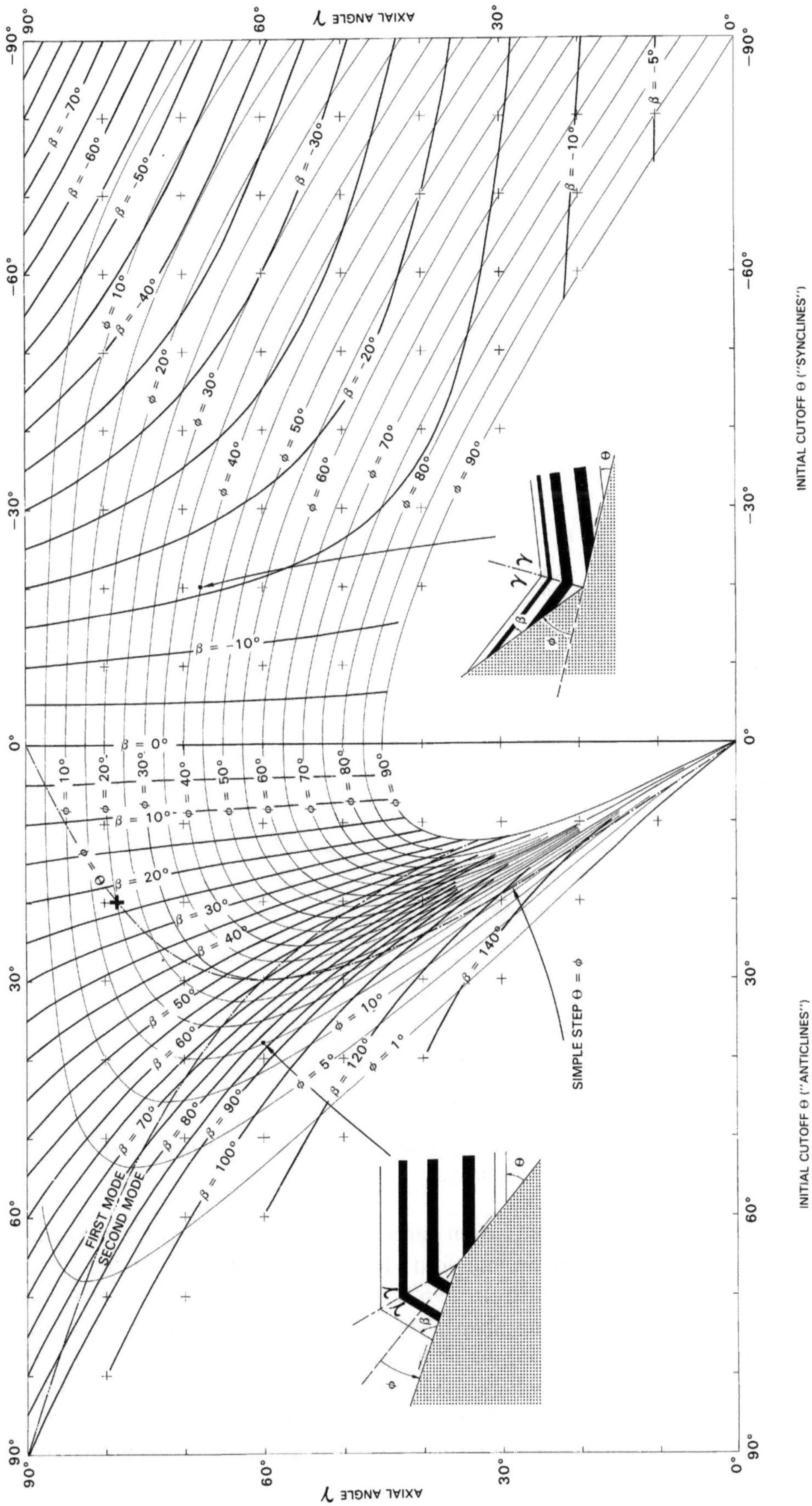

Figure 10-38 Fault bend fold graph showing angular relationships between the initial cutoff angle (θ), the frontal dip panel (β), and the axial surface angle (γ). (From Suppe 1983. Published by permission of the American Journal of Science.)

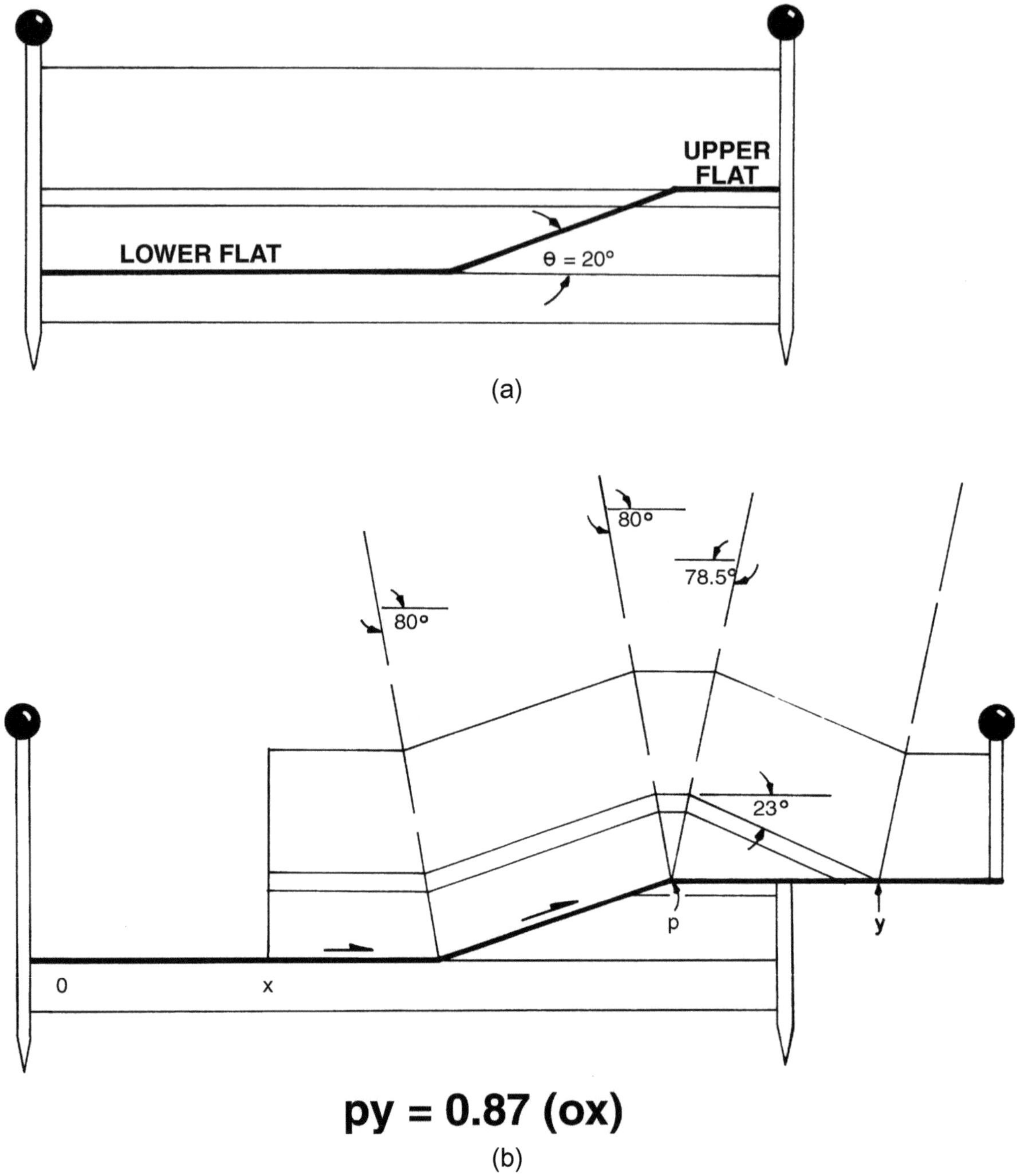

Figure 10-39 Fault bend fold exercise for beds ramping up a fault with a 20-deg cutoff angle and with slip on the lower decollement equal to the ramp length.

We can now present a major conclusion concerning fold geometry. Perhaps you have noticed the relationship between fault shape and fold shape. The shape of the fault is related to the shape of the fold (Rich 1934; Dahlstrom 1969) and indeed, *fault shape determines fold shape* (Fig. 10-36). Thus, if you know fault shape, you can predict fold shape, and conversely, if you know fold shape, you can infer something about fault shape. For example, on Figure 10-37b and c, notice how the thrust ramps up where the fault flat intersects the synclinal axial surface BY. So, as the fault ramp parallels the back limb of the upturned hanging wall beds. The location where the thrust intersects the active axial surfaces AX or B′X, the thrust forms the upper flat.

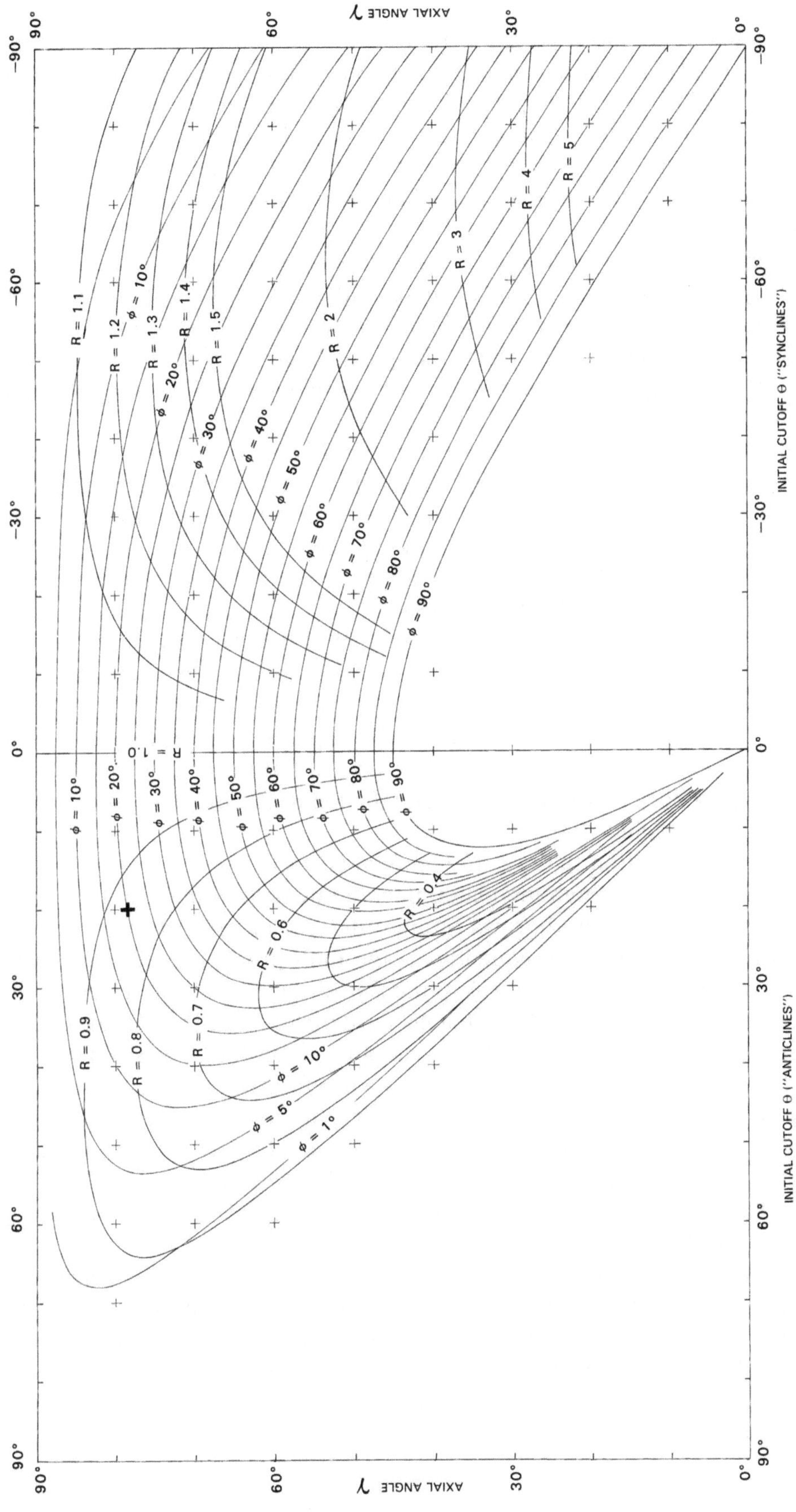

Figure 10-40 Fault bend fold graph showing the amount of slip to be expected along different portions of a fault surface. The R lines indicate the ratio of slip along the upper flat relative to slip along the lower flat. (From Suppe 1983. Published by permission of the American Journal of Science.)

Fault Propagation Folds

Fault propagation folds are a common fold type observed in outcrop and on seismic data (Figs. 10-41 and 10-42), and like fault bend folds, they are known to be good hydrocarbon producers. Fault propagation folds possess the particular characteristic that as the fold grows, the deformation advances at the tip of a propagating thrust fault (Fig. 10-43), hence the name "fault propagation fold" (Suppe 1985). As long as the structure has not been faulted through (i.e., been subject to breakthrough), the slip is consumed by bedding plane slip within the frontal limb of the fold (Fig. 10-43c).

Fault propagation folds typically have higher cutoff angles than fault bend folds, in the range of about 20 deg to 40 deg, which causes these fold types to possess steeply dipping to overturned frontal limbs that commonly do *not* image on seismic sections, along with a characteristic *asymmetry* (Figs. 10-41 and 10-44). This striking asymmetry, where imaged on seismic sections across folds with less dip, has the appearance of a striking snake, giving rise to the expression *snakehead structure* (Fig. 10-41).

The kinematics of simple fault propagation folds are as follows. Suppe and Medwedeff (1990) discuss more complex models in their paper. A fault, propagating upward from a decollement, causes beds at the front of the propagating fault tip to bend forward as material moves up the ramp (Fig. 10-43a). As in fault bend folding, the beds will also bend up the ramp created by the propagating thrust fault as they move through axial surface B, creating the back dip panel outlined by axial surfaces B and B′ (Fig. 10-43a).

In this style of folding, an increase in the amount of deformation within the core of the fold accommodates the slip on the fault. Therefore, the beds near the tip of the thrust fault *bend forward*, commonly at steep angles (Fig. 10-43a). The rotation of these steeply dipping beds, and bedding plane slip between the beds, *consumes* the slip along the thrust fault. Thus, the slip on the fault dies out within the core of the fold. This type of fault is referred to as a *blind thrust*. The more steeply dipping beds between the front and the top of the structure form two axial surfaces, A and A′, as shown in Figure 10-43a.

During fault propagation folding, *all* the axial surfaces are active and, with the exception of axial surface B, move through the material as the beds deform (Fig. 10-43). As the propagating fault extends and the fold grows in amplitude, it incorporates more material into the frontal limb of the structure. Consequently, as the fault propagates forward and as axial surface A′ moves away from axial surface A, point 2 of Figure 10-43a and b rolls through A′ into the steeply dipping frontal limb. With increasing deformation, the dip panels, as defined by axial surfaces A and A′ and by B and B′, broaden (Fig. 10-43c). Axial surface B′ is an *active* surface, as beds roll through it from the crest of the fold into the back limb. Axial surfaces A and B′ form a branch point at the same stratigraphic horizon as the fault tip. As the fault propagates, the loci of deformation, and thus the axial surfaces and the branch point, move forward and upward. The structurally lower beds fold more tightly, and the back limb of the fold widens. As the fold grows, the deformation also fractures the rock, which can affect the porosity and permeability.

Fault propagation folding can exhibit a variety of structural styles, depending on the cutoff angle (Fig. 10-44) and the amount of slip. As the cutoff angle increases, and for the same amount of slip, the folding will appear to be more symmetric on seismic sections even though the amount of slip remains unchanged. If the fold forms according to the processes described in Figure 10-43, the cutoff angle can be determined directly from the dip of the beds within the back dip panel, as these beds parallel the ramp.

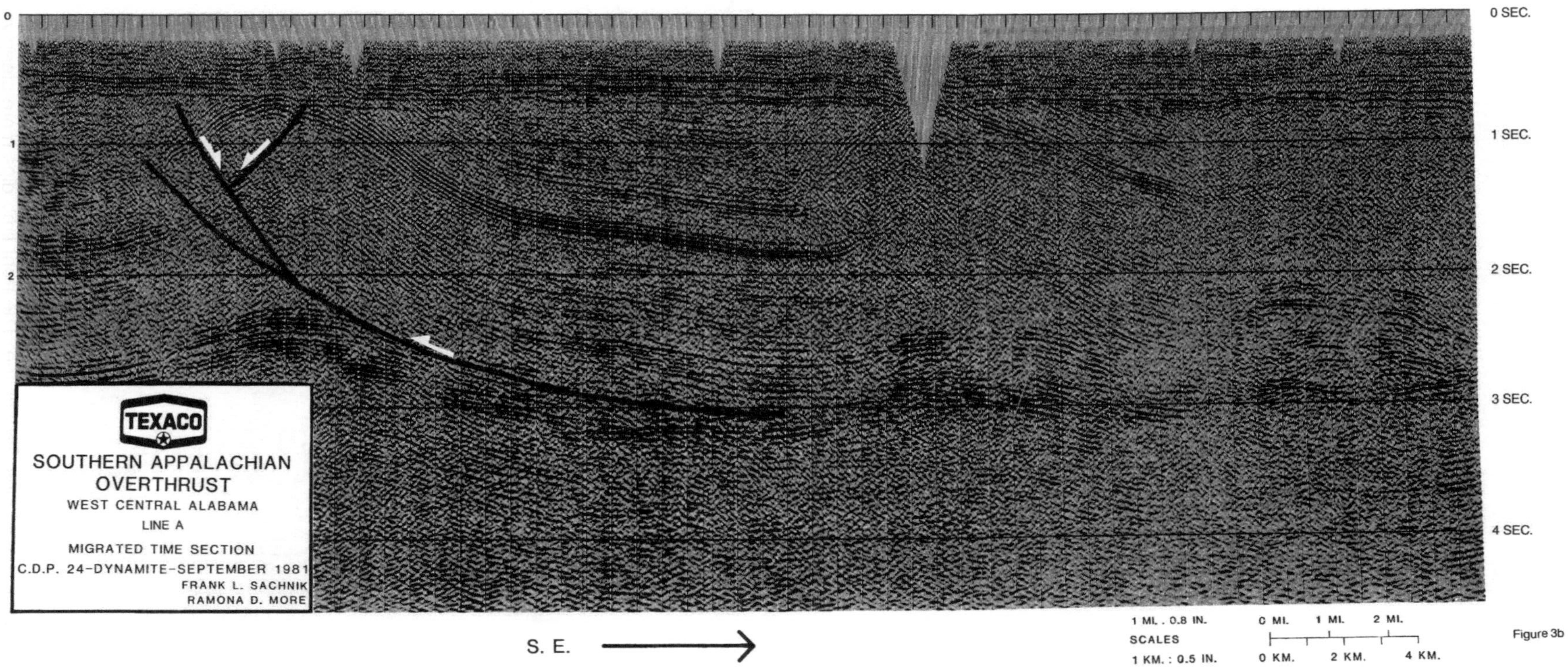

(a)

Figure 10-41 (a) Two-dimensional seismic section imaging asymmetric fault propagation folds, Southern Appalachians, Alabama. (Interpretation by R. Bischke. After Sachnik and More in Bally 1983 AAPG©1983, reprinted by permission of the APPG whose permission is required for further use.) (b) A 3D seismic example of a fault propagation fold from Northern Apennines, Italy (Storti et al. 2005. Reprinted by permission of the AAPG whose permission is required for further use.)

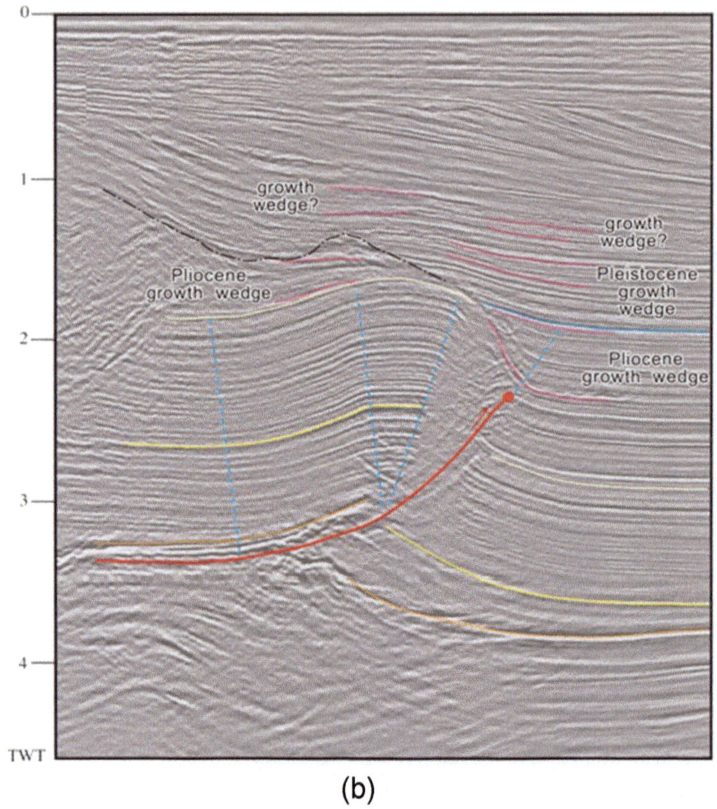

Figure 10-41 *(continued)*

Figure 10-42 Fault propagation fold, Appalachians, Tennessee. The front limb dips more steeply than the back limb. Thrust fault dies near synclinal axial surface at the front of the fold. (From Suppe 1985.)

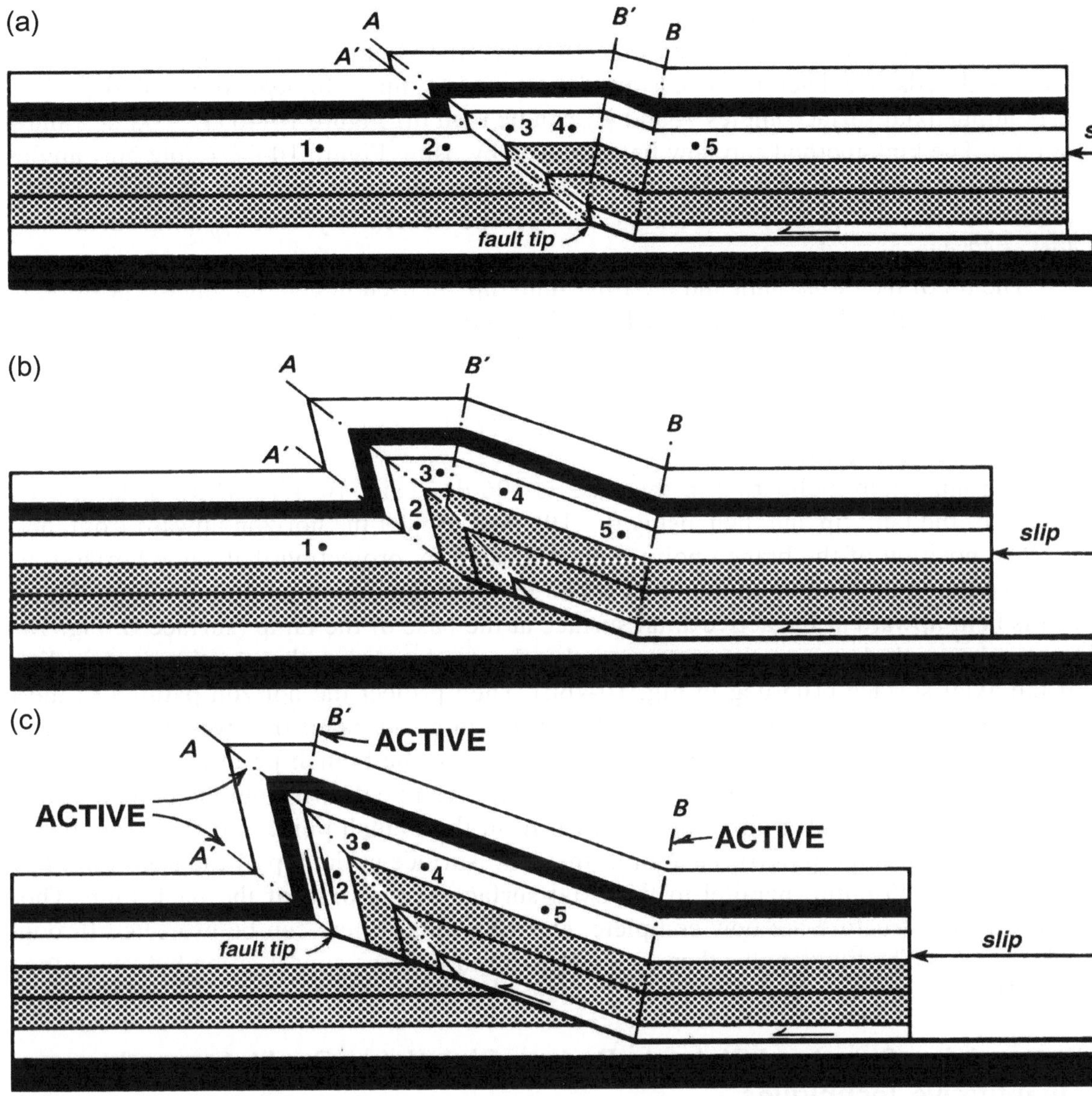

Figure 10-43 Fault propagation fold kinematics, illustrating the progressive development of beds deforming at the tip of a propagating thrust fault. (Modified after Suppe 1985.)

Given additional amounts of slip, the fault propagation may find a weak or incompetent horizon that parallels bedding and becomes a hybrid fault bend fold (Fig. 10-45b, decollement breakthrough). Alternatively, the structure can break through the anticlinal, the synclinal, or the overturned limb portions of the fold, creating more complex geometries (Fig. 10-45a, c, and d).

As with fault bend folds, fault propagation folds can be balanced using formulas or graphs, as in Figure 10-43 (Suppe 1988; Suppe and Medwedeff 1990). In outcrop or on seismic sections, simple fault propagation folds can be balanced by observing either the ramp angle (θ) or the back limb dip of the fold. When using seismic data, remember to depth-correct the seismic or choose sections that are roughly on a scale of one to one. This can be readily accomplished on the workstation over a given interval by using checkshot data or velocity information. We have found this procedure to be adequate for most cases.

We study a simple case for balancing fault propagation folds, using Figure 10-47. For example, you may observe beds on a depth-converted seismic section dipping and overlying horizontal beds. The back limb beds are determined to dip at 30 deg, so $\theta = 30$ deg. The corresponding axial surface angles γ_p and γ^*_p can be read off Figure 10-46 ($\gamma_p = 53$ deg and $\gamma^*_p = 38$ deg). The kink method can now be employed by using Figure 10-47. (Note how angles are measured in the figure.) First, construct a 30-deg dipping ramp, with the tip of the thrust fault as best determined from seismic data. Construct the structurally lower γ_p axial surface, which dips at 53 deg.

The tip of the thrust fault and the front limb dip are used in determining the position of the branch point defined by the upper γ_p, B′, and γ^*_p axial surfaces (Fig. 10-47). The inclination of the front limb (β) is defined by

$$\beta = \theta + 2(\gamma^*_p)$$

so the frontal limb inclination in this case is 106 deg (Fig. 10-47b). Draft a flat horizon to the left of the fault tip and then, using $\beta = 106$ deg, project the horizon upward from the fault tip. The position of the branch point is then located by projecting that same horizon, which is above the fault flat and on the level of the fault tip, across the fold's back limb. This horizon is bent upward at the active axial surface at the base of the ramp (surface B, Fig. 10-43). Bisect the angle between the ramp and the flat to determine the position and inclination of the axial surface (105 deg in Fig. 10-46b). Then project the horizon parallel to the fault ramp, upward to the intersection with the front limb dip panel that was projected upward from the fault tip. The projected horizons intersect at the branch point (Fig. 10-47b). Next, the structurally higher γ_p axial surface (53-deg dip) can be drawn from the branch point. The γ^*_p axial surface (38-deg dip) can be drawn from the branch point into the core of the fold and to the fault (Fig.10-47b). Then draft the axial surface at the top of the back limb upward from the branch point, parallel to the axial surface at the base of the back limb. The elements of the structure are now complete, and additional layers can be projected throughout the structure. The line lengths then can be measured for area conservation between pinpoints (Fig. 10-47c).

Determining Structural Styles to Reduce Risk Using Qualitative and Quantitative Techniques

As structural style affects wellbore position and reservoir size, the correct identification of structural styles can increase production and reduce costs and risks (Kostenko et al. 2008). There are two major end-member structure styles: (1) fault bend folds and the related shear fault bend folds and (2) simple fault propagation folds (Suppe 1983; Suppe and Medwedeff 1990; Bischke 1994a; Shaw et al. 2005) (see Figs. 10-37 and 10-43). In particular, the recognition of an optimal wellbore plan, relative to the steep limb associated with fault propagation fold styles, is likely to increase production and reduce costs (Kostenko et al. 2008). Variations on these two major structural styles exists and are described in detail in Shaw et al. (2005) and Hughes and Shaw (2014). Here we concentrate on the end-member structures, which should help geoscientists recognize general structural styles and, in particular, fault propagation folds (Fig. 10-41) from other structural styles.

When examining these two major fold styles, geoscientists should recognize that each fold style has *dependent geometric properties*. For example, **the amount of repeated section is related to the amount of throw, fault dip, and bed dip** (Chapter 7, or Fig. 10-55), **and structural relief is related to the amount of displacement**. Also, *these geometric*

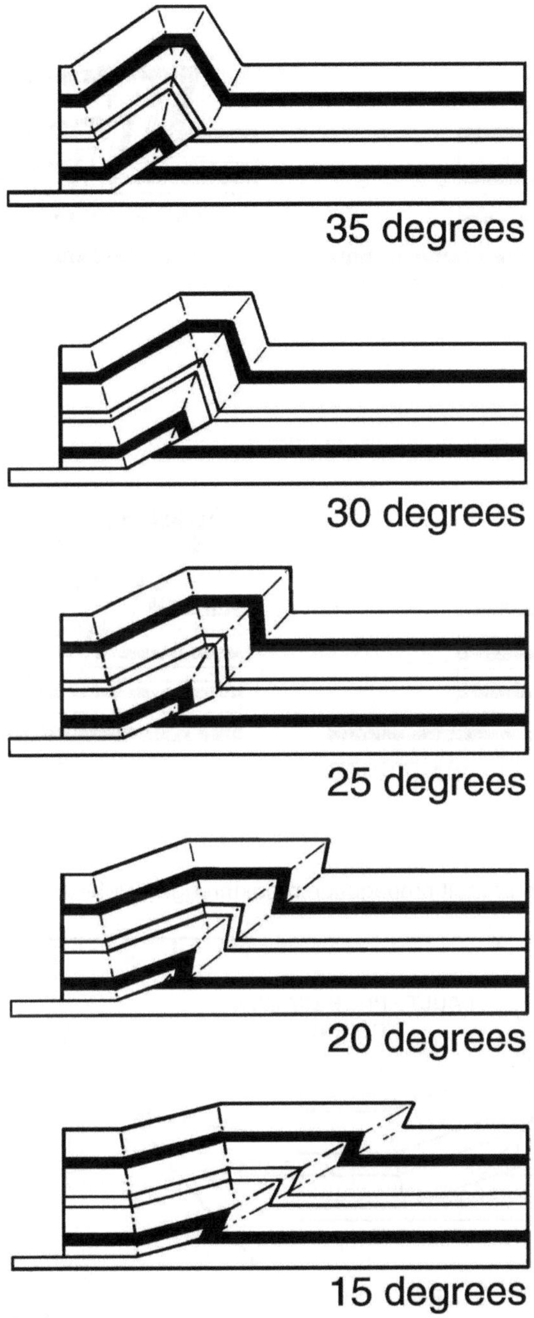

Figure 10-44 Fault propagation folds at different cutoff angles. Frontal limb dips increase as the cutoff angle decreases. At low cutoff angles, the frontal limb dips are too high to be imaged on conventional seismic sections. (Published by permission of John Suppe.)

parameters differ for the two major fold styles, and these differences are discussed in detail in the following sections. However, and regardless of structural style, all of the measurements of the geometric components should be consistent with each other. The measurements proposed are meant to be precise and to detect minor structural-style differences. The measurements can also be used to confirm interpretations, detect miscorrelations, and have properties similar to quick-look techniques proposed by Tearpock et al. (1994) and Wilkerson and Dicken (2001).

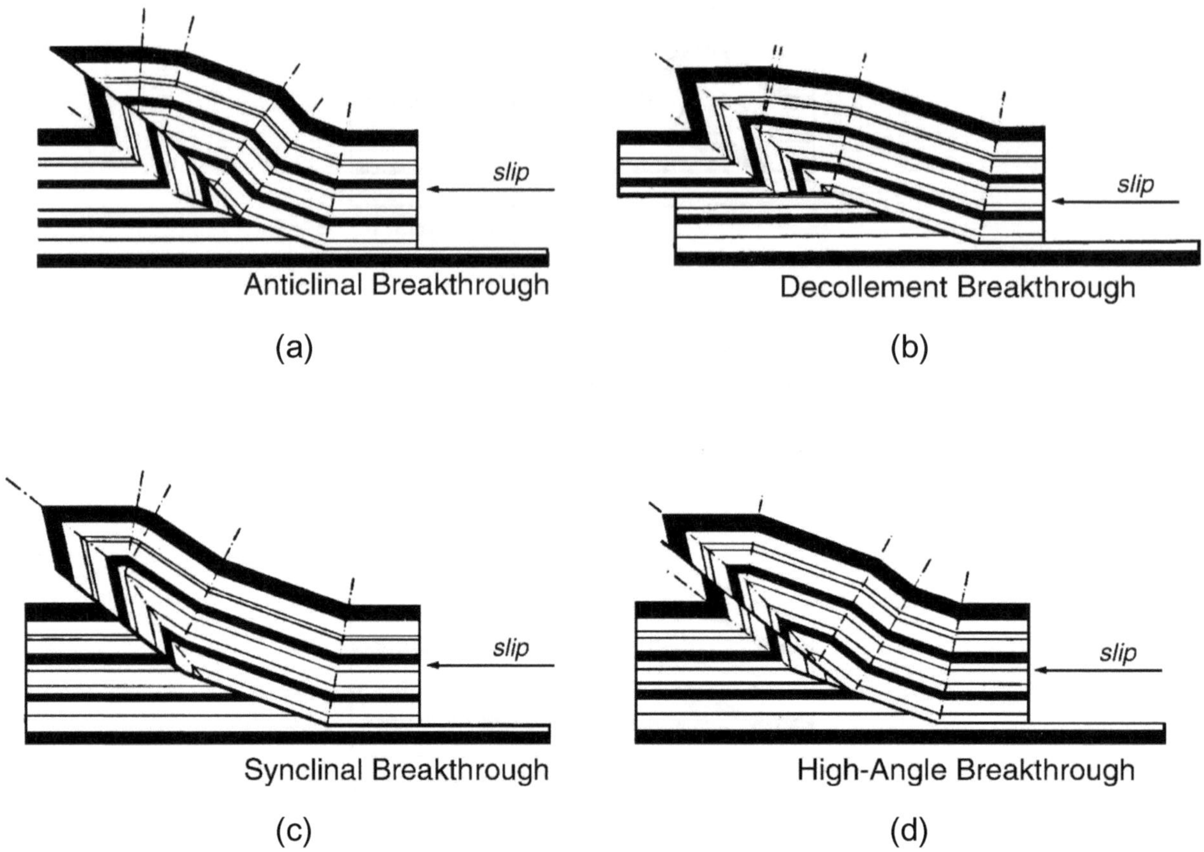

Figure 10-45 Different types of fault propagation breakthrough. (Published by permission of John Suppe.)

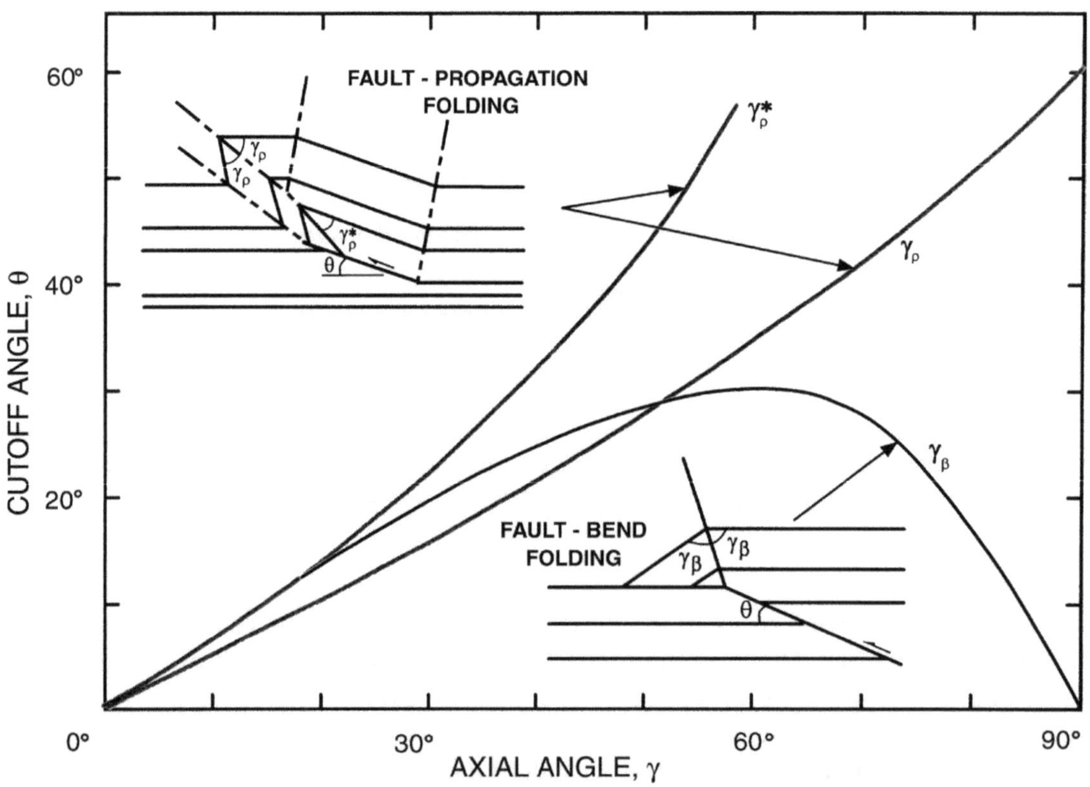

Figure 10-46 Fault propagation fold graph for a simple step-up from a decollement surface. (Modified from Suppe 1985.)

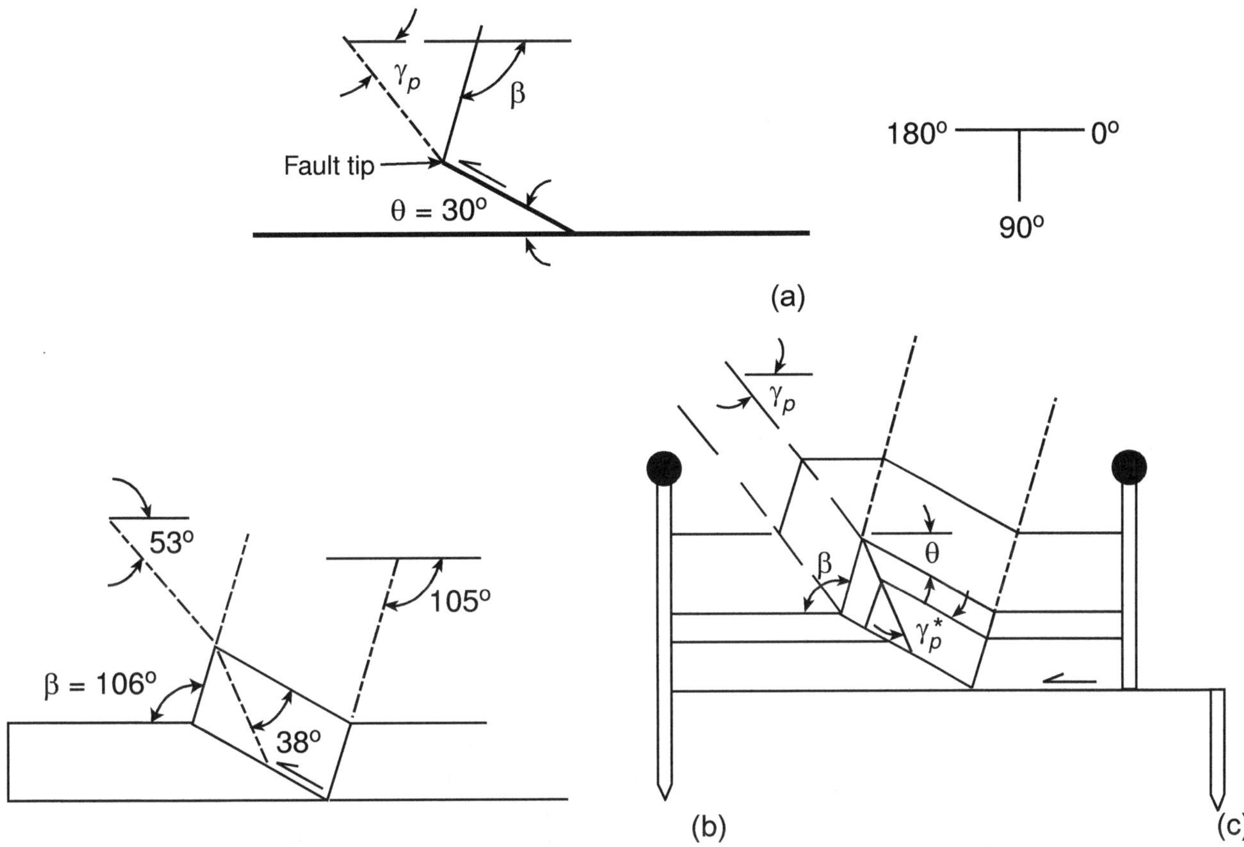

Figure 10-47 Fault propagation fold exercise.

Qualitative Observations for Determining Structural Styles

The two main structural styles are geometrically different and can often be distinguished on interpreted and well-imaged seismic sections. However, real data often complicate interpretations, and a good understanding of structural styles will result in optimal well locations. Thus, we shall study both qualitative and quantitative methods for distinguishing structural style. We discuss qualitative methods first.

In general, fault propagation fold structural styles are tight *asymmetric* structures (Figs. 10-41 to 10-43) that can have steep frontal limb dips that often do not image on seismic (consult Fault Propagation Folds section). This nonimaged region is often interpreted as being faulted (Kostenko et al. 2008), whereas the main fault is often present at the base of the nonimaged region. The fault associated with these structures dies upward into the core of the fold, which means that the thrust fault decreases its displacement toward the surface. This has the effect of *increasing the structural relief at shallower depths, while increasing reservoir size.*

In contrast, fault bend fold structural styles are more *symmetric* than fault propagation fold styles and typically have wider crests, which may result in larger reservoir sizes than fault propagation folds. In fault bend folds, the fault typically bends downward before surfacing, and the displacements on the flatter dipping portions of the fault do not die out (Hughes and Shaw 2014) (Figs. 10-34 to 10-37). This has the effect of *maintaining the amount of the structural relief.* In other words, on fault bend folds, *the structural relief remains about constant* from the crest of structure to the base of the limb. On fault bend folds, which have horizons that are horizontal or restored to the horizontal, the structural relief equals the throw and the repeated section (Fig. 10-48). Repeated section is *the vertical distance or vertical separation* from the hanging wall horizon to its corresponding footwall horizon (Chapter 7, Fig. 7-28).

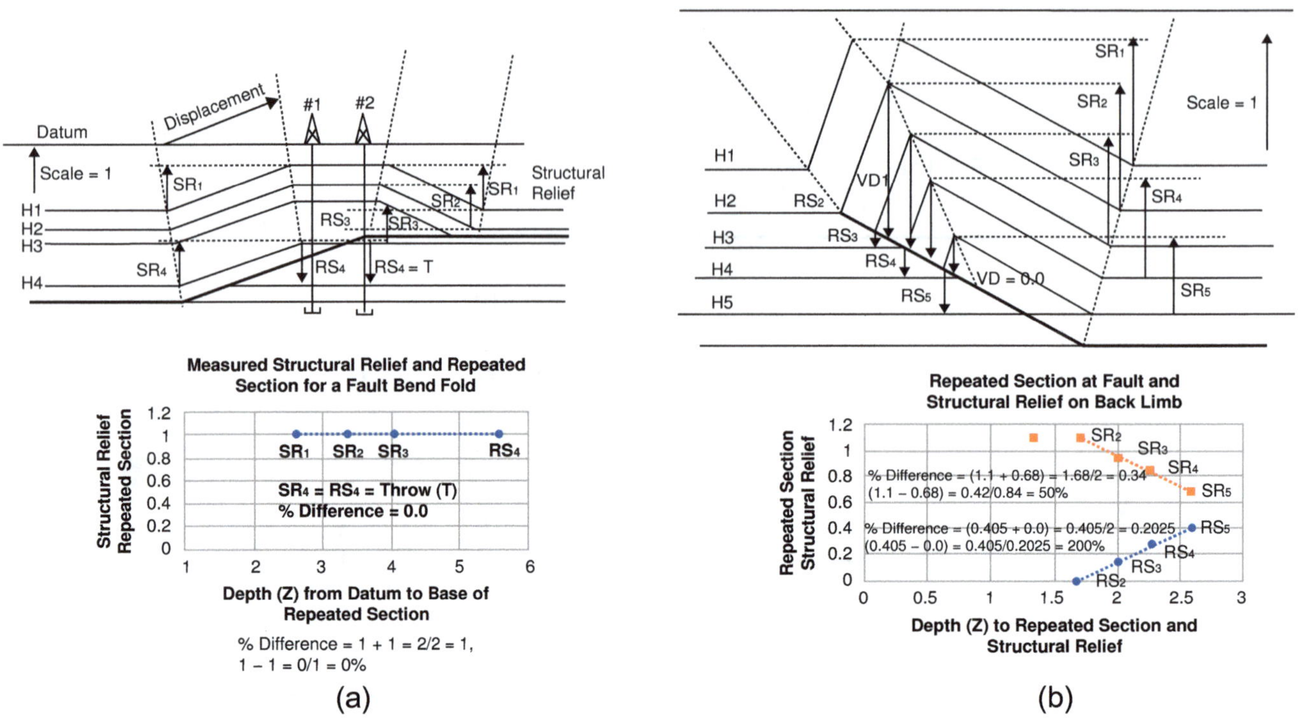

Figure 10-48 (a) A balanced end member, fault bend fold model that ramps off of the horizontal horizons. In this most basic of models of horizontal and uniform thickness beds, the structural relief (SR) equates to the throw (T) and repeated section (RS). Horizons can be projected from the hanging wall into the footwall to measure the projected repeated section. Also, the displacement or dip slip = SR/sin (bed dip). Real structures can be compared to the modeled SR and RS data on an SR/RS versus depth (Z) plot (see Bischke 1994b). On this end-member fault bend fold model, the measurements plot as a horizontal line and the % difference = 0.0. (b) A balanced end member, fault propagation fold model that ramps off of horizontal beds. In this end member model, the structural relief at the back and front of the structure decreases with increasing depth (Z), while the repeated section increases with increasing depth. This geometry contrasts with fault bend folds (Fig. 10-45a), as shown in the SR/RS versus depth (Z) plot. Furthermore, the % differences are 50% to 200% and not 0.0%, and thus % differences help determine structural styles. (Published by permission of R. Bischke.)

Quantitative Methods for Determining Structural Styles

There are several quantitative methods for determining structure style, which must be consistent. One is to measure the displacements (vertical separations) of the horizons across the fault surface versus the distances along the fault surface (Hughes and Shaw 2014). The vertical separations are a measure of the dip slip on the thrust. On fault bend folds, the displacements on the ramping fault are about **constant from the base of the ramp up to where the fault bends downward** (Hughes and Shaw 2014). The displacements then decrease at the downward bend in the fault surface (see Figs. 3 and 6 in Hughes and Shaw 2014). In contrast to fault bend folds, on fault propagation folds, displacements on the ramping fault **decrease upward** and often die out before reaching the surface (see Figs. 9, 10, and 12 in Hughes and Shaw 2014).

Hughes and Shaw conclude that structural deformation mechanisms can transition along the fault surface and contain components of the end-member fold types (their Fig. 14). For example, on the same structure, shear fault bend folds (see Fig. 10-53 for an example) can transition to fault bend folds. On shear fault bend folds, the shear parallels the bed surfaces (Corredor, Shaw, and Suppe 2005, in Shaw et al. 2005). Furthermore, fault bend folds can exist in combination with fault propagation fold mechanisms. Our study supports their conclusions that many structures have different mechanisms.

The second method, described here, is to measure the vertical separation (Reid 1909 for definition, or see Fig. 7-28) in the form of *repeated section* (RS) from the hanging wall to the footwall, or the amount of the *structural relief* (SR) for each significant horizon. The structural relief is the vertical distance from the top to the base of the limb of the structure (Fig. 10-48a and b), often defined by an inflection point. Another method is to measure the throw (T).

Next, plot these measurements on a structural relief or on a repeated section versus depth (Z) graph (Fig. 10-48a) (also see Chapter 13). Measurements of structural relief can be obtained from either limb of the fault bend fold, although the back, or the more *gently dipping limb, may image better* than the frontal limb. This procedure is different from studying the slip on the fold and complements Hughes and Shaw's (2014) methods of measuring and confirming the displacements on reverse faults. Furthermore, in the hanging wall, the structural relief measurements can be applied to uninterpreted horizons. This is particularly valuable if the footwall horizons do not image or if velocity problems exist that distort geometry (Sheriff 1977, 1980), thus making *displacement* measurements possible. In addition, the core in the *frontal limbs of fault propagation folds often do not image*, as the front limb can be near vertical, thus containing incoherent data (see Storti et al. 2005) (Fig. 10-41).

A number of structural geometries or components on folds are geometrically related to each other and must be consistent with each other. In Chapter 7, we discuss the geometric relationships of repeated section and throw. For example, using Figures 7-27 and 7-28, we derive equations for repeated section (Eqs. [7-4] and [7-5]) that relate repeated section to throw using bed and fault dips. In turn, **throw is also related to displacement or slip on the fault and to fault dip** (Figs. 7-27 and 7-28). In addition, **structural relief is related to fault dip, shape, and displacement** (Figs. 7-27, 10-34, and 10-48a and b). We start with ideal, generic end-member examples of fault bend (Fig. 10-48a) and fault propagation (Figs. 10-48b and 10-49a) folds before studying well-constrained real examples.

A review of existing depth-corrected, 3D contractional structures indicates that there is value in measuring more that one of the structural components to determine the precise structural style. In other words, different conclusions concerning fault interactions or mechanisms are possible if geoscientists measure a combination of dip-slip displacements, throw, structural relief, and repeated section. Multiple procedures, such as quick-look techniques (Tearpock et al. 1994; Wilkerson and Dicken 2001), can rapidly detect horizon miscorrelations and retrodeformability. Also, 3D vertical separations are *simpler to measure than 2D displacements on fault surfaces*. This is particularly true when interpreting vertical well log data.

Figure 10-48a is a balanced, end-member fault bend fold on a scale of one to one. The vertical change in depth of the limbs of the fold define the structural relief, labeled SR1 to SR3 for the top three horizons (H1 to H3). One simply measures the vertical distance of the SR, as indicated by the vertical arrows. The SR measurements are taken at an inflection point on the dipping limb to the crest of structure. Often, the more gently dipping back limb images better that the front limb, and therefore measurement are often better conducted on the back limb of folds. Furthermore, when measuring SR, the beds need not be horizontal.

Two wells penetrate the generic crest of structure, with Well No. 1 penetrating the top of H4. The generic well thus encounters repeated section in the footwall, labeled RS4. The top or crest of H4 in the hanging wall is projected to Well No. 2 in order to measure the projected repeated section, RS4. In addition, Well No. 2 intersects the top of H3 in the hanging wall and the footwall, creating RS3. In this case, as the beds are horizontal, the repeated section equals the throw. *Notice that, for* ***uniform thicknesses*** *and horizontal beds, SR1, SR2, and SR3 equals RS3, RS4, and throw (T)*. Thus, when graphed on a repeated section/structural relief versus total depth (Z) correlations graph, the result is a horizontal line, indicating that the structural relief and repeated section are constant (lower Fig.10-48a). To determine structural style, these measurements can replace or complement measuring the displacements on the fault surface. Hughes and Shaw (2014) concluded that on the ramp, the displacements on end-member fault bend folds are constant.

In the section on real examples, and in the generic examples presented in the next paragraph, we present the % difference of the data. The % difference is the difference between the largest and smallest length measurements divided by the averaged sum of the difference between the largest and smallest length measurements. These values quantitatively define the percentage change of the measurements. On real fault bend folds, the % differences may approach zero (Figs. 10-48a, 10-51a, and 10-52). The % difference measurements help define structural style and also detect complex deformation histories (Hughes and Shaw 2014).

When measuring a repeated section on structures that exhibit large displacements and repeated section (and ramp off the footwall horizon), *measure the repeated section at about the same point where you measure the structural relief* (Figs. 10-50 and 10-51). This procedure is similar to drilling a vertical well. However, in some cases, it may be more convenient to measure the repeated section on either the forward or trailing side of the dip panel (Fig. 10-48a and b). Using an axial surface inflection point works best. In real structures, the hanging wall beds need not parallel the footwall beds as the fault ramps through the footwall, and thus the structural relief need not equal the repeated section. In the examples shown in Figures 10-50 and 10-51, the repeated section is larger than the structural relief.

In contrast to fault bend folds, fault propagation folds tend to behave differently, and the structural relief will not remain constant with increasing depth (Figs. 10-48b and 10-49a). While at shallow depths, the structural relief can be about constant, with increasing depths the structural relief decreases as the repeated section increases (Fig. 10-49b). Notice in Figure 10-48b that *in the frontal portions of the hanging wall, the vertical distance (VD), measured from the back limb to the fault surface, decreases to zero*. This vertical distance is typically measurable even if the seismic data in the hanging wall are incoherent. In addition, where seismic data are lacking, geoscientists can measure the vertical distance from the back limb to its corresponding *projected* correlation in the footwall (Fig. 10-49a).

In Figure 10-48b, measuring repeated section is identical to measuring the vertical separation at the fault surface to the corresponding horizon in the footwall (Figs. 6-35, 7-26, 7-27, and 13-19). In contrast to the fault bend folds, in the fault propagation fold structural styles, the % difference will be high and can exceed 50%. Thus, % difference measurements should help distinguish structural style.

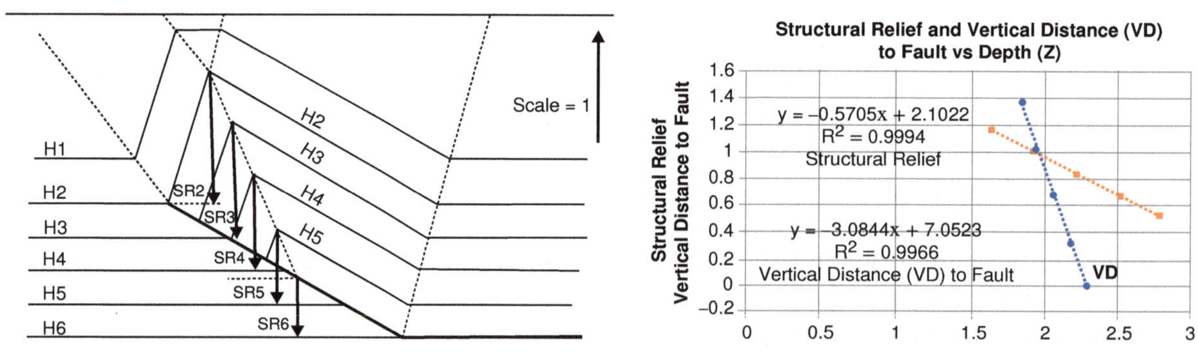

SR % Difference = (1.159 + 0.51) = 1.669/2 = 0.8345
(1.159 − 0.51) = 0.649/0.8345 = 78%
(a)

VD to Fault % Difference = (1.365 + 0.0) = 1.365/2 = 0.6825
(1.365 − 0.0) = 1.365/0.6835 = 200%
(b)

Figure 10-49 (a) Fault propagation fold displaying the structural relief (SR) in the front limb from the hanging wall to the footwall. (b) On this SR and vertical distance (VD) versus depth (Z) plot, the SR decreases with increasing depth (Z), and the SR % difference is large at 78%, indicating that the structure is a fault propagation fold. The structural style is confirmed as the VD from the hanging wall fault propagation folds horizon to the fault surface also decreases with increasing depth (Z) on the plot. At depth, the VD on the fault surface = 0.0, and thus the VD % difference = 200%. (Published by permission of R. Bischke.)

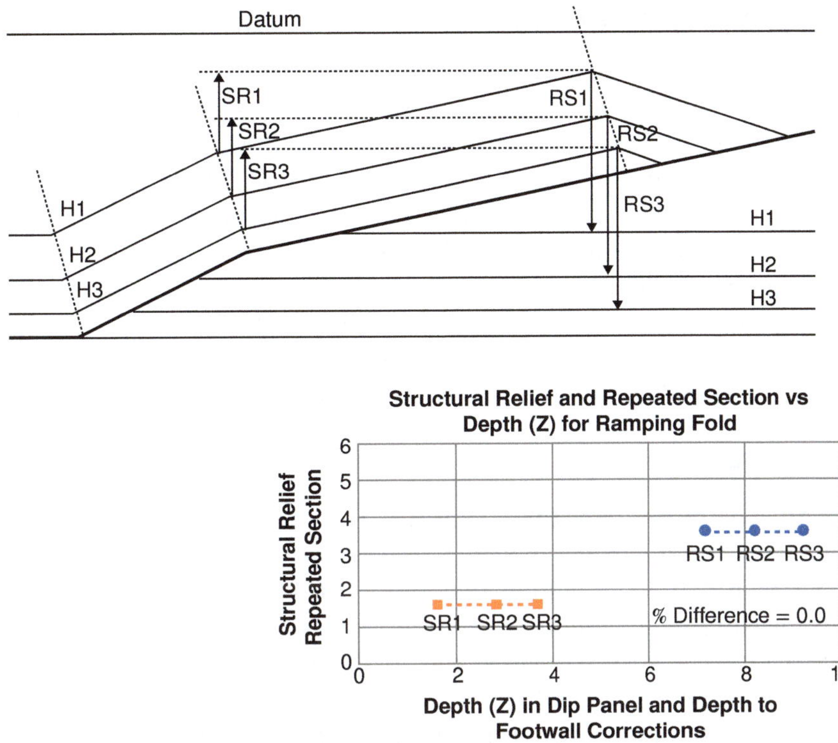

Figure 10-50 Thrust fault ramps through footwall and bends downward. This bend causes the structural relief to be different from the repeated section. The % differences are, however, about equal to zero, indicating that this structure is a fault bend fold. (Published by permission of R. Bischke.)

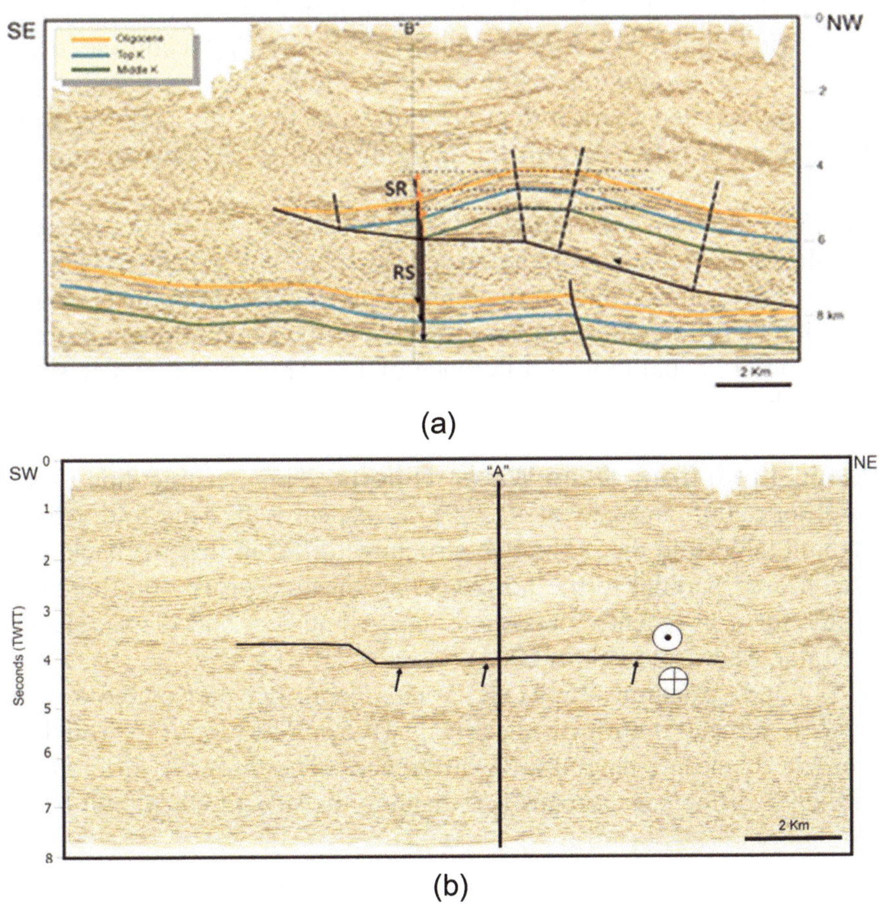

Figure 10-51 (a) A depth-corrected 3D seismic line is a dip line of the well-imaged El Furrial Fault-Bend Fold, Venezuela. The repeated section and structural relief data were measured at about the same position (Morales et al. 2005). (Reprinted by permission of the AAPG whose permission is required for further use.) (b) Strike line through the frontal panel of the dip line shown in Figure 10-51a is plane "B." On Figure 10-51b, the dip line shown in Figure 10-51a crosses the strike line on plane "A." The thrust fault is located where dipping beds terminate over horizontal beds.

Real Fault Bend Fold and Fault Propagation Data

Repeated section can exist in both well log and seismic section data. Here we study outcrop data, depth-corrected seismic data, and well-constrained well log examples that use correlations, surface dip, and dipmeter data. We start with constrained fault bend fold data before addressing fault bend fold data. In all of the real data calculations, the % difference measurements include all interpretation measurements, which includes existing structural interpretations, miscalculations or data problems, such as lateral velocity changes and poorly imaged footwalls. After increasing the size of the figures to 400% and 200%, the ability to measure with a magnifying glass is about 1 part in 60, or about 1.7%. In Chapter 7, the repeated section is defined as AE, and the throw as AC, and this convention is followed in the equations. The repeated section (AE)/throw (AC) ratio (Fig. 10-55) will be calculated when possible for each structure to confirm the interpretation.

Fault Bend Folds

Tiehchanshan Anticline, Western Taiwan. The Tiehchanshan Anticline is constrained by well log correlations and surface dip data. Based on reasonable correlations and lithologies, Suppe and Namson (1979) conclude that the TCS-34 well has 700 m of repeated section. According to measurements conducted on their cross section, the repeated section at the well is about 685 ft (1 m = 3.28 ft). Thus, the % difference is 2.2%. The measured AE/AC is 690/1150 = 0.6, and the AE/AC value calculated is (1 − tan(5 deg))/tan (average dip 13.5 deg) = 0.62. The calculated values are within the calculated values of the sine function and measurement errors.

Depth Corrected El Furrial 3D Seismic, Monagas Venezuela. The data shown in Figure 10-51 are from measurements collected from the well-imaged giant El Furrial Venezuela structure (Morales et al. 2005, in Shaw et al. 2005). The structural relief and repeated section measurements are from the prolific Oligocene to the less prolific Middle Cretaceous section. The structural relief is measured from the southeast side of the southeast dipping panel adjacent to the crest of structure. The repeated section data are collected from the hanging wall reflections at the *base* of the structural relief data in the hanging wall to the corresponding footwall reflections.

Figure 10-52 shows repeated section and structural relief calculations in kilometers (1 km = 3281 ft) along with the % differences for the structural relief and repeated section. Structural relief % differences for the southeast side of the dipping panel adjacent to the crest are equal to zero. The repeated section % differences data are collected at the same horizontal position as the structural relief data and are equal to a % difference of 1%. Thus, along this seismic line, the structure is a classic fault bend fold.

Depth-Corrected 3D Seismic Line from Deep-Water Nigeria. This seismic line from Nigeria (Corredor, Shaw, and Bilotti 2005, in Shaw et al. 2005) is a shear fault bend fold, which is related to the end-member fault bend fold data. The two structures differ in that the back and front limbs of shear folds are sheared parallel to the horizons (see sheared profile on the back limb of Fig. 10-53). Shear folds form in recently deposited, unconsolidated, and plastic shales subject to gravity sliding (Corredor, Shaw, and Bilotti 2005).

Figure 10-54 is a throw (T) and structural relief (SR) versus depth (Z) plot of the a to b back limb of the structure that was subject to uniform shear (Fig. 10-53). The plot of the roughly uniform-thickness, red to green horizon data have low % differences. The throw % differences are 5.7%, and the structural relief % differences are 4.8%, indicating the structure is a fault bend type of fold.

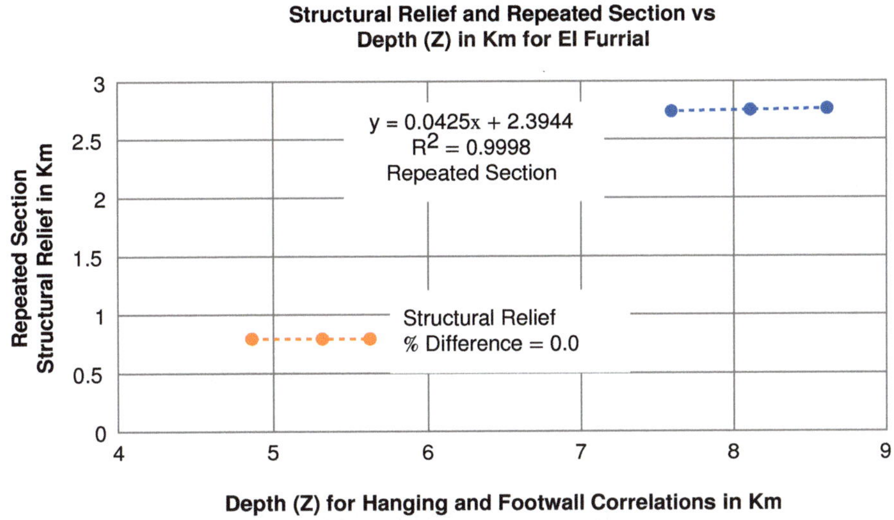

Repeated Section % Difference = (2.717 km + 2.76 km) = 5.477/2 = 2.739, (2.76 km – 2.717 km) = 0.43/2.739 = 1.57%

Figure 10-52 Structural relief/repeated section plot for the super-giant El Furrial, Venezuela, structure (Morales et al. 2005). Measurements from a 3D line indicate that at this location the structure is a classic fault bend fold. The R^2 repeated section data are close to 1.0, and structural relief data are equal to zero. (Published by permission of R. Bischke.)

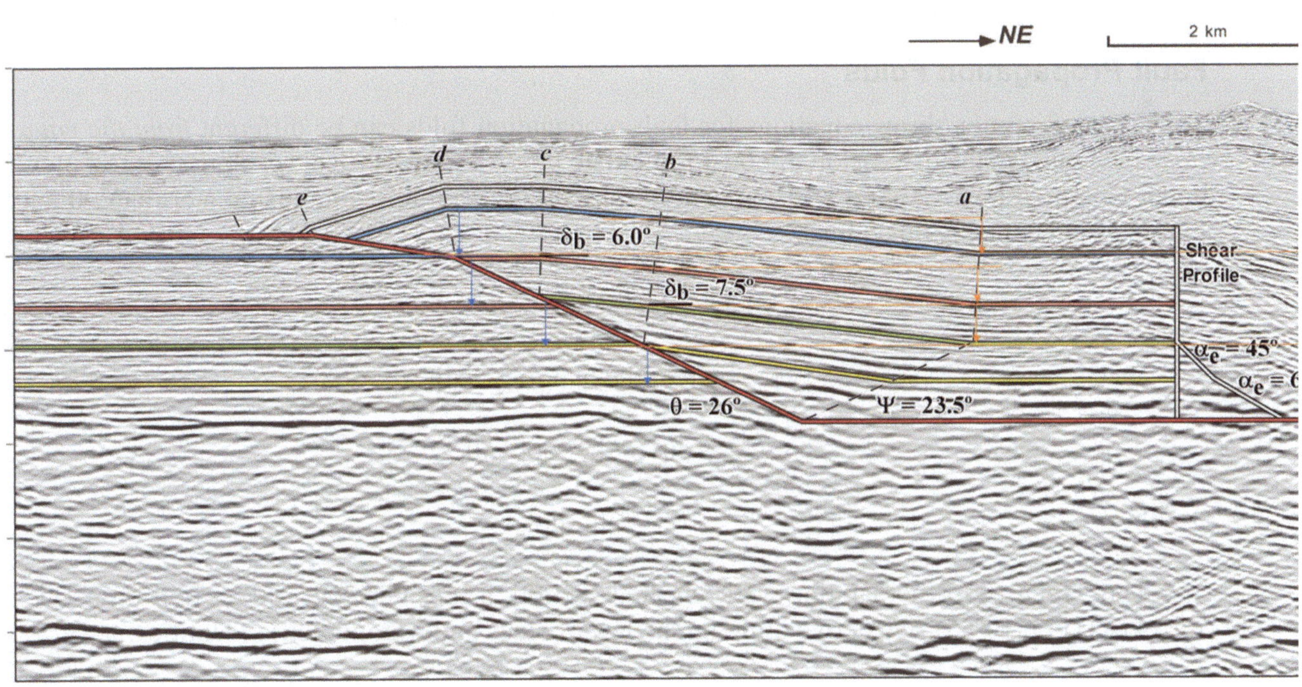

Figure 10-53 A well-imaged 3D and depth-corrected line of a shear fault bend from the deep-water Nigeria fold-thrust belt (in Corredor, Shaw, and Suppe 2005, in Shaw et al. 2005; reprinted by permission of the AAPG, whose permission is required for future use.)

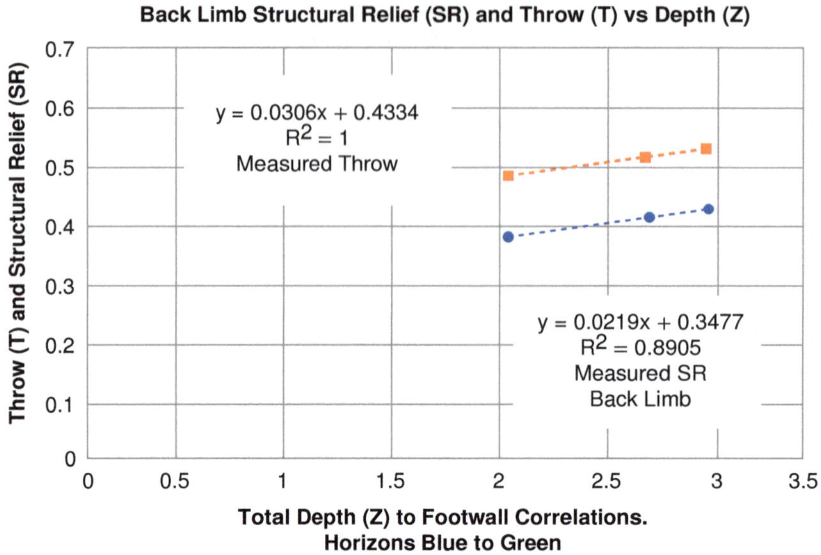

Throw % Difference 0.524 – 0.495 = 0.029
(0.524 + 0.495) = 1.019/2 = 0.5095
0.029/0.5095 = 5.7%

SR Back Limb % Difference = 0.4095 – 0.3904 = 0.019
(0.4095 + 0.3904)/2 = 0.4 or 0.019/0.4 = 4.8%

Figure 10-54 A throw and structural relief plot of the shear fault bend fold imaged in Figure 10-53. The data come from the uniformly sheared back limb of the structure and have low % differences. This indicates that the structure is related to fault bend folding. (Published by permission of R. Bischke.)

Fault Propagation Folds

The repeated section/throw equations for fault propagation folds can be different *from the equations* described in the discussion of how to calculate throw versus vertical separation in Chapter 7 under Compression. Also, Eq. (10-1) changes for hanging wall horizon dips greater than 90 deg. Figure 10-55 contains three dips.

$$\phi 1, \phi 2, \text{ and } \theta, \text{ where}$$

$$\phi 1 = \text{dip in the hanging wall}$$

$$\phi 2 = \text{dip in the footwall}$$

$$\theta = \text{fault dip}$$

Thus, relative to the hanging wall repeated section, a repeated footwall section ratio B′C′/AC exists (Fig. 10-55). We derive utilizing the law of sines and trigonometric identities:

$$(\sin(\phi 1 - \pi/2))/S = (\sin(\pi - \phi 1 + \theta))/B'C'$$

$$S = \sin(\pi/2)AC/\sin(\theta)$$

$$(\sin(\phi 1 - \pi/2))\sin(\theta)/(\sin(\pi/2)AC = (\sin(\pi - \phi 1 + \theta))/B'C'$$

$$B'C'/AC = (\sin(\pi - \phi 1 + \theta))/(-\cos(\phi 1))(\sin(\theta)) \qquad (10\text{-}1)$$

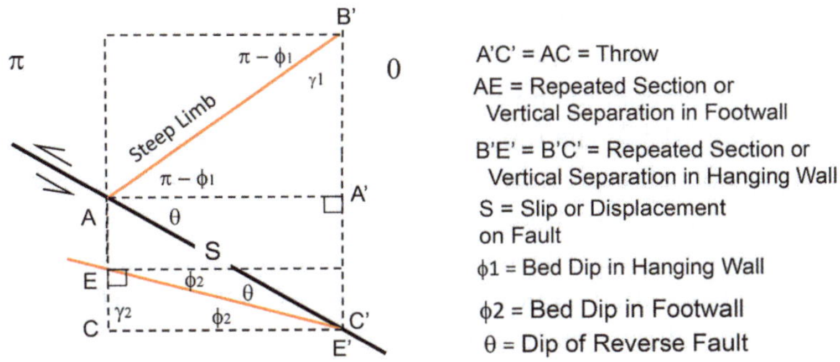

$AE/AC = \sin(\theta - \phi 2)/\sin(\pi/2 + \phi 2)\sin(\theta) = 1.0 - \tan(\phi 2)/\tan(\theta)$ for the hanging wall and top footwall cutoff.
For $\theta = \phi 1$, also use
$AE/AC = 1.0 - \tan(\phi 2)/\tan(\theta)$ (i.e., for Melin footwall 1)

Figure 10-55 As fault propagation folds are not similar to fault bend folds, different repeated section formulae are required to predict repeated section/throw calculations. The figure allows for a calculation of fault propagation folds that have a difference in dip between the hanging wall and footwall. Consult text for details. (Published by permission of R. Bischke.)

Outcrop of Fault Propagation Fold Dunlap, Tennessee. As stated previously, fault propagation folds are geometrically different from fault propagation folds. The first example is a fault propagation folds from an outcrop in the Appalachian fold-thrust belt. We measure a photograph interpreted by Boyer (1986; see his dashed lines or, in this book, Fig. 10-42). The structural relief on the horizon above the shale section establishes the scale on Figure 10-56. The datum for the depth (Z) measurements are relative to the top of the Boyer's Figure 6.

Relative to the fault surface, the measurements taken from Boyer's figure and plotted on Figure 10-56 show that the structural relief on the front limb of this fault propagation fold decreases to zero at the thrust fault (e.g., examine the arrows in the hanging wall of the front limb of Fig. 10-48b). This geometry is typical of fault propagation folds. In contrast to fault bend folds, the % difference calculation is large at 200%. The back limb structural relief also corresponds to fault propagation folding, with the % differences decreasing with depth. The % difference is high at 69%.

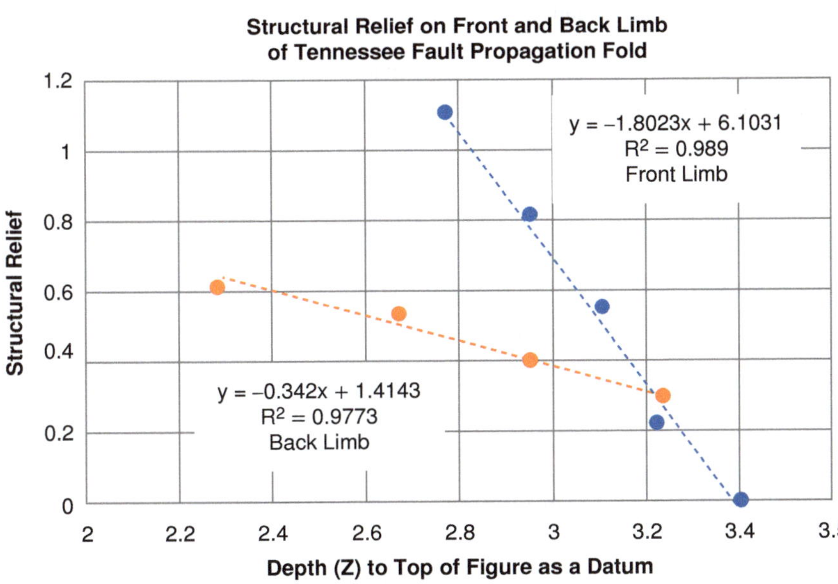

Front Limb % Difference = (1.1 + 0.0) = 1.1/2 = 0.55
(1.1 − 0.0) = 1.1/0.55 = 200%
Back Limb % Difference = (0.617 + 0.3) = 0.917/2 = 0.459
(0.617 − 0.3) = 0.317/0.459 = 69%

Figure 10-56 A back limb and front limb structural relief plot for an outcrop in Tennessee (Boyer 1986). The structural relief measurements exceed 50%, indicating that the structure is a fault propagation fold style. (Published by permission of R. Bischke.)

Melin Anticline, Western Taiwan. The Melin Anticline, interpreted by Suppe and Medwedeff (1990), is constrained by surface dips, dipmeter well logs, repeated section, and well log formation correlations. Our measurements of the back and front limb of the structural relief/repeated section data are shown in Figure 10-57. The repeated section data are measurements taken from the hanging wall to the equivalent footwall correlation. Again, the % differences are large at 53% for the structural relief and 142% for the repeated section.

For the top Chinshui Shale, the hanging wall and footwall measured repeated section/throw ratio = 0.882, and the calculated values are AE/B′E′ = 0.895 (Fig. 10-55 and Eq. 10-1). The footwall wall repeated section/throw measurements can be compared to the calculated values, and thus B′E′/A′C′ = 0.833. For the top Chinshui Shale, calculated values for the hanging wall = 0.835.

Fault Propagation Seismic Fold from Nigeria. This depth-corrected and well-imaged seismic line is from Nigeria (Hughes and Shaw 2014) (Fig. 10-58) and was interpreted as a fault propagation fold. The front limb structural relief was measured at the same position as the repeated section. As both data sets contain zeros, the % differences are large (Fig. 10-59), and both the structural relief % difference and the repeated section % difference are 200%. This fold has all of the characteristics of fault propagation folding (Hughes and Shaw 2014), although some interpreters may conclude after examining the structure that the structure is a fault bend fold!

Conclusions

This study of structural styles concludes that several or all of the geometric components should be measured to more precisely define structural styles. These measurements include fault displacements, throw, structural relief, repeated section, and repeated section/throw measurements and calculations. These geometric properties will detect problem interpretations or correlations and will establish a sound basis for confirming structural styles when the structural style is not obvious. This problem particularly occurs in the frontal limb of fault propagation style folds (Bischke 1994a; Kostenko et al. 2008), which can result in costly errors.

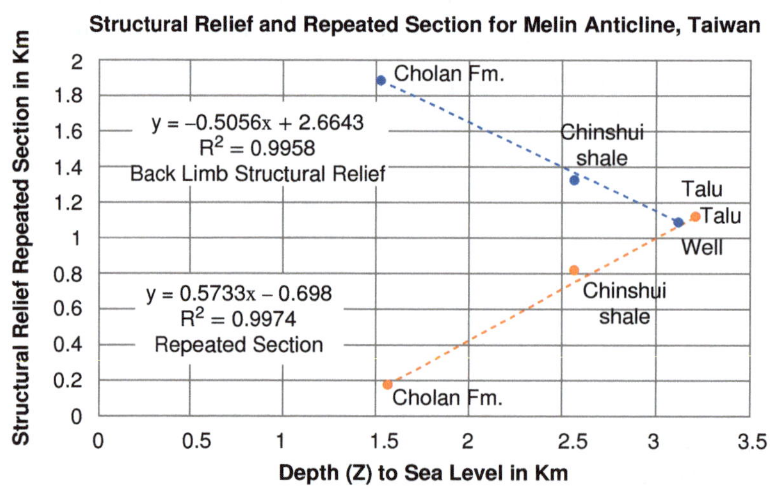

Back Limb Structural Relief % Difference = (1.9 km + 1.1 km) = 3.0/2 = 1.5
(1.9 km − 1.1 km) = 0.8/1.5 = 53%

Repeated Section % Difference = (1.12 km + 0.19 km) = 1.31/2 = 0.655
(1.12 km − 0.19 km) = 0.93/0.655 = 142%

Figure 10-57 The Melin structure is constrained by surface and well log dips, repeated section, and well log correlation data (Suppe and Medwedeff 1990). This balanced structure has % differences in excess of 50%, consistent with FPF. (Published by permission of R. Bischke.)

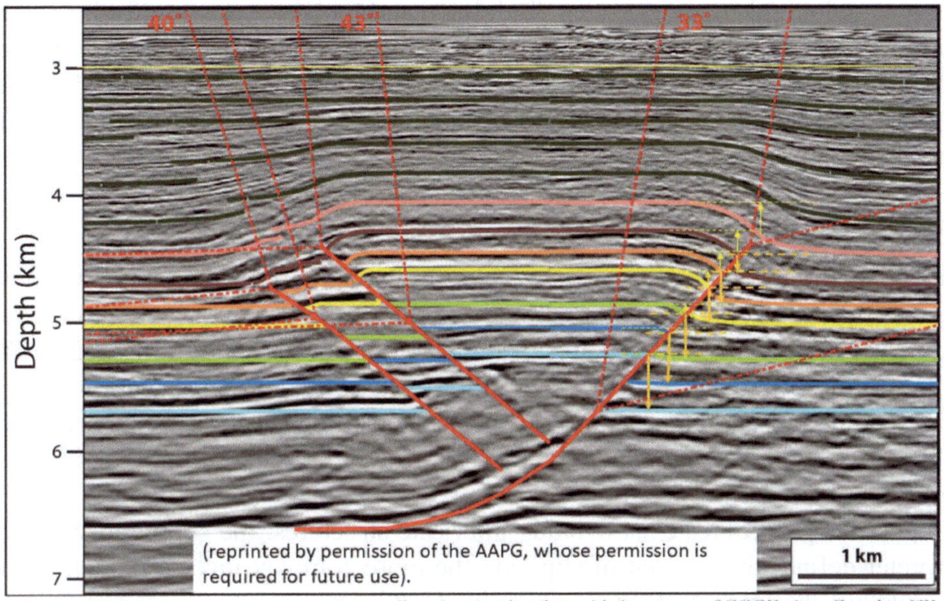

Data is owned and provided courtesy of CGGVeritas, Crawley, UK.

Figure 10-58 A well-imaged and depth-corrected 3D seismic line from Nigeria (Hughes and Shaw 2014). The yellow upward arrows depict structural relief, and the downward orange arrows depict repeated section. (Reprinted by permission of the AAPG whose permission is required for further use.)

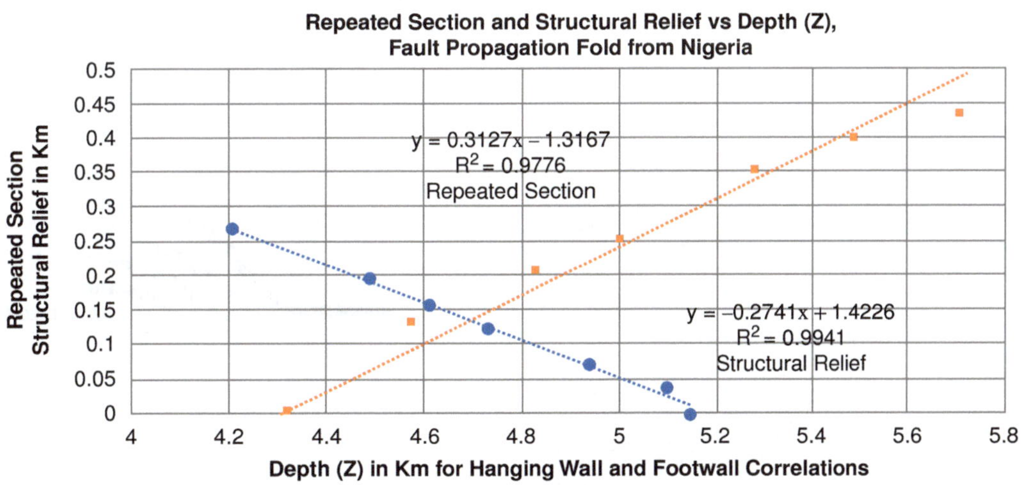

Back Limb % Difference = (0.269 + 0.0) = 0.269/2 = 0.1345
(0.269 – 0.0) = 0.269/0.1345 = 200%

Repeated Section % Difference = (0.44 – 0.0) = 0.44/2 = 0.22
(0.44 – 0.0) = 0.44/0.22 = 200%

Figure 10-59 A structural relief/repeated section plot from a 3D seismic line from deep-water Nigeria (Hughes and Shaw 2014). Measurements of the repeated section and structural relief data are taken from the same point on the fault and along the fold. As the structural relief and repeated section % differences are 200%, this structure appears to be a classic fault propagation fold.

The two major end-member styles, fault bend and fault propagation folding, exhibit distinctly different geometric properties. Fault bend folds from reasonably well-constrained structures contained % difference calculations of approximately 0.0 and did not exceed 6%. Conversely, fault propagation folds had structural relief % differences greater than 50% and repeated section % differences that approached or equaled 200%.

Imbricate Structures

As the thrust belt moves progressively over the foreland, there is a tendency for new thrust faults to form near the toe (front) of the thrust belt and for these thrusts to seek a lower structural level. Where a thrust fault forms below a pre-existing fault(s), motion along the deeper fault will cause the shallow fault(s) and its overlying structure(s) to fold. The deformation can produce some rather interesting and complex geometries of stacked folds (Fig. 10-60). Thrust faults that form at a higher structural level and above the newly formed imbricate thrust faults have been, perhaps inappropriately, called *out-of-sequence* thrusts.

Two possible procedures exist to study imbricate structures: the simple forward model in which the interpreter defines the amount of slip, and the more complex inverse model. In the more useful inverse modeling, the amount of slip is constrained but is unknown to the interpreter, which makes the problem more difficult. We address the forward modeling before proceeding to the inverse model.

This complex process is best described through example. We first assume that a fault bend fold formed near the front of a thrust belt, as shown in Figure 10-61a. In this example, we assume that faulting formed ramp AB and that the ramp angle is 20 deg. We can now determine the frontal dip panel angles by using the methods developed in the section on fault bend folds. We also assume for purposes of demonstration that, at a particular time, the fault breaks through at the lower level and another ramp forms in front of the ramp that formed the fault bend fold. The new ramp along the lower decollement is ramp CD in Figure 10-61b.

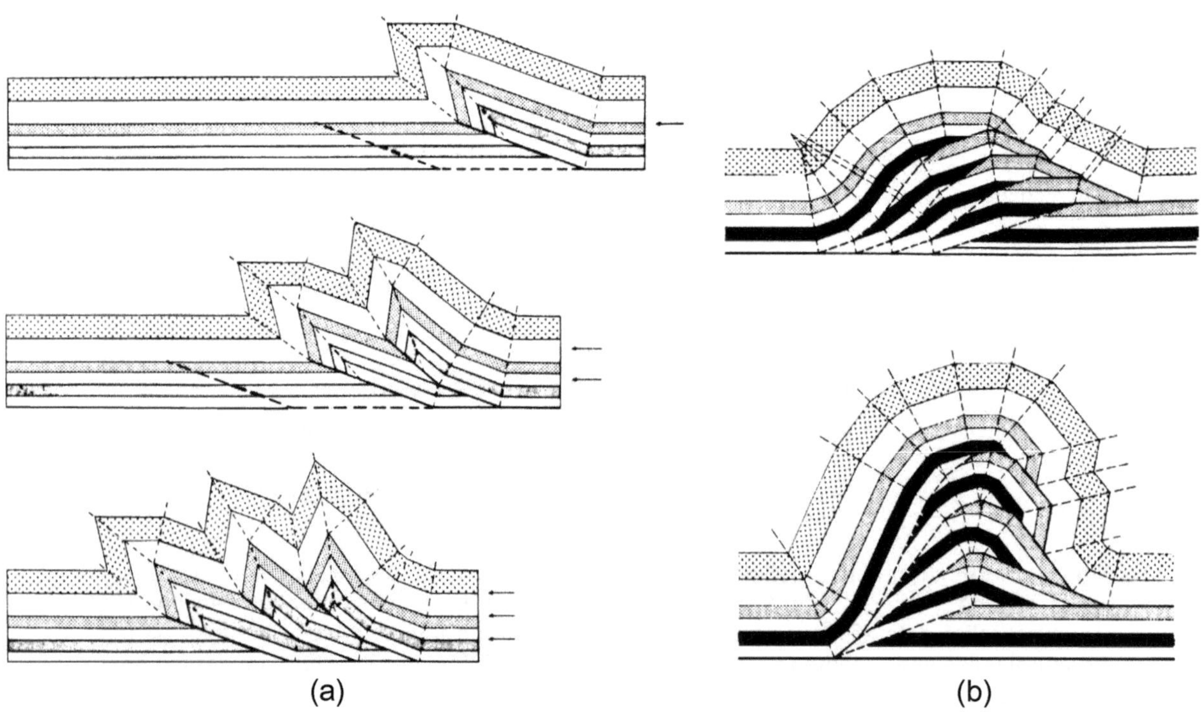

Figure 10-60 Diagram showing different types of duplexes. (a) Stacked fault propagation folds. (b) Foreland and anticlinal stacked duplexes. (From Mitra 1986; AAPG©1986, reprinted by permission of the AAPG whose permission is required for further use.)

This wedge-shaped structure, which is completely surrounded by ramps AB and CD and flats BD and AC, is called a **horse** (Boyer and Elliott 1982). If several horses move up their ramps, then they form a **duplex** of folded imbricate thrusts (Fig. 10-60b).

We assume for purposes of demonstration that the thrusting on the lower fault has progressed to a stage such that the original distance between points A and C has been exactly halved, which means that only a portion of the rhomb-shaped horse has moved onto the upper flat (Fig. 10-61c). As the frontal part of the rhomb-shaped horse moves up the ramp, it will bend in the same manner as layers deform during fault bend folding. Thus, for this part of the deformation, we can determine the deformed shape of the horse using the kink method and the techniques developed in the section on fault bend folds. It now follows that as the horse block moves up a 20-deg ramp and onto an upper flat, it will have the same frontal dip angle (β) as the fault bend fold had when it moved up the 20-deg ramp AB in Figure 10-61a. Therefore, frontal dip angle can be determined to be 23 deg from Figure 10-38 (or from Table 10-1). The amount of slip consumed by the bending of the layers within the frontal dip panel of the horse can be determined from Figure 10-40, which in this case is about 0.87 of the total slip. Therefore, after deformation, the distance DD will be equal to 0.87 of the distance 1/2 AC in Figure 10-61b.

Next, we bisect the angle between the part of the horse that was bent (where it rode up over the top of the ramp) and its undeformed portion. This produces an axial surface with a dip of 78.5 deg. Then we project the uppermost portions of the horse to the left. Bed length consistency requires that the layers be the same length before and after deformation, so length BD before deformation (Fig. 10-61b) should be equal to length BD′ after deformation (Fig. 10-61d).

The problem of the deformed horse can now be resolved. The upper deformed part of the horse located near point B (Fig. 10-61d) rode up ramp CD without being deformed. As the cutoff angle of ramp AB is 20 deg, the upper segment of ramp AB can be projected downward at a 20-deg angle in Figure 10-61d (i.e., line segment FB). Similarly, the lower portion of ramp AB (located near point A of Fig. 10-61b) slid along the lower flat without being deformed, and thus the lower part of the ramp AB can be projected upward at a 20-deg angle (Fig. 10-61c). The central part of the horse, however, has been subject to deformation as the wedge-shaped horse moved up ramp CD and through axial surface EC. Axial surface EC is pinned to the lower footwall cutoff at point C and is an active axial surface. Ramp CD has a 20-deg initial cutoff angle, so the bisecting axial surface EC dips at 80 deg (Fig. 10-61d and e). It also defines the extent of line AE.

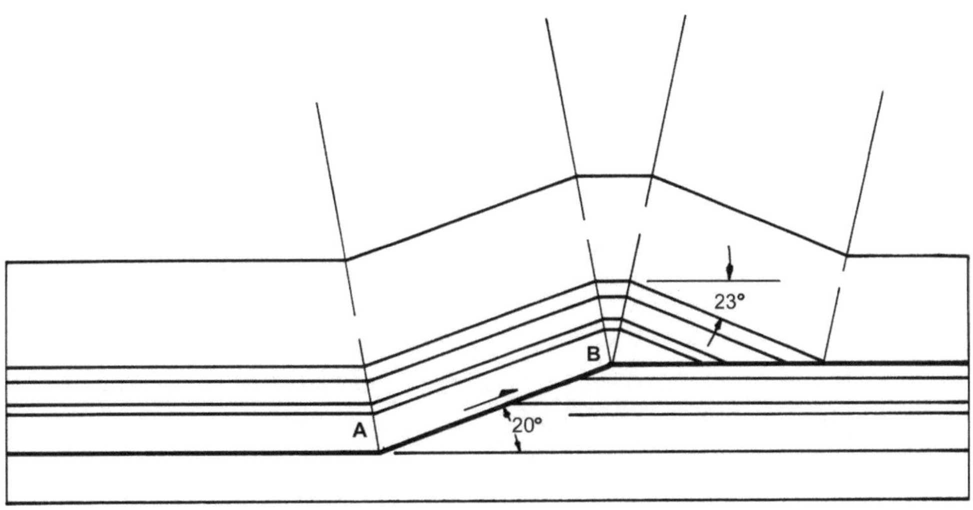

(a)

Figure 10-61 (a)–(e) Duplex exercise, forward model.

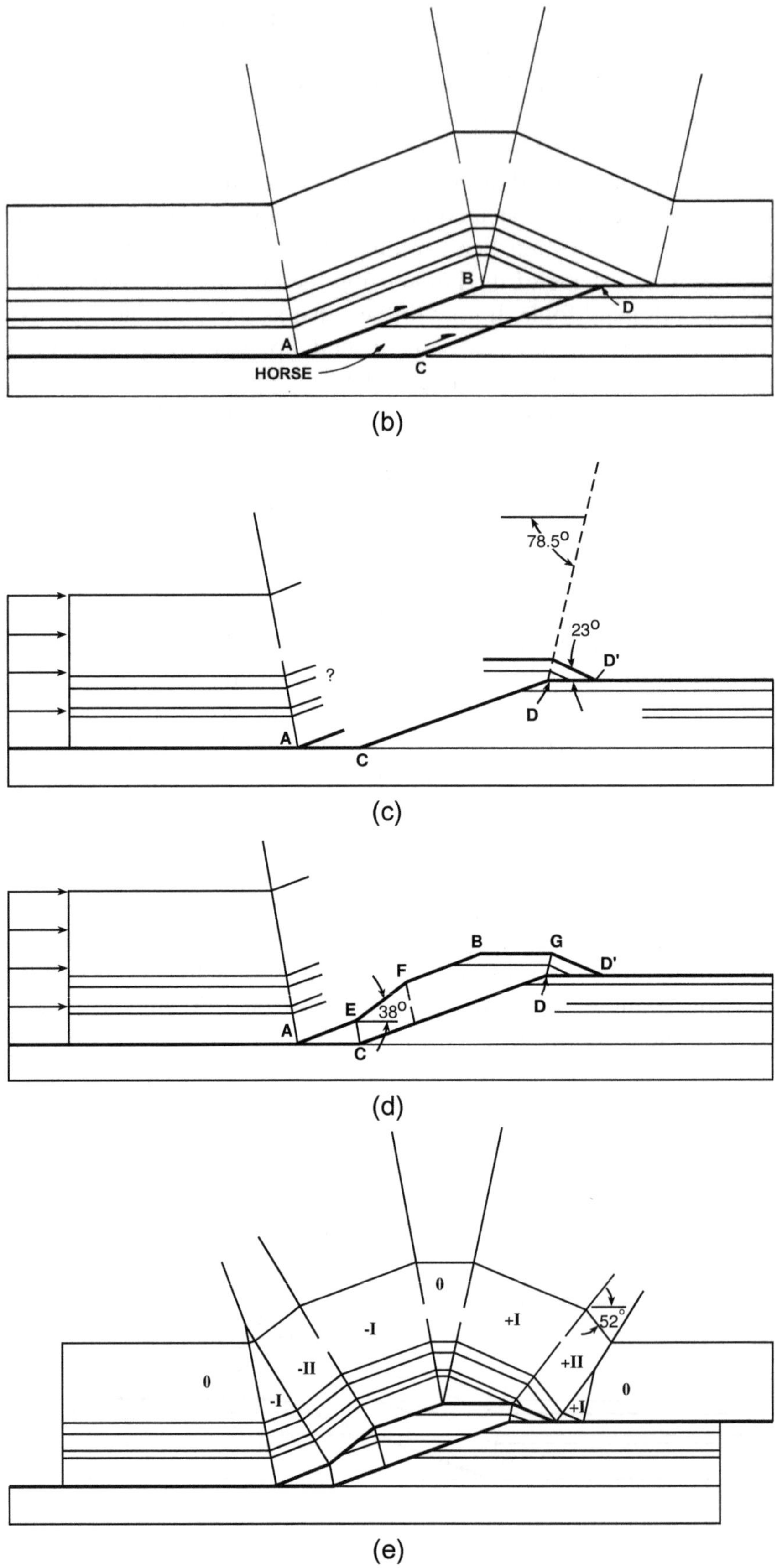

Figure 10-61 *(continued)*

We have now determined that as the deformed fault ramp AB moves through axial surface EC, it must deform (bend upward). Ramp AB initially dipped at 20 deg before deformation, so it must dip at an *even higher* angle *after* deformation. If a 20-deg dipping line (line AB) moved up a 20-deg dipping ramp (line CD of Fig. 10-61d), one might *incorrectly* conclude that the central deformed portion of fault ramp AB (line segment EF) would dip at 40 deg. We use Table 10-1 to provide the correct dip angle for line EF.

The following is an example of the rationale for Table 10-1. In the section on fault bend folding, we learned that in order to maintain line lengths, the angle β, the dip of the frontal limb, must be greater than the (initial) cutoff angle θ. This relationship must be maintained on every structural level within imbricate structures. Thus, as the frontal part of the horse rides over the top of the ramp, it rotates forward to a dip angle of 23 deg (Fig. 10-38, or Table 10-1). This rotation causes the overlying beds to dip forward at a higher angle (dip panel +II in Fig. 10-61e). The dips at the higher structural level experience a *quantum increase* in dip (Suppe 1980, 1983). In other words, the insertion of the horse onto the upper flat will cause the beds above it (panel +II) to dip at an angle that is greater than twice 23 deg, or in this case 52 deg, as determined from Table 10-1. In Table 10-1, the central column contains values for the cutoff angle θ, and the other columns provide the calculated dips within each panel of the fold. Thus, for a 20-deg cutoff angle, the front limb dips at 23 deg and at 52 deg in panels +I and +II respectively. In order to maintain line length and formation thickness, the strata above the horse will shear in such a manner that an increase in dip in the frontal panel is accommodated by a decrease in the expected dip in the back panel. Thus, the compensating dip of line segment EF (panel −II) in Figure 10-61d and e is 38 deg rather than 40 deg (Table 10-1). The dip panels III through VII in Table 10-1 relate to higher-order duplex structures.

The length of line AE was defined previously by constructing the axial surface at point C in Figure 10-61d. Point F is determined from the amount of slip on the lower fault. Measure that slip up the ramp and draw the inactive axial surface parallel to the one at point C; then draw line EF at a dip of 38 deg (Table 10-1) upward to the axial surface. Complete the deformed fault surface by drawing line

Table 10-1 Dip Spectral Analysis

Forward dips (+)							***Fundamental***	***Back dips (—)***						
VII	***VI***	***V***	***IV***	***III***	***II***	***I***	***cutoff angle θ***	***I***	***II***	***III***	***IV***	***V***	***VI***	***VII***
61.6°	52.5°	43.0°	34.0°	25.2°	16.6°	8.2°	8°	8°	15.9°	23.4°	30.6°	37.3°	43.5°	49.3°
70.2°	59.2°	48.6°	38.3°	28.3°	18.6°	9.2°	9°	9°	17.8°	26.2°	34.0°	41.3°	47.9°	53.9°
80.6°	67.6°	55.2°	43.3°	31.9°	20.9°	10.3°	10°	10°	19.7°	28.9°	37.4°	45.1°	52.0°	58.2°
93.1°	77.3°	62.6°	48.8°	35.7°	23.3°	11.4°	11°	11°	21.6°	31.5°	40.6°	48.7°	55.9°	62.2°
109°	88.8°	71.0°	54.8°	39.8°	25.8°	12.6°	12°	12°	23.5°	34.1°	43.7°	52.1°	59.5°	65.9°
128°	102°	80.5°	61.5°	44.3°	28.5°	13.8°	13°	13°	25.4°	36.7°	46.7°	55.4°	62.9°	69.4°
160°	119°	91.3°	68.6°	48.9°	31.2°	15.0°	14°	14°	27.2°	39.1°	49.5°	58.4°	66.1°	72.5°
—	146°	104°	76.3°	53.6°	33.9°	16.2°	15°	15°	29.1°	41.5°	52.3°	61.4°	69.0°	75.5°
	—	124°	85.9°	59.0°	36.8°	17.4°	16°	16°	30.9°	43.9°	54.9°	64.1°	—	
		—	99.2°	65.6°	40.2°	18.8°	17°	17°	32.7°	46.2°	57.5°	—		
		—	123°	73.1°	43.7°	20.2°	18°	18°	34.4°	48.4°	59.9°	—		
			—	82.2°	47.4°	21.6°	19°	19°	36.2°	50.6°	—			
			—	97.6°	52.0°	23.2°	20°	20°	37.9°	52.7°	—			
				—	57.0°	24.8°	21°	21°	39.6°	—				
				—	63.6°	26.6°	22°	22°	41.3°	—				
				—	72.0°	28.4°	23°	23°	42.9°	—				
					—	30.4°	24°	24°	—					

Published by permission of the American Journal of Science (Suppe 1983).

FB (point B was determined above). The inactive axial surface beneath F will be the line at which you project horizons within the horse downward and parallel to the lower ramp, as in Figure 10-61e.

The strata above deformed ramp AB will parallel that surface and bend at axial surfaces located at points E and F. Bisect the angles along the deformed upper ramp (angles AEF and EFB in Fig. 10-61d) and complete the dip domains toward the surface (Fig. 10-61e). The crest and the front limb (with a 23-deg dip) of the original fold were above the upper fault flat, to the right of point B in Figure 10-61a. As described previously, part of this frontal dip panel of the original fault bend fold was subsequently deformed by the frontal portions of the horse, and it dips at 52 deg (Fig. 10-61e). However, the part of the horse between points F and G (Fig. 10-61d) is undeformed, so the dips of the overlying strata and the axial surfaces at B are maintained from the original fold (dip panels −I, 0, and +I in Fig. 10-61e). A small dip panel (+I) exists to the right of panel +II and has a dip equal to that of the original front panel (23 deg), as it is a part of the frontal limb of the original fold that was not deformed (Fig. 10-61b). The axial surface on the right of small dip panel +I has an inclination of 78.5 deg, maintained from the original fold. The axial surfaces bounding panel +II are placed at bends in the upper fault surface and drafted at inclinations that bisect the adjacent dips. Complete the cross section by using dip data in Table 10-1 to draw the horizons in panels −I, +I, and 0 (flat). The finished, balanced cross section in Figure 10-61e depicts a *duplex* structure.

Although this exercise may at first seem to be an unnecessary complication, we use these small changes of dip to our advantage in what is called *dip spectral analysis* (Suppe 1980, 1983). Dip spectral analysis can be used to interpret poorly imaged subthrust plays.

Consider the geometry present in Figure 10-61e, which has several implications concerning petroleum exploration. Assume that the thin horizon above the lower flat is a productive reservoir horizon. Notice that this reservoir can be intersected on two structural levels, resulting in two potential plays. The first play is the closure associated with the original fault bend fold. The second play is a partial closure located within the horse, and thus its prospectivity would depend on the trapping mechanism or the permeability of the beds above the thin reservoir horizon.

Dip Spectral Analysis. This section discusses the method to locate potential subthrust plays in practice. We learned from the previous imbricate structure exercise that for uniform step-up angles, and as one moves up the structural pile, the dips at the front of the imbricate structures increase at an increasing rate. Furthermore, the back dips, whereas exhibiting a corresponding increase, do so at a *decreasing* rate. Therefore, the frontal dips exhibit a *unique quantum* increase in dip (in our case, 52 deg is greater than twice 23 deg), whereas the back dips exhibit a *unique quantum rate decrease* in dip (Table 10-1). These unique changes in dip allow us to estimate the number of subthrusts and their approximate position. For example, in Figure 10-60b, the duplex structure in the lower figure has four different back dips relative to the flat regional dip. This means that the duplex has four or more thrusts and four or more potential repeated sections.

Notice that in Figure 10-61e the final structure exhibits three frontal dip panels or domains (labeled +I, +II, and +I) and three back dip panels (labeled −I, −II, and −I) that are separated by a region of initial dip (panel 0). In nature, detailed surface mapping across the structure shown in Figure 10-61e could result in the topographic section shown in Figure 10-62a. In this figure, the following dips occur from left to right: 0, −38, −20, 0, 23, 52, 0, −20, −20, −20, and 0 deg (designating leftward dips as negative). If a regional dip of 5 deg to the left were to exist, then the corresponding dips would be −5, −43, −25, −5, 18, 47, −5, −25, −25, −25, and −5 deg. Regional dip should be removed prior to dip spectral analysis. All angles are determined relative to regional dip, but by removing regional dip, we can conveniently use angles measured from the horizontal, as given in Table 10-1. The resultant dips corrected for 5-deg regional dip are 0, −38, −20, 0, 23, 52, 0, −20, −20, −20, and 0 deg.

We have shown that higher dips exist at the front of the structure, whereas lower dips occur at the back (Fig. 10-61e). Thus, the 52-deg dip and its associated 23-deg dip are *forward dips*, whereas the −38-deg dip and its associated −20-deg dip are *backward dips* (see Table 10-1). As there are two forward dips, the 52-deg dip represents a +II domain, and the 23-deg dip represents a +I dip domain. These numbers, each of which represents an individual dip domain, can be compared to Table 10-1 (line 13) to indicate that there are *two thrust faults* ramping at 20 deg, causing second-order frontal and back dips of 52 deg and −38 deg respectively, and first-order frontal and back dips of 23 deg and −20 deg respectively, in the upper structure. If *three faults* are present, then we would expect an additional forward dip of about 98 deg and a back dip of about 53 deg (Table 10-1). As only two forward dips exist in our example, Table 10-1 suggests that only two thrust faults are present. Therefore, surface and subsurface data can be compared to Table 10-1 in order to *determine the number of imbrications that may exist in the area of study.* The data could be surface bed dips, subsurface dipmeters, and/or depth-corrected seismic data.

Let us proceed to solve the structure presented in Figure 10-62a, which is based on the forward model completed in Figure 10-61e, but uses only surface outcrop data. This inverse model is more complex and requires inverted thinking; thus, the interpretation process is more difficult.

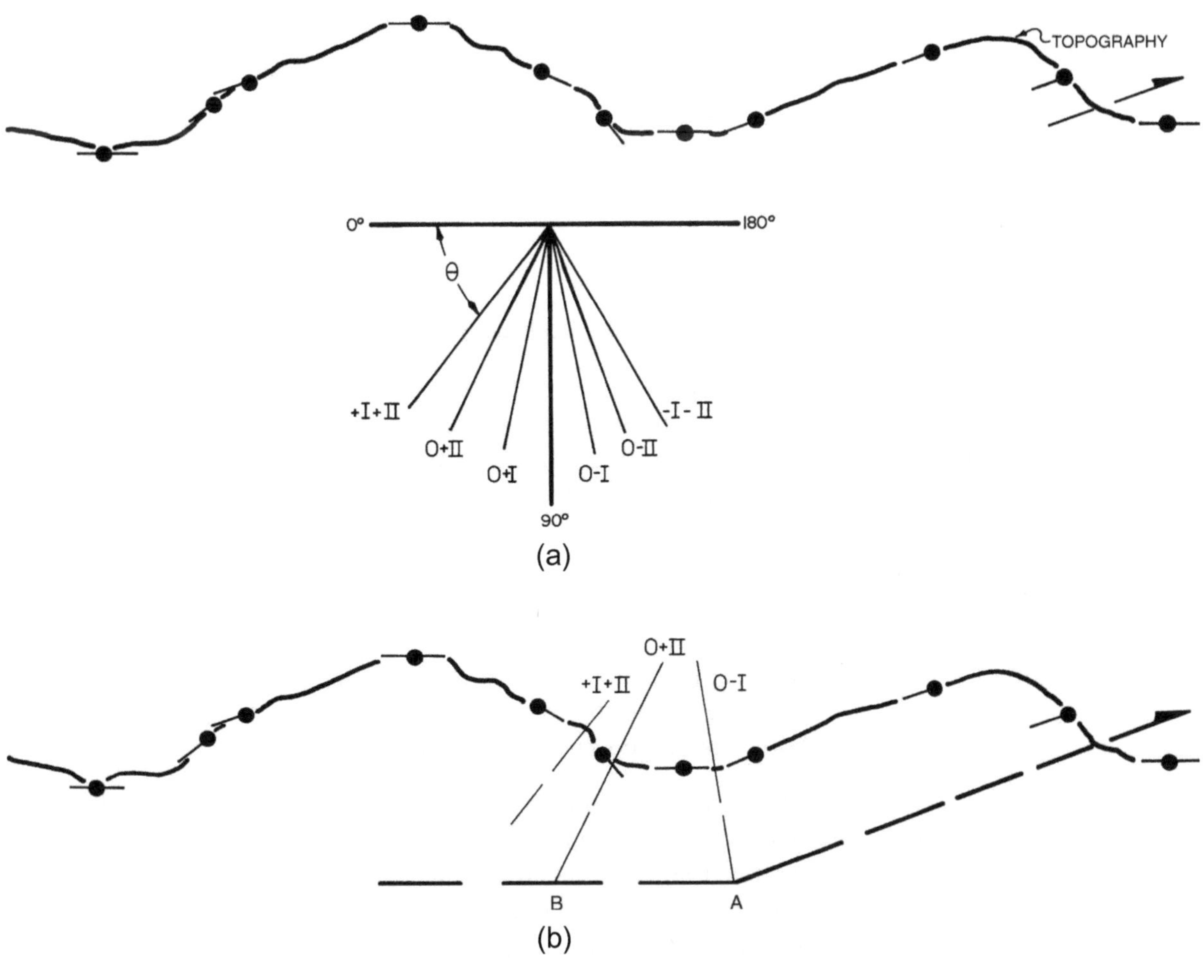

Figure 10-62 (a)–(d) Duplex exercise, inverse model.

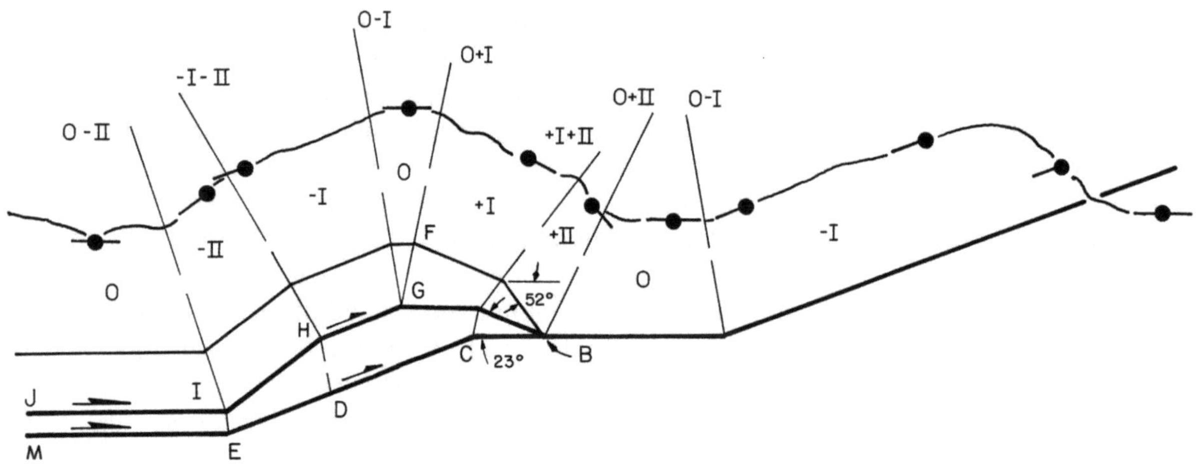

(c)

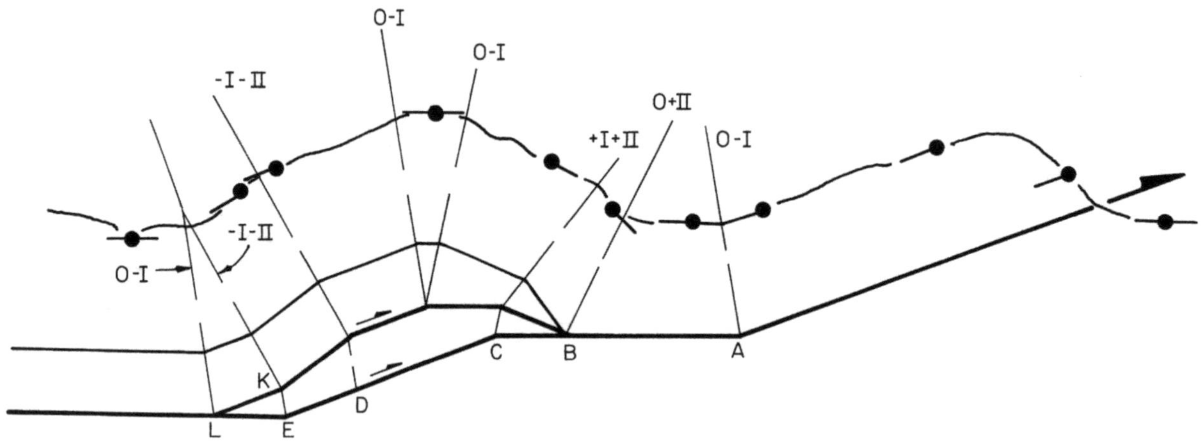

(d)

Figure 10-62 *(continued)*

After analysis of the surface dip data, the first task is to determine the dips of the axial surfaces, which separate the observed dip domains. Once the *related* dips within each of the dip domains have been averaged, as was described in the section on the kink method, the dip of the axial surface between given adjacent dip domains can be determined from the following formula:

$$\theta = (\mathrm{Dip}_1 + \mathrm{Dip}_2 + 180°)/2$$

where

θ = dip of axial surface taken counterclockwise from the horizontal

Dip_1 = average structural dip in dip domain 1

Dip_2 = average structural dip in dip domain 2

For this equation only, structural dip is taken to be *negative* if it is to the right. It would fall within the 90-deg to 180-deg quadrant, as defined in Figure 10-62a. If dip is to the left, it falls within the 0-deg to 90-deg quadrant and is assigned a positive value in the equation. The dips of the axial surfaces for the appropriate dip domains presented in Figures 10-62a and 10-61e were calculated and are presented in Table 10-2. For example, at the top of the hill on the left, a 0-deg dip exists adjacent to a 23-deg dip to the right (Fig. 10-62a). These two dips represent a 0 dip domain and a +I dip domain (Fig. 10-61e and Table 10-1). They are separated by a dip domain boundary, which is the 0 +I axial surface. Thus, the 0 +I dip domain boundary dips at (0 − 23 + 180)/2 = 78.5 deg (Table 10-2).

Table 10-2 Axial Surface Calculations

Dip domain boundary*	***(Dip$_1$* + Dip$_2$* + 180°)/2***	***Dip of axial surface (in degrees)***
–I – II	(38 + 20 + 180)/2	119.0
0 – II	(0 + 38 + 180)/2	109.0
0 – I	(0 + 20 + 180)/2	100.0
0 + I	(0 – 23 + 180)/2	78.5
0 +II	(0 – 52 + 180)/2	64.0
+ I + II	(–23 – 52 + 180)/2	52.5

*Dips and dip domains are taken from Fig. 10-46a

These axial surface dip calculations are best applied to the solution of problems using a method that was suggested by John Suppe and is shown in the lower portion of Figure 10-62a. First, calculate the dips of all axial surfaces for the adjacent pairs of dip domains. These are shown in Table 10-2. Then create a reference set of axial surfaces, as shown in Figure 10-62a. The dip of each axial surface is projected downward from a central point and labeled. The dip of each axial surface is measured off in a *counterclockwise* direction. Two triangles can now be aligned with the axial surface dip data and slid into any position on the cross section that the interpreter desires. This procedure makes the interpretation process more rapid.

Possessing the dip data set presented along the topographic profile in Figure 10-62a and the knowledge gained from the solution to the problem presented in Figure 10-61e will not allow us to arrive at a unique solution to our problem. Trial and error and some guessing are required to solve the problem presented in Figure 10-62a. The advantage of solving this problem is obvious for areas where seismic data fails to image imbricate structures. It may result in the identification of duplex structures and generation of additional prospects and perhaps additional oil and gas discoveries.

Before proceeding further, notice that Figure 10-62a does not include the −20-deg dip panel above the lower flat, which is present in Figure 10-61e and which is to be expected from Table 10-2. This should make our problem more interesting and realistic. The 0 and −II data are included in Table 10-2 for reasons that will soon become apparent.

The first step in the solution of our problem is to examine the data from a geometric point of view. Two observations are critical to an accurate interpretation. First, the observed dips tend to follow the topographic slope, which is often the case in nature. Second, the beds above the thrust fault at the outcrop dip at about the same angle as the thrust fault, suggesting that a ramp is responsible for the 20-deg tilt to these beds. Therefore, the thrust fault can be projected from the outcrop downward to where it intersects the adjacent (0-deg) dip domain. But where is this point? It exists where the 0 −I axial surface intersects the projected thrust fault. The change in topography is used to position and project the 0–I axial surface with a 100-deg dip (from Table 10-2) downward to the point where it intersects the thrust fault, thus determining the base of the upper ramp (Fig. 10-62b, point A). In addition, we have predicted the structural level of the upper flat or decollement (Fig. 10-62b).

Looking at the dip data further, we see the 0-deg dip at the hilltop on the left and the 23-deg and 52-deg dips to the right. Using Table 10-1, we can infer that two 20-deg ramping thrusts create the observed 23-deg and 52-deg forward dips, and thus these two thrusts must be imbricated (i.e., stacked) in order to produce the observed quantum increase in dip. The data suggest that a 52-deg dip domain adjoins a 0-deg dip domain, and therefore, a 64-deg axial surface (between the 0 and +II domains of Table 10-2) is positioned at the appropriate change in topographic slope

(Fig. 10-62b). We know that two dips exist to the left of the intersection of the 0 +II axial surface with the upper flat at point B. First, the structurally higher beds dip at 52 deg to the right, and second, a 23-deg deformed thrust must exist beneath the 52-deg dipping beds. This follows from a direct application of Table 10-1 and the theory that we presented earlier.

Consequently, at point B we interpret and construct a 52-deg dipping bed and the deformed 23-deg dipping fault (Fig. 10-62c). The 52-deg dipping bed and the deformed 23-deg fault are then projected up to the +I +II axial surface, which has a dip of 53 deg (Table 10-2) and is positioned at the break in topographic slope. To the left of this axial surface, the fault will be flat like that in an undeformed part of the older fault. Now the kink method is applied to the deformed horse block in order to map the forward dip within it, beneath and parallel to the deformed fault. An axial surface must extend downward from the kink in the fault. Given the adjacent dips of 0 deg and −23 deg, the dip of the axial surface calculates to be 78.5 deg (Table 10-2). A (0 +I) axial surface is drawn downward from the kink in the deformed fault to where it intersects the flat, thus locating point C in Figure 10-62c. The frontal portions of the *deformed* horse block have now been properly defined (i.e., we have separated the undeformed and the deformed regions of the horse block). Point C not only marks the position where the horse block has ridden onto the upper flat but also determines where the structurally lower ramp can be projected downward at a 20-deg angle (Fig. 10-62c, line EC).

Proceeding with the construction, the 52-deg dipping beds in panel +II above the horse can be projected from the +I +II axial surface at a 23-deg angle up to the 0 +I axial surface (Fig. 10-62c, point F). The axial surface was positioned from topographic data. The flat of the upper thrust is projected to the 0 +I axial surface to establish point G, which defines the edge of the horse block, or the top of the ramp on the original fault (Fig. 10-61a and d, point B). From point G, the upper fault can be projected downward at a 20-deg angle (defined by the dip data in panel −I to the −I −II axial surface (Fig. 10-62c, point H). This axial surface was positioned at a break in the topographic slope. The structural and fault dip must change at this axial surface, so the deformed fault is projected downward at a 38-deg angle, as defined by the dip data in panel −II, until the upper fault intersects the 0 −II axial surface (Fig. 10-62c, point I). This axial surface, with an inclination of 109 deg (Table 10-2), is located at a break in the topographic slope. Using the 0 −II axial surface as a bisector, project the upper fault to level out on the IJ structural level parallel to the overlying flat structural dip. The lower flat on the structurally lower fault still must be interpreted. An axial surface must exist between the −20-deg ramp and the flat. This surface calculates to have a 100-deg dip, and it must extend downward from point I. Thus, point E is located at the base of the ramp. The fault flat is then drawn to the left of point E.

Finally, the axial surface between points H and D is constructed parallel to active axial surface IE (Fig. 10-62c). It is the inactive axial surface within the horse. The length of the line ED should be the amount of slip on the younger fault (Fig. 10-61, distance EF), according to this interpretation.

What is the final result of our interpretation process? We have seemingly followed the methods with a result that is somewhat complex and confusing! *Notice that the upper and lower faults do not merge on the same structural level* (compare I to E, Fig. 10-62c). How can we fix this solution, and what features should we look for when attempting to arrive at a more satisfactory solution? First, check the slip on the upper flat (CB) of the younger fault to the slip up the ramp (ED). They are incompatible. Using Figure 10-40 to determine the slip correction factor for a step-up angle of 20 deg., *line CB should be 0.87 of ED, or CB/0.87.* However, in Figure 10-62c, the distance ED is too large. Thus, an error was made in this portion of our analysis that represents not only a clue to the proper solution but also the approximate position of our difficulties. Second, returning to Table 10-1, we reexamine the possible dip domains that are associated with two 20-deg ramping thrusts, as there are obvious problems at the back of the structure, in that

JI does not coincide with ME. Thus, a problem may exist in this region. The data in Table 10-1 suggest that a −I dip domain can exist between a 0 dip domain and a −II back dip domain of 38 deg. Examining the dip panels on our solution reveals that a −I-deg dip domain was not included in the back part of our solution.

We now backtrack, modifying the first solution by *inserting a −I back dip domain between the 0 and −II and the back dip panels* (Fig.10-62d). Using Figure 10-40, we know that the distance ED in Figure 10-62c and d should be equal to CB/0.87, and this distance is entered on the figure by measuring down the ramp from point D to E. From the newly calculated position of point E, we construct a newly positioned fault flat. Then, from point E, we project a 0 −I axial surface upward at 100 deg (Table 10-2) to the deformed upper thrust (point K on Fig. 10-62d). This portion of the deformed upper thrust dips at 38 deg, downward from point H (compare Fig. 10-62c to d). From point K, we next project a −I −II axial surface upward at 119 deg (Table 10-2). We can also conclude that to the left of point K exists a 20-deg −I back dip panel and another 0 −I axial surface that intersects the lower decollement at point L (Fig. 10-62d), which is established by projecting the upper fault at a 20-deg dip to the combined lower fault flat. This solution, although only slightly different from Figure 10-62c, creates a *deformed horse block* and a more reasonable interpretation.

From this complex exercise, we conclude that (1) duplexes produce *more rounded structures;* (2) structural balancing can be nonunique, even under ideal situations; and (3) the interpretation process can be rigorous, but in this age of global energy economics and price variations, the rewards could be substantial. The better we understand the detailed geometry of structures, the more likely we are to find additional reserves of oil and gas.

Box and Lift-Off Structures

Box and **lift-off structures** represent a particular class of folds that, when viewed relative to the regional dip, are roughly *symmetric* but angular structures that contain steeply dipping limbs (Figs. 10-63 and 10-64). Both structural types form along a zone of weak detachment located at depth and possess the characteristic that the decollement is isoclinally folded into the hanging wall (Laubscher 1961; Namson 1981). In the Jura Mountains, this zone of weakness consists of evaporites (i.e., gypsum), although overpressured shales are likely to produce a similar deformational style. Box and lift-off structures differ from diapiric structures in that there is less mass transport or flow into the cores of these folds. This causes the box and lift-off structures to have almost vertically dipping limbs at the lower structural level. In addition, diapiric structures typically result from a gravity instability, whereas box and lift-off folds result from compression.

The box fold structural style was once thought to be a relatively rare structural style. Today the structures have been observed in many compressional environments and can be productive of hydrocarbons. Box folds can be distinguished from other structural styles in that the width of the box fold, across the crest of the fold, maps as a region of constant width (Fig. 10-63a). No other fold style exhibits this geometry in map view.

Box and lift-off structures can be recognized in outcrop from their bilateral symmetry and also from their angular geometry. If broad zones of vertically dipping beds are encountered in outcrop (e.g., 70 deg to 80 deg), then these structural styles are suspect (Fig. 10-64).

Box folds have nearly flat tops, vertically dipping limbs, and axial surfaces that dip at about 45 deg (Fig. 10-63a). If in a region of vertically dipping beds two axial surfaces intersect at nearly right angles, then you should consider the possibility that box folds are present. On seismic sections, vertically dipping beds *do not image*, and thus a pattern of gently dipping reflectors separated by two zones of noncoherent reflectors, representing the almost vertically dipping beds, may be an indication of this style of deformation. However, zones of noncoherent reflectors on seismic sections can result from other causes, such as strike-slip faulting, rock type, or data acquisition problems. Seismic reflection analysis (Payton 1977; Sheriff 1980) could resolve the correlation problem because the sedimentary sequences on the flanks of box and lift-off folds are elevated within the cores of these structures.

Lift-off folds differ from box folds in that the shallow limbs of the lift-off fold style tend to dip in the 45-deg to 62-deg range relative to the regional dip. At depth, the shallow, steeply dipping limbs merge into a zone of nearly vertical-dipping limbs (Fig. 10-63). If you observe nearly vertical beds just above a decollement, it may be impossible to determine whether you are observing a lift-off or box fold, as both structural types possess 45-deg dipping axial surfaces on this structural level. In practice, however, this difference may be academic.

If the lift-off fold is not subject to bedding plane shear, then the limbs of the fold at a higher structural level dip at about 53 deg (Fig. 10-63b). This dip angle changes with increasing bedding plane shear (Namson 1981), and the amount of shear can be calculated from the dip of the fold limbs (Mitra and Namson 1989). If there is bedding plane shear within the structure, then Mitra and Namson (1989) show that this shear affects the depth to detachment (as presented in the section Depth-to-Detachment Calculations in this chapter). However, the difference is not major for small amounts of shear. Mitra and Namson (1989) should be consulted for more accurate depth-to-detachment calculations.

Box and lift-off folds are commonly found in association with each other. In the Prealps, Mosar and Suppe (1988) observed that lift-off structures form in the leading or the trailing position relative to fault propagation folds. In addition, they observed that fault propagation folds may transform laterally into lift-off structures and that the two structural styles may be related to each other as the local cutoff angle steepens. Thus, at low cutoff angles (less than about 18 deg to 20 deg), fault bend folds may form in an area, whereas if the cutoff angle is greater than about 20 deg to 25 deg, fault propagation folds usually form instead of fault bend folds. If the cutoff angle increases to over 60 deg along the strike of a structure, then the structure may transform into a lift-off or box fold.

When mapping box or lift-off structures, apply the kink method in your mapping. When applying this method to these symmetric structures, remember that the hanging wall decollement is assumed to rise vertically above the basal detachment and to *fold back upon itself* (Figs. 10-63

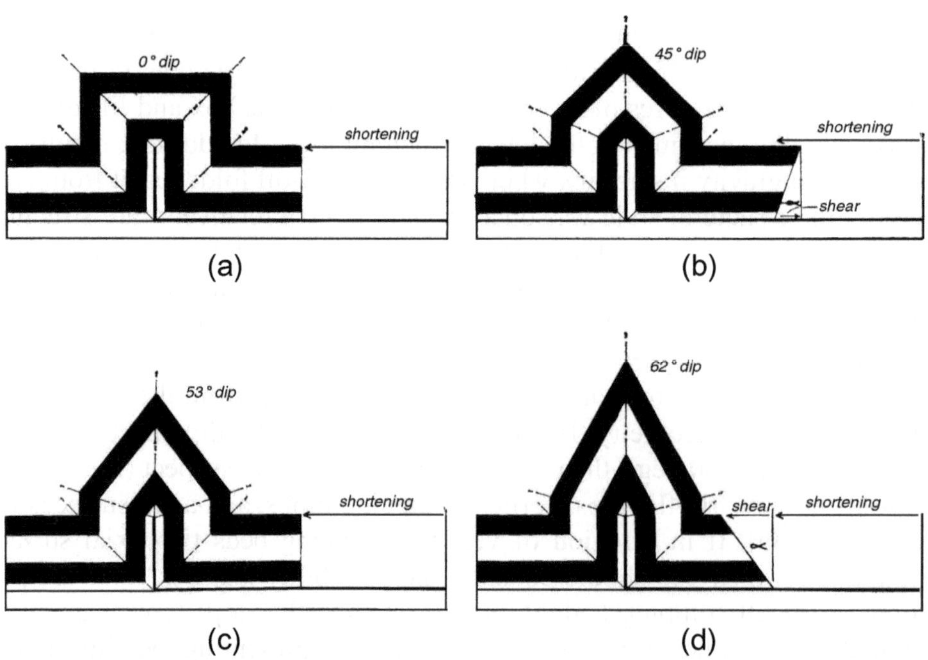

Figure 10-63 (a) Box and (b–d) lift-off structures. (From Namson 1981. Published by permission of the Chinese Petroleum Institute.)

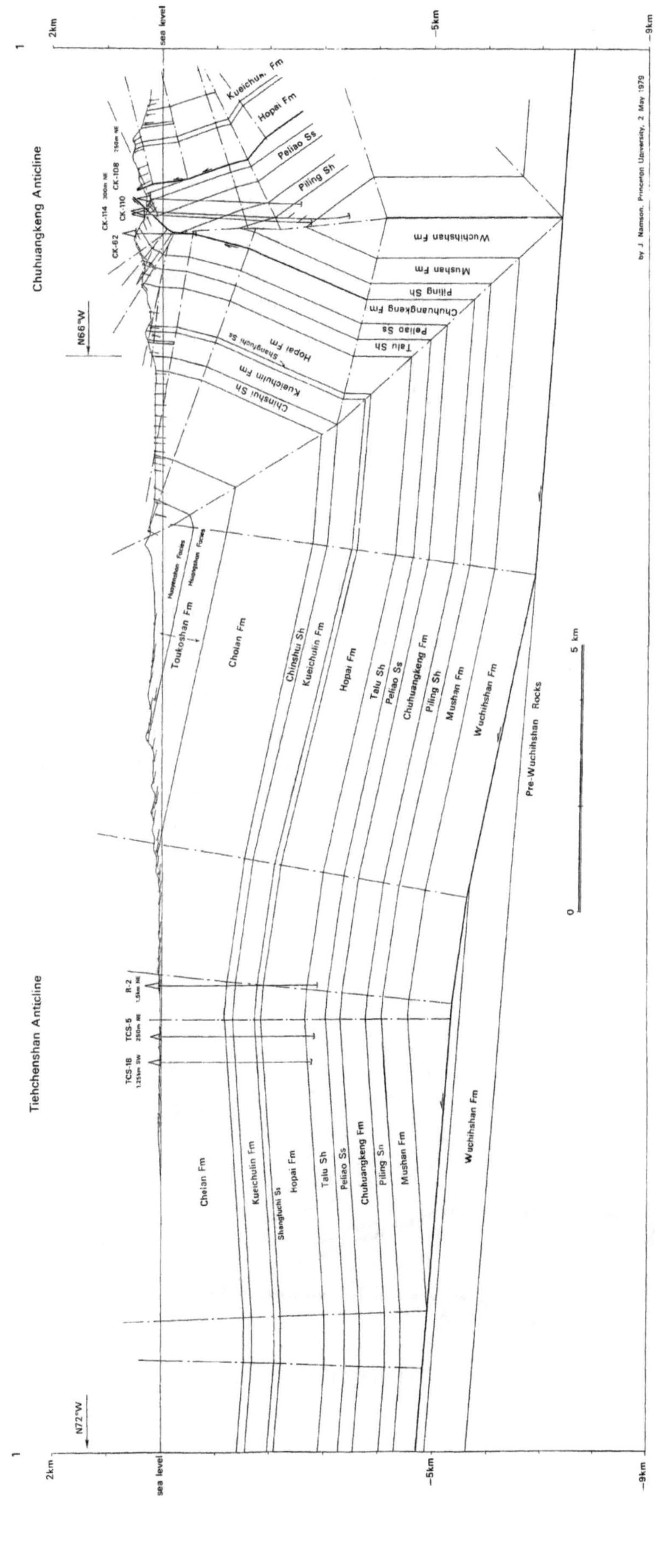

Figure 10-64 Cross section of the Chuhuangkeng Anticline, Taiwan, showing a broad region of near-vertical dips. (From Namson 1981. Published by permission of the Chinese Petroleum Institute.)

and 10-64). Box folds can be distinguished from other structural styles in that the width of the box fold, across the crest of the fold, maps as a region of constant width (Fig. 10-63). No other fold style exhibits this geometry in map view.

Triangle Zones and Wedge Structures

Triangle zones and wedge structures are complex structures that exhibit both a lower and an upper detachment. The basal detachment is often called the *sole thrust*, or *floor thrust*, whereas the uppermost thrust is called the *roof thrust* (Fig. 10-65) (Boyer and Elliott 1982). In the case of triangle zones, the roof thrust is a *passive* back thrust. The wedge moves above the sole thrust and beneath the roof thrust, peeling off the shallow portions of the cover.

Gordy and Frey (1975) and Gordy et al. (1977) initially used the concept of a **triangle zone** to explain the complex relationships associated with an anticlinorium located at the front of the Canadian Rockies. Jones (1982) refined the concept and showed that the structure contained a duplex and that it was responsible for the termination of the eastern-directed thrusting along the Rocky Mountain thrust front. We have learned that during the orogenic process, the deformation progresses (advances) toward the foreland. Therefore, a frontal portion that existed at a previous time during the formation of the thrust belt would exist today hinterland of the thrust front. This implies that *fossil triangle zones can exist within the cores of mountain ranges,* perhaps representing the frontal edge of the deformation at a previous time.

A simple triangle zone that uses the concept of a ramping monocline is illustrated in Figure 10-65. Notice that the deformation terminates where the roof thrust meets the sole thrust, creating a half-syncline. This monoclinally shaped syncline with only one limb lies foreland of the thrust belt. Jones (1982) mapped a half-syncline along the Rocky Mountain front and concluded that a wedge-shaped body of material must be thrust underneath the dipping beds of the half-syncline. For an explanation of the details of the complex structure of triangle zones, consult Sterne (2006).

A seismic section of a complex triangle zone is imaged in Figure. 10-66. Notice the wedge-shaped body represented by the duplication of the reflection located between sp 190 to sp 240 and at 1.2 sec to 1.5 sec. That reflection appears to correlate to the flat reflection at 1.5 sec in

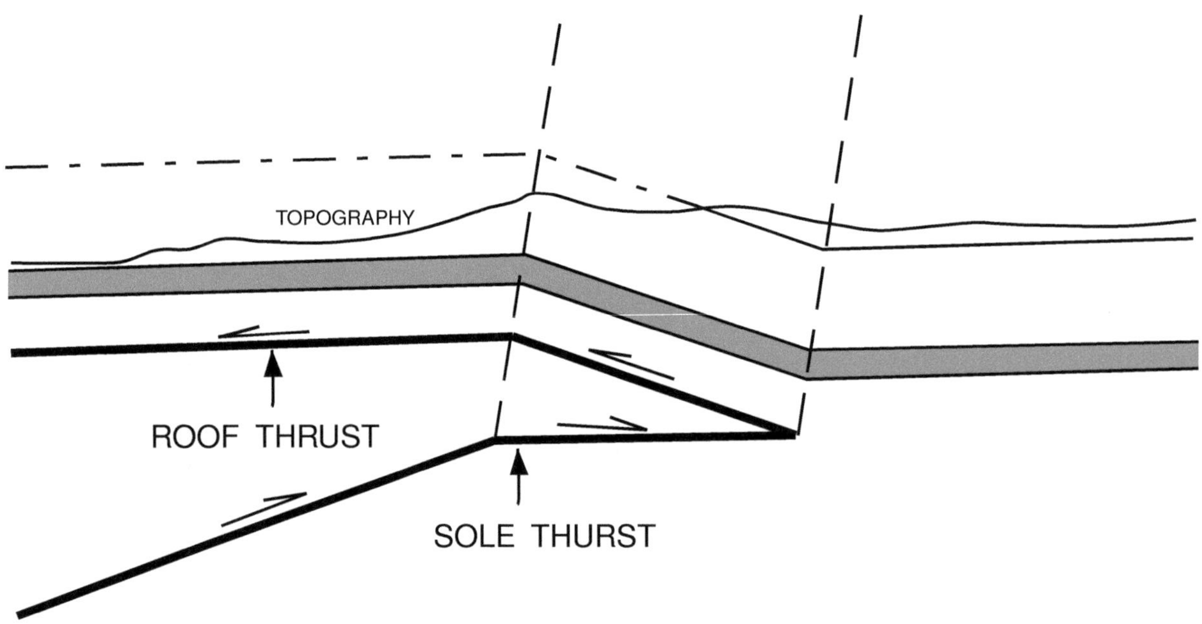

Figure 10-65 Simple triangle zone with passive roof backthrust.

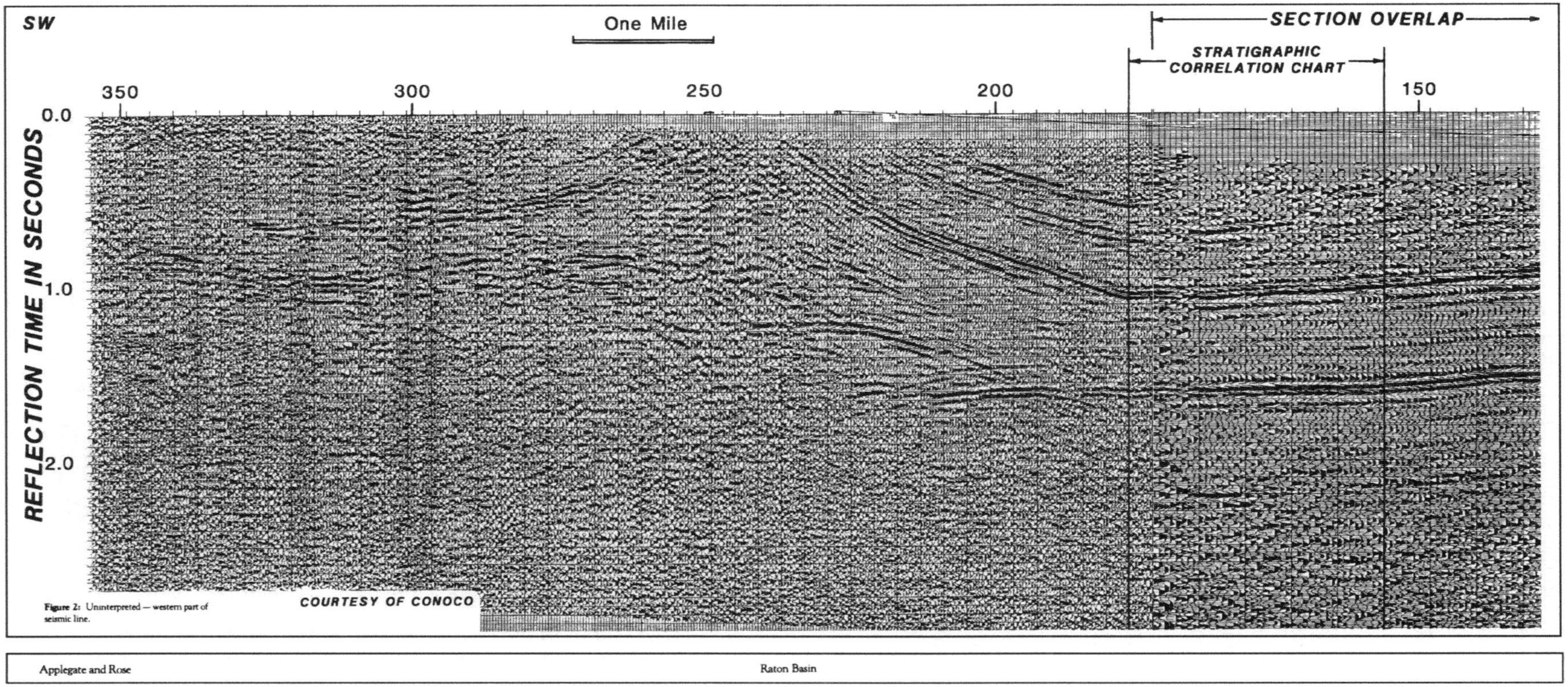

Figure 10-66 Seismic section imaging a triangle zone to the right of sp 250, Raton Basin, Colorado. (After Applegate and Rose 1985 in Gries and Dyer 1985. Published by permission of the Rocky Mountain Association of Geologists and the Denver Geophysical Society.)

front of the structure. On the seismic line, use the following procedures to locate the backthrust. First, project the synclinal axial surface at the front of the triangle zone downward to where the axial surface intersects the sole thrust, as in Figure 10-66. At the point where the two surfaces intersect, construct a line that is parallel to the monoclinally dipping beds. Project this line that represents the backthrust toward the surface. The backthrust conforms to the shape of the hanging wall beds. Notice at sp 220 and at 1000 ms how the backthrust separates dipping beds in its hanging wall from flatter beds in the footwall. This change in bed dips across the backthrust is indicative of a decollement or faulting.

Medwedeff (1988, 1989) extends the concept of interactive sole and roof thrusts to single structures, and he calls these interactive thrusts **wedge structures**. Figure 10-67a is an example of an incipient wedge structure that has two bends on its sole thrust and a single bend in its roof thrust. As the deformation progresses (Fig. 10-67b), motion along the sole thrust deforms the roof thrust. Back and frontal dip panels form over what is essentially a fault bend fold that also has an upper roof detachment. Notice, however, that *the overlying beds in effect ride up the roof thrust*, and they will also form fold panels above the upper detachment. This structurally higher fold is caused by the bends in the roof thrust, so its dip panels terminate at the upper detachment (Fig. 10-67b). The result is two folds for the price of one, which are slightly offset from each other. As the deformation progresses (Fig. 10-67c), the axial surfaces interfere and annihilate each other as they form branch points. This example illustrates that the deformation process can be very transient and that the introduction of additional fault bends results in folds that have more rounded tops (Fig. 10-67c).

(a)
(b)
(c)

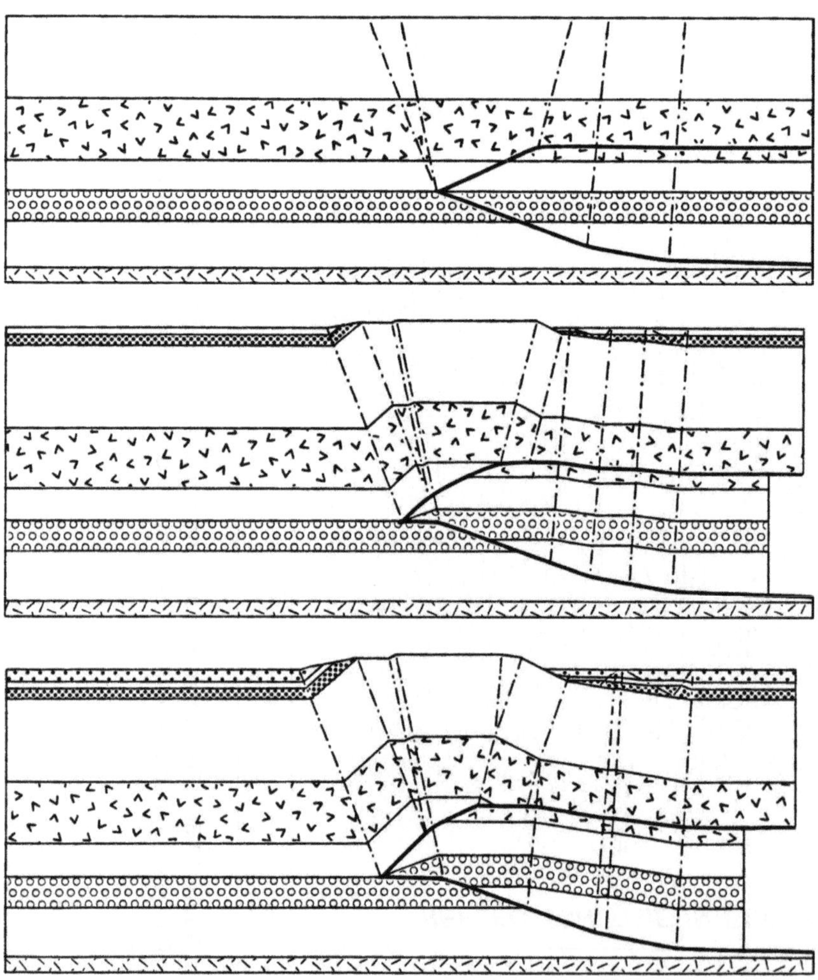

Figure 10-67 (a)–(c) Wedge structure showing progressive stages of development. (Published by permission of Don Medwedeff 1988.)

Medwedeff (1988) also uses wedge structures to model the complex stratigraphic relationships present at Wheeler Ridge, California. The restored structure and the present structure, with the corresponding positions of the wells (with their well logs), are shown in Figures 10-68 and 10-69 respectively. These figures demonstrate how well logs can be used to define the complex relationships that exist within some structures. *Precise correlations and balancing can be effectively integrated to locate prospects that may not be recognized by normal mapping techniques.*

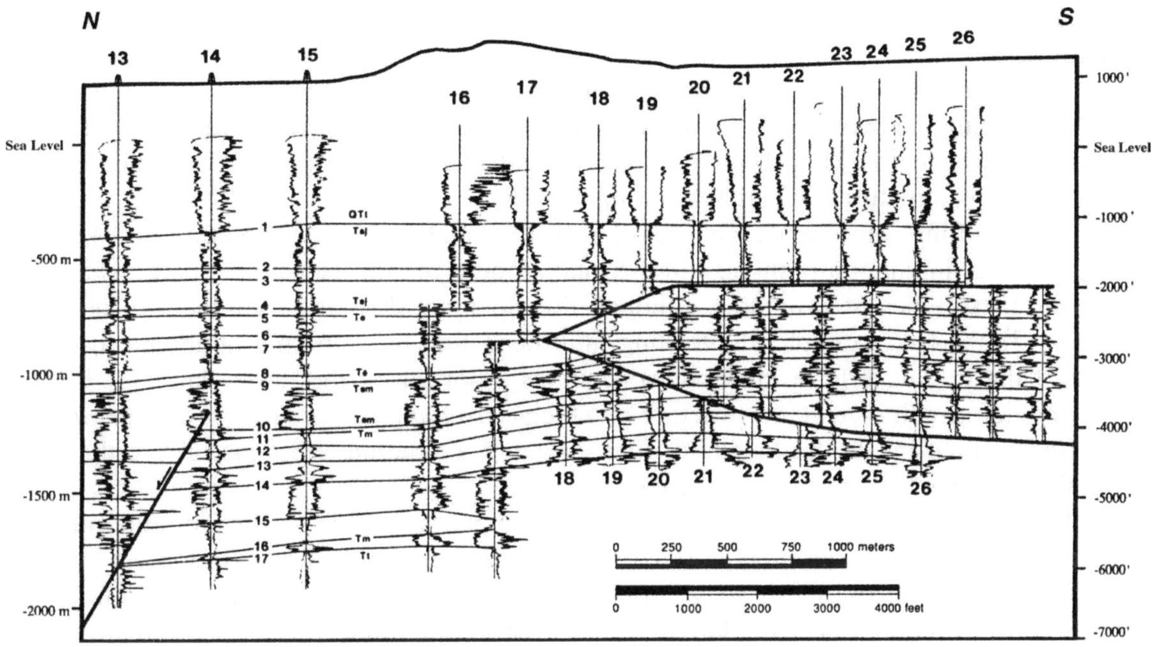

Figure 10-68 Wedge structure in its initial, or restored, state as defined by well logs; Wheeler Ridge, California, USA. (Published by permission of Don Medwedeff 1988.)

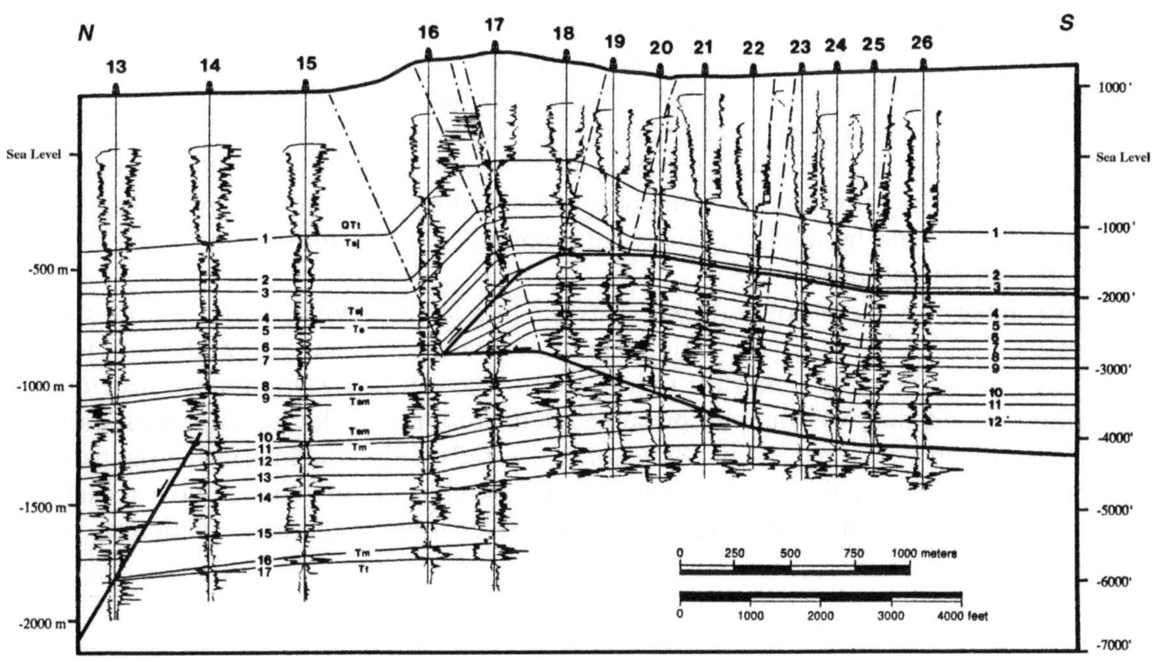

Figure 10-69 Wedge structure in its present state as defined by well logs, Wheeler Ridge, California, USA. (Published by permission of Don Medwedeff 1988.)

Interference Structures

Would you believe that anticlines can form over synclines with no evidence of an intervening fault or evidence of more than one deformation? Nevertheless, clear evidence for this seemingly contradictory relationship can be seen on seismic sections and in outcrops. In the previous section on wedge structures, we saw that deformation on a lower level can modify the shape and the form of dip panels of structures located on a higher structural level. Where structural modification of this type results from a single deformation along one thrust surface, as illustrated in Figure 10-70a, the resulting structures are called *interference structures* (Suppe 1988).

Interference structures are commonly present where the spacing between ramps is relatively narrow, causing the back dip panel of the leading structure to interfere with the frontal dip panel of the trailing structure (Fig. 10-70a and b). The interference tends to produce chevron folds and conjugate kink structures (Weiss 1972; Suppe 1988).

The resulting interference patterns that are created by the deformation are dependent on ramp spacing, initial cutoff angle, and the total amount of slip. Two model patterns are useful in mapping these types of structures, although this does not deny the usefulness of other types of patterns. In the first example, the leading fault bend fold has run up a ramp and the frontal dip panel of the trailing fold (a monocline) occupies a portion of the lower flat (Fig. 10-70a). The resulting deformation creates a structure in which *a frontal fold lies beneath a structurally higher anticline* formed by the trailing fold. Flat dips and a syncline exist just above the lower flat, directly in front of the trailing monocline.

As the deformation progresses, the frontal portions of the trailing monocline will start to run up the leading ramp (Fig. 10-70b). If both ramps have about the same cutoff angle, then as the trailing monocline runs up the second ramp, the beds in the frontal dip panel of the trailing monocline will flatten. One result of the deformation is to create a region of nearly flat dips and a narrow syncline over the leading ramp as the monocline unfolds. This example once again stresses the progressive nature of the deformation. *Structures were not cast in their present-day positions;* they move, and thus strata bend and rebend. Knowledge of which regions of a fold have been subject to refolding should aid in the prediction of *fracture porosities* and better well site locations, which can result in greater productivity. In our example, some of the strata in the back dip panel of the leading fault bend fold were first bent backward, and then forward, by the advancing monocline. As the active axial surfaces sweep through the structure, particular regions within these folds will be subject to repeated deformation and bending that enhances fracture porosity. One can study the refolding by *applying the kink method* and by modeling increasing amounts of slip into the structure to study how the deformation progresses.

We make two more points before leaving this subject. First, as the initial cutoff angle decreases, the structurally higher anticline will move vertically away from the structurally lower syncline, but at the same time shift to a position where it is located almost *directly above the syncline.* Second, the two examples presented here are for a clockwise shear within the interfering frontal and back dip panels. In other words, the beds within the interfering dip panels exhibit a *Z* vergence (Suppe 1988). An example of an *S* vergence (counterclockwise shear), in which the frontal dip panel of the trailing monocline passes through the upper anticline, is shown in Figure 10-70c.

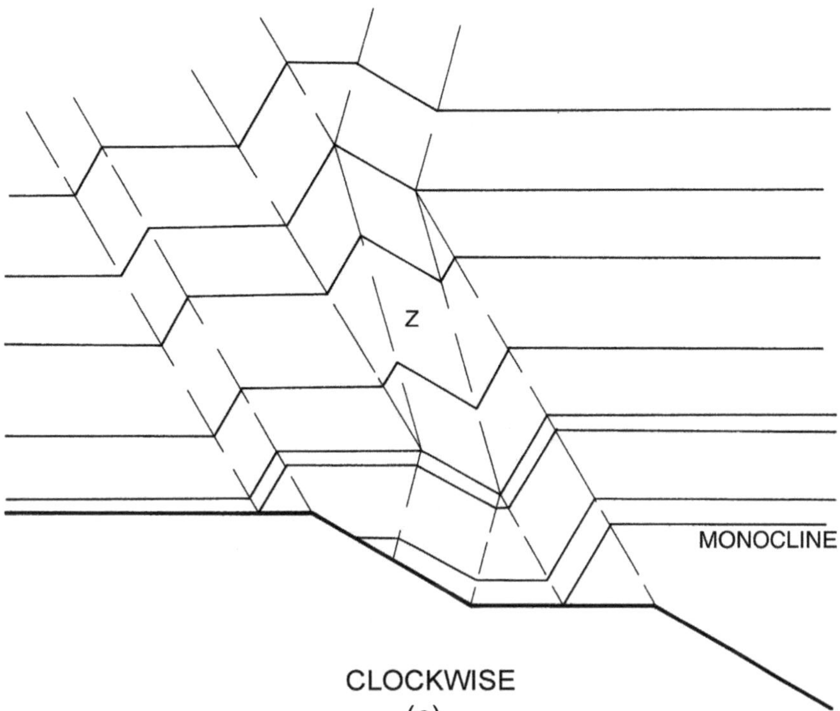

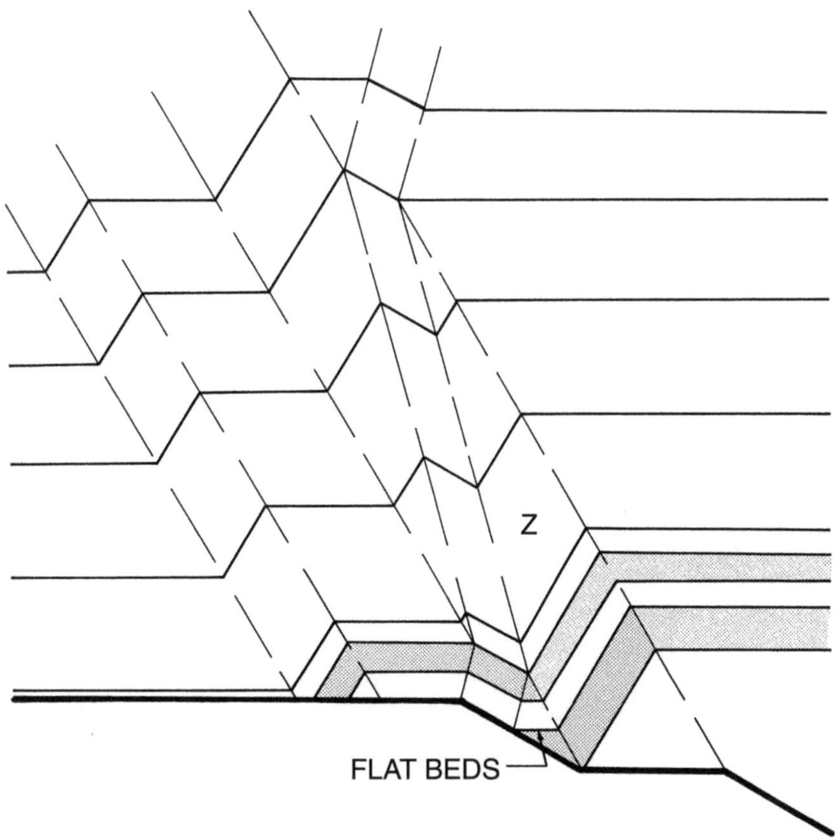

Figure 10-70 Interference structures (a) and (b) for clockwise deformation and increasing slip (c) for counterclockwise shear. (Modified after Suppe 1988. Published by permission of John Suppe.)

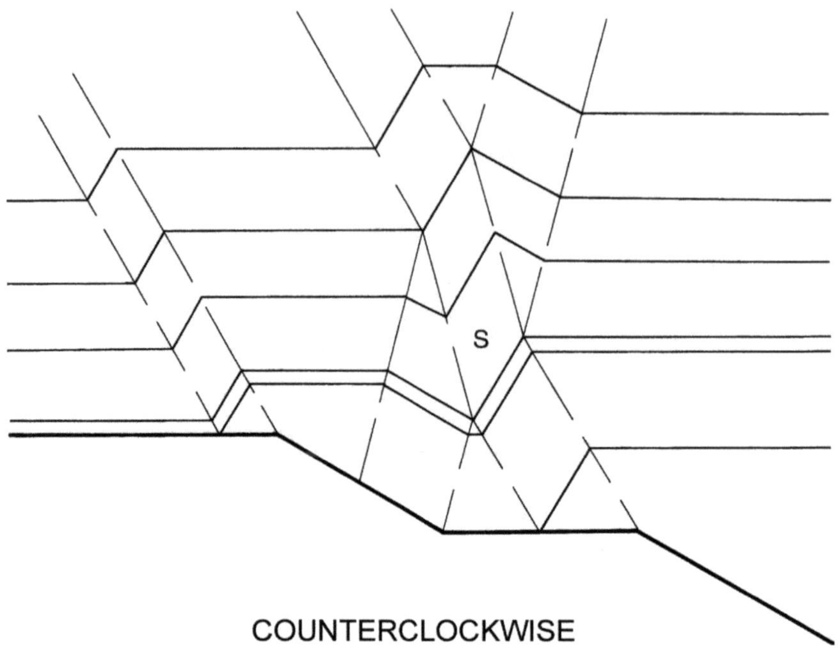

COUNTERCLOCKWISE
(c)

Figure 10-70 *(continued)*

CHAPTER 11

EXTENSIONAL STRUCTURES: BALANCING AND INTERPRETATION*

INTRODUCTION

In the first edition of this text, we mentioned that the balancing of extensional structures was in the initial stages of development. Since then, several major oil companies have successfully applied extensional balancing concepts in the northern Gulf of Mexico, Nigeria, Indonesia, Brazil, and elsewhere (Dula 1991; Nunns 1991; Withjack et al. 1995; Shaw et al. 1997; Shaw et al. 2005). Accordingly, we now have a higher degree of confidence in extensional structural and balancing methods and techniques. As with compressional structural geology and balancing, a common theme to this chapter is the relationship between hanging wall fold geometry and fault geometry. Thus, we introduce a new method in the section Using Structural Relief and Throw to Predict Downthrown Stratigraphic Section. In this section, we describe how to project footwall sand and its associated downthrown depth into the hanging wall. Similarly, in shale units, footwall porosity or hydrocarbon content information can also be projected downthrown into the corresponding hanging wall correlations. The method has high square correlations and is based on closed form solutions to the extensional balancing problem.

This chapter is divided into two parts. The first is the origin of hanging wall anticlines (commonly called *rollover anticlines* or *rollovers*), antithetic and synthetic faults, and keystone structures, and how knowledge of the genesis of a structure can help find additional hydrocarbons (Tearpock et al. 1994). We address such subjects as (1) why some growth faults die in both the upward and downward directions and what this means to exploration, (2) where rollover

* For all figures in this chapter (in the printed book only), see the preface for information about registering your copy on the InformIT site for access to the electronic versions in color.

anticlines are likely to *exist or be positioned along major listric normal faults* (i.e., faults that exhibit large vertical expansion across the fault surface), and (3) how faults form prospects. The second part of the chapter is the study of compaction effects along growth normal faults and how sandstone/shale ratios are utilized to project growth faults into poor data areas or, conversely, how fault shape is used to predict *percent sand in the footwall of the fault*. The chapter ends with projecting footwall correlations into the corresponding downthrown and undrilled hanging wall section.

ORIGIN OF HANGING WALL (ROLLOVER) ANTICLINES

Often, geoscientists work on the edge of coherent data or in areas where seismic data deteriorates. In this section, we present methods for extrapolating known data into poor data areas. Rollover anticlines have been successfully drilled for many years, yet questions remain concerning their origin and how rollovers form prospects. A major insight into the origin of rollover was initiated by Hamblin (1965) when he recognized that these strange "reverse drag folds" are the natural consequence of motion along listric normal faults. He reasoned that if the hanging wall block separates from the footwall block, a small void opens between the two blocks. Collapse due to gravity would instantaneously close the void and, as the hanging wall block collapses onto the footwall block, a reverse drag structure forms. However, Hamblin did not exactly specify how this gravitational collapse occurs. Gibbs (1984) later recognized from North Sea data that extensional structures (as an analogy with compressional structures) seemed to form duplexes, complete with horses, transfer zones, and so on. Yet a number of questions remain to be fully answered, such as:

1. Can the shallow portion of rollover structures be used (a) to predict the deeper structure (e.g., subfault structure) or (b) to extrapolate into regions of poor or nonexistent data (e.g., into poorly imaged seismic zones)?
2. What is the precise origin of rollover structures, and is it possible to predict the amplitude, style, and position of a rollover from observable and mappable geologic features?
3. What are the origins of the antithetic, or compensating, faults and the keystone structures, and how do these structures influence rollover geometry?
4. What causes some (perhaps most) antithetic growth faults to exhibit displacements along their surfaces that die in both the upward and downward directions in a seemingly contradictory relationship?

Answers to at least some of these questions should aid geoscientists in reducing the time and expense in (1) locating rollovers; (2) generating better interpretations, maps, and prospects; and (3) isolating trapping mechanisms. Our studies and those of others suggest that compaction, fault shape, antithetic and synthetic faulting, and cross structures affect the geometry of rollover structures.

When developing a theory for a structure that is as complex as a rollover, the initial theories are likely to be overly simplistic (Eichelberger et al. 2017). Our position is that a theory is only as good as its ability to quantitatively explain the observations. Therefore, and although less work has been conducted on extensional balancing in recent years, we expect that in the future, modifications and refinements are likely to make these theories more exacting.

Coulomb Collapse Theory

A theory of rollover formation based on Coulomb failure, or breakup (Suppe 1985) of the hanging wall onto the footwall block, has been advanced by Xiao and Suppe (1988, 1992), Groshong

(1989), and others. According to this theory, rocks fail along Coulomb fracture surfaces oriented at about 20 deg to 30 deg to the maximum principal stress (σ_1) (Billings 1972), which in the case of extensional deformation is subvertical. This theory describes many of the features observed on seismic sections, and the model commonly mimics observed rollover geometries subject to depth correction (Fig. 11-1). An understanding of this theory of genesis of rollovers can help you better interpret rollover geometries in poor data areas and thereby generate better prospects (Tearpock et al. 1994).

The elements of the Coulomb theory are shown in Figure 11-2a through e. Figure 11-2 represents the initial prefaulted or pregrowth state. This simple example shows an incipient major fault that has, for purposes of demonstration, a single concave upward bend. In nature, major listric normal faults are normally curved, and this more realistic case can be duplicated by introducing several concave upward bends into a model.

As the hanging wall block moves over the footwall block, a small void opens between the hanging wall and the footwall blocks, as described by Hamblin (1965) (Fig. 11-2b). Gravitational forces cause the hanging wall block to instantaneously collapse into the hole (created by the sliding) along the Coulomb failure surfaces. In Figure 11-2b the Coulomb collapse angle is 70 deg

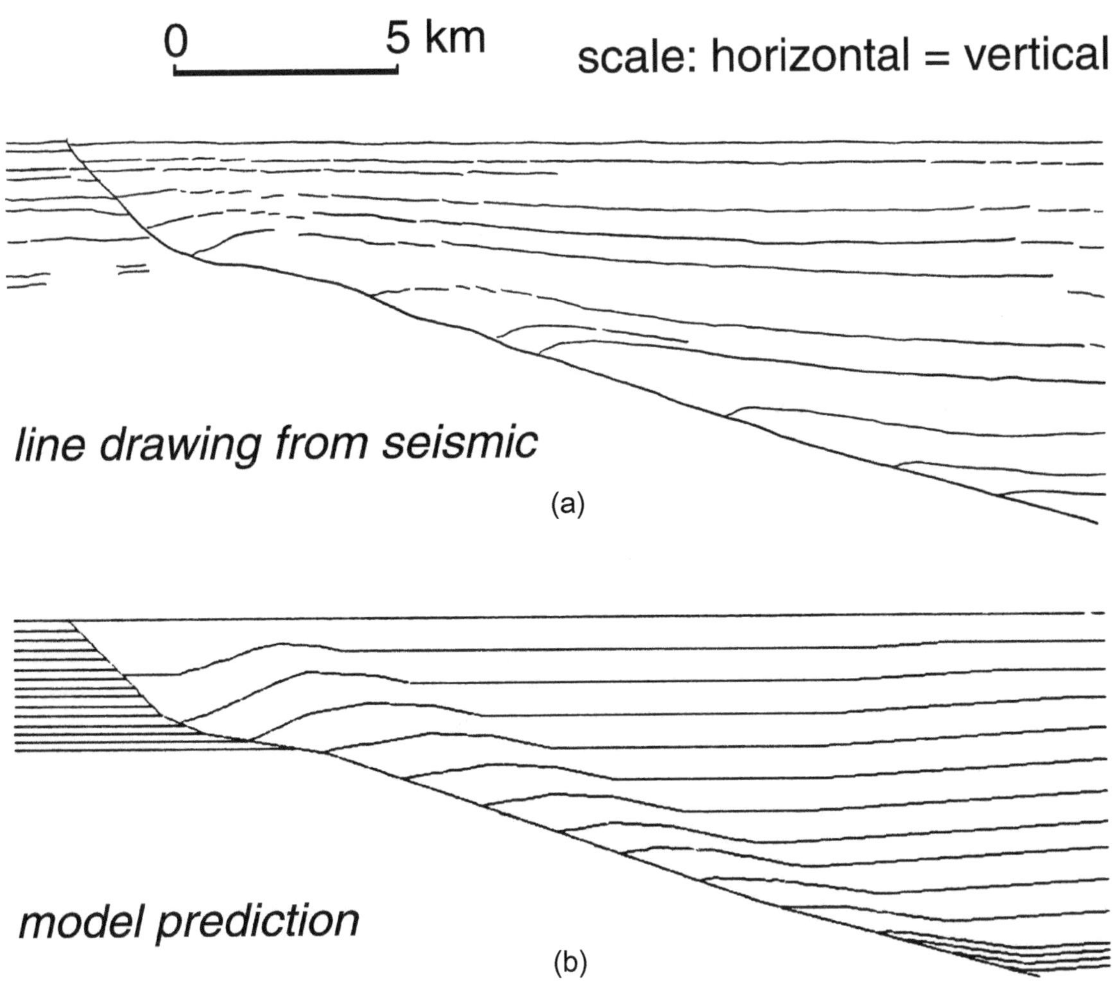

Figure 11-1 Comparison of (a) Brazos Ridge seismic data to (b) computer model, utilizing the Coulomb breakup theory. (Published by permission of H. Xiao and J. Suppe.)

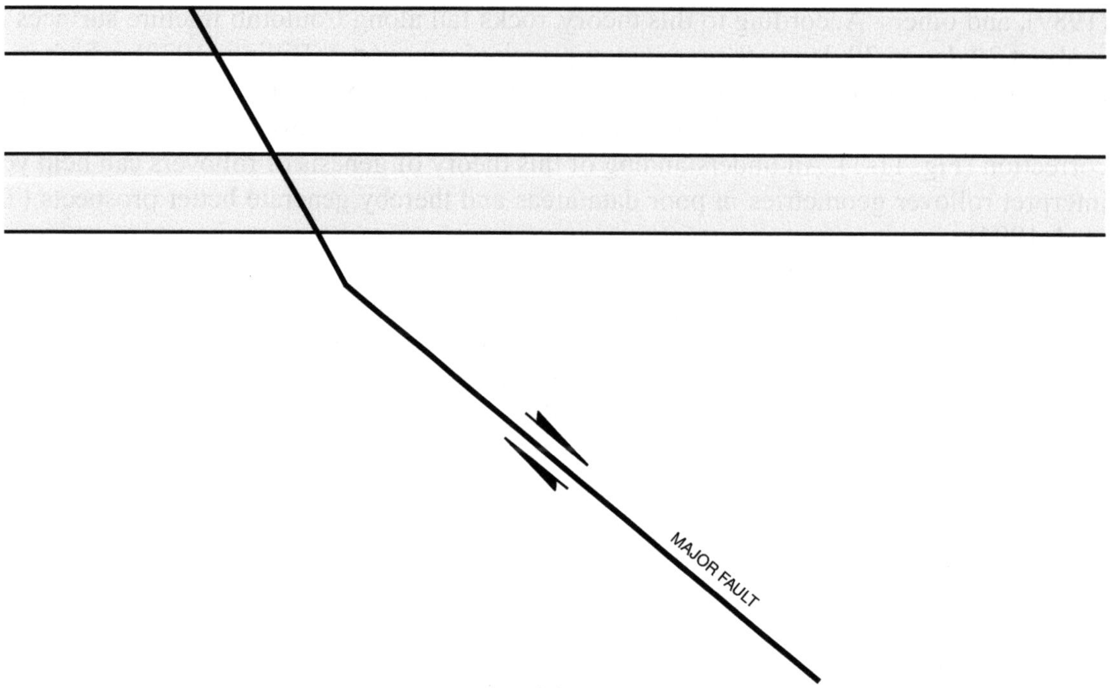

(a)

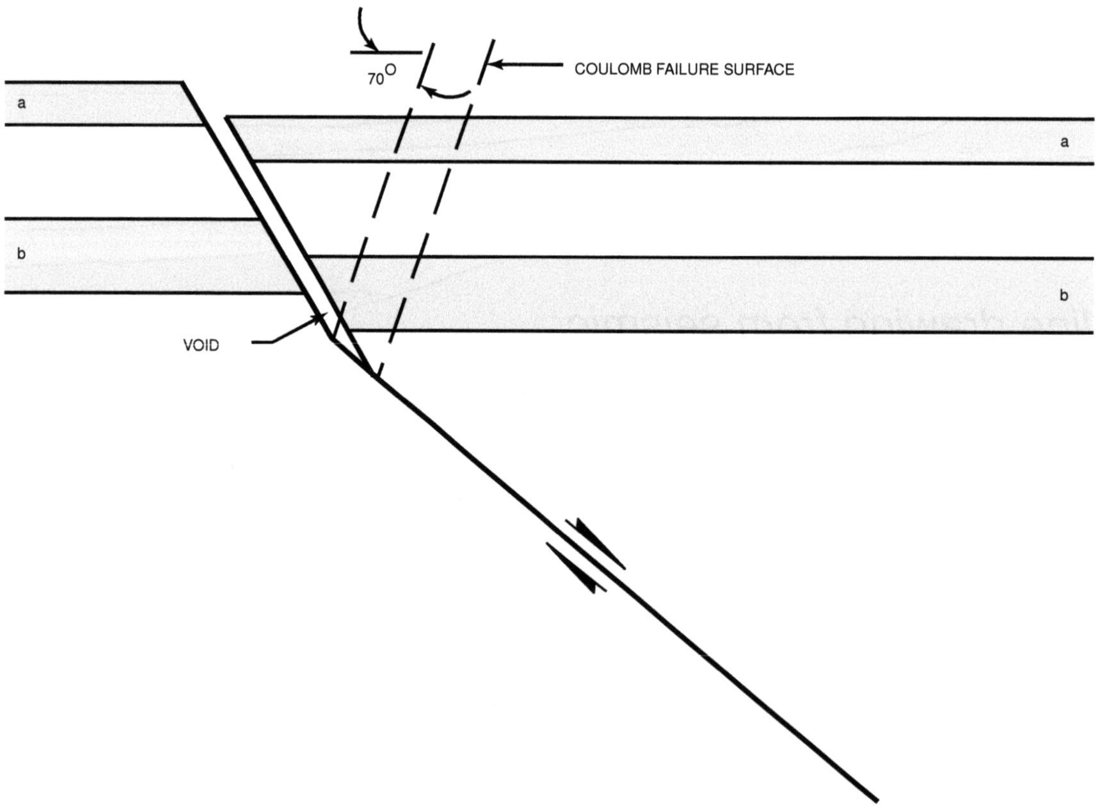

(b)

Figure 11-2 (a)–(e) Coulomb failure or shear collapse model showing different stages in the development of a simple rollover. A bend in the major fault subjects the hanging wall to deformation at the active axial surface, which is fixed to the bend in the footwall. Increased slip causes the inactive axial surface to migrate away from the surface of active deformation. (Modified after Suppe 1988. Published by permission of John Suppe.)

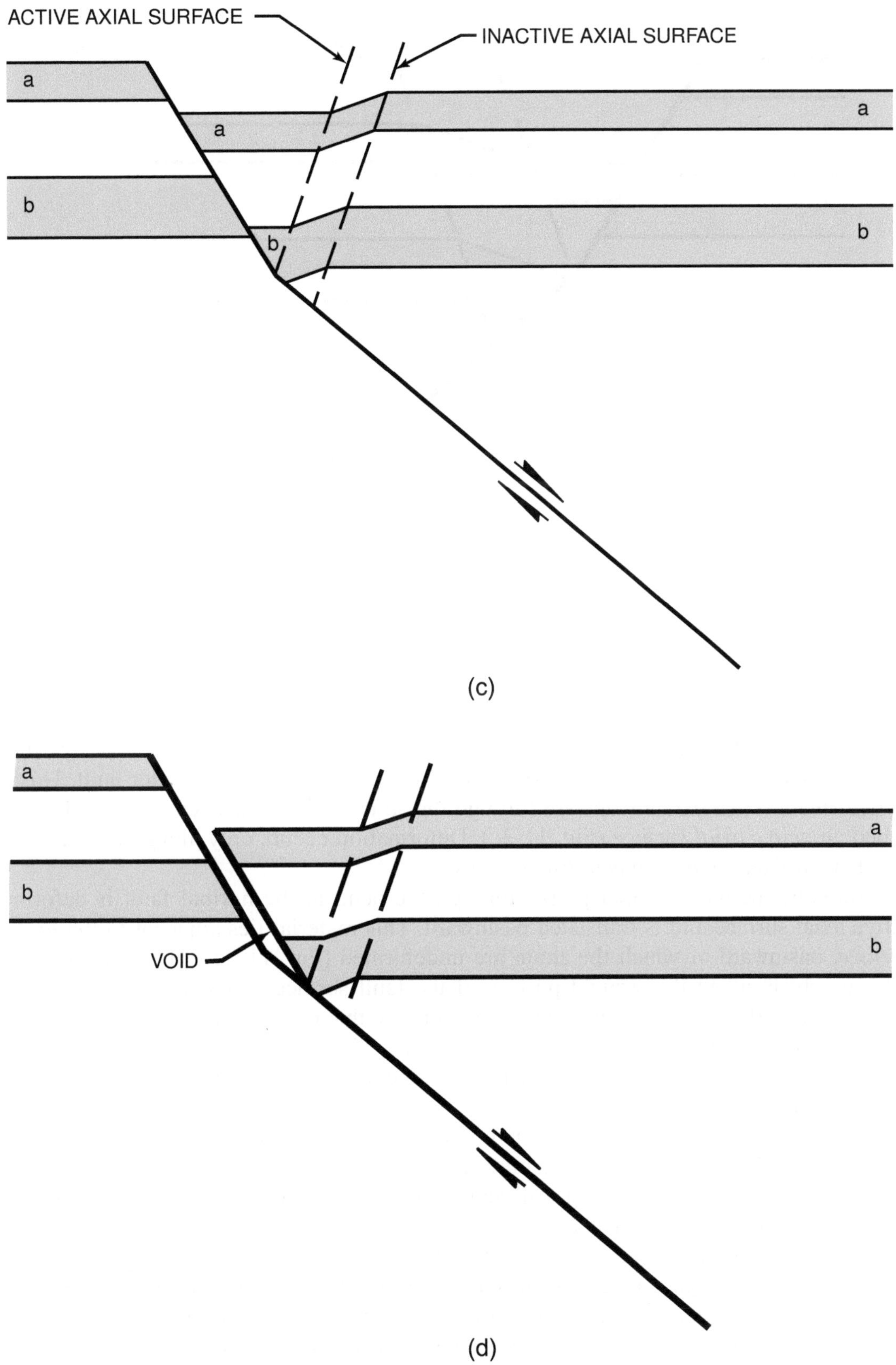

Figure 11-2 *(continued)*

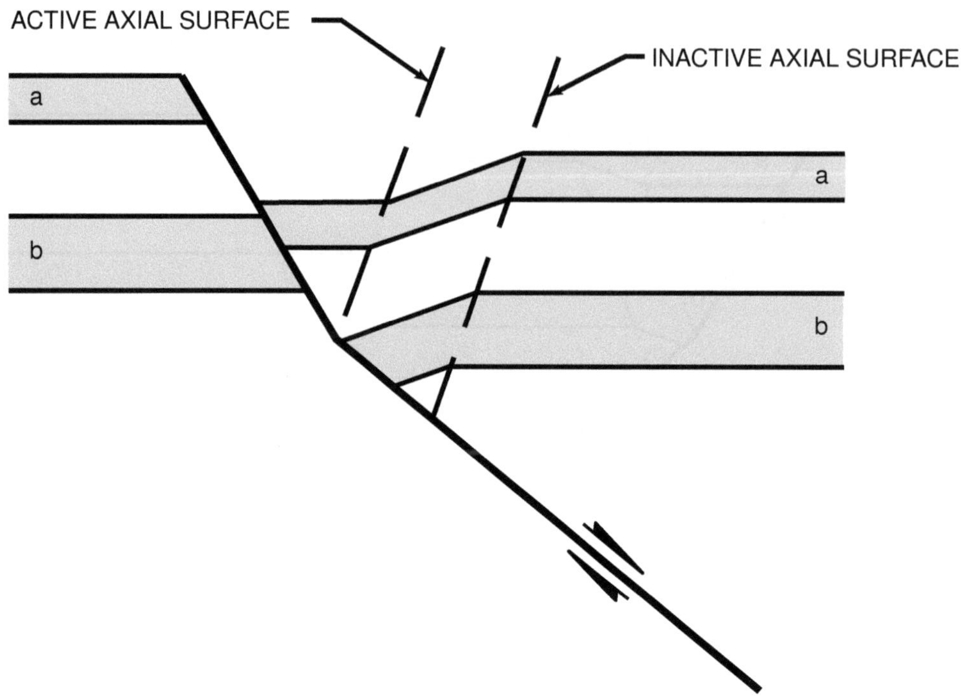

Figure 11-2 *(continued)*

with respect to the horizontal. Beds in the hanging wall shear parallel to the Coulomb failure surfaces, and the material fills the hole, causing the beds to extend (Fig. 11-2c).

Observe in Figure 11-2b and c that material experiences rotation as it passes through the Coulomb shear surface that is fixed to the concave upward bend in the major fault. This is a locus of deformation that extends upward through the strata and, for that reason, this shear surface is called an *active axial surface* (Fig. 11-2c). Deformation occurs only along active axial surfaces, which are *affixed to the bends in listric faults.*

Initially, the material that passes through the bend in the normal fault is deformed at the active axial surface and is translated basinward. This material lies adjacent to the *inactive axial surface*, basinward of which the strata are undeformed (Fig. 11-2c). The inactive axial surface rides passively along the straight portion of the fault surface as movement progresses. As this surface is not affixed to the bend in the fault surface that creates the void, the inactive surface is passive and does not cause the hanging wall beds to dip toward the fault surface. Notice that the slip on the fault surface is the distance between the active and inactive fault surfaces, parallel to the fault surface.

This process is more readily understood if the fault model is subject to another increment of sliding (Fig. 11-2d). Of course, in nature these increments are infinitesimal. The sliding opens another void between the hanging wall and the footwall blocks, and the active axial surface that formed in Figure 11-2c translates basinward (Fig. 11-2d). However, the void instantaneously closes by gravitational shear failure, pinning the active axial surface to the bend. The resultant structure shown in Figure 11-2e contains a graben-like feature adjacent to the steepest portion of the major fault and is a monoclinally shaped rollover structure. Sedimentary compaction within the basin creates basinward dip and closes the structure in the direction to the right in the figure.

You can better understand these processes by redrawing Figure 11-2e and cutting the hanging wall block from the footwall block with scissors. Then subject the major fault to another increment of motion and collapse the suspended material onto the footwall parallel to the Coulomb failure surfaces.

Growth Sedimentation

In areas like the Gulf of Mexico, most major normal faults are active growth faults. This means that sedimentation occurs simultaneously with fault slip, and thus the stratigraphic intervals are subject to vertical expansion across the fault surface (i.e., the intervals thicken on the downthrown side of the growth fault). How does this syndepositional sedimentation affect the structure, and can an analysis of this growth aid us in finding hydrocarbons?

Referring back to Figure 11-2c, let us assume that a layer of sediment is deposited across the hanging wall and footwall portions of our structure (Fig. 11-3a). The graben-like feature contains more accommodation space than the top of the rollover, so the sediments will be thickest over the graben and thinner over the footwall and the crestal portions of the monoclinal rollover. We saw in Figure 11-2b through e that the *active axial surface* is fixed to the bend in the major fault and that it represents the locus of deformation of the hanging wall beds. The *inactive axial surface* is passive and migrates basinward. Notice in Figure 11-3a that the inactive surface does not extend upward into the recently deposited growth sediments. The active axial surface, however, is a locus of active deformation that affects the entire body of growth sediments. When deformation is viewed as an incremental process, a *growth axial surface* must connect the active axial surface to the inactive axial surface, as shown in Figure 11-3a. You can convince yourself of this statement by visualizing a thin layer of additional growth sediments deposited above layer 1. This layer will be horizontal. Another small increment of sliding along the major fault causes Coulomb collapse that deforms this recently deposited layer in the region where the active and the growth axial surfaces converge (Fig. 11-3a). An additional increment of sliding, combined with growth sedimentation, produces the geometry observed in Figure 11-3b. The growth wedge *expands* as the growth sediments move through the active axial surface.

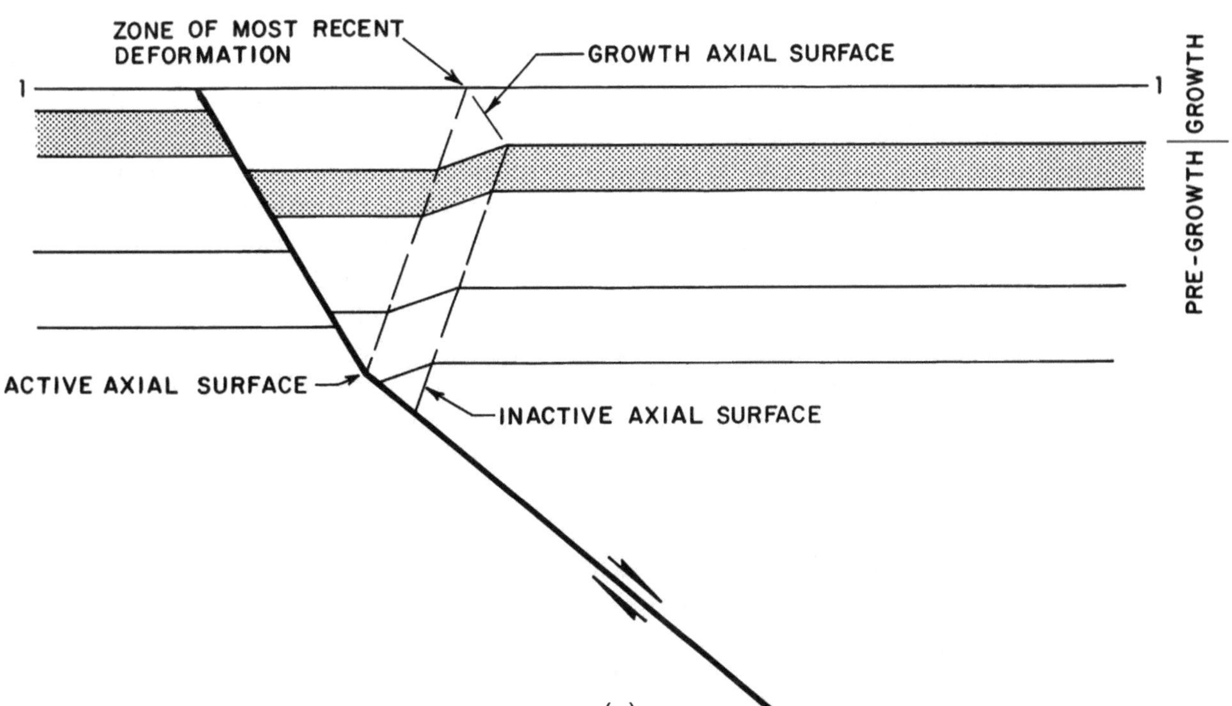

(a)

Figure 11-3 (a)–(b) Rollover development showing deformation during growth sedimentation. The most recently deposited sediments are deformed at the point where the active and growth axial surfaces converge. (Modified after Suppe 1988. Published by permission of John Suppe.)

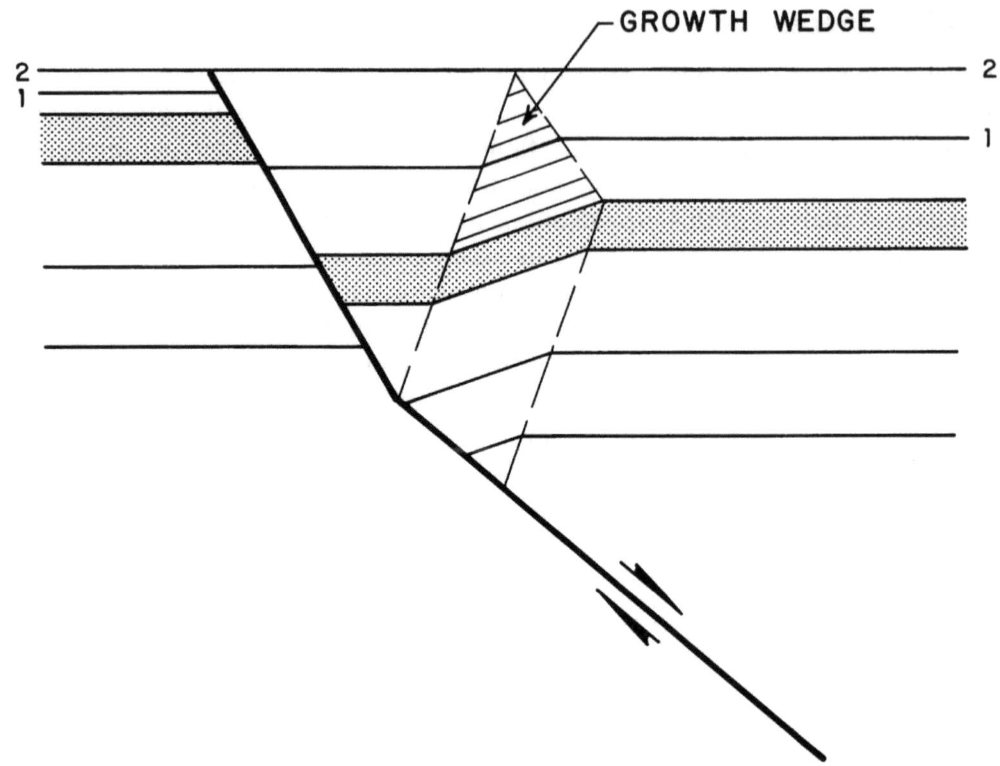

(b)

Figure 11-3 *(continued)*

For larger rollover structures, sands are more likely to be deposited in the graben, whereas suspended sediments are more likely to be dominant across the top of the rollover. Therefore, facies changes and stratigraphic traps are predicted to occur in the vicinity of the growth axial surface. A more realistic example of growth faulting is presented in the following section on projecting large growth faults to depth.

The growth axial surface can be located on a seismic section by using dip domain analysis (Tearpock et al. 1994). This surface is located between the steeper-dipping beds at the front of the structure and the flatter-dipping beds located at the crest. After you locate the growth axial surface, you can determine the stratigraphic interval that was deposited as the structure *grew*. This is important, as no structure exists to trap hydrocarbons prior to the growth phase. To determine the growth phase of the structure, follow the growth axial surface downward. The growth axial surface subparallels the main fault during the time of growth, thus delimiting the history of the structural growth. The growth axial surface terminates at the inactive axial surface, and this point indicates the top of the pregrowth strata (Fig. 11-3a and b). No structure existed during the pregrowth phase of sedimentation to trap hydrocarbons (Fig. 11-2a).

Example of a Rollover Structure. Some listric faults are observed to dip more gently with depth and then, at greater depth, to increase in dip (Fig. 11-12). This creates a concave downward bend in the major fault. Above this bend, the Coulomb collapse theory predicts that the deformation takes place along basinward-dipping conjugate shear surfaces. Shear can occur along two conjugate surfaces (Billings 1972) and, in our example, this shear occurs in the clockwise direction (Fig. 11-4a).

The Brazos Ridge, in the northern Gulf of Mexico, can be used to demonstrate the Coulomb breakup theory for a more realistic case. The Corsair fault is the major listric normal fault in the Brazos Ridge area. In the most general sense, the Corsair fault dips at about 45 deg, flattens to about 10 deg, and then steepens to approximately 20 deg, as represented in a computer model in Figure 11-4c. The Brazos region is subject to synchronous deformation and sedimentation, and thus the sediments deposited in the hanging wall block are growth sediments. If the simple Corsair fault model described in Figure 11-4c were subject to gravitational (basinward) sliding, two voids would be created, one above the concave-up portion and the other adjacent to the concave-down portion of the master fault. Clockwise collapse above the deeper concave downward bend (Fig. 11-4a) and counterclockwise collapse above the concave upward bend (Fig. 11-4b) would produce the geometry observed in Figure 11-4c. Notice that there is a relationship between the shape of the fault and the shape of the rollover in the hanging wall.

If deposition rates are high, as in the Brazos Ridge area, the sediments will fill the low area adjacent to the shallowest portions of the major fault, thin over the top of the rollover structure, and then thicken basinward (Fig. 11-4c). The vertical expansion of stratigraphic intervals, as observed in this example, demonstrates why the sediments are referred to as growth sediments. Also notice the growth axial surface in Figure 11-4c. It subparallels the growth fault, and it terminates with depth at an inactive axial surface that dips toward the fault (compare to Fig. 11-3b). The model in Figure 11-4c contains rollover amplitudes that are higher than observed along the Brazos Ridge (Fig. 11-1a). However, compaction combined with synthetic and antithetic crestal faulting reduces the amplitudes of rollovers.

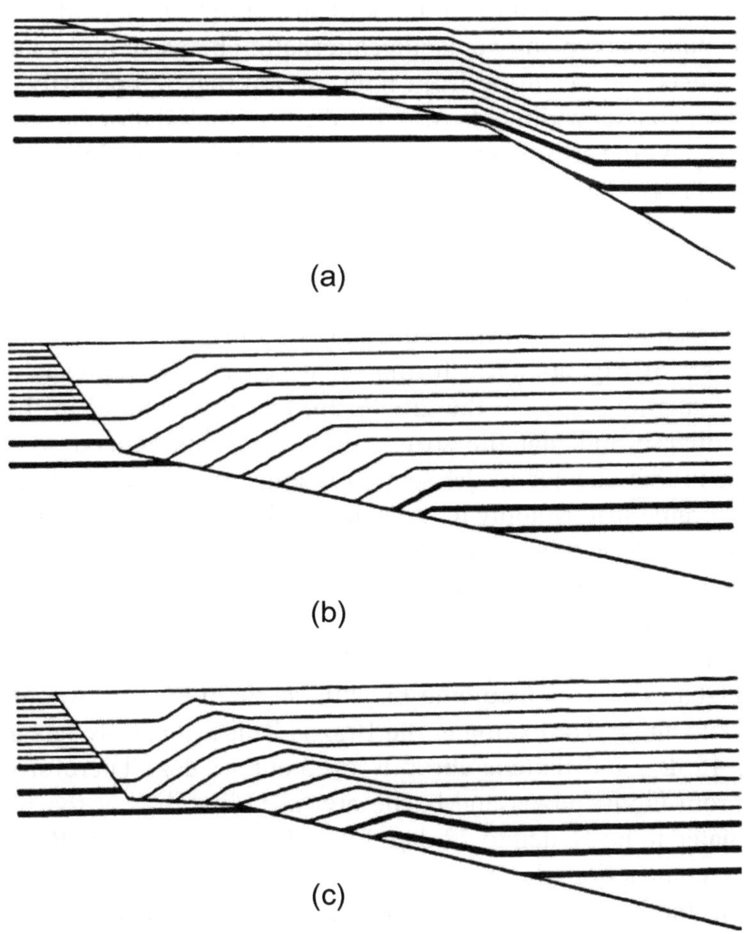

Figure 11-4 Deformation of growth sediments at the bends in normal faults. (a) A downward bend in the fault activates basinward dipping or synthetic shear. (b) An upward bend activates landward or antithetic shear. (c) Simple generic model of Brazos Ridge rollover development, not including compaction. (Published by permission of H. Xiao and J. Suppe.)

The thickness of the crestal growth section can vary with the shape of the fault. In the case of the Brazos Ridge, the large expansion fault that creates the structure dips at about a 20-deg angle at depth. Thus, as the hanging wall moves over the footwall, the hanging wall slides down a 20-deg dipping fault surface. This slope causes the hanging wall to drop in elevation relative to the footwall, and the downward motion creates *ample space for the accumulation of thick growth section over the crest of the structure* (Figs. 11-3a and 11-12). Later we show that this fault geometry creates productive *four-way closures* (Fig. 11-21).

However, if a large expansion fault flattens at depth and is subhorizontal, then as the hanging wall moves over the footwall, the crest of the rollover structure does not change elevation with respect to the footwall (Fig. 11-5). This fault geometry creates the classic *half-graben structure*, with accommodation space developing between the fault and the flank of the half-graben structure but not over the crest of the structure. In this case the growth sediments are more likely to thin onto the flank of structure, creating stratigraphic traps. However, on actively growing structures, this *thinning results from an interaction of stratigraphic and structural processes*, classified under the heading of tectonostratigraphy (Chapter 13). Figure 11-5 shows a seismic profile of a half-graben structure from the Central Sumatra Basin, Indonesia. As the structure grows, growth sediments deposited in the half-graben structural low will be subject to axial surface deformation, as shown in Figure 11-5. The growth sedimentary section thins dramatically across the growth axial surfaces and onto the flank of structure (Suppe et al. 1992; Shaw et al. 1997). As growth sedimentation is typically episodic, alternating between high and low sedimentation rates, the local subsidence rate resulting from fault motion may temporarily exceed the sedimentation rate. In this case, coarser sediments may fill the accommodation space in the half-graben (Xiao and Suppe 1992). As the sedimentation versus subsidence rates fluctuate, the tectonostratigraphic process creates facies changes and unconformities along strike and down-dip of the growth axial surfaces.

Provided that information exists on the shape of a fault that has created a rollover, the Coulomb breakup theory can be used to predict the rollover angle (θ) or bed dips. Based on Coulomb deformation, the rollover angle (θ) is related to the change in fault dip (ϕ) through a complicated set of trigonometric formulae, which can be represented by graphs (Fig. 11-6a and b). The assumptions used in the graphs are that the material in the hanging wall is subject to Coulomb collapse and that the structure has *not* experienced sedimentary compaction, as described later in this chapter in the section Compaction Effects along Growth Normal Faults. Thus, these diagrams strictly apply to the nongrowth phase of the sedimentation and will approximate the rollover angle within growth sediments. Let us assume that a seismic section images the shallow and deeper portions of a listric fault but does not image the bed dips. If the above conditions are met and if the initial fault dip (β) and the fault dip at a deeper level (α_1) can be observed on a *depth-converted* seismic section, then the rollover dip can be estimated from Figure 11-6a or b. Figure 11-6a assumes a Coulomb breakup angle of 60 deg, whereas Figure 11-6b assumes a 70-deg breakup angle. For example, the Brazos Ridge generally has a major fault that initially dips at about 45 deg and flattens abruptly to about 15 deg. Therefore, the change in fault dip $\phi = 30$ deg. From Figure 11-6a and b, the predicted bed dips at the front of the rollover structures are estimated to be 18 deg to 22 deg. Even though the Brazos Ridge growth sediments have been subject to compaction, these results compare favorably to observed bed dips, which average about 20 deg at the bend in the fault (see Fig. 11-1a and b). For presentation purposes, Figure 11-6a and b can also be used to generate generic or idealized models of real rollover structures. The generic structure shows how the rollover structure formed.

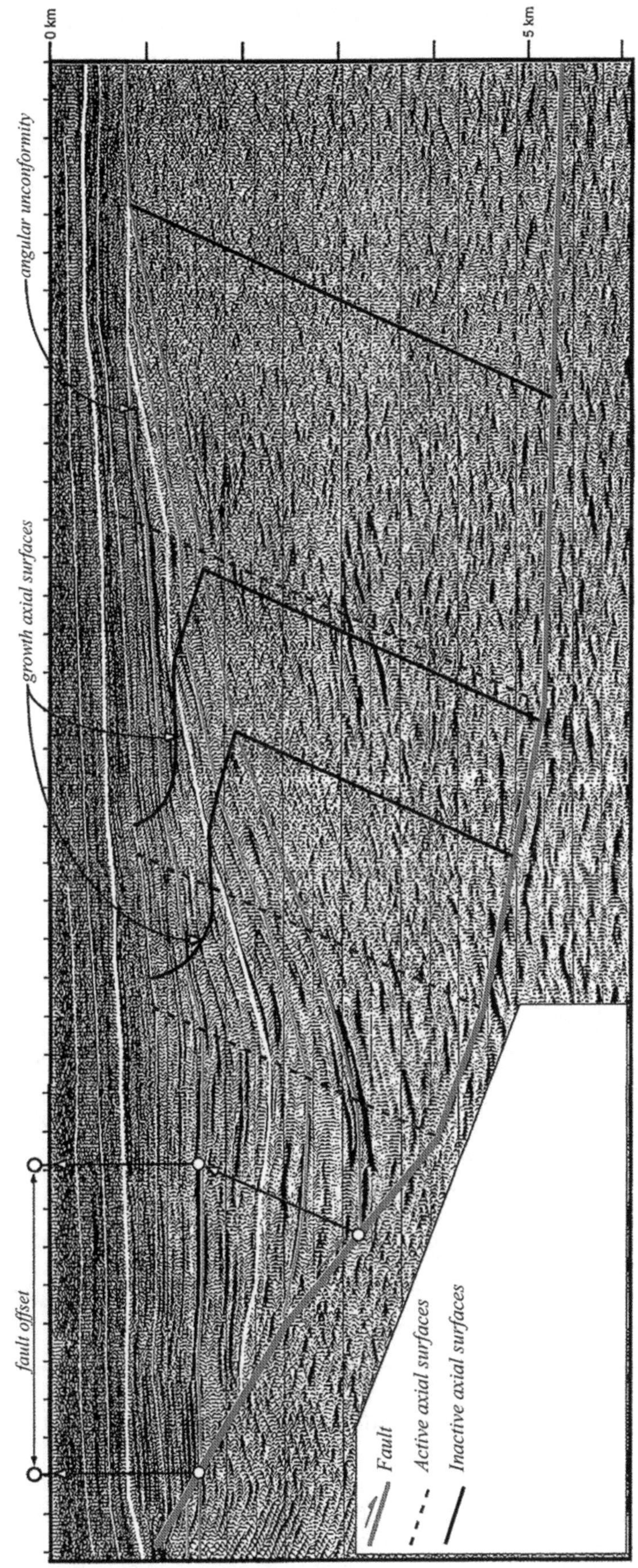

Figure 11-5 Seismic profile of a half-graben structure from Central Sumatra Basin, Indonesia, showing axial surfaces emanating from bends in the fault surface. Growth axial surfaces on the flank of the structure (subhorizontal curved lines) define the position where growth section thins onto structure. The syntectonic sedimentation created potential facies changes and unconformities at the position of growth axial surfaces. (From Shaw et al. 1997. AAPG©1997, reprinted by permission of the AAPG whose permission is required for further use.)

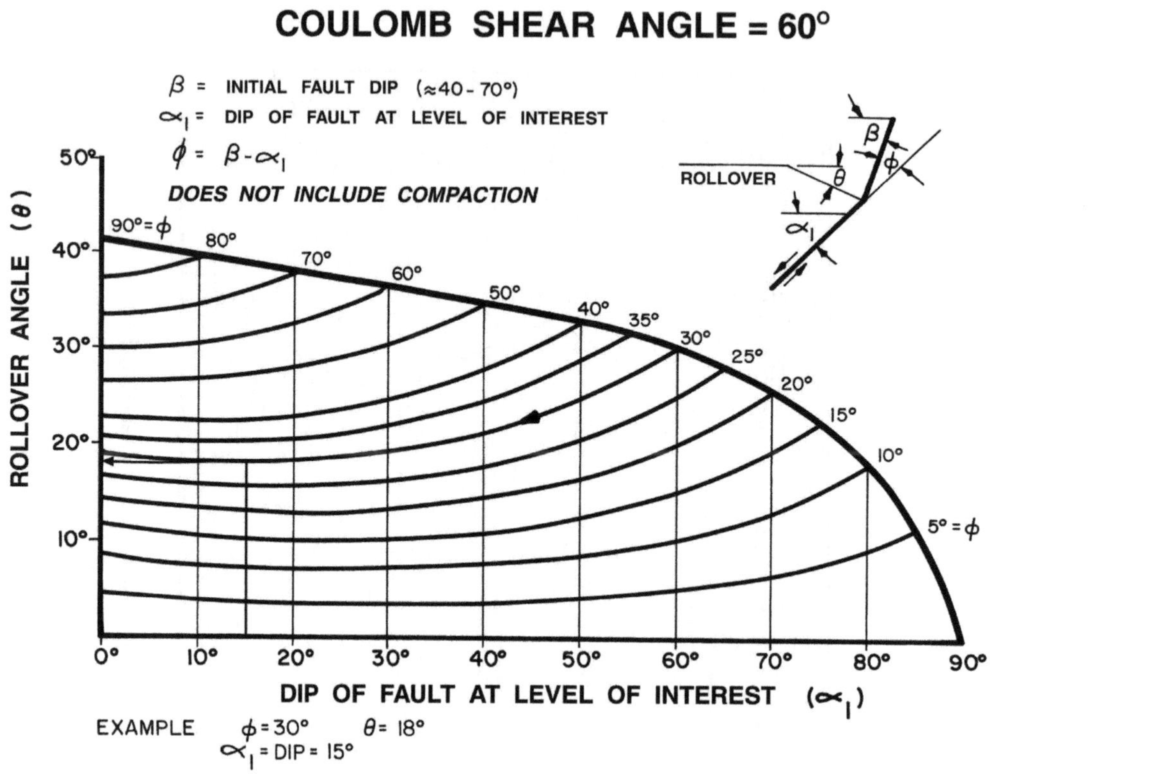

(a)

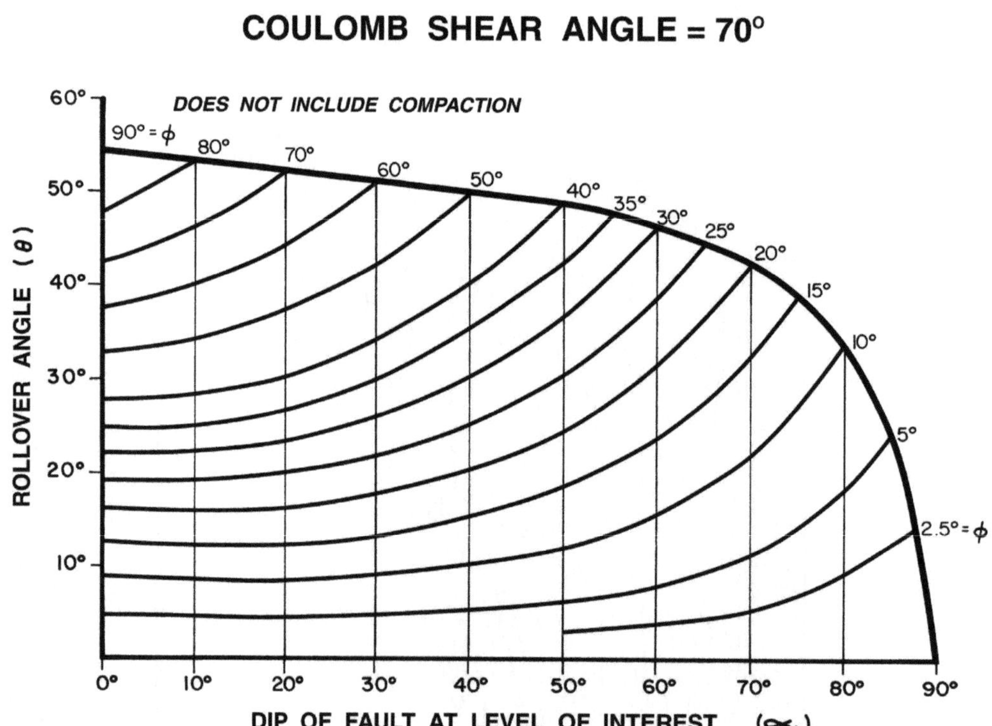

(b)

Figure 11-6 (a) Theoretical prediction of rollover angle (θ) from fault dips (ϕ and α1) for a Coulomb shear angle of 60 deg. Diagram does not include compaction. (b) Same as (a), but for a Coulomb shear angle of 70 deg. (Published by permission of R. Bischke.)

A GRAPHICAL DIP DOMAIN TECHNIQUE FOR PROJECTING LARGE GROWTH FAULTS TO DEPTH

Often, geoscientists work on the edge of, or below the level of, coherent seismic data. Large listric normal faults may be present in the poor data zone. If a proposed well crosses a large growth fault and the well encounters an unexpected stratigraphic section, the result is often disappointment and confusion regarding the geology. Also, large growth faults commonly are pressure-sealing faults, so the unexpected penetration of a large fault by a drilling well could be disastrous.

On the Gulf of Mexico shelf, in Indonesia, China, Nigeria, and elsewhere, rollover structures form downthrown to listric growth faults. Growth sequences of sediments that accumulate downthrown to normal faults are common throughout the world. We use these growth sections to formulate a technique for projecting normal faults into poor data areas using the well-imaged portions of rollover structures.

In the section on the origin of rollover, we show how rollover structures form as the hanging wall conforms to the shape of the footwall. Thus, there is a causative relationship between the shape of the fault and the shape of the fold. We use this causative relationship to present a graphical technique for projecting large growth faults into poor data areas. This inclined shear collapse technique is subject to measurement errors. Therefore, the method can project a fault through only a limited number of bends (two or three) within poor data zones. In order to compare theory to observation, we present a well-constrained example of projecting a fault surface to depth using the shallow parts of a rollover structure.

We base our graphical method on the kinematics developed by Hamblin (1965) and make no assumption as to the shape of the fault surface. For example, the fault surface can contain both concave and convex bends, and the fault surface need not flatten at depth. Bischke and Tearpock (1999) present formulas for projecting faults to depth and for calculating Coulomb collapse angles. The Coulomb collapse angle formula requires knowledge of fault throw and has application to rollover structures in extensional environments anywhere in the world. We also present a graphical method for predicting Coulomb collapse angles in this section.

Rollover Geometry Features

A major advancement in the understanding of structures occurred when Hamblin (1965) recognized that rollover anticlines form as hanging wall beds slip along listric normal faults. As the hanging wall slides along the fault surface, *the hanging wall conforms to the shape of the fault surface* by collapsing onto the fault surface. Hamblin indicated that the process could result from vertical or Coulomb collapse (Fig. 11-7). Numerous authors tested the two mechanisms and favor Coulomb collapse over the vertical collapse mechanism (White et al. 1986; Worrall and Snelson 1989; Groshong 1989; Bischke and Suppe 1990b; Dula 1991; Nunns 1991; Withjack et al. 1995). Xiao and Suppe (1992) show, from empirical observations using forward models, that many rollovers form as a result of Coulomb collapse at angles of 65 deg to 70 deg with respect to the horizontal. Industrial software programs are available that incorporate the vertical and inclined collapse algorithms to model rollover structures (Rowan and Kligfield 1989).

In a listric normal fault system, as the hanging wall beds move through bends in the fault surface, small voids open between the hanging wall and the footwall (Fig. 11-7). Large gravitational forces instantaneously close these voids, forcing the hanging wall to conform to the shape of the footwall. This collapse mechanism causes the hanging wall beds to dip, or diverge, toward the fault surface (Fig. 11-8). Furthermore, the strata steepen in dip as they move across bends in the fault, so the *more listric the fault surface, the steeper the dip of the beds* (Fig. 11-8).

Motion of the hanging wall down the fault surface creates *accommodation space for growth sediments*. The recently deposited sediments, having existed for only a short period of time, have not moved through many fault bends and therefore exhibit gentle bed dips (Fig. 11-8). The deeper and older sediments have moved through numerous fault bends (Xiao and Suppe 1992). Thus, older growth sediments dip at higher angles than shallow growth sediments (Fig. 11-8).

Several observations can be made about rollover fold shapes. First, the more listric the fault, the steeper the dip of the growth section. Second, there is a relationship between fault shape and rollover shape. If fault shape is known, then rollover geometry can be predicted with a high degree of certainty (Xiao and Suppe 1992; Shaw et al. 1997). Conversely, if rollover shape is known, then fault shape can be predicted, but with less certainty. Predicting fault shape from rollover shape is an inverse technique and calculation errors accumulate. These errors originate from uncertainties resulting from depth correction, geological and stratigraphic variations, and simple measurement errors. However, both techniques are important tools in extensional structural interpretation.

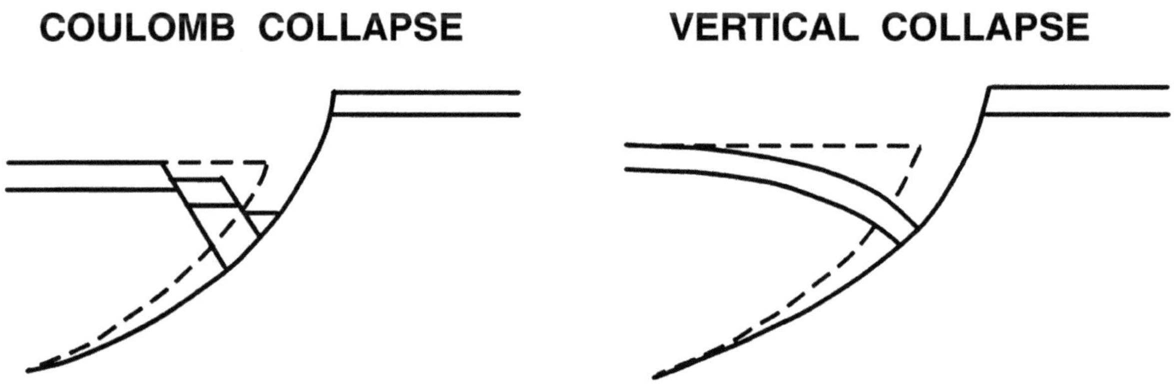

Figure 11-7 Rollovers form as the hanging wall collapses onto the footwall. Hanging wall collapse could occur as inclined (Coulomb) collapse or as vertical collapse. (Modified after Hamblin 1965. Published by permission of the Geological Society of America.)

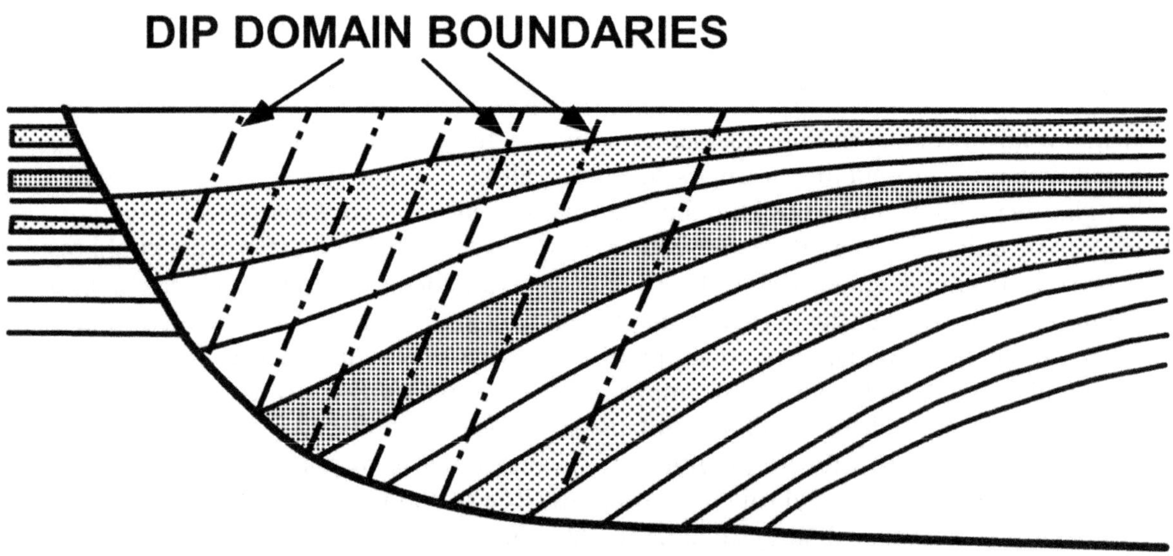

Figure 11-8 Listric normal faults cause the hanging wall beds to diverge toward the fault as the beds move through bends in fault surfaces. The fault bends define the dip domain boundaries. Young growth sediments dip at gentle angles relative to the deeper beds, as the shallow beds have not moved through many fault bends. The older and deeper beds, having moved through many fault bends, dip at higher angles. (Modified from Xiao and Suppe 1992. Published by permission of the Gulf Coast Association of Petroleum Geologists.)

Projecting Large Normal Faults to Depth

We present a graphical method for projecting large normal faults to depth using hanging wall beds that have been subject to noticeable rollover. We also present methods for estimating fault shape and position from the shape of a well-constrained rollover structure.

In order to apply this method, two features must be known: (1) a marker bed that exhibits noticeable rollover and (2) the true dip of the fault at the hanging wall cutoff level of the marker beds. A profile is constructed that must be a dip profile, so that it is perpendicular (as close as possible) to strike of the fault and strike of the beds. A fault surface map can help you determine the dip direction of the profile. The structural information can come from well log data, depth-corrected seismic profiles, or cross sections. If possible, start the projection process by locating a marker bed above the known level of the main expansion fault. Then use the known portions of the fault to calibrate and test the projection procedure. If satisfied with the results, apply the method to the unknown levels of the fault.

Procedures for Projecting Large Normal Faults to Depth

1. *If using a seismic profile, depth-correct the seismic section over the interval of concern.* Depth-correct the section on the workstation using local checkshot or other more modern depth conversion methods such as employing formation tops to calibrate models derived from time surfaces and processing velocities.
2. If secondary faulting offsets the marker bed, restore the bed to its unfaulted position before proceeding further. This is readily done by scanning or digitizing the marker bed into a graphics program and restoring the upthrown and downthrown cutoffs of the marker bed (Bischke and Tearpock 1999).
3. In the graphics program, locate a *deep marker bed* on a level that intersects the main fault. The bed should exhibit obvious rollover and intersect the imaged portion of the main fault (Fig. 11-9a).
4. Fit tangents to the marker bed in order to segment the rollover into *dip domains* (see Cross-Section Construction and Kink Method Approximation sections in Chapter 10). Each dip domain contains a region of roughly uniform bed dips (Shaw et al. 1997). Curved rollovers will possess several dip domains (Figs. 11-8 and 11-9a). Our generic example has four important dip domains: dip domains 1, 2, 3, and 4 (Fig. 11-9b). Where the tangent lines intersect, construct five dip domain boundary lines (S_1, S_2, S_3, S_4, and S_5) to each of the dip domain panels (Fig. 11-9b). The dip domain boundaries are drawn at an angle of either 68 deg or 112 deg. In Figure 11-9b, dip is measured clockwise, with 0 deg to the right and 180 deg to the left. If the marker bed dips toward the main fault, then these domain boundaries dip at about 112 deg with respect to the horizontal, and they dip in the *antithetic direction* toward the main fault. If the marker bed dips away from the main fault, then these domain boundaries dip at about 68 deg in the *synthetic direction*, or in the same direction as the main fault. As an alternative to the above assumed angles, you can derive the collapse angle from the geometry of rollover structures, as described later in the chapter (Bischke and Tearpock 1999).
5. Project the first dip domain boundary S_1 downward to where it intersects the straight-line projection of the main fault surface (Fig. 11-9c). Where the two surfaces intersect at point (1), the fault dip decreases and the fault becomes more listric.
6. As the first step to determine the fault position beyond the first bend in the main fault, *construct horizontal lines between the right and the left dip domain boundaries*, or between boundaries S_1 and S_2, S_2 and S_3, and so on (Fig. 11-9b and c). Then, *parallel* to the dip

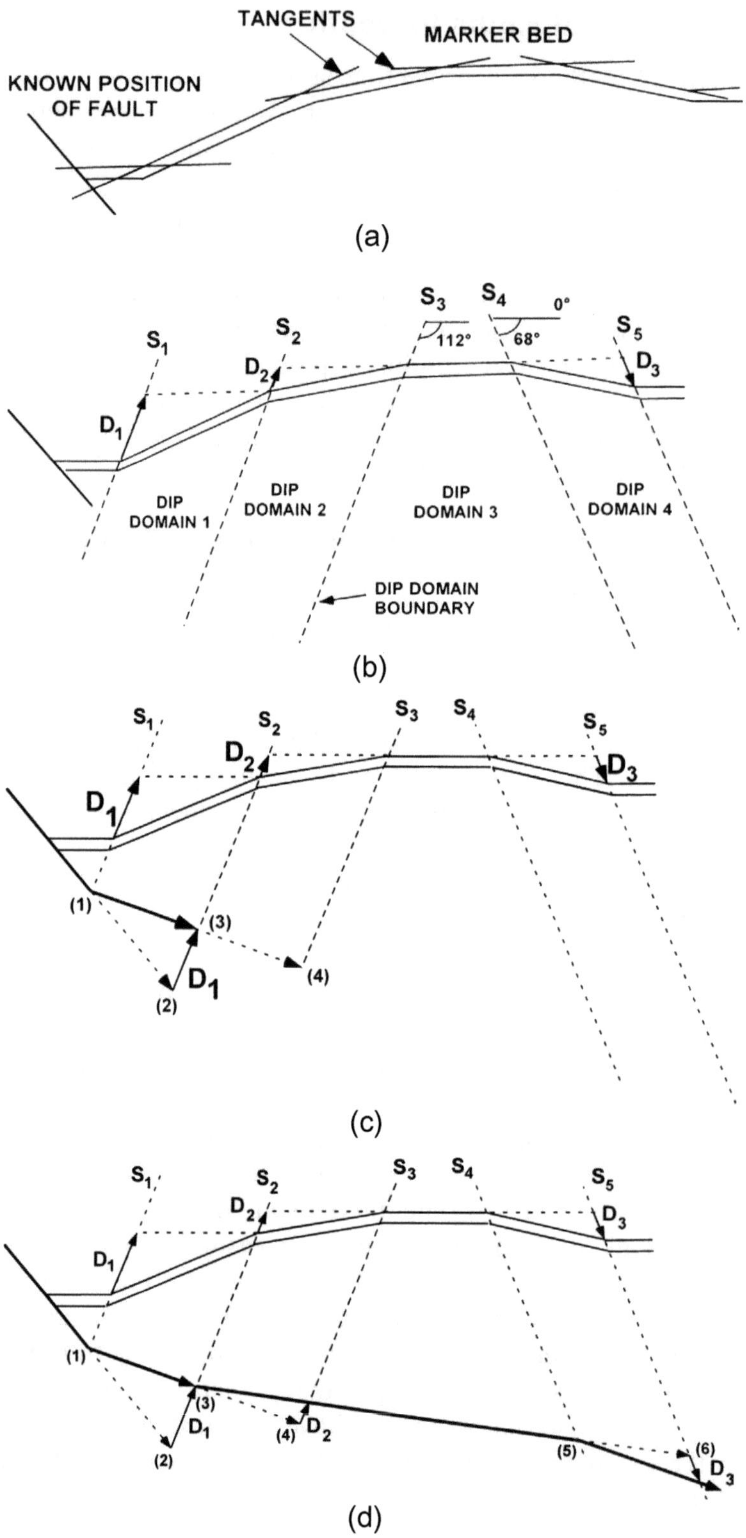

Figure 11-9 (a)–(d) Graphical procedures for projecting a normal fault to depth, using the shape of a generic rollover. For explanation, see text.

domain boundaries and between the marker bed and the horizontal line, measure the distances D_1, D_2, and D_3. In other words, measure distances D_1 and D_2 along the 112-deg inclined dip domain boundaries. Similarly, measure distance D_3 along the 68-deg inclined dip domain boundary S_5 (Fig. 11-9c).

7. Next, project the known portion of the main fault forward, from the left domain boundary at point (1) to the projection of the right dip domain boundary at point (2) (Fig. 11-9c). Where the straight-line projection of the main fault intersects the right domain boundary S_2, measure the distance D_1 from point (2) upward and in line with the right domain boundary S_2. This procedure determines the position of point (3). The position of the main fault is now known at point (1) and at its projected position at point (3), defined by distance D_1. Next, construct the projected portion of the fault surface between points (1) and (3), establishing the position of the fault beneath dip domain 1 (Fig. 11-9c).

8. Project the main fault forward from point (3) to point (4) (Fig. 11-9c) and repeat step 7 for the next fault bend (Fig. 11-9d). For this fault bend, measure distance D_2 from point (4) at the right domain boundary S_3 (Fig. 11-9d).

9. If the rollover has a *horizontal crest*, this means that the *fault does not change dip* beneath the flat crest of the rollover. Thus, the main fault dips at the same angle beneath the crest of the rollover (Fig. 11-9d). In other words, the fault is straight beneath dip domain 3 in Figure 11-9b. Project the fault beneath the flat crest to dip domain boundary S_4 at point (5). At point (5) the fault turns down and becomes *convex* where the beds turn down, but what is the dip of the fault beyond dip domain boundary S_4?

10. Repeat step 7 again, but in this case, the fault turns down at the axial surface where the marker bed turns down. So subtract distance D_3 from point (6). Lastly, we construct the convex projection of the fault surface beneath dip domain 4 (Fig. 11-9d).

This technique is very precise on ideal or generic rollovers, but it suffers from measurement errors and geologic variations on real rollover structures, where these measurement errors add through each fault bend. Therefore, the projection technique may deteriorate after projecting the fault surface through three or four fault bends.

Field Example. The method has application to all extensional regimes. A field example of a rollover structure from the Gulf of Mexico basin illustrates the procedures used to employ the fault projection method.

The example is from a 3D depth-corrected profile from the Burgentine Lake Field, onshore Texas (Fig. 11-10) (Bischke and Tearpock 1999). The horizons and fault surface were tied to well log data and mapped. The profile was depth-corrected using checkshot data. A graphical technique for determining the Coulomb collapse angle (described in the next section) derives a collapse angle ψ of 63 deg for the structure (Fig. 11-10a). This angle is slightly less than the 67-deg to 68-deg collapse angles observed on some Gulf of Mexico rollovers (Xiao and Suppe 1992). Employing the steps described in the preceding section on procedures, and using a ψ of 63 deg, we generate the cross section shown in Figure 11-10b. *The projected portion of the fault surface lies over the depth-corrected portion of the fault surface*. In this example, the agreement between theory and observation is very good. Therefore, we have shown that the method has application in working with real rollover structures.

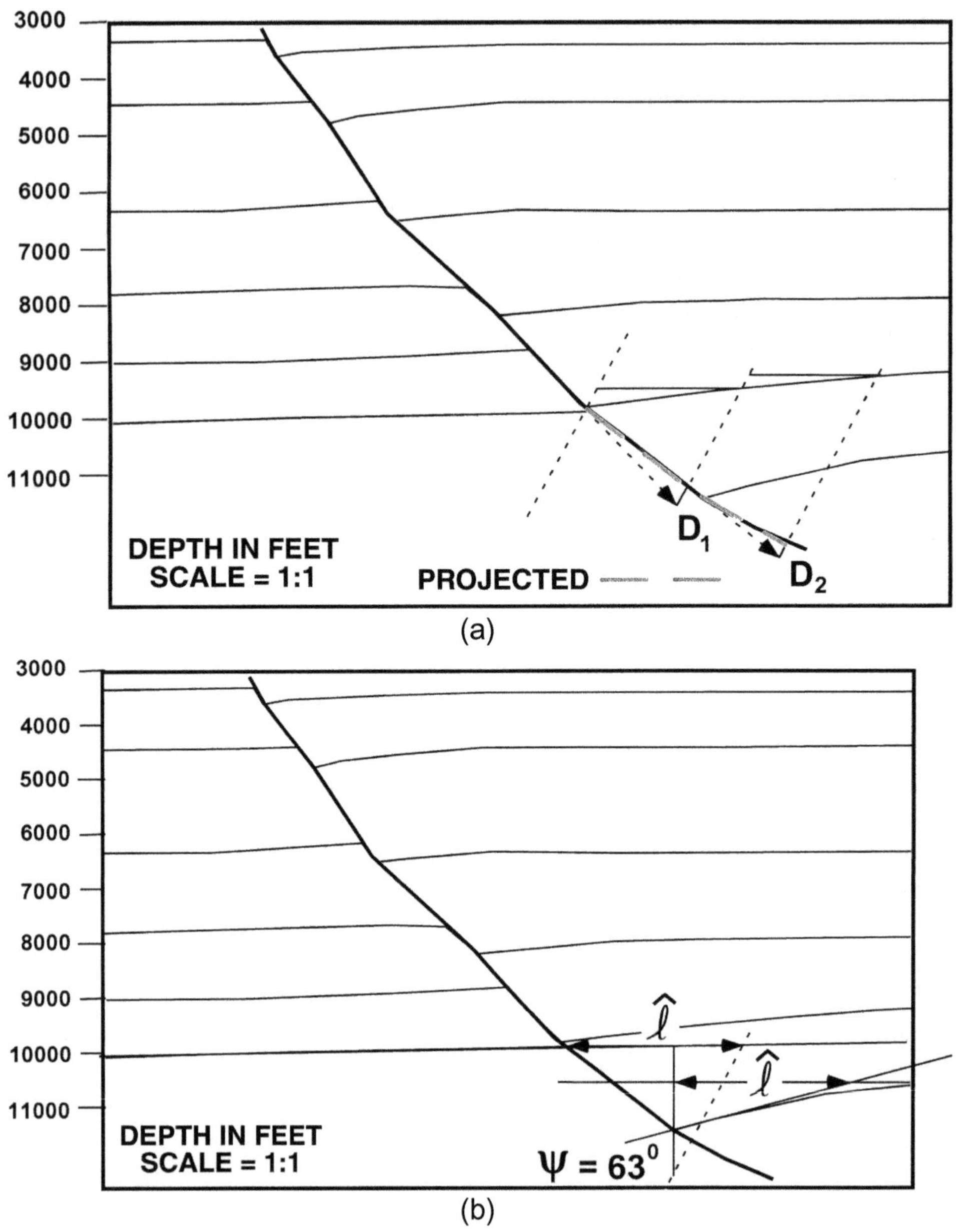

Figure 11-10 (a) Graphical procedures for determining the Coulomb collapse angle on a rollover anticline at Burgentine Lake, Texas, USA. For detailed explanations, see text. (b) Predicted fault surface conforms to depth-corrected fault surface [dashed in (a)].

Determining the Coulomb Collapse Angles from Rollover Structures

We conclude this section by describing a graphical technique for determining the Coulomb collapse angle. In Bischke and Tearpock (1999), we present the mathematical basis for these graphical methods. Here we describe the graphical method.

Using a dip-oriented, depth-corrected seismic profile, begin by locating a deep dip domain that lies adjacent to a large expansion fault, as outlined by the dashed lines in the hanging wall block in Figure 11-11a. The fault is represented by the lines L and S. A tangent line ℓ is fit to the *steepest dipping beds* adjacent to the fault surface, and another tangent line is fit to the flat crest of the rollover. The technique requires the determination of two displacement parameters: (1) the throw T on the fault and (2) the vertical distance V between the marker bed's footwall position and the position of the marker bed at the crest of the rollover (Fig. 11-11b).

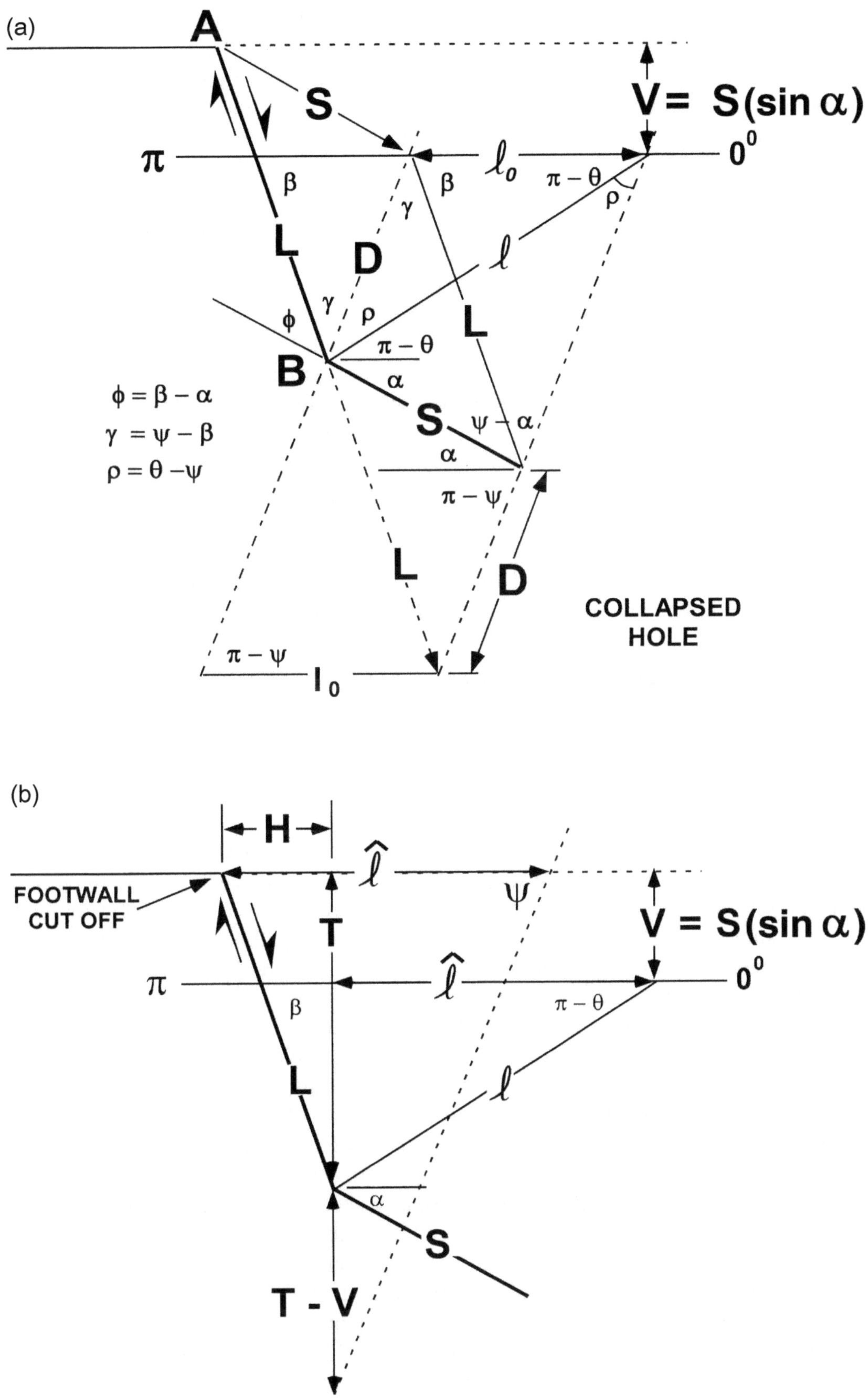

Figure 11-11 (a)–(b) Graphical method for predicting Coulomb collapse angles from rollover structures. For detailed explanations, see text.

At the position of the marker bed's hanging wall cutoff, construct the vertical throw vector line T on the depth-corrected profile (Fig. 11-11b). Determine the vertical distance V. Next, extend the throw line T by the distance T-V (Fig. 11-11b). Determine length ($\hat{\ell}$) by first locating the point of intersection between tangent line ℓ, at the front, and the tangent line at the crest of the rollover. Length $\hat{\ell}$ is the horizontal distance between this intersection point at the crest and the vertical line T (Fig. 11-11b). Measure length $\hat{\ell}$ at the same structural level as the tangent line drawn at the flat crest of the rollover structure. Next, position a line of length ($\hat{\ell}$) adjacent to the marker bed's footwall cutoff position (Fig. 11-11b). The final step in the process is to determine the value of the Coulomb collapse angle ψ (Fig. 11-11b). The angle ψ is equal to the inclination of the dip domain boundary (Fig. 11-11a). From the right-hand termination of line ($\hat{\ell}$), positioned adjacent to the marker bed's footwall cut off, construct a dashed line downward to the lower extension of the T-V line (Fig. 11-11b). Lastly, measure the Coulomb collapse angle ψ from the horizontal, using a protractor. At Burgentine Lake, the Coulomb collapse angle ψ is 63 deg if measured counterclockwise, as dip (Fig. 11-10b).

In conclusion, Hamblin's (1965) inclined collapse mechanism of rollover formation combined with dip domain analysis provides a graphical method for projecting normal faults into poor data regions. The fault projection method, when compared to depth-corrected profiles of fault shape, *yields reasonable results*. The technique is sensitive to measurement errors and may deteriorate after projecting a fault through more than a few fault bends. The method is also dependent on the collapse angle (Bischke and Tearpock 1999), so we presented a graphical method for calculating the Coulomb collapse angle for rollover structures.

ORIGIN OF SYNTHETIC AND ANTITHETIC FAULTS, KEYSTONE STRUCTURES, AND DOWNWARD DYING GROWTH FAULTS

In this section, we describe in more detail how rollover structures and secondary faults form in relation to the main structures. Some of the secondary features observed on rollover structures are inconsistent with some elementary textbook explanations.

In many growth fault areas, most of the antithetic or compensating faults do not connect to the concave upward bend in the major normal fault (e.g., Fig. 11-12). Surprisingly, few antithetic faults are present in the region directly above the concave upward bend in the major fault. As this region is subject to extension, as was discussed previously, this is contrary to what might be expected. Instead, many of the *antithetic faults terminate* at, and lie above and basinward of, *a synthetic fault* that appears to be associated with the concave *downward* bend in the major fault (Fig. 11-12) (Bischke and Suppe 1990b). For lack of a better term, this synthetic fault is called the *master synthetic fault*. In Figure 11-12, a master synthetic fault is imaged between shotpoint (sp) B and sp C at 0.2 sec to 2.0 sec. Figure 11-12 also shows antithetic faults terminating along the master synthetic fault. Antithetic faults generally do not exist beneath this synthetic fault.

From Figures 11-4c and 11-12, and our discussion of simple rollover structures, it appears likely that the master synthetic fault is related to the active axial surface that forms at the concave downward bend in the master fault. More importantly, the master synthetic fault can be observed to offset reflectors between the 1.0-sec to 2.0-sec level, but *not* reflectors at 2.1 sec to 2.4 sec near sp C. Notice that the bold reflector at sp C and at 2.1 sec below sp C is coherent (i.e., not displaced). A closer inspection of the master synthetic fault shows not only that the *displacements decrease downward* to zero but also that the displacements die upward toward the sea floor. The fact that the displacements are negligible near the sea floor is not surprising, as little time has lapsed since the most recent sediments were deposited. Our examination of the seismic lines in this area has resulted in two seemingly confusing relationships concerning the slip along the

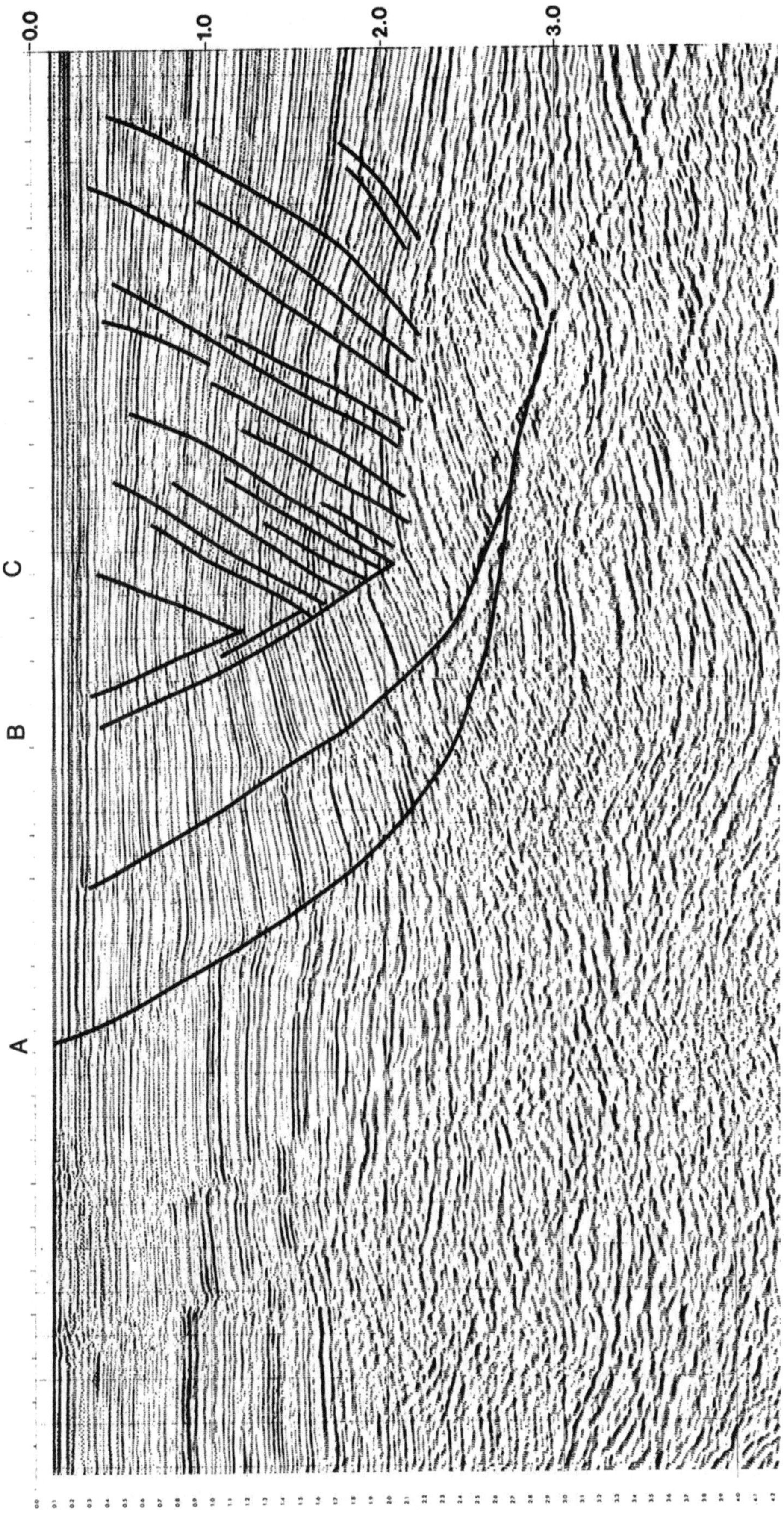

Figure 11-12 Jebco Seismic Inc. line from Brazos Ridge imaging Corsair fault (sp A), rollover structure, and downward dying antithetic and synthetic faults. Some antithetics terminate along a master synthetic fault. (Interpretation by R. Bischke. From Xiao and Suppe 1990. Published by permission of John Suppe.)

antithetics and the master synthetic fault: (1) the slip along the antithetics generally terminates at the master synthetic fault; and (2) the slip along the master synthetic fault in Figure 11-12 at first increases from shallow depths but then dies at greater depths. Where does all this slip go, and does it really go to zero? A close examination of Figure 11-12 shows that these relationships of downward-decreasing slip continue along the deep portions of the main listric growth fault shown at sp A. However, the conservation of fault displacement and the linking of faults, discussed in Chapter 10, suggest otherwise. In Chapter 10, we make the point that slip along a thrust fault need not remain constant and that the slip could totally die in the core of a fault propagation fold. We do, however, account at all times for the changes in slip and how the slip is consumed, or dissipated. In extensional tectonic areas such as shown here, this slip, which appears to vanish so abruptly, apparently follows the *bedding planes*. How does this process work?

Backsliding Process

One can envision these large listric growth faults as being within a large, slowly moving landslide (Xiao and Suppe 1992). As the hanging wall fault block slips, a void opens between the hanging wall and the footwall, and the overlying material instantaneously collapses along Coulomb shear surfaces, producing the rollover (Figs. 11-2 and 11-4). We learned from Figure 11-6a and b that the greater the concave bend (ϕ) in the major normal fault, the greater the rollover angle (θ). If the dip along the major fault decreases with depth, then eventually the rollover angle may increase to an angle such that the bed dip (θ) exceeds the dip on the major fault (Fig. 11-13). Frictional failure could then occur along the bedding surfaces that form the front limb of the rollover structure, and not just along the Coulomb shear surfaces. The higher the dip of the rollover structure, the closer the sedimentary beds come into parallelism with the antithetic Coulomb

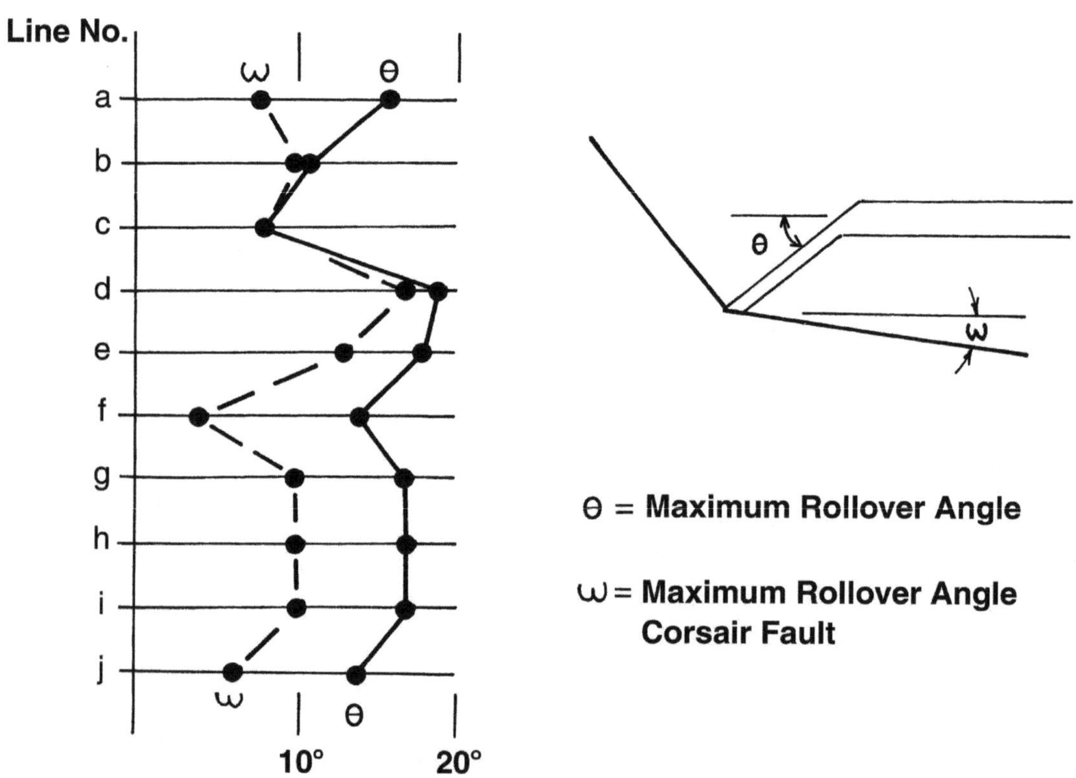

Figure 11-13 Minimum dip (ω) on Corsair fault (U.S. Gulf of Mexico) versus maximum bed dips (θ) at the front of the rollover and above its base. The frontal limb dip on the rollover structures generally exceeds the gentle dips on the Corsair fault. (Published by permission of R. Bischke.)

failure direction. At some angle of inclination, perhaps less than the Coulomb shear angle, the bedding planes are likely to present less frictional resistance than a sequence of sedimentary layers. Thus, frictional failure along the bedding surfaces would be favored over the Coulomb shear surface (of 60 deg to 70 deg), which cuts across the layering at a high angle (Fig. 11-2b). These potential bedding plane slip surfaces are likely to occur in weak layers, such as *overpressured shale zones*, which would present the least possible frictional resistance.

We therefore conclude that **backsliding** along bedding surfaces is a mechanism that can account for the apparent termination of the slip along the downward dying synthetic faults and antithetic faults. Using the examples of the Brazos Ridge, the backsliding appears to initiate where the rollover angle exceeds about 10 deg. The backsliding is a second-order effect relative to the total amount of slip along the major normal fault, about one part of bedding plane slip to about every seven parts of the slip along the Corsair fault. Although the bedding plane slip within the frontal limb of the rollover structure is small compared to the total slip, we shall see that it can have a marked effect on the amplitude of the rollover structure.

Backsliding Model. Exactly how does this backsliding mechanism operate? We present two examples: the first outlines the backsliding process, and the second, a more realistic model, assumes that the backsliding consumes about 1 part in 10 of the amount of slip that occurs along the major detachment fault. Depending on geological conditions, however, backsliding may at times be the dominant process.

Figure 11-14 illustrates the backsliding mechanism. Initially, slip along the master fault must occur to an extent such that a *critical rollover failure angle* is exceeded (Fig. 11-14a). This critical failure angle depends on local geological conditions, such as the pore pressures, and is likely to vary from area to area. As the pore pressure is unlikely to be large in pregrowth sediments, backsliding is most likely to occur within the growth sedimentary package. At some time, a weak rock, such as an overpressured shale, moves through the concave upward bend in the normal fault surface. When this occurs, an increment of backsliding along the bedding surfaces that comprise the frontal portions of the rollover structure fills the lower part of the void created by the basinward sliding (Fig. 11-14b). Initially, the plane of detachment propagates along the bedding plane surface and, in the case of the Brazos Ridge, up to the master synthetic fault. At this location, the synthetic shear associated with the *concave downward* bend in the major fault turns the strata downward in the basinward direction and forms the upper portions of the rollover (Figs. 11-4c and 11-12).

At the position of the master synthetic fault, the shear within the weak horizon can follow one of four possible paths. First, the propagating bedding plane slip (overpressured) could also turn downward and follow the bedding surfaces across the top of the rollover structure. This possibility seems unlikely, as large amounts of energy would be required. Second, the propagating slip could cut across the more gently dipping bedding surfaces located near the top of the rollover to a pre-existing antithetic fault (plane of weakness), and then follow the older antithetic fault to the surface. This would create small, triangular-shaped structures. Third, at the position of the master synthetic, the growing bedding plane fault could follow a Coulomb failure surface up to the sea floor, creating a new antithetic fault (Fig. 11-14b). This deformation path appears to be the mechanism of least resistance and the path most favored by many compensating faults. This mechanism would also result in antithetic faults that appear to terminate along the master synthetic, which are normally observed on our Brazos Ridge seismic section example. Fourth, the hanging wall block above the bedding plane detachment and the major fault and master synthetic fault could detach and slump toward the major fault (not shown in Fig. 11-14). This mechanism would produce a surface of detachment along the master synthetic. This mechanism could be a dominant process in certain areas. We favor the third, and to a lesser extent, the second mechanism to explain the observed antithetic faulting that we have studied in examples such as the Brazos Ridge.

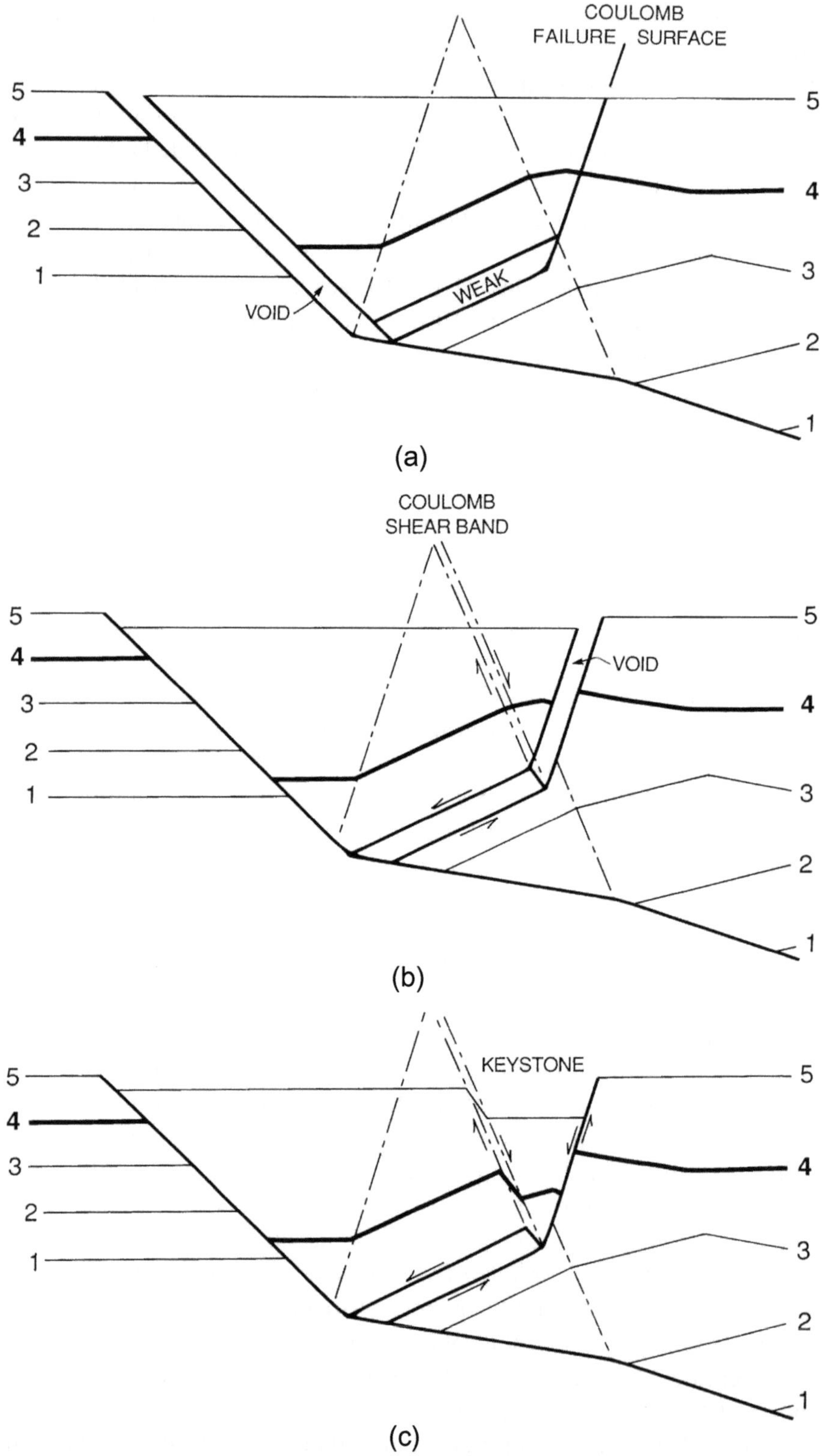

Figure 11-14 Backsliding model in growth sedimentary section. (a) Uniform shear along bedding surfaces causes the hanging wall to detach from the undeformed footwall wedge. (b) The backsliding forms an antithetic fault and opens a void along the newly formed fault surface. (c) The upper void is filled by synthetic Coulomb collapse along a basinward-dipping shear band. The balanced model forms a keystone and downward dying antithetic and synthetic faults. (Published by permission of R. Bischke.)

The third mechanism would cause another void to open along the newly formed antithetic fault (Fig. 11-14b). This hole can be filled only by collapse along the basinward-dipping Coulomb failure surfaces (Fig. 11-14c), forming a keystone structure. According to the balanced deformation model just described, another increment of backsliding will result in the formation of another antithetic, which forms above the master synthetic and the previously formed antithetic.

Another consequence of this model is that the bedding plane slip is likely to occur along shale horizons that could be related to fluctuations in sea level (Payton 1977). As a result, the intervening horizons are more likely to contain *sands*, or *reservoir rock*. Notice in Figure 11-12 that the antithetic faults die downward into a unit of semicoherent reflections between 2.1 sec and 2.4 sec. This unit is an overpressured shale, and we have observed normal faults that die into overpressured shale intervals in other structures.

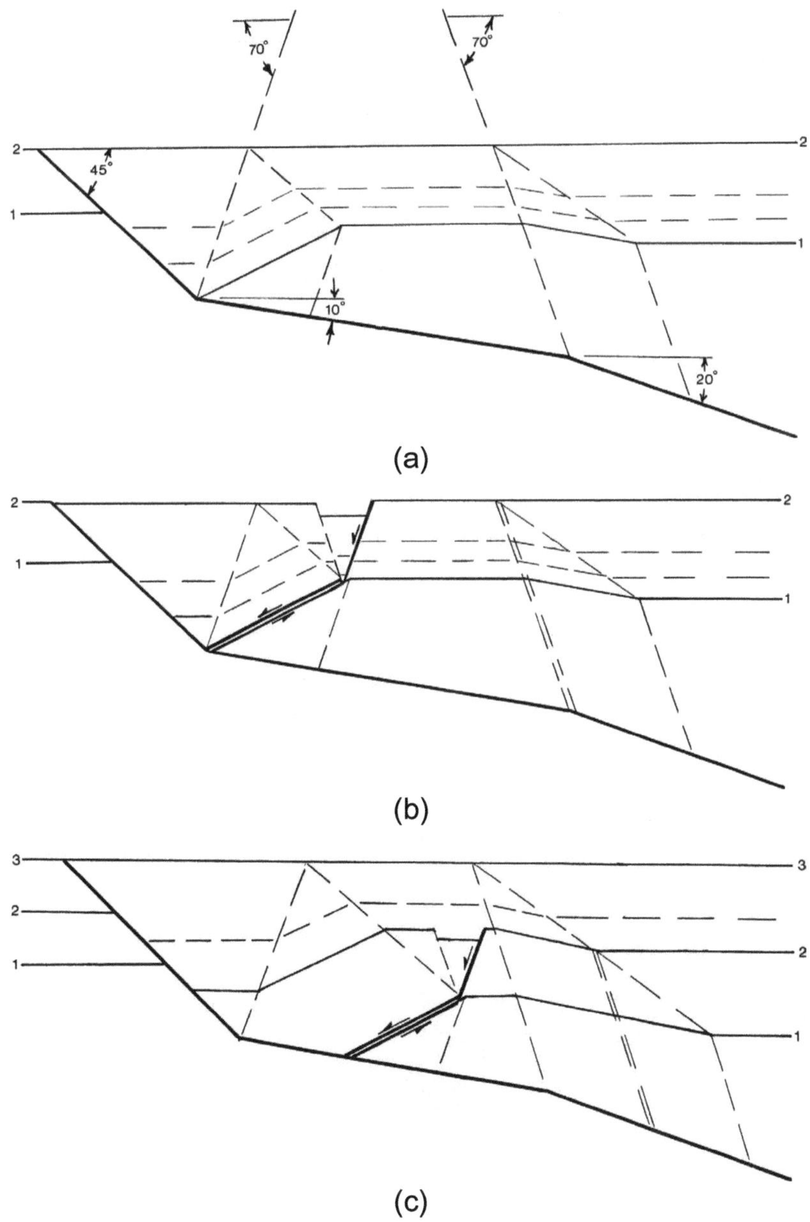

Figure 11-15 (a)–(i) Generic model of the Corsair fault illustrating the progressive development of downward dying antithetic and synthetic faults through growth stages 1 to 7. (Published by permission of R. Bischke.)

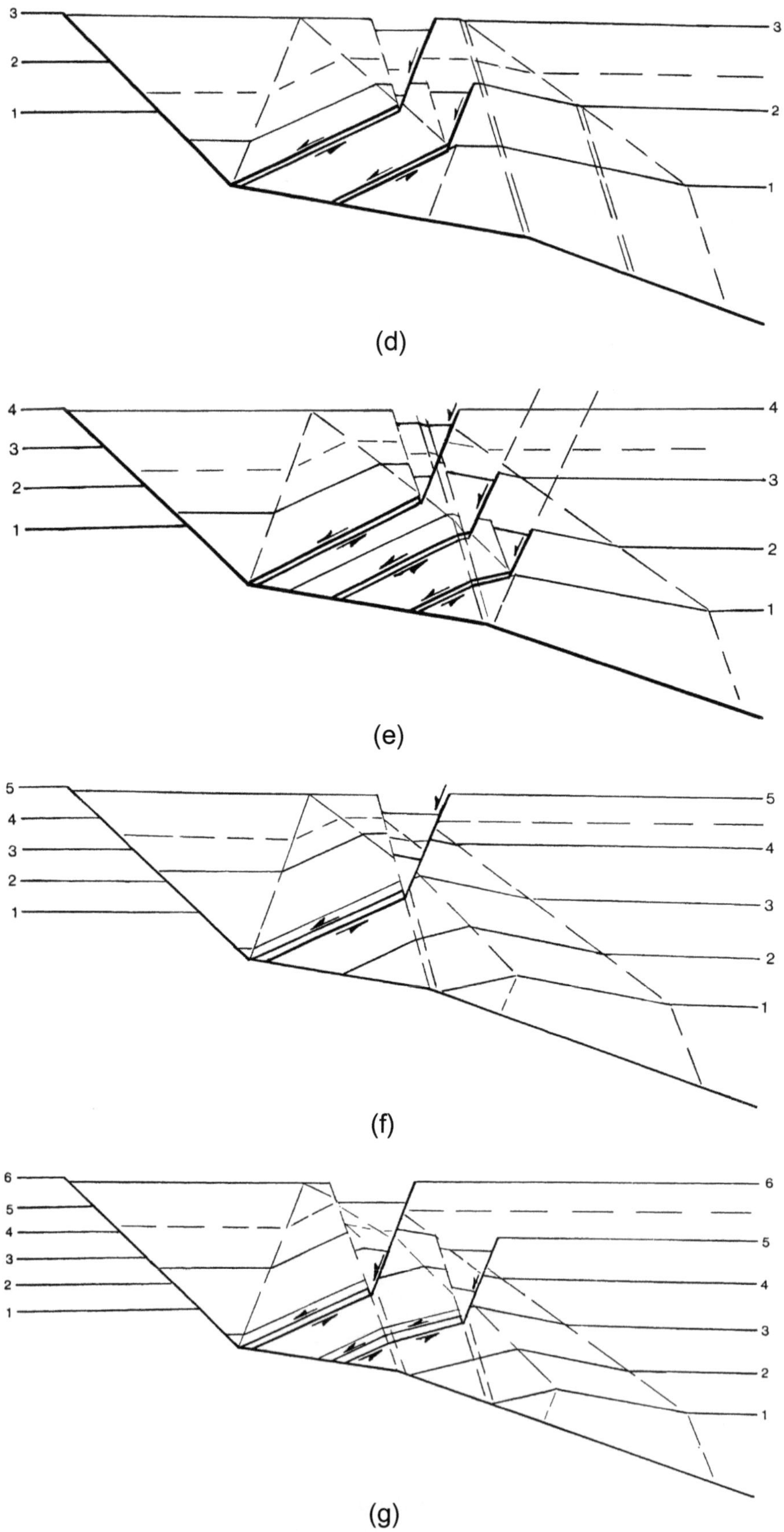

Figure 11-15 *(continued)*

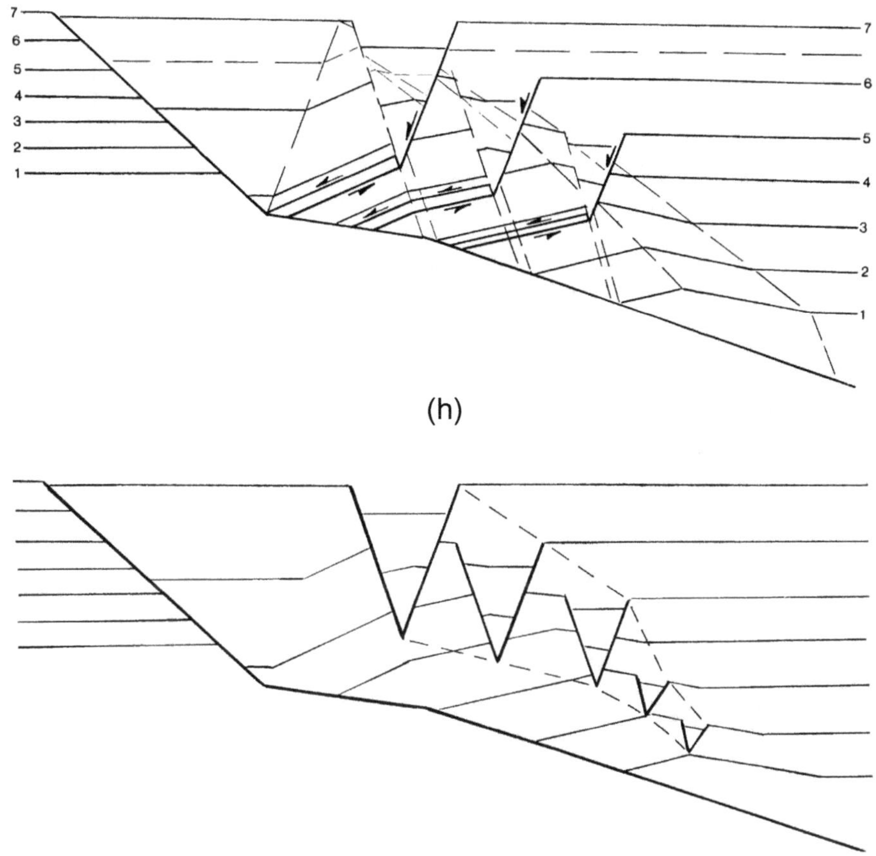

(h)

(i)

Figure 11-15 *(continued)*

Example from Corsair Trend. A more realistic, generic example is illustrated in Figure 11-15. Here the major fault initially dips at 45 deg, decreases to 10 deg on the flat, and then steepens basinward to 20 deg. This fault configuration is similar to the shape of the Corsair fault in Figure 11-12. Also, the Coulomb failure angle is assumed to be 70 deg and the growth phase of the sedimentation to begin at Horizon 1 (Fig. 11-15a). In Chapter 13, we describe how to distinguish the growth interval from the nongrowth interval using growth plots. Since the sediments deposited beneath Horizon 1 are pregrowth sediments, and since the backsliding mechanism is likely to initiate in an overpressured horizon, the backsliding process is not likely to begin until a critical dip angle is reached within the growth sediments. At this stage, the first increment of backsliding occurs and a keystone structure forms (Fig. 11-15b) at Horizon 1. Notice in this example that the distance between the concave upward bend and the concave downward bend in the fault is greater than in the example in Figure 11-14.

Additional sliding along the major fault produces the geometry present in Figure 11-15c, and another increment of backsliding results in Figure 11-15d. The backsliding has the effect of creating keystone blocks that become younger upward as the blocks grow larger. The basinward sliding along the major fault has deactivated the antithetic fault that formed in Figure 11-15b, and this antithetic fault stopped growing after Horizon 2 was deposited (Fig. 11-15c) and was buried by the more recent growth sediments. The backsliding also has the effect of reducing the rollover amplitude. After the deposition of Horizon 4, the geometry appears as

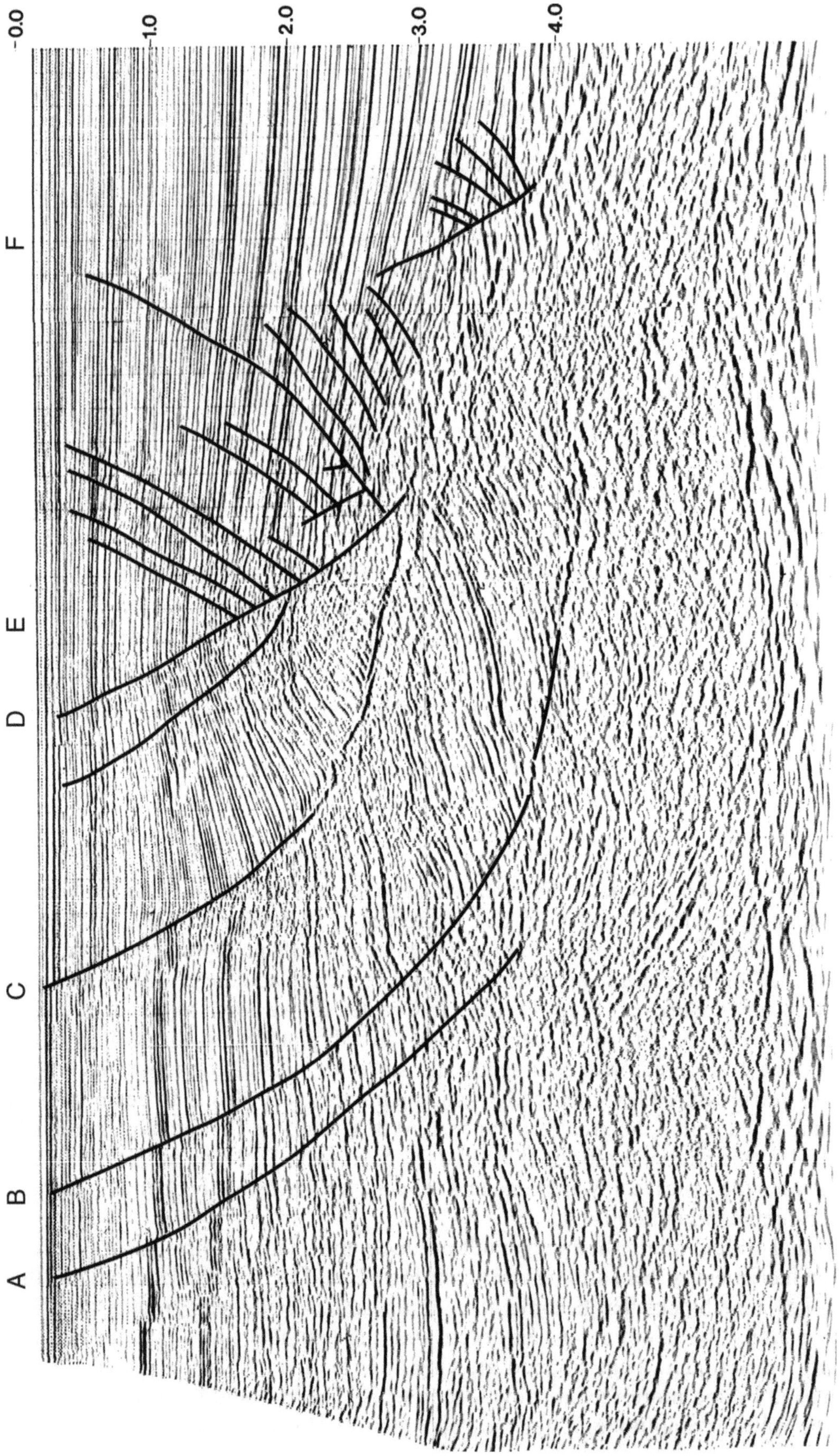

Figure 11-16 Jebco Seismic Inc. migrated line of Brazos Ridge showing antithetic faults that die with depth into the sedimentary basin. Corsair fault is imaged at sp C and two other less active faults at sp A and B. (Based on unpublished work of Xiao and Suppe. Published by permission of John Suppe.)

shown in Figure 11-15e. At this stage, the lower two antithetic faults, which formed during the interval of time that Horizons 1 to 3 were deposited, have moved through the locus of deformation (active axial surface) that is associated with the concave downward bend in the major fault. The clockwise shear associated with this deformation has the effect of slightly rotating the antithetic faults clockwise (Fig. 11-15e). During the interval of time that Horizon 5 is deposited (Fig. 11-15f), slip along the major fault has advanced to the stage that the newly forming antithetics initiate to the right of the clockwise shear-active axial surface. Thus, the antithetics that form after the deposition of Horizon 4 will not be rotated clockwise by the concave downward bend in the major fault (Fig. 11-15g). Additional increments of backsliding are shown in Figure 11-15h and i.

Figure 11-15i is a generalization of the deformation that occurred during the deposition of Horizons 1 through 7. Dashed lines are drawn at the base of the antithetic faults where the slip entered the bedding planes and at the top of the antithetic faults where they ceased to grow and became inactive (Fig. 11-15i). These lines can be considered *axial surfaces* that are associated with the growth phase of the antithetic crestal faulting and the formation of the keystones (Tearpock et al. 1994).

Again, notice that the more recent antithetic faulting and keystones form to the left of the older rollovers. If this model is correct, then it can be tested by data. Figures 11-16 and 11-17 are two seismic lines from the northern Gulf of Mexico and Brunei (Southeast Asia) respectively. On both of these lines, the antithetic faults have a tendency to be younger upward and to be older in the deeper sediments. The pattern of downward dying faults and keystone blocks is similar to our example in Figure 11-15.

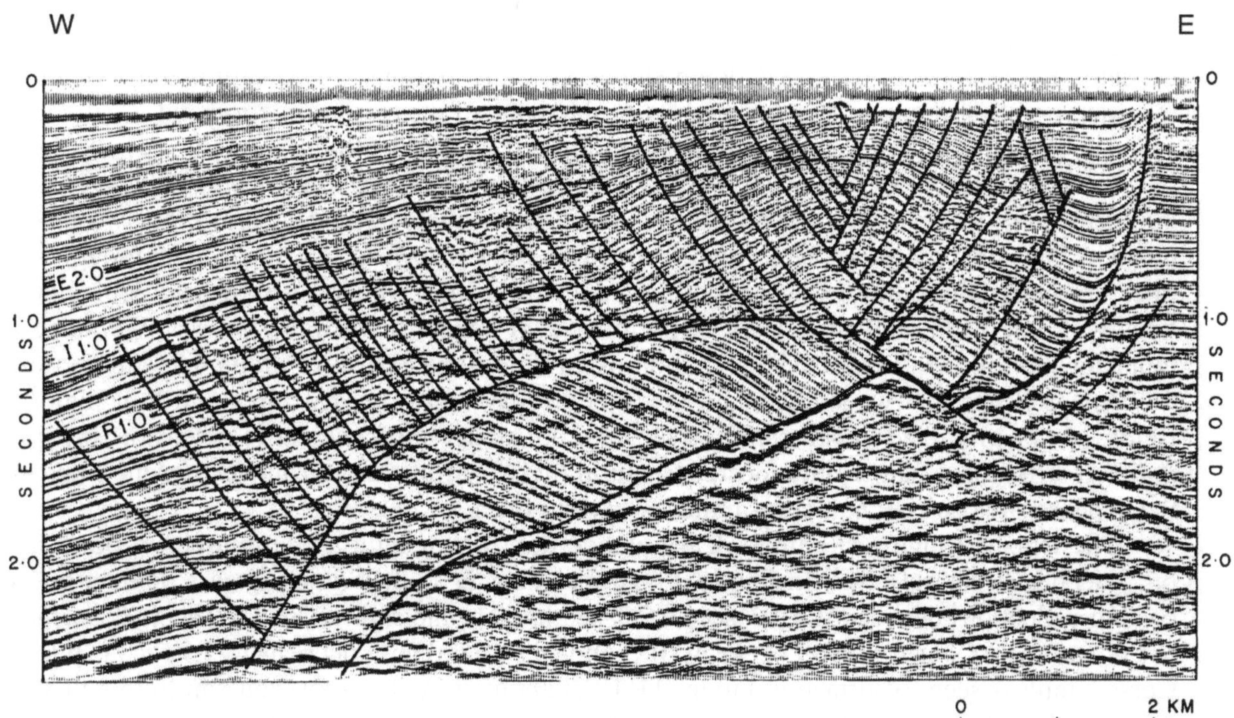

Figure 11-17 Seismic line from Brunei (Borneo) showing antithetic faults that appear to terminate along bedding surfaces. The antithetic faulting also dies with depth into the basin. (Published by permission of Muzium Brunei.)

Let us briefly review the consequences of the above-described deformation.

1. Growth antithetic faults form basinward and above synthetic faults located near the crests of rollovers.
2. The antithetics become younger upward with the older antithetics being positioned basinward and terminating at a deeper level.
3. Slip along the master synthetic may die with depth, and slip along the antithetic faults appears to terminate at the position of the master synthetic.
4. The deformation mechanism forms a keystone structure, or a graben, which with the antithetic faults has the effect of reducing the amplitude of the rollover structure.
5. The deformation and sedimentary mechanisms form faults that die in both the upward and downward directions.
6. The process may be controlled by the sedimentary cycle.

THREE-DIMENSIONAL EFFECTS AND CROSS STRUCTURES

Our observation of the Brazos Ridge tie lines revealed the existence of previously unreported **cross structures**, or **transverse structures**, within the rollover structure. As these cross structures *determine the position of structural highs (closures), cross structures may play a fundamental role in locating closures in other rollover structures*. In fact, all the Brazos Ridge closures that we studied could be located directly from *strike lines (tie-lines)* or from *fault surface maps*.

Many geoscientists, however, may mistrust 2D regional strike lines, and this is most unfortunate. They correctly reason that collecting data parallel to strike of a steeply dipping surface, such as the Corsair fault, is like collecting data from a roof. For example, seismic energy from the up-slope portion of the roof returns to the receiver, while the energy returning from directly beneath the receiver is deflected down-slope. Thus, in regions of structural dip, the energy recorded on 2D strike lines comes from out of the plane of the seismic section and is called *sideswipe*. Consequently, reflections on strike lines are difficult to depth-correct.

Nevertheless, our experience has led us to conclude that *strike lines can be used to correctly interpret data*. Strike lines may not image structures in their correct location, but we take the position that often more can be learned from strike lines than from dip lines, even though strike lines are generally more poorly imaged relative to dip lines. Indeed, strike lines look like dip lines with one additional advantage. Strike lines also image the *cross structures*. We make this point for the following reason. The location of all the rollover closures along a 25-mi section of the Brazos Ridge that we studied, and this includes five rollover structures and four major gas fields, *can be predicted from a single strike line*. In frontier areas subject to extension, the use of regional strike lines can rapidly locate structural highs within a large rollover structure and save the expense of acquiring numerous unnecessary dip lines.

The Brazos Ridge strike lines clearly image the Corsair fault and a number of other faults on a deeper structural level. In Figure 11-18, the Corsair fault is imaged as a distinct boomer (strong reflection) between the 2.3-sec to 2.7-sec level to the right of sp C. *The fault surface is not planar*. The Corsair is level between sp C and sp D, but then it deepens to 2.7 sec at sp E. The Corsair fault is seen to surface in Figure 11-18 at about sp A (0.2 sec). You can follow this fault downward to about the sp B area, where another *deeper* fault is observed to *deform*, or roll over, the Corsair fault, analogous to the geometry in a compressional duplex (Chapter 10). This deeper fault continues beneath the Corsair to perhaps the 3.55-sec level, where it levels off (at sp D). On the dip lines, this and other deeper faults can be observed to surface to the north or landward of the Corsair fault, to dip beneath the Corsair fault, and to be presently active, although at much lower slip rates (Fig. 11-16, sp A and B). In Figure 11-16, the Corsair fault is seen to surface

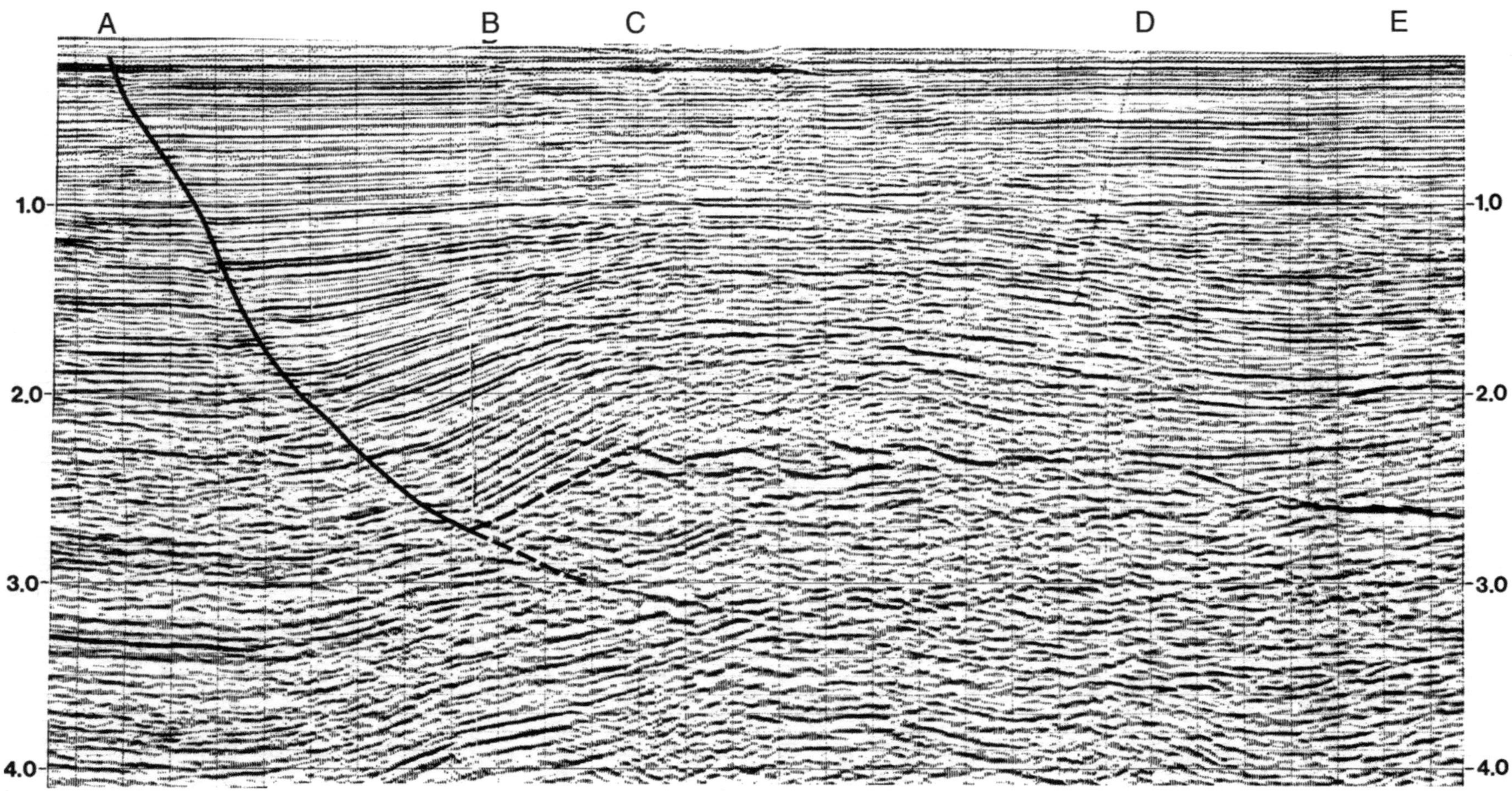

Figure 11-18 Jebco Seismic Inc. strike line along Corsair fault. Deeper faults have deformed the Corsair into a series of highs and lows that create cross structures on the fault surface map (Fig. 11-19). The low on the Corsair fault at sp B creates Chute 1 on Figure 11-20. (Published by permission of Jebco Seismic Inc., Houston, Texas, USA.)

at sp C. On the strike lines, these deeper faults can be observed to intersect the Corsair fault (Fig. 11-18, sp B), to offset the fault, or to form cross structures.

These cross structures trend subperpendicular to the Corsair fault *and deform and partition* the footwall into a series of high-gradient and low-gradient areas (Fig. 11-19). On a regional scale, the series of highs and lows on the Corsair fault, which are caused by the *major subfault deformation*, are bounded by the cross structures (Fig. 11-20). The low-gradient, shelf-like, or "flat" areas on the Corsair fault produce what we term **bows** on the fault surface. Bows are known in the U.S. Gulf of Mexico to be a key indicator of a petroleum trap. We propose the term **chute** for the higher gradient regions of indentation into the fault surface. The strike line (Fig. 11-18) images chute 1 (Fig. 11-20) at sp B.

Finally, we demonstrate that the general position of the rollover structures can be predicted from the fault surface map of the Corsair fault (Fig. 11-19), with the bows (lowest gradient areas) on the fault surface corresponding to the position of the structural highs within the rollover structure and the chutes (higher gradient areas) corresponding to the position of the saddles between the structural highs. These relationships are shown in Figure 11-21, which is a restored time map of a horizon within the rollover structure. The map was restored to a common level by simply closing the faults. A comparison of Figure 11-21 to Figures 11-19 and 11-20 demonstrates that the *bows correlate to the position of the structural highs, whereas the chutes correlate to the saddles*. Thus, rollover closures can be located from fault surface maps, which further demonstrates the value of constructing these maps (Chapter 7).

What does the saddle geometry look like in 3D? Figure 11-22 is a 3D perspective view of the flanks of two structures within a large rollover structure, based on a study of axial surface deformation in the Central Sumatra Basin, Indonesia (Shaw et al. 1997). The fault surface contains a central low, or chute, and the chute is bordered on its flanks by two fault surface highs. As the hanging wall block moves down the fault surface, the displaced horizons conform to the shape of the 3D low in the fault surface, forming a saddle and the flanks of the adjoining structures, as depicted in the inset in Figure 11-22. The geometry resembles that observed on the strike line (Fig. 11-18) between sp D and sp E, where the flank of a closure in the hanging wall block is imaged. The structures flanking the low dip toward the chute in the fault surface (Fig. 11-22).

STRIKE-RAMP PITFALL

We discuss the necessity of constructing accurate maps throughout this book, and we emphasize that here by presenting examples of the pitfall in interpreting strike-ramps between en echelon normal faults, which are common in extensional terranes (Tearpock et al. 1994). We demonstrate how to recognize, locate, and avoid this costly pitfall. We have observed this problem and its resultant dry wells on a number of 2D and 3D data sets. The examples illustrate the necessity of constructing accurate maps of subsurface faults.

Strike-ramps form along arrays of normal faults that contain en echelon offsets, or stepovers (Morley et al. 1990; Peacock and Sanderson 1991). Strike-ramps are commonly ubiquitous features on extensional data sets. Examine the air photograph shown in Figure 11-23, taken from the Yucca Mountains, Nevada (Ferrill et al. 1999). Notice how the Boomerang fault almost connects to the uppermost extension of the Fatigue Wash fault, but no connecting fault links the two fault systems to form a two-way fault closure. On 3D data sets, the seismic data may become semicoherent, or deteriorate, where two faults overlap, perhaps due to the fault shadow effect. If this occurs, then *structural horizon mapping may result in the mapping of two fault surfaces as a single fault surface*. In addition, on 2D and 3D seismic data sets, structural aliasing can result in mapping the two overlapping faults as one fault (Fig. 11-24a and b). Furthermore, this overlap has a limited lateral zone. Even if the data are good, en echelon faults may be overlooked if the interpretation is based on insufficiently close seismic lines.

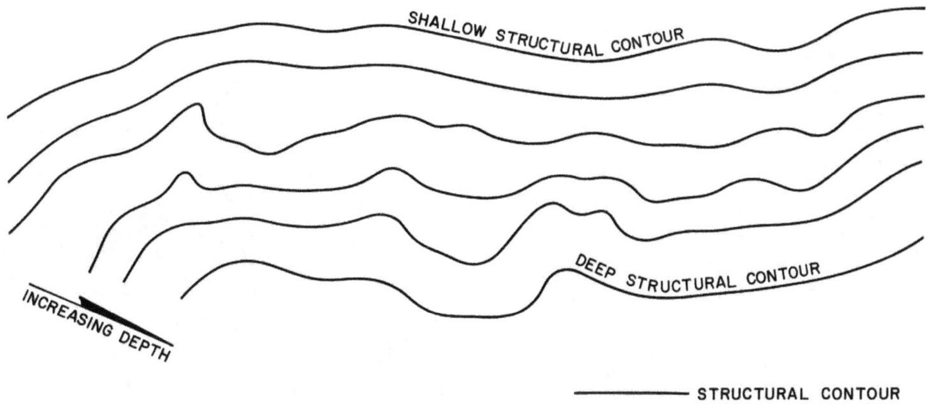

Figure 11-19 Fault surface map of a portion of the Brazos Ridge. The structural contours deepen toward the lower portion of the diagram. The cross structures segment the fault surface map into a series of low-gradient areas (bows) and high-gradient areas (chutes). (Prepared by W. L. Keyser. Published by permission of Texaco USA, Eastern E & P Region.)

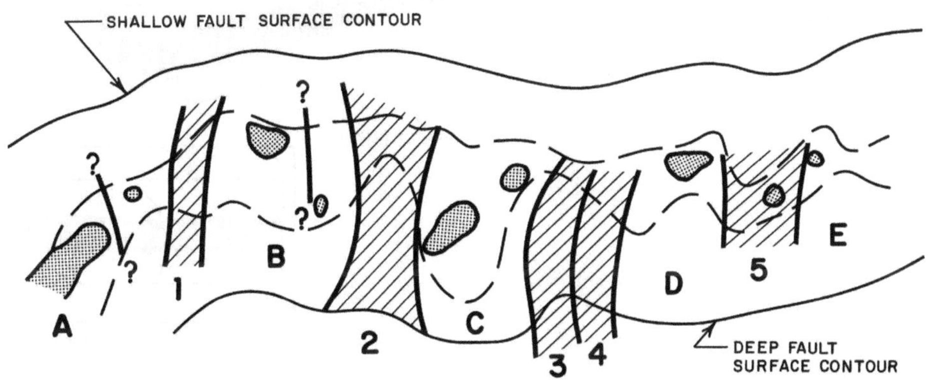

Figure 11-20 Location of chutes and bows. The chutes exist as intrusions into, and the bows exist as protrusions upon, the fault surface map (Fig. 11-19). Five bows and chutes exist along trend, with chutes 3 and 4 being a double chute. (Published by permission of Texaco USA, Eastern E & P Region.)

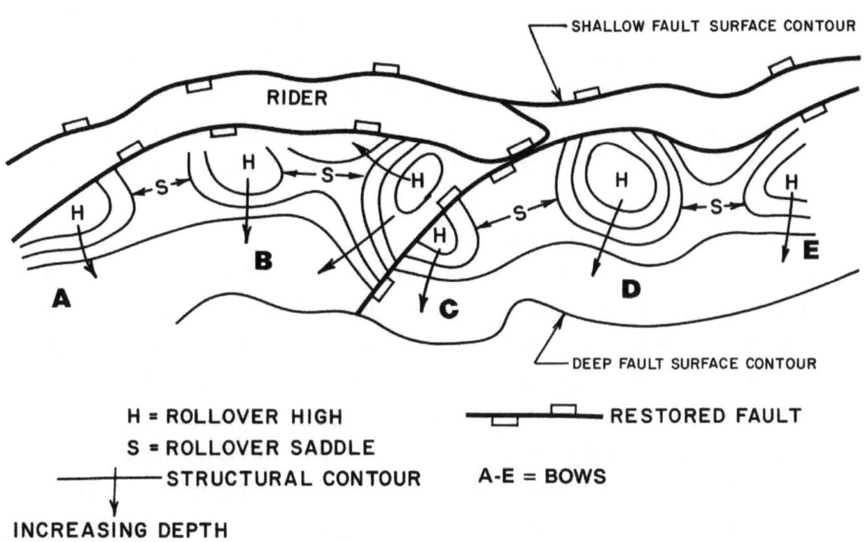

Figure 11-21 Partially restored horizon time map. The structural highs correspond to the position of bows, whereas the saddles correlate to the chutes. (Published by permission of Texaco USA, Eastern E & P Region.)

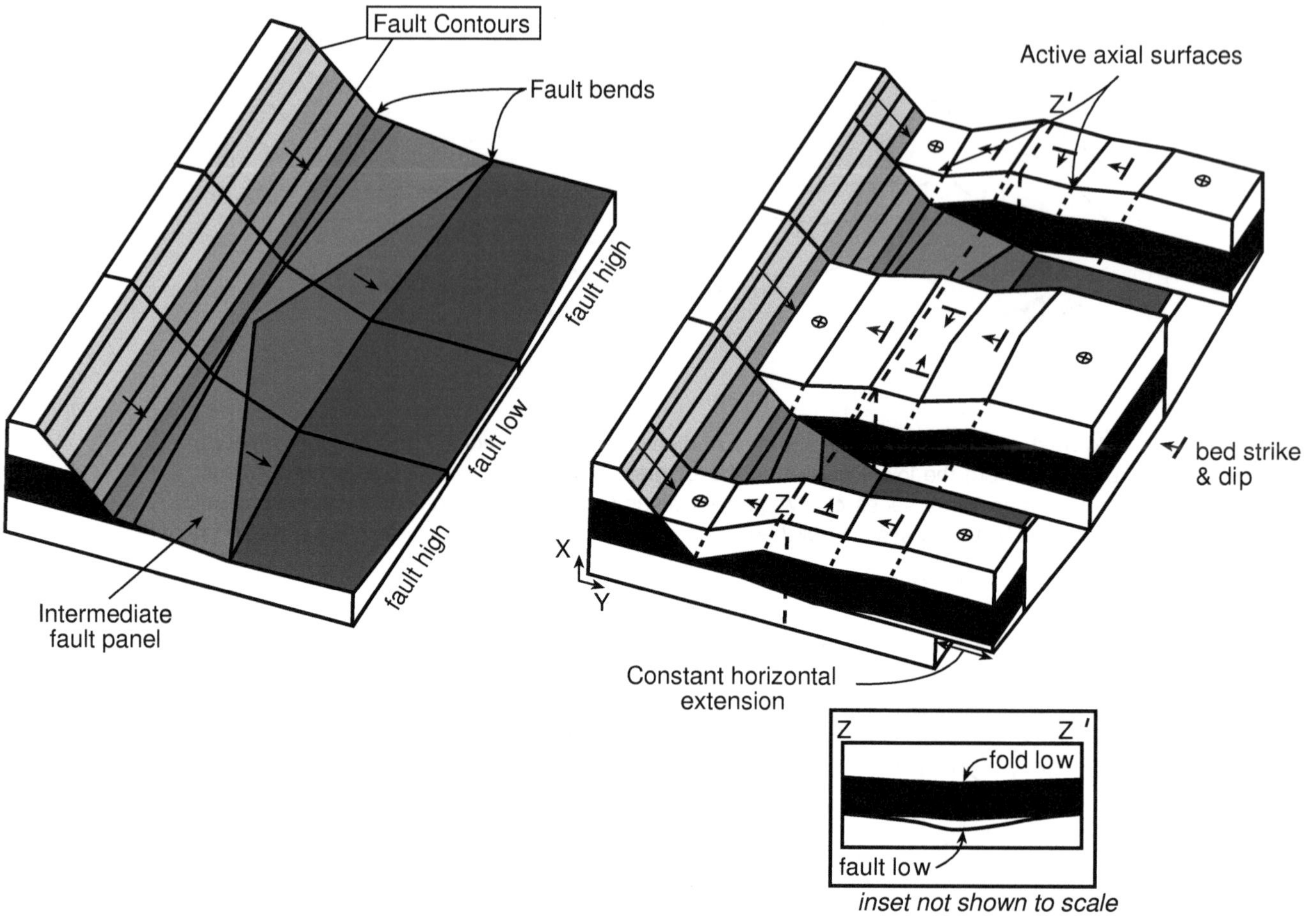

Figure 11-22 Perspective view of a 3D structural saddle based on a fault surface map from Central Sumatra Basin, Indonesia. The low area on the fault surface generates a structural low. (Modified after Shaw et al. 1997; AAPG©1997, reprinted by permission of the APPG whose permission is required for further use.)

Notice on Figure 11-24a that the strata downthrown to the overlapping faults form a structural low relative to the upthrown strata. Where Fault 1 dies out, or *loses displacement to the east*, Fault 2 *increases in displacement to the east*. This reversal of displacements between the offsetting fault surfaces causes the strata within the overlap area to dip to the west, forming a strike-ramp. The pitfall is that *this ramp may or may not contain a fault* that would form a three-way fault closure (Brenneke 1995).

Let us assume that the strike-ramp shown on Figure 11-24a is within a 3D survey area, that a fault shadow effect causes the data to deteriorate between the offset fault surfaces, and that the data were interpreted along equally spaced lines. If on the 3D data set the horizontal distance between the two overlapping fault surfaces is large, then seismic horizon mapping should detect the strike-ramp. On narrower strike-ramps, however, structural horizon mapping of the offset fault surfaces is subject to possible fault miscorrelation problems. Geoscientists may assume that the fault surfaces are throughgoing and, in this case, the deterioration of the seismic data set may not resolve that two faults are present. If the geoscientist connects Fault 1 to Fault 2, then the result is a nonexistent prospect based on a three-way closure (Fig. 11-24b). In practice, the data are commonly incoherent between the offset fault surfaces, so in such a case, *interpreting every line on a 3D data set may not even resolve this problem.*

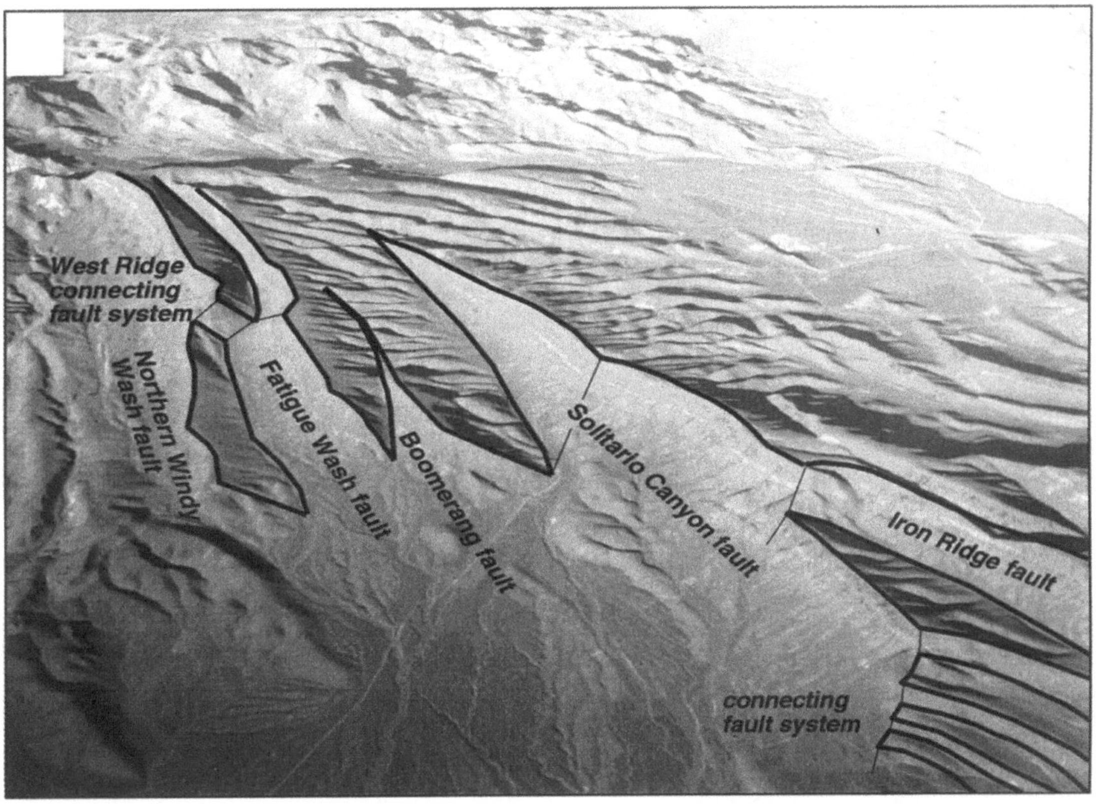

Figure 11-23 En echelon normal fault surfaces from Yucca Mountains, Nevada, USA, showing faults and fault offsets in bold lines. An unfailed block, or strike-ramp, separates the Boomerang fault from the Fatigue Wash fault. Failure of the block between the Fatigue Wash and Northern Windy Wash faults creates the West Ridge cross fault. In the subsurface, a structural geometry identical to the Yucca Mountains normal fault system would trap hydrocarbons in the block downthrown to the West Ridge connecting fault system, but not in the block between the Boomerang and the Fatigue Wash Faults. (From Ferrill et al. 1999 Published by permission of Elsevier Science, Journal of Structural Geology.)

How does one locate, recognize, and resolve fault correlation problems? Figure 11-25 shows a fault surface map that contains two en echelon faults, which are incorrectly mapped as a single fault. Can you detect the strike-ramp between the two faults? The strike-ramp exists as a *minor bend* or *curve* on the fault surface map (Brenneke 1995).

We are aware of two methods to prevent this costly pitfall, both of which involve fault surface mapping. Geoscientists can analyze 3D **coherency** data over many time slices (Fig. 11-26). Coherency data *commonly image the en echelon offsets associated with strike-ramps*, particularly on the shallow structural levels. When interpreting seismic data, assign a fault segment to every fault that exists on the multilevel time slices. These interpreted fault segments on the time slices represent strike (trace) segments relative to the dip profiles. Thus, faults observed on time slices easily tie to faults observed on dip profiles.

Lacking coherency data on 3D data sets, or when using 2D data, you can locate en echelon offsets by constructing fault surface maps, as described in Chapter 9. If the fault surface map contains a **kink** or a **bend**, as in Figure 11-25, then en echelon faults may be present. Two overlapping fault surfaces, separated by a significant horizontal distance, will contain a contouring bend if mapped as a single fault surface. However, two closely spaced faults may exhibit only a subtle kink or bend in the fault surface. When using 3D data, we can readily solve the problem of

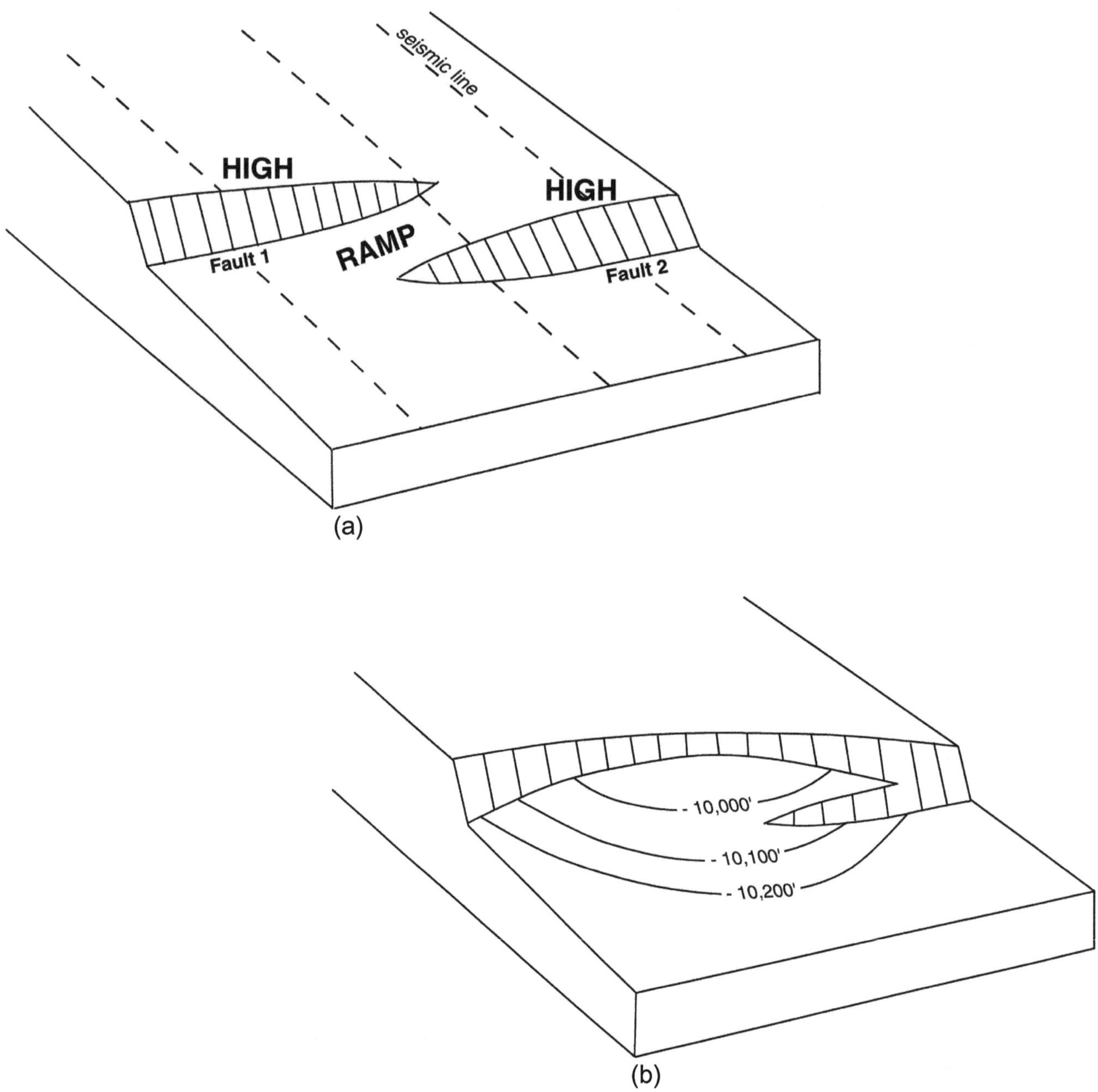

Figure 11-24 (a) Two en echelon faults die out in opposite directions, creating a strike-ramp. The unfaulted ramp is incapable of trapping hydrocarbons. An improperly spaced 2D data set, structural aliasing, or poor 3D seismic data in the area of fault overlap could result in the prospect shown in (b). (b) If fault surface maps are not constructed, then a combination of incomplete seismic control, structural aliasing, or poor 3D seismic data can create a nonexistent prospect. (Published by permission of R. Bischke.)

distinguishing between a gentle bend in a single fault from an en echelon offset of two fault surfaces. Construct several arbitrary lines across the kink or the bend in the fault surface, at oblique angles to the kink. Where strike-ramps are present, these arbitrary lines will commonly encounter regions of coherent data in the areas where *the two fault surfaces do not overlap*. Accordingly, *two faults, and not one fault, will image on the arbitrary lines*.

In order to solve the strike-ramp problem on 2D data sets, properly oriented oblique or dip profiles must exist that cross both faults. These lines may not exist. Thus, be suspicious of any bend, kink, or curve in a normal fault surface that acts as a potential trapping fault for a prospect. Risk this bend in the fault accordingly. An additional line or two may need to be acquired to resolve the problem.

Alternatively, additional motions along the strike-ramp may generate faults that link two originally separate faults, creating a fault trap (Tearpock et al. 1994). The probability of an

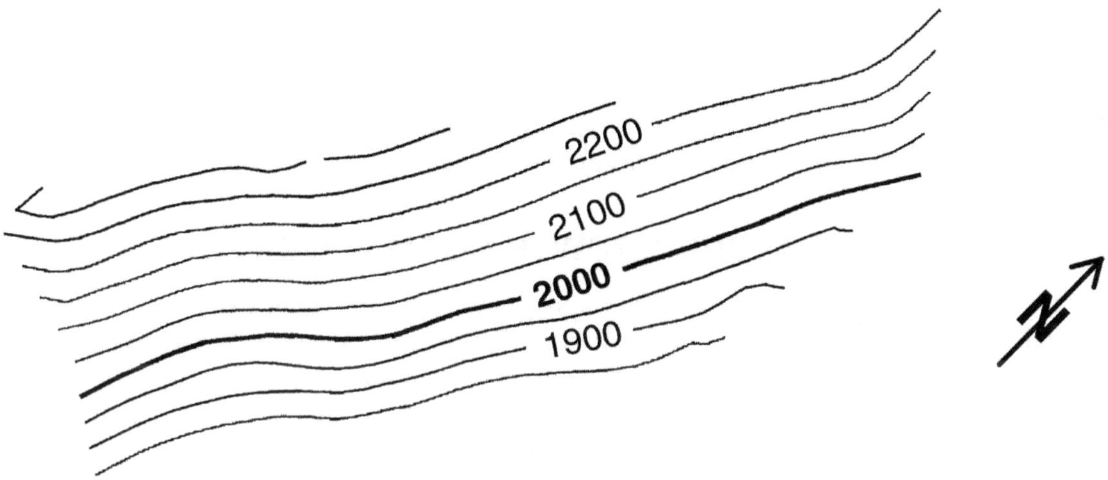

Figure 11-25 Fault surface map of two en echelon fault surfaces mapped as a single fault surface. Bend in fault surface map suggests the possible presence of two faults that do not intersect rather than one fault. Map is in two-way time (msec). It was generated using a triangulation algorithm that does not smooth through bends in fault surfaces. Compare to the coherency cube display in Figure 11-26 that clearly shows the presence of two separated faults. (Published by permission of R. Bischke.)

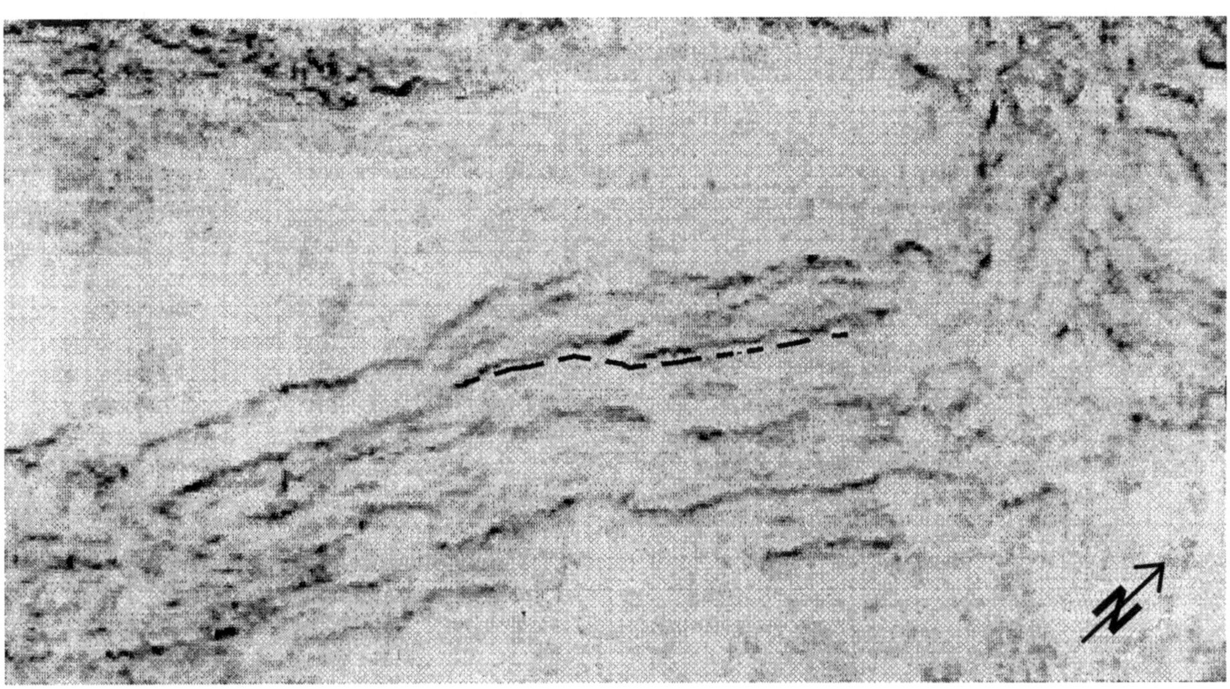

Figure 11-26 Strike-ramp on coherency cube data taken from St. Charles Ranch, Texas. Just above the dashed line, the en echelon pattern of the faults is evident. The dashed line represents a fault surface contour on Figure 11-25, where the curvature of the contours is a clue to possible en echelon faults. (Published by permission of R. Bischke.)

accumulation in this type of trap depends on the timing of these younger faults. Chapter 13 in this book and Tearpock et al. (1994) present methods for studying fault timing.

Geoscientists should exercise care when interpreting faults and constructing fault surface maps. Geoscientists *loop-tie* fault surfaces maps for the same reason that they loop-tie horizon surfaces: to make sure they remain on the same fault. Also, consider that on 3D data sets, randomly placed arbitrary lines may or may not locate the strike-ramps. Strike-ramps that exist along faults that constitute the same fault system are often *subtle* features on 3D data sets. Geoscientists must know where to place the arbitrary lines in order to demonstrate that a trapping fault exists. Correctly interpreted and loop-tied fault surface maps show geoscientists where to place the arbitrary lines. Fault surface mapping also reduces the horizon mis-tie problem, often saving time. The final product is a higher quality interpretation and prospects.

COMPACTION EFFECTS ALONG GROWTH NORMAL FAULTS

We present a theory developed by Xiao and Suppe (1989), which explains why some growth normal faults are listric (concave upward) or antilistric (concave downward). When using depth-corrected seismic data, it is apparent that faults are commonly listric in shape through predominately shale sections and turn antilistric into thick sand packages. Generally, about 20% to 30% of the faults in an area like the northern Gulf of Mexico coastal region are mapped as antilistric, whereas in specific areas like Brazos Ridge, the majority of the antithetic faults are antilistric. We stress that the following techniques for projecting faults into poor data areas apply only to *growth faults*, which were active when the sediments were being deposited, but not to the sections of those faults within pregrowth sediments deposited prior to the normal faulting. In other words, these techniques apply only to those *portions* of a fault that were active during deposition.

Furthermore, this method uses the true dip of a fault, so any seismic profile used for data must be dip-oriented (perpendicular to the strike of the fault), depth-corrected, and as close as possible to true scale. Similarly, cross-sections must be dip-oriented and at true scale.

An understanding of the change in fault shape with depth is critical in applying the fault projection technique. Consider two small columns of material, column A located at the sea floor and column B located at depth, as illustrated in Figure 11-27. Column A is just recently deposited and not subject to significant compaction, whereas column B is buried and is compacted. Initially, column A has width ΔX and an initial porosity of ϕ_0 If we define the height of column A as ΔY, then the initial dip of the growth normal fault relative to the *footwall* is defined by

$$\tan\theta_0 = \frac{\Delta Y}{\Delta X} \qquad (11\text{-}1)$$

θ^0 is the initial fault dip

After burial, column A will dewater and compact and take on the shape of column B (Fig. 11-27).

Column B has a buried width, height, fault dip, and porosity of $\Delta X'$, $\Delta Y'$, θ, and ϕ respectively. Therefore, the fault dip at B is defined as

$$\tan\theta = \frac{\Delta Y'}{\Delta X'} \qquad (11\text{-}2)$$

However, the compaction primarily affects the height and not the width; therefore,

$$\Delta X = \Delta X' \qquad (11\text{-}3)$$

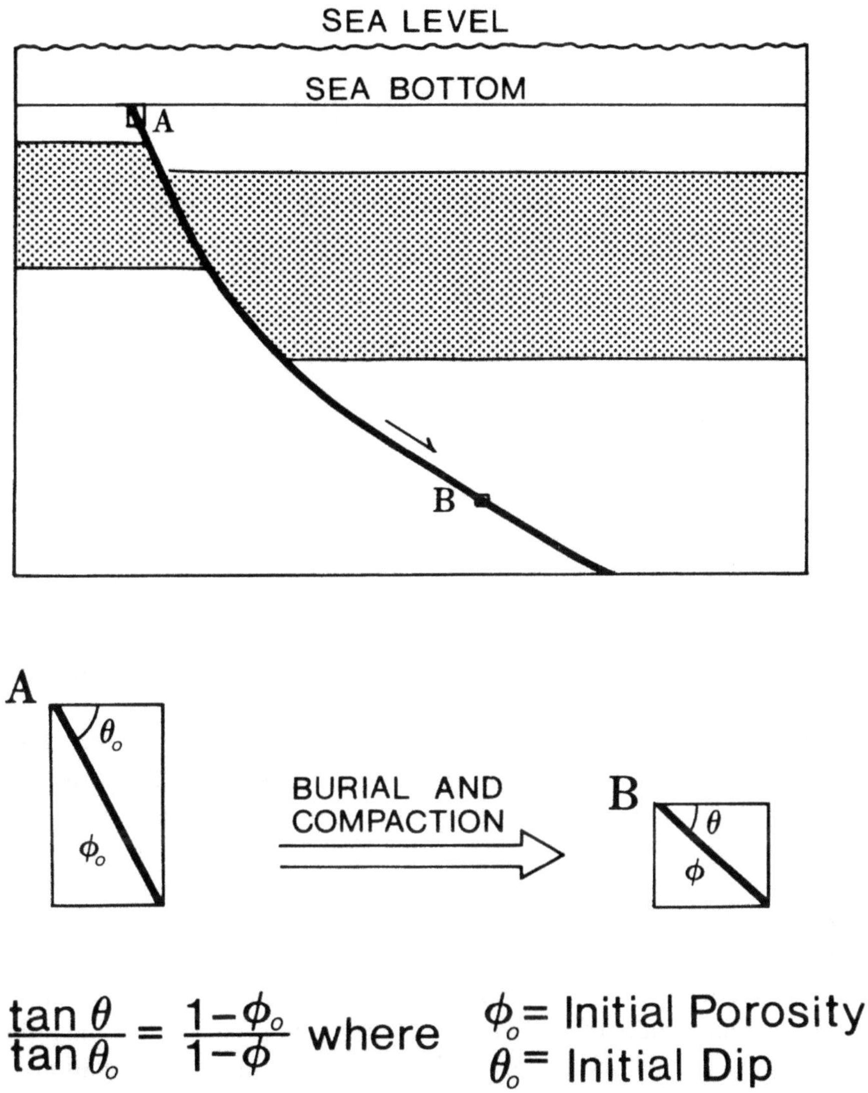

Figure 11-27 Compaction and burial along a growth normal fault. Element A at the sea floor compacts to the geometry present at B. These relationships are taken relative to the compacted footwall. (From Xiao and Suppe 1989. AAPG©1989, reprinted by permission of the AAPG whose permission is required for further use.)

and thus, using Eqs. (11-1) and (11-2),

$$\frac{\tan\theta}{\tan\theta_0} = \frac{\Delta Y'}{\Delta Y} \tag{11-4}$$

where θ = bed dip

As porosity controls the extent of compaction (Baldwin and Butler 1985), this mass conservation formula written in terms of solid volume is

$$(\Delta X')(\Delta Y')(1-\phi) = (\Delta X)(\Delta Y)(1-\phi_0) \tag{11-5}$$

and from Eqs. (11-4) and (11-5), we derive

$$\frac{\tan\theta}{\tan\theta_0} = \frac{(1-\phi_0)}{(1-\phi)}, \text{ or}$$

$$\theta\text{dip} = \text{Arctan}((\tan(\theta^0)(1-\phi^0))/(1-\phi_{cal})) \quad (11\text{-}6)$$

where

$$\phi_{cal} = \phi_{ss} \text{ or } \phi_{sh} \text{ depending on rock type}$$

[see Eqs. (11-7) to (11-10)]

$$\phi_{sh} = 0.68$$

$$\tan(\theta^0) = 68 \text{ deg}$$

Therefore, if we are able to predict the initial fault dip (θ_0) and initial porosity (ϕ_0) along with the porosity (ϕ) of the rock at any given depth level, we can determine the approximate fault dip (θ) at any level of interest (Xiao and Suppe 1989) (Fig. 11-27).

The process of dewatering and compaction is rapid, occurring within about the upper 700 ft (Baldwin and Butler 1985), so any compaction equation is controlled primarily by the porosity, which is in turn related to the sandstone/shale ratio. Thus, if the depths of a normal fault are known in each of several neighboring wells, and if the sandstone/shale ratios are known from local or adjoining footwall well logs, then the fault shape and its location can be extrapolated between the adjoining wells using Eq. (11-6). Case histories of the process are presented later in this section.

Before applying Eq. (11-6) to an area, we must be able to calculate the amount of porosity to be expected in sandstone and shale horizons at any given depth. Porosity/depth equations are dependent on the region being studied (Baldwin and Butler 1985). These equations represent the average porosity of sandstone or shale at any given depth, so determine these relationships by using *local data*. The porosity/depth equations represent averages, so results will also represent averages.

For the U.S. Gulf Coast, the following empirically derived sandstone and shale porosity/depth equations can be applied (Xiao and Suppe 1989). For shale,

$$\phi\text{sh} = 0.2684 - 0.1972\, x \,\log 10\ (z/3280 \text{ ft}) \quad (11\text{-}7)$$

where

$$\phi_{sh} = \text{shale porosity}$$

$$z = \text{depth in ft}$$

For sandstone, Eqs. (11-8) through (11-10) apply:

$$\phi_{ss} = 0.3198 - 0.0327(z/3280 \text{ ft}) \quad (11\text{-}8)$$

below 8000 ft

and

$$\phi_{ss} = 0.3933 - 0.0635(z/3280 \text{ ft}) \quad (11\text{-}9)$$

above 8000 ft

Alternatively, you may use the following equation (Atwater and Miller 1965).

$$\phi_{ss} = \phi^0 - 0.01265\, x\, (\text{depth}/1000 \text{ ft}) \quad (11\text{-}10)$$

where

ϕ_{ss} = sand stone porosity

ϕ^0 = the initial porosity

From Eq. (11-9) we can determine that the initial sandstone porosity is about 39% by setting the depth to zero. Porosity/depth equations for areas other than the Gulf of Mexico are presented by Baldwin and Butler (1985) and Sclater and Christie (1980).

The technique can now be applied. First, the sandstone portions of a well are determined by locating the sandstone and the shale base lines from spontaneous potential (SP) logs. Common industry practice suggests that sandstones exist wherever the SP log exceeds 25% of the distance between the sandstone and shale baselines (Fig. 11-28). Values of less than 25% on the divided SP log are taken to be shale (Schlumberger 1987). Second, a "splicing method" is employed, using Figure 11-29. In those depth intervals that have been determined from the SP log to be sandstone, use Eq. (11-8), (11-9), or (11-10), and use Eq. (11-7) in those depth intervals that represent shale.

Equation (11-6) is also dependent on the initial fault dip and initial porosities. As an example, for the northern Gulf of Mexico these values have been empirically determined to be 67.5 deg for initial fault dip, and 39% and 68% porosity for sandstone and shale, respectively (Xiao and Suppe 1989). You may wish to confirm these initial values from local data prior to applying Eq. (11-6) to your area of interest. We have now determined the sandstone and the shale portions of the sedimentary section from the SP logs and can calculate their average porosity at any given depth from Eqs. (11-7) to (11-10). Therefore, we can now calculate the average fault dip (θ) at any depth from Eq. (11-6) and the splicing method. These formulae were tested with good results in a number of examples in the Gulf of Mexico, such as the one shown in Figure 11-30.

In areas of poor seismic data, or in areas with well log data and limited seismic data, the technique can be used to help interpret the shape of important growth faults. The shape of a fault with depth may be important in generating a prospect, designing a prospective well location, or the determination of reserves based on the areal extent of a reservoir.

Perhaps a prospect is under evaluation and the location of the trapping fault is of critical importance. The method can be used as a "quick-look" technique (Tearpock et al. 1994) to evaluate the presented interpretation and, in particular, the shape and location of the trapping fault.

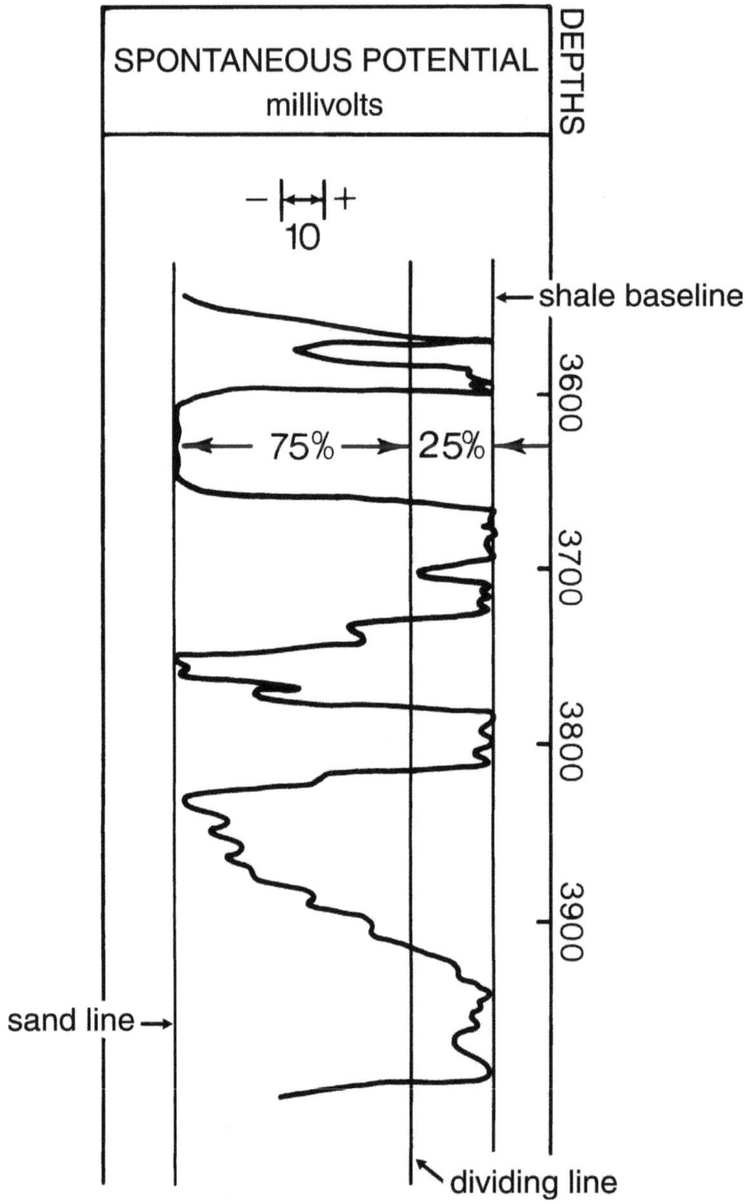

Figure 11-28 Relationships utilized to determine sand/shale ratios. Sand is present wherever the SP log deviates more than 25% off the shale base line. (From Xiao and Suppe 1989. AAPG©1989, reprinted by permission of the AAPG whose permission is required for further use.)

Prospect Example

A good application of this method arises when analyzing a prospect. Examine Figure 11-31, showing Well No. 1, which penetrates a large growth fault, and Well No. 2, which has a discovery at the D Sand level on a rollover structure. Seismic data from this area are not of the best quality, but *they image gentle south dips*. The proposed well, to be drilled up-dip of the **discovery well**, *looks like a lead-pipe cinch!* Would you participate in the proposed well? *If not, why not?*

Let's do a quick examination of the cross section in Figure 11-31. We first observe that the normal fault is planar in shape, dipping at an angle of about 55 deg. The discovery and proposed wells appear to be in the hanging wall rollover anticline formed by the fault, which is penetrated by Well No. 1.

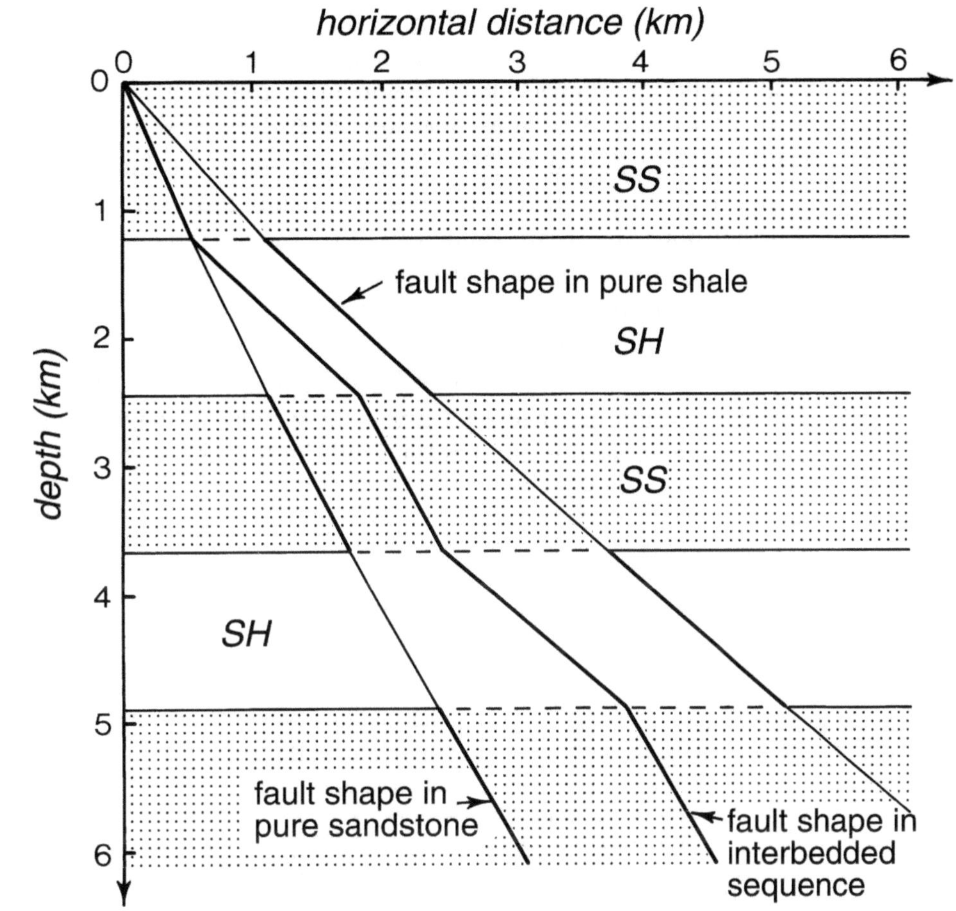

Figure 11-29 The splicing method for estimating fault dips from sand/shale horizons determined from SP logs. (From Xiao and Suppe 1989. AAPG©1989, reprinted by permission of the AAPG whose permission is required for further use.)

Several questions come to mind. Why is the fault in Figure 11-31 not listric? Why do the beds dip or roll toward a fault surface that is a planar surface? Geoscientists know that rollover structures are associated with listric fault surfaces; that is, rollover structures form above fault surfaces that are not planar. In the section on the origin of rollover in this chapter, we demonstrate that *only above listric fault surfaces* do hanging wall strata form a rollover anticline.

Let's take a look at the fault surface shown in Figure 11-31 and refer back to Figure 11-29. Figure 11-29 allows us to estimate fault dips from lithology. The lithology in the footwall of the fault in Well No. 1 allows us to predict the dip and shape of the fault. Well No. 1, which penetrates the fault near −4000 ft, contains about 50% sand in the footwall block between the −4000-ft and −8000-ft levels (−1200 m to −2400 m). The average depth of the sandy section is −6000 ft (−1800 m). On Figure 11-29, a fault at a depth of −6000 ft (−1800 m) in a 50% sand section will dip at an angle that lies *halfway* between the fault shape in pure sandstone and the fault shape in pure shale. We can fit a protractor to this figure to obtain a fault dip of about 55 deg. The fault shown on Figure 11-31 dips about 55 deg between −4000 ft and −8000 ft, exhibiting good agreement between observation and theory.

Notice that Well No. 1 penetrates a predominantly shale section below −8000 ft (−2400 m). Thus, the footwall section below −8000 ft should be subject to significant compaction. *Faults in compacting shales will dip at gentler angles than faults traversing sand-rich sections.* Let's take another look at Figure 11-29, but this time we assume a pure shale section between −8000 ft and −20,000 ft (−2400 m to −6100 m). The *average* depth of this pure shale section is about −14,000 ft (−4250 m). Fit a protractor to the *fault shape on the pure shale curve* (Fig. 11-29) at an average depth of −14,000 ft. Figure 11-29 predicts that the fault will dip at an *average of 40 deg* below −8000 ft. If the fault dips at an average of 40 deg below −8000 ft, then the fault should follow the path shown in Figure 11-32. The listric bend in the fault causes the strata to roll toward the fault and to form a rollover anticline in the hanging wall block (see section on origin of rollover). Furthermore, the listric shape of the fault causes the strata to dip at a higher angle than those shown in Figure 11-31 and causes the fault to be farther to the south than previously interpreted. The result of our quick analysis is to predict the possibility that *the productive D Sand will be faulted out in the proposed well* (Fig. 11-32). What appeared at first to be a sure thing may not be. An alternative to the quick-look method is to use a spreadsheet and the procedures outlined in this section to write a program using Eqs. (11-6), (11-7), (11-8), (11-9), and (11-10). The program can be used to predict the fault dip through a varying sandstone/shale section.

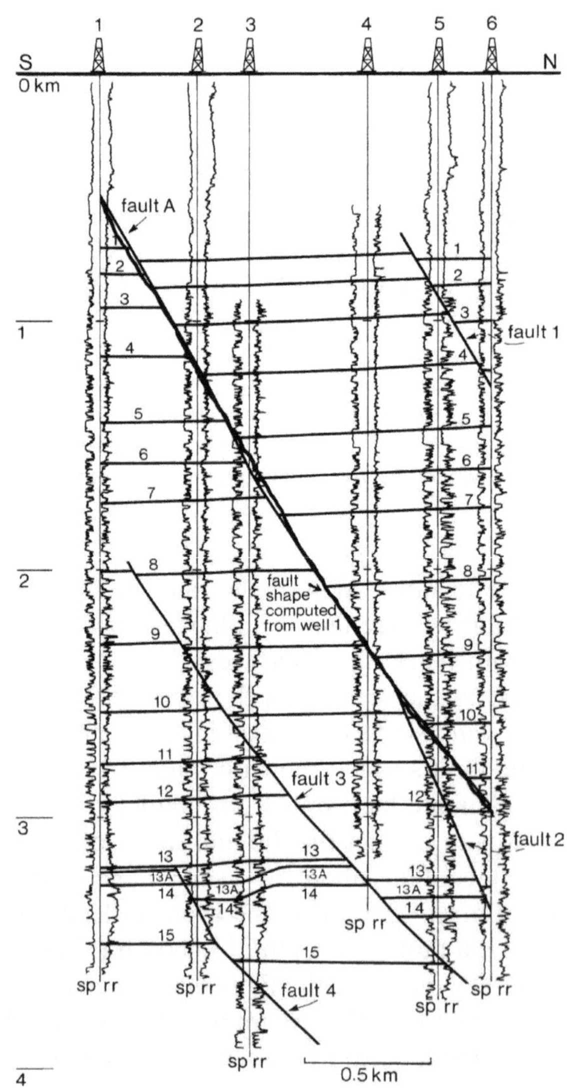

Figure 11-30 Test example for projecting fault dips between wells. The SP log from Well No. 1, the fault dip (Fig. 11-27), and the porosity/depth equations are utilized to predict the shape and position of Fault A with depth. (From Xiao and Suppe 1989. AAPG©1989, reprinted by permission of the AAPG whose permission is required for further use.)

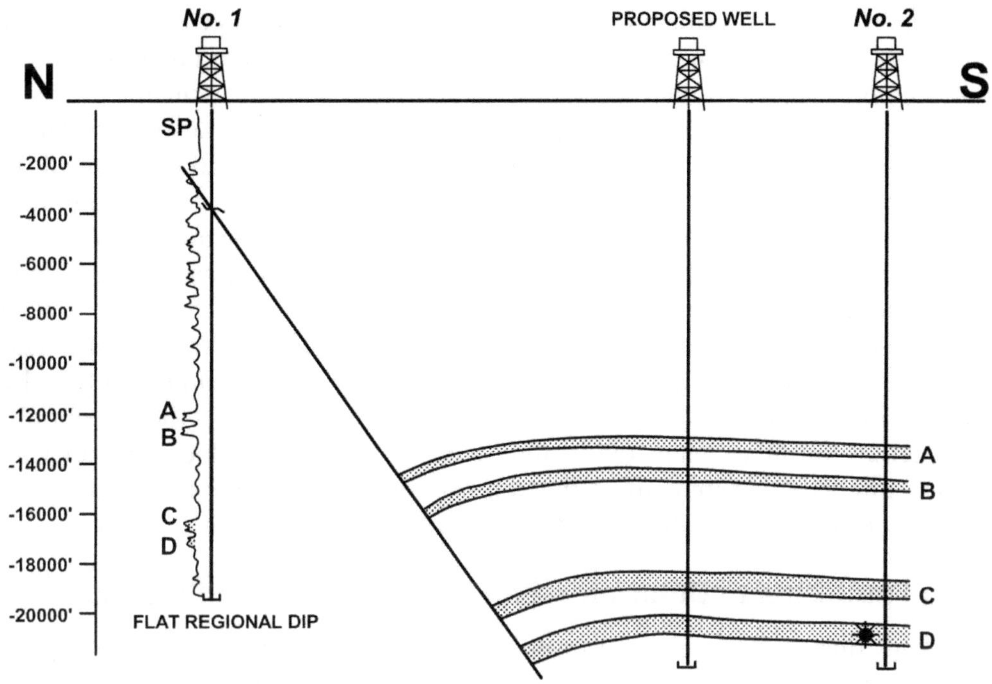

Figure 11-31 Proposed well position located up-dip of the producing Well No. 2, and downthrown to a planar fault surface that intersects Well No. 1 at −4000 ft. Quick-look techniques can determine structural viability of the proposed production well. (Published by permission of R. Bischke.)

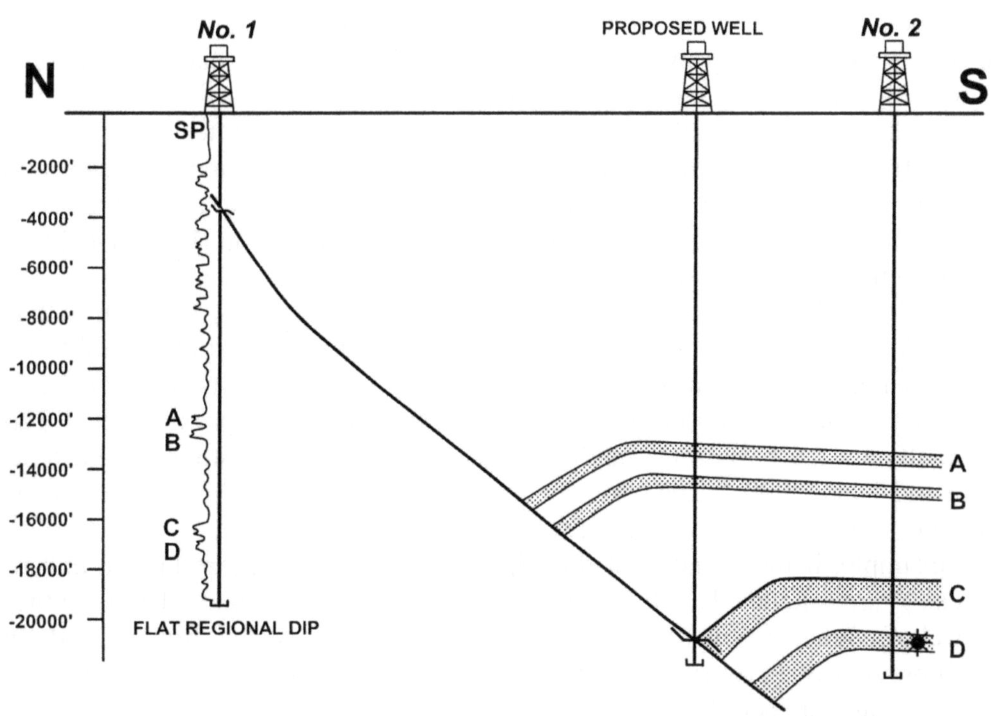

Figure 11-32 The SP log in Well No. 1 and the compaction formula result in a listric fault surface that dips at a lower angle below −8,000 ft, causing the interpreted fault to displace to the south. Listric shape of the fault surface causes the beds to roll toward the fault surface. Given this structure and the relocated fault, a good probability exists for the reservoir being faulted out in the proposed well. (Published by permission of R. Bischke.)

The importance of this example is to stress that there is often a geometric relationship of fault shape to fold shape, and it can be used when generating geologic interpretations. The cross section shown in Figure 11-31 does not conform to certain basic geologic principles. Therefore, a good possibility exists that the interpretation is in error. The error may be small or large. The quick-look technique indicates that the error, in this case, may be large enough to result in a dry hole. This quick analysis should direct the prospect generator to additional work to either substantiate the interpretation, as shown in Figure 11-32, or modify the interpretation based on the principles shown in this section on the effects of compaction on the shape of growth normal faults.

The compaction work presented by Xiao and Suppe (1989) can result in accurate predictions of fault shape where compaction is the predominant process. The method is stable and tends to *average* the well log data. As the method predicts the *average fault dip* based on the *average lithology over an interval*, the method averages through intervals that do not obey the idealized compaction formulae. Thus, the method tends to average out local and aberrant secondary cementation intervals that may affect the compaction curves. *Deformation of the fault surface*, either by salt or by another fault, may affect the results of the method.

Inverting Fault Dips to Determine Sand/Shale Ratios or Percent Sand

Inverting fault dips is a technique used to estimate the percent sand in a prospective interval, which is valuable information for a prospect. When well log data are not available on the level of interest, and provided that a normal fault is present, it may be possible to estimate the amount of sand off of a seismic line. The seismic line must be depth-corrected. With an estimate of the percent sand over the prospect area, a certain amount of risk can be mitigated, better volumetrics can be estimated, and overall economic evaluation can be improved.

Normal faults can be *listric* in shape and dip at lower angles with increasing depth, or the faults can be *antilistric* and dip at higher angles with increasing depth (Fig. 11-29). Observation by industry indicated that growth normal faults commonly dip at shallower angles in shale sections and at higher angles in sand sections (Gow 1962; Roux 1978). Industry uses this relationship between fault shape and lithology (Roux 1978) to predict sand intervals from depth-corrected seismic profiles. If on a depth-corrected, dip-oriented seismic section a growth fault steepens at depth, then the fault may be entering a sand-rich interval (Bischke and Tearpock 1993). In this section, we develop a technique based on the Xiao-Suppe equation [Eq. (11-6)] to estimate percent sand from the shape of faults on depth-corrected seismic sections.

In the Brazos Ridge area in the northern Gulf of Mexico, many of the growth normal faults at shallow depths, and within the crests of rollover structures, dip at angles of 40 deg to 45 deg, and with greater depths dip at higher angles (50 deg to 60 deg). These *antilistric normal faults steepen into a known section of higher sandstone/shale ratios* (approximately 1.0 to 1.5). These normal faults are the basis for a method of using fault dips to estimate sandstone/shale ratios (Bischke and Suppe 1990a). The basic concept for accomplishing this task has been presented in the previous section and is implicit in Eq. (11-6).

For example, if the initial fault dip (θ_0) and porosities (ϕ_0) are known and if the average porosities (ϕ) are calculated at all depth levels using Eqs. (11-7) to (11-10), then the fault dip (θ) at the level of interest can be calculated from Eq. (11-6). This is the forward modeling procedure described previously. Forward modeling is more accurate than inverse modeling, and ratios such as sand/shale ratio are sensitive.

If, however, the fault dip (θ) is constrained by good depth-corrected seismic sections, then the average porosities (ϕ) for sand and for shale can be calculated from Eqs. (11-7 to 11-9) at any depth level. This is called *inverse modeling*, which is sensitive to measurement errors. As porosities are dependent on lithology [see Eqs. (11-7) to (11-10)], we can estimate the (sand or shale/

depth curve in any area, provided that we have information on local porosity depth curves. This information can be derived from the nearest wells, which in frontier areas could be many miles from the area of interest.

First, we must initiate a sandstone/shale ratio formula. This can be accomplished using Figure 11-33. In practice, the observed height of the figure is the top and the bottom of a portion of an interpreted growth normal fault on a depth-corrected seismic section. This distance represents a column of interbedded sandstone and shale. Using a deck-of-cards analogy, the sandstone layers are removed from the deck and placed at its bottom.

We now calculate from Figure 11-33:

$$h_2 = hsh/\sin(\theta_{sh})$$
$$h_3 = hss/\sin(\theta_{ss})$$

where

hsh = thickness of shale
hss = thickness of sandstone
θ_{sh} = calculated fault dip in shale interval
θ_{ss} = calculated fault dip in sandstone interval
θ_{obs} = observed fault dip on a depth corrected seismic at the working level
h_2 = length of fault in shale interval
h_3 = length of fault in sandstone interval

Using the law of sines,

$$\sin(\theta_{ss} - \theta_{obs})/h_2 = \sin(\theta_{obs} - \theta_{sh})/h_3$$

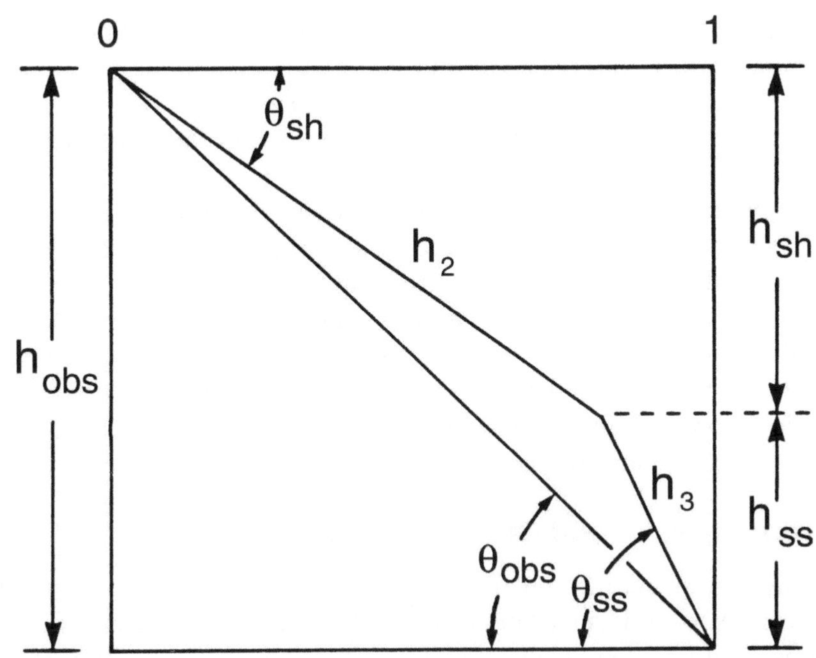

Figure 11-33 An alternating sequence of sand and shale beds can be generalized into a shale interval of thickness hsh and a sandstone layer of thickness hss. As shales have a higher initial porosity than sandstones, shales compact more than sandstones. (Published by permission of R. Bischke.)

and substituting and rearranging yields

$$r_{calc} = hss/hsh = [\sin(\theta_{obs} - \theta_{sh}) \sin(\theta_{ss})]/[\sin(\theta_{ss} - \theta_{obs})\sin(\theta_{sh})] \tag{11-11}$$

or

$$r_{calc} = [(\tan\theta_{ob}/\tan\theta_{sh}) - 1.0]/[1.0 - (\tan\theta_{ob}/\tan\theta_{ss}) - 1] \tag{11-12}$$

where

r_{calc} = average estimated sandstone/shale ratio

θ_{obs} = observed fault dip calculated from an interpreted and depth-corrected section

Sandstone/shale ratios can be converted to percent sand by the following equation:

$$\% \text{ SAND} = [1.0 - (1.0/(hss/hsh + 1.0))] \times 100 \tag{11-13}$$

As θ_{sh} and θ_{ss} can be obtained from Eq. (11-6) and the porosity from Eqs. (11-7) to (11-10), we can calculate the average porosity, or sand/shale, curve with increasing depth.

Before proceeding to a test case, we outline a number of factors that affect the accuracy of our methods.

1. Inversion methods are sensitive, and errors occur when measurements are not accurate. The method is most accurate when penetrating thick sand sections interbedded with thick shale formations, such as exist in the Brazos Ridge and Segno Field examples presented in the next section (Figs. 11-34, 11-36, and 11-37).
2. *Accordingly, accurate measurements are necessary.*
3. *The method applies only to growth normal faults* and thus to areas that are experiencing both active sedimentation and extension.
4. Although the method can be applied wherever velocity information is available, the velocity function is critical to obtaining reliable results because it is used to determine depth of the fault. Velocity determinations are discussed in Chapter 5.
5. The method in its present stage of development involves only compaction, and thus *deformation of normal faults by other processes*, such as salt flow, *may invalidate the method.*
6. The method is restricted to the depth to which the seismic reflections can be resolved.
7. The method applies only to sandstone/shale lithologies, and not to carbonates. However, faults also steepen into carbonate reservoirs.

Example from Corsair Fault Offshore Texas

In the following test of the method from data located on the Corsair fault, Texas, we compare a depth-corrected growth normal faults to well log data (Bischke and Suppe 1900a). In order to compare observed well log results, which resolves horizons to within feet to seismic data, in which the resolution is a function of depth, the well log data must be converted to longer wavelengths. This is accomplished by first determining the sand and shale horizons from SP logs, as discussed earlier (Fig. 11-28), and then measuring the wavelength of coherent reflectors from the seismic section. The frequency content of the seismic section is thus determined. As the frequency content decreases with depth, the wavelength of the reflections increases with depth. This wavelength is then depth-corrected and passed over the SP data as a moving average, with the length of the operator increasing with depth. The result of this averaging process from an example well is the sandstone/shale depth curve shown in Figure 11-34. The processed data are near several antilithic faults located on a rollover above the Corsair fault, Texas (Bischke and Suppe 1990a, their Figs. 8 and 9). The data are plotted at a frequency of two seismic wavelengths. This long-wavelength well log data can be directly compared to normal fault calculations based on seismic data.

Typical results, calculated using Eq. (11-13) and based on a growth fault located near Well No. 9, are shown in Figure 9 of Bischke and Suppe (1990a). This fault is the closest fault to Well No. 9. The results approximate the well log data. The seismic data was processed utilizing stacking velocities. The normal fault, which terminates at a depth of −7000 ft, has been used to predict the higher sandstone/shale ratios present in the −4800-ft to −6300-ft level. Other examples, located closer to the main fault, encounter a sedimentary dumb or more precisely a prograding wedge, which has higher sandstone/shale ratios ranging from 2 to 4. The prograding wedge thins and dies out near the crest of structure. A typical calculated example is shown in Fig. 11-35, in which high sandstone/shale ratios occur near 1 between a depth of 5500 ft to 6000 ft.

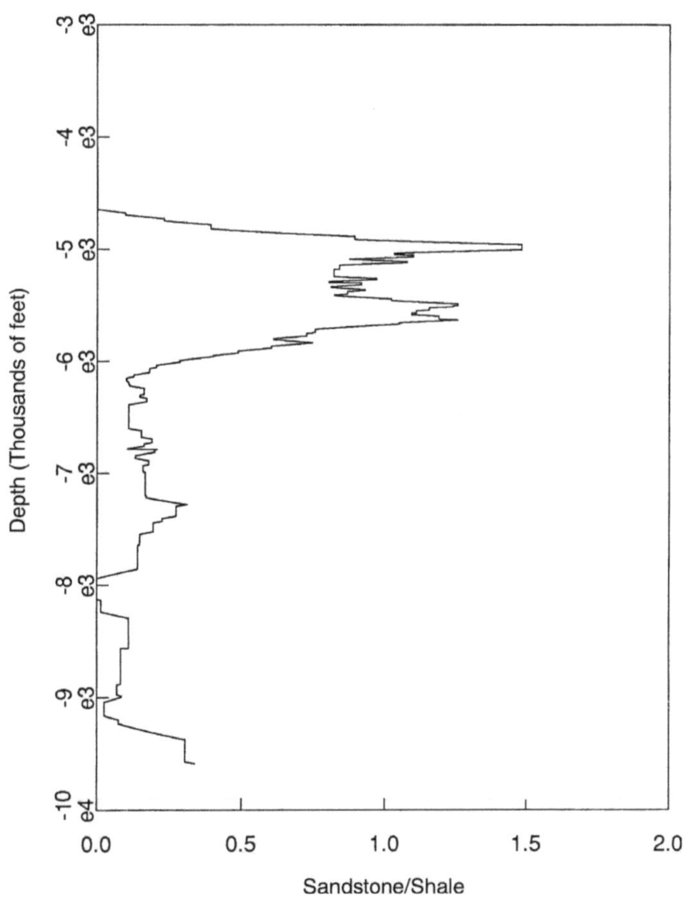

Figure 11-34 Processed sandstone/shale depth curve obtained by averaging sand/shale horizons taken from an SP log from Well No. 9 (Bischke and Suppe 1990a). The data are plotted at a frequency of two seismic wavelengths and range in depth from −3000 ft to −10,000 ft. (Published by permission of R. Bischke.)

Elements of the sandstone-shale theory were known to industry for many years and were applied worldwide. As a final example, we examine a field in Texas where an early form of the method was applied in the 1960s.

Example from Segno Field, Polk County, Texas. The Segno Field rollover structure formed downthrown to an east-west trending normal fault (Fig. 11-36a). Examine the footwall portion of cross section A-A′ constructed from dense well control on the eastern flank of the field (Fig. 11-36b). This section from Gow (1962) through the Tertiary Jackson, Cook Mountain, Sparta, and Weches shale sections shows that the normal fault dips at angles of 30 deg to 40 deg. Gow wrote on the profile, "Note that fault plane dip steepens in sand formations" (Gow 1962). In the Yegua and Wilcox sand sections the fault surface steepens to about 50 deg to 55 deg (Fig. 11-36b).

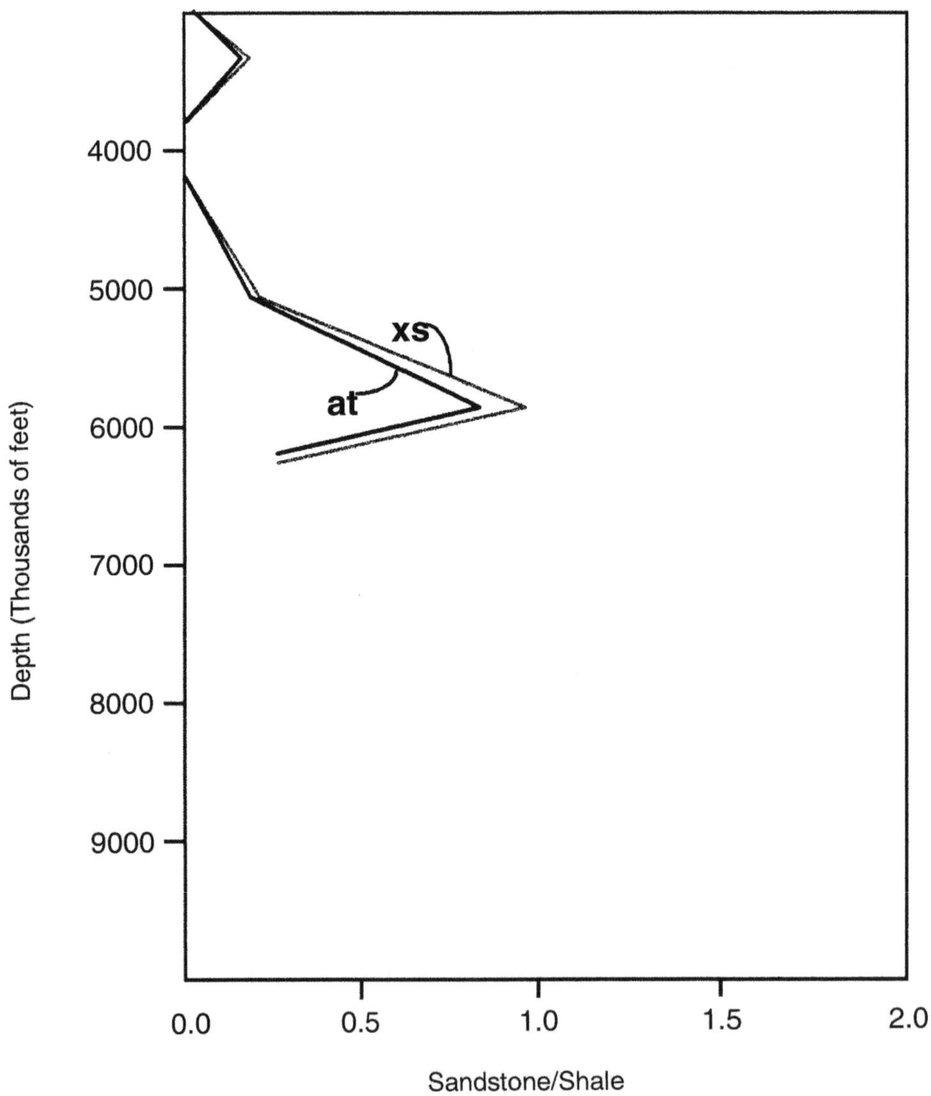

Figure 11-35 Plot showing predicted sandstone/shale depth curve derived from fault dips interpreted on seismic profile. The fault is adjacent to the well data shown in Figure 11-9 in Bischke and Suppe (1990a). The xs curve uses the Xiao-Suppe compaction equation, whereas the at curve denotes the Atwater equation. (Published by permission of R. Bischke.)

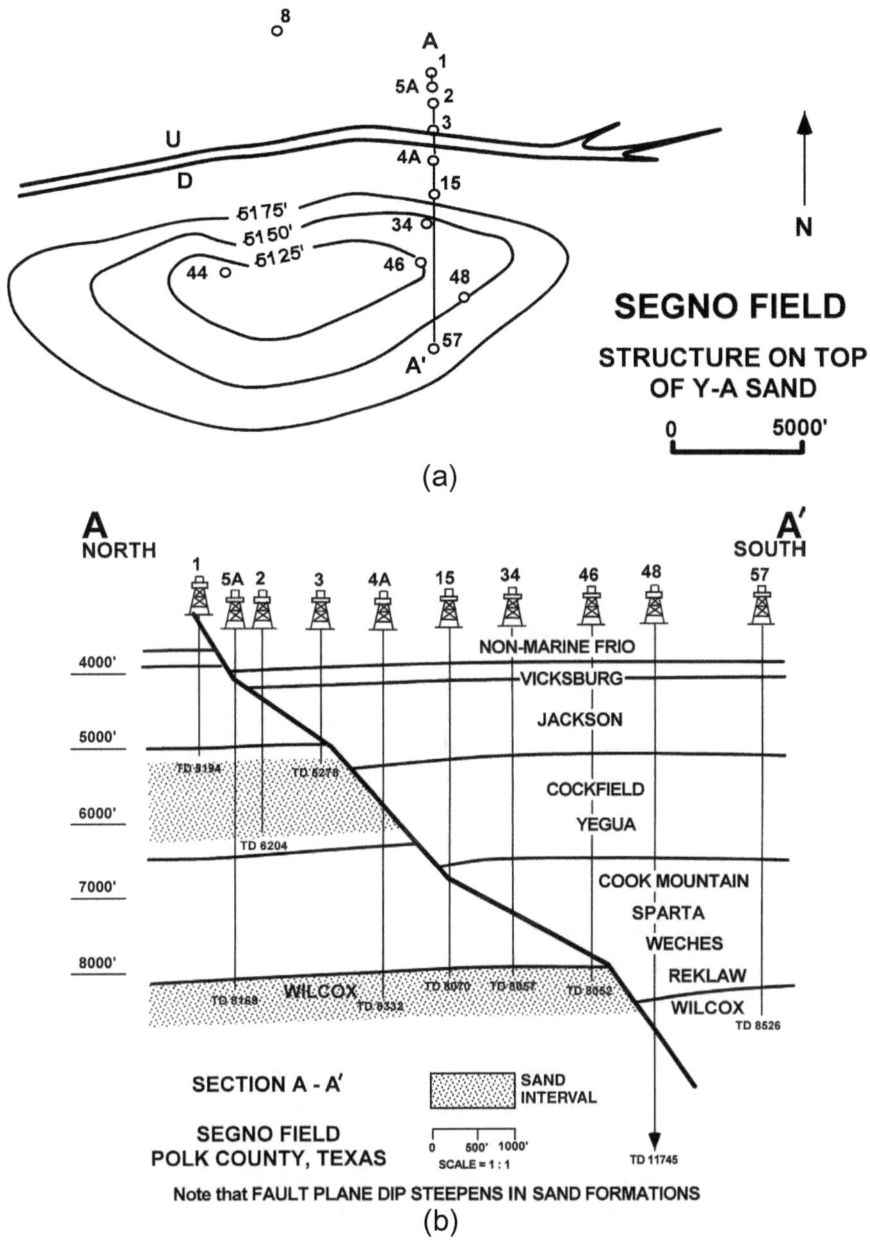

Figure 11-36 (a) Map of Segno Field, Texas, USA, with location of Cross Section A-A . (b) Section A-A shows growth fault steepening in sand intervals. As Gow (1962) noted on the section, "Note that fault plane steepens in sand formations." (From Bischke and Tearpock 1993. Redrawn from Gow 1962. Published by permission of the Houston Geological Society.)

A change in fault dip angle from 30–40 deg to 50–55 deg is typically apparent on depth-corrected seismic profiles. Indeed, Roux (1978) presents an example of a fault that changes dip through a thick sand and shale section on a vintage seismic time profile. This profile is in Lowell's book on structural styles (Lowell 1985). Industry recognized, and at times used, this "change in fault dip technique" to locate sand sections during the 1960s and 1970s. Subsequent bright spot analysis apparently replaced the technique, but geoscientists are again using the technique where direct hydrocarbon indicators are not applicable or not in support of amplitude versus offset (AVO) analysis.

We obtained a seismic line that crosses the crest of the structure. The line was digitized and depth-corrected using interval velocities. Interval velocities approximate true velocities and are always available to the geoscientist. On the profile (Fig. 11-36b), the fault dips at high

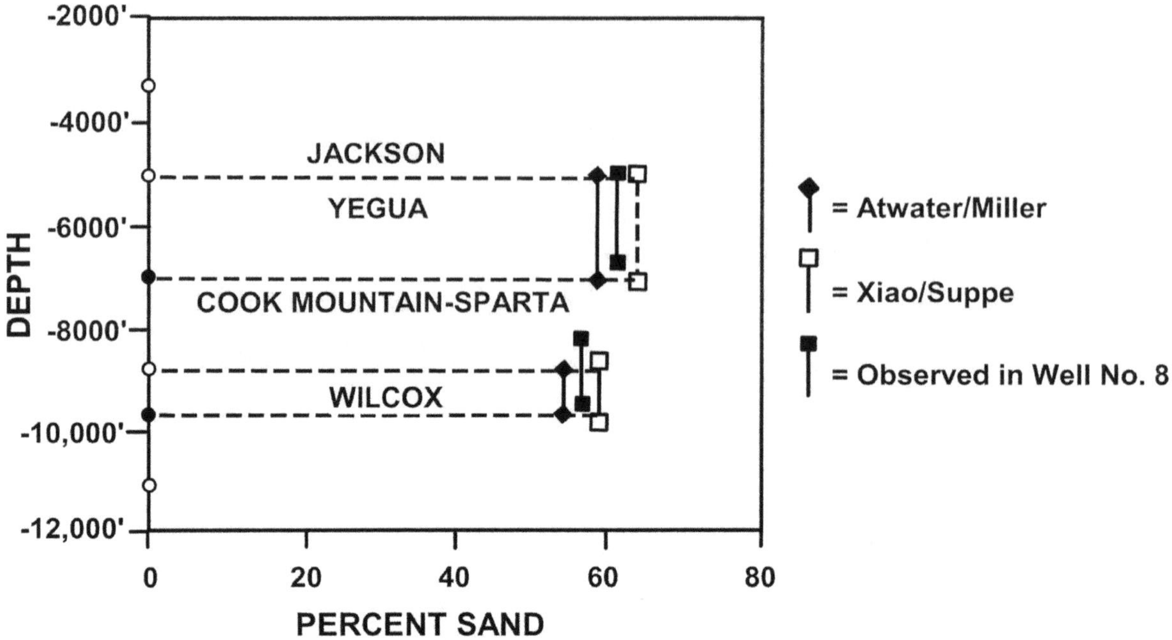

Figure 11-37 Percent sand plot for Segno Field growth fault, based on changes in dip of fault. Fault steepens through Yegua and Wilcox sand sections, generating a high sand percentage response from about −5000 ft to −7000 ft and below about −9000 ft. Compare calculated sand percentages to sand percentages based on log of footwall Well No. 8 (closed box bar on graph). (From Bischke and Tearpock 1993. Published by permission of the Houston Geological Society.)

angles through the Yegua and Wilcox sands and at less steep angles through the Jackson, Cook Mountain, and Sparta shale sections (Bischke and Tearpock 1993). After picking the fault on the vintage seismic line, the inflection points on the fault surface were depth-corrected for input into a sand percentage prediction program. This program uses formulas presented in the previous section. Program results shown in Figure 11-37 are comparable to Well No. 8 in the footwall block (Fig. 11-36a), located adjacent to the seismic line (Bischke and Tearpock 1993). In Figure 11-37, the diamond-shaped bar on the left shows the resulting computations from the Atwater-Miller sand compaction formula [Eq. (11-10)], whereas the open box bar on the right defines the Xiao-Suppe compaction curve results [Eqs. (11-8) and (11-9)]. The Atwater-Miller sand compaction formula is more conservative than the Xiao-Suppe formula. The closed box bar (middle) is the percent sand based on the Well No. 8 log.

When subject to computation, the interpreted fault trace predicts a sand response on the level of the Yegua sand between the depths of about −5000 ft and −7000 ft. The Atwater-Miller and Xiao-Suppe bars bracket the sand percentage in Well No. 8 and predict a sand percentage of about 60% (Fig. 11-37). Between the −5000-ft and −6700-ft depths in Well No. 8, the sand percentage is about 60% (Bischke and Tearpock 1993). An examination of Figure 11-37 indicates that a shale section should exist at about the −7000-ft to −8800-ft level and that sand should exist below −8800 ft. Well No. 8 contains shale between −6700 ft and −8200 ft before encountering the Wilcox sand (Fig. 11-36b). These estimates, which are a good match to the well data, are based on imprecise interval velocities, and better velocity estimates and seismic data should improve the depth percent sand and sandstone/shale predictions.

The method works best on seismic sections where you can pick the fault trace accurately and where thick sand and shale sections are present. Fault surfaces also respond to interbedded sand/shale sections, and the section need not be predominantly sand (Bischke and Tearpock 1993;

Tearpock and Bischke 1991). Why drill a well to a depth of 20,000 ft if there is no direct evidence for sand at that level, without first using fault dip analysis to help reduce your sand risk?

This method has application to petroleum exploration, as was recognized by industry years ago (Gow 1962; Roux 1978). Normally, when a well is drilled, *one of* the *least known parameters is the sand/shale ratio of the reservoir*. As the section in the Segno Field example contains shale from −7000 ft to −8800 ft (Fig. 11-37), the method also produces estimates for seal thicknesses. Such information can be used by a geoscientist or engineer to better determine the *target depths* for wells and to improve the estimates of potential reserves prior to drilling. The method can be applied to the third dimension, so 3D percent sand maps can be constructed to better select well locations.

We have seen examples of exploration well locations being moved by as much as 1000 ft on the basis of this analysis. If a prospect has good structural closure over a large lateral area, the specific choice of the actual well location may be based on where the prospective intervals have the highest percent sand. This technique can be used, under the circumstances discussed, to make such determinations. All that is required to apply the method is an array of *small antilistric, growth normal faults*. The usefulness of the method increases as well control becomes limited and the distance between wells increases.

USING STRUCTURAL RELIEF AND THROW TO PREDICT THE DOWNTHROWN STRATIGRAPHIC SECTION

This section discusses several methods that may be used in some cases to predict the downthrown stratigraphic section on deep rollover structures. The requirements are a depth-corrected seismic section that images the master (major) fault surface, the general geometry of the structural relief of the hanging wall rolled over, anticlinal beds, and the stratigraphy of the upthrown beds. These beds are typically at the depths to known sand intervals, or sandy shale horizons that contain a high percentage of total organic carbon content (%TOC). These intervals need not be drilled downthrown to the fault. One method shows how the expanded downthrown stratigraphic section may be predicted at greater depths by measuring the R/T ratio, which is the downthrown structural relief (R) divided by the throw (T) on the interpreted portions of the fault using the imaged upthrown and downthrown portions of a growth rollover structure. A second method is to calculate the ratio of the structural relief (R) divided by the entire thickness of the growth stratigraphic section (Gh). In this case, only downthrown data are required.

The depth of any important horizon, expressed as a mapped horizon on seismic, is dependent on the fault slip or sedimentation rate, which is dependent on *time*. However, this important horizon, which exists adjacent to the normal fault, exists at a particular *depth*. Depth, like the R/T ratio, is a length measurement and is *independent of time*. This results as both the changes in the depth of the structural relief (R) and the changes in throw (T) cancel out in the R/T ratio, and thus the R/T ratio measure is *dimensionless*.

Provided that the growth is near linear (which is often the case), the methods described here can then be used, from closed form mathematic derivations, to project known data into the undrilled portions of the stratigraphic section. Models of generic structures are developed in order to better understand how real structures behave. The seismic line must be in the fault dip direction or corrected for apparent dip using a well-known formula. The subject of accuracy or robustness of the techniques is addressed with %difference and R-square calculations of linear regression taken from the R/T measurements and calculations. The methods often have a high degree of accuracy on both modeled and real rollover structures, even though the growth intervals may contain large variations. Near-linear growth is often present on real growth faulted structures, which enables the projection equations to estimate the deeper stratigraphic section.

Nonlinear results occur when R-square measurements are large. Nonlinear results exist at mis-correlations, large unconformities, or large changes in growth.

Topics

During a field study of rollover structures, Hamblin (1965) realized that rollovers form when the hanging wall beds move through bends in the causative listric fault. Typically, at shallow depths, the beds dip at shallow angles and *bed dips steepen* where the master normal fault increases its curvature (ϕ) (Fig. 11-38). This means that the greater the curvature of the large master fault, the more the beds diverge or dip toward the master fault. As growth faults typically increase curvature with increasing depth, the *structural relief* (R) also increases with increasing depth (Fig. 11-38). A comparison of the orange, yellow, and blue horizons shows how R and ϕ increase with depth (see Table 11-1 for a glossary of terms).

The mechanics of the rollover formation process is well understood and has been discussed by many geoscientists (e.g., Dula 1991; Nunns 1991; Xiao and Suppe 1992; Withjack et al. 1995; Tearpock and Bischke (2003) but has been subject to few studies in recent years. This section reviews only the most basic aspects of rollover formation. Xiao and Suppe (1992) showed, using theoretical analysis and empirical data, that if three primary parameters are known, then the geometry of rollover structures could be accurately modeled. These primary parameters are the Coulomb collapse angle (ψ), the shape of the listric normal fault, or the change in normal fault dip angle (ϕ), and the growth or expansion index (Ei) (see Fig. 11-39a to c). Ei = Gn/Gf is calculated as the ratio between the corresponding hanging wall (Gf) and footwall horizon thicknesses (Gn). Thorson (1963) was first to define the concept of growth, which does not involve time (consult Chapter 13 for more about growth structures).

For the Gulf of Mexico and Indonesia, the Coulomb collapse angles are about 67 deg to 70 deg (Xiao and Suppe 1992; Shaw et al. 1997; Bischke and Tearpock 2003) (see Fig.11-11). The change in fault angle (ϕ) can be obtained from depth-corrected seismic (Tearpock and Bischke 2003) and the expansion index from depth-corrected seismic or well log data. Thus, the basic theory and procedures exist to accurately predict hanging wall rollover geometries from using structural balancing software packages. Nomograms for predicting the shape of the bed dips in the hanging wall structure are presented in Tearpock and Bischke (1991, 2003). Prospectors typically know that the greater the change of the listric fault angle (ϕ), the steeper the bed dips on the flank of the rollover and the greater the structural relief (R) to the crest of the rollover structure (Fig. 11-38).

This section addresses both pregrowth and synsedimentary or growth structures. On pregrowth structures there is a general mathematical relationship between the structural relief (R), the throw (T) on the fault, and the change in fault dip (ϕ) [see Eq. (11-18) in the section Pregrowth Structure and Calculations], or

$$\text{Structural relief (R)/Throw (T)} = K \qquad (11\text{-}14)$$

where K is a different function for pregrowth or for growth faulted structures.

In pregrowth modeled structures, the closed form equation derivations show that the K function is primarily influenced by the change in fault dip angle (ϕ) (see Pregrowth Structure and Calculations). However, other parameters also influence the shape of pregrowth rollover structures, and in general, K is a function of

$$K(\beta, \alpha, \phi, \psi)$$

[Fig. 11-39b; also see Eq. (11-18) in Pregrowth Structure and Calculations], where (β) = the initial or steepest dipping portion of the master fault and (α) = the dip of the deep portion of

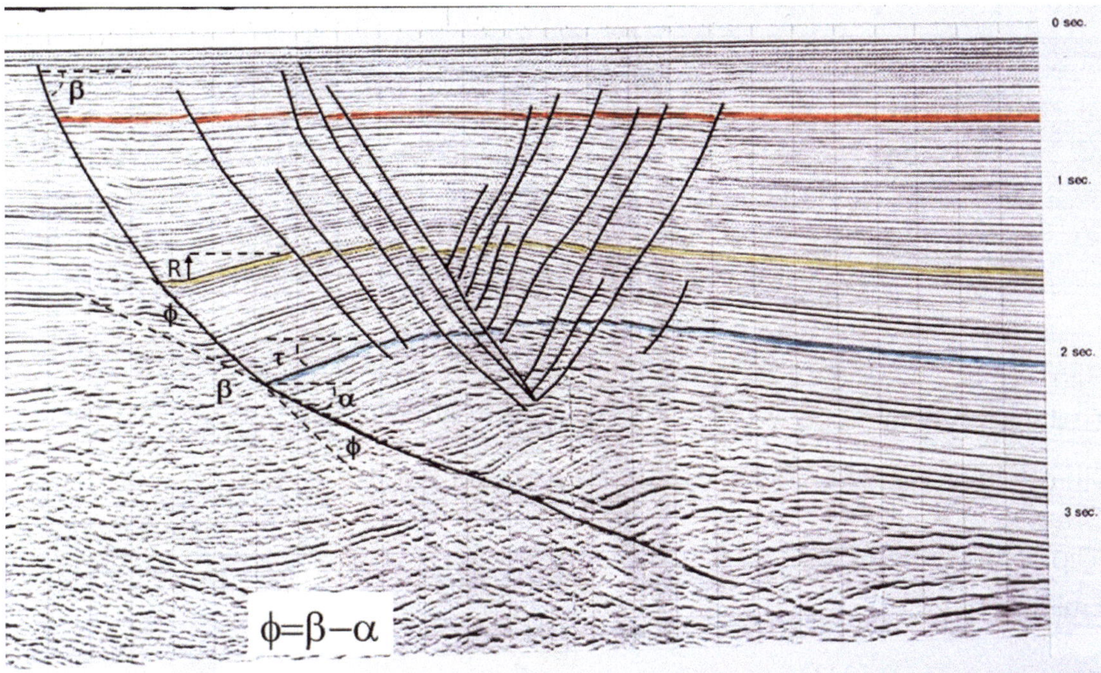

Figure 11-38 An expansion fault from offshore Texas. As the master fault decreases in dip (ϕ), the stratigraphic units diverge toward the listric fault as the structural relief (R) increases at depth. Also defined are the shallow (β) and deep (α) dips on the master fault. (Published by permission of John Suppe.)

the fault beneath the crest of the rollover (Fig. 11-38). Except for the Coulomb collapse angle (ψ), the remaining parameters, such as the structural relief (R)/throw (T), can be measured from depth-corrected seismic sections. All calculations apply to the top of the horizons. However, the base of Horizon H1 is the top of Horizon H2.

General Eq. (11-14) may be used to determine the deep rollover structure and stratigraphy of pregrowth horizons, where the fault surface and seismic reflections image but where the seismic resolution may be of poor quality and/or the sequence stratigraphic relationships are uncertain. Thus, *the seismic does not need to be perfectly coherent.*

On the more common growth or synsedimentary faults, K is a function of

$$K(\psi, \rho, \tau)$$

[see Eq. (11-18) in Pregrowth Structure and Calculations], where (ψ) = the Coulomb collapse angle, (ρ) = the growth axial surface (Xiao and Suppe 1992), and (τ) = angle that the rollover limb dips toward the master fault, defined by inflection points located near the crest and base of the limb (Fig. 11-39b or 11-41).

On pregrowth and growth structures, the method and its equation may also be used to check interpretations for consistency and structural validity where structure is complex or where the correlations are uncertain. *Again, the seismic section must be depth-corrected.* Following Xiao and Suppe (1992), we introduce simple, modeled derived structures to better understand the properties of real models based on seismic data. They classified models into pregrowth structures and growth structures. *The pregrowth, or no-growth, structures are simpler than growth structures and are studied first.* The application of these models to existing Gulf of Mexico rollover structures follows. At first, the models are simple, approximating curvilinear surfaces as straight lines, and become increasingly complex.

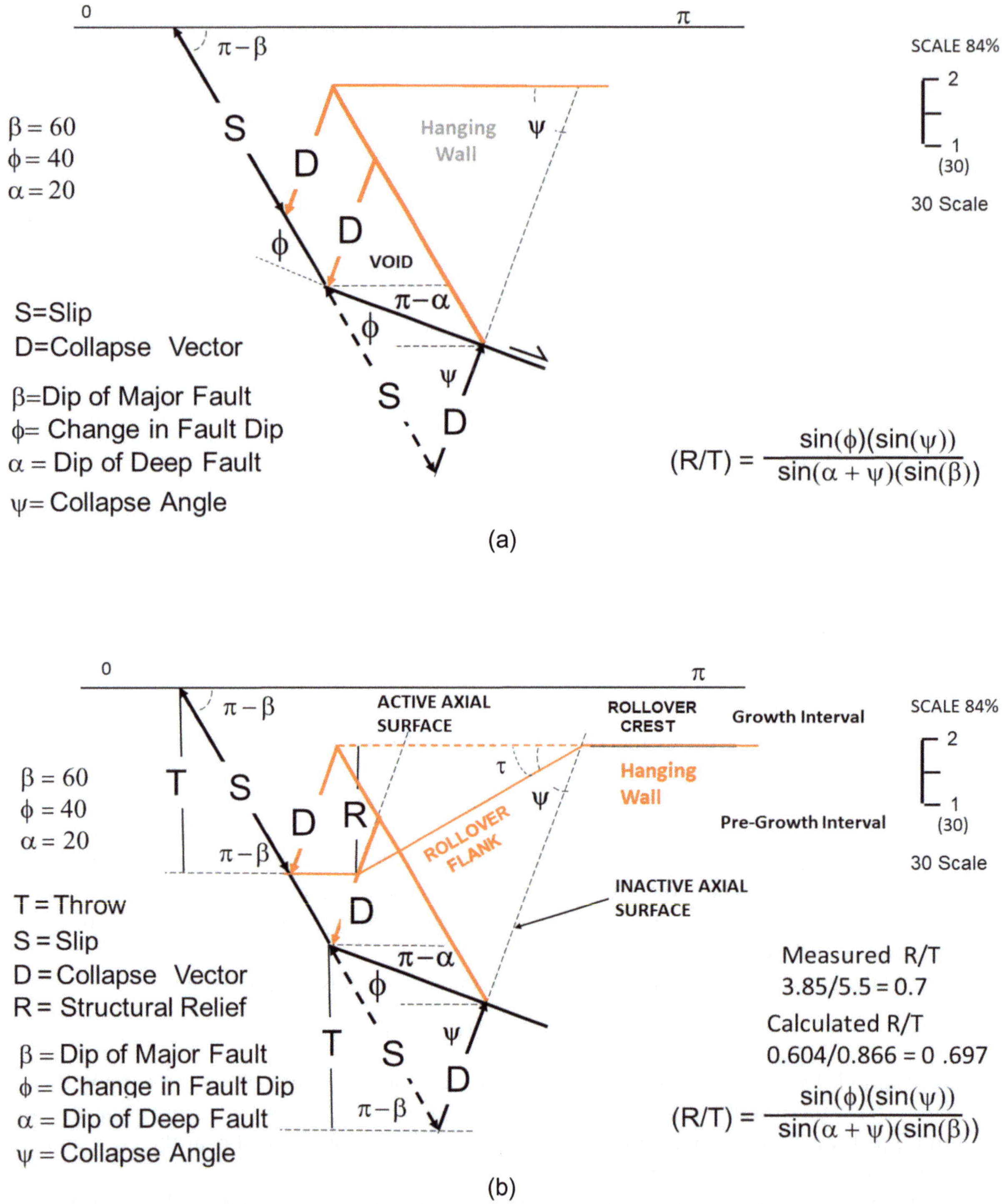

Figure 11-39 (a) Following Hamblin (1965), the hanging wall (colored orange) slips an amount S along the fault surface, creating a void. This void collapses an amount D onto the footwall, generating a rollover structure. (b) The resultant rollover has a structural relief (R) and a throw (T), and thus an R/T ratio can be measured and calculated. This graphic representation shows how the projection of the dashed line S vector beneath the fault bend and the addition of the collapse vector D along the collapse vector (ψ) determines the position of the master fault below the bend. Thus, (b) geometrically defines the ϕ and α angles for pregrowth structures. As the amount of slip increases, the throw also increases along with the D vector, which determines the values of the ϕ and α angles. In addition, and employing the angles shown in (b), the law of sines allows for a calculation of the Rn/Tn ratios. (c) The results of Eq. (11-20) when b = ψ. (Published by permission of R. Bischke.)

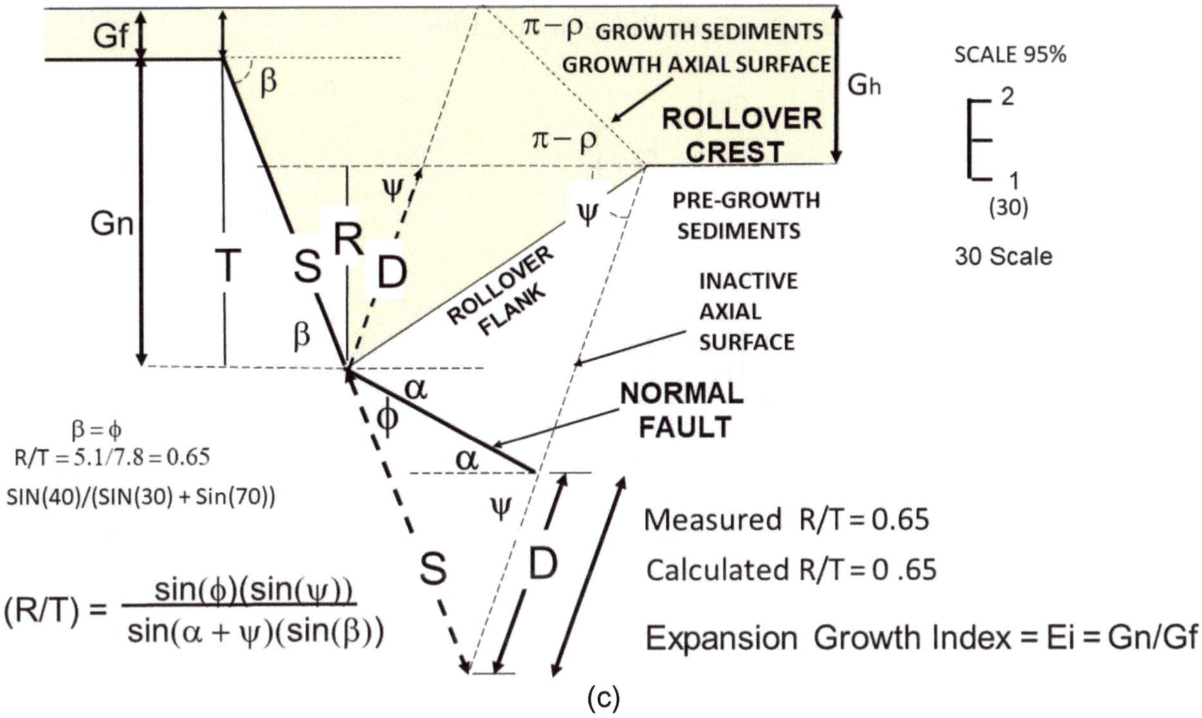

Figure 11-39 *(continued)*

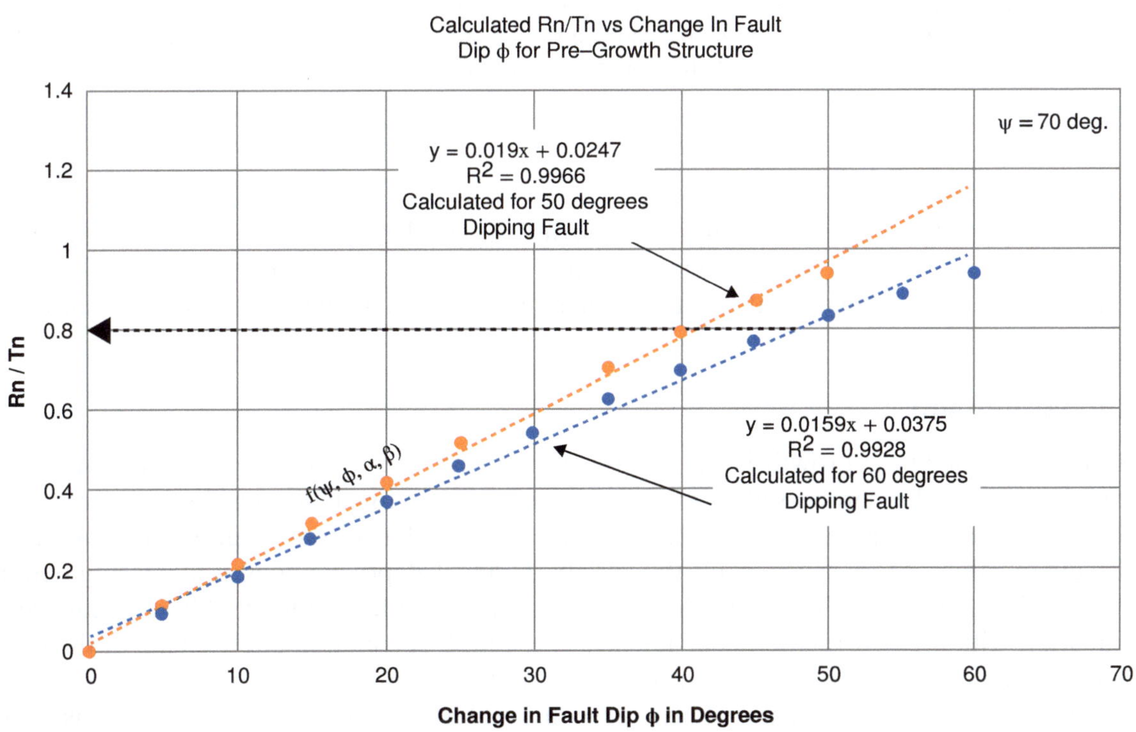

Figure 11-40 A graph of Rn/Tn values versus the change in the dip (ϕ) of the fault for a 50-deg and a 60-deg dipping master fault. These values are generated from Eq. (11-18). For pregrowth structures, the graph can quickly check R/T values. For example, using this figure, a change in fault dip of ϕ = 40 deg on a fault initially dipping at β = 50 deg yields an Rn/Tn ratio of 0.8. (Published by permission of R. Bischke.)

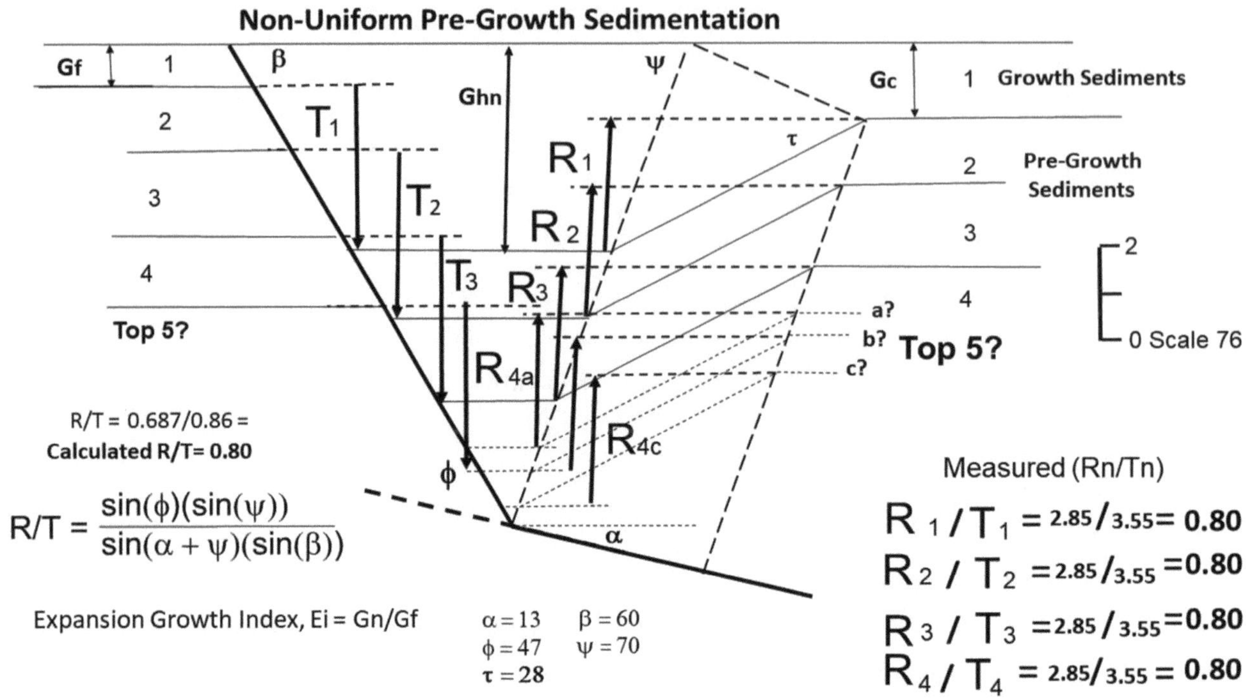

Figure 11-41 A simple modeled pregrowth structure that exhibits uniform sedimentation in the hanging wall and footwall for horizons H1 to H4. The amount of throw (Tn) and structural relief (Rn) vectors are shown for each horizon. Unlike growth structures, pregrowth structures dip in a uniform horizontal width toward the largest bend (ϕ) in the normal fault. Hand-measured values of Rn/Tn are posted on the figure along with the calculated value from Eq. (11-18), and the definition of the expansion index (Ei). Horizon 5 is assumed to be in a poor data area, but its hanging wall depth can be predicted by the Rn/Tn plot proposed in this study. (Published by permission of R. Bischke.)

Lastly, real structures are analyzed. Xiao and Suppe (1992) realized that real structures could be modeled with smaller straight-line segments. In fact, real listric fault surfaces are often curved and have multiple active axial surfaces emanating from small bends in the fault surface (Xiao and Suppe 1992).

Linear Growth

Linear or near-linear growth is not always present but can occur when the R/T picks of seismic horizons, when processed in a spreadsheet, generate linear or near-linear regression coefficients in an x/y equation. The PF (see Table 11-1 for definitions) regressions may have high R-square values. More specifically, the variable R and T depth changes at increasing depths, and the depths of the R and T values adjacent to the master fault are modeled by a linear regression equation with high correlation coefficients. These high correlation coefficients of the measured and calculated depths result in a near linear (T/R) versus depth graph (e.g., Fig. 11-40, and see Chapter 13). Linear or near-linear growth is interesting in that linear functions with high R-square calculations are able to make predictions. For example, if the position of sands or a high %TOC horizon exists in the footwall, and these identical beds have not been drilled in the downthrown block, then the approximate depth of the downthrown beds maybe estimated utilizing the near-linear formula.

The correlations were conducted using well log parasequences and on high seismic acoustic impedance contrast data. Of course, the parasequence data had more correlations than the seismic data. The zero crossings on the seismic were not included in order to simplify the interpretation of the seismic profile. It is generally accepted that if random error is not included, then fewer independent variable correlations are required to justify linearity, even as low as 2 to 3 (Warner 2013). Later in the discussion of syntectonic models, we visually explain, using seismic data, how to identify near-linear growth using growth axial surfaces (Fig. 11-39c).

The cause of this linear or near-linear growth is not well studied but may result from uniform sedimentary loading, causing the master fault to generate linear growth. Linear growth was first recognized in paleontological data (Shaw 1964), then in contractional (compressional) setting using seismic, well logs and outcrop data (Suppe et al. 1992; Suppe et al. 1997; and Shaw et al. 2005). The latter case is interesting in that in order to get linear growth, the sediments must flood over the uplifted structure at a steady state. Linear growth was also recognized in extensional well log data, and segments of linear growth exist between some sequence boundaries (Bischke 1994b; Sanchez et al.1997; Chatellier and Porras 2001, 2004) and was also recognized in field studies (Bischke et al. 1999). Linear growth is also confirmed in well log data collected along strike-slip faults (Shaw et al. 1994). When using well logs, the parasequence correlations can be typically determined within less than a meter (several feet), so the growth data graphs have a high degree of accuracy (Bischke 1994b).

Furthermore, linear or near-linear growth is not uncommon and exists in all environments (Tearpock and Bischke 2003) (for examples, see Figures 13-8, 13-9, 13-11, 13-12, 13-14, 13-17, and 13-21). For example, seven parasequences on depth-corrected seismic had an R-square = 0.982. Knowing beforehand from the seven parasequence examples presented above that *only one of the examples was linear* suggests that linear growth may be more common than realized. Linear growth just exists within sedimentary sequences. We do not suggest that linear growth is always present, as published examples in the preceding references also contain nonlinear sedimentary growth. Despite the evidence of linear or near-linear growth along normal faults, no studies have been published on how linear growth occurs or on its extent. On the other hand, linear growth associated with contractional structures has been subject to numerous studies (see preceding references).

Unlike contractional structures in which the growth sediments must flood over the uplifted structures, as long as the normal fault is slipping, extensional normal faults create accommodation space to trap growth sediments. It is important to understand that *growth measurements involve thickness changes and are not directly related to time* (Suppe et al.1992). Unlike growth sedimentation, empirical determinations of sedimentation rates are known to scale—or obey fractal mathematical relationships, which plot as a negative sloping power law on a log/log scale (Sadler 1999). In other words, mean sedimentation rates are high over short time periods, and sedimentation rates are smaller over long time periods. In effect, longer time periods allow for more and/or larger hiatus, which include nondeposition or erosion (Sadler 1999). The important point is that sedimentation rates are strongly dependent on the time frame of reference.

Industry acknowledges that thin growth sections on normal faults (let us assume a growth section of 100 feet, or 33 m) were deposited over shorter time frames than growth sections measuring several miles thick (assume 2 miles, or 3.2 km) (for extensional examples, see Tearpock and Bischke 2003; for contractional examples, see Echavarria et al. 2003). Thus, in analogy with sedimentary rate data, thin growth sections are unlikely to contain many large or numerous hiatus (sequence boundaries, unconformities, or erosion), while thick growth sections are likely to contain more and larger hiatus. We are not familiar with any study that addresses these issues for normally faulted sections.

Table 11-1 Glossary of Terms

(β)	Initial or shallow dipping portion of the master fault (Fig. 11-38).
(α)	Dip of the deep portion of the fault beneath the crest of the rollover (Fig. 11-38).
(ϕ)	Change in normal fault dip angle.
(ρ)	Dip of Growth axial surface (Xiao and Suppe 1992).
(τ)	Angle that the rollover limb dips toward the master fault, defined by inflection points located near the crest and base of the limb.
R	Structural relief of the limb of the rollover, defined by inflection points and the crest and base of the limb of the structure.
T	Throw defined by the vertical displacement between the footwall and its equivalent hanging wall horizon.
(ψ)	Coulomb collapse angle (typically 67 to 70 deg).
Hn	Horizon number n = 1, 2, 3, . . .m.
Ghn	Total thickness of growth sediments for the base of Horizon Hn, which is equal the total depth(Zn) of the horizon being studied. The Ghn depth is taken relative to a datum depth, which can exist below the surface or sea floor.
Gn	Hanging wall horizon thicknesses.
Gf	Footwall horizon thicknesses.
Ei	Expansion index, the growth calculated from the hanging wall horizon thicknesses divided by the equivalent footwall horizon thickness (Gn/Gf).
FP	Parameter correlation coefficient of a linear regression through the calculated or measured thickness change ratios displayed on (Tn/Rn) vs depth(Zn) or Rn/Ghn vs depth(Zn) graphs.

Calculations and Examples

This section shows how to calculate and describe equations for pregrowth and growth structures. Pregrowth structures are discussed first. The following pregrowth and growth models contain both measured and calculated values (see definitions described in the next section, Measured and Calculated Values on Graphs). Only the most salient aspects of rollover formation are reviewed, and the details of rollover formation are covered by Xiao and Suppe (1992) and in the first five sections of Chapter 11 in Tearpock and Bischke (2003).

Pregrowth Structure and Calculations. Rolled-over beds (hanging wall anticlines) are caused by a *void* that develops between the hanging wall and foot wall of a normal fault, as the hanging wall beds slip (S) along the footwall fault surface (Hamblin 1965) (Fig. 11-39a). The weight of the hanging wall instantaneously closes the void caused by Coulomb shear collapse (ψ). The shear, which parallels the collapse angle, produces the rolled-over bed geometry (Hamblin 1965; Xiao and Suppe 1992) (Fig.11-39b).

In Figure 11-39b, the amount of the collapse required to fill the void is shown by the vector D, which parallels the collapse angle (ψ). In other words, the magnitude of vector D represents the amount of displacement required to cause the hanging wall to collapse onto the footwall (Hamblin 1965), closing the void. This collapse process forms the flank of the rollover, which dips toward the fault between the active and inactive axial surfaces and forms the structural relief (R) (Fig. 11-39b). The vector D is the vector Ax in Xiao and Suppe (1992). The structural relief is defined by the inflection points on the flank of the structure. The vector D vector is determined by the amount of slip (S) or throw (T), the Coulomb collapse angle (ψ), the change in dip (ϕ), and the dip of the fault (α) at depth as the hanging wall moves over the footwall (Fig. 11-39b). (Consult Table 11-1 for the definition of symbols.)

A graphical or inverse model solution to the structure consists of several steps. First, a copy of the modeled hanging wall is constructed from the fault surface, and the hanging wall slips over the footwall. Next, parallel to the dip of the shallow portion of the fault surface (β), project the amount of slip (S) below the bend in the major fault surface. Then, parallel to the inactive axial surface, add the D vector to the tip of the slip vector (S) (Fig. 11-39b). The D vector determines the position of the deep fault surface (α) and also determines the flank dip angle (τ) of the rollover.

Using the law of sines, a mathematical solution from Figure 11-39b is as follows:

$$(\sin(\pi - \alpha) + \psi)/S = (\sin(\phi))/D \qquad (11\text{-}15)$$

where

$$\sin(\pi - \beta) = Tn/S \qquad (11\text{-}16)$$

$\sin(\psi) = Rn/D$ (Figs. 11-39b and c) (11-17)

where

Tn = throw for Horizon Hn
Rn = structural relief of the flank from the base to the crest of the rollover structure for Horizon Hn, where n = 1, 2, 3, . . . m

Substituting Eqs. (11-16) and (11-17) into Eq. (11-15) and employing trigonometric identities eliminates π from Eqs. (11-15) and (11-16), and all of the angles are first-quadrant angles. This formulation yields

$$((\sin(\alpha + \psi)\ (\sin(\beta))/Tn) = ((\sin(\phi))\ (\sin(\psi))/Rn)$$

and rearranging yields

$$Rn/Tn = ((\sin(\phi))\ (\sin(\psi)))/((\sin(\alpha + \psi)\ (\sin(\beta)) \qquad (11\text{-}18)$$

or

$$Rn/Tn = f(\psi, \phi, \alpha, \beta) \qquad (11\text{-}19)$$

If $\psi = \beta$, then Eq. (11-18) simplifies to

$$Rn/Tn = (\sin(\phi))/(\sin(\alpha + \psi)) \qquad (11\text{-}20)$$

$$Tn = Rn/f(\psi, \phi, \alpha, \beta) \text{ or} \qquad (11\text{-}21)$$

$$Rn = (Tn)\ f(\psi, \phi, \alpha, \beta) \qquad (11\text{-}22)$$

where

α = dip of normal fault at depth
ψ = Coulomb collapse angle, or in this case, the active axial surface that emanates from the bend (ϕ) in the fault
ϕ = change in fault dip
β = dip of the normal fault at shallow depth

Equation (11-22) yields the Rn values if the Tn values are measured. Equation (11-20) is plausible, as the major fault dip (β) could initiate or fracture at a 67-deg to 70-deg angle, or for dips approximating the Coulomb collapse angle ψ [see Fig. 11-39c to calculate R/T for Eq. (11-20)]. A study of the shallow portions of large expansion faults using depth-corrected data had a range of values between 60 and 45 deg for β. Figure 11-40 was created using Eq. (11-18) for Figures 11-39b and 11-41, assuming 50-deg and 60-deg angles for β, $\alpha = 20$ deg, and $\phi = 40$ deg. Even with four sine equations, the results are nearly linear. The accuracy of Figure 11-40 can be checked against Figures 11-39b and 11-41. In addition, from depth-corrected data and knowing the footwall stratigraphy and fault shapes, from well control or seismic data, Eq. (11-18) can be used to predict the throw or structural relief across large pregrowth faults and thus the downthrown stratigraphy (Fig. 11-41).

A Pregrowth Structure

A pregrowth structure (Fig. 11-41) is distinguished from a growth structure in that the rolled-over beds form *a band or dip domain of uniform horizontal width* that typically converges toward a large bend in the listric fault. However, if the growth is a syntectonic or growth fault and the growth increases with depth, then the growth-faulted rolled-over beds grow wider with increasing depth (Xiao and Suppe 1992) (Figs. 11-38 and 11-44). In Figure 11-39b, notice that the bed is rolled over at the bend in the fault surface (ϕ). The active axial surface emanates from this bend and dips at angle of about 67 deg. The (ψ) surface is *active as shear along the* (ψ) surface propagates *into the hanging wall* and defines where the beds roll over (Xiao and Suppe 1992).

In the hanging wall, the vertical distance between the top, or crest, of the structure and the depth at which the dipping beds are near horizontal defines the structural relief (Rn). Let us assume in Figure 11-41 that (1) sands are present in the footwall at the top of horizon 5, (2) in the hanging wall and in the vicinity of horizon 5, the data are semicoherent and **cannot be correlated from the hanging wall into the footwall using reflection character**, and (3) all of the wells in the **hanging** wall penetrate only as deep as horizon 4. Which is the correct solution to the depth of horizon 5? To determine the correct horizon, geoscientists can measure the Rn/Tn ratios for horizons 5a, 5b, or 5c (Fig. 11-42).

A convenient way of illustrating the correct horizon position—5a, 5b, or 5c—is by projection and then displaying the Rn/Tn results on an Rn/Tn versus depth(Zn) plot (Fig. 11-42). Depth(Zn) is the total depth to the **base** of the Rn/Tn or Hn values, typically the subsea depths to the horizons or to a datum below the surface. These plots have been called Δd/d or throw (T) versus depth (Z) plots in the literature (Bischke 1994b; Pochat et al. 2004). Again, depth(Zn) is the total depths to the base of the throw(Tn) and structural relief(Rn) vectors. When applied to well log data, these plots have an accuracy of better than 5 ft/10,000 ft, or a 0.05% error, at a depth of 10,000 ft (3049 m) (Bischke 1994b). Seismic resolution accuracy at this depth is substantially lower than well log data (Sheriff 1980), and thus the error is higher. Geoscientists can determine seismic resolution from vertical Fresnel zone nomograms that are a function of frequency, that decreases with depth, velocity, and wavelength (Sheriff 1980).

Figure 11-42 is a Δd/d or structural relief(Rn)/throw(Tn) versus depth(Zn) graph for the measured data in Figure 11-41. Also, note that the *base of Horizon H1 is the top of Horizon H2.* Projecting the base of horizons H1 to H4 and the R2/T2 to R5/T5 data and to depth leads to the obvious conclusion that the Horizon H5b data point on the graph is the correct answer. Notice that the H5b horizon R5/T5 data matches Rn/Tn data for the top of horizons H2, H3, and H4. Also, notice that the R5b/T5b data point has a unique depth(Zn) position on the graph of 9.05 units (Fig. 11-42). This follows, as the Rn/Tn measurements for the top of horizons H2 to 5b all have the same values of Rn/Tn = 0.80 (Figs. 11-41 and 11-42). The Rn-square or FP correlation values (see Table 11-1 for definition) for these horizons is 1.0. This makes sense, as the *pregrowth sequence was deposited prior to the faulting and the faulting is post-depositional* (the Tn values in Fig. 11-41 all have the same value). Thus, all of the horizons must have the same throw(Tn) values, in this case equal to 3.55 units. A plot of the throw(Tn) values versus depth(Zn) would confirm these conclusions. The calculated Rn/Tn value is 0.80, which is calculated from Eq. (11-18).

Figure 11-40 is a Rn/Tn versus depth(Zn) graph, which has properties similar to those of Δd/d plots in that *outlying points constitute an anomaly*. The cause of an anomalous point must be investigated, and this technique is discussed in detail in Chapter 13. The possibility that the horizon points H5a and H5c are the result of variable geology are unlikely. This conclusion follows, as the horizons were continuous when deposited and were deposited in a near horizontal position prior to the faulting. Excluding the possibility of a channel, stratigraphy does not change across a nonexistent fault. The possibility of a channel can be checked on adjoining lines. A possible but less likely explanation is that a major unconformity terminates by coincidence at the fault surface. Again, this explanation is unlikely, as the nonexistent fault was not moving and could not represent a buttress to onlapping sediments. Also, adjacent lines may resolve this possibility. Another option is to create a figure similar to Figure 11-42 using different depth(Z) values, such as points H5a and H5c, which do not match shallower Rn/Tn = 0.8. If the problem is an *improper depth correction and a velocity error*, then the velocities can be checked and *the graph will help correct the incorrect velocities* to a proper value. The *correct depth* will correspond to the depth in which Rn/Tn = 0.8. Lastly, another option exists when confronted with an anomalous situation in which it is not possible to correlate from the footwall into the hanging wall using seismic reflection character. In pregrowth structures, as the hanging wall along the fault surface must fit back into the footwall, structural balancing or, more specifically, retrodeformation may help (see Chapter 10, Fig. 10-10).

The next plot (Fig. 11-43) is based on Figure 11-41 and contains a deep single bend (ϕ) below Horizon H5. As the beds have not moved through the bend in the fault, the data points shown in Figure 11-43 are *impossible in uniform thickness pregrowth strata*, as the downthrown and upthrown beds must have the same thicknesses and thus the same Rn/Tn values. This interpretation of the data is possible as a result of poor seismic data or poor stratigraphic correlations. The Rn/Tn versus depth(Zn) graph (Fig. 11-43) looks reasonable except for the FP calculation of 0.357. In addition to the low FP value, the Rn/Tn plot in Figure 11-43 looks more random than the data in Figure 11-42. Assuming that the best linear fit is the most probable answer, possible problems exist in horizon correlations H4 and H5. These horizons appear more anomalous than correlations H2 and H3, which could be better. If there are any questions as to data anomalies, the following ratios can be graphed. Ratios, such as Rn/Tn ratios, are more sensitive than direct depth measurements such as the throw (T), which in this case, because the faulting is post-depositional, has a uniform value of 3.55 units (Fig. 11-41).

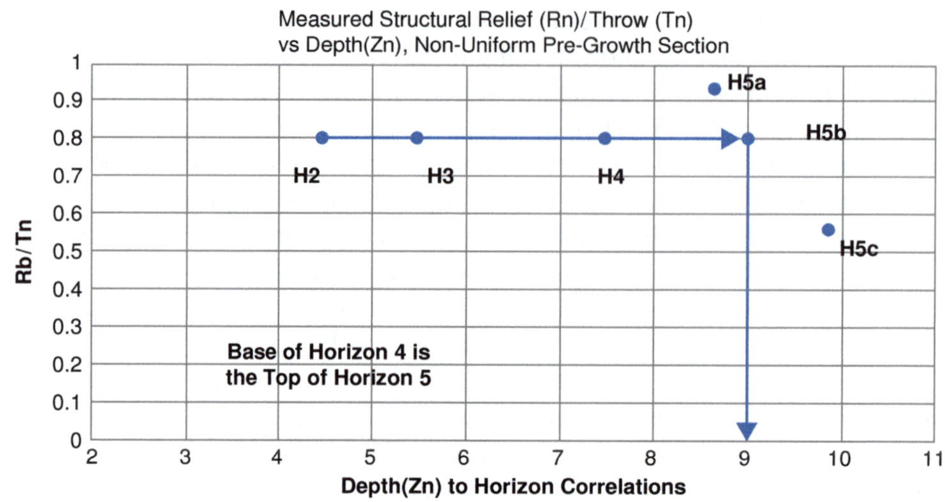

Figure 11-42 A structural relief(Rn)/throw(Tn) versus depth(Zn) plot for the uniform R/T data posted on Figure 11-41. The top of horizons H2 to H5b have the same R-square value of 1.0. From this figure, horizon 5 has a unique depth(Zn) value of 9.05 units. Horizons H5a and H5c represent potential miscorrelations. (Published by permission of R. Bischke.)

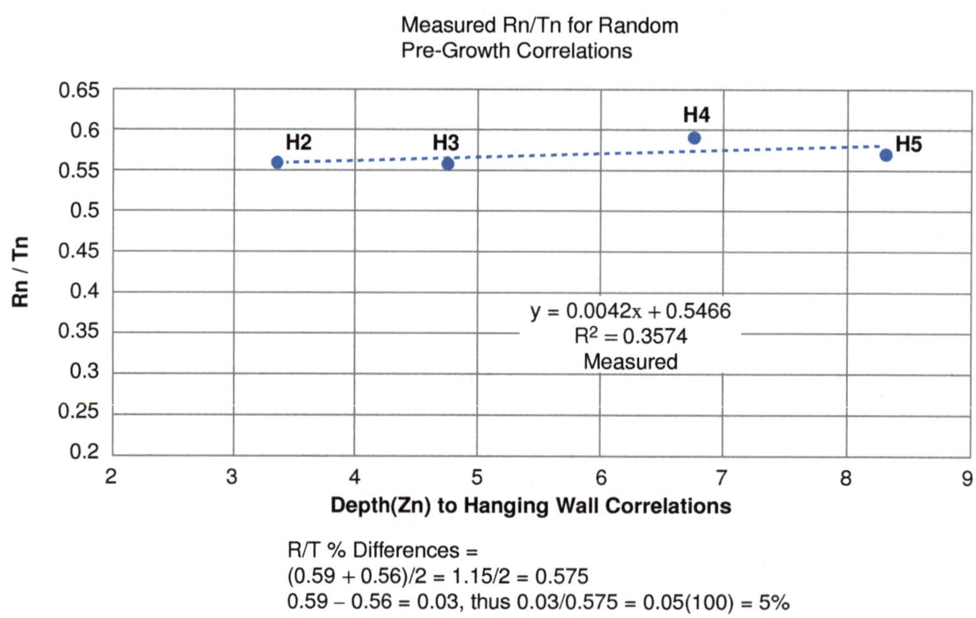

Figure 11-43 Nonuniform sedimentation between the hanging wall and footwall cannot exist in pregrowth structures. This follows, as the marine sediments are typically deposited at low dip angles, and the faulting is post-depositional. The low linear regression R-squared result of 0.357 indicates that there is a problem with the interpretation. (Published by permission of R. Bischke.)

Accuracy of Methods

One way to address accuracy or correctness is to calculate the FP and R-squares of the data. Another is to calculate the % differences of the data. Percentage differences have the property that the order of the numbers do not matter. The % difference takes the averaged sum of the largest and smallest horizon thicknesses (Gn) (Fig. 11-44), then divides this number into the horizon thickness differences between the largest value minus the smallest value multiplied by 100. As there are no horizon thickness differences in the correct values measured on Figure 11-42, the

% difference = 0.0, which means that the data have zero variations. The following discussion on growth and real structures contains % differences that are much larger. Large % differences can affect the accuracy of any projection method, but methods exist to recognize, address, and mitigate large % differences in real structures. (See Fig. 11-45 for the general equation).

The calculated values for graphs use Eqs. (11-18) and (11-28) to (11-30), and the measured values often used higher-imaged scales (200% to 400%), a magnifier, and typically a 30 scale (1/30 of an inch). Thus, the measured values have errors of up to 0.0167 in. = $(1/(30 \times 2))$ = (0.042 cm) (the viewing scale is shown as *a percentage above the measured scale at 1/60 inches*).

Growth Structure and Calculations

Figure 11-44 shows a modeled growth structure labeled to calculate growth on a simple fault (0 is on the left, and π is on the right). Measurements are taken from the horizontal, and ρ can exceed 90 deg, but in the formula $\pi - \rho$ is less than 90 deg. As described in the section Details for Constructing Triangles, the object of the following calculations is to create a triangle often, but not always, bounded by an axial surface dip ψ. In this case, the triangle is formed by the vector Mn, flank dip vector Nn, and ρn (see Fig. 11-44).

Down to the Right Faults. The following formula is for a down to the right fault, or one that flattens to the right. Thus, if by convention, 0 is next to the shallow portion of the master fault, and π is above the deeper portion, then down to the right faults satisfy Eq. (11-28) (see Table 11-1 for definitions of terms). In this case, for $(\rho n > \psi)$ and from the law of sines and Figure 11-44,

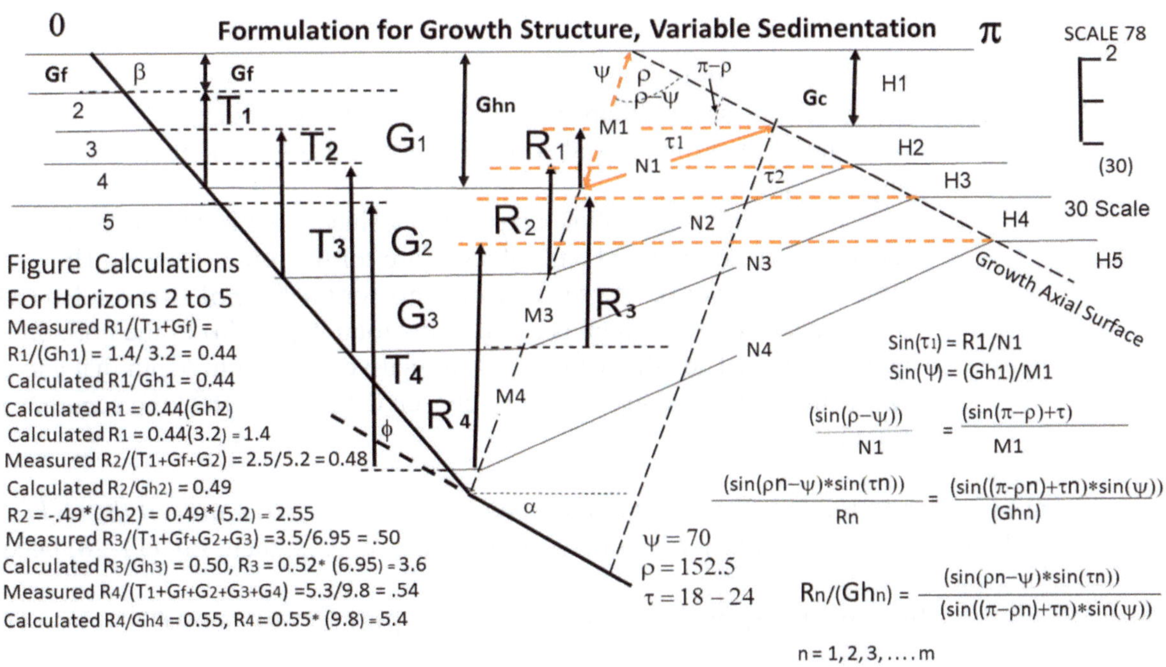

Figure 11-44 Figure used to calculate growth formula for variable sedimentation. The M1, N1 portions of the triangle are colored in orange, and the law of sines formulation for growth structures is shown for the M1 and N1 vectors on the right side of the figure [Eq. (11-28)]. For clarity, the rounded measured and calculated values for the M1–N1 to M4–N4 vectors are posted on the left side of figure. (Published by permission of R. Bischke.)

$$(\sin(\rho - \psi))/N1 = (\sin(\pi - \rho) + \tau)/M1 \tag{11-23}$$

where

$$\sin(\tau 1) = R1/N1 \tag{11-24}$$

$$\sin(\psi) = (T1 + Gf)/M1 = (Gh1)/M1 \tag{11-25}$$

where, as before, R1 is the structural relief for Horizon H1, but (T1 + Gf) is the *throw plus the sum of the growth section* from the base of the horizon being calculated to some datum located at or below the sea floor (Fig. 11-62a). Similarly, Gh1 is the total sedimentary or growth section for the horizon being studied to a datum, which in this case is from the base to the top of Horizon H1 (Fig. 11-44).

Substituting Eqs. (11-24) and (11-25) into Eq. (11-23) yields

$$R1/Gh1 = (\sin(\rho - \psi)*\sin(\tau 1))/(\sin((\pi - \rho) + \tau 1)*\sin(\psi)) \tag{11-26}$$

$$R1/Gh1 = f(\psi, \rho, \tau) \tag{11-27}$$

However, in general and as the horizon deepens and the flank dip often increases, the total Depth(Zn) of the **base** of Horizon Hn increases and thus, in general, from Figure 11-44,

$$Rn/Ghn = (\sin(\rho n - \psi)*\sin(\tau \cdot n))/(\sin((\pi - \rho n) + (\tau n))*\sin(\psi)) \tag{11-28}$$

where

n = 2, 3, 4, . . . m for top of Horizon Hn
ψ = Coulomb shear collapse angle, typically 67 to 70 deg
ρn = dip of the growth axial surface in Figure 11-44, which in this case is the angle $\pi - \rho$
τn = frontal dip of the rollover structure defined from inflection points, located near the crest and base of structure for the top of Horizon Hn
Ghn = total thickness of growth sediments for the base of Horizon Hn (which has a thickness Gn), which is equal to the total depth(Zn) of the horizon being studied. The Ghn depth is taken relative to a datum depth that can exist below the surface or sea floor. The chosen datum can affect the accuracy of Eq. (11-28).

One exception to Eq. (11-28) is that *in multibend or nonlinear structures, if the right side of the triangle is not* ρn *and is replaced by* ρm, *where m = a, b, c, . . .z,* then, Eq. (11-28) becomes

$$Rn/Ghn = (\sin(\rho m - \psi)*\sin(\tau \cdot n))/(\sin((\pi - \rho m) + (\tau\, n))*\sin(\psi)) \tag{11-29}$$

(See sections The Growth Axial Surface Increases Dip and The Growth Axial Surface Decreases.)

Formula for Down to the Left Faults. This equation follows the same formulation as Eq. (11-28), and its derivation is also

$$Rn/Ghn = (\sin(\rho n - \psi)*\sin(\tau\, n))/(\sin((\pi - \rho n) + \tau\, n)*\sin(\psi)) \tag{11-30}$$

but with *0 on the right.*

Accuracy of Eqs. (11-28), (11-29), and (11-30). Sine functions can be sensitive, and the datum can be adjusted to more stable angles. The ψ, or the Coulomb collapse angle, is a constant, and the dip of the limb dip (τn) range is not as large as the ρn angle range. Thus, the calculation accuracy of Eqs. (11-28) to (11-30) is mainly dependent on ρn. If the ρn − ψ dip angles are less than about 5 to 10 degrees, then if possible, use a datum that results in ρn − ψ angles that are between about 45 to 130 degrees. This procedure will result in lower calculated ρn − ψ errors and thus in better Rn/Ghn values.

Details for Constructing Triangles. In Figure 11-44, vector M1, which is an active axial surface ψ for Horizon H1, the growth axial surface ρ1, and the dip of the flank N1 form a triangle. Vector M1 in Eq. (11-25) defines Gh1/sin(ψ). Also, N1 = R1/sin(τ1) [Eq. (11-24)]. As long as the growth axial surface, ρn, and Nn values do not change dip, the angles do not change. According to the general Eq. (11-28), R3/Gh3 is defined by angle (ρ − ψ), and length of vectors M3 and N3 (line N3 dips at τ3 degrees) (Fig. 11-44). It is important to remember that when the horizons or the growth axial surface (ρn) changes dip, the *calculations require the formation of new triangles, with new angles for each horizon studied.* In this case, the τ1 angle dips more gently at the crest of Horizon H2, and the dip increases with increasing depth.

When analyzing a seismic line, geoscientists can predict whether the depth-corrected structure is near linear by observing the dip of the growth axial surface ρ. If ρ appears to be near linear, then the calculated and measured data points are likely to generate a near-linear line when processed by a linear regression option in a spreadsheet. Thus, along some seismic lines a straight line, or a nearly straight line ρ, can be drawn on the seismic line at the crest of structure (red lines on Fig. 11-62a). The ρ lines, although not directly entered into the equations, are **part of the visual data**.

As before, Rn is the structural relief, and Ghn includes the total thickness of growth sediments for Horizon Hn to a datum located at or below the sea floor as described previously. Figure

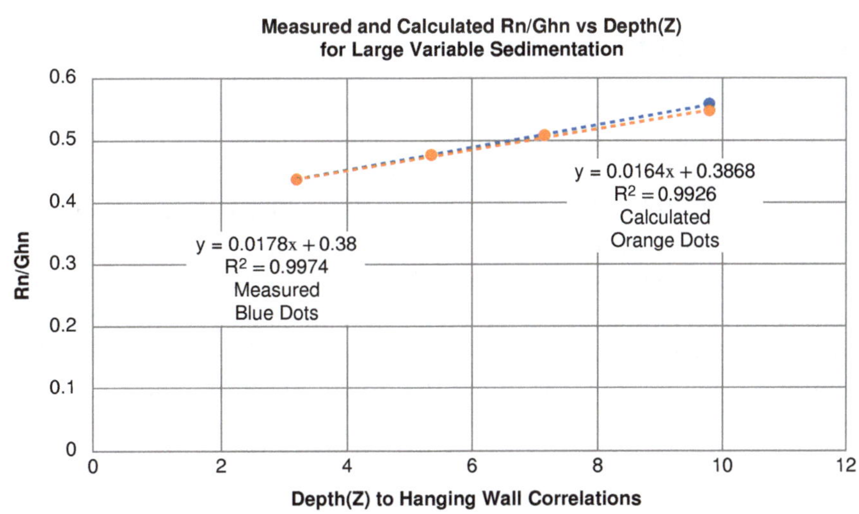

Gn % Difference = (Largest horizon Gn thickness + Smallest horizon thickness)2.0, next (Largest horizon Gn thickness − smallest horizon thickness). Final value ((Largest horizon Gn thickness − smallest horizon thickness)/ (Largest horizon Gn thickness + Smallest horizon thickness)2.0) × 100

Gn % Difference = (3.25 + 1.75) = 5.0/2 = 2.5
(3.25 − 1.75) = 1.5/2.5 = 60%

Figure 11-45 A Rn/Ghn plot versus total depth(Zn) for the Figure 11-44. The measured and calculated values for variable sedimentation show near-linear growth. The % difference values are high with a 60% variation. Thus, the R-squared result for the measured values is 0.997, but the calculated result is only near linear at 0.993. (Published by permission of R. Bischke.)

11-44 is a high-growth rollover type of model resulting from changing Ghn and Gn thicknesses, and changes in the frontal rollover dip angles τn. The changing τn dip values could be caused by sedimentary onlap. The variable sedimentation Rn/Ghn plot (Fig. 11-45), generated from Figure 11-44 data, nevertheless exhibits a near-linear calculated and measured growth versus depth plot. Every point on the plot on Figure 11-45 represents a new formula formed by new values in the equation. On this plot, the Gn % difference is 60%. Calculated values, using Eq. (11-28) may generate larger R-square values than measured values, but not always. PF coefficients (see Table 11-1 for definition) typically get better at lower % differences calculations ranging from below about a 50% to 70% difference. Thus, R-square variations reflect the linearity of the equations.

Measured and Calculated Values on Graphs. For measured values, the figures were made larger at a scale of 200% to 400%, and the Rn/Tn values are measured directly with a scale ruler and a magnifying glass, resulting in a measurement error of about 1/60. For syntectonic structures, defined by Eqs. (11-28) to (11-30), the calculated and measured values are compared to each other on the graphs. The posted **calculated values** on all of the graphs use angular measurements defined from the right-hand side of Eqs. (11-28) to (11-30) (calculated values on Fig. 11-45). Notice that from the left-hand side of Eqs. (11-28) to (11-30) is the ratio

$$\text{Rn/Ghn} \tag{11-31}$$

This ratio is measurable. Thus, the **measured values** for Eq. (11-31) are first obtained by measuring the lengths of Rn and Ghn. The left-hand side of Eqs. (11-28) to (11-30) is then determined from Eq. (11-31) by dividing Rn by the measured value of the total growth (Ghn), or depth(Zn), from a chosen datum to the horizons being studied. The Ghn = depth(Zn) values are much larger than the Rn values.

Syndepositional Growth Models

In nature, syndepositional growth structure is more common than pregrowth structure and is structurally more complex. For example, geoscientists have recorded growth on normal faults using expansion indices (Thorsen 1963) or throw (Tn) versus depth(Zn) diagrams (Bischke 1994b; Sanchez et al. 1997; Tearpock and Bischke 2003). Expansion indices (Ei) typically compare downthrown horizon or sequence package thicknesses (Gn) divided by the equivalent upthrown sequence thicknesses, here defined as Gn/Gfn (see Fig. 11-46 for the definition of depth(Zn) and Gn/Gf). An alternative is to correlate thickness on the crest of the rollover structure to upthrown thicknesses, here defined as Gc/Gf (Fig. 11-46, right-hand side). If the growth is roughly uniform, then Eq. (11-18) in the Pregrowth Structure and Calculations section can be used to confirm the correlations. If the growth axial surface changes rapidly, then Eq. (11-28) is modified to correct for the change in growth (Fig. 11-52).

This discussion starts with the simplest models first, by developing models that have uniform or a small amount of growth. For this low-growth analysis, the late and rapidly growing sediments infilling the most recently deposited sediments are not included in any calculation. For example, if the thickness of Gf is about 0.0 in the models, then the expansion index calculations Gn/Gf would be near infinite, making the % difference evaluations useless. Thus, Horizon H1 (Fig. 11-46) will be included only when investigating large-growth sedimentation models. The initial models will concentrate on simple uniform growth. These models have a straight-line growth axial surface (ρ) (Xiao and Suppe 1992) and where no variations in sedimentary thicknesses exist. This assumption permits for a simple understanding of the methods, which are then made more realistic. *Typically, lower R-square values indicate a more complex model, and thus R-square values are reported in the text.* Understand that the R-square values reflect the correlation calculations!

Uniform Growth

In the simple uniform sedimentation model (Fig. 11-46), the measurements and calculations start at the top of Horizon H2. In this model, the downthrown stratigraphic horizon thicknesses do not change, and the growth and corresponding Rn/Tn measurements are linear. The geometry of syndepositional structures expand, and the horizontal width of the structure grows wider with depth. The growth axial surface (ρ) defines the top or crest of structure (Xiao and Suppe 1992) and is a constant in this model (Fig. 11-46). If (ρ) appears to be linear, as in Figure 11-46, then the linear regression results will exhibit high R-square values, and the growth will be linear. This occurs in real or modeled data. If the growth axial surface does not change dip, and the horizons (Gn) do not change thickness, then the expansion index Gn/Gf does not change and the growth is uniform. Again, the vertical distance between the top, or crest, of the structure and its flank, which is the depth at which the beds no longer dip, define the structural relief (Rn). Thus, R1 is the vertical distance between a horizontal line drawn at the crest of structure (that terminates at the position of the growth axial surface) and a point on the flank of the structure where the horizon is roughly horizontal. Again, we assume that the position of the top of Horizon H6 is not clearly known (Fig. 11-46).

Figures 11-47a and b are an Rn/Tn and Rn/Ghn versus depth(Zn) plot. The mathematical formulation of expansion faults is covered in the Pregrowth Structure and Calculations section. *Notice that the Rn/Ghn plot (Fig. 11-47b) does not require footwall data, as Ghn exists only in the hanging wall.* The measured structural relief(Rn)/throw(Tn) and calculated Rn/Ghn values are near linear with an R-square very close to 1. Measurement errors are the cause of the imperfections. The misinterpreted horizons 6a and 6c clearly deviate from the more linear trend on Figure 11-47a. Ratios, such as Rn/Tn versus depth(Zn) graphs, are more sensitive than direct measurements of Rn and Tn, both of which can contain minor errors. Studies of the models generally

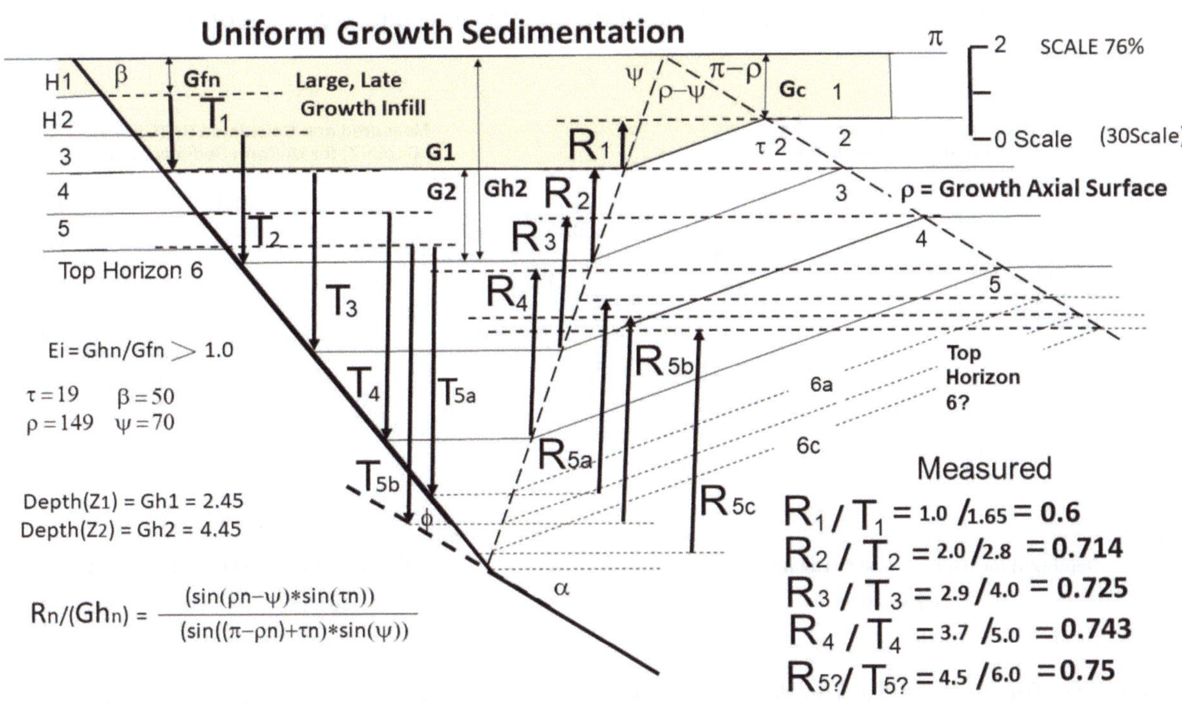

Figure 11-46 This figure is a growth or syndepositional structure that exhibits uniform growth. On growth structures, the throw(Tn) and structural relief(Rn) values increase with depth as the limb of the structure widens with increasing depth(Zn). Accordingly, the Rn/Tn values also increase with increasing depth. The equation on the figure is the growth equation for a uniform growth axial surface (ρ). (Published by permission of R. Bischke.)

show that the *structural relief(Rn)* versus depth(Zn) graphs tend to have the highest R-square results from the linear regression. Figure 11-47b contains the same data plotted in Figure 11-46, but the data are *calculated* using Eq. (11-16) and angles. *Growth measurements* (Ghn) is always measured from the datum for horizon Gn to the base of the horizon being studied [Fig.11-50 and Eq. (11-13)]. In this data set, the direct measurements result in an R-squared of 0.9919 (Fig. 11-47a). The calculated data from Eq. (11-18) result in an R-squared = 1.0 (Fig. 11-47b). The result is constant, as the angles do not change between horizons. The measured values result in an R-square of 0.9, which results from direct measurement errors (consult Measured and Calculated Values on Graphs for definitions of measured versus calculated values). Data with a positive slope on these Rn/Tn diagrams means that the structure is growing larger with depth(Zn). As a result of this growth, the sedimentary wedge grows thicker. On Figure 11-47a, the total depth to the top of horizon 6b is 10.3 units. The position of the R5/T5/base depth(Z5) (which defines the top of horizon 6) value has a *unique position on the linear, calculated uniform growth graphs* (see the Gh1 = 2.45 and Ghn = 4.45 calculated depths on Figure 11-46). Thus, by projection, the top of Horizon H6b is the correct horizon top for uniform growth, and not horizons H6a or H6c.The data set need not be perfectly linear to conduct this projection exercise. An R-square near or at 1.0 means a high correlation is involved in the projection process.

In normally faulted structures, delta (Δd)/d (change in depth/depth) plots of stratigraphic sequences are also often near linear, indicating near-linear stratigraphic growth within sedimentary sequences (Bischke 1994b; Sanchez et al. 1997; and Bischke et al. 2006). Thus, near-linear growth is not uncommon (Tearpock and Bischke 2003) (Figs. 13-8, 13-9, 13-11, 13-12 13-14, 13-17, and 13-21).

The graph represented in Figure 11-48 is similar to the graph shown in Figure 11-46, except that the displacements are larger and horizons H7, H8, and H9 have moved through the active axial surface into an inactive zone. In this case, these horizons have moved through the bend in the fault bend defined by (ϕ), and these horizons dip onto the deep portion of the fault defined

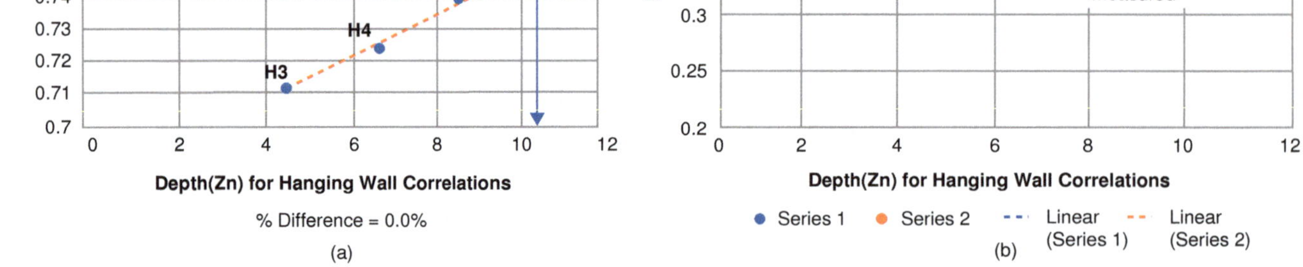

Figure 11-47 (a) The measured Rn/Tn values are linear correlation for horizons H3 to H6b and have a high R-squared value. The depth(Zn) value is measured from the base of Horizon Hn to the top of the growth. (b) The measured Rn/Ghn values are not as accurate, which is not typical. However, the calculated linear regression R-squared results, using Eq. (11-28), are 1.0 for H3 to H6b. (Published by permission of R. Bischke.)

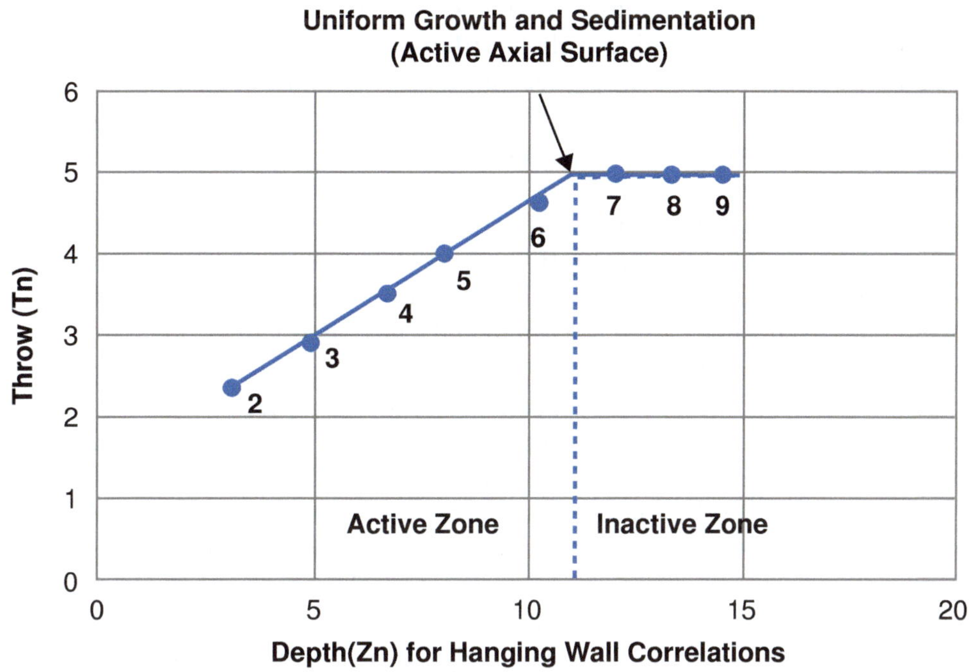

Figure 11-48 This model is very similar to the model shown in Figure 11-46, but the strata pass through the active axial surface into the inactive zone to the right of the active axial surface. No deformation occurs in the inactive zone. (Published by permission of R. Bischke.)

by α (Fig. 11-46). Notice that the calculations for throw(Tn) on Figure 11-48 for horizons H7 to H9 do not change and have the same values. The constant values mean that the region to the right of the active axial surface is *inactive and lacks any deformation.*

Low Variances in Growth Sedimentation

Similarly, if the sedimentation differences have relatively low slopes, then the growth plots are nearly linear (Fig. 11-49) and are similar to Figure 11-47a. As the growth approximates uniform growth, the calculated Rn/Tn values have an R-square = 0.99, and thus this model, like the previous model, is relatively simple. Hereafter both the growth and sedimentation variances are posted on Figure 11-49 and on all of the remaining higher-growth figures. The following discussion on growth contains % differences that are much larger than low-growth sedimentation differences.

The Rn or Tn % differences have no meaning on growth structures, as the Rn or Tn values often increase and have a larger value with increasing depth(Zn). However, the horizon thicknesses Gn or the Ghn/Gfn values can define differences in sedimentation thicknesses that affect growth and its linearity. Thus, Gn % differences will define sedimentation variations on growth structures. On Figure 11-49, the Gn % differences have a 27% variation.

High Variations in Growth

Thorsen (1963) showed from correlations of well log data that expansion index Ghn/Gfn defined displacements on listric growth faults. The growth typically increases with increasing depth, reaches a maximum, and then decreases (Fig. 13-3). Thus, growth on listric faults is *typically not random*. Nevertheless, we test a nonuniform, or random, sedimentation model on a relatively high horizon thickness Gn, hanging wall % difference data (Figs. 11-50 and 11-51a and b) scenario. As a result of the large % difference values of 113%, the Rn/Tn data are less reliable with a measured R-square of 0.9335 (Fig. 11-51a) that indicates lower correlations, but this R-square value could indicate a less certain interpretation.

However, the calculated Rn/Ghn plot (Fig. 11-51b) is near linear with a correlated R-square result = 0.988 even though the Gn % difference variance is 113%. When the % difference data exceed about 50% to 70%, the correlation of the data may or may not significantly decrease. Also, posted on Figure 11-50 are the rounded measured and calculated values for R2/Gh2 and R2/Gh3 and the Rn/Tn values. (See Measured and Calculated Values on Graphs for the definition of calculated and measured values).

Figure 11-51c tests the accuracy of the correlation, projection method assuming that the position of the top of Horizon H6 is unknown. This is accomplished by excluding Horizon H6 data from the data set and using only the H2 to H5 data on the plot. The recalculated Rn/Ghn values are shown in Figure 11-51c (notice that a comparison of Fig. 11-51b and c show slight differences in the y, Rn/Ghn, and R-square values). Thus, it is possible to predict the top Horizon H6 by evaluating the two y/x equations in Figure 11-51c. Simply insert the x = depth value for the top of Horizon H6, which is 9.25, into the two equations (Fig. 11-51c). The measured and calculated solutions are 0.49 and 0.476 respectively. Next, the *relative % errors* can be determined from the results of the two plots by (1) taking the measured and calculated values from Figure 11-51b, then (2) subtracting the values from the Figure 11-51c results, and (3) determining the % relative error by multiplying the difference by 100. The measured and calculated relative % errors are 1.8% and 0.085% respectively (Fig. 11-51c).

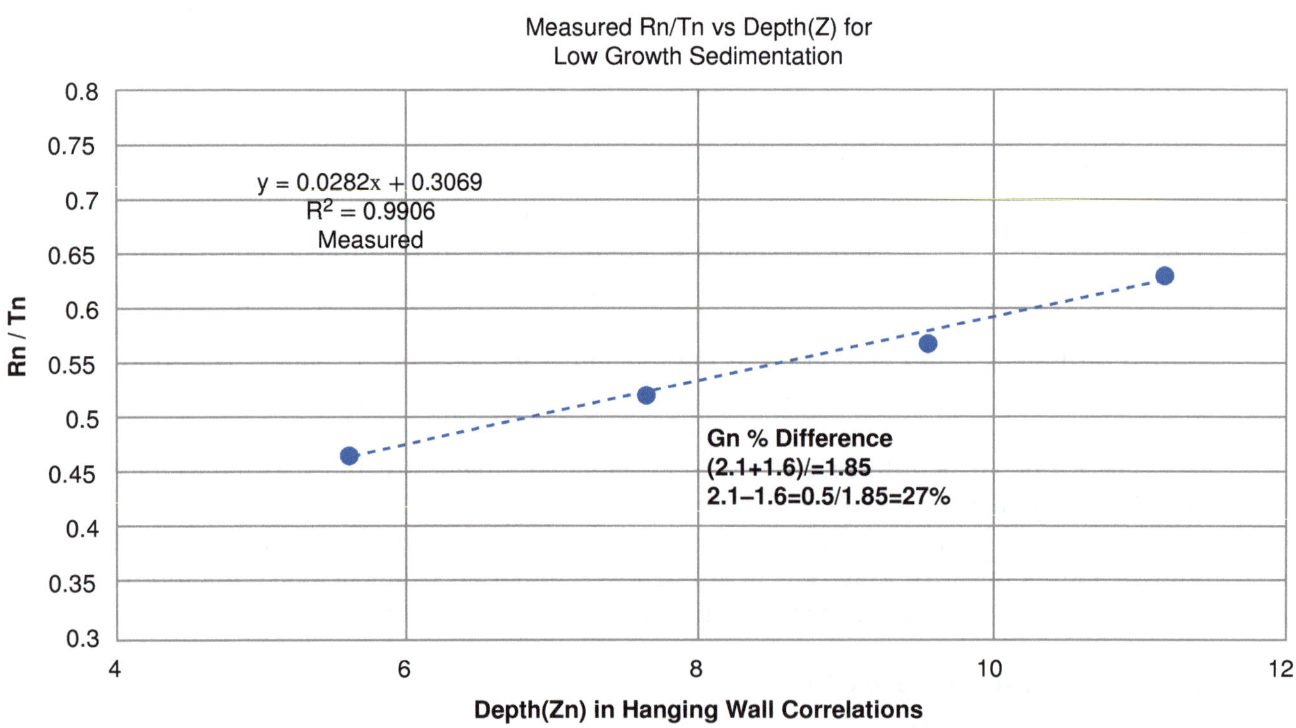

Figure 11-49 If the growth does not vary significantly, the results are similar to uniform growth, but the % differences in this low-growth model increase to 27%. (Published by permission of R. Bischke.)

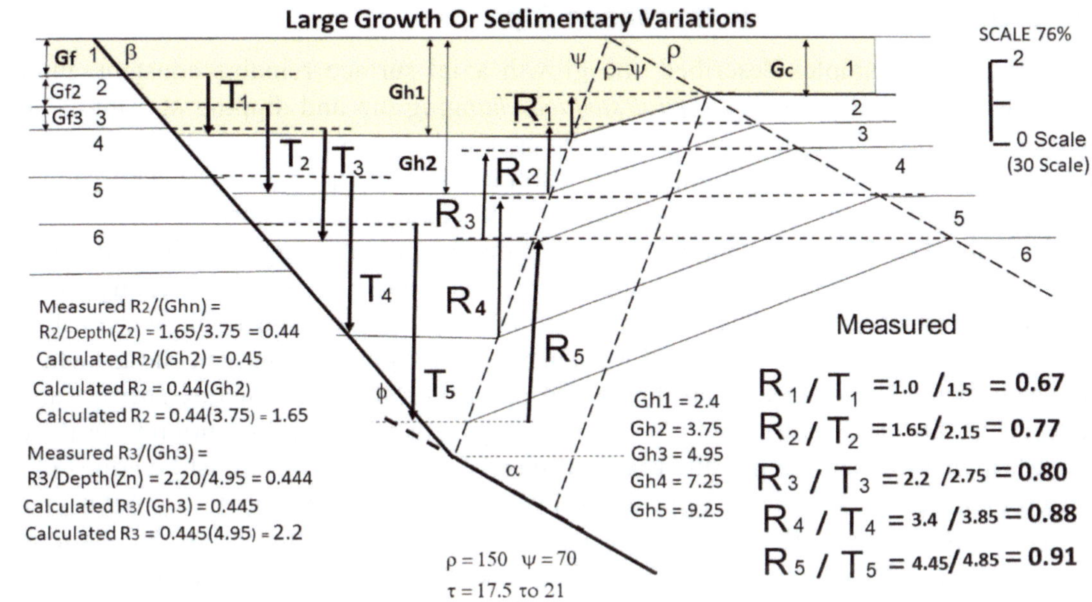

Figure 11-50 A model showing nonuniform growth sedimentation, which is not typical of many structures. Measured and calculated values for horizons H3 and H4 (R2/Gh2 and R3/Gh3) are posted on the left-hand side of the figure. (Published by permission of R. Bischke.)

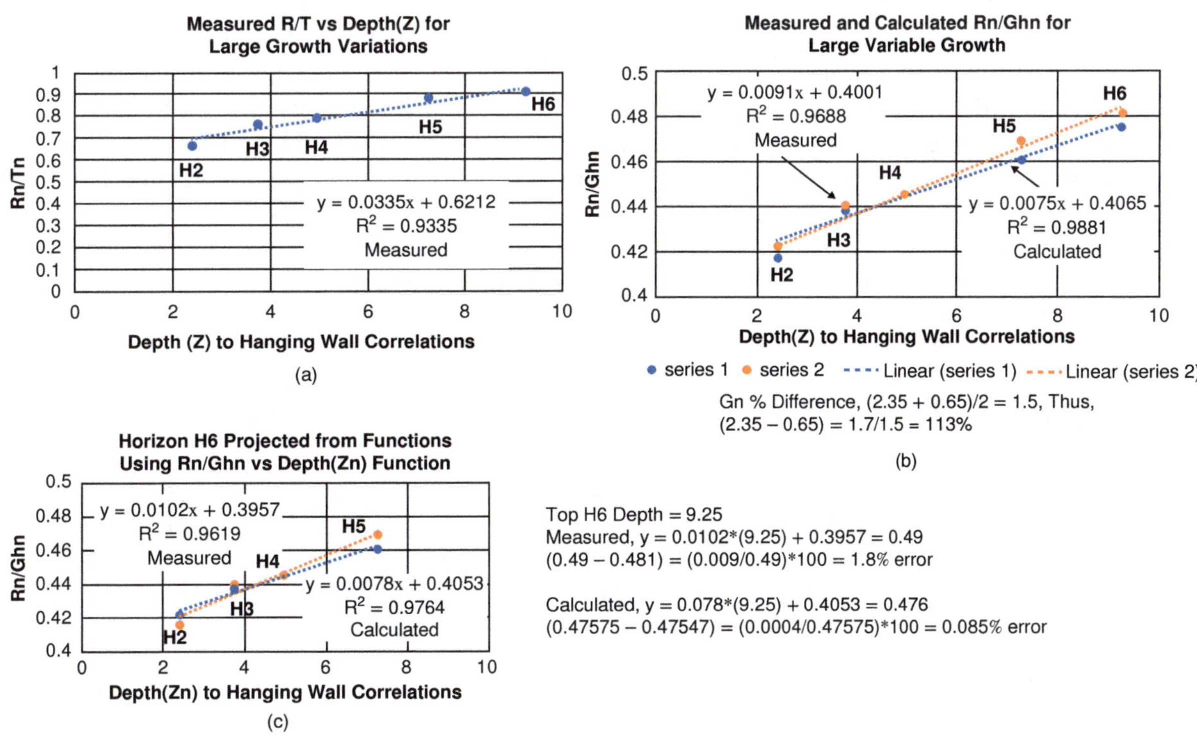

Figure 11-51 (a) As a result of more random sedimentation, the measured Rn/Tn data are less reliable and have a lower R-squared than more uniform growth. (b) However, the calculated data using Eq. (11-28) is more reliable with a linear regression result of R-square 0.988. When the % difference is around 50% to 70% or greater, with the exception of increasing growth, the linear projection assumption maybe less reliable. (c) Test of the projection method assuming that the location of the top of Horizon H6 is unknown. Measured and calculated R6/Gh6 values are shown along with the relative errors. (Published by permission of R. Bischke.)

The Growth Axial Surface Increases Dip

In the modeled examples described, the growth axial surface remains constant. Figure 11-52 shows an example of the *growth axial surface changing* dip and dipping at a lower angle, thus increasing the growth-angle (ρ1) dips at 139 deg and then dips at ρ2 = 153 deg at greater depths. In Eq. (11-28), the ρ1 and ρ2 angles are subtracted from (ψ) = 70 deg. Thus, there are two sources of changing growth in Figure 11-52: the change in the horizon thicknesses and the change in the dip of the growth axial surface. A more gently dipping growth axial surface means that the growth decreases (Xiao and Suppe 1992). On seismic sections, this change in the growth axial surface is recognized by a *shift in the position of the crest* and flank of the rollover structure. In this case, the crest of the rollover shifts to the right. If, however, the growth increases, then the growth axial surface bends downward, and the crest of the structure narrows and shifts to the left (Fig. 11-54). Furthermore, this change in growth creates a discontinuity in the data (Fig. 11-53a), indicating that the **problem is nonlinear**. This nonlinearity must be adjusted for in the measurements and calculations. Typically, solutions to nonlinear problems are not trivial.

This topic is also briefly discussed in the Pregrowth Structure and Calculations section, where a detailed derivation of Eq. (11-18) is presented. The following is a description of how to calculate the right-hand side of Eq. (11-28) to adjust for nonlinearity. Notice on Figure 11-52 that the growth axial surface (ρ1) for the crest of horizons H1 and H2 does not change position. Thus, (ρ1) does not change in the calculations for the first two horizons. The solution for the top of Horizon H3 in Eq. (11-28) is to form a triangle that includes the angles (ρ1), ψ, and the value of the bed dip τ2 [use radians in Eq. (11-28) for a computer-generated solution].

At the crest of Horizon H4 (or the bottom of Horizon G3), the growth axial surface abruptly changes dip from ρ1 = 139 to ρ2 = 153 deg, and this change in the growth axial surface shifts the position of the crests of horizons H4 and H5 to the right. In order to adjust for this nonlinear behavior, the triangle formed does not include the growth axial surface (ρ2) but rather the angle ρa = 142.5 deg. Thus, in Eq. (11-28), the *ρm replaces the angle ρn*, and Eq. (11-29) contains the term ρm − ψ in the numerator (consult Fig. 11-52 for variable *ρ values*). The lengths M3 and

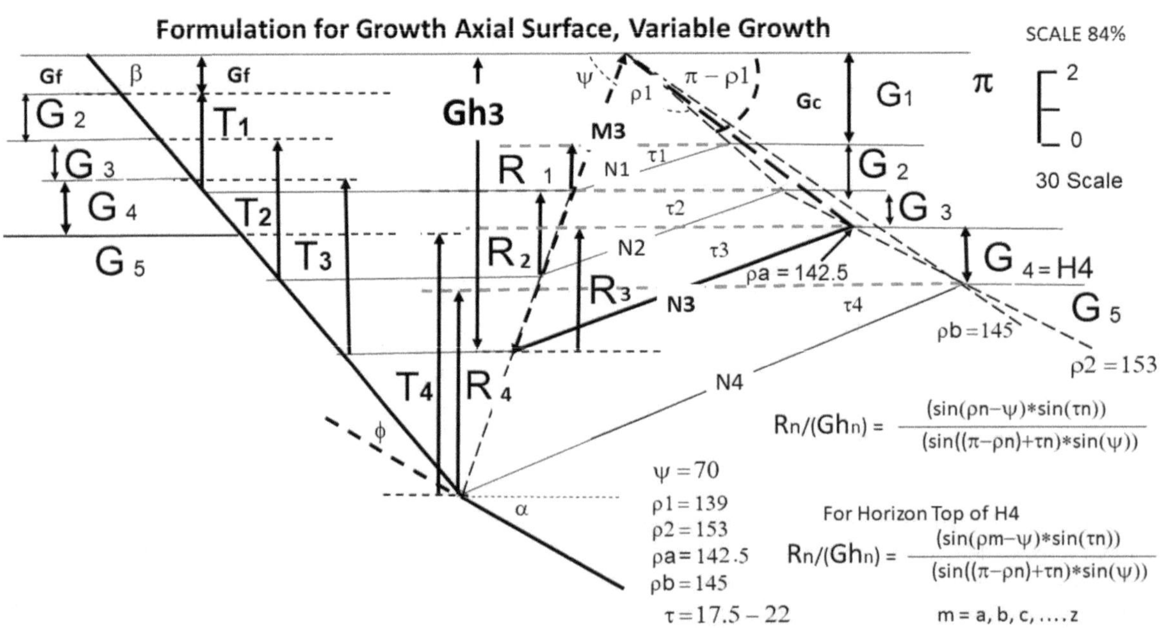

Figure 11-52 As the growth axial surface (ρn) changes, the problem is nonlinear. (Published by permission of R. Bischke.)

N3 and the angle $\rho a = 142.5$ deg are shown in bold on Figure 11-52. Also, notice that for the base of Horizon H4, the value of the growth axial surface changes to $\rho b = 145$ deg, both N4 and M4 lengthen, and $\tau 4$ increases. This change must also be incorporated into Eq. (11-28) [see Eq. (11-29)], and the values can change with every horizon. However, this procedure corrects for the nonlinearity of the growth.

Figure 11-53a and b shows the measured and calculated results from Figure 11-52 for an Rn/Tn and Rn/Ghn versus depth(Zn) graphs. Figure 11-53 has a relatively large Gn % difference of 61% and a growth axial surface that changes from 139 deg to 153 deg, which has the result of decreasing sedimentary growth. These large variability changes have the effect of creating a distinct discontinuity in the tops of horizons H3 and H5 on the Rn/Tn graph (compare the tops of H2 to H4 and H4 to H5 on Fig. 11-53a). In spite of the discontinuity on Figure 11-53a, the Rn/Ghn R-square valves are near unity.

Increasing or decreasing variable growth is present on many rollovers that exhibit a divergent bed dip pattern and is not uncommon (Fig. 11-38). However, real structures can have a high degree of correlation similar to the modeled values. The number of different-sized triangles can vary for each sedimentary sequence on some rollover profiles. *To solve for nonlinear growth, one simply recognizes where the growth axial surface is located on the seismic section.* The growth can increase or decrease several times, and this changing growth can be calculated *by adjusting the sides and angles of the triangle*. Also recognize that even if the structure is nonlinear in its formation, precise information on model-derived Rn/Tn data are available in the nonlinear portion of the curve between imaged horizons. For example, on Figure 11-53a, an equation exists for horizons H4 to H5. This follows as two points determine a straight line!

The Growth Axial Surface Decreases Dip

Figure 11-54 is an example of the effects of increasing horizon growth on linearity of the regression analysis. This increasing growth axial surface model was created by decreasing the dip of the growth axial surface in Figure 11-54. In this new model, the growth axial surface $\rho 1$, which is 139 deg, is the same as in Figure 11-52. However, $\rho 2$ is decreased in dip to 130 deg from 153 deg in Figure 11-52. This change in the dip of the growth axial surface also makes the problem nonlinear like Figure 11-52, and the problem is solved as described in the prior section, The Growth Axial Surface Increases Dip. The equation shown on Figure 11-52 determines the values of the plot in Figure 11-55b. Also, this change in the dip of the growth axial surface causes the crest of the structure to narrow, and the crest of structure shifts to the left.

Relative to Figure 11-52, Figure 11-54 has a lower Gn % difference = 46% and has a relatively high calculated linear regression R-square of 0.9516 (Fig. 11-55a). Some seismic lines image a decreasing growth axial surface. When the % difference values approach about 50%, the R-squares tend to be lower, as seen in Figure 11-55a. Nevertheless, one could argue that a 5% error is better than guessing. In Figure 11-55b, the calculated and measured linear regression R-square values are 0.9975 and 0.9821 respectively; these values are not significantly different. The similar R square-linear regression results are characteristic of the Rn/Ghn graphs.

Figure 11-52 shows how to correct for nonlinear data measurements. In the model, the growth axial surface changes from $\rho 1 = 139$ deg degrees above Horizon H3 to $\rho 2 = 153$ deg degrees at depth. The change in the dip of the growth axial surface causes the limb of the structure to widen with increasing depth. A triangle must be formed in order to employ Eq. (11-29) and solve for one of the data points. This triangle is shown in bold for the top of Horizon H4 and is defined by the growth axial surface $\rho a = 142.5$ deg, N3, M3, and Gh3 in Figure 11-52.

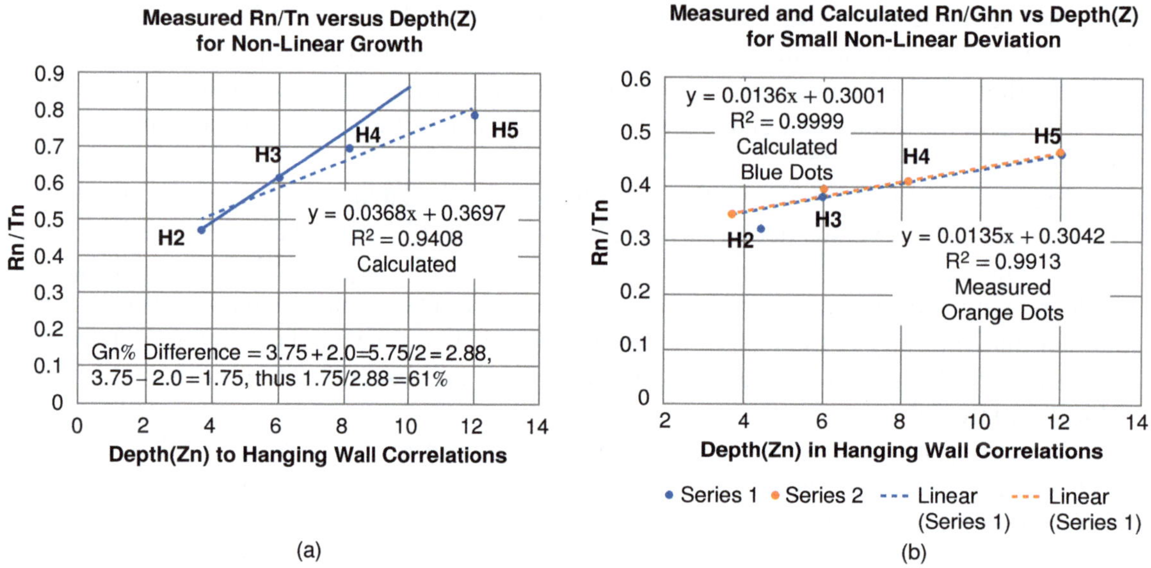

Figure 11-53 (a) The large changes in growth on nonlinear structures cause the Rn/Tn values to have a lower R-square value of 0.94. The high Gn % difference = 61% results in a lower R-square result. (b) However, the values for the calculated linear regression results are R-square = 0.999 and 0.99 for the measured Rn/Ghn values. (Published by permission of R. Bischke.)

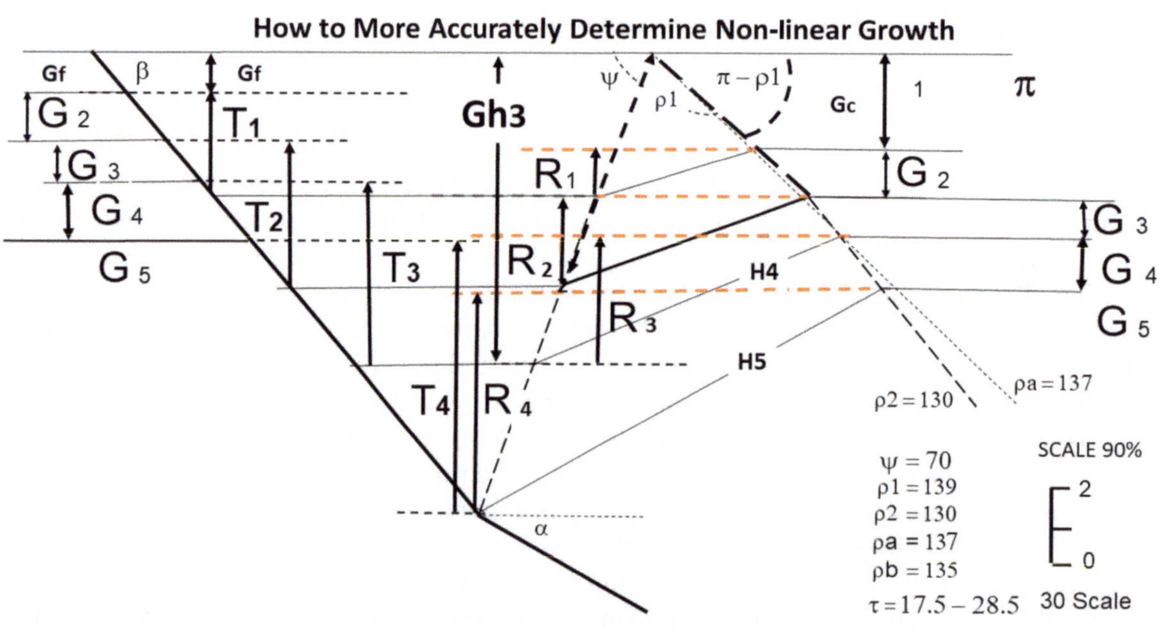

Figure 11-54 A growth model, modified from Figure 11-52. Steeply dipping growth axial surfaces may image on seismic, and in this model decrease from ρ1 = 139 to ρ2 = 130 deg; if recognized on seismic, decreasing growth can be corrected on the Rn/Tn and Rn/Ghn plots (Fig. 11-55a and b). (Published by permission of R. Bischke.)

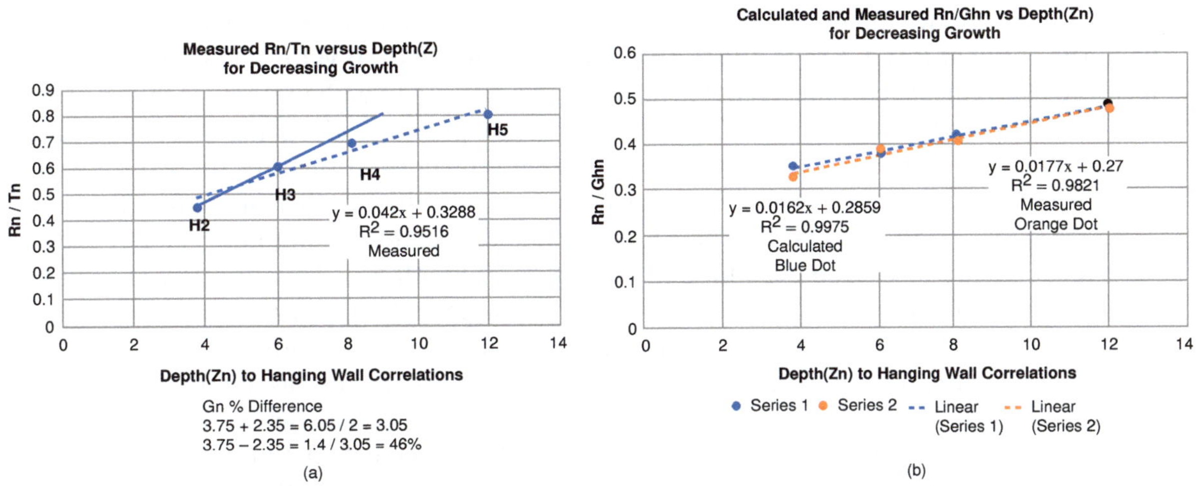

Figure 11-55 (a) In this model, with a decreasing growth axial surface, ρ2 turns down at Horizon H3. The Rn/Tn graph has an R-square of 0.9516, which is caused by higher growth. (b) However, the calculated and measured R-square values are higher for the Rn/Ghn plots at 0.9975 and 0.9821 respectively. (Published by permission of R. Bischke.)

Large Growth, Multibend Structure

This multibend model contains large growth variations as the frontal limb dip angles change from 12 to 32 deg (Fig. 11-56). In this model, there are two fault bends and two active axial surfaces emanating from the bends. Also, for each horizon, there are two dips that define the structural relief and a growth axial surface equal to 150 deg. Consequently, the structure has two distinct dip domains. Therefore, each domain produces two different structural reliefs (Rnm) and (Tnm) for each throw. The two structural reliefs are defined as Rna for the shallow dip domain and Rnb for the domain near the fault that is deeper. This means that Rna/Ghna depths are shallower than the Rnb/Ghnb depths (consult Fig. 11-56 for the posted, measured, and calculated values for R1a/Gh1a and R1b/Gh1b). Notice that the Rna dips nearer to the crest of structure are smaller than the Rnb dips. Thus, there are two groups of problems to be solved in Figure 11-56. This geometry is similar to some large growth faults (Fig. 11-16).

Domain 1. The first domain group of solutions involves the structural relief of domain R1a. This involves the growth axial surface ρ1a = 150 deg, N1a, and the active surface (ψ) to form the first triangle. The second problem also involves the growth axial surface related to ρ1a dipping at 150 deg, the active axial surface, and N2a. The third set of solutions involves the triangle formed by ρ1a and ρ3b dipping at 110 deg and N3a. and not the angle (ψ) (Fig. 11-56). *As the left-hand side of the triangle is not the angle (ψ) but ρ3b = 110 deg for this triangle*, (ψ) **is replaced by ρ3b = 110 deg when formulating Eq. (11-29)**. Also, for the measured values of Rna/Ghna, the Ghna values are measured at 1.5, 3.5, and 5.95 units.

Domain 2. The second domain Rnb/Ghnb are solved next. The three triangles are formed by ρ1b = 158 deg, N1b, and the shallowest active axial surface (ψ) at the first bend in the fault. Next, ρ2b = 131, Nb2, and the active axial surface form a triangle. The last problem involves ρb3 dipping at 110 deg, Nb3, and the active axial surface located at the second bend in the fault (Fig. 11-56). Horizon H4 is in the *inactive zone of no deformation* and has the same values as H3 (Fig. 11-56). In this case, the Ghnb values are larger than Ghna values in the first domain.

As there are two dip domains for each horizon, in order to determine the correct values of Rn/Ghn, **the value of R1a/Gh1a is added to the value of R1b/Gh1b values**. Figure 11-56 shows the measured and calculated values R1a/Gh1a and R1b/Gh1b along with the sum of the two values.

The Pregrowth Structure and Calculations section provides additional discussion on changing growth axial angles and how to model changing growth axial surfaces on real rollover structures. The large Gn variations in this modeled structure test the viability of the Rn/Ghn projection methods (Figs. 11-56 and 11-57). The Rn/Ghn value calculations using Eqs. (11-28) and (11-29) have an R-square of 0.9986 even though Gn = 82% (Fig. 10-57). This linearity at high Gn values is probably the result of a constant growth axial surface. In this case, the measured R-square is equal to 0.9999 and is a direct linear measurement, measured with a scale. Real structures (e.g., Figs. 11-61, 11-62b and c) also can have high R-squares, as do modeled structures. However, the growth model study, which includes other models not shown, suggests that the projection methods are often less reliable if the Gn % difference calculations exceed about 50% to 70%. The study of model structures predicts that the only structures that cannot be projected to greater depths are structures in which random horizon thicknesses (Gn) are present or in which the growth axial surface (ρn) changes significantly.

Thorsen (1963), and other work on extensional growth structures (Bischke et al. 2006), suggests that random structures are not common. An exception would be structures that rapidly increase or decrease in growth with increasing depth or contain large unconformities. However, many real structures, described in the following Real Structures section, have near-linear R-squares.

In summary, as many real rollovers have curved frontal limbs, the two-bend models look more realistic [Fig. 11-56 has two fault bends (ϕ) of 20 deg each]. Each bend requires a τn and ρn calculation or measurement. How to calculate this type of structure is also covered in the Pregrowth Structure and Calculations section, but the calculations are relatively simple when

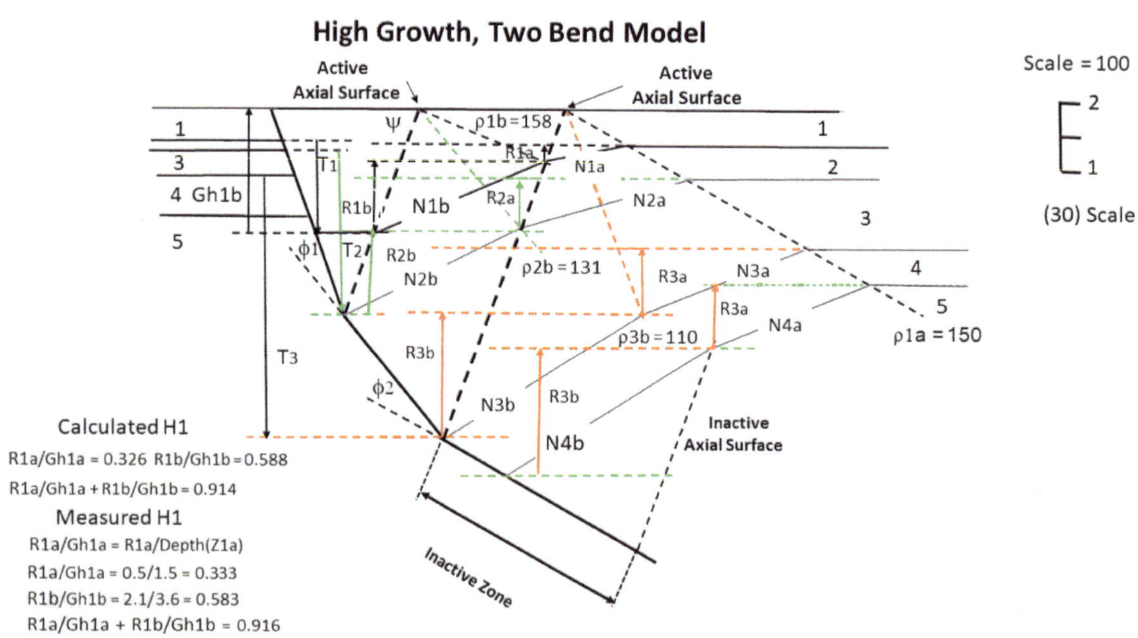

Figure 11-56 This multibend model may look more realistic than the previous models and tends to look more like some real structures. The model employs two active axial surfaces emanating from two bends in the fault. In order to solve Eq. (11-29), three inactive axial surfaces, R1a, R2a, and N1a, are used to form the triangles (see text for details). Horizon H5 has moved through the deepest active axial surface in the major fault, and thus the shape of the top of Horizon H4 is the same as Horizon H5. (Published by permission of R. Bischke.)

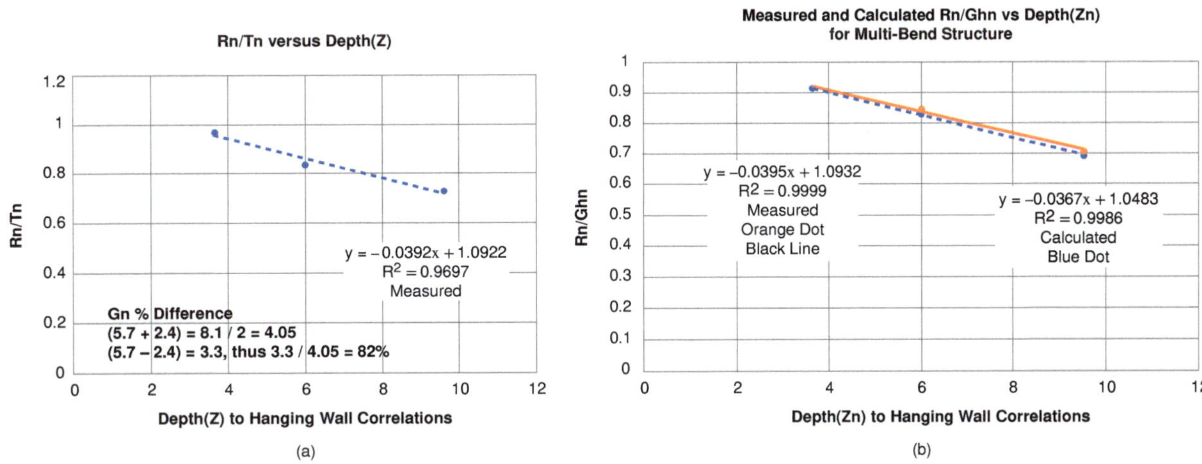

Figure 11-57 (a) The Rn/Ghn measured and calculated values versus depth(Z) graph for multibend model in Figure 11-56. This graph has a relatively high % difference = 82%, but nevertheless, the linear regression R-square results are about 1.0, in this case probably due to a linear growth axial surface. Horizon H5 has moved into the inactive zone, having passed through the deepest active axial surface. (Published by permission of R. Bischke.)

using a spreadsheet. Every change in dip requires an axial surface (ψ). Again, *if the left-hand side of the triangle is not the angle (ψ), but the angle $\rho\mu$, then for this triangle,* **$\rho\mu$ replaces (ψ) in Eq. (11-28); in this case, use Eq. (11-29).** Notice that on Figure 11-56, the structure moved through the *deepest active axial surface and entered the inactive zone.*

Real Structures

Real structures can be observed in both seismic and well log data. Seismic data is described first. *Published public domain 3D, depth-corrected seismic data are not common, probably for several reasons, including proprietary issues.* First, vertically expanded data, such as time data, expand the vertical component of the time section, making correlations and sequence boundaries more obvious. Second, depth correction takes more time. In contrast, in compressional regimes, non-depth-corrected seismic data can lead to major interpretation problems (Chapter 10), although this distortion of geometry is less problematic in extensional regimes. Nevertheless, simply converting a depth-corrected seismic line with a local or regional velocity/depth function permits geoscientists to view the prospect on a scale that is closer to one to one and presents a less distorted view of the actual geometry. Chapter 9 discusses in detail the modern depth-correction process, mapping methods, and techniques involving several 3D shale basin data sets.

Seismic

Southern Louisiana Rollover. Figure 11-58a is a measured Rn and throw(Tn) diagram from an older depth-corrected 2D line taken from Xiao and Suppe's (1992) paper on the origin of rollover. This seismic line was one of the first balanced lines for extensional structures. Horizon H6 was not included in this calculation, as its base is a pregrowth horizon. Horizon 5, which is a multibend horizon, was divided into two-line segments with different dips and depths (Zn) (see Fig. 11-56 for an example). Four active axial surfaces, $\psi = 67$ deg, were constructed at the position where the **base** of divergence of horizons H2 to H5 intersect the fault surface. The throw(Tn) correlations of y to x is high, and thus the R-squares results are high at 0.973, indicating roughly linear slip with increasing depth (Fig. 11-58a).

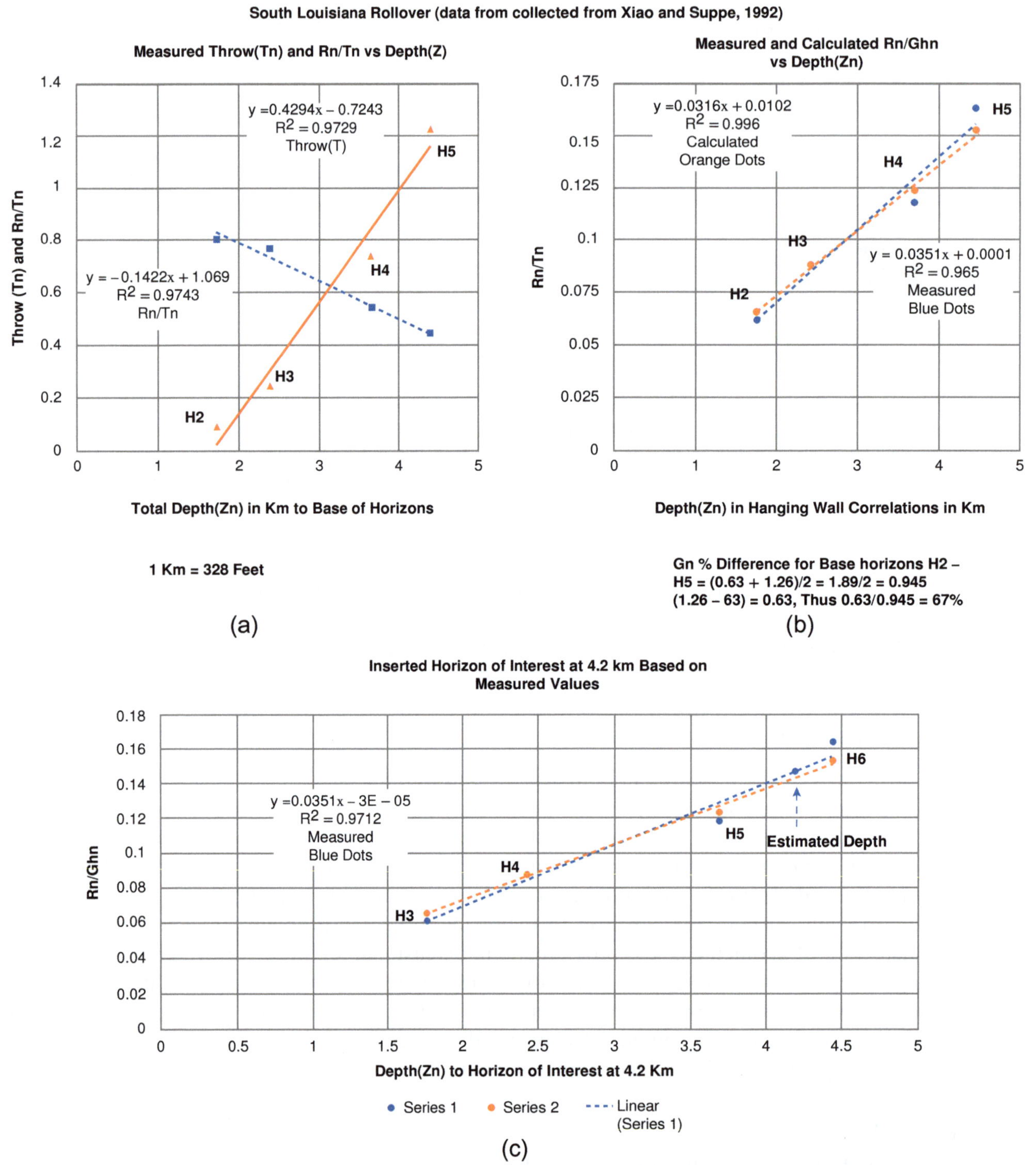

Figure 11-58 (a) The throw (T) correlation has a high R-square and thus the slip on the fold is near constant. (b) The depth-corrected balanced structure from southern Louisiana is published in Xiao and Suppe (1992). On this rollover, the growth axial surface for the first three horizons have about a constant growth axial surface at the base with ρ1 = 139 deg, which then decreases its dips to a lower angle with ρ2 = 164 deg at the top of H4. Thus, the crest of structure widens at depth. Both the measured and calculated R-squares are near linear, even though the Gn % difference = 67%. (c) The anomalous horizon must lie on or near the near-linear formula at a depth of about 4.2 km. This calculated depth can then be mapped onto the crest of structure. (Published by permission of R. Bischke.)

The base of horizons H2 to H4 have a growth axial surface with $\rho = 139$ deg, although the base of Horizon H5 shifts to the right and the value for $\rho 2 = 164$ deg. This makes the rollover slightly nonlinear. However, both the measured and calculated Rn/Ghn data have a high R-squared even though the horizon thickness Gn % difference = 67% (Fig. 11-58b). This high variance is caused by the large change in the growth axial surfaces (ρn) and the horizon thicknesses (Gn) that change with depth. The high R-squares result from a near-uniform growth axial surface (ρn) for the base of horizons H2 to H4. In Figure 11-58b, the calculated and measured linear regression values are 0.996 and 0.965 respectively. Remember that for the left-hand side of the shallow triangle on the base of Horizon H5, which bends from $\tau = 5$ to $\tau = 12$ deg,ψ must be replaced by $\rho a = 98$ deg in Eq. (11-29). This rollover expands and the horizons diverge or thicken with depth (consult the Gn numbers in Fig. 11-58b).

Next, let us assume in the case of this near-linear structure that a sand or high TOC shale exists in a footwall well below the top of Horizon H4 at a depth of 3.18 km (see Xiao and Suppe 1992, their Fig. 20, at 3.18 km) but has not been penetrated in the hanging wall. As the measured data are near linear, and the base of horizons H2 to H5 were mapped in the hanging wall and footwall, it is possible to estimate the depth of this high TOC horizon in the hanging wall within some degree of accuracy. This accuracy can be estimated using the determined R-square values and then comparing these values to a perfect R-square value of 1.0.

First, the *% difference of the footwall* depths are calculated for Horizon H4 at 2.92 km and the horizon of interest by calculating the depth to the top of Horizon H4 and subtracting the depth to the horizon of interest. This value is then divided by the Horizon H4 thickness (as shown in Fig. 11-59), which equates to 0.704% for the footwall. Thus, in the *hanging wall*, the high TOC zone is located at about at a thickness of 70.4% below the top of Horizon H4. Based on near-linear growth, Figure 11-59 also shows the estimated depth to where the high TOC horizon is located in the hanging wall. This calculated depth near the fault is at about 4.2 km (Fig. 11-59).

There is a check on this work that involves the linear regression y-x formula. As the measured data are near linear, the best depth estimate of 4.2 km for the TOC horizon is inserted into the growth formula. Next, the Rn/Ghn value times its depth should plot as a point along

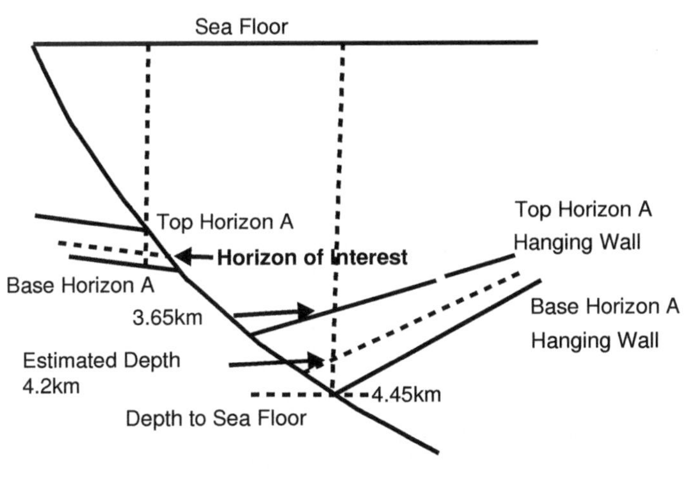

South Lousiana Rollover Example

For Footwall:

Depth to (**Horizon of Interest** – Top Horizon A)
/ Thickness of Horizon A in Footwall

(3.11 – 2.92)km = 0.19km/(3.19 – 2.92)km =
0.19/0.27 = 0.704

For Hanging Wall:

Depth to (Base Horizon A – Top of Horizon A)

(4.45 – 3.65)km = (0.8km) × (0.704) = 0.56km
3.65km + 0.56km = 4.2km Estimate for Horizon of Interest

Estimate for Point between Base H4 and H5

Measured Formula Rn/(Ghn) Value
Y = 0.0351(4.2km depth estimate) – 0.0001 =
Rn/(Ghn) = 0.1474 – 0.0001 = 0.1473

Figure. 11-59 Profile showing how to calculate depth values for a horizon of interest located in the footwall and then calculate its approximate depth in the hanging wall. (Published by permission of R. Bischke.)

the measured near-linear trend (estimated depth on Fig. 11-58c). The y-x formula values were calculated before the insertion of the depth of the TOC horizon at 4.2 km in the hanging wall. The estimated value for Rn/Ghn = 0.1473 should lie along the measured data trend line (Fig. 11-58c).

Corsair Trend/Brazos Ridge Offshore Texas

Figure 11-60 is a tracing of a depth-corrected seismic line drawing of horizons construction by Xiao and Suppe (1992) with the help of a major oil company. A good discussion of the major producing Corsair trend, offshore Texas, is found in Vogler and Robison (1987). The 2D line is imaged in Figure 11-58a, b, and c.

Figure 11-1a is from the large Corsair fault, which exhibits obvious bed dip divergence associated with large changes in the dip of the Corsair fault and large throws exceeding 10,000 ft (3 km) (Christiansen 1983). The beds dip toward the fault with maximum dip of 24 deg at a depth of 2.5 km. The depth-corrected line tracing was constrained by numerous deep wells that were drilled into the Big Hum, producing zone to depths exceeding 17,000 ft (5 km) (Vogler and Robison 1987). The line drawing generated by Xiao and Suppe is published in Tearpock and Bischke (2003) (Fig. 11-1). In Figure 11-60, the line drawing was slightly restored by removing crestal faults with small displacements (Fig. 11-12). The growth on this large rollover decreases with depth, as ρ1 dips at 135 deg at shallow depths, and ρ2 dips at 161 deg at deeper depths (the dashed lines on Fig. 11-60).

In Figure 11-61, measurements taken from this two-bend depth-corrected line is presented in the form of as Rn/Ghn graph. In spite of the changes in the growth axial surfaces and large thickness changes (the Gn % difference = 100%), the R-squares are near 1.0 (Fig. 11-60).

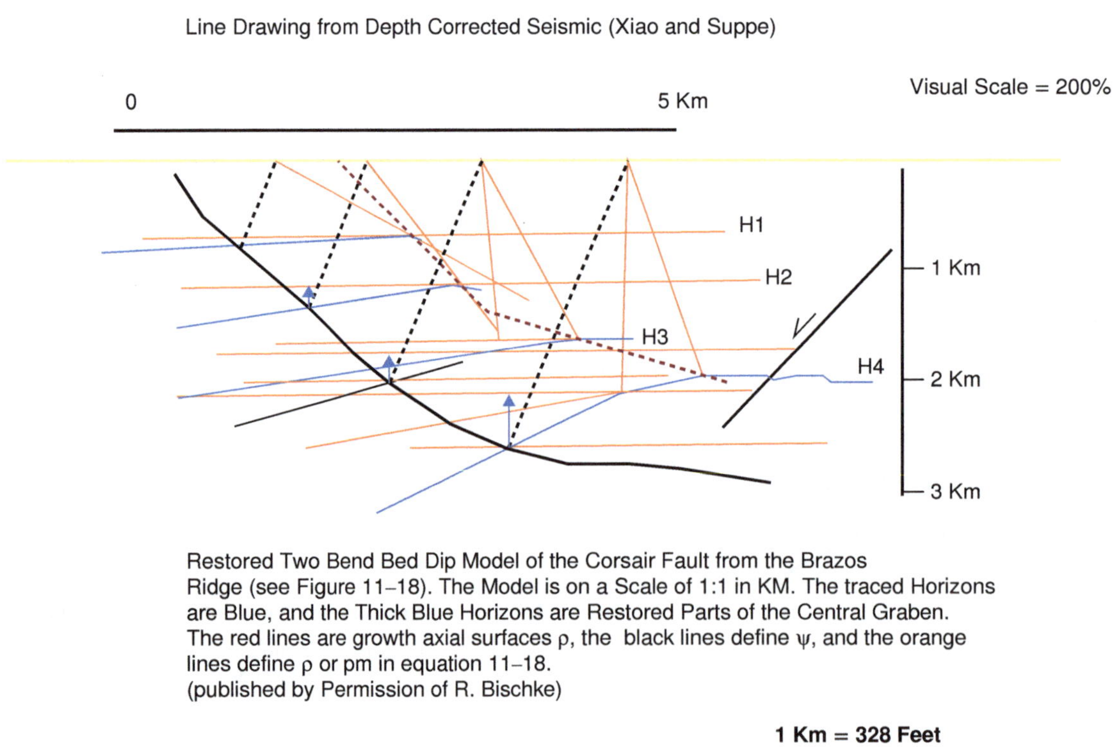

Figure 11-60 A tracing of Figure 11-1 for a depth-corrected line drawing by Xiao and Suppe. The structure was fit to the line tracing. (Published by permission of R. Bischke.)

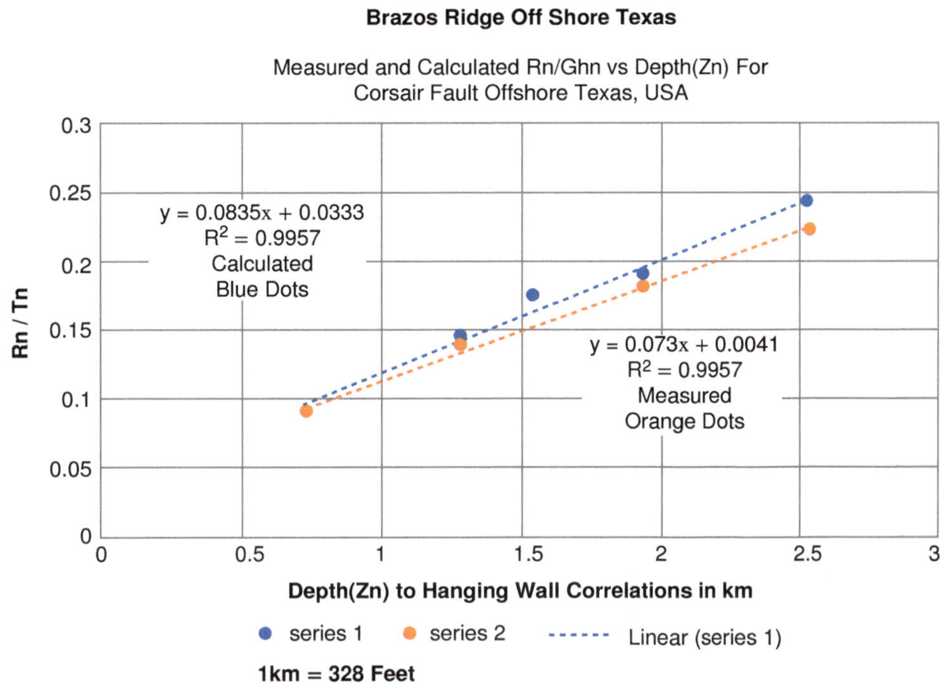

Figure 11-61 The measured and calculated Rn/Ghn values for a depth-corrected rollover from the Corsair fault, offshore Texas. This profile is a line tracing generated by Xiao and Suppe that was published in Tearpock and Bischke (2003) (Fig. 11-1). A restored model of Figure 11-1 is shown in Figure 11-60. Both the calculated and measured points on the figure were generated from a linear regression. These high R-square results from the linear regression occur in spite of large changes in horizon thicknesses with a Gn % difference = 100%. (Published by permission of R. Bischke.)

Haynesville Shale Play, Texas/Louisiana

The next example is a 3D depth-corrected line from the Haynesville shale play from Texas/Louisiana (Fig. 11-62a). This salt-related example images a fault that bottoms in a salt weld at about 15,500 ft (4.7 km). The first three horizons exhibit linear growth with the beds dipping from 4.5 to 7.5 deg. The measured and calculated Rn/Ghn values have R-squares at about 1.0, with horizon thicknesses of Gn % difference = 40% (Fig. 11-62b). The slope of the graph is positive, indicating increasing growth with depth (flat to dipping beds).

The three horizons above the salt weld also exhibit linear growth, but in this case with a negative slope indicative of decreasing growth with increasing depth (Fig. 11-62c). Above the salt weld, the bed dips range from 9.5 to 10.5 deg. The measured R-square values are also at about 1.0, with a Gn % difference = 58%. Horizon H4 is a probable unconformity, does not fit the two linear trends, and exhibits decreasing growth.

Starting at the base of Figure 11-62c and working toward the surface, an interpretation of the tectono-stratigraphy is as follows. At first, the sediments loaded the salt as the sediments thickened upward to Horizon H5 (increasing growth, growth is in dimensionless time; see Chapter 13). The thicken upward sequence was linear and ended at Horizon H5. There was a change in growth resulting in the unconformity at Horizon H4. Then the growth decreased linearly (Fig. 11-62b) until the sedimentary loading expelled all of the salt, creating the weld. This event happens at the flat horizon above Horizon H1 (see Fig. 11-62a). This sequence style was first described by Thorsen (1963) using the expansion index (Ei) for the Norco, Bonnet Carre faults and in particular on the LaPlace fault (Fig. 13-3).

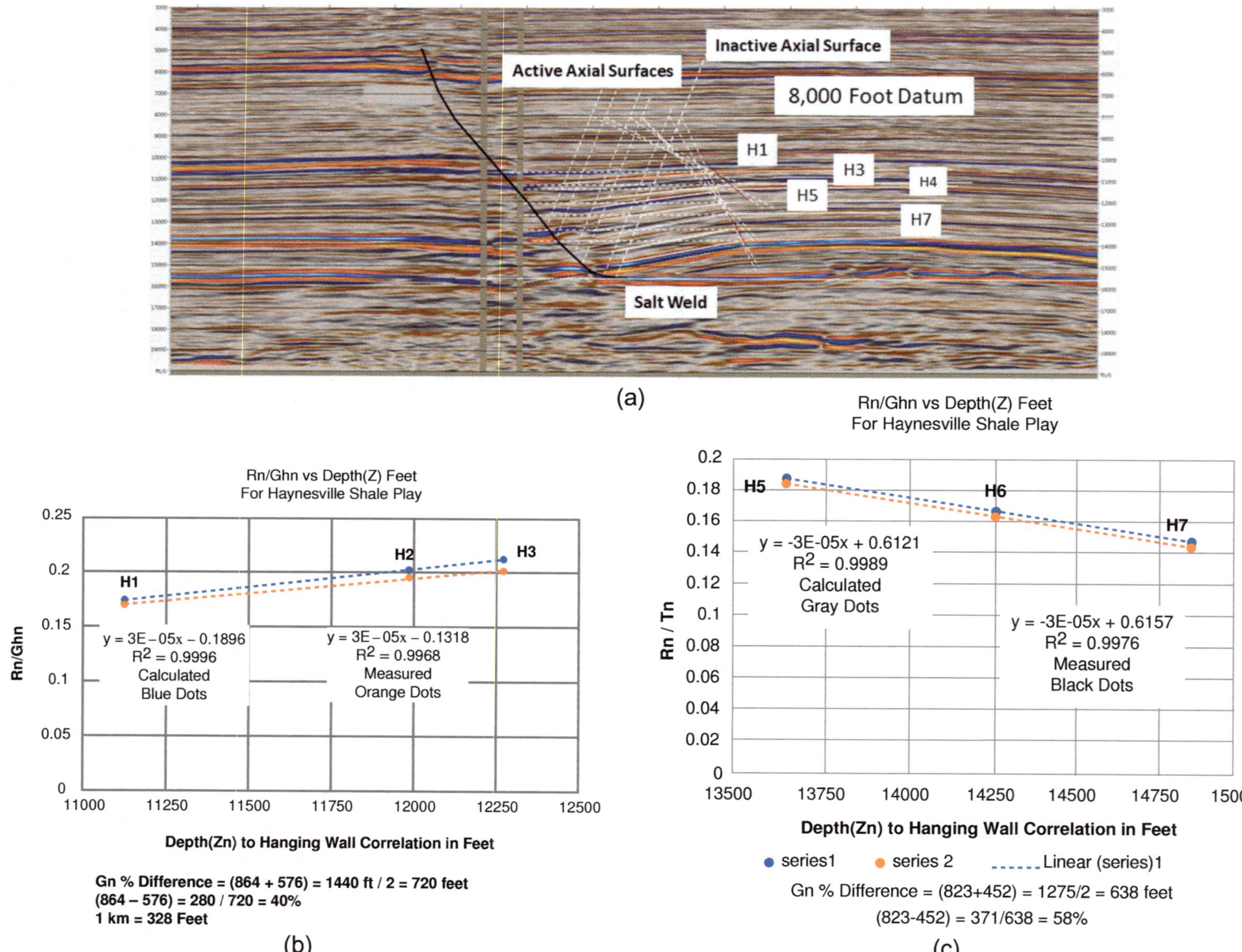

Figure 11-62 (a) This depth-corrected 3D line is from the Haynesville shale play trend in Texas/Louisiana. The large fault bottoms in a salt weld. Horizons from H1 to H3 diverge between 10,000 to 13,000 ft (3 to 4 km). (b) Measured and calculated Rn/Ghn values for horizons H1 to H3 between 10,000 ft on the crest of structure to 13,000 ft (3 to 4 km). Calculated and measured Rn/Ghn values are near linear. (c) R-square values = 0.9988 for a decreasing growth of horizons H5 to H7 directly above the salt weld. (Published by permission of R. Bischke.)

Well Log Data

The last example is a near-linear example of growth using well log data. This example uses missing section data from the Rabbit Island Field, Louisiana (Fig. 11-63a, interpreted by Dan Tearpock). The wells cut a downward dying fault and its *fault surface map along with a seismic line* is published in Tearpock and Bischke (2003) (see Fig. 13-27). At shallow depths, the missing section values increase, and then decrease below a depth of about 5600 ft (1 ft = 0.3048 m). The seismic line exhibits *coherent data* at about 11,000 ft, and a well at this level contains a complete stratigraphic section with no faulting to trap hydrocarbons (Tearpock and Bischke 2003). The missing section data are not precisely linear (mostly the result of the deepest correlation point) but are readily fit to a second-order polynomial with an X-square coefficient of only 7xE-06. However, if the missing section data are analyzed from the 5700- to 9700-ft level, the data are nearly linear (Fig. 11-63b). This portion of the Rabbit fault slipped at a near constant distance.

Unconformity in Shale Basin

Figure 11-64a shows a shale basin 3D seismic line with a large unconformity between the H3 and H4 horizons, indicating that not all growth is near linear and that unconformities can cause nonlinear growth. Figure 11-64b is a plot of the growth discontinuity (Bischke 1994b) in which Horizon H3 is clearly offset from the averaged R-square general trend using all four data points.

Most of the generic and real examples shown exhibited near-linear coefficient correlations of the crestal data. The example in Figure 11-64a is more complex and exhibits a changing growth axial surface (red lines), and thus this example has high R-squares (Fig. 11-64b). A large unconformity offsets the structure along the near horizontal portion of the dashed red line, as the unconformity shifts the crest of structure to the left.

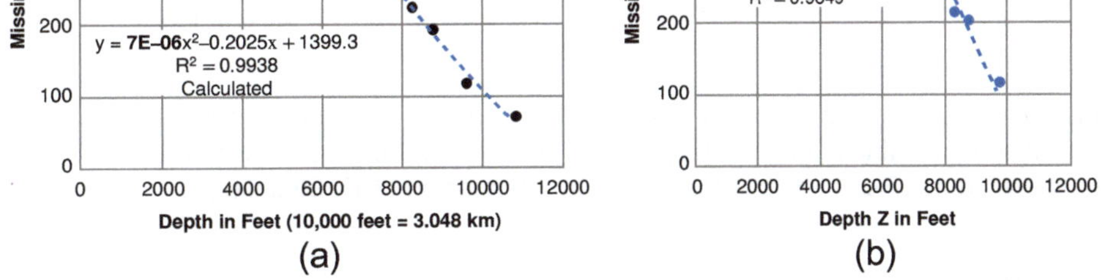

Figure 11-63 (a) This figure uses missing section well log data interpreted by Dan Tearpock. This example is constrained by a fault surface map and a seismic dip line that crosses the fault (see Fig. 13-27). The wells cut a downward dying fault from the Rabbit Island Field, Louisiana, that is published in Tearpock and Bischke (2003). Below 5500 ft, the missing section decreases from 490 ft to 70 ft (Fig. 13-27). The missing section data are best fit to a second-degree polynomial that contains a small coefficient of 7E-06. Thus, if the deepest data point is eliminated, the data are almost linear. (Published by permission of R. Bischke.)

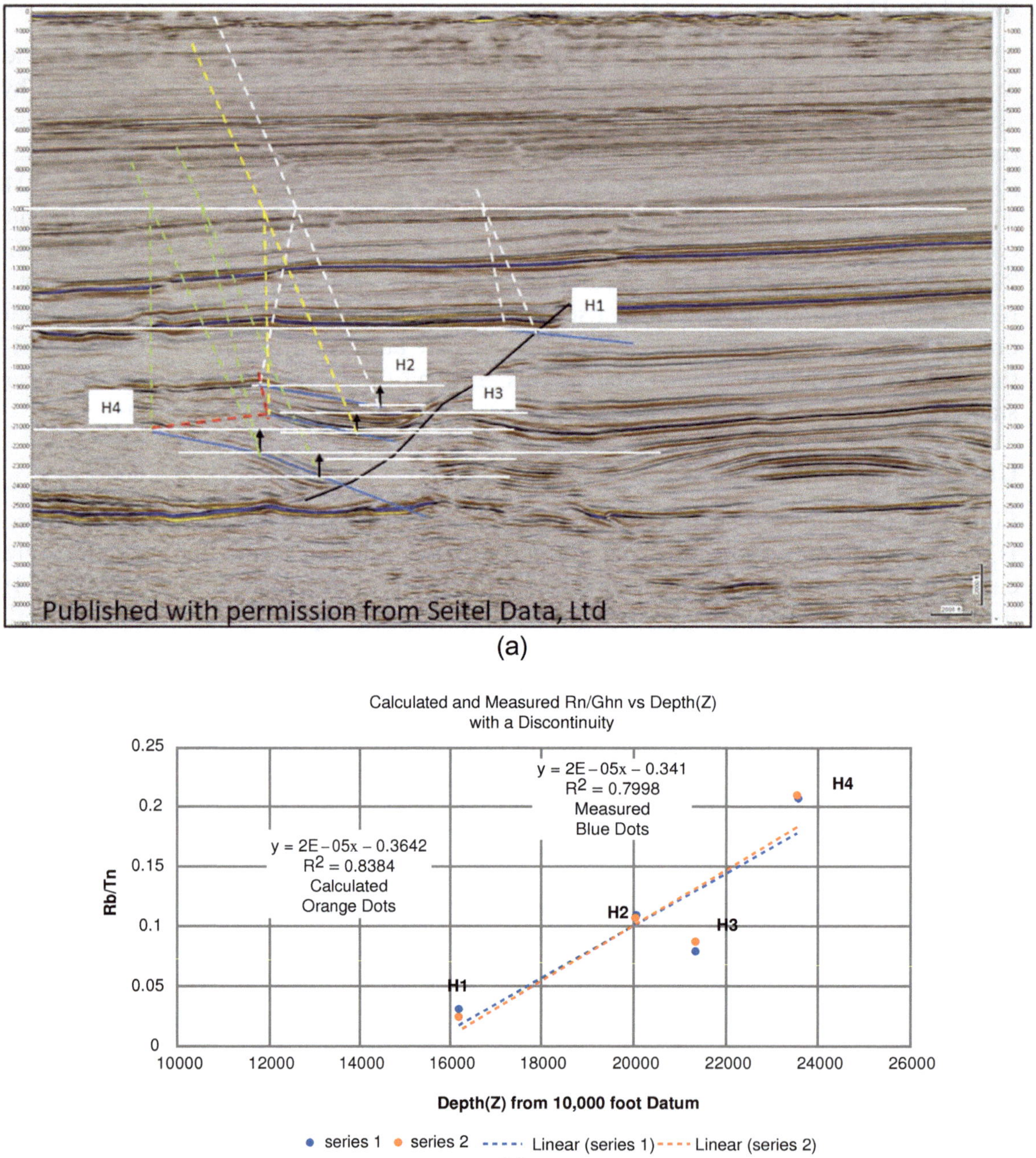

Figure 11-64 This figure is a shale basin example that contains an unconformity and therefore nonlinear growth. The existing discontinuity in growth is clearly present on the growth diagram, as Horizon H3 is offset from the general trend (Fig. 11-64b). (Published by permission of R. Bischke.)

Conclusions

This study started by discussing linear growth in general and then concentrated on which factors affect linear growth on generic models, finally applying these results to several real data. Many real examples were subject to the linearity tests. Growth becomes less linear if the % difference calculations exceed about 50% to 70%. Data become less linear if the growth is large and variable, alternating between increasing then decreasing growth, or if the growth axial surface (ρ) changes dip. Many seismic sections exhibit nonlinear growth, particularly if the section contains a large unconformity (Fig. 11-64).

Public domain, depth-corrected seismic data are not common in the literature, and thus firm conclusions regarding near-linear growth on numerous real structures is unavailable. The section Real Structures and examples presented in Chapter 13 on growth structures show that near-linear growth exists in well log data and on some older, depth-corrected 2D seismic data. Our study of 3D seismic and depth-corrected seismic show that episodes of near-linear data, obtained from Chapter 9, exist in shale play basins.

Linear growth may have applicability for the following reasons. First assume that linear growth is present in both the footwall and the hanging wall. If an interesting horizon, for example, either a hard shale or a high TOC section, is present in the footwall but has not been penetrated in the hanging wall, then projecting the footwall interval to depth in the hanging wall is possible. Next, run a linear regression analysis of the depths in the known footwall intervals. This analysis results in a y-x equation. The depth of the interesting interval in the hanging wall will lie on a Ghn/depth (Z) graph of the linear equation.

CHAPTER 12

STRIKE-SLIP FAULTS AND ASSOCIATED STRUCTURES*

INTRODUCTION

Strike-slip displacements occur along near-vertical faults that offset basement (Harding 1990). Displacements along strike-slip faults are predominantly in the strike direction of the fault, and the vertical separations of the horizons along strike-slip faults may alternate between normal and reverse separations. Along active strike-slip faults, dip-slip components of displacement are also common (Clark et al. 1984). As they move, the crustal blocks typically encounter curves or bends in the near-vertical fault surfaces. Material moving into these fault bends may generate structures that can trap hydrocarbons. Commonly, areas may be subject to strike-slip, compressional, and extensional displacements, and the compressional or extensional displacements *need not be contemporaneous* with the strike-slip faulting (Wright 1991; Shaw and Suppe 1996). This complex style of displacements, combined with the 3D structural development and the progressive nature of the deformation, which changes with time, makes the interpretation of strike-slip faults and their related structures difficult. These complexities often result in the misidentification of strike-slip faults and the misinterpretation of structural styles (Harding 1985, 1990). Strike-slip deformation is truly a *four-dimensional problem*, which requires an understanding of how the predominantly horizontal displacements occur through time. In this chapter and in Chapter 13, we address the 3D strike-slip problem and propose methods to solve the complexities associated with strike-slip deformation, thus providing ways to improve interpretations used to explore for and develop hydrocarbon resources.

Problems concerning the interpretation of strike-slip faulted structures involve the issue of recognizing empirical evidence for strike-slip faulting (Harding 1985, 1990). According to Harding

* For all figures in this chapter (in the printed book only), see the preface for information about registering your copy on the InformIT site for access to the electronic versions in color.

(1990), "Many workers do not provide evidence for their assertions of strike-slip deformation." More specifically, some strike-slip interpretations *lack direct evidence in support of horizontal displacements. Evidence for horizontal displacements fundamental to strike-slip faulting is critical* and, therefore, strike-slip fault interpretations that lack direct evidence for horizontal displacements are questionable. Unfortunately, other structural styles are readily confused with strike-slip styles (Harding 1990), and strike-slip interpretations are often the cause of numerous discussions among working groups. For example, strike-slip interpretations made from poor quality 2D seismic data are often attributed to a different structural style after 3D data are acquired. Once data quality improves, preliminary strike-slip fault interpretations may become (1) inversion structures (Mitra 1993; Link et al. 1996), (2) duplex folding (Mitra 1986; see the Imbricate Structures section in Chapter 10 in this book), (3) growth compressional folding (Suppe et al. 1992; Shaw and Suppe 1994) (Chapter 13), (4) basement structures (Narr and Suppe 1994), or (5) areas of high bed dip. Poor-quality data and an inappropriate structural model can easily result in the misinterpretation of structural style. Our experience suggests that, at times, interpretations may incorporate numerous strike-slip faults into an area that contains minor amounts of strike-slip faulting. We make the case that strike-slip faulting, when viewed in a regional context, is often no more difficult to confirm than other types of faulting. We believe that the **releasing bends** and **restraining bends**, common to known strike-slip faults, *are fundamental to the understanding and recognition of strike-slip faults and their related structures* (Crowell 1974a, 1974b). Their presence is supportive and is often the only direct evidence for strike-slip faults present in subsurface data sets.

In this chapter, we briefly review several misconceptions concerning strike-slip faulting and problems encountered when working with these fault systems. We present methods and techniques for recognizing, restoring, and balancing strike-slip structures. Our goal is to provide the methods with which to generate interpretations and prospects based on *observational evidence* for lateral displacement, and *not on the absence of data.* This chapter concentrates on several models and interpretation methods used to interpret strike-slip faulting, and we briefly present the strengths and weaknesses of each model. The manner in which we apply models to geological and geophysical data affect our understanding of the petroleum system and the ultimate success of a prospect or of field development. These models are the stress/strain, surface feature, restoration, releasing bend/restraining bend, and balanced cross-section models. A common theme to this discussion is the application of the standard mapping techniques, methods, and philosophy that we employ when interpreting extensional and compressional regimes. Geoscientists seem best served by establishing 3D structural validity to their interpretations and prospects, as described in Chapters 5, 7, 8, and 9, rather than relying on theoretical models that involve stress or strain. We must get the strike-slip fault geometry correct before embarking on the regional geological model and prospect generation.

All too often, discussions concerning the interpretation of strike-slip faults seem to result in theoretical arguments involving stress or strain. Stress is not commonly measured in subsurface data and are rarely measured in outcrop. Accordingly, the application of stress or strain concepts to prospect generation or regional studies can easily result in disappointment and confusion. For this reason, we begin the discussion of geological models by briefly reviewing the application of stress and strain as applied to prospect generation.

MAPPING STRIKE-SLIP FAULTS

The procedures and methods for recognizing and mapping strike-slip faults are no different from mapping and recognizing other styles of faulting. As discussed in Chapters 1 and 7, the interpretation process starts by constructing viable maps of the large, or structure-forming, fault surfaces. In the interpretation of faults from seismic data, the interpreted profiles are *loop-tied* in order not to map two faults as a single fault. Viable fault surface maps are typically smooth surfaces devoid of major kinks. Kinks in fault surface maps typically indicate a more complex interpretation, such as the presence of two faults rather than one fault or intersecting faults.

In a typical exploration or development study, after mapping the faults, we integrate chosen horizons into the fault surface maps, as discussed in Chapter 8. We are unaware of any techniques, other than those already discussed in this book, required to integrate strike-slip faults into horizon maps. Thus, if a strike-slip fault exhibits normal separations, then employ the techniques described in the section Techniques for Contouring across Normal Faults in Chapter 8. If a strike-slip fault exhibits reverse separations, then consult the discussion on reverse faults and compressional tectonics in Chapter 8. A discussion of other geometries concerning strike-slip faults and their associated features and styles are in the section on strike-slip faults in Chapter 8 and in the bulk of this chapter.

The Problem of Strike-Slip Fault Interpretation

Our review of strike-slip fault interpretations and the associated structures and prospects from around the world suggests that some interpretations of strike-slip faults do not present direct evidence for *horizontal displacements*. Often, geoscientists *do not generate fault surface maps* in support of strike-slip faulting. Instead, the working group may rely too heavily on stress/strain models or may interpret strike-slip faults to be within no-data zones on seismic sections. Some of the observed geometries associated with strike-slip faults are *inconsistent with simple models of stress or strain* (Sylvester 1988). For example, some textbooks on structural geology teach that strike-slip faults are oriented at about a 30-deg to 45-deg angle to the direction of maximum principal stress or strain (the compressional direction), as in Figure 12-1. This simple model of strike-slip faulting assumes a *continuous (unfractured), isotropic, and homogeneous body.* However, faults and joints introduce *discontinuities* into rock that invalidate the continuous material body

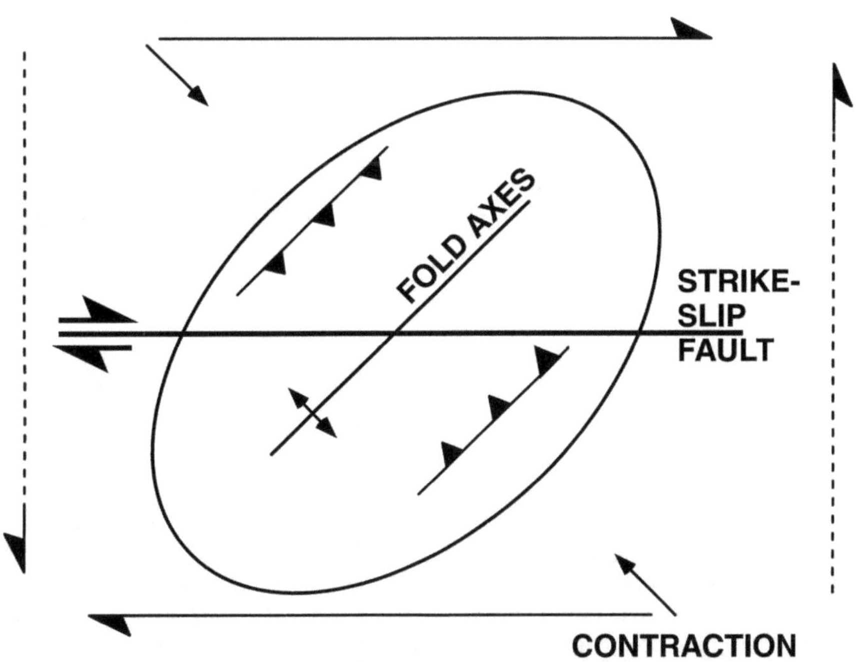

Figure 12-1 Simple model for compression and strain. Strike-slip faults lie at a 45-deg angle, and reverse faults and fold axes form at about a 45-deg angle, from the contraction direction (the direction of maximum principal stress). The simple double-couple model for stress and strain assumes a continuous (unfractured) material body. This model is inconsistent with observed geometrics associated with strike-slip faults.

assumption. The crust contains numerous fractures and is not a continuous body (Pollard and Segall 1987). In the following section, we review evidence that this simple model of deformation is inconsistent with the regional orientation and distribution of Neogene fold axes relative to strike-slip faults (Mount and Suppe 1987, 1992).

Strain Ellipse Model

A common model used in prospect generation is one in which strike-slip displacements *cause contractional folds to form along and near strike-slip faults*. Some structural geology texts teach that these folds form along strike-slip faults according to a simple model of strain, as shown in Figure 12-1. The folds and the throughgoing faults are thought to form as a result of the *maximum shear strain oriented parallel to the fault surface*. When applying the simple model of strain to strike-slip faults, geoscientists may orient the model so that a shear couple parallels the direction of the throughgoing fault (Fig. 12-1). This assumes a continuous and homogenous medium and that the direction of maximum contractional strain or stress is at a 45-deg angle to the throughgoing fault. However, rock fractures at about a 30-deg angle from the maximum principal stress (Ramsay 1967). Some texts infer from an analysis of the simple shear model that (1) the model accounts for the origin of folds in the vicinity of strike-slip faults, and (2) the model requires the fold axes to trend at high angles (about 30 deg to 45 deg) to the strike-slip faults. This analysis assumes that *the plane of maximum shear stress or strain lies near, or in the plane of, the throughgoing strike-slip fault* (Fig. 12-1). However, actual folds trend at much lower angles to strike-slip faults (Sylvester 1988), perhaps as a result of rotational displacements.

In the section on balancing strike-slip fault interpretations, we discuss how folds form adjacent to and along strike-slip faults. Folds adjacent to strike-slip faults may or may not exhibit fold axes that trend at 30-deg to 45-deg to the fault surfaces. Figures 12-2, 12-3, and 12-4 show

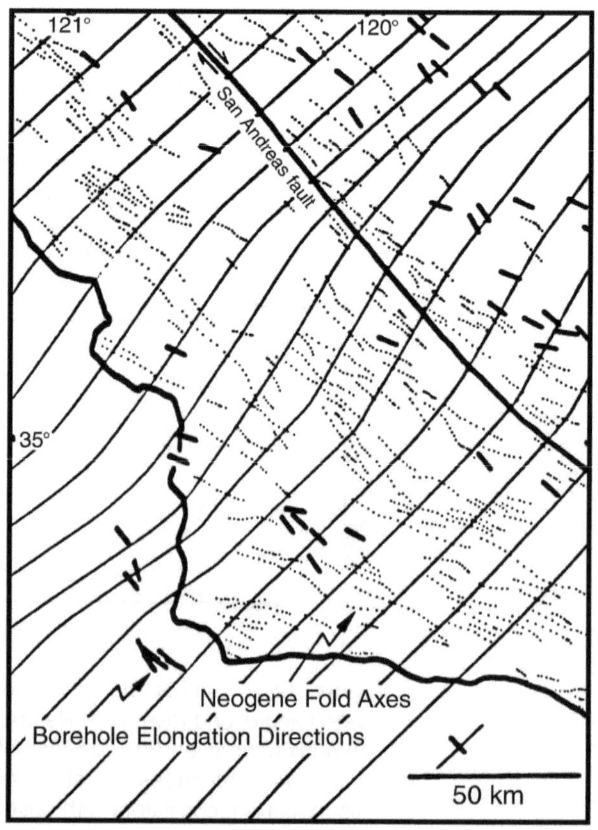

Figure 12-2 Borehole elongations measure the direction of maximum principal stress across the San Andreas Fault, California. Borehole breakout directions (short bold lines), Neogene fold axes (dotted lines), and predicted maximum compressive stress trajectory direction (long thin lines) from breakout data, using a statistical smoothing technique developed by Hanson and Mount, are shown. Maximum compressive stress direction is subnormal to, and Neogene fold axes are subparallel to, the San Andreas Fault. Neogene fold axes extend about 100 km from fault zone. (Published by permission of Van Mount.)

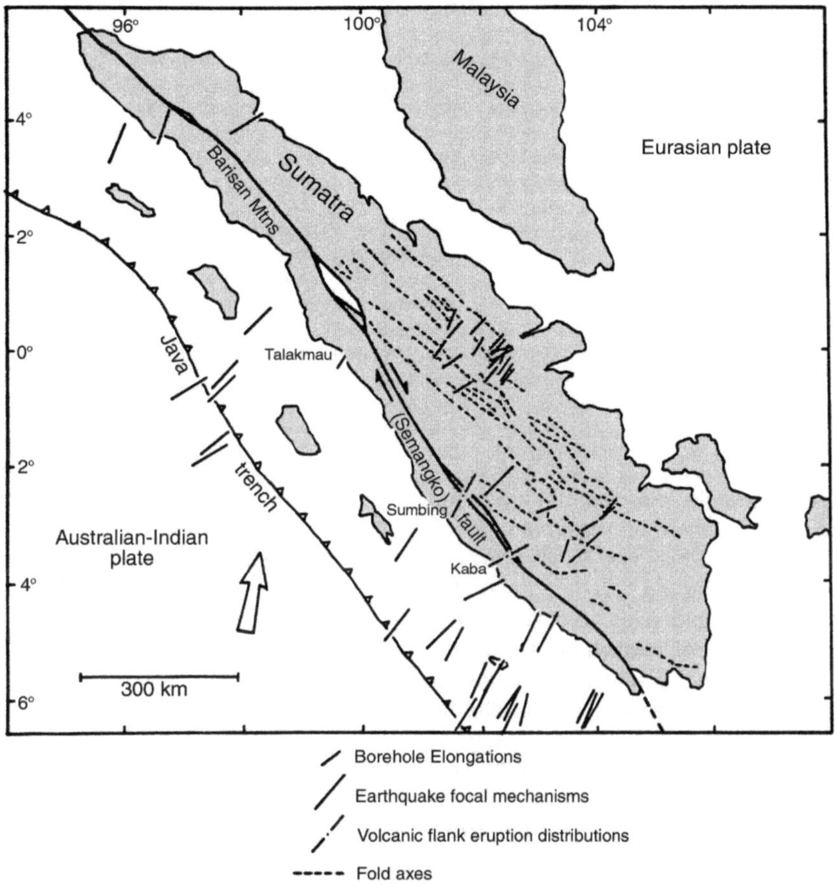

Figure 12-3 Borehole elongations measure the direction of maximum principal stress across the Semangko Fault, Indonesia. Borehole breakout directions rotated 90 deg (short thin lines), Neogene fold axes (dashed lines), and thrust fault earthquake focal mechanism solutions (long thin lines) are shown. Arrows show direction of relative plate motion. Maximum compressive stress direction is subnormal to, and Neogene fold axes are subparallel to, the Semangko Fault. Neogene fold axes extend about 300 km from fault zone. (From Mount and Suppe 1992. Published by permission of the American Geophysical Union.)

regional Neogene fold axes relative to the San Andreas, Semangko (Indonesia), and Philippine Faults. Notice on the figures that most of the fold axes trend subparallel to the strike-slip fault zones. One would assume that these fold axes lie in a plane that is subnormal to the maximum contraction direction, which in these examples would be *subnormal to the surface trace of the strike-slip faults*. This orientation of the maximum contraction direction is inconsistent with the simple strain model, and it suggests that the compression subnormal to the faults may be independent of the strike-slip displacements. Furthermore, the neotectonic folds are *not* concentrated along the fault zones, but rather they exist as far as *50 km to 300 km* from the fault zones. Thus, empirical data obtained from areas such as the San Andreas, Semangko, and Philippine Faults do not support the simple model of stress and strain, resulting in the "stress paradox" (Sylvester 1988). The simple theory, if applied to geological or geophysical data, could affect regional and prospect interpretations and your understanding of the petroleum system under study.

The fold geometry present on Figures 12-2, 12-3, and 12-4 seems to be in conflict with deductive reasoning concerning continuous material behavior taught in many texts on structural geology. However, some of these observations are consistent with or predicted by more advanced

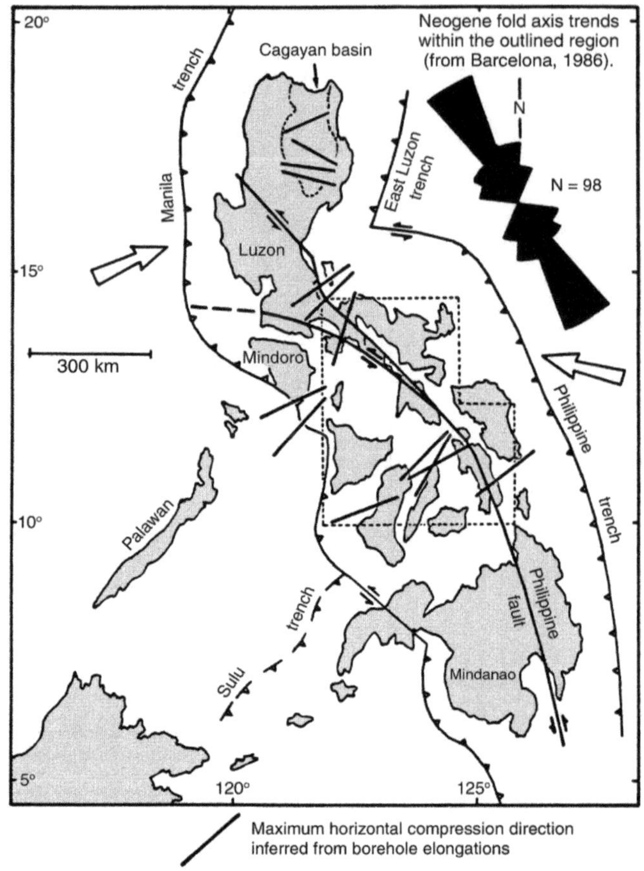

Figure 12-4 Borehole elongations are used to measure the direction of maximum principal stress across the Philippine Fault. Borehole breakout directions rotated 90 deg (long thin lines) and Neogene fold axes (rose diagram). Arrows show direction of relative plate motions. Maximum compressive direction is subnormal to, and Neogene fold axes are subparallel to, the Philippine Fault and its branches. Neogene fold axes do not concentrate near the fault zone, but rather extend about 200 km from the fault zone. (From Mount and Suppe 1987 and 1992. Published by permission of Van Mount.)

theories on *discontinuous* material behavior, as presented in texts on strength of materials and rock mechanics.

How are the observations present on Figures 12-2, 12-3, and 12-4 consistent with known principles of mechanics? Displacements along strike-slip faults often create a zone of rubble in the "fault zone." Broken material is incapable of supporting large stresses or strains (Billings 1972), thus releasing shear strain in the vicinity of the fault zone. The fault gouge also reduces the frictional stress in the fault zone. If the shear strain in the rubble zone approaches zero, then the fault surface lies near a principal plane of stress (Ramsay 1967), and the maximum principal stress could rotate into a position that is subnormal to the fault zone.

Mount and Suppe (1987) propose that the strike-slip motions decouple from the compressional motions, and that plate tectonics may use the transform faults as the "weak link in the chain." Motion along the weak and low-friction transform faults has the effect of minimizing the work (strain energy) required to drive the global tectonic system. Physics teaches us that the least work solution is the correct solution.

Problems Interpreting Stress

Misconceptions concerning stress can result in incorrect geological interpretations. First, *stress is a mathematical and not a physical concept* (Jaeger 1962) and is by definition a measure of the intensity of the state of a reaction. *Stresses are invisible*, so one cannot see a stress; one sees only the results of stress. Alternatively, the force vector is by definition a *directed line segment*. Forces can be visualized as a load or as a weight. Second, stress is a ***second-order tensor*** (Jaeger 1966). The stress tensor exhibits both rotational components and invariant components that are independent of the coordinate system. The tensor components of stress or strain interact with discontinuities to cause the state of stress or strain within a body to be complicated. Rock mechanics textbooks present numerous examples of complicated stress trajectory patterns, related to simple structures, that are certainly not intuitive (e.g., Obert and Duvall 1967). Chinnery (1963) solved the problem of the state of stress along a strike-slip fault using elastic dislocation models. Elastic dislocation models show complicated stress trajectory patterns involving simple structures (Fig. 12-5), particularly at the ends of fractures (Bischke 1974; Xiaohan 1983). The simple model of stress or strain cannot predict the complicated stress patterns associated with faults and fractures.

Geoscientists sometimes attempt to infer the state of stress from geological features or structures. Inferences concerning the state of stress are complicated by several factors, including the rotational property of the stress and strain ellipsoid. As the stress tensor may rotate through time, the *finite strain* observed in rocks need not result from a unique stress direction (Flinn 1962; Ramsay 1967). Furthermore, geoscientists *must measure the state of stress; they cannot directly observe stress*. No one has ever seen a stress, certainly not a stress in the distant past. If a structure formed in the distant past, then the stresses that formed the structure may not be recoverable. Thus, we believe that inferences, deductions, and speculations concerning stress often lead to incorrect conclusions concerning structures and prospects. It is not our intent to discuss all the ramifications of stress or strain or their field measurements, which are presented in texts on rock mechanics (e.g., Jaeger 1962; Obert and Duval 1967; Jaeger and Cook 1969). Our point is that speculations concerning stress have little value during the interpretation and prospect-generation process and that these speculations may cause more harm than good. Accordingly, we concentrate on interpretation methods that involve the *displacement of stratigraphic units*. This approach has

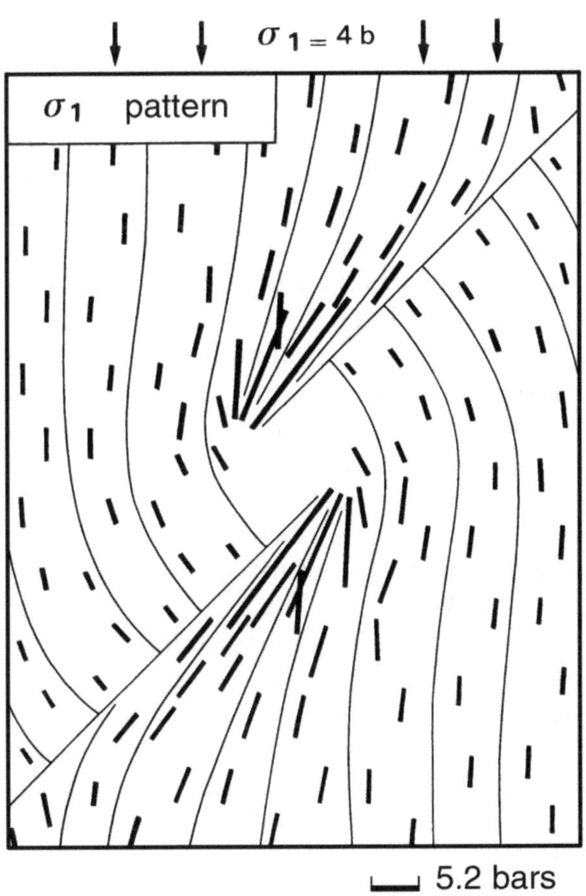

Figure 12-5 Stress trajectories (thin lines) around an en echelon offset, or stepover, in a strike-slip fault. Near the ends of the stepover, the stress trajectories of the maximum principal stresses rotate into the fault as a result of the discontinuity. Stresses are discontinuous across the fault surface and change intensity according to the length of the short bold lines. (From Xiaohan 1983; Guiraud and Seguret 1985. Published by permission of the Society for Sedimentary Geology.)

an advantage in that *displaced horizons are subject to direct observation*, as opposed to stresses that are *at best measured.*

Stress Measurements across Strike-Slip Faults

When attempts were made to measure the stress on the San Andreas (California) and other large strike-slip faults, the results were puzzling (Zoback et al. 1987; Mount and Suppe 1987, 1992). Figures 12-2, 12-3, and 12-4 show the direction of the maximum horizontal stress trajectories, as determined from borehole breakout measurements and earthquake focal mechanisms solutions, for the San Andreas (California), Semangko (Indonesia), and Philippine Faults (Mount and Suppe 1992). The maximum horizontal stress lies in the same plane as the maximum principal stress σ_1. These measurements record the state of stress during the Neogene. The borehole breakouts obtained from deep wells react to stresses at depth, below the influence of surface topographic effects. Hydro-fracturing experiments, conducted during enhanced recovery efforts, show that wellbore breakout data record the direction of the minimum principal stress σ_3 (Zoback et al. 1985; Zheng et al. 1989; Dart and Swolfs 1992). The direction of maximum principle stress σ_1 is 90 deg from the σ_3 direction and lies in a plane oriented at right angles to the borehole breakout direction. For three of the world's major strike-slip faults, stress directions derived from borehole breakout data, as shown on Figures 12-2, 12-3, and 12-4, suggest that the maximum horizontal stress is predominantly *subnormal to the strike-slip faults.* This is in contrast to the 30-deg to 45-deg angles as predicted by the simple model of stress. This may suggest that major strike-slip faults are low shear stress, or weak, faults. Their fault surfaces apparently lie near a *principal plane of stress* (Ramsay 1967), where the shear stress, and thus the frictional stress, is low. These stress measurements are consistent with the lack of a heat flow anomaly on the San Andreas Fault (Brune et al. 1969).

From this discussion, we conclude that the simple models of stress and strain are inconsistent with the contractional direction implied by the regional Neotectonic fold axes and borehole breakout directions (Figs. 12-2, 12-3, and 12-4). The simple models of stress and strain do not include material discontinuities in the analysis, so these simple models often fail to predict the trends and the distribution of the observed tectonic features. The fact that the simple theory fails to predict natural features is consistent with advanced theories on discontinuous material behavior and elastic deformation (Pollard and Segall 1987; Obert and Duval 1967). The basic problem of identifying or mapping strike-slip faults is not a mechanical problem, but rather a *geometric problem*. Perhaps it is time to rethink the strike-slip fault problem, particularly as strike-slip faulting has been difficult to document when interpreting subsurface data.

Applying the appropriate model to describe the stress and strain along fault surfaces is important. Often when confronted with possible strike-slip motions, some geoscientists may model the observed fold axes *at about 30-deg to 45-deg angles to the throughgoing strike-slip fault* (Fig. 12-1). This approach, which conforms to the assumption of high shear stress, can cause several mapping and interpretation problems. We have seen maps derived from 2D data in which faults and structures *were forced* to conform to the simple strain assumption shown in Figure 12-1. An incorrect conclusion derived from the application of the simple model may cause several interpretation problems, which are discussed below. It is not our intent to single out specific interpretations, but to improve the understanding of strike-slip structures in order to help geoscientists generate high-quality prospects. Thus, we treat strike-slip faulting as a purely geometric problem that involves standard mapping techniques and methods.

One example of interpretation problems is fitting the simple strain model to nonexistent faults and thereby forcing these nonexistent faults into interpretations. Also, little or no consideration may be given to the exploration potential of structures that exist at a distance from a strike-slip fault. Alternatively, confusion may exist on the part of management as to why the structures do not exist at 30-deg angles to the fault zone, or why structures exist at a distance from the fault zone. Management may assign a higher risk to the area due to these apparent "strange structural complications."

Geoscientists may believe that folding must have been accompanied by strike-slip faulting in an area under study. Figure 12-1 implies that folds and strike-slip faults exist in common association, which occurs in many areas (Figs. 12-2 and 12-3). However, if strike-slip faults do not exist where folds are present, which is common to many folded terrains, then geoscientists *may force strike-slip faults through data in order to satisfy a 30-deg to 45-deg angle assumption*. Nonexistent faults may be drawn downward through poor seismic data zones to converge into a deep master basement fault, creating a *thick-skinned* strike-slip environment, where a *thin-skinned* compressional environment and hydrocarbon migration model is the appropriate model for the area. The interpretation of an intensely fractured structure may cause management to abandon a good prospect (Tearpock et al. 1994).

In cases concerning *en echelon* folded structures, interpreters may force strike-slip faults through coherent seismic data that constitute lateral ramps (Chapter 10), or force faults along axial surfaces in an attempt to satisfy the strain ellipse assumption (Fig. 12-1). These practices may result in incorrect seismic correlations, incorrect interpretations and maps, incorrect models of faulting and the petroleum system, and in numerous disappointments and dry holes. This may lead to further confusion concerning strike-slip faulting when evaluating the results of a drilling program and the local petroleum system.

CRITERIA FOR STRIKE-SLIP FAULTING

How does one recognize that strike-slip displacements exist in an area? Harding (1985, 1990) discusses seismic criteria for the recognition of strike-slip faults. His checklist includes the first three of the following main criteria, to which we add two other important and definitive criteria recognizable in petroleum-related data sets. We also provide additional criteria that directly follow from Harding's analysis.

1. *During strike-slip faulting, a large, near-vertical master fault offsets basement and cover rocks* (Harding 1990). In many cases, magnetic deconvolution can help resolve the depth to magnetic basement (Hartman et al. 1971). These relationships are shown on Figure 12-6 near sp 820 where a *near-vertical branch of the Philippine Fault displaces magnetic basement and cover rocks* (Bischke et al. 1990). The fault cannot rotate into a vertical position as a result of imbricate faulting.
2. *The seismic stratigraphy of the sediments on the opposite sides of the near-vertical fault should be fundamentally different* (Harding 1990) (Fig. 12-6 at sp 980). The juxtaposed seismic stratigraphy should be shown as not resulting from *inversion structure* or growth reverse faulting (see Chapter 13).
3. *The structure and seismic reflections are discontinuous across a high-angle fault surface* (Harding 1990) (Fig. 12-6 at sp 980). If you can easily correlate the seismic reflection character and geology across a possible strike-slip fault, then strike-slip faulting is probably not present. In this case other types of faulting should be considered.
4. *Fault surface maps depict a steeply dipping, high-angle surface* (Shaw et al. 1994) *that may contain en echelon offsets, or stepovers* (Aydin and Nur 1982). Fault surface mapping introduces 3D structural validity into interpretations. Surprisingly, in the literature we have seen no viable map of a fault surface, generated from seismic profiles, presented in support of strike-slip faulting. Figure 12-18 is a map of a restraining bend on the San Andreas Fault based on aftershock activity rather than on seismic or well data.

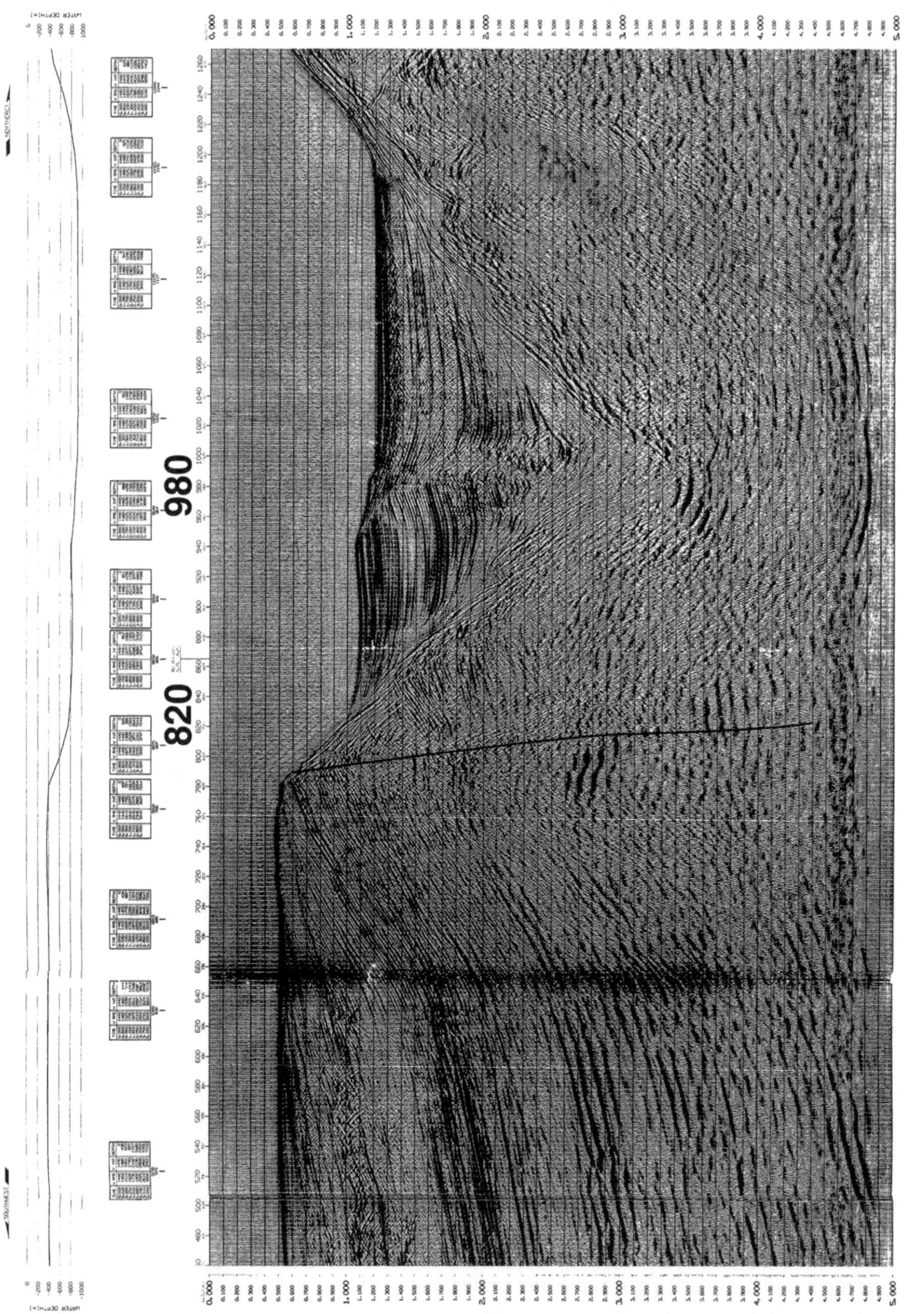

Figure 12-6 A large branch of the Philippine Fault, projected in from subsurface data and nearby surface maps from the island of Masbate. Large fault offsets magnetic basement represented by the bold reflections at sp 820 at 2.70 sec. The fault surface is nearly vertical. A minor splay images near sp 980, where the seismic stratigraphy and bed dips change across the fault surface. (From Bischke et al. 1990. Published by permission of the Philippine National Oil Company.)

5. The *en echelon* offsets are the sites of the *compressional restraining bends* and the *extensional releasing* bends, such as the rhombochasms and tipped wedge basins (Crowell 1974a, 1974b). These bends are recognizable on fault surface maps constructed along strike-slip faults, where they appear as bends in the fault surface. Criteria 4 and 5 are discussed in detail in the following sections.

If the preceding criteria are met within an area, then the application of a strike-slip fault model is warranted (Harding 1990) but *not proven*. Across normal and reverse faults, the hanging wall and footwall cutoffs of correlatable beds record the sense and the amount of the vertical and horizontal separations. If the cutoffs are recognized, *then vertical and horizontal displacements are proven*. However, across strike-slip faults, commonly no cutoffs exist to record the horizontal separations along straight-line segments of strike-slip faults. Unfortunately, 2D fault patterns or sets of divergent or convergent fault patterns, as interpreted on seismic profiles, *do not provide* direct evidence for lateral displacements. These divergent or convergent fault patterns provide evidence for *vertical separations,* but not actual *horizontal* displacements. Thus, *strike-slip faulting requires a 3D analysis in support of horizontal displacements.* The problem with these criteria, which are suggestive of strike-slip faulting, is the inability to directly address the issue of *horizontal displacements*. Economics obligates geoscientists to present viable interpretations rather than concepts "drawn on a seismic background," as Bally (1983) correctly recognized.

ANALYSIS OF LATERAL DISPLACEMENTS

In this section, we review a number of geological features used to document horizontal and vertical displacements on strike-slip faults. Documenting vertical displacements is important in the extensional releasing bends and the compressional restraining bends (Crowell 1974a, 1974b). We place particular emphasis on the local and regional restoration of geological features. These methods are capable of documenting the amount of lateral displacements, which are important when projecting sand trends or when analyzing the petroleum system. We begin with a discussion of surface geological features.

Surface Features

A number of surface features exist in association with strike-slip faults that support lateral or horizontal displacements in the Holocene. These geomorphic features include sag ponds, shutter ridges and pressure ridges (Allen 1962), offset river channels (Wallace 1968), and surface fractures (Wallace 1973). Topographic and geological maps present evidence of Holocene strike-slip motions.

Piercing Point or Piercing Line Evidence

Piercing point or piercing line evidence involves the displacement of geological features that were *initially intact, or unbroken,* prior to faulting. A piercing line can be some linear feature that is offset by faulting, and each offset end of the piercing line is a piercing point. Piercing points can also represent a nonlinear feature offset by faulting. These reference features can be seen in outcrop, imaged on a seismic profile, or constructed within a map or cross section.

Pregrowth strata, which are intact prior to faulting, constitute the best piercing line evidence. In growth strata, if stratigraphic intervals correlate across the structure or fault surface, then these syntectonic strata can provide good piercing line evidence. In this case, the sedimentation rate exceeds the tectonic uplift or fault slip rate. If, however, the fault slip rate temporarily exceeds the sedimentation rate, then the syntectonic sediments may contain an initial offset across the fault surface. Alternatively, reconstructions based on sediments deposited in a starved basin environment will contain a displacement error commensurate with the size of the initial offset. If the strata correlate across the fault surfaces, then these errors are likely to be small.

Faults displace distinctive stratigraphic horizons, diapirs, dike swarms, mountain ranges or basins. If one of these features is cut by a fault, then the feature may be *restorable* to its initial position. To determine the approximate slip on a fault, simply move the strata back along the fault until the displaced feature restores to its approximate initial, or intact, position (Sylvester 1988). The feature restores back to its initial configuration at a corresponding *piercing point or line*. For example, if a normal fault displaces a horizon, then the hanging wall cutoff restores back into its corresponding footwall cutoff. A 2D seismic profile would intersect the hanging wall and footwall cutoffs at two points. If the profile aligns in the direction of fault slip, then the structure restores back to its corresponding piercing points.

Some features provide better piercing point evidence than others. For example, stratigraphic pinchouts or subcrops provide better piercing point evidence than isopach or isochore maps constructed from syntectonic sedimentary intervals. The stratigraphic pinchout or subcrop information represents a "line in space," whereas an isopach map represents thickness information. Thus, problems concerning thickness changes related to syntectonic sedimentation often arise where the *stratigraphic units change thickness across active fault surfaces*. Stratigraphic thickness changes can result from a variety of causes, such as growth faulting and its associated syntectonic sedimentation, or from paleotopographic slopes that cause changes in basin configuration and environments. Geoscientists have documented growth sedimentation across all known geological structural styles, including compressional features and strike-slip faults (see Chapter 13).

Isopach or isochore information, if based on *pregrowth strata* that change stratigraphic thickness, provides good piercing point evidence. A discussion of methods for distinguishing between pregrowth and growth strata are in Chapter 13.

Other features that represent good piercing point evidence include offset zoned diapirs, mountain ranges, volcanic belts, and basins. Zoned diapirs and basins contain walls and flanks that represent offset surfaces where faulted. These features are useful for recognizing strike-slip faulting if the contacts are nearly vertical and correlatable. Mountain ranges and volcanic belts contain structures and trends that may be restorable, although mountain ranges and basins could contain pre-existing offsets such as the *salients* and *en echelon folds* common to fold-thrust belts. Offsets along salients and en echelon folds can be large and can introduce large errors into reconstructions. Subsequent fault motion may occur along these potentially weak, pre-existing offsets.

We discuss two types of piercing point evidence: first, the less definitive regional restoration process, then, the more definitive local and balanced restoration process.

Regional Restoration. Geoscientists use regional features to determine the approximate amount of slip along strike-slip faults. The Great Glen Fault in Scotland displaced a zoned granite batholith an apparent distance of 65 miles (105 km) (Kennedy 1946). In California, the San Andreas Fault offsets rocks to the north of the Salton Sea and those flanking the Soledad Basin by 250 km (Crowell 1962). The Garlock Fault apparently displaces a Mesozoic dike swarm and other geological features over a horizontal distance of 65 km (Fig. 12-7). The Philippine Fault system offsets Oligocene-Miocene intermediate and siliceous igneous rocks, ophiolite belts, and the intervening Central Luzon Valley-Llocos Basins up to 200 km to 300 km (Fig. 12-8). Gravity, isopach, and isochron maps can be subject to similar regional restoration.

Regional restorations of long-wavelength features such as volcanic chains and mountain belts, gravity and magnetic anomalies, unconformity intersections, and isopach thicknesses provide insight into the approximate amount of horizontal slip on strike-slip faults. These

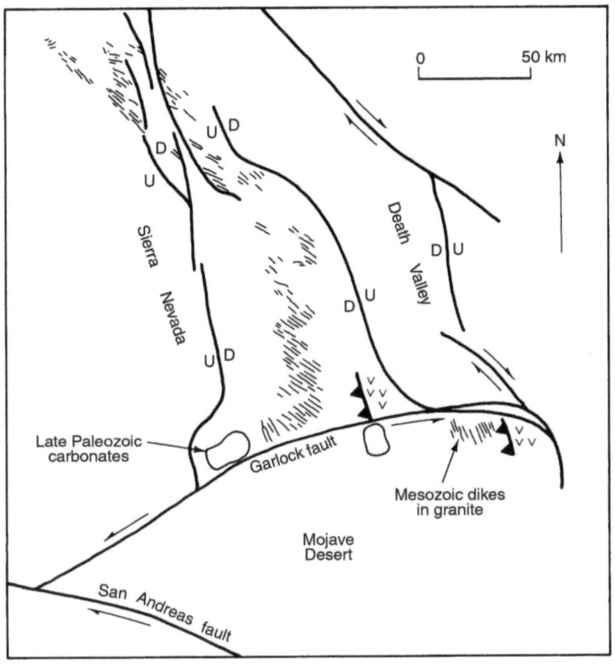

Figure 12-7 Offset features along Garlock Fault, California (Suppe 1985).

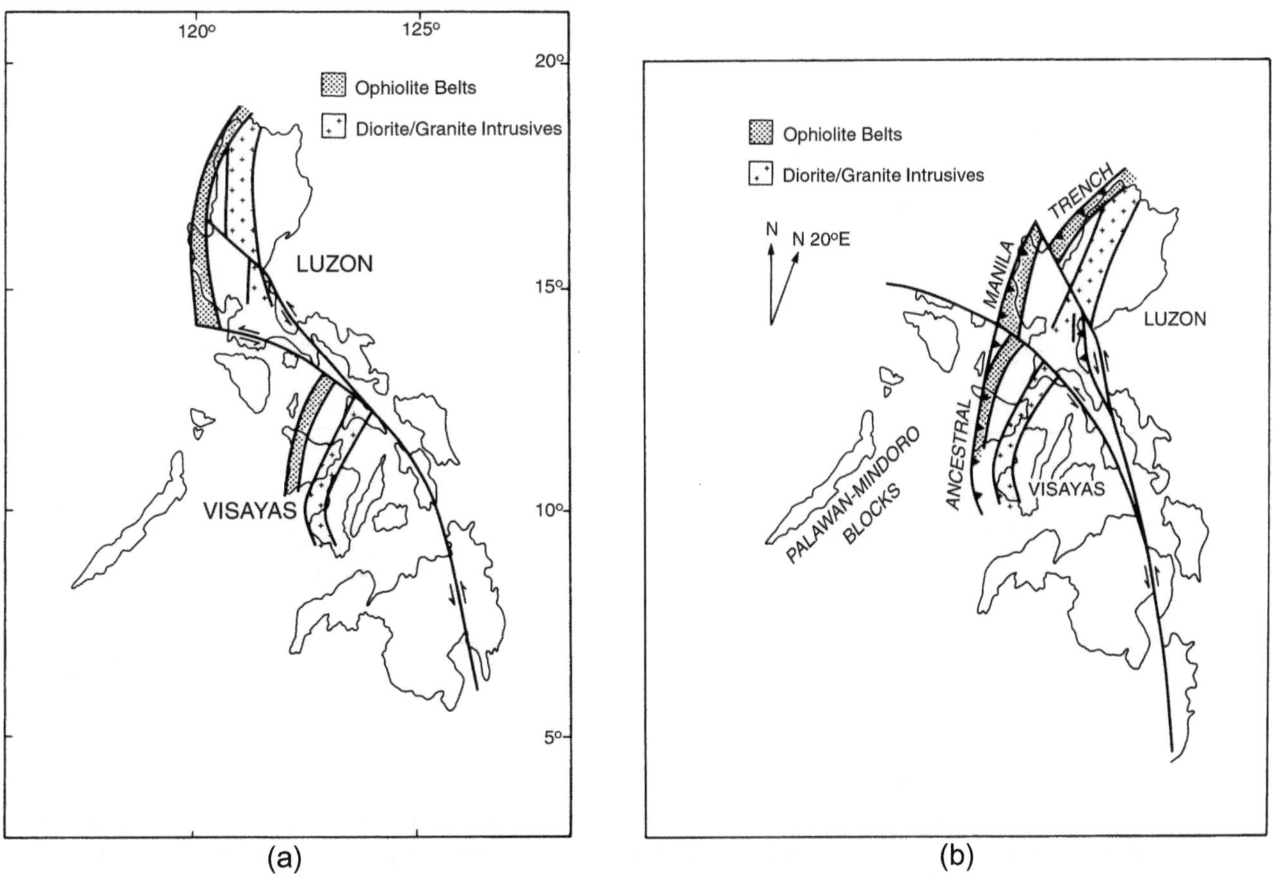

Figure 12-8 (a)–(b) Restoration of Philippine Fault, using long-wavelength features such as ophiolite belts, sedimentary basins, and volcanic chains. (From Bischke et al. 1990. Published by permission of Tectonophysics.)

approximate restorations, in support of strike-slip motion, are more convincing when supported by the criteria discussed in the section on criteria for strike-slip faulting.

Local Restoration. Commonly, the best available evidence in support of strike-slip faulting is the presence of the ubiquitous **restraining bends** and **releasing bends** (Crowell 1974a, 1974b). Restraining and releasing bends document horizontal displacements, and the restoration of these features, using piercing point and piercing line evidence, can determine the *approximate slip* along fault surfaces (Hill and Dibblee 1953).

The known strike-slip faults of the world are not perfectly straight faults, but contain *small to large* en echelon offsets, called stepovers, that form restraining and releasing bends (Aydin and Nur 1985) (Fig. 12-9). Table 12-1, taken from Aydin and Nur (1982), lists constraining and releasing bends from various areas of the world.

Alternatively, restraining and releasing bends form at bends in strike-slip fault surfaces (Crowell 1982). If slip along the linked strike-slip fault system moves material *away from* the stepover or bend in a fault surface, then the resulting *extension* forms *releasing bends* (Fig. 12-9a). If, however, slip along the stepover or bend moves material *into* the stepover or bend in the fault surface, then the resulting *compression* causes *restraining bends* (Fig. 12-9b) (Crowell 1974a, 1974b; McClay and Bonora 2001).

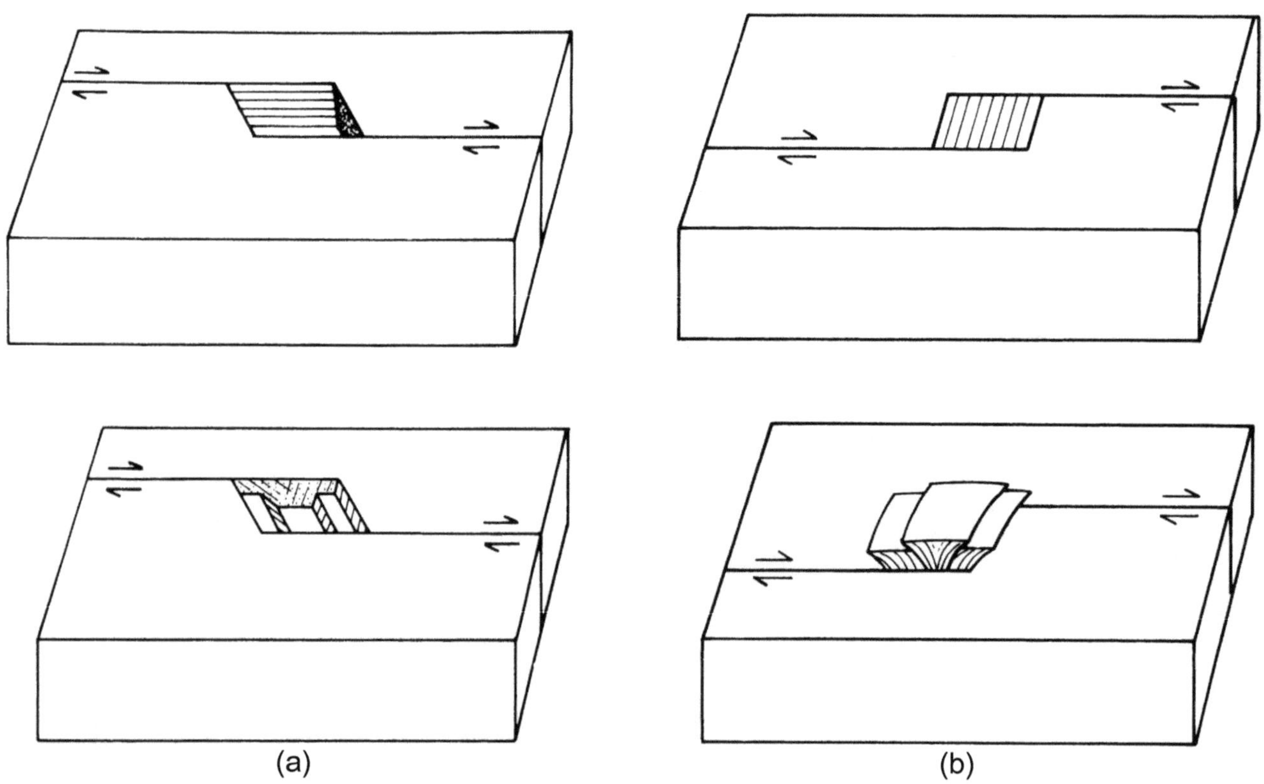

Figure 12-9 (a)–(b) Geometry of releasing and restraining bends (Aydin and Nur 1985). (a) If material moves away from a stepover, then extension and resultant normal faults develop a releasing bend. Structural lows exist in the area of the stepover. (b) If material moves into a stepover, then compression and resultant reverse and thrust faults develop a restraining bend. Structural highs exist in the area of the stepover. (From Aydin and Nur 1985. Published by permission of the Society for Sedimentary Geology.)

Table 12-1 Size of restraining and releasing bends along strike-slip faults. (Aydin and Nur 1982. Published by permission of the American Geophysical Union.)

Fault and/or Location	***Basin or Mountain Range***	***Graben (G) or Horst (H)***	***Dimension (M)***		***Reference***
			Length	***Width***	
Motagua, Guatemala	Motagua Valley	G	50,000	20,000	Schwartz et al. [1979]
	Rio El Tambor	G	25	8	
Polochic	Lago de Izabal	G	80,000	30,000	Bonis et al. [1970]; Plafker [1976]; this study
Dead Sea Rift, Israel	Hula	G	20,000	7,000	Freund et al. [1968]
	Lake Kineret	G	17,000	5,000	
	Ayun	G	6,600	1,600	
	East of Timna	G	1,000	250	
	North of Ayun	G	1,200	400	Garfunkel et al. [1982]
		G	1,200	400	
		G	1,600	450	
		G	5,000	1,200	
		G	2,000	500	
	South of Timna	G	8,800	3,000	
		G	20,000	6,000	
	West of the Dead Sea	G	3,500	750	Garfunkel [1982]
		G	3,000	750	
		G	3,000	800	
		G	6,000	1,500	
		G	7,500	1,800	
		G	3,000	750	
	East of the Dead Sea	G	4,500	1,500	
Paran	Karkom	G	18,000	6,000	Bartov [1979]
		G	6,000	1,500	
Bir Zrir, Sinai		G	5,000	2,000	Eyal et al. [1980]
Gulf of Elat	Elat	G	45,000	10,000	Ben-Avraham et al. [1979]
	Aragonese	G	40,000	9,000	
	Tiran-Dakor	G	65,000	8,000	
Dasht-e Bayaz, Iran		G	1,200	500	Freund [1974]
Hope, New Zealand	Medway-Karaka	G	700	230	Freund [1971]
	Glynnwye	G	980	210	
	Glynnwye Lake	G	1,800	550	
	Polars Station	G	2,300	900	
	Hanmer Plains	G	13,000	3,500	Freund [1974]

(*continued*)

Table 12-1 Size of restraining and releasing bends along strike-slip faults. (Aydin and Nur 1982. Published by permission of the American Geophysical Union.) (*continued*)

Fault and/or Location	Basin or Mountain Range	Graben (G) or Horst (H)	Dimension (M)		Reference
			Length	Width	
Hope, New Zealand	Medway-Karaka	H	90	30	Freund [1971]
	Glynnwye Lake	H	300	90	
	Poplars Station	H	300	150	
	Hanmer Plains	H	4,500	2,700	
North Anatolian, Turkey	Niksar	G	25,000	10,000	Seymen [1975]; this study
	Erzincan	G	40,000	12,000	Ketin [1969]
	Susehri	G	23,000	6,000	
San Andreas, California, USA	Cholame Valley	G	17,000	3,000	Jennings [1959]; Brown [1970]
	San Bernardino Mountains	H	32,000	14,000	Dibblee [1975]
Imperial	Brawley	G	10,000	7,000	Johnson and Hadley [1976]; Sharp [1976, 1977]
Elsinore	Elsinore Lake	G	12,000	3,000	Rogers [1965]
Garlock	Koehn lake	G	40,000	11,000	Jennings et al. [1969];
		G	300	150	Smith [1964]; Clark [1973]; this study
		G	600	110	
	West of Quail Mountain	G	600	100	
		G	240	90	Clark [1973]
		G	900	220	
	Searleys Valley	G	1,600	380	
	East of Christmas Canyon		1,250	250	
San Jacinto,	Hog Lake	G	680	170	Sharp [1972]
	Hemet	G	22,000	5,000	Sharp [1975]
Buck Ridge	Santa Rosa Mountain	G	6,000	1,700	Sharp [1972]
Coyote Creek	Ocotillo Badlands	H	5,500	1,800	Sharp and Clark [1972]
	Borrega Mountain	H	4,000	1,600	
	Bailey's Well	G	500	200	Clark [1972]
		G	190	80	

Fault and/or Location	Basin or Mountain Range	Graben (G) or Horst (H)	Dimension (M)		Reference
			Length	Width	
Olinghouse, Nevada	Tracy-Clark Station	G	70	40	Sanders and Slemmons [1979]; this study
		G	160	90	
		G	450	175	
		G	980	250	
Bocono, Venezuela	La Gonzales	G	23,000	6,200	Schubert [1980a]
	Merida-Mucuchies	G	6,200	1,700	Schubert [1980b]
		G	700	200	
		G	280	70	
Valencia	Lake Valencia	G	30,000	11,500	Schubert and Laredo [1979]
El Pilar	Casanay	H	3,000	1,200	Schubert [1979]

Releasing Bends. If motion at a stepover or bend in a strike-slip fault creates extension, then a basin bounded by normal faults develops (Fig. 12-9a and b). Two examples of this type of motion are segments of the Hope Fault, New Zealand, and the San Jacinto Fault, California, USA (Suppe 1985) (Fig. 12-10b and c). The San Jacinto Fault is part of the San Andreas Fault system.

Releasing bends form as material within a stepover is subjected to extension. The amount of extension is related to the amount of slip on the strike-slip fault system. According to the releasing bend model, *strike-slip faults bound the basin on two parallel sides and normal faults bound the basin on the other two sides.* The strike-slip faults form the walls of the basin *between* the stepover, and normal faults form basin margins at the *ends* of the stepover (Fig. 12-9a). As the basin extends, material slumps into the void created by the extension parallel to the direction of strike-slip fault motion, and thus the extension *records the amount of displacement that formed the basin*. To determine the approximate amount of motion that formed the bend, restore the bend by moving the correlatable strata back in a direction that is opposite to the direction of strike-slip displacements.

If the subsidence rate in the basin exceeds the sedimentation rate, then the syntectonic sediments deposited concurrent with strike-slip displacements may contain initial offsets across the normal fault surfaces. Although *releasing bends provide direct evidence* for horizontal displacements, the restoration of releasing bends *may overstate* the amount of the strike-slip displacements, if the upthrown block does not contain growth sediments. Sequence stratigraphic evidence, based on high-stand or low-stand evidence, can minimize the amount of error encountered during the restoration process. However, if the stratigraphy easily correlates across fault surfaces, then any error in restoration is likely to be small.

The normal faults that form the margins of the basin contain hanging wall and footwall cutoffs that form piercing lines (Fig. 12-11a and b). These piercing lines represent the hanging wall and footwall cutoffs mapped in three dimensions. The piercing lines form as the blocks at the edge of the extensional basin slump into the basin subnormal to the surface trace of the

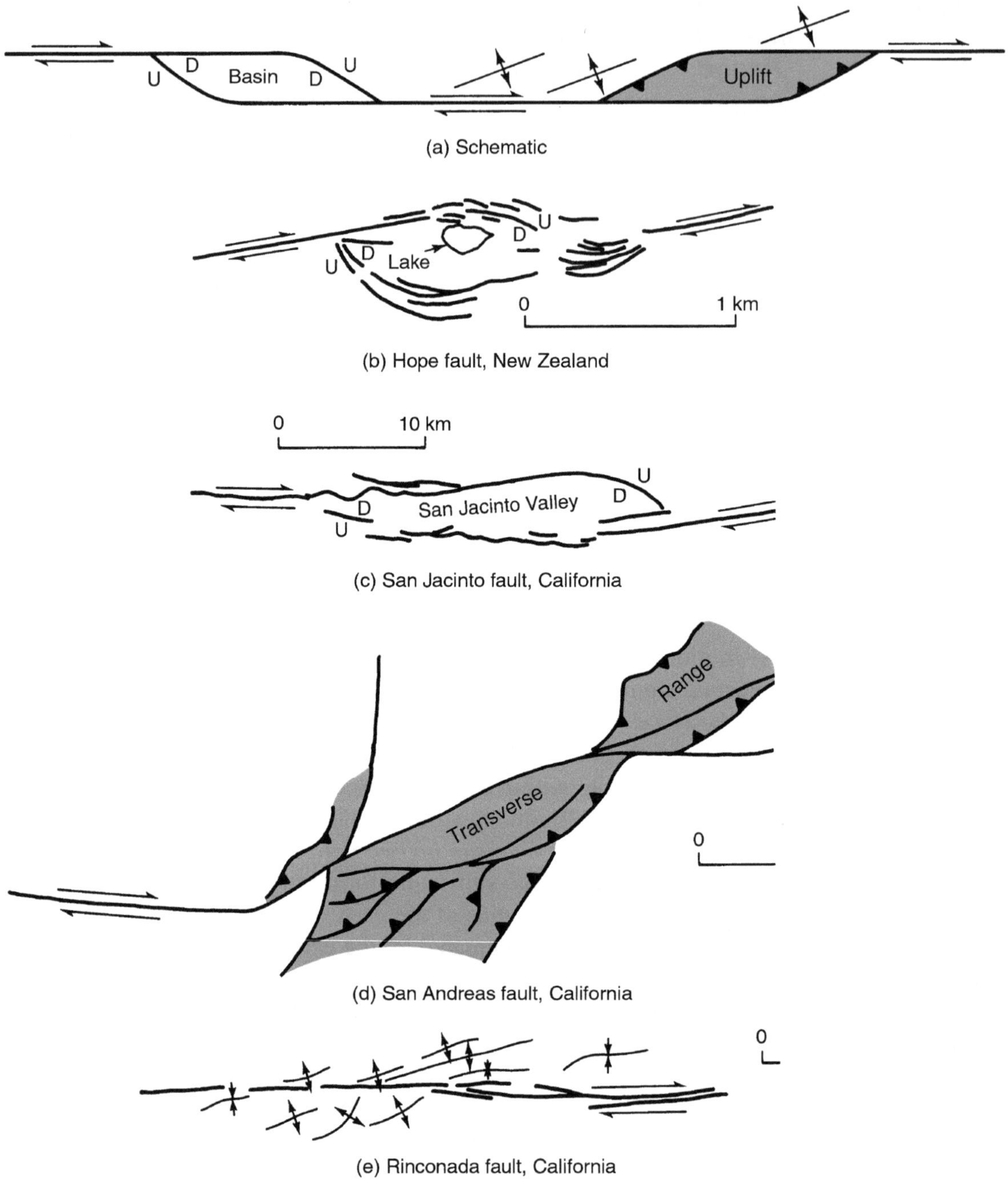

Figure 12-10 (a)–(e) Examples of releasing and restraining bends (Suppe 1985).

strike-slip fault. Thus, if we can assume that *the direction of strike-slip motion is parallel to the surface trace* of the strike-slip fault (i.e., in the strike direction of the mapped fault surface), then we can restore, or close, the basin by moving the strata in the opposite direction of fault displacements and along any number of profiles that parallel the surface trace of the strike-slip fault (Fig. 12-11b and c). The hanging wall cutoffs restore back into the footwall cutoffs at their corresponding piercing points located along the piercing line (Fig. 12-11c).

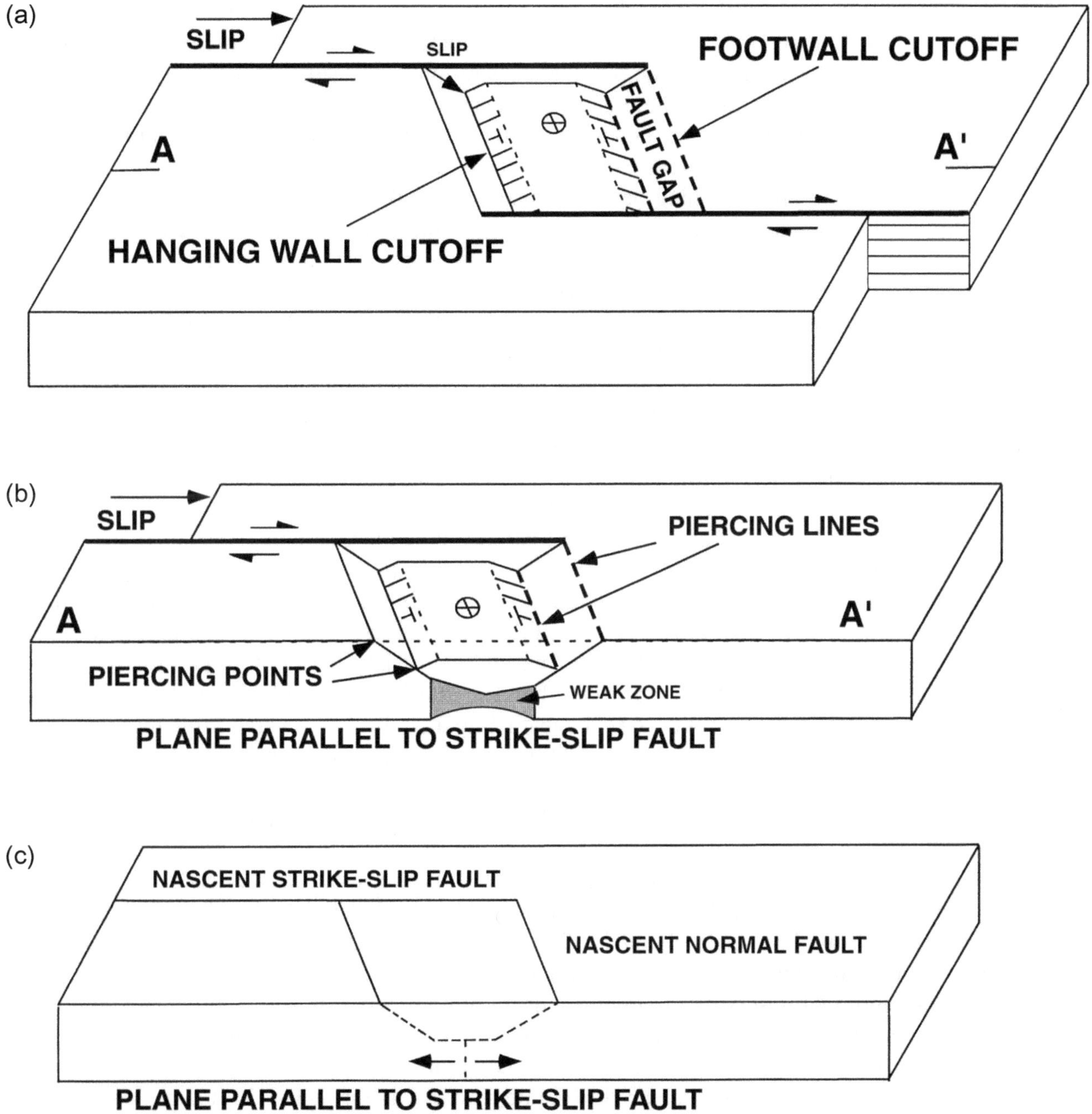

Figure 12-11 Motion along a strike-slip fault can be restored by (a) mapping the hanging wall and footwall cutoffs. (b) A profile taken parallel to the surface trace of the strike-slip fault cuts the piercing lines, creating piercing points. (c) The piercing points are restored back to an undeformed position. (Published by permission of R. Bischke.)

Restraining Bends. If material moves into a stepover or bend, compression occurs in the restraining bend (Crowell 1974a, 1974b; Aydin and Nur 1985; Cunningham and Mann 2007). The compression forms reverse faults, thrust faults, and pressure ridges, or pop-ups (Fig. 12-9b). A description of the complex deformation occurring on a pop-up from 3D seismic data is in Durand-Riard et al. (2013). This example is on a clear restraining bend. McClay and Bonora

(2001) present several examples of restraining stepovers from Nevada, Wyoming, Chile, and the Netherlands. The Transverse Ranges of California, which exist on the Great Bend, a stepover in the San Andreas Fault, are an example of a large restraining bend (Fig. 12-10d). Contraction also occurs at restraining bends on continuously curved strike-slip fault surfaces. Modern 3D seismic time seismic data image these offset bends (Benesh et al. 2014). We discuss an example along the Loma Prieta bend in the San Andreas Fault in a later section in this chapter.

Let's restore a small restraining bend to illustrate how the process confirms the presence of strike-slip displacements, restores the initial Pliocene stratigraphic trends, and allows us to estimate the amount of Pliocene displacements. This knowledge will allow interpreters to converge on the geometry and history of the Pliocene structures and the associated petroleum system. The bend is located on the flank of the Long Beach Anticline in the Newport-Inglewood Trend, southern California, USA. The Newport-Inglewood Trend is a classic zone of "transpressional" deformation (Harding 1973). The strike-slip faults along the trend form left-stepping, en echelon offsets, and the Cherry Hill and Northeast Flank Faults are major faults in the Newport-Inglewood Trend. Along the southern flank of the Long Beach Anticline, the Northeast Flank Fault steps over to the Cherry Hill Fault (Wright 1991) (Fig. 12-12a). Trenching indicates that the Cherry Hill Fault dies out to the southeast of the map area. Motion along the Newport-Inglewood Trend is right-lateral and, therefore, the bend should be subject to compression. If we assume that material enters the bend parallel to the surface trace of the Northeast Flank Fault or the Cherry Hill Fault, then the resulting contraction is restorable. This contraction forms the Signal Hill Promontory, or pressure ridge (Fig. 12-12b), and a reverse fault that dips at about 50 deg to the southeast (Taylor 1973) (Fig. 12-13). The method of implied fault strike (Tearpock et al. 1994), when applied to Taylor's map of the bend in the Northeast Flank Fault (Fig. 12-12a), shows that the fault strikes about N65E beneath Signal Hill. The reverse fault motions cause the hanging wall beds and footwall beds to form piercing lines that strike northeast-southwest beneath Signal Hill.

Some textbooks seem to imply that strike-slip and compressional displacements are causatively related, and that transpressional strike-slip displacements commonly generate compressional folds adjacent to strike-slip faults (Fig. 12-1). The Signal Hill pressure ridge formed on the flank of the Long Beach Anticline, but strike-slip displacements need not be the cause of the Long Beach Anticline itself (Wright 1991). In this case, a decoupling of the strike-slip displacements from the compressional displacements may be more appropriate.

Several empirical observations support a decoupling process, which could change the interpretation of the Newport-Inglewood Trend. For example, according to the transpressional explanation shown in Figure 12-1, the axis of the folds would initiate at 30-deg to 45-deg angles to the strike-slip fault. The strains required to rotate a fold axis through a 30-deg to 45-deg angle suggest large amounts of strike-slip displacements. As the axis of the Long Beach Anticline is parallel to the Cherry Hill Fault (Fig. 12-12a), the strike-slip displacement on the Cherry Hill Fault should be substantial. Harding (1973) makes the observation that fold axes are presently offset only by 200 m to 800 m, but some or most of this displacement could be an initial offset of the axes of compartmentalized folds (Chapter 8, Fig. 8-74). The front limb of the Long Beach Anticline is not offset from its crest (Wright 1991) (Figs. 12-13 and 12-14). A second observation is that the structural contours in the hanging wall of the Northeast Flank Fault (Fig. 12-12a) are compatible with the footwall structural contours. This structural compatibility also implies small Plio-Pleistocene displacements (see Chapter 8). Thus, some or all of the displacements on the Northeast Flank Fault appear to be younger than the Long Beach Anticline, which supports Wright's (1991) observations that the Newport-Inglewood Trend exhibits a complex structural and stratigraphic history not easily reconciled with a simple strike-slip origin.

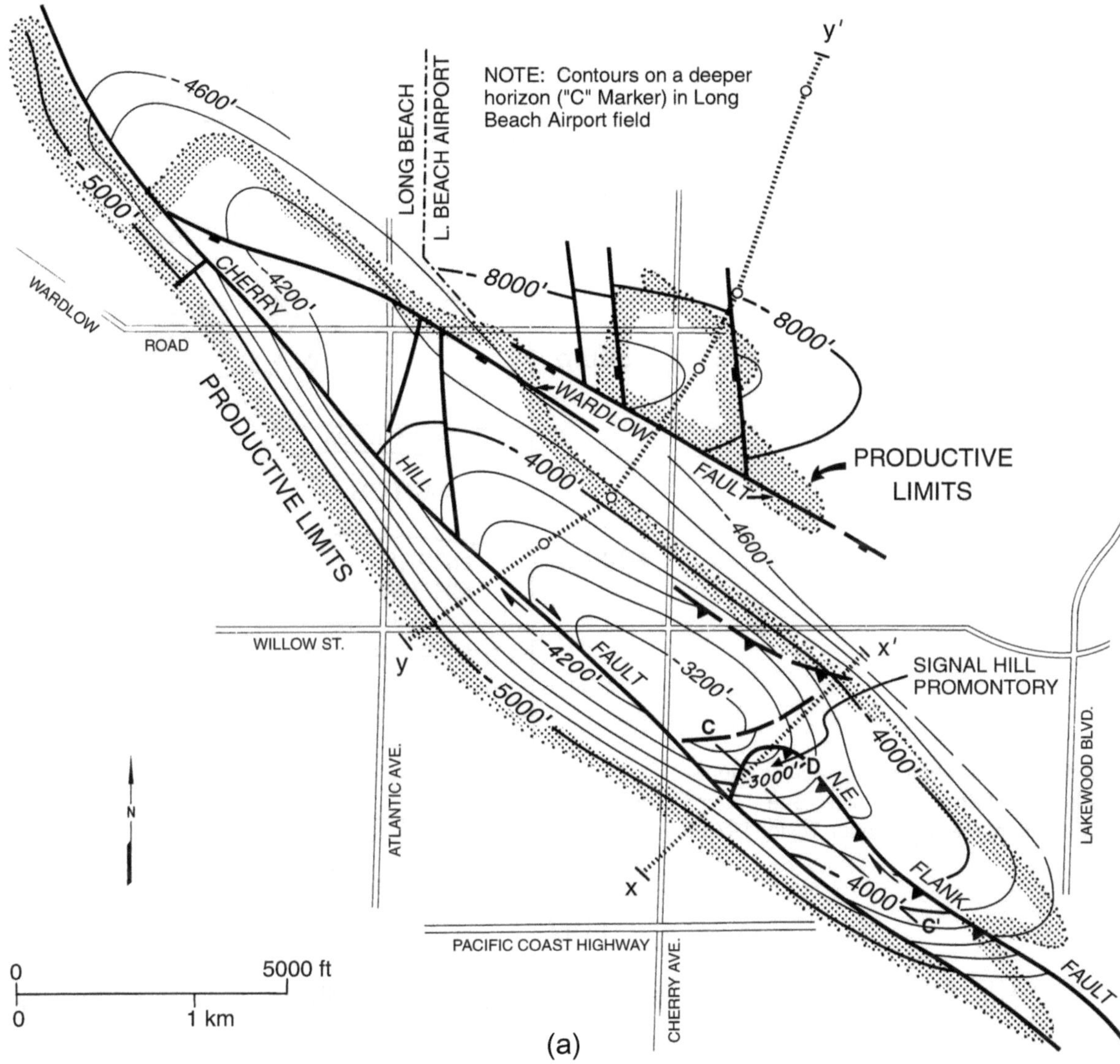

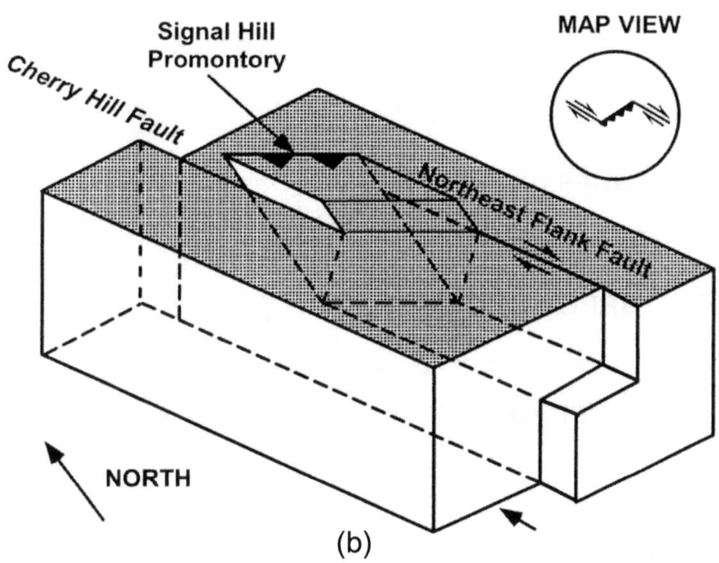

Figure 12-12 (a) Structural map of Long Beach Anticline showing Cherry Hill and Northeast Flank faults that locally define the Newport-Inglewood Trend. Profile C-C′, which trends NW-SE across the Signal Hill pressure ridge, is parallel to the surface trace of the Northeast Flank and Cherry Hill Faults. (Modified from Wright 1991. AAPG©1991, reprinted by permission of the AAPG whose permission is required for further use). (b) Structural model for Signal Hill restraining bend. Northeast Flank Fault bends to the southwest linking to Cherry Hill fault, forming a restraining bend. (Published by permission of J. Shaw and R. Bischke.)

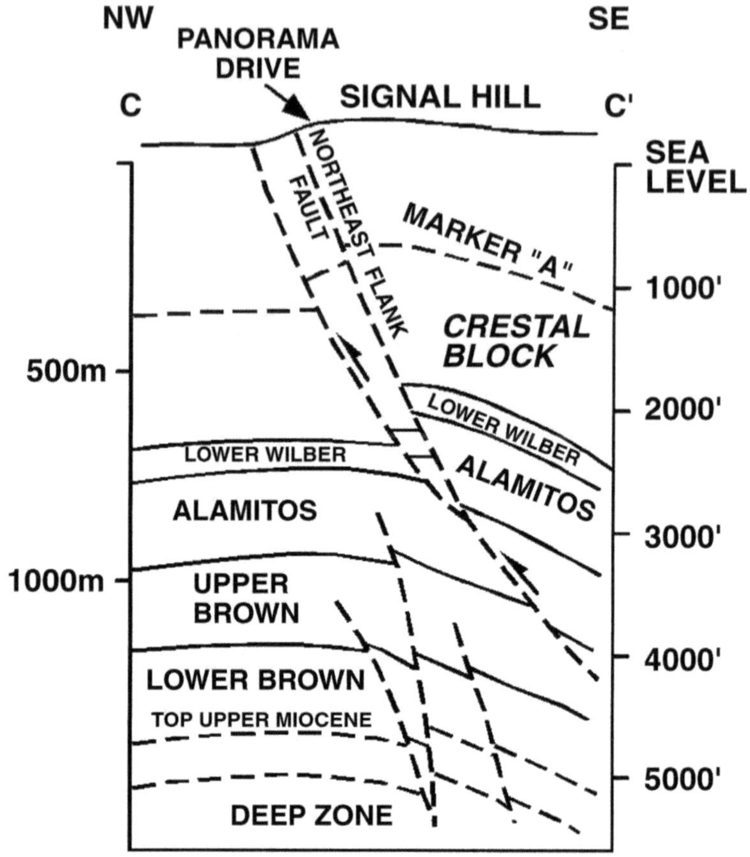

Figure 12-13 Cross section C-C′ across Signal Hill, parallel to Cherry Hill Fault. See Figure 12-12a for location. (Redrawn after Taylor 1973.)

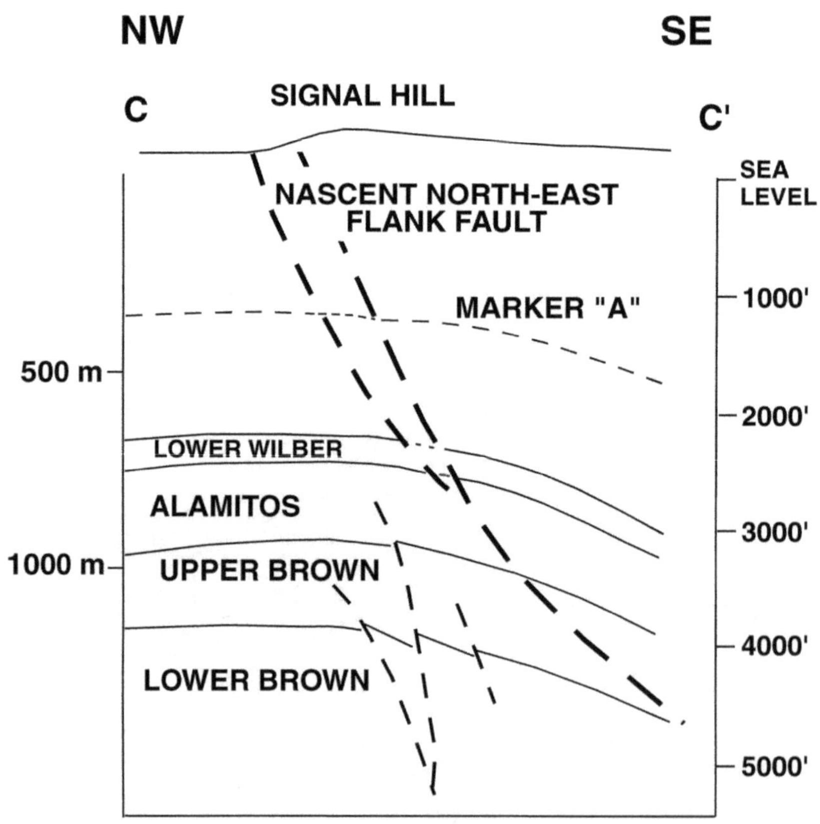

Figure 12-14 Restored Signal Hill block that was thrust to the northwest. Block restores by moving strata back along the Northeast Flank Fault parallel to the surface trace of the Cherry Hill and Northeast Flank Faults. Restoration shows a pre-existing Long Beach folded anticline. (Published by permission of R. Bischke.)

If you construct a fault surface map for the Northeast Flank Fault along the southeastern flank of the Long Beach Anticline, it shows that the fault surface curves beneath Signal Hill (Taylor 1973), where the fault strikes at about N65E (Fig. 12-12). To the southeast of the anticline, the fault dips at high angles and strikes at about N60W (Fig. 12-12). A seismic or geological profile, such as cross section C-C′ taken in the NW-SE direction (and across the curved portion of the fault surface), will confirm whether the beds are thrust, reverse, or normally faulted. In this case they are reverse faulted (Fig. 12-13) and, referring to Figure 12-9, we can deduce the correct sense of strike-slip motion. That confirms the Signal Hill restraining bend to be a small but obvious restraining bend.

It is obvious from this discussion that in order to locate bends in fault surfaces, *it is necessary to construct accurate maps of the fault surfaces*, as described in Chapter 7. Typically, a 2D seismic grid is sufficient for constructing general fault surface maps and permits the detection of gentle bends in fault surfaces. Bends or offsets in fault surfaces have important consequences concerning the correct interpretation of data, prospect generation, or the absence of prospects (Tearpock et al. 1994). This is another reason for constructing *loop-tied fault surface maps*. As curved fault surfaces create releasing and restraining bends along strike-slip faults, maps of these fault surfaces may readily solve difficult structure problems.

The Signal Hill restraining bend is restorable along any number of profiles aligned subparallel to the surface trace of the Cherry Hill or Northeast Flank Faults (Figs. 12-13 and 12-14). The Northeast Flank Fault exhibits about 150 m of vertical separation and about 170 m of horizontal separation in the Pliocene Lower Wilber and Alamitos horizons. These displacements are in general agreement with Harding's (1973) estimates. The Northeast Flank Fault cuts the southeastern limb of a pre-existing Long Beach anticline (Figs. 12-12 and 12-14) and appears *younger than the compressional folding* that formed the Long Beach anticline. The strike-slip motion may decouple from the compressional motions that formed the Long Beach Anticline (Shaw and Suppe 1996). Therefore, the compressional and strike-slip motions may not be directly related.

A misconception concerning piercing points is that offset fold axes, once thought to represent a continuous line, are good piercing point evidence because these offset fold axes restore to a single line. However, fold compartmentalization caused by tear faults are known to exist (Chapter 8, Fig. 8-74), and *associated folds may form with an initial offset*, and not along a single unbroken fold axis. These initial offsets can be large. In addition, folds forming along tear faults may grow during the deformation process, contributing to the offset. Unfortunately, we find that most types of piercing point evidence are rare or absent from most subsurface data sets.

In summary, the releasing and restraining bends, which are common to the known strike-slip faults throughout the world, often provide the best evidence in support of strike-slip faulting (Aydin and Nur 1985). If you are working a suspected strike-slip fault, locate a bend to confirm the strike-slip motions. Partially linked strike-slip fault systems typically contain stepovers. Often, a quick look at a structure map containing a strike-slip fault interpretation can resolve the presence or absence of strike-slip faulting in a matter of minutes. If strike-slip faults are thought to be in the area, examine a structure map for the presence of en echelon stepovers. Next, inquire as to the direction of fault motion. If fault motion moves material into the bend, then *a structural high should exist on the map* in the area of the restraining bend (Figs. 12-9b and 12-12). If, however, fault motion moves material out of the bend, then *a low should exist on the map* in the area of the releasing bend (Figs. 12-9a and 12-10b and c). An exception to this *quick-look technique* is structural inversion, which can also produce high and low areas along en echelon inversion faults. If a regional strike-slip fault interpretation does not contain restraining and releasing bends, perhaps style of faulting is present.

The documentation of strike-slip motion may be no more difficult than the documentation of other types of fault motion, but specific technical work must be done. First, fault surface maps, constructed in support of the deformation, should contain geometries that are consistent with the highs and lows on horizon structure maps. Second, in other tectonic regimes, whether it be compressional, extensional, or salt-related deformation, geoscientists and management require *direct evidence* of the type and style of the deformation. Strike-slip faults *are not two-dimensional, and thus they require a 3D analysis.* In many cases, the answer lies in a 4D analysis of the problem (Wright 1991). We presented techniques that can rapidly resolve the interpretation of strike-slip faulting that are no more difficult than techniques required to confirm other types of faulting. These techniques are supported by empirical evidence that establishes 3D structural validity.

Modern explorationists place emphasis on the petroleum system and its associated risk factors. To better understand how structures form and how faulting affects structural development and hydrocarbon migration, a good understanding of fault timing and geometry is necessary. We know of no way to address these issues concerning risk without fault surface maps, *correctly interpreted from loop-tied data.* If fault surface maps do not exist, then interpreters working an area have imprecise knowledge as to how the local structures formed and how the recognized faults may have channeled hydrocarbons. How can geoscientists correctly identify and understand the type and style of faulting, and generate viable prospects, without constructing viable maps based on loop-tied data? Furthermore, these maps of the fault surfaces may contain subtle bends that indicate small restraining and releasing curves, thus generating additional prospective structures. Fault surface maps also may record the strike-slip motions, as we show in a following section.

If accurate fault surface maps are not available, then it is possible to *misidentify* the structural style. On 2D seismic data, one may factor flower structure fault geometries (Harding 1985) into the risk analysis, where in reality inversion is the correct structural style. The petroleum system model may erroneously contain *thick-skinned vertical faults*, whereas in reality the faults *are thin-skinned, low-angle faults.* This can have a major effect on the interpretation of how hydrocarbons migrate and enter structures. Thus, differences in structural style can impact the understanding of the petroleum system, the interpretation model, the determination of risk, and the ultimate success of an exploration or development program.

Strike-slip faulting is sometimes over-interpreted and confused with other structural styles (Harding 1985), and some interpretations may lack *positive evidence in support of horizontal displacements.* Too often, strike-slip fault interpretation and the related structures are supported by a *lack* of data, no-seismic-data zones, high bed dips, axial surfaces, misapplied seismic interpretation rules, and so on (Tearpock et al. 1994). Certainly, strike-slip faulting should be subject to the *same scientific methods and subsurface mapping standards* that apply to other styles of faulting. Your success as a geoscientist depends on high-quality interpretations, maps, and prospects. Only solid scientific work, as described here, can ensure the best chance of success.

SCALING FACTORS FOR STRIKE-SLIP DISPLACEMENTS

Scaling laws for faults allow us to predict the approximate length of all faults, including strike-slip faults. These empirical laws can keep interpretations focused and accurate in order to improve the success rate for finding hydrocarbons. Aydin and Nur (1982, 1985) conducted two interesting studies of restraining and releasing bends and developed a scaling law for the length of strike-slip fault bends. Their study has implications regarding the development of strike-slip faults that may not be totally understood. Aydin and Nur investigated 11 major strike-slip faults in various areas around the world, and they measured the length and the width of 70 restraining and releasing bends associated with those faults (Table 12-1). The width of bends was not independent of the length of bends, as one would assume from the model shown in Figure 12-15. Instead, the lengths (L) of bends are proportional to the widths (W) by the relation

$$L \cong 3.2\ W \tag{12-1}$$

Equation 12.1 indicates that L is approximately equal to 3.2 W, at the 95% confidence level. The formula implies that the widths of bends widen as the slip on the faults increases. This implies that smaller bends may coalesce and cluster into larger bends (Aydin and Nur 1982), or that fault zones grow wider as the faults grow in length.

The consequences of their study are multifold. Strike-slip bend widths can be used to estimate the lengths of bends. Large strike-slip faults have wide bends (Fig. 12-10d), whereas small strike-slip faults have narrow bends. It is possible, however, that a large strike-slip fault may deactivate and a new fault may replace a major fault, which would start the process over again. As *large* strike-slip faults have *wide* bends, these wide bends should be *easy to locate* during a regional analysis, and help to confirm the strike-slip interpretation. Narrow bends on small strike-slip faults may be more difficult to locate. In this case, however, the *strike-slip process is less important* and other processes may dominate, such as extensional or compressional faulting or folding.

We know of no *well-documented* example by which large amplitude anticlines, domes, structural highs, or basins form as a result of small amounts of strike-slip motion. If the amount of strike-slip deformation is small, the exploration emphasis must shift from processes associated with strike-slip faulting to processes associated with other types of faulting. Another consequence of the study of releasing and restraining bends is that the bends are relatively common along strike-slip faults (Table 12-1). Lastly, as releasing bends grow wider with increasing slip, smaller basins that form within each bend may widen to form larger basins.

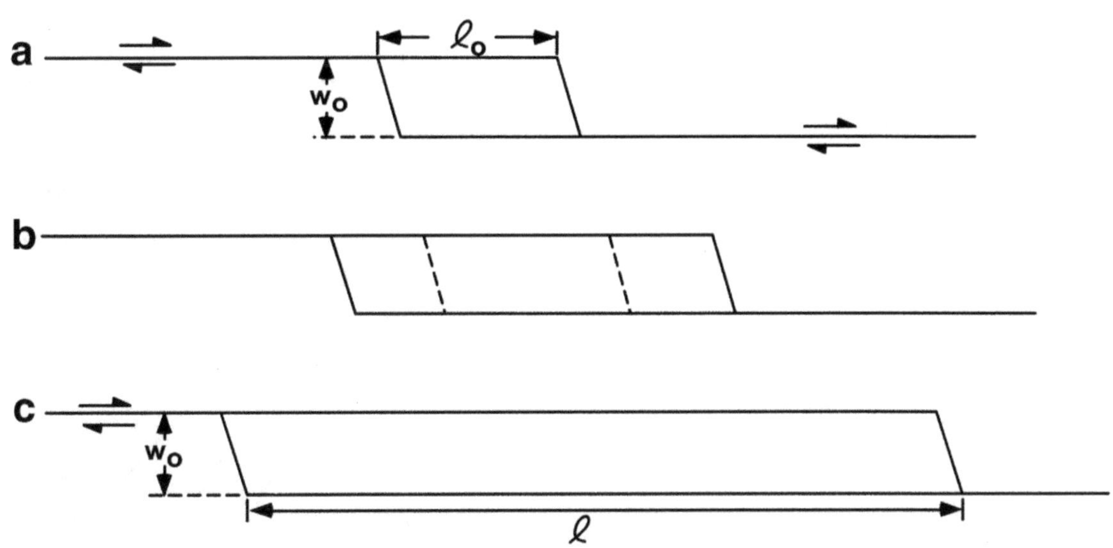

Figure 12-15 A simple model of a strike-slip fault that has a releasing bend that grows in length as the slip on the fault increases. According to this model, the width (w) of the bend should be independent of the length (l) of the bend. Strike-slip faults do not behave according to this simple model, and real strike-slip faults exhibit a relationship in which length is proportional to width. (From Aydin and Nur 1982. Published by permission of the American Geophysical Union.)

BALANCING STRIKE-SLIP FAULTS

The subject of balancing strike-slip faults is in its infancy. We have heard arguments to the effect that large amounts of displacement along some strike-slip faults preclude any attempt to balance or restore the displacements. In the Local Restoration section, we show that large strike-slip faults are locally restorable at their restraining and releasing bends. The concept of piercing lines or points enables geoscientists to restore the section under the assumption that material either enters or leaves the bend parallel to the surface trace of the strike-slip fault. Similarly, in the Regional Restoration section, long-wavelength restorations of major geological or geophysical features document the amount of strike-slip displacements on several well-known faults. Geoscientists have made considerable progress in restoring and documenting (Durand-Riard et al. 2013) strike-slip deformation in three dimensions.

Compressional restraining bends balance if we use the methods outlined in the section on compressional faulting in Chapter 10. We can balance extensional releasing bends using the methods described for extensional faulting in Chapter 11.

Compressional Folding along Strike-Slip Faults

Surface geological information allows us to generate a balanced cross section that is comparable to the subsurface geometry along a portion of the San Andreas Fault in California, as defined by earthquake hypocenters. The Loma Prieta Earthquake occurred on a restraining bend of the San Andreas Fault (Shaw et al. 1994; Schwartz et al. 1994) (Fig. 12-16). In the Loma Prieta epicentral zone, the surface of the San Andreas Fault changes its strike from N40W (320 deg) to N50W (310 deg) (Fig. 12-17). Material entering this restraining bend should therefore be subject to compressional as well as strike-slip motions. Accordingly, the focal mechanism solution for the earthquake inferred from first-motion studies is oblique-reverse, right-lateral motion (Oppenheimer 1990). Geodetic data indicate 1.6 m ± 0.3 m of strike-slip and 1.2 m ± 0.3 m of reverse slip during the earthquake (Lisowski et al. 1990). A fault surface map derived from aftershock hypocentral locations is shown in Figure 12-18a. The fault surface changes strike and

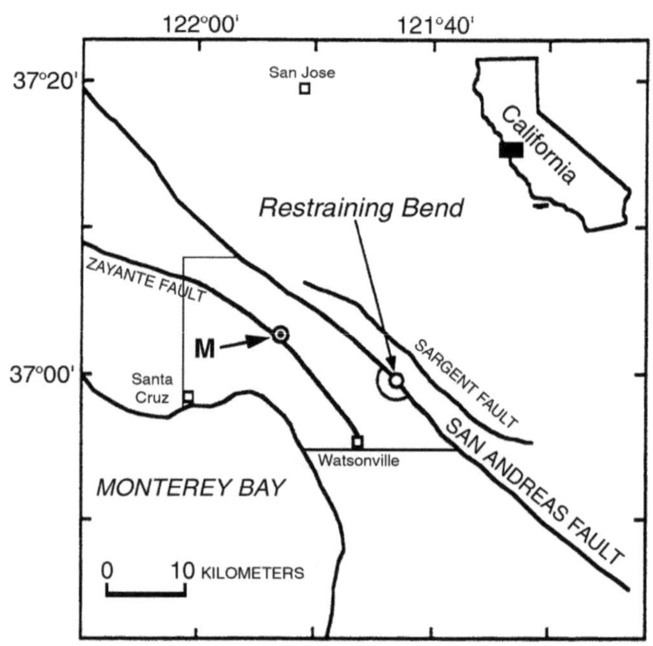

Figure 12-16 Location map of Loma Prieta earthquake and restraining bend on San Andreas Fault, with 1989 earthquake epicenter at M. (From Shaw et al. 1994, United States Geological Survey Publication.)

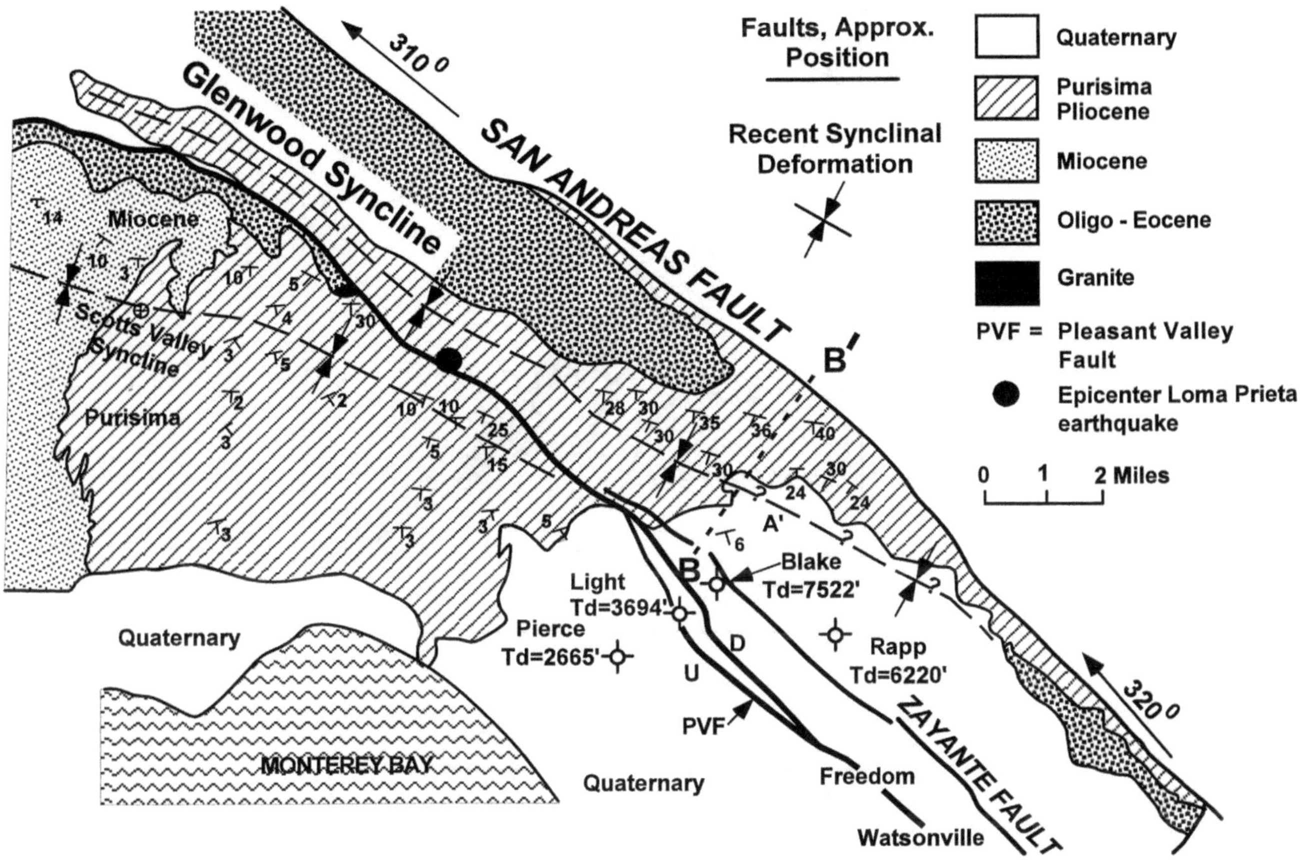

Figure 12-17 Geological map of Loma Prieta epicentral area showing Glenwood syncline emanating from bend in surface trace of San Andreas Fault. Balanced cross section B-B' is shown in Figure 12-19. San Andreas Fault turns from 320 deg to 310 deg, forming a restraining bend. (From Shaw et al. 1994, United States Geological Survey Publication.)

dip along its trend. The curved shape of the fault surface, as defined by the hypocentral data, generates a restraining bend. Other strike-slip faults should exhibit similar geometries (Durand-Riard et al. 2013; Benesh et al. 2014).

Before entering the bend, the Pacific Plate moves in the N40W (320 deg) direction parallel to the surface trace of the San Andreas Fault. As material enters the bend to the north of Watsonville/Freedom (Fig. 12-17), the Pacific Plate moves up the ramp formed by the southwesterly dipping, high-angle reverse-strike-slip fault surface (Fig. 12-18a). The upward motion generates compressional synclines, similar to synclines that form at the base of compressional thrust fault ramps (Chapter 10). In this example, the syncline should emanate from where the fault surface bends and departs from its general N40W (320 deg) trend. In the south, hypocentral solutions define a San Andreas Fault that dips at 82 deg and strikes N40W (320 deg) (Profile 3 of Fig. 12-18b). In the restraining bend and at Profile 2 of Figure 12-18b, the fault changes in dip to 65-70 deg at 8 km depth and strikes at N50W (310 deg). As predicted, material entering this bend in the fault surface generates the Glenwood syncline (Fig. 12-17). This syncline terminates at the southern bend in the fault surface, northeast of the Rapp Well, where the San Andreas Fault departs from its general N40W (320 deg) trend (Fig. 12-17). Ground-surface dip data present on the surface geological map and well log data constrain the geometry of the Glenwood Syncline (Figs. 12-17 and 12-19).

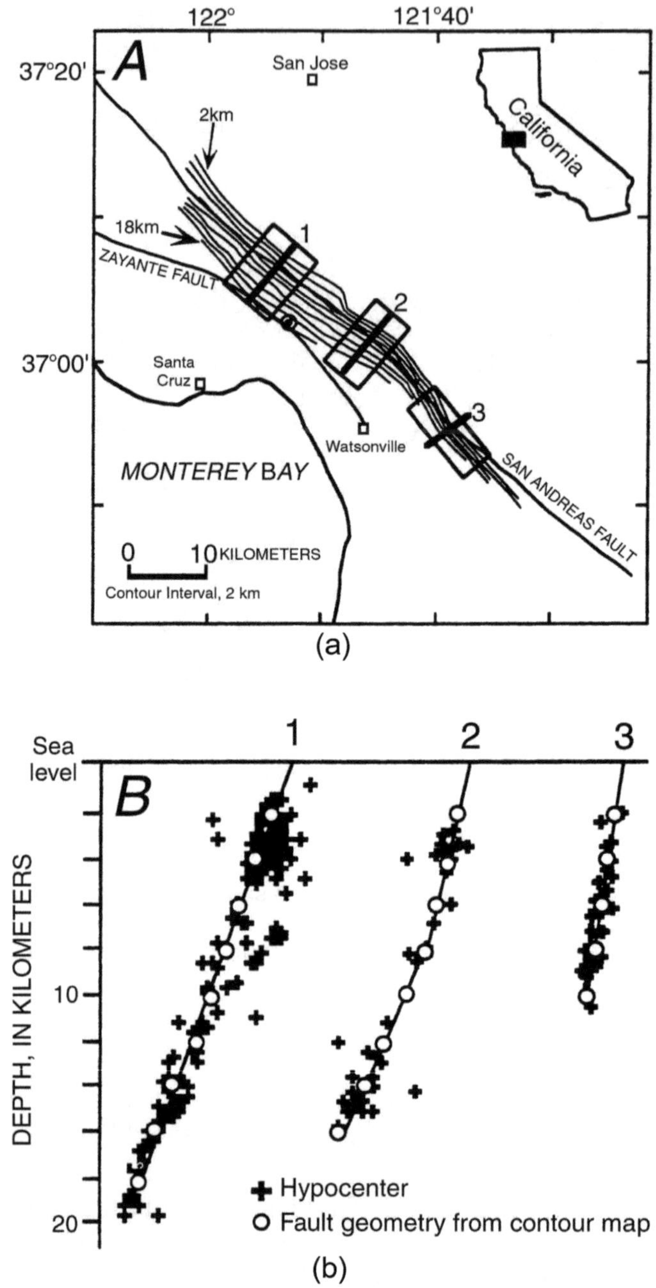

Figure 12-18 (a) Fault surface map for Loma Prieta restraining bend showing locations of cross sections 1, 2, and 3. (b) Cross sections of San Andreas Fault as defined by hypocentral activity. The fault bends at cross section 2 at a depth of 8 km and it dips at a higher angle at cross section 3. (From Shaw et al. 1994.)

In the brittle regime of the earth's crust, folds form as hanging wall beds move over nonplanar fault surfaces (Bally et al. 1966; Suppe 1983). Thus, we can use surface and well data related to the Glenwood Syncline to generate balanced models of strike-slip compressional folding along the San Andreas Fault. Cross section balancing allows us to predict the subsurface fault geometry from the surface and well data. In practice, this exercise allows geoscientists to predict the vertical and lateral dimensions of the hanging wall geometry (Tearpock et al. 1994). Hanging wall geometry is important in positioning wells, in understanding the size of a prospect, and in

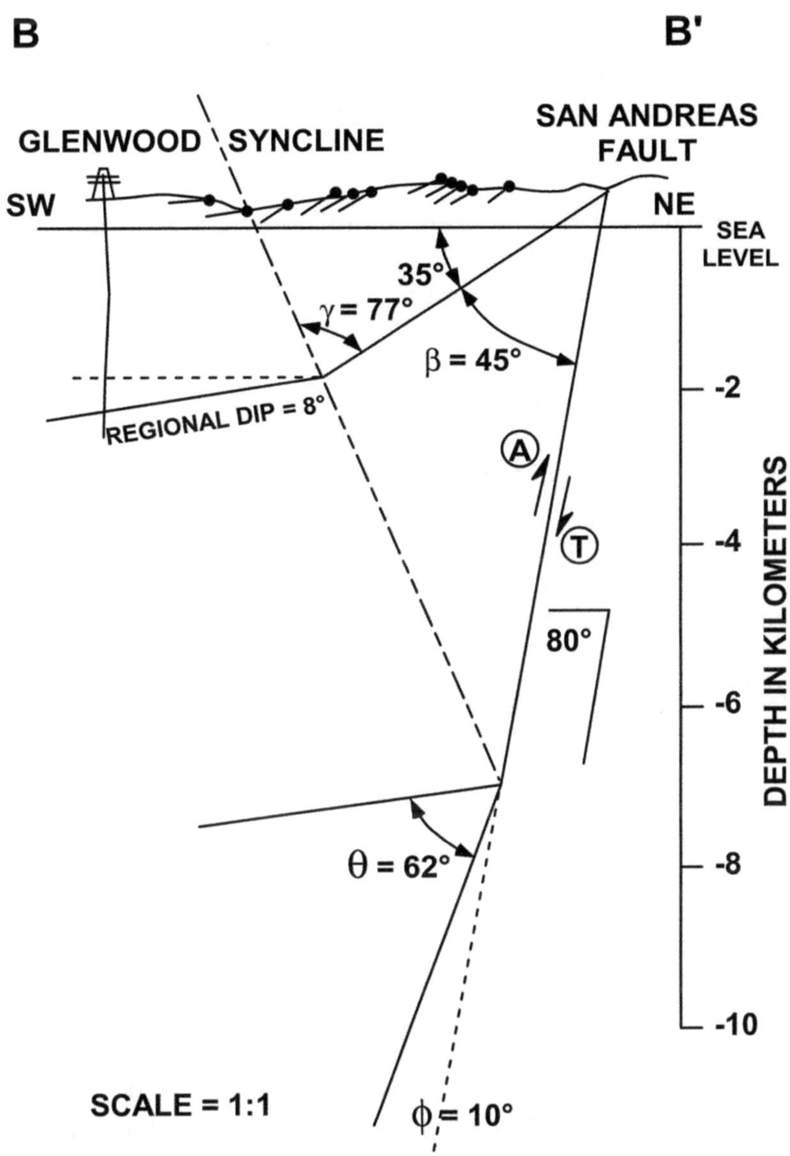

Figure 12-19 Example of a balanced strike-slip fault model for the Loma Prieta restraining bend, San Andreas Fault, California, simplified from Shaw et al. (1994). Model uses ground-surface bed dip data, well control, and dip of San Andreas Fault at the ground surface. The model contains a fault bend at 7 km, in good agreement with hypocentral data along the fault during the Loma Prieta Earthquake. See Figure 12-17 for location.

generating admissible interpretations of strike-slip faults (see the generic example in this section). We use the balanced model to predict subsurface fault geometry and compare the model predictions against the observed fault surface geometry as defined by hypocentral earthquake activity.

Sometimes fault surfaces image on seismic data sets, but the hanging wall structure does not clearly image. Balancing techniques can predict fold shape from fault shape (Tearpock et al. 1994; Shaw et al. 1994) and help constrain interpretations. We balance profile B-B′ on Figure 12-17 (profile 2 in Fig. 12-18b) and present it as Figure 12-19. One of two approaches can be employed in order to balance a profile. If the subsurface fault geometry is known from depth-corrected seismic sections and well control or, in this case, hypocentral aftershock activity, then geoscientists can generate a balanced, or generic, model of the hanging wall structure. Alternatively, if the shallow hanging wall geometry variables γ and β are known from depth-corrected seismic sections, outcrop data, or well control, then geoscientists can estimate the fault surface geometry variables θ and ϕ (Fig. 12-19), where

γ = angle between synclinal axial surface and the adjacent strata
β = angle between shallower portion of strike-slip fault and bedding
θ = angle between deeper portion of strike-slip fault and bedding
ϕ = difference in dip angle between shallower and deeper portion of strike-slip fault

In this case, we use surface dip data and shallow well control to determine γ, the angle between the axial surface and the adjacent beds of the Glenwood Syncline (Fig. 12-19). The hanging wall cutoff angle β, in Figure 12-19, can be determined from local bed dips and fault geometry. By definition,

$$(\beta) = \text{dip of shallow fault} - \text{dip of beds adjacent to fault surface} \qquad (12\text{-}2)$$

The flank of the Glenwood Syncline adjacent to the fault dips at an average of 35 deg to the southwest (Figs. 12-17 and 12-19), whereas regional dip is 8 deg southwest in the area west of the synclinal axis. Well log and surface dip data from along the trend of the Glenwood Syncline constrain the regional dip. Thus, to determine β, subtract the dip of the flank of the Glenwood Syncline from the observed surface dip of the San Andreas Fault within the restraining bend. The surface dip of the fault is about 80 deg (Brabb 1989), and thus β = 80 deg − 35 deg = 45 deg. From inspection of Figure 12-19, the axial surface angle γ can be determined from the kink law (Chapter 10), or from the following equation.

$$2\,\gamma = 180 \text{ deg} - \text{dip of synclinal limb adjacent to fault} + \text{regional dip}$$

Therefore,

$$\gamma = (180 \text{ deg} + \text{regional dip} - \text{dip of synclinal limb adjacent to fault})/2 \qquad (12\text{-}3)$$
$$\gamma = (180 \text{ deg} + 8 \text{ deg} - 35 \text{ deg})/2$$
$$\gamma = 77 \text{ deg}$$

Dip data at the ground surface is used to position the axial surface (Fig. 12-19). The axial surface is drawn downward at an angle of 69 deg (77 deg − 8 deg) until it reaches the fault. This point determines the position of the bend in the fault. Horizons on the synclinal limbs can now be constructed, with the fold hinge bisected by the axial surface.

We next consult the fault-bend fold graph (Fig. 10-38) to generate a balanced solution to the fault surface problem. The correct graph to use is the diagram for synclines (right graph). The values required are γ = 77 deg, recorded on the y-axis of the graph, and β = 45 deg. The values for β are recorded by the right-sloping bold diagonal lines. Project a horizontal line into the graph from the value of γ = 77 deg. Where the γ = 77 deg line intersects the β = 45 deg curve, the change in fault dip angle ϕ is read off the graph from the thin, near-horizontal and downward-sloping curve on the graph. The value of ϕ is slightly less than 10 deg.

Therefore, we conclude that the Glenwood Syncline is created by about a 10-deg subsurface bend in the San Andreas Fault. The axial surface of the Glenwood Syncline emanates from this subsurface bend in the San Andreas Fault, where the previously undeformed beds moved over the bend in the fault surface (Fig. 12-19). As described previously, the depth to the bend in the fault surface was determined by projecting the axial surface of the Glenwood Syncline downward to where it intersects the 80-deg dipping fault. The axial surface intersects the San Andreas fault at a depth of about 7 km (Fig. 12-19). Below this depth, the fault takes a 10-deg bend and the balanced model predicts a fault dip of about 70 deg below 7 km (Fig. 12-19).

We can now compare the model-generated values and the predicted depth of the fault bend to profile 2 in Figure 12-18b. The strike-slip fault bend fold model predicts that the San Andreas Fault makes about a 10-deg angle bend at a depth of about 7 km and that the fault dips at about 70 deg below 7 km (Fig. 12-19). These values compare favorably with the hypocentral data that suggests about a 10-deg fault bend at a depth of about 8 km. Below this depth, the fault dips at about 70 deg, as shown in Figure 12-18b, Profile 2. Given the good agreement between theory and observation, strike-slip fault bend fold theory may allow geoscientists to make viable predictions of the structural geometry along strike-slip faults (Shaw et al. 1994). Balanced cross sections of compartmentalized displacements along strike-slip faults may help geoscientists generate higher quality prospects in a tectonic environment that has proven to be difficult to quantify. When millions of dollars are at stake, deterministic models may add value to conceptual interpretations of prospects generated in strike-slip faulted tectonic regimes.

The San Andreas Fault is subject to hundreds of kilometers of slip, which created the Glenwood Syncline with a broad south-dipping limb (Fig. 12-17). This limb could form a three-way closure against the San Andreas Fault. Other strike-slip faults that contain smaller amounts of slip should succumb to an analysis similar to what we have presented here. Balanced solutions of these structures can improve prospect viability, perhaps solve complex structural problems, and thereby reduce prospect risk.

Generic Example of Strike-Slip Compressional Folding. If fault surface maps are obtainable from well log or seismic data, then balanced models of hanging wall geometries are possible. Balanced models of hanging wall geometries lead to viable, low-risk prospects. A generic example is shown in the cross section presented in Figure 12-20. Perhaps your working group

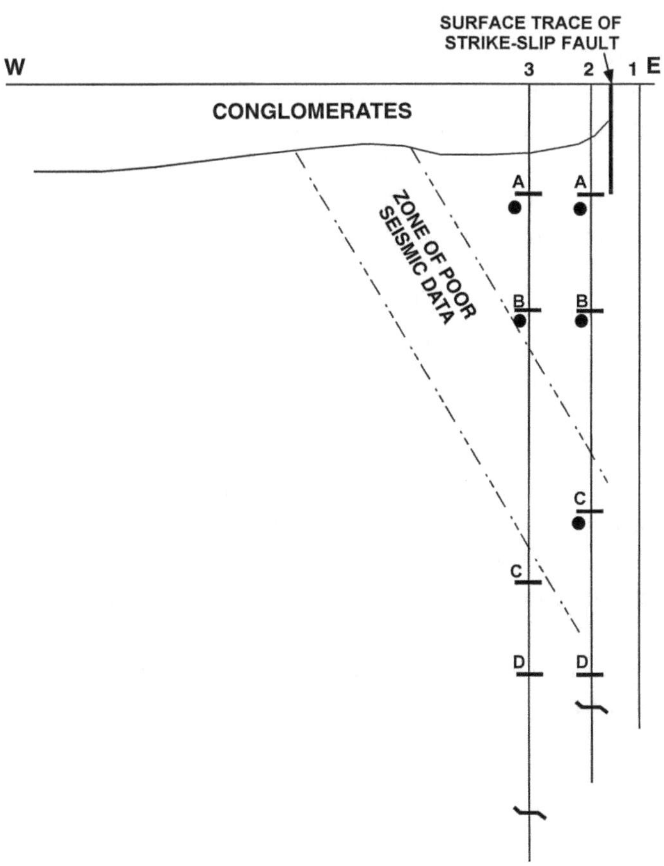

Figure 12-20 Well log data along a known strike-slip fault. The seismic data as collected are incoherent between dashed lines. Structural data are not subject to unique interpretations, which can result in dramatically different solutions to well-constrained seismic and well log data. (Published by permission of R. Bischke.)

was subject to the following type of problem that employed well log and seismic data. Let us assume that two companies find hydrocarbons in Wells No. 2 and 3, in the A, B, and C Horizons adjacent to a vertical fault. Outcrop and seismic data suggest that the near-vertical fault, containing restraining and releasing bends, is located between Wells No. 1 and 2 (Fig. 12-20). The two companies construct cross sections of the new field in order to propose additional development wells. In Figure 12-20, well log data constrain the fault geometry, but what is the limit of the field, and is it necessary to drill additional wells?

We consider two interpretations of the data: a qualitative interpretation presented by Company A and a quantitative interpretation presented by Company B. We discuss the Company A interpretation first. We make the assumption that geoscientists from Company A do not have a solid background in structural geology and therefore have not been trained in volume conservation or structural balancing concepts. They construct a cross section through the field that employs conceptual, but not volume, conservation concepts. On the other hand, geoscientists from Company B construct a balanced cross section of the existing data. How may these two working groups and their interpretations differ? Can the difference affect future exploration and success? We will assume that the 3D seismic data that crosses the area is of reasonable quality but suffers from the usual problems, such as surface statics and the inability to image steeply dipping beds.

After examining the data present in Figure 12-20, the Company A geoscientist interprets a secondary fault along the western flank of the structure. Two independent sources of evidence exist for this fault. The first piece of evidence is the no-data seismic zone on the flank of the structure (Fig. 12-20). The second piece of evidence is the change in thickness of the stratigraphic units above the D Horizon, between Well No. 2 and Well No. 3. The interpretation is shown in Figure 12-21. The proposed secondary fault exhibits normal, strike-slip, and reverse separations, and it explains the bed dip and thickness variations between the B and C Horizons and the C and D Horizons in Wells No. 2 and 3. This fault geometry also accounts for the slip reversal between the B and D Horizons. Strike-slip faults can exhibit normal, reverse, and lateral separations. The interpreted fault turns down and merges with the large, master strike-slip fault at depth.

Lastly, Company A completes the seismic interpretation by correlating the reflections from the crest of structure into the off-structure flank position located to the west of the no-data zone on the seismic data set. In the low dip area to the west of the no-data zone, the seismic images a gently folded structure in which the reflectors turn up beneath the secondary fault (Fig. 12-21). The turned-up beds are interpreted to be caused by fault drag on the secondary fault, as is shown in many structural geology textbooks (Billings 1972).

The Company A team present the cross section of the folded and faulted structure shown in Figure 12-21. Using cross sections and maps of the field, they propose two additional wells to define the western limits of the field. These wells will test the upturned A and B Horizons located beneath the fault.

The Company A interpretation makes three basic assumptions: first, that the no-data zone on the seismic data set results from faulting; second, that faulting causes changes in thickness of stratigraphic intervals as seen in well logs, perhaps associated with repeated or missing section; third, that folds tend to be gently curved, rounded features, as shown in some textbooks on structural-metamorphic geology (see Chapter 10). Does another interpretation of the data exist that applies to the low temperature portions of mobile belts?

The Company B geoscientist team, having knowledge of structural geology concepts and techniques, attempts to balance the data shown in Figure 12-20 using procedures developed in Chapter 10 and in this chapter. They notice that the D Horizon is on the same structural level in Wells No. 2 and 3, and thus this horizon is not subject to possible faulting or folding between these two wells. Procedures outlined in the section on compressional folding along strike-slip faults suggest that the D Horizon may be near a bend in the vertical fault. Using the fault cuts

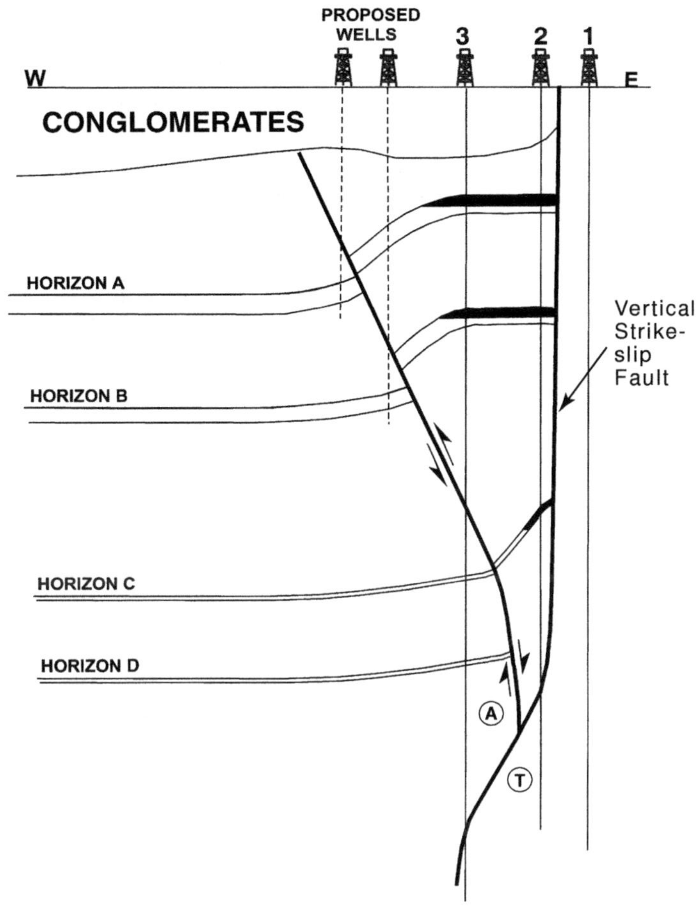

Figure 12-21 Company A geoscientists solve the folded-structure problem by using the concept of a secondary fault, related to strike-slip displacements, to explain the apparently high bed dips and vertical thickness changes interpreted from the wells. They propose two additional wells to test the A and B Horizons. (Published by permission of R. Bischke.)

located below the D Horizon in Wells No. 2 and 3, the Company B geoscience team reasons that the cutoff angle (θ) between the D Horizon and the deeper part of the fault is 60 deg (Fig. 12-22). Thus, the difference (ϕ) in dip angle of the fault is 30 deg.

The geoscientists can now complete the cross section shown in Figure 12-22, using the methods outlined in the previous section on compressional folding along strike-slip faults. Strata are deformed above the bend in the fault surface, so an axial surface emanates from this fault bend. They determine the dip of the axial surface by using the method described in the preceding section. Knowing that $\theta = 60$ deg and $\phi = 30$ deg, they use Figure 11-34 (right graph) to determine the axial surface angle γ to be 60 deg. As the regional dip is 0 deg, the actual dip of the axial surface is 60 deg. The axial surface projects upward at 60 deg from the bend in the fault (Fig. 12-22). The axial surface bisects the hinges of a fold, so by construction the A, B, and C Horizons dip at 60 deg *through the no-seismic-data zone*. The axial surface correlates to the western limit of the no-data zone on the seismic, and thus the no-data zone can result from changing bed dips and not from faulting, as assumed in the first interpretation.

Using similar reasoning with respect to Horizons A and B, the up-dip limit of the no-data zone can define a second axial surface. The axial surfaces are positioned as shown in Figure 12-22. As described earlier, seismic character correlation suggests the structural position of Horizons A and B are to the west (Fig. 12-20).

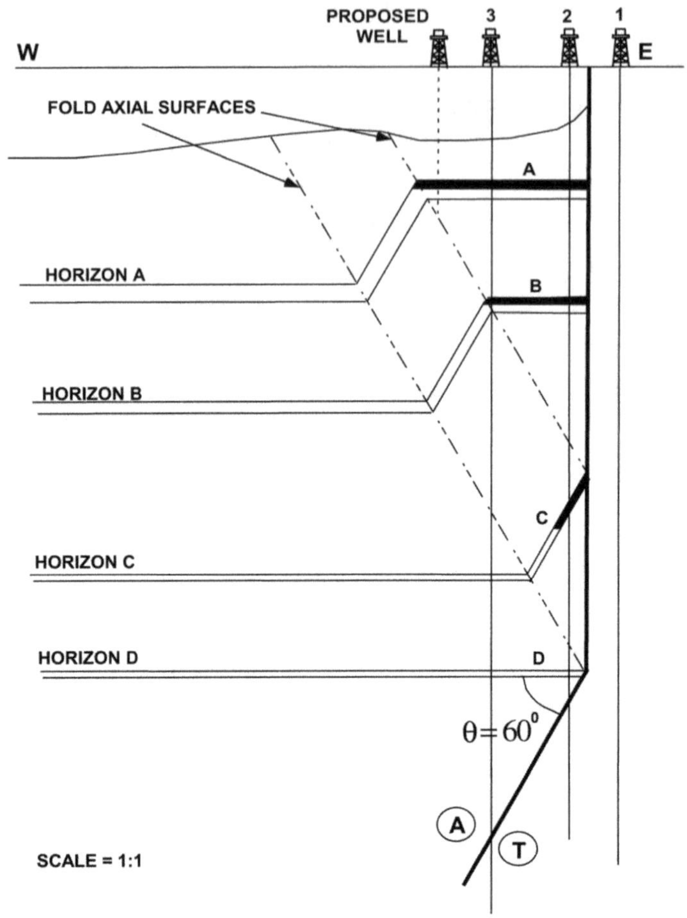

Figure 12-22 Company B geoscientists attempt to balance the structure and generate a strike-slip fault bend fold model that explains the anomalous bed dips and interval thickness changes at and below horizon C. In this model, the high bed dips occur at axial surfaces that define the flank of a monocline. Geoscientists propose a single shallow well at the A Sand to define the western limits of the field. This model shares properties with box and lift-off structures. (Published by permission of R. Bischke.)

The Company B team assumes that the structure can be modeled using the Kink Method described in Chapter 10, and thus their cross section has more angular folds than the Company A cross section shown in Figure 12-21. Petroleum-scale structures, exhibiting a kink fold structural style, are common to the low temperature portions of fold and thrust belts (Suppe 1985; Boyer 1986). Furthermore, the Company B team are suspicious of the upturned beds imaged below the no-data seismic zone. The folding elevates high density and high velocity strata to a structural level that is higher than the equivalent beds located to the west of the no-seismic-data zone. Thus, the area below the region of high bed dips may be exhibiting velocity pull-up. In addition, as Company B concluded that the no-data zones result from folding and not faulting, there is no reason to postulate a fault-drag effect. They model the beds to the west of the no-data zone to be nearly flat surfaces.

The balanced interpretation presented by Company B also makes three assumptions. First, the no-data zone on the seismic data set results from high bed dips caused by folding. Second, folding causes changes in bed dips as well as in the different well log thicknesses. Third, folds in the low-temperature portions of mobile belts commonly exhibit kink geometry. The interpretation presented in Figure 12-22 has properties similar to lift-off or box folds (Chapter 10). The two interpretations contrast fundamentally. The interpretation presented by Company A emphasizes faulting, whereas the interpretation presented by Company B stresses folding.

The interpretation by Company B may not require any additional wells, depending on the areal extent of the reservoir and the drive mechanisms. However, to maximize development potential, one well is proposed to define the western limit of the accumulation in Horizon A. No wells are proposed to test the footwall plays in Horizons A and B (as presented by Company A), since the interpretation set has no closure in the footwall.

These two different interpretations will have a major impact on the exploration and development program. The interpretation shown in Figure 12-21 requires more wells and higher costs. It is based on negative evidence, limited structural knowledge, and a misunderstanding of the fold deformation in the footwall of the proposed secondary fault. Referring back to Chapter 1, we emphasize that a strong background in structural geology, not to be confused with an understanding of seismic interpretation, is a major component of a successful exploration or development program. This is particularly true in complex structural areas such as those discussed in this section.

The differences in the two interpretations result primarily from philosophical differences as to the approach used to resolve no-data zone problems and as to what represents real data. Rather than repeat many of the structural principles discussed in Tearpock et al. (1994) and this and other texts, we propose the following approach to evaluating conflicting or questionable structural interpretations. This approach is suggested by our experience in reviewing the results of many wells drilled in similar structural settings around the world.

1. Begin by examining interpretations to determine whether *they honor all the data*. Then look for obvious problems. Tearpock et al. (1994) present many quick-look techniques that aid in this analysis.
2. Certainly, geology can be complex, but where alternative solutions exist that honor all the data, the probable solution is commonly the *less complex solution.*
3. Valid interpretations of geological structures should be based on *positive empirical evidence* that confirms the presence of observed features. This requires an understanding of valid structural principles. The fault in Figure 12-21 may exist, *but this interpretation is of high risk and is less likely*. What constitutes *negative evidence*? A no-data zone on seismic, if used to support an argument, can qualify as an argument based on negative evidence. Thus, if faulting is proposed that is based on a no-data zone, then this argument is based on negative evidence. Again, this does not mean that the fault is not present; it means that *this fault is a high-risk fault*. Another way of approaching this problem is to consider what feature(s) on the data set will lower risk.

   If an argument is based on negative evidence, ask yourself what positive evidence is required to support the argument. Throughout the world, most normal faults that dip at 70 deg image on seismic sections. Does a proposed fault in a no-data zone have a surface reflection or exhibit *discontinuous reflections* across the surface, and if not, why not? If the fault surface images, can a viable map of the fault surface be constructed that supports the faulting? Other observable features, such as large amounts of missing or repeated section in well log data, are direct evidence for faulting. These features lower fault risk. However, a change in dip on a dipmeter log can result from either faulting or the well *crossing an axial surface.*
4. Can another model better explain the observed features? This is where experience and training are important. Perhaps the no-data zone in our example results *from high bed dips caused by folding* and not by faulting. Geoscientists who work fold-thrust belts know that high bed dips are common to fold belts. Folding creates high bed dips that do not image on many seismic data sets. Furthermore, our experience shows that if faulting is present, then

dip domain analysis (Chapter 10) typically locates direct evidence for the faulting in the form of *bed dip discontinuities*. If another explanation is as probable, or more probable, then caution is prudent. Risk the uncertainty accordingly. Good scientific work should yield better interpretations, resulting in more success at lower costs.

A structure similar to Figure 12-22 is shown in Figure 12-23 from Pecos Country, New Mexico, USA (Kelley 1971). The geological map of the region shows several long, arcuate northeast-southwest trending faults attributed to strike-slip displacements. This structure, called the Y-O Buckle, is bound by a near-vertical fault that contains a bend. A monoclinal fold forms above the bend in the fault, and below the bend the beds are not folded. The folded structures along these faults are subparallel to the surface trace of the faults, and they tend to exist at gentle bends in the faults. Thus, some of the folds appear to form at restraining bends. Field mapping demonstrates that the faults contain vertical separations that commonly change from normal to reverse separations along strike (Kelley 1971). The strata are deformed adjacent to the fault surfaces, often as disharmonic folds with near-vertical limbs (Figs. 12-22 and 12-23).

Extensional Folding along Strike-Slip Faults

The balancing of releasing bends is similar to the balancing of restraining bends, except that the displacements are in the opposite direction. Rather than employing the techniques used in Chapter 10 on compressional structures, we use the techniques previously discussed for extensional structures (Chapter 11). We apply these ideas to a possible releasing bend in the classic Ridge Basin, southern California, that was carefully mapped by Crowell and his colleagues and

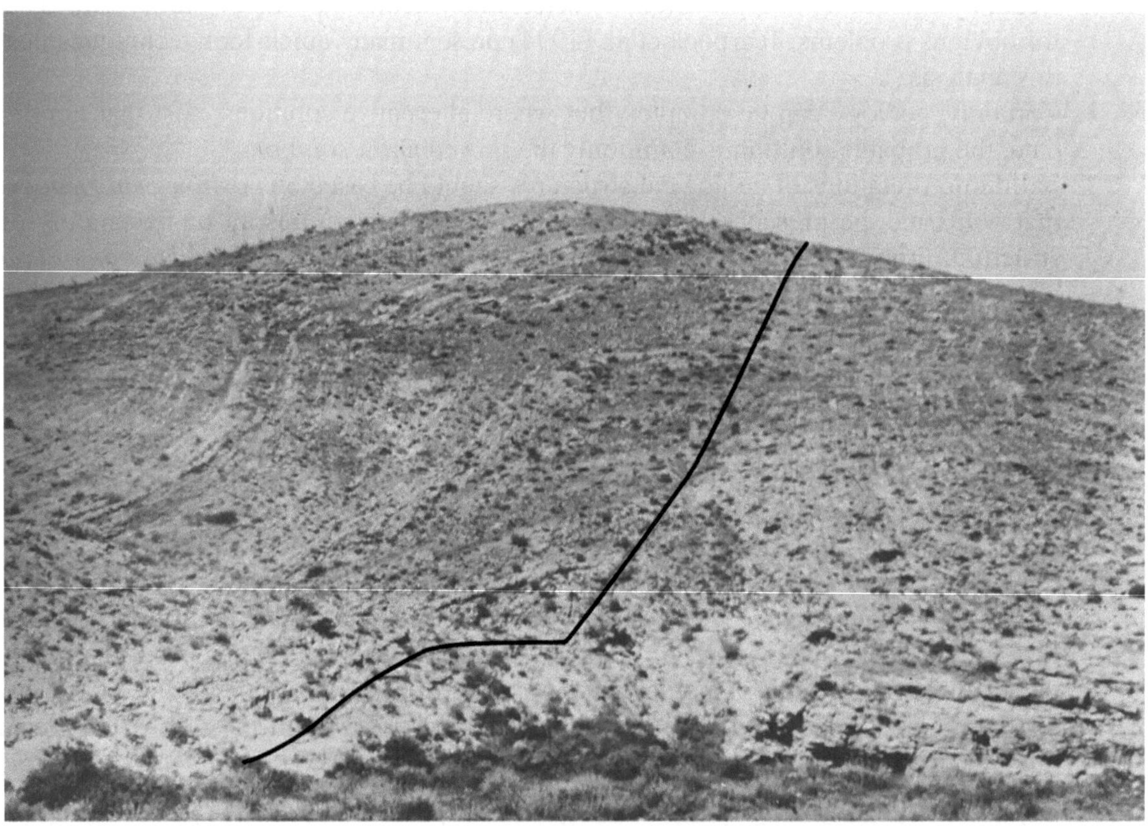

Figure 12-23 Box-like detachment fold in Pecos County, New Mexico, USA, along a long, arcuate fault system that appears to contain restraining and releasing bends. In the photograph, the strata are folded above the bend in the fault surface, similar to Figure 12-22. (From Kelly 1971. Published by permission of the New Mexico Bureau of Mines and Mineral Resources.)

students (Crowell 1950, 1982, 2003a, 2003b, 2003c). Crowell (2003c) and Link (2003) describe the detailed structure and stratigraphy of the basin along with the geological complications. In this section, we attempt to integrate surface dip data, seismic data, well log data, sedimentary data, and structural data into a model for Ridge Basin.

Ridge Basin Geology. Ridge Basin formed along the northeastern boundary of the large right lateral San Gabriel strike-slip fault (Fig. 12-24), which was active during the late Miocene. The San Gabriel Fault is an inclined fault surface that strikes at about N40W and dips 50 deg to 70 deg to the northeast (Crowell 2003c). Piercing point evidence, consisting of offset belts of pre–Late Miocene rocks and structures, suggests that the San Gabriel Fault experienced about 60 to 80 km of strike-slip displacement (Crowell 1982, 2003c).

Crowell (1974a, 1974b, 2003c) proposed that Ridge Basin formed as a result of *a restraining bend* located to the northwest of Ridge Basin and beneath the Frazier Mountain and Dry Creek Thrusts, that overthrust the northwestern part of Ridge Basin (Fig. 12-24). In his more recent model (Crowell 2003c) removed the extensional deformation from the basement that floors Ridge Basin and proposes squeezing and uplift at the stepover. As the evidence supports that 60 to 80 km of material moved through this restraining bend, a question arises as to why the active east-west-trending *compressional Transverse Ranges terminate at the San Gabriel fault* (Crowell 2003a) (Fig. 12-24); or why this east-west thrust faulting trend in the Transverse Ranges does not exist within domains 1 and 2 of Ridge Basin (Crowell, 1982, 2003c) (Fig 12-24; consult Restraining Bends section). An alternative interpretation of the regional data is that the right lateral San Gabriel Fault was originally part of the San Andreas Fault, and that *modern strand* of the San Andreas Fault has about 240 km of displacements (Ehlert 2003).

A seismic line presented by May et al. (1993) trends across the northwestern portions of Ridge Basin (Figs. 12-24 and 12-25). This line contains reflections that diverge toward an apparent listric normal fault that branches (splays) off the San Gabriel Fault. We call this proposed branch fault the Hungry Valley Fault. The deepest coherent reflections at about −10,000 feet on this depth-corrected seismic line dip at about 30 degrees, similar to the surface dips on the Geologic Map of Ridge Basin (Crowell 1982, 2003a, 2003b, 2003c). Most important, the line appears to image a *rollover structure* (May et al. 1993) (Fig. 12-26; also see Figs. 11-16 and 11-17). Typically, divergent dipping reflections dip toward listric normal faults and dip at

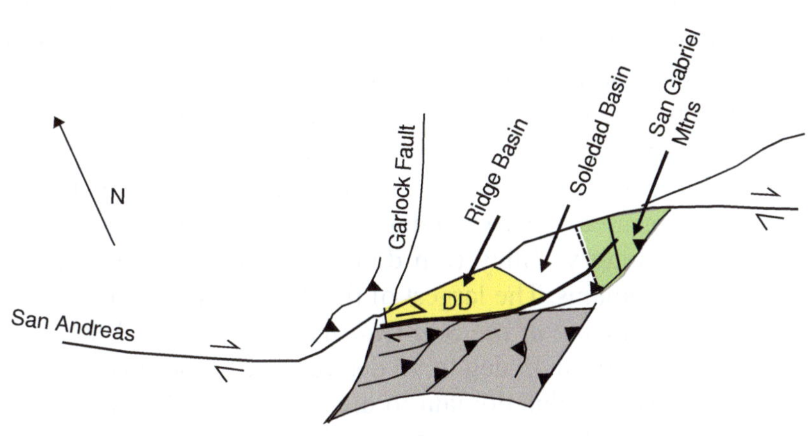

Figure 12-24 The compressional Transverse Range terminates at the right lateral San Gabriel Fault. (From Crowell 2003a. Published by permission of the Geological Society of America.)

about right angles to the normal fault. Therefore, the rolled-over reflections may dip toward a north-striking, branch normal fault that splays off the San Gabriel Fault (Link 2003).

This structural configuration favors an extensional releasing bend (Figs.12-9a and 12-11), *subjecting Ridge Basin to a component of both dip-slip and strike-slip deformation.* Furthermore, the Hungry Valley Fault, as shown in Figure 12-25, has normal separations that create accommodation space for the accumulation of the Ridge Basin sediments. This depocenter beneath Hungry Valley trends about north-south (Link 2003), so the Hungry Valley Fault also trends in a north-south direction (Link 2003). The Hungry Valley Fault was apparently responsible for the accumulation, rotation, and shingling of a section of Ridge Basin syntectonic sediments that are 14 km thick. The basin, however, was only about 5 km to 6 km deep (Crowell 2003c; Link 2003, 2003). Some deformational process was responsible for the rotation of these strata to dip at about 30 deg to the west (Crowell 2003a, 2003b, 2003c, Geologic Map of Ridge Basin and Geologic Cross-Section of Ridge Basin, profile A).

That this rotation is tectonic in nature is supported by the type of sedimentation in Ridge Basin, which involves the Peace Valley formation, and is about 9.0 km thick (Crowell, Geologic Cross-Section of Ridge Basin, profile A; Link 2003). Within the clastic units of the Peace Valley Formation the paleocurrent directions are to the southwest (Link 2003). According to Link et al. (1978) and Smyth (1982) the Peace Valley formation sediments are largely lacustrine, and 40% to 50% of the Violin Breccia consists of lacustrine fan deltas (Link 2003). These lacustrine deposits within the Peace Valley formations start with the shallow, freshwater Osito Canyon and Cereza Peak members and end with the Alamos Canyon and Posey Canyon members, which consist of brackish, deep-water lake sediments (Wood 1981; Link et al. 1978; Smith 1982). According to Smith (1982), the shallow fresh water deposits contain "wave-ripples, mudcrack casts, mammal tracks and arthropod burrows" and that these features are common. On Crowell's Cross-Section A, these sediments extend for 15 km along the axis of the basin and downlap a thin veneer of Violin Breccia at high angles directly above horizontal basement. Thus, any model for Ridge Basin should account for these 30-deg dips through a 9.0-km thick lacustrine section.

A geological map of the basin published by Crowell (1982, 2003a, 2003b, 2003c) shows the following major features (Fig. 12-24). The geology adjacent to the San Gabriel Fault is dominated by the *syntectonic* Violin Breccia, deposited throughout the Late Miocene (Crowell 1982). Paralleling the breccia is the Ridge Basin syncline. The close association of the Violin Breccia with the Ridge Basin syncline suggests that the two features are related and that Ridge Basin syncline is a syntectonic feature. As Ridge Basin Group sediments onlap the flanks of the syncline, the stratigraphy supports the interpretation that the syncline is syntectonic (Link 2003). In the vicinity of Hungry Valley, the Pliocene sediments dip at gentle angles of 5 deg to 10 deg. These gently warped sediments were subject to late-stage Pliocene to Holocene folding that postdated the structure in areas north and northwest of dip domain 1 (May et al. 1993). Thus, the area surrounding Hungry Valley represents a gently dipping domain that was subsequently folded (Fig. 12-24).

Surface bed dips measured by Crowell define two large dip domains that dominate the structure to the southeast of Hungry Valley (dip domains 1 and 2 in Fig. 12-24). The Ridge Basin syncline partitions these domains. The largest of the domains (domain 1) dips to the west and *lies about 1 to 3 km to the northeast of the San Gabriel Fault* (Wood 1981; Crowell 1982, 2003c). The beds in domain 1 strike from 45 deg to 90 deg relative to the surface trace of the San Gabriel Fault (Fig, 12-24). We interpret this domain to dip toward the Hungry Valley Fault. The average bed dips, partitioned into smaller subdomains, are shown on Figure 12-24.

Dip domain 2 lies adjacent to the San Gabriel Fault and is *syncline-separated* from domain 1 (Fig. 12-25). In the northwest, this domain extends about 3 km to the northeast of the San Gabriel Fault, but to the southeast of Fisher Spring, the domain narrows to within 1 km of the fault zone. In domain 2 and in map view, the beds generally *dip at high angles to the north and uniformly dip away from the San Gabriel Fault at oblique angles* to the fault zone. Bed dips in this domain *average*

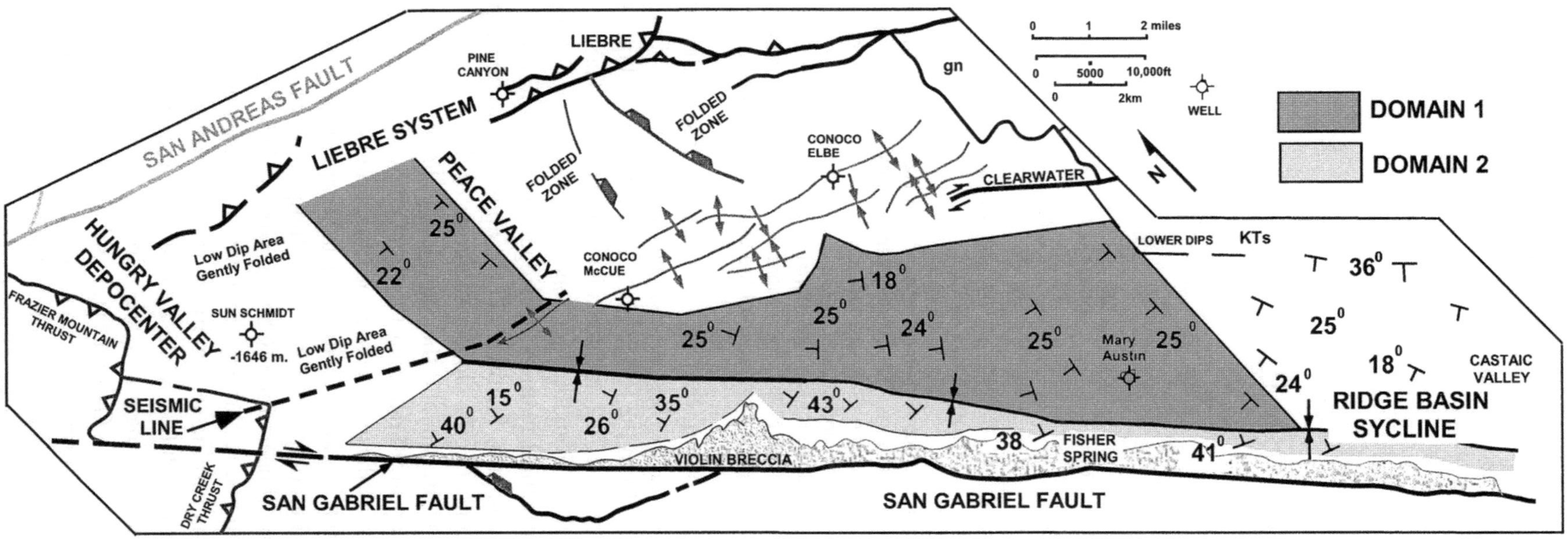

Figure 12-25 Generalized geologic map of Ridge Basin, California, modified after Crowell (1982, 2003a, Plate 1, a fold-out detailed map). Dip domain 1, located 1 to 3 km from the San Gabriel Fault, strikes north-south and dips to the west. Dip domain 2, adjacent to the San Gabriel Fault, dips to the north and is syncline-separated from dip domain 1. The two domains form the Ridge Basin syncline that subparallels the San Gabriel Fault. The gentle dips beneath Hungry Valley (in the west) are slightly folded by later compressional forces. Posted on the figure are the average values of the bed dips in the individual subdomains, along with the dip and strike directions. Average bed dips and strikes in domain 1 are 25W and N10E, and plunge of the synclinal axis is 24NW. Bed dips in domain 2 generally dip at high angles to the north, typically between the 25-deg to 45-deg range. (Published by permission of R. Bischke.)

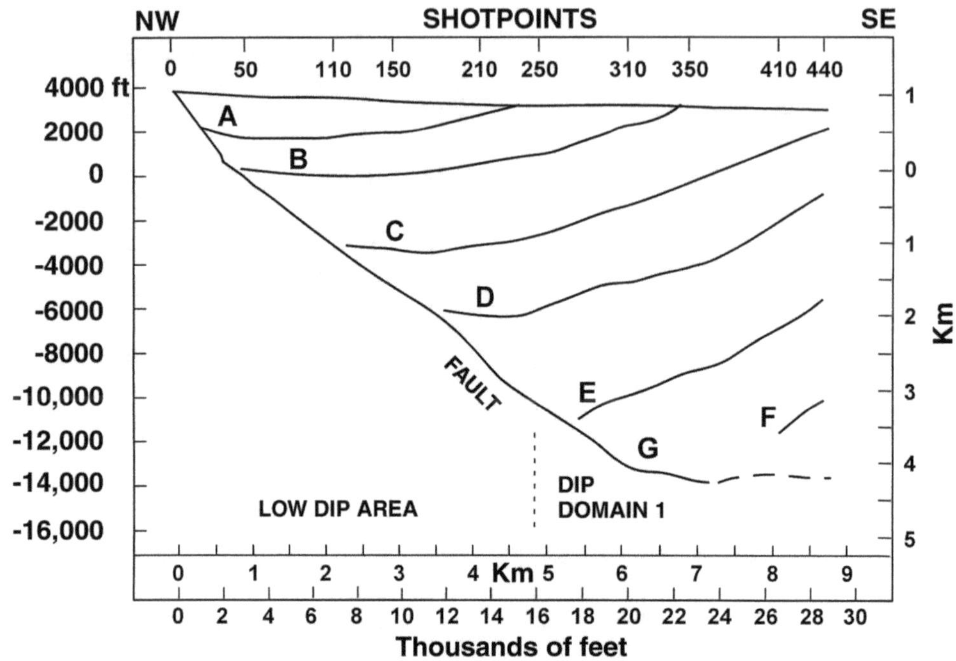

Figure 12-26 Tracing of depth-corrected seismic line across Ridge Basin that images an apparent half-graben rollover structure. The oblique seismic profile images reflections that contain apparent dips to the northwest. Where the profile crosses the low dip area of Hungry Valley, the hanging wall reflections directly above and adjacent to the fault surface exhibit low apparent dips. Seismic line crosses into domain 1 at about sp 250. See Figure 12-25 for the line location. Consult Link (2003) for a more detailed drawing of the seismic line. (After May et al. 1993. Published by permission of the Geological Society of America.)

34 deg to the north. We emphasize, on typical rollover structures, from different areas of the world, *beds dip toward normal faults at about right angles, not away from the fault surface at oblique angles*. Thus, dip domains 1 and 2 form the Ridge Basin syncline. The oblique seismic line extends near the northwest portions of dip domain 2, located adjacent to the San Gabriel Fault, before entering dip domain 1, located about 4 km from the fault zone to the seismic line (Fig. 12-25).

Any model of the gross structural features in Ridge Basin should qualitatively and quantitatively explain the following structural features: (1) the approximate bed dips and the direction of the dips within the two domains; (2) the close association of the syntectonic Ridge Basin syncline with the syntectonic Violin Breccia and the San Gabriel Fault; (3) the processes that cause the strike of the beds to rotate up to 90 deg near the San Gabriel Fault zone, forming the Ridge Basin synclinal kink fold, and how these processes operate; and (4) how Ridge Basin was filled with at least 14 km of sediments, 7.5 km of which are lacustrine that presently dip at 30 deg to the west (the model also must be compatible with the stratigraphy of Ridge Basin [Link 2003] and with Goguel's law of volume conservation [Chapter 10]); and (5) the interpretation should be restorable in that the cover should be restorable along with the basement (Chapter 10). Thus, one problem is how to **restore the lacustrine sediments to the horizontal along with the basement without leaving a hole** (see Crowell 2003a, Plate 1, Geologic Cross-Section of Ridge Basin, profile A). The dip domains and the syncline mapped by Crowell provide insight into the major tectonic processes associated with the San Gabriel Fault and the formation of Ridge Basin. As the subsurface geology is not well-constrained, we consider a solution to the Ridge Basin geology that assumes an extensional origin. This model will test the viability of an extensional origin for Ridge Basin based primarily on surface data. Other explanations may be possible, especially if more seismic data are acquired (consult Link 2003 for a review of the existing proposed models).

Geometry of Strike-Slip Extensional Folding. The deformation of the hanging wall beds along a releasing bend differs from the deformation along a normal fault, as is seen in the perspective view diagram presented in Figure 12-27. An inclined strike-slip fault and a branch normal fault are shown. The dip of the branch normal fault surface decreases with depth. For purposes of convenience only, in Figure 12-27 we show this deep fault surface to be virtually horizontal. Displacement of the hanging wall block over the footwall causes the beds to be downthrown to the branch normal fault. These displacements also create accommodation space for growth sediments. According to the model, the beds adjacent to the strike-slip fault *rotate downward* at high and acute angles toward the branch fault. The beds forming to the right of the synclinal axis and located above the deep branch fault surface *rotate toward* the branch normal fault at right angles. The process generates a synclinally folded structure downthrown to the branch normal fault, with a synclinal axis that subparallels the strike of the strike-slip fault, mimicking the Ridge Basin geologic map (compare Fig. 12-27 to Crowell 2003a, Geologic Map of Ridge Basin, or Fig. 12-25).

The detailed kinematics of the process are best shown in Figure 12-28, which is drawn such that the left-hand portion of the block is in the same vertical plane as the *synclinal axis* **GH** in Figure 12-27 (or plane GCDEH in Fig. 12-28b). In Figure 12-28a, slip on the strike-slip fault forms an inclined strike-slip fault surface, ABCD, and forms a nascent right-stepping releasing

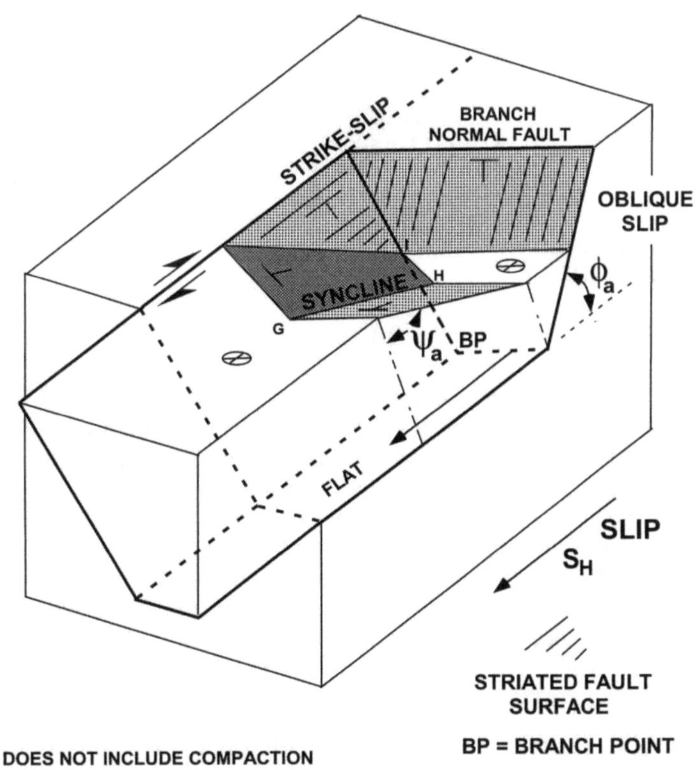

Figure 12-27 Perspective view of a synclinal, rollover structure forming along an inclined strike-slip fault. The strike-slip and branch normal faults form a releasing bend, and the strike-slip fault may be throughgoing. Slip vector is parallel to the strike direction of the strike-slip fault surface. The dip domain panel adjacent to the strike-slip fault dips away from the strike-slip fault, as the dip domain situated across the synclinal fold axis dips toward the branch fault surface. This fold style occurs along the San Gabriel Fault as the Ridge Basin syncline (Fig. 12-25 and Crowell 2003a, Plate 1, Geologic Map of Ridge Basin). Synclines may emanate from inclined releasing bends subject to large components of strike-slip motion. Figure is not drawn to scale. (Published by permission of R. Bischke.)

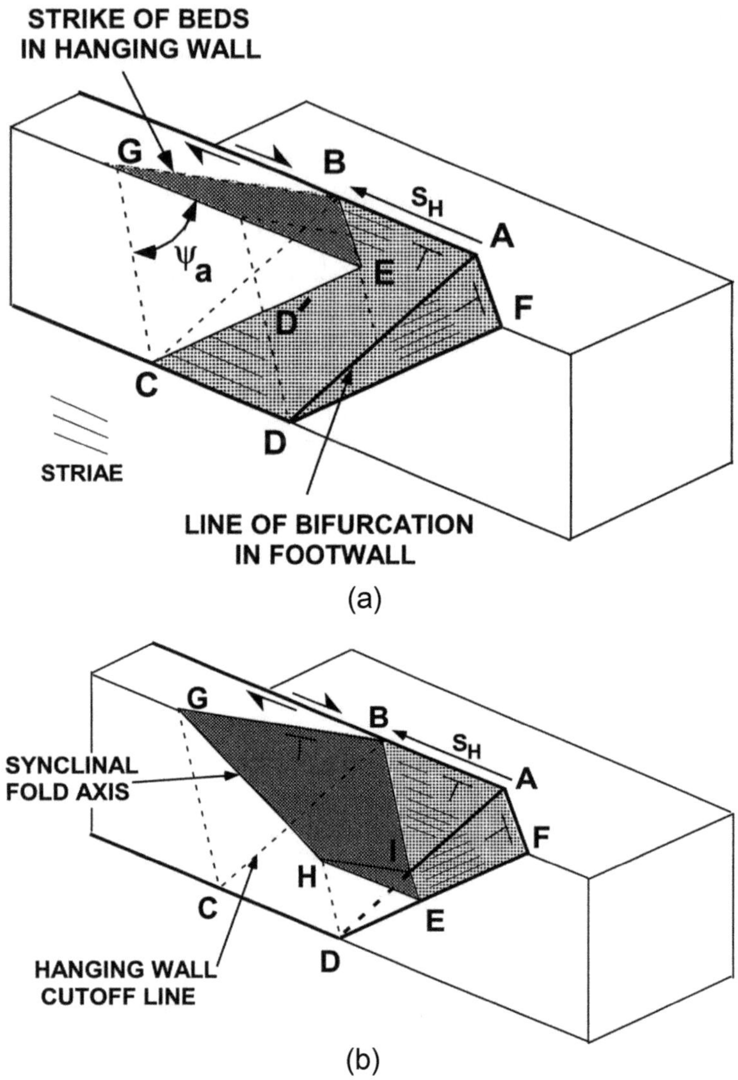

Figure 12-28 The kinematics of rollover folding above a releasing bend in a strike-slip fault. Slip component S_H parallels the strike of the fault surface. (a) Slip from point A to point B causes the hanging wall cutoff BC to separate from footwall cutoff AD. The resultant motion opens a void beneath surface EBC, and the hanging wall will collapse by Coulomb shear onto the footwall. Point D' collapses onto point D along the apparent collapse angle ψ_a. (b) The resultant deformation causes the beds to dip away from the strike-slip fault surface in a direction perpendicular to the surface trace of the Coulomb collapse surface BG. Stereographic projection methods can determine the plunge of the line of bifurcation AD (branch normal fault in Fig. 12-29), based on knowledge of the strike and dip of the beds adjacent to the strike-slip fault and the apparent collapse angle ψ_a. Figure is not drawn to scale. (Published by permission of R. Bischke.)

bend, fault surface ADF (Fig 12-28b) (also consult Figs. 11-2 and 12-30). The hanging wall in Figure 12-28 exhibits normal separations. In other words, the hanging wall block slips a small distance S_H over the inclined strike-slip fault from point A to B, parallel to the strike of the strike-slip fault. This motion opens a small void beneath the hanging wall, a void which extends to the hidden, dashed line BC (Fig. 12-28a). The line BC was originally coincident with the line of bifurcation, AD, of the two fault surfaces (Fig. 12-28a). Thus, this small void exists above the surface ABCDF. As fault ADF is a normal fault subject to oblique-slip, this void is instantaneously filled by gravitational collapse (along Coulomb failure surface ψ_a in Figs. 12-28a and 12-30) of the hanging wall material onto the surface ABCDF (Fig. 11-2).

In Figures 12-27 and 12-28b, line GH defines the plunge of the synclinal axial surface. Volume conservation therefore dictates that point D′ collapses onto the footwall in the plane containing line GH. As points D and D′ lie in the plane defined by the **apparent** Coulomb collapse angle ψ_a, point D′ collapses onto point D, forming a rollover. The collapse of the hanging wall onto the footwall causes the hanging wall beds to rotate and displace toward point E in Figure 12-28b, by bending along the *active* axial surface trace line BG. The active axial surface trace line BG is the surface expression of the *inactive* Coulomb collapse surface, BGC. The strike of the beds adjacent to the strike-slip fault define this deformation, which in the case of Ridge Basin is dip domain 2 in Figure 12-25. This deformation causes the beds to rotate away from the inclined strike-slip fault surface, *forming one-half of a synclinal fold.*

The remaining half of the synclinal fold is shown in Figure 12-27 as the surface above the fault flat that dips toward the branch normal fault. For clarity, that surface was not shown in Figure 12-28. Collapse along Coulomb shear surfaces causes the surface to the right of the synclinal axis to dip toward the branch fault. In the case of Ridge Basin, this surface is in **dip domain 1** (Fig. 12-25).

If a dipping strike-slip fault and branch fault form a fault flat, as shown in Figure 12-27, then two dip domains form as the hanging wall block slides over the footwall. The beds adjacent to the strike-slip fault dip away from the strike-slip fault, and the beds above the fault flat dip toward the branch normal fault. This deformation results in *a synclinal structure with an axial surface that trends subparallel to the surface trace of the strike-slip fault* (Fig. 12-27). The Ridge Basin syncline exhibits this type of geometry (Fig. 12-25).

Geometry of the Hungry Valley Fault. After determining the strike and dip of the proposed Hungry Valley Fault, which branches off the San Gabriel Fault, we model the dip of the deep normal fault beneath Ridge Basin. The modeling uses stereographic projection methods and extensional inverse balancing techniques, described in Chapter 11. As stated previously, near Hungry Valley, the San Gabriel Fault strikes at about N40W and dips to the northeast (Crowell 2003c) (Fig. 12-25). In addition, the trend of the Hungry Valley depocenter of about N10E (Link 2003), and the measured average N10E strike of the beds in domain 1, constrain the strike of the proposed Hungry Valley branch normal fault.

Notice on Figure 12-25 or on Crowell's Geologic Map of Ridge Basin, that the synclinal axial surface near the seismic line subparallels the surface trace of the San Gabriel Fault. This allows the kinematics of strike-slip extensional folding to simply construct a balanced model for Ridge Basin (Fig.12-28b). Notice on Figure 12-28b that any vertical profile, *oriented parallel to the surface trace* of a dipping master strike-slip fault, contains a *zero-deg* ($\alpha_a = 0.0$) apparent dip for the deep fault surface (Fig. 12-30).

In other words, Figure 12-28b contains a vertical surface GCDEH that intersects the dipping master fault along dipping surface ABCD. The line of intersection between the two dipping surfaces GCDEH and ABCD, define the angle (α_a). This *apparent dip angle* (α_a) *is independent of the dip on the dipping portion ABCD of the major strike-slip fault surface.* This means that a general solution to the problem of strike-slip extensional folding is independent of the dip on the San Gabriel Fault surface. Also notice that from a comparison of Figure 12-28b to 12-30, the surface GHE (Fig 12-28b) allows a calculation of *five apparent angles:* (1) angle (α_a), (2) angle (θ_a) = the apparent dip of the Ridge Basin syncline, (3) angle (ψ_{a1}) = the apparent dip of the Coulomb collapse angle, (4) angle (β_a) = the apparent dip of the proposed Hungry Valley Fault, and angle, (5) (ϕ_a) = the apparent change in dip in the apparent fault surface. Also from Fig. 12-25, measurements exists for the calculation of (1) the strike of the San Gabriel Fault trending at N40W,

(2) the strike of the branch fault from the averaged strikes of dip domain 1 located to the west of Peace Valley, and (3) the averaged dip and strike of domain 2 adjacent to the San Gabriel fault.

Lastly, the Coulomb collapse angle typically varies from 60 deg to 70 deg (Xiao and Suppe 1992; Bischke and Tearpock 1999). Here we assume a 65-deg Coulomb collapse angle. Next, we propose two balanced extensional folding models for Ridge Basin, one for dip domain 1, which traverses the deeper portions of the basin, and the second for dip domain 2, which parallels and cuts the San Gabriel Fault. Both domains are portrayed on a stereographic projection as thin lines (Fig. 12-29). We start with dip domain 2. These procedures may apply to other strike-slip faults (Fig. 12-33).

Model for Dip Domain 2. We start with the best constrained data, the strike of the San Gabriel Fault and the Ridge Basin syncline, which is N40W on Figures 12-25 and 12-29. Next, the kinematics of pure strike-slip folding shown in Figures 12-27 and 12-28 permit a balanced and retrodeformable model of Ridge Basin along a vertical plane defined by line GCD (Fig. 12-28a). A *profile in dip domain 2, constructed parallel to the surface trace* of the San Gabriel Fault, can

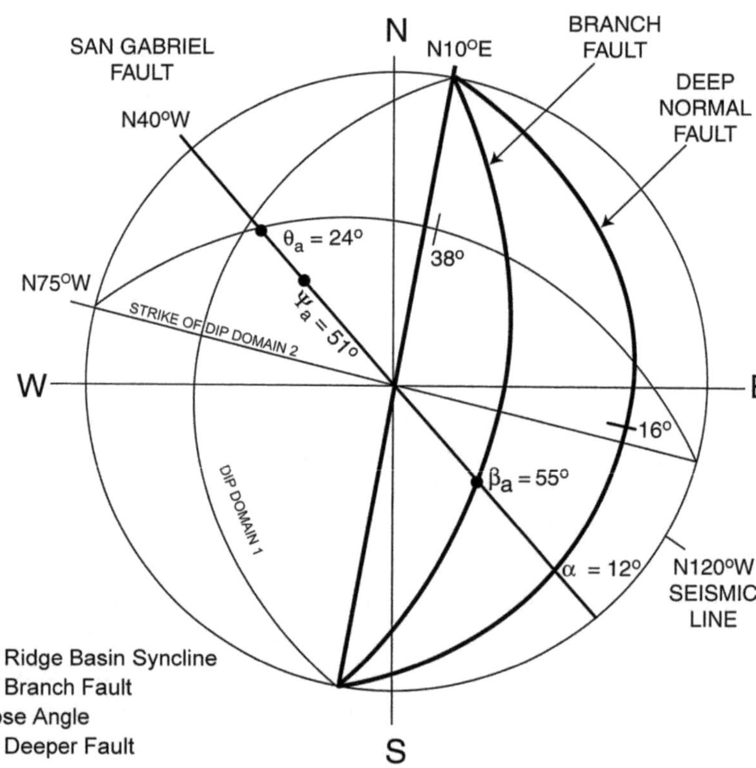

θ_a = Apparent Dip of Ridge Basin Syncline
β_a = Apparent Dip of Branch Fault
Ψ_a = Apparent Collapse Angle
α_a = Apparent Dip of Deeper Fault

Figure 12-29 Stereoplot showing interpreted surface and subsurface structure at the Hungry Valley branch point (Fig. 12-32). Fault surfaces are bold lines and dip domains are thin lines on the projection. Constraints on the interpretation include the structural cross sections of Crowell (2003c), a seismic line, the surface trace of the San Gabriel Fault that trends at N40W, the measured average dip and strike of beds in dip domains adjacent to Hungry Valley, the N10W trend of the Hungry Valley depocenter (Link 2003), and the Sun Schmidt Well. The plunge of dip domain 2 and the Ridge Basin syncline (θ_a = 24 deg at N40W) help constrain the apparent dip (β_a = 55 deg at N40W) of the Hungry Valley Fault that strikes at about N10E. The Hungry Valley Fault flattens at depth to form a listric fault surface. Measured dips in dip domain 1 (25W/N10E) and dip domain 2 (36NE/N75W) intersect to form the Ridge Basin syncline that trends subparallel to the San Gabriel Fault. Apparent bed dip of θ_a = 24NW constrains the bend in the normal fault surface, as does ϕ_a = 55 deg. (Published by permission of R. Bischke.)

resolve the rollover structural geometry (Fig. 12-30). The problem is solvable if methods exist to *measure* (1) the apparent collapse angle ψ_{a1} (Bischke and Tearpock 1999), (2) the apparent bed dip (θ_a), and (3) the apparent dip (α_a), of the fault at depth. It then follows that the apparent fault bend (ϕ_a) in the fault surface is obtainable using either forward or inverse modeling techniques described in Chapter 11, in this case, the angles are measurable as $\phi_a = \beta_a$.

Notice in Figure 12-28b that line GHE lies in the plane that parallels the direction of fault slip and the Ridge Basin synclinal plunge GH (Fig. 12-27). Furthermore, material does not cross this plane. Also, notice that the active Coulomb collapse plane BGC intersects line GHE at an acute angle BG, located at the top of the hanging wall (Fig. 12-28a). This means that (1) not only will a balanced profile lie in the plane defined by the direction of fault slip, but (2) the Coulomb collapse angle (ψ_a), *must be corrected for apparent dip* (along with the other appropriate angles, such as apparent bed dip θ_a, β_a, ψ_{a1} (Fig. 12-30).

Assuming a Coulomb collapse angle of 65 deg, we first calculate the apparent collapse angle (ψ_{a1}), from the strike direction of the Coulomb collapse surface. We obtain this angle by measuring the strike direction of the beds in domain 2 adjacent to the San Gabriel Fault (Fig. 12-25). In the model in Figure 12-28, the calculated strike of the beds in domain 2 is parallel to the active axial surface trace line BG. The measured strike of the beds on dip domain 2 from Figure 12-25 averages N75W and is labeled on stereoplot in Figure 12-29. When θ_a is measured, then the bed dips in dip domain 1 and 2 can be predicted from the stereograph. Accordingly, and using stereographic projection methods, the true Coulomb collapse plane ψ dips at 65 deg and dips perpendicular to the active axial surface line BG, striking at N75W. However, on plane GCDEH (Fig. 12-28b), the apparent Coulomb collapse angle measurement is $\psi_{a1} = 51$ deg and lies in a profile trending at N40W (Figs. 12-29 and 12-30). Next, using the same procedures to determine ψ_{a1} (Fig. 12-30), we measure the apparent plunge angle θ_a of the Ridge Basin syncline, and the apparent dip of the branch fault, β_a.

The plunge angle of the Ridge Basin syncline on plane GCDEH of Figure 12-28b is $\theta_a =$ 24 deg toward N40W (Fig. 12-30b). This value constrains the model as the great circles of dip domain 1, with dip domain 2, define the Ridge Basin syncline (Figs. 12-27 and 12-29). This 24-deg dip results in two conclusions. First, the great circle defining dip domain 2 determines the maximum dip for dip domain 2, which is about 38 deg to the north (Fig 12-29). Measured values from Figure 12-25 are 34N, striking at N85W, and thus calculated values from Figure 12-29 are close to measured values for dip domain 2. Second, in this solution, the branch fault dips at an apparent dip of 55 deg. SE (β_a on Figs. 12-29 and 12-30).

In generating the model, we will later describe that the strike of the beds in dip domain 1 will be consistent with the strike of the Hungry Valley Branch Fault at about N10E. As stated in the previous paragraph, the apparent bed dip θ_a at the intersection of domains 1 and 2 defines *the plunge angle of the Ridge Basin syncline* (Figs. 12-29 and 12-30), which means that this 24-deg dip trending at N40W dip in domain 2 must be *similar to the apparent dip in domain 1*. This is required in order to maintain continuity at and across the synclinal axial surface. This apparent dip lies in the same vertical profile as the strike direction of the strike-slip fault and the Ridge Basin syncline, which in this case is the San Gabriel Fault treading at N40W (Figs. 12-25 and 12-29). If a profile lies in the strike direction of an inclined strike-slip fault, then the apparent fault dip $\alpha_a = 0.0$ deg (Fig 12-30a), and the apparent bend ϕ_a in the fault surface is equal to β_a, the apparent dip of the Branch Fault (Fig. 12-30). Thus, the value for the apparent dip of the Hungry Valley Branch Fault can also be plotted on the stereographic projection with an apparent dip $\beta_a = 55$ deg southeast (Fig. 12-29), and trending at S140E. We discuss in the next section that the trend of the branch fault is about N10E. Thus, the great circle through the apparent dip $\beta_a =$ 55 deg southeast defines the maximum dip of the branch fault at 59E/N10E (Fig. 12-29).

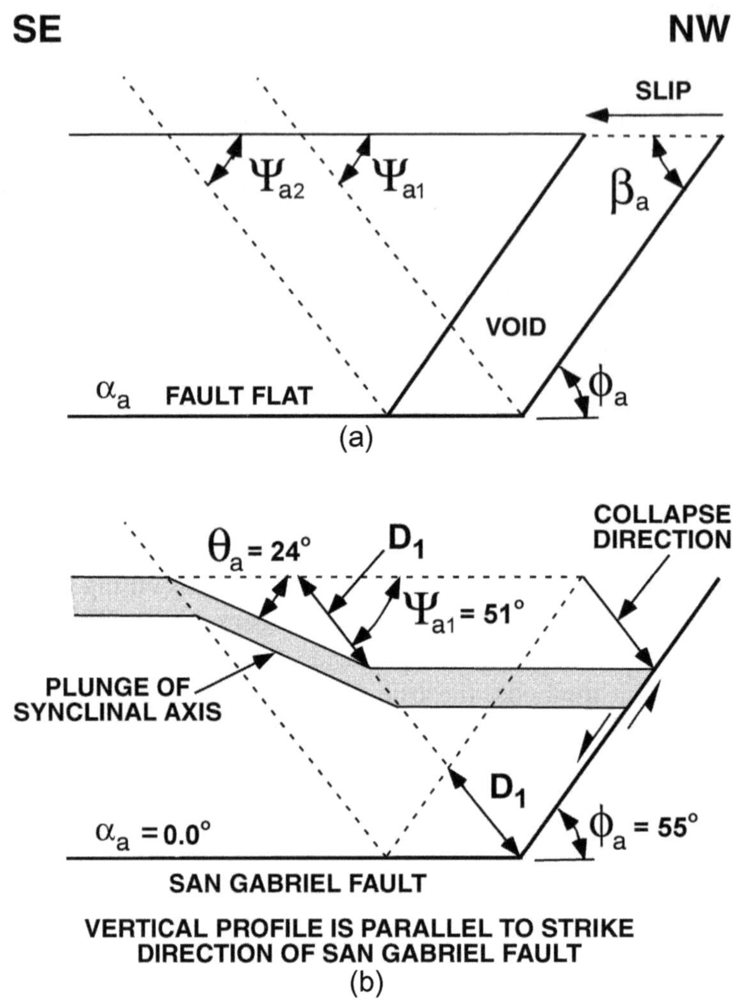

Figure 12-30 (a)–(b) Model for dip domain 2 adjacent and parallel to the San Gabriel Fault. (Published by permission of R. Bischke.)

Inverse Modeling of Dip Domain 1 and the Branch Normal Fault. Lastly, using the extension inverse theory (Chapter 11), and briefly described in this section, we first justify the shallow geometry of the Hungry Valley Fault from the geological data and then determine the dip of the *deeper portion* of the Hungry Valley Branch Fault, which forms the basal portions of Ridge Basin. We briefly describe the sedimentary evidence first.

According to the extensional model, the branch fault forms the western flank of the Hungry Valley depocenter. This assumption seems reasonable, as the stratigraphy of the basin (Link and Osborne 1982; Link 2003) predicts that a depocenter exists in the area of Hungry Valley (Figs. 12-25 and 12-26). A normal fault would form a depocenter. In addition, the direction of sedimentary transport into the Hungry Valley depocenter is from the north at about N10E (Link 2003). If we assume that the shallow portions of the proposed Hungry Valley Fault parallels the strike direction of dip domain 1, then the Hungry Valley Branch Fault may subparallel the present trend of the Hungry Valley basin. Using the observation that dipping beds, related to normal faulting, typically dipping *at about a right angles* to the strike of the *normal fault* (typical of rollover structures), measurements of strike of domain 1 (Fig. 12-25) predict a strike for the proposed Hungry Valley Fault is N10E, which is consistent with the sedimentary source direction (Link 2003).

Next, we estimate the strike of the branch fault using (1) the plunge angle of the Ridge Basin syncline of θ_a = deg toward the proposed Hungry Valley Fault, (2) the stereographic plot of the measured domain 1 bed dips, and (3) inverse modeling (Chapter 11). Inverse modeling, which is the reverse of forward modeling, employs surface bed dips or depth-corrected seismic data to predict deep subsurface fault geometry. The consequence of decoupling known surface data from the unknown deep fault geometry introduces uncertainty into the subsurface depth predictions. This uncertainty results primarily from measurement errors that are inherent in all inversion methods, including seismic and potential field methods. Although inversion methods contain error, they have the advantage of being able to make *predictions*.

The inversion of the surface dip data and the 55-deg southeast apparent dip on the Hungry Valley normal fault predicts the gross crustal structure beneath Ridge Basin. This exercise results in a generic and retrodeformable model of Ridge Basin (Chapter 11). Additional subsurface data, such as seismic profiles, are required to confirm these predictions.

Inverse theory requires a number of assumptions that include knowledge of the Coulomb collapse angle (typically between 60 deg and 70 deg), taken here to be 65 deg. This inversion uses a profile constructed parallel to the strike of the San Gabriel Fault using domain 1 dips in the direction of the San Gabriel Fault slip, which is N40W.

Retrodeformable Figure 12-31 represents any profile that trends in a direction of N40W across dip domain 1. This profile can trend adjacent to the Ridge Basin syncline or across Hungry and Peace Valleys in the northern portions of Ridge Basin (Fig. 12-25). As the profile parallels the slip direction of the San Gabriel Fault, the bed dip angle θ_a and deep normal fault dip angle α_a are apparent dip angles. Domain 1 has an average bed dip θ of about 25 deg west, and it is separated from the gently dipping and folded Hungry Valley domain by an axial surface (Fig. 12-25). Correcting the bed dips in domain 1 to the profile direction of N40W results in an apparent dip θ_a =20 deg. This profile intersects the Hungry Valley Fault at an apparent dip of β_a = 55 deg (Figs. 12-29 and 12-31).

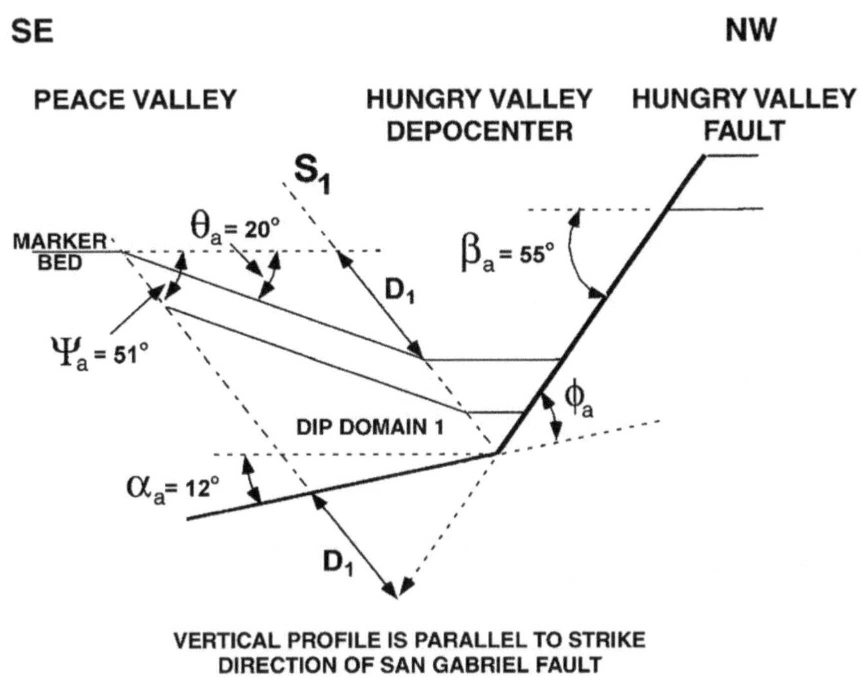

Figure 12-31 Inversion model for dip domain 1, used to calculate dip α_a of the normal fault at depth. The listric normal fault causes the beds between Hungry and Peace Valleys to dip toward that fault, thus forming the Hungry Valley depocenter. The interpretation based on the surface dip data suggests that the 16-deg dipping normal fault may intersect the brittle-ductile transition beneath the volcanic Soledad Basin. (Published by permission of R. Bischke.)

To determine the apparent dip of the deeper part of the fault α_a, use the methods described in the section Procedures for Projecting Large Normal Faults to Depth, in Chapter 11. As described in that section, simply project the distance D_1 (Fig. 12-31) from the active to the inactive axial surface at the projection of the shallow fault surface. This procedure establishes the position and dip of the deeper part of the branch fault α_a. In this case, the dip angle of the deep fault $\alpha_a = 12$ deg (Fig. 12-31). Using the stereonet (Fig. 12-29), this 12-deg apparent dip of the deeper fault, in a S140E direction, establishes the plane or great circle representing the deep normal fault. The true dip of the fault is 16 deg in a S100E direction (Fig. 12-29).

A normal fault dip of 16 deg projects to the east to intersect the brittle-ductile transition at a depth of about 20 km beneath the Soledad Basin that contains Tertiary volcanic rocks (Ehlert, 2003). This prior volcanic activity may have weakened the crust, facilitating the crustal extension. A generalized fault surface map for the Ridge Basin basement is shown in Figure 12-32. According to the model, hanging wall beds moving over the nonplanar fault surfaces *deform at the bends in the fault surfaces*. This motion replicates the main structural features and bed dips observed in Ridge Basin. Thus, the fault surface map represents a 3D fault surface model for the Ridge Basin structure (Fig. 12-25). The faults are tied to cross sections constructed by Crowell (2003c) and to the Sun Schmidt Well (Fig. 12-32). On Figure 12-32, all the fault surfaces are planar and the lines of fault intersection are determined with stereoplots. Where well control and seismic data are sparse, this procedure allows for the projection of fault surfaces over long

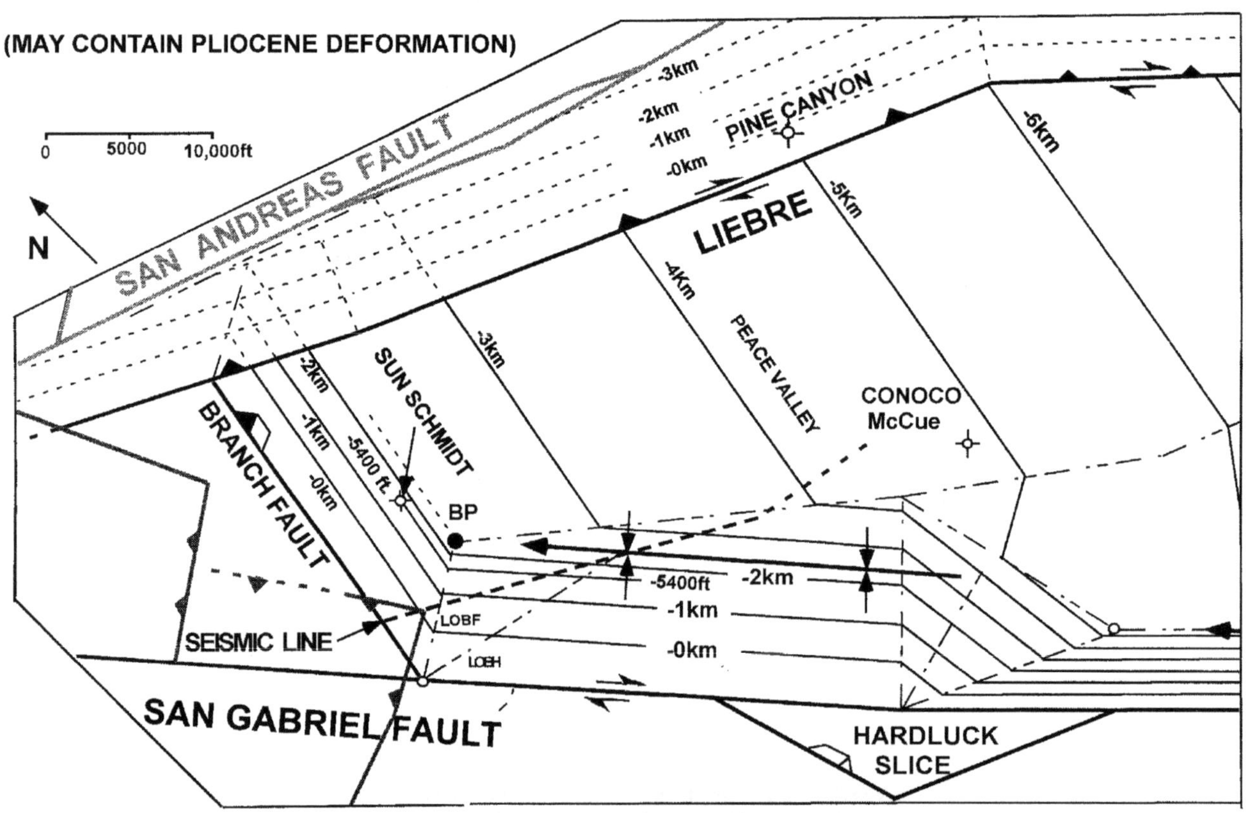

Figure 12-32 Generic fault surface map for the Ridge Basin basement, California. The interpretation of dip domain, seismic, and well data suggest that a branch fault forms a releasing bend in the area of Hungry Valley. This fault geometry, when combined with Coulomb collapse theory, models the general dip and strike pattern shown in Figure 12-25. The synclinal fold structure emanates from branch points (BP) located beneath Hungry Valley. Stereographic projection methods determine the fault plane intersections. (Published by permission of R. Bischke.)

distances and provides strong constraints on an interpretation surface dip data (Suppe 1983). Methods involving curved line projection techniques require dense data control and are not applicable to underconstrained data sets.

The generalized fault surface map for the basement shown in Figure 12-32 is consistent with the major structural features and the bed dips observed in the Ridge Basin surface dips (Fig. 12-25). The generalized model shows that *the Ridge Basin syncline emanates from the branch point* (BP) of the San Gabriel Fault with the Hungry Valley Branch Fault. As the San Gabriel Fault dips at a *higher angle* in the southeast area of the map, the intersection of the deep normal fault with the San Gabriel Fault migrates to the south. This map may explain why *domain 2 narrows to the southeast near Fisher Spring*. The geometry of the San Gabriel Fault controls the position of the Ridge Basin syncline.

Tests of Model. Measurements and stereographic projection can determine the true bed dip θ for dip domain 2, and measurements on profiles can determine apparent dip of the beds θ_a parallel to the San Gabriel Fault. This apparent dip angle θ_a lies in the same *plane, defined by the great circle, as the true bed dip* (Fig. 12-29). Near Hungry Valley, the strike of the beds in dip domain 2 is about N75W (Figs. 12-25 and 12-30). As the apparent dip of the beds (θ_a = 24 deg) lies in a profile that trends at N40W and in a plane that strikes at N75W, the maximum bed dip from the stereogram is determined to be 38 deg in a N15E strike direction (Fig. 12-29). This calculated value of 38N/N75W for the dip and strike of the beds in domain 2 is close to the observed average dips of 34N/N85W in dip domain 2 (Fig. 12-25).

Plotted on Figure 12-29 is dip domain 1, which dips and strikes at 25W/N10E. This plane intersects dip domain 2 to form the Ridge Basin syncline, which subparallels the surface trace of the San Gabriel Fault (compare Figs. 12-25 and 12-29). In Figure 12-29, the plane representing domain 1 (25W/N10E) intersects the plane representing dip domain 2 (36N/N75W) in a direction *that is subparallel to the surface trace of the San Gabriel Fault and the Ridge Basin syncline at a direction of N43W*. Thus, there is a good correspondence between the model and the observed surface bed dip data (Fig. 12-25 and Fig, 12-29).

SUMMARY FOR SECTION

The balancing of strike-slip fault displacements is in its infancy and requires additional work involving restorable kinematic models that describe the structural styles observed along strike-slip fault zones. In Ridge Basin, the good correspondence between model-based predictions and observation is encouraging. However, these kinematic models should be subject to additional tests utilizing well-constrained data sets and viable fault surface maps. Nilsen and McLaughlin (1985) report that Ridge Basin contains some structures that are similar to Hornelen Basin in Norway and to portions of Little Sulphur Creek Basin in northern California (Fig. 12-33). Two of these basins contain a throughgoing strike-slip fault that bounds one flank of the basin, and talus and debris flow deposits exist adjacent to the strike-slip fault. These syntectonic breccias demonstrate growth through time. Paralleling the fault along the talus and debris flow deposits is a synclinal fold structure (Fig. 12-33). Nilsen and McLaughlin (1985) conclude that many of the world's strike-slip extensional basins share similar tectonostratigraphic histories.

When examining a strike-slip structure, we should ask ourselves how the structure and bed dips moved into their present positions. In this case, the question is, *How did 25- to 30-deg lacustrine sediments downlap basement, along a N40W profile, across the deepest portions of the basin?* A kinematic description of displacement of the structure is essential prior to embarking on the more theoretical analysis of how the structure reacted to stress and strain. The theory of elasticity is clear on this subject: kinematics is followed by dynamics (Obert and Duval 1967;

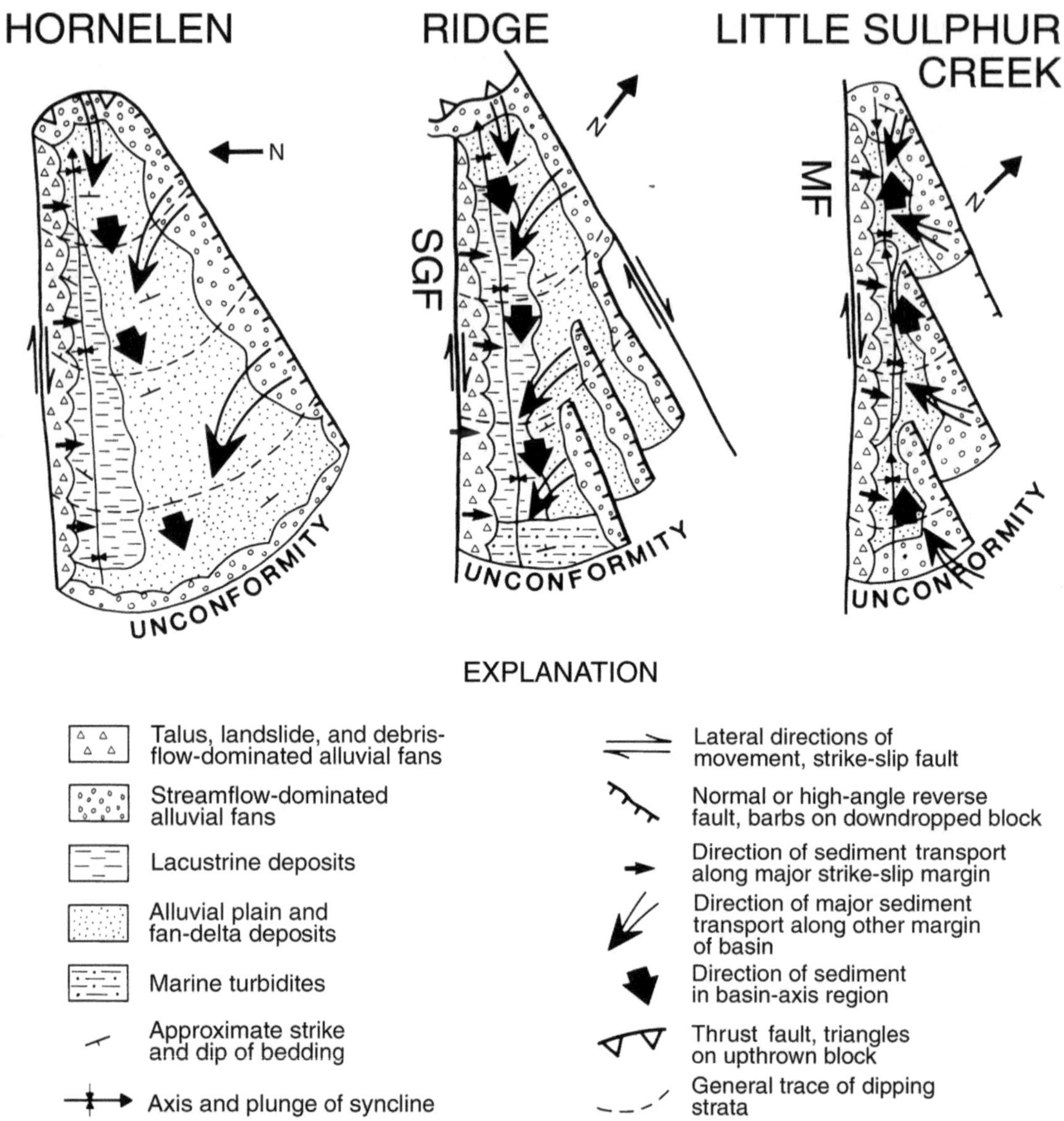

Figure 12-33 Hornelen Basin (Norway), Ridge Basin (southern California), and Little Sulphur Creek Basin (northern California), showing strike-slip fault and associated breccia and synclinal structure that parallel the strike-slip fault zone. (From Nilsen and McLaughlin 1985. Published by permission of the Society for Sedimentary Geology.)

Ramsay 1967). Clearly, scientists must develop the kinematic equations prior to formulating the tensor equations for stress and strain or differential equations. Structural studies that restrict the analysis to stress, while ignoring displacement analysis, seem questionable and unrealistic.

GENERAL CONCLUSIONS FOR CHAPTER 12

We are unaware of any method obtained from seismic or well log correlation data, or from models concerning *stress* theory (or the strain ellipsoid), that leads to resolving the state of stress on fault surfaces. Deductions concerning stress *are not a viable exploration or prospect-generation tool.* Stress is a mathematical concept, is invisible, and its direction cannot be deduced from neotectonic or older faulting data. Furthermore, clay analog models of strike-slip faulting contain a number of assumptions that affect the understanding of the structures and the styles of faulting that form along strike-slip faults. Clay models do not scale to the real earth (Hubbert 1937) and suffer from near-field boundary condition problems that impart contractional motions across a fault surface. The clay models do not correctly model the orientation or the regional distribution of real structures (Figs. 12-3, 12-4, and 12-5). Geoscientists are best served by applying correct interpretation procedures and mapping techniques to their data and prospects rather than relying on theoretical models involving stress or strain.

Early in an exploration project, geoscientists should concentrate their interpretation on observable and reliable data and should construct viable maps of recognizable fault surfaces. During the interpretation and construction of admissible fault surface maps, geoscientists will gain insight into the appropriate tectonic style affecting their area. When the mapped data lead geoscientists to a viable tectonic style, then geoscientists may employ a model to better understand and interpret the area. If interpreters approach a data set with *preconceived ideas* of the structure or bias toward a particular model, then the results of the exploration project are predictable—the interpretation and maps will conform to the preconceived ideas based on the model. A ***data-first***, *models-second* approach is a more objective approach to any data set involving subsurface structure. Geoscientists then develop the appropriate model from the data rather than interpret the data using a convenient model. This approach is more likely to result in admissible interpretations and maps of the subsurface that are valid in 3D. This approach will minimize structural risks where exploration prospects exist.

Strike-slip fault interpretations should present direct evidence for horizontal displacements. As strike-slip faulting is a 3D problem, 2D seismic profiles can, at best, provide only suggestions of strike-slip faulting. *However, the construction of viable maps of the faults is the first step in providing direct evidence for the horizontal displacements that are required for strike-slip interpretations.* These viable maps of fault surfaces should show evidence of curved, or bent, fault surfaces. Strike-slip displacements at the bends will create restraining and releasing bends. Thus, restraining and releasing bends seem fundamental to the existence and understanding of strike-slip faults (Crowell 1974a, 1974b). Their ubiquitous presence along the major strike-slip fault zones presents direct evidence for strike-slip faulting. However, restraining and releasing bends may not record the total amount of strike-slip displacements. The bends may be obvious, such as the large bend along the Transverse Ranges of California (Fig. 12-10b). Alternatively, the bends may be subtle, like the bend at Loma Prieta along the San Andreas Fault (Fig. 12-18). Most important, these bends contain information that can confirm the presence of the horizontal displacements that accompany strike-slip faulting. If strike-slip faulting is present on a subsurface map, one only needs to look for the presence of restraining and releasing bends. Restraining bends should map as structural highs and releasing bends as structural lows. If evidence for restraining and releasing bends is not present on a regional scale subsurface map, then another interpretation may be warranted. If the strike-slip fault breaks the surface, then look for offset rock types on the surface geologic map (Crowell 1974a, 1974b).

We discussed several published techniques for recognizing strike-slip displacements. Three-dimensional regional and local restorations of strike-slip faulting may be supportive of horizontal displacements. Fault surface maps may show curved fault surfaces that form restraining and releasing bends. Folds form as the hanging wall beds move over curved fault surfaces (Bally et al. 1966; Rich 1934). Fault bend folding along strike-slip faults may prove useful in describing and analyzing structures in fault bends and along mapped fault surfaces. Balanced cross sections of restraining and releasing bends provide constraints on structural interpretation problems that lead to 3D structural validity and to admissible interpretations of subsurface data. Balanced cross sections of restraining and releasing bends should assist industry to generate high-quality prospects with reduced risk. In short, strike-slip fault interpretations should be subject to the same *high-quality and rigorous interpretation techniques required of other styles of faulting.*

CHAPTER 13

GROWTH STRUCTURES*

INTRODUCTION

Tectonic and sedimentary growth is important to petroleum exploration. For example, growth along faults creates both structural traps and seals (Brenneke 1995), and this growth can be coincident with the deposition of reservoir units. Growth faults may trap hydrocarbons during the early stages of a hydrocarbon cycle. The expanded sections that typically form downthrown to down-to-the-basin normal faults are the *highest growth* sections found in the petroleum regime (Thorsen 1963). These expanded downthrown sections often contain large accumulations of hydrocarbons (Tearpock et al. 1994). Hydrocarbon accumulations correlate not only to high growth intervals but often to the highest growth intervals within the petroleum system (Fisher and McGowen 1967; Woodbury et al. 1973; Branson 1991; Pacht et al. 1992). Therefore, in order to find large accumulations of hydrocarbons, we must understand and analyze growth. Also, growth defines the history of prospect formation.

We emphasized the importance of constructing a 3D structural framework in Chapters 7, 8, and 9. In this chapter we address the *4D time factor*, which is important when exploring for and developing hydrocarbons. We consider the time variable through the *superposition*, or *relative age, principle*. In other words, the oldest rocks are the deepest rocks, except for the inverted sequences discussed in Chapter 10. Rocks record cycles of changing sedimentary influx that relate to systems tract and sequence boundaries. For example, if we define the thickness of an interval of sediments as d, and time as t, then the sedimentation rate is d/t. The sedimentation rate for that interval may change between two locations, and the change in rate is $\Delta d/t$. We can take the ratio of the change in sedimentation rate to the sedimentation rate, or $(\Delta d/t)/(d/t)$. In this ratio, *time cancels out*. This dimensionless growth (Suppe et al. 1992), or relative age, ratio $\Delta d/d$ is a measure of **growth** (Bischke 1994b). *When using subsurface data, information on growth and relative age is always available to evaluate growth and its impact on exploration.*

* For all figures in this chapter (in the printed book only), see the preface for information about registering your copy on the InformIT site for access to the electronic versions in color.

We demonstrate from correlation data that each sedimentary sequence, being a genetically related unit, experiences different patterns of growth. Furthermore, sedimentary sequences in all tectonically active regimes exhibit growth, not just in the extensional growth fault regime. Typically, *linear or monotonic patterns of growth characterize sedimentary sequences*. The linear growth recorded in a sequence is typically punctuated by abrupt changes in growth rate. These punctuated growth patterns may aid in defining systems tract and sequence boundaries, and they can contribute to the resolution of a variety of practical problems involving changes in growth.

When geoscientists think of growth faulting, they are often referring to the large listric, growth normal faults. However, growth faults can be *small*, with only tens of feet of vertical displacements. Any normal fault is a growth fault if the stratigraphic section expands across the fault (Thorsen 1963; Dawers and Underhill 2000). Furthermore, all structural styles can exhibit growth, including **growth reverse** and **growth thrust faults** (Suppe et al. 1992) as well as **growth strike-slip faults** (Shaw et al. 1994; see Chapter 12) and **growth inversion structures** (Mitra 1993; Link et al. 1996). In this chapter, we discuss these and other complex structural styles that are sometimes misidentified with well log and seismic data. Structural styles are important because the recognition of the correct style for a region affects our interpretations and our understanding of the petroleum system. We discuss techniques for recognizing structural styles using growth analysis, with examples of growth structures from the four major tectonic regimes: extensional, compressional, strike-slip, and salt. Large accumulations of hydrocarbons correlate to timing of migration and fault movement, so it is important to understand the timing related to fault growth and deposition.

The methods presented in this chapter, the expansion index method (Thorsen 1963) and the high-resolution $\Delta d/d$ technique and its related methods (Bischke 1994b; Sanchez et al. 1997), can have a high resolution, about one part in a thousand, and are robust (Chatellier and Porras 2001, 2004) when integrated with other geological and geophysical data, thus providing a check on existing interpretations. Growth analysis has helped resolve a number of existing or potential problems that include (1) rapidly distinguishing faults from unconformities on the flanks of salt bodies (Bischke et al. 1999) and confirming small faults or uncertain fault interpretations; (2) locating sequence boundaries (Bischke 1994b; Sanchez et al. 1997) and subtle stratigraphic traps and predicting erosional surfaces and potential bald structures on the crests of anticlines; (3) solving general correlation problems (discussed in this chapter); (4) calculating sand/shale ratios from seismic sections downthrown to normal faults (Pochat et al. 2004) and predicting major lithology from growth sections (Castelltort et al. 2004); (5) locating the highest growth or highest petroleum potential intervals (i.e., the thickest sands) (Fisher and McGowen 1967; Woodbury et al. 1973; Branson 1991; Pacht et al. 1992); (6) determining the time of structural growth and timing of faults (Cartwright et al. 1998, Echavarria and Allmendinger 2003); (7) checking interpretations for consistency with the existing growth history; (8) recognizing unidentified problems; and (9) quality-controlling the *input values in a well log data base*. The method has also been applied to determine timing on strike-slip faults (Shaw et al. 1994) and reverse and thrust faults (Echavarria et al. 2003). More recently, Chatellier and Rueda (2010) used these techniques to recognize missing section resulting from miscorrelations and for locating missing section on about the *same structural level* on a contractional folded structure. The missing section data contour into a shallow-dipping fault surface. This is consistent with a folded anticline subsequently faulted by another period of thrusting.

In this chapter, we describe practical examples pertaining to sequence stratigraphic interpretations, to strike-slip and compressional deformation, to unconformities and correlation problems related to salt, and to the extensional regime that typically contains growth sediments.

Expansion Index for Growth Faults

Thorsen (1963) proposed the **expansion index method** to categorize growth along normal faults. Indeed, the term *expansion fault* comes from Thorsen's classic work on growth faulting. The expansion index is a ratio calculated by comparing a downthrown stratigraphic thickness to its correlative upthrown thickness. Evaluation of the indices for a sequence of stratigraphic units determines the timing and activity on growth faults, and a plot of the indices provides a record of the movement of a fault throughout its history. Although Thorsen (1963) developed the technique by studying growth faults in southeast Louisiana, USA, his method for analyzing growth faults *applies to any region of the world* that contains syntectonic faults.

For normal faults, the procedure starts by determining the correlative upthrown and downthrown stratigraphic units, and then comparing the thickness of each unit in the upthrown (footwall) and downthrown (hanging wall) blocks. By definition,

$$\text{Expansion Index} = \frac{\text{Thickness Downthrown}}{\text{Thickness Upthrown}}$$

Where growth is recognized, an expansion index greater than 1.0 typifies growth. If the expansion index ratio is less than 1.0, then a correlation error has been made or an unconformity is present in the hanging wall fault block.

Figure 13-1 is a simple example to illustrate the determination and application of the expansion index for a generic growth fault. The left side of the figure contains a cross section of a growth fault with one well in the upthrown block and one in the expanded, downthrown block. For simplicity, the correlative horizons, *a* through *k*, in the upthrown fault block bound units that are all of equal thickness ($t = 1.0$). The thickness of each unit in the downthrown block is used to calculate the expansion index for each.

Thorsen presents the results in the form of a bar graph on the right side of Figure 13-1, and the fault movement with time is readily apparent. By analyzing the graph of the expansion indices, we conclude that movement on the fault began at time *j* and reached a maximum period of growth activity between times *f* and *g*. Fault movement ceased at time *b*. The maximum period of fault growth between times *f* and *g* involved a 2.1 to 1 expansion of the downthrown stratigraphic section. If the interval *f* to *g* contains thick sands, then this thick interval has the highest potential for hydrocarbon accumulation.

Each growth fault appears to have its own unique *fingerprint* in reference to its expansion indices. The rate of movement on a growth fault increases from the initial movement to some maximum rate, and then decreases again until fault movement ceases. During its life, a growth fault may move sporadically, resulting in an irregular growth history. Each growth fault has its own unique growth history as a result of fault movement, the rate of subsidence, and the rate of sediment supply across the fault. Even over long horizontal distances, the fingerprint for a particular fault can usually be recognized along strike, as shown in Figure 13-2, even though the values of the expansion indices may not be constant along strike. The plots of the expansion indices for this fault, calculated at separate locations two miles apart, are similar. A review of the individual expansion indices and the overall pattern at each location shows that they are similar enough to be identified as representing the same fault.

Because of the unique characteristics of the plot of the expansion indices for each growth fault, the technique can be used in complexly faulted areas as a lateral correlation tool for the identification of individual growth faults over long distances. The expansion index technique is applicable for use with either electric well logs or seismic sections. Because of the greater stratigraphic detail that is found in well logs versus seismic sections, expansion indices calculated

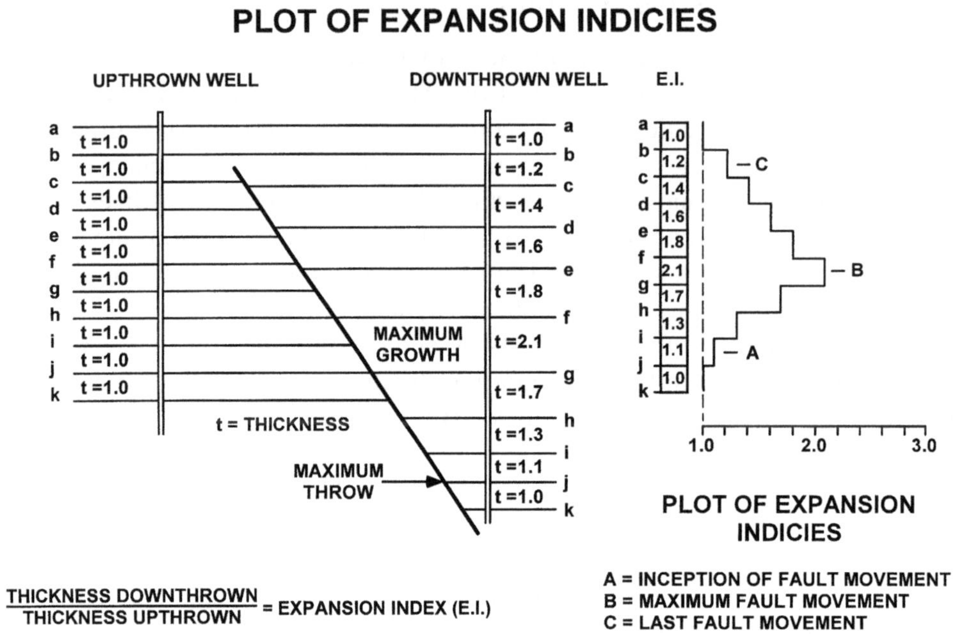

Figure 13-1 Expansion index can be used to measure the timing and rate of movement along a growth fault. Expansion indices are commonly illustrated in bar graph form (right side of figure). (From Thorsen 1963. Published by permission of the Gulf Coast Association of Geological Societies.))

using well logs are more accurate. One important note needs to be made with regard to this technique and seismic sections. As time distorts vertical distance on seismic sections, always *depth-convert* the seismic sections before calculating the expansion index.

The expansion index also serves as a vertical correlation aid for well log and seismic interpretations. The thickness of the stratigraphic section in the upthrown block should almost always be equal to or less than the thickness for the equivalent section in the downthrown block, so the expansion index that is calculated for any specific unit is typically greater than 1.0. Therefore, a calculated expansion index less than 1.0 using well logs or seismic data indicates either a bust in the correlation of the individual units across the fault or an unconformity in the hanging wall fault block.

Figure 13-3 shows the analysis of three separate growth faults in southern Louisiana. The plots of the expansion indices indicate that the age of each fault is different and that they become progressively younger in a basinward direction (see map insert). The time of maximum fault growth, indicated by dark arrows, corresponds to the age of maximum sedimentation in the area across each fault. In certain geological settings, such information is vital to the exploration for hydrocarbons. In the Miocene trend onshore and offshore Louisiana, for example, many of the major structural features relate to the genesis of diapirs of the Jurassic Louann Salt, which directly ties to the timing of the sediment load across growth faults.

The analysis of a growth fault, based on the pattern of its expansion indices, provides information important to the exploration for hydrocarbons. If time information is available, then at least eight benefits derive from analyzing the expansion index plot of a growth fault.

1. The plot shows the time of fault inception.
2. It indicates the time of maximum fault movement (maximum expansion).
3. It shows the time of cessation of fault movement.
4. It provides a complete history of fault movement.

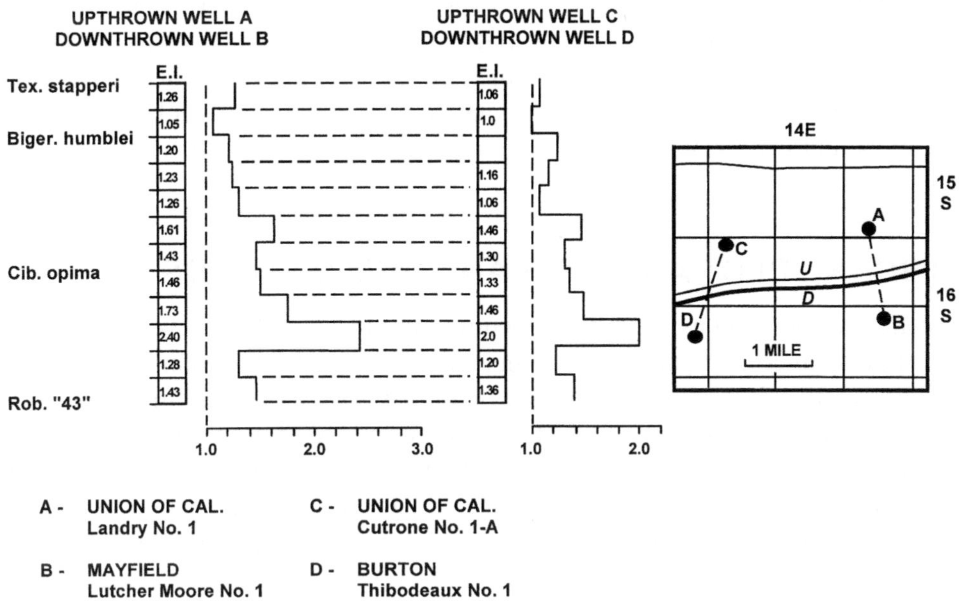

Figure 13-2 Comparison of expansion indices, taken over 2 mi apart, indicates that a growth fault's fingerprint can be recognized over long distances. (From Thorsen 1963. Published by permission of the Gulf Coast Association of Geological Societies.)

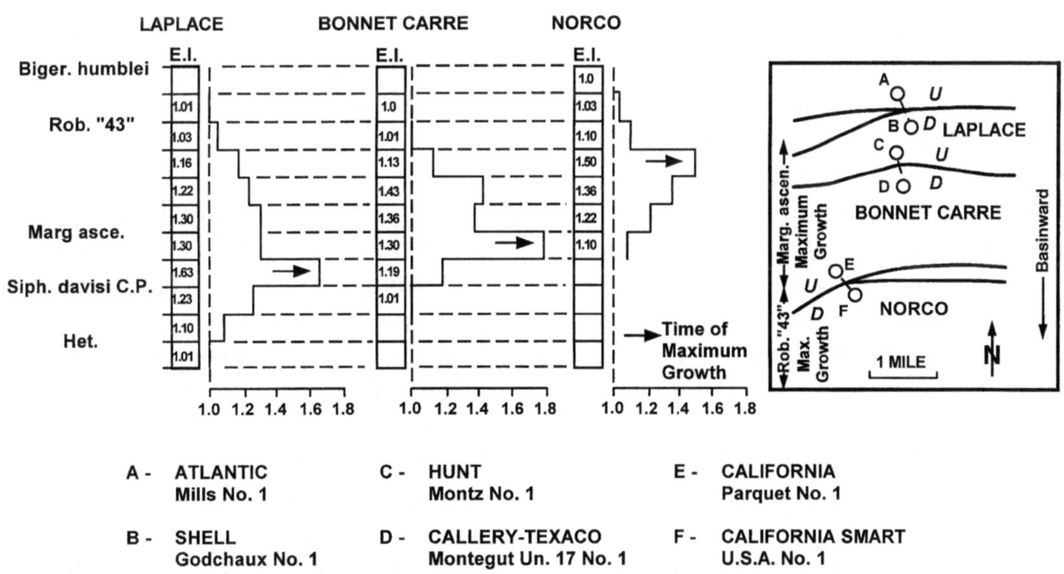

Figure 13-3 Expansion indices can be used to date growth faults with regard to their initial movement, time of maximum growth, and last movement. (From Thorsen 1963. Published by permission of the Gulf Coast Association of Geological Societies.)

5. Regionally, it can correlate stratigraphic units associated with maximum fault growth.
6. It can be used with structural growth indices to compare fault growth with structural growth of uplifted areas.
7. It can serve as a vertical correlation tool for seismic sections and well logs.
8. It serves as a lateral correlation tool for recognizing specific faults over long distances using well logs or separate seismic lines.

Some disadvantages to using the expansion index method are that, because it is a ratio, it has a low resolution and may contain mathematical ranges that can be significant (Bischke 1994b). In the next section, we describe the related, but high-resolution, Δd/d technique, also referred to as the Multiple Bischke Plot Analysis (MBPA). This technique minimizes mathematical ranges and errors, with maximum errors on the order of 0.1 percent (see Accuracy of Method section).

MULTIPLE BISCHKE PLOT ANALYSIS AND Δd/d METHODS

Growth structures develop as sediments accumulate during tectonic events and sea level fluctuations, and they form coincident with faulting, subsidence, and diapirism. By analyzing the subsea depths of correlative horizons, we can glean information about sedimentary and tectonic history from the syntectonic sediments. Indeed, the geometry of a growth section contains the *only records of the dynamic processes* that are available to geoscientists. If we are to provide answers to tectonostratigraphic problems, then we must consult the growth section to decipher the details. In many areas, reliable correlation data exist to constrain interpretations, but the data may be ambiguous over problem intervals. These problem areas may result from facies changes, high bed dips, or rapidly thinning horizons. An analysis of the reliable portions of the growth section can help resolve practical problems that conventional methods, such as 3D seismic, cannot resolve.

In this section, we describe two methods for analyzing growth: the Δd/d method proposed by Bischke (1994b) and a related but more powerful method that Sanchez et al. (1997) and Chatellier and Porras (2001, 2004) call the Multiple Bischke Plot Analysis (MBPA). A single graphical plot of *time-stratigraphic correlation markers* forms the basis for the Δd/d method of growth analysis. Alternatively, MBPA is an enhanced graphical technique based on *multiple Δd/d plots* of correlation markers. Sanchez et al. (1997) show that comparing several Δd/d plots to each other often provides additional information for rapidly understanding the growth history or resolving problems related to growth.

These methods display, in graphical form, the growth history of a sedimentary section. Because the plots record *changes in growth*, the methods have a very high resolution, typically about *one part in one thousand*, when using well log correlations (Bischke 1994b). The technique appears capable on well log data of resolving subtle stratigraphic details that occur on sequence boundaries that have only several feet of missing section (see Accuracy of Method). On good-quality seismic profiles, the methods can identify disconformities having only 20 ft to 50 ft of missing section. The methods may help solve problems that result from changing sedimentary, eustatic, or tectonic conditions. These include (1) correlation problems in general and, specifically, on the flanks of salt bodies; (2) questions about fault timing and trapping of hydrocarbons; (3) locating subtle stratigraphic traps; and (4) identifying a variety of stratigraphic and structural problems not readily recognized by conventional interpretation techniques. The methods have application in *all* tectonic regimes, whether extensional, compressional, strike-slip, or diapiric. Geoscientists often encounter problems adjacent to salt diapirs if they cannot correlate well logs due to sands thinning dramatically toward or onto salt or to a thick multiple sand section, and they cannot correlate the seismic data because of poor imaging of high bed dips. In these situations, the *MBPA method* may be the only remaining avenue available to the solution of correlation problems. Furthermore, if correlation data is entered into a spreadsheet that contains a graphics program, then these methods can be very fast.

Growth can be recognized by a change in the true stratigraphic thickness of a unit within an area, and the Thorsen method calculates and compares thicknesses of a number of units. The MBPA method depicts growth by using a number of horizons and plotting the *changes in depth of each horizon* within an area. Unlike the Thorsen method, the interval thicknesses are not determined. Therefore, the MBPA method is easier to use, involves fewer measurement errors, and is more accurate. The method allows for the recognition of faults, unconformities, and incorrect correlations in the form of discontinuities, or anomalies, on the plots.

Method

In this section we describe the high-resolution Δd/d and MBPA methods and present a number of practical examples of these methods. *The methods have application* ***from the producing-field scale****, with excellent well control,* ***to the regional basin scale*** *with minimal well control.* We present several examples of salt-related problems that are not resolvable using conventional techniques, but are resolvable using growth plots. We first describe the Δd/d method, which is the foundation for MBPA.

To begin the method in an area of study, locate two wells in a general dip direction, so that one well is structurally higher than the other. Consider first a stable, or nongrowth, tectonic environment in which the correlative intervals between two wells have about the same thickness (Fig. 13-4a left). Correlative horizons are represented by correlative markers in the well logs. For simplicity in Figure 13-4a, the top correlative marker is at the same depth in both wells. The difference in depth of the lower marker reflects growth. Plot this depth difference (Δd) of each marker against its subsea true vertical depth (SSTVD) d in the structurally higher well (Fig. 13-4a right). In a stable tectonic environment, the slope on the Δd/d plot is *approximately flat*.

In an unstable, or growth, tectonic environment, differences in the thickness of correlative stratigraphic intervals are greater, and thus larger vertical distances (Δd) separate the horizons in the wells (Fig. 13-4b left). Typically, only two wells establish this relationship, although the wells need to be in the dip direction. The resulting Δd/d plot (Fig. 13-4b right) has a *higher*, or *steeper, slope* than the one for a stable environment.

Using a number of correlation markers generates more points for the Δd/d plot. *The slope of the curve on the plot reflects the growth history of the interval plotted.* In applying the technique, use many correlation markers that represent time-stratigraphic horizons. Such horizons include parasequence boundaries, which are valuable in MBPA. Determine the difference in depth of each marker in the pair of wells, and then generate a plot of the data. If you enter the subsea depths of the correlations in a spreadsheet, simply calculate the difference in depths of the correlations in the structurally higher well and those in the lower well to generate Δd values. Place the subsea depths of the markers in the higher well on the x-axis, and then plot the Δd value corresponding to each marker on the y-axis. This is best done in a spreadsheet program that contains a graphics package.

A pair of wells is always used in the analysis, but one of the wells can then be compared with a third well, a fourth well, and so on. This comparison forms the basis of the MBPA described in the next paragraph. A plot is generated for each pair of wells. To provide more analytical power to the method, when a given up-dip well is compared with a number of other wells, place all the plots generated on a single graph (Fig. 13-17b). This type of plot provides significant information on the overall area of study in terms of growth, faults, and unconformities. We provide several examples of the Δd/d analysis and MBPA later in this chapter.

When conducting an MBPA, compare a reference well to any number of adjacent wells by using a *set of Δd/d plots* of the reference well versus each of the other wells (Sanchez et al. 1997). The reference well on the cross plots could be *a type well that has a complete stratigraphic section*. Alternatively, the reference well can be *any other well positioned on any structural level*. Unlike the Δd/d method, plot the depths in the reference well on the x-axis, even though it may not be the structurally higher well. *Thereby the reference well is present on all the plots*.

The advent of 3D statistical programs enhances MBPA by allowing the comparison of a reference well to any number of other wells. The procedure rapidly generates Δd/d plots, printed on a single page. A visual comparison of these plots rapidly locates *tectonostratigraphic anomalies. If the anomaly is present in the reference well, then the anomaly exists on all plots. However, if the anomaly is present in only one well, then the anomaly is in the well used for comparison, rather than in the reference well.* This process allows interpreters *to rapidly identify problem wells and to better understand the cause of the anomaly* (Sanchez et al. 1997; Chatellier and Porras 2001).

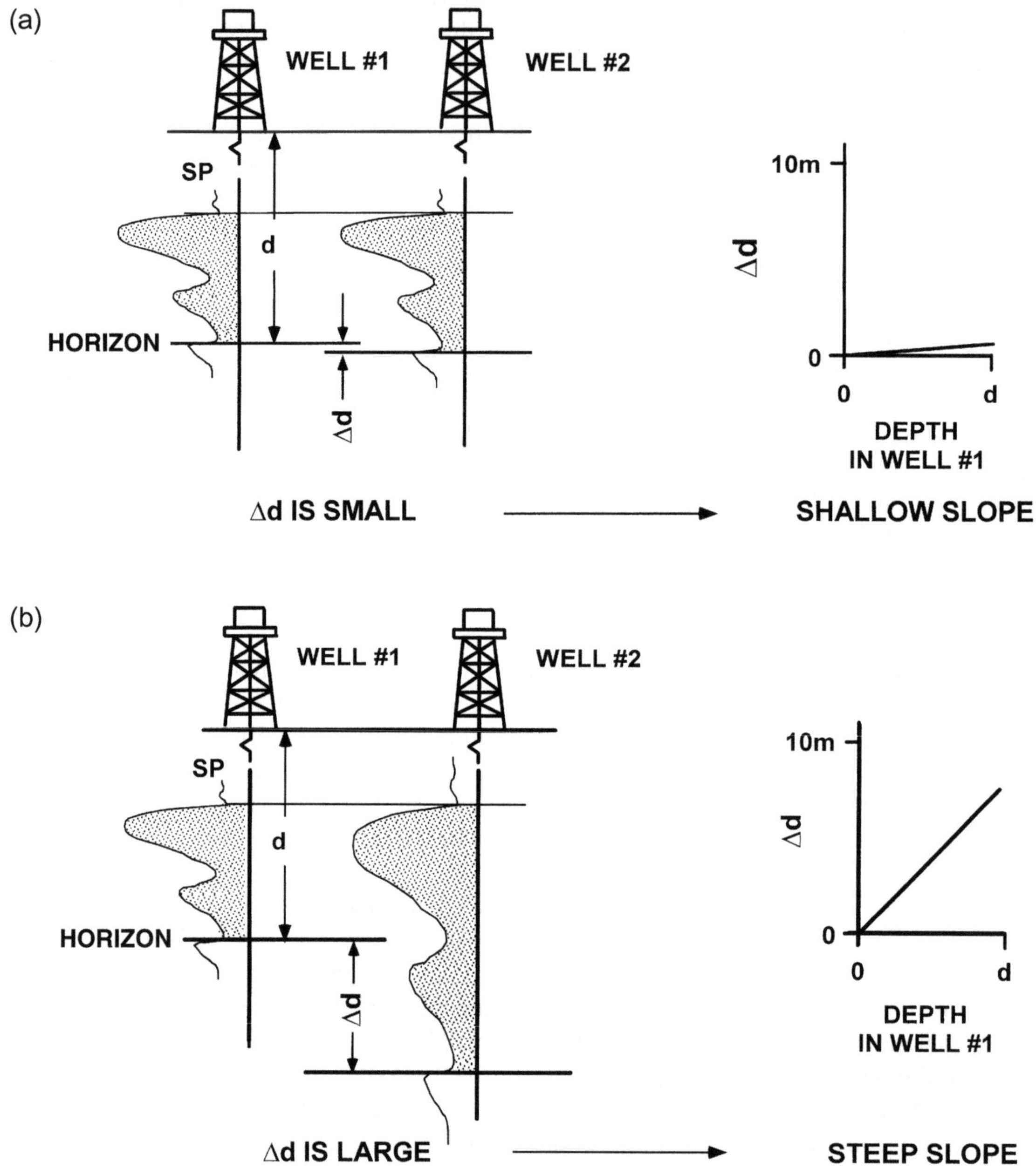

Figure 13-4 Sedimentary environments can be categorized into stable and unstable growth environments. (a) In a stable, nongrowth environment, the vertical distance between correlative markers is small. If the markers are plotted on Δd/d plots, then the curves will have a gentle slope. (b) In an unstable, growth environment, the vertical distance between correlative markers is large, and the resulting Δd/d curves are more steeply sloping. (Modified from Bischke 1994b, AAPG©1994. published by permission of the AAPG whose permission is required for further use.)

Structural and stratigraphic analysis, based on comparison of Δd/d plots, can be enhanced if stratigraphic data are posted on the plots, as in Figures 13-14 and 13-17b. Then MBPA can be applied with respect to discrete stratigraphic intervals and thereby provide information directly relevant to the history of the study area.

Sanchez et al. (1997) describe their experience with previously correlated well logs. Should interpreters use pre-existing correlations, and perhaps be prejudiced by these correlations? Often this results in purely cosmetic changes to the correlations, adding little to the understanding of the field. A second, more time-consuming but more objective approach is to recorrelate the logs, ignoring the existing correlations. A third, less time-consuming approach is to subject the existing correlations to MBPA in order to rapidly identify correlation problem areas.

Consider the purchase of an older, producing field. The task at hand, besides producing the proven reserves, is to find any additional potential. How do we, as geoscientists and engineers, find additional potential in previously worked areas? New potential is found by disagreeing with, and changing, the previous correlations—in other words, finding error in the older work. We must remember that oil and gas reserves are not found by making maps. It is correlation work that identifies the potential, and then the mapping based on the correlations that graphically displays that potential.

If we take previous geoscience and engineering work, in the form of correlations of electric logs or correlations on seismic sections, and submit these correlations to a test of validity (MBPAs), we can rapidly identify problem correlations, previous correlation errors, faults and even unconformities. We can define the growth history of the area, providing us guidance as to the stratigraphic intervals that might hold the best potential for hydrocarbons. And *most important*, we can do this while *reducing the cycle-time of the correlation and mapping process* by as much as 50 percent (Bischke and Tearpock 1999).

This methodology should be applied to all correlation data sets during the initial stages of any project. Numerous field studies show that the MBPA method, when applied at the beginning of a study, results in a higher-quality product and can aid in the rapid identification of new hydrocarbon potential.

Common Extensional Growth Patterns

In extensional regimes, growth plots detect *missing section* due to normal faults or unconformities. *The missing section occurs as* ***discontinuities*** *on the plots*. In extensional regimes, three types of growth patterns commonly occur and can be used to interpret structural or stratigraphic history (Fig. 13-5).

Normal faulting typically results in missing section in correlated well logs. On the Δd/d plots, large faults produce large discontinuities, or offsets, in the plotted data, whereas small faults produce small offsets in the plots. If faulting is present, two types of patterns exist: a downward, or *negative displacement pattern*, in which Δd becomes smaller with increasing depth (Fig. 13-5a), and an upward, or *positive displacement*, in which Δd becomes larger with increasing depth (Fig. 13-5b). If a *negative discontinuity* is present, then *the identified fault cuts the structurally lower well* within the intervals being correlated (Fig. 13-5a). A *positive displacement* pattern means that *the fault cuts the structurally higher well* (Fig. 13-5b). Each of the faults in Figure 13-5a and b is interpreted as a growth fault because the growth rate is higher within the stratigraphic interval that is downthrown to the fault in the faulted well.

An unconformity is a surface of erosion and/or nondeposition caused by changing environmental conditions that affect growth. Erosion or downlap will create thickness changes within stratigraphic units. Unconformities, as do normal faults and sequence boundaries commonly, eliminate stratigraphic section and produce missing section in correlated logs, and so they

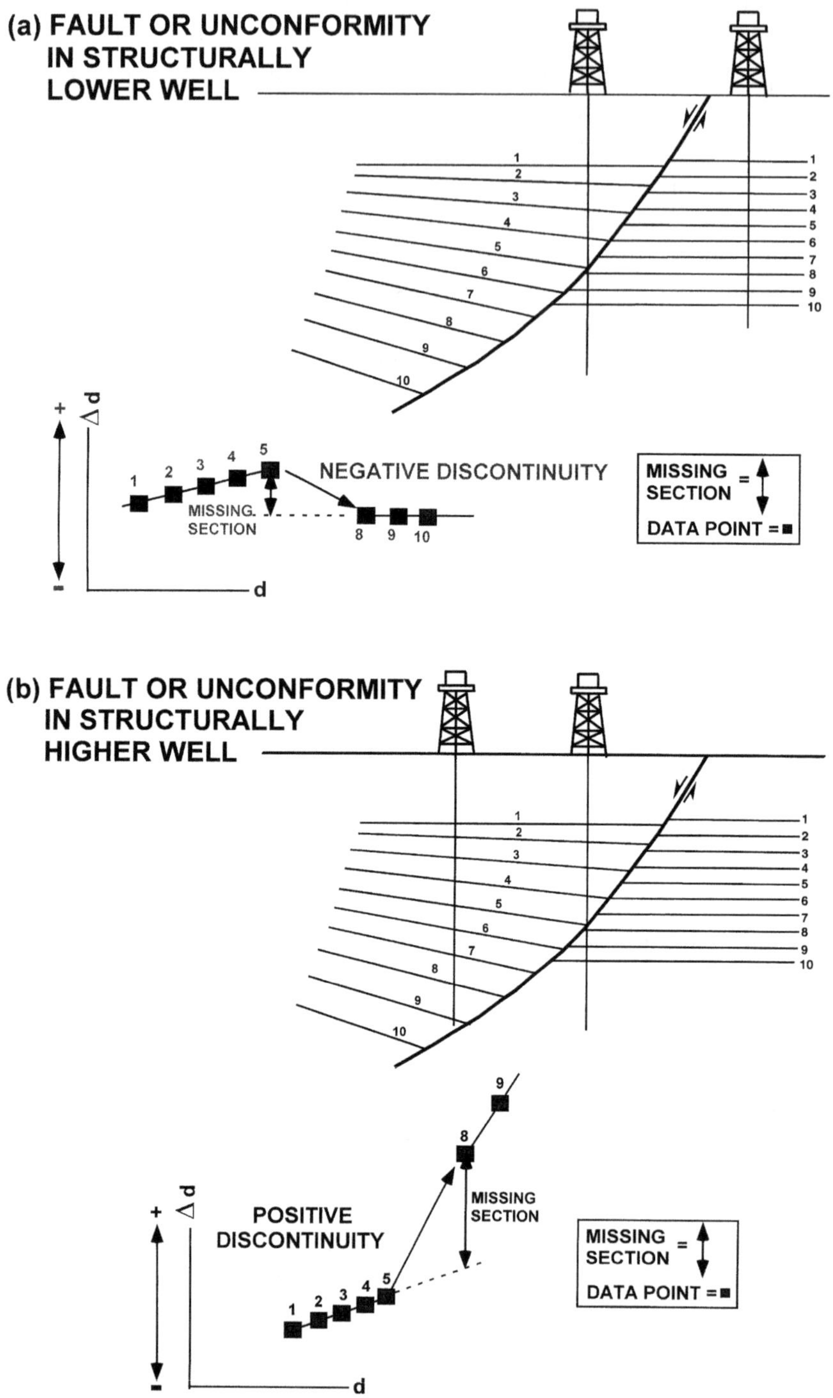

Figure 13-5 In extensional regimes, three patterns of discontinuities are observed on Δd/d plots. The discontinuities on the plots are caused by missing section that may be due to a fault or an unconformity. (a) If the missing section is in the structurally lower well, then the plot of Δd/d contains a negative, or downward, discontinuity. We use an example of a fault. (Published by permission of R. Bischke and D. Tearpock.) (b) If the missing section is in the structurally higher well, then the plot contains a positive, or upward discontinuity. We use an example of a fault. These patterns can help distinguish faults from unconformities in areas of high bed dip. (Published by permission of R. Bischke and D. Tearpock.) (c) A hiatus causes a change in slope on the plot. (From Bischke 1994b. AAPG©1994, published by permission of the AAPG whose permission is required for further use.)

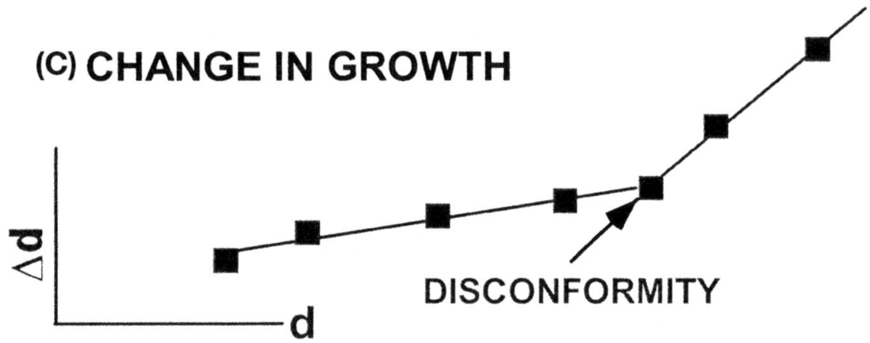

Figure 13-5 *(continued)*

generate discontinuity patterns on the growth plots. Thus, large unconformities resemble faults. As a rule, however, unconformities *tend to remain at or near the same stratigraphic level over large areas and in many wells*. Geologists recognize that normal faults usually dip at high angles of 40 deg to 60 deg (Ocamb 1961) and that a fault intersects different wells at different stratigraphic levels. An unconformity, however, if it covers a large area, appears at about the same stratigraphic level in the wells in the area and typically dips at a lower angle. For example, a so-called fault, interpreted to be the cause of the same stratigraphic section missing in many wells in a study area, actually would be an unconformity rather than a fault. The existence of missing section at the same stratigraphic level is obvious on a set of Δd/d plots. Accordingly, geoscientists employ the method to rapidly locate unconformities and to rapidly distinguish faults from unconformities in well log data. If many wells exist in an area of study, then the plots aid in keeping track of the missing section and its approximate value and stratigraphic level in each well (Fig. 13-17b). The plots represent a record of the missing section in the form of a visual display that is readily apparent. Refer to Figure 13-17b and review Fault J and the recognized unconformity.

Often a more difficult problem develops for geoscientists where structure was growing very slowly relative to changing sedimentation rates or where sedimentation rates were high relative to deformation rates. In these cases, it is difficult to detect disconformities in well log data, and downlap may not be obvious or present on seismic sections. If there were a hiatus in deposition (a condensed section), then when sedimentation resumes, the rate of deposition would likely change. Tobias (1990) stresses this point. Subtle unconformities, caused by gradual processes, occur as *changes in slope* of the Δd/d curves (Fig. 13-5c). Alternatively, if there is the slightest change in the sedimentation rate, regardless of its cause (e.g., sea level fluctuations), then changing environmental conditions are likely to produce changes in the slope of a Δd/d curve displayed on the plot. These plots are very sensitive to changes in the tectonically or eustatically controlled sedimentation rate, so disconformities and/or subtle sequence boundaries commonly occur as *breaks in slope* of the curves generated on the Δd/d plots (Fig. 13-5c).

Unconformity Patterns

Two unconformity growth patterns exist, one for onlap and the other for downlap. These patterns may help define systems tract boundaries.

If beds **onlap** an unconformity, then at the time of deposition the beds above the unconformity dip at a lower angle than the beds below the unconformity. Figure 13-6a shows the typical occurrence of onlap above an unconformity, with the missing section being greater in the well located at the structurally higher position at the time of deposition. Erosion typically removes more section in the up-dip direction than in the down-dip direction. The missing section caused by the unconformity occurs as a *positive displacement* on the Δd/d plot (Fig. 13-6a), even if strata were later structurally rotated. Thus, *if a positive discontinuity is present on a Δd/d plot* (Fig. 13-5b), the missing section results either from *a fault in the structurally higher well or from erosion, with subsequent onlap of sediments above the unconformity.*

If beds **downlap** an unconformity, then at the time of deposition, the beds above the unconformity dipped at a higher angle than the beds below the unconformity. Downlap causes a negative discontinuity on Δd/d plots (Fig. 13-6b), even if strata were structurally rotated. Again, more section is missing in the area that was up-structure at the time of deposition. Thus, *if a negative discontinuity exists on a plot*, then the missing section responsible for the discontinuity results *either from a fault in the structurally lower well or from erosion, with subsequent downlap of sediments above the unconformity.*

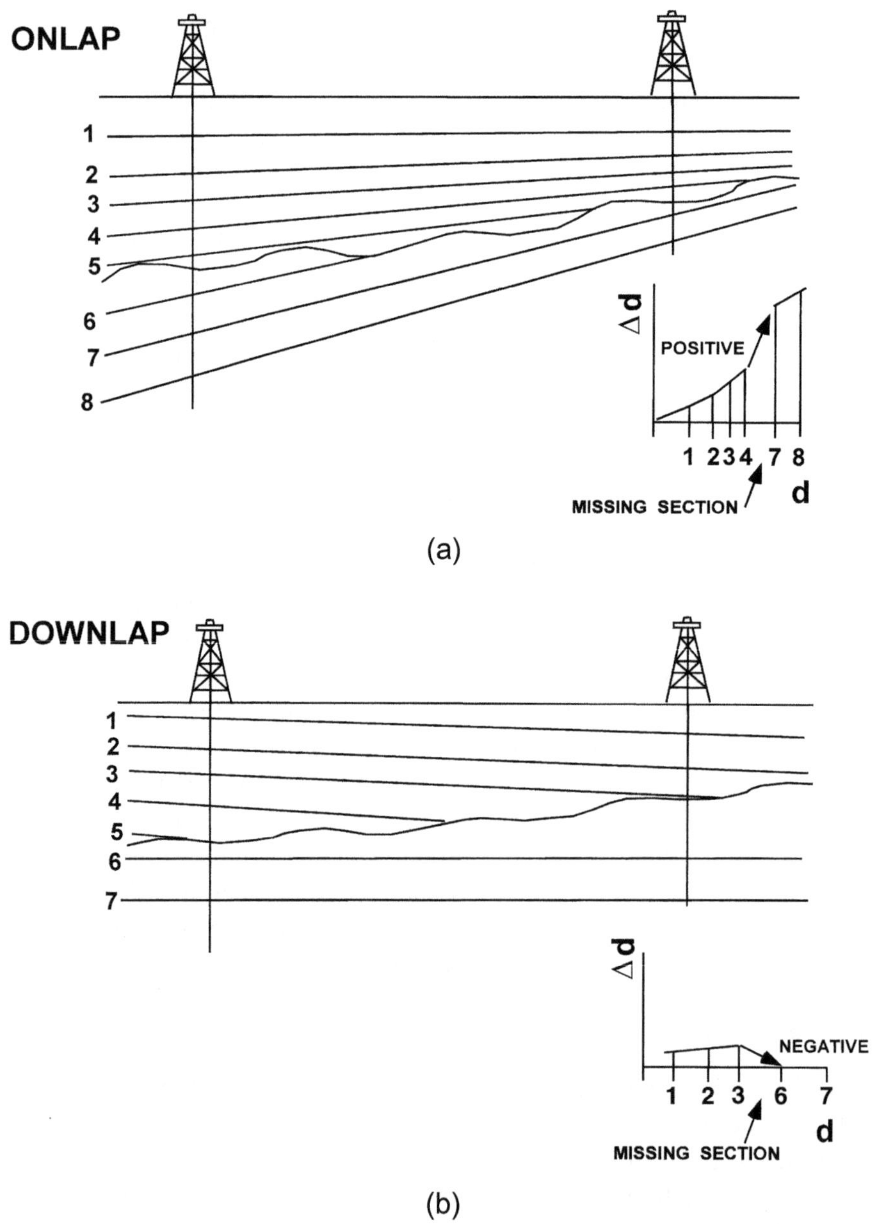

Figure 13-6 (a) Onlap produces a positive discontinuity on the Δd/d plot, whereas (b) downlap creates a negative discontinuity. (From Bischke 1994b; AAPG©1994. Published by permission of the AAPG whose permission is required for further use.)

These patterns enable us to distinguish faults from unconformities on the steeply dipping flanks of salt structures, where bed dips are commonly too steep to image on seismic data. For example, on the flanks of salt diapirs the sedimentary units typically onlap, rather than downlap, an unconformity. Therefore, unconformities on the flanks of salt domes produce positive, rather than negative, discontinuity patterns (Fig. 13-6a). This concept can be used in comparing an up-dip well to one or more structurally lower wells. If missing section causes a negative displacement on the plots, as shown in Figure 13-5a, then the missing section in the structurally lower well *must be due to a fault rather than to an unconformity*.

The larger the missing section, the larger the discontinuity in the Δd/d plots. However, consider that an unconformity removes section in both of the wells used in a Δd/d plot (Fig. 13-6). Then the amount of missing section, estimated from that plot, is the difference between the amounts of the two missing sections, rather than the actual amount of section missing in one of the wells.

ACCURACY OF METHOD

Well log data has a higher resolution than 3D seismic data. At a depth of −15,000 ft, geologists can commonly determine the depths between two correlation markers to within a few feet (e.g., 5 ft). A Δd/d plot is a plot of change-in-depth versus total depth. Therefore, if we define error as the precision in correlation between two tops, divided by the subsea depths to the correlations, then the error in well log data at −15,000 ft is 5/15,000, or 1 part in 3,000 (0.03%). This is a *very high degree of accuracy*. Thus, Δd/d plots constructed from well log data typically have errors on the order of less than 0.1%.

Seismic data at a depth of −15,000 ft may have a velocity of about 10,000 ft/sec one-way time, and a frequency of 30 Hz. These data would have a wavelength of about 500 ft and a resolution of about 100 ft (Sheriff 1980). At a depth of −15,000 ft, seismic data has an error of 100/15,000, or about 1 part in 150, or about 0.7%. Well log data are at least one order of magnitude more precise than seismic data. The intent of this discussion is not to degrade seismic data that we use on a daily basis but rather to point out the accuracy of the methods when using well log versus seismic data.

The plots will contain *errors in the slope of the growth curves* if a well encounters *changes in bed dips or changes in wellbore deviation angle* (Chatellier and Porras 2001). This error affects only the *slope* of the growth curves, and *the error* increases as *the tangent of the change in bed dip angle or wellbore deviation angle* (Bischke 1994b). The plots of (1) vertical wells against the ramp portion of deviated wells or (2) two subparallel deviated wells, compared to each other, contain fewer errors than plots based on highly deviated wells and horizontal wells, and these latter types of plots should be avoided. Thus, during correlation work, if vertical wells, the ramp portion of deviated wells, or two subparallel deviated wells are available for study, then it is generally not necessary to correct the data. If, however, the intent of the study is to distinguish high-growth intervals from low-growth intervals, and the dipmeter data indicate rapidly changing bed dips, then the data can be corrected for changing bed dips by using the correction factors presented by Bischke (1994b). The errors present in uncorrected data accumulate as the tangent of the change in bed dip angle; for example, if the bed dip changes by 10 degrees, then the error introduced in the *slope* of the curves is 18%.

EXAMPLES OF THE Δd/d METHOD

Generic Example of a Delta

In order to understand the application of Δd/d plots and Multiple Bischke Plots, let's review a generic example prior to discussing practical examples of the two methods. We use a generic delta that contains six parasequences to demonstrate how the technique works (Fig. 13-7).

The stratigraphic intervals between horizons 3 and 6 are of constant thickness, so these units represent a nongrowth section. These units downwarp with the deposition of prograding units, between horizons 0 and 3, upon the older sequences. For this example, we study three wells advantageously positioned on the delta (Fig. 13-7).

From the correlation data obtained from the wells, we construct two Δd/d plots in the form of bar graphs. We generate the data by correlating the logs and noting depth differences for each correlation marker, as illustrated in Figure 13-7, by drawing a horizontal dashed line between the wells for each correlation marker. Horizontal lines are drawn between Wells No. 1 and 2 to illustrate the depth of correlation markers 1 through 6 in Well No. 1. Measure the vertical distance (difference in depth) for each of the correlative markers, 1 through 6, in each pair of wells. Plot these Δd values on the y-axis against the subsea depth of the correlations in the structurally higher well, plotted on the x-axis (Fig. 13-7). In practice, calculate the Δd measurements on a spreadsheet by entering the subsea depths as positive values and subtracting the depths in the structurally higher well from the depths in the structurally lower well. The plot for Wells No. 1 and 2 are shown in Figure 13-7. These data can be *from anywhere in the world* and *from any tectonostratigraphic environment*, yet the following interpretation will be correct.

The interval between correlations 1 to 3 is an expanded, *growth* interval, and the interval between correlations 3 through 6 is a *nongrowth* interval. We can see that the interval 1 to 3 has

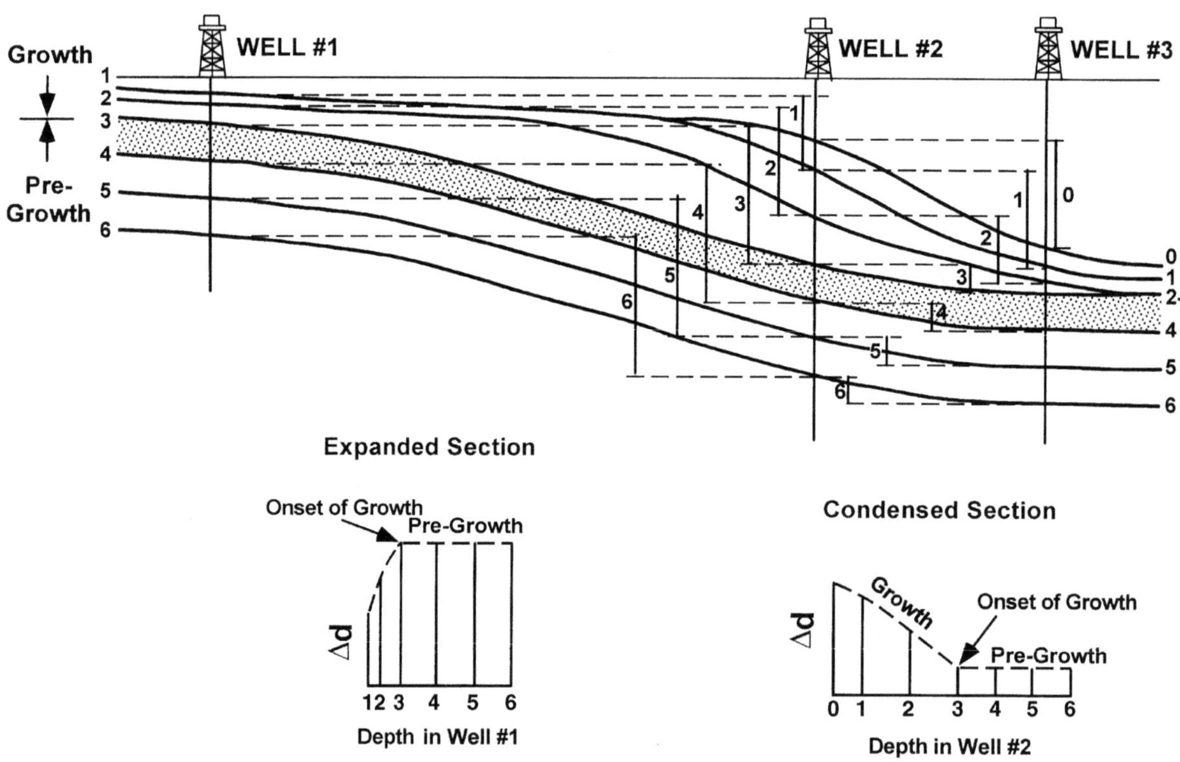

Figure 13-7 Stratigraphic interpretation of a generic delta based on well control. Parasequence correlations define a nongrowth (pregrowth) section in the wells beneath a disconformity located at correlative horizon 3. Growth section above the disconformity contains an expanded section between Wells No. 1 and 2, and a condensed section between Wells No. 2 and 3. Expanded and condensed sections are characteristic of wells located across a delta, and thus the method may define sedimentary environments in areas that lack dense well control. A seismic line located between Wells No. 1 and 3 would not detect the hiatus. (From Bischke 1994b. AAPG©1994, published by permission of the AAPG whose permission is required for further use.)

a *positive slope* on the plot, demonstrating expansion from Well No. 1 toward Well No. 2. The slope is flat between correlations 3 through 6, indicating no growth. As no onlap is present, the break in slope at correlation 3 is a possible disconformity that would not image on seismic profiles between Wells No. 1 and 2.

To continue the analysis, we conduct the same procedure between Wells No. 2 and 3. In this case, a sequence boundary, labeled 0, is present in the two wells (Fig. 13-7). Again, a Δd/d plot shows the growth history relative to the two wells. Well No. 2 is the structurally higher well, so it is plotted on the x-axis (Fig. 13-7).

A unique interpretation of the plot of Wells No. 2 and 3 is made. The interval between correlations 1 and 3 is a *condensed section in Well No. 3*, and the interval between correlations 3 through 6 is a nongrowth interval. The interval 1 to 3 has a *negative slope*, which means that the section thins from Well No. 2 toward Well No. 3, indicating a condensed section in Well No. 3. The slope is flat between markers 3 through 6, confirming the nongrowth interval. The break in slope at correlation 3 is probably a time-transgressive disconformity that exists in all three wells. We can now conclude that the growth section expands between Wells No. 1 and 2 and contracts between Wells No. 2 and 3, which is characteristic of only a few sedimentary environments, including a delta. When integrated with other geological information, the technique can identify or constrain the number of possible structural styles or sedimentary environments in the area of study, even where data are limited.

Applying the Δd/d Method to Seismic Data

We can construct Δd/d plots from either well logs or depth-converted seismic sections. As discussed previously, the seismic sections used *must* be depth-converted, since time distorts vertical distance on seismic sections. The entire section need not be depth-converted. As long as we have reasonable time–depth data, the correlation markers picked in time from the section can be converted to depth before input of the data into a Δd/d plot.

For this example, we use a seismic section (Fig. 13-8a) from the offshore Gulf of Mexico, across the area called the Brazos Ridge. The seismic profile illustrates a large, listric growth fault, which nearly flattens with depth. A faulted rollover anticline exists in the hanging wall block. The hanging wall anticline has one master synthetic fault, several downward-dying synthetic faults, and a number of downward-dying antithetic faults.

In this example, we use the correlated seismic data between the two major synthetic faults to look for unconformities or sequence boundaries and to help us understand the growth history of the area. The Δd/d plots, which record changes in the sedimentation rate, are very sensitive to changing structural and sedimentary conditions.

The two dashed lines between sp C and sp D define the area used to obtain the data for analysis. Since this area is free of any visible faults, the resulting plot will not be complicated with fault patterns. The two dashed lines can be considered for simplicity as two fictitious directionally drilled wells. These dashed lines are essentially parallel to each other and to the bounding faults. We can obtain correlation data on each event that has been correlated on the seismic section. The more events (markers) correlated, the better the results.

By observation of this single seismic section, can you see the sequence boundaries? Can you see an angular unconformity that might set up traps? Can you visualize the growth history of this area? Most likely, the answer to each question is no. However, once we subject the correlation data to Δd/d analysis, the questions should be easily answered.

Figure 13-8b is the Δd/d plot generated from the correlation data obtained from interpretations of the seismic section. More than 26 markers were correlated in the master synthetic fault block (Fig. 13-8a) and carried through the two dashed lines. Since the beds dip into the fault near sp C, the correlation data obtained from the dashed line near sp D is considered as the data that would be equivalent to that obtained from a higher structural well (it is in the higher structural

(a)

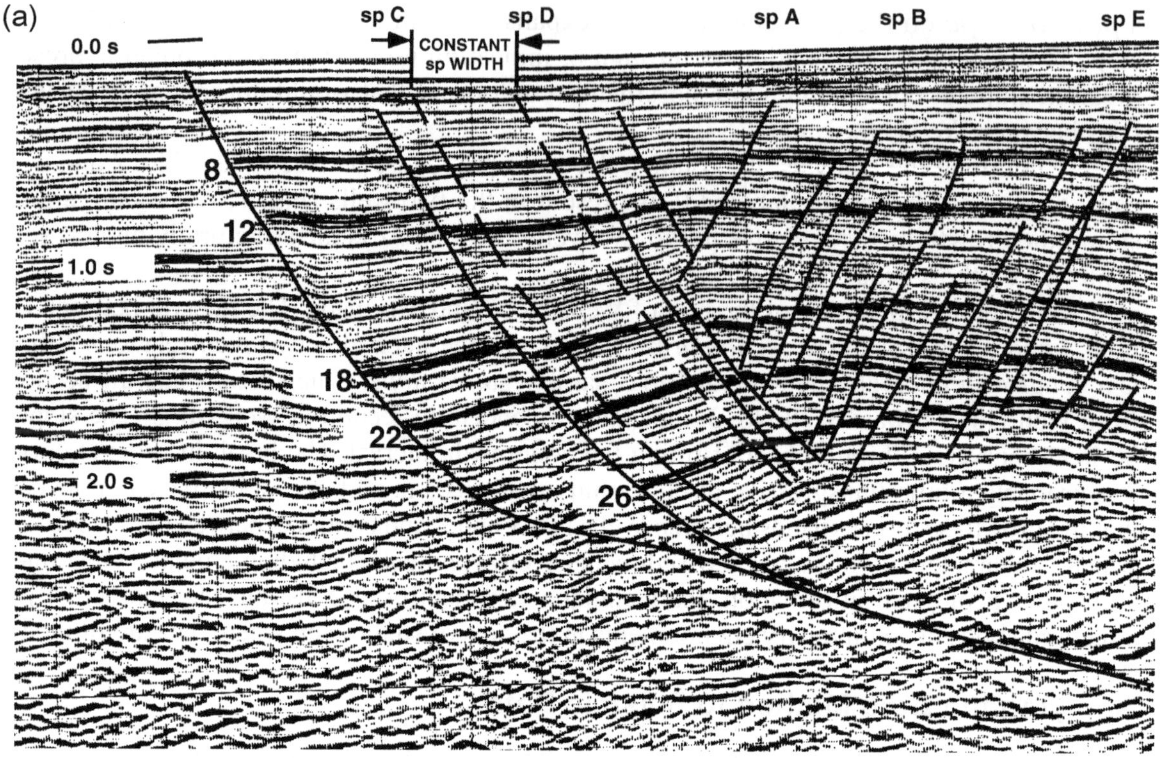

(b)

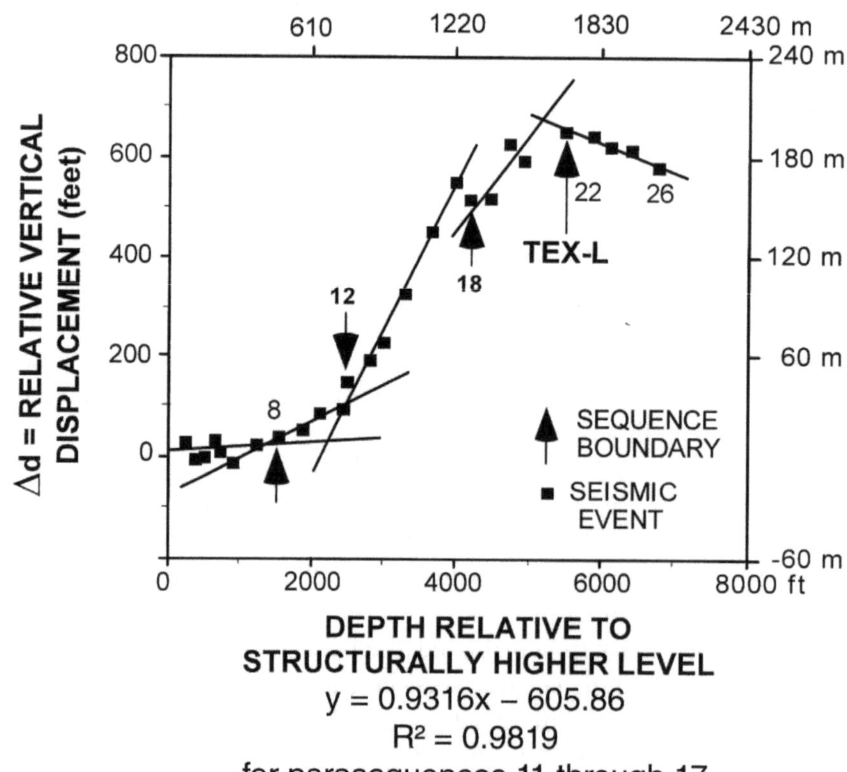

Figure 13-8 (a) Seismic section in the Brazos Ridge area, northern Gulf of Mexico. Inclined dashed lines are locations for correlation data in a Δd/d plot. (b) A Δd/d plot of data, derived from (a), reflects changes in growth history of the structure. The breaks in slope, at data points 8, 12, 18, and 22, mark major sequence boundaries. (Bischke 1994; AAPG©1994 reprinted by permission of the AAPG whose permission is required for further use.)

position). The correlation data, which are in time, are first depth-converted using local time–depth information, and then plotted as shown in Figure 13-8b.

We see various changes in the growth history of the area and at least five distinctive changes in the slope of the plotted data. The plot illustrates at least four major breaks in the growth trends, from a condensed section between Δd/d points 22 to 26 to a nongrowth section above −1000 ft. The maximum growth occurred in this area between correlation points 12 and 18. And finally, the breaks at correlation points 8, 12, 18, and 22 appear to mark major sequence boundaries.

This one Δd/d plot, from one seismic section from a 3D seismic survey, provides significant information about the stratigraphic and growth history of this area of the Brazos Ridge. Plots from other seismic sections also exist. The combined use of multiple Δd/d plots, using seismic sections like the one shown in Figure 13-8a, provide significant information on the tectonostratigraphic history of an area and begin to define its hydrocarbons potential.

Resolving a Log Correlation Problem

We now turn to an example of the Δd/d method and consider how it aids in the resolution of difficult correlations. Such difficulties are common between on-structure and off-structure wells in complex tectonic settings, such as diapiric areas. The first example is from the northern Gulf of Mexico, offshore USA. Well No. 1 is on top of a salt dome, and it has a section that is difficult to correlate between 7000 ft and 9000 ft. 3D seismic data are incoherent in this area.

We begin by successfully correlating an off-structure type well (Well No. 5) with several surrounding wells, as well as with up-dip wells. Based on these correlations, the off-structure type log of Well No. 5 has a complete stratigraphic section. We then correlate Well No. 5 with the problem Well No. 1 on top of the salt diapir. A total of 31 parasequence boundaries are correlated between the two wells, resulting in the Δd/d plot shown in Figure 13-9a. Notice that the x-axis has an expanded scale relative to the y-axis, which is typical for Δd/d plots.

Although Wells No. 1 and 5 are not close to each other, the growth plot shows near-linear growth through the entire 9000-ft section. Notice on Figure 13-9a that parasequence boundary correlation No. 27, located in the difficult-to-correlate portion of Well No. 1, lies about 230 ft off the general linear trend. This data point is based on the initial correlation shown in Figure 13-9b.

Two possible interpretations of the data are (1) a miscorrelation exists in Well No. 1 as correlated with Well No. 5, and correlation marker No. 27 *is about 230 ft too high*; or (2) the anomalous correlation marker No. 27 is, by coincidence, *faulted up* in Well No. 1 by 230 ft relative to Well No. 5, and then *faulted down* in Well No. 5 by the same amount. By comparison to Figure 13-5a, *the possible 230-ft fault, responsible for down-faulting correlation marker No. 27, must exist in the off-structure Well No. 5 (at about the 8600-ft level)*. However, correlation with wells adjacent to Well No. 5, together with the 3D seismic data, indicate that *Well No. 5 does not penetrate a 230-ft fault near the 8600-ft level*. Therefore, interpretation (2) cannot be correct. So a miscorrelation exists in Well No. 1.

Notice on Figure 13-9b that another possible correlation to the 8600-ft sand in Type Well No. 5 exists below the 8000-ft level in Well No. 1. This sand is about *230 ft deeper* than the initial correlation at 7800 ft in Well No. 1. The Δd/d method, when integrated with the existing geological information, strongly suggests that the lower sand correlation, below 8000 ft, is the correct correlation. This change in correlation affects the interpretation, including the fault pattern, reservoir delineations, volumetrics, and additional potential production.

When used in conjunction with other geological information, the Δd/d method may "uniquely" resolve correlation problems. The method also determines whether correlations of horizons are either too high or too low, and it specifies the approximate vertical distance between the correct and incorrect correlation picks. In Figure 13-9, this vertical distance is 230 ft.

The method is most useful in areas where reliable correlation data exists but nevertheless the data deteriorate within a problem interval. This commonly occurs around salt diapirs or on other structural features that exhibit steep bed dips, such as compressional structures. If a near-linear growth trend is identifiable on the plot, then the method directs the interpreter to an alternative correlation. The method uses a *nonarbitrary process*.

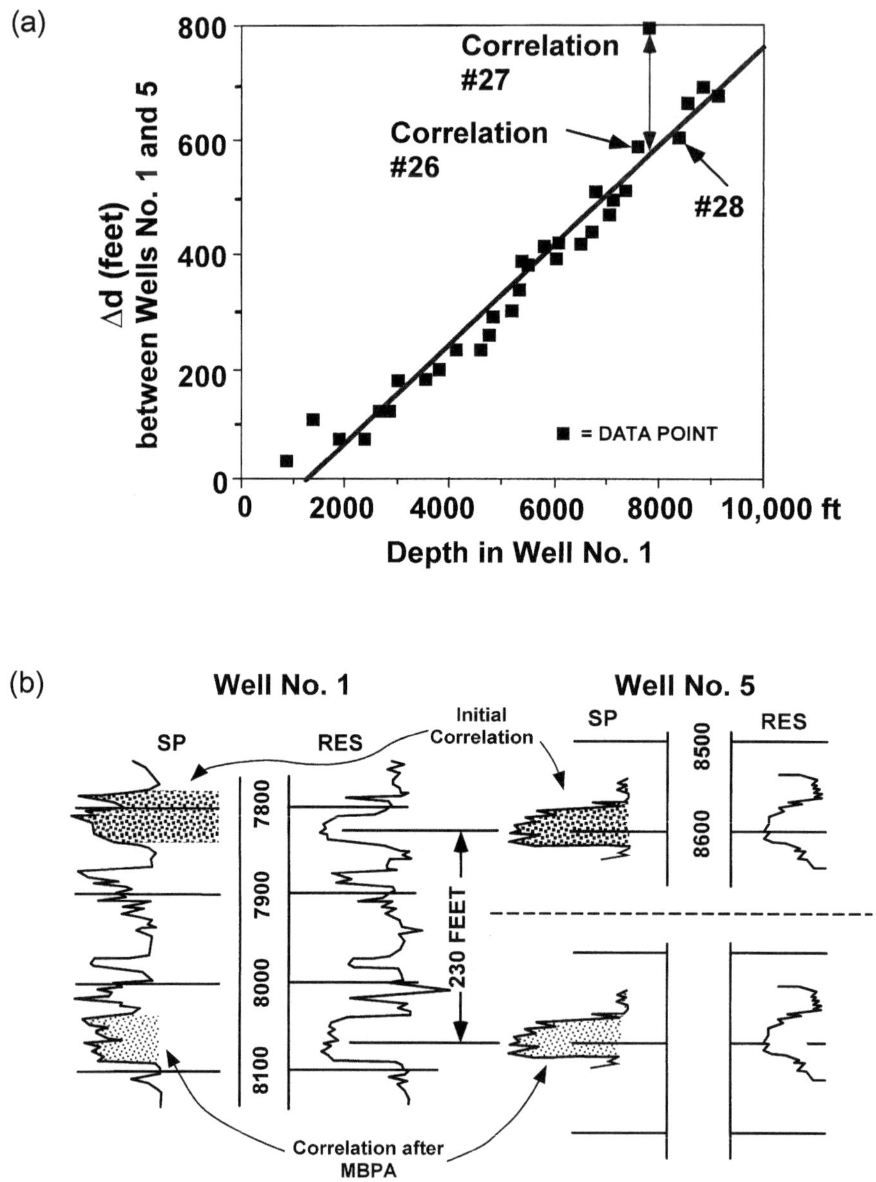

Figure 13-9 (a) Growth plot of the type Well No. 5 and the difficult-to-correlate Well No. 1 in the Eugene Island 208/215 field. Correlation marker No. 27 lies 230 ft off the general growth trend, suggesting that this anomaly is due to miscorrelation or to faults in the two wells. See text for explanation that marker No. 27 in Well No. 1 is miscorrelated to the type log by about 230 ft. (b) The 8600-ft sand in Well No. 5 was initially incorrectly correlated to the 7800-ft sand in Well No. 1. The Δd/d method, when integrated with other geological information, demonstrates that the 8600-ft sand in Well No. 5 correlates to the sand below 8000 ft in Well No. 1. (From Bischke et al. 1999. Published by permission of the Gulf Coast Association of Geological Societies.)

An Example of Stratigraphic Interpretation

The Δd/d method is capable of locating disconformities and hiatuses from well log data and on seismic profiles, along with associated sequence boundaries (Fig. 13-8). In this section, we show how the method can distinguish between lithostratigraphic units and genetically related stratigraphic units, which may aid in locating sequence boundaries.

The example is from a carbonate section in the Guadalupe Mountains, New Mexico, USA (Fig. 13-10). The Grayburg Formation formed in a shallow marine-to-supertidal environment within a sigmoidally shaped carbonate ramp that contains fifth-order parasequences that shallow upward (Lindsay 1991). The Grayburg Formation is subdivided into five stratigraphic units called zones. The overlying Queen Formation contains shallow marine sandstones, carbonates, and evaporites (Vanderhill 1991). Kerans and Nance (1991) interpret the Grayburg Formation to be in the high stand systems tract (HST) below a probable Type 1 sequence boundary. They conclude that the base of the Queen Formation reflects a rise in sea level.

Using core and well log correlations from Lindsay (1991), we generate a growth plot from Wells No. EMSU 458 and 457 (Fig. 13-10). The data points include the tops of Lindsay's Zones 1 to 5 within the Grayburg Formation and tops of several parasequences in the Queen Formation. Zone 1, at the top of the Grayburg Formation, is within a low-growth section (Fig. 13-10). This linear low-growth pattern continues into the Queen Formation, and thus Zone 1, in the Grayburg Formation, appears to be genetically related to the Queen Formation. Contrastingly, a high-growth section is seen in Zones 2 through 5. Growth rate was highest during deposition of Zone 5, then gradually declined through Zone 2. Zones 2 to 5 in the Grayburg Formation *define the carbonate ramp*.

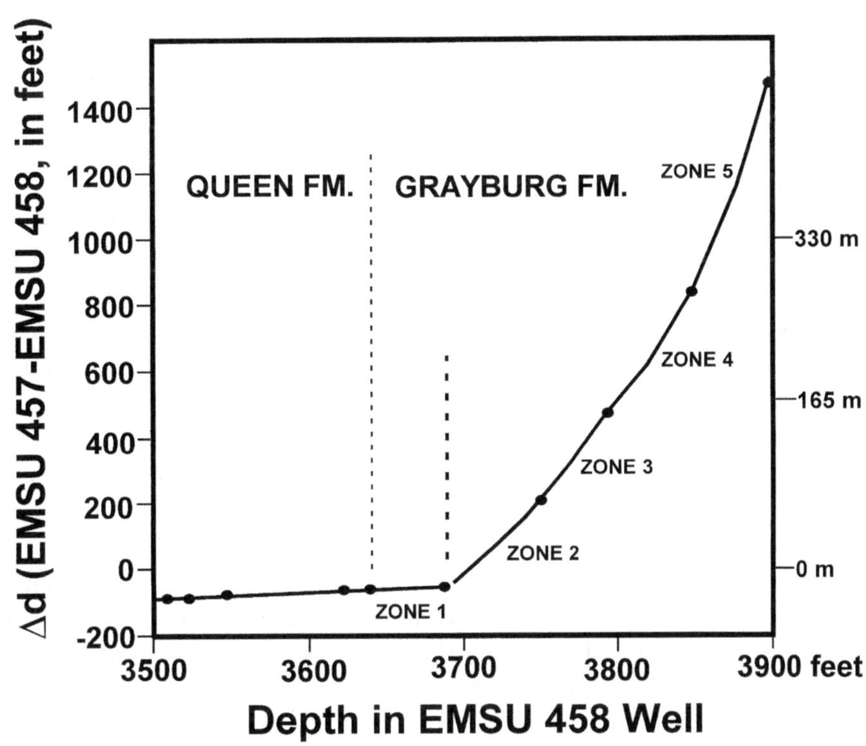

Figure 13-10 Growth plot illustrating carbonate ramp in Grayburg formation, Guadalupe Mountains, New Mexico, USA. The change in slope at the top of Zone 2 partitions the Grayburg Formation into high-growth and low-growth sequences, and indicates a probable tectonostratigraphic boundary. A genetic relationship exists between Zone 1 in the Grayburg Formation and the overlying Queen Formation. (Published by permission of R. Bischke.)

A disconformity appears to exist at the top of Zone 2 within the Grayburg Formation (compare Figs. 13-5c and 13-10). We interpret this hiatus to result from a change in sea level. The growth plot, constructed from correlations based on well log and core data, is compatible with a genetically related boundary at, or very near, the top of Zone 2 in the Grayburg Formation. Zone 1 in the Grayburg is genetically related to the Queen Formation. A Δd/d plot of Wells EMSU 458 and 247, not shown in this book, yields similar results with the carbonate ramp represented beneath the top of Zone 2. If sedimentary sequences exist as genetically related units, regardless of the cause, then these units are likely to be related to sediment supply. Changes in sediment supply should occur at or near system tract boundaries, where changes in sea level affect growth. We conclude that changes in growth may occur at, or near, major sequence boundaries and may aid in locating the position of sequence boundaries (Tobias 1990; Sanchez et al. 1997). The method can help distinguish between lithostratigraphic and tectonostratigraphic boundaries.

Locating Sequence Boundaries in a Compressional Growth Structure

The Newport-Inglewood Trend, in the western Los Angeles Basin, California, USA (Harding 1973; Wright 1991), is a classic zone of compressive and strike-slip deformation that involves growth. Producing fields exist in folds within the trend, and the trend was mapped by the California Division of Oil and Gas beginning in the late 1920s. The Division geologists typically mapped the unconformities and were able to carry the upper and lower Pliocene and Miocene sequence boundaries from growth fold to growth fold.

Growth plots that we made support the ability to correlate sequence boundaries between fields. As an example, Figure 13-11 shows a Δd/d plot for the Huntington Beach Field Anticline, located in the southern part of the Newport-Inglewood Trend. The plot uses data from an off-structure well and a well located near the crest of structure, as presented by Harding (1973). The plot consists of near-linear increments of growth punctuated by discontinuities. Correlation to adjoining wells shows that the wells are not faulted, so these discontinuities represent missing section due to disconformities. These disconformities represent several tens of feet of missing section and are regional time-stratigraphic boundaries (Wright 1991; Blake 1991). The hiatus at point 1 is near the top of the Pliocene, and the Top Reppeto, at point 2, is a late Pliocene sequence boundary in the Huntington Beach Field Anticline. Points 4 and 5 are early Pliocene and Top Miocene sequence boundaries. The California Division of Oil and Gas mapped every disconformity except the anomaly at point 3, and were thus able to readily correlate between adjoining fields. The California Division of Oil and Gas recognized the value of sequence stratigraphy in the late 1920s.

The slopes on the plot indicate intervals of growth of the Huntington Beach Anticline, which may have implications concerning hydrocarbon generation and migration. This anticline grew during the Pliocene (Fig. 13-11). If hydrocarbon generation and migration occurred before the Pliocene, then there was no structure to trap the hydrocarbons. The low-growth interval, shallower than point 1 on the plot, represents the onlap of the Pleistocene Pico Formation across the crest of the anticline. These Pleistocene sediments are cut by the Inglewood Fault System and were not subject to folding (Harding 1973). The conclusion is that the migration of hydrocarbons into the Huntington Beach Anticline was post-Miocene.

Lastly, the plot helps to interpret the structural style. The Newport-Inglewood Trend is subject to several interpretations. Harding (1973) proposes that strike-slip displacements caused, and were coincident with, the compressional folding. Wright (1991) employs structural maps and well log data from fields along the trend to conclude that the folding and faulting exhibit a complex structural and stratigraphic history, not readily reconciled with a simple strike-slip origin. From a seismic study of the Los Angeles Basin, Shaw and Suppe (1996) conclude that the

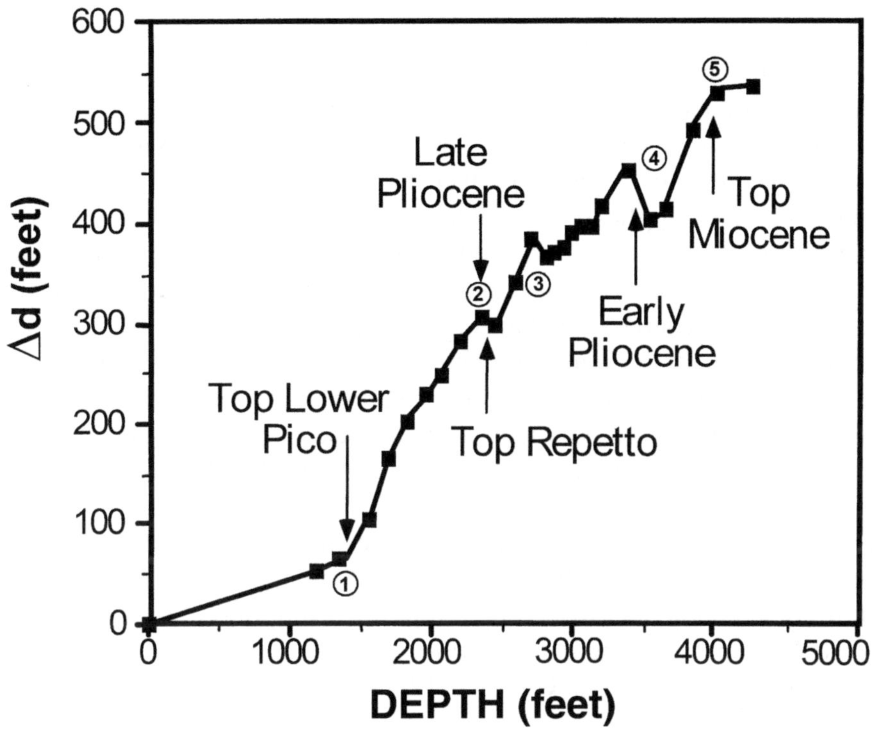

Figure 13-11 Growth plot for Huntington Beach Field Anticline, Newport-Inglewood Trend, California, USA. The compressional fold grew during the Pliocene. The highest growth interval is between the Top Repetto and Top Lower Pico section. Disconformities exist at discontinuities on the plot, indicated by the circled numbers. The California Division of Oil and Gas mapped these disconformities between fields in the Newport-Inglewood Trend.

Los Angeles Basin is subject to large-scale, blind-thrust faulting, and that the folding is not coeval with recent strike-slip displacements on the Newport-Inglewood fault system. Thus different interpretations are possible for the structural formation of the local folds. Which is correct? Can the growth plot suggest the correct solution? The Δd/d plot seems to support the interpretation of Shaw and Suppe (1996). In the same Newport-Inglewood trend, a study of several growth plots and a restoration of the Signal Hill restraining bend, at the Long Beach Anticline, provide evidence that the strike-slip faulting postdates the compressional folding. The Signal Hill analysis is presented in Chapter 12.

A Δd/d plot of the correlations presented in Harding (1973), when integrated with the local geological data (Wright 1991), suggests that the Huntington Beach field is a *growth anticline* (Fig. 13-11). A **growth compressional style** may be new to many geoscientists and engineers alike. A lack of knowledge of growth compressional structures can cause misinterpretations that may affect regional interpretations, the interpretation of the local petroleum system, and ultimately the prospects generated. *Any interpretation based on existing correlations must be consistent with the correlations, regardless of the structural style under study.*

Analysis of the Timing of a Strike-Slip Growth Structure

The Δd/d method can be applied to strike-slip faults as well, whereby we can determine when a strike-slip fault was active. This information may provide support as to when the fault system provided a hydrocarbon conduit into associated structures. Many strike-slip faults cut through the sedimentary cover into basement (Harding 1990), and thus some strike-slip faults may tap deep source rocks. Motion on active strike-slip faults may create deep conduits for hydrocarbons, particularly at extensional, releasing bends (Chapter 12). Therefore, active strike-slip faults may have a greater potential for hydrocarbon accumulation in associated traps, than inactive faults. In addition, the trapping structures along strike-slip faults must exist prior to the migration of hydrocarbons along the nearly vertical fault surfaces.

The Δd/d plots have application for determining the growth history of strike-slip displacements. Contrary to intuition, the Δd/d method applies also to faults that exhibit predominantly horizontal displacements, for several reasons. First, prior to the movement along the fault, the area affected by the later strike-slip faulting may contain paleogeographic or paleobathymetric relief. Once the motion on the fault commences, the nonuniform topographic relief may lead to progressive juxtaposition of topographic highs against lows. Growth sediments deposited across the paleosurfaces are thinner across the highs and thicker across the lows.

Most strike-slip faults contain components of dip-slip displacements, and growth sedimentation may occur across the faults during fault activity. The Δd/d plots readily record the dip-slip component of growth in the same manner as recorded across growth normal faults. The assumption is that the dip-slip component of displacement correlates to the strike-slip component of displacement. However, a dip-slip component of motion need not accompany every strike-slip component of motion. The Δd/d method cannot distinguish between horizontal and dip-slip motions. Chapter 12 presents methods for recognizing horizontal displacements.

As an example of strike-slip fault analysis, we use the Zayante Fault, a large fault located about 5 km south of the surface trace of the San Andreas Fault in southern California (Clark and Reitman 1973). Deformed Pliocene and Quaternary sediments along the fault suggest recent activity on the Zayante Fault. We apply the Δd/d plot to wells separated by a splay of the Zayante Fault in order to determine when the fault was active (Shaw et al. 1994).

We select two wells for studying displacements on the Zayante Fault, and the two well locations are shown on the map in Chapter 12, Figure 12-17. The structurally higher Pierce Well is about 2 km to the south of the fault zone. The Light Well is the structurally lower well, and it is located northeast of the Pierce Well and on the opposite side of a splay off the main Zayante Fault (Shaw et al. 1994). The Light Well contains a thicker stratigraphic section than the Pierce Well, and the thicker section is shown to be a growth section by Δd/d analysis. Sixteen parasequence boundaries were correlated within the Pliocene and Pleistocene sections of the two wells. The Δd/d plot in Figure 13-12, generated from Pierce and Light well log data, shows that growth initiated in the section between the deposition of parasequence boundaries 5 and 6 (late Pliocene). The plot exhibits a pregrowth, or low-growth, interval below 200 m. Between parasequence boundaries 1 and 5, the stratigraphic interval continues to expand into the Pleistocene section. This linear expansion continues into the Recent, as evidenced by offset alluvial terraces (Fig. 13-12). Thus, displacements on the splay of the Zayante Fault began in the late Pliocene, between parasequence boundaries 5 and 6, and continued into the Recent, as evidenced by the offset terraces (Fig. 13-12). Other Δd/d plots from the area show similar growth histories on different splays of the Zayante Fault. Therefore, in some cases Δd/d plots can detect time of movement on strike-slip faults. If hydrocarbon migration occurred within the fault zones during periods of movement, the Δd/d analysis of fault timing provides valuable information regarding petroleum potential. The structures formed by movement on the fault(s) would necessarily predate hydrocarbon migration in order for accumulation to occur.

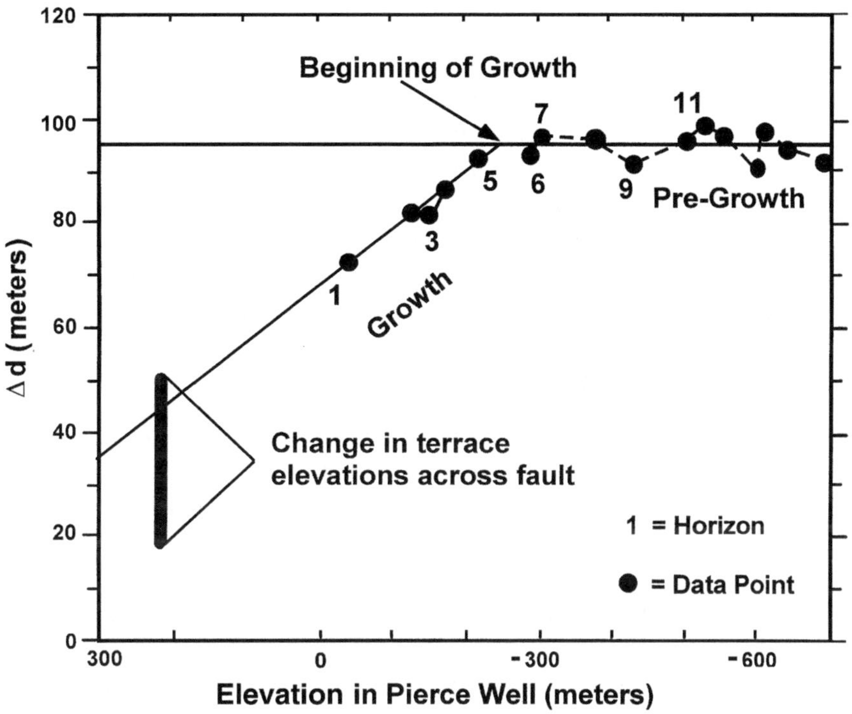

Figure 13-12 Growth plot for a splay of the Zayante Fault, California. Growth began between the deposition of parasequence boundary correlations 5 and 6 in the late Pliocene. Growth continues into the Recent, as evidenced by offset terraces. (From Shaw et al. 1994. Published by permission of the United States Geological Survey.)

THE MULTIPLE BISCHKE PLOT ANALYSIS

Sanchez et al. (1997) developed a significant enhancement to the Δd/d method. They use well log data from Venezuela to generate *multiple cross-plots* of correlation data, using the Δd/d method. They call this multiple well log correlation technique the **Multiple Bischke Plot Analysis (MBPA)**.

The MBPA can be more powerful than that of a single Δd/d plot. First, the method compares many wells to each other, increasing the probability that an *anomalous* correlation marker in a problem well will not be overlooked. Second, a quick perusal of several Δd/d plots may rapidly direct geoscientists to the well that contains an anomalous correlation marker or problem. Additional well log correlation or seismic analysis can then be directed at the anomalous correlation to rapidly identify the cause of the anomaly. MBPA is versatile and does not require knowledge of which well is in the structurally higher position. The wells can be in any structural position and can even be along strike. Lastly, different types of plots can be constructed to emphasize or to demonstrate a concept or observation. Several Δd/d plots, all using the same reference well, can be grouped on a single plot to form a **stacked Multiple Bischke Plot** (Chatellier and Porras 2001). Stacked MBPs allow geoscientists to compare thickness changes between wells, or to locate an anomaly or missing section that exists in every well. As an example of a stacked MBP, see Fig 13-17b.

Another type of analysis is the **inverted MBP** (Chatellier and Porras 2001). For this analysis, the axes on the standard Δd plot are inverted, with Δd plotted on the x-axis and d plotted on the y-axis. Inverted MBPs contain properties similar to conventional stratigraphic cross sections, in that the y-axis values correspond to different depths in a reference well. These plots help to identify decollement levels and stratigraphic intervals that are partially faulted out of wells (Chatellier and Porras 2001, 2004).

In the format presented by Sanchez et al. (1997), MBPA is based on three or four Δd/d plots placed on a single page. Common to all these Δd/d plots is the same reference well. On the left side of each of the plots is the number of the reference well (top number), along with the number of the well used for comparison (bottom number), as in (Fig. 13-13). The Δd values are on the y-axis, but in this case, the depths in the *reference well* are on the x-axis, regardless of its structural position relative to the comparison well. Thus, the reference well is present on all the plots.

MBPA to Recognize Correlation Problems

Using an example from eastern Venezuela, we employ MBPA to locate an interpretation anomaly that is due to a miscorrelation in a particular well. Figure 13-14 is Multiple Bischke Plot showing a stratigraphic miscorrelation from a producing field in eastern Venezuela (Sanchez et al. 1997, Chatellier and Porras, 2004). The correlation data points on the plots are flooding surfaces interpreted from the field data. The numbers C4 through C7 represent productive units within the field. The Multiple Bischke Plots compare Wells No. 485, 910, and 834 to reference Well No. 365. The vertical lines on the plots indicate known faults. Notice on all three plots that a correlation point for a flooding surface within the C6 unit lies about 70 ft below the straight-line portion of the curves (see arrows). As this anomaly exists on all three plots, the authors attribute the anomaly to a miscorrelation in reference Well No. 365.

The next example is from a salt-cored structure and involves a well that bottoms in salt near the crest of the structure (Bischke et al. 1999). Steep bed dips caused the 3D seismic data to

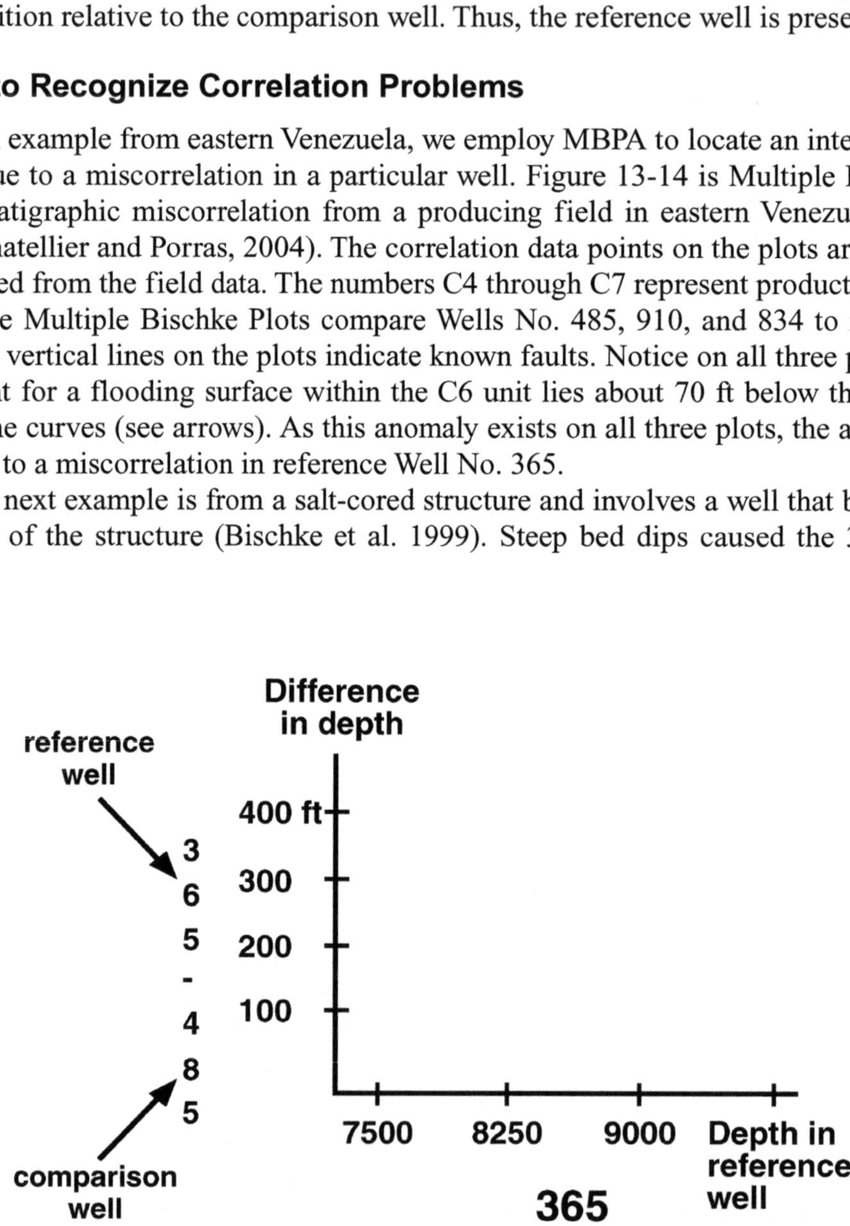

Figure 13-13 Definitions for a Multiple Bischke Plot. The comparison of a reference well (top number on left), with depths plotted on the x-axis, to a comparison well (bottom number on left), forms the basis for the Multiple Bischke Plots. Unlike the Δd/d method, the reference well can be on any structural level and need not be the structurally higher well.

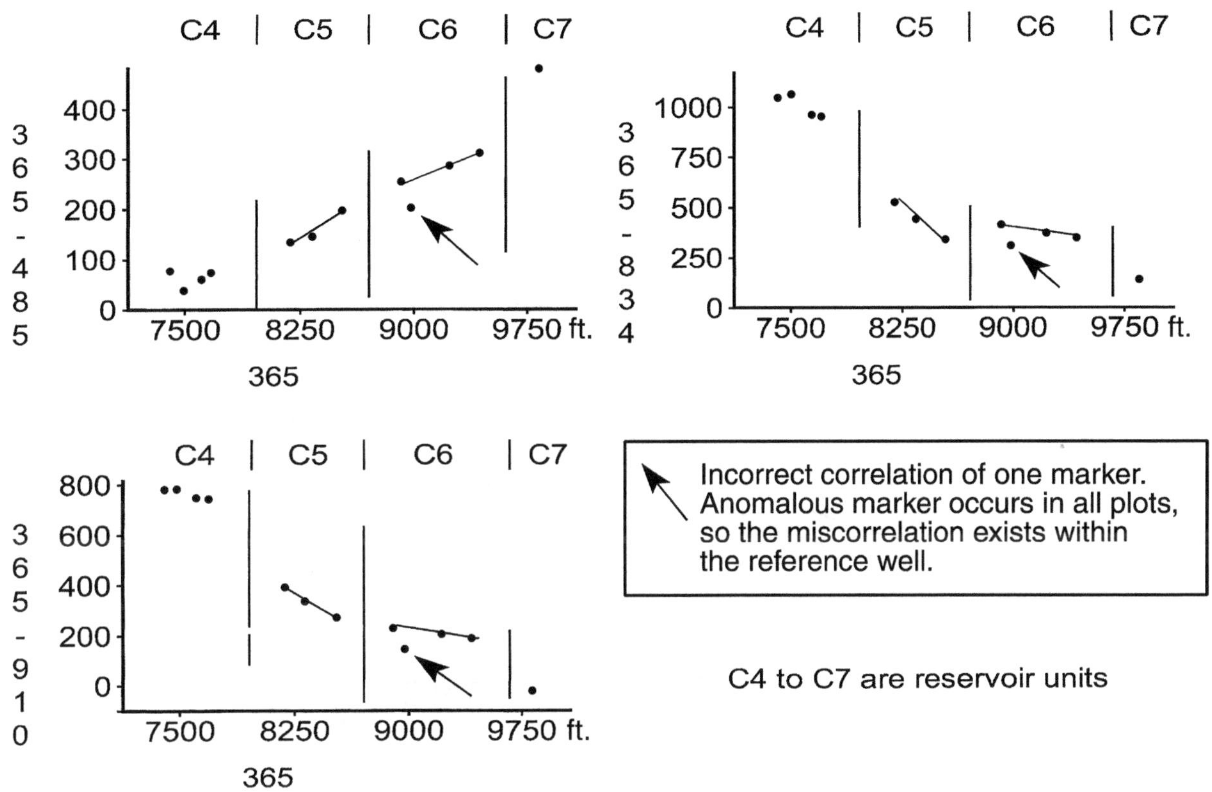

Figure 13-14 A Multiple Bischke Plot of flooding surfaces from a field in Venezuela. Anomalous point in the C6 unit, in all plots, suggests a possible stratigraphic miscorrelation in the reference well (From Sanchez et al. 1997. Published by permission of Latinoamericano de Sedimentologia, Sociedad Venezolana de Geologos.)

become incoherent, and a rapidly thinning stratigraphic section on the flanks of the dome caused reduced confidence in the well log correlations. The 3D seismic data were tied to off-structure wells. The initial mapping, based on a computer-mapping program, indicated that two horizons crossed each other in the up-dip direction, near the top of salt, which is impossible (Fig. 13-15). How do you resolve a mapping and correlation problem if you cannot correlate the 3D seismic data or the well logs with confidence?

The high bed dips and a rapidly thinning stratigraphic section along the flank of the diapir and near Wells No. 1, 2, and 3 made the seismic and well log correlations inconclusive (Fig. 13-15). A three-point problem, constructed from other wells that flank the diapir, results in an average bed dip of 42 deg. Higher bed dips are likely to be encountered up-dip, as the strata conform to the salt face. Best-guess correlations of the well log data resulted in an inaccurate pick for what is called the 20 Horizon in the up-dip Well No. 1. Seismic picks were tied to the well control and maps were generated. The computer-mapping program projected mapping horizons into the area of incoherent seismic data. The result was that the deeper 20 Horizon crosses the shallower 18 Horizon up-dip of Wells No. 2 and 3. This is impossible. There is a significant distance between where the seismic data deteriorate and where the log data for Wells No. 1, 2, and 3 are used to generate ties to the seismic data, which were used in the final interpretation and maps.

Recognizing the problem, a solution must be determined. There are three questions to answer: (1) Is the 20 Horizon correlated too high, or is the 18 Horizon correlated too low? (2) Could there be unrecognized faulting that is not imaged in the 3D data set? (3) What does one do when both the seismic and well log data cannot be correlated with confidence? When confronted with

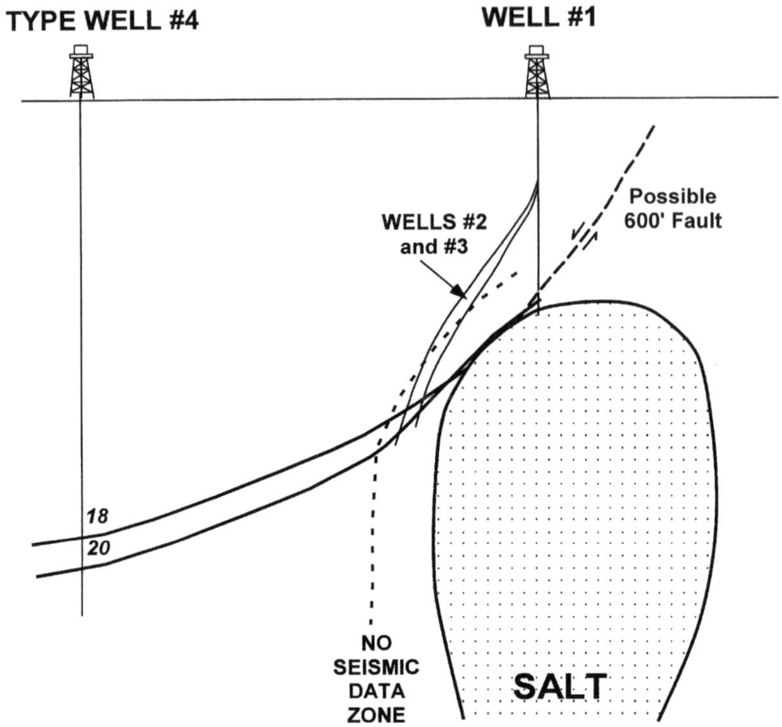

Figure 13-15 High bed dips on the flanks of diapirs may cause 3D seismic data to deteriorate, as in this example. Inability to correlate the well logs may result in incorrectly correlated horizons in the structurally higher wells. A projection, by a mapping program, of the 18 and 20 Horizons up-dip into the no-seismic-data zone caused the 18 and 20 Horizons to cross, which is impossible. MBPA provides the means to resolve the structural and stratigraphic problems.

difficult problems of this kind, the MBPA may converge on a viable structural solution to the problem (Bischke et al. 1999). Given their associated correlation problems, we want to compare Well No. 1 with nearby Wells No. 2 and 3. These wells are down-dip of Well No. 1, have thicker stratigraphic sections, are close to each other, and are easier to correlate than Well No. 1.

When using MBPA, experience has shown that comparing wells that have correlation problems to a *type well* can help resolve those problems. Wells No. 1, 2, and 3 are compared to a reference Well No. 4, which is an off-structure well. Figure 13-16a and b are two Δd/d plots constructed for Wells No. 2 and 3, using Well No. 4 as the reference well. These two plots indicate that the growth rate gradually increases from the 1 Horizon down through the 20 Horizon in both Wells No. 2 and 3. Gradual (monotonic) changes in growth on Δd/d plots are a common pattern observed in different tectonic environments (Bischke 1994b).

Figure 13-16c is a Δd/d plot for Wells No. 1 and 4, using the original correlations. Well log correlations are difficult below the 11 Horizon in Well No. 1. Notice where the correlation for the 20 Horizon appears in the Δd/d plot. The plot is subject to one of three possible interpretations:

1. The growth increases dramatically between the 11 and 20 Horizons around Well No. 1;
2. Based on the discontinuity in the plot, a normal fault or unconformity, with a missing section of about 600 ft, is present in Well No. 1 between the 11 and 20 Horizons; or
3. The 20 Horizon correlation in Well No. 1 is about 600 ft too high for a possible position of the 20 Horizon. We obtain the value of about 600 ft by projecting the growth curve, defined by the 1 to 11 data points, to beneath the 20 Horizon data point, and then determining the difference in depths.

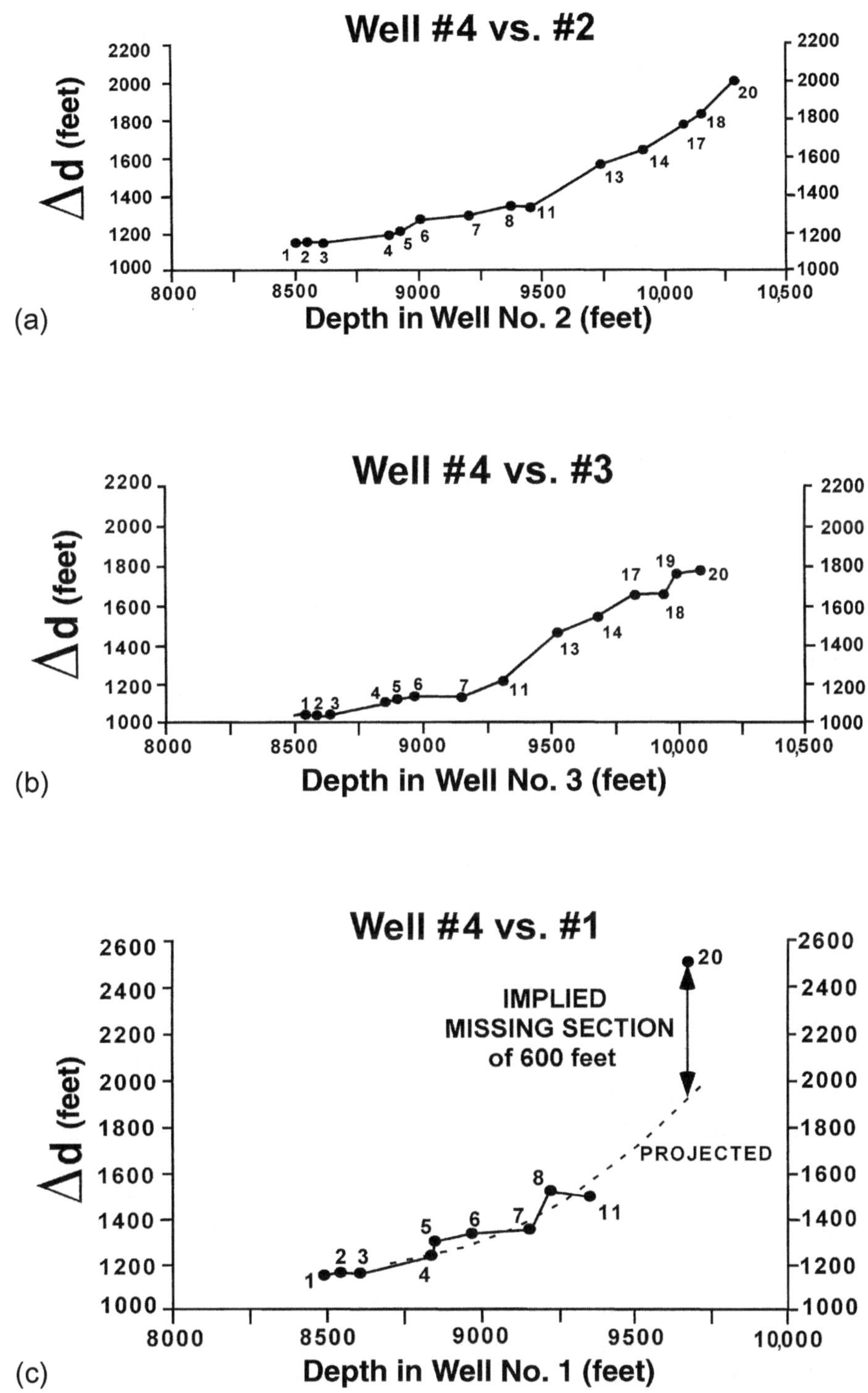

Figure 13-16 MBPA using Well No. 4 in Figure 13-15 as a reference well. (a) Growth plot using Well No. 2 for comparison. Growth increases monotonically between the 1 and 20 Horizons. (b) Growth plot using Well No. 3. Growth increases monotonically between the 1 and 20 Horizons. A comparison to Figure 13-5 shows no significant apparent missing section in either Well No. 2 or 3. (c) Growth plot for Well No. 1 located near the crest of the diapir. The plot suggests that about 600 ft of missing section exists at the level of the 20 Horizon. As 600 ft of section is not missing in Wells No. 2 and 3, or in other surrounding wells, the missing section in Well No. 1 is likely to result from a data miscorrelation, rather than a fault. (From Bischke et al. 1999. Published by permission of the Gulf Coast Association of Geological Societies.)

Using the data obtained from the Multiple Bischke Plots and knowledge of the local geology, two of these interpretations can be eliminated and we can resolve the problem. We first consider the increased growth interpretation. Although dramatic increases in growth are possible, this increased growth is not observed in the nearby Wells No. 2 and 3 or in other nearby wells, thus making the high growth interpretation very unlikely.

The interpretation based on a 600-ft fault or unconformity can be rejected for the following reasons. A 600-ft unconformity should be observable on the down-dip portions of seismic profiles where bed dips are gentle or in growth plots constructed from other nearby, on-structure wells. Furthermore, Wells No. 1, 2, and 3 *are drilled from the same platform*, and Wells No. 2 and 3 are deviated more to the south than Well No. 1. If a large 600-ft fault cuts Well No. 1, then it could also cut Wells No. 2 and 3, or other nearby wells. Plots of the data shown in Figure 13-16a and b do not contain a large 600-ft discontinuity, nor do plots with other nearby wells. This means that the Δd/d plots do not indicate a large fault or unconformity below the 11 Horizon, making interpretation No. 2 also highly unlikely. Also, the seismic data over the structure is of reasonable quality. No large faults or unconformities are recognized on the seismic data near Well No. 1.

Reexamination of the well logs suggests that the correlation of the 20 Horizon, near the bottom of Well No. 1, is probably an incorrect correlation and that interpretation No. 3 is the likely solution. Additional correlation work indicates that the miscorrelated horizon, which is about 550 ft below the 11 Horizon in Well No. 1, is not the 20 Horizon, but rather it is most likely the 14 Horizon. This change in correlation results in a completely new interpretation from the 14 Horizon through the 20 Horizon. The changes affect the fault interpretation, structure maps, reservoir maps, and the upside prospect potential. The method results in a reasonable solution to a difficult problem in an area where well log correlation and 3D seismic data alone could not resolve the problem.

The Use of a Stacked Multiple Bischke Plot

This example, from the northern Gulf of Mexico, involves a salt-related structure. During the growth of salt diapirs, stratigraphic units onlap the flanks of the structures, creating large unconformities. Numerous faults associated with the growth sedimentation and developing structure may also be present, causing problems for a geoscientist *who has to distinguish between missing section that results from faults and missing section that results from unconformities*. Complicating the problem is the fact that the missing section caused by unconformities can exceed the missing section caused by growth faulting. Although this problem relates to salt, the general principles outlined in this section apply to locating unconformities in any tectonic setting.

Figure 13-17a shows well locations on the flank of a salt dome and the measured bed dips in the wells at a stratum called the Rob L Horizon. Figure 13-17b shows a Multiple Bischke Plot using the structurally higher Well No. 6 as a reference well. All the data from the off-structure wells can be plotted on the same diagram because each well is being compared to Well No. 6. This display, which emphasizes growth between the wells, is called a stacked plot (Chatellier and Porras 2001). Referenced along the x-axis are the Horizons B through Q in each well. Using the Multiple Bischke Plot, we can interpret the data. The structure experienced two growth phases, between the B and M Horizons and between the O and Q Horizons. The interval from the M to O Horizons exhibits little growth. Negative discontinuities are present in the plots of Wells No. 1 and 3. The discontinuities represent *missing section*, either due to a fault in the structurally lower well (compare to Fig. 13-5a) or due to downlap (compare to Fig. 13-6b). As *downlap is rare on the flanks of salt diapirs*, it is more likely that these discontinuities result from faulting in the structurally lower Wells No. 1 and 3. Correlation to other off-structure wells confirms that the missing sections in Wells No. 1 and 3 on the MBP are due to Fault J, which is noted on Figure 13-17b.

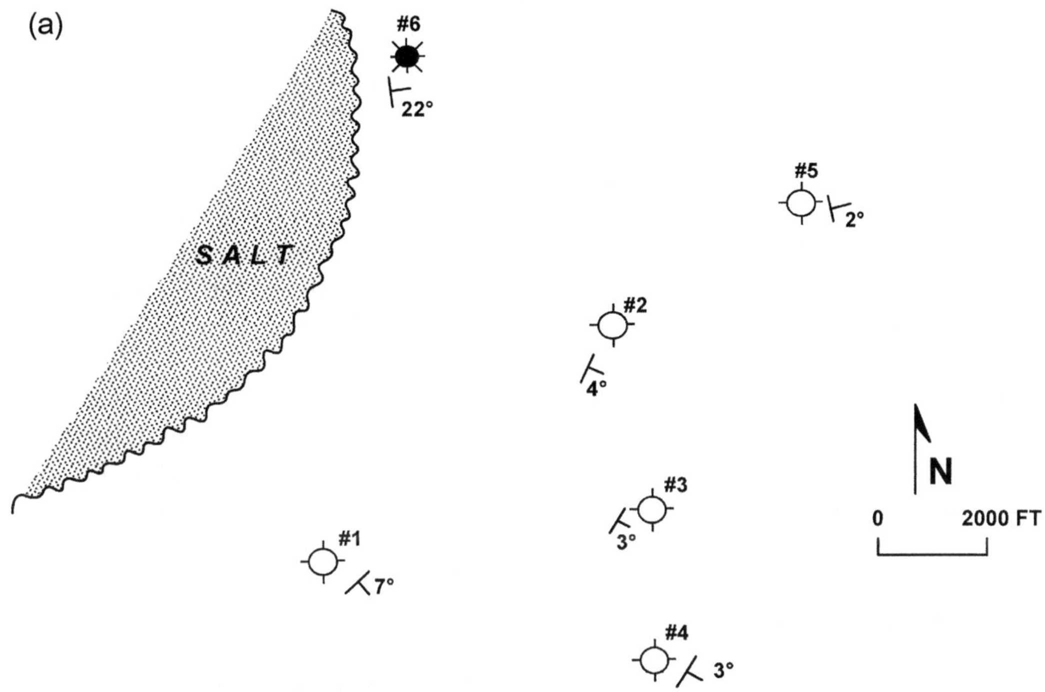

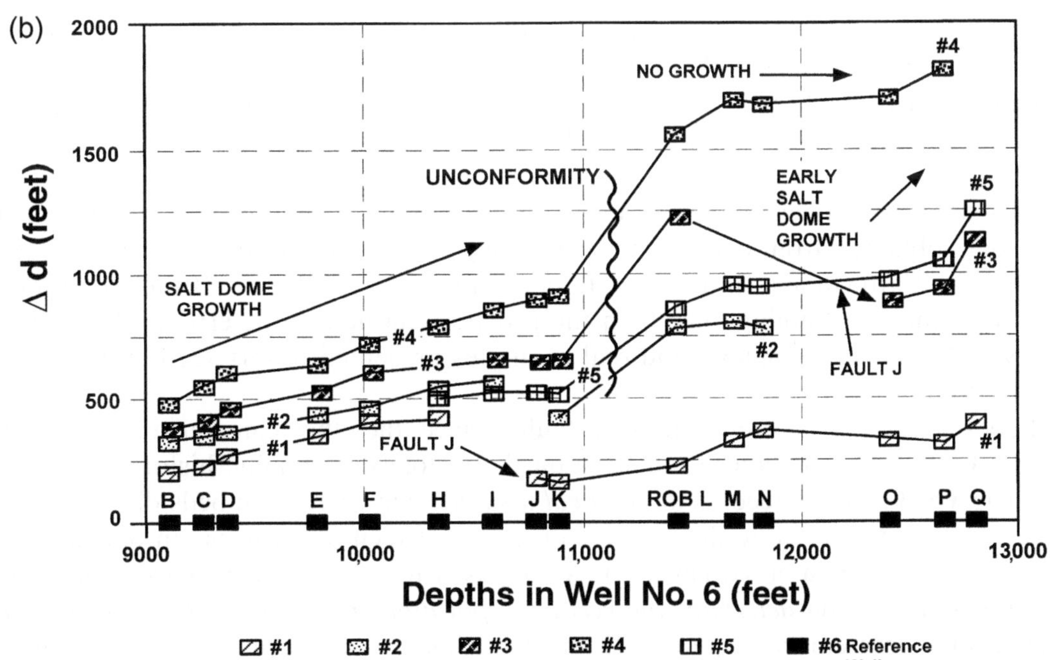

Figure 13-17 (a) Map showing strike and dip on the Rob L horizon at locations on the flank of a salt diapir in the northern Gulf of Mexico, USA. High bed dips on the flanks of salt structures cause stratigraphic thinning and deterioration of the seismic data. In this environment, geoscientists have difficulties attributing missing section to faults or to large unconformities. (b) The Multiple Bischke Plot for the structurally higher reference Well No. 6 versus the five off-structure comparison wells. The missing section above the Rob L sand in every well is interpreted to be due to a large unconformity. Fault J produces about 250 and 340 ft of missing section in Wells No. 1 and 3 respectively, based on the plot. (Published by permission of R. Bischke.)

On the flanks of salt diapirs, the amount of missing section typically increases up-dip as the horizontal distance between the on-structure and off-structure wells increases (Chapter 4, Fig. 4-44a). The Δd/d plots detect the amount of missing section caused by unconformities, as measured relative to the two wells used for a given plot. If both wells penetrate the same unconformity, then the amount of missing section determined from the plot will be the *difference* in the amounts of sections missing in the wells.

A positive discontinuity exists in every well plot in Figure 13-17b. Furthermore, this missing section occurs above the Rob L Horizon in every well plot. Thus, we attribute this missing section above the Rob L to represent an *unconformity*. The results of the MBPA guided additional correlation work and the interpretation when working with other structurally higher wells that flank the diapir.

Therefore, the MBPA, using six wells on this structure, provides quantitative graphical evidence for the following interpretation. The MBPA method (1) locates and defines the size (relative to certain wells) of the unconformity above the Rob L horizon, (2) defines the growth history of the structure, (3) provides initial evidence for Fault J, (4) aids in distinguishing faults from unconformities, and (5) helps keep the interpretation focused.

VERTICAL SEPARATION VERSUS DEPTH METHOD

In Chapters 7 and 8, we discuss missing or repeated section resulting from normal or reverse faults as being equal to the fault component vertical separation (VS). The vertical separation can also be analyzed through the use of Multiple Bischke Plots. In the case of analyzing the vertical separation, the plots are now set up as **VS/d** rather than the normal Δd/d.

A plot of vertical separation versus depth can be used to analyze growth structures. In previous sections, we showed that information about tectonostratigraphic history is available by comparing two wells (Bischke 1994b). Outcrop, well log, depth-corrected seismic data, or well-constrained cross sections, provide the data required to construct the profiles. Now we apply vertical separation (**VS**) versus depth (**d**) plots to a variety of structural uses. The vertical separation data can be derived directly from seismic data or from missing or repeated section data obtained from correlation of well logs. These VS/d plots of data can be used to document fault growth and timing, and to determine *structural style*. If the structural style in an area is complex or ambiguous, then the profiles can help to determine the correct structural style. A number of generic and real examples presented in this section illustrate how to apply the method.

If geoscientists can correlate across a fault surface, then they can construct maps of faulted horizons. These maps contain information on the tectonostratigraphic history (Yeats and Beall 1991). The **VS/d method** records the *vertical distance* that correlative beds displace across fault surfaces. Geoscientists define this vertical distance between offset beds, on opposite sides of a fault, as vertical separation (Chapter 7). As opposed to other fault displacement components, geoscientists *can always determine the vertical separation component of fault displacement*. For example, slip measurements require a 3D understanding of fault displacements. *One-dimensional well logs and two-dimensional seismic profiles do not contain 3D information*, and they cannot independently provide slip measurements (Chapter 7). Fault throw may be obtainable only *after* the construction of 3D maps, or from seismic sections that are perpendicular to the strike of the fault surfaces (Chapter 8).

The VS/d plots measure the vertical component of fault displacement. We typically think of vertical separation for normal and reverse faults. However, strike-slip faults also commonly exhibit vertical displacements (Harding 1973; Wright 1991; Shaw et al. 1994), particularly at restraining and releasing bends (Crowell 1974a and b) (Chapter 12). Also, because paleotopographic surfaces are not planar and strike-slip faults exhibit vertical slip components, the method can apply to growth sedimentation associated with strike-slip faults (Shaw et al. 1994).

Where subsurface data are sparse, the construction of subsurface maps and cross sections typically involves the introduction of assumptions and interpretation. *Fault slip components are rarely obtainable in outcrop data*, making fault slip determinations questionable even where faults are subject to direct observation. However, *VS is always obtainable* directly from missing or repeated section in well logs or from a single seismic profile (Chapter 7). Thus, we present a method for studying fault displacements based on VS.

Method

Information on fault history and structural style is obtainable by plotting VS against the depth of the faulted horizons at the footwall cutoff *(Fig. 13-18). Every displaced horizon has a value of VS (VS_1, VS_2, VS_3, . . . VS_n). For a VS/d plot, the VS is measured from the footwall horizon to the hanging wall horizon. For a normal fault, the footwall horizon is projected across the fault (Fig. 13-18). The measurement of VS in these cases is* downward *(increasing depth), and the amount of VS is taken to be a* positive *value for purposes of the VS/d plot (Fig. 13-18). For a reverse fault, the measurement of VS is* upward *(decreasing depth), and the amount of VS is taken to be a* negative *value (Fig. 13-19).*

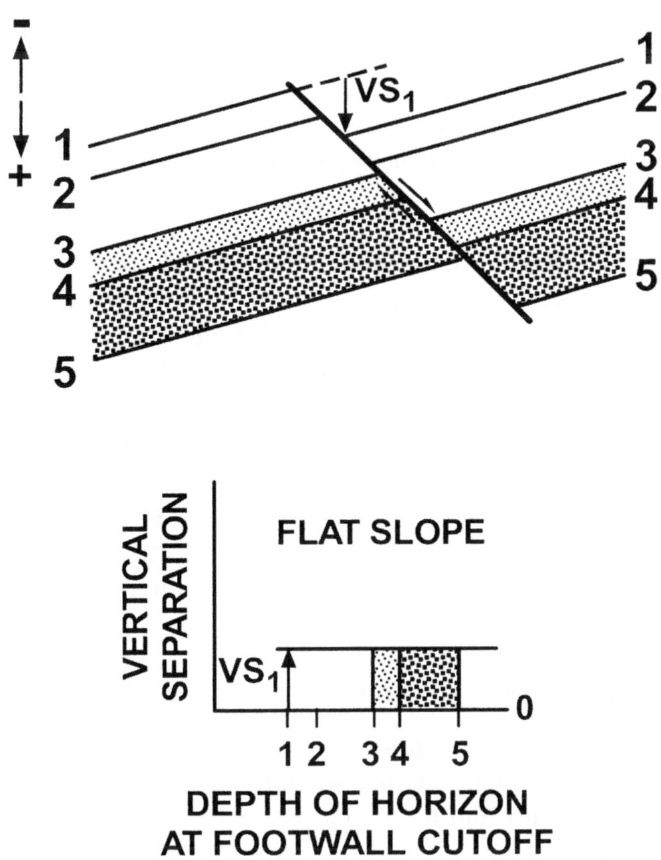

Figure 13-18 A generic example of a post-depositional normal fault and its VS/d plot. A post-depositional normal fault plots in the positive quadrant and has a flat slope. (Published by permission of R. Bischke.)

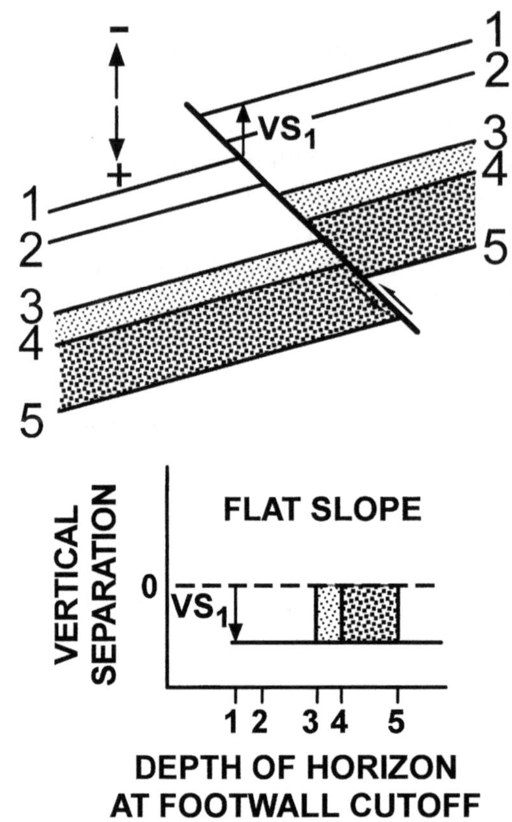

Figure 13-19 A generic example of a post-depositional reverse fault and its VS/d plot. A post-depositional reverse fault plots in the negative quadrant and has a flat slope. (Published by permission of R. Bischke.)

If the fault exhibits growth, then sediments deposited coeval with fault motion will change thickness across the fault surface, and the values of VS will change with depth. The slope of the VS/d plots can be used to define fault growth, with higher slopes indicating higher growth rates. A zero slope indicates a post-depositional (nongrowth) fault (Figs. 13-18 and 13-19). So, the method readily defines the timing of fault movements.

The plots of VS versus depth document vertical displacements. If age data are available, then it is possible to determine the displacement history within a time framework. However, the plots cannot independently detect large rotational motions (see Chapter 12). Where applicable with seismic data, use the workstation to convert the seismic data to depth before constructing the plots. Typically, interval velocities are adequate, although better velocity versus depth functions improve the results.

Generic and Real Examples of Analysis of VS/d Plots

Vertical separation data taken along *fault surfaces* provides information on fault displacement history and structural style. Thus, in this section we concentrate on fault surfaces. The data required are VS measurements and the depth to the footwall cutoff of each correlated horizon. We present a number of generic and real examples of several common, and some less known, structural styles. There are two basic types of faults, growth (expansion) faults and post-depositional (nongrowth) faults (Thorsen 1963). The plots of VS versus depth document growth faulting or post-depositional faulting. Our discussion starts with simple examples, followed by more complex ones.

Post-Depositional (Nongrowth) Faults. Correlative stratigraphic intervals on opposite sides of post-depositional faults do not exhibit significant thickness changes across fault surfaces. As VS does not change with increasing depth, post-depositional faults exhibit zero (flat) slopes on VS/d plots (Figs. 13-18 and 13-19). As a matter of convention, we define *normal faults* to contain positive displacements. Thus, post-depositional normal faults plot *above zero VS and in the positive quadrant* (Fig. 13-18). The reason for distinguishing between positive and negative quadrants will become clear in the Complex Growth Structures section.

In contrast to the normal fault, the sign of VS along a post-depositional reverse fault is negative, and thus *reverse faults plot in the negative quadrant* (Fig. 13-19). So, on VS/d plots, reverse and normal faults plot in different quadrants. Post-depositional reverse faults exhibit flat slopes on VS/d plots, as do the post-depositional normal faults (compare Figs. 13-18 and 13-19).

Growth Normal Faults. Sedimentary units change thickness across faults when the sedimentation is concurrent with faults that have dip-slip components of displacement. Growth associated with faulting is common in areas that have rapid sedimentation rates (Suppe et al. 1992; Shaw et al. 1994), and thus normal, reverse, and strike-slip faults may exhibit growth. One can distinguish growth faults from post-depositional faults in that the growth faults have *nonzero slopes* on the VS/d plots. The VS/d method has a high resolution and can detect growth where growth *is not obvious* in surface or subsurface data. The expansion faults common to deltas exhibit positive slopes on the VS versus depth plots, and the data are in the positive quadrant (Fig. 13-20).

Figure 13-21 is a VS/d plot for a growth fault in the northern Gulf of Mexico. Growth Faults tend to seal through an interval of rapid growth, and any faults with large displacement tend to seal. The larger the fault, the greater the probability that the fault will seal. Large displacements on a fault can create a clay smear or barrier in the fault zone or can juxtapose permeable rocks against

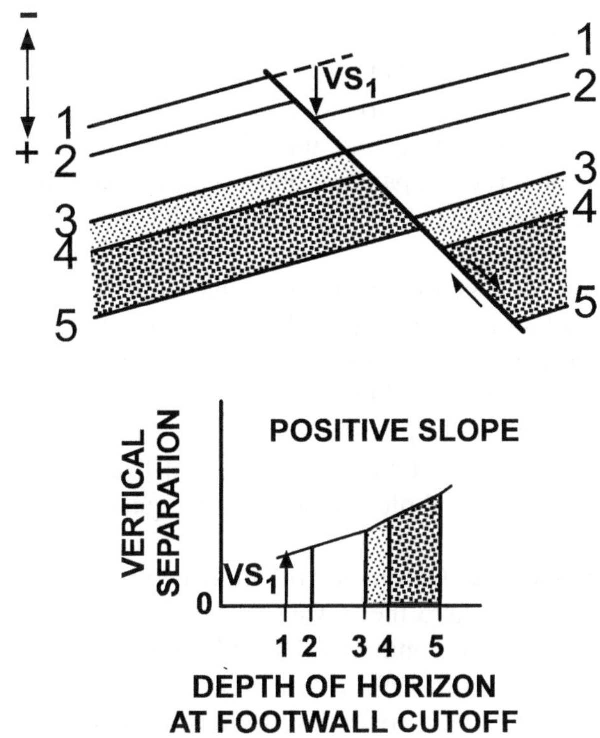

Figure 13-20 A generic example of a growth normal fault and its VS/d plot. A growth normal fault plots in the positive quadrant and has a positive slope. (Published by permission of R. Bischke.)

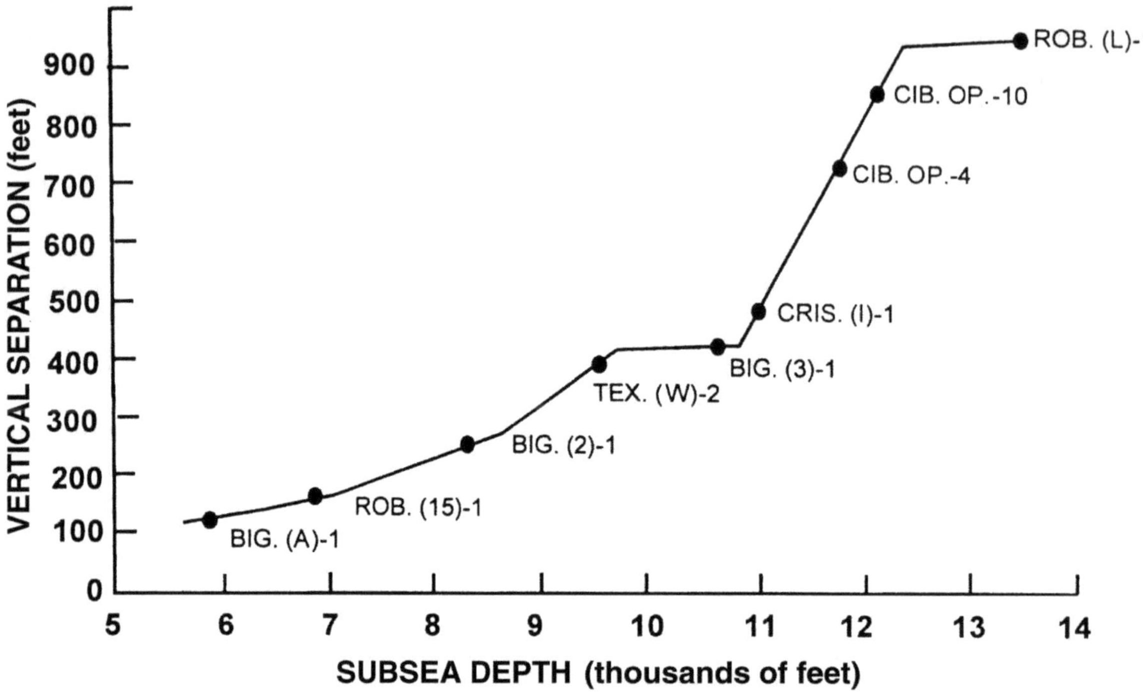

Figure 13-21 VS/d plot illustrating fault timing on a normal fault from the northern Gulf of Mexico, offshore USA. Rapid growth on the fault, between 11,000 ft and 12,400 ft, enhances its trapping capabilities within that depth range. (Published by permission of R. Bischke.)

shale (Brenneke 1995). In the United States, strata are typically dated by index fossil taxa, such as *Cibicides optima* (abbreviated CIB.OP.). A review of Figure 13-21 shows that the earliest growth on the fault occurred just prior to CIB.OP.-10 time, with the older faulted interval being pre-growth strata. The fault grew rapidly from CIB.OP.-10 time to about Bigenerina (3)-1 [BIG.(3)-1] time, and then became inactive. Just before Textularia (W)-2 [TEX.(W)-2] time, the fault became active again and continued growing, at a decreasing rate, until Bigenerina (A)-1 [BIG.(A)-1] time. As the fault exhibits 400 ft to 900 ft of vertical separation and was growing rapidly during CIB.OP.-10 time to BIG.(3)-1 time, the fault has a high probability of sealing below the 11,000-ft level.

Growth Reverse Fault. Most geoscientists know that growth faulting is common to extensional environments. Less well known is the case where deposition is concurrent with folding and/or reverse faulting, and **compressional growth structures** form (Suppe et al. 1992; Tearpock et al. 1994; Chapter 10 of this book). If, during compressional folding, the sediments flood across and over the rising structure, then a compressional growth structure forms (Fig. 13-22). Plots of VS versus depth for growth reverse faults are distinguished from those of growth normal faults in that they exhibit *negative slopes* and plot in the negative quadrant (Fig. 13-22).

Complex Growth Structures. Complex growth structures exist where more than one fault, or slip surface, is present, or where there was more than one period of deformation. Complex structures occur in all tectonic environments. The hybrid inversion structures that occur in Brunei (James 1984), the North Sea (Mitra 1993), Venezuela (Link et al. 1996), and elsewhere are readily recognizable on the VS/d plots. **Inversion structures** *originate as normal fault structures that are later subject to reverse fault displacements.* Growth across the faults may be noticeable during both the extensional and compressional phases. On growth plots, the VS data initially plot in the negative and then in the positive quadrant (Fig. 13-23). As these structures

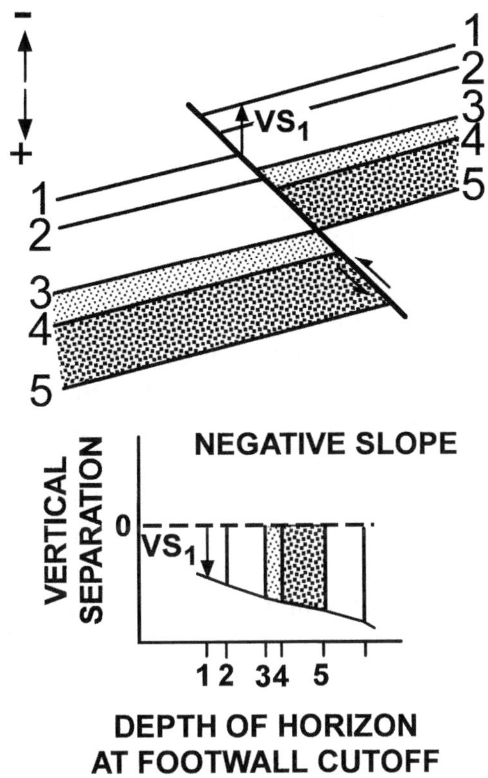

Figure 13-22 A generic example of a growth reverse fault and its VS/d plot. A growth reverse fault plots in the negative quadrant, with a negative slope. (Published by permission of R. Bischke.)

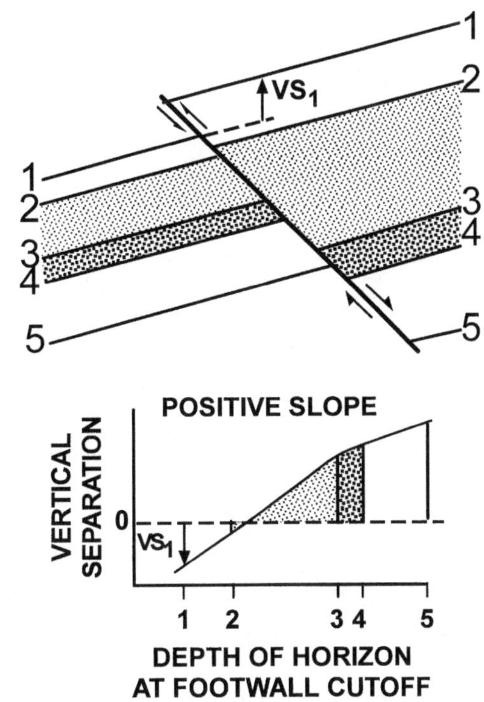

Figure 13-23 A generic example of an extensional, growth inversion fault and its VS/d plot. An extensional growth inversion fault plots in both the positive and negative quadrants, with a positive slope. (Published by permission of R. Bischke.)

plot in different quadrants and have positive slopes, *inversion structures are distinct* and easily recognized from other fault styles.

Inversion structural styles have been easily confused with strike-slip fault structural styles. In Chapter 12, we discuss methods and procedures for documenting horizontal displacements, which are critical in determining strike-slip faulting. Plots of VS versus depth aid in recognizing inversion structures. In the hanging wall portions of inversion structures, the shallow stratigraphy is thin relative to its correlative footwall units, whereas the deeper stratigraphy is thick relative to its correlative footwall units (Fig. 13-24 and Mitra 1993). Before the recognition of inversion structures, and on older seismic data that poorly imaged the listric normal fault, geoscientists commonly interpreted well log and seismic data incorrectly. We use a depiction of an inversion structure in Venezuela as an example. First, at the shallow structural level of the stratigraphic unit, geoscientists recognized that the thin, shallow-water environment, footwall stratigraphy correlated to the thicker, deep-water environment strata in the hanging wall block (Fig. 13-24b). Second, the thickness of the sediments and the environmental conditions *reverse* at the shallower structural level, with the footwall section being thicker relative to the hanging wall section. The deeper, and once downthrown, units were subsequently thrust back up, along the pre-existing normal faults. This reverse motion thrusts the deeper water units to a structurally higher level than the correlative thin shallow water units (Fig. 13-24b). Before the recognition of inverse

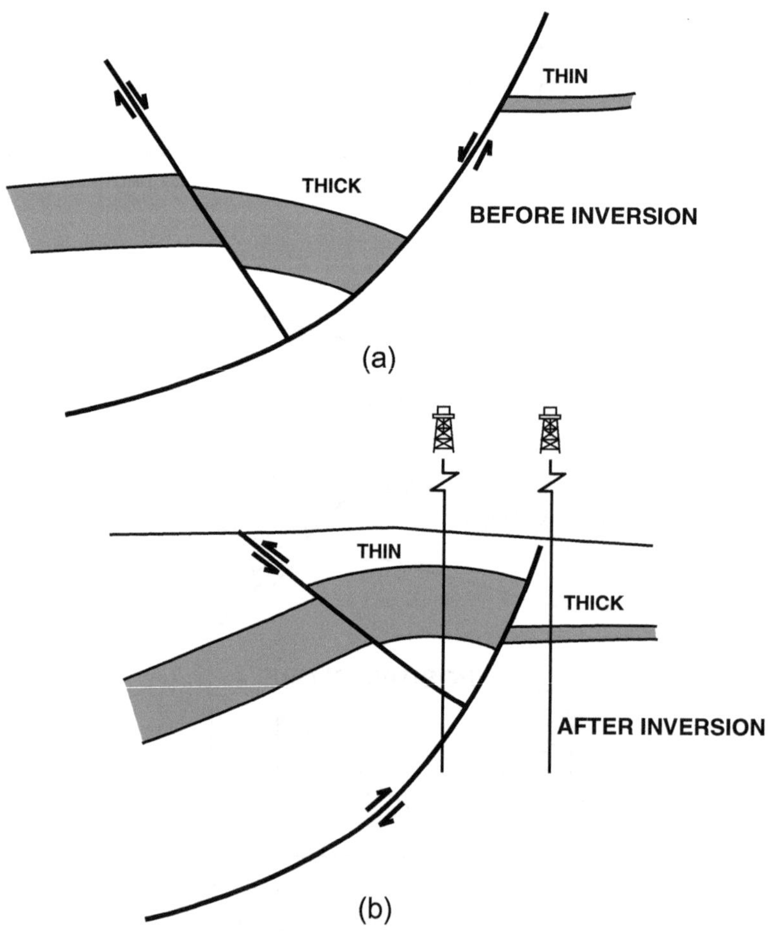

Figure 13-24 Schematic example of inversion. As the displacements reverse on an inversion structure, the thick, originally downthrown section is subsequently thrust back over a thin correlative section located in the footwall. Geoscientists often interpreted these confusing stratigraphic and structural relationships to result from strike-slip faulting. (Published by permission of R. Bischke.)

structures, these complex and confusing relationships were interpreted to result from large strike-slip displacements. Only strike-slip displacements were considered capable of creating this confusing geometric style. Thus, inversion structures may be misidentified as "flower" structures. To help distinguish inversion structures from flower structures, examine the correlation markers on VS/d plots. If the markers plot in both the negative and positive quadrants and have positive slopes, then the structure is most likely an inversion structure.

We can distinguish between an inversion structure and another structural style. **Compressional inversion structures** are present *where the displacements on reverse faults invert and become normal.* The positive and negative data points, combined with negative slopes, indicate inversion on a VS/d plot (Fig. 13-25).

Downward-Dying Growth Faults. Another important but problematic structure is the **downward-dying growth fault**, which commonly forms in the extensional environment (Bischke and Suppe 1990b; Tearpock and Bischke 1991). Some geoscientists err in being unaware that a normal fault may die downward. An extension of a fault with depth is unreasonable if supporting data does not exist. However, this fault style can be recognized because it plots on a VS/d diagram in the positive quadrant and exhibits positive slopes at shallow depths and negative slopes at greater depths (Fig. 13-26).

Some faults in deltaic areas may decrease in displacement in all directions away from a central point. If the vertical separations on these faults surfaces are contoured into a map of fault displacements, then these faults exhibit a more or less elliptical displacement pattern (Barnett et al. 1987; Bergen and Shaw 2010). The maximum displacements occur at the center of the ellipse

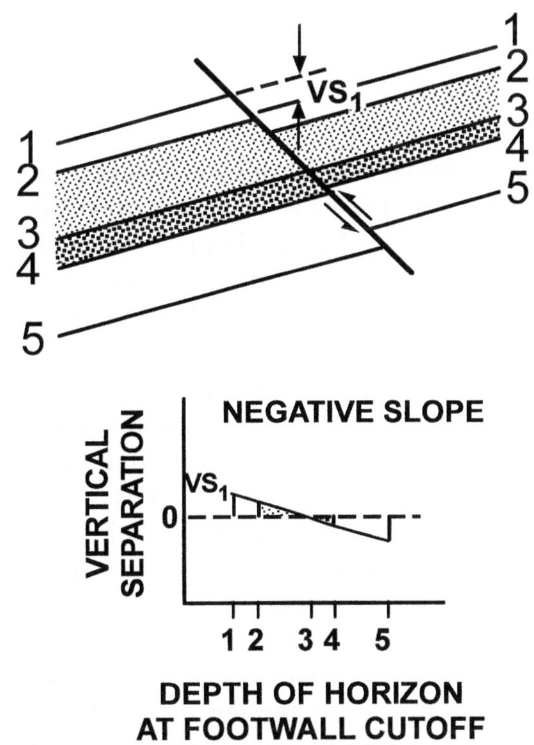

Figure 13-25 A generic example of a compressional, growth inversion fault and its VS/d plot. On a compressional, growth inversion structure, the thin upthrown section is later downthrown, juxtaposing thin sections in the hanging wall against thicker sedimentary sections in the footwall. A compressional inversion fault plots in both the positive and negative quadrants and has a negative slope. (Published by permission of R. Bischke.)

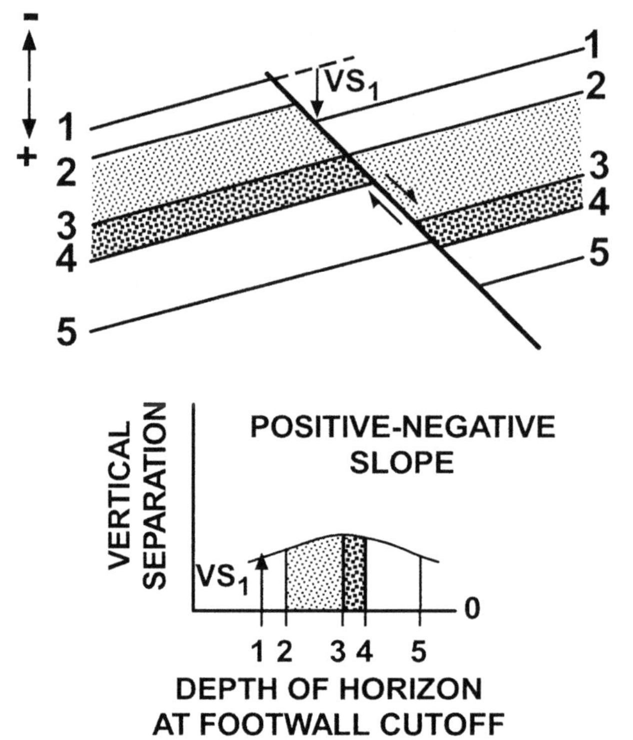

Figure 13-26 A generic example of a downward-dying, growth normal fault and its VS/d plot. The displacements on downward-dying faults decrease with increasing depths. A downward-dying growth fault plots in the positive quadrant. It has a positive slope at shallower depths and a negative slope at greater depths. (Published by permission of R. Bischke.)

and the fault dies out in every direction, upward, downward, and laterally. In Chapter 11, we present a model to explain these unusual relationships in a syntectonic depositional environment.

Figure 13-27 shows a map of a downward-dying growth fault, with accompanying seismic section, and a VS/d plot (Rabbit Island, coastal Gulf of Mexico, USA). The Rabbit Island fault exhibits 450 ft of missing section in a well at the 5500-ft level and about 70 ft of missing section in a well at about the 11,000-ft level. The structural question is, what happens to the fault below 11,000 ft? In order to determine the downward extent of the fault, based only on well control, there must be wells that go deeper than 11,000 ft in the area where the fault could reach this depth. With seismic data, one must approach the interpretation with caution. At times, interpreters do not pay attention to the displacements when picking faults, but rather they concentrate on interpreting the fault location on the seismic sections. If an interpreter does not realize that he or she is dealing with a downward-dying fault, an interpretation can be generated of a throughgoing fault extending to depth. This apparent throughgoing fault may set up prospects or perhaps divide producing reservoirs. If the fault in our example is absent beneath the 11,000-ft level, then no fault exists to trap hydrocarbons beneath that level. Or, if a reservoir exists just beneath 11,000 ft, then the partitioning of the reservoir is incorrect. In either case the interpretation is wrong and can have serious implications to any exploration or development activities. Geoscientists can construct VS/d plots to predict fault displacement and to evaluate whether they are dealing with a downward-dying fault, and to determine at which level the downward-dying fault loses its displacement (Tearpock et al. 1994). The Rabbit Island fault apparently dies out near the 11,000-ft level, as shown on the seismic line in Figure 13-27c, and it cannot trap hydrocarbons beneath that depth.

(a)

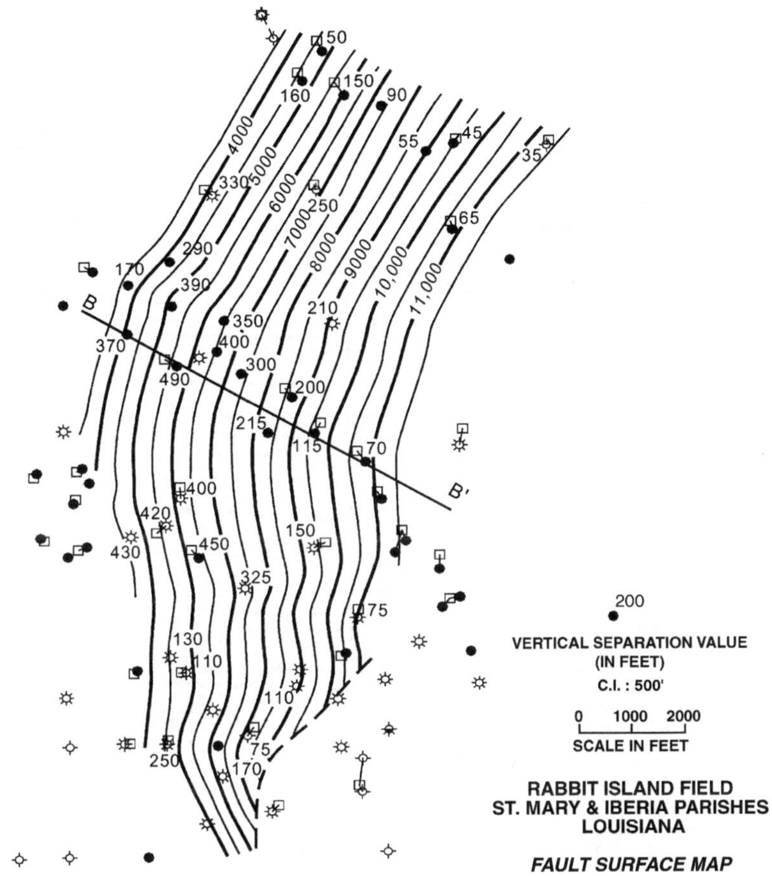

(b)

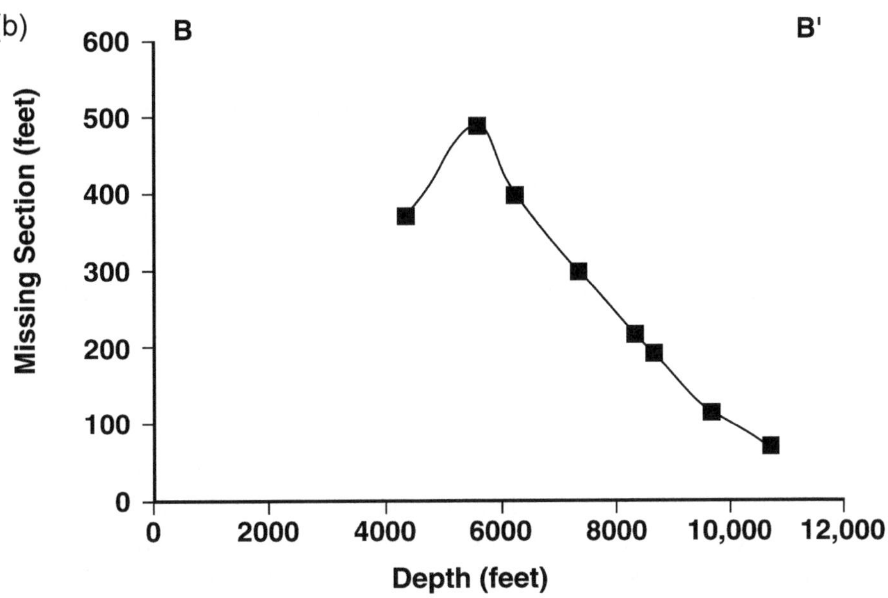

Figure 13-27 (a) Map, (b) VS/d plot, and (c) seismic line showing a downward-dying growth fault in Rabbit Island Field, coastal Gulf of Mexico, USA. Vertical separation decreases with increasing depth. The trapping capabilities along the fault surface diminish at about the 11,000-ft level. Figures 13-27a and b interpreted by D. Tearpock. (Maps and seismic line published by permission of Texaco, Inc.; VS/d plot published by permission of R. Bischke.)

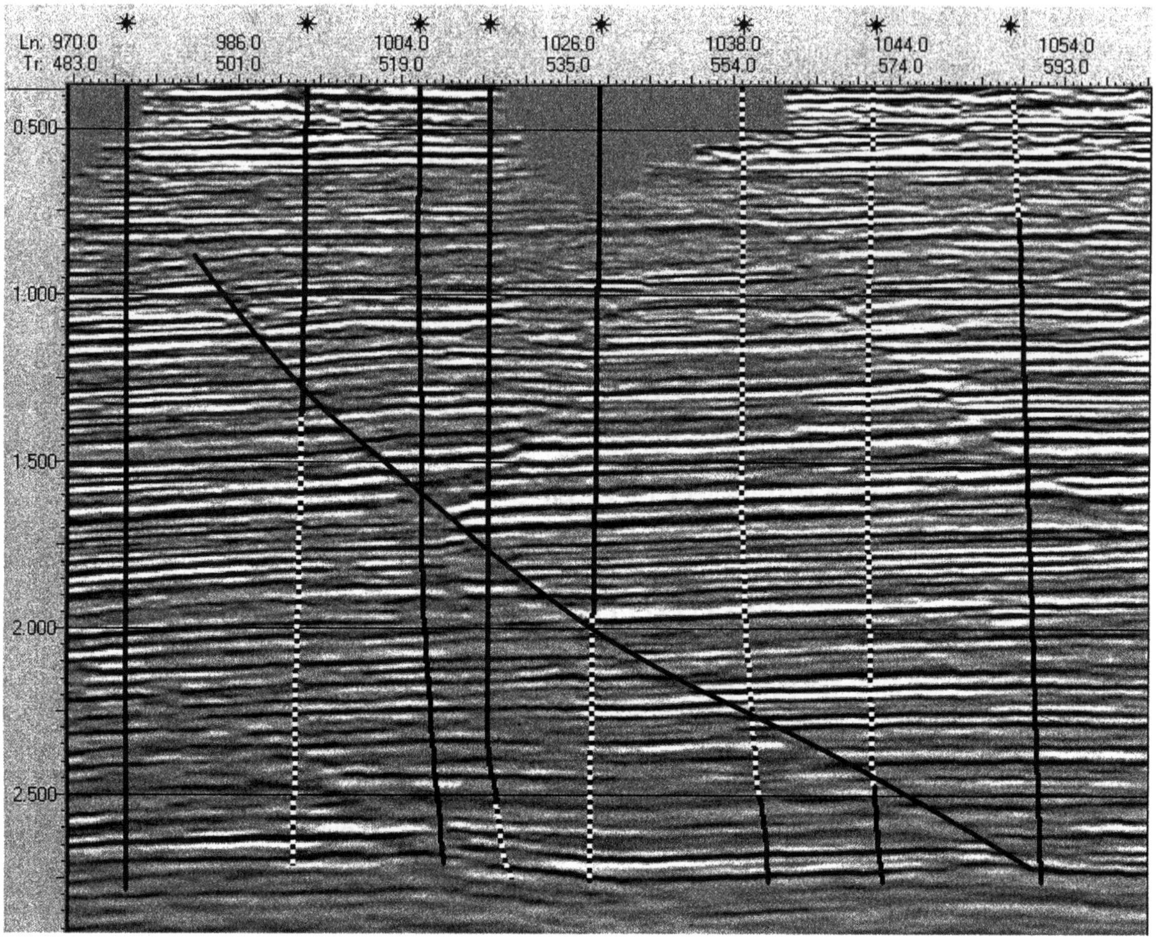

(c)

Figure 13-27 *(continued)*

Downward-dying growth faults are common in many areas, and seismic displays of them from the northern Gulf of Mexico and Brunei are in Chapter 11. In other areas, some faults may extend to depth, whereas other faults die downward. When dealing with potential hydrocarbon traps on downward-dying faults, care must be exercised to be sure that the potentially trapping fault will offset the objective horizon (Tearpock et al. 1994). The VS/d method or MBPA can aid in the recognition and documentation of the trapping capability of downward-dying growth faults.

CONCLUSIONS

Reliable correlation data are obtainable in many areas, but the data may deteriorate in key areas. The Δd/d, MBPA, or VS/d methods, when integrated with geological and geophysical data, can resolve difficult structural and stratigraphic problems where 3D seismic and well log correlation data are ambiguous (Sanchez et al. 1997). The methods are easy to use, require little additional effort, and can eliminate the guesswork that could lead to incorrect structural interpretations and maps that could result in dry holes. Where correlation problems result from *complicated structural features, the Multiple Bischke Plot Analysis can resolve these problems during the initial stages of a project*, prior to constructing subsurface maps and generating prospects. This can help eliminate the need to remap the structure and/or reinterpret the seismic and well log data at a later stage in a project.

Additionally, as the methods are easy to use and rapid to apply, they serve as excellent checks on any interpretation involving structures or correlations. Often, after finishing a project, interpreters assume that their interpretations and maps are correct and do not subject their interpretations to *consistency checks* (Tearpock et al. 1994). This can result in interpretations that are inconsistent with the interpreter's own correlation data, maps, or cross sections. Growth techniques are an excellent consistency check and introduce a relative age factor into interpretations. Managers may find these techniques useful during prospect reviews and when analyzing prospects and the petroleum system.

The Δd/d and MBPA methods have a resolution of about one part in a thousand when using well log data, so the methods can often identify disconformities, subtle systems tract boundaries, or small growth faults (Bischke 1994b). The techniques can distinguish faults from unconformities where bed dips are high and where the seismic data is incoherent. If a well log correlation is the probable cause of the problem, then the method can direct geoscientists to the correct correlation, thus eliminating guesswork. Analysis of growth history often leads to viable structural solutions to complex problems.

Syntectonic sedimentation can occur in any tectonostratigraphic regime, not solely in the extensional, normal fault regime. In areas where the sedimentation rates keep pace with the tectonic uplift rates, structures formed by compressional and strike-slip faults can exhibit growth (Suppe et al. 1992; Shaw et al. 1994). These growth environments allow for the determination of fault timing that influences migration and petroleum system studies. Lastly, in areas where structure is complex or ambiguous, growth plots provide the results to determine structural style and resolve interpretation problems. Any interpretation of a stratigraphic sequence or structure must be consistent with the correlations.

CHAPTER 14

ISOCHORE AND ISOPACH MAPS

INTRODUCTION

Two key terms, **isochore** and **isopach**, are often used synonymously in the petroleum industry as measures of thickness, but they are different. An ***isochore map*** (Fig. 14-1a) delineates the ***true vertical thickness*** of a stratigraphic interval, whereas an ***isopach map*** (Fig. 14-1b) illustrates the ***true stratigraphic thickness*** of a stratigraphic interval. These two terms are often confused with respect to their geological meaning. It is vital for both exploration and development work that the correct meaning and, more important, the correct application of these two thicknesses be understood.

An isochored or isopached unit may be as small as an individual sand only a few feet thick or as large as several thousand feet thick and encompassing a number of stratigraphic units. An isopach map is extremely useful in determining the stratigraphic framework or the structural relationship responsible for a given type of sedimentation and for recognizing paleohigh areas. The shape of a basin, the position of the shoreline, areas of uplift, and under some circumstances the amount of vertical uplift and erosion can be recognized by mapping the variations in thickness of a given stratigraphic interval (Bishop 1960).

Isochore and isopach maps are used for a number of purposes by the petroleum geologist, including (1) depositional environment studies, (2) genetic sand studies, (3) growth history analyses, (4) depositional fairway studies, (5) derivative mapping, (6) the history of fault movement, and (7) calculation of hydrocarbon volumes.

In this chapter, we discuss several different types of isochore and isopach maps important to the evaluation of petroleum potential. These include interval isopach maps and net sand and net pay isochore maps. An interval isopach map delineates the *true stratigraphic thickness* of a specific unit or units. Interval isopach maps are discussed briefly near the end of this chapter.

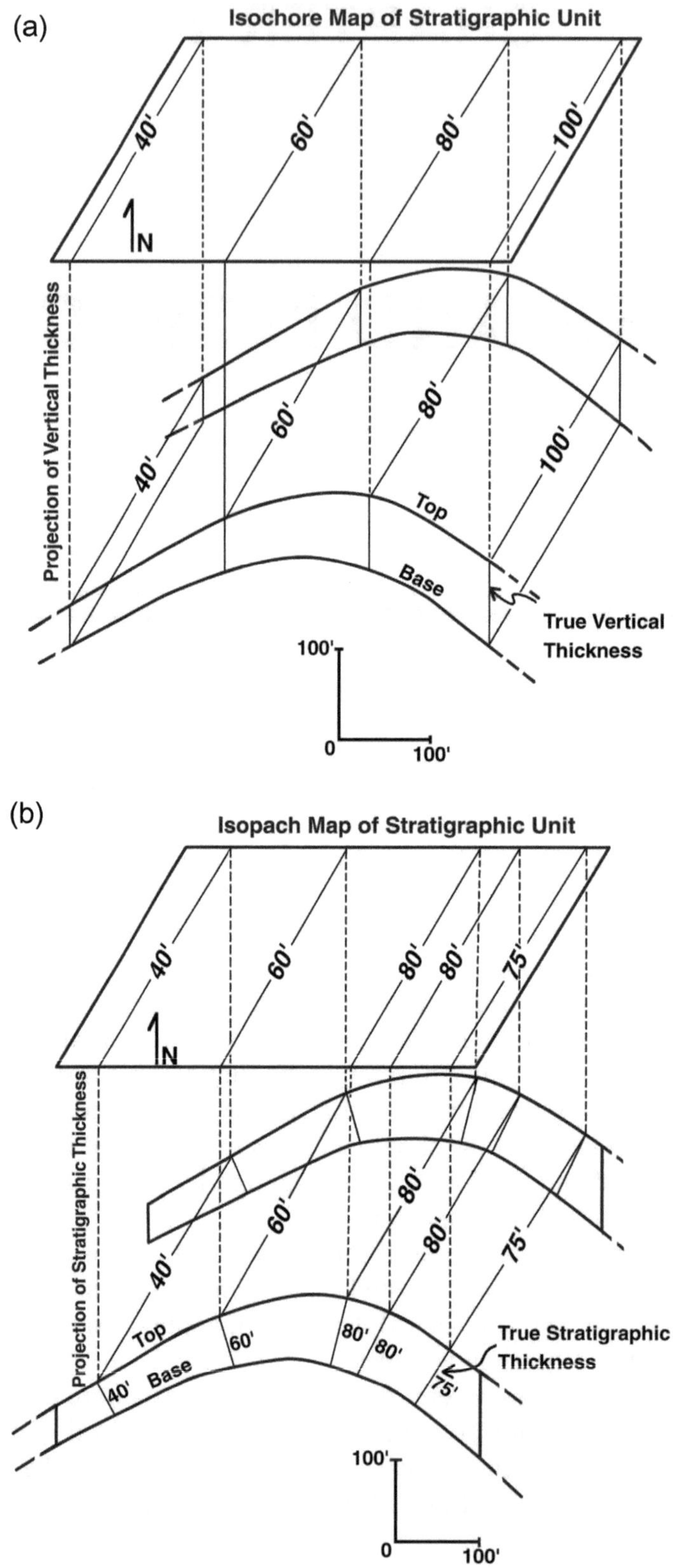

Figure 14-1 (a) An isochore map delineates the true vertical thickness of a stratigraphic interval, such as a rock unit containing a reservoir. (b) An isopach map delineates the true stratigraphic thickness of a stratigraphic interval. The same dipping stratigraphic unit is used in both (a) and (b), with the same edge-of-map boundaries. Note different thickness values assigned to the isochore map versus the isopach map of the same unit.

An isochore map may be based on the total vertical thickness of a specific unit or based on an aggregate vertical thickness of a particular rock type within a stratigraphic unit. For example, a given unit may consist of interbedded permeable and impermeable strata. An isochore map can be made for the total unit consisting of both permeable and impermeable strata, or an isochore map can represent the aggregate vertical thickness of only the reservoir-quality rock within that stratigraphic unit. For simplicity and brevity in this chapter, we use the term **net sand isochore map** to refer to a map of only the *reservoir-quality rock, although the term* **net reservoir isochore** *might be more precise*. Therefore, a net sand isochore map represents the *aggregate vertical thickness of reservoir-quality rock* present in a particular stratigraphic interval, which is illustrated in Figure 14-2. The techniques and calculations to derive vertical thickness are explained in detail in this chapter. The fluid contained in a stratigraphic unit may be hydrocarbons or water, or any combination of the two. Figure 14-3 shows a net sand isochore map of the 10,500-ft Sand in Golden Meadow Field, Lafourche Parish, Louisiana, USA. A net pay isochore map delineates the *aggregate vertical thickness of reservoir-quality rock that contains hydrocarbons (gas, oil, or both)*. An example is shown in Figure 14-5.

Net sand and net pay isochore maps of subsurface units are usually prepared from well log data, although seismic data may also be used. Interval isopach maps may also be constructed from well log data and seismic data, where coverage is adequate. As with structure maps, the completeness and accuracy of an isochore or isopach map depend on the amount and accuracy of data available. Even in isochore and isopach mapping, we cannot get away from the importance of log correlation work. Well log data, particularly correlations, should be studied very carefully in order to prepare an accurate and precise isochore map.

For volumetric reserve calculations, we are interested in obtaining the volume of a reservoir (solid material plus fluid-filled pore space). In this book, we use the volume unit acre-foot, which is standard practice in the United States. One acre-foot is equivalent to 1233.48 cubic meters and to 0.123348 hectare-meters. To many people, an acre-foot is an abstract measurement, but the concept is relatively simple. One acre-foot can be defined as that volume of rock plus fluid contained in an area one acre in size, with a thickness of one foot. How big is an acre? There is a

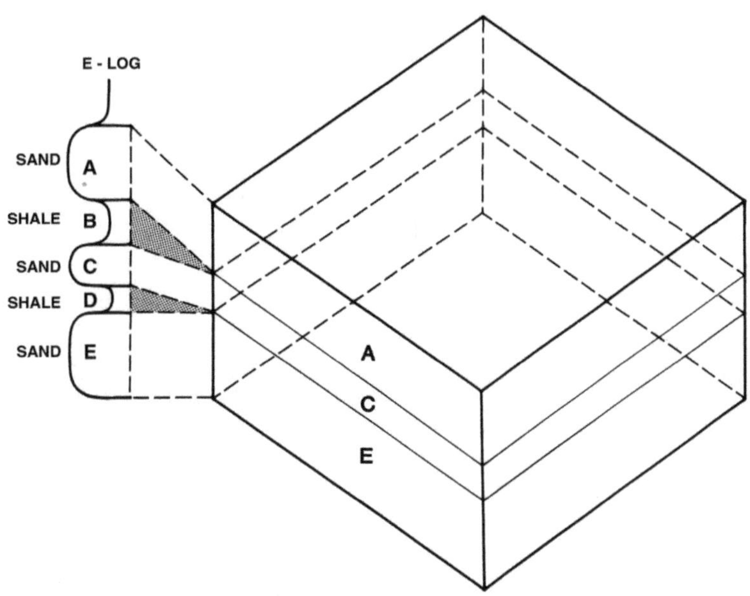

Figure 14-2 Net sand consists of porous reservoir-quality rock. All nonreservoir-quality rock is ignored. (From Tearpock and Harris 1987. Published by permission of Tenneco Oil Company.)

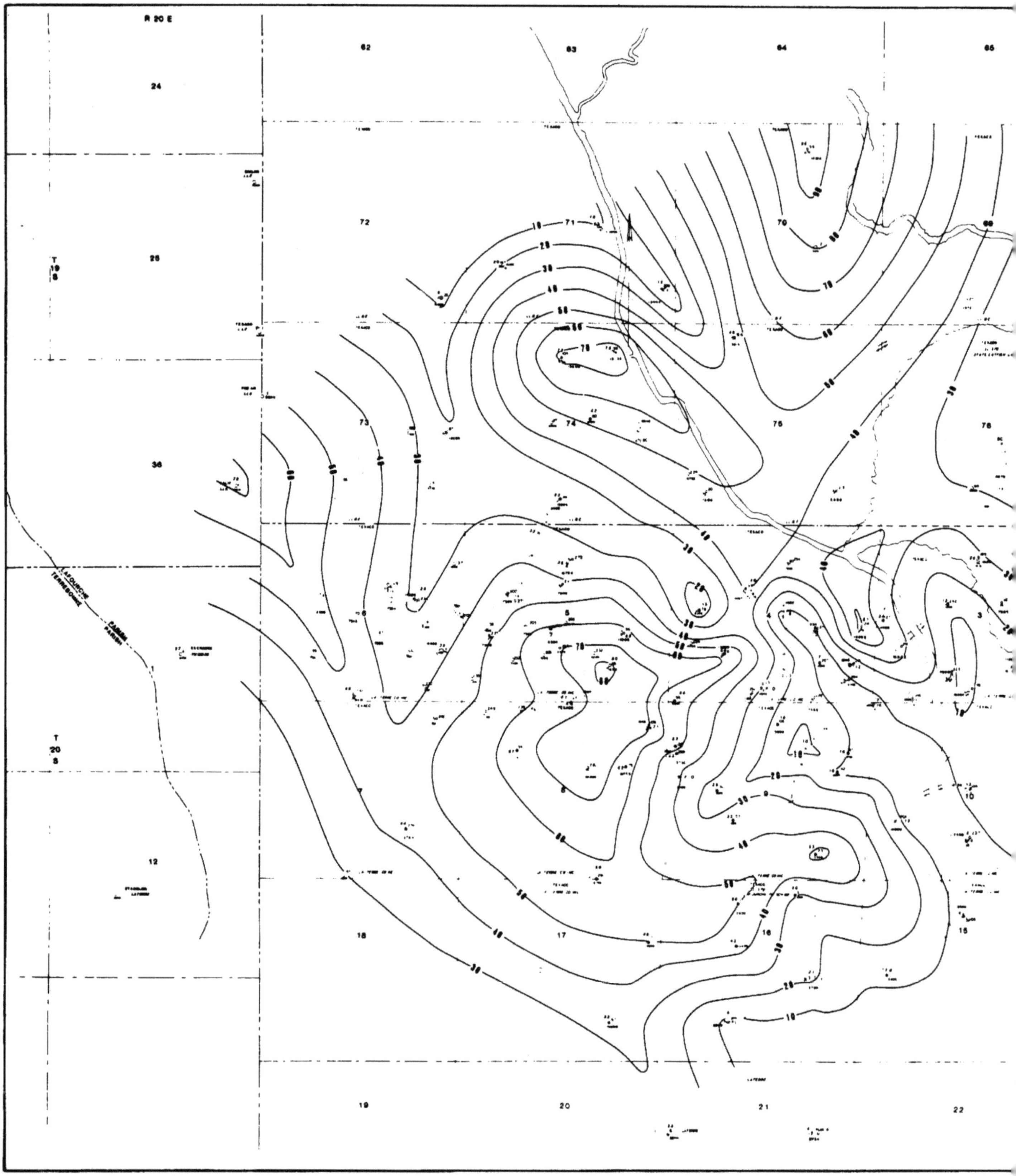

Figure 14-3 Portion of the net sand isochore map of the 10,500-ft Sand in Golden Meadow Field, Lafourche Parish, Louisiana, USA. (Published by permission of Texaco, USA.)

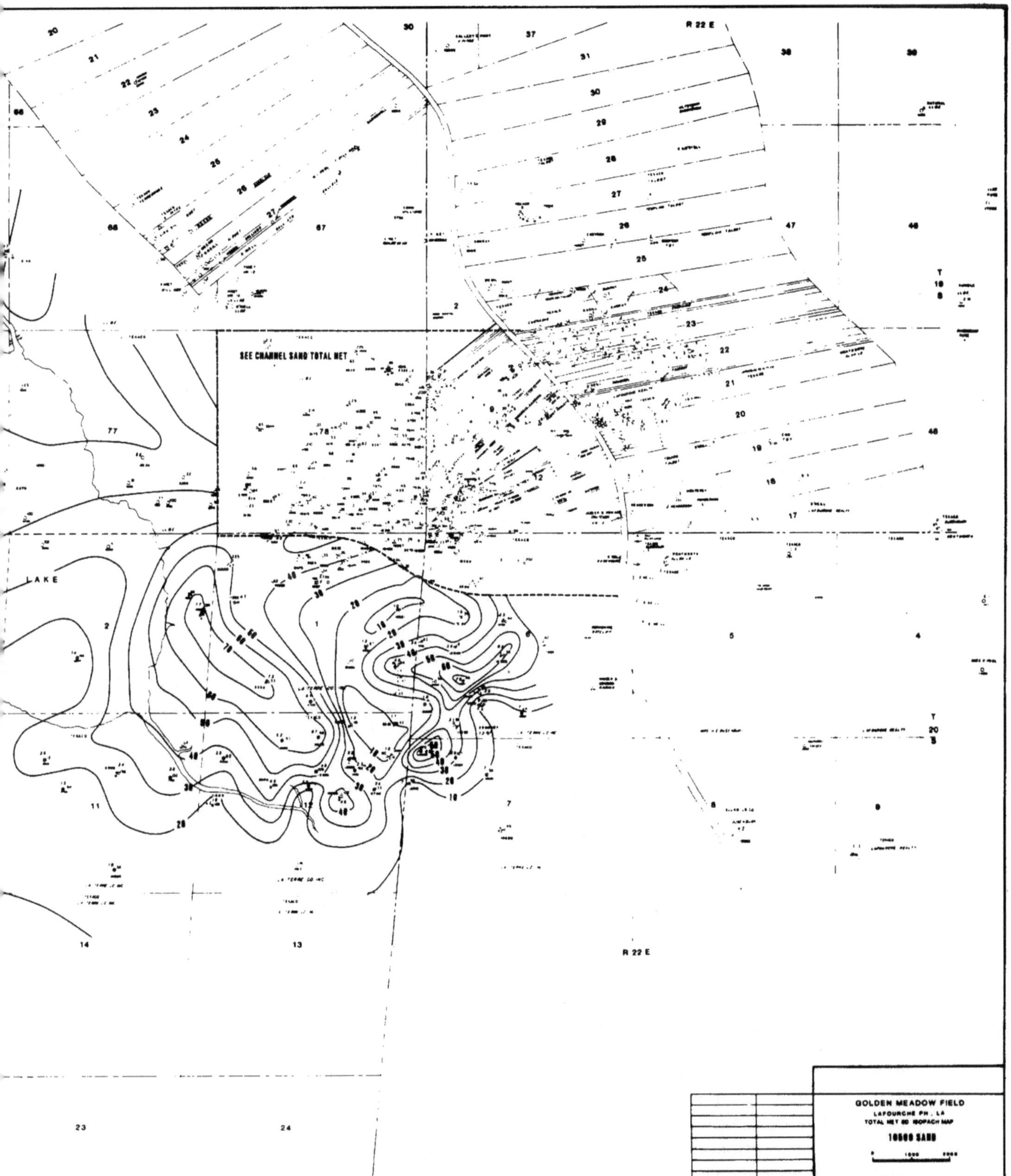
R 22 E
SEE CHANNEL SAND TOTAL NET
LAKE
T 20 S
R 22 E
GOLDEN MEADOW FIELD
LAFOURCHE PH., LA

very easy way for some people to visualize the size of an acre. One acre contains just about the same area as an American football field from goal line to goal line. A football field 300 ft long and 160 ft wide is equal to 48,000 sq ft, whereas one acre is equal to 43,560 sq ft. If we fill, with oil, a box that is one foot deep and the size of an American football field, the total volume of oil is just about equal to one acre-foot. In terms of barrels of oil, there are 7758 barrels (1233 cubic meters) of oil in one acre-foot. However, a reservoir volume of one acre-foot actually comprises solid material plus fluids that occupy the pore space. A calculation of oil-in-place in a reservoir must take into account pore space and water saturation as well as other factors. In creating a net pay map, we assume that within a net sand interval that contains hydro-carbons, all pore spaces are filled with hydrocarbons. The later calculation of hydrocarbons-in-place takes account of water saturation. In most cases, the structurally lower limit of a reservoir is a hydrocarbon/water contact, typically defined on the basis of parameters related to economic productivity, such as porosity and water saturation.

SAND–SHALE DISTRIBUTION

Most individual rock bodies do not consist exclusively of permeable rock; shale and other impermeable rock material are commonly distributed throughout the rock body as interbedded shale or impervious (tight) zones. The percentage and distribution of shale members or impervious zones varies widely within rock units. Net sand and net pay isochore maps are drawn on net effective sand (porous and permeable sand) only; therefore, shale and other nonreservoir-quality rock must be subtracted from the total interval to determine the net effective sand for isochore mapping.

In a given oil or gas well, the net effective sand to be used for isochore mapping is normally determined by computer analysis of digital log curves or detailed analysis of 5-in. well logs, supplemented by available core analysis. In Chapter 4, in the section Annotation and Documentation, we outline a method for annotating the percentage and distribution of net sand and impermeable layers that are present within a particular productive unit (Fig. 4-47). Once the net sand is determined for each well, a net sand map can be prepared for that sand. The aggregate of net sand for any particular well may contain water or hydrocarbons; net pay is that portion of the net sand that contains hydrocarbons in sufficient quantities to produce hydrocarbons at economic rates.

For a net sand or net pay map, the gross, or overall, interval to be mapped extends from the top of porosity to the base of porosity within the productive unit. Within that interval, the amounts of net sand and net pay vary with location. The net/gross ratio is the amount of net sand divided by the thickness of the gross interval, as determined from well logs and core analyses. The net/gross ratio within a productive unit can be mapped, based on log and core data, and used to estimate net sand and net pay at selected locations within the unit, as described later in this chapter.

BASIC CONSTRUCTION OF ISOCHORE MAPS

The procedure used in constructing an isochore map depends on whether the reservoir being mapped is a bottomwater or an edgewater reservoir. A bottomwater reservoir is a reservoir that is completely underlain by water (Fig. 14-4). An edgewater reservoir is one not completely underlain by water; some portion(s) of the productive unit is completely filled with hydrocarbons, and water underlies the remainder of the reservoir (Fig. 14-6).

It is important to be able to visualize a hydrocarbon reservoir in three dimensions. Your ability to understand the configuration of a reservoir can affect the location of development wells, completion practices, and planned production.

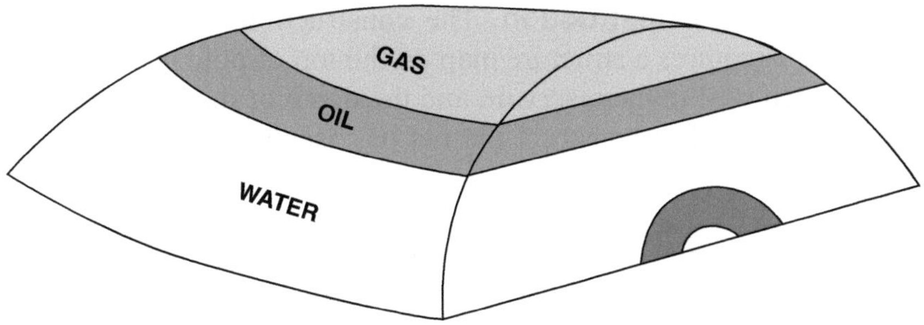

Figure 14-4 Three-dimensional model of a bottomwater reservoir.

Bottomwater Reservoir

Figure 14-4 is a 3D model of a bottomwater reservoir. This hydrocarbon accumulation, consisting of oil and gas, is trapped within an anticlinal structure. Notice how the oil and gas are segmented within the reservoir and completely underlain by water.

Figure 14-5 illustrates, in both map and cross-sectional views, a bottomwater oil reservoir. Looking at the cross section in the center of the figure, we can see the oil/water contact. The hydrocarbons are completely underlain by water; therefore, nowhere in the productive unit is there a full-thickness of oil. At any location within the reservoir, the amount of net pay equals the amount of net sand *above the oil/water contact*.

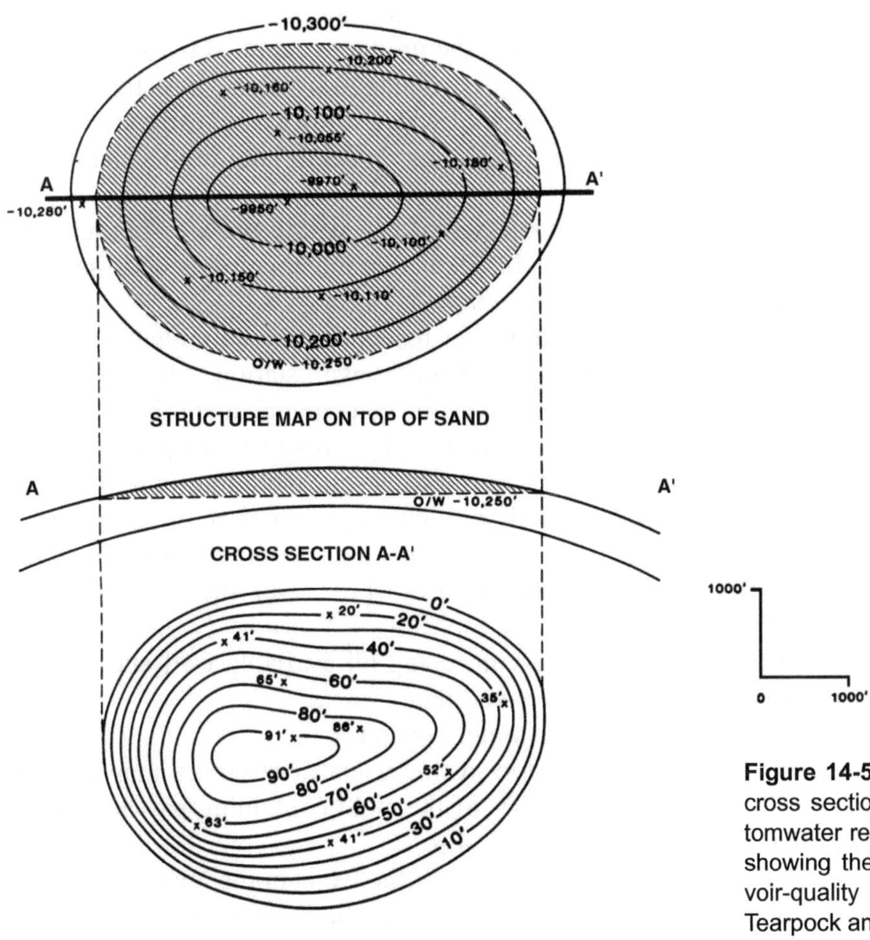

Figure 14-5 Structure map of the top of porosity, cross section, and net oil isochore map for a bottomwater reservoir. A net oil isochore map is a map showing the vertical aggregate thickness of reservoir-quality rock containing hydrocarbons. (From Tearpock and Harris 1987. Published by permission of Tenneco Oil Company.)

Net Pay Isochore Map Construction. The construction of a net pay isochore map for a bottomwater reservoir requires a structure map on the top of porosity for a given productive unit, net pay values for each well in the reservoir, and the depth of the hydrocarbon/water contact. The following procedure is used to construct the net pay isochore map for a bottomwater reservoir (Fig. 14-5).

1. Post the net pay values for each well on a basemap. If deviated wells are included, the net pay values must be corrected to true vertical thickness.
2. Overlay the isochore basemap on the top-of-porosity structure map for the reservoir, and draw the outer limit of the hydrocarbon-bearing reservoir. The outer limit may be any boundary, or combination of boundaries, such as an oil/water contact, fault, pinchout, or permeability barrier. Where this outer limit of the productive reservoir area is a hydrocarbon/water contact, it becomes the *zero contour line* on the net pay isochore map, as shown in Figure 14-5. In this case, the outer limit of the reservoir is an oil/water contact at a depth of −10,250 ft. However, if a boundary of part of a reservoir is a more or less vertical surface, such as a vertical fault or a vertical salt face, then the amount of net pay adjacent to this surface will exceed zero.
3. Contour the net pay isochore map, which is contained within the area outlined by the zero line on the basemap. Be sure to honor all posted net pay values. The net pay contours may generally conform to the structure contours. Net pay contours are commonly drawn as being equally spaced. However, due to variations in net sand thickness within the reservoir, the isochore contours need not be equally spaced. Figure 14-5 is a net pay map of a reservoir in which the net oil contours reflect the structure in a general way, yet the contours are proportionally spaced due to variation in net sand within the productive unit. If the well control is limited, additional points of contour control may be obtained by using a method called *walking wells* or by using a net/gross ratio map, each of which allows us to estimate net pay at chosen points in a reservoir. These methods are explained in detail later in this chapter.

Always indicate the maximum or minimum thicknesses of pay within the area bounded by a maximum or closed minimum net pay contour (Fig. 14-5), as this information is necessary for volumetric calculation after planimetering of the net pay map.

If you are constructing a net pay map for a bottomwater reservoir on a computer, you have two choices. You can follow the preceding steps 1, 2, and 3 on the workstation. Alternatively, if the net sand is evenly distributed vertically within the reservoir, you can calculate the bulk rock volume (BRV) of the reservoir and multiply the resulting volume by a net/gross map to construct a net pay map. The BRV is simply the volume of rock in the reservoir, including both reservoir and nonreservoir lithologies. To use this second method, you will require a structure map on the top of the reservoir, knowledge of the depth of the hydrocarbon/water contact, a net/gross ratio map, and net pay values for every well within the productive area of the reservoir.

1. Construct a structure map of the top of the reservoir, using all available well and seismic data and making sure that the seismic data ties the well data.
2. Construct a surface representing the hydrocarbon/water contact.
3. By computer, compute the volume between the top of the reservoir structure map and the hydrocarbon/water contact. This volume is the BRV of the field containing pay and corresponds to the BRV of the oil and gas slices in Figure 14-4.
4. Determine the net/gross ratio for the reservoir unit in all of the wells in the area of the accumulation. The wells used in the net/gross mapping should not be limited to the accumulation itself but should extend beyond it as far as necessary to delineate any important trends.

5. Construct a net/gross ratio map of the gross reservoir unit (see Fig. 14-14a). This map should honor the net/gross data from the wells and reflect your understanding of the geology of the reservoir unit.
6. Multiply the BRV by the net/gross map to determine the net pay volume. Post the net pay for every well in the field, and construct a net pay isochore map of the net pay volume, honoring all well control.

This procedure may seem complicated, but with the use of a computer, a net pay isochore map can be constructed quickly once a top-of-reservoir structure map is constructed and net/gross and net pay counts have been calculated for the relevant wells. Some software packages allow various steps to be programmed so that a macro can be constructed, essentially automating this process. It should be emphasized that this method is applicable only if the nonreservoir rock is uniformly, or relatively uniformly, distributed vertically within the gross reservoir. It should also be emphasized that the net/gross described here is net sand divided by gross reservoir interval, not net sand divided by gross sand.

Edgewater Reservoir

Figure 14-6 is a 3D model of an edgewater reservoir containing oil and gas. The hydrocarbons are trapped within an anticlinal structure. Compare the configuration of this reservoir with the bottomwater reservoir model in Figure 14-4. In this example of an edgewater reservoir, part of the productive unit is filled with gas, part is filled with oil, and wedge-shaped volumes of gas and oil are also present. It is obvious that this type of reservoir is more complex and more difficult to visualize in three dimensions.

Figure 14-7 illustrates in map and cross-sectional views an edgewater reservoir containing oil. The cross section shows an oil-filled part of the productive unit and a wedge of oil that is underlain by a wedge of water. Between the oil/water contact on the top of the unit (top of porosity) and the oil/water contact on the base of the unit (base of porosity), an oil wedge overlies a water wedge. Up-dip of the oil/water contact on the base of porosity, the reservoir is full of oil from the top to base of porosity. That area is the full-thickness area of the reservoir. In the full-thickness area, net pay equals net sand. In the wedge, net pay equals the amount of net sand *above the water level*.

Edgewater reservoirs are obviously more complex than bottomwater reservoirs. An edgewater reservoir can be extremely complex if it contains both oil and gas and is cut by one or more faults. When mapping a reservoir cut by faults, consideration may be given to mapping one or more fault wedges in addition to mapping the hydrocarbon wedges above water. The result may be a complex isochore map. Fault wedges are discussed in detail later in this chapter.

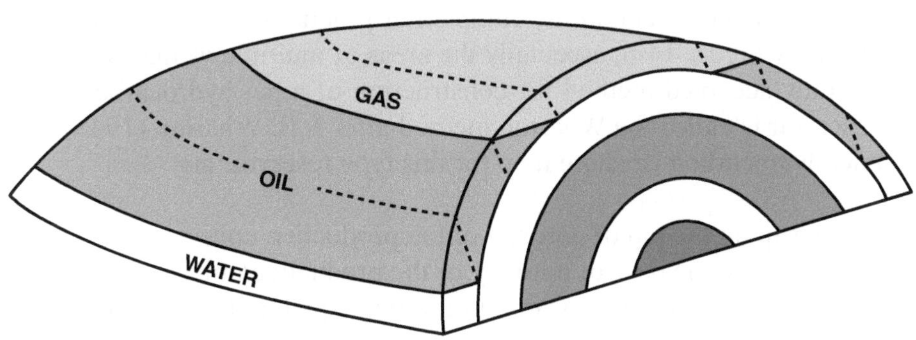

Figure 14-6 Three-dimensional model of an edgewater reservoir.

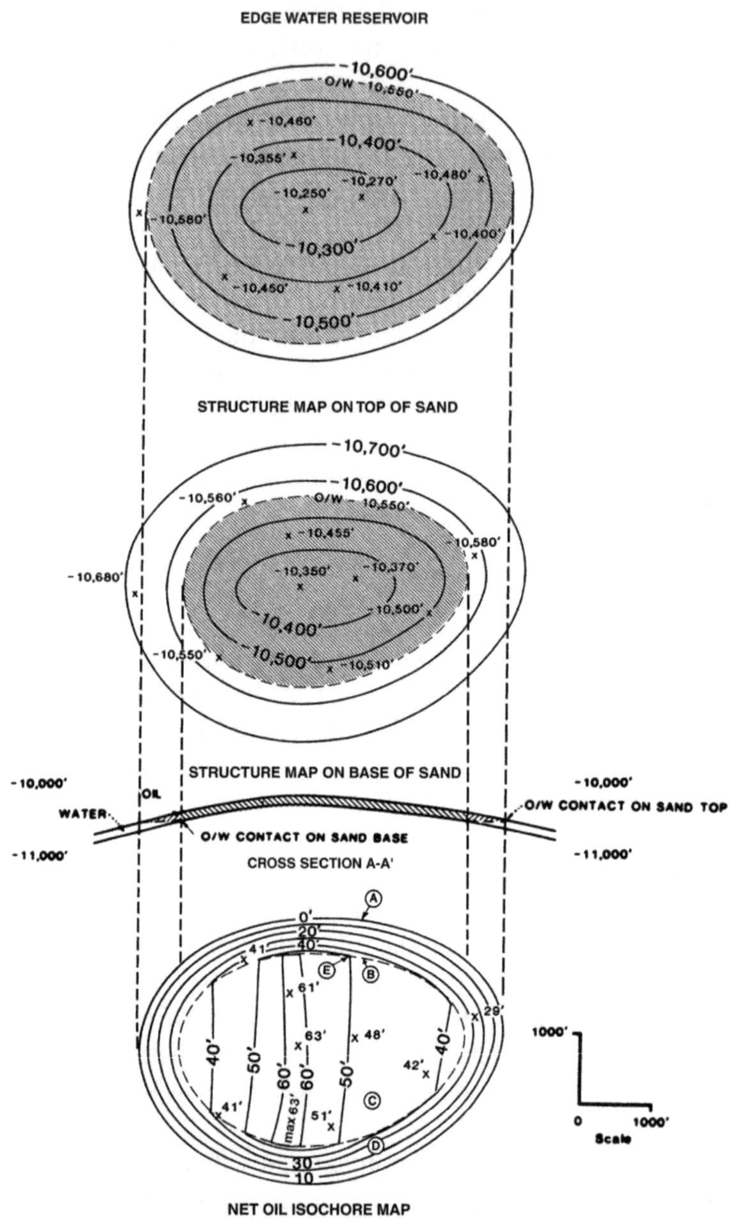

Figure 14-7 Structure maps, cross section, and net oil isochore map for an edgewater reservoir.

An understanding of the reservoir type and configuration is very important in such decisions as the location of development wells, completion practices, and production plans. Take a few minutes and review Figure 14-6, especially the areas of multiple wedge zones.

The generally accepted method for construction of a net hydrocarbon isochore map for an edgewater reservoir is called the Wharton method after J. B. Wharton (1948). The data needed to construct a net hydrocarbon isochore map for this type reservoir are

1. a structure map on the top of porosity of the productive unit;
2. a structure map on the base of porosity of the productive unit;
3. a net sand isochore map (with a contour interval equal to that to be used in the net pay map);
4. net pay values for all available wells; and
5. depth or elevation of all fluid contacts (oil/water, gas/water, oil/gas).

Net Pay Isochore Map Construction for a Single-Phase Reservoir. We first outline the procedure for construction of a net hydrocarbon isochore map for an edgewater reservoir containing a single type of hydrocarbon. For this example, we consider the rock type to be sandstone and the hydrocarbon to be oil. The procedure is illustrated in Figure 14-8.

1. Start with a basemap with all the well control spotted.
2. Place the basemap over the structure map on the top of the unit (top of porosity) (Fig. 14-8a), and trace the outer limit of the productive reservoir. In this example, the limit is the oil/water contact on the top of the unit, and it becomes the *zero line* on the net oil isochore map, as in the previous example. The zero line is shown in Figure 14-8b. From this point forward, we refer to the base map as the net oil isochore map.
3. Place the net oil isochore map over the structure map on the base of the unit (base of porosity) (Fig. 14-8c), and trace the oil/water contact on the isochore map using a *dashed line*. This dashed line represents the inner limit of water (ILW) for the reservoir. Within the area inside this dashed line, oil fills the productive unit (Fig. 14-8d), and this area is referred to as the *full-thickness area*. The intersection of the oil/water contact with the top and base of the unit outline the wedge zone.
4. Post net pay values for all wells within the reservoir, corrected to true vertical thickness, on the net oil isochore map.
5. In the full-thickness area, the entire net sand is filled with oil, so the net oil in this area equals the amount of net sand, as interpreted on the net sand isochore map. Notice that the net sand contours are based on *all* wells within and outside the reservoir. Therefore, to contour the full-thickness area, place the net oil isochore map over the net sand isochore map as shown in Figure 14-8e, and trace the contours within the dashed-line area onto the net oil isochore map. Label the maximum thickness of oil (63 ft) within the area enclosed by the maximum net pay contour (60 ft). The full-thickness area of the net oil isochore map is now finished, as illustrated in Figure 14-8f.
6. The next step is to contour the oil wedge. The wedge zone is the area between the oil/water contact on the top of the unit and the oil/water contact on the base of the unit, as shown in Figure 14-8d and f. The oil wedge overlies a water wedge (see cross section in Fig. 14-10). All well data in the wedge must be honored. In the full-thickness area, net sand distribution controls the net pay contours. However, in the oil wedge, the structural attitude of the productive unit, and the distribution of impermeable rock within the unit influence the position of net pay contours. The contours conform in general to the structure contours, but they are not necessarily equally spaced (variations in contour spacing are discussed below). As the first step in contouring the wedge, draw the net pay contours from the full-thickness area into the wedge. Notice in Figure 14-8g that the full-thickness contour lines turn abruptly at the ILW, in the direction of increasing sand thickness (in the direction of the next net sand contour of higher value). They connect through the wedge with contours of the same value elsewhere in the full-thickness area. Lastly, draw the net pay contours in the oil wedge that do not correspond to contours within the full-thickness area. Honor the well data within the wedge, and more or less equally space contours where data is lacking (Fig. 14-8h). If the dip of the productive unit is more irregular than in this example, then that variation in structure may be considered in contouring the wedge.

The completed net oil isochore map is shown in Figure 14-8h. This figure highlights five important points of the net pay isochore map construction. We recommend that you leave the dashed ILW on the net pay isochore map. It allows others to review and verify that you have used the water contact on the base of porosity correctly in creating the net pay map.

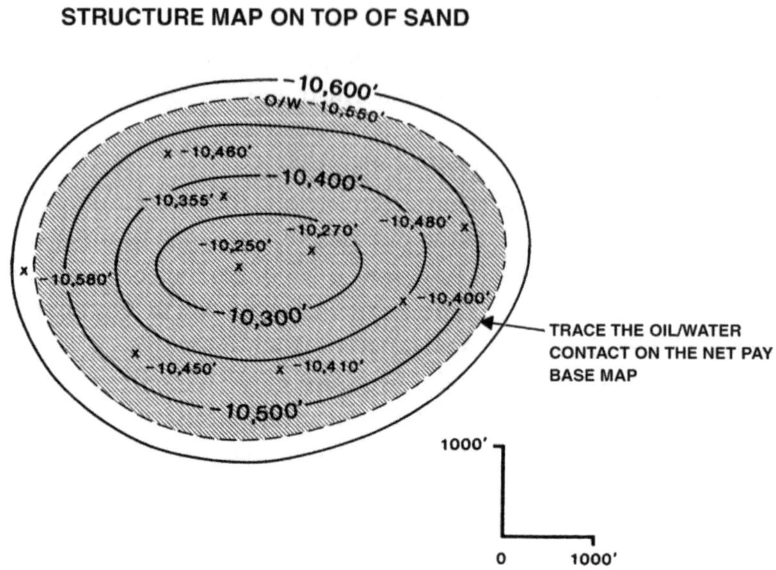

(a)

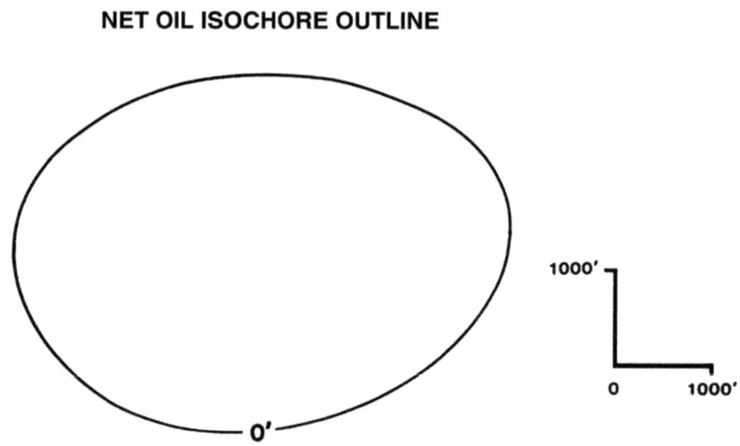

(b)

Figure 14-8 (a) Overlay a net oil basemap on the structure map of the top of sand (top of porosity). The oil/water contact is traced on the overlay. (b) The oil/water contact becomes the zero contour line on the net oil isochore map. (c) Overlay the net oil isochore map on the structure map of the base of sand (base of porosity), and trace the oil/water contact as a dashed line. (d) Isochore basemap delineating the two major areas comprising the net oil isochore map: (1) oil wedge from the zero line to the inner limit of water (ILW; oil/water contact on the base of the sand), and (2) the area of full hydrocarbon thickness. (e) To contour the full-thickness area, superimpose the net oil isochore map on a net sand isochore map and trace the net sand contours inside the ILW (dashed line). (f) Net oil isochore map with contours drawn for the oil-filled area. (g) All full-thickness area contours that intersect the ILW must connect through the wedge with contours of equal value (see text for procedure). (h) Completed net oil isochore map with important points of isochore construction listed. (From Tearpock and Harris 1987. Published by permission of Tenneco Oil Company.)

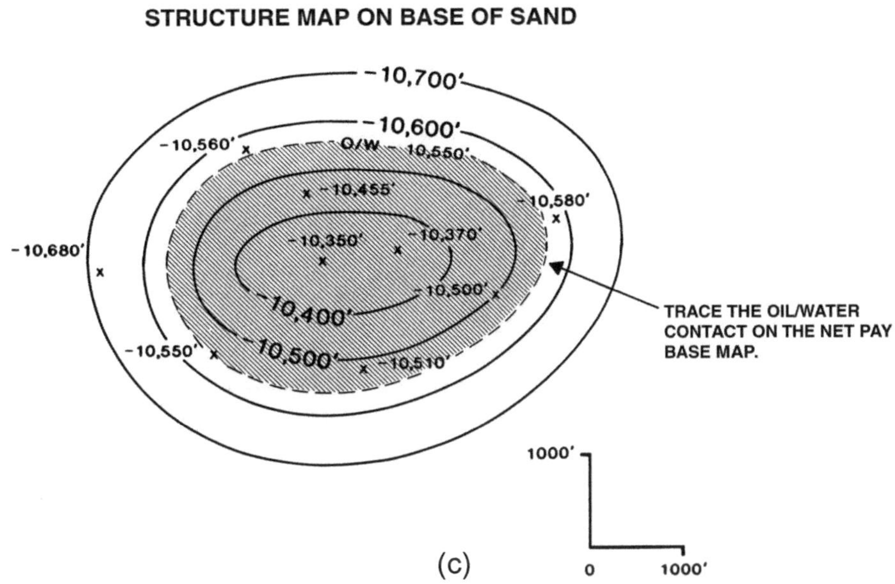

(c)

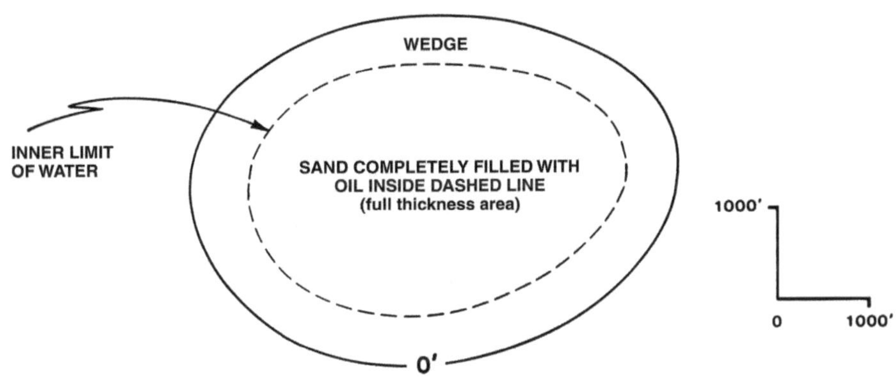

(d)

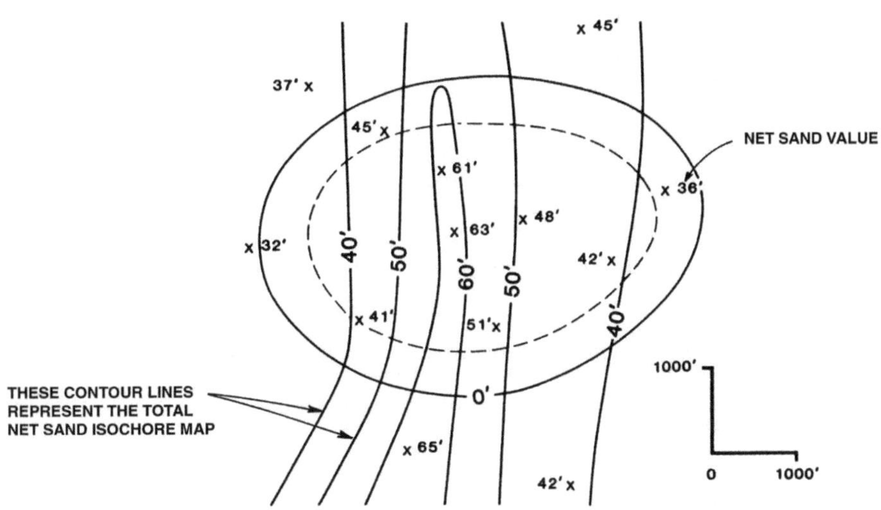

(e)

Figure 14-8 *(continued)*

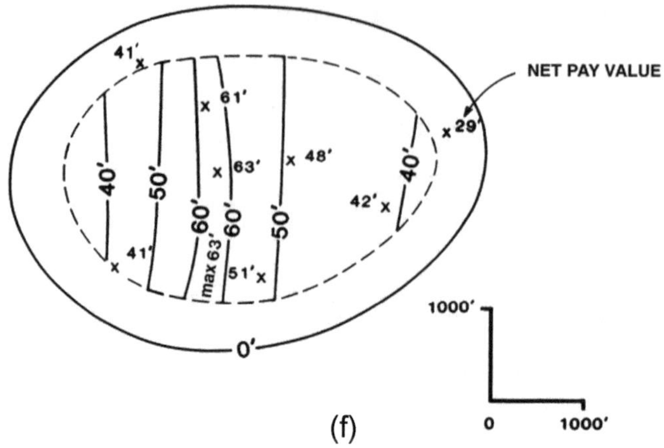

(f)

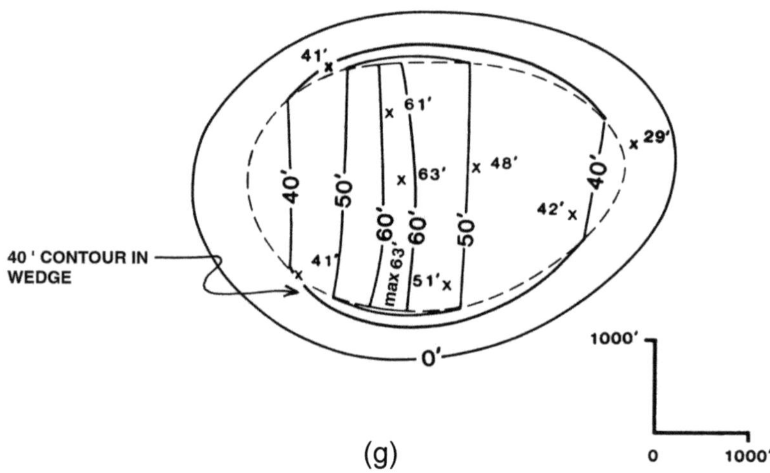

(g)

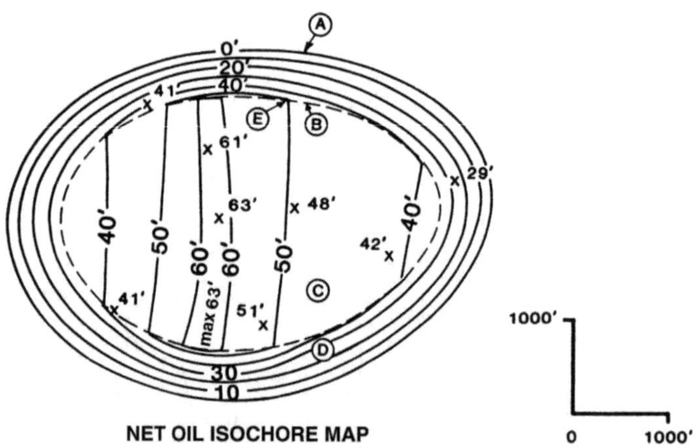

A. "0" LINE, OR OUTER LIMIT OF HYDROCARBON / WATER CONTACT FROM STRUCTURE MAP ON TOP OF SAND.

B. INNER LIMIT OF WATER IS HYDROCARBON / WATER CONTACT ON BASE OF SAND.

C. AREA UPDIP OF "B" IS TOTALLY FILLED WITH HYDROCARBONS.

D. HYDROCARBON WEDGE.

E. ABRUPT ANGLE TOWARD NEXT NUMERICALLY LARGER NET SAND CONTOUR.

(h)

Figure 14-8 *(continued)*

The method of connecting the full-thickness contours to those in the wedge zone is extremely important and deserves special attention. Why can we not extend the net pay contours in the full-thickness area on trend, past the ILW and into the wedge? Look at Figure 14-9, which is similar to Figure 14-8g. Let's discuss the construction of the easternmost 50-ft contour in the figure. Why must this contour line sharply change direction at the ILW on the net oil isochore map rather than continue straight into the wedge zone? Figure 14-10 is a diagrammatic cross section along the 50-ft net sand contour. The net sand is exactly 50 ft everywhere along the cross section. The net sand is exactly 50 ft everywhere along the cross section. In the portion of the reservoir that is up-dip to the oil/water contact on the base of porosity (ILW), oil fills the entire 50 ft of net sand. However, a point one foot down-dip of the ILW is within the wedge zone, where the sand contains both oil and water. Therefore, anywhere outside the ILW, both oil and water are present and, since the total net sand is still 50 ft thick, there must be less than 50 ft of oil. The 50-ft net pay contour, therefore, cannot continue along the 50-ft net sand contour down-dip of the ILW.

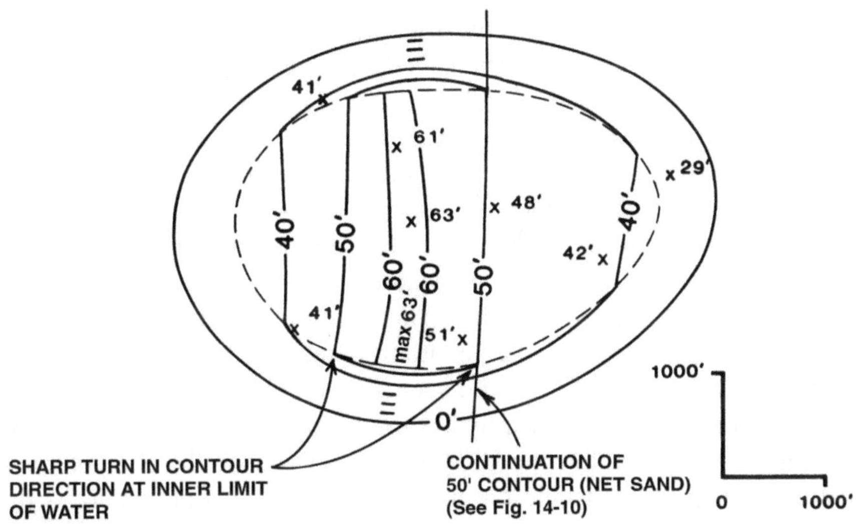

Figure 14-9 Full-thickness net pay contours make an abrupt turn at the inner limit of water toward the next net sand contour of higher value. (Modified from Tearpock and Harris 1987. Published by permission of Tenneco Oil Company.)

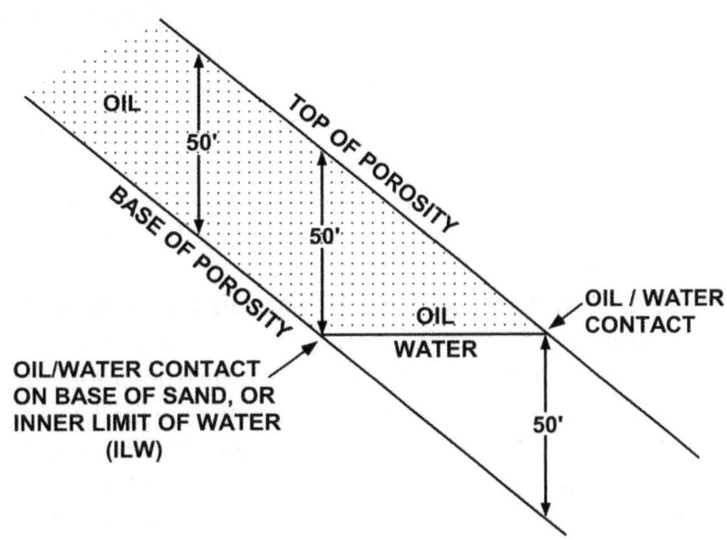

Figure 14-10 Cross section along the 50-ft net sand contour line in Figure 14-9.

Where must the 50-ft contour from the full-thickness area be drawn in the wedge zone? It must be drawn through an area of 50 ft of net oil in the wedge zone. This area exists only where the net sand is *greater* than 50 ft. In Figure 14-8e, notice that the net sand increases in thickness west of the 50-ft contour to a maximum of 63 ft. Therefore, the 50-ft contour, at its intersection with the ILW, must turn sharply toward the area of thicker sand. Since contour lines must close, the contour connects to the west to connect with the other 50-ft contour in the full-thickness area (Fig. 14-8g).

This procedure must be undertaken for all contour lines contained within the full-thickness area of the net pay isochore map. The correct application of this technique is most important. If the full-thickness area net pay contours are carried incorrectly into the wedge zone, the volume of hydrocarbons determined for the reservoir will be *overestimated*.

Figure 14-11a and b present a summary of the foregoing method for preparing a net pay isochore map for an edgewater reservoir containing one type of hydrocarbon.

Comprehension of, and commitment to, the foregoing procedure will help you avoid the most common pitfall in net pay mapping, which is the mapping of more net pay than net sand that is filled with hydrocarbons. In the full-thickness area, a net pay contour corresponds exactly to a net sand contour. In the hydrocarbon wedge, a net pay contour corresponds to that same amount of net sand *above the water level*. Therefore, in the wedge, a net pay contour of a given value must be within an area where the *total* net sand is of greater value. *Always* overlay and compare the net pay contour map to the net sand contour map in order to check that the positions of the net pay contours are reasonable.

As with a bottomwater reservoir, you have two options if constructing a net pay isochore map of an edgewater reservoir on a computer workstation. First, you can follow the foregoing procedure on the workstation. The complication with this approach is that few if any software packages understand the implication of an ILW line when contouring net pay. In order to force the software package to duplicate the Wharton method, it may be necessary to add a significant number of digital control points to force the software to turn abruptly toward thicker net sand contours at the ILW.

A second option is to calculate the BRV of the reservoir on the computer and multiply this volume by a net/gross map to determine the net pay volume. This procedure is directly analogous to using the computer to construct a net pay map on a bottomwater reservoir except that a base of reservoir structure map is required in addition to a top of structure map and knowledge of the depth of the hydrocarbon/water contact. In this case, the base of reservoir structure map is combined with the surface representing the depth of the hydrocarbon/water contact. The volume between the top of structure and the combined base of structure/hydrocarbon/water contact is the BRV of the reservoir. Multiplying this volume by the net/gross map yields the net pay volume, which may then be contoured. As with a bottomwater reservoir, make sure to plot the net pay in each well within the accumulation to help in contouring the net pay volume. Make sure that all well control within the accumulation is honored because failure to honor well control is a red flag for anyone reviewing a net pay isochore map. It is still best practice to plot the ILW on the net pay isochore map. It is also best practice to construct a net sand map to compare with the net pay isochore contours in the full-thickness area and wedge zone.

Net Pay Isochore Map Construction for a Reservoir Containing Oil and Gas. Two methods may be used to estimate the volumes of oil and gas in a reservoir containing both types of hydrocarbons. The simplest and quickest method is to construct a total net hydrocarbon isochore map and a net gas isochore map, calculate the volumes of each, and subtract the gas volume from the total hydrocarbon volume to determine the oil volume. This method is appropriate when only one lease owner is involved or when only the total estimated volumes of oil and gas are required (there being no interest in the actual distribution of oil and gas within the reservoir). Where a reservoir underlies two or more separate ownerships, it is very important to know the estimated volumes of gas and oil under each lease. In this case, net gas and net oil isochore maps must be constructed, preferably using the procedure outlined in this section.

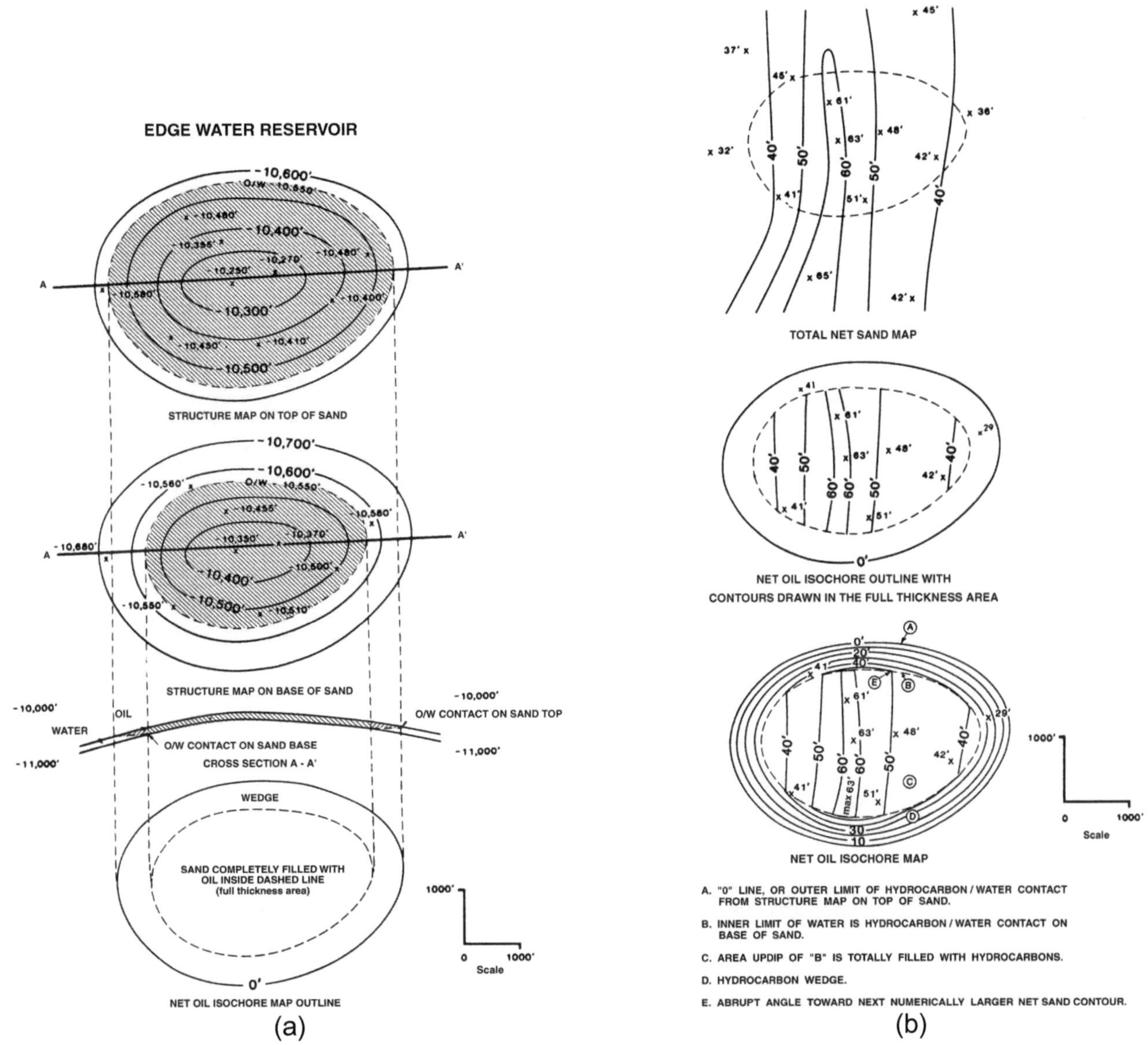

Figure 14-11 (a)–(b) Summary of method for constructing a net hydrocarbon isochore map for an edgewater reservoir containing one hydrocarbon (oil or gas). (Modified from Tearpock and Harris 1987. Published by permission of Tenneco Oil Company.)

Construct the net gas isochore map first. Draw the basic maps used in the Wharton method, which are the structure map on top of porosity, structure map on base of porosity, and net sand isochore map. Using these maps, construct the net gas isochore map as shown in Figure 14-12a and b. The net gas isochore map is constructed using the same procedure explained in the previous section on edgewater reservoirs containing one type of hydrocarbon. The only difference in this case is that the gas/oil contact defines the down-dip, outer limit of the gas reservoir, whereas in the previous cases the limit was determined by a hydrocarbon/water contact.

The last map to be constructed is the net oil isochore map. This map differs from the previous maps in that it has two wedge zones, an inner wedge zone (gas/oil) and an outer wedge zone (oil/water), as shown on the cross section in Figure 14-12a. The outer oil wedge and any full-thickness areas are constructed using the Wharton method, as previously discussed; the inner oil wedge requires additional steps.

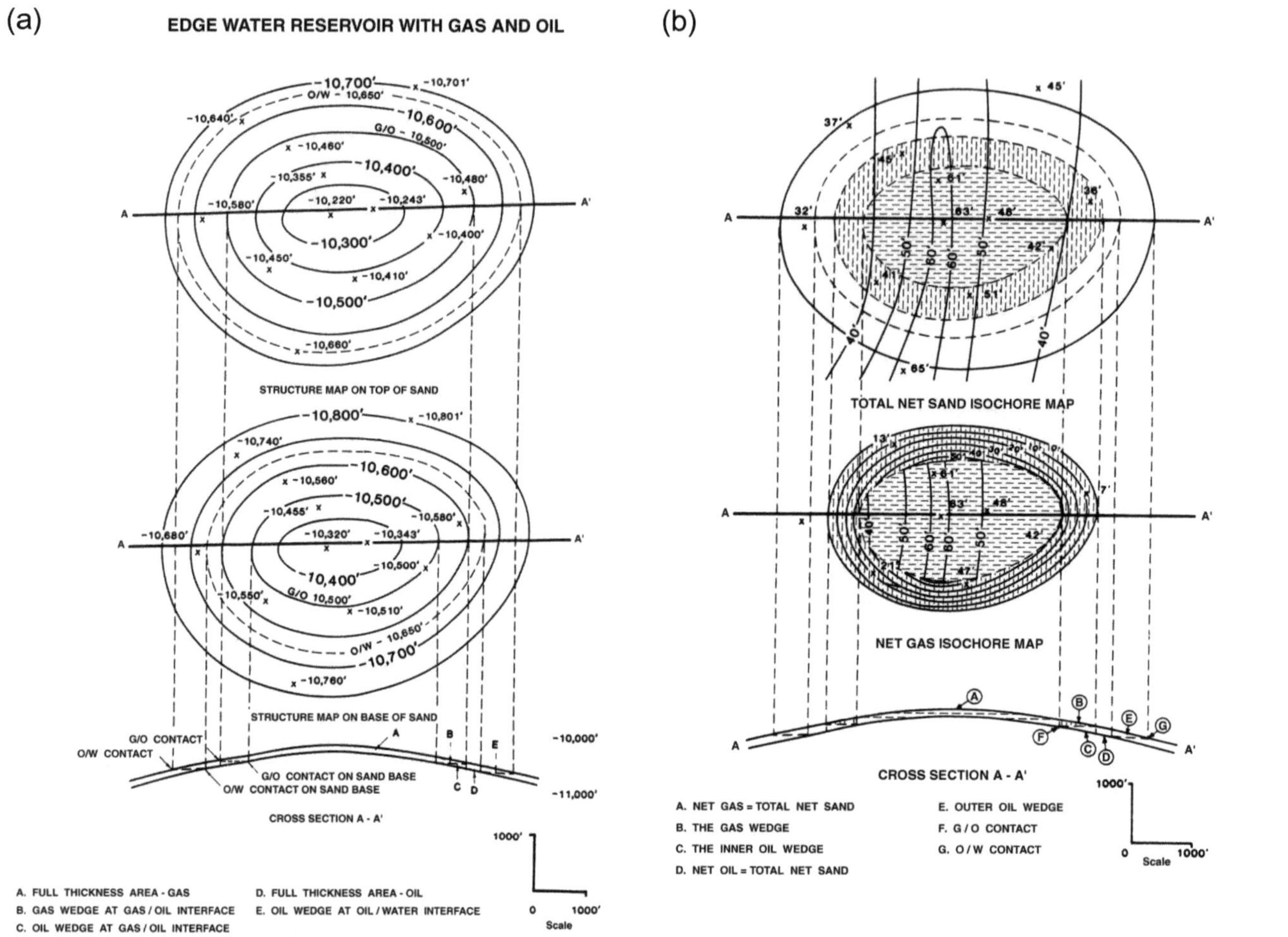

Figure 14-12 (a)–(b) Summary of procedure to construct a net gas isochore map for a reservoir containing both oil and gas. (Modified from Tearpock and Harris 1987. Published by permission of Tenneco Oil Company.)

Figure 14-13a shows a partitioned map (without wells) for the oil reservoir, with an inner wedge and an outer wedge of oil, and an area of full-thickness in between. The fluid contacts on the structure maps are used to determine the zero net pay limits and dashed wedge limits by tracing them onto the net oil map.

First, contour the area containing a full-thickness of oil. Post the net oil values next to each well and overlay the isochore basemap onto the net sand isochore map, as shown in Figure 14-13b. Trace the net sand contours on the net oil map only within the full-thickness area of oil, as shown in Figure 14-13c. Indicate the maximum thickness of oil in each of the areas enclosed by the 60-ft net pay contours. This portion of the isochore map is now complete.

As with edgewater reservoirs containing a single phase, edgewater reservoirs that consist of a gas cap and an oil rim can easily be isochored on a computer workstation. Using the procedures outlined previously for bottomwater and edgewater reservoirs, as appropriate, determine the bulk rock volume of the gas reservoir and the bulk rock volume of the oil reservoir. Multiply each volume by the net/gross map to determine the net pay volume for both the gas reservoir and the oil reservoir. Alternatively, you can determine the net pay volume of the gas reservoir and the net pay volume of the entire hydrocarbon reservoir and determine the net pay volume of the oil reservoir by difference.

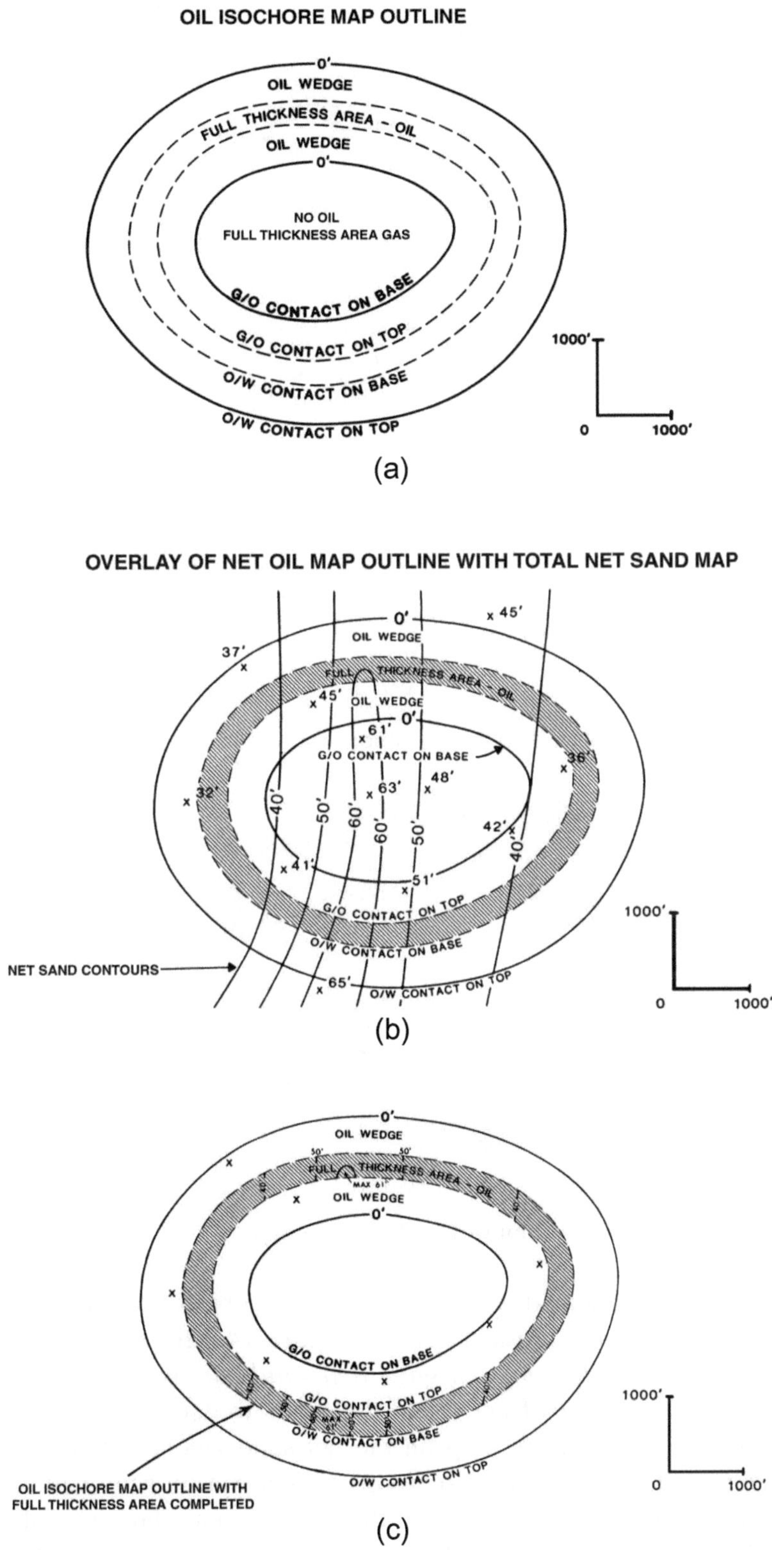

Figure 14-13 (a) Outline of oil isochore map showing the inner and outer wedge zones and the full-thickness area. (b) Overlay of net oil isochore map on the net sand isochore map. The contours in the area of full oil thickness are equal to the net sand contours. (c) Full-thickness area is contoured. (d) Overlay of net gas and net sand isochore maps used to aid in the construction of the inner oil wedge contours. (e) Completed net oil isochore map. (From Tearpock and Harris 1987. Published by permission of Tenneco Oil Company.)

OVERLAY OF NET GAS MAP AND TOTAL NET SAND MAP

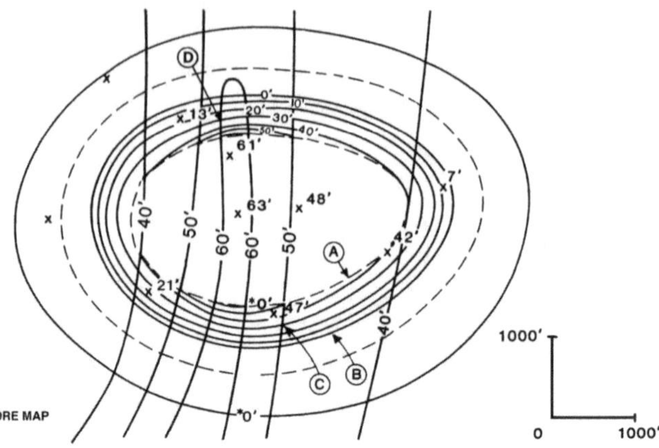

A. - GAS/OIL CONTACT ON BASE OF SAND - INNER LIMIT OF OIL

B. - GAS/OIL CONTACT ON SAND TOP - OUTER LIMIT OF GAS

C. - 30' GAS CONTOUR CROSSES 50' TOTAL NET SAND CONTOUR. THIS POINT IS IN THE AREA THAT IS TOTALLY FILLED WITH HYDROCARBONS. IF TOTAL HYDROCARBON IS 50' AND ONLY 30' IS GAS, THEN THE OTHER 20' IS OIL.

D. - 30' GAS CONTOUR CROSSES 60' TOTAL NET SAND CONTOUR. NET OIL IS 30' IN OIL WEDGE.

(d)

COMPLETED NET OIL ISOCHORE MAP

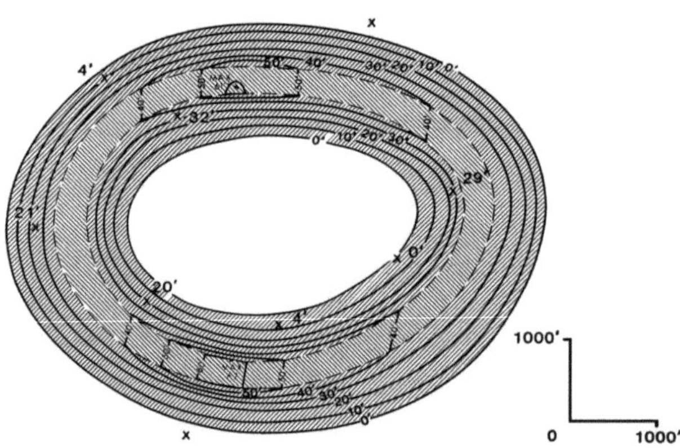

(e)

Figure 14-13 (*continued*)

We now contour the inner oil wedge on the map, to be followed by contouring of the outer oil wedge. By referring to the cross section in Figure 14-12b, we see that within the inner oil wedge, *the sum of net oil-filled and net gas-filled sand equals the total net sand*. Therefore, in order to determine the amount of net oil in the inner wedge zone, the following procedure is used. Overlay the net gas isochore map on the net sand isochore map, and note each location where the contours on the two separate maps cross. The net oil sand value at each contour intersection is equal to the difference in values of the two contours. For example, at point C in Figure 14-13d, the 30-ft contour line on the net gas isochore map crosses the 50-ft contour line on the net sand isochore map. By subtracting the 30 ft of net gas from the 50 ft of net sand, a value of 20 ft of net oil is obtained for this point. Take a minute and review the data at point D. As indicated, a known value is established wherever a contour line on the net gas isochore crosses a contour line on the net sand isochore. The net oil sand value at each intersection is the difference in values of the two contours. Figure 14-13d shows 48 points of control, plus data from four wells, to aid in contouring the inner wedge zone of the oil isochore map. Overlay the net oil map on the net gas and net sand maps, mark selected points with the corresponding value of oil, and then contour the inner oil wedge. The net pay contours at the edge of the full-thickness area turn abruptly towards the thicker net sand values (Fig. 14-13e), just as described in the contouring of a single-phase reservoir.

When contouring the inner oil wedge, the net oil isochore map should be overlain on both the net gas and net sand isochore maps. This allows mapping the inner oil wedge with the constraint, and thus the assurance, that the sum of the net gas and net oil does not exceed the total net sand. This is one of the most complex, and therefore most difficult, areas to construct within an isochore map.

Finally, contour the outer wedge precisely as described in the previous section on contouring the wedge in a single-phase reservoir (Fig. 14-8h). We now have constructed the inner and outer wedges and the full-thickness area for the oil isochore. The completed net oil isochore map is shown in Figure 14-13e.

METHODS OF CONTOURING THE HYDROCARBON WEDGE

Limited Well Control and Evenly Distributed Impermeable Rock

With limited well control in the wedge zone of a reservoir and a fairly even distribution of impermeable rock within the productive unit, the most common method for contouring the wedge is to combine proportional and equal spacing of the net pay contours, while honoring the available well control. This is how the outer oil wedge and gas wedge in the last example were constructed. Typically, you may equally space contours between the outermost wells and the zero net pay contour. Within the area of well control, proportionally space the contours according to the net pay values and the well locations. Contouring in this manner assumes that the vertical distribution of net sand is uniform in the reservoir. On the other hand, the horizontal distribution of net sand may not be uniform and, if so, the net pay contours will reflect that stratigraphy only to the extent of the density of control points. The configuration of the contours within a wedge is primarily controlled by the structural attitude of the productive unit and the distribution of net sand and impermeable rock within the overall unit. If the distribution of nonreservoir-quality rock (e.g., shale) is fairly uniform both vertically and horizontally, the primary influence on the contours in the wedge is the structural attitude of the unit. In such a case, the contours should more or less conform to the structure contours and be more or less proportionally and equally spaced.

If the well control in the wedge is limited and sand distribution varies significantly horizontally, we can use the ratio of net sand to gross interval (net/gross ratio) within the productive unit to approximate net pay within the wedge. This allows us to more accurately contour in the wedge than if we simply equally and proportionally space the contours. Using the wells within and outside the reservoir, first calculate the net/gross ratio in each of the wells. Then map the net/gross ratio in the area of the reservoir, using contours with a typical contour interval of 5% or 10% (0.05 or 0.10). When we map the net/gross ratio, we make the assumption that the distribution of net sand is uniform vertically. This means that, at any point in the reservoir, the net/gross ratio is the same *above the water level* as it is for the entire thickness. Once the map is complete, estimate the net pay at selected points in the wedge. At a given point, calculate the gross interval thickness above the water level as the difference in depth between the top of porosity and the water level. Multiply this calculated gross interval by the net/gross ratio to estimate the net sand above the water level, which is equal to net pay at the selected point.

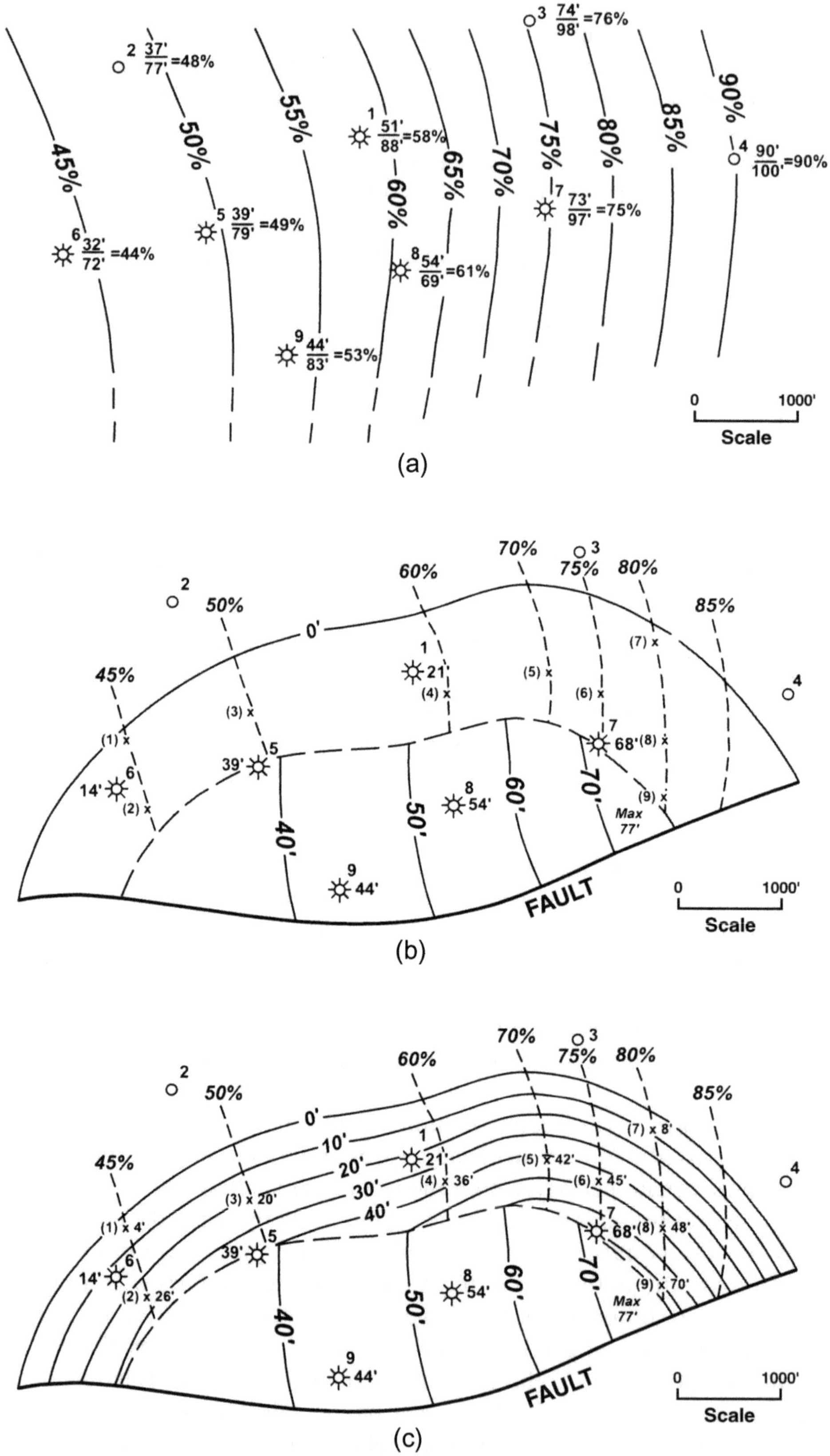

Figure 14-14 (a) Net/gross ratio map based on well control. (b) Overlay of partially completed net gas map on the net/gross ratio map. (c) Completed net gas isochore map. (Published by permission of D. Tearpock.)

This process is done automatically if you use the workstation to compute the BRV of the reservoir and multiply it by a net/gross map. Lateral changes in gross thicknesses are captured by the bulk rock volume. Lateral changes in the net/gross ratio are captured in the net/gross map.

We use the maps in Figure 14-14 as an example of the technique, applied to an edgewater gas reservoir with a gas/water contact at −8350 ft. The net/gross mapping technique begins after construction of the outline of the net pay map and contouring of net pay within the full-thickness area (Fig. 14-14b). As the first step in constructing a net/gross ratio map, calculate the net/gross ratio for wells *within* and *outside* the reservoir. Figure 14-14a shows the net/gross ratio calculation beside each well in the area of study. Contour the values using a contour interval of 5%, or 0.05. Then estimate net pay at selected points within the wedge. Figure 14-14b is a partially completed net pay isochore map of the reservoir, shown as an overlay on some contours in the net/gross map. For efficiency, select estimated net pay points to lie directly on the net/gross contours, such as points (1) through (9). As an example, use point (3) on the 50% contour. Calculate the gross interval thickness *above the water level* at point (3) by taking the difference in depth between the gas/water contact at −8350 ft and the top of porosity at point (3), which is −8310 ft on the structure map (not shown). The gross interval thus equals 40 ft. Then estimate the net pay by using the following calculation.

$$
\begin{aligned}
\text{gross interval above the water level} &= 40\text{ ft}\\
\text{net/gross} &= 0.50\\
\text{net} &= (0.50)(40\text{ ft})\\
&= 20\text{ ft net sand above the water level}\\
\text{net pay} &= 20\text{ ft}
\end{aligned}
$$

Using the same procedure, create as many points of estimated pay as necessary for adequate control. Figure 14-14c shows the completed net pay map for the reservoir. If using the BRV multiplied by net/gross map method to calculate net pay using a workstation, this process will be utilized automatically.

The net/gross mapping method provides added control for contouring net pay in the hydrocarbon wedge. The method is reasonably accurate if net sand is more or less uniform in vertical distribution. However, the method is significantly less accurate if the reservoir includes intervals of impermeable rock that vary considerably in thickness. For a reservoir with this erratic vertical distribution of net sand and impervious rock, we recommend estimating net pay by using the walking wells method, described in the next section.

Walking Wells—Unevenly Distributed Impermeable Rock

A technique called **walking wells** can be used to improve the accuracy of the contouring in the hydrocarbon wedge. At times, we may wish to better define the distribution of net pay sand within the wedge instead of just using the proportionally spaced method. For example, a dispute exists as to the equity of various leases overlying a reservoir, and the reservoir is being mapped for equity determination between various companies. Furthermore, the distribution of net sand versus impermeable rock varies within the reservoir, such as both thick and thin shale intervals occurring within the productive unit. In such cases, a more detailed estimate of the reserves in the wedge zone may be required.

Our purpose in walking a well is to estimate the net pay at selected points within the wedge. To accomplish this, we assume that the stratigraphy seen in a given well log is representative of the stratigraphy along a designated net sand contour line extending through the wedge. In our imagination, we will move, or walk, the well along this line and estimate net pay at selected points, as described below.

Let's say we wish to construct a net gas isochore map, with a 10-ft contour interval, for a sandstone reservoir with limited well control. We want additional control in the gas wedge, so we decide to walk a well through the wedge and estimate the amount of net gas at selected points. Any well to be walked can be located in the reservoir itself, or even down-dip of the hydrocarbon/water contact and thus outside the reservoir. *The key point in walking a well is to choose a well that can be walked parallel to the nearest contour line on the net sand isochore map.* This is a critical point because when walking a well through the wedge, the assumption is made that the amount and distribution of net sand, as seen in the well log, are the same within the reservoir in a direction parallel to the net sand contours. For example, if a well has 50 ft of net sand, it will fall on the 50-ft total net sand contour line. Construction of this 50-ft contour line assumes that exactly 50 ft of net sand exists all along this contour line, not just at the well location (Fig. 14-15). If a series of wells were drilled along this contour line, each well should encounter exactly 50 ft of net sand. We further make the assumption that along the net sand contour line (at least for any limited distance), the vertical *distribution* of net sand and impermeable rock, as seen in the well, remains constant within the unit being mapped. This assumption regarding the rock distribution may not be true along the 50-ft contour over long distances, or on opposite limbs of a fold, because of changes in depositional environments, variable structural growth histories, and other factors. However, it is a reasonable assumption to make for a limited distance from the well, parallel to the nearest net sand contour line.

Procedure for Walking a Well. Use three maps in walking a well through the wedge zone. First, lay the structure map of the top of porosity over the net sand isochore map. Then overlay the net gas base map on the two maps, as shown in Figure 14-16a. We wish to walk Well No. 2, which is located near the crest of the structure, through the wedge zone in order to estimate net pay in part of the reservoir. We walk the well along the dashed line, parallel to the 50-ft net sand contour.

Notice that the ILW is shown on the structure map of the top of porosity as a means to delimit the full-thickness area on this map. From its position on the base-of-porosity structure map, the ILW was simply traced on the top-of-porosity structure map. Now we have the data necessary to make a net pay isochore map after walking the well, as described in the following procedure.

1. Preparatory to walking a well, use a computer analysis of the digital log curves or an analysis of a detailed 5-in. electric log to determine the net sand in the well and the distribution of sand and impermeable rock. The 5-in detailed electric log for Well No. 2 is shown in Figure 14-l6b. The well contains 48 ft of net sand, and the log shows increments of 10 ft of net sand per gross feet of interval, so chosen because we use a 10-ft contour interval in the net pay map. The productive unit is full of gas in Well No. 2, at its position in the full-thickness area.
2. Overlay the top-of-porosity structure map on the net sand map, then overlay the net gas basemap. Trace the gas/water contact as a zero net gas contour on the basemap, and trace the limit of the full-thickness area as a dashed line.
3. On the net gas map, lightly mark a dashed line for the well path, parallel to the nearest net sand contour. The net sand contour map may be removed for now; use it later in net pay mapping of the full-thickness area and for checking positions of net pay contours in the wedge.
4. To begin walking Well No. 2, move the well, from its actual structural position, along the well path and place it such that the top of porosity in the productive unit is at the gas/water contact at −9298 ft, which corresponds to the zero contour line on the net gas isochore map

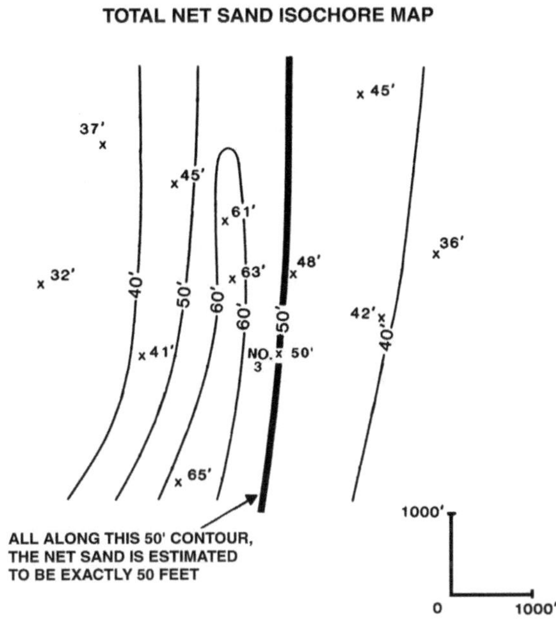

Figure 14-15 Net sand contour map. Each net sand contour line is constructed with the assumption that the amount of net sand is constant along the contour line.

(Point A in Fig. 14-16a). A well drilled at this location would (1) encounter the top of porosity of the productive unit at the gas/water contact, (2) contain 48 ft of net sand, and (3) have no pay.

5. Starting at the top of the sand on the electric log, determine the number of vertical gross feet of section containing 10 ft of net sand. In Figure 14-16b, a total of 16 ft of gross section includes the 10 ft of net sand. From Point A, move the well up-structure 16 ft, along the dashed line to locate Point B at −9282 ft. This Point B becomes a 10-ft net gas data point for contouring the gas wedge, because a well drilled at Point B would encounter the top of the sand 16 ft above the gas/water contact, and it would contain 10 ft of net gas sand.
6. To determine the location of the 20-ft net gas point, start at the base of the previous 10-ft net sand section in the log and repeat the procedure. In this example (Fig. 14-16b), it requires 21 vertical gross ft to obtain the next 10 ft of reservoir-quality sand. Move the well up-structure 21 ft from Point B, along the dashed line, to locate Point C at −9261 ft. Point C is the 20-ft net gas data point at 37 ft (16 ft + 21 ft) above the water level.
7. Continue the same procedure until the well is back at its original structural position. Erase the dashed well path on the net gas map.

Using this method, the well may be walked completely across the wedge, resulting in a more accurate contour spacing than using the method incorporating equal spaced and proportionally spaced contours. We make no assumption about the vertical distribution of net sand being uniform. We use the real distribution of net sand in an actual well to estimate net sand distribution in a part of the reservoir. Figure 14-16c shows the completed net gas isochore map constructed for this reservoir, incorporating the data obtained from walking Well No. 2.

We can walk as many wells as we deem necessary. However, we emphasize strongly that a well *must be walked parallel to the nearest contour line on the net sand isochore map*. Significant net pay contouring errors can occur if this procedure is not followed. Consider Well No. 5 in the western portion of the reservoir (Fig. 14-17). If we wish to develop a more accurate contour

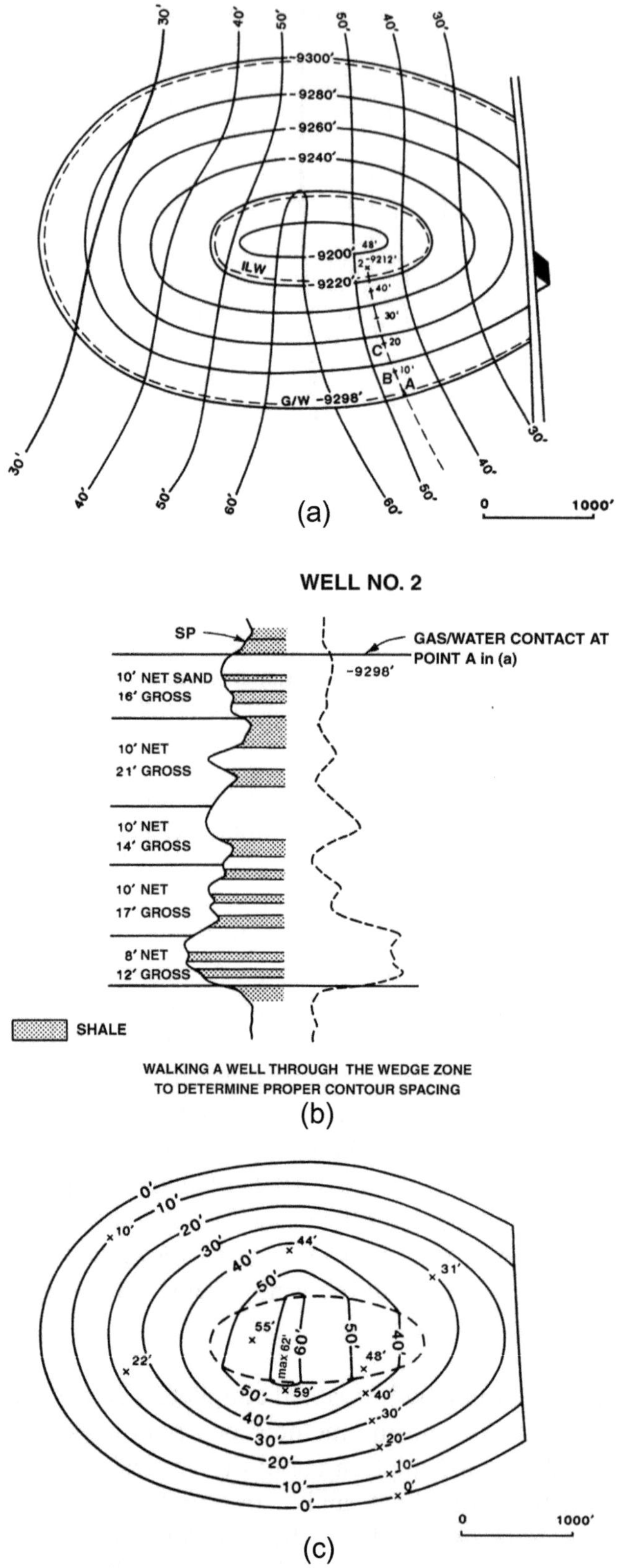

Figure 14-16 (a) Net gas basemap overlain on top-of-porosity structure map and net sand isochore map. Walk Well No. 2 through the wedge zone along the dashed line, parallel to the nearest net sand contour line (50 ft). (b) Five-inch detailed log for the 9200-ft reservoir, showing increments of 10 ft of net sand per gross feet of interval. (c) Completed gas isochore map for the 9200-ft reservoir. The contour spacing in the southeastern portion of the wedge zone was improved by walking Well No. 2. (From Tearpock and Harris 1987. Published by permission of Tenneco Oil Company.)

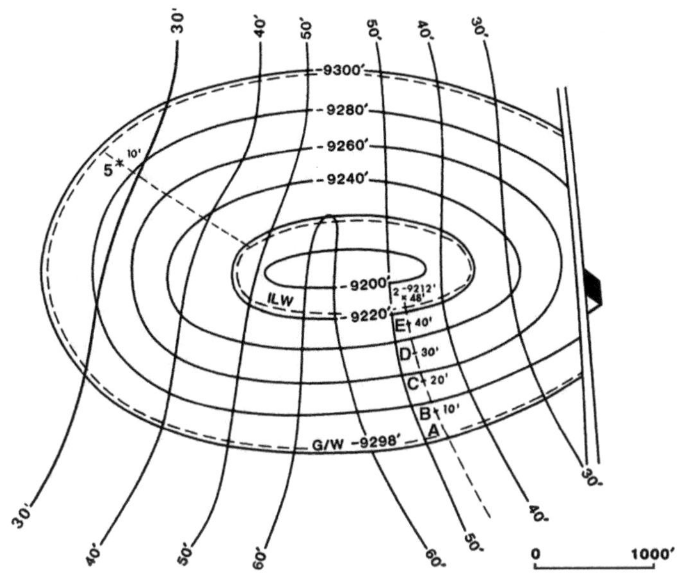

Figure 14-17 Structure map superimposed onto the net sand isochore map. Well No. 5 in the western portion of the reservoir cannot be walked through the wedge zone to improve the spacing of net pay contours. (Modified from Tearpock and Harris 1987. Published by permission of Tenneco Oil Company.)

spacing in this area of the reservoir, can Well No. 5 be walked from the water level up-dip to the inner limit of water, along the dashed line? The answer is no. Well No. 5 has 28 ft of net sand. If the well is walked from the gas/water contact up-dip, it will be walked across net sand contours into an area of greater net sand than is actually present in the well, as seen on the net sand map. Therefore, Well No. 5 cannot be walked to improve the contour spacing in this area of the wedge. Caution must be taken when walking wells to be sure that the assumptions made in choosing a well to walk are geologically reasonable and can be supported by a net sand isochore map or, if necessary, additional study of sand distribution.

Reservoirs with Significant Shale Intervals. If a reservoir is encountered with one or more significant shale intervals between net sand, such as that shown in Well No. 2 in Figure 14-18a, the accurate construction of the net hydrocarbon isochore map within the wedge may depend on the walking of wells through the wedge. It is obvious from a review of the 5-in. detail log for Well No. 2 that the net sand and impervious rock are not evenly distributed throughout the gross interval. Therefore, the use of the equally spaced contour method for contouring the wedge would result in significant error.

The map in Figure 14-18b shows the location of Well No. 2, in its position on structure, and the placement of the 10-ft and 20-ft net gas values used for contouring the net gas isochore map based on walking Well No. 2 through the wedge. Notice that the first 10 ft of net sand are obtained in 16 ft of gross interval; however, it takes another 52 ft of gross interval to obtain another 10 ft of net pay sand. Therefore, the estimated 10-ft net gas point is at −9282 ft (16 ft above the water level of −9298 ft), and the estimated 20-ft net gas point is at −9230 ft (68 ft above the water level). Well No. 3 was also walked to generate the 10-ft and 20-ft net gas points shown in Figure 14-18b.

Figure 14-19a shows a net gas isochore map prepared for this reservoir using a combination of equally spaced and proportionally spaced contours within the wedge, which honor the net pay values assigned to each well. Figure 14-19b is a net gas isochore map prepared by walking Wells

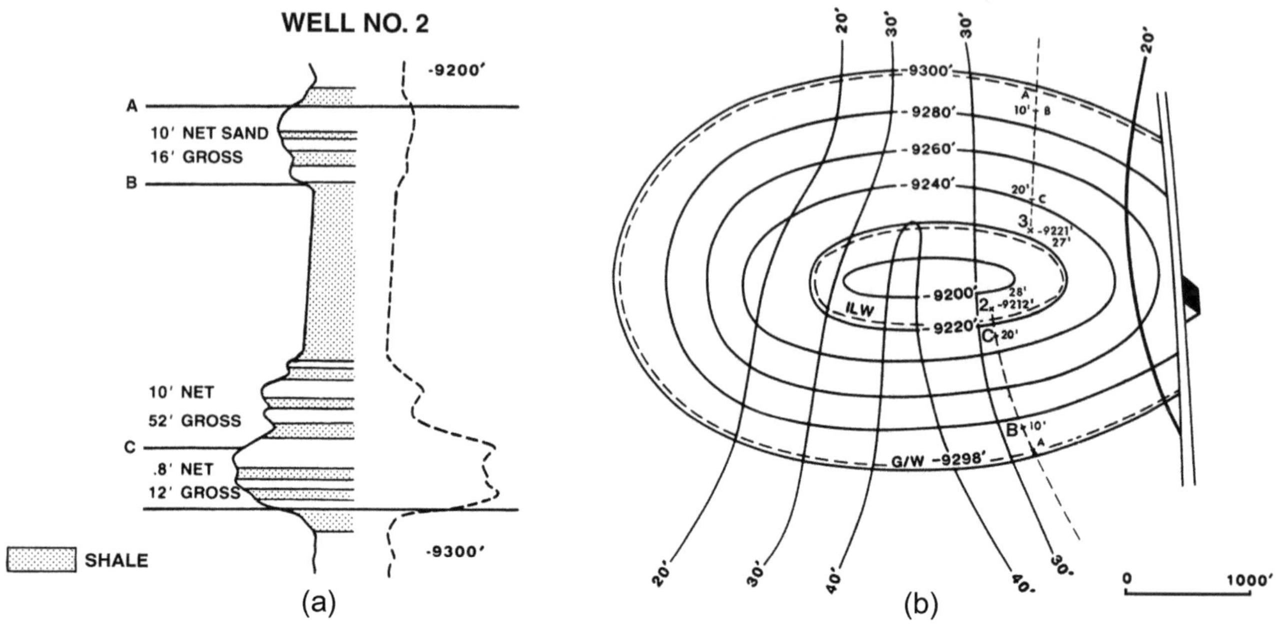

Figure 14-18 (a) Five-inch detailed electric log for Well No. 2. The sand and impermeable rock are not evenly distributed throughout this productive unit. (b) Top-of-porosity structure map superimposed on the net sand isochore map. The dashed lines indicate paths along which Wells No. 2 and 3 were walked through the wedge zone to improve the net gas contour spacing. (From Tearpock and Harris 1987. Published by permission of Tenneco Oil Company.)

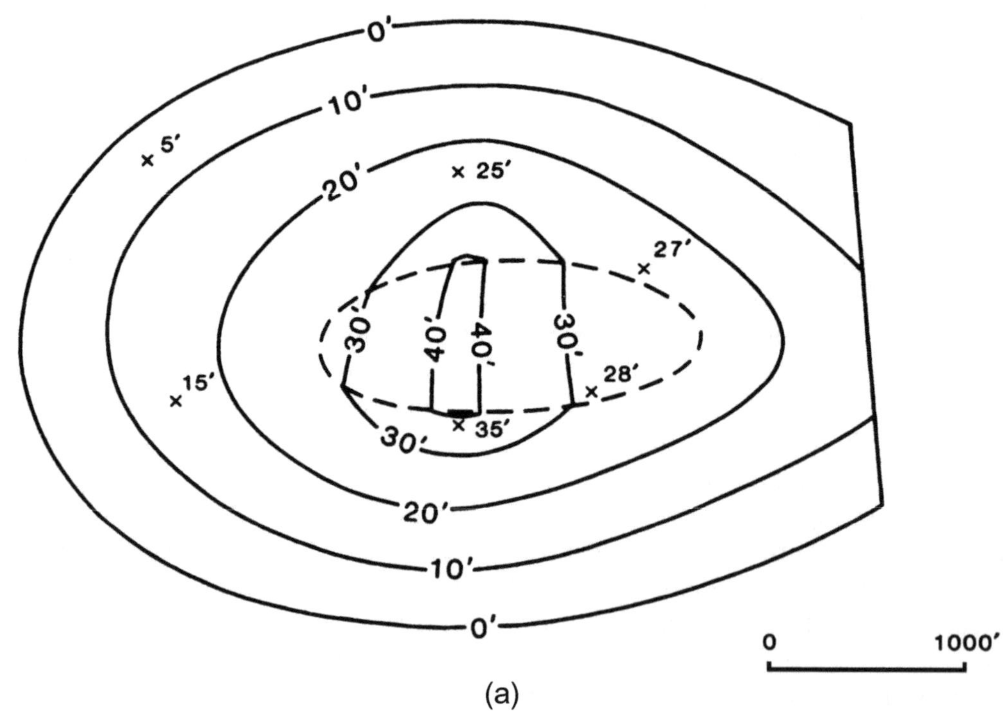

Figure 14-19 (a) Net gas isochore map based on equally and proportionally spaced contours. (b) Net gas isochore map with the contour spacing based on walking Wells No. 2 and 3 through the wedge. Compare this map to that shown in Figure 14-19a. (From Tearpock and Harris 1987. Published by permission of Tenneco Oil Company.)

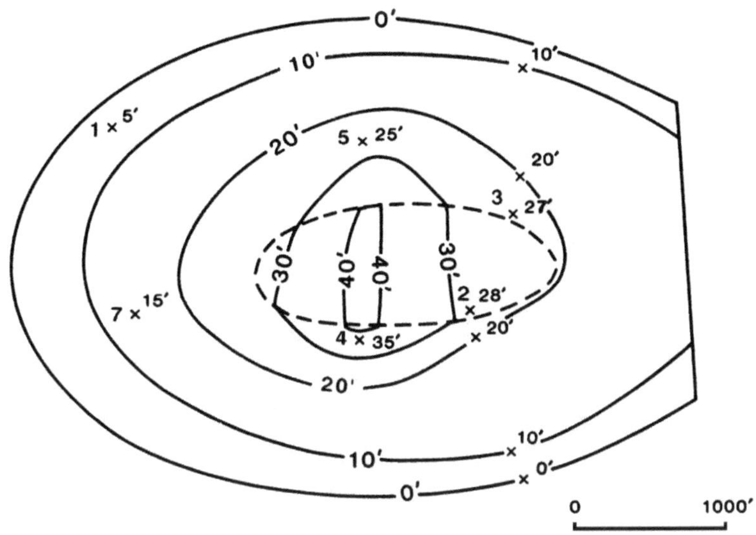

(b)

Figure 14-19 *(continued)*

No. 2 and 3 through the wedge. Observe the significant difference between the two net gas isochore maps. If there were several leases involved in this gas reservoir, the net gas isochore map prepared by walking the wells would provide a more accurate map for assigning equities to each lease.

Several other methods exist for constructing accurate net pay isochore maps. For example, there is a more accurate method of estimating pay when we walk wells. Also, using the method that employs the construction of a net-to-gross ratio map for the entire productive unit provides estimates of net pay at points within the wedge. In this section of the chapter, we review the more detailed method of walking wells.

Using the same reservoir as in Figure 14-18, we illustrate the more detailed technique of walking wells. Figure 14-20 shows a north-south diagrammatic cross section along the path used to walk Well No. 2. On the right side of the figure, we position Well No. 2 so that the top of the sand is at the gas/water contact (−9298 ft). On the left side of the figure, we position the well at the inner limit of water (−9218 ft) where the base of porosity is at the gas/water contact. From the right, Well No. 2 must be walked 16 ft up-structure from the gas/water contact to −9282 ft, to obtain 10 ft of net pay sand. From this point up-dip to −9248 ft, the well gains no additional pay because the section being raised above the water level contains nothing but shale. At −9248 ft, the net pay remains 10 ft. At this point, the gas/water contact intersects the top of the lower sand member; therefore, moving up-structure from this position, the well gains additional pay. Continuing to walk the well up-dip, it must reach −9230 ft before an additional 10 ft of net sand, in the lower member, lies above the water contact. We estimate 20 ft of net pay at −9230 ft. At the edge of the full-thickness area, with the gas/water contact on the base of sand, all the net sand (28 ft) lies above the water contact. From this point to the actual well location, shown on the far left in the figure, the reservoir contains 28 ft of estimated pay.

On the cross section, two locations have 10 ft of net gas sand (−9282 ft and −9248 ft), and the area between these points has a constant 10 ft of net gas sand. The accuracy of the net gas isochore map can be improved by constructing a map honoring the two 10-ft net pay values. Well No. 3 was also walked through the wedge zone, as indicated in Figure 14-18b, to aid in the construction of the net pay isochore map.

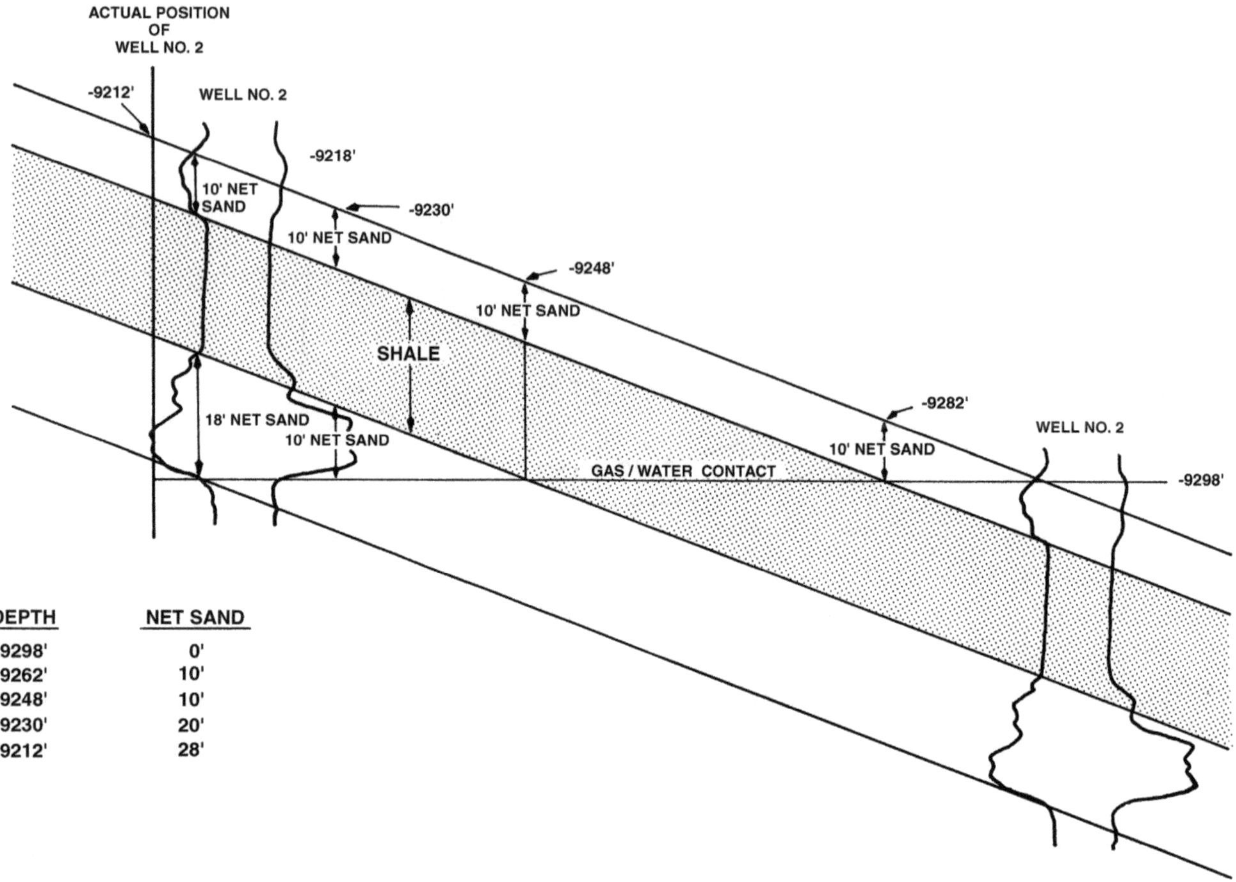

DEPTH	NET SAND
-9298'	0'
-9262'	10'
-9248'	10'
-9230'	20'
-9212'	28'

Figure 14-20 Diagrammatic cross section along the path taken to walk Well No. 2.

Figure 14-21 shows the resulting net gas isochore map. At first glance, it may appear as if an important contouring rule was broken in the construction of the isochore map: contours cannot merge or split. However, no rules are broken. The two 10-ft contour lines that appear to merge represent the limits of a *very wide* 10-ft contour line. Everywhere within the area of the wide contour, the net gas has a constant value of 10 ft.

One may ask, since the sands are so far apart and separated by such a thick shale break, why we would not map each sand separately and construct two isochore maps. In the western part of the reservoir, some of the sands within the two sand members shale-out, and sands develop within the thick shale interval. The result is a reservoir consisting of interfingering shales and sands in which there is productive communication throughout the reservoir, with the sands acting as a single reservoir. Therefore, the thick shale wedge is localized in the eastern section of the reservoir. The rapid decrease in width of the 10-ft contour line reflects the fact that the shale interval decreases to the west. If such a shale interval were known to be continuous over the entire reservoir, it would be necessary to prepare a structure map for each sand member and construct a separate net pay isochore map for each sand member.

The procedure outlined in this section is more involved than the two previous methods shown, but the technique provides further accuracy in the construction of a net hydrocarbon isochore map. The method chosen to prepare a net hydrocarbon isochore map depends on a number of factors, including the available time, detail, and accuracy required.

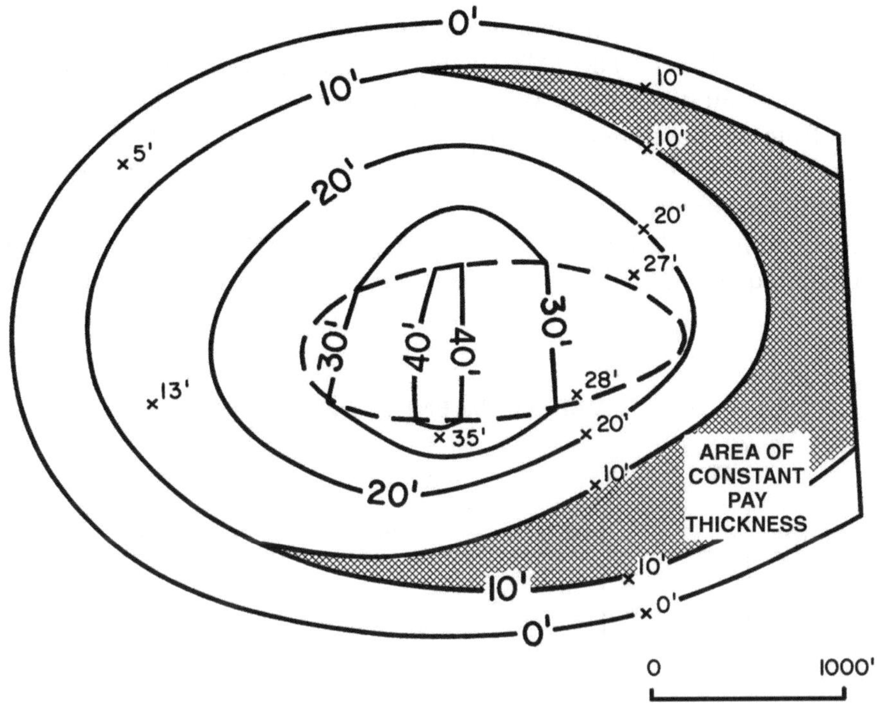

Figure 14-21 Net gas isochore map using a more accurate method of contouring the wedge based on the results of walking Wells No. 2 and 3.

VERTICAL THICKNESS DETERMINATIONS

True vertical thickness (TVT) is the thickness of an interval measured in a vertical direction. As mentioned in several sections of the text (see Chapter 4), vertical thickness is a very important measurement. It is this vertical thickness that is required to measure the vertical separation (equal to missing or repeated section in a well) of a fault. It is also the thickness required to accurately count net sand and net pay from 5-in. detailed logs, and it is the thickness used to construct net sand and net pay isochore maps. As described in the introduction to this chapter, the vertical thicknesses of net sand and net pay typically represent aggregate vertical thicknesses.

In a vertical well, the actual thickness measured on the electric log is the TVT. In the case of a directionally drilled well, however, a correction factor must be applied to correct the exaggerated or diminished measured log thickness (MLT) to TVT.

For a horizontal reservoir (zero bed dip), the thickness that is used for net sand or net pay isochore mapping equals the true stratigraphic thickness (TST) (Fig. 14-22). However, if the same reservoir is rotated to some angle, such as 20 deg, the thickness of the reservoir required for net sand and net pay isochore mapping does not equal the stratigraphic thickness. Figure 14-22 illustrates the cross-sectional area of a reservoir with a fixed width in the third dimension, and we use the cross section to represent the volume of the reservoir. The horizontal reservoir (zero bed dip) in the lower portion of the figure has a cross-sectional area of 50,000 sq ft. The reservoir has a length of 500 ft, as seen in map view, and a thickness of 100 ft. Since the dip of the reservoir is zero, the TVT equals the TST (100 ft). If the same reservoir rotates to an angle of 45 deg, as shown in the upper portion of the figure, the length of the reservoir shortens to 354 ft in map view. The cross-sectional area of the reservoir does not change, as the TST remains 100 ft. In order to map the reservoir and maintain a cross-sectional area of 50,000 sq ft, the thickness to be used must exceed 100 ft. The TVT of the dipping reservoir measures to be 141.25 ft, and 141.25 ft × 354 ft = 50,002.5 sq ft. We conclude from this example that as a reservoir of fixed length rotates from the horizontal, the areal extent of the reservoir decreases in map view; therefore, in order to maintain the same cross-sectional area or volume of the reservoir, the shortened length must be multiplied by the TVT.

For directionally drilled wells, the log thickness of a given stratigraphic interval can either be thicker, equal to, or thinner than that seen in a vertical well drilled through the same

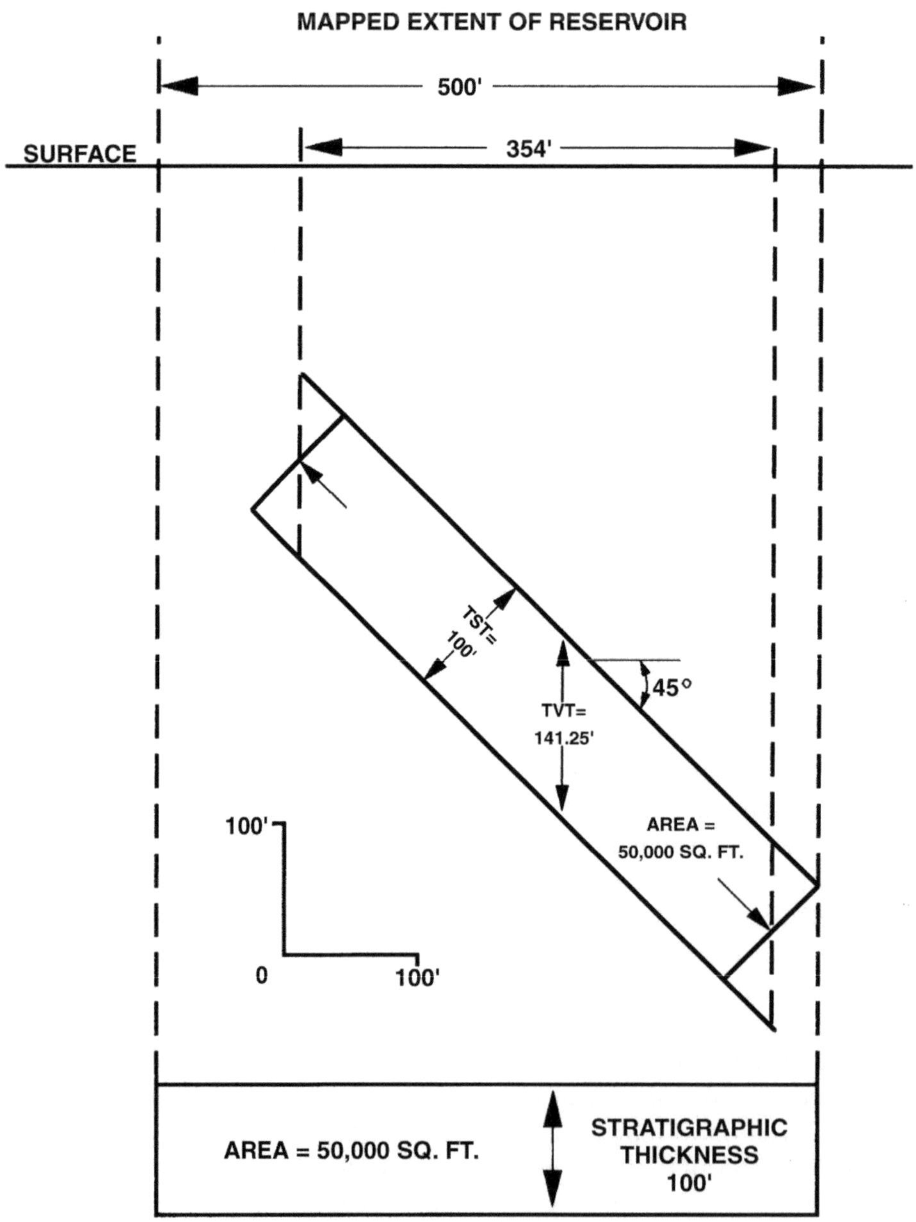

Figure 14-22 Cross-sectional area of two reservoirs of equal volume and a constant TST of 100 ft. One reservoir is horizontal, the other is dipping at 45 deg.

stratigraphic section. A correction factor must be applied to the MLT in most deviated wells to convert the borehole thickness to TVT. The correction factor consists of two parts: (1) the correction for wellbore deviation angle within the interval of interest, and (2) the correction for bed dip. Any one of Eqs. (4-3), (4-4), (4-5), or (4-6) in Chapter 4 may be used to calculate this correction factor. In Chapter 4, we used the equations to estimate the TVT of missing or repeated section in a well as the result of a fault. Remember, the vertical separation of a fault at a well equals the TVT of the stratigraphic section missing or repeated in the wellbore. In this chapter, we look at the same correction factor equations in order to convert deviated wellbore thickness to TVT, for use in net sand and net pay isochore mapping.

For convenience, we repeat the correction factor given in Eq. (4-6), presented here as Eq. (14-1). Equation (14-1), which is a 3D equation, is the preferred equation to use for correction factors because this one equation can be used to calculate the thickness correction factor regardless of the direction of wellbore deviation, and the true dip of the beds is used instead of the apparent dip required in the 2D equations. Recall from Chapter 4 that we refer to this equation as Setchell's equation:

$$\text{TVT} = \text{MLT}[\cos\psi - (\sin\psi\cos\alpha\tan\phi)] \tag{14-1}$$

where

TVT = true vertical thickness
MLT = measured log thickness
ψ = wellbore deviation angle
ϕ = true bed dip
α = Δ azimuth (acute angle between the wellbore azimuth and the azimuth of true bed dip)

If the beds are horizontal, then Setchell's equation reduces to the simple correction factor equation given in Chapter 4, Eq. (4-3):

$$\text{TVT} = \text{MLT}\,(\cos\psi)$$

which is equivalent to correcting for wellbore deviation only.

Figure 14-23 illustrates the measurement of azimuth and Δ azimuth for use in Eq. (14-1). For convenience, the Δ azimuth used in the equation is typically the acute angle between the azimuth of the wellbore and the azimuth of the true bed dip.

In order to more closely examine the two directionally drilled wells shown in Figure 14-24, look first at the well drilled to the east in a down-dip direction (Fig. 14-24a). Consider the interval to be a reservoir filled with gas or oil. The reservoir has an MLT of 476 ft. We apply the correction factor for wellbore deviation only, using Eq. (4-3) from Chapter 4. The MLT reduces to 357 ft, shown in the figure as the TVD thickness, or the true vertical depth thickness (TVDT). This thickness also exceeds the TVT of the interval, because the correction for only wellbore deviation does not take into account the dip of the beds. The TVDT is that thickness of an interval obtained from a true vertical depth (TVD) log and, for dipping beds, TVDT does *not* equal TVT (see Chapter 4). With the final correction for bed dip, the MLT converts to the TVT of 150 ft, shown in Figure 14-24a at the penetration point of the wellbore in the top of the reservoir. Note that the TST is 123 ft. The TST is calculated by multiplying the TVT by the cosine of the angle of bed dip (35 deg in this example).

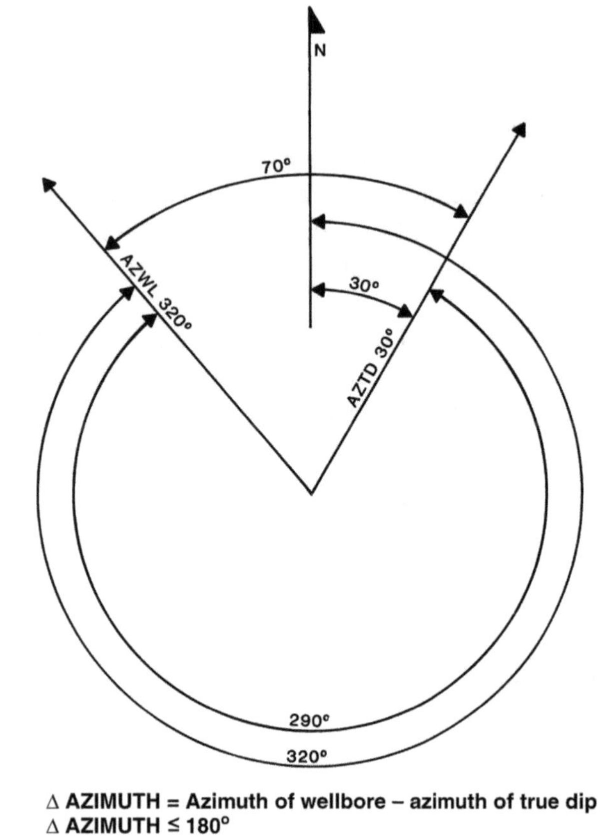

Figure 14-23 Azimuth is measured from 0 deg to 360 deg in a clockwise direction from true north. A Δ azimuth is the difference in azimuth of the wellbore and the azimuth of true bed dip. For convenience, the minimum Δ azimuth is typically used.

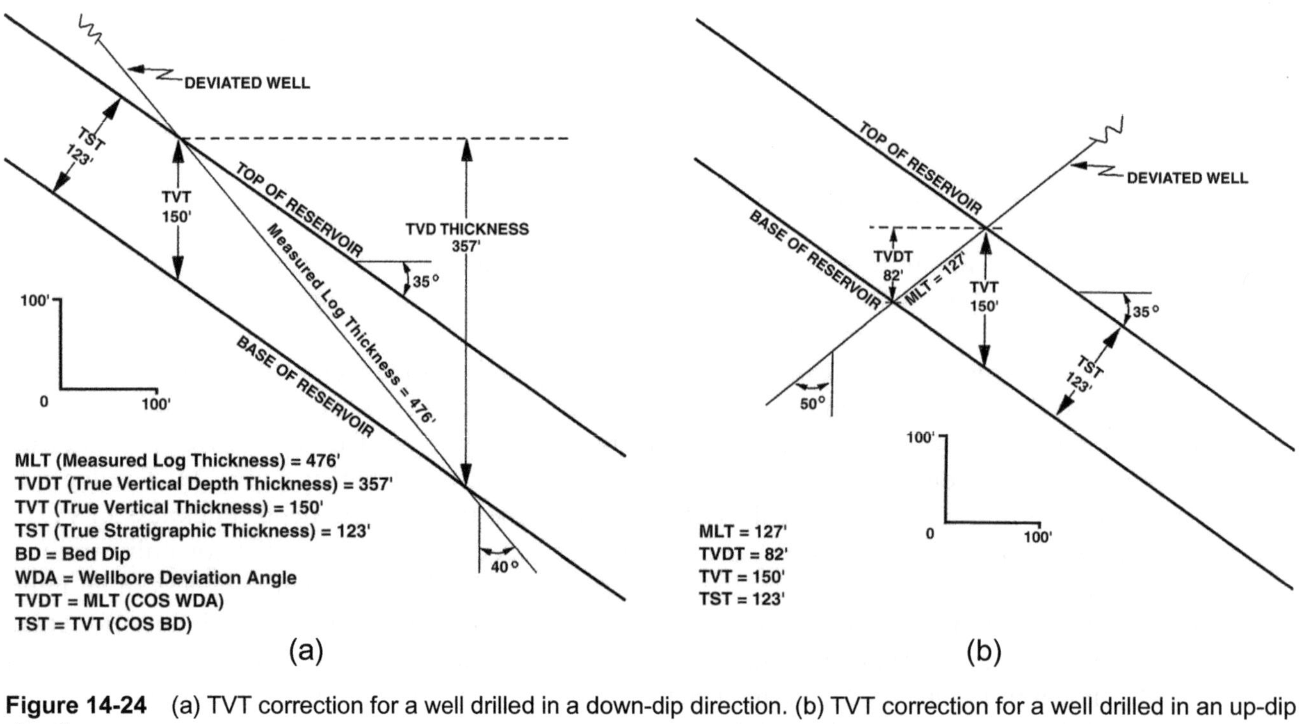

Figure 14-24 (a) TVT correction for a well drilled in a down-dip direction. (b) TVT correction for a well drilled in an up-dip direction.

The well in Figure 14-24b deviates up-dip, to the west. The MLT of 127 ft measures less than the TVT. Applying a correction factor for the well deviation angle alone, which is equivalent to the correction to TVDT, provides an even smaller thickness of 82 ft. When Eq. (14-1), the correction factor equation for both bed dip and wellbore deviation, is applied, the MLT converts to the TVT of 150 ft, the value needed for net sand and net pay mapping. As an exercise, use Eq. (14-1) to calculate the TVT for the two wells in Figure 14-24 and to confirm the results shown.

Various computer programs can be used to create TVD, TVT, and TST logs from measured depth (MD) logs for use in mapping. The deviated well log data, the directional survey for the well, and bed dip information are necessary as input data. The log data are obtained from the logging company tapes or digitized from the actual log. The directional survey data are furnished by the directional company that worked the well. The bed dip information can be obtained either from completed structure maps or from a dipmeter. The output logs can be in standard presentation or at any scale desired. The logs in Figure 14-25 were created by using IEPS (Integrated Exploration and Production System). The log sections for Well No. MP-D5, shown from left to right in the figure, represent the (1) measured depth log, (2) true vertical depth log, (3) true vertical thickness log, and (4) true stratigraphic thickness log. This well is in an area of significant bed dip. Notice the similarity of the MD log and the TVD log. This is so because the TVD log thickness is equivalent to a correction for wellbore deviation only, and not bed dip. However, the TVT log shows a considerable reduction in thickness from the MD log because the TVT log reflects corrections for both wellbore deviation and bed dip.

We caution here that TVD logs, which are a standard part of the log suite for a deviated well, are used too often for purposes that are not applicable with this log. A *widespread misunderstanding* exists that a TVD log prepared from a MD log can be used to (1) correlate with other well logs, (2) determine the vertical separation for a fault, and (3) count net sand and net pay for isochore mapping. Remember, a TVD log is corrected only for wellbore deviation, and not bed dip. In areas of flat-lying beds, a TVD log is equivalent to a TVT log because the only correction factor required is for wellbore deviation (Fig. 14-26). However, if the beds are dipping (particularly over 10 deg), a TVD log typically does not represent the log thickness required to aid in correlation work, to determine the vertical separation for a fault, or for use in net sand and net pay counting. For these purposes, we *must* correct a deviated well log so that the log thickness represents the TVT. Look again at Figure 14-25 and observe the significant difference in thickness between the TVD and the TVT logs. To determine the vertical separation of a fault by correlation of a faulted well with a deviated well, and to count all net sand and net pay from a deviated well log, we *must* use a TVT log or its equivalent. By the equivalent of a TVT log, we mean calculating and using correction factors if a TVT log is unavailable, which is commonly the case. Therefore, for each interval on the deviated well log requiring the conversion of MLT to TVT, determine correction factors and apply them to the MLTs of the intervals of interest.

The Impact of Correction Factors

Figure 14-27 presents an example of two separate net pay isochore maps prepared for a reservoir on the flank of a salt dome in the offshore Gulf of Mexico. Notice that there are two platforms from which wells were drilled. Platform D is located in the up-structure position, with the D Platform wells directionally drilled down-dip. The A platform is located on the flank of the structure, with most of the A Platform wells directionally drilled up-dip.

Figure 14-27a shows a net pay isochore map prepared for the T-1 Sand, Reservoir A. The net pay values posted on the net pay isochore map were corrected for borehole deviation, but not for bed dip of about 35 deg at this location on the flank of the structure. In addition to the error of failing to correct the net pay values for bed dip, several other serious isochoring problems exist but are not discussed here.

Figure 14-27b presents a net pay isochore map for the same reservoir with new net pay values, which were generated after correction for borehole deviation and bed dip. This new isochore map was

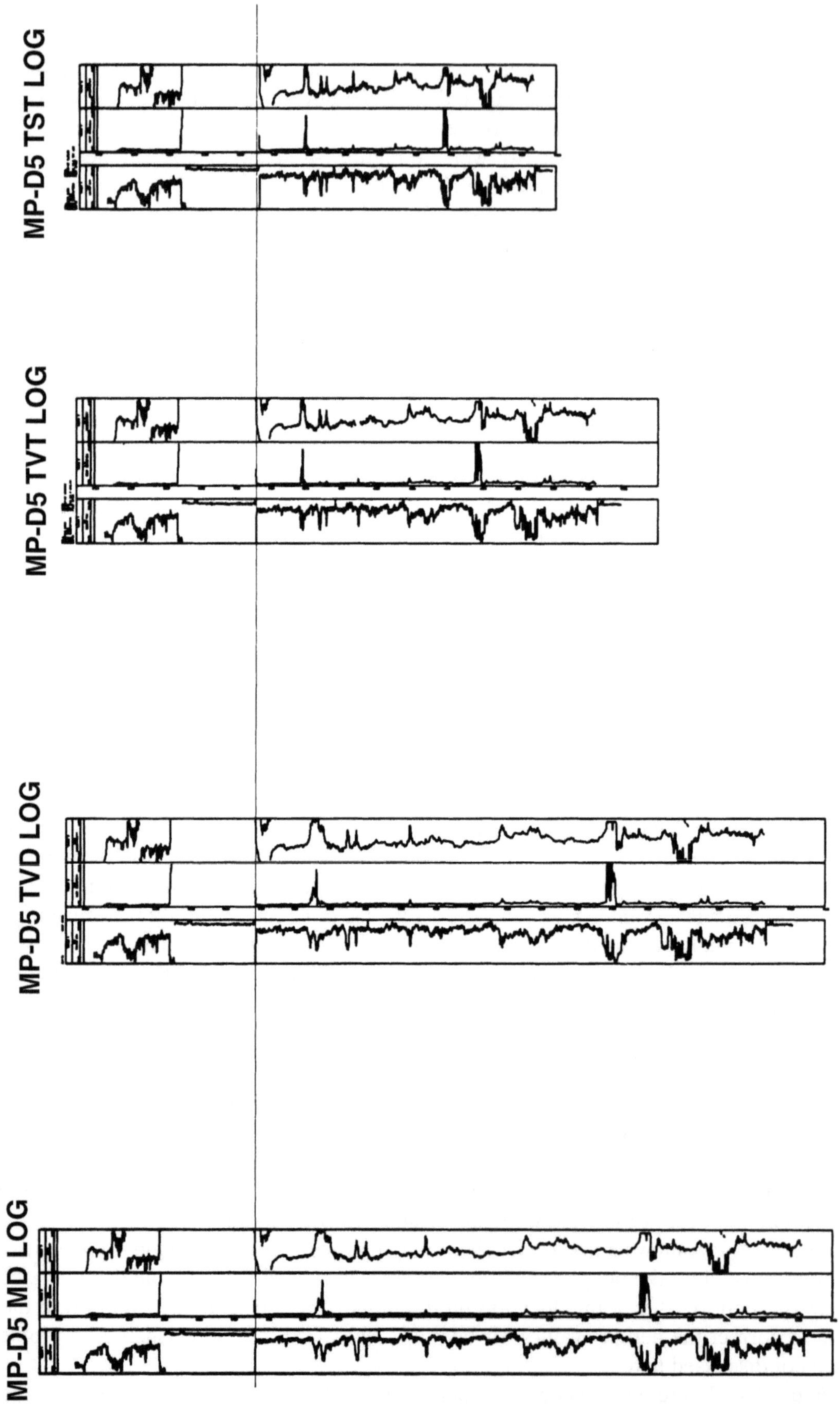

Figure 14-25 Computer-generated electric logs illustrating the difference in thickness between measured depth, true vertical depth, true vertical thickness, and true stratigraphic thickness logs, generated for the same well. (From Tearpock and Harris 1987. Published by permission of Tenneco Oil Company.)

prepared solely to compare the effect of the correction factor for bed dip on the ultimate calculation of the total volume of the reservoir. Therefore, the isochore errors made in Figure 14-27a (not discussed here) are incorporated in the net pay isochore map in Figure 14-27b. The planimetered volume for the isochore map in Figure 14-27a, with MLTs corrected for bed dip, is 524 acre-feet, whereas the map in Figure 14-27b with uncorrected MLTs provides a volume of 617 acre-feet. Therefore, the estimated hydrocarbon volume for the map prepared without correcting the net pay thickness values for bed dip is 18% greater than the isochore map prepared by taking into account the correction factor for bed dip. So, based on the incorrect map, the volumetric calculation overstates the reserves by 18%.

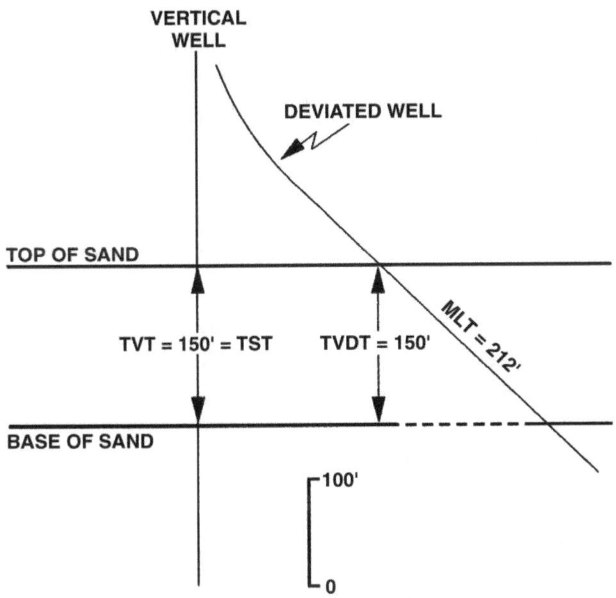

Figure 14-26 True vertical depth thickness equals true vertical thickness where the stratigraphic unit is horizontal.

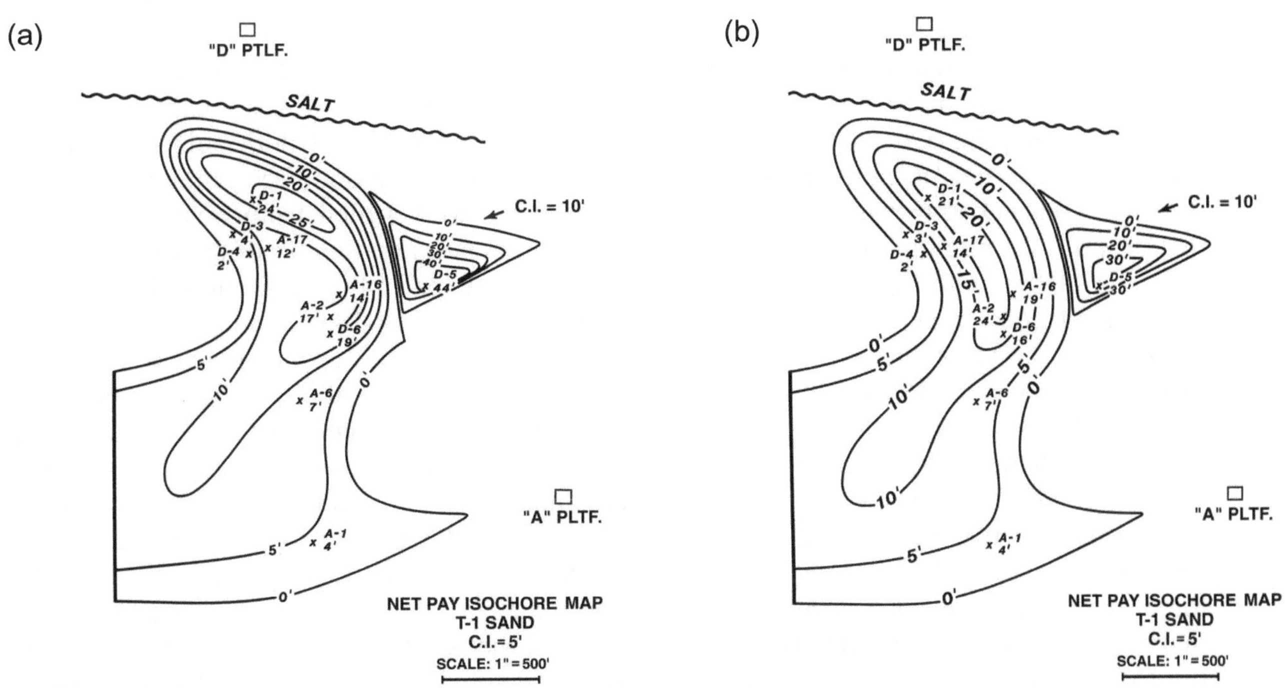

Figure 14-27 (a) Net pay isochore map for the T-1 Sand, Reservoir A. The net pay values for the deviated wells from Platforms A and D were corrected only for wellbore deviation. (b) Net pay isochore map for the T-1 Sand, Reservoir A. The net pay values for the deviated wells from Platform A and D were corrected for wellbore deviation and bed dip. Compare the net pay value for each well with those shown in Figure 14-27a. (From Tearpock and Harris 1987. Published by permission of Tenneco Oil Company.)

In a situation like this, we would expect the error factor to be larger than 18%, and it would be in most cases. However, look at Wells No. A-2 and D-5. Well No. A-2 was corrected upward from 17 ft net pay to 24 ft net pay, whereas Well No. D-5 was corrected downward from 44 ft net pay to 30 ft net pay. It so happens that the D Platform wells, drilled down-dip, result in a reduction in net pay values when the correction factor for bed dip is considered, whereas the A Platform wells, drilled up-dip, result in an increase in net pay. Therefore, a significant part of the error is negated because of the directions in which the wells were drilled.

This reservoir is only one of a number of oil and gas reservoirs that are productive within this field. A complete remapping of the field was undertaken when several major mapping errors, such as the one shown here, were identified. The remapping of the field resulted in a significant write-down of hydrocarbon reserves that were overestimated because of mapping errors, such as the failure to incorporate the proper correction factors in determining net pay values. Several development wells planned for the field based on the overestimated reserves were not required, saving the company millions of dollars in unnecessary wells. This example illustrates the impact that correction factors can have on estimated hydrocarbon volumes determined from volumetric calculations using net pay isochore maps.

We have seen error factors of several hundred percent. In one example, the net pay used for the reserve estimate from a deviated well was 750 ft, when in fact the actual TVT value was about 150 ft. This is a major error, resulting in a significant overestimation of reserves.

TVT Correction (Setchell's Equation) Discussion

If one closely examines Figure 4-22, the TVT value that is being calculated is the TVT at the point where the well exits the bed and the bed dip that is used in this derivation is the dip of the top of the bed. The top and the base of the bed are parallel in Figure 4-22, so the TVT at the point where the well exits the bed is equal to the TVT where the well enters the bed. Indeed, the TVT is constant along the entire length of the well within the bed if the bed boundaries are parallel. If the bed boundaries are parallel, the TVT value of the bed can be posted at both the point where the well enters the bed and the point where it exits the bed or anywhere along the path of the well within the bed. If the values at both points and along the well path do not fit the isochore map, it indicates that the top and base of the bed are not parallel. Consider, for example, a hypothetical deviated well drilled from west to east near the crest of the structure in Figure 14-14c. If this well penetrated the top of the reservoir just west of Well No. 9, the gross interval thickness would be expected to be about 80 ft, based on the 83 ft of gross interval thickness seen in Well No. 9 (Fig. 14-14a). If the well penetrated the base of the reservoir near the 70-ft net gas contour line, the gross thickness would be expected to be about 100 ft (70 ft of net sand divided by NTG of about 0.73). Obviously, in the case of this reservoir, the top and base of the reservoir are not parallel.

Although the derivation of the Setchell equation shown in Figure 4-22 uses the dip of the top of the bed and calculates the TVT of the bed at the point where the well exits the bed, we could equally easily derive the equation for the TVT where the well enters the top of the bed using the dip of the base of the bed. The fact that the TVT at the point of exit of the bed uses the dip of the top of the bed and the TVT at the point of entry into the bed uses the dip of the base of the bed makes intuitive sense. We know the depth of the base of the bed at the point of exit. To determine the TVT of the bed at this point, we need to know the depth of the top of the bed at the point of entry and the dip of the bed to a point directly above the exit point. The dip of the base of the bed is irrelevant to this determination. An analogous argument can be made for determining the TVT at the point of entry to the bed.

If the top and base of the bed being isochored are not parallel, the Setchell equation, Eq. (14-1), can be used to determine the TVT at both the point of entry into the bed and the point of exit out of the bed.

$$TVT_{exit} = MLT \cos \psi - (\sin \psi \cos \alpha \tan \phi_{bedtop}) \quad (14\text{-}1a)$$

$$TVT_{entrance} = MLT \cos \psi - (\sin \psi \cos \alpha \tan \phi_{bedbase}) \quad (14\text{-}1b)$$

Consider the well illustrated in Figure 14-24a, which is drilled directly down-dip at an inclination of 40 deg through a bed with the top of the bed dipping at 35 deg. We have shown that if the well has an MLT of 476 ft, the TVT where the well exits the bed is 150 ft. Consider if the base of the bed is not parallel to the top of the bed but dips at 40 deg rather than 35 deg. In this case, we can use Eq. (14-1b) to determine that the TVT of the bed is 108 ft at the point the well enters the bed [476 × (cos 40 − (sin 40 cos 0 tan 40) = 108]. This result makes intuitive sense. If the base of the bed dips more than the top of the bed, the bed must be expanding down-dip. If we know that the TVT of the bed is 150 ft where the well exits the bed, the TVT up-dip where the well enters the bed must be less than 150 ft.

If the azimuth of the dip of the top of the bed differs from the azimuth of the base of the bed, α will have to be adjusted in the two equations.

Few if any beds have parallel tops and bases over their entire extent. This situation would yield a constant isochore thickness over the bed's entire extent if the bed dip is constant. Many beds, however, have nearly parallel tops and bases, and thus essentially constant isochore thicknesses, over moderate areas. The assumption that bed boundaries were essentially parallel is reasonable for many conventional deviated wells. As very-high-angle deviated wells become more common, deviated wells travel greater distances within beds, and assuming that bed boundaries are parallel is less likely to be reasonable. In these cases, it is important to post the TVT of the bed calculated using the dip of the top of the bed directly above the point where the well exits the bed. The TVT of the bed at the point where the well enters the bed can be calculated separately if the dip of the base of the bed is also known.

One of the best ways to determine the dip of a bed is from a structure map. A structure map can also be used to determine the apparent dip of a bed along the path of a deviated well. This is done by measuring the distance between the well's point of entry into the bed and its point of exit from the bed. This distance is the run along the path of the well. The difference between the depth of the bed at the point at which the well penetrates the top of the formation and the depth of the top of the bed, read from the map, directly above the point where the well exits the base of the bed. This depth difference is the rise along the well path. As we saw in Chapter 2, Eq. (2-1),

$$\tan \phi = (C.I.)/(C.S.)$$

where the contour interval (*C.I.*) is frequently called the *rise* and the contour spacing (*C.S.*) is frequently called the *run*. If we measure the rise and run along the path of a well, we can get an analogous equation:

$$\tan \phi_{alongwell} = rise_{alongwell}/run_{alongwell}$$

Using the dip along the well path means that the difference in azimuth between the well path and the dip is zero, allowing the Setchell equation to be modified to

$$TVT = MLT \cos \psi - (\sin \psi \tan \phi_{alongwell})$$

The preceding discussion of the Setchell equation is only completely accurate for cases in which a well first intersects the top of a bed and subsequently exits through the base of a bed. These cases comprise the vast majority of all wells. It is possible, however, for a well to drill up through stratigraphy, that is, to drill first through the base of a bed and subsequently exit through the top of a bed. This situation can occur if the sum of the deviation of the wellbore plus the apparent dip of the bed along the path of the wellbore exceeds 90 deg. This situation is directly analogous to drilling a normal fault backwards and encountering repeated section on a normal fault (see Fig. 4-36 and Fig. 4-37).

Using the Setchell equation [Eq. (14-1)] as written for such wells yields unexpected results (Sia Agah, personal communication, 2016). Consider the case illustrated in Figure 14-28 where a wellbore drilled directly down-dip and deviated at 55 deg encounters a bed dipping at 38 deg. Figure 14-28 shows that the well intersects the base of the bed and exits through the top of the bed. Assuming an MLT of 106 ft and applying Eq. (14-1) as written, with $\psi = 55$ deg, $\alpha = 0$ deg, and $\phi = 38$ deg, yields a TVT of the bed of −6 feet. Obviously, a bed thickness cannot be a negative number, so Eq. (14-1) should be revised:

$$TVT = |MLT\ [\cos \psi - (\sin \psi \cos \alpha \tan \phi)]| \qquad \text{(14-1 revised)}$$

where TVT is the absolute value of the MLT multiplied by the Setchell correction factor.

A second outcome of drilling a well through the base of the bed and exiting the top of the bed is that the TVT of the bed is calculated at the point where the well enters the base of the bed if the dip of the top of the bed is used in Eq. (14-1 revised). The TVT of the bed is calculated at the point where the well exits the top of the bed if the dip on the base of the bed is used in the calculation. If the bed boundaries are parallel, the two TVT values are the same.

We do not know if most, or any, computer workstations plot TVT values in the correct location. This uncertainty is not an issue if bed boundaries are parallel or nearly so. If bed boundaries are not parallel, it is incumbent on the subsurface mapper to make sure that the TVT value is plotted in the correct location. If the dip of both the top and base of the bed in question are available, the TVT of the bed at both the point of intersection of the top of the bed and the base of the bed should be plotted and used in contouring. This may require the use of digital control points such as pseudo or dummy wells. Some computer programs may calculate TVT at both the entry and exit points of the bed. One program plots a single TVT value midway between the entry and exit points of the well, although this may be adjustable within the software. It is critical that geoscientists making isochore maps on a workstation understand what their computer software is doing with data from deviated wells so that they can ensure that their isochore maps are accurate.

There is a second important assumption in the Setchell equation (Eq. 14-1): that the well is roughly straight between the entry and exit points of the bed. If the well changes the deviation angle or azimuth significantly within the bed in question, additional calculations are required for accurate TVT calculations. The straight-line distance between the entry and exit points needs to be calculated and used in place of the MLT. Similarly, the average azimuth of the well needs to be determined from a map showing the entry and exit points. Finally, the average inclination of the well may need to be calculated from the TVD difference between the entry and exit points and the straight-line distance between these points.

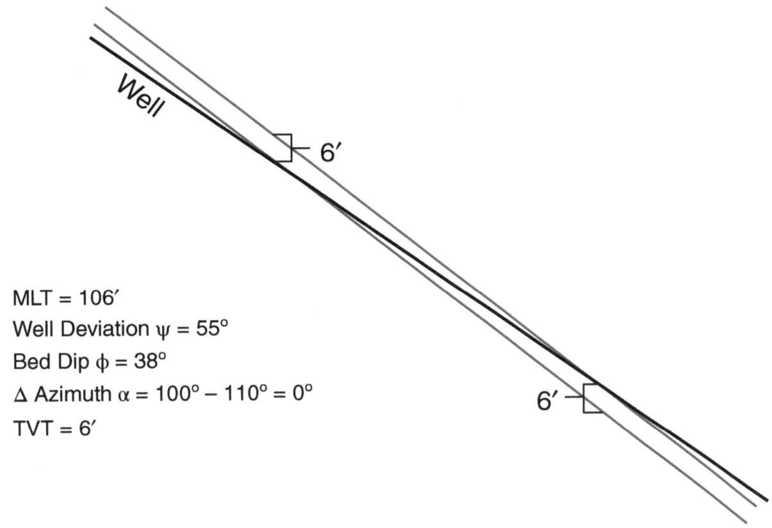

Figure 14-28 A well drilled through the base of a dipping bed and out the top of the bed. Using the original Eq. (14-1), the thickness where the well penetrates the base of the bed is calculated to be −6 ft thick.

VERTICAL THICKNESS AND FLUID CONTACTS IN DEVIATED WELLS

The mathematical equations that are used for converting log thicknesses to true vertical thicknesses were reviewed in Chapter 4 and in the previous section in this chapter, but some additional discussion is required regarding these correction factors when dealing with deviated wells that penetrate fluid contacts. The mathematical treatment in these situations is not as straightforward as for deviated wellbores within the full-thickness area of a reservoir.

Normally, we determine net sand and net pay values for isochore mapping at the position where a well penetrates the top of the reservoir. For a vertical well, the penetration points of all horizons are at the same location in map view, directly under the surface location of the well (Fig. 14-29). However, for a directionally drilled well, the intersections of the well with the top and base of a stratigraphic unit occur in different locations with respect to the surface location of the well, as seen in map view along the path of the well (Fig. 14-29).

In many cases, due to the low angle of wellbore deviation or the minimal thickness of a sand, the calculation and positioning of the net sand or net pay values at the point where the well penetrates the top of porosity of the productive unit provides sufficient accuracy. There are situations involving highly deviated wells, fluid contacts, dipping beds, or thick sands, however, where a single data point for net sand or net pay at the penetration point of the well at the top of the reservoir may be insufficient, as well as being incorrectly calculated or posted for isochore mapping. These special conditions are discussed here.

Equation (14-1) can be used to calculate the correction factor for reservoir TVT within the full-thickness area, where no fluid contacts exist in the well (Fig. 14-30, well on the right). In this figure, the deviated well penetrates a reservoir with a TVT of 150 ft. The log thickness measures 219 ft from the depth where the well penetrates the top of the reservoir to its penetration at the reservoir base. Using Eq. (14-1) and the data given in Figure 14-29, the 219 ft of MLT convert to 150 ft TVT. Observe an oil/water contact just down-dip from the well's penetration of the base of the reservoir. The entire wellbore penetration is therefore confined to the full-thickness area of the reservoir, with no fluid contacts present in the well.

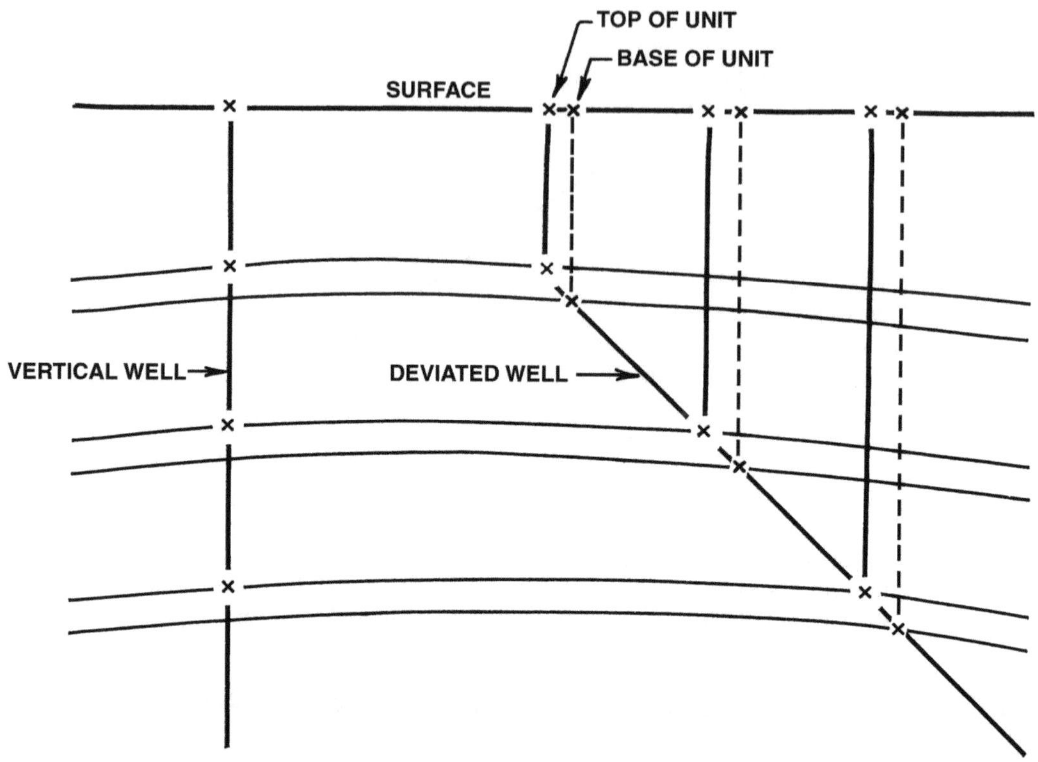

Figure 14-29 A deviated well penetrates the top and base of a stratigraphic unit at different positions relative to the surface location of the well.

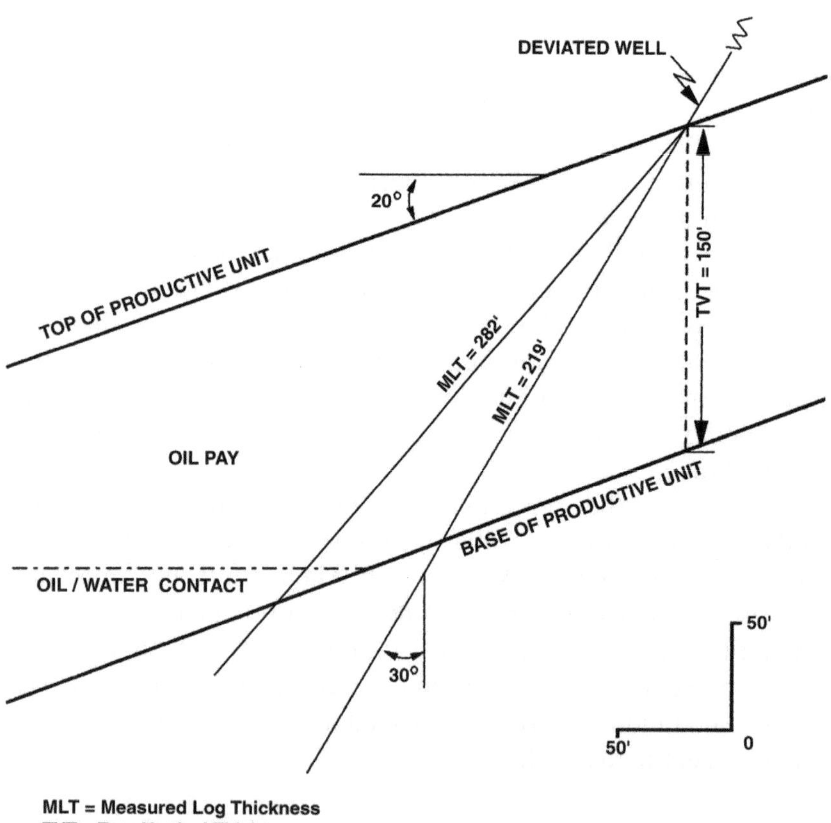

Figure 14-30 Cross section illustrates a deviated well penetrating an oil reservoir within the full-thickness area and a second deviated well penetrating the oil/water contact. See text for explanation of conversion of MLT to TVT for each well.

The well on the left in Figure 14-30 penetrates an oil/water contact. The MLT of the entire stratigraphic interval is 282 ft. What correction factor equation would you use for this MLT to calculate TVT of the oil column, at the penetration point of the top of the reservoir? In this particular situation where the penetration point is within the full-thickness area, Setchell's equation, applied to the MLT of 282 ft, will provide the TVT of the oil at the penetration point in the top of the reservoir.

Application of Setchell's equation to the MLT of the net oil (less than 282 ft) would result in an incorrect TVT at the penetration point in the top of the reservoir. However, if you wish to obtain the value of the vertical feet of oil directly above the penetration point of the wellbore *at the oil/water contact*, the use of the Setchell equation provides this value of net pay in the wedge. We will discuss this situation next.

Now consider the following situation. A well penetrates the top of an oil reservoir within the oil wedge (i.e., directly above the oil/water contact). The following data apply to the well and well log:

1. Wellbore deviation is 30 deg due west;
2. Bed dip is 20 deg due west;
3. MLT of net sand within the productive unit, from top to base, is 219 ft; and
4. MLT of net sand from the top of the unit to the oil/water contact is 115 ft.

What is the vertical thickness of the oil column under the penetration point of the well at the top of the reservoir? Draw a cross section showing the relationship of the well to the productive unit and, using Eq. (14-1), calculate the TVT of the oil under the penetration point at the top of the reservoir. Is the correct answer 150 ft, 100 ft, or 79 ft? If you calculated 150 ft, this thickness is equal to the *total* true vertical thickness of net sand directly under the penetration point of the well at the top of the sand (Fig. 14-30). If you calculated 79 ft, this is the vertical thickness of net oil pay directly above the point where the well penetrates the oil/water contact. A review of Figure 14-31 shows that the oil/water contact is a horizontal surface. Therefore, we do not have to consider any bed dip correction factor to calculate the net pay directly under the penetration point. No bed dip affects the TVT of the net oil sand, since the oil/water contact is a horizontal surface, and the penetration point at the top of the sand is a point *directly over* the oil/water contact. Therefore, Eq. (14-1), as applied to calculation of net oil sand directly beneath the penetration point of the well at the top of the sand, reduces to the correction factor for wellbore deviation multiplied by the MLT from the top of the sand to the oil/water contact, which is 115 ft. The net oil pay calculates as

$$\text{TVT} = \text{MLT}\ (\cos\psi)$$
$$\text{TVT} = 115\ (0.866)$$
$$\text{TVT} = 100 \text{ ft net oil sand at the position where the well penetrates the top of the sand.}$$

Notice the result if we consider bed dip, as well as wellbore deviation, to determine the correction factor for this case; that is, if we incorporate bed dip equal to 20 deg in Setchell's equation, Eq. (14-1). Then we are calculating the correction factor used to determine the TVT of total *net sand* at the penetration point at the top of the reservoir, as well as the correction factor for the TVT of net oil sand *directly above the penetration point of the well at the oil/water contact* (Fig. 14-31). Calculating the correction factor (CF) in Setchell's equation,

$$\text{CF} = [\cos\psi - (\sin\psi \cos\alpha \tan\phi)]$$
$$\text{TVT} = \text{MLT} \times \text{CF}$$

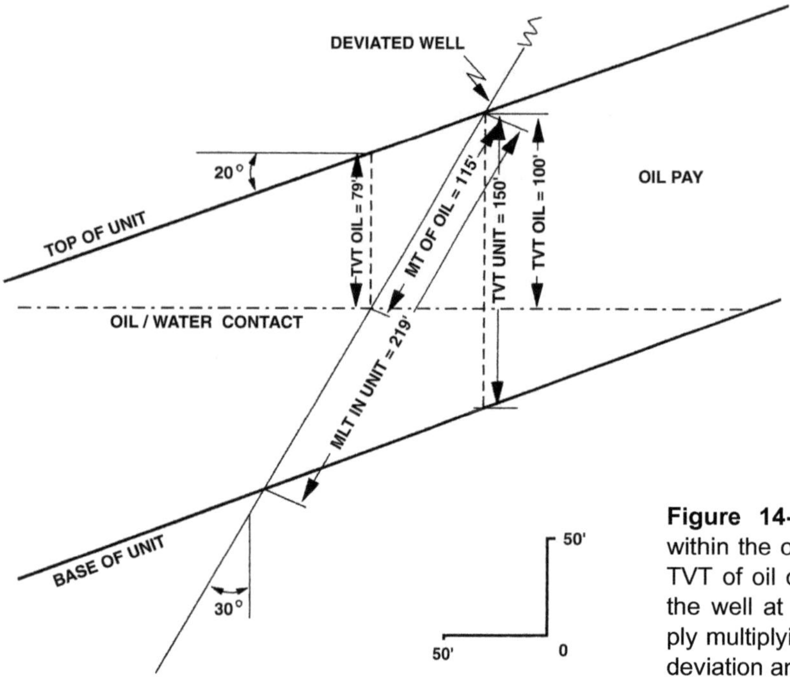

Figure 14-31 Deviated well penetrates the reservoir within the oil wedge (directly above the oil/water contact). TVT of oil column, directly below the penetration point of the well at the top of the reservoir, is calculated by simply multiplying the MLT of oil by the cosine of the wellbore deviation angle.

Using the data provided in Figure 14-30, the correction factor is based on

$$\text{Wellbore deviation } (\psi) = 30 \text{ deg}$$
$$\text{True bed dip } (\phi) = 20 \text{ deg}$$
$$\text{Delta azimuth } (\alpha) = 0 \text{ deg}$$

Therefore,

$$CF = [\cos 30° - (\sin 30° \cos 0° \tan 20°)]$$
$$CF = 0.866 - (0.5)(1)(0.364)$$
$$CF = 0.684$$

1. The true vertical thickness of the net sand at the penetration point of the well at the top of the sand is

$$TVT = MLT \times CF$$
$$TVT = (219 \text{ ft})(0.684)$$
$$TVT = 150 \text{ ft}$$

2. The true vertical thickness of the net oil pay sand above the penetration point of the well at the oil/water contact is

$$TVT = (\text{MLT from top of sand to the oil/water contact})(CF)$$
$$TVT = (115 \text{ ft})(0.684)$$
$$TVT = 79 \text{ ft}$$

The preceding discussion shows that a fluid contact in a deviated well is a special condition for which only the correction factor for wellbore deviation may be required to convert a MLT to TVT, even though the beds are dipping at a significant angle. We make an important conclusion here: *Wherever the penetration point of a well at the top of a reservoir is above a fluid contact (in the wedge), as in Figure 14-31, the dip of the beds can be considered zero in the equation for determining the net pay TVT at the penetration point at the top of the reservoir.* In other words, use the simple equation TVT = MLT cos ψ. Furthermore, as shown in Figure 14-31, if the actual reservoir bed dip and the MLT of the net pay sand are used in Eq. (14-1), Setchell's equation, then the TVT of the net pay is calculated directly above the *point of penetration of the oil/water contact*. Spot this point on the net oil map and use its TVT net pay value in the contouring of net oil, in addition to using the net pay value at the penetration point at the top of the reservoir.

Figure 14-32 illustrates a situation in which a reservoir containing oil and gas is penetrated by a deviated well. In this case, the vertical thicknesses of the oil and gas columns must be determined for isochore mapping. Using the data provided in the figure, verify the TVT for the oil and for the gas at two separate locations: (1) directly beneath the penetration point of the well at the top of the reservoir, and (2) the penetration point of the gas/oil contact. Finally, calculate the TVT of water, oil, and gas at the point where the well penetrates the oil/water contact (dashed line).

Figure 14-33 illustrates a well deviated in an up-dip direction, penetrating an oil reservoir in the oil wedge. The same procedures as previously discussed are used to calculate the TVT of oil at the locations where the well penetrates the top of the reservoir and the oil/water contact. If desired, the TVT of oil can also be calculated at the position where the well penetrates the base of the productive unit. First, calculate the TVT of the water column at this location, and then subtract this value from the TVT of the total net sand to arrive at the vertical thickness for oil.

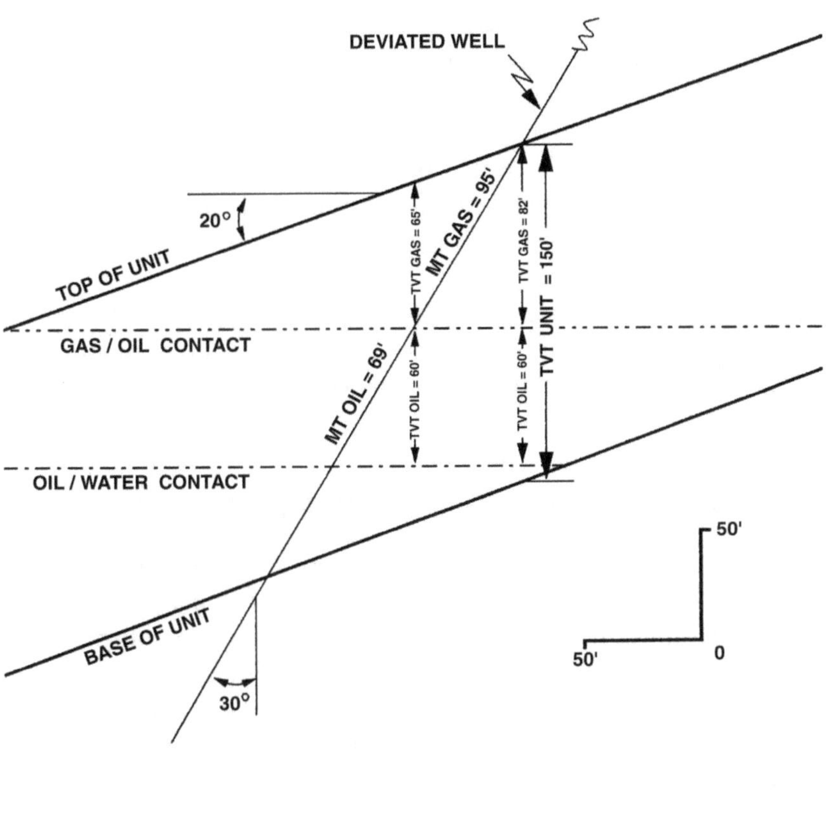

Figure 14-32 The deviated wellbore penetrates the reservoir directly above two separate fluid contacts (gas/oil and oil/water).

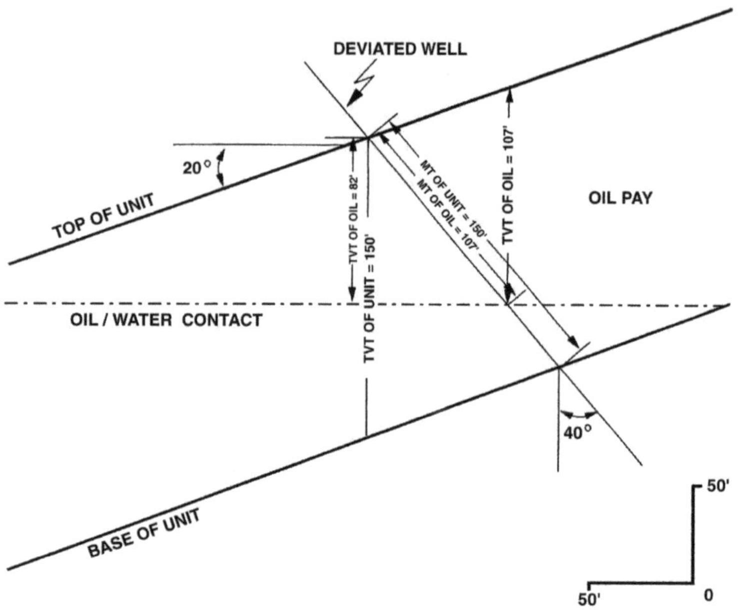

Figure 14-33 Deviated well drilled in an up-dip direction, penetrating a reservoir within the oil wedge. True vertical thickness calculation methods are the same as those for a well drilled in a down-dip direction, penetrating the top of a reservoir within the hydrocarbon wedge.

Consider the case if the well in Figure 14-33 had been drilled slightly to the right and penetrated the top of the reservoir within the wedge zone but penetrated the base of the reservoir above the oil/water contact. Use of Setchell's equation allows calculation of the TVT of the net pay at the point where the well penetrates the base of the reservoir. If the depth of the oil/water contact is not known, the TVT of net pay at the point the well penetrates the top of the reservoir is not known, although a minimum amount of net pay can be determined by assuming the oil/water contact is immediately below the penetration point of the base of the reservoir.

If a well penetrates a reservoir up-dip of the ILW, the TVT of the pay in the well at the point of penetration of the reservoir is given by Setchell's equation. If a well penetrates a reservoir down-dip of the ILW, the TVT of the pay in the well at the point of penetration of the reservoir is given by the vertical distance between the penetration point and the hydrocarbon/water contact.

By having a good understanding of the geometric relationship of the productive unit, wellbore, and fluid contacts, we can use this knowledge to our advantage. The calculation of the net gas or net oil at various points in a well, such as the well's penetration point at the top of the productive unit, base of the unit, or at fluid contacts, can provide additional net pay values. These values can be used in the preparation of the net gas or net oil isochore maps, providing additional control in the wedge zones.

The detailed calculations shown in this section are not always required or justified. However, the use of these techniques may prove to be very important where detailed, accurate mapping is needed for some specific reserve estimate, development plan, enhanced recovery program, unitization, equity determination, or litigation.

MAPPING THE TOP OF STRUCTURE VERSUS THE TOP OF POROSITY

We discuss the effect of mapping on a structure top versus a porosity top with regard to structure mapping in Chapter 8, and here we review this special condition as it relates to net pay isochore maps. We mention in Chapter 8 that the upper portion of a reservoir unit may be composed of nonreservoir-quality rock. The terms tight zone and tight streak refer to an interval of impermeable rock. Although the top of the unit may represent the actual stratigraphically equivalent top, it does not constitute the top of reservoir-quality rock. Therefore, the structure map prepared to interpret the structure may not be useful to map and evaluate the reservoir itself.

Once the structure mapping is complete, the question arises as to whether a separate map on the *top of porosity* is required for accurate delineation of the reservoir and for use in the construction of net hydrocarbon isochore maps, from which estimates of reserves are made. Two parameters are considered in evaluating the importance of the nonreservoir-quality rock: (1) the thickness of the tight zone, and (2) the relief of the structure. A thick tight zone has a greater effect on ultimate reserve estimates than one that is thin. Low-relief structures introduce greater error in delineating the limits of a reservoir than steeply dipping structures, particularly if the low-relief structure contains a reservoir with bottomwater. This is true because a steeply dipping reservoir is associated with a relatively small wedge zone when compared to the total area of the reservoir.

On a low-relief structure, the wedge zone of a reservoir can represent a significant portion of the total reservoir area (Fig. 14-34). Figure 14-34a shows in map and cross-sectional views a low-relief bottomwater reservoir mapped on the top of a productive unit, which consists of nonreservoir-quality rock in the upper 75 ft. The same reservoir is mapped on the top of porosity in Figure 14-34b. The net oil isochore map prepared from each structure map is shown in Figure 14-34c. The same net pay values are assigned to each well in both isochore maps. In this case, because the reservoir is on a low-relief structure, the difference in reservoir volume between the isochore map based on the top of structure and the isochore map based on top of porosity is 637 acre-feet, or a significant overestimate of *32%* (Fig. 14-34d). Consequently, the volume of recoverable hydrocarbons is overestimated by 32%.

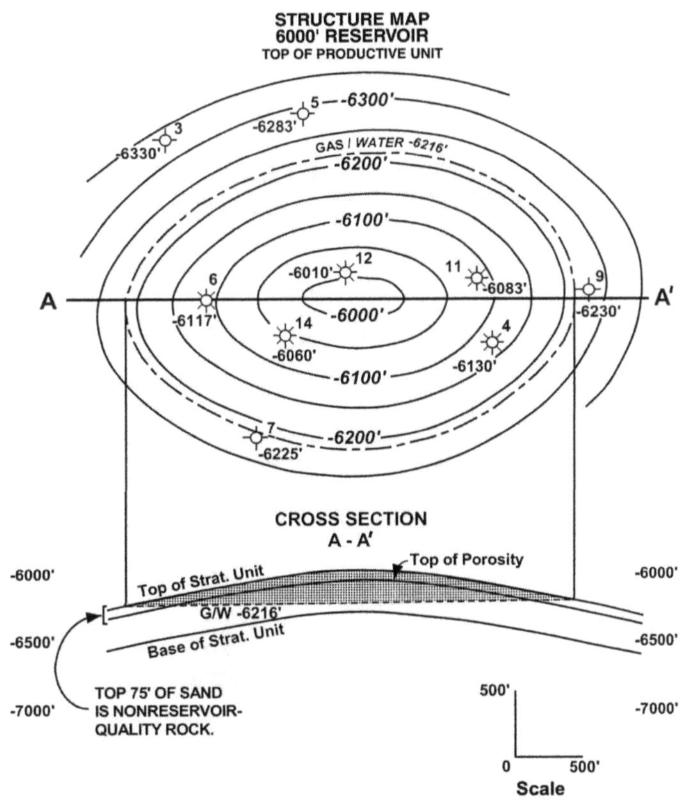

(a)

Figure 14-34 (a) The structure map on the top of the stratigraphic unit containing the 6000-ft reservoir, and the cross section A-A′. Upper 75 ft of the unit contains nonreservoir-quality rock. (b) The structure map on top of porosity (6000-ft reservoir) and the cross section A-A′. (c) Two separate net pay isochore maps: (1) the upper isochore map is based on the structure map on the top of the stratigraphic unit, and (2) the lower isochore map is based on the top-of-porosity structure map. Net pay values for all the wells are the same for each map. (d) There is a 32% reduction in reservoir volume for the net pay isochore map constructed from the top of porosity map versus the net pay isochore map constructed from the top of the stratigraphic unit. This is a significant reduction in volume.

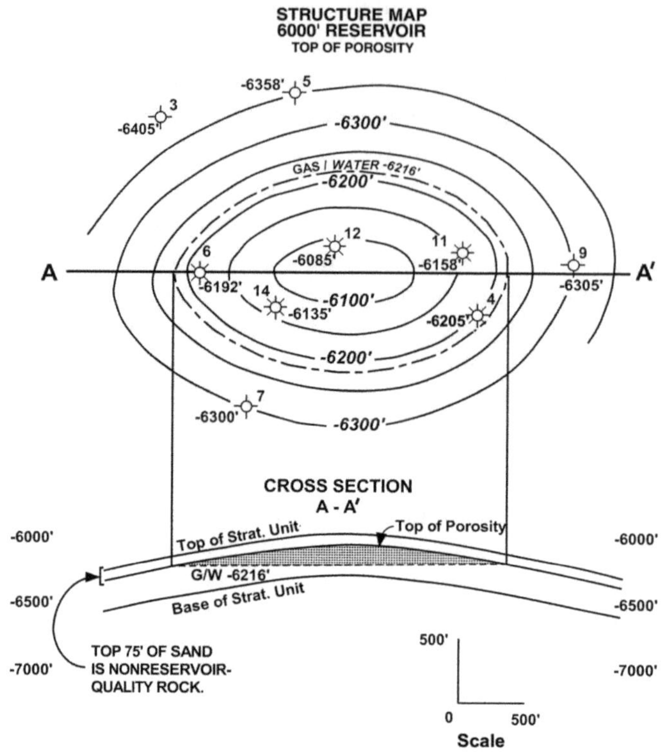

(b)

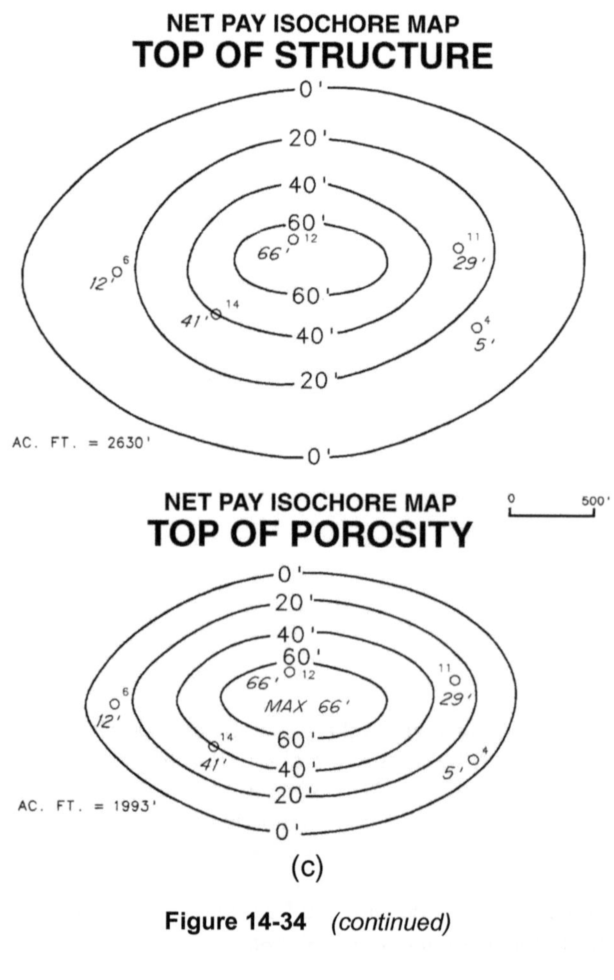

(c)

Figure 14-34 *(continued)*

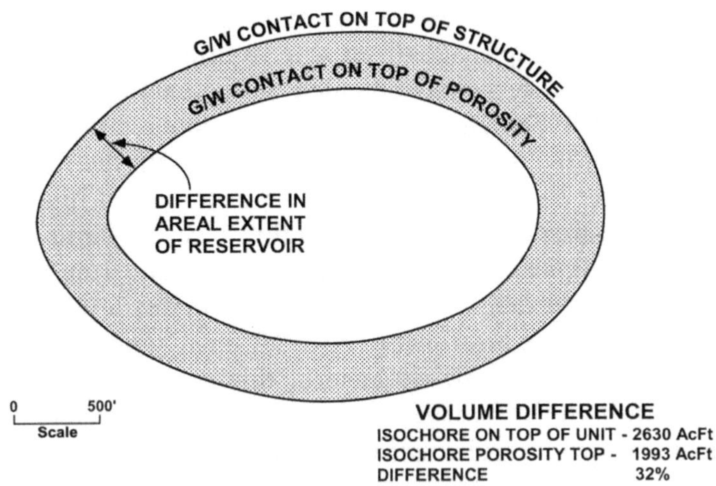

(d)

Figure 14-34 *(continued)*

The decision to prepare a separate map on the top of porosity, where the upper portion of a productive unit is tight, needs to be made on a reservoir-by-reservoir basis. Depending on the geometry of the reservoir and thickness of the tight zone, the difference in volume between a map on the top of the unit and one on the top of porosity may be too insignificant to warrant additional mapping.

If, for some reason, it is not possible to create a separate top-of-porosity map, it may be possible to walk wells through the wedge zone, as described earlier in this chapter. In the case illustrated by Figure 14-34, this procedure would result in a very wide zero contour, with the first net pay occurring 75 ft above the oil/water contact. Creating a separate top-of-porosity map allows you to use the BRV times net/gross method to determine the net pay. In this case, the net/gross refers to the interval from the top of the porosity to the base of the reservoir.

FAULT WEDGES

A **fault wedge** within a reservoir is defined as a wedge-shaped section of strata bounded by a fault. It is commonly just as important to map net pay within the fault wedge of a productive reservoir as it is to map net pay within the hydrocarbon wedge above water. If a productive unit is thin or the dip of the bounding fault is a high angle, the reservoir volume affected by the fault wedge may be insignificant and can be ignored for all practical purposes. In cases where the reservoir is thick or the fault dips at a low angle, the reservoir volume within the fault wedge may be significant and must therefore be included when constructing the net pay isochore map. There are several ways to handle the mapping of a fault wedge. It may be contoured in the conventional way, or a mid-trace (mid-point) method can be used.

Conventional Method for Mapping a Fault Wedge

The conventional method of contouring the fault wedge is the most accurate and should be employed whenever possible. With this method, the wedge is actually contoured using all control points in the same manner as contouring a water wedge. As with a water wedge, where the impermeable zones (shale, tight rock, etc.) are fairly evenly distributed throughout the reservoir section, the fault wedge net pay contours may be evenly spaced in the wedge zone. All the control points must be honored, however, even though it may cause an uneven spacing of the contours. Figure 14-35 shows an example of a structure map on the top of a productive unit. We use a simple, idealized map in order to provide a clear concept of mapping the fault wedge. The reservoir is bounded on the east by a west-dipping fault, with the remainder of the reservoir bounded by an oil/water contact at −8000 ft. The structure map shows the intersection of the top of the productive unit with the fault, as well as the intersection of the fault with the base of the unit (shown as a dashed line). The area between these two intersections is the area of fault wedge. In this case, the fault dips to the west at 45 deg and the productive unit dips at 30 deg. It is readily apparent from the fault intersections with the top and base of the unit that the fault wedge comprises a large portion of this reservoir. For simplicity, we assume the reservoir to have 50% net sand and 50% shale, evenly distributed throughout the gross interval. Because of this even shale distribution, we can evenly space the contours in both the hydrocarbon and fault wedges. The reservoir is in the downthrown fault block, so the control required to map the fault wedge in this example includes the intersections of the top and base of the productive unit with the fault. The upthrown trace of the fault, and therefore the fault gap as seen on a structure map, plays no part in mapping the fault wedge.

Figure 14-36a illustrates the net oil isochore map for this reservoir with both the water and fault wedges conventionally contoured. The cross section A-A′, drawn below the net oil isochore map, depicts certain key control points in the reservoir, including the oil/water contact on the top of the productive unit, the inner limit of water (the oil/water contact on the base of the unit), the intersection of the base of the reservoir with the fault, and the intersection of the top of the reservoir with the fault. These key control points play an important part in the construction of the wedges for this net oil isochore map.

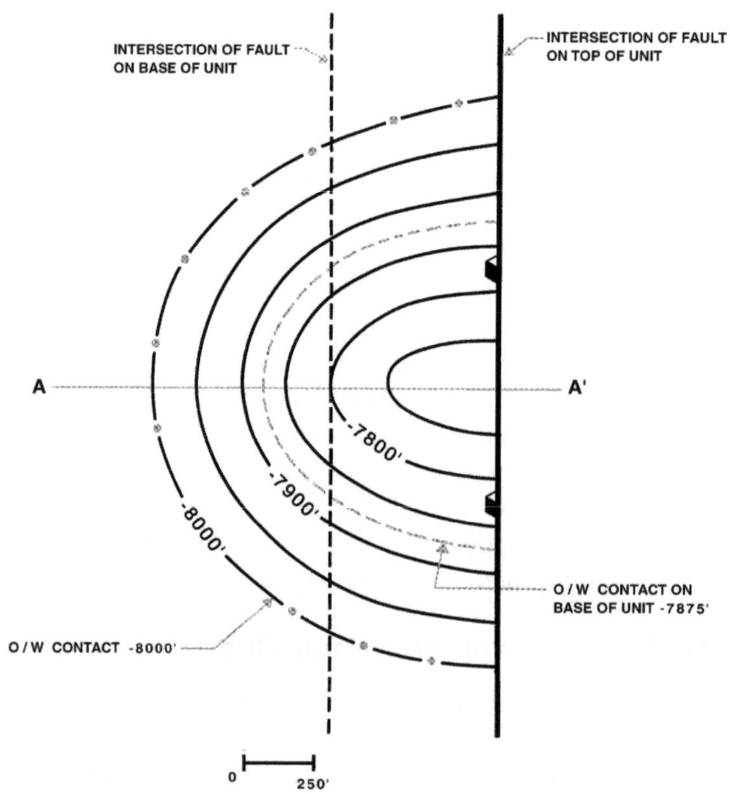

Figure 14-35 Structure map on top of a reservoir shows the intersections of a fault with the top and base of the unit. These two intersections are required in mapping the fault wedge.

If mapping the net pay using the BRV times net/gross method, the fault surface becomes one of the bounding surfaces used in determining the BRV. If the reservoir is a bottomwater reservoir, the BRV is determined from the top of the reservoir surface, the fault surface, and the hydrocarbon/water contact. If the reservoir is an edgewater reservoir, the BRV is determined from the top of the reservoir surface, the fault surface, the hydrocarbon/water contact, and the base of the reservoir surface.

Mid-Trace Method

The use of the conventional method of contouring the fault wedge can be tedious and at times unjustified. In such cases, we can use a shortcut method for contouring the fault wedge, referred to as the **mid-trace method**, or **mid-point method** (Fig. 14-36b). To construct an isochore map using this method, first draw a line that is halfway between the intersections of the top and base of the reservoir with the fault. This line is the mid-trace. Draw the line to stop at the ILW, shown as points a and a′ in Figure 14-36b. Next, extend this line from the ILW straight to the intersection of the fault with the oil/water contact on the top of the reservoir, as indicated by a-b and a′-b′ on the figure. This line b-a-a′-b′ becomes a boundary of the reservoir. Any part of the reservoir in the fault wedge, but outside this boundary line, is considered as being folded under to convert the wedge zone inside this line to an area of full-thickness. This is illustrated in the cross section below the net oil isochore map in Figure 14-36b. Notice that this procedure extends the full-thickness area to the mid-trace. Next, extend all contours in the hydrocarbon wedge to intersect with line segments a-b and a′-b′, as illustrated in the figure. The finished map includes an enlarged full-thickness area (125 ft of net pay) plus a wedge.

(a)

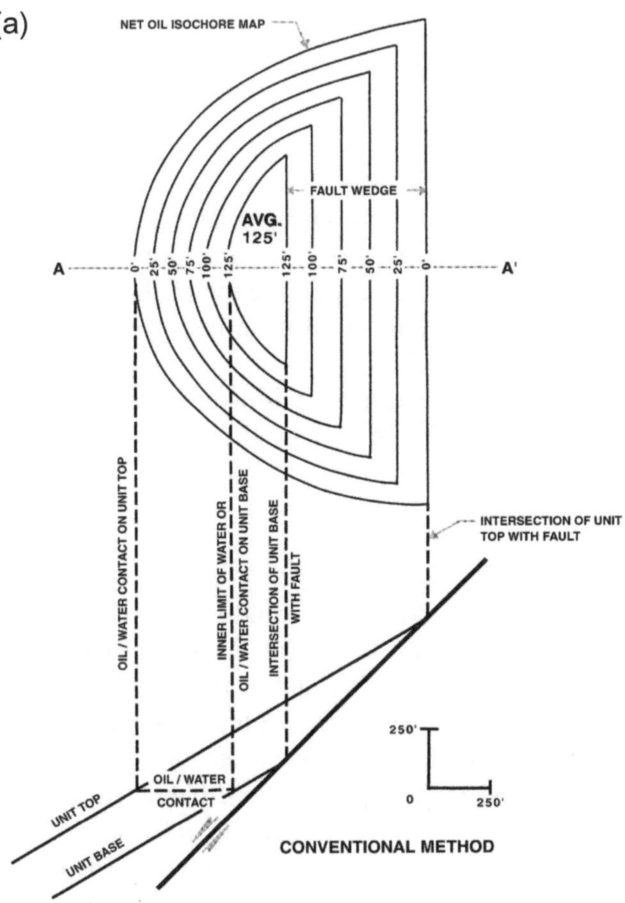

Figure 14-36 (a) The fault wedge is mapped using the conventional method. See cross section below map. (b) Mid-trace method for mapping the fault wedge. See cross section.

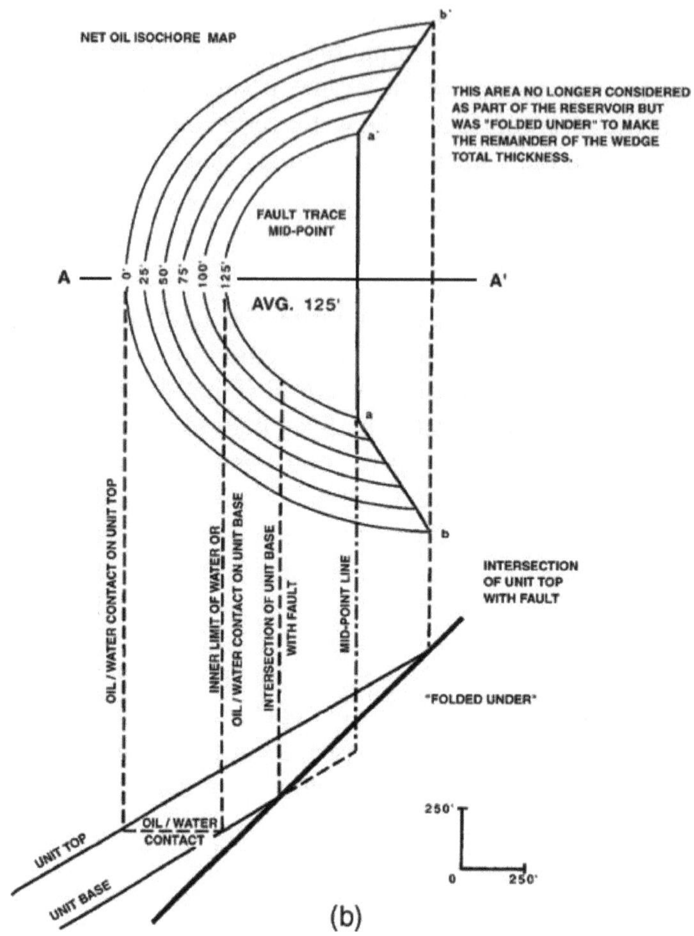

Figure 14-36 *(continued)*

Figure 14-37 illustrates the use of the mid-trace method of contouring a fault wedge with a reservoir bounded by two intersecting faults. Note that the mid-trace extension, from the inner limit of water to the intersection of the fault with the oil/water contact on the top of the unit, is in opposite directions on the individual fault wedges. Remember, always draw a line from the mid-trace to the intersection of the zero net pay contour with the fault.

It should be pointed out that Figures 14-36 and 14-37 are drawn to illustrate the mid-trace method. The fault shape implied by the fault traces (see Figs. 7-42b and 8-54b) in these figures is unlikely to occur in normal geological settings.

NONSEALING FAULTS

Although faults play a very important role in the trapping of hydrocarbons, studies have shown that some faults are nonsealing, thereby permitting the migration of hydrocarbons from one fault block to the next. One of the most common situations resulting in a nonsealing fault is the juxtaposition of permeable stratigraphic units across a fault, with capillary properties of the fault zone being insufficient to create a seal. This can occur where a fault does not have sufficient displacement to separate an entire permeable unit from one fault block to the next, or where juxtaposition of different permeable units exists. The concept of sealing and nonsealing faults is discussed at more length in Chapter 6.

Hydrocarbon accumulations can occur on both sides of a nonsealing fault if a trap exists in the up-structure fault block. The accumulations would have a common water level. With nonsealing faults, it is important to map fault wedges because they can contain significant amounts of hydrocarbons. We contour the fault wedge basically the same as presented in the previous section, with one exception. Since there is hydrocarbon pay upthrown and downthrown to the fault, two reservoirs and two fault wedges must be mapped to account for all the hydrocarbon volume. The isochore maps for the two reservoirs can be constructed individually or contoured as one map.

Figures 14-38 and 14-39 illustrate an example of reservoirs juxtaposed across a nonsealing fault. A productive sandstone unit contains two fault-separated reservoirs on the southern flank of a salt structure. Figure 14-38a shows the structure map on the top of porosity in Reservoirs A and B. The reservoirs are limited to the north by a vertical salt face, to the east and west by faults, and in the down-dip direction by a common gas/water contact at −10,550 ft. The interval from top to base is in excess of 200 ft thick, and it is all reservoir-quality sand. Fault B, with a vertical separation of 200 ft, has insufficient displacement to completely separate the sand, so part of the sand is juxtaposed across the fault. The same gas/water contact in both Reservoirs A and B indicates that the fault is nonsealing and that the two reservoirs are in communication. Figure 14-38b

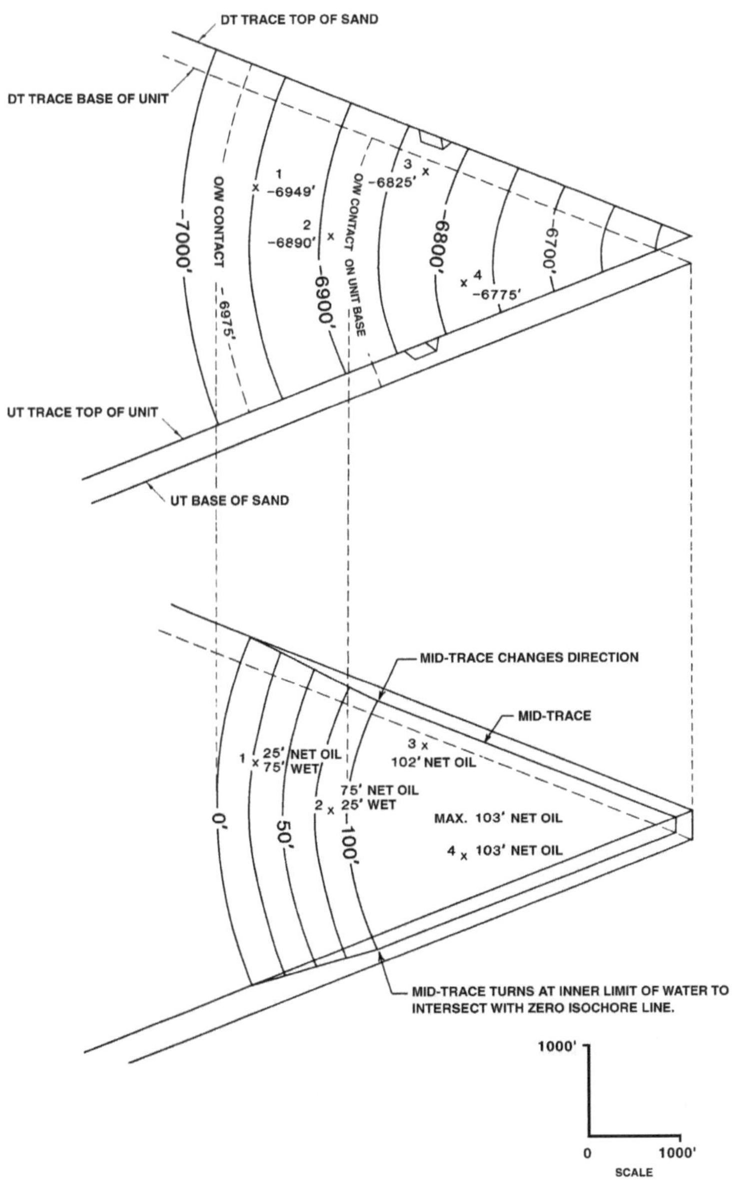

Figure 14-37 Construction of two fault wedges involving intersecting faults. Note the change in direction of the mid-traces at the inner limit of water.

through d are the structure map on the base of the 9500-ft Sand, the fault surface map, and the net sand map respectively. These maps are required for isochore construction.

The net gas isochore maps for the Reservoirs A and B can be constructed separately or as one single net gas isochore map. Figure 14-39a and b show the individually constructed net gas isochore maps for Reservoirs A and B respectively. Notice that the fault wedges for Faults A and B comprise a significant portion of each of the reservoirs. Figure 14-39c illustrates how the two reservoirs can be mapped together by combining the wedges for Fault B.

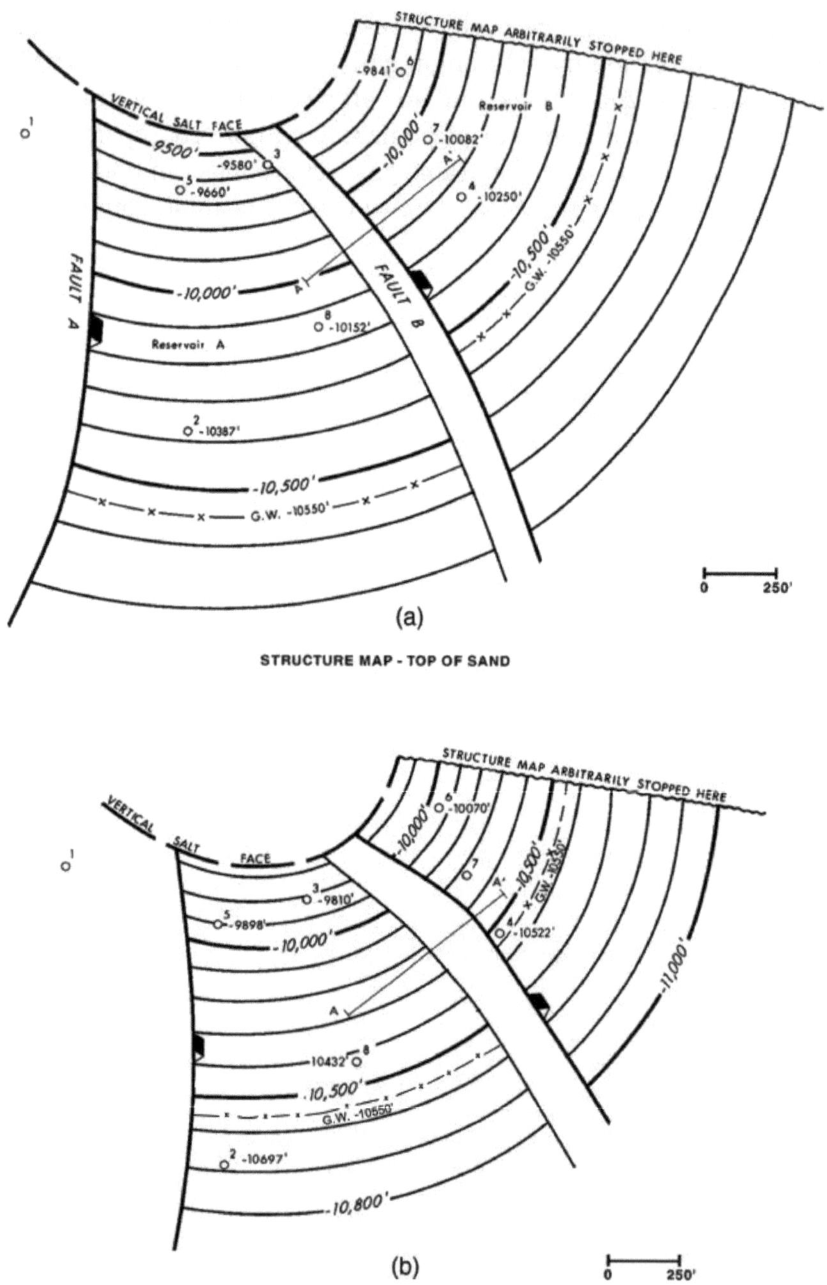

Figure 14-38 (a) Structure map on top of Reservoirs A and B. Salt face is vertical at this depth.(b) Structure map on base of Reservoirs A and B. (c) Fault surface map of Faults A and B. (d) Net sand isochore map for 9500-ft Sand.

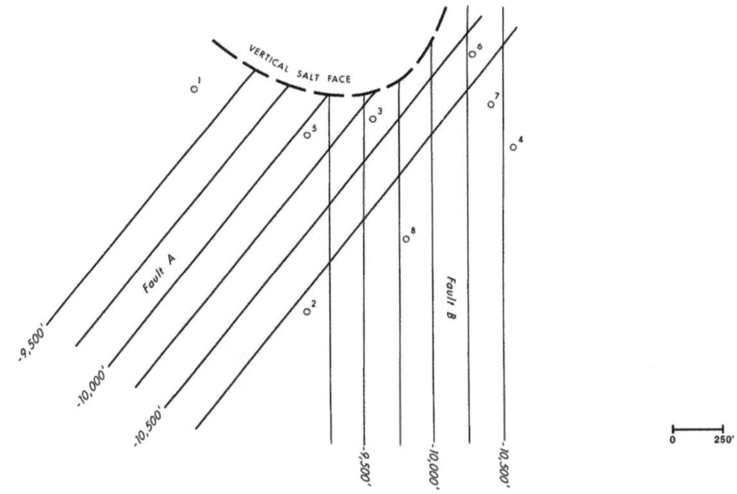

(c)

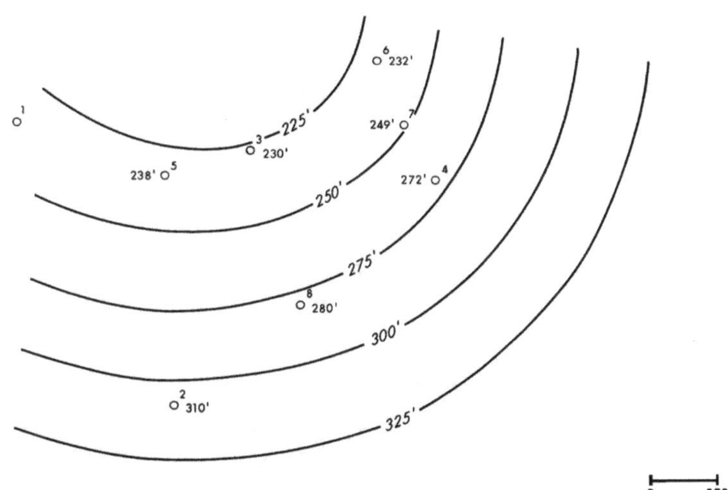

(d)

Figure 14-38 *(continued)*

Cross section A-A′ shows about 75 ft of sand juxtaposed across the fault at this position in the reservoir.

Isochore maps in which the fault wedges are combined, such as the one shown in Figure 14-39c, are difficult to construct and can easily result in errors. We recommend for simplicity of construction and planimetering that, even with nonsealing faults, each reservoir and fault wedge be mapped separately, as shown in Figure 14-39a and b. Constructing net pay isochore maps for each productive fault block can also be important in developing field development plans, as even nonsealing faults can act as baffles during production time scales. Thus, a well in one fault block may not access reserves in an adjacent fault block, even if the intervening fault is nonsealing and the two fault blocks share a common water level.

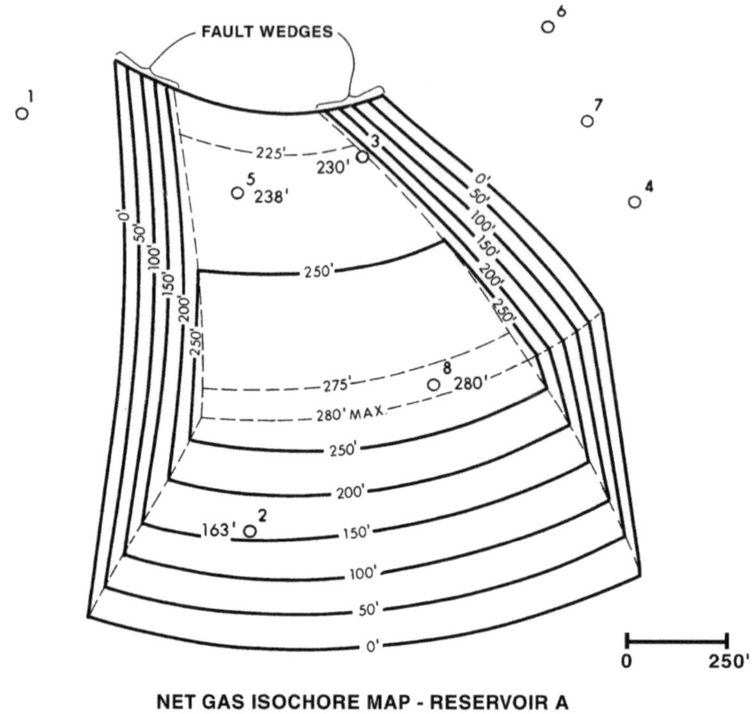

(a)

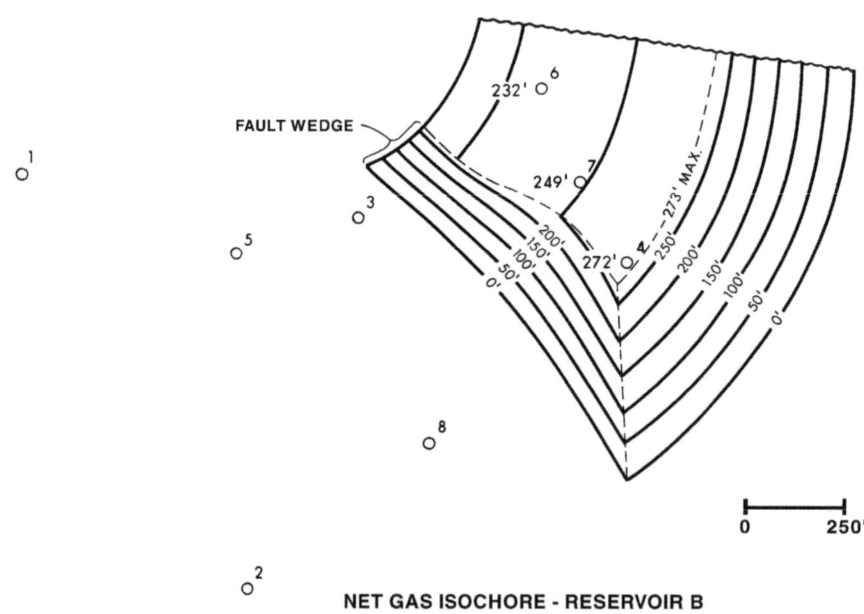

(b)

Figure 14-39 (a) Net gas isochore map for Reservoir A. The map includes two fault wedges and a water wedge. (b) Net gas isochore map for Reservoir B. Fault wedge and water wedge are significant portions of the reservoir volume. (c) Composite net gas isochore map for Reservoirs A and B. Notice the complexity in the mapping of the two fault wedges created by Fault B when both reservoirs are mapped together.

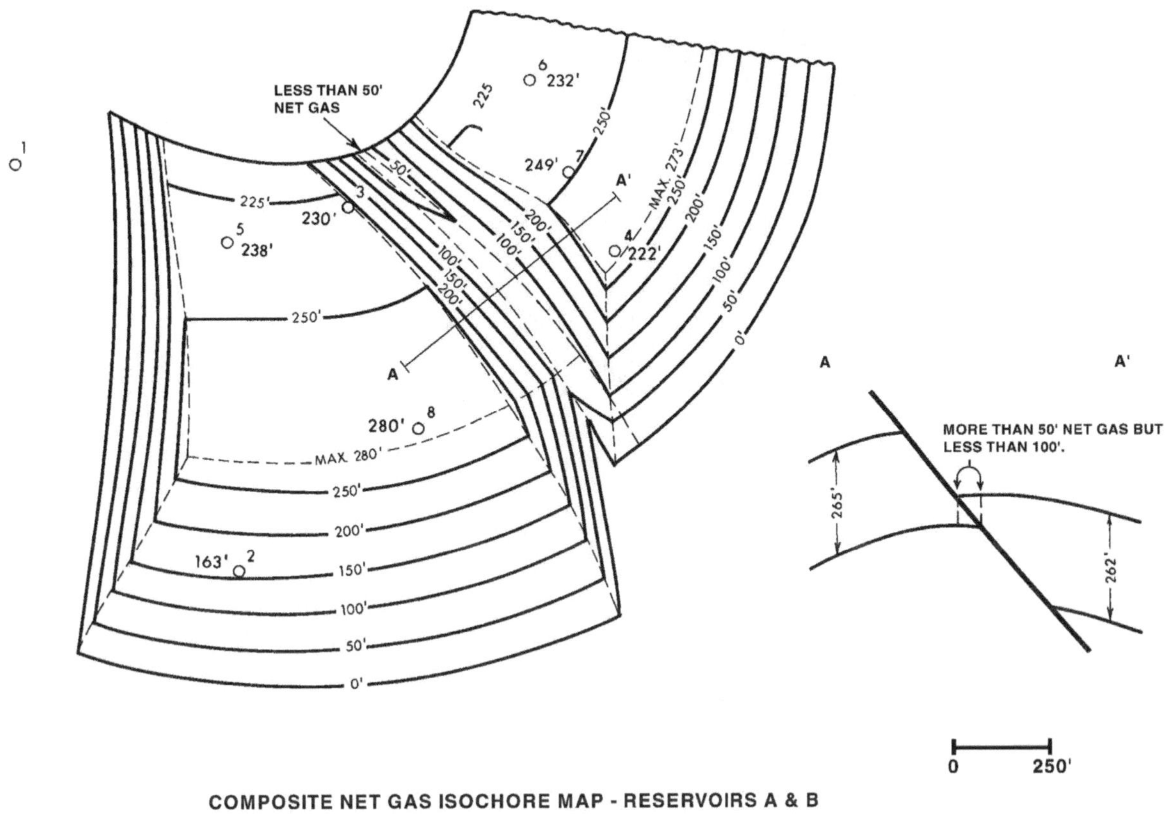

(c)

Figure 14-39 *(continued)*

VOLUMETRIC CONFIGURATION OF A RESERVOIR

We use a net pay map to determine the volume of a reservoir, so we want to consider how this volume, as depicted by the map, corresponds to the actual structural configuration of the reservoir. Figure 14-40 relates the net pay map depiction of volume of a gas-on-oil accumulation to the structural configuration of the productive unit. The cross section in the center of the figure shows the actual structural configurations of the oil reservoir and the gas reservoir. The cross sections in the lower portion of the figure show the varying thickness of each reservoir, with each represented as a polygon with a flat base. If we consider possible changes in thickness in the third dimension, then each polygon represents an irregular solid that has a volume equal to that of the reservoir. It is important to notice that the depicted volume is not a true representation of the structural configuration of the reservoir. The reservoir has been artificially flattened, or piled up (referred to as isochore piling), to represent the same volume that exists within the structure. The net pay map will be planimetered, and then the depicted volumetric solid is used in the calculation of the reservoir volume.

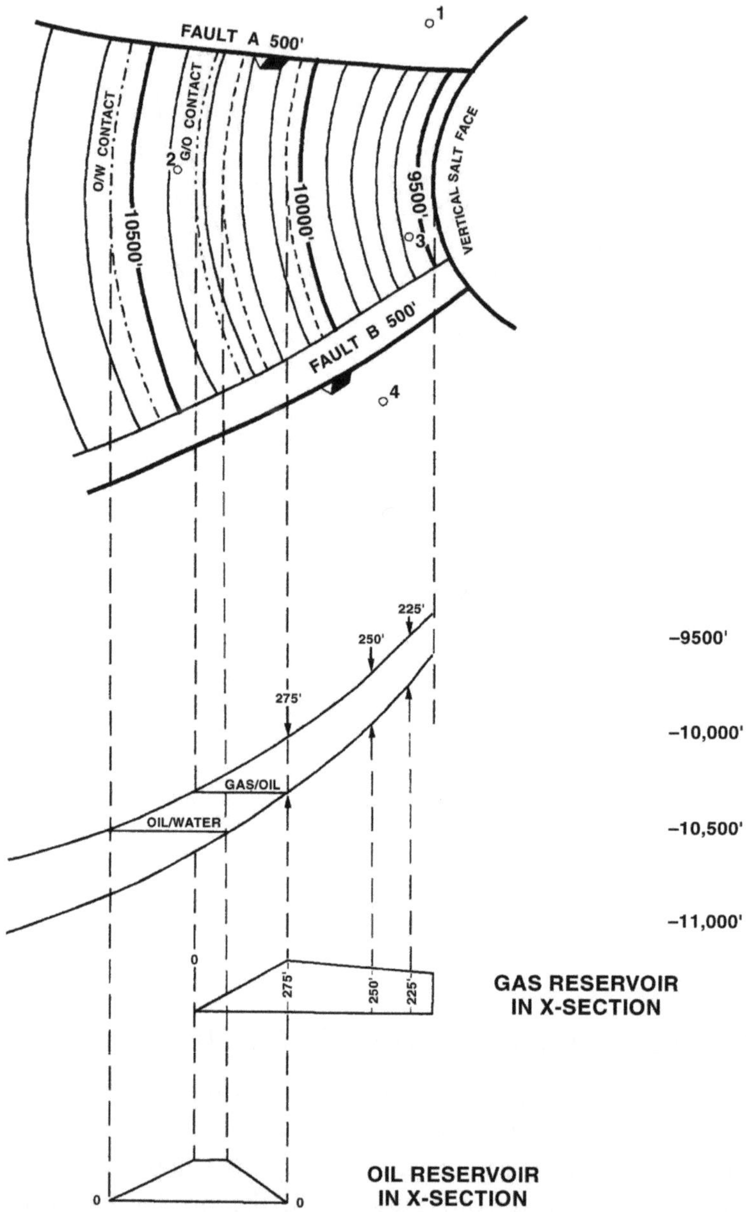

Figure 14-40 In the preparation of a net hydrocarbon isochore map, the configuration of the reservoir is completely rearranged and the base is artificially flattened. Compare the configuration of the net gas and net oil cross sections in the lower portion of the figure to that of the structure cross section in the center.

RESERVOIR VOLUME DETERMINATIONS FROM ISOCHORE MAPS

Two methods are commonly used to determine reservoir volume from net pay isochore maps using a mechanical planimeter, the **horizontal slice method** and the **vertical slice method**. Previous editions of this book describe these methods in significant detail (Tearpock and Bischke 1991, 2003). Mechanical planimeters are rarely used today. Most reservoir volumes are calculated by digitizing maps using a planimetering program or are calculated within the software that created the maps. Individuals interested in calculating reservoir volumes using mechanical planimeters are referred to the earlier editions of this book.

TILTED HYDROCARBON/WATER CONTACTS

So far in this chapter, we have assumed that hydrocarbon/water contacts are flat. This is an excellent assumption and accurate in most cases, but there are exceptions. Hubbert (1953) recognized that hydrodynamic flow in an aquifer causes a hydrocarbon/water contact to tilt in the direction of flow. A comprehensive discussion of tilted hydrocarbon/water contacts is beyond the scope of this book, but a basic introduction to the concept is worthwhile because of the potential impact on net pay isochore mapping.

The inclination of hydrocarbon/water contacts is described by the following equation (Hubbert 1953):

$$dz/dx = (\rho w/(\rho w - \rho h)) * dh/dx \quad (14\text{-}2)$$

where

dz/dx = inclination of the hydrocarbon/water contact in, for example, feet per mile
ρw = density of the formation water
ρh = density of the hydrocarbon
dh/dx = hydraulic head of the aquifer in, for example, feet per mile

Equation (14-2) shows that for any given hydraulic head, the inclination of the hydrocarbon/water contact is a function of the density difference between the hydrocarbon and water. The greater the density difference, the less the tilt on the contact. Lower-gravity, higher-density oil in an aquifer of fresh water will have a greater inclination in the hydrocarbon/water contact than a high-gravity, lower-density oil in an aquifer of high-salinity water for any given hydraulic head. Gas/water contacts show even less inclination for a given hydrodynamic head because of the greater density difference in the two fluids.

Tilted hydrocarbon contacts are not caused by the viscous drag of aquifer water on trapped hydrocarbons. They are caused by pressure gradients, or technically hydraulic potential gradients, within the reservoir. Both inclined contacts and aquifer flow are responses to the same force but do not cause each other.

Most cases of hydrodynamic flow are driven by elevation differences between the recharge area of an aquifer and its eventual discharge region. In the case of the U.S. Gulf Coast, the hydraulic head of shallow, low-salinity aquifers may be two or three feet per mile (Hubbert 1953). In the U.S. Great Plains, the hydraulic head may be ten feet per mile, whereas in intermontane basins, the hydraulic head may be hundreds of feet per mile (Hubbert 1953). The greater the hydraulic head, the greater the inclination of the hydrocarbon/water contact for a given density contrast between hydrocarbons and aquifer water. For gases, $\rho w/(\rho w - \rho h)$ is between one and two, so a gas/water contact is typically inclined between one and two times the hydraulic gradient. Thus, an aquifer gradient of 10 ft/mi would result in an inclined gas/water contact of between 10 and 20 ft/mi (Hubbert 1953). For oils, the term $\rho w/(\rho w - \rho h)$ ranges from two to infinity but is generally between seven and fifteen with ten being a reasonable average number (Hubbert 1953).

Many aquifers are hydrodynamic. If the formation water is fresh, it must be hydrodynamic, and a hydrocarbon/water contact will be tilted (Hubbert 1953). If formation water is saline, it may be flowing (hydrodynamic), and a hydrocarbon/water contact will be tilted; or the formation water may be static, and the hydrocarbon/water contact will be horizontal. Most hydrodynamic aquifers are onshore and associated with significant topographic relief. Work in the North Sea, however, suggests that some deep, offshore aquifers are hydrodynamic due to dewatering of overpressured shales into laterally extensive permeable units that subcrop normally pressured permeable units (Dennis et al. 2000).

Even in areas with hydrodynamic flow, the inclination on an oil/water contact may be small, a maximum of a foot or two from one side of the accumulation to another. Small inclinations on oil/water contacts can be very difficult to resolve from uncertainties in depth measurements, as discussed in Chapter 3. For a very large field covering several miles laterally, however, the same inclination of the contact can result in tens of feet of elevation difference from across the field in the direction of hydrodynamic flow (Dennis et al. 2000). The inclination of the contact can be differentiated conclusively from depth measurement uncertainties if good-quality formation pressure is available (Fig. 14-41).

Oil fields with gas caps in hydrodynamic units behave differently depending on how much inclination there is on the oil/water contact. In a hydrostatic environment, both the gas/oil contact and the oil/water contact in an oil field are horizontal (Fig. 14-42a). If the oil/water contact is inclined such that the gas cap is still underlain everywhere by oil, the oil/water contact is inclined but the gas/oil contact is horizontal. This is because the oil is in contact with the hydrodynamic aquifer, whereas the gas is only in contact with the hydrostatic oil (Fig. 14-42b). If the oil/water contact is inclined further, part of the gas cap may be underlain by the hydrodynamic aquifer. It this case, the gas/oil contact will be horizontal but the gas/water contact will be inclined, although less so than the oil/water contact (Fig. 14-42c). Finally, if the oil/water contact is sufficiently inclined, the oil accumulation may be totally separated from the gas cap, so that the gas will be underlain everywhere by water and the gas/water contact will be inclined (Fig. 14-42d).

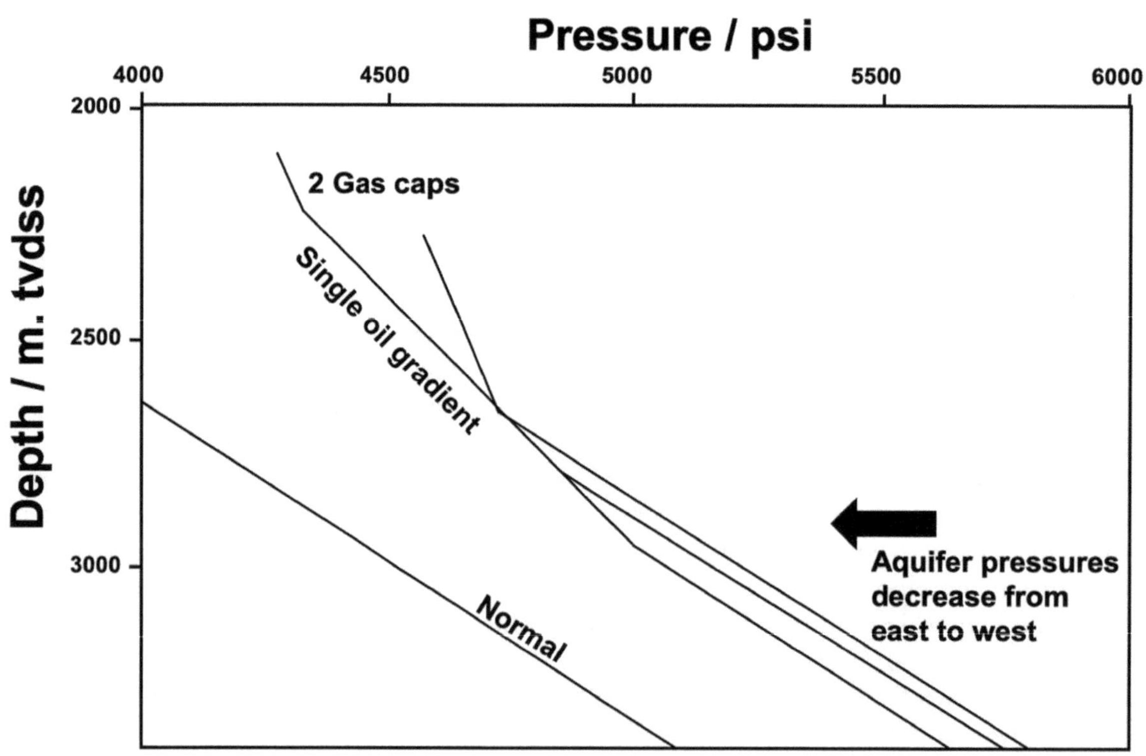

Figure 14-41 Formation pressures for Pierce Field in the central North Sea. Pierce Field is associated with two salt domes and has two separate culminations with two separate gas caps with different gas/water contacts. All of the oil lies on a single pressure gradient. Aquifer pressures on different sides of the field have similar gradients but are slightly offset from one another with the aquifer pressure at a constant depth decreasing from east to west. The oil/water contact at Pierce Field dips to the west at about 5 deg, or almost 500 ft/mi (90 m/km). (Modified from Dennis et al. 2000. Published with permission from Norwegian Petroleum Society.)

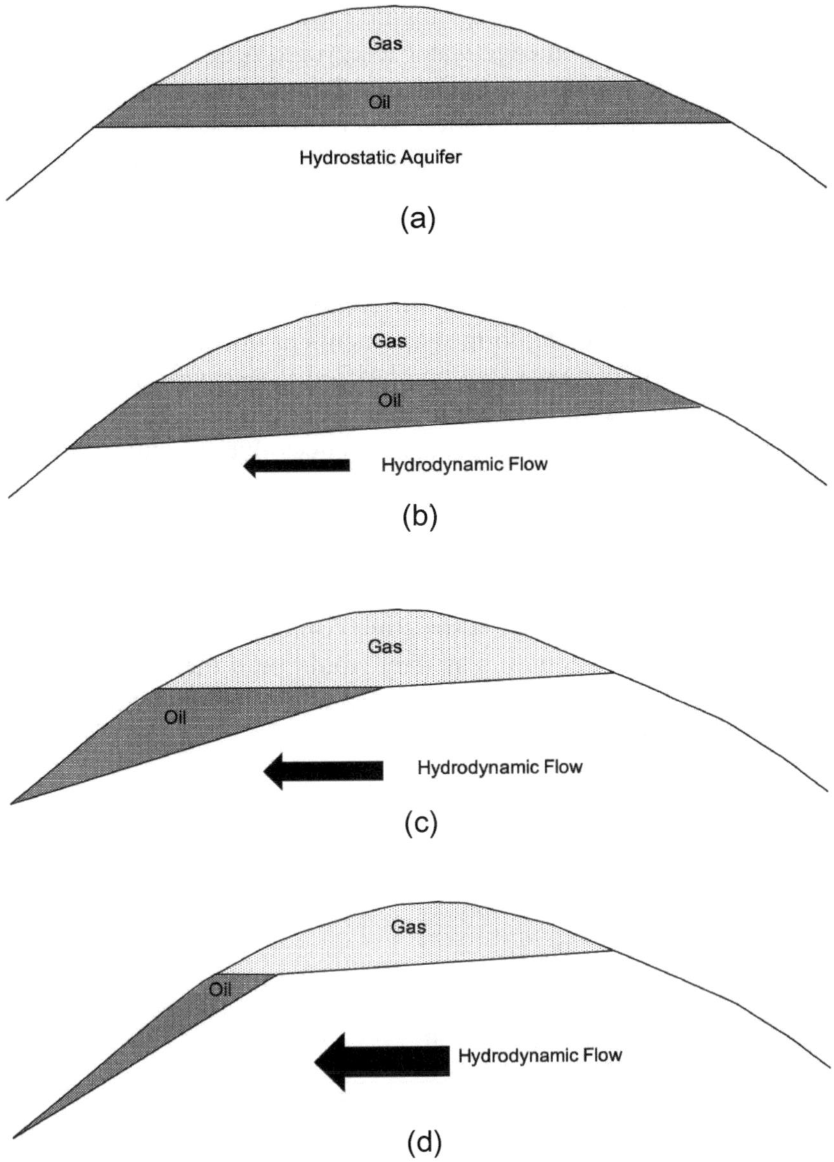

Figure 14-42 (a) Oil field with a gas cap in a hydrostatic aquifer. Both the gas/oil and oil/water contacts are horizontal. (b) Oil field with a gas cap in a hydrodynamic aquifer with a small hydraulic head. The oil/water contact is inclined, as the oil is in contact with the hydrodynamic aquifer. The gas/oil contact is horizontal, as the gas is only in contact with the hydrostatic oil. (c) Oil field with a gas cap in a hydrodynamic aquifer that has a moderate hydraulic head. The oil column has been inclined sufficiently that part of the gas cap is underlain by hydrostatic oil and part of the gas cap is underlain by hydrodynamic water. The gas/oil contact is horizontal and the gas/water contact is slightly inclined. (d) Oil field with a gas gap in a hydrodynamic aquifer with a significant hydraulic head. The oil column has been inclined to the point that it is almost completely separated from the gas cap. The oil/water contact is strongly inclined, and most of the gas/water contact is inclined. (Modified from Hubbert 1953. Published by permission of J. Brenneke.)

To properly calculate the volumetrics of an oil field, it is important to distinguish inclined hydrocarbon/water contacts from perched hydrocarbon/water contacts because inclined contacts tend to have much greater impact on net pay volumes than do perched contacts. Both cases have hydrocarbon/water contacts at different elevations in different parts of the field. In both cases, reservoir formation pressures of the hydrocarbon phase will lie on a single line, reflecting the in situ density of the hydrocarbon. The water pressures will intersect the hydrocarbon pressure line at various depths, reflecting the different oil/water contacts (compare Figs. 8-86 and 14-41). One key difference is that perched contacts can occur only in edge water reservoirs. A second difference is that perched contacts are associated with closed lows on the base of the reservoir, whereas tilted contacts are not. For perched water to be an acceptable explanation, you must be able to map a closed low on the base of the reservoir. A third difference is that the chemistry of the formation water beneath a perched contact may be significantly different that the chemistry of the main aquifer, whereas there are not significant differences in formation water chemistry in hydrodynamic reservoirs over short distances, although there may be chemistry trends on a basinal scale (e.g., Bond 1972).

Hydrocarbon/water contacts at slightly different depths may reflect actual different contacts, or they may reflect difference in depth measurements or uncertainties in directional surveys (see Chapter 3). Much of this uncertainty can be clarified if reservoir pressures are available, as discussed in Chapter 3 (Fig. 3-27). One additional uncertainty is in the measurement of reservoir pressures (Fig. 14-43). Very small pressure differences may be due to gauge differences in the repeat formation tester (RFT) tools.

Tilted hydrocarbon/water contacts can affect both bottomwater and edgewater reservoirs. The procedure for mapping net pay for reservoirs with inclined hydrocarbon/water contacts is very similar to mapping net pay for reservoirs with horizontal contacts. With inclined contacts, the zero net pay contour does not track a structure contour, and the isochore contours may not be as similar in shape to structure contours as they would be if the contact were horizontal. If mapping these reservoirs on a computer, the only difference is that hydrocarbon/water surface used in the gross reservoir determination is not horizontal.

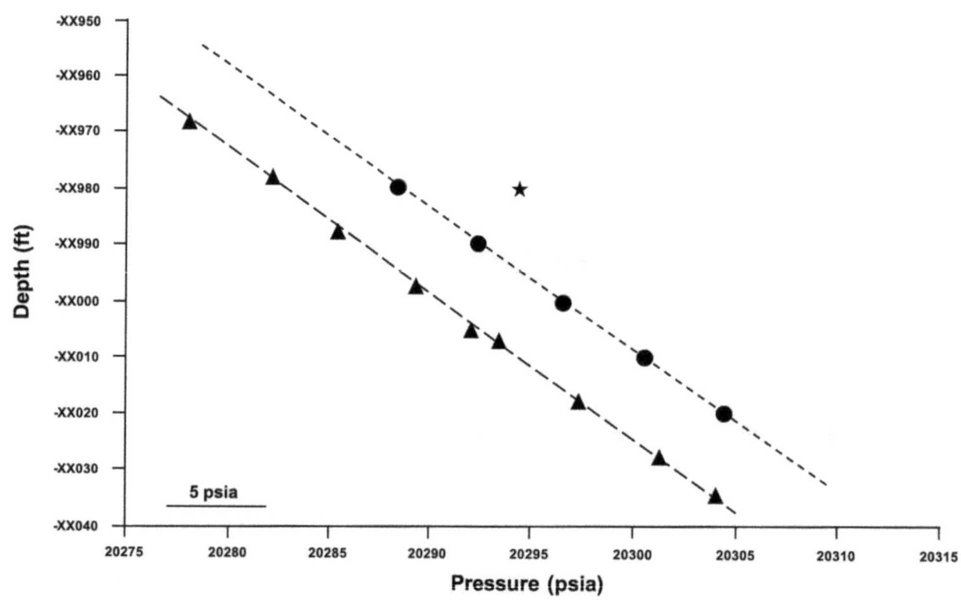

Figure 14-43 Repeat formation test pressures from two gauges (triangle and circle) in a single wireline tool in a deep water Gulf of Mexico well. As the two gauges were made up into a single tool, their relative depth is known. The difference between the two gauges is a measure of the accuracy of the gauges. The difference between the gauges, about 6 psia, is 0.03% of the pressure measurement, well within the specs quoted by the service provider. The single star is a logging-while-drilling (LWD) pressure from the same well that has been depth-shifted to tie the log measurements. The difference between the LWD pressure and the wireline pressures, again about 0.03% of the total gauge pressure, may be due to differences in gauges, or it may be due to differences in tool design and operation. (Published by permission of J. Brenneke.)

INTRODUCTORY RESERVOIR ENGINEERING

Reservoir engineering must be involved in all phases of oil and gas prospect evaluation from the initial phases of exploration through field development. Prior to any wells being drilled on a prospect, the most likely reservoir engineering parameters must be assumed in a preliminary evaluation to estimate the potential hydrocarbons. Initial drilling provides basic reservoir parameters, but offset information from analogous fields is used to predict the performance of the new discovery. Finally, during the field's productive life, performance data needs to be constantly evaluated to allow revisions to the original assumptions and to develop and modify the field's depletion plan. The last phase is especially critical to optimize drainage of recoverable reserves yielding maximum productivity and revenue.

An excellent review of the history of reservoir engineering is found in Craft and Hawkins (1959). One of the major benefits to the geologist of becoming familiar with basic reservoir engineering is to understand what a reservoir engineer requires for the evaluation of hydrocarbon reserves. A list of the symbols, nomenclature, and other pertinent information adopted by the Society of Petroleum Engineers for use in reservoir engineering can be found at the Society's website (www.spe.org). It is essential for a geologist to become familiar with the basics of petroleum engineering and to work closely with the petroleum engineer to obtain the necessary data to evaluate hydrocarbon reserves.

Reservoir Characterization

After the identification of hydrocarbons in a permeable stratigraphic unit, a plan must be established to estimate the in-place volumes of these hydrocarbons and what might be expected to be recovered. Recoverable hydrocarbons are referred to as *hydrocarbon reserves*. Because the productive characteristics of a reservoir or field may not be fully known until it is maturely developed, offset analogous reservoirs need to be considered as models to the development of any new discovery. With a model to follow, various reservoir-engineering parameters such as reservoir porosity, permeability, and water saturation can be estimated to allow volumetric analyses of the reservoir or field to estimate potentially recoverable reserves.

Estimation of Reserves

Reservoir net volume is calculated from net hydrocarbon isochore maps, as earlier discussed in this chapter. From detailed log analysis, rock properties can be identified. As an example, interstitial water saturation must be estimated for reserve determinations. Detailed reservoir petrophysics for use in calculating reserves is beyond the scope of this book; however, a presentation of general reservoir engineering may be helpful, and it is outlined here.

Using the letter symbols G and N, initial in-place volumes of gas and oil can be determined by the following equations.

$$G = (43{,}560)(\phi)(1 - S_{wi})(B_{gi})(\text{reservoir volume in acre-feet}) \tag{14-3}$$

and

$$\text{N} = (7758)(\phi)(1 - S_{wi})(\text{reservoir volume in acre-feet})/B_{oi} \tag{14-4}$$

where

G = original gas-in-place, in cu ft
N = original oil-in-place, in barrels
ϕ = effective porosity, fraction
S_{wi} = interstitial water saturation, fraction
B = formation volume factor, dimensionless
Reservoir volume = net reservoir volume
43,560 = cu ft per acre-foot
7758 = barrels per acre-foot
subscript o = oil-bearing zone
subscript g = gas-bearing zone
B_{gi} = standard cu ft/reservoir cu ft
B_{oi} = reservoir barrels/stock tank barrels

Gas Reservoirs. Equation (14-3) is used to estimate original gas-in-place. There are several unknown factors in the equation that must be determined. The formation volume factor (FVF) or B_{gi} is defined as the relationship of gas volumes from surface conditions to reservoir conditions. For gas reservoirs, B_{gi} is expressed in standard cubic feet per reservoir cubic foot, SCF/cu ft. As defined here, B_{gi} is greater than 1 as gas expands from reservoir conditions to standard conditions, which approximate surface conditions. B_{gi} can also be defined as reservoir cu ft/standard cu ft. In that case, B_{gi} is less than 1 and moves from the numerator to the denominator in Eq. (14-3), and Eq. (14-3) becomes directly analogous to Eq. (14-4). The porosity is expressed as a fraction of the net volume, and the interstitial water (S_{wi}) is a fraction of the pore volume.

To determine the unit recovery for a gas reservoir, the final reserve volume per acre-foot is determined on the basis of the reservoir drive mechanism. This fractional recovery, or *recovery factor*, represents the difference between the initial unit-in-place gas and the final, or abandonment, unit-in-place gas.

$$\text{Recovery Factor}\,(RF) = \frac{100(G - Ga)}{G}$$

or

$$RF = \frac{100(B_{gi} - B_{ga})}{B_{gi}}\,\text{Percent} \qquad (14\text{-}5)$$

where

B_{gi} = formation volume factor at initial conditions
B_{ga} = formation volume factor at abandonment conditions

This recovery factor is indicative of depletion drive reservoirs, where interstitial water saturation remains unchanged and, conversely, gas saturation remains constant. The other end of the spectrum with regard to drive mechanisms is a strong water drive, where produced gas is being replaced by encroaching water (there is no appreciable pressure loss and $B_{gi} = B_{ga}$). The recovery factor for a water-drive gas reservoir, which is representative of the change in gas and water saturations in the reservoir due to production, is shown in Eq. (14-6).

$$RF = \frac{100(1 - S_{wi} - S_{gr})}{(1 - S_{wi})} \text{Percent} \tag{14-6}$$

where

S_{wi} = interstitial water saturation, fraction
S_{gr} = residual gas saturation, fraction

Oil Reservoirs. Oil reservoirs are often more difficult to analyze for a number of reasons, including the presence in some reservoirs of both oil and gas. If an oil reservoir is found without free gas, the oil is said to be undersaturated. An oil reservoir with a free gas cap is indicative of a saturated oil reservoir.

Equation (14-4) is used to volumetrically determine original oil-in-place. The variables in Eq. (14-4) are very similar to those discussed for Eq. (14-3). If we consider an undersaturated oil reservoir under a strong water drive, the recovery factor is based on the following equation.

$$RF = \frac{100(1 - S_{wi} - S_{or})}{(1 - S_{wi})} \text{Percent} \tag{14-7}$$

where

S_{or} = residual oil saturation, decimal

Where there is an initial gas cap, the oil is saturated. In such cases, the reservoir can be produced under drive mechanisms other than water drive. These include dissolved gas, gas cap, or a combination drive. The opposite end of the recovery factor spectrum from a water drive reservoir is that of a dissolved gas drive reservoir.

$$RF = \frac{(1 - S_{wi} - S_{ga})(B_{oi})}{(1 - S_{wi})(B_{oa})} \text{Percent} \tag{14-8}$$

where

S_{wi} = interstitial water saturation, fraction
S_{ga} = gas saturation at abandonment, fraction
B_{oi} = initial formation volume factor
B_{oa} = formation volume factor at abandonment
RF = recovery factory, percent

Field Production History

Once reservoir or field production has been established and sufficient quantities of hydrocarbons produced, performance data can be used to estimate original and remaining reserves, and to forecast future performance. It is good practice to monitor both production and reservoir pressure data. These points of reference are invaluable in evaluating reservoir heterogeneities and are applicable to material balance equations.

Various performance curves can be used to evaluate reserves and forecast future production trends. The most common performance curves are those which plot the production of oil, gas, and water versus time. In many instances, performance evaluation is by far the most accurate method for estimating the original in-place volume of hydrocarbons, estimating recoverable reserves, and forecasting future performance. The following is a partial list of the types of performance curves that can be plotted.

1. Monthly production (oil or gas) versus time
2. Flowing tubing pressure versus time
3. Bottom-hole pressure data versus cumulative production
4. Percent oil cut versus cumulative production
5. Water yield versus cumulative gas production

A detailed discussion of these performance curves is beyond the scope of this book. We again refer you to Craft and Hawkins (1959).

INTERVAL ISOPACH MAPS

As discussed in the beginning of this chapter, an isopach map is one in which the true stratigraphic thickness of a single stratigraphic interval is contoured, such as a stratigraphic unit or an interval between two stratigraphic markers. Such a map is referred to as an **interval isopach map**. Interval isopach maps are particularly useful in determining the history of movement along faults and of development of growth structures, such as salt dome structures and folds, as well as being useful in the interpretation of depositional environments and paleostructures.

The full application of interval isopach maps is beyond the scope of this text. More specifically, we address the determination and use of the proper interval thicknesses obtained from well logs or seismic data to be used for construction of interval isopach maps. Interval isopach maps reveal the *true stratigraphic thicknesses of units* rather than vertical thicknesses. The TST is the thickness measured perpendicular to the top and base of the stratigraphic interval. Where strata have no dip or have only gentle dips, TST equals or approximates TVT. But with increasing dip, the differences become substantial and a correction factor is applied to TVT to obtain TST. The relationship is given as

$$\text{TST} = (\text{TVT})(\cos \phi) \tag{14-9}$$

where

$$\text{TST} = \text{true stratigraphic thickness}$$
$$\text{TVT} = \text{true vertical thickness}$$
$$\phi = \text{true bed dip}$$

For example, with a 10-deg bed dip, an interval with a TST of 100 ft has a vertical thickness of 101.5 ft; at a 20-deg dip, the vertical thickness is 106.5 ft. At a 45-deg dip, however, the vertical thickness becomes 141 ft, which is significantly different than the stratigraphic thickness. If the TVT is the value contoured, the map is more correctly referred to as an *isochore map*. Therefore, net sand and net pay maps are isochore maps.

True Stratigraphic Thickness from Well Logs

The determination of stratigraphic thickness from well logs presents a few complications. In areas of nearly flat-lying beds, the TVT approximates TST. The determination of stratigraphic thickness does, however, become more complicated around steeply dipping structures. Figure 14-44 shows the effect of changing bed dip on vertical well log thickness of a stratigraphic unit whose TST is constant. With zero bed dip, TVT equals TST. At a 20-deg bed dip, the TVT is only 1.06 times TST. At a 40-deg bed dip, the TVT is equal to 1.30 times TST, and at 60 deg, the TVT is twice as thick as the TST. Assuming vertical wells, if the upper and lower markers chosen for interval isopaching are parallel or nearly so (that is, they are at or near the same dip), Eq. (14-9) can be used to convert vertical thickness to stratigraphic thickness.

If the thickness of the interval changes, and thus the dips of the upper and lower surfaces of the interval are different, then the correction factor based on the cosine of a single bed dip, either at the upper surface or lower surface, will not be accurate. For such cases, another equation is required (Tearpock and Harris 1987).

$$\mathrm{TST} = \mathrm{TVT}/((\sin\alpha \tan\phi) + \cos\alpha) \qquad (14\text{-}10)$$

where

TST = true stratigraphic thickness
TVT = true vertical thickness
α = dip of upper horizon
ϕ = dip of lower horizon

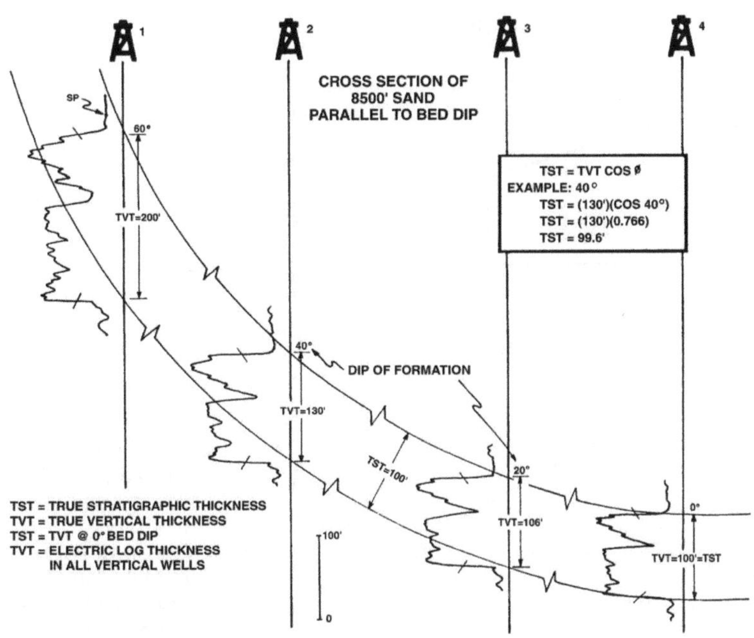

Figure 14-44 Effect of changing bed dip on TVT of a stratigraphic unit, where TST is constant.

Figure 14-45 shows the equation and an example problem. This equation takes into account the dip of the upper and lower surfaces and the vertical thickness of the interval.

Finally, in the case of deviated wells, the measured log thickness must be first corrected to TVT, and then corrected to TST. The procedure for vertical thickness conversions is discussed in this chapter and in Chapter 4.

To avoid making laborious stratigraphic thickness calculations, the graph in Figure 14-46 can be used to calculate TST if the vertical thickness and the dips of the upper and lower surfaces are known. The horizontal axis represents the dip of the upper surface, and the vertical axis represents the correction factor. The curves within the graph represent the values obtained by subtracting the dip of the upper surface from the dip of the lower surface (note positive and negative values). Consider the following example.

Data:

$$\text{Dip of upper surface} = 20 \text{ deg}$$
$$\text{Dip of lower surface} = 30 \text{ deg}$$
$$\text{TVT} = 1000 \text{ ft}$$

1. To use the graph, first subtract the dip of the upper surface from the lower surface. This value determines which of the curved lines to use for the correction factor.
2. Enter the graph on the horizontal axis at 20 deg and project vertically until the line intersects the curve with value equal to the dip of the lower surface minus the dip of the upper surface. In this case, it is the +10-deg curve.
3. From the intersection with the curve, project horizontally to the left to intersect the vertical axis and determine the correction factor. In this case, it is 0.88.

Therefore,

$$\text{TST} = (1000 \text{ ft})(0.88)$$
$$\text{TST} = 880 \text{ ft}$$

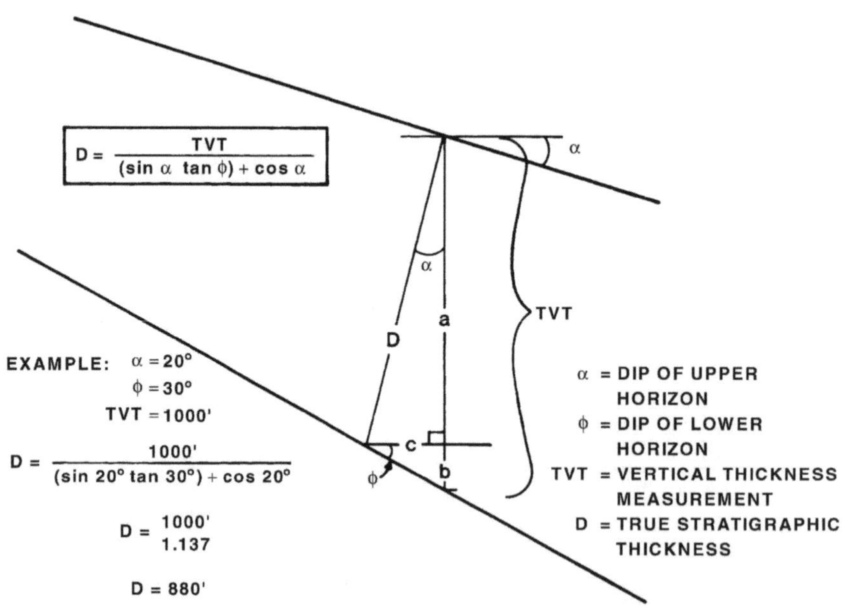

Figure 14-45 Cross section showing the geometric relationship between two horizons that have different angles of dip. Eq. (14-10) is used in this type of situation to convert TVT to TST. (Prepared by C. Harmon. Modified from Tearpock and Harris 1987.Published by permission of Tenneco Oil Company.)

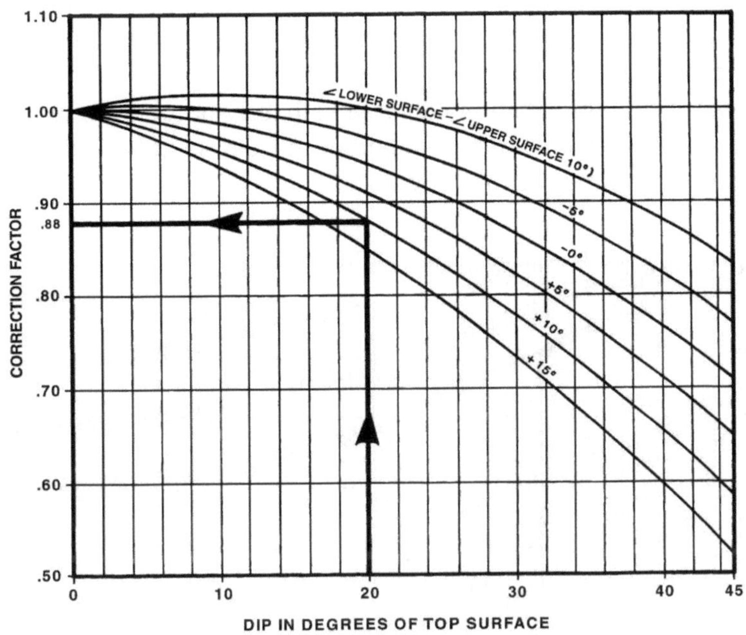

Figure 14-46 Graph derived from Eq. (14-10) and used to determine the correction factor for converting TVT of a stratigraphic interval to TST, where the upper and lower surfaces dip at different angles. (Prepared by C. Harmon. From Tearpock and Harris 1987. Published by permission of Tenneco Oil Company.)

Interval Isopach Construction Using Seismic Data

Using seismic data for interval isopach construction can give you many additional data points between well control. In areas of relatively low dip (10 deg or less) and parallel horizons, the TVT calculated from seismic is a close approximation of the TST. If you have seismic data that has been converted to depth, the process is straightforward. If the seismic data is in time, the procedure in this type of area is almost as straightforward: The time-converted depth of the upper horizon is subtracted from the time-converted depth of the lower horizon to arrive at an interval thickness. The basic requirement is an accurate time–depth function.

In areas of steeper dip and nonparallel horizons, you should be aware of some visual pitfalls inherent in time seismic sections. The basic point to remember is that *a time section is not a cross section*. It is distorted due to the two very different dimensions displayed on a section: time along the vertical axis and distance along the horizontal axis. These dimensional differences often introduce some very pronounced vertical exaggeration.

To illustrate this, observe the two horizons indicated in Figure 14-47a. They obviously diverge from one another as the interval thickens into a fault. How do we determine the TST of the interval? The first inclination is to draw a perpendicular line, shown as A on the figure, from the top horizon to intersect the bottom horizon, and calculate trigonometrically the stratigraphic thickness using the time-converted depths at both points, along with the lateral distance between the two points. *THIS IS WRONG.*

To see graphically what is actually present, look at Figure 14-47b, which shows the horizons converted to depth and displayed at a true 1:1 scale. Line A, drawn earlier as perpendicular to the top horizon on the seismic line, is in fact a longer segment than the true perpendicular, which is line B on the seismic line and cross section. The reason for this pitfall is that the time seismic line, at this depth, has about a 2:1 vertical exaggeration. In this case, you would post a larger thickness for the interval than is actually present. To obtain corrected data points, you need to apply Eq. (14-10), which uses the dip of the top and bottom horizons, and the vertical thickness of the interval. The graph in Figure 14-46 can also be used to determine the stratigraphic thickness.

In summary, seismic information can be a valuable source of interval thickness data, as long as you are aware of the visual distortion inherent in time seismic data and properly account for it in the calculation of stratigraphic thicknesses.

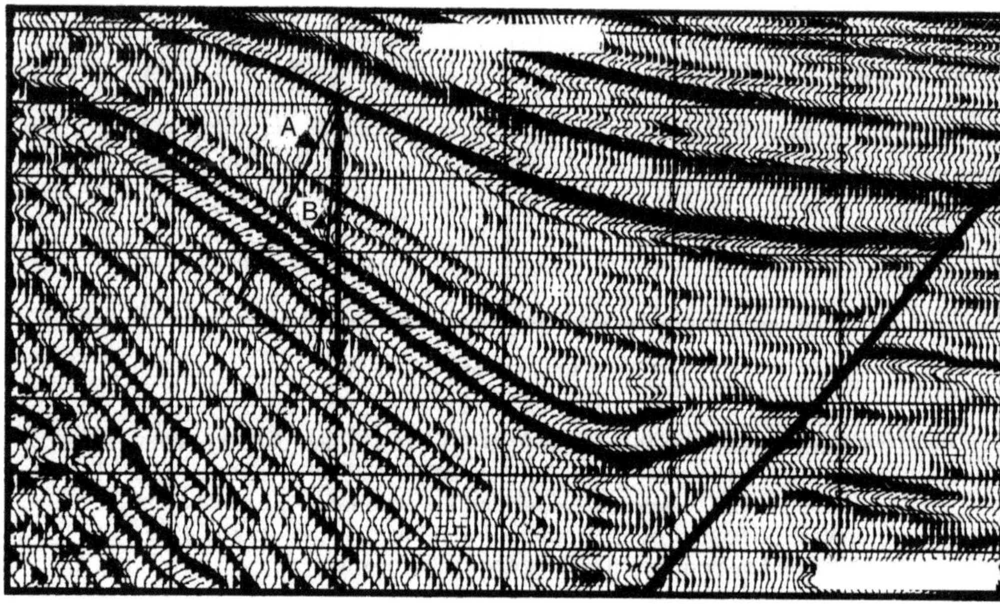

(a)

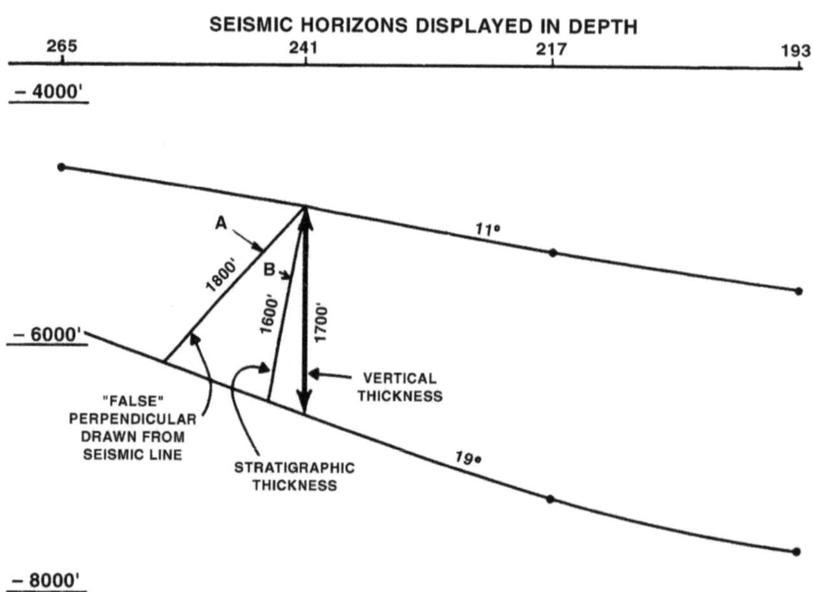

(b)

Figure 14-47 (a) Which line, A or B, represents the true stratigraphic thickness of the designated interval? (b) True (1:1 scale) cross section of the seismic interval shown in (a). (Prepared by C. Harmon. From Tearpock and Harris 1987. Published by permission of Tenneco Oil Company.)

APPENDIX

GENERAL MAP SYMBOLS

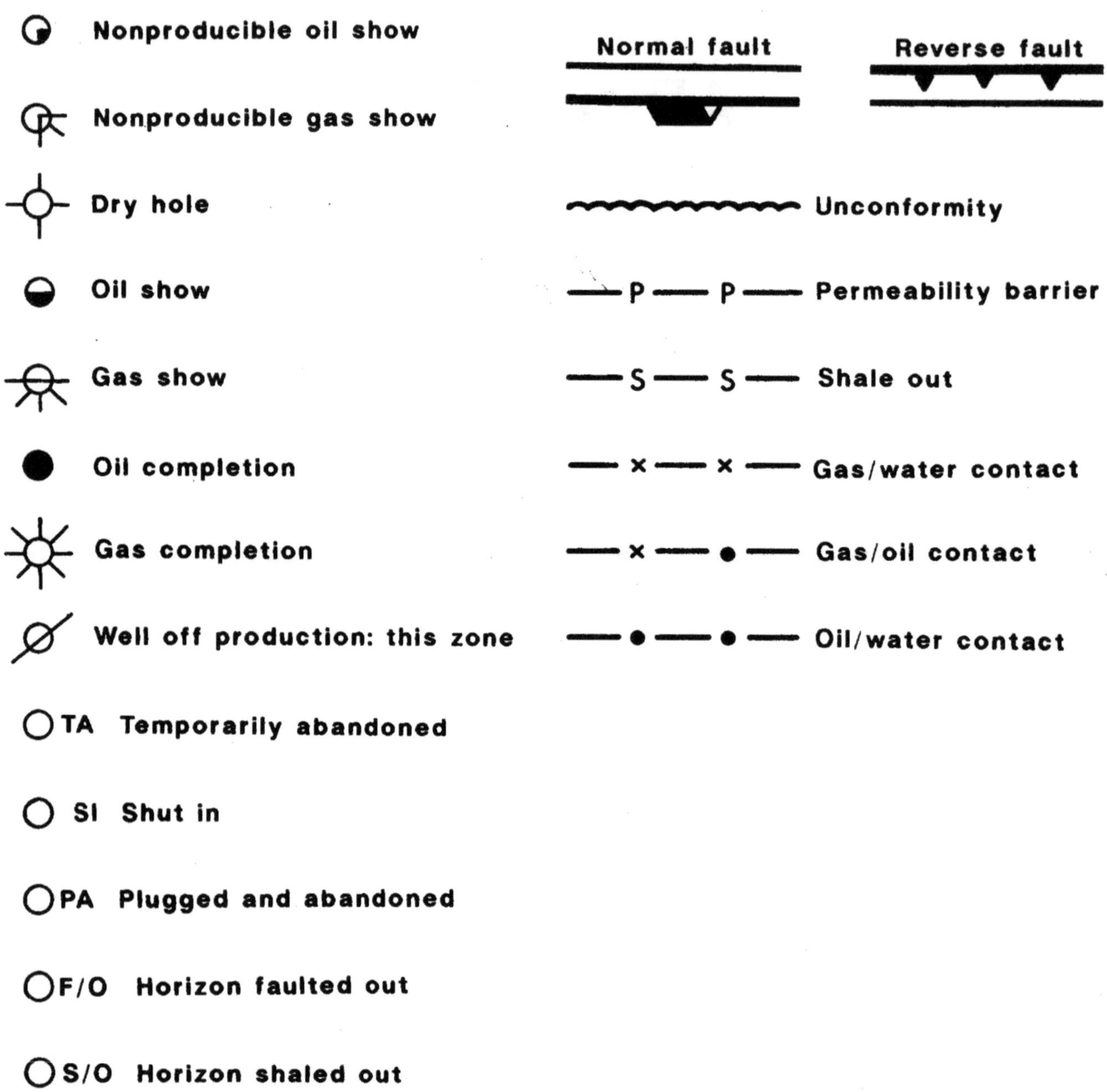

MAP SYMBOLS USED THROUGHOUT THE TEXT ARE NOT ALWAYS CONSISTENT WITH THOSE SHOWN HERE. A WIDE VARIETY OF SYMBOLS ARE USED BY DIFFERENT COMPANIES, SO NO CONSISTENT USE OF SYMBOLS EXISTS THROUGHOUT THE INDUSTRY.

Figure A-1 *Map symbols.*

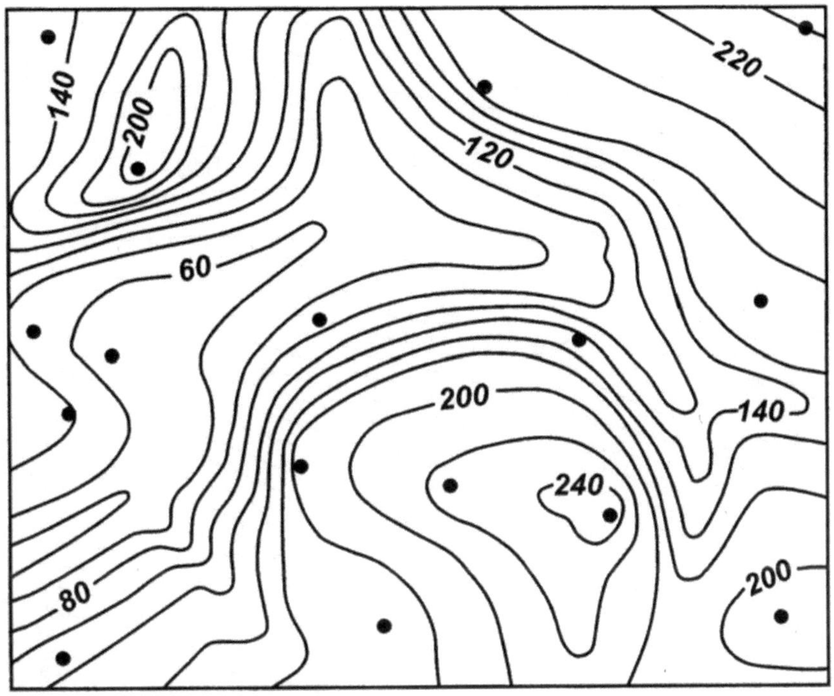

Figure A-2 *Map constructed using the interpretive contouring technique. (Reproduced from Analysis of Geologic Structures by John M. Dennison, by permission of W. W. Norton & Company, Inc. Copyright © 1968 by W. W. Norton & Company, Inc.)*

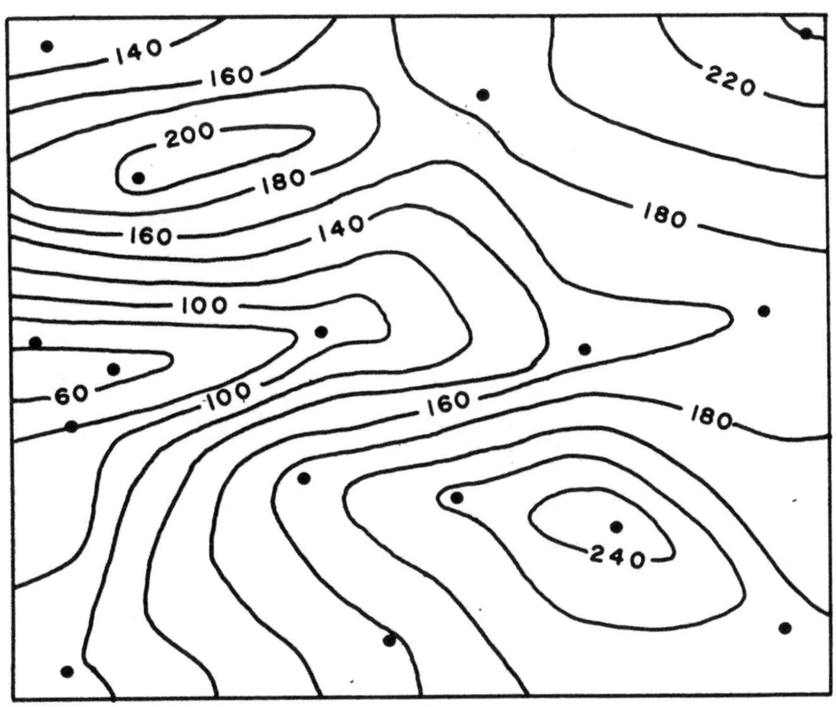

Figure A-3 *Map constructed using the mechanical contouring technique. (Reproduced from Analysis of Geologic Structures by John M. Dennison, by permission of W. W. Norton & Company, Inc. Copyright © 1968 by W. W. Norton & Company, Inc.)*

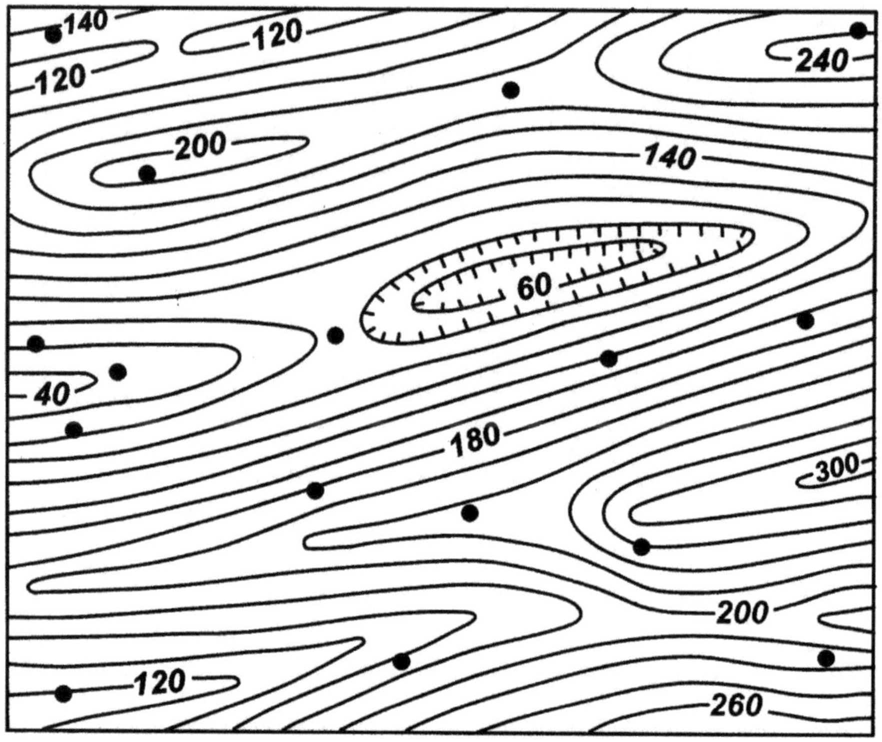

Figure A-4 *Map constructed using the equal-spaced contouring technique. (Reproduced from Analysis of Geologic Structures by John M. Dennison, by permission of W. W. Norton & Company, Inc. Copyright © 1968 by W. W. Norton & Company, Inc.)*

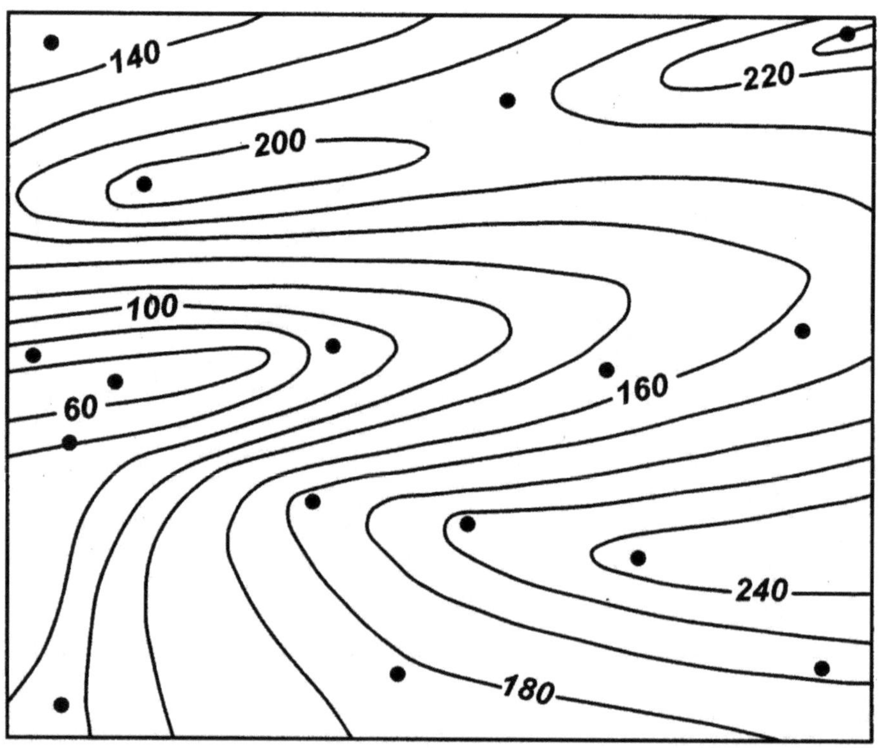

Figure A-5 *Map constructed using the parallel contouring technique. (Reproduced from Analysis of Geologic Structures by John M. Dennison, by permission of W. W. Norton & Company, Inc. Copyright © 1968 by W. W. Norton & Company, Inc.)*

REFERENCES

Adams, G. F., 1957, Block diagrams from perspective grids: Journal of Geoscience Education, v. 5, no. 2, p. 10–19.

Allan, U. S., 1980, A model for the migration and entrapment of hydrocarbons (abs.): AAPG Research Conference on Seals for Hydrocarbons, Unpublished Abstracts, 16 p.

______, 1989, Model for hydrocarbon migration and entrapment within faulted structures: American Association of Petroleum Geologists Bulletin, v. 73, p. 803–811.

Allaud, L., 1976, Vertical net sandstone determination for isopach mapping of hydrocarbon reservoirs: American Association of Petroleum Geologists Bulletin, v. 60, p. 2150–2153.

Allen, C. R., 1962, Circum-Pacific faulting in the Philippines-Taiwan region: Journal of Geophysical Research, v. 67, p. 4795–4812.

Amer, A., F. Chinellato, S. Collins, J.-M. Denichou, I. Dubourg, R. Griffiths, R. Koepsell, S. Lyngra, P. Marza, D. Murray, and I. Roberts, 2013, Structural steering—A path to productivity: Oilfield Review, v. 25, no. 1, p. 14–31.

Anderson, A., 1929, Underground surveys of oil wells, SPE-929030-G: Transactions of the AIME, v. 82, no. 1, 11 p.

Anspach, D. H., S. E. Tripp, R. E. Berlitz, and J. A. Gilreath, 1987, Postdevelopment analysis of producing shelf-slope environments of deposition, High Island area: Transactions Gulf Coast Association of Geological Societies, v. 37, p. 1–10.

Anstey, N. A., 1977, Seismic Interpretation: The Physical Aspects: Boston, International Human Resources Development Corporation Press, 637 p.

______, 1982, Simple Seismics for the Petroleum Geologist, the Reservoir Engineer, the Well-Log Analyst, the Processing Technician, and the Man in the Field: Boston, International Human Resources Development Corporation Press, 168 p.

Applegate, J. K., and Rose, P. R., 1985, Structure of the Raton basin from a regional seismic line, *in* Gries, R. R., and R. C. Dyer, eds., Seismic Exploration of the Rocky Mountain Region: Denver, CO, Rocky Mountain Association of Geologists and Denver Geophysical Society, p. 259–265.

Archie, G. E., 1942, The electrical resistivity log as an aid in determining some reservoir characteristics: Transactions American Institute of Metallurgical Engineers, v. 146, p. 54–62.

______, 1947, Electrical resistivity an aid in core-analysis interpretation: American Association of Petroleum Geologists Bulletin, v. 31, p. 350–366.

Atwater, G. I., and E. E. Miller, 1965, The effect of decrease in porosity with depth on future development of oil and gas reserves in South Louisiana (Abs.): American Association of Petroleum Geologists Bulletin, v. 49, p. 334.

Atwater, G. I., and M. J. Foreman, 1959, Nature of growth of southern Louisiana salt domes and its effect on petroleum accumulation: American Association of Petroleum Geologists Bulletin, v. 43, p. 2592–2622.

Aydin, A., and A. Nur, 1982, Evolution of pull-apart basins and their scale independence: Tectonics, v. 1, p. 91–105.

______, 1985, The type and role of stepovers in strike-slip tectonics, *in* Biddle, K. T., and N. Christie-Blick, eds., Strike-Slip Deformation, Basin Formation, and Sedimentation: Society of Economic Paleontologists and Mineralogists, Special Publication No. 37, p. 35–44.

Badgley, P. C., 1959, Structural Methods for the Exploration Geologist and a Series of Problems for Structural Geology Students: New York, Harper & Bros., 280 p.

Badley, M. E., 1985, Practical Seismic Interpretation: Boston, International Human Resources Development Corporation Press, 266 p.

Baldwin, B., and C. O. Butler, 1985, Compaction curves: American Association of Petroleum Geologists Bulletin, v. 69, p. 622–626.

Ball, S. M., 1982, Exploration application of temperatures recorded on log-headings—An up-the-odds method of hydrocarbon-charged porosity prediction: American Association of Petroleum Geologists Bulletin, v. 66, p. 1108–1123.

Bally, A. W., 1983, Seismic Expression of Structural Styles: American Association of Petroleum Geologists Bulletin, Studies in Geology Series No. 15, v. 3., p. 167–172.

Bally, A. W., P. L. Gordy, and G. A. Stewart, 1966, Structure, seismic data, and orogenic evolution of southern Canadian Rocky Mountains: Canadian Society of Petroleum Geologists Bulletin, v. 14, p. 337–381.

Banks, C. J., and J. Wharburton, 1986, "Passive-roof" duplex geometry in the frontal structures of the Kirthar and Sulaiman Mountain Belts, Pakistan: Journal of Structural Geology, v. 8, p. 229–237.

Barker, T. B., and C. E. Guzman, 1992, Defining the salt with walkaway VSPs and salt proximity survey, Auger field, Gulf of Mexico: Expanded Abstracts 62nd Annual International Meeting, Society of Exploration Geophysicists, 25–29 October, New Orleans, p. 284–287.

Barnes, M., 2005, High resolution satellite imagery applied to oil and gas projects: SEG 2005 Annual Meeting, 6–11 November, Houston, Texas.

Barnett, J. A. M., J. Mortimer, J. H. Rippon, J. J. Walsh, and J. Waterson, 1987, Displacement geometry in the volume containing a single normal fault: American Association of Petroleum Geologists Bulletin, v. 71, p. 925–937.

Bartel, D. E., M. Busby, J. Nealon, and J. Zaske, 2006, Time to depth conversion and uncertainty assessment using average velocity modeling: SEG Annual Meeting, 1–6 October, New Orleans, Louisiana.

Beckwith, R. H., 1941, Trace-slip faults: American Association of Petroleum Geologists Bulletin, v. 25, no. 12, p. 2181–2193.

______, 1947, Fault problems in fault planes: Geological Society of America Bulletin, v. 58, p. 79–108.

Bell, W. G., 1956, Tectonic setting of Happy Springs and nearby structures in the Sweetwater Uplift area, central Wyoming, *in* Geological Record of the American Association of Petroleum Geologists, Rocky Mountain Section.

Benesh, N. P., A. Plesch, and J. Shaw, 2014, Geometry, kinematics, and displacement characteristics: An example from the deep-water Niger Delta: American Association of Petroleum Geologists Bulletin, v. 98, p. 465–482.

Bengston, C. A., 1980, Statistical curvature analysis methods for interpretation of dipmeter data: Oil and Gas Journal, June 23 issue, p. 172–190.

Bennison, G. M., 1975, An Introduction to Geological Structures and Maps, third ed.: London, Edward Arnold Ltd., 144 p.

Berg, O. R., and D. G. Woolverton, 1985, Seismic Stratigraphy II—An Integrated Approach to Hydrocarbon Exploration: American Association of Petroleum Geologists Memoir 39, 276 p.

Berg, R. R., 1967, Point bar origin of Fall River sandstone reservoirs, northeastern Wyoming: Society of Petroleum Engineers, Paper 1953, 6 p.

Bergen, J. B., and J. H. Shaw, 2010, Displacement profiles and displacement-length scaling relationships of thrust faults constrained by seismic-reflection data, Geological Society of America Bulletin, v. 122, p. 1209–1219.

Billings, M. P., 1954, Structural Geology, second ed.: New York, Pearson, 514 p.

______, 1972, Structural Geology, third ed.: Englewood Cliffs, NJ, Pearson, 606 p.

Biot, M. A., 1961, Theory of folding of stratified viscoelastic media and its implication in tectonics and orogenesis: Geological Society of America Bulletin, v. 72, p. 1595–1620.

Bischke, R. E., 1974, A model of convergent plate margins based on the Recent tectonics of Shikoku, Japan: Journal of Geophysical Research, v. 79, p. 4845–4857.

______, 1990, Applied Structural Balancing: Gulf Coast Association of Geological Societies Short Course, 40th Annual Convention, Lafayette, LA.

______, 1994a, The compressional off-structure problem: Houston Geological Society Bulletin, May issue, p. 29–34.

______, 1994b, Interpreting sedimentary growth structures from well log and seismic data (with examples): American Association of Petroleum Geologists Bulletin, v. 7, p. 873–892.

Bischke, R. E., and D. J. Tearpock, 1993, A method for estimating gross sand percentage and reservoir thickness from seismic section: Example from Segno Field, Polk Co., Texas: Houston Geological Society Bulletin, May issue, p. 22–23, 40–43.

Bischke, R. E., and D. J. Tearpock, 1999, A graphical dip domain technique for projecting large growth faults to depth using imaged hanging wall structure: Transactions Gulf Coast Association of Geological Societies, v. 49, p. 112–120.

Bischke, R. E., and J. Suppe, 1990a, Calculating sand/shale ratios from growth normal fault dips on seismic profiles: Transactions Gulf Coast Association of Geological Societies, v. 40, p. 39–49.

______, 1990b, Geometry of rollover: Origin of complex arrays of antithetic and synthetic crestal faults (Abs.): American Association of Petroleum Geologists Bulletin, v. 74, p. 611.

Bischke, R. E., J. B. Brewton, and D. J. Tearpock, 2006, A robust technique for resolving faulting problems on the steeply dipping flanks of salt domes using the multiple Bischke plot analysis: Gulf of Mexico Association of Geological Societies, v. 55, p. 5–22.

Bischke, R. E., J. Suppe, and R. del Pilar, 1990, A new branch of the Philippine Fault system, as observed from aeromagnetic and seismic data: Tectonophysics, v. 183, p. 243–264.

Bischke, R. E., W. Finley, and D. J. Tearpock, 1999, Growth analysis (Δd/d): Case histories of the resolution of correlation problems as encountered while mapping around salt: Transactions Gulf Coast Association of Geological Societies, v. 49, p. 102–110.

Bishop, M. S., 1960, Subsurface Mapping: New York, John Wiley & Sons, 198 p.

Blake, G. H., 1991, Review of the Neogene biostratigraphy of the Los Angeles Basin and implications for basin evolution, *in* Biddle, K. T., ed., Active Margin Basins: American Association of Petroleum Geologists Memoir 52, p. 135–184.

Blomquist, P. K., 2016, Wolfcamp Horizontal Play, Midland Basin, West Texas: Search and Discovery Article #10890, p. 1–34.

Bloomer, R. R., 1977, Depositional environments of a reservoir sandstone in west-central Texas: American Association of Petroleum Geologists Bulletin, v. 61, no. 3, p. 344–359.

Bolt, H., 2015, Wireline Depth Determination, Rev. 4.0, 110 p.

______, 2016, Wireline logging depth quality improvement: Methodology review and elastic-stretch correction: Petrophysics, v. 57, no. 3, p. 294–310.

Bond, D. C., 1972. Hydrodynamics in deep aquifers of the Illinois Basin, Circular 470, Illinois State Geological Survey, 72 p.

Boos, C. M., and M. F. Boos, 1957, Tectonics of eastern flank and foothills of Front Range, Colorado: American Association of Petroleum Geologists Bulletin, v. 41, no. 12, p. 2603–2676

Bott, W. F., Jr., and T. T. Tieh, 1987, Diagenesis and high fluid pressures in the Frio and Vicksburg shales, Brooks County, Texas: Transactions Gulf Coast Association of Geological Societies, v. 37, p. 323–334.

Bouvier, J. D., C. H. Kaars-Sijpesteijn, D. F. Kleusner, C. C. Onyejekwe, and R. C. Van der Pal, 1989, Three-dimensional seismic interpretation and fault sealing investigations, Nun River Field, Nigeria: American Association of Petroleum Geologists Bulletin, v. 73, p. 1397–1414.

Bower, T. H., 1947, Log map, new type of subsurface map: American Association of Petroleum Geologists Bulletin, v. 127, no. 7, p. 340–349.

Boyer, S. E., 1986, Styles of folding within thrust sheets: Examples from Appalachian and Rocky Mountains of the U.S.A. and Canada: Journal of Structural Geology, v. 8, p. 325–339.

Boyer, S. E., and D. Elliott 1982, Thrust systems: American Association of Petroleum Geologists Bulletin, v. 66, no. 9, p. 1196–1230.

Brabb, E. E., 1989, Geologic map of Santa Cruz County, California: U.S. Geological Survey Miscellaneous Investigations Series Map I-1905, scale 1:62,500.

Branson, R. B., 1991, Productive trends and production history, South Louisiana and adjacent offshore, *in* Goldthwaite, D., ed., An Introduction to Central Gulf Coast Geology: New Orleans Geological Society, p. 61–70.

Brenneke, J. C., 1995, Analysis of fault traps: World Oil, December issue, p. 63–71.

Breyer, J. A., ed., 2016, The Eagle Ford Shale: A renaissance in U.S. oil production: American Association of Petroleum Geologists Memoir 110, 389 p.

Bristol, H. M., 1975, Structural geology and oil production of northern Gallatin County and southernmost White County, Illinois: Illinois State Geological Survey, Illinois Petroleum 105, 20 p.

Bristol, H. M., and J. D. Treworgy, 1979, The Wabash Valley fault system in southeastern Illinois: Illinois State Geological Survey, Circular 509, 19 p.

Brooks, A., H. Wilson, A. Jamieson, D. McRobbie, and S. G. Holehouse, 2005, Quantification of depth accuracy, SPE 95611, 2005 SPE Annual Technical Conference and Exhibition, 9–12 October, Dallas, Texas, 5 p.

Broussard, M. J., and B. E. Lock, 1995, Modern analytical techniques for fault surface seal analysis: A Gulf Coast case history: Transactions Gulf Coast Association of Geological Societies, v. 45, p. 87–93.

Brown, A. R., 1999, Interpretation of Three-Dimensional Seismic Data, fifth ed.: American Association of Petroleum Geologists Memoir 42, 510 p.

______, 2011, Interpretation of Three-Dimensional Seismic Data, seventh ed.: American Association of Petroleum Geologists Memoir 42, 648 p.

Brown, Alton, 2003, Improved interpretation of wireline pressure data, American Association of Petroleum Geologists Bulletin, v. 87, p. 295–311.

Brown, G. G., D. L. Katz, R. C. Alden, and Natural Gas Processors Association, 1948, Natural Gasoline and Volatile Hydrocarbons: Tulsa, OK, National Gasoline Association of America.

Brown, L. F., and W. L. Fisher, 1984, Stratigraphic Interpretation and Petroleum Exploration: American Association of Petroleum Geologists Bulletin, Continuing Education Course Note Series No. 16.

Brown, R. L., 1981, Thickness Corrections for Deviated Wells—A Simplified Technique Utilizing the Apparent Dip along a Well Path: Contribution to the 1981 Geological/Petrophysical Conference, Houston, TX.

Brown, R. N., 1980, History of exploration and discovery of Morgan, Ramadan and July Oil Fields, Gulf of Suez, Egypt, *in* Miall, A. D., ed., Facts and Principles of World Petroleum Occurrence: Canadian Society of Petroleum Geologists Memoir 6, p. 733–764.

Brown, W. G., 1982, New Tricks for Old Dogs—A Short Course in Structural Geology: American Association of Petroleum Geologists, Southwest Section Meeting, Wichita Falls, TX, 79 p.

______, 1984a, Working with Folds: American Association of Petroleum Geologists, Structural Geology School Course Notes.

______, 1984b, Basement Involved Tectonics, Foreland Areas: American Association of Petroleum Geologists, Continuing Education Course Series No. 26, 92 p.

Bruce, C. H., 1973, Pressured shale and related sediment deformation: Mechanism for development of regional contemporaneous faults: American Association of Petroleum Geologists Bulletin, v. 57, p. 878–886.

Brune, J. N., T. L. Henyey, and R. F. Roy, 1969, Heat flow, stress and rate of slip on the San Andreas Fault, California: Journal of Geophysical Research, v. 74, p. 3821–3827.

Bucher, W. H., 1933, The Deformation of the Earth's Crust: Princeton, NJ, Hafner Publishing Company, 518 p.

Busch, D. A., 1959, Prospecting for stratigraphic traps: American Association of Petroleum Geologists Bulletin, v. 43, no. 12, p. 2829–2843.

______, 1974, Stratigraphic Traps in Sandstones—Exploration Techniques: American Association of Petroleum Geologists Memoir 21, 174 p.

Busch, D. A., and D. A. Link, 1985, Exploration Methods for Sandstone Reservoirs: Tulsa, OK, Oil and Gas Consultants, Inc., 327 p.

Busk, H. G., 1929, Earth Flexures, Their Geometry and Their Representation and Analysis in Geological Section, with Special Reference to the Problem of Oil Finding: Cambridge, Cambridge University Press, 106 p.

______, 1957, Earth Flexures, Their Geometry and Their Representation and Analysis in Geological Section, with Special Reference to the Problem of Oil Finding: New York, William Trussell, 106 p.

Cape, D. C., S. McGeary, and G. A. Thompson, 1983, Cenozoic normal faulting and the shallow structure of the Rio Grande Rift near Socorro, New Mexico: Geological Society of America Bulletin, v. 94, p. 3–14.

Cartwright, J. A., R. Bouroullec, D. James, and H. D. Johnson, 1998, Polycyclic motion history of some Gulf Coast growth faults from high-resolution displacement analysis, Geology, v. 26, no. 9, p. 819–822.

Castelltort, S., S. Pochat, and J. Van den Driessche, 2004, Using T–Z plots as a graphical method to infer lithological variations from growth strata, Journal of Structural Geology, v. 26, p. 1425–1432.

Challinor, J., 1933, The “throw” of a fault: Geological Magazine, v. 70, p. 384–393.
Chamberlin, R. T., 1910, The Appalachian folds of central Pennsylvania: Journal of Geology, v. 18, p. 228–251.
Chamberlin, T. C., 1897, The method of multiple working hypothesis: Journal of Geology, v. 5, p. 837–848.
Chatellier, J-Y., and C. Porras, 2001, The Multiple Bischke Plot Analysis, a simple and powerful graphic tool to integrate stratigraphic studies: American Association of Petroleum Geologists, Annual Convention Program Book, p. A34–A35.
______, 2004, The Multiple Bischke Plot Analysis: A Simple and Powerful Graphic Tool for Integrated Stratigraphic Studies: Search and Discovery Article #40110.
Chatellier, J-Y., and S. Olave-Hoces, 2009, Successful Exploration in a Thrust Belt, Lessons Learned from the Giant Fields of Eastern Venezuela (the Furrial Trend): Search and Discovery Article #10202.
Chatellier, J-Y., and M. Rueda, 2010, Challenging the Paradigm “Missing Section-Normal Fault”—Implications for Hydrocarbon Exploration, Search and Discovery Article #40550.
Chen, A., 2014, The practice of graphical fluid-gradient interpretations of formation tester pressure data, American Association of Petroleum Geologists Bulletin, v. 98, p. 1431–1448.
Chenoweth, P. A., 1972, Unconformity traps, *in* King, R. E., ed., Stratigraphic Oil and Gas Fields—Classification, Exploration Methods, and Case Histories: American Association of Petroleum Geologists Memoir 16, p. 42–46.
Chia, C. R., H. Laastad, A. Kostin, F. Hjortland, and G. Bordakov, 2006, A new method for improving LWD logging depth, SPE 102175, SPE Annual Technical Conference and Exhibition, San Antonio, Texas, 24–27 September, 9 p.
Chilingarian, G. V., and P. Vorabutr, 1981, Drilling and Drilling Fluids, Developments in Petroleum Science, book 11, New York, Elsevier, 767 p.
Chinnery, M. A., 1963, The stress changes that accompany strike-slip faulting: Bulletin of the Seismological Society of America, v. 53, p. 921–932.
Chopra, S., D. Lubo-Robles, and K. J. Marfurt, 2018, Some Machine Learning Applications in Seismic Interpretation: Search and Discovery Article #42270, p. 1–7.
Christensen, A. F., 1983, An example of a major syndepositional listric fault, *in* Bally, A. W., ed., Seismic Expression of Structural Styles: American Association of Petroleum Geologists Bulletin, Studies in Geology No. 15, v. 2, p. 2.3.1–36 to 2.3.1–40.
Christie-Blick, N., and K. T. Biddle, 1985. Deformation and basin formation along strike-slip faults: Strike-Slip Deformation, Basin Formation, and Sedimentation, Society for Sedimentary Geology, Special publication, v. 37, 1–34.
Clark, J. C., 1959, Problems of fault nomenclature: American Association of Petroleum Geologists Bulletin, v. 43, no. 11, p. 2653–2674.
Clark, J. C., and J. D. Reitman, 1973, Oligocene stratigraphy, tectonics, and paleogeography southwest of the San Andreas Fault, Santa Cruz Mountains and Gabilan Range, California Coastal Ranges: United States Geological Survey, Professional Paper 783, 18 p.
Clark, M. M., K. K. Lienkaemper, J. J. Harwood, D. S. Lajoie, K. R. Matti, J. C. Perkins, J. A. Rymer, A. M. Sarna-Wojcicki, A. M. Sharp, J. D. Sims, J. C. Tinsley, and J. I. Ziony, 1984, Preliminary slip-rate table and map of late-Quaternary faults of California: United States Geological Survey Open-File Report 84–106, 12 p.
Clark, S. K., 1943, Classification of faults: American Association of Petroleum Geologists Bulletin, v. 27, p. 1245–1265.
Clemons, K., H. Bello, R. Bodziak, M. McChesney, R. Meek, and A. Stephens, 2016, Seismic attributes: Exploiting seismic data to understand heterogeneous reservoir performance in the Eagle Ford Shale, south Texas, U.S.A., *in* J. A. Breyer, ed., The Eagle Ford Shale: A renaissance in U.S. oil production: American Association of Petroleum Geologists Memoir 110, p. 241–258.
Cloos, E., 1942, Distortion of stratigraphic thickness due to folding: Proceedings of the National Academy of Science, v. 28, no. 10, p. 401–407.
______, 1968, Experimental analysis of Gulf Coast fracture patterns: American Association of Petroleum Geologists Bulletin, v. 52, p. 420–444.
Coates, J., 1945, The construction of geological sections: Quarterly Journal, Geological Mining and Metallurgical Society of India, v. 17, p. 1–11.
Cole, W., 1969, Reservoir Engineering Manual, second ed.: Houston, Gulf Publishing Company, 385 p.
Coleman, J. M., 1982, Deltas: Processes of Deposition and Models for Exploration: Boston, International Human Resources Development Corporation, 124 p.
______, 1988, Dynamic changes and processes in the Mississippi River Delta: Geological Society of America Bulletin, v. 100, p. 999–1015.

Corredor F., J. H. Shaw, and F. Bilotti, 2005, Structural styles in the deep-water fold and thrust belts of the Niger Delta, American Association of Petroleum Geologists Bulletin, v. 89, p. 753–780.

Corredor, F., J. H. Shaw, and J. Suppe, 2005, Shear fault-bend fold, deep water Niger Delta, *in* Shaw, J., C. Connors, and J. Suppe, eds., Seismic Interpretation of Contractional Fault-Related Folds, American Association of Petroleum Geologists, p. 87–92.

Craft, B. C., and M. F. Hawkins, 1959, Applied Petroleum Reservoir Engineering: Englewood Cliffs, NJ, Pearson, 437 p.

Crans, W., G. Mandl, and J. Haremboure, 1980, On the theory of growth faulting: A geomechanical delta model based on gravity sliding: Journal of Petroleum Geology, v. 2, p. 265–307.

Cribbs, B., 2009, Practical Wellbore Formation Test Interpretation: Search and Discovery Article #120009, 20 p.

Crowell, J. C., 1948, Template for spacing structure contours: American Association of Petroleum Geologists Bulletin, v. 32, p. 2290–2294.

______, 1950, Geology of Hungry Valley area, southern California: American Association of Petroleum Geologists Bulletin, v. 34, p. 1623–1646.

______, 1959, Problems in fault nomenclature: American Association of Petroleum Geologists Bulletin, v. 43, p. 2653–2674.

______, 1962, Displacement along the San Andreas fault, California: Geological Society of America Special Paper 71, 61 p.

______, 1974a, Origin of late Cenozoic basins in southern California, *in* Dickinson, W. R., ed., Tectonics and Sedimentation: Society of Economic Paleontologists and Mineralogists, Special Publication No. 22, p. 190–204.

______, 1974b, Sedimentation along the San Andreas Fault, California, *in* Dott, R. H., Jr., and R. H. Shaver, eds., Modern and Ancient Geosynclinal Sedimentation: Society of Economic Paleontologists and Mineralogists, Special Publication No. 19, p. 292–303.

______, 1982, The tectonics of Ridge Basin, Southern California, *in* Crowell, J. C., and M. Link, eds., Geologic History of Ridge Basin, Southern California: Pacific Section, Society of Economic Paleontologists and Mineralogists, p. 25–42.

______, 2003a, Introduction to geology of Ridge Basin, southern California, *in* Crowell, J. C., ed., Evolution of Ridge Basin, Southern California: An Interplay of Sedimentation and Tectonics, Special Paper 37: Geological Society of America, p. 1–16.

______, 2003b, Overview of rocks bordering Ridge Basin, southern California, *in* Crowell, J. C., ed., Evolution of Ridge Basin, Southern California: An Interplay of Sedimentation and Tectonics, Special Paper 37: Geological Society of America, p. 89–112.

______, 2003c, Tectonics of Ridge Basin, southern California, *in* Crowell, J. C., ed., Evolution of Ridge Basin, Southern California: An Interplay of Sedimentation and Tectonics, Special Paper 37: Geological Society of America, p. 157–204.

Crowell, J. C., and M. H. Link, eds., 1982, Geologic History of Ridge Basin, Southern California: Los Angeles, CA, Pacific Section, Society of Economic Paleontologists and Mineralogists, 304 p.

Cunningham, W. D., and P. Mann, 2007, Tectonics of strike-slip restraining and releasing bends, Geological Society, London, Special Publications, 290, 482 p.

Currie, J. B., 1952, Three-dimensional method for solution of oil field structures: American Association of Petroleum Geologists Bulletin, v. 36, p. 889–890.

______, 1956, Role of concurrent deposition and deformation of sediments in development of salt-dome graben structures: American Association of Petroleum Geologists Bulletin, v. 40, p. 1–16.

Dahlen, F. A., and J. Suppe, 1988, Mechanics, growth, and erosion of mountain belts: Geological Society of America, Special Paper 218, p. 161–178.

Dahlen, F. A., J. Suppe, and D. Davis, 1984, Mechanics of fold-and-thrust belts and accretionary wedges: Cohesive Coulomb theory: Journal of Geophysical Research, v. 89, p. 10,087–10,101.

Dahlstrom, C. D. A., 1969, Balanced cross-sections: Canadian Journal of Earth Sciences, v. 6, p. 743–757.

______, 1970, Structural geology in the eastern margin of the Canadian Rocky Mountains: Bulletin of Canadian Petroleum Geology, v. 18, p. 332–406.

Dart, R. L., and H. S. Swolfs, 1992, Subparallel faults and horizontal-stress orientations: An evaluation of in-situ stresses inferred from elliptical wellbore enlargements, *in* Larsen, R. M., H. Brekke, and B. T. Larson, eds., Structural and Tectonic Modeling and Its Applications to Petroleum Geology: Norwegian Petroleum Society, Special Publication No. 1, p. 519–529.

Davis, D., and T. Engelder, 1985, The role of salt in fold-and-thrust belts: Tectonophysics, v. 119, p. 67–88.

Davis, D., J. Suppe, and F. A. Dahlen, 1983, Mechanics of fold-and-thrust belts and accretionary wedges: Journal of Geophysical Research, v. 88, p. 1153–1172.

Davis K., M. H. Burbank, D. Fisher, S. Wallace, and D. Nobes, 2005, Thrust fault growth and segment linkage in the active Ostler fault zone, New Zealand, Journal of Structural Geology, v. 27, p. 1528–1546.

Davis, T. L., 1987, Seismic facies analysis: pitfalls and applications in cratonic basins: The Leading Edge, v. 6, p. 18–23, 37.

Davison, I., 1986, Listric normal fault profiles: Calculation using bed-length balance and fault displacement: Journal of Structural Geology, v. 8, no. 2, p. 209–210.

______, 1987, Normal fault geometry related to sediment compaction and burial: Journal of Structural Geology, v. 9, p. 393–401.

Dawers, N. H., and J. R. Underhill, 2000, The role of fault interaction and linkage in controlling synrift stratigraphic sequences: Late Jurassic, Statfjord East Area, North Sea: American Association of Petroleum Geologists Bulletin, v. 84, p. 45–64.

De Paor, D. G., and G. Eisenstadt, 1987, Stratigraphy and structural consequences of fault reversal: An example from the Franklinian Basin, Ellesmere Island: Geology, v. 15, p. 948–949.

Dennis, H., J. Baillie, T. Holt, and D. Wessel-Berg, 2000, Hydrodynamic activity and tilted oil-water contacts in the North Sea, *in* Ofstad, K., J. E. Kittilsen, and P. Alexander-Marrack, eds., Improving the Exploration Process by Learning from the Past, Norwegian Petroleum Society Special Publication 9, Amsterdam, Elsevier, p. 171–185.

Dennis, J. G., 1972, Structural Geology: New York, Ronald Press, 532 p.

Dennison, J. M., 1968, Analysis of Geologic Structures: New York, W. W. Norton & Co., 209 p.

Dickey, P. A., 1981, Petroleum Development Geology, second ed.: Tulsa, OK, PennWell Publishing Co., 428 p.

______, 1986, Petroleum Development Geology, third ed.: Tulsa, OK, PennWell Publishing Co., 530 p.

Dickinson, G., 1954, Subsurface interpretation of intersecting faults and their effects upon stratigraphic horizons: American Association of Petroleum Geologists Bulletin, v. 38, no. 5, p. 854–877.

Dickinson, W. R., and D. R. Seely, 1979, Structure and stratigraphy of forearc regions: American Association of Petroleum Geologists Bulletin, v. 63, p. 2–31.

Diegel, F. A., J. F. Karlo, D. C. Schuster, R. C. Shoup, and P. R. Tauvers, 1995, Cenozoic structural evolution and tectono-stratigraphic framework of the northern Gulf Coast continental margin, *in* Jackson, M. P. A., D. G. Roberts, and S. Snelson, eds., Salt Tectonics: A Global Perspective: American Association of Petroleum Geologists Memoir 65, p. 109–151.

Dischinger, J. D., and S. Mitra, 2006, Three-dimensional structural model of Painter and East Painter reservoir structures, Wyoming Fold and Thrust Belt: American Association of Petroleum Geologists Bulletin, v. 90, no. 8, p. 1171–1185.

Dixon, J. M., 1975, Finite strain and progressive deformation in models of diapiric structures: Tectonophysics, v. 28, p. 89–124.

Dobrovolny, J. S., 1951, Descriptive geometry for geologists: American Association of Petroleum Geologists Bulletin, v. 35, p. 1674–1686.

Doebl, F., and R. Teichmuller, 1979, Zur Geologie und heutigen Geothermik mittleren Oberrhein-Graben: Fortschritte in der Geologie von Rheinland und Westfalen, v. 27, p. 1–17.

Dolly, E. D., 1979, Geological techniques utilized in Trap Spring Field discovery, Railroad Valley, Nye County, Nevada, *in* Newman, G. W., and H. D. Goode, eds., Basin and Range Symposium and Great Basin Field Conference: Denver, CO, Rocky Mountain Association of Geologists and Utah Geological Association, p. 455–467.

Donath, F. A., and R. B. Parker, 1964, Folds and folding: Geological Society of America Bulletin, v. 75, p. 45–62.

Donn, W. L., and J. A. Shimer, 1958, Graphic Methods in Structural Geology, New York, Appleton-Century-Crofts, 180 p.

Doveton, J. H., 1986, Log Analysis of Subsurface Geology: New York, John Wiley & Sons, 273 p.

Downey, M. W., 1984, Evaluating seals for hydrocarbon accumulation: American Association of Petroleum Geologists Bulletin, v. 68, p. 1752–1763.

Dula, W. F., Jr., 1991, Geometric models of listric normal faults and rollover folds: American Association of Petroleum Geologists Bulletin, v. 75, p. 1609–1625.

Duran, L. G., 1951, Trigonometric and graphical solution of problems in structural mapping: World Oil, v. 133, no. 6, p. 94–98.

Durand-Riard, P., Shaw, J., Plesch, A., and Lufadeju, G., 2013, Enabling 3D geomechanical restoration of strike- and oblique-slip faults using geological constraints, with applications to the deep-water Niger Delta: Journal of Structural Geology, v. 48, p. 33–44.

Dury, G. H., 1952, Map Interpretation: London, Pittman, 203 p.

Eardley, A. J., 1938, Graphic treatment of folds in three dimensions: American Association of Petroleum Geologists Bulletin, v. 22, p. 483–489.

Eastman Christensen, n.d., Magnetic Single-Shot Operations Manual: Salt Lake City, UT, Eastman Christensen, 38 p.

Eastman Whipstock, n.d., Introduction to Directional Drilling: Houston, TX, Eastman Whipstock, 38 p.

Ebanks, W. J., Jr., and J. F. Weber, 1982, Development of a shallow heavy-oil deposit in Missouri: Oil and Gas Journal, September 27 issue, p. 222–234.

Echavarria, L., R. Hernandez, R. Allmendinger, and J. Reynolds, 2003, Subandean thrust and fold belt of northwestern Argentina: Geometry and timing of the Andean evolution, AAPG Bulletin, v. 87, p. 965–985.

Edwards, M. B., 2001, Wilcox Regional Study—South Texas Zapata Phase: Unpublished multiclient report.

Ehlert, K. W., 2003, Tectonic significance of the middle Miocene Mint Canyon and Caliente Formations, southern California: *in* Crowell, J. C., ed., Evolution of Ridge Basin, Southern California: An Interplay of Sedimentation and Tectonics, Special Paper 367, p. 113–130.

Eichelberger, N. W., A. G. Nunns, R. H. Groshong Jr., and A. N. Hughe, Direct estimation of fault trajectory from structural relief: American Association of Petroleum Geologists Bulletin, v. 101, p. 635–653.

Elliott, D., 1976, The energy balance and deformation mechanisms of thrust sheets: Philosophical Transactions of the Royal Society of London, Series A, v. 283, p. 289–312.

______, 1983, The construction of balanced cross sections: Journal of Structural Geology, v. 5, p. 101.

Emery, K. O., 1980, Continental margins—Classification of petroleum prospects: American Association of Petroleum Geologists Bulletin, v. 64, p. 297–315.

Engelder, T., and R. Engelder, 1977, Fossil distortion and decollement tectonics of the Appalachian Plateau: Geology, v. 5, p. 457–460.

Ewing, T. E., 1983, Growth faults and salt tectonics in the Houston Diapir Province—Relative timing and exploration significance: Transactions Gulf Coast Association of Geological Societies, v. 33, p. 83–90.

Fagin, R. A., J. E. Trusty, L. R. Emmet, and M. K. Mayo, 1991, MWD resistivity tool guides bit horizontally in thin bed: Oil & Gas Journal, December 9 issue, p. 62–65.

Faill, R. T., 1969, Kink band structure in the Valley and Ridge Province, central Pennsylvania: Geological Society of America Bulletin, v. 80, p. 2539–2550.

______, 1973, Kink band folding, Valley and Ridge Province, Pennsylvania: Geological Society of America Bulletin, v. 84, p. 1289–1314.

Fallaw, W. C., 1973, Grabens on anticlines in Gulf Coastal Plain, and thinning of sedimentary section in downthrown block: American Association of Petroleum Geologists Bulletin, v. 57, p. 198–203.

FEMA (Federal Emergency Management Agency), 2014, Vertical Datum Conversion, Guidance Document 24, 11 pages.

Ferrill, D. A., A. P. Morris, and R. N. McGinnis, 2009, Crossing normal conjugate faults in field exposures and seismic data, American Association of Petroleum Geologists Bulletin, v. 93, no. 11, pp. 1471–1488.

Ferrill, D. A., A. P. Morris, J. A. Stamatakos, and D. W. Sims, 2000, Crossing conjugate normal faults, American Association of Petroleum Geologists Bulletin, v. 84, pp. 1543–1549.

Ferrill, D. A., J. A. Stamatakos, and D. Simmons, 1999, Normal fault corrugation: Implications for growth and seismicity of active normal faults: Journal of Structural Geology, v. 21, p. 1027–1038.

Fisher, W. L., and J. H. McGowen, 1967, Depositional systems in the Wilcox Group of Texas and their relationship to occurrence of oil and gas: Transactions Gulf Coast Association of Geological Societies, v. 27, p. 105–125.

Fitzgerald, P., and B. K. Pedersen, 2007, A technique for improving the accuracy of wireline depth measurements, SPE 110318, 2007 SPE Annual Technical Conference and Exhibition, 11–14 November, Anaheim, California.

Flinn, D., 1962, On folding during three-dimensional progressive deformation: Quarterly Journal of the Geological Society, v. 118, p. 385–433.

Forgotson, J. M., Jr., 1960, Review and classification of quantitative mapping techniques: American Association of Petroleum Geologists Bulletin, v. 44, p. 83–100.

Forsyth, D., H. Bolt, and T. Loermans, 2013. Improved depth quality measurement: Where old theory should meet (near) future practice, SPWLA (Society of Petrophysicists and Well Log Analysts) 54th Annual Logging Symposium, June 22–26, New Orleans, 13 p.

Fox, F. G., 1959, Structure and accumulation of hydrocarbons in southern Foothills, Alberta, Canada: American Association of Petroleum Geologists Bulletin, v. 43, p. 992–1025.

Frey, M. G., and W. H. Grimes, 1970, Bay Marchand-Timbalier Bay-Callou Island Salt Complex, Louisiana, *in* Halbouty, M. T., ed., Giant Oil and Gas Fields of the Decade 1968–1978: American Association of Petroleum Geologists Memoir 14, p. 277–291.

Frischknecht, F. C., R. Grette, P. V. Raab, and J. Meredith, 1985, Location of abandoned wells by magnetic surveys—Acquisition and interpretation of aeromagnetic data for five test areas: U.S. Geological Survey Open-File Report 85-614-A, 64 p.

Fugitt, D. S., J. E. Florstedt, G. J. Herricks, M. R. Wise, C. E. Stelting, and W. J. Schweller, 2000, Production characteristics of sheet and channelized turbidite reservoirs: Garden Banks191, Gulf of Mexico: The Leading Edge, v. 19, p. 356–369.

Gaafar, G. R., M. M., Altunbay, and A. A. Shaharudin, 2016, Perched-water, the concept and its effects on exploration and field development plans in sandstone and carbonate reservoirs, OTC-26653-MS, Offshore Technology Conference, Kuala Lumpur, Malaysia, 12 p.

Gatewood, L. E., 1970, Oklahoma City Field—Anatomy of a giant, *in* Halbouty, M. T., ed., Giant Oil and Gas Fields of the Decade 1968–1978: American Association of Petroleum Geologists Memoir 14, p. 223–254.

Geikie, A., 1903, Text-Book of Geology, fourth ed.: New York, Macmillan and Company, v. 1.

Geiser, P. A., 1988, The role of kinematics in the construction and analysis of geological cross sections in deformed terrains, *in* Mitra, G., and S. Wojtal, eds., Geometrics and Mechanisms of Thrusting, with Special Reference to the Appalachians: Geological Society of America, Special Paper 222, p. 47–76.

Ghignone, J. I., and G. Deandrade, 1970, General geology and major oilfields of Reconcavo Basin, Brazil, *in* Halbouty, M. T., ed., Giant Oil and Gas Fields of the Decade 1968–1978: American Association of Petroleum Geologists Memoir 14, p. 337–358.

Gibbs, A. D., 1983, Balanced cross section construction from seismic sections in areas of extensional tectonics: Journal of Structural Geology, v. 5, p. 153–160.

______, 1984, Structural elevation of extensional basin margins: Quarterly Journal of the Geological Society, v. 141, p. 609–620.

Giles, A. B., and D. H. Wood, 1983, Oakwood salt dome, east Texas: Geologic framework, growth history, and hydrocarbon production: Texas Bureau of Economic Geology, Geological Circular 83–1, 55 p.

Gill, W. D., 1953, Construction of geological sections of folds with steep-limb attenuation: American Association of Petroleum Geologists Bulletin, v. 37, p. 2389–2406.

Goguel, J., 1962, Tectonics, second ed.: San Francisco, W. H. Freeman and Co., 384 p.

Gordy, P. L., and F. R. Frey, 1975, Geological cross sections through the Foothills: Foothills Fieldtrip Guide Book, Canadian Society of Petroleum Geologists/Canadian Society of Exploration Geophysicists.

Gordy, P. L., F. R. Frey, and D. K. Norris, 1977, Geological guide for the Waterton-Glacier Park Field Conference: Canadian Society of Petroleum Geologists, 93 p.

Gow, K. L., 1962, Segno Field, Polk County, Texas, *in* Denham, R. L., ed., Typical Oil and Gas Fields of Southeast Texas: Houston, TX, Houston Geological Society, p. 187–191.

Gries, R. R., and R. C. Dyer, eds., 1985, Seismic Exploration of the Rocky Mountain Region: Denver, CO, Rocky Mountain Association of Geologists and Denver Geophysical Society, 300 p.

Groshong, R. H., Jr., 1975, "Slip" cleavage caused by pressure solution in a bucklefold: Geology, v. 3, p. 411–413.

______, 1989, Half graben structures: Balanced models of extensional fault-bend folds: Geological Society of America Bulletin, v. 101, p. 96–105.

Guiraud, M., and M. Seguret, 1985, A releasing solitary overstep model for the Late Jurassic—Early Cretaceous Soria strike-slip basin (northern Spain), *in* Biddle, K. T., and N. Christie-Blick, Strike-Slip Deformation, Basin Formation, and Sedimentation: Society of Economic Paleontologists and Mineralogists, Special Publication No. 37, p. 159–175.

Gwinn, V. E., 1964, Thin-skinned tectonics in the Plateau and northwestern Valley and Ridge province of the Central Appalachians: Geological Society of America Bulletin, v. 75, p. 863–900.

Halbouty, M. T., 1967, Salt Domes: Gulf Region, United States and Mexico: Houston, Gulf Publishing Company, 425 p.

______, 1979, Salt Domes: Gulf Region, United States and Mexico, second ed.: Houston, Gulf Publishing Company, 561 p.

______, 1982, The deliberate search for the subtle trap: American Association of Petroleum Geologists Memoir 32, 351 p.

Hamblin, W. K., 1965, Origin of "reverse drag" on the downthrown side of normal faults: Geological Society of America Bulletin, v. 76, p. 1145–1164.

Hammack, R. W., G. A. Veloski, D. G. Hodges, and C. M. White, 2016, Methods for Finding Legacy Wells in Large Areas, NETL-TRS-6-2016, EPAct Technical Report Series: Pittsburgh, PA, U.S. Department of Energy, National Energy Technology Laboratory, 28 p.

Hammes, U., and J. F. W. Gale, eds., 2013, Geology of the Haynesville Gas Shale in East Texas and West Louisiana, U.S.A.: American Association of Petroleum Geologists Memoir 105, 236 p.

Handley, E. J., 1954, Contouring is important: World Oil, v. 138, p. 106–107.

Hardage, B. A., R. P. Major, and T. F. Hentz, eds., 1998, Integrated strategies for carbonate reservoir growth: An example from the Ellenburger Group, Permian Basin, West Texas: University of Texas at Austin, Bureau of Economic Geology, Contract Report No. GRI-98/0070, 275 p.

Harding, T. P., 1973, Newport-Inglewood Trend, California—An example of wrenching style of deformation: American Association of Petroleum Geologists Bulletin, v. 57, p. 97–116.

______, 1974, Petroleum traps associated with wrench faults: American Association of Petroleum Geologists Bulletin, v. 58, p. 1290–1304.

______, 1976a, Predicting productive trends related to wrench faults: World Oil, v. 182, no. 7, p. 64–69.

______, 1976b, Tectonic significance and hydrocarbon trapping consequences of sequential folding synchronous with San Andreas faulting, San Joaquin Valley, California: American Association of Petroleum Geologists Bulletin, v. 60, p. 356–378.

______, 1984, Graben hydrocarbon occurrences and structural style: American Association of Petroleum Geologists Bulletin, v. 68, p. 333–362.

______, 1985, Seismic characteristics and identification of negative flower structures, positive flower structures, and positive structural inversion: American Association of Petroleum Geologists Bulletin, v. 69, p. 582–600.

______, 1990, Identification of wrench faults using subsurface structural data: Criteria and pitfalls: American Association of Petroleum Geologists Bulletin, v. 74, p. 1590–1609.

Harding, T. P., and J. D. Lowell, 1979, Structural styles, their plate-tectonic habitats, and hydrocarbon traps in petroleum provinces: American Association of Petroleum Geologists Bulletin, v. 63, p. 1016–1058.

Harrington, J. W., 1951, The elementary theory of subsurface structural contouring: Transactions American Geophysical Union, v. 32, p. 77–80.

Hart, B. S., 2011, Introduction to Seismic Interpretation: American Association of Petroleum Geologists Discovery Series No. 16 digital version.

Harth P., P. Ensign, and B. Davis, 2000, Rapidly and automatically tying time to depth using sequential volume calibration: SEG Technical Program Expanded Abstracts, p. 681–684.

Hartman, K., C. Joy, A. Reitz, L. Chetel, and S. Chakraborty, 2017, Immediate business impact from fast-track processing of long offset ocean-bottom nodes survey, Thunder Horse field, Gulf of Mexico: SEG Technical Program Expanded Abstracts 2017, p. 2215–2219.

Hartman, R. R., D. J. Teskey, and J. L. Friedberg, 1971, A system for rapid digital aeromagnetic interpretation: Geophysics, v. 36, p. 891–918.

Hay, J. T. C., 1978, Structural development in the northern North Sea: Journal of Petroleum Geology, v. 1, p. 65–77.

Heisey, E. L., ed., 1977, Rocky Mountain Thrust Belt Geology and Resources: Casper, WY, Wyoming Geological Society, 787 p.

Herold, S. C., 1933, Projection of dip angle on profile section: American Association of Petroleum Geologists Bulletin, v. 17, p. 740–742.

Herron, D. A., 2011, First Steps in Seismic Interpretation, Geophysical Monograph Series Number 16, Society of Exploration Geophysics, 203 p.

Heybroek, P., U. Haanstra, and D. A. Erdman, 1967, Observations on the geology of the North Sea area: Proceedings of the 7th World Petroleum Congress, v. 2, p. 905–916.

Hill, M. L., 1942, Graphic method for some geologic calculations: American Association of Petroleum Geologists Bulletin, v. 26, p. 1155–1159.

______, 1947, Classification of faults: American Association of Petroleum Geologists Bulletin, v. 31, p. 1669–1673.

______, 1959, Dual classification of faults: American Association of Petroleum Geologists Bulletin, v. 43, p. 217–237.

Hill, M. L., and T. W. Dibblee, 1953, San Andreas, Garlock, and Big Pine faults, California (a study of the character, history, and tectonic significance of their displacements): Geological Society of America Bulletin, v. 64, p. 443–458.

Hills, E. S., 1963, Elements of Structural Geology: London, Methuen & Co., 483 p.

Hooper, N. J., J. G. M. Raven, and M. J. Kilpatrick, 1992, Computer modeling of multiple surfaces with faults: The Ivanhoe Field, Outer Moray Firth Basin, U.K. North Sea, *in* Hamilton, D. E., and T. A. Jones, eds., Computer Modeling of Geologic Surfaces and Volumes: Tulsa, OK, American Association of Petroleum Geologists Bulletin, p. 161–174.

Horsfield, W. T., 1977, An experimental approach to basement-controlled faulting: Geologie en Mijnbouw, v. 56, p. 363–370.

______, 1980, Contemporaneous movement along crossing conjugate normal faults: Journal of Structural Geology, v. 2, p. 305–310.
Hossack, J. R., 1983, A cross-section through the Scandinavian Caledonides constructed with the aid of branch-line maps: Journal of Structural Geology, v. 5, p. 103–111.
Hubbert, M. K., 1937, Theory of scale models as applied to the study of geologic structures: Geological Society of America Bulletin, v. 77, p. 1247–1264.
______, 1953, Entrapment of petroleum under hydrodynamic conditions, American Association of Petroleum Geologists Bulletin, v. 37, pp. 1954–2026
Hubbert, M. K., and W. W. Rubey, 1959, Role of fluid pressure in mechanics of overthrust faulting: Geological Society of America Bulletin, v. 70, p. 115–166.
Hudec, M. R., and M. P. A. Jackson, 2007, Terra infirma: Understanding salt tectonics, Earth-Science Reviews, v. 82, p. 1–28.
Hughes, A. N., and J. H. Shaw, 2014, Fault displacement-distance relationships as indicators of contractional fault-related folding style, American Association of Petroleum Geologists Bulletin, v. 98, no. 2, p. 227–251.
Hull, C. E., and H. R. Warman, 1968, Asmari Oil Fields of Iran: London, British Petroleum Company, p. 428–437.
Illies, J. H., 1981, Mechanism of graben formation: Tectonophysics, v. 73, p. 249–266.
Ivanhoe, L. F., 1956, Integration of geological data on seismic sections: American Association of Petroleum Geologists Bulletin, v. 40, p. 1016–1023.
Jackson, M. P. A., and C. J. Talbot, 1991, A glossary of salt tectonics: Geological Circular, v. 91–94, University of Texas at Austin, Bureau of Economic Geology, 44 p.
Jackson, M. P. A., and S. J. Seni, 1983, Geometry and evolution of salt structures in a marginal rift basin of the Gulf of Mexico, east Texas: Geology, v. 11, p. 131–135.
Jackson, M. P. A., and W. E. Galloway, 1984, Structural and depositional styles of Gulf Coast Tertiary continental margins: Application to hydrocarbon exploration: American Association of Petroleum Geologists Bulletin, Continuing Education Course Note Series No. 25, 226 p.
Jaeger, J. C., 1962, Elasticity, Fracture and Flow with Engineering and Geological Application: London, Methuen, 268 p.
Jaeger, J. C., and N. G. W. Cook, 1969, Fundamentals of rock mechanics: London, Methuen, 513 p.
Jaeger, L. G., 1966, Cartesian Tensors in Engineering Science: New York, Pergamon Press, 166 p.
Jageler, A. H., and D. R. Matuszak, 1972, Use of well logs and dipmeters in stratigraphic-trap exploration: American Association of Petroleum Geologists Memoir 16, p. 107–135.
James, D. M. D., ed., 1984, The Geology and Hydrocarbon Resources of Negora Brunei Darussalam: Bandar Seri Begawan, Muzium Brunei, 169 p.
Jamieson, A., 2012, Introduction to Wellbore Positioning, V09.10.17, 247 p.
Jenyon, M. K., 1988, Fault-salt wall relationships, southern North Sea: Oil & Gas Journal, September 5 issue, p. 76–81.
Johanson, D. B., 1987, Structural evolution of Grand Lake Field, Cameron Parish, Louisiana: Transactions Gulf Coast Association of Geological Societies, v. 37, p. 113–122.
Johnson, H. A., and D. H. Bredeson, 1971, Structural development of some shallow salt domes in Louisiana Miocene Productive Belt, American Association of Petroleum Geologist Bulletin, v. 55, p. 204–226.
Jones, P. B., 1971, Folded faults and sequence of thrusting in Alberta Foothills: American Association of Petroleum Geologists Bulletin, v. 55, p. 292–306.
______, 1982, Oil and gas beneath east-dipping underthrust faults in the Alberta Foothills: Rocky Mountain Association of Geologists, v. 1, p. 61–74.
______, 1988, Balanced cross-sections – an aid to structural interpretation: The Leading Edge, v. 7, no. 8, p. 29–31.
Joyce, S., 2016, Abandoned Wells—Living on Top of Forgotten Oil and Gas Wells: Wyoming Public Radio broadcast, insideenergy.org.
Katz, D. L., M. W. Legatski, M. R. Tek, L. Gorring, and R. L. Neilsen, 1966, How water displaces gas from porous media: Oil and Gas Journal, January 10 issue, p. 55–60.
Kay, M., 1945, Paleogeographic and palinspastic maps: American Association of Petroleum Geologists Bulletin, v. 29, p. 426–450.
______, 1954, Isolith, isopach, and palinspastic maps (Geological Note): American Association of Petroleum Geologists Bulletin, v. 38, p. 916–917.
Kelley, V., 1971, Geology of Pecos Country, southern New Mexico: New Mexico Bureau of Mines and Mineral Resources, Memoir 24, 79 p.

Kendrick, J., 2000, Turbidite reservoir architecture in the Northern Gulf of Mexico deepwater: Insights from the development of Auger, Tahoe, and Ram/Powell fields, *in* Weimer P., et al. (eds.) Deep-Water Reservoirs of the World, Proceedings of the GCSSEPM Foundation 20th Annual Research Conference: Houston: Write Enterprise, p. 450–468.

Kennedy, W. Q., 1946, The Great Glen Fault: Quarterly Journal of the Geological Society, v. 102, p. 41–76.

Kerans, C., and H. S. Nance, 1991, High frequency cyclicity and regional depositional patterns of the Grayburg Formation, Guadalupe Mountains, New Mexico, *in* Meader-Roberts, S. J., M. P. Candelaria, and G. E. Moore, eds., Sequence Stratigraphy, Facies, and Reservoir Geometries of the San Andres, Grayburg and Queen Formations, Guadalupe Mountains, New Mexico: Society of Economic Paleontologists and Mineralogists, Special Publication 91–32, p. 53–70.

Klein, G. D., 1985, Sandstone Depositional Models for Exploration for Fossil Fuels, third ed.: Boston, International Human Resources Development Corporation, 209 p.

Kostenko O. V, S. J. Naruk, W. Hack, M. Poupon, H. Jurg Meyer, M. Mora-Glukstad, C. Anowai, and M. Mordi, 2008, Structural evaluation of column-height controls at a toe-thrust discovery, deep-water Niger Delta, American Association of Petroleum Geologists Bulletin, v. 92, no. 12, p. 1615–1638.

Kosters, E. C., D. G. Bebout, S. J. Seni, C. M. Garrett Jr., L. F. Brown Jr., H. S. Hamlin, S. P. Dutton, S. C. Ruppel, R. J. Finley, and N. Tyler, 1989, Atlas of Major Texas Gas Reservoirs: University of Texas at Austin, Bureau of Economic Geology, Atlas of Major Texas Gas Reservoirs, 161 p.

Krantz, B., C. Ormond, and B. Freeman, eds., 2016, 3-D Structural Interpretation: Earth, Mind, and Machine: American Association of Petroleum Geologists Memoir 111, 237 p.

Krumbein, W. C., 1942, Criteria for subsurface recognition of unconformities: American Association of Petroleum Geologists Bulletin, v. 26, p. 36–62.

______, 1948, Lithofacies maps and regional sedimentary stratigraphic analysis: American Association of Petroleum Geologists Bulletin, v. 32, p. 1909–1923.

______, 1952, Principles of facies map interpretation: Journal of Sedimentary Petrology, v. 22, p. 200–211.

Kuhme, A. K., 1987, Seismic interpretation of reefs: The Leading Edge, v. 6, no. 8, p. 60–65.

Kyte, D. G., D. N. Meehan, and T. R. Svor, 1994, Method of maintaining a borehole in a stratigraphic zone during drilling: United States Patent 5,311,951 (May 17, 1994).

Lafayette Geological Society, 1964, Typical oil and gas fields of Southwestern Louisiana: Lafayette, LA, Lafayette Geological Society, v. 1, 41 p.

______, 1970, Typical oil and gas fields of Southwestern Louisiana: Lafayette, LA, Lafayette Geological Society, v. 2, 30 p.

______, 1973, Offshore Louisiana Oil & Gas Fields: Lafayette, LA, Lafayette Geological Society, v. 3, 30 p.

Lamb, C. F., 1980, Painter Reservoir Field—Giant in Wyoming Thrust Belt: American Association of Petroleum Geologists Bulletin, v. 64, p. 638–373.

Lamerson, P. R., 1982, The Fossil Basin and its relationship to the Absaroka thrust system, Wyoming and Utah, *in* Powers, R. B., ed., Geologic Studies of the Cordilleran Thrust Belt: Denver, CO, Rocky Mountain Association of Geologists, p. 279–340.

Langstaff, C. S., and D. Morrill, 1981, Geologic Cross Sections: Boston, International Human Resources Development Corporation Press, 108 p.

Laubscher, H. P., 1961, Die Fernschilbhypothese der Jurafaltilng: Eclogae Geologicae Helvetiae, v. 54, p. 222–282.

______, 1977, Fold Development in the Jura: Tectonophysics, v. 37, p. 337–362.

Leake, J., and F. Shray, 1991, Logging while drilling keeps horizontal well on small target: Oil & Gas Journal, September 23 issue, p. 53–59.

LeBlanc, R. J., 1972, Geometry of sandstone reservoir bodies, *in* Cook, T. D., ed., Symposium on Underground Waste Management and Environmental Implications: American Association of Petroleum Geologists Memoir 18, p. 133–190.

Lelek, J. J., 1982, Geologic factors affecting reservoir analysis, Anschutz Ranch East Field, Utah-Wyoming: Society of Petroleum Engineers 15th Annual Offshore Technology Conference, Society of Petroleum Engineers Paper 10992.

Leroy, L. W., and J. W. Low, 1954, Graphic Problems in Petroleum Geology: New York, Harper & Bros., 238 p.

Leroy, L. W., D. O. Leroy, and J. W. Raese, eds., 1977, Subsurface Geology, fourth ed.: Golden, CO, Colorado School of Mines, 941 p.

Leroy, L. W., D. O. Leroy, J. W. Raese, and S. D. Schwochow, eds., 1987, Subsurface Geology, fifth ed.: Golden, CO, Colorado School of Mines, 1002 p.

Leroy, L. W., ed., 1950, Subsurface Geologic Methods—A Symposium, second ed.: Golden, CO, Colorado School of Mines, 1156 p.

Levorsen, A. I., 1943, Discovery thinking: American Association of Petroleum Geologists Bulletin, v. 27, p. 887–928.

______, 1954, Geology of Petroleum: San Francisco, W. H. Freeman and Co., 703 p.

Ley, H. H., 1930, Structure contouring: American Association of Petroleum Geologists Bulletin, v. 14, p. 103–105.

Li, Y., J. Jackson, S. Espinosa, and D. Faw, 2003, Uncertainties in mapping salt flanks with 3D Salt Proximity Survey, SPE 84574: SPE Annual Technical Conference and Exhibition, 5–8 October, Denver, Colorado, 16 p.

Lindsay, R. F., 1991, Grayburg Formation (Permian-Guadalpian): Comparison of reservoir characteristics and sequence stratigraphy in the Northwestern Central Basin Platform with outcrops in the Guadalupe Mountains, New Mexico, *in* Meader-Roberts, S. J., M. P. Candelaria, and G. E. Moore eds., Sequence Stratigraphy, Facies, and Reservoir Geometries of the San Andres, Grayburg and Queen Formations, Guadalupe Mountains, New Mexico: Society of Economic Paleontologists and Mineralogists, Special Publication 91–32, p. 111–118.

Link, M. H., and R. H. Osborne, 1982, Sedimentary facies of Ridge Basin, southern California, *in* Crowell, J. C., and M. H. Link, eds., Geologic History of Ridge Basin, Southern California: Society of Economic Paleontologists and Mineralogists, Pacific Section, p. 63–78.

Link, M. H., C. K. Taylor, N. G. Muñoz, J. E. Bueno, and P. J. Muñoz, 1996, 3-D seismic examples from Central Lake Maracaibo, Maraven's Block 1 Field, Venezuela, *in* Weimer, P., and T. L. Davis, eds., Application of 3-D Seismic Data to Exploration and Production, American Association of Petroleum Geologists, Studies in Geology No. 42 and Society of Exploration Geophysicists Geophysical Developments Series No. 5: Tulsa, OK, American Association of Petroleum Geologists and Society of Exploration Geophysicists, p. 69–82.

Link, M., 2003, Sedimentation of Ridge Basin, *in* Crowell, J. C., ed., Evolution of Ridge Basin, Southern California: An Interplay of Sedimentation and Tectonics, Special Paper 37: Geological Society of America, p. 17–88.

Link, P. K., 1982, Basic Petroleum Geology: Tulsa, OK, Oil and Gas Consultants International, 235 p.

Link, T. A., 1949, Interpretations of foothills structures, Alberta, Canada: American Association of Petroleum Geologists Bulletin, v. 33, p. 1475–1501.

Lisowski, M., W. H. Prescott, J. C. Savage, and M. J. S. Johnston, 1990, Geodetic estimate of coseismic slip during the 1989 Loma Prieta, California, earthquake: Geophysical Research Letters, v. 17, p. 1437–1440.

Lister, G. S., M. A. Etheridge, and P. A. Symonds, 1986, Detachment faulting and the evolution of passive continental margins: Geology, v. 14, p. 246–250.

Lock, B. E., 1989, Subsurface Geological Investigations, unpublished course notes.

Lock, B. E., and S. L. Voorhies, 1988, Sequence stratigraphy as a tool for interpretation of the Cockfield/Yegua in Southwest Louisiana: Transactions Gulf Coast Association of Geological Societies, v. 39, p. 123–131.

Lou, M., D. Cheng, and F. Doherty, 2012, 3D anisotropic calculation of salt-face exit points for VSP salt proximity surveys, SEG 2012 Annual Meeting, 4–9 November, Las Vegas, 5 p.

Low, J. W., 1951, Subsurface maps and illustrations, *in* Leroy, L. W., ed., Subsurface Geologic Methods—A Symposium, second ed.: Golden, CO, Colorado School of Mines, p. 894–969.

Lowell, J. D., 1985, Structural Styles in Petroleum Exploration: Golden, CO, Oil & Gas Consultants International, 460 p.

Lyle, H. N., 1951, Southwest Texas faults: Oil & Gas Journal, v. 49, no. 41, p. 108–112.

Mackenzie, D. B., 1972, Primary stratigraphic traps in sandstone, *in* King, R. E., ed., Stratigraphic Oil and Gas Fields—Classification, Exploration Methods, and Case Histories: American Association of Petroleum Geologists Memoir 16, p. 47–66.

Mackin, J. H., 1950, The down-structure method of viewing geologic maps: Journal of Geology, v. 58, p. 55–77.

Macurda, D. B., Jr., 1987, Seismic interpretation of transgressive and progradational sequence: The Leading Edge, v. 6, no. 4, p. 18–21.

Malhase, J., 1927, Constructing geologic sections with unequal scales: American Association of Petroleum Geologists Bulletin, v. 11, p. 755–757.

Mannhard, G. W., and D. A. Busch, 1974, Stratigraphic trap accumulation in southwestern Kansas and northwestern Oklahoma: American Association of Petroleum Geologists Bulletin, v. 58, p. 447–463.

Marshak, S., and G. Mitra, 1988, Basic Methods of Structural Geology: Englewood Cliffs, NJ, Pearson, 446 p.

Martin, R. G., 1980, Distribution of salt structures in the Gulf of Mexico: Map and descriptive text: U.S. Geological Survey Map MF-1213.

May, R. S., K. D. Ehman, G. G. Gray, and J. C. Crowell, 1993, A new angle on the tectonic evolution of the Ridge Basin, a "strike-slip" basin in Southern California: Geological Society of America Bulletin, v. 105, p. 1357–1372.

Mayuga, M. N., 1970, Geology and development of California's giant—Wilmington Oil Field, *in* Halbouty, M. T., ed., Geology of Giant Petroleum Fields: American Association of Petroleum Geologists Memoir 14, p. 158–184.

McClay, K., and M. Bonora, 2001. Analog models of restraining stepovers in strike-slip fault systems: American Association of Petroleum Geologists Bulletin, v. 85, 233–260.

McClay, K., and M. Bonora, 2001, Analog models of restraining stepovers in strike-slip fault systems: American Association of Petroleum Geologists Bulletin, v. 85, p. 233–260.

Medwedeff, D. A., 1988, Structural analysis and tectonic significance of Late-Tertiary and Quaternary compressive-growth folding, San Joaquin Valley, California [Ph.D. thesis]: Princeton University, 184 p.

______, 1989, Growth fault-bend folding at southeast Lost Hills, San Joaquin Valley, California: American Association of Petroleum Geologists Bulletin, v. 73, p. 54–67.

Medwedeff, D. A., and J. Suppe, 1986, Growth-fault bend folding—Precise deformation kinematics, timing and rates of folding and faulting from syntectonic sediments: Geological Society of America, Abstracts with Programs, v. 18, p. 692.

Merret, R., and R. W. Almendinger, 1991, Estimates of strain due to brittle faulting: Sample of fault populations: Journal of Structural Geology, v. 13, p. 735–738.

Merritt, J. W., 1946, Geotechniques of oil exploration: Oil Weekly, v. 121, no. 5, p. 17–26.

Mertie, J. B., Jr., 1947a, Calculation of thickness in parallel folds: Geological Society of America Bulletin, v. 58, p. 779–802.

______, 1947b, Delineation of parallel folds and measurement of stratigraphic dimensions: Geological Society of America Bulletin, v. 58, p. 779–802.

______, 1948, Application of Brianchon's theorem to construction of geologic profiles: Geological Society of America Bulletin, v. 59, p. 767–786.

Mertosono, S., 1975, Geology of Pungut and Tandun Oil Fields, Central Sumatra: Indonesian Petroleum Association (June issue), p. 165–179.

Meyer, H. J., and H. W. McGee, 1985, Oil and gas fields accompanied by geothermal anomalies in Rocky Mountain region: American Association of Petroleum Geologists Bulletin, v. 69, p. 933–945.

Mitra, S., 1986, Duplex structures and imbricate thrust systems: Geometry, structural position, and hydrocarbon potential: American Association of Petroleum Geologists Bulletin, v. 70, p. 1087–1112.

______, 1988, Three-dimensional geometry and kinematic evolution of the Pine Mountain thrust system, southern Appalachians: Geological Society of America Bulletin, v. 100, p. 72–95.

______, 1993, Geometry and kinematic evolution of inversion structures: American Association of Petroleum Geologists Bulletin, v. 77, p. 1159–1191.

Mitra, S., and V. S. Mount, 1998, Foreland basement-involved structures: American Association of Petroleum Geologists Bulletin, v. 82, no. 1, p. 70–109.

Mitra, S., and J. Namson, 1989, Equal-area balancing: American Journal of Science, v. 289, p. 563–599.

Montgomery, S. L., 1997a, Raster logs may be basis for a geologic workstation: Oil and Gas Journal, April 7 issue, p. 84–88.

Montgomery, S. L., 1997b, A thing of beauty: Oil and Gas Investor, August issue, p. 49–51.

Moody, J. D., 1973, Petroleum exploration aspects of wrench-fault tectonics: American Association of Petroleum Geologists Bulletin, v. 57, p. 449–476.

Morales, M., E. Hung, and R. Bischke, 2005, Three-dimensional interpretation of the El Furrial trend, Eastern Venezuela Basin, Venezuela, *in* American Association of Petroleum Geologists Seismic Interpretation of Contractional Fault-Related Folds, AAPG Seismic Atlas Studies in Geology #53, Part 2, p. 83–86.

Morley, C. K., 1999, Patterns of displacement along large normal faults: Implications for basin evolution and fault propagation, based on examples from East Africa: American Association of Petroleum Geologists Bulletin, v. 83, p. 613–634.

Morley, C. K., R. A. Nelson, T. L. Patton, and S. G. Munn, 1990, Transfer zones in the East African Rift System and their relevance to hydrocarbon exploration in rifts: American Association of Petroleum Geologists Bulletin, v. 74, p. 1234–1253.

Mosar, J., and J. Suppe, 1988, Fault propagation folds: models and examples from the Pre-Alps and the Jura: 6ieme Reunion du Groupe Tectonique Suisse, Neuchatel, 8/9.

Motrenko, A., V. Strijov, G.-W. Weber, 2014, Bayesian sample size estimation for logistic regression: Journal of Computational Applied Mathematics, v. 255, p 743–752.

Mount, V. S., and J. Suppe, 1987, State of stress near the San Andreas fault: Implications for wrench tectonics: Geology, v. 15, p. 1143–1146.

______, 1992, Present-day stress orientations to active strike-slip faults: California and Sumatra: Journal of Geophysical Research, v. 97, p. 11,995–12,013.
Mount, V. S., J. Suppe, and S. C. Hook, 1990, A forward modeling strategy for balancing cross sections: American Association of Petroleum Geologists Bulletin, v. 74, p. 521–531.
Murray, G. E., 1968, Salt structures of Gulf of Mexico basin—A review, *in* Braunstein, J., and G. D. O'Brien, eds., Diapirism and Diapirs: American Association of Petroleum Geologists Memoir 8, p. 99–121.
Namson, J., 1981, Structure of the western Foothills Belt, Miaoli-Hsinchu area, Taiwan, (1) Southern Part: Petroleum Geology of Taiwan, no. 18, p. 31–51.
Narr, W., and J. Suppe, 1994, Kinematics of basement-involved compressive structures: American Journal of Science, v. 294, p. 802–860.
Needham, D. T., G. Yielding, and B. Freeman, 1996, Analysis of fault geometry and displacement patterns, *in* Buchanan, P. G., and D. A. Nieuwland, eds., Modern developments in structural interpretation, validation and modelling: Geological Society Special Publication 99, p. 189–199.
Nelson, P. H. H., 1980, Role of reflection seismic in development of Nembe Creek Field, Nigeria, *in* Halbouty, M. T., ed., Giant Oil and Gas Fields of the Decade 1968–1978: American Association of Petroleum Geologists Memoir 30, p. 565–576.
New Orleans Geological Society, 1965, Oil & Gas Fields of Southeast Louisiana, v. 1: New Orleans, LA, New Orleans Geological Society.
______, 1967, Oil & Gas Fields of Southeast Louisiana, v. 2: New Orleans, LA, as above.
______, 1983, Oil & Gas Fields of Southeast Louisiana, v. 3: New Orleans, LA, as above.
______, 1988, Offshore Louisiana Oil & Gas Fields, v. 2: New Orleans, LA, as above.
Nilsen, T. H., and R. J. McLaughlin, 1985, Comparison of tectonic framework and depositional patterns of the Hornelen strike-slip basins of Norway and the Ridge and Little Sulphur Creek strike-slip basins of California, *in* Biddle, K. T., and N. Christie-Blick., eds., Strike-slip Deformation, Basin Formation and Sedimentation: Society of Economic Paleontologists and Mineralogists, Special Publication No. 37, p. 79–103.
Norvick, M.S., 2001. Chronostratigraphic sections of the northern margins of the Australian plate: Final report. Unpublished report for AGSO, 3 September.
Nunns, A. G., 1991, Structural restoration of seismic and geological sections in extensional regimes: American Association of Petroleum Geologists Bulletin, v. 75, p. 278–297.
O'Brien, C., 1988, Pragmatic migration: A method for interpreting a grid of 2D migrated seismic data: The Leading Edge, v. 7, no. 2, p. 24–29.
O'Brien, J. O., 2004, Salt Proximity Survey, SEG Int'l Exposition and 74th Annual Meeting, 10–15 October, Denver, Colorado, 4 p.
Obert, L., and W. I. Duval, 1967, Rock Mechanics and the Design of Structures in Rock: New York, John Wiley & Sons, Inc., 650 p.
Ocamb, R. D., 1961, Growth faults of south Louisiana: Transactions Gulf Coast Association of Geological Societies, v. 11, p. 139–175.
Oppenheimer, D. H., 1990, Aftershock slip behavior of the 1989 Loma Prieta earthquake: Geophysical Research Letters, v. 17, p. 1199–1202.
Pacht, J. A., B. Bowen, B. L. Shaffer, and W. R. Pottorf, 1992, Systems tracts, seismic facies, and attribute analysis within a sequence stratigraphic framework—Example from offshore Louisiana Gulf Coast, *in* Rhodes, E. G., and T. F. Moslow, eds., Marine Clastic Reservoirs: New York, Springer-Verlag, p. 21–38.
Page, D., 1859, Handbook of Geological Terms and Geology: Edinburgh, William Blackwood & Sons, 415 p.
Payton, C. E., ed., 1977, Seismic Stratigraphy—Applications to Hydrocarbon Exploration: American Association of Petroleum Geologists Memoir 26, 516 p.
Peacock, D. C. P., and D. J. Sanderson, 1991, Displacements, segment linkage, and relay ramps in normal fault zones: Journal of Structural Geology, v. 13, p. 721–733.
Pedersen, B. K., and M. V. Constable, 2006, Operational procedures and methodology for improving LWD and wireline depth control, Kristen Field, Norwegian Sea: SPWLA 47th Annual Logging Symposium, June 4–7, Veracruz, Mexico.
Pelto, C. R., 1954, Mapping of multicomponent systems: Journal of Geology, v. 62, p. 501–511.
Pennebaker, P. E., 1972, Vertical net sandstone determination for isopach mapping of hydrocarbon reservoirs: American Association of Petroleum Geologists Bulletin, v. 56, p. 1520–1529.
Perrier, R., and J. Quiblier, 1974, Thickness changes in sedimentary layers during compaction history: Methods for quantitative evaluation: American Association of Petroleum Geologists Bulletin, v. 58, p. 507–520.
Perry, W. J., D. H. Roeder, and D. R. Lageson, 1984North American thrust-faulted terranes: American Association of Petroleum Geologists Bulletin, Reprint Series No. 27, 466 p.

Peters, W. C., 1987, Exploration and Mining Geology, fourth ed.: Golden, CO, Colorado School of Mines, 696 p.

Pettijohn, F. J., P. E. Potter, and R. Siever, 1972, Sand and Sandstone: New York, Springer-Verlag, 618 p.

______, 1987, Sand and Sandstone, second ed.: New York, Springer-Verlag, 553 p.

Pineda, W., and Bergeron, J., 2017, Understanding wireline depth control in wells with high cable tension and best practices to account for wireline stretch while performing stationary measurements: SPWLA 58th Annual Logging Symposium, June 17–21, Oklahoma City, Oklahoma.

Pochat, S., S. Castelltort, J. Van Den Driessche, K. Besnard, and C. Gumiaux, 2004, A simple method of determining sand/shale ratios from seismic analysis of growth faults: An example from upper Oligocene to lower Miocene Niger Delta deposits: American Association of Petroleum Geologists Bulletin., v. 88, p. 1357–1367.

Pollard, D. D., and P. Segall, 1987, Theoretical displacements and stresses near fractures in rock, with applications to faults, joints, veins, dikes and solution surfaces, *in* Atkinson, B. K., ed., Fracture Mechanics of Rock: New York, Academic Press, p. 277–350.

Posamentier, H. W., and P. R. Vail, 1988, Eustatic controls on clastic deposition II, *in* Wilgus, C. K., B. J. Hastings, H. Posamentier, J. C. Wagoner, C. A. Ross, and C. G. St. C. Kendall, eds., Sea Level Changes: An Integrated Approach: Society of Economic Paleontologists and Mineralogists, Special Publication 42, p. 125–154.

Posamentier, H. W., M. T. Jervey, and P. R. Vail, 1988, Eustatic controls on clastic deposition I, *in* Wilgus, C. K., B. J. Hastings, H. Posamentier, J. C. Van Wagoner, C. A. Ross, and C. G. St. C. Kendall, eds., Sea Level Changes: An Integrated Approach: Society of Economic Paleontologists and Mineralogists, Special Publication 42, p. 109–124.

Potter, P. E., 1963, Late Paleozoic sandstones of the Illinois Basin: Illinois State Geological Survey, Report of Investigations 217, 92 p.

Powers, R. B., ed., 1982, Geologic Studies of the Cordilleran Thrust Belt, 3 volumes: Denver, CO, Rocky Mountain Association of Geologists.

Price, R. C., 1986, The southeastern Cordillera: Thrust faulting, tectonic wedging and delamination of the lithosphere: Journal of Structural Geology, v. 8, p. 239–254.

Projects, SEG 2005 Annual Meeting, 6–11 November, Houston, Texas.

Ragan, D. M., 1985, Structural Geology: An Introduction to Geometrical Techniques: New York, John Wiley & Sons, 393 p.

Ramsay, J. G., 1967, Folding and fracturing of rocks: New York, McGraw-Hill, 568 p.

Rees, F. B., 1972, Methods of mapping and illustrating stratigraphic traps, *in* King, R. E., Stratigraphic Oil and Gas Fields—Classification, Exploration Methods, and Case Histories: American Association of Petroleum Geologists Memoir 16, p. 168–221.

Reid, H. F., 1909, Geometry of faults: Geological Society of America Bulletin, v. 20, p. 171–196.

Reid, H. F., W. M. Davis, A. C. Lawson, F. L. Ransome, and Committee, 1913, Report of the committee on the nomenclature of faults: Geological Society of America Bulletin, v. 24, p. 163–186.

Reiter, W. A., 1947, Contouring fault planes: World Oil, v. 126, no. 7, p. 34–35.

Reitz, A., and Hartman, K., 2017, Demonstrating the value of subsalt 3D VSPs; A case study at Thunder Horse North, Gulf of Mexico, SEG Technical Program Expanded Abstracts, p. 2210–2214.

Remmelts, G., 1995, Fault-related salt tectonics in the southern North Sea, The Netherlands, *in* Jackson, M. P. A., D. G. Roberts, and S. Snelson, eds., Salt tectonics: A Global Perspective: American Association of Petroleum Geologists Memoir 65, p. 261–272.

Rettger, R. E., 1929, On specifying the type of structural contouring: American Association of Petroleum Geologists Bulletin, v. 13, p. 1559–1560.

Rich, J. L., 1932, Simple graphical method for determining true dip from two components and for constructing contoured maps from dip observations: American Association of Petroleum Geologists Bulletin, v. 16, p. 92–94.

______, 1934, Mechanics of low-angle overthrust faulting as illustrated by Cumberland thrust block, Virginia, Kentucky, and Tennessee: American Association of Petroleum Geologists Bulletin, v. 18, p. 1584–1596.

______, 1935, Graphical method for eliminating regional dip: American Association of Petroleum Geologists Bulletin, v. 19, p. 1538–1540.

______, 1951, Three critical environments of deposition and criteria for recognition of rocks deposited in each of them: Geological Society of America Bulletin, v. 62, p. 1–20.

Rider, M. H., 1978, Growth faults in Carboniferous of Western Ireland: American Association of Petroleum Geologists Bulletin, v. 62, p. 2191–2213.

Rippon, J. H., 1985. Contoured patterns of the throw and hade of normal faults in the Coal Measures (Wesphalian) of the north-east Derbyshire. Proceedings of the Yorkshire Geological Society 45 (3), p. 147–161.

Robinson, J. P., 1982, Petroleum exploration in southeastern Arizona: Anatomy of an overthrust play, *in* Powers, R. B., ed., Geologic Studies of the Cordilleran Thrust Belt: Denver, CO, Rocky Mountain Association of Geologists, p. 665–674.

Roeder, D. H., 1973, Subduction and orogeny: Journal of Geophysical Research, v. 78, p. 5005–5024.

______, 1983, Hydrocarbons and Geodynamics of Folding Belts: Rocky Mountain Association of Geologists, Continuing Education Short Course Notes.

Rolfsvag, T. A., and Danielsen, T. M., 2016, Perched water static model, SPE-180000-MS, *in* Proceedings of the SPE Bergen One Day Seminar, 20 April, Bergen, Norway, 10 p.

Roux, W. F., Jr., 1978, The development of growth fault structures: American Association of Petroleum Geologists Bulletin, Structural Geology School Notes.

Rowan, M. G., and R. Kligfield, 1989, Cross-section restoration and balancing as aid to seismic interpretation in extensional terrains: American Association of Petroleum Geologists Bulletin, v. 73, p. 955–966.

Rowan, M. G., and R. Linares, 2000, Fold evolution matrices and axial surface analysis of fault bend folds: Application to the Medina Anticline, eastern Cordillera, Colombia: American Association of Petroleum Geologists Bulletin, v. 84, p. 741–764.

Rowan, M. G., M. P. A. Jackson, and B. D. Trudgill, 1999, Salt-related fault families and fault welds in the northern Gulf of Mexico, American Association of Petroleum Geologists Bulletin, v. 83, no. 9, p. 1454–1484.

Royse, F., Jr., M. A. Warner, and D. C. Reese, 1975, Thrust belt structural geometry and related stratigraphic problems, Wyoming-Idaho-northern Utah, *in* Bolyard, D. W., ed., Symposium on Deep Drilling Frontiers in the Central Rocky Mountains: Denver, CO, Rocky Mountain Association of Geologists, p. 41–54.

Sachnik, F. L., and R. D. More, 1983, Southern Appalachian faulting and folding, *in* Bally, A. W., ed., 1983, Seismic Expression of Structural Styles: American Association of Petroleum Geologists, Studies in Geology No. 15, v. 3, p. 3.4.1–79 to 3.4.1–82.

Sadler, P. M., 1999, The influence of hiatus on sedimentary accumulation rates, GeoResearch Forum, v. 5, p. 15–40.

Sanchez, R., J. Y. Chatellier, R. de Sifontes, N. Parra, and P. Muñoz, 1997, Multiple Bischke Plot Analysis, a powerful method to distinguish between tectonic or sedimentary complexity and miscorrelations: Memoras del Primero Congreso Latinoamericano de Sedimentologia, Sociedad Venezolana de Geologos, Tomo II, p. 257–264.

Sanford, A. R., 1959, Analytical and experimental study of simple geologic structures: Geological Society of America Bulletin, v. 70, p. 19–52.

Santiago-Acevedo, J., 1980, Giant fields of the Southern Zone, Mexico, *in* Halbouty, M. T., ed., Giant Oil and Gas Fields of the Decade 1968–1978: American Association of Petroleum Geologists Memoir 30, p. 339–385.

Schlumberger, 1987, Log Interpretation Principles/Applications: Schlumberger Educational Services.

Schmid, C. F., and E. H. Maccannell, 1955, Basic problems, techniques and theory of isopleth mapping: Journal of the American Statistical Association, v. 50, no. 269, p. 220–239.

Scholle, P. A., and D. Spearing, 1982, Sandstone Depositional Environments: American Association of Petroleum Geologists Memoir 31, 410 p.

Schwartz, S. Y., D. L. Orange, and R. S. Anderson, 1994, Complex fault interactions in a restraining bend on the San Andreas Fault, Southern Santa Cruz Mountains, California, *in* Simpson, R. W., ed., The Loma Prieta, California, Earthquake of October 17, 1989: Tectonic Processes and Models: United States Geological Survey Professional Paper 1550-F, p. F49–54.

Sclater, J. G., and P. A. F. Christie, 1980, Continental stretching: An explanation of the post-mid-Cretaceous subsidence of the central North Sea Basin: Journal of Geophysical Research, v. 85, p. 3711–3739.

Sebring, L., Jr., 1958a, Chief tool of the petroleum exploration geologist: The subsurface structural map: American Association of Petroleum Geologists Bulletin, v. 42, p. 561–587.

______, 1958b, Subsurface map: Underground guide for oil men: Oil and Gas Journal, v. 56, no. 27, p. 186–189.

Sengbush, R. L., 1986, Petroleum Exploration: A Quantitative Introduction: Boston, International Human Resources Development Corporation Press, 233 p.

Seni, S. J., and M. P. A. Jackson, 1983, Evolution of salt structures, east Texas diapir province, part 1: Sedimentary record of halokinesis: American Association of Petroleum Geologists Bulletin, v. 67, p. 1219–1244.

Serra, O., 1985, Sedimentary Environments from Wireline Logs: Schlumberger, 211 p.

Setchell, J., 1958, A nomogram for determining true stratum thickness: Shell Trinidad EP 28884, Abstract in PA Bulletin, No. 127/128: N. V. DeBataafache Petroleum Maatschappij, The Hague, Production Department, p. 8.

Shaw, A. B., 1964, Time in Stratigraphy: New York, McGraw Hill, 365 p.

Shaw, J. H., and J. Suppe, 1994, Active faulting and growth folding in the eastern Santa Barbara Channel, California: Geological Society of America Bulletin, v. 106, p. 607–626.

______, 1996, Earthquake hazards of active blind-thrust faults under the central Los Angeles basin, California: Journal of Geophysical Research, v. 101, p. 8623–8642.

Shaw, J. H., R. E. Bischke, and J. Suppe, 1994, Relations between folding and faulting in the Loma Prieta epicentral zone: Strike-slip fault bend folding, *in* Simpson, R. W., ed., The Loma Prieta, California, Earthquake of October 17, 1989: Tectonic Processes and Models: United States Geological Survey Professional Paper 1550-F, p. F3–21.

Shaw, J. H., C. C. Connors, and J. Suppe, 2005, Seismic interpretation of fault related contractional fault-related folds, American Association of Petroleum Geologists Seismic Atlas, Studies in Geology #53, 156 p.

Shaw, J. H., S. C. Hook, and E. P. Sitohang, 1997, Extensional fault-bend folding and synrift deposition: An example from the Central Sumatra Basin, Indonesia: American Association of Petroleum Geologists Bulletin, v. 81, p. 367–379.

Shelton, J. W., 1984, Listric normal faults: An illustrated summary: American Association of Petroleum Geologists Bulletin, v. 68, p. 801–815.

Sheriff, R. E., 1973, Encyclopedic Dictionary of Exploration Geophysics: Tulsa, OK, Society of Exploration Geophysicists, 266 p.

______, 1977, Limitations on resolution of seismic reflections and geologic detail derivable from them, *in* C. E. Payton, ed., Seismic stratigraphy—Applications to hydrocarbon exploration: American Association of Petroleum Geologists Memoir 26, p. 3–14.

______, 1980, Seismic Stratigraphy: Boston, International Human Resources Development Corporation Press, 227 p.

______, 1989, Geophysical Methods: Englewood Cliffs, NJ, Pearson, 605 p.

Shoup, R., 2017, Geological cross sections: Combining geologic models and best practices to finding and exploiting reserves, hidden potential in mature basins: Play Analogs and Best Practices, Geosciences Technology Workshop, 13–14 September, Bandung, Indonesia.

Shumaker, R. C., 1992, Paleozoic structure of the Central Basin Uplift and adjacent Delaware Basin, West Texas, American Association of Petroleum Geologists Bulletin, v. 76, no.11, p. 1804–1824.

Silver, A., 1982, Techniques of Using Geologic Data: Oklahoma City, OK, IED Exploration.

Sinclair, S., L. Crespo, L. Waite, K. Smith, and C. Leslie, 2017, Resource assessment in the northern Midland Basin: Detailed mapping of Late Pennsylvanian, Wolfcampian, and Early Leonardian margins and flooding surfaces using well logs and seismic data: URTeC paper 2692102, p. 1–13.

Singer, J. M., 1992, An example of log interpretation in horizontal wells: The Log Analyst, March–April issue, p. 85–95.

Situmorang, B., E. T. Siswoyo, and F. Paltrinieri, 1976, Wrench fault tectonics and aspects of hydrocarbon accumulation in Java: Indonesian Petroleum Association, June issue, p. 53–67.

Skilbeck, C. G., and Lennox, eds., 1984, Seismic Atlas of Australia and New Zealand Sedimentary Basins: Sydney, Earth Resources Foundation, 301 p.

Sloss, L. L., 1984. Comparative anatomy of cratonic unconformities, *in* Schlee, J. S., ed., Interregional Unconformities and Hydrocarbon Accumulations, American Association of Petroleum Geologists Memoir, v. 36, p. 1–6.

Smith, D. A., 1966, Theoretical consideration of sealing and non-sealing faults: American Association of Petroleum Geologists Bulletin, v. 50, p. 363–374.

______, 1980, Sealing and nonsealing faults in Louisiana Gulf Coast Salt Basin: American Association of Petroleum Geologists Bulletin, v. 64, p. 145–172.

Smith, D. A., and F. A. E. Reeve, 1970, Salt piercement in shallow Gulf Coast salt structures: American Association of Petroleum Geologists Bulletin, v. 54, p. 1271–1289.

Smith, P. R., 1982, Sedimentology and diagenesis of the Miocene Peace Valley Formation, Ridge Basin, southern California, *in* Crowell, J. C, and M. H. Link eds., A Geologic History of Ridge Basin, Southern California: Los Angeles, CA, Pacific Section, Society of Economic Paleontologists and Mineralogist, Pacific Section, p. 151–158.

Smoluchowski, M., 1909, Folding of the earth's surface: Information of mountain chains: Academy of Science Cracovie Bulletin, v. 6, p. 3–20.

Sneider, R. M., F. H. Richardson, D. D. Paytner, R. E. Eddy, and I. A. Wyant, 1977, Predicting reservoir rock geometry and continuity in Pennsylvanian reservoirs, Elk City Field, Oklahoma: Journal of Petroleum Technology, v. 29, p. 851–866.

Spencer, E. W., 1977, Introduction to the Structure of the Earth, second ed.: New York, McGraw-Hill, 640 p.

Stephenson, E. A., and D. D. Haines, 1946a, Preparation of contour maps: Oil and Gas Journal, v. 45, no. 15, p. 115.

______, 1946b, Use of contour maps: Oil and Gas Journal, v. 45, no. 16, p. 131.

Sterne, E. J., 2006, Stacked, "evolved" triangle zones along the southeastern flank of the Colorado Front Range: Mountain Geologist, v 43, p. 65–92.

Stewart, W. A., 1950, Unconformities, *in* Leroy, L. W., ed., Subsurface Geologic Methods—A Symposium, second ed.: Golden, CO, Colorado School of Mines.

Stigant, J., 2012, Geospatial Data Quality and Mapping in Shale Plays—Technical and Business Impact: Search and Discovery Article #40940.

Stockwell, C. H., 1947, The use of plunge in the construction of cross-sections of folds: Geological Association of Canada, v. 3, p. 97–121.

Stone, D. S., 1969, Wrench faulting and Rocky Mountain tectonics: Mountain Geologist, v. 6, no. 2, p. 67–79.

Stoneburner, R. K., 2017, The Evolution of the American Shale Plays: Where We Are and How We Got There: Search and Discovery Article #110228, p. 1–30.

Storey, M., 2013, Is well depth underestimated?: Oilfield Technology, v.6, no. 8, p. 29–32.

______, 2014, 10 Things to know about well logs: Foundations, Journal of the Professional Petroleum Data Management Association, v.1, no. 3, p. 19–21.

Storti, F., S Tavani, S. Merlini, A. Mosconi, and F. Salvini, 2005, Basil anticline, Northern Apennines, Italy, *in* American Association of Petroleum Geologists Seismic Interpretation of Contractional Fault-Related Folds, AAPG Seismic Atlas Studies in Geology #53, Part 2, p. 93–95.

Strahler, A. N., 1948, Geomorphology and structure of the West Kaibab fault zone and Kaibab Plateau, Arizona: Geological Society of America Bulletin, v. 59, p. 513–540.

Straley, W. H., III, 1932, Some notes on the nomenclature of faults: Journal of Geology (a University of Chicago Press journal), v. 42, no. 7 (Oct.–Nov.), p. 756–763.

Stude, G. R., 1978, Depositional environments of Gulf of Mexico South Timbalier Block 54 salt dome and salt dome growth models: Transactions Gulf Coast Association of Geological Societies, v. 28, p. 627–646.

Suppe, J. G., G. T. Chou, and S. C. Hook, 1992, Rates of folding and faulting from growth strata, *in* McClay, K. R., ed., Thrust Tectonics, New York, Chapman and Hall, p. 105–121.

Suppe, J., 1980, Imbricate structure of western Foothills Belts, south-central Taiwan: Petroleum Geology of Taiwan, no. 17, p. 1–16.

______, 1983, Geometry and kinematics of fault-bend folding: American Journal of Science, v. 283, p. 684–721.

______, 1985, Principles of Structural Geology: Englewood Cliffs, NJ, Pearson, 537 p.

______, 1988, Short Course on Cross-Section Balancing in Petroleum Structural Geology: Los Angeles, CA, Pacific Section, American Association of Petroleum Geologists.

Suppe, J., and D. A. Medwedeff, 1984, Fault-propagation folding: Abstracts with Programs, Geological Society of America Bulletin, v. 16, p. 670.

Suppe, J., and D. A. Medwedeff, 1990, Geometry and kinematics of fault-propagation folding: Eclogae Geologicae Helvetiae, v. 83, p. 409–454.

Suppe, J., and J. Namson, 1979, Fault-bend origin of frontal folds in the western Taiwan fold-and-thrust belt: Petroleum Geology of Taiwan, no. 6, p. 1–18.

Suppe, J., and Y. L. Chang, 1983, Kink method applied to structural interpretation of seismic sections, western Taiwan: Petroleum Geology of Taiwan, no. 19, p. 29–47.

Suppe, J., F. Sabat, J. A Munoz, J. Poblet, E. Roca, and J. Verges, Bed-by-bed fold growth by kink-band migration: Sant Llorens de Morunys, eastern Pyrenees, 1997, Journal of Structural Geology, v. 19, p. 443–461.

Suter, H. H., 1947, Exaggeration of vertical scale of geologic sections: American Association of Petroleum Geologists Bulletin, v. 31, p. 318–339.

Swiecicki, T., and K. Maynard, 2009, Geology and sequence, stratigraphy of Block 06/94, Nam Con Son Basin, Offshore Vietnam, *in* Proceedings of the 2009 South East Asia Petroleum Exploration Society (SEAPEX) Conference: Singapore, p. 1–17.

Sylvester, A. G., 1988, Strike-slip faults: Geological Society of America Bulletin, v. 100, p. 1666–1703.

Tanner, J. H., III, 1967, Wrench fault movements along Washita Valley Fault, Arbuckle Mountain area, Oklahoma: American Association of Petroleum Geologists Bulletin, v. 51, p. 126–141.

Taylor, J. C., 1973, Recent developments at Signal Hill, Long Beach oil field, *in* Guidebook for the Society of Economic Paleontologists and Mineralogists: Annual Meeting with the American Association of Petroleum Geologists and the Society of Economic Geophysicists, Field Trip 1, p. 16–25.

Tearpock, D. J., 1997, Project Management and Synergistic Team Development: Short Course Training Manual, Subsurface Consultants & Associates, LLC.

______, 1998, The ten commandments of quality interpretation and mapping: GeoLOGIC, a quarterly technical newsletter of Subsurface Consultants & Associates, Fall issue.

______, 1999, Rebuttal to expert witness reports of John C. Davis, Michael J. Economides, Jack C. Goldstein, and Harry F. Manbeck, Jr.: Union Pacific Resources Company, Plaintiff v. Chesapeake Energy Corporation and Chesapeake Operating, Inc., Defendants, C.A. No. 4–96CV-726-Y, Deposition Exhibit 10, Tearpock.

______, 2000, Outsourcing: A key to success in the twenty-first century: Energy Houston, v. 2.

Tearpock, D. J., and R. E. Bischke, 1980, The structural analysis of the Wissahickon Schist near Philadelphia, Pennsylvania: Geological Society of America Bulletin, v. 91, p. 644–647.

______, 1990, Mapping throw in place of vertical separation: A costly subsurface mapping misconception: Oil and Gas Journal, July 16, v. 88, no. 29, p. 74–78.

______, 1991, Applied Subsurface Geological Mapping: Upper Saddle River, NJ, Pearson, 648 p.

______, 2003, Applied Subsurface Geological Mapping: With Structural Methods, second ed.: Upper Saddle River, NJ, Pearson, 822 p.

Tearpock, D. J., R. E. Bischke, and J. L. Brewton, 1994, Quick Look Techniques for Prospect Evaluations: Lafayette, LA, Subsurface Consultants & Associates, 286 p.

Tearpock, D. J., and J. C. Brenneke, 2001a, Shared earth modeling: A true multidisciplinary approach to E&P: Energy Houston, v. 4, no. 1, p. 40–45.

______, 2001b, Multidisciplinary teams, integrated software for shared-earth modeling key E&P success: Oil and Gas Journal, December 10, p. 84–88.

Tearpock, D. J., and J. Harris, 1987, Subsurface Geological Mapping Techniques—A Training Manual: Tenneco Oil Co.

Tearpock, D. J., and J. Harris, 1990a, Applied Subsurface Mapping Techniques: Gulf Coast Association of Geological Societies Short Course, Annual Convention, Lafayette, LA.

Tearpock, D. J., and J. Harris, 1990b, Isopach Maps and Their Application in Subsurface Mapping: Lafayette Geological Society, Continuing Education Short Course Notes.

Tearpock, D. J., and J. Harris, 1990c, Quantitative Mapping Techniques: Houston Geological Society, Continuing Education Short Course Notes.

Tearpock, D. J., and H. Pousson, 1990, A three-dimensional correction factor equation for directionally drilled wells: Transactions Gulf Coast Association of Geological Societies, v. 40.

Thomas, W. A., 1968, Contemporaneous normal faults on flanks of Birmingham anticlinorium, central Alabama: American Association of Petroleum Geologists Bulletin, v. 52, p. 2123–2136.

Thompson, T., J. Vogt, and Y. Zaretskiy, 2017, Designing and validating 2D reservoir models, SPE-188066-MS, in Proceedings of the SPE Kingdom of Saudi Arabia Annual Technical Symposium and Exhibition, 24-27 April, Dammam, Saudi Arabia, p. 1–13.

Thorogood, J. L., 1986, Well surveying: Past progress, current status and future needs: World Oil, January issue, p. 87–91.

______, 1990, Instrument performance models and their application to directional surveying operations: Society of Petroleum Engineers, Paper 18051.

Thorsen, C. E., 1963, Age of growth faulting in southeast Louisiana: Transactions Gulf Coast Association of Geological Societies, v. 13, p. 103–110.

Tobias, S., 1990, Expansion profiles and sequence stratigraphy: A new way to identify systems tracts, sequence boundaries and eustatic histories, *in* Armentrout, J. M., and B. F. Perkins, Sequence Stratigraphy as an Exploration Tool—Concepts and Practices in the Gulf Coast; Eleventh Annual Research Conference, Gulf Coast Section, Society of Economic Paleontologists and Mineralogists: Austin, TX, Earth Enterprises, p. 351–361.

Todd, R. G., and R. M. Mitchum, 1977, Seismic stratigraphy and global changes of sea level, Part 8: Identification of Upper Triassic, Jurassic, and Lower Cretaceous seismic sequences in Gulf of Mexico and offshore West Africa, *in* Payton, C. E., ed., Seismic Stratigraphy—Applications to Hydrocarbon Exploration: American Association of Petroleum Geologists Memoir 26, p. 145–163.

Travis, R. B., 1978, Graphic determination of stratigraphic and vertical thicknesses in deviated wells: American Association of Petroleum Geologists Bulletin, v. 63, p. 845–866.

Trusheim, F., 1960, Mechanism of salt migration in northern Germany: American Association of Petroleum Geologists Bulletin, v. 44, p. 1519–1541.

Tucker, P. M., 1982, Pitfalls Revisited: Tulsa, OK, Society of Exploration Geophysicists, 19 p.

______, 1988, Seismic contouring: A unique skill: Society of Exploration Geophysicists, v. 53, no. 6, p. 741–749.

Tucker, P. M., and H. J. Yarston, 1973, Pitfalls in seismic interpretation: Society of Exploration Geophysicists, Monograph No. 2, 50 p.

Turk, L. B., 1950, Significance and use of lap-out maps in prospecting for oil and gas (Abs.): American Association of Petroleum Geologists Bulletin, v. 34, p. 625.

Usdansky, S. I., and R. H. Groshong, Jr., 1984a, Analytical extrapolation of cross sections of vertical drape folds by digital computer: Geological Society of America, Abstract with Programs, Part A, v. 16, p. 258.

______, 1984b, Comparison of analytical models for dip-domain and fault-bend folding: Geological Society of America, Abstract with Programs, Part B, v. 16, p. 680.

Vail, P. R., and R. M. Mitchum Jr., 1977, Seismic stratigraphy and global changes of sea level, Part I: Overview, *in* Payton, C. E., ed., Seismic Stratigraphy—Applications to Hydrocarbon Exploration, American Association of Petroleum Geologists Memoir 26, 516 p.

Vail, P. R., and W. Wornardt Jr., 1991, An integrated approach to exploration and development in the 90's: Well log-seismic sequence stratigraphy analysis: Transactions Gulf Coast Association of Geological Societies, v. 41, p. 630–650.

Van Wagoner, J. C., H. W. Posamentier, R. M. Mitchum, P. R. Vail, J. F. Sarg, T. S., Loutit, and J. Hardenbol, 1988, An overview of the fundamentals of sequence stratigraphy and key definitions, *in* Wilgus, C. K., B. S. Hastings, C. G. St. C. Kendall, H. W. Posamentier, C. A. Ross, and J. C. Van Wagoner, eds., Sea Level Changes: An Integrated Approach: Society of Economic Paleontologists and Mineralogists, Special Publication No. 42, p. 39–45.

Van Wagoner, J. C., R. M. Mitchum, K. M. Campion, and V. D. Rahmanian, 1990, Siliciclastic sequence stratigraphy in well logs, cores, and outcrops: Concepts for high-resolution correlation of time and facies: American Association of Petroleum Geologists, Methods in Exploration 7, 55 p.

Vanderhill, J. B., 1991, Depositional setting and reservoir characteristics of Lower Queen (Permian Guadalupian) sandstones, Keystone (Colby) Field, Winkler County, Texas, *in* Meader-Roberts S., M. P. Candelaria, and G. E. Moore, eds., Sequence stratigraphy, facies, and reservoir geometries of the San Andres, Grayburg and Queen Formations, Guadalupe Mountains, New Mexico: Society of Economic Paleontologists and Mineralogists, Special Publication 91–32, p. 119–130.

Vogler, H. A., and B. A. Robison, 1987, Exploration for deep geopressure gas: Corsair Trend, offshore Texas: American Association of Petroleum Geologists Bulletin, v. 71, p. 777–787.

Wadsworth, A. H., Jr., 1953a, Percentage of thinning chart—New techniques in subsurface geology: American Association of Petroleum Geologists Bulletin, v. 37, no. 1, p. 158–162.

______, 1953b, The percentage of thinning chart: Oil and Gas Journal, v. 51, no. 43, p. 72–73.

Wallace, R. E., 1968, Notes on stream channel offsets by the San Andreas Fault, Southern Coastal Ranges, California, *in* Dickinson, W. R., and A. Grantz, eds., Proceedings of Conference on Geologic Problems of San Andreas Fault System: Palo Alto, CA, Stanford University Press, v. 11, p. 6–21.

______, 1973, Surface fracture patterns along San Andreas fault, *in* Kovach, R. L., and A. Nur, eds., Proceedings of Conference on Geologic Problems of San Andreas Fault System: Palo Alto, CA, Stanford University Press, v. 13, p. 248–258.

Warner, A., 2013, Applied Statistics for Bivariate through Multivariate Techniques, Thousand Oaks, CA, Sage, 1172 p.

Weber, K. J., 1995, Perched hydrocarbon-water contacts—A common but poorly understood phenomenon: Abstracts of the 57th EAGE Conference, Glasgow, F035.

Weber, K. J., and E. Daukora, 1976, Petroleum geology of the Niger Delta: Proceedings of the Ninth World Petroleum-Congress, v. 2, p. 209–221.

Weber, K. J., G. Mandl, W. F. Pilaar, F. Lehner, and R. G. Precious, 1978, The role of faults in hydrocarbon migration and trapping in Nigerian growth fault structures: 10th Annual Society of Petroleum Engineers Offshore Technological Conference, Preprint no. OTC-3356, p. 2643–2653.

Weiss, L. E., 1972, The Minor Structures of Deformed Rocks; A Photographic Atlas: New York, Springer-Verlag, 431 p.

Wernicke, B., 1985, Uniform-sense normal simple shear of the continental lithosphere: Canadian Journal of Earth Sciences, v. 22, p. 108–125.

Wernicke, B., P. L. Guth, and G. L. Axen, 1984, Tertiary extensional tectonics in the Sevier thrust belt of southern Nevada, *in* Lintz, J., ed., Western Geological Excursions, v. 4: Guidebook for the Annual Meeting of the Geological Society of America, Reno, NV, Mackay School of Mines, p. 473–510.

West, J., and H. Lewis, 1982, Structure and palinspastic reconstruction of the Absaroka Thrust, Anschutz Ranch area, Utah and Wyoming: Rocky Mountain Association of Geologists, 1982 Annual Symposium, p. 633–639.

Westaway, R., 1995. Deformation around stepovers in strike-slip fault zones, Journal of Structural Geology, v. 17, p. 831–846.

Weston, J. L., A. G. Ledroz, and R. Ekseth, 2014, New gyro while drilling technology delivers accurate azimuth and real-time quality control for all well trajectories: SPE Drilling & Completions, v. 29, no. 3, 295–303.

Wethe, D., 2015, Walking oil rigs spur cheaper, quicker drilling in supply glut: Rigzone, August 31.

Wharton, J. B., Jr., 1948, Isopachous maps of sand reservoirs: American Association of Petroleum Geologists Bulletin, v. 32, p. 1331–1339.

Wheeler, H. E., 1958, Time and stratigraphy: American Association of Petroleum Geologists Bulletin, v. 42, p. 1047–1063.

Wheeler, R. L., 1980, Cross-strike structural discontinuities: Possible exploration tool for natural gas in Appalachian Overthrust Belt: American Association of Petroleum Geologists Bulletin, v. 64, p. 2166–2178.

White, N. J., J. A. Jackson, and D. P. McKenzie, 1986, The relationship between the geometry of normal faults and that of the sedimentary layers in their hanging walls: Journal of Structural Geology, v. 8, p. 897–909.

Wiggins, G. B., and D. J. Tearpock, 1985, Methods in oil and gas reserves estimation: Prospects, newly discovered and developed properties: American Association of Petroleum Geologists, Annual Convention Short Course.

Wilcox, R. E., T. P. Harding, and D. R. Seely, 1973, Basic wrench tectonics: American Association of Petroleum Geologists Bulletin, v. 57, p. 74–96.

Wilgus, C. K., H. Posamentier, C. A. Ross, and C. G. St. C. Kendall, eds., 1988, Sea Level Changes: An Integrated Approach: Society of Economic Paleontologists and Mineralogists Special Publication No. 42, p. 39–45.

Wilkerson, S., and C. L. Dicken, 2001, Quick-look techniques for evaluating two-dimensional cross sections in detached contractional settings, American Association of Petroleum Geologists Bulletin, v. 85, p. 1759–1770.

Williamson, H., 1999, BP Amoco Upstream Technology Group Directional Survey Handbook (BPA-D-004), 260 p.

Wilson, C. W., and R. G. Stearns, 1958, Structure of the Cumberland Plateau, Tennessee: Geological Society of America Bulletin, v. 69, p. 1283–1296.

Wilson, H., J. C. Lofts, G. C. Page, A. G. Brooks, and D. Walder, 2004, Depth control: Reconciliation of LWD and Wireline Depths, Standard Practice and an alternative simple but effective method, SPE 89899: SPE Annual Technical Conference and Exhibition, 26–29 September, Houston, Texas, 11 p.

Winker, C. D., and M. B. Edwards, 1983, Unstable progradational clastic shelf margins, *in* Stanley, D. J., and G. T. Moore, eds., The Shelfbreak: Critical Interface on Continental Margins: Society of Economic Paleontologists and Mineralogists Special Publication No. 33, p. 139–157.

Winker, C. D., R. A. Morton, T. E. Ewing, and D. D. Garcia, 1983, Depositional setting, structural style, and sandstone distribution in three geopressured geothermal areas, Texas Gulf Coast: University of Texas, Bureau of Economic Geology, Report of Investigations 134, 60 p.

Winn, R. D., Jr., M.G. Bishop, and P. S. Gardner, 1985, Shallow-water and sub-storm-base deposition of Lewis Shale in Cretaceous Western Interior seaway, south-central Wyoming: American Association of Petroleum Geologists Bulletin, v. 71, p. 859–881.

Withjack, M. O., Q. T. Islam, and P. R. La Pointe, 1995, Normal faults and their hanging wall deformation: An experimental study: American Association of Petroleum Geologists Bulletin, v. 79, p. 1–18.

Wolff, C. J. M., and J. P. De Wardt, 1981, Borehole position uncertainty—Analysis of measuring methods and deviation of systematic error model: Journal of Petroleum Technology, v. 33, p. 2339–2350.

Wood, M. F., 1981, Depositional environments of the Apple Canyon Sandstone, Ridge Basin, central Transverse Range, California [MS thesis]: Los Angeles, University of Southern California, 266 p.

Woodbury, H. O., I. B. Murray Jr., and R. E. Osbourne, 1980, Diapirs and their relation to hydrocarbon accumulation, *in* Miall, A. D., ed., Facts and Principles of World Petroleum Occurrence: Canadian Society of Petroleum Geologists Memoir 6, p. 119–142.

Woodbury, H. O., I. B. Murray Jr., P. J. Pickford, and W. H. Akers, 1973, Pliocene and Pleistocene depocenters, outer continental shelf, Louisiana and Texas: American Association of Petroleum Geologists Bulletin, v. 57, p. 2428–2439.

Woodward, N. B., 1987, Stratigraphic separation diagrams and thrust belt structural analysis: Thirty-Eighth Field Conference, 1987 Wyoming Geological Association Guidebook, p. 69–77.

Woodward, N. B., S. E. Boyer, and J. Suppe, 1985, An Outline of Balanced Cross Sections: University of Tennessee Department of Geological Sciences, Studies in Geology 11, second ed.: Knoxville, University of Tennessee, 170 p.

Worrall, D. M., and S. Snelson, 1989, Evolution of the Gulf of Mexico, with emphasis on Cenozoic growth faulting and the role of salt, *in* Bally, A. W., and A. R. Palmer, The Geology of North America, v. A: Boulder, CO, Geological Society of America, p. 97–137.

Wright, T. L., 1991, Structural geology and tectonic evolution of the Los Angeles basin, California, *in* Biddle, K. T., ed., Active Margin Basins: American Association of Petroleum Geologists Memoir 52, p. 35–134.

Wyoming Geological Association, 1987, The Thrust Belt Revisited, Thirty-Eighth Field Conference Guidebook: Casper, WY, Wyoming Geological Association, 403 p.

Xiao, H., and J. Suppe, 1988, Origin of rollover: Geological Society of America, Annual Convention, Abstracts with Programs, p. A109.

______, 1989, Role of compaction in the listric shape of growth normal faults: American Association of Petroleum Geologists Bulletin, v. 73, p. 777–786.

______, 1992, Origin of rollover: American Association of Petroleum Geologists Bulletin, v. 76, p. 509–529.

Xiao, H., F. A. Dahlen, and J. Suppe, 1991, Mechanics of extensional wedges: Journal of Geophysical Research, v. 96, p. 10,301–10,318.

Xiaohan, L., 1983, Perturbations de contraintes liées aux structures cassantes dans les calcaires fins du Languedoc. Observations et simulations mathématiques [unpubl. 3 eme cycle thesis]: Montpellier, Université des Sciences et Techniques du Languedoc, 130 p.

Yeats, R. S., and J. M. Beall, 1991, Stratigraphic controls of oil fields in the Los Angeles Basin, *in* Biddle, K. T., ed., Active Margin Basins: American Association of Petroleum Geologists Memoir 52, p. 221–235.

Yielding, G., B. Freeman, and T. Needham, 1997, Quantitative fault seal prediction, American Association of Petroleum Geologists Bulletin, v. 81, 897–917.

Yielding, G., P. Bretan, and B. Freeman, 2010, Fault seal calibration: A brief review, *in* Jolley, S. J., Q. J. Fisher, R. B. Ainsworth, P. J. Vrolijk, and S. Delisle, eds., Reservoir Compartmentalization: London, The Geological Society, p. 243–255.

Zagorski, W. A., G. R. Wrightstone, and D. C. Bowman, 2012, The Appalachian Basin Marcellus gas play: Its history of development, geologic controls on production, and future potential as a world-class reservoir, *in* J. A. Breyer, ed., Shale Reservoirs—Giant Resources for the 21st Century: American Association of Petroleum Geologists Memoir 97, p. 172-200.

Zheng, Z., J. Kemeny, and N. G. Cook, 1989, Analysis of borehole breakouts: Journal of Geophysical Research, v. 94, p. 7171–7182.

Zoback , M. D., D. Moos, L. G. Mastin, and R. N. Anderson, 1985, Well bore breakouts and in situ stress: Journal of Geophysical Research, v. 90, p. 5523–5530.

Zoback, M. D., M. L. Zoback, V. S. Mount, J. Suppe, J. P. Eaton, J. H. Healy, D. Oppenheimer, P. Reasenberg, L. Jones, C. B. Raleigh, I. G. Wong, O. Scotti, C. Wentworth, 1987, New evidence on the state of stress of the San Andreas fault system: Science, v. 238, p. 1105–1111.

INDEX

C

D

E

F

G

H

I

M

N

O

P

Q

R

S

T

U

W

X

Y

Z

CREDITS

Page xxiii: "The petroleum geologist is vital to the economic security of … experience of both, they have combined to write a text that will quickly become the 'Bible' of subsurface mapping." Frank W. Harrison, Jr., Consulting Geologist, Past President AAPG (1981-82)

Page xxiv: "The daily life of geoscientists includes only … fortunate to have the skills of Dan Tearpock and Richard Bischke because both of these authors are passionate about their subject and dedicated to sharing their ideas with others." Dr. Richard S. Bishop, Past President AAPG (1998-1999)

Page xxiv: "Geoscientists must visualize the unknown … celebrated good science serving human needs in a business context." Dr. Charles A. Sternbach, Consulting Geologist, Past President AAPG

Page 4: "If something can go wrong it will," Murphy's Law.

Page 11, Figure 2-1: Modified from Appelbaum. Geological & Engineering Mapping of Subsurface: A workshop course by Robert Appelbaum. Published by permission of Prentice-Hall, Inc.

Page 11, Figure 2-2: Modified from Appelbaum. Geological & Engineering Mapping of Subsurface: A workshop course by Robert Appelbaum. Published by permission of Prentice-Hall, Inc.

Page 15, Figure 2-5: From Tearpock and Harris 1987. Published by permission of Tenneco Oil Company.

Page 15, Figure 2-6: From Tearpock and Harris 1987. Published by permission of Tenneco Oil Company.

Page 19, Figure 2-10: Modified from Bishop 1960, Subsurface Mapping: New York, John Wiley & Sons, 198 p. Published by permission of author.

Page 20, Figure 2-12: Modified from Bishop 1960, Subsurface Mapping: New York, John Wiley & Sons, 198 p. Published by permission of author.

Page 22, Figure 2-14: Modified from Bishop 1960, Subsurface Mapping: New York, John Wiley & Sons, 198 p. Published by permission of author.

Page 22, Figure 2-15: Modified from Bishop 1960, Subsurface Mapping: New York, John Wiley & Sons, 198 p. Published by permission of author.

Page 23, Figure 2-16: Reproduced from Analysis of Geologic Structures by John M. Dennison, by permission of W. W. Norton & Company, Inc. Copyright 1968 by W. W. Norton & Company, Inc.

Page 25, Figure 2-17: AAPG©1991, reprinted by permission of the AAPG whose permission is required for further use.

Page 26, Figure 2-18: AAPG©1991, reprinted by permission of the AAPG whose permission is required for further use.

Page 28, Figure 2-19: AAPG©1991, reprinted by permission of the AAPG whose permission is required for further use.

Page 28, Figure 2-20: Published by permission of Scientific Computer Applications, Inc.

Page 29, Figure 2-21: AAPG©1991, reprinted by permission of the AAPG whose permission is required for further use.

Page 30, Figure 2-22: AAPG©1991, reprinted by permission of the AAPG whose permission is required for further use.

Page 31, Figure 2-23: Published by permission of Scientific Computer Applications, Inc.

Page 32, Figure 2-24: AAPG©1991, reprinted by permission of the AAPG whose permission is required for further use.

Page 33, Figure 2-25: AAPG©1991, reprinted by permission of the AAPG whose permission is required for further use.

Page 34, Figure 2-26: Published by permission of Scientific Computer Applications, Inc.

Page 34, Figure 2-27: Published by permission of Scientific Computer Applications, Inc.

Page 35, Figure 2-28: Published by permission of Scientific Computer Applications, Inc.

Page 35, Figure 2-29: Published by permission of Scientific Computer Applications, Inc.

Page 36, Figure 2-30: Published by permission of Scientific Computer Applications, Inc.

Page 37, Figure 2-31: Published by permission of Scientific Computer Applications, Inc.

Page 38, Figure 2-32: From Thompson et al. 2017. Published by permission of Society of Petroleum Engineers.

Page 38, Figure 2-33: From Thompson et al. 2017. Published by permission of Society of Petroleum Engineers.

Page 39, Figure 2-34: From Thompson et al. 2017. Published by permission of Society of Petroleum Engineers.

Page 40, Figure 2-35: Published by permission of Scientific Computer Applications, Inc.

Page 41, Figure 2-36: Published by permission of Scientific Computer Applications, Inc.

Page 41, Figure 2-37: Published by permission of Scientific Computer Applications, Inc.

Page 42, Figure 2-38: Published by permission of Scientific Computer Applications, Inc.

Page 44, Figure 2-39: Published by permission of Subsurface Consultants & Associates, LLC.

Page 44, Figure 2-40: Published by permission of Subsurface Consultants & Associates, LLC.

Page 45, Figure 2-41a: Published by permission of Subsurface Consultants & Associates, LLC.

Page 45, Figure 2-41b: Published by permission of Subsurface Consultants & Associates, LLC.

Page 46, Figure 2-42a: Published by permission of Subsurface Consultants & Associates, LLC.

Page 46, Figure 2-42b: Published by permission of Subsurface Consultants & Associates, LLC.

Page 47, Figure 2-43a: Published by permission of Subsurface Consultants & Associates, LLC.

Page 47, Figure 2-43b: Published by permission of Subsurface Consultants & Associates, LLC.

Page 48, Figure 2-44: Published by permission of Subsurface Consultants & Associates, LLC.

Page 51, Table 3-1: Bolt, H., 2015, Wireline Depth Determination, Rev. 4.0, 110 p.

Page 52, Figure 3-1: Bolt, H., 2015, Wireline Depth Determination, Rev. 4.0, 110 p. Published with permission from Harald Bolt.

Page 53, Figure 3-2: Published with permission of J. Brenneke.

Page 55, Figure 3-3: Published with permission of J. Brenneke.

Page 56, Figure 3-4: Storey, M., 2013, Is well depth underestimated?: Oilfield Technology, v.6, no. 8, p. 29–32. Published by Oilfield Technology

Page 57, Figure 3-5: Jamieson, A., 2012, Introduction to Wellbore Positioning, V09.10.17, 247 p, image courtesy of Pegasus Vertex, Inc.

Page 58, Figure 3-6a: Pineda, W., and Bergeron, J., 2017, Understanding wireline depth control in wells with high cable tension and best practices to account for wireline stretch while performing stationary measurements: SPWLA 58th Annual Logging Symposium, June 17–21, Oklahoma City, Oklahoma. Published with permission of Society of Petrophysicists and Well Log Analysts.

Page 58, Figure 3-6b: Pineda, W., and Bergeron, J., 2017, Understanding wireline depth control in wells with high cable tension and best practices to account for wireline stretch while performing stationary measurements: SPWLA 58th Annual Logging Symposium, June 17–21, Oklahoma City, Oklahoma. Published with permission of Society of Petrophysicists and Well Log Analysts.

Page 58, Figure 3-7: Pedersen, B. K., and M. V. Constable, 2006, Operational procedures and methodology for improving LWD and wireline depth control, Kristen Field, Norwegian Sea: SPWLA 47th Annual Logging Symposium, June 4–7, Veracruz, Mexico. Published with permission of Society of Petrophysicists and Well Log Analysts.

Page 59, Figure 3-8a: Chia, C. R., Laastad, H., Kostin, A., Hjortland, F. and Bordakov, G., 2006, A New Method for Improving LWD Logging Depth, SPE 102175, 2006 SPE Annual Technical Conference and Exhibition. Published with permission of Society of Petroleum Engineers.

Page 59, Figure 3-8b: Chia, C. R., Laastad, H., Kostin, A., Hjortland, F. and Bordakov, G., 2006, A New Method for Improving LWD Logging Depth, SPE 102175, 2006 SPE Annual Technical Conference and Exhibition. Published with permission of Society of Petroleum Engineers.

Page 61, Figure 3-9a: Published by permission of Tenneco Oil Company

Page 61, Figure 3-9b: Published by permission of Tenneco Oil Company

Page 61, Figure 3-9c: published by permission of J. Brewton.

Page 63: , Figure 3-10: Published by permission of Eastman Christensen.

Page 63, Figure 3-11: From Leroy, L. W., D. O. Leroy, and J. W. Raese, eds., 1977, Subsurface Geology, fourth ed.: Golden, CO, Colorado School of Mines, 941 p.. Published by permission of the Colorado School of Mines.
Page 64, Figure 3-12: Data from Baker Hughes website. (published with permission of J. Brenneke).
Page 65, Figure 3-13a: Published by permission of Gardes Directional Drilling.
Page 65, Figure 3-13b: Published by permission of Gardes Directional Drilling.
Page 66, Figure 3-14: Published by permission of Gardes Directional Drilling.
Page 67, Figure 3-15: Published by permission of Eastman Christensen
Page 70, Figure 3-17: Published by permission of Gardes Directional Drilling.
Page 72, Figure 3-18: Wolff, C. J. M., and J. P. De Wardt, 1981, Borehole position uncertainty – analysis of measuring methods and deviation of systematic error model: Journal of Petroleum Technology, v. 33, p. 2339–2350. Published by permission of the Journal of Petroleum Technology and Gyrodata, Inc.
Page 74, Figure 3-20: Wolff, C. J. M., and J. P. De Wardt, 1981, Borehole position uncertainty – analysis of measuring methods and deviation of systematic error model: Journal of Petroleum Technology, v. 33, p. 2339–2350. Published by permission of the Journal of Petroleum Technology and Gyrodata, Inc.
Page 75, Figure 3-21: Jamieson, A., 2012, Introduction to Wellbore Positioning, V09.10.17, 247 p.
Page 78, Figure 3-25: Published by permission of Gardes Directional Drilling.
Page 79, Figure 3-26: Modified after Stigant, J., 2012. Geospatial Data Quality and Mapping in Shale Plays – Technical and Business Impact, Search and Discovery Article #40940., Used with permission of American Association of Petroleum Geologists [AAPG].
Page 80, Figure 3-27: Cribbs, B., 2009, Practical Wellbore Formation Test Interpretation, Search and Discovery Article 120009, 20 pages.
Page 87, Figure 4-4: Published by permission of Texaco USA.
Page 88, Figure 4-5: Published by permission of the New Orleans Geological Society.
Page 103, Figure 4-15a: Published by permission of Tenneco Oil Company
Page 103, Figure 4-15b: Published by permission of Tenneco Oil Company
Page 103, Figure 4-15c: Published by permission of J. Brewton.
Page 114, Figure 4-21: Published by permission of D. Tearpock and R. Bischke.
Page 116: Figure 4-22: Published by permission of D. Tearpock.
Page 124, Figure 4-25: From Tearpock 1999. Published by permission of Anadarko Petroleum Corporation.
Page 125, Figure 4-26: From Tearpock 1999, Rebuttal to Expert Witness Reports of John C. Davis, Michael J. Economides, Jack C. Goldstein, and Harry F. Manbeck, Jr.: Union Pacific Resources Company, Plaintiff v. Chesapeake Energy Corporation and Chesapeake Operating, Inc., Defendants. C.A. No. 4–96CV-726-Y, Deposition Exhibit 10, Tearpock.Published by permission of Anadarko Petroleum Corporation.
Page 126, Figure 4-27: From Tearpock 1999, Rebuttal to Expert Witness Reports of John C. Davis, Michael J. Economides, Jack C. Goldstein, and Harry F. Manbeck, Jr.: Union Pacific Resources Company, Plaintiff v. Chesapeake Energy Corporation and Chesapeake Operating, Inc., Defendants. C.A. No. 4–96CV-726-Y, Deposition Exhibit 10, Tearpock. Published by permission of Anadarko Petroleum Corporation.
Page 127, Figure 4-28: From Tearpock 1999, Rebuttal to Expert Witness Reports of John C. Davis, Michael J. Economides, Jack C. Goldstein, and Harry F. Manbeck, Jr.: Union Pacific Resources Company, Plaintiff v. Chesapeake Energy Corporation and Chesapeake Operating, Inc., Defendants. C.A. No. 4–96CV-726-Y, Deposition Exhibit 10, Tearpock. From Tearpock 1999. Published by permission of Anadarko Petroleum Corporation.
Page 128, Figure 4-29: From Kyte, D. G., D. N. Meehan, and T. R. Svor, 1994, Method of maintaining a borehole in a stratigraphic zone during drilling: United States Patent 5,311,951 (May 17, 1994). Published by permission of Anadarko Petroleum Corporation.
Page 130, Figure 4-30: Published by permission of W. C. Ross and A2D Technologies.
Page 133, Figure 4-31: Published by permission of W. C. Ross and A2D Technologies.
Page 135, Figure 4-32: Published by permission of W. C. Ross and A2D Technologies.
Page 137, Figure 4-33: Published by permission of W. C. Ross and A2D Technologies.
Page 141, Figure 4-38: Published by permission of Texaco, USA.
Page 142, Figure 4-39: From Tearpock, D. J., and J. Harris, 1987, Subsurface Geological Mapping Techniques – A Training Manual: Tenneco Oil Co. Published by permission of Tenneco Oil Company.
Page 152, Figure 4-45: Published by permission of Tenneco Oil Company.
Page 153, Figure 4-46: From Tearpock, D. J., and J. Harris, 1987, Subsurface Geological Mapping Techniques – A Training Manual: Tenneco Oil Co. Published by permission of Tenneco Oil Company.
Page 154, Figure 4-47: Modified from Tearpock, D. J., and J. Harris, 1987, Subsurface Geological Mapping Techniques – A Training Manual: Tenneco Oil Co. Published by permission of Tenneco Oil Company.
Page 159, Figure 5-3: Herron, D. A., 2011, First Steps in Seismic Interpretation, Geophysical Monograph Series Number 16, Society of Exploration Geophysics, 203 p. Published by permission of Society of Economic Geophysicists.
Page 161, Figure 5-5: Published by permission of TGS/GECO.
Page 162, Figure 5-7: After Anstey 1982. Simple Seismics for the Petroleum Geologist, the Reservoir Engineer, the Well-Log Analyst, the Processing Technician, and the Man in the Field: Boston, International Human Resources Development Corporation Press, 168 p. Published by permission of the International Human Resources Development Corporation Press.
Page 164, Figure 5-8: Published by permission of TGS/GECO.
Page 172, Figure 5-15a: Seismic line published by permission of TGS/GECO.
Page 174, Figure 5-16: From Badley, M. E., 1985, Practical Seismic Interpretation: Boston, International Human Resources Development Corporation Press, 266 p, provided by Merlin Profilers, Ltd.
Page 178, Figure 5-19: Seismic data published by permission of TGS/GECO.
Page 179, Figure 5-20: Seismic data published by permission of TGS/GECO.
Page 180, Figure 5-21: Seismic data published by permission of TGS/GECO.
Page 186, Figure 5-26: Seismic data published by permission of TGS/GECO.
Page 190, Figure 5-29: Published by permission of J. Brenneke.
Page 199, Figure 6-1: Modified from Oil and Gas Fields of Southeast Louisiana, v. 3, 1983. Published by permission of the New Orleans Geological Society.
Page 200, Figure 6-2: Modified from Oil and Gas Fields of Southeast Louisiana, v. 3, 1983. Published by permission of the New Orleans Geological Society.
Page 203, Figure 6-5: Reprinted from Oil and Gas Fields of Southeast Louisiana, v. 3, 1983. Published by permission of the New Orleans Geological Society.
Page 205, Figure 6-6: From Mannhard and Busch 1974; AAPG©1974, reprinted by permission of the American Association of Petroleum Geologists (AAPG) whose permission is required for further use.
Page 207, Figure 6-7: Potter, P. E., 1963, Late Paleozoic sandstones of the Illinois Basin: Illinois State Geological Survey, Report of Investigations 217, 92 p.
Page 222, Figure 6-16: Modified from Lamerson 1982. Published by permission of the Rocky Mountain Association of Geologists.
Page 225, Figure 6-18: Modified from Suter 1947; AAPG©1947, reprinted by permission of the AAPG whose permission is required for further use.
Page 227, Figure 6-19: From Langstaff and Morrill 1981. Published by permission of the International Human Resources Development Corporation Press.
Page 227, Figure 6-20: Modified from Langstaff and Morrill 1981. Published by permission of the International Human Resources Development Corporation Press.
Page 228, Figure 6-21: Modified from Brown 1984a.reprinted by permission of the AAPG whose permission is required for further use.
Page 229, Figure 6-22: Modified from Brown 1984a. AAPG©1984a reprinted by permission of the AAPG whose permission is required for further use.
Page 230, Figure 6-23: From W. Brown 1984a; AAPG © 1984. reprinted by permission of the AAPG whose permission is required for further use.

Page 231, Figure 6-24: Modified from W. Brown 1984a; AAPG©1984. reprinted by permission of the AAPG whose permission is required for further use.
Page 232, Figure 6-25: Modified from W. Brown 1984a; AAPG©1984. reprinted by permission of the AAPG whose permission is required for further use.
Page 233, Figure 6-27: Published by permission of Tenneco Oil Company.
Page 234, Figure 6-28: Published by permission of Tenneco Oil Company.
Page 235, Figure 6-29: Published by permission of Tenneco Oil Company.
Page 238, Figure 6-32: From Offshore Louisiana Oil and Gas Fields, v. II, 1988. Published by permission of the New Orleans Geological Society.
Page 246, Figure 6-36: Modified after Bower 1947; AAPG©1947, reprinted by permission of the AAPG whose permission is required for further use.
Page 247, Figure 6-37: Published by permission of Texaco USA.
Page 248, Figure 6-38: Modified from Boeckelman, unpublished. Published by permission of author.
Page 249, Figure 6-39: From Langstaff and Morrill 1981. Published by permission of the International Human Resources Development Corporation Press.
Page 249, Figure 6-40: From Bishop 1960. Published by permission of author.
Page 251, Figure 6-41: From Langstaff and Morrill 1981. Published by permission of the International Human Resources Development Corporation Press.
Page 252, Figure 6-42: From Boeckelman, unpublished. Published by permission of author.
Page 253, Figure 6-43: From Boeckelman, unpublished. Published by permission of author.
Page 255, Figure 6-44: From Boeckelman, unpublished. Published by permission of author.
Page 257, Figure 6-46: Published by permission of W. C. Ross and A2D Technologies.
Page 258, Figure 6-47: Published by permission of W. C. Ross and A2D Technologies.
Page 260, Figure 6-48: Published by permission of W. C. Ross and A2D Technologies.
Page 261, Figure 6-49: Published by permission of W. C. Ross and A2D Technologies.
Page 262, Figure 6-50: Published by permission of W. C. Ross and A2D Technologies.
Page 263, Figure 6-51: Published by permission of W. C. Ross and A2D Technologies.
Page 264, Figure 6-52: Published by permission of W. C. Ross and A2D Technologies.
Page 266, Figure 6-53: Broussard, M. J., and B. E. Lock, 1995, Modern analytical techniques for fault surface seal analysis: A Gulf Coast case history: Transactions Gulf Coast Association of Geological Societies, v. 45, p. 87–93. Published by permission of the Gulf Coast Association of Geological Societies.
Page 266, Figure 6-54: Broussard, M. J., and B. E. Lock, 1995, Modern analytical techniques for fault surface seal analysis: A Gulf Coast case history: Transactions Gulf Coast Association of Geological Societies, v. 45, p. 87–93. Published by permission of the Gulf Coast Association of Geological Societies.
Page 267, Figure 6-55: Broussard, M. J., and B. E. Lock, 1995, Modern analytical techniques for fault surface seal analysis: A Gulf Coast case history: Transactions Gulf Coast Association of Geological Societies, v. 45, p. 87–93. Published by permission of the Gulf Coast Association of Geological Societies.
Page 268, Figure 6-56: After Broussard and Lock 1995. Published by permission of the Gulf Coast Association of Geological Societies.
Page 269, Figure 6-57: Published by permission of Badley Earth Sciences.
Page 270, Figure 6-58: Published by permission of Badley Earth Sciences.
Page 271, Figure 6-59: Published by permission of Badley Earth Sciences.
Page 271, Figure 6-60: Published by permission of Badley Earth Sciences.
Page 272, Figure 6-61: Modified from Shoup 2017, personal communication. Used with permission of R. Shoup.
Page 273, Figure 6-62: Shoup, R., 2017. Geological Cross Sections; Combining Geologic Models and Best Practices to Finding and Exploiting Reserves, Hidden Potential in Mature Basins: Play Analogs and Best Practices, Geosciences Technology Workshop, 13-14 September 2017, Bandung, Indonesia, used with permission of Dr. Tony Swiecicki.
Page 274, Figure 6-63: From Norvick 2001. Published with permission from M. S. Norvick.
Page 276, Figure 7-1: Harding, T. P., and J. D. Lowell, 1979, Structural styles, their plate-tectonic habitats, and hydrocarbon traps in petroleum provinces: American Association of Petroleum Geologists Bulletin, v. 63, p. 1016–1058. reprinted by permission of the AAPG whose permission is required for further use.
Page 277: "Probably no portion of geologic literature has a more confused terminology than that dealing with faults." Straley, W. H., III, 1932, Some notes on the nomenclature of faults: Studies for Students: Chicago, University of Chicago, p. 756–763
Page 278, Figure 7-2: Tearpock, D. J., and J. Harris, 1987, Subsurface Geological Mapping Techniques – A Training Manual: Tenneco Oil Co. Published by permission of Tenneco Oil Company.
Page 281, Figure 7-3: Modified from Billings 1972, Structural Geology, third ed.: Englewood Cliffs, NJ, Prentice-Hall, p. 606. Published by permission of Pearson Education, Inc.
Page 281, Figure 7-4: Published by permission of D. Tearpock and R. Bischke.
Page 282, Figure 7-5: Published by permission of D. Tearpock and R. Bischke.
Page 284, Figure 7-6: Published by permission of D. Tearpock and R. Bischke.
Page 286, Figure 7-7: Published by permission of D. Tearpock.
Page 287, Figure 7-8: Published by permission of D. Tearpock.
Page 289, Figure 7-9: Modified from Bishop 1960. Published by permission of author.
Page 296, Figure 7-12: Published by permission of Sandefer Oil and Gas, Inc
Page 297, Figure 7-13: Modified from Hooper, N. J., J. G. M. Raven, and M. J. Kilpatrick, 1992, Computer modeling of multiple surfaces with faults: The Ivanhoe Field, Outer Moray Firth Basin, U.K. North Sea, in Hamilton, D. E., and T. A. Jones, eds., American Association of Petroleum Geologists Bulletin, Computer Applications in Geology No. 1, p. 161–174. reprinted by permission of the AAPG whose permission is required for further use.
Page 301, Figure 7-16: Published by permission of Texaco, USA.
Page 306: "zone of combined throw", Dickinson, G., 1954, Subsurface interpretation of intersecting faults and their effects upon stratigraphic horizons: American Association of Petroleum Geologists Bulletin, v. 38, no. 5, p. 854–877
Page 308, Figure 7-22: Modified from Tearpock and Harris 1987. Published with permission of Tenneco Oil Company.
Page 309, Figure 7-23: Published by permission of D. Tearpock and J. Brewton.
Page 310, Figure 7-24: Published by permission of Texaco, USA.
Page 311, Figure 7-25: Dickinson, G., 1954, Subsurface interpretation of intersecting faults and their effects upon stratigraphic horizons: American Association of Petroleum Geologists Bulletin, v. 38, no. 5, p. 854–877. reprinted by permission of the AAPG whose permission is required for further use.
Page 314, Figure 7-26: Modified from Tearpock and Harris 1987. Published by permission of Tenneco Oil Company.
Page 314, Figure 7-27: Published by permission of R. Bischke.
Page 315, Figure 7-28: Published by permission of R. Bischke.
Page 321, Figure 7-32: From Lamerson 1982. Published by permission of the Rocky Mountain Association of Geologists.
Page 323, Figure 7-33: Lamerson 1982. Published by permission of the Rocky Mountain Association of Geologists.
Page 324, Figure 7-34: Modified after Rowan and Linares 2000; AAPG©2000. reprinted by permission of the AAPG whose permission is required for further use.
Page 325, Figure 7-35: Prepared by C. Harmon. From Tearpock and Harris 1987. Published by permission of Tenneco Oil Company.
Page 326, Figure 7-36: Prepared by C. Harmon. Modified from Tearpock and Harris 1987. Published by permission of Tenneco Oil Company.
Page 328, Figure 7-37: Modified from Tearpock and Harris 1987. Published by permission of TGS Offshore Geophysical Company and Tenneco Oil Company.
Page 330, Figure 7-38: From Tearpock and Harris 1987. Published by permission of Tenneco Oil Company.
Page 331, Figure 7-39: Modified from Tearpock and Harris 1987. Published by permission of Tenneco Oil Company.

Page 332, Figure 7-40a: Modified from Tearpock and Harris 1987. Published by permission of Tenneco Oil Company.
Page 332, Figure 7-40b: From Tearpock and Harris 1987. Published by permission of Tenneco Oil Company.
Page 334, Figure 7-41: From Tearpock and Harris 1987. Published by permission of Tenneco Oil Company.
Page 336, Figure 7-42: From Tearpock and Harris 1987. Published by permission of Tenneco Oil Company.
Page 337, Figure 7-43: From Tearpock and Harris 1987. Published by permission of Tenneco Oil Company.
Page 344, Figure 7-47: From Tearpock and Harris 1987. Published by permission of Tenneco Oil Company.
Page 350, Figure 7-50: Published by permission of Texaco, USA.
Page 352, Figure 7-51: Published by permission of Texaco, USA.
Page 355, Figure 8-1: From Tearpock and Harris 1987. Published by Tenneco Oil Company
Page 356, Figure 8-2: From Bishop 1960. Published by permission of Tenneco Oil Company.
Page 359, Figure 8-5: From Bishop 1960. Published by permission of Tenneco Oil Company.
Page 360, Figure 8-6: Modified after Bishop 1960. Published by permission of author.
Page 361, Figure 8-7: From Bishop 1960. Published by permission of author.
Page 364, Figure 8-9: Modified after Bishop 1960. Published by permission of author.
Page 367, Figure 8-11: Published by permission of D. Tearpock.
Page 369, Figure 8-12: Published by permission of D. Tearpock.
Page 370, Figure 8-13: Published by permission of D. Tearpock.
Page 372, Figure 8-14: Published by permission of D. Tearpock and R. Bischke.
Page 373, Figure 8-15: Published by permission of Tearpock and Bischke.
Page 374, Figure 8-16: Published by permission of Tearpock and Bischke.
Page 386, Figure 8-23d: published by permission of J. Brenneke.
Page 392, Figure 8-27: Prepared by J. Bollick. Published by permission of Tenneco Oil Company.
Page 395, Figure 8-29: Published by permission of D. Tearpock.
Page 397, Figure 8-30a: Modified after Bishop 1960; Published by permission of author
Page 397, Figure 8-30c: Modified after Bishop 1960; Published by permission of author
Page 397, Figure 8-30f: From Bishop 1960. Published by permission of author
Page 401, Figure 8-32: Published by permission of Tenneco Oil Company.
Page 404, Figure 8-33: Tearpock and Harris 1987. Published by Tenneco Oil Company
Page 405, Figure 8-36: Published by permission of Tenneco Oil Company.
Page 407, Figure 8-37: Published by permission of Tenneco Oil Company.
Page 409, Figure 8-38: Published by permission of Tearpock et al. 1994.
Page 410, Figure 8-39: Modified from Tearpock, D. J., and J. Harris, 1987, Subsurface Geological Mapping Techniques – A Training Manual: Tenneco Oil Co. Published by permission of Tenneco Oil Company.
Page 412, Figure 8-40: Prepared by C. Harmon. Modified from Tearpock and Harris 1987. Published by permission of Tenneco Oil Company.
Page 415, Figure 8-42a: Herron, D. A., Latimer, R. B., & Society of Exploration Geophysicists. (2011). First steps in seismic interpretation. Tulsa, Okla: Society of Exploration Geophysicists. Published with permission from Society of Economic Geophysicists, seismic courtesy of PGS.
Page 415, Figure 8-42b: From Herron 2011, First steps in Seismic Interpretation. Published with permission from Society of Economic Geophysicists, seismic courtesy of PGS
Page 416, Figure 8-43: From Geology of Petroleum, first ed. By A. I. Levorsen, Copyright 1954 by W. H. Freeman and Company. Reprinted by permission.)
Page 425, Figure 8-50: Modified from Tearpock and Harris 1987. Published by permission of Tenneco Oil Company
Page 426, Figure 8-51: Published by permission of the New Orleans Geological Society.
Page 427: "map each fault block separately and independently using the available well or seismic control. Each contoured fault block must be individually integrated with the fault surface map for the displacing fault to determine the upthrown and downthrown fault traces". From Tearpock and Harris 1987. Published by Tenneco Oil Company
Page 428, Figure 8-52: From Santiago-Acevedo, J., 1980, Giant fields of the Southern Zone, Mexico: in Halbouty, M. T., ed., Giant Oil and Gas Fields of the decade 1968–1978: American Association of Petroleum Geologists Memoir 30, p. 339–385.; AAPG©1980, reprinted by permission of the AAPG whose permission is required for further use.
Page 440, Figure 8-56: Published by permission of the Lafayette Geological Society.
Page 441, Figure 8-57: Published by permission of the New Orleans Geological Society.
Page 446, Figure 8-59: From Dickinson 1954; AAPG ©1954. reprinted by permission of the APPG whose permission is required for further use.
Page 447, Figure 8-60: Modified from Dickinson 1954; AAPG©1954. reprinted by permission of the APPG whose permission is required for further use.
Page 450, Figure 8-63: Published by permission of the New Orleans Geological Society.
Page 452, Figure 8-64: Published by permission of the New Orleans Geological Society.
Page 453, Figure 8-65: From Atwater and Forman 1959; AAPG©1959, reprinted by permission of the APPG whose permission is required for further use.
Page 455, Figure 8-66: From Seni and Jackson 1983; AAPG©1983, reprinted by permission of the AAPG
Page 456, Figure 8-67: From Remmelts 1995; AAPG©1995, published by permission of the AAPG whose permission is required for further use
Page 456, Figure 8-68: From Salt Domes, by Michel T. Halbouty. Copyright 1979 by Gulf Publishing Company, Houston, Texas. Used with permission. All rights reserved
Page 457, Figure 8-69: Published by permission of the New Orleans Geological Society.
Page 459, Figure 8-71: Published by permission of the Lafayette Geological Society.
Page 460, Figure 8-72: From Salt Domes, by Michel T. Halbouty. Copyright 1979 by Gulf Publishing Company, Houston, Texas. Used with permission. All rights reserved
Page 461, Figure 8-74a: From Stone 1969. Reprinted by permission of the Rocky Mountain Association of Geologists.
Page 461, Figure 8-74b: From Bell, W. G., 1956, Tectonic setting of Happy Springs and nearby structures in the Sweetwater Uplift area, central Wyoming, in Geological Record of the American Association of Petroleum Geologists, Rocky Mountain Section.; AAPG©1956. reprinted by permission of the AAPG whose permission is required for further use.
Page 461, Figure 8-74c: From Brown 1982; AAPG©1982. reprinted by permission of the AAPG whose permission is required for further use.
Page 464, Figure 8-76: Modified from Harding 1973; AAPG©1973. reprinted by permission of the APPG whose permission is required for further use
Page 465, Figure 8-77: After Mayuga, M. N., 1970, Geology and development of California's giant – Wilmington Oil Field, in Halbouty, M. T., ed., Geology of Giant Petroleum Fields: American Association of Petroleum Geologists Memoir 14, p. 158–184.; AAPG©1970. reprinted by permission of the AAPG whose permission is required for further use.
Page 466, Figure 8-78: From Lamerson, P. R., 1982, The Fossil Basin and its relationship to the Absaroka thrust system, Wyoming and Utah, in Powers, R. B., ed., Geologic Studies of the Cordilleran Thrust Belt: Denver, CO, Rocky Mountain Association of Geologists, p. 279–340. Reprinted by permission of the Rocky Mountain Association of Geologists.
Page 470, Figure 8-80a: Modified from Bishop 1960. From Bishop 1960. Published by permission of author.
Page 471, Figure 8-80c: From Bishop 1960. Published by permission of author.

Page 472, Figure 8-81: From Rowan, M. G., and R. Linares, 2000, Fold evolution matrices and axial surface analysis of fault bend folds: Application to the Medina Anticline, eastern Cordillera, Colombia: American Association of Petroleum Geologists Bulletin, v. 84, p. 741–764. reprinted by permission of the AAPG whose permission is required for further use.
Page 474, Figure 8-82: From Lamerson, P. R., 1982, The Fossil Basin and its relationship to the Absaroka thrust system, Wyoming and Utah, in Powers, R. B., ed., Geologic Studies of the Cordilleran Thrust Belt: Denver, CO, Rocky Mountain Association of Geologists, p. 279–340. Reprinted by permission of the Rocky Mountain Association of Geologists.
Page 475: "movable water above the transition zone that is not directly connected to the actual free-water table of a given hydrocarbon column", Gaafar et al. (2016)
Page 476, Figure 8-83: Modified after Weber, 1995
Page 476, Figure 8-84: Modified after Weber 1995
Page 477, Figure 8-85: Modified after Kendrick 2000
Page 477, Figure 8-86: Modified after Fugitt et al 2000
Page 482, Figure 9-1: Data published with permission from Seitel Data Ltd.
Page 482, Figure 9-2: Data published with permission from Seitel Data Ltd.
Page 483, Figure 9-3: Data published with permission from Seitel Data Ltd.
Page 483, Figure 9-4: Data published with permission from Seitel Data Ltd.
Page 484, Figure 9-5: Data published with permission from Seitel Data Ltd.
Page 484, Figure 9-6: Data published with permission from Seitel Data Ltd.
Page 484, Figure 9-7: Data published with permission from Seitel Data Ltd.
Page 484, Figure 9-8: Data published with permission from Seitel Data Ltd.
Page 485, Figure 9-9: Data published with permission from Seitel Data Ltd.
Page 485, Figure 9-10: Data published with permission from Seitel Data Ltd.
Page 485, Figure 9-11: Pioneer Natural Resources Company, Micah Reasnor. ©2020 Pioneer Natural Resources Company.
Page 486, Figure 9-12a: Pioneer Natural Resources Company, Micah Reasnor. ©2020 Pioneer Natural Resources Company.
Page 486, Figure 9-12b: Pioneer Natural Resources Company, Micah Reasnor. ©2020 Pioneer Natural Resources Company.
Page 487, Figure 9-13: Pioneer Natural Resources Company, Micah Reasnor. ©2020 Pioneer Natural Resources Company.
Page 489: "Teams don't plan to fail, they fail to plan", Tearpock, D. J., 1997, Project Management and Synergistic Team Development: Short Course Training Manual, Subsurface Consultants & Associates, LLC.
Page 491, Figure 9-14: Data published with permission from Seitel Data Ltd
Page 492, Figure 9-15: Data published with permission from Seitel Data Ltd
Page 493, Figure 9-16: Data published with permission from Seitel Data Ltd
Page 493, Figure 9-17: Data published with permission from Seitel Data Ltd
Page 494, Figure 9-18: Data published with permission from CGG Land
Page 495, Figure 9-19: Data published with permission from CGG Land
Page 496, Figure 9-20: Data published with permission from CGG Land
Page 496, Figure 9-21: Data published with permission from CGG Land
Page 497, Figure 9-22: Data published with permission from CGG Land
Page 498, Figure 9-23: Data published with permission from CGG Land
Page 500, Figure 9-24: Pioneer Natural Resources Company, Micah Reasnor. ©2020 Pioneer Natural Resources Company.
Page 502, Figure 9-25: Data published with permission from Seitel Data Ltd.
Page 503, Figure 9-26: Data published with permission from Seitel Data Ltd.
Page 504, Figure 9-27: Data published with permission from Seitel Data Ltd.
Page 505, Figure 9-28: Data published with permission from Seitel Data Ltd.
Page 506, Figure 9-29: Data published with permission from Seitel Data Ltd.
Page 506, Figure 9-30: Data published with permission from Seitel Data Ltd.
Page 506, Figure 9-31: Data published with permission from Seitel Data Ltd.
Page 506, Figure 9-32: Data published with permission from Seitel Data Ltd.
Page 507, Figure 9-33: Data published with permission from Seitel Data Ltd.
Page 508, Figure 9-34: Data published with permission from Seitel Data Ltd.
Page 508, Figure 9-35: Data published with permission from Seitel Data Ltd.
Page 509, Figure 9-36: Data published with permission from Seitel Data Ltd.
Page 510, Figure 9-37: Data published with permission from Seitel Data Ltd.
Page 511, Figure 9-38: Data published with permission from Seitel Data Ltd.
Page 512, Figure 9-39: Data published with permission from CGG Land.
Page 512, Figure 9-40: Data published with permission from CGG Land.
Page 512, Figure 9-41: Data published with permission from CGG Land.
Page 513, Figure 9-42: Data published with permission from CGG Land.
Page 513, Figure 9-43: Data published with permission from CGG Land.
Page 514, Figure 9-44: Data published with permission from CGG Land.
Page 514, Figure 9-45: Data published with permission from CGG Land.
Page 515, Figure 9-46: Data published with permission from CGG Land.
Page 515, Figure 9-47: Data published with permission from CGG Land.
Page 516, Figure 9-48: Data published with permission from CGG Land.
Page 517, Figure 9-49: Seismic data published with permission from Seitel Data Ltd.
Page 518, Figure 9-50: Published with permission from Seitel Data Ltd.
Page 518, Figure 9-51: Published with permission from Seitel Data Ltd.
Page 519, Figure 9-52: Published with permission from Seitel Data Ltd.
Page 519, Figure 9-53: Published with permission from Seitel Data Ltd.
Page 520, Figure 9-54: Published with permission from Seitel Data Ltd.
Page 521, Figure 9-55: Published with permission from Seitel Data Ltd.
Page 522, Figure 9-56: Published with permission from Seitel Data Ltd.
Page 522, Figure 9-57: Published with permission from Seitel Data Ltd.
Page 523, Figure 9-58: Published with permission from Seitel Data Ltd.
Page 524, Figure 9-59: Published with permission from Seitel Data Ltd.
Page 524, Figure 9-60: Published with permission from Seitel Data Ltd.
Page 526, Figure 9-61: Published with permission from Seitel Data Ltd.
Page 526, Figure 9-62: Published with permission from Seitel Data Ltd.

Page 527, Figure 9-63: Published with permission from Seitel Data Ltd.
Page 528, Figure 9-64: Published with permission from Seitel Data Ltd.
Page 529, Figure 9-65: Published with permission from Seitel Data Ltd.
Page 529, Figure 9-66: Published with permission from Seitel Data Ltd.
Page 530, Figure 9-67: Published with permission from Seitel Data Ltd.
Page 531, Figure 9-68: Pioneer Natural Resources Company, Micah Reasnor. ©2020 Pioneer Natural Resources Company.
Page 532, Figure 9-69: Pioneer Natural Resources Company, Micah Reasnor. ©2020 Pioneer Natural Resources Company.
Page 532, Figure 9-70: Pioneer Natural Resources Company, Micah Reasnor. ©2020 Pioneer Natural Resources Company.
Page 533, Figure 9-71: Pioneer Natural Resources Company, Micah Reasnor. ©2020 Pioneer Natural Resources Company.
Page 534, Figure 9-72: Pioneer Natural Resources Company, Micah Reasnor. ©2020 Pioneer Natural Resources Company.
Page 534, Figure 9-73: Pioneer Natural Resources Company, Micah Reasnor. ©2020 Pioneer Natural Resources Company.
Page 535, Figure 9-74: Pioneer Natural Resources Company, Micah Reasnor. ©2020 Pioneer Natural Resources Company.
Page 535, Figure 9-75: Pioneer Natural Resources Company, Micah Reasnor. ©2020 Pioneer Natural Resources Company.
Page 535, Figure 9-76: Pioneer Natural Resources Company, Micah Reasnor. ©2020 Pioneer Natural Resources Company.
Page 536, Figure 9-77: Pioneer Natural Resources Company, Micah Reasnor. ©2020 Pioneer Natural Resources Company.
Page 537, Figure 9-78: Pioneer Natural Resources Company, Micah Reasnor. ©2020 Pioneer Natural Resources Company.
Page 538, Figure 9-79: Pioneer Natural Resources Company, Micah Reasnor. ©2020 Pioneer Natural Resources Company.
Page 542, Figure 10-2: Modified after Rich 1934; AAPG©1934. reprinted by permission of the American Association of Petroleum Geologists (AAPG) whose permission is required for further use.
Page 547, Figure 10-5: Modified after Dahlstrom 1969. Published by permission of the National Research Council of Canada.)
Page 549, Figure 10-6: Published by permission of R. Bischke.
Page 550, Figure 10-7: Published by permission of R. Bischke.
Page 552, Figure 10-8: Published by permission of D. Roeder 1983.
Page 554: "As to the interpretations presented, the reader … a seismic background that illustrate an author's concept", Bally, A. W., 1983, Seismic Expression of Structural Styles: American Association of Petroleum Geologists Bulletin, Studies in Geology Series No. 15, v. 3., p. 167–172
Page 555, Figure 10-9: From Bally, A. W., 1983, Seismic Expression of Structural Styles: American Association of Petroleum Geologists Bulletin, Studies in Geology Series No. 15, v. 3., p. 167–172; AAPG©1983, reprinted by permission of the AAPG whose permission is required for further use.
Page 556, Figure 10-10: Published by permission of R. Bischke.
Page 559, Figure 10-11: Bally, A. W., P. L. Gordy, and G. A. Stewart, 1966, Structure, seismic data, and orogenic evolution of southern Canadian Rocky Mountains: Canadian Society of Petroleum Geologists Bulletin, v. 14, p. 337–381. Published by permission of the Canadian Society of Petroleum Geologists.
Page 560, Figure 10-12: From Boyer, S. E., 1986, Styles of folding within thrust sheets: examples from Appalachian and Rocky Mountains of the U.S.A. and Canada: Journal of Structural Geology, v. 8, p. 325–339. Published by permission of the Journal of Structural Geology.
Page 561, Figure 10-13: Published by permission of the Bureau of Economic Geology, University of Texas at Austin; edited by Hardage et al. 1998. (Interpretation by R. Bischke)
Page 562, Figure 10-14a: Published by permission of R. Bischke.
Page 563, Figure 10-14b: Published by permission of R. Bischke.
Page 563, Figure 10-15: Published by permission of the Bureau of Economic Geology, University of Texas at Austin; edited by Hardage et al. 1998. (Modified by R. Bischke)
Page 564, Figure 10-16: Published by Bureau of Economic Geology, University of Texas at Austin, Edited by Hardage, Major and Hentz, 1998, (Modified by R. Bischke)
Page 566, Figure 10-17a: Published by Permission of R. E. Bischke
Page 566, Figure 10-17b: Published by Permission of R. E. Bischke
Page 568: "folding and fault linkage [are] the probable causes of much of the variation in displacement- length scaling ratios observed in our data sets","fault interactions play a major role in the ultimate displacement profile geometries for thrust faults." Bergen and Shaw 2010
Page 569, Figure 10-18: Published by permission of Ted Snedden.
Page 572, Figure 10-20: Modified after Elliott, D., 1976, The energy balance and deformation mechanisms of thrust sheets: Philosophical Transactions of the Royal Society of London, Series A, v. 283, p. 289–312. Published by permission of the Royal Society of London.
Page 573, Figure 10-21: Modified from Marshak, S., and G. Mitra, 1988, Basic Methods of Structural Geology: Englewood Cliffs, NJ, Prentice-Hall, p. 446.
Page 577, Figure 10-23a: Modified after Suppe 1983, 1985. Published by American Journal of Science.
Page 577, Figure 10-23b: Published by permission of R. Bischke
Page 579, Figure 10-25: Published by permission of R. Bischke.
Page 580, Figure 10-26: Modified from Tearpock, D. J., R. E. Bischke, and J. L. Brewton, 1994, Quick Look Techniques for Prospect Evaluations: Lafayette, LA, Subsurface Consultants & Associates, 286 p.
Page 581, Figure 10-27: Modified from Tearpock, D. J., R. E. Bischke, and J. L. Brewton, 1994, Quick Look Techniques for Prospect Evaluations: Lafayette, LA, Subsurface Consultants & Associates, 286 p.
Page 582, Figure 10-28: From Bischke 1994. Published by permission of the Houston Geological Society.
Page 583, Figure 10-29: From Bischke 1994. Published by permission of the Houston Geological Society.
Page 585, Figure 10-30: Modified after Laubscher 1961; Suppe 1985. Published by Swiss Geological Society.
Page 586, Figure 10-31: Modified after Laubscher 1961
Page 590, Figure 10-32a: From Seismic Atlas of Australian and New Zealand Basins, Skilbeck and Lennox 1984. Reprinted by permission of the AAPG whose permission is required for further use.
Page 590, Figure 10-32b: Reprinted by permission of the AAPG whose permission is required for further use.
Page 591, Figure 10-33: Compliments of Jon Mosar.
Page 591, Figure 10-34: Published by permission of John Suppe.
Page 592, Figure 10-35: From Rich 1934; AAPG©1934. reprinted by permission of the AAPG whose permission is required for further use.
Page 592, Figure 10-36: From Rich 1934; AAPG©1934, reprinted by permission of the AAPG whose permission is required for further use.
Page 593, Figure 10-37: Modified after Suppe Geometry and kinematics of fault-bend folding: American Journal of Science, v. 283, p. 684–721. Published by permission of the American Journal of Science.
Page 595, Figure 10-38: From Suppe 1983, Geometry and kinematics of fault-bend folding: American Journal of Science, v. 283, p. 684–721. Published by permission of the American Journal of Science.
Page 597, Figure 10-40: From Suppe 1983, Geometry and kinematics of fault-bend folding: American Journal of Science, v. 283, p. 684–721. Published by permission of the American Journal of Science.
Page 599, Figure 10-41a: Interpretation by R. Bischke. After Sachnik and More in Bally (1988); AAPG©1988, Reprinted by permission of the AAPG whose permission is required for further use.
Page 599, Figure 10-41b: Storti et al. 2005. Reprinted by permission of the AAPG whose permission is required for further use.
Page 600, Figure 10-42: Suppe, J., 1985, Principles of Structural Geology: Englewood Cliffs, NJ, Prentice-Hall, p. 537.
Page 601, Figure 10-43: Modified after Suppe 1985 Principles of Structural Geology: Englewood Cliffs, NJ, Prentice-Hall, p. 537.

Page 603, Figure 10-44: Published by permission of John Suppe
Page 604, Figure 10-45: Published by permission of John Suppe
Page 604, Figure 10-46: Modified from Suppe 1985 Principles of Structural Geology: Englewood Cliffs, NJ, Prentice-Hall, p. 537.
Page 606, Figure 10-48: Published by permission of R. Bischke)(see the Preface for information about accessing the color versions of these figures
Page 608, Figure 10-49: Published by permission of R. Bischke
Page 609, Figure 10-50: Published by permission of R. Bischke
Page 609, Figure 10-51a: Morales M., E. Hung, and R. Bischke, 2005, Three-dimensional interpretation of the El Furrial trend, Eastern Venezuela Basin, Venezuela, in AAPG Seismic Interpretation of Contractional Fault-Related Folds, AAPG Seismic Atlas Studies in Geology #53, Part 2, p. 83-86.
Page 609, Figure 10-51b: Morales M., E. Hung, and R. Bischke, 2005, Three-dimensional interpretation of the El Furrial trend, Eastern Venezuela Basin, Venezuela, in AAPG Seismic Interpretation of Contractional Fault-Related Folds, AAPG Seismic Atlas Studies in Geology, p. 85.
Page 611, Figure 10-52: Morales et al. 2005, Published by permission of R. Bischke
Page 611, Figure 10-53: Corredor et al., in Shaw et al. 2005, reprinted by permission of the AAPG whose permission is required for further use.
Page 612, Figure 10-54: Published by permission of R. Bischke
Page 613, Figure 10-55: Published by permission of R. Bischke
Page 613, Figure 10-56: Boyer, 1986, Published by permission of R. Bischke
Page 614, Figure 10-57: Suppe, J., and D. A. Medwedeff, 1990, Geometry and kinematics of fault-propagation folding: Eclogae Geologicae Helvetiae, v. 83, p. 409–454. Published by permission of R. Bischke.
Page 615, Figure 10-58: Hughes and Shaw, 2014. Reprinted by permission of the AAPG whose permission is required for further use.
Page 615, Figure 10-59: Hughes and Shaw, 2014. Published by permission of R. Bischke.
Page 616, Figure 10-60: From Mitra, S., 1986, Duplex structures and imbricate thrust systems: Geometry, structural position, and hydrocarbon potential: American Association of Petroleum Geologists Bulletin, v. 70, p. 1087–1112.; AAPG©1986. reprinted by permission of the AAPG whose permission is required for further use.
Page 619: TBL10-1: Published by permission of the American Journal of Science (Suppe 1983, Geometry and kinematics of fault-bend folding: American Journal of Science, v. 283, p. 684–721.).
Page 626, Figure 10-63: From Namson, J., 1981, Structure of the western Foothills Belt, Miaoli-Hsinchu area, Taiwan, (1) Southern Part: Petroleum Geology of Taiwan, no. 18, p. 31–51. Published by permission of the Chinese Petroleum Institute.
Page 627, Figure 10-64: From Namson, J., 1981, Structure of the western Foothills Belt, Miaoli-Hsinchu area, Taiwan, (1) Southern Part: Petroleum Geology of Taiwan, no. 18, p. 31–51. Published by permission of the Chinese Petroleum Institute.
Page 629, Figure 10-66: After Applegate and Ross, in Gries, R. R., and R. C. Dyer, eds., 1985, Seismic Exploration of the Rocky Mountain Region: Rocky Mountain Association of Geologists and Denver Geophysical Society, 300. Published by permission of the Rocky Mountain Association of Geologists and the Denver Geophysical Society.
Page 630, Figure 10-67: Published by permission of Don Medwedeff, D. A., 1988, Structural analysis and tectonic significance of Late-Tertiary and Quaternary compressive- growth folding, San Joaquin Valley, California [Ph.D. thesis]: Princeton University, 184 p
Page 631, Figure 10-68: Published by permission of Don Medwedeff, D. A., 1988, Structural analysis and tectonic significance of Late-Tertiary and Quaternary compressive- growth folding, San Joaquin Valley, California [Ph.D. thesis]: Princeton University, 184 p
Page 631, Figure 10-69: Published by permission of Don Medwedeff, D. A., 1988, Structural analysis and tectonic significance of Late-Tertiary and Quaternary compressive- growth folding, San Joaquin Valley, California [Ph.D. thesis]: Princeton University, 184 p
Page 633, Figure 10-70: Modified after 1988, Short course on cross-section balancing in petroleum structural geology: Pacific Section, American Association of Petroleum Geologists. Published by permission of John Suppe.
Page 637, Figure 11-1: Published by permission of H. Xiao and J. Suppe.
Page 638, Figure 11-2: Modified after Suppe 1988, Short course on cross-section balancing in petroleum structural geology: Pacific Section, American Association of Petroleum Geologists. Published by permission of John Suppe.
Page 641, Figure 11-3: Modified after Suppe 1988, Short course on cross-section balancing in petroleum structural geology: Pacific Section, American Association of Petroleum Geologists. Published by permission of John Suppe.
Page 643, Figure 11-4: Published by permission of H. Xiao and J. Suppe.
Page 645, Figure 11-5: From Shaw, J. H., S. C. Hook, and E. P. Sitohang, 1997, Extensional fault-bend folding and synrift deposition: An example from the Central Sumatra Basin, Indonesia: American Association of Petroleum Geologists Bulletin, v. 81, p. 367–379; AAPG©1997, reprinted by permission of the AAPG whose permission is required for further use.
Page 646, Figure 11-6: Published by permission of R. Bischke.
Page 648, Figure 11-7: Modified after Hamblin, W. K., 1965, Origin of "reverse drag" on the downthrown side of normal faults: Geological Society of America Bulletin, v. 76, p. 1145–1164. Published by permission of the Geological Society of America.
Page 648, Figure 11-8: Modified from Xiao and Suppe 1992, Origin of rollover: American Association of Petroleum Geologists Bulletin, v. 76, p. 509–529. Published by permission of the Gulf Coast Association of Petroleum Geologists.
Page 655, Figure 11-12: Interpretation by R. Bischke. From Xiao and Suppe 1990. Published by permission of John Suppe
Page 656, Figure 11-13: Published by permission of R. Bischke.
Page 658, Figure 11-14: Published by permission of R. Bischke.
Page 659, Figure 11-15: Published by permission of R. Bischke.
Page 662, Figure 11-16: Interpretation by R. Bischke. From Xiao and Suppe 1990. Unpublished work of Xiao and Suppe. Published by permission of John Suppe.
Page 663, Figure 11-17: Published by permission of Muzium Brunei.
Page 665, Figure 11-18: Published by permission of Jebco Seismic Inc., Houston, Texas, USA.
Page 667, Figure 11-19: Prepared by W. L. Keyser. Published by permission of Texaco USA, Eastern E & P Region.
Page 667, Figure 11-20: Published by permission of Texaco USA, Eastern E & P Region.
Page 667, Figure 11-21: Published by permission of Texaco USA, Eastern E & P Region.
Page 668, Figure 11-22: Modified after Shaw, J. H., S. C. Hook, and E. P. Sitohang, 1997, Extensional fault-bend folding and synrift deposition: An example from the Central Sumatra Basin, Indonesia: American Association of Petroleum Geologists Bulletin, v. 81, p. 367–379.; AAPG©1997. reprinted by permission of the APPG whose permission is required for further use.
Page 669, Figure 11-23: From Ferrill, Stamatakos, and Simms 1998. Published by permission of Elsevier Science, Journal of Structural Geology.
Page 670, Figure 11-24: Published by permission of R. Bischke.
Page 671, Figure 11-25: Published by permission of R. Bischke.
Page 671, Figure 11-26: Published by permission of R. Bischke.
Page 673, Figure 11-27: From Xiao and Suppe 1989, Role of compaction in the listric shape of growth normal faults: American Association of Petroleum Geologists Bulletin, v. 73, p. 777–786.; AAPG©1989. reprinted by permission of the AAPG whose permission is required for further use.
Page 676, Figure 11-28: From Xiao and Suppe 1989, Role of compaction in the listric shape of growth normal faults: American Association of Petroleum Geologists Bulletin, v. 73, p. 777–786.; AAPG©1989. reprinted by permission of the AAPG whose permission is required for further use.
Page 677, Figure 11-29: From Xiao and Suppe 1989, Role of compaction in the listric shape of growth normal faults: American Association of Petroleum Geologists Bulletin, v. 73, p. 777–786.; AAPG©1989. reprinted by permission of the AAPG whose permission is required for further use.

Page 678, Figure 11-30: From Xiao and Suppe 1989, Role of compaction in the listric shape of growth normal faults: American Association of Petroleum Geologists Bulletin, v. 73, p. 777–786.; AAPG©1989. reprinted by permission of the AAPG whose permission is required for further use.
Page 679, Figure 11-31: Published by permission of R. Bischke.
Page 679, Figure 11-32: Published by permission of R. Bischke.
Page 681, Figure 11-33: Published by permission of R. Bischke.
Page 683, Figure 11-34: Bischke, R. E., and J. Suppe, 1990a, Calculating sand/shale ratios from growth normal fault dips on seismic profiles: Transactions Gulf Coast Association of Geological Societies, v. 40, p. 39–49, Published by permission of R. Bischke.
Page 684: "Note that fault plane dip steepens in sand formations", Gow, K. L., 1962, Segno Field, Polk County, Texas, in Denham, R. L., ed., Typical Oil and Gas Fields of Southeast Texas: Houston, TX, Houston Geological Society, p. 187–191.
Page 684, Figure 11-35: Published by permission of R. Bischke.
Page 685, Figure 11-36b: From Bischke and Tearpock 1993. Redrawn from Gow 1962. Published by permission of the Houston Geological Society.
Page 686, Figure 11-37: From Bischke and Tearpock 1993. Published by permission of the Houston Geological Society.
Page 689, Figure 11-38: Published by permission of John Suppe.
Page 690, Figure 11-39: Published by permission of R. Bischke.
Page 691, Figure 11-40: Published by permission of R. Bischke.
Page 692, Figure 11-41: Published by permission of R. Bischke.
Page 698, Figure 11-42: Published by permission of R. Bischke.
Page 698, Figure 11-43: Published by permission of R. Bischke.
Page 699, Figure 11-44: Published by permission of R. Bischke.
Page 701, Figure 11-45: Published by permission of R. Bischke.
Page 703, Figure 11-46: Published by permission of R. Bischke.
Page 704, Figure 11-47: Published by permission of R. Bischke.
Page 705, Figure 11-48: Published by permission of R. Bischke.
Page 706, Figure 11-49: Published by permission of R. Bischke.
Page 707, Figure 11-50: Published by permission of R. Bischke.
Page 707, Figure 11-51: Published by permission of R. Bischke.
Page 708, Figure 11-52: Published by permission of R. Bischke.
Page 710, Figure 11-53: Published by permission of R. Bischke.
Page 710, Figure 11-54: Published by permission of R. Bischke.
Page 711, Figure 11-55: Published by permission of R. Bischke.
Page 712, Figure 11-56: Published by permission of R. Bischke.
Page 713, Figure 11-57: Published by permission of R. Bischke.
Page 714, Figure 11-58: Published by permission of R. Bischke.
Page 715, Figure 11-59: Published by permission of R. Bischke.
Page 716, Figure 11-60: Published by permission of R. Bischke.
Page 717, Figure 11-61: Published by permission of R. Bischke.
Page 718, Figure 11-62: Published by permission of R. Bischke.
Page 719, Figure 11-63: Published by permission of R. Bischke.
Page 720, Figure 11-64a: Published with permission from Seitel Data Ltd.
Page 720, Figure 11-64b: Published by permission of R. Bischke.
Page 726, Figure 12-2: Published by permission of Van Mount.
Page 727, Figure 12-3: From Mount, V. S. and J. Suppe 1992, Present-day stress orientations to active strike-slip faults: California and Sumatra: Journal of Geophysical Research, v. 97, p. 11,995–12,013. Published by permission of the American Geophysical Union.
Page 728, Figure 12-4: From Mount and Suppe 1987 and 1992. Published by permission of Van Mount.
Page 729, Figure 12-5: From Xiaohan, L., 1983, Perturbations de contraintes liées aux structures cassantes dans les calcaires fins du Languedoc. Observations et simulations mathématiques [unpubl. 3 eme cycle thesis]: Montpellier, Université des Sciences et Techniques du Languedoc, 130 p.; Guiraud, M., and M. Seguret, 1985, A releasing solitary overstep model for the Late Jurassic – Early Cretaceous Soria strike-slip basin (northern Spain), in Biddle, K. T., and N. Christie-Blick, Strike-Slip Deformation, Basin Formation, and Sedimentation: Society of Economic Paleontologists and Mineralogists, Special Publication No. 37, p. 159–175. Published by permission of the Society for Sedimentary Geology.
Page 729, Figure 12-5: From Xiaohan, L., 1983, Perturbations de contraintes liées aux structures cassantes dans les calcaires fins du Languedoc. Observations et simulations mathématiques [unpubl. 3 eme cycle thesis]: Montpellier, Université des Sciences et Techniques du Languedoc, 130 p.
Page 729, Figure 12-5: Guiraud, M., and M. Seguret, 1985, A releasing solitary overstep model for the Late Jurassic – Early Cretaceous Soria strike-slip basin (northern Spain), in Biddle, K. T., and N. Christie-Blick, Strike-Slip Deformation, Basin Formation, and Sedimentation: Society of Economic Paleontologists and Mineralogists, Special Publication No. 37, p. 159–175. Published by permission of the Society for Sedimentary Geology.
Page 732, Figure 12-6: Bischke, R. E., J. Suppe, and R. del Pilar, 1990, A new branch of the Philippine Fault system, as observed from aeromagnetic and seismic data: Tectonophysics, v. 183, p. 243–264. Published by permission of the Philippine National Oil Company.
Page 733: "Drawn on a seismic background", Bally, A. W., 1983, Seismic Expression of Structural Styles: American Association of Petroleum Geologists Bulletin, Studies in Geology Series No. 15, v. 3., p. 167–172.
Page 735, Figure 12-7: Suppe 1985, Principles of Structural Geology: Englewood Cliffs, NJ, Prentice-Hall, p. 537.
Page 735, Figure 12-8: Bischke, R. E., J. Suppe, and R. del Pilar, 1990, A new branch of the Philippine Fault system, as observed from aeromagnetic and seismic data: Tectonophysics, v. 183, p. 243–264. Published by permission of Tectonophysics.
Page 736, Figure 12-9: From Aydin and Nur 1985, The type and role of stepovers in strike-slip tectonics: in Biddle, K. T. and N. Christie-Blick, eds., Strike-Slip Deformation, Basin Formation, and Sedimentation: Society of Economic Paleontologists and Mineralogists, Special Publication No. 37, p. 35–44. Published by permission of the Society for Sedimentary Geology.
Page 737: TBL12-1: Aydin, A., and A. Nur, 1982, Evolution of pull-apart basins and their scale independence: Tectonics, v. 1, p. 91–105. Published by permission of the American Geophysical Union
Page 740, Figure 12-10: Suppe 1985, Principles of Structural Geology: Englewood Cliffs, NJ, Prentice-Hall, p. 537.
Page 741, Figure 12-11: Published by permission of R. Bischke.
Page 743, Figure 12-12a: Modified from Wright, T. L., 1991, Structural geology and tectonic evolution of the Los Angeles basin, California, in Biddle, K. T., Active Margin Basins: American Association of Petroleum Geologists Memoir 52, p. 35–134.; AAPG"1991. reprinted by permission of the AAPG whose permission is required for further use
Page 743, Figure 12-12b: Published by permission of J. Shaw and R. Bischke.
Page 744, Figure 12-13: Redrawn after Taylor, J. C., 1973, Recent developments at Signal Hill, Long Beach oil field: in Guidebook for the Society of Economic Paleontologists and Mineralogists, Annual Meeting with the American Association of Petroleum Geologists and the Society of Economic Geophysicists, Field Trip 1, p. 16–25.
Page 744, Figure 12-14: Published by permission of R. Bischke.

Page 747, Figure 12-15: Aydin, A., and A. Nur, 1982, Evolution of pull-apart basins and their scale independence: Tectonics, v. 1, p. 91–105. Published by permission of the American Geophysical Union.
Page 753, Figure 12-20: Published by permission of R. Bischke.
Page 755, Figure 12-21: Published by permission of R. Bischke.
Page 756, Figure 12-22: Published by permission of R. Bischke.
Page 758, Figure 12-23: From Kelly 1971. Published by permission of the New Mexico Bureau of Mines and Mineral Resources.
Page 760: "wave-ripples, mudcrack casts, mammal tracks and arthropod burrows", Smith (1982).
Page 759, Figure 12-24: From Crowell 2003a. Published by permission of the Geological Society of America.
Page 761, Figure 12-25: Published by permission of R. Bischke.
Page 762, Figure 12-26: After May, R. S., K. D. Ehman, G. G. Gray and J. C. Crowell, 1993, A new angle on the tectonic evolution of the Ridge Basin, a "strike-slip" basin in Southern California: Geological Society of America Bulletin, v. 105, p. 1357–1372. Published by permission of the Geological Society of America.
Page 763, Figure 12-27: Published by permission of R. Bischke.
Page 764, Figure 12-28: Published by permission of R. Bischke.
Page 766, Figure 12-29: Published by permission of R. Bischke.
Page 768, Figure 12-30: Published by permission of R. Bischke.
Page 769, Figure 12-31: Published by permission of R. Bischke.
Page 770, Figure 12-32: Published by permission of R. Bischke.
Page 772, Figure 12-33: From Nilsen, T. H., and R. J. McLaughlin, 1985, Comparison of tectonic framework and depositional patterns of the Hornelen strike-slip basins of Norway and the Ridge and Little Sulphur Creek strike-slip basins of California, in Biddle, K. T., and N. Christie-Blick., eds., Strike-slip Deformation, Basin Formation and Sedimentation: Society of Economic Paleontologists and Mineralogists, Special Publication No. 37, p. 79–103. Published by permission of the Society for Sedimentary Geology.
Page 778, Figure 13-1: From Thorsen, C. E., 1963, Age of growth faulting in southeast Louisiana: Transactions Gulf Coast Association of Geological Societies, v. 13, p. 103–110. Published by permission of the Gulf Coast Association of Geological Societies.
Page 779, Figure 13-2: From Thorsen, C. E., 1963, Age of growth faulting in southeast Louisiana: Transactions Gulf Coast Association of Geological Societies, v. 13, p. 103–110. Published by permission of the Gulf Coast Association of Geological Societies.
Page 779, Figure 13-3: From Thorsen, C. E., 1963, Age of growth faulting in southeast Louisiana: Transactions Gulf Coast Association of Geological Societies, v. 13, p. 103–110. Published by permission of the Gulf Coast Association of Geological Societies.
Page 782, Figure 13-4: Modified from Bischke, R. E., and J. Suppe, 1990b, Geometry of rollover: Origin of complex arrays of antithetic and synthetic crestal faults (Abs.): American Association of Petroleum Geologists Bulletin, v. 74, p. 611. 1994b; AAPG©1994. published by permission of the AAPG whose permission is required for further use.
Page 784, Figure 13-5a: Published by permission of R. Bischke and D. Tearpock.
Page 784, Figure 13-5b: Published by permission of R. Bischke and D. Tearpock.
Page 784, Figure 13-5c: From Bischke 1994b; AAPG"1994, published by permission of the AAPG whose permission is required for further use.
Page 786, Figure 13-6: From Bischke, R. E., and J. Suppe 1994b, Interpreting sedimentary growth structures from well log and seismic data (with examples): American Association of Petroleum Geologists Bulletin, v. 7, p. 873–892. ; AAPG"1994, published by permission of the AAPG whose permission is required for further use.
Page 788, Figure 13-7: From Bischke, R. E., and J. Suppe, 1990b, Geometry of rollover: Origin of complex arrays of antithetic and synthetic crestal faults (Abs.): American Association of Petroleum Geologists Bulletin, v. 74, p. 611. 1994b; AAPG"1994, published by permission of the AAPG whose permission is required for further use.
Page 790, Figure 13-8: Bischke 1994; AAPG"1994, reprinted by permission of the AAPG whose permission is required for further use.
Page 792, Figure 13-9: From Bischke, R. E., and D. J. Tearpock, 1999, A graphical dip domain technique for projecting large growth faults to depth using imaged hanging wall structure: Transactions Gulf Coast Association of Geological Societies, v. 49, p. 112–120. Published by permission of the Gulf Coast Association of Geological Societies.
Page 793, Figure 13-10: Published by permission of R. Bischke.
Page 799, Figure 13-14: From Sanchez, R., J. Y. Chatellier, R. de Sifontes, N. Parra, and P. Muñoz, 1997, Multiple Bischke Plot Analysis, a powerful method to distinguish between tectonic or sedimentary complexity and miscorrelations: Memoras del Primero Congreso Latinoamericano de Sedimentologia, Sociedad Venezolana de Geologos, Tomo II, p. 257–264. Published by permission of Latinoamericano de Sedimentologia, Sociedad Venezolana de Geologos.
Page 801, Figure 13-16: From Bischke, R. E., W. Finley and D. J. Tearpock, 1999, Growth analysis (Δd/d): Case histories of the resolution of correlation problems as encountered while mapping around salt: Transactions Gulf Coast Association of Geological Societies, v. 49, p. 102–110. Published by permission of the Gulf Coast Association of Geological Societies.
Page 819, Figure 14-2: From Tearpock, D. J., and J. Harris, 1987, Subsurface Geological Page Mapping Techniques – A Training Manual: Tenneco Oil Co. Published by permission of Tenneco Oil Company.
Page 820, Figure 14-3: Published by permission of Texaco, USA.
Page 823, Figure 14-5: From Tearpock, D. J., and J. Harris, 1987, Subsurface Geological Mapping Techniques – A Training Manual: Tenneco Oil Co. Published by permission of Tenneco Oil Company.
Page 828, Figure 14-8: From Tearpock, D. J., and J. Harris, 1987, Subsurface Geological Mapping Techniques – A Training Manual: Tenneco Oil Co. Published by permission of Tenneco Oil Company.
Page 831, Figure 14-9: Modified from Tearpock, D. J., and J. Harris, 1987, Subsurface Geological Mapping Techniques – A Training Manual: Tenneco Oil Co. Published by permission of Tenneco Oil Company.
Page 833, Figure 14-11: Modified from Tearpock, D. J., and J. Harris, 1987, Subsurface Geological Mapping Techniques – A Training Manual: Tenneco Oil Co. Published by permission of Tenneco Oil Company.
Page 834, Figure 14-12: Modified from Tearpock, D. J., and J. Harris, 1987, Subsurface Geological Mapping Techniques – A Training Manual: Tenneco Oil Co. Published by permission of Tenneco Oil Company.
Page 835, Figure 14-13: From Tearpock, D. J., and J. Harris, 1987, Subsurface Geological Mapping Techniques – A Training Manual: Tenneco Oil Co. Published by permission of Tenneco Oil Company.
Page 838, Figure 14-14: Published by permission of D. Tearpock.
Page 842, Figure 14-16: From Tearpock, D. J., and J. Harris, 1987, Subsurface Geological Mapping Techniques – A Training Manual: Tenneco Oil Co. Published by permission of Tenneco Oil Company.
Page 843, Figure 14-17: Modified from Tearpock, D. J., and J. Harris, 1987, Subsurface Geological Mapping Techniques – A Training Manual: Tenneco Oil Co. Published by permission of Tenneco Oil Company.
Page 844, Figure 14-18: From Tearpock, D. J., and J. Harris, 1987, Subsurface Geological Mapping Techniques – A Training Manual: Tenneco Oil Co. Published by permission of Tenneco Oil Company
Page 844, Figure 14-19: From Tearpock, D. J., and J. Harris, 1987, Subsurface Geological Mapping Techniques – A Training Manual: Tenneco Oil Co. Published by permission of Tenneco Oil Company
Page 852, Figure 14-25: From Tearpock, D. J., and J. Harris, 1987, Subsurface Geological Mapping Techniques – A Training Manual: Tenneco Oil Co. Published by permission of Tenneco Oil Company
Page 853, Figure 14-27: From Tearpock, D. J., and J. Harris, 1987, Subsurface Geological Mapping Techniques – A Training Manual: Tenneco Oil Co. Published by permission of Tenneco Oil Company

Page 876, Figure 14-41: Modified from Dennis et al. 2000. Published with permission from Norwegian Petroleum Society.
Page 877, Figure 14-42: Modified from Hubbert 1953. Published by permission of J. Brenneke.
Page 878, Figure 14-43: Published by permission of J. Brenneke.
Page 884, Figure 14-45: Prepared by C. Harmon. From Tearpock, D. J., and J. Harris, 1987, Subsurface Geological Mapping Techniques – A Training Manual: Tenneco Oil Co. Published by permission of Tenneco Oil Company.
Page 885, Figure 14-46: Prepared by C. Harmon. From Tearpock, D. J., and J. Harris, 1987, Subsurface Geological Mapping Techniques – A Training Manual: Tenneco Oil Co. Published by permission of Tenneco Oil Company.
Page 886, Figure 14-47: Prepared by C. Harmon. From Tearpock, D. J., and J. Harris, 1987, Subsurface Geological Mapping Techniques – A Training Manual: Tenneco Oil Co. Published by permission of Tenneco Oil Company.
Page 803, Figure 13-17: Published by permission of R. Bischke.
Page 805, Figure 13-18: Published by permission of R. Bischke.
Page 806, Figure 13-19: Published by permission of R. Bischke.
Page 807, Figure 13-20: Published by permission of R. Bischke.
Page 808, Figure 13-21: Published by permission of R. Bischke.
Page 809, Figure 13-22: Published by permission of R. Bischke.
Page 809, Figure 13-23: Published by permission of R. Bischke.
Page 810, Figure 13-24: Published by permission of R. Bischke.
Page 811, Figure 13-25: Published by permission of R. Bischke.
Page 812, Figure 13-26: Published by permission of R. Bischke.
Page 813, Figure 13-27: Maps and seismic line published by permission of Texaco, Inc.; VS/d plot published by permission of R. Bischke.
Page 889, Figure APP-2: Reproduced from Analysis of Geologic Structures by John M. Dennison, by W. W. Norton & Company, Inc. Copyright©1968 by W. W. Norton & Company, Inc.
Page 889, Figure APP-3: Reproduced from Analysis of Geologic Structures by John M. Dennison, by W. W. Norton & Company, Inc. Copyright©1968 by W. W. Norton & Company, Inc.
Page 890, Figure APP-4: Reproduced from Analysis of Geologic Structures by John M. Dennison, by W. W. Norton & Company, Inc. Copyright©1968 by W. W. Norton & Company, Inc.
Page 890, Figure APP-5: Reproduced from Analysis of Geologic Structures by John M. Dennison, by W. W. Norton & Company, Inc. Copyright©1968 by W. W. Norton & Company, Inc.
Cover images: David C. Metzner
Cover: Data published with permission from CGG Land.
Cover: Data published with permission from Seitel Data Ltd.
Cover: Original figure drawn by Jim Brenneke.
Cover: Pioneer Natural Resources Company, Micah Reasnor. ©2020 Pioneer Natural Resources Company.
Cover: Data published with permission from CGG Land.

Photo by izusek/gettyimages

Where are the color figures?

Register your purchase of
Applied Three-Dimensional Subsurface Geological Mapping, Third Edition,
to access color figures.

Students—follow these steps:

1. Go to **informit.com/register** and log in or create a new account.
2. Enter the ISBN: **9780134859613**
3. Answer the challenge question as proof of purchase.
4. Click on the "Access Bonus Content" link in the Registered Products section of your account page, to be taken to the page where the downloadable PowerPoint files are available.

Professors—follow these steps:

1. Go to **https://www.pearson.com/us/higher-education/subject-catalog/download-instructor-resources.html** and log in or register to request access to the Instructor Resource Center.
2. Locate this product.
3. Download the PowerPoints containing the color figures.